9995

MECHANICAL AND ELECTRICAL EQUIPMENT FOR BUILDINGS

MECHANICAL AND ELECTRICAL EQUIPMENT FOR BUILDINGS

Ninth Edition

Benjamin Stein
Consulting Architectural Engineer

John S. Reynolds
Professor of Architecture
University of Oregon

John Wiley & Sons, Inc.

New York   Chichester   Weinheim   Brisbane   Singapore   Toronto

DISCLAIMER

The information in this book has been derived and extracted from a multitude of sources including building codes, fire codes, industry codes and standards, manufacturer's literature, engineering reference works and personal professional experience. It is presented in good faith, but although the authors and the publisher have made every reasonable effort to make the information presented accurate and authoritative, they do not warrant, and assume no liability for its accuracy or completeness or its fitness for any specific purpose. It is primarily intended as a learning and teaching aid, and not as a source of information for the actual final design of building systems. It is the responsibility of users to apply their professional knowledge in the use of the information presented in this book, and to consult original sources for current detailed information as needed, in actual design situations.

This publication is designed to provide accurate and authoritative information in regard to the subject matter covered. It is sold with the understanding that the publisher is not engaged in rendering professional services. If professional advice or other expert assistance is required, the services of a competent professional person should be sought.

Library of Congress Cataloging-in-Publication Data:

Stein, Benjamin.
    Mechanical and electrical equipment for buildings / Benjamin Stein, John S. Reynolds. — 9th ed.
      p.   cm.
    Includes index.
    ISBN 0-471-15696-5 (cloth : alk. paper)
    1. Buildings—Mechanical equipment. 2. Buildings—Electric equipment. 3. Buildings—Environmental engineering. I. Reynolds, John, 1938– . II. Title.
    TH6010.S74 1999
    696—dc21   99-24973

Printed in the United States of America

10 9 8 7 6 5 4 3 2 1

PREFACE

Approximately four generations have passed since the first edition of *Mechanical and Electrical Equipment for Buildings* was printed in 1935. At birth, this book was 5.7 × 8.9 inches (145 × 226 mm) with 429 pages. Now in our 9th edition, we are 7.5 × 10 inches (191 × 254 mm) with almost 1800 pages, a volumetric increase of 420%. Many topics have been added, yet a few have disappeared; computers are now routinely used in system design; equipment and distribution systems have undergone huge changes; mechanical cooling has become commonplace; fuel sources have shifted (coal provided half of the U.S.'s fuel in 1935). And in recent editions, the book has increasingly added "why" to its long list of "how to".

Most of the topics in this text involve energy consumption. As North America has moved from its initial reliance on renewable energy sources (wind, water and horse power) to today's nonrenewable fossil fuels, it has also added vastly to its population, and increased its per capita energy use. The resulting environmental degradation (primarily evident in air and water quality) has spurred efforts to reverse this decline. Governmental regulations are examples of such efforts, but this text emphasizes the investigations of alternative fuels and design approaches.

It is increasingly clear that a gradual global warming is underway. It is less clear to what extent our hugely increased energy consumption is responsible, with its associated heat release and gaseous additions to the atmosphere. But it is very clear that the world's supply of fossil fuel is diminishing, with future consequences for all buildings that today rely so thoroughly on nonrenewable energy sources.

Our buildings are today contributing to problems with global consequences tomorrow, and our approach to mechanical and electrical systems must consider how best to respond. Thus, on-site resources—daylighting, passive solar heating, passive cooling, solar water heating, rainwater, waste water treatment, photovoltaic electricity— share chapters with the traditional off-site networks (natural gas, oil, the electrical grid, water and sewer lines). On-site processes can be area-intensive, labor-intensive, and can involve increased first costs that require years to recover. On-site energy use requires us to look beyond the building, to pay as much attention to our building's context as to the mechanical and electrical spaces and systems within.

Over the many editions of this book, another trend has emerged. We have moved from systems that centralize all sources of heating, cooling, water and electricity toward those that encourage more localized production and control. Increased sophistication of digital control systems have encouraged this trend. Further encouragement comes from multi-purpose buildings whose schedules of occupancy are fragmented, and corporations with varying work schedules that result in partial occupancy on weekends. Another factor in this decentralization is worker satisfaction; we find increasing evidence that productivity increases with a sense of individual control of one's work environment. To this end, residences are increasingly used as office work environments. Our communications networks have made this possible. As our residential designs thus become more complex (office lighting, zones for heating/cooling, added raceways, sound isolation), our work environments become more attractive and individual.

Air and water pollution from buildings and their energy sources are widely recognized and condemned, but perhaps our most pervasive pollutant is noise (unwanted sound). This impacts building siting, space planning, choice of exterior and interior materials; even choice of cooling systems such as natural ventilation. Air and water pollution can result in physical illness, but so can noise pollution, along with its additional disadvantage of mental stress.

This book is written for the North American building design community, and has always emphasized examples from this region. Yet other areas of the world, some with similar traditions and fuel sources, have worthy examples of new strategies for building design utilizing on-site energy and energy conservation. Thus some buildings from Europe appear in this 9th edition, along with many North American examples. We have added listings for buildings, researchers and architects/engineers to the index of this edition.

Building system design is now widely performed on computers, often using proprietary software that includes hundreds of built-in assumptions. This book is intended to encourage the designer to take a dual approach to system design: use computers as a tool to check the designer's decisions, and use this book to check that computer output. Rough hand-calculated results should point toward the same results obtained with the computer; the greater the disparity, the greater the need to check both approaches.

We write this book mindful of the student, the architect-in-training, and the practicing professional. Basic theory, preliminary design guidelines, and detailed design procedures allow this book to serve both as an introductory text, and as a more advanced reference. We intend this to be a textbook for multiple courses in architecture, architectural engineering and building/construction management.

We are excited to announce that, for the first time ever, an Instructor's Manual has been developed in conjunction with the ninth edition (**ISBN 0-471-35769-3**). The manual, prepared by Walter Grondzik of Florida A&M University, will outline the contents and terminology in each chapter; highlight concepts of special interest or difficulty; and feature sample quiz and exam questions. The manual is available to instructors who are using this book in their courses. It is our hope that both full-time and adjunct professors will find the Instructor's Manual to be a valuable aid when planning their courses, and a helpful tool when tailoring this comprehensive volume for classroom use. It continues to serve as a reference for the architectural registration examinations in the U.S. and Canada. We also intend to provide a useful reference book for the offices of architects, engineers, and building managers.

Benjamin Stein
John S. Reynolds

ACKNOWLEDGMENTS

Many people and organizations have contributed to this book. We begin with those from whose work we have borrowed at length: J. Douglas Balcomb, Baruch Givoni, John Tillman Lyle, Murray Milne, William McGuinness, and Victor Olgyay; ASHRAE (American Society of Heating, Refrigerating and Air-Conditioning Engineers), the American Solar Energy Society, the National Drinking Water Clearinghouse, the National Small Flows Clearinghouse, the National Fire Protection Association, the Illuminating Engineering Society of North America; and the many equipment manufacturers whose product information and photographs are used to illustrate the text discussion.

We are grateful to several reviewers and others who made helpful and clarifying suggestions, some of whom carefully read drafts of various chapters. These include Carl Bovill, University of Maryland; Paul Buel, Southeast Community College-Milford; Mark DeKay, Washington University; and Jeff S. Haberl, Texas A&M University.

Several professionals provided valuable assistance in assembling materials, and clarifying ideas and details. These include Craig Christiansen (NREL research reports), William Lowry (climate of cities), Jason McLennan ("life of X" cartoons), Daniel Panetta (AIWPS and recycling at California Polytechnic, San Luis Obispo), Dr. Jonathan Stein (computer application), John A. Van Deusen (vertical transportation), and Martin Yoklic (cooltower performance analysis).

We are pleased to include many more new drawings by Michael Cockram, whose engaging work first appeared in the eighth edition. We here acknowledge also the many architects and engineers who provided illustrations of their buildings and processes throughout this book.

Testing in the classroom is a particularly valuable way to find needed improvements in any textbook. Students at the University of Oregon have, over several years, raised many questions whose answers have resulted in changes. Valuable suggestions have come from the many graduate teaching fellows at the University of Oregon, particularly Alfredo Fernandez, Susie Harriman and Therese Peffer. Former Oregon students who helped with research include Troy Anderson, Daniel Irurah, Reza Javandel, and Jeff Joslin.

We also extend a special word of thanks to Walter Grondzik of Florida A&M University who created the instructor's manual to accompany the ninth edition.

A large portion of the work involved in producing a manuscript is accomplished by supporting personnel. Among these, we wish particularly to thank Lila Stein for handling correspondence, typing a large portion of the manuscript, assisting in proofreading and other production items, and offering many helpful suggestions. We also wish to thank Stephanie Thompson for typing and editing a portion of the manuscript.

Finally, we are indebted to the staff at John Wiley & Sons for their diligent and highly professional work, especially Amanda Miller, Senior Editor; Matt Van Hattem, Associate Developmental Editor; Robert J. Fletcher, Project Manager; and Nyshie Perkinson, Editorial Assistant.

CONTENTS

PART I   ENERGY OVERVIEW

CHAPTER 1
CONTEXT FOR BUILDING
SYSTEMS DESIGN   3

1.1 Buildings and Sites *3*
1.2 Sources of Materials *4*
1.3 Sources of Energy *9*
1.4 Transportation Systems *17*
1.5 Information Systems *18*
1.6 Design Strategies for
     Sustainability *21*
1.7 Design Challenges *35*

CHAPTER 2
CLIMATE, COMFORT AND
DESIGN STRATEGIES   37

2.1 The Body *37*

2.2 Comfort *41*
2.3 Climates *52*
2.4 Design Strategies for Cooling *55*
2.5 Design Strategies for Heating *58*
2.6 Combining Strategies *63*

CHAPTER 3
SITES AND RESOURCES   67

3.1 Climates within Climates *67*
3.2 Analyzing the Site *71*
3.3 Site Design Strategies *73*
3.4 Direct Sun and Daylight *75*
3.5 Sound and Air *83*
3.6 Rain and Groundwater *104*
3.7 Plants *107*

PART II   THERMAL CONTROL

CHAPTER 4
HEAT FLOW   115

4.1 The Building Envelope *115*
4.2 Connectors, Filters, and Barriers *115*
4.3 Switches and User's Choice *116*
4.4 Heat Flow, Seasons and the
     Building *120*
4.5 Heat Flow Through Solids *124*
4.6 Heat Flow Through Air *127*
4.7 Heat Flow Through the Opaque
     Building Envelope *128*
4.8 Heat Flow Through Windows
     and Skylights *137*
4.9 Moisture and Infiltration *143*

CHAPTER 5
DESIGNING FOR HEATING
AND COOLING   191

5.1 Organizing the Problem *191*
5.2 Zoning *195*
5.3 Daylighting Guidelines *197*
5.4 Passive Solar Heating
     Guidelines *205*
5.5 Summer Heat Gain Guidelines *225*
5.6 Passive Cooling Guidelines *225*
5.7 Re-Integrating Daylighting, Passive
     Solar Heating, and Cooling *248*
5.8 Calculating Worst-Hourly Heat
     Loss *249*

5.9 Calculations for Heating-Season Fuel Consumption (Conventional Buildings) *251*

5.10 Passive Solar Heating Performance *255*

5.11 Approximate Method for Calculating Heat Gain (Cooling Load) *276*

5.12 Psychrometry *288*

5.13 Detailed Hourly Heat Gain (Cooling Load) Calculations *293*

5.14 Passive Cooling Calculation Procedures *295*

CHAPTER 6
INTERIOR AIR QUALITY   **331**

6.1 Sources and Impacts *331*
6.2 Predicting IAQ *332*
6.3 Zoning for IAQ *337*
6.4 Passive and Low-Energy Approaches to Ventilation *341*
6.5 Equipment for IAQ *351*

CHAPTER 7
HVAC FOR SMALLER BUILDINGS   **371**

7.1 Review of the Need for Mechanical Equipment *371*

7.2 Heating, Ventilating, and Air Conditioning (HVAC): typical Design Processes *372*

7.3 Location and Service Distribution *373*

7.4 Controls for Smaller Building Systems *375*

7.5 Refrigeration Cycles *375*

7.6 Cooling-Only Systems *377*

7.7 Heating-Only Systems *385*

7.8 Heating-Cooling Systems *412*

7.9 Psychrometrics and Refrigeration *423*

CHAPTER 8
LARGER BUILDING HVAC SYSTEMS   **427**

8.1 HVAC and Building Organization *427*

8.2 Heating, Ventilating, Air-Conditioning (HVAC) System Types *444*

8.3 Central Equipment *453*

8.4 Air Distribution Within Spaces *484*

8.5 All-Air HVAC Systems *495*

8.6 Air & Water Systems *500*

8.7 All-Water Systems *511*

8.8 District Heating and Cooling *516*

8.9 Cogeneration *519*

PART III   WATER AND WASTE

CHAPTER 9
WATER AND BASIC DESIGN   **531**

9.1 Water in Architecture *531*
9.2 The Hydrologic Cycle *534*
9.3 Basic Planning *536*
9.4 Rainwater *541*
9.5 Collection and Storage *546*
9.6 Rainwater and Site Planning *551*
9.7 Components *561*

CHAPTER 10
WATER SUPPLY   **573**

10.1 Water Quality *573*

10.2 Filtration *577*
10.3 Disinfection *579*
10.4 Other Water Treatments *582*
10.5 Water Sources *585*
10.6 Hot Water Systems and Equipment *597*
10.7 Fixtures and Water Conservation *625*
10.8 Fixture Accessibility and Privacy *638*
10.9 Water Distribution *640*
10.10 Piping, Tubing, Fittings, and Controls *651*
10.11 Sizing of Water Pipes *656*
10.12 Irrigation *663*

CHAPTER 11
WATER AND WASTE 669

11.1 Waterless Toilets and Urinals *669*
11.2 Principles of Drainage *678*
11.3 Piping, Fittings, and Accessories *681*
11.4 Design of Residential Waste Piping *686*
11.5 Design of Larger-Building Waste Piping *697*
11.6 On-Site Individual Building Sewage Treatment *704*
11.7 On-Site Multiple-Building Sewage Treatment *714*
11.8 Larger-Scale Sewage Treatment Systems *728*
11.9 Recycling and Graywater *735*
11.10 Stormwater Treatment *741*

CHAPTER 12
SOLID WASTE 747

12.1 Waste and Resources *747*
12.2 Resource Recovery: Central or Local? *755*
12.3 Solid Waste in Small Buildings *757*
12.4 Solid Waste in Large Buildings *759*
12.5 Equipment for the Handling of Solid Waste *763*
12.6 The Service Core *764*

PART IV FIRE PROTECTION

CHAPTER 13
FIRE PROTECTION 769

13.1 Design for Fire Resistance *769*
13.2 Smoke Management *781*
13.3 Water for Fire Suppression *786*
13.4 Other Fire Suppression Methods *808*
13.5 Lightning Protection *815*

FIRE ALARM SYSTEMS 820

13.6 General Considerations *820*
13.7 Fire Codes, Authorities, and Standards *820*
13.8 Fire Alarm Definitions and Terms *822*
13.9 Types of Fire Alarm Systems *823*
13.10 Circuit Supervision *825*
13.11 Conventional Systems *825*
13.12(a) System Coding *827*
13.12(b) Signal Processing *828*
13.13 Addressable Fire Alarm Systems *828*
13.14 Addressable Analog (Intelligent) Systems *829*
13.15 Automatic Fire Detection: Incipient Stage *830*
13.16 Automatic Fire Detection: Smoldering Stage *831*
13.17 Automatic Fire Detection: Flame Stage *835*
13.18 Automatic Fire Detection: Heat Stage *836*
13.19 Special Types of Fire Detectors *840*
13.20 False Alarm Mitigation *841*
13.21 Manual Stations *842*
13.22 Sprinkler Alarms *844*
13.23 Audible and Visual Alarm Devices *844*
13.24 General Recommendations *845*
13.25 Residential Fire Alarms *845*
13.26 Multiple-Dwelling Alarm Systems *846*
13.27 Commercial and Institutional Building Alarm Systems *846*
13.28 High-Rise Office Building Fire Alarm System *847*
13.29 Industrial Facilities *849*

PART V   ELECTRICITY

CHAPTER 14
PRINCIPLES OF ELECTRICITY   853

14.1   Electric Energy *853*
14.2   Unit of Electric Current—The Ampere *853*
14.3   Unit of Electric Potential—The Volt *853*
14.4   Unit of Electric Resistance—The Ohm *854*
14.5   Ohm's Law *854*
14.6   Circuit Arrangements *855*
14.7   Direct Current and Alternating Current (dc and ac) *857*
14.8   Electric Power Generation—dc *858*

PHOTOVOLTAIC SYSTEMS   858

14.9   Photovoltaic Power—General *858*
14.10 PV Systems—Types and Application *860*
14.11 Codes and Standards *863*
14.12 Terminology and Definitions *864*
14.13 Solar PV Cells *864*
14.14 PV System Arrays *866*
14.15 PV System Batteries *866*
14.16 BOS *869*
14.17 Design of Stand-Alone PV System *869*
14.18 Design of a Grid-connected (Interactive) PV System *874*
14.19 PV System—Conclusion *875*

ELECTRIC POWER AND ENERGY   877

14.20 Electric Power Generation—ac *877*
14.21 Power and Energy *878*
14.22 Power in Electric Circuits *878*
14.23 Energy in Electric Circuits *880*
14.24 Electric Demand Charges *882*
14.25 Electric Demand Control *883*
14.26 Electrical Measurements *887*

CHAPTER 15
ELECTRICAL SYSTEMS AND MATERIALS: WIRING AND RACEWAYS   893

15.1   System Components *893*
15.2   National Electrical Code *893*

15.3   Economics of Material Selection *894*
15.4   Energy Considerations *896*
15.5   Electrical Equipment Ratings *896*
15.6   Interior Wiring Systems *896*
15.7   Conductors *897*
15.8   Conductor Ampacity *897*
15.9   Conductor Insulation and Jackets *898*
15.10 Copper and Aluminum Conductors *898*
15.11 Flexible Armored Cable *901*
15.12 Nonmetallic Sheathed Cable (Romex) *901*
15.13 Conductors for General Wiring *903*
15.14 Special Cable Types *903*
15.15 Busway/Busduct/Cablebus *903*
15.16 Light-Duty Busway, Flat Cable Assemblies, and Lighting Track *905*
15.17 Cable Tray *906*
15.18 Design Considerations for Raceway Systems *907*
15.19 Steel Conduit *909*
15.20 Aluminum Conduit *911*
15.21 Flexible Metal Conduit *912*
15.22 Nonmetallic Conduit *913*
15.23 Surface Metal Raceways (Metallic and Nonmetallic) *913*
15.24 Outlet and Device Boxes *913*
15.25 Floor Raceways *916*
15.26 Underfloor Duct *917*
15.27 Cellular Metal Floor Raceway *922*
15.28 Precast Cellular Concrete Floor Raceways *922*
15.29 Full Access Floor *922*
15.30 Under-Carpet Wiring System *924*
15.31 Ceiling Raceways and Manufactured Wiring Systems *924*

CHAPTER 16
ELECTRICAL SYSTEMS AND MATERIALS: SERVICE AND UTILIZATION   933

16.1   Electric Service *933*
16.2   Overhead Service *933*

16.3 Underground Service *934*
16.4 Underground Wiring *934*
16.5 Service Equipment *937*
16.6 Transformers *937*
16.7 Transformers Outdoors *940*
16.8 Transformers Indoors: Heat Loss *941*
16.9 Transformers Indoors: Selection *942*
16.10 Transformer Vaults *943*
16.11 Service Equipment Arrangements and Metering *943*
16.12 Service Switch(es) *944*
16.13 Switches *944*
16.14 Contactors *947*
16.15 Special Switches *948*
16.16 Solid-State Switches, Programmable Switches, Microprocessors, and Programmable Controllers *949*
16.17 Equipment Enclosures *952*
16.18 Circuit-Protective Devices *953*
16.19 Switchboards and Switchgear *956*
16.20 Unit Substations (Transformer Load Centers) *959*
16.21 Panelboards *959*
16.22 Principles of Electric Load Control *960*
16.23 Intelligent Panelboards *962*
16.24 Electric Motors *963*
16.25 Motor Control Standards *966*
16.26 Motor Control *967*
16.27 Motor Control Equipment *968*
16.28 Wiring Devices: General Description *970*
16.29 Wiring Devices: Receptacles *972*
16.30 Wiring Devices: Switches *973*
16.31 Wiring Devices: Specialties *975*
16.32 Low-Voltage Switching *976*
16.33 Wireless Switching and Control *979*
16.34 Power Line Carrier Systems *979*

16.35 Power Conditioning *982*
16.36 Power Conditioning Equipment *983*
16.37 Surge Suppression *984*
16.38 Uninterruptible Power Supply *989*
16.39 Emergency/Standby Power Equipment *994*
16.40 System Inspection *996*

CHAPTER 17
ELECTRIC WIRING DESIGN 997

17.1 General Considerations *997*
17.2 Load Estimating *998*
17.3 System Voltage *1001*
17.4 Grounding and Ground-Fault Protection *1007*
17.5 Energy Conservation Considerations *1010*
17.6 Design Procedure *1011*
17.7 Electric Spaces *1012*
17.8 Electric Closets *1015*
17.9 Equipment Layout *1015*
17.10 Application of Overcurrent Equipment *1016*
17.11 Branch Circuit Design *1020*
17.12 Branch Circuit Design Guidelines: Residential *1023*
17.13 Branch Circuit Design Guidelines: Nonresidential *1025*
17.14 Load Tabulation *1032*
17.15 Spare Capacity *1033*
17.16 Feeder Capacity *1034*
17.17 Panel Feeder Load Calculation *1037*
17.18 Harmonic Currents *1038*
17.19 Riser Diagrams *1039*
17.20 Service Equipment and Switchboard Design *1041*
17.21 Emergency Systems *1042*

PART VI ILLUMINATION

CHAPTER 18
LIGHTING FUNDAMENTALS 1049

18.1 Introductory Remarks *1049*

PHYSICS OF LIGHT 1050

18.2 Light as Radiant Energy *1050*
18.3 Transmittance and Reflectance *1050*

18.4   Terminology and Definitions *1051*
18.5   Luminous Intensity *1053*
18.6   Luminous Flux *1053*
18.7   Illuminance *1054*
18.8   Luminance, Exitance, and
        Brightness *1055*
18.9   Illuminance Measurement *1058*
18.10   Luminance Measurement *1060*
18.11   Reflectance Measurements *1061*
18.12   Inverse Square Law *1062*
18.13   Luminous Intensity: Candela
        Measurements *1063*
18.14   Intensity Distribution Curves *1063*

LIGHT AND SIGHT   *1065*

18.15   The Eye *1065*
18.16   Factors in Visual Acuity *1067*
18.17   Size of Visual Object *1068*
18.18   Subjective Brightness *1068*
18.19   Contrast and Adaptation *1069*
18.20   Exposure Time *1072*
18.21   Secondary Task-Related
        Factors *1073*
18.22   Observer-Related Visibility
        Factors *1073*
18.23   The Aging Eye *1074*

QUANTITY OF LIGHT   *1075*

18.24   Illuminance Levels *1075*
18.25   Illuminance Category *1076*
18.26   Illuminance
        Recommendations *1077*

QUALITY OF LIGHTING   *1081*

18.27   Considerations of Lighting
        Quality *1081*
18.28   Direct (Discomfort) Glare *1084*
18.29   Veiling Reflections and Reflected
        Glare *1086*
18.30   Equivalent Spherical Illumination
        (ESI) and Relative Visual
        Performance (RVP) *1092*
18.31   Control of Reflected Glare *1094*
18.32   Luminance Ratios *1097*
18.33   Patterns of Luminance: Subjective
        Reactions to Lighting *1098*

FUNDAMENTALS OF COLOR   *1100*

18.34   Color Temperature *1100*
18.35   Object Color *1101*
18.36   Reactions to Color *1103*

CHAPTER 19
**LIGHT SOURCES: THEIR CHARACTERISTICS
AND APPLICATION**   **1107**

19.1   General Remarks *1107*

DAYLIGHTING   *1107*

19.2   Daylighting as a Lighting Design
        Factor *1107*
19.3   Permanent Supplementary Artificial
        Lighting Interiors (PSALI) *1109*
19.4   Characteristics of Outdoor
        Illumination *1111*
19.5   Solid Overcast Sky *1114*
19.6   Clear Sky *1116*
19.7   Partly-Cloudy Sky *1118*
19.8   Concept and Characteristics of
        Daylight Factor *1120*
19.9   Daylight Analysis: CIE
        Method *1125*
19.10   Graphic Daylighting Design
        Method (GDDM) *1132*
19.11   Daylight Design: IESNA Lumen
        Method *1133*
19.12   Daylight Analysis: Computer
        Use *1145*
19.13   Physical Modeling in Daylight
        Design *1148*
19.14   Additional Factors in Interior
        Daylighting *1151*
19.15   Advanced Daylighting
        Techniques *1153*

INCANDESCENT LAMPS   *1157*

19.16   The Incandescent Filament
        Lamp *1157*
19.17   Special Incandescent Lamps,
        Including Energy-Saving
        Lamps *1163*
19.18   Tungsten–Halogen
        (Quartz–Iodine) Lamps *1164*
19.19   Halogen Lamp Types *1165*

FLUORESCENT LAMPS   *1166*

19.20   Fluorescent Lamp
        Construction *1166*
19.21   Federal Standards for Fluorescent
        Lamps *1171*
19.22   The Fluorescent Lamp:
        Characteristics of Operation *1171*
19.23   Special Fluorescent Lamps *1177*
19.24   Compact Fluorescent Lamps *1179*

19.25   Magnetic Core-and-Coil
         Ballasts *1180*
19.26   Electronic Ballasts for Fluorescent
         Lamps *1183*

**HIGH-INTENSITY DISCHARGE
(HID) LAMPS**   *1184*

19.27   Mercury-Vapor Lamps *1185*
19.28   Metal–Halide Lamps *1187*
19.29   High-Pressure Sodium (HPS)
         Lamps *1190*
19.30   Low-Pressure Sodium (SOX)
         Lamps *1193*
19.31   Induction Lamps *1193*
19.32   Sulphur Lamps *1194*
19.33   Special Purpose Lamps *1195*

COLOR OF LIGHT   *1195*

19.34   Chromaticity *1195*
19.35   Spectral Distribution of Light
         Sources *1196*
19.36   Color Rendering (Index)
         CRI *1199*

**CHAPTER 20
LIGHTING DESIGN**   *1203*

20.1   General *1203*
20.2   Goals of a Lighting Design *1203*
20.3   Lighting Design Procedure *1204*
20.4   Cost Factors *1206*
20.5   Power Budgets *1206*
20.6   Task Analysis *1207*

PRELIMINARY DESIGN   *1209*

20.7   Energy Considerations *1209*
20.8   Preliminary Design *1212*
20.9   Illumination Methods *1213*
20.10   Types of Lighting Systems *1213*
20.11   Indirect Lighting *1213*
20.12   Semi-Indirect Lighting *1216*
20.13   Direct–Indirect and General
         Diffuse Lighting *1216*
20.14   Semi-Direct Lighting *1218*
20.15   Direct Lighting *1218*
20.16   Size and Pattern of
         Luminaires *1220*
20.17   Other Design Considerations *1224*

LUMINAIRES   *1226*

20.18   Design Considerations *1226*
20.19   Lighting Fixture Distribution
         Characteristics *1227*

20.20   Luminaire Light Control *1229*
20.21   Luminaire Diffusers *1233*
20.22   Uniformity of Illumination *1236*
20.23   Luminaire Mounting Height *1242*
20.24   Lighting Fixtures *1243*
20.25   Lighting Fixture Construction *1244*
20.26   Lighting Fixture Structural
         Support *1245*
20.27   Lighting Fixture Appraisal *1245*
20.28   Luminaire–Room System
         Efficiency: Coefficient of
         Utilization *1246*
20.29   Luminaire Efficacy Rating
         (LER) *1246*

DETAILED DESIGN PROCEDURES   *1247*

20.30   Calculation of Average
         Illuminance *1247*
20.31   Calculation of Horizontal
         Illuminance by the Lumen (Flux)
         Method *1247*
20.32   Calculation of Light-Loss Factor
         (LLF) *1248*
20.33   Determination of Coefficient of
         Utilization (CU) by the Zonal
         Cavity Method *1250*
20.34   Zonal Cavity Calculations:
         Illustrative Examples *1253*
20.35   Zonal Cavity Calculation by
         Approximation *1259*
20.36   Effect of Cavity Reflectances on
         Illuminance *1261*
20.37   Modular Lighting Design *1261*
20.38   Calculating Illuminance at a
         Point *1262*
20.39   Design Aids *1264*
20.40   Calculating Illuminance from a
         Point Source *1265*
20.41   Calculating Illuminance from
         Linear and Area Sources *1266*
20.42   Computer-Aided Design *1267*
20.43   Computer-Aided Design:
         Illustrative Example *1267*
20.44   Average Luminance
         Calculations *1273*

EVALUATION STAGE   *1276*

20.45   Design Evaluation *1276*

**CHAPTER 21
LIGHTING APPLICATION**   *1279*

21.1   Introduction *1279*

LIGHTING CONTROL   *1279*

21.2   Requirement for Lighting
       Control *1279*
21.3   Lighting Control: Switching *1280*
21.4   Lighting Control: Dimming *1281*
21.5   Lighting Control: Control
       Initiation *1281*
21.6   Lighting Control Strategy *1284*

RESIDENTIAL OCCUPANCIES   *1292*

21.7   Residential Lighting: General *1292*
21.8   Residential Lighting: Energy
       Factors *1292*
21.9   Residential Lighting Sources *1292*
21.10 Residential Lighting: Design
       Suggestions *1293*
21.11 Residential Lighting: Luminaires
       and Architectural Lighting
       Elements *1295*
21.12 Residential Lighting: Control *1295*

EDUCATIONAL FACILITIES   *1295*

21.13 Institutional and Educational
       Buildings *1295*
21.14 General Classrooms *1299*
21.15 Special-Purpose Classrooms *1301*
21.16 Assembly Rooms, Auditoriums,
       and Multipurpose Spaces *1302*
21.17 Gymnasium Lighting *1303*
21.18 Lecture Hall Lighting *1303*
21.19 Laboratory Lighting *1303*
21.20 Library Lighting *1303*
21.21 Special Areas *1305*
21.22 Other Considerations in School
       Lighting *1306*

COMMERCIAL INTERIORS   *1306*

21.23 Office Lighting: General *1306*

21.24 Lighting for Areas with Visual
       Display Terminals (VDTs) *1307*
21.25 Office Lighting Guidelines *1313*
21.26 Task-Ambient Office Lighting
       Design Using Ceiling-Mounted
       Units *1316*
21.27 Task-Ambient Office Lighting
       Using Furniture-Integrated
       Luminaires *1316*
21.28 Integrated and Modular
       Ceilings *1317*
21.29 Lighting and Air
       Conditioning *1317*

INDUSTRIAL LIGHTING   *1317*

21.30 General *1317*
21.31 Levels and Sources *1318*
21.32 Industrial Luminance Ratios *1319*
21.33 Industrial Lighting Glare *1319*
21.34 Industrial Lighting
       Equipment *1319*
21.35 Vertical-Surface Illumination *1321*

**SPECIAL LIGHTING: APPLICATION
TOPICS**   *1321*

21.36 Emergency Lighting *1321*
21.37 Floodlighting *1326*
21.38 Street Lighting *1326*
21.39 Light Pollution *1326*
21.40 Remote Source Lighting *1328*
21.41 Fiber-Optic (FO) Lighting *1329*
21.42 Fiber-Optic Terminology *1330*
21.43 Fiber Optic
       Lighting—Arrangements and
       Applications *1332*
21.44 Hollow Light Guides *1333*
21.45 Prismatic Light Guides *1334*
21.46 Prismatic Film Light Guide *1336*
21.47 Standards and Nomenclature *1336*

PART VII   SIGNAL EQUIPMENT

**CHAPTER 22
SIGNAL SYSTEMS**   *1345*

22.1   Introduction *1345*
22.2   Principles of Intrusion
       Detection *1345*

PRIVATE RESIDENTIAL SYSTEMS   *1347*

22.3   General *1347*
22.4   Residential Intrusion Alarm
       Systems *1352*
22.5   Residential Intercom Systems *1352*

22.6 Residential Telecommunication and Data Systems *1352*

22.7 Premise Wiring *1353*

MULTIPLE-DWELLING SYSTEMS   1354

22.8 Multiple-Dwelling Entry and Security Systems *1354*

22.9 Multiple-Dwelling Television Systems *1355*

22.10 Multiple-Dwelling Telephone Systems *1355*

22.11 Hotels and Motels *1356*

SCHOOL SYSTEMS   1357

22.12 General *1357*

22.13 School Security Systems *1357*

22.14 School Clock and Program Systems *1358*

22.15 School Intercom Systems *1360*

22.16 School Sound Systems *1360*

22.17 School Electronic Teaching Equipment *1361*

OFFICE BUILDING SYSTEMS   1362

22.18 General *1362*

22.19 Office Building Security Systems *1362*

22.20 Office Building Communications Systems *1363*

22.21 Office Building Communications Planning *1364*

22.22 Office Building Control and Automation Systems *1366*

INDUSTRIAL BUILDING SYSTEMS   1366

22.23 General *1366*

22.24 Industrial Building Personnel Access Control *1367*

22.25 Industrial Building Sound and Paging Systems *1369*

AUTOMATION   1371

22.26 General *1371*

22.27 Stand-Alone Lighting Control Systems *1372*

22.28 Building Automation System (BAS) *1374*

22.29 Glossary of Computer and Control Terminology *1374*

22.30 BAS System Arrangement *1376*

22.31 Intelligent Buildings *1380*

22.32 Intelligent Residences *1381*

PART VIII   TRANSPORTATION

CHAPTER 23
VERTICAL TRANSPORTATION:
PASSENGER ELEVATORS   1385

GENERAL INFORMATION   1385

23.1 Introduction *1385*

23.2 Passenger Elevators *1385*

23.3 Codes and Standards *1385*

ELEVATOR EQUIPMENT   1386

23.4 Principal Components *1386*

23.5 Gearless Traction Machines *1388*

23.6 Geared Traction Machines *1388*

23.7 Arrangement of Elevator Machines, Sheaves, and Ropes *1389*

23.8 Safety Devices *1390*

23.9 Elevator Doors *1390*

23.10 Cabs and Signals *1393*

23.11 Requirements for the Handicapped *1394*

ELEVATOR CAR CONTROL   1398

23.12 Drive Control *1398*

23.13 Thyristor Control, ac and dc *1398*

23.14 Variable-Voltage dc Motor Control *1400*

23.15 Variable-Voltage, Variable-Frequency (VVVF) ac Motor Control *1400*

23.16 Elevator Operating Control *1401*

23.17 System Control Requirements *1402*

23.18 Single Automatic Pushbutton Control *1402*

23.19 Collective Control *1402*

23.20 Selective Collective Operation *1402*

23.21 Computerized System
        Control *1402*
23.22 Rehabilitation Work: Performance
        Prediction *1405*
23.23 Lobby Elevator Panel *1405*
23.24 Car Operating Panel *1406*

ELEVATOR SELECTION   *1406*

23.25 General Considerations *1406*
23.26 Definitions *1407*
23.27 Interval or Lobby Dispatch Time
        and Average Lobby Waiting
        Time *1407*
23.28 Handling Capacity *1408*
23.29 Travel Time or Average Trip
        Time *1409*
23.30 Round-Trip Time *1410*
23.31 System Relationships *1418*
23.32 Car Speed *1419*
23.33 Single-Zone Systems *1419*
23.34 Multizone Systems *1421*
23.35 Other Elevator Selection
        Recommendations *1422*

**THE PHYSICAL PROPERTIES AND SPATIAL
REQUIREMENTS OF ELEVATORS**   *1424*

23.36 Shafts and Lobbies *1424*
23.37 Dimensions and Weights *1425*
23.38 Structural Stresses *1428*

POWER AND ENERGY   *1431*

23.39 Power Requirements *1431*
23.40 Energy Requirements *1432*
23.41 Energy Conservation *1433*
23.42 Emergency Power *1434*

SPECIAL CONSIDERATIONS   *1434*

23.43 Fire Safety *1434*
23.44 Elevator Security *1434*
23.45 Elevator Noise *1435*
23.46 Elevator Specifications *1435*
23.47 Novel Designs *1436*

**CHAPTER 24:
VERTICAL TRANSPORTATION:
SPECIAL TOPICS**   *1441*

SPECIAL SHAFT ARRANGEMENTS   *1441*

24.1   Sky Lobby Elevator System *1441*
24.2   Double-Deck Elevators *1442*

HYDRAULIC ELEVATORS   *1442*

24.3   Conventional Plunger-Type
        Hydraulic Elevators *1442*
24.4   Hole-less Hydraulic Elevators *1445*
24.5   Roped Hydraulic Elevators *1445*

FREIGHT ELEVATORS   *1445*

24.6   General *1445*
24.7   Freight Car Capacity *1447*
24.8   Freight Elevator Description *1448*
24.9   Freight Elevator Cabs, Gates,
        and Doors *1450*
24.10 Freight Elevator Cost Data *1450*

SPECIAL ELEVATOR DESIGNS   *1452*

24.11 Observation Cars *1452*
24.12 Inclined Elevators *1452*
24.13 Rack and Pinion Elevators *1452*
25.14 Residential Elevators and
        Chair Lifts *1452*
24.15 Linear Elevator Motor Drive *1457*

MATERIAL-HANDLING EQUIPMENT   *1459*

24.16 General *1459*
24.17 Manual Load/Unload
        Dumbwaiters *1459*
24.18 Automated Dumbwaiters *1462*
24.19 Horizontal Conveyors *1462*
24.20 Selective Vertical Conveyors *1462*
24.21 Pneumatic Tubes *1467*
24.22 Pneumatic Trash and Linen
        Systems *1467*
24.23 Automated Container Delivery
        Systems *1468*
24.24 Automated Self-Propelled
        Vehicles *1468*
24.25 Conclusion *1470*

**CHAPTER 25
MOVING STAIRWAYS AND WALKS**   1471

MOVING ELECTRIC STAIRWAYS   *1471*

25.1   General *1471*
25.2   Parallel and Crisscross
        Arrangements *1471*
25.3   Location *1476*
25.4   Size, Speed, Capacity, and
        Rise *1478*
25.5   Components *1479*

25.6   Safety Features *1479*
25.7   Fire Protection *1481*
25.8   Lighting *1481*
25.9   Application *1481*
25.10   Elevators and Escalators *1484*
25.11   Electric Power Requirements *1485*
25.12   Special-Design Escalators *1485*
25.13   Preliminary Design Data and Installation Drawings *1486*

25.14   Budget Estimating for Escalators *1487*

MOVING WALKS AND RAMPS   1487

25.15   General *1487*
25.16   Application of Moving Walks *1487*
25.17   Application of Moving Ramps *1488*
25.18   Size, Capacity, and Speed *1489*
25.19   Components *1490*

PART IX   ACOUSTICS

CHAPTER 26
FUNDAMENTALS OF ARCHITECTURAL
ACOUSTICS   1493

SOUND THEORY AND HEARING
PHENOMENA   1493

26.1   General *1493*
26.2   Sound Definition and Generation *1493*
26.3   Frequency *1494*
26.4   Velocity of Propagation *1496*
26.5   Wavelength and Types of Propagation *1497*
26.6   Sound Magnitude *1497*
26.7   Sound Intensity *1497*
26.8   Intensity Level (IL) and the Decibel (dB) *1499*
26.9   Human Hearing *1503*
26.10   Sound Pressure and Sound Pressure Level (SPL) *1504*
26.11   Loudness Level: Fletcher-Munson Contours *1506*
26.12   Measurement of Sound Pressure Level *1508*
26.13   Other Factors in Hearing *1510*
26.14   Speech *1510*
26.15   Other Sounds *1511*

NOISE CRITERIA   1511

26.16   Negative Effects of Noise *1511*
26.17   Noise and Annoyance *1511*
26.18   Noise Criterion Curves *1512*
26.19   Room Criterion Curves *1513*
26.20   High Noise Levels *1516*

CHAPTER 27
SOUND IN ENCLOSED SPACES   1519

27.1   Sound in Enclosures *1519*

ABSORPTION   1519

27.2   Sound Absorption *1519*
27.3   Mechanics of Absorption *1520*
27.4   Absorptive Materials *1520*
27.5   Installation of Absorptive Materials *1525*

ROOM ACOUSTICS   1526

27.6   Reverberation *1526*
27.7   Sound Fields in an Enclosed Space *1528*
27.8   Sound Power Level (PWL) and Sound Pressure Level (SPL) *1528*
27.9   Noise Reduction by Absorption *1530*
27.10   Noise Reduction Coefficient *1534*

ROOM DESIGN   1535

27.11   Reverberation Criteria for Speech Rooms *1535*
27.12   Criteria for Music Performance *1536*
27.13   Sound Paths *1538*
27.14   Ray Diagrams *1541*
27.15   Auditorium Design *1542*

SOUND REINFORCEMENT SYSTEMS   1544

27.16   Objectives and Criteria *1544*
27.17   Components and Specifications *1545*
27.18   Loudspeaker Considerations *1546*

CHAPTER 28
BUILDING NOISE CONTROL 1549

NOISE REDUCTION *1549*

28.1 Principles of Noise Reduction *1549*

ABSORPTION *1549*

28.2 Panel and Cavity Resonators *1549*
28.3 Acoustically Transparent Surfaces *1550*
28.4 Absorption Recommendations *1552*
28.5 Characteristics of Absorptive Materials *1553*

SOUND ISOLATION *1556*

28.6 Airborne and Structure-Borne Sound *1556*

AIRBORNE SOUND *1559*

28.7 Transmission Loss (TL) and Noise Reduction (NR) *1559*
28.8 Barrier Mass *1560*
28.9 Stiffness and Resonance *1561*
28.10 Compound Barriers (Cavity Walls) *1562*
28.11 Sound Transmission Class (STC) *1564*
28.12 Composite Walls and Leaks *1567*
28.13 Doors and Windows *1570*
28.14 Diffraction: Barriers *1574*
28.15 Flanking *1577*

SPEECH PRIVACY *1577*

28.16 Principles of Speech Privacy between Enclosed Spaces *1577*
28.17 Sound Isolation Descriptors *1580*
28.18 Speech Privacy Design for Enclosed Spaces *1582*
28.19 Principles of Speech Privacy in Open Area Offices *1585*
28.20 Open Office Speech Privacy Levels and Descriptors *1589*

28.21 Design Recommendations for Speech Privacy in Open Offices *1592*

STRUCTURE-BORNE SOUND *1595*

28.22 Structure-Borne Impact Noise *1595*
28.23 Control of Impact Noise *1595*
28.24 Impact Isolation Class (IIC) *1597*

MECHANICAL SYSTEM NOISE CONTROL *1597*

28.25 Mechanical Noise Sources *1597*
28.26 Quieting of Machines *1598*
28.27 Duct System Noise Reduction *1599*
28.28 Active Noise Cancellation *1602*
28.29 Piping System Noise Reduction *1603*
28.30 Electrical Equipment Noise *1605*
28.31 Noise Problems Due to Equipment Location *1606*
28.32 Sound Isolation Enclosures, Barriers and Damping *1607*

STC AND IIC RECOMMENDATIONS AND CRITERIA *1607*

28.33 Multiple Occupancy Residential STC/IIC Criteria *1607*
28.34 Specific Occupancies *1610*

OUTDOOR ACOUSTIC CONSIDERATIONS *1612*

28.35 Sound Power and Pressure Levels in Free Space (Outdoors) *1612*
28.36 Building Siting *1613*

REFERENCE MATERIAL *1613*

28.37 Glossary *1613*
28.38 Reference Standards *1618*
28.39 Units and Conversions *1618*
28.40 Symbols *1618*

PART X   APPENDICES

APPENDIX A
CLIMATIC CONDITIONS FOR THE UNITED STATES, CANADA, AND MEXICO 1623

APPENDIX B
SOLAR DATA 1635

APPENDIX C
ANNUAL SOLAR PERFORMANCE   1661

APPENDIX D
SUNSHADING   1693

D.1 Introduction *1693*
D.2 Solar Position: Altitude and
        Azimuth *1693*
D.3 Sun Time and Clock Time *1693*
D.4 Sunpath Projection *1695*
D.5 Shadow Angles and Shading
        Masks *1701*
D.6 Use of Shading Masks *1702*
D.7 Finite Horizontal Elements *1705*

APPENDIX E
ECONOMIC ANALYSIS   1711

E.1 Economic Decision *1711*
E.2 Life-Cycle Cost (LCC) *1711*
E.3 Initial (Simple) Rate of Return *1712*
E.4 Cost-Effectiveness Comparison *1713*
E.5 Internal Rate of Return (IRR) *1715*
E.6 Payback Period *1715*

APPENDIX F
SOUND TRANSMISSION DATA
FOR WALLS   1717

APPENDIX G
SOUND TRANSMISSION AND
IMPACT INSULATION DATA FOR
FLOOR/CEILING CONSTRUCTION   1729

APPENDIX H
OPTIONS FOR COMPUTATION
ASSISTANCE   1739

APPENDIX I
METRICATION; SI UNITS;
CONVERSIONS   1741

I.1 General Comments on
       Metrication *1741*
I.2 SI Nomenclature, Symbols *1741*
I.3 Common Usage *1742*
I.4 Conversion Factors *1743*

APPENDIX J
ASHRAE STANDARDS FOR ENERGY
EFFICIENT BUILDING DESIGN   1747

INDEX   1763

TABLES

Table 1.1
Approximate Total Embodied Energy in
Building Materials   **6**

Table 1.2
Residential Salvage for Reuse   **9**

Table 1.3
Energy End-Use in U.S. Buildings by Fuel
Type (Quads)   **12**

Table 1.4
Daily Arrival of Solar Energy on Earth
Compared to Other Energy Quantities   **18**

Table 2.1
Metabolic Rates for Typical Tasks   **38**

Table 2.2
Typical Insulation Values for Clothing
Ensembles   **40**

Table 2.3
Adaptive Behavior for
Thermal Comfort   **47**

Table 2.4
Indoor Air Velocity and Comfort   **48**

Table 2.5
Normal Data from Annual Summary
of Local Climatological Data for
Dodge City, Kansas   **56**

Table 3.1
Average Changes in Climate Effects
Caused by Urbanization   **68**

Table 3.2
Heat Generated Within Cities   **70**

Table 3.3
Air Pollution: Sources and Effects   **95**

Table 3.4
Beaufort Scale (Lower Speeds Only)   **100**

Table 3.5
Deciduous Trees for Summer Shading,
Winter Solar Collection   **111**

Table 4.1
Heat Symbols, Terms, Definitions   **124**

Table 4.2
Thermal Properties of Typical Building
and Insulating Materials   **147**

Table 4.3
Properties of Surface Air Films and
Air Spaces   **158**

Table 4.4
Thermal Resistances of Plane
Air Spaces   **160**

Table 4.5
U-Values for Some Whole-Wall,
Whole-Roof, and Whole-Floor Systems **163**

Table 4.6
Corrected R-Values for Wall Systems
with Steel Studs   **164**

Table 4.7
U-Values Comparing Whole-Wall
Framing and SIP (Structural Insulated
Panels)   **164**

Table 4.8
Heat Loss Coefficient of Floor Slabs, F_2,
per Unit Length of Perimeter   **165**

Table 4.9
Heat Flow Coefficients of Floor Slabs,
F_2, for Various Values of R **166**

Table 4.10
Heat Flow Below Grade   **167**

Table 4.11
Corrected R-Values for Roof Systems
with Metal Trusses   **168**

Table 4.12
Coefficients of Transmission, U, for
Wood and Steel Doors   **169**

Table 4.13
Window Characteristics   **171**

Table 4.14
Skylight U-Factors   **173**

Table 4.15
Shading Coefficients for
Domed Skylights   174

Table 4.16
Solar Optical Properties of
Several Glazings   175

Table 4.17
Approximate Shading Coefficients of
External Shading Devices   175

Table 4.18
Shading Coefficients for Louvered
Sun Screens   176

Table 4.19
Adjustable Sunshading   177

Table 4.20
Shading Coefficients (SC)   178

Table 4.21
Shading Coefficients for Single and
Insulating Glass with Draperies   179

Table 4.22
Recommended Outdoor Air Requirements for Ventilation   180

Table 4.23
Estimated Overall Infiltration Rates for
Small Buildings   188

Table 4.24
Approximate Infiltration Through Doors
and Windows of Small Buildings   189

Table 5.1
Proceeding Through Chapter 5   192

Table 5.2
Recommended Daylight Factors   200

Table 5.3
Daylight Factor (DF) Design Guidelines
for Overcast Sky Conditions   201

Table 5.4
Design Data for a Typical Second-Floor
Bay of the Office Building (Fig. 5.7f)   209

Table 5.5
Overall Heat Loss Criteria for Solar
Guidelines   209

Table 5.6
Design Guidelines for Passive Solar
Glazing Area   212

Table 5.7
Design Guidelines for Passive Solar
Thermal Mass   217

Table 5.8
Estimating Summer Heat Gains   226

Table 5.9
Comparing Passive Cooling Design
Guidelines with Detailed Calculation
Procedures   231

Table 5.10
Approximate Heat Values of Fuels   251

Table 5.11
Typical Residential Daily Internal Heat
Gains from Appliances and Lighting   253

Table 5.12
Some Typical Values of Annual Fuel
Utilization Efficiency (AFUE)   255

Table 5.13
Passive Solar Heating Systems
Compared   256

Table 5.14
Transparent Insulation Materials
(TIM)   260

Table 5.15
Design Data for Some Early Passive
Solar Buildings   261

Table 5.16
Time Lag Through Homogeneous
Walls   275

Table 5.17
Indoor Temperature Swing, ΔT Swing 276

Table 5.18
Design Equivalent Temperature
Differences (DETD)   278

Table 5.19
Design Cooling Load Factors Through
Glass   280

Table 5.20
Sensible Cooling Load Factors Due to
Infiltration and Ventilation   282

Table 5.21
Rates of Heat Gain from Occupants of
Conditioned Spaces   283

Table 5.22
Rates of Heat Gain from Office
Equipment   284

Table 5.23
Rate of Heat Gain from Miscellaneous
Appliances 285

Table 5.24
Cooling Load Through Building
Envelope 287

Table 5.25
Comparison of Methods of Heat Gain
Calculation 295

Table 5.26
Cooling Climate Data for Some North
American Cities 296

Table 5.27
Night Ventilation of Thermal Mass 304

Table 5.28
Common Mass Heat Capacities 304

Table 5.29
Oregon Office Building Performance:
Cooling by Night Ventilation of
Thermal Mass for a Typical
Second-Floor Bay 307

Table 5.30
Cooltower Performance Hourly
Simulations 316

Table 5.31
Roof Pond Design Data 318

Table 5.32
Earth Tube Applications 323

Table 6.1
Common Air Pollutants 333

Table 6.2
Characteristics of Particles and Particle
Dispersoids 336

Table 6.3
Estimating Indoor Air Quality 338

Table 6.4
Air Replacance Compared to Input Air
Changes per Hour (ACH) 339

Table 6.5
Air Quality Standards 340

Table 6.6
Design Decisions and Impacts: EPA
Campus, Research Triangle Park 342

Table 6.7
Turbine Ventilator Performance 347

Table 6.8
Representative Heat Exchanger Data
for Smaller Buildings 354

Table 6.9
Air Filter Characteristics 363

Table 7.1
Solar Performance of Roofing
Materials 382

Table 7.2
Approximate Average Wood Density 388

Table 7.3
Hydronic Baseboard Convectors 399

Table 7.4
Air Velocities for Ducts and Grilles 408

Table 7.5
Typical Residential Forced-Air Register,
$2\frac{1}{4} \times 12$ in. (60×300 mm) 412

Table 8.1
Basic HVAC Systems: Tasks and
Components 429

Table 8.2
Procedure for Matching Zones and
Systems 434

Table 8.3
Approximate Space Sizes for Major
Heating and Cooling Equipment 451

Table 8.4
Approximate Space Sizes for Air-
Handling Equipment 452

Table 8.5
Tons of Refrigeration: Design
Guidelines 453

Table 8.6
Cooling Storage Comparison 478

Table 8.7
Air Handling Equipment Rooms 480

Table 9.1
Water Measurement Terms 537

Table 9.2
Planning Guide for Water Supply 538

Table 9.3
Minimum Number of Plumbing
Facilities 542

Table 9.4
Estimated Sewage Flow Rates   **544**

Table 9.5
Rainfall Cistern Sizing Procedure   **551**

Table 9.6
Some Constituents of Stream Water   **552**

Table 9.7
Size of Gutters   **565**

Table 9.8
Sizing Roof Drains, Leaders, and
Vertical Rainwater Piping   **568**

Table 9.9
Sizing of Horizontal Rainwater Piping **570**

Table 10.1
Water Use and Quality in Buildings   **574**

Table 10.2
Example of Chemical Analysis   **575**

Table 10.3
Common Water Quality Problems and
Treatment in Small Systems   **577**

Table 10.4
Pumps for Water Supply   **590**

Table 10.5
Calculations to Establish Capacity of
Water Storage Tanks at Kinloch   **595**

Table 10.6
Representative Hot Water
Temperatures   **599**

Table 10.7
Domestic Hot Water
Consumption—Residences   **601**

Table 10.8
HUD–FHA Minimum Water Heater
Capacities, Residential   **606**

Table 10.9
Procedure for Sizing a Residential
Water Heater Storage Tank   **607**

Table 10.10
Domestic Hot Water, Commercial,
Institutional   **609**

Table 10.11
Solar Domestic Hot Water (DHW)
Systems   **616**

Table 10.12
Thermal Expansion of Pipe   **649**

Table 10.13
Water Supply Piping Materials   **652**

Table 10.14
Flow and Pressure to Typical Plumbing
Fixtures   **656**

Table 10.15
Water Supply Fixture Units (WSFU)   **661**

Table 10.16
Allowance in Equivalent Length of Pipe
for Friction Loss in Valves and
Threaded Fittings   **662**

Table 11.1
Materials for Waste Piping   **682**

Table 11.2
Drainage Fixture Units (dfu)   **691**

Table 11.3
Horizontal Fixture Branches and
Stacks   **693**

Table 11.4
Size and Developed Length of Stack
Vents and Vent Stacks   **694**

Table 11.5
Building Drains and Sewers   **695**

Table 11.6
Drainage Fixture Units (dfu) for
Example 11.1   **697**

Table 11.7
Septic Tank Capacity   **707**

Table 11.8
Septic Tank and Leaching Area Design
Criteria for Five Typical Soils   **708**

Table 11.9
Location of On-Site Sewage Disposal
Systems   **710**

Table 11.10
Disposal Field Trenches   **711**

Table 11.11
Sand Filter Design Values   **713**

Table 11.12
Residential Graywater Sources for
Irrigation   **737**

Table 11.13
Heat Recovery from Graywater   **738**

Table 11.14
Location of Residential Graywater
System   **739**

Table 11.15
Graywater Design Criteria, Six Typical Soils **741**

Table 12.1
Typical Recyclable Materials in Building Operation **748**

Table 12.2
Solid Waste Sources in Building Operation **750**

Table 12.3
Space Planning for Solid Waste **756**

Table 12.4
Building Characteristics and Core Placement **766**

Table 13.1
Major Factors Influencing Fire Growth **771**

Table 13.2
Summary of Life Safety Code Requirements for Interior Finish **773**

Table 13.3
Common Path, Dead End, and Travel Distance Limits **776**

Table 13.4
Summary of NFPA 101®, Life Safety Code®, Provisions for Occupant Load and Exit Capacity **778**

Table 13.5
Requirements for Exit Stairs **779**

Table 13.6
Common Failure Modes for Automatic Sprinklers **794**

Table 13.7
Sprinkler Heads **796**

Table 13.8
Relative Fire Hazard for Various Occupancies, as Related to Sprinkler Installations **798**

Table 13.9
Protected Area, Maximum Spacing, and Distances Below Ceiling for Sprinklers **800**

Table 13.10
Maximum Floor Areas for Sprinkler Systems **802**

Table 13.11
Class A Fire Extinguishers **816**

Table 14.1
Physical Characteristics of Typical 12-V Batteries **870**

Table 15.1
Physical Properties of Bare Copper Conductors **898**

Table 15.2
Allowable Ampacities of Insulated Conductors Rated 0 through 2000 Volts, 60° to 90°C (140° to 194°F) **899**

Table 15.3
Current-Carrying Capacity Derating Factors **900**

Table 15.4
Typical Ambient Temperatures **900**

Table 15.5
Conductor Insulation and Application **901**

Table 15.6
Life-Cycle Relative Cost Comparison of 2000-A, 208-V Feeder Installation **905**

Table 15.7
Comparative Dimensions and Weights of Metallic Conduit **910**

Table 15.8
Metric Equivalents of Rigid Conduit Sizes **911**

Table 16.1
Typical Transformer Data **938**

Table 16.2
Air-Cooled Transformer Electrical Insulation Temperature Ratings (Based on 40°C Ambient) **939**

Table 16.3
Typical Dry-Type Transformer: Dimensions and Weights **939**

Table 16.4
Relative Installation Costs for a 300 to 1000 kVA Transformer **943**

Table 16.5
Control Equipment Enclosures **947**

Table 16.6
Rating and Approximate Dimensions of ac Full-Voltage Conventional Single-Speed Motor Controllers, 3-Phase Combination Circuit Breaker Type **970**

Table 16.7
Power Conditioning Equipment **984**

Table 16.8
Typical Dimensional Data for UPS
Equipment **993**

Table 17.1
Electric Load Estimating **999**

Table 17.2
Nominal Service Size in Amperes **1001**

Table 17.3
System and Utilization Voltages **1006**

Table 17.4
Effects of Undervoltage on Utilization
Equipment **1007**

Table 17.5
Recommended Branch Circuit Loads **1022**

Table 17.6
Branch Circuit Requirements **1023**

Table 17.7
Load, Circuit, and Receptacle Chart for
Residential Electrical Equipment **1026**

Table 17.8
Panel Initial and Expanded Loads **1034**

Table 17.9
Maximum Number of Conductors in
Trade Sizes of Rigid Metal Conduit **1035**

Table 17.10
Maximum Wire and Ampacity of a
Conduit, and Ampacity Gain on
Rewiring **1036**

Table 17.11
Current and Volt-Ampere
Relationships **1038**

Table 18.1
Lighting Units—Conversion Factors **1055**

Table 18.2
Typical Luminance Values **1069**

Table 18.3
Preferred and Permissible
Luminances **1069**

Table 18.4
Typical Illuminance Recommendations,
CIBSE (UK) **1078**

Table 18.5
Examples of Activities/Interiors
Appropriate for Each Maintained
Illuminance **1079**

Table 18.6
Design Maintained Illuminance Flow
Chart **1081**

Table 18.7
Illuminance Categories and Illuminance
Values for Generic Types of Activities
in Interiors **1082**

Table 18.8
Effect of Weighting Factors on General
Lighting Illuminance Values, for
Illuminance Categories A, B, and C **1082**

Table 18.9
Effect of Weighting Factors on Task
Lighting Illuminance Value, for
Illuminance Categories D through I **1083**

Table 18.10
Typical Reflectances **1089**

Table 18.11
Recommended Maximum Luminance
Ratios **1098**

Table 19.1
Efficacy of Various Light Sources **1108**

Table 19.2
Typical Solar Altitude and Azimuth
Data as a Function of Date and Time of
Day **1112**

Table 19.3
Effect of Wall Reflectance Factor on
Proportion of Internally Reflected
Component (IRC) in Daylight Factor **1122**

Table 19.4
Horizontal Illuminances (E_H) from
Overcast Sky **1124**

Table 19.5
Correction Factors to Be Used in CIE
Daylight Calculations **1128**

Table 19.6
Transmittance Data of Glass and Plastic
Materials **1137**

Table 19.7
Typical Light-Loss Factors for Daylighting Design **1137**

Table 19.8
Coefficient of Utilization from Window
Without Blinds; Sky Component
$E_{xvk}/E_{xhk} = 0.75$ **1138**

Table 19.9
Coefficient of Utilization from Window
Without Blinds; Sky Component
$E_{xvk}/E_{xhk} = 1.00$ **1139**

Table 19.10
Coefficient of Utilization from Window
Without Blinds; Sky Component
$E_{xvk}/E_{xhk} = 1.25$ **1140**

Table 19.11
Coefficient of Utilization from Window
Without Blinds; Sky Component
$E_{xvk}/E_{xhk} = 1.50$ **1141**

Table 19.12
Coefficient of Utilization from Window
Without Blinds; Sky Component
$E_{xvk}/E_{xhk} = 1.75$ **1142**

Table 19.13
Coefficient of Utilization from Window
Without Blinds; Ground Component   **1143**

Table 19.14
Reflectances of Building Materials and
Outside Surfaces   **1144**

Table 19.15
Typical Incandescent Lamp Data   **1162**

Table 19.16
Minimum Required Efficacy for R and
PAR Lamps   **1163**

Table 19.17
Typical Fluorescent Lamp Data   **1172**

Table 19.18
EPACT Minimum Standards for
Fluorescent Lamps   **1172**

Table 19.19
Comparative Characteristics of 48 in.
Tubular Fluorescent Lamps   **1178**

Table 19.20
Cross-Reference of Trade Names for
Major CFL Products   **1179**

Table 19.21
Minimum Required Ballast Efficacy
Factors (BEF)   **1181**

Table 19.22
Typical Data for Mercury-Vapor
Lamps   **1186**

Table 19.23
Typical Data for Clear High-Pressure
Sodium (SON) Lamps   **1191**

Table 19.24
Effect of Illuminant on Object Colors   **1200**

Table 20.1
Approximate Reflection Factors   **1208**

Table 20.2
Coefficients of Utilization for Typical
Luminaires with Suggested Maximum
Spacing Ratios   **1238**

Table 20.3
Percent Effective Ceiling or Floor
Cavity Reflectance (ρ_{CC}, ρ_{FC}) for Various
Reflectance Combinations   **1252**

Table 20.4
Factors for Effective Floor Cavity
Reflectances Other Than 20% (Any
Wall Reflectance)   **1255**

Table 20.5
Wall Luminance Coefficients and Ceiling
Cavity Luminance Coefficients for
Typical Luminaires   **1274**

Table 21.1
Illuminance Recommendations for
General Lighting   **1293**

Table 21.2
Current Illuminance Recommendations
for Specific Residential Visual Tasks   **1294**

Table 21.3
Illuminance Levels for Safety   **1300**

Table 21.4
Lighting Application Guide   **1327**

Table 21.5
Typical Characteristics of Plastic Optical
Fiber Cables   **1331**

Table 22.1
Elements of Residential Signal
Systems   **1348**

Table 23.1
Elevator Capacity and Speed;
Approximate Metric Equivalents   **1389**

Table 23.2
Comparative Energy Use of Various
Motion Control Arrangements   **1401**

Table 23.3
Comparative Characteristics of Elevator
Drive Systems   **1401**

Table 23.4
Recommended Elevator Intervals and
Related Lobby Waiting Time   **1408**

Table 23.5
Car Passenger Capacity (p)   **1409**

Table 23.6
Minimum Percentage Handling Capacities
(*PHC*)   **1409**

Table 23.7
Population of Typical Buildings for
Estimating Elevator and Escalator
Requirements   **1410**

Table 23.8
Office Building Efficiency   **1410**

Table 23.9
Elevator Equipment
Recommendations   **1419**

Table 23.10
Relative First Cost Figures for Passenger
Elevators at Various Speeds (fpm) (for
a Given Drive System)   **1422**

Table 23.11
Office Buildings: Cost of Elevator and
Electric Work   **1422**

Table 23.12
Typical Machine Room Dimensions   **1425**

Table 25.1
Escalator Passenger Capacity   **1478**

Table 25.2
Approximate Maximum Escalator
Rise   **1479**

Table 25.3
Typical Standard-Duty Escalator Motor
Sizes   **1485**

Table 25.4
Relative Escalator Prices   **1487**

Table 25.5
Maximum Permissible Operating Speeds
of Moving Ramps   **1489**

Table 25.6
Moving Walks/Ramps: Commercial
Ratings   **1489**

Table 26.1
Sound Propagation Velocity in Various
Media   **1497**

Table 26.2
Comparison of Decimal, Exponential,
and Logarithmic Statements of Various
Acoustic Intensities   **1499**

Table 26.3
Subjective Loudness Changes and
Corresponding Intensity Level
Changes   **1499**

Table 26.4
Addition of Uncorrelated Sound Pressure Levels   **1502**

Table 26.5
Common Noise Levels   **1505**

Table 26.6
Typical Industrial Noise Levels   **1516**

Table 26.7
Mean Attenuation Data (dBA) For
Ear-Protectors at Test Frequencies (Hz)
Shown   **1518**

Table 27.1
Octave Band Average Sound Absorption Coefficients   **1522**

Table 28.1a
Improvements in STC Rating of Stud
Partitions   **1567**

Table 281.b
STC Ratings of Masonry Walls   **1568**

Table 28.2
Typical STC Values for Doors   **1571**

Table 28.3
Sound Transmission Class (STC) of
Window Construction   **1574**

Table 28.4
Relative Quality of Sound Isolation   **1580**

Table 28.5
Relation Between Barrier STC and
Hearing Condition on the Receiving
Side, Background Noise Level at
NC-25   **1581**

Table 28.6
Typical STC Ratings of Interior
Partitions   **1583**

Table 28.7
Recommended Category Classification
and Suggested Noise Criteria Range for
Steady Background Noise   **1584**

Table 28.8
Articulation Index (AI) and Speech
Privacy   **1591**

Table 28.9
Maximum Air Speeds in Ducts to Yield
NC-15 or NC-25 Background Levels   **1600**

Table 28.10
Maximum Sound Levels: Dry-Type
Transformers   **1606**

Table 28.11
Acoustic Criteria for Selection of
Conventional Core-and-Coil Fluorescent
Lamp Ballasts   **1606**

Table 28.12
Recommended STC for Partitions;
Specific Occupancies   **1609**

Table 28.13
Criteria for Airborne Sound Insulation
of Wall Partitions Between Dwelling
Units   **1610**

Table 28.14
Criteria for Airborne and Impact Sound
Insulation of Floor–Ceiling Assemblies
Between Dwelling Units   **1611**

Table 28.15
Acoustic Units and Conversions   **1616**

Table 28.16
Symbols and Abbreviations Used
Commonly in Acoustics   **1617**

Table A.1
Outside Design Conditions: United
States and Canada   **1624**

Table A.2
Outdoor Design Conditions: Mexico and
Puerto Rico   **1633**

Table B.1
Solar Intensity and Solar Heat Gain
Factors for 16° N Latitude (Conven-
tional Units)   **1636**

Table B.2
Solar Intensity and Solar Heat Gain
Factors for 32° N Latitude (Conven-
tional Units)   **1637**

Table B.3
Solar Intensity and Solar Heat Gain
Factors for 40° N Latitude (Conven-
tional Units)   **1638**

Table B.4
Solar Intensity and Solar Heat Gain
Factors for 48° N Latitude (Conven-
tional Units)   **1639**

Table B.5
Solar Intensity and Solar Heat Gain
Factors for 64° N Latitude (Conven-
tional Units)   **1640**

Table B.6
Solar Intensity and Solar Heat Gain
Factors for 16° N Latitude (SI Units) **1641**

Table B.7
Solar Intensity and Solar Heat Gain
Factors for 32° N Latitude (SI Units) **1642**

Table B.8
Solar Intensity and Solar Heat Gain
Factors for 40° N Latitude (SI Units) **1643**

Table B.9
Solar Intensity and Solar Heat Gain
Factors for 48° N Latitude (SI Units) **1644**

Table B.10
Solar Intensity and Solar Heat Gain
Factors for 64° N Latitude (SI Units) **1645**

Table B.11
Solar Position and Clear Day Insolation,
32° N Latitude   **1646**

Table B.12
Solar Position and Clear Day Insolation,
40° N Latitude   **1648**

Table B.13
Solar Position and Clear Day Insolation,
48° N Latitude   **1650**

Table B.14
Solar Position and Clear Day Insolation,
64° N Latitude   **1652**

Table B.15
Average Insolation, Temperature, and
DD Data   **1655**

Table C.1
Characteristics of Selected Passive Solar
Heating Systems   **1662**

Table C.2
Thermal Properties of Various
Materials   **1665**

Table C.3
Annual Passive Heating Performance:
SSF   **1667**

Table E.1a
Present Value of *n* Future Payments
Beginning at the End of the First
Period   **1713**

Table E.1b
Present Value of *n* Future Payments
Beginning at the End of the First Period
and Escalating at *K* per Period   **1714**

Table E.1c
Present Value of *n* Future Payments
Beginning at the End of the First Period
and Escalating at *K* per Period   **1714**

Table E.1d
Present Value of *n* Future Payments
Beginning at the End of the First Period
and Escalating at *K* per Period   **1714**

Index F.1
Sound Transmission Class: Walls   **1718**

Index F.2
Sound Transmission Loss: Walls   **1719**

Table F.1
Acoustic Characteristics of Walls   **1720**

Index G.1
Sound Transmission Class: Floor/Ceiling
Constructions   **1729**

Index G.2
Impact Insulation Class: Floor/Ceiling
Constructions   **1730**

Index G.3
Floor/Ceiling Sound Transmission and
Construction Data   **1731**

Table G.1
Acoustic Characteristics of Floors   **1732**

Table I.1
These Prefixes May Be Applied to All
SI Units   **1741**

Table I.2
Typical Abbreviations: All Systems of
Units   **1743**

Table I.3
Common Approximations   **1744**

Table I.4
Useful Conversion Factors:
Alphabetized   **1744**

Table J.1
Residential Fenestration *U* Values
and SC   **1748**

Table J.2
Residential Envelope Requirements
Above 10,000 HDD65   **1749**

PART I
ENERGY OVERVIEW

Often mechanical and electrical equipment for buildings is not considered until many important design decisions have already been made. Often, too, such equipment is considered to have a corrective function, which permits a building to "work" in a climate it essentially ignores.

Part I is intended to encourage designers to include both climate and comfort in their early decisions. Chapter 1 discusses the fuel and resource relationship to buildings, from design through demolition. It includes a set of principles to guide regenerative design. Chapter 2 discusses human comfort, the variety of conditions that seem "comfortable," and some implications of a more broadly defined "comfort zone." It includes an introduction to design strategies for lighting, heating, and cooling. Chapter 3 encourages a view of a building site as a collection of renewable resources, to be used and shared in the lighting, heating, and cooling of buildings.

1

CONTEXT FOR BUILDING SYSTEMS DESIGN

The designer of today's buildings will add (or reshape) built spaces on a planet that has, for eons, evolved with a fixed arrival rate of renewable energy from the sun. Suddenly (in geologic time) our planet is experiencing population growth, nonrenewable resource depletion, and apparent global warming. We who design today have available a vast, yet declining, reservoir of material resources and fossil-fuel energy; we design for a rapidly increasing population, each of whom is consuming more resources than did her/his grandparents. Eventually, our planet must live within a fixed budget of renewable energy and material resources. Therefore, how can we best move toward this sustainable goal with our present building designs?

Population growth is both the fundamental source of our work as architects and the fundamental source of our greatest problems. Our planet did not sustain a human population of 1 billion until about 1830, at which time the United States depended almost entirely on the renewable energy sources of fuel wood and work animals; interior lighting was accomplished by burning oil or gas. In less than 200 years, 4 billion more people were added to our planet, whereas the United States shifted to depend almost entirely on nonrenewable fuels: coal, oil, and natural gas (Fig. 1.1). Another 1 billion people are expected before the year 2010. We are involved in the decisions about where they will live and work and how many of what kinds of resources they will use. The mechanical and electrical systems that support our new buildings can be a part of the problem or the solution.

1.1 Buildings and Sites

Our buildings are, after all, temporary occupants of their sites. The arrival of a building usually produces rapid and dramatic changes to the biological systems that evolved on the site. Buildings are guests, sites are hosts; how might they most productively coexist?

The site offers the building earth for support, for a potential heat source and heat sink, and for the growth of plants where building density permits. At some level, an ecosystem of life is already established. Sounds on the site depend on the context, urban or rural. Water is somewhere below the site, flows across it, falls on it as rain, perhaps collects on its surface. Wind moves erratically across the site. Solar energy arrives in diurnal and seasonal cycles. The overall long-term arrival rate of sun, wind, and water is steady, although great variations can occur over shorter time spans.

The building arrives, bringing with it people and vehicles, a flow of materials into and out of the building, sounds generated by activity, and imported utility services such as electricity, water, and natural gas. The building offers the site electric light by night, a continuous flow of heat, and water containing waste or nutrients, depending on one's viewpoint. In our society, this water outflow is most often whisked off for treatment elsewhere.

When the site is larger than the building, this outflow of water might interact positively with the site. Large sites also offer smaller buildings an opportunity to filter sun, wind, sound, and rainwater, making selective use of on-site resources.

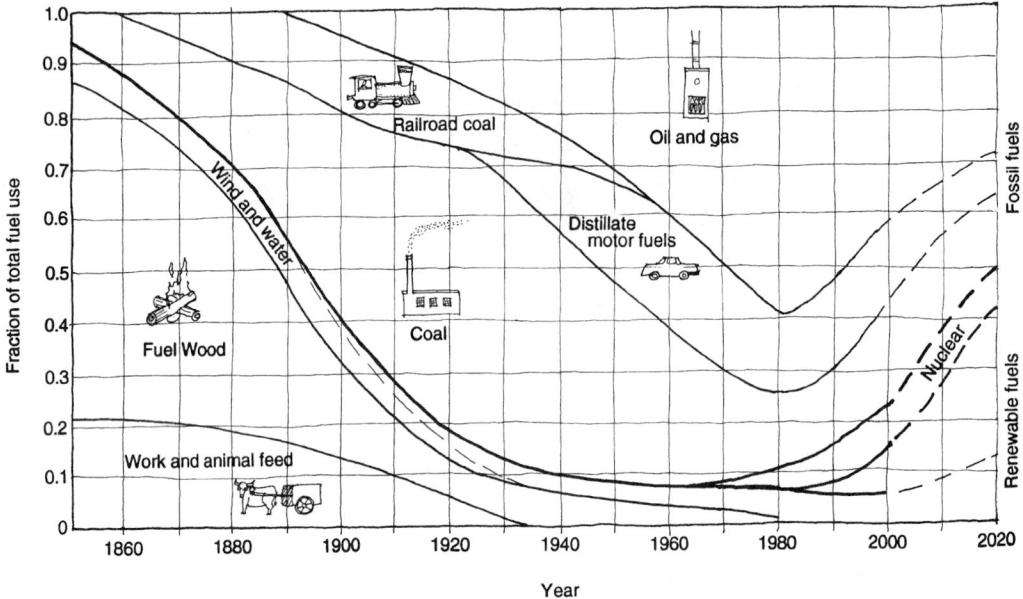

Fig. 1.1 *U.S. fuel input since 1850. This shows our progression from dependence on renewable fuels (wood and work animals) to fossil fuels (coal, then oil and gas). Wind and water power changed their form of use from "mills" to hydroelectricity between 1890 and 1940. Similarly, although not shown here, much fossil fuel is now converted to electricity before use. In the future, a return toward reliance on renewable sources is possible if we invest sufficiently in research and development. [Data 1850–1970 are from Fisher (1974); data 1970–1980 are from Meyers (1983); future projections from Brower (1990). Drawing by Michael Cockram, © 1998 by John S. Reynolds, A.I.A., all rights reserved.]*

Roofs, walls, and floors can play a role in this interaction.

When the building fills the site, on-site resource use is more limited. Because buildings that completely fill sites are so often in densely built-up areas, less wind is available around the walls. The earth is less able to act as a heat sink and treat water. The roof thus becomes the major opportunity to receive the sun and rain and discharge heat to the wind. It may be the only opportunity to grow plants on the site. In dense urban areas, the future roofscape will be dominated by climate–building interactions: solar collection, rain collection, heat rejection, water treatment, and gardens. Compared to today's urban roofscape dominated by horizontal, black or gray surfaces behind parapet walls, this represents a dramatic opportunity for esthetic, social, and technical change.

The site and its building are a part of several much larger systems. In the design of mechanical and electrical systems, the natural systems (climate) and several manufactured systems (building material supply, energy and utilities, transportation, and information) are of primary interest. A detailed look at climate begins in Chapter 2.

1.2 Sources of Materials

On-site sources of building materials are very limited: wood, straw, and earthen construction require huge areas to supply materials for even a small building. The more common choices involve materials brought to the site from some distance: between renewable and nonrenewable materials and between virgin and recycled materials.

Wood is the only renewable material in wide use in North America today. It is easily worked, supports a wide variety of finishes, has moderate structural strength, requires regular maintenance for long life, burns easily, and has a moderate value either as thermal mass or as insulation. This common building material illustrates the impact a rapidly increasing population has on a fixed, even if renewable, resource base. As huge, old trees are harvested to the point of disappearance, growing demand for wood outstrips the supply available from younger, smaller trees. New methods are devised, such as laminates, to allow larger wood members to be constructed of smaller components. The value of salvaged, older wood members increases.

Nonrenewable materials are by far the most commonly used in mechanical and electrical systems; metals and plastics predominate, with some clay products. Their advantages include strength, durability, fire resistance, and either conductivity or resistivity. However, most such materials are obtained at a significant energy cost to mine, transport, and shape them for our use. With rapidly increasing demand for such materials, how is the designer on a planet with fixed resources to respond?

(a) Embodied Energy. One measure is "embodied energy," a complex and therefore elusive indicator of how much energy must be invested to mine or harvest, transport, and refine a unit of building material. Table 1.1 summarizes information on such embodied energy for common units of today's prevalent construction materials. Such apparently simple numbers are complicated by variations in availability of the raw resource (more work to extract requires more energy), variations in distance from raw resource to refining stations, and variations in the fuels used (and their efficiency) in the refining or reshaping processes.

Consider two alternatives for exterior wall surfaces: wood siding and aluminum siding. Wood has embodied energy from a renewable source—the sun. It takes the energy of humans and chain saws to cut the trees, and fuel to haul them, perhaps 100 miles (160 km), to a mill. At the mill, more energy is invested as logs become lumber, and still more energy as lumber becomes finished siding. Aluminum begins as bauxite, which requires energy for its removal from deposits, then more energy to ship great distances to smelters, which use large quantities of electricity in their process. Cheap electricity (as in the Pacific northwest, with its once-surplus hydropower) draws shipped bauxite from thousands of miles away.

Once the aluminum ingots are formed at the smelter, they are then shipped—again, sometimes thousands of miles—to factories that make products such as siding from the ingots. For a given area of finished siding, the aluminum alternative represents about 100 times as much embodied energy as the wood.

(b) Recycled or Virgin Material. It seems paradoxical that while the world's population is increasing and its raw materials are decreasing, the costs of labor are growing so much faster than the costs of raw materials. One consequence is that labor-intensive practices become less economically attractive. Recycling is one labor-intensive option (see examples in Chapter 12), whether at the scale of a household, an office building, or an entire industry.

Building construction, renovation, and demolition demonstrates many opportunities for recycling, but is at present a major waste problem. The U.S. Environmental Protection Agency's 1998 report, "Characterization of Building-Related Construction and Demolition Debris," estimated that 136 million tons per year (123 Mg) of such material is produced, and that 65% to 85% of that ends in landfills.

Consider building demolition. If more of a demolished building can be recycled, more of the energy embodied in its material can be recovered. At present, the recovery of usable materials from demolition is limited because the cost of labor is high and the cost of energy relatively low. Therefore, it is currently easier, quicker, and cheaper to reduce a building to rubble and haul it to a landfill. As landfill capacity becomes scarce, design regulations concerning recycled materials can be expected.

An Atlantic City, New Jersey, demolition was able to recycle 90% of its demolition waste. Of the total of 1,583 tons of demolition waste, only 152 tons were nonrecyclable. Concrete and masonry became crushed aggregate for road building. Glass became "glasphalt" embedded in road surfaces as reflectors. Wood waste became mulch. At Fort Ord in California, four buildings totaling about 11,000 ft² (1,022 m²) were dismantled rather than demolished, saving roof boards, framing lumber, and tongue-and-groove wood flooring. Unpainted drywall was reclaimed for composting.

TABLE 1.1 **Approximate Total Embodied Energy in Building Materials**[a]

Building Material	Volume Btu	Volume MJ	Weight Btu/lb	Weight MJ/kg	Area Btu/ft²	Area MJ/m²	Per Unit Btu	Per Unit MJ
Masonry								
Brick: clay fired[b]			4,000	9.3			14,000/brick	14.8/brick
ceramic glazed[c]							33,400/brick	35.2/brick
Adobe: semistabilized[b]			123	0.3			3,700/block	3.9/block
Concrete block[b]			730–960	1.7–2.24			24,100– 31,800/block	25.5– 33.6/block
Quarry tile[c]					51,000	580		
Ceramic tile[b]					25,160	285		
Concrete								
Ready-mix concrete[d]								
3,000 psi	1,500,000/yd³	2,070/m³						
4,000 psi	1,700,000/yd³	2,346/m³						
5,000 psi	2,000,000/yd³	2,760/m³						
Ingredients:								
Portland cement[b]			2,400–4,000	5.6–9.3				
Sand for concrete, unprocessed[d]	43,200/yd³	60/m³						
Sand, washed	67,500/yd³	93/m³						
Crushed stone, dry, for concrete[d]	180,225/yd³	250/m³						
Slaked lime[e]			2.2	0.005				
Reinforcing steel: 25 mm[e]			12,680	29.5				
Reinforcing bars[c]: #2							2,600/ft	0.84/m
#8							41,800/ft	13.4/m
Welded wire mesh[c]: 2 × 4, 14/14			3,900	9				
2 × 12, 8/8			25,400	59				
Metal Framing								
Steel framing[b]			19,200	44.7				
Steel shapes[c]: W12 × 65, carbon							1,217,800/ft	390/m
W12 × 65, alloy							1,749,200/ft	560/m
WT6 × 27, carbon							543,350/ft	175/m
WT6 × 27, alloy							780,400/ft	250/m
Angles, 50 × 8 mm[e]			14,060	32.7				
Joists, 203 × 152 × 52 mm (1 kg/m)[e]			16,830	39.1				
Aluminum shapes[c]: 8 I 8.81							811,800/ft	260/m
6 I 5.10							469,900/ft	150/m

Material							
Wood Framing							
Wood framing[b]	91,620/ft³	3,400/m³					
2 × 4[b]						3,750/ft	1.2/m
Lumber[c]	160,800/ft³	6,000/m³	5,720	13.3		7,600–9,800/bd ft	(8–10.3/bd ft)
Glue laminated timbers[b]						13,400/bd ft	(14.2/bd ft)
Metal Sheets							
Steel[c]: 22 guage					29,400	333	
16 guage					58,800	667	
Galvanized[c]: 22 guage					49,800	565	
16 guage					98,500	1,118	
Corrugated galvanized iron (0.6 mm)[e]			40,330	93.8			
Aluminum plate[c]: $\frac{1}{4}$ in.					420,700	4,775	
1 in.					1,680,300	19,070	
Stainless steel[c]: cold rolled			138,300	321.7			
hot rolled			80,800	188			
Copper[e]			265,140	617			
Lead[e]			105,620	245			
Wood Products							
Shingles[c]					5,000–5,800	56.8–65.8	7,300/bd ft (7.7/bd ft)
Plywood: $\frac{3}{8}$ in. softwood[c]							10,300–14,300/bd ft (10.9–15.1/bd ft)
Flooring[c]							17,900 bd ft (18.9/bd ft)
Mouldings[c]					67,500	766	
Mineral surface insulating board[c]							
Roofing (See also Wood Products)							
Asphalt shingle[b]: self-sealing					29,730	337	
regular strip					25,330	288	
Rolled roofing[c]					7,800–11,000	89–125	
Saturated felt[b] 15 lb					1,840	21	
30 lb					3,680	42	
Plaster and Lath							
Lath board[b]					2,600	29.5	
Steel lath[b]					2,600	29.5	
Gypsum board[b] $\frac{3}{8}$ in.[c]			12,000	27.9	5,800	60.2	

(continued)

7

TABLE 1.1 Approximate Total Embodied Energy in Building Materials[a] (continued)

Building Material	Volume Btu	Volume MJ	Weight Btu/lb	Weight MJ/kg	Area Btu/ft²	Area MJ/m²	Per Unit Btu	Per Unit MJ
Glass[b,g]			6,750–7,500	15.7–17.4				
Flat glass: double strength					15,430	175		
tempered					72,600	824		
Plate and float glass[c]: $\frac{1}{8}$ to $\frac{1}{4}$ in.					48,000	545		
Laminated plate glass[c]: $\frac{1}{4}$ in.					212,500	2,412		
Thermal Insulation								
Fiberglass[b]			13,000	30.2				
Polystyrene[b]			50,400	117				
Polyurethane[b]			31,000	72.1				
Mineral wool, $4\frac{1}{2}$ in.[c]					8,300	94.2		
Acoustical Ceiling Systems								
Steel suspension systems[b]			19,200	44.7				
Aluminum suspension systems[b]			103,500	241				
Flooring (see also Wood Products)								
Linoleum[b]			7,350	17.1				
Vinyl[b]			22,560–27,730	52.2–64.5				
Vinyl composition tile[b]			5,900	13.7				
Modified resin vinyl tile[e]					68,370	776		
Finishes								
Water-based paint[b]	503,670/gal	1,400/L	33,303	76.8				
Stains and varnishes[b]	437,000–508,500/gal	1,220–1,420/L						
Paints[c]								

[a] These numbers are rounded in most cases.
[b] From Environmental Resource Guide (1996).
[c] From Hannon et al. (1977).
[d] From Construction Technology Laboratories, Martha Van Geem, P.E. (correspondence, 1997). Metric conversions by the author.
[e] From D. K. Irurah (1997) An embodied energy algorithm for energy conservation in building construction as applied to South Africa, Ph. D. thesis, February 1997, University of Pretoria.
[f] Plaster information not yet available.
[g] Glass products currently available with 30% less embodied energy; under development with 60% less embodied energy.

Opportunities thus include crushed wallboard as a replacement for lime as an agricultural land conditioner, carpet ground up for attic fill insulation, plate glass crushed for use in fiberglass insulation, and pulverized wood as an aid to composting at sewage sludge treatment facilities. Used acoustic ceiling tiles can become part of the slurry from which new acoustic tiles are made. Building materials are now increasingly made from recycled materials: reinforcing bars from ferrous scrap metal; cellulose insulation from newsprint; parking lot bumper strips, fence posts, and park benches from recycled plastics; and nonstructural concrete from incinerator ash. Even plastic yogurt containers, complete with their scraps of aluminum foil, are made into a terrazzo-like floor tile.

Architect Pliny Fisk has developed a wide array of such applications. His Advanced Green Builder home near Austin, Texas, shows several applications of "ashcrete," made with coal fly ash and bottom ash, producing a 97% recycled-content concrete. This is used as ferro-cement for columns and beams and is foamed for hollow wall infill. There are a wide variety of other innovative recycled-material applications as well.

The purest recycling involves the reuse of a building component "as is." A residential demolition-by-hand salvage project in Portland, Oregon, recovered doors, windows, bathroom fixtures, framing lumber, plywood, siding, flooring, and bricks; the energy and economic summary is shown in Table 1.2.

1.3   Sources of Energy

Renewable fuel sources are those that are available indefinitely but arrive at a relatively fixed rate. For example, the influx of solar energy varies from day to day, but on average it continues at a steady rate. A woodlot produces a limited amount of wood per year, but does so for centuries if properly managed. A popular analogy for renewable fuel sources is a fixed-but-steady monetary income, such as a salary with no raises.

Using the same analogy, nonrenewable fuel sources (such as coal, oil, and natural gas) are like savings accounts that draw no interest—once spent, they are gone.

The United States, like other industrialized countries, has spent the time since the 1840s in an energy transition (see Fig. 1.1). We began with renewable energy sources, obtained locally, that did relatively low-grade work: animals pulled or pushed, and wood was burned to yield heated air, water, or steam. Buildings and people were directly affected by energy sources; work animals were fed, tended, and housed; fuel wood was cut nearby and stacked in large sheds (Fig. 1.2); fireplaces were social centers of the buildings they warmed; and smoke from chimneys indicated activity inside (and fouled the air outside). Fuel use depended on human labor. Architects of the era responded to the visual appeal of fireplaces and chimneys (Fig. 1.3).

We are now almost entirely dependent on nonrenewable energy sources, an increasing portion

TABLE 1.2 **Residential Salvage for Reuse**

	Embodied Energy,[a] BTU/ft^2 floor area	Value, US \$/ft^2 floor area
Total for reusable salvage[b]	46,890	4.90
Demolition energy consumed[c]	−3,380	
Value of energy embodied in salvage[d]		+.50
Value of avoided dumping fees		+2.70
Total energy savings and value	43,510	8.10

Source: Joslin, J. et al. "The Waste Papers: Analysis and Discussion of the Potential for Salvage and Re-Use of Construction Materials from Residential Demolition," Metro Solid Waste Department Publications, Portland, OR, 1993.

[a]Based on Stein, R. et al. *Handbook of Energy Use for Building Construction,* U.S. Department of Energy, 1981.

[b]Framing lumber alone constituted 38% of this embodied energy.

[c]Gasoline for transportation and hauling, plus human labor at 254.6 BTU/h.

[d]Assumed at \$.04/kWh, very low rate typical of the Pacific Northwest.

ENERGY OVERVIEW

(a)

(b)

Fig. 1.2 *Residential heating, past and present. (a) The house dependent on fireplaces or wood stoves also depends on someone to tend the fire. The warmer area near the fire in this early Oregon farmhouse was used for social purposes; the colder extremities served for sleeping areas and for storage of food and fuel. [Based on a plan drawn by Philip Dole.] (b) The contemporary suburban home has either a small furnace area or electric heat built into each room. Heating equipment is no longer a major influence on building form.*

Fig. 1.3 *The fireplace and the more efficient wood stove can inspire architectural form. This chimney symbolizes permanence as well as protection from the cold. The major social space of the house is represented both by the arched window and by the fireplace chimney. (Photo by William Johnston.)*

of which is imported oil and natural gas. This is partly due to rapid growth of both population and per capita energy consumption. It is also due to the highly concentrated energy available in fossil fuels, which encourages high-grade uses such as generating electricity and other industrial processes. Buildings and people are now largely oblivious to their energy sources: electricity is generated in far-off power plants, natural gas arrives from buried pipelines. Fuel use is regulated by automatic controls, and heating and cooling equipment is stored out-of-sight. If we neither personally control the thermostat nor pay the utility bill, we have no direct contact with energy sources. Architects pay engineers to design, and

then to successfully hide, mechanical and electrical equipment.

Our buildings designed today are likely to rely heavily on electricity (Fig. 1.4), which carries serious implications about resource depletion and environmental quality, for several reasons:

1. Consumption of electricity is expected to rise about twice as fast as the overall energy demand, and more often we are expected to use electricity in place of other energy forms. Part of the reason for this is that, for some energy sources such as coal, conversion to electricity before distribution to buildings is the most convenient option.

2. Other than daylight, electricity is almost the only source of illumination in many buildings. Furthermore, the heat produced by electric lights reduces the building's need for space heating, but increases its need for space cooling, and mechanical cooling is typically provided by electric air conditioning equipment.

34.4 Quads in U.S. Buildings[1]

Electricity wasted in production

15.77%

Electricity delivered

7.05%

2.24% Other[3]

0.65% Renewables[2]

Natural Gas

8.68%

Industry

34.9%

Transportation

24.5%

Fig. 1.4 Sources of energy as consumed by end-use sector in the United States, 1996. Total: 94.0 Quads (1 Quad = 10^{15} Btu). Not included are raw materials used in manufacturing. (1) Both residential and commercial buildings are included. (2) Renewables do not include passive solar energy use; hydroelectricity and PV are included in "electricity." (3) "Other" includes fuel oil, liquefied natural gas, coal, kerosene, and other petroleum products. (Data from U.S. Office of Building Technologies, Core Databook, *1998. Drawing by Michael Cockram, © 1998 by John S. Reynolds, A.I.A., all rights reserved.)*

3. Electricity is a convenient and versatile energy form; it not only serves such high-temperature and highly concentrated (or "high-grade") tasks as lighting and motive power, but it can also serve the low-temperature simpler (or "low-grade") tasks, such as cooking, heating water, and space heating (Table 1.3). "All-electric" buildings have become commonplace, even though they are subject to paralysis in blackouts, as is any building dependent on a single energy source. In Table 1.3, out of a total of 34.40 quads used for all building energy sources, the primary energy for electricity generation used 22.82 quads, equal to two-thirds of the total.

4. Electricity generated by thermal processes (except for cogeneration) delivers less than one-third the total energy that goes into its production; more than two-thirds is usually lost as waste heat at the generating plant. (In Table 1.3, for every 1 quad of electricity delivered, 3.24 quads were assumed used in generation.)

As we remove our planet's raw materials, including fossil fuels, we look to a future return to renewable and sustainable energy. This time we will be aided by solar energy available directly as electricity (photovoltaics) produced on or very near the building it serves. Solar collectors (some for water heating, others for electricity) will shape the roofs and therefore the silhouettes of buildings. Architects can choose to hide or celebrate such opportunities. We will store and distribute hydrogen as a high-grade fuel, produced from water by using electricity from renewable sources such as solar thermal and wind. The mechanical and electrical equipment we specify today will probably operate in such a future within their lifetimes. In this near future, "lower-tech" processes such as biomass (combustion of wood and waste products) may develop faster than "higher-tech" processes such as photovoltaics (PV). However, PV is growing very rapidly; its growth curve is as steep as that of computer technology during its first 15 years. One major oil company's energy scenario (in the late 1990s) anticipates 50% of the world energy to be met by alternatives to fossil fuels by the year 2050.

The designers of the Montana State University EPICenter (Fig. 1.5) in Bozeman, Montana, chose to both recognize and celebrate a close building–site relationship with a renewable-fuel future. This ambitious building, a combination of classrooms and laboratories, presents a wide variety of settings for the exchange of information and ideas, a building "where sustainable design principles are evident and delightful; the building itself as a teacher." The "cartoons" presented here, "Life of Electricity" and "Life of Heat," are part of a series of such analyses, more of which is seen later in this chapter (Fig. 1.8), and again in Chapters 6 (Fig. 6.25) and 11 (Fig. 11.45).

Our fossil-fueled economy seems so entrenched that it defies a transition to renewable energy. A common question: Is solar energy ade-

TABLE 1.3 Energy End-Use in U.S. Buildings, by Fuel Type (Quads)[a]

End-Use	Natural Gas	Fuel Oil[b]	LPG	Other Fuel[c]	Renewable Energy[d]	On-Site Electric	On-Site Total	On-Site %	Primary Electric[e]	Primary Total	Primary %
Space heating[f]	5.38	1.20	0.31	0.24	0.62	0.73	8.49	45.6	2.38	10.14	29.5
Space cooling[g]	0.02					0.97	0.99	5.3	3.13	3.16	9.2
Ventilation[h]						0.17	0.17	0.9	0.54	0.54	1.6
Water heating[i]	1.88	0.17	0.07		0.02	0.54	2.68	14.4	1.75	3.89	11.3
Lighting						1.49	1.49	8.0	4.83	4.83	14.0
Refrigeration						0.68	0.68	3.6	2.19	2.19	6.4
Cooking	0.34		0.03			0.16	0.54	2.9	0.53	0.90	2.6
Wet clean[j]	0.05					0.27	0.31	1.7	0.86	0.91	2.6
Office equipment						0.27	0.27	1.4	0.87	0.87	2.5
Color television						0.21	0.21	1.1	0.67	0.67	2.0
Miscellaneous[k]	0.26	0.02	0.08	0.03		1.02	1.41	7.6	3.29	3.68	10.7
Other[l]	0.75	0.09				0.55	1.40	7.5	1.77	2.62	7.6
Total	8.68	1.48	0.50	0.27	0.65	7.05	18.63	100	22.82	34.40	100

Source: Office of Building Technologies, Core Databook, June 1998.

[a]Quad = 10^15 Btu. Data are for the year 1996.

[b]Includes distillate fuel oil (1.34 quads) and residual fuel oil (0.15 quad).

[c]Includes distillate fuel oil (1.34 quads) and residual fuel oil (0.15 quad).

[c]Kerosene (0.11 quad) and coal (0.13 quad) are attributed to space heating; motor gasoline (0.03 quad) attributed to miscellaneous end-uses.

[d]Passive solar space heating is not included. Includes wood (space heating, 0.61 quad), geothermal (including space heating, 0.01 quad), solar (water heating, 0.02 quad) and biomass (electric generation, 0.003 quad).

[e]Site-to-source electricity conversion = 3.24, due to generation and transmission losses.

[f]Includes electric furnace fans (0.12 quad) and district services of 0.28 quad natural gas, 0.06 quad distillate oil, and 0.03 quad electric.

[g]Includes district services of 0.001 quad natural gas and less than 0.001 electric.

[h]Commercial only (residential fan and pump energy use included proportionally in space heating and cooling).

[i]Includes district services of 0.11 quad natural gas, 0.02 quad distillate fuel oil, and 0.008 quad electric.

[j]Includes clothes washers (0.03 quad), natural gas clothes dryers (0.05 quad), electric clothes dryers (0.19 quad), and dishwashers (0.05 quad).

[k]Includes commercial service station equipment, emergency electric generators, fuel oil cooking, natural gas-driven pumps, natural gas lighting, automated teller machines, telecommunications equipment, medical equipment, residential pool/hot tub heating, outdoor grilles, and outdoor natural gas lighting.

[l]Attributable to commercial buildings, but not directly to building end-uses.

Fig. 1.5 The proposed EPICenter at Montana State University (Bozeman) is an ambitious integration of systems that support a building of classrooms and research laboratories. (a) Section looking west; the central commons are flanked by the solar aquatic facility on the south and the mechanically conditioned research labs on the north. (b) "Life of Electricity" and (c) Section describe the building's interaction with the utility grid as it generates a portion of its own electricity. (d) "Life of Heat" and (e) heating section, and (f) cooling section describe the building's emphasis on renewable energy and on reclaiming otherwise "waste" heat. (Courtesy of Place Architecture LLC, Bozeman, Montana, and Berkebile Nelson Immenschuh McDowell, Architects, Kansas City, Missouri.)

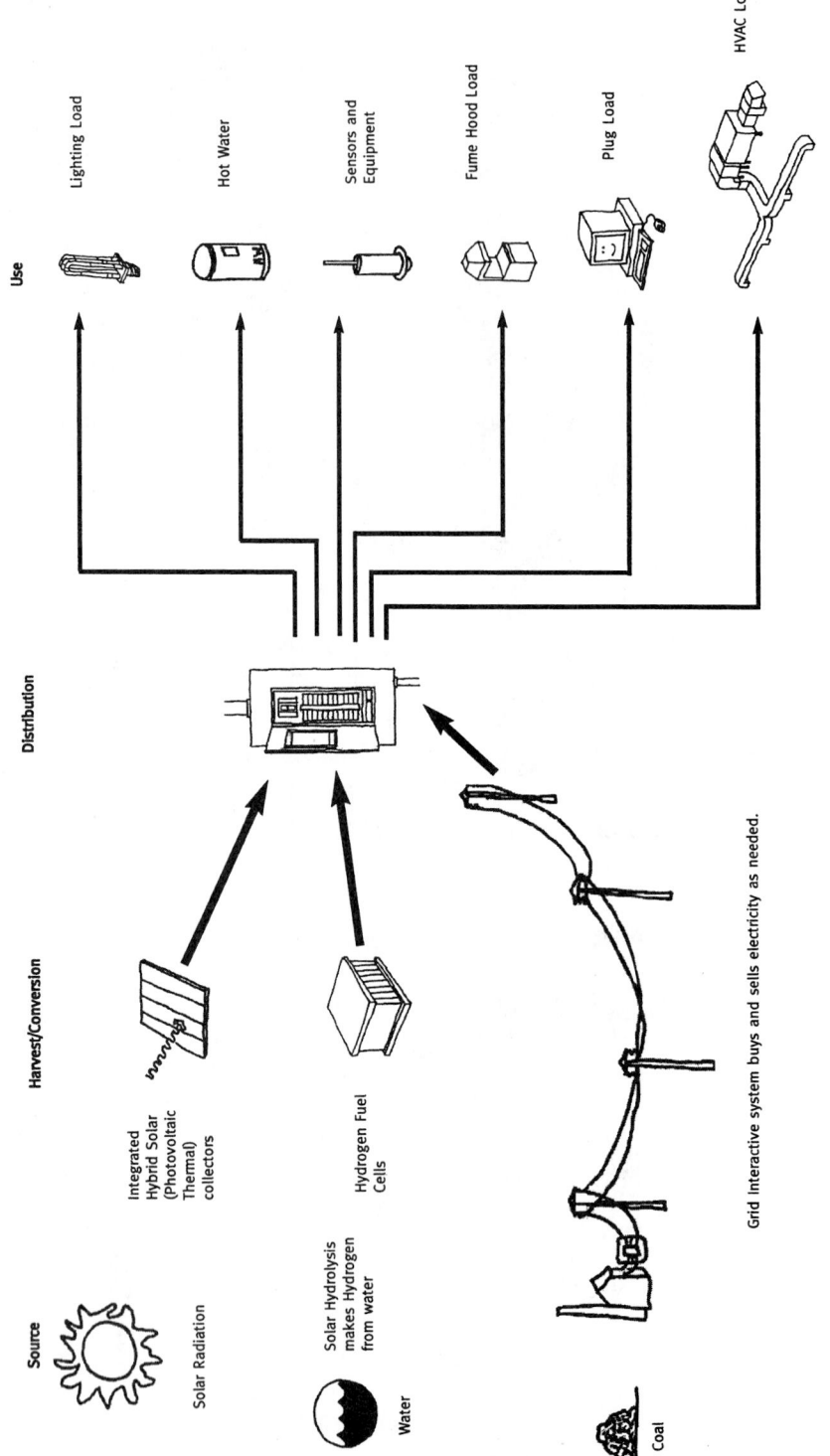

Source **Harvest/Conversion** **Distribution** **Use**

Solar Radiation

Integrated Hybrid Solar (Photovoltaic Thermal) collectors

Lighting Load

Hot Water

Sensors and Equipment

Fume Hood Load

Plug Load

HVAC Load

Water

Solar Hydrolysis makes Hydrogen from water

Hydrogen Fuel Cells

Coal

Grid Interactive system buys and sells electricity as needed.

(b)

Fig. 1.5 *(continued)*

14

SLOPE SKYLIGHT PV

ELECTRICITY FOR LIVING MACHINE

ROOF INTEGRATED PV

INVERTER

POWER DISTRIBUTOR

ELECTRICITY FOR BUILDING

WATER

ELECTROLIZER

PHOTON EXCHANGE MEMBRANE FUEL CELL

HYDROGEN STORAGE

HYDROGEN FUEL CELLS (PHOSPHORIC ACID)

ELECTRICITY GENERATION

(c)

Fig. 1.5 *(continued)*

Fig. 1.5 (continued)

(e)

(f)

Fig. 1.5 (continued)

quate for our energy needs? Table 1.4 compares the earth's receipt of solar energy at the surface in a single day with some other energy phenomena.

1.4 Transportation Systems

The link between transportation systems and mechanical/electrical equipment may seem ob-

scure, but consider the impact on buildings of the automobile and its internal combustion engine. Fresh air intakes at street level face significant pollution from engine exhausts. Parking below buildings, often at several levels and necessitating either long ramps or sloped floors, competes for basement space with heavy mechanical and electrical equipment such as boilers, chillers, and ice storage tanks. Sloped floors make future use for

TABLE 1.4 **Daily Arrival of Solar Energy on Earth Compared to Other Energy Quantities**

Solar energy received each day	1
Melting of an average winter's snow during the spring	1/10
A monsoon circulation between ocean and continent	1/100
Use of energy by all mankind in a year	1/100
A midlatitude cyclone	1/1,000
A tropical cyclone	1/10,000
Kinetic energy of motion in earth's general circulation	1/100,000
The first H bomb	1/100,000
A squall line containing thunderstorms and perhaps tornados	1/1,000,000
A thunderstorm	1/100,000,000
The first A bomb	1/100,000,000
The daily output of Boulder Dam	1/100,000,000
A typical local rain shower	1/10,000,000,000
A tornado	1/100,000,000,000
Lighting New York City for one night	1/100,000,000,000

Source: Reprinted by permission from Lowry (1988).

other purposes quite difficult. Fumes from automobiles require large fans, and energy to serve them, to ventilate parking levels. Unheated subterranean parking areas break much of the thermal link between basement floors and the earth, enhancing earth's value as a heat sink for mechanical equipment discharge.

In a future of electrically powered vehicles, PV arrays over rooftop parking areas can provide electricity for the building and recharge the batteries of the parked cars of workers throughout the day (Fig. 1.6). In a future of hydrogen fuel cell–powered vehicles, engine discharge consists simply of water. No fumes are emitted from such future cars, saving fan space and energy. To the extent that cars become smaller, or are replaced by public transit or bicycles, significant built space may be reclaimed from parking for other uses—providing the floors are not sloped.

Where entire buildings are now built as single-purpose parking garages, placing all ramps in the center of such a structure, serving flat parking floors, might allow for the future removal of the ramps and their replacement with a daylight atrium. Thus, the building can be renovated rather than demolished to serve new functions. However, this assumes that sufficient head room is available for building services, which in turn might cause the original parking garage to have longer ramps to serve greater floor-to-floor heights.

1.5 Information Systems

Control systems for mechanical and electrical equipment have become much more sophisticated, thanks to developments in information systems. With the advent of "smart houses," "intelligent buildings," and "smart appliances," it is now possible to regulate an array of building systems from great distances, allowing an unoccupied building to save energy by being minimally heated or cooled. With advance remote signals from its absent occupants, the building can reach thermal comfort levels by the time they arrive. For a fully occupied building, with some zones requiring heating and other zones cooling, some zones with available daylight and others without, an automatic central operating system can, without human intervention, integrate the control of fresh air quantities, sunlight through movable shading devices, and intensity of electric lighting to achieve the maximum amount of on-site, renewable energy used by such a building.

The Albany, New York, County Airport uses automated controls to regulate solar gain through its large central skylight (Fig. 1.7). The computer monitors indoor and outdoor temperatures, keeps track of solar altitude and azimuth, then regulates the insulated shading louver position according to the building's need for heating or cooling. Solar gain is stored in the masonry wall that supports

Fig. 1.6 *The winning entry to a competition (sponsored by the New York Power Authority and the New York State Association of Architects) features a canopy of PV (solar) cells over a 19-acre (7.7 hectare) parking lot. In a future of electric-powered automobiles, such arrays can charge the batteries of parked cars, trucks, and busses while keeping them cool. (Courtesy of Kiss Cathcart Anders Architects, New York, New York.)*

(a)

① The operable louvers are open to allow
   sunlight into the building.

② The sunlight warms a brick wall at the back of
   the skylight.

③ Air is drawn through a plenum behind the wall,
   and is heated during its passage over the
   warm bricks.

④ The heated air is drawn into the building
   heating system for distribution.

⑤ Natural light is provided to the space below.

① The operable louvers are closed to provide an
   insulated "ceiling" below the glass and reduce
   heat loss through the skylight.

② The louvers are insulated, blocking the flow of
   heat from the building.

③ The brick wall remains warm for some time
   after sunset due to its "thermal mass".

④ Air is drawn through a plenum behind the wall,
   and is heated during its passage over the
   warm bricks. When the heat stored in the brick
   wall is used up, air flow through the plenum is
   stopped.

⑤ The heated air is drawn into the building
   heating system for distribution.

① The operable louvers are partially open.

② Direct sunlight is reflected by the louvers.

③ Diffuse northern light is provided to the spaces
   below.

④ The area below the skylight serves as a heat
   banking space. Warm air from the building
   below will collect beneath the skylight.
   Exhaust fans will then discharge the warm air
   from the building.

Sunny Winter Day

The sun is used to
provide heat and light.

Winter Night

Loss of heat through the
skylight is reduced.

Sunny Summer Day

The sun is used to
provide light, but is not
allowed to penetrate into
the building and
generate excessive heat.

(b)

Fig. 1.7 *The Albany, New York, County Airport features a central skylight* (a) *that provides 40% of
the light and 20% of the heat for the 57,000-ft² (5,296-m²) building.* (b) *The insulated louvers are
controlled by computer to admit or block the sun and to store winter night heat within the building.
(Courtesy of Einhorn Yaffee Prescott, Architects, Albany, New York.)*

the skylight. A plenum behind the wall then heats air to be supplied to the vestibule areas of the airport, where winter heat losses are greatest. Photoelectric controls turn off electric lights when daylight is adequate. The skylight meets 40% of the lighting and 20% of the heating needs of this 57,000-ft^2 (5295-m^2) building.

The EPICenter building at Montana State (see Fig. 1.5) recognizes the role of information systems in the "Life of a Building" cartoon (Fig. 1.8). Because of the complexity of the systems interacting in this project, computer monitoring is particularly helpful. The post-occupancy evaluation (POE) aspects reinforce the design idea of "the building as a teacher"; with time and experience, operations become more efficient.

Information systems promise enormous energy conservation achievements while using very small amounts of energy themselves. They also have one of the least space-consuming distribution systems, or "distribution trees," of all building service systems, especially compared to air ducts and plumbing distribution trees [see Chapters 17 (Electrical Wiring Design) and 22 (Signal Systems)]. In return for such agreeable characteristics, architects must recognize that developments in information technology are so rapid that the nature of these distribution systems is likely to undergo frequent and dramatic change. The distribution tree may be quite small, but it should be highly accessible. Where information can be transmitted without wires or cables, the impact on building service space is even smaller. Such information systems can make more feasible the renovation, rather than demolition, of older buildings for new tenants.

1.6 Design Strategies for Sustainability

A rapid transition to renewable materials and energy sources provides architects with varied opportunities to integrate site and building design with components and processes, many of which are quite new. Some strategies to help designers achieve this site/building/component integration were proposed by architect and landscape architecture professor John Tillman Lyle in his book *Regenerative Design for Sustainable Development* (1994). Those strategies most applicable to the

design of mechanical and electrical systems are presented here, along with a guide to the chapters in this book where such strategies may be applied.

(a) Let Nature Do the Work. Designers can often find ways to use natural processes on-site, where they occur, in place of dependence on (or supplementary to) services that are remote. Smaller buildings on larger sites are particularly good candidates for this strategy. Chapter 3 contains many examples of on-site energy resource usage. Chapter 9 has the example of a California residential development that retains rainwater, rather than dumping it in a storm sewer, designed in the early 1980s and still functioning successfully (see Fig. 9.19).

This strategy, and the nine that follow, were guides to the design of the Center for Regenerative Studies, at the California Polytechnic State University at Pomona (near Los Angeles), California (Fig. 1.9). [Extensive descriptions can be found in Lyle (1994).] The first stage of construction was completed in 1993. A 16-acre (6.5-hectare) site supports facilities for integrated studies of shelter, energy, food, water, and waste. Students study and live in buildings that are arrayed along a south slope, with terraces that support vegetables, vines, and fruit trees; these in turn provide shading for south windows in summer. Cross and stack ventilation (see Chapter 5) provide fresh air and night cooling.

Water and its waste from the kitchen and bathrooms is taken to treatment ponds, where it is taken up by microorganisms and aquatic plants (see Chapter 11 for similar systems). The enriched water in these ponds is used for irrigation crops on this site, in turn providing food for the kitchen. Fish from these ponds also become food. Aquatic plants grow rapidly, and are harvested for two purposes: to be digested for biogas (eventually to be used for cooking) and to feed animals, whose meat, eggs, and milk feed the residents. Manure from the animals also becomes fertilizer for the crops.

(b) Consider Nature as Both Model and Context. A look at present and past editions of this textbook reveals a strong reliance on physics to guide design: heat flow, water flow, electricity, light, and sound follow rules established by phys-

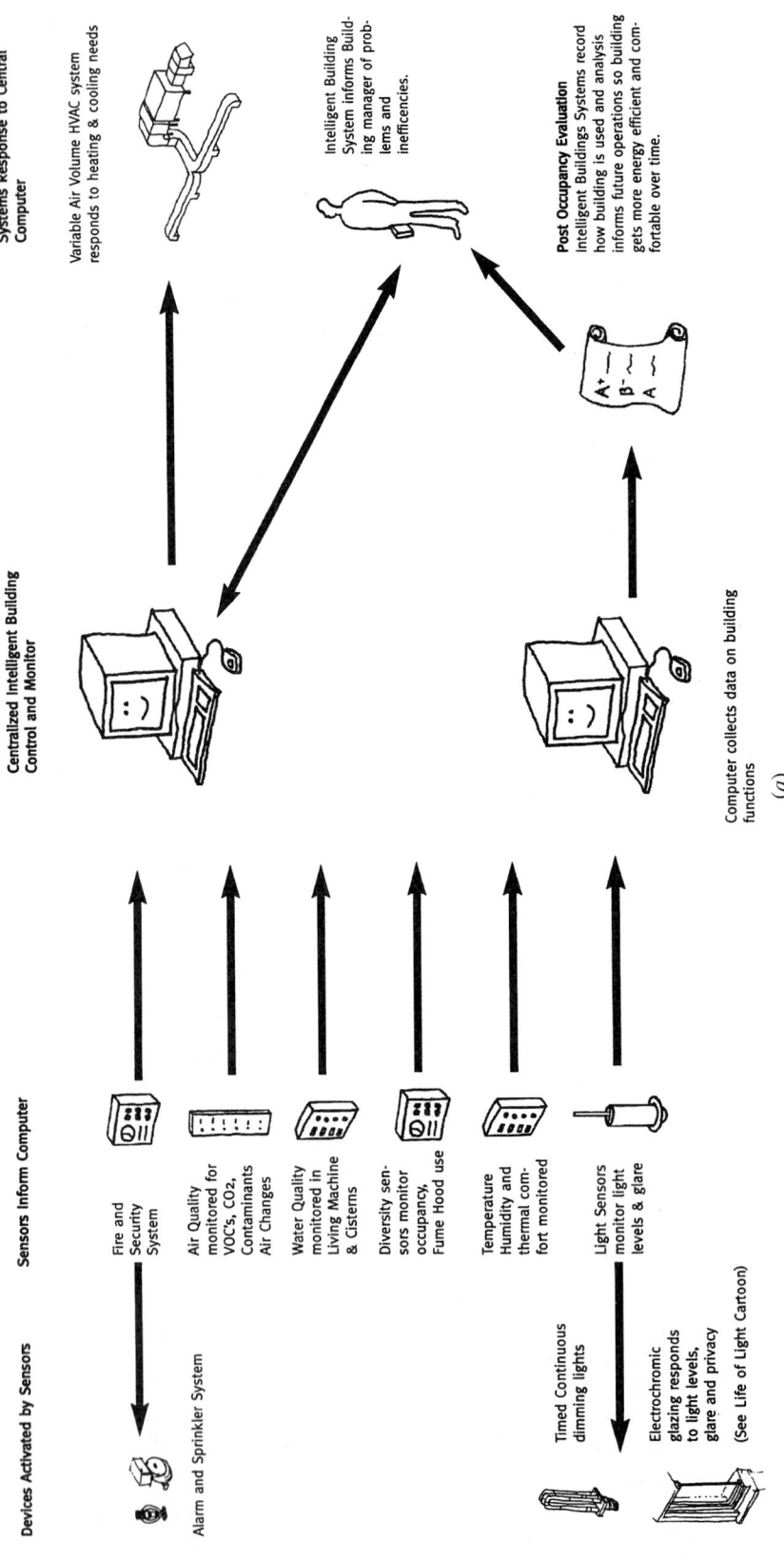

Fig. 1.8 The proposed EPICenter at Montana State University will rely on an extensive monitoring system to optimize the performance of its many innovative systems. (a) "Life of Building" diagram and (b) section describe the controls and equipment involved. (Courtesy of Place Architecture LLC, Bozeman, Montana, and Berkebile Nelson Immenschuh McDowell, Architects, Kansas City, Missouri.)

(b)

Fig. 1.8 *(continued)*

- - - POTABLE WATER SUPPLY
━━━ RECLAIMED WATER SUPPLY
PUMPING TO KNOLLTOP
● HOLDING TANK
← IRRIGATION DISTRIBUTION
●━ AMMONIA TREATMENT

WATER FLOW PLAN

⬙ RACEWAYS
◆ NURSERY PONDS
▬ GROWOUT PONDS
HAND-TECH PONDS
RESERVOIR
DRAIN TO STREAM

••ǀ• PRETREATMENT
•••••• SEWAGE
POND DRAIN
BYPASS TO SEWER
SEWAGE TREATMENT

(a)

Fig. 1.9 *The Center for Regenerative Studies at California State Polytechnic University, Pomona.* (a) *Site plan, tracing the path of imported fresh water, imported reclaimed water (from Pomona Water Reclamation Plant), as well as sewage from the buildings, and the integrated treatment systems.* (b) *Aerial sketch of eventual build-out, seen from the west.* (c) *Detail plan of first phase buildings and adjacent ponds.* (d) *Looking north across a pond to the vine-shaded balconies of living units; ducks and chickens live in the shade below the building.* (e) *Sections through pond-side and hillside living units. [Drawings from Lyle, John Tillman (1994).* Regenerative Design for Sustainable Development. *John Wiley & Sons, Inc., New York.]*

23

(b)

(c)

Fig. 1.9 *(continued)*

ics. This strategy looks to nature for biological, more than physical, models of integral design, as in the case of the Center for Regenerative Studies. A particularly revealing example is the Santee Water Reclamation Plant in arid southern California (see Fig. 11.50). Also in Chapter 11, new on-site waste water treatments are shown that significantly reduce the pressure on central municipal sewage treatment systems while returning nutrients to the site. "Waste" water is diverted from a problem to be solved elsewhere to a local opportunity.

(c) Aggregate Rather than Isolate. In today's conventional designs for nonresidential buildings, those with high expectations of thermal comfort are kept within a narrow range of air temperatures, best achieved by isolating the build-

(d)

(e)

Fig. 1.9 (continued)

ing from the highly variable exterior climate. Windows are sealed, and fresh air is filtered and/or humidified or dehumidified at a central location, then ducted to each space. In contrast, this strategy asks designers to allow open windows to admit light, sun, and fresh air directly when appropriate; to make the building–site relationship an obvious one. Larger sites with smaller buildings are easier targets for this strategy, yet new and large buildings are turning "back" to this approach [see England's Inland Revenue Centre, Fig. 5.48, and Berry, et al. (1995)]. Control of humidity and dust or pollen may have to be exchanged for a wider personal choice of temperature and access to outdoor air, although very high humidity and particulates may intervene. Many designs that rely on site-based resources, such as passive solar heating and passive cooling, are found in Chapter 5.

In England's cold and cloudy winter, a solar-heated school has performed for decades with back-up heating only from its occupants and electric lights (Fig. 1.10). Classrooms have high ceilings to better distribute daylight and are constructed of thermally massive materials. Fresh air is occupant controlled, with cross ventilation entering through the multiple-position window in the south wall and leaving through simple high openings in the classroom's north wall to the single-loaded north corridor, then through similar openings in the corridor's outside wall. In its most

(*a*)

Fig. 1.10 *St. George's School at Wallasey, England, is an early example of a passive solar building.* (a) *The south wall admits light, solar heat, and fresh air; a single-loaded corridor on the north side completes the air flow through the building.* (b) *Movable panels behind the solar wall allow sun to either pass through or be reflected and help to retain stored heat by night. (Photos by Reg McDonald.) The operable portions of the south wall pivot on the horizontal axis at the center of the window.* (c) *The most-closed position, when no air flow directly through the window is desired (very cold outside).* (d) *Minimum fresh air enters at waist and head levels with the window in the vertical position.* (e) *More fresh air enters as the window top tilts in toward the room. The window can be locked in this position for night ventilation with security.* (f) *Maximum ventilation is achieved with the window nearly horizontal.*

(*b*)

(c) *(d)*

(e) *(f)*

Fig. 1.10 *(continued)*

"closed" position, the classroom can receive ventilation air that has been prewarmed in the large cavity between the south glass walls. (A contemporary variation on the double glass wall, but one that isolates more than aggregates, can be seen in the Occidental Chemicals office building, Fig. 6.11.)

(d) Match Technology to the Need. Here we are asked to avoid using high-grade resources for low-grade tasks. It is obviously wasteful to flush toilets with purified water, but how impure can water be that is acceptable for irrigation? See Chapter 11 for some variations. Under what circumstances might we justify using the high-grade energy of electricity to do a simple, low-grade task such as space heating (Fig. 1.11)? In this example, the high-grade energy is more appropriate for a low-grade task when used in a higher-

technology device—the heat pump—rather than the much more simple electric resistance heater. See Chapter 7 for details on such heating and cooling components.

(e) Seek Common Solutions to Disparate Problems. Do special design efforts and devices for sustainability always cost more initially, even if they save some costs of energy later in operation? The strategy of making a "special" design feature serve multiple tasks is one way to counteract the problem of high first cost. This can be as simple and low technology as using the heat from garden composting to help warm a greenhouse; it can be complex and high technology, like the office building in Chapter 8 that uses the fire sprinkler system to exchange heat among the heat pumps serving many thermal zones (see Fig. 8.72). A broadly integrated example is the office build-

ENERGY OVERVIEW

Fig. 1.11 Variations on higher-grade energy and lower-grade tasks. (a) *The fossil fuel natural gas is often burned in furnaces to provide low-grade space heating. With today's high-efficiency furnaces, about 80% of the gas' energy is delivered to the building as space heat.* (b) *However, when that natural gas is instead used to generate (higher-grade) electricity, and then the electricity is used for resistance space heat, the inefficiencies at the electric power plant cut deeply into the available energy: only about 27% is delivered to the building.* (c) *However, when the electricity generated by natural gas is used to drive a heat pump and the outdoor air is above freezing, about 71% of the gas' energy is delivered as space heat. (Drawing by Michael Cockram, © 1998 by John S. Reynolds, A.I.A., all rights reserved.)*

ing in Chapter 5 that arranges its structural system to enhance daylighting and passive cooling by night ventilation, adding solar heating, improved indoor air quality, electric light dimming, and acoustic absorption in the process (see Fig. 5.4).

(f) Shape the Form to Guide the Flow.

The most obvious examples of this strategy are the solar heated buildings that are shaped to gather winter sun, many of which are found in Chapter 5. In Fig. 1.12, Connecticut's snowy landscape acts as a large reflector of winter sun into the building's south-facing windows; so does the flat roof in front of the upper story windows. A skylight along the north wall adds more sun while washing an otherwise dark north wall with daylight that serves to reduce the "cave effect" for both floors. The landscape adds sunlight in front and earth warmth in back, as the home is set into the hillside.

Daylighting is another obvious place to apply this "form follows flow" strategy; the library in Fig. 1.13 reaches up at its center for north skylight in the Pacific northwest. The clerestory's interior surface is a curved reflector, shaping the flow of diffused daylight to the central light well below.

Supplementing the "low-grade" general-purpose daylighting are individually controlled, "high-grade" electric lights for reading stations. The on-site energy resource is generally and plentifully used, whereas the imported energy is used sparingly for special purposes. The psychologically "cool" (blue) north daylight is warmed by the "warm" (yellow) light of the task light; see Chapter 18 for light quality details.

(g) Shape the Form to Manifest the Process.

This is more than a variation on "if you've got it, flaunt it"; this strategy asks for a building to inform its users as to how it works inside and out. In passive solar heated and passively cooled buildings, much of the thermal performance is evident in the very form of the exterior envelope and the interior space, rather than hidden in a closet or mechanical penthouse. Again, Chapter 5 contains many such examples, such as the Agua Blanca ranch and its prominent cooltowers (see Fig. 5.53).

The restaurant in Fig. 1.14 is stretched out along the east–west axis, allowing plentiful south light to enter both for solar heating and for daylighting. The lower windows, angled to face due

(a)

(b)

Fig. 1.12 *Passive solar-heated, balanced-daylight residence of architect Donald Watson, FAIA, near New Canaan, Connecticut. (a) South facade in winter, with hillside providing earth sheltering. (b) Daylight model shows balanced light between south windows and a skylight washing the north wall with daylight down through two stories. (c) First floor plan, showing elongation along east–west axis and an art gallery to take advantage of subdued natural light at the north wall. (d) Upper floor plan. (Courtesy of Donald Watson, FAIA, architect.)*

south, receive added sunlight reflected from the parking in front; the upper clerestory windows (facing southeast) receive added sunlight reflected from a shiny-surfaced flat roof. The floor is a concrete slab, darker in color, to absorb solar energy. Additional thermal mass is desirable, both for winter solar heating and summer night ventilation of mass, so 30-gallon (113-L) barrels of water

First Floor Plan

(c)

Second Floor Plan

(d)

Fig. 1.12 *(continued)*

are painted a dark, intense blue. Some of these barrels support tables, others support plants at south windows, but most are arrayed on a high slotted shelf inside the clerestory windows. Here they intercept winter sun, releasing their stored heat by night. Here also, they are washed by cool night air in summer. Thus they act as storage in both seasons: accepting solar heat by day in winter, soaking up building heat by day in summer. To help flush the building on summer nights, a culvert, open at either end, runs below the building, with frequent openings up through the floor slab. Cool night air thus enters the building, rises around the barrels, and is drawn out through five large stacks topped by wind-gravity ventilators (see Chapter 6). On summer evenings, outdoor dining occurs below the grapevines; along the saw-tooth plan, the segments of insulated east wall (between the south glass windows segments) swing open, giving plentiful access from inside to outside.

Because summer sun is unwelcome indoors (even in Oregon), the clerestory is covered in summer by exterior shading cloth, admitting light but very little direct sun. Downstairs, grapevines provide dense shade, leafing out bright-green in May, dropping their yellow leaves in October. The grapes in this sunny south-facing microclimate, protected from winds, are quite sweet, and garnish each meal served in the fall. In winter, the downstairs windows are protected by interior roll-down insulated shades, whereas the clerestory windows have thick insulated interior shutters controlled by rope and pulley systems that resemble those of a sailboat. The back-up heating system is one wood stove. We might expect that kitchen heat would be more than enough to warm the restaurant, but the requirements of ventilation air for the large kitchen grill are so high that most of the kitchen heat is unavailable for the eating area.

(h) Use Information to Replace Power.
The sophisticated control systems discussed in Section 1.5 are examples of this strategy. There is also a role for the users of buildings; to make the decisions when to do what in order to maintain the desired conditions. This is often called "thermal sailing" (see Chapter 4), where controls are placed in the hands of users. These controls' operation—and their consequences—are then made

(a)

(b) Section through library

Fig. 1.13 *The library designed by Alvar Aalto at Mt. Angel Abbey, Oregon, combines general day-lighting with task electric lighting. (a) The central north-facing skylight diffuses light over a two-story central opening, shown in section (b). The individual electric lights (c), each with a pull chain, can be used for reading when more light is desirable.*

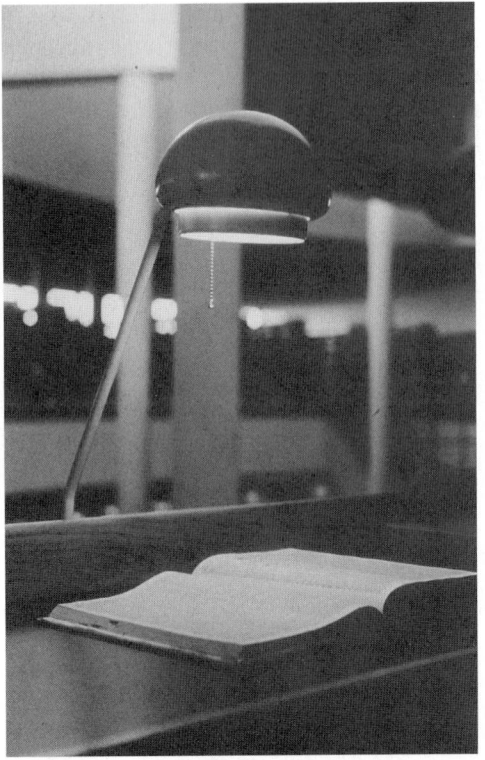

as evident as possible. The school in Fig. 1.10 gives each classroom control over access to outdoor air. The restaurant in Fig. 1.14 allows users to manipulate both the shutters above, via ropes and pulleys, and the rolling shades below; the stacks of the wind-gravity ventilators overhead can be opened or closed with pull-chains attached to dampers; fires in the wood stove can be controlled with air-intake dampers.

This reliance on a building's users is not so much an energy saver—most controls need very little power—as it is an education. A user who understands how a building gets and conserves its heat in cold weather is more likely to respond by lowering the indoor temperature and tightening up any heat leaks. Furthermore, some studies of worker comfort indicate that with more personal control (such as operable windows), workers express feelings of comfort in a wider range of temperatures than with centrally controlled air conditioning.

As we face a continued increase in world population and continued depletion of nonrenewable resources, it seems obvious that labor-intensive

Fig. 1.13 *(continued)*

Fig. 1.14 *The Cottage Restaurant, Cottage Grove, Oregon. (a) Northeast view, showing the wind-gravity (turbine) ventilators that eject heat by day and night in summer, and the lack of windows in either the north or east wall. (b) Southeast view, with generous south-facing windows and a southeast-facing clerestory. An extensive grape-covered trellis provides summer shading; the clerestory is seasonally protected with exterior shading cloth. (c) Interior, showing the thermally massive concrete floor, and many of the 30-gallon (113-L) barrels (on high shelf) that help to store solar heat by winter day and store coolth by summer night. (d) Section and (e) plan explain the relationship of the thermal elements. For alternatives to turbine ventilators, see Fig. 6.8. (Photos by G. Z. Brown. Drawings by Michael Cockram, © 1998 by John S. Reynolds, A.I.A., all rights reserved.)*

Fig. 1.14 *(continued)*

alternatives must be made available. The designer's challenge is to make these alternatives attractive and rewarding to use; to engage the users in indoor environmental control without making them slaves to a building's thermal regulation.

(i) Provide Multiple Pathways. This strategy shows redundancy as a virtue. Within this book's many topics, the most obvious need for multiple pathways may be two fire stairs in addi-

tion to elevators. There are many other examples, from back-up heating and cooling systems to multiple water reservoirs and piping pathways for fire sprinklers to emergency electrical and lighting systems. However, this strategy also applies to climate–site–building interactions, when one site-based resource may temporarily weaken and can be replaced by another.

In a future of PV arrays that are cheaper, more efficient, and integrated into walls and roofs,

(d) Section

(e) Plan

Fig. 1.14 (continued)

buildings will be able to generate considerable amounts of electricity on-site. At times, such as sunny mild weather weekend days when a commercial building is unoccupied, more electricity may be generated than can be immediately used on-site. When such an electrical system is also tied into the utility grid, the excess PV electricity may be used to "run the meter backwards" and supply other users elsewhere. Conversely, at night, this building's electrical needs can be served by the utility grid. See Section 14.9 for details on grid-connected PV systems.

(j) Manage Storage as a Key to Sustainability. In the previous example, a PV system used the utility grid instead of a bank of batteries

as storage for its temporary excess production. Storage appears repeatedly as an issue throughout this book. The greater the magnitude of the supply cycle, the more urgent becomes storage management. The high variability of wind-generated electrical output could be teamed with hydrogen storage; momentary excess wind power is used to split water into its oxygen and hydrogen components, and is thus converted to a combustible fuel that can be drawn on at a rate and time independent of wind speed.

On sunny winter days, a room's excess solar energy can be stored in its thermally massive surfaces, to be released by night. On cool summer nights, "coolth" (the opposite of heat) can be stored in these same surfaces and used to cool the room by day. This kind of storage is strongly dependent on the architecture. Storage more dependent on mechanical equipment is described in Chapter 8, in which off-peak (cheaper) electricity is used to make ice in a large thermal storage unit, to provide on-peak cooling next day.

Rainwater can be stored in cisterns, involving a balancing of normal daily demands for water against variable monthly supplies. Some hot water systems are now managed by varying the storage temperature in anticipation of demand; during a few hours of heavier demand, the water is kept hotter (see Chapter 10). In this manner, many hours of lower storage temperature results in less heat loss from the tank, conserving energy.

1.7 Design Challenges

The buildings we design today are very likely to experience major changes in the way that they are used and in their sources of energy supply. This poses some overall challenges to the designer.

(a) Design for Building Recycling. Designing for the recycling of buildings is a two-part balancing act. First, the designer should provide enough flexibility to prolong the useful life of the building by enabling it to adapt easily to changed usage. Flexibility can be expensive to implement physically and can result in a bland "sameness" throughout the structure. The latter characteristic is easier for the designer to change than the former. Second, the design can allow for demounting parts so that the structure can remain intact safely while reusable materials and components are removed. Yet this can result in heavier buildings in which floor systems are not structurally integral with beams. It also discourages integration of mechanical and structural systems, as is discussed in Chapter 8. Furthermore, a demountable building is perhaps especially subject to energy leaks, such as cracks widening around self-contained components of the facade.

Some initial guidelines for recyclable buildings are as follows:

1. Design the structure to be separable from everything else and, of itself, to be easily disassembled. Extensive remodeling is then possible without major structural modifications, and at the end of a building's life, elements of its structure can be reused elsewhere.
2. Design for "breathing room" where possible: between a building and its neighbors, or between major spaces within a building. Some expansion is thus possible without rebuilding. This could include designing the columns and footings to support an extra floor or two, for vertical expansion.
3. Maximize the utilization of on-site (or natural) forces such as sun and wind. The less sophisticated the mechanical and electrical equipment, the less obvious will be the obsolescence of such equipment with the passing of time.
4. Use the materials and components distinctly: avoid combinations that make recycling of these elements difficult. A steel or plastic pipe embedded in a concrete slab is neither repaired nor recycled easily; some "sandwiches," or manufactured panels of building materials, do not allow metals, plastics, and other products they contain to be separated for reuse at the end of the panel's life.

Although maximum savings of energy can be realized when a building component can be reused "as is," even the crushing and reprocessing of some (separated) building materials will save energy when compared to their original manufacture from virgin material (see Chapter 12).

(b) Design for Energy Transition. Two more challenges to designers arise:

1. To design our buildings not only to save energy, but also so that they can eventually be weaned away from dependence on nonrenewable fuels.

   The transition from "imported" electricity off the grid, to site-generated PV electricity, seems easy enough. Interestingly, because PV cells produce direct-current (DC) electrical current, and today's recreational vehicle industry has spawned a host of DC appliances such as televisions, lights, and motors, buildings that anticipate some DC distribution may be expected.

2. To use energy wisely; to expect only a "fair share" of locally available renewable fuels, recognizing that such sources are limited even though they are continuously available.

For example, with supplies stretched tight by increasing density, it is tempting to erect a larger solar collector to intercept sunlight that would otherwise be utilized by the neighboring building. This temptation grows stronger as the building is designed to rely more heavily on the sun. The concept of a "solar envelope" to protect each building's fair share is discussed in Section 3.4.

REFERENCES

American Institute of Architects (1996). *Environmental Resource Guide.* John Wiley & Sons, New York.

Berry, B., Bown, J., Thornton, J., Turzynski, J., and Walton, M. (1995). "The New Inland Revenue Centre, Nottingham," in *The Arup Journal,* April 1995. Ove Arup Partnership, London.

Brower, D. (1990). *Cool Energy.* Union of Concerned Scientists, Cambridge, Mass.

Fisher, J. (1974). *The Energy Crisis in Perspective.* John Wiley & Sons, New York.

Hannon, B., Stein, R., Segal, B., and Serber, D. (1977). *Energy Use for Building Construction.* Center for Advanced Computation, University of Illinois, Champaign-Urbana, Illinois.

Joslin, J., et al. (1993). *The Waste Papers: Analysis and Discussion of the Potential for Salvage and Re-Use of Construction Materials from Residential Demolition.* Metro Solid Waste Department Publications, Portland, OR.

Lyle, J. T. (1994). *Regenerative Design for Sustainable Development.* John Wiley & Sons, Inc., New York.

Meyers, R. (1983). *Handbook of Energy Technology and Economics.* John Wiley & Sons, Inc., New York.

Stein, R., et al. (1981). *Handbook of Energy Use for Building Construction.* U.S. Department of Energy, Washington, D.C.

U.S. Environmental Protection Agency. (1998). *Characterization of Building-Related Construction and Demolition Debris in the United States.* EPA530-R-98-010. Washington, D.C.

U.S. Office of Building Technologies. (1998). *Core Databook* (Version 2.2, June 1998). Washington, D.C.

2

CLIMATE, COMFORT, AND DESIGN STRATEGIES

NOTE: The following convention is observed throughout this text: Temperatures of air, liquids, and solids are shown as °F, or as °C, or as K. Temperature *differences* are shown as F°, C°, or K. This is intended to help distinguish between actual temperatures and *changes* in temperature.

One of the earliest reasons for building was to create shelter from the climate and to enhance thermal comfort. This chapter is about the interrelationship among bodies, buildings, and climate; we begin by discussing body heat flow, then thermal comfort, and then the design strategies that are appropriate to various climates.

2.1 The Body

Because we are alive, we are always generating body heat. Because the body's core must stay within a narrow temperature range, we nearly always need to lose this internally generated heat to our environment. The rate at which we produce heat changes frequently, as does the environment's ability to accept or reject heat. To regulate our body heat loss, we have available three common layers between our body cores and our environment: our first skin, which is our own; the second skin, or our clothing; and the third skin, in this case, a building.

(a) Metabolism. The rate at which we generate heat (our metabolic rate) depends mostly on our level of muscular activity, partly on what we ate and drank (and when), and partly on where we are in our normal daily cycle. Our heat production is measured in metabolic (*met*) units (Table 2.1). One met is defined as 50 kcal/h m², equal to 18.4 Btu/h ft², or 58.2 W/m². One met is the energy produced per unit of surface area by a seated person at rest. Under these conditions, the total heat produced by a normal adult is about 360 Btu/h, or 106 W. The more active we are, the more heat we produce, and our own "first" skin is the most important regulator of heat flow.

Of the many interactions between our skin and the rest of our body, consider these three: touch, blood, and water. The sensation of touch includes pressure and pain as well as heat and cold. The sensations of heat and cold are produced by contact with surfaces or moving air as well as by radiant heat. They are frequently our signals for shifts in body heat regulation, which is controlled by the thermostat in our brains, called the *hypothalamus.*

In response to signals from our surface and to changes in our core temperature, the hypothalamus calls for changes in our blood distribution system. If we are too cold, we need to decrease our rate of heat loss, so the flow of blood from our core toward the surface of our skin decreases. Blood carries heat (in HVAC terms, we are an "all-water" system): the less exposure to cool air at the surface, the less heat is lost. This decreased blood flow toward the surface is called *vasoconstriction,* and is triggered in part by cold signals from our skin. In this condition, less water is forced to the skin surface by our sweat glands, which reduces evaporation and therefore reduces heat loss.

Note the implications of this zoning within the body: We strive to maintain, at all costs, a nearly constant core temperature for our vital organs. This most-protected zone takes thermal precedence over the less-vital zone of our extremities, such as arms and legs; next in priority are our fingers and toes. The farther from our central body mass (fingers and toes) and the greater the

TABLE 2.1 **Metabolic Rates for Typical Tasks**

Activity	Metabolic Rate[a]		
	met units[b]	Btu/h ft^2	W/m^2
Resting			
Sleeping	0.7	13	40
Reclining	0.8	15	45
Seated, quiet	1.0	18	60
Standing, relaxed	1.2	22	70
Walking (on the level)			
2 mph (0.9 m/s)	2.0	37	115
3 mph (1.2 m/s)	2.6	48	150
4 mph (1.8 m/s)	3.8	70	220
Office activities			
Reading, seated	1.0	18	60
Writing	1.0	18	60
Typing	1.1	20	65
Filing, seated	1.2	22	70
Filing, standing	1.4	26	80
Walking about	1.7	31	100
Lifting, packing	2.1	39	120
Driving/flying			
Car	1.0–2.0	18–37	60–115
Aircraft, routine	1.2	22	70
Aircraft, instrument landing	1.8	33	105
Aircraft, combat	2.4	44	140
Heavy vehicle	3.2	59	185
Miscellaneous occupational activities			
Cooking	1.6–2.0	29–37	95–115
House cleaning	2.0–3.4	37–63	115–200
Seated, heavy limb movement	2.2	41	130
Handling 110-lb (50-k) bags	4.0	74	235
Pick and shovel work	4.0–4.8	74–88	235–280
Machine work			
Sawing (table saw)	1.8	33	105
Light (electrical industry)	2.0–2.4	37–44	115–140
Heavy	4.0	74	235
Miscellaneous leisure activities			
Dancing, social	2.4–4.4	44–81	140–255
Calisthenics/exercise	3.0–4.0	55–74	175–235
Tennis, singles	3.6–4.0	66–74	210–270
Basketball	5.0–7.6	90–140	290–440
Wrestling, competitive	7.0–8.7	130–160	410–505

Source: Reprinted with permission of the American Society of Heating, Refrigerating, and Air-Conditioning Engineers, Inc. From 1997 *ASHRAE Handbook of Fundamentals.*

[a]For average adult with body surface area of 19.6 ft^2 (1.8 m^2). For whole-body average heat production, see also Table 5.21.

[b]One met = 18.4 Btu/h ft^2 = 58.2 W/m^2.

surface area (ears), the faster the surface temperature will drop in cold conditions. The most variable thermal zone of all is our skin surface.

When cold conditions worsen, we get "goose flesh," symptoms of our skin's unsuccessful attempt to create insulation by fluffing up our body hair. Because we cannot add insulation this way, we soon increase our metabolic rate, or burn more fuel, by shivering, muscular tension, or increased muscular activity. At the point that shivering incapacitates us, we may reach 6 mets. Before this point, we seek help from our second and then third skins of clothing and buildings.

The opposite occurs when we are too hot: first, blood flow toward the skin surface increases (*vasodilatation*), triggered primarily by warm signals from our core. The sweat glands greatly increase their secretion of water and salt to the skin surfaces. This increases heat loss by evaporation (although salt accumulations impede evaporation by lowering the vapor pressure of water).

(b) Heat Flow. Once the blood and water get our surplus heat to the skin surface, we have four ways to pass it to the environment: *convection* (air molecules pass by our surface, absorbing heat), *conduction* (we touch cooler surfaces, and heat is transferred), *radiation* (when our skin surface is hotter than other surfaces "seen" but not touched, heat is radiated to these cooler surfaces, and vice-versa when other surfaces are hotter than our skin surface), and *evaporation* (a liquid can evaporate only by removing large quantities of heat from the surface it is leaving). The amount of heat we lose by each of these four methods depends on the interaction of our metabolism, our clothing, and our environment. Figure 2.1 illustrates the typical situation of a person at rest as environmental conditions change.

As air and surface temperatures approach our own body temperature, we lose the options of convection, conduction, and radiation. Evaporation becomes essential, so access to dry, moving air is greatly appreciated. As air and surface temperatures fall, evaporation decreases while convection, conduction, and, particularly, radiation increase. Under the "normally comfortable" conditions in Fig. 2.1 of about 70°F (18°C), the

Heat generated, Btuh	400	400	400	400	400 (curve 1)
Heat lost by:					
Radiation and convection	350	300	200	100	0 (curve 2)
Evaporation	50	100	200	300	400 (curve 3)
Total, Btuh	400	400	400	400	400 Total
					(curve 1)

Fig. 2.1 Heat generated and lost (approximate) by a person at rest (rh fixed at 45%).

proportions of body heat loss per hour are as follows:

Radiation, convection, and conduction	72%
Evaporation	
From skin surface	15%
From lungs (exhaled air)	7%
Warming of air inhaled to lungs	3%
Heat expelled in feces and urine	3%

(c) Clothing. Usually clothing acts as an insulating layer and is particularly effective at retarding radiation, convection, and conduction. As air and surface temperatures in our environment fall well below our body's temperature, we adjust this second skin to provide increased insulation. However, in a hot, humid environment, our first skin needs exposure to moving air to encourage heat loss, yet needs protection from the sun's radiant heat: We need a simple second skin that acts as a sunshade. In a hot, arid environment, our second skin may keep us from losing too much valuable water while also playing the vital role of shading.

The insulating value of clothing is measured in "clo" units, 1.0 clo being equivalent to the typical American man's business suit in 1941, when clo was born (1 clo is equal to 0.88 ft^2 h F/Btu, or to 0.155 m^2 K/W). The total clo of what you are now wearing can be estimated by assuming 0.35 clo per kilogram (0.15 clo per pound) of your clothing's weight. The total clo of your attire may also be estimated by simply adding the clo of each individual item from Table 2.2; this total will be a bit higher than the actual clo value of the ensemble.

Sometimes the position of clothing is as important as its clo value; consider the role of socks as they separate our feet from contact with a cold floor.

Our second skin is just as likely as our third skin to be dominated by considerations of style more than of thermal regulation; we cannot always count on our clothing—or buildings—to increase our thermal comfort.

TABLE 2.2 **Typical Insulation Values for Clothing Ensembles**

Ensemble Description[a]	clo[b]
Walking shorts, short-sleeve shirt	0.36
Trousers, short-sleeve shirt	0.57
Trousers, long-sleeve shirt	0.61
Same as above, plus suit jacket	0.96
Same as above, plus vest and T-shirt	1.14
Trousers, long-sleeve shirt, long-sleeve sweater, T-shirt	1.01
Same as above, plus suit jacket and long underwear bottoms	1.30
Sweat pants, sweat shirt	0.74
Long-sleeve pajama top, long pajama trousers, short ¾ sleeve robe, slippers (no socks)	0.96
Knee-length skirt, short-sleeve shirt, pantyhose, sandals	0.54
Knee-length skirt, long-sleeve shirt, full slip, pantyhose	0.67
Knee-length skirt, long-sleeve shirt, half slip, pantyhose, long-sleeve sweater	1.10
Same as above, replace sweater with suit jacket	1.04
Ankle-length skirt, long-sleeve shirt, suit jacket, pantyhose	1.10
Long-sleeve coveralls, T-shirt	0.72
Overalls, long-sleeve shirt, T-shirt	0.89
Insulated coveralls, long-sleeve thermal underwear, long underwear bottoms	1.37

Source: Based on 1997 *ASHRAE Handbook of Fundamentals,* published by the American Society of Heating, Refrigerating and Air-Conditioning Engineers, Inc., Atlanta, Ga.

[a]All ensembles include shoes and briefs or panties. All ensembles except those with pantyhose include socks unless otherwise noted.

[b]One clo = 0.88 ft^2 h °F/Btu = 0.155 m^{2}K/W.

2.2 Comfort

A positive definition of comfort is "a feeling of well-being." The more common experience of thermal comfort is a lack of discomfort—or of being unconscious of how you are losing heat to your environment. There are three categories of factors that affect comfort: personal, measurable environmental, and psychological. Most *personal* (or physical) factors are under your control: they are your metabolism and your clothing, as well as various adaptations such as migration to a more comfortable place, or drinking or eating warm or cold foods. *Measurable environmental* factors are the familiar tools of the designer and engineer: air temperature, surface temperature, air motion, and humidity. *Psychological* factors are also familiar designers' tools, but they are much harder to measure for comfort: color, texture, sound, light, movement, and aroma. These psychological factors are often overlooked as we strive to meet the numerical (therefore calculable) physical criteria for thermal comfort. However, our primary objective is to heat or cool people, and buildings are our means to that end.

Consider the courtyard in a hot, dry climate (Fig. 2.2). Its fountain suggests coolness in the color and texture of its water; running water provides splashing sounds and sparkles of light, and may generate some air motion. Vines provide shade and their leaves sway in the slightest breeze, evidence of at least some air motion. Blossoms of flowering plants yield a cool fragrance that blends with the aroma of moistened surfaces in hot, dry surroundings. The measured coolness of such a courtyard may be but a slight improvement over the environment beyond, but it seems successfully cool.

Then consider the fireplace in a cold climate (Fig. 2.3). The fire's color is intensely warm, and it dances and casts a flickering light; it crackles, and it yields a smoky aroma (the latter not appreciated by many). Few textures seem hotter than that of glowing coals. Yet a fireplace might add but slightly to the overall warmth of the room, because it draws in cold air to replace the air that passes up the chimney. We seek a place near the fire for the "real" radiant warmth, but also for the psychological comfort the fire offers.

In our society, designers are encouraged to take numerical data quite literally. The measur-

Fig. 2.3 *The open fire seems the very spirit of warmth, despite the heat lost by large quantities of exhaust air up the chimney. Senses of sight, sound, touch, and smell all may be involved in a perception of warmth, wrapped in this red brick environment. (Courtesy of the architect and photographer, Edward Allen.)*

Fig. 2.2 *Indicators of coolness in a courtyard include running water and shade from vines that move with the breeze. Senses of sight, sound, touch, smell, and taste all may be involved in a perception of coolness.*

able environmental factors have been tested extensively in laboratories—but such testing is done by excluding other factors. A holistic view of designing for comfort considers numbers as a nonabsolute guide; common sense and a designer's own thermal experience play important roles as well. Lisa Heschong's *Thermal Delight in Architecture* (1979) is an excellent extension of these points of view.

Ultimately, our buildings will be expected to demonstrate success with regard to the measurable environmental factors of comfort, so it is necessary to understand how air and surface temperatures, air motion, and humidity are related to heat transfer.

Heat Transferred by:	Is Primarily Dependent on:
Conduction	Surface temperature
Convection	Air temperature, then air motion, then humidity
Radiation	Surface temperature (and orientation to the body)
Evaporation	Humidity, air motion, air temperature

From this comparison, it is evident that humidity is of relatively low importance in cold conditions, where heat loss by convection, radiation, and conduction is dominant. However, humidity is of primary importance in hot conditions, dominated by evaporative heat loss. This is further evident in comfort studies, which show that skin temperature is an important factor in cold conditions, whereas skin wettedness (percent covered by water) is of most importance in hot conditions.

A detailed analysis of a space's surrounding surface temperatures can be performed using the *mean radiant temperature* (MRT) measurement. Such a calculation involves determining the temperatures of all the surfaces and the position in the space at which to measure the MRT, then determining the solid angle that is formed between the measuring position and the outer edges of each surface. These values are averaged to find the MRT. However, a more useful term—both for impact on comfort and because it can be measured physically instead of laboriously calculated—is *operative temperature* (t_{op}), defined later. (For de-

tails on MRT calculations, as well as t_{op}, see Chapter 8 of the 1997 *ASHRAE Handbook of Fundamentals* published by the American Society of Heating, Refrigerating, and Air-Conditioning Engineers.)

(a) Comfort Standards. The interaction between comfort and important environmental factors can be generally summarized in Fig. 2.4. The "comfort zone" represents combinations of air temperature and relative humidity that most often produce comfort for a seated North American adult in shirtsleeves (total 0.6 clo) in the shade and without noticeable air motion. Surface temperatures are assumed to be not markedly different from air temperatures. As we move from this zone, at air temperatures below (to the left of) this comfort zone, comfort is still attainable if added radiant heat (increased surface temperatures), such as increased exposure to the sun, is

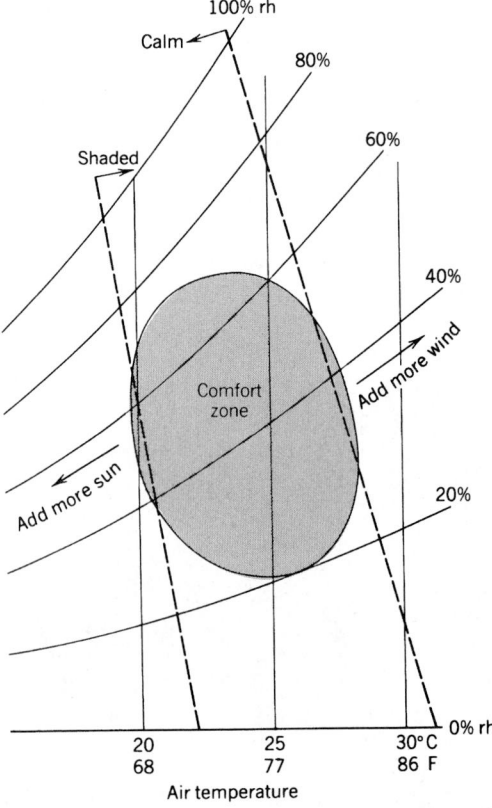

Fig. 2.4 *Comfort zones defined by relative humidity and air temperature.*

provided. (Also, more activity and more clothing would help.) Similarly, at air temperatures above (to the right of) the comfort zone, comfort is still attainable if added air motion is provided, such as exposure to wind. (Less activity and less clothing would also help.) In both cases, limits are soon reached, but the important point is that the basic comfort zone can be expanded by utilizing more sun or more wind and, in very dry climates, more moisture to the air.

Designers of buildings can aim for more specific comfort goals, especially where measurable environmental factors are concerned. Figure 2.5(a,b) shows the acceptable ranges of air and surface temperatures (t_{op}) and relative humidity for sedentary persons wearing typical clothing for both summer and winter. Air motion is assumed

to be unnoticeable. The terms used here are defined as follows:

Operative Temperature (t_{op}). Essentially an average of the air temperature of a space and the average of the various surface temperatures surrounding the space (or MRT). (To get a working grasp of operative temperature, get a 6-in.–diameter metal toilet float, paint it flat black, and drill a small hole in it. Then through a rubber stopper, insert an ordinary air-temperature thermometer about 5 in. long, so that its bulb is near the center of the toilet float interior. The thermometer will read approximate operative temperature; it will rise noticeably when passed from shade to sun; it will fall slightly in a breeze. Such a device is called a *globe thermometer.*) Ordinary

(a) *(b)*

Fig. 2.5 *Variations on the "comfort zone." The winter and summer comfort zones for light activity (1.2 met) that conform to ANSI/ASHRAE Standard 55a-1995 are shown in Fahrenheit (a) and Celsius (b). Operative temperature includes the influences of both air and surface temperatures. (Reprinted with permission of the American Society of Heating, Refrigerating, and Air-Conditioning Engineers, Inc., from ANSI/ASHRAE Standard 55a-1995, Thermal Environmental Conditions for Human Occupancy.) (c) Relationships among preferred (optimal) operative temperature, clothing, and activity. The shaded bands represent the degree to which departures from the optimal are tolerable; that is, nearly all people will find the departures only "slightly cool" or "slightly warm." (P. O. Fanger, Laboratory of Heating and Air Conditioning, Technical University of Denmark, Copenhagen.)*

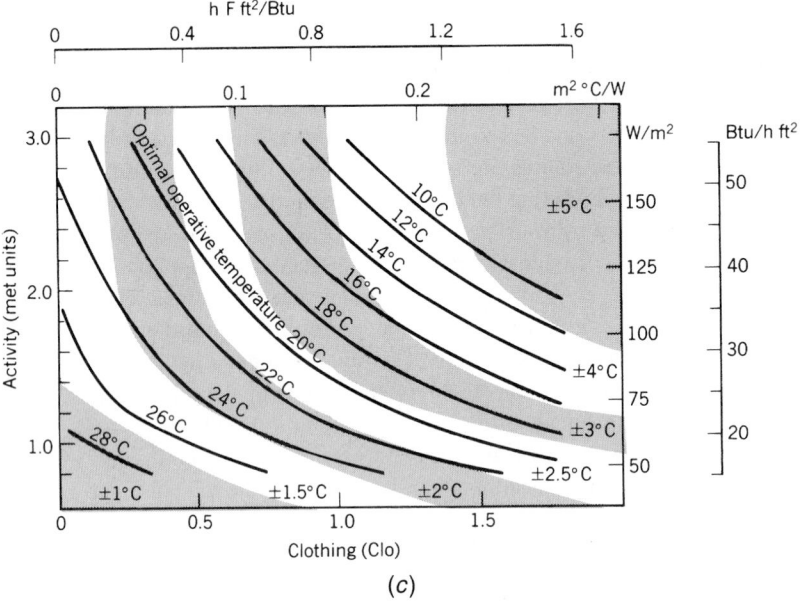

Fig. 2.5 (*continued*)

air temperature by itself is often not an adequate comfort indicator, especially in passively solar heated or passively cooled spaces where radiant temperature or air motion may be more influential than air temperature.

Relative Humidity (rh). This familiar weather report term is approximately the ratio of the actual density of water vapor in air to the maximum density of water vapor that such air could contain, at the same temperature, if it were saturated (at 100% rh).

Wet-Bulb Temperature (WB). The temperature shown by a thermometer with a wetted bulb rotated rapidly in the air to cause evaporation of its moisture. This term is derived from the instrument shown in Fig. 2.6, the sling psychrometer. In dry air the moisture evaporates and draws heat out of the thermometer to produce a large *wet-bulb depression* (difference between dry- and wet-bulb temperatures). This is an index of low relative humidity. Slow evaporation when the air is already moisture laden results in a small wet-bulb depression and indicates a high relative humidity. Note that at 100% rh, DB, and WB are equal.

Fig. 2.6 *Sling psychrometer and its usage. Air motion encourages evaporation from the moist cloth, lowering the wet-bulb temperature below the surrounding air temperature, whereas the dry-bulb temperature stays constant at the surrounding air temperature. (At 100% rh, wet-bulb and dry-bulb temperatures will be equal.)*

Dry-Bulb Temperature (DB). The temperature of the ambient mixture of air and water vapor measured in the normal way with a simple thermometer.

The criteria for operative temperature–relative humidity combinations should be specific for combinations of activity and clothing, because such personal factors are just as important in determining comfort. Figure 2.5c shows not only that much lower operative temperatures are comfortable at high activity–high clo levels (such as heavy work in heavy clothing outdoors), but also that greater tolerance (wider zone borders) can be expected under such conditions.

The standards in Fig. 2.5a,b purport to apply to any sex, age, or national origin; research cited by ASHRAE supports such a view. High humidity is avoided (to avoid mold and mildew as well as discomfort) and air motion is assumed to be unnoticeable. However, occupants of naturally ventilated buildings seem to find comfortable a higher combination of temperature and relative humidity than do occupants of sealed, air-conditioned buildings. Figure 2.7a shows a much higher incidence of these warmer, more humid environments in Hawaiian naturally ventilated classrooms than in similar air-conditioned classrooms. Yet, the majority of these Hawaii students and teachers voted these conditions acceptable. These observations were made in both "hot" and "cool" seasons (see Kwok, 1998).

Does long-term *acclimatization* influence the

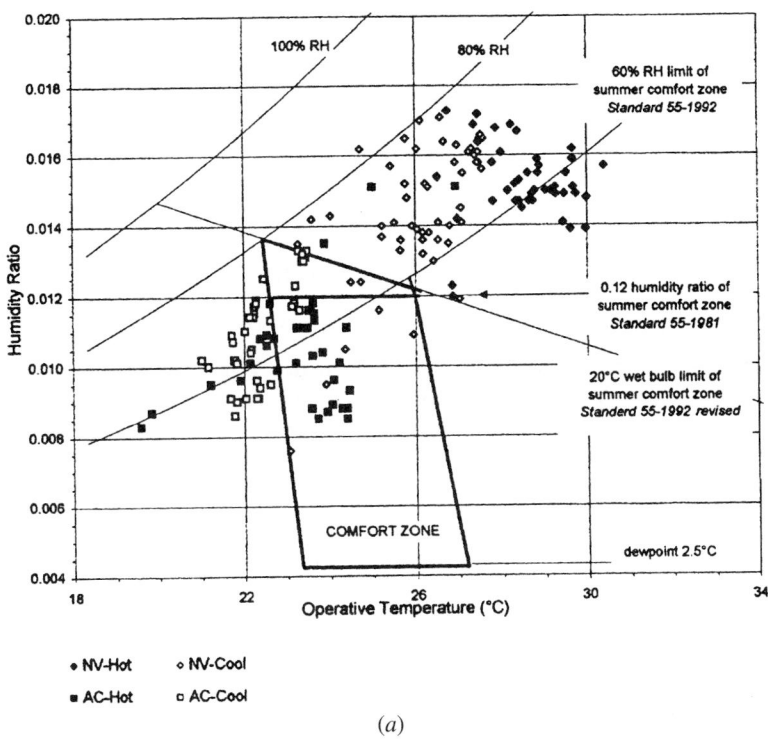

(a)

Fig. 2.7 *Natural ventilation and comfort.* (a) *Measured conditions compared to the ASHRAE summer comfort zone (see Fig. 2.5b) for naturally ventilated and air-conditioned classrooms in Hawaii. The majority of the occupants voted these conditions acceptable. A higher temperature and humidity comfort zone for naturally ventilated buildings is supported by studies such as this (Kwok, 1998). (Reprinted with permission of the American Society of Heating, Refrigerating, and Air-Conditioning Engineers, Inc. from ASHRAE Transactions 1998, Vol. 104, Number 1.)* (b) *Suggested boundaries of outdoor air temperature and humidity within which indoor comfort can be provided by natural ventilation. Assumed air speed is 2 m/s (4.5 mph). The higher limits for "hot-developing countries" assume acclimatization by those cultures. (Based on Givoni, 1998.)*

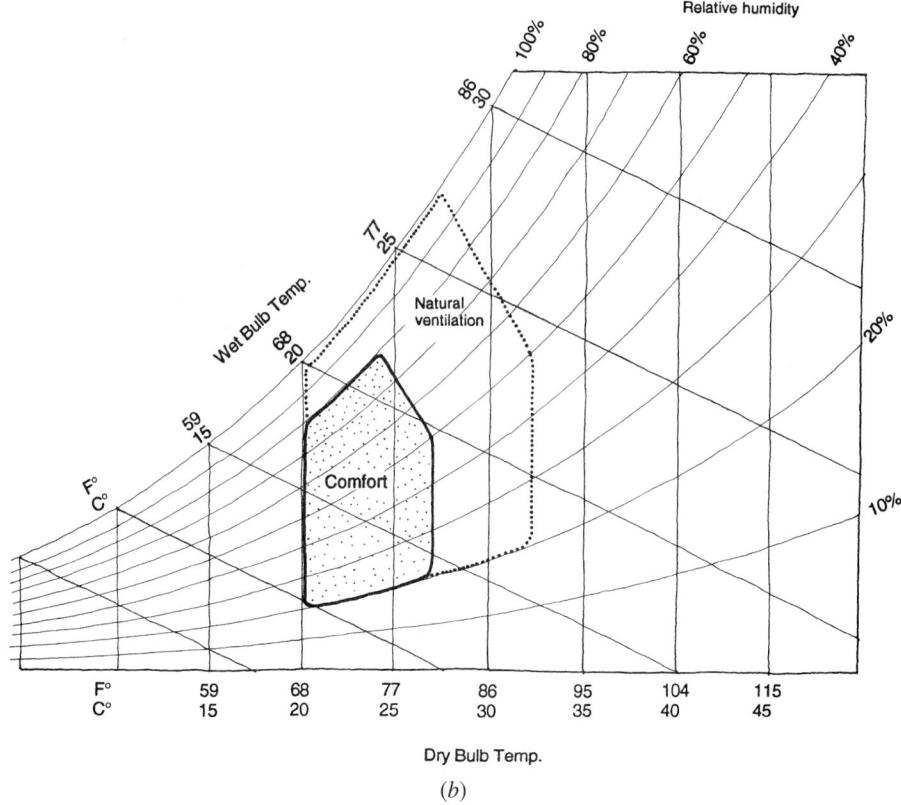

(b)

Fig. 2.7 (*continued*)

sensation of thermal comfort? Researchers disagree. ASHRAE Standard 55 recognizes but two seasonal comfort zones, influenced by activity and clothing. Givoni (1998) advocates a higher temperature, higher humidity limit for "hot-developing" countries, Fig. 2.7*b,* based on a combination of traditional and economic factors. In this case, an interior air speed of 2 m/s (about 5 mph) is assumed.

An *adaptive model* of comfort recognizes acclimatization's influence as well as personal abilities to influence one's comfort. Humphreys and Nicol (1998) elaborate on this approach, with occupant behaviors summarized in Table 2.3. They present a formula to determine the "indoor comfort" temperature, T_n, relative to an exponentially weighted running average of outdoor temperature, and applicable to "free-running" buildings (without mechanically narrowly controlled indoor temperatures):

$$T_n = 0.534\,T_{rmo} + 12.9\,(°C)$$

where T_{rmo} is in °C and represents the mean outdoor air temperature during a period of hot weather. Note that relative humidity is not included in this equation. Using this equation, with Phoenix, Arizona's average July–August temperature (to approximate an exponentially weighted running average) of 32.3°C (90.2°F):

$$T_n = 0.534\,(32.3) + 12.9 = 17.25 + 12.9$$
$$= 30.1°C\ (86.2°F)$$

In contrast, the ASHRAE summer comfort zone (from Fig. 2.5) is about 23°C to 26°C (at 50% rh).

Returning to Fig. 2.5*a,* based on ASHRAE (1995), some details are:

Activity: 1.2 met (mainly sedentary)

Winter: 0.9 clo, heavy slacks, long-sleeved shirt and sweater. Air motion at maximum = 0.15 m/s (30 fpm). The lower the temperature, the less air motion is desirable. Note: Optimum winter operative temperature is 22.7°C (71°F).

TABLE 2.3 **Adaptive Behavior for Thermal Comfort**

Response Category	Actions in Response to Cold	Actions in Response to Heat
Regulating the rate of internal heat generation	Increasing muscle tension and shivering	Reducing one's level of activity Drinking cold liquids (induces sweating) Drinking hot liquids (induces sweating) Eating less Adopting the siesta routine (matching activity to thermal environment)
Regulating the rate of body heat loss	Vasoconstriction (reduces blood flow to the surface tissues) Curling or cuddling up (reduces exposed surface area) Adding some clothing	Vasodilation (increases blood flow to the surface tissues) Adopting an open posture (increases exposed surface area) Taking off some clothing Sweating
Regulating the thermal environment	Turning up the thermostat Lighting a fire Complain to management (so someone else will raise the temperature) Insulate loft or wall cavities Improving windows or doors, weather-stripping	Turning on the air conditioner Switching on a fan Opening a window Shading a window from sun
Selecting a different thermal environment	Finding a warmer spot (such as going to bed) Visit a friend (with a warmer place) Visit a heated public building Building a new home Emigrating: a long-term solution	Finding a cooler spot Visit a friend (with a cooler place) Visit a cooled public building (or go swimming) Building a new home Emigrating: a long-term solution
Modifying the body's physiological comfort conditions	Acclimatizing, letting body and mind become more resistant to cold stress	Acclimatizing, letting body and mind adjust so that heat is less stressful

Source: Based on Humphreys, M., and Nicol, J. F., "Understanding the Adaptive Approach to Thermal Comfort," in *ASHRAE Transactions 1998,* Vol. 104, Number 1.

For each 0.1 clo increase in clothing, a decrease of 0.6C° (1F°) is allowed in the borders of the winter comfort zone; for each 1 met increase in activity, a decrease of 1.4C° (2.5F°) is allowed in the borders of the winter comfort zone.

Summer: 0.5 clo, light slacks and short-sleeved shirt or blouse. Air motion is assumed at 0.25 m/s (50 fpm). The borders of the summer comfort zone may be raised 1C° for each 0.275 m/s increase in air motion (1F° for each 30 f/m); up to a limit of 28°C at 0.8 m/s (82.45° F at 160 f/m). At this point, loose paper, hair, and other objects might start to be blown about (however, see Table 2.4). Note: Optimum summer operative temperature is 24.4°C

ENERGY OVERVIEW

TABLE 2.4 **Indoor Air Velocity and Comfort**

Air Velocity	Possible "Lower Temperature" Comfort Sensation (between 80° and 90°F; larger numbers correspond to high-humidity areas)	Probable Impact
Up to 50 fpm (0.25 m/s)	No change in comfort sensation	Unnoticed
50–100 fpm (0.25–0.51 m/s)	2–3F° lower (1.1–1.7C°)	Pleasant
100–200 fpm (0.51–1.02 m/s)	4–5F° lower (2.2–2.8C°)	Generally pleasant, but causing a constant awareness of air movement
200–300 fpm (1.02–1.52 m/s)	5–7F° lower (2.8–3.9C°)	From slightly drafty to annoyingly drafty
Above 300 fpm (1.52 m/s)	More than 5–7F° lower (2.8–3.9C°)	Requires corrective measures if work is to be efficient and health secured

Source: Adapted from Victor Olgyay, *Design with Climate: Bioclimatic Approach to Architectural Regionalism,* Copyright © 1963, Princeton University Press. Reprinted by permission.

(76°F). For each 0.1 clo decrease in clothing, an increase of 0.6C° (1F°) is allowed in the borders of the summer comfort zone.

The upper and lower borders for relative humidity are based on less precise data, but in general, the lower limit seeks to avoid problems such as coughs, nosebleeds, static electricity, and dust mites from excessively dry air; the upper limits try to keep skin wettedness within acceptable levels, as well as discouraging the growth of mold and mildew.

(b) Passive Building Comfort Standards.
A somewhat different approach to comfort standards has been proposed for buildings that take a more passive approach to heating and cooling in areas where acceptable humidity can be maintained. In these buildings, direct sun might add significantly to body warming in winter; strong air currents might be expected as a part of summer body cooling. These tend to be the kind of "free-running" buildings mentioned earlier as candidates for the adaptive model of comfort prediction.

Arens et al. (1980) demonstrated a wider range of comfort conditions and considerably more tolerance for summer air motion than ASHRAE (1995). A summary of these proposals in shown in Fig. 2.8. Such graphs are called "bioclimatic" charts because of their interrelation of climate and human comfort factors. Where the users of

buildings are expected to adjust to the wider temperature swings associated with passive buildings, these standards can be used. Some differences in the basic assumptions of Arens et al. from ASHRAE (1995) should be noted:

Activity: 1.3 met (ASHRAE, 1.2 met)

Winter: 0.8 clo (ASHRAE, 0.9 clo)

Summer: 0.4 clo (ASHRAE, 0.5 clo)

Note particularly that rather than operative temperature, ordinary (DB) air temperature is charted here. Added radiant heat below the shading line can be utilized to extend the comfort zone to lower temperatures; added air motion above the nearly still air line can extend the comfort zone to higher temperatures. *Radiant heat* to be added to a lower-extended comfort zone in Fig. 2.8 is shown in two quantities: effective radiant field (ERF) is a measure of the net radiant heat flux to the body from all surfaces at temperatures other than air temperature. A more convenient figure for designers may be the total solar radiation (or *insolation*) on a horizontal surface, termed I_{TH}. Such values are more readily available, and the added-radiation lines of Fig. 2.8 represents I_{TH} converted to its approximate radiant impact on a human body's surface area, when the sun is at an altitude of 45°. Arens et al. (1980) give further details on the effects of solar radiation on comfort at other sun altitudes, activity levels, and clothing combinations.

ENERGY OVERVIEW

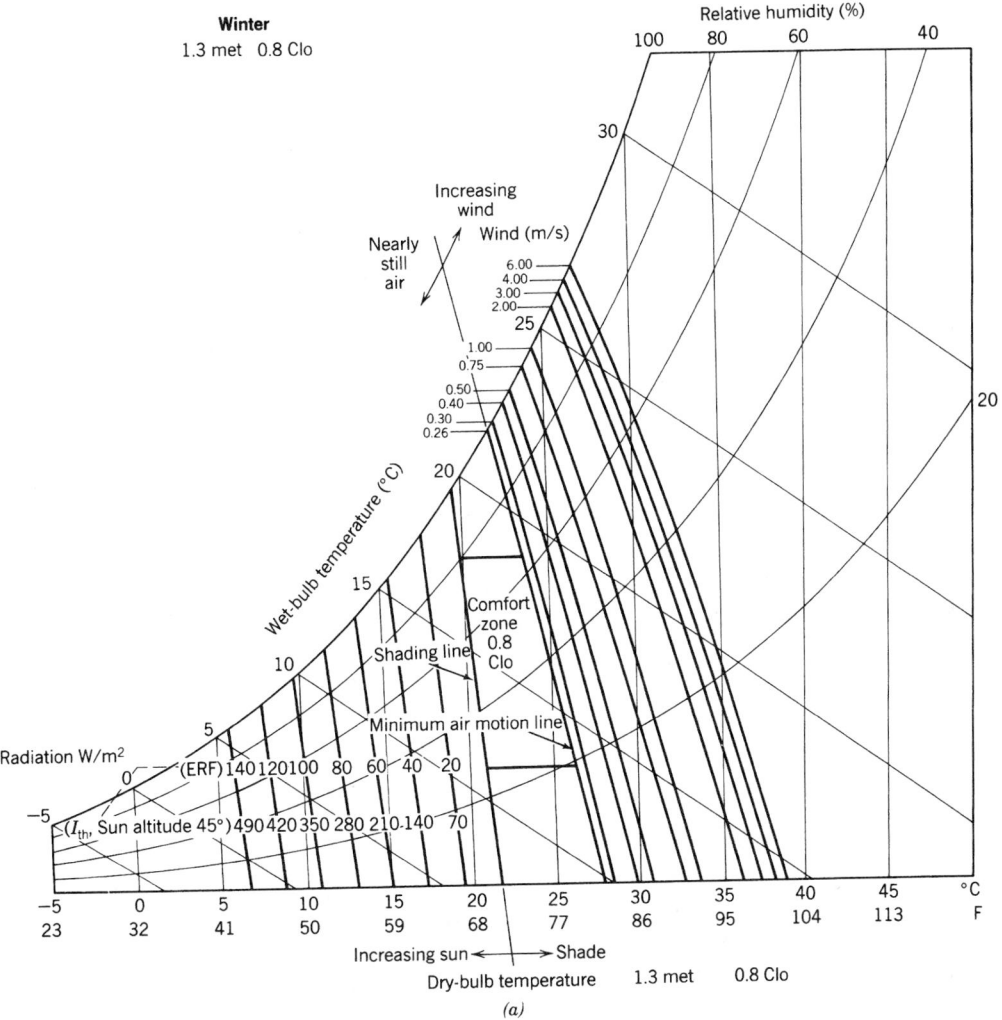

Fig. 2.8 *Comfort zones that encourage passively heated and cooled buildings. The winter comfort zone
(a) and the summer comfort zone (b) are combined for the year-round zone shown in (c). (Based on
Arens et al., 1980.)*

Added air motion in Fig. 2.8 is shown for quite a wide range of velocities. Table 2.4 indicates that people easily tolerate air motion outdoors up to about 3 m/s (about 600 f/m), but indoor studies have indicated that about 2 m/s (400 f/m) is the maximum tolerance from overhead fans.

Caution is still advised when relative humidity is above 70% or below 20%, for reasons cited earlier.

(c) Localized Comfort. What about assuring comfort at each work station, or wherever people spend longer time periods? The location

of heating, cooling, and ventilating components is an important detail. The human body is most affected by its thermal environment in thermally sensitive places. Although we sweat and are sensitive to heat or cold over nearly all our skin surface, we are most thermally sensitive due to concentrations in these places:

Heat receptors: Fingertips, nose, elbows

Cold receptors: Upper lip, nose, chin, chest, fingers

So, on a hot and humid day, cool air moving across the face is a particularly strong promise of

ENERGY OVERVIEW

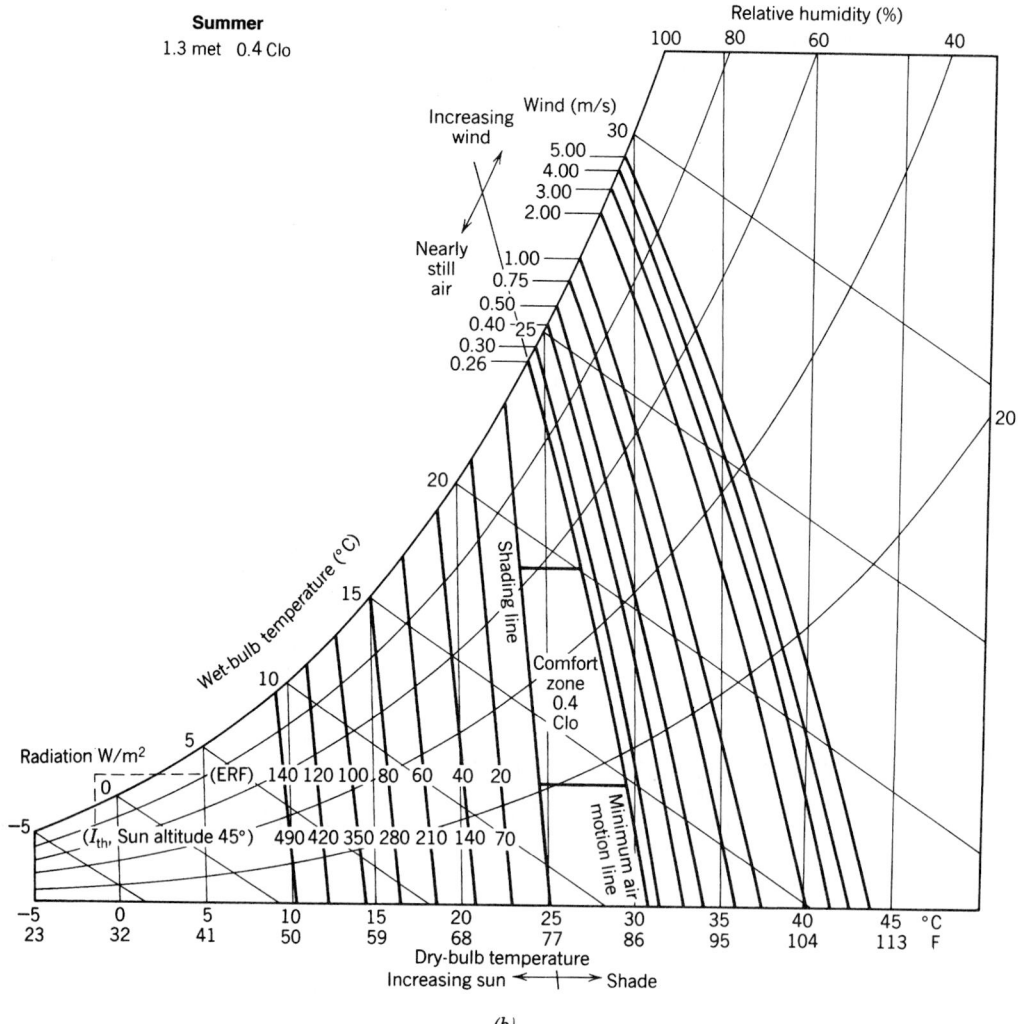

(b)

Fig. 2.8 *(continued)*

comfort, whereas on a cold day, a burst of heat (such as radiant heat from an electric light, a heater, or a cup of coffee) to the face and fingers is quickly effective.

Regarding our sense of touch: our fingertips are really most sensitive to the *rate at which heat is being conducted* to colder objects, or from hotter objects. The temperature of our skin at fingertips under ordinary conditions is in the high 20s C (high 80s F), so our sense of touch works against many passively heated surfaces in winter, which feel "cool" even though they are successfully warmer than room air temperature. These sur-

faces are made of materials that conduct heat rapidly, so that they can soak up solar radiation without overheating the room. This high *conductivity* makes them eager to accept human warmth as well, and persuades us as we touch them that they are cooler than, in fact, they are. Everyone who has walked barefoot from a rug to an unheated tile floor will sense that the tile is colder, even when both surfaces are at exactly the same surface temperature. This same characteristic thus works for passively cooled surfaces in summer.

Air motion in excess is called a *draft,* and results in an undesirable local cooling of the body.

Combined seasons
1.3 met

Fig. 2.8 (*continued*)

It is a particularly serious threat in winter, yet summer drafts from very cool conditioned air are also bothersome. Figure 2.9 shows the percentage of people dissatisfied due to perceptible draft on their head region (head, neck, shoulders, and back). Again, this applies to sedentary people wearing normal indoor clothing. Note the influence of air temperature: the lower the temperature, the higher the dissatisfaction. Many air-conditioning systems deliver summer air as low as 13°C (55°F), well below the lowest temperature shown here. The standard summer 0.25 m/s (50 fpm) air motion at such supply air temperatures seems to threaten more than 40% dissatisfaction

from drafts. However, ceiling fans with the same air motion, for comfort at higher temperatures such as 27°C (80°F), should cause less than 15% dissatisfaction.

Another comfort factor is *radiant asymmetry,* a difference between two opposite surfaces' temperatures as experienced by a seated body between them, when one of these surfaces is markedly different in temperature from all other surfaces within a space. Figure 2.10 shows that the greatest dissatisfaction arises from "warm ceilings" (typical of winter radiantly heated ceilings), in which almost half the experimental subjects experienced discomfort with ceilings 15C° (27F°)

ENERGY OVERVIEW

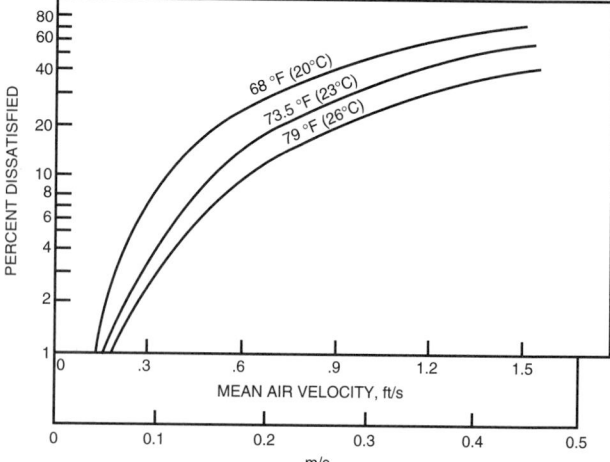

Fig. 2.9 *Percentage of people dissatisfied as a function of mean air velocity. The warmer the air, the greater the tolerance. (Reprinted with permission of the American Society of Heating, Refrigerating, and Air-Conditioning Engineers, Inc. from 1997* ASHRAE Handbook of Fundamentals.)

warmer than the space's average temperature. Less serious were "cool walls" (typical of a cold window in winter). Less than 10% dissatisfaction resulted from "cool ceilings" (such as radiant cooling panels in summer), and even less from "warm walls" (typical of a sunny window in summer, or a passively heated mass wall in winter).

Because warm air is less dense than cold air, it rises; therefore, in most building spaces air temperature is somewhat higher at the ceiling than at the floor, regardless of season. Figure 2.11 shows experimental results of dissatisfaction with a verti-

cal air temperature difference between the head at 1.1 m (43 in.) and the feet at 0.1 m (4 in.). (When the head level was cooler than the floor, much greater temperature differences were tolerated.)

2.3   Climates

Our most familiar names for climates describe their most severe season, as shown in Fig. 2.12a. This is a convenient description, but can be mis-

Fig. 2.10 *Percentage of people expressing discomfort due to asymmetric radiation. A warmer ceiling produces the highest discomfort; a warmer wall, the least discomfort. (Reprinted with permission of the American Society of Heating, Refrigerating, and Air-Conditioning Engineers, Inc. from 1997* ASHRAE Handbook of Fundamentals.)

Fig. 2.11 *Percentage of people dissatisfied as a function of vertical air temperature difference between head (higher temperature) and ankles (lower temperature). A cold floor proved uncomfortable. (Reprinted with permission of the American Society of Heating, Refrigerating, and Air-Conditioning Engineers, Inc. from 1997* ASHRAE Handbook of Fundamentals.*)*

leading for designers. "Cold" climates can have very hot, sometimes humid summer days; hot-arid climates can have bitterly cold winter conditions. Before designing the buildings that will modify exterior conditions to provide indoor comfort, we should know when and how much of this modifying is appropriate.

One useful method of analyzing outdoor conditions is shown in Fig. 2.12*b–e.* This "timetable" view of daily cycles across the entire year was devised by Olgyay (1963). A building's schedule of occupancy can be overlaid on such a graph, with resulting insights about climate/building interaction (see Fig. 5.6 for an example). On the climate timetable graph for each location, the isolines correspond roughly to the comfort lines shown in Fig. 2.8*a;* those labeled "100, 200, and 300" are Btu/h-ft^2 (315, 630, and 960 W/m^2) of radiation needed to remain comfortable outdoors without wind. These correspond roughly to 61°, 53°, and 45°F (16°, 12°, and 7°C), respectively, although the higher the relative humidity, the lower the air temperature to which these isolines correspond. The "shading line" corresponds to about 70°F (21°C) at lowest rh, and to about 68°F (20°C) at highest rh. The darker shaded ("overheated") zone is outlined by an isoline that corresponds to about 82°F (28°C) at lowest rh, and to about 78°F (26°C) at highest rh. At temperatures above these, there is a need for increased air motion as well as shade to remain comfortable outdoors. Further isolines within this "overheated" zone correspond to thermal comfort need for 300-

and 700-fpm (1.5- and 3.6-m/s) air velocities. From Table 2.4, however, it is evident that even at 300 fpm (1.5 m/s), excessive air motion for most work situations is likely.

Cool area

Temperate area

Hot–arid area

Hot–humid area

(a)

Fig. 2.12 *Climate zones and timetables.* (a) *Regional climate zones of the North American continent.* (b) *Basis for the isolines on the subsequent timetables (c, d, and e). (From Victor Olgyay,* Design with Climate: Bioclimatic Approach to Architectural Regionalism, *Copyright © 1963 by Princeton University Press. Reprinted by permission.)*

(b) The isolines on the timetables roughly correspond to these lines on the comfort zone (see Fig. 2.8)

(c) Timetable for New York, NY

(d) Timetable for Phoenix, AZ

(e) Timetable for Miami, FL

Fig. 2.12 (*continued*)

The timetable for New York City shows that about one-fifth of the year needs shade outdoors, with most hours of mid-July to mid-August in the overheating period; little relief can be expected by night in summer. Perhaps one-third of the year is "too cold," with temperatures too low for shirtsleeve comfort outdoors and requiring more than 300 Btu/h-ft² (960 W/m²) of radiation; an unlikely solar occurrence in New York winters.

The timetable for Phoenix, Arizona, shows that about one-half the year needs shade outdoors, and most of that time is in the overheated period. Air temperatures are so extreme that over 700-fpm (3.6-m/s) wind is needed about one-eighth of the year. Like New York, there is little temperature relief on summer nights, although the daily temperature range is much greater in arid Phoenix. This high "daily range" is evident throughout the year. Perhaps surprising to northerners is that about one-tenth of the year is "too cold," needing more than 300 Btu/h-ft² (960 W/m²) of radiation; by day, the Phoenix sun can help, but most of these colder hours occur after dark.

The timetable for Miami, Florida, shows that about four-fifths of the year needs shade outdoors, and about three-fifths is in the overheated period.

Again, there is a large portion needing over 700-fpm (3.6-m/s) wind, although unlike Phoenix, these hours tend to occur also by night. The added dotted line indicates a need for wind to counteract high vapor pressure, independent of high temperature. Clearly, wind is important to comfort in Miami in any unconditioned building. There is no such zone as "too cold" here.

These climatic timetables give us a graphic view of outdoor conditions in a particular location. In New York, the timetable suggests an emphasis on space heating, but needs for cooling as well. In Phoenix, cooling is paramount, but space heating is still a minor issue. In Miami, the concern is all about shading and cooling. Several strategies for both heating and cooling are available that utilize the climate as heat source or sink. In Section 2.4, cooling strategies are outlined; in Section 2.5, solar heating strategies. Combinations of day-

lighting, heating, and cooling are considered in Section 2.6. (Solar water heating is discussed in Chapter 10.)

2.4 Design Strategies for Cooling

The chart in Fig. 2.13 is a guide to the various approaches to passive cooling in temperate climates. It is based on the "building bioclimatic chart" developed by Milne and Givoni (1979). It helps to assess the suitability of a summer climate for four approaches to passive cooling. The average monthly data (as in Table 2.5) is now available in software and at various websites (see Appendix H). It can be plotted on this chart and, if not exceeding the "boundaries" of a strategy zone, then a match is indicated between the climate and that strategy. Although the edges of each strategy

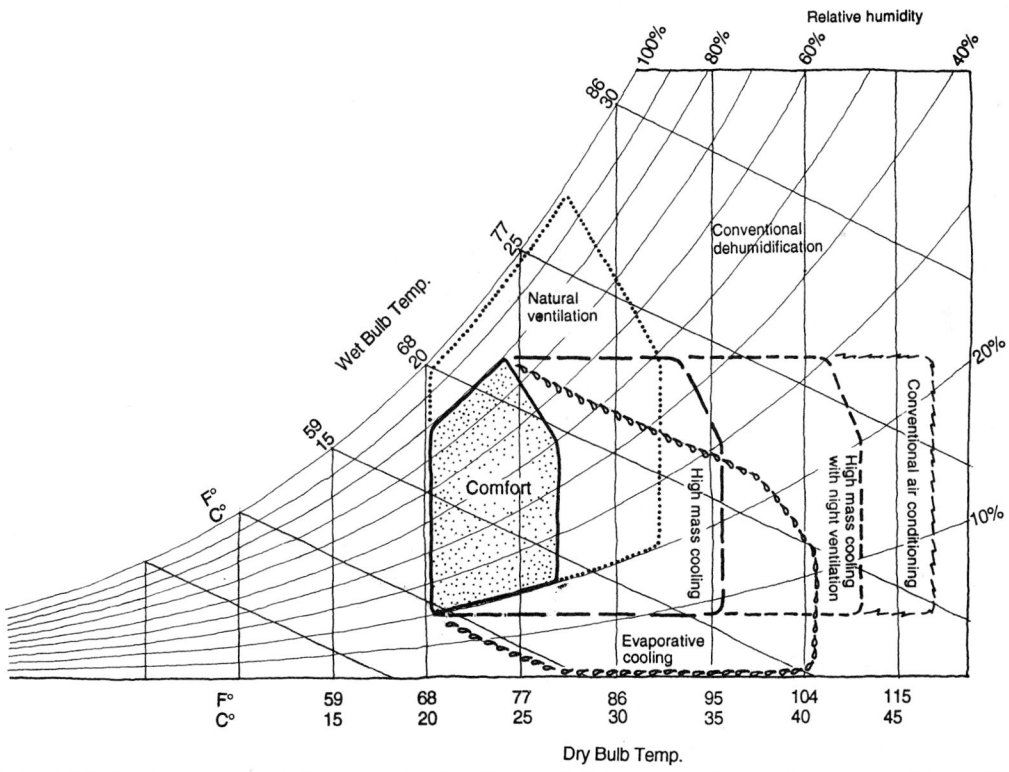

Fig. 2.13 *Passive cooling design strategies by climate. Buildings usually contain sources of heat. The more heat that is generated within the building, the more an* artificially warmer *climate is created. Thus, after plotting the outdoor climate data on this chart, consider how* shifting *these plots would affect your design strategy. The more* solar gains allowed inside, electric lights, and business machines, etc., *the further your plots shift to the right. The more* people, cooking, bathing, and other heat-plus-moisture sources, *the further your plots move both upward and to the right. (Based on Milne and Givoni, 1979.)*

TABLE 2.5 **Normal Data from Annual Summary of Local Climatological Data for Dodge City, Kansas**[a]

| Month | Daily Temperatures (°F) | | | | Relative Humidity (%) at Hour | | | | Wind | |
| | Normal | | Extreme[b] | | | | | | | |
	Max	Min	Max	Min	00	06	12	18	Mean Speed (mph)	Prevailing Direction
June	86.0	61.4	108	41	63	75	44	38	14.4	South
July	91.4	66.9	109	47	66	78	46	42	12.9	South
August	90.4	65.7	107	47	71	80	50	46	12.7	South

[a]Records as of 1979.

[b]Extremes do not occur on the same day or series of days.

zone are drawn as lines, these boundaries are more broad and vague than the lines suggest. When a climate surpasses the boundaries of all four strategy zones, conventional air-conditioning is almost certainly desirable.

The similarity between Figs. 2.8 and 2.13 is evident: the "comfort zones" include somewhat warmer conditions for naturally ventilated buildings.

To plot a climate's average monthly conditions on this strategy chart, assemble the information such as that shown in Table 2.5. This is available at every climatological station maintained by the National Oceanic and Atmospheric Administration (NOAA) of the U.S. Department of Commerce. These climate summaries, often called LCDs, are available from the National Climatic Data Center (Federal Building, Asheville, NC 28801). Many libraries also carry such summaries in publications such as "Climates of the States," available in the QC section.

In addition to a typical hot summer day, it is useful to check for the "design condition," that hottest moment for which engineers would select a building's mechanical cooling equipment. The design condition shows the climate near its "worst" while avoiding the freakish conditions of "extreme" temperatures listed in the NOAA data. This design data is available in Appendix A as Summer Design Dry-Bulb and Mean Coincident Wet-Bulb (2.5%), meaning that such a combination will, in a typical summer, only be exceeded by 2.5% of the total hours in the 4-month period of June through September.

EXAMPLE 2.1. Plot the typical day of the hottest month for Dodge City, Kansas.

SOLUTION. By inspection of Table 2.5, the hottest month is July, but August is close, and sun angles are lower (more gain through windows). To chart the typical August day, find the approximate rh for the coldest and the hottest hours. Because rh is listed only at four times, first select the highest rh, occurring here at 6 AM: this will coincide approximately with the coldest hour. Thus, one end of the linear climate plot for August in Dodge City will be at the combination of 80% rh and 65.7°F. The other end of the plot will be at the lowest rh and the hottest hour, in this case at 6 PM, a combination of 46% rh and 90.4°F. The line between these points in shown in Fig. 2.14.

We see from Fig. 2.14 that one passive cooling strategy—high mass with night ventilation—appears clearly adequate to serve for the typical August day in Dodge City. (In this strategy, the *low temperature* end of the line is also important and should be below the comfort zone for best results.) Another strategy, high mass, appears just barely adequate. Two other strategies, natural ventilation and evaporative cooling, appear inadequate because the highest temperature/lowest rh combination falls outside their boundaries; slightly so for natural ventilation, but greatly so for evaporation.

Now, check on the "design condition" from Appendix A. The summer dry bulb and mean coincident wet bulb is 97/69; this combination is shown as a circle around the point in Fig. 2.14. The 97°F DB temperature is plotted on the corresponding vertical line; the 69°F WB temperature is obtained by first locating the 69°F DB vertical line, and following that line up to 100% rh, where 69°F DB and WB are coincident. Then, follow

Fig. 2.14 *Hot-month daily ranges and the summer design condition for Dodge City, Kansas, superimposed on the design strategy chart of Fig. 2.13.*

downward to the right along the constant wet-bulb line to its intersection with 97°F DB.

The design condition point falls just outside the zone of high mass cooling, and just within the zone of evaporative cooling. Because the design condition falls well within the zone of high mass with night ventilation, this strategy is thus confirmed as clearly the best of the passive cooling strategies for Dodge City. ■

(a) Natural Ventilation Cooling. This is the most obvious strategy suggested by the comfort charts presented earlier, in which higher air temperatures were offset by increased air motion. It may be the only passive strategy available in more humid, hot climates in which temperatures are only slightly lower by night than by day (yet very high humidities may only be avoided by sealing and conditioning buildings). Buildings should be very open to breezes while simultaneously closed to direct sun. They may be thermally lightweight as well, because night air is not cool enough to remove much stored daytime heat.

Natural ventilation has two variations: cross ventilation and stack ventilation. *Cross ventilation* is driven by wind and is accomplished with windows. It relies on rather narrow plans with large ventilation openings on either side. Thus, it is naturally compatible with daylighting. *Stack ventilation* depends on very low openings to admit outside air and very high openings to exhaust air; it is driven by the principle that hot air rises. Stack ventilation is generally weaker than cross ventilation—except when there is no wind at all. Design guidelines for sizing apertures for ventilation are found in Section 5.6a,b.

(b) High-Mass Cooling. This strategy is for warm, dry summers, when the extremes of hot days are tempered by the still-cool thermal mass of the building. Cool nights then slowly drain away the heat that such mass accumulates during the day. The thermal mass can be on floors, walls, or roofs, but will need a "sink" to which it can reject its heat by night. The roof has the advantage of radiating to the cold night sky, but should be pro-

tected from exposure to sun by day. The masonry courtyard-type buildings of the Mediterranean are indigenous examples of this passive cooling strategy; their courtyard floors and roofs can be protected with movable shading devices (*toldos*) by day, then opened to night-sky radiation.

"Roof ponds" are a form of high-mass cooling for one- and two-story buildings. Because they require only the roof to be massive, they allow for considerable design freedom below, in walls and fenestration. Where cooling is the only objective, this approach uses water that is stored between the metal ceiling and the roof insulation; by night, it is pumped (and/or sprayed) over the exposed roof surface and allowed to trickle back through the insulation to the storage pond. At lower latitudes (with high winter sun altitudes), roof ponds can be used for passive solar heating as well. This strategy uses sliding panels of insulation over bags of water; panels slide open on winter days to collect sun, open on summer nights to radiate heat to the sky. A design guideline sizing procedure is found in Section 5.6*f*.

Another variation on high-mass cooling depends on earth contact. The earth acts as a heat sink, keeping walls and floors (even roofs where soil is the surface) cool. However, if the earth is allowed to continue to act as a heat sink in winter, heating needs could be greatly increased. Thus, a strategy for summer contact and winter isolation might be appropriate.

(c) High-Mass Cooling with Night Ventilation.

This hot–dry summer design strategy must use the air at subcomfortable nighttime temperatures to flush away the heat stored in daytime. The fewer the subcomfortable night hours, the more area of thermally massive surfaces must be provided to store the day's heat. Also, because there are more hours of daylight and few of night, the ventilation must occur more quickly and thoroughly, probably using fans. (Nightly wind velocities are typically lower than daily velocities, because summer wind is so often driven by regional solar overheating of the earth surface.) The building switches from a thermally closed condition by day (to exclude sun and hot outdoor air) to an open condition at night (to allow ventilation to cool the mass). Note: Nighttime temperatures must be cooler than the comfort zone if this strat-

egy is to be effective. Figure 2.15 shows this need to inspect both ends of the typical day's plot for this strategy.

This cooling strategy is highly compatible with passive solar heating strategies that rely on large areas of thermal mass, such as *direct gain* (Section 2.6*a*). It is suitable for large, high buildings, particularly buildings that have concrete (thermally massive) structural systems. An example of this strategy is shown in Fig. 5.7. A design guideline sizing procedure is found in Section 5.6*c*.

(d) Evaporative Cooling.

This design strategy relies on the principle that when moisture is added to air, the air increases in relative humidity while decreasing in dry-bulb temperature. (On the bioclimatic chart, this pattern exactly follows the WB line, upward and to the left.) In conditions that are more uncomfortably dry than uncomfortably hot, higher humidity is gladly exchanged for lower air temperature. However, large quantities of both water and outdoor air are needed; fan-driven evaporative coolers (see Fig. 7.8) are the most common way to provide this kind of cooling.

Evaporative cooling imposes few constraints on designers because the equipment resembles conventional HVAC systems. Rather high indoor air velocities and their associated sounds are typical of these systems, and the aroma of the wetted material of the cooler is often noticeable. The coolest air will be in the vicinity of the air inlet to the space from the cooler, the warmest air at the outlet from the space. Design guideline sizing procedures are found in Section 5.6*d* for fan-driven coolers, and Section 5.6*e* for passive "cool-towers."

Indirect evaporative cooling, involving more mechanical equipment, is first discussed in Chapter 6 and Fig. 6.16.

2.5 Design Strategies for Heating

Figure 2.16 is an indication of how greatly solar space heating might contribute to a building in the winter months. To use these charts, refer to Appendix B, Table B.15. For your location, find the average January ambient (DB) air temperature, called "TA," and the average January daily

Fig. 2.15 *Strategy chart segment for night ventilation of thermal mass. This strategy is most successful when cool nighttime temperatures are available, preferably below the comfort zone for several hours.*

solar radiation on a vertical south-facing surface, called "VS." Enter each chart with these two data points to determine the approximate percent contribution that solar energy can make to your building's winter seasonal fuel needs. In Fig. 2.16*a*, these points of TA and VS are plotted for seven North American cities. Sunny and cold Denver, Colorado, contrasts sharply with cloudy and cool Portland, Oregon. Yet, the subsequent charts show that even in Portland, solar can contribute to space heating.

The remainder of Fig. 2.16 compares two building envelope types with three common solar heating strategies. The left column assumes a building with insulation at code-required minimums and some south-facing glass (unshaded in winter). For this modestly solar building type, the seasonal fuel bill will be reduced, but rarely more than 40%, by using solar energy. The right column assumes a building that exceeds minimum insulation and is designed with much greater areas of south-facing

glass unshaded in winter. For this seriously solar building type, the seasonal fuel bill will be reduced by about twice as much as its modest counterpart. [For more details, see Section 5.10; the left column assumes a load-collector ratio (LCR) of 100, the right column, an LCR of 30.]

The three solar heating strategies are *direct gain* (let sunlight and heat into the space, where it warms exposed thermally massive surfaces), *indirect gain* (sun strikes thermal mass first, is then passed only as heat to the space behind), and *isolated gain* (a sunspace or greenhouse heated greatly by the sun that then passes some of its heat to the space behind). Direct-gain systems carry the symbol DG; indirect gain, TW (for Trombe wall); isolated gain, SS (for sun space). For more details, see Appendix C; all the systems in this figure assume double-glazed windows (and for the direct and indirect ones, low-ε, argon-filled windows), but none use night insulation over windows. The direct gain is system type DGB2, the

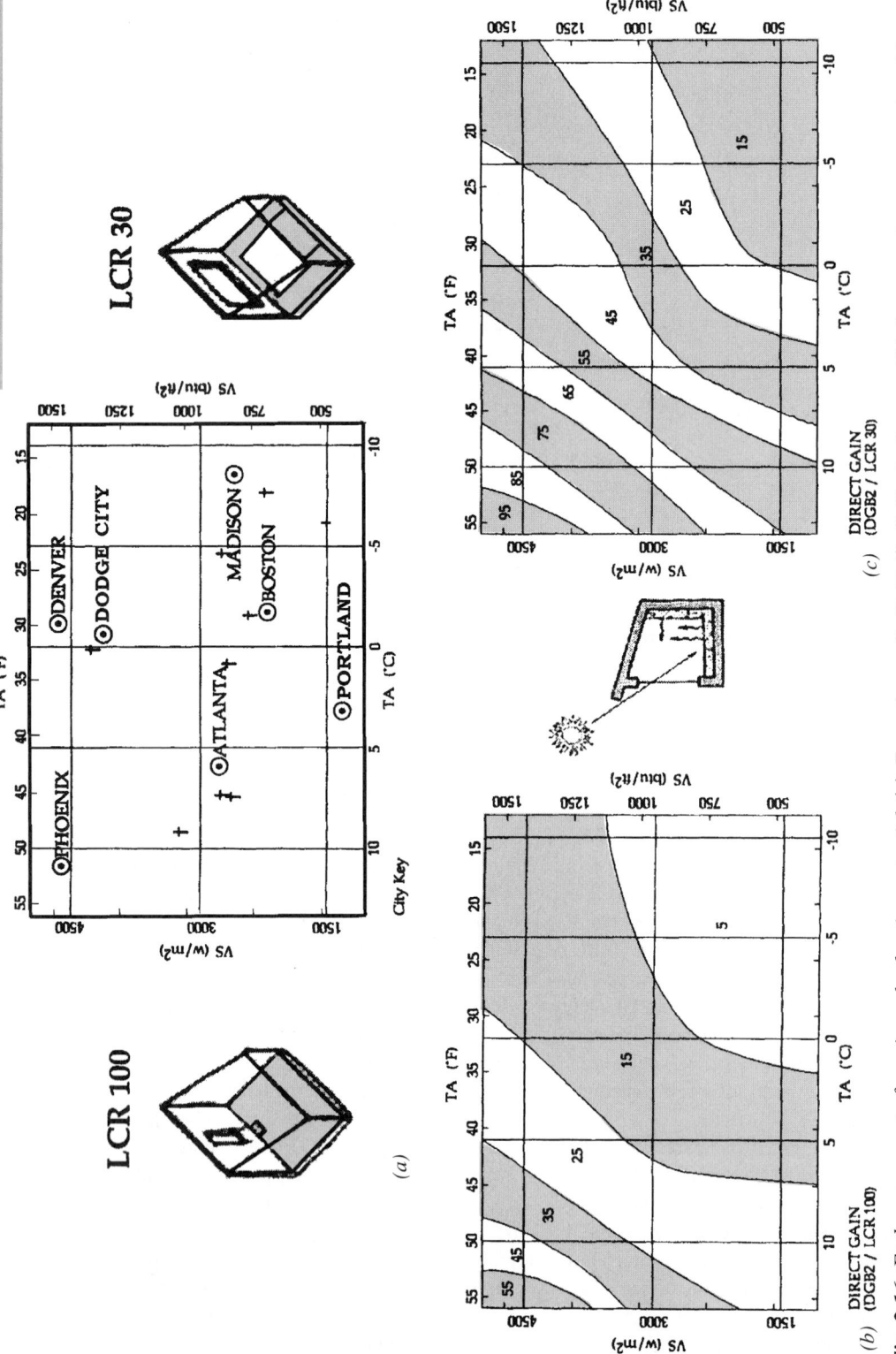

Fig. 2.16 Early assessment of passive solar heating potential. (a) From Appendix B, Table B.15, the January TA (average air temperature) and January VS (daily solar radiation on a vertical surface) are plotted for Atlanta, Georgia; Boston, Massachusetts; Denver, Colorado; Dodge City, Kansas; Madison, Wisconsin; Phoenix, Arizona; and Portland, Oregon. The Solar Savings Fraction (SSF), the percentage reduction in heating fuel achieved by passive solar design, is approximated for modestly solar "LCR 100" (smaller south windows, moderately well insulated) and seriously solar "LCR 30" (larger south windows, very well insulated buildings) († Indicates where SSF is approximately 6% higher than shown on graph for these locations: in California: Fresno, Mt. Shasta, Sacramento, and San Francisco; in Idaho: Boise and Pocatello; in Montana: Helena and Missoula; and in Reno, Nevada.) (b) Direct gain systems in modestly solar and (c) seriously solar buildings "LCR 30".

60

Fig. 2.16 (*continued*) (d) *Indirect gain systems in modestly solar and* (e) *seriously solar buildings; and* (f) *Isolated gain systems in modestly solar and* (g) *seriously solar buildings.*

ENERGY OVERVIEW

indirect gain is a vented TWD2, and the isolated gain is SSE1.

(a) Direct Gain. Almost any building with some south-facing glass could claim to be a direct-gain building. The sun is admitted to the space to be heated, striking furnishings and room surfaces. In a well-designed direct-gain space, there are ample thermally massive surfaces (such as concrete slab; concrete block; or brick, quarry, or ceramic tile) that directly receive much of this incoming sun. Such massive surfaces should be of at least three times the area of south-facing glass in order to keep the space from overheating in sunny hours. The mass should be thick enough, typically at least 4 in. (100 mm) to absorb and later reradiate a winter day's dosage of direct sun. In Chapter 5 and Appendix C, variations on direct gain approaches involve glazing types, relative mass areas and thickness, and whether night insulation is used over windows.

Direct gain is popular because of its simplicity and its ample daylight and view to the south. Relative to the other passive solar approaches, it has problems of glare, overheating on sunny days, and large radiant heat losses to glass areas by night; thus, a large diurnal difference in interior temperature and fading of furnishings in direct sunlight. This winter heating approach is well matched with the summer cooling strategy of night ventilation of mass; both systems depend on large areas of internal exposed thermal mass, so such an investment pays off in both winter and summer.

(b) Indirect Gain. This is the passive solar approach encountered least often, perhaps because a sheet of glass covering an opaque wall seems such a denial of "window." The mass wall behind the glass is usually 8 to 12 in. (200 to 300 mm) thick, and should be of a dense and highly conductive material such as standard-weight concrete or dense brick. Water in containers is also an option. An air space between the glass and the mass allows for an option to "vent" the mass wall, where cool air from the heated space behind the wall enters at the bottom, rises with solar heating, then exits into the space behind near the ceiling.

Sometimes this air space is made wide enough to admit a person for cleaning the inside surface of the glass, but operable casement or awning-type glazing are other cleanable options. The back surface of the mass (facing the heated space) should be kept clear of hangings or large furniture so as to facilitate radiant heat transfer to the space. In Chapter 5 and Appendix C, variations on indirect gain approaches involve mass characteristics (water or masonry), thickness, and surface treatments; whether masonry walls are vented; glazing types; and whether night insulation is used behind the glazing.

Indirect gain is less popular because it admits much less daylight and lacks a view to the south. Yet, relative to the other passive solar approaches, it has advantages of less glare, significantly less overheating on sunny days, and no fading of furnishings in direct sunlight. Another advantage is its large radiant heat contributions in the evening after sunny days. Many applications of indirect gain incorporate a smaller area of direct gain (window) within the larger indirect gain wall surface.

(b) Isolated Gain. This is a popular passive solar approach because it provides a uniquely sunny inhabitable space, alternately called "sunspace," "greenhouse," "sun room," or "winter garden." It experiences great variations in temperature—hot in the afternoon, cold before dawn—so that the space behind it can be kept reasonably comfortable with solar heating delivered as needed. The sunspace usually has both south-facing vertical glazing and inclined glazing, increasing insolation in both winter and summer. In Chapter 5 and Appendix C, variations on isolated gain approaches involve configurations of the glazings, whether the sunspace is surrounded by the building or is "added on," whether the common wall between sunspace and building is masonry or insulated frame, whether the sunspace's end walls (facing east and west) are glazed or insulated; glazing types; and whether night insulation is used behind the glazing.

Chapter 5 presents much more detail about building insulation standards, south-facing window areas, thermal mass, and options for movable window insulation. These passive solar systems are described and compared in Table 5.13, and a design guideline sizing procedure is given in Section 5.4. This introduction to passive solar heating design is intended to help the designer find combinations of daylighting, passive cooling, and passive solar heating design approaches that are feasible in a given climate.

2.6 Combining Strategies

Renewable sources of on-site energy for lighting, heating, and cooling allow a building to function with less need for imported energy sources. As building designs are shaped by the use of on-site sources, they become more regional in appearance. This works against the one-design-fits-all approach common to chain retailing (such as video stores or fast-food outlets), but works for the subtle integration of a building into its regional climate and culture while reducing the economic and environmental costs of imported energy.

(a) Daylighting. This may be the most obvious and commonly used of all on-site energy sources. Two approaches are *sidelighting,* using windows in walls, and *toplighting,* using skylights in roofs. A hybrid approach is the clerestory window in very high walls; depending on the clerestory window's relation to the floor plan, it could be considered either side- or toplighting.

Sidelighting not only admits light but allows for view to the exterior (and of the interior, especially at night). Sidelighting varies substantially by orientation and season, and direct sun can pour large quantities of heat through unshaded windows. With windows that face east and west, daylighting problems can occur due to the low sun altitudes in morning and evening that result in glare and unwanted summer heat gain.

Toplighting admits more light per unit area than does sidelighting, as skylights are directly exposed to the bright sky dome as well as direct sunlight. Toplighting less often results in glare problems because the light source is high in the space, unlike a sunny window that is in one's field of view. However, direct sun passing through a clear skylight may strike interior surfaces, with glare and fading consequences. Moreover, unshaded skylights (more than unshaded windows) are exposed to summer sun by day and to the cold winter sky. Although more efficient as admitters of light, they can represent a significant thermal penalty. Therefore, seasonal controls on skylights are even more important than on windows.

Design guidelines for the sizing of windows and skylights are found in Section 5.3. More extensive daylight design methods are found in Chapter 19.

(b) Daylighting, Cooling, and Heating. Although the relationships between these three aspects of environmental control are quite complex, in general the decision to extensively daylight a building tends to decrease its need for electricity and cooling and increase its need for heating. This is due to the influence of the large glass areas associated with daylighting. In summer, properly shaded daylighting openings contribute less heat gain than the electric lights that the daylight is replacing. This reduces the need for both electricity and cooling. In winter, some solar gain through daylight openings helps with heating (especially with south-facing windows), but the daylight still replaces electric lights that would otherwise be additional sources of heat. Thus, compared to a conventional building of the same size and function, the daylit building is likely to use more heating energy (unless passive solar heating is emphasized), less electricity for lighting, and less cooling energy.

The matrix of design strategies in Fig. 2.17 illustrates some opportunities for a simple one-space building across a range of on-site energy sources. "Ventilative" cooling strategies depend on the air as a heat sink; "evaporative/radiative" cooling strategies depend on both sky and air as heat sinks; "earth coupling" cooling strategies depend on the earth as a heat sink. For a more extensive treatment of combined strategies by region for daylighting, cooling, and heating, see Lechner (1991) for 17 U.S. climate zones with bioclimatic charts of monthly conditions, vernacular design approaches, and recommended climate design priorities.

PASSIVE COOLING STRATEGIES

EARTH COUPLING   RADIATIVE/EVAPORATIVE   VENTILATIVE

PASSIVE SOLAR
DIRECT GAIN
TOPLIGHTING   SIDELIGHTING

CROSS
VENTILATION

STACK
VENTILATION

NIGHT
VENTILATION
OF THERMAL MASS

COURTYARD

ROOF POND

EARTH
CONTACT

EARTH
TUBE

- Cooling System Driven by Sun   - Systems Cools by Night Only

Fig. 2.17 Combining daylight, passive cooling, and passive solar heating opportunities. These schematic sections of simple one-cell buildings remind designers of the potential for site-based, renewable resources and how they might interact. (Sharon Shoshani and Alan Rutherford, University of Oregon class project.)

HEATING STRATEGIES

ISOLATED GAIN
TOPLIGHTING   SIDELIGHTING

INDIRECT GAIN
TOPLIGHTING   SIDELIGHTING

Fig. 2.17 (*continued*)

ENERGY OVERVIEW

REFERENCES

Arens, E., Gonzalez, R., Berglund, L., McNall, P., and Zeren, L. (1980). "A New Bioclimatic Chart for Passive Solar Design," in Vol. 5.2, *Proceedings of the Fifth National Passive Solar Conference,* American Section of the International Solar Energy Society, Boulder, Colo., pp. 1202–1206.

ASHRAE (1995). ANSI/ASHRAE Standard 55a-1995, addendum to ANSI/ASHRAE 55-1992, *Thermal Environmental Conditions for Human Occupancy.* American Society of Heating, Refrigerating, and Air-Conditioning Engineers, Atlanta, GA.

ASHRAE (1997). *Handbook of Fundamentals.* American Society of Heating, Refrigerating, and Air-Conditioning Engineers, Atlanta, GA.

Givoni, B. (1998). *Climate Considerations in Building and Urban Design.* Van Nostrand Reinhold, New York.

Humphreys, M., and Nicol, J. F. (1998). "Understanding the Adaptive Approach to Thermal Comfort," in *ASHRAE Transactions 1998,* Vol. 104, Number 1.

Kwok, A. (1998). "Thermal Comfort in Tropical Classrooms," in *ASHRAE Transactions 1998,* Vol. 104, Number 1.

Lechner, N. (1991). *Heating, Cooling, Lighting Design Methods for Architects.* John Wiley & Sons, New York.

Milne, M., and Givoni, B. (1979). "Architectural Design Based on Climate," in *Energy Conservation Through Building Design,* D. Watson, Editor, McGraw-Hill, New York.

Olgyay, V. (1963). *Design with Climate: Bioclimatic Approach to Architectural Regionalism,* Princeton University Press, Princeton, NJ.

3
SITES AND RESOURCES

In Chapter 1, designers are encouraged to consider the site as a partner with the building in using renewable energy and conserving resources. In Chapter 2, climate and indoor comfort are analyzed in ways to encourage, where appropriate, a close relationship between building and site. Such buildings can open themselves often to the free energy available on-site, rather than retreating permanently within an indoor climate dependent on purchased energy. This chapter looks at microclimates: the localized climate near buildings where inside meets outside.

A designer's early decisions in site planning will influence later choices of both the building's mechanical and electrical equipment and its overall consumption of energy. When the site is seen as a collection of resources (sun, wind, water, plants) and also as part of the environment we all share, the buildings we design can approach self-sufficiency of energy supply. They can do this without limiting the availability of such local energy resources for neighboring buildings. In addition to saving energy, the use of on-site resources can create outdoor spaces that become especially pleasant to use. Such spaces can direct winter sun to a glass wall while blocking the wind, or funnel summer breezes through shade to an open window. Site planning is greatly influenced by economic considerations, zoning regulations, and adjacent developments, all of which can interfere with the design of the site to utilize the sun and the wind. Integration of all these concerns at the site-planning stage is the first step in adapting a building to its climate.

The microclimate is influenced by the interaction of the characteristics of both site and climate:

Site	Climate
Soil type	Sun
Ground surface	Air temperature
Topography (profile)	Humidity
Vegetation	Precipitation
Water presence	Air motion
View	Air quality
Human activity (heat, noise, etc.)	

This chapter looks briefly at some results of site–climate interactions.

3.1 Climates within Climates

The statistics describing the climate at a particular site can be quite different from the statistics that are published as being representative of the entire region. This is particularly evident where a neighboring hill blocks wind or winter sun, or an adjacent lake cools summer breezes or adds a damp chill to the winter air. It may be less obvious that urban sites are under the influence of an urban subclimate—probably the best known of whose features is a "heat island" —that differs from the conditions of the surrounding countryside unaffected by urbanization. The cities' climatological stations are often at a more rural site, such as an outlying airport, further masking the effects of the heat island. The effects of this urban heat island are summarized in Table 3.1 and Fig. 3.1.

The most obvious reason for the city's relative warmth year-around is its concentration of *heat sources:* the air conditioners, furnaces, electric lights, and internal combustion engines in cars and buildings. This "internal" urban and industrial heat production is shown in Table 3.2 and Fig. 3.2. It appears that cities and industrial regions of the world release less internal heat per capita as people live and work closer together. Not surprisingly, commercial and industrial cities release more heat for a given population density, and tropical cities release less, while still conforming

TABLE 3.1 **Average Changes in Climate Effects Caused by Urbanization**[a]

Effect	Comparison with Rural Environment
Contaminants	
Condensation nuclei and particulates	10 times more
Gaseous admixtures	5 to 25 times more
Cloudiness	
Cover	5 to 10% more
Fog, winter	100% more
Fog, summer	30% more
Precipitation[b]	
Totals	5 to 10% more
Days with less than 5 mm (2 in.)	10% more
Snowfall	5% less
Relative humidity	
Winter	2% less
Summer	8% less
Radiation	
Global	15 to 20% less
Ultraviolet, winter	30% less
Ultraviolet, summer	5% less
Sunshine duration	5 to 15% less
Temperature	
Annual mean	0.5 to 1.0 C° more
Winter minima (average)	1 to 2 C° more
Heating degree days	10% less
Wind speed	
Annual mean	20 to 30% less
Extreme gusts	10 to 20% less
Calms	5 to 20% more

Source: Landsberg (1970).

[a]These effects vary from city to city and from day to day.

[b]Research since 1970 has shown it is not at all certain that urbanization causes increases in precipitation amount within a city.

to the idea that greater compactness permits less energy use per capita.

The *rain* that falls on the city and countryside can be an effective evaporative cooling mechanism, especially as water evaporates from wet surfaces. Yet the materials of streets and buildings are usually designed to shed water quickly and thoroughly, sending rainwater into storm sewers instead of evaporating slowly in the wind and sun. Thus, evaporative cooling for these surfaces is minimized.

The city also changes the overall cooling action of the *wind* by channeling it into narrow streets. The geometry of high vertical walls and narrow streets also increases the summer heat in cities as the high *sun* is reflected downward and is absorbed and then reradiated by the often rocklike streets and building surfaces. In winter, however, this geometry puts most urban surfaces at a solar disadvantage because the low sun strikes only the upper portion of south-facing walls. The reduction in radiant heat loss at night, caused by this lack of access to the sky, is a key element in the formation of the urban heat island, as summarized in Fig. 3.3. Sky access is discussed in more detail in Section 3.4.

A more subtle urban climate influence is *contaminated air.* Small particles in the city's air can

Fig. 3.1 *Urban heat island; a nighttime difference between the original countryside climate and that of the new urban area within it. (a) Direct solar radiation is likely to be reflected within the city, thereby increasing solar gain to urban areas. (b) Idealized profile of the air temperature difference between urban and rural, at times of peak differences: calm, clear nights. (c) Based on (b), typical isotherms of equal temperature provide a "contour map" of the urban heat island. (d) Temperature records at a rural site (solid line) and in the central city (dashed line) during a typical night and day. The city's heat-conducting materials and thin cloud of polluted air acting alone would not change the average air temperature, but would reduce the day–night difference (producing the dotted line). Adding to that the heat from increased solar gain and internal heat sources warms at all hours, producing the observed urban record (dashed line). (e) The urban heat island can affect the countryside "downstream." [Reprinted by permission from Lowry (1988).]*

keep some sunlight from reaching the city, yet it can help to keep the city's heat from radiating outward. These particles also form additional nuclei for fog droplets. Table 3.1 suggests that fog may occur in a city in winter up to twice as often as in the surrounding countryside. Trees and greenery are not as available in the city to act as crude filters of airborne dust.

The city thus changes its climate from that of its surroundings. In winter, the city's factories, vehicles, surface materials, and geometry combine to reduce the amount of energy needed for heating buildings. The typical means of providing additional winter heating (by fossil fuel–burning furnaces and power plants) contribute to airborne particles and urban fog. Solar-assisted heating would diminish this pollution—and with less air pollution, more sun would reach a solar collector. However, the greater the population density, the more difficult is access to winter sun, especially at latitudes farther from the equator, where winter heating is most needed. High buildings readily block low sun altitudes.

In summer, the city's internal heat makes things worse. Air conditioners add their own mechanical heat to the building heat that they pour into the air; their appetite for electricity taxes the power plants, in turn adding vastly more waste

TABLE 3.2 **Heat Generated Within Cities**

City or Region[a]	Population (millions)	Area (km²)	Population Density (10³/km²)	Energy Use per Capita (kW/capita)	Energy Use per Unit Area[b] (W/m²)
Cities and Industrialized Regions					
3 Fairbanks[c]	0.045	80	0.6	10.91	6.55
11 Vancouver	0.6	111	5.4	3.55	19.2
12 Brussels	1.3	163	8.0	3.5	28.0
14 West Berlin	2.3	235	9.8	2.14	21.0
9 St. Louis	0.75	250	3.0	5.3	15.9
10 Munich	0.9	300	3.0	3.0	9.0
5 New Jersey suburbs of NYC	4.7	6,100	0.8	9.1	7.3
1 Los Angeles County	7.0	10,000	0.07	10.5	0.74
4 Nordheim–Westfalen[d]	16.9	33,800	0.5	8.0	4.0
2 BosNyWash[e]	33.0	87,000	0.38	11.2	4.3
Commercial/Industrial Cities					
15 Sheffield	0.5	48	10.4	1.83	19.0
17 Manhattan	1.7	59	28.8	5.52	159.0
16 Montreal	1.1	78	14.1	7.02	99.0
13 Budapest	1.3	113	11.5	3.74	43.0
8 Cincinnati (summer)	0.6	222	2.7	9.3	25.1
7 Hamburg	1.83	763	2.4	5.3	12.7
6 Chicago	3.5	1,842	1.9	27.2	51.7
Tropical					
18 Hong Kong	3.9	105	37.1	0.88	32.6
19 Singapore	2.1	140	15.0	0.81	12.2

Source: Lowry (1995), by permission.

[a]Index numbers refer to Fig. 3.2.

[b]0.317 W/m² = 1 Btu/ft².

[c]Data for Fairbanks, Alaska, include a sparsely populated incorporated area surrounding the CBD.

[d]Nordheim–Westfalen is the heavily urbanized and industrialized region in the lower Rhine valley near Düsseldorf–Dortmund.

[e]*BosNyWash* is the term often used by urbanists to refer to the megalopolis that stretches from Boston through New York to Washington.

heat. Again, solar energy can help, this time by generating photovoltaic (PV) electricity to run the air conditioners. High summer sun angles make solar access much less problematic. The stronger the sun, the more electricity is produced for the air conditioners. PV arrays are currently high in first cost, but are noiseless, emit no exhaust gases, and add no internally generated heat to the summer air.

Site- and urban-planning responses to these urban climate characteristics can sometimes lead in contradictory directions. For example, the provision of greenways within cities would bring softer ground surfaces cooled in summer by shading, breezes, and evaporation. However, its winter impact might be an increase in fog via evaporation of retained storm water; or perhaps fog would be discouraged locally by increased speeds in wind

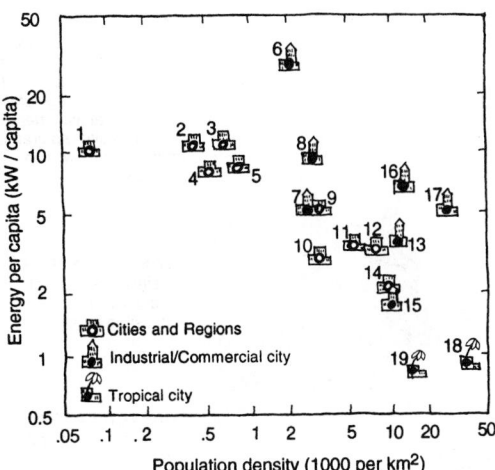

Fig. 3.2 *Heat released from cities: population density and energy use per capita. Numbers refer to locations in Table 3.2. [Data with permission from Lowry (1995).]*

channels formed by the greenways. The summer impact of the greenway is also mixed; changes in comfort would depend on the particular combination of increased humidity and cooler temperature the greenways produced. On balance, the

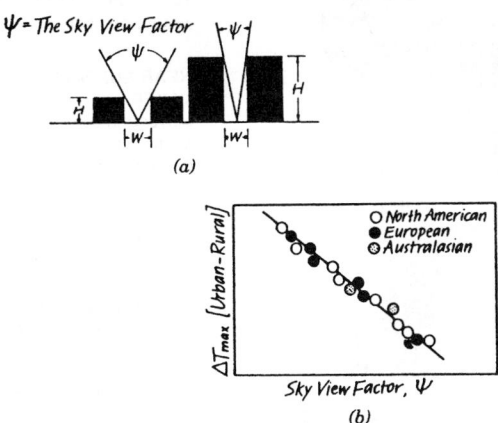

Fig. 3.3 *The urban heat island is particularly strong on calm, clear nights.* (a) *With a greatly reduced "sky view factor" (Ψ) to the cold night sky, the walls and floors of urban canyons—streets—cannot lose heat as readily as can the open countryside.* (b) *The more narrow the Ψ, the more pronounced is the effect of the urban heat island, ΔT, in cities throughout the world. [Reprinted by permission from Lowry (1988).]*

greenway within a city seems to be a positive and esthetic step in ameliorating the urban climate.

While considering how the sun, wind, and other climate elements can be utilized on a site for the benefit of a building, it is important to remember the need to protect the access of others to these same, shared resources. In "Tragedy of the Commons" (1968), Garrett Hardin writes about the commons as publicly owned meadows shared by many herders. Each herder realizes that his personal wealth will increase as animals are added to his herd, so all herders increase their livestock. However, the meadow capacity does not increase; overgrazing occurs, and as a result the commons become unable to support any animals. The following discussions of the resources of sun, air, and water found on a site are influenced both by the "private" needs of a building and the "public" patterns of resource availability, which should remain accessible to all.

3.2 Analyzing the Site

After recognizing the resources that exist on and around a site, the designer then decides how best to integrate these resources into the building design while making this building and site design a successful addition to the larger patterns of the surroundings. Schematic site plans are typically used as a kind of inventory; overlaid "bubble diagrams" can test possible design arrangements that relate rooms or functions to their surroundings in the plan. Sun and wind conditions (in both summer and winter), noise sources, and water runoff patterns are often included in this schematic plan. It is particularly important to identify microclimates on the site, the places that have special characteristics differing from the regional climate. Microclimates can present opportunities for an "expanded" comfort zone; more sun in cool air, more wind in warm air. Microclimates can sometimes reveal building sites where less energy is consumed because the winter is warmer or the summer cooler. Microclimates can also be problem areas to be avoided for building or outdoor activity, if possible, or where special design measures need to be taken to correct their difficulties.

ENERGY OVERVIEW

ENERGY OVERVIEW

Increasing long range view, increasing privacy and increasing exposure to the wind, sun, daylight, and rain.

Maximum exposure to public, to surface activities and sounds, to distribution and to a variety of microclimates.

Increasing privacy, thermal stability, and exposure to groundwater.

SKY LAYER
Isolation by height; too far from surface to see or hear its activity in detail, too high to climb stairs regularly. Extensive activity here places heavy requirements on layers below (see Fig.12.9)

NEAR-SURFACE LAYER
Detailed overview of surface activities, accessible by stairs.

SURFACE LAYER
The most varied and public level.

SUBSURFACE LAYER
Isolation by enclosure; often plays a supporting services role for structure, mechanical and electrical equipment.

(a)

Increasing long range view, increasing privacy and increasing exposure to the wind, sun, daylight, and rain.

Maximum exposure to public, to surface activities and sounds, to distribution and to a variety of microclimates.

Increasing privacy, thermal stability, and exposure to groundwater.

SKY LAYER
Least frequent public contact public works, housing, health, administration parks and recreation, building, and redevelopment.

NEAR-SURFACE LAYER
Mayor and council offices, council chambers, reference library, news conferences, exhibits.

SURFACE LAYER
Most frequent public contact; entry and lobby, complaints, elections, licensing, assessing, health registration

SUBSURFACE LAYER
Parking, mechanical equipment, data processing, inactive files, and storage.

(b)

Fig. 3.4 (a) *Characteristics of horizontal layers for site analysis.* (b) *Layers and form: Boston City Hall, 1969. (Kallman, McKinnell and Knowles, Architects.)*

The site's collection of microclimates is not limited by those visible in plan, on the surface. Conditions of privacy and accessibility, view, heat, light, air motion, sound, and water all change with vertical distance from the surface (Fig. 3.4*a*).

To minimize energy consumption in constructing and using buildings and to integrate them with their surroundings, the conditions best suited to various functions should approach the characteristics of the "layer" of the site in which they are located. Consequently, both vertical and horizontal site analyses are needed. A lecture hall, needing both an acoustically isolated and a closely controlled environment, is an obvious candidate for the subsurface layer. Electrical equipment, which benefits from cool environments, is also suitable to the subsurface layer.

One building whose form responds to these layers is Boston's City Hall (Fig. 3.4*b*). Activities with the most frequent public interaction are located near the surface, such as the skylit, high-ceilinged lobby, whereas special ceremonial functions are elevated to distinctive forms in the near-surface layer above. The city offices with less frequent public contact occupy several floors in the sky layer, where daylight is plentiful. Storage and mechanical functions, as well as parking, are in the subsurface layer. An esthetic equivalent of this horizontal layering is architect Louis Sullivan's concept of a facade as a "base, body and capital."

3.3 Site Design Strategies

In Chapter 2, appropriate building design strategies were identified by analyzing climate data. Sites can also be organized to aid in the heating, lighting, cooling, and noise control characteristics of buildings; a few of the most common site strategies are compared with their seasonal roles by Watson and Labs (1983) in Fig. 3.5. These graphics suggest housing but also apply to any function that can utilize the extended comfort zone. Later in this chapter, several of these strategies are discussed in more detail.

One example of how the information in Figs. 3.4 and 3.5 can be applied (and manipulated) in the planning of a building and its site is a house designed by Frank Lloyd Wright in the 1940s (Fig. 3.6). The direct gain of solar heat through its south-facing windows in winter makes this an early example of "passive solar heating" (by contrast, "active solar heating" includes collectors and a separate storage volume for solar heat, such as a tank of water or a bed of heated rocks). This

ENERGY OVERVIEW

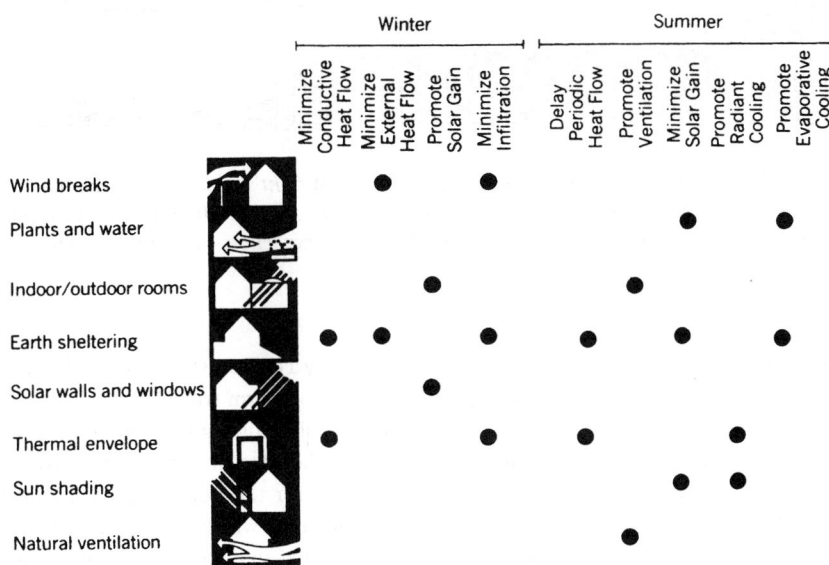

Fig. 3.5 *Generic bioclimatic site design concepts. (Reprinted from* Passive Cooling *by permission of the publisher, American Solar Energy Society, Inc.)*

North
↑

Balcony Level

Ground Fl.

Entry Tunnel

Clo.

Sunken Lawn

(a)

(b)

Fig. 3.6 *An early passive solar-heated home, Frank Lloyd Wright's Solar Hemicycle (Jacobs House II) near Madison, Wisconsin (between the cool and temperate zones). The house was designed in the early 1940s and built in 1948. (a) Floor plans. (b) Section-perspective, looking east toward the entry tunnel in the berm wall.*

house, known as the Solar Hemicycle, was built in 1948 near Madison, Wisconsin, which lies between the cool and the temperate climate zones. Winter heating is the dominant thermal influence in this area. The house stands on a rise in the prairie, particularly vulnerable to winter winds. During construction, earth was scooped from in front of the south face of the house and bermed against the entire curved north wall almost to roof

level. Only a narrow strip of second-floor windows separate the berm from the roof. This berm and curve combination gives winter-wind protection from northeast to northwest and provides further insulation for the north wall. This north wall is made of stone, which absorbs and stores the winter solar heat that comes in through the floor-to-ceiling, southeast-to-southwest–facing glass. The concrete slab-on-grade first floor also stores solar heat in winter. The impression that this house and site are sun collectors is heightened by an entrance tunnel in the northeast end of the house and berm that leads from the parking area through the berm and onto the sunny, protected south terrace overlooking a sunken "sun-trap" circular lawn. Passage from the cold north side through an even colder tunnel to the sunny south side prepares the visitor for the solar-heated interior.

This house plan is shorter in the north–south and longer in the east–west directions (about 1:3). This elongation is typical of passive solar designs, which are able to store and use the winter solar heat gain, and thus profit from having long south-facing glass walls to act as passive collectors. Had these large windows been well insulated at night, and the roof and walls insulated up to the present standards, this house would be a very up-to-date example of passive solar heating.

In the Solar Hemicycle house, protection from summer overheating is provided by an overhang along the south glass walls, as well as the now-shaded and cool thermal mass of the north stone/berm wall combination. The high windows in both north and south walls allow warmed air to rise and escape.

Almost all of today's energy sources have the sun as an ancestor. The earth receives a very small percentage of the sun's daily energy output; the amount of solar energy available to each site varies both seasonally and daily. Typically, the closer the sun to a position directly overhead, the more solar energy reaches a horizontal site. Fortunately, direct sunlight's radiant energy is not its only resource; direct sun is our most intense light source, and its use is complicated by its slow but constant change of position. Indirect sun, such as on an overcast day or from the clear north sky, is a diffuse and readily utilized light source. Daylighting was the dominant method of lighting buildings until the middle of the 20th century.

3.4 Direct Sun and Daylight

(a) Access to Light and Sun. The value of daylight (and air) to buildings has long been recognized in zoning laws, which require that minimum distances (setbacks) be maintained between a building and the property line in lower-density areas. Height restrictions often accompany these setbacks, defining a maximum "buildable volume" in which a building can grow (Fig. 3.7). As buildings become taller and density increases, daylight reflecting down between them is diminished; in response, the maximum buildable volume becomes more narrow as it rises.

When *direct sun* in winter is desirable at the ground floor of each site, this buildable volume is sharply reduced in height due to the low angle of the winter sun at northern latitudes. This most-restricted buildable envelope, called the *solar envelope,* is at present uncommon, but solar access ordinances have been enacted for residential zones in some cities. The most restrictive feature of the solar envelope is the low slope of its northern face, usually corresponding to the altitude of the sun above the horizontal at about 2 hours from noon on December 21. This allows 4 hours of access to direct sun for your northern neighbor, on even the shortest day.

A brief history of daylight and solar access

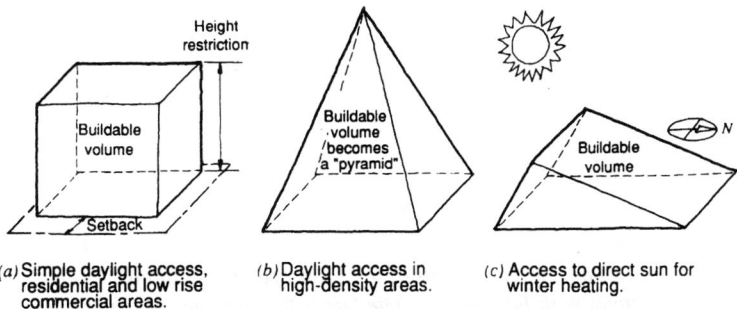

(a) Simple daylight access, residential and low rise commercial areas.   *(b)* Daylight access in high-density areas.   *(c)* Access to direct sun for winter heating.

Fig. 3.7 *Access to light. Some regulatory methods that compromise between private optima (e.g., maximum rentable floor space) and public optima (e.g., daylight at street level).*

Daylight   Daylight   Atkinson,
Spacing Angles   Indicators   1912

Fig. 3.8 *Maximum allowable building envelopes for daylight access. These envelopes are applied to a 200' × 400' (61 × 122 m) block at 40° N. Latitude. The east–west streets (along the longer side) are 65' (20 m) wide; the north–south streets are 45' (14 m) wide. In this case, "daylight spacing angles" and "daylight indicators" produce nearly identical envelopes. [From DeKay (1992), with permission of the American Solar Energy Society.]*

Solar   New York,   New York,
Envelope   1916   1980

(a)

(b)

(c)

Fig. 3.9 *These solar envelopes are refinements of the solar access "pyramid" of Fig. 3.7. (a) The slope of the solar envelope changes with latitude. (b) The larger the site, the greater the buildable volume of the solar envelope. (c) Solar envelopes for various orientations of individual sites. (d) Solar envelopes for east–west elongated blocks. (e) Solar envelopes for north–south elongated blocks. (Reprinted by permission of R. Knowles,* Sun, Rhythm, Form, *© 1981, MIT Press.)*

(d)   (e)

Fig. 3.9 (continued)

regulations and current design guidelines is discussed in DeKay (1992). Figure 3.8 shows the application of six variations of regulations on an urban block at 40° N. latitude, elongated east–west. The east–west streets were wider than the north–south streets. Further development of the solar envelope is shown in Fig. 3.9, based on the work of Knowles (1981). Clearly, winter solar access is closely related to maximum density of development.

For daylight access, building surfaces can be almost as important as building geometry. The importance of reflected light from vertical and horizontal surfaces is apparent in the daylighting calculations in Chapter 19. Lighter-colored surfaces produce more internal light, especially in crowded urban conditions where a view of the sky from windows is not common. An increase in daylight results for all surfaces on an urban street if the building and site surfaces are light colors. This can conflict with the use of planting for shading and evaporative cooling, although lighter-colored groundcover plants are available. Although daylight reflected in a diffused way from building and ground surfaces is a potential benefit, the harsh specular reflections of direct sun in mirror surfaces are often unwelcome (see Section 3.4e).

(b) Charting the Sun. For site planning, one of the earliest tasks is to determine when direct sun reaches a building or activity—such as a playground—on a site. The details of how a building utilizes the sun are found in later chapters on solar heating (Chapter 5, and for water heating, Chapter 10) and on daylighting (Chapter 19). Numerical data on sun position and intensity are found in Appendices B and C.

The charts in Fig. 3.10 show the sun's position for three latitudes that cross the United States and also show, for the 6 hours of greatest insolation, what percentage of clear-day insolation is gained each hour by an unshaded *south-facing* vertical window. This emphasis on south orientation calls attention to its useful characteristic of receiving *more* sun in winter and *less* sun in summer than any other orientation. To convert these percentages to actual heat gains, see Appendix Tables B.1 through B.10 (for *clear* days at a given latitude), or Appendix Table B.15 for *average* day data by climate station (January and July only).

These sun-position charts are actually graphs superimposed on the eye of an observer. At the vertical center of each chart, the observer is looking due south. The horizon (a horizontal plane at observer eye level) is the line at the bottom of the chart.

(c) The "Band of Sun." We return to the section as a site analysis tool. A north–south section is drawn through the proposed building and its site. At three times of year (each solstice and the equinox), the noon solar altitude is determined, as in Fig. 3.11. Then the "band of sun" that strikes the building and site is drawn, incoming at the noon altitude. The portion of this available direct sun that is used by the building and site can then be analyzed. The more of the band that is unused in each season, the higher the potential for redesign in order to better utilize this solar resource. This analysis helps to show how wintertime solar usages depend on lower sun altitudes; summertime uses depend on higher altitudes.

(d) Skylines and Winter Sun. The actual skyline as seen from a given place can be charted to determine access to direct sun at any time of year (Fig. 3.12). Such an analysis should precede the planning of the location of a solar collector,

(a) Altitude

Two coordinates are used to locate the position of the sun in the sky. They are called the <u>altitude</u> and the <u>azimuth</u> (also called the bearing angle).

Solar altitude is the angle, above the horizon, of the position of the sun measured from the horizon. The horizontal lines on the chart represent altitude angles in 10° increments above the horizon.

(b) Azimuth

Solar azimuth is the angle, along the horizon, of the position of the sun measured to the east or west from true south.

(c) Skydome

The grid on the chart thus represents vertical and horizontal angles of the whole skydome. It is as if there was a clear dome around the observer, and then the chart was peeled off of this dome and laid flat.

(d) Sun Position

With the altitude and azimuth angles, the sun can be located at any point in the sky.

Fig. 3.10 *Sun charts. Illustrations* (a) *through* (g) *describe the development of the charts shown for three latitudes that cross the United States, map* (h). *(From Edward Mazria and David Winitsky,* Solar Guide and Calculator, *Center for Environmental Research, University of Oregon, 1976.) Charts* (i) *through* (k) *include the approximate percentage of clear day insolation **for south-facing windows,** for each of the 6 maximum hours of sun each month.*

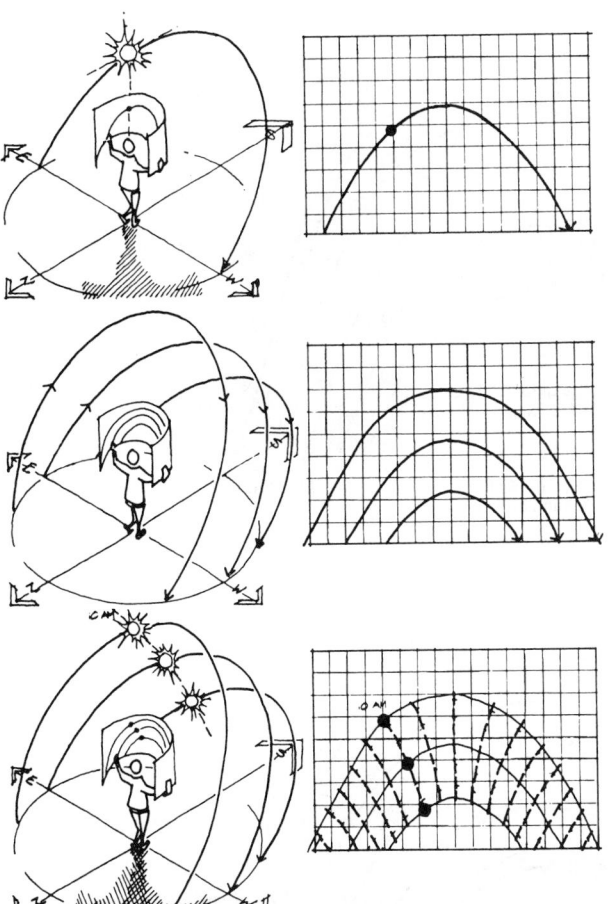

(e) Sun Path

By connecting the points of the location of the sun throughout various times of the day, the sun's path for that day can be drawn.

(f) Seasonal Paths

Thus, we can plot the sun's path for any day of the year. Those paths shown represent summer, fall/spring, and winter. The sun path is greatest during the summer when it reaches its highest altitude and rises and sets with the widest azimuth angle from true south. During winter, the sun is much lower in the sky and rises and sets with the narrowest azimuth angles from true south.

(g) Times of Day

Finally, if we connect the hour lines on each path, we get the heavy dotted line which represents the hours of the day. This completes the sun chart. NOTE: The times on the sun chart are for solar time. This may vary from standard time as much as an hour and fifteen minutes for different locations and different times of the year. This is adequate for most practical uses of the sun chart. It is important to remember to at least use standard time when using the charts.

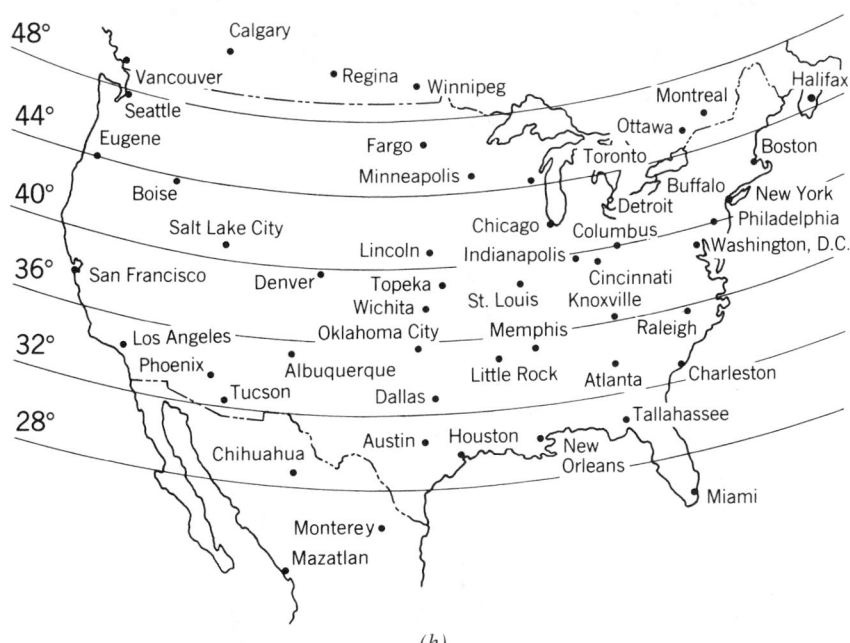

(h)

Fig. 3.10 *(continued)*

ENERGY OVERVIEW

Fig. 3.10 (continued)

(a)

(b)

(c)

Fig. 3.11 *The Band of Sun available for a proposed building and its site is charted at solar noon on a north–south section.* (a) *The summer solstice, where collecting surfaces are most optimum at near-horizontal tilt angles.* (b) *The equinox.* (c) *The winter solstice, where collecting surfaces are most optimum at near-vertical south-facing tilt angles.*

Fig. 3.12 *Charting the skyline from a specific location. (From Edward Mazria and David Winitsky,* Solar Guide and Calculator, *Center for Environmental Research, University of Oregon, 1976.)*

whether it be a south-facing window or a mechanical device. Seasonal obstructions within the six "best" hours (9 AM to 3 PM) are particularly serious and should be minimized in collector siting.

At the same time, consideration of the neighbors' access to direct sun is appropriate. This can be checked by charting another skyline somewhere along the northern boundary of your site.

The outline of the building being designed should be included on this skyline, and can then be modified, if necessary, to preserve solar access for its neighbor.

The location charted in Fig. 3.12 is quite a good one for a summer solar usage, such as a swimming pool heater, or for a PV array that serves summer loads such as air conditioners. It is also a possible location for a solar collector for heating of domestic water, which does most of its work in the 6 months from March through September; obstructions are few within this period. However, it is quite poor for winter solar collection, allowing only about one-fifth of the potential solar gain in the November to January period.

(e) Sun and Shadows: Model Technique.
These graphic techniques have a limitation: each graph applies only to *one* particular location. To study multiple locations, multiple graphs must be constructed. By contrast, a three-dimensional model used in conjunction with a sun shadow plot (Fig. 3.13) can yield the three-dimensional sun penetration patterns as they change over time, for *any* location, indoors or out, on the model. Models are initially time-consuming to build but can save time when considering alternate locations on a site, alternate window and space combinations, alternate shading devices, and so on. Perhaps most important to you, the designer, models suggest three-dimensional alternative solutions, because you are testing *space* rather than merely plan or section.

However, if obstructions to needed sun exist far from the site (such as nearby mountains), you might want to rely on graphs rather than try to include such large obstructions on your model.

In Fig. 3.13, "sun peg shadow plots" are found for seven latitudes. A copy of such plots, correctly attached to a model of *any scale,* allows the designer to quickly determine exact sun penetration and shadow patterns at many times, for any date. These plots are one of the most important tools for the solar designer, both early and late in the design process.

In Fig. 3.13*h*, a model is observed at a sun position corresponding to 3 PM on December 21. At this end of the 6-hour period of best collection, we see that sun still fills the south side glazed open-frame circulation space and enters spaces beyond through south openings. We can also see that passing from these spaces into the circulation space might be an experience of considerable glare because the sun is so low in the sky. In such a case, vertical fins inside the south glazing would help intercept and diffuse direct sun while still permitting solar heat gain.

(f) Controlling Solar Reflections.
The use of highly reflective (or "mirror") glass to reduce heat gain in office buildings has increased the frequency of annoying solar reflections from buildings (Fig. 3.14). Large areas of non–mirror glazing, such as on passive solar heated buildings, can also cause reflection problems. The farther the sun's rays are from being perpendicular (or normal) to any surface, the more of the sun's light is reflected, rather than absorbed, by that surface. Thus, the intensity of the reflection is greatest (and the transmission of solar gain is the least) when the sun's rays are nearly parallel to a reflecting surface.

Because these most intense reflections occur when the sun's rays are nearly parallel to a wall, they are fairly easily blocked or intercepted. For example, foliage (Fig. 3.15) can intercept reflected sunlight from a south-facing window without blocking solar collection during the best hours. Another common approach is to use external projections around windows, as in Fig. 3.16. Again, the use of the model and sun peg shadow chart is a good way to explore reflections. By using mirrored surfaces to represent reflective glass on the model, the resulting reflections will be evident along their three-dimensional path.

3.5 Sound and Air

Sound and air are considered together because they are so difficult to separate. Many buildings that could be opened to ventilation or cooling by breezes rely instead on forced ventilation because of the noise that would accompany the breeze through an opened window. Polluted air is another potential deterrent to "natural" ventilation. Almost any device that reduces sound will also reduce the velocity of the breeze, as is true of most filtering devices to remove dust particles.

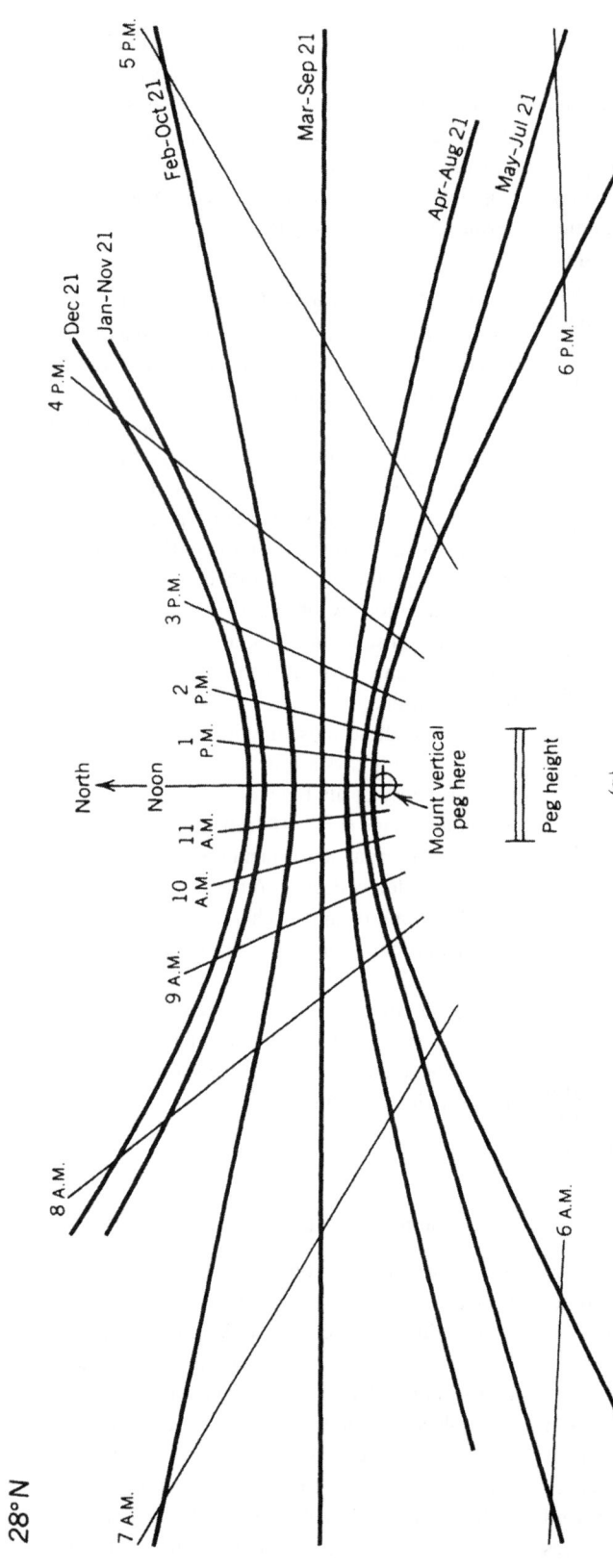

28°N

(a)

Fig. 3.13 (a–g) *Sun peg shadow charts. These charts will show the exact position of sunlight and shadow on a model of any scale, on any date, at any time of day between shortly after sunrise and shortly before sunset.*

1. *Find the chart nearest your latitude [map, Fig. 3.10 (h)].*
2. *Make a copy of this chart. (If the copier changes the chart size, the peg height line will also change; there should be no problem.)*
3. *Construct a "peg" whose finished height above the chart surface corresponds exactly to the "peg height" shown on your copy of the chart. (This peg must stand and remain perfectly vertical relative to the model.)*
4. *Mount your copy of the chart on the model to be tested. The chart must be perfectly horizontal over its entire surface, and the north arrow on the chart must correspond to true north on the model.*
5. *Mount your vertical peg at the location shown on the chart. Be sure it is vertical relative to the model and will not be bumped out of vertical alignment.*
6. *Choose a test time and date. Take your model out into direct sunlight. (With a small model, a camera tripod might serve as a mount.) Then tilt the model until the shadow of the peg points toward the intersection of the chosen time's line and the chosen date's curve. When the end of the peg's shadow touches this intersection, your model will show the same sun-shadow patterns as would occur on the time and date at the intersection you chose. (Reprinted by permission from Brown, Reynolds, and Ubbelohde, Inside Out: Design Procedures for Passive Environmental Technologies, John Wiley & Sons, © 1982.)*

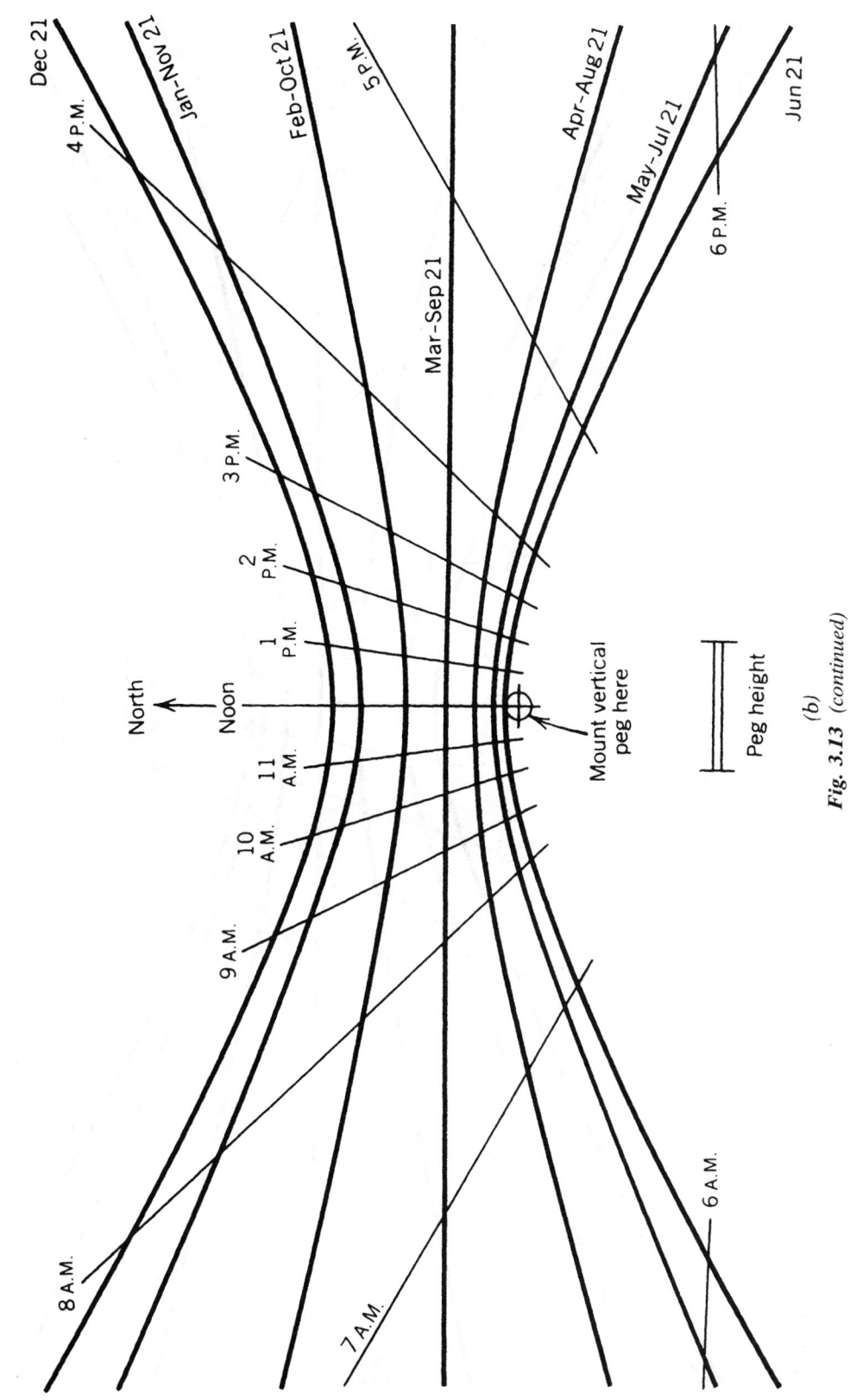

32°N

Dec 21
Jan–Nov 21
Feb–Oct 21
5 P.M.
Mar–Sep 21
Apr–Aug 21
May–Jul 21
Jun 21

4 P.M.
6 P.M.

North
Noon

3 P.M.
2 P.M.
1 P.M.
11 A.M.
10 A.M.
9 A.M.
8 A.M.
7 A.M.
6 A.M.

Mount vertical
peg here

Peg height

(b)

Fig. 3.13 (continued)

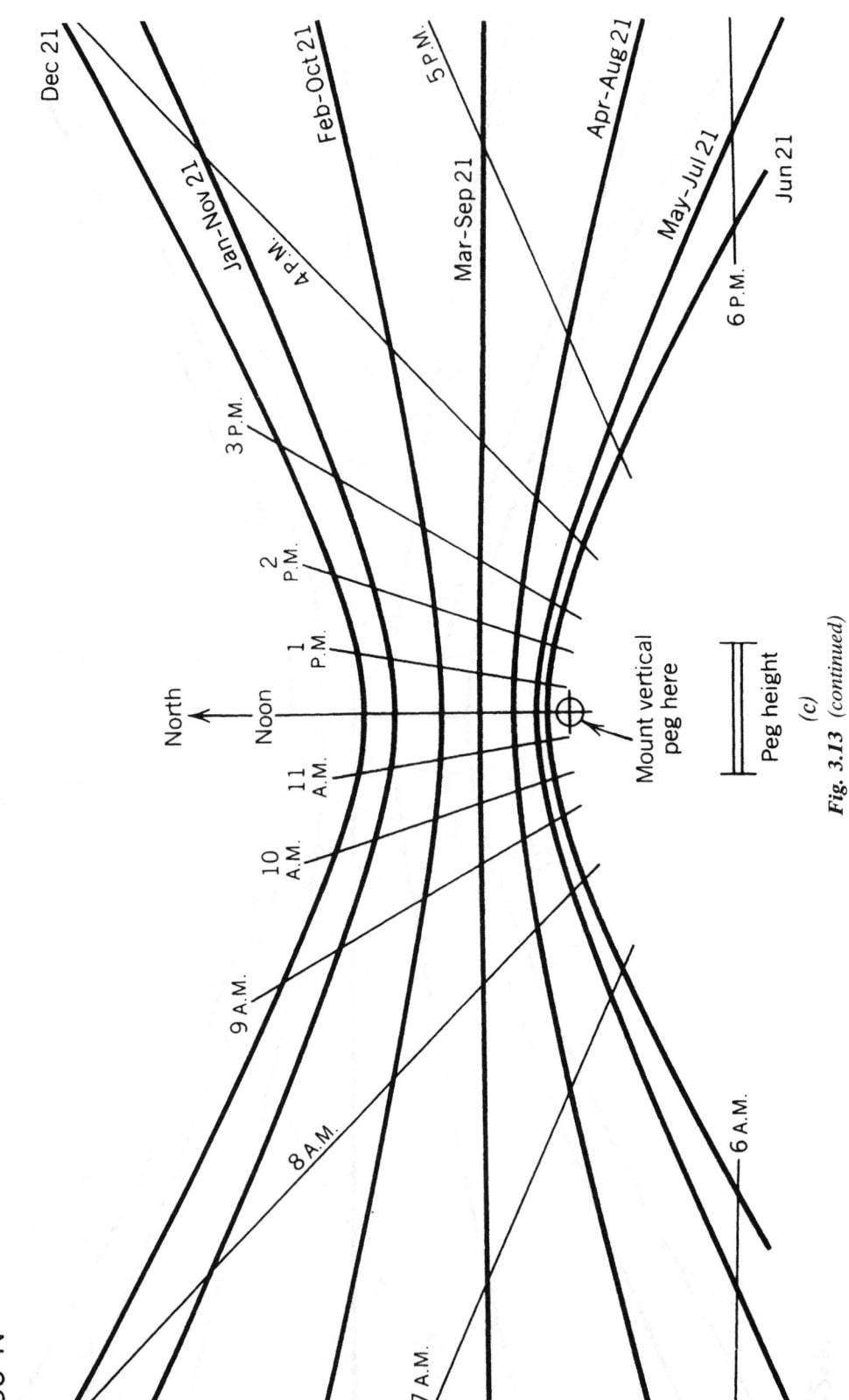

Fig. 3.13 (continued)

40° N

Fig. 3.13 (continued)

(d)

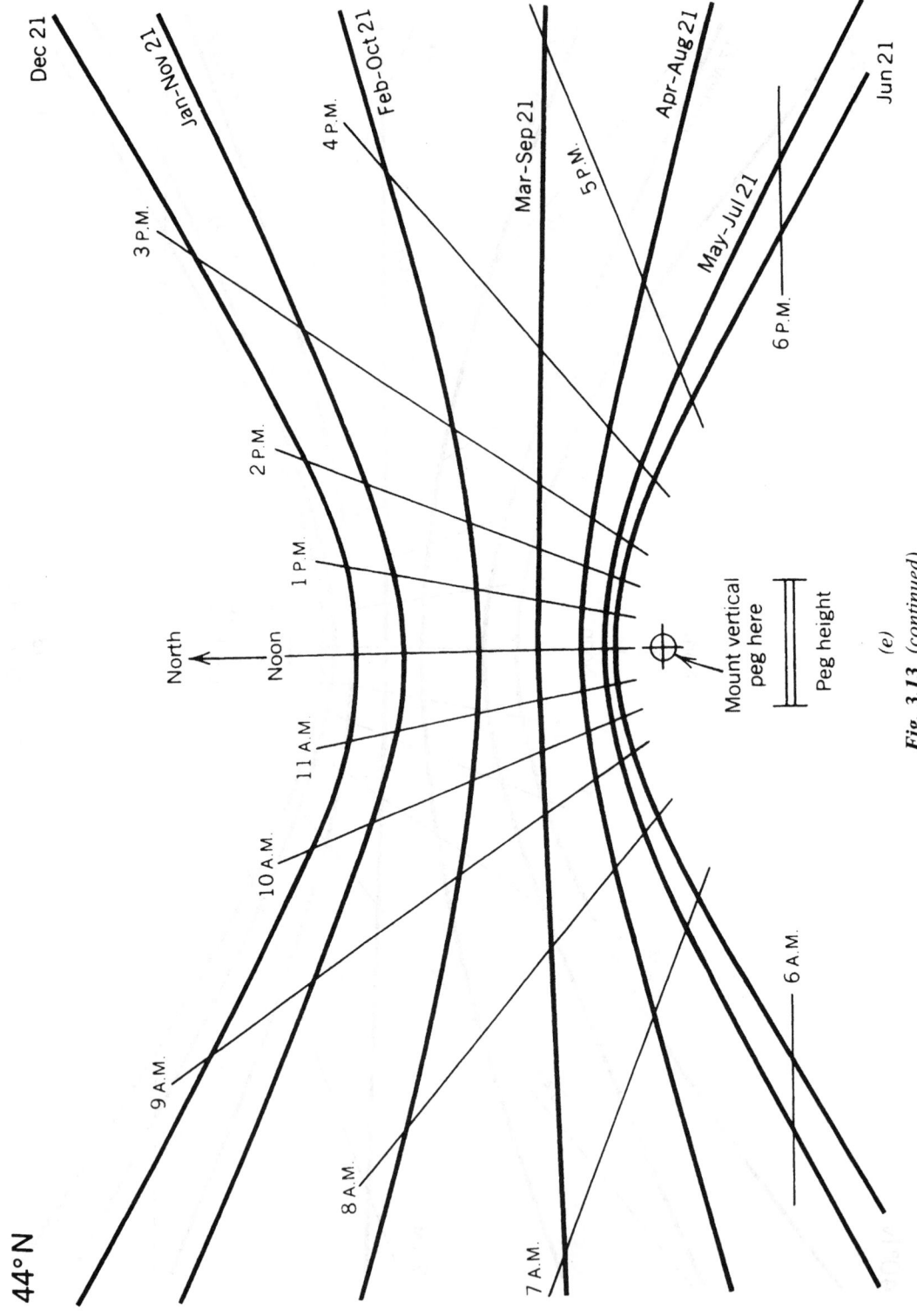

44°N

Dec 21

Jan-Nov 21

Feb-Oct 21

Mar-Sep 21

Apr-Aug 21

May-Jul 21

Jun 21

3 P.M.

4 P.M.

5 P.M.

6 P.M.

2 P.M.

1 P.M.

North

Noon

11 A.M.

10 A.M.

9 A.M.

8 A.M.

7 A.M.

6 A.M.

Mount vertical peg here

Peg height

(e) (continued)

Fig. 3.13 *(continued)*

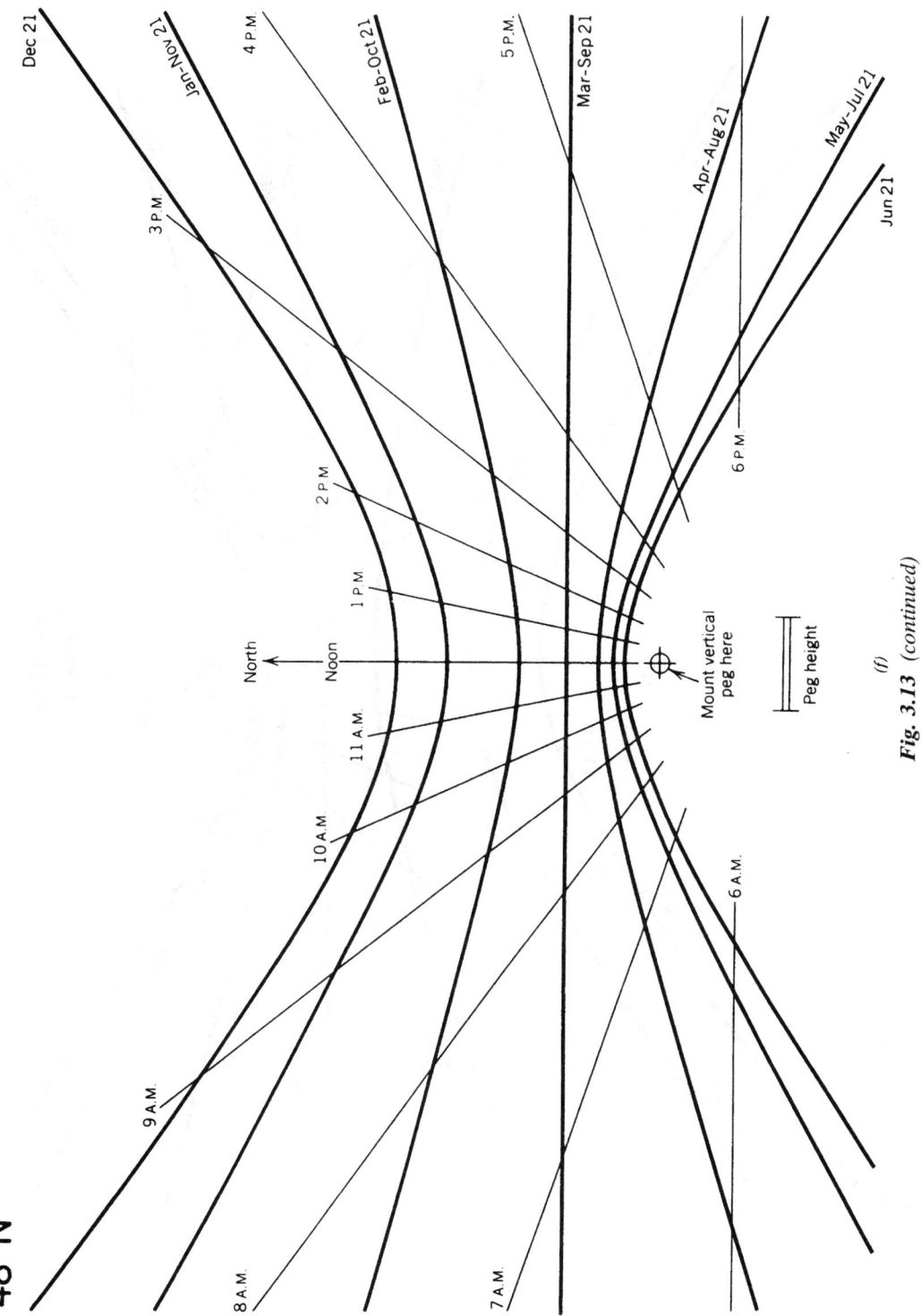

48° N

ENERGY OVERVIEW

Fig. 3.13 *(continued)*

(f)

Mount vertical peg here

Peg height

Dec 21
Jan-Nov 21
4 P.M.
Feb-Oct 21
5 P.M.
Mar-Sep 21
Apr-Aug 21
May-Jul 21
Jun 21

3 P.M.
2 P.M.
1 P.M.
North
Noon
11 A.M.
10 A.M.
9 A.M.
8 A.M.
7 A.M.
6 A.M.
6 P.M.

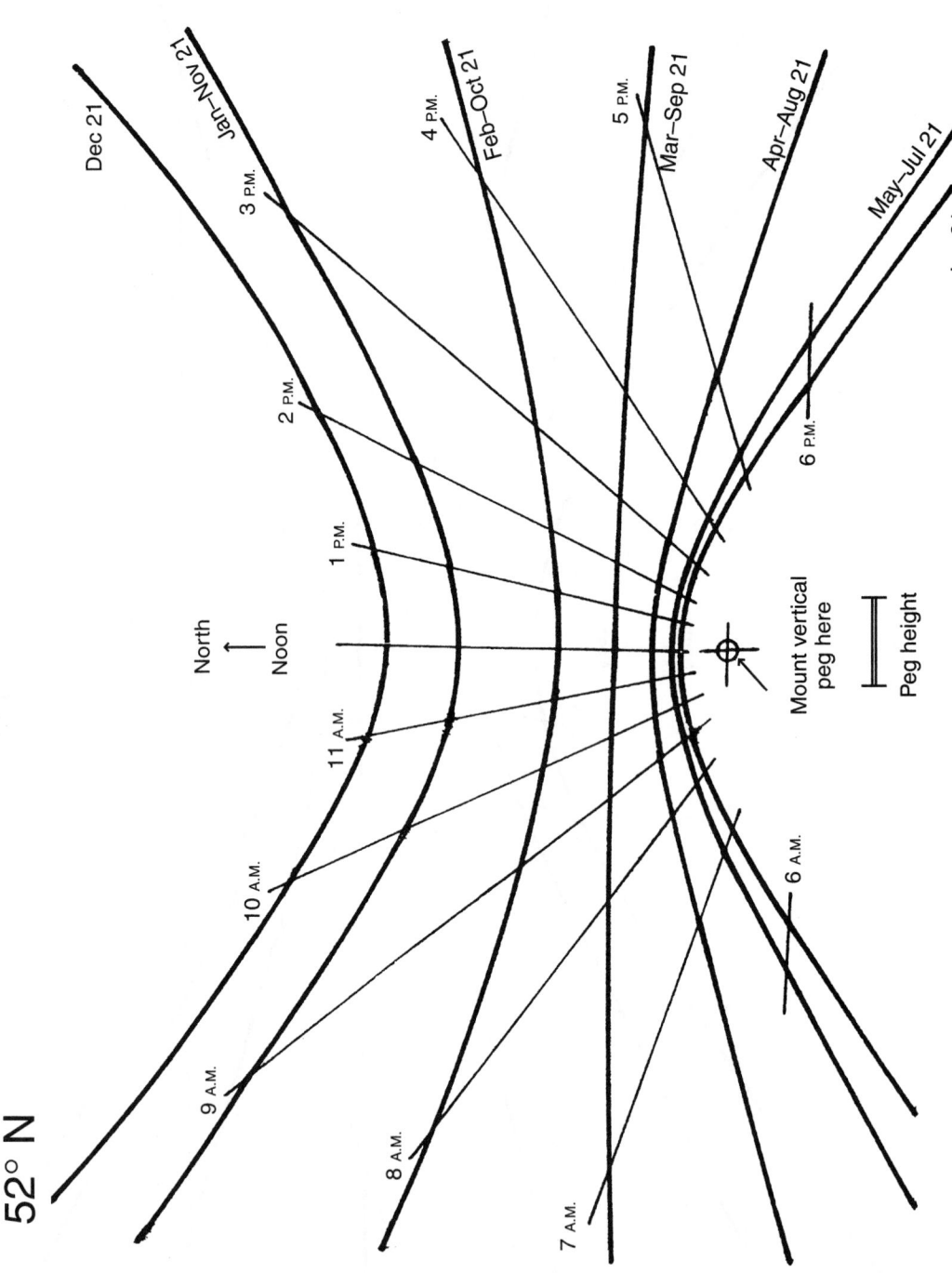

52° N

North

Noon

Dec 21

Jan–Nov 21

3 P.M.

Feb–Oct 21

4 P.M.

5 P.M.

Mar–Sep 21

Apr–Aug 21

May–Jul 21

Jun 21

2 P.M.

1 P.M.

11 A.M.

10 A.M.

9 A.M.

8 A.M.

7 A.M.

6 A.M.

6 P.M.

Mount vertical peg here

Peg height

(g)

Fig. 3.13 (continued)

Fig. 3.13 (h) *The model of a small building with a glazed open-frame circulation space on the south side is observed at a sun position of 3 PM on December 21.*

(a) Noise. Any sound that is unwanted becomes "noise." The urban building in Fig. 3.17 is unusual both in providing an opportunity for the wind ventilation—air moves freely below it as well as around it—and in the extraordinary intensity of traffic noise that is carried through such air.

Two characteristics of cities (see Section 3.1) contribute to increased noise at street level: *hard surfaces* reflect rather than absorb sound, and *parallel walls* intensify sound between them rather than dissipating it. Increasing horizontal and vertical distances from urban noise sources affects the

outdoor noise level in various ways, shown in the graphs of Fig. 3.18. Unfortunately, obtaining much distance in dense urban areas is quite difficult.

Although building surfaces are generally of hard materials for durability in weathering, softer and multiplaned materials (such as plants) are desirable from a public noise–reduction viewpoint. Their impact on measured sound levels may be slight, but visually softer surfaces reinforce our perception of acoustically softer environments (much as the sound of running water reinforces our perception of cool environments). Further-

Fig. 3.14 *Reflections. Mirror-glass windows in a newer office building (left) in Milwaukee, Wisconsin, cast strong reflections on the north-facing wall of the older hotel next door. Although this reflected heat might be welcome in winter, the glare is intense. In summer, the older building is particularly disadvantaged.*

Fig. 3.15 *Selective protection from reflections.* (a) *The tree standing west of this south window wall does not interfere with solar access during the best hours for solar collection (around noon), nor does it prevent early morning sun from entering the windows. The reflections of early morning are intercepted by the tree, before they can escape to annoy elsewhere.* (b) *The late-afternoon sun is blocked by the tree, before either solar gain or reflections can occur.*

more, the sounds made by plants as the wind moves through their leaves can help mask the unwanted noises of the street beyond. Fountains are especially useful sources of this *masking sound*; they can be kept flowing as long as the

Fig. 3.16 *The "eggcrate" shading devices shown on the southwest corner of this University of Oregon building prevent reflections by blocking the sun from either side of the glass. However, different shading devices are appropriate for different orientations of facades.*

noise persists, and they enhance the cooling function of natural ventilation, especially in drier climates. Not all spaces benefit from masking sound; where a single sound source is expected to predominate over all others (as in an amphitheater), masking sound interferes and becomes unwanted "masking noise."

Where site conditions allow, barriers to street noise can be installed that cast "sound shadows" on the surface of a site (Fig. 3.19). Such barriers do little to reduce noise levels at upper windows, but surface activities can be given much lower noise levels, especially near the barrier. Many cities now require such barriers between new housing developments and highways or railroads.

Another urban noise source is the mechanical equipment of buildings themselves. Many of the noise complaints against buildings involve air-conditioning equipment (the compressive refrigeration cycle and its year-round utilization as a heat pump are described in Chapter 7). Noise is generated both by the compressor and by the great quantities of outdoor air that must be rapidly pushed through outdoor coils (further heating outdoor air in summer so that the indoors may be cooled). When densely packed buildings are forced to rely on mechanical cooling, such closeness makes the noise of the systems even more annoying. The machine's appetite for outdoor air is so enormous that attempts to surround it with noise shields can hinder its efficient operation and shorten its life. Fortunately, more recent energy-efficient equipment is also more quiet, although not silent.

Fig. 3.17 *Apartment buildings in a series straddle the approach ramps to New York's George Washington Bridge. (a) Section along the freeway. (b) Looking down to freeway. These buildings were the scene of a study linking noise level with reading disabilities for occupants of the apartments (Cohen et al., 1973).*

Fig. 3.18 *Predicting noise levels outdoors. (a) Distance as a factor in noise intensity. (b) Building height as a factor in noise propagation. (From Clifford R. Bragdon,* Noise Pollution: The Unquiet Crisis, *University of Pennsylvania Press, 1971. Reprinted by permission.)*

(a)

(b)

(c)

Fig. 3.19 *Barriers to outdoor noise sources.* (a) *To determine the approximate reduction in noise in decibels (dBA) due to an outdoor barrier, construct a section locating the noise source (N), the solid barrier (B), and the receiver's location (R). On this section, determine the effective height (H) of the barrier, and the diffraction angle (β) with the resulting "noise shadow." Enter the graph (b) with H and β; where their lines intersect determines the noise reduction in dBA (left margin). A reduction of 10 dBA is perceived as "half as loud" as the original source. Note perceptible noise reduction from simply breaking the line of sight (β = 1°). (From Leslie Doelle,* Environmental Acoustics, *McGraw-Hill Book Company, 1972. Reprinted by permission.) For a more detailed procedure, see Fig. 27.33.* (c) *A noise barrier along a freeway near Washington, DC. Columns fill the space between the giant slabs, creating a solid wall.*

TABLE 3.3 **Air Pollution: Sources and Effects**

Gas or Pollutant	Sources	Effects
Carbon monoxide (CO)	Gasoline-powered vehicles; industry using oil and gas; building heating using oil and gas; biomass burning	Enters human bloodstream rapidly, causing nervous system dysfunction and death at high concentrations; interferes with self-cleansing of atmosphere
Carbon dioxide (CO_2)	Fossil-fuel combustion, deforestation	Contributes to greenhouse effect
Methane (CH_4)	Rice fields, cattle, landfills, fossil-fuel production	Contributes to greenhouse effect
Sulfur oxides [sulfur dioxide (SO_2) and sulfur trioxide]	Industry using coal and oil, heating using coal and oil, power plants using coal, oil and gas; ore smelting	Acid rain, damaging plants and attacking building skin materials; irritates human respiratory tract and complicates cardiovascular disease; decreases visibility in atmosphere
Nitrogen oxides, NO_x [nitric oxide (NO) and nitrogen dioxide (NO_2)]	Gasoline-powered vehicles; building heating using oil and gas; industry and power plants; biomass burning	Acid rain, damaging plants and attacking building skin materials; irritates human eyes, nose, and upper respiratory tract; triggers development of smog; decreases visibility in atmosphere
Nitrous oxide (N_2O)	Nitrogenous fertilizers, deforestation, biomass burning	Contributes to greenhouse effect
Hydrocarbons (compounds of hydrogen and carbon)	Petroleum-powered vehicles, petroleum refineries, general burning	Promotes smog; toxic to humans at high concentrations
Chlorofluorocarbons	Aerosol sprays, refrigerants, foams	Contributes to greenhouse effect and to stratospheric ozone depletion
Particulates (liquid or solid particles smaller than 500 micrometers)	Vehicle exhausts, industry, building heating, general burning, spore- and pollen-bearing vegetation	Promotes precipitation formation; some are toxic to humans; some pollens and spores cause allergic reactions in humans

Sources: Adapted from Marsh (1991) and Graedel and Crutzen (1989).

In residential neighborhoods, the greater distance between buildings might be expected to lessen these difficulties. Yet the much lower ambient (or background) sound level of residential areas is one of their more appealing characteristics, and an intruding compressor on a hot summer night can trigger lawsuits from neighbors who formerly enjoyed cool—and quiet—night breezes.

(b) Air Pollution. Problems of global importance are resulting from air pollution; these are summarized in Table 3.3. Building construction and operation are among the contributors to air pollution. The "greenhouse effect" (Fig. 3.20) threatens global warming, a trend many now believe is underway. The greenhouse effect occurs because gases that absorb long-wave radiation (heat) from the earth's surface are accumulating in the atmosphere. Energy production and use (especially of fossil fuels) are contributing heavily to these greenhouse gases, which include carbon dioxide, methane, nitrous oxide, ozone, and

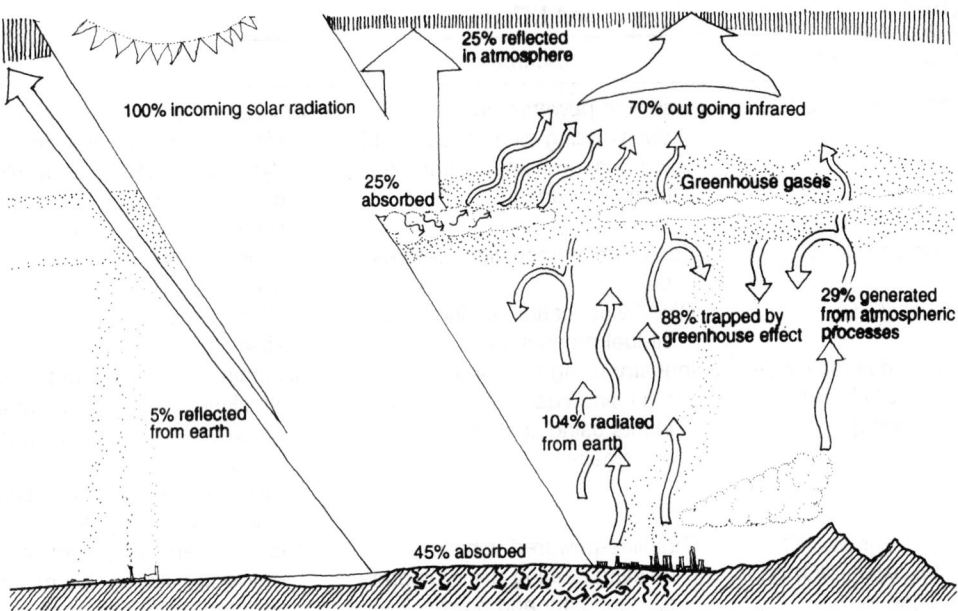

Fig. 3.20 *"Greenhouse effect" traps heat in the earth's upper atmosphere. Clouds and particles in the atmosphere reflect about one fourth of incoming solar radiation, while they are returning about two thirds of the heat that the earth would otherwise lose to outer space. Currently, the atmosphere keeps earth about 33C° (60F°) warmer than it would be without this heat trapping. Increases in greenhouse gases theoretically will reflect more incoming solar radiation, but reflect even more outgoing radiation, resulting in global warming. (Based on "The Changing Climate" by Stephen F. Schneider,* in Scientific American, *September 1989.)*

chlorofluorocarbons (CFCs). Another serious threat is stratospheric ozone depletion, with potentially devastating consequences to ecosystems due to increased ultraviolet radiation received at the earth's surface.

The designers of buildings are affected in several ways. First, we can help to greatly reduce the air pollution caused by electric power plants and by fuel burning in buildings, by designing both for *greater energy conservation* and by utilizing *clean and renewable energy sources* within our buildings. Chapter 5 is concerned largely with this topic. Second, we can specify materials and equipment that, through their manufacture or operation, lessen air pollution. This suggests avoiding fuel combustion (coal, oil, trash, wood, and natural gas, roughly in descending order of air pollution threat). It encourages specifying that refrigeration equipment use *non-CFC refrigerants,* and that urethane foam insulation or upholstery products be made with non-CFC blowing agents. Many of these practices are mandated, but the designer

can seek out the most environmentally benign products.

Our buildings are substantial contributors to air pollution; the fuel combustion within, the electric power plants that supply, the incinerators and landfills that receive waste from buildings. Buildings and power plants are major contributors to the greenhouse effect and the primary causes of acid rain (sulfur oxides) and smog (nitrogen oxides). The transportation that takes people to and from buildings is another major air pollution source. We design buildings to utilize "fresh" air, whether by natural or forced ventilation; we must then also design buildings to preserve our fresh air resource. The less energy they require, the cleaner their outdoor air will be.

At the site itself, local sources of air pollution must be minimized and, insofar as possible, isolated. Combustion gases pose a threat, especially where vehicles approach buildings (Fig. 3.21). Idling truck motors at loading zones or automobile motors at drive-up service windows can

Fig. 3.21 Protecting outdoor air intakes. This loading dock was a source of indoor air pollution from truck motor fumes, prompting the warning sign.

threaten building openings such as doors and windows; mechanical system fresh air intakes are particularly vulnerable, because outdoor air is being drawn rapidly into them. Again, consider the neighboring buildings: Will the location of some activity on your site threaten a neighbor's fresh air?

(c) Wind Control. For most buildings, wind (like sun) changes from friend to enemy with the change of seasons. In many locations, wind also changes prevailing direction with the seasons. Control of wind often means utilizing wind-sheltered areas in winter while encouraging increased wind speeds in summer. From spring through fall, outdoor spaces might benefit from this block-to-admit changeover on a daily basis; less wind in cool mornings, more in hot afternoons. Fortunately, winds are generally weakest in the early mornings and strongest in the afternoons because of the effect of the sun's heating on the surrounding environment.

The generalized patterns of wind flow around thin windbreaks and thicker buildings (Fig. 3.22) help us to understand where shelter and increased flows occur. However, they are much more complicated than they first appear and are highly influenced by objects upstream, to the sides, and downstream of the wind-directing object being analyzed. Wind-tunnel tests are far more reliable

than these patterns; unfortunately, such tests are expensive and still fraught with opportunity for misprediction. With these warnings in mind, the site can be analyzed graphically for seasonal wind utilization.

Wind ultimately returns to its original flow pattern after encountering an obstacle such as a windbreak or a building. Before it reaches the obstacle, it slows, builds (positive) pressure, and turns upward or sideways. As it passes the obstacle, it increases its speed, and reduced (negative) pressure results at the sides of and behind the obstacle. These pressure differences, flow patterns, and the size and shape of the wind-protected areas behind the obstacle are all utilizable for control of air motion, inside and out.

Windbreaks are commonly used to protect outdoor areas; these can be either fences or plants. Figure 3.23 shows the relative reduction in wind velocity at the level of the windbreak as wind approaches then passes. The distances (horizontal axis) are in units of the height of the windbreak. Note that the most dense windbreak produces the greatest reduction in wind speed behind it—but that the wind recovers its full velocity closer to such a barrier, compared to a less dense windbreak. Thus, the more dense the windbreak, the greater reduction in wind speed, but the smaller the area so affected.

Gaps in windbreaks can produce increased wind speeds through such gaps. Figure 3.24 shows a small area of wind speeds even greater than the unmolested wind speed above a windbreak. Although a gap is a threat to winter wind protection, it is an opportunity for summer wind speed enhancement. With constant prevailing summer wind directions, a windbreak gap could provide a small area with above-average air motion.

Wind around buildings is a complex matter: some rules of thumb for shelter areas were shown in Fig. 3.22, whereas Figs. 3.25 and 3.26 show some wind behavior to be expected in typical building combinations. Beranek (1980) discussed methods for charting the shelter areas in these building groups. Localized effects of wind turbulence between buildings can have a particular impact on entryways. People leaving the controlled air motion of the interior lobby are often unprepared for the speed and turbulence of channeled wind just beyond the doorway.

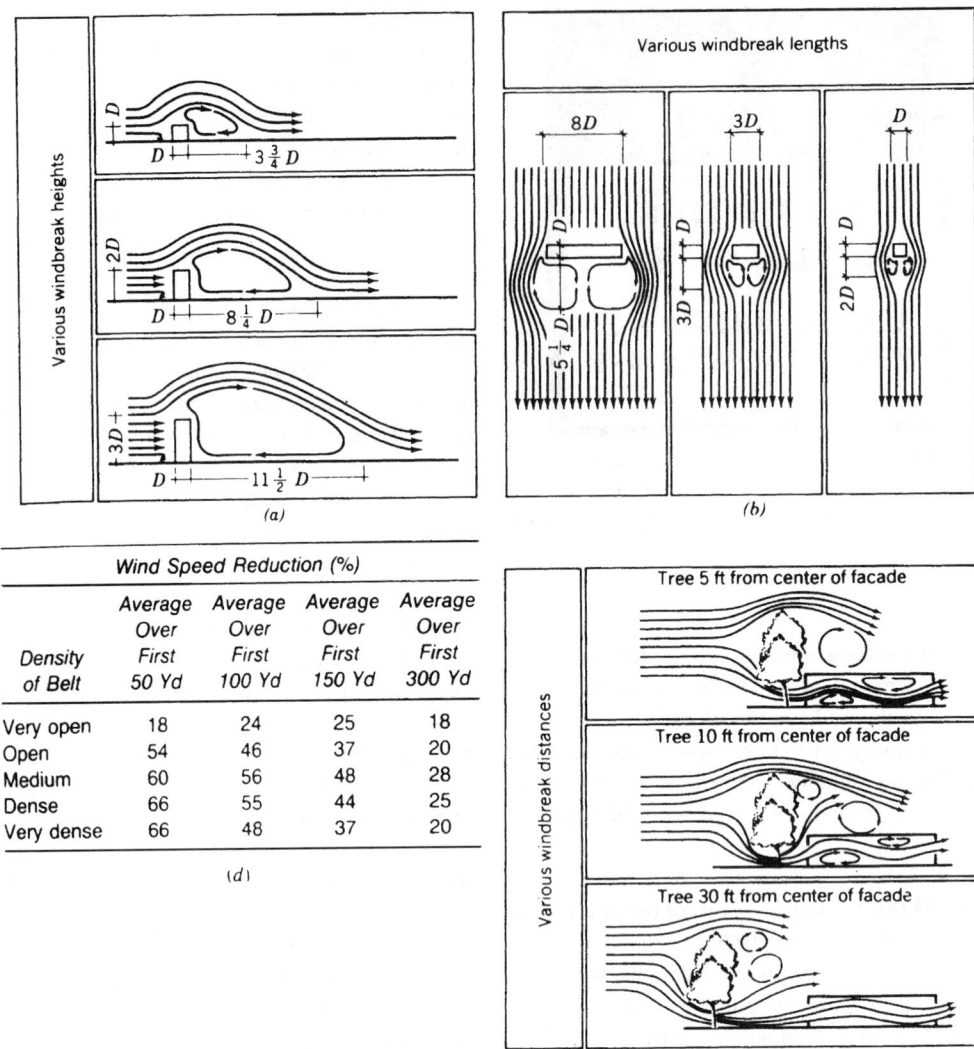

Fig. 3.22 *Approximate patterns of wind around objects. (a) Difference in barrier height. (b) Difference in barrier distance. (c) Wind through trees and buildings. (d) Reduction in wind speed due to windbreak density. (Reproduced with the permission of the American Institute of Architects, © 1981, AIA.)*

Wind used to cool buildings is discussed in detail in Chapter 5.

(d) Ventilation and Cooling. We introduce outdoor air to buildings for two distinct reasons. *Ventilation* involves the provision of "fresh air" to interiors to replenish the oxygen used by people and help carry away their byproducts of carbon dioxide and body odors. Ventilation is desirable all year round; recommended minimum rates of fresh air are found in Chapter 4 (Table 4.22). *Cooling* (with outdoor air) replaces heated indoor air with cooler outdoor air. Thus, cooling by breezes is a seasonal opportunity, limited to times when the outdoor air temperature is lower than the indoor air temperature and outdoor humidity is at or below that desired indoors. When outdoor air temperatures are about the same as interior temperatures, breezes might still be useful to increase interior air motion, thus "extending" the comfort zone, as discussed in Chapter 2.

Cooling can require far greater quantities of air than ventilation, and its influence on building siting and window size and placement is consider-

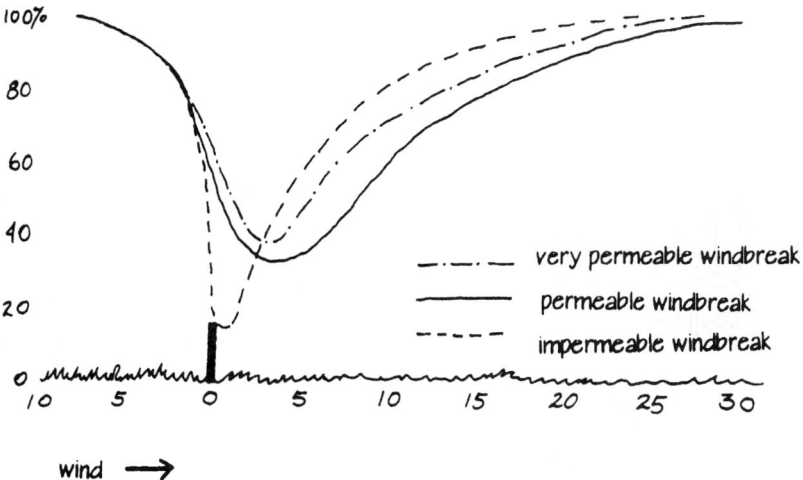

Fig. 3.23. *Wind speed reduction behind windbreaks of varying permeability. Solid (impermeable) barriers produce the lowest wind speeds, but these are effective over the shortest distance beyond the windbreak. Units of distance = heights of windbreak. [From Brown and Gillespie (1995).]*

able. Figure 3.27 illustrates the differing approaches to these two uses and suggests that window position and human occupants be considered together in design.

Two examples of contemporary buildings that use the wind appear in Figs. 3.28 and 3.29. The first served as a summer exhibition space in Montreal, Canada, and relied on prevailing winds (and on rising heated air) to remove the air heated by display lighting and crowds of people. The second is a chapel in a cemetery in the hot and humid climate of Houston, Texas. The architects in-

tended this shaded, open-air space as a respite from the sealed, air-conditioned buildings and cars prevalent in Houston. In this building, the funeral is a time and place for reunion with physical phenomena; breezes and shade trees provide cooling, rainwater collection into a pool reflects skylight, and a narrow shaft of direct sun marks its daily path along the rough stone interior north wall.

The building that relies on prevailing wind for cooling must be sited with attention to wind direction. One building should not be erected to ob-

Fig. 3.24 *Wind speeds accelerate through a gap in a windbreak. Numbers are percentage of the wind speed unaffected by the windbreak. [From Caborn, J. M. (1957). Shelterbelts and microclimate. Edinburgh, Scotland: H. M. Stationery Office; cited in McPherson (1984).]*

Fig. 3.25 *Wind patterns around single buildings. (a) Tall, slender buildings: height greater than 2.5 times width. (b) Tall, rather wide buildings; height between 2.5 and 0.6 times width. (c) Long buildings; height less than 0.6 times width. (From W. J. Beranek, "General Rules of the Determination of Wind Environment" Wind Engineering, J. E. Cermak, ed., Vol. 1, © 1980, Pergamon Press Ltd., reprinted by permission.)*

struct the next building's access to breeze. As seen previously, obstacles upstream from intake openings, or downstream near the outlets, can substantially reduce the velocity—and thereby the cooling effect—of the wind. Although wind cools people in hot weather primarily by speeding evaporation from skin, it can itself become an irritant at high velocity (Tables 2.4 and 3.4). Manual controls of openings are a necessary part of natural ventilation equipment. Finally, the proper

TABLE 3.4 **Beaufort Scale (Lower Speeds Only)**

Beaufort Number	Speed 6 m (19.7 ft) Above Ground			Description of Effects Outdoors	
	m/s	*fpm*	*mph*	*On Land*	*Over Water*
0	0.3	Less than 88	Less than 1	Smoke rises; no perceptible movement	Smooth sea
1	0.6–1.7	88–264	1–3	Smoke drift shows wind direction; tree leaves barely move	Scale-like ripples
2	1.8–3.3	352–616	4–7	Wind felt on face; leaves rustle	Small wavelets
3	3.4–5.2	704–968	8–11	Leaves, twigs in constant motion; hair is disturbed; wind extends light flag	Large wavelets; occasional white foam crests
4	5.3–7.4	1056–1408	12–16	Small branches move; dust rises; hair disarranged	Small waves become longer

Source: Reprinted from *Passive Cooling* by permission of the publisher, American Solar Energy Society, Inc.

Fig. 3.26 Wind patterns among building clusters. (a) Increased wind speeds can occur, especially where wind is at 45° to the face. With the bar effect, the downward spinning wind behind these buildings can reach 1.4 times the speed of the average wind. (b) Venturi effect: Where few obstructions occur upwind

ENERGY OVERVIEW

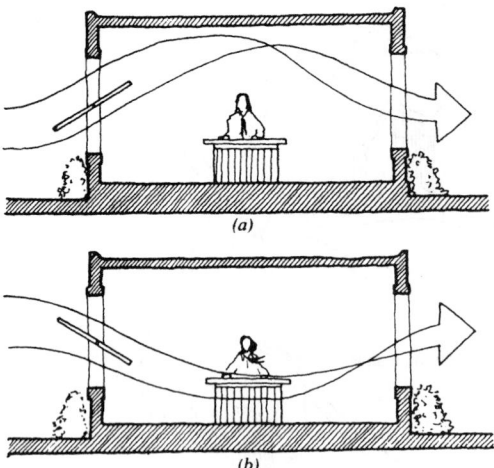

Fig. 3.27 *Ventilation, with and without cooling. The window's size can influence the stream of air within a space, providing either ventilation alone, or adding a cooling function to moving air. (a) Ventilation: the pivoting window directs the breeze upward, where it replaces hot inside air at the ceiling. It has minimum contact with people below. (b) Ventilation and cooling: this position of the window directs the breeze toward the floor, where it encounters people and provides a direct cooling effect from air motion.*

Fig. 3.28 *Cooling with the wind. The air inside this exhibition hall is heated by lights and crowds of people. It thus rises and is sucked out of the large clerestory opening by the prevailing winds. Note that winds do not blow* into *this high opening, but create a negative pressure outside this opening, encouraging outward flow of the interior's air. (African Place, at Expo '67 Montreal, Canada; John Andrews, Architect.)*

size and placement of openings, with an unobstructed ventilation path through the building, must be provided.

(e) Wind, Daylight, and Sun. When daylight and wind ventilation are desired, they combine to limit the width of buildings. Multistory hotels are an example; where such hotels fill entire blocks, they are often arranged around a central atrium, and the surrounding building is but two hotel rooms and one corridor in thickness. Multistory office buildings used to have a similar

---

or downwind from the narrow neck of this building, wind speeds through the neck can reach 1.3 times the average, up to heights of 30 m, and 1.6 times the average at about 50 m in height. (c) The gap effect begins to occur with perpendicular winds at buildings with more than 5 stories (15 m) height; by 7 stories, wind speeds 1.2 times the average can occur through the gaps; by 60 stories, gap wind speeds can be 1.5 times the average. (d) For higher buildings, increased wind speeds occur at the corners (localized within a radius, equal to the width of the building, d, around the corner). Where height is 15 m, wind speed can reach 1.2 times the average; for heights above 35 m, wind speed can be 1.5 times the average. Where two towers approach each other, increased wind around corners and between the towers can go as high as 2.2 times the average for towers 100 m high. (e) Increased wind speed and turbulence within the wake of buildings can be especially serious for towers at heights from 16 to 30 stories, where wind speeds can reach 1.4 to 2.2 times the average. (From J. Gandemer, "Wind Environments Around Buildings," Wind Effects on Buildings and Structures, K. J. Eaton, ed., © 1977, Cambridge University Press. Reprinted by permission.)

(a)

(b)

Fig. 3.29 *The Beth Israel Chapel and Memorial Garden, Houston, Texas. Open to the breeze, shaded by trees and a double-membrane roof, this chapel reunites people with natural forces during a funeral. (a) Plan. Curved walls visually separate the open south courtyard from the roofed chapel, but allow breezes to pass. The narrow triangular roof opening allows a shaft of direct sun to fall along the interior north wall, marking the daily passing of time. (b) Section, south to north. The curved roof sheds its rainwater to an oversize gutter above the curved walls. When breezes are weak, suspended paddle fans provide air motion. (c) View from the west. The oversize gutter delivers rainwater to a pond that reflects daylight. Trees on islands help with evening shade. (Photo by Timothy Hursley. Courtesy of Solomon Inc., Architecture and Urban Design, San Francisco.)*

(c)

Fig. 3.29 (continued)

form, but increasing urban density and reliance on electric lighting and cooling have changed their form considerably (Fig. 3.30). There is now a trend back to arranging office workspaces nearer daylight and, in Europe, to including operable windows for fresh air as well.

When wind and sun are combined, they tend to be influential in the opposite seasons. Still, fresh air is desirable year round, so in winter, a supply of tempered fresh air is necessary. In Fig. 3.31, a designer weighs the decision whether to face a clerestory to the north or south for a given set of seasonal prevailing winds.

3.6   Rain and Groundwater

Most buildings interact with water in four forms: rainwater and groundwater are less controlled and exterior interactions, whereas potable and wastewater are more controlled and interior. A detailed treatment of water within buildings is found in Part III, Water and Waste, and a closer look at rainwater utilization is found in Chapter 9.

Rainwater, like solar energy, is a diffuse, intermittent, and often seasonal resource. As a source of water, it is most often collected and used where other water sources are scarce or of poor quality. Rain, like sun, has an influence on building design:

heavy rains and pitched roofs have long been found in the same locales. Overhangs may extend further beyond walls exposed to storm winds; even gutter and downspout details can become a design feature, as shown in the chapel of Fig. 3.29 and several examples in Chapter 9. A building that reflects the combined influences of daylight, wind, and rain is shown in Fig. 3.32.

Rainwater's impact on site design can be thought of as similar to that of wind. Once on the surface, it will flow downhill in a wide path (on a wide plane), and buildings that obstruct this path must make provisions for diverting it. Slight, shallow ditches called *swales* are frequently used in such diversion; orientation of the building to the slope can affect surface water diversion, as shown in Fig. 3.33.

Surface water can be used to advantage in thermal, acoustic, and daylighting roles. Hot, dry breezes that pass over water surfaces (and especially through misty sprays above ponds) gain substantial moisture while undergoing a drop in dry-bulb temperature, as seen in the previous discussion of evaporative cooling (Section 2.4). Such air can provide improved comfort in hot, dry conditions. If the water body also provides the sound of running water, it can serve to mask unwanted sounds such as freeway traffic or conversations in adjacent rooms.

ENERGY OVERVIEW

(a)

(b)

Fig. 3.30 *In contrast to office building plan* (a), *which uses daylight and wind ventilation in each office, office building* (b) *receives cooled, filtered air, is less subject to the noise of the city, typically provides both constant light and temperature throughout, and provides for more rentable floor space on its site. It also allows less daylight to reach the street level, requires much more electricity (though probably less heating fuel), and thus contributes more heat (and possibly more noise from mechanical equipment) to the city all year round.*

Surface water has a more complex role in daylighting due to the reflection characteristics of water. The surface of water is highly reflective to light that strikes it at *low angles of incidence;* the reflections of the setting sun from the small pond will illuminate the ceiling and east wall of the chapel in Fig. 3.29. Off the west coast, reflected sun from an infinitely huge surface can throw a blinding sheet of light across buildings at the beach. Conversely, sunlight near noon in summer

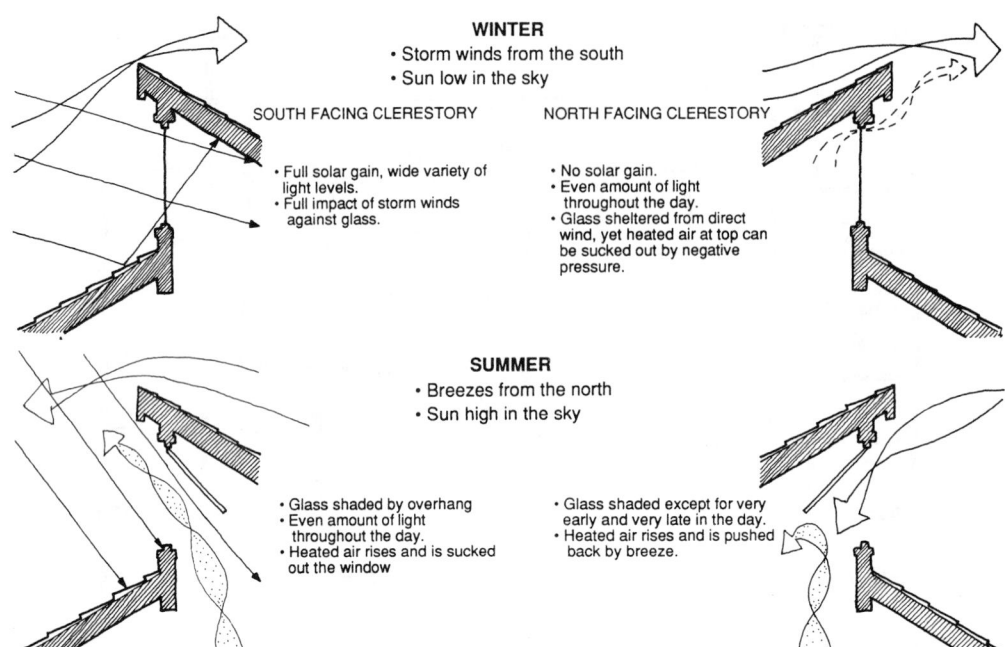

WINTER
- Storm winds from the south
- Sun low in the sky

SOUTH FACING CLERESTORY   NORTH FACING CLERESTORY

- Full solar gain, wide variety of light levels.
- Full impact of storm winds against glass.

- No solar gain.
- Even amount of light throughout the day.
- Glass sheltered from direct wind, yet heated air at top can be sucked out by negative pressure.

SUMMER
- Breezes from the north
- Sun high in the sky

- Glass shaded by overhang
- Even amount of light throughout the day.
- Heated air rises and is sucked out the window

- Glass shaded except for very early and very late in the day.
- Heated air rises and is pushed back by breeze.

Fig. 3.31 *Which way to orient a clerestory? Some relative advantages of north versus south orientation for a clerestory window/shed roof combination. (Seasonal wind directions shown prevail in the Pacific northwest.)*

Fig. 3.32 *Rain, wind, sun, daylight, and design. This outdoor covered tennis facility at the University of Oregon, Eugene, was designed to ward off the rain-bearing south winds in winter. Direct sun's glare is also unwelcome; instead, north skylight is admitted along with reflected light from roof surfaces. Cool outdoor temperatures are maintained, appropriate to the strenuous activity. Gutters at the lower edge of each roof plane carry away rain; the courts stay dry for all but a few days each year. (Unthank Seder Poticha Architects.)*

Fig. 3.33 *Rain as surface flow.* (a) *Where buildings intercept surface water, provisions for diversion are necessary. The building oriented as in* (b) *needs less elaborate provisions.*

strikes a water body at *high angles of incidence,* nearly perpendicular at southern U.S. latitudes. Water is highly absorptive to light at these angles, reflecting relatively little light. Therefore, water bodies east, south, and west of our buildings can provide increased reflected light on sunny winter days and somewhat decreased light in summer, relative to grass surfaces. However, on heavily *overcast* days, when the sky is uniformly gray (and the least but most glare-free daylight is available), water surfaces will not be particularly helpful.

The reflection of sun off water tends to be in sparkling patches of always-changing light. This can be fascinating or it can become annoying, either as glare to the eye or as distraction from a visual task. Reflected off a matte-finish ceiling, such dancing light might be welcome; directly into eyes or onto a work surface, it easily becomes a problem. Sparkle for one viewer may be glare for another.

Groundwater is generally avoided by designers where possible, as a threat to foundations and below-ground spaces. This avoidance carries over into site planning, and marshy places are usually unwelcome near buildings. Urban dryness is intensified as ponds are drained and streams are piped away. However, architects have used groundwater as a heat sink, discharging building heat to the groundwater in summer and removing the groundwater heat for the building's winter benefit. The quantity of groundwater available to act as a heat sink thus varies with subsurface conditions, and little information is yet available to indicate its limits. More buildings discharging more heat will eventually raise groundwater temperatures; but how many degrees and for how long remains generally unknown. Another risk is that of pollution, through increased handling of groundwater, or by using refrigerants to move the heat from building to sink. Many localities now strictly regulate the use of groundwater for heat sinks as a result.

The ability of water to conduct and store heat, which encourages its use as a heat sink, also makes it a thermal enemy of heat storage tanks or bins in solar buildings. There is little point in collecting solar energy to store in an underground tank for later use if groundwater is allowed to rob the tank of its heat. However, groundwater helps those storage tanks that are to be kept as cool as possible. Such tanks provide cool water to buildings at peak-heat hours of hot days and can greatly lessen the demand for electricity to run conventional air conditioners (see Section 8.3*f*).

Snow has special implications: it delays runoff, provides a blanket of thermal insulation, absorbs sound, and reflects more light than almost any other naturally occurring surface. Wind patterns can deposit snow much more thickly in some places, for better or worse. Snow hampers the moving of external devices such as awnings or thermal shutters, and can create a disabling glare when it reflects the low winter sun into windows at eye level.

3.7 Plants

Plants have several roles: they affect the absorptivity and emissivity of the earth's surface; they are part of both the food and water cycles; they turn carbon dioxide into oxygen by day; they provide organic matter suitable for building materials; and they help us tell time both by growth and by change with the seasons. Our associations with plants are mostly pleasant ones, and they contribute to our enjoyment of the places where they grow.

Plants are also of immediate practical value to buildings because they enhance privacy, slow the winter wind, reduce the glare of strong daylight, or prevent summer sun from entering and overheating buildings. In this latter role, plants are particularly noteworthy, because they enhance a feeling of coolness when breezes rustle or sway their leaves. Most importantly, they respond more to cycles of *outdoor temperature* than to the cycles

December 21

June 21

September 21/ March 21

Fig. 3.34 *Fixed overhang sun control. This south-facing corridor in Oregon is open-air all year round. The low winter sun fills the space, and some of its heat is stored by the tile floor. However, in cold weather the offices that it connects are disadvantaged by exposure to cold air. In summer, little of the corridor is exposed to sun. Sun control is identical in the (cool) spring and (warm) fall.*

of *sun position.* Unlike fixed-position sunscreens on buildings, plants thus provide deepest shade in the hottest weather.

To illustrate this contrast, a fixed sunscreen (here, an overhang for south-facing windows) is shown in Fig. 3.34. Such sunscreens block the sun for some portion of the year, centered on June 21; that is, maximum shade is provided at the summer solstice. A typical approach to such sunscreens in the U.S. temperate zone is to shade at least half a south window in a residence (or all the window in an office with internal heat sources) from March 21 to September 21 (vernal equinox to autumnal equinox). Yet, March is on the average a colder month than September; March 21 is the last day of winter, whereas September 21 is the last day of summer. (Indeed, average monthly temperatures for June and September can be quite similar.) Full solar radiation is more welcome in early spring than in early fall, yet sun position is identical at these times. In contrast to fixed sunscreens, deciduous plants do most of their shading from the middle of June to early October, giving windows access to solar radiation through much of the spring (Figs. 3.35 and 3.36).

Deciduous trees have a potential solar heating disadvantage in that certain species (such as sweet gum and some oaks) hold on to their leaves well into the heating season, a tendency increased by fertilizing or irrigating near the tree. Other trees have a dense branch structure, blocking a surprisingly high percentage of solar radiation even when bare (Table 3.5). Agricultural extension services in most areas can provide information on the hardiness and dates of defoliation for various tree species. For solar collection, avoiding trees or large shrubs within the area shown in Fig. 3.37 is recommended. In this "protected zone," a deciduous vine on a trellis above the south-facing windows is the better planting strategy.

Many large buildings have a relatively short heating season because they have constant internal heat sources. For such internal heat–dominated buildings, a tree or vine that foliates early in spring and defoliates late in fall and has a low transmissivity in summer is advantageous. So, the choice of trees, shrubs, and vines for the improvement of a building's microclimate can best be made after an analysis of building heating and cooling needs by month.

(a)

(b)

(c)

(d)

Fig. 3.35 *A deciduous tree as a naturally "smart" shading device. (a) In hot August, the tree is in full, dense leaf, providing shade to sidewalk and shops beyond. (b) In cold January, the bare branches admit winter sun. The following photographs were taken on the equinox at 4:45 PM sun time. (c) On September 21, the local average temperature is 75°F (24°C), and the tree continues its role as a shading device. (d) On March 21, the local average temperature is 53°F (12°C), and the tree has not yet begun to develop its leaves. [From Reynolds (1976).]*

(a) May: 44—68F   (b) July 50—83F
Average Daily Temperature Range for Eugene, Oregon

(c) November: 38—53F   (d) January 33—46F

Fig. 3.36 *Deciduous vines, temperature, and sun position. The sun's path through the sky is identical in late May (a) and late July (b) (see Fig. 3.10). Identical but lower sun paths occur in late November (c) and late January (d). This deciduous vine responds more to the temperature of its Oregon climate than to the sun's position at 44° N latitude, which makes it particularly useful as a potential sun control device. (For termites and wall longevity, it is best to keep vines on a trellis rather than on the wall surface.) [From Reynolds (1976).]*

TABLE 3.5 **Deciduous Trees for Summer Shading, Winter Solar Collection**

Botanical Name	Common Name	Transmissivity[a] Range %		Foliation[b]	Defoliation[c]	Mature Height	
		Summer	Winter			m	ft
Acer platanoides	Norway Maple	5–14	60–75	E	M	15–25	
Acer rubrum	Red Maple	8–22	63–82	M	E	20–35	
Acer saccharinum	Silver Maple	10–28	60–87	M	M	20–35	
Acer saccharum	Sugar Maple	16–27	60–80	M	E	20–35	
Aesculus hippocastanum	Horse Chestnut	8–27	73	M	L	22–30	
Amelanchier canadensis	Serviceberry	20–25	57	L	M		
Betula pendula	European Birch	14–24	48–88	M	M–L	15–30	
Carcis canadensis[d]	Red Bud	62	74	L	M	12	40
Carya ovata	Shagbark Hickory	15–28	66			24–30	
Catalpa speciosa	Western Catalpa	24–30	52–83			18–30	
Cornus florida[d]	Dogwood	43	53	L	E	11	35
Fagus sylvatica	European Birch	7–15	83	L	L	18–30	
Fraxinus pennsylvanica	Green Ash	10–29	70–71	M–L	M	18–25	
Gleditsia tricanthos inermis	Honey Locust	25–50	50–85	M	E	20–30	
Juglans nigra	Black Walnut	9	55–72	L	E–M	23–45	
Liquidamber styraciflua[d]	Sweet Gum	33	47	M	L	24	80
Liriodendron tulipifera	Tulip Tree	10	69–78	M–L	M	27–45	
Platanus acerifolia	London Plane Tree	11–17	46–64	M–L		30–35	
Populus detoides	Cottonwood	10–20	68	E	M	23–30	
Populus tremuloides	Trembling Aspen	20–33		E	M	12–15	
Quercus alba	White Oak	13–38				24–30	
Quercus palustris[d]	Pin Oak	45	47	L	L	23	75
Quercus rubra	Red Oak	12–23	70–81		M	23–30	
Robinia pseudoacacia[d]	Black Locust	38	40		E	21	70
Tilia cordata	Littleleaf Linden	7–22	46–70	L	E	18–21	70
Ulmus americana	American Elm	13	63–89	M	M	18–24	

Source: Brown and Gillespie (1995), except as noted.

[a]Transmissivity to solar radiation; varies by instruments used by various researchers.

[b]Foliation: E = Early (before April 30); M = Middle (May 1–15); L = Late (after May 15).

[c]Defoliation: E = Early (before November 1); M = Middle (Nov. 1–30); L = Late (after November 30).

[d]From Montgomery (1987); transmissivity % = (100% − blockage %).

This distance is equal to at least twice the ultimate height of the trees of shrubs beyond.

Fig. 3.37 Protecting access to winter sun, given a lawn or terrace of minimum size and shape to the south of solar collecting surfaces. Coniferous and even deciduous plants within this "protected zone" should be avoided, unless they are very low growing or are a reliably early defoliating species (Table 3.5). Summer sun protection for these south windows is best provided by flexible architectural controls, such as awnings or hanging screens.

REFERENCES

American Solar Energy Society. (1981). *Passive Cooling,* ASES, Boulder, Colo.

Beranek, W. J. (1980). "General Rules for the Determination of Wind Environment," in *Wind Engineering,* J. E. Cermak, Editor, Pergamon Press, Elmsford, N.Y.

Bragdon, C. (1970). *Noise Pollution, The Unquiet Crisis,* University of Pennsylvania Press, Philadelphia.

Brown, G. Z., Reynolds, J., and Ubbelohde, M. S. (1982). *Inside/Out: Design Procedures for Environmental Technologies,* Wiley, New York.

Brown, R. D., and Gillespie, T. J. (1995). *Microclimatic Landscape Design: Creating Thermal Comfort and Energy Efficiency,* Wiley, New York.

Cohen, S., Glass, D. C., and Singer, J. E. (1973). "Apartment Noise, Auditory Discrimination, and Reading Ability in Children," *Journal of Experimental Social Psychology,* Vol. 9.

DeKay, M. (1992). "A Comparative Review of Daylight Planning Tools and a Rule-of-Thumb for Street Width to Building Height Ratio," in *Proceedings of the 17th National Passive Solar Conference,* Cocoa Beach, Florida, 1992, American Solar Energy Society, Boulder, CO.

Doelle, L. (1972). *Environmental Acoustics,* McGraw-Hill, New York.

Gandemer, J. (1977). "Wind Environment Around Buildings: Aerodynamic Concepts," in *Wind Effects on Buildings and Structures,* K. J. Eaton, Editor, Cambridge University Press, New York.

Gandemer, J. (1981). "The Aerodynamic Characteristics of Windbreaks, Resulting in Empirical Design Rules," *Journal of Wind Engineering and Industrial Aerodynamics,* Vol. 1, July. Elsevier Scientific Publishing Co., Amsterdam.

Graedel, T. E., and Crutzen, P. J. (1989). "The Changing Atmosphere," in *Scientific American,* September.

Knowles, R. (1981). *Sun, Rhythm, Form.* MIT Press, Cambridge, Mass.

Landsberg, H. E. (1970). "Climates and Urban Planning," in *Urban Climates,* World Meteorological Organization Technical Note 108, Geneva.

Lowry, W. P., and Lowry, P. P. (1995). *Fundamentals of Biometeorology, Volume 2: The Biological Environment.* Peavine Publications, McMinnville, Ore.

Lowry, W. P., (1988). *Atmospheric Ecology for Designers and Planners,* Van Nostrand Reinhold, New York.

Marsh, W. M. (1991). *Landscape Planning, Environmental Applications, 2nd Edition.* John Wiley & Sons, New York.

McPherson, E. G. (Editor) (1984). *Energy-Conserving Site Design,* American Society of Landscape Architects, Washington, DC.

Montgomery, D. (1987). "Landscaping as a Passive Solar Strategy," in *Passive Solar Journal,* Vol. 4, No. 1. American Solar Energy Society, Boulder, Colo.

Reynolds, J. S. (1976). *Solar Energy for Pacific Northwest Buildings,* Center for Environmental Research, University of Oregon, Eugene, Ore.

Watson, D., and Labs, K. (1983). *Climatic Design.* McGraw-Hill, New York.

PART II
THERMAL CONTROL

The next five chapters deal with the topics of heating, ventilating, and cooling. The objective continues to be the comfort of people, for whom we heat, ventilate, and cool buildings: Thermal comfort criteria were discussed in Chapter 2. In Chapter 4, the basic theory of heat flow is discussed; whether we are calculating heat gain in summer or heat loss in winter, the process is based on the same principles of heat flow. Chapter 5 presents criteria for building performance. Initially, the strategies are abbreviated as "design guidelines," later followed by detailed heat loss and heat gain procedures. Chapter 6 has interior air quality (IAQ) as its focus, including some equipment for heat recovery associated with adequate building ventilation. Chapter 7 deals with heating and cooling equipment for smaller buildings, while Chapter 8 deals with larger buildings, and groups of buildings.

It is important to keep our perspective during the process of calculating thermal loads for buildings. Throughout these lengthy chapters, keep the following points in mind:

1. Human comfort is the ultimate objective of our calculations and resulting architectural manipulations; this comfort is not attainable by numbers alone. Most of us have experienced situations in which psychological factors seemed to have at least as much influence on comfort as temperature, humidity, or air motion. Achieving numerical adequacy is important, but it is not by itself a guarantee of comfort. We find increasing evidence that occupants are more likely to be satisfied with a building's performance to the extent that they can control their personal environment within that building.

2. These calculations are a rare opportunity to numerically evaluate a building design. When the numbers are all cranked out, sit back and take a larger view. What design strategies do they suggest? Smaller or larger windows? More insula-

tion or thermal mass? We have worked hard to get these numbers; now let them work for us.

3. There are two specific objectives to our calculations, which yield different kinds of answers and require different amounts of work.

(a) How big should a device be to furnish enough heating (or cooling) during the coldest (or hottest) conditions we are likely to encounter for our building?

(b) About how much energy will be used by our building in a typical season?

For the first question, we deal with extreme "design" conditions, as discussed in Chapter 2 (and detailed in Appendix A). We calculate our hourly winter heat loss (or summer heat gain) under these special conditions, then adjust our buildings, probably recalculate, and finally size our equipment according to these calculations. We have, in effect, a snapshot of the building's performance at a special high-stress time—one that occurs rarely in a building's lifetime.

By contrast, the second question looks at normal conditions, as they change over a season. Our calculations for a design-hour performance can be used, but must be adjusted for normal and continuous-change conditions. These calculations take much longer, and are even more suited to automatic computation (Appendix H) than simple design-hour answers. Once completed, we again have the opportunity to adjust our buildings. This time, adjustments are aimed more at reducing long-term energy use than at equipment sizing. We have, in effect, a series of photos that become a movie of our building's performance over time, under normal circumstances. We get another perspective on architecture as a performing art.

4. Buildings, like our bodies, exchange heat with the outside environment in two distinct ways: Through the envelope, akin to heat exchange through human skin, and with incoming fresh air, akin to our heat exchange with the air we inhale.

When calculating envelope (skin) losses or gains, materials, areas, and rates of heat flow through the envelope are used. When calculating fresh air (lung) losses or gains, volume of space and rates of fresh air exchange are used. For heavily insulated buildings in cold weather, it is usual for "lung" losses to exceed "skin" heat losses. Skin and lung losses (or gains) must be added together to determine the total rate of a building's heat exchanges with its environment; both can be manipulated through design to improve a building's thermal performance.

5. Buildings nearly always experience internal heat production; this is helpful in meeting cold weather demands, but harmful in hot weather. For most buildings, it will be helpful to manipulate the envelope to minimize internal gains (e.g., to reduce the need for electric lights by specifying larger, better-placed, summer-shaded windows), since the reduction in internal heat is so helpful in summer, and can so often be replaced by an increased reliance on passive solar heating in winter.

6. An occupied building can be very different thermally from an unoccupied one. Many buildings (such as offices and schools) experience high internal gains and rates of fresh air for only about a third of a typical day, and only 5 days per week; this amounts to less than a quarter of a year. Setback thermostats promise energy savings for such occupancy patterns. However, it is important to remember that we cannot depart too far from comfortable interior temperatures during the night or weekend without straining the heating and cooling systems early in the morning on the following working day. Said another way, one hour's surplus heat can be another hour's needed warmth.

4
HEAT FLOW

This chapter is about heat flow in buildings, both through a building's skin and by way of outdoor air brought in to replace "stale" indoor air. Basic calculations of heat flow are in this chapter, whereas most applications (passive solar heating, passive cooling, equipment sizing, seasonal energy usage) are found in Chapter 5.

NOTE: Except for the introductory Table 4.1 (terms and definitions), the tables that accompany this chapter are grouped together at the chapter's end. They begin on page 147.

4.1 The Building Envelope

The envelope of a building is not merely a set of two-dimensional exterior surfaces, it is a transition space—a theater where the interaction between outdoor forces and indoor conditions can be experienced. Sun and daylight are admitted or redirected to the interior; breezes and sounds are channeled; and rain is deflected. This transition space is a place where people indoors experience something of what the outdoors is like at that moment, as well as where people outside get a glimpse of the functions within. The more suited the outdoors to comfort, the more easily indoor activity can move into this transition space. At entries, a person will be especially aware of the difference between indoors and outdoors during passage between them. An example of an entry as a space, not merely a surface, is shown in Fig. 4.1.

The envelope also has a fourth dimension: it changes with time. Seasonal changes have a marked effect on the transition space in Fig. 4.1, and a more subtle effect on the east-facing balconies of the apartments in Fig. 4.2. Here, the year-round usable volume of these apartments is increased by making the balcony into a sun porch. This allows sun to enter while blocking the wind, in response to Oregon's long but mildly cold winter. Not all buildings encourage such flexibility

and individual expression; an unchanging envelope can be a symbol of stability and is considered appropriate for some governmental and religious monuments. The more that users are involved in decisions about how much of the outside to bring inside, the more changeable the building envelope will be.

4.2 Connectors, Filters, and Barriers

Called by their familiar names, the basic components of the envelope are windows, doors, floors, walls, and roofs. On closer inspection, windows can include skylights, clerestories, screens, shutters, drapes, blinds, diffusing glass, and reflecting glass—an array of components that determines more exactly how the envelope does its job of making the transition between inside and outside. Norberg-Schulz (1965, p. 113) suggests that a component can also be thought of by its function in the exchange of energies: as a *filter, connector, barrier* or *switch*.

> In general, we define a *connector* as a means to establish a direct connection, a *filter* as a means to make the connection indirect (controlled), a switch as a regulating connector, and a *barrier* as a separating element. . . . An opaque wall thus serves as a filter to heat and cold, and as a barrier to light. Doors and windows have the character of switches, because they can stop or connect at will.

This approach to the choice of envelope components can be illustrated by two opposite concepts of design: the *open frame* and the *closed shell*. In harsh climates (or where unwanted external influences such as noise or intruding activities abound), the designer frequently conceives of a building's envelope as a closed shell and proceeds to selectively punch holes in it to make limited and special contacts with the outdoors. Such an approach might be called "barrier-dominated."

Fig. 4.1 *The envelope is more than a surface. This south-facing office facade in Oregon forms a microclimate that buffers the transition between indoors and outdoors. Groundcover plants at seated eye level and deciduous vines overhead give a seasonally changing frame to the outdoor view that also admits winter sun, year-round daylight, and summer breeze. (Photo © John S. Reynolds, AIA.)*

When external conditions are very close to the desired internal ones, the envelope begins as an open structural frame, with pieces of building skin selectively added to modify only a few outdoor forces. Such an approach might be called "connector-dominated." The open-frame or the closed-shell approach to envelope design, when combined with material availability and the influence of local culture, can produce a distinct regional architecture (Fig. 4.3).

With a wide range of energy sources, building materials, and mechanical equipment available, it is possible to design connector-dominated buildings anywhere, despite the climate. The consequences of the resulting energy consumption can be severe. By contrast, if defending against outdoor conditions becomes an overriding consideration, barrier-dominated envelopes can occur in any climate. The potential of using natural energy sources can thus be greatly reduced. The designer's combination of connectors, filters, and barriers are basic to the design of building exteriors. With the addition of switches that allow these elements to respond to changing conditions, a liveliness can result that makes a building an attractive addition to its neighborhood. (Note: A connector, filter, or barrier for one natural force

may change its role for another: glazing may be a connector to daylight but a barrier to wind.)

4.3 Switches and User's Choice

Switches are a designer's way to give building occupants some control of their own environment. Seasons and functions change; people are unpredictable. If the designer has carefully considered the range of choices that a switch will provide, successful user control of that switch is possible. Buildings whose skins are plentifully supplied with switches become continuing demonstrations that architecture is a performing art, not simply sculpture.

In Chapter 1, the designer's response to both users and site was discussed in Section 1.6c, "Aggregate Rather than Isolate." The varied positions of a ventilating window switch were shown (Fig. 1.10) at a passively heated school in England. Passive solar heating systems also rely on switches to throttle down the incoming sun on warmer days, seen in Fig. 4.2b. Ventilating switches may be separate from "windows," and Fig. 4.4 shows a particularly integrated concept of a switch-dominated envelope.

(a)

(b)

(c)

Fig. 4.2 *Envelopes change with time. This east-facing building changes by season and over years. (a) December morning sun floods the open balconies in the building's early years. (b) Months later, June morning sun is blocked by shading devices on some balconies. (c) Years later, many balconies have been enclosed for greater winter comfort in this Oregon climate. (Cascade Manor, Eugene, Oregon; John Graham & Associates, Architects.)*

Open Frame

Seminole

Hot Humid Climate: To the <u>open frame</u>, a barrier roof of local plant materials is added to reject rain and sun. A raised floor avoids damp earth and its creatures and allows breezes to pass over and under its users.

Crow

Temperate Climate: This <u>open frame</u> is wrapped in light—filtering animal <u>skin, doubled</u> near the ground. Wind and rain are rejected; protection against cold is provided by users' clothing (blankets) more than by the envelope. The switch at the crown controls smoke (see also Fig. 2.20). Portability of shelter is a cultural factor here.

Closed Shell

Pueblo

Arid Climate: The <u>closed shell</u> of mud block is a barrier to wind and sunlight; it filters heat by both delaying and reducing its impact on the interior. Some light and heat are admitted directly by small connectors: the door and window, typically south—facing.
By early morning, the cold interiors are abandoned in favor of rapidly warming south terraces.

Eskimo

Cold Climate: The igloo's <u>closed shell</u> of ice is a filter to light and heat, a barrier to wind. Holes for entry and for smoke are allowed, but sparingly. Fur—bearing hides hung inside can increase thermal comfort for users.

Fig. 4.3 *Open frame and closed shell; climate, materials, and culture. The influence of climate is clear, but material availability and cultural patterns are also strong determinants in the choice of envelope design in these examples.*

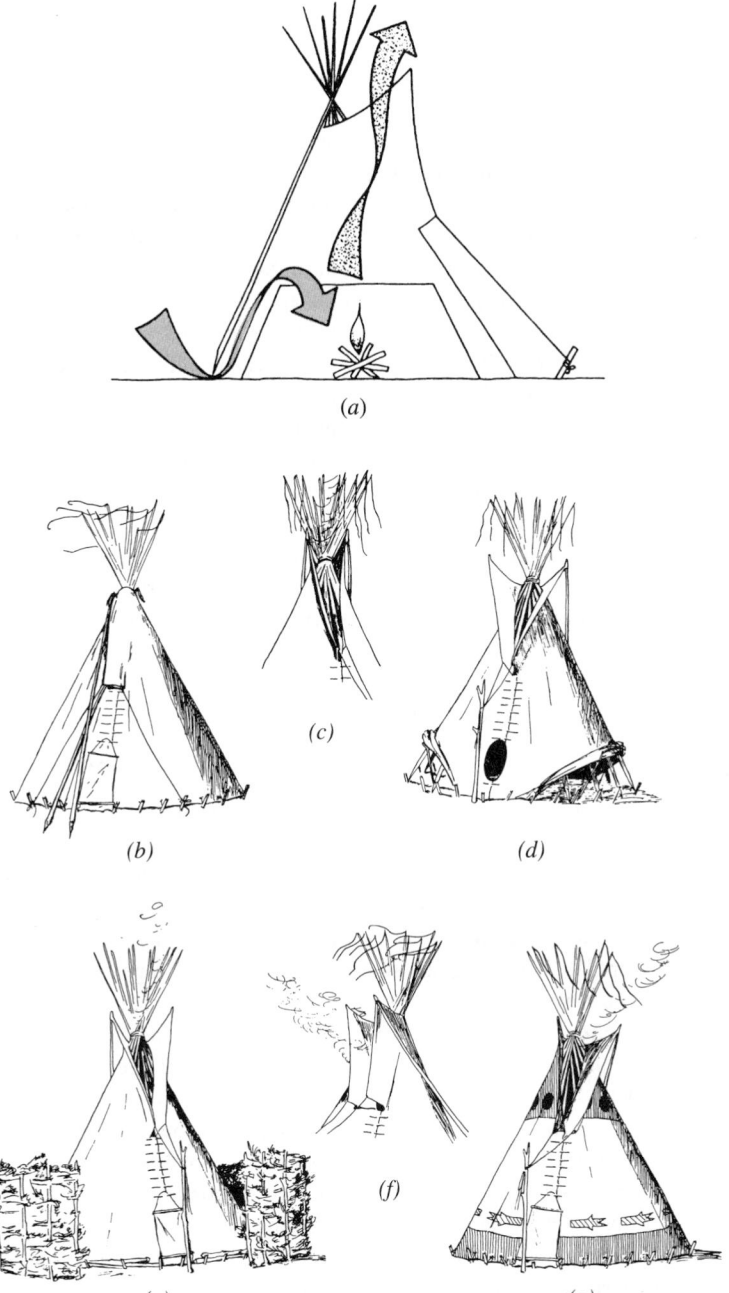

Fig. 4.4 *Thermal switches on the Great Plains.* (a) *With an interior fire, smoke flaps at top are adjusted to vent the smoke, and an interior liner forces the cold combustion air entering at ground level to first rise along the sides of the tipi, gaining some warmth, before it moves across the occupants on its way to the fire. Six adjustments to the lightweight, translucent, and portable tipi are shown: the tipis in these diagrams are facing east, with their backs to the prevailing westerly winds.* (b) *In severe rainstorms, the smoke flaps at the top can be closed.* (c) *For ordinary conditions of west wind, the smoke flaps block the wind, thus creating a suction to draw out the smoke.* (d) *For hot weather, breezes are admitted under the cover at the ground.* (e) *For extremely cold weather, a temporary windbreak can be added.* (f) *For the unusual wind, or* (g) *southwest wind, the smoke flaps are manipulated to block wind, again encouraging smoke draw-out, as in* (c). [(b)–(g) *From* The Indian Tipi: Its History, Construction, and Use, *by Reginald and Gladys Laubin. Copyright 1957 by the University of Oklahoma Press.*]

THERMAL CONTROL

Daylighting switches are perhaps our most common and visible ones. Awnings block direct sun at some times, and admit it at others. Opaque drapes can expose all the window to incoming daylight on dark days, or block out daylight entirely when desired. Translucent curtains can change bright sun into a diffuse light for the interior, or be drawn back to allow strong direct sun and shadow to be cast inside a room.

Visually, switches are a particularly promising source of three- and four-dimensional interest on a building's exterior. For the office buildings in Fig. 4.5, on which facade would the daily and seasonal changes be most visually evident? If it is easier to imagine human beings working behind the windows of one building than of the other, might that also suggest a more satisfactory work environment? Might more personal control of a window promote greater comfort?

Switches encourage interaction between users and their environments. This is usually satisfying to the users, who are able to select the desired exposure to climate at that moment. Yet without automation, supervision, or training in their uses, switches can be detrimental to system performance. Examples are a thermal shade left to cover a passive solar collecting window on a cold, sunny morning; a vent left open during the hottest hours in a high-mass, night-ventilated building; an awning rolled up to expose a window to summer sun. People in buildings often use switches in unexpected ways, as in Fig. 4.6. Conventionally air-conditioned buildings typically do away with user-operated ventilating switches (operable windows) so that the system will instead function with a closely controlled, filtered flow of outdoor air. Thermally efficient as this practice may be, it can also be the source of widespread dissatisfaction with air-conditioned spaces. Sealed windows greatly curtail people's contact with sounds, smells, and breezes of the outdoors. This is frequently beneficial in urban areas, yet on beautiful days it can be very frustrating. A lack of switches can contribute to a feeling of helplessness about one's own environment.

Passive heating and cooling techniques are especially reliant on switches, hence on the understanding and cooperation of the people within these spaces. These users often must base their actions of the moment on what will be needed later, by manipulating thermal switches. This practice, called *thermal sailing,* is similar to the actions of outdoor workers in the far north, who learn to unbutton their coats in the cold early hours of the workday *before* they begin to sweat. Sweat would soak their insulating clothing, with harmful results later in the day as temperatures fall rapidly near dusk.

Misjudgments in passively solar-heated residences can result in extraordinarily high temperatures on a sunny day or nights without stored solar heat. For the building closely related to its climate, the design of switches is also the design of an educational process for the users. The challenge is to involve, but not enslave, the users in the management of their environment. Automated controls are a partial answer to this challenge; switches that are easy, fun, and obvious to use are another.

4.4 Heat Flow, Seasons, and the Building

The flow of heat through a building envelope varies both by season (heat from the building in winter and to the building in summer) and by the path of the heat (through the building's skin, or by way of outdoor air entering the interior). This variation can change the focus of the designer who strives for energy conservation. In the 1970s, designers experienced first an emphasis on much tighter control of the building skin to conserve energy (especially under winter conditions) and then a greatly increased desire for outdoor air in response to "sick building syndrome." Thus we find that today's typical new North American building loses somewhat more winter heat to its incoming fresh air than it loses through its skin. Yet in summer it gains somewhat more heat through the skin than it does to its incoming fresh air. This is due in part to the influence of direct sun being so plentiful in summer, and to the outdoor–indoor temperature differences being so much greater in winter than in summer in most North American locations.

The following discussion of heat flow will focus first on the building skin, then on air exchange.

(a) Convection, Conduction, and Radiation. Whenever an object is at a temperature

(a)

(b)

(c)

Fig. 4.5 *Sun control for Boston offices. (a) Shading switches both inside and outside the glass are evident, and the windows themselves are operable. Awnings are more likely to be fully extended as the windows sit higher up the canyon of the street. (New England Merchants' Bank: Shepley Rutan & Coolidge, Architects, demolished 1966.) (b) Three-dimensional filters (overhangs) dominate the south facade (right) of this building at Boston University; the west windows (where overhangs are less effective) have internal switches (blinds), shown here in a variety of positions. (Law and Education Building; Sert, Jackson and Gourley, Architects). (c) The two-dimensional filter of reflective glass sheathes all faces of this office tower, sending a sheet of reflected sunlight to the neighborhood below. The switches in this case are thermostats; users are not involved in the outside–inside relationship. (John Hancock Headquarters; I. M. Pei and Partners, Architects. Photo by Stephen Tang.)*

THERMAL CONTROL

(a)

(b)

Fig. 4.6 *Which is spring, which is summer? (a) The awnings for this south-facing office in Eugene, Oregon, are joined by overhangs and side walls (or fins) to provide sun control for windows. The spring sun altitude is low enough to threaten glare in the lower windows. (b) The summer sun, at higher altitude, is readily blocked by the overhangs, and the awnings can be rolled up with less risk of glare. (Moreland-Unruh-Smith, Architects.)*

different from its surroundings, heat flows from the hotter object to the colder object. Chapter 2 discussed the human body's means of disposing its surplus heat to the cooler environment. Buildings, like bodies, experience heat loss to, and gain from, the environment in three principal ways. In *convection*, molecules of cooler air absorb heat from a warmer surface, expand in volume, rise, and carry it away. In *conduction*, heat is transferred directly from the molecules of warmer building

Any solid material
wall, floor or roof

A single solid material illustrates the transfer of heat from the warmer to the cooler particles by conduction (1).

An air space in a wall

As air is warmed by the warmer side of the air space it rises. As it falls down along the cooler side it transfers heat to this surface (2). Radiant energy (3) is transferred from the warmer to the cooler surface. The rate depends upon the relative temperature of the surfaces and upon their emissive and absorptive qualities. Direction is always from the warmer to the cooler surface.

The convective action (2) in the air space of a roof is similar to that in a wall although the height through which the air rises and falls is usually less. The radiant transfer is up in this case because its direction is always to the cooler surface.

An air space in a roof or floor

When the higher temperature is at the top of a horizontal air space the warm air is trapped at the top and, being less dense than the cooler air at the bottom, will not flow down to transfer its heat to the cooler surface. This results in little flow by convection. The radiant transfer in this case is down because that is the direction from the warmer surface to the cooler.

Inside air
motion

Outside wind

A composite assembly of building
materials in place at the site

This example of a wall in place illustrates the several methods by which heat is lost through a composite assembly of materials. Conduction at varying rates in different materials is accounted for in 1a, 1b, 1c. Convection currents (2) and radiation (3) carry the heat across the air space.

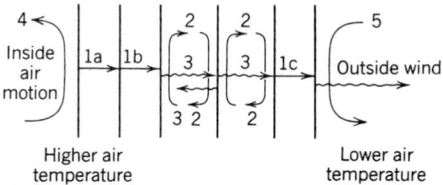

Inside
air
motion

Outside wind

Higher air
temperature

Lower air
temperature

Heat is conducted from the room air by warm air currents that strike the inside wall. Heat is conducted away from the exterior surface of the wall by the action of the wind.

Fig. 4.7 *Nature of heat flow through materials, air spaces, and assembled structures. Thermal action is identified by the following numbers: 1, conduction; 2, convection; 3, radiation; 4, inside surface conductance; and 5, outside surface conductance.*

THERMAL CONTROL

surfaces to the molecules of cooler solids (such as earth) in contact with the building. In *radiation,* heat flows in electromagnetic waves from hotter surfaces to detached, distant colder ones, through any transport medium, even empty space. *Evaporation* can also be involved, carrying heat away from wet surfaces, but this is much less influential for most buildings than for our bodies.

This combination of heat flow by convection, conduction, and radiation is shown in Fig. 4.7 through some typical combinations of materials. Notice that multiple air spaces and reflective surfaces are an inexpensive but useful way to slow the flow of radiative heat in hot spaces where temperature differences are large. Some of the most effective insulating materials therefore combine dead-air spaces and layers of reflective films. Heat flow through the various components of a building's skin involves both heat flow through solids and heat flow through layers of air.

4.5 Heat Flow Through Solids

Certain terms are fundamental to any discussion of heat flow; Table 4.1 lists terms of energy, terms of power (energy used over a given time span), and terms of heat flow rate.

(a) Conductivity. Each material has a characteristic rate at which heat will flow through it. For such individual, or homogeneous, solids, this rate is called its *conductivity,* and in the inch–pound (I-P) system is designated as k; the number of British thermal units per hour (Btu/h) that flow through 1 square foot (ft^2) of material that is 1 in. thick, when the temperature drop through that material is $1F°$ (under conditions of steady heat flow). Thus the I-P units of conductivity are Btu-in./h ft^2 °F.

The System International (SI) equivalent is the rate at which Watts flow through 1 square meter

TABLE 4.1 Heat Symbols, Terms, Definitions[a]

I-P (Inch-Pound) Units			SI (Systéme International) Units		
Symbol	Term	Definition or Usage; Conversions	Symbol	Term	Definition or Usage; Conversions
Part A. **Energy**					
Btu	British thermal unit	The amount of heat required to raise 1 pound of water by 1F°. (It roughly corresponds to the energy released in burning an ordinary wooden match.) 1 Btu = 1.055 kilojoules	J	joule	Newton-meter, or a force of 1 newton (1 N) acting through a distance of 1 meter. In terms of heat, a joule is 1/4.184 of the amount of heat required to raise 1 gram of water by 1C°.
			kJ	kilojoule	1000 joules; 1 kJ = 0.9478 Btu
Part B. **Power**					
Btu/h	British thermal units per hour	Commonly used to express the total heat loss or gain of a building, and to express the size of heating or cooling equipment. 1 Btu/h = 0.2929 W	W	watt	The power required to produce energy at the rate of 1 joule per second. Commonly used to specify size for light bulbs, furnaces, air conditioners, and other heat producers. 1 W = 3.412 Btu/h
hp	horsepower	1 hp = 746 W			

TABLE 4.1 **Heat Symbols, Terms, Definitions**[a] (*continued*)

I-P (Inch-Pound) Units			SI (Systéme International) Units		
Symbol	Term	Definition or Usage; Conversions	Symbol	Term	Definition or Usage; Conversions

Part C. ***Rate of Heat Flow***

I-P (Inch-Pound) Units			SI (Systéme International) Units		
Btu/h ft² °F		(Btu/h ft² °F) $\times$ 5.678 = W/m²K (W/m² K) $\times$ 0.176 = Btu/h ft² °F These are the units of conductance and of overall heat transmission for the following terms:	W/m² K (or W/m² °C)		
	C	(conductance). The rate of heat flow through a homogeneous material (or combination of materials) of a stated thickness.			
	a	(air space conductance). The rate of heat flow through an air space bounded by two surfaces.			
	h	(film or surface conductance coefficient). The rate of heat flow from a surface due to air (or other fluid) motion against the surface. (Identical to f, the film conductance.)			
	U	(overall coefficient of heat transmission). The overall rate of heat flow through any combination of materials, air layers, and air spaces. It is equal to the reciprocal of the sum of all resistances R (see below) that are involved in this combination. This is the term used directly in building envelope heat loss or gain calculations.			
Btu-in./h-ft²-°F	k	(conductivity). The rate of heat flow through a homogeneous material, per unit of thickness.	W/m K (or W/m °C)		
h ft² °F/Btu	R	(resistance). A measure of resistance to the passage of heat; the reciprocal of conductance.	m² K/W		
	ε	(emittance). The ratio of the radiant flux from a given surface to that of a blackbody (a "perfect" emitter) at the same temperature.			
	E	(effective emittance). The combined effect of the emittance of parallel surfaces bounding an air space.			

[a]A more complete listing of conventional–SI conversion units is found in Appendix I.

(m²) of material, 1 meter thick, when the temperature drop through that material is 1K (equal to 1C°) under conditions of steady heat flow. Thus, the SI units of conductivity are W/m K (or W/m °C; these two terms are used interchangeably in the tables of this chapter).

Conductivity is an important factor in passive heating or cooling design that depends heavily on the rate at which heat is conducted into a material from its surface. When we touch a solid's surface, we sense its conductivity (rather than its tempera-

ture). Conductivity for each solid is established by tests and is published as a basic rating.

(b) Conductance. Many solids such as common brick, wood siding, thermal batt or board insulation, gypsum board, and so on are widely available in standard thicknesses. For such common materials, it is useful to know the rate of heat flow for that standard thickness instead of the rate per inch. *Conductance,* designated as *C,* is the number of Btu per hour that flow through

THERMAL CONTROL

1 square foot of a given thickness of material when the temperature difference is 1F°. The units are Btu/h ft² °F. (In SI, the units are W/m² K.)

(c) Resistance. Conductivity and conductance are compared in Table 4.1 and Fig. 4.8, which include another useful term, *resistance.* Designated as *R,* resistance tells us how effective any solid is as an insulator. The reciprocal of conductivity (or conductance), *R* is measured in *hours* needed for 1 Btu to flow through 1 ft² of a given thickness of a solid when, again, the temperature difference is 1F°. In the I-P system, the units are h-ft²-°F/Btu. In SI, units are m²-K/W. Resistances and other important thermal properties are listed for many materials in Table 4.2 (page 147).

Resistance is especially useful when comparing insulating materials, because the greater the *R*-value, the more effective the insulator. For this purpose, *R* is sometimes listed as "per inch" of thickness, in which case the units are h-ft²-°F/

Btu-in. (In SI, units are m-K/W, or "per meter" of thickness.)

(d) Insulators. Architectural materials interact with heat flow either as *insulators* to retard the flow of heat, or as *conductors* to encourage heat flow, as in the case of thermal storage materials. It is common to find both insulators and conductors in the same construction. For example, a wall can have an inner layer that is highly conductive and thermally massive (for thermal storage), with an outer layer that is also highly conductive and thermally massive (for durability and weathering), whereas between inner and outer surfaces it has a highly insulative, low mass substance.

Materials used for insulation fall into three categories: (1) *inorganic* fibrous or cellular (such as glass, rock wool, slag wool, perlite, or vermiculite), (2) *organic* fibrous or cellular (such as cotton, synthetic fibers, cork, foamed rubber, or polystyrene), and (3) *metallic* or metalized organic *reflec-*

Glass Fiber Insulation Board

Sand and Gravel Concrete

Fig. 4.8 *Sample conductivity* (k) *for 1-in. thickness, conductance* (C) *for a specified thickness (4 in. in this example), and resistance* (R) *for two materials. Glass fiber is a material of low conductivity; concrete is a material of high conductivity. Note: Standard unit of area is 1 square foot; standard air temperature differential is 1 degree Fahrenheit (shown as 1F°).*

tive membranes (which must face an air space to be effective).

Insulating materials are available in a wide variety of forms. Most adaptable are *loose fill* (as above a ceiling on the floor of an attic); *insulating cement,* a loose material mixed with a binder and troweled onto a surface; and *formed-in-place* materials such as expanded pellets or liquid-fiber mixtures that are poured, frothed, sprayed, or blown in place. Less adaptable and more common are the batts and blankets of *flexible, semirigid* insulation, with varying degrees of compressibility and adaptability.

Rigid insulation is applied in blocks, boards, or sheets, and can be preformed to nonplanar surfaces such as pipes. Exterior insulation and finish systems have become very popular, both for retrofit and for new construction. Utilizing expanded polystyrene rigid boards applied to exterior gypsum, plywood, or cementitious substrates, then covered with fabric-reinforced acrylic, this method achieves slightly more than *R*-4 per inch of thickness (RSI about *R*-28 per meter).

Reflective materials are found in sheets and rolls of either single or multiple layers, sometimes as preformed shapes with integral airspaces. When used without attachment to blanket or batt insulation, a reflective layer is called a *radiant barrier* and is especially applicable to roofs in the warmer climates nearer the equator. Radiant barriers are also useful in east- and west-facing walls in such climates; details of applications in the southern United States are found in Chandra et al. (1986).

For a summary of contemporary insulation materials, including environmental impacts and life-cycle considerations, see Wilkinson (1999).

(d) Conductors. Materials used as conductors are typically dense, durable, and diffuse heat readily. Table 4.2 lists density, conductivity, or conductance, and specific heat for many common materials. For a given material, the higher the numerical values of these three characteristics, the more successful will be that material's performance as a conductor. Some specifics on thermal mass are found in Section 5.14c.

The previous paragraphs concentrated on heat flow by conduction, but convection and radiation are also involved. Both these latter flows become more evident when looking at heat flow through air.

4.6 Heat Flow Through Air

At the exposed surfaces of solids, heat transfer takes place both by convection and by radiation. Evaporation also can occur, occasionally with thermally significant results; see Fig. 5.46 (British Pavilion, Seville) for an example of an entire wall so used. Convection is highly dependent on air motion, so wind outdoors must be considered. Also, because warm air rises and cold air falls, vertical surfaces that encourage this kind of airflow exchange heat faster than the same surfaces placed horizontally, unless the direction of the heat flow is *upward* through this horizontal layer, as was evident in Fig. 4.7.

(a) Resistance. When air motion along surfaces is minimal, an *insulating layer* of air is created. The resistance of this layer of still air along a vertical surface is numerically equal to that of a thickness of ½-in. (12.7-mm) plywood. However, when this air layer is disturbed, its resistance drops quickly; with a 15-mph (6.7-m/s) wind, resistance drops to about one quarter of the still-air value (Table 4.3). Similar drops in resistance occur when forced-air grilles are located immediately above or below windows.

(b) Conductance. These surface layers of air are often evaluated by their conductance, the reciprocal of resistance, because we often wish to *encourage* heat transfer between solids and air. For example, passive heating and cooling systems are dependent on large interior surfaces as heat exchangers. By contrast, active (conventional) systems typically concentrate on heat exchange within mechanical equipment.

Surface conductances are termed h_i for interior air layers and h_o for exterior or outside air layers (sometimes f_i and f_o are used instead). Like other conductances, they are expressed in Btu/h ft² °F (in SI units, W/m² K). The variations in surface resistances and conductances are shown in Table 4.3.

THERMAL CONTROL

(c) Emittance. Radiation is also highly influenced by surface characteristics; shiny materials are much less able to radiate than are common rough building materials. This characteristic is called *emittance,* the *ratio* of the radiation emitted by a given material to that emitted by a blackbody at the same temperature. The impact of shiny versus ordinary surfaces is seen in Tables 4.3 and 4.4, which deal with air layers and air spaces within construction; the lower the emittance, the lower the radiative heat exchange. For most materials, emittance is related to absorptance: a highly absorptive (low reflectance) material will usually have a high emittance as well. ("Selective surfaces" are sometimes used in passive solar collection; these manage to be highly absorptive yet have very low emittance.)

(d) Insulation. The combination of dead (still) air spaces and reflective surfaces produces some of our most effective insulating products, especially when they are made of lightweight materials of low conductivity. Glass fiber, cellular glass, expanded styrenes (foamed plastics), and mineral fibers all exhibit the characteristics of enclosing vast numbers of dead-air spaces per unit volume. When bonded to reflective films and properly installed (the shiny film facing a dead air space), high resistance to heat flow is achieved. Resistances to heat flow can be compared between various building materials, including insulation products (Table 4.2) and air films and air spaces (Tables 4.3 and 4.4)

4.7 Heat Flow Through the Opaque Building Envelope

When the process of heat flow is understood, the calculations can begin. Initially, the *hourly* heat loss or heat gain is calculated, because these rates can later be used either to establish equipment size requirements (design conditions) or approximate energy consumption (average conditions).

To calculate the hourly flow through a building's envelope, these factors are necessary:

1. The *rate* at which heat flows through the various assemblies of materials that make up the envelope.

2. The *area* of each of these assemblies.
3. The *temperature difference* between inside and outside for the hour being calculated.

Often, only the first two factors are calculated, which allows a comparison of one building with any other, anywhere. The third factor, temperature difference, is strongly climate related.

The *areas* of walls, roofs, windows, doors, and floors of various kinds are determined from preliminary architectural design drawings; final versions of these drawings should be subject to changes suggested by these calculations. Such changes are usually in roof, wall, and other construction (affecting the rate of heat flow) rather than in area. The *temperature differences* for design conditions are listed in Appendix A; for energy consumption, see Section 5.9. The *rates* at which heat flows must now be determined.

(a) U-Value. The variety of terms used so far to express heat flow is potentially bewildering. These terms are, however, but part of a larger picture; what is needed is *one* final, overall expression of the steady-state rate at which heat flows through architectural skin elements (wall, roofs, etc.). This is provided by the *U value* (also called the *U*-factor), where *U* is thermal transmittance, again expressed in the now-familiar terms of Btu/h ft^2 °F (in SI units, W/m^2 K). Because *U* values are both common and rather complex, several pages are devoted to listing them for familiar constructions of walls, floors, roofs, doors, and windows (Tables 4.5 to 4.15 at the end of this chapter). *U* values are also used in design criteria, presented in Appendix J and Chapter 5.

These *U* values are calculated for a particular element (roof, wall, etc.) by finding the resistances of each of its component materials, including air layers and internal air spaces, then adding all these resistances to obtain ΣR. The *U* value is the reciprocal of this sum of resistances:

$$U = \frac{1}{\Sigma R}$$

This procedure is illustrated in Fig. 4.9. Unfortunately, many common constructions are not so simple, as shown in the sections that follow. "Precalculated" *U* values for many common constructions are found in Tables 4.5 through 4.15, and

Component	R	Reference Table
Inside air layer	0.68	Table 4.3
Common brick	0.20	Table 4.2
Nominal 6-in. batt fiberglass	19.00	Table 4.2
½-in. Plywood	0.62	Table 4.2
1-in. Wood siding	0.79	Table 4.2
Outside air layer	0.17	Table 4.3
ΣR	21.46	

$$U = \frac{1}{\Sigma R} = \frac{1}{21.46} = 0.046$$

Fig. 4.9 *Procedure for determining the* U-*value (overall thermal transmittance) for a wall with interior thermal mass.*

are based on an outside wind velocity of 15 mph (6.7 m/s), except for summer conditions as noted.

Many codes and standards specify maximum *U*-values (or, for insulation alone, minimum R- or maximum *C*-values) for various components of the envelope. An example of such requirements for a typical residence is found at the end of Section 4.8, after all envelope components have been discussed.

(b) Trends in U- and R-Values. Before looking individually at each component of the envelope, some history and prediction may be useful. As a result of the oil embargo of the early 1970s, there have been huge increases in component *R* (therefore huge decreases in component *U*), sometimes because of increased thickness and insulation (as with walls and roofs), sometimes because of substantially changed materials (as with windows). What might the next quarter century bring, and how might the designer of today's building envelope anticipate—or participate in—these advances?

Structural systems are changing fundamentally due to advent of structural insulated panels (SIP). They promise greatly improved insulation and airtightness, compared to the site-built wood or steel framing systems they replace. They can incorporate the latest developments in both insulated cores and thermal storage surfaces, and do so with less thickness and lighter weight. Since the early 1970s, walls have gone from about *R*-7 (RSI 1.23) batts between typical 2-in. × 4-in. studs to *R*-26 (RSI 4.58) batts between 2-in. × 6-in. studs plus insulating sheathing, a three-fold reduction in heat flow. With SIP insulated cores approaching *R*-25 per inch (RSI 174 per meter), the next quarter century could produce yet another three-fold reduction while decreasing wall thickness. Floors and roofs can similarly benefit from SIP rather than framed construction.

Insulation developments include aerogel, gas-filled panels, powder-evacuated panels, and compact vacuum. Some of these products are suitable for retrofit applications in existing construction. Aerogel is both transparent and porous, one of the lightest existing solid materials. Silica aerogel can be foamed into cavities, and done so without ozone-depleting chlorofluorocarbons (CFCs). In a 90% vacuum, silica aerogel has *R*-20 per inch (RSI 139 per meter). By adding carbon to absorb infrared radiation, R-32 per inch (RSI 222 per meter) is possible.

Gas-filled panels are hermetically sealed plastic bags enclosing honeycomb baffles of thin polymer films and low-conductivity gas (argon, krypton, or xenon). Relative resistance values of this insulating core: argon, *R*-5.2 per inch; krypton, *R*-13.4 per inch; and xenon, *R*-19.3 per inch (RSI 36, 93, and 134 per meter, respectively).

Powder-evacuated panels contain a vacuum and compacted silica-based powder sealed within a multilayer gas barrier. Expensive to produce and threatened by punctures, this may be a construction technique more suited for appliances than for buildings. The resistance is about *R*-20 to *R*-25 per inch (RSI 139 to 174 per meter).

Windows have undergone the most dramatic change. Resistances for the unit (glazing plus frame) have evolved from single-glazed windows of about *R*-1 to late-1990s values of *R*-6.5 or more. Prototypes as high as *R*-20 (RSI 3.53) exist. This has been accomplished with a series of developments: a second layer of glass; a wider air space between glazing layers; tinted, reflective, photochromic, and low-emissivity coatings; low-conductivity gas between glazings; intermediate films between glazings; operable blinds between glazings (for better control); lower-conductivity glazing spacers and frames; and more air-tight weather stripping. Future developments are likely to include "smart windows" with variable light transmission on demand. For example, in electrochromic glazing, an applied electric field switches the window from a clear state to one with a deep coloration; intermediate states of coloration are possible. The more likely application is in cooling-dominated climates where control of glare, solar gain control, and visual privacy are paramount.

Before the oil embargo of the early 1970s, windows represented the greatest heat flow rate through the envelope of the typical building. Today, the highest rate of heat flow is more likely to be outside air infiltration (or deliberate ventilation). Windows do so many things in addition to weather protection—admit daylight, allow views outdoors and in, admit breezes, and admit warming winter sun—that their relative thermal weakness continues to challenge manufacturers. Over the next quarter century, windows may well continue to exhibit the most dramatically improved resistance to heat flow of all the components of the envelope.

(c) Walls. Compared to other surfaces of the building envelope, wall U values are quite straightforward. There are few complications such as ground contact or crawl spaces with floors, or intermediary attic spaces with roofs. There is, however, the matter of thermal bridging: where framing interrupts the insulation, an averaged U-value must be found. An example of a wood frame wall is shown in Fig. 4.10. With metal studs, the detrimental effects of thermal bridging are more serious; Table 4.6 is a first approximation of the resulting R-values.

The latest edition of ASHRAE/IES Standard 90.1 *Energy Efficient Design of New Buildings Except New Low-Rise Residential Buildings* presents many tables of whole-wall U-values that account for thermal bridging. To illustrate the complexity, consider this variety of proposed values, all for nominal 4-in. (100 mm) thick framed walls

with R-13 (RSI 2.3) insulation only within the stud cavity:

Metal building walls, nominal 4 in.	$U = 0.14$
(100 mm	$U = 0.74$)
Steel frame walls, 3.5 in. at 16-in. o.c.	$U = 0.134$
(90 mm at 400 mm o.c.	$U = 0.76$)
3.5 in. at 24-in. o.c.	$U = 0.115$
(90 mm at 600 mm o.c.	$U = 0.65$)
Wood frame walls, 3.5 in. at 16-in. o.c.	$U = 0.094$
(90 mm at 400 mm o.c.	$U = 0.53$)
3.5 in. at 24-in. o.c.	$U = 0.091$
(90 mm at 600 mm o.c.	$U = 0.52$)

If a detailed calculation of thermal bridging is desired, consult 1997 *ASHRAE Handbook of Fundamentals,* Chapter 24.

Several new developments are promising to improve the thermal performance of wall systems.

Structural insulated panels (SIP) are now available in a wide variety of structural surfaces and interior insulation type and thicknesses. Because a single factory-built panel replaces site-built framing, savings in labor can be substantial. Thermal performance is considerably improved because no framing members penetrate the insulated core. The typical SIP consists of two structural surfaces (often oriented-strand board, or OSB) that enclose a core of either expanded polystyrene (EPS) (R-4 per inch; RSI 27.8 per meter) or polyisocyanurate foam (R-6.5 per inch; RSI 45.1 per meter). The panels are connected with plywood splines that connect the adjacent panels' structural surfaces, in a manner that reduces the thickness, but does not interrupt, the insulated core. Another joining method, shiplap joints, maintains full-thickness core insulation. Air-tightness is generally tighter than with stud framing. Relative U-values of framed walls and SIP are found in Table 4.7.

Some potential developments with SIP include insulated cores consisting of the fore-mentioned products under development: aerogel, gas-filled panels, and perhaps powder-evacuated panels. Another potential development is an interior panel surface of phase-change wallboard, that stores then releases heat at room temperatures.

Resistance between framing	h ft² °F / Btu	m² K / W
Outside air surface, 15 mph wind	0.17	
(24 km/h)		(0.03)
Wood siding, 0.5-in. x 8-in. lapped	0.81	
(13 x 200 mm)		(0.14)
Sheathing, 0.5-in. vegetable fiber board	1.32	
(12.7 mm)		(0.23)
Mineral fiber, approx. 5.5-in. batt insulation	21.00	
(approx. 140 mm)		(3.67)
Gypsum wallboard, 0.5-in.	0.45	
(12.7 mm)		(0.079)
Inside air surface, still air	0.68	
		(0.12)
Total thermal resistance between framing	24.43	(4.27)

Resistance at framing	h ft² °F/ Btu	m² K / W
Outside air surface, 15 mph wind	0.17	
(24 km/h)		(0.03)
Wood siding, 0.5-in. x 8-in. lapped	0.81	
(13 x 200 mm)		(0.14)
Sheathing, 0.5-in. vegetable fiber board	1.32	
(12.7 mm)		(0.23)
Wood studs, nominal 2-in. x 6-in.	6.82	
(approx. 140 mm)		(1.20)
Gypsum wallboard, 0.5-in.	0.45	
(12.7 mm)		(0.079)
Inside air surface, still air	0.68	
		(0.12)
Total thermal resistance at framing	10.25	(1.80)

Average Resistance, R_{av}

With 12%[*] framing, typical of 2-in. 6-in. studs at 24-in. on center:

R_{av} = 0.88 (24.43) + 0.12 (10.25) = 21.50 + 1.23 = 22.73

[RSI_{av} = 0.88 (4.27) + 0.12 (1.80) = 3.76 + 0.22 = 3.98]

U_{av} = 1 / 22.73 = 0.044 Btu/h ft² °F

[U_{av} = 1 / 3.98 = 0.25 W/ m² K]

Fig. 4.10 *Determining the average* U-*value in a wood frame wall. The 12% framing is based on the face area of studs as well as the sill plates, compared to the face area of insulation.*

This could increase thermal mass considerably with very little increase in weight, with advantages for winter daily solar heating and summer night ventilation of mass.

Insulated masonry systems offer lower *U*-values while preserving a masonry exterior (for weathering and appearance) and a masonry interior (for thermal mass). Past approaches (filling concrete block or tile hollow cores with insulation) left considerable thermal bridging through the solid masonry, with rather high resulting *U*-values. Even in cavity walls, the layer of insulation between wythes of block is usually penetrated by ties with resulting thermal bridging. The most elementary improvement here are ties with less thermal conductivity, such as fiber composite materials in place of steel. Another approach is to manufacture precast blocks that can be stacked integrally with preformed rigid insulation; to the extent that the insulation layer can both be thick and unbroken by ties, thermal bridging is greatly reduced. Manufacturers' data contain whole-wall *U*-values; below is one sample, for medium-density (120 lb/ft³) concrete block:

8-in. two-core concrete block, uninsulated $U = 0.39$

8-in. two-core concrete block with insulated cores $U = 0.19$

8-in. proprietary two-wythe concrete block and core $U = 0.11$

If one of the masonry surfaces is not to be exposed, then a continuous layer of insulation can be applied to that face, with much lower whole-wall U-values resulting.

Transparent insulation can be combined with exterior glazing on the outer surface of south-facing masonry or concrete walls to add solar heat gain to mass walls. This system is discussed in Chapter 5, for passive solar heating; see Table 5.15 and Fig. 5.25.

Where poured concrete is not to be a surface material, insulating concrete forms (ICFs) use preformed rigid insulation as the formwork for poured concrete, remaining in place after the wall is poured. Rigid insulation thus protects (but isolates as thermal mass) both faces of the structural wall; exterior and interior finishes are then applied.

Appendix J presents guidelines for choosing wall U-values for residential construction, based on *ANSI/ASHRAE Standard 90.2-1993, Energy Efficient Design of New Low-Rise Residential Buildings.* See Figs. J.3 through J.7 for single family residences, and Figs. J.19 through J.25 for multifamily structures.

(d) Floors. When a floor is exposed to outdoor air flowing beneath it, the U-value is calculated in a similar manner to walls and roofs. Again, where framing cavities are the location of the insulation, the effects of thermal bridging must be included. The latest edition of *ASHRAE/IES Standard 90.1 Energy Efficient Design of New Buildings Except New Low-Rise Residential Buildings* presents several tables of whole-floor U-values that account for thermal bridging. In the case of floors, it is increasingly common to provide a continuous layer of insulation below the framing cavity, thus greatly reducing the detrimental effects of thermal bridging.

Over an enclosed crawl space, the most simple procedure is to insulate the floor above the crawl space and assume that the crawl space is at outdoor temperature. Many codes require vented crawl spaces. This assumption overestimates heat flow rate, because a vented crawl space will rarely be as cold as outdoor air under winter design conditions.

If a more accurate crawl space design temperature is desired, or if a warmer crawl space is desired and its walls are to be insulated, then a more complex procedure is used, as presented in the 1997 *ASHRAE Handbook of Fundamentals,* Chapter 27. Heat losses to the ground must be considered, as well as potentially significant moisture problems resulting from the contact of the earth's surface with warmer crawl space air. It is rarely energy conserving, especially in colder climates, to insulate the walls of the crawl space instead of insulating the floor above it.

When the floor is in direct contact with the ground as well as with outdoor air, the heat flow calculation changes. The ground is often at a temperature different from outdoor air, and earth is more conductive than air. Table 4.8 shows the heat flow rate through the edge of a concrete slab on grade, with this heat flow rate in F_2 units (per foot or meter of perimeter length) rather than U-values (per unit area). Four edge insulation conditions are shown, with corresponding values of F_2 for three climate zones. Interpolation can be used to estimate F_2 for other climates (represented by degree days). This procedure assumes that all the heat loss from the floor occurs through the perimeter.

Table 4.8 lists a maximum slab-edge R-value of 5.4 h ft^2 °F/Btu (0.95 K m^2/W). If more slab-edge insulation seems desirable, consult Table 4.9. Note how much more effective is vertical insulation at the exposed perimeter, compared to insulation beneath the slab.

At this point we encounter the principle of diminishing returns: much higher R-value does not produce as much less heat flow. An insulation truism is: The first few inches (mm) of insulation make a much greater impact on the rate of heat flow than do the last few inches (mm). Indeed, the highest F_2 value listed in Table 4.9 is for insulation at R-10 (RSI-1.8) not only around the perimeter, but also under the entire area of the floor slab. This lowest rate of heat flow is only a reduction of 25% from that same insulation employed only around the perimeter to a depth of 4 ft (1.2 m).

EXAMPLE 4.1. A warehouse will be built with a slab on grade. The warehouse is 80 ft square in plan. What is the relationship between total slab insulation and rate of heat flow through the slab?

SOLUTION. From Table 4.9, F_2 is related to total insulation as follows: Building slab perimeter: 4 sides at 80 ft each = 320 ft
Building slab area: 80 ft $\times$ 80 ft = 6,400 ft^2

Construction	Total Volume of Slab Insulation	F_2 (Btu/h ft °F)	$F_2 \times$ Perimeter = Rate of Heat Flow (Btu/h °F)
Uninsulated slab	None	0.73	233
R-5 vertical, 2 ft	2 ft $\times$ 320 ft = 640 ft^2 @ 1 in.	0.58	186
R-10 vertical, 4 ft	4 ft $\times$ 320 ft = 1,280 ft^2 @ 2 in.	0.48	154
R-10 fully insulated	2 ft $\times$ 320 ft = 640 ft^2 @ 2 in. plus 6,400 ft^2 @ 2 in.		
	7040 ft^2 @ 2 in.	0.36	115

Thus, for the minimal initial investment of 640 ft^2 @ 1-in. thick insulation, we cut the uninsulated slab heat flow by one fifth. However, for the vastly greater investment of 7,040 ft^2 @ 2-in. thick (22 times as much insulation!), we cut the lesser-insulated heat flow by one fourth. ■

Appendix J presents guidelines for choosing floor *U*-values and slab edge *R*-values for residential construction, based on *ANSI/ASHRAE Standard 90.2-1993, Energy Efficient Design of New Low-Rise Residential Buildings.* See Figs. J.8, J.9, and J.12 for single family residences, and Figs. J.26 through J.28 and J.31 and J.32 for multifamily structures.

(e) Basements. Heat flow through basement walls and floors is complicated by an increasing length of path of heat flow with increasing depth, as shown in Fig. 4.11. A further complication is that the surface temperature of the earth varies seasonally. To obtain a ''design temperature'' of the earth surface, first approximate the mean winter temperature of your location. This can be done from Appendix B, Table B.15: take the average between ambient temperature (TA) for January and for the year. Then from this average temperature, *subtract* the value of the constant amplitude, Fig. 4.11c. The result is the design temperature. Table 4.10 is then used to determine the heat flow rate.

EXAMPLE 4.2. Calculate the heat loss for a basement in Minneapolis, Minnesota, which is 28 ft wide by 30 ft long, sunk 6 ft below grade. An insulation of *R*-8.34 is applied to this wall to a depth of 2 ft below grade. An internal temperature of 70°F is to be maintained.

SOLUTION. The earth surface design temperature is estimated from Table B.15; for Minneapolis, the TA in January is 12°F; TA for the year is 44°F; the average of these is 28°F. From Fig. 4.11c, the amplitude is about 24F° at Minneapolis. The design temperature at the earth surface is therefore 28 − 24 = 4°F.

The heat loss through the basement wall is estimated using Table 4.10, Part A:

1st ft below grade (insulated)	0.093 Btu/h ft °F
2nd ft below grade (insulated)	0.079
3rd ft below grade (uninsulated)	0.155
4th ft below grade (uninsulated)	0.119
5th ft below grade (uninsulated)	0.096
6th ft below grade (uninsulated)	0.079
Total per 1-ft length of wall	0.621 Btu/h ft °F

Basement perimeter = 2 (28 + 30 ft) = 116 ft
Total wall heat loss = 0.621 Btu/h ft °F $\times$ 116 ft = 72 Btu/h °F

Fig. 4.11 *Heat flow below grade. (a) The paths of heat flow through basement walls and floor follow radial isotherms as shown. (b) Heat flow paths for a partially insulated basement wall. (c) Lines of constant amplitude of the ground temperature; just below the earth's surface, the yearly average ground temperature will be this much higher in hottest summer and this much lower in coldest winter. (Reprinted with permission of the American Society of Heating, Refrigerating, and Air-Conditioning Engineers, Inc. from 1997 ASHRAE Handbook of Fundamentals.)*

The heat loss through the basement floor is calculated using Table 4.10 Part B;

Average heat loss per ft² floor = 0.025 Btu/h ft² °F (for 28 ft width, 6 ft depth);

Floor area = 28 ft × 30 ft = 840 ft²

Total floor heat loss = 0.025 Btu/h ft² °F × 840 ft² = 21 Btu/h °F

Total heat loss for the basement, below grade: walls 72 + floor 21 = 93 Btu/h °F

Design temperature difference : 70°F inside − 4°F at earth surface = 66F°

Maximum rate of heat loss below grade = 93 Btu/h °F × 66F° = 6,138 Btu/h. ∎

Appendix J presents guidelines for choosing basement wall insulation's C-values for residential construction, based on *ANSI/ASHRAE Standard 90.2-1993, Energy Efficient Design of New Low-Rise Residential Buildings.* See Figs. J.10 and J.11 for single family residences, and Figs. J.29 and J.30 for multifamily structures.

(f) Roofs. In the most simple case of a roof directly above a ceiling in a conditioned space, the U-value is calculated in a similar manner to walls and floors. Once again, where framing cavities are the location of the insulation, the effects of thermal bridging must be included. Table 4.11 shows the first-approximation correction factors to be applied to such insulation between metal trusses; the latest edition of *ASHRAE/IES Standard 90.1 Energy Efficient Design of New Buildings Except New Low-Rise Residential Buildings* presents several detailed tables of whole-roof U-values that account for thermal bridging. However, insulation is now commonly placed entirely above (or entirely below) the supporting structure, greatly reducing the effects of thermal bridg-

ing. (If a detailed calculation of thermal bridging is desired, consult 1997 *ASHRAE Handbook of Fundamentals,* Chapter 24.)

Where an insulated ceiling is below an uninsulated attic roof, the most simple procedure is to assume that the attic is at outdoor temperature. [An attic is a space having an average distance of 1 ft (0.3 m) or more between ceiling and the underside of a roof. A vented attic is often desired to carry away moisture that may have migrated through the insulated ceiling.] This assumption probably overestimates heat flow rate because a vented attic may approach the lowest temperature of outdoor air by night under winter design conditions, but by day the attic is warmed by solar radiation. Some stored solar heat probably keeps the nightly temperature above the design condition.

If a more accurate attic temperature is desired, then a more complex procedure is used, as presented in the 1997 *ASHRAE Handbook of Fundamentals,* Chapter 27.

There is growing variety of above-the-structure insulation materials, some penetrated by fastenings, others essentially unbroken. Among the latter are *spray polyurethane foam* (SPF), made up of the two components polyol and isocyanate, forming the polyisocyanurate foam that is also used in the cores of some SIPs. The aged *R*-value is 6.2 to 6.5 per inch (RSI 43 to 45 per meter), depending on the manufacturer. These materials are rigid yet lightweight, adhering fully to the roof substrate and therefore requiring no penetration by fasteners. They can be sloped and are self-flashing. They also require a protective coating, which is renewed every 5 to 10 years, again depending on the coating chosen. Although polyurethane foam commonly contains ozone-depleting HCFC, a newer process uses hydrocarbons instead to form rigid panels. If this process can be applied to on-site blown roofing, it will be environmentally more friendly.

Appendix J presents guidelines for choosing roof *U*-values for residential construction, based on *ANSI/ASHRAE Standard 90.2-1993, Energy Efficient Design of New Low-Rise Residential Buildings.* See Figs. J.1 and J.2 for single family residences, and Figs. J.17 and J.18 for multifamily structures.

Roofs are of particular interest to heat gains and losses because they are subjected to such extremes. On a cold clear night, radiant losses to the sky can lower the roof's surface temperature below that of the outdoor air; to be celebrated in summer, but a more serious matter in winter. In summer, the elevated temperature in sunlight is even more dramatic.

Sol-air temperature is the higher temperature of outdoor air just above a roof surface that is produced by solar radiation (and radiation exchange with the sky). The simplified formula for roofs (in I-P units) is:

$$t_e = t_o + \frac{\alpha I}{h_o} - 7F°$$

where t_e = sol-air temperature
   t_o = outdoor (ambient) DB temperature
   α = absorptance of surface for solar radiation (for light-colored roofs, usually assumed = 0.45) (for dark-colored roofs, usually assumed = 0.90) (detailed values are listed in Table 7.1, page 382)
   I = total solar radiation incident on surface, Btu/h ft² (see Appendix B for solar heat gain factors, approximately equivalent to I, for horizontal surfaces)
   h_o = coefficient of heat transfer by longwave radiation and convection at the roof surface (usually assumed to be = 3.0 Btu/h ft² °F)

The simplified formula for roofs (in SI units) is:

$$t_e = t_o + \frac{\alpha I}{h_o} - 3.9C°$$

where the differences from I-P units are as follows:
   I = total solar radiation incident on surface, W/m²
   h_o = usually assumed to be = 17.0 W/m² K

EXAMPLE 4.3.   What is the sol-air temperature above a horizontal white roof, compared to a dark roof, on a clear July 21 at noon, at a location at 40° N latitude? Assume outdoor air temperature at 90°F.

SOLUTION. From Appendix B, Table B.3, at July 21 noon at 40° NL, the solar heat gain factor on a horizontal surface = 262 Btu/h ft².

$$\text{Sol-air temperature, } t_e = t_o + \frac{\alpha I}{h_o} - 7\text{F}°$$

For a white roof:

$$t_e = 90°\text{F} + \frac{0.45 \times 262 \text{ Btu/h sf}}{3.0 \text{ Btu/h sf °F}} - 7\text{F}°$$

$$= 90 + 39.3 - 7 = 122.3°\text{F}$$

For a dark roof:

$$t_e = 90°\text{F} + \frac{0.90 \times 262 \text{ Btu/h sf}}{3.0 \text{ Btu/h sf °F}} - 7\text{F}°$$

$$= 90 + 78.6 - 7 = 161.6°\text{F}$$

This elevated temperature in full sun will drive considerably more heat through the roof; total hourly heat gain = $U \times$ Area $\times \Delta T$. The dark roof has a ΔT about 1.3 greater than the white roof at this time and place. ∎

(g) Doors. Table 4.12 lists U-values for solid (no glazing) doors in common use in North America. These are convenient, but somewhat suspect, because the resistances of air films are rather large factors in their thermal performance. Many doors have either forced-air supply or return grilles in close proximity, reducing the resistance of the indoor air layer. Conversely, exterior air layers may be less disturbed because doors are typically protected somewhat by overhangs, porches, and so on. Also, the wind conditions in this table are typical of winter rather than summer. Where significant door glazing is involved, U-values for windows should be substituted.

Appendix J presents guidelines for choosing door U-values for residential construction, based on *ANSI/ASHRAE Standard 90.2-1993, Energy Efficient Design of New Low-Rise Residential Buildings.* See Fig. J.14 for single family residences and Fig. J.33 for multifamily structures.

(h) Overall Envelope Heat Flow. Once the U-value has been found for each element of the envelope (including windows and skylights, next section), the hourly conductive heat loss (or gain) through the building's skin can be calculated as follows,

$$q = \Sigma(U \cdot A)\Delta t$$

I-P units: $\text{Btu/h} = \Sigma[(\text{Btu/h ft}^2 \, °\text{F}) \, (\text{ft}^2)] \, (\text{F}°)$

SI units: $\text{W} = \Sigma[(\text{W/m}^2 \, \text{K}) \, (\text{m}^2)] \, (\text{K}).$

where q = hourly heat flow through the envelope, by conduction, due to temperature difference between inside and outside.
U and A (U-values and areas) are specific to each skin element.
Σ indicates that all UA for these elements are to be added together.
Δt is the temperature difference between inside and outside. For *design* conditions, use outside temperature from Appendix A. For *average* conditions, use average temperatures (such as from Appendix B, Part 3, or local climatological data).

Once q has been determined, it is combined with several other factors to arrive at a building's total hourly need for heating (or cooling). These other factors include heat loss (or gain) from outside air infiltration or ventilation, and (in summer) heat gains from solar radiation through glazing, and from internal sources of heat. This process of arriving at the total building heat loss (or gain) is discussed in Chapter 5.

Meanwhile, another useful procedure becomes available with two of the above quantities, U (using its component resistances, R) and Δt. The surface temperatures and the internal temperature at any point within walls, roofs, or floors can be determined given the inside and outside temperatures. The gradual change of temperature through construction is called the *thermal gradient,* and it can be charted by proportioning the amount of temperature change to the amount of thermal resistance at any point. This procedure is illustrated in Fig. 4.12.

Surface temperatures are useful in predicting comfort: Just how cold in winter, or hot in summer, will that interior surface be? Internal temperatures are useful in establishing where water vapor might condense within a wall, roof, or floor; this is discussed further in Section 4.9.

R	Component	ΣR from Interior	Temperature Drop from Interior (F°)	Temperature at Outer Edge of Component (F°)
0.68	Inside air layer	0.68	0.68/21.46 × 36 = 1.1	66.9
0.20	Common brick	0.88	0.88/21.46 × 36 = 1.5	66.5
19.00	Nominal 6-in. insulation	19.88	19.88/21.46 × 36 = 33.3	34.7
0.62	½-in. Plywood	20.5	20.5/21.46 × 36 = 34.3	33.7
0.79	1-in. Wood siding	21.29	21.29/21.46 × 36 = 35.7	32.3
0.17	Outside air	21.46		32

Fig. 4.12 *Procedure for calculating the thermal gradient through construction. In this example, the outside and inside temperatures are 32 and 68°F (0 and 20°C), respectively. Thus, the Δt = 68 − 20 = 36F°.*

4.8 Heat Flow Through Windows and Skylights

Windows and skylights get special attention for several reasons. Despite dramatic improvements (Section 4.7c), they still usually have the lowest *R* (highest *U*) of all the components of the envelope. Also, they are major contributors to the infiltration of outdoor air that adds to the need for winter heating or summer cooling. Perhaps thermally most important, they admit solar heat, for better (winter) or worse (summer). They also are the admitters of daylight to buildings, and often of desired ventilation. The huge variety of available windows and the variety of important roles they play have led to a standardized approach to rating windows. This approach, still in development, is coordinated by the National Fenestration Ratings Council (NFRC). Each window or skylight manufactured in the United States bears an NFRC label, certifying that the window has been independently rated. Each label carries two columns: one for *residential* (smaller size) windows and one for *commercial* (larger size) windows. The difference is due to the smaller window having a relatively greater effect of heat flow through the frame, as compared to the glass; therefore, the overall heat flow rate through the residential window will usually be slightly higher, whereas the commercial window likely has a slightly larger solar heat gain and air leakage rate. (Characteristics of some residential-sized windows are found in Table 4.13.) The label also carries a brief description of the product itself, such as "Model #, Casement, Low-e = 0.2, 0.5″ gap, Argon Filled." Each label lists (or will soon list) the following information, for both residential and commercial sizes: *U*-factor, solar heat gain coefficient, visible light transmittance, air leakage, fenestration heating rating, and fenestration cooling rating.

(a) U-factor. As with opaque components of the envelope, heat flow due to temperature difference through windows and skylights is calculated as

$$\text{Btu/h} = U\text{-factor (Btu/h ft}^2\,°\text{F)} \times \text{Area (ft}^2)$$
$$\times \Delta\text{T (F°)}$$
$$[\text{W} = U\text{-factor (W/m}^2\,\text{K)}$$
$$\times \text{Area (m}^2) \times \Delta\text{T (K)}]$$

The value of U for windows and skylights is complicated by potentially large differences in heat flow rates between the center-of-glass, edge-of-glass, and frame. The "U-factor" averages these into one value for the entire window or skylight. Again, the lower the U-factor, the less the heat flow for a given temperature difference inside to outside. The size of the air gap between glazings, the coatings on the glazings, the gas fill between glazings, and the frame construction are all influential here. Tables 4.13 and 4.14 list U-factors for various windows and skylights.

Appendix J presents guidelines for window U-values and shading coefficients for residential construction, based on *ANSI/ASHRAE Standard 90.2-1993, Energy Efficient Design of New Low-Rise Residential Buildings.* See Figs. J.15 and J.16 for single family residences and Figs. J.34 and J.35 for multifamily structures.

(b) Solar Heat Gain Coefficient (SHGC).
This is also based on the performance of the entire frame, not just the glass. It represents that portion of the incident solar radiation (at all spectra) that is passed through the entire window or skylight, at normal incidence angle, with 1.0 as the maximum. Thus, a higher SHGC is desirable for solar heating applications, whereas a lower SHGC is better for windows where cooling is the dominant thermal issue. The SHGC depends on the type of glass, as well as tinting, reflective coatings, blinds, even some shading by the window or skylight frame.

The heat flow through a window or skylight that results from solar radiation is calculated as

$$\text{Btu/h} = \text{SHGC} \times \text{Area (ft}^2) \times \text{SHGF (Btu/h ft}^2)$$

$$[\text{W} = \text{SHGC} \times \text{Area (m}^2) \times \text{SHGF (W/m}^2)]$$

where SHGF is the *solar heat gain factor,* in Btu/h ft^2 (or W/m^2), found in Appendix B for several latitudes.

Until the late 1990s, the term *shading coefficient* (SC) was used for a similar purpose; it too had a maximum value of 1.0. However, this was based only on the glass, not the frame, and was expressed relative to single clear glazing at ⅛-in. (3-mm) thickness. The SC is a useful, first-approximation estimate, and may still be found when discussing skylights (Table 4.15) or basic

properties of glazing (Table 4.16). The shading coefficient is also used when assessing the effects of external or internal shading devices (see Section 4.8*k*).

(c) Visible Transmittance (VT).
This represents the fraction of the incident amount of light in the visible spectrum (at a normal incidence angle) that passes through the glazing. The higher the VT, the better for daylight quantities indoors. It is influenced by the color of the glazing (clear has the highest VT) as well as by coatings and number of glazings.

We might expect that any glazing or coating that reduces the SHGC will also reduce the VT. However, spectrally selective glazings and selective coatings are available that greatly reduce SHGC with little reduction in VT. This relationship can be expressed by the *light-to-gain ratio, LSG,* obtained by dividing the VT by the SHGC. Although not listed on the NFRC label, these values (including the effects of the frame) are included in Tables 4.13 and 4.16. The greater the LSG, the more suitable for daylighting in hot climates (or wherever cooling is the dominant thermal condition).

(d) Air Leakage.
This is the tested rate of outdoor air infiltration between the window and its frame, under pressure equivalent to a 25-mph (40-km/h, or 11-m/s) wind, with the window locked. ("High-performance" windows may be tested at even higher pressures.) As weather-stripping deteriorates with age, higher rates of infiltration may be expected.

(e) Fenestration Heating Rating (FHR).
These express the approximate seasonal percent reduction in energy used for residential space heating, as a result of using this window compared to the use of a "reference window" (single-glazed aluminum frame, equivalent to window #1 in Table 4.13). While intended to give the layperson a "bottom-line" comparison, such a simplistic rating ignores orientation (and thus useful winter solar gain) and assumes a ratio of window area : floor area that may be markedly different from a passively solar-heated house, or one in which more plentiful daylight is desired. A rating of FHR 25 means a seasonal heating

energy use of 25% less than the "reference window" version of the same house.

(f) Fenestration Cooling Rating (FCR).

These express the approximate seasonal percent reduction in energy used for residential space cooling as a result of using this window compared to the use of a "reference window" (single-glazed aluminum frame, equivalent to window #1 in Table 4.13). Again, such a simplistic rating ignores orientation and assumes a ratio of window area : floor area that may be markedly different from a passively solar-heated house, or one in which more plentiful daylight is desired. A rating of FCR 25 means a seasonal cooling energy use of 25% less than the "reference window" version of the same house.

(g) Low-emittance (low-ε) Coatings.

These are typically applied to one glass surface facing into the air gap between glazings. Such a coating blocks a great deal of the radiant transfer between the glazings, reducing the flow of heat through the window. Indeed, one such coating is almost as effective as adding another layer of glazing; in Table 4.13, compare the U-factor for window #5 with those of #7 and #12. Two approaches to these low-ε films are *hard-coat* (durable, less expensive, but less thermally effective) and *soft-coat* (better thermal performance but more expensive and subject to degradation by oxidation in the manufacturing stage). An important added benefit of these films is their reduction of ultraviolet transmission, thus reducing fading of objects and surface materials in rooms.

Three common types of low-ε coatings are:

1. High-transmission low-ε: for passive solar heating applications, where a low U-factor is combined with a high SHGC; window #7 is an example. The coating is on the inner glazing, where it traps outgoing infrared radiation that otherwise would be lost. Summer overheating can be avoided with external shading devices.

2. Selective-transmission low-ε: where winter heating and summer cooling are both important, with low U-factor and low SHGC, but with a relatively high VT for daylighting. Window #9 is an example, with a LSG ratio of 1.65. The coating is on the outer glazing, where it traps incoming infrared radiation, which is then convected away by outdoor air. Future spectrally selective coatings should provide even higher VT with lower SHGC.

3. Low-transmission low-ε: where the sun is the enemy, low U-factor, low SHGC, and even low VT seem warranted. Window #10 is an example, again with the coating on the outer glazing, where it rejects more of the solar gain. With a tinted exterior glazing, even lower SHGC and VT could result.

(h) Inert Gas in the Air Gap.

We have already seen (Section 4.7c) the relative benefits of filling an enclosed air space with argon or krypton. Their less-conductive properties greatly reduce heat transfer by convective currents within the air gap between glazings, producing lower U-factors. As a result, the inner surface of the glass is maintained at a temperature closer to that of indoors, with greater comfort (because radiant heat to or from the window surface is reduced) and less chance of condensation on the inside surface. To preserve this gas over the life of the window, a very reliable edge seal is required.

For argon, the optimum air gap width is about ½ in. (12 mm). When a less thick window is needed, more expensive krypton can fill an air gap of only ¼ in. (6 mm). Because the combination of inert gas and low-ε coatings is so effective at lowering the U-factor, most manufacturers offer them together (as with the windows #7 through #11 in Table 4.13) rather than separately.

A much wider variety of window and skylight U-factors, SHGC, and VT can be found in Chapter 29 of the 1997 *ASHRAE Handbook of Fundamentals*. A small sample of U-factors for skylights is shown in Table 4.14. Note in this table the great thermal penalty paid when using smaller units: the "manufactured skylights" are 2 ft × 4 ft (600 mm × 1200 mm), whereas the "site-assembled glazing" unit size is 4 ft (1200 mm) square. For otherwise identical construction, the whole-component U- factors are almost twice as high for the smaller units. With these smaller units, a smart investment is thus in the frame with the better thermal performance.

Important wavelengths in energy-efficient windows

Transmission curves of Heat Mirror insulations

Fig. 4.13 *Transmission and reflection performance for a selective transmitter film* (Heat Mirror). *Incoming solar radiation (both visible and near-infrared) is mostly transmitted, whereas radiant heat from room-temperature objects is reflected and thereby kept within heated spaces. (Courtesy of Southwall Corporation.)*

(i) Selective Transmission Films. Heat flow due to radiation can be greatly reduced by the introduction of selective transmitter (or "low-emittance") film somewhere within the glazing cavity. As shown in Fig. 4.13, these films admit most of the incoming solar radiation in both visible and near-infrared (short) wavelengths. As objects become warmed within the room, however, they begin to emit far-infrared (long-wave) radiation that is reflected back into the room by the selective film. These selective films typically are available as separate sheets that can be inserted between sheets of glazing as a window is fabri-

cated. As a separate sheet, a selective film could also be applied to existing windows—for instance, between storm windows and the ordinary windows they protect.

Now that such products have gained acceptance, an increasing range of choices are offered. For example, a range from Heat Mirror 77 (low reflectance, recommended for vertical glass) to Heat Mirror 33 (low transmission, recommended for sloping glass) is available, and includes seven colors. (A clear Heat Mirror 88 has the lowest reflectance of all.) Some skylight and greenhouse manufacturers offer this low-emittance product as a standard option.

(j) Superwindows. When all of the currently available conservation options are combined in one product, it is called a *superwindow.* The combination of multiple glazings and/or suspended films, coatings, inert gas fill, and sealed, thermally broken frame construction yields a lower heat flow rate and a higher price. In Table 4.13, window #11 is one example. Fig. 4.14 compares the heating and cooling season performance, in a residential application, of several windows in three locations in the United States with quite different seasons. A double-glazed window with two selective films, yielding three gas-filled cavities, is shown in Fig. 4.15. With their combination of superior thermal resistance and solar gain potential, superwindows can conceivably have a *better* heating season thermal performance than that of an insulated opaque wall. The potential design consequences are enormous.

(k) Shading. Perhaps the single most important energy-related component for passively cooled buildings is the sunshade. Because many solar-heated buildings will experience overheating in hot weather, sunshading is critical for passively heated buildings as well. If correctly implemented, sunshading rejects most solar heat gains yet aids in distributing daylighting deep into buildings, where it replaces or reduces the internal gains due to electric lights.

If a building is arranged to intercept the intense rays of the sun *before* they pass through its glass walls instead of afterward, the air-conditioning cooling load can often be cut in half. In approximate terms, external shading rejects about 80% of

(a) Heating season

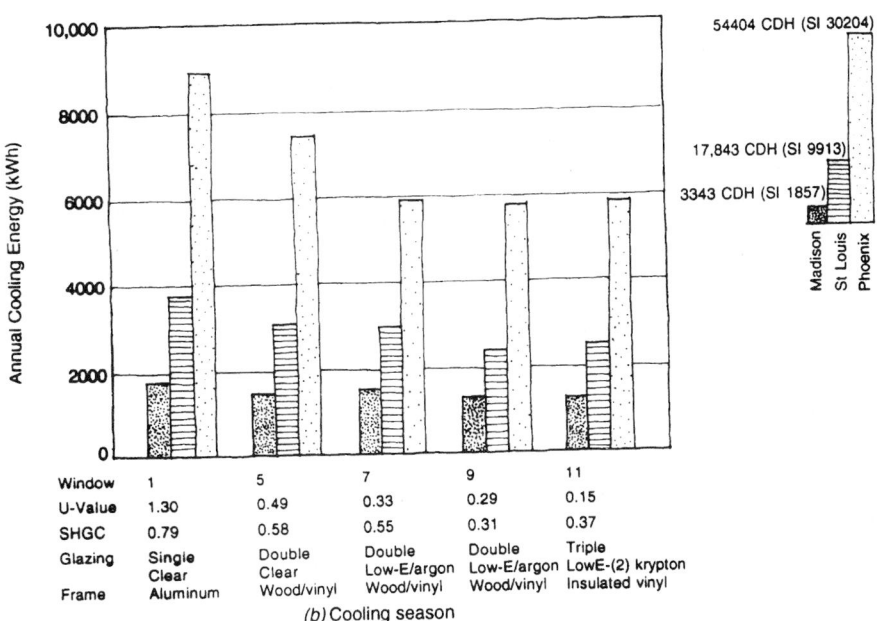

(b) Cooling season

Fig. 4.14 *Seasonal energy performance comparisons of various windows in a "typical" 1,540 ft² (143 m²) residence. Glazing area was assumed at 15% of floor area, with equal glazing area at all four orientations, and no external or internal shading. Window numbers correspond to those from Table 4.13. MBtu = millions of Btu; HDD = heating degree days, base 65°F; CDH = cooling degree hours, base 74°F. (From* Residential Windows: A Guide to New Technologies and Energy Performance, *by John Carmody, Stephen Selkowitz, and Lisa Heschong. Copyright © 1996 by John Carmody, Stephen Selkowitz, and Lisa Heschong. Adapted by permission of W. W. Norton & Company, New York.)*

THERMAL CONTROL

THERMAL CONTROL

Two sheets of
Heat Mirror™

Three gas-filled
spaces

Non-conductive
spacer

Unique gas
retention system

Fig. 4.15 The Superglass *window system utilizes two films of* Heat Mirror *between two outer panes of glass. Three air spaces are thus created and filled with a nonreactive gas mixture to retard convection. Nonmetallic spacers reduce heat flow at edges. The resulting window has a center-of-glass insulating value of R-8.1 (RSI 1.43). [ASHRAE lists for a similar window, nonoperable, 4 ft (1220 mm) square, an overall window* U = 0.19.] *Very low transmission of ultraviolet light is another characteristic. (Courtesy of Hurd Millwork Company, Medford, Wis.)*

the fierce attack of solar energy, whereas internal shading absorbs and reradiates 80% of it. Outside louvers have a chance to cool off in an occasional breeze, but inside draperies are part of a heat trap, and they constitute a system of hot-weather radiant heating that discomforts those who work near perimeter surfaces.

To properly reject direct sun yet allow for view and daylight, many sunshades project out from the windows they protect. These exterior projections become highly visible elements of facades, and they tempt some designers into superimposing formal esthetic criteria that can be damaging to the solar control functions. A frequent example is the application of the same sunshade geometry to all elevations of a building. When the sunshades are fixed in position, this tends to help one facade but not the others (see Fig. 3.16). Where they are movable, as with awnings, this same-sunshade approach is not so serious.

Fixed sunshading devices are very common, partly because they lack the moving parts and controls that can be expensive and troublesome.

As explained in Section 3.4, they pose a dilemma in the spring and fall, because in order to block the sun on any elevation in September, they must also block it in March. For many buildings in temperate climates, March is a heating-need month, whereas September is a cooling-need month. A procedure for evaluating fixed shading device performance is presented in Appendix D; some approximate shading effects, expressed as SC (shading coefficients) are found in Tables 4.17 and 4.18.

Adjustable sunshading, once enormously common (it seemed as if every 1930s shop and office window had an awning) but later considered "old-fashioned," has made a comeback with the advent of passive cooling. Some basic approaches to adjustable devices are shown in Table 4.19. The first type shown, made of durable materials, has proved effective against both theft and hurricanes.

How are movable shading devices adjusted to the desired position? Most manufacturers offer three types of controls: manual, motorized, and automatic. Manual systems are cheap and relatively trouble-free, but they require thoughtful, timely action on the part of the users of the building. So do motorized controls, but for large or heavy devices in remote places (clerestories, for example), the motorized control provides a steady hand for smooth operation. Automatic systems have the advantages of freeing the users from adjustment tasks and of taking into account the thermal needs of the building as a whole when setting the sunshade's position. With computerized controls becoming widespread for larger buildings, the added costs of automatic sunshading control can easily be incorporated into the overall cost of controls.

Interior shading devices are less effective than exterior devices, but far more common. They are not subject to weathering or dirt accumulation, and generally are easier to adjust manually. The designer who prefers an unchanging facade will rely on such interior devices. Tables 4.20 and 4.21 list SC for several combinations of interior shading devices.

EXAMPLE 4.4. A residence of one story is being designed in Billings, Montana. The ceiling is the underside of the roof (no attic), and heating/cooling ductwork will be within the conditioned space, treated as a low furred-down ceiling at the perimeter walls. Some of these walls will be frame, others masonry with exterior insulation (that is, the thermal mass is exposed to the interior). There will be a vented crawl space below a wood frame floor.

What are the prescribed U-values (and other energy conservation characteristics) for such a building?

SOLUTION. Refer to Appendix A for climate data, and to Appendix J, for the provisions of *ANSI/ASHRAE Standard 90.2-1993, Energy Efficient Design of New Low-Rise Residential Buildings.*

Climate: Billings, Montana: HDD65: 7,220; CDH74: 6,167 (Table A.1).

Roof: Maximum U-value: 0.033 (Fig. J.2A).
This will require insulation of at least $R = 1/0.033 = 30$ before the effects of thermal bridging are considered.

Frame walls: Maximum U-value: 0.063 (Fig. J.3A). Table 4.7 cites an overall $U = 0.073$ for an insulated 2 in. $\times$ 6 in. stud wall. This will require an additional layer of insulation outside the frame: overall $R = 1/0.063 = 16$. Overall R of insulated frame is 13.7 (again, from Table 4.7), so an outer layer of about R-3 will be needed.

Masonry walls: Maximum U-value: 0.08 (Fig. J.4A).
This will require insulation of at least $R = 1/0.08 = 12.5$ outside the masonry.

Floor (over vented crawl space): Maximum U-value: 0.047 (Fig. J.9A).
This will require insulation of at least $R = 1/0.047 = 21$ before the effects of thermal bridging are considered.

Doors: If non-wood, maximum U-value: 0.19 (Fig. J.14A).
From Table 4.12, Part B, an insulated steel slab, double glazed 6% of door area, with low-ε, argon filled, has $U = 0.18$.

Windows: If 15% of floor area or less, maximum U-value: 0.36 (Fig. J.15A), and maximum SC = 0.7 (Fig. J.16A).
From Table 4.13, windows #7 through #12 qualify for both maximum U and SC (considered equivalent to SHGC for this purpose).

We revisit this example in Chapter 5, Section 5.7, after considering daylight and window size. ∎

4.9 Moisture and Infiltration

Buildings, like our bodies, exchange moisture and air with the environment, as well as exchanging heat. Although most of this moisture exchange occurs during the exchange of fresh air, some moisture is exchanged through a building's skin. This can cause problems, either in hot, humid climates or very cold ones.

In hot-humid conditions, cool inside surfaces are often encountered—for example, a "radiant cooling" panel containing chilled water. If hot and humid air contacts such a surface, *condensation* can occur: the moisture vapor in the air condenses to form visible droplets of water on the cool surface. The result can be mildly annoying if droplets of condensation fall on one's body, or serious as water stains occur, and, eventually, mold grows on damp surfaces. For these reasons, the fresh (ventilation) air is dehumidified; even then, large cooled surfaces are often avoided in hot-humid climates. Even in less humid but hot conditions, condensation can occur within the air stream through cooling equipment, potentially creating problems of bacterial growth and odor.

In cold climates, cold interior surfaces also occur, especially at windows. Although the air indoors may not be particularly humid (40% to 50% rh is common), it contains enough moisture to permit condensation on cold surfaces. (For details, see the discussion on psychrometry in Section 5.12.) Again, mild annoyance or more serious damage can result. A much less visible moisture threat occurs within walls, ceilings, or floors. Almost all common building materials, including gypsum board, concrete, clay masonry, wood, and fiberglass insulation, are easily permeated by moisture. Most surface materials are also permeable. In cold climates, the air outside contains relatively little moisture, even though the rh may be high. By contrast, inside air contains much more moisture per unit of volume, despite its probably lower rh. (This is evident from the psychrometric chart, Fig. 5.35.) The result is a flow of vapor from high vapor pressure to low vapor pressure (typically warm to cold).

(a) Cold Climate Moisture Control. The problem with such a flow occurs when the temperature within the wall (floor, ceiling, etc.) drops low enough for this vapor to condense. Insulation can then become wet and thereby less effective, because water conducts heat far better than the air pockets it has filled. If wet insulation compacts, these air pockets are then permanently lost. Worse yet, moisture damage can occur, such as dry rot in wood structural members.

The usual remedy for such a potential problem in cold climates is to install a *vapor barrier* within the building envelope; because very low permeability is desired, these barriers are commonly of plastic film installed with as few holes as possible. Because the moister air on the warm side is the source of the problem, this barrier needs to be installed *as close to the warm side as possible*— typically, just behind the interior surfaces (gypsum board, wood floors, etc.). However, with higher *R*-values of insulation, it is becoming common to install the vapor barrier within the insulation, at a point about one third the distance from the interior to the exterior. This allows the inner one third of the wall to be used as a chase for wiring or plumbing without penetrating the barrier, yet with enough insulation beyond this point to prevent condensation.

Another approach is to use vinyl wallpaper or vapor barrier paints on interior surfaces. However, these give less protection than the around-the-corners wrap that can be achieved with properly installed plastic films. This disadvantage is shared by aluminum foil–faced insulation; it is very effective thermally, but less effective as a vapor barrier. (Also, the aluminum foil must face an air space if it is to be thermally effective; see Table 4.4.)

A substantial benefit of plastic films is that they reduce airflow through construction. Outdoor air is always infiltrating a building, gradually replacing the indoor air. This unintentional source of fresh air becomes a problem when temperatures outside are very different from those inside, especially when strong winds force outdoor air indoors fast enough to produce noticeably cold (or hot) drafts. Some fresh air is always desirable in buildings, but so is user control of how and where it is admitted. Therefore, the moisture- and infiltra-

tion-tight characteristics of plastic film vapor barriers are usually beneficial. When effective vapor barriers are installed, the smaller air-change values that accompany "tight" construction may be assumed in the following calculations of heat flow due to infiltration.

(b) Hot Humid Climate Moisture Control. Here, the object is to keep the moisture in the warmer outside air from penetrating to the cooler (and hopefully less humid) interior. One method is to use a *drainage plane,* installed just inside the exterior surface material. The drainage plane can consist of simple tar paper (building felt). If any moisture succeeds in penetrating this drainage plane, it should be allowed to wick to the interior; therefore, no vapor barrier such as plastic, vinyl wallpaper, or vapor barrier paints on interior surfaces should be used.

(c) Heat Flow by Infiltration. Two common terms for outdoor air that enters a building are infiltration and ventilation. *Infiltration* is the "accidental" influx of outdoor air due to air leakage through the building skin. *Ventilation* is the deliberate, designed introduction of outdoor air.

Winter heat loss (and summer heat gain in cooled, closed buildings) occurs when "fresh" outdoor air enters the building to replace "stale" indoor air. This heat exchange, somewhat analogous to human lung heat losses (or gains), must also be calculated when sizing heating or cooling equipment or when estimating energy use per season.

It is important to provide at least a minimum amount of fresh air indoors, both for comfort and for health. Odors and a sense of staleness can be uncomfortable, and dangerous buildups of radon gas and other pollutants such as formaldehyde can be produced within buildings. These pollutants are most effectively removed with air changes through rooms. Table 4.22 lists the recommended design outdoor air flow rates to produce acceptable indoor environments. A more thorough discussion of indoor air quality is found in Chapter 6.

The calculation of sensible heat lost (or gained) by the flow of outdoor air into spaces is as follows (for sensible and latent heat, see Section 5.11):

$$q_v = (V) (1.08) (\Delta t)$$

where q_v = sensible heat exchange due to ventilation (Btu/h)

  V = volume flow rate, in cubic feet per minute (cfm) of outdoor air introduced

  Δt = temperature difference between outdoor and indoor (F°)

  1.08 = a constant derived from the density of air at 0.075 lb/ft³ under "average" conditions, multiplied by the specific heat of air (heat required to raise one lb of air 1F°), which is 0.24 Btu/lb °F, and by 60 min/h. The units of this frequently encountered constant are Btu-min/ft³ h °F.

In SI units, this becomes

$$q_v = (V) (1.2) (\Delta t)$$

where q_v is in watts, V is in liters per second, 1.2 is the density of air [at 1.20 kg/m³ multiplied by the specific heat of air (at 1.0 kJ/kg K)], and Δt is in degrees K.

The next task is to determine how much air, V (in cfm or L/s) should be assumed for this heat flow calculation. Two basic methods are detailed here; one assumes that natural infiltration will meet the needs for fresh air indoors; the other deliberately introduces outdoor air, via fans, as required by building codes for many occupancies. (The passive options of windows and stack effect are detailed in Chapter 5.)

(d) Infiltration by Approximation.

For residences and small commercial buildings without significant internal heat sources, some building codes do not require mechanical (or forced) ventilation. In some codes, if the operable window area is 5% of the floor area of each room, adequate natural ventilation is assumed to be achievable. The following methods are typically used with design condition calculations; that is, to find the worst-hour winter loss (or summer gain) to the building when the windows are closed.

The *air-change method* is very quick (but tends to overestimate); a glance at Table 4.23 is suffi-

cient to find an estimated number of air changes per hour (ACH) for a space; then

$$V = \frac{(ACH)(\text{volume, ft}^3)}{60 \text{ min/h}}$$

where V is in cfm.

In SI units:

$$V = \frac{(ACH)(\text{volume, m}^3)}{3600 \text{ seconds/h}}$$

where V is in L/s.

The *crack method* takes longer and assumes that data on window and door construction and wind velocities are known. It also assumes that all infiltration under worst-hour conditions is due to cracks around doors and windows. If the building's windows are NFRC-rated, use the air leakage rate (see Table 4.13) for all windows *on the windward exposure(s) only.* In this case:

$$V = [\text{total lin. ft of window edge}] \times [\text{air leakage rate, cfm per lin. ft}]$$

where V is in cfm.

In SI units:

$$V = [\text{total lin. m of window edge}] \times [\text{air leakage rate, L/s per lin. m}]$$

where V is in L/s.

If the air leakage rates are unknown, Table 4.24 shows how many cfm per foot of crack should be assumed, again *on the windward exposure(s) only,* to arrive at the total cfm due to infiltration. Multiply the total lineal feet of crack by the infiltration rate from part B of Table 4.24.

With this approximate total V due to infiltration from either the air change or crack method, the resulting winter sensible heat loss (or summer heat sensible gain) can then be calculated as previously:

$$q_v = (V) (1.08) (\Delta t)$$

[In SI units:

$$q_v = (V) (1.2) (\Delta t)]$$

To these sensible gains must be added some adjustment for latent gains when mechanical cooling is used. These simple and quick methods are in sharp contrast to much more detailed procedures, outlined in the 1997 *ASHRAE Handbook of Fundamentals,* Chapter 25.

(e) Forced Ventilation. Because many codes require a given rate of outdoor air based on either a building's population or floor area, we can calculate the resulting winter sensible heat loss (or summer sensible heat gain). In this case, V is provided by fans. Where a building's required outdoor air (Table 4.22) is based on population:

$$V = [\text{cfm (or L/s) outdoor air per person}] \times [\text{number of people}].$$

Where the required outdoor air is based on floor area,

$$V = [\text{cfm (or L/s) outdoor air per unit floor area}] \times [\text{total floor area}]$$

This quantity, V, is then used to select a fan to force the required/desired amount of outdoor air directly into spaces. It is often forced through equipment that might filter it or alter its temperature or humidity before introduction to the space. Often, the fans are used instead to suck air out of the space.

A *heat exchanger* in either case is an excellent opportunity to save energy. Outgoing and incoming air are kept as separate but nearby airstreams, and heat can be transferred from one stream to another. Thus, incoming very cold winter air can be given much of the heat, but none of the pollutants, of outgoing warm air. (The flow of heat is reversed in summer, cooling the incoming air.) Our bodies do the same thing with arteries and veins, where arms and legs intersect the trunk. In cold weather, blood from the extremities is warmed before it enters the thermally protected core of our body. The equipment involved in this exchange is discussed in Section 6.5.

REFERENCES

Publications of the American Society of Heating, Refrigerating, and Air-Conditioning Engineers, Inc., Atlanta, Ga.:
ASHRAE Handbook of Fundamentals, 1997 and 1981
ASHRAE Standard 62-1989, Ventilation for Acceptable Indoor Air Quality, and Addenda.
ASHRAE Standard 90.1, Energy Efficient Design of New Buildings Except New Low-Rise Residential Buildings (latest edition).
ASHRAE Standard 90.2-1993, Energy Efficient Design of New Low-Rise Residential Buildings, and Addenda.
Carmody, J., Selkowitz, S., and Heschong, L. (1996). *Residential Windows: A Guide to New Technologies and Energy Performance,* W. W. Norton and Company, New York.
Chandra, S., Fairey, P. W., and Houston, M. (1986). *Cooling with Ventilation,* Solar Energy Research Institute, Golden, Col.
Laubin, R., and Laubin, G. (1975). *The Indian Tipi: Its History, Construction and Use,* University of Oklahoma Press, Norman, Okla.
National Fenestration Ratings Council (NFRC), *Certified Products Directory.* NFRC Incorporated, Silver Spring, Md.
Norberg-Schulz, C. (1965). *Intentions in Architecture,* MIT Press, Cambridge, Mass.
Wilkinson, G. (1999). "Beyond R-Value—Insulating for the Environment," in *Environmental Design and Construction,* January–February 1999.

TABLE 4.2 **Thermal Properties of Typical Building and Insulating Materials**[a]

		Density, lb/ft³	Conductivity[b] (k), Btu-in / h ft² °F	Conductance (C), Btu / h ft² °F	I-P Resistance[c] (R)		Specific Heat, Btu / lb °F	SI Resistance[c] (R)	
					Per Inch Thickness (1/k), °F ft² h / Btu-in	For Thickness Listed (1/C), °F ft² h / Btu		Per Meter Thickness (1/k), k m / W	For Thickness Listed (1/C), k m² / W
Description									
Building Board									
Asbestos-cement board	0.125 in.	120	4.0	—	0.25	—	0.24	1.73	—
Asbestos-cement board	0.25 in.	120	—	33.00	—	0.03		—	0.005
Asbestos-cement board	0.375 in.	120	—	16.50	—	0.06		—	0.011
Gypsum or plaster board	0.375 in.	50	—	3.10	—	0.32	0.26	—	0.056
Gypsum or plaster board	0.5 in.	50	—	2.22	—	0.45		—	0.079
Gypsum or plaster board	0.625 in.	50	—	1.78	—	0.56		—	0.099
Plywood (Douglas Fir)[d]		34	0.80	—	1.25	—	0.29	8.66	—
Plywood (Douglas Fir)	0.25 in.	34	—	3.20	—	0.31		—	0.055
Plywood (Douglas Fir)	0.375 in.	34	—	2.13	—	0.47		—	0.083
Plywood (Douglas Fir)	0.5 in.	34	—	1.60	—	0.62		—	0.11
Plywood (Douglas Fir)	0.625 in.	34	—	1.29	—	0.77		—	0.14
Plywood or wood panels	0.75 in.	34	—	1.07	—	0.93	0.29	—	0.16
Vegetable fiber board									
Sheathing, regular density[e]	0.5 in.	18	—	0.76	—	1.32	0.31	—	0.23
	0.78125 in.	18	—	0.49	—	2.06		—	0.36
Sheathing intermediate density[e]	0.5 in.	22	—	0.92	—	1.09	0.31	—	0.19
Nail-base sheathing[e]	0.5 in.	25	—	0.94	—	1.06	0.31	—	0.19
Shingle backer	0.375 in.	18	—	1.06	—	0.94	0.31	—	0.17
Shingle backer	0.3125 in.	18	—	1.28	—	0.78		—	0.14
Sound deadening board	0.5 in.	15	—	0.74	—	1.35	0.30	—	0.24
Tile and lay-in panels, plain or acoustic		18	0.40	—	2.50	—	0.14	17.0	—
	0.5 in.	18	—	0.80	—	1.25		—	0.22
	0.75 in.	18	—	0.53	—	1.89		—	0.33
Laminated paperboard		30	0.50	—	2.00	—	0.33	13.9	—
Homogeneous board from repulped paper		30	0.50	—	2.00	—	0.28	13.9	—

(continued)

TABLE 4.2 **Thermal Properties of Typical Building and Insulating Materials**[a] (*continued*)

Description	Density, lb/ft³	Conductivity[b] (k), Btu-in / h ft² °F	Conductance (C), Btu / h ft² °F	I-P Resistance[c] (R) Per Inch Thickness (1/k), °F ft² h / Btu-in	I-P Resistance[c] (R) For Thickness Listed (1/C), °F ft² h / Btu	Specific Heat, Btu / lb °F	SI Resistance[c] (R) Per Meter Thickness (1/k), k m / W	SI Resistance[c] (R) For Thickness Listed (1/C), k m² / W
Building Board								
Hardboard[e]								
Medium density	50	0.73	—	1.37	—	0.31	9.50	—
High density, service-tempered grade and service grade	55	0.82	—	1.22	—	0.32	8.46	—
High density, standard-tempered grade	63	1.00	—	1.00	—	0.32	6.93	—
Particleboard[e]								
Low density	37	0.71	—	1.41	—	0.31	9.77	—
Medium density	50	0.94	—	1.06	—	0.31	7.35	—
High density	62	.5	1.18	—	0.85	—	5.90	—
Underlayment 0.625 in.	40	—	1.22	—	0.82	0.29	—	0.14
Waferboard	37	0.63	—	1.59	—	—	11.0	—
Wood subfloor 0.75 in.	—	—	1.06	—	0.94	0.33	—	0.17
Building Membrane								
Vapor—permeable felt	—	—	16.70	—	0.06		—	0.011
Vapor—seal, 2 layers of mopped 15-lb felt	—	—	8.35	—	0.12		—	0.21
Vapor—seal, plastic film	—	—	—	—	Negl.		—	Negl.
Finish Flooring Materials								
Carpet and fibrous pad	—	—	0.48	—	2.08	0.34	—	0.37
Carpet and rubber pad	—	—	0.81	—	1.23	0.33	—	0.22
Cork tile 0.125 in.	—	—	3.60	—	0.28	0.48	—	0.049
Terrazzo 1 in.	—	—	12.50	—	0.08	0.19	—	0.014
Tile—asphalt, linoleum, vinyl, rubber	—	—	20.00	—	0.05	0.30	—	0.009
Vinyl asbestos						0.24		
Ceramic						0.19		
Wood, hardwood finish 0.75 in.	—	—	1.47	—	0.68	—	—	0.12

Insulating Materials

Blanket and Batt[f,g]

Description								
Mineral fiber, fibrous form processed from rock, slag, or glass								
Approx. 3–4 in.	0.4–2.0	—	0.091	—	11	—	—	1.94
Approx. 3.5 in.	0.4–2.0	—	0.077	—	13	—	—	2.29
Approx. 3.5 in.	1.2–1.6	—	0.067	—	15	—	—	2.63
Approx. 5.5–6.5 in.	0.4–2.0	—	0.053	—	19	—	—	3.32
Approx. 5.5 in.	0.6–1.0	—	0.048	—	21	—	—	3.67
Approx. 6–7.5 in.	0.4–2.0	—	0.045	—	22	—	—	3.91
Approx. 8.25–10 in.	0.4–2.0	—	0.033	—	30	—	—	5.34
Approx. 10–13 in.	0.4–2.0	—	0.026	—	38	—	—	6.77
Board and Slabs								
Cellular glass	8.0	0.33	—	3.03	—	0.18	19.8	—
Glass fiber, organic bonded	4.0–9.0	0.25	—	4.00	—	0.23	27.7	—
Expanded perlite, organic bonded	1.0	0.36	—	2.78	—	0.30	19.3	—
Expanded rubber (rigid)	4.5	0.22	—	4.55	—	0.40	31.6	—
Expanded polystyrene, extruded (smooth skin surface) (CFC-12 exp.)	1.8–3.5	0.20	—	5.00	—	0.29	—	—
Expanded polystyrene, extruded (smooth skin surface) (HCFC-142b exp.)[h]	1.8–3.5	0.20	—	5.00	—	0.29	34.7	—
Expanded polystyrene, molded beads	1.0	0.26	—	3.85	—	—	26.7	—
	1.25	0.25	—	4.00	—	—	27.7	—
	1.5	0.24	—	4.17	—	—	28.9	—
	1.75	0.24	—	4.17	—	—	28.9	—
	2.0	0.23	—	4.35	—	—	30.2	—
Cellular polyurethane/polyisocyanurate[d] (CFC-11 exp.) (unfaced)	1.5	0.16–0.18	—	6.25–5.56	—	0.38	43.3–38.5	—
Cellular polyisocyanurate[i] (CFC-11 exp.) (gas-permeable facers)	1.5–2.5	0.16–0.18	—	6.25–5.56	—	0.22	43.3–38.5	—
Cellular polyisocyanurate[i] (CFC-11 exp.) (gas-impermeable facers)	2.0	0.14	—	7.04	—	0.22	48.8	—
Cellular phenolic (closed cell) (CFC-11, CFC-113 exp.)[k]	3.0	0.12	—	8.20	—	—	56.8	—
Cellular phenolic (open cell)	1.8–2.2	0.23	—	4.40	—	—	30.5	—
Mineral fiber with resin binder	15.0	0.29	—	3.45	—	0.17	23.9	—

(*continued*)

TABLE 4.2 **Thermal Properties of Typical Building and Insulating Materials**[a] *(continued)*

Description	Density, lb/ft³	Conductivity[b] (k), Btu-in / h ft² °F	Conductance (C), Btu / h ft² °F	I-P Resistance[c] (R) Per Inch Thickness (1/k), °F ft² h / Btu-in	I-P Resistance[c] (R) For Thickness Listed (1/C), °F ft² h / Btu	Specific Heat, Btu / lb °F	SI Resistance[c] (R) Per Meter Thickness (1/k), k m / W	SI Resistance[c] (R) For Thickness Listed (1/C), k m² / W
Insulating Materials								
Mineral fiberboard, wet felted								
Core or roof insulation	16–17	0.34	—	2.94	—	—	20.4	—
Acoustical tile	18.0	0.35	—	2.86	—	0.19	19.8	—
Acoustical tile	21.0	0.37	—	2.70	—	—	18.7	—
Mineral fiberboard, wet molded								
Acoustical tile[f]	23.0	0.42	—	2.38	—	0.14	16.5	—
Wood or cane fiberboard								
Acoustical tile[f] 0.5 in.	—	—	0.80	—	1.25	0.31	—	0.22
Acoustical tile[f] 0.75 in.	—	—	0.53	—	1.89	—	—	0.33
Interior finish (plank, tile)	15.0	0.35	—	2.86	—	0.32	19.8	—
Cement fiber slabs (shredded wood with Portland cement binder)	25–27.0	0.50–0.53	—	2.0–1.89	—	—	13.9–13.1	—
Cement fiber slabs (shredded wood with magnesia oxysulfide binder)	22.0	0.57	—	1.75	—	0.31	12.1	—
Loose Fill								
Cellulosic insulation (milled paper or wood pulp)	2.3–3.2	0.27–0.32	—	3.70–3.13	—	0.33	25.6–21.7	—
Perlite, expanded	2.0–4.1	0.27–0.31	—	3.7–3.3	—	0.26	25.6–22.9	—
	4.1–7.4	0.31–0.36	—	3.3–2.8	—	—	22.9–19.4	—
	7.4–11.0	0.36–0.42	—	2.8–2.4	—	—	19.4–16.6	—
Mineral fiber (rock, slag, or glass)[g]								
Approx. 3.75–5 in.	0.6–2.0	—	—	—	11.0	0.17	—	1.94
Approx. 6.5–8.75 in.	0.6–2.0	—	—	—	19.0	—	—	3.35
Approx. 7.5–10 in.	0.6–2.0	—	—	—	22.0	—	—	3.87
Approx. 10.25–13.75 in.	0.6–2.0	—	—	—	30.0	—	—	5.28
Mineral fiber (rock, slag, or glass)[g] Approx. 3.5 in. (closed sidewall application)	2.0–3.5	—	—	—	12.0–14.0	—	—	2.1–2.5

Material								
Vermiculite, exfoliated	7.0–8.2	0.47	—	2.13	—	0.32	14.8	—
	4.0–6.0	0.44	—	2.27	—	—	15.7	—
Spray Applied								
Polyurethane foam	1.5–2.5	0.16–0.18	—	6.25–5.56	—	—	43.3–38.5	—
Ureaformaldehyde foam	0.7–1.6	0.22–0.28	—	4.55–3.57	—	—	31.5–24.7	—
Cellulosic fiber	3.5–6.0	0.29–0.34	—	3.45–2.94	—	—	23.9–20.4	—
Glass fiber	3.5–4.5	0.26–0.27	—	3.85–3.70	—	—	26.7–25.6	—
Reflective Insulation								
Reflective material ($\varepsilon < 0.5$) in center of ¾ in. cavity forms two ³⁄₈ in. vertical air spaces[m]	—	—	0.31	—	3.2	—	—	0.57
Metals (See 1997 *ASHRAE Handbook of Fundamentals*)								
Roofing								
Asbestos-cement shingles	120	—	4.76	—	0.21	0.24	—	0.037
Asphalt roll roofing	70	—	6.50	—	0.15	0.36	—	0.026
Asphalt shingles	70	—	2.27	—	0.44	0.30	—	0.077
Built-up roofing 0.375 in.	70	—	3.00	—	0.33	0.35	—	0.058
Slate 0.5 in.	—	—	20.00	—	0.05	0.30	—	0.009
Wood shingles, plain and plastic film faced	—	—	1.06	—	0.94	0.31	—	0.166
Plastering Materials								
Cement plaster, sand aggregate	116	5.0	—	0.20	—	0.20	1.39	—
Sand aggregate 0.375 in.	—	—	13.3	—	0.08	0.20	—	0.013
Sand aggregate 0.75 in.	—	—	6.66	—	0.15	0.20	—	0.026
Gypsum plaster:								
Lightweight aggregate 0.5 in.	45	—	3.12	—	0.32	—	—	0.056
Lightweight aggregate 0.625 in.	45	—	2.67	—	0.39	—	—	0.066
Lightweight aggregate on metal lath 0.75 in.	—	—	2.13	—	0.47	—	—	0.083
Perlite aggregate	45	1.5	—	0.67	—	0.32	4.64	—
Sand aggregate	105	5.6	—	0.18	—	0.20	1.25	—
Sand aggregate 0.5 in.	105	—	11.10	—	0.09	—	—	0.016
Sand aggregate 0.625 in.	105	—	9.10	—	0.11	—	—	0.019
Sand aggregate on metal lath 0.75 in.	—	—	7.70	—	0.13	—	—	0.023
Vermiculite aggregate	45	1.7	—	0.59	—	—	4.09	—

(continued)

TABLE 4.2 **Thermal Properties of Typical Building and Insulating Materials**[a] (*continued*)

Description	Density, lb/ft³	Conductivity[b] (k), Btu-in, h ft² °F	Conductance (C), Btu, h ft² °F	Resistance[c] (R) Per Inch Thickness (1/k), °F ft² h, Btu-in	Resistance[c] (R) For Thickness Listed (1/C), °F ft² h, Btu	Specific Heat, Btu, lb °F	SI Resistance[c] (R) Per Meter Thickness (1/k), k m, W	SI Resistance[c] (R) For Thickness Listed (1/C), k m², W
Masonry Materials								
Masonry Units								
Brick, fired clay	150	8.4–10.2	—	0.12–0.10	—	—	0.83–0.68	—
	140	7.4–9.0	—	0.14–0.11	—	—	0.94–0.77	—
	130	6.4–7.8	—	0.16–0.12	—	—	1.08–0.89	—
	120	5.6–6.8	—	0.18–0.15	—	0.19	1.24–1.02	—
	110	4.9–5.9	—	0.20–0.17	—	—	1.42–1.18	—
	100	4.2–5.1	—	0.24–0.20	—	—	1.65–1.36	—
	90	3.6–4.3	—	0.28–0.24	—	—	1.93–1.61	—
	80	3.0–3.7	—	0.33–0.27	—	—	2.31–1.87	—
	70	2.5–3.1	—	0.40–0.33	—	—	2.77–2.23	—
Clay tile, hollow								
1 cell deep — 3 in.	—	—	1.25	—	0.80	0.21	—	0.14
1 cell deep — 4 in.	—	—	0.90	—	1.11	—	—	0.20
2 cells deep — 6 in.	—	—	0.66	—	1.52	—	—	0.27
2 cells deep — 8 in.	—	—	0.54	—	1.85	—	—	0.33
2 cells deep — 10 in.	—	—	0.45	—	2.22	—	—	0.39
3 cells deep — 12 in.	—	—	0.40	—	2.50	—	—	0.44
Concrete blocks[n,o]								
Limestone aggregate								
8 in., 36 lb, 138 lb/ft³ concrete, 2 cores	—	—	—	—	—	—	—	—
Same with perlite filled cores	—	—	0.48	—	2.1	—	—	0.37
12 in., 55 lb, 138 lb/ft³ concrete, 2 cores	—	—	—	—	—	—	—	—
Same with perlite filled cores	—	—	0.27	—	3.7	—	—	0.65
Normal weight aggregate (sand and gravel)								
8 in., 33–36 lb, 126–136 lb/ft³ concrete, 2 or 3 cores	—	—	0.90–1.03	—	1.11–0.97	0.22	—	0.20–0.17

Description								
Same with perlite filled cores	—	0.50	—	2.0	—	—	—	0.35
Same with vermiculite filled cores	—	0.52–0.73	—	1.92–1.37	—	—	—	0.34–0.24
12 in., 50 lb, 125 lb/ft³ concrete, 2 cores	—	0.81	—	1.23	0.22	—	—	0.217
Medium weight aggregate (combinations of normal weight and lightweight aggregate)								
8 in., 26–29 lb, 97–112 lb/ft³ concrete, 2 or 3 cores	—	0.58–0.78	—	1.71–1.28	—	—	—	0.30–0.22
Same with perlite filled cores	—	0.27–0.44	—	3.7–2.3	—	—	—	0.65–0.41
Same with vermiculite filled cores	—	0.30	—	3.3	—	—	—	0.58
Same with molded EPS (beads) filled cores	—	0.32	—	3.2	—	—	—	0.56
Same with molded EPS inserts in cores	—	0.37	—	2.7	—	—	—	0.47
Lightweight aggregate (expanded shale, clay, slate or slag, pumice)								
6 in., 16–17 lb, 85–87 lb/ft³ concrete, 2 or 3 cores	—	0.52–0.61	—	1.93–1.65	—	—	—	0.34–0.29
Same with perlite filled cores	—	0.24	—	4.2	—	—	—	0.74
Same with vermiculite filled cores	—	0.33	—	3.0	—	—	—	0.53
8 in., 19–22 lb, 72–86 lb/ft³ concrete	—	0.32–0.54	—	3.2–1.90	0.21	—	—	0.56–0.33
Same with perlite filled cores	—	0.15–0.23	—	6.8–4.4	—	—	—	1.20–0.77
Same with vermiculite filled cores	—	0.19–0.26	—	5.3–3.9	—	—	—	0.93–0.69
Same with molded EPS (beads) filled cores	—	0.21	—	4.8	—	—	—	0.85
Same with UF foam filled cores	—	0.22	—	4.5	—	—	—	0.79
Same with molded EPS inserts in cores	—	0.29	—	3.5	—	—	—	0.62
12 in., 32–36 lb, 80–90 lb/ft³ concrete, 2 or 3 cores	—	0.38–0.44	—	2.6–2.3	—	—	—	0.46–0.40
Same with perlite filled cores	—	0.11–0.16	—	9.2–6.3	—	—	—	1.6–1.1
Same with vermiculite filled cores	—	0.17	—	5.8	—	—	—	1.0

(continued)

TABLE 4.2 **Thermal Properties of Typical Building and Insulating Materials**[a] (*continued*)

Description	Density, lb/ft³	Conductivity[b] (k), Btu-in / h ft² °F	Conductance (C), Btu / h ft² °F	I-P Resistance[c] (R) Per Inch Thickness (1/k), °F ft² h / Btu-in	I-P Resistance[c] (R) For Thickness Listed (1/C), °F ft² h / Btu	Specific Heat, Btu / lb °F	SI Resistance[c] (R) Per Meter Thickness (1/k), k m / W	SI Resistance[c] (R) For Thickness Listed (1/C), k m² / W
Masonary Materials								
Stone, lime, or sand								
Quartzitic and sandstone	180	72	—	0.01	—	—	0.10	—
	160	43	—	0.02	—	—	0.16	—
	140	24	—	0.04	—	—	0.29	—
	120	13	—	0.08	—	0.19	0.53	—
Calcitic, dolomitic, limestone, marble, and granite	180	30	—	0.03	—	—	0.23	—
	160	22	—	0.05	—	—	0.32	—
	140	16	—	0.06	—	0.19	0.43	—
	120	11	—	0.09	—	—	0.63	—
	100	8	—	0.13	—	—	0.90	—
Gypsum partition tile								
3 × 12 × 30 in., solid	—	—	0.79	—	1.26	0.19	—	0.222
3 × 12 × 30 in., 4 cells	—	—	0.74	—	1.35	—	—	0.238
4 × 12 × 30 in., 3 cells	—	—	0.60	—	1.67	—	—	0.294
Concretes[o]								
Sand and gravel or stone aggregate concretes	150	10.0–20.0	—	0.10–0.05	—	—	0.69–0.35	—
(concretes with more than 50%	140	9.0–18.0	—	0.11–0.06	—	0.19–0.24	0.77–0.39	—
quartz or quartzite sand have conductivities in the higher end	130	7.0–13.0	—	0.14–0.08	—	—	0.99–0.53	—
of the range)								
Limestone concretes	140	11.1	—	0.09	—	—	0.62	—
	120	7.9	—	0.13	—	—	0.88	—
	100	5.5	—	0.18	—	—	1.26	—
Gypsum-fiber concrete (87.5% gypsum, 12.5% wood chips)	51	1.66	—	0.60	—	0.21	4.18	—
Cement/lime, mortar, and stucco	120	9.7	—	0.10	—	—	0.71	—
	100	6.7	—	0.15	—	—	1.04	—
	80	4.5	—	0.22	—	—	1.54	—

Material	Density (lb/ft³)	k	C	R (1/k)	R (1/C)	Specific Heat	
Lightweight aggregate concretes							
Expanded shale, clay, or slate; expanded slags; cinders; pumice (with density up to 100 lb/ft³); and scoria (sanded concretes have conductivities in the higher end of the range)	120	6.4–9.1	—	0.16–0.11	—	—	1.08–0.76
	100	4.7–6.2	—	0.21–0.16	—	0.20	1.48–1.12
	80	3.3–4.1	—	0.30–0.24	—	0.20	2.10–1.69
	60	2.1–2.5	—	0.48–0.40	—	—	3.30–2.77
	40	1.3	—	0.78	—	—	5.40
Perlite, vermiculite, and polystyrene beads	50	1.8–1.9	—	0.55–0.53	—	—	3.81–3.68
	40	1.4–1.5	—	0.71–0.67	—	0.15–0.23	4.92–4.65
	30	1.1	—	0.91	—	—	6.31
	20	0.8	—	1.25	—	—	8.67
Foam concretes	120	5.4	—	0.19	—	—	1.32
	100	4.1	—	0.24	—	—	1.66
	80	3.0	—	0.33	—	—	2.29
	70	2.5	—	0.40	—	—	2.77
Foam concretes and cellular concretes	60	2.1	—	0.48	—	—	3.33
	40	1.4	—	0.71	—	—	4.92
	20	0.8	—	1.25	—	—	8.67

Siding Materials (on flat surface)

Shingles

Material	Density (lb/ft³)	k	C	R (1/k)	R (1/C)	Specific Heat	
Asbestos-cement	120	—	4.75	—	0.21	0.21	0.037
Wood, 16 in., 7.5 exposure	—	—	1.15	—	0.87	0.31	0.15
Wood, double, 16-in., 12-in. exposure	—	—	0.84	—	1.19	0.28	0.21
Wood, plus ins. backer board, 0.312 in.	—	—	0.71	—	1.40	0.31	0.25

Siding

Material	Density (lb/ft³)	k	C	R (1/k)	R (1/C)	Specific Heat	
Asbestos-cement, 0.25 in., lapped	—	—	4.76	—	0.21	0.24	0.037
Asphalt roll siding	—	—	6.50	—	0.15	0.35	0.026
Asphalt insulating siding (0.5 in. bed.)	—	—	0.69	—	1.46	0.35	0.26
Hardboard siding, 0.4375 in.	—	—	1.49	—	0.67	0.28	0.12
Wood, drop, 1 × 8 in.	—	—	1.27	—	0.79	0.28	0.14
Wood, bevel, 0.5 × 8 in., lapped	—	—	1.23	—	0.81	0.28	0.14
Wood, bevel, 0.75 × 10 in., lapped	—	—	0.95	—	1.05	0.28	0.18
Wood, plywood, 0.375 in., lapped	—	—	1.69	—	0.59	0.29	0.10

(continued)

TABLE 4.2 Thermal Properties of Typical Building and Insulating Materials[a] (*continued*)

Description	Density, lb/ft³	Conductivity[b] (k), Btu-in / h ft² °F	Conductance (C), Btu / h ft² °F	I-P Resistance[c] (R) Per Inch Thickness (1/k), °F ft² h / Btu-in	For Thickness Listed (1/C), °F ft² h / Btu	Specific Heat, Btu / lb °F	SI Resistance[c] (R) Per Meter Thickness (1/k), k m / W	For Thickness Listed (1/C), k m² / W
Siding Materials								
Aluminum, steel, or vinyl[p,q] over sheathing								
Hollow-backed	—	—	1.64	—	0.61	0.29[q]	—	0.11
Insulating-board backed nominal 0.375 in.	—	—	0.55	—	1.82	0.32	—	0.32
Insulating-board backed nominal 0.375 in., foil backed	—	—	0.34	—	2.96	—	—	0.52
Architectural (soda-lime float) glass	158	6.9	—	—	—	0.21	—	0.018
Woods (12% moisture content)[e,r]								
Hardwoods						0.39[s]		
Oak	41.2–46.8	1.12–1.25	—	0.89–0.80	—		6.2–5.5	—
Birch	42.6–45.4	1.16–1.22	—	0.87–0.82	—		6.0–5.7	—
Maple	39.8–44.0	1.09–1.19	—	0.92–0.84	—		6.4–5.8	—
Ash	38.4–41.9	1.06–1.14	—	0.94–0.88	—		6.5–6.1	—
Softwoods						0.39[s]		
Southern Pine	35.6–41.2	1.00–1.12	—	1.00–0.89	—		6.9–6.2	—
Douglas Fir-Larch	33.5–36.3	0.95–1.01	—	1.06–0.99	—		7.3–6.9	—
Southern Cypress	31.4–32.1	0.90–0.92	—	1.11–1.09	—		7.7–7.6	—
Hem-Fir, Spruce-Pine-Fir	24.5–31.4	0.74–0.90	—	1.35–1.11	—		9.3–7.7	—
West Coast Woods, Cedars	21.7–31.4	0.68–0.90	—	1.48–1.11	—		10.3–7.7	—
California Redwood	24.5–28.0	0.74–0.82	—	1.35–1.22	—		9.4–8.5	—

Source: Reprinted with permission of the American Society of Heating, Refrigerating, and Air-Conditioning Engineers, Inc. from 1997 *ASHRAE Handbook of Fundamentals.*

Note: The I-P units for resistance (R), either per inch (1/k) or for thickness stated (1/C), are given in Table 4.1. The SI units for resistance are shown in the last two columns

[a]Values are for a mean temperature of 75°F (24°C). Representative values for dry materials are intended as design (not specification) values for materials in normal use. Thermal values of insulating materials may differ from design values depending on their *in-situ* properties (density, moisture content, orientation, etc.) and variability experienced during manufacture.

[b]To obtain thermal conductivities in Btu/h ft °F, divide the *k*-factor by 12 in./ft. The symbol λ is also used to represent thermal conductivity.

[c]Resistance values are the reciprocals of *C* before rounding off *C* to two decimal places.

[d]From Lewis (1967), *Thermal Conductivity of Wood-Base Fiber and Particle Panel Materials*. Forest Products Laboratory, Research Paper FPL 77, June.

[e]U.S. Dept. of Agriculture (1974), *Wood Handbook*, Handbook No. 72.

[f]Does not include paper backing and facing, if any. Where insulation forms a boundary (reflective or otherwise) of an airspace, see Tables 4.3 and 4.4 for properties of an airspace.

[g]Conductivity varies with fiber diameter. Batt, blanket, and loose-fill mineral fiber insulations are manufactured to achieve specified *R*-values, the most common of which are listed here. Due to differences in manufacturing processes and materials, the product thicknesses, densities and thermal conductivities may vary over considerable ranges for a specified *R*-value.

[h]This material is relatively new and data are based on limited testing.

[i]For additional information, see Society of Plastics Engineers (SPI) *Bulletin* U108. Values are for aged, unfaced board stock. For change in conductivity with age of expanded polyurethane/polyisocyanurate, see Chapter 22 of the 1997 *ASHRAE Handbook of Fundamentals*.

[j]Values are for aged products with gas-impermeable facers on the two major surfaces. An aluminum foil facer of 0.001 in. (25 μm) thickness or greater is generally considered impermeable to gases. For change in conductivity with age of expanded polyurethane/polyisocyanurate, see Chapter 22 of the 1997 *ASHRAE Handbook of Fundamentals*.

[k]Cellular phenolic insulation may no longer be manufactured. These thermal conductivity and resistance values do not represent aged insulation, which may have a higher thermal conductivity and lower thermal resistance.

[l]Insulating values of acoustical tile may vary, depending on density of the board and on type, size, and depth of perforations.

[m]Cavity is framed with 0.75 in. (20 mm) wood furring strips. Caution should be used in applying this value for other framing materials. The reported value was derived from tests and applies to the reflective path only. The effect of studs or furring strips must be included in determining the overall performance of the wall.

[n]Values for fully grouted block may be approximated using values for concrete with a similar unit weight.

[o]Values for concrete block and concrete are at moisture contents representative of normal use.

[p]Values for metal or vinyl siding applied over flat surfaces vary widely depending on the amount of ventilation of the airspace beneath the siding; whether airspace is reflective or nonreflective; and on thickness, type, and application of insulating backing used. Departures of ±50% or more from these values may occur.

[q]Vinyl specific heat = 0.25 Btu/lb °F (1.0 kJ/kg K).

[r]The conductivity values listed are for heat transfer across the grain. The thermal conductivity of wood varies linearly with the density, and the density ranges listed are for those normally found for the wood species given. For extrapolation to other moisture contents, see footnotes for Table 4, Chapter 24, 1997 *ASHRAE Handbook of Fundamentals*.

[s]For an empirical equation for the specific heat of moist wood at 75°F (24°C), see footnotes for Table 4, Chapter 24, 1997 *ASHRAE Handbook of Fundamentals*.

TABLE 4.3 **Properties of Surface Air Films and Air Spaces**

Part A. Surface Conductances^a and Resistances^b for Surface Air Films

		Surface Emittance, ε											
		I-P Units^c						SI Units^d					
		Nonreflective $\varepsilon = 0.90$		Reflective $\varepsilon = 0.20$		$\varepsilon = 0.05$		Nonreflective $\varepsilon = 0.90$		Reflective $\varepsilon = 0.20$		$\varepsilon = 0.05$	
Position of Surface	Direction of Heat Flow	h_i	R	h_i	R	h_i	R	h_i	R	h_i	R	h_i	R
Still Air													
Horizontal	Upward	1.63	0.61	0.91	1.10	0.76	1.32	9.26	0.11	5.17	0.19	4.32	0.23
Sloping–45°	Upward	1.60	0.62	0.88	1.14	0.73	1.37	9.09	0.11	5.00	0.20	4.15	0.24
Vertical	Horizontal	1.46	0.68	0.74	1.35	0.59	1.70	8.29	0.12	4.20	0.24	3.35	0.30
Sloping–45°	Downward	1.32	0.76	0.60	1.67	0.45	2.22	7.50	0.13	3.41	0.29	2.56	0.39
Horizontal	Downward	1.08	0.92	0.37	2.70	0.22	4.55	6.13	0.16	2.10	0.48	1.25	0.80
Moving Air (any position)		h_o	R					h_o	R				
Winter Wind 15-mph (6.7-m/s)	Any	6.00	0.17					34.0	0.030				
Summer Wind 7.5-mph (3.4-m/s)	Any	4.00	0.25					22.7	0.044				

Part B. Emittance Values of Various Surfaces and Effective Emittances of Air Spaces

	Average Emittance ε	Effective Emittance ε_{eff} of Air Space	
		One Surface Emittance ε; Other Surface 0.9	Both Surfaces Emittance ε
Aluminum foil, bright	0.05	0.05	0.03
Aluminum foil with condensate just visible	0.30	0.29	—
Aluminum foil with condensate clearly visible	0.70	0.65	—
Aluminum sheet	0.12	0.12	0.06
Aluminum-coated paper, polished	0.20	0.20	0.11
Steel, galvanized, bright	0.25	0.24	0.15
Aluminum paint	0.50	0.47	0.35
Building materials: wood, paper, masonry, nonmetallic paints	0.90	0.82	0.82
Regular glass	0.84	0.77	0.72

Source: Reprinted with permission of the American Society of Heating, Refrigerating, and Air-Conditioning Engineers, Inc. from 1997 *ASHRAE Handbook of Fundamentals.*

[a]Conductances are for surfaces of the stated emittance facing virtual blackbody surroundings at the same temperature as the ambient air. Values are based on a surface–air temperature difference of 10F° (5.5C°) and for surface temperatures of 70°F (21°C).

[b]No surface has both an air space resistance and a surface resistance value.

[c]I-P units: surface conductance h_i and h_o measured in Btu/h ft^2 °F; resistance R measured in h ft^2 °F/Btu.

[d]SI units: surface conductance h_i and h_o measured in W/m^2 K; resistance R measured in m^2 K/W.

THERMAL CONTROL

TABLE 4.4 Thermal Resistance of Plane[a] Air Spaces

All resistance values expressed in ft² · °F · h/Btu.

Values apply only to air spaces of uniform thickness bounded by plane, smooth, parallel surfaces with no leakage of air to or from the space. These conditions are not normally present in standard building construction. When accurate values are required, use overall U-factors determined for your particular construction. Thermal resistance values for multiple air spaces must be based on careful estimates of mean temperature differences for each air space.

Position of Air Space	Direction of Heat Flow	Air Space Mean Temp,[b] (F°)	Temp Diff,[b] (F°)	0.5-in. Air Space[d] Value of E[b,c]					0.75-in. Air Space[d] Value of E[b,c]				
				0.03	0.05	0.2	0.5	0.82	0.03	0.05	0.2	0.5	0.82
Horiz.	Up ←	90	10	2.13	2.03	1.51	0.99	0.73	2.34	2.22	1.61	1.04	0.75
		50	30	1.62	1.57	1.29	0.96	0.75	1.71	1.66	1.35	0.99	0.77
		50	10	2.13	2.05	1.60	1.11	0.84	2.30	2.21	1.70	1.16	0.87
		0	20	1.73	1.70	1.45	1.12	0.91	1.83	1.79	1.52	1.16	0.93
		0	10	2.10	2.04	1.70	1.27	1.00	2.23	2.16	1.78	1.31	1.02
		−50	20	1.69	1.66	1.49	1.23	1.04	1.77	1.74	1.55	1.27	1.07
		−50	10	2.04	2.00	1.75	1.40	1.16	2.16	2.11	1.84	1.46	1.20
45° Slope	Up ↗	90	10	2.44	2.31	1.65	1.06	0.76	2.96	2.78	1.88	1.15	0.81
		50	30	2.06	1.98	1.56	1.10	0.83	1.99	1.92	1.52	1.08	0.82
		50	10	2.55	2.44	1.83	1.22	0.90	2.90	2.75	2.00	1.29	0.94
		0	20	2.20	2.14	1.76	1.30	1.02	2.13	2.07	1.72	1.28	1.00
		0	10	2.63	2.54	2.03	1.44	1.10	2.72	2.62	2.08	1.47	1.12
		−50	20	2.08	2.04	1.78	1.42	1.17	2.05	2.01	1.76	1.41	1.16
		−50	10	2.62	2.56	2.17	1.66	1.33	2.53	2.47	2.10	1.62	1.30
Vertical	Horiz. →	90	10	2.47	2.34	1.67	1.06	0.77	3.50	3.24	2.08	1.22	0.84
		50	30	2.57	2.46	1.84	1.23	0.90	2.91	2.77	2.01	1.30	0.94
		50	10	2.66	2.54	1.88	1.24	0.91	3.70	3.46	2.35	1.43	1.01
		0	20	2.82	2.72	2.14	1.50	1.13	3.14	3.02	2.32	1.58	1.18
		0	10	2.93	2.82	2.20	1.53	1.15	3.77	3.59	2.64	1.73	1.26
		−50	20	2.90	2.82	2.35	1.76	1.39	2.90	2.83	2.36	1.77	1.39
		−50	10	3.20	3.10	2.54	1.87	1.46	3.72	3.60	2.87	2.04	1.56
45° Slope	Down ↗	90	10	2.48	2.34	1.67	1.06	0.77	3.53	3.27	2.10	1.22	0.84
		50	30	2.64	2.52	1.87	1.24	0.91	3.43	3.23	2.24	1.39	0.99
		50	10	2.67	2.55	1.89	1.25	0.92	3.81	3.57	2.40	1.45	1.02
		0	20	2.91	2.80	2.19	1.52	1.15	3.75	3.57	2.63	1.72	1.26
		0	10	2.94	2.83	2.21	1.53	1.15	4.12	3.91	2.81	1.80	1.30
		−50	20	3.16	3.07	2.52	1.86	1.45	3.78	3.65	2.90	2.05	1.57
		−50	10	3.26	3.16	2.58	1.89	1.47	4.35	4.18	3.22	2.21	1.66

Position of Air Space	Direction of Heat Flow		Mean Temp (F°)	Temp Diff (F°)	1.5-in. Air Space Value of $E^{b,c}$					3.5-in. Air Space Value of $E^{b,c}$				
					0.03	0.05	0.2	0.5	0.82	0.03	0.05	0.2	0.5	0.82
Horiz.	Down	→	90	10	2.48	2.34	1.67	1.06	0.77	3.55	3.29	2.10	1.22	0.85
			50	30	2.66	2.54	1.88	1.24	0.91	3.77	3.52	2.38	1.44	1.02
			50	10	2.67	2.55	1.89	1.25	0.92	3.84	3.59	2.41	1.45	1.02
			0	20	2.94	2.83	2.20	1.53	1.15	4.18	3.96	2.83	1.81	1.30
			0	10	2.96	2.85	2.22	1.53	1.16	4.25	4.02	2.87	1.82	1.31
			−50	20	3.25	3.15	2.58	1.89	1.47	4.60	4.41	3.36	2.28	1.69
			−50	10	3.28	3.18	2.60	1.90	1.47	4.71	4.51	3.42	2.30	1.71

Position of Air Space	Direction of Heat Flow		Air Space Mean Temp[b] (F°)	Temp Diff[b] (F°)	1.5-in. Air Space Value of $E^{b,c}$					3.5-in. Air Space Value of $E^{b,c}$				
					0.03	0.05	0.2	0.5	0.82	0.03	0.05	0.2	0.5	0.82
Horiz.	Up	↑	90	10	2.55	2.41	1.71	1.08	0.77	2.84	2.66	1.83	1.13	0.80
			50	30	1.87	1.81	1.45	1.04	0.80	2.09	2.01	1.58	1.10	0.84
			50	10	2.50	2.40	1.81	1.21	0.89	2.80	2.66	1.95	1.28	0.93
			0	20	2.01	1.95	1.63	1.23	0.97	2.25	2.18	1.79	1.32	1.03
			0	10	2.43	2.35	1.90	1.38	1.06	2.71	2.62	2.07	1.47	1.12
			−50	20	1.94	1.91	1.68	1.36	1.13	2.19	2.14	1.86	1.47	1.20
			−50	10	2.37	2.31	1.99	1.55	1.26	2.65	2.58	2.18	1.67	1.33
45° Slope	Up	↗	90	10	2.92	2.73	1.86	1.14	0.80	3.18	2.96	1.97	1.18	0.82
			50	30	2.14	2.06	1.61	1.12	0.84	2.26	2.17	1.67	1.15	0.86
			50	10	2.88	2.74	1.99	1.29	0.94	3.12	2.95	2.10	1.34	0.96
			0	20	2.30	2.23	1.82	1.34	1.04	2.42	2.35	1.90	1.38	1.06
			0	10	2.79	2.69	2.12	1.49	1.13	2.98	2.87	2.23	1.54	1.16
			−50	20	2.22	2.17	1.88	1.49	1.21	2.34	2.29	1.97	1.54	1.25
			−50	10	2.71	2.64	2.23	1.69	1.35	2.87	2.79	2.33	1.75	1.39
Vertical	Horiz.	→	90	10	3.99	3.66	2.25	1.27	0.87	3.69	3.40	2.15	1.24	0.85
			50	30	2.58	2.46	1.84	1.23	0.90	2.67	2.55	1.89	1.25	0.91
			50	10	3.79	3.55	2.39	1.45	1.02	3.63	3.40	2.32	1.42	1.01
			0	20	2.76	2.66	2.10	1.48	1.12	2.88	2.78	2.17	1.51	1.14
			0	10	3.51	3.35	2.51	1.67	1.23	3.49	3.33	2.50	1.67	1.23
			−50	20	2.64	2.58	2.18	1.66	1.33	2.82	2.75	2.30	1.73	1.37
			−50	10	3.31	3.21	2.62	1.91	1.48	3.40	3.30	2.67	1.94	1.50

(continued)

TABLE 4.4 **Thermal Resistance of Plane**[a] **Air Spaces** (*continued*)

All resistance values expressed in ft² · °F · h/Btu.

Values apply only to air spaces of uniform thickness bounded by plane, smooth, parallel surfaces with no leakage of air to or from the space. These conditions are not normally present in standard building construction. When accurate values are required, use overall U-factors determined for your particular construction. Thermal resistance values for multiple air spaces must be based on careful estimates of mean temperature differences for each air space.

Position of Air Space	Direction of Heat Flow	Air Space Mean Temp,[b] (F°)	Temp Diff,[b] (F°)	1.5-in. Air Space[d] Value of E[b,c]					3.5-in. Air Space[d] Value of E[b,c]				
				0.03	0.05	0.2	0.5	0.82	0.03	0.05	0.2	0.5	0.82
45° Slope	Down ↗	90	10	5.07	4.55	2.56	1.36	0.91	4.81	4.33	2.49	1.34	0.90
		50	30	3.58	3.36	2.31	1.42	1.00	3.51	3.30	2.28	1.40	1.00
		50	10	5.10	4.66	2.85	1.60	1.09	4.74	4.36	2.73	1.57	1.08
		0	20	3.85	3.66	2.68	1.74	1.27	3.81	3.63	2.66	1.74	1.27
		0	10	4.92	4.62	3.16	1.94	1.37	4.59	4.32	3.02	1.88	1.34
		−50	20	3.62	3.50	2.80	2.01	1.54	3.77	3.64	2.90	2.05	1.57
		−50	10	4.67	4.47	3.40	2.29	1.70	4.50	4.32	3.31	2.25	1.68
Horiz.	Down →	90	10	6.09	5.35	2.79	1.43	0.94	10.07	8.19	3.41	1.57	1.00
		50	30	6.27	5.63	3.18	1.70	1.14	9.60	8.17	3.86	1.88	1.22
		50	10	6.61	5.90	3.27	1.73	1.15	11.15	9.27	4.09	1.93	1.24
		0	20	7.03	6.43	3.91	2.19	1.49	10.90	9.52	4.87	2.47	1.62
		0	10	7.31	6.66	4.00	2.22	1.51	11.97	10.32	5.08	2.52	1.64
		−50	20	7.73	7.20	4.77	2.85	1.99	11.64	10.49	6.02	3.25	2.18
		−50	10	8.09	7.52	4.91	2.89	2.01	12.98	11.56	6.36	3.34	2.22

Source: Reprinted with permission of the American Society of Heating, Refrigerating, and Air-Conditioning Engineers, Inc. from 1997 *ASHRAE Handbook of Fundamentals.*

Note: To convert to RSI, multiply above values of R by 0.176.

[a]Thermal resistance values were determined from the relation, $R = 1/C$, where $C = h_c + \varepsilon_{eff} h_r$; h_c is the conduction–convection coefficient; ε_{eff} h_r is the radiation coefficient $= 0.0068\varepsilon_{eff}[(t_m + 460)/100]^3$; and t_m is the mean temperature of the space. For extrapolation from this table to air spaces less than 0.5 in. (as in insulating window glass), assume $h_c = 0.159 (1 + 0.0016t_m/l$, where l is the air space thickness in inches, and h_c is heat transfer through the air space only.

[b]Interpolation is permissible for other values of mean temperature, temperature difference, and effective emittance ε_{eff}. Interpolation and moderate extrapolation for air spaces greater than 3.5 in. are also permissible.

[c]Effective emittance, ε_{eff}, of the air space is given by $1/\varepsilon_{eff} = 1/\varepsilon_1 + 1/\varepsilon_2 - 1$, where ε_1 and ε_2 are emittances of the surfaces of the air space (from Table 4.3).

[d]A single resistance value cannot account for multiple air spaces. Each air space requires a separate resistance calculation that applies only for the established boundary conditions.

[e]Resistances of horizontal spaces with heat flow downward are substantially independent of temperature difference.

TABLE 4.5 *U*-Values for Some Whole-Wall, Whole-Roof, and Whole-Floor Systems

Part A. Conventional Construction

Component	Construction, from Exterior Inward	Insulation R-Value I-P °F ft² h Btu	Insulation R-Value SI k m² W	Whole-Component U-Value, Including Framing I-P °F ft² h Btu	Whole-Component U-Value, Including Framing SI k m² W
Wall[a]	Wood siding, fiber board sheathing, 2-in. × 4-in. studs @ 16-in. o.c. and R-11 blanket insulation, gypsum board	R-11	1.94	0.081	0.52
Roof[a,b] (flat)	Built-up roof, plywood deck, 2-in. × 8-in. rafters @ 24-in. o.c. and R-19 blanket insulation with nonreflective air space, gypsum board, acoustic tile	R-19	3.35	0.046	0.3
Floor[a,c]	Metal lath and lightweight aggregate plaster, 2-in. × 8-in. floor joist with R-19 blanket insulation and nonreflective air space, wood subfloor, plywood, and floor tile	R-19	3.35	0.052	0.3

Part B. Passive Solar Collecting Walls

Component	Whole-Component U-Value, Winter[d] No Night Insulation °F ft² h Btu	No Night Insulation k m² W	With Night Insulation[e] of Specified (R) (R-4) °F ft² h Btu	(RSI-0.7) k m² W	(R-9) °F ft²h Btu	(RSI-1.58) k m² W
Direct gain[f]	0.55	3.1	0.33	1.9	0.28	1.6
Trombe wall, 18-in. (457-mm) thick[f]	0.22	1.2	0.16	0.9	0.13	0.7
Water wall[f]	0.33	1.9	0.22	1.2	0.19	1.1

[a]Based on 1981 *ASHRAE Handbook of Fundamentals.*

[b]Heat flow upward, winter conditions.

[c]Heat flow downward, winter conditions; open to outdoor conditions below.

[d]Summer-day *U*-value can be developed from the designer's assumptions about the hours for which the insulation is in place.

[e]Insulation is assumed to be in place from 5:30 PM to 7:30 AM, solar time. These *U*-values are thus average, rather than instantaneous, values, unlike the others in this table.

[f]From Balcomb et al. (1980).

THERMAL CONTROL

THERMAL CONTROL

TABLE 4.6 **Corrected *R*-Values for Wall Systems with Steel Studs**

Nominal Studs Size[a]	Spacing of Framing	Cavity Insulation R-value I-P °F ft² h / Btu	SI K m² / W	Correction Factor	Effective Framing/ Cavity R-value °F ft² h / Btu	K m² / W
2 in. × 4 in. (50 mm × 101 mm)	16-in. o.c. (406 mm)	R-11	1.94	0.50	R-5.5	0.96
		R-13	2.29	0.46	R-6.0	1.06
		R-15	2.64	0.43	R-6.4	1.13
2 in. × 4 in. (50 mm × 101 mm)	24-in. o.c. (610 mm)	R-11	1.94	0.60	R-6.6	1.16
		R-13	2.29	0.55	R-7.2	1.27
		R-15	2.64	0.52	R-7.8	1.37
2 in. × 6 in. (50 mm × 152 mm)	16-in. o.c. (406 mm)	R-19	3.35	0.37	R-7.1	1.25
		R-21	3.70	0.35	R-7.4	1.30
2 in. × 6 in. (50 mm × 152 mm)	24-in. o.c. (610 mm)	R-19	3.35	0.45	R-8.6	1.52
		R-21	3.70	0.43	R-9.0	1.59
2 in. × 8 in. (50 mm × 203 mm)	16-in. o.c. (406 mm)	R-25	4.41	0.31	R-7.8	1.37
2 in. × 8 in. (50 mm × 203 mm)	24-in. o.c. (610 mm)	R-25	4.41	0.38	R-9.6	1.69

Source: Reprinted with permission of the American Society of Heating, Refrigerating, and Air-Conditioning Engineers, Inc. from *ANSI/ASHRAE Standard 90.2-1993, Energy Efficient Design of New Low-Rise Residential Buildings.*
[a]Steel studs up to a maximum uncoated thickness of 0.064 in. (64 mils) (16 gauge) (1.6 mm).

TABLE 4.7 ***U*-Values Comparing Whole-Wall Framing and SIP (Structural Insulated Panels)**

Component	Center-of-Insulated-Cavity R-value[a] I-P °F ft² h / Btu	SI k m² / W	Whole-Wall R-value[b] I-P °F ft² h / Btu	SI k m² / W	Whole-Wall U-value I-P °F ft² h / Btu	SI k m² / W
2-in. × 4-in. insulated stud wall at 16-in. o.c.	13.6	2.4	9.6	1.69	0.104	0.59
2-in. × 6-in. insulated stud wall at 24-in. o.c.	19.6	3.46	13.7	2.41	0.073	0.41
6 ½-in. SIP	25.1	4.35	21.6	3.8	0.046	0.26

Source: Oak Ridge National Laboratory (in *Environmental Building News*).
[a]Center-of-insulated-cavity R-value is the maximum total R-value of the wall system, including insulation, sheathing, drywall, etc.
[b]Whole-wall R-value is an average value that includes the framing as well as some openings in a standardized wall system.

TABLE 4.8 **Heat Loss Coefficient of Floor Slabs, F_2, per Unit Length of Perimeter**

Construction	Insulation	Btu/h °F ft Perimeter Degree-Days[a] (65° F base)			W/Km Perimeter Kelvin Days[a] (18° C base)		
		2950	5350	7433	1640	2970	4130
Block wall, 8 in. (200 mm), brick facing	Uninsulated	0.62	0.68	0.72	1.07	1.17	1.24
	Insulated from edge to footer: R-5.4 h ft² °F/Btu (RSI-0.95 m² K/W)	0.48	0.50	0.56	0.83	0.86	0.97
Block wall, 4 in. (100 mm), brick facing	Uninsulated	0.80	0.84	0.93	1.38	1.45	1.61
	Insulated from edge to footer: R.5.4 h ft² °F/Btu (RSI-0.95 m² K/W)	0.47	0.49	0.54	0.81	0.85	0.93
Metal stud wall, stucco	Uninsulated	1.15	1.20	1.34	1.99	2.07	2.32
	Insulated from edge to footer: R-5.4 h ft² °F/Btu (RSI-0.95 m² K/W)	0.51	0.53	0.58	0.88	0.92	1.00
Poured concrete wall with duct near perimeter[b]	Uninsulated	1.84	2.12	2.73	3.18	3.67	4.72
	Insulated from edge to footer, and 3 ft. (910 mm) under floor slab R-5.4 h ft² °F/Btu (RSI-0.95 m² K/W)	0.64	0.72	0.90	1.11	1.24	1.56

(a) 8-IN. BLOCK WALL   (b) 4-IN. BLOCK WALL   (c) METAL STUD WALL   (d) CONCRETE WALL

TABLE 4.8 Heat Loss Coefficient of Floor Slabs, F_2, per Unit Length of Perimeter (*continued*)

Source: Reprinted with permission of the American Society of Heating, Refrigerating, and Air-Conditioning Engineers, Inc. from 1997 *ASHRAE Handbook of Fundamentals*.

[a]Approximate degree-day/design temperature correlation:

2950 HDD: greater than 5° F	1640 HDD: greater than −15° C
5350 HDD: from 5° to −5° F	2970 HDD: from −15° to −31° C
7433 HDD: less than −5° F	4130 HDD: less than −31° C

[b]Weighted average temperature of heating duct was assumed at 110°F (43° C) during the heating season, with outdoor temperature less than 65° F (18° C).

NOTE: To use this table:

$$q = F_2 P \, \Delta t$$

where q = heat loss through perimeter [Btu or W].
   F_2 = heat loss coefficients from above.
   P = perimeter of exposed slab edge [ft or m].
   Δt = temperature difference between indoor and outdoor air [F° or C°]

Do not assume additional losses from the slab to the earth below. Heat gains are assumed nonexistent.

TABLE 4.9 Heat Flow Coefficients of Floor Slabs, F_2, for Various Values of R

Insulation Strategy R-Value, Position, and Width (or Depth) of Insulation	F_2	
	(Btu/h ft °F)	(W/m K)
Uninsulated slab	0.73	1.26
R-5 Horizontal insulation, 2 ft, no thermal break	0.70	1.21
R-10 Horizontal insulation, 2 ft, no thermal break	0.70	1.21
R-15 Horizontal insulation, 2 ft, no thermal break	0.69	1.19
R-5 Horizontal insulation, 4 ft, no thermal break	0.67	1.16
R-10 Horizontal insulation, 4 ft, no thermal break	0.64	1.11
R-15 Horizontal insulation, 4 ft, no thermal break	0.63	1.09
R-5 Vertical insulation, 2 ft	0.58	1.0
R-10 Vertical insulation, 2 ft	0.54	0.93
R-15 Vertical insulation, 2 ft	0.52	0.9
R-5 Vertical insulation, 4 ft	0.54	0.93
R-10 Vertical insulation, 4 ft	0.48	0.83
R-15 Vertical insulation, 4 ft	0.45	0.78
R-10 Fully insulated slab (insulated under entire slab as well as around edge)	0.36	0.62

Source: "*Super Good Cents Heat Loss Reference, Vol. 1,*" Ecotope Group, for Bonneville Power Administration, 1988.

Insulation is extruded polystyrene, R = 5.0 h ft^2 °F/Btu-in. (RSI = 34.7 m k/W).

Soil conductivity is 0.75 Btu/h ft °F (1.30 W/Km).

No thermal break at edge of slab, where so indicated. If a thermal break is provided with horizontal insulation, use the corresponding value for vertical insulation.

TABLE 4.10 **Heat Flow Below Grade**[a]

Part A. Basement Walls

SI Heat Loss[b] (W/m²K)				Depth		Path Length Through Soil		I-P Heat Loss[b] (Btu/ft² °F)			
Uninsulated	R = 0.73	R = 1.47	R = 2.20	m	(ft)	m	(ft)	Uninsulated	R = 4.17	R = 8.34	R = 12.5
2.33	0.86	0.53	0.38	0–0.3	(0–1)	0.2	(0.68)	0.410	0.152	0.093	0.067
1.26	0.66	0.45	0.36	0.3–0.6	(1–2)	0.69	(2.27)	0.222	0.116	0.079	0.059
0.88	0.53	0.38	0.30	0.6–0.9	(2–3)	1.18	(3.88)	0.155	0.094	0.068	0.053
0.67	0.45	0.34	0.27	0.9–1.2	(3–4)	1.68	(5.52)	0.119	0.079	0.060	0.048
0.54	0.39	0.30	0.25	1.2–1.5	(4–5)	2.15	(7.05)	0.096	0.069	0.053	0.044
0.45	0.34	0.27	0.23	1.5–1.8	(5–6)	2.64	(8.65)	0.079	0.060	0.048	0.040
0.39	0.30	0.25	0.21	1.8–2.1	(6–7)	3.13	(10.28)	0.069	0.054	0.044	0.037

Part B. Basement Floors

Depth of Foundation Wall Below Grade		SI Heat Loss[b] (W/m²K), Width of House in Meters				I-P Heat Loss[b] (Btu/ft² °F), Width of House in Feet			
m	(ft)	6	7.3	8.5	9.7	20	24	28	32
1.5	(5)	0.18	0.16	0.15	0.13	0.032	0.029	0.026	0.023
1.8	(6)	0.17	0.15	0.14	0.12	0.030	0.027	0.025	0.022
2.1	(7)	0.16	0.15	0.13	0.12	0.029	0.026	0.023	0.021

Source: Reprinted with permission of the American Society of Heating, Refrigerating, and Air-Conditioning Engineers, Inc. from 1997 *ASHRAE Handbook of Fundamentals.*

[a]Only heat losses are assumed because the interior temperature is normally higher than the ground temperature.

[b]Soil conductivity assumed at 1.38 W/m K (9.6 Btu-in/h ft² °F).

TABLE 4.11 Corrected *R*-Values for Roof Systems with Metal Trusses[a]

Insulation *R*-value (h ft² °F/Btu)	5	10	15	20	25	30	35	40	45	50	55
RSI (K m²/W)	0.88	1.76	2.64	3.53	4.41	5.29	6.17	7.05	7.93	8.81	9.70
Correction Factor	0.96	0.92	0.88	0.85	0.81	0.79	0.76	0.73	0.71	0.69	0.67

Source: Reprinted with permission of the American Society of Heating, Refrigerating, and Air-Conditioning Engineers, Inc., from *ASHRAE/IES Standard 90.1-1989 Energy Efficient Design of New Buildings Except New Low-Rise Residential Buildings.*

[a]Based on 4-ft (1.2-m) o.c. metal trusses that penetrate the insulation, and 0.66-in. (16.8-mm) crossbars every 1 ft (0.3 m).

TABLE 4.12 Coefficients of Transmission, U, for Wood and Steel Doors

Nominal Door Thickness		Description	Wood Storm Door[a]		Metal Storm Door[b]		No Storm Door	
in.	mm		Btu/h ft² °F	W/m² K	Btu/h ft² °F	W/m²K	Btu/h ft² °F	W/m²K

Part A. Unglazed Doors

Unglazed Wood Doors[c,d]

in.	mm	Description	Btu/h ft² °F	W/m² K	Btu/h ft² °F	W/m²K	Btu/h ft² °F	W/m²K
1⅜	35	Panel door with 7/16-in. (11-mm) panels[e]	0.33	1.87	0.37	2.10	0.57	3.24
1⅜	35	Hollow core flush door	0.30	1.70	0.32	1.82	0.47	2.67
1⅜	35	Solid core flush door	0.26	1.48	0.28	1.59	0.39	2.21
1¾	45	Panel door with 7/16-in. (11-mm) panels	0.32	1.82	0.36	2.04	0.54	3.07
1¾	45	Hollow core flush door	0.29	1.65	0.32	1.82	0.46	2.61
1¾	45	Panel door with 1⅛-in. (29-mm) panels	0.26	1.48	0.28	1.59	0.39	2.21
1¾	45	Solid core flush door	—	—	0.26	1.48	0.40	2.27
2¼	57	Solid core flush door	0.20	1.14	0.21	1.19	0.27	1.53

Unglazed Steel Doors[d]

in.	mm	Description	Btu/h ft² °F	W/m² K	Btu/h ft² °F	W/m²K	Btu/h ft² °F	W/m²K
1¾	45	Fiberglass or mineral wool core with steel stiffeners, no thermal break[f]			—	—	0.60	3.41
1¾	45	Paper honeycomb core without thermal break[f]			—	—	0.56	3.18
1¾	45	Solid urethane foam core without thermal break[c]			—	—	0.40	2.27
1¾	45	Solid fire-rated mineral fiberboard core without thermal break[f]			—	—	0.38	2.16
1¾	45	Polystyrene core without thermal break[f] [18-gage (1.31-mm) commercial steel]			—	—	0.35	1.99
1¾	45	Polyurethane core without thermal break[f] [18-gage (1.31-mm) commercial steel]			—	—	0.29	1.65
1¾	45	Polyurethane core without thermal break[f] [24-gage (0.70-mm) residential steel]			—	—	0.29	1.65
1¾	45	Polyurethane core with thermal break and wood perimeter[f] [24-gage (0.70-mm) residential steel]			—	—	0.20	1.14
1¾	45	Solid urethane foam core with thermal break[c]			0.16	0.91	0.20	1.14

(*continued*)

TABLE 4.12 **Coefficients of Transmission, *U*, for Wood and Steel Doors** (*continued*)

Part B. Glazed Doors^g

| | Single Glazing | | Double Glazing | | | |
| | | | Airspace ½ in. (12.7 mm) | | Low-ε 0.10, Argon ½ in. (12.7 mm) | |
	Btu/h ft² °F	W/m²K	Btu/h ft² °F	W/m²K	Btu/h ft² °F	W/m²K
Wood slab in wood frame^h						
6% glazing	0.48	2.73	0.46	2.61	0.44	2.50
25% glazing	0.58	3.29	0.46	2.61	0.42	2.38
45% glazing	0.69	3.92	0.46	2.61	0.39	2.21
Insulated steel slab with wood edge in wood frame^h						
6% glazing	0.21	1.19	0.19	1.08	0.18	1.02
25% glazing	0.39	2.21	0.26	1.48	0.23	1.31
45% glazing	0.58	3.29	0.35	1.99	0.26	1.48
Foam insulated steel slab with metal edge in steel frame^i						
6% glazing	0.44	2.50	0.41	2.33	0.39	2.21
25% glazing	0.55	3.12	0.48	2.73	0.44	2.50
45% glazing	0.71	4.03	0.56	3.18	0.48	2.73

Source: Reprinted with permission of the American Society of Heating, Refrigerating, and Air-Conditioning Engineers, Inc. from 1997 *ASHRAE Handbook of Fundamentals*.

Note: All *U*-values listed are for exterior doors with less than 50% glazing, except for the storm doors, which are in addition to the main exterior door. Glazing area greater than 50% in exterior doors should be included with the appropriate glass type and analyzed as windows (see Table 4.13). Interpolation and moderate extrapolation may be made for door thicknesses other than those specified.

^aValues for wood storm door are approximately 50% glass.

^bValues for metal storm door are for any percent glass area.

^cValues are based on a nominal 32-in. × 80-in. (810-mm × 2030-mm) door size.

^dOutside air conditions: 15 mph (24 km/h) wind speed, 0° F (−18°C) air temperature; inside air conditions: natural convection, 70° F (21 °C) air temperature.

^e55% panel area.

^fNominal 3 ft × 7 ft (910-mm × 2130-mm) door size.

^gRough opening 38 in. × 82 in. (970 mm × 2080 mm).

^hThermally broken sill [if not thermally broken, add 0.03 Btu/h ft² °F (0.17 W/m²K)].

^iNonthermally broken sill.

TABLE 4.13 **Window Characteristics**[a]

Glazing Description and Reference Number	Layers of Glazing and Spaces (outside to inside)	Frame (and spacer)	U-Factor		Total Window					Air Leakage			
			Btu/h ft² °F	W/m²K	SHGC[b]	VT[c]	LSG[d]	FHR[e]	FCR[e]	cfm/lin. ft.	L/s m	cfm/ft²	L/s m²
1. Single-glazed clear	1/8 in. (3 mm) clear	Aluminum, no thermal break	1.30	7.37	0.79	0.69	0.87	0	0	0.65	1.01	0.98	4.98
2. Single-glazed bronze	1/8 in. (3 mm) bronze	Aluminum, no thermal break	1.30	7.37	0.69	0.52	0.75	−2	8	0.65	1.01	0.98	4.98
3. Double-glazed clear	1/8 in. (3 mm) clear 1/2 in. (13 mm) air 1/8 in. (3 mm) clear	Aluminum, thermal break (aluminum)	0.64	3.63	0.65	0.62	0.95	19	12	0.37	0.57	0.56	2.84
4. Double-glazed bronze	1/8 in. (3 mm) bronze 1/2 in. (13 mm) air 1/8 in. (3 mm) clear	Aluminum, thermal break (aluminum)	0.64	3.63	0.55	0.47	0.85	17	20	0.37	0.57	0.56	2.84
5. Double-glazed clear	1/8 in. (3 mm) clear 1/2 in. (13 mm) air 1/8 in. (3 mm) clear	Wood or vinyl (aluminum)	0.49	2.78	0.58	0.57	0.98	24	18	0.37	0.57	0.56	2.84
6. Double-glazed bronze	1/8 in. (3 mm) bronze 1/2 in. (13 mm) air 1/8 in. (3 mm) clear	Wood or vinyl (aluminum)	0.49	2.78	0.48	0.43	0.90	22	25	0.37	0.57	0.56	2.84
7. Double-glazed low-ε[f]	1/8 in. (3 mm) clear 1/2 in. (13 mm) argon 1/8 in. (3 mm) low-ε 0.20	Wood or vinyl (stainless)	0.33	1.87	0.55	0.52	0.95	32	19	0.10	0.15	0.15	0.76

(continued)

TABLE 4.13 **Window Characteristics**^{a} (*continued*)

Glazing Description and Reference Number	Layers of Glazing and Spaces (outside to inside)	Frame (and spacer)	U-Factor Btu/h ft² °F	W/m²K	SHGC^b	VT^c	LSG^d	FHR^e	FCR^e	cfm/lin. ft.	L/s m	cfm/ft²	L/s m²
					Total Window					Air Leakage			
8. Double-glazed low-ε^f	⅛ in. (3 mm) low-ε 0.08 ½ in. (13 mm) argon ⅛ in. (3 mm) clear	Wood or vinyl (stainless)	**0.30**	*1.70*	0.44	0.56	1.27	32	27	0.10	*0.15*	0.15	*0.76*
9. Double-glazed spectrally selective^f	⅛ in. (3 mm) low-ε 0.04 ½ in. (13 mm) argon ⅛ in. (3 mm) clear	Wood or vinyl (stainless)	**0.29**	*1.65*	0.31	0.51	1.65	30	36	0.10	*0.15*	0.15	*0.76*
10. Double-glazed spectrally selective^f	⅛ in. (3 mm) low-ε 0.10 ½ in. (13 mm) argon ⅛ in. (3 mm) clear	Wood or vinyl (stainless)	**0.31**	*1.76*	0.26	0.31	1.19	27	40	0.10	*0.15*	0.15	*0.76*
11. Triple-glazed low-ε^f superwindow	⅛ in. (3 mm) low-ε 0.08 ½ in. (13 mm) krypton ⅛ in. (3 mm) clear ½ in. (13 mm) krypton ⅛ in. (3 mm) low-ε 0.08	Insulated vinyl (insulated)	**0.15**	*0.85*	0.37	0.48	1.30	38	33	0.05	*0.08*	0.08	*0.41*
12. Triple-glazed clear	⅛ in. (3 mm) clear ½ in. (13 mm) air ⅛ in. (3 mm) clear ½ in. (13 mm) air ⅛ in. (3 mm) clear	Wood or vinyl (stainless)	**0.34**	*1.93*	0.52	0.53	1.02	32	22	0.10	*0.15*	0.15	*0.76*

Source: Residential Windows: A Guide to New Technologies and Energy Performance, by John Carmody, Stephen Selkowitz, and Lisa Heschong. Copyright © 1996 by John Carmody, Stephen Selkowitz and Lisa Heschong. Adapted by permission of W. W. Norton & Company, New York. (Metric conversions by the author.)

^{a}Based on a casement window, 2 ft × 4 ft (610 mm × 1220 mm).

^{b}Solar heat gain coefficient (higher numbers mean more solar heat flow).

^{c}Visible transmittance (higher numbers mean more light transmitted).

^{d}Light-to-solar gain ratio, LSG = VT/SHGC. For center-of-glass values, see Table 4.16.

^{e}FHR and FCR are heating season and cooling season estimates of percent energy saved in a typical application compared to using window #1.

^{f}Low-ε ratings indicate what percentage of the long-wavelength radiant energy is admitted; 0.20 therefore admits 20% and reflects 80%.

TABLE 4.14 **Skylight *U*-Factors**[a]

Part A. *I-P units (Btu/h ft² °F)*

| Glazing | Manufactured Skylight[b] | | | | Site-Assembled Glazing[c] | | |
| | Aluminum, Thermal Break | | Reinforced Vinyl/ Aluminum-Clad Wood | Wood/ Vinyl | Aluminum, Thermal Break | | Structural Glazing |
	No	Yes			No	Yes	
Single glazed	1.98	1.89	1.75	1.47	1.36	1.25	1.25
Double glazed, ½-in. gap	1.30	1.10	1.04	0.84	0.81	0.69	0.65
Double glazed, ½-in. gap, argon low-ε 0.20 on surface[d] 2 or 3	1.15	0.95	0.89	0.68	0.66	0.55	0.51
Double glazed, ½-in. gap, argon low-ε 0.10 on surface[d] 2 or 3	1.13	0.93	0.87	0.67	0.65	0.53	0.49
Double glazed, ½-in. gap, low-ε 0.05 on surface[d] 2 or 3	1.11	0.91	0.85	0.65	0.63	0.52	0.47
Triple glazed, ½-in. gaps	1.10	0.87	0.81	0.61	0.62	0.51	0.45
Triple glazed, ½-in. gaps, argon low-ε 0.10 on surfaces[d] 2 or 3 and 4 or 5	0.95	0.72	0.67	0.47	0.48	0.37	0.31

Part B. *SI units (W/m²K)*

| Glazing | Manufactured Skylight[b] | | | | Site-Assembled Glazing[c] | | |
| | Aluminum, Thermal Break | | Reinforced Vinyl/ Aluminum-Clad Wood | Wood/ Vinyl | Aluminum, Thermal Break | | Structural Glazing |
	No	Yes			No	Yes	
Single glazed	11.24	10.73	9.96	8.34	7.73	7.09	7.09
Double glazed, 12.7-mm gap	7.39	6.27	5.90	4.74	4.59	3.95	3.70
Double glazed, 12.7-mm gap, argon, low-ε 0.20 on surface[d] 2 or 3	6.52	5.39	5.04	3.89	3.77	3.12	2.87
Double glazed, 12.7-mm gap, argon, low-ε 0.10 on surface[d] 2 or 3	6.41	5.29	4.94	3.79	3.67	3.03	2.78
Double glazed, 12.7-mm gap, low-ε 0.05 on surface[d] 2 or 3	6.31	5.18	4.84	3.69	3.57	2.93	2.68

(*continued*)

TABLE 4.14 **Skylight** *U*-**Factors**[a] (*continued*)

	Part B. **SI units (W/m²K)**						
	Manufactured Skylight[b]				Site-Assembled Glazing[c]		
	Aluminum, Thermal Break		Reinforced Vinyl/ Aluminum-Clad Wood	Wood/ Vinyl	Aluminum, Thermal Break		Structural Glazing
Glazing	No	Yes			No	Yes	
Triple glazed, 12.7-mm gaps	6.22	4.92	4.62	3.48	3.51	2.88	2.56
Triple glazed, 12.7-mm gaps, argon, low-ε 0.10 on surfaces[d] 2 or 3 and 4 or 5	5.38	4.07	3.79	2.66	2.72	2.09	1.78

Source: Excerpted with permission of the American Society of Heating, Refrigerating, and Air-Conditioning Engineers, Inc. from 1997 *ASHRAE Handbook of Fundamentals.*

[a] All glazings are ⅛-in. (3-mm) thickness. Winter conditions, 15 mph (24 km/h) wind with 0° F (−18° C) outside, 70° F (21° C) inside, no solar radiation. Glazing is sloped at 20° from the horizontal.

[b] Skylight is 2 ft × 4 ft (600 mm × 1200 mm) with frame width as follows: aluminum, 0.7 in. (18 mm); wood/vinyl/aluminum-clad, 0.9 in. (23 mm).

[c] Site-assembled glazing size is 4 ft × 4 ft (1200 mm × 1200 mm) with aluminum frame width of 2.25 in. (57 mm), structural glazing width of 2.5 in. (64 mm).

[d] Glazing layer surfaces are numbered from outside to inside.

TABLE 4.15 **Shading Coefficients for Domed Skylights**

		Curb (See Below)			
Dome	Light Diffuser (Translucent)	Height, [mm (in.)]	Width-to-Height Ratio	Shading Coefficient	U-Factor at Center of Skylight
Clear $\tau = 0.86$[a]	Yes $\tau = 0.58$	0 (0)	∞	0.61	2.6 (0.46)
		230 (9)	5	0.58	2.4 (0.43)
		460 (18)	2.5	0.50	2.3 (0.40)
Clear $\tau = 0.86$	None	0 (0)	∞	0.99	4.5 (0.80)
		230 (9)	5	0.88	4.3 (0.75)
		460 (18)	2.5	0.80	4.0 (0.70)
Translucent $\tau = 0.52$	None	0 (0)	∞	0.57	4.5 (0.80)
		460 (18)	2.5	0.46	4.0 (0.70)
Translucent $\tau = 0.27$	None	0 (0)	∞	0.34	4.5 (0.80)
		230 (9)	5	0.30	4.3 (0.75)
		460 (18)	2.5	0.28	4.0 (0.70)

Source: Reprinted with permission of the American Society of Heating, Refrigerating, and Air-Conditioning Engineers, Inc. from 1981 and 1997 *ASHRAE Handbook of Fundamentals.*

TABLE 4.16 **Solar Optical Properties of Several Glazings**

Part A. **Glass**

Type of Glazing	SHGC[a]	VT[b]	LSG[c]
Reflective blue-green	0.38	0.33	0.87
Film on clear glass	0.22	0.19	0.86
Green tinted, medium	0.69	0.75	1.09
Green low-ε	0.49	0.71	1.45
Sun-control low-ε + green	0.23	0.36	1.56
Super low-ε + clear	0.40	0.71	1.77
Super low-ε + green	0.30	0.60	2.00

Part B. **Transparent Plastic**

	Transmittance		
Type of Plastic	Visible	Solar	SC
Acrylic			
Clear	0.92	0.85	0.98
Gray tint	0.16	0.27	0.52
Gray tint	0.33	0.41	0.63
Gray tint	0.45	0.55	0.74
Gray tint	0.59	0.62	0.80
Gray tint	0.76	0.74	0.89
Bronze tint	0.10	0.20	0.46
Bronze tint	0.27	0.35	0.58
Bronze tint	0.49	0.56	0.75
Bronze tint	0.61	0.62	0.80
Bronze tint	0.75	0.75	0.90
Reflective[d]	0.14	0.12	0.21
Polycarbonate			
Clear, ⅛ in. (3 mm)	0.88	0.82	0.98
Gray or bronze, ⅛ in. (3 mm)	0.50	0.57	0.74

Source: Reprinted with permission of the American Society of Heating, Refrigerating, and Air-Conditioning Engineers, Inc. from 1997 *ASHRAE Handbook of Fundamentals.*

[a]Solar heat gain coefficient.

[b]Visible transmittance.

[c]Light-to-solar gain ratio, LSG = VT/SHGC. For whole-window values, see Table 4.13.

[d]Aluminum metalized polyester film on plastic.

TABLE 4.17 **Approximate Shading Coefficients of External Shading Devices**

Awnings	
Of venetian blind type, ⅔ drawn	0.43
fully drawn	0.15
Dark or medium canvas	0.25
Shading screens	0.28–0.23
Louvres, movable	0.15–0.10
Overhang: continuous, completely shading window	0.25
Dense tree casting heavy shade	0.25

Source: Adapted from Olgyay and Olgyay (1957).

TABLE 4.18 **Shading Coefficients for Louvered Sun Screens**[a]

Profile Angle, (deg)	Group 1		Group 2		Group 3		Group 4		Group 5		Group 6	
	Trans-mittance	SC	Trans-mittance	SC	Trans-mittance	SC	Trans-mittance	SC	Trans-mittance	SC	Trans-mittance	SC
10	0.23	0.35	0.25	0.33	0.40	0.51	0.48	0.59	0.15	0.27	0.26	0.45
20	0.06	0.17	0.14	0.23	0.32	0.42	0.39	0.50	0.04	0.11	0.20	0.35
30	0.04	0.15	0.12	0.21	0.21	0.31	0.28	0.38	0.03	0.10	0.13	0.26
≥40	0.04	0.15	0.11	0.20	0.07	0.18	0.20	0.30	0.03	0.10	0.04	0.13

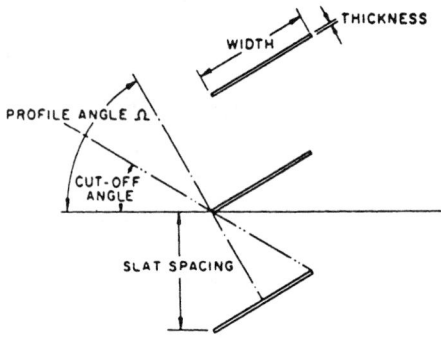

Source: Reprinted with permission of the American Society of Heating, Refrigerating, and Air-Conditioning Engineers, Inc., from 1997 *ASHRAE Handbook of Fundamentals.*

[a]**Group 1.** Black, width over spacing ratio 1.15/1; 23 louvers/in. (1.1 mm between louvers).

Group 2. Light color; high reflectance, otherwise same as group 1.

Group 3. Black or dark color; w/s ratio 0.85/1; 17 louvers/in. (1.5 mm between louvers).

Group 4. Light color or unpainted aluminum; high reflectance; otherwise same as group 3.

Group 5. Same as group 1, except two lights of 0.25-in. clear glass with 0.5-in. airspace (6 mm glass, 13 mm airspace).

Group 6. Same as group 3, except two lights of 0.25-in. clear glass with 0.5-in. airspace (6 mm glass, 13 mm airspace).

U **value = 0.85 Btu/h-ft²-F (4.83 W/m²K)** for all groups when used with single glazing.

TABLE 4.19 **Adjustable Sunshading**

	Cooling and Daylighting Performance While Shading					
	Solar Heat Rejection	Daylight Distribution	View	Ventilation	Typical Orientation	Heating: Night Insulation
Roll-down Shades Solid	Blocks both direct and ground-reflected light. Heat builds up behind shade, unless vented.	Translucent shades diffuse sunlight evenly but may themselves become sources of glare, due to their brightness. If colored, they will color incoming diffuse light.	Blocks view.	Blocks ventilation.	East and west, where it is rolled down daily during hours of threat from direct sun.	Can be effective if infiltration above and around the shade is blocked.
Slated	Blocks most direct sun and ground-reflected light.	Thinly striped pattern of direct sun is admitted.	Heavily filters view.	Greatly reduces ventilation.		No value, unless slats can fit together to eliminate openings.
Fold-out Retractable Awning	Blocks direct sun, admits ground-reflected light. Heat on awning easily dissipated by breezes.	Translucent awnings diffuse sunlight evenly. (Colored translucent awnings color the diffused light.) Awnings admit ground-reflected light toward ceiling, where white surfaces diffuse it deeper into space.	Blocks sky view. Allows ground view.	Reduces ventilation somewhat; tends to direct wind upward into space.	Any orientation (on south, awnings can be deep and can be left in position for weeks at a time).	If awning folds back against window, it provides some insulation, but infiltration reduces its value.
Pivoting Louvers Horizontal	Blocks direct sun, admits ground-reflected light.	Sunlight and ground-reflected light directed toward ceiling, where white surfaces diffuse the light deeper into space.	Blocks sky view. Filters ground view.	Reduces ventilation somewhat; directs wind upward into space.	South; also used on other elevations.	
Vertical	Blocks direct sun, admits some sky and ground-reflected light.	Diffuse and ground-reflected light directed sideways into space. Does not usually help with deeper daylight penetration.	Filters view.	Reduces ventilation somewhat; directs wind sideways into space.	East and west; sometimes south and north.	Closed louvers provide some insulation, but infiltration reduces its value.

TABLE 4.20 **Shading Coefficients (SC)**

Part A. **SC for Insulating Glass[a] with Indoor Shading by Venetian Blinds or Roller Shades**

Type of Glass	Nominal Thickness, Each Light	Solar Trans.[b] Outer Pane	Solar Trans.[b] Inner Pane	Venetian Blinds[c] Medium	Venetian Blinds[c] Light	Roller Shade Opaque Dark	Roller Shade Opaque White	Roller Shade Translucent Light
Clear out	2.5, 3 mm ($\frac{3}{32}$, $\frac{1}{8}$ in.)	0.87	0.87					
Clear in				0.62	0.58 (45° open)	0.71	0.35	0.40
Clear out				0.63	0.58 (closed)[d]			
Clear in	6 mm ($\frac{1}{4}$ in.)	0.80	0.80					
Heat-absorbing[e] out	6 mm ($\frac{1}{4}$ in.)	0.46	0.80	0.39	0.36	0.40	0.22	0.30
Clear in								
Reflective coated glass								
SC[f] = 0.20				0.19	0.18			
0.30				0.27	0.26			
0.40				0.34	0.33			

Part B. **Shading Coefficients for Double Glazing with Between-Glass Shading**

Type of Glass	Nominal Thickness, Each Pane	Solar Trans.[b] Outer Pane	Solar Trans.[b] Inner Pane	Description of Air Space	Venetian Blinds Light	Venetian Blinds Medium	Louvered Sun Screen
Clear out Clear in	2.5, 3 mm ($\frac{3}{32}$, $\frac{1}{8}$ in.)	0.87	0.87	Shade in contact with glass or shade separated from glass by airspace	0.33	0.36	0.43
Clear out Clear in	6 mm ($\frac{1}{4}$ in.)	0.80	0.80	Shade in contact with glass; voids filled with plastic	—	—	0.49
Heat-absorbing[e] out				Shade in contact with glass or shade separated from glass by airspace	0.28	0.30	0.37
Clear in	6 mm ($\frac{1}{4}$ in.)	0.46	0.80	Shade in contact with glass; voids filled with plastic	—	—	0.41

Source: Reprinted with permission of the American Society of Heating, Refrigerating, and Air-Conditioning Engineers, Inc. from 1997 *ASHRAE Handbook of Fundamentals.*

[a]Factory-manufactured units with 5-, 6-, or 13-mm ($\frac{3}{16}$-, $\frac{1}{4}$-, or $\frac{1}{2}$-in.) airspace.

[b]Refer to manufacturer's literature for exact values.

[c]For vertical blinds with opaque, white, or beige louvers, tightly closed, SC is approximately the same as for opaque white roller shades.

[d]Use these values only when operation is automated for solar gain reduction (as opposed to daylight use).

[e]Refers to bronze or green-tinted heat-absorbing glass.

[f]SC for glass with no shading device.

TABLE 4.21 **Shading Coefficients for Single and Insulating Glass with Draperies**

Glazing	Glass Trans.	Glass SC[a]	Range of Shading Coefficients Drapery Fabrics[b] High Transmittance, Low Reflectance[c]	Low Transmittance, High Reflectance[c]
Single Glass				
3 mm (⅛ in.) clear	0.86	1.00	0.87	0.37
6 mm (¼ in.) clear	0.80	0.95	0.80	0.35
12 mm (½ in.) clear	0.71	0.88	0.74	0.35
6 mm (¼ in.) heat abs.	0.46	0.67	0.57	0.33
12 mm (½ in.) heat abs.	0.24	0.50	0.43	0.30
Reflective coated (see manufactur-	—	0.60	0.57	0.33
er's literature for exact values)	—	0.50	0.46	0.31
	—	0.40	0.36	0.26
	—	0.30	0.25	0.20
Insulating Glass, 12-mm (½-in.) air space clear out and clear in	0.64	0.83	0.66	0.35
Heat abs. out and clear in	0.37	0.55	0.49	0.32
Reflective coated (see manufac-	—	0.40	0.38	0.28
turers' literature for exact	—	0.30	0.29	0.24
values)	—	0.20	0.19	0.15

Source: Based on 1997 *ASHRAE Handbook of Fundamentals,* published by the American Society of Heating, Refrigerating and Air-Conditioning Engineers, Inc., Atlanta, GA.

[a]For glass alone, with no drapery.

[b]Draperies of 100% fullness, loose hanging.

[c]See 1997 *ASHRAE Handbook of Fundamentals,* Chapter 27, Table 29, for more detailed listings.

TABLE 4.22 **Recommended Outdoor Air Requirements for Ventilation**[a]

Part A. **Commercial Facilities (Offices, Stores, Shops, Hotels, Sports Facilities, etc.)**

Application	Estimated Occupancy[b] (Persons per 1000 ft² or 100 m² Floor Area): Use Only When Design Occupancy is Not Known	Outdoor Air Requirements				Comments
		cfm per person	L/s per person	cfm per ft²	L/s per m²	
Dry Cleaners, Laundries						
Commercial laundries	10	25	13			Dry cleaning processes may require more air.
Commercial dry cleaners	30	30	15			
Storage, pickup	30	35	18			
Coin-op laundries	20	15	8			
Coin-op dry cleaners	20	15	8			
Food and Beverage Service						
Dining rooms	70	20	10			Supplementary smoke removal equipment may be required. Makeup air for hood exhaust may require more ventilating air. The sum of the outdoor air and transfer air of acceptable quality from adjacent spaces shall be sufficient to provide an exhaust rate of not less than 1.5 cfm/ft² (7.5 L/s per m²).
Cafeterias, fast food	100	20	10			
Bars, cocktail lounges	100	30	15			
Kitchens (cooking)	20	15	8			

Garages, Repair Service Stations

Application	Occupancy	cfm	L/s	Comments
Enclosed parking garages		1.50	7.50	Distribution among people must consider worker location and concentration of running engines; stands where engines are run must incorporate systems for positive engine exhaust withdrawal. Contaminant sensors may be used to control ventilation.
Auto repair rooms		1.50	7.50	

Hotels, Motels, Resorts, Dormitories

Application	Occupancy	cfm	L/s	Comments
Dormitory sleeping areas	20	15	8	See also food and beverage services, merchandising, barber and beauty shops, garages.
		cfm/room	L/s/room	Independent of room size.
Bedrooms		30	15	
Living rooms		30	15	
Bathrooms		35	18	Installed capacity for intermittent use.
Lobbies	30	15	8	
Conference rooms	50	20	10	
Assembly rooms	120	15	8	
Gambling casinos	120	30	15	Supplementary smoke removal equipment may be required.

Offices

Application	Occupancy	cfm	L/s	Comments
Office space	7	20	10	Some office equipment may require local exhaust.
Reception areas	60	15	8	Supplementary smoke removal equipment may be required
Telecommunication centers and data entry areas	60	20	10	
Conference rooms	50	20	10	

(continued)

THERMAL CONTROL

181

TABLE 4.22 **Recommended Outdoor Air Requirements for Ventilation**[a] (*continued*)

Part B. Institutional Facilities

Application	Estimated Max. Occupancy[b] (Persons per 1000 ft² or 100 m²)	Outdoor Air Requirements				Comments
		cfm per person	L/s per person	cfm per ft²	L/s per m²	
Public Spaces						
Corridors and utility rooms				0.50	0.25	
Public rest rooms cfm/wc or urinal		50	25			Mechanical exhaust with no recirculation is recommended
Locker and dressing rooms				0.5	2.5	
Smoking lounge	70	60	30			Normally supplied by transfer air, local mechanical exhaust. Exhaust with no recirculation recommended.
Elevators				1.00	5.00	Normally supplied by transfer air.
Retail Stores Sales Floors and Showroom Floors						
Basement and street	30			0.30	1.50	
Upper floors	20			0.20	1.00	
Storage rooms	15			0.15	0.75	
Dressing rooms				0.20	1.00	
Malls and arcades	20			0.20	1.00	
Shipping and receiving	10			0.15	0.75	
Warehouses	5			0.05	0.25	
Smoking lounge	70	60	30			Normally supplied by transfer air, local mechanical exhaust. Exhaust with no recirculation recommended.

Specialty Shops

	Occ.	cfm/person	L/s·person	cfm/ft²	L/s·m²	Comments
Barber	25	15	8			
Beauty	25	25	13			
Reducing salons	20	15	8			
Florists	8	15	8			Ventilation to optimize plant growth may dictate requirements.
Clothiers, furniture				0.30	1.50	
Hardware, drugs, fabric	8	15	8			
Supermarkets	8	15	8			
Pet shops				1.00	5.00	

Sports and Amusement

	Occ.	cfm/person	L/s·person	cfm/ft²	L/s·m²	Comments
Spectator areas	150	15	8			
Game rooms	70	25	13			When internal combustion engines are operated for maintenance of playing surfaces, increased ventilation rates may be required.
Ice arenas (playing area)				0.50	2.50	
Swimming pools (pool and deck area)				0.50	2.50	Higher values may be required for humidity control.
Playing floors (gymnasium)	30	20	10			
Ballrooms and discos	100	25	13			
Bowling alleys (seating areas)	70	25	13			

Theaters

	Occ.	cfm/person	L/s·person	cfm/ft²	L/s·m²	Comments
Ticket booths	60	20	10			
Lobbies	150	20	10			Special ventilation will be needed to eliminate special stage effects (e.g., dry ice vapors, mists, etc.)
Auditorium	150	15	8			
Stages, studios	70	15	8			

Transportation

	Occ.	cfm/person	L/s·person	cfm/ft²	L/s·m²	Comments
Waiting rooms	100	15	8			
Platforms	100	15	8			
Vehicles	150	15	8			Ventilation within vehicles may require special consideration.

(*continued*)

TABLE 4.22 **Recommended Outdoor Air Requirements for Ventilation**[a] (*continued*)

Part B. *Institutional Facilities*

Application	Estimated Max. Occupancy[b] (Persons per 1000 ft² or 100 m²)	Outdoor Air Requirements				Comments
		cfm per person	L/s per person	cfm per ft²	L/s per m²	
Workrooms						
Meat processing	10	15	8			Spaces maintained at low temperatures, −10° to +50°F (−23° to +10° C) are not covered by these requirements unless the occupancy is continuous. Ventilation from adjoining spaces is permissible. When the occupancy is intermittent, infiltration will normally exceed the ventilation requirement.
Photo studios	10	15	8			
Darkrooms	10			0.50	2.50	
Pharmacy	20	15	8			
Bank vaults	5	15	8			
Duplication printing				0.50	2.50	Installed equipment must incorporate positive exhaust and control (as required) of undesirable contaminants (toxic or otherwise).

Part B. *Institutional Facilities*

Application	Estimated Max. Occupancy[b] (Persons per 1000 ft² or 100 m²)	Outdoor Air Requirements				Comments
		cfm per person	L/s per person	cfm per ft²	L/s per m²	
Education						
Classroom	50	15	8			Special contaminant control systems may be required for processes or functions including laboratory animal occupancy.
Laboratories	30	20	10			
Training shop	30	20	10			
Music rooms	50	15	8			
Libraries	20	15	8			
Locker rooms				0.50	2.50	
Corridors				0.10	0.50	
Auditoriums	150	15	8			
Smoking lounges	70	60	30			Normally supplied by transfer air, local mechanical exhaust. Exhaust with no recirculation recommended.
Hospitals, Nursing and Convalescent Homes						
Patient rooms	10	25	13			Special requirements or codes and pressure relationships may determine minimum ventilation rates and filter efficiency.
Medical procedure	20	15	8			
Operating rooms	20	30	15			Procedures generating contaminants may require higher rates.
Recovery and ICU	20	15	8			
Autopsy rooms				0.50	2.50	Air shall not be recirculated into other spaces.
Physical therapy	20	15	8			

(continued)

185

TABLE 4.22 **Recommended Outdoor Air Requirements for Ventilation**[a] (*continued*)

Part B. Institutional Facilities

Application	Estimated Max. Occupancy[b] (Persons per 1000 ft² or 100 m²)	Outdoor Air Requirements				Comments
		cfm per person	L/s per person	cfm per ft²	L/s per m²	
Correctional Facilities						
Cells	20	20	10			
Dining halls	100	15	8			
Guard stations	40	15	8			

Part C. Residential Facilities (Private Dwellings, Single, Multiple)

Application	Outdoor Air Requirements[c]	Comments
Living areas	0.35 air changes per hour but not less than 15 cfm (7.5 L/s) per person	For calculating the air changes per hour, the volume of the living areas shall include all areas within the conditioned space. The ventilation is normally satisfied by infiltration and natural ventilation. Dwellings with tight enclosures may require supplemental ventilation supply for fuel-burning appliances, including fireplaces and mechanically exhausted appliances. Occupant loading shall be based on the number of bedrooms as follows: first bedroom, two persons; each additional bedroom, one person. Where higher occupancies are known, they should be used.
Kitchens[d]	100 cfm (50 L/s) intermittent —or— 25 cfm (12 L/s) continuous —or— Openable windows	Installed mechanical exhaust capacity.[e]
Baths, toilets[d]	50 cfm (25 L/s) intermittent —or— 25 cfm (12 L/s) continuous —or— Openable windows	Installed mechanical exhaust capacity.[e]

Garages

Separate for each dwelling unit	100 cfm (50 L/s) per car	Normally satisfied by infiltration or natural ventilation.
Common for several units	1.5 cfm/sq ft (7.5 L/s/m^2)	See "Enclosed Parking Garages," Part A.

Source: Reprinted with permission of the American Society of Heating, Refrigerating, and Air-Conditioning Engineers, Inc. from ASHRAE Standard 62-1989, *Ventilation for Acceptable Indoor Air Quality.*

[a]This table prescribes supply rates of acceptable outdoor air required for acceptable indoor air quality. These values have been chosen to control carbon dioxide and other contaminants with an adequate margin of safety, and to account for health variations among people, varied activity levels, and a moderate amount of smoking.

[b]Net occupiable spaces.

[c]The outdoor air is assumed to be acceptable.

[d]Climatic conditions may affect choice of ventilation option chosen.

[e]The air exhausted from kitchens, bath and toilet rooms may utilize air supplied through adjacent living areas to compensate for the air exhausted. The air supplied shall be of sufficient quantity to meet the requirements of this table.

TABLE 4.23 **Estimated Overall Infiltration Rates for Small Buildings**

Part A. **Construction Types**

Construction Type	Description
Tight	Good multifamily residential construction with close-fitting doors, windows, and framing; also, one-story, under 1500 ft² (140 m²) with full vapor retarder, no fireplace, well-fit windows, and weather-stripped doors.
Medium	Below-average multifamily construction; also, new two-story frame houses (or one-story houses more than 10 years old with average maintenance), floor area >1500 ft² (140 m²), average-fit windows and doors, and a fireplace with damper and glass closure.
Loose	Average manufactured homes; poorly constructed multifamily and single family residences with poorly fit windows and doors, more than 20 years old, of average maintenance, fireplace with no damper or glass closure, or having more than an average number of vented appliances.

Part B. **Design Infiltration Rate (ACH) for Winter: Indoors 68°F (20°C); Wind Speed = 15 mph (6.7 m/s)**

Construction Type	Winter Outdoor Design Temperature									
	°F: 50 / °C: 10	40 / 4	30 / −1	20 / −7	10 / −12	0 / −18	−10 / −23	−20 / −29	−30 / −34	−40 / −40
Tight	0.41	0.43	0.45	0.47	0.49	0.51	0.53	0.55	0.57	0.59
Medium	0.69	0.73	0.77	0.81	0.85	0.89	0.93	0.97	1.00	1.05
Loose	1.11	1.15	1.20	1.23	1.27	1.30	1.35	1.40	1.43	1.47

Part C. **Design Infiltration Rate (ACH) for Summer: Indoors 75°F (24°C); Wind Speed = 7.5 mph (3.4 m/s)**

Construction Type	Summer Outdoor Design Temperature					
	°F: 85 / °C: 29	90 / 32	95 / 35	100 / 38	105 / 41	110 / 43
Tight	0.33	0.34	0.35	0.36	0.37	0.38
Medium	0.46	0.48	0.50	0.52	0.54	0.56
Loose	0.68	0.70	0.72	0.74	0.76	0.78

Part D. **Infiltration Rates per Unit Floor Area**

Air Changes per Hour

Ceiling Height ft	m	0.3	0.4	0.5	0.6	0.7	0.8	0.9	1.0	1.1	1.2	1.3	1.4	1.5	1.6	1.7	1.8	1.9	2.0
									cfm/ft²										
7.5	2.29	0.04	0.05	0.06	0.08	0.09	0.10	0.11	0.13	0.14	0.15	0.16	0.18	0.19	0.20	0.21	0.23	0.24	0.25
8	2.44	0.04	0.05	0.07	0.08	0.09	0.11	0.12	0.13	0.15	0.16	0.17	0.19	0.20	0.21	0.23	0.24	0.26	0.27
8.5	2.59	0.04	0.06	0.07	0.09	0.10	0.11	0.13	0.14	0.16	0.17	0.18	0.20	0.21	0.23	0.24	0.26	0.27	0.28
9	2.74	0.05	0.06	0.08	0.09	0.11	0.12	0.14	0.15	0.17	0.18	0.20	0.21	0.23	0.24	0.26	0.27	0.29	0.30
									Btu/h ft² °F										
7.5	2.29	0.04	0.05	0.07	0.08	0.09	0.11	0.12	0.14	0.15	0.16	0.18	0.20	0.20	0.22	0.23	0.24	0.26	0.27
8	2.44	0.04	0.06	0.07	0.09	0.10	0.12	0.13	0.14	0.16	0.17	0.19	0.22	0.22	0.23	0.24	0.26	0.27	0.29
8.5	2.59	0.05	0.06	0.08	0.09	0.11	0.12	0.14	0.15	0.17	0.18	0.20	0.23	0.23	0.24	0.26	0.28	0.29	0.30
9	2.74	0.05	0.06	0.08	0.10	0.11	0.13	0.15	0.16	0.18	0.19	0.21	0.24	0.24	0.26	0.28	0.29	0.31	0.32

Sources: Parts A through C, reprinted with permission of the American Society of Heating, Refrigerating, and Air-Conditioning Engineers, Inc. from 1997 *ASHRAE Handbook of Fundamentals;* Part D, from ASHRAE *Cooling and Heating Load Calculation Manual,* 1979.

Notes: Multiply cfm/ft² by **5.08** to obtain L/s m².
Multiply Btu/h ft² °F by **5.673** to obtain W/m² K.

TABLE 4.24 **Approximate Infiltration Through Doors and Windows of Small Buildings**

Part A. *Converting Wind Speed to Velocity Head Factor*

NOTE: Typical design assumptions:
Winter wind $V_w = 15$ mph = velocity head factor of 0.105
Summer wind $V_w = 7.5$ mph = velocity head factor of 0.028

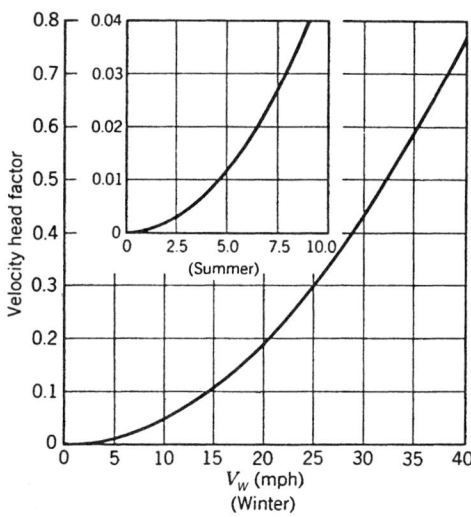

Part B. *Infiltration Rates for Velocity Head Factors*

NOTE: Enter this graph with velocity head factor (from Part A) to find infiltration rate in cfm/ft of crack (using values of k found in Part C or D).

(*continued*)

THERMAL CONTROL

TABLE 4.24 **Approximate Infiltration Through Doors and Windows of Small Buildings** (*continued*)

Part C. Classifications of Windows for Infiltration

Window Fit	Wood Double-Hung (Locked)	Other Types
Tight, $k = 1.0$	Weather stripped; average gap ($\frac{1}{64}$-in. crack)	Wood casement and awning windows; weather stripped.
		Metal casement windows; weather stripped.
Average, $k = 2.0$	Non–weather stripped; average gap ($\frac{1}{64}$-in. crack) or weather stripped; large gap ($\frac{3}{32}$-in. crack)	All types of vertical and horizontal sliding windows; weather stripped. If average gap ($\frac{1}{64}$-in. crack), this could be tight-fitting window.
		Metal casement windows; non–weather stripped. If large gap ($\frac{3}{32}$-in. crack), this could be a loose-fitting window.
Loose, $k = 6.0$	Non–weather stripped; large gap ($\frac{3}{32}$-in. crack)	Vertical and horizontal sliding windows; non–weather stripped.

Part D. Classification of Residential-type Doors for Infiltration

Door Fit	Comments
Tight, $k = 1.0$	Very small perimeter gap and perfect fit weather stripping—often characteristic of new doors
Average, $k = 2.0$	Small perimeter gap having stop trim fitting properly around door; weather stripped
Loose, $k = 6.0$	Large perimeter gap having poor fitting stop trim; weather stripped or Small perimeter gap; no weather stripping

Source: Copyright © American Society of Heating, Refrigerating and Air-Conditioning Engineers, Inc., Atlanta, GA. Reprinted by permission from *Cooling and Heating Load Calculation Manual,* 1979.

THERMAL CONTROL

5

DESIGNING FOR HEATING AND COOLING

Chapters 1 to 4 provide some preparation for design: a perspective on national energy resources, an understanding of human comfort, an analysis of the climate resources available on site, a list of the general design strategies appropriate to various climates, a discussion of the basics of heat transfer calculation, and some attention to requirements for the components of the building envelope. Hence, there is much we now know about the variables outside a building, the desired conditions inside, and the individual components of a building's skin. In this chapter, all these variables are integrated into the process of designing for heating and cooling with the important related factor of daylighting. The design of mechanical support systems is introduced in Chapter 7.

The information in Chapter 5 is organized from the general to the particular. It also serves two distinct kinds of buildings—buildings using on-site energy resources and more conventional buildings using imported energy. The sections that pertain to one or the other building type—and to both—are outlined in Table 5.1.

5.1 Organizing the Problem

How should your building's skin respond to the sometimes conflicting needs for heating, cooling, and daylighting? Heating and cooling design strategies were related to climate in Sections 2.3 through 2.6, in which it became evident that internal heat gains can shift the appropriate design strategies for a building toward cooling, perhaps eliminating heating needs entirely. Typically, much of this internal heat is provided by electric lighting. Daylighting can replace electric lighting for most of the typical working day in most building types—*if* the building is designed to allow daylight to reach most of the interior.

(a) Fenestration. Codes and standards typically prescribe a relationship between floor area and fenestration area (residential buildings) or total wall area and fenestration area (nonresidential buildings). These prescriptions assume that a building will be designed "conventionally"; that is, to rely on imported energy for lighting, heating, and cooling. Thus, prescribed areas of fenestration tend to be rather small. If a designer wishes to rely on daylighting to a greater extent, then some proof of benefit will be needed.

Daylighting is accompanied by large glass areas, which increase a building's heating needs in winter; yet in such buildings less heat from electric lighting is available to fill those needs. Another complication is that adequate daylight requires much larger glass areas under dim winter skies than under bright summer skies. If the glass area is sized for winter daylighting, then excessive daylight—and along with it, excessive heat—might be admitted in summer. With proper controls, daylighting can reduce summer cooling loads, relative to electric lights, but it will usually increase winter heating loads. Where passive solar heating or surplus heat from another source is readily available, this trade-off is attractive.

One of the earliest and most difficult questions for the designer is how much fenestration is optimum for a building. In Chapter 1, Table 1.3 showed national energy end-use in buildings: space heating accounted for about 30%; space cooling and ventilation, about 11%; and electric lighting, about 14%. By building type and location, however, this proportion of energy use can change substantially. In Section 5.3, daylighting is considered first because it so largely determines whether

191

THERMAL CONTROL

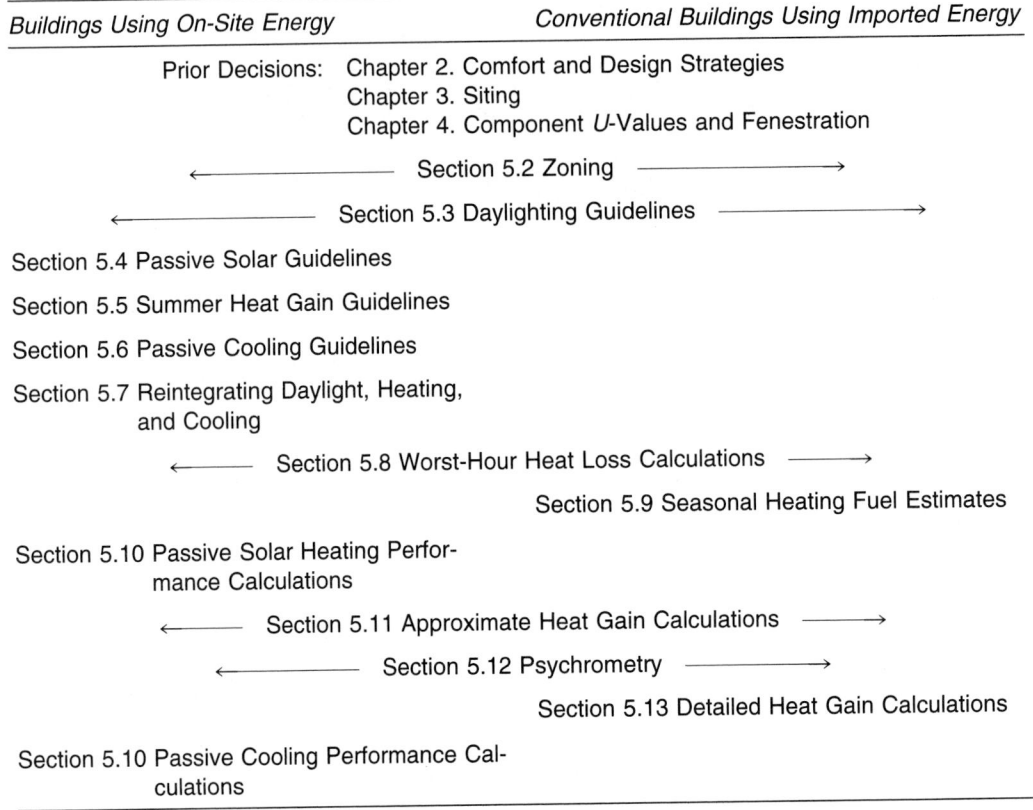

TABLE 5.1 **Proceeding Through Chapter 5**

Buildings Using On-Site Energy	*Conventional Buildings Using Imported Energy*

Prior Decisions: Chapter 2. Comfort and Design Strategies
Chapter 3. Siting
Chapter 4. Component *U*-Values and Fenestration

←——————————— Section 5.2 Zoning ————————→

←——————————— Section 5.3 Daylighting Guidelines ————————→

Section 5.4 Passive Solar Guidelines

Section 5.5 Summer Heat Gain Guidelines

Section 5.6 Passive Cooling Guidelines

Section 5.7 Reintegrating Daylight, Heating, and Cooling

←———— Section 5.8 Worst-Hour Heat Loss Calculations ————→

Section 5.9 Seasonal Heating Fuel Estimates

Section 5.10 Passive Solar Heating Performance Calculations

←———— Section 5.11 Approximate Heat Gain Calculations ————→

←———— Section 5.12 Psychrometry ————→

Section 5.13 Detailed Heat Gain Calculations

Section 5.10 Passive Cooling Performance Calculations

space heating or space cooling will be the dominant need within the building.

(b) Building Form. At its simplest, form can be reduced to questions of tall or short, thick or thin. Figures 5.1 and 5.2 compare these form variations to the relative importance of heating, cooling, and daylighting. It is apparent that thicker, taller buildings have more floor space away from climate influences; being electrically lit rather than daylit, they generate heat and need cooling all year. These buildings are called *internal load dominated* (ILD) (Fig. 5.1). In contrast, thinner buildings in which nearly all spaces have an exterior wall need heating in cold weather and cooling in hot weather; electric lights by day are largely unnecessary. These buildings are called *skin load dominated* (SLD) (Fig. 5.2).

The ultimate choice of building form is determined by a combination of design issues; the energy use issue can help in the selection process.

Once the building form has been chosen, the functions can be distributed according to typical architectural criteria, including the thermal zoning considerations mentioned in Section 5.2.

In the selection of a building form, some particularly important questions accompany each energy use. Because the question of daylighting versus electric lighting frequently is so influential in determining whether heating or cooling will be the dominant concern, we begin with daylighting.

Daylighting issues.

1. What will be the relative emphasis on side-lighting (characterized by uneven distribution and glare in the visual field but little glare on horizontal surfaces) and toplighting (the reverse characteristics)?

2. What role will direct sun play in daylighting? In winter, can solar heat without glare be admitted?

THERMAL CONTROL

Fig. 5.1 *Internal-load–dominated (ILD) buildings: the relative importance of cooling, daylighting, and heating varies with building form (and with climate), as do limited opportunities for the use of on-site or natural energy sources.*

Fig. 5.2 *Skin-load–dominated (SLD) buildings: the relative importance of cooling, daylighting, and heating varies with climate and building, as do plentiful opportunities for the use of on-site or natural energy sources.*

3. How can seasonal adjustments be made in the size of daylight openings?
4. To what extent will daily changes in daylighting control be necessary?
5. How can adequate daylight be admitted in an even way, such that unwelcome dark-appearing places are avoided?

Heating issues.

1. Can the sun be used to heat spaces? If so, how will south-wall design be affected?
2. How can openings in walls facing other directions be kept to a minimum without daylight being shut out? Where such openings are desirable, how low a fenestration U-value can be justified?
3. What role will direct sun through south glass or skylights play in daylighting?
4. How can daylight be admitted but the chilling effects of large, cold glass surfaces be minimized?
5. How can incoming fresh air be warmed be-

fore it chills the people sitting near the fresh air opening?
6. Is there surplus heat elsewhere in the building that can be used to help warm perimeter spaces?

Cooling issues.

1. Will the strategy be to open the building to breeze or close the building for "coolth" retention, or use a combination of these alternatives (open by night, closed by day)?
2. How can direct sun be kept out of the building? Can east and west windows be minimized and adequate daylight still be provided?
3. How can adequate daylight be admitted for winter conditions without overlighting (and thus overheating) for summer conditions?
4. When can cooling be provided by outdoor air, rather than by a refrigeration cycle?
5. Can the operation of refrigeration machinery be concentrated during the coldest

(nighttime) hours, when electric power is cheapest?

6. How can incoming fresh air be cooled before it warms the people sitting near the fresh air opening?

7. Can the structure of the building be used to absorb heat by day, then be flushed with night air in climates with cool nights?

(c) Building Envelope. The next design step involves relating the climate to the design of the building's skin. Each skin element provides an opportunity for thermal and luminous exchange between inside and out; heating, cooling, ventilating, and daylighting devices can be mixed as needed. Figure 5.3 shows some of the most common of these devices for varying orientations. Sections 5.3, 5.4, 5.5 and 5.6 give numerical criteria for sizing these skin elements with an emphasis on the use of on-site, renewable energy resources. These criteria may conflict with the size relationships that may be prescribed by codes and standards for conventional buildings.

An example of how lighting, heating, and cooling considerations can combine with other criteria (structural and acoustic, in this example) is shown in Fig. 5.4. These design diagrams explore opportunities in an office building in a cool-winter, dry-summer climate; the actual resulting building is shown in Fig. 5.7.

5.2 Zoning

Before calculating winter heat loss and summer heat gain, zoning should be considered. Dividing the calculations into zones facilitates later decisions about sizes of equipment as well as fenestration and areas of thermal mass. The thermal and luminous zoning of a building recognizes that different envelopes and support systems may be required around and within the building. The more carefully zoning is considered in these early design stages, the better will be the lighting and thermal performance and the lower will be the annual energy consumption. (Also, the less likely it will be that all sides of a building will have an identical appearance.)

Zoning is most often influenced by the following factors:

1. *Function.* Particularly important because of the variations in internal heat gains between functions, function may also influence the vertical organization of a building, as in Fig. 5.5 (and was shown in Fig. 3.4). Comfort conditions may vary considerably between functions; air temperatures can be lower for a strenuous activity than for a sedentary activity, or heat tolerance may be greater for some activities (restaurant kitchens) than for others. Some functions thrive in daylight, others shun it. Some functions adversely affect the indoor air quality of other functions.

2. *Schedule.* Closely related to function, scheduling can influence both the envelope and the support system. An activity scheduled only between 9 AM and 4 PM can often be entirely daylit at a time when the outside temperatures are the warmest of the daily cycle. By contrast, activity that takes place only from 9 PM to 4 AM will be entirely dependent on electric lighting, whose heat can be used to overcome the chill of the outside temperatures during these hours in winter. (In the summer, such heat can be flushed away with the cool outside night air in many U.S. locations.) Support systems are often divided by scheduling considerations: If one activity has operating hours different from those of the remainder of the building, a separate mechanical system is often provided. This saves energy, because large equipment will not be underused to provide heating or cooling for only one zone.

3. *Orientation.* The degree of exposure to daylight, direct sun, and wind is obviously important to zoning. Consider the block-square office building floors, Fig. 5.5, on a cold, sunny, and windy day. Perimeter spaces with direct sun through the windows may gain more heat than is lost, and thus need cooling. This might be done by the opening of windows, but too much cold air (especially on the windy side of a building) may make the workers near the windows uncomfortable. Perimeter spaces without direct sun may have a net heat loss due to heat loss through glass, infiltration, and a lack of electric lights (because daylight is adequate). These spaces will need heat from a

THERMAL CONTROL

Fig. 5.3 *The components of a building's envelope can be used both to conserve energy and to admit on-site or natural energy sources. Some of the many opportunities for heating, ventilating, cooling, and daylighting are shown here.*

mechanical support system. Interior (no-daylight) spaces are overheated by electric lights because they cannot lose heat. These spaces will need cooling from the support system.

Zoning considerations are among the most im-portant influences on building form and external design, along with the familiar esthetic, social, legal, economic, and technical influences that combine in a tug-of-war familiar to the building designer.

	North walls	East walls	West walls

Fig. 5.3 (continued)

5.3 Daylighting Guidelines

The information in this section can be used either before preliminary design, as guidelines to window sizing, or after preliminary design, as criteria for predicted daylighting performance. In the previous chapters, windows and skylights have been seen in a somewhat negative light, as threats to winter heat loss and summer heat gain. It is time to numerically consider their benefits.

When a building is designed to rely on daylighting, a prime design concern is the *daylight factor* (DF), which is expressed as a percentage of the outdoor light *under overcast skies* that is available indoors.

$$DF = \frac{\text{indoor illumination from daylight}}{\text{outdoor illumination}} \times 100\%$$

Chapter 19 presents detailed procedures for determining the DF for various points within a build-

Fig. 5.4 *Design diagrams explore opportunities in an office building in western Oregon (see Fig. 5.7 for the resulting building). (a) Choosing a one-way structure allows windows to be placed as high as possible for deeper daylight penetration. (b) Maximizing exposed interior surfaces that are thermally massive facilitates both winter solar heat storage and summer night cooling. Carpet is at the insistence of the client. (c) Shaping the window with more area above a lightshelf helps reduce contrast between light near the window and dark further in.*

(d) Indirect electric lighting

(e) Sound absorbtion

Fig. 5.4 *(continued)* (d) *Indirect electric lighting is facilitated when higher ceilings are available; less glare on computer screens results.* (e) *With so much exposed mass surface, acoustic absorption is a must. This solution continues to expose the concrete ceiling, and does not cast shadows from incoming daylight. (Drawings by Michael Cockram, © 1998 by John S. Reynolds, A.I.A, all rights reserved.)*

ing and for clear and partly cloudy sky conditions. For now, some simple target overcast sky DFs are presented in Table 5.2, while Table 5.3, parts A and B, give the simplest design guidelines that provide these target DFs. These design guidelines consider two factors: *How high* is the window in the wall, and *how large* is the window or skylight area, compared to the floor area for each daylit space?

For sidelighting (from windows), the first critical factor is to determine H, the height of the top of the window above the floor. The floor area served by this window will then extend to a maxi-mum distance (for DF calculation purposes) of $2.5H$ perpendicular to the window wall.

Remember that the DF is a ratio that is based on how much light is available outdoors; the resulting light indoors will vary with the light outdoors. The target daylight factors listed in Table 5.2 provide sufficient light during most of the daylight hours on overcast winter days. The relationship between office working hours, daylight, and outdoor temperature is explored in the *climatic timetables* of Fig. 5.6, for the building whose design diagrams appeared earlier (see Fig. 5.4). Obviously, much more light will be available on sum-

Offices–5 zones

Stores–5 zones

Fig. 5.5 *Zoning for a mixed-use building with perimeter and internal zones as well as varying types of use. Scheduling and/or internal load differences within any one of these zones could require division into additional zones.*

TABLE 5.2 **Recommended Daylight Factors**[a]

Task	DF
Ordinary seeing tasks, such as reading, filing, and easy office work	1.5–2.5%
Moderately difficult tasks, such as prolonged reading, stenographic work, normal machine tool work	2.5–4.0%
Difficult, prolonged tasks, such as drafting, proofreading poor copy, fine machine work, and fine inspection	4.0–8.0%

Source: Millet and Bedrick (1980).

[a]Use the smaller DF values for southern latitudes with plentiful winter daylight.

mer days—probably more light than is needed, bringing heat along with it. When sizing windows and skylights, remember that controlling direct sun is necessary and that less opening area is needed in summer than in winter. For one example of such controls, we now look at the office building of Figs. 5.4 and 5.6.

EXAMPLE 5.1. Many of the design guidelines in this chapter will be illustrated by applying them to a typical bay of the building in Fig. 5.7. (Example 5.1 will continue to appear in parts throughout this chapter.) This 24,000-ft^2 (2,230 m^2) office building for an electric utility is located in Eugene, in western Oregon's wet-winter, dry-summer climate typical of the U.S. Pacific northwest. Energy conservation and daylighting considerations played a major role in this building's design and were integrated with solar heating, night ventilation of mass, structure, and acoustics (Fig. 5.4). Table 5.4 presents the types of design statis-

tics that will be needed for any building to which this chapter's design guidelines will be applied.

The various applications of design guidelines to this building can be found as follows:

Example 5.1, Part B. Overall rate of Btu/DD ft^2 heat loss, page 210.

Example 5.1, Part C. Approximate SSF (solar savings fraction), page 219.

Example 5.1, Part D. Approximate heat gain, page 225.

Example 5.1, Part E. Cross ventilation guidelines, page 232.

Example 5.1, Part F. Night ventilation of thermal mass guidelines, page 237.

Example 5.1, Part G. January balance point temperature, page 253.

Example 5.1, Part H. Annual SSF based upon LCR, page 267.

Example 5.1, Part I. Clear January day indoor temperature swing, page 275.

Example 5.1, Part J. Detailed night-cooling of mass calculation, page 306.

EXAMPLE 5.1, PART A. Daylighting design for this office building began with design diagrams (Fig. 5.4) and an assessment of daylighting potential (Fig. 5.6). After establishing that daylighting was available during almost all normal working hours, the building plan was then orga-

TABLE 5.3 Daylight Factor (DF) Design Guidelines for Overcast Sky Conditions

Part A. Sidelighting[a,b]

$$DF_{av} = 0.2 \left(\frac{\text{window area}}{\text{floor area}} \right)$$

$$DF_{min} = 0.1 \left(\frac{\text{window area}}{\text{floor area}} \right)$$

Part B. Toplighting[c]

Vertical monitors:

$$DF_{av} = 0.2 \left(\frac{\text{skylight glazing area}}{\text{floor area}} \right)$$

North-facing sawtooth:

$$DF_{av} = 0.33 \left(\frac{\text{skylight glazing area}}{\text{floor area}} \right)$$

Horizontal skylights:

$$DF_{av} = 0.5 \left(\frac{\text{skylight glazing area}}{\text{floor area}} \right)$$

Part C. Buildings with Atrium[d]

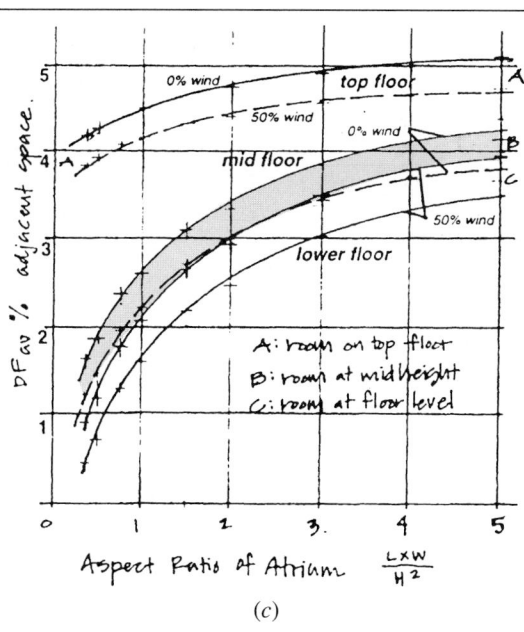

(c)

Source: Parts A and B: Millet and Bedrick (1980). Part C: Brown and DeKay (2000).

[a]Assumes windows in one wall of a room with relatively light-colored surfaces.

[b]Window height/room depth relationships based on the works of R. G. Hopkinson and others at the British Research Station.

[c]Assumes an even distribution of such skylights in the roof so that an even distribution of light results in the room below: thus, only average DF, no minimum, is listed.

[d]Based on model tests of a square atrium with white walls, open to the sky. "No windows" average atrium wall reflectance = 70%. "50% windows" average atrium wall reflectance = 40%. DF are for an office of 9 × 9 × 3 m (30 × 30 × 10 ft). The window opening to atrium is 1.5 m high × 9 m long (5 ft high × 30 ft long) sill height 0.85 m (2.8 ft).

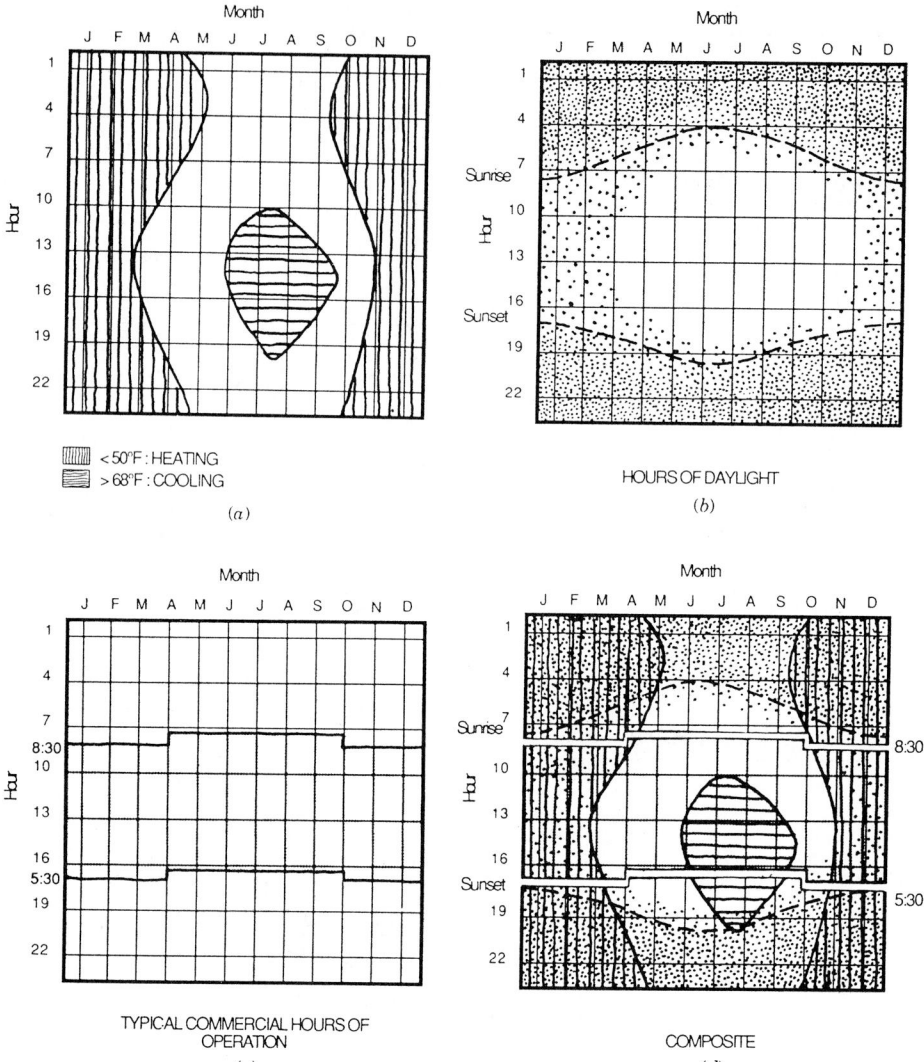

Fig. 5.6 *A series of climate "timetables" (see Fig. 2.12) were prepared while designing the office building for the Emerald People's Utility District near Eugene, Oregon (Fig. 5.7). (a) Average outdoor temperatures show little daily variation in winter, large daily range in summer. (b) Daylight peaks around noon near the summer solstice. (c) Work schedule shows the impact of summer's daylight savings time. (d) A composite timetable indicates that daylight is available during almost all working hours, and that half the hottest hours occur after the building closes on summer afternoons—thanks to daylight savings time. Cool nighttime temperatures make night ventilation of thermal mass an attractive cooling option. (Courtesy of Equinox Design, Inc., and WEGroup, PC, Architects, Eugene, Oregon.)*

nized so that almost all windows faced either south or north (Fig. 5.7b,c) to avoid the problems of low-altitude sun (year-round glare and summer heat gain) that accompany east- and west-facing windows. Then the building section was designed so that the height of the windows (H) was related to the depth of the floor plan served by those windows ($2.5H$), shown in Fig. 5.7d.

After choosing a generous target $DF_{av} = 4.0\%$ from Table 5.2, the next step was to size the windows and clerestories (as skylights), using the DF guidelines from Table 5.3. Applicable formulas

THERMAL CONTROL

for both sidelighting and for vertical monitor skylights:

$$DF_{av} = 0.2 \frac{\text{window (or skylight) area}}{\text{floor area}}$$

Applied to the entire typical bay,

$$DF_{av} = 0.2 \frac{\begin{array}{c}97 \text{ ft}^2 \text{ north} + 97 \text{ ft}^2 \text{ south} \\ + 132 \text{ ft}^2 \text{ clerestory}\end{array}}{1440 \text{ ft}^2}$$

$$= 0.045, \text{ or } 4.5\%$$

slightly above the target DF_{av} of 4.0%.

Note that in this example, DF_{min} from the sidelighting occurs near the center of the building, about the point (on the second floor) where the most light is available from the skylight. Therefore, relatively even daylighting distribution is expected on the upper floor. This is helped by the use of lightshelves and the T-shaped windows, shown in Fig. 5.7g and 5.7h.

The seasonal window size question (need more in winter, less in summer) was answered in this example by the use of deciduous vines outside the south windows. In winter and cool spring, the vines are bare of leaves. Warm weather brings

(a) Site plan

Fig. 5.7 *The office building for the Emerald People's Utility District near Eugene, Oregon, is elongated east–west for maximum window areas facing north–south (a). The ground floor plan (b), second floor plan (c), and diagrammatic daylight section (d) show how almost all workstations are daylit. The section (e) shows relation between windows, trellis for shading, and exposed concrete construction. The cutaway axonometric (f) of a typical open-office two-story bay shows air flush through hollow cores of exposed precast concrete slabs, windows and shading, and suspended sound-absorbing baffles. Lightshelves (g) reduce the contrast between lots of daylight near windows and too little daylight farther in. The resulting windows are T-shaped (h) with more glass above the lighshelves, less glass below. The deciduous vines on trellises at south facade change the building's appearance by season; (i) at summer solstice, the shading stripes will soften as vines leaf out with age; (j) at fall equinox, the leaves will still shade, whereas at spring equinox the branches will be bare to allow sun to reach the windows; (k) at winter solstice most of the window is exposed to the sun. (Drawings c and g courtesy of Virginia Cartwright.) (Courtesy of Equinox Design, Inc., and WEGroup, PC, Architects, Eugene, Oregon.)*

(b) Ground floor plan

(c) Second floor plan
Fig. 5.7 (continued)

THERMAL CONTROL

(d) Daylighting cross section (south to right)

(e) Section through two-story wing

Fig. 5.7 *(continued)*

leafy shade, lasting well into the warm fall (Fig. 5.7 *i–k*). ■

Table 5.3, Part C, has design guidelines for buildings with an atrium to provide daylighting to surrounding offices. Because the lowest DF will result in the offices on the lowest floor (deepest within the atrium), the designer might find the required atrium "aspect ratio" first for the lowest floor, and then size the atrium on that basis. (The atrium aspect ratio = length × width/height²).

This procedure would, of course, provide higher DF in all the offices higher in the atrium. For a detailed discussion of the relationship atrium size, rentable office floor area, and latitude, see De-Kay (1992).

5.4 Passive Solar Heating Guidelines

When the building function, climate, and site are favorable, winter solar gains can contribute

THERMAL CONTROL

Suspended acoustic panels

Precast double beam

Trellis with deciduous vines above south clerestory

Air enters cores

Precast hollow core slab - exposed to below

Core flush intake air

Trellis with deciduous vines

VAV Supply duct

Core flush return duct

Carpeted floors

Concrete block fin walls

Light shelf with indirect lights

Conventional return air duct

Upper windows for daylight

Lower windows for view

(*f*) Axonometric of a typical two-story bay

Fig. 5.7 (continued)

(*g*) Lightshelves

(*h*) North Facade
Fig. 5.7 *(continued)*

greatly to a properly designed building. One of the first criteria to be considered is that of energy conservation; there is little benefit of pouring solar heat into a leaky building. For solar designers, a motto should be "insulate before you insolate."

(a) Whole-Building Heat Loss Criteria. The recommended maximum rates of heat loss [in Btu per degree days (DD) per square foot] shown in Table 5.5 are the basis for development of the following passive solar heating guidelines. They are also useful as a quick check on overall envelope performance in residential or small commercial buildings. (Some states have adopted similar criteria as part of their building codes; be sure to check the applicable code before relying on these numbers for code compliance.) The heat loss rates are shown for two conditions, as follows.

1. *Conventional   (Non-Passively   Solar–Heated) Small Buildings.* The overall rate of Btu/DD-ft^2 is based on *total* heat loss, including all portions of the envelope *and* infiltration. To determine the whole-building heat loss rate, list for each envelope component (roof, walls, floor,

(*i*) South Facade at Summer Solstice

(*j*) South Facade at the Spring Equinox

(*k*) South Facade Near the Winter Solstice
Fig. 5.7 (*continued*)

TABLE 5.4 **Design Data for a Typical Second-Floor Bay of the Office Building (Fig. 5.7*f*)**

Component of Envelope	U-Value (Btu/h-ft²-°F)	Area (ft²)	U × A (Btu/h-°F)
Floor Area *(basis for guidelines)*		1440	
Exposed to outside:			
Roof[a]	0.028	1512	42
Exterior opaque walls[b] (north and south)	0.084	296	25
North window[c]	0.74	97	72
Total for passive solar guidelines[d] UA_{ns}			139
South window[c]	0.74	97	
South clerestory[c]	0.74	132	
Total for passive solar guidelines A_p		229	
Interior thermal mass:			
Roof underside		1512	
Bearing walls, beams and columns		720	
Total for passive heating and cooling		2232	

Ventilation
Volume of enclosed space, 22,900 ft³
Occupancy is 8 persons
Ventilation is provided at the rate of 20 cfm per person
Openable window area: 16 ft² on north, 16 ft² on south

[a]*R*-40 over precast slab. Assume corrected *R*-35, therefore *U* = 0.28.
[b]4-in. concrete block outside, 1-in. airspace, *R*-19 batt with metal studs at 24 in. and gypsum inside; approximate corrected *R*-11.9, therefore *U* = 0.084.
[c]Clear double glazed, ⅜-in. airspace. nonthermal break aluminum, overall *U* = 0.74.
[d]Passive solar heating calculations do not directly include losses from south-facing glass. Also, there is no floor heat loss or gain, because this is a second-floor space.

TABLE 5.5 **Overall Heat Loss Criteria for Solar Guidelines**

		Maximum Overall Heat Loss			
		Btu/DDF ft²		W/DDK m²	
Annual Heating Degree Days (base 65°F)	(base 18°C)	Conventional Buildings	Passively Solar Heated Buildings, Excluding Solar Wall[a]	Conventional Buildings	Passively Solar Heated Buildings, Excluding Solar Wall[a]
Less than 1000	Less than 556	9	7.6	51	43
1000–3000	556–1667	8	6.6	45	37
3000–5000	1667–2778	7	5.6	40	32
5000–7000	2778–3889	6	4.6	34	26
Over 7000	Over 3889	5	3.6	28	20

Source: Balcomb et al. (1980). SI conversions approximated by the author.
[a]The guidelines in Table 5.6 assume a solar building that meets this criterion.

THERMAL CONTROL

windows, etc.) the *U*-value (Chapter 4) and the total exposed area *A,* then simply multiply $U \times A$.
(For slab floors on grade, see Chapter 4 for determining perimeter heat losses.) For the special
cases of walls below grade, such as berm walls, an
approximation is needed. During the coldest
weather, the temperature outside such walls will
nearly always be higher than the outdoor air temperature, so any procedure based on Btu/DD will
overpredict the heat loss through these walls. As a
result, designers often calculate the *UA* of below-
grade walls by using their actual *U*-value, but using
only *half* of their actual area; this lesser *UA* roughly
compensates for the lesser ΔT through these walls.

For infiltration, determine the number of air
changes per hour (ACH) under winter design conditions (Section 4.9) and multiply this infiltration
(or fresh air) rate by a constant that accounts for
density and specific heat:

$$or \quad \left.\begin{array}{l} \text{ACH (volume, ft}^3) \times 0.018 = \\ \text{ACH (volume, m}^3) \times 0.33 = \end{array}\right\} UA \text{ for infiltration}$$

Add the envelope *UA* values to those for infiltration,
multiply by 24 h/day to account for degree days
(DD), and divide by the building's total heated
floor area:

$$\frac{(UA_{\text{envelope}} + UA_{\text{infiltration}}) \times 24\text{ h}}{\text{total heated floor area (ft}^2)} = \text{Btu/DD-ft}^2$$

2. *Passively Solar–Heated Buildings.* Here the
overall rate of Btu/DD-ft² *excludes* the solar collecting portion(s) of the envelope; otherwise, it is also
based on total heat loss from all other portions of
the envelope, and it includes infiltration. The equation used to determine the overall rate is as follows:

$$\frac{(UA_{\text{envelope, except south glass}} + UA_{\text{infiltration}}) \times 24\text{ h}}{\text{total heated floor area (ft}^2)} = \text{Btu/DD-ft}^2$$

One of the biggest unknowns in this procedure is
the assumed rate of infiltration. A carefully designed
and constructed small building can easily achieve a
rate of 0.75 ACH; with increased attention to infiltration (vapor) barrier installation, caulking of all
cracks, and so on, rates below 0.33 ACH have
been demonstrated.

EXAMPLE 5.1, PART B. Find the overall
rate of heat loss, Btu/DD-ft², for the Oregon office
building presented in Fig. 5.7.

SOLUTION. This office building is passively
solar heated, so the overall rate of heat loss, Btu/
DD-ft², is determined by using the total $U \times A$
of the non-south–glass envelope. From Table 5.4,
this is $42 + 25 + 72 = 139$ Btu/h-°F.

To this must be added the effects of ventilation.
From the same table, outdoor air is shown to be
supplied at the rate of

$$20\text{ cfm/person} \times 8\text{ persons} = 160\text{ cfm}$$

which can also be expressed as 160 cfm × 60 min/
h = 9600 cfh. Comparing this hourly rate to the
volume of the typical bay, we have

$$\frac{9600\text{ ft}^3/\text{h}}{22,000\text{ ft}^3} = 0.44\text{ ACH}$$

The $U \times A$ for ventilation is therefore

$$0.44\text{ ACH} \times 22,000\text{ ft}^3 \times 0.018 = 174\text{ Btu/h-}°\text{F}$$

and the total rate of heat loss is

$$\frac{(139\text{ Btu/h-}°\text{F} + 174\text{ Btu/h-}°\text{F}) \times 24\text{ h/day}}{1440\text{ ft}^2} = 5.2\text{ Btu/DD-ft}^2$$

This building is located in Eugene, Oregon, which
corresponds most closely to the location of Salem,
Oregon, as listed in Appendix B, Table B.15,
where we find DD65 in Salem = 4852 per year.
From Table 5.5, the maximum recommended rate
of heat loss for a passively solar heated building
in a 3000- to 5000-DD climate is 5.6 Btu/DD-ft².
This building's heat loss rate is under this maximum, thus enhancing energy conservation. ■

Passive solar heating and energy conservation
have a complex relationship. Relative to "conventional" buildings, passively solar–heated buildings usually conserve purchased energy; yet,
buildings that aim at very high percentages of
solar heating can use more *total* heating energy
than is used by buildings with smaller window
areas. Designers interested primarily in saving
purchased energy may aim at lower solar percentages and more insulation; designers interested in
buildings that closely relate to climate and climatic
changes may aim at higher solar percentages (and
more daylighting), along with higher thermal
masses and, probably, greater ranges of indoor
temperature.

(b) Solar Savings Fraction (SSF). The term *solar savings fraction* is used to evaluate a building's solar heating performance. The solar savings fraction is the extent to which a solar design *reduces a building's auxiliary heat requirement* relative to a "reference" building—one that has, instead of a solar wall, an energy-neutral wall that experiences neither solar gain nor heat loss; otherwise, the solar building and the reference building are identical. The solar savings fraction compares the auxiliary energy needed by the solar building to the auxiliary energy needed by the reference building, as illustrated in Fig. 5.8. Remember that the SSF is *not* the percentage of the solar building's heat supplied by the sun; typically, the sun provides a much *higher* fraction of a building's total space heat than does the SSF. Rather, the SSF is more a measure of the solar building's *conservation* advantage.

A starting point for passive solar preliminary design is Table 5.6. For your location, both a range of SSF values and a range of ratios of areas of south glass/floor area can be determined. The table shows the SSF ranges for both "standard performance" (simple double-glazed windows) and "superior performance" (for either night-insulated or "superwindow") solar openings. This same information is shown graphically for six geographically diverse locations in Fig. 5.9. (You are encouraged to add your own location's information from Table 5.6 to a copy of this graph.)

The difference between "standard" and "superior performance" is explored in Fig. 5.10 for the September–June period in Boston, Madison, and Seattle. The solid line charts the lower performance of simple double glazing (listed as $U = 0.55$ Btu/h ft^2 °F). The other two higher-performance alternatives (movable insulation and superwindows) show why the newer superwindows have made movable insulation nearly obsolete: Without the need for any actions by the user, superwindows have a higher net heat gain than movable insulation in all three locations.

(c) Thermal Mass. Another early design question involves the amount of thermal mass area necessary to store the solar heat admitted each day. This reveals the simplistic, early-design nature of Table 5.6, because it does not distinguish between the various approaches to passive solar heating (*direct gain, indirect gain,* and *isolated gain,* from Chapter 2). Table 5.7 details the simple relationship between SSF and the area (also the weight) of water or masonry that should be provided in direct-gain designs.

The *distribution* of the thermal mass is also important, however. Indirect gain systems (Trombe wall and water wall systems) usually place the thermal mass in full sun for the entire day, often just inside the glazing. In direct-gain systems, this thermal mass should be within (or should enclose) the direct-gain–heated space, and

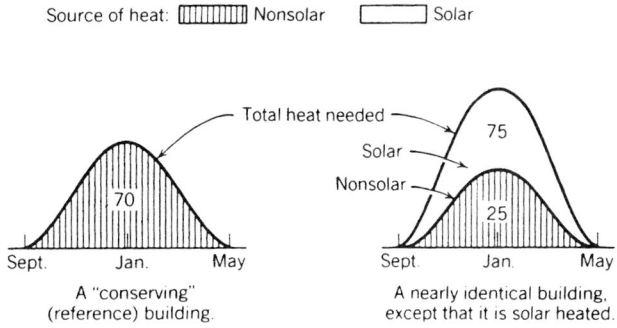

Fig. 5.8 *The solar savings fraction (SSF) compares the auxiliary heat needed by solar-heated buildings to that needed by a nonsolar but energy-conserving building that is otherwise similar, called the "reference" building. In this example, the solar building needs 25 units of auxiliary heat per year, whereas the reference building needs 70 units. The difference is 70 − 25 = 45 units, or 64% of the reference 70 units. Therefore, SSF = 64%. (Note, however, that the solar building is 75% solar heated.) (From Brown, Reynolds and Ubbelodhe,* InsideOut, Design Procedures for Passive Environmental Technologies, *© 1982, John Wiley & Sons, Inc. Reprinted by permission.)*

TABLE 5.6 **Design Guidelines for Passive Solar Glazing Area**

| | Area of Solar Glazing[a] as Ratio of Floor Area | | Approximate SSF Values | | | |
| | | | Standard Performance[b] | | Superior Performance[c] | |
Location	Low	High	Low	High	Low	High
Birmingham, Alabama	0.09	0.18	22	37	34	58
Mobile, Alabama	0.06	0.12	26	44	34	60
Montgomery, Alabama	0.07	0.15	24	41	34	59
Phoenix, Arizona	0.06	0.12	37	60	48	75
Prescott, Arizona	0.10	0.20	29	48	44	72
Tucson, Arizona	0.06	0.12	35	57	45	73
Winslow, Arizona	0.12	0.24	30	47	48	74
Yuma, Arizona	0.04	0.09	43	66	51	78
Fort Smith, Arkansas	0.10	0.20	24	39	38	64
Little Rock, Arkansas	0.10	0.19	23	38	37	62
Bakersfield, California	0.08	0.15	31	50	42	67
Daggett, California	0.07	0.15	35	56	46	73
Fresno, California	0.09	0.17	29	46	41	65
Long Beach, California	0.05	0.10	35	58	44	72
Los Angeles, California	0.05	0.09	36	58	44	72
Mount Shasta, California	0.11	0.21	24	38	42	67
Needles, California	0.06	0.12	39	61	49	76
Oakland, California	0.07	0.15	35	55	46	72
Red Bluff, California	0.09	0.18	29	46	41	65
Sacramento, California	0.09	0.18	29	47	41	66
San Diego, California	0.04	0.09	37	61	46	74
San Francisco, California	0.06	0.13	34	54	45	71
Santa Maria, California	0.05	0.11	31	53	42	69
Colorado Springs, Colorado	0.12	0.24	27	42	47	74
Denver, Colorado	0.12	0.23	27	43	47	74
Eagle, Colorado	0.14	0.29	25	35	53	77
Grand Junction, Colorado	0.13	0.27	29	43	50	76
Pueblo, Colorado	0.11	0.23	29	45	48	75
Hartford, Connecticut	0.17	0.35	14	19	40	64
Wilmington, Delaware	0.15	0.29	19	30	39	63
Washington, DC	0.12	0.23	18	28	37	61
Apalachicola, Florida	0.05	0.10	28	47	36	61
Daytona Beach, Florida	0.04	0.08	30	51	36	63
Jacksonville, Florida	0.05	0.09	27	47	35	62
Miami, Florida	0.01	0.02	27	48	31	54
Orlando, Florida	0.03	0.06	30	52	37	63
Tallahassee, Florida	0.05	0.11	26	45	35	60
Tampa, Florida	0.03	0.06	30	52	36	63
West Palm Beach, Florida	0.01	0.03	30	51	34	59
Atlanta, Georgia	0.08	0.17	22	36	34	58
Augusta, Georgia	0.08	0.16	24	40	35	60
Macon, Georgia	0.07	0.15	25	41	35	59
Savannah, Georgia	0.06	0.13	25	43	35	60
Boise, Idaho	0.14	0.28	27	38	48	71
Lewiston, Idaho	0.15	0.29	22	29	44	65
Pocatello, Idaho	0.13	0.26	25	35	51	74

TABLE 5.6 **Design Guidelines for Passive Solar Glazing Area** (*continued*)

Location	Area of Solar Glazing[a] as Ratio of Floor Area		Approximate SSF Values			
			Standard Performance[b]		Superior Performance[c]	
	Low	High	Low	High	Low	High
Chicago, Illinois	0.17	0.35	17	23	43	67
Moline, Illinois	0.20	0.39	17	22	46	70
Springfield, Illinois	0.15	0.30	19	28	42	67
Evansville, Indiana	0.14	0.27	19	29	37	61
Fort Wayne, Indiana	0.16	0.33	13	17	37	60
Indianapolis, Indiana	0.14	0.28	15	21	37	60
South Bend, Indiana	0.18	0.35	12	15	39	61
Burlington, Iowa	0.18	0.36	20	27	47	71
Des Moines, Iowa	0.21	0.43	19	25	50	75
Mason City, Iowa	0.22	0.44	18	19	56	79
Sioux City, Iowa	0.23	0.46	20	24	53	76
Dodge City, Kansas	0.12	0.23	27	42	46	73
Goodland, Kansas	0.13	0.27	26	39	47	74
Topeka, Kansas	0.14	0.28	24	35	45	71
Wichita, Kansas	0.14	0.28	26	41	45	72
Lexington, Kentucky	0.13	0.27	17	26	35	58
Louisville, Kentucky	0.13	0.27	18	27	35	59
Baton Rouge, Louisiana	0.06	0.12	26	43	34	59
Lake Charles, Louisiana	0.06	0.11	24	41	32	57
New Orleans, Louisiana	0.05	0.11	27	46	35	61
Shreveport, Louisiana	0.08	0.15	26	43	36	61
Caribou, Maine	0.25	0.50	— NR[c] —		53	74
Portland, Maine	0.17	0.34	14	17	45	69
Baltimore, Maryland	0.14	0.27	19	30	38	62
Boston, Massachusetts	0.15	0.29	17	25	40	64
Alpena, Michigan	0.21	0.42	— NR —		47	69
Detroit, Michigan	0.17	0.34	13	17	39	61
Flint, Michigan	0.15	0.31	11	12	40	62
Grand Rapids, Michigan	0.19	0.38	12	13	39	61
Sault Ste. Marie, Michigan	0.25	0.50	— NR —		50	70
Traverse City, Michigan	0.18	0.36	— NR —		42	62
Duluth, Minnesota	0.25	0.50	— NR —		50	70
International Falls, Minnesota	0.25	0.50	— NR —		47	66
Minneapolis–St. Paul, Minnesota	0.25	0.50	— NR —		55	76
Rochester, Minnesota	0.24	0.49	— NR —		54	76
Jackson, Mississippi	0.08	0.15	24	40	34	59
Meridian, Mississippi	0.08	0.15	23	39	34	58
Columbia, Missouri	0.13	0.26	20	30	41	66
Kansas City, Missouri	0.14	0.29	22	32	44	70
Saint Louis, Missouri	0.15	0.29	21	33	41	65
Springfield, Missouri	0.13	0.26	22	34	40	65
Billings, Montana	0.16	0.32	24	31	53	76
Cut Bank, Montana	0.24	0.49	22	23	62	81
Dillon, Montana	0.16	0.32	24	32	54	77
Glasgow, Montana	0.25	0.50	— NR —		55	75
Great Falls, Montana	0.18	0.37	23	28	56	77

(*continued*)

THERMAL CONTROL

TABLE 5.6 **Design Guidelines for Passive Solar Glazing Area (*continued*)**

Location	Area of Solar Glazing[a] as Ratio of Floor Area		Approximate SSF Values			
			Standard Performance[b]		Superior Performance[c]	
	Low	High	Low	High	Low	High
Helena, Montana	0.20	0.39	21	25	55	77
Lewistown, Montana	0.19	0.38	21	25	54	76
Miles City, Montana	0.23	0.47	21	23	60	80
Missoula, Montana	0.18	0.36	15	16	47	68
Grand Island, Nebraska	0.18	0.36	24	33	51	76
North Omaha, Nebraska	0.20	0.40	21	29	51	76
North Platte, Nebraska	0.17	0.34	25	36	50	76
Scotts Bluff, Nebraska	0.16	0.31	24	36	49	74
Elko, Nevada	0.12	0.25	27	39	52	76
Ely, Nevada	0.12	0.23	27	41	50	77
Las Vegas, Nevada	0.09	0.18	35	56	48	75
Lovelock, Nevada	0.13	0.25	32	48	53	78
Reno, Nevada	0.11	0.22	31	48	49	76
Tonopah, Nevada	0.11	0.23	31	48	51	77
Winnemucca, Nevada	0.13	0.26	28	42	49	75
Concord, New Hampshire	0.17	0.34	13	15	45	68
Newark, New Jersey	0.13	0.25	19	29	39	64
Albuquerque, New Mexico	0.11	0.22	29	47	46	73
Clayton, New Mexico	0.10	0.20	28	45	45	73
Farmington, New Mexico	0.12	0.24	29	45	49	76
Los Alamos, New Mexico	0.11	0.22	25	40	44	72
Roswell, New Mexico	0.10	0.19	30	49	45	73
Truth or Consequences, New Mexico	0.09	0.17	32	51	46	73
Tucumcari, New Mexico	0.10	0.20	30	48	45	73
Zuñi, New Mexico	0.11	0.21	27	43	45	73
Albany, New York	0.21	0.41	13	15	43	66
Binghamton, New York	0.15	0.30	— NR —		35	56
Buffalo, New York	0.19	0.37	— NR —		36	57
Massena, New York	0.25	0.50	— NR —		50	71
New York (Central Park), New York	0.15	0.30	16	25	36	59
Rochester, New York	0.18	0.37	— NR —		37	58
Syracuse, New York	0.19	0.38	— NR —		37	59
Asheville, North Carolina	0.10	0.20	21	35	36	61
Cape Hatteras, North Carolina	0.09	0.17	24	40	36	60
Charlotte, North Carolina	0.08	0.17	23	38	36	60
Greensboro, North Carolina	0.10	0.20	23	37	37	63
Raleigh–Durham, North Carolina	0.09	0.19	22	37	36	61
Bismarck, North Dakota	0.25	0.50	— NR —		56	77
Fargo, North Dakota	0.25	0.50	— NR —		51	72
Minot, North Dakota	0.25	0.50	— NR —		52	72
Akron–Canton, Ohio	0.15	0.31	12	16	35	57
Cincinnati, Ohio	0.12	0.24	15	23	35	57
Cleveland, Ohio	0.15	0.31	11	14	34	55
Columbus, Ohio	0.14	0.28	13	18	35	57
Dayton, Ohio	0.14	0.28	14	20	36	59
Toledo, Ohio	0.17	0.34	13	17	38	61

TABLE 5.6 **Design Guidelines for Passive Solar Glazing Area** (*continued*)

Location	Area of Solar Glazing[a] as Ratio of Floor Area		Approximate SSF Values			
			Standard Performance[b]		Superior Performance[c]	
	Low	High	Low	High	Low	High
Youngstown, Ohio	0.16	0.32	— NR —		34	54
Oklahoma City, Oklahoma	0.11	0.22	25	41	41	67
Tulsa, Oklahoma	0.11	0.22	24	38	40	65
Astoria, Oregon	0.09	0.19	21	34	37	60
Burns, Oregon	0.13	0.25	23	32	47	71
Medford, Oregon	0.12	0.24	21	32	38	60
North Bend, Oregon	0.09	0.17	25	42	38	64
Pendleton, Oregon	0.14	0.27	22	30	43	64
Portland, Oregon	0.13	0.26	21	31	38	60
Redmond, Oregon	0.13	0.27	26	38	47	71
Salem, Oregon	0.12	0.24	21	32	37	59
Allentown, Pennsylvania	0.15	0.29	16	24	39	63
Erie, Pennsylvania	0.17	0.34	— NR —		35	55
Harrisburg, Pennsylvania	0.13	0.26	17	26	38	62
Philadelphia, Pennsylvania	0.15	0.29	19	29	38	62
Pittsburgh, Pennsylvania	0.14	0.28	12	16	33	55
Wilkes Barre–Scranton, Pennsylvania	0.16	0.32	13	18	37	60
Providence, Rhode Island	0.15	0.30	17	24	40	64
Charleston, South Carolina	0.07	0.14	25	41	34	59
Columbia, South Carolina	0.08	0.17	25	41	36	61
Greenville–Spartanburg, South Carolina	0.08	0.17	23	38	36	60
Huron, South Dakota	0.25	0.50	— NR —		58	79
Pierre, South Dakota	0.22	0.43	21	23	58	80
Rapid City, South Dakota	0.15	0.30	23	32	51	76
Sioux Falls, South Dakota	0.22	0.45	18	19	57	79
Chattanooga, Tennessee	0.09	0.19	19	32	33	56
Knoxville, Tennessee	0.09	0.18	20	33	33	56
Memphis, Tennessee	0.09	0.19	22	36	36	60
Nashville, Tennessee	0.10	0.21	19	30	33	55
Abilene, Texas	0.09	0.18	29	47	41	68
Amarillo, Texas	0.11	0.22	29	46	45	72
Austin, Texas	0.06	0.13	27	46	37	63
Brownsville, Texas	0.03	0.06	27	46	32	57
Corpus Christi, Texas	0.05	0.09	29	49	36	63
Dallas, Texas	0.08	0.17	27	44	38	64
Del Rio, Texas	0.06	0.12	30	50	39	66
El Paso, Texas	0.09	0.17	32	53	45	72
Fort Worth, Texas	0.09	0.17	26	44	38	64
Houston, Texas	0.06	0.11	25	43	34	59
Laredo, Texas	0.05	0.09	31	52	39	64
Lubbock, Texas	0.09	0.19	30	49	44	72
Lufkin, Texas	0.07	0.14	26	43	35	61
Midland–Odessa, Texas	0.09	0.18	32	52	44	72
Port Arthur, Texas	0.06	0.11	26	44	34	60
San Angelo, Texas	0.08	0.15	29	48	40	67
San Antonio, Texas	0.06	0.12	28	48	38	64

(*continued*)

THERMAL CONTROL

THERMAL CONTROL

TABLE 5.6 **Design Guidelines for Passive Solar Glazing Area** (*continued*)

Location	Area of Solar Glazing[a] as Ratio of Floor Area		Approximate SSF Values			
			Standard Performance[b]		Superior Performance[c]	
	Low	High	Low	High	Low	High
Sherman, Texas	0.10	0.20	25	41	38	64
Waco, Texas	0.08	0.15	27	45	38	64
Wichita Falls, Texas	0.10	0.20	27	45	41	67
Bryce Canyon, Utah	0.13	0.25	26	39	52	78
Cedar City, Utah	0.12	0.24	28	43	48	75
Salt Lake City, Utah	0.13	0.26	27	39	48	72
Burlington, Vermont	0.22	0.43	— NR —		46	68
Norfolk, Virginia	0.09	0.19	23	38	37	62
Richmond, Virginia	0.11	0.22	21	34	37	61
Roanoke, Virginia	0.11	0.23	21	34	37	61
Olympia, Washington	0.12	0.23	20	29	38	59
Seattle–Tacoma, Washington	0.11	0.22	21	30	39	59
Spokane, Washington	0.20	0.39	20	24	48	68
Yakima, Washington	0.18	0.36	24	31	49	70
Charleston, West Virginia	0.13	0.25	16	24	32	54
Huntington, West Virginia	0.13	0.25	17	27	34	57
Eau Claire, Wisconsin	0.25	0.50	— NR —		53	75
Green Bay, Wisconsin	0.23	0.46	— NR —		53	75
La Crosse, Wisconsin	0.21	0.43	— NR —		52	75
Madison, Wisconsin	0.20	0.40	15	17	51	74
Milwaukee, Wisconsin	0.18	0.35	15	18	48	71
Casper, Wyoming	0.13	0.26	27	39	53	78
Cheyenne, Wyoming	0.11	0.21	25	39	47	74
Rock Springs, Wyoming	0.14	0.28	26	38	54	79
Sheridan, Wyoming	0.16	0.31	22	30	52	75
Canada						
Edmonton, Alberta	0.25	0.50	— NR —		54	72
Suffield, Alberta	0.25	0.50	28	30	67	85
Nanaimo, British Columbia	0.13	0.26	26	35	45	66
Vancouver, British Columbia	0.13	0.26	20	28	40	60
Winnipeg, Manitoba	0.25	0.50	— NR —		54	74
Dartmouth, Nova Scotia	0.14	0.28	17	24	45	70
Moosonee, Ontario	0.25	0.50	— NR —		48	67
Ottawa, Ontario	0.25	0.50	— NR —		59	80
Toronto, Ontario	0.18	0.36	17	23	44	68
Normandin, Quebec	0.25	0.50	— NR —		54	74

Source: Adapted from Balcomb et al. (1980).

NR = Not recommended.

[a]Due south–facing openings are assumed.

[b]Double-glazed, clear glass (approximately equal to window #3, Table 4.13)

[c]Either movable window insulation of R-9, in place from 5:30 PM to 7:30 AM, solar time; or superwindows with overall U-value near 0.30 (approximately equal to windows #7 or #12, Table 4.13).

TABLE 5.7 **Design Guidelines for Passive Solar Thermal Mass**

Expected Solar Savings Fraction (SSF), %	Thermal Storage by Weight/Collector Area				Recommended Effective[a] Thermal Storage Area Per Unit Area of Solar Collection Area	
	Water		Masonry		Water Surface Area[b]	Masonry Surface Area[c]
	lb/ft²	kg/m²	lb/ft²	kg/m²	Collector Surface Area	Collector Surface Area
10	6	29	30	147	0.1	0.7
20	12	59	60	293	0.2	1.5
30	18	88	90	440	0.3	2.2
40	24	117	120	586	0.4	2.9
50	30	147	150	733	0.5	3.7
60	36	176	180	879	0.6	4.4
70	42	205	210	1026	0.7	5.1
80	48	234	240	1172	0.8	5.9
90	54	264	270	1319	0.9	6.6

Source: Adapted from Balcomb et al. (1980).

[a]Effective area is that area exposed at some point to direct sun during a clear winter day.

[b]For a water container 12-in (300-mm) thick.

[c]For a 4-in. (100-mm) thick brick, density 123 lb/ft³ (1970 kg/m³).

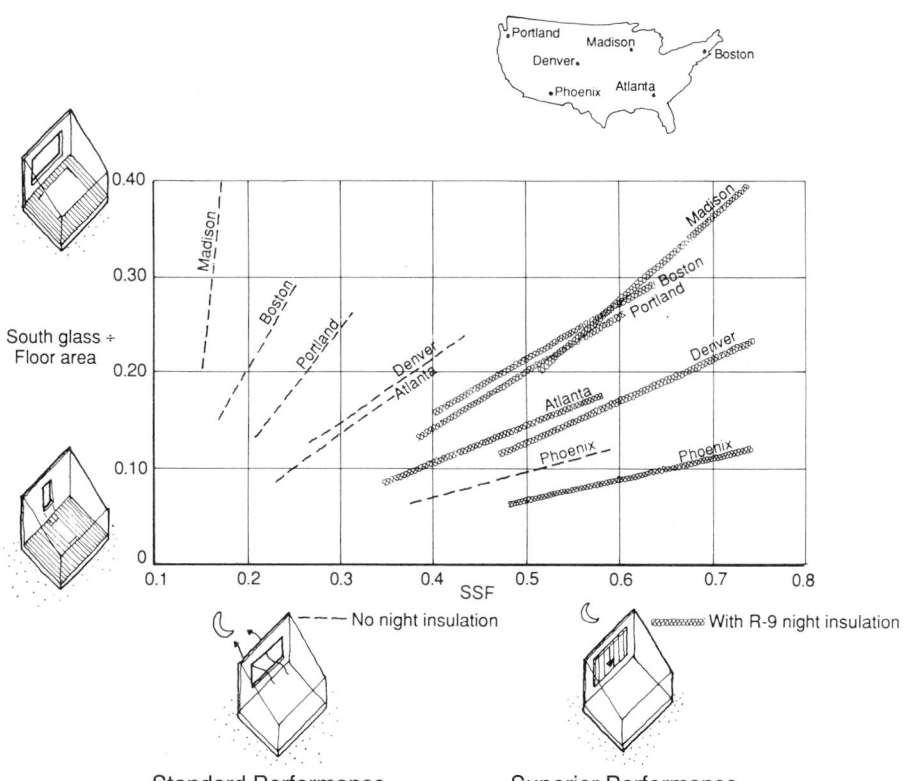

Fig. 5.9 *Design guidelines for passive solar heating: ratio of south glass to floor area. The information of Table 5.6 is presented graphically for six U.S. cities. "No night insulation" = standard performance; "With R-9 night insulation" = superior performance.*

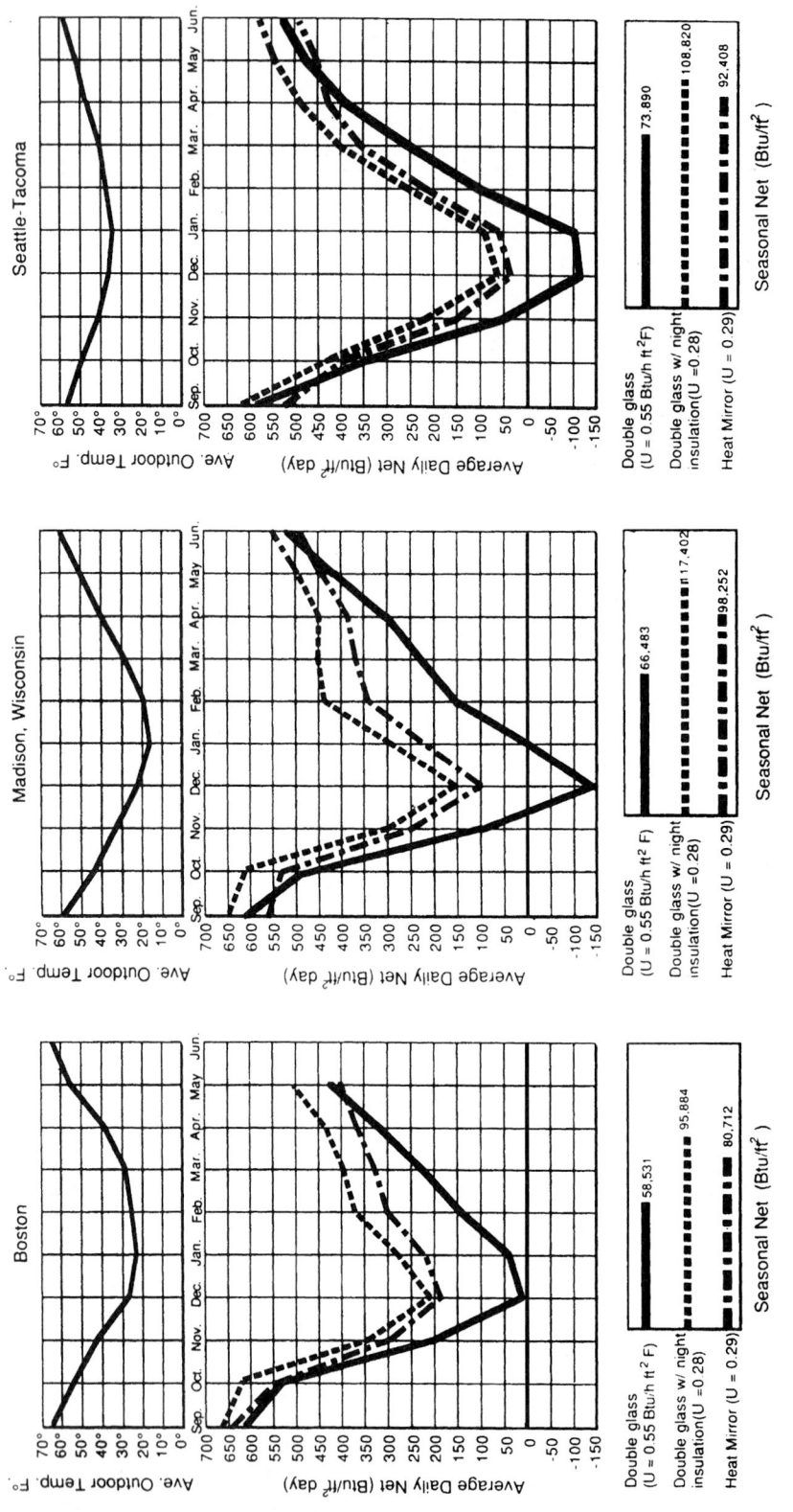

Fig. 5.10 *Average daily net heating, per square foot of south window, for three glazing assemblies in three climates. (a) In Boston, ordinary double glazing is a slight energy plus even in December. (b) In Madison, ordinary double glazing is a net loser for almost 2 months. (c) In Seattle, ordinary double glazing is a net loser for more than 2 months. (From Johnson, 1981.)*

the exposed surface area of the mass should be *at least three times the glazing area.* Masonry surfaces are less thermally effective (on a daily basis) beyond a depth of about 4 to 6 in. (100 to 150 mm). Note that thermal storage area is relatively unimportant at low SSF values, but as the SSF increases, so does the relative proportion of thermal mass area to solar glazing.

EXAMPLE 5.1, PART C. Find the approximate SSF of a typical bay of the Oregon office building (Fig. 5.7).

SOLUTION. "Standard performance" windows (simple double glazed) are used in this building, and the approximate SSF can be found from Tables 5.4 (building data) and 5.6 (SSF guidelines) as follows:

$$\frac{\text{south glass area } 97 \text{ ft}^2 + 132 \text{ ft}^2}{\text{floor area } 1440 \text{ ft}^2} = 0.16$$

For Salem, Oregon, with standard performance windows, a range of 0.12 to 0.24 ratio glass/floor yields SSF 21 to 32%. So, with an actual ratio of 0.16, the approximate SSF is determined as follows:

$$\text{Salem ratio range } 0.24 - 0.12 = \Delta\, 0.12$$

$$\text{Salem SSF range } 32\% - 21\% = \Delta\, 11\%$$

$$(\text{Actual ratio } 0.16)$$

$$- (\text{Salem minimum ratio } 0.12) = 0.04,$$

$$\text{therefore} = \frac{0.04\,(\Delta\,11\%)}{\Delta\,0.12}$$

$$= 3.7\%, \text{ to be}$$

$$\text{added to the minimum SSF;}$$

$$\text{Actual SSF} = \text{minimum } 21\% + 3.7\% = 24.7\%.$$

This glass area/floor area ratio (0.16) is consistent with the previously calculated daylighting design considerations, although there is somewhat more south wall area that could be utilized for additional solar heat collection. (If more solar heat but not more daylight is desired, the additional collection area could be provided as Trombe wall.)

Table 5.7 shows that direct gain thermal mass appropriate to the SSF = 25% would be about 75 lb of masonry per square foot of south glass, or about 2 ft² of mass surface per square foot of south glass. From Table 5.4, the total exposed area of thermal mass in this typical bay is 2232 ft²:

$$2232 \text{ ft}^2 \text{ mass}/229 \text{ ft}^2 \text{ south glass}$$

$$= \text{almost } 10 \text{ ft}^2 \text{ mass/ft}^2 \text{ glass}$$

This is much more mass area than the minimum recommended. This excess of mass area will help prevent overheating on sunny winter days and contribute to thermal stability the year around. For a more detailed analysis of this building's SSF, see Example 5.1, Part H, page 267.

At this point, one must wonder about why standard rather than superior performance windows were used. For Salem, the approximate SSF would have increased to about 44%, with more than adequate internal areas of thermal mass. The simple answer is that the client did not want moveable insulation, and superwindows were not available at an attractive cost when the building was being designed. ∎

Phase-change materials are an alternative to simple thermally massive masonry surfaces. Flat bags of eutectic salts are shown in Fig. 5.11. Thin, horizontal tiles packed with these phase-change materials can store great quantities of heat with the phase change from solid to liquid. This change can be formulated to occur in the low 70s °F (20s °C) to prevent overheating of the space. As flat tiles enclosing bags of salt, they can form the finished surface in any horizontal application—floors, ceilings, or counter- and tabletops, for example. Tubes and trays of salts can be arranged as desired. Because the function of these materials is to keep room temperatures steady, their perfor-

Insulation

Eutectic salt bags

Dense building board

Venetian blinds

Fig. 5.11 *Using bags of phase-change material (eutectic salt) for thermal storage above a flat ceiling. Mirrored-surface venetian blinds reflect direct sun to the ceiling, where it strikes a heavy [at least 90 lb/ft² (440 kg/m²)] board product with high conductivity. Metal ceilings or plaster are also suitable; ordinary gypsum board is not sufficiently conductive. Bags of salt weighing 5 lb/ft² (25 kg/ m²) are in contact with the ceiling material and must be installed in a horizontal position. They are usually formulated to melt between 70 and 75°F (21 and 24°C). (Adapted by permission from Johnson, 1981.)*

mance in preventing overheating on a sunny winter day will also be appreciated on hot summer days, provided they are taken below the phase change or melting temperature at night. For U.S. locations with large daily temperature ranges in summer (MDR in Appendix A), thermal storage surfaces for passive solar heating in winter are potentially useful for night ventilation cooling in summer.

As a preliminary guide, *the phase-change tile surface area = 1 to 3 times the area of solar opening*. Additional information on these materials can be found in Johnson (1981).

The Society for the Preservation of New Hampshire Forests building utilizes phase-change storage above a metal ceiling (Fig. 5.12). It also uses water-filled tubes as an interior partition, masonry interior walls, and a concrete floor slab to help store the winter daily solar gain.

The New Canaan Nature Center (Fig. 5.13) is another innovative New England example using many passive/low energy components. In addition to eutectic salts used in the south-facing railing on the upper level, the 4000-ft² (372 m²) building utilizes extensive clear, double-glazed south glass, movable insulating shades, operable vents at the skylight for stack-effect ventilation, ceiling fans, a woodstove, solar collectors placed inside the skylight monitor (their water is used for warming planting beds), warm air heat recovery ducts, a well-insulated envelope, and even a rainwater collection system. The manually operated switches for ventilation, insulation, and shading represent an unusual degree of user–building interaction. There is also unusual attention to microclimates and transition between inside and outside in this award-winning building.

Rock beds are sometimes used to store the excess heat that indirect gain (sun space) systems can generate. They are typically placed directly beneath a concrete floor slab. The disadvantage of this approach is that any below-grade location raises suspicions of condensation and/or groundwater that may facilitate mold growth, yet cleaning of the rock bed is difficult.

The general guidelines cited by Mazria (1979) are as follows:

Rock bed *volume*, ft³, per ft² of solar opening:

Cold climates: ¾ to 1½

Temperate climates: 1½ to 3

Rock bed *surface area,* in contact with floor above:

Cold climates: 75% to 100% of floor area above

Temperate climates: 50% to 75% of floor area above

(d) Orientation. How important is it that the passive solar opening face due south? The general recommendation is that this orientation be *within 30° of south*. In Volume 2 of *The Passive Solar Design Handbook*, the average penalties for off-south orientation are listed as follows.

5% decrease in SSF at 18° east or 30° west of true south.

10% decrease in SSF at 28° east or 40° west of true south.

Fig. 5.12 *The Conservation Center for the Society for the Protection of New Hampshire Forests, Concord, NH, Banwell White Arnold Hemberger & Partners, Architects.* (a) *The plan is elongated on the east–west axis to maximize south exposure and facilitate daylighting. Direct gain serves the reception area, workroom, and offices, and a sunspace double envelope combination warms the lecture room. A wood-fired boiler provides backup heat.* (b) *Section with south glazing and thermal storage materials: translucent water tubes, masonry walls, and phase-change materials in ceiling and windowsill positions. The circulation of hot air that collects at the clerestory is also shown. In summer, the hot air is vented and an awning will shade the clerestory. Daylighting is diffused through translucent tubes to the spaces on the north side.* (c) *At the workroom, mirror-finish venetian blinds reflect direct sun to the phase-change materials in bags above the dark metal deck ceiling. (Photo by C. Stuart White, Jr.)*

221

Summer vent

Continuous louvre

Dec. 22

Vented roof

Structure & insulation

Continuous ceiling plenum

Intake

Summer awning

Mar. 21

Base of phase change material (P.C.M.)
above ⅝" gypsum board ceiling

Solar heat storage
in water filled tube

Corridor

Exhaust
duct

12" concrete blk.

Office   Secretary

Jan. 21
12:00

Berm

"MIT System" minimizes
glare and temperature
swings at work space

Workroom

P.C.M. tiles

4" conc. floor
6" deep hypocaust

Fan units

Stratified direct gain heat
moved through masonry
wall & floor for low temp.
radiant heat

Duct/mech. tunnel

Light directing louvres
(adjustable) between double
glazed windows

A/A   Section A-A

(b)

(c)

Fig. 5.12 (continued)

(a)

(b)

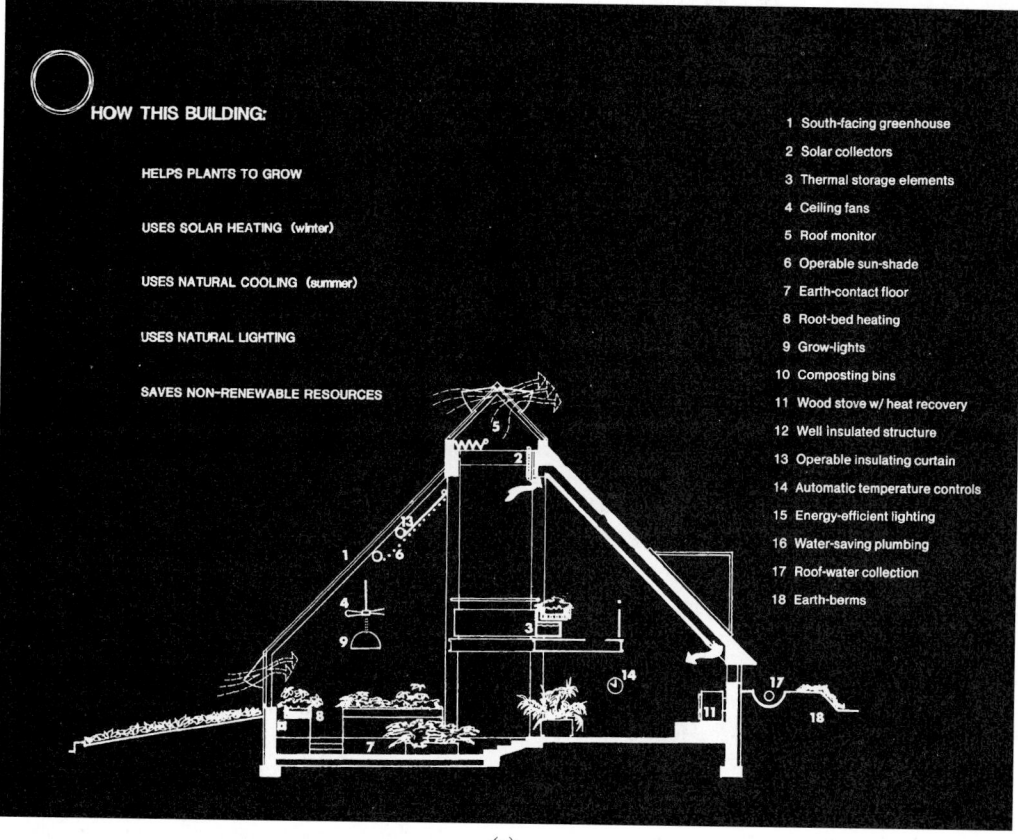

(c)

Fig. 5.13 *The New Canaan (Connecticut) Nature Center (a,b) illustrates many passive and low-energy components. (c) Section illustrates the design features that utilize renewable energy and enhance energy conservation. The "thermal storage elements" are filled with eutectic salts that absorb excess solar heat via phase change (from solid to liquid) by day, then again change phase (from liquid to solid) at night to release their stored heat to the space. (Buchanan/Watson, Architects. Photo by Robert Perron. Courtesy of Donald Watson FAIA, Troy, New York.)*

20% decrease in SSF at 42° east or 54° west of true south.

See Section 5.10 for more detailed coverage of passive heating performance that distinguishes between direct, indirect, and isolated gain systems and includes the expected annual auxiliary energy needed by a passively solar–heated building.

EXAMPLE 5.2. A small office building in Omaha, Nebraska, is to be passively solar heated. How much "superwindow" south glass area should be provided?

SOLUTION. Table 5.6 shows that an Omaha building with a south glass area equal to 20% of its floor area can expect an SSF of 51 if its solar openings are "superior performance."

Select 20% of floor area in south glass: SSF = 51. Thermal mass can be estimated from Table 5.7; for SSF = 50, either 30 lb of water or 150 lb of masonry (per square foot of south glass) should be provided. Using brick as a thermally massive surface, 3.7 ft² of brick should be provided for each square foot of south glass. ■

(e) Roof Ponds. This is yet another approach to passive solar heating, one not covered by the preceding guidelines. In general, roof ponds are used in warmer, less humid areas of the southern United States, where snow will not impede the movement of roof insulating panels and the winter sun is higher in the sky than at northern latitudes. They are frequently sized for their *summer cooling* performance. In the U.S. southern latitudes, a pond sized for cooling will usually be adequate to absorb the needed winter sun. As a check for the pond's heating capacity, Mazria (1979) recommends the following guidelines.

Roof pond area = 85% to 100% of floor area for winter average outdoor temperatures of 25° to 35°F (−4° to +2°C).

Roof pond area = 60% to 90% of floor area for winter average outdoor temperatures of 35° to 45°F (2° to 7°C).

These guidelines are for roof ponds that have two layers of enclosing material between the water and the sky (i.e., are "double glazed") and movable night insulation.

The roof-pond–cooling design guideline, which usually governs, is discussed in Section 5.6.

(f) Active Solar Heating. In contrast to passive systems, which incorporate sun collection and storage as part of a building's walls, floors, or ceilings, active solar heating uses mechanical equipment to collect and store solar energy. The most common early design questions for such systems involve the area of the solar collectors, their tilt and azimuth, and the size of the thermal storage. [Domestic hot water (DHW) solar heating information is given in Section 10.6*i*.]

For active solar space-heating (SH) systems, the rules of thumb are more complex. Building heating needs vary by function and climate. The percentage of space heating that can economically be provided with active solar systems is another big variable. Nevertheless, a rough guide is desirable as a design starting point.

Collector/floor area ratio = the smaller of the two window/floor area ratios listed in Table 5.6

This should provide a portion of the annual heating load somewhere in the range of high performance SSF from Table 5.6. Larger arrays of collectors can be designed, of course, but they rarely will be economically attractive.

For collector tilt and azimuth optima:

Optimum tilt = latitude plus 10° to 15°; optimum azimuth is from due south to 15° W of south

where the tilt angle is measured up from horizontal. The orientation somewhat west of south is attractive in climates with frequent morning fog. Also, because air temperatures are higher in the afternoon, collectors lose less heat then and therefore operate more efficiently.

The design guidelines for storage size are as follows:

2 gal of water storage per ft² (81 L per m²) of collector

or

0.5 to 0.75 ft³ of rock bed per ft² (or 0.15 to .23 m³ per m²) of collector

The large arrays of collectors necessary for space heating must be served by correspondingly large pipes or ducts. Whereas pipe size rarely influences design, the air ducts for air-type collectors can consume large amounts of space. Therefore, the following typical flow rates are:

water flow rate of 0.25 to 0.5 gpm per ft^2 (0.17 to 0.34 L/s per m^2) of collector

or

airflow rate of 2 cfm per ft^2 (10 L/s per m^2) of collector

Approximate pipe sizes can be determined from Chapter 10, and approximate duct sizing is shown in Section 8.4.

5.5 Summer Heat Gain Guidelines

Design guidelines for cooling are complicated by the fact that cooling loads frequently are more related to individual building characteristics than to climate; sunshading and internal heat gains are particularly influential on cooling loads. Because passive cooling guidelines (Section 5.6) are expressed in heat to be removed per unit of floor area, it is first necessary to estimate the extent of the heat gain problem. Later (in Section 5.13), more detailed passive cooling procedures are shown. A comparison of the guidelines with these more detailed procedures is given in Table 5.9, page 231.

Detailed calculations for heat gain are presented in Sections 5.11 and 5.13. For now, Table 5.8 gives a quick *approximation*. Many buildings (restaurants, factories, stores selling heating appliances, etc.) have special heat sources within. For these unusually heat-loaded situations, Table 5.8 will be inadequate. As a starting point for preliminary passive-cooling sizing for typical buildings, however, it should be helpful.

Note that "open" buildings, such as those that are naturally ventilated, do not have heat gains from infiltration, because they assumedly maintain internal temperatures that are slightly *above* exterior temperatures. However, these buildings do experience heat gains through windows, walls, and roofs, due to solar impacts on these surfaces. For "closed" buildings, heat gain from infiltration

or ventilation must be added, because these structures maintain internal temperatures lower than outside temperatures.

EXAMPLE 5.1, PART D.   Find the approximate heat gain for the Oregon office building in Fig. 5.7.

SOLUTION.   From Table 5.8:

A. People and equipment (using the high end for Europe office buildings, based on assumptions about computer use)   5.8 Btu/h-ft^2

B. Electric lighting   0.5 Btu/h-ft^2
(with a 4% DF, from Example 5.1, Part A; this procedure assumes most electric lights are off when daylight is available)

C. Envelope (Eugene's design temperature, Appendix A, is 89°F)
South glass, shaded by vines:

$$\frac{(97\ ft^2 + 132\ ft^2) \times 16}{1440\ ft^2} \qquad 2.54\ Btu/h\ ft^2$$

North glass, unshaded: (Table 5.19 Part B; regular double glass, venetian blinds):

$$\frac{97\ ft^2 \times 14\ Btu/h\text{-}ft^2}{1440\ ft^2} \qquad 0.94\ Btu/h\text{-}ft^2$$

Walls:

$$\frac{296\ ft^2 \times 0.084\ Btu/h\text{-}ft^2\text{-}°F \times 15}{1440\ ft^2} \qquad 0.26\ Btu/h\text{-}ft^2$$

Roof:

$$\frac{1512\ ft^2 \times 0.028\ Btu/h\text{-}ft^2\text{-}°F \times 35}{1440\ ft^2} \qquad 1.03\ Btu/h\text{-}ft^2$$

D. Total heat gains, internal and through building skin: 11.07 Btu/h-ft^2

E. Heat gains from ventilation:

$$\frac{20\ cfm/person \times 8\ persons \times 16}{1440\ ft^2} \qquad 1.8\ Btu/h\text{-}ft^2$$

Total approximate heat gains: 12.87 Btu/h-ft^2
■

5.6 Passive Cooling Guidelines

The design guidelines for passive cooling are much newer and less tested than are those for

TABLE 5.8 **Estimating Summer Heat Gains**

Part A. Internal Heat Sources—People and Equipment

Function	Area per Person[a]		Sensible Heat Gain (Btu/h ft² of Floor Area)			Sensible Heat Gain (W/m² of Floor Area)		
	(ft²)	(m²)	People[b]	Equipment[c]	Total	People[b]	Equipment[c]	Total
Office, U.S.[c]	180–100	16.7–9.3	1.3–2.3	0.4–1.1	1.7–3.4	4.1–7.3	1.2–3.4	5.3–10.7
Office, Europe[e]			1–1.6	2.2–4.2	3.2–5.8	3–5	7–13.1	10–18.1
School: elementary, U.S.	100–20	9.3–1.9	2.3–11.5	0–0.6	2.3–12.1	7.3–36.3	0–2.0	7.3–38.3
Schools, Europe[e]			3.8–8.0	0–0.6	3.8–8.6	12–25.2	0–2.0	12.0–27.2
School: secondary, college	150–100	13.9–9.3	1.7–2.6	0–0.6	1.7–3.2	5.4–8.2	0–2.0	5.4–10.2
Healthcare:								
Sleeping (hospital)	240	22.3	0.9	0.6[e]	1.5	2.8	2.0[e]	4.8
In-patient (clinic)	120	11.1	1.9	Varies	1.9+	6.0	Varies	6.0+
Assembly: fixed seats	15	1.4	14.0	—	14.0	44.2	—	44.2
standing space, concentrated use	15–7	1.4–0.7	21.0–45.0	0–0.5	21.0–45.5	66.3–142.0	0–1.6	66.3–143.6
Restaurant:[f]								
Fast food: dining area	15	1.4	17	3.4	20.4	53.6	10.7	64.3
Kitchen, refrigeration				17.1	17.1	54.0	54.0	54.0
Sit-down: dining area	25	2.3	10.2	5.1	15.3	32.2	16.1	48.3
Kitchen, refrigeration				7.2	7.2		22.7	22.7
Mercantile: street floor	50–30	4.7–2.8	6.3–10.5	3.4	9.7–13.9	19.9–33.1	10.7	30.6–43.8
Other sales floors	60–50	5.6–4.7	5.3–6.3	3.4	8.7–9.7	16.7–19.9	10.7	27.4–30.6
Shopping center, Europe[e]			3.2	0.3–1.3	3.5–4.5	10	1.0–4.0	11.0–14.0
Warehouse	1000–300	92.9–27.9	0.4–1.2	—	0.4–1.2	1.3–3.8	—	1.3–3.8
Hotels, nursing homes	300–200	27.9–18.6	0.8–1.2	3.4	4.2–4.6	2.5–3.8	10.7	13.2–14.5
Apartments[g]	300–200	27.9–18.6	0.8–1.2	See note g	See note g	2.5–3.8	See note g	See note g

Part B. Internal Heat Sources—Electric Lighting

Function	Sensible Heat Gain[h] (Btu/h ft² of Floor Area)			Sensible Heat Gain[h] (W/m² of Floor Area)		
	DF<1	1<DF<4[h]	DF>4[h]	DF<1	1<DF<4[h]	DF>4[h]
Office	5.1	2.0	0.5	16.1	6.3	1.6
School: elementary	6.3–6.8	2.5–2.7	0.6–0.7	19.9–21.5	7.9–8.5	1.9–2.2
School: secondary, college	6.3–6.8	2.5–2.7	0.6–0.7	19.9–21.5	7.9–8.5	1.9–2.2
Healthcare:						
Sleeping (hospital)	6.8	2.7	0.7	21.5	8.5	2.2
In-patient (clinic)	6.8	2.7	0.7	21.5	8.5	2.2
Assembly	3.8	1.5	0.4	12.0	4.7	1.3
Restaurants[j]	6.3	2.5	0.6	19.9	7.9	1.9
Mercantile	5.1–6.8	2.0–2.7	0.5–0.7	16.1–21.5	6.3–8.5	1.6–2.2
Warehouse	2.4	1.0	0.2	7.6	3.2	0.6
Hotels, nursing homes	6.8	2.7	0.7	21.5	8.5	2.2
Apartments[g]	Up to 6.8	Up to 2.7	Up to 0.7	Up to 21.5	Up to 8.5	Up to 2.2

Part C. Heat Gain through Envelope[i]

	(Btu/h ft² of Floor Area) Outdoor Design Temperature		(W/m² of Floor Area) Outdoor Design Temperature	
	90°F	100°F	32°C	38°C
I. Gains through externally shaded windows[k]:				
Find ratio, $\dfrac{\text{total window area}}{\text{total floor area}}$, then multiply by	16	21	50	66
II. Gains through opaque walls:				
Find ratio, $\dfrac{\text{total opaque wall area}}{\text{total floor area}} \times (U_{\text{wall}})$, then multiply by	15	25	8	14
III. Gains through roofs:				
Find ratio, $\dfrac{\text{total opaque roof area}}{\text{total floor area}} \times (U_{\text{roof}})$, then multiply by	35	45	19	25

(*continued*)

TABLE 5.8 **Estimating Summer Heat Gains** (*continued*)

Part D. Summary Gains

I. Passive cooling systems for (thermally) "open" buildings:
   Cross ventilation
   Stack ventilation
   Nighttime or "open" hours of thermal mass/night ventilation
   Total gains: Add Parts A, B, and C to obtain cooling load, Btu/h ft² [W/m²] of floor area.
II. Passive cooling systems for (thermally) "closed" buildings:
   Roof ponds
   Evaporative cooling
   Daytime or "closed" hours of thermal mass/night ventilation
   Total gains: Add Parts A, B, C, and E to obtain cooling load, Btu/h ft² [W/m²] of floor area.

Part E. Gains from Infiltration/Ventilation of "Closed" Buildings

	Outdoor Design Temperature		Outdoor Design Temperature	
	90°F	100°F	32°C	38°C
Infiltration:				
Find ratio, $\dfrac{\text{total window + opaque wall area}}{\text{total floor area}}$, then multiply by	1.0	1.9	3.2	6.0
Ventilation:				
Find $\left[\dfrac{\text{known total cfm of outdoor air}}{\text{total floor area, ft}^2}\right]$, then multiply by	16	27		
Find $\left[\dfrac{\text{known total L/s of outdoor air}}{\text{total floor area, m}^2}\right]$, then multiply by			9.9	16.8

[a]Lower density from elevator population estimates, Chapter 23; higher density from fire population estimates, Chapter 13. Lower density for whole-building estimates; higher density for single-space estimates.

[b]Sensible gains per adult from Table 5.21 for activities as stated.

[c]The typical "miscellaneous" load of 1 watt/ft^2 produces 3.4 Btu/h ft^2.

[d]P. Komor, "Space Cooling Demand from Office Plug Loads," *ASHRAE Journal*, December 1997. This is considerably lower than the "typical" 2 to 4 W/ft^2 for offices, used by many designers.

[e]Adapted from Hastings (1994).

[f]Based on total area of restaurant + kitchen. From ACSA (1994) *Design with PV*, data from Electric Power Research Institute.

[g]Residential internal gains are often assumed at 230 Btu/h per occupant plus 1200 to 1600 Btu/h total from appliances [67.3 W per person, plus 351.4 to 468.5 W total from appliances].

[h]Adapted from Northwest Power Planning Council, *Maximum Lighting Standards*, 1983. Values shown for DF $\geq$ 1 assume automatic dimming in the presence of daylight.

[i]Lighting is often much lower in sit-down than in fast-food restaurants.

[j]Averaged from the more specific data found in Table 5.18.

[k]If windows are not externally shaded, see Table 5.19 for multipliers, which will vary by orientation.

passive heating. It is important that the designer *first* check the match between the climate and the cooling strategy, as was done in Chapter 2 using Fig. 2.13, before applying these guidelines to a building design.

(a) Cross Ventilation. One of the oldest methods known, this strategy provides plentiful fresh air, but maintains a building at temperatures slightly *above* those outdoors. The cross ventilation inlet (window) areas, expressed as percentage of total floor area, are related to wind speed and resulting heat removal in Fig. 5.14*a*. Remember that *an equal (or greater) area of outlet openings* must also be provided. Furthermore, any internal obstructions (such as partitions) must have a total of openings at least equal to this required inlet area. The assumptions about wind direction and indoor–outdoor temperature differences that were used to produce these guidelines are explained in the figure. For more detailed wind speed and direction information, see Section 5.14.

The ΔT of 3F° used in Fig. 5.14 is deliberately kept small to encourage "open" strategies in

milder summer climates. Thus, an interior temperature of 83°F, which is comfortable with sufficient air motion, lower-percentage relative humidity, and comfortable surface temperatures, would be obtainable with an outside temperature of 80°F. However, a greater ΔT is often appropriate—for example, for spring or fall cooling of office buildings, or for summer cooling of factories or kitchens where internal temperatures may remain in the low 90s. In such cases, find the percentage of inlet area required, then multiply by the ratio

$$\frac{3F°}{\text{actual } \Delta T}$$

to obtain required cross ventilation areas for any specific temperature difference (see Example 5.1, Part E).

See Table 5.9 for a comparison of these design guidelines with the more detailed calculations of Section 5.14, for both cross and stack ventilation.

EXAMPLE 5.1, PART E. Find the required inlet areas for cross ventilation of a typical bay of the office building in Fig. 5.7.

Fig. 5.14 (a) *Cross-ventilation design guidelines for heat removed per unit floor area and relationship of inlet openings and wind speed. Total inlet opening area is expressed as a percent of total floor area served by cross ventilation. Note: Outlet areas must also be at least equal to inlet areas. This graph is based on an internal temperature 3F° (1.7C°) above the exterior temperature, and assuming that the wind is not quite perpendicular to the window openings, for a wind effectiveness factor of 0.4 (see Section 5.14).*
Fig. 5.14 (b) *Stack-ventilation design guidelines for heat removed per unit floor area and relationship of stack height and inlet openings. Area of stack inlet/throat/outlet is expressed as a percentage of total floor area served by stack ventilation. Note: Outlet areas and stack throat area must also be at least equal to inlet areas. This graph is based on an internal temperature of 83°F, exterior temperature 80°F, for a 3F° differential (28.3°C internal, 26.7°C internal, for a 1.6C° differential).*

TABLE 5.9 **Comparing Passive Cooling Design Guidelines with Detailed Calculation Procedures**

Design Guideline	Detailed Analysis Procedure
Part A. Cross Ventilation	
Section 5.6(a) • Assume 3F° (1.67C°) ΔT • Assume window orientation to wind	Section 5.14(a) • Use actual ΔT • Use actual window orientation to wind
Part B. Stack Ventilation	
Section 5.6(b) • Assume 3F° (1.67C°) ΔT	Section 5.14(b) • Use actual ΔT
Part C. Night Ventilation of Thermal Mass	
Section 5.6(c) • Assume ratio of mass area/floor area • Assume cooling during hour of max. ΔT • Assume max. ΔT for natural ventilation estimation	Section 5.14(c) • Use actual exposed mass area • Use actual mass heat capacity • Use actual hourly chart of air and mass temperatures • Find total cooling and required air flow rate
Part D. Evaporative Cooling (Active)	
Section 5.6(d) • Assume 2.67 cfm/ft² floor area • Assume 83°F exhaust air	Section 5.14(d) • Use actual outdoor temperature • Find actual indoor supply temperature • Find allowable ΔT as air passes through indoors • Then cfm = (Btu/h)/(1.08)(ΔT)
Part E. Cooltowers (Passive, Evaporative)	
Section 5.6(e) • Find approximate exit air temperature • Find approximate flow rate • Then Btu/h = (cfm) (1.08)(ΔT)	Section 5.14(e) • Use actual outdoor temperature and wet-bulb depression • Find actual exit air (supply) temperature • Find actual exit air flow rate • Then Btu/h = (cfm) (1.08)(ΔT)
Part F. Roof Ponds	
Section 5.6(f) • Assume pond max. temperature 80°F • Estimate pond minimum temperature • Assume 30% gain through roof insulation • Assume pond depth from 3 in. to 6 in. • Find area of pond	Section 5.14(f) • Actual outdoor temperature, resulting pond temperature • Actual heat gain through roof insulation • Actual hours of internal heat gains • Choice of fan-forced or still-air layer at ceiling • Evaporation or not, on upper pond surface • Find actual pond depth and area • Find size of backup cooling

(continued)

THERMAL CONTROL

TABLE 5.9 **Comparing Passive Cooling Design Guidelines with Detailed Calculation Procedures (*continued*)**

Design Guideline	Detailed Analysis Procedure
Part G. **Earth Tubes**	
Section 5.6(g) • Assume 65°F soil temperature • Assume soil conductivity • Does not specify depth • Assume 85°F outdoor air • Assume 500 fpm velocity • Choose diameter and length to match cooling load	*Section 5.14(g)* • Actual underground temperature, assume resulting tube temperature is within 4F° • Actual soil conductivity • Actual depth • Actual outdoor air temperature • Assume 500 fpm velocity • Calculate actual cooling (assume 1.3 Btu/h F ft length × ΔT)

SOLUTION. This office building is located in Eugene, Oregon, whose monthly average wind velocities may be obtained from NOAA Local Climatological Data. For July, the hottest month, the average wind velocity is about 8 mph, from a nearly due-north average direction. From Table 5.4 the inlet area (=outlet area also, in this case) is 16 ft². The inlet area, as a percentage of floor area, is therefore

$$16\ \text{ft}^2 / 1440\ \text{ft}^2 = 1\%$$

Entering Fig. 5.14a with 8-mph wind velocity and 1% inlet area, we find that somewhat less than 15 Btu/h-ft² heat gain can be removed by cross ventilation. In Example 5.1, Part D, we approximated a heat gain (excluding ventilation) of about 11 Btu/h-ft². Therefore, under average July conditions, the window area is probably adequate to remove heat gains and maintain an indoor temperature 3F° *above* outdoor temperature.

Because Eugene's average July daily high is 83°F, this would produce an interior daily high of 86°F. From Fig. 2.4b, such a temperature would require somewhat more than minimum air motion in order to feel comfortable. At Eugene's summer *design* temperature of 89°F, the interior temperature would be 92°F. For these reasons this building was designed to be night ventilated, relying on thermal mass rather than on cross ventilation for hot-day comfort.

In milder weather, these windows can achieve substantial cooling because the temperature difference, ΔT, is more than the 3F° assumed in Fig. 5.14. For example, when the temperature indoors is 75°F, outdoors 65°F, the fully opened windows with the 8-mph wind would result in a cooling rate of

$$\frac{\text{about 15 Btu/h-ft}^2 \times 10\text{F}°}{3\text{F}°} = \text{about 50 Btu/h-ft}^2$$

which is considerably more than the actual gain of 11 Btu/h-ft². Therefore, under such conditions the windows are only partially opened. ∎

(b) Stack Ventilation. Another historically useful strategy, like cross ventilation, it provides plentiful fresh air, but maintains a building at temperatures slightly *above* outdoor temperatures. The stack inlet areas, expressed as percentage of total floor area, are related to stack height and the resulting heat removal in Fig. 5.14b. Remember that *an equal (or greater) area of stack outlet openings as well as at least an equal cross-sectional area through the vertical stack* are also required. Again, consider internal obstructions: partitions must have a total of openings at least equal to this required inlet area. The assumptions about indoor–outdoor temperature differences that were used to produce this guideline are explained in the figure.

Adjustment of the ΔT for this guideline is similar to the procedure for cross ventilation. It requires multiplication of the required percentage of stack area by the ratio

$$\sqrt{\frac{3\text{F}°}{\text{actual}\ \Delta T}}$$

to obtain the required stack ventilation areas for any other specific temperature difference.

Until the advent of air conditioning, these methods of cross and stack ventilation were used in virtually all commercial buildings. In some climates with mild summers, newer buildings are turning to these older methods (see Fig. 5.48 in Section 5.14). Figure 5.15 shows the Thoreau Center for Sustainability in San Francisco's Presidio National Park. An older military hospital was renovated for offices for nonprofit environmental organizations. To maximize daylighting and passive cooling, the original operable windows are utilized. Interior partitions are kept low enough to facilitate cross ventilation, with integral indirect electric lighting to the higher white ceiling. The section indicates that both cross ventilation and stack effect are anticipated. However, the stack openings to the attic and, subsequently, through the roof are less than 1% of the total floor area.

The most likely benefit of the stack effect here is to keep the attic at cooler summer temperatures (with air drawn in at the eave vents, not flowing outward as shown).

(c) Night Ventilation of Thermal Mass.

This strategy maintains a building at temperatures lower than the outside by day and flushes the building with plentiful fresh air by night. The sizing procedure is shown in Fig. 5.16, in which climate data are related to two representative types of thermally massive building. For each type, the graphs show the *daily* Btu per square foot of floor area that can be stored. The "average" mass building is represented by a building with an exposed concrete floor 4 in. thick (or an exposed ceiling of equivalent construction). There is one unit area of exposed mass for each unit floor area. The "high" mass building is similar to the typical passively solar–heated, direct-gain building, or to

(a)

Fig. 5.15 The Thoreau Center for Sustainability occupies renovated hospital buildings (a) *in San Francisco's Presidio National Park.* (b) *Relatively narrow floor plans facilitate daylighting and natural ventilation.* (c) *Interior at center corridor.* (d) *Section illustrates daylighting and cross ventilation. Very small stack openings relative to floor area served limit the stack effect to a minor role. (Drawings courtesy of Tanner Leddy Maytum Stacy Architects, San Francisco.)*

THERMAL CONTROL

SECOND FLOOR PLAN

FIRST FLOOR PLAN

(b)

(c)

Fig. 5.15 *(continued)*

a multistory building with an exposed concrete structure, in which *both sides* of the floor slab are available for thermal storage, or an equivalent exposed mass area in walls, and so on. For this building, there are two units area of exposed mass for each unit floor area.

The climate data needed for Fig. 5.16 (maxi-

mum summer design DB temperature and mean daily range) are given in Appendix A. These data also allow calculation of the *minimum* summer design DB temperature; that is, maximum design DB temperature minus mean daily range. This minimum temperature is of interest here because the thermal mass of the building will be lowered

TYPICAL SECTION AT BUILDING 1013

PERSPECTIVE AT BUILDING 1013

(d)

Fig. 5.15 (continued)

toward (but not quite to) it during night ventilation. For high daily range climates, the lowest temperature obtained by the thermal mass will be *about one fourth of the mean daily range above the minimum air temperature* (for lower daily range climates—30F° or less—*one fifth* of the mean daily range).

EXAMPLE 5.3. Compare the potential for night ventilation of mass in Sacramento, California, and Oklahoma City, Oklahoma.

SOLUTION. Sacramento, California:

Highest DB = 98°F (Appendix A)

Mean daily range = 36F° (Appendix A)

Lowest DB = 98 − 36 = 62°F

Approximate lowest mass temperature: 1/4 of 36 = 9F°; 62 + 9 = 71°F.

Oklahoma City, Oklahoma:

Highest DB = 97°F

Mean daily range = 23F°

Lowest DB = 97 − 23 = 74°F

Approximate lowest mass temperature: 1/5 of 23 = 4.6F°; 74 + 4.6 = 78.6°F.

From these data, it appears that Sacramento, with a comfortably low 71°F mass temperature, is a likely site for night ventilation of heat stored in the building's mass. However, Oklahoma City's 78.6°F lowest mass temperature is at the middle of the comfort zone, a climate in which this passive strategy will be harder to achieve. In either location, performance studies can best be carried out with the details of this mass cooling procedure, found in Section 5.14.

We can also compare the daily heat removed.

Fig. 5.16 *Night ventilation of thermal mass design guidelines. Parts a and b show Btu/day stored and removed per square foot of floor area (W/day per square meter of floor area). Enter these graphs with your location's summer design outdoor DB temperature and the MDR (mean daily range). (a) "Average mass" structure has 1 ft² of mass surface exposed per square foot of floor area, where the mass is a 4-in. (100-mm) ordinary-density concrete slab. (b) "High-mass structure" has 2 ft² of mass surface exposed per square foot of floor area, where the mass is a 3-in. (75-mm) ordinary-density concrete slab (or both sides of a 6-in. concrete wall or slab exposed). Both average- and high-mass structures go into "open" mode when outdoor temperature drops below 80°F(26.7°C); therefore, the highest mass temperature is assumed to be 80°F (26.7°C). Part (c) shows the approximate percentage of the daily total heat stored that is removed during the night-time hour of maximum cooling. Part (d) shows the ΔT between the outside air and inside mass that exists during that hour of maximum cooling. Required ventilation rates can then be determined from (c) and (d).*

For a high-mass building, from Fig. 5.16*b:*

Sacramento (DB = 98°F, MDR = 36F°) storage is somewhat more than 100 Btu/ft² day.

Oklahoma City (DB = 97°F, MDR = 23F°) storage is much less than 80 Btu/ft² day. ∎

Before these rules of thumb can be fully utilized, the number of hours for which the building must be "closed" must be determined. During these hours, heat will be stored in the structure,

up to the maximum storage indicated in Fig. 5.16. Typically, these buildings will go into the thermally "closed" mode (allowing minimum ventilation only) at about 6 AM for a day with 100°F maximum, or about 8 AM for a day with 85°F maximum, remaining in the closed mode until the outdoor temperature drops below 80°F. (To approximate this hour, assume that the midpoint outdoor temperature, between daily high and daily low, occurs around 10 PM.) Thus, the typical office building remains thermally "closed" during the 8 or 9 hours of summer occupancy. All the heat generated during the "closed" mode—8 to 9 daytime hours—must be stored in the structure: Heat to be stored/ft² floor area =

$$\begin{bmatrix} \text{hours occupied in} \\ \text{"closed" mode} \end{bmatrix}$$
$$\times \begin{bmatrix} \text{heat gain, Btu/h ft}^2 \text{ floor} \\ \text{(from Table 5.8)} \end{bmatrix}$$

Having found the amount of heat that can be stored each day, the designer next must solve the problem of removing the heat by night. Either natural or forced ventilation can be used, but in many locations wind speed is very low on summer nights. The ventilation rate is determined by the "best hour" of cooling during the night: that is, the hour during which the temperature difference, ΔT, between inside mass and outside air is greatest, hence the most heat is removed. This "best hour" information can be found in Fig. 5.16c and d. Then the cross or stack ventilation rules of thumb can be used to size the openings.

See Table 5.9 for a comparison of these guidelines and more detailed calculation procedures for night ventilation of thermal mass.

EXAMPLE 5.1, PART F. Find the potential for night ventilation of mass in the office building in Fig. 5.7.

SOLUTION. For Eugene, Oregon, from Appendix A, the summer design temperature is 89°F and MDR is 31F°. From Table 5.4, the typical bay has a ratio of exposed thermal mass/floor area of

$$\frac{1512 \text{ ft}^2 + 720 \text{ ft}^2}{1440 \text{ ft}^2} = 1.55$$

which places this example halfway between "average" mass (ratio 1.0) and "high" mass (ratio 2.0). Entering Fig. 5.16a, with Eugene's climate data, we find that an average mass building can store about 100 Btu/ft²-day, whereas a high mass building (Fig. 5.16b) can store about 170 Btu/ft²-day. This building can therefore store about 135 Btu/ft²-day.

Assuming a 9-h typical working day (including noon hour), at the average approximate heat gain rate of 13 Btu/h-ft² (Example 5.1 Part D), the daily heat gain in this building is

$$9 \text{ h} \times 13 \text{ Btu/h ft}^2 = 117 \text{ Btu/ft}^2\text{-day}.$$

This average heat gain is less than the storage capacity of 135 Btu/ft²-day, so the building should perform satisfactorily with this method of cooling.

Although this building was designed for nighttime forced ventilation because of summer night calm, we will check to see whether the openable windows are adequate for natural night cross ventilation. In Fig. 5.16c, we find that during the hour of maximum night cooling, something less than 14% of the total stored heat gains must be removed:

$$14\% \times 117 \text{ Btu/ft}^2\text{-day} = 16 \text{ Btu/h-ft}^2$$

From Fig. 5.16d, the resulting maximum hourly ΔT is about 13F°.

In Example 5.1, Part E, we determine that less than 15 Btu/h-ft² would be removed by cross ventilation under average July conditions, at a 3F° ΔT. Because during this hour of maximum cooling the indoor–outdoor ΔT is 13F°, cross ventilation could remove

$$\frac{13\text{F}° \times 15 \text{ Btu/h-ft}^2}{3\text{F}°} = 65 \text{ Btu/h-ft}^2$$

Cross ventilation by night appears feasible for this building, because excess stored heat could be removed during this hour of maximum cooling. Unfortunately, Eugene's summer *nighttime* average wind velocity is nearly zero. Forced ventilation was therefore chosen, also in order to flush all interior mass surfaces thoroughly with cool air.

The rate of forced ventilation during the hour of maximum cooling can be estimated from Section 4.6:

$$q_v = \frac{V(1.08 \text{ Btu-min/ft}^3\text{-°F-h})(\Delta T)}{60 \text{ min/h}}$$

Expressed per square foot floor area in this building,

$$16 \text{ Btu/h-ft}^2 = V \text{ cfh } [0.018 \text{ Btu-min./ ft}^3\text{-°F-h}] 13\text{F}°$$

$$V = \frac{16}{0.018 \times 13}$$

$$= 68.4 \text{ cfh outdoor air/ft}^2 \text{ floor area}$$

This rate of forced outdoor air ventilation produces this air change: Average ceiling height is 15.3 ft, so

$$\frac{68.4 \text{ cfh/ft}^2 \text{ floor}}{15.3 \text{ ft}^3/\text{ft}^2 \text{ floor}} = 4.47 \text{ ACH}$$

a rate considerably greater than that required for minimum outdoor air per person by day. This promises an attractive indoor air quality by next morning. ∎

The administration building for the Fetzer Winery sits among grapevines in northern California's Mendocino Valley (Fig. 5.17). The summer design condition is approximately 96/68, with MDR 30F° (35.6/20, MDR 16.7C°). As an "average mass" structure, this building could store and remove about 60 Btu/ft² day (190 W/m² day); as a "high-mass structure," it could store and remove about 100 Btu/ft² day (315 W/m² day).

(d) Evaporative Cooling. The most common evaporative cooler is not strictly "passive" cooling, as it depends on a fan to force large quantities of outdoor air through a wet filter, thereby lowering the air's temperature and raising its relative humidity before delivering the air to the space to be cooled. In hot and arid climates, the energy used by the fan in evaporative systems is less than the energy needed to achieve conventional cooling based on the compressive refrigeration cycle. Although this process requires quantities of water, it does not use refrigerants that pose a threat to the earth's ozone layer.

Before using these guidelines, be sure to check the cooling strategy chart in Fig. 2.13 to determine whether evaporative cooling is appropriate for your climate. It is unlikely that evaporative cooling will be helpful in the humid southeast United States.

The rates of evaporative cooling presented in Fig. 5.18 are based on a rather high airflow of 2.67 cfm per ft² of floor area. (In conventional cooling, an airflow rate closer to 1 cfm per ft² is more common.) With this amount of air motion, a highest indoor air temperature of 83°F is assumed; so the evaporatively cooled air, after absorbing the heat from the space, exits at 83°F. (Note: this may still be well below the outdoor temperature: might it do some useful cooling?) To use Fig. 5.18, first

(a)

Fig. 5.17 *The Fetzer Winery administration building* (a) *in its rural Mendocino County, California, setting. The south-facing clerestory and a deciduous vine and trellis are the more visible parts of the daylighting, passive solar heating, and passive cooling strategies.* (b) *Plan shows relation of workstation to north and south daylight.* (c) *Section, with ducts for the night air flush as well as for conventional heating and cooling.* (d) *North side dormers are for night air intake and exhaust. (Drawings courtesy of Valley Architects, St. Helena, California.)*

(b)

(c)

(d)

Fig. 5.17 (continued)

Fig. 5.18 *Fan-forced evaporative cooling design guideline for Btu/h removed per square foot of floor area (W per m². A forced airflow rate of 2.67 cfm per square foot of floor area (13.56 L/s per m² of floor area) is assumed, as is an exiting (exhaust) air temperature of 83°F (28.3°C). The dotted lines toward the left side of the graph represent increasingly humid indoor air, which may cause discomfort.*

find in Appendix A, Table A.1, the summer design dry-bulb and mean coincident wet-bulb temperatures for your location. Enter the graph at these two data points, and at their intersection, find the approximate amount of heat in Btu/h-ft² that evaporative cooling can remove. As wet-bulb temperatures surpass 68°F (represented by dotted lines in the graph), indoor conditions become increasingly humid, reaching almost certain discomfort indoors by wet-bulb 75°F.

A more thorough method for evaluating evaporative cooling potential is presented in Section 5.14, where various airflow rates and temperatures of supply air and exit air can be examined. See Table 5.9 for a comparison of this guideline and the more thorough method.

EXAMPLE 5.4. Evaluate the potential for evaporative cooling for a 3000-ft² retail store in Denver, Colorado. Due to large electric lighting and equipment display loads, the approximate heat gain is 30 Btu/h-ft².

SOLUTION. From Table A.1, Denver has a summer design dry-bulb temperature of 91°F, with a coincident wet-bulb temperature of 59°F. Checking back with Fig. 2.14, these design conditions fall well within the climate served by evapo-

rative cooling. Next, enter Fig. 5.18 at the points of 91° DB and 59° WB; the intersection of these data lines shows that at such conditions, heat gains can be removed by evaporative cooling at a rate of about 37 Btu/h-ft² of floor area.

37 Btu/h-ft² capacity

> 30 Btu/h-ft² approximate heat gains

Therefore, evaporative coolers can meet the need. Because Fig. 5.18 assumes an air flow rate of 2.67 cfm per ft², the cooler's approximate size will be

$$3000 \text{ ft}^2 \times 2.67 \text{ cfm/ft}^2 = 8000 \text{ cfm}$$

This cooler size could be reduced by the ratio of cooling capacity to heat removal need:

$$8000 \text{ cfm} \times \frac{30 \text{ Btu/h-ft}^2 \text{ need}}{37 \text{ Btu/h-ft}^2 \text{ capacity}} = \text{about 6500 cfm}$$

However, the larger capacity will allow a lower indoor air exit temperature than the 83°F assumed in Fig. 5.18, so the larger capacity is a safer approximation. ∎

(e) Cooltowers. A more passive approach to evaporative cooling appears as a tower on the residence in Fig. 5.19. This University of Arizona experimental building near the Tucson Airport

Fig. 5.19 An experimental residence in Tucson, Arizona, is passively and evaporatively cooled by a "cooltower" (at the right in the photograph). Hot dry air passes through the wetted pads at the tower's top and undergo a drop in DB temperature that causes it to fall, entering the residence at the bottom of the tower. The tower at the left is a "solar chimney" to help speed the flow of warmed exhaust air.

has a "cooltower" with wetted pads on all four faces at the top. Hot, dry air is cooled as it passes through the pads, dropping to the base of the tower then into the house. Analysis by Givoni (1994) indicates that such a tower's delivery of wetter, cooler air is almost independent of wind speed, and also is not dependent on the second tower at the opposite end of this building, a "solar chimney" through which air from the house is discharged. Temperature and flow rate of air delivered to a building by a passive cooling tower can be approximated from Fig. 5.20, but such data are based on very little experimental work. A newer building using this principle, as well as Trombe wall passive solar heating, is the Zion National Park Visitor Center (Fig. 5.21). More details are found in Section 5.14 (*e*).

EXAMPLE 5.5 A 3000 ft² (279 m²) retail store in Las Vegas, Nevada, has a summer heat gain of 14 Btu/h ft² (44.2 W m²). How large a passive cooling tower would be needed under summer design conditions?

SOLUTION. From Appendix A, Las Vegas has a summer DB = 106°F with a mean coincident WB = 65°F. This represents a "wet-bulb depression" (DB − WB) of 41F°. From Fig. 5.20*a*, at 106°F DB and 41F° (DB − WB), the air exiting a passive cooling tower will be about 69°F.

Next, determine the amount of supply air at 69°F needed to maintain this building at an indoor temperature of 82°F (probably a reasonable indoor temperature when it is 106°F outdoors, given the air motion provided by this passive cooling tower). From Chapter 4, Section 4.9*a*:

$$Cfm = (Btu/h)/(1.08)(\Delta T)$$
$$= (3000\ ft^2 \times 14\ Btu/h\ ft^2)/(1.08)(82 - 69)$$
$$= about\ 3000\ cfm$$

At the Las Vegas design condition of 106 DB, 41F° DB − WB, from Fig. 5.20*c*, it appears that a passive cooling tower of 40 ft (12 m) in height, with a total wetted pad area of 32 ft² (3 m²), will provide about 3300 cfm (1.67 m²/s), somewhat more than the minimum just calculated. ∎

(f) Roof Ponds. This promising strategy has rarely been implemented, possibly due to water phobia among architects and clients alike. Yet it demonstrates the most stable interior temperatures of any of these techniques, needing electricity only to open and close the sliding insulation panels on the roof. Roof ponds sized for cooling will likely be nearly equal in area to the floors of the buildings they cool. Average pond depth is between 3 and 6 in. (75 and 150 mm). A more detailed and precise procedure for determining

(a) Exit air temperatures

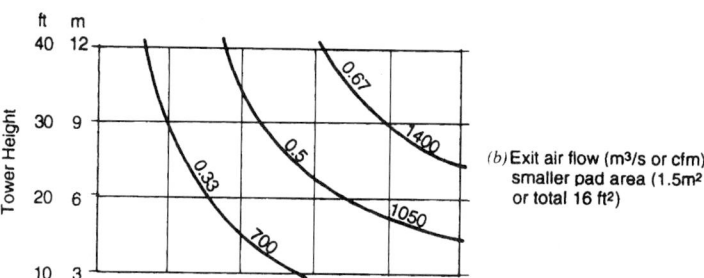

(b) Exit air flow (m³/s or cfm), smaller pad area (1.5m² or total 16 ft²)

(c) Exit air flow (m³/s or cfm) larger pad area (3m² or 32 ft² total)

(Design DB) - (mean coincident WB)

Fig. 5.20 *Design guidelines for cooltowers. (a) The temperature of the air as it leaves the cooltower (enters the building) depends on the outdoor DB, as well as the difference between the DB and WB temperatures. The flow rate of the exit air depends on the tower height, as well as the difference between the DB and WB temperatures. (b) Flow rate for a smaller area of wetted pads, 1.5 m² (16 ft²). (c) Flow rate for a larger area of wetted pads, 3 m² (32 ft²). Based on work of Givoni (1994).*

SOUTH ELEVATION
VISITOR TRANSIT CENTER
ZION NATIONAL PARK

(a)

SECTION THROUGH COOL TOWER
VISITOR TRANSIT CENTER
ZION NATIONAL PARK

(b)

Fig. 5.21 *The new Visitor Center at Zion National Park (Utah) features both cooling by cooltowers and solar heating* (a) *by Trombe walls and a clerestory. Because many of the building's functions can be performed outside, each cooltower* (b) *has both an indoor and an outdoor outlet, either of which can be closed off when not needed. Air enters through wet media pads at the top of the tower* (c). *An X-baffle then diverts the air downward. The building also features PV and water harvesting. (Courtesy of the National Park Service, Denver Service Center.)*

(c)

Fig. 5.21 (continued)

pond area is presented in Section 5.14, but here is the quick approximation:

- pond maximum temperature: 80°F

- pond minimum temperature = minimum nighttime DB (= max. daytime temp − mean daily range, from Appendix A)

- pond ΔT = pond maximum of 80°F − pond minimum

- pond's allowable daily heat stored (from building), Btu/day-ft^2 = (0.7) (pond ΔT) (pond depth, feet*) (62.5 lb./ft^3 water)(1.0 Btu/lb. °F)

- (this equation assumes 70% of the pond's daily heat gain from the building below and 30% through the insulated panels above the pond)

- required size of pond (ft^2 pond per ft^2 floor area) =

$$\frac{\text{building total heat gain per day (Btu/day-ft}^2\text{ floor area)}}{\text{pond allowable heat stored per day (Btu/day-ft}^2\text{ floor area)}}$$

See Table 5.9 for a comparison of the design guideline and the more detailed calculations for roof ponds.

*Depth not to exceed 0.75 ft.

EXAMPLE 5.6, PART A. An office building in Albuquerque, New Mexico, is to be cooled by a roof pond 4 in. deep. An hourly heat gain of 15 Btu/h-ft^2 is assumed (excluding heat gain through the roof), somewhat more than the Oregon office building of Example 5.1.

SOLUTION. For Albuquerque:

Maximum temperature = 94°F (Appendix A)

Mean daily range = 27F° (Appendix A)

Minimum (night) temperature = 94 − 27 = 67°F

Therefore:

Pond ΔT = 80 maximum − 67 minimum = 13F°

Pond storage capacity for building heat =

0.7(13°)(0.33 ft)(62.5 lb./ft^3)(1 Btu/lb. °F)
$$= 188 \text{ Btu/day-ft}^2$$

The required pond size, then, is

$$\frac{15 \text{ Btu/h-ft}^2 \times 9 \text{ h/day}}{188 \text{ Btu/day-ft}^2} = 0.72 \text{ ft}^2/\text{ft}^2 \text{ floor area}$$

Therefore, a 4-in. pond covering 72% of the one-story building's floor area will be approximately large enough to cool the building. (If the entire ceiling is desirable as a cooled surface, then a slightly more shallow pond of greater area could be used. However, the sliding insulation panels must be stacked somewhere beyond the edge of the roof pond.)

We revisit this building as Example 5.6 Part B, in Section 5.14. ■

The residence in Fig. 5.22 is one of the first roof-pond structures. Located in Atascadero, California, at about 35° N. latitude, it has a remarkable history of thermal stability, serving both as passive solar heating in winter and passive cooling in summer (nearby Paso Robles has 2976 HDD65 and 5349 CDH74). Roof-pond inventor Harold Hay donated the building to California State Polytechnic at San Luis Obispo in the late 1990s. To emphasize the pond as the solar heating device, this house has almost no south-facing glass (most windows face east toward a small lake). The sliding insulation panels are stacked when open (by day in winter, by night in summer) over the carport on the north end of the building.

(a)

(b) Roof pond cycle in summer

(b)

Fig. 5.22 *One of the first roof pond buildings, this Atascadero, California, residence is now a research building for California State Polytechnic University at San Luis Obispo. (a) The exterior from the west. (b) Sections showing summer day roof pond insulation and night exposure. (c) Sections showing winter day roof pond exposure and night insulation. (d) Year-long record of indoor temperature range, compared to outdoor. (e) Plan. South exposure is minimal; the roof is the surface of interest for heating and cooling. (f) The roof insulating panels in closed position (as on a summer day). (g) The roof insulating panels in open position (as on a summer night). Water is contained within plastic bags laid directly on the metal deck ceiling. (Drawings based in part on Sandia National Laboratory (1977)* Passive Solar Buildings: A Compilation of Data and Results, *Sand 77-1204.)*

Roof pond cycle in winter

(c)

(d)

(e)

(f)   (g)

Fig. 5.22 (continued)

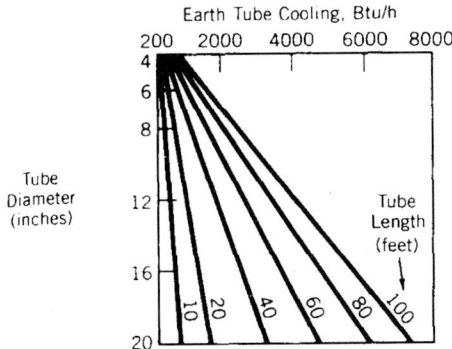

Earth Tube Cooling, Btu/h

Tube Diameter (inches)

Tube Length (feet)

Fig. 5.23 *Cooling of intake air provided by an earth tube, in Btu/h. This tube is set in heavy, damp soil (k = 0.75 Btu/h-°F-ft) whose temperature around the tube is 65°F. The outdoor air entering the tube is at 85°F and is moved at a flow velocity of 500 fpm. (Adapted by permission from Donald W. Abrams,* Low Energy Cooling, *© 1986, Van Nostrand Reinhold Company.)*

(g) Earth Tubes. These provide a way to cool outdoor air before it enters a building. A fan is used to force sufficient quantities of air through these long tubes. Because earth tubes need to be well underground as well as of rather great length in order to cool outdoor air, it is rarely economical to install enough earth tubes to completely meet a building's need for cooling. If long trenches are needed for another purpose (underground water lines, for example), an earth tube is more feasible. Where earth tubes are contemplated, that component of building heat gain that is represented by

cooling fresh air (Table 5.8, Part E) can sometimes be met using this strategy.

The estimate of cooling for the earth tubes described in Fig. 5.23 is based on Abrams (1986), who assumed a soil conductivity equal to that of heavy, damp earth. One of the most influential variables in earth tube performance is soil conductivity, so Fig. 5.24 (also from Abrams) compares cooling performance of 6-in. (152-mm) diameter earth tubes when soil conductivity changes. Note that this design guideline presents total Btu/h cooling of *the outdoor air in the tube,* rather than Btu/h-ft² (for a typical ft² of building floor area). Therefore, two adjustments must be made: first, find the Btu/h cooling load of the entire building, and second, adjust for the difference between hotter air outdoors and cooler air desired within the building.

See Section 5.14(g) for a more detailed calculation procedure for earth tubes in which tube depth, soil temperature, and soil conductivity are variable. See Table 5.9 for a comparison of the design guideline with the more detailed calculation procedure.

EXAMPLE 5.7, PART A. A partially underground nature center of 4000 ft² in northern Pennsylvania, design DB = 91°F, has an hourly heat gain of 9.3 Btu/h-ft², of which 2.2 Btu/h-ft² are due to outdoor air required ventilation. If earth tubes are used only for cooling the fresh air, how many tubes at what length will be needed?

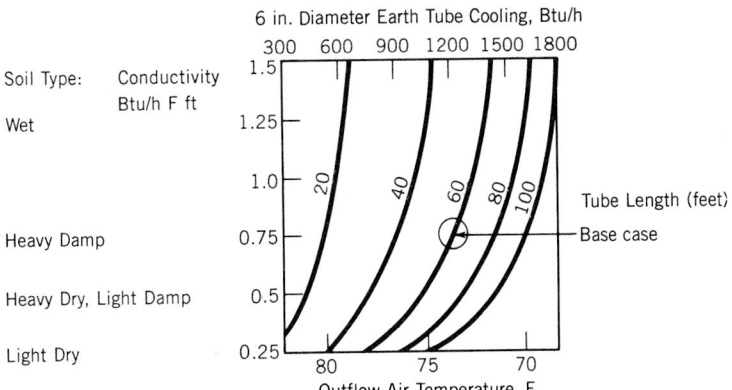

6 in. Diameter Earth Tube Cooling, Btu/h

Soil Type: Conductivity Btu/h F ft

Wet

Heavy Damp

Heavy Dry, Light Damp

Light Dry

Tube Length (feet)

Base case

Outflow Air Temperature, F

Fig. 5.24 *Six-inch diameter earth tube cooling capacities as soil conductivity varies. The "base case" is a 6-in. diameter earth tube, 60 ft long, shown in Fig. 5.23 to deliver 1200 Btu/h of cooling at soil conductivity 0.75, with other conditions as specified in Fig. 5.23. (Adapted by permission from Donald W. Abrams,* Low Energy Cooling, *© 1986, Van Nostrand Reinhold Company.)*

THERMAL CONTROL

SOLUTION. First, the total building cooling needed from the earth tubes is

$$(4000\ \text{ft}^2)(2.2\ \text{Btu/h ft}^2) = 8800\ \text{Btu/h}.$$

Next, adjust this figure to account for hot outdoor air versus cooler indoor air. If we accept 82°F indoors (the walls are earth sheltered and therefore cooler), then the total earth tube cooling needed is (8800 Btu/h)(91°/82°), or about 9770 Btu/h.

From Fig. 5.23, assuming heavy damp soil, a 60-ft long tube of 20 in. diameter will deliver about 4500 Btu/h; two such tubes will deliver about 9000 Btu/h.

We return to this example in Section 5.14(*g*), Example 5.7, Part B. ■

Table 5.9 compares the design guuuidelines for passive cooling with the more detailed calculation procedures that are given in Chapter 4 (basement walls and floor) and later in Section 5.14.

5.7 Reintegrating Daylighting, Passive Solar Heating, and Cooling

At this point, preliminary investigations of envelope component *U*-values and fenestration/floor area ratios have proceeded through various codes, standards, and design guidelines. It is quite likely that conflicting advice has resulted. Depending on the location, building type, and client's objectives, a very different mix of glass, insulation, and thermal mass could result from prescriptive conservation approaches, compared to passive solar heating, or any of several approaches to passive cooling.

When prescriptive standards (such as presented in Appendix J) appear to preclude either passive solar heating or passive cooling strategies in a building, the designer should remember that there are procedures for either window area tradeoffs (in which more insulation in walls or roofs might allow more glass areas) or methods to compare whole-building annual energy consumption that can demonstrate the superiority of using well-designed passive strategies over a simple conventional prescribed approach.

EXAMPLE 5.8. We revisit the one story residence in Billings, Montana, analyzed in Example 4.4. How does the prescribed fenestration area for conventional thermal performance compare with those for daylight, solar heating and ventilation cooling?

SOLUTION. In Example 4.4, windows that were ≥15% of floor area needed a maximum *U*-value of 0.36 (Fig. J.15A), and maximum SC = 0.7 (Fig. J.16A). From Table 4.13, windows #7 through #12 qualified for both maximum *U* and SC (considered equivalent to SHGC for this purpose).

We now need to assume the floor area and dimensions in order to proceed. A plan 32 ft wide by 64 ft long (long walls to face north/south) yields a floor area of 2048 ft². Most windows can be placed in the long walls for both daylight distribution and cross ventilation. However, the south wall will need the most window area to facilitate solar heating. Tentatively, assign window area as follows:

South window area = 8% of floor area
North window area = 5% of floor area
East and west window area = 2% (1% each) of
                                              floor area
                                    ____
                                    15%

Daylight: For sidelighting, DF_{av} = 0.2 [window area/floor area], where all the floor area is within 2.5 *H* of the window wall. If we assume that *H* = 7.5 ft (ceiling at wall 8 ft high), then the daylight area is 2.5 × 7.5 = 18.75 deep. Light from opposite exterior walls produces the maximum daylight width of 2 × 18.75 = 37 ft. With our assumed maximum 32-ft width, all floor area is within the daylight zone. For our total of 15% of floor area in fenestration, DF_{av} = 0.2 [0.15] = .03, or 3%. This seems to be sufficient according to the target DFs in Table 5.2.

Passive Solar Heating: From Example 4.3, windows #7 through #12, Table 4.13, are under consideration. From Table 5.6, footnote *c*, windows #7 and #12 qualify as "superior performance" glazings. For Billings, Montana, a "low" ratio of 0.16 south glass/floor area would produce a SSF = 53%. From Table 5.7, such an SSF would require at least 3.8 ft² interior masonry surface for each square foot of south glass. However, we have allocated only 8% south window area/floor area. The designer can:

· reallocate window area so that essentially all windows face south (with daylight and cross-ventilation penalties), or

· increase the total window area to more than 15%, with correspondingly higher fenestration U-values required (from Table J.1, up to 21% of floor area would require $U = 0.29$), or

· settle for a lower SSF with only 8% south window area/floor area.

Cross Ventilation: First we need a quick estimate of heat gains. From Table 5.8:

A. Internal gains people and equipment: assume 4 people at 230 Btu/h plus 1200 Btu/h appliances:

$$(4 \times 230) + 1200 = 2120 \text{ Btu/h}$$

$$2120 \text{ Btu/h}/2048 \text{ ft}^2 = \textbf{1.04 Btu/h ft}^2$$

B. Lights: assume with 3% DF a gain of about **2.0 Btu/h ft^2**

C. Gain through windows: Billings' design temperature (Appendix A) is 91°F. Externally shaded window gains = $(0.15) \times 16 = \textbf{2.4 Btu/h ft}^2$ Gain through walls: total vertical area = $[(2 \times 32 \text{ ft}) + (2 \times 64 \text{ ft})] = 1536 \text{ ft}^2$. Of this, window area = $(0.15 \times 2048) = 307 \text{ ft}^2$; opaque wall area = $1536 - 307 = 1229$.
From Example 4.3, frame walls must have maximum $U = 0.073$, whereas masonry walls have maximum U-value: 0.08. Therefore,

gain through walls = [1229 sq ft wall area/

2048 sq ft floor area] $\times 0.08 \times 15$

$$= \textbf{0.72 Btu/h ft}^2$$

Gain through roof: assume roof area equal to floor area, and from Example 4.3, maximum U-value: 0.033. Therefore,

gain through roof = $(1.0) \times 0.033 \times 35$

$$= \textbf{1.16 Btu/h ft}^2$$

D. For an "open" building (using cross ventilation),

total heat gain = $1.04 + 2.0 + 2.4 + 0.72 + 1.16$

$$= \textbf{7.32 Btu/h ft}^2$$

We can assume a total openable area of perhaps 90% of the glazed area, using fully openable casement windows. With 8% floor area in south windows and 5% in north windows, the maximum available ventilation area = $0.9 \times 5\% = 4.5\%$ of the floor area. From Fig. 5.14*a*, it appears that this rate of heat gain can be removed through this inlet area by a wind velocity of only about 2 mph. (However, when it is 91°F outdoors, that suggests 94°F indoors! Another cooling strategy is highly likely.)

Summary: Probably increase south glass area (and overall ratio window/floor area) and use windows with lower U-values, with summer external shading. Floor and many interior wall surfaces will be faced with thermally massive materials. ■

5.8 Calculating Worst-Hourly Heat Loss

Sections 4.7 to 4.9 showed how heat is exchanged by conductance through a building's envelope (q) and through the ventilation of a building (q_v). By combining these rates of heat exchange, we can obtain a building's total hourly heat loss in winter. (To calculate the total hourly heat gain in summer, other gains must be added; this more complicated procedure is discussed in Sections 5.11 and 5.13). The total hourly heat loss of a building can be calculated under several different assumptions reflecting different purposes.

(a) Maximum Hourly Loss: Sizing Conventional Heating Equipment. The most typical use of $q + q_v$ is to determine the maximum amount of heat per hour that heaters must provide. Two important assumptions are usually made:

1. No internal heat gains (lights, people, etc.) and no solar gains are present in the building.
2. The lowest outdoor temperature ("design condition"—see Appendix A) is occurring.

These are conservative assumptions that lead to the installation of heating equipment that is rarely used to capacity. Such equipment does provide a

THERMAL CONTROL

safety margin for those rare tin...
lower than design temperatures...
windows are inadvertently left op...
porary and unexpected heat lea...
cold weather.

Thus, to obtain design hou...
culate

$$q_{\text{total}} = q + c$$

where

$$q = (\Sigma UA)\,\Delta T$$

with ΣUA the sum, for all exposed components of the building's envelope, of $U \times A$; $\Delta T =$ (interior–exterior design temperature); and

$$q_v = 1.08\,V\,\Delta T \qquad (1.2V\,\Delta T;\,\text{SI units})$$

where V is the volume in cfm (L/s) of outdoor air introduced.

(b) Maximum Hourly Heat Loss: Sizing Auxiliary Heating for Passive Solar Buildings.
The one important difference between the maximum heat loss calculations for conventional and passive solar buildings is the following assumption:

No internal heat gains (lights, people, etc.) are present, but there is sufficient stored solar energy to at least cancel out the heat losses through the south solar collection area.

Otherwise, the procedure is the same as that for conventional buildings:

$$q_{\text{total}} = q + q_v$$

where $q = \Sigma UA_{\text{ns}}\,\Delta T$
$q_v = 0.018\,V\,\Delta T$ (ΔT again based on the design condition, Appendix A)
$(1.2\,\Delta T;\,\text{SI units})$
ΣUA_{ns} *excludes* the solar collector area

This can lead to occasional chilly interiors, as when several days of heavily overcast skies coincide with design-condition outdoor temperatures. In some locations, therefore, designers use the more conservative, conventional procedure for sizing the auxiliary heaters for passive solar buildings.

...cking De-
...produce q
...a building's
...e heat loss is
...opportunities
...ased comfort.
...s is occurring
...wall, for exam-
...s:

...chitectural and
...may override.)
2. Go to ... -value to reduce heat loss and increase ...e winter surface temperature of this large glass area. (Cost and detailing considerations may override.)
3. Add thermal shades or shutters to dramatically reduce heat loss and increase winter surface temperature. (Architectural, view, cost, and detailing considerations may override.)

If much of the building's heat loss occurs through ventilation air, consider these options:

1. Reduce infiltration, by tighter construction (or reduce mechanical ventilation toward the code-required minimum).
2. Add a heat exchanger between outgoing and incoming air.

These calculations can also be used to check your building against published criteria for thermal performance (see Appendix J). Redesign of building envelopes to meet such criteria is fairly common in the early stages of building design.

(d) Hourly Rates of Fuel Consumption.
When outdoor temperatures in winter drop below the building balance point (see Section 5.9), heating systems usually begin to operate. The hourly rate of fuel consumption depends on the hourly heat loss from the interior space. If the boiler (or furnace) is selected to run continuously at the outdoor, critical winter design temperature, then it will cycle (run intermittently) at higher outdoor temperatures. The equipment, however, *is* selected on the basis of the maximum winter demand rate and therefore relates to the calculated heat loss at the design temperature. Sometimes your energy supplier will ask for the maximum hourly rate.

TABLE 5.10 **Approximate Heat Values of Fuels**

Fuel	Heat Value		Typical Seasonal Efficiency, Percent
	Conventional Units	SI Units	
Anthracite coal	14,600 Btu/lb	33,980 kJ/kg	65–75
No. 2 oil	141,000 Btu/gal	39,300 kJ/L	70–80[c]
Natural gas[a]	1,050 Btu/ft³	39,100 kJ/m³	70–80[c]
Propane	2,500 Btu/ft³	93,150 kJ/m³	70–80[c]
Electricity	3,413 Btu/kW	1 kW	95–100
Wood[b]	7,000 Btu/lb	16,290 kJ/kg	30–50

NOTE: This table includes the thermal value of electricity used on-site (but not losses in fuel energy at the electrical generating plant). Approximate seasonal efficiencies of typical burner–boiler equipment are also shown.

[a]Natural gas is frequently sold in "therms"; one therm = 100,000 Btu/h (SI units, 29.3 kW).

[b]At 20% moisture content.

[c]Higher efficiency furnaces are now available.

EXAMPLE 5.9. Calculate the rates of burning for several fuels (or the rate of using electricity) to make up the hourly heat loss, under design conditions, of a mercantile store. Its maximum hourly heat loss is 159,840 Btu/h. For fuel values, refer to the data in Table 5.10.

SOLUTION. If, for instance, oil were used, the statement would be

gal/h × Btu/gal heat value × efficiency
$$= \text{Btu/h heat loss}$$

Transposing, we have

$$\text{gal/h} = \frac{\text{Btu/h heat loss}}{\text{Btu/gal} \times \text{efficiency}}$$

(Other efficiency statements are similar.) Applying values to this and to statements for the other fuels, we obtain the rates

Coal: 159,840/14,600 × 0.70 = 15.6 lb/h

Oil: 159,840/141,000 × 0.75 = 1.51 gal/h

Gas (older, less efficient equipment): 159,840/ 1052 × 0.75 = 203 cfh

Electricity: 159,840/3.41 × 1.00 = 46,600 W (46.6 kW)

These results are based on the assumption that the boiler and its piping are enclosed within the useful volume of the store. If they were in cold basements, or if the ducts or pipes ran through unheated space, more fuel would be used and system efficiency would decline. The rates estab-lished set the values by which the fuel-burning apparatus is selected. For instance, if oil were used, a nozzle that would inject oil at the rate of about 1 gph should be tried. The above rates are for design (extreme) conditions and are not typical of the lower average rate of operation throughout the winter. ∎

5.9 Calculations for Heating-Season Fuel Consumption (Conventional Buildings)

The following method of estimating the fuel used for space heating in a typical season best applies to residences and small commercial buildings that are skin-load dominated and not passively solar heated beyond SSF = 10%. To the extent that the combination of internal and solar gains can be predicted accurately, this method yields a reasonable estimate of annual fuel consumption for any building. (For passively solar heated buildings, use instead the method given in Section 5.10.)

Internal and solar gains make almost any building warmer than the outdoors during the heating season. The furnace (or other space-heating device) is not needed until the outdoor temperature drops to the point at which these internal and solar gains are insufficient to heat the building by themselves; that is, when the heat lost through the building's skin and infiltration matches the heat gained through solar plus internal loads. This particular outdoor temperature is called the *bal-*

ance point; it represents the beginning of the need for space-heating equipment.

To estimate the annual energy needed for a building's space heating, it is necessary to know the following:

- The building's heat-loss rate (envelope and infiltration).
- The building's internal plus solar gain rate.
- The building's balance point temperature.
- The time period during which the outside temperature falls below the building's balance point temperature ("degree days").

(a) The Balance Point Temperature.

When a building needs neither heating or cooling, the internal gains equal the external losses:

$$Q_i = \text{balance point } q_{\text{total}}$$

where Q_i = internal gains plus solar gains (Btu/h, or W), and

balance point
$$q_{\text{total}} = UA_{\text{total}} (T_i - T_b)$$

T_b = balance point temperature

T_i = average interior temperature over 24 h, winter

UA_{total} = total heat loss rate (envelope plus infiltration)(Btu/h-°F, or W/°C)

Rewriting this equation to solve for the balance point temperature:

$$T_b = T_i - \frac{Q_i}{UA_{\text{total}}}$$

To determine the total heat loss rate UA_{total}, combine the envelope (or skin) losses ΣUA (as was done in Sections 4.7 and 5.8) and the infiltration (or ventilation) losses $1.08V$ (or, in SI units, $1.2V$), where V is the volume in cfm (or L/s).

The quantity Q_i cannot be determined so straightforwardly. The internal gains can be estimated as shown in Table 5.8 (remember, these are summertime gains) or calculated more precisely from known building population, lighting, and equipment data. For residences, the following *daily total* internal gains are considered typical:

People: Two adults and two children (average times of occupancy): 23,000 to 24,500 Btu/day (6.7 to 7.2 kWh/day)

Lights and equipment: See Table 5.11 for individual heat sources, but if actual appliances are unknown, then:

53,000 Btu/day (15.5 kWh/day) for energy efficient (as of 1980) equipment

100,000 Btu/day (29.3 kWh/day) for typical (as of 1970) equipment.

For more details on internal gains in residences, see the 1997 *ASHRAE Handbook of Fundamentals,* Chapter 27.

The solar gains are elusive; each month has a different average gain. For simplicity in the heating season, use the average January daily insolation on a vertical surface (found in Appendix B) with this approach.

During the heating season:

$$Q_i = \frac{\text{internal gains (Btu/day)}}{24 \text{ h}}$$
$$+ \frac{\left[\begin{array}{c} \text{January insolation} \qquad \text{area (ft}^2\text{),} \\ \text{(Btu/ft}^2 \text{ day average)}, \quad \times \quad \text{south} \\ \text{vertical surface} \qquad \text{glass} \end{array} \right]}{24 \text{ h}}$$

The balance point temperature, T_b, can be used to do several things besides predict fuel consumption. By noting the T_b on a graph of monthly outdoor temperatures, the designer can quickly see the relative importance of heating versus cooling for a specific building in a given climate. Also, it can be used to gain a better understanding of how zones in a building interrelate. Once the T_b is calculated for each zone, it can be determined when the entire building needs heating (outdoor temperature below any zone's T_b) or cooling (outdoor temperature above any zone's T_b), or when one zone's surplus heat can be another zone's space heating source (outdoor temperature higher than one zone's T_b, but lower than that of another). If thermal exchange between zones occurs for a major portion of the typical year, the choice of either a heating or a cooling strategy could be influenced.

TABLE 5.11 **Typical Residential Daily Internal Heat Gains from Appliances and Lighting**

*Part A. **Electric Appliances***

Heat Source	Btu/day	kWh/day
Frost-free refrigerator	16,000	4.7
Freezer	10,900–15,700	3.2–4.6
Dryer[a]	8,900–11,600	2.6–3.4
Range[a]	6,500–11,300	1.9–3.3
Television[b]	3,400–4,400	1.0–1.3
Dishwasher	2,400	0.7
Lights and miscellaneous[c]	9,500–18,800	2.8–5.5
Water heater[d]	13,700–27,400	4.0–8.0

*Part B. **Gas Appliances***

Heat Source	Btu/day
Water heater	13,700–27,400
Dryer[a]	12,000–19,000
Range[a]	12,000–27,000

Source: A New Prosperity: The SERI Solar Conservation Study, Brick House Publishing, Andover, Mass., 1981. Adapted by permission from the 1989 *Handbook of Fundamentals,* published by the American Society of Heating, Refrigerating, and Air-Conditioning Engineers, Inc., Atlanta, Ga.

[a]These are for appliance's consumption per day; heat gain to house is less, depending on the amount of heated exhaust air.

[b]Total daily use of TV per household.

[c]These figures are for an average 1350-ft^2 house; rate per square foot of floor area may be extrapolated.

[d]Standby heat loss from water heater to house.

EXAMPLE 5.1, PART G. Calculate the January balance point temperature for a typical bay of the office building in Fig. 5.7. Assume that the average indoor January temperature over a 24-h period is an energy-conserving 68°F.

SOLUTION. The (summer) internal gains from Example 5.1, Part D, may need some modifications. In darker winter, assume the gain from electric lights are closer to the 2<DF<4 value of 2.0 Btu/h ft^2. The internal gain is then 5.8 + 2.0 = 7.8 Btu/h-ft^2 from people, equipment, and electric lights. The average internal gains are as follows:

$$\frac{7.9\,\text{Btu/h-ft}^2 \times 9\,\text{h}}{24\,\text{h}} = 2.96\,\text{Btu/h-ft}^2$$

The January solar gain, *VS*, from Appendix B, Table B.15, for Salem, Oregon, is 471 Btu/ft^2 -

day per ft^2 of south glass:

$$\frac{471\,\text{Btu/h-ft}^2 \times (97\,\text{ft}^2 + 132\,\text{ft}^2)\,\text{south glass}}{24\,\text{h} \times 1440\,\text{ft}^2}$$

$$= 3.1\,\text{Btu/h-ft}^2$$

Therefore,

$$Q_i = 2.96 + 3.1 = 6.06\,\text{Btu/h-ft}^2$$

UA for the typical bay (from Table 5.4 and Example 5.1, Part B) includes, for this calculation, all elements of the building envelope:

$$(42 + 25 + 72 + 72 + 98 + 174\text{ infiltration})$$
$$= 483\,\text{Btu/h-°F},$$

expressed per square foot of floor area,

$$483/1440 = 0.34\,\text{Btu/h-°F-ft}^2$$

THERMAL CONTROL

Therefore,

$$T_b = 68°F - \frac{6.06 \text{ Btu/h-ft}^2}{0.34 \text{ Btu/h-°F-ft}^2} = 68 - 17.8$$

$$= 50.2°F$$

Consulting NOAA data for Salem, Oregon, 50.2°F, although based on January insolation, is closest to the daily average temperature for the months of April and October. ■

(b) Degree Days. These data are published for each climate station and are calculated to various *base* temperatures (found in Appendix B). Heating degree days to a specified base temperature, such as HDD65, refer to average temperatures *below* the base temperature of 65°F. Cooling degree days to a specified base temperature, such as CDD65, refer to average temperatures *above* the base temperature of 65°F. (Cooling degree hours such as CDH74 similarly refer to the average number of hours above a base temperature of 74°F.) For smaller buildings, such as residences, heating degree days are in much wider usage than are cooling degree days (or hours), and many data sources simply list DD65; this notation is interchangeable with HDD65, and is used for the remainder of this section. Until recently, degree days (DD) were always based on 65°F, because older, indifferently insulated buildings, with low internal gains, had a typical balance point of about 65°F. The combination of much higher levels of insulation and much more electric equipment has shoved the average building's balance point temperature downward; hence DD50, DD55, and DD60 are included with the traditional DD65 in Appendix B.

To derive heating DD for a particular climate and X base temperature DDX, each day's mean temperature (halfway between high and low) is subtracted from the base temperature; the result is the number of DDX for that day. If the mean temperature equals or exceeds the base temperature, no DDX are recorded. Then the DDX are totaled for an average year.

For example, assume that a day in Troy, New York, had a high of 60°F and a low of 34°F. The mean temperature was $[60-34/2] + 34 = 47°F$.

$$65 - 47 = 18 \text{ DD65}$$

$$60 - 47 = 13 \text{ DD60}$$

$$55 - 47 = 8 \text{ DD55}$$

$$50 - 47 = 3 \text{ DD50}$$

Clearly, a building with a 65°F balance point will need more heat that day than will a building with a 50°F balance point temperature.

To convert DD I-P (inch-pound) to DD SI, simply multiply DD I-P by ⁵⁄₉:

$$DDSI = 0.56 \text{ DDI-P}$$

To obtain the "DD balance point" needed to estimate your particular building's heating needs, interpolate between the various base DDs as required. (If your balance point is below 50°F, get lower DD base figures from your local weather station. Do not extrapolate!)

(c) Yearly Space Heating Energy. To estimate the energy needed over an average year for a building, E, calculate

$$E = \frac{(UA)(\text{DD balance point})(24 \text{ h})}{(AFUE)(V)}$$

where E is in units of fuel consumed per year (therms of gas or kWh of electricity).

UA is the total heat loss rate, envelope + infiltration (Btu/h-°F, or W/°C).

DD balance point is obtained as just described.

$AFUE$ is the annual fuel utilization efficiency, displayed on all furnaces manufactured within the United States (Table 5.12).

V is the heating value of the fuel, from Table 5.10.

EXAMPLE 5.10. A residence in Springfield, Illinois, has a total heat loss rate UA of 544 Btu/h-°F and a balance point temperature of 55°F. It will have a natural gas condensing furnace, located indoors, for which $AFUE = 0.93$. For Springfield, DD55 = 3434. The approximate average annual energy used for space heating is

$$E_{\text{therms}} = \frac{(544 \text{ Btu/h-°F})(3434 \text{ DD})(24 \text{ h/day})}{0.93 \times 100,000 \text{ Btu/therm}}$$

$$= 482 \text{ therms} \quad ■$$

TABLE 5.12 **Some Typical Values of Annual Fuel Utilization Efficiency (AFUE)**

Type of Furnace	AFUE (%)	
	Indoor	ICS[a]
Natural Gas		
Natural-draft with standing pilot[b]	64.5	63.9
Natural-draft with intermittent ignition[b]	69.0	68.5
Natural-draft with intermittent ignition and auto vent damper[b]	78.0	68.5
Fan-assisted combustion with standing pilot or intermittent ignition	80.0	78.0
Fan-assisted combustion with improved heat transfer, with standing pilot or intermittent ignition	82.0	80.0
Direct vent with standing pilot, preheat[b]	66.0	64.5
Direct vent, fan-assisted combustion, and intermittent ignition	80.0	78.0
Fan-assisted combustion (induced draft)	80.0	78.0
Condensing	93.0+	91.0+
Oil		
Standard[b]	71.0	69.0
Standard with improved heat transfer[b]	76.0	74.0
Standard, with improved heat transfer and auto vent damper[b]	83.0	74.0
Condensing	91.0	89.0

Source: 1996 ASHRAE Systems and Equipment Handbook, copyright © by the American Society of Heating, Refrigerating, and Air-Conditioning Engineers, Inc., Atlanta, GA.

[a]Isolated combustion system: combustion air is drawn from outdoors, not indoors.

[b]Since 1992, all new furnaces are required to have an AFUE (ICS) level of at least 78.0%.

5.10 Passive Solar Heating Performance

Section 5.4 presents the design guidelines for determining solar opening size, thermal mass, and the solar savings fraction (SSF). As a building design takes shape, more detailed information becomes useful: Which passive system matches the architectural program? Which performs better thermally? For a given solar opening, what exactly is the resulting SSF? How much auxiliary fuel consumption per year must accompany that SSF? If a building overheats on sunny winter days, how hot will it get?

To this point, passive solar heating has been treated as a single approach; the design guidelines for SSF distinguished only between "standard" and "superior" system performance. Important architectural differences, however, characterize the various passive solar heating approaches, as summarized in Table 5.13. On the basis of the wider architectural implications presented in Table 5.13 and the detailed sizing information found in Appendix C, the designer can select a passive solar heating approach with some confidence in its applicability and its yearly need for auxiliary space heating.

(a) Glazing Performance.

In the reference systems of Appendix C, Table C.1, the glazing is assumed to face due south. The choices of glazing conditions include single, double, and triple glazing, and night insulation "no" or "yes." Single, double, and triple glazing with no night insulation are common approaches. Moveable night insulation, once more common (Fig. 5.25), has largely been supplanted by "superwindows." Yet what equivalent of the current superwindows might provide a performance approximately equivalent to "double glazing with R-9 night insulation"?

The 24-hour averaged U-value of the double-glazed fixed window with R-9 night insulation is found as follows: The ordinary double-glazed window, wood/vinyl frame, corresponds to window #5, Table 4.13, with an overall $U = 0.49$. "Night insulation" assumes adding R-9 insulation, in place from 5:30 PM to 7:30 AM. This extra-insulated window should at night then have R-overall of

TABLE 5.13 **Passive Solar Heating Systems Compared**

	Influence on Plan	Heating Characteristics
Direct gain (DG) 	Sun can enter through south windows or skylights; open plan can allow sun and stored heat to serve entire top floor of building. Large areas of thermal mass surface should be darker colored and free of rugs, wall hangings, etc. Light-colored surfaces near glass reduce glare. Outdoor view and access to south are encouraged.	Quick to warm up in the morning; fast response to sun. Tendency to overheat at midday; large temperature swings. Much radiant loss to bare window by night; movable insulation encouraged (or triple glazing or selective film). Warmth spread throughout space along with thermal mass.
Thermal storage wall **Trombe (masonry)** **Wall (TW)** **Water wall (WW)** 	Needs to be on south wall of building. Inner wall of TW or WW should be kept clear of hangings, furniture, etc., but rest of space is unrestricted. Not much solar impact beyond about 25 ft from TW or WW. Outdoor view and access to south are discouraged.	Unvented TW, WW are slow to warm by day, slow to cool by night; small temperature swings. Most radiant heat arrives in evening; comfort is most likely near the TW or WW surface. (Vented TW behave midway between unvented TW and DG systems.)
Sunspace (SS) 	Same influences as TW and WW above; or sunspace can be insulated from building, becoming less efficient heat source for a rock bed. Floor above rock bed should be kept free of rugs. Sunspace becomes a special place with different characteristics from rest of building. Access to sunspace thereby encouraged. Access to south encouraged, view to south filtered.	Sunspace thermally like DG, but with extreme temperature swings and accentuated radiant loss by night. Movable insulation often omitted and night use curtailed. Building beyond is thermally like TW or WW, depending on its connection with sunspace. Warm floor above rock bed in months with SS surplus heat.
Roof pond 	Flat or nearly flat roof is desirable. Skylight is discouraged. Plan is completely unrestricted; sidelighting and views, access to outdoors encouraged.	Low temperature swings; steady temperatures both summer and winter. Winter air stratification possible, since warmest surface is also highest surface in the space.

TABLE 5.13 **Passive Solar Heating Systems Compared** (*continued*)

	Daylighting	Cooling
DG	Very high DF possible, with high glare potential at lower windows. Possible conflict between light-colored surfaces for glare reduction and dark-colored surfaces for solar absorption. Summer shading greatly reduces DF. Encourages both side and top lighting.	Cross ventilation: encouraged by large windows to south. Stack ventilation: helped when clerestories are used. Night ventilation/thermal mass: excellent potential, much mass surface and capacity. Other closed-building cooling: large south windows are big threat. Shading (and lower DF) necessary in overheating months.
TW, WW	No daylighting through TW, unless interrupted by windows (high glare potential). Diffuse light possible through translucent WW. Discourages sidelighting.	Cross ventilation: discouraged by solid TW, WW. Stack ventilation: TW or WW can itself produce stack effect, but with risk of evening overheating. Night ventilation/thermal mass: limited interior mass surface exposure, but lots of mass capacity. Other closed-building cooling: with summer shading of TW or WW, good match with controlled cooling; thermal mass delays and reduces peak heat gains.
SS	Very high DF within sunspace; little or no daylight through common wall, except as encouraged by view and access to surface. Summer shading of sunspace reduces DF.	Cross ventilation: only to extent that common wall is penetrated for view and access. Stack ventilation: SS can become moderately effective stack. Night ventilation/thermal mass: both surfaces of common mass wall are available, but two spaces must be cooled. Other closed-building cooling: with summer shading of SS, good match with controlled cooling (except in SS itself).
Roof pond	Discourages toplighting, encourages sidelighting, with very light color on underside of roof.	Cross ventilation: excellent potential. Stack ventilation: discouraged. Night ventilation/thermal mass: easily achieved with roof undersurface. Other closed-building cooling: roof pond night sky cooling is often sufficient by itself, making other cooling unnecessary.

THERMAL CONTROL

Source: Illustrations from AIA, Ramsey/Sleeper, *Architectural Graphic Standards* 7th ed., © 1981 John Wiley & Sons, Inc., New York. Reprinted by permission.

(a) Absorber parallel   *(b)* Absorber perpendicular

(c) Cavity structure   *(d)* Quasi-homogeneous

Absorber

Temperature graph

Transparent insulation

Air cavity

(e) Trombe Wall

(f)

(g)

Fig. 5.25 *Variations on insulation for solar apertures. Transparent insulation materials (TIMs) in most applications are really more translucent than transparent. (a) Absorber-parallel systems utilize multiple films, as in "superwindows." The more the layers, the greater the optical losses, however. (b) Absorber-vertical structures include honeycomb or capillary materials. Some scattering and absorption of light occurs, but optical losses are lower than with type (a). (c) Cavity structures combine both absorber-parallel and absorber-vertical structures. (d) Quasi-homogenous layers are similar to cavity structures and include aerogel. (e) These translucent materials are particularly effective as insulation for Trombe walls, where the heated mass can be insulated from the cold winter air temperatures yet still receive strong sunlight. (Courtesy of Fraunhofer Institute, Freiburg, Germany and* Solar Today, *American Solar Energy Society.) (f)The Duette window shade, when installed over double glazing, has winter R-values as follows: at ⅜-in. (9.5-mm) thick: translucent, R 3.23; ¾-in. (19-mm) thick: translucent R 3.57; opaque, R 4.2. (g) The shades can move either vertically or horizontally, cover both planar and curved surfaces, and be either motor or manually operated. (Courtesy of Hunter Douglas, Inc., Broomfield, Colorado.)*

at least

$$[1/0.49] + 9 = \text{R-11};$$

then nighttime $U = 1/11 = 0.09$.

$$U_{24\,av} = \left[\frac{(14h)(0.09)] + [(10h)(.49)}{24\ h} \right] = 0.257$$

The equivalent "superwindow" would then have these characteristics, based on window #5's solar gain:

$$U = 0.26, \text{SHGC} = 0.58, VT = 0.57$$

Such a combination may be difficult to obtain. Here are some choices for "superwindow" substitutes for the "double/yes n.i." option within the Appendix C reference types:

· Where listed, choose "triple glazing" option instead of the double/yes n.i.

· Use window #7, Table 4.13, with $U = 0.33$, SHGC = 0.55, $VT = 0.52$, with nearly identical solar gain characteristics, and interpolate between double/yes n.i., and double/no n.i., on the basis of U-values:
  double/no n.i., $U = 0.49$
  double/yes n.i., $U = 0.26$
  window #7, $U = 0.33$
  This places window #7 about one third the difference, closer to double/yes n.i.

· Use the clear triple-glazed window (#12, Table 4.13) in place of double/yes n.i., and given its $U = 0.34$, SHGC = 0.52, $VT = 0.53$, interpolate for U-value as with window #7.

Another option is *transparent insulation materials* (TIM), discussed earlier in Section 4.7c as aerogel (one example). These materials transmit diffuse light and solar radiation, but are opaque to thermal radiation. The four generic types of TIM are shown in Fig. 5.25a–e. The absorber-parallel structures (Fig. 5.25a) are similar to the sheets of spectrally selective materials in "superwindows"; the more layers, however, the less the transmission of solar gain. Absorber-vertical structures (Fig. 5.25b) include honeycomb or capillary materials that reflect and transmit the incoming solar radiation with fewer transmission losses. Combining these structures (Fig. 5.25c) can result in either transparent rectangular cross sections or transparent foam with bubble sizes of a

few millimeters. Quasi-homogeneous layers (Fig. 5.25d) include aerogel. These scatter the incoming sunlight differently than do the type structures shown in Fig. 5.25c.

For passive space heating, TIM use may be limited by the lack of *visual* transparency: these materials diffuse sunlight. They are thus appealing primarily for Trombe wall and water wall applications (Fig. 5.25e), where view through the glazing is unimportant. Some information on U-values and diffuse transmittance is shown in Table 5.14. Again, if we seek a substitute for "moveable insulation over double glazing," we would like $U = 0.26$, SHGC = 0.58, $VT = 0.57$; from Table 5.14, any of the honeycomb or capillary materials in Part A will be close, if we remember to reduce the τ diffuse and slightly increase the U-value, to account for a single cover sheet of glass.

(b) Direct Gain (DG) Systems. These are the most commonly encountered systems because nearly all south-facing spaces have windows, and at least some have areas of internal thermal mass. In Table 5.15, three of the four early examples utilize some DG collection. Direct gain buildings are shown in Figs. 1.12 (Watson residence) and 1.14 (Cottage Restaurant), as well as the Oregon office example in Fig. 5.7. The most common problems are overheating on sunny winter days and inadequate area of thermal mass.

For DG spaces, the thermal mass should be widely distributed around the room so that direct sun can strike the mass surface and/or be reflected to the mass surface as soon as possible on entering the window. Table 5.7 gave design guidelines for mass; the DG reference systems (Appendix C) list either a 3:1 or a 6:1 mass:glass area ratio. However, even more mass area will give a more thermally stable performance, as will soon become clear in Section 5.10(i); a common recommendation for DG spaces above 50% SSF is for a thermal mass area five to seven times the area of glass (for no more than the optimum 4-in. masonry thickness). There are many common ways to provide such mass surfaces, including brick veneer and clay tile over a bed of grout. These surfaces can be applied to frame construction. Floors (such as slab on grade) are easy ways to achieve large mass areas, but carpet and rugs are very popular with occupants!

THERMAL CONTROL

TABLE 5.14 **Transparent Insulation Materials (TIM)**

Part A. *Honeycomb and Capillary Structures*

		U-Values [a,b]	
TIM	τ diffuse[b]	Btu/h ft² °F	W/m² K
Honeycomb polycarbonate,			
thickness 1.97 in. (50 mm)	0.85	0.35	2.0
thickness 3.94 in. (100 mm)	0.78	0.19	1.07
Capillaries, polycarbonate,			
thickness 3.94 in. (100 mm)	0.73	0.17	0.98
Capillaries, PMMA (acrylic glass),			
thickness 3.94 in. (100 mm)	0.80	0.16	0.91

Part B. *Aerogel Between Double Glazing*

		U-Values [a]	
Aerogel Granule Diameter	τ diffuse[b]	Btu/h ft² °F	W/m² K
<0.079 in. (<2 mm)	0.22	0.17	0.98
0.118–0.157 in. (3–4 mm)	0.40	0.18	1.03
0.157–0.236 in. (4–6 mm)	0.42	0.20	1.13
0.236–0.315 in. (6–8 mm)	0.43	0.20	1.15

Source: Wittwer, V., Platzer, W. and Rommell, M. "Transparent Insulation," *Solar Today,* January–February 1991.
[a]Mean temperature 50°F (10°C), ΔT 18F° (10 K)
[b]Part A values are based on the TIM only, without the glass cover necessary in the actual application.

(c) Sunspaces (SS). These are the next most common systems, especially the prefabricated "add-on" type shown in the Kelbaugh house (Table 5.15). Perhaps the most common problem is the expectation that a sunspace will be a "greenhouse" where exotic plants will thrive. But the sunspace is a "thermal slave" to the "thermal master" space behind it; comfortable temperatures in the master space are achieved at the cost of *very wide temperature swings* in the slave space. Consequently, while there are many times in a day when the sunspace is at a comfortable temperature, there are other times when it is not. In the First Village example (Table 5.15 and Fig. 5.26), the sunspace contains the stairs. Although not always thermally comfortable, the passage between floors is at least rapid and visually pleasant. The "master" spaces beyond, on both floors, can choose the degree of interaction with the sunspace by opening doors and windows into it.

When the sunspace's common wall is masonry, the conductive flow of heat from slave to master

is built in. Yet many of the SS systems in Appendix C show an insulated, lightweight frame common wall. It is important to remember that, with such insulated common walls, the sunspace is assumed to contain *a row of water containers,* extending across the full east–west width of the sunspace. These theoretical containers are twice as high as they are wide and sit adjacent to (but not on) the floor and the common wall. They have a volume of 1 ft³ for every 1 ft² of common wall. This takes quite a bit of sunspace floor area.

Both masonry and insulated common wall SS systems assume available thermo-circulation vents in the common wall; the top vents are 8 ft (2.4 m) above the bottom vents, and the top vents constitute 3% of the area of the common wall, as do the bottom vents. All sunspace systems are assumed to have a thermally massive, perimeter-insulated floor slab on grade.

The SS system descriptions are quite explicit as to dimensions (Fig. C.1), but variations from these dimensions are not a problem *if the propor-*

TABLE 5.15 **Design Data for Some Early Passive Solar Buildings**

Name, Function, Location	Floor Area (ft²)	Area Ratio, S. glass / Floor	System Type (Appx. C)	Approx. LCR: ACH = 1.0	ACH = 0.5	Approx. SSF: ACH = 1.0	ACH = 0.5
Dove Publications, Pecos, N.M.							
Office	2660	0.37	DGB1 and WWB4	37[a]	25[a]	43%[a]	51%[a]
Warehouse	5040	0.11	DGC1				
Karen Terry House, Santa Fe, N.M.	850	0.45	DGC1	22	17	52%	60%
Kelbaugh House, Princeton, N.J.	1640[b]	0.49[c]	TWE2, DGB1, and SSB4	13	9	62%	72%
First Village, Unit 1, Santa Fe, N.M.	1800[b]	0.22[c]	SSE1	49	31	38%	52%

[a]Treating office and warehouse as one large zone.
[b]Not including sunspace floor area.
[c]Includes sunspace south glazing, but not sunspace floor area.

(b) Second floor plan

8" insulated
frame wall

14" adobe wall

up

Green house

(a) Ground floor plan

Warm Air vented in cooling mode

Direct gain

Mass in adobe and floor

Fan

Rock bed

(c) Section

Fig. 5.26 *An early and well-known example of sunspace passive solar heating, the Unit 1, First Village residence borders the northeast and northwest sides of the enclosed sunspace. (a) Main floor plan. (b) Upper floor plan. (c) Section showing how hottest air from sunspace (midday to late afternoon) can be forced into a rockbed below the main floor for nighttime radiant heat. (Drawings based in part on Sandia National Laboratory (1977)* Passive Solar Buildings: A Compilation of Data and Results, *Sand 77-1204.)*

tions of length-width-height are maintained. It is the shape rather than the dimensions of each SS type that influences its performance.

(d) Trombe Walls (TW). These systems are considerably less common. When used, they often have rather large DG openings within the Trombe wall for daylight and view. This is the case with the Kelbaugh house (Table 5.15), an unusual example that mixes three passive systems. The main advantage of TW systems is thermal stability; the diurnal temperature swings are less than with most other passive systems. They deliver a large portion of their heat by radiation to the space. The main disadvantages seem to be the loss of view and daylight, and keeping the air space clean between the Trombe mass wall and the glass. Objects that interfere with radiant heat transfer from the interior surface must also be minimized.

Two major choices are *vented* and *unvented.* The vented TW systems provide naturally moving air (via the stack effect between the mass and the glass) in addition to heat conducted through the mass wall. They deliver warmer air sooner than an identical-but-unvented mass wall, thanks to the flow of warmed air. They also introduce dust and dirt to the space between the mass and the glass.

The unvented TW systems deliver heat quite a bit later, with the result of quite low daily temperature variations and warmth arriving in the evening. They are somewhat less efficient than vented TW because the very high temperature at midday between the mass and glass results in more heat flow out through the glass.

The warehouse in Fig. 5.27 utilizes a very high vented TW in the New England winter. The detail shows a simple device to prevent reverse air flow at night. This Vermont warehouse utilizes ceiling fans to prevent stratification of warm air at the ceiling. Daylighting is provided by roof monitors; the TW is vented to the outside in summer.

(e) Water Walls (WW). These are the least common systems of them all. Perhaps, as with roof ponds, it is another case of water phobia? They can be specially made, or made of corrugated galvanized steel culverts, steel drums, or fiberglass-reinforced plastic tubes (for which man-

ufacturers provide suggested installation details). Within water wall containers, some air space should be provided, as water expands when it heats. Either a rust inhibitor or a sacrificial anode should be added to the water within steel containers. Remember that WW systems are opaque, not transparent; the more transparent the tubes, the closer to DG will be their performance. This is discussed later in Section 5.10(*g*).

Specially made WW can fit neatly below windows or anywhere else within exterior wall framing. The San Luis Solar Group Complex includes a house and architectural studio (Fig. 5.28) near Santa Margarita, California. Both buildings utilize daylight (including clerestories and skylights), DG, PV, and domestic water panel collectors; there is also a small hydropower system adjacent to the complex. Of special interest here are the WW steel panels, 9-in. (230-mm) thick and painted on the outer surface with a black selective paint, behind double glazing. The selective surface paint is an excellent absorber of short-wave solar radiation, but a poor emitter of long-wave radiation from the heated black surface. The inside steel surface is painted and exposed to the space to be heated. These water walls are used in the conference room of the studio, the dining area, and the upstairs bedroom. They are unobtrusive and correspond approximately to WWC2, Appendix A.

(f) Load Collector Ratio (LCR) Annual Performance. The following procedure is based on Balcomb et al. (1984), *Passive Solar Heating Analysis,* published by ASHRAE, and reprinted by permission. The reference offers a much wider variety of passive systems and a wider network of location listings than can be presented in this book. Along with more "sensitivity curves" to allow prediction for nonstandard passive systems, the reference also provides a much more time-consuming and detailed method for calculating the *monthly* SSF and auxiliary energy needs; the method presented here gives annual results only. Thus, Balcomb et al. (1984) is an important and perhaps indispensable reference for the serious passive solar designer.

The method presented here, called the *load collector ratio* (LCR), yields the annual SSF and

SUMMER VENT
6" WHITE ELBOW

6" φ PIPE
PLUG

EXPANDED
METAL

HINGED DOOR
OPEN WINTER-
CLOSED SUMMER

2 1/4" AIR
SPACE

SPACER DISCS
TYP. 5"φ x 1 1/2"
THICK

PLASTIC FILM
CHECK VALVE

3/4" ± SPACE
FOR PLASTIC
EXPANSION
DEFLECTION

EXPANDED
METAL

16 x 16 GALV
SLEEVE

DOUBLE WALL
PLASTIC PANELS

DRILLED FOR
WEEPS

SCHEMATIC THROUGH TROMBE WALL

(a)

(b)

(c)

Fig. 5.27 *The warehouse for the Famolare shoe company in Brattleboro, Vermont* (a) *uses a huge vented Trombe wall, 20 ft by 184 ft (6.1 m by 56.1 m). The Trombe wall faces 10° east of south. The pond in foreground is an absorption basin for rainwater from the roof and pavement. (Photo by Robert Perron.)* (b) *Interior shows lower and upper vents.* (c) *Simple switches at bottom prevent reverse thermosiphoning on cold nights; seasonal switches at top provide building ventilation, via Trombe wall, in summer. Spacer disks prevent plastic double glazing from deflecting too far inward when its inner surface overheats. (Reprinted from* Architectural Record, *November 1979, © 1979 by McGraw-Hill, Inc. All rights reserved. Reproduced with the permission of the publisher. Courtesy of Banwell White Arnold Hemberger & Partners, Architects, Hanover, NH)*

(a)

Trout Creek
Studio/Office
Research (Photovoltaics above)
Conference
Living

(b)

Fig. 5.28 *The San Luis Solar Group Complex near Santa Margarita, California, includes a house and architectural studio. PV and micro-hydro electricity generation joins daylighting, passive solar heating, and passive cooling in this innovative installation. (a) South facade of studio. (b) Site plan. (c) Section north–south. (d) Exterior of a specially built water wall below a window, serving the dining room. (Drawings courtesy of San Luis Solar Group, Santa Margarita, California.)*

THERMAL CONTROL

(c)

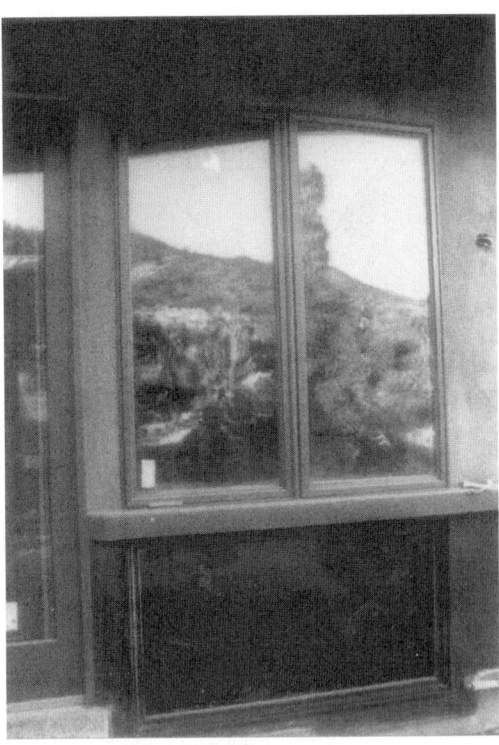

(d)

Fig. 5.28 *(continued)*

auxiliary energy needs for a building. This method has the following steps:

STEP 1. Choose the location and the reference passive system that most closely coincide with your building and its site. The locations listed in Appendices B and C are shown in Fig. B.1; the reference passive systems for which performance data are available (Appendix C) are summarized in Table C.1. If your system differs significantly from the closest reference system, see Section 5.10(g), "Variations on Reference Systems."

STEP 2. Tentatively select a size for the solar openings, balancing the design guidelines for SSF (Table 5.6) with that for daylighting (Table 5.3) and, if applicable, for ventilation (Fig. 5.14).

STEP 3. Calculate the "non-south" envelope heat loss rate, UA_{ns}, for the building design—one that *excludes* the solar openings but *includes* all other envelope losses, as well as the infiltration loss. Then multiply UA_{ns} by 24 h to obtain Btu/DD; this is called the *building load coefficient* (BLC).

$$BLC = 24 \times UA_{ns}$$

STEP 4. Check your building's overall loss rate against the criteria from Table 5.5:

$$Btu/DD\ ft^2 = \frac{BLC}{\text{floor area (ft}^2)}$$

Does your building envelope conserve energy sufficiently, or do you need more insulation (or less non-south glass, or less infiltration)?

STEP 5. Determine the *vertical projection* of the solar opening area A_P. (For a vertical solar opening, A_P is identical to the actual opening area; for a 45°-inclined solar opening, $A_P = 0.707$ actual area.)

STEP 6. Find the LCR, expressed in Btu/DD-ft^2:

$$LCR = \frac{BLC}{A_p}$$

STEP 7. For the reference system that most closely approaches your design, consult Table C.3 (Appendix C) at the appropriate location. By interpolation, find the annual SSF that corresponds to your passive system's listed LCR. Note also the annual DD65 listed in this table.

STEP 8. Finally, determine the approximate annual auxiliary heating Q required:

$$Q = (1 - SSF) \times BLC \times DD$$

Although this quick annual results method is based on the DD65 listed in Table C.3, it is possible to adjust Q to *approximately* account for higher internal gains or better-insulated envelopes. In this adjustment, use DD based on the balance point instead of DD65 in the previous equation.

STEP 9. Now compare the design guideline–predicted relationship between collector size and SSF to the actual one you have just calculated. If the SSF is *smaller* than you had hoped, can you decrease BLC (improve conservation) or increase collector size, or switch to another passive system with a more favorable SSF for the same LCR? If the SSF is *larger*, will you be happy with the increased fuel savings, or will you reduce the collector size, or consider another, less-efficient passive system that has some architectural advantage over the one for which you calculated SSF? Table 5.15 shows four early and well-known passive solar buildings, with system type and approximate LCR.

EXAMPLE 5.1, PART H. What is the annual SSF for the Oregon office building of Fig. 5.7?

SOLUTION. The reference passive system most closely corresponding to the direct gain in a typical bay is found in Appendix C, Table C.1, Part D: DGC1 has a high thermal storage capacity, based on an average thermal mass thickness of 4 in., and a ratio of mass to south glass area of 6:1. It has "standard performance" double-glazed windows with no night insulation. (From Table 5.4, the actual building's mass to south glass area ratio is 2232 ft^2/229 ft^2 = almost 10:1.) The BLC of the typical bay is

24 h/day $\times$ (42 + 25 + 72 + 174 infiltration)
$$Btu/h\text{-}°F = 7512\ Btu/DD$$

The LCR of the typical bay is therefore

$$\frac{7512\ BLC}{229\ ft^2\ \text{collector area}} = 32.8$$

From Appendix C, Table C.3, for Salem, Oregon, with LCR = 33, system DGC1, SSF = about 27%, compared to the design guideline (Example 5.1, Part C) of SSF = 25%.

Note that if "superior performance" windows had been used throughout the typical bay, the BLC would be reduced. The north window UA would become 97 ft^2 $\times$ 0.30 Btu/h-ft^2 °F = 29. Thus

$$BLC = 24\ h/day$$
$$\times\ (42 + 25 + 29 + 174\ \text{infiltration})$$
$$Btu/h\text{-}°F = 6480\ Btu/DD$$

The passive system now closest to the actual building is DGC2, with triple glazed, thus lower U-value openings. The new LCR is

$$\frac{6480\ \text{BLC}}{229\ \text{ft}^2\ \text{collector}} = 28.3$$

Therefore, DGC23 in Salem, Oregon, with LCR of 28.3, find SSF as follows:

at LCR 25, SSF = 41; at LCR 30, SSF = 37 (from Table C.3)

Salem SSF range 41 − 37 = Δ 4

Salem LCR range 30 − 25 = Δ 5

(Actual LCR) − (Salem minimum LCR) = 28.3-25 = 3.3,

therefore = 3.3 (Δ 4)/Δ 5 = 2.6, to be added to the minimum SSF;

Actual SSF = 37 + 2.6% = 39.67%.

This is a substantial improvement that increases the energy saved by about 50%. ∎

EXAMPLE 5.11. We now take a more detailed look at the solar collecting area required for the passively solar heated building in Omaha, Nebraska, discussed in Example 5.2.

SOLUTION. From the design guidelines used in Example 5.2, we expected that 20% of the floor area in south glass with superior performance would yield a 51% SSF. More detailed characteristics of this building are shown in Fig. 5.29. From a program requirement of 2900 ft^2 of floor area, the 20% south glass area equals 0.20 × 2900 = 580 ft^2. For daylighting by sidelight only, DF_{av} = 0.2 (window area/floor area); if all the south glass area is available for daylighting (as with DG systems), DF_{av} = 0.2 × 20%, or 4%. This is adequate for office work. However, because you want to avoid dark areas near the rear walls of these spaces and investigate Trombe walls, and so on, some north light is desirable. Choose about 3% of the floor area in north glass (say, 90 ft^2 glass); added DF_{av} = 0.2 × 3% = 0.6% additional.

(Wall and roof insulation and overall percentage wall area in fenestration should be checked against local codes and the current ASHRAE Standard 90.1. The fenestration area:

south glass 580 ft^2 + north glass 90 ft^2 = 670 ft^2

The total wall area:

north 1100 ft^2 + south 1100 ft^2 + east 600 ft^2
+ west 600 ft^2 = 3400 ft^2

Fenestration/wall area ratio = 670 ft^2/3400 ft^2 = 20%.

Coincidentally, this is also the percentage floor area in south glazing.)

We now assume that the following maximum U-values meet codes and standards and that ventilation at the rate of 20 cfm per person is provided. The building will house 2900 ft^2/180 ft^2 per person = 16 people. Outdoor air required (in place of infiltration) = 20 cfm per person × 16 people = 320 cfm.

STEP 1. Given the emphasis on daylight and the original assumption of night ventilation on the windows, choose system DGC2. This has triple-glazed windows and will require six times as much thermally massive surface area (minimum 4-in. thick) as south glass area:

6 × 580 ft^2 south glass
= 3480 ft^2 exposed thermal mass, minimum

If all interior surfaces of exterior walls are exposed concrete block, we obtain

3400 ft^2 wall total − 670 ft^2 window
= 2730 ft^2 interior mass area

which is not sufficient. So add at least 750 ft^2 exposed concrete slab floor area, such as a 10-ft–wide strip just inside the south windows, for the entire length of the building.

STEP 2. The solar opening was selected at 20% floor area, or 580 ft^2.

STEP 3. Calculate UA_{ns}.

North, east, and west
  opaque (mass) walls: 2210 ft^2 × 0.08 = 177

South opaque (mass) wall: 520 ft^2 × 0.08 = 42

Roof (insulation
  above deck):   1974 ft^2 × 0.048 = 95

South elevation: Glass, 580 ft² Opaque wall, 520 ft²

East: Opaque, 600 ft²

North elevation: Glass, 90 ft² Opaque wall, 1010 ft²

West elevation: Opaque wall, 600 ft²

Main floor plan: 2100 ft²

10 ft

N

2nd floor: 800 ft²

Fig. 5.29 *Plans and elevations of the Omaha, Nebraska, office building discussed in Example 5.11.*

Slab perimeter:

$$200 \text{ lin. ft} \times 0.54^* \text{ Btu/h-°F-lin. ft} = 108$$

North glass

(triple glazed, $U = 0.33$): $90 \text{ ft}^2 \times 0.33 = 30$

Ventilation:

$$320 \text{ cfm} \times 1.08 \text{ Btu min./h ft}^3\text{°F} = \underline{346}$$

Total 798 Btu/h-°F

Then

$$BLC = 798 \text{ Btu/h-°F} \times 24 \text{ h/DD}$$

$$= 19,152 \text{ Btu/DD}$$

STEP 4. Calculate overall heat loss rate:

$$\frac{19,152 \text{ Btu/DD}}{2900 \text{ ft}^2 \text{ floor}} = 6.6 \text{ Btu/DD-ft}^2$$

This is considerably in excess of the Table 5.5 criterion of 4.6 Btu/DD-ft² for this 6601 DD65 climate. Unless further conservation steps are taken, the annual SSF is likely to be much lower than was predicted by the design guideline. Promising targets: more insulation on wall, slab edge, and roof; a heat exchanger for that 320 cfm of outdoor air.

STEP 5. Determine the vertical projection of the solar opening. The south windows are vertical; therefore, A_p is identical to the actual area, 580 ft².

STEP 6. Find the LCR:

$$\frac{19,152 \text{ Btu/DD}}{580 \text{ ft}^2} = 33.0$$

STEP 7. Enter Table C.3 for Omaha, Nebraska, LCR = 33, and DGC2. Interpolate between LCR 30 and 40. The annual SSF is found to be about 29.2%.

STEP 8. The approximate annual auxiliary heating required is

$$Q = (1 - 0.292) \times 19,152 \times 6601 \text{ DD}$$

$$= \text{about } 89,500,000 \text{ Btu}$$

STEP 9. Compare actual SSF with predicted: SSF = 29.2% is only about three fifths of the 51% SSF predicted by the guideline, due to inadequate conservation measures as explained in Step 4. Obviously, a lower LCR yields a higher SSF; a larger

*From Table 4.8, assuming insulated construction type (a), $R = 5.4$, and interpolating for 6601 DD65.

collector area would lower the LCR. However, Step 4 indicated that lower BLC, not higher A_p, is the better way to a lower LCR in this case.

Table C.3 can be used to compare the relative efficiencies of many passive heating systems. Using the column of LCR = 30, for instance, the best-performing systems are:

Omaha, Nebraska, LCR = 30

WWB4	47% SSF
WWC2	45%
SSE2	44%
DGC3	42% (movable insulation)
TWD4	42%
TWE2	42%

If any of these systems are more architecturally compatible with the building program and/or design intent, they could be used instead of DGC2. In this case, DGC2 gives substantial daylighting with little impact on the exterior appearance; the large area of thermal mass necessary was relatively easily obtained. DGC2 appears to be a reasonable choice for this building in this climate, if the client prefers not to use movable insulation over windows. ∎

What if two or more passive systems are used in the same building? In that case, first do calculations for the entire building assuming that *one* system has *all* the solar area, and find the SSF. Next, do the calculations assuming that the second system has all the solar area, and find its SSF. Then average the SSF values according to the relative solar areas of the two systems.

EXAMPLE 5.12. A veterinary clinic in Buffalo, New York, is using two systems: water walls for examining rooms and direct gain for the waiting/reception area. The building's characteristics are

balance point = 50°F

$UA_{ns} = 356 \text{ Btu/h-°F}$

A_p water wall = 240 ft² (reference system WWB4)

A_p direct gain = 150 ft² (reference system DGA3)

total floor area = 1900 ft²

predicted SSF from design guideline = about 36%, superior performance

Checking the overall heat loss criteria (Table 5.5),

$$\frac{356 \times 24}{1900 \text{ ft}^2} = 4.5 \text{ BTU/DD ft}^2$$

This is less than 4.6 (for 5000 to 7000 DD), so it is acceptable.

First, calculate as though all A_p (390 ft²) were system WWB4:

$$BLC = 24 \times 356 = 8544 \text{ Btu/DD}$$

$$LCR = \frac{8544}{390} = 22$$

For Buffalo, LCR 20 yields SSF = 0.37, and LCR 25 yields SSF = 0.31, so SSF = 0.35, by interpolation.

Next, calculate as though all A_p were system DGA3. For LCR 22, Buffalo SSF = 0.28, by interpolation.

Now calculate the average SSF, given that water wall comprises 240/390 = 62% of total solar opening, and direct gain comprises 150/390 = 38%:

$$\frac{0.35(62\%) + 0.28(38\%)}{100\%} = 0.217 + 0.106$$

$$= 0.323; SSF = 0.32$$

Result: The SSF is about 90% of what was predicted (0.32 as opposed to 0.36); because the glass area is already large, program requirements probably prevent further increases in it. The building already seems to conserve energy well, because it meets the Table 5.5 heat loss criteria.

The approximate annual auxiliary heating energy required is

$$Q = (1 - 0.323) \times 8544 \text{ Btu/DD}$$

$$\times 3322 \text{ DD50 (balance point)}$$

$$= 19 \text{ million Btu}$$

(This is equivalent to about 5600 kWh, or about 220 therms of natural gas burned at 85% efficiency.) ∎

(g) Variations on Reference Systems. A particularly wide set of choices faces the designer of direct-gain and sunspace systems, in which mass distribution and glass orientation can assume thousands of different combinations. The "sensitivity curves" furnished in Balcomb et al. (1984) give some guidance on how a predicted SSF might vary as an actual passive system departs from a reference system.

Sensitivity curves can serve as early general design guidelines. Looking at the curves for your location, which design changes yield dramatic results and which make little difference? The curves may also be used to adjust the SSF you found for a reference design.

One passive solar example that departs radically from any reference case is the Class of 1959 Chapel at the Harvard Business School in Cambridge, Massachussetts (Fig. 5.30). The glazed "sunspace" resembles system type SSA in that it is attached to (rather than set into) the building behind it. However, this is manifestly *not* the simple rectangular sunspace plan, and the solar aperture is divided into southeast- and southwest-facing halves. The aperture A_p is doubly complicated: the vertical dimension will be the vertical projection of the sloping wall, whereas the horizontal dimension will be the due-south projection of the plan. From Table C.1, because of the glazed end walls and masonry common wall, it comes closest to either SSA3 (standard performance) or SSA4 (superior performance). This depends on the glazing chosen; in this case, clear double glazing with low-ε, airspace approximately ½-in. (13-mm) thickness. In Cambridge (Boston), for a small LCR (a huge solar aperture relative to the building behind) at 3320 DD65, the Balcomb reference shows:

If LCR =	25	20	15
If SSA3, then SSF =	37	41	47
If SSA4, then SSF =	55	61	69

Given the complex geometry of this sunspace, the safer assumption is "standard performance."

Water walls are rather uncommon in buildings, despite their relatively high performance throughout Table C.3. These WW systems are based on opaque containers, but some designers are attracted by the idea of daylight filtering through translucent water containers. Algae growth in transparent plastic tubes, encouraged by the daylight that passes through them, is usually controlled by the addition of an algaecide to the water. Dyes may be added to change the color of daylight through tubes or water-filled glass block, as in Fig. 5.31. The "water wall" of this Long Island library is actually a hybrid between *direct gain* (light passes through the glass to strike massive surfaces beyond) and *water wall* (all light is converted to heat within the opaque water containers, then passed to the space beyond). To assess the performance of such a system, interpolate between the entire aperture considered as the relevant DG system and same entire aperture considered as the relevant WW system. The higher the visible transmittance, the closer to DG performance will be the result.

(h) Thermal Lag Through Mass Walls.

The time necessary for solar heat to pass through various thermally massive materials is shown in Table 5.16. This time lag can be put to use when the time of maximum solar heat is different from the time of maximum internal heat need. A typical example is for a residence's living room in winter, in which the late evening sedentary entertainment hours occur with cold temperatures outside, yet maximum warmth is desired inside. A Trombe wall, for instance, made of solid grouted concrete block and 12-in. thick will delay the arrival of maximum solar gain to the interior by almost 8 h. If maximum solar heat gain occurs at about 1 PM (maximum sun at noon, but highest temperatures at about 2 PM), then such a Trombe wall would deliver the maximum heat to the inner surface at about 9 PM.

(i) Internal Temperatures.

Two quantities are of particular interest to passive solar designers: How much higher, compared to outdoors, will the average indoor temperature be from solar heating alone? Also, how widely will this internal temperature vary (swing) on a clear winter day?

The approximate temperature difference between inside and outside on a *clear* January day, called ΔT *solar*, can be estimated from Fig. 5.32; it varies with latitude and with the LCR. Sunspaces are not shown in the figure. Although the temperature within the sunspace cannot be easily approximated, the ΔT solar for the room beyond the sunspace can be approximated by using Fig. 5.32a if these spaces have an insulated common wall, or Fig. 5.32b if they have a masonry common wall.

To determine the average winter indoor temperature,

1. Find the average January ambient (outdoor) temperature, TA, from Appendix B.

(a)

(b)

Fig. 5.30 *The Class of 1959 Chapel at the Harvard Business School* (a) *is entered through a sunspace with a sunken garden* (b). (c) *Plan; south is about 30° east of the point of the sunspace.* (d) *Section. This is not a "typical" sunspace type from Appendix C!* (e) *South-facing glazing of the sunspace. (Drawings courtesy of Moshe Safdi and Associates, Inc., Somerville, Mass.)*

2. Find ΔT solar for the building and its site (Fig. 5.32).

3. Find the ΔT due to internal heat sources:

$$\Delta T \text{internal} = \frac{\text{total internal gains (Btu/day)}}{\text{BLC} + (UA_s \times 24)}$$

0   10 ft.

(c)

Fig. 5.30 (continued)

where UA_s is for the solar area only. *Note:* ΔT internal averages 5 to 7F° for residences.

4. Add the quantities from the first three steps to find the average January clear-day indoor temperature.

When internal gains are high, there is less need for ΔT solar; the building is mostly "heating itself" (becoming an ILD rather than an SLD building). If the average indoor temperature is too high, smaller solar openings should be considered unless your climate is predominantly cloudy (clear days rare) in November through January.

The other important comfort question is the size of *temperature swing* due to passive solar heating. Controlled by the sun and by the actions of users rather than by a thermostat, passive solar buildings typically experience larger daily variations (swings) in indoor temperature than do conventional buildings, especially on clear days. To

(d)

(e)

Fig. 5.30 (continued)

(a)

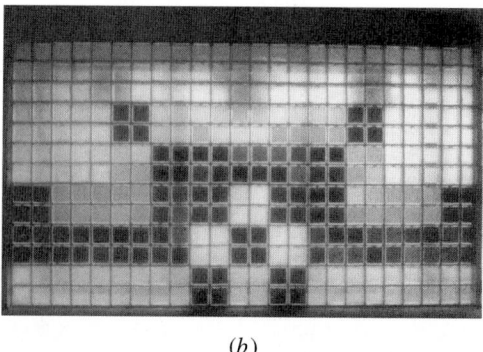

(b)

Fig. 5.31 *The Center Moriches Free Public Library (Long Island, New York) utilized water-filled glass block as a heat storage and daylight-diffusing south wall (a). Because the water storage is translucent, it functions somewhere between a DG and a WW system. Dyes add color to selected blocks, forming the pattern shown (b). Lightshelves and stack ventilators add to its energy-conserving performance. (Courtesy of Banwell White Arnold Hemberger & Partners, Architects, Hanover, NH).*

THERMAL CONTROL

TABLE 5.16 **Time Lag Through Homogeneous[a] Walls**

Material	Thickness (in.)	U-Value[b] (Btu/h-ft²)	Time Lag (h)
Stone	8	0.67	5.5
	12	0.55	8.0
	16	0.47	10.5
	24	0.36	15.5
Solid concrete	2	0.98	1.1
	4	0.84	2.5
	6	0.74	3.8
	8	0.66	5.1
	12	0.54	7.8
	16	0.46	10.2
Common brick	4	0.60	2.3
	8	0.41	5.5
	12	0.31	8.5
	16	0.25	12.0
Face brick	4	0.77	2.4
Wood	$\frac{1}{2}$	0.68	0.17
	1	0.48	0.45
	2	0.30	1.3
Insulating board	$\frac{1}{2}$	0.42	0.08
	1	0.26	0.23
	2	0.14	0.77
	4	0.08	2.7
	6	0.05	5.0

Source: Victor Olgyay, *Design with Climate: Bioclimatic Approach to Architectural Regionalism,* Copyright © 1963 by Princeton University Press. Reprinted by permission.

[a]For composite constructions, add an estimated additional time lag to the sum of the individual materials' time lags, as follows:

Two-layer, light construction: $\frac{1}{2}$ h more
Three or more layers: 1 h more
Very heavy construction: 1 h more

[b]The U-value is based on outdoor surface conductance of 4.0 and an indoor surface conductance of 1.65 Btu/h-ft²-°F.

estimate your building's clear-day January temperature swing, or ΔT *swing,* see Table 5.17. The average indoor temperature determined previously will fall in the middle of this ΔT swing.

EXAMPLE 5.1, PART I. Find the clear January day temperature swing in the typical bay of the office building in Fig. 5.7.

SOLUTION.

1. (Appendix B) Salem's average January outdoor temperature = 39°F.

2. (Fig. 5.32a) At 44° NL, using a direct gain system, with LCR = 33, ΔT solar = 24F°.
3. ΔT internal = (from Example 5.1, Part G, winter, and Part H)

$$\frac{7.9 \text{ Btu/h-ft}^2 \times 1440 \text{ ft}^2 \times 9 \text{ h/day}}{7512 \text{ Btu/DD}}$$
$$+ (229 \text{ ft}^2 \times 0.74 \text{ Btu/h-ft}^2\text{-°F} \times 24 \text{ h/day})$$

$$= \frac{102,384}{11,579} = 8.8\text{F}°$$

4. Average January clear-day temperature indoors is 39°F + 24F° + 8.8F° = 72.8°F (within the comfort zone)

TABLE 5.17 **Indoor Temperature Swing, ΔT Swing**

NOTE: These swings are based on a thermal storage mass capacity of 45 Btu/ft² °F; ΔT solar can be found in Fig. 5.32.

Passive Solar System	ΔT Swing
Direct gain: $\dfrac{\text{mass area}}{\text{glass area}} = 1.5$	$1.11 \times \Delta T$ solar
$= 3$	$0.74 \times \Delta T$ solar
$= 9$	$0.37 \times \Delta T$ solar
Water wall	$0.39 \times \Delta T$ solar
Trombe wall, vented for 3% of wall area	$0.65 \times \Delta T$ solar
Trombe wall, unvented	$0.13 \times \Delta T$ solar

Source: Balcomb et al. (1980).

5. (Table 5.17) Indoor temperature swing, for direct gain with a 9 : 1 ratio of mass to south glass area = $0.37 \times \Delta T$ solar:

$$0.37 \times 24\text{F}° = 8.9\text{F}° \ \Delta T \text{ swing}$$

Therefore, the clear January day interior will vary by 8.9/2 on either side of the average temperature:

$$72.8 + 8.9/2 = 77.3°\text{F high}$$

$$72.8 - 8.9/2 = 68.4°\text{F low}$$

This appears to be a quite comfortable range. Because the Pacific Northwest is largely overcast in winter, this daily range typically will be less wide, but the ΔT solar will also be smaller. Thus, the average January day temperature indoors will be lower. ■

5.11 Approximate Method for Calculating Heat Gain (Cooling Load)

Unlike the calculations for winter worst-hour heat loss, which simply assume nighttime conditions and few if any internal gains, summer worst-hour heat gain calculations are very complex. The difference in air temperature between inside and outside, which was so influential in winter, is much less important in summer. Solar gains and internal gains from lights, people, and equipment *must* be included. Summer calculations are complicated further by the fact that the hourly change in summer load can be very great, both from changing

Fig. 5.32 *Graphs of "ΔT solar," the temperature difference to be expected between the average inside temperature and the average outside temperature on a clear January day. The curve marked "40°/20°" applies to both 40° latitude and 20° latitude. (a) ΔT solar for direct gain, water wall, vented Trombe wall or sunspace (insulated common wall) systems. (b) ΔT solar for unvented Trombe wall or sunspace (masonry common wall) systems. (From Balcomb et al., 1980.)*

sun position and changing internal loads. Also, in summer the thermal mass of the building becomes influential, delaying the impact of the radiant component of heat gains from all sources. The design guideline approach represented in Table

5.8 cannot respond to hourly changes or to thermally massive construction. Furthermore, Table 5.8 deals only with sensible heat gains and ignores latent heat gains.

Sensible heat is the kind of heat that *increases the DB temperature of air;* the glowing coil of an electric range is adding sensible heat to the air of the kitchen.

Latent heat is the kind of heat that is present within *increased moisture* in air; boiling water in a pan on top of the electric range is evaporating, and as it evaporates it increases the latent heat in the kitchen air. This latent heat will increase the WB temperature, but not the DB temperature, of the air. In conventionally cooled buildings, the mechanical cooling equipment must be sized so that latent as well as sensible heat will be removed.

A simplified heat gain procedure has been developed for residential buildings; with some risk and some judgment, it can also be used for a quick *approximation* of the conditions in commercial buildings. This simplified method was devised to be used with buildings that, like residences,

1. Are occupied and air conditioned (internal temperatures closely controlled) for 24 h/day. (Separate calculations can be done for weekdays and weekends.)
2. Derive much of their gains through the building envelope and ventilation rather than internally.
3. Are tolerant of undersized cooling equipment, with the result that unusually hot weather means noticeably higher indoor temperatures (and that interior temperatures will vary, or "swing," during a typical summer day).

Because many commercial buildings do *not* have these characteristics, this method should not be applied if very accurate estimates are desired for such buildings. However, the method is rapid, if risky. Compare the results from this procedure (Example 5.13) with those from the more exact procedure for commercial buildings in Table 5.25. (*Note:* In Sections 5.11 and 5.12, we use the order SI (I-P).

(a) Gains Through Roof and Walls. The

sensible heat gains through opaque parts of a building's envelope are calculated with the equation

$$q = U \times A \times \text{DETD}$$

where U-values are for summer, and
 A = area of the roof or wall, and
DETD (design equivalent temperature differences); values are listed for broad categories of construction in Table 5.18.

The DETD values are based on an average indoor temperature of 23.8°C (75°F) and the outdoor conditions listed. Note that for lightweight wall construction (and doors) the DETD varies by orientation. A means of correcting DETD for other temperatures is as follows.

Where the design temperature difference [outdoor design temperature minus assumed indoor 23.8°C (75°F)] is not an even increment of 2.8C° (5F°), the equivalent temperature difference should be corrected 0.5C° (1F°) for each 0.5C° (1F°) difference from the tabulated values.

For rapid approximation, however, select the DETD directly from the table for the conditions nearest to your building/climate combination. The design temperature for your climate is listed in Appendix A.

Note that Table 5.18 lists considerably higher DETD for roofs than for any other component. This is due to the *sol-air temperature,* the effective temperature of outdoor air just above a solar-heated surface. When sun strikes surface, it adds its heat to that from the outdoor air. The darker the surface, the greater the resulting effective (sol-air) temperature. This elevated outdoor temperature thus increases the ΔT through the roof, resulting in greater heat gains. Sol-air temperature is discussed in the 1997 *ASHRAE Handbook of Fundamentals,* Chapter 28, with some tabulated values for July 21 at 40° N latitude. Uninsulated or poorly insulated roofs of darker colors are particularly impacted by sol-air temperature. The sol-air formula was discussed in Section 4.7*f.*

(b) Gains Through Glass. The quick way

to approximate these gains is

$$q = A \times \text{DCLF}$$

DCLF (design cooling load factors) values are listed in Table 5.19 and *include the U-values* as

TABLE 5.18 **Design Equivalent Temperature Differences (DETD)**

Part A. Roofs, Mass Walls, and Floors

Outdoor Design Temperature — Daily Temperature Range[a] (L = low, M = medium, H = high)

Walls and Doors / Ceilings and Roofs[b]	29.4°C L	29.4 M	32.2 L	32.2 M	32.2 H	35.0 L	35.0 M	35.0 H	37.7 M	37.7 H	40.5 H	43.3 H	85°F L	85 M	90 L	90 M	90 H	95 L	95 M	95 H	100 M	100 H	105 H	110 H
Walls and Doors																								
1. Masonry walls, 200-mm (8-in.) block or brick	5.7	3.5	8.5	6.3	3.5	11.3	9.1	6.3	11.8	9.1	11.8	14.6	10.3	6.3	15.3	11.3	6.3	20.3	16.3	11.3	21.3	16.3	21.3	26.3
2. Partitions, frame	5.0	2.7	7.7	5.5	2.7	10.5	8.3	5.5	11.1	8.3	11.1	13.8	9.0	5.0	14.0	10.0	5.0	19.0	15.0	10.0	20.0	15.0	20.0	25.0
masonry	1.4	0.0	4.2	1.9	0.0	6.9	4.7	1.9	7.5	4.7	7.5	10.3	2.5	0	7.5	3.5	0	12.5	8.5	3.5	13.5	8.5	13.5	18.5
Ceilings and Roofs[b]																								
1. Ceilings under naturally vented attic or vented flat roof—dark	21.1	18.8	23.8	21.6	18.8	26.6	24.4	21.6	27.2	24.4	27.2	30.0	38.0	34.0	43.0	39.0	34.0	48.0	44.0	39.0	49.0	44.0	49.0	54.0
—light	16.6	14.9	19.4	17.2	14.4	22.2	20.0	17.2	22.7	20.0	22.7	25.5	30.0	26.0	35.0	31.0	26.0	40.0	36.0	31.0	41.0	36.0	41.0	46.0
2. Built-up roof, no ceiling—dark	21.1	18.8	23.3	21.6	18.8	26.6	24.4	21.6	27.2	24.4	27.2	30.0	38.0	34.0	43.0	39.0	34.0	48.0	44.0	39.0	49.0	44.0	49.0	54.0
—light	16.6	14.4	19.4	17.2	14.4	22.2	20.0	17.2	22.7	20.0	22.7	25.5	30.0	26.0	35.0	31.0	26.0	40.0	36.0	31.0	41.0	36.0	41.0	46.0
3. Ceilings under unconditioned rooms	5.0	2.7	7.7	5.5	2.7	10.5	8.3	5.5	11.1	8.3	11.1	13.8	9.0	5.0	14.0	10.0	5.0	19.0	15.0	10.0	20.0	15.0	20.0	25.0

Floors

Floors	Values (29°C … 43)	Values (85°F … 110)
1. Over unconditioned rooms	5.0 2.7 7.7 5.5 10.5 8.3 5.5 11.1 8.3 11.1 13.8	9.0 5.0 14.0 10.0 5.0 19.0 15.0 10.0 20.0 15.0 20.0 25.0
2. Over basement, enclosed crawl space or concrete slab on ground	0.0 0.0 0.0 0.0 0.0 0.0 0.0 0.0 0.0 0.0 0.0	0 0 0 0 0 0 0 0 0 0 0 0
3. Over open crawl space	5.0 2.7 7.7 5.5 10.5 8.3 5.5 11.1 8.3 11.1 13.8	9.0 5.0 14.0 10.0 5.0 19.0 15.0 10.0 20.0 15.0 20.0 25.0

Part B. *Frame Walls, and Doors*

Outdoor Design Temperature

Daily Temperature Range[a]

Orientation	29°C L	29°C M	29°C H	32 L	32 M	32 H	35 L	35 M	35 H	38 L	38 M	38 H	41 L	41 M	41 H	43 L	43 M	43 H	85°F L	85°F M	85°F H	90 L	90 M	90 H	95 L	95 M	95 H	100 L	100 M	100 H	105 L	105 M	105 H	110 L	110 M	110 H
Frame walls, Doors:																																				
North	4	2		7	4	2	10	7	4		10	7		13	10			13	8	3		13	8	3	18	13	8		18	13		23	18			23
NE and NW	8	5		11	8	5	13	11	8		13	11		16	13			16	14	9		19	14	9	24	19	14		24	19		29	24			29
East and West	10	7		13	10	7	16	13	10		16	13		18	16			18	18	13		23	18	13	28	23	18		28	23		33	28			33
SE and SW	9	6		12	9	6	14	12	9		14	12		17	14			17	16	11		21	16	11	26	21	16		26	21		31	26			31
South	6	3		9	6	3	12	9	6		12	9		14	12			14	11	6		16	11	6	21	16	11		21	16		26	21			26

Source: Part A, 1981 *ASHRAE Handbook of Fundamentals*; Part B, 1997 *ASHRAE Handbook of Fundamentals*; both copyright © by the American Society of Heating, Refrigerating, and Air-Conditioning Engineers, Inc., Atlanta, GA.

[a]Daily temperature range:

L (Low) less than 9 K (16 F°)

M (Medium) 9 to 14 K (16 to 25 F°)

H (High) greater than 14 K (25 F°)

(MDR, Appendix A)

[b]For roofs in shade, 18-h average DETD = 6.1 C° (11 F°). At 32° C (90° F) design temperature and medium daily range, DETD for light-colored roof = 6.1 + (0.71) (21.6 − 6.1) = 17.1 C° [11 + (0.71)(39 − 11)] = 31 F°].

TABLE 5.19 **Design Cooling Load Factors Through Glass**

Part A. W/m²

	Regular Single Glass						Regular Double Glass						Heat-Absorbing Double Glass						Clear Triple Glass		
Outdoor Design Temp.ª	29.4	32.2	35.0	37.7	40.5	43.3	29.4	32.2	35.0	37.7	40.5	43.3	29.4	32.2	35.0	37.7	40.5	43.3	29.4	32.2	35.0
No Awnings or Inside Shading																					
North	72.6	85.2	97.8	110.4	123.0	138.8	59.9	66.2	75.8	82.0	88.3	94.7	37.9	44.1	53.7	60.0	66.2	72.6	53.7	60.0	63.1
NE and NW	176.6	189.3	202.0	214.6	227.1	243.0	145.1	151.4	161.0	167.2	173.6	179.9	85.1	91.4	101.0	107.2	113.6	119.9	132.6	135.7	138.9
East and west	255.5	268.1	280.8	293.4	306.0	321.9	214.6	220.9	230.3	236.7	243.0	249.2	132.6	138.9	148.2	154.6	160.9	167.2	195.7	198.8	202.6
SE and SW	220.9	233.4	246.0	258.8	271.3	287.1	186.1	192.4	202.0	208.2	214.6	220.9	110.4	116.8	126.2	132.6	138.9	145.1	167.2	173.6	176.7
South	126.2	138.9	151.4	164.0	176.7	192.4	104.1	110.4	119.9	126.2	132.6	138.9	66.0	75.8	75.8	82.0	88.3	94.7	94.7	97.9	104.1
Horiz. skylight	504.8	517.4	530.0	542.7	555.2	571.0	438.6	444.9	454.3	460.7	467.0	473.2	280.8	296.6	296.6	302.9	309.1	315.6	397.6	400.7	407.0
Draperies or Venetian Blinds																					
North	47.3	60.0	72.6	85.1	97.9	113.6	37.9	44.1	53.7	60.0	66.2	72.6	28.3	34.7	44.1	50.4	56.8	63.1	34.7	37.9	44.2
NE and NW	101.0	113.6	126.2	138.9	151.4	167.2	85.1	91.4	101.0	107.2	113.6	119.9	63.1	69.4	78.9	85.1	91.4	97.9	75.8	82.0	85.1
East and west	151.4	164.0	176.7	189.3	202.0	217.7	132.6	138.9	148.2	154.6	161.0	167.2	94.7	101.0	110.4	116.8	123.0	129.3	119.9	123.0	129.3
SE and SW	126.2	138.9	151.4	164.0	176.7	192.4	110.4	116.8	126.2	132.6	138.9	145.1	75.8	82.0	91.4	97.9	104.1	110.4	101.0	104.1	107.2
South	72.5	85.1	97.9	110.4	123.0	138.9	63.1	69.4	78.9	85.1	91.4	97.9	47.3	53.7	63.1	75.8	75.8	82.0	56.8	60.0	66.2
Roller Shades Half-Drawn																					
North	56.8	69.4	82.0	92.7	102.2	123.0	47.3	53.7	63.1	69.4	75.8	82.0	31.6	37.9	47.3	53.7	60.0	66.2	41.0	44.2	47.3
NE and NW	126.2	138.9	151.4	164.0	176.7	192.4	119.9	126.2	135.7	142.0	148.2	154.6	75.8	82.0	91.4	97.9	104.1	110.4	107.2	110.4	110.4
East and west	192.5	205.0	217.7	230.3	243.0	258.8	170.3	176.7	186.1	192.4	198.8	205.0	110.4	116.8	126.2	132.5	138.9	145.1	154.6	154.6	157.8
SE and SW	164.0	176.7	189.3	202.0	214.6	230.3	145.1	151.4	161.0	167.2	173.6	179.9	94.7	101.0	110.4	123.0	123.0	129.3	129.3	132.6	135.7
South	91.4	104.1	116.8	129.3	142.0	157.8	85.1	91.4	101.0	107.2	113.6	119.9	56.8	63.1	72.6	78.9	85.1	91.4	78.9	82.0	82.0
Awnings^b																					
North	63.1	75.8	88.3	101.0	113.6	129.3	41.0	47.3	56.8	63.1	69.4	75.8	31.6	37.9	47.3	53.7	60.0	66.2	34.7	37.9	41.0
NE and NW	66.2	78.9	91.4	104.1	116.8	132.6	44.2	50.4	60.0	66.2	72.6	78.9	34.7	41.0	50.4	56.8	63.1	69.4	37.9	41.0	44.2
East and west	69.4	82.0	94.7	107.2	119.9	135.7	44.2	50.4	60.0	66.2	72.6	78.9	37.9	44.2	53.6	60.0	66.2	72.6	37.9	41.0	44.2
SE and SW	66.2	78.9	91.4	104.1	116.8	132.6	44.2	50.4	60.0	66.2	72.6	78.9	34.7	41.0	50.4	56.8	63.1	69.4	37.9	41.0	44.2
South	66.2	75.8	88.3	101.0	113.6	129.3	41.0	47.3	56.8	63.1	69.4	75.8	34.7	41.0	50.4	56.8	56.8	75.8	34.7	37.9	41.0

Part B. *Btu/h ft²*

Outdoor Design Temp.[a]	Regular Single Glass						Regular Double Glass						Heat-Absorbing Double Glass						Clear Triple Glass		
	85	90	95	100	105	110	85	90	95	100	105	110	85	90	95	100	105	110	85	90	95
No Awnings or Inside Shading																					
North	23	27	31	35	39	44	19	21	24	26	28	30	12	14	17	19	21	23	17	19	20
NE and NW	56	60	64	68	72	77	46	48	51	53	55	57	27	29	32	34	36	38	42	43	44
East and west	81	85	89	93	97	102	68	70	73	75	77	79	42	44	47	49	51	53	62	63	64
SE and SW	70	74	78	82	86	91	59	61	64	66	68	70	35	37	40	42	44	46	53	55	56
South	40	44	48	52	56	61	33	35	38	40	42	44	19	21	24	26	28	30	30	31	33
Horiz. skylight	160	164	168	172	176	181	139	141	144	146	148	150	89	91	94	96	98	100	126	127	129
Draperies or Venetian Blinds																					
North	15	19	23	27	31	36	12	14	17	19	21	23	9	11	14	16	18	20	11	12	14
NE and NW	32	36	40	44	48	53	27	29	32	34	36	38	20	22	25	27	29	31	24	26	27
East and west	48	52	56	60	64	69	42	44	47	49	51	53	30	32	35	37	39	41	38	39	41
SE and SW	40	44	48	52	56	61	35	37	40	42	44	46	24	26	29	31	33	35	32	33	34
South	23	27	31	35	39	44	20	22	25	27	29	31	15	17	20	22	24	26	18	19	21
Roller Shades Half-Drawn																					
North	18	22	26	30	34	39	15	17	20	22	24	26	10	12	15	17	19	21	13	14	15
NE and NW	40	44	48	52	56	61	38	40	43	45	47	49	24	26	29	31	33	35	34	35	35
East and west	61	65	69	73	77	82	54	56	59	61	63	65	35	37	40	42	44	46	49	49	50
SE and SW	52	56	60	64	68	73	46	48	51	53	55	57	30	32	35	37	39	41	41	42	43
South	29	33	37	41	45	50	27	29	32	34	36	38	18	20	23	25	27	29	25	26	26
Awnings[b]																					
North	20	24	28	32	36	41	13	15	18	20	22	24	10	12	15	17	19	21	11	12	13
NE and NW	21	25	29	33	37	42	14	16	19	21	23	25	11	13	16	18	20	22	12	13	14
East and west	22	26	30	34	38	43	14	16	19	21	23	25	12	14	17	19	21	23	12	13	14
SE and SW	21	25	29	33	37	42	14	16	19	21	23	25	11	13	16	18	20	22	12	13	14
South	21	24	28	32	36	41	13	15	18	20	22	24	11	13	16	18	20	22	11	12	13

Source: Copyright © by the American Society of Heating, Refrigerating and Air-Conditioning Engineers, Inc., Atlanta, GA. Reprinted by permission from the *ASHRAE Handbook of Fundamentals*, 1981.

[a]Based on indoor design temperature of 23.8°C (75°F) and outdoor design temperatures as indicated. Interpolate to obtain factors for outdoor design temperatures other than those given.

[b]For other external shading devices which completely shade the glass at any orientation, use the values for "Awnings, north."

well as the equivalent temperature differences. (These DCLF values do *not* correspond to the worst-hour gains; they were obtained by averaging the hours from 5:30 AM to 6:30 PM at both 30° N and 40° N latitudes.) Again, the DCLF values were based on an inside temperature of 23.8°C (75°F) and outside temperatures as listed in Table 4.19. (For *rapid* approximation, ignore the corrections procedure shown in note *a.*) Glass protected by exterior shading devices that exclude all direct sun may be considered equal to the values listed in Table 5.19 for "north glass protected by awnings."

(c) Gains from Outdoor Air. In residences, outdoor air often enters by infiltration. In many other buildings, codes require the deliberate introduction of outdoor air by mechanical ventilation. Whichever way outdoor air enters your building in summer, its sensible heat gain can be calculated by either

$$q_{\text{infiltration}} = (A_{\text{exposed}})(\text{infiltration factor})$$

where A_{exposed} is the total area of exposed wall surface, including windows and doors and the infiltration factor is found from Table 5.20; or

$$q_{\text{mechanical ventilation}} = (Q)(\text{ventilation factor})$$

where Q is the volume of outdoor air (cfm or L/ s; see Section 4.9), and the ventilation factor is found from Table 5.20.

(d) Gains from People. Only the sensible gains are tabulated, because a simple overall factor for latent gains is included later. The rate of heat gain from people in various activities is shown in Table 5.21; values in the "Sensible Heat"

column are generally used. [For residences, sensible heat gain per occupant is often assumed at 67.3 W (230 Btu/h)].

$$q_{\text{people}} = (\text{number of occupants})$$
$$(\text{sensible gain per occupant})$$

(e) Gains from Lights. The power supplied to electric lights (those that normally are on while cooling equipment is functioning) can be added directly to the sensible heat gain. Be sure to include ballast heat gains along with fluorescent lights, usually done by taking *from 1.12 to 1.2 times the total bulb wattage* of such lights (use the lower figure with energy-efficient ballasts).

(f) Gains from Equipment. In residences, a standard assumption is that 350 to 470 W (1200 to 1600 Btu/h) of sensible heat gain is produced by appliances. (Other residential heat loads are assumed to be vented.) For other buildings, see the gains of specific pieces of equipment given in Tables 5.22 and 5.23.

(g) Latent Heat Gains. These vary a great deal with the type of occupancy, but this simple method assumes that the latent gains are closely associated with outdoor air infiltration. Figure 5.33 is a method to estimate the percentage of *total sensible gains* that are *additional latent heat.* The design DB and mean coincident WB temperatures from Appendix A, as well as the relative tightness of building construction, determine the latent percentage of total sensible gain. A minimum of 10% additional latent, and probable maximum of 30% additional latent, are recommended limits.

TABLE 5.20 **Sensible Cooling Load Factors Due to Infiltration and Ventilation**

°C:	29.4	32.2	35.0	37.7	41.5	43.3	(Units)	Design Temperature	(Units) °F:	85	90	95	100	105	110
	2.2	3.5	4.7	6.0	6.9	8.2	W/m²	Infiltration, per gross exposed wall area	Btu/h ft²	0.7	1.1	1.5	1.9	2.2	2.6
	6.8	9.9	13.6	16.7	19.8	23.6	W per L/s	Mechanical ventilation	Btu/h cfm	11.0	16.0	22.0	27.0	32.0	38.0

Source: Copyright © by the American Society of Heating, Refrigerating and Air-Conditioning Engineers, Inc., Atlanta, GA. Reprinted by Permission from the *ASHRAE Handbook of Fundamentals,* 1981.

TABLE 5.21 Rates of Heat Gain from Occupants of Conditioned Spaces

Activity	Location	Total Heat Gain							
		W				Btu/h			
		Adult Male	Adjusted M/F[b]	Sensible[a] Heat	Latent[a] Heat	Adult Male	Adjusted M/F[b]	Sensible[a] Heat	Latent[a] Heat
Seated at theater	Theater, matinee	115	95	65	30	390	330	225	105
Seated at theater, night	Theater, night	115	105	70	35	390	350	245	105
Seated, very light work	Offices, hotels, apartments	130	115	70	45	450	400	245	155
Moderately active office work	Offices, hotels, apartments	140	130	75	55	475	450	250	200
Standing, light work; walking	Department or retail store	160	130	75	55	550	450	250	200
Walking, standing	Drug store, bank	160	145	75	70	550	500	250	250
Sedentary work	Restaurant[c]	170	160	80	80	590	550	275	275
Light bench work	Factory	235	220	80	140	800	750	275	475
Moderate dancing	Dance hall	265	250	90	160	900	850	305	545
Walking 4.8 km/h (3 mph), light machine work	Factory	295	295	110	185	1000	1000	375	625
Bowling[d]	Bowling alley	440	425	170	255	1500	1450	580	870
Heavy work	Factory	440	425	170	255	1500	1450	580	870
Heavy machine work, lifting	Factory	470	470	185	285	1600	1600	635	965
Athletics	Gymnasium	585	525	210	315	2000	1800	710	1090

Source: Reprinted with permission of the American Society of Heating, Refrigerating, and Air-Conditioning Engineers, Inc. from 1997 *ASHRAE Handbook of Fundamentals.*

[a]All values are rounded to the nearest 5 W (and 5 Btu/h). Based on 24° C (75° F) room DB temperature. For 27° C (80° F) room DB, the total heat remains the same but the sensible heat values should be *decreased* by approximately 20%, and the latent heat values *increased* accordingly.

[b]Adjusted heat gain based on normal percentage of men, women, and children for the application listed, assuming gain from an adult female is 85%, and from children 75%, of that from an adult male.

[c]Adjusted heat gain includes 18 W (60 Btu/h) for food per individual; 50% sensible, 50% latent.

[d]Calculate but one person per alley actually bowling, and all others as sitting, standing, or walking slowly.

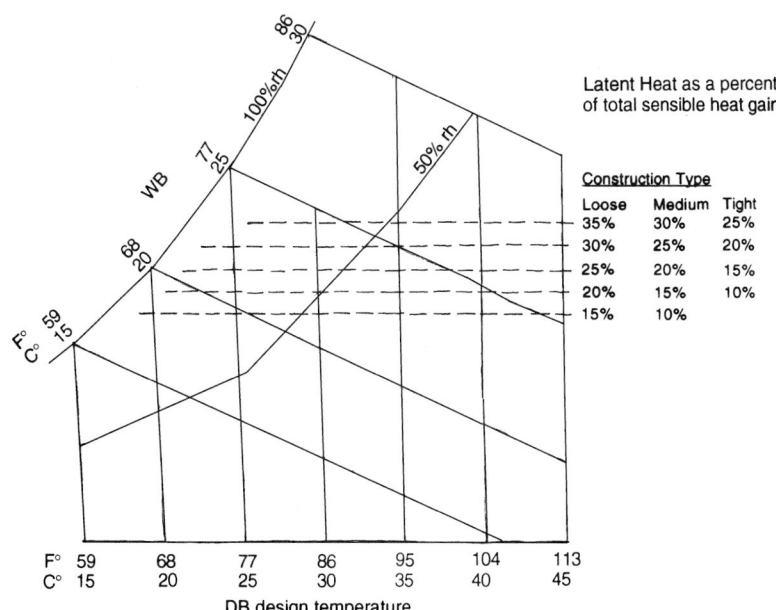

Fig. 5.33 *Estimating the latent heat gain as a percent of total sensible heat gain, assuming these additional latent gains are closely associated with outdoor air infiltration. The design DB and mean coincident WB temperatures are from Appendix A. Relative tightness of building construction is also included.*

TABLE 5.22 **Rates of Heat Gain from Office Equipment**

Appliance	Maximum Input Rating (W)	Recommended Rate of Heat Gain	
		W	Btu/h
Check processing workstation, 12 pockets	4,800	2,460	8,410
Computer Devices			
Communication/transmission	1,800–4,600	1,640–2,810	5,600–9,600
Disk drives/mass storage	1,000–10,000	1,000–6,570	3,412–22,420
Microcomputer	100–600	90–530	300–1,800
Minicomputer	2,200–6,600	2,200–6,600	7,500–15,000
Optical reader	3,000–6,000	2,350–4,980	8,000–17,000
Plotters	75	63	214
Printers			
Letter quality, 30–45 characters/min.	350	292	1,000
Line, high speed, 5000 or more lines/min	1,000–5,300	730–3,810	2,500–13,000
Line, low speed 300–600 lines/min	450	376	1,280
Tape drives	1,200–6,500	1,000–4,700	3,500–15,000
Terminal	90–200	80–180	270–600
Copiers/Duplicators			
Copiers, large; 30–67[a] copies/min	1,700–6,600	1,700–6,600	5,800–22,500
Copiers, small; 6–30[a] copies/min	460–1,700	460–1,700	1,570–5,800
Feeder	30	30	100
Microfilm printer	450	450	1,540
Sorter/collator	60–600	60–600	200–2,050
Audio Equipment			
Cassette recorders/players	60	60	200
Receiver/tuner	100	100	340
Signal analyzer	60–650	60–650	90–2,220
Mail Processing			
Folding machine	125	80	270
Inserting machine, 3600–6800 pieces/h	600–3,300	390–2,150	1,330–7,340
Labelling machine, 1500–30000 pieces/h	600–6,600	390–4,300	1,330–14,700
Postage meter	230	150	510
Vending Machines			
Cigarette	72	72	250
Cold food/beverage	1,150–1,920	575–960	1,960–3,280
Hot beverage	1,725	862	2,940
Snack	240–275	240–275	820–940
Miscellaneous			
Barcode printer	440	370	1,260
Cash registers	60	48	160
Coffee maker, 10 cups	1,500	1,050 sensible / 450 latent	3,580 sensible / 1,540 latent
Microfiche reader/printer	1,150	1,150	3,920
Microwave oven, 28 L (1 ft^3)	600	400	1,360
Paper shredder	250–3,000	200–2,420	680–8,250
Water cooler, 30 L/h (32 qt/h)	1,750	1,750	5,970

Source: Excerpted with permission of the American Society of Heating, Refrigerating, and Air-Conditioning Engineers, Inc. from 1997 *ASHRAE Handbook of Fundamentals.*

[a]Input of power is not proportional to capacity.

TABLE 5.23 **Rate of Heat Gain from Miscellaneous Appliances**[a]

Miscellaneous Data	Manufacturer's Rating Watts	Recommended Rate of Heat Gain (W)			Appliance	Miscellaneous Data	Manufacturer's Rating Btu/h	Recommended Rate of Heat Gain (Btu/h)		
		Sensible	Latent	Total				Sensible	Latent	Total
Electrical Appliances										
Blower type	1580	675	120	785	Hair dryer	Blower type	5,400	2,300	400	2,700
Helmet type 60 heaters @ 25 W	700	550	100	650	Hair dryer	Helmet type	2,400	1,870	330	2,200
91.44-cm normal use	1500	250	50	300	Permanent wave machine	60 heaters @ 25 W 36-in. normal use	5,000	850	150	1,000
1.27-cm diameter		28		28	Neon sign, per linear meter of tube[b]	0.5-in. diameter		30		30
0.95-cm diameter		56		56		0.375-in. diameter		60		60
	1100	190	350	540	Sterilizer, instrument		3,750	650	1200	1,850
	202	102		102	Magnetic card type-writer		690	350		350
Running	1760	1760		1760	Small copier	Running	6,000	6,000		6,000
Standby	880	880		880		Standby	3,000	3,000		3,000
Running	3515	3515		3515	Large copier	Running	12,000	12,000		12,000
Standby	1760	1760		1760		Standby	6,000	6,000		6,000
Gas-Burning Appliances					Lab burners					
1.1-cm barrel	880	495	125	620	Bunsen	0.4375-in. barrel	3,000	1,680	430	2,100
3.8-cm wide	1465	820	205	1025	Fishtail	1.5-in. wide	5,000	2,800	700	3,500
2.54-cm diameter	1760	985	245	1230	Meeker	1-in. diameter	6,000	3,360	840	4,200
Mantle type	585	530	60	590	Gas light, per burner	Mantle type	2,000	1,800	200	2,000
Continuous flame	730	265	30	295	Cigar lighter	Continuous flame	2,500	900	100	1,000

Source: Copyright © by the American Society of Heating, Refrigerating and Air-Conditioning Engineers, Inc., Atlanta, GA. Reprinted by permission from the *ASHRAE Handbook of Fundamentals,* 1981.

[a]For Residential appliances, see Table 5.11.

[b]Conventional (Btu/h) values are per linear foot of tube.

Fig. 5.34 *Plan of the office building for which heat gain calculations are shown in Example 5.13 (also Table 5.25).* (Reprinted with permission of the American Society of Heating, Refrigerating, and Air-Conditioning Engineers, Inc. from 1997 *ASHRAE Handbook of Fundamentals.*)

EXAMPLE 5.13. (This example is adapted from the 1997 *ASHRAE Handbook of Fundamentals.* Reprinted with permission of the American Society of Heating, Refrigerating, and Air-Conditioning Engineers, Inc., Atlanta, GA.)

A one-story office building (Fig. 5.34) is located "in the eastern United States near 40° N latitude." The adjoining buildings on the north and west are not conditioned, and their inside air temperatures are, for simplicity, assumed equal to the outdoor air temperature at any time of day. This is an unusual building both for its uninsulated walls and for very high lighting loads.

Roof construction: 115-mm (4.5-in.) flat roof deck of 50-mm (2-in.) gypsum slab on metal roof deck, 50-mm (2-in.) rigid above-deck roof insulation, surfaced with two layers of mopped felt vapor seal built-up roofing having dark-colored gravel surface, no false ceiling. Summer $U = 0.51$ W/m^2 -°C (0.09 Btu/h-ft^2 -°F).

South wall construction: 100-mm (4-in.) face brick, 200-mm (8-in.) common brick, 16-mm (0.625-in.) plaster, 6-mm (0.25-in.) plywood panel glued on plaster. Summer $U = 1.36$W/m^2 -°C (0.24 Btu/h-ft^2 -°F).

West wall and adjoining north party wall construction: 325-mm (13-in.) solid brick, no plaster.

Interior application $U = 1.40$ W/m^2 -°C (0.25 Btu/h-ft^2 -°F).

North exposed wall and east wall construction: 200-mm (8-in.) concrete block and 16-mm (0.625-in.) plaster. Summer $U = 2.73$ W/m^2 -°C (0.48 Btu/h-ft^2-°F).

Floor construction: 100-mm (4-in.) concrete on ground.

Fenestration: 1-m × 1.5-m (3-ft × 5-ft) windows of regular plate glass with light-colored venetian blinds, not openable. Summer $U = 4.6$ W/m^2 -°C (0.81 Btu/h-ft^2 -°F).

Front doors: Two, 1.5 m × 2 m (2.5 × 7 ft).
Side doors: Two, 1.5 m × 2 m (2.5 × 7 ft).
Rear doors: Two, 1.5 m × 2 m (2.5 × 7 ft), interior.

Door construction: Light colored 45-mm (1.75-in.) steel door with solid urethane core and thermal break. Summer $U = 1.08$ W/m^2 -°C (0.18 Btu/h-ft^2 -°F).

(*U*-values for all doors and outside walls were calculated assuming a wind speed of 12 km/h (7.5 mph). For party and inside walls, still air was assumed.)

Outdoor design conditions: Dry-bulb temperature, 35°C (94°F); daily range, 11C° (20F°). Wet-bulb temperature, 25°C (77°F).

Indoor design conditions: Dry-bulb temperature, 24°C (75°F); wet-bulb temperature, 18°C (62.5°F).

Occupancy: 85 office workers from 8 AM to 5 PM.

Lights: 17,500 W, fluorescent, from 8 AM to 5 PM; and 4000 W, tungsten, continuous.

Equipment: This example does *not* include any office equipment.

Ventilation: The ventilation rate is 7 L/s (15 cfm) per person, for a total of 595 L/s (1275 cfm).

The conditioning equipment is located in the adjoining structure to the north.

Determine the sensible, latent, and total space cooling load at design conditions. (*Note:* The results of this example are compared to much more detailed calculation procedures in Table 5.25.)

SOLUTION. Before beginning the calculations, *estimate* the heat gain using the design guideline procedure of Table 5.8.

TABLE 5.24 Cooling Load Through Building Envelope (Example 5.13)

Net Area (m²)	U-value (W/m² K)	ΔT (C°)	DETD (C°)	DCLF	Cooling Load (W)	Section	Reference	Net Area (ft²)	U-value (Btu/h ft² °F)	ΔT (F°)	DETD (F°)	DCLF	Cooling Load (Btu/h)
371.6	0.51		24.4		4620	Roof	Table 5.18	4000	0.09		44.0		15,840
37.6[a]	1.36		9.1		470	South wall	Table 5.18	405[a]	0.24		16.3		1,580
71.1[a]	2.7		9.1		1750	East wall	Table 5.18	765[a]	0.48		16.3		5,990
15.8[a]	2.7		9.1		390	North wall exposed	Table 5.18	170[a]	0.48		16.3		1,330
98.9[a]	1.41	11.2[b]			1560	Party walls		1065[a]	0.25	20[b]			5,330
3.25	1.08		13.1		50	Doors: S	Table 5.18	35	0.19		23.6		160
3.25	1.08	11.2[b]			40	N	Table 5.18	35	0.19	20[b]			130
3.25	1.08		13.1		50	E	Table 5.18	35	0.19		23.6		160
5.6	(4.59)[c]			97.9	550	Windows: S	Table 5.19	60	(0.81)[c]			31	1,860
2.8	(4.59)[c]			72.6	200	N	Table 5.19	30	(0.81)[c]			23	690
				Total:	9680							Total:	33,070

[a]Calculated from gross wall area, less windows and door areas.
[b]Design temperature difference, inside to outside.
[c]DCLF for glass *includes* the U-value.

People (omit equipment in this example):
$$2.3 \times 2^a = 4.6 \text{ Btu/h-ft}^2$$

Lighting[b]:
$$21,500 \text{ W}/4000 \text{ ft}^2$$
$$= 5.4 \text{ W/ft}^2 \times 3.41 \text{ Btu/h-W} = 18.4$$

Envelope gains:
Windows: $90 \text{ ft}^2/4000 \text{ ft}^2 \times 18^c = 0.4$
Walls:
 South: $405 \text{ ft}^2/4000 \text{ ft}^2 \times 0.24 \times 19^c = 0.5$
 East, north exposed:
 $935 \text{ ft}^2/4000 \text{ ft}^2 \times 0.48 \times 19^c = 2.1$
 Party: $1065 \text{ ft}^2/4000 \text{ ft}^2 \times 0.25 \times 19^c = 1.3$
 Roof: $4000 \text{ ft}^2/4000 \text{ ft}^2 \times 0.09 \times 39^c = 3.5$

Ventilation: $1275 \text{ cfm}/4000 \text{ ft}^2 \times 21^c = \underline{6.7}$

Total 37.5 Btu/h-ft²

[a]4000 ft²/85 people = 47 ft² per person, more than double the heat gain at a lower-range density of 100 ft²/person assumed in this approximate procedure, so increase people gain by factor of 2.
[b]This is considerably in excess of the lighting limitations on which this approximate procedure was based. Without the detailed knowledge of installed electric lighting, the design guideline would have suggested, based on

$$\text{DF} = 0.2 \times \frac{90 \text{ ft}^2 \text{ windows}}{4000 \text{ ft}^2 \text{ floor}} = 0.5\% \text{ DF},$$
$$< 1, \text{ a gain of only 5.1 Btu/h-ft}^2$$

[c]Interpolate for 94° F, between 90° and 100° F outdoor temperature.

Cooling load due to heat gain through building envelope: The heat gains through roof, exposed walls, and doors shown in Table 5.24 were calculated by

$$q = U \times A \times \text{DETD}$$

for which DETD values are taken from Table 5.18. In line with the approximate (and therefore rapid) nature of this calculation method, *no corrections* were made to DETD values to adjust for either outside or inside design temperatures differing from those listed, and actual gains were rounded to nearest 10. The climate's daily temperature range is medium; the roof is dark (typical of commercial buildings in air-polluted areas). Outside design temperatures of 35°C (95°F) were used. To keep things simple, the temperature difference through party walls was made equal to that on which the table was based; in this case, $35 - 23.8 = 11.2\text{C}°$ ($95 - 75 = 20\text{F}°$).

Cooling load due to heat gain through glass: These heat gains were calculated by

$$q = A \times \text{DCLF}$$

for which the DCLF values (which include U-value) were taken from Table 5.19. Again, *no corrections* were made, and gains were rounded to nearest 10.

Cooling load due to heat gain from lights, people, and equipment: The rate of heat gain from the lighting can be approximated simply by taking the energy input (including extra power required by ballasts for fluorescents). For fluorescent lights,

17,500 W × 1.2 ballast factor

$$= 21,000 \text{ W } (71,650 \text{ Btu/h})$$

For incandescent lights,

$$4000 \text{ W } (13,650 \text{ Btu/h})$$

The sensible gains from people can be determined from Table 5.21. For office work, assume 75 W (250 Btu/h):

$$85 \text{ people} \times 75 = 6380 \text{ W } (21,250 \text{ Btu/h})$$

Heat gains from office equipment are ignored in this example. (See Table 5.8 for typical ranges of such gains.) The sensible gains from lights, people, and equipment thus total 31,380 W (106,550 Btu/h).

Cooling load due to ventilation or infiltration: Because this is a commercial building incorporating deliberate introduction of outdoor air, infiltration will be ignored. [Although the mechanical ventilation suggested in Table 4.22 is 9.3 L/s (20 cfm) per person, this example assumes 7 L/s (15 cfm) per person.] Total mechanical ventilation is 595 L/s (1275 cfm). Sensible heat gains (from Table 5.20):

mechanical ventilation

$$13.6 \times 595 \text{ L/s} = 8090 \text{ W}$$

$$[22.0 \times 1275 \text{ cfm} = 28,050 \text{ Btu/h}]$$

Latent heat gains: The climate description suggests a typical northeastern U.S. climate at 40° NL. Newark, NJ (91 DB /73 WB), and Philadelphia, PA (90 DB, 74 WB), are examples. From Fig. 5.33, estimate latent gain as about 20% of the total sensible load. ∎

	W	Btu/h
Sensible gains, envelope	9,680	33,070
Sensible gains, lights and people	31,380	106,550
Sensible gains, ventilation	8,090	28,050
Sensible gains, total	49,150	167,670
Additional latent gains (20%, this example)	9,830	33,530
Total latent and sensible heat gains	58,980	201,200

5.12 Psychrometry

Moisture, air, and heat interact with some consequences that are threats to and other consequences that are opportunities for building performance. In winter, condensation within insulation due to falling air temperatures can be disastrous. In summer, adding moisture to hot dry air can lower its DB temperature while raising its humidity to more comfortable levels.

These moisture, air, and heat interactions are complex. As air temperature rises, its capacity to hold moisture rises also, and the warmer air becomes less dense. These combined interactions are described by *psychrometry,* the study of moist air. Fortunately, these interactions can be combined within a single chart (Fig. 5.35 with I-P units and Fig. 5.36 with SI units).

Already, in Chapters 2, 4, and 5, we have encountered the terms *dry-bulb* (DB), *wet-bulb* (WB), and *relative humidity* (rh). These elements are combined in the basic chart of Fig. 5.37, where the term *saturation line,* at 100% relative humidity, is introduced. This is also called the *dew point* (DP) because dew forms (water vapor condenses) when saturated air touches any surface at or below the air's dew point temperature. This saturation is sometimes undesirable, as within walls or roofs, or on ceiling, air duct, or glass surfaces. However, it is often desirable, as on air-conditioner coils, where the resulting reduction of the moisture content in the air is deliberate.

The psychrometric chart may be used to graph a wide variety of processes, which are summarized in Fig. 5.38. To understand these processes, we must add to the basic chart of Fig. 5.37. The first addition is *humidity ratio,* which indicates the amount of moisture by weight within a given weight of dry air. Air treatment processes that travel along these horizontal lines of constant humidity ratio (Fig. 5.39) are the familiar processes of simple heating (air passing through the heating coil of a furnace or through a solar collector) and simple (sensible) cooling (air passing through the cooling coil of an air conditioner, before saturation). The humidity ratio is used in Section 5.12 in calculating latent heat gains from outdoor air.

The next addition shows how the *density* of air varies as its temperature and moisture content vary. These lines are those of *specific volume,*

Fig. 5.35 *Psychrometric chart, I-P units. Conditions at sea level, standard atmospheric pressure (29.921 in. Hg). (Courtesy of the Carrier Corporation.)*

THERMAL CONTROL

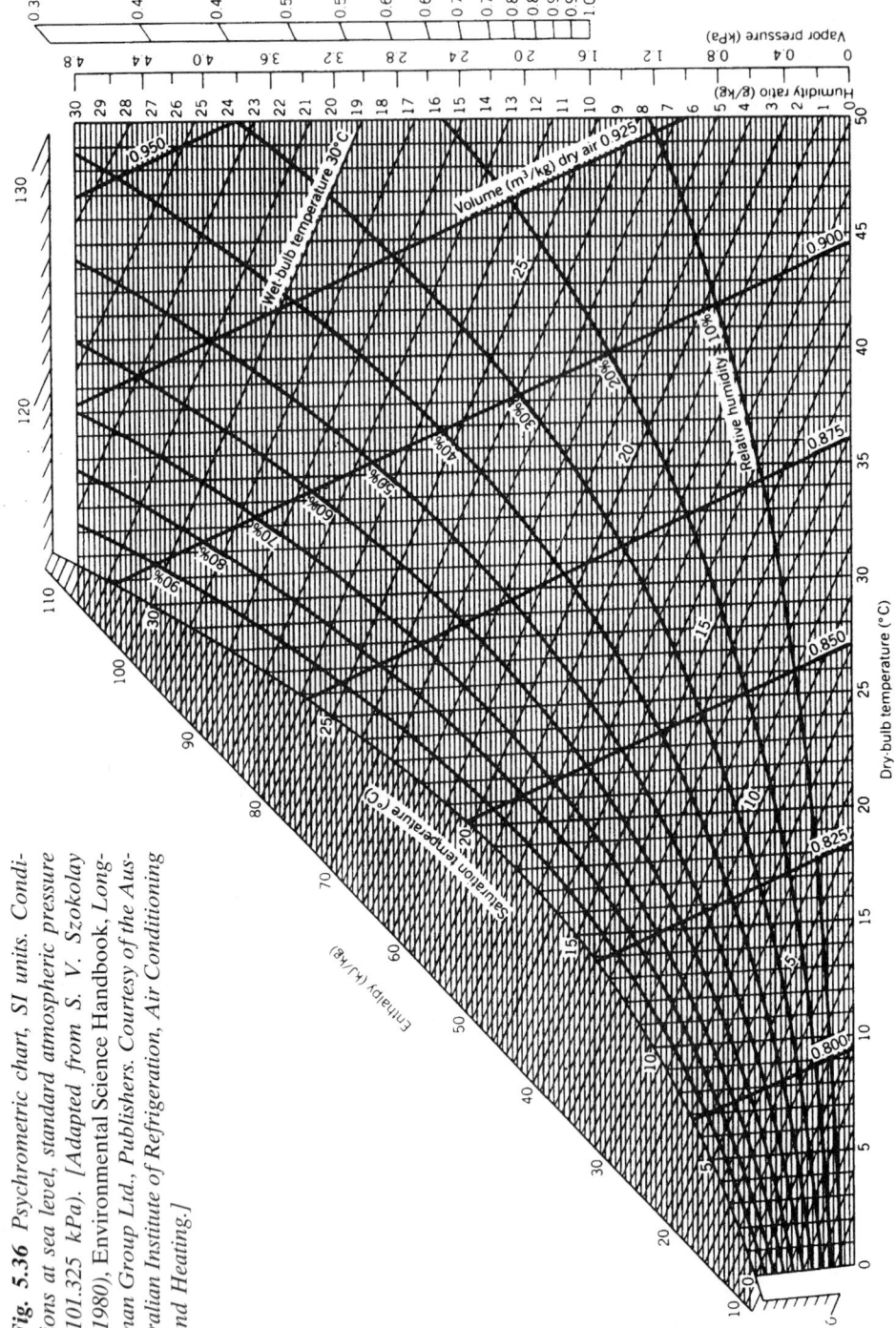

Fig. 5.36 *Psychrometric chart, SI units. Conditions at sea level, standard atmospheric pressure (101.325 kPa). [Adapted from S. V. Szokolay (1980), Environmental Science Handbook, Longman Group Ltd., Publishers. Courtesy of the Australian Institute of Refrigeration, Air Conditioning and Heating.]*

Fig. 5.37 *Some basic components of the psychrometric chart: DB, WB, and rh.*

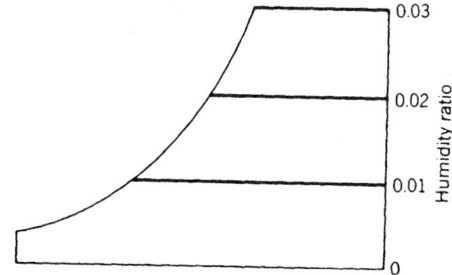

Fig. 5.39 *Humidity ratio on the psychrometric chart: I-P units are lb moisture/lb of dry air; SI units are kg moisture/kg dry air.*

the reciprocal of density, a useful quantity in air-conditioning calculations and helpful in understanding the stack effect [Section 5.14(*b*)]. The specific volume is given in ft³/lb (m³/kg) of dry air. It is evident from these lines in Fig. 5.40 that a pound of hot air is larger (has more volume) than a pound of cold air. This larger volume per unit of weight increases buoyancy; thus, hot air rises, whereas cold air sinks.

The next addition (Fig. 5.41) contains *enthalpy,* the sum of the sensible and latent heat content of an air–moisture mixture, relative to the sum of the sensible and latent heat in air at 0°F (0°C in SI units) at standard atmospheric pressure. Enthalpy units are Btu/lb (kJ/kg) of dry air. Enthalpy lines are almost parallel to those of WB temperature. Perhaps the most familiar process

to travel along the lines of constant enthalpy is evaporative cooling, where increased moisture and lower air DB temperature are obtained *without changing the enthalpy* (total heat content) of the air. There is indeed a drop in sensible heat as the temperature drops, but this is matched by an increase in latent heat as the moisture content increases. The opposite process is chemical (desiccant) dehumidifying, where decreased moisture

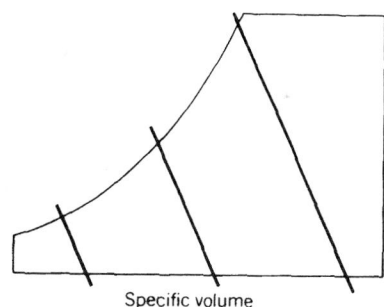

Fig. 5.40 *Specific volume on the psychrometric chart: I-P units are ft³/lb dry air; SI units are m³/kg dry air.*

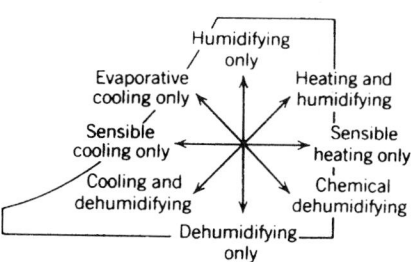

Fig. 5.38 *Climatic-conditioning processes, expressed on the psychrometric chart. [Adapted from "Architectural Design Based on Climate," by M. Milne and B. Givoni, in Watson (ed.),* Energy Conservation in Building Design. *Reprinted with the permission of the publisher, McGraw-Hill, Inc.]*

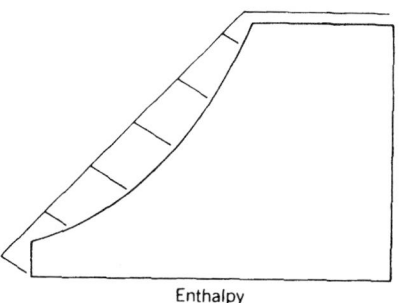

Fig. 5.41 *Enthalpy on the psychrometric chart: I-P units are Btu/lb; SI units are kJ/kg.*

content is obtained at the price of increased air DB temperature; again, no change in enthalpy (total heat) occurs.

(a) Cooling Process.

The work to be done by mechanical air-conditioning equipment is measured by the total change in enthalpy that must occur within the air that is treated by such equipment. The psychrometric chart is used to accurately size an air conditioner.

First, consider the problem of determining the total change in enthalpy. Assume outdoor conditions of 90°F DB, and 76°F WB, with desired indoor conditions at 75°F DB and 50% rh. From Fig. 5.35, these indoor conditions are 75 DB/62.7 WB. In the simple (but relatively rare) case of cooling 100% outdoor air, the total heat to be removed is determined as shown in Fig. 5.42. (Once-through systems cooling 100% outdoor air require substantial energy. They may be found in hospital surgical rooms and laboratories with many fume hoods.) How much total heat will be removed in taking air from 90 DB/76 WB to 75 DB/62.7 WB?

For every pound of "dry" air (based on the weight of the air alone) that is cooled and dehumidified, $39.6 - 28.3 = 11.3$ Btu must be extracted. Similarly, for every pound of air so treated, $0.0162 - 0.0093 = 0.0069$ lb of condensed moisture must be disposed of.

The actual cooling process is more complex, as shown in Fig. 5.43. The conditioned air must be introduced to the space at both a lower DB temperature and lower rh than the desired indoor conditions, so that such supply air might "soak up" heat and moisture, then leave through the return air grilles no worse than the desired indoor conditions (75 DB/50% rh). So another set of conditions, lower in DB and rh, will be established, depending on the rates at which air is introduced and the amounts of heat and moisture to be absorbed by the air passing through the space.

Within the cooling equipment, the outdoor air follows a complex path; the lines labeled 1, 2, and

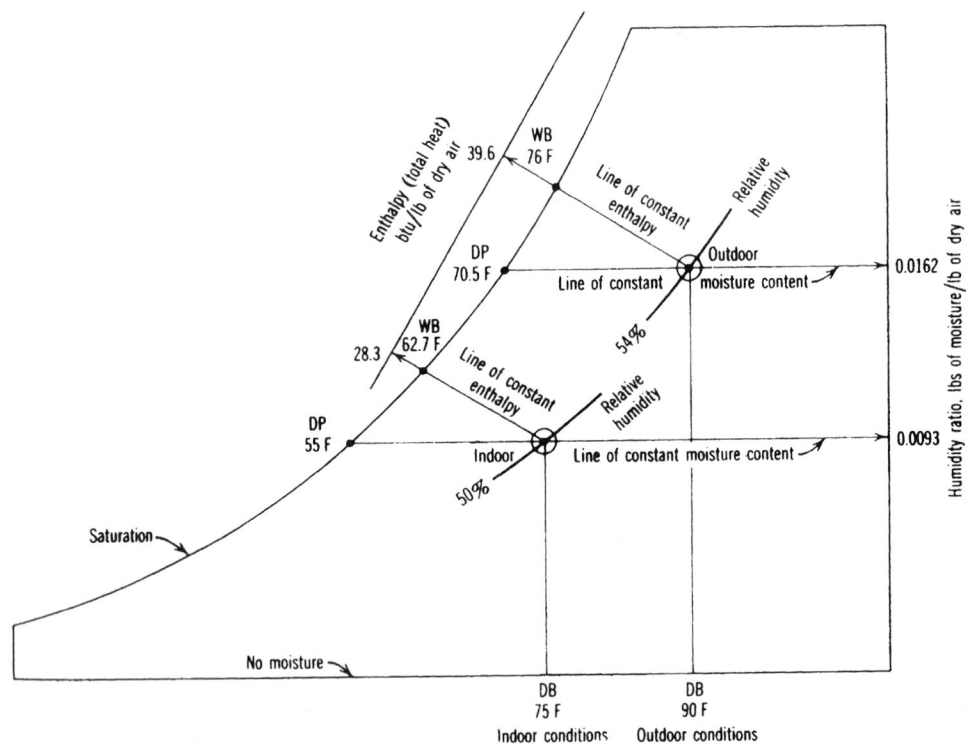

Fig. 5.42 *Use of the psychrometric chart to determine change in enthalpy between given outdoor and indoor conditions.*

Fig. 5.43 *Process of cooling and dehumidifying outdoor air, summer conditions.*

3 in Fig. 5.43 trace the cooling and the dehumidifying steps. Outdoor air is cooled without a loss of moisture (step 1) until it reaches saturation (its dew point). The cooling then continues; the air continues to lose sensible heat in step 2, and now loses moisture as well, extracting more heat. When step 3 begins (known as the *reheating process*), the remaining smaller amount of moisture and the air are both heated to 65°F, 40% rh—slightly below the conditions to be maintained in the space. The changes in the heat and moisture content of the air at the various stages are measured along the enthalpy scale.

Obviously, deliberately adding heat to just-cooled air sounds wasteful! Many heat-recovery processes can be incorporated in cooling equipment so that the reheat is provided by some of that just removed. Heat recovery equipment is discussed in Chapter 6.

An example of the sizing of an air conditioner for a dance hall is shown in Section 7.9, Example 7.4.

(b) Heating Process. The process of heating outdoor air in winter is charted in Fig. 5.44. Air at low temperatures in winter often has a humidity

ratio so low that this moisture content would be unacceptably low when the outdoor air is warmed (step 1). Therefore, moisture must be added. Often this is in the form of a warm water spray. However, in Fig. 5.44 an adiabatic spray is used—that is, no change in enthalpy—shown in step 2. Then, in step 3, the saturated air is warmed to a higher temperature (105°F in this case) than the desired indoor conditions of 70 DB and 40% rh. This supply air then loses its heat to the space. Note that in this process, we start at desired indoor conditions and work "backwards" (along step 3) to saturation, then downwards along the constant enthalpy line (along step 2) until we reach the line of constant humidity ratio of the warming outdoor air—in this case, 67°F DB. As with cooling, the changes in heat can be read on the enthalpy scale.

5.13 Detailed Hourly Heat Gain (Cooling Load) Calculations

Ordinarily, the approximate method of calculating the peak cooling load (Section 5.11) is sufficient for a designer's preliminary estimates of

THERMAL CONTROL

Fig. 5.44 *Process of heating and humidifying outdoor air, winter conditions.*

cooling equipment size for a small building. However, much more detailed procedures are used by engineers to actually size the equipment and to assess the peak-load impact of various design options such as shading devices.

The 1997 *ASHRAE Handbook of Fundamentals* (Chapter 28) presents three detailed methods for calculating heat gains and equipment cooling loads. The transfer function method (TFM) and the total equivalent temperature differential, time-averaging method (TETD/TA) each yield

hourly values over a 24-h day. They account for thermal storage by building mass, which can significantly shift impact of instantaneous heat gains on the actual cooling load for the HVAC equipment, as shown in Fig. 5.45. These detailed methods are so tedious and/or complicated as to require the use of a computer. Another more direct, one-step method, cooling load temperature differential/cooling load factor (CLTD/CLF), more quickly yields a 1-h value that also accounts for the effects of thermal mass; however, consider-

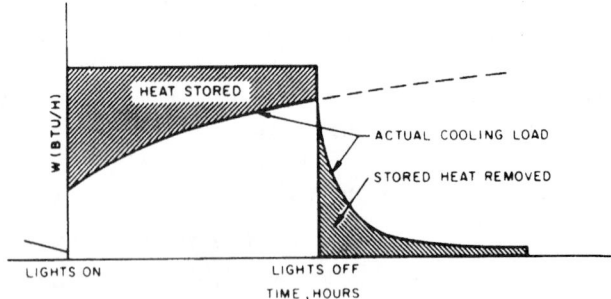

Fig. 5.45 *Thermal storage effect in the cooling load caused by electric lights.* (Reprinted with permission of the American Society of Heating, Refrigerating, and Air-Conditioning Engineers, Inc. from 1997 *ASHRAE Handbook of Fundamentals.*)

TABLE 5.25 **Comparison of Methods of Heat Gain Calculation (Example 5.13)**

Method	Maximum Hourly Gain (W)			24-h Average Gain (W/m²)			Maximum Hourly Gain (Btu/h)			24-h Average Gain (Btu/h ft²)		
	Sensible	Latent	Total	Sensible	Latent	Total	Sensible	Latent	Total	Sensible	Latent	Total
Design Guideline (Table 5.8)	(118/m²)						(37.5/ft²)					
Approximate[a] (Section 5.11)	49,150 (131/m²)	9,830	58,980				167,670 (41.9/ft²)	33,530	201,200			
Transfer Function[b] (worst hour 4 PM)	41,415 (110/m²)	15,040	56,455	68.3	32.0	100.3	149,623 (37.4/ft²)	61,168	210,791	21.3	12.5	33.8
CLTD/CLF[b] (for 4 PM only)	53,499 (143/m²)	15,040	68,489				179,140 (44.8/ft²)	63,208	242,348			
TETD/TA[b] (worst hour 4 PM)	50,096 (134/m²)	15,040	65,136	68.3	32.0	100.3	169,082 (42.3/ft²)	61,168	230,250	21.3	12.5	33.8

[a]See Table 5.24 and Example 5.13 for calculations.
[b]From 1997 *ASHRAE Handbook of Fundamentals,* Chapter 28.

able judgment is necessary to choose the "correct" hour for analysis. These methods are used by engineers, rarely by architects, and are not presented here. However, a summary of the results of the calculations by each method for the office building of Example 5.13 is shown in Table 5.25.

The quick and uncomplicated heat gain methods shown in Table 5.8 and in Section 5.11 served the simple building in Example 5.13 rather well. Had the design guideline (Table 5.8) been used to size an air conditioner, it matched the TFM worst-hour sensible gains, but was only about 84% of the other methods' worst-hour sensible gains. Latent gain estimates would have been necessary as well. The approximate method underpredicted latent gains, but the total (latent and sensible) gains of this quick, by-hand method ranged from 83% to 95% of the computer-based methods' results.

As a quick approximation for estimating overall equipment size or passive cooling approach, these by-hand methods are adequate for smaller and more simple buildings. When utilizing these quicker methods, keep in mind that some judgments may require subdividing the calculations or even changing some of the multipliers to reflect unusual conditions. A designer's checklist of considerations for heat gain should include:

1. Characteristics of the building's envelope: materials, sizes, external surface colors, and shapes.
2. Building location and orientation, as well as the extent of external shading of the building by trees or adjacent structures.
3. Outdoor design conditions (see Appendix A).
4. Indoor design conditions: DB, WB, and ventilation rate.
5. The schedule of lighting, occupancy, equipment, and any other processes that contribute to internal heat gain.
6. Thermal zoning requirements.

5.14 Passive Cooling Calculation Procedures

When outdoor air is 85°F (30°C) or below, it is possible to cool buildings by simple ventilation, maintaining conditions indoors that are within the "comfort zone." Design guidelines for sizing windows and stacks for either cross ventilation or stack ventilation were presented in Section 5.6.

In the many climates in which the outdoor temperature is above 85°F (30°C) for a large number of working-day hours, buildings that are closed to the hot exterior during those hours are practical. Between 80° and 85°F (27° to 30°C), outdoor air can keep people within the comfort zone if it moves across the body fast enough (see Fig. 2.8). In areas with reliable winds, open buildings are feasible up to 85°F (30°C). Between 55° and 69°F (13° and 21°C), outdoor air is cool enough to use in place of, or to greatly supplement, mechanically cooled air, if the humidity is sufficiently low. Passive cooling is even easier under these conditions; see Table 5.26 for some cities and related cooling data. Appendix A also shows CDD74 for a wider variety of locations.

From the advent of air conditioning to recent years, the standard response of designers was to turn to mechanical cooling, utilizing the refrigera-

tion cycle and forced air. More recently, some passive cooling alternatives have proved to be effective in climates that have clear or cool summer nights. The following procedures go beyond the quicker design guidelines that appeared in Section 5.6 and allow the designer to better adjust

TABLE 5.26 **Cooling Climate Data**[a] **for Some North American Cities**

| City, State, or Province | I-P Units, °F | | SI Units, °C |
	CDD50	No. Hours 8 AM–4 PM (°F) 55 < DB < 69 (°C) 13 < DB < 21	CDD10
Albuquerque, New Mexico	3908	703	2171
Albany, New York	2525	605	1403
Amarillo, Texas	4128	680	2293
Atlanta, Georgia	5038	749	2799
Baltimore, Maryland	3683	593	2046
Birmingham, Alabama	5206	760	2892
Brownsville, Texas	8777	422	4876
Boise, Idaho	2807	647	1559
Boston, Massachusetts	2897	713	1609
Buffalo, New York	2468	697	1371
Burlington, Vermont	2228	637	1238
Calgary, Alberta	1167	NA	648
Charleston, South Carolina	5722	767	3179
Cheyenne, Wyoming	1886	608	1048
Chicago, Illinois (O'Hare)	2941	613	1634
Columbia, Missouri	3752	633	2084
Columbus, Ohio	3119	708	1733
Denver, Colorado	2732	739	1518
Detroit, Michigan	3199	632	1777
Dodge City, Kansas	5001	637	2272
El Paso, Texas	5488	735	3049
Fort Wayne, Indiana	3077	601	1709
Fort Worth, Texas	6174	772	3430
Fresno, California	5350	785	2972
Great Falls, Montana	1993	641	1107
Honolulu, Hawaii	9949	69	5527
Houston, Texas	7215	703	4008
Jackson, Mississippi	5900	640	3278
Jacksonville, Florida	6847	674	3804
Lake Charles, Louisiana	6813	668	3785
Las Vegas, Nevada	6745	719	3747
Los Angeles, California	4777	1849	2654
Louisville, Kentucky	4000	636	2222
Lubbock, Texas	4833	743	2685
Madison, Wisconsin	2389	658	1327
Medford, Oregon	2989	749	1661
Memphis, Tennessee	5467	851	3037
Miami, Florida	9474	259	5263
Minneapolis, Minnesota	2680	566	1489
Montreal, Quebec	2146	NA	1192

TABLE 5.26 **Cooling Climate Data**[a] **for Some North American Cities (***continued***)**

City, State, or Province	I-P Units, °F		SI Units, °C
	CDD50	No. Hours 8 AM–4 PM (°F) 55 < DB < 69 (°C) 13 < DB < 21	CDD10
Nashville, Tennessee	4689	749	2605
New Orleans, Louisiana	6910	789	3893
New York (Central Park), New York	3643	790	2019
Norfolk, Virginia	4478	685	2488
Oklahoma City, Oklahoma	4972	733	2762
Omaha, Nebraska	3618	586	2010
Ottawa, Ontario	2045	NA	1136
Philadelphia, Pennsylvania	3623	646	2013
Phoenix, Arizona	8425	746	4681
Pittsburgh, Pennsylvania	2836	700	1576
Portland, Maine	1943	665	1079
Portland, Oregon	2517	1060	1398
Raleigh, North Carolina	4499	740	2499
Regina, Saskatchewan	1620	NA	900
Richmond, Virginia	4223	716	2346
Sacramento, California	4474	990	2486
Salt Lake City, Utah	3276	586	1820
San Antonio, Texas	7170	690	3983
San Diego, California	5223	1911	2902
San Francisco, California	2883	1796	1602
Saint Louis, Missouri	4193	614	2329
Seattle–Tacoma, Washington	2021	982	1123
Sioux City, Iowa	3149	602	1749
Tampa, Florida	8239	592	4577
Topeka, Kansas	3880	608	2156
Toronto, Ontario	2370	NA	1317
Tucson, Arizona	6921	716	3845
Tulsa, Oklahoma	5150	591	2861
Vancouver, British Columbia	1536	NA	853
Washington, DC	3734	657	2074
Winnipeg, Manitoba	1784	NA	991

Source: ANSI/ASHRAE Standard 90.1-1989, *Energy Efficient Design of New Buildings Except Low-Rise Residential Buildings,* copyright © by the American Society of Heating, Refrigerating, and Air-Conditioning Engineers, Inc., Atlanta, GA.

NA = not available

[a]CDH74 (cooling degree hours base 74° F) is found in Appendix A.

a preliminary design to the program and the climate.

To introduce these more detailed methods, consider the United Kingdom Pavilion at the 1992 Seville World Expo (Fig. 5.46). This (now dismantled) was an unusually large passive building [65 m × 32 m × 25 m high (213 ft × 105 ft × 82 ft high)], with an unusually wide collection of passive cooling techniques, inspired by the traditional cooling methods of hot-and-dry southern Spain:

shading, water, and mass. The Expo was an April to October event, with the greatest challenge in July–August, when average daily temperatures range from about 18° to 39°C (64° to 102°F).

Shading began with the flat roof, where standing wave-form racks carry both translucent fabric shades and PV cells facing up to the south. Outside the south facade, projecting sail-like "fly-sheets" keep off direct sun. The east and west facades took different strategies.

THERMAL CONTROL

Water became a theme of the exhibit, beginning with the wavelike silhouette on the roof and featured as a continuous sheet of water streaming down the east facade, dripping into a pool. The water is pumped with some power (about 50%) provided by the PV cells on the roof racks. Waiting visitors standing in line outside are treated to the sight and sound of this controlled waterfall, then cross a bridge over the pond, flanked by spray fountains, to enter the Pavilion. The pond continues below the east wall, exposed to the interior. Water also plays a much less evident role on the west facade.

Mass was a particular challenge, as the building was designed to be a lightweight structure to be dismantled and largely reused elsewhere. The west wall consisted of white metal containers filled with water to reflect the greater part of the direct sun and store the remainder during the day, radiating by night to cooler outdoor air.

(a)

Fig. 5.46 *The United Kingdom Pavilion at the Seville (Spain) World Expo, 1992.* (a) *Wave-form shading devices on roof also carry PV cells, whose electricity helps pump water from the pool that flows continuously over the east glass wall.* (b) *Section (E–W) with a combination of cooling strategies.* (c) *Section (E–W) with estimated maximum temperatures. (Sections are courtesy of Ove Arup Partnership, London, England.)*

(b)

(c)

Fig. 5.46 *(continued)*

Thermal zoning was a vital part of this concept, recognizing that people were entering from very hot, unshaded conditions, and that most visitors would then move through rather quickly. Passive cooling was given the task of keeping visitors cooler than the outside; mechanical refrigeration was responsible for maintaining even lower temperatures only where people gathered over extended time, as in the "pods" that contain the restaurant and the cafeteria. Gardner and Hadden (1992) describe how the typical July visitor left a sunny outdoor environment at about 38°C (100°F) for the cooler interior at about 32°C (90°F) supple-

mented by radiant "coolth" from the huge east glass wall at a maximum of 24°C (75°F) thanks to evaporation from the running water. Exhaust air from the pods lowered temperatures in their vicinity to about 28°C (82°F). The pods themselves, maintained at about 23°C (73°F), were entered only after undergoing a lengthy transition through the Pavilion from the hot outdoors.

To some extent, both cross and stack ventilation were also involved, despite the outdoor heat; the hottest air indoors rose to the ceiling, replaced by outdoor air that is cooled slightly as it passed through the watery gap between the east glass

THERMAL CONTROL

wall and the pond. Very high vents at the top of both east and west walls admitted wind to sweep across the ceiling and remove the hottest air.

(a) Cross Ventilation. Once again, a reminder that cross ventilation cooling only works when the *outside is cooler than the inside.* Otherwise, why bring in air that is warmer to try to cool a space? For ventilation using windows, the quantity of outdoor air admitted is termed V. The flow of air from inlet windows through the building to outlet windows should have a minimum of obstacles. In I-P units,

$$V = C_v A v$$

where V = volume flow rate of air, cfm
   A = area of openable windows on inlet side or sides, ft^2
   C_v = effectiveness factor (dimensionless) that adjusts for different wind orientations: 0.5 to 0.6 for winds perpendicular to the window openings; 0.25 to 0.35 for wind diagonal to the window openings.
   v = velocity of wind, feet per minute (= mph × 88).

In SI units,

$$V = 1000C_v A v$$

where V is in L/s (= m^3/s, multiplied by 1000)
   A is in m^2
   v is in m/s

The sensible heat removed by this flow of outdoor air through indoor spaces was presented in Chapter 4:

$$q_v = (V)(1.08)(\Delta t)$$

where q_v = sensible heat exchange due to ventilation (Btu/h)
   V = volume flow rate, in cubic feet per minute (cfm) of outdoor air introduced
   Δt = temperature difference between outdoor and indoor (F°)
   1.08 = a constant (the density of air, multiplied by the specific heat of air); units are Btu-min/ft^{3}h °F.

In SI units, this becomes

$$q_v = (V)(1.2)(\Delta t)$$

where q_v is in watts, V is in liters per second, 1.2 is the density of air (at 1.20 kg/m^3 multiplied by the specific heat of air (at 1.0 kJ/kg K), and Δt is in K.

A further modification to these cooling formulas is usually necessary because wind data are usually taken at airports at a height of about 10 m (30 ft) above open, unobstructed ground. Rarely will buildings that rely on cross ventilation be so favorably situated. Figure 5.47 shows correction factors to be applied to such airport wind velocities for variations in terrain and height above the ground. The dimensionless correction factor should then be applied to the quantity V, usually resulting in a reduced volume flow rate.

A more detailed procedure for calculating cross ventilation in residences given in Chandra et al. (1986) accounts for such factors as neighboring buildings and terrain, height of opening above grade, insect screens, and framing members within openings. To analyze natural ventilation in even more detail, wind tunnel tests are more likely to be useful than are detailed calculations.

(b) Stack Ventilation. Yet again, the reminder that stack ventilation cooling only works when the *outside is cooler than the inside.* The stack effect needs several conditions: warmer air

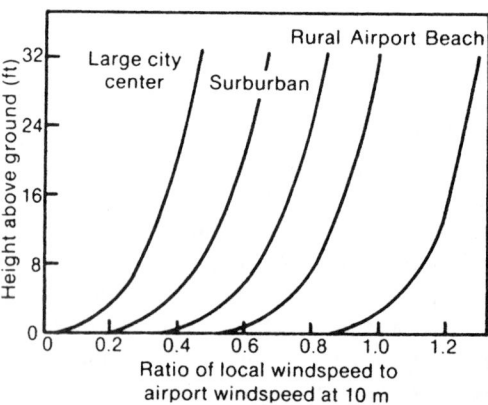

Fig. 5.47 *Windspeed variation with height for various terrains. (From* Cooling with Natural Ventilation, *Florida Solar Energy Center, published by the Solar Energy Research Institute.)*

indoors that can enter the bottom of the stack, cooler air outdoors, and low inlets to admit that cooler outdoor air to the building. This cooler outdoor air picks up heat from the building and enters the bottom of the stack. Within the stack, this now-warm air rises, because it is less dense—therefore lighter—than the cooler outdoor air that surrounds the top of the stack.

The following equation applies when there is no significant resistance to air flow within the building. To calculate ventilation cooling by the stack effect:

$$V = 60 \, KA \sqrt{g \, h \, (T_i - T_o)/T_i}$$

where V = airflow rate, cfm

K = discharge coefficient for opening, assume = 0.65 for multiple inlet openings

A = in square feet, the smaller of either total free area of inlet or outlet openings or horizontal cross-sectional area (throat area) of the stack

g = gravitational constant, 32.2 ft/s²

h = height of stack from inlet to outlet, ft

T_i = temperature indoors (must be greater than outdoors), °F

T_o = temperature outdoors (must be less than indoors), °F

In SI units,

$$V = 60 \, KA \sqrt{g \, h \, (T_i - T_o)/T_i}$$

where V = airflow rate, m³/s (× 1000 = L/s)

K = discharge coefficient for opening, assume = 0.65 for multiple inlet openings

A = in m², the smaller of total free area of inlet or outlet openings or horizontal cross-sectional area (throat area) of the stack

g = gravitational constant, 9.81 m/s²

h = height of stack from inlet to outlet, m

T_i = temperature indoors (must be greater than outdoors), K

T_o = temperature outdoors (must be less than indoors), K

The height of the stack and its cross-sectional area present an unusual opportunity for building form. The Inland Revenue Centre at Nottingham, England (Fig. 5.48) took advantage of this opportunity to provide cylindrical stair towers at corners (or at ends) of its buildings as stack ventilators. These are three and four-story buildings, with a width of 13.6 m (44.6 ft) to encourage both daylight penetration and natural ventilation. The tower stacks serve all floors except the top. The fabric-covered top of each tower is 7 m (23 ft) above the highest floor served, and can be raised up to 1 m (3.3 ft) or lowered (closed) to vary the flow of exhaust air. The stacks are enclosed in glass block, so that solar gain might assist the heating—and thus the flow—of air by day within the stack. Berry et al. (1995) report that the air flow increased from 4.8 to 6.2 ACH with the tower operating in solar mode.

Each of these lower floors' openings to each stack is 4 m² (43 ft²), with the doors held open magnetically unless fire alarms close them. This rather small opening to the stack (relative to the floor area served) is the limiting factor in the outward flow of heated air; to supplement the supply of outdoor air, small three-speed fans are installed in the floor below each window to draw in fresh air, then distribute it through a wide floor grille. The triple-glazed windows are also operable, both as sliding "doors" and as tilt-ins from the top.

This fan-assisted stack effect is used on summer nights to help pre-cool the building for the following day; the lower floors feature very large areas of exposed structural mass in their precast concrete ceilings.

The top floor is a lighter-weight structure, with the stack effect culminating in a ridge vent. Monitoring reveals that this floor operates at a few degrees warmer than the massive, night-ventilated lower floors.

(c) Night Ventilation of Thermal Mass.
Before doing these detailed calculations, be sure that (1) you have checked your summer climate against the passive cooling design strategies (Fig. 2.13) and that night ventilation of thermal mass is appropriate, and (2) your building and climate have been checked for approximate performance

(a)   (b)

(c)

Fig. 5.48 *The Inland Revenue Centre at Nottingham, England. (a) Cylindrical stair towers at corners and ends of three- and four-story office blocks act as stacks; solar gain through glass helps speed the flow of exhaust air. (b) Tops of the towers are fabric and can be raised to increase the stack aperture, and therefore increase the flow. (Photos courtesy of Vaughan Reynolds.) (c) Section of three-story wing with stack-ventilation stair tower. (Drawing courtesy of Ove Arup Partnership, London.) (d) Section perspective with fan-assisted intakes below each window and exposed precast concrete ceiling that helps store heat for night ventilation cooling.*

Light colored exposed concrete
acts as a heat sink and relector

Lightshelf/shade

Deep piers help
shade glazing
Openable triple glazed
window with solar blinds

User controls for
blinds, heat & fan

Balcony/shade

Night ventilation

Void

Air inlet

Fan/heat

(d)

Fig. 5.48 (continued)

based on the design guidelines (Fig. 5.16) for this cooling strategy. Also have in mind a positive night ventilation strategy; Fig. 5.49 shows an example of a forced-air system integral with a concrete joist and girder structural system. The following procedure is adapted from one developed by Karen Crowther for a workshop at the Fifth National Passive Solar Conference in Amherst, Massachusetts, as presented in Miller (1980).

STEP 1. In column II of Table 5.27, list the hourly outdoor temperatures for the design condition (these may be approximated from the summer DB temperature and mean daily range given in Appendix A). This will give the worst-day performance. (To get average-day performance, list average hourly temperatures, which are available from local weather service.) You need not list temperatures above 80°F (27°C) because outdoor air will not be used for cooling above that temperature.

STEP 2. Calculate the 24-h heat gain for the building, in Btu. Find the sum of all the hourly heat gains through the envelope, the minimum ventilation while "closed," and the internal gains while operating. List on line H of Table 5.27.

STEP 3. Find the total area of the thermal mass surface that is *exposed* (no rugs, etc.) both to the

space to be cooled *and* to moving night air during the ventilation ("open") cycle; list on line A of Table 5.27. *Note:* The larger the mass area exposed, the better the performance. This is why direct-gain solar-heated buildings, with plentiful exposed mass, often make such good candidates for night ventilation cooling. Two additional comments on mass surface: place it where people "see" it, so that it can readily receive their radiant heat; and keep direct sun off the mass (and out of the building) during the cooling season.

STEP 4. Find the mass heat capacity for the entire space to be cooled: mass volume × density × specific heat. Table 4.2 lists both den-

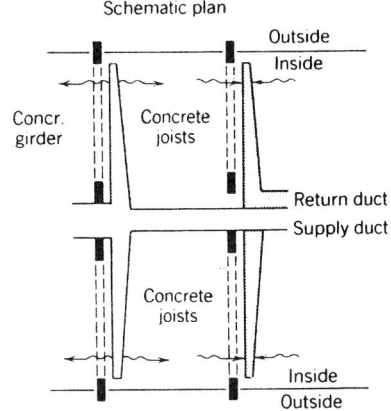

Schematic plan

Outside
Inside

Concr.
girder

Concrete
joists

Return duct
Supply duct

Concrete
joists

Inside
Outside

Schematic axonometric

Inside

Outside

Fig. 5.49 *Example of ductwork for night ventilation and the thermal mass of a concrete structural system.*

TABLE 5.27 **Night Ventilation of Thermal Mass**

Calculation Procedure
A. Mass surface area (from step 3) .... ft²
B. Mass heat capacity (from step 4) .... Btu/°F
C. Floor area (supplementary cooling, step 5), and .... ft²
    Total building volume .... ft³

(I) Hour	(II) Outside Air Temperature (°F)	(III) Cooling (Btu)	(IV) Mass Temperature (°F)
8 PM			
9 PM			
10 PM			
11 PM			
Midnight			
1 AM			
2 AM			
3 AM			
4 AM			
5 AM			
6 AM			
7 AM			
8 AM			
9 AM			

D. Total mass cooling .... Btu
E. Final mass temperature .... °F
F. Supplementary cooling (see steps 5 and 11) .... Btu
G. Total cooling, D + F .... Btu
H. 24-h heat gain, from step 2 .... Btu
I. Flow rate required for night ventilation .... cfh   or .... ACH

sity and specific heat for most common building materials; Table 5.28 shows a quick way to get mass heat capacities for the most common thermal mass materials. Enter this total mass heat capacity on line B of Table 5.27.

STEP 5. For "supplementary" cooling due to surfaces *other* than those of the principal thermal mass, list the space's *floor area*. This step should only be taken for spaces with a significant amount of roof, wall, or floor area *in addition* to the thermal mass areas counted in step 3. For example:

· If your space has exposed concrete ceilings, walls, and floors—all counted already as thermal mass—skip this step.

TABLE 5.28 **Common Mass Heat Capacities**

Conventional Units				S.I. Units
ft³ ×	(Btu/ft³ °F)	=	Btu/°F	m³ × (kJ/m³K) = kJ/K
Volume ×	(62.4)		Water	Volume × (4181)
Volume ×	(22.5)		Ordinary concrete	Volume × (1507)
Volume ×	(18.7)		Masonry, grout-filled	Volume × (1253)
Volume ×	(15.6)		Brick	Volume × (1045)

Source: Adapted from Crowther, "Night Ventilation Cooling of Mass," in Miller (1980).

- If all your thermal mass is in the form of free-standing water containers, enter the entire floor area.
- If the entire ceiling is thermal mass, but the walls or the floors are not, enter half the floor area.
- If the entire ceiling is thermal mass, but the floor is not, and there are few or no walls (e.g., open office plan), enter one-third to one-fourth of the floor area.

STEP 6. Complete column III hour by hour, after determining the mass temperature (column IV for the preceding hour):

$$\frac{\text{cooling}}{\text{Btu/h}} = \left[\begin{array}{c}\text{previous} \\ \text{hour} \quad\quad \text{outside} \\ \text{mass} \quad - \text{temp., °F} \\ \text{temp., °F} \quad (\text{col. II}) \\ (\text{col. IV})\end{array}\right]$$

$$\times \begin{array}{c}\text{mass} \\ \text{surface} \\ \text{area, ft}^2 \\ (\text{line A})\end{array} \times \begin{array}{c}\text{surface} \\ \text{conductance,} \\ \text{Btu/h} - \text{ft}^2 - \text{°F}\end{array}$$

For the first hour, assume that the mass temperature is 80°F. The surface conductance is usually assumed as 1.0 Btu/h-ft²-°F. (See Table 4.3 for other surface conductances under various conditions, many of which are considerably more than 1.0.)

STEP 7. Complete column IV hour by hour, after calculating the cooling Btu/h (column III):

$$\text{mass temp.} = \frac{\text{previous hour}}{\text{mass temp. (col. IV)}}$$

$$- \frac{\text{cooling Btu/h (col. III)}}{\text{mass heat capacity, Btu/°F (line B)}}$$

STEP 8. Continue this hourly process using columns III and IV *until* the falling temperature of the mass equals the rising temperature of the outdoor air. At that point, continuing with plentiful ventilation will only rob the mass of its "coolth"; the building therefore switches to (thermally) closed mode, with minimal ventilation.

STEP 9. Add all the hourly cooling Btu/h values (column III) to obtain the total mass cooling, Btu (line D).

STEP 10. Note the final mass temperature from column IV. This is probably at least 5F° *above*

the *lowest* air temperature of the night (column II). If this lowest mass temperature is significantly higher, consider redesigning for more exposed mass surface area.

STEP 11. If supplementary cooling is appropriate (see step 5), calculate it as follows.

$$\begin{array}{c}\text{supplementary} \\ \text{cooling Btu}\end{array} = \left[80\,\text{F} - \begin{array}{c}\text{Final mass} \\ \text{temperature} \\ (\text{line E})\end{array}\right]$$

$$\times 2.25 \times \left[\begin{array}{c}\text{Floor area} \\ (\text{line C})\end{array}\right]$$

The 2.25 factor assumes a modest role for the other, less thermally massive surfaces. Enter this supplementary cooling on line F.

STEP 12. Obtain total cooling by adding lines D and F; enter total on line G.

STEP 13. In step 2, you entered the 24-h heat gain for the building on line H. Compare this cooling Btu needed to the total cooling provided (line G). If you have not provided enough cooling, and the final mass temperature is more than 7F° above the lowest nighttime outdoor temperature, consider redistributing the building mass over a wider surface (e.g., 3000 ft² of 4-in. slab rather than 2000 ft² of 6-in. slab) and trying again. If you do not have enough cooling, and the final mass temperature is 5 to 7F° above the lowest nighttime outdoor temperature, consider providing *both* more mass and more surface area (e.g., 3000 ft² of 4-in. slab rather than 2000 ft² of 4-in. slab) and trying again.

STEP 14. Determine and enter on line 1 the approximate flow required for night ventilating air. Use the following formula:

$$\text{cfh} = \frac{\text{Btu/h}}{0.018\,\Delta T}$$

where cfh (ft³/h) = the minimum required flow rate of night air

Btu/h = the cooling Btu for the hour of maximum cooling during the night (column III)

ΔT = the temperature difference between the final mass temperature (column IV) and the outdoor air (column II) for that same hour of maximum cooling

It is often useful to express this night ventilation flow rate in terms of air changes per hour (ACH).

$$\text{ACH} = \frac{\text{cfh required}}{\text{building volume (ft}^3)}$$

EXAMPLE 5.1, PART J. How much heat will be removed, and what is the final thermal mass temperature, in a typical bay of the Oregon office building in Fig. 5.7? (See pages 203–209.) In this procedure, assume that only the exposed surfaces of the thermal mass participate in night flush cooling. (In the actual building, the interior surfaces of the continuous cores in the precast floor slabs are also flushed with night air, yielding added cooling.)

SOLUTION.

STEP 1. The design temperature for Eugene, Oregon, is 89°F, with a mean daily range of 31F°. This is distributed by hour (using the sine curve equation) as shown in Table 5.29, with the minimum daily temperature assumed at 4 AM, maximum at 4 PM.

STEP 2. The 24-h heat gain is estimated from Example 5.1, Part D. During the 9-h workday, the heat gains total 13 Btu/h-ft². During the unoccupied hours, there are negligible gains from electric lights, people and equipment, windows, or ventilation. However, the *opaque* envelope gains remain because this envelope stores daily heat and releases it gradually. These unoccupied gains (from Table 5.4) total 0.26 Btu/h-ft² from the walls, plus 1.03 Btu/h-ft² from the roof, totaling 1.29 Btu/h-ft².

Occupied: 9 h × 13 Btu/h-ft²
$$= 117 \text{ Btu/ft}^2$$
Unoccupied: 15 h × 1.29 Btu/h-ft²
$$= \underline{19.35 \text{ Btu/ft}^2}$$
$$136.35 \text{ Btu/day-ft}^2$$

For the entire typical bay,

136.35 Btu/day-ft² × 1440 ft² = 196,344 Btu/day

Enter this total on line H.

STEP 3. The area of exposed mass, from Table 5.4, is a total of 2232 ft². The concrete floor is covered by carpet and is therefore assumed disconnected from thermal storage for this space. The wallboard interior surfaces of the exterior walls are also ignored.

STEP 4. The mass heat capacity is estimated assuming that all elements are solid concrete at 150 lb/ft³ and 4-in. deep. The bearing walls are 8-in. thick, exposed on both sides, with concrete fill in the cores of all concrete blocks. The beams are 8-in. thick, but their inner faces are unexposed (they form the air duct for the night flush system). The precast ceiling slabs are 10-in. thick, but much of this thickness consists of air-filled cores; therefore, only the lower 4 in. is assumed to participate as "exposed" mass. From Table 5.28:

Mass heat capacity = 2232 ft² × 0.33 ft × 22.5 Btu/ft³-°F = 16,573 Btu/°F

STEP 5. Little supplementary cooling is assumed, although the carpeted floor provides some (albeit insulated) mass, as does the wallboard surface of the exterior walls. Assuming that *only one fourth* of the floor area contributes to cooling,

$$0.25 \times 1440 \text{ ft}^2 = 360 \text{ ft}^2$$

STEPS 6–10. See Table 5.29.

STEP 11. Determine supplementary cooling:

$$(80°F - 67.2°F) \times 2.25 \times 360 \text{ ft}^2 = 10,368 \text{ Btu}$$

STEP 12. See Table 5.29.

STEP 13. The building has more than adequate cooling capacity to meet this 89°F design condition; on *typical* summer days, with a high of only 83°F, there will be even more excess mass capacity. (A more conservative and very detailed analysis, assuming more electric lights left on, extreme high temperatures, and a less favorable coefficient of heat transfer between mass and air, led to the decision to also flush the precast cores at night.)

Because of this excess capacity, it appears advisable to begin the night flush a few hours *later,* when the ΔT between indoors and outdoors is greater and therefore more cooling is achieved per hour of fan operation. (Repeat the procedure above, beginning the flush at midnight with a ΔT of 80 − 68 = 12F°, to see how similar results can be obtained with 3 fewer hours of fan operation.) The disadvantage is that the interior will remain hotter for a longer period in the evening, which

TABLE 5.29 **Oregon Office Building (Fig. 5.7) Performance: Cooling by Night Ventilation of Thermal Mass for a Typical Second-Floor Bay**

A. Mass surface area	2,232 ft^2
B. Mass heat capacity	16,573 Btu/°F
C. Floor area (supplementary cooling)	360 ft^2
Total building (bay) volume	22,900 ft^3

(I) Hour	(II) Outside Air Temperature (°F)	(III) Cooling[a] (Btu/h)	(IV) Mass Temperature (°F)
PM			
8	80	No heat removed	80
9	77	(80 − 77) 2232 (line A) = 6,696	$80 - \dfrac{6696}{16{,}573\ (\text{line B})} = 79.6$
10	74	(79.6 − 74) 2232 = 12,499	$79.6 - \dfrac{12{,}499}{16{,}573} = 78.8$
11	71	(78.8 − 71) 2232 = 17,410	$78.8 - \dfrac{17{,}410}{16{,}573} = 77.7$
12	68	(77.7 − 68) 2232 = 21,650	$77.7 - \dfrac{21{,}650}{16{,}573} = 76.4$
AM			
1	65	(76.4 − 65) 2232 = 25,445	$76.4 - \dfrac{25{,}445}{16{,}573} = 74.9$
2	62	(74.9 − 62) 2232 = 28,793	$74.9 - \dfrac{28{,}793}{16{,}573} = 73.2$
3	60	(73.2 − 60) 2232 = 29,462	$73.2 - \dfrac{29{,}462}{16{,}573} = 71.4$
4	58	(71.4 − 58) 2232 = 29,909	$71.4 - \dfrac{29{,}909}{16{,}573} = 69.6$
5	60	(69.6 − 60) 2232 = 21,427[b]	$69.6 - \dfrac{21{,}427}{16{,}573} = 68.3$
6	62	(68.3 − 62) 2232 = 14,062	$68.3 - \dfrac{14{,}062}{16{,}573} = 67.5$
7	65	(67.5 − 65) 2232 = 5,580	$67.5 - \dfrac{5580}{16{,}573} = 67.2$
8	68	Stop flush: mass temperature is now below outdoor temperature	

D. Total mass cooling	212,933 Btu
E. Final mass temperature	67.2°F
F. Supplementary cooling	10,368 Btu
G. Total cooling (212,933 + 10,368)	223,301 Btu
H. Compare to 24-h heat gain	196,344 Btu
I. Flow rate required for night ventilation	About 5.5 ACH

[a]A surface conductance of 1.0 Btu/h-ft^2-°F is assumed in this calculation.

[b]At this point, enough heat has been removed to meet the typical bay's design-day heat gain.

could discomfort a late worker. Alternatively, the flush could be stopped after 5 AM, by which time the stored heat has been removed.

STEP 14. The airflow required will be checked at 4 AM, the hour of maximum cooling:

$$\frac{29{,}909 \text{ Btu/h}}{0.018 \text{ Btu/ft}^3\text{-}°\text{F-h} \times (71.4 - 58)\text{F}°} = 124{,}000 \text{ cfh}$$

$$\text{ACH} = \frac{124{,}000 \text{ cfh}}{22{,}900 \text{ ft}^3} = 5.41$$

which is somewhat more than that predicted by the design guideline in Exercise 5.1, Part F, of 4.47 ACH. ■

What might be the advantages of this approach, compared to conventional mechanical cooling? For this Oregon building, detailed records of electricity end-use are available (see Ash-

Fig. 5.50 *Comparing night ventilation of mass with conventional compressive cooling. The Emerald People's Utility District offices (Fig. 5.7 and Example 5.1, Part J) are monitored for end-use of electricity. In August 1991 [65 CDD18 (117 CDD65)], with compressive cooling only, 160 kW/ day of electricity was used, nearly all during periods of heavy use by day. Compare this to the hotter August 1990 [83 CDD18 (150 CDD65)], when 126 kW/day was used for compressive cooling by day, plus 34 kW/day for night-ventilation fan-forced cooling. (From Ashley and Reynolds, 1994.)*

ley and Reynolds, 1994). Figure 5.50 compares several August monthly records for the (conventional) compressors and the night-flush fans. In the first two Augusts, the building operated as a night-ventilated structure. In the third August (1991), due to a building-wide computer control changeover, the building operated only as a conventionally cooled structure. The sun's impact on the building should be relatively equal for these 3 months, but outdoor temperature varied as shown by the CDD18°C (= CDD65°F × 5/9). Not only was as much electricity necessary to cool the conventionally operated building, despite lower CDD than the previous August; but also, all the fan energy use and almost all the compressor energy occurred during the daytime hours of high consumption. In many locations, utilities charge more for electricity used during the daytime. With night ventilation, the fans deliver huge quantities of fresh air off-peak by night, unlike the conventional daytime operation that recycles some air to save energy.

Another night-ventilation example is the California Highway Patrol offices at Gilroy, California, shown in Fig. 5.51. This one-story building sits near a freeway interchange south of San Jose (81°F DB/65°F WB, MDR = 26°F). In this building, the nighttime air is drawn in as far from the freeway as possible, taken underground a short distance (with just a bit of earth tube cooling), then distributed throughout the central corridor. At either end of a continuous skylight over the central corridor, large fans exhaust air by night. Corridor walls are concrete block, and the floor is a concrete slab on grade. The cool night air may carry a seasonal scent; Gilroy calls itself the garlic capital of the world.

(d) Fan-Assisted Evaporative Cooling.
Before beginning the following calculations, be sure that:

1. You have checked your summer climate against the passive cooling design strategies (Fig. 2.13) and the "evaporative cooling" strategy is appropriate.
2. Your building and climate have been checked for approximate performance based on the design guidelines for evaporative cooling [Section 5.6(d)].

First find the total sensible heat gain in Btu/h that is to be removed from your building by evaporative cooling. Generally, this is calculated at your climate's summer design DB and mean coincident WB temperatures (Table A.1). The psychrometric chart is then used to plot the progress of evaporatively cooled air, as shown in Fig. 5.52. (The complete and complex psychrometric chart, I-P units, is found in Fig. 5.35.)

STEP 1. *Determine outdoor air conditions.* Enter the psychrometric chart at the summer design DB and mean coincident WB temperatures for your climate (point *A* in Fig. 5.52). As the air is blown through the evaporative cooler, it proceeds along the constant WB line toward saturation or 100% rh (from point *A* toward point *B* in Fig. 5.52).

STEP 2. *Determine supply air temperature.* The most efficient evaporative cooler will be able to cool and humidify the outdoor air to a point *about three fourths* of the total distance on the chart from entering air conditions to saturation (the three-fourths point is reached at point *B* in Fig. 5.52); a more reasonable operating assumption may be two thirds of that total distance.

STEP 3. *Determine ΔT indoors.* At conditions of point *B,* the air leaves the evaporative cooler and enters the building. It immediately begins to

pick up sensible heat (moving from point *B* toward point *C* in Fig. 5.52). When the air has picked up enough heat to reach an uncomfortably warm temperature, it is exhausted to the outdoors (point *C* in Fig. 5.52). Note, however, that this exhaust air is still well below outdoor temperatures, and therefore capable of some cooling just outside the building.

STEP 4. *Calculate heat removed and airflow rate indoors.* The heat removed from the building by this exhaust air depends on two factors: (1) the DB temperature difference (ΔT) between the supply air and the exhaust air (point *B* to point *C* in Fig. 5.52), and (2) the airflow rate usually expressed in cubic feet per minute:

Btu/h removed = (cfm airflow)

$$\times (1.08 \text{ Btu min/ft}^3\text{-}°\text{F-h})$$

$$\times (\Delta T, F°)$$

The designer can vary the cfm airflow rate by choosing larger or smaller evaporative coolers; any point less than three fourths of the total distance from outdoor air to saturation may also be chosen.

EXAMPLE 5.14. A 4000-ft² retail store near Tucson, Arizona, has been calculated to have a total sensible heat gain of 100,000 Btu/h at sum-

(a)

Fig. 5.51 *The California Highway Patrol building near Gilroy, California* (a) *is flushed with night air for cooling.* (b) *Axonometric shows intake air through a very short earth tube; night flush exhausts are at either end of the central skylit corridor with thermally massive walls and floor.* (c) *Site Plan.* (d) *Looking up at one night flush exhaust grille.* (e) *Exterior with night flush exhaust grille. (Drawings courtesy of The Colyer/Freeman Group, San Francisco, California.)*

Fan exhausts hot air

Concrete block thermal mass radiates accumulated heat

Cool air distributed throughout corridor

Fan exhausts hot air

Cool night air drawn into corridor from landscaped area

(b)

Fig. 5.51 (continued)

(c)

(d)   (e)

Fig. 5.51 *(continued)*

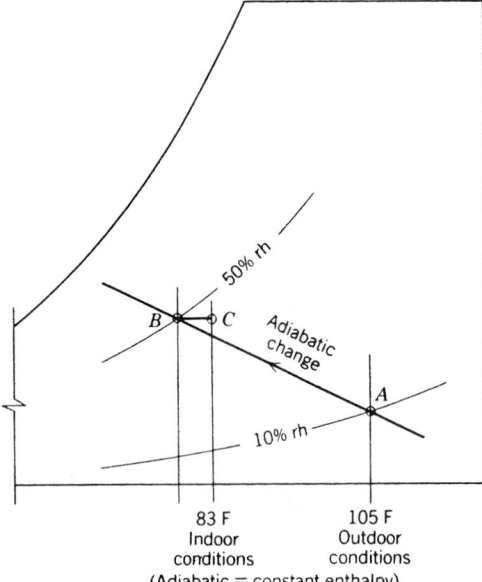

Fig. 5.52 *Once-through cycle of outdoor air through a fan-forced evaporative cooler. Values are transcribed from the psychrometric chart (Fig. 5.35). Hot-dry air can be humidified adiabatically (no change in total heat content) from point A to point B, reaching indoor conditions that fall within, or close to, the optimum summer comfort zone shown in Figs. 2.7 and 2.8. This is accomplished without the use of the high-energy–consuming compressive refrigeration cycle. As the humidified air picks up heat from the space, it moves from point B to point C.*

mer design conditions (105 DB, 66 WB for this location).

SOLUTION.

STEP 1. Enter Fig 5.52 at 105 DB, 66 WB (point *A*).

STEP 2. Measure along the constant 66-WB line to saturation, and determine that three fourths of this distance corresponds to 76 DB, 66 WB (point *B*).

STEP 3. ΔT indoors depends on the assumed exhaust air DB temperature; assume 83°F in this example (point *C*, corresponding to 83° DB, 68° WB). Then, $\Delta T = 83 - 76 = 7$F°.

STEP 4. Calculate heat removed and airflow rate indoors.

100,000 Btu/h removed

$$= (\text{cfm airflow})$$
$$\times (1.08 \text{ Btu-min/ft}^3\text{-°F-h})$$
$$\times (7\text{F°})$$

Rearranging the equation yields

$$\text{cfm} = \frac{100,000 \text{ Btu/h}}{1.08 \text{ Btu-min/ft}^3\text{-°F-h} \times 7\text{F°}} = 13,277 \text{ cfm}$$

Note that this is a rate of 13,277/ 4000 = 3.3 cfm/ft² of floor area.

This airflow rate can be provided by one larger or several smaller cooling units, depending on possible zones within the building. For instance, two evaporative coolers at 7000 cfm each would meet the requirements for this retail store; one might serve a smaller-area zone in front with show windows and lots of solar gain, another serve a larger-area interior zone with less solar gain. ∎

Some typical evaporative coolers are shown in Section 7.6.

(e) Cooltowers. One of the first North American cooltowers was shown in Fig. 5.19, along with estimated cooling design guidelines in Fig. 5.20. To more closely estimate the performance of such a passive, evaporative device with calculations, some very complex equations are involved [for example, see Thompson et al. (1994)]. For a shorter but less exact approach, consider this formula proposed by Givoni (1994), in either °F or °C:

$$T_{\text{exit}} = \text{DB} - 0.87[\text{DB} - \text{WB}]$$

where T_{exit} = air entering the space (exiting the cooltower)
   DB = dry-bulb temperature, outside
   WB = coincident wet-bulb temperature, outside

This simple equation ignores the potential influences of both wind (driving more air through the evaporative cooling pads) and the optional solar chimney exhaust tower (seen in Fig. 5.19).

Cooling achieved by the tower depends both on the temperature of this exit air and its flow rate. Givoni (1994) proposes, in SI units:

$$V = 0.033 \, A_{\text{evap}} \sqrt{H \, (\text{DB} - \text{WB})}$$

where V = flow, in m³/s ($\times$ 1000 = L/s)

A_{evap} = total area of wetted pads, m², in cooltower

H = overall height of tower, m

DB = dry-bulb temperature, outside

WB = coincident wet-bulb temperature, outside

0.033 = accounts for pressure drop through the tower and the building

Note that this equation ignores the area both of the cooltower and the outlet. Knowing both the temperature and flow rate, cooling achieved is

$$W = 1.2V \left[T_{int} - T_{exit} \right]$$

where W = Watts removed from space by cooltower

V = flow from cooltower, in m³/s ($\times$ 1000 = L/s)

T_{int} = DB temperature to be maintained indoors

T_{exit} = air entering the space (exiting the cooltower)

In I-P units, this process becomes:

$$V = 2.7\, A_{evap} \sqrt{H\,(DB - WB)}$$

where V = flow, in cfm

A_{evap} = total area of wetted pads, ft², in cooltower

H = overall height of tower, ft

DB = dry-bulb temperature, outside

WB = coincident wet-bulb temperature, outside

2.7 = accounts for pressure drop through the tower and the building

Note that this equation ignores the area both of the cooltower and the outlet. Knowing both the temperature and flow rate, cooling achieved is

$$Btu/h = 1.08\, V \left[T_{int} - T_{exit} \right]$$

where Btu/h = heat removed from space by cooltower

V = flow from cooltower, in cfm

T_{int} = DB temperature to be maintained indoors

T_{exit} = air entering the space (exiting the cooltower)

In a remote desert west of Tucson, the Agua Blanca ranch uses several cooltowers to serve a building around a T-shaped courtyard (Fig. 5.53). This off-grid building complex uses PV for lights and water pumping, propane for a refrigerator, one wall heater and stand-by generator, and collects rainwater in a cistern below the courtyard. Wood stoves and fireplaces supplement direct solar gain in winter. The courtyard connects to the surrounding ranch by three passages, one covered with a roof and small open skylights, one covered with a deciduous vine on a trellis, and one open to the sky. At the outer end of each passage are large rolling barn doors that provide both security and the ability to trap cool air from the cooltowers (or via the indoor spaces), thus increasing hot-weather comfort outdoors.

Three cooltowers, 22-ft (6-m) high, each with a total evaporative pad area of 64 ft² (5.9 m²), serve three units around the T-shaped courtyard. At the outlet of each cooltower, switches allow the flow to be shared, sent to either of two locations, or closed off. One of the cooltowers can deliver air directly to the covered passage off the courtyard.

Design simulations (Yoklic and Layseca, 1998) for a typical June in Tucson predict a flow from each cooltower of about 5000 cfm (2.36 m³/s), maintaining indoors at 74°F (23.3°C). In these quite small buildings, this translates to an ACH of about 60, considering the volume to include a height up to 6 ft (1.8 m), above which stratification of warm air negates the space.

These computer-generated design simulations are summarized in Table 5.30. These results differ somewhat from the results of Givoni's equations, suggesting that perhaps in this case,

$$T_{exit} = DB - 0.70[DB - WB],$$

and that flow rate (in SI units) may be closer to

$$V = 0.04\, A_{evap} \sqrt{2\,z\,(DB - WB)}$$

where z = effective height of the tower: that is, from the bottom of evaporative pads to the top of outlet to the building.

(f) Roof Pond Cooling. Before doing the following detailed calculations, be sure that (1) you have checked your summer climate against the passive cooling design strategies (Fig 2.13)

(a)

(b)

1 KITCHEN
2 DINING
3 MULTIPURPOSE
4 BEDROOM
5 LIVING RM
6 N-S ALLEY
7 E-W ALLEY
8 COURTYARD
9 VERANDA
10 PATIO
▦ COOL TOWER
▭ CISTERN

N

FLOOR PLAN

SECTION: WEST VERANDA

SECTION: E-W ALLEY

SECTION: N-S ALLEY

0   10

(c)

Fig. 5.53 *Cooltowers dominate the silhouette of the Agua Blanca ranch in a remote desert near Tucson, Arizona.* (a) *Exterior, showing north and west sides.* (b) *Plan of three units, each with a cooltower, around a T-shaped courtyard/covered passage outdoor space.* (c) *Sections through the three cooltowers.* (d) *West side, with one of the large rolling barn doors that allow the courtyard/outdoor spaces to be connected with cooler breezes or to be protected from hot winds. The cooltowers' air, after use within the units, can be discharged to the courtyard, while (under the hottest conditions) it is still cooler than outdoor air. Thus it helps to form a "cool pool" of air in the protected courtyard. (Courtesy of Martin Yoklic, Environmental Research Laboratory, University of Arizona.)*

(d)

Fig. 5.53 *(continued)*

and that either the "evaporative" or the "high thermal mass" strategy is appropriate, and (2) your building and climate have been checked for approximate performance based on the design guidelines. Note, however, that the design guideline is based *only* on summer night DB temperature. With evaporative cooling (such as a light spray onto the surface of the roof pond containers), better performance can be expected, shown by the following calculation procedure, based on Fleischhacker et al. (1982). This procedure can be used to check the size and depth required for your building's roof pond. It assumes an *optimum pond depth of 4 in. (100 mm)*, but allows for other depths as well.

STEP 1. First, assemble the following data about your climate.

Maximum DB temperature (Appendix A).

Mean daily range (Appendix A).

Minimum DB temperature (= max DB − mean daily range).

Design wet-bulb (2½%) (Appendix A).

Average maximum rh for July (from local climatological data, such as those given in Table 2.4).

July average temperature (TA July, in Appendix B).

From these data, determine two further characteristics for your climate: (1) minimum WB tempera-

ture, and (2) average July operating hours for residential air conditioning, or N.

Minimum WB temperature can be determined from the psychrometric chart (shown in Figs. 5.35 or 5.36). Enter the chart with minimum DB and move vertically along the constant DB line until you reach the average maximum rh for July. At that intersection, refer to the diagonal WB temperature lines, from which minimum WB temperature can be determined (Fig. 5.54).

Average July operating hours, N, can be estimated from Fig. 5.55. Enter at July TA for your climate. Your climate's N will fall somewhere between the maximum N and minimum N lines; as a guide to estimating N for your climate compare your climate's mean daily range and design WB to those of the cities shown in Fig. 5.55. The higher the mean daily range and the lower the design WB, the lower the N.

STEP 2. Calculate the peak hourly heat gain, *but exclude internal gains.* (The method given in Section 5.11 is appropriate for this roof-pond–sizing procedure.)

STEP 3. Approximate the daily total heat (excluding internal heat) to be stored in the roof pond, or Q_E, in Btu.

$$Q_E = \frac{\text{(Btu/h, peak hourly gain)} \times \text{(N, hours for July)}}{31, \text{days in July}}$$

TABLE 5.30 Cooltower Performance Hourly Simulations[a]

Part A. Tower Effective Height[b] = 10 ft   (3 m)

Ambient (°F)			Tower Outflow					Ambient (°C)			Tower Outflow				
DB	WB	RH (%)	DB	RH (%)	Velocity (fpm)	Flow Rate (cfm)	Water Consumed (gal/h)	DB	WB	RH (%)	DB	RH (%)	Velocity (m/s)	Flow Rate[c] (m³/s)	Water Consumed (L/h)
82	56	18.2	61.6	72.0	170	4254	9.8	27.8	13.3	18.2	16.4	72.0	0.86	2.00	37.1
87	59	18.0	65.1	71.0	175	4384	10.8	30.6	15	18.0	18.4	71.0	0.89	2.07	40.9
93	62	16.7	68.9	69.1	183	4572	12.3	33.9	16.7	16.7	20.5	69.1	0.93	2.16	46.6
98	64	15.0	71.8	67.0	190	4752	13.9	36.7	17.8	15.0	22.1	67.0	0.97	2.24	52.6
102	64	11.4	72.9	63.2	199	4987	16.1	38.9	17.8	11.4	22.7	63.2	1.01	2.35	60.9

Part B. Tower Effective Height[b] = 15 ft   (4.6 m)

Ambient (°F)			Tower Outflow					Ambient (°C)			Tower Outflow				
DB	WB	RH (%)	DB	RH (%)	Velocity (fpm)	Flow Rate (cfm)	Water Consumed (gal/h)	DB	WB	RH (%)	DB	RH (%)	Velocity (m/s)	Flow Rate[c] (m³/s)	Water Consumed (L/h)
82	56	18.2	62.2	69.6	205	5133	11.5	27.8	13.3	18.2	16.8	69.6	1.04	2.42	43.5
87	59	18.0	65.7	68.7	212	5289	12.7	30.6	15	18.0	18.7	68.7	1.08	2.50	48.1
93	62	16.7	69.6	66.7	221	5516	14.4	33.9	16.7	16.7	20.9	66.7	1.12	2.60	54.5
98	64	15.0	72.5	64.5	229	5734	16.3	36.7	17.8	15.0	22.5	64.5	1.16	2.71	61.7
102	64	11.4	73.7	60.5	241	6017	18.9	38.9	17.8	11.4	23.2	60.5	1.22	2.84	71.5

Part C. Tower Effective Height[b] = 20 ft   (6.1 m)

Ambient (°F)			Tower Outflow					Ambient (°C)			Tower Outflow				
DB	WB	RH (%)	DB	RH (%)	Velocity (fpm)	Flow Rate (cfm)	Water Consumed (gal/h)	DB	WB	RH (%)	DB	RH (%)	Velocity (m/s)	Flow Rate[c] (m³/s)	Water Consumed (L/h)
82	56	18.2	62.6	68.0	235	5864	12.9	27.8	13.3	18.2	17.0	68.0	1.19	2.77	48.8
87	59	18.0	66.2	67.0	242	6043	14.2	30.6	15	18.0	19.0	67.0	1.23	2.85	53.7
93	62	16.7	70.1	65.0	252	6302	16.1	33.9	16.7	16.7	21.2	65.0	1.28	2.97	60.9
98	64	15.0	73.0	62.7	262	6550	18.2	36.7	17.8	15.0	22.8	62.7	1.33	3.09	68.9
102	64	11.4	74.3	58.7	275	6874	21.1	38.9	17.8	11.4	23.5	58.7	1.40	3.24	79.9

Source: Excerpted from Yoklic, M. and Layseca, M. "Earth Construction, Courtyards, Verandahs, and Cooltowers: An Arizona Ranch House with Middle East Vernacular," in *Proceedings of the 23rd National Passive Solar Conference,* American Solar Energy Society, Boulder, Colo. Additional simulations courtesy of Martin Yoklic. Metric conversions by the author.

[a]Tower has a total evaporative pad area of 64 ft² (5.95 m²); pad thickness is 4 in. (100 mm); tower outlet area is 25 ft² (2.32 m²); building outlet area is 50 ft² (4.65 m²). Ambient data is selected from Tucson, Arizona, typical June.

[b]Effective height is measured from the bottom of the evaporative pads to the top of the outlet to the building.

[c]Flow rate (m³/s) × 1000 = L/s.

STEP 4. Determine the rate of the internal gain, in Btu/h, while the building is occupied.

STEP 5. Approximate the daily total internal heat to be stored in the roof pond, or, Q_I, in Btu.

Q_I = hourly internal gains, in Btu/h

× daily hours of building occupation

STEP 6. Calculate the daily heat gain directly to the roof pond through its insulated covers, or Q_P, in Btu. This formula assumes R16 (RSI − 2.8)

insulating panels with white upper surfaces and foil faces on the under surface.

$$Q_P = 0.4(A_c)(4 \times DB \text{ max, °F} - DB \text{ min, °F} - 200)$$

where A_c is the horizontal surface area of pond in square feet.

STEP 7. Consider whether fans will be used to stir the room air below the roof pond (to help the heat exchange between the pond and the room, as well as to aid comfort) and determine the value

Dodge City, Kansas: Max DB   97 F
                     Mean daily range − 25
                     Min DB =   72 F
          Average July max rh =   75%

WB
67 F

3. The WB temperature
at intersection *A*
is the minimum WB
temperature.

75% rh

A

2. Move along constant
DB line until average
July max rh
is reached.

1. Enter chart at min DB : 72 F
                               DB

Fig. 5.54 Finding the minimum (nightly) WB temperature, summer design conditions, for Dodge City, Kansas.

to be used for h, the overall heat transfer coefficient. A corrugated steel deck ceiling is assumed.

Air velocity	fpm (m/s)	0 (0)	44 (1.22)	73 (0.37)	115 (0.58)
Value of h		1.25	1.47	1.53	1.70

STEP 8. Determine the highest comfortable internal air temperature, T_{op}, from the comfort criteria provided in Fig. 2.8.

STEP 9. Calculate the maximum allowable internal air temperature, T_{imax}, which will be higher than T_{op} by a factor F that is related to your climate's characteristics (see Table 5.31 Part A for values of F).

$$T_{imax} = T_{op} + F$$

STEP 10. Determine the allowable temperature swing of the roof pond.

(a) Calculate the maximum pond T:

max. pond T

$$= T_{imax} - \frac{\left(\begin{array}{c}\text{peak total hourly heat gain,}\\\text{including internal gains, Btu/h}\end{array}\right)}{h(A_c)}$$

where T_{imax} is from step 9, h is from step 7, and A_c is the horizontal surface area of the pond in square feet.

(b) Calculate the minimum pond T, which depends on whether evaporative cooling will be used to help lower the pond's tempera-

ture and on several characteristics of the pond and the building.

1. For a *dry* pond surface:

min. pond T_{dry}

$$= DB_{min} + 1.5F° \pm \text{corrections } F°$$

(if any, from Table 5.31 Part B)

Fig. 5.55 Range of N, *average July operating hours, for residential air conditioners. (Based on Fleischhacker et al., 1982.)*

TABLE 5.31 **Roof Pond Design Data**

Part A. ***Values of Factor "F"***

Location	Design DB/ Mean Coincident WB (°F)	MDR (F°)	"F," Based on Interior Air Motion	
			None	Fans: 115 fpm (0.58 m/s)
Miami, Florida	90/77	15	2.0	1.0
San Antonio, Texas	97/73	19	3.0	1.5
Phoenix, Arizona	107/71	27	4.0	2.0

Part B. ***Corrections to Minimum (DB) Pond Temperature, F°***

Pond Depth (in.)				Night Internal Load (Btu/h ft²)				Pond Portion Exposed to Sky			
2	4	6	10	0	2	4	8	⅓	½	⅔	Fully
−1.5	+0	+0.7	+1.0	+0	+0	+0.8	+2.5	+3.3	+2.1	+1.2	+0

Source: Fleischhacker, P., Bentley, D. and Clark, G. (1982). "A Simple Verified Methodology for Thermal Design of Roof Pond Cooled Buildings," in *Proceedings of the 7th National Passive Solar Conference."* American Solar Energy Society, Boulder, Colo.

2. For a *wet* pond surface:

$$\text{min. pond } T_{wet} = DB_{min} - \frac{DB_{min} - WB_{min}}{2}$$

(c) Find the pond temperature swing ΔT_p:

$$\Delta T_{pdry} = \text{max. pond } T - \text{min. pond } T_{dry}$$

$$\Delta T_{pwet} = \text{max. pond } T - \text{min. pond } T_{wet}$$

(Obviously, if neither minimum pond T dry nor minimum pond T wet is lower than maximum pond T, a roof pond cannot be used for cooling!)

STEP 11. Determine the required pond depth D (in inches).

$$D = \frac{(0.19)(Q_E + Q_I + Q_p)}{(\Delta T_p)(A_c)}$$

where Q_E, Q_I, and Q_P are the daily pond heat gains, from steps 3, 5, and 6; ΔT_p is from step 10; and A_c is the horizontal surface area of the pond in square feet.

Note: If D is less than 4 in., consider reducing the pond size and recalculating. If D is much more than 4 in., consider a larger pond area, more air motion indoors, or the use of a wet pond surface to more closely approach the optimum 4-in. depth, or see optional step 12.

STEP 12. (optional) Auxiliary mechanical air conditioning may offer a more economical alternative than increased pond size. The size, in "tons" (12,000 Btu/h) of air conditioning required, is determined as follows.

(a) Desired D is 4 in., optimum.

(b) $\Delta T_p = \left(\frac{0.19}{D}\right)\frac{(Q_E + Q_I + Q_p)}{A_c}$

(c) so

$$\text{max. pond } T = \text{min. pond } T + \Delta T_p$$

(d) And

tons of AC

$$= \frac{\text{peak total hourly heat gain}}{6000} - \frac{(\text{Btu/h} - [h(A_c)](T_{imax} - \text{max pond } T)}{6000}$$

where the peak total hourly heat gain *includes* internal gains, h is from step 7, A_c is the horizontal surface area of the pond in square feet, T_{imax} is from step 9, and max. pond T is from step 10.

EXAMPLE 5.6, PART B. Size the roof pond for the Albuquerque office building for which we predicted that a 4-in.–deep pond equal to three fourths of the building's floor area would be sufficient. Assume that the one story office is about 4000 ft² in area. Try a 4-in.–deep pond of 3000 ft².

SOLUTION.

STEP 1. Albuquerque, design data:

DB_{max} is 94°F (Appendix A).

MDR is 27F° (Appendix A).

DB min. is $94 - 27 = 67$°F.

July rh *maximum* (nighttime) is approximately 50% (local data).

July TA is 79°F (Appendix B).

WB_{min} is from Fig. 5.35 at the intersection of the 67 F DB line, and 50% rh is 56°F.

July operating hours N for this dry, high area should be close to the minimum for TA = 79°; from Fig. 5.55, this is about 160 h.

STEP 2. Determine the peak hourly gain. Earlier in this exercise, a total gain of 15 Btu/h-ft² was assumed as an average. Assume that 9 Btu/h-ft² of this total represents the load from electric light, people, and equipment, and that 6 Btu/h-ft² is due to envelope and ventilation gains. (No heat gains to the interior are assumed through the roof pond.) Then,

peak hourly gains (excluding internal)
$$= 6 \times 4000 \text{ ft}^2 = 4,000 \text{ Btu/h}$$

STEP 3.

$$Q_E = \frac{24,000 \text{ Btu/h} \times 160 \text{ h}}{31 \text{ days}} = 123,870 \text{ Btu}$$

STEP 4.

internal gains $= 9 \text{ Btu/h-ft}^2 \times 4000 \text{ ft}^2$

$$= 36,000 \text{ Btu/ft}^2$$

STEP 5.

$Q_I = (36,000 \text{ Btu/h})(9 \text{ h operation})$

$$= 324,000 \text{ Btu}$$

STEP 6.

$Q_p = 0.4(3000 \text{ ft}^2)(4 \times 94 \text{ F} - 67 \text{ F} - 200)$

$$= 1200 \times (376 - 67 - 200)$$

$$= 1200 \times 109 = 130,800 \text{ Btu}$$

STEP 7. Fans will be used, at 115 fpm, for added comfort as well as for heat transfer. Therefore, $h = 1.7$.

STEP 8. From Fig. 2.4, with 115 fpm (0.58 m/s) airspeed, it appears that 83°F is within the comfort zone. So $T_{op} = 83$°F.

STEP 9. $T_{imax} = T_{op} + F$. In Table 4.30, Part A, Albuquerque's conditions appear to be somewhere between those of San Antonio and those of Phoenix, so assume F to be 1.75.

$$T_{imax} = 83 + 1.75 = 84.75\text{°F}$$

STEP 10.

(a) max. pond $T = T_{imax} - \dfrac{\text{total hourly gain}}{h(A_c)}$

$$= 84.75 - \frac{24,000 + 36,000}{1.7 \times 3000}$$

$$= 84.75 - 11.75 = 73\text{°F}$$

(b) Assume a dry pond surface. Because this is a 4-in.–deep pond that is fully exposed to sky, with no night internal load, there is no correction factor from Table 4.30 Part B.

min. pond $T_{dry} = $ DB min. $+ 1.5$F°

$$= 67 + 1.5 = 68.5\text{°F}$$

For a wet pond,

min. pond $T_{wet} = DB_{min} - \dfrac{DB_{min} - WB_{min}}{2}$

$$= 67 - \frac{67 - 56}{2}$$

$$= 61.5\text{°F}$$

(c) the pond temperature swing is therefore

$$\Delta T_p \text{ dry} = 73 - 68.5 = 4.5\text{F°}$$

$$(\Delta T_p \text{ wet} = 73 - 61.5 = 11.5\text{F°})$$

STEP 11.

$$D_{rqd} = \frac{0.19(123,870 + 324,000 + 130,800)}{4.5 \times 3000}$$

$$= \frac{109,947}{13,500} = 8.1 \text{ in., for a dry pond}$$

(For a wet pond, $D_{rqd} = 3.2$ in.)

It appears that a 4-in. deep pond with a wet surface of somewhat less than 3000 ft² would be sufficient for this building in this climate. ∎

Once the area and the depth of roof ponds have been determined, questions arise about the architectural integration of such large horizontal surfaces and their relative emphasis on heating or cooling. Figure 5.56 shows variations in the treatment of containers and the insulating panels to match the most important thermal role of roof ponds; Fig. 5.57 shows three typical approaches to the placement of roof ponds on buildings. A valuable source of information about the container bags, sliding insulation, roof decks, and other components needed for roof ponds is the *California Passive Solar Handbook* by Phillip Niles and Ken Haggard, written for the California Energy Commission in 1980. Commercially available products include the Skytherm® system, developed and pioneered by Harold R. Hay.

A variation of the roof pond (applying primarily to the cooling mode) is to move the water rather than move the insulation. Water-impervious insulation (such as Styrofoam) is secured in place above the metal ceiling at a distance sufficient to be a reservoir for the cooling water. At night, this water is pumped onto the top of the insulation, where it cools by radiation and evaporation, then trickles slowly back into the reservoir. This variation is now being tested at the School of Engineering Technology, University of Nebraska at Omaha campus. In winter, the reservoir of water above the metal ceiling could act as thermal mass for a DG passive system. However, cold winter rain would need to be excluded from that reservoir.

(g) Earth Tubes. The use of the earth as a heat sink, in our culture at any rate, is still in an early development stage. The most direct application would be an underground building with uninsulated concrete walls set against the soil. The problem, of course, is that winter heat loss will likely exceed summer loss; if solar heat can be admitted (e.g., through skylights) to counterbalance the increased winter loss, uninsulated walls sized for the desired summer loss become more attractive. (However, except in dry climates, condensation on walls may pose a seasonal or even year-round problem.) A more thorough discussion of this option can be found in Watson and Labs (1983).

The long-term potential of the earth as a heat sink is lessened by the fact that soils are relatively slow heat conductors. For example, we ignore winter heat loss through concrete slabs to the earth below, calculating instead the heat losses through the slab's exposed perimeter.

Earth tubes [as discussed under design guidelines, Section 5.6 (g)] are devices for cooling the incoming ventilating air through earth contact, before it enters the building. A small amount of air (perhaps equal to the minimum fresh air requirements—see Table 4.22) is usually brought in through several tubes. These tubes are usually 8 to 20 in. in diameter, are buried at a depth of 5 to 10 ft, and are up to 200 ft long. Table 5.32 summarizes several tested summertime applications.

The lowest temperature to which air can be cooled in such tubes will *approach* earth temperature; the slower the air moves through the tube, the more time there will be for cooling. The air temperature might come within 4F° (about 2C°) of the earth temperature, under good conditions.

(a)

Fig. 5.56 (a) *Prinicipal components of roof ponds with moveable insulation.* (b–g) *Variations on roof ponds for optimization of heating or cooling performance. (Reprinted from Niles and Haggard, 1980, by permission of the California Energy Commission.)*

Day   Night

(b) Cooling Only

The ponds are shaded from the sun so that cooling can be accomplished by evaporation on a 24-hour basis. Removal of heat from the interior space can be done by radiation and convection to the bottom of the ponds, or by thermosiphoning the cooled water to an interior tank in contact with the room, as shown.

(c) Cooling and Some Heating

Flooding the roof ponds adds evaporative cooling to radiation and convection losses. During the day, forcing air over flooded ponds below closed insulation creates additional evaporative cooling.

(d) Cooling Emphasis

Flooding the ponds increases cooling through evaporation. Sealed but nonglazed ponds emphasize radiant cooling and are still capable of heating at lower latitudes.

(e) Balanced Heating and Cooling

The use of an inflated air cell over the ponds increases heating capability by increasing insulation. During the cooling season, the cell is deflated to allow maximum night sky radiation and convection cooling.

(f) Heating Emphasis

The use of lift or bi-fold insulation allows the insulation to also act as a reflector when in the open position, thus increasing solar radiation to the ponds.

(g) Maximum Heating and Some Cooling

The roof pond is enclosed under a roof that slopes to the north, protecting the ponds from snow and providing permanent insulation on the north side. In the open position, the movable insulation acts as a reflector for low winter sun.

Fig. 5.56 *(continued)*

THERMAL CONTROL

THERMAL CONTROL

(a)

(b)

(c)

Fig. 5.57 *Three examples of roof pond integration with building form. (a) Insulating panels slide open to stack over exterior (or untempered) space. See Fig. 5.22 for this system in use. (b) Hinged insulating panels with reflective undersurfaces can act as reflector to increase winter insolation. However, some interference with summer night sky radiation can result, because the pond surface cannot "see" the entire sky dome. (c) Bifold insulating panels also act as reflectors, interfering less with diffuse solar radiation and allowing somewhat more exposure to the night sky than (b). (Adapted from Niles and Haggard, 1980, by permission of the California Energy Commission.)*

To approximate earth temperature around the tube:

1. Determine T_{GW}, the groundwater temperature (in °F) (assume equal to deep-underground earth temperature) from the map in Fig. 5.58.
2. Determine T_{AMP}, the average amplitude of surface temperature (in F°), using the map in Fig. 4.11, page 134.
3. Determine CF, amplitude correction factor, depending on whether the soil outside your

earth tube is dry, average, or wet during the cooling period. The amplitudes, or seasonal variations in earth temperature, *diminish to near zero* at these depths (also called CF):

	CF	
Dry soil	14 ft	(4.25 m)
Average soil	18 ft	(5.5 m)
Wet soil	22 ft	(6.7 m)

4. Determine D, the depth (ft) at which the earth tube will be installed.
5. Determine T_{SG}, the late summer ground temperature (°F) around the earth tube:

$$T_{SG} = (T_{GW}) + (T_{AMP})\,[1 - (D/CF)]$$

(Note that when the tube is very deep and $D \geq CF$, $T_{SG} = T_{GW}$.)
6. Determine the temperature difference that makes cooling possible:

$$\Delta T = \text{desired indoor temperature} - T_{SG}$$

7. Determine the long-term cooling rate:

Btu/h ft of tube length

$$= (\Delta T)(C)$$

(Area, tube surface, ft²/ft of length)

where C is a factor that accounts for the long-term effects of heat flow from the tube to the soil:

$C = 0.11$ for dry soil (soil conductivity $k = 0.25$ Btu/h-ft-°F)

$C = 0.28$ for average soil ($k = 0.75$)

$C = 0.44$ for wet soil ($k = 1.5$)

In SI units,

$C = 0.62$ for dry soil

$C = 1.59$ for average soil

$C = 2.50$ for wet soil

This represents an attempt to simplify a complex problem. The air flow rate is ignored, even though with very slow flow, there is more time for heat exchange in the tube. The longer the tube is flooded with hot air, the higher will be the earth temperature. For more detailed information about and other approaches to earth tubes, see Givoni (1994), C. E. Frances, "Earth Cooling

TABLE 5.32 **Earth Tube Applications**[a]

Part A. I-P Units

Soil, No. of Tubes, Arrangement	Tube Diameter (in.)	Length (ft.)	Depth[b] (ft)	Temperatures (°F) Soil	Intake Air[c]	Air Velocity (ft/min)	Total Flow Rate (ft³/h)	Flow per Tube (ft³/h)	Peak Sensible Cooling (Btu/h ft)
Sandy soil; 8 tubes, radial	10	100	8	73°	67–105°	220	57,600	7,200	**20**
Silt loam, wet; 9 tubes, radial (3 trenches)	5	150	8–11.5	73°	62–95°	500	60,000	4,000	**11**
Silt loam; 22 tubes, parallel @ 4 ft	10	180	8–9	72°	70–94°	500	359,700	16,350	**30**
Silt loam; 5 tubes, parallel	12	260	10	63°	60–92°	530	123,600	24,720	**40**

Part B. SI Units

Soil, No. of Tubes, Arrangement	Tube Diameter (cm.)	Length (m)	Depth[b] (m)	Temperatures (°C) Soil	Intake Air[c]	Air Velocity (m/s)	Total Flow Rate (m³/h)	Flow per Tube (m³/h)	Peak Sensible Cooling (W/m)
Sandy soil; 8 tubes, radial	25	30.5	2.4	23°	19.4–40.6°	1.1	1,632	204	**19.7**
Silt loam, wet; 9 tubes, radial (3 trenches)	13	46	2.4–3.5	23°	16.7–35°	2.5	1,080	120	**10.6**
Silt loam; 22 tubes, parallel @ 1.2 m	25	55	2.4–2.7	22°	21.1–34.4°	2.5	10,200	463	**29.5**
Silt loam; 5 tubes, parallel	30	79	3	17°	15.6–33.3°	2.7	3,500	696	**38.8**

Source: Givoni, B. (1994). *Passive and Low Energy Cooling of Buildings.* Van Nostrand Reinhold, New York. I-P conversions by the author.

[a]Central Illinois farm applications, summertime.

[b]Tubes often slope to drain the condensation that occurs as air reaches the dewpoint within the tube.

[c]Intake air temperature range during the period of study.

Tubes, Case Studies . . ." in Bowen et al. (1981), Abrams (1980), and Brown et al. (1992).

EXAMPLE 5.7, PART B. A partially underground nature center of 4000 ft² (372 m²) in northern Pennsylvania has an hourly heat gain of 9.3 Btu/h-ft², of which 2.2 Btu/h-ft² are due to outdoor air–required ventilation. A long trench at a depth of 6 ft (1.8 m) in heavy, damp soil must be dug for a utility connection. If one earth tube, used only for cooling the fresh air, is installed at the bottom of the trench, what length and diameter will be needed?

SOLUTION. The total cooling needed from the earth tube is

$$(4000 \text{ ft}^2)(2.2 \text{ Btu/h ft}^2) = 8800 \text{ Btu/h}$$

(In Part A using design guidelines, assuming heavy damp soil, a 60-ft–long tube of 20-in. diameter promised to deliver about 4500 Btu/h; two such tubes, about 9000 Btu/h.)

1. Determine T_{GW}: from Fig. 5.58, northern Pennsylvania has an average underground temperature of 52°F (11.1°C).
2. Determine T_{AMP}: the seasonal amplitude (from Fig. 4.11) is about 18F°.
3. Determine CF: in wet soil, this amplitude extends to 22 ft.
4. Determine D: given trench depth is 6 ft.
5. Determine T_{SG}, the late summer ground temperature (°F) around the earth tube:

$$T_{SG} = (T_{GW}) + (T_{AMP})\left[1 - (D/CF)\right]$$

$$T_{SG} = (52) + (18)\left[1 - (6/22)\right]$$

$$T_{SG} = (52) + (13) = \text{about } 65°F \text{ at most.}$$

6. Determine the temperature difference that makes cooling possible:

$$\Delta T = \text{desired indoor temperature} - T_{SG}$$

In this case we accept maximum 82°F air temperature (the walls will be cooler,

Fig. 5.58 *Distribution of well-water temperature (°F) in the United States, adapted from the National Water Well Association. These temperatures may be assumed to be the yearly average ground temperature. For seasonal temperature change amplitude, see Fig. 4.11, page 134. (Reprinted from* Passive Cooling, *published by the American Solar Energy Society, Boulder, Colo.)*

THERMAL CONTROL

thanks to underground location), so 82 − 65 = 17F°.

7. Determine the long-term cooling rate:

Btu/h ft of tube length

$$= (\Delta T)(C)$$

(Area, tube surface, ft²/ft of length);

Here, C = about 0.44 for wet soil. Try first the 20-in. diameter tube from the earlier design guidelines:

Btu/h ft of tube length

$$= (17F°)(0.44)\left[\frac{20 \text{ in.} \times \pi \times 1 \text{ ft length}}{12 \text{ in./ft}}\right]$$

$$= 39 \text{ Btu/h ft}$$

To deliver the full 8800 Btu/h, the tube would need to be about 8800/39 = 225 ft in length. The design guideline promised considerably more, in this case, than could be delivered!

In early summer, the earth temperature around the tube would be at its yearly average temperature of 52°F. With this larger ΔT, 82 − 52 = 30F°, the performance improves:

Btu/h ft of tube length

$$= (30F°)(0.44)\left[\frac{20 \text{ in.} \times \pi \times 1 \text{ ft length}}{12 \text{ in./ft}}\right]$$

$$= 69 \text{ Btu/h ft}$$

and the tube length under those conditions would be about 8000/69 = 127 ft. ∎

(h) Passive Cooling Summary. As a summary example of a passively cooled and solar heated building, consider the Visitor Center for the Antelope Valley, California, Poppy Reserve, about 85 miles northeast of Los Angeles (Fig. 5.59). Set in a high desert where winter temperatures sometimes fall below 20°F, the building is 100% passively solar heated by a combination of direct-gain and Trombe wall strategies, interior thermal shades, and thermally massive ceiling, walls, and floor. An earth-sheltered, thermally massive building was chosen for its suitability to

the environmentally sensitive site as well as for its all-year thermal advantage in a desert climate. On summer days when air temperatures frequently exceed 100°F, a combination of an earth tube and evaporative cooling provides a modest supply of fresh air without overheating. Continuous power ventilating of the building at night draws the cool outdoor air through the tube, then over the thermally massive surfaces to provide added cooling capacity for the following day. The 2100-ft² building also is served by an 8-kW wind generator and has its own well.

EXAMPLE 5.12. Use the psychrometric chart format to investigate the cooling process at the Antelope Valley Visitor Center. The earth tube is 24-in. (610-mm) diameter, 150-ft (45.7-m) long, at an average depth of 6 ft (1.8 m).

SOLUTION. First, determine the design conditions. Barstow, California, is a nearby, similar climate; the design condition (°F) is 104/68, with MDR 37 (and thus a nightly low of 104 − 37 = 67°F, even at design conditions). Enter the psychrometric chart diagram (Fig. 5.60) at slightly more severe conditions, such as 105 DB, 70 WB (point *A*). Note that rh = about 17% in this desert location.

Next, examine the earth tube conditions:

1. Determine T_{GW}: from Fig. 5.58, the deep-earth temperature is about 64°F.
2. Determine T_{AMP}: the seasonal amplitude (from Fig. 4.11) is about 19F°.
3. Determine CF: in dry (desert) soil, this amplitude extends to 14 ft.
4. Determine D: given trench depth is 6 ft.
5. Determine T_{SG}, the late summer ground temperature (°F) around the earth tube:

$$T_{SG} = (T_{GW}) + (T_{AMP})[1 − (D/CF)]$$

$$T_{SG} = (64) + (19)[1 − (6/14)]$$

$$T_{SG} = (64) + (10.9) = \text{about } 75°F \text{ at most.}$$

(Note that, during April poppy season, the earth temperature is likely to be at the yearly average of 64°F.)

The air in the earth tube will approach to about 4F° above this late-summer temperature, or

FLOOR PLAN

(a)

(b)

Fig. 5.59 *Visitor Center for the Antelope Valley California Poppy Reserve, near Palmdale, California.*
(a) *Plan shows earth sheltering for an elongated east–west axis, served both by direct gain and Trombe wall solar heating and by earth tube/evaporative cooling.* (b) *Solar energy enters mostly through the south glass; the glass block skylights and north clerestory are mostly for even distribution of daylighting. At rest rooms, a vented Trombe wall is used. Thermal shades on the interior protect the south glass and north clerestory on winter nights; a roll-down exterior sunscreen greatly reduces direct solar gain through south glass in summer. Concrete block walls and concrete roof and floor provide thermal mass, as useful for summer cooling as for winter solar heating.* (c) *In summer, the desert air first passes through an earth tube (assisted by a fan), then is evaporatively cooled within the building, in sight of visitors. This cooled air picks up the building's heat and is then exhausted from the stack ventilator.* (d) *The Visitor Center is set into the field of poppies; an 8-kW wind generator and the earth tube intake can be seen on the hillside above. (Courtesy of The Colyer/Freeman Group, San Francisco, California.)*

PASSIVE COOLING
Earth tube, evaporative cooling fountain, passive stack ventilator
Season: June, July, August, September

(c)

(d)

Fig. 5.59 (continued)

Fig. 5.60 Following the path of passively cooled air at the Antelope Valley Visitor Center. At summer design conditions, air enters the earth tube at Point A It is sensibly cooled within the tube, perhaps to the extent shown at Point B. It then enters the building through a small fountain, perhaps reaching an evaporatively cooled state as far as Point C. As the cooled air then picks up heat from the building, the air heads toward Point D, at which point it is exhausted through the stack ventilator.

74.8 + 4 = approximately 79°F. (This may not sound very cool, but compare it to 104°F outside.) To follow what happens to the air in the tube, move horizontally on the chart in Fig. 5.60, from point *A* toward 79 DB (point *B*): at 79 DB, about 62 WB, the rh in the air has risen to about 34%.

At this point, the air enters the building through a small fountain, consisting of water droplets. To chart this evaporative cooling, move from point *B* along the constant 62 WB line. If this fountain were a really good heat exchanger, we might achieve 80% of the distance from point *B* to saturation; call that point *C* (about 66 DB, 62 WB, rh now = 80%). This combination (of pessimistic air/earth temperatures and optimistic evaporative cooling expectations) is actually below the summer comfort zone!

The air now begins to pick up heat from the building's interior. With few sources of latent

heat, we follow sensible heat gain horizontally to the right; if we are to maintain an interior temperature of, say, 82°F (with air motion and cool underground surfaces), the air can be heated to point *D:* 82 DB, 67 WB, rh = about 47%. At this condition, it is discharged through the stack ventilator at the west end of the building. (Note that it is much cooler than the outdoor air at point *A,* even at the point of discharge.)

We have avoided quantifying so far, in favor of graphically following the air/moisture conditions. A typical velocity in earth tubes is 500 fpm (2.5 m/s); in a tube of 24-in. (or 2-ft) (610-mm) diameter, the flow rate is

$$500 \text{ ft/min} \times \pi \, (2 \text{ ft}/2)^2 = 1570 \text{ cfm}$$

In its passage through the interior, the air from point *C* to point *D* gains 82 − 66 = 16F°. If actually achievable with such a flow rate, this represents cooling equal to

$$\text{Btu/h} = 1570 \text{ cfm } (1.08)(16\text{F}°)$$

$$= 27,130 \text{ Btu/h}$$

(per unit floor area, 27,130 Btu/h/2,100 ft² = 12.9 Btu/ft² floor area, a considerable contribution). ∎

REFERENCES

Abrams, D. W. (1986). *Low Energy Cooling,* Van Nostrand Reinhold, New York.

Ashley, R. and Reynolds, J. (1994). "Overall and Zonal Energy End Use in an Energy Conscious Office Building," in *Solar Energy,* Vol. 52, No. 1. 1993, Pergamon Press, Ltd.

ASHRAE (American Society of Heating, Refrigerating, and Air-Conditioning Engineers, Inc.), Atlanta, GA. *1997 Handbook of Fundamentals,* and *Cooling and Heating Load Calculation Manual, 2nd edition,* 1992.

Balcomb, J. D., Barley, D., McFarland, R., Perry, J. Jr., and Wray, W. (1980). *Passive Solar Design Handbook,* Vol. 2; *Passive Solar Design Analysis,* U.S. Department of Energy, Washington, DC.

Balcomb, J. D., Jones, R. (ed.), McFarland, R., and Wray, W. (1984). *Passive Solar Heating Analysis.* American Society of Heating, Refrigerating, and Air-Conditioning Engineers, Inc., Atlanta, GA.

Berry, B., Bown, J., Thornton, J., Turzynski, J. and Walton, M. (1995). "The New Inland Revenue Centre, Nottingham," in *The Arup Journal,* April. Ove Arup Partnership, London.

Bowen, A., Clark, E., and Labs, K. (1981). *Passive Cooling.* American Solar Energy Society, Boulder, Colo.

Brown, G. Z., and DeKay, M. (2000). *Sun, Wind and Light,* 2nd Edition. John Wiley & Sons, Inc., New York.

Brown, G. Z., Haglund, B., Loveland, J., Reynolds, J., and Ubbelohde, S. (1992). *Inside Out, Design Procedures for Passive Environmental Technologies,* 2nd Edition. John Wiley & Sons, Inc., New York.

Chandra, S., Fairey, P. W., and Houston, M. (1986). *Cooling with Ventilation,* Solar Energy Research Institute, Golden, Colo.

DeKay, M. (1992). "Volumetric Implications and a Rule-of-Thumb for Thickness of Atria Buildings," in *Proceedings of the 17th National Passive Solar Conference.* American Solar Energy Society, Boulder, Colo.

Fleischhacker, P., Bentley, D., and Clark, G. (1982). "A Simple Verified Methodology for Thermal Design of Roof Pond Cooled Buildings," in *Proceedings of the 7th National Passive Solar Conference.* American Solar Energy Society, Boulder, Colo.

Gardner, I. and Hadden, D. (1992). "UK Pavilion, Expo '92, Seville," from *The Arup Journal,* Autumn. Ove Arup Partnership, London.

Givoni, B. (1994). "Passive and Low Energy Cooling of Buildings." Van Nostrand Reinhold, New York.

Hastings, S. R. (Ed.), (1994). *Passive Solar Commercial and Institutional Buildings.* International Energy Agency. John Wiley & Sons, Chichester.

Johnson, T. (1981). *Solar Architecture, The Direct Gain Approach.* McGraw-Hill, New York.

Mazria, E. (1979). *The Passive Solar Energy Book.* Rodale Press, Emmaus, Pa.

Miller, H. (Ed.) (1980). *Proceedings of the Passive Cooling Workshop,* 5th National Passive Solar Conference, American Solar Energy Society, Boulder, Colo.

Millet, M. and Bedrick, J. (1980). *Manual Graphic Daylighting Design Method,* Department of Architecture, University of Washington, Seattle, Wash.; sponsored by the U.S. Department of Energy.

Niles, P. and Haggard, K. (1980). "California Passive Solar Handbook," California Energy Commission, Sacramento.

Olgyay, V. and Olgyay, A. (1957). *Solar Control and Shading Devices,* Princeton University Press, Princeton, N.J.

Thompson, T., Chalfoun, N., and Yoklic, M. (1994). "Estimating the Performance of Natural Draft Evaporative Coolers," in *Energy Conversion Management,* Vol. 35, No. 11, Elsevier Science Ltd., London.

Watson, D. and Labs, K. (1983). *Climatic Design; Energy Efficient Building Principles and Practices,* McGraw-Hill, New York.

Yoklic, M. and Layseca, M. (1998). "Earth Construction, Courtyards, Verandahs, and Cooltowers: An Arizona Ranch House with Middle East Vernacular," in *Proceedings of the 23rd National Passive Solar Conference,* American Solar Energy Society, Boulder, Colo.

6

INTERIOR AIR QUALITY

Although ventilation is the "V" in HVAC, this textbook has not previously had a separate chapter on either ventilation or interior air quality (IAQ). Ventilation has been a subissue within heating and cooling system design. However, in centuries past, there were special systems to deal with providing outside air to buildings, even on a residential scale. Banham (1969) describes both large and small buildings where outside air was deliberately introduced. In the home that Dr. John Hayward built for his family in 1867, the Octagon in Liverpool, England (Fig. 6.1), outdoor air was brought into the basement, slowed to precipitate some particulates, then heated to help it rise throughout the four-story building. Ceiling vents just above gas lights drew "vitiated" air from each room, which was then vented to a "foul air chamber" in the attic. From there, a large shaft utilized a combination of a siphon and the stack effect. Powered at its low point by ever-present heat from the kitchen cooking range, it drew the foul air down, then up a very high chimney to discharge.

Several trends have combined to bring IAQ back into prominence. First, an increasingly large percentage of people's time is now spent indoors, and in more tightly controlled environments, as a service-based economy overtakes a manufacturing-based one. Second, the oil embargo of 1973 raised the world consciousness of finite energy sources, producing a sudden and powerful rush toward energy-conserving design. This in turn encouraged designers to limit the introduction of outdoor air that demanded cooling in summer and heating in winter. Third, a proliferation of chemicals has produced a vast array of potential air pollutants, from synthetic products permanently installed within buildings, to equipment used indoors, to cleaning fluids used in maintenance. With more time spent in less fresh air and

surrounded by more pollution sources, we experienced increasing numbers of buildings with *sick building syndrome* (SBS). One definition of SBS is that more than 20% of the occupants complain of symptoms associated with SBS, such as headaches, upper respiratory irritation, and irritations of the eyes, among others. If these symptoms disappear after leaving the workplace (weekends are especially good periods of contrast), SBS is even more strongly indicated.

The building designer has an elusive task when air quality is at issue, because so little can be predicted. Heat flow rates, occupancy schedules, and typical weather patterns can be combined to predict with some confidence how much energy will be consumed by a building; construction types can then be altered in the design stage to yield predictably different results. As yet, designers have few tables that yield rates of outgassing for various materials at given temperatures, nor have we the averages and the design conditions for the quality of outdoor ("fresh") air, even though we do have such information for its temperature and humidity.

Yet controlling the quality of indoor air may be as important as controlling its temperature and humidity. Buildings that are thermally comfortable can still suffer SBS when pollutants are sufficiently numerous. Designers know that saving energy for heating and cooling lowers the cost of maintaining a building; employers know that lost productivity from either on-the-job illness or sick leave can result in much greater costs. One estimate for a large office building compared the cost of increasing ventilation and improving air filtration to the value of projected health and productivity benefits. Initial improvements yielded estimated benefit-to-cost ratios of 50 to 1 (increased ventilation) and 20 to 1 (improved filtration).

Fig. 6.1 *Fresh air intake, stale air exhaust at the Octagon, Grove Street, Liverpool, 1867. Dr. John Hayward built this system of basement intake, attic-to-cooking range-to-chimney exhaust. Dirty air outside (coal-fired industries and buildings were common) and inside (gas lamps) probably produced an IAQ considerably worse than today's norms. 1. Fresh air intake; 2. Settling chamber in basement; 3. Heating coils; 4. Air passages in lobby floors; 5. Air passages in cornice; 6. Extract above gas lamp; 7. Foul air chamber; 8. Foul air down duct; 9. Foul air chimney; 10. Flue from kitchen range. (Reprinted by permission from Banham, Reyner (1969)* The Architecture of the Well-Tempered Environment, *The Architectural Press, London.)*

IAQ depends on four major considerations, three of which depend largely on the designer:

1. Limit pollution at the source (choose materials and equipment carefully).
2. Isolate unavoidable sources of pollution.
3. Provide for adequate supply and filtering of fresh air (and return air).
4. Maintain the building and its equipment in clean condition.

The designer can thus provide for improved IAQ by careful location on site; choosing of materials and (where possible) equipment; zoning to isolate pollutant sources; and providing clean, adequate, and well-distributed outdoor air, air-cleaning devices, and building commissioning (often including a "flush mode" following completion of construction). It is up to the building managers to maintain the building's IAQ health by a regular equipment maintenance program, regular interior cleaning, and a careful selection of cleaning agents. Furthermore, a flush of the building after every unoccupied weekend or holiday period will be helpful in removing accumulated pollutants from finishes and furnishings.

6.1   Sources and Impacts

Indoor air pollution can be described both in terms of the types of contaminants (gaseous, organic, or particulate) and the types of effects (odors, irritants, toxic substances). We not only inhale contaminants, we also absorb some and ingest some; not only the nose is involved in sensing IAQ. For some contaminants, the only method of avoidance is to design for their exclusion; equipment will not remove them, although increased ventilation can reduce their impact. Examples are asbestos, radon, and pesticides. Table 6.1 summarizes some common indoor air pollutants, their effects, and simple strategies to ameliorate them.

(a) Odors. One of the most immediate indicators of IAQ problems is odor detection. People are sensitive to odors over an extraordinary range, while equipment to detect and classify odors is woefully lacking. Odors are perceived most strongly on initial encounter, then "fatigue" oc-

TABLE 6.1 **Common Air Pollutants**

Pollutant	Sources	Effects	Control Strategies
Excess moisture[a]	Cooking (heating open liquids), washing, exhaling	Increases growth of fungus, bacteria and dust mites	Exhaust ventilation at source; dehumidification
Carbon dioxide (CO_2)	Human respiration	Minor discomfort at high concentrations; "stuffiness"	CO_2 is a good indicator of ventilation rate in tightly enclosed spaces or where occupancy is high
Carbon monoxide (CO)	Incomplete combustion: furnaces, stoves, fireplaces; motor vehicle exhaust	Headaches, dizziness, sleepiness, muscle weakness, potentially lethal	Sealed combustion burners, adequate combustion air, safe exhaust flues
Nitrogen oxides	High-temperature combustion	Irritation, possible immune suppression	Safe exhaust flues, sealed combustion burners
Sulfur oxides	Combustion fuels containing sulfur (oil, coal)	Potential irritant, burning eyes, reduces lung function	Alternative fuels, safe exhaust flues, sealed combustion burners
Polynuclear aromatic hydrocarbons	Smoking, combustion of wood or coal, barbecuing, burnt food	Irritants and carcinogens	Prohibit smoking, lower-temperature cooking, use clean fuels, burn wood in enclosed firebox with adequate oxygen supply
Ozone	Laser printers, photocopiers, small motors, electronic air cleaners	Inflammation of bronchia, wheezing and shortness of breath, dizziness, asthma attacks	Remove sources or exhaust at source, maintain electronic air cleaners
Volatile organic compounds (VOCs)			
Formaldehyde	Particle board, interior laminated panels, glues, fabric treatments, paints	Burning eyes and nose, skin rash, shortness of breath, headaches, nausea, dizziness, fatigue	Use alternative materials, seal particle board if used, ventilate
Others	Paints, solvents, carpets, soft plastics, adhesives, caulkings, softwoods, paper products, cleaning and maintenance products	Intoxication, burning eyes and nose, shortness of breath, headaches, nausea, dizziness, loss of judgment, panic	Use alternative materials, age materials before installing, ventilate
Lead	Pre-1970s paint, pre-1985 pipes and solder, dust and soil near roads (residue from leaded gas)	Neurotoxic, especially if ingested by young children; learning disabilities, nausea, trembling, numbness of extremities	Identify and remove or seal old paint, replace pipes and solder, avoid foods grown by roadside

(continued)

TABLE 6.1 **Common Air Pollutants** (*continued*)

Pollutant	Sources	Effects	Control Strategies
Pesticide residues[b]	Treated basements and foundations, treated ceiling and wall cavities, treated cabinets and closets, treated soil outside foundation	Neurotoxic or long-term risk of liver, kidney, and other disease, including cancers	Identification and removal by expert if history known, sealing in pesticide if possible
Asbestos fiber	Pre-1975 steam pipe and duct insulation, furnace and furnace parts, pre-1980 reinforced vinyl floor tile and fiber cement shingles and siding	Long-term cancer risk from inhaling fibers	Leave material undisturbed, get expert identification and removal if required, seal with special sealant and cover with sheet metal if not crumbling
Mineral and glass fiber	Thermal insulation, pipe insulation, fire-resistant acoustic tile and fabrics	Potential irritant, burning eyes, itching skin, long-term risk of lung damage and cancer	Handle only with respirator and gloves, seal and enclose, do not disturb in place
Fungus particles, dust mites	Grow in basements, damp carpet, bedding, fabrics, walls and ceilings, closets	Very allergenic, burning eyes and nose, sneezing, skin rash, congestion, and shortness of breath	Keep surfaces dry and clean, cover bedding and upholstery with barrier cloth, ventilate, use borax treatments to retard fungus
Hazardous bacteria (e.g., *Legionella*)	Standing warm water, untreated hot tubs, air-conditioning drain pans, humidifier reservoirs	Severe respiratory illness, potentially lethal	Prevent standing water, clean and treat tubs and reservoirs
Radon gas	Natural radioactivity in soils	Increased lifetime lung cancer risk	Seal foundation and floor drains, ventilate subsoil
Methane and other soil gases	Decomposing garbage in landfills, leaking sewage lines, toxic waste	Possibly explosive or toxic, nuisance odors	Know site history before building, remove soil if necessary, seal foundation and floor drains, ventilate subsoil

Source: Adapted from Rousseau, D. and Wasley, J. (1997). *Healthy by Design.* Hartley and Marks, Point Roberts, Washington.

[a]Too little moisture also adversely affects health.

[b]Pre-1980s treatments are more likely to leave residues and be toxic.

curs and perception fades; thus, visitors are more likely to detect odors than are the long-term inhabitants of a space. Odors may be simply unpleasant, or may be indicators of a more serious IAQ problem that has physiological consequences. When an unfamiliar odor is detected, our reactions are either positive, neutral, or negative, depending on whether we perceive a threat associated with the odor.

Sometimes odors are directly traceable to a source, but in office environments the odor is usually more complex. A typical blend for an office environment's odor may be body odor, grooming products (perfumes, colognes), copy machines, cleaning fluids, and outgassing from materials. More rarely now, tobacco smoke may also be present. This complexity produces an interesting reaction; people tend to become less sensitive to each of the component odors, with the result of overall masking. However, an architectural masking approach—the deliberate introduction of a "perfume" to cover offending odors—is rarely successful. (Sweet-and-sour may work nicely for the palate, but olfactorily, it can be uniquely nauseating.) Conversely, as the indoor environment is freed of multiple odors, people become more sensitive to the one or two odors that remain.

Sometimes a simple measurement of carbon dioxide concentration is used as a first indicator of potential body odor and/or inadequate ventilation problems, because the CO_2 concentration indoors is generally proportional to human concentration.

Filtering out the odors from indoor air is usually achieved with electronic or with activated charcoal filters, described in Section 6.5.

(b) Irritants.

Unlike odors, which are immediately perceived and fade with prolonged exposure, irritants are often imperceptible at first but cause increasing distress over time. Symptoms of irritants include itching or burning eyes, sneezing, coughing, dry nose and throat, sore throat, and tightness of the chest. Most irritants are present in the form of particles and gas dispersoids (Table 6.2).

Sources of irritants include the building itself and the equipment and occupants within. New and newly renovated buildings are particularly prone to problems from outgassing of paints, adhesives, sealants, office furniture, carpeting, and vinyl wall coverings. *Volatile organic compounds* (VOCs) are chemicals containing carbon molecules that are volatile; that is, they off-gas or evaporate from the surface of materials at room temperatures. The list is long: methane, ethane, methylene chloride, trichloroethane, CFCs, HCFCs, HFCs, formaldehyde, and hydrocarbons such as styrene, benzene, and alcohols. All are now found more frequently in these new spaces.

Long-term occupancy brings other irritants. Ozone, a friend in the upper atmosphere but a smog component below, is produced from copy machines, high-voltage electrical equipment, and—ironically—from electrostatic air cleaners. Mineral fibrous particles can result from the breakdown of interior duct insulation and fireproofing. Hydrocarbon compounds come from copy machines and their papers. Tobacco smoke is a mixture of gases and fine particles especially irritating to sensitive individuals. Low humidity can exacerbate problems with irritants, producing symptoms similar to those from chemicals. Carpet shampooing yields organic solvents and ammonia; nighttime cleaning coinciding with reduced or nonexistent ventilation is especially problematic. In contrast, night maintenance with *increased* ventilation rates—as with cooling by night ventilation of thermal mass—can reduce this threat.

As with odor control, the impacts of irritants can be reduced with increased outdoor air. Filtering for the removal of irritants usually consists of particulate filters; less common are gaseous-removal filters, air washers, and electronic air cleaners. These are also discussed later in Section 6.5.

(c) Toxic Particulate Substances.

At the top of this list is asbestos, widely used in buildings until its toxicity was realized in the 1970s. We encounter asbestos in tightly bound form as in asbestos-cement and in vinyl-asbestos floor tiles, and in loosely bound form as in sprayed-on asbestos insulation. The latter form is particularly dangerous, readily releasing toxic asbestos fibers over the life of the material. With asbestos, neither increased ventilation nor filtering is acceptable; it must be either removed under stringent controls to isolate the air within the affected space, or sealed and left in place.

TABLE 6.2 **Characteristics of Particles and Particle Dispersoids**

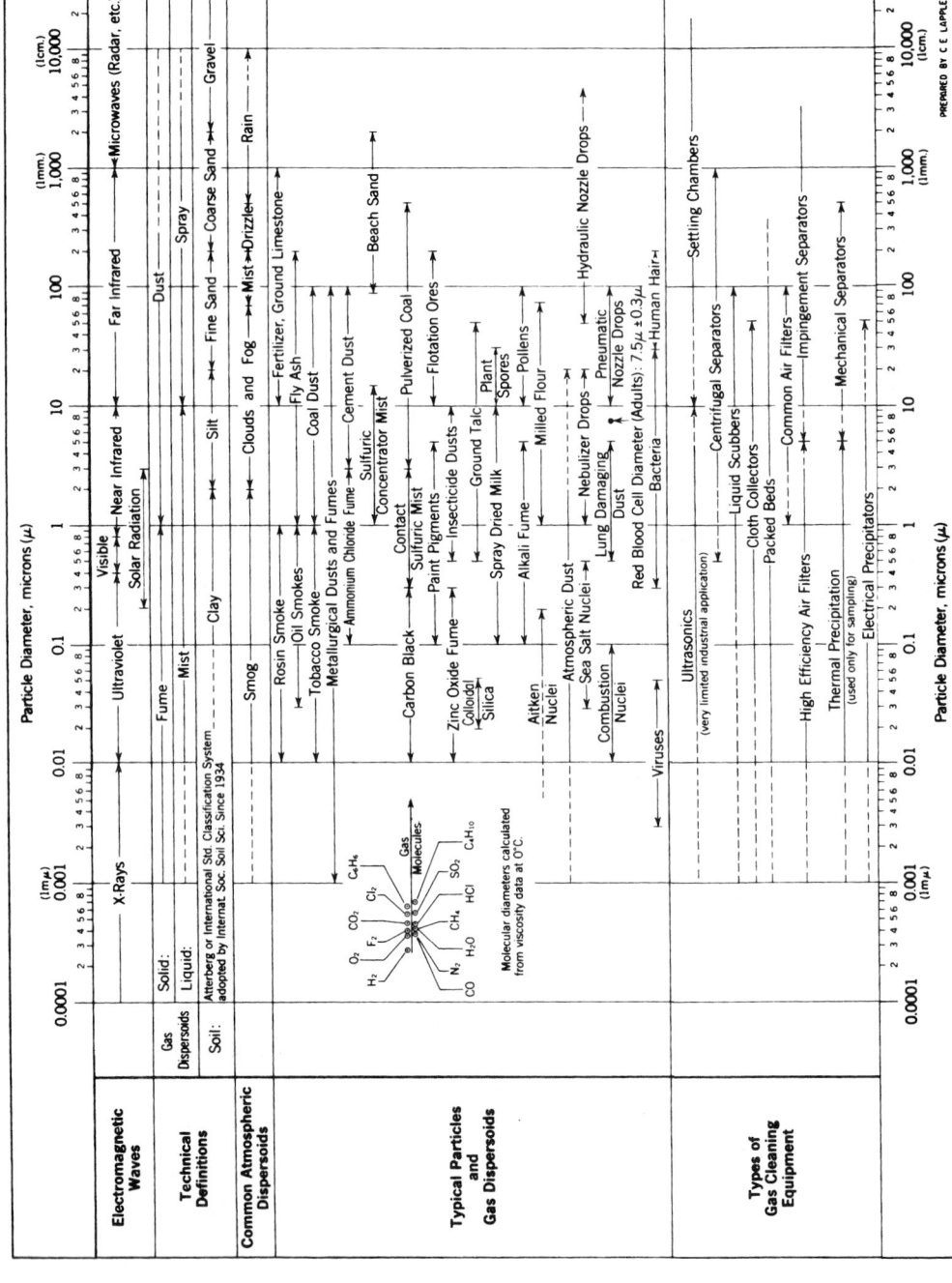

Source: Reprinted with permission of SRI International (formerly the Stanford Research Institute), Menlo Park, CA.

Some of the respirable particles (see Table 6.2) that result from incomplete combustion are toxic. Incomplete combustion can occur from smoking, in woodstoves, fireplaces, gas ranges, and unvented gas or kerosene space heaters. Lacking control of combustion at its source, the remedies are to isolate the source insofar as possible, exhaust air from the immediate vicinity, increase outdoor air to the area, and utilize particle filtering.

(d) Biological Contaminants.

Because living things inhabit both buildings and outdoor air, there will be biological contaminants such as bacteria, fungi, viruses, algae, insect parts, and dust within buildings. Moisture encourages both the retention and growth of these contaminants; standing water (in HVAC system components) and moist interior surfaces are likely trouble sites. Allergic reactions and infectious and noninfectious diseases can result. Outbreaks of Legionnaire's disease occurred when improperly maintained HVAC systems within large buildings incubated then distributed the disease-causing microorganisms. Now, residential humidifiers, dehumidifiers, and air conditioners' drain pans are suspect.

Remedies for biological contaminants begin with good design and end with vigilant maintenance. Although exposure to ultraviolet light is sometimes employed, equipment such as filters are rarely an effective solution in this case.

(e) Radon and Soil Gases.

Radon is a radioactive gas that decays rapidly, releasing radiation at each stage. It is colorless and odorless, thus undetectable by humans. If we inhale radon, radiation release in the lungs can cause lung cancer. Other soil gases include methane (usually odiferous), and some pesticides that can volatize and enter buildings with soil gases. Effects on humans are not likely to be beneficial.

In many of the buildings with high levels of radon, the problem has been traced to exposed earth. Radon penetrates through cracks and openings around plumbing, and below-grade spaces are particularly at risk. Penetrations of below-grade walls and floors should be both minimized and well sealed; under-slab ventilation (Section 6.4c) may be appropriate, especially in areas of high radon risk.

(f) Multiple Chemical Sensitivity.

This is an unusual and sometimes controversial condition, also known as *environmental illness,* in which IAQ becomes the prime indoor environmental issue. People with this affliction avoid environments with any of the foregoing risk factors. The controversy arises because the causes of this condition are poorly understood; if causes are mysterious, so might be the remedies.

In San Rafael, California, "Ecology House" is an apartment complex for low-income people with multiple chemical sensitivity. Its construction avoided plywood, using Douglas fir sheathing instead; the floors are tile instead of wall-to-wall carpet; cabinets are metal, not plywood or OSB; the heating system is radiant hot water, not forced air. Barbecues and fireplaces are absent; painted surfaces minimized; any window coverings are alternatives to curtains. There is even an "airing room" where, for example, newspapers can be hung to evaporate ink odors before they are read.

6.2 Predicting IAQ

Designers want to know how much outdoor air and how much filtering will produce an acceptable indoor environment. These questions are elusive.

(a) Ventilation Rate.

The most common remedy for SBS (after controlling the pollution source) is to increase the rate of outdoor air ventilation. Recommended rates of ventilation are found in Table 4.22. Although very small amounts of outdoor air will provide sufficient oxygen, and human body odor control is usually achievable at a rate of from 6 to 9 cfm (3 to 4.5 L/s) of outdoor air per occupant, outdoor air has more work to do than provide oxygen and control odors. The current minimum rate for offices (Table 4.22) is 20 cfm (10 L/s) of outdoor air per occupant. Because this rate is based on a density of 7 persons per 1000 ft^2 (per 100 m^2), it corresponds to an outdoor air rate of 0.14 cfm/ft^2 (0.1 L/s m^2) of floor area. In a typical office with an 8-ft (2.4-m) ceiling, this corresponds to slightly more than 1 ACH (air change per hour).

The less volume of space per occupant (i.e., lower ceilings or greater density), the higher the rate of ventilation is appropriate. An analogy:

One fish will pollute a small pond more rapidly than it will a larger pond.

Two units have been proposed to integrate the various indoor air pollutants in the same way as they are perceived by human beings. The *olf* is a unit of pollution (1 olf = the bioeffluents produced by the average person); the *decipol* is a unit of perceived air quality. These are related in this proposed comfort formula:

$$Q = 10 \frac{G}{C_i - C_o}$$

where Q = ventilation rate, L/s
  G = total pollution sources, olf
  C_i = perceived indoor air quality, decipol
  C_o = perceived outdoor air quality, decipol

At present, C_i is recommended to be set at 1.4 decipol, which represents an expectation of 80% of occupants satisfied with indoor air quality. C_o and G may be roughly estimated from Table 6.3. The greatly increased ventilation rates for existing buildings in Part C [compared to ASHRAE's 0.14 cfm/ft^2 (0.1 L/s m^2) of floor area] suggest dealing with indoor pollution at its source rather than increasing outdoor air flushing—and related energy consumption—to such high levels. These proposals are discussed in more detail in Fanger (1989).

One reason for greatly increased ventilation rates above the guidelines given in Table 4.22 is the concept of *replacance*. Table 6.4 shows that at the rate of 1 ACH of outdoor air, an indoor

TABLE 6.3 **Estimating Indoor Air Quality**

Part A. *Perceived Outdoor Air Quality, C_o*

During smog episodes	>1 decipol
In cities with moderate air pollution	0.05–0.3 decipol
On mountains or at sea	0.01 decipol

Part B. *Estimated olf Loads in Offices per m^2 Floor Area*

Pollution Source	olf/m^2
Occupants (10 m^2 per person)	
Bioeffluents	0.1
Additional load from 20% smokers	0.1
40%	0.2
60%	0.3
Materials and ventilation system	
Average in existing buildings[a]	0.4
Low-olf buildings[b]	0.1
Total load in office buildings	
Average in existing buildings (40% smokers)	0.7
Low-olf buildings (nonsmoking)	0.2

Part C. *Ventilation Requirements in Office Buildings[c]*

Based on 1.4 decipol indoor air quality	
Existing buildings[a] (0.7 olf/m^2) smoking	5 L/s-m^2
Low-olf buildings[b] (0.2 olf/m^2)	1.4 L/s-m^2
ASHRAE Standard 62-1989[d]	0.8 L/s-m^2

Source: Fanger (1989). Reprinted by permission from the *ASHRAE Journal,* © 1989 by the American Society of Heating, Refrigerating, and Air-Conditioning Engineers, Inc., Atlanta, Ga.

[a]Based on field studies of 15 randomly selected buildings in Copenhagen.

[b]Designed to contain low-outgassing materials and a frequently maintained ventilating system.

[c]Assumes 10 m^2 per person and negligible outdoor air pollution.

[d]See Table 4.22.

TABLE 6.4 **Air Replacance Compared to Input Air Changes per Hour (ACH)**

NOTE: Mixing is considered continuous and vigorous—as would be obtained in a forced-ventilation system.

Time from Start of Run (h)	Rate of Air Input (ACH):	Replacance (%)								
		0.06	0.12	0.25	0.5	1	2	4	8	16
1/16		0.4	0.8	1.6	3.1	6.1	11.7	22.1	39.3	63.2
1/8		0.8	1.6	3.1	6.1	11.7	22.1	39.3	63.2	86.5
1/4		1.6	3.1	6.1	11.7	22.1	39.3	63.2	86.5	98.2
1/2		3.1	6.1	11.7	22.1	39.3	63.2	86.5	98.2	99.9
1		6.1	11.7	22.1	39.3	63.2	86.5	98.2	99.9	100
2		11.7	22.1	39.3	63.2	86.5	98.2	99.9	100	100
4		22.1	39.3	63.2	86.5	90.2	99.9	100	100	100
8		39.3	63.2	86.5	98.2	99.9	100	100	100	100
16		63.2	86.5	98.2	99.9	100	100	100	100	100

Source: Shurcliff (1981). Reprinted by permission.

THERMAL CONTROL

space would have only 63% "new" air after 1 hour; about 8 hours at this rate is required for all the "old" air to be exhausted. There is, then, a difference between the fresh air input rate (ACH) and the *replacance*—the fraction of air molecules at one specified time that was *not* in the indoor space at an earlier, reference time. This relationship, along with details of air pollutants (and of heat exchanger design, for energy conservation), is thoroughly discussed in Shurcliff (1981).

The new campus for the Environmental Protection Agency (EPA) in Research Triangle Park, North Carolina, was designed with special emphasis on IAQ (Fig. 6.2). The designers considered several alternatives for fresh air provisions, deciding that a simple variable air volume (VAV) sys-

Fig. 6.2 *The new campus of the U.S. EPA (Environmental Protection Agency) at Research Triangle Park, North Carolina, features concentrated parking separate from the buildings. High exhaust stacks disperse air from laboratories; intakes are kept well away from exhaust. (Courtesy of Hellmuth, Obata + Kassabaum, Washington, DC.)*

tem, set at a minimum of 3 ACH (of combined fresh and recycled air) would be acceptable. If the system were designed with a typical minimum, 1 ACH would have resulted in periods when neither heating nor cooling were required (spring and fall typical conditions). However, a 6 ACH alternative would have had dramatically increased energy consumption.

(b) Material Safety Data Sheets. Manufacturers of building materials furnish these reports, which list all chemical constituents that

make up at least 1% of the material (and are not deemed "proprietary"). Unfortunately, this does not predict emission rates. The designer is left with the suspicion that the higher the percentage content of a chemical, the more likely its outgassing.

(c) Testing. When a client is especially interested in IAQ, full-scale time tests can be used. At the new EPA campus (Fig. 6.2), the contractor was given a target for allowable air concentrations (Table 6.5) for IAQ. Any material assemblies

TABLE 6.5 **Air Quality Standards**

Part A. **National Ambient-Air Quality Standards for Outdoor Air**

Contaminant	Long-Term Concentration Averaging			Short-Term Concentration Averaging		
	$\mu g/m^3$	ppm		$\mu g/m^3$	ppm	
Sulfur dioxide	80	0.03	1 year	365	0.14	24 hours
Total particulate	75[a]	—	1 year	260	—	24 hours
Carbon monoxide				40,000	35	1 hour
Carbon monoxide				10,000	9	8 hours
Oxidants (Ozone)				235[b]	0.12[b]	1 hour
Nitrogen dioxide	100	0.055	1 year			
Lead	1.5	—	3 months[c]			

Source: U.S. Environmental Protection Agency.

Part B. **Maximum Indoor Air Concentration Standards at EPA Campus**

Indoor Contaminants	Allowable Air Concentration Levels[d]
Carbon monoxide (CO)	<9 ppm
Carbon dioxide (CO₂)	<800 ppm
Airborne mold and mildew	Simultaneous indoor and outdoor readings
Formaldehyde	<20 $\mu g/m^3$ above outside air
Total VOC	<200 $\mu g/m^3$ above outside air
4 Phenyl cyclohexene (4-PC)[e]	<3 $\mu g/m^3$
Total particulates	<20 $\mu g/m^3$
Regulated pollutants	<National Ambient-Air Quality Standards (see Part A)
Other pollutants	<5% of TLV-TWA[f]

Source: Hellmuth, Obata + Kassabaum, P.C. The Greening Curve: Lessons Learned in the Design of the New EPA Campus.

[a]Arithmetic mean.

[b]Standard is attained when expected number of days per calendar year with maximum hourly average concentrations above 0.12 ppm (235 $\mu g/m^3$) is equal to or less than 1.

[c]Three-month period is a calendar quarter.

[d]These levels must be achieved prior to acceptance of building. These levels do not account for contributions from office furniture, occupants, and occupant activity.

[e]4-PC is an odorous contaminant constituent in carpets with styrene-butadiene-latex rubber (SBR).

[f]TLV-TWA is threshold limit value–time weighted average.

deemed likely to contribute more than one-third of these allowable air concentrations, and used in large quantities, were to be tested as assemblies before acceptance. One desirable outcome of such testing is the possible avoidance (or shortening) of an anticipated flush-out of the completed building with outdoor air for 90 days before occupancy.

6.3 Zoning for IAQ

After implementing pollution control at the source (cleaner equipment, prohibiting smoking, careful material choices, etc.), remaining unavoidable pollutant sources should be identified. Then isolate the more sensitive areas from the contaminators. This is sometimes difficult, as in "open offices" where walls are unwelcome but copying machines are essential. In such cases, erect as much of a barrier as is possible around the offender, then "task ventilate" to remove the contaminated air immediately. Sometimes air pollution sources also produce unwanted sound, in which case the argument for a more complete barrier may become more compelling.

Many health care and laboratory buildings have "clean" and "dirty" zones, even separate circulation pathways. Often, differential air pressures are maintained to discourage air flow from dirty to clean; higher pressure in clean areas, lower in dirty areas. Lower-pressure areas can be created simply by installing exhaust fans from such spaces, as well as limiting the volume of supply air. Higher-pressure areas can be created by installing make-up air equipment, as well as increasing the volume of supply air from the HVAC system.

On a site-planning scale, try to locate air intakes upwind from pollution sources. Because winds frequently change direction, this may be more a matter of adequate separation distance than direction. The most obvious example is a major air intake for a central HVAC system, which should be as far as possible from parking areas, delivery docks, and streets—and from the exhaust outlets from that same HVAC system or outlets from other buildings' systems. Even exhaust outlets should be located carefully, because there is a possibility that at times outdoor air can be drawn into these "exhaust" grilles. The mechanical equipment room is the typical loca-

tion for both intake and exhaust; energy conservation devices such as heat exchangers benefit from close proximity of intake and exhaust. Most animals use the same "ducts" to breathe in and exhale, obviously inviting such air re-entrainment. However, for a building, separation of these openings is prudent design.

The new campus for the EPA (Fig. 6.2) is an example of predesign planning for IAQ. This is a 1,000,000 ft^2 (92,900 m^2) building complex, serving a population of more than 2000 on 133 acres (54 hectares) of farm land that has reverted to second-growth hardwoods. Table 6.6 is a summary of design decisions and their impact on IAQ. Some of the more visible design consequences are the separation of parking and building, the concentration of parking in a structure (less impact on the existing landscape, more control of vehicle fumes), and the height of the exhaust stacks from the laboratories. Many other design decisions are hidden within the building's materials and HVAC system, as detailed in Table 6.6.

The topic of zoning includes decisions about local versus central equipment. Should individual exhaust fans be installed (creating selective lower-pressure areas) or a central exhaust fan (that can discharge up a very tall stack)? Should air cleaners be installed locally, where they may be selected according to the degree of pollution, or central cleaning, where it can be more easily and regularly maintained? What about heat exchangers for tempering incoming air: lots of smaller ones, or one large one? The larger and more complex the building, the more likely that a mix will result of local, specialized zones and a large, more general zone centrally served. Figure 6.3 explores the issue of the location of the office copier; at the edge, where task ventilation is easy, but plentiful daylight and view may be "wasted" on this function; or away from the edge, where a central exhaust system is more likely to be utilized.

6.4 Passive and Low-Energy Approaches to Ventilation

This section deals with ventilation, the approach that assumes "the solution to pollution is dilution." The other major approach, air cleaning, almost always involves forcing air through various

THERMAL CONTROL

TABLE 6.6 **Design Decisions and Impacts: EPA Campus, Research Triangle Park**

Item	Decision	Impact on IAQ
Siting of building	Locate exhaust downwind from air intakes, separate by >100 ft (30 m)	Minimizes re-entrainment of laboratory exhaust air at air intakes
	Maximize separation between parking areas and air intakes	Reduces the potential for vehicular exhaust entering building
Location of parking garages	Locate parking structure away from the building	Reduces the potential for vehicular exhaust entering building
Laboratory exhaust stacks	Stack height increased to 30 ft (9 m) based on wind tunnel testing	Minimizes re-entrainment of laboratory exhausts into air intakes
Radon	Site-specific testing confirmed low levels of radon	Confirmed that radon levels are safe
Delivery/loading zone	Maintain negative pressure in loading area, positive pressure in building	Eliminates entrainment of delivery vehicle exhaust
Landscaping	Low maintenance and nonsporulating plants selected	Intake of spores, fertilizer, or chemicals entering building is reduced
	Plants used as a barrier to vehicle exhaust	Minimizes entrainment of delivery vehicle exhaust
Laboratory fume hoods	Install flow gauges and alarms	Provides warning of air contaminants present in laboratory areas due to loss of flow
Acoustic insulation of ducts	Ductwork increased in size to reduce need for acoustical insulation; in select areas, mylar-coated duct silencers are used as ductwork transitions out of equipment rooms	Minimizes release of fibers into the air stream and possible contamination of the HVAC system (duct liners are difficult to monitor or to clean and can be sites of microbial growth)
Moisture accumulation	Install drain pans pitched toward drain pipe	Reduces moisture, which could result in introduction of bacterial contamination into HVAC system
Humidity control	No moisture carry over into system	Minimizes moisture in HVAC system and resultant bacterial contaminants due to moisture
Corrosion inhibitors	Inhibitors do not contain volatile amines	Eliminates exposure to certain air contaminants
System maintenance	Provide access panels at ductwork appurtenances and ample clearance around equipment	Maximizes ease of maintaining HVAC system
Outside ventilation rate	100% outdoor air in laboratories; 20 cfm (10 L/s) per person in offices	Maximizes occupant comfort and removal of air contaminants
	Flexibility to increase ventilation rate 20% for unexpected sources	Minimizes possible occupant exposure to contaminants
Airflow efficiency	Minimum airflow rate set at 3 ACH for VAV system in office areas	Increases air movement and ventilation effectiveness
	Flexible connections to room diffusers in open office areas	Facilitates modifications to enhance airflow as necessary

TABLE 6.6 **Design Decisions and Impacts: EPA Campus, Research Triangle Park** (*continued*)

Item	Decision	Impact on IAQ
Air cleaning	ASHRAE 30% efficiency pre-filters with 85% final filters	Minimizes dust and other aerosols entering indoor air via the HVAC system
	Flexibility to add scrubbers to laboratory exhaust	Minimizes release of contaminants to ambient air
	Use bird-proofing mesh screen	Eliminates bird droppings and possible microorganism infection in HVAC system
Thermal control	Building Automated Control system	Provides optimum control of temperature and pressure
	Fixed windows	Eliminates unconditioned air from entering building; maintains positive pressure in laboratories
Exhaust system	100% exhaust for photocopying rooms, laboratories, food preparation areas	Eliminates potential to recirculate contaminants and odors through building via HVAC system
	Photocopiers located within 10 ft (3 m) of exhaust vent	Control potential source of air contaminants; recirculated air is filtered prior to its return
Smoking	Designate building as non-smoking	Eliminates exposure to second-hand smoke and recirculation of tobacco smoke via the HVAC system
Building materials, finishes, furnishings	Materials selected to minimize release of contaminants from products	Minimize occupant exposure to contaminants as a result of off-gassing from building material, finishes, furnishings

Source: Hellmuth, Obata + Kassabaum, Architects, Washington DC. This material was produced with U.S. Government sponsorship through order #70-2124-NTLX.

filtering devices. Equipment that controls both ventilation and air cleaning is discussed in the section following.

(a) Windows. Operable windows are one of the oldest and most common "switches" of all. Passive ventilation through windows and skylights is influenced by the position of the open window; if wind strikes the glass surface in its open position, it will be deflected. The direction of the wind approaching the window is generally unpredictable. Also, whereas for simple ventilation (without cooling) it is usually desirable to keep wind *away* from people, for cooling at temperatures above the standard comfort zone, wind *across* the body is helpful (Fig. 3.27). For these reasons, a window that can be opened in a variety of positions can be useful; some examples are shown in Fig. 6.4. A multiple-position window in the Wallasey school was shown in Fig. 1.10.

Perhaps the best aspect of operable windows is that they give the building occupant some control over the source of outdoor air. Perhaps the worst aspect is that they rarely offer any means of filtering this incoming air. They also can confound attempts by a central HVAC system to regulate air flow and resulting pressure. Sometimes they admit air (windward side); at other times, they

(a)

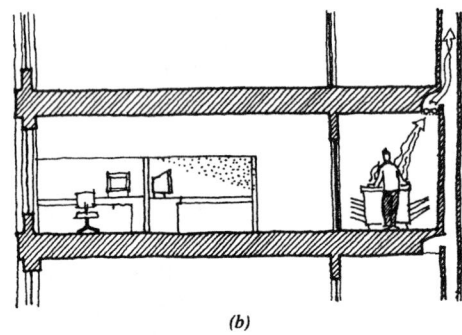

(b)

Fig. 6.3 *Zoning for IAQ: the office copier. This notorious source of VOCs can be located at the perimeter, where access to fresh air and direct exhaust is simple; however, perimeter space is premium for daylight and view. If located in less desirable interior space, exhaust air can create a negative air pressure, drawing air from adjacent offices and containing the VOCs; longer runs of exhaust ducts are required.*

exhaust it (leeward side). Note that the EPA Campus (Fig. 6.2) elected fixed, not operable, windows.

Some windows offer more "free area" of opening than others of the same size. Figure 6.5 compares some common window types. The pattern of incoming flow is also highly influenced by the way in which these windows open. The outward flow is somewhat affected as well. Insect screens will reduce the flow of air. Details of windows and screens are discussed in Chandra et al. (1986). Estimates of air flow driven by wind through window openings are found in Section 5.6a, detailed calculations in Section 5.14a.

Windows work best in the presence of wind. In calm conditions, they may still admit—or exhaust—air due to the stack effect. The taller the building, the more pronounced this effect. Operable windows in very tall buildings have been shunned by designers until recently. The Commerzbank in Frankfurt, Germany, is a 56-story tower. Each office's exterior window (Fig. 6.6) is operable in temperate weather; the lock is controlled by a building management system (BMS), although the occupant decides the degree of openness. The full-height office window's outer skin is fixed single-glazed safety glass, with 5-in. (125-mm) ventilation slots all across the top and bottom. These serve the 8-in. (200-mm) wide cavity between the outer and inner window skins. The

Fig. 6.4 *A window that can open in more than one position can enhance ventilation performance.* (a) *The window tilts from the top, directing the incoming air towards the ceiling; fresh air does not directly encounter workers within the space.* (b) *The window swings inward, allowing incoming air to move across people, enhancing warm-weather cooling. (Courtesy of Three Rivers Aluminum Company, Inc.)*

(a)

(b)

Fig. 6.5 *The percent of actual openable area varies with window type. (Reprinted by permission from Moore, Fuller (1994).* Environmental Control Systems: Heating, Cooling, Lighting. *McGraw-Hill Company, New York.)*

User controlled window motor

Single glazed panel

Double glazed in-swing window

Internal blinds

Fig. 6.6 *The Commerzbank is a recent addition to Frankfurt, Germany's skyline. Operable windows in this 57-story tower are tilted in from the top by occupants while a fixed outer pane with continuous venting slots, top and bottom, keeps air currents under control. Shown here in exhaust mode, the window can also supply fresh air to the office. Under adverse outdoor conditions, the building management system (BMS) locks the inner window in the closed position.*

inner window is double glazed, hinged at floor level, and has a motor-operated tilt-in mechanism at the top. Motorized blinds for solar shading are located within the cavity between window skins. A central atrium provides a stack effect, so that an open window is usually a source of incoming rather than exhaust air, although the top and bottom slots allow for a slight stack effect at each window.

(b) Stack Effect. Several applications of the principle that hot air rises are applicable to IAQ. For the simple stack effect, estimates of air flow are found in Section 5.6b, detailed calculations in Section 5.14b. Devices can be used to enhance the stack effect by creating suction when wind blows across the top of the stack. Probably the most common (available in chain-store catalogs) are wind gravity or turbine ventilators (Fig. 6.7); their performance characteristics are listed in Table 6.7. This is probably not the most effective topping device, however. Figure 6.8 compares the volumetric airflow results for the turbine and several other ventilators with those for a simple open stack. The tests were done in a wind tunnel at Virginia Polytechnic Institute.

Because the stack effect works more forcefully with increased height, intakes should be as low as possible. When these openings are near the earth surface, precooling of summer intake air is possible. The Cottage Restaurant in Oregon (Fig. 1.14) took some advantage of this. The Olivier Theatre at the Bedales School in rural Hampshire, England (Fig. 6.9) uses a gently sloping site to similar advantage. The tightly packed audience of a theater generates considerable heat, as do the lights. For this theater, the maximum acceptable temperature around the audience is about 25°C (77°F); the heat produced by people and lights provides an uplift of about 7C° (12.6F°). Therefore, whenever the outdoor temperature is about 18°C (64.4°F), cooling of the incoming air is needed.

At this theater, air is introduced to an "undercroft" (crawl space) with a concrete floor, on which are built many concrete block walls, forming an indirect path for the incoming air. This undercroft is cooled by night ventilation, thus made ready for the next event's heat gains. The inlet openings are 5% of the theater floor area; the surface area of the undercroft is 3 m² (32 ft²) per person. The audience of 270 people is ventilated and cooled by the air rising from this undercroft, through openings that total 3.5% of the floor area. Gaining heat, the air rises toward the central cupola, aided when necessary by a "punkah" fan (see later Fig. 7.4), and exits through four louvered sides of the cupola, with a total outlet area at 6% of the floor area. The overall height of this stack is 15.5 m (51 ft); a maximum of 15 ACH is expected at 3C° difference, inside warmer than outside. This ventilation–cooling system is silent and utilizes no refrigerant. In the heating season at partial occupancy, the stack outlets are closed, and the fan can be run in reverse to send collected warm air downward to the floor.

(c) Underslab Ventilation. Although the theater used the ground's coolth to advantage, there are some places where the ground contains radon (or other soil gases). A county map of the United States (Fig. 6.10) shows relative risk from radon, a long-term harmful gas. Buildings on for-

Fig. 6.7 Two wind-gravity (turbine) ventilators accent the skyline of the addition to Barton Hall at the University of New Hampshire. These draw hot exhaust air from the auditorium; cool outdoor air is admitted through low windows and shutters in the auditorium's north wall. (Courtesy of Banwell White Arnold Hemberger & Partners, Inc., Hanover, New Hampshire.)

TABLE 6.7 **Turbine Ventilator Performance**

NOTE: The combination of wind suction and stack effect produces the following exhaust capacities (cfm) for various throat diameters and stack heights of turbine ventilators. Recommended spacing between ventilators is 20 ft (6 m). Cfm × 0.472 = L/s.

Outdoor Wind Velocity: mph (m/s)		2 (0.9)			4 (1.8)			6 (2.7)			8 (3.6)			10 (4.5)		
ΔT, indoors-outdoors F°		10	20	30	10	20	30	10	20	30	10	20	30	10	20	30
C°		5.6	11.1	16.7	5.6	11.1	16.7	5.6	11.1	16.7	5.6	11.1	16.7	5.6	11.1	16.7
Turbine Throat Diameter	Height Above Intake ft (m)							Exhaust Capacity (cfm)								
6" (150 mm)	10 (3)	114	125	130	210	221	226	314	325	330	426	437	442	534	545	550
	20 (6)	122	135	144	218	231	240	322	335	344	434	447	456	542	555	564
	30 (9)	129	144	156	225	240	252	329	344	356	441	456	468	549	564	576
	40 (12)	135	152	166	231	248	262	335	352	366	447	464	478	555	572	586
10" (250 mm)	10 (3)	209	222	274	370	383	435	545	558	610	728	741	793	915	928	980
	20 (6)	234	269	301	395	430	462	570	605	637	753	788	820	940	975	1007
	30 (9)	254	301	328	415	462	489	590	637	664	773	820	847	960	1007	1034
	40 (12)	269	318	355	430	479	516	605	654	691	788	837	874	975	1024	1061
14" (350 mm)	10 (3)	333	383	422	558	608	647	804	854	893	1062	1112	1151	1324	1374	1413
	20 (6)	376	444	496	601	669	721	847	915	967	1105	1173	1225	1367	1435	1487
	30 (9)	413	496	560	638	721	785	884	967	1031	1142	1225	1289	1404	1487	1551
	40 (12)	444	539	614	669	764	839	915	1010	1085	1173	1268	1343	1435	1530	1605
18" (450 mm)	10 (3)	476	564	623	755	843	902	1071	1159	1218	1399	1487	1546	1737	1825	1884
	20 (6)	549	662	747	828	941	1026	1144	1257	1342	1472	1585	1670	1810	1923	2008
	30 (9)	611	747	853	890	1026	1132	1206	1342	1448	1534	1670	1776	1872	2008	2114
	40 (12)	662	819	941	941	1098	1220	1257	1414	1536	1585	1742	1864	1923	2080	2202
24" (610 mm)	10 (3)	716	874	978	1101	1259	1363	1522	1680	1784	1963	2121	2225	2412	2570	2674
	20 (6)	844	1046	1196	1229	1431	1581	1650	1852	2002	2091	2293	2443	2540	2742	2892
	30 (9)	954	1196	1384	1339	1581	1769	1760	2002	2190	2201	2443	2631	2650	2892	3080
	40 (12)	1046	1324	1542	1431	1709	1927	1852	2130	2348	2293	2571	2789	2742	3020	3238
30" (760 mm)	10 (3)	1139	1385	1545	1719	1965	2125	2379	2625	2785	3068	3314	3474	3769	4015	4175
	20 (6)	1342	1655	1890	1922	2235	2470	2582	2895	3130	3271	3584	3819	3972	4285	4520
	30 (9)	1514	1890	2185	2094	2470	2765	2754	3130	3425	3443	3819	4114	4144	4520	4815
	40 (12)	1655	2090	2430	2235	2670	3010	2895	3330	3670	3584	4019	4359	4285	4720	5060
36" (910 mm)	10 (3)	1613	1967	2201	2475	2829	3063	3418	3772	4006	4414	4768	5002	5428	5782	6016
	20 (6)	1901	2354	2692	2763	3216	3554	3706	4159	4497	4702	5155	5493	5716	6169	6507
	30 (9)	2148	2692	3115	3010	3554	3977	3953	4497	4920	4949	5493	5916	5963	6507	6930
	40 (12)	2354	2981	3470	3216	3843	4332	4159	4786	5275	5155	5782	6271	6169	6796	7285
42" (1070 mm)	10 (3)	2183	2663	2998	3350	3835	4170	4645	5125	5460	6000	6480	6815	7365	7845	8180
	20 (6)	2588	3203	3668	3760	4375	4840	5050	5665	6130	6405	7020	7485	7770	8385	8850
	30 (9)	2928	3668	4243	4100	4840	5415	5390	6130	6705	6745	7485	8060	8110	8850	9425
	40 (12)	3203	4058	4723	4375	5230	5895	5665	6520	7185	7020	7875	8540	8385	9240	9905
48" (1220 mm)	10 (3)	2868	3500	3925	4412	5044	5469	6078	6710	7135	7843	8475	8900	9638	10270	10695
	20 (6)	3378	4185	4785	4922	5729	6329	6588	7395	7995	8353	9160	9760	10148	10955	11555
	30 (9)	3817	4785	5535	5361	6329	7079	7027	7995	8745	8792	9760	10510	10587	11555	12305
	40 (12)	4185	5300	6175	5729	6844	7719	7395	8510	9385	9160	10275	11150	10955	12070	12945

Source: Reprinted courtesy of Western Ventilating Equipment, Inc., Los Angeles, California.

mer industrial sites or landfills could be threatened by other dangerous soil gases. Even ordinary sites can be threatened with methane gas from a leaking sewer line. A precaution against soil gas is to design for a *passive subslab depressurization system*. This involves at least one 4-in. (100-mm) pipe, open at both ends. The lower end is set into a layer of clean, crushed rock at least 4-in. (100-mm) thick that lies immediately below the floor slab. The object is to allow air within this rock layer to enter the open end of the pipe. The slab is poured and carefully sealed around the pipe,

THERMAL CONTROL

THERMAL CONTROL

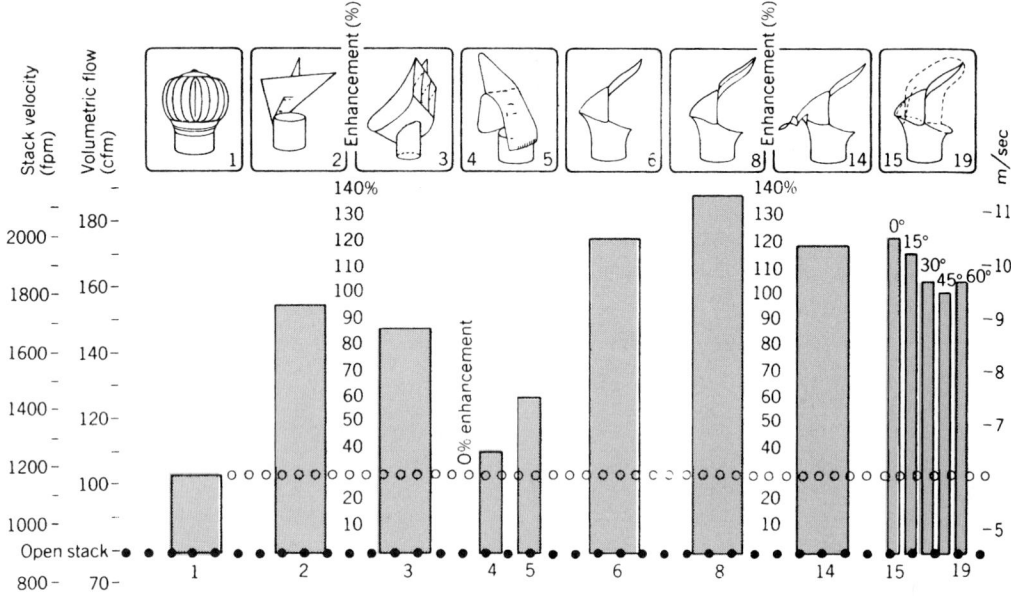

Fig. 6.8 *Performance of some passive ventilators, compared to a simple open stack. The popular turbine ventilator is shown as #1, but is outperformed by many other topping devices. (From the work of R. P. Schubert and Philip Hahn; reprinted from* Progress in Passive Solar Energy Systems, © *1983, by permission of the publisher, American Solar Energy Society, Boulder, Colorado.)*

and the pipe is extended (through interior walls) through the roof, where it can vent radon and other soil gases at a safer place. The heat from the building drives the stack effect within this pipe.

(d) Preheating Ventilation Air. Winter fresh air brought directly into a space will improve IAQ, but at the expense of thermal comfort. Several passive or low-energy strategies are available. The office building in Fig. 6.11 is surrounded by a 4-ft (1.2-m) wide cavity between inner and outer glass surfaces. Air within this cavity is heated, both by the sun and by indoor heat sources, and rises out a damper-controlled opening. Although this building does not utilize such heated air for ventilation, it demonstrates the possibility of such an approach. See the Wallasey school (Fig. 1.10) and the Comstock Building (Fig. 8.72) for examples.

The south-facing wall is used as a winter preheat opportunity in the (unglazed) *transpired collector,* available as Solarwall (Fig. 6.12). Aluminum sheeting, specially finished for solar absorption and penetrated by thousands of tiny

holes, is the exterior surface. Behind this is a cavity, kept under negative pressure by a fan. Outdoor air is drawn through the holes, heated by the dark outer surface, and drawn up the cavity to the fan and then on to the space. Insulation and the interior surface complete the south wall. Thus, heat loss from the building through the south wall is recaptured by the inflowing outside air. In summer, the fan draws directly from the outside, bypassing the solar cavity. The holes at the top of the wall serve as outlets for a stack effect produced by the solar gains through the outer surface. There are numerous installations around the world, one of the largest, at 108,000 ft² (10,034 m²), for an aircraft manufacturer in Quebec, Canada. For a design procedure, see Federal Technology Alert (April 1988).

Another approach to both residential winter ventilation and heat exchange is the "breathable wall" combined with an exhaust air heat pump. This system depends on the house being under negative pressure, assured by installing a heat pump that takes heat from forced exhaust air and invests that heat either in space heating or domestic water heating. The fresh air to replace that

(a)

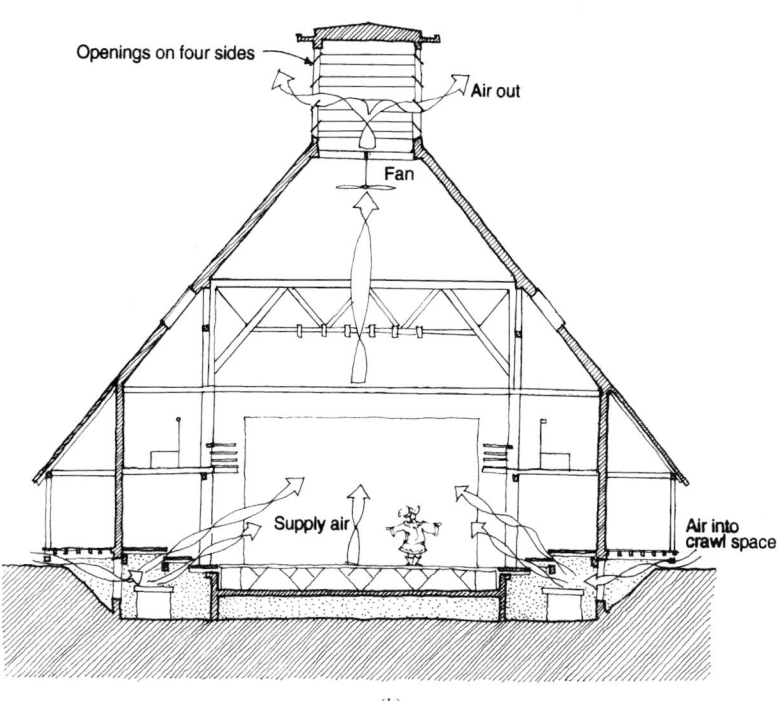

(b)

Fig. 6.9 *Stack ventilation caps the Olivier Theatre at the Bedales School in rural Hampshire, England. (a) View from the east. (Photo by VIEW/Dennis Gilbert.) (b) Section shows the openings that admit outdoor air to the concrete "undercroft" where it is cooled, then admitted below seats to the auditorium; gaining heat, it rises (assisted when necessary by a "punkah" fan) and out the four-sided dampered openings in the cupola. (Designed by Fielden Clegg Architects, Bath, Avon; engineering by Max Fordham and Partners, London.)*

being expelled is drawn in through the outside walls, in a unique combination of fiberglass lap siding board, fiberglass insulation batts, a breathable sheathing, and *no* vapor barrier. This allows a slow, steady stream of cold air to enter and be warmed by the insulation, then enter the house. More information is available from the Canadian National Research Council.

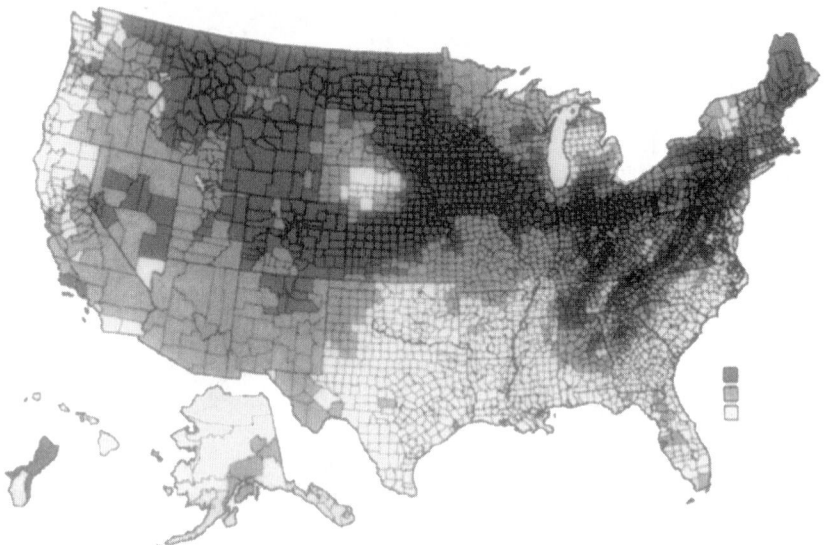

Fig. 6.10 *Relative risks of high radon levels in the soil by county in the United States. The darker the zone, the higher the risk. (U.S. Environmental Protection Agency, courtesy of* Environmental Building News.*)*

(a)   (b)

Fig. 6.11 *The Occidental Chemical Corporate Office Building, Niagara Falls, New York. Architects: Cannon Design, Inc.* (a) *Fully exposed to sun and wind despite its downtown site, the building appears as a conventional curtainwall box.* (b) *An ordinary-looking plan with a central core and suspended ceilings.* (c) *At all four sides, a 4-ft (1.2-m) cavity allows for maintenance of movable daylight louvers and window washing, as well as for ample natural ventilation by the stack effect. Although this air is released at the roof, it could be utilized during winter as tempered fresh air to the interior.*

Motorized Damper

Motorized Damper

Wall Section Detail

(c)

Fig. 6.11 *(continued)*

6.5 Equipment for IAQ

This section is about equipment that moves, heats or cools, humidifies or dehumidifies, and cleans air. A large range of capacities is involved, from room-sized to central whole-building air handling. A general note about heating and cooling system choice: IAQ will be easier to achieve if the heating and cooling systems utilize forced air motion, because some filtering is built in to the air-handling

equipment. However, separate air-cleaning systems are increasingly common, so radiant-only heating systems with added forced-air cleaning can yield high indoor air quality. For cooling, the economizer cycle (see Fig. 8.35) provides up to 100% outdoor air at times, and cooling by night ventilation of thermal mass provides many complete air changes, during the nightly building maintenance activities that are so fume producing. Evaporative cooling (sections 5.6 *d,e* and 5.14 *d,e*) provides a continuous flow of outdoor air, more humid than outdoor conditions.

(a) Exhaust fans. This is one device we all have used. Exhaust fans remove air that is odorous and/or excessively humid before it can spread beyond bathrooms or kitchens, creating a negatively pressured area in the process that further limits the spread of undesirable air. In buildings that have non–air-motion heating systems (radiant heat), exhaust fans are often the only built-in devices for moving air. They are often very noisy, sometimes useful for covering other sounds associated with toilets, but also noisy enough to discourage their use over long periods of time. In its most simple application, the exhaust fan is a stand-alone device, with no thought about from where the replacement air will be drawn, and rarely much concern about where this unwanted air will be discharged outdoors. (Discharge into attics, basements, or crawl spaces is prohibited.)

ANSI/ASHRAE 90.2-1993, *Energy-Efficient Design of New Low-Rise Residential Buildings,* requires intermittent (user-controlled) exhaust fans of at least 50-cfm (25-L/s) capacity for bathrooms, and 100 cfm (50 L/s) for kitchens. The intake for these fans should be as near to the source of polluted air as possible so as not to drag such air across other locations before it leaves the space. In kitchens, this is directly above the range (grease, odors, and water vapor); in bathrooms, in the ceiling above the toilet (and shower) in order to remove the warmest (therefore most moist) air. Some options are shown in Fig. 6.13.

A more comprehensive approach in a residence includes the addition of a "principal exhaust fan," which should be centrally located (draw from the greatest space), quiet, and suitable for continuous use. Its exhaust capacity should be at least 50% of the entire system capacity. In turn,

(a)

Fig. 6.12 *Fresh air winter pre-heating on the south wall is accomplished with this unglazed transpired collector* (a). *Tiny holes in the aluminum skin admit air to a cavity outside the wall's insulation, where it is preheated by sun on the dark aluminum, as well as heat escaping through the insulation. A fan draws this air to a supply plenum, after which it is distributed to the space. In summer* (b), *the fan draws directly from the outside, and the collector is self-ventilated by the stack effect. (Courtesy of the National Renewable Energy Laboratory.)*

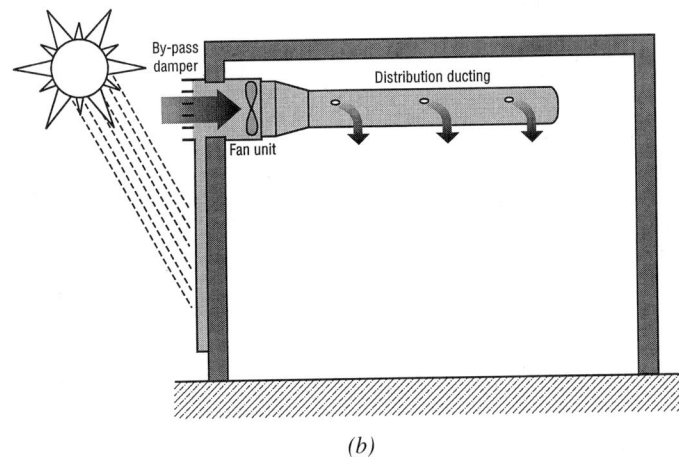

(b)

this entire capacity is typically no less than 0.3 ACH. When this principal exhaust fan is operating, outdoor air must be brought in, tempered, and circulated throughout the residence. The tempering can be by mixing with indoor air, or by heating. Figure 6.14 shows two approaches to whole-house exhaust, one for forced-air systems, one without forced air motion.

This additional, continuous exhaust capacity might cause problems of inadequate air for or spillage of fumes from combustion equipment. This equipment (furnaces, stove-top barbecues, etc.) with a net exhaust greater than 150 cfm (75 L/s) must be provided with a make-up air fan that turns on/off at the same time.

(b) Heating/Cooling of Make-Up Air.
Where climates are mild and/or energy inexpensive, special equipment (Fig. 6.15) other than heat exchangers can be used to heat and/or cool a particularly large quantity of make-up air. Especially common in factories or laboratory buildings with high exhaust air requirements, these simple devices often supplement the building's main heating/cooling system, which deals primarily with heat gains/losses through the building's skin. In warm dry climates, evaporative coolers are often used for make-up air because they are already designed to utilize 100% outdoor air. Even in hot and more humid climates, *indirect evaporative cooling* can help lower the temperature of make-

(a)

(b)

(c)

Exhaust
soffit
above
WC

Houses
duct

Infrared
light

Activated as
needed

Odors
exhausted

(d)

Fig. 6.13 *Options for exhaust air from toilet rooms:* (a) *Daylight enters from the sides; incoming air from the lower window picks up heat and moisture near the ceiling, exits through high clerestory.* (b) *Two skylights admit daylight; one exhausts hot moist air at the ceiling. Make-up air will be needed from the adjacent rooms.* (c) *Adding an exhaust fan will insure negative pressure in the bathroom, isolating moisture and odors from the adjacent rooms.* (d) *In public toilet rooms, ventilation from directly above toilets is desirable. In cold weather, small radiant heat lamps with timer switches could add task heat and light.*

up air. In Fig. 6.16, outdoor air is at 104°F (40°C) and 10% rh (Point *A*). Two streams of this outdoor air are involved. An evaporative cooler cools one air stream along a constant WB temperature line to Point *B*, where it is now 70% rh but considerably cooler at 73°F (23°C): in the comfort zone, but rather hot and humid. At this point, it enters an air-to-air heat exchanger (see Section 6.5c). The other outdoor air stream enters the other side of this heat exchanger, again at Point *A*. As the two streams exchange heat, they move toward the same temperature: the evaporatively cooled

stream moves from Point *B* to Point *C*, about 86°F (30°C) and 42% rh, and is then exhausted. On the other side, outdoor air moves from Point *A* to Point *D*, about 91°F (33°C) and 16% rh. This second air stream is "indirectly evaporatively cooled." Although it is still well above the comfort zone, it can then be either cooled by typical refrigerant systems, or evaporatively cooled until it reaches the comfort zone.

(c) Heat Exchangers. As the tightness of construction increases and fewer air changes per

Fig. 6.14 *Assured ventilation for residences. (Based on Canadian Research Council studies.)*

TABLE 6.8 **Representative Heat Exchanger Data for Smaller Buildings**

Part A. *Air-to-Air Sensible Heat Exchangers*

Approximate Size[a] L × W × H		Airflow Range		Temperature Recovery at 32°F (0°C)
in.	mm	cfm	L/s	
21 × 13 × 25	535 × 330 × 630	50–120	25–60	70%
34 × 17 × 21	865 × 435 × 530	75–225	35–110	80%
53 × 17 × 21	1335 × 435 × 530	75–225	35–110	93%
47 × 20 × 37	1195 × 505 × 940	350–700	165–330	65%

Source: Conservation Energy Systems, Inc., Saskatoon and Minneapolis.

Part B. *Air-to-Air Sensible + Latent Heat Exchangers*

Approximate Size[b] L × W × H		Airflow Range		Wheel Efficiency
in.	mm	cfm	L/s	
46 × 32 × 24	1170 × 815 × 610	500–1000	235–470	84–82%
54 × 48 × 34	1375 × 1220 × 865	600–1600	285–760	87–80%
		1600–2300	760–1090	75–70%
67 × 67 × 43	1705 × 1705 × 1095	1750–3250	830–1535	85–80%
		3000–4500	1420–2125	76–71%
124 × 84 × 60[c]	3150 × 2135 × 1525	3500–6500	1660–3070	85–80%
		6000–9000	2835–4250	76–71%

Source: Greenheck Fan Corporation, Schofield, Wisconsin.

[a]Access to one face (L × W) is needed for servicing.

[b]Several arrangements (interior and rooftop) are available; dimensions do not include service clearance access, supply hoods, or exhaust hoods.

[c]This unit contains two wheels.

Fig. 6.15 *Make-up air heaters/coolers.* (a) *Distributed on a factory roof, these units prevent negative indoor air pressures by forcing replacement (make-up) air for the air that is exhausted from processes indoors. Often, a separate heating/cooling system will be used for gains or losses through the building skin.* (b) *Heat recovery ventilator, utilizing a cross-flow core, serves from 200 to 900 cfm (100 to 450 L/s) with supply air entering at a temperature 0.6 to 0.8 of exhaust air temperature. (© Conservation Energy Systems, Minneapolis, Minnesota.)*

hour (ACH) occur from natural infiltration, forced ventilation becomes more attractive as a means of reducing indoor air pollution. When a heat exchanger is used, it is possible to maintain an adequate supply of fresh air without severe energy consumption consequences. Figure 6.17 illustrates the basic principle of a simple air-to-air heat exchanger that is becoming increasingly common for tightly built small buildings. Note

that the outgoing and incoming air streams must be adjacent.

Some commercially available heat exchangers are capable of extracting 70% or more of the heat from exhaust air. The lower the volume of airflow, the higher the efficiency. Table 6.8, Part A shows representative sizes, airflows, and efficiencies. For the best diffusion of incoming fresh air through a building, the heat exchanger should be

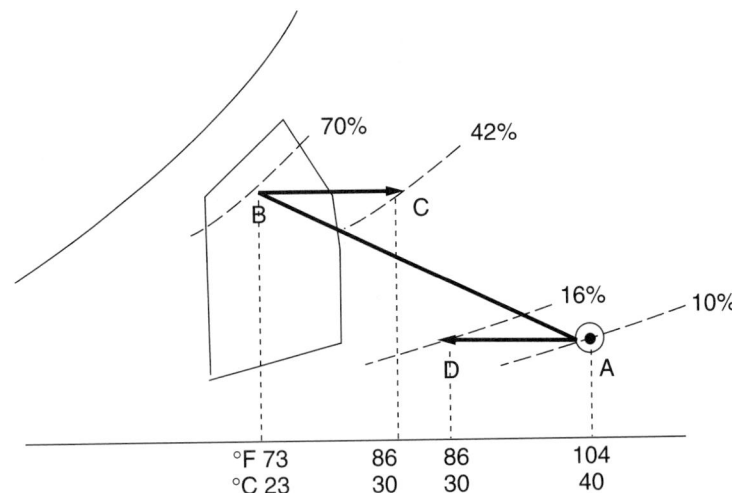

Fig. 6.16 *Indirect evaporative cooling can precool fresh air. A similar process is shown in Fig. 7.10.*

(a)

(b)

(c)

Fig. 6.17 *Air-to-air heat exchangers are particularly helpful in cold weather.* (a) *The basic principle of operation.* (b) *A super-insulated home is heated by coils fed from the domestic water heater, plus a heat exchanger to preheat incoming cold air. (Reprinted by permission of Alberta Agriculture, Home and Community Design Branch,* Low Energy Home Designs, © 1983 by Brick House Publishing Co.) (c) *Fan-powered heat exchanger, utilizing a heat/moisture exchange wheel (see also Fig. 6.20). This model is adjustable from 70 to 200 cfm (35 to 100 L/s), can be ceiling, wall, or floor mount, and operates at 75% to 80% thermal efficiency. (Courtesy of Air Xchange, Inc., Rockland, Massachusetts.)*

incorporated at the central forced-air fan (Fig. 6.17b). When a central forced-air system is not available, heat exchangers can be placed at various points in the building; typically, each heat exchanger is then equipped with its own fan. These may serve as make-up air units that are more energy-efficient than the devices discussed in Section 6.5(b).

Some cautions about air-to-air heat exchangers:

1. Avoid using them on exhaust airstreams that are contaminated with grease, lint, or excessive moisture (through cooking and clothes drying in particular) because clogging, frosting, and fire hazard problems can develop.

2. In colder winter conditions, a built-in defroster, which will require energy, will be needed.

3. Carefully locate the outdoor fresh air intake. Keep this air intake as far as possible from the exhaust air outlet to avoid drawing indoor-contaminated air back into the building. Keep it away from pollution sources such as vehicle exhaust, furnace flues, dryer and exhaust fan vents, and plumbing vents.

A student housing complex in Greensboro, North Carolina, utilizes heat exchangers on the exhaust air from bathrooms (Fig. 6.18). These exhaust devices are called *energy recovery ventilators* (ERV). Each floor of the three-story complex

(a)

Fig. 6.18 *Energy recovery ventilator (ERV) serves two small bathrooms in a student housing complex in Greensboro, North Carolina. The ERV is activated by switches in either bathroom. Air is drawn out beside the toilet and exchanges heat (at about 85% efficiency) with incoming fresh air. This tempered fresh air is then mixed with some return air and fed directly to the heat pump's indoor unit, above the ERV. The supply air is then fed to other rooms, insuring a negative pressure in the bathroom while the ERV operates. Each of 16 apartments has its own ERV and heat pump. (Design by Harry John Boody, Jamestown, N.C.)*

EAO = EXHAUST AIR OUT (TO OUTSIDE)

SAO = SUPPLY AIR OUT (TO R/A PLENUM)

EAI = EXHAUST AIR IN (FROM BATHROOM)

SAI = SUPPLY AIR IN (FRESH AIR FROM OUTSIDE)

ERV DETAIL

#3 FRESH AIR TO RETURN AIR

#1 STALE AIR INTAKE FROM BATHROOM

#2 FRESH AIR INTAKE (HIGH MOUNT)

ERV ACCESS SIDE

#4 STALE AIR EXHAUST (LOW MOUNT)

ENERGY RECOVERY VENTILATOR
MODEL#:   160 CFM
BDD = BACKDRAFT DAMPER IN DUCT.

18 × 6 TO (2) 6⌀

2 1/2" LIN

1/2" LIN

CC

FAN

COIL

R/A

2" ISO PADS

(b)

Fig. 6.18 *(continued)*

has four apartments, each with 1078 ft² (100 m²) floor area and 8-ft (2-m) ceilings. Each apartment has its own air-to-air ERV that accepts air from the two small bathrooms in the apartment. Control is by individual switches in the bathrooms. Prefiltered outdoor air is drawn into the ERV, exchanging heat with the outgoing exhaust air. This fresh air is then fed directly into the air handler for each apartment's heat pump, adjacent to the ERV. Thus the bathrooms are under negative pressure relative to the rest of the apartment.

Outside, the fresh air intake is located high on the wall, the exhaust outlet lower. All intakes and outlets are separated by a minimum of 8 ft (2 m) on the exterior walls.

The *heat pipe* (Fig. 6.19) also transfers sensible heat between adjacent airstreams. Within the heat pipe, a charge of refrigerant spends its life alternately evaporating, condensing, and migrating by capillary action through the porous wick. Because the only thing that moves is refrigerant and it is self-contained, the ideal of no maintenance and

Fig. 6.19 *The heat pipe is a self-contained, no-moving-parts device.* (a) *It silently helps transfer sensible heat between adjacent fresh-air intake and stale-air exhaust airstreams. (Reprinted by permission from AIA, Ramsey, and Sleeper,* Architectural Graphic Standards, *9th ed., © 1994 by John Wiley and Sons.)* (b) *The heat pipe can increase moisture removal by first precooling incoming hot air before it reaches the cooling coils. Then, the heat is returned to the cooled air, bringing it to a temperature–humidity combination acceptable for supply air to a space. (Courtesy of Heat Pipe Technologies, Inc. and Environmental Building News.)*

long life is obtained. Efficiency is from 50% to 70%; modular sizes are available to 54 in. × 138 in. × 8 rows deep.

Heat pipes can assist in the dehumidification and cooling of incoming air. The typical cooling process using cooling coils was shown in Fig. 5.43. This potentially energy-intensive process first overcools the air to "wring out" (condense) water, then reheats the air. The heat pipe precools the air (subtracting heat) before the cooling coil, then warms it (adding heat) after the cooling coil. No energy input is required. A configuration of the heat pipe in this task is shown in Fig. 6.19b.

Energy transfer wheels (Fig. 6.20) go further than the two preceding devices, in that they transfer latent as well as sensible heat. In winter, they recover both sensible and latent heat from exhaust air; in summer, they can both cool and dehumidify the incoming fresh air. Seals and laminar flow of air through the wheels prevent mixing of exhaust air and incoming air. A further precaution in the process purges each sector of the wheel briefly, using fresh air to blow away any unpleasant residual effects of the exhaust air on the wheel surfaces. Carryover of exhaust air qualities, except

those of heat and moisture, is between 4% and 8% without purging, and less than 1% with purging. Efficiency is from 70% to 80%, and available sizes range up to 144 in. in diameter. Table 6.8, Part B shows some representative sizes, airflows, and efficiencies. An smaller example of this device was shown in Fig. 6.17c.

(d) Dessicant Cooling. Another rotating-wheel process involves *dessicant cooling*. Dessicant cooling systems (Fig. 6.21) are attractive because they use no refrigerants that contain CFCs and they lower humidity without having to resort to overcooling air. The dessicants (such as silica gel, activated alumina, or synthetic polymers) in an "active" system must be heated to drive out the moisture they remove from the incoming air; at present, natural gas is typically used, but solar energy is a promising source due to its plentiful summer availability. Waste heat from other mechanical systems may also be used. In a "passive" dessicant system, the heat from the building's exhaust air is enough to release and vent the moisture removed from incoming air.

Research into materials suitable for dessicant

THERMAL CONTROL

Fig. 6.20 *Energy transfer wheels. The wheel surface is impregnated with lithium chloride (or other similar materials), which absorbs moisture and transfers it to the other airstream. (a,b) The wheel delivers moist air in winter and dry air in summer. (c) Cross section through the wheel and the two airstreams it serves. Exhaust air may be filtered to help keep the wheel clean. (d) Multiple-unit installation. Room exhaust air passes through the upper chambers and incoming fresh air passes through the lower chambers. Wheels rotate at 8 to 10 rpm. (e) Rooftop unit supplies up to 1000 cfm (500 L/s). (Courtesy of Air Xchange, Inc., Rockland, Massachusetts.)*

Exhaust air
to outdoors

Air exhausted
from room

Energy exchange
wheels

Room
exhaust

Supply fan
to room

To cooling coil,
reheat, etc.

Fresh air in

(d)

EXHAUST AIR

SUPPLY
AIR

ROOF CURB

STALE AIR

Return from ceiling
Plenum or ducted from
specific building
locations. (Install
balancing damper)

Supply fresh air to
return ductwork of cooling and
heating system or directly to space.
(Install balancing damper)

FRESH AIR
SUPPLY

(e)

Fig. 6.20 *(continued)*

Fig. 6.21 *Dessicant cooling utilizes the rotating wheel, impregnated with either silica gel or Type 1M material. The wheel removes moisture from incoming hot outdoor air, delivers some to outgoing exhaust air. The remaining moisture must be driven from the wheel by a heat source; natural gas, or eventually solar energy. (National Renewable Energy Laboratory.)*

cooling and solar-driven regeneration should produce improvements in the types available as well as their performance. For an extended discussion of both solid and liquid-based dessicant systems, see Lorsch (Ed.) (1993).

(e) Task Dehumidification and Humidification.

Humidity affects comfort; for a sedentary person, a 30% rh change will produce about the same comfort sensation as a 2F° (1C°) change in temperature. Higher humidity adds to hot air discomfort, making evaporation more difficult, increasing skin wettedness, and increasing the friction between skin and clothing or furniture surfaces. As rh exceeds 60%, problems with IAQ increase due to mold and mildew growth. Lower humidity produces sensations of cooler and drier; but too-low conditions can irritate the skin.

For spaces that need only dehumidification rather than mechanical cooling, *refrigerant dehumidifiers* are commercially available (Fig. 6.22). Their advantage over dessicant dehumidifiers is that the air temperature remains essentially unchanged during dehumidification (dessicant dehumidifiers raise the temperature of the dried air). However, these devices consume energy to run the refrigeration cycle, and this is added, as heat, to the space. Accumulated water must periodically be removed from the unit, and if untended

could become a source of disease. Most refrigerant dehumidifiers encounter operating difficulties at air temperatures below 65°F (18°C), at which point frost forms on their cooling coils. This could cause problems in a tightly enclosed residence in winter.

Task humidifiers are widely available, often used to relieve symptoms of respiratory illnesses. Again, the problem of bacterial and mold growth in water reservoirs arises; some products add a ultraviolet light (UV) bulb to counteract this threat.

(f) Filters.

There is a wide variety of particle air pollutants, described in Table 6.2. The particulates of larger size are the easiest to remove, but much smaller respirable particles pose a greater threat to health. Not all pollutants can be removed by filters. Figure 6.23 compares filter groups and testing methods with associated filter efficiencies. (A new ASHRAE Standard method is under development, and will be issued under 52.2P.) Groups I, II, and III are widely used and are generally illustrated in Fig. 6.24. The highest efficiency filter, the high-efficiency particulate arrestance (HEPA), is most often found in special air cleaners for unusually polluted or IAQ-demanding environments. Air filter characteristics are summarized in Table 6.9.

Typical Performance Data

Watts	290	450	465	630
Pints per day[a]	14	20	25	30
Unit dimensions (in.)				
Height	20½	20½	21½	21½
Width, 11¾				
Depth, 16¾				
"Average" room size served	20′ × 32′	35′ × 40′	40′ × 48′	40′ × 60′

[a]For air at 80 F, 60% rh.

Fig. 6.22 *Refrigerant dehumidifiers typically are installed as free-standing units in smaller buildings.*

Again, it is wise to remove pollutants at the source. ASHRAE Standard 62-1989, *Ventilation for Acceptable IAQ,* recommends dust collectors rather than filters when the dust loading equals or exceeds 10 mg/m³. This reference offers considerable useful information about filters.

Particulate filters. Particulate filters are very common; *panel filters* are furnished with HVAC equipment and function mainly to protect the fans from large particles of lint or dust. Because they are relatively crude, they are not properly considered to be air cleaning equipment. *Media filters*

TABLE 6.9 **Air Filter Characteristics**

Media and Type	Percent Efficiency Range		Dust-Holding Capacity	Airflow Resistance (In. Water)[a]
	Atmospheric Dust	Small Particles		
Dry panel, throwaway	15–30	NA	Excellent	0.1–0.5
Viscous panel, throwaway	20–35	NA	Good	0.1–0.5
Dry panel, cleanable	15–20	NA	Superior	0.08–0.5
Viscous panel, cleanable	15–25	NA	Superior	0.08–0.5
Mat panel, renewable	10–90	0–60	Good to superior	0.15–1.0
Roll mat, renewable	10–90	0–55	Good to superior	0.15–0.65
Roll oil bath	15–25	NA	Superior	0.3–0.5
Close pleat mat panel	NA	85–95	Varies	0.4–1.0
High-efficiency particulate	NA	95–99.9	Varies	1.0–3.0
Membrane	NA	to 100	NA	NA
Electrostatic with mat	80–98	NA	Varies	0.15–1.25

Source: Reprinted by permission from AIA, Ramsey, and Sleeper, *Architectural Graphic Standards,* 9th ed., © 1994 by John Wiley & Sons.

[a]Higher airflow resistance values will require increased fan energy.

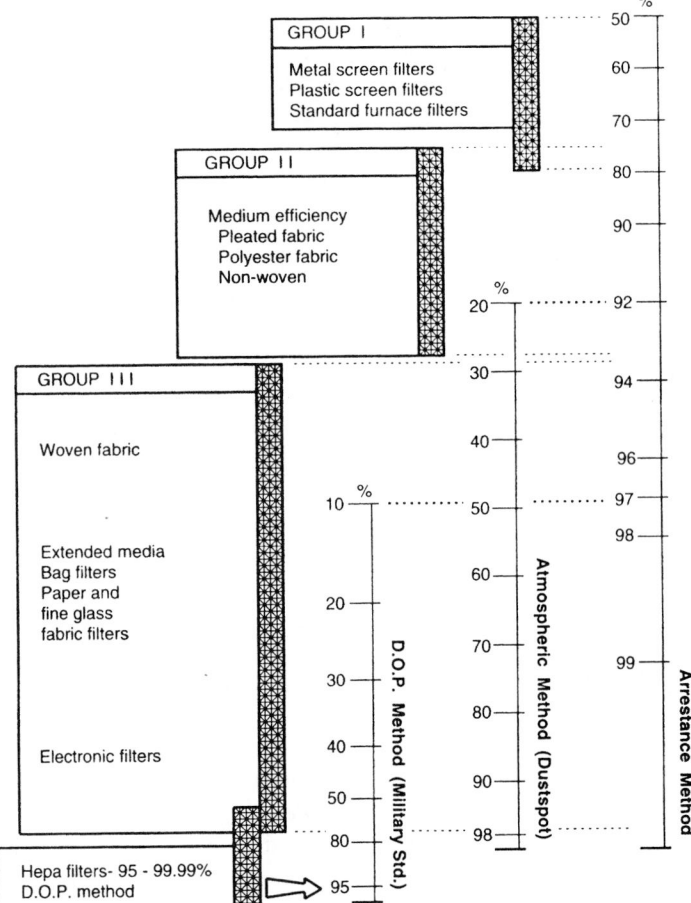

Fig. 6.23 *Air filter efficiencies. Methods of determining efficiency vary; the Arrestance method can be used across all groups, while the Atmospheric (Dust Spot) method should be used for medium- and high-performance filters. The detection operational program (DOP) method is a military standard. (From Rousseau, D. and Wasley, J. (1997). Healthy by Design, Hartley & Marks, Point Roberts, Wash.*

are much finer, using highly efficient filter paper in pleats within a frame. They work both by straining and impaction. The larger particles are strained out by the closely spaced filter fibers while some of the smaller particles that would otherwise pass through are pushed into the fibers due to the air turbulence. Particulate filters need regular maintenance, especially the media filters, which can become blocked and cause damage to HVAC equipment if not replaced frequently enough.

Media filters of high quality are expected to perform at an efficiency of 90% (minimum) and are typically at least 6 in. (150 mm) deep; this is for a 6-month minimum life-cycle.

Adsorption filters. Adsorption filters are for gaseous removal and vary according to the pollut-

ant in question. *Activated-charcoal filters* are the more common of these types, absorbing materials with high molecular weights but allowing those of lower weights to pass. Other adsorption filters use porous pellets impregnated with active chemicals such as potassium permanganate; the chemicals react with contaminants, reducing their harmful effects. Adsorption filters must regularly be regenerated or replaced.

High-quality adsorption filters contain gas adsorbers and/or oxidizers and are of sufficient capacity to remain active over a full service cycle of 6 months at 24 hours per day, 7 days per week. The air velocity should allow the air to remain in the filter about 0.06 second.

Air washers. Air washers are sometimes used to control humidity and bacteria growth. The

Fig. 6.24 *Air filter types. (Not shown are pleated and roll types.)*

moisture involved can pose a threat if these devices are not well maintained.

Electronic air cleaners. Electronic air cleaners can pose a different threat due to ozone production but have the advantage of demanding less maintenance. Static electricity is produced in the *self-charging mechanical filter* by air rushing through it; larger particles thus cling to the filter. The more humid and/or higher velocity the air, the less the filtering efficiency. In the *charged media filter,* an electrostatic field is created by applying a high dc voltage to the dielectric material of the filter. Many particles are not polarized, however, due to an insufficiently strong field. A *two-stage electronic air cleaner* first passes dirty air between ionizing wires of a high-voltage power supply. Electrons are stripped from the particulate contaminants, leaving them positively charged. Then, these ionized particles pass between collector plates that are closely spaced and oppositely charged. The particles are simultaneously repelled by the positive plates and attracted to the negative plates, where they are collected.

(g) Locating Air Cleaning Equipment.

Before the advent of concern about IAQ, buildings were often designed with rather crude panel filters only at the HVAC equipment itself; they were primarily intended to intercept materials that might adversely affect combustion or heat exchange. In addition to these equipment-protecting devices, a building deserving a high IAQ will now have a combination of high-efficiency particle filters and adsorption filters.

The panel filters provided with the HVAC equipment are usually located *upstream* from the unit's fan. Subsequently, the high-efficiency particle and adsorption filtering system should be located *downstream* from the HVAC cooling coils and their drain pan, to ensure that any microbiological contaminants living on those wet surfaces are removed rather than being distributed throughout the building.

In buildings where the filtering system must be integrated with the central HVAC system (typical of small buildings), that system should be capable of continuously circulating at the rate of 6 to 10 times per hour, and of operating against

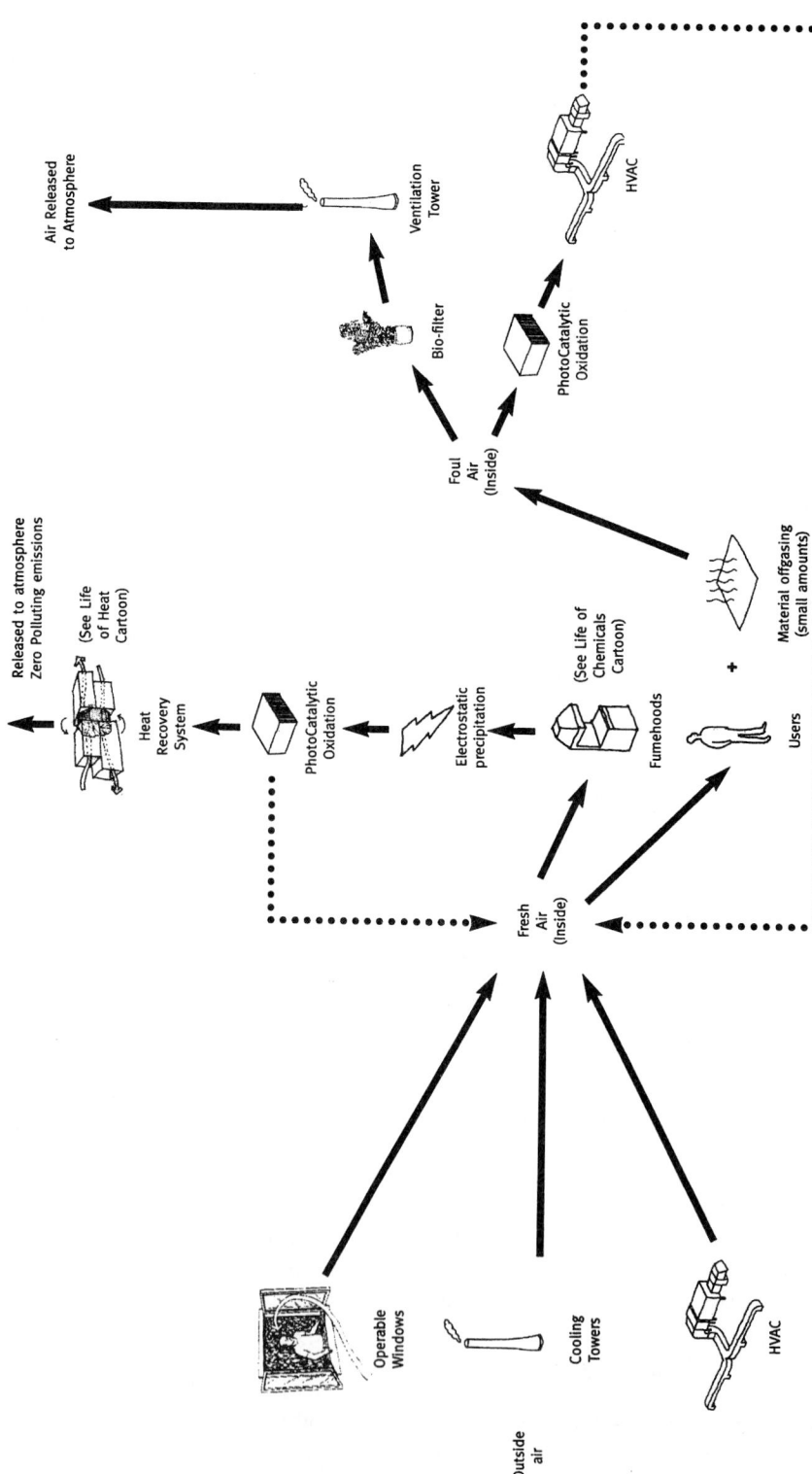

Fig. 6.25 *The EPICenter at Montana State University (see again Fig. 1.5) uses a combination of strategies for IAQ. (a) "Life of Air" cartoon describes a zoning process for treatment, separating the methods for fume hoods from those for classroom/commons air. (b) Section describes how some biofiltering in the large south-facing sunspace will be employed for further air processing of classroom/commons air, before release. (Courtesy of Place Architecture LLC, Bozeman, Montana, and Berkebile Nelson Immenschuh McDowell, Architects, Kansas City, Missouri.)*

CONTROLLED VENTILATION

MECHANICALLY VENTILATED ZONE

AIR AND VENTILATION STRATEGIES

(*b*)

Fig. 6.25 *(continued)*

the rather considerable static pressure that results from high-efficiency filters.

(h) Ultraviolet Light (UV). Since early in the 20th century, ultraviolet light in the C band (200–280 nm wavelength) has been used to kill harmful microorganisms, but under tightly controlled conditions. Now there are UV lamp devices that work within HVAC systems, promising to control fungi, prevent the development and spread of bacteria, and reduce the spread of viruses. As an additional benefit, cooling coils and drain pans stay cleaner. The germicidal output of these lamps is somewhat higher at room temperatures (and higher) than at colder temperatures. These devices take up very little space (they work within ductwork) and generate no ozone or other chemicals. They are even more effective when installed in a UV-reflective duct interior: aluminum seems the best UV reflector commonly available. Tube life is from 5000 to 7500 hours.

UV light also looks promising as treatment for VOCs. The National Renewable Energy Laboratory is helping develop a process that bombards polluted air with UV in the presence of special catalysts. Pollutants including cigarette smoke, formaldehyde, and toluene are quickly broken down into molecules of water and carbon dioxide.

The EPICenter at Montana State University was introduced in Fig. 1.5. This building contains both chemistry laboratories and classrooms, and is considering *photocatalytic oxidation* as a process for both laboratory fume hoods and general indoor air quality. UV light is focused on a catalyst in the presence of water vapor, to destroy airborne microorganisms and VOCs. Hydroxy radicals are generated in this process, more powerful than chlorine in their ability to oxidize VOCs and neutralize microbes. The combination of IAQ strategies in this project is shown in Fig. 6.25.

(i) Individual Space Air Cleansing. Energy conservation considerations have reduced the air circulation rate in many central air handling systems; moving less air saves energy used by fans. One result can be a poor distribution efficiency, which means poorly mixed air within the occupied spaces. With local (individual space) air-filtering equipment, both a high circulation rate and proper air mixing are achievable. Each such unit has its own fan that can operate either with or without the central HVAC fan. An example of an independent electrostatic air filter is shown in Fig. 6.26.

Air circulation through these filters should occur at rates between 6 and 10 times per hour. The

Fig. 6.26 *A stand-alone electrostatic air filter operates independently of the central HVAC system; no ductwork is required. Airflow rates vary; the slower the flow, the more efficient the filtering. Flush mount (recessed in suspended ceiling), ceiling surface mount, wall mount, and portable models are available. (Courtesy of Tectronic Products Co. Inc., East Syracuse, New York.)*

air is then ducted to diffusers, hence circulating in a sweeping pattern across a space to return air intakes on the *opposite side.*

A variation on electrostatic air cleaning is shown in Fig. 6.27. In this equipment, a mixture of outdoor and indoor air is passed through a complex electrical field produced by both a high-voltage dc and a high-frequency ac. This supply air emerges with greatly reduced submicron particles that have coagulated into larger particles more

Fig. 6.27 *This air-cleaning system uses a combination of high-voltage dc and a high-frequency ac. The supply air emerges with greatly reduced submicron particles, which have coagulated into larger particles more easily carried by air currents. As this air moves through the occupied space, it picks up submicron particles and is either exhausted or returned to the equipment, where it is filtered to remove the larger particles. (Courtesy of CRS Industries Inc., Tampa, Florida.)*

easily carried by air currents. As this air moves through the occupied space, it picks up submicron particles and is returned to the equipment (some is exhaust air), where it is filtered to remove the larger particles.

Portable air cleaners abound. One more elaborate model combines ultraviolet germicidal light with carbon/oxidizing media and HEPA. It is a counter-height rolling device about 15-in. (380-mm) square that emits cleansed air from the top. It claims to purify air in a space of about 12,000 ft³ (340 m³).

(j) Controls for IAQ. There are now a large number of air quality monitoring devices, some of which control the operation of related equipment. One of the oldest and most simple devices measures the concentration of carbon dioxide (CO_2) in parts per million (ppm). Wherever there are indoor concentrations of people, elevated levels of CO_2 can be expected. Thus, CO_2 becomes a kind of "canary in the coal mine," an early indicator of pollutant build-ups due to dense occupancy. In the Bedales school theater (Fig. 6.9), the operation of the exhaust air dampers in the cupola is controlled by a BMS. Under ordinary conditions, the damper openings are regulated by information from the CO_2 monitor. This can be overridden by indoor and outdoor temperatures (summer night ventilation is encouraged), air velocity monitors below the seats (too much velocity could produce discomfort), the presence of wind and rain (openings limited), or by a fire alarm (fully open to aid in smoke extraction).

Where air pollution is expected from sources in addition to humans, monitors may be installed for carbon monoxide, or for combinations of VOCs, or fuels such as propane, butane, or natural gas; even for a depletion of oxygen. These monitors can be installed as stand-alone alarms or with additional relays that activate equipment. They are about the size of a programmable thermostat; mounting height depends on which gas is to be monitored.

Such devices can regulate ventilating heat exchangers, such as the ERV (shown in Fig. 6.18). This could be especially useful in unoccupied periods, holidays, and weekends. Many HVAC systems are shut off during such extended periods; VOCs from finishes and furnishings continue to accumulate, however. Periodic flushing from ERVs, controlled by VOC monitors, will help maintain acceptable air quality and could eliminate (or greatly reduce) the need for a Monday morning pre-flush.

REFERENCES

American Institute of Architects, Ramsey, C. G., and Sleeper, H. R. (1994). *Architectural Graphic Standards,* 9th ed., Wiley, New York.

American Society of Heating, Refrigerating, and Air-Conditioning Engineers, Inc. (1981). *Handbook of Fundamentals,* and ASHRAE/IES 90.1-1989, *Energy Efficient Design of New Buildings Except New Low-Rise Residential Buildings,* ASHRAE, Atlanta, Ga.

Banham, R. (1969). *The Architecture of the Well-Tempered Environment.* University of Chicago Press, and the Architectural Press, London.

Chandra, S., Fairey, P. W., and Houston, M. (1986). *Cooling with Ventilation,* Solar Energy Research Institute, Golden, Colo.

Fanger, P. O. (1989). "The New Comfort Equation for Indoor Air Quality," *ASHRAE Journal,* October 1989, pp. 33–36.

Federal Technology Alert (April 1988). *Transpired Collectors (Solar Preheaters for Outdoor Ventilation Air).* National Renewable Energy Laboratory, U.S. Department of Energy.

Lorsch, H. (Ed.) (1993). *Air-Conditioning Systems Design Manual.* ASHRAE, Atlanta, Ga.

Shurcliff, W. A. (1981). *Air-to-Air Heat Exchanges for Houses.* Brick House, Andover, Mass.

SMACNA (1993). *Indoor Air Quality,* 2nd edition. Sheet Metal and Air Conditioning Contractors National Association, Inc., Chantilly, Va.

THERMAL CONTROL

7

HVAC FOR SMALLER BUILDINGS

The preceding chapters have dealt with most of the early steps in the design process, such as the choice of thermal design strategy (Chapter 2), the choice of siting (Chapter 3), and the choice of the components and initial sizing of envelopes (Chapters 4 and 5). Chapter 6 dealt with strategies and equipment for interior air quality. Chapters 7 and 8 carry the design process into the specifics of mechanical systems and equipment for heating and cooling.

There are various ways to organize this large collection of components and systems. The next two chapters are organized as follows:

Chapter 7:

1. A review of the relative thermal role of building envelopes versus their internal heating/cooling equipment.
2. A look at the heating/cooling *system design process* begun by the architect and completed by the engineer.
3. Equipment location and sizing for smaller buildings (one or a few thermal zones).
4. Controls for smaller buildings.
5. Refrigeration cycles.
6. Cooling-only systems.
7. Heating-only systems.
8. Heating and cooling systems.
9. Psychrometry and refrigeration.

Chapter 8:

10. The basics of the process by which mechanical systems are integrated into building design.
11. The four generic HVAC system types for large buildings.
12. Larger, centralized HVAC equipment.
13. Air distribution considerations within spaces.

14–17. Details of the four generic HVAC systems.
18, 19. Large-scale system opportunities.

Items 1, 2, and 10 are a more theoretical introduction to systems and equipment; hence, rather than reading the chapter in order, you could read those items first, then go on to the component-and-equipment sections.

7.1 Review of the Need for Mechanical Equipment

One of the most basic functions of buildings is to provide shelter from weather. In a carefully designed building, the roofs, walls, windows, and interior surfaces alone can maintain comfortable interior temperatures for most of the year in most North American climates. With appropriate scheduling, the most uncomfortable hours within buildings can often be avoided; for example, the siesta avoids the hottest afternoon hours within stores and office buildings. Several aspects of comfort and climate, however, pose difficult challenges for ordinary building forms and materials.

A building surface primarily influences comfort through its *surface temperature;* secondarily, it slowly changes *air temperature* (as when cool air moves across a warm surface). Important as these two determinants of comfort are, they sometimes are not sufficient by themselves. Especially in cooling situations, *air motion* and *relative humidity* are significant comfort determinants; and for many indoor activities, *air quality* becomes an important issue in both heating and cooling.

Building form can work with climate to produce air motion for cooling, although the faster air speeds that extend the human comfort zone above 83°F (28°C) may be difficult to provide without mechanical assistance. Relative humidity is still controlled by mechanical (or chemical) means. Building form and materials may be able to keep spaces surprisingly cool, but without de-humidification, surfaces in many North American summer climates can become clammy and covered with mold.

It is difficult to filter air without a fan to force the air through the filtering media. Electrostatic filtering, of course, is done within mechanical equipment.

Thus, whereas the desired air and surface temperatures can often be achieved by passive means (a combination of building form, surface material, and occasional user response), the comfort determinants of air motion, relative humidity, and air quality often require mechanical devices. As the control of air properties—motion, moisture, particulate content—becomes more critical to comfort, the designer becomes more likely to respond with a sealed building, excluding outdoor air except through carefully controlled mechanical equipment intakes. In the recent past, this exclusion of outdoor air has often been accompanied by the exclusion of daylight, of view, of solar heat on cold days—in sum, by a general rejection of all aspects of the exterior environment. As designers come to terms with the role of mechanical equipment, we should also clarify the role of these devices in relation to the climate: Are they occasional modifiers, permanent interpreters, or permanent excluders of the outdoors?

7.2 Heating, Ventilating, and Air Conditioning (HVAC): Typical Design Processes

HVAC systems usually involve a minimum of three design stages. In the *preliminary design* phase, the most general combinations of comfort needs and climate characteristics are considered:

Activity comfort needs are listed.

Activity schedule is listed.

Site energy resources are analyzed.

Climate design strategies are listed.

Building form alternatives are considered.

Combinations of passive and active systems are considered.

One or several alternatives are sized by general design guidelines.

This level of analysis covers the process discussed in Chapters 1 to 4 and Sections 5.1 to 5.6. For smaller buildings, this analysis is often done by the architect alone. For innovative or unusual systems in smaller buildings, and especially for larger, multiple-zone buildings, consultants such as engineers and landscape architects often are included. The team approach is particularly valuable in assessing the strengths of various design alternatives. The architect and the consultants have very different perspectives, and when mutual goals can be clearly agreed on early in the design process, these perspectives are not only mutually supporting but can produce striking innovations whose benefits extend far beyond services to the clients of a particular building. By setting an example, the team can make available better environments for less energy for hundreds of subsequent buildings. With inspired teamwork, the distribution of HVAC services can enhance building form, as many examples in this chapter show.

By the time the *design development* phase is reached, one of the design alternatives has probably been chosen as the most promising combination of esthetic, social, and technical solutions for the program. The consulting engineer (or architect, on a smaller job) is furnished with the latest set of drawings and the program. Typically, the architectural or mechanical engineer then:

I. Establishes design conditions.
    A. By activity, lists the range of acceptable air and surface temperatures, air motions, relative humidities, lighting levels, and background noise levels.
    B. Establishes the schedule of operations.

II. Determines the HVAC zones, considering:
    A. Activities.
    B. Schedule.
    C. Orientation.
    D. Internal heat gains.

III. Estimates the thermal loads on each zone:
   A. For worst winter conditions.
   B. For worst summer conditions.
   C. For the average condition or conditions that represent the great majority of the building's operating hours.
   D. Frequently, an estimate of annual energy consumption is made.
IV. Selects the HVAC systems. Often, several systems will be used within one large building because orientation, activity, or scheduling differences may dictate different mechanical solutions. Especially common is one system for the all-interior zones of large buildings and another system for the perimeter zones.
V. Identifies the HVAC components and their locations.
   A. Mechanical rooms.
   B. Distributions trees—vertical chases, horizontal runs.
   C. Typical in-space components, such as under-window fan-coil units, air grilles, and so on.
VI. Sizes the components.
VII. Lays out the system. At this stage, conflicts with other systems (structure, plumbing, fire safety, circulation, etc.) are most likely to become evident. Because insufficient vertical clearance is one of the most common building coordination problems with HVAC systems (especially all-air systems—see Sections 8.2 and 8.5), the layouts must include sections as well as plans. Opportunities for integration with other systems also become more apparent at this stage: air ducts can also help distribute daylighting, act as sunshading devices, or fulfill other functions.

After the architect and the other consultants hold conferences in which HVAC system layout drawings are compared to those for other primary systems (structure, plumbing, electrical, etc.), *design finalizing* occurs. At this final stage, the HVAC system designer verifies the match between the loads on each component and the component's capacity to meet the load. Final layout drawings then are completed.

7.3 Equipment Location and Service Distribution

This chapter considers smaller buildings with one or a few thermal zones. For larger buildings, turn now to Chapter 8.

Smaller buildings are typically skin-load dominated: that is, for them, the climate dictates whether heating or cooling is the major concern. In some climates, only heating systems are needed; the building can "keep itself cool" during hot weather without mechanical assistance. In other climates, only cooling is needed. In still others, both heating and cooling are required.

(a) Central or Local? A skin-load–dominated building may have such differing but simultaneous needs that a room-by-room solution ("local") to heating, ventilating, and cooling is desirable. Consider the building with both north- and south-facing spaces on a cold sunny winter day; one side gets ample solar heat, while the other side needs additional heat. The advantage of local systems is an ability to respond quickly to individual rooms' needs.

The central system also has advantages: the equipment is contained within its own space rather than taking up space within each room, and maintenance can be carried out without disrupting activities within those rooms.

(b) Central Heating or Cooling Equipment. In the early design stages, an approximate size for the largest equipment is sometimes useful. Once the heating or cooling capacities are known, manufacturers' catalogs can be consulted for the dimensions of the heating and cooling equipment.

The critical decision in sizing the heating equipment is the *design temperature:* What is the lowest reasonable outdoor temperature for which a heating device can be sized if the desired interior temperature is to be maintained? These winter design temperatures are listed in Appendix A.

THERMAL CONTROL

When this is known, the next step is

design ΔT = inside temperature

$\qquad$ − outside design temperature

In Section 5.4, the criteria for Btu/DD ft² were listed in Table 5.5. To convert to the required capacity of a building's heating equipment, calculate

$$\frac{\text{Btu/DD ft}^2}{24 \text{ h}} \times \Delta T \times \text{ft}^2 \text{ floor area}$$

$$= \text{Btu/h heating capacity}$$

[Note: For passively solar heated buildings, the backup heating unit is sometimes sized for the heat loss of the *entire* envelope and sometimes for the envelope *minus the solar wall,* just as the criteria were defined in Table 5.5. Similarly, buildings with *reliable* internal gains (lights, equipment, people, etc.) are sometimes designed with smaller heating units because internal gains supply a constant portion of the space heating needs. For a more complete discussion, see Section 5.8.]

The sizing of cooling (mechanical refrigeration) units is not so straightforward, as was evident when detailed hourly heat gain procedures were presented (Sections 5.11 and 5.13). However, a *very approximate* early estimate can be obtained from the estimated hourly gains in Table 5.8. (Warning: This estimate is likely to be *lower* than that obtained using the peak heat gain hour, for which cooling equipment is often sized.)

sensible Btu/h cooling capacity

$$= [\text{approx. heat gain (Btu/h ft}^2)][\text{floor area (ft}^2)]$$

Another common unit used for sizing mechanical refrigeration is "tons" of cooling capacity, one ton being equivalent to the useful cooling effect of a ton of ice, or 12,000 Btu/h (3516 W). Therefore, the required capacity in tons is

$$\frac{\text{heat gain (Btu/h)}}{12,000} = \text{tons of cooling}$$

With this information about needed Btu/h (or Watts), equipment is then selected; physical size and service access areas are specified in catalogs. For a more general floor area estimate, see the sizing nomographs in Tables 8.3 and 8.4. Even more generally, equipment size may be estimated using these guidelines:

For ordinary equipment: 500 ft²/ton (46.5 m²/ton)

High-efficiency chillers: 1,000 ft²/ton (93 m²/ton)

(c) Distribution Trees. Central heating/cooling systems produce heating and cooling in one place, then distribute them to other building spaces according to respective needs. The distribution tree is the means for delivering heating and cooling: the "roots" are the machines that provide heat and cold, the "trunk" is the main duct or pipe from the mechanical equipment to the zone to be served, and the "branches" are the many smaller ducts or pipes that lead to individual spaces.

For now, the questions to be answered about distribution trees for buildings are: How many? What kind? and Where? A building can have one giant distribution tree, several medium-sized trees, or an orchard of much-smaller trees. At one extreme, a large mechanical room is the scene of all heating and cooling production; leading from this room is a very large trunk duct with perhaps hundreds of branches. At the other extreme, each zone has its own mechanical equipment (such as a rooftop heat pump), with short trunks and relatively few branches on each tree.

What kind of distribution tree? Most simply, air (ducts) or water (pipes). Air distribution trees are bulky and therefore likely to have major visual impacts unless they are concealed above ceilings, below floors, or within vertical chases. Water distribution trees consume much less space (a given volume of water carries vastly more heat than does the same volume of air at the same temperature) and can be easily integrated within structural members, such as columns. Both air and water trees can be sources of noise.

Where does the distribution tree fit in? On the exterior, it can lend a three-dimensional organizational structure to a facade. Exterior trees can take up a smaller amount of rentable floor space but require expensive cladding and are subject to considerable heat losses and gains, which could increase energy usage. Interior trees are often combined with other continuous vertical spaces, such as elevator shafts and stairways. If the choice is an exterior distribution tree, its potential contribution to facade performance should be considered. For example, the distribution tree might act as a sunshade or as a light reflector.

To carry the tree analogy to its logical conclusion, consider the "leaves," the points of interchange between the piped or ducted heating or cooling and the spaces served. One example is a large, bulky device such as a fan-coil unit on the exterior wall, below windows. In contrast, a perforated ceiling system with thousands of small holes acting as a widely spread grille is essentially invisible.

7.4 Controls for Smaller Building Systems

In the past, when buildings had mechanical equipment for heating only, thermostats were simple on–off devices; when they dropped below a setpoint, the heat was turned on. With the advent of mechanical cooling and then concerns about energy conservation, thermostats added a "deadband" separating heating need temperature from cooling need temperature, and automatic setback provisions that allow buildings to change internal temperature between occupied and unoccupied hours. ANSI/ASHRAE Standard 90.2-1993, *Energy Efficient Design of New Low-Rise Residential Buildings,* calls for thermostats capable of being set from 55°F to 85°F (13°C to 29°C) as well as an adjustable deadband, the range of which includes a setting of 10 F° (5.6 C°).

Building management systems (BMS) are now capable of regulating far more than temperature. They are capable of remote control, allowing systems to be activated in advance of the occupants' arrival; this is particularly useful for weekend and vacation homes, where heating/cooling and domestic hot water systems, if left on, could waste much energy during unoccupied periods. Door and window locks, security cameras, lighting, and appliances are potential partners in a comprehensive control system.

Future developments include "neural networks," where automation systems are capable of learning while being used. They thus predict usage patterns, adjusting in advance without needing specific commands from occupants. When building use pattern is highly predictable, as with many retail and commercial occupancies, these self-programming systems should learn very quickly how to anticipate needs while conserving energy.

In residences, usage patterns are likely more varied and less predictable. Experiments at the University of Colorado have revealed that even here, patterns may be more predictable than first thought; for details, see Mozer (1998). The Adaptive Control of Home Environments (ACHE) system has two objectives: to anticipate inhabitants' needs and to save energy. Lighting, air temperature, ventilating, and water heating are controlled so that just enough energy, just when needed, might be provided. Lights will be set to the minimum required, hot water maintained at the minimum temperature to meet demands, only occupied rooms will be kept at optimum comfortable temperature, and so on.

The control framework compares energy costs to "misery" costs. Minimum settings and their time patterns are developed as the building is occupied over time. Whenever a preset minimum is overruled by the occupant, the system "learns" and readjusts accordingly; however it occasionally "tests" lower minimum settings to be sure that energy conservation is not being unduly sacrificed to past desires for more.

7.5 Refrigeration Cycles

Unlike the slow and diffuse heat transfer processes that characterize passive heating and cooling approaches, mechanical equipment can rapidly concentrate heating or cooling, on demand. The refrigeration cycle is a particularly useful mechanical process in heating as well as cooling applications. The two types of heat transfer process commonly used in mechanical equipment for buildings are the *compressive* and the *absorption* refrigeration cycles. (Equipment that utilizes these cycles is discussed in the remainder of this chapter and Chapter 8.)

(a) Compressive Refrigeration. As applied in Fig. 7.1, the compressive refrigeration

Fig. 7.1 *Schematic arrangement of the compressive refrigeration cycle, providing chilled water for a building cooling system.*

cycle is a scheme for transferring heat from one circulated water system (chilled water) to another (condenser water). The means for doing this is the liquefaction and evaporation of a refrigerant, during which processes it gives off and takes on heat, respectively. The heat it gives off must be disposed of (except in the heat pump), but the heat that it acquires is drawn out of the circulated water known as the chilled water, which is the medium for subsequent cooling processes.

Refrigerants are a gas at normal temperatures and pressures, so must be compressed and liquefied to be of service later as a heat absorber. To be liquefied (see Fig. 7.1), the refrigerant must first be compressed to a high-pressure vapor; then, by means of cool water, latent heat is extracted from the refrigerant, which condenses it to a liquid. This high-pressure liquid is a potential heat absorber because, when it is released through an expansion valve, it springs back mechanically to gaseous form. In this change of state it must take on latent heat, by drawing heat out of the circulated water of the chilled water system.

It may be said that the refrigeration cycle pumps the heat out of the chilled water system into the condenser water system. Indeed, by special (reverse cycle) arrangements of the water systems, a *heat pump* is the result. (The compressive refrigeration cycle can be used to transfer heat between almost any media; whereas Fig. 7.1 illustrates a water–water cycle, Fig. 7.38 shows the heat pump in both air–air and water–air applications.)

The piston-type compressor in Fig. 7.1 can instead be one of several other types: *rotary, scroll,* or *screw* compressors, each with characteristics suitable to particular applications. See Section 8.3*b* for further discussion.

(b) Alternative Refrigerants. Unfortunately, the most commonly used refrigerants of the early 1990s are a threat to our atmosphere, because they can escape from the equipment as chlorofluorocarbon (CFC) gases. The threats are stratospheric ozone depletion and global warming. A related global warming threat is the CO_2 released from the production of electricity that powers chillers and other refrigeration machines. There is a potential conflict if lower-threat refrigerants are less efficient and therefore produce higher energy consumption. This conflict assumes continued power generation by fossil fuels; when photovoltaic (PV) is used to power these cooling devices, this aspect of the threat to global warming disappears.

Production of CFC refrigerants was banned in the United States by the middle of the 1990s. (A black market quickly developed for the widely used CFC refrigerants such as Freon.) Fears that replacement refrigerants would be less efficient have largely disappeared; combinations of better chillers and new refrigerants produce energy savings. The U.S. EPA estimated in 1998 that when 44% of the CFC chillers are replaced or converted in 2000, the resulting U.S. energy use will be reduced by 7 billion kWh/year.

Considerable efforts are under way both to reduce the likelihood of escaping refrigerant and to develop more efficient non-CFC refrigerants. The first alternative was hydrochlorofluorocarbons (HCFC) refrigerants—still a threat to our atmosphere, but better than CFC. HCFC still contains chlorine, a major influence on ozone depletion. Thus, HCFCs themselves are due to be phased out in the first decades after 2000. A longer-term replacement is hydrofluorocarbons (HFCs); even this improvement still threatens our atmosphere. Yet another possibility is natural hydrocarbons (HC). In general, comparing these two alternatives, HFCs have higher global warming potential and long atmospheric lifetimes, but low toxicity and are nonflammable. HCs have negligible global warming potential and short atmospheric lifetimes, but are flammable and explosive.

One alternative is ammonia, used in the early years of compressive refrigeration, but discontinued because of its acute toxicity and (relative to CFCs) its flammability. It is now returning in some absorption cooling cycles. Ammonia does have the advantage of its strong and unmistakable odor; it warns of its leakage, unlike the other refrigerants. Use of ammonia will require regular maintenance, good ventilation, and good access/escape routes. It will encourage systems that keep the refrigerant loop within controlled spaces such as mechanical rooms.

The search for the ideal refrigerant continues and may never end. This is another area in which rapid developments may be expected in the near future. For a comprehensive review of refrigerants, see James E. Calm, "Refrigerant Safety," in the *ASHRAE Journal,* July 1994.

(c) Absorption Refrigeration Cycle. This process is illustrated in Fig. 7.2. There are no CFCs or HCFCs used here; the process illustrated uses distilled water as the refrigerant, and lithium bromide as the absorber (or "salt solution"). In order to remove heat from chilled water, this cycle uses still more heat in regenerating the salt solu-

tion and is typically somewhat less efficient than the simple compressive cycle, and will need about twice the capacity for rejecting heat. However, the heat for salt solution regeneration may be provided by solar energy, or by relatively high-temperature waste heat from another source, such as steam or hot water, such as from a fuel cell. Because the high-grade energy (electricity) needed to run a compressor is replaced by the lower-grade heat needed to run the generator, the absorption cycle can enjoy an energy advantage over the compressive cycle, even though it is less efficient.

There are several variations on the absorption cycle. When a fuel such as natural gas is used as the heat for the generator, it is called "direct fired." When another heat source (such as "waste heat") is used, it is called "indirect fired." The relatively simple (!) cycle shown in Fig. 7.2 is a "single-effect" absorption cycle, using one heat exchanger between the strong and weak salt solution. A "double-effect" absorption cycle (powered by steam in Fig. 7.3) adds a second generator and condenser that operate at a higher temperature. It approximately doubles the efficiency of the single-effect cycle. A "triple-effect" cycle will, in turn, provide a 50% efficiency improvement over the double-effect cycle.

7.6 Cooling-Only Systems

(a) Fans. Before the advent of mechanical air conditioning, cooling was commonly achieved with simple air motion provided by fans. The summer comfort chart shown in Fig. 2.8 encourages increased air motion as a way to extend comfort into air temperatures in the mid-80s °F. As a general rule, people will perceive a 1 F° decrease in air temperature for every 15 fpm increase in that air's speed past the body (about a 1 C° decrease for every 1 m/s increase). Ceiling fans are often installed and run at slow speed to destratify warm air at the ceiling in winter; they can be run at higher speed in summer to provide added comfort through increased air motion. The air motion produced by ceiling fans will vary with fan height above the floor, the number of fans in a space, and the fan's power, speed and blade size. Figure

1 Evaporator and absorber

Consider two connected, closed tanks with a salt solution (lithium bromide) in one and water in the other. Just as common table salt absorbs water on a damp day, the salt solution in the absorber soaks up some of the water in the evaporator. The water remaining is thereby cooled by evaporation.

2 Evaporator coil and pump added

This refrigeration effect is utilized by putting a coil in the evaporator tank. Water from this tank is pumped to a spray header which wets the coil. The spray's evaporation chills water in the coil as it circulates to the refrigeration load. Solution pumped to spray in absorber raises efficiency.

3 Solution pumps and generator added

In an actual operating cycle, the salt solution is continuously absorbing water vapor. To keep the salt solution at proper concentration, part of it is pumped directly to a generator where excess water vapor is boiled off. The reconcentrated salt solution is returned to the absorber tank where it mixes with the solution sprayed to absorber in step 2.

4 Condenser and heat exchanger added

Water vapor boiled off from the weak solution is condensed and returned to the evaporator. A heat exchanger uses the hot, concentrated salt solution leaving the generator to preheat the cooler, weak solution coming from the absorber. Finally, condensing water circulating through the absorber and condenser coils removes the waste heat.

Fig. 7.2 *Absorption–refrigeration cycle. Refrigeration load is a building's chilled water supply system. Heat source for the generator can be "indirect fired" (steam, hot water, waste heat, or solar energy) or "direct-fired" (natural gas). (Adapted, courtesy of the Carrier Corporation.)*

7.4 shows expected air speeds with one 48-in. (1220-mm) ceiling fan in a typical residential living room.

In the hot summer climate of Davis, California, an experimental house sponsored by Pacific Gas & Electric, designed by the Rocky Mountain Institute, has eliminated a conventional cooling system through a series of alternative approaches. Fans play a major role: In addition to ceiling fans, a whole-house fan removes the hottest air from

Fig. 7.3 *Indirect-fired double-effect absorption cycle. Lithium bromide and water are used in this cycle. (Courtesy of the Carrier Corporation.)*

the central hallway, exhausting to the ventilated attic. More thermal mass (tile floors, double dry-wall), an attic radiant barrier, and low-ε, gas-filled windows are also used.

(b) Unit Air Conditioners. The device shown in Fig. 7.5 is perhaps the most commonly seen piece of mechanical equipment in the United States. Perched in windows in full view of pas-sersby, these window-box air conditioners noisily remind us that many of our buildings still are *not* centrally mechanically cooled. Mechanical cool-ing was considered a luxury until well after World War II.

Built-in, through-wall air conditioners offer a low first-cost way to provide separate zones for individual apartments, motel rooms, and so on (see Section 7.8). In noisy cities, the drone of these units masks street noise for the interior, thus potentially helping to promote relaxation. Unfortunately, such scattered units rarely afford the chance to conserve energy through the ex-change of waste heat or the higher efficiencies that can accompany larger equipment. However, if turned on only when cooling is needed (i.e., when people are present), they can provide sub-stantial savings over the larger always-on systems.

(c) Evaporative Cooling: Misting. As ex-plained in Section 5.12, the net effect of evapora-tive cooling is *no total change* in heat content (enthalpy) of the treated air; its DB temperature is *lowered,* but there is an *increase* in relative hu-midity. *People* feel cooler, although no change in total heat has occurred. One of the most direct approaches is a misting or fogging system, where a fine spray of water droplets is blown into the air. A common application is for small outdoor areas; the team benches of football stadiums, or refreshment pavilions. However, very large spaces in hot dry climates can also benefit; Fig. 7.6 shows mist descending from a skylight diffuser in the Atocha railroad station in Madrid, Spain. A large space using fogging (in a Michigan conservatory) is shown later in Fig. 8.52.

(d) Evaporative Cooling: Roof Spray. In the past, when roofs were poorly insulated, roof sprays were rather common ways to reduce heat gains. New variations promise new energy savings.

Roof color is the first consideration; white or near-white roofs are the first step toward energy savings, as was shown in Example 4.3, sol-air tem-perature. However, emissivity is also involved; the higher the emissivity, the faster a roof surface reradiates its heat to the sky. Table 7.1 compares solar absorptance, albedo (overall solar re-flectance), and emissivity for some common roofing materials. The combination of high albedo

THERMAL CONTROL

(a)

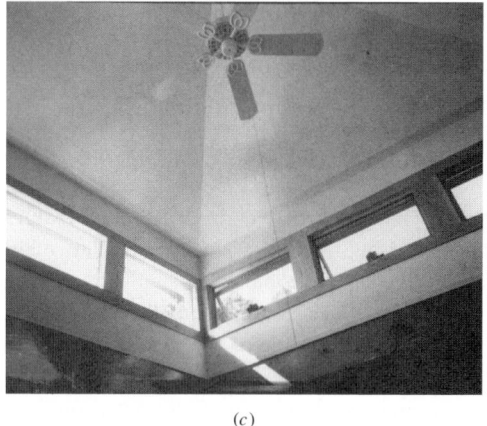

(c)   (b)

Fig. 7.4 *Ceiling fans are useful in heating and cooling. In winter, at slow speeds, they destratify warm air at the ceiling (see again Fig. 6.9, the Bedales School theater.) In summer (a) they extend the comfort zone by providing increased air motion. Room size is typical of residential living rooms. (Reprinted with permission of the American Society of Heating, Refrigerating, and Air-Conditioning Engineers, Inc. from 1997* ASHRAE Handbook of Fundamentals.) *(b) and (c) A ceiling fan is a visual feature in this North Carolina residence. (Christopher C. Morgan, AIA, Architect, Charlotte, North Carolina.)*

and high emissivity resists solar heat gain most effectively. (It also promises to reduce the heat island effect in urban areas; see Chapter 3.) A solar reflectance index (SRI) is under development to allow quick comparisons between roofing products. Currently, the SRI scale = 0 at the combination of albedo 5% and emittance 90% (about equal to black asphalt shingles). The SRI scale = 100 at albedo 80%, emittance 90% (about equal to T-EPDM).

The *night roof spray thermal storage system* (NRSTS) cools water on a roof by night, using both night sky radiation and evaporation. The water is then stored for use the next day in build-

ing cooling. The water can be stored either on the roof (below floating insulation, above a structural ceiling) or in a tank below the roof. When stored on the roof, it becomes a variant of the roof pond cooling system (Section 5.14f). When stored in a tank, the water can be circulated through a cooling coil to precool air before it enters the chiller-fed cooling coils. It is thus an assistance to mechanical cooling. Three variations are shown in Fig. 7.7.

At a Nogales, Arizona, border patrol station, a 6500 ft^2 (604 m^2) flat white roof was retrofitted with a NRSTS, utilizing a 10,000-gal (37,850 L) above-ground tank. Over the summer of 1997, it averaged 250 Btu/ft^2 day of cooling. Water use at

Fig. 7.5 *Diagram of a typical through-the-wall air conditioning (cooling) unit. Direct heat exchange occurs between air and the process of evaporation (on one side) and condensation (on the other) of the refrigerant. The unit is self-contained, requiring access to outside air and electricity to power the compressor and the two fans. Typical capacities are 1 to 2 tons (12,000 to 24,000 Btu/h, 3.5 to 7 kW).*

Fig. 7.6 *Mist is sprayed from a diffusing system located high in the Atocha railroad station in Madrid, Spain. It is psychological reinforcement of evaporative cooling in a hot, dry summer.*

TABLE 7.1 **Solar Performance of Roofing Materials**

Material	Absorptance (%)	Albedo (%)	Emittance (%)
White asphalt shingles	79	21	91
Black asphalt shingles	95	5	91
White granular-surface bitumen	74	26	92
Red clay tile	67	33	90
Red concrete tile	82	18	91
Unpainted concrete tile	75	25	90
White concrete tile	27	73	90
Galvanized steel (unpainted)	39	61	4
Aluminum	39	61	25
Siliconized white polyester over metal	41	59	85
Kynar white over metal	33	67	85
Gray EPDM	77	23	87
White EPDM	31	69	87
T-EPDM	19	81	92
Hypalon	24	76	91

Source: Lawrence Berkeley National Laboratory.

various sites has averaged 4 to 5 gallons/h (15 to 19 L/h) per 1000 ft^2 (per 93 m^2) of roof area. Details are described in Bourne and Hoeschele (1998).

(e) Evaporative Coolers. These are also affectionately termed "swamp coolers" and "desert coolers" and are familiar devices in hot, arid climates (Fig. 7.8). (They are used in other climates for special, high-heat applications such as restaurant kitchens.) They require a small amount of electricity to run a fan, and some water to increase the relative humidity of the air they supply to the building. Their process of cooling is explained in Section 5.14d.

The typical evaporative cooler shown in Fig. 7.8 needs full access to outdoor air and is thus often set on the roof; through-the-wall units are also available. Great quantities of dry, hot outdoor air are blown through pads kept moist by recirculated and makeup water. The "cooled" air is then delivered to the indoor space. The effect of the gently moving cool air is to cool the body and also to produce further cooling by evaporation of body moisture.

Air introduced into the indoor space must then be exhausted for the system to operate properly. By selecting the room through which the air is exhausted, one can route cool air as desired in any chosen path from unit to relief opening. The closer to the relief opening, the warmer the indoor air.

(f) Indirect Evaporative Cooling. The preceding discussion concerns the most simple application of evaporative cooling, known as the *direct* process (once-treated outdoor air directly introduced to the space). Unfortunately, some areas have such hot summer daytime conditions that such simple evaporative approaches cannot produce comfort indoors. *Direct and indirect* processes have been combined to achieve more "real" cooling and better indoor comfort conditions; a psychrometric process diagram was shown in Fig. 6.16.

One of several such approaches is shown in Figs. 7.9 and 7.10. Warm, rather dry night outside air is evaporatively cooled and fed into a rock bed. The air's temperature is low enough to cool the rock bed, and its relative humidity is moderate. (At the same time, the house is directly evaporatively cooled by a second cooling unit.) Figure 7.10 traces the process by day. Extremely hot, dry outdoor air (A) is drawn into the rock bed, where it is cooled by contact (D). It can then be passed through an evaporative cooler, to achieve a better combination of rh and DB temperature (E). After picking up both sensible and latent heat, the air

(a)

(b)

(c)

Fig. 7.7 *Night roof spray thermal storage systems (NRSTS).* (a) *This version approximates the perfor-mance of the roof pond (see Section 5.14f).* (b) *Remote water storage allows use of the cooled water at any time.* (c) *The floor slab is used as thermal storage in this version. [From Technical Installation Review (December 1997):* WhiteCap Roof Spray Cooling System. *Federal Energy Management Program, U.S. Department of Energy.]*

Model	Dimensions (in.)			hp	Air Delivery	
	H	W	D		cfm at 0.1 in.	cfm at 0.5 in.
AS50, AD50	28	42	45	1/3	2650	1600
				1/2	3110	2130
				3/4	3630	2806
AS70, AD70	35	42	48	1/2	3750	2610
				3/4	4350	3580
				1	4820	4100

Fig. 7.8 *The UltraCool evaporative cooler has a low profile and uses internal components of plastic or stainless steel to reduce corrosion. Where ducts are short, use friction of "cfm at 0.1 in." (Courtesy of Champion Cooler Corporation, El Paso, Texas.)*

Fig. 7.9 *Direct–indirect evaporative cooling, utilizing a rock bed for storage and heat exchange. (Reprinted by permission of the Environmental Research Laboratory, University of Arizona.)*

Fig. 7.10 *Comparing direct (A to B to C) and indirect cooling (A to D to E to F) on the psychrometric chart.*

is exhausted (at approximately temperature *F*). Note that at condition *F*, it is still much cooler than outdoor air.

By comparison, simple direct evaporative cooling by day would have produced indoor supply air too hot and humid for comfort (*B*). Again, upon exhaust (*C*), it is still cooler than outdoor air.

Indirect evaporative cooling is combined with a direct refrigerant system in an innovative tent structure over a San Francisco department store (Fig. 7.11). Two layers of fiberglass, Teflon-coated fabric, separated by an average of 12 in. (300 mm), are supported by a network of cables hung from eight masts. The tent roof covers about 70,000 ft² (6,500 m²) of sales floor; its 7% translucency to sunlight provides 450 to 550 fc of daylight. This greatly reduces the need for electric lighting, although some, clipped onto exposed fire sprinkler pipes, is still used as accent lighting. About 3.5 W/ft² of solar gain, mostly in the form of this diffused daylight, penetrates the tent cover. When this solar gain is combined with heat gains from people and electric accent lights, a cooling load is always generated in San Francisco's mild climate, which rarely falls below 45°F (7°C). The roof's relatively high *U*-value is thus advantageous in helping lose heat. Because San Francisco overheats even more rarely, such a low resistance to heat flow is not seriously disadvantageous in summer.

To remove this heat, four sets of direct refrigerant equipment are provided at the tent perimeter, on grade. These feed into a perimeter plenum,

from which the entire store is supplied with cooled air. The exhaust air is collected at the center and returned to help with the task of cooling. Indirect evaporative cooling (also called sensible evaporative refrigeration) units are used to cool air (see Fig. 7.11*b*). In this application, two units work in tandem. Air to be cooled (supply air) enters the heat exchanger of the first unit, which is cooled by evaporative cooling of outside air. During peak temperature periods, the supply air is only somewhat cooled by this process; sufficiently low temperatures for use on the interior are obtained by passing it through the second unit's heat exchanger. This second unit is cooled by evaporative cooling of the *exhaust* air from the store, which is cooler than outside air under summer conditions. Thus the exhaust air does some work beyond the direct cooling of the tent's interior. This two-stage, indirect evaporative cooling process allows the supply air to be cooled without the relative humidity being raised, as would be the case if direct evaporative cooling were used.

7.7 Heating-Only Systems

There are substantial areas of North America in which the summers are so mild, but winters so cold, that heating systems are installed, but not cooling systems. This is most common for residences and small commercial buildings.

Where should heating devices be placed within a space? The relative comfort in heated rooms from entire ceilings or floors as a heat source is discussed in Chapter 2; Fig. 2.9 shows dissatisfaction with increased air velocity (the warmer the air, the higher velocity was tolerated), Fig. 2.10 shows dissatisfaction with asymmetric radiation (warmer ceilings less tolerated), and Fig. 2.11 shows dissatisfaction with vertical air temperature difference (cold floors were less tolerated). Figure 7.12 shows why designers usually locate heat sources below windows, despite the fact that warmer temperatures just inside an exterior wall will drive more heat through the wall in cold weather (heat loss = $U \times A \times \Delta T$). As windows and walls become better insulated, their interior surface temperatures rise, and the need for heat at the edge grows less. Indeed, with super-insulated components, the need for space heat disappears

(a)

(b)

Fig. 7.11 *Bullock's Department Store, San Mateo, California.* (a) *The eight-masted white fabric roof, highly reflective to ward off solar heat, transmits about 7% of daylight to provide ample diffuse daylight for sales areas. (Photo by Steve Proehl. Environmental Planning and Research, Inc., architects; Giampolo and Associates, Inc., mechanical and electrical engineers.)* (b) *Cooling is provided by a two-stage, indirect evaporative cooling system, which uses much less energy than does conventional compressive refrigeration.*

because internal gains from the sun, lights, appliances, and occupants can heat the space.

(a) Wood Heating Devices.

After the sun, the most ancient method of heating is the radiant effect of fire. With each step from campfire to

fireplace to wood stove, more of the fuel's heat was captured for the room rather than wasted to the outdoors (Fig. 7.13). Although many people enjoy the sight, sound, and smell of the open fireplace, the tightly enclosed wood stove with catalytic combustor is a substantially more efficient

(a)   (b)

Fig. 7.12 *Locating a heat source near an interior wall (a) encourages a cold draft along the floor in winter. Below a window (b) it evens the temperature throughout the room, but also loses more heat through the window.*

and less polluting approach to heating. Indoor as well as outdoor air pollution is a serious issue with fireplaces and stoves. Combustion generates carbon monoxide, breathable particulates, and at times, nitrogen dioxide. Wood smoke can cause nose and throat irritation; it can remain in the lungs, it can trigger asthmatic attacks. Keeping a clean chimney, burning small and hot fires rather than large and smoky ones, using seasoned wood, and insuring adequate ventilation to the wood-burning device are strategies to minimize pollution and risk.

Open fireplaces may be lovely to look at, but the amount of air exhausted up the chimney can quickly cause more heat losses than heat gained from the fire. The colder the outside air, the greater the net heat loss. Masonry mass around fireplaces can store and release some heat; for energy conservation, the mass should be surrounded by the building, rather than located on an exterior wall. ANSI/ASHRAE Standard 90.2-1993, *Energy Efficient Design of New Low-Rise Residential Buildings,* requires fireplaces to have a tight-fitting damper, firebox doors, and a source of outside combustion air within the firebox.

Wood stoves are available in a wide variety of styles and are made of several materials. The sizes of such stoves are often difficult to determine. Manufacturers rarely specify the Btu/h output, which depends on the density, moisture content, and burn time of the wood fuel. Wood that has been split, loosely stacked, and covered from rain for at least 6 months should achieve a moisture content of about 20% by weight. The following sizing procedure assumes no more than this 20% moisture content. For more details, see *Alternate Sources of Energy Magazine* 35 (1978).

Fireplace
Large front opening

Efficiency factor:
0.1 - Poor

Nonairtight stove
Door over opening but air leaks in around body parts

Efficiency factor:
0.2 - Low

Almost airtight stove
Air leaks in around door only

Efficiency factor:
0.35 - Better

Catalytic Converter reduces pollutants

Airtight stove
Air enters through a controlled damper only; door is gasketed

Efficiency factor:
0.5 - Best

Fig. 7.13 *Wood-burning devices have substantially increased in efficiency since the open fireplace. (Copyright © 1978 by* Alternate Sources of Energy, *Issue 35.)*

The formula for the hourly heat output to a room from a wood stove is

$$\text{Btu/h} = \frac{(V)(E)(D)(7000)}{T}$$

where

V = useful (loadable) volume of the stove (ft³)

E = percent efficiency, expressed as a decimal (<1.0); see Fig. 7.13

D = density of the wood fuel (Table 7.2)

T = burn time (h) for a complete load of firewood; usually assumed at 8-h minimum

7000 = Btu/lb of firewood, 20% moisture content

("Bone-dry" wood can be assumed to have 8600 Btu/lb.) Note that a *drop* in burn time *increases* the Btu/h output; it is evident that when the air supply is increased to the stove, the fire burns hotter, consuming the wood more quickly. To meet the design heat loss (worst condition) for a room, burn times of 8 to 10 h should be assumed. Stoves rarely need relighting with a 10-h burn time.

Pellet stoves were introduced in 1984 and have several desirable characteristics. The pellets are made from densified quality sawdust, a manufac-

TABLE 7.2 Approximate Average Wood Density

Type	Density (lb/ft³)[a]
Shagbark hickory	40.5
White oak	37.4
Red oak	36.2
Beech	36.2
Sugar maple	34.9
Yellow birch	34.3
White ash	33.7
Black walnut	31.2
American elm	28.7
Spruce	25.6
Hemlock	23.7
Aspen	23.1
Red cedar	18.7
White pine	17.5

Source: Alternate Sources of Energy Magazine, #35, © 1978. Reprinted by permission.

[a]These values are approximate; they vary a great deal.

turing by-product. The form and content of this fuel produces a highly efficient burn with less pollution emitted. The fuel is cleaner and takes less storage space than cordwood; an electric auger automatically feeds fuel into the burnplace to maintain a fire. From 10,000 to 50,000 Btu/h (2930 to 14,650 W) are produced, depending on the model and operating settings.

Wood stoves are frequently used as the sole mechanical heat source for an entire building, such as a residence or a small commercial building that is passively solar heated. Because radiant heat is the dominant form of heat output, the areas that "see" the stove will get most of the benefit. However, "circulating" stoves convert a larger portion of their heat to convected heat, which produces a layer of hot air at ceiling level. By providing a path between rooms at the ceiling, this hot air will slowly spread throughout a building; it also easily finds its way upstairs, because warm air rises. The more thermally massive the ceiling construction, the longer it will store and reradiate the heat from this warm air mass.

The flue leading from a wood stove carries very hot gases that are a potential source of heat (and pollution). The flue can be exposed to a space, making its radiant heat available, or simple heat exchangers can be constructed (such as the preheating of domestic hot water). More elaborate heat recovery devices—for boiler flue heat recovery—are discussed in Section 8.3.

Catalytic combustors, a recent development, reduce the air pollution from wood burning. These devices are honeycomb-shaped, chemically treated disks, as much as 6 in. in diameter and 3-in. thick (150-mm diameter, 75-mm thick). They are either inserted in the flue or built into the stove itself. When wood smoke passes through the combustor, it reacts with the chemical and ignites at a much lower temperature; this causes gases to burn that otherwise would have gone up the flue. The result is more heat produced, less creosote buildup in the flue, and fewer pollutants in the atmosphere. Like the catalytic converters in autos, these devices impose limits on the fuel: plastic, colored newsprint, metallic substances, and sulfur are ruinous to combustors, which means that the stove must be used as a wood burner, not a trash incinerator.

Wood stoves have a larger impact on building

design than do most other heating devices. Either noncombustible materials must be placed below and around them, or minimum clearance to ordinary combustible building materials must be provided. Furniture arrangements and circulation paths must be designed with the very hot stove surfaces in mind. Hot spots occur near the stove; cold spots occur whenever visual access to the stove is blocked. Thermally massive materials near the stove are advantageous in leveling the large temperature swings that can accompany the on–off cycle of the stove; this affinity for thermal mass has made the wood stove a popular choice for backup heat in passively solar-heated buildings. Finally, the amount of space required for wood storage should not be overlooked; recall the impact of the wood storage space on the house shown in Fig. 1.2. A covered, well-ventilated, easily accessible, and quite large space is optimum.

Masonry heaters overcome many of the metal wood stoves' disadvantages. Their footprint is rather small compared to their height; typically, they are used to heat the entire building (such as a residence). An inner vertical firebox supports a hot, clean burn, resulting in efficient combustion; combustion gases then flow downward in outer chambers, transferring heat to exterior masonry surfaces. In Finnish masonry heaters, this is termed *contraflow* (Fig. 7.14). Cool air at the floor of the room flows upward as it is heated by contact with this masonry; the temperature difference between masonry and air remains fairly constant, with highest temperatures at the top. Heat is gentle and even; dangerously hot surfaces are avoided. Fires may be built at 6 PM, and combustion is completed by a family's bedtime; the heat continues to radiate all night, but no fire is burning while people sleep. Research at Finland's Tampere University of Technology has resulted in optimum masonry heater designs, described in Barden and Hyytiainen (1993).

(b) Electric Resistance Heaters. These common devices carry the disadvantage of using high-grade energy to do a low-grade task, as shown in Fig. 1.11. Their advantages, however, are impressive: low first cost and individual thermostatic control that can easily be used to make each room a separate heating zone. Thus, the energy wasted at the electricity-generating plant

Warmed room air rises

Heating smoke sinks as it cools   To the chimney

Fig. 7.14 *The Finnish "contraflow" masonry heater.*

(usually, 60% to 70%) can be partially "recovered" at the building, where unused rooms can remain unheated. A few of the many types of electric resistance heaters are shown in Fig. 7.15. As in the case of metal wood stoves, surfaces can sometimes reach high temperatures, requiring care in the location of heaters relative to furniture placement, draperies, and traffic flow. Electric heaters are sized by their capacity in kilowatts (1 kW = 3413 Btu/h). The maximum watt density allowed is 250 W per linear foot of heater (820 W per linear meter).

(c) Gas-Fired Heaters. These are often found in semi-outdoor locations such as loading docks and repair shops. They are fired with either natural gas or propane. When vented, they can be used in more traditional environments such as the retail store in a remodeled warehouse (Fig. 7.16). Their advantage is that they heat surfaces first rather than air, so that comfort is obtained without the need for such high air temperatures.

NATURAL CONVECTION UNITS
Heating units for wall mounting, recessed placement or surface placement are made with elements of incandescent bare wire or lower temperature bare wire or sheathed elements. An inner liner or reflector is usually placed between elements so that part of the heat is distributed by convection and part by radiation. Electric convectors should be located so that air movement across the elements is not impeded. Small units with ratings up to 1650 W operate at 120 V. Higher wattage units are made for 208 or higher voltages and require heavy duty receptacles.

BASEBOARD HEATER (Wall Mounted)
L 24″ to 120″
D 2″ to 8″
H 4″ to 12″
CAP 300 W to 4000 W

CABINET CONVECTOR (Surface Mounted or Recessed)
L 24″ to 96″
D 3″ to 8″
H 11″ to 32″
CAP 1000 W to 4000 W

FLOOR HEATER (Recessed)
L 14″ to 108″
W 5″ to 8″
H 8″ to 11″
CAP 300 W to 2000 W

HYDRONIC BASEBOARD (Floor Mounted)
L 23″ to 107″
D 3″ to 6″
H 9″ to 12″
CAP 300 W to 2000 W

NATURAL CONVECTION UNITS

RADIANT HEATING
Heat is produced by a current that flows in a high resistance wire or ribbon and is then transferred by radiation to a heat absorbing body. Manufacturer's recommendations for clearance between a radiant fixture and combustible materials or occupants should be followed.

INFRARED HEATER (Pendant Mounted) Circular heat lamp is available
L 14″ to 86″
W 4″ to 12″
H 3″ to 16″
CAP 500 W to 7000 W

RADIANT HEAT PANEL (Surface Mounted or Recessed) Decorative murals are available
L 48″ to 144″
W 24″ to 48″
D 1″
CAP 500 W to 1000 W

RADIANT CEILING WITH EMBEDDED CONDUCTORS
Dimensions and capacity vary with coverage

RADIANT HEATING UNITS

FORCED AIR UNITS
Unit ventilators and heaters combine common convective heating with controlled natural ventilation.

Unit ventilators are most often mounted on an outside wall for air intake and at windowsills to prevent the down draft of cold air.

CEILING HEATER (Recessed) Circular unit with light is available
L 10″ to 14″
W 8″ to 14″
D 4″ to 8″
CAP 500 W to 1500 W

UNIT HEATER (Bracket Mounted)
W 12″ to 52″
D 6″ to 22″
H 12″ to 26″
CAP 1.5 KW to 50 KW

UNIT VENTILATOR (Surface Mounted or Recessed)
L 48″ to 104″
D 11″ to 26″
H 26″ to 32″
CAP 1 KW to 36 KW

WALL HEATER (Recessed)
W 10″ to 18″
D 2″ to 6″
H 9″ to 24″
CAP 750 W to 4000 W

PORTABLE HEATER
W 10″ to 72″
D 2″ to 12″
H 7″ to 24″
CAP 500 W to 5000 W

FORCED AIR UNITS

Fig. 7.15 *Varieties of electric resistance heating units. (Reprinted by permission from AIA, Ramsey, and Sleeper,* Architectural Graphic Standards, *9th ed., © 1994 by John Wiley and Sons.)*

When high rates of air exchange are expected, high-intensity radiant heaters are often used.

Radiant heaters should be sized by the surface temperature change they produce. Many manufacturers specify this surface ΔT for specified mounting heights and angles relative to the surface to be heated.

Gas fired baseboard heaters are also available, using either natural gas or propane. They heat both by convection and radiation, as do their electric counterparts. At a steady-state efficiency of 80%, they use a lower grade resource (than electricity) to do this lower-grade task. They are direct

vented (using a built-in fan) to the outside, and therefore must be installed on or near an exterior wall; vents are 1½-in. diameter, maximum 19-in. length (38-mm diameter, 483-mm length). With a cross section of 9-in. high by 5-in. deep (230-mm × 130-mm), a 48-in. (1220-mm) length will deliver 5800 Btu/h (1700 W); a 72-in. (1830-mm) length will deliver 9400 Btu/h (2755 W).

(d) Ceiling Electric Resistance Heat.
Ceilings can be constructed to include electric resistance in wiring (Fig. 7.17). Because the ceiling is not touched, it can be safely heated to a rather

(a)

(b)

Fig. 7.16 *A high-intensity infrared heater adds to the "historic" atmosphere of this retail store in an old warehouse* (a). *Exposed mechanical equipment includes the chains that operate the clerestory windows. The vented gas heater has an adjustable reflector that enables its radiant heat to be directed. (Clark-Ditton Architects, Eugene, Oregon.)* (b) *Vented gas high-intensity infrared heaters are available in both straight-line and U-shaped units. (Courtesy of Solaronics, Inc., Rochester, Michigan.)*

high temperature. The primary disadvantage of ceiling heat is that hot air stratifies just below the ceiling, so that air motion is discouraged; remember also that Figs. 2.10 and 2.11 predict more discomfort with the warmer ceiling. Finally, the wires are hidden within the ceiling surface, and unwary occupants can puncture wires while installing hooks or additional luminaires.

(e) Hot Water Boilers. The remaining choices for whole-building heating systems are discussed in the following order: (1) hot water and (2) forced air. Such systems include a fuel, a heat source, a "mover" (such as pump or fan), a distribution system, a heat exchanger or terminal within the space, and a control system.

Hot water boilers are rated according to heating capacity by several different categories. *Heating capacity* is the rate of useful heat output with the boiler operating under steady-state conditions, often expressed in *MBh* (1000 Btu/h). This "useful heat" assumes that the boiler is within the heated envelope of the building; thus the heat that escapes from the boiler walls is available to help heat the building. *AFUE,* the annual fuel utilization efficiency, is defined as 100% minus the losses up the stack during both the on and off cycles, and the losses due to infiltration of outdoor air to replace the air used for combustion and for draft control. Finally, the *net I=B=R rating* is published by the Hydronics Institute Division of GAMA. The net *I=B=R rating* load is lower than the heating capacity rating, because it consists only of the heating to be delivered to the spaces, and excludes the heat loss of the boiler itself.

Select a boiler whose rating matches the calculated critical heat loss of the house or building; too small a boiler will result in lower indoor temperatures at design conditions; too large a boiler costs more and is a waste of space. When using AFUE to select an efficient boiler, take care to see that the assumptions about "inches of water draft" and percentage CO_2 are similar for the boilers being compared. Minimum AFUE are specified both in ANSI/ASHRAE Standard 90.2-1993, *Energy Efficient Design of New Low-Rise Residential Buildings,* and the latest version of BSR/ASHRAE/IESNA 90.1, *Energy Standard for Buildings Except Low-Rise Residential Buildings.*

Boilers and their accessories comprise a wide inventory. A few selected types are discussed as follows:

1. *Oil-fired steel boiler.* A refractory chamber receives the hot flame of the oil fire. Combustion continues within the chamber and the fire tubes. Smoke leaves through the breeching at the rear. Water, *outside* the chamber, receives the heat generated in the combustion chamber. If a domestic hot water coil is connected for use, a larger capacity boiler is selected. An aquastat (water thermo-

Fig. 7.17 *Radiant heating electrical cable is installed prior to completion of a plaster ceiling.*

Fig. 7.18 *Gas-fired cast iron sectional boiler for hot water heating. Very high operating efficiency is possible; no chimney is required, as lower temperature exhaust gases can be vented through the wall to the exterior.*

stat) turns on the burner whenever the boiler water cools off, thereby maintaining a reservoir of hot water ready for heating the building.

2. *Gas-fired cast iron hot water boiler* (Fig. 7.18). Cast iron sections contain water that is heated by hot gases rising through these sections. Output is related to the number of sections. Additional heat is gained from a heat extractor in the flue. With induced draft combustion, condensing unit in the flue, and intermittent electronic ignition instead of a pilot light, up to 90% AFUE is attainable. The American Gas Association (AGA) sets standards for gas-fired equipment.

3. *Oil-fired, cast iron hot water boiler.* Primary and secondary air for combustion may be regulated at the burner unit. Flame enters the refractory chamber and continues around the outside of the water-filled cast iron sections.

(f) Hot Water Baseboard and Radiator Systems.

Hot water heating circuits that serve baseboards or radiators come in four principal arrangements. Figure 7.19a shows the series loop system, usually run at the building's perimeter. The water flows to and *through* each baseboard or fintube in turn. Obviously, the water at the end of the circuit is a little cooler, but because in all hot water systems the water temperature drop seldom exceeds 20 F° (11 C°) in residences, the *average* temperature can usually be used to select the baseboard or other elements. Valves at each heating element are not possible, because any valve would shut off the entire loop. Adjustment is by a damper at each baseboard, which reduces the natural convection of air over the fins. This is a "one-zone" system—all elements on, or off, together. There is no general rule about the maximum allowable length of water circuit, but for long runs, the pipe size can be increased, or *several* loops used in parallel to create more than one thermal zone.

The one-pipe system shown in Figs. 7.19b and 7.20 is a very popular choice. Special fittings act to divert part of the flow into each baseboard. A valve may be used at each one to allow for reduced heat or for a complete shutoff, to conserve energy—an advantage that the loop system does not provide. The one-pipe system uses a little more piping and thus is not as economical to install as the loop system, in which piping is minimal. Again, the supply water temperature will be lower at the end of the run than at the beginning.

The two-pipe reverse–return shown in Fig. 7.19c provides the same supply water temperature to each baseboard or radiator, because it is not cooled either by passing through a previous baseboard or accepting the cooler return water. Equal friction, resulting in equal flow, is achieved through all baseboards (numbers 1 to 5) by *reversing* the return instead of running it directly back to the boiler. This equality is effected by equal lengths of water flow through any baseboard together with its lengths of supply-and-return main. More pipe is required for this system than for the systems shown in Fig. 7.19a or 7.19b.

Figure 7.19d shows an arrangement that is not usually favored because the path of water through baseboard number 1 is much shorter than that through the others, and especially number 5. Baseboard number 5 could easily be undesirably cool, because it is short-circuited by the others.

Pipe expansion requires expansion joints in long runs of pipe and clearance around all pipes passing through walls and floors. Each time a hydronic system changes from room temperature to a heated condition, the piping will undergo the following expansion, assuming 70°F (21°C) initial temperature:

Water Temperature (F)	Pipe Expansion per 100 ft	
	Iron Pipe (in.)	Copper Tubing (in.)
160	0.7	1.0
180	0.9	1.3
200	1.0	1.5
220	1.2	1.7

An *air cushion tank,* compression tank, or expansion tank is a closed tank containing air, usually located above the boiler. When the water in the system is heated, it expands, compressing the air trapped in the tank. This tank allows for the usual range of temperatures within the system, including temperatures above the usual boiling point of water, without the frequent opening of the pressure relief valve. One type of air cushion tank, called a *diaphragm tank,* separates the air and water with an inert flexible material; this prevents reabsorption of the air by the water.

THERMAL CONTROL

Fig. 7.19 *Diagrams of hydronic distribution options, seen in plan. Baseboard convectors are shown here. Controls are not shown.*

For a conventional (unpressurized) air cushion tank, allow 1 gallon of tank capacity for every 5000 Btu/h (1 L for every 385 W) of the total system heat loss. For a pressurized tank of at least 8 lb/in² (55 kPa), allow 1 gallon of tank capacity for each 7000 Btu/h (1 L for each 540 W); or see manufacturer's recommendations.

Air vents and water drains are part of the distri-bution system. Except for the necessary air cush-ion in the upper part of the compression tank above the boiler, air must not be allowed to accu-mulate at high points in the piping or at the con-vector branches. Air vents at all high points re-lieve these possible air pockets that would otherwise make the system air-bound and inoper-ative.

Fig. 7.20 *One-pipe hydronic system—type* (b) *of Fig. 7.19.*

If a system is to be drained and left idle in a cold house, water trapped in low points could freeze and burst the tubing or fittings. Operable drain valves must be provided at such locations and, of course, at the bottom of the boiler, as shown in Fig. 7.21.

Hydronic and electrical controls allow automatic operation, described in Fig. 7.21. There are two options for system control:

1. As in Fig. 7.21, the thermostat controls the circulating pump and the boiler. In colder weather, the system operates almost continuously, and the average temperature in the system gradually rises.
2. The thermostat controls only the boiler, and the circulating pump operates continuously. This uses more energy for the pump but minimizes system temperature variation, and thus the possibility of expansion noises.

Makeup water is added as required, the air level in the tank is regulated by the air control fittings, and the circulator and burner operate as controlled by the aquastat and thermostat. If air vents in the *piping* are not automatic, they will require periodic manual "bleeding" of unwanted air.

Circulating pumps are used to overcome the friction of flow in the piping and fittings and to deliver water at a rate sufficient to offset the hourly heat loss of the house or building.

Pipe insulation is required whenever the pipes are outside the heated envelope of the building; ANSI/ASHRAE Standard 90.2-1993, *Energy Efficient Design of New Low-Rise Residential Buildings,* specifies insulation appropriate to the temperature of the heated water.

(g) Radiant Panels. Radiant floors have several comfort advantages over radiant ceilings (see again the introduction to Section 7.7). The components are largely the same as in the baseboard/radiator systems, except that now whole coils of pipes replace the individual radiators and baseboards. A balancing valve should be installed on each coil. Where uninsulated spaces underlie floors (or are above ceilings), special attention should be paid to adequate insulation, because the radiant panel will generate especially high temperatures, thus a very high ΔT through the floor (or ceiling).

THERMAL CONTROL

Fig. 7.21 *An oil-fired boiler and its hydronic and electrical controls.*

Thanks to higher insulation, today's buildings often require panels smaller in area than the floor (or ceiling) area available. In a conventional radiant panel system, the panel is placed nearest the exterior walls, where the heat loss is greatest. In a solar-heated building, this is not so clear. If the panel heats the floor surface just inside south-facing glass, how much warming will be left to the sun? A preheated slab will absorb much less solar radiation. However, in cloudy cold weather, the area closest to the south glass could become uncomfortably cool.

In the past, copper tubing was widely used for the heating coils. This typically involved a number of connections within a coil; each represented a

potential point of failure over the life of the panel. Today, coils are typically one-piece, and made of synthetic materials such as cross-linked polyethylene tubing. When the floor is concrete or other cast-in-place material, the coil is either directly embedded in the slab (tied down to resist floating during the pour) or the coil is stapled to an underfloor, over which the slab is poured. Radiant floors with coils underneath wood floors are increasingly popular.

Rugs or carpets over radiant floors are a mixed blessing; they are soft on the feet, but interfere with the exchange of heat. Special undercarpet pads can help in heat transfer; higher water temperatures might be used in the coils, because

(A) *Compression tank.* Accommodates the expansion of the water in the system.

(B) *Air control fittings.* Vent out unwanted air in the boiler and maintain the level in the compression tank.

(C) *Pressure relief valve.* Usually set for 30 psi. Initial cold pressure about 12 psi. Relieves excessive system pressure.

(D) *Oil burner.* Responds to aquastat or thermostat.

(E) *Stack temperature control.* Senses stack temperature and stops oil injection if ignition has not occurred.

(F) *Drain valve.* At low point in the water system.

(G) *Aquastat.* Maintains temperature of boiler water by starting the oil burner when temperature of water drops below the aquastat's setting. Sometimes set at about 180 F.

(H) *Remote switch.* At a safe distance from the boiler so that the plant can be turned off in case of trouble during which the boiler cannot be approached.

(I) *Junction box and relays.* General control center.

(J) *Thermostat.* When the room temperature drops below its setting, it turns on both the oil fire and the circulating pump.

(K) *Electrical power source.* Operates from a separate individual circuit at the power panel.

(L) *Hot water supply.* Copper tubing to convectors or baseboards.

(M) *Hot water return.* Copper tubing from convectors or baseboards.

(N) *Draft adjuster.* Regulates the draft (combustion air) over the fire.

(O) *House cold water main.* From which water is fed automatically into boiler.

(P) *Flow control valves.* Prevent casual flow of water by gravity when the circulator is not running.

(Q) *Temperature pressure gauge.* Indicates water temperature and pressure. Sometimes supplemented by immersion thermometers in supply and return mains.

(R) *Pressure-reducing valve.* Admits water into the system when the pressure there drops below about 12 psi. Has a built-in check valve to prevent backflow of boiler water into the water main.

(S) *Shutoff valves.* Normally open. Can be closed to isolate the system and permit servicing of components.

(T) *Circulator.* Centrifugal circulating pump that moves the water through the tubing and heating elements.

Fig. 7.21 *(continued)*

skin contact with the floor is prevented by the carpet.

(h) Hydronic Heating Sizing. The calculations for the sizing of a water distribution system are presented in Section 10.11; they are based on the required flow and the friction in the piping. (For domestic water supply, another factor is the vertical distance through which the water must be raised. In these closed-loop heating systems, however, the weight of the cooler water falling back to the boiler essentially counterbalances the weight of the hot water being raised. Furthermore, gravity is helping the hot—lighter—water rise, and the cooler water fall.)

The key to pipe sizing is the overall required flow rate. Ordinarily, the temperature drop that occurs as the hot supply water gives up heat to the space (through the convector) is about 20 F° (11 C°) in residences; in commercial applications,

30, 40, or 50 F° (17, 22, or 28 C°) are also common, as recommended by the manufacturers of unit heaters and convectors. Because the entire building's design heat loss is overcome by this system, in I-P units,

total flow rate, gpm

$$= \frac{\text{design heat loss, Btu/h}}{20\,\text{F°} \times 60\,\text{min/h} \times 8\,\text{lb/gal} \times 1\,\text{Btu/lb F°}}$$

$$= \frac{\text{design heat loss}}{9600}$$

in SI units,

total flow rate, L/s

$$= \frac{\text{design heat loss, W}}{11\,\text{C°} \times 1\,\text{kg/L} \times 4180\,\text{W sec/kg C°}}$$

$$= \frac{\text{design heat loss}}{45980}$$

Fig. 7.22 *Typical pump capacity (or performance) curves, for four pumps used in hydronic heating systems. One foot of water = 2.99 kPa; 1 gpm = 0.0631 L/s. (Courtesy of the Hydronics Institute Division of GAMA, Berkeley Heights, New Jersey.)*

Then, by using Section 10.11, we can account for friction through piping and fittings and size the main supply and return pipes. The same procedure can be applied to branches, proportioned to the heat they must deliver. The total friction to be overcome in the most distant run is then converted from total psi to "feet of head," with each foot of head = 0.433 psi (the pressure exerted by a foot-high column of water). In SI, 1 ft of water = 2.99 kPa.

With both the friction of the system expressed in feet of water, or head, and the flow rate established, a pump can be selected. Figure 7.22 shows typical performance curves for four pumps. The designer enters the curve with desired flow rate, then selects the pump with a head capacity greater than or equal to the head required. Pump performance curves are provided by the manufacturer.

The critical choice, however, is not pipe size: It is relatively easy to distribute such small-diameter pipes within wall and floor/ceiling construction. Rather, the critical choice is *hot-water supply temperature:* the higher this temperature, the smaller the convector units that discharge the heat to each space. However, higher temperatures endanger occupants, who may suffer skin burns if they touch exposed parts of the convectors or distribution tree. Higher temperatures also can lead to steam within the boiler/distribution tree, although the system is under pressure and therefore the boiling

point is greater than 212°F (100°C). These systems are not designed to accommodate steam, and serious injury can sometimes result. A safer choice of average water distribution temperature is 180°F, even though these temperatures result in larger convectors. A slightly lower annual average energy consumption should accompany the lower distribution temperature.

Baseboard convector selection is then made from manufacturer's data tables, such as the one shown in Table 7.3. The two common baseboard types are *RC,* usually cast iron with a water-backed front surface and extended rear heating surface, and *finned tube,* a metal tube with extended surface in the form of fins, usually placed behind a metal enclosure.

EXAMPLE 7.1. What length of baseboard convector (Table 7.3) is necessary along a 20-ft (6-m) long living room wall? From heat loss calculations, the living room requires 9000 Btu/h (2635 W). The average water temperature is 180°F, and water flow is about 1 gpm (0.6 L/s).

SOLUTION. With 180°F supply water and 1 gpm flow rate, Table 7.3 shows that 580 Btu/h will be delivered for each lineal foot of baseboard convector. 9000 Btu/h ÷ 580 Btu/h ft = 15.5 ft. Choose a baseboard combination of 16-ft overall length, consisting of two 8-ft (finished length) pieces. Total active finned length will be about 15.5 ft. ∎

Panel (radiant ceiling or floor) heating design usually depends on a water temperature of 120°F (49°C) for heated floors and 140°F (60°C) for heated ceilings. An uncarpeted concrete floor slab using ¾-in. diameter pipe or tube on 12-in. centers (20-mm diameter on 300-mm centers) with an average water temperature of 120°F, will deliver *50 Btu/h-ft² (158 W/m²) of floor panel.*

A ceiling panel with nominal ⅜-in. tube on 6-in. centers (10-mm diameter on 150-mm centers), with an average water temperature of 140°F, will deliver *60 Btu/h-ft² (189 W/m²) of ceiling panel.*

To determine the area of heated panel, divide the room's design heat loss by the rate of panel heat delivery. Although radiant ceilings deliver more heat per unit area, they also discourage air motion because the warmest air rises to lie against the warmest surface. In contrast, at a radiant floor

TABLE 7.3 **Hydronic Baseboard Convectors**

Water Supply				Heat Delivered per Unit Length[b]	
Temperature		Flow Rate[a]			
°C	°F	L/s	gpm	Watts/m	Btu/h ft
60	140	0.06	1	307	320
		0.25	4	326	340
66	150	0.06	1	365	380
		0.25	4	384	400
71	160	0.06	1	432	450
		0.25	4	461	480
77	170	0.06	1	490	510
		0.25	4	518	540
82	180	0.06	1	557	580
		0.25	4	586	610
88	190	0.06	1	614	640
		0.25	4	653	680
93	200	0.06	1	682	710
		0.25	4	720	750
99	210	0.06	1	739	770
		0.25	4	778	810

Source: Slant/Fin Ltd. Mississauga, Ontario.
[a]Use flow rate of 0.6 L/s (1 gpm) unless flow rate is known to be ≥0.25 L/s (4 gpm).
Pressure drop is as follows: flow 0.6 L/s (1 gpm) = 3.76 mm/m (47 mil.in./ft).
flow 0.25 L/s (4 gpm) = 42 mm/m (525 mil.in./ft).
[b]Baseboard finished length will be 76 mm (3 in.) longer than the length required to meet the heating need.

the coolest air drops to contact the warmest surface, is then warmed, and rises to be cooled at the ceiling, drops to the floor, and repeats this cycle continually.

Each panel will contain one or more coils. In floor panels, each coil delivers 10,000 Btu/h (2930 W) and should be no longer than about 200 linear feet (60 m). In ceiling panels (with smaller-diameter tubes), each coil delivers 3000 Btu/h (880 W) and should be no longer than about 100 linear feet (30 m).

Sizing of the main pipes and pump is based on the longest circuit, measured along the length of the supply pipe from the boiler to the coil, and back along the return line to the boiler. Two guidelines apply:

1. Do not include the length of the coil itself in this circuit length.
2. No section of a floor panel main should be less than a diameter of ¾ in. (20 mm). ■

EXAMPLE 7.2. Determine the panel area and coils required for a radiant floor in a living room 15 ft × 25 ft (= 375 ft²) with a design heat loss of 12,000 Btu/h (4.6 m × 7.6 m = 34.8 m², heat loss 3514 W).

SOLUTION. The panel area required is

$$\frac{12,000 \text{ Btu/h}}{50 \text{ Btu/h ft}^2} = 240 \text{ ft}^2 \text{ of floor panel}$$

The room has 375 ft² of floor available, so the heated panel is usually placed along the exterior walls where the room heat loss is greatest (unless solar heat through south windows is involved). The number of coils in the panel is

$$\frac{12,000 \text{ Btu/h}}{10,000 \text{ Btu/h per coil}} = 1.2 \text{ coils}$$

Use one coil in this panel. ■

(i) Hydronic Zoning. Figure 7.23 shows that zoning is relatively easy to accomplish with hydronic systems. The installation shown in the figure is made up of three separately heated areas—the first, second, and third floors. Each

(a)

(b)

Fig. 7.23 *Three-zone, multicircuit, one-pipe system.* (a) *Each convector has connections to the one pipe.* (b) *Boiler, piping, and water controls suitable for this three-zone, one-pipe system. Each one-pipe circuit should be provided with two flow-control valves and a circulator (also called booster, or pump) on the supply (or return) pipe.*

can be heated to different temperatures as called for by thermostats in each separate apartment. For example, if only the thermostat serving the second floor (zone *B*) calls for heat, it turns on pump *B*. Flow-control valves *B* open, admitting hot water from the boiler header to main *B*. Flow-control valves *A* and *C* remain closed, preventing flow in mains *A* and *C*. Any or all of the zones may operate at one time. The boiler keeps a supply of hot water continually ready to supply any zone on demand. This is achieved by an aquastat immersed in the boiler water. When the boiler water drops below the prescribed temperature, it turns on the firing device, such as an oil burner or a gas burner, which brings the water up to temperature. If an overhead main supplies downfeed, as in the first floor of this installation, special downfeed supply and return fittings are necessary. For the second- and third-floor zones, one special return tee is sufficient. If the designer also elects to use a special upfeed *supply* tee of the venturi type, higher outputs of the convectors will result.

Two of the more famous residences in the Midwest utilize hydronic systems. Frank Lloyd Wright's Robie House (Chicago) has wall radiators integrated below the north windows in the living room. Underfloor radiators with grilles in the floor were provided for below the full-height south windows, but were apparently never installed. The boiler sits in a basement room. Mies van der Rohe's Farnsworth House (Fox River, Illinois) preserves its four walls of ceiling-to-floor glass by concealing radiant heating pipes in the floor slab. The boiler sits within the central utility "closet."

Today's radiators are designed to reflect the sleek and simple lines of contemporary architecture. The radiator is getting new exposure with some colorful and pleasing products (Fig. 7.24). The Mayer Art Center at Phillips Exeter Academy in New Hampshire (Fig. 7.25) features new exposed radiators in some older buildings. These radiators are based on simple components [typically, 2¾ in. (70 mm) wide] that can be combined in many heights and widths, inviting the designer to feature them rather than to hide them in metal cabinets.

(j) Heating Equipment Efficiency, Combustion, and Fuel Storage. As fuels burn to produce heat, they require oxygen to support the combustion. Because oxygen constitutes only about one-fifth of the volume of air, reasonably large rates of airflow are required. The air should be drawn in from outdoors at a position close to the fuel burner or (preferably) led to this location by a duct. ANSI/ASHRAE Standard 90.2-1993, *Energy Efficient Design of New Low-Rise Residential Buildings,* calls for 0.5 cfm/1000 Btu/h (1 L/s per 1242 W). This supply duct should be arranged to remain open at all times. Although permitted by Standard 90.2, this combustion air should *not* be drawn from the general building space—it is a waste of energy, and contemporary "tight" construction inhibits such airflow. A dangerous condition is created whenever stack flow is restricted.

High-efficiency boilers and furnaces manage to remove so much heat from the exhaust gases that smaller flues at much lower temperatures result. These relatively small pipes can be vented through a wall to the exterior. Eliminating a chimney has lessened the impact on building design of such boilers and furnaces.

For older or less-efficient fuel burning equipment, it is important that when chimneys carry high-temperature flue gases, they be safely isolated from combustible construction, to prevent the possibility of fire. The size of flue will depend upon the boiler or furnace selected. Flue height (Fig. 7.26*a*) had traditionally been 35 to 40 ft (11 to 12 m). The function of providing a draft, for which chimney height was an important consideration, is now provided by fans. For example, oil is injected under pressure, accompanied by air, and forced in by a fan. Often a draft adjuster in the breeching (smoke pipe) that carries the flue gases to the chimney is arranged to open slightly to *reduce* the normal stack draft. If increased draft should ever be required, an induced draft fan that puts a suction on the flue side of the fire is usually chosen instead of greater stack height. Draft hoods above gas burners prevent downdraft from blowing out the flames.

Prefabricated chimneys (Fig. 7.26*b*) are replacing with increasing frequency the bulkier and heavier field-built masonry. They offer a number of advantages and may be easily supported on a normal structure.

The storage space to be allowed for fuel oil

THERMAL CONTROL

TUBES
HEADER

"V" PANEL RADIATOR
LAMELLAE (OR FINS)
HEADER
TUBES

"C" CONVECTOR
LAMELLAE
TUBES

"C" CONVECTOR

TUBES
HEADERS

"H" PANEL RADIATOR
TUBES
HEADERS

"R" COLUMN RADIATOR
HEADER
ROUND TUBES (HORIZONTAL)
FLAT STRAPS (VERTICAL)

"G" GRILLE RADIATOR

RUNTAL STEEL RADIATORS

CUSTOM CURVED ENCLOSURE
FLOOR-MOUNTED ENCLOSURE ON PEDESTALS

LOW PROFILE FLOOR-MOUNTED ENCLOSURE
DOUBLE SLOPE ENCLOSURE

VULCAN FIN TUBE RADIATORS

Fig. 7.24 *Hot water radiators are available in bright colors and are based on simple components. (Reprinted by permission from AIA, Ramsey, and Sleeper,* Architectural Graphic Standards, *9th ed., © 1994 by John Wiley and Sons.)*

Fig. 7.25 *Hot water radiators fit below a window* (a) *at the Mayer Art Center, Philips Exeter Academy, Exeter, New Hampshire. Amsler Hagenah MacLean, Architects. (Photo by Alex Beatty.)* (b) *Radiator formed by flat tubes. (Courtesy of Runtal/ North American Energy Systems.)*

Fig. 7.26 *The need for 40-ft (12-m) chimneys* (a) *has been eliminated by controlled draft in burners* (b), *also by the use of high-efficiency heating equipment that can be directly vented to the exterior. Note the buried fuel oil tank, a potential environmental threat.*

depends on the proximity of the supplier and the space available at the building. For oil, when more than 275 gal (1040 L) was stored, it was common practice to use an outside tank buried in the ground. This practice eventually led to leaking tanks and contaminated soil and groundwater. Thus, several factors converged to discourage the use of oil: a cleaner and more efficient alternative (natural gas), few basements in new construction where oil tanks might be stored, and unsightly outdoor above-ground oil tanks.

Both ANSI/ASHRAE Standard 90.2, *Energy Efficient Design of New Low-Rise Residential Buildings,* and BSR/ASHRAE/IESNA 90.1, *Energy Standard for Buildings Except Low-Rise Residential Buildings,* specify minimum efficiency ratings for heating and cooling equipment. Depending on the size and the type, one of these terms will apply:

Annual fuel utilization efficiency (AFUE) is the ratio of annual fuel output energy to annual input energy, which includes any nonseason pilot input loss.

Coefficient of performance (COP) is defined slightly differently depending on the task: for *cooling,* it is the ratio of the rate of heat removal to the rate of energy input in consistent units, for a complete cooling system (or factory-assembled equipment) as tested under a nationally recognized standard or designated operating conditions. For *heating* (heat pump), it is the ratio of the rate of heat delivered to the rate of energy input in consistent units, for a complete heat pump system as tested under designated operating conditions. Supplemental heat is not included in this definition.

Energy efficiency ratio (EER) is the ratio of net equipment cooling capacity in Btu/h to the total rate of electric input in watts under designated operating conditions. (When consistent units are used, this ratio is the same as COP.)

Integrated part load value (IPLV) is a single-number figure of merit based on part-load EER or COP expressing part-load efficiency for air-conditioning and heat pump equipment on the basis of weighted operation at various load capacities for the equipment.

Seasonal energy efficiency ratio (SEER) is the total cooling output of an air conditioner during its normal annual usage period for cooling, in Btu/h, divided by the total electric energy input during the same period, in watt-hours.

(k) **Warm Air Heating Systems.** These systems began to supersede the open fireplace in about 1900. Originally, an iron furnace that stood in the middle of the basement was hand-fired by coal. Surrounding it was a sheet metal enclosure. An opening in its side near the bottom admitted cool combustion air that gravitated to the basement. A short duct from the top of the enclosure delivered the warm air by gravity to a large grille in the middle of the floor of the parlor. Other rooms, including those in upper stories, shared a little of this warmth when doors were left open.

Very gradual changes had culminated by middle of the 20th century in systems essentially like the ones described in Fig. 7.27. The improvements included:

Automatic firing of oil or gas.

Operational and safety controls.

Ducted air to and from each room.

Blowers to replace gravity.

Filters.

Adjustable registers.

By the 1960s, the basement was beginning to disappear, as subslab perimeter systems became popular for basementless houses (Fig. 7.28). The heat source was located centrally and fully *within* the insulated volume of the house; heat escaping from the unit itself merely helped heat the building. In general, air was delivered from below, upward across windows, to be taken back at a central high-return grille.

When electricity was used instead of oil and gas for space heating, provisions for combustion, chimneys, and fuel storage were unnecessary. Horizontal electric furnaces began to appear in shallow attics or above furred ceilings. Air was delivered down from ceilings across windows and taken back through door-grilles and open plenum space. Electric resistance furnaces use more electricity than do heat pumps to do the same heating task (see again Fig. 1.11), so heat pumps have largely supplanted such furnaces.

THERMAL CONTROL

FLOOR AREA REQUIRED BY WARM AIR FURNACE

OUTPUT CAPACITY (BTU/HR)	FURNACE FLOOR AREA (SQ FT)*
Up to 52,000	2.4
52,000–84,000	4.2
84,000–120,000	6.6
120,000–200,000	13.1

*Based on net floor area occupied by the upflow or downflow furnace. Low boy unit requires 50% more floor area. Space for combustion air should be added as required by local codes. Adequate space should be provided for service.

Fig. 7.27 *Typical furnace types and duct distribution arrangements. (Reprinted by permission from AIA, Ramsey, and Sleeper,* Architectural Graphic Standards, *9th ed., © 1994 by John Wiley and Sons.)*

As energy-saving design gained strength, insulated windows and well-insulated roofs, walls, and floors lessened the need for space heating. From a central furnace or heat pump, short ducts could deliver warm air to the inner side of each room, because warming at the insulated windows was less essential. Air returned to the unit through open grilles in doors and at the furnace or heat pump enclosure.

Comfort is one of warm air heating's advantages. The motion of air in the space helps to assure uniform conditions and reasonably equal temperatures in all parts of a building. The building can quickly be warmed with a forced-air system. It is possible to clean both the recirculated air and the outdoor air by means of filters and other special air-cleaning equipment. Air may be

circulated in nonheating seasons. Fresh air may be introduced to reduce odors and to make up the air exhausted by fans in kitchens, laundries, and bathrooms. Central cooling can be incorporated or introduced if ducts are designed originally to do so; cooling often calls for greater rates of air circulation. Humidification can be achieved by a humidifier in the air stream, and, if cooling is included in the design, dehumidification can be accomplished in summer. For both heating and cooling, a common arrangement is to place the supply registers in the floor, below areas of glass. This is important for winter operation. With adequate attention to supply register placement, return grilles can be located so as to minimize return air ductwork. High return grilles pick up the warmer air for reheating at the equipment. In

Fig. 7.28 *Forced-warm-air, perimeter loop system, adaptable for a cooling coil at the furnace. No returns are taken from kitchen, baths, or garage.*

many systems, air circulates at all times and is warmed or cooled as required.

Planning for warm air systems begins with the attempt to locate the furnace reasonably close to the center of the building. After the system is designed, a furnace must be selected. It should be capable of burning fuel at a rate suitable to make up the building's hourly heat loss. The rate of air delivery depends on the air temperature *rise* that is planned. Finally, the motor and blower must be powerful enough to overcome the friction of air against metal in both the supply and return duct systems, as well as the friction of air flowing through the furnace, filters, registers, and grilles (see later Fig. 7.31). Minor adjustments can be made at the furnace to adapt to the demands of the system and the building.

Some of the system components are discussed below.

Furnaces have become much more efficient in recent years, thanks to forced-draft chimneys and heat exchangers, as shown in Fig. 7.29. Seasonal efficiencies of up to 95% are possible, in contrast to about 62% for older furnaces. The AFUE rat-

ings for furnaces are based on an "isolated combustion system" that requires that all combustion and dilution air be drawn from outside.

Figure 7.30 shows the relationship between a furnace, the duct distribution tree, and some elements of the spaces they serve.

Ducts are constructed of sheet metal or glass fiber and are either round or rectangular. Ductwork will conduct noise unless these suggestions are followed:

Do not place the blower too close to a return grille.

Select quiet motors and cushioned mountings.

Do not permit connection or contact of conduits or water piping with the blower housing.

Use a flexible connection between bonnet and ductwork.

Ducts also can be lined with sound-absorbing material to further discourage noise transfer, but beware of materials that encourage mold and mildew growth.

Duct sizes may be selected on the basis of

(a)

(b)

Fig. 7.29 *Furnaces with greatly increased operating efficiencies are now available.* (a) *An Amana gas furnace.* (b) *The small, high-efficiency heat exchanger utilized by this furnace, which recovers heat from exhaust gases. (Courtesy of Amana Refrigeration, Inc.)*

permissible air velocity in the duct (Table 7.4). (See Section 8.4 for more detailed duct sizing procedures.)

EXAMPLE 7.3. The main duct in the low-velocity, warm air system of a residence delivers 1600 cfm (755 L/s). It is located above a drywall ceiling, and has a rectangular cross section. Select a size for this duct.

SOLUTION. Table 7.4 indicates that for a residence, RC(N) should be between 25 and 35. Choose the quieter RC(N) = 25; then 1700 fpm would be a maximum acceptable velocity. The area of the duct in square inches would be

$$\frac{1600 \text{ cfm} \times 144 \text{ (in.}^2/\text{ft}^2)}{1700 \text{ fpm}} = 135 \text{ in.}^2$$

A 10- $\times$ 14-in. (140-in.^{2}) duct is acceptable; a larger duct would produce less noise and require less fan power to overcome friction. ∎

Figure 7.31 illustrates a device that simplifies duct sizing. At a glance, it will show many duct cross-sectional configurations that will satisfy the combined requirements of friction, airflow, and air velocity.

Dampers will be necessary to balance the system and adjust it to the desires of the occupants (Fig. 7.32). Splitter dampers are used where branch ducts leave the larger trunk ducts. The flow of each riser can be controlled by an adjustable damper in the basement at the foot of the riser. Labels should indicate the rooms served. Some codes require dampers of fire-resistant material actuated by fusible links, to prevent the possible spread of fire through a duct system (see Fig. 13.6). Figure 7.32*d* shows how turning vanes can be used to assist airflow at sharp turns in ductwork. Such assistance reduces friction within the ductwork, thus reducing the total static head (Fig. 7.33) against which the supply fan must work.

Supply registers (Fig. 7.34) should be equipped

TABLE 7.4 **Air Velocities for Ducts and Grilles**

*Part A. **Main Ducts**[a]*

Duct Location	Design RC(N)	Maximum Airflow Velocity			
		fpm		m/s	
		Rectangular	Circular	Rectangular	Circular
In shaft, or above	45	3500	5000	17.8	25.4
drywall ceiling	35	2500	3500	12.7	17.8
	25	1700	2500	8.6	12.7
Above suspended	45	2500	4500	12.7	22.9
acoustic ceiling	35	1750	3000	8.9	15.2
	25	1200	2000	6.1	10.2
Duct within occupied	45	2000	3900	10.2	19.8
space	35	1450	2600	7.4	13.2
	25	950	1700	4.8	8.6

*Part B. **Face Velocities at Supply and Return Openings**[b]*

Type of Opening	Design RC(N)	Maximum Airflow Velocity	
		fpm	m/s
Supply air outlet	45	630	3.2
	40	550	2.8
	35	490	2.5
	30	430	2.2
	25	350	1.8
Return air opening	45	750	3.8
	40	670	3.4
	35	590	3.0
	30	490	2.5
	25	430	2.2

*Part C. **Ranges of RC (N)**[c]*

Function	HVAC System Noise in Unoccupied Spaces, RC (N)
Residential, private	25–35
Hotels, individual rooms, meeting rooms	25–35
Lobbies, corridors, service areas	35–45
Office buildings, private offices, conference rooms	25–35
Teleconference rooms	25 (max.)
Open-plan offices	30–40
Lobbies, circulation	40–45
Hospitals and clinics, private rooms, operating rooms	25–35
Wards, corridors, public spaces	30–40
Performing Arts, drama theaters, music teaching rooms	25 (max.)
Music practice rooms	35 (max.)
Churches, mosques, synagogues	25–35
Schools, lecture halls	35 (max.)
Classrooms over 750 ft² (70 m²)	35 (max.)
Classrooms up to 750 ft² (70 m²)	40 (max.)
Libraries	30–40

TABLE 7.4 **Air Velocities for Ducts and Grilles** (*continued*)

Part C. **Ranges of RC (N)**[c] (*continued*)

Function	HVAC System Noise in Unoccupied Spaces, RC (N)
Laboratories (with fume hoods), group teaching	35–45
Research, telephone use, speech communication	40–50
Testing/research, minimal speech communication	45–55
Courtrooms, unamplified speech	25–35
Amplified speech	30–40
Sports indoors, school gymnasiums and pools	40–50
Large capacity, with amplified speech	45–55

Source: ASHRAE 1995 *Applications Handbook,* copyright © by the American Society of Heating, Refrigerating, and Air-Conditioning Engineers, Inc., Atlanta, Ga.

[a]Branch duct velocities should be about 80% of those listed, whereas velocities in final runouts to outlets should be 50% or less. Elbows and other fittings can increase noise substantially, so airflows should be reduced accordingly.

[b]These are "free" opening velocities. Diffusers and grilles can increase sound levels, in which case these velocities should be reduced accordingly.

[c]See also Table 28.7 for NC recommended levels.

Fig. 7.30 *Conventional warm-air furnace and ducts. Low supply registers under windows and high return grilles at interior walls help assure distribution within the spaces.*

Fig. 7.31 *The Ductulator is a duct-sizing device available from the Trane Company. The designer selects any two factors (e.g., friction and air-flow volume) and the device yields all the other factors (e.g., air velocity, diameter of the round duct required, or combinations of rectangular-duct cross sections required).*

(27° and 35°C) is selected for the fan switch in the furnace bonnet. After the burner starts, the fan switch turns on the blower when the furnace air reaches the selected cut-in temperature. Burner and blower then continue to run while heat is needed. When the burner turns off, the blower continues to run until the temperature in the furnace drops to a level a little below the cut-in temperature of the fan switch. If, during operation, the temperature unexpectedly exceeds 200°F (93°C), a high-limit switch turns off the burner in the interest of comfort and safety. As in all automatically fired heating units, a stack temperature control in the breeching cuts off the fuel if ignition fails.

with dampers and their vanes should be arranged to disperse the air and to reduce its velocity as soon as possible after it enters the room. This is commonly done by providing vanes that divert the air, half to the right and half to the left. When a supply register is in the corner of a room, it is best if the vanes deflect all the air away from the corner. Return grilles are of the slotted type in walls and of the grid type in floors. All registers and grills should be made tight at the duct connection. See Tables 7.4 and 7.5 for selection of registers based on output and recommended face velocity.

Controls: The burner is started and stopped by a thermostat, which is usually placed in or near the living room at a thermally stable location that is protected from cold drafts, direct sunlight, and the warming effects of nearby warm air registers. A cut-in temperature of between 80° and 95°F

Fig. 7.32 *Air controls in ducts.* (a) *Air adjustment by opposed-blade dampers.* (b) *Air adjustment by splitter damper.* (c) *Conventional turns in ducts.* (d) *Right-angle turns with turning vanes—a more compact method.*

Furnace and blower

A + B = Static head, inches of water

Fig. 7.33 *The static head is the pressure, measured in inches of water, available to overcome friction in the entire system.*

Damper adjustment

(a)

Section

Spread

Throw

Aspiration

Return register

(b)

Fig. 7.34 *The typical floor register (diffuser), in its 2¼-in. x 12-in. (60- x 300-mm) size* (a). *It has diverting vanes for "spread" and an adjustable damper. See Table 7.5 for characteristics.* (b) *"Spread" and "throw"; cooler room air is induced by aspiration to join the stream of warm air, resulting in a bland and pleasant airstream that crosses the room.*

TABLE 7.5 **Typical Residential Forced-Air Register, 2¼ × 12 in. (60 × 300 mm)**

Part A. **I-P Units**

Heating (Btu/h)	3045	4565	6090	7610	9515	11,415	13,320	15,220
Cooling (Btu/h)	855	1280	1710	2135	2670	3,200	3,735	4,270
Cfm	40	60	80	100	125	150	175	200
Vertical throw (ft)	3	4	5	6	8	10	12	14
Vertical spread (ft)	6	8	10	11	14	17	22	25
Face velocity (fpm)	280	420	565	705	880	1,050	1,230	1,400

Source: Lima Register Company.

Part B. **SI Units**

Heating (W)	890	1340	1780	2230	2790	3340	3900	4460
Cooling (W)	250	375	500	625	780	940	1090	1250
L/s	19	28	38	48	59	71	83	95
Vertical throw (m)	0.9	1.2	1.5	1.8	2.4	3.0	3.7	4.3
Vertical spread (m)	1.8	2.4	3.0	3.4	4.3	5.2	6.7	7.6
Face velocity (m/s)	1.4	2.1	2.9	3.6	4.5	5.3	6.2	7.1

SI conversions by the author.

7.8 Heating/Cooling Systems

Mechanical cooling is a much more recent development than heating; in its early days, it was widely adopted as a retrofit to existing heating systems. The first two approaches to mechanical cooling reflect this early attitude.

(a) Cooling Coils Added to Warm Air Furnaces.

This common system utilizes a rather simple arrangement of the refrigeration cycle. Figure 7.1 illustrated the circuit of a refrigerant in compression, condensing, and evaporation, in which the condenser heat is carried away by water and the evaporation process draws heat out of water in another circuit to produce *chilled* water. Thus, the heat is *moved* to a heat rejection location outdoors. Figure 7.35 is a schematic diagram of an air-to-air (in contrast to a water-to-water) refrigeration device. Air instead of water can be used to cool the condenser, and indoor air can be cooled directly by being passed over the evaporator coil in which the refrigerant is expanding from a liquid to a gas. Thus, heat is moved from the indoor air to the outdoor air by the step-up action or heat-pumping nature of the refrigeration cycle. When indoor air is cooled directly in this manner, by the expanding refrigerant, the process is usu-

ally known as *direct expansion*. The cooling coils therefore are often referred to as *DX coils*. Figure 7.36 shows another popular arrangement, in which the air-flow through the furnace and coils is horizontal.

Meanwhile, the compressor–condenser unit is placed outdoors, on a concrete slab or on the roof. The unit creates a noisy, hot microclimate in summer—an influence on both site and building planning.

(b) Hydronic and Coils.

This system combines a perimeter hot water heating pipe with an overhead air-handling system. A boiler with a tankless coil supplies domestic hot water. The boiler's heat output supplies both the perimeter loop and a coil in the air-handling unit of the duct system. The total heating load is met by the combination of radiant heat generated by the perimeter loop and heated air from the overhead air-handling system.

As installed in the Levittown standardized houses, the perimeter loop consists only of ½- or ¾-in. (13- or 19-mm) tubing embedded 4 in. (100 mm) below the top of the floor slab to overcome the cold slab effect. It has the capacity to maintain a 35 F° (19 C°) differential between the inside and outside temperatures at the perimeter.

(a)

Air to system
warm or cool

Cooling
coil

Return air

(b)

Fig. 7.35 *The compressive refrigeration cycle* (a) *as used in an exterior air-cooled heat rejection unit, and an interior cooling coil, suitable for placement in a central airstream.* (b) *Cooling/heating air-handling unit. Cutaway view shows an upflow air pattern. At lower left is return air intake and filter; adjacent are fan and motor. Above these components are the gas-burning elements. At the top is the direct-expansion cooling coil. (Courtesy of American Furnace Division, Singer Company.)*

The air-handling unit and overhead duct system, incorporating supply outlets in each room and central return, is used throughout the year. Its cooling coil is connected to an adjacent outdoor condensing unit (Fig. 7.37).

Because the heating load is shared by the slab loop and by warm air, the winter indoor temperature remains more constant. The air can be distributed at no more than about 120°F (49°C), or 20 F° (11 C°) less than with a conventionally ducted system.

(c) Air–Air Heat Pumps. These use the refrigeration cycle to both heat *and* cool, thus eliminating the distinction between furnace (or boiler) and DX cooling coils. As shown in Fig. 7.38, heat is "pumped" from indoors to outdoors in summer (Fig. 7.38a) and from outdoors to indoors in winter (Fig. 7.38b). Heat pumps can transfer heat air–air, air–water, and water–water. The most common application for smaller buildings is the air–air heat pump, shown in Figs. 7.38 and 7.39. In a *single-package* (also called *unitary*) system (Fig. 7.39a), only one piece of equipment is involved. A single-package air–air heat pump moves heat between an outdoor airstream and an indoor airstream; although kept separate, both streams pass through a *single* outdoor unit. A system with both outside and inside components is called a *split* system (Fig. 7.39b). A split-system air–air heat pump moves heat via a refrigerant loop, between the outdoor unit (which also contains the compressor), through which outdoor air passes, and the indoor unit (which usually contains backup heating coils) for the treatment and circulation of indoor air.

Single-package heat pumps are commonly located on roofs, where they have unlimited access to outdoor air and where their noise is less likely to annoy—provided they are sufficiently isolated from the building's structure. This approach is shown in the daylighted, passively solar heated Mount Airy, North Carolina, library (Fig. 7.40). This 14,000-ft² (1300-m²) building also has a solar preheating system for its hot water. The five air–air heat pumps utilize economizer cycles (up to 100% outside air when temperatures are favorable). The average annual building energy consumption has been monitored at about 17,000

THERMAL CONTROL

Fig. 7.36 *Compact, horizontal flow combinations for forced-air heating and cooling. The small pipe between the two refrigerant pipes of the cooling unit is a water drain that carries away the condensed moisture from the recirculated and outdoor air. Heating unit will require gas and flue or exhaust gas connections. Refrigerant pipes connect to the outdoor compressor–condenser unit.*

Btu/ft² (53,635 W/m²)—approximately one-third of nearby similar-function buildings.

Individual air–air heat pump units are especially common in building types with all-perimeter spaces with varying orientations and numerous thermal zones. Motels are a prime example. In Fig. 7.41, separate air–air heat pumps serve each motel room; at best, their constant noise helps mask the intermittent sounds from the adjacent parking lot/circulation space. Opportunities for heat exchange between these heat pumps are scarce; if a central water loop (see Section 7.8e) were substituted for outdoor air as the heat source/sink, energy costs would go down, although first cost would rise.

Split systems are popular because the noise of the compressor and the outdoor air fan are removed from the interior, and the size of the indoor unit can be quite small. This indoor ele-

ment is often mounted either high on the wall or on the ceiling. Such indoor units are available with automatically changing louvers; when it is the cooling mode, it delivers cool air along the ceiling, from where it sinks to the level of occupancy; cold air blowing directly on people is avoided. In the heating mode, the louvers shift to direct hot air steeply downwards. The greater the distance between the indoor and outdoor units, the greater the strain on the refrigerant loop.

Heat pumps have a high initial cost, and they have shown a relatively high frequency of compressor failure. Noise from the compressor and the outdoor air fan may affect site planning, especially for residences.

One of the primary attractions of the heat pump is that in its heating mode it can give more energy than it receives (electrically). Although energy (usually electricity) is required to run the

(a)

(b)

Fig. 7.37 *Combination hydronic and forced air system* (a), *using a perimeter hot water loop in the floor slab, and an overhead air supply. (Courtesy of Levitt and Sons. Design by mechanical engineer John Liebl.)* (b) *Close-up of the heating and cooling components. The boiler heats both the slab perimeter and the coils in the airstream.*

(a)

(c)

(b)

Fig. 7.38 *Heat pump applications of the compressive refrigeration cycle. The simple air–air heat pump provides cooled air* (a) *or heated air* (b). *Teamed with a solar collector and water storage tank* (c), *the heat pump can yield usefully warm temperatures in the airstream while increasing the collector efficiency by lowering the storage tank temperature. (Reprinted with permission from* Popular Science © *1978 by Times Mirror Magazines, Inc.)*

(a)

(b)

Fig. 7.39 *The package unit* (a) *and the split-system* (b) *are popular applications of the air–air heat pump. The package unit and the outdoor unit of the split system are also typically placed outside walls, as well as on roofs as shown here.*

(a)

(b)

Fig. 7.40 *Mount Airy Library, Mount Airy, North Carolina. Mazria/Schiff & Associates, Architects.* (a) *View from the southwest. (Photo by Gordon H. Schenck, Jr.)* (b) *North–South sections relate sunlight and natural heat flow. Individual package heat pumps are set on the flat-roof sections.* (c) *Plan.*

(c)

Fig. 7.40 *(continued)*

cycle, the pump draws "free" heat from a source such as outdoor air. The total heat delivered to the building is more than the heat (electricity) required to run the cycle. The measure of this heat advantage is called the *coefficient of performance* (COP), defined as

$$\text{COP} = \frac{\text{heat delivered to space}}{\text{necessary work input}}$$

(See earlier discussion in Section 7.7*j*.)

In typical space-heating applications, a seasonal COP of 2 or more is common in mild-winter areas. Because the COP changes with outdoor conditions and indoor load, a *seasonal energy efficiency ratio* (SEER) rating system has been established. SEER measures the number of Btu/h removed for each watt of energy input, averaged over the cooling season. The higher the SEER, the more efficient the heat pump's seasonal performance. SEER ranges are roughly as low as 5

and as high as 15. The heating cycle of the heat pump has a similar rating system, called the *heating seasonal performance factor* (HSPF).

As might be expected from a device that draws heat from winter outdoor air, there are limitations to its heating performance. As outdoor temperatures approach 32°F (0°C), the COP drops and the outdoor coil tends to ice over. Built-in electric resistance coils must then be used; this, of course, ends the efficiency advantage that made the heat pump attractive. See Fig. 7.42 for a demonstration of falling performance with falling temperatures. Because of this characteristic, air–air heat pumps are less frequently used in cold-winter climates. They also generally make questionable backup choices for passively solar–heated buildings in colder climates, because backup sources are typically needed only in the coldest weather.

Heat pumps that pump heat from *water* sources, such as wells or solar-heated storage

(a)

(b)

Fig. 7.41 *Air–air heat pumps serve a motel. (a) A panel set just forward of the exterior wall allows the heat pump to inhale and exhale outdoor air around the panel edges. In summer it discharges warm air; in winter, cool air. (b) Interior cabinet contains the heat pump. Room air is taken in as shown, then discharged (after cooling or heating) upward across the glass surface—the likely point of maximum heat loss or gain.*

tanks, or from the *ground*, are much more dependable cold weather performers. Water–air heat pumps (shown in Fig. 7.38c) and solar collectors make an effective team. As these rather high-COP heat pumps remove heat from the solar storage tank (to deliver it to the indoor air), the resulting lower temperature of the solar-heated water *increases* the solar collector's performance. (See Fig. 10.36 for solar collector efficiencies.)

Assume that on a cold, partly sunny day, the collector being fed water from the tank at 90°F is able to raise its temperature by 4 F° to 94°F (from 32° to 35°C). This improvement is slight because of the rather high heat loss that a 94°F collector experiences when surrounded by cold air. If, however, the collector were to be fed 59°F (15°C) water, its heat loss would be greatly reduced. The heat that the collector *does not* lose to the cold

Fig. 7.42 *Typical air–air heat pump operating characteristics; heat delivery falls with falling outdoor temperature.*

air can be invested in the 59°F water, which will thus leave at a considerably higher temperature than 59°F + 4 = 63°F. Thus, more solar energy is collected, and more is available for transfer to the building via the heat pump. The heat pump, in turn, can heat the 59°F tank water to much higher temperatures, to serve the coils in the air handling unit.

Although most of today's heat pumps utilize electricity to drive the compressive refrigeration cycle, there are also absorption cycle heat pumps that utilize natural gas. The *GAX heat pump* (Fig. 7.43) not only uses this lower-grade resource for heating and cooling, it avoids CFCs and HCFCs and functions at outdoor temperatures much lower than the electric air–air heat pump. Eventually, solar energy can be used to drive the absorption cycle. Solar-driven refrigeration is a particularly elegant blend of energy source and task—the hotter the sun, the higher the cooling capacity.

(d) Ground Source Heat Pumps. Ground–air heat pumps, also called *geothermal heat pumps* or *GeoExchange systems,* are found in several configurations throughout North America. They often provide domestic hot water in addition to heating and cooling. An "environmentally safe" refrigerant is circulated through a loop installed underground (or in a pond or lake), taking heat from the soil in winter, discharging

Fig. 7.43 *An air-cooled advanced GAX (generator-absorber heat exchange) heat pump uses aqua ammonia as the absorbent, and is operational down to −10°F (−23°C). The target cooling COP is 0.95, heating COP 1.55. Designed for light commercial applications, it will range from 5 to 25 tons capacity. (RHX = refrigerant heat exchanger; IP ABS = intermediate pressure absorber.) (Courtesy of Energy Concepts Company, Annapolis, Maryland.)*

(a) Horizontal ground source closed loop heat pump

(b) Vertical ground source closed loop heat pump

(c) Pond or lake closed loop heat pump

(d) Ground water source heat pump

(e) Standing column ground source heat pump

Fig. 7.44 *Configurations of ground source heat pump (also called GeoExchange systems, or geothermal heat pumps). (a) Horizontal ground source closed loop heat pump, laid in trenches. (b) Vertical ground source closed loop heat pump, placed in boreholes. (c) Pond or lake closed loop heat pump. (d) Groundwater-source heat pump, taking water from one well, discharging to another. (e) Standing column groundwater-source heat pump, taking water from, then discharging to, the same well.*

heat to the soil in summer. The loop is often high-density polyethylene (HDPE). Below the surface, soil temperatures are more stable year-round than outdoor air temperatures, thus raising the COP relative to air–air heat pumps. Annual well-water (deep soil) temperatures were shown in Fig. 5.58. This system is almost completely out-of-sight, with no maintenance or weathering of exterior equipment. Noise is confined to the compressor in an indoor, small mechanical room.

Some common configurations are shown in Fig. 7.44. In the closed systems (*a,b,c*) the flow rate is typically 2 to 3 gpm/ton of refrigeration

(0.3 to 0.5 mL/J), with lower flows in the open loop systems.

The horizontal ground source closed loop heat pump (Fig. 7.44*a*) requires trenches 3 to 6 ft (1 to 2 m) deep; typically, 400 to 650 ft (120 to 200 m) of pipe are installed per ton (12,000 Btu/h, or 3.5 kW) of heating and cooling capacity. To squeeze more pipe length into a trench, a "slinky" coiled pipe is sometimes used. The trenches can be placed below parking lots or lawns and gardens.

The vertical ground source closed loop heat pump (Fig. 7.44*b*) is particularly applicable where

THERMAL CONTROL

site area is limited. Vertical holes are bored from 150 to 450 ft (46 to 137 m) deep. Each hole contains a single, full-depth loop, and is backfilled (or grouted) after the loop is installed. Because the temperature is much lower at greater depths, less pipe length is required than for horizontal loops. Distance between boreholes varies from a minimum of 15 ft (4.6 m) with high water table and low building cooling loads to as much as 25 ft (7.6 m) for buildings with high cooling loads. A minimum distance of 20 ft (6 m) is usually recommended.

The pond or lake closed loop heat pump (Fig. 7.44c) is sometimes used, when a building is close to an adequately large body of water. The loop is submerged, and the surrounding water conducts heat far more rapidly than does soil. The resulting shorter length required, and the low cost of placing the coils in water, can make this attractive. However, the water level in the pond should never drop below a minimum of 8 ft (2.5 m), and must have sufficient surface area for heat exchange.

The groundwater–source open loop heat pump ("pump and dump") is only suitable where groundwater is plentiful, and may be prohibited by local codes and environmental regulators. One variant of this system (Fig. 7.44d) takes water from one well, through a heat exchanger within the building, then discharges it to a second well. Another variant (Fig. 7.44e) takes water from the bottom, and discharges back into the top, of the same (standing) well, typically 6-in. (150-mm) diameter, and as deep as 1500 ft (460 m).

The Wildlife Center of Virginia at Waynesboro is a teaching–research hospital for wildlife. Its 5700 ft^2 (530 m^2) floor area served by four geothermal heat pumps (two at 4 tons, two at 5 tons), connected to 11,350 ft (3460 m) of underground horizontal pipe, laid in the "slinky" configuration, and thus fitting within about 2500 ft (760 m) of trench. The trench was dug around existing trees and placed under future roadways in its forest setting. Energy simulations estimate that annual heating, cooling, and hot water will use about 35,000 kWh. Had air–air heat pumps and an electric water heater been used instead, the estimate is 66,000 kWh, a 47% yearly savings advantage for the geothermal system.

Ground source heat pumps are often used in retrofits, especially in schools where site areas are plentiful, or historic structures where small-size interior mechanical equipment is highly desirable. The Daniel Boone High School near Johnson City, Tennessee, was built in 1971 with a two-pipe chilled water system and electric resistance heat. A 1998 ASHRAE Technology Award was won when this 160,000 ft^2 (14,864 m^2) school was retrofitted with a ground source vertical closed loop heat exchanger that is fed by 320 boreholes, each 150 ft (46 m) deep. Each borehole loop is 300 ft (91 m) of ¾-in. (19-mm) polyethylene pipe. The boreholes are arranged in sections of 20 holes each. The holes are 15 ft (5 m) on center. The 20-hole sections are separated by 20 ft (6 m). The loops all connect to an 8-in. (203-mm) supply and same size return line to the new heat exchanger within the existing mechanical room. Within the building, a new water loop heat pump system is installed; one heat pump in each zone.

(e) Water Source Heat Pumps.

The school described previously uses an interior closed water loop that connects all the heat pumps of all thermal zones. This system facilitates heating in one

Fig. 7.45 *The Water Loop Heat Pump system, also called water source heat pumps. Each water–air heat pump either deposits heat into the loop (while cooling), or withdraws heat (while heating). This system is particularly suited to buildings where simultaneous heating and cooling needs occur. Supplementary heat sources (boilers) and heat rejectors (cooling towers) are usually provided. (Reprinted by permission from AIA, Ramsey, and Sleeper,* Architectural Graphic Standards, *9th ed., © 1994 by John Wiley and Sons.)*

zone while another zone is being cooled, because simultaneous heat "deposits" and heat "withdrawals" actually help the system to function most efficiently. Figure 7.45 diagrams the typical water loop system, where a supplementary heat source (such as a boiler) in cold weather, and a supplementary heat rejector (such as a cooling tower) in hot weather, are installed to maintain useable water temperatures within the loop. A large office building in Pittsburgh that uses such a system is shown in Fig. 8.72.

Motels are good candidates for water loop heat pumps, because some rooms face sunny conditions, others are shaded; some are occupied, some unoccupied; and substantial domestic hot water needs all combine to make a heat-sharing water loop attractive.

7.9 Psychrometrics and Refrigeration

The procedure for sizing cooling equipment is more complicated than for heating because latent as well as sensible heat must be considered. The example following refers to the psychrometric chart, introduced in Fig. 5.35; a review of Figs. 5.35 to 5.43 may be helpful.

EXAMPLE 7.4. Find the total heat to be removed, and thus the refrigeration capacity required, for a dance hall. The design conditions are:

Room conditions (summer)	75°F DB (24°C), 50% rh
Number of occupants	80 people
Activity	Dancing
Ventilation provided	35 cfm (18 L/s) per person [Table 4.22, Part A requires a minimum of 25 cfm (13 L/s) per person]
Conditions, outdoor air	90°F DB, 75°F WB (32.2°C , 23.9°C)

Heat Gains in the Room	Sensible Heat, SH (Btu/h)	Latent Heat LH (Btu/h)
80 people dancing (see Table 5.21)		
80 × 305 Btu/h	24,400	
80 × 545 Btu/h		43,600
Total transmission and solar gain, and lights, equipment, etc.	67,600	(none)
	Room sensible heat (RSH) = 92,000	Room latent heat (RLH) = 43,600
Total heat gains in room:		135,600 Btu/h (RSH + RLH)

SOLUTION. First, determine the portion of the heat gain that is due to sensible heat gain, called the *sensible heat factor,* SHF:

$$\text{SHF} = \frac{\text{RSH}}{\text{RSH} + \text{RLH}} = \frac{92,000}{135,600} = 0.68$$

On the psychrometric chart (simplified in Fig. 7.46a), draw a line between the fixed "bull's-eye" (80°F DB, 50% rh) and the value of 0.68 on the SHF scale, at the upper right edge of the chart. This is called the "SHF line."

Point A is the room condition (of the "used" air) within the dance hall, as it is returned for reprocessing: 75°F DB, 50% rh (62.5°F WB). Next, decide how much cooler the supply air should be than the return air. To avoid uncomfortable drafts, this supply temperature is usually 20 F° (or less) below the space's air temperature. In this case, with such well-stirred air, choose 20 F°. Then the quantity of air required to cool the room will be:

$$\text{cfm} = \frac{\text{RSH}}{1.08\Delta t} = \frac{92,000 \text{ Btu/h}}{1.08\,(20\text{ F}°)} = 4260 \text{ cfm}$$

Fig. 7.46 Sizing cooling equipment using the psychrometric chart (detailed charts in Figs. 5.35 and 5.36). (a) Finding the conditions for the supply air. (b) Finding the conditions for the return air–outdoor air mixture. (c) Points A, B, C, and D, shown within the building and its cooling equipment.

(The factor 1.08 is the constant explained in the ventilation formulas in Section 4.9.)

The portion of this supply air that is *outdoor* air is as follows:

$$80 \text{ people} \times 35 \text{ cfm/person} = 2800 \text{ cfm}$$

So the percentage of outdoor air is

$$\frac{2800}{4260} = 66\%$$

Now, several important points can be located on the chart. *Point B* is the condition of the air entering the rooms; it has been decided that it will be 20 F° cooler than 75°F, which places point *B* somewhere on the 55° DB line. To determine exactly where, draw a dashed line through point *A, parallel* to the "SHF line," and extend it until it crosses the vertical 55° DB line. This is *Point B,* and occurs at 55° DB, 51.3° WB; enthalpy (h_B) = 21.0 Btu/lb. *Point D* is the condition of the outdoor air, given (under design conditions) at 90°F DB, 75°F WB.

Point C (Fig. 7.46*b*) represents the mixture of 66% outdoor air and 34% return air that is brought to the cooling equipment for treatment and distribution back to the dance hall. First, connect points *A* (the return air) and *D* (outdoor air); then plot *C* at 66% of the distance from *A* to *D*. This occurs at 85°F DB, 71.3°F WB; enthalpy (h_C) = 35.2 Btu/lb.

The cooling equipment must remove the grand total heat (GTH) according to the formula

$$\text{GTH} = 4.5 \times \text{cfm} \times (h_C - h_B)$$

(where 4.5 is a constant = 60 min/h × 0.075 lb/ft³ average air density). So, in this example:

$$\text{GTH} = 4.5 \times 4260 \text{ cfm} \times (35.2 - 21.0)$$
$$= 272,214 \text{ Btu/h}$$

The size of the required refrigeration unit is specified in "tons," where 1 ton = 12,000 Btu/h.

refrigeration required

$$= \frac{272,214 \text{ Btu/h}}{12,000 \text{ Btu/h ton}} = 22.7 \text{ tons}$$

Note: If the *minimum* outdoor air requirements of 25 cfm per person were provided,

$$80 \text{ people} \times 25 \text{ cfm/person} = 2000 \text{ cfm}$$

The percentage of outdoor air becomes 2000/4260 = 47%.

Point *C* then moves to about 82°F DB, 68.8°F WB, at which point h_C = about 33 Btu/lb.

The refrigeration required then becomes

$$4.5 \times 4260 \times (33 - 21)$$

$$= \frac{230,040 \text{ Btu/h}}{12,000 \text{ Btu/h ton}} = \text{about 19 tons}$$

This represents a first-cost saving in equipment size and of course energy savings over the life of the dance hall. (Dancers would smell more sweat, however. It is that trade-off again—energy conservation vs. IAQ.) ∎

REFERENCES

American Institute of Architects, Ramsey, C. G., and Sleeper, H. R. (1994). *Architectural Graphic Standards,* 9th ed., Wiley, New York.

American Society of Heating, Refrigerating, and Air-Conditioning Engineers, Inc. (1997). *Handbook of Fundamentals,* and ASHRAE/IES 90.1-1989, *Energy Efficient Design of New Buildings Except New Low-Rise Residential Buildings,* and ANSI/ASHRAE Standard 90.2-1993, *Energy Efficient Design of New Low-Rise Residential Buildings,* ASHRAE, Atlanta, Ga.

Barden, A. A., and Hyytiainen, H. (1993). *Finnish fireplaces: Heart of the home,* 2nd ed. Finnish Building Centre, Ltd., Helsinki.

Bourne, R. and Hoeschele, M. (1998). "Performance Results for a Night Roof Spray Storage Cooling System," in *Proceedings of the 23rd National Passive Solar Conference,* American Solar Energy Society, Boulder, Colo.

Hydronics Institute Division of GAMA (1996). *Installation Guide for Residential Hydronic Heating Systems, No. 200.* Berkeley Heights N. J.

Lorsch, H. (Ed.) (1993). *Air-Conditioning Systems Design Manual.* ASHRAE, Atlanta, Ga.

Mozer, M. (1998). "The Neural Network House: An Environment that Adapts to its Inhabitants," in *Proceedings of the American Association for Artificial Intelligence, Spring Symposium on Intelligent Environments,* March 1998, Stanford University, Palo Alto, Calif.

Natural Gas Cooling Equipment Guide (4th edition, 1996). American Gas Cooling Center, Arlington, Va.

8

LARGE BUILDING HVAC SYSTEMS

This chapter continues the discussion that began in Chapter 7, with focus now on more complex systems for larger buildings with many thermal zones.

Before Selecting an HVAC System for a Large Building: For those using this book for reference more than for a text, you may have opened this book to this chapter. Your choice of an HVAC system should follow preliminary design decisions that are discussed in these preceding sections:

Chapter 6—Indoor Air Quality discusses the relationship between comfort, zoning, and equipment that helps maintain acceptable air quality in buildings.

7.1 Review of the Need for Mechanical Equipment reviews the relative thermal role of building envelopes compared to their internal heating, cooling, and ventilating systems.

7.2 Typical Design Process outlines the decisions made by the architect and the engineer.

7.5 Refrigeration Cycles reviews both the compressive and the absorptive refrigeration processes.

At the onset of the 21st century, large building HVAC is showing several trends. One is the increasing willingness to let mechanical equipment share its tasks with natural ventilation and daylighting. Building automation has made this easier to manage. Another trend is toward underfloor plenum air supply, also known as *displacement ventilation,* rather than using ducts to diffusers and return grilles, both on the ceiling. Concern about air quality indoors and environment outdoors is producing a variety of approaches to increased ventilation that avoid refrigerants containing CFCs and HCFCs. Fuel cells and photovoltaics (PVs) are promising increased energy autonomy to larger buildings.

8.1 HVAC and Building Organization

By this time, many decisions about a building design have been made: design strategies appropriate to the climate and the building's activities have been identified, and the basic siting and overall form of the building have been determined from daylighting and thermal considerations, among others. This section begins by considering the internal yet broad issues of zoning and system choice and ends with a discussion of the more detailed consequences of system choice. A general guide to estimation of a building's thermal zone requirements was developed in Section 5.2, which discussed the importance of differences in function, schedule, and orientation.

(a) Zoning. The minimum number of thermal zones for a conventionally designed multipurpose building is shown in Fig. 8.1; more than these 16 zones could be produced by differences in scheduling within a zone, such as between "offices" and "stores." As is true of the other occupied floors, "apartments" have a minimum of five zones (based on orientation); however, the emphasis on individual controls—and the variation in usage patterns—often produces as many zones as there are apartments. When the details of zoning are added to the other preliminary design decisions, the details of HVAC systems can be considered.

(b) System Anatomy. Table 8.1 describes the basic organization of any HVAC system. Three kinds of common tasks (heating, cooling, and ventilating) are done by production components; usually, they require distribution and deliv-

Fig. 8.1 *The minimum number of thermal zones for a rather large, conventionally designed, multi-purpose building.*

ery components. Intake supplies and exhaust by-products accompany each task. Although the eventual choice of HVAC system should follow an analysis of the zone's needs, some early concepts underlie system choices.

(c) Central versus Local Systems. This was explored in Sections 6.5(*g*) for IAQ and 7.3(*a*) for smaller buildings; it is a question even more applicable to larger buildings. Central systems require one or several large mechanical spaces (often in basements and/or on roofs), sizable distribution trees, and complex control systems. The noise, heat, and other characteristics of such mechanical rooms can be controlled fairly easily, because the machinery is concentrated at a few locations. Similarly, maintenance is easy to perform without interrupting normal activities, although breakdowns in central equipment can paralyze the entire building. Air quality can be controlled by location of the air intakes high above the pollution at street level and by regular maintenance of the centralized air filtering equipment. Longer equipment life can be expected with regular maintenance. Energy conservation can be served by the recovery of one machine's heat by-product for a nearby machine's heat input. Although there are many ways to provide for the differing thermal needs of the many zones served

by central systems, one important drawback of central systems is the size and length of the distribution trees necessary to carry centralized services to many local receivers. Another drawback is a difference in zone scheduling: When the entire system must be activated to serve one zone (such as computer operations in an office building on a weekend), energy is wasted.

Local systems therefore become increasingly attractive as scheduling differences multiply. Also, pronounced differences in other factors—function (with resulting comfort expectations) or placement within the building, for example—can lead to the choice of local systems. Large and centralized equipment spaces are not required with local systems; rather, production equipment is distributed throughout the building (or over the roof of low-rise structures). Dispersal of equipment minimizes the size of distribution trees and greatly simplifies control systems. Moreover, system breakdowns affect only small portions of the building. However, noise and other by-products of multiple machines pose numerous potential threats to occupied spaces, and maintenance is demanding, because access to so many separate locations is often disruptive or constricted. Then, too, air quality depends on the regular cleaning of many filters scattered over the building, often within occupied spaces. The potential for energy conservation seems promising because heating or cooling is produced only as locally needed, but there is little chance to use one zone's waste heat as another's needed source.

Central heat/cool, local air distribution has become a popular way to take advantage of favorable characteristics of both central and local approaches. This is shown in Fig. 8.2*b*, with a central boiler/chiller space remotely located, and fan rooms on each floor. This minimizes the bulky distribution tree for air; although the distribution tree for heated and chilled water is extensive, it is also of much smaller diameter, therefore relatively easily accommodated. The central equipment room makes energy recovery systems from boilers and chillers more feasible.

(d) Uniformity versus Diversity. How similar should be the interior environments of buildings? This question encompasses not only thermal experiences, but visual and acoustical ones as well.

TABLE 8.1 **Basic HVAC Systems: Tasks and Components**

	Production/motion Movers, converters, processors	Distribution Supply and return trees, delivery and control components	Results
Heat:	Boilers Furnaces Pumps Fans Filters Heat pumps	Pipes Ducts Electricity conduits Diffusers Grilles Radiators Thermostats Valves, dampers	Warm air or surfaces Air motion often controlled Humidity control sometimes needed
Cool:	Evaporative coolers Heat pumps Chillers and cooling towers Coils Pumps Fans Filters	Pipes Ducts Diffusers Grilles Radiators Thermostats Valves Dampers	Cool air or surfaces Air motion usually controlled Humidity control usually provided
Vent:	Fans Filters	Ducts Diffusers Grilles Switches Dampers	Fresh air Air motion usually controlled Air quality control often needed

Source: Class notes developed by G. Z. Brown, University of Oregon.

THERMAL CONTROL

The advantages of uniformity are most evident in a rapidity of design and construction that, through mass production and speed, often brings lower first costs. As mentioned in Section 1.7 (designing for recycling), uniform ceiling heights, light placement, grille locations, and so on promote a flexibility in office arrangements that can extend a building's usable life span. However, there are at least four types of offices, which may need to be interchangeable within such "flexible" space. The typical *enclosed office* has the privacy of four walls and a door. The *bullpen office* has repeated, identical workstations, with low dividers at about the height of the desk surface. The *uniform open plan office* resembles the bull pen, but with higher divider partitions for added privacy. The *free-form open plan office* has some individually designed workstations with divider partitions of varying heights (sometimes reflecting varying status of workers). In the bullpen and uniform open plan office, the resulting uniformity is not always attractive to users, and diversity is often encouraged at a more personal level—with office furnishings, for example. A more thorough approach to diversity can provide stimulus to the user who spends many hours away from the variability of the exterior climate.

If offices must be uniform in ceiling lighting, air handling, and size, the corridors that connect them and the lounges or other supporting service

Fig. 8.2 Fan rooms (F) *can either be combined with, or separated from, boiler/ chiller "equipment" rooms (*). (a) Common location for a central, combined equipment room. (b) Increasingly common arrangement of a small fan room on each floor, equipment room in the basement. (c) An intermediate floor may be able to give space to a central fan room, while the heavier and noisier equipment remains in basement. (d) With a top floor central fan room, the equipment may be located either on the roof or in a mechanical penthouse, or remain in the basement. (Adapted by permission from Edward Allen,* The Architect's Studio Companion, *2nd edition, © 1994, John Wiley & Sons, New York.)*

* BOILER AND CHILLER EQUIPMENT ROOM

spaces can deliberately be made different. Diversity requires a complete and detailed design of places; it gives the builder a more complex and interesting task; and it can provide orientation and interest to the users. The attractiveness of diversity is evident in most collections of retail shops, in which light and sound—and sometimes heat and aroma—are used to distinguish one shop from the next.

Diversity in the thermal conditions to be maintained, such as warmer offices and cooler circulation spaces in the winter, can be used to enhance the comfort of the office users. Designers have long recognized that a space can be made to seem brighter and higher if it is preceded by a dark, low transition space. Thermal comfort impressions can be manipulated similarly. Less-than-comfortable conditions in circulation spaces or other less critical zones not only make the critical spaces seem more comfortable by contrast, but also save significant amounts of energy over the

life of a building. Furthermore, such conditions can make passive strategies more attractive.

A large-scale demonstration of diversity in thermal zones is shown in Fig. 8.3. Passive solar heating can make a significant contribution, even through a shallow-sloped, single-glazed cover in cloudy Glasgow, Scotland, largely because the mall area and leisure areas are allowed a much wider thermal range than would be permitted in stores and offices. The overcast skies are quite suitable for daylight, and the addition of summer sunshading makes natural ventilation (through stack effect, assisted by fans) possible during the cool summers. U.S. Pacific Northwest climate conditions are similar.

(e) Comparing Systems and Zones. In the process of selecting systems from the wide variety available, it is helpful to consider the match between the zones' characteristics and those of various systems. Among the considerations are zone

Fig. 8.3 *St. Enoch's Square, Glasgow, Scotland, a proposal to use passive solar heating, daylighting, and natural ventilation. Reiach & Hall and GMW Partnership, architects (joint venture); Cosentini Associates, energy consultants; Princeton Energy Group, daylighting consultants. (a) Schematic section showing winter operation; the mall temperature varies around 63°F (17°C) during operating hours, while offices are kept near 70°F (21°C). (b) Schematic section showing summer operation; the mall temperature swings from about 68° to 74°F (20 to 23°C) during operating hours. (c) Estimates of annual energy consumption, for a conventional-design "base case" and several alternative configurations. Note the significantly lower heating energy requirements, resulting in part from the lower winter temperatures allowed for the less-critical zones such as the mall and the leisure areas, in configuration A to E.*

placement (near to or away from the building skin), the zone's thermal loads, the comfort determinants based on the zone's activities, the space available for system components within the zone, and the life-cycle costs of various system alternatives.

Zone placement will sometimes preclude local systems, which depend on easy access to outdoor air both for fresh air and for a heat source or sink. Local systems for interior (away-from-skin) zones are awkward. Relationships between zone placement and building forms are shown in Fig. 8.4.

The thermal loads on each zone determine the extent to which heating or cooling is the dominant problem—which, in turn, can influence the choice of system. A zone with little cooling load and low moisture production may be well served by a simple system of fresh air plus heating, with no humidity control. Zones that require cooling will usually also require more complete control of air motion and relative humidity. Although it is risky to generalize about which comfort determinants are most important (given differences between activities and between individuals), it can generally be assumed that comfort and thermal tasks are related.

THERMAL CONTROL

Fig. 8.4 *Zone placement and building form are related to heating, cooling, and ventilating tasks; some applications take on typical building forms. (From class notes developed by G. Z. Brown, University of Oregon.)*

	For Heating of Spaces	For Cooling of Spaces
More important	Surface, air temperatures	Air motion
↓	Air motion	Relative humidity
	Relative humidity	Air, surface temperatures
Less important	Air quality	Air quality

Thus, the choice of systems can be based partly on whether the system provides good control of the more important comfort determinants.

(f) Distribution Trees. Section 5.1 raised the preliminary questions about distribution trees: how many, what kind (air or water), and where to place them within buildings. HVAC system choice is influenced by the amount of space the system requires. In some cases, it is easy to provide small equipment rooms at regular intervals throughout a building, such that little or nothing in the way of a distribution tree will be required. In other cases, a network of distribution trees and central, large equipment spaces are easier to accommodate. These central systems typically fall into one of three classifications:

All-air (the largest distribution trees).

Air & water.

All-water (the smallest distribution trees, with local control of fresh air).

The details of these systems can be found in Sections 8.5 to 8.7 along with typical applications and space requirements. Figure 8.5 shows the matrix of central–local, air–water influences on distribution trees.

A simplified procedure for matching zones and systems is shown in Table 8.2, in which preliminary system choices are made for a building such as the multipurpose structure shown in Fig. 8.1. In this process, the original 16 zones are translated into three local systems and three central systems: one all-air, one air and water, and one all-water.

The Fox Plaza Building in San Francisco, which illustrates many of the matches between systems and zones, is shown in Fig. 8.6. This project includes four major building types in one structure:

1. Underground garage for the storage of cars.
2. Commercial center at ground level, includ-

ing a bank, a women's specialty store, and other commercial establishments.
3. Ten floors of offices.
4. Sixteen floors of apartments.

The mechanical level is located between the office portion of the building and the apartments above. The distribution trees—heating, air conditioning, electrical, and so on—are thus directed both upward and downward, resulting in two shorter trees rather than one longer tree. The spatial requirements of offices and those of apartments are quite different; thus the floor-to-floor heights; window treatment; and the heating, cooling, electrical, elevators, and other services are different. The placement of the mechanical level between the offices and the apartments also provides for a definite visual separation between the two functions.

Quite unusual is the placement of the steam boilers on the 13th floor instead of in the conventional basement location. Only a small amount of auxiliary equipment is located on the roof and in a small portion of the garage. Residential areas have hot-water heating (residential cooling being rarely needed in San Francisco), offices have dual-duct, high-velocity heating/cooling, and commercial (ground-floor) tenants are supplied with hot and chilled water for individual climate control requirements.

(g) Central Equipment Location. The Fox Plaza Building has an intermediate location for the heating and cooling production equipment—one that separates floors of apartments from floors of offices. Other typical locations for central equipment are in the basement (where machine noise is most easily isolated, utilities are easily accessed, and machine weight is little problem) and on the roof, where access to air as a sink for reject heat is easiest of all and headroom is unlimited. Very tall buildings may require several intermediate mechanical floors. Examples of these approaches are found throughout the rest of this chapter.

THERMAL CONTROL

TABLE 8.2 **Procedure for Matching Zones and Systems**

Capsule Description

A multipurpose building (similar to that shown in Fig. 8.1) is situated in cold winter–mild summer climate.

Apartments are on upper floors, surrounding an open-air central court; they have adequate daylight and cross ventilation. Floor heights are quite low.

Offices are rented to various tenants. Exterior offices have high ceilings, to facilitate daylighting, and therefore have low internal gains but restricted clearance for horizontal ducts. Interior offices have lower ceilings; vertical chase space is limited because it reduces rentable area.

Shops are located around the perimeter on the high-ceilinged ground floor; some smaller shops are located on the mezzanine and ground floors in the interior zone. Space for vertical chases is severely limited on these highest rental floors.

The parking area is below grade, surrounded by air and light wells. Floor heights are very limited, to reduce ramp length.

Activities (Program)	Apartments	Offices			Shops			Parking
		Computer Center	Interior	Exterior	Interior	Exterior	Restaurant	
Schedule	24 h	24 h	9 h		9 h		12 h	
Placement	Exterior	Exterior for access	Interior	Exterior	Interior	Exterior	Exterior for access	Entire floor
Internal gains	Vary	High	Medium	Low	High	Medium	High plus moisture	Low with exhaust gas
Dominant HVAC task(s)	Heat-ventilate	Cool	Cool	Heat/cool	Cool	Cool	Cool	Ventilate

HVAC space available

Vert. (in plan)	Medium	Medium	Medium	Tight	Tight	Tight	Tight	Medium
Horiz. (in section)	Tight	Medium	Tight	Medium	Tight	Ample	Ample	Tight
System choices								
Local		A						
Central								
All-air							B	C
Air & water			(D)	D	D	(D)		
All-water	F		E			E		

Summary

A. The computer center's unique schedule and rate of internal gain usually requires a separate system, equipped with humidity and air quality controls to protect the equipment. Some heat recovery for use in *E* and *F* seems advisable.

B. The restaurant's special problems of heat, moisture, and aroma, as well as its schedule, require a separate system.

C. The parking area needs only plenty of fresh air; it requires no tie with the other zones at all.

D. The always-cooling loads of interior zones are best served by all-air systems offering control of humidity and air quality. However, vertical chase size is tight, and high-velocity distribution may be required. (The exterior zones could also be served by all-air. But the need for heating, plus the likelihood of fresh air infiltration and the tight clearance for horizontal ductwork, all suggest that the system for exterior zones be separated from the system for interior zones.)

E. Quick changes from heating to cooling are best handled by water; some central air-quality control is offered by air & water systems.

F. A central all-water system offers energy conservation advantages, recovering waste heat from system *D* (and potentially from *A*). Fresh air is easily and cheaply handled on a local basis, which also provides cooling.

Mechanical Space

Probably best located on the top office floor, or on a floor of its own between offices and apartments. Distribution tree sizes will thereby be minimized on the high-rent ground floor.

435

Fig. 8.5 *Matrix of distribution trees. (© 1998 John S. Reynolds, drawing by Michael Cockram.)*

The equipment's considerable heat, moisture, air motion, noise, and vibration potentially annoy nearby floors (or even neighboring buildings). As shown in the Fox Plaza example, the equipment can be expressive of building services and can serve a useful demarcation role between vertical layers of high-rise buildings. Moving the equipment off the roof also frees this prestigious view location for high-rent occupancy, and a roof form much more expressive of great height than a flat roof with a cooling tower.

(h) Concealment and Exposure. The pipes, ducts, and conduits that take the necessary resources to and from the interior are often carried within a network of spaces unseen by anyone except builders and repair people. The advantages of concealment include less noise from moving water and air, fewer surfaces requiring cleaning, less care necessary in construction (leaks, not looks, are important), and more control over the appearance of the interior ceiling and wall surfaces. Although maintenance access to such hidden supply lines is more difficult, a variety of readily removable covers is available, particularly in suspended ceilings.

However, the exposure of these supply networks provides an honest and direct source of visual (and occasionally acoustical) interest. Exposure in corridors and service areas and concealment in offices constitute an approach used in many office buildings. Flexibility is usually encouraged by exposure; changes can be easily made when there is no need for neatly cut holes in concealing surfaces. However, flexibility from full-height movable partitions requires constant ceiling heights—a feature of the suspended-ceiling approach.

One of the more spectacular examples of exposed mechanical (and structural) systems is shown in Fig. 8.7—the result of a design competition for a museum of modern art, reference li-

(a)

(b)

Fig. 8.6 *The Fox Plaza Building, San Francisco, Victor Gruen Associates, Inc., architects and engineers. (a) Elongated facade facing northeast shows the 16 floors of apartments above, 10 floors of offices below, the 13th floor mechanical space. In that space are chillers, and pumps for cooling tower water and chilled water, as well as boilers and converters (steam to hot water) for the fan-coil units in the residential stories above, and hot water coils in the office stories below. Air handling for the offices is also located here, downfed by high-velocity ducts. Residences are heated, offices are heated and cooled. (The roof has the cooling tower and the domestic hot water generator–storage units for the residential stories.) (Courtesy of* Progressive Architecture.*) (b) Construction photo with air-handling units visible on the thirteenth floor, and downfeed ducts that supply high-velocity hot and cold air to the office floors. (Photo by Morley Baer).*

Fig. 8.7 *Centre Georges Pompidou, Paris. A view of the mechanical support systems. Piano + Rogers, architects. (Photo by John Tingley.)*

brary, center for industrial design, center for music and acoustic research, and supporting services, in downtown Paris.

When users are invited to play an active role in adjusting conditions inside, exposure of the switches they manipulate is helpful. Visible mechanisms not only remind users of their opportunities, but also encourage user interaction. In this way, adjustments are sometimes discovered that the designer had not anticipated.

(i) Mechanical–Structural Integration or Separation.

The similarity of these two technical support systems—structures and environmental controls—has intrigued designers ever since mechanical systems began to require substantial volume for distribution, as in air-duct systems. As the complexity and size of the mechanical distribution systems was *increasing* with technological development (typically, more air is required to cool a space than simply to heat it, because of a lower ΔT), the increased strength of materials was *reducing* the size of the structural system. The

"uncluttered" floor areas between the more widely spaced columns became desirable for flexibility in spatial layout. With the mechanical systems at or within these columns, floor areas remained clear, thus giving mechanical–structural integration further impetus. With the new expectations for cooling, the refrigeration cycle's cooling tower often moved to the roof, often taking the air-handling machinery with it. This further encouraged the merging of systems, for one system was growing wider as the other diminished (Fig. 8.8). Thus a fixed-column cross section, consisting mostly of the structural column at the base and the air duct at the top, became possible.

Yet the functions of these systems differ widely: compared to the dynamic on-off air, water, and electrical distribution systems, the structural system is static—gravity never ceases. The moving parts in mechanical systems need maintenance far more frequently than the connections of structural components. Changes in occupancy can mean enormous changes in mechanical systems, requiring entirely different equipment; structural changes of such magnitude usually occur only at demolition. Mechanical systems can invite user adjustment; structural systems rarely do.

Thus, although it is possible to wrap the mechanical systems in a structural envelope, it is of questionable long-term value, given the differing life spans and characteristics of these systems. The probability of future change suggests that the *mechanical* system be the exposed one, despite the

Fig. 8.8 *Distribution trees: With rooftop centralized air handling, the supply and return air duct sizes decrease as they approach the ground. Conversely, the structural load increases toward the ground.*

appeal to many designers of the structural system's cleaner lines.

(j) Distribution Tree Placement Options.

These options are summarized in Fig. 8.9. Vertical placement options are important because they affect floor space, influencing the flexibility of spatial layout and the availability of usable (or rentable) floor space. Horizontal placement options affect ceiling height—a particular issue in daylighting design, and sometimes a critical factor when overall height limits are imposed yet a maximum of usable floor space is desired. (In Washington, DC, for instance, no building can rise higher than the Capitol.) Both vertical and horizontal distribution at the edges can have a dramatic impact on building appearance.

The history of distribution trees and high-rise buildings is one of trends and countertrends. Initially, a multistory building relied on daylight and cross ventilation, so a thin, relatively high-ceiling plan with much perimeter was favored (refer to Fig. 3.30). The heat gain and loss was all at the perimeter, so perimeter distribution trees (carrying only steam or heated water, and of quite small diameter) were generally used. As electric lighting and thus the need for air conditioning increased, so did the thickness of floor plans; large central internal areas needed lots of forced, cooled air. Central boilers, chillers, and fan rooms were the norm. Thus, bulky air distribution trees appeared. At about the same time, the glass curtain wall and its slick, two-dimensional look of modernity became fashionable. The air distribution trees were so visually intrusive on the facades that they were pushed to the core, where cooling needs were relatively steady. However, the thin glass perimeter experienced extreme needs for both heating and cooling; getting from vertical trees at the core to the perimeter required larger cavities above suspended ceilings. This pushed the ceiling in the offices down, to keep floor-to-floor

Fig. 8.9 Distribution tree placement options, both vertical (with impact on plan) and horizontal (with impact on section). (From class notes developed by G. Z. Brown, University of Oregon.)

distances economical. Vast office areas resulted that were visually dull, low ceilinged, and without daylight.

Now, countertrends include decentralized air handling, with small fan rooms on each floor. Vertical air distribution trees are shrinking; horizontal ones more common. At the same time, daylighting is pushing office ceilings higher; so is a preference for indirect lighting and its compatibility with computer screen visual comfort. Night cooling utilizing thermal mass is encouraging the exposure of concrete structure, and favoring raised-floor displacement ventilation systems. A renewed interest in sun control is encouraging three-dimensional facades, replacing two-dimensional reflective glass facades (which merely redirect the sun toward someone else). With increased three-dimensionality at the facade, perimeter distribution trees are once again conceivable.

It is logical to place at the perimeter the parts of the system that deal with the effects of sun, shade, and temperature change in the several perimeter zones, leaving at the core a separate network to handle the more stable interior areas. The disadvantages of perimeter distribution include (usually) higher construction costs and an environment that is more thermally hostile, due to the extremes of outdoor temperature.

Vertical distribution within internal circulation cores is very common, as it leaves a maximum of plan flexibility for the rest of each floor and does not disturb the prized floor areas nearest windows. However, one centralized vertical distribution trunk will require large horizontal branches near the core, so with this choice early thought must be given to the horizontal placement options.

An unusual example of vertical air distribution at the core is shown in Fig. 8.10. The Fox Plaza, Los Angeles, office building's unique features include both fan rooms on each floor *and* a large central vertical air shaft. This air shaft begins at bottom as a fresh air intake to each floor, and tapers to become, at top, an exhaust (heated) air outlet from each floor. Thus, the stack effect is utilized to help supply fresh and exhaust stale air from a large building, with help from small supply fans at each floor.

Vertical distribution integrated with structure has some intriguing possibilities where the structure–HVAC integration concept is suitable. Mul-

tiple HVAC trees are implied (because there are multiple columns with which they are integrated), so the horizontal branches tend to be small. However, these branches often join the vertical trunk at the same place that critical column-to-girder structural connections need to be made; interference is common and can be costly to solve. Vertical distribution at the *edges* is potentially dramatic in form but costly to enclose (if outside) or wasteful of prime floor space (if inside).

Horizontal distribution above corridors is very common, since reduced headroom here is more acceptable than in the main activity areas. Furthermore, corridors tend to be away from windows, so their lower ceilings do not interfere with daylight penetration. Because corridors connect nearly all spaces, horizontal service distribution to such spaces is also provided. Furthermore, exposure of these services above corridors can heighten the contrast between such serving spaces and the uncluttered, higher ceiling offices that are served. Horizontal distribution at the *structure* is sometimes chosen, particularly where U-shaped beams or box beams provide ready channels for HVAC distribution. However, the penetration of horizontal structure members by these continuous service runs must be coordinated. Horizontal distribution at the *edges* can be integrated usefully with sun-shading devices and light shelves; it can also act as a spandrel element that contrasts with the window strips. Horizontal distribution within *whole layers* below floors (or above ceilings) is often utilized, now increasingly common with displacement ventilation systems.

An example of supply at the edge for both vertical and horizontal distribution can be found in the International Building in San Francisco (Fig. 8.11). Here the vertical shafts are prominently exposed at the corners; these shafts carry supply and return ducts serving the four perimeter air conditioning zones. Air-handling equipment and a 750-ton refrigeration plant are located on the floors just below the terrace level (those least desirable for renting). Each corner duct branches to serve two zones, separately controlled. Pressure reduction and blending are done by equipment in the hung ceiling, and from these points air flows to strip-grille diffusers directly above the glass on the four sides of the building. Local controls offer comfort to personnel in each area.

(a)

(b)

Fig. 8.10 *The Fox Plaza, Los Angeles office building is a 34-story, 800,000 ft² (74,320 m²) granite and glass tower* (a) *with an unusual vertical distribution tree.* (b) *Typical lower floor plan (floors 6 to 16) shows both a fan room and a large vertical air shaft. At this lower level most of the shaft area is supplying outdoor air (from an intake in the bluff face below the building); the remainder is exhausting stale air toward the roof. Note the lack of columns between the core and perimeter, contributing to office layout flexibility.*

OLYMPIC BOULEVARD

AIR SHAFT

FAN ROOM

TEL. RM.

STAIR

ELECT. RM.

ELEVATOR LOBBY

STAIR

SERV. ELEVATOR

SHAFT

MEN

WOMEN

N

(c)

Fig. 8.10 (c) *Typical upper floor plan (floors 31 to 33) shows fewer elevators; by this level, most of the shaft area is exhausting stale air toward the roof.* (d) *Section shows the tapered interior of the constant-cross-section central air shaft, that relies upon the stack effect to bring in (usually cooler) outdoor air at the base, and expels hotter exhaust air at the top. (Courtesy of Johnson Fain Pereira Associates, Architects, Los Angeles: and Kim, Casey and Harase, Inc., Engineers, Los Angeles. Photo by Wolfgang Simon.)*

(d)

Interior zones on each floor are supplied by a riser duct in the building core, which branches at each floor to a loop just outside the line of elevators. The loop serves ceiling diffusers.

Between the perimeter loop and the interior loop, a return loop collects air for return to the central station (second and third floors). These return loops on the 11th to 21st floors are picked up by external return risers on alternate exterior corners. From the 10th floor down, the loops are picked up (as shown in Fig. 8.11) by an interior return riser that extends down through the core

in front of the blank faces of the high-rise elevators. To provide a clear space between the elevator banks on the main floor (4th or terrace), the two core ducts' risers are offset at the ceiling of that story.

In summary, perimeter air for all stories is supplied through corner ducts. Central air for all stories is supplied through a core duct. All return air above the 10th floor is carried down through the return ducts at the *other* two corners. Return air from the 10th floor and below is carried down by a return duct in the core.

(a)

Fig. 8.11 *The International Building, San Francisco. Anshen and Allen, architects; Eagelson, Engineers (Charles Krieger, E.E.) mechanical designers. (Courtesy of* Progressive Architecture.*) (a) One of the four corner main duct enclosures.*

THERMAL CONTROL

Supply ducts to all stories
Cold–air duct
Hot–air duct

Air returns thru these ducts to central vertical return 10th and below; to external vertical returns 11th thru 21st

16′ cantilever, all four facades

Return–air duct 11th thru 21st

Conditioned air to interior areas

Continuous strip ceiling outlet

Low–rise elevators 1–11

Pressure reduction and mixing (typical)

Supply duct, conditioned–air to interior areas all stories

Hot–air duct
Cold–air duct

Return–air duct 10th and below

Continuous strip ceiling outlet

24′–6″

High–rise elevators 11–22

Above the 10th floor these return ducts connect to the external–corner vertical return–ducts

Hot–air duct
Cold–air duct

Supply ducts to all stories

Return–air duct 11th thru 21st

Plan, tenth floor

(b)

Fig. 8.11 (b) *The major supply ducts (both hot and cold) to all 21 floors are located in two opposite corners. Each of these supply distribution trees serves two adjacent sides of perimeter offices. The conditioned air is supplied from a third-floor mechanical space. In the opposite two corners, return air from the upper 11 floors is collected and taken down to the mechanical space. The remainder of the return air is taken down through the core.*

Further examples of the relationships among HVAC systems, their distribution trees, and buildings is given in examples that accompany the more detailed descriptions of large-building HVAC systems in Section 8.4 to 8.7.

8.2 Heating, Ventilating, Air-Conditioning (HVAC) System Types

Large buildings have so many thermal zones, and there are so many ways to move heat from one place to another, that hundreds of HVAC system variations have been devised. A few of the most typical are introduced in this section; the following

section treats in detail the major components of HVAC production and delivery. Finally, some common variations on each of the four main system classifications are presented.

One way to classify HVAC systems is by the media used to transfer heat. Although thousands of liquids and gases can be used as carriers of heat, the three most common in building applications are air, water, and refrigerant. Traditionally, there are four main system classifications:

Direct refrigerant systems.

All-air systems.

Air & water systems.

All-water systems.

In the last three cases, the heating/cooling production equipment typically is located centrally in a large building, often rather far from the thermal zones it serves. The air-handling components may be either centrally served, or served floor-by-floor. Distribution tree size and placement thus become important issues when those systems are selected. In direct refrigerant systems, the heating/cooling machine usually is located adjacent to the zone(s) it serves; thus, the machine's environment—the microclimate it creates, and its needs—relative to the zone's environment becomes an important consideration.

(a) Direct Refrigerant Systems.

These systems nearly eliminate the distribution trees of air or water, relying instead on a heating/cooling device adjacent to or within the space to be served. Thus they are prevalent in skin-dominated buildings with extensive perimeter zones; these tend to be smaller buildings, so these systems are discussed in Chapter 7, Section 7.8.

(b) All-Air Systems.

The more common variations on all-air systems are shown in Fig. 8.12. Because air is the only heat transfer media used between the mechanical room (central station) and the zones it serves, and because air holds much less heat per unit volume than water, the distribution trees for this class are quite thick. Sometimes to reduce duct sizes, higher velocities are used for supply air. This generates more noise and higher friction, resulting in more energy used by fans; higher velocity should be used only sparingly, where space limitations are extreme.

For comfort, however, these systems are, overall, the best. The quantities of air moved through the central station(s) are heated or cooled, humidity-controlled, filtered, and freshened with outdoor air—all under controlled conditions. Within the zones, supply registers and return grilles allow a well-planned stream of conditioned air to thoroughly permeate all work areas. More details on air distribution are found in Section 8.4 and on this HVAC class in Section 8.5.

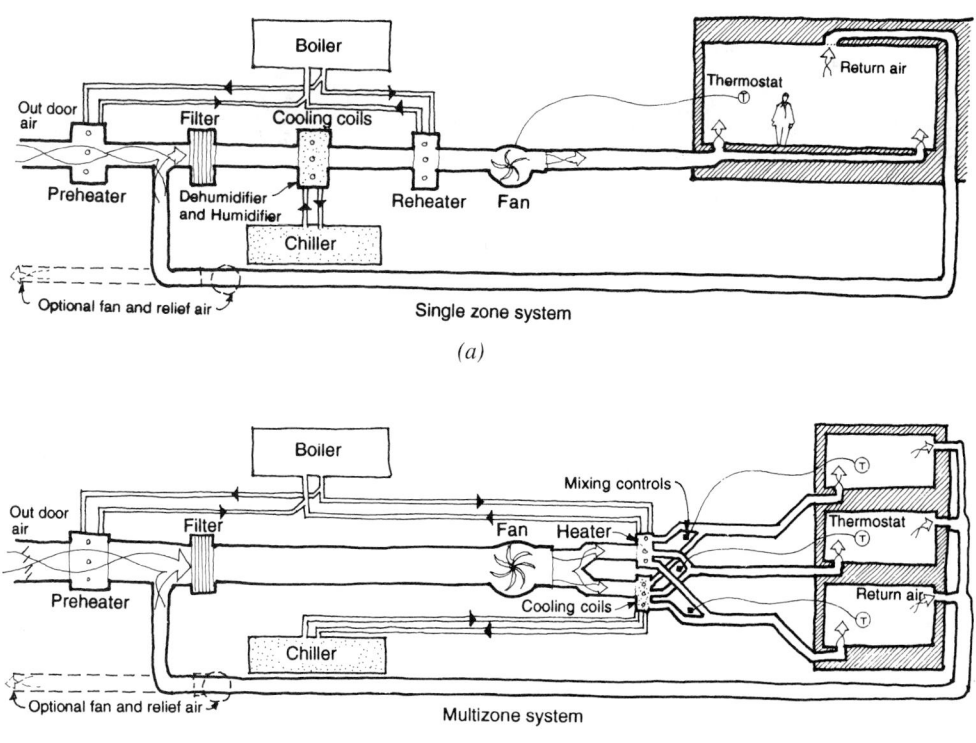

Fig. 8.12 *All-air HVAC systems. Underfloor air supply is shown here, but ceiling supply is also common.*

THERMAL CONTROL

(c)

Single duct variable volume system

(d)

Double duct system

(e)

Single duct reheat system

(f)

Fan powered variable air volume

Fig. 8.12 *(continued)*

Single-zone system (Fig. 8.12a). This is the common, small-building forced air system, controlled by a single thermostat.

Single-duct, variable-air-volume (VAV) systems (Fig. 8.12c). The most popular large building system of the 1990s, its single duct requires less building volume for distribution than multiduct systems, and the variation of air *volume* flow rate (rather than of air temperature) saves energy relative to the single duct with reheat (Fig. 8.12e). Depending on outdoor conditions and prevailing indoor needs, the central station supplies at normal velocity either a heated or a cooled stream of air. Automatic volume controls (linked to each zone's thermostat) adjust the volume admitted to that zone, within an air terminal diffuser (often located above a suspended ceiling). When the central station is supplying cold air, a zone that needs more cooling will get more air; an unoccupied room with no internal gains, or a space with heat loss through an exterior wall, will get less air. Clearly, such a system is well suited to serve the interior, always-hot zones of internal-load-dominated buildings. Less clear is its suitability for the perimeter zones of buildings in cold, cloudy conditions.

Fan-powered VAV systems (Fig. 8.12f). This variation allows individual units to heat when the main supply system is cooling; it might therefore serve perimeter zones. In such case, the cool air is reduced to minimum for IAQ, and the unit's fan draws additional air from a ceiling (or floor) plenum, heating it as required.

Multizone systems (Fig. 8.12b). Because each zone has an individual, centrally conditioned airstream, the total distribution tree volume grows to astonishing size with only a few zones. The central station produces both warm and cool airstreams, which are mixed at the central location to suit each zone. These systems are more likely to be found on medium-size buildings, or on larger buildings in which smaller central stations are located on each floor. The single-return airstream collects air from all zones (as is the case for the other systems in this class). Energy savings result when a "bypass" deck is added, allowing each zone to choose some un-thermally treated return air as part of the supply air.

Single duct with reheat (Fig. 8.12e). This system (along with VAV) has the smallest distribution tree of this class, because at each zone the only object added to the duct is a small reheat coil (heat provided by steam, hot water, or electric resistance). (Technically, this could be also called an air and water system.) The central station provides a single stream of cold air that must be cold enough to meet the maximum cooling demand of any one zone. All other zones *reheat* this air as needed. In cold weather, outdoor air at temperatures as low as 38°F (3°C) can be used; the colder this single central airstream, the less air need be circulated (and the smaller the ducts). For buildings with large interior zones in most U.S. climates, however, the central airstream must be *cooled* most of the time; then, more energy must be spent to reheat the airstream at most zones. These systems thus are notorious for energy wastage, although careful engineering can make them attractive for some climates.

Double-duct, constant-volume systems (Fig. 8.12d). Two complete distribution trees are required; at the height of summer the cooling airstream does all the work, whereas in the coldest winter conditions the heating airstream carries the load. Most of the time, air from these two streams is mixed to order at each zone's air terminals. Because both temperature and volume can be controlled, this system offers better comfort under reduced load conditions than does the single-duct, variable-volume system. (Example: an only partially occupied room.) However, it is much more expensive to install, consumes much building volume for the two distribution trees, and is usually more energy consuming than the single-duct, variable-volume system that has largely replaced it.

(c) Air & Water Systems. Several variations on air & water systems are shown in Fig. 8.13. Most of the heating and cooling of each zone is accomplished via the water distribution tree, which is much thinner than the tree needed by air. For air quality—filtering, humidity, freshness—a small, centrally conditioned airstream, equal to the total fresh air required, is provided. Thus, several distribution trees are involved, yet the total space they require is almost always less than that required by all-air systems.

Fig. 8.13 *Air & water HVAC systems. Where underfloor air supply is shown here, ceiling supply is also common. In* (b) *the supplementary air is often delivered directly to the fan coil unit.*

Exhaust air may be gathered in a return air duct system, making heat recovery possible. Or (a cheaper alternative) air can be exhausted locally, to avoid the construction of yet another distribution tree. If the water distribution provides either heating *or* cooling only, it is called a two-pipe system (shown throughout Fig. 8.13). If it provides simultaneous heating *and* cooling, it is a four-pipe system, or yet another heating

and cooling variation utilizing heat pumps (see Fig. 7.45). [Three-pipe systems were a first-cost saving alternative that allowed simultaneous heating and cooling (two supply pipes) but wasted energy by mixing the very warm hot return and the mildly warm cold return in only one return pipe. They are no longer permitted in most locales.]

This class of system frequently serves the

perimeter zones of large buildings, whereas all-air systems (commonly, single-duct, variable-volume) are used for the interior zones. More details on this HVAC class can be found in Section 8.6.

Induction systems (Fig. 8.13a). This familiar system's air terminal may be found below windows throughout the United States. A high-velocity (and high-pressure), constant-volume fresh air supply is brought to each terminal, where it is forced through an opening in such a way that air already within the room ("bypass," or secondary, air) is *induced* to join the incoming jet of air. A fairly thorough circulation of room air is thus accomplished with only a little centrally treated air. Air then passes over finned tubes for heating or cooling. Thermostats control the unit's output by controlling either the flow of the water or the flow of secondary air.

Fan-coil with supplementary air (Fig. 8.13b). Another familiar piece of below-window equipment is the fan coil, which moves the room air as it provides either heating or cooling. Centrally conditioned, tempered fresh air is brought to the space in a constant-volume stream; the fan moves both fresh and room air across a coil that either heats or cools the air, as required.

Radiant panels with supplementary air (Fig. 8.13c). Either ceiling or wall panels contain the heated or cooled water to provide a large surface for radiant heat exchange. Centrally conditioned, tempered fresh air is brought to the space in a constant-volume stream. The "piece of equipment" within the space is replaced by a large surface, which must be kept clear of obstructions to radiant heat exchange.

Water loop heat pumps. This variation on the two-pipe distribution system was shown in Fig. 7.45. Heat pumps (water–air) either draw heat from the loop (heating mode) or discharge heat to it (cooling mode). For a large building in cold weather, the excess heat of the interior zones is thus used to warm the perimeter zones. The loop's temperature ranges between 65° and 90°F (18° and 32°C); in hot weather, a central cooling tower disposes of the loop's excess heat, whereas in cold weather a central boiler adds the loop's needed heat. The loop is sized to carry 2 to 3 gpm (0.13 to 0.19 L/s) per ton, where the total tonnage equals the sum of all the individual units (often greater than the actual load).

Because individual heat pumps are used, this system is closely related to direct refrigerant systems; it is often considered an all-water system.

(d) All-Water Systems. The more simple-appearing all-water systems are shown in Fig. 8.14. These systems only heat and cool; the distribution trees are indeed slim. Air quality is dealt with elsewhere—either locally, by means of infiltration or windows; or by a separate fresh air supply system; or simply by fresh air from an adjacent system, such as a ventilated interior zone. This ambiguity about fresh air leads to similar ambigu-

Fig. 8.14 *All-water HVAC systems.* (a, b, c) *Four-pipe distribution trees need smaller volumes than do those for air systems; however, less-thorough conditioning of air is provided.*

ities about whether a system is "air & water" or "all-water." A fan-coil terminal is often employed so that air motion occurs along with heating or cooling. (Sometimes the fan-coil unit is located against the exterior wall, so that fresh air may be brought in and mixed with the room air through the fan.) Both baseboard and valence (above-window) units are also commonly available.

Because air is handled so locally, there is very little mixing of air from one zone to another, making this attractive where potential air contamination (or smoke from a fire) is an especial concern. It is also an easy system to retrofit. However, maintenance is high; filters in each fan-coil must be cleaned, and drain pans are potentially problematic.

More details on this HVAC class can be found in Section 8.7.

Two-pipe water distribution systems were shown throughout Fig. 8.13. They provide either heating or cooling. One pipe is for supply, the other for return. In a typical large-building application, they are used for heating in winter, cooling in summer. This raises the question of what changeover period will be required, a problem made much easier with the following alternative system.

Four-pipe systems are shown in Fig. 8.14*a, b,* and *c*. They allow quick changeover heating and cooling, utilizing two supply and two return pipes. When simultaneous cooling and heating is required within the same fan-coil unit (Fig. 8.14*c*), a split or double coil is used. In the coldest winter and hottest summer weather, both supply pipes can carry hot or chilled water.

(e) Equipment Space Allocations.
An important early design decision is whether to integrate or separate the *heating/cooling* equipment and the *air handling* equipment (see Fig. 8.2). If integrated, one or a few central mechanical room(s) can serve many floors and each mechanical room will need area and height sufficient for both heating/cooling and air handling equipment. If separated, one (or several, in tall buildings) large space for heating/cooling equipment is typically located in the basement or the penthouse, with a smaller fan room on each floor.

Each mechanical room should have both a central location relative to the area it serves and

direct access to the outside—contradictory requirements in many cases. Central locations within the area served to minimize distribution tree size; access to the outdoors facilitates the use of outdoor air as a heat source (winter) or sink (summer) and allows equipment to be installed or removed in later remodeling. Mechanical rooms serving both heating/cooling and air handling equipment need relatively high ceilings; 12 ft (3.7 m) clear is a typical minimum, 20 ft (6 m) clear a typical maximum.

Tables 8.3 and 8.4 present the approximate space requirements for conventional mechanical systems. (For a more detailed look at equipment room space requirements, see Figs. 8.18, 8.29, and 8.41.)

The sizing graphs shown in Tables 8.3 and 8.4 are generous if buildings are designed with energy conservation in mind. For buildings with large heat gains or losses, these graphs may slightly undersize the areas needed. These graphs are intended to give a very fast approximation of areas; more detailed procedures for sizing are presented in Section 8.3 (for boilers, chillers, fan rooms) and Section 8.4 (for air ducts).

EXAMPLE 8.1. Approximate the central equipment room needed to serve the Oregon office building (Fig. 5.7, page 203) that was analyzed for passive performance throughout Chapter 5. The total floor area of this office building is 24,000 ft^2 (2230 m^2).

SOLUTION. Enter Table 8.3, spaces for heating and cooling, for an office building at 24,000 ft^2. The space for the boiler and chiller is estimated at about 500 ft^2. Enter Table 8.4, spaces for air handling, for an office building at 24,000 ft^2. The space for the fan room is estimated at about 800 ft^2. Thus a central mechanical room (boiler, chiller, and main fans) is estimated at 500 + 800 = 1300 ft^2 (121 m^2). (The actual size of the mechanical room in this building is 1200 ft^2, a very close prediction.)

Now, approximate the largest duct sizes required. On the building's upper floor are five bays, each 1440 ft^2, for a total of 7200 ft^2 (669 m^2). The building's mechanical room is beyond the east end of this upper floor, so all the HVAC system supply air must enter through one end of the

TABLE 8.3 Approximate Space Sizes for Major Heating and Cooling Equipment

10'-10" (3.30)	17'-1" (5.21)	20'-6" (6.25)	25'-0" (7.62)	25'-0" (7.62)	36'-3" (11.05)	39'-3" (11.96)	Length
7'-3" (2.21)	7'-3" (2.21)	7'-3" (2.21)	7'-3" (2.21)	7'-3" (2.21)	7'-8" (2.34)	7'-8" (2.34)	Width
4'-11" (1.50)	4'-11" (1.50)	4'-11" (1.50)	4'-11" (1.50)	4'-11" (1.50)	7'-9" (2.36)	7'-9" (2.36)	Height

Example: Approximate sizes for all-air system in a 150,000-ft² department store:

Cooling capacity: about 450 tons

Total space boiler/chiller room: about 3200 ft²

Cooling towers: about 560 ft² of roof area (or other outdoor space)

Packaged unit sizes: no single unit large enough; largest such unit (39 ft 3 in. × 7 ft 8 in. × 7 ft 9 in.) serves about 33,000 ft² of this building, so five such units could be arranged on rooftop, to meet total load.

Source: Reprinted by permission from Edward Allen, *The Architect's Studio Companion,* © 1989 by John Wiley & Sons, New York.

TABLE 8.4 **Approximate Space Sizes for Air-Handling Equipment**

Example: Approximate sizes for all-air system in a 150,000-ft² department store:
Central fan: moves a volume of about 220,000 cfm
Main supply/return ducts: each, 120 ft² total cross-sectional area
Branch supply/return ducts: each, 300 ft² total cross-sectional area
Fan rooms: 5200 ft² total area
Fresh air louvers: 500 ft² total area
Exhaust air louvers: 400 ft² total area
If rooftop packaged units are used, each unit needs both supply and return main duct area of
   about 21 ft²; and both supply and return branch duct area of about 35 ft².

Source: Reprinted by permission from Edward Allen, *The Architect's Studio Companion,* © 1989 by John Wiley &
Sons, New York.

TABLE 8.5 **Tons of Refrigeration: Design Guidelines**

Type of Occupancy	Floor Area Served		Assumptions
	ft²/ton	m²/ton	
General occupancy:	350–550	32.5–51	400 cfm/ton (189 L/s · ton)
Perimeter spaces			1.5 cfm/ft² (0.7 L/s · m²)
Interior spaces			0.6 cfm/ft² (0.3 L/s · m²)
Offices	500	46.5	200 ft²/person (18.6 m²/person)
			3.5 W/ft² (38 W/m²) for lights, equipment
High-rise apartments			
North-facing	1000	93	
Other orientations	500	46.5	
Hospitals	333	31	1000 ft²/bed (93 m²/bed)
Shopping centers	400	37	
Department stores			2 W/ft² (21.5 W/m²)
Specialty stores			5 W/ft² (53.8 W/m²)
Hotels	350	32.5	
Restaurants	150	14	
Central Plants			
Urban districts	380	35.5	
College campuses	320	29.5	
Commercial centers	475	44	
Residential centers	500	46.5	

NOTE: 1 ton = 12,000 Btu/h (= 3.514 kW)

Source: Lorsch, Harold (ed.) (1993). *Air-Conditioning System Design Manual.* Reprinted with permission of American Society of Heating, Refrigerating and Air-Conditioning Engineers, Atlanta, Georgia.

central duct. Enter Table 8.4, spaces for air handling, for an office building at 7200 ft². The following ranges of sizes are indicated:

Volume of air: 7,000–11,000 cfm

Area, main supply or return ducts: 4 to 6.5 ft²

(The actual central supply duct maximum size in this building's upper floor is 2.6 ft². The table thus overestimated duct size by about 150% in this case of a well-insulated, shaded, and daylit building.) ∎

Engineers often refer to cooling equipment sizes in "tons" of refrigeration. The relationship between tons and floor area served is explored in Table 8.5.

For the thermally well-designed detached residence of today, a design guideline is

1000 ft² of floor area/ton.

8.3 Central Equipment

The many HVAC systems that are included in the categories of all-air, air & water, and all-water have in common a dependence on *central equipment* for the generation of heating and cooling, and/or air quality control. Figure 8.15 shows the basic relationships between some of the major pieces of central equipment and the spaces they serve. This section offers a general guide to some central equipment options and sizes. The consulting engineer chooses such equipment based upon a much more detailed analysis.

(a) Boilers. These devices heat the recirculating hot water system used for building heating. The type of boiler selected depends on the size of the heating load, the heating fuels available, the desired efficiency of operation, and whether single or modular boilers are to be installed. Boiler sizes are commonly stated either in Btu/h of net output or in (gross) horsepower, where, in I-P units,

boiler horsepower

$$= \frac{\text{heating load (Btu/h)}}{\% \text{ boiler efficiency}}$$
$$\times 33,470 \text{ Btu/h per horsepower}$$

(a)

(b)

Fig. 8.15 *Some basic components of HVAC central equipment. (a) A simplified diagram of a cooling cycle, in which chilled water is circulated to air-handling coils and heat is disposed of through a cooling tower. (b) Schematic diagram of major components of central equipment for both heating and cooling.*

In SI units,

boiler horsepower

$$= \frac{\text{heating load (kW)}}{\% \text{ boiler efficiency}}$$
$$\times 9.81 \text{ kW per horsepower}$$

Efficiency depends partly on the number of passes that the hot gases make through the water—the more passes, the higher the efficiency. It also depends on burner efficiency and on regular maintenance. Finally, efficiency is best when the equipment is operating near its capacity. Figure 8.16 compares typical boiler types, including two- and three-pass boilers.

Fire tube boilers. The hot gases of the fire are taken through tubes that are surrounded by the water to be heated. Firebox boilers place the boiler shell on top of the combustion chamber. Scotch-marine boilers feature multiple passes of the combustion gas through tubes. Fire tube boilers can be either dryback or wetback. Dryback designs have chambers outside the vessel to take combustion gases from the furnace to the tank. Wetback designs have water-cooled chambers that conduct the combustion gases.

Water tube boilers. The water to be heated is taken through tubes that are surrounded by the boiler's fire. They hold less water than the fire tube models, and so respond faster and can generate steam (where desired) at higher pressures.

Cast-iron boilers. Often used in residential and light-commercial applications, these are lower-pressure and lower-efficiency boilers. They do have the advantage of being modular.

In addition to the boilers themselves, there are choices of burner types [depending on fuel(s) used], burner controls, and boiler feedwater systems. Consult the latest ASHRAE *HVAC Systems and Equipment Handbook* for details.

Fossil-fuel–burning boilers need flues for exhaust gases, fresh air for combustion, and required air pollution control equipment. The exhaust gas is usually first taken *horizontally* from the boiler; this horizontal enclosure, or flue, is called the *breeching*. The *vertical* flue section is called the *stack.* Guidelines for sizes and arrangements of breeching and stacks are shown in Fig. 8.17. Local codes determine the quantity of air required for combustion; local air pollution authorities set pollution control requirements. As a general rule, combustion air can be supplied in a duct to the boiler at an average velocity of 1000 fpm (0.01 m/s). The duct should be large enough to carry at least 2 cfm (1 L/s) per boiler horsepower. Furthermore, ventilation air to the boiler room should be provided; preferably, the inlet and outlet should be on opposite sides of the room. Minimum sizes: enough for 2 cfm (1 L/s) per boiler horsepower, at a velocity of about 500 fpm (.005 m/s).

Space requirements for boilers are summarized in Fig. 8.18, in which multiple boilers are shown. Note that clear space within the room must be provided, so that the tubes of the boiler can be pulled when they must be replaced. Access for eventually replacing entire boilers must be considered.

Several types of single boilers are discussed as follows. The final boiler type discussed, the modular boiler, is preferred for energy conservation.

1. *High-Output, Package-Type Steel Boiler.* For large buildings that use steam as a primary heating medium, one or several such boilers may be used. Direct use of steam can be seen in Fig. 8.15*b*, supplying preheat and reheat coils and also a humidifying unit. The relative lightness of this boiler type, compared to the older styles with ponderous masonry bases (boiler settings), makes it suitable for use on upper floors of tall buildings. Figure 8.6 shows two such boilers on the thirteenth floor of the Fox Plaza Building.

2. *Converter, Steam to Hot Water (Fig. 8.19).* When, in a building that uses primary steam boilers, secondary circuits that use hot water for heating are required, a converter is used. It is considered a heat exchanger. In Fig. 8.6, there is downfeed steam supply for the two boilers on the 13th floor to two such converters, one for hot water heating in the apartments and one below the garage ceiling for hot water heating in the commercial area. A converter may also be used to transfer heat from steam to *domestic* (service) water. Converters are fre-

Fig. 8.16 *Comparisons of boiler types.* (a) *Cast iron sectional type.* (b) *Two-pass fire tube.* (c) *Three-pass wet-back Scotch marine.* (*Reprinted by permission from AIA, Ramsey, and Sleeper,* Architectural Graphic Standards, *9th ed., © 1994 by John Wiley and Sons.*)

VERTICAL
CONNECTION

OFFSET CONNECTION
(RECOMMENDED)

STACK DIAMETER—SINGLE BOILER
VENT OR STACK

BOILER HORSE-POWER	STACK DIAMETER (IN.)	A (IN.)	B (IN.)	C (IN.)
15-20	6	15	15	12
25-40	8	20	20	16
50-60	10	25	25	20
70-100	12	30	30	24
125-200	16	40	40	32
250-350	20	50	50	40
400-800	24	60	60	48

STACK DIAMETER—MULTIPLE BOILERS:
COMMON BREECHING AND STACK

BOILER HORSE-POWER	MINIMUM STACK DIAMETER (IN.)					
	NUMBER OF BOILERS					
	2		3		4	
	100 FT	200 FT	100 FT	200 FT	100 FT	200 FT
25-40	11	12	13	14	14	16
50-60	13	14	15	16	17	18
70-100	16	17	19	20	21	23
125-200	21	22	24	26	28	30
250-350	26	28	32	34	34	40
400-600	32	34	38	40	42	46

MULTIPLE BOILERS WITH COMMON BREECHING

BREECHING DIAMETER—
SINGLE AND MULTIPLE BOILERS

BOILER HORSE-POWER	MINIMUM BREECHING DIAMETER (IN. OD)			
	A (IN.) 1 BOIL-ER	B (IN.) 2 BOIL-ERS	C (IN.) 3 BOIL-ERS	D (IN.) 4 BOIL-ERS
15-20	6	8	9	9
25-40	8	10	11	12
50-60	10	12	14	15
70-100	12	15	17	18
125-200	16	20	22	24
250-350	20	25	28	30
400-600	24	30	33	36
700-800	24	34	38	42

Note: Stack diameter should be larger than breeching diameter.

Fig. 8.17 *Breeching and stack size guidelines for fossil-fuel-fired boilers. (Reprinted by permission from AIA, Ramsey, and Sleeper,* Architectural Graphic Standards, *9th ed., © 1994 by John Wiley and Sons.)*

THERMAL CONTROL

BOILER ROOM SPACE REQUIREMENTS

BOILER HP	15-40	50-100	125-200	250-350	400-800
Dimension A	5'-9"	6'-6"	6'-10"	7'-9"	8'-6"
Dimension B	7'-5"	8'-9"	9'-7"	11'-9"	14'-3"

Fig. 8.18 *Boiler room space requirements. Dimension* A *includes an aisle of 3'–6" (1 m) between the boiler and the wall. Dimension* B *between boilers includes an aisle of at least 3'–6" (1 m), up to 5 ft (1.5 m) for the largest boilers. (Reprinted by permission from AIA, Ramsey, and Sleeper,* Architectural Graphic Standards, *8th ed., © 1988 by John Wiley and Sons.)*

quently used where central steam supply systems are available, as in large-city downtown areas. The easier, quieter distribution of heat by hot water has largely replaced steam heating distribution trees within buildings.

3. Electric Boilers. Where electric rates are competitive with those of fossil fuels, electric boilers are sometimes used. Both hot water and steam electric boilers are available. The advantage of electric boilers is the elimination of combustion air, the flue, and air pollution at the building. The disadvantages are the use of a high-grade energy source for a relatively low-grade task and the pollution impact at the electric generating plant. In order to protect against high electric demand charges, a large number of control steps is desirable.

4. Compact Boilers (Fig. 8.20). Newer, smaller-dimension boilers with high thermal efficiencies are available, with rather high thermal efficiencies. In addition to their space-saving footprint, they feature a variety of venting options that make them easily adaptable to smaller equipment rooms.

5. Modular Boilers (Fig. 8.21). The primary advantage of modular boilers is efficiency. Boilers achieve maximum efficiency when they are operated continuously, at their full rated fuel input. The single boilers discussed previously operate this way only under outside design conditions, which by definition occur, at most, during 5% of a normal winter. In a modular boiler design, each section is run independently. Therefore, only one section need be fired for the mildest heating needs; as the weather gets colder, more sections are gradually added. Because each section operates continuously at full-rated fuel input, efficiency is greatly increased (Fig. 8.22). Each module, being rather small, requires little time to reach useful temperature and (unlike the larger, single boilers) does not waste a lot of heat as it cools down. Thus, modular boilers usually produce a 15% to 20% fuel savings for the heating season, relative to single boilers. Their other advantages include ease of maintenance (one module can be cleaned while others carry the heating load) and small size (allowing (*a*) Section illustrating the principle of heat transfer from steam to water.

(*b*) A converter connected to steam supply and equipped with all
devices necessary for a complete hot water heating system.

Fig. 8.19 *Conversion unit that transfers heat from steam to hot water. (Courtesy of ITT Bell and Gossett.)*

easy installation and replacement in existing buildings).

Modular boilers also eliminate the initial cost of *oversizing* heating equipment. In cold climates, conventional boiler systems often use two or three large boilers to ensure that heat is available even if one large boiler fails. When two such boilers are used, it is common practice to size each boiler at two-thirds of the total heating load; an oversize of one-third results. When three such boilers are used, it is common

practice to size each boiler at 40% of the total heating load; an oversize of 20% results. However, when a minimum of five modular boilers are used, oversizing can be eliminated because the failure of a single module will not have a crippling impact on the overall heat output.

Gas-fired pulse boilers are an even smaller and more energy-efficient choice for modular boilers. Pulse boilers utilize a series of 60 to 70 small explosions per second, making the hot flue gases pulse as they pass through the

Copper Finned Tubes
High heat transfer low volume, low mass adds BTUS to the system fast not the boiler.

Field Adjustable Top To Rear Intake & Exhaust
(Patent Pending Adjusts for the installation reat configuration makes stacking easy.

High Temperature Ceramic Fiber Combustion Chamber
Rated to 2,300°.

ASME Relief Valve

Combustion Air Blower With Air Shutter
Adjucts to the application for maximum efficiency.

Flow Switch

Factory Sized & Installed Pump
(Domestic hot water supply boilers only)

Removable Bronze Headsers
Easy inspection & service of all water-ways.

Exclusive New Dia-Scam II
Microprocessor-based control. Monitors every electrical boiler function. Pinpoint diagnostic capabilities. Digital LED temperature display. Function status and fault LEDs. Plus or minus 1° accuracy. Optional: Remote control panel and or modem with Software.

Separate Remote Flame Sensor
Senses flame at the most remote burners. Provides maximum safety and extends ignitor life

Hot Surface Ignition

Baked-on Enamel Finish

Small Footprint
Up to 1/3 smaller than most standard boilers.

Premixed Pressurized Combustion System
Low flame temperatures provide clean combustion & low NOx.

Stainless Steel Burner With Titanium Added
For long life & high temperature resistance.

Fig. 8.20 Burkay Genesis hot water boiler, fueled by either natural gas or propane, is available in ratings from 200,000 to 750,000 Btu/h. All units are 30 in. high × 24 in. deep (762 mm × 610 mm); the smallest boiler is 23 in (584 mm) wide, the largest is 57 in. (1,454 mm) wide. The copper heat exchanger has an 83.7% thermal efficiency rating and a variety of venting options are available. (Courtesy of A.O. Smith Water Products Company, Irving, Texas.)

firetube. This makes for very efficient heat transfer. Pulse boilers are available up to about 300,000 Btu/h (88 kW).

Pulse boilers operate with lower water temperatures so water vapor in the flue gas can condense and drain. This change of state liberates additional heat, allowing these pulse boil-ers to achieve efficiencies up to 90%. They exhaust moist air, not hot smoke, so flues can be small-diameter plastic pipe rather than large-diameter heat-resistant materials.

(b) Chillers. These devices remove the heat gathered by the recirculating chilled water system

Fig. 8.21 Modular boilers. (a) *A bank of four modules—total input 1.5 million Btu/h (439 kW).* (b) *Details of one module.* (c) *Schematic of flow conditions in mild weather, with only one module in operation.*

(a)

One-piece refractory combustion chamber

Horizontal cast-iron sections

Burner control

Burner control

Cutaway of Heating Module
385,000 Btu/h input
Factory assembled with burners and all oil controls
20" × 32" × 48 ½" high

Shipping weight:	644 lb	Floor loading	179 lb/ft.^{2}
Water content:	7.25 gal	Pressure drop ($\Delta T = 20F°$):	1.7' WC
Fire side surface:	49 ft.^{2}	Water side surface:	42 ft.^{2}
Horsepower:	9.1 hp	Pressure rating	100 psi ASME

(b)

SCHEMATIC FLOW DIAGRAM

For 4-step 1,540,000 Btu/h Input Multi-Temp (MO-1540)

(c)

Fig. 8.21 *(continued)*

TYPICAL HEAT LOAD DISTRIBUTION IN
5600 DEGREE DAY AREA

HOW SHORT CYCLING REDUCES BOILER
EFFICIENCY

Fig. 8.22 *One large boiler versus many smaller ones. Whether in mild weather with few modules working, in cooler weather with several working, or in extreme weather with all working, the modules actually operating contribute with minimal short-cycling (lower diagram). Short-cycling of a single large boiler could drop the general efficiency to the lower levels of the 66% to 75% efficiency range.*

as it cools the building. The selection of chillers depends largely on the fuel source and the total cooling load. Chillers include both absorption and compressive refrigeration processes, in a wide range of sizes.

New developments in chillers continue to result from a combination of concerns about the role of CFCs and HCFCs in global climate change, and from changes in utility regulations that are producing unstable energy prices in many areas. Chillers capable of changing quickly between electricity and natural gas are becoming available as a result.

Where central steam is available, or high-temperature water (from solar collectors, as waste heat from an industrial process, a fuel cell, etc.), the single-effect indirect-fired *absorption chiller* (Fig. 8.23) is attractive. This device uses the absorptive refrigeration cycle (explained in Fig. 7.2). Direct-fired absorption chillers use natural gas to power the cycle. In general, absorption equipment is less efficient than compressive refrigeration cycle equipment, although a "cheap" or even "free" heat source to power the cycle can rapidly overcome efficiency disadvantages. Absorption machines have fewer moving parts (and therefore require less maintenance) and are generally quieter than compressive cycle equipment. They are environmentally attractive, despite their much higher waste heat output (about 31,000 Btu/ton, compared to at most 15,000 Btu/ton for compressive cycle equipment), because they do not use CFCs or HCFCs, and because they require far less electricity to operate. Newer developments include the double-effect absorption chiller (see

(a)   (b)

Fig. 8.23 (a) *An absorption chiller, driven by heat to produce chilled water. The Carrier Corporation (Courtesy of Ingersoll-Rand).* (b) *Two-stage absorption chiller utilizing steam, producing from 200 to 800 tons of cooling. (Reprinted by permission from AIA, Ramsey, and Sleeper,* Architectural Graphic Standards, *9th ed., © 1994 by John Wiley and Sons.)*

Fig. 7.3) and now the triple-effect chiller, each accompanied by an increase in efficiency.

The compressive refrigeration cycle (explained in Fig. 7.1) is used in the other types of chillers. Larger units are *centrifugal chillers* (Fig. 8.24), whose compressors either can be driven by an electric motor or can utilize a turbine driven by steam or gas. (When a steam-driven turbine is used, the exhaust steam is often used to run an auxiliary absorption cycle machine. These two devices make an efficient combination, and the steam plant that supplied them in summer can supply heating in winter.) Centrifugal chillers usually require about 1 hp/ton (0.57 kW, or 10 ft³ gas, or about 15 lb of steam, per ton). These large chillers usually require a cooling tower. *Dual-condenser chillers* (Fig. 8.25) can choose whether to reject their heat to a cooling tower (via the heat rejection condenser) or to building heating (via the heat recovery condenser).

Somewhat smaller chillers use either twin screws or a scroll in place of a piston in the compressor. The screw compressor (Fig. 8.26) has a pair of helical screws; as they rotate they mesh and thus compress the volume of the gas refrigerant. They are small and quiet with little vibration. The scroll compressor (Fig. 8.27) uses two interfitting spiral-shaped scrolls. Again, the refrigerant gas is compressed as one scroll rotates against the other fixed one. Gas is brought in at one end

while the compressed gas is released at the other. Quiet and low-maintenance, they are also more efficient than reciprocating compressors.

Even smaller compressive-cycle machines are called *reciprocating chillers* (Fig. 8.28). Usually electrically driven, they are often combined with an air-cooled heat rejection process, rather than a cooling tower. This makes them a closer relative of the smaller direct refrigerant machines discussed in Section 7.8.

Chilled water is usually supplied at between 40° and 48°F (4° to 9°C). When the chilled water is supplied cold and returns much warmer, the large rise in temperature reduces the initial size (cost) of equipment and increases the efficiency (thereby reducing operating cost as well). Water treatment may be needed for chilled water to control corrosion or scaling.

Typical cooling capacities and space requirements of chillers are shown in Fig. 8.29. Each refrigeration machine in this illustration requires two pumps—one for the chilled water (to cool the building) and one for condenser water (to deal with reject heat). Typically, space is provided for future chiller additions, which may be required by building expansion and/or by higher internal gains from as-yet-uninstalled equipment, such as computer terminals within offices. Improved-efficiency chillers may replace older ones when energy costs and environmental regulations be-

THERMAL CONTROL

Shell and tube
type condenser

Electric motor.
Could be steam
turbine or diesel

Condenser
water in

Condenser
water out

Chilled water
out

Chilled water
in

Centrifugal
compressor

(a)

(b)

Fig. 8.24 (a) *A centrifugal chiller—a machine of large capacity using the compressive refrigeration cycle. (Courtesy of the Carrier Corporation.) (b) Centrifugal chiller with a flooded cooler and condenser within a single outer shell. This low-pressure unit typically produces 100 to 400 tons of cooling. (Reprinted by permission from AIA, Ramsey, and Sleeper,* Architectural Graphic Standards, *7th ed., © 1981 by John Wiley and Sons.)*

Fig. 8.25 *Dual-condenser chiller. Heat drawn from the chilled water system is either rejected to the cooling tower, or recovered for use in building heating. (Reprinted by permission from AIA, Ramsey, and Sleeper,* Architectural Graphic Standards, *9th ed., © 1994 by John Wiley and Sons.)*

come compelling. Adequate clearance access to the equipment room is a major design issue.

(c) Condensing Water Equipment. With chillers, there must be a way to reject the heat that is removed from the recirculating chilled water system. Reject heat is handled by the condensing water system, which serves the condensing process within refrigeration cycles. For larger buildings, the condensing water requirement is most likely to be met by a *cooling tower.*

The cooling tower's place within the overall equipment layout was shown in Fig. 8.15*b*; a more detailed guide to sizes and types is given in Fig. 8.30. The object is to maximize the surface area contact between outdoor air and the heat condensing water. In crossflow towers, fans move air horizontally through water droplets and wet layers of fill (or packing), whereas in counterflow towers (prevalent in larger buildings), fans move the air up as the water moves down.

Cooling towers create a special—and usually unpleasant—microclimate. They demand huge quantities of outdoor air [approximately 300 cfm (142 L/s)], which they make considerably more humid. In cold weather, they can produce fog. They are typically very noisy—a natural consequence of forced-air motion. The condensing water flows are about 2.8 gpm (0.18 L/s) per ton of compressive refrigeration, and about 3.5 gpm (0.22 L/s) per ton of absorption refrigeration.

The water that escapes as vapor from the tower is between 1.6 and 2 gph (1.7 and 2.1 mL/s). This water must be replaced, which is done automatically. The steady evaporation and exposure to the outdoors under hot and humid conditions spells trouble for the condensing water: Controls for scaling, corrosion, bacterial, and algae growth are especially important. Ozone treatment systems have the advantage of reliable biological control, and leave no chemical residue. Since the discovery of the link between Legionnaires' disease and cooling towers, biological control has assumed greater importance.

The vapor that escapes the cooling tower should be kept from the vicinity of fresh air intakes, and from neighboring buildings or parked cars, where feasible. The floor space requirements

Fig. 8.26 *Screw, or helical, compressor is a quieter, smaller machine with little vibration. (Reprinted by permission from Bobenhausen,* Simplified Design of HVAC Systems, *© 1994 by John Wiley and Sons.)*

THERMAL CONTROL

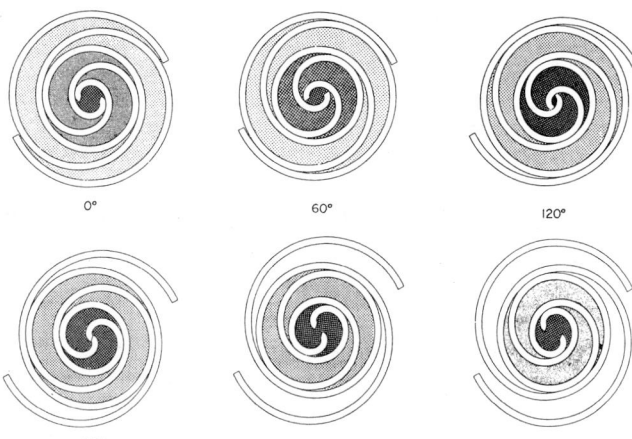

0°   60°   120°

180°   240°   300°

Fig. 8.27 Scroll compressor rotates one scroll form against another, with a quiet and efficient compression of the refrigerant. (Reprinted by permission from Bobenhausen, Simplified Design of HVAC Systems, *© 1994 by John Wiley and Sons.)*

can be approximated from Table 8.3, or use the average of 1/500 of the building gross floor area [for towers up to 8 ft (2.4 m) high], or 1/400 of the building gross floor area (for higher towers).

Although it is tempting to try to block the noise of cooling towers with solid barriers, it is critical that noise control not interfere with air circulation. The manufacturer's recommended clearances to solid objects near cooling towers must be consulted before a tower is enclosed in any way. The roof is thus a favorite location for cooling towers, where wind can disperse the vapor and the noise and odor are remote from the street. However, the cooling tower can sometimes be featured; near downtown Denver, the cooling tower for the large performing arts complex sits in a forlorn stretch of grass bordered by arterial streets, and away from pedestrians (Fig. 8.31). Its plume adds visual interest as it twists ghostlike above the equipment.

When fouling of the condensing water system cannot be tolerated, an alternative approach, called the *closed-circuit evaporative cooler,* is taken. Its schematic operation is described in Fig. 8.32. Usually used to cool the refrigerant directly, it can also be used for the condenser water, as well as used on water loop heat pump systems (Fig. 7.45). Either refrigerant or condenser water is protected within an always-closed loop, while a separate body of water is recirculated through the cooler, with steady evaporation and attendant problems. It requires much less make-up water than the cooling towers.

```
L    7'-10"
W    3'-0"
H    5'-0"
WT.  3500 LB
```

Fig. 8.28 A reciprocating chiller—a small-capacity machine that uses the compressive refrigeration cycle. Typical at sizes of less than 200 tons of cooling. (Reprinted by permission from AIA, Ramsey, and Sleeper, Architectural Graphic Standards, *9th ed., © 1994 by John Wiley and Sons.)*

(d) Energy Conservation Equipment. One big advantage of central equipment rooms is the opportunity they present for energy conservation. Regular maintenance is simplified when all the equipment lives in a generous space kept at optimum conditions; with regular maintenance comes increased efficiency of operation. Another conservation opportunity is that of heat transfer between various machines, or between distribution trees, where one's waste meets another's need.

Boiler flue economizers achieve heat transfer by passing the hot gases in a boiler's stack through a heat exchanger, thus preheating the incoming boiler water (Fig. 8.33).

Runaround coils (Fig. 8.34) can be used for

Plan

ROOM DIMENSIONS

Dimension A allows for a minimum 3 ft 6 in. aisle between the water column on the boiler and the wall. Dimension B between boilers allows for a clear aisle of:

    3'6"— 15–200 hp
    4'0"—250–350 hp
    5'0"—400–800 hp

The shortest boiler room length is obtained by allowing for possible future tube replacement (from front or rear of boiler) through a window or doorway. Allowance is only made for minimum door swing at each end of the boiler.

AIR SUPPLY

Two permanent air supply openings on opposite walls of the boiler room are recommended. These openings should be located below a height of 7 ft with a total clear area of at least 1 sq ft. Air supply openings can be louvered for weather protection.

Size the openings by using the following formula:

$$\text{area (sq ft)} = \frac{\text{CFM}}{\text{FPM}}$$

Amount of air required (CFM):

    Combustion air—max. boiler HP × 2 CMF/BHP
    Ventilation air—max. boiler HP × 2 CFM/BHP
    NOTE: a total of 10 CFM/BHP applies up to 1000 ft elevation. Add 3% more per 1000 ft of added elevation.

Air velocity required (FPM):   Up to 7 ft height—   250 FPM
                               Above 7 ft height—   500 FPM
                               Supply air duct to boiler—1000 FPM

REFRIGERATION EQUIPMENT ROOM SPACE REQUIREMENTS

EQUIPMENT (TONS)	DIMENSIONS								MINIMUM ROOM HEIGHT
	L	W	HEIGHT	T	A	B	C	D	
RECIPROCATING MACHINES									
Up to 50	10'-0"	3'-0"	6'-0"	8'-6"	3'-6"	3'-6"	4'-0"	3'-0"	11'-0"
50 to 100	12'-0"	3'-0"	6'-0"	9'-0"	3'-6"	3'-6"	4'-0"	3'-6"	11'-0"
CENTRIFUGAL MACHINES									
Up to 120	17'-0"	6'-0"	7'-0"	16'-6"	3'-6"	3'-6"	4'-6"	4'-0"	11'-6"
120 to 225	17'-0"	6'-0"	7'-0"	16'-6"	3'-6"	3'-6"	4'-6"	4'-0"	11'-6"
225 to 350	17'-0"	6'-6"	7'-6"	16'-6"	3'-6"	3'-6"	5'-0"	5'-0"	11'-6"
350 to 550	17'-0"	8'-0"	8'-0"	16'-6"	3'-6"	3'-6"	6'-0"	5'-6"	12'-0"
550 to 750	17'-6"	9'-0"	10'-6"	17'-0"	3'-6"	3'-6"	6'-0"	5'-6"	14'-0"
750 to 1500	21'-0"	15'-0"	11'-0"	20'-0"	3'-6"	3'-6"	7'-6"	6'-0"	15'-0"
ABSORPTION MACHINES									
Up to 200	18'-6"	9'-6"	12'-0"	18'-0"	3'-6"	3'-6"	4'-6"	4'-0"	15'-0"
200 to 450	21'-6"	9'-6"	12'-0"	21'-0"	3'-6"	3'-6"	5'-0"	5'-0"	15'-0"
450 to 550	23'-6"	9'-6"	12'-0"	23'-0"	3'-6"	3'-6"	6'-0"	5'-6"	15'-0"
550 to 750	26'-0"	10'-6"	13'-0"	25'-6"	3'-6"	3'-6"	6'-0"	5'-6"	16'-0"
750 to 1000	30'-0"	11'-0"	14'-0"	29'-6"	3'-6"	3'-6"	7'-0"	6'-0"	17'-6"

***Fig. 8.29** Chiller room space requirements. Each refrigeration machine is served by two pumps (chilled water and condenser water). (Reprinted by permission from AIA, Ramsey, and Sleeper, Architectural Graphic Standards, 9th ed., © 1994 by John Wiley and Sons.)*

(a)

SMALL CROSSFLOW INDUCED DRAFT PACKAGE COOLING TOWER

TONS 3GPM/TON 95-85-78	OVERALL DIMENSIONS (IN.)			OPERATING WEIGHT (LB)	MOTOR (HP)
	L	W	HT.		
5	69	33	60	940	1/4
25	75	46	80	1600	1
50	84	64	92	2500	3
100	93	100	92	4200	5
150	100	144	112	8000	7 1/2

(b)

COUNTERFLOW FORCED DRAFT PACKAGE COOLING TOWER

TONS 3GPM/TON 95-85-78	OVERALL DIMENSIONS (IN.)			OPERATING WEIGHT (LB)	MOTOR (HP)
	L	W	HT.		
20	36	36	78	950	2
50	72	36	96	1700	7 1/2
150	144	56	122	4800	20
400	140	118	192	14,000	50

(c)

Fig. 8.30 *Cooling towers that serve the condensing water system for large buildings. (a) Cutaway view of a large capacity (200 to 700 tons) crossflow induced-draft package cooling tower. (b) Size ranges for crossflow induced-draft package cooling towers. (c) Size ranges for counterflow induced-draft package cooling towers. Note: "3 gpm/ton 95-85-78" refers to the cooling capacity in tons, with condensing water flow at 3 gpm per ton, the condensing water entering the tower at 95°F and leaving at 85°F, with outside air at no more than 78° WB.*

(d)

Fig. 8.30 (d) *The more wall clearance, the better the operation. A = maximum height of enclosure above tower outlet; minimize this dimension. B = as large as possible, especially if walls have no air openings. (Consult manufacturer for minimum dimensions.) (Reprinted by permission from AIA, Ramsey, and Sleeper,* Architectural Graphic Standards, *9th ed., © 1994 by John Wiley and Sons.)*

heat transfer between intake and exhaust air ducts when these two airstreams are rather far apart. This circulating heat-transfer fluid usually contains antifreeze; it provides simple sensible heat transfer, with no restrictions on exhaust and intake location. No contamination of intake air by exhaust air is caused by this arrangement. The efficiency of such coils runs between 50% and 70%, and they are available in modular sizes up to 20,000 cfm (9440 L/s).

Heat exchange between incoming and exhaust air streams was discussed in Section 6.5(*c, d*), where heat pipes (Fig. 6.19) and thermal transfer wheels (Fig. 6.20) can play major roles in energy conservation. The thermal transfer wheels can add to the central equipment space requirements. Other dessicant systems were discussed in Section 6.5(*d*).

Economizer cycles (Fig. 8.35) use cool outdoor air, as available, to ease the burden on a refrigeration cycle as it cools the recirculated indoor air. The economizer cycle can thus be thought of as a central mechanical substitute for the open window; when it is cool enough (below the supply air temperature), 100% outside air can be provided, and no chilled water is needed. When the outdoor air temperature is higher than the supply air temperature, but lower than the return air temperature, 100% outdoor air is still brought in, but chilled water is used to lower its temperature. Above the return air temperature, outdoor air is reduced to that volume required for IAQ.

Relative to open windows, this cycle has several advantages: energy-optimizing automatic thermal control, filtering of the fresh air, tempering of the cool outdoor air to avoid unpleasant drafts, and an orderly diffusion of fresh air throughout the building. Its disadvantages are the loss of personal control that windows offer and thus loss of awareness of exterior-interior interaction. In hot humid climates, the moisture brought by 100% outdoor air may be unwelcome.

Economizer cycles are available as options on most direct refrigerant machines (such as single-package rooftop units) and are typically installed for large-building central air supply systems. Buildings with high internal gains (internal load dominated) are particularly good targets for economizer cycles because they need cooling even when the outside temperature is chilly. Economizer cycles lend themselves readily to a cooling strategy of night ventilation of thermally massive structures because they have a built-in option for 100% outdoor air.

(e) GeoExchange Systems. Using the earth as a heat source and sink for small buildings was explored in Section 7.8*d*. Four typical applications were shown in Fig. 7.44. Larger buildings can also utilize such systems. A late 1990s building just east of Central Park, Manhattan, utilizes two wells 1500 ft (457 m) deep; all but the top 50 ft (15 m) are lined by bedrock. Heat will be taken from (or discharged to) water, that will be pumped from one well and discharged to the other; a "groundwater source" system. The average year-round temperature in these deep wells (in an intensely urban area) is estimated at 56°F (13°C), about the temperature at which chilled air is delivered to a space in summer. Going to such depths is perhaps the only geothermal option in densely built-up areas.

The Cambridge (Massachusetts) Cohousing Project (Fig. 8.36) is also in a densely settled urban neighborhood. The 41 living units feature passive solar heating, underground parking [this reserves

THERMAL CONTROL

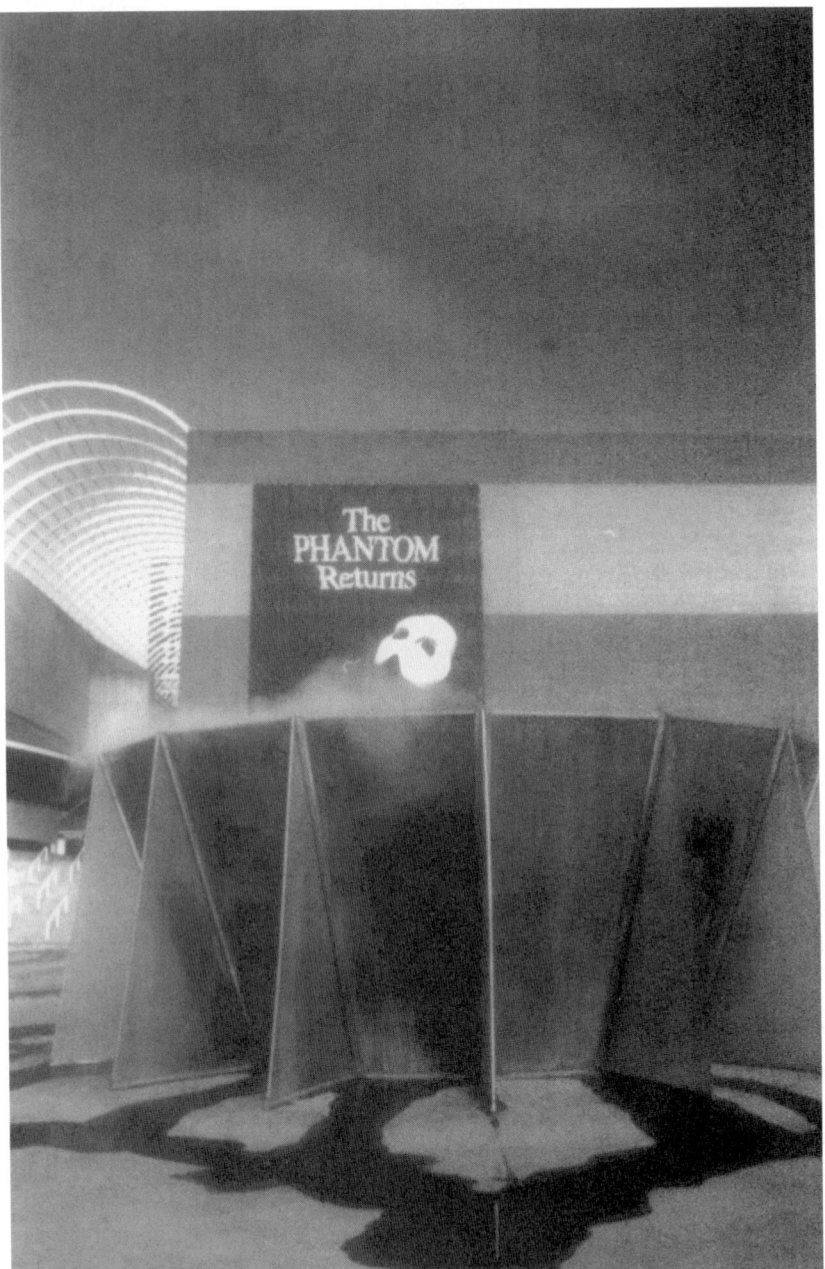

Fig. 8.31 *A plume of mist hovers ghost-like above a cooling tower in full public view near the Denver performing arts complex.*

some 20,000 ft² (1860 m²) of the surface for open green space], and centralized heating/cooling utilizing ground-source heat pumps, with locally controlled thermal zones. In this urban setting, the relative quiet of the indoor heat pump's compres-

sors and the lack of hot air (summer) or cold air (winter) noisy discharges are welcome amenities.

In England, the Hyndburn Borough Council decided that their new headquarters building should set an example as a "zero energy" building:

Fig. 8.32 *Closed-circuit evaporative coolers, which cool the condensing water system while protecting it from contact with outside air. A self-contained water system is circulated through the evaporative cooler; steady evaporation losses are replaced by makeup water. (Based on AIA, Ramsey, and Sleeper,* Architectural Graphic Standards, *8th ed., © 1988 by John Wiley and Sons.)*

Fig. 8.34 *Runaround coils for heat transfer between fresh intake air and stale exhaust air, used where the airstreams are in separate locations. (Reprinted by permission from AIA, Ramsey, and Sleeper,* Architectural Graphic Standards, *9th ed., © 1994 by John Wiley and Sons.)*

that is, over a typical year, it should generate as much energy as it imports. The 38,750 ft^2 (3600 m^2) building (Fig. 8.37) is elongated east–west and boasts an ambitious section that combines daylighting, PVs, a well-insulated skin, and even

Fig. 8.33 *Heat recovery for boilers. Flue gas entering at 500°F (260°C) leaves the "economizer" at 325°F (163°C), a temperature still high enough to prevent condensation in the stack. The heat recovered here is added to incoming boiler water, raising its temperature from 200° to 248°F (93° to 120°C). (Reprinted by permission from AIA, Ramsey, and Sleeper,* Architectural Graphic Standards, *9th ed., © 1994 by John Wiley and Sons.)*

rainwater collection to use for flushing toilets. Windpower will add to electricity generation. Summer cooling will be by ventilation (assisted by night ventilation of mass) through a raised "Termodeck" system. This is a prefabricated hollow-core slab through which cool night air circulates, storing coolth to assist the next day's hottest hours. (This is a concept explored by the Oregon office building of Fig. 5.7). The raised floor then provides a distribution plenum, with a design rate of 4 ACH.

For more extreme, future summer cooling, but now mostly for winter heating (beyond that provided by passive solar heating), an adjacent reservoir acts as a thermal sink. A "lake closed source heat pump" will serve the mechanical ventilation system. The lake is south of the building, providing some local evaporative cooling in summer and inviting a very climate-oriented south facade, behind which are open-office areas. The north facade faces the city and is more traditionally institutional in character, as are the individual offices behind it.

(f) Energy Storage. We commonly experience daily changes from warmer to colder conditions, both in winter and in a hot-dry summer.

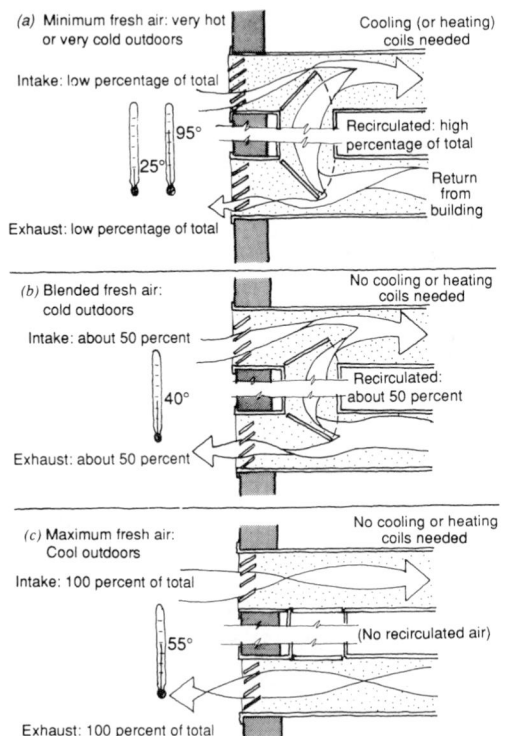

(a) Minimum fresh air: very hot
or very cold outdoors

Cooling (or heating)
coils needed

Intake: low percentage of total

95°

25°

Recirculated: high
percentage of total

Return
from
building

Exhaust: low percentage of total

(b) Blended fresh air:
cold outdoors

No cooling or heating
coils needed

Intake: about 50 percent

40°

Recirculated:
about 50 percent

Exhaust: about 50 percent

(c) Maximum fresh air:
Cool outdoors

No cooling or heating
coils needed

Intake: 100 percent of total

55°

(No recirculated air)

Exhaust: 100 percent of total

Fig. 8.35 *The economizer cycle controls the relationships between fresh, exhaust and recirculated air. (a) When outside air is hot (or very cold), the economizer cycle is inactive, and minimum fresh air is introduced. (b) As very cold outside air gets warmer, it can be blended with recirculated air, and neither heating nor cooling coils are needed. (c) When outside air is cool, it can completely replace circulated air, making mechanical cooling inactive.*

Central storage equipment for large buildings can take advantage of this cycle to increase operating efficiency, save energy, and significantly reduce electricity demand charges. Some electric utilities offer incentives to install thermal storage, in order to reduce the peak strain on their generating facilities.

Water storage tanks are one common approach to storage, such as those shown in Fig. 8.38. On typical winter days, the total internal heat generated by a large building can be somewhat greater than its total need for heating at the perimeter zones. Instead of being thrown away as exhaust air, this surplus heat is captured and stored in large water tanks, from which it can be withdrawn

and used on cold winter nights and weekends. In the summer, chillers can work at night, when efficiency is high because cool outdoor air helps the refrigeration cycle reject its heat. By storing the coolth produced, less work need be done by chillers during the next day's peak, when electric rates are highest and operating efficiency lowest.

Design guidelines for water storage are from 0.5 to 1 gal/ft² (20 to 40 L/m²) of conditioned space. Such large and heavy tanks are frequently located in basements and underground parking facilities. Water tanks require additional floor space for equipment, although they may allow somewhat smaller chillers to be installed. They may also contribute to lower fire insurance premiums because considerable water is stored and ready to use.

Ice storage tanks are another storage approach, as shown in Fig. 8.39. Because they take advantage of the latent heat of fusion [143.5 Btu/lb of water at 32°F (334 kJ/kg at 0°C)], such units can store far more energy in a given-size tank than can water. Design guidelines for ice storage are from 0.13 to 0.25 gal/ft² (5 to 10 L/m²) of conditioned space. A comparison of the relative sizes of storage tanks needed for an 18-story office building is shown in Table 8.6. Also, with ice storage there is less undesired mixing of hot and cold water within the tank.

Ice can be made in several ways. A common method is to form ice as a layer around a pipe that carries refrigerant in a closed circuit through the tank. Control of the thickness of this layer is important, because if too little ice is made at night, too little cooling will be available the following day. In another method, ice is formed and thawed by circulating a brine through coils in cylindrical water tanks. Other methods form ice on plates, then harvest it in an insulated bin.

The Iowa Public Service Building (Fig. 8.40) uses ice storage in six tanks that occupy about half the floor space in the ground floor mechanical room. This installation serves a five-story, 167,000-ft² (15,515 m²) utility office building that utilizes solar collectors supplemented by small backup boilers. The six ice-making machines serve a 75,000-gal (283,900 L) ice storage pit with 90 million Btu (26,355 kWh) capacity. Winter heat exchange opportunities include reject heat from the ice-making machines, heat from the central

(a)

(b)

Fig. 8.36 *Cambridge (Massachusetts) CoHousing development takes advantage of three boreholes on its urban site (a) to provide central heating and cooling, serving individual residential fan-coil units. It also preheats water (b) for the DHW (domestic hot water) system. The residences are sited to provide winter solar access, and a positive fresh air intake is located on the side sheltered from adjacent railroad tracks. Underground parking preserves open space for gardens and recreation. (Courtesy of Building Science Engineering, Harvard, Massachusetts.)*

(a)

Fig. 8.37 *The Hyndburn Borough Council (England) headquarters faces south toward a reservoir* (a) *that provides an evaporatively cooled microclimate, and also acts as a heat source/sink for a water-water heat pump.* (b) *Daylighting, passive solar heating, photovoltaics, and a well-insulated shell are featured. Termodeck is a precast hollow-core slab that stores coolth on summer nights, and precools ventilation air. The supply air then rises from the plenum created by raised floor. (Courtesy of Jestico + Whiles + Associates, architects, London, England.)*

Fig. 8.37 *(continued)*

toilet exhaust air, and heat from return air taken through luminaires. Solar collectors on the roof preheat the ventilation air, which is fed into the ceiling plenum at each floor. Fan-coil units then draw from this fresh air supply.

Water and ice storage tanks are sometimes located at the top of a building, especially when other related mechanical equipment is there also. Despite the weight of such tanks, it can become a structural advantage. The "tuned mass damper" method of reducing lateral vibration (or sway) in high-rise buildings utilizes a heavy, moving mass at the top; when the building begins to sway, the mass is moved in the opposite direction. In the case of the Crystal Tower in Osaka, Japan, this mass is provided by ice storage tanks. This 515 ft (157 m), 37-story building has an ice thermal storage total of 25,400 ft³ (720 m³) divided into nine tanks. Six of the ice storage tanks provide

the movable structural mass, suspended from roof girders, weighing 540 tons (489,880 kg), including steel framing. The chiller and condenser are also in the equipment penthouse.

(g) Air-Handling Equipment. A given HVAC system may have many variations. In some variations, all the air is passed through one central equipment room. In others, air handling may be done in many separate and smaller rooms, whereas central heating and cooling will require only one equipment room. Some detailed design guidelines for air-handling equipment for either case are shown in Fig. 8.41. Air handling equipment room dimensions are shown in Table 8.7. Total air quantities (cfm) may be estimated from Table 8.4 or obtained more precisely from cooling load calculations.

Fig. 8.38 *Water tank heat storage. The 870,000 ft² (80,825 m²) Transportation office building (Park Plaza, Boston) population 2000, uses a three-compartment insulated concrete tank storing 750,000 gal (2,838,990 L) of water. (a) At outside air 40° to 50°F (4.4 to 10°C), the surplus internal heat is stored in the tanks, rather than rejected as exhaust air. (b) By the time the outside air is about 50°F (10°C), the tanks are fully charged; up to outside 60°F (15.5°C), the economizer cycle provides energy-conserving cooling. (c) At about 60°F (15.5°C) chillers must operate; but working all night when the outside air is cooler enables them to work less by day. Their nightly production is stored as cold water, available to help with the following day's peak. Smaller machines and more efficient operation are the result. (Courtesy of Shooshanian Engineering Associates, Inc.)*

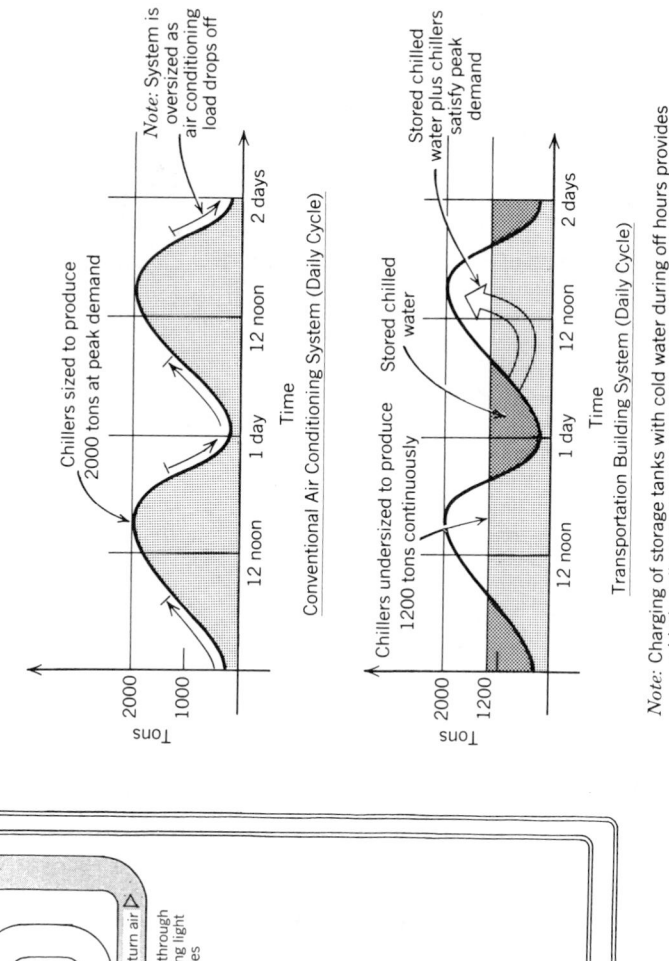

Fig. 8.38 (*continued*)

477

THERMAL CONTROL

Fig. 8.39 Ice tank heat storage. Chilled water in an ice tank forms a layer of ice around the refrigerant pipe (supplied by the refrigeration machine). As chilled water at about 35°F (1.7°C) is removed from the tank, it enters a heat exchanger, so that its temperature is closer to 45°F (7.2°C) for distribution to cooling coils throughout the building.

Some common fan types include:

Panel: the most simple of this list; for high efficiency air delivery; the motor is generally mounted in the center of the propellers. Likely the most noisy. Not designed for much static pressure from ductwork, filters, etc.

Fixed pitch vaneaxial: capable of working against somewhat more static pressure than the panel fan, more common in industrial applications.

Centrifugal: with an airfoil bladed wheel, it has high efficiency over a wide operating range, and is quieter than the previous two. Major changes in pressure result in only minor changes in volume of air delivered.

Vane axial adjustable pitch: takes less space than centrifugal fans and can work against more static pressure. The pitch is adjustable occasionally, not continuously; for system balancing or seasonal changes in air volume.

TABLE 8.6 **Cooling Storage Comparison**

Eighteen-story office building: 375,000 gross ft² (34,840 m²), 800 tons peak load, 6250 ton-h/day at design, 75 million Btu (21,900 kWh) storage needed.

	Water Storage	*Ice Storage*
Btu/lb[a] (kJ/kg)	15 (34.9)	164 (381.5)
Pounds storage	5,000,000	457,000
(Kilograms)	(2,267,960)	(207,290)
Gallons	599,500	54,800
(Liters)	(2,269,300)	(207,435)
Storage efficiency	0.90	1.0
Percent ice		66%
Net gallons	666,000	83,030
(Liters)	(252,100)	(314,295)
Cubic feet	89,046	11,101
(Cubic meters)	(2,522)	(314)
Floor area of tank, approx. 8 ft deep	80- by 150-ft	30- by 50-ft
(2.5 m deep)	(24- by 46-m)	(9- by 15-m)
Storage ratio, water to ice	8 : 1	

[a]Btu/lb based on $\Delta T = 15$ F° for water storage and on $\Delta T = 20.5$ F° in the melted ice water (added to 143.5 Btu/lb at fusion).

Source: Reprinted by permission from *Specifying Engineer*, January 1983. Metric conversions by the author.

FIRST FLOOR PLAN

Fig. 8.40 *The ground floor of the Iowa Public Service offices (Sioux City) contains six ice machines within the mechanical space adjacent to the loading dock. (Courtesy of Rossetti Associates and Foss, Engelstad, Heil Associates, joint venture architects.)*

TABLE 8.7 Air Handling Equipment Rooms

Part A. I-P Units

Cfm Range	Approximate Overall Dimension of Supply Air Equipment			Recommended Room Dimensions		
	Width ×	Height ×	Length	Width ×	Height ×	Length
1000–1,800	4′–9″	2′–9″	14′–9″	12′–6″	9′–0″	18′–9″
1,801–3,000	5′–0″	3′–6″	16′–0″	13′–9″	9′–0″	20′–0″
3,001–4,000	6′–9″	4′–6″	16′–0″	17′–6″	9′–0″	20′–0″
4,001–6,000	7′–6″	4′–6″	16′–9″	18′–0″	9′–0″	20′–9″
6,001–7,000	7′–6″	4′–9″	18′–3″	18′–6″	9′–6″	22′–3″
7,001–9,000	8′–0″	5′–0″	18′–9″	19′–0″	10′–0″	22′–9″
9,001–12,000	10′–0″	5′–6″	21′–0″	23′–0″	11′–0″	25′–0″
12,001–16,000	10′–3″	6′–0″	22′–0″	13′–6″	12′–6″	26′–0″
16,001–19,000	10′–6″	6′–6″	23′–9″	24′–0″	13′–0″	27′–9″
19,001–22,000	11′–9″	7′–3″	25′–0″	26′–9″	15′–0″	29′–0″
22,001–27,000	11′–9″	8′–6″	26″–0″	27′–0″	16′–0″	30′–0″
27,001–32,000	13′–0″	9′–9″	27′–9″	29′–0″	18′–0″	31′–9″

Part B. SI Units

L/s Range	Approximate Overall Dimension of Supply Air Equipment (m)			Recommended Room Dimensions (m)		
	Width ×	Height ×	Length	Width ×	Height ×	Length
470–849	1.4	0.8	4.5	3.8	2.7	5.7
850–1,415	1.5	1.1	4.9	4.2	2.7	6.1
1,416–1,887	2.1	1.3	4.9	5.3	2.7	6.1
1,888–2,831	2.3	1.3	5.1	5.5	2.7	6.3
2,832–3,303	2.4	1.4	5.6	5.6	2.9	6.8
3,304–4,247	2.4	1.5	5.7	5.8	3.0	6.9
4,248–5,662	3.0	1.7	6.4	7.0	3.4	7.6
5,663–7,550	3.1	1.8	6.7	7.2	3.8	7.9
7,551–8,966	3.2	2.0	7.2	7.3	4.0	8.5
8,967–10,381	3.6	2.2	7.6	8.1	4.6	8.8
10,381–12,741	3.6	2.6	7.9	8.2	4.9	9.1
12,742–15,100	4.0	3.0	8.5	8.8	5.5	9.7

Source: AIA, Ramsey and Sleeper, *Architectural Graphic Standards*, 9th ed., © 1994 by John Wiley and Sons. Metric conversions by the author.

Vane axial controllable pitch: the automatically controlled pitch responds to changes in temperature, humidity, air flow, etc., depending on the sensors used. Commonly used in VAV systems.

Filters and other methods of IAQ control were discussed in Chapter 6. There are a huge variety of combinations of air handlers, filters, and coils for air heating/cooling, selected by the consulting engineer. For the architect, the primary issues include adequate space, adequate access (for present maintenance and future equipment replacement), and noise isolation.

(h) Controls. Although the most obvious HVAC control function is to maintain desired comfort conditions, controls also increase fuel economy by promoting optimum operation, and

(a) **Equipment Room Plan**

(b) **Equipment Room Plan**

(c) **Equipment Room Section A–A**

Fig. 8.41 *Some variations on a huge variety of HVAC equipment rooms.* (a) *Plan, with floor-mounted air-conditioning unit. Outdoor air is drawn from an adjacent shaft; return air ducts are above the equipment, and a remote exhaust fan expels what is not recirculated.* (b) *Plan, on an exterior wall. Outdoor air is taken through the wall, and a return air fan is overhead.* (c) *Section through Plan* b. *(Reprinted by permission from AIA, Ramsey, and Sleeper,* Architectural Graphic Standards, *9th ed., © 1994 by John Wiley and Sons.)*

act as safety devices, limiting or overriding mechanical equipment. They also eliminate human error: controls fall asleep only during a power failure.

Although precise control of temperature and humidity everywhere in a building may be a tempting thought, controls can usually maintain only a *range* of conditions, not a setpoint. This range (within which neither heating nor cooling is called for) is called the *deadband.* Temperature variations occur vertically within a space—the higher the space, the greater the variation. Variations between horizontal positions within a zone are highly likely, especially where one or more walls are exterior and where different rooms share a zone controlled by a single thermostat. Variations in time also occur: a building will always be warmer at the moment the furnace fan turns off than at the moment the fan turns on.

Individual controls can be classified as follows: *controllers,* which measure, analyze, and initiate action; *actuators,* which are the controller's slaves and in turn become the masters of pieces of equipment; *limit and safety controls,* which may function only infrequently, preventing damage to equipment or buildings; and *accessories,* a miscellaneous collection.

Systems of controls can be classified by power source: electric (both analog and direct digital control); pneumatic (in which compressed air is the motivating force); and self-contained, including "passive" controls such as those motivated by thermal expansion of liquids or metals. Another way to classify control systems is by the motion of the controller equipment: *two-position* systems are of the simple on–off type; *multiposition* systems have several varieties of ON position, commonly used for separate operation of more than one machine; *floating* controls can assume any position in the range between minimum and maximum; *central logic control* systems can be programmed to integrate the many aspects of building control into one decision-making unit, and are now the prevalent approach to building management.

Control diagrams for two common HVAC applications are shown in Fig. 8.42. Single-duct, variable-air-volume systems (Fig. 8.42*a*) are further described in Section 8.5; with the constantly varying flow rate, the fan must be regulated so as to

THERMAL CONTROL

Fig. 8.42 *Controls for some common HVAC applications.* (a) *Single air duct with variable air volume (VAV).* (b) *Economizer cycle.*

maintain the minimum pressure (and therefore, flow) needed at the most demanding outlet. This outlet may be either the one most remote from the fan or the one needing the greatest flow (because it has the highest gain) at the moment. Economizer cycles (Fig. 8.42*b*) compare outside to inside positions, and vary the proportion of fresh (outdoor) air to return air, to provide "free" cooling (see again Fig. 8.35).

Most of today's large buildings are regulated by central logic control systems, usually called building management systems (BMS). The goal is a productive, cost-effective environment achieved by optimizing the interrelationships between the building's structure, systems, services, and management. The HVAC system is centrally regulated and inter-tied with lighting, electric power (such as load shedding), elevators, service hot water, access control and security, telecommunications, and information management. Such systems not only maintain comfort with energy conservation, they also sound the alarm about malfunctions, can learn from past practice, and keep records of performance.

This integration and automation of the many building control systems is made possible through

direct digital control (DDC), which can be applied to a wide variety of elements. They provide "dynamic control" in the case of HVAC anticipating time-based changes in heat flow patterns or in occupancy schedules. These actions depend on direct digital microcontrollers, located on each piece of regulated equipment throughout the building. Three building types show applications of this comprehensive automated control opportunity.

Laboratories have proven to be especially difficult HVAC control problems due to their fume hoods. Conditioned air is provided from the central HVAC system, often by a VAV supply. Whenever a fume hood is exhausting air (frequently in huge quantities), the VAV supply and return systems are affected. Complicating this relationship is the nature of the laboratory work; where the experiment could be damaged by outside contaminants, the lab should be positively pressurized to minimize infiltration (and the VAV must therefore supply and return slightly *more* air than the fume hood exhausts). However, where the experiments involve diseased, toxic, or other hazardous substances, the lab must be negatively pressurized (and the VAV must supply

Office cross-section

(a)

Floor construction

(b)

Fig. 8.43 *Natural ventilation and daylight dominate the north–south section* (a) *of the British Research Establishment building for fire research, "Building 16," at Garston.* (1) Stack ventilation (hot, calm); (2) clerestory BMS-controlled ventilation; (3) night ventilation through slab, BMS-controlled; (4) cross ventilation by-pass over enclosed offices; (5) enclosed office single-sided ventilation; (6) corridor cross-over zone; (7) manually operated lower-level windows; (8) high-level BMS-controlled windows; (9) motorized external shading louvers, also BMS-controlled. (b) *East–west section detail of precast floor structure.* (1) Luminaire with integral photosensors; (2) heated/cooled screed, using a geothermal source; (3) raised access floor for wiring; (4) cross ventilation duct (night ventilation and/or cross-over ventilation for enclosed offices); (5) wave-form precast concrete ceiling with poured-in-place topping slab. *(Courtesy of Feilden Clegg Architects, Bath, England.)*

and return slightly *less* air than the fume hood exhausts). DDCs tied to a central system can balance energy conservation, lab worker safety, and safety for the nonlab environment.

Hotel rooms can present serious energy loss problems, from heating/cooling either an unoccupied room or a room with open windows. With DDCs inter-tied with the registration desk, an "unoccupied" mode of operating can be remotely controlled, with bare-minimum heating or cooling. When the room is occupied, the supply of either hot water heating or cooled air can be throttled back whenever the window is open. Also, a "purge" mode can enable a new arrival (or the front desk, in anticipation) to select a greatly increased flow of outdoor air for a limited time

period, to dilute cigarette smoke or other odors. Chapter 22 presents more detailed information on "intelligent buildings."

Offices might be provided with DDC not only for the VAV supply units, but inter-tied DDCs for a ventilating window (preventing simultaneous open windows and treated forced-air delivery), daylight reflectors (mini-lightshelves), venetian blinds, radiant heater valve, electric light switch, and an insulating shade. A control panel or "dashboard" gives the worker an opportunity to interact with the central control in operating these devices.

Office buildings that use more passive strategies also benefit from BMS. The British Research Establishment building for fire research, "Building 16" (Fig. 8.43), depends on cross and stack ventilation for its cooling, and moveable louvers on the south windows for sunshading and daylight penetration. A common network links this building's control systems; each worker has a TV-like controller that regulates lights and can override the programmed settings of the nearest high-level windows (for ventilation) and south-window louvers (for sun control and daylighting). This innovative office building uses a ground heat pump ("borehole") to supplement solar and internal winter heat gains, or to provide supplementary cooling to the night ventilation system. Note the provisions for cross ventilation for individually enclosed offices on the ground floor, utilizing a cavity above the ceiling to carry the ventilation air on to exhaust on the other side of the building.

Rapid changes in centralized control systems are continuing. Initially, DDC systems were developed by individual companies, with little or no opportunity for system communication between systems. Such proprietary systems left designers frustrated by an inability to specify a wide variety of products, all controlled by a single BMS. In the late 1990s, two competing systems emerged that promise integration of products and systems from different manufacturers. BACnet was developed by ASHRAE committee members as a nonproprietary communication standard. LonMark was developed by the LonMark Interoperability Association, a user-funded organization of building owners, specifiers, system integrators, and product suppliers. "Open control system architecture" is an opportunity for a BMS in which components from several vendors interoperate over a BACnet-adapted Ethernet LAN (local area network). It makes possible a system combining BACnet, LonMark, and proprietary subsystems (Fig. 8.44).

8.4   Air Distribution Within Spaces

When large office buildings had far greater interior (core) areas than perimeter areas and were filled with less efficient lighting at high luminance

Fig. 8.44 *Building Management System (BMS) that is open to various protocol standards. (Courtesy of Honeywell, Inc.)*

levels, they were considered to always need cooling. Because cool air supplied at the ceiling would naturally fall toward the level of occupants and because suspended ceilings were ubiquitous, supply air from the ceiling was almost universal. The suspended ceiling was also tempting for the return air provisions, whether as a plenum or another ducted system. With both supply and return air at the ceiling, the danger of short-circuiting arose; supply air heading quickly for the return opening, with resulting shortages of both cooling and IAQ. In this section, we look first at approximate duct sizing, then consider three air distribution systems for multistory office buildings.

(a) Air Ducts. Duct sizes (in cross section) are frequently of interest early in the design process. Duct depths can help determine floor heights; duct cross sections influence the sizes and shapes of the vertical cores that serve multistory buildings. An approximation of duct size can be obtained as follows.

1. Determine the quantity of air to be distributed through the largest duct, using Table 8.5, or the air changes per hour, from a calculation of night ventilation of thermal mass (e.g., Example 5.1, Part J). This will usually be expressed in cubic feet per hour (cfh).
2. If necessary, convert cfh to cfm:

$$\frac{\text{cfh} \times 1\,\text{h}}{60\,\text{min}} = \text{cfm}$$

3. Find the maximum velocity of this air within the duct from Table 7.4 (page 408), expressed in feet per minute (fpm).
4. The approximate *minimum* required cross-sectional area of the duct A is then

$$A_{\text{in.}^2} = \frac{\text{volume of air (cfm)}}{\text{velocity (fpm)}}$$

$$\times\ 144\ \text{in}^2/\text{ft}^2 \times \text{friction allowance}$$

where the friction allowances are

- round ducts = 1.0 (may be neglected)
- nearly square ducts (ratio of width to depth, 1 : 1):

    small (<1000 cfm) = 1.10

    large (>1000 cfm) = 1.05

- thin rectangular ducts (ratio of width to depth, 1 : 5) = 1.25.

Then check this against the *recommended* duct cross-sectional area from Table 8.4. Remember that the *minimum* duct area will carry a penalty of increased noise and friction. Techniques for noise suppression in ductwork are discussed in Chapter 28.

(b) Ceiling Air Supply. This approach is so widespread that many lighting fixtures are made to either serve as diffusers for supply air or as intakes for return air. As return air fixtures, they are especially effective because they remove much of the heat from electric lighting before it can contribute to overheating the office space below. This makes less work for the cooling equipment.

When individual offices are enclosed, a supply diffuser and return grille insures air circulation. However, as open office space proliferated, concerns grew about whether the workspace cubicles were being adequately served with conditioned air. Air diffusers and grilles (Fig. 8.45) on the ceilings are uncluttered by furniture and independent of rearrangements of workstation cubicles. Canadian research of the early 1990s indicated that the most important variable was the quantity of the airflow, not the location of the diffusers relative to workstations or whether the cubicle partitions have a small gap where they meet the floor.

Are higher ceilings better for air distribution? There seems to be a tradeoff between more vertical clear space (between ceiling and cubicle partitions) that would allow a wider range of air distribution and the increased distance from the actual occupant who should be receiving the conditioned air. The increased distance would allow colder air to be distributed, resulting in smaller ducts, and could also allow higher velocity of supply air, further reducing duct size. Higher ceilings allow deeper daylight penetration and represent a larger "pool" of air, slower to grow polluted by occupants or office equipment. Higher ceilings also encourage underfloor distribution, as seen in the next subsection.

When the noise from forced air diffusers is critical, as in telecasting and recording studios, pin-hole perforated diffusers (Fig. 8.46) can pro-

THERMAL CONTROL

L   24" to 72"
W   24" to 72"
Capacity varies with air
velocity

(b)

AIR DISTRIBUTION OUTLETS

KEY

1 RECTANGULAR LOUVERED FACE DIF-
   FUSER: Available in 1, 2, 3, or 4-way pattern,
   steel or aluminum. Flanged overlap frame or inserted in 2 × 2 ft or 2 × 4 ft baked enamel steel
   panel to fit tile modules of lay-in ceilings. Supply
   or return.

2 ROUND LOUVERED FACE DIFFUSER: Normal 360° air pattern with blank-off plate for
   other air patterns. Surface mounting for all type
   ceilings. Normally of steel with baked enamel
   finish. Supply or return.

3 RECTANGULAR PERFORATED FACE DIF-
   FUSER: Available in 1, 2, 3, or 4-way pattern,
   steel or aluminum. Flanged overlap frame or
   2 × 2 ft and 2 × 4 ft for replacing tile of lay-in
   ceiling can be used for supply or return air.

4 ROUND PERFORATED FACE DIFFUSER:
   Normal 360° air pattern with blank-off plate for
   other air patterns. Steel or aluminum. Flanged
   overlap frame for all type ceilings. Can be used
   for supply or return air.

5 LATTICE TYPE RETURN: All aluminum square
   grid type return grille for ceiling installation with
   flanged overlap frame or of correct size to replace
   tile.

6 SADDLE TYPE LUMINAIRE AIR BOOT: Provides air supply from both sides of standard size
   luminaires. Maximum air delivery (total both
   sides) approximately 150 to 170 cfm for 4 ft long
   luminaire.

7 SINGLE SIDE TYPE LUMINAIRE AIR BOOT:
   Provides air supply from one side of standard size
   luminaires. Maximum air delivery approximately
   75 cfm for 4 ft long luminaire.

8 LINEAR DIFFUSER: Extruded aluminum, anodized, duranodic, or special finishes, one way or
   opposite direction or vertical down air pattern.
   Any length with one to eight slots. Can be used
   for supply or return and for ceiling, sidewall, or
   cabinet top application.

9 INTEGRATED PLENUM TYPE OUTLET FOR
   "T" BAR CEILINGS: Slot type outlet, one way
   or two way opposite direction air pattern. Available in 24, 36, 48, and 60 in. lengths. Replaces or
   integrates with "T" bar. Approximately 150 to
   175 cfm for 4 ft long, two-slot unit.

10 SIDEWALL OR DUCT MOUNTED REGISTER:
   Steel or aluminum for supply or return. Adjustable horizontal and vertical deflection. Plaster
   frame available. Suitable for long throw and high
   air volume.

(a)

Fig. 8.45 *Common air distribution outlets.* (a) *Air diffusers of types 1 through 5 are typical ceiling application; types 6 and 7 integrate with luminaires; types 8 and 10 may be used in other-than-ceiling applications; type 9 is a nearly "invisible" hung-ceiling application.* (b) *Typical heat-of-light luminaires that include both supply and return air. (Reprinted by permission from AIA, Ramsey, and Sleeper,* Architectural Graphic Standards, *9th ed., © 1994 by John Wiley and Sons.)*

vide large quantities of low-velocity air. This is an especially challenging combination of high heat gains (from lights) and low background noise requirements. Again, ceiling distribution leaves the floor unencumbered.

(c) Underfloor Supply with Displacement Ventilation. As open office spaces grow larger in area and their ceilings higher (for daylight penetration and indirect lighting), and concerns about IAQ deepen, the floor has gained in popularity as the air source location. Displacement ventilation provides fresh air, cooled to just below the design room temperature, at a low velocity. Furthermore, fresh air supply rates are typically near 45 cfm (22 L/s), compared to about 30 cfm (15 L/s) often used in ceiling supply systems. This high-volume, low-ΔT, low-velocity combination requires greatly increased duct sizes, provided by using the plenum below a raised floor system. The fresh air rises quickly to the occupant level, drawn upwards by the stack effect of heat gain from lights, people, and office equipment. It continues to the ceiling, where it stratifies, and where various approaches to return air collection are located. Freshest and coolest at the floor, hottest and most stale at the ceiling, this system promises better IAQ and thermal comfort than would be available from ceiling supply–return systems. Furthermore, the outlets at the floor are adjustable and easily reached by the office workers.

The raised floor is typically supported on a 2 × 2 ft (600 × 600 mm) module. The module that contains the diffusing outlet often also contains an outlet for power and cable. This diffuser is placed off-center within the modular unit, further expanding the variability of the floor opening relative to furniture placement because each plan rotation of the square module places the outlet slightly differently. Some codes restrict the height of the raised floor plenum; many codes require specially coated wiring within the plenum because moving air will wash over the wires for the life of the building.

The overall floor-to-floor height needs to be great enough to accommodate both the raised floor and the higher office ceiling [at least 9 ft (2.7 m) to accommodate warm air stratification]. However, the elimination of the suspended ceiling and the supply air ductwork also eliminates the suspended ceiling cavity. This tends to expose the

Fig. 8.46 *Pin-hole perforated diffusers (above lighting grid) provide large quantities of low-velocity air to the television studio at the Community Media Center of Santa Rosa, California. The background noise from more typical HVAC systems would be amplified by sound recording equipment, hence unacceptable. (© TLCD Architecture, Santa Rosa, California.)*

THERMAL CONTROL

(a)

(b)

Fig. 8.47 *Library Square, Vancouver, British Columbia, supplies conditioned air to the central book stacks (a) via an underfloor plenum (b); this displacement ventilation results in return air taken through openings in the exposed structural ceiling. (Plan courtesy of Moshie Safdie and Associates; section adapted from information supplied by Blair McCarry, Keen Engineering, North Vancouver, B.C.)*

lighting, fire sprinkler, and public address systems below the structural ceiling, although the latter can be distributed in the raised floor cavity above, and puncture the structural ceiling as required.

The structural slab below the raised floor can be of concrete, and this thermal mass can be night ventilated to store coolth to assist the ventilation air for the next day. The Inland Revenue Building (see Fig. 5.48) incorporates a night-ventilated raised floor section within its precast wave-form floor sections.

Library Square in Vancouver, British Columbia (Fig. 8.47), utilizes a raised floor to serve its interior zones of book stacks. The library is seven floors, totals 390,000 ft² (36,320 m²), and exposes its concrete structure as a finished ceiling. The floor-to-floor height is 16.4 ft (5 m). The underfloor supply is fed by VAV boxes and floor fan units (FFU) that blend low-temperature supply air (about one-third) with local filtered return air (about two-thirds) to provide 63°F (17°C) air to the pressurized plenum. Flush to the floor, high induction swirl floor diffusers [8 in. (200 mm) in diameter] produce an upward air motion, taking the fresh-return air blend directly into the occupancy zone [the first 6.5 ft (2 m) of height on each floor]. The stack effect of heat from people, computers, and lights takes hot air on to the strati-

fication zone, above. Thus, people-generated pollutants tend to rise to this hot, stale air zone, then into return air slots in the precast concrete ceiling. Return air is taken away in insulated ducts within the supply air plenum of the floor above. The plenum height is an unusually generous 2 ft (600 mm) high; it also contains the sprinkler distribution and all wiring systems. The warm-dry summer climate supports a night-ventilation system, flushing the plenum and storing coolth in the exposed concrete structure. This has reduced the use of an ice-storage system.

(d) Workstation Delivery Systems. The air delivery system shown in Fig. 8.48 allows a variety of individual controls at each workstation. As developed by Johnson Controls, Inc., these systems are called Personal Environments® systems. A mixture of outdoor and recirculated indoor air, called *primary air,* is brought from the

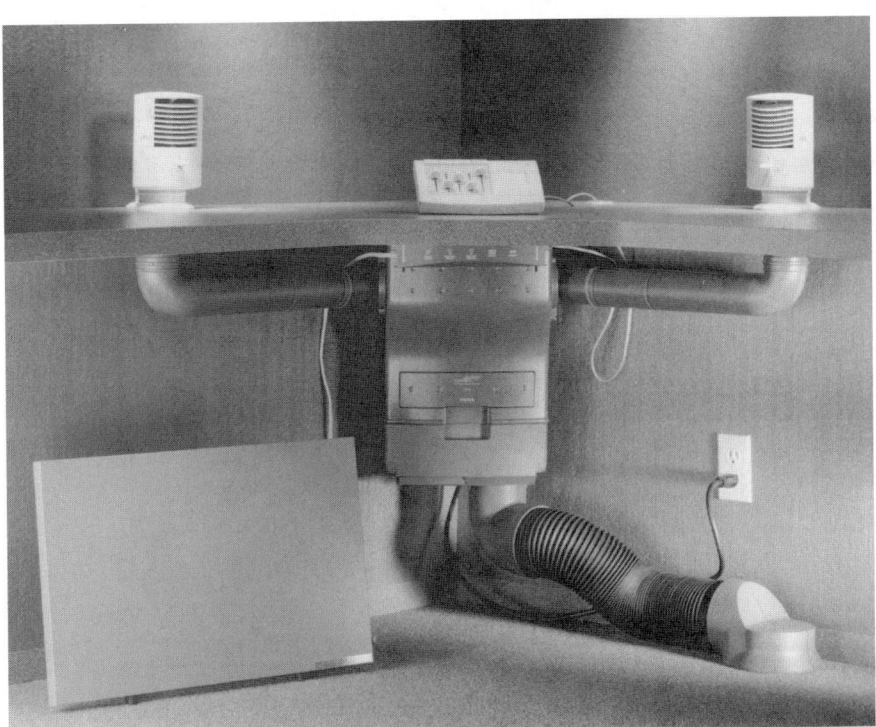

Fig. 8.48 *The Personal Environments® system provides each workstation with a fan, air filters, air mixing box, and a background sound ("white noise") generator. The control panel allows adjustment of task lighting, background sound, fan speed, primary/recirculated air mixture, and radiant heating. Two diffusers distribute both air and background sound. Diffusers are adjustable about both horizontal and vertical axes. The below-desk radiant heating panel warms the lower body. (Courtesy of Johnson Controls, Inc., Milwaukee, Wisconsin.)*

main duct (or floor plenum) to each Personal Environments system's mixing box, in a duct carrying at maximum 120 cfm (56 L/s), but typically less. Each worker can adjust supply air temperature, the mixture of primary and locally recirculated air, air velocity (and therefore volume) and direction, and radiant supplementary heat (below desk level). Task lights can be dimmed and the back-

ground (masking) sound level is adjustable. The worker has almost as much environmental control at each office workstation as we have in our automobile's front seat. When the workstation is unoccupied, an occupancy sensor shuts the system down, maintaining a minimum air flow of 12 cfm (5.7 L/s).

The S.C. Johnson Commercial Markets, Inc.,

(a)

Fig. 8.49 *The S.C. Johnson Commercial Markets, Inc., building in Sturtevant, Wisconsin, takes advantage of high ceilings to provide daylighting and underfloor distribution to offer personal control for each workstation. (a) An upper floor plan shows a mix of individual laboratories and open office areas. Each office workstation has a Personal Environments system. (b) Section at office perimeter, with 6 ft (1.8 m) wide lightshelf. From the suspended ceiling, a rolling translucent shade desends when needed, just inside the upper window. Radiant heating units sit below the lower window. A workstation served by a Personal Environments system is visible at the left side. The raised floor above the structural slab forms a supply plenum, fed by the main duct below the structural slab. (Courtesy of the Zimmerman Design Group, Milwaukee, Wisconsin.)*

(b)

Fig. 8.49 *(continued)*

building in Sturtevant, Wisconsin (Fig. 8.49) gives each workstation a Personal Environments system within a large open-office floor area. Total gross floor area is 277,000 ft^2 (25,734 m^2). Each floor contains both laboratories and office areas, so the ceiling heights in the office areas are a generous 12 ft (3.7 m). This allows for deeper daylight penetration from the large glass areas on the south and west sides, supplemented by daylight from the central atrium. A lightshelf 6-ft (1.8-m) wide helps distribute the perimeter daylight more evenly. The Personal Environments systems are fed with primary air from the floor plenum, which is in turn fed by a main duct

that lies between the structural slab and the suspended ceiling of the floor below. The individual controls at each workstation result in the overall ambient indoor temperature being slightly higher in summer, slightly lower in winter, saving energy in both seasons.

Air from the workstations rises to the high ceiling and some is reintroduced to the ceiling plenum, then recirculated to the Personal Environments systems. Most "used" air is expelled via toilet rooms, coffee nooks, and copy rooms, helping to maintain IAQ throughout the building. When the economizer cycle is operating, additional air is exhausted from the top of the atrium.

(e) Alternative Supply/Return Systems.
So far, the supply/return systems have been either both from the ceiling or supply from the floor and return at the ceiling. Supply from the ceiling, return through the floor is another possibility. The fresh air gets to the occupancy level later, and individual control is much more difficult, but debris from the floor falls into a return, not a supply, plenum. There seems less danger of debris falling into a floor supply diffuser while the system operates; it is during fan-off periods, that may include custodial maintenance, that this possibility increases.

An alternative approach with return at the perimeter is shown in Fig. 8.50. It is variously called an *air-extract window, air curtain window,* or a *climate window.* Developed in Scandinavia in the 1950s, this is a triple-glazed window that passes room air between a typical outer double-glazed window and an inner, single pane. The inner pane thus is kept at very nearly the same temperature as the room air, which greatly increases comfort near windows on very cold (or very hot) days. Venetian blinds are often inserted in this cavity, where they can intercept direct sun and redirect its light toward the ceiling. The solar heat intercepted by the blind is carried off by the room air to a plenum above the ceiling, where such air can be either exhausted or recirculated, and its heat content either reclaimed or rejected. The U-value of these windows is dependent on the rate of airflow between the glazing (Fig. 8.50c); typical flow rates are 4 to 6 cfm per foot (6 to 9 L/s per meter) of window width.

The Ocosta Junior/Senior High School at Westport, Washington (Fig. 8.51), uses these windows on the south facade. (For an example of such windows in a larger U.S. building, see Fig. 8.72, which shows the Comstock Center in Pittsburgh.) The classrooms have photocells that switch off some lights nearest the windows when daylight is adequate, and water loop heat pumps help to transfer passive solar heat from the south side to the sunless north side rooms. Excess heat in the Ocosta school is stored in an underground water tank. The compact plan, highly insulated envelope, and mass capacity within the school reduce the need for a boiler to supplement the winter temperature of the water loop. During consistently warm weather, the BMS exhausts the air from the window to the outdoors.

Supply openings in vertical risers are relatively rare, but at times function effectively. The Frederick Meijer Gardens near Grand Rapids, Michigan, is a huge indoor tropical rain forest. This glass structure is more than 65 ft (20 m) high; it has more glass area than floor area—great for a high daylight factor, but imagine the heating loads. High humidity must be maintained, yet water condensation on leaves is undesirable. A strong horizontal air flow could be helpful to the plants, acting as wind would to develop a stronger root structure.

The solutions here begin with a perimeter tunnel serving as an air supply duct, delivering air evenly around the glass walls, heated by finned tubes at planting level that are largely hidden by foliage. The tunnel insulates the planting beds

Fig. 8.50 Air extract window (or air curtain, or climate window). (a) Cross section showing the window as a solar collector. (b) The window as a "solar chimney" exhausting hot air. (c) The U-value of these windows varies with the rate of airflow within the glazing: typical flow rate is 4 to 6 cfm per foot (6 to 9 L/s per meter) of window width. (From articles by D. Aitken and O. Seppanen in Proceedings of the Sixth National Passive Solar Conference. *© 1981 by the American Solar Energy Society, Boulder, Colorado.)*

Fig. 8.51 *The Ocosta Junior/Senior High School at Westport, Washington, uses air extract (air curtain) windows to help distribute solar energy from the south side [to the right in section (a)] to the north classrooms. (b) Detail through the south facing windows. Photocells control the electric lights during daylight hours; venetian blinds direct sunlight toward the ceiling and deeper into the classroom. (Courtesy of Burr Lawrence Rising + Bates Architects, Tacoma, Washington.)*

THERMAL CONTROL

(a)

(b)

Fig. 8.52 *With more glass area than floor area, the Frederick Meijer Conservatory near Grand Rapids maintains a tropical environment in Michigan's winter. (a) A perimeter tunnel delivers air evenly around the walls, heated by finned tubes just above planting level. (b) A separate "wind" system provides a continuous breeze from diffusers in the vertical ducts, to strengthen plants' root systems. (c) A fogging system maintains tropical humidity while adding evaporative cooling under summer conditions. (Photos by Roger Van Vleck, courtesy of Fishbeck, Thompson, Carr & Huber, Ada, Michigan.)*

(c)

Fig. 8.52 (continued)

from the extremes of the exterior, maintaining constant ground temperatures year round. A separate "wind" system continuously feeds air to vertical ducts with several diffusers (Fig. 8.52b), making leaves and branches sway. Humidity is provided by a high-pressure water system that creates fog (very small water particles). This is located high in the space (but also near waterfalls and a stream for special effects). This is also an evaporative cooling system to counteract summer overheating. Peak air flow is needed only for cooling; return air is taken through a large opening hidden behind a waterfall. About one-third of the peak supply air is returned; the remainder is exhausted at the top of the structure, through operable vents high in the glass roof. A BMS controls the roof vents, the fogging system, the finned tube heating system, rolling shading screens, and the heating/cooling equipment in a basement, all in response to changing conditions outside this enormous glass house.

8.5 All-Air HVAC Systems

This large and complex family of systems was introduced in Section 8.2. All-air HVAC systems require thick distribution trees—therefore more building volume—but can promise comfortable results because they effectively regulate IAQ, temperature, and humidity. This approach concentrates the mechanical equipment and its maintenance within a central mechanical room, freeing the rest of the building from repair interruptions and worries about water leaks from heating/cooling pipes. However, this family does less well with perimeter zones in cold climates.

With all-air systems, air quality can be manipulated through pressure control: *negative pressures* in odorous or excessively humid locations (kitchens, toilet rooms, pet shops in shopping centers, etc.), *positive pressures* in shopping malls, corridors of apartment houses, stair towers, and so on. The difference in pressure sets up an overall direction of air flow that helps prevent the spread of odorous or otherwise contaminated air and can help manage smoke in a fire (see Chapter 13). Positive pressures in connecting spaces help each adjacent space keep its air to itself.

(a) Single-Zone Systems (Fig. 8.53). Smaller buildings are often served by this least-complicated all-air example (see examples in Section 7.8). Where an entire building is essentially

Fig. 8.53 *Single-zone, single-duct HVAC system. Underfloor air supply is shown here, but ceiling supply is also common.*

one thermal zone, uncomplicated enough to be served by only one thermostat, choice of this system offers very low first cost. A closely related system is the multizone (see later this chapter), which is a combination of single zones.

(b) Single-Duct VAV Systems (Fig. 8.54).

As described in Section 8.2, this system is an energy-saving option ideally suited to the internal zones of large buildings, where cooling is always needed. However, it can also be adapted to serve the entire building, with important savings over the constant-volume (CV) systems described later [paragraph (*e*)]. The air-handling units (fans, etc.) for each thermal zone are sized to meet the peak demand on that zone. In most CV systems, the fans always run at this peak condition speed, even though peaks are often of rather short duration. In VAV systems, fans only run at peak speed during peak hours. This obviously saves considerable energy needed to run fans. When the building uses only one or a few central fans, a VAV system will require smaller fans because the fans need meet only one peak condition at a time and do not have to be sized for all zones' peak flows. The variation in demands on the fan can be met either by selecting variable-pitch blades or (less expensively) by varying the speed of the fan.

Where a VAV system is used, provision must be made for at least code-minimum fresh air levels, sometimes resulting in overcooling or in wasteful reheating. Although VAV systems are typically less noisy than CV systems (because less air is moving), the air motion noise *varies* with the volume, and variable noise sources are inherently more noticeable than are steady ones.

Within (or near) the spaces it serves, the VAV system typically needs a mixing box or terminal; this is often placed above a suspended ceiling or below a raised floor. Of the several terminal types available, the standard and most simple one is shown in Fig. 8.55. This terminal serves not only to vary the quantity of air, but to both attenuate

Fig. 8.54 *Variable-air-volume (VAV) HVAC system. Underfloor air supply is shown here, but ceiling supply is also common.*

RANGE OF DIMENSIONS

CFM	HEIGHT	LENGTH	WIDTH
400	8″- 9″	24″-39″	14″-30″
800	10″-11″	24″-53″	18″-42″
1600	14″	30″-48″	22″-44″
2400	16″	42″-60″	26″-54″
3200	18″	42″-67″	33″-54″

Fig. 8.55 The most simple type of VAV terminal, which pinches back the volume of incoming air as thermal loads decrease. (Reprinted by permission from AIA, Ramsey, and Sleeper, Architectural Graphic Standards, *9th ed., © 1994 by John Wiley and Sons.)*

the noise and reduce the velocity of air from the main trunk of the distribution tree. High velocity is commonly used in main ducts because it reduces the size (thickness) of these critical, large portions of the tree.

The Utah Department of Natural Resources in Salt Lake City (Fig. 8.56) utilizes VAV throughout its three-story, 105,000 ft² (9755 m²) office building. Air is supplied from linear ceiling diffusers located between the rows of indirect luminaires, parallel to the windows. Return air is taken into the suspended ceiling plenum. Highly insulated walls, windows, and roof lessen the need for perimeter heating, although a supplementary warm water heating system is built in below the windows to offset perimeter heat losses. The lower 1 ft (300 mm) is a slot where room air is drawn in by convection, and a perforated metal grille at the top forms the window sill.

The building is elongated east–west for best daylight control and winter solar gain; lightshelves even the daylight penetration. With almost no east or west facing glass, the building relies on 4200 ft² (390 m²) of south glass and 3800 ft² (353 m²) of north glass to serve its width of

Fig. 8.56 The Utah Department of Natural Resources building features lightshelves, direct and indirect evaporative cooling, and the economizer cycle. Economized, evaporatively cooled air is distributed by a VAV system supplied and returned between luminaires at the ceiling. (Based on the design by Gillies Stransky Brems Smith Architects, Salt Lake City, Utah.)

125 ft (38 m). Supplementary indirect lighting is controlled by photocells and dimming ballasts.

Utah's hot dry summers permit a four-stage approach to cooling. First is the economizer cycle (see Fig. 8.35); at higher temperatures, direct evaporative cooling is used; then indirect evaporative cooling (see Figs. 7.10 and 7.11) utilizing an oversized cooling tower; and finally (estimated at about 10 days per year), a conventional chiller.

There are several variations on the basic VAV system, some of which respond to the problem of minimum airflows for rooms with little thermal load or to the desire to serve both interior and perimeter areas with the same HVAC system.

(c) Fan-Powered VAV Systems (Fig. 8.57).

To maintain minimum fresh air, VAV terminals are usually set so that they cannot be entirely closed off, ensuring some outdoor air at all times. If this fresh air minimum, entering at low velocity, does not provide the desired air motion and mixing within the room, VAV terminals can be equipped with fans, which are activated as needed with decreasing incoming air volume. These self-contained fans mix room air with incoming air to provide an airstream of the right temperature and velocity to maintain comfort.

This approach to VAV is sometimes taken when simultaneous heating and cooling are needed, such as at perimeter zones, which can generate sizable heating needs while the rest of the building needs cooling. Another approach is to utilize an induction-type VAV terminal, with which air heated by electric lights is induced to join the incoming cool airstream. Greater heating needs often require the use of reheat terminals

supplied by a circulating hot water system or by electric resistance heating. This reheat application is more energy efficient than the standard—CV—reheat system described in subsection (e), because a much smaller volume of air is first cooled, then reheated. The water or electric coils can be incorporated either in the VAV terminal or in the ductwork between the terminal and the space it serves.

An example of a floor-by-floor VAV system (Fig. 8.58) is a 1-million ft^2 (92,900 m^2) 28-floor Chicago office building designed by Skidmore Owings & Merrill. This mid-rise approach to office towers utilizes three "stacked" atriums to relieve the monotony of the wide interior floors. Another result is lower structural and energy costs per square foot relative to conventional high-rise structures. The cubelike shape of the building exposes less skin area (38% of which is in insulating glass) to Chicago's cold winters; electric lighting at about 1.8 W/ft^2 holds down internal gains. To accommodate the differing schedules and comfort needs of a variety of tenants, each floor is provided with two VAV supply fans that can be operated nights and weekends, independently of the rest of the building. Each floor's mechanical core has one exterior wall (on an alley) to facilitate fresh air intake/stale air exhaust. The perimeter heating system is electric resistance fin radiation; an economizer cycle provides cooling with outdoor air below 55°F (13°C).

(d) Multizone Systems (Fig. 8.59).

A multizone system is a collection of single-zone systems served by a single supply fan; such systems rarely exceed eight zones per air-handling unit.

Fig. 8.57 *Fan-powered variable-volume (VAV) HVAC system. Underfloor air supply is shown here, but ceiling supply is also common.*

Simultaneous heating of some zones and cooling of others is possible, but leakage between zones at the decks of hot and cold coils is common. Return air from all zones is mixed within one return duct; a by-pass at the heating and cooling decks requires more space but allows such air to be mixed with supply without undergoing unnecessary heating or cooling. One multizone system per floor of medium- to high-rise buildings is an increasingly common application.

(e) Single Duct with Reheat (Fig. 8.60).
Formerly a widespread system and usually supplied at constant volume (CV), single-duct reheat systems now are severely restricted by many codes and ASHRAE standards. For high-pressure

and high-velocity main ducts, CV reheat boxes (Fig. 8.61a) are used to control the noise, temperature, pressure, and velocity of the supply air. For more simple all-low-velocity systems, simple duct insert heaters (Fig. 8.61b) can be used. These are sized to fit the duct, which needs to be enlarged only slightly to accommodate them.

(f) Double-Duct Systems (Fig. 8.62).
This is still considered the "Cadillac" of HVAC systems, not only because of its superior comfort control and flexibility for simultaneous heating/cooling zones, but also its high initial cost, large size, and high energy usage. The double-duct system is more rarely installed now, except in hospitals. Building volumes needed for a double-duct

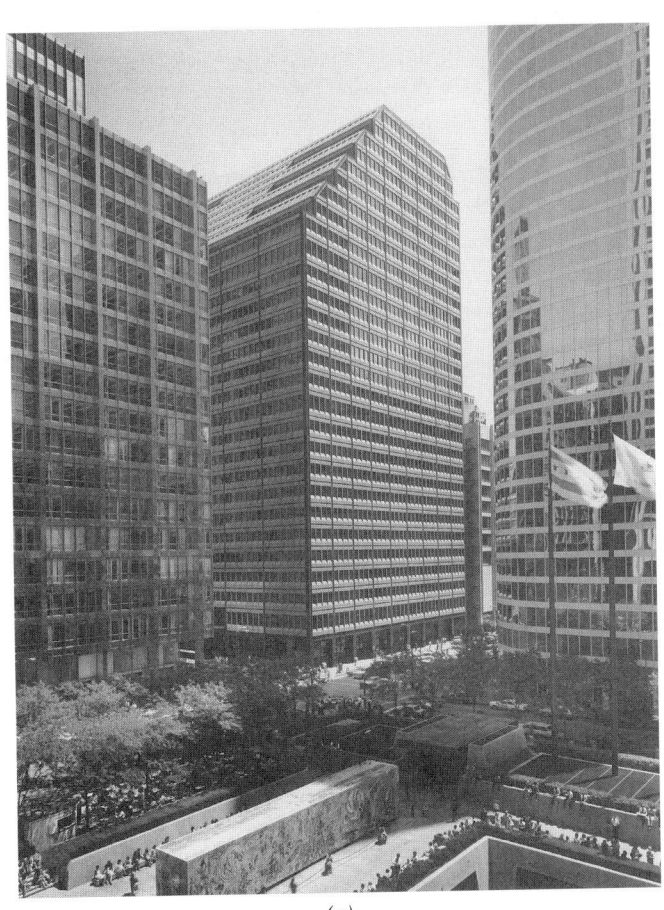

(a)

Fig. 8.58 *Floor-by-floor fan rooms supply VAV for this Chicago office building.* (a) *Exterior view of 33 West Monroe (photo by Merrick, Hedrick-Blessing).* (b) *Section perspective showing "stacked atriums."* (c) *Plan of a typical lower floor; two fan rooms occupy the space along the exterior wall. (Courtesy of Skidmore, Owings & Merrill, Architects-Engineers, Chicago.)*

THERMAL CONTROL

(b)

Fig. 8.58 *(continued)*

system's three (two supply, one return) full-sized air distribution trees are harder to justify, given that VAV systems can provide acceptable comfort for most common spaces. An example of a double duct office building was seen in Fig. 8.11, San Francisco's International Building.

The mixing boxes (terminals) of double-duct systems are similar to those of other all-air systems (Fig. 8.63), but are generally larger for the same air-flow capacity. This adds still more to their impact on building volume. They are expensive and may require maintenance. Although most double-duct systems are CV, they can be VAV when the reduction in airflow is no more than 50% below the maximum. For details on other double-duct variations, see Lorsch (1993).

8.6 Air & Water Systems

These systems, introduced in Fig. 8.13 and Section 8.2, have the design complexity—and first

33 WEST MONROE   FLOOR 6

(c)
Fig. 8.58 *(continued)*

Fig. 8.59 *Multizone HVAC system. Underfloor air supply is shown here, but ceiling supply is also common.*

Fig. 8.60 *Single duct with reheat HVAC system. Underfloor air supply is shown here, but ceiling supply is also common.*

RANGE OF DIMENSIONS

CFM	HEIGHT	LENGTH	WIDTH
200	9″-11″	30″-50″	16″-22″
400	9″-11″	30″-51″	18″-30″
800	9″-11″	30″-51″	22″-42″
1600	14″-16″	48″-51″	40″-44″
2400	16″-18″	60″-55″	40″-54″
3200	16″-18″	60″-55″	16″-66″
5000	20″-18″	60″-55″	20″-80″

(a)

(b)

Fig. 8.61 *Single-duct with reheat coils* (a) *require a simple terminal where velocity and pressure are reduced.* (b) *Electric coil within a duct. (Reprinted by permission from AIA, Ramsey, and Sleeper,* Architectural Graphic Standards, *9th ed., © 1994 by John Wiley and Sons.)*

Fig. 8.62 *Double-duct HVAC system. Underfloor air supply is shown here, but ceiling supply is also common.*

(a)

MIXED AIR OUTLET TO SUPPLY
DUCT AND ROOM OUTLETS

W

L

I

MIXING
DAMPER MOTOR
(PNEU. OR ELEC.)

ROUND OR
RECTANGULAR
CONDITIONED COLD
AIR INLET

ROUND OR
RECTANGULAR
CONDITIONED
HOT AIR INLET

High, medium, or low velocity systems. Inlet pressure
¼ to 1½ in. W.C. Capacity range from 150 to 2000
cfm per box (low velocity) to 5000 cfm (high velocity).
Box serves as converter from high to low velocity air
system, noise attenuator, and control device by mixing
hot and cold air streams.

RANGE OF DIMENSIONS

CFM	HEIGHT	LENGTH	WIDTH
400	6″-10″	40″-51″	30″-19″
800	8″-11″	50″-51″	42″-24″
1600	12″-14″	48″-51″	44″-40″
2400	14″-18″	60″-55″	54″-44″
3200	14″-18″	60″-55″	54″-44″
5000	16″-18″	60″-55″	54″-66″

(b)

Fig. 8.63 *High-velocity double-duct terminal, providing mixing and attenuation (pressure and sound reduction). (a) Pneumatically controlled from a thermostat, this unit blends and delivers air at the selected temperature. (Courtesy of Anemostat.) (b) Typical mixing box dimensions. (Reprinted by permission from AIA, Ramsey, and Sleeper,* Architectural Graphic Standards, *9th ed., © 1994 by John Wiley and Sons.)*

cost—of supply-and-return distribution trees for both water and air. This disadvantage is offset by the space-saving advantages of thinner water trees, the possibility of only one air tree (supply), and still having the superior comfort characteristics offered by air. When only fresh air is centrally treated and distributed, only an equal quantity of air need be exhausted, which rarely needs an extensive distribution tree. Because this exhaust air is not recirculated, these systems are attractive for hospitals and other buildings in which the mixing of air between zones is undesirable. When return air is used rather than exhausted, it forms a small percentage of the supply air.

These systems are most often used in perimeter zones, where extensive extra heating (winter) and

cooling (summer) is readily provided by water. If only one water supply tree is used (two-pipe system), there is a problem of deciding when to changeover between heating and cooling. In-space maintenance is required (filters in particular), and humidity is less tightly controlled than in all-air systems. In most air & water systems, the air is circulated in a cooled and dehumidified condition, and heated by the water for most of the year.

Air & water systems can often be found in the perimeter zones of office buildings, hospitals, schools, apartments, and laboratories.

(a) Induction (Fig. 8.64). Centrally conditioned fresh air is supplied (at either high or me-

Fig. 8.64 *High-velocity induction HVAC system.*

dium pressures and velocities) to each induction terminal. Each terminal then mixes 20% to 40% incoming fresh air with the 80% to 60% room air that is induced to flow along with the fresh air, all passing over finned tubes for heating or cooling and circulating this mixture of air to the space.

In two-pipe systems, either hot or cold water—not both—is available to temper this air mixture ordered by the thermostat linked to each terminal. Because cool air is distributed for most of the year, the two-pipe system is largely in heating mode (in colder climates).

In four-pipe systems, the availability of both hot and chilled water makes it possible to switch instantly from heating to cooling for excellent thermal control.

The induction terminals typically are located either below perimeter windows (Fig. 8.65) or above a suspended ceiling. Condensation from the cooling coil in summer and the need to clean filters of the induced room air make under-window locations much easier to maintain, even if they intrude on the floor area available.

The CBS Tower in New York City (Fig. 8.66) uses a high-velocity induction system for its perimeter zones. The triangular black granite-faced exterior columns give the facade a three dimensionality quite unlike the slick glass boxes of its contemporaries (it was designed by Eero Saarinen, and built in 1962). These thick columns enclose the perimeter's distribution trees; every other column contains the high velocity air supply; in between, every fourth column contains the supply water and every alternate fourth column the return water trees. These constant column sizes belie the thicker air, thinner water distribution

trees. The interior zone of this tower is served by a VAV system; return air from both perimeter and interior zones is collected at the core.

(b) Fan Coil with Supplementary Air (Fig. 8.67).

This system is closely related to the preceding induction system; however, this system uses a fan at each unit, rather than high-velocity primary air, to move the mixture of supply and room air through the unit. A typical arrangement is a supply air plenum over a corridor, feeding horizontal fan-coils above the suspended ceilings of spaces on either side. These fan-coils draw some supply air and mix it with room return air to maintain desired temperatures. If condensation and resulting pans of standing water are to be avoided, the supply air must be sufficiently dehumidified.

Fan-coils are very widely used, both in conjunction with conditioned supply air (air & water) or as stand-alone units that take in their own fresh air at the perimeter (all-water), or even as room-air-only units. A building that uses both fan-coil units (at entries) and radiant panels (in offices) is shown in Fig. 8.68.

(c) Radiant Panels with Supplementary Air (Fig. 8.69).

Large areas of radiant surface can be used to offset large losses of bodily radiant heat, as in the case of large areas of cold glass on a winter day, or when users are both scantily clothed and sedentary. In summer, such panels can help offset radiant gain from electric lights or large glass areas. The ceiling is often favored for the panel location, because it is uncluttered by the furniture, tackboards, and other items that

1 Wall mounting strip
2 Base unit, air plenum
3 Base unit, coil section
4 Base unit enclosure
5 Enclosure end–frames
6 Run–out enclosures
7 Shelf end–panels
8 Shelving
9 Enclosure panel
10 Base unit removable panel
11 Lint screen
12 Base unit air plug
13 Air transition fitting
14 Discharge grille section

Assembly

Fig. 8.65 Installation of a high-velocity induction unit. Conditioned 100% outdoor air is brought in through a high-velocity duct, to provide ventilation and to induce the circulation of room air. It is attenuated and silenced in the chamber (2), and then, through jets in the front of the plenum, it induces the flow of room air, which is heated or cooled at the fin coil (3). The lint screen (11) requires periodic maintenance for proper airflow. (Courtesy of Carrier Corporation.)

(a)

(b)

(c)

Fig. 8.66 *The identical triangular exterior columns of the CBS Tower in New York City* (a) *enclose a variey of distribution trees.* (b) *Plan of a typical floor; perimeter zones are served by a high-induction air & water system, interior by VAV. All return air is collected at the core.* (c) *At the perimeter, column type A contains an air supply tree, feeding the high-velocity induction unit on either side of the column. Column type S contains the supply water tree; hot in winter, cold in summer. Column type R contains the return water tree. (Reprinted by permission from Guise, David,* Design and Technology in Architecture, *© 1985, John Wiley and Sons.)*

Fan coil with supplementary air

Fig. 8.67 *Fan coil with supplementary air HVAC system. Underfloor air supply is shown here, but ceiling supply is also common. Supplementary air is often delivered directly to the fan coil unit.*

Cooling system

(a)

Fig. 8.68 *An innovative Canadian office building combines a radiant panel system* (a) *with a dehumidified summer ventilation system* (b). *Green on the Grand at Kitchener, Ontario, also features daylighting, gas absorption chiller/heater, rainwater pond heat rejection, and two-stage heat recovery. (Courtesy of Enermodal Engineering, Ltd., Kitchener, Ontario.)*

Ventilation system

(b)

Fig. 8.68 (continued)

Radiant panels with supplementary air

Fig. 8.69 *Radiant panels with supplementary air HVAC system.*

cover floors and walls. At the 56-story Commerz-bank in Frankfort, Germany, chilled ceilings serve all office floors in conjunction with the ventilation system.

Green on the Grand (Fig. 8.68) is a 23,465 ft^2 (2180 m^2) two-story office complex in Kitchener, Ontario. Its design combines a very highly insulated envelope, daylighting, water conservation, a gas absorption (non-CFC, non-HCFC) chiller/heater with a heat rejection pond, a separate IAQ system, and heat recovery of exhaust air. It also features a dedicated air system for ventilation and a hydronic system for heating and cooling. The hydronic system uses fan coils for the entryways, and radiant panels on ceilings of office spaces. (The conditioned air in this example is provided by the ventilating system.) Fan coils were better able to handle the extremes of entryway heat losses and gains. The radiant panels in the offices are used for both heating and cooling, but sized to meet cooling loads and cover 30% of the ceiling area. Tenants were given their choice of two designs: steel panels suspended below a drywall ceiling (painted to match the ceiling color), or extruded aluminum panels fit into a suspended ceiling system. To prevent condensation on these radiant panels while cooling, the ventilation air is dehumidified.

The ventilation system supplies low on the walls, exhausts high on the walls, a form of displacement ventilation. There are two rates of fresh air: 20 cfm (10 L/s) per person, and (when economizer cycles operate, or when more fresh air or more cooling is required) 40 cfm (20 L/s) per person. The exhaust air then flows through two heat exchangers: the first reheats fresh air that has been greatly cooled for dehumidification, the second preconditions the incoming fresh air at the point of entry. The latter heat exchanger is a rotary-wheel ERV, capable of transferring both heat and moisture.

Several approaches to the radiant cooling (and heating) panel, independent of a dehumidified air supply, are shown in Fig. 8.70. These approaches assume a minimal floor/ceiling thickness, which could contribute either to reducing the floor-to-floor height, or to greater ceiling height (for daylight penetration or displacement ventilation). If a raised floor is used above these concrete slabs,

even higher percentage heat gains would come from the ceiling.

Radiant cooled floors are sometimes used, as in the Bleshman Regional Day School (Fig. 8.71). The designers of this school for the handicapped recognized that the majority of its users would spend much of their time quite close to the floor, and that the colder air near the floor could be uncomfortable, especially in the New Jersey winter. The entire floor is warmed by the supply (ventilation) air in winter, which enters just below windows to counteract the downdraft off cold glass. In summer, the cool supply air first cools the floor, then cools the air in front of the warm glass. The concrete cellular "air floor" provides a thermal mass that helps maintain steady temperatures. Heated or cooled air is provided by rooftop air–air heat pumps, one of which is provided for each cluster of three to six classrooms. This HVAC example is therefore related to the direct refrigerant family, discussed in Section 7.8.

(d) Water Loop Heat Pump. This approach was diagrammed in Fig. 7.45 and used in the school in Fig. 8.51. In the 175,000-ft^2 (16,260 m^2) Comstock Center in Pittsburgh (Fig. 8.72), there are six to eight small heat pumps on each of 10 floors, located above the suspended ceiling. Their connecting loop doubles as the building's wet-pipe sprinkler system supply; this is possible because the heat pumps keep the loop between 65° and 85°F (18° and 29°C). Because neither hot nor chilled water-supply and water-return distribution trees are needed in addition to the sprinklers, there is a substantial first-cost saving, and relatively little building volume is consumed. A 23,000-gal (87,060 L) water storage tank allows daytime heat, rejected to the loop by the heat pumps, to be stored and recalled for nighttime heating. When necessary, penthouse equipment is used: two small boilers to maintain the 65°F minimum temperature in the water loop, or an evaporative condenser to hold temperatures at 85°F.

Another feature of the Comstock Center is its use of air extract windows (see Fig. 8.50) to control infiltration and to moderate the perimeter zone's temperatures. After the return air passes up inside this window and arrives in the ceiling plenum, it

Fig. 8.70 *Approaches to radiant cooling systems. (a) Concrete core system: water is circulated through plastic tubes embedded in the concrete floor/ceiling slab. (b) With carpet, pad, and insulation on the floor, most of the heat radiates to the ceiling. (c) Panel system, usually using aluminum facing, connected to metal tubes. (d) Panel system has the highest percent heat to the ceiling. (e) Cooling grid made of capillary tubes embedded in ceiling plaster (or in gypsum board, or mounted on ceiling panels). This is the most even surface temperature distribution. (f) Again, a high percentage of the heat is radiated to the cooled ceiling, depending on the use of insulation below the carpet and pad. (From* Center for Building Science News, *Fall 1994, Lawrence Berkeley Laboratory, California.)*

DISCHARGED AIR

INSULATED GLASS

CONVECTION HEATING-COOLING

ALUMINUM DIFFUSER

PLASTIC LAMINATE
WINDOW STOOL

"OUTSULATION"

RADIANT FLOOR

INSULATION

AIR
SPACE

2" COVER SLAB

AIR FLOOR DOMES

WARMED OR COOLED AIR
FROM AIR SOURCE HEAT PUMP

2" GROUND SLAB

CONCRETE AIR
SUPPLY TRENCH

HVAC—ENERGY
CONSERVATION SYSTEM

(a)

Fig. 8.71 *Underfloor air distribution achieved with cellular concrete "air floor" (a) serves the Bleshman Regional Day School (for the multiply handicapped), Paramus, New Jersey. (b) Plan, showing courtyards that bring daylight between clusters of classrooms and into the central circulation spine. (Courtesy of Rothe-Johnson, architects.)*

is mixed with fresh air, then tempered and recirculated by the heat pumps.

The Comstock Center also utilizes a large daylighting atrium, whose temperature control is provided largely by exhaust air from the offices; the stack effect is utilized to provide natural ventilation on its west face in summer conditions.

8.7 All-Water Systems

These systems (see Fig. 8.14 and Section 8.2*d*) typically deal only with temperature control; air quality is left to separate systems. An especially familiar all-water application is the simple *fan-coil unit* (Fig. 8.73). (Because units with the same

THERMAL CONTROL

KEY
1 CLASSROOM
2 FOOD SERVICE
3 PHYSICAL EDUCATION
4 ADMINISTRATION
5 EDUCATIONA SUPPORT
6 HEALTH SUITE
7 THERAPY
8 HOME CARE FACILITY
9 SUPPORT

(b)

Fig. 8.71 (continued)

name are used in air & water systems, this further blurs the distinction between these two HVAC families.) These units, which may be found above ceilings, below windows, or in corners, simply control the temperature (and, to a limited extent, the relative humidity) of the air already in the room, which is blown through the coils. Because water is often condensed from the room air when cooling is in progress, a drain line is required; water standing in drain pans is unhealthy. Exterior air

(a)

(b)

Fig. 8.72 *The Comstock Center, Pittsburgh, Burt Hill Kosar Rittelmann Associates, architects.* (a) *The north and west faces.* (b) *Ground floor plan.* (c) *Water loop heat pumps are linked via the sprinkler system on each floor. This is also called a "tri-water system."* (d) *Extract-air windows (see again Fig. 8.50) control infiltration and moderate the perimeter zone temperatures. Fresh air is ducted to the plenum, where it mixes with return air before being treated and recirculated through the heat pumps.* (e) *The central daylighting atrium is tempered by exhaust air from the offices; exhaust air then leaves via the stack effect.* (f) *The stack effect also controls summer overheating.*

PENTHOUSE

HEAT RECOVERY
WATER STORAGE TANK

PUMP   PUMP

EVAPORATIVE COOLER

HEAT
EXCHANGER

BOILERS(2)

RETURN →

← SUPPLY

TYPICAL FLOOR

SUPPLY →

RETURN AIR

SUPPLY AIR

← RETURN

REVERSE-CYCLE
HEAT PUMP UNIT

SPRINKLER SYSTEM

(c)

(d)

Fig. 8.72 *(continued)*

OFFICE EXHAUST AIR INTO ATRIUM

VERTICAL BLINDS TO CONTROL SOLAR GAIN

SINGLE GLAZING

DOUBLE GLAZING

METAL GRATE TO PROVIDE OPENINGS FOR
NATURAL UPWARD AIR-FLOW

(e)

DURING SUMMER OPERATION OUTDOOR AIR ENTERS
THE SOLAR CHIMNEY THROUGH MOTORIZED AIR-
INTAKE DAMPERS AT THE BASE. SOLAR HEATED
AIR INDUCES A NATURAL UPWARD AIR FLOW
RISING APPROXIMATELY 100 FEET, WHICH IS
THEN EXHAUSED OUT THROUGH THE TOP OF THE
SHAFT.   DURING WINTER OPERATION THE AIR
SHAFT IS SEALED TO CREATE A DEAD-AIR SPACE
BETWEEN THE INNER SINGLE AND OUTER DOUBLE
GLAZINGS.

(f)

Fig. 8.72 *(continued)*

intake grilles can easily be added when fan-coil units are located below windows, to allow the local provision and tempering of ventilation air; this is especially common in motels, hotels, and apartments.

Fan-coil units are inherently noisy due to the fan, so careful attention to sound ratings is required when background sound levels must be low. In offices, a relatively high background sound level can help to maintain acoustic privacy at the workstation; the sound of the fan-coil can be reassuring that some kind of air treatment system is at work.

All-water perimeter systems with local fresh air can also take the simple form of *operable windows with hot water finned-tube radiation.* An expressive variation on this approach appears in the reading rooms and staff workrooms for the Seeley

Fig. 8.73 *Simple fan-coil units, without fresh air.* (a) *Standard below-window unit.* (b) *"High-rise" unit for corners or cabinet locations.* (c) *Above-ceiling unit.* (d) *Low-profile unit, such as below large windows. (Reprinted by permission from AIA, Ramsey, and Sleeper,* Architectural Graphic Standards, *9th ed., © 1994 by John Wiley and Sons.)*

G. Mudd Library at Yale University (Fig. 8.74). Limestone spandrels are curved in to allow fresh air to enter these smaller perimeter rooms just below the windows, where hot water finned-tube radiators are available when needed. The incoming fresh air replaces exhaust air, which flows out the upper operable windows. The remainder of the building has conventional forced-air heating and cooling.

8.8 District Heating and Cooling

Often, large projects made up of many large buildings are well served by *one* central station heating/cooling plant. The familiar "economies of scale" apply here; very large, efficient, and well-maintained boilers and chillers encourage energy recovery through heat exchange, reduce air pollution, and remove the noise and other nuisances associated with heating and cooling from the other buildings. This approach is called *district heating and cooling.*

District heating for residences and small commercial buildings is common in northern Europe; district steam systems serve the central areas of many U.S. cities. Most often electrical generating plants are the heat (or steam) source; *cogeneration* (discussed in Section 8.9) allows the waste heat from the generating plant to be put to use either as

(a)

Fig. 8.74 *Local fresh air and finned-tube radiation at the perimeter. The Seeley G. Mudd Library at Yale University; Roth and Moore, architects.* (a) *Exterior, showing curved limestone spandrels, along which flows incoming fresh air. (Photo © Steve Rosenthal.)* (b) *Section, showing the fresh air intake, finned-tube radiation, and upper operable sash for exhaust air. This sytem is used for the smaller reading room and staff workrooms at the perimeter (at right of plan).* (c) *Plan.*

(b)

01 5 10
Third Floor Plan

(c)

Fig. 8.74 *(continued)*

space heating or steam (indirect-fired) absorption cooling. Ironically, trends toward better-insulated buildings reduce the "market" for distributed heat, making the installation less cost-effective. This discourages the supplier both from investing in a district system, and from encouraging technical innovations in energy conservation on the part of their customers. Smaller, densely built-up central heating districts tend to be more successful than large, sprawling ones; lengthy lines cost more to install and maintain, and lose more energy.

(a) High-Temperature Water and Chilled Water. Long-distance steam distribution has been used for more than a century; the development of *high-temperature water* (HTW) and chilled water distribution among buildings is a more recent development. Offering many advantages (although steam is still frequently chosen for city distribution), circulated high-temperature, high-pressure hot water and chilled water in closed systems is widely used in Air Force bases

and airports, and for groups of buildings such as hospital complexes and college campuses. Now there are increased efforts to install new district heating/cooling networks served by existing fossil-fueled electricity generating plants. Such plants "waste" more than half their fossil fuel input (see Fig. 1.11), and district heating/cooling could intercept much of that waste; see also the next section on cogeneration.

Water will not flash into steam if kept at sufficiently high pressure. It may then be circulated by pumps through supply and return mains and through branches to heat exchangers, which operate conventional low-pressure hot water systems, generate steam, and perform numerous other thermal tasks. Pressures are on the order of 400 psig (pounds per square inch, gauge) and temperatures are about 300°F. During its circuit the water will sometimes lose up to 150 F° and 60 psig in pressure. A section shown in Fig. 8.75 illustrates a common arrangement.

High-temperature water has a number of ad-

Fig. 8.75 *Typical arrangement of a high-temperature water system. (Reprinted from* High-Temperature Water Systems, *Industrial Press. By courtesy of author Owen S. Lieberg, consulting engineer.)*

vantages over steam for certain installations. It is a two-pipe system, and the temperature drop in the *supply* main is often as little as 10 F°. With reasonably high water velocities, mains can be reduced to almost half the size of those required for steam distribution, with no need for steam traps and pressure-reducing valves. The pipes need not pitch to low points, as in the case of steam (to accommodate condensation), but may follow the contours of the ground. Although installation costs are greater, operational costs are less than for steam. Feed water treatment is negligible and corrosion is minimal. Underground problems of expansion and insulation are the same as in other subterranean systems. Large sweep-type loops accommodate expansion between fixed points, and underground piping is embedded in special, thermally efficient insulative fill.

Chilled water systems also offer advantages. The remote central chillers are likely candidates to use waste heat in a non-CFC absorption cooling cycle. Natural sources are possible; the Toronto (Canada) District Heating Corporation will use Lake Ontario water, drawn from a 1.6-mile (2.6-km) intake at a depth of 200 ft (61 m). This water is expected to be at 40°F (4.5°C) year around. Passed through a heat exchanger (with the district chilled water) and then treated, the lake water then joins the city water supply.

With district heating/cooling, all facilities except air handling and ducts are located together in a remote central plant. This frees the buildings from the space requirements and visible impacts of stacks, boilers, fuel storage, water chillers, and cooling towers; the associated heat, humidity, fouled air, and noise is as remote as the central plant. When such a system serves individual customers, the heated or chilled water is metered. When it is owned by the group of buildings served (such as college campuses), it is usually not metered; this can become a problem when efforts to identify building energy waste and subsequent savings are being investigated.

8.9 Cogeneration

In the course of these eight chapters, electricity has repeatedly been called a "high-grade" source, in reference to the high temperatures needed to produce electricity by conventional (fossil-fuel) means and to the large amount of waste heat (often twice the fraction of electricity) produced in the process. Cogeneration (also called "total energy") is an attempt to recover some of the otherwise-wasted, lower-grade heat that accompanies the generation of electricity by steam turbines. Industrial cogeneration facilities are especially cost-effective in the pulp and paper, petroleum, and chemical industries. Building-scale cogeneration is less obvious an energy bargain.

(a) Electrical Power Generation at the Site. Where conditions are favorable, electricity for power and light can be generated economically by a system that also supplies the building with heating in winter and cooling in summer. Such a system utilizes a fuel such as gas or oil and is often supplemental to the local electric utility company. Although installation costs are greater than for the more conventional systems that use separate services of electricity and fuel, the savings in annual operating cost can sometimes pay for the excess installation cost in a reasonable time. Operational savings continue thereafter. This method developed largely in the 1960s and is used in hundreds of commercial and industrial buildings and in many schools. Cogeneration is particularly attractive for district heating/cooling plants.

(b) Early On-Site Power Generation.
Before 1900 and for several years thereafter, nearly all large buildings and groups of buildings were supplied with direct current generated on or near the premises. The motive power was usually in the form of steam-driven reciprocating engines with belt connections to direct-current generators. Direct current cannot be transformed to different voltages and must be generated and distributed at the voltage used in the building. At these relatively low voltages, power loss in the distribution system is great, and distance adds greatly to the loss. This tended to keep electricity usage within a building.

With the development and use of alternating-current machines, utility companies were able to establish central power stations from which electricity could be transmitted great distances to the user at high voltage. There it was transformed down to domestic voltages for use. Because, at high voltage, power losses are very low, this system became universal. During the 1920s and 1930s, owners removed their private power generators and accepted utility service, with its savings in operating expense.

(c) How Cogeneration Developed.
Older buildings used steam produced from coal-burning boilers from which little or no energy salvage was possible. Today's fuels, used directly in reciprocating engines or turbines (including those that generate electricity), have residual heat value that can be recovered for purposes of heating or cooling. For cogeneration to be successful, there should be a reasonably steady demand in the building for the power generated and also for the heat recovered. Lighting and the demand for power by computers, electrical business machines, and other devices can create nearly constant demand for power throughout the year. Similarly, the exhaust heat recovery from the engines or turbines that power the generators is in demand for either heating or indirect-fired absorption cooling at most times of the year.

(d) Turbines and Reciprocating Engines.
Figure 8.76 shows two principal systems for total energy. In both systems—one using a turbine and the other a reciprocating engine to operate the generator that supplies electric power—heat is reclaimed to produce steam or hot water. The steam or hot water is then used for heating or, by use of an absorption chiller, for cooling. When a turbine is used, the fuels are natural gas or fuel oil. The heat is recovered by passing the hot turbine exhaust through a waste heat boiler to produce steam. Fuel for a reciprocating engine is natural gas or diesel fuel. Both the jacket cooling water and the hot engine exhaust are passed through heat exchangers that utilize the heat to produce steam or hot water for heating or cooling. In both systems, an auxiliary boiler, fired directly by gas or oil, stands ready to help maintain a balance in the system.

Cogeneration offers a degree of electrical independence and a way to use otherwise wasted heat. The latter is essentially free energy because it would otherwise have to be purchased and paid for separately in the form of electricity or other fuels.

(e) Cogeneration for Housing.
The Harbortown apartment and townhouse complex in Detroit, Michigan, utilizes a year-around natural-gas fired primary energy and cogeneration/waste recovery system (Fig. 8.77). Each apartment has one or two stackable, upright fan coil units, using either chilled or heated water. This water is provided by two direct-fired chiller–heaters: each incorporates a natural gas boiler and a two-stage absorption chiller in the same unit, taking less space in the mechanical room. The natural gas–driven cogeneration electric plant serves rather constant loads such as corridor and outdoor lighting. The generator's waste heat is passed to a

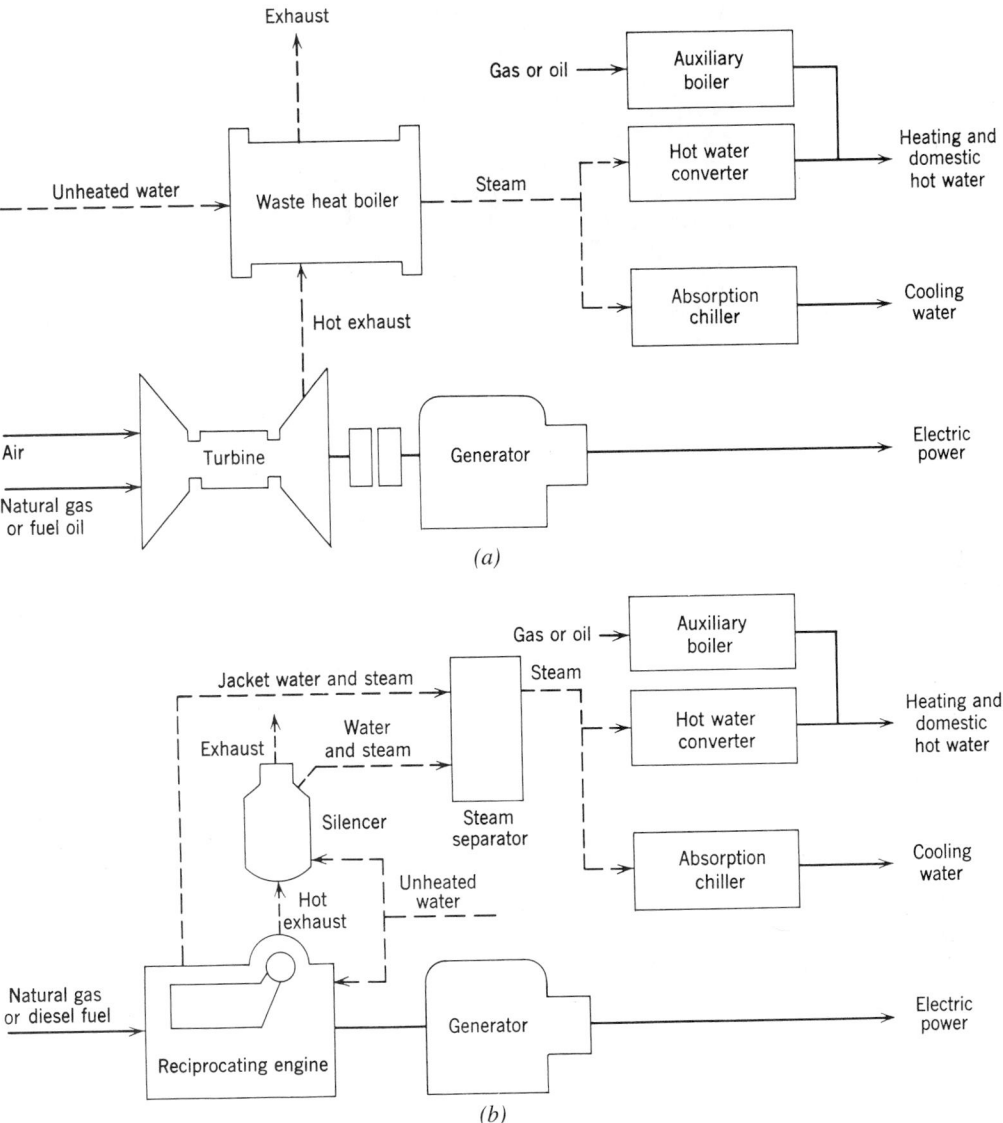

Fig. 8.76 *Cogeneration (total energy) systems.* (a) *Using turbine.* (b) *Using reciprocating engine.* (Reprinted by permission from Total Energy, Educational Facilities Laboratories.)

storage tank, where it preheats domestic hot water for the tenants. This meets about 87% of the hot water energy usage.

(f) Fuel Cells. We first encountered fuel cells in connection with the explorations of outer space. Spaceships use stored oxygen and hydrogen to feed the fuel cell, yielding—seemingly miraculously—electricity, heat, and pure water. In effect, this reverses the process of electrolysis: using electricity to split water into its two compo-

nents, hydrogen and oxygen. This theory dates back to Sir William Grove in 1839, long before the needs of space travel.

A fuel cell generates direct current power by converting the *chemical energy* of hydrogen and oxygen into electricity and heat. Because no combustion is involved, nitrogen oxides and carbon monoxide are nearly eliminated. For a building's fuel cell power plant, several components are added: first, a fuel processor (or fuel reformer) to prepare a hydrogen-rich stream of fuel, then

(a)

(b)

(c)

FLOW DIAGRAM OF HEAT RECOVERY SYSTEM

(d)

Fig. 8.77 *Harbortown is a 120,000 ft² (11,150 m²) apartment and townhouse complex* (a) *in Detroit that uses natural gas as a primary fuel year-around. Photo © 1989, William Kildow. Chiller-heaters (Hitachi) combine boiler and two-stage absorption chiller in one compact unit.* (b) *Comparison of headspace and* (c) *floor space required by chiller-heaters versus conventional units of similar capacities.* (d) *Cogeneration plant sends its waste heat to preheat domestic hot water.* (e) *Typical partial plan shows the space advantages of stackable two-pipe fan coil (FCR) units. (Courtesy of Skidmore, Owings and Merrill, Architects-Engineers, Chicago.)*

(e)

Fig. 8.77 (continued)

the fuel cell stack, then a power conditioner (or inverter) to convert the fuel cell's dc to ac current.

Fuel cells are not, by themselves, that much more efficient than combustion processes; they run about 40% efficient for power production. When combined with a way to use that 60% waste heat, they climb toward a total of 90% efficiency, with very few harmful emissions.

Hydrogen is the ultimate fuel; already hydrogen-rich, it needs no processor and produces the least environmental impacts from the fuel cell. Although hydrogen supply networks are not yet commonplace, scenarios exist for a "hydrogen economy" (see Section 1.6*j*), in which surplus electricity from windpower is used to split water into hydrogen and oxygen atoms, and the hydrogen stored for later distribution. Windpower's extreme variability often presents power grid distribution challenges; storage solves this problem. Solar energy or biomass are other renewable sources for electrical generation. However, other hydrocarbon fuels also can feed fuel cells: natural gas (most common for building fuel cell power plants), propane, methanol, and, with additional pretreatment, methane from landfills and anaerobic digester gas from sewage treatment plants.

Even with these less hydrogen-rich fuels, the emissions from fuel cells are well below those of combustion processes; the fuel processor emits carbon dioxide and a trace of carbon monoxide.

The proton-exchange membrane (PEM) fuel cells (Fig. 8.78) are promising for transportation when zero-emission vehicles are the objective and hydrogen is available as the fuel. This process has building applications as well. It works at a temperature below the boiling point of water. Relative to other fuel cell types, it has a smaller size, lighter weight, and lower noise levels. The PEM fuel cell stack involves several subsystems, typically including water purifiers and pumps, air compressors, and heat rejection components (coolant pumps and a radiator). This complicates performance at subfreezing temperatures and adds maintenance requirements. But the smaller size promises smaller building applications in the future.

The phosphoric acid fuel cell power plant (Fig. 8.79) is at work in (or outside) a number of buildings and municipalities around the world. The ONSI Corporation's 200-kW fuel cell power plant measures 10 ft × 18 ft, 10 ft high (3 m × 5.5 m, 3 m high), weighs 40,000 lb (18,144 kg),

Fig. 8.78 *The proton-exchange membrane (PEM) fuel cells use a special sheet that only protons—the nuclei of hydrogen atoms—can cross. Hydrogen pumped along one side of the membrane loses its electrons when it comes into contact with the platinum catalyst. The protons travel through the membrane, and on the other side combine with oxygen atoms (in air) to form water. The electrons lost by the hydrogen travel through a circuit, producing an electrical direct current. (Courtesy of Nucleus, Fall 1994, © Union of Concerned Scientists, Cambridge, Massachusetts.)*

and needs a clearance of 8 ft (2.5 m) around the module for maintenance. It also needs a cooling module (or a building's cooling tower) for discharge of heat in excess of that recovered for building use. The acid electrolyte in the fuel cell stack works at about 390°F (200°C). At its rated power, it also produces 700,000 Btu/h (205 kW) to heat a water stream to 140°F (60°C). The rated sound level is 62 dBA at 30 ft (9 m) from the module.

Fig. 8.79 *A fuel cell power plant, using natural gas (or other hydrocarbon fuel). The fuel processor (reformer) converts the natural gas to a hydrogen rich stream, that enters the fuel cell power section (or stack). The output is electricity (dc), water, carbon dioxide and considerable heat. Some heat is used by the fuel processor, some is rejected, but most is usable in a building or industry. A power conditioner (inverter) converts the dc to ac power. (Courtesy of ONSI Corporation.)*

(a)

Fig. 8.80 *The 48-story office building at 4 Times Square, New York City* (a), *under construction in 1998.* (b) *Many environmentally friendly features are included, such as facade-integrated PV and two fuel cell power plants like those shown in Fig. 8.79. Fed by natural gas, the fuel cells are expected to work without maintenance interruptions for up to 5 years and provide considerable hot water as well as electricity. (The fuel cells are actually installed at the fourth floor level rather than at the penthouse.) (Courtesy of Fox and Fowle, Architects, P.C., New York. Consentini Associates, Mechanical Engineers; the Durst Organization, Developer.)*

The Durst Organization's high-rise office building at 4 Times Square, Manhattan (Fig. 8.80), is a demonstration of several future-oriented strategies: PV cells integral to the facade, direct-fired (natural gas) absorption chillers, increased fresh air for IAQ, a dedicated exhaust air shaft (for smoking and other polluting activities), waste chutes to facilitate recycling, among others. It also utilizes two 200-kW fuel cell power packages, fed by natural gas. The electricity is about 80% destined for external lighting by night; huge electrical signs will be a prominent part of the building's Times Square facades. By day, about 80% of the power goes to the building's base load. The hot water will be used in winter for perimeter heating. In summer, the water is wasted. Although there was hope that one of the absorption chillers could be fed by this water, a variety of considerations precluded this: the fuel cells are located on the fourth floor (accessible to maintenance and close to the huge signs) whereas the chillers are located in the penthouse.

The 200-kW fuel cell of Fig. 8.79 is usually fed by natural gas, but other examples include hydrogen fuel (Hamburg, Germany: electricity to the grid, hot water to heat an apartment building), anaerobic digester gas (Yonkers, New York: waste-water treatment plant; electricity and hot water used within the facility), and landfill gas (Groton, Connecticut: closed landfill; electricity to the grid, plans for a hydroponic-tomato–growing greenhouse to utilize the hot water). Plentiful hot water is a byproduct of the fuel cell; it seems that finding a year-round use for this water is often a problem of excess supply.

Other fuel cell types include alkaline [operates at 480°F (250°C)], molten carbonate [works faster at 1110–1300°F (600–700°C)] and solid oxide

THERMAL CONTROL

FUEL CELLS
ALTERNATIVE ENERGY SOURCE

GAS-FIRED ABSORPTION HVAC SYSTEM
NO CFCs OR HCFCs

HIGH EFFICIENCY VARIABLE SPEED PUMPS, MOTORS, FANS

PHOTOVOLTAIC PANELS
IN CURTAINWALL

ENVIRONMENTALLY FRIENDLY BUILDING MATERIALS

FLOOR-BY-FLOOR AIR QUALITY MONITORING, CONTROL & PURGE SYSTEM

50% MORE FRESH AIR PROVIDED

FILTRATION SYSTEM FOR AIR POLLUTANTS

ADDITIONAL EXHAUST SHAFT
FOR SMOKING, FUMES & HEAT

ENVIRONMENTALLY FRIENDLY BUILDING MAINTENANCE

EXISTING FOOTINGS REUSED

CENTRALIZED, AUTOMATED BUILDING MANAGEMENT

INTEGRATED COMMUNICATIONS TOWER

REMOVABLE CRANE
FOR BUILDING MAINTENANCE

HAT TRUSS STRUCTURE
REDUCES STRUCT. STEEL USE

HIGH PEFORMANCE LOW-E GLASS CURTAINWALL
WITH LARGER WINDOW AREAS

EFFICIENT LIGHTING
OCCUPANCY SENSORS & CONTROLS

RECYCLED & RECYCLABLE BUILDING MATERIALS

RECYCLABLE WASTE CHUTES

CONCRETE CORE STRUCTURE
REDUCES STRUCT. STEEL USE

COMMISSIONING

RECYCLABLE WASTE STORAGE FACILITIES

ENERGY

INDOOR AIR QUALITY

RECYCLING

MANAGEMENT

(b)

Fig. 8.80 (continued)

[1200 to 1830°F (650 to 1000°C)] with a higher efficiency, and waste heat sufficient to run gas turbines. These higher temperature processes are more likely at central station power plants.

Clearly, cogeneration and fuel cells have a promising future.

REFERENCES

American Institute of Architects, Ramsey, C. G., and Sleeper, H. R. (1994). *Architectural Graphic Standards,* 9th ed., John Wiley and Sons, New York.

American Society of Heating, Refrigerating, and Air-Conditioning Engineers, Inc. (1997). *Handbook of Fundamentals,* ASHRAE, Atlanta, Ga.

American Society of Heating, Refrigerating, and Air-Conditioning Engineers, Inc. (1989). ASHRAE Standard 90.1-1989: *Energy Efficient Design of New Buildings Except New Low-Rise Residential Buildings,* ASHRAE, Atlanta, Ga.

Bobenhausen, W. (1994). *Simplified Design of HVAC Systems,* John Wiley and Sons, New York.

Lorsch, H. (ed.) (1993). *Air-Conditioning Systems Design Manual.* ASHRAE, Atlanta, Ga.

THERMAL CONTROL

PART III
WATER AND WASTE

Almost every building designed today in North America is supplied with potable (drinkable) water. In the vast majority of these buildings, *most* of this clean water is used to carry away organic waste. The result is a wide range of design impacts covering everything from the detailed arrangement of bathroom fixtures and interior surfaces to the overall plans of very large and complex water and sewage treatment facilities.

This section of the book presents the following topics: basic planning information and rainwater (Chapter 9); water supply (Chapter 10); organic waste disposal, both water and waterless (Chapter 11); and solid waste design issues (Chapter 12). Fire protection, which often requires the largest amount of water supply piping within a building, is covered in Chapter 13.

9

WATER AND BASIC DESIGN

"The next great world crisis will be water supply." This prediction is increasingly frequent as countries downstream threaten war over water rights with countries upstream, and water withdrawals in coastal areas allow saltwater intrusions into aquifers. State against state, county against county, water rights continue to cause problems. With a finite planetary water supply pitted against an increasing population and, worse yet, an increasing *per capita* consumption of water, we see again (as with fossil fuels) the problem of limited resources versus growing demand. At least in this case the amount of water is fixed, not diminishing. However, the problem of fair allocation remains (Fig. 9.1). Countries have fought wars over oil; must they also wage war over water?

Agricultural and industrial use of water may dwarf that of buildings, but designers still have a role in this dilemma. We can use water beautifully where it is appropriate and avoid its use when possible.

9.1 Water in Architecture

Throughout history, in nearly all climates and cultures, the designer's major concern about water was how to keep it *out* of a building. Only since the end of the 19th century has water supply *within* a building become commonplace in industrialized countries. In the rest of the world today, running water is still not available within most buildings. Water's potential contributions to life-style and architecture are as numerous and varied as are appropriate design responses to the supply, use, and return of such a versatile commodity.

(a) Nourishment. Much of the human body is water, the most abundant chemical in our bodies as well as in our diet. The amount of really "pure" (potable) water that we need for drinking and cooking is very small—only about 3 g/cd (gallons per capita per day) (11.4 L/cd) in the United States. The most common supply system throughout history has been the central municipal fountain or well (Fig. 9.2), whose technical importance to the community has often been emphasized by the esthetics of both the fountain's overall sculptural composition and the elegance of detail in water spouts, basins, and other elements. The social importance is evident from its location in the central plaza, from which site small amounts of water are carried daily by townspeople to homes or workplaces. Opportunities abound for conversations around the fountain.

As daily potable water became available within buildings, water-related social opportunities diminished. However, employees still converse around the drinking fountain—for example, when it is located in a place that invites lingering or relaxing.

(b) Cleansing and Hygiene. Water is a nearly ideal medium for the dissolution and transport of organic waste, and its high heat-storage capacity easily attains comfortable temperatures for bathing. Much larger quantities of water are used for cleaning than for nourishment: in the average U.S. home, about 14 g/cd (53 L/cd) is used for clothes washing and dishwashing and another 21 g/cd (79.5 L/cd) is used for bathing and personal hygiene.

In the past, water for cleaning was carried to the home less frequently; the Saturday night (only) bath was typical well into the 20th century in the United States. Bathing vessels were usually portable and sometimes were combined with other pieces of furniture (a couch that sat over a tub, a metal tub that folded up inside a tall wooden cabinet, etc.). Thus, a *bathplace* rather than a bathroom was the common design response.

Although the physical constraints of water carrying were important design influences on bathing, cultural attitudes were at least as strong an

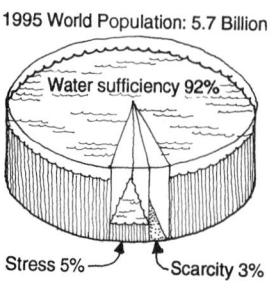

1995 World Population: 5.7 Billion

Water sufficiency 92%

Stress 5% — Scarcity 3%

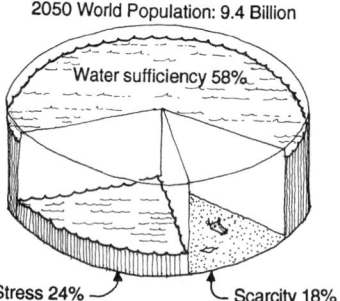

2050 World Population: 9.4 Billion

Water sufficiency 58%

Stress 24% — Scarcity 18%

Fig. 9.1 *Earth's present and projected future annual per capita fresh water supply for all purposes. The relative size of the graphs reflects population growth. "Sufficient" supply is 1700 m³ (449,000 gal) per person; "stress" supply is between 1000 and 1700 m³ (264,180 and 449,000 gal) per person; "scarcity" is less than 1000 m³ (264,180 gal) per person. Currently, worldwide water use is 70% irrigation, 20% commercial-industrial, and 10% residential. (Worldwatch, State of the World, 1999.) (Courtesy of Population Action International.)*

Fig. 9.2 *The fountain in the central piazza at Tarquinia, Italy, still serves as a gathering place and as a symbol of the town's provision for a basic need.*

influence throughout history. Perhaps the most startling contrast is between the medieval concept of bathing as an almost sinful indulgence and the earlier Roman attitude toward public baths as the social centers of cities.

Now, bathing facilities are commonly designed to be used on a very personal scale, in privacy. There also are welcome opportunities to design more social bathing (recreational, rather than cleansing) places, such as swimming pools, bath houses, and hot tubs. The characteristics of the water supply (spouts, jets, cascades) can be matched with those of the water body (mirror-smooth, gently flowing, rippled, rolling, foaming) to obtain the desired atmosphere.

(c) Ceremonial Uses. Largely through its associations with cleaning, water acquired a ceremonial significance that remains particularly evident in religious services. Examples of the ceremonial use of water include vessels containing holy water at entrances to Catholic churches, pools in the forecourts of mosques, and full-immersion baptismal fonts at the altars of some Protestant churches. The opportunities for esthetic expression are particularly rich in these ceremonial applications.

(d) Transportation Uses. In stark contrast to its uses in nourishing, cleansing, and celebrating, water is used in our buildings principally to transport organic waste. There is perhaps no more flagrant example of a mismatch in architecture than the high-grade resource of pure water being used for the low-grade task of carrying away a cigarette butt. The typical U.S. home uses 32 g/cd (121 L/cd) just to flush toilets.

In the past, table scraps were commonly fed to animals or composted, and human waste thrown out of windows (accompanied by warning cries) or deposited in holes below outhouses. Organic waste disposal was thus dependent on either portable vessels or special structures set apart from the typical building.

As water supplies were developed, water's advantages over the foul smell and inconvenience of these methods became irresistible. A typical sequence of events unfolded on Manhattan Island. In the 1700s, Manhattan was farm country that, like all other areas that later developed into large cities, had minimal water needs. Potable water was available in shallow wells and from some springs and streams. These sources were largely unaffected by the minor ground pollution from widely separated dry-pit privies (outhouses) that received human wastes. Paved city streets appeared in the 1800s, at which time the natural streams were enclosed in pipes, called *storm sewers*. These pipes led the rainfall to the many waterways surrounding the island. Then in the later 1800s, flush toilets appeared. It seemed natural to connect the toilets to the already established "storm sewers" and to rename the pipeways "combined" sewers, which now carried both storm water and so-called "sanitary" drainage to the rivers (sanitary for the building, but not for the rivers). Fast-flowing rivers are natural sewage treatment plants, and, surprisingly, for many decades they did a fair job of keeping pollution reasonably in check. With the prospect of future sewage treatment plants, separate "sanitary" sewers were built. Also, there were some remaining (and some newly built) storm sewers that did not carry the wastes from toilets.

In cities where this confused pattern of sewer systems still exists (this includes most larger and older cities), it is now extremely difficult and expensive to sort out and reroute sewers so that *only* sanitary drainage goes to treatment plants and *all* storm drainage goes to waterways or into the ground. It seems particularly ironic that in most U.S. locations the rainwater that falls on a home's roof is adequate in both quality and quantity to supply a family's cleansing needs [21 + 14 = 35 g/cd (132 L/cd)]. In Chapters 9 and 10, these possibilities for rainwater are developed further.

As the human waste disposal place became a room within the building, the design issues grew more complex. Physically, there was the need for running water, and for large-diameter pipes that sloped downward continuously from the toilet to a sewer or septic tank. As sewer gas became a recognized problem, an elaborate system of traps and vents became necessary. Again, cultural attitudes were also influential: How private an activity was this elimination from the body? To what extent could/should one plumbing fixture accommodate both body cleansing and waste elimination? If males insisted on standing rather than sitting while urinating, how to devise a toilet that would also properly accommodate defecation, which requires a low seat? Some designer responses to these questions can be found in Chapter 10.

(e) Cooling. Water has a remarkable cooling potential: it stores heat readily, removes large quantities of heat when it evaporates, and vaporizes readily at temperatures commonly found at the human skin surface. In hot-dry climates, designers can place water surfaces (or sprays) upwind from the place to be cooled, or resort to the evaporative coolers discussed in Sections 2.4, 5.6, 5.14, and 7.6. Cooling towers (Section 8.3) are familiar components of large-building cooling systems.

Because all of us have experienced the physical cooling of the skin by water, we all carry psychological associations between water and cooling that can enhance our comfort on hot days. The sight of sunlight reflected on a water surface, with its characteristic "dancing" quality, connotes coolness, as does the sound of running or splashing water. Thus, even when water does not physically cool people, it can make an important psychological contribution to human comfort (Fig. 9.3).

(f) Ornamental Uses. In almost any landscaping application, indoors or out, water becomes a center of interest. Our association of water with nourishing, cleansing, and cooling make water a very powerful design element—a fact recognized by landscape designers throughout history. In arid regions, water is often used sparingly, in small, tightly controlled channels and at lower

WATER AND WASTE

Fig. 9.3 *Water as an aesthetic feature of a court-yard during the hot-arid season in Colima, Mexico. The sound and sight of running water add to a psychological impression of coolness.*

flow rates. The gardens of Islamic architecture are especially effective demonstrations of such design restraint. Where water is more plentiful, it has been used lavishly, as at the Villa d'Este in Tivoli, outside Rome, where much of a river's flow is diverted through the gardens.

Especially useful design characteristics of water include its *reflectivity,* which sets it apart from most plant and ground materials in a garden; its *liquidity,* which attracts attention to its motion and creates unique sounds wherever it is moved; and its *life-sustaining* potential, which allows the addition of both water plants and animals to a garden.

(g) Protective Uses. Every designer dreads water's ability to penetrate a roof and damage a building and its contents. However, we all depend on water as the best fire protection available in most buildings. The vast quantities of water po-

tentially required for firefighting must be delivered quickly; the result is pipes of enormous diameter regulated by very large valves. Because this system's distribution tree must be immediately obvious to firefighters, some degree of exposure is prudent. Despite its size and guarantee of at least partial exposure in public places, a fire protection water supply system is rarely treated as a visually integral design element. This mismatch of potential and actuality is discussed further in Chapter 13.

Another protective use has been as a means to control circulation; moats around castles may seem quaint today, but designers still sometimes use water as a means of directing traffic over a bridge to an entry.

9.2 The Hydrologic Cycle

There is a finite quantity of water in the earth and its atmosphere. [This statement is challenged by a controversial theory of a steady rain of small water-bearing comets from outer space; see Gleick (1998).] The process whereby this water constantly circulates, powered by about one-fourth of earth's solar energy, is called the *hydrologic cycle* (Fig. 9.4). More than 99% of this water is "inaccessible"—either because it is saltwater or because it is frozen in glaciers on polar ice caps. The most accessible sources of water are precipitation and runoff.

Precipitation has the advantage of relative purity, although acid rain is a growing threat in many parts of the world, including much of the United States and Canada. Like solar energy, precipitation is a very large but very thinly spread resource; its capture is therefore likely to take place on an individual basis. Until we experience a "water crisis" similar to the "energy crisis" that began in the 1970s, rainwater capture will remain a mostly untapped freshwater source in the United States.

Runoff enjoys the advantage of a concentrated flow of water, which permits easy capture of large quantities. Its most serious disadvantage is the possibility of pollution—organic, chemical, and radioactive—depending on what is upstream from the point of capture. In some regions of North America, river water is reused 50 times on

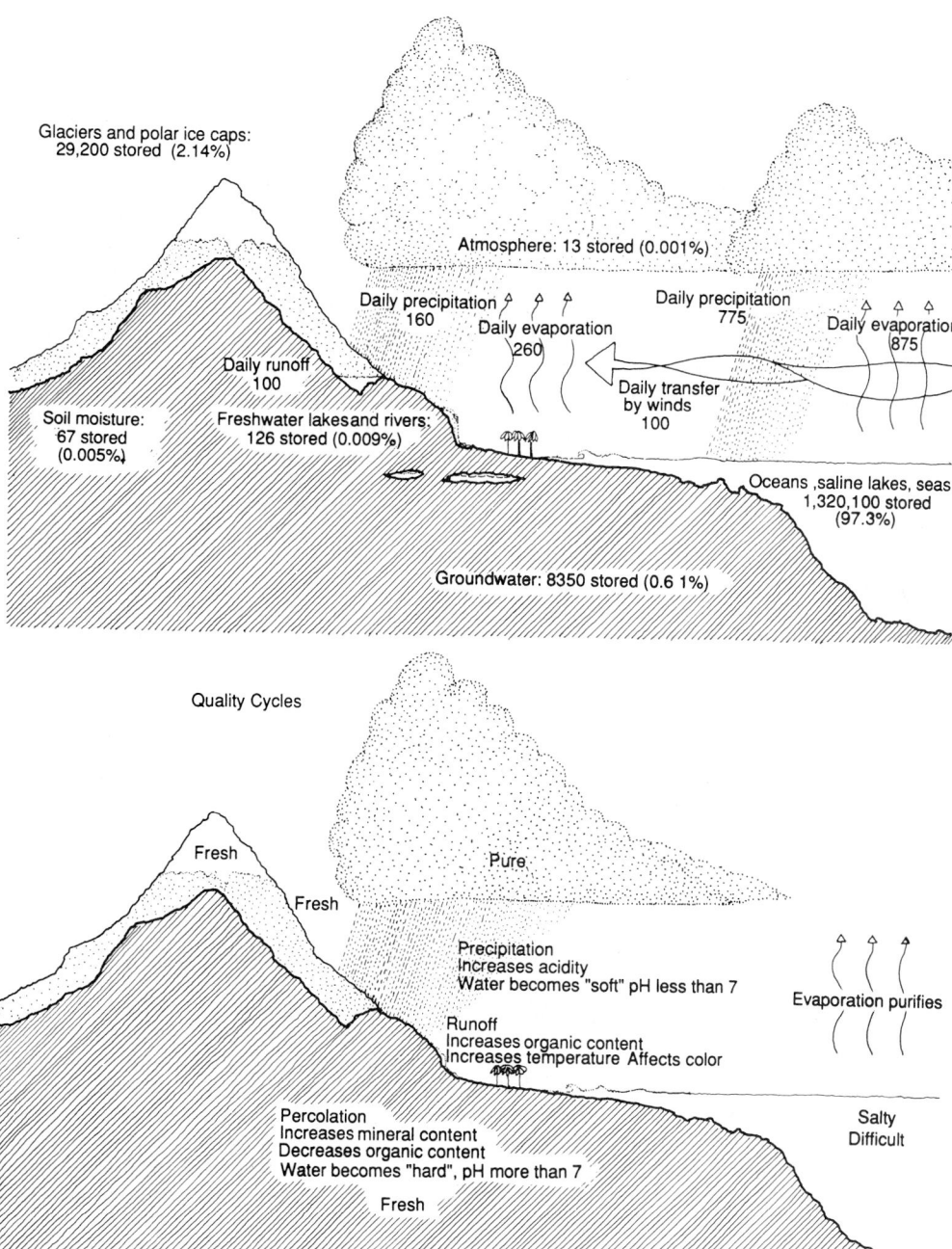

Fig. 9.4 (a) *Hydrologic cycle. The figures given are in cubic kilometers. "Evaporation" includes transpiration from plants as well as evaporation from surfaces. "Precipitation" can be rain, hail, sleet, or snow. The vast majority of stored water is in the ocean.* (b) *Quality of water at various stages within this cycle. See also Chapter 10 for quality and treatment considerations.*

its way to the ocean. Further discussion of water sources and treatments occurs in Chapters 10 and 11.

That part of daily precipitation which neither evaporates nor joins the runoff becomes part of *soil moisture*. Much soil moisture is used by growing plants and is soon transpired (evaporated) by the plant to the atmosphere. As water works downward below the root zone of plants, it eventually reaches a zone of saturation, where all voids in the earth's material are filled with water. This zone of saturation is called *groundwater;* the upper surface of groundwater is called the *water table.* Wells are commonly sunk to a point well below the water table, so that the latter's seasonal fluctuations will not interrupt the well's access to groundwater.

9.3 Basic Planning

After considering the relationship between a building and the roles that water play within it, the designer must do basic sizing—of the quantities of water needed, of the areas in which water will be used, and of the areas and equipment associated with water's return to the hydrologic cycle.

However, before discussing basic planning for the amounts of water used daily within buildings and for the treatment of that water, we should note that water is also an important component of building construction. The production of 1 ton (907 kg) of bricks requires 580 gal (2200 L) of water; 1 ton of steel requires 43,600 gal (165,000 L); 1 ton of plastic requires 348,750 gal (1.32 million L). Water is one of the main ingredients of concrete; a typical 94-lb (42.6 kg) bag of cement requires about 6 gal (23 L) of water.

Another way in which buildings contribute to water consumption is through electricity consumption. Most power plants require very large quantities of water, which they quickly return to the hydrologic cycle warmer in temperature (and perhaps as vapor). A large nuclear power plant that utilizes a cooling tower can evaporate daily the approximate equivalent transpiration from about 9 mi^2 (23 km^2) of forest.

(a) Water Supply. The task of rough sizing for water needs is complicated by the conflict be-

tween *current practice* and *conservation.* Current practice tends toward the use of large amounts of water for very low-grade tasks. Conservation reserves high-quality water for high-grade tasks and emphasizes recycling as well as diminished overall usage of water.

Water supply often is first estimated in terms of gallons per capita per day (liters per capita per day). Table 9.1 shows some common terms used in water supply. Typical quantities were matched with nourishing, cleansing, and other usages in the United States in Section 9.1. To gain an appreciation of how the rate of urban water usage has changed over time, consider the following figures [from Milne (1976)]:

Imperial Rome	38 g/cd (144 L/cd)
London, 1912	40 g/cd (151 L/cd)
American cities just before World War II	115 g/cd (435 L/cd)
Los Angeles, mid-1970s	182 g/cd (689 L/cd)

Compare these usage rates to the "Basic Water Requirement" for four domestic needs—drinking, sanitation, bathing, and cooking—of about 13 g/cd (50 L/cd) proposed by Gleick (1998).

Although the historical trend has clearly been toward higher per capita use of water, the recent emphasis on conservation has resulted in significant changes in this pattern. In Section 11.1, the influences of various conservation practices will be considered in more detail. Table 9.2 can be used to estimate the daily indoor usage of water in various facilities, if *current practices* are anticipated. For very rough approximations of *conservation* effects:

· Reduce Table 9.2 values by 25%, assuming simple conservation measures such as flow controls.

· Reduce Table 9.2 values by 50%, assuming partial recycling.

The more water used for flushing toilets within the building types listed in Table 9.2, the greater the potential for savings through conservation.

In urban areas, public water mains usually provide the necessary quantities of water at the pressures and rates of flow required to operate typical plumbing fixtures. For isolated systems, or those independent of public networks, water supply can

TABLE 9.1 **Water Measurement Terms**

Quantity

1 gal	= 8.33 lb
	= 231 in.^{3}
	= 0.134 ft^3
	= 3.785 L
1 liter	= .0263 gal
1 cubic foot	= 7.48 gal
	= 62.4 lb
	= 0.028 m^3
1 cubic meter	= 1000 L
	= 35.32 ft^3
1 ton	= 240 gal
1 acre	= 0.4 hectare
	= 40 km^2
1 acre-foot (a-f)	= 325,851 gal
	= 43,560 ft^3
	= 12 acre-inches (ac-in.)
1 million gallons	= 3.07 a-f

Flow

1 cfs	= 1.98 a-f/day
	= 0.028 m^3/s
1 a-f/year	= annual water supply for five people at 180 gal/cd
1000 gpm	= 2.23 cfs
	= 4.42 a-f/day
1 million gpd	= 694.4 gpm
	= 1.55 cfs
	= 1120 a-f/yr

Leaks

Slow drip	= 170 gpd
	= 62,050 gal/yr
⅛-in. (3.2-mm) diameter stream	= 3600 gpd
	= 4 a-f/yr

Cost

10¢ per 1000 gal	= 7.48¢ per 100 ft^3
	= $32.59 per a-f
10¢ per 100 ft^3	= $43.56 per a-f
	= $13.40 per 1000 gal
10¢ per ton	= $136 per a-f

Sources: Milne (1976) and Ferguson (1998).

come from individual sources: wells, springs, cisterns, lakes, and so forth. For the minimum pressures and flow rates necessary from these sources (and/or the storage vessels associated with them), see Table 10.14.

(b) Cisterns. Where rainwater is to be utilized, a rough approximation of catchment area and cistern storage volume is initially needed. This procedure is detailed later in this chapter. For now,

1. From Table 9.2, find the quantity of rainwater to be used daily:

$$\text{g/cd} \times \text{population} = \text{gpd}$$

$$(\text{L/cd} \times \text{population} = \text{L/d})$$

2. Convert this quantity to the yearly need for water:

$$\text{gpd} \times 365 \text{ days} = \text{gal/yr}$$

$$(\text{L/d} \times 365 \text{ days} = \text{L/yr})$$

TABLE 9.2 **Planning Guide for Water Supply**[a]

Building Usage	Per Capita (as Listed) Daily Usage	
	Gallons	(Liters)
Airports (per passenger)	3–5	(11–19)
Apartments, multiple family (per resident)	60	(227)
Bath houses (per bather)	10	(38)
Camps		
Construction, semipermanent (per worker)	50	(189)
Day with no meals served (per camper)	15	(57)
Luxury (per camper)	100–150	(378–568)
Resorts, day and night, with limited plumbing (per camper)	50	(189)
Tourist, with central bath and toilet facilities (per person)	35	(132)
Cottages with seasonal occupancy (per resident)	50	(189)
Courts, tourist, with individual bath units (per person)	50	(189)
Clubs		
Country (per resident member)	100	(378)
Country (per nonresident member present)	25	(95)
Dwellings		
Boardinghouses (per boarder)	50	(189)
Additional kitchen requirements for nonresident boarders	10	(38)
Luxury (per person)	100–150	(378–568)
Multiple-family apartments (per resident)	40	(151)
Rooming houses (per resident)	60	(227)
Single family (per resident)	50–75	(189–284)
Estates (per resident)	100–150	(378–568)
Factories (per person per shift)	15–35	(57–132)
Highway rest area (per person)	5	(19)
Hotels with private baths (two persons per room)	60	(227)
Hotels without private baths (per person)	50	(189)
Institutions other than hospitals (per person)	75–125	(284–473)
Hospitals (per bed)	250–400	(946–1514)
Laundries, self-service (per washing)	50	(189)
Livestock (per animal)		
Cattle (drinking)	12	(45)
Dairy (drinking and servicing)	35	(132)
Goat (drinking)	2	(8)
Hog (drinking)	4	(15)
Horse (drinking)	12	(45)
Mule (drinking)	12	(45)
Sheep (drinking)	2	(8)
Steer (drinking)	12	(45)
Motels with bath, toilet, and kitchen facilities (per bed space)	50	(189)
With bed and toilet (per bed space)	40	(151)
Parks		
Overnight, with flush toilets (per camper)	25	(95)
Trailer, with individual bath units, no sewer connection (per trailer)	25	(95)
Trailer, with individual baths, connected to sewer (per person)	50	(189)
Picnic		
With bathhouses, showers, and flush toilets (per picnicker)	20	(76)
With toilet facilities only (per picnicker)	10	(38)

TABLE 9.2 **Planning Guide for Water Supply**[a] (*continued*)

Building Usage	Per Capita (as Listed) Daily Usage	
	Gallons	(Liters)
Poultry		
Chickens (per 100)	5–10	(19–38)
Turkeys (per 100)	10–18	(38–68)
Restaurants with toilet facilities (per patron)	7–10	(26–38)
Without toilet facilities (per patron)	2½–3	(9–11)
With bar/cocktail lounge (additional quantity per patron)	2	(8)
Schools		
Boarding (per pupil)	75–100	(284–378)
Day, with cafeteria, gymnasium, and showers (per pupil)	25	(95)
Day with cafeteria but no gymnasiums or showers (per pupil)	20	(76)
Day without cafeteria, gymnasiums, or showers (per pupil)	15	(57)
Service stations (per vehicle)	10	(38)
Stores (per toilet room)	400	(1514)
Swimming pools (per swimmer)	10	(38)
Theaters		
Drive-in (per car space)	5	(19)
Movie (per auditorium seat)	5	(19)
Workers		
Construction (per person per shift)	50	(189)
Day (school or office, per person per shift)	15	(57)

Source: Manual of Individual Water Supply Systems (1975).
[a]These values may be reduced as follows: with flow controls, up to 25% reduction
: with water recycling, up to 50% reduction.

3. Assume, conservatively, that a "dry" year will have two-thirds the precipitation of an average year; this measurement is the "design precipitation." (Average annual precipitation is available from NOAA annual summaries.)

Average annual precipitation

$$\times \, \tfrac{2}{3} = \text{design precipitation}$$

4. From Fig. 9.5, determine the catchment area required.
5. Roughly size the cistern (storage) capacity by finding the longest dry period (in days of negligible rainfall, from NOAA local climatological data):

cistern capacity = gpd × days of dry period

6. Convert capacity to volume by the formula

1 ft³ stores 7.48 gal of water

(1 m³ stores 1000 L water)

EXAMPLE 9.1, PART A. A 20,000-ft² one-story factory near Salem, Oregon, will use roof-collected rainwater to flush its toilets. Water-conserving toilets using 3 gal/flush and serving 20 workers are to be used. Table 9.2 shows that factory workers use between 15 and 35 g/cd. Because usage at this factory will be low—no showers, for example—assume 15 gal/cd:

15 g/cd × 20 workers = 300 gpd, for *all* usages

Because low-flush toilets are to be used, reduce this figure by 25%:

0.75 × 300 = 225 gpd

Toilets will probably account for most of this 225 gpd: for example, at three flushes per day per worker,

3 flushes/day × 3 gal/flush

× 20 workers = 180 gpd

Fig. 9.5 *Yields of rainfall catchment areas (roofs) in terms of total precipitation. In these graphs, 75% of the total precipitation is assumed to be catchable; the remainder is lost to evaporation or spillage.* (a) *Conventional units, from the U.S. Environmental Protection Agency's* Manual of Individual Water Supply Systems, *1975.* (b) *SI units (1 m³ = 1000 L).*

Assume, then, that up to 200 gpd of rainwater will be utilized.

Catchment Area: Salem's average annual rainfall is 41 in. Design rainfall is 2/3 × 41 = 27.3 in., in a "dry" year; yearly need is 200 gpd × 365 days = 73,000 gal. (The combination of 73,000 gal and 27.3 in. is off the chart in Fig. 9.5a, so divide 73,000 gal by 2, to obtain 36,500 gal, and then double the resulting catch-

ment area.) At 27.3 in. of precipitation, 36,500 gal will be caught by 2800 ft². The catchment area for 73,000 gal is, therefore, 2 × 2800 = 5600 ft². (So about 28% of the 20,000-ft² factory's roof area will suffice for a catchment area.)

Cistern Capacity: Salem normally has very dry summers; average monthly rainfall is as follows:

May	2.1 in.
June	1.4 in.
July	0.4 in.
August	0.6 in.
September	1.5 in.
October	4.0 in.

The dry period, then, runs from mid-June to mid-September—about 90 days. Thus

$$capacity = 200 \text{ gpd} \times 90 \text{ days} = 18,000 \text{ gal}$$

$$volume = \frac{18,000 \text{ gal}}{7.48 \text{ gal/ft}^3} = 2406 \text{ ft}^3$$

(For example, 5-ft deep × 22-ft square = 2420 ft³.) A more detailed sizing procedure for this building's cistern is presented later in this chapter. ■

(c) Required Facilities. Another important early design question is that of how many plumbing fixtures should be provided. Table 9.3 lists the minimum plumbing facilities required for various types and sizes of building occupancies. (Note that local requirements may differ somewhat from this particular guide, which is taken from the 1997 International Plumbing Code. There are many plumbing codes in use in North America, and they sometimes disagree.) Because these are considered "minimal" requirements, more generous provisions are sometimes appropriate. Chapter 10 discusses the design of the spaces in which these facilities are used.

(d) Sewage. Where public sewers are available (as in most urban areas), the designer usually is not concerned with estimating the flow of sewage. However, where private or on-site sewage treatment is required, or where public sewage treatment facilities are overtaxed, total sewage flow is an early design concern. Table 9.4 lists these sewage flows by type of occupancy, again in g/cd (L/cd). Note that sewage flow may differ from supply flow, especially where supply water is used for irrigation, car washing, or in evaporative processes.

Once the daily flow of sewage is established, some early guidelines are needed for determining the suitability of and the area required by some treatment processes. One of the most common reasons for the rejection of a potential rural building site is a lack of suitability for sewage disposal. A geologic analysis of structural and sewage disposal potential is one of the first documents needed by the designer. Chapter 11 details the sizing of some common treatment methods. For this earlier stage, the following design guidelines can be useful:

Septic tank drainfields. In conventional units (minimum 750-ft^2 area for any system):

- For shallow trenches in poorly draining soil: drainfield area = total sewage flow in gpd $\times$ 3.6 ft^2/gal.
- For deep trenches in well-draining soil: drainage area = total sewage flow in gpd $\times$ 0.4 ft^2/gal.

In SI units (minimum 70 m^2 area for any system):

- For shallow trenches, poor soil: L/day $\times$ 0.087 m^2/L
- For deep trenches, good soil: L/day $\times$ 0.01 m^2/L

(These guidelines allow for an expansion area equal to the original size of the drainage field, in case of field failure.)

Mounds. These are built-up leaching fields on top of an existing grade (see Fig. 11.39). For a single-family dwelling, allow for a 4-ft high (1.2 m) mound whose bottom area is a square 44 ft (13.4 m) on each side and whose sides slope at a 1:3 vertical-to-horizontal ratio.

Package sewage plant drainfields. In these, sewage is treated to a much greater extent than in septic tanks, and effluent is filtered:

- In poorly draining soil: total sewage flow in gpd $\times$ 0.49 ft^2/gal (L/day $\times$ 0.012 m^2/L).
- In well-draining soil: total sewage flow in gpd $\times$ 0.23 ft^2/gal (L/day $\times$ 0.006 m^2/L).

Sewage lagoons. These consist of at least two open treatment ponds (primary and secondary) and are sized on the basis of pounds of biochemical oxygen demand (BOD) rather than gallons of sewage flow. A typical assumption for estimating BOD is

- 0.2 lb BOD/person for "ordinary" domestic sewage.
- 0.3 lb BOD/person where garbage grinders or other devices contribute added organic material to domestic sewage.

The *total* acreage (ac) needed for the two ponds can be estimated as

- 20 lb BOD/ac for the (colder) northern United States.
- 35 lb BOD/ac for the (drier, warmer) southern United States.

(The primary pond is usually sized for 50 lb BOD/ac.)

9.4 Rainwater

There is a striking similarity between rainwater and solar energy. Both are essential for agriculture and both have been well understood and utilized by farmers since agrarian societies first emerged. Both can be very beneficial to architecture and were utilized as needed by anonymous builders for centuries. Both fell out of favor with designers as plentiful supplies of pure, centrally treated water and concentrated, centrally controlled fuels became commonplace. Both are thinly yet relatively evenly spread over the world's population, so they are at least seasonally available to help meet nearly every building's needs for water and heat. Yet both are difficult to utilize in industrialized societies, because they require *individual, preoccupancy* expenditures.

Consider the typical public water and electric utility in a U.S. city. It can raise the funds to build large water treatment plants, electricity generating plants, and the network of pipes and wires that bring these commodities to every building. The utility's costs, including interest on its construction debts, will be passed on to its consumers on a monthly basis, along with a margin of profit that is usually controlled by state governments.

WATER AND WASTE

TABLE 9.3 **Minimum Number of Plumbing Facilities**[a]

Occupancy	Water Closets, Urinals[b]		Lavatories	Bathtubs/Showers	Drinking Fountains[c]	Others
	Male	Female				
Assembly						
Theaters	1 per 125	1 per 65	1 per 200	—	1 per 1000	1 service sink
Nightclubs	1 per 40	1 per 40	1 per 75	—	1 per 500	1 service sink
Restaurants	1 per 75	1 per 75	1 per 200	—	1 per 500	1 service sink
Halls, museums, etc.	1 per 125	1 per 65	1 per 200	—	1 per 1000	1 service sink
Coliseums, arenas	1 per 75	1 per 40	1 per 150	—	1 per 1000	1 service sink
Churches[d]	1 per 150	1 per 75	1 per 200	—	1 per 1000	1 service sink
Stadiums, pools, etc.	1 per 100	1 per 50	1 per 150	—	1 per 1000	1 service sink
Business[e,f,g]	1 per 25		1 per 40	—	1 per 100	1 service sink
Educational	1 per 50		1 per 50	—	1 per 100	1 service sink
Factory and industrial	1 per 100	1 per 100	1 per 100	[h]	1 per 400	1 service sink
High hazard[e,f]	1 per 100	1 per 100	1 per 100	[h]	1 per 1000	1 service sink
Institutional						
Residential care	1 per 10	1 per 10	1 per 10	1 per 8	1 per 100	1 service sink
Hospitals, ambulatory nursing home patients[i]	1 per 50	1 per 50	1 per room[j]	1 per 15	1 per 100	1 service sink per floor
Day nurseries, sanitariums, nonambulatory nursing home patients, etc.[i]	1 per 15	1 per 15	1 per 15	1 per 15[k]	1 per 100	1 service sink
Employees, other than residential care[i]	1 per 25	1 per 25	1 per 35	—	1 per 100	—
Visitors, other than residential care	1 per 75	1 per 75	1 per 100	—	1 per 500	—
Prisons[i]	1 per cell	1 per cell	1 per cell	1 per 15	1 per 100	1 service sink
Asylums, reformatories, etc.[i]	1 per 15	1 per 15	1 per 15	1 per 15	1 per 100	1 service sink
Mercantile[e,f,g]	1 per 500	1 per 500	1 per 750	—	1 per 1000	1 service sink
Residential						
Hotels, motels	1 per guestroom	1 per guestroom	1 per guestroom	1 per guestroom	—	1 service sink
Lodges	1 per 10	1 per 10	1 per 10	1 per 8	1 per 100	1 service sink
Multiple family	1 per dwelling unit	1 per dwelling unit	1 per dwelling unit	1 per dwelling unit	—	1 kitchen sink per dwelling unit; 1 automatic clothes washer connection per 20 dwelling units

Occupancy					
Dormitories	1 per 10	1 per 10	1 per 8	—	1 service sink
One- and two-family dwellings	1 per dwelling unit	1 per dwelling unit	1 per dwelling unit		1 kitchen sink per dwelling unit; 1 automatic clothes washer connection per dwelling unit[l]; 1 service sink
Storage[e,f]	1 per 100	1 per 100	h	1 per 1000	1 service sink

Source: Copyright © 1997, International Code Council, Inc., Falls Church, Virginia. *International Plumbing Code.* Reprinted with permission of author. All rights reserved.

[a]Based on one fixture being the minimum required for the number of persons indicated or any fraction of the number of persons indicated. The number of occupants shall be determined by the building code. Unless otherwise shown, the required water closets, lavatories, and showers or bathtubs shall be distributed evenly between the sexes based on the percentage of each sex anticipated in the occupant load. The occupant load shall be composed of 50% of each sex, unless statistical data approved by the code official indicate a different distribution.

[b]In each bathroom or toilet room, urinals shall not be substituted for more than 50% of the required water closets.

[c]Drinking fountains shall not be installed in public restrooms. Where water is served in restaurants or where bottled water coolers are provided in other occupancies, drinking fountains shall not be required.

[d]Fixtures located in adjacent buildings under the ownership or control of the church shall be made available during the periods the church is occupied.

[e]Separate employee facilities shall not be required in occupancies in which 15 or less people are employed. Separate facilities for each sex shall not be required in structures or tenant spaces with a total occupant load, including both employees and customers, of 15 or less in which food or beverage is served for consumption within the structure or tenant space.

[f]Access to toilet facilities in occupancies other than assembly or mercantile shall be from within the employees' regular working area. The required toilet facilities shall be located not more than one story above or below the employees' regular work area, and the path of travel to such facilities shall not exceed a distance of 500 ft (152 m).

[g]In mercantile and assembly occupancies, employee toilet facilities shall be either separate facilities or public customer facilities. Separate employee facilities shall not be required in tenant spaces of 900 ft² (84 m²) or less where the travel distance from the main entrance to a central toilet area does not exceed 500 ft (152 m) and where such central facilities are located not more than one story above or below the tenant space.

[h]Emergency showers and eyewash stations shall be provided with a supply of cold water as required by the manufacturer. Waste connection shall not be required for emergency showers and eyewash stations.

[i]Toilet facilities for employees shall be separate from facilities for inmates or patients.

[j]A single-occupant room with one water closet and one lavatory serving not more than two adjacent patient rooms shall be permitted where such room is provided with direct access from each patient room and with provisions for privacy.

[k]For day nurseries, a maximum of one bathtub shall be required.

[l]For attached one- and two-family dwellings, one automatic clothes washer connection shall be required per 20 dwelling units.

TABLE 9.4 **Estimated Sewage Flow Rates**

Type of Occupancy	Unit Gallons (Liters) per Day
Airports	15 (56.8) per employee
	5 (18.9) per passenger
Auto washers	Check with equipment manufacturer
Bowling alleys (snack bar only)	75 (283.9) per lane
Camps, campgrounds with central comfort station	35 (132.5) per person
Campgrounds with flush toilets, no showers	25 (94.6) per person
Day camps (no meals served)	15 (56.8) per person
Summer and seasonal	50 (189.3) per person
Churches (sanctuary)	5 (18.9) per seat
With kitchen waste	7 (26.5) per seat
Dance halls	5 (18.9) per person
Factories, no showers	25 (94.6) per employee
With showers	35 (132.5) per employee
Cafeteria, add	5 (18.9) per employee
Hospitals	250 (946.3) per bed
Kitchen waste only	25 (94.6) per bed
Laundry waste only	40 (151.4) per bed
Hotels (no kitchen waste)	60 (227.1) per bed (2 person)
Institutions (resident)	75 (283.9) per person
Nursing home	125 (473.1) per person
Rest home	125 (473.1) per person
Laundries, self-service	50 (189.3) per wash cycle
Commercial	Per manufacturer's specifications
Motel	50 (189.3) per bed space
With kitchen	60 (227.1) per bed space
Offices	20 (75.7) per employee
Parks, mobile homes	250 (946.3) per space
Picnic parks (toilets only)	20 (75.7) per parking space
Recreational vehicles, without water hook-up	75 (283.9) per space
With water and sewer hook-up	100 (378.5) per space
Restaurants/cafeterias	20 (75.7) per employee
Toilet	7 (26.5) per customer
Kitchen waste	6 (22.7) per meal
Add for garbage disposal	1 (3.8) per meal
Add for cocktail lounge	2 (7.6) per customer
Kitchen waste—disposable service	2 (7.6) per meal
Schools—staff and office	20 (75.7) per person
Elementary students	15 (56.8) per person
Intermediate and high	20 (75.7) per student
With gym and showers, add	5 (75.7) per student
With cafeteria, add	3 (11.4) per student
Boarding, total waste	100 (378.5) per person
Service station, toilets	1000 (3785) for 1st bay
	500 (1892.5) for each additional bay
Stores	20 (75.7) per employee
Public restrooms, add (per unit of floor space)	1 per 10 ft^2 (4.1 per m^2)
Swimming pools, public	10 (37.9) per person
Theaters, auditoriums	5 (18.9) per seat
Drive-in	10 (37.9) per space

Source: Uniform Plumbing Code, copyright © 1997. Printed by permission of the International Association of Plumbing and Mechanical Officials.

Fig. 9.6 *This "low-entropy kindergarten" competition entry for Frankfort, Germany, features sun, adjustable shade, and a roof that collects rainwater and leads it to an interior stone-filled catchment. The rainwater downspout can be seen in the center of the drawing; the stone-filled catchment is below the stair landing. Overflow then goes outside, through a layer of stones and into a below-ground cistern. (Courtesy of Fernau & Hartman, Architects, Berkeley, California.)*

Thus, our society has a well-established method for encouraging central suppliers of water and heat.

Now consider the individual building owner. To build a cistern and a solar-heated building, she must borrow money at an interest rate higher than that which the utility pays; and both options cost more initially than simple connections to the utility's pipes and lines. Even though she is willing to flush her toilets with rainwater rather than with chlorinated and filtered potable water from the utility, and even though she is willing to heat with lower-grade solar energy than with higher-grade electricity, she must pay a substantial first-cost penalty—with interest—to do so.

The overall public good could be well served by a mixture of public networks of pure water and electricity and individual cisterns and solar applications. The environmental benefits would be substantial—less water withdrawn from rivers, lakes, and underground aquifers; less energy and chemicals used to treat and deliver such water; less stormwater discharged to pollute rivers; less fuel used to generate electricity; and less environmental damage from power plants. Yet we continue to economically discourage the individual who uses the rain and the sun.

Architects Fernau and Hartman's design for an invited competition for a "low-entropy kindergarten" in Frankfort, Germany (Fig. 9.6) celebrates sun and rainwater as teaching elements. In their words, "The building reveals natural phenomena such as the play of light and shadow over a wall during the course of a year, the warmth of a wintergarden in January, the cooling shade from deciduous vines on a trellis overhead, water pouring off the roof into an interior stone catchment during a downpour."

9.5 Collection and Storage

In terms of both quality and quantity, rainwater is an attractive alternative. Figure 9.4*b* shows that rainwater is near to the purest state in the hydrologic cycle. More recently, it is true, air pollution began to threaten the quality of rainwater in some areas, as "acid rain" has become widespread in the northeastern section of the North American continent and in Europe. In some particularly air-polluted locations, lead poses a threat to rainwater quality. Also, on any catchment surface, dust and bird droppings are common pollutants that must be considered. Other factors that bear on rainwater quality are roofing materials and the form of the roof. The appropriate health authority should be consulted for a list of roofing materials that will have no toxic effects on rainwater. Steeper roofs are scoured by winds and thus collect less dust and give cleaner runoffs. Devices to discourage roosting birds are strongly recommended, as are periodic checks of cistern water for bacteria. Fungicides (for moss control) should be scrupulously avoided, as should roofing paints containing lead. For these reasons, urban rainwater commonly is not used for drinking and cooking. (Bottled water or on-site water distilling can supply potable water.) For the typical residence, however, that still leaves about 95% of indoor water usage that could be provided by rainwater. Also, in those cities that have very "hard" public water supplies, rainwater's "soft" characteristics make it particularly attractive.

The quantity of rainwater available in most U.S. locations could meet a high percentage of typical home or business needs. Milne (1976) pointed out that the 42 in. (1067 mm) of rain that falls annually on the streets and roofs of Manhattan Island could, if collected and stored, provide 148% of the residential needs of the 1.7 million inhabitants. For the typical U.S. suburban house [roof area of 1500 ft^2 (139 m^2)], the annual catchment can be estimated by combining the rainfall quantities (Fig. 9.7) with the resulting catchment yield (Fig. 9.5). Even at a "dry" rate of only 20 in. annually, a 1500-ft^2 roof would yield about 12,000 gal, or 33 gpd—nearly enough to meet the clothes cleaning and dishwashing needs of the family. (In SI units, a 140-m^2 roof with annual 500 mm precipitation will yield about 50,000 L, or 137 L/d.)

Unfiltered rainwater seems particularly well-

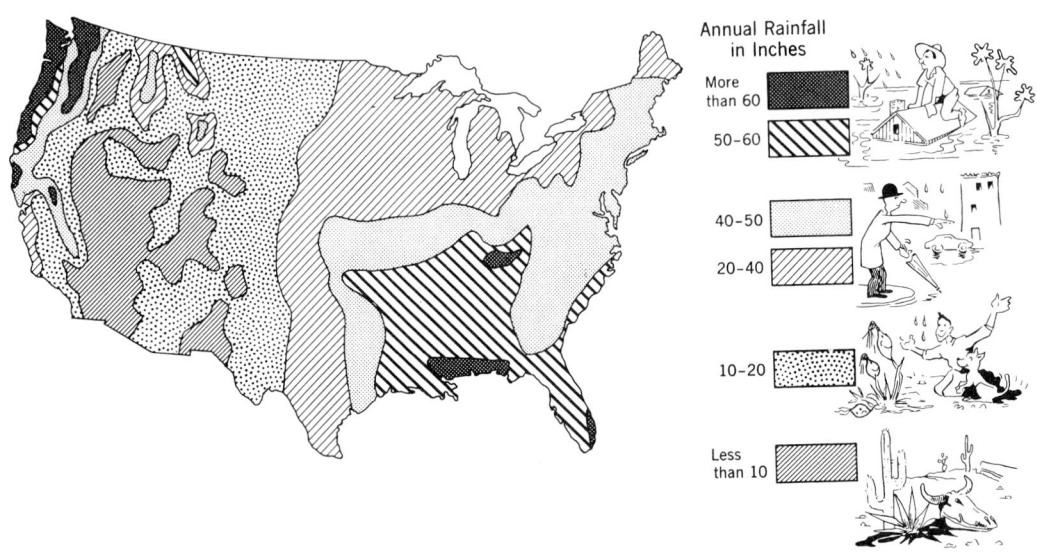

Fig. 9.7 *Annual rainfall in the United States. Cistern volume conversion factors:*

$$1 ft^3 = 7.48 \ gal = 0.02832 \ m^3$$
$$1 m^3 = 1000 \ L = 264.2 \ gal$$

suited to the irrigation of small lawns or gardens, both because it lacks additives unneeded by plants, such as chlorine or sodium fluoride, and because it can reduce the user's demand on the public water supply on the hottest summer days. Cisterns located above the irrigated area have the advantage of replacing a pump with simple gravity flow. The rainbarrel at the bottom of a downspout was an example of this approach—although one of limited capacity.

In many of the world's drier areas, smaller cisterns within the home are common. Such cisterns, which can be fed both by rainwater and by the public supply, are frequently used for all domestic purposes, including drinking. The presence of such a large water volume also can be advantageous in the event of fire. Sometimes these storage cisterns are required because the public supply is diminished or cut off at peak-usage hours, due to insufficient water main capacity. Figure 9.8 shows a cistern in the typical outdoor location, along with various options by which pollution from dirty roofs can be minimized.

When cisterns are taken seriously as water storage devices, their function and size can become strong design-form determinants. The country home of architect John Andrews in the dry ranchland of New South Wales, Australia, offers a particularly striking example of cisterns and form (Fig. 9.9). These demonstrate the design image of "pregnant downspouts": storing the water above where it will be used, rather than below ground. The corner rainwater collectors also help deflect cooling summer breezes into the house, along the diagonal walls at each corner. The central skylight then vents the breezes, along with the house's heat. From the corner cisterns, rainwater is pumped up to the central tower's storage tank. It then can be heated by solar collectors (to be installed on the sloping top of the tower), or used to supply the house's plumbing fixtures, or even used to sprinkle the metal roof surface, whose surface temperature could quickly be lowered by evaporation. Unevaporated roof water simply runs back into the cisterns. As a final design-with-water gesture, the shower (off bedroom 1) has been made into a true celebration place for cleansing and refreshing, with a sweeping view of the ranch.

In this passively solar heated home, the living areas are placed on the elongated north side—the warmer side in the cold season of this southern hemisphere house. Passive cooling is aided by pergolas on the west and south, which are covered with vines for hot season shading of windows.

Sizing. The procedures described in Section 9.3 are for rough sizing of both the catchment area and the storage capacity for cisterns. When rainwater is to be a primary, as opposed to merely a supplementary source, a closer look must be taken at rainfall deposits and user withdrawals from a cistern.

This procedure depends on the monthly average rainfall (from NOAA Local Climatological Data), the monthly water usage, and the catchment area yield (from Fig. 9.5).

EXAMPLE 9.1, PART B. Take a closer look at the cistern that was approximately sized in Part A. Daily usage for this cistern, to be used for toilet flushing in a factory near Salem, Oregon, was estimated at 200 gal and catchment area of 5600 ft². The cistern capacity was estimated at 18,000 gal. Begin the process in the midst of the wettest months.

From Table 9.5, the following conclusions can be drawn:

1. For an 18,000-gal cistern, when the end of December cumulative capacity is added to January's surplus, the cistern will be at capacity from November through April.
2. With no allowances for abnormally dry months, we could reduce the cistern size by about 3000 gal (the size of the smallest cumulative capacity, in September).
3. A larger cistern—one that could utilize everything from the catchment area—could be built. Its approximate size would be the year-end surplus of 38,510 gal plus the maximum spring monthly cumulative capacity of 25,410 gal, in April.
4. Or the surplus could be devoted to additional usage of rainwater, beyond mere toilet flushing, from November through April. ∎

Fig. 9.8 *Section through a typical outdoor cistern. (a) A "roof washer" gets the dirtiest, first runoff from the roof. It can later be emptied either by opening the faucet wide or by leaving the faucet slightly open so that it will slowly drain. (b) In place of the "roof washer," a sand filter may be used. The "flapper valve" is then used rarely, perhaps to divert the first rainfall after a prolonged dry spell. After this, the valve is left in a position to divert all rainwater to the sand filter. (Based on the U.S. Environmental Protection Agency's* Manual of Individual Water Supply Systems, *1975.) (c) Another roof-washing option is a "tipping valve," which dumps the first X gallons of each rainfall. After each rainfall, the valve must be manually (or spring) reset to "dry" position if it is to again intercept dirty water. In an extended wet period, it would probably be left in the "wet" position.*

(a)

(b)

Fig. 9.9 *Country home of architect John Andrews, Eugowra, NSW, Australia. (a) View from the southwest, showing corner cisterns, central daylight/ventilation barrel vault, and "energy tower" with water storage tank. (b) The view from the south, showing shading pergola to be covered by vines. (c) Floor plan, showing fireplace at center and cisterns at all corners. Insect screens protect the openings to summer breezes. (d) Section. (Courtesy of John Andrews International Pty, Ltd.)*

entry

water tank

water tank

outdoor dining area

outdoor living area

breakfast

dress

pergola

dining area

energy tower over fireplace

COATS

living area

bedroom 1

BAR

office

laundry

dress

water tank

water tank

pergola

pergola

bedroom 2

court

water tank

garage

water tank

water tank

FLOOR PLAN

0 1M 2M 3M 4M 5M 6M

(c)

GARAGE   PERGOLA   BEDROOM 2   LIVING AREA   ENERGY TOWER   OUTDOOR LIVING   PERGOLA

SECTION THRU LIVING AREA

0 1 2 3 4 5

(d)

Fig. 9.9 *(continued)*

TABLE 9.5 **Rainfall Cistern Sizing Procedure**

I		II	III	IV	V	VI
						Cumulative Capacity Adjusted for
		Catchment			Cumulative	Actual Size[f]
Month and Rainfall[a] (in.)		Yield[b] (gal)	Usage[c] (gal)	Net[d] (gal)	Capacity[e] (gal)	(gal)
January	6.9	18,630	6,200	12,430	12,430	12,430
February	4.8	12,960	5,600	7,360	19,790[g]	18,000
March	4.3	11,610	6,200	5,410	25,200[g]	18,000
April	2.3	6,210	6,000	210	25,410[g]	18,000
May	2.1	5,670	6,200	−530	24,880[g]	17,470
June	1.4	3,780	6,000	−2,220	22,660[g]	15,250
July	0.4	1,080	6,200	−5,120	17,540	10,130
August	0.6	1,620	6,200	−4,580	12,960	5,550
September	1.5	4,050	6,000	−1,950	11,010	3,600
October	4.0	10,800	6,200	4,600	15,610	8,200
November	6.1	16,470	6,000	10,470	26,080[g]	18,000
December	6.9	18,630	6,200	12,430	38,510[g]	18,000

Source: Based upon procedure from Brown, Reynolds, Ubbelohde, *InsideOut: Design Procedures for Passive Environmental Technologies,* copyright © 1982 by John Wiley & Sons.

[a]NOAA data for Salem, Oregon.

[b]From Fig. 9.5, according to which 10 in. precipitation yields about 27,000 gallons for this 5600-ft^2 catchment area.

[c]The factory uses 200 gpd times days in each month.

[d]Col. II minus col. III.

[e]Values in col. IV, added month by month.

[f]Set at 18,000 gallons for the factory in Example 9.1.

[g]The cumulative capacity exceeds the actual storage capacity of 18,000 gallons.

9.6 Rainwater and Site Planning

Prior to the spread to rural areas of buildings, streets, roads, and paved parking lots, water from rainfall and melting snow found its own way to natural destinations. Surface flow to creeks, streams, and rivers accounted for part of this drainage. Underground flow aided the general runoff. Outcropping of flowing groundwater created springs and artesian wells. Low, dished areas formed lakes that in turn overflowed to outlet streams. Flat areas sometimes developed into swamps or marshes.

At a time when there was a choice of locations for towns and villages, sites next to rivers were usually chosen. The waterways provided transportation and water was supplied from the river or from adjacent wells. As streets and roads were built, slopes could be arranged whereby the rain falling on these areas and flowing onto them from roofs of buildings could run to the river. At interior parts of the country, high ground was favored for building sites and growing communities. Obviously, swampy or marshy ground would not be chosen, but it did provide terminal locations for the stormwater that ran off the high ground. In the course of this natural flow, much of the water was drawn by evaporation to the clouds. The rest, conforming to topographic river basins, continued along to seek its way to the sea.

As building increased, desirable locations grew scarce. The possibility of selecting high, dry ground diminished. Great areas, formerly low and marshy, were filled in and buildings constructed, often on piles. From such locations stormwater could not be disposed of by drainage to some adjacent lower area, or even recharged to the earth through dry wells. Moreover, extensive grids of paved streets and sidewalks in these level developments caught and held the water, resulting in considerable "ponding." Storm sewers had to be built and the water transported great distances, often having to be lifted at intermediate pumping stations before reaching its destination, which might be a remote river.

This emphasis on the removal of stormwater has led to an expensive and elaborate system based on quick "disposal" of rainfall. By decreasing the time between precipitation and runoff, quick disposal increases the peak flows within such systems, thereby increasing flooding of rivers during storms but reducing the rivers' flow between storms. It also contributes pollutants to wa-

terways that otherwise would have been filtered by the soil, as detailed in Table 9.6.

The overloading of storm sewers not only causes minor flooding, but also can influence building design. The designers of the New Orleans Convention and Exhibition Center, a building with 610,000 ft^2 (56,670 m^2) of roof area, spared the city's storm sewers from the impact of a 14-acre (5.7-hectare) runoff by carrying the rainwater over the roofs of adjoining wharfs, and thus discharging it directly into the Mississippi River.

As urban storm sewers reach capacity and suburban groundwater levels drop, designers have

TABLE 9.6 **Some Constituents of Stream Water**

Constituent	Source in Nature	Role in Natural Ecosystem	Source of Urban Excess	Role of Excess
Sediment	Banks of meandering channels	Maintain stream profile and energy gradient; store nutrients	Construction sites; eroding stream banks	Abrade fish gills; carry excess nutrients and chemicals in adsorption; block sunlight; cover gravel bottom habitats
Organic compounds	Decomposing organic matter	Store nutrients	Herbicides; pesticides; fertilizers	Deprive water of oxygen by decomposition
Nutrients	Decomposing organic matter	Support ecosystems	Organic compounds; organic litter; fertilizers; food waste; sewage	Unbalance ecosystem; produce algae blooms; deprive water of oxygen by decomposition
Trace metals	Mineral weathering	Support ecosystem	Cars; construction materials; all kinds of foreign chemicals	Reduce resistance to disease; reduce reproductive capacity; alter behavior
Chloride	Mineral weathering	Support ecosystems	Pavement deicing salts	Sterilize soil and reduce biotic growth
Bacteria	Native animals	Participate in ecosystems	Pet animals; dumpsters; trash handling areas	Cause risk of disease
Oil	Decomposing organic matter	Store nutrients	Cars	Deoxygenate water

Source: Ferguson, B. (1998). *Introduction to Stormwater.* © John Wiley and Sons, New York.

Fig. 9.10 *The Eco-roof Pavilion is a demonstration structure at the confluence of the Columbia and Willamette rivers. Overlooking a wetlands that further treats outflows from Portland's sewage treatment, this three-segmented roof compares the rain retention characteristics of a sod roof, a conventional roof, and an artificial soil roof. Waterfall pylons will offer direct runoff comparisons. (Courtesy of Energy Studies in Buildings Laboratory, University of Oregon.)*

begun to emphasize "stormwater infiltration" (or recharge of groundwater), rather than "quick runoff." Three design strategies for encouraging such recharge have emerged: roofs that will retain water and slowly release it, porous pavement, and onsite infiltration of runoff.

(a) Roof Retention. If stormwater is to be sent to storm sewers or to soak into the ground, a *slow flow* from roofs will help by diminishing peak flows in sewers and giving soaked soil a longer time to absorb still more runoff. A nearly flat roof with specially designed drains (see later Fig. 9.24) forms a temporary pond that permits slower discharge, yet eventually drains completely dry (to discourage mosquito breeding, etc.). (For cisterns, however, sloped roofs should be used, as they stay cleaner than flat roofs.) Another possibility is a sod roof that retains water for longer periods than ponded flat roofs (Fig. 9.10). Another option is to store the cistern water on the roof itself (Fig. 9.11).

However achieved, this temporary pond on top of a building will clearly add to structural requirements. Another problem could be posed by high winds blowing sheets of water onto people below. In summer, however, a flooded roof can provide the significant thermal advantage of greatly lowering daytime roof surface temperatures. Rainfall-retaining roofs are now required in some urban areas with overtaxed storm sewers. Sewer districts frequently charge for stormwater

runoff; by ponding stormwater on site, such charges can be reduced or avoided.

(b) Porous Pavement. To retain rainfall on-site, many builders use either porous asphalt (Fig. 9.12), porous concrete, "incremental" paving, or open-celled pavers (alternation of paving materials with grass or ground-cover plants) for parking lots and roadways.

Porous concrete has been used for many years in building construction as a low-strength, high-porosity material that has some insulating properties (R-5 for 10-in. thickness). Patented porous concrete pavement now in use in Florida has a strength of more than 3800 psi and a permeability of 2.3 gallons of water per minute per square foot. (A 2500-psi mix has a permeability of 18.5 gallons per minute per square foot.) In cold weather areas, the freeze–thaw cycle could be destructive to porous concrete.

Incremental paving features many small, adjacent paving units; the resulting many joints allow water to pass through more readily than solid, unbroken surfaces.

Open-celled pavers alternate small concrete or plastic paving units with grass or ground cover, as shown in Fig. 9.13. These are particularly applicable to shorter-term parking, as for a sports stadium, or for more remote areas of retail mall parking lots. They are also possible for roadway shoulders, and emergency vehicle–only access areas.

Fig. 9.11 *Retaining rainwater so that gravity flow can encourage its usage.*

(a)

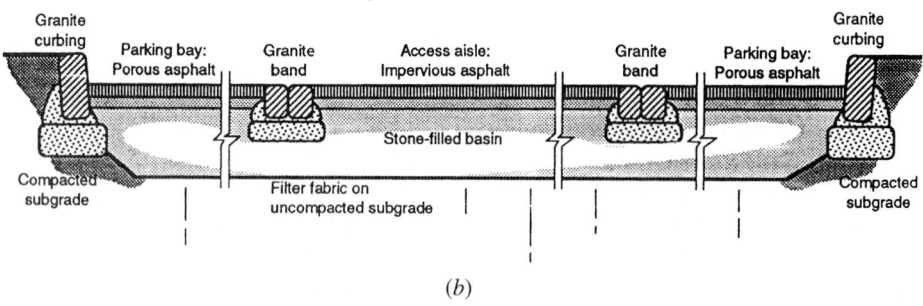

(b)

Fig. 9.12 *Porous paving encourages groundwater recharge rather than storm runoff.* (a) *Standard porous asphalt pavement as used in Rockville, Maryland.* (b) *Subsurface basin allows required retention ponds to serve as parking lots; this example is at the Morris Arboretum, Philadelphia. (Reprinted by permission from Ferguson, B. (1998)* Introduction to Stormwater, *© John Wiley and Sons, New York.)*

(a)   (b)

Fig. 9.13 *Open-celled pavers present a surface of grass or hardy ground cover, reinforced by small units of concrete or plastic. (a) An Oregon office building uses Grasscrete's concrete paving units to reinforce its parking area off an alley. Unthank Seder Poticha, architects, Eugene, Oregon. (Photo by Jane Lidz.) (b) Miami, Florida's Orange Bowl features a huge parking area, where the driveways (aisles) are asphalt paved, and the parking areas reinforced by Grasspave² plastic reinforcing open cells. (Courtesy of Invisible Structures, Inc., Aurora, Colorado.)*

If stormwater is not retained, its runoff can be polluted to the extent that treatment is required. Using strips of planting for such treatment (phytoremediation) is discussed at the end of Chapter 11.

(c) Site Design for Recharging. This tactic is especially advisable for suburban-density developments in drier climates with absorptive soil (sand, gravel, etc.). The first option is at each house; design of the entire subdivisions can also be influenced.

The simplest design approach is the gutterless sloped roof illustrated in Fig. 9.14, which is applicable to one-story, basementless homes with wide, overhanging roofs. A gravel-filled trench skirting the perimeter directly below the edge of the eaves catches the water flowing off the roof.

Some designers do not like the appearance of conventional gutters and leaders; other designers

Gravel filled trench

Fig. 9.14 *Gutters and leaders are not always essential, provided doorways, walls, foundations, and landscaped lawns are not subjected to rain concentrations.*

(a)

(b)

Fig. 9.15 *Falling rain is featured at this Oregon residence, where the gutter is held below the metal flashing at the roof's edge. Runoff is seen from indoors as it falls into the suspended gutter. Unthank Poticha Waterbury, Architects, Eugene, Oregon.*

celebrate them (Fig. 9.15). There are many ingenious ways to avoid or modify their use and yet provide proper drainage. In many cases, however, gutters and leaders will be required, either to collect rainwater for cisterns or to control conditions around the perimeter of the house. Several op-

tions for stormwater recharge can be used with the gutter–leader combination.

A splash pan at the foot of each leader (Fig. 9.16a) offers the simplest method. It will lead the water a few feet from the house but will accommodate only a relatively low rate of flow.

(a) Concrete splash pan   (b) Gravel filled pipe, vitrified   (c) Dry well
                               clay or reinforced concrete

Fig. 9.16 *Roof drainage for houses. Method* (a) *is suitable for low rates of flow introduced into very pervious soil. When denser soil is encountered,* (b) *is used to get the water into the ground and thus avoid surface erosion. For heavy flow or to lead the water farther from the structure,* (c) *may be used with one or several dry wells.*

A gravel-filled pipe is somewhat more effective (Fig. 9.16*b*). When the soil is not very permeable (as, for instance, with clay), it is best to use a dry well with extended area and many perforations through which the water can be discharged to the ground.

Footing drains are often used to collect and lead away groundwater that accumulates around the foundations. This reduces the likelihood of basement wall leakage. These drains are most necessary when higher ground near the buildings increases the flow of groundwater against underground walls. Figure 9.17 illustrates this and also shows how stormwater from drains and roofs may be led to a surface absorption area of rock and gravel beyond a head wall where the general storm drain outcrops. This method can be chosen where there is sufficient property area and slope.

It has the advantages of easy maintenance and service. Also, one can observe whether it is functioning correctly.

In new suburban developments, for which there are no storm sewers, recharge basins are sometimes required to deliver stormwater to the ground. Water from numerous roofs, paved areas, and curb catch basins is collected and piped to an open, unpaved pit, where it sinks into the earth. This method is not recommended in areas of dense, impervious clay soil (Fig. 9.18). A particularly effective example of this approach is offered by the community of passively solar heated residences known as Village Homes, outside Davis, California. This area receives only about 20 in. (about 500 mm) of rain annually, so the recharge of groundwater was to this garden-oriented community a far more attractive (and less-expensive)

Fig. 9.17 *Disposal of stormwater on the site but remote from the house or building. When a wall is against a hill, it is usually subjected to pressure of groundwater during storms. Open-joint clay, plastic, or fiber tile accepts this water and carries it away. Footing drains are tight-joint clay tile or bituminous fiber pipe. Flow through stone and gravel returns the water to the earth. Head wall is appropriate in lieu of a dry well if the site permits.*

option than loss of the rainwater to the storm sewer. Stormwater flows from leaders to dry, rockbed channels, along which are gardens and bicycle paths (Figs. 9.19 and 9.20). Occasionally, small dams across these channels create temporary holding ponds, in case the runoff has not yet soaked through the channel bottom. In the event of extraordinarily heavy rainfall, an inlet to the public storm sewer is available beyond the final holding pond; this inlet is needed approximately once every 5 years.

The National Wildflower Research Center (Fig. 9.21) is built over an endangered aquifer southwest of Austin, Texas. This complex of small buildings and larger gardens includes two cisterns, one of which forms the base of a central observation tower. Water gardens and irrigation are both served by the cisterns. The buildings' roofs are designed for rainwater collection, and a small channel (aqueduct) carries water over a wall to the outer cistern.

The planning of the landscape around buildings may closely follow such considerations as irrigation. The "hydrozone" concept of landscape planning is shown in Fig. 9.22. To minimize water consumption, exotic plant species are kept to a minimum and located near the house, where storm runoff and irrigating water is readily available. Native and adapted plantings, which can survive on normal rainfall, are utilized elsewhere.

Fig. 9.18 *Recharge basins in suburban communities. When topography, groundwater level, and porosity of the soil permit, developers are sometimes required to install systems that collect stormwater and carry it from catch basins at street curbs, the roofs of all houses, and paved areas to a recharge basin that receives the water and returns it to the ground. For the safety of children, a fence is sometimes required to prevent unauthorized access to the basin.*

(a)

(b)

Fig. 9.19 *Village Homes, a northern California subdivision of solar homes and stormwater recharge areas.* (a) *Plan shows the emphasis on bicycle paths, narrow streets, and widespread community-maintained garden space through which the recharge streambeds are led.* (b) *Site section explains the gradual drainage from leaders at houses to recharge stream. (Reprinted by permission from M. Corbett,* A Better Place to Live, *copyright © 1983 by the Rodale Press.)*

Fig. 9.20 *The open drainage ways and pedestrian paths of Village Homes. Photo by Alan Butler.*

(a)

Fig. 9.21 The National Wildflower Research Center is designed to capture and utilize rainwater. (a) Site plan shows two cisterns, one at far right (1) the other in the central observation tower (9). (b) Section through the observation tower/cistern. (Courtesy of Overland Partners, architects, San Antonio, Texas.) 1. Cistern; 2. Aqueduct; 3. Auditorium; 4. Gallery; 5. Main plaza; 6. Gift shop; 7. Children's discovery room; 8. Kitchen; 9. Observation tower/cistern; 10. Library/board room; 11. Administration; 12. Research laboratories; 13. Seed silo; 14. Pergola; 15. Comparative garden; 16. Pavilion; 17. Display greenhouse; 18. Research greenhouse; 19. Shade house; 20. Demonstration garden.

1 WATER CISTERN
2 STAIRWELL
3 CISTERN ROTUNDA
4 OCULUS
5 OBSERVATION
   PLATFORM

TOWER SECTION

(b)

Fig. 9.22 *The Hydrozone concept of landscape planning restricts exotic planting to special, easily watered areas. Native and adapted plantings are used elsewhere. Hydrozonics are here shown on a typical suburban lot. (Reprinted by permission from* Energy Conserving Site Design, *E. G. McPherson, ed., copyright © 1984 by the American Society of Landscape Architects.)*

Landscape that requires no additional water is often called a xeriscape; see Fig. 10.68, and Section 10.12 for irrigation systems and an example in California's Napa Valley.

9.7 Components

The first stormwater system design decision to be made involves the establishment of "watersheds" on a building's roof. To what edges, or at what points, will runoff be directed? To what depths will it accumulate before it leaves the roof? Because the answers to these questions depend on the intensity of storms as well as on the roof's geometry, it is necessary to find the maximum hourly rainfall for each location. This figure is available from local building code officials, or from Fig. 9.23.

Gutters and *leaders* (downspouts) can be sized through the use of Tables 9.7 and 9.8. The sizing of both gutters and leaders depends both on the horizontal projected area of the roof, as shown in Fig. 9.16, and the maximum hourly rainfall.

EXAMPLE 9.2. Select a gutter and two leaders for the front half of a house, as shown in

Fig. 9.16. Rainfall rate is 4 in./h. Projected roof area is 700 ft^2, and the slope of the gutter is $\frac{1}{16}$ in. in 1 ft of length (0.5%).

SOLUTION. From Table 9.7, choose a semicircular gutter with a 6-in. diameter. (Note that if a steeper gutter slope were designed, then at $\frac{1}{2}$-in. slope per foot (4%), only a 4-in. diameter gutter would be required.)

Because two leaders will be used, each will drain 350 ft^2. Table 9.8 shows that a 2-in. leader can be used. For this gutter–leader combination, specify the detail of Fig. 9.24*a*. ■

Storm gutters and leaders can have an important impact on a building's appearance (Fig. 9.25). Alternatively, leaders can be set within buildings and gutters can be built into a roof's surface, to minimize the visual impact.

Where routing of stormwater inside a building is preferable, drains and leaders can be sized from Table 9.8, and *horizontal piping* can be sized from Table 9.9. Care should be taken to insulate such lines; cold rainwater inside pipes can cause condensation to form on the outside, sometimes resulting in staining and other water damage.

Fig. 9.23 *Maximum 100-year, 1-hour rainfall (in.):* (a) *Eastern U. S.* (b) *Central U. S.* (c) *Western U. S. [Note: Alaska, range 0.4 to 1.4 in.; Hawaii, range 1.5 to 8 in.]* (International Plumbing Code. *Reprinted with permission. All rights reserved. Copyright © 1997, International Code Council Inc., Falls Church, Virginia.*)

(b)

Fig. 9.23   (continued)

(*c*)

Fig. 9.23 (*continued*)

TABLE 9.7 **Size of Gutters**

Part A. *I-P Units*

Diameter of Gutter (in.) 1/16 in./ft Slope	Maximum Rainfall (in. per h)				
	2	3	4	5	6
3	340	226	170	136	113
4	720	480	360	288	240
5	1250	834	625	500	416
6	1920	1280	960	768	640
7	2760	1840	1380	1100	918
8	3980	2655	1990	1590	1325
10	7200	4800	3600	2880	2400

Diameter of Gutter (in.) 1/8 in./ft Slope	Maximum Rainfall (in. per h)				
	2	3	4	5	6
3	480	320	240	192	160
4	1020	681	510	408	340
5	1760	1172	880	704	587
6	2720	1815	1360	1085	905
7	3900	2600	1950	1560	1300
8	5600	3740	2800	2240	1870
10	10,200	6800	5100	4080	3400

Diameter of Gutter (in.) 1/4 in./ft Slope	Maximum Rainfall (in. per h)				
	2	3	4	5	6
3	680	454	340	272	226
4	1440	960	720	576	480
5	2500	1668	1250	1000	834
6	3840	2560	1920	1536	1280
7	5520	3680	2760	2205	1840
8	7960	5310	3980	3180	2655
10	14,400	9600	7200	5750	4800

Diameter of Gutter (in.) 1/2 in./ft Slope	Maximum Rainfall (in. per h)				
	2	3	4	5	6
3	960	640	480	384	320
4	2040	1360	1020	816	680
5	3540	2360	1770	1415	1180
6	5540	3695	2770	2220	1850
7	7800	5200	3900	3120	2600
8	11,200	7460	5600	4480	3730
10	20,000	13,330	10,000	8000	6660

TABLE 9.7 **Size of Gutters** (*continued*)

Part B. **SI Units**

Diameter of Gutter (mm) 5.2 mm/m Slope	Maximum Rainfall (mm per h)				
	50.8	76.2	101.6	127.0	152.4
76.2	31.6	21.0	15.8	12.6	10.5
101.6	66.9	44.6	33.4	26.8	22.3
127.0	116.1	77.5	58.1	46.5	38.7
152.4	178.4	119.1	89.2	71.4	59.5
177.8	256.4	170.9	128.2	102.2	85.3
203.2	369.7	246.7	184.9	147.7	123.1
254.0	668.9	445.9	334.4	267.6	223.0

Diameter of Gutter (mm) 10.4 mm/m Slope	Maximum Rainfall (mm per h)				
	50.8	76.2	101.6	127.0	152.4
76.2	44.6	29.7	22.3	17.8	14.9
101.6	94.8	63.3	47.4	37.9	31.6
127.0	163.5	108.9	81.8	65.4	54.5
152.4	252.7	168.6	126.3	100.8	84.1
177.8	362.3	241.5	181.2	144.9	120.8
203.2	520.2	347.5	260.1	208.1	173.7
254.0	947.6	631.7	473.8	379.0	315.9

Diameter of Gutter (mm) 20.9 mm/m Slope	Maximum Rainfall (mm per h)				
	50.8	76.2	101.6	127.0	152.4
76.2	63.2	42.2	31.6	25.3	21.0
101.6	133.8	89.2	66.9	53.5	44.6
127.0	232.3	155.0	116.1	92.9	77.5
152.4	356.7	237.8	178.4	142.7	118.9
177.8	512.8	341.9	256.4	204.9	170.9
203.2	739.5	493.3	369.4	295.4	246.7
254.0	133.8	891.8	668.9	534.2	445.9

Diameter of Gutter (mm) 41.7 mm/m Slope	Maximum Rainfall (mm per h)				
	50.8	76.2	101.6	127.0	152.4
76.2	89.2	59.5	44.6	35.7	29.7
101.6	189.5	126.3	94.8	75.8	63.2
127.0	328.9	219.2	164.4	131.5	109.6
152.4	514.7	343.3	257.3	206.2	171.9
177.8	724.6	483.1	362.3	289.9	241.4
203.2	1040.5	693.0	520.2	416.2	346.5
254.0	1858.0	1238.4	929.0	743.2	618.7

Source: Reprinted by permission from the 1997 Uniform Plumbing Code. Copyright © by the International Association of Plumbing and Mechanical Officials, Walnut, California.

Fig. 9.24 *Storm drainage components.* (a) *Conventional gutter and leader for houses; sizes vary by manufacturer.* (b) *Ordinary roof drain (Josam Manufacturing Co.).* (c) *Roof drain for controlled flow.* (From Specifying Engineer, *November 1982.*)

TABLE 9.8 **Sizing Roof Drains, Leaders, and Vertical Rainwater Piping**

NOTES: The sizing data for vertical conductors, leaders and drains is based on the pipes flowing $7/24$ full. For rainfall rates other than those listed, determine the allowable roof area by dividing the area given in the 1 inch/hour (25 mm/hour) column by the desired rainfall rate. Vertical piping may be round, square, or rectangular. Square pipe shall be sized to enclose its equivalent round pipe. Rectangular pipe shall have at least the same cross-sectional area as its equivalent round pipe, except that the ratio of its side dimensions shall not exceed 3 to 1.

Part A. ***I-P Units***

Size of Drain, Leader or Pipe (in.)	Flow (gpm)	Maximum Allowable Horizontal Projected Roof Areas Square Feet at Various Rainfall Rates					
		1″/h	2″/h	3″/h	4″/h	5″/h	6″/h
2	23	2176	1088	725	544	435	363
3	67	6440	3220	2147	1610	1288	1073
4	144	13,840	6920	4613	3460	2768	2307
5	261	25,120	12,560	8373	6280	5024	4187
6	424	40,800	20,400	13,600	10,200	8160	6800
8	913	88,000	44,000	29,333	22,000	17,600	14,667

Part B. ***SI Units***

Size of Drain, Leader or Pipe (mm)	Flow (L/s)	Maximum Allowable Horizontal Projected Roof Areas Square Meters at Various Rainfall Rates					
		25 mm/h	50 mm/h	75 mm/h	100 mm/h	125 mm/h	150 mm/h
50	1.5	202	101	67	51	40	34
75	4.2	600	300	200	150	120	100
100	9.1	1286	643	429	321	257	214
125	16.5	2334	1117	778	583	467	389
150	26.8	3790	1895	1263	948	758	632
200	57.6	8175	4088	2725	2044	1635	1363

Source: Reprinted by permission from the 1997 Uniform Plumbing Code. Copyright © by the International Association of Plumbing and Mechanical Officials, Walnut, California.

Fig. 9.25 *A downspout terminates the roof with flair at a BART (Bay Area Rapid Transit) station at Pleasanton, near San Francisco. (Photo by Jane Lidz.)*

Fig. 9.26 *(Example 9.3.) Separate storm drainage. Areas drained and corresponding sizes of vertical leaders and horizontal drains are from Tables 9.8 and 9.9. Storm drain piping within a building needs insulative covering with a vapor barrier on the outside. This prevents condensation (sweating) on the pipes when in winter, warm and moisture-laden air in the building could otherwise reach the pipe surface (which would be cold from carrying icy water), condense there, and lead to wet, dripping conditions on the pipes. Each roof has two drains, in case one is temporarily blocked.*

TABLE 9.9 **Sizing of Horizontal Rainwater Piping**

NOTES: The sizing data for horizontal piping is based on the pipes flowing full. For rainfall rates other than those listed, determine the allowable roof area by dividing the area given in the 1 inch/hour (25 mm/hour) column by the desired rainfall rate.

Part A. ***I-P Units***

Size of Pipe (in.)	Flow at ⅛ in./ft Slope (gpm)	Maximum Allowable Horizontal Projected Roof Areas Square Feet at Various Rainfall Rates					
		1″/h	2″/h	3″/h	4″/h	5″/h	6″/h
3	34	3288	1644	1096	822	657	548
4	78	7520	3760	2506	1880	1504	1253
5	139	13,360	6680	4453	3340	2672	2227
6	222	21,400	10,700	7133	5350	4280	3566
8	478	46,000	23,000	15,330	11,500	9200	7670
10	860	82,800	41,400	27,600	20,700	16,580	13,800
12	1384	133,200	66,600	44,400	33,300	26,650	22,200
15	2473	238,000	119,000	79,333	59,500	47,600	39,650

Size of Pipe (in.)	Flow at ¼ in./ft Slope (gpm)	Maximum Allowable Horizontal Projected Roof Areas Square Feet at Various Rainfall Rates					
		1″/h	2″/h	3″/h	4″/h	5″/h	6″/h
3	48	4640	2320	1546	1160	928	773
4	110	10,600	5300	3533	2650	2120	1766
5	196	18,880	9440	6293	4720	3776	3146
6	314	30,200	15,100	10,066	7550	6040	5033
8	677	65,200	32,600	21,733	16,300	13,040	10,866
10	1214	116,800	58,400	38,950	29,200	23,350	19,450
12	1953	188,000	94,000	62,600	47,000	37,600	31,350
15	3491	336,000	168,000	112,000	84,000	67,250	56,000

Size of Pipe (in.)	Flow at ½ in./ft Slope (gpm)	Maximum Allowable Horizontal Projected Roof Areas Square Feet at Varous Rainfall Rates					
		1″/h	2″/h	3″/h	4″/h	5″/h	6″/h
3	68	6576	3288	2192	1644	1310	1096
4	156	15,040	7520	5010	3760	3010	2500
5	278	26,720	13,360	8900	6680	5320	4450
6	445	42,800	21,400	14,267	10,700	8580	7140
8	956	92,000	46,000	30,650	23,000	18,400	15,320
10	1721	165,600	82,800	55,200	41,400	33,150	27,600
12	2768	266,400	133,200	88,800	66,600	53,200	44,400
15	4946	476,000	238,000	158,700	119,000	95,200	79,300

TABLE 9.9 **Sizing of Horizontal Rainwater Piping** (*continued*)

Part B. ***SI Units***

Size of Pipe (mm)	Flow at 10 mm/m Slope (L/s)	Maximum Allowable Horizontal Projected Roof Areas Square Meters at Various Rainfall Rates					
		25 mm/h	50 mm/h	75 mm/h	100 mm/h	125 mm/h	150 mm/h
75	2.1	305	153	102	76	61	51
100	4.9	700	350	233	175	140	116
125	8.8	1241	621	414	310	248	207
150	14.0	1988	994	663	497	398	331
200	30.2	4273	2137	1424	1068	855	713
250	54.3	7692	3846	2564	1923	1540	1282
300	87.3	12,375	6187	4125	3094	2476	2062
375	156.0	22,110	11,055	7370	5528	4422	3683

Size of Pipe (mm)	Flow at 20 mm/m Slope (L/s)	Maximum Allowable Horizontal Projected Roof Areas Square Meters at Various Rainfall Rates					
		25 mm/h	50 mm/h	75 mm/h	100 mm/h	125 mm/h	150 mm/h
75	3.0	431	216	144	108	86	72
100	6.9	985	492	328	246	197	164
125	12.4	1754	877	585	438	351	292
150	19.8	2806	1403	935	701	561	468
200	42.7	6057	3029	2019	1514	1211	1009
250	76.6	10,851	5425	3618	2713	2169	1807
300	123.2	17,465	8733	5816	4366	3493	2912
375	220.2	31,214	15,607	10,405	7804	6248	5202

Size of Pipe (mm)	Flow at 40 mm/m Slope (L/s)	Maximum Allowable Horizontal Projected Roof Areas Square Meters at Various Rainfall Rates					
		25 mm/h	50 mm/h	75 mm/h	100 mm/h	125 mm/h	150 mm/h
75	4.3	611	305	204	153	122	102
100	9.8	1400	700	465	350	280	232
125	17.5	2482	1241	827	621	494	413
150	28.1	3976	1988	1325	994	797	663
200	60.3	8547	4273	2847	2137	1709	1423
250	108.6	15,390	7695	5128	3846	3080	2564
300	174.6	24,749	12,374	8250	6187	4942	4125
375	312.0	44,220	22,110	14,753	11,055	8853	7367

Source: Reprinted by permission from the 1997 Uniform Plumbing Code. Copyright © by the International Association of Plumbing and Mechanical Officials, Walnut, California.

EXAMPLE 9.3. Select sizes for vertical conductors and horizontal storm drains for the building shown in Fig. 9.26. Roof, balcony, and courtyard areas are as shown, rainfall rate is 4 in./h, and the pitch of horizontal drains is ¼-in. slope in 1 ft of run (2%).

SOLUTION. Sizes selected and shown in Fig. 9.26 may be verified in Tables 9.8 and 9.9. ∎

REFERENCES

Ferguson, B. (1998). *Introduction to Stormwater.* John Wiley & Sons, New York.

Gleick, Peter (1998). *The World's Water.* Island Press, Washington DC.

International Code Council (1997). *International Plumbing Code.* (The complete *International Plumbing Code* may be obtained by contacting the following organizations: Building Officials and Code Administrators, Inc.; 4051 West Flossmoor Road, Country Club Hills, IL 60478-5795; International Conference of Building Officials; 5360 Workman Mill Road, Whittier, CA 90601-2298; or Southern Building Code Congress International, Inc.; 900 Montclair Road, Birmingham, AL 35213-1206.)

Milne, M. (1976). *Residential Water Conservation,* U.S. Office of Water Research and Technology, Department of Commerce, NTIS.

U.S. Environmental Protection Agency (1975). *Manual of Individual Water Supply Systems,* U.S. EPA, Washington, DC.

10
WATER SUPPLY

With water, one of the designer's first concerns is to match the quality of the water to the task it performs. As this becomes a more serious design issue, designers will provide for the recycling of water within and around buildings, as well as to specify plumbing fixtures that use less water. Table 10.1 shows typical relationships between quality and usage. Water recycling and conservation are discussed in detail in Chapter 11. In this chapter we deal primarily with *potable* (drinkable) water: first with issues of water quality, and then with the matter of assuring an adequate supply of water throughout the building.

10.1 Water Quality

A summary of water quality at the various stages of the hydrologic cycle was shown in Fig. 9.4. As precipitation, water contains few impurities: almost no bacterial content is present, and only small amounts of minerals and gases can be expected. To collect this nearly pure water, surfaces are needed—and at this point, foreign substances can readily contaminate the water. These pollutants can affect water's physical (mostly organic), chemical (mostly inorganic), biological, or radiological characteristics. Both surface water and groundwater are subject to pollution.

The U.S. National Drinking Water Clearinghouse is a source of information on water supply and waste-water treatment systems. Their brief history of waterborne diseases worldwide begins with a major cholera epidemic that began in Calcutta, India, in 1817, killing thousands, and spread by 1832 to New York, emptying the streets in panic over the disease. In 1854, a London physician demonstrated that local cases of cholera could be traced to one pump contaminated with sewage from a nearby house, but how the disease was transmitted remained a mystery. In summer 1859, London's Thames River, carrying combined storm and sanitary wastes, stank so badly that Parliament was suspended. In 1892, the bacteria causing cholera was identified during an epidemic in Hamburg, Germany, proving the relationship between contaminated water and the disease. In 1939, an outbreak of typhoid fever killed 60 people at an Illinois mental hospital; typhoid and gastroenteritis resulted from an accidental pumping of polluted river water into supply mains in Rochester, New York, in 1940. Another worldwide cholera epidemic began in Indonesia in 1961, eventually reaching Latin America by 1991. In 1993, an outbreak of cryptosporidiosis in the public water supply of Milwaukee, Wisconsin, shook the public's faith in water systems across the United States; this outbreak killed 104 and infected more than 400,000 people.

In response to such history, we disinfect huge quantities of water, expending energy and adding chemicals so that every drop of water entering North American buildings is potable. Again, the high-grade resource/low-grade use question arises; recycled water for lower-grade use is discussed in Chapter 11.

(a) Physical Characteristics. Some of the most noticeable alterations of water quality fall within this category. Water from surface sources (roof runoff, streams, rivers, lakes, ponds) is particularly subject to physical pollutants.

Turbidity is easy to see, and thus a likely source of dissatisfaction for the would-be consumer. It is caused by the presence of suspended material such as clay, silt, other inorganic material, plankton, or finely divided organic material. Even those materials that do not adversely affect health are usually esthetically objectionable.

Color, another visible alteration, is often caused by dissolved organic matter, as from decaying vegetation. Some inorganic materials also color water, as do growths of microorganisms. Like turbidity, such color changes usually do not

TABLE 10.1 **Water Use and Quality in Buildings**

Use	Desired Quality
A. Consumed	
1. Drinking and cooking	Potable
2. Bathing	Potable
3. Laundering	Soft
4. Irrigation and watering of livestock	Unpolluted
5. Industrial processes	As required
6. Vapor to increase the relative humidity of air	
B. Circulated	
1. Hot water for heating	Note: Make-up
2. Chilled water for cooling	water should be
	soft or neutral
3. Condenser cooling water	and, for swimming,
	potable
4. Swimming pool water	
5. Steam for heating, later condensed	
C. Generally static	
1. Water stored for fire protection	No special requirement
2. Water in fire stand-pipes	
3. Water in sprinkler piping	
D. Controlled	
1. Vapor condensed to reduce relative humidity of air	

NOTE: For water uses in Section A, flow is often continuous. Section B comprises uses for which flow other than circulation is intermittent or at a relatively low rate, the water added to the systems being known as "make-up water." Items C2 and C3 call for piping to provide adequate though infrequent flow in emergencies. Obviously, item D1 relates only to moisture condensed out of the air and involves no design for supply.

threaten health, but are often psychologically undesirable.

Taste and *odor* can be caused by organic compounds, inorganic salts, or dissolved gases. This condition can be treated only after a chemical analysis has identified which source is responsible.

Temperature is another characteristic of psychological importance—we expect drinking water to be cool. In general, water supplied between 50° and 60°F (10° to 16°C) is preferred.

Foamability is usually caused by concentrations of detergents. The foam itself does not pose a serious health threat, but it may indicate that other, more dangerous pollutants associated with domestic waste are also present. Because of increased foaming in water in the 1960s, today's detergents must use biodegradable linear alkylate sulfonate (LAS), which biodegrades rapidly—except in the absence of oxygen. Because this lack of oxygen is characteristic of some septic tank drainage fields, foam in drinking water should be investigated promptly.

(b) Chemical Characteristics. Groundwater is particularly subject to chemical alteration because as it moves downward from the surface it slowly dissolves some minerals contained in rocks and soils. A chemical analysis (such as that given in Table 10.2) is usually necessary for individual water supply sources. These analyses will indicate (1) the possible presence of harmful or objectionable substances, (2) the potential for corrosion within the water supply system, and (3) the tendency for the water to stain fixtures and clothing. Concentrations are expressed in milligrams per liter (mg/L), which is essentially equivalent to parts per million (ppm).

Some general terms commonly used to describe chemical characteristics of water are defined as follows.

Alkalinity is caused by bicarbonate, carbonate, or hydroxide components. Testing for these components of water's alkalinity is a key to determining which treatments to use.

Hardness is a relative term (see Fig. 10.1). "Hard" water inhibits the cleaning action of soaps and detergents, and it deposits scale on the inside of hot water pipes and cooking utensils, thus wasting heating fuel and making utensils unusable. Hardness, which is caused by calcium and magnesium salts, can be classified as temporary (carbonate) or permanent (noncarbonate). Temporary hardness is largely removed when the water is heated—it forms the scale just described. Permanent hardness cannot be removed by simple heating (see Section 10.4c).

pH is a measure of the water's hydrogen ion concentration, as well as its relative acidity or alkalinity (Fig. 10.1). A pH of 7 is neutral. Measurements below 7 indicate increasing acidity (and

TABLE 10.2 **Example of Chemical Analysis**

Quality		Parts per Million (ppm)[a]
Total hardness	as $CaCO_3$	30
Calcium hardness	as $CaCO_3$	20
Alkalinity (methyl orange)	as $CaCO_3$	27
Alkalinity (Phenolphthalein)	as $CaCO_3$	0
Free carbon dioxide	as CO_2	13.5
Chlorides	as Cl	6
Sulfates	as SO_4	4
Silica	as SiO_2	19
Phosphates—normal	as PO_4	0
Phosphates—total	as PO_4	0.5
Iron—total	as Fe	1.6
Total dissolved solids		66
Turbidity or sediment		present

Source: A report by Olin Water Service for a private well in Virginia.
[a]Note that "ppm" and "mg/L" are essentially equivalent terms for concentration in water.

corrosiveness); water in its natural state can have a pH as low as 5.5, with 0 the ultimate acidity. Measurements higher than 7 indicate increasing alkalinity; a pH as high as 9 can be found in water in its natural state, with 14 the ultimate alkalinity. The pH value is the starting point for determining treatments for corrosion control, chemical dosages, and disinfection.

Unintentional chemical additions to water supplies most commonly include the following elements.

Toxic substances are occasionally present in water supplies. Local health authorities can provide information about acceptable concentrations of such substances as arsenic (As), barium (Ba), cadmium (Cd), chromium (Cr^{6+}), cyanides (CN), fluoride (F), lead (Pb), selenium (Se), and silver (Ag). Although limited amounts of *fluoride* are frequently added to water supplies to help prevent tooth decay, fluorides in excess of such "optimum concentrations" can produce mottling of teeth. *Lead* poses a dangerous threat, even in relatively

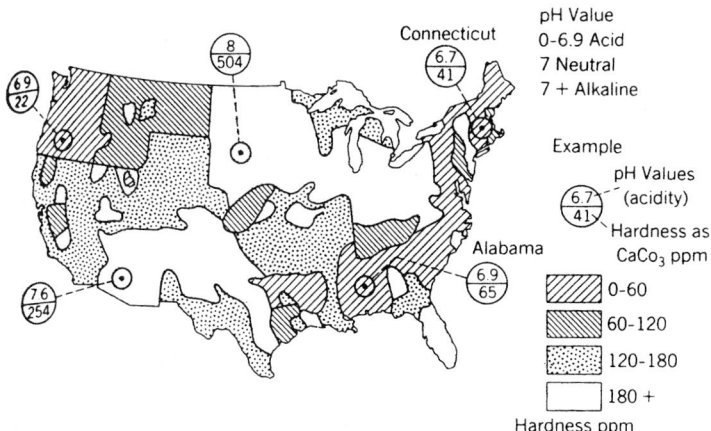

Fig. 10.1 *Approximate groundwater chemical characteristics across the United States. Treatment may be needed when pH is less than 7.0 (acidity results in corrosion) or when hardness as $CaCo_3$ exceeds 65 ppm (essentially, ppm and mg/L are identical measures). (Courtesy of* Progressive Architecture.*)*

Fig. 10.2 *How groundwater becomes contaminated. The "plume" formed by contaminants can often go undetected. (Copyright © 1982 by Newsweek, Inc., all rights reserved. Reprinted by permission.)*

small amounts, because it is a cumulative poison. Lead in water usually comes from lead piping (in older buildings or cities) or corrosive water on lead-painted roofs. A maximum recommended concentration is 0.05 mg/L.

Unfortunately, the list of toxic chemicals is expanding, and as chemical waste dumps have been abandoned or mismanaged, groundwater has become contaminated. The U.S. Environmental Protection Agency (EPA) has estimated that 75% of both active and abandoned chemical waste dumps—some 51,000 in all—are leaking. In addition to the following list of inorganic chemicals, we are becoming aware of many new organic chemicals as well, some of which are suspected of causing 5% to 20% of U.S. cancers. The threat to our groundwater supplies is illustrated in Fig. 10.2. Once polluted, aquifers are extremely difficult to clean. This is one reason why ground source heat pumps are tightly regulated.

Chlorides can enter water as it passes through geologic deposits formed by marine sediment, or because of pollution from seawater, brine, or industrial or domestic wastes. A noticeable taste results from chloride in excess of 250 mg/L.

Copper can enter water from natural copper deposits or from copper piping that contains corrosive water. Concentrations of copper in excess of 1.0 mg/L can result in an undesirable taste.

Iron is frequently present in groundwater. Corrosive water in iron pipes will also add iron to water. At concentrations above 0.3 mg/L, iron can lend a brownish color in washed clothes and can affect the taste of the water.

Manganese can both pose a physiological threat (it is a natural laxative) and produce color and taste effects similar to those produced by iron. The recommended limit is 0.05 mg/L.

Nitrates in high concentrations pose a threat to infants, in whom it can cause "blue baby" disease. In shallow wells, nitrate concentrations can indicate seepage from deposits of livestock manure.

Pesticides, a growing threat to water supplies, are particularly common in wells near homes that have been treated for termite control. Avoid using pesticides near wells.

Sodium is primarily dangerous for people with heart, kidney, or circulatory ailments. For a low-sodium diet, the sodium in drinking water should not exceed 20 mg/L. Salts spread on roadways for ice control can leach into the soil and enter groundwater. Note that some "water softeners" (discussed in Section 10.4c) can raise sodium concentrations in water.

Sulfates, which have laxative effects, can enter groundwater from natural deposits of Epsom salts (magnesium sulfate) or Glauber's salt (sodium sulfate). Concentrations should not exceed 250 mg/L.

Zinc sometimes enters groundwater in areas where it is found in abundance. Although not a health threat, it can cause an undesirable taste at concentrations above 5 mg/L.

(c) Biological Characteristics. Potable water should be kept as free as possible from disease-producing organisms—bacteria, protozoa, and viruses. These organisms are not easily identified; a thorough biological water test is complex and time-consuming. For this reason, the standard test is for *one* kind of bacteria—the coliform group (*E. coli*), which is always present in the fecal wastes of humans (as well as those of many animals and birds) and which outnumbers all other disease-producing organisms in water. The recommended maximum concentration for coliform bacteria is one organism per 100 mL (about ½ cup) water.

For biological activity to be kept to a minimum in drinking water, a water source should be chosen that does not normally support much plant or animal life; hence the popularity of groundwater, rather than surface water, as a source. In addition, the supply should be protected from subsequent biological contamination. Where cities depend on small lakes for water, humans are frequently excluded from the watersheds. Organic fertilizers and nutrient minerals should also be kept out of the water supply, to further discourage biological activity. For the same reason, stored water should be kept dark and at low temperatures. Finally, organisms (or their by-products) are commonly destroyed at treatment facilities.

(d) Radiological Characteristics. The mining of radioactive materials and the use of such material in industry and power plants have produced radiological pollution in some water supplies. Because radiological effects are cumulative, concentrations of radioactive materials should be low indeed. "Safe" minimum concentrations have continually been revised downward for other radiation exposures; consult the local public health service for current recommendations.

10.2 Filtration

In the preceding section, water pollution was broken down into physical, chemical, biological, and radiological categories. The various forms of treatment for such pollutants do not necessarily fall into the same categories, as one treatment may be effective for several different polluted conditions. A general look at common domestic water quality problems is provided in Table 10.3.

TABLE 10.3 **Common Water Quality Problems and Treatment in Small Systems**

Item	Cause	Bad Effect	Correction
Hardness	Calcium and magnesium salts from underground flow	Clogging of pipes by scale, burning out of boilers, and impaired laundering and food preparation	Ion-exchanger (Zeolite process)
Corrosion	Acidity, entrained oxygen and carbon dioxide (low pH)	Closing of iron pipe by rust, leaking connections, destruction of brass pipe	Raising the alkaline content (Neutralizer)
Biological pollution	Contamination by organic matter or sewage	Disease	Chlorination by sodium hypochlorite or chlorine gas; or ozonation
Color	Iron and manganese	Discoloration of fixtures and laundry	Chlorination or ozonation and fine filtration
Taste and odor[a]	Organic matter	Unpleasantness	Filtration through activated carbon (Purifier); aeration
Turbidity[a]	Silt or suspended matter picked up in surface or near-surface flow	Unpleasantness	Filtration

[a]These problems are not common in private systems that use deep wells.

We begin with filtration, because it so often is the first treatment in a series; it is also one of the oldest and simplest methods.

This very common treatment removes suspended particles, some bacteria, and color or taste, by passing water through a permeable fabric or a porous bed of materials. The more common approaches are listed below, beginning with filtering to remove suspended particles then moving on to more specialized applications for the removal of iron and/or manganese, tastes, and odors.

(a) Sedimentation.

Before entering a filter, this process removes some suspended matter from water simply by allowing time and the inactivity of the water to do the work of settling out heavier suspended particles. Simple basins, ponds, or tanks constructed for this purpose are large enough to retain the water for at least 24 h and are equipped with baffles to slow the water flow. To clean out the sediment, water usually is diverted to an identical second basin while the first is being cleaned.

(b) Coagulation.

This process also removes suspended matter, along with some coloration. A chemical such as alum (hydrated aluminum sulfate) is added to water made turbulent by baffles or static mixers to distribute the chemicals evenly.

(c) Flocculation.

The water is then held in a quiet condition, where the suspended particles will combine with the alum to form floc. These heavy particles then settle out, in a process similar to sedimentation. Some adjustment of the pH may be necessary.

See the National Drinking Water Clearinghouse, *Tech Brief* (September 1996) *Filtration* for details on the following methods of filtering.

(d) Slow Sand Filters.

These are common in small-scale water supply systems (Fig. 10.3; also see Fig. 9.8 for rainwater application). Not suitable for water with high turbidity, they do not usually require coagulation/flocculation, and may not even require sedimentation. Water should not be chlorinated before entering this filter, because it will interfere with the subsequent biological activity. These filters are able to remove up to 99.9% of *Giardia* cysts.

These are low-maintenance, easily constructed devices that should be cleaned as often as the turbidity of the water demands—from once a day to perhaps once a month. They are cleaned by removal and replacement of about the top 1 in. (25 mm) of sand, which has formed a layer of biological slime called (descriptively) the *schmutzdecke,* which traps small particles and degrades organic material in the water. This sand is then either washed for reuse or discarded.

The approximate rate of flow is slow, requiring a rather large surface area: between 0.03 to 0.10 gpm per ft^2 (0.02 to 0.07 L/s m^2) of filter bed surface; in other units, 40 and 140 gallons per day per square foot (1630 to 5720 liters per day per

Fig. 10.3 *Slow sand filter, effective in removing cysts of* Giardia. *Water should not be chlorinated before the water enters the filter. From National Drinking Water Clearinghouse,* Tech Brief *(September 1996)* Filtration. *U.S. Environmental Protection Agency, Washington, DC.*

square meter) of filter bed surface. Overall thickness is usually 30 to 48 in. of sand over 12 in. of gravel (900 to 1200 mm of sand over 300 mm of gravel) with an underdrain system.

In freezing temperatures, they must be housed; if they develop an ice layer, this prevents cleaning.

(e) Diatomaceous Earth Filters. Also known as precoat or diatomite filtration, these can be of either the vacuum or the pressure type. They rely on a layer of diatomaceous earth, a minimum of ⅛-in. (3-mm) thick placed on a septum or filter element [for *Giardia* removal, thickness should be increased to about ⅕-in. (5-mm)]. They are most suitable for water with low bacterial counts and low turbidity. Simple to operate and effective in removing cysts, algae, and asbestos, they require periodic attention to remain effective, including backwashing every 1 to 4 days.

(f) Direct Filtration. Intended for water supplies of high quality and seasonally consistent flow, these systems omit sedimentation, but should include coagulation for most effective *Giardia* removal. These are often used with steel pressure tanks, to maintain pressure in the water supply line.

Packaged filtration combines features such as chemical addition, flocculation, and sedimentation, along with filtration, in one compact unit. Most often used for small community water supplies, they treat surface water to remove turbidity, color, and coliform organisms.

(g) Membrane Filtration. Also called *microfiltration* or *ultrafiltration* this rapidly developing technique can remove bacteria, *Giardia,* and some viruses. It does not require coagulation as pretreatment. Using hollow fiber or spiral-wrapped membranes, it is able to exclude all particles greater than 0.2 microns from the water stream. It is best used on water supplies of low turbidity, because of the fouling of the fibers or membranes.

Water is forced at high pressure through these filters. The contaminants trapped on the inflow side must be frequently removed by reversing the flow and flushing the waste; calcium and other persistent contaminants must periodically be removed with chemical cleaning.

(Nanofiltration, using much smaller pores, is discussed in Section 10.3.)

(h) Cartridge Filtration. Increasingly popular on lavatory faucets as well as on small supply systems, these are easy to operate and maintain. They require water of low turbidity and last longer when some prefiltering by more crude means is added upstream. They can exclude particles of 0.2 microns (or even smaller). A disinfectant can prevent surface-fouling microbial growth on the cartridge filters; some periodic chemical cleaning will likely be required.

(i) Other Filters. *Activated carbon filters* are particularly effective for removing tastes and odors. The water is passed through granular carbon, which attracts large quantities of dissolved gases, soluble organics, and fine solids.

Porous stone, ceramic, or unglazed porcelain filters (also called *Pasteur filters*) are usually made in small sizes so that they can be attached to water faucets. They are used widely in some countries, such as Mexico, but poor maintenance or hairline cracks often lead to bacterial infiltration, complicating the filtration process. A more positive approach to the disinfection of drinking water thus is desirable.

10.3 Disinfection

Disinfection is the most important health-related water treatment, because it destroys microorganisms that can cause disease in humans. Disinfection is required of water supply systems that rely on surface water or groundwater sources under the influence of surface water. Initially, primary disinfection achieves the desired level of microorganism kill (inactivation), then secondary disinfection maintains a disinfectant residual in the treated water that prevents microorganism regrowth.

Although chlorination has become the standard approach to removing harmful organisms from water, there are alternatives: nanofiltration, ultraviolet (uv) light (unsuitable for water with high turbidity because the light cannot easily penetrate), bromine, iodine, ozone, and heat treatment, among others. Chlorine continues to disin-

Fig. 10.4 Cylinder-mounted chlorinator for automatic disinfection of supply water. From National Drinking Water Clearinghouse, Tech Brief *(June 1996)* Disinfection. *U.S. Environmental Protection Agency, Washington, DC.*

fect after the initial application. It is this continuing secondary disinfection that has made it so universally relied on, despite dangers such as that posed by deadly chlorine gas. Although chlorine affects the taste and odor of water, it is also effective in removing less-desirable tastes and odors. Unfortunately, chlorine can react with organic materials in water to form halogenated by-products. It is easier to either remove the organic materials before treatment, or to use another disinfectant strategy, than to try to remove these halogenated by-products after chlorine treatment.

See the National Drinking Water Clearinghouse, *Tech Brief* (June 1996), *Disinfection* for details on the following methods (*a*) through (*d*).

(a) Chlorination. Factors that affect chlorine's ability to disinfect include

1. *Chlorine concentration.* The higher the concentration, the faster and more complete the rate of disinfection.
2. *Contact time.* The longer the chlorine contacts the organisms in water, the more complete the disinfection. At a minimum, 0.4 mg/L of chlorine should contact water 30 minutes before use.
3. *Water temperature.* The higher the temperature during contact, the more complete the disinfection.
4. *pH.* The lower the pH, the more effective the disinfection.

There are three common forms in which chlorine is used to disinfect water supplies. *Chlorine gas* is stored in a cylinder (Fig. 10.4) as a liquid under high pressure and is released (as a gas) by a regulator to an injector attached to a water pipe or tank. The injector passes highly pressurized water through a venturi orifice, creating a vacuum that draws the chlorine into the water stream.

Sodium hypochlorite solution is easier to handle than deadly chlorine gas but is very corrosive and decomposes rather quickly. It should be stored in a cool, dark, dry area for no more than a month. *Hypochlorinators* automatically pump (or inject) a sodium hypochlorite solution into water. They are usually no larger than the pumps used in small water systems (see Fig. 10.13). Some hypochlorinators are specially designed for low and fluctuating water pressures or for use where electricity is not available.

Solid calcium hypochlorite is a white solid containing 65% available chlorine that dissolves easily in water. It is corrosive, with a strong odor, but very stable and can be stored for up to a year. However, it readily absorbs moisture, forming chlorine gas; also, reactions between calcium hypochlorite and organic materials (wood, cloth, petroleum products) can generate enough heat to cause a fire or explosion. Again, hypochlorinators are used to deliver the disinfectant to water.

(b) Chloramine. This is generated on-site by adding ammonia to water containing chlorine, or when water containing ammonia is chlorinated. This is a weaker disinfectant against viruses or protozoa than the chlorination processes but produces fewer disinfection by-products. It is most often used as a secondary rather than primary disinfectant. Again, hypochlorinators are used to inject chlorine, after which ammonia is added.

(c) Ozonation. This was first used in full-scale drinking water treatment in 1906. It is a powerful oxidizing and disinfecting agent, destroying most bacteria, viruses, and other pathogenic organisms. It requires shorter contact time and dosage than chlorine and leaves no chlorine taste. Ozone is formed by passing dry air (or pure oxygen) through a system of high voltage elec-

Fig. 10.5 *Recycled cooling tower water is treated by an ozonator, magnetic descaler, and a filtration system. This controls scale formation, algae and slime, corrosion, and sludge buildup. (Courtesy of Aqua-Flo, Inc., Baltimore.)*

trodes; about 80% of the energy is converted to heat. It is an unstable gas and must be generated onsite. When ozone reacts with an organic, it produces oxygen and an oxidized form of the organic. Ozone not used in this process quickly decays to oxygen.

In the United States, ozone is commonly used in cooling tower water treatment (Fig. 10.5), where its effectiveness against *Legionella pneumophila* is especially appreciated, as well as its control of algae and scale that can greatly reduce cooling efficiency. Ozone is also used in food processing, waste water cleanup, smoke removal, swimming pools and spas, bottled water, and pulp and paper bleaching.

Equipment includes an ozone generator, contactor, and destruction unit, plus instrumentation and controls. Operation and maintenance are relatively complex; electricity is 26% to 43% of the operating costs for small systems. Because it acts only as a primary disinfectant, a secondary disinfectant (often chlorine) is usually required.

(d) Ultraviolet Light. Special lamps are used within a reactor (Fig. 10.6), whose light disrupts the genetic material of the cells of organisms, making them unable to reproduce. Although effective against bacteria and viruses, UV radiation does not inactivate either *Giardia* or *Cryptosporidium* cysts. It is otherwise an effective primary disinfectant system, requiring a short contact time, and without halogenated by-products. Yet again, a secondary disinfectant system is usually necessary. This system is not suitable for water that contains high levels of suspended solids, turbidity, color, or soluble organic matter.

(e) Nanofiltration. These membranes start with pore sizes of 0.2 to 0.3 microns, and are then dipped into a polymer that leaves a thin film, decreasing the pore size to 1 nanometer. This removes bacteria, viruses, pesticides and organic material. It also gives the membranes an affinity for calcium, contributing to water "softening." However, it also means that the membranes need periodic acid cleaning to remove the calcium deposits. Adding phosphates to nanofiltered water reduces its capacity to dissolve lead.

With such extremely small pore sizes, this process requires very high water pressures, in turn requiring energy. Yet again, a secondary disinfectant system is usually necessary.

WATER AND WASTE

Fig. 10.6 *Ultraviolet water purifier, effective against bacteria and viruses (but not against* Giardia *or* Cryptosporidium *cysts). Courtesy of Atlantic Ultraviolet Corporation, Hauppauge, New York.)*

10.4   Other Water Treatments

(a) Aeration (Oxidation). This simple process can improve the taste and color of water and help remove iron and manganese. In aeration, as much of the water's surface as possible is exposed to air. The methods used are rich in esthetic possibilities—the spraying of water into air, the fall of a turbulent stream of water over a spillway, and *flowforms,* sculptural waterfalls designed to carry water in a rhythmical, pulsating, figure-eight pattern. The Real Goods headquarters in Hopland, California, features these as a waterfall (Fig. 10.7) within a recycled water irrigation system.

To guard against contamination, these processes are often enclosed; if exposed, they must be kept clean. For aeration within tanks, water is passed through a series of perforated plates in streams or droplets.

Aeration improves the flat taste of distilled water and cistern water by adding oxygen. It also oxidizes iron or manganese, which then can more easily be removed by filtration and removes odors caused by hydrogen sulfide and algae.

Because aeration raises the level of dissolved oxygen in water, it should be avoided as a treatment when corrosion is a threat.

(b) Corrosion Control. It is important to control corrosion in order both to keep water systems operating freely and to prevent corrosive water from increasing the concentration of hazardous materials (as from copper pipes). It also imparts a taste and/or odor to water that is objectionable. Corrosion is a slow degradation of a metal by a flow of electric current from the metal to its surroundings. Some factors involved in corrosion control are:

1. *Acidity.* The more acid (low pH, less than 6.0), the more corrosive the water.
2. *Conductivity.* As dissolved mineral salts increase the water's conductivity, they encourage the flow of the electrical current of corrosion.
3. *Oxygen content.* Dissolved oxygen destroys the thin protective hydrogen film on immersed metals, thus promoting corrosion.

(b)

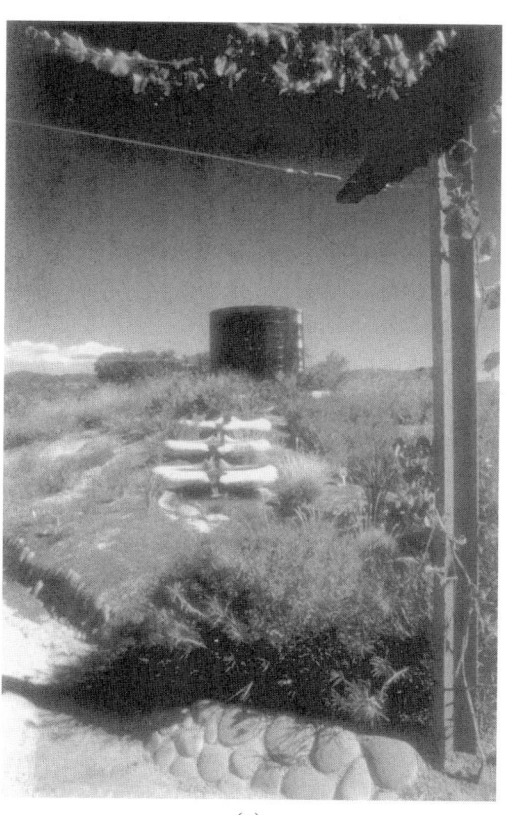

(a)

Fig. 10.7 *The Real Goods Solar Living Center at Hopland, California illustrates photovoltaic (PV) power, passive solar heating, daylighting, and passive cooling, as well as creative water recycling. Water from on-site irrigation ponds is pumped by PV to a tank on a small hill* (a) *where, as flow varies with strong sunlight, it overflows and splashes down a series of* flowforms *to oxygenate the water. It passes through a plaza in front of the retail store, eventually* (b) *returning to the ponds. Landscape design by Land and Place.*

4. *Carbon dioxide content.* Carbon dioxide forms carbonic acid, which attacks metal surfaces.
5. *Water temperature.* Increased temperature increases corrosion.
6. *Lower flow rates.* Reduced turbulence means reduced erosion of the protective layers that form on the inner surfaces of pipes.

The products of corrosion often contribute to scale formation. Scale then lines surfaces, eventually clogging openings.

See the National Drinking Water Clearinghouse, *Tech Brief* (February 1997) *Corrosion Control* for details on this subject.

Acid neutralizers can be installed on water supplies with low pH; their function is often combined with those of hypochlorinators. Typically, neutralizing solutions are mixtures of lime, soda ash, and water. However, pH adjustment should be done just before water delivery, *after* treatment processes such as coagulation and disinfection.

Corrosion inhibitors cause protective coatings to form on pipes. They are commonly fed into the water as are other chemicals. Inorganic phosphates, sodium silicates, and mixtures of phosphates and silicates are the more commonly used additives.

Other corrosion control strategies include commercial pipe coatings/linings, installing dielectric or insulating unions (to avoid complications from dissimilar pipe metals), and avoiding metal piping and fixtures altogether.

(c) Softening. Water "hardness" is caused primarily by calcium and magnesium deposits; when they are removed, water will be "soft." Where water hardness is mild enough to affect only laundering, cisterns may be used to collect "soft" rainwater to use for washing clothes. Where water hardness produces scale in pipes and within water heating appliances and cisterns are not feasible, water softening equipment is used. Demineralization of water is accomplished with

one of three methods: ion exchange, reverse osmosis, or electrodialysis.

See the National Drinking Water Clearinghouse, *Tech Brief* (May 1997) *Ion Exchange and Demineralization* for details on the following methods of treatment.

Ion exchange is popular for small systems, and is effective not only with hardness ions but with radionuclides. It can be used with fluctuating flow rates, but needs pretreatment of most surface waters, and its waste is highly concentrated (requiring careful disposal). There are a large variety of specific resins used for the exchange medium, each effective for specific contaminants.

On the exchange medium's charged surface, one (contaminant) ion is exchanged for another (regenerant) ion. Eventually saturation occurs; the contaminants are flushed and the medium is regenerated—once per day is the common shortest cycle. Sodium chloride is often used to regenerate the exchange medium, resulting in a rather high sodium residual—an undesirable development for people on low-sodium diets. Other regenerant materials include potassium chloride.

Equipment includes prefiltration, ion exchange, disinfection, storage, and distribution; in smaller systems, single package units incorporate all these.

Reverse osmosis (RO), like ion exchange, is popular for small systems and can be used with fluctuating flow rates, but needs pretreatment of most surface waters and its waste is also highly concentrated. RO is effective not only with hardness ions but also with radium, natural organic substances, pesticides, and microbiological contaminants. Used in series, RO units can remove an even higher percentage of contaminants.

An inert, semipermeable membrane has high-pressure supply water on one side; as the pressure slowly forces water through this filtering membrane, most of the contaminants are removed. Water must be used to flush the semipermeable membrane so that mineral buildup is avoided; this produces a brine requiring careful disposal. Membranes are available in varying types and pore sizes; membranes are also prone to fouling.

Commercial reverse-osmosis units are available in sizes ranging from a 1 gpm (3.9 L/m) water delivery rate [using two membranes and a 3-hp motor, requiring 4.5 gpm (17 L/m) feedwater for

a 22% recovery rate] to a water delivery rate of 12.5 gpm (47.3 L/m) [using 12 membranes and a 15-hp motor, requiring 19.2 gpm (72.8 L/m) for a 65% recovery rate].

Electrodialysis effectively removes fluoride and nitrate, and can also remove barium, cadmium and selenium. It is relatively insensitive to levels of flow and of total dissolved solids, but has an enormous appetite for water; from 10% to 80% of the total water supplied to an RO filtering unit is delivered to the user; the remainder is a waste stream. It also requires a higher level of pretreatment.

In this process, membranes adjacent to the inflowing stream are charged (either positively or negatively), attracting counter-ions to these membranes. The membranes allow either positively or negatively charged ions to pass through, thus they leave the inflow stream and enter the waste streams (one on the other side of each membrane). High water pressure and a source of dc power are needed in this process.

Membranes can become fouled when the pores are clogged by salt precipitation, or blocked by suspended particulates. For either of these water conditions, pretreatment is essential.

Reversing the charge on the membranes, "electrodialysis reversal," helps to flush the attached ions from the membrane surface, and can extend the time between membrane cleanings.

(d) Nuisance Control. Some organisms may not be injurious to health but can multiply so rapidly that piping or filters become clogged or the water's appearance, odor, and taste are affected. This process is most common in surface water sources, and it is within surface reservoirs that these treatments are most often applied. Algae growths, the most prevalent nuisance, can usually be controlled by applying copper sulfate (blue stone or blue vitriol) to the water body. Sudden and massive algae kills can have adverse impacts on other life forms within the water, because the decomposing algae rob the water of oxygen. As a further precaution, stored water should be shielded from sunlight whenever possible.

Cooling towers are an especially difficult water treatment problem. The murky water with high turbidity and bacterial count leads to clogged passages (thus inefficient performance), deteriorated surfaces, and to the growth of potentially lethal bacteria (*L. pneumophila*). As a result, enormous quantities of water are commonly used once-through the cooling tower, then dumped to storm sewers, rather than recycled in a closed system. (As late as 1989, San Francisco City Hall was reported to be using 96,000 gallons per summer day for once-through cooling—in a year of drought.) To treat cooling tower water successfully, a method is needed for microbial control, removing organics, and precipitating inorganics. Ozonation (Fig. 10.5) is a common answer to this widespread treatment problem.

Fig. 10.8 *A solar still can be used to provide a small daily quantity of pure water; this installation serves a laboratory. (Courtesy of McCracken Solar Co., Alturas, California.)*

(e) Fluoridation. A heated controversy continues over the addition of fluoride to drinking water. The advantage of fluoridation is that children who drink fluoridated water have lower rates of tooth decay; and because everyone drinks water, all children benefit, not just those who can afford fluoride pills. Its disadvantages are that only children need the fluoride, not adults. Also, in amounts *above* those used in water treatment, fluoride is toxic and can cause mottled teeth. Opponents of fluoridation suggest that because sugar is a leading cause of tooth decay in children, sugar, rather than water, should be fluoridated.

Small water systems can be equipped with fluoridation units. However, fluoride levels in the water supply must be carefully monitored.

(f) Distillation. This is a simple, low-technology approach to purification that produces the equivalent of bottled water for drinking, cooking, and laboratory uses. In one process, it promises the removal of suspended solids, salts, bacteria, and (apparently) halogenated hydrocarbons. When water pollution is extreme, as in the case of sea (salt) water, distillation may be the best treatment. Water is heated to encourage evaporation. As the water turns to vapor, virtually all pollutants are left behind. When this vapor encounters a cooler surface, it condenses, and pure water (although flat in taste) can be collected from this surface.

Any heat source can be used in the distilling of water; solar stills (Fig. 10.8) are gaining in popularity because the energy used is free. Solar distillation of cistern water is an autonomous approach. In semiarid rather sunny climates, a solar still should produce about $\frac{1}{12}$ gallon per square foot of collector surface area (4 liters per square meter) per day. This rate of production suggests that only the water used directly for drinking or other specialized purposes is usually feasible for distillation. Another factor to consider is that the cleaning of the still is generally accomplished by flushing it with twice as much water as was delivered. If not excessively brackish, this flush water could be used for irrigation or other nonpotable-quality tasks.

10.5 Water Sources

This section focuses on the equipment used to capture and store groundwater from wells. Other water sources are less often used for smaller systems, either because they require much more extensive treatment (surface water from lakes or rivers) or because they provide water intermittently (cisterns). A multistage treatment (flocculation, sedimentation, etc.) may be inappropriate for small water systems that receive only occasional maintenance. Cisterns were discussed in Chapter 9. Another increasingly important source of city water, the treated effluent of sewage treatment plants, is discussed in Chapter 11.

(a) Wells. Farms and remote housing developments usually have private water systems. In rural and suburban areas where the progress of building is faster than the development of municipal water supplies, private sources must be sought. Driven or drilled wells are preferable; water from these sources usually has at least the advantages of purity, coolness, and freedom from turbidity, odor, and unpleasant taste—any of which *may* be encountered in addition to either acidity or hardness.

Bored wells, which are dug with earth augers, are usually less than 100 ft (30 m) deep. They are used when the earth to be bored through is boulder-free and will not cave in. The diameter range is 2 to 30 in. (50 to 760 mm). The bored well is then cased with metal, vitrified tile, or concrete.

Driven wells are the simplest and, usually, the least expensive type. A steel drive-well point is fitted on the end of pipe sections and driven into the earth. The drive point is usually 1¼ to 2 in. (32 to 50 mm) in diameter. The materials and design of drive-well points vary according to the expected characteristics of the earth in which the well is driven. First, a pilot hole is bored (frequently with a simple hand auger) and the drivewell point and pipe sections lowered into it. Then the well is driven to a point well below the water table.

Jetted wells require a source of water and a pressure pump. A washing well point is supplied with water under pressure; this loosens the earth and allows the point and pipe to penetrate.

Drilled wells require more elaborate equipment of several types, depending on the geology of the site. The percussion (or cable-tool) method involves the raising and dropping of a heavy drill bit and stem. Having thus been pulverized, the earth being drilled is mixed with water to form a slurry, which is periodically removed. As drilling proceeds, a casing is also lowered (except when drilling through rock).

Rotary drilling methods (either hydraulic or pneumatic) utilize a cutting bit at the lower end of the drill pipe; a drilling fluid (or pressurized air) is constantly pumped to the cutting bit to aid in the removal of particles of earth, which are then brought to the surface. After the drill pipe is withdrawn, a casing is lowered into position.

Another method is the down-the-hole pneumatic (air) hammer, which combines the percussion effect with the rotary drill bit.

Local well drillers, who usually use the method most suited to the prevailing geology of the region, can offer useful advice about well construction methods. When clients plan to build in a remote location, the architect and engineer will want to advise them about water problems. Quality-corrective measures can always be taken and pumping equipment purchased, but the amount of water that can be obtained from the ground and the depth and cost of wells are important considerations. There are some problem areas where wells several hundred feet in depth will yield as little as 5 gpm (0.3 L/s), or nothing. The cost of drilling a number of exploratory wells may be excessive. Unfortunately, when such difficulties occur, there is often no easy solution. Conferences with neighboring owners, state and federal geologists, and local well drillers can all be helpful.

A low-yield well can be combined with storage tanks, so that the pump can run all night, slowly filling the tanks to meet the next day's demands.

(b) Pumps. The Hydraulic Institute, a non-profit trade association of pump manufacturers, has several recommendations for reducing the amount of electricity consumed by pumps. Some of these apply to small water supply systems.

1. Design systems with lower capacity and total head requirements. (Larger pipe sizes, minimizing elevation of tanks are examples.)
2. Avoid excessive capacity. It is typically less expensive to add pumping capacity later on if needs increase. Operating a smaller pump nearer its capacity saves energy, compared to a larger pump operating well under its capacity.
3. Select the most efficient pump type and size, even if its first cost is greater; life-cycle costs are likely to be lower.
4. Use two or more smaller pumps instead of one larger one so that excess pump capacity can be turned off.
5. Maintain pumps and system components in virtually new condition to avoid efficiency loss.

Three common types of pumps used in well water supply are the positive displacement, the centrifugal, and the jet.

Positive displacement pumps. There are two principal types of positive displacement pumps. In a *reciprocating pump,* a plunger moves back and forth within a cylinder equipped with check valves. The cylinder is best located near or below the groundwater level. Water enters the cylinder through an initial check valve (which allows flow in only one direction). As the plunger moves toward this check valve, the water is forced through the second check valve, located within the plunger itself. Then, as the piston returns to its original position, the water is forced upward toward the surface.

A *rotary pump* has a helical or spiral rotor—a turning vertical shaft within a rubber sleeve. As the rotor turns, it traps water between it and the sleeve, thus forcing the water to the upper end of the rotor.

Centrifugal pumps. This type of pump contains an impeller mounted on a rotating shaft. The rotating impeller increases the water's velocity while forcing the water into a casing, thus converting the water's velocity into higher pressure. Each impeller and casing is called a stage; many stages can be combined in a multistage pump. The number of stages depends on the pressure needed to operate the water supply system, as well as the height to which the water must be raised. The most common centrifugal pumps are those used in deep wells.

The *turbine pump* has a vertical turbine located below groundwater levels and a driving motor located higher up, usually over the well casing at grade. A long shaft is thus required between the motor and the turbine. Substantial head clearances for this shaft's removal can be required. Figure 10.9 shows a turbine pump installation for a small community on Long Island, New York. The water is taken from the ground by multistage turbine pumps at depths of several hundred feet. The water is delivered to submerged hydropneumatic tanks at a pressure of about 80 psi (about 550 kPa). As water is demanded in the houses, the air under pressure in the upper part of the tanks forces it through the mains.

Submersible pumps are designed so that the motor can be submerged along with the turbine (Fig. 10.10). The lengthy pump shaft is thus eliminated.

Jet (or ejector) pumps. In a jet pump, a venturi tube is added to the centrifugal pump. A portion of the water that is discharged from a centrifugal pump at the wellhead is forced down to a nozzle and venturi tube (Fig. 10.11). The lower pressure within the venturi tube induces well water to flow in, and the velocity of the water from the nozzle pushes it upward toward the centrifugal pump, which can then more easily lift it by suction.

Pump selection. Pump variations and characteristics are summarized in Table 10.4. The type of pump selected depends on many factors, including the rate of yield of the well, the daily flow (and maximum instantaneous flow rate) needed by the users, the size of storage or pressure tank used, and the total operating pressure against which the pump works (including the height to which water must be raised within the well). First cost, maintenance, and reliability are also factors, as is the energy used by the pump. In cold climates, the pump and the water supply system must be protected from freezing.

Of these factors, the two critical determinants are the flow rate (gallons per minute or per hour to be delivered) and the total pressure (or "head"). The flow rate depends on the number of fixtures to be served (Fig. 10.12), and the total pressure (Fig. 10.13) includes the suction lift, static head, and friction loss plus pressure head. This relationship will be explained in detail in Section 10.11.

(c) Pressure Tanks. Serving also for water storage, these are frequently used both to maintain a constant pressure on a pump-supplied water system and to allow for temporary peaks in water supply rates that exceed the capacity of the pump. (Elevated tanks offer one alternative to pressure tanks, cisterns another—although the latter usually are not located high enough to provide pressure to the supply system.)

Pressure tanks are often housed in outbuildings, along with the pump and any water treatment equipment (Fig. 10.13). The temperature of the outbuilding must be kept above freezing and its roof or walls should be removable to allow for replacement of parts over time. One type of pressure-storage tank is shown in Fig. 10.14.

The capacity of pressure tanks usually is small in comparison to the daily *total* water consumption; they provide short-term responses to peak flow demands. As a general rule, the pressure tank should be sized to deliver about 10 times

WATER AND WASTE

Fig. 10.9 (a) *A turbine well-pump.* (b) *Its operation.* (c) *Measurement of its capacity.* (d, e) *Its use in supplying a small community with groundwater.* (f) *A Jacuzzi multistage lineshaft turbine well-pump. (Jacuzzi Bros., Inc.) Capacities of turbine pumps range from 50 to 16,000 gpm (3 to 1010 L/s). (By permission of* Progressive Architecture.)

Fig. 10.10 *Submersible pump (centrifugal type), in exploded view. (From the U.S. Environmental Protection Agency's* Manual of Individual Water Supply Systems, *1975.)*

(*f*)

Fig. 10.9 (*continued*)

the pump's capacity in gpm (L/m). For a typical residence, allow 10 to 15 gal (38–57 L) tank capacity per person served.

For larger installations, the size of a pressure storage tank can be calculated by

$$Q = \frac{Qm}{1 - \dfrac{P_1}{P_2}}$$

where Q = tank volume (gal)
Qm = 15 minutes of storage at peak usage rate (gal)
P_1 = minimum allowable operating pressure (psi) plus atmospheric pressure (14.7 psi)

P_2 = maximum allowable operating pressure (psi) plus atmospheric pressure (14.7 psi)

The ranges of allowable pressures are discussed in Section 10.9. (An alternative tank-sizing procedure is shown in Table 10.5.)

EXAMPLE 10.1. An office building in a remote location has a water supply system served by a pump and well. The peak demand is 50 gpm. The fixtures will operate at a minimum of 50 psi; a maximum of 70 psi should not be exceeded. Therefore,

$$Qm = 15 \text{ min} \times 50 \text{ gpm} = 750 \text{ gal}$$

and

$$Q = \frac{Qm}{1 - \dfrac{P_1}{P_2}} = \frac{750}{1 - \dfrac{50 + 14.7}{70 + 14.7}} = \frac{750}{1 - 0.76} = 3125 \text{ gal}$$

∎

TABLE 10.4 **Pumps for Water Supply**

Type of Pump	Practical Suction Lift[a]	Usual Well-Pumping Depth	Usual Pressure Heads	Advantages	Disadvantages	Remarks
Positive Displacement						
1. Reciprocating				1. Positive action. 2. Discharge against variable heads. 3. Pumps water containing sand and silt. 4. Especially adapted to low capacity and high lifts.	1. Pulsating discharge. 2. Subject to vibration and noise. 3. Maintenance cost may be high. 4. May cause destructive pressure if operated against closed valve.	1. Best suited for capacities of 5–25 gpm against moderate to high heads. 2. Adaptable to hand operation. 3. Can be installed in very small diameter wells (2" casing). 4. Pump must be set directly over well (deep well only).
(a) Shallow well	22–25 ft	22–25 ft	100–200 ft			
(b) Deep well	22–25 ft	Up to 600 ft	Up to 600 ft above cylinder			
2. Rotary				1. Positive action 2. Discharge constant under variable heads. 3. Efficient operation.	1. Subject to rapid wear if water contains sand or silt. 2. Wear of gears reduces efficiency.	
(a) Shallow well (gear type)	22 ft	22 ft	50–250 ft			
(b) Deep well (helical rotary type)	Usually submerged	50–500 ft	100–500 ft	1. Same as shallow well rotary. 2. Only one moving pump device in well.	1. Same as shallow well rotary except no gear wear.	1. A cutless rubber stator increases life of pump. Flexible drive coupling has been weak point in pump. Best adapted for low capacity and high heads.
Centrifugal						
1. Shallow well				1. Smooth, even flow. 2. Pumps water containing sand and silt. 3. Pressure on system is even and free from shock. 4. Low-starting torque. 5. Usually reliable and good service life	1. Loses prime easily. 2. Efficiency depends on operating under design heads and speed.	1. Very efficient pump for capacities above 60 gpm and heads up to about 150 ft.
(a) Straight centrifugal (single stage)	20 ft max.	10–20 ft	100–150 ft			

Type	Practical suction lift[a]	Usual well-pumping depths	Usual pressure heads	Advantages	Disadvantages	Remarks
(b) Regenerative vane turbine type (single impeller)	28 ft max.	28 ft	100–200 ft	1. Same as straight centrifugal except maintains priming easily. 2. They are self-priming.	1. Same as straight centrifugal not suitable for pumping water containing sand or silt.	1. Reduction in pressure with increased capacity not as severe as straight centrifugal.
2. Deep well						
(a) Vertical line shaft turbine (multistage)	Impellers submerged	50–300 ft	100–800 ft	1. Same as shallow well turbine. 2. All electrical components are accessible, above ground.	1. Efficiency depends on operating under design head and speed. 2. Requires straight well large enough for turbine bowls and housing. 3. Lubrication and alignment of shaft critical. 4. Abrasion from sand.	
(b) Submersible turbine (multistage)	Pump and motor submerged	50–400 ft	50–400 ft	1. Same as shallow well turbine. 2. Easy to frost-proof installation. 3. Short pump shaft to motor. 4. Quiet operation. 5. Well straightness not critical.	1. Repair to motor or pump requires pulling from well. 2. Scaling of electrical equipment from water vapor critical. 3. Abrasion from sand.	1. 3500 RPM models, although popular because of smaller diameters or greater capacities, are more vulnerable to wear and failure from sand and other causes.
Jet						
1. Shallow well	15–20 ft below ejector	Up to 15–20 ft below ejector	80–150 ft	1. High capacity at low heads. 2. Simple in operation. 3. Does not have to be installed over the well. 4. No moving parts in the well.	1. Capacity reduces as lift increases. 2. Air in suction or return line will stop pumping.	
2. Deep well	15–20 ft below ejector	25–120 ft 200 ft max.	80–150 ft	1. Same as shallow well jet. 2. Well straightness not critical.	1. Same as shallow well jet. 2. Lower efficiency, especially at greater lifts.	1. The amount of water returned to ejector increases with increased lift—50% of total water pumped at 50-ft lift and 75% at 100-ft lift.

Source: U.S. Environmental Protection Agency, *Manual of Individual Water Supply Systems*, 1975.

[a]Practical suction lift at sea level. Reduce lift 1 ft for each 1000 ft above sea level.

(a)

(c)

(b)

CHOOSE THE CORRECT JH FROM THESE CHARTS DEEP WELL (Down to 120 feet)				
The JH in the ½, ¾, 1 and 1½ horsepower rating, matched with the appropriate injector and pipe sizes, will pump this amount of water at the indicated lift or depth to the water in the well:				
If suction lift or depth to water is	JH ½ HORSEPOWER	JH ¾ HORSEPOWER	JH 1 HORSEPOWER	JH 1½ HORSEPOWER
	Produces these gallons per hour between 20-50 lbs. discharge pressure		Produces these gallons per hour between 30-60 lbs. discharge pressure	
30 feet	795 G.P.H.	990 G.P.H.	1140 G.P.H.	1620 G.P.H.
40	680	875	1000	1470
50	575	735	875	1300
60	445	630	745	1200
70	360	495	620	920
80	310	385	530	820
90	255	315	435	700
100	220	275	340	590
110	195	240	295	540
120		205	250	480

(d)

Fig. 10.11 *Details of the deep-well jet pump. (a) Photograph of a multistage jet pump housing and equipment. At the top is the on/off electrical switch activated by pressure settings. It controls the direct-connected electric motor at the left. Impellers are enclosed in the pump housing at the right. Circulating connections to and from the jet can be seen at the right, the pump discharge at the top. Note the pressure gauge readings up to 100 psi, within which range the pump can be set to operate. (b) Well casing and circulating lines. Jet element can be seen at the bottom of the left-hand (larger) pipe. (c) Cutaway section of the pump. (d) Pumping capacity gph under various conditions. Note the two discharge pressure ranges, 20 to 50 psi and 30 to 60 psi. (Jacuzzi Bros., Inc.) (e) Jet-type (also known as venturi or ejector) deep-well pump and storage tank for a house or small building [for well lifts greater than 25 ft (7.6 m)]. Reduced pressure at (f), the jet nozzle, induces flow of groundwater into the circulated flow.*

Air volume control

Pressure gauge

Electric
motor

Centrifugal
impeller

(f)

Detail of
venturi ejector

To the house or building

Well
lift

Water table

Venturi
ejector *(a)*

Foot valve

Strainer

(e)

Fig. 10.11 *(continued)*

Fig. 10.12 *The relationship between the recommended flow-rate capacity of a pump and the number of domestic plumbing fixtures it supplies. See details in Section 10.9a. (From the U.S. Environmental Protection Agency's* Manual of Individual Water Supply Systems, *1976.)*

The capacity of elevated tanks usually is equal to at least 2 days of average water usage. For firefighting or other special requirements, the capacity may have to be even greater.

A summary of the quality, treatment, and supply issues raised thus far is provided in the example of a rural estate shown in Fig. 10.15 and Table 10.5. On this large estate, water was required for an estimated demand of 30,000 gpd (113,560 Lpd). This was for domestic use only; irrigation was served by a separate installation pumping from a lake. Wells were dug for the domestic supply. It was quickly evident that despite the great depths of the wells drilled, the available flow would be small. Four wells yielded a total rate of only 25 gpm.

Fig. 10.13 *The components of the total operating pressure (or head), a critical determinant of pump size. (For details, see Section 10.9a.) The pumphouse is usually a separate structure with water treatment and storage components. (From the U.S. Environmental Protection Agency's* Manual of Individual Water Supply Systems, *1975.)*

Fig. 10.14 *Small pressure-storage tanks are installed primarily to keep a water supply system at constant pressure. These Hydrocel models are small [8½ in. in diameter, 27 in. long (216 mm diameter, 686 mm long)], so they can be installed almost anywhere along the supply system. (Courtesy of Jacuzzi Bros., Inc.)*

Calculations for the amount of water to be stored to supplement this meager supply are shown in Table 10.5. The table shows that during the 14 daytime hours under conditions of peak demand, the well pumps would run continuously. Concurrently, an additional 9000 gal would be drawn from the tanks. At night, the well pumps would run for 6 h to restore the 9000 gal drawn from the tanks during the day.

The operation of the system is illustrated in the diagram and notes of Fig. 10.15*a*. The pressure of 75 psi in the hydropneumatic tank is sufficient to raise the water to the greatest height in the distribution sytem, overcome friction in the piping, and leave a residual pressure available at each fixture of about 10 to 15 psi, as needed. Excessive pressure can always be moderated by a valve in the branch supply of the fixture.

Pressure in this tank is assured by the air compressor, activated by the pressure switch. The float switch starts the centrifugal pump to deliver water from the storage tanks as needed. The storage tanks, piped together and acting as a single reservoir, are fed by the four wells. The wells operate in unison, singly, or in groups, depending on the level in the storage tanks. If the level drops rapidly, all well pumps can run. Although they are arranged to operate all day when needed, in periods of minimum demand only one or two may be called upon. Each has its own power supply, but controls emanate from panel *N*.

WATER AND WASTE

TABLE 10.5 **Calculations to Establish Capacity of Water Storage Tanks at Kinloch (Fig. 10.15)**

1. Potable domestic water demand per day	30,000 gpd
2. Assumed hours of use	7 AM to 9 PM 14 hours
3. Assumed hours of virtual nonuse	9 PM to 7 AM 10 hours
4. Well-yield rate per hour	25 gpm × 60 min = 1500 gph
5. Total well yield in 14 hours	15,000 gph × 14 hr = 21,000 gal
6. Total water needed in 14 hours	30,000 gal
7. Well yield in 14 hours	21,000 gal (see 5 above)
8. Net-tank capacity required, minimum	30,000 − 21,000 = 9000 gal
9. Net-tank capacity as designed	
20,000 × 0.80 (80% full) = 16,000 (2 tanks at 10,000 gal each)	
2000 × 0.70 (70% full) = 1,400 (1 tank at 2,000 gal)	
17,400 gal, O.K., greater than 9000	
10. Well operation at night to restore 9000 gal to tanks	9000 ÷ 1500 = 6 hours of the 10 night hours available.

WATER AND WASTE

Legend

(A)	Wells
(B)	Well water header for 4 wells
(C)	Well water storage, vented tanks, (atmospheric pressure)
(D)	Pump supply pipe
(E)	Electric centrifugal pumps (one is a standby)
(F)	Pump discharge to hydro-pneumatic tank
(G)	Hydro-pneumatic tank, 75 lbs. per sq. in.
(H1) (H2)	Discharge pipes, pressurized well water
(I)	Water softeners, (ion exchangers)
(J)	Treated well water under pressure
(K1) (K2)	Branch mains to buildings (treated well water)
(L)	Air compressor maintains 75 psi in tank
(M)	Sodium chloride for regeneration
(N)	Electrical control panel
(O)	Float switches
(P)	Pressure switch (actuates air compressor)
(Q)	Control wiring to well pumps
(R)	Control wiring, float switch to panel
(S)	Control wiring, pressure switch to compressor
(T)	Compressed air piping
(U)	Control wiring, float switch to centrifugal pump
(V)	Gauge glass indicates water level

This schematic diagram shows system components and connections. For clarity, details (valves, drains, check-valves etc.)
are not shown. All control wiring, here simply indicated, operates through control panel (N). See text for operation.

(a)

Fig. 10.15

(b)

Fig. 10.15 *(continued)(a) Water control and distribution center for the Kinloch Estate. (Bentel & Bentel, Architects, FAIA.) The plant is located in a hillside, below the bathing and dressing pavilion. (b) One of two pavilions adjacent to the swimming pool that serves residents and guests at Kinloch. Below this pavilion is the control center. Exterior tubing to and from the control center is all below grade.*

Swimming pools should be supplied with pure, potable water. Biological tests showed that this well water was safe. Thus, the fill-line to the swimming pool is connected from *this* control center rather than from the supply of lake water used for irrigation. Because swimming pool water is separately recirculated and provided with its own purification treatment at a location adjacent to the pool, this supply line for make-up water to the pool would have infrequent and off-hour use. Water for domestic use in the buildings is passed through one of three ion-exchangers to provide softening and to make the water more suitable for washing, bathing, and cooking. Periodically, the calcium precipitate can be flushed out and the tanks regenerated by sodium chloride.

10.6 Hot Water Systems and Equipment

There are many ways to provide the "domestic" or "service" hot water needs within a building—the hot water used *not* for space heating, but for bathing, clothes washing, dishwashing and other related functions. Whereas much of the world either heats such water on a cookstove or does without it, North Americans enjoy an array of choices for domestic hot water (DHW) supply systems. In this section, rough supply estimates

are considered first, then basic design choices, conventional heater tanks, solar water heaters, and finally, energy recovery heaters.

With today's better-insulated buildings, the demand for space heating can fall to a level about equal to DHW. As a result, a trend is developing toward using one rather larger water heater, designed to meet both the need for space heating and for DHW. This saves both floor space and equipment cost, compared to the more conventional boiler for space heating, water heater for DHW. Some restrictions apply; the 1997 International Plumbing Code restricts the maximum water outlet temperature to 160°F (71°C), and requires water in such a heating system to remain potable.

(a) Water temperature. Higher water supply temperatures have several advantages, but carry risks of scalding. Table 10.6 shows that although water becomes uncomfortably hot to the touch above 110°F (43°C), much higher temperatures are often used for some commercial processes. When determining the temperature at which DHW is to be supplied, consider the following factors.

High temperatures

· Allow the installation of smaller storage tanks (less hot water is mixed with cold to achieve final usage temperature at the shower, sink, or lavatory)—but require larger heating units.

· Can be achieved at the point of use (rather than in the tank) by heaters built into equipment such as dishwashers.

· Can cause scale to form on heating coils and within piping [above 140°F (60°C) in areas of "hard" water quality].

· May be required by code for some applications, limited by code for others.

· Limit the potential for growth of *Legionella pneumophila* bacteria [above 140°F (60°C)].

Lower temperatures

· Are less likely to cause burns, but may not achieve desired sanitation.

· Mean less energy consumed, because storage and pipe heat losses are lower.

(a)

(b)

Fig. 10.16 (a) *Example of indirect, tankless heater for DHW, utilizing a boiler that provides hot water for space heating. Because no storage is used, the point of use for DHW should be very close to the boiler.* (b) *Internal tankless heating coil for domestic hot water immersed in the jacket water of a gas-fired hot water heating boiler. Approximate capacity range 3 to 15 gpm at 100F° rise (0.2 to 0.9 L/s at 55C° rise).* (c) *External-type, tankless heater for domestic hot water. Boiler water is piped to the unit and circulates by gravity, transferring heat to the coil. Approximate capacity range 3 to 15 gpm at 100F° rise (0.2 to 0.9 L/s at 55C° rise). Note: Because it can be highly inefficient to provide summertime DHW with a boiler designed for winter-time space-heating loads, codes may alter or prohibit this arrangement.*

(c)

Fig. 10.16 *(continued)*

- Allow the installation of smaller heating units—but require larger storage tanks.
- Make possible the use of lower-grade heat sources for DHW, such as solar energy or waste heat recovery. Table 1.3 showed that 14% of annual U.S. energy usage is for DHW, which requires the lowest-grade heat source of all the common usages.

(b) Heat Sources and Methods. These include the familiar concentrated-energy, high-grade sources such as natural gas and electricity. Oil- and coal-fired boilers are frequently equipped with DHW coils as well. Buildings served by steam use steam as a DHW heat source. Cogeneration provides heat for DHW, as in Fig. 8.79. Because of the relatively low-grade final temperatures needed, DHW can also be readily provided by wood-burning equipment, incinerators, solar energy equipment, heat pumps, and heat recovery devices (as in commercial ice-making machines whose discharged heat is contributed to DHW).

Heating methods. There are two basic methods:

1. *Direct* heating brings water against directly heated surfaces: electric-resistance elements within tanks, or other electrically warmed surfaces, or surfaces directly exposed to fire or hot gases.
2. *Indirect* heating can be accomplished in sev-

TABLE 10.6 **Representative Hot Water Temperatures**

Use	Temperature	
	°F	°C
Lavatory		
Hand washing	105	40
Shaving	115	45
Showers and tubs	110	43
Therapeutic baths	95	35
Commercial and institutional laundry (based on fabric)	up to 180	up to 82
Residential dish washing and laundry	140	60
Surgical scrubbing	110	43
Commercial spray-type dishwashing[a]		
Wash	150 minimum	65 minimum
Final rinse	180 to 195	82 to 90

Source: Reprinted with permission of the American Society of Heating, Refrigerating, and Air-Conditioning Engineers, Inc., from the 1995 *ASHRAE Applications Handbook.*

[a]For other types of commercial dishwashers, see the source listed above.

eral ways. Coils containing steam or fluids can be submerged within water tanks or (Fig. 10.16) set within boilers, whose primary function usually is to provide space or industrial process heating. Alternatively, the coils containing DHW can be placed outside the boiler but within a casing containing steam, hot exhaust gases, or very hot water.

Both direct or indirect methods can be utilized in a variety of equipment:

Storage tank water heaters, the type most commonly used for residential and small commercial purposes (see Section 10.6h).

Circulating storage water heaters, in which the water is first heated by a coil, then circulated through the storage tank (as in some solar heaters—see Section 10.6i).

Tankless (instantaneous) heaters, in which the water is very quickly raised to a desired temperature within a heating coil and immediately sent to the point of usage.

(c) Tankless Water Heaters. Also called instantaneous water heaters, they are available in larger sizes for central hot water systems (Fig. 10.16) or in smaller sizes for wall mounting adjacent to remote plumbing fixtures that occasionally need hot water, or for isolated bathrooms, laundry rooms, and so forth. Decentralized tankless units can be as small as "instant hot water taps" for kitchen or bar sinks—electric-resistance heaters capable of generating perhaps 3 gph at up to 200°F (2 L/h at up to 93°C). Bathroom groups can require either electric resistance or gas-fired units (Fig. 10.17); electric heaters require very high amperage and convert only about one-third of the primary energy into electricity; gas heaters should be vented and are most efficient without continuously burning pilots.

These tankless heaters can be very small; one unit of but 2.65 ft³ produces up to 4 gpm (0.24 L/s) with a temperature rise of 45°F (20°C). With a greater temperature rise of 100°F (55.5°C), the flow rate drops to somewhat less than half. Because tankless heaters often consist of fairly long coils through which the water passes as it is heated, they may add considerable friction to the total for which the supply system is designed (see Section 10.11).

To gain a better understanding of hot water used per activity and the role of instantaneous heaters, see Table 10.7. Domestic water system design usually concentrates on the storage tank size [see subsection (h)].

(d) Energy Factors (EF). The U.S. Department of Energy developed this standardized measure of annual overall efficiency. Standard storage

(a)

(b)

Fig. 10.17 *Instantaneous or tankless water heater.* (a) *In this "instant-flow" water heater, a series of coils heats water as it flows through.* (b) *Installation below a typical lavatory. A wide range of sizes is available; the 4.6 kW unit raises water by 31F° at 1 gpm; the 9 kW unit raises water by 61F° at 1 gpm. (Courtesy of Chronomite Laboratories, Inc.)*

TABLE 10.7 **Domestic Hot Water Consumption—Residences**

	Gallons (Liters) per Use Hot Water Required	
	14-lb (6.4-kg) Machine	18-lb (8.2-kg) Machine
Clothes Washing Machine		
Hot wash/hot rinse	38 gal (144 L)	48 gal (182 L)
Hot wash/warm rinse	28 gal (106 L)	36 gal (136 L)
Hot wash/cold rinse	19 gal (72 L)	24 gal (91 L)
Warm wash/cold rinse	10 gal (38 L)	12 gal (45 L)
Dishwashing	*Small*	*Large*
Dishwashing machine	10 gal (38 L)	15 gal (57 L)
Sink washing	4–8 gal (15–30 L)	
Personal Hygiene		
Tub bathing	12–30 gal (45–134 L)	
Wet shaving/hair washing	2–4 gal (8–15 L)	
Showering	2–6 gpm (13–38 L/s)	

Source: Reprinted by permission from Russell Plante, *Solar Domestic Hot Water*, copyright © 1983 by John Wiley & Sons.

tank water heaters that are gas-fired may reach EF of 0.60 to 0.64. Because they have virtually none of the stand-by losses of storage tank heaters, gas-fired tankless water heaters offer EF up to 0.69 with continuous pilots, and as high as 0.93 with electronic ignition. (However, this minor usage of electricity is not included in the EF rating.)

Some gas-fired instantaneous heaters are designed to operate with battery-powered pilot light ignition. These are useful in remote installations with propane but without electricity.

(e) Central versus Distributed Equipment.

This choice can be particularly complex. In the United States a central water-heater storage tank is standard equipment in homes and small stores. However, the growing use of solar water heating, along with awareness of the heat losses from water-heater storage tanks, has created new interest in *combinations* of central and distributed DHW.

Consider a large residential DHW application such as that shown in Fig. 10.18. Here, two areas in which hot water is used are separated by some 50 ft (15 m). If the simple, centralized water-heater storage tank is used (Fig. 10.18*a*), it would probably be placed nearest the maximum-use fixtures—dishwashers and clothes washers. (Also, floor space for the water-heater storage tank is

usually more plentiful there.) Codes typically require that at least the first 8 ft (2.4 m) of the hot water pipe be insulated, as it leaves the water heater. However, each time hot water is needed in the remote bathroom serving the bedrooms, the previously heated water in at least 50 ft (15 m) of supply pipe—which, despite insulation even over its entire length, will almost certainly have cooled—must be drained off before hot water finally arrives. (A ¾-in. diameter pipe, for example, will hold 4.6 gal in 50 ft of pipe. If the water heater is set at an energy-conserving level of 120°F, and the incoming city water is 50°F, the 4.6 gal of wasted water will also waste the 2682 Btu invested when it is heated.)

Another way to conserve water is with a recirculating hot water system (Fig. 10.18*b*). The primary disadvantages of this system are the increased heat loss through the hot water pipe [now kept at 120°F (49°C) for 24 h per day, rather than for just a few minutes] and the energy required to run the circulating pump (again for 24 h daily). Codes typically require a conveniently located cut-off switch for the circulating pump, and pipe insulation with a minimum $k = 0.3$ Btu-in/h ft²°F (0.04 W/m K) over the entire length of the circulating hot water system.

A decentralizing option (Fig. 10.18*c*) is to install two water-heater storage tanks, one for each

Fig. 10.18 *Options for DHW in a larger residence.* (a) *Typical centralized water-heater storage tank; water and energy is wasted in the 50-ft (15 m) run.* (b) *Recirculating pump added; this saves water but requires continuous energy to operate pump and make up the pipe's heat losses.* (c) *Decentralized approach using two water-heater storage tanks; this roughly doubles the energy lost from storage but eliminates water and energy loss from the 50-ft (15 m) run.* (d) *Central solar water heater/decentralized tankless heating coil combination; saves on water and energy, but not initial cost.*

group. First cost will be greater, and more floor space required. The daily waste heat from *two* heaters is clearly a disadvantage that will at least partially offset the energy saved by eliminating the 50-ft (15-m) run of pipe. (Furthermore, in warm weather this waste heat will add to discomfort within the house.)

A decentralized-centralized mix (Fig. 10.18*d*) will save energy, water, and floor space, but it will almost certainly be costlier to install. A central solar water heater—either passive (shown) or active—brings the water up to warm (winter) or very hot (summer). At each fixture group, a tankless heating coil instantaneously heats the water, as needed. Almost no water is wasted, and the energy lost in the 50 ft (15 m) of pipe will be solar energy, not purchased energy from natural gas or electricity.

Another option would be to omit the solar water heater and simply install two tankless heaters. The primary disadvantage of this setup would be the need for larger-capacity heaters, whose instantaneous demand for energy could lead to electric cost penalties in areas where peak-load rates are high. Compared to central storage tanks, however, less heat is lost.

(f) Distribution Trees. The choice of a distribution tree associated with central water heaters must take into account the gradual cooling of water within the distribution system once it has left the central heater. If the heat losses associated with constantly circulating hot water (looped trees) are preferable to the heat and water wastage associated with simple, single hot water distribution trees, such loop systems can be achieved in either of two ways.

Thermosiphon hot water circulation depends on the fact that water expands and becomes lighter when heated, as may be seen in Fig. 10.19*a*. If heat is applied to the lower loop of a glass tube, both ends of which have been inserted in an inverted bottle containing water, the water will move from *A* to *B* and will rise through the tube *BC* into the bottle. It here becomes cooled and drops through the tube *DA* to *A*, is again heated, and rises in the tube *BC*—thus completing the circulation. Because the movement depends on the difference in weight between the two columns of water, the velocity and consequent efficiency

(a)   *(b)*

Fig. 10.19 (a) *Principle of hot water circulation by gravity (thermosiphon).* (b) *Its application to hot water service. During periods of no demand there is sufficient incidental cooling between C and D so that dense (less warm) water at A forces the lighter (hot) water at B to rise for speedy availability at each faucet.*

of the circulating system increase as both the temperature of the water and the height of the circuit increase. Hot water supply systems therefore usually consist of a heater with a storage tank, piping to carry the heated water to the farthest fixture, and a continuation of this piping to return the unused cooled water back to the heater. A constant circulation is thereby maintained, and hot water may be drawn at once from a fixture without first draining off through the faucet the cooled water that would be standing in the supply pipe if there were no recirculation. Because heat increases the corrosive action of water on metals, copper tubing or rated PVC pipe is often chosen for use in hot water and hot water circulating systems. Thermosiphon circulating systems are particularly effective in multistory buildings, because of the beneficial effect of increased circuit height on circulation efficiency.

Forced circulation of hot water is often utilized where a height advantage is unavailable. Low, long, rambling buildings, such as some large one-story residences, schools, and factories, lack the height needed to set up good hot water circulation

Fig. 10.20 *Forced circulation of DHW, for a long, low building. (Note that the use of space-heating boilers for DHW may cause system inefficiencies in warm weather.)*

by gravity. Also, the flow is diminished by friction in long pipe runs. For such buildings the forced-circulation scheme shown in Fig. 10.20 offers one option. Three independent aquastats—devices that create an electrical signal impulse when water temperature drops—control this system. Aquastats *A, B,* and *C,* respectively, sense the temperatures of the water in the heater, the tank, and the end of the circulation-return main. As needed, they turn on the oil or gas burner, the tank-circulating pump, and the system-circulating pump. Fixtures remote from the tank are as close to hot water as the length of their hot water runout pipes. Water is usually available at full temperature in 5 to 10 seconds. Trial aquastat settings in °F could be *A* 180, *B* 160, *C* 120 (°C: *A* 82, *B* 71, *C* 49).

(g) Variable Storage Temperature.

An energy-saving computer control is available for hotels, motels, apartments, and larger commercial buildings. This device (Fig. 10.21) varies the supply temperatures of hot water, so that the hottest water is supplied at the busiest hours. In hours of low usage, much lower supply temperatures mean that more "hot" water will be mixed with less "cold" water at showers, lavatories, and sinks. This increased "hot" water quantity poses no problem at off-peak hours, and the lower temper-

ature means significant decreases in heat loss from the recirculating supply water system. A typical payback period is 1 year. The device stores a memory (adjusted weekly) of the typical daily patterns of usage, and varies the supply temperature accordingly.

(h) Conventional Water Heater Selection.

One of the most common appliances used in the United States is the water-heater storage tank, an energy-conserving model of which is shown in Fig. 10.22.

For residences, the capacity of hot water heating/storage equipment can be taken from the values in Table 10.8. Note that for the familiar tank-type direct water heaters (gas, electric, or oil), there is a stated relationship among storage size, rate of hourly heat input, rate of draw (demand) over 1 h, and recovery rate. Example 10.2 uses Table 10.8 for the sizing of a residential water heater.

EXAMPLE 10.2. Select a natural gas water heater for a five-bedroom house with 3 baths. The *minimum* requirements of HUD-FHA are shown in Table 10.8: 50 gallons, 47,000 Btu/h, 90-gal draw per hour, and 40 gph recovery.

Fig. 10.21 *Variable temperature supply.* (a) *Energy savings possible when the supply temperature of hot water is varied. Heat losses from supply pipes are greatly reduced for most of the typical day. A computer controls the supply temperature.* (b) *Schematic of computer controlled variable temperature DHW system. (Courtesy of Fluidmaster, Inc., Anaheim, California.)*

SOLUTION. From Fig. 10.22 and Table 10.9, BTH 120 is selected; it exceeds all the minimums. ∎

Estimation of the demand for commercial and institutional buildings is not so simple, and design guidelines are less reliable. For larger buildings, there is a trade-off between quick recovery (high heating capacity) and storage size; big tanks have small heaters, and vice versa. Another variable is storage temperature, as was discussed earlier in subsection (*a*).

WATER AND WASTE

TABLE 10.8 HUD–FHA Minimum Water Heater Capacities, Residential

Part A. *I-P Units*

Number of Baths	1–1.5			2–2.5				3–3.5			
Number of Bedrooms	1	2	3	2	3	4	5	3	4	5	6
Gas[a]											
Storage, gal	20	30	30	30	40	40	50	40	50	50	50
Input, 1000 Btu/h	27	36	36	36	36	38	47	38	38	47	50
1-h draw, gal	43	60	60	60	70	72	90	72	82	90	92
Recovery, gph	23	30	30	30	30	32	40	32	32	40	42
Electric[a]											
Storage, gal	20	30	40	40	50	50	66	50	66	66	80
Input, kW	2.5	3.5	4.5	4.5	5.5	5.5	5.5	5.5	5.5	5.5	5.5
1-h draw, gal	30	44	58	58	72	72	88	72	88	88	102
Recovery, gph	10	14	18	18	22	22	22	22	22	22	22
Oil[a]											
Storage, gal	30	30	30	30	30	30	30	30	30	30	30
Input, kW	70	70	70	70	70	70	70	70	70	70	70
1-h draw, gal	89	89	89	89	89	89	89	89	89	89	89
Recovery, gph	59	59	59	59	59	59	59	59	59	59	59
Tank-type indirect[b,c]											
I-W-H-rated draw, gal in 3 h, 100F° rise		40	66		66	66[e]	66	66	66	66	66
Manufacturer-rated draw, gal in 3 h, 100F° rise		49	49		75	75[e]	75	75	75	75	75
Tank capacity, gal		66	66		66	66[e]	83	66	82	82	82
Tankless-type indirect[c,d]											
I-W-H-rated draw, gpm, 100F° rise		2.75	2.75		3.25	3.25[e]	3.75	3.75	3.75	3.75	3.75
Manufacturer-rated draw, gal in 5 min, 100F° rise		15	15		25	25[e]	35	25	35	35	35

Part B. *SI Units*

Number of Baths	1–1.5			2–2.5				3–3.5			
Number of Bedrooms	1	2	3	2	3	4	5	3	4	5	6
Gas[a]											
Storage, L	76	114	114	114	150	150	190	150	190	190	190
Input, kW	7.9	10.5	10.5	10.5	10.5	11.1	13.8	11.1	11.1	13.8	14.6
1-h draw, L	163	227	227	227	265	273	341	273	311	341	350
Recovery, mL/s	24	32	32	32	32	36	42	34	34	42	44
Electric[a]											
Storage, L	76	114	150	150	190	190	250	190	250	250	300
Input, kW	2.5	3.5	4.5	4.5	5.5	5.5	5.5	5.5	5.5	5.5	5.5
1-h draw, L	114	167	220	220	273	273	334	273	334	334	387
Recovery, mL/s	10	15	19	19	23	23	23	23	23	23	23
Oil[a]											
Storage, L	114	114	114	114	114	114	114	114	114	114	114
Input, kW	20.5	20.5	20.5	20.5	20.5	20.5	20.5	20.5	20.5	20.5	20.5
1-h draw, L	337	337	337	337	337	337	337	337	337	337	337
Recovery, mL/s	62	62	62	62	62	62	62	62	62	62	62
Tank-type indirect[f,g]											
I-W-H-rated draw, L in 3 h, 55 K rise		150	150		250	250[e]	250	250	250	250	250
Manufacturer-rated draw, L in 3 h, 55 K rise		186	186		284	284[e]	284	284	284	284	284
Tank capacity, L		250	250		250	250[e]	310	250	310	310	310

TABLE 10.8 **HUD–FHA Minimum Water Heater Capacities, Residential (*continued*)**

Part B. SI Units

Number of Baths	1–1.5			2–2.5				3–3.5			
Number of Bedrooms	1	2	3	2	3	4	5	3	4	5	6
Tankless-type indirect[g,h] I-W-H-rated draw, mL/s, 55 K rise		170	170		200	200[e]	240	200	240	240	240
Manufacturer-rated draw, L in 5 min, 55 K rise		57	57		95	95[e]	133	95	133	133	133

Source: Reprinted with permission of the American Society of Heating, Refrigerating, and Air-Conditioning Engineers, Inc. from 1995 *ASHRAE Applications Handbook.*

[a]Storage capacity, input, and recovery requirements are typical and may vary with individual manufacturers. Any combination of these requirements to produce the stated 1-h draw will be satisfactory.

[b]Boiler-connected water heater capacities (180°F boiler water, internal, or external connection).

[c]Heater capacities and inputs are minimum allowable. Variations in tank size are permitted when recovery is based on 4 gph/kW at 100F° rise for electrical, AGA recovery ratings for gas, and IBR ratings for steam and hot water heaters.

[d]Boiler-connected heater capacities (200°F boiler water, internal, or external connection).

[e]Also for 1.5 baths and 4 bedrooms for indirect water heaters.

[f]Boiler-connected water heater capacities (82°C boiler water, internal, or external connection).

[g]Heater capacities and inputs are minimum allowable. Variations in tank size are permitted when recovery is based on 4.2 mL/s kW at 55 K rise for electrical, AGA recovery ratings for gas, and IBR ratings for steam and hot water heaters.

[h]Boiler-connected heater capacities (93°C boiler water, internal, or external connection).

Small shops and very small office buildings can be treated like residences for hot water sizing. For larger buildings, the trade-off between heaters and storage tanks should be explored. To begin this process, consult Table 10.10, which shows the maximum hourly and daily demands and contrasts them to the average day (which should be used to estimate monthly energy consumption).

The relationship between heater size and tank size is graphed in Fig. 10.23. The advantage of a *larger heater* is its smaller tank, which consumes less floor space and volume, weighs less, and probably has a lower first cost. The advantage of a *larger tank* is that its smaller heater tends to work steadily rather than in spurts; this lower, steadier demand for energy will lead to lower utility rates in the case of electric heating, and thus to money savings over the life of the system. When the fuel

TABLE 10.9 **Procedure for Sizing a Residential Water Heater Storage Tank**

Characteristic	Table 10.8: HUD–FHA 5 Bedrooms, 3 Baths	Fig. 10.22 Data: Cyclone XHE Model BTH-120
Storage (gal)	50	60
Btuh input	47,000	120,000
1-hr draw (= tank capacity + 1-hr recovery)	90	231 @ 80F°
Recovery	40	171

WATER AND WASTE

Fig. 10.22 *"Cyclone XHE" water-heater storage tanks, using natural gas (or propane). This model achieves 94% thermal efficiency with a burner located at the top, forcing blower-driven air into a counter-current contact with rising fuel. A precise mix of air and fuel at the point of ignition results in high combustion efficiency, and the swirling flame continues downward in a submerged central combustion chamber. The heated gas is then forced at high velocity through the spiral heat exchanger coil, with a swirling action that maximizes heat exchange with the water. (Courtesy of the A.O. Smith Water Products Company, Irving, Texas.)*

Cyclone XHE Data:

[all models are 27¾ × 30 in. (705 × 762 mm) with 3-in. (76 mm) PVC vents]

Model	Storage Capacity Gallons (L)	Btuh Input/Output	Recovery (gal/h) @ 94% Thermal Efficiency		Height	
			80F° ΔT	100F° ΔT	in.	mm
BTH-120	60 (227)	120,000/112,800	171	137	55½	(1,410)
BTH-150	100 (378)	150,000/141,000	214	171	74½	(1,892)

TABLE 10.10 **Domestic Hot Water, Commercial, Institutional**

Type of Building	Maximum Hour	Maximum Day	Average Day
Men's dormitories	3.8 gal (14.4 L)/student	22.0 gal (83.4 L)/student	13.1 gal (49.7 L)/student
Women's dormitories	5.0 gal (19 L)/student	26.5 gal (100 L)/student	12.3 gal (46.6 L)/student
Motels: no. of units[a]			
20 or less	6.0 gal (23 L)/unit	35.0 gal (132.6 L)/unit	20.0 gal (75.8 L)/unit
60	5.0 gal (20 L)/unit	25.0 gal (94.8 L)/unit	14.0 gal (53.1 L)/unit
100 or more	4.0 gal (15 L)/unit	15.0 gal (56.8 L)/unit	10.0 gal (37.9 L)/unit
Nursing homes	4.5 gal (17 L)/bed	30.0 (114 L)/bed	18.4 gal (69.7 L)/bed
Office buildings	0.4 gal (1.5 L)/person	2.0 gal (7.6 L)/person	1.0 gal (3.8 L)/person
Food service establishments:			
Type A—full meal restaurants and cafeterias	1.5 gal (5.7 L)/max meals/h	11.0 gal (41.7 L)/max meals/h	2.4 gal (9.1 L)/avg meals/day[b]
Type B—drive-ins, grilles, luncheonettes, sandwich and snack shops	0.7 gal (2.6 L)/max meals/h	6.0 gal (22.7 L)/max meals/h	0.7 gal (2.6 L)/avg meals/day[b]
Apartment houses: no. of apartments			
20 or less	12.0 gal (45.5 L)/apt.	80.0 gal (303.2 L)/apt.	42.0 gal (159.2 L)/apt.
50	10.0 gal (37.9 L)/apt.	73.0 gal (276.7 L)/apt.	40.0 gal (151.6 L)/apt.
75	8.5 gal (32.2 L)/apt.	66.0 gal (250 L)/apt.	38.0 gal (144 L)/apt.
100	7.0 gal (26.5 L)/apt.	60.0 gal (227.4 L)/apt.	37.0 gal (140.2 L)/apt.
200 or more	5.0 gal (19 L)	50.0 gal (195 L)/apt.	35.0 gal (132.7 L)/apt.
Elementary schools	0.6 gal (2.3 L)/student	1.5 gal (5.7 L)/student	0.6 gal (2.3 L)/student[b]
Junior and senior high schools	1.0 gal (3.8 L)/student	3.6 gal (13.6 L)/student	1.8 gal (6.8 L)/student[b]

Reprinted with permission of the American Society of Heating, Refrigerating, and Air-Conditioning Engineers, Inc. from 1995 *ASHRAE Applications Handbook*.

[a]Interpolate for intermediate values.

[b]Per day of operation.

supply is solar energy or waste heat, the larger tank again is usually better suited to the characteristics of the fuel supply.

EXAMPLE 10.3. (Adapted from the 1995 *ASHRAE Applications Handbook*). A women's dormitory housing 300 students, with a cafeteria serving 300 meals in 1 h, is to be built. Find the required storage size for two conditions: (1) assuming a minimum recovery rate for both dorm and cafeteria; and (2) assuming a dorm recovery rate of 2.5 gph (2.6 mL/s), which is half the maximum hourly value given in Table 10.10, and a cafeteria recovery rate of 1.0 gph (1.1 mL/s), which is two-thirds the maximum hourly value given in Table 10.10.

SOLUTION. *Minimum recovery:* From Fig. 10.23a, the minimum recovery rate for women's dormitories is 1.1 gph. For 300 students,

$$300 \times 1.1 = 330 \text{ gph recovery}$$

At this rate, again from Fig. 10.23a, the minimum usable storage capacity is 12 gal/student. Assume that 70% of the total size is this "usable" capacity. This means that after 70% of the stored hot water is withdrawn, the remaining water has been cooled (by incoming water) to an unusably low temperature. Storage size must be increased by

$$\frac{100\%}{70\%} = 1.43$$

Thus,

$$12 \times 300 \times 1.43 = 5150 \text{ gal storage}$$

From Fig. 10.23e, the minimum recovery rate for the cafeteria (serving full meals, type *A*) is 0.45 gph. For 300 meals,

$$300 \times 0.45 = 135 \text{ gph recovery}$$

At this rate, the minimum usable storage capacity is 7 gal/meal. Thus

$$300 \times 7 \times 1.43 = 3000 \text{ gal storage}$$

WATER AND WASTE

Fig. 10.23 *Domestic hot water sizing; the tradeoff between recovery (heater) capacity and storage capacity. "Usable" storage capacity is usually considered to be between 60 and 80% of the total tank capacity. For (b) Motels, and (f) Apartments, curves are based upon the number of dwelling units. For (e) Food service, Types A, B, see Table 10.10. (Reprinted by permission of the American Society of Heating, Refrigerating, and Air-Conditioning Engineers, Inc. from the* 1995 *ASHRAE Applications Handbook.)*

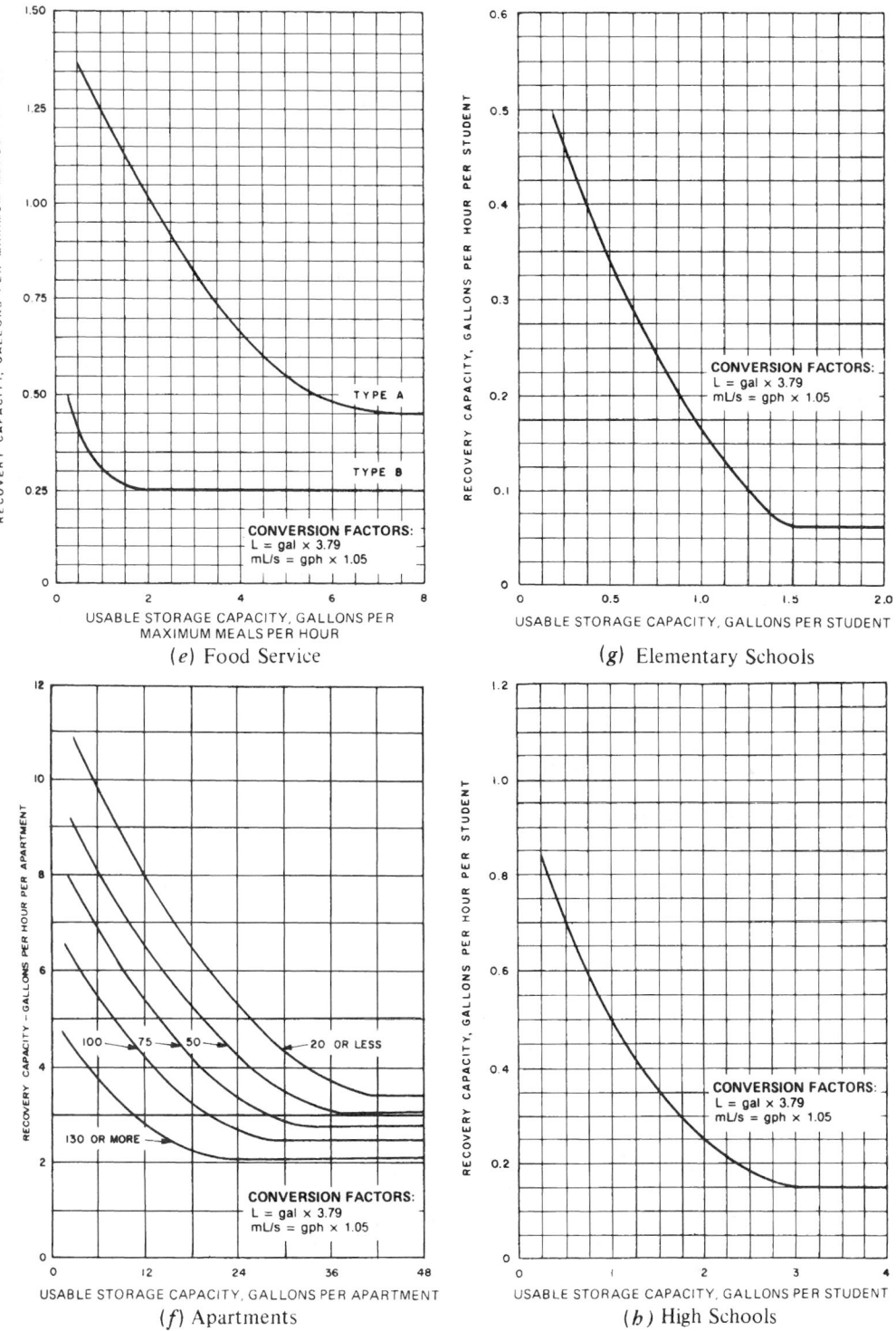

(e) Food Service

(g) Elementary Schools

(f) Apartments

(h) High Schools

Fig. 10.23 (continued)

Combining these requirements for dorm and cafeteria,

$$recovery = 300 + 135 = 465 \text{ gph}$$

$$storage = 5150 + 3000 = 8150 \text{ gal}$$

Faster recovery: At the specified dorm recovery rate of 2.5 gph,

$$300 \times 2.5 = 750 \text{ gph}$$

and the minimum usable storage required is 5 gal/student. Thus

$$300 \times 5 \times 1.43 = 2150 \text{ gal}$$

At the specified cafeteria recovery rate of 1.0 gph,

$$300 \times 1.0 = 300 \text{ gph}$$

and the minimum usable storage required is 2 gal/meal. Thus

$$300 \times 2.0 \times 1.43 = 860 \text{ gal}$$

Combining these requirements for dorm and cafeteria,

$$recovery = 750 + 300 = 1050 \text{ gph}$$

$$storage = 2150 + 860 = 3010 \text{ gal}$$

Note: This sizing procedure does not account for *system losses,* that is, heat lost from the storage tanks and from the hot water piping. Recovery capacities are usually increased because of these losses, which are simple to calculate when the *U*-values of the tank and pipe insulation are known: $Btu/h = U \times A \times \Delta T$. ■

For this example, an increase in heater size of 225% for faster recovery allows the size of tank to be reduced to only 37% of original size. For these larger DHW applications, central steam is often used as a heat source. Figure 10.24 shows an indirect storage tank system. Such systems may cost more than tankless systems, but they are usually cheaper to operate, both because the peak prices for the instantaneous fuel demands of tankless heaters are avoided and because low-grade steam or waste high-temperature water sources can be utilized.

In any system that heats water under pressure, safety precautions are necessary. To minimize the dangers of excessive pressures (which can damage the system) and of superheated water (which can severely injure people), most codes require *pressure and temperature (P/T) relief valves* to be installed on top of all water heaters. Because these valves are designed to release hot water whenever dangerous pressure or temperature levels are reached, they should be attached to a length of pipe that will conduct the released water to a drain.

(i) Solar Water Heating. Solar energy is most attractive to a designer when it will do most of its work during the time that it is most available. On a seasonal basis (Fig. 10.25), solar energy is

Fig. 10.24 *Storage tank and heater for domestic hot water for large-demand applications. Steam coil submerged in tank (see Fig. 10.54). Capacities 100 to 10,000 gph (380 to 37,850 L/h) varying by length of coil, for 140F° (40° to 180°F) temperature rise (78C° rise).*

Hot water outlet

Relief valve

Thermometer

Manhole

Steam inlet

Steam coil

Blow off

Cold water inlet

Drip

Air valve

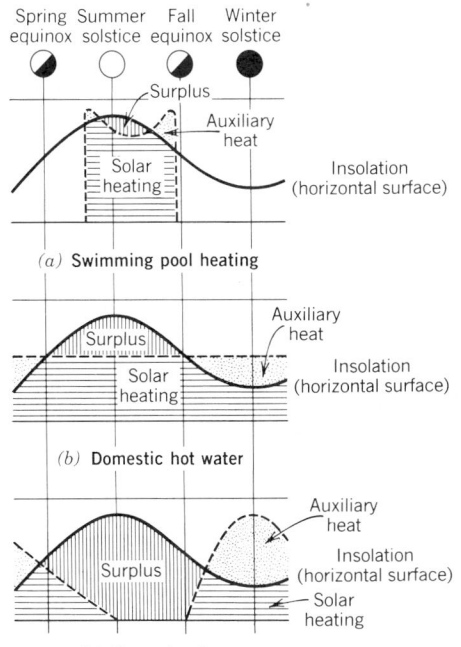

Spring Summer Fall Winter
equinox solstice equinox solstice

Surplus

Auxiliary
heat

Solar
heating

Insolation
(horizontal surface)

(a) Swimming pool heating

Auxiliary
heat

Surplus

Insolation
(horizontal surface)

Solar
heating

(b) Domestic hot water

Auxiliary
heat

Surplus

Insolation
(horizontal surface)

Solar
heating

(c) Space heating

Fig. 10.25 *Comparison of the pattern of supply of solar energy (on a horizontal surface) to various patterns of heating needs.*

especially attractive for outdoor swimming pool heating, as it will be used only in sunny, warm months. Throughout the United States, solar energy can easily meet most of the summertime demand for DHW as well. In the northern part of the country, a much smaller solar contribution can be expected in winter. An especially difficult problem is the winter space-heat mismatch between greatest need and least solar supply. For both solar DHW and space-heating systems, winter brings the added complication of the need to protect against freezing.

Again, there are many ways in which solar energy can be used to heat water. Solar water heating systems are usually classified by the means of fluid circulation ("passive" or "active"), the means by which heat from the collector piping is transferred to the DHW itself ("direct," "indirect," or closed loop), and the means of protection against freezing.

Passive systems rely on gravity for circulation, as was explained in Section 10.6f. Hence, the storage tank usually must be placed *above* the collector (Fig. 10.26) and the number of bends in the

WATER AND WASTE

Fig. 10.26 *Solar water heaters are widely used. This house in Miami, Florida, incorporates a storage tank above the collector, which is enclosed in the chimney-like form. This permits passive (thermosiphon) circulation, without a pump. (Photo by M. Steven Baker.)*

collector's supply and return piping minimized, to minimize friction. Heavy storage tanks located high in a building can cause structural problems. The advantages of this passive approach include lower cost of components, high mechanical reliability (no pump, etc.), and low operational costs.

Active systems use pumps to force the fluid to the collector. Although this setup allows the collector and storage to be located anywhere that is convenient for the designer, it introduces the complications of mechanical breakdown, increased maintenance, and the cost of the energy needed to run the pump. Active DHW systems are widespread in North America; see Figs. 10.27 and 10.28 for typical systems.

Direct systems utilize only one fluid: the water to be heated for use in the building is circulated through the solar collector. Such a system has the advantages of simplicity and efficiency, as it does not require a separate loop of fluid and the attendant piping complications and inefficiencies of heat exchange.

Indirect systems use a closed loop containing a fluid that circulates through the collector and storage tank. The fluid is not mixed with the DHW itself; rather, heat is passed from one fluid to the other through a heat exchanger. One advantage of this system is that it allows nonfreezing solutions to be used in the collector loop. It also allows collectors to be operated at low pressure, rather than at the high pressures typical of urban public water systems. The choice of fluid is determined by its freezing and boiling points, its specific heat, and its level of toxicity.

(a)

(b)

Fig. 10.27 Active solar water heating utilizes differential controllers (a) *that compare the collector temperature to that of the storage tank.* (b) *Typical DHW system with separate solar and conventional heater tanks. (The alternative of only one storage tank produces less stand-by heat loss, but may reduce solar collection by feeding the collector with gas-heated water.)*

Fig. 10.28 (a) *Cutaway section of a typical active solar collector.* (b) *Cross section at header.* (c) *Method of connecting manifolds of adjacent solar collectors.*

As Table 10.11 shows, these design elements (passive and active, direct and indirect) are combined in a number of typical solar DHW systems. Brief descriptions of some of the more common systems follow.

1. *Batch systems.* In these, the simplest of all such systems, a black-painted storage tank is exposed to the sun within a glazed collector box (Fig. 10.29). The price of simplicity is the inefficiency of exposing a tank of hot water to the cold nighttime conditions of the collector. Movable insulation, however inconvenient, can be used to reduce this problem. These heaters are also referred to as "breadbox" water heaters and "integral passive solar water heaters." The batch system was used at the Antelope Valley California Poppy Reserve (Fig. 5.59), as shown in Fig. 10.30. Because of their simplicity, they are favored by do-it-yourself builders, especially in mild-winter climates.

A more recent batch collector utilizes eight copper tubes in a well-insulated, double-glazed (glass out, Teflon in) frame. The water flows in series, allowing the colder replacement water to enter at the bottom of the collector, with the hottest water at the top, ready for use. Insulation between the tubes helps maintain this temperature difference. Collectors are available in storage sizes of approximately 30, 35, 40, and 50 gallons, weighing when filled 425 to 664 pounds (114, 133, 152, and 190 L, weighing 193 to 301 kg).

2. *Thermosiphon systems.* The sun acts as both the "pump" and the heat source for these systems (Fig. 10.31). With no moving parts, maintenance needs are low. The collector, however, must be lower than the tank, and piping must be kept as simple as possible. Because the coldest water remains in the lower collector at night, the hot water in the upper tank is not threatened with

TABLE 10.11 **Solar Domestic Hot Water (DHW) Systems**

Type	Main Features	Advantages	Disadvantages
Batch system	Batch tank inside collector box. Potable water within collector/tank.	No external power. Few components. Collector/tank at any location.	Seasonal; dependent on freezing locations. Heat loss at night from storage.
Thermosiphon system[a]	Flat plate liquid collectors. Normally open loop but no pump or external power (passive DHW system). Storage tank higher than collector.	No external power. Few components. High performance.	Seasonal; dependent on freezing locations (if water is collector fluid). Needs structural support for high storage tank.
Geyser pumping system	Flat-plate liquid collectors. Methanol-water solution passively pumped by solar heat to a heat exchanger below.	No freezing damage. No mechanical or electric parts. No liquid service.	Temporarily stops operation at subzero temperatures, as solution freezes to a slush.
Closed-loop freeze-resistant system[a]	Flat plate liquid collectors. Closed loop of piping from collectors to storage tank. Uses external energy (circulator and differential controller). Uses nonfreezing collector fluid. Pressurized stonelined storage tank.	Can be used in coldest climates. More and better-established competition. Circulator; small consumption of external energy. High performance.	Liquid; service/maintenance required. More components required.
Drain-back system[a]	Flat plate liquid collectors. Water is collector fluid (open loop). Potable water circulates through the heat exchanger in the storage tank (not through the collectors). Large heat exchanger. Pitched headers.	Can be used in the coldest climates. No antifreeze used. Most simple of active flat plate systems (no valves).	Larger pump; larger consumption of external energy. System must drain thoroughly. Use of corrosion inhibitor recommended.

System	Characteristics	Disadvantages
Drain-down system[a]	Flat plate liquid collectors. Potable water circulated through collectors. Line pressure feeds collectors (open loop). Has automatic drainage valves. Pitched headers.	No heat exchanger or extra storage tank needed. High performance.	In some instances a larger pump; larger consumption of external energy. System must drain thoroughly. No corrosion inhibitor possible. Freeze danger with valve failure.
Air-to-liquid system[a]	Flat plate air collectors. Air-to-water heat exchanger. Ductwork and blower. Pipes and circulator. Larger collector area than liquid system.	Won't freeze (dependent on exchanger location). Air leaks won't cause damage. Integrates well with space heating.	Hard to detect leaks. More space required for ducts. Blower and circulator required. More carpentry involved. Less efficient than other systems.
Phase-change system[a]	Flat plate liquid collectors. Freon 114 or R12.[b] Storage tank higher than collectors (passive type). Closed-loop refrigerant-grade piping from collectors to storage tank (passive type) or to condenser (subambient type).	No external power (passive type). Can be mounted at any location (subambient type).	Very hard to detect leaks. Special equipment to install. More components required (subambient type).

[a]Reprinted by permission from Russell H. Plante, *Solar Domestic Hot Water: A Practical Guide*, Copyright © 1983 by John Wiley & Sons.
[b]HCFC or HC refrigerants must be used in new systems.

Hot water pipe
P/T relief valve
Drain valves
Dip tube
Three-valve bypass
Batch tank
Vacuum breaker
Auxiliary water heater
Pressure/temperature relief pipe
Cold water

Fig. 10.29 *Batch solar water-heating system. (Reprinted by permission from Daniel K. Reif,* Passive Solar Water Heaters, *copyright © 1983 by Brick House Publishing Company, Inc.)*

Hot water from tank
Cold water to tank
Hot water to fixture
Mixing valve Max. output 120°F
Water supply from well

Fig. 10.30 *A batch solar water heater supplies the hot water for the Antelope Valley California Poppy Reserve (also described in Fig. 5.59). (Courtesy of the Colyer/Freeman Group, Architects, San Francisco.)*

Fig. 10.31 *Thermosiphon solar water-heating system. (Reprinted by permission from Daniel K. Reif,* Passive Solar Water Heaters, *copyright © 1983 by Brick House Publishing Company, Inc.)*

undue heat loss. However, freezing conditions pose a severe threat to the collector. Accordingly, indirect (closed-loop) systems containing a non-freezing fluid are frequently used; *phase-change systems* are a promising cold-winter option for this type of passive system.

3. *Closed-loop freeze-resistant systems.* In addition to being used in thermosiphoning systems, these are commonly used in active systems (Fig. 10.32). A small pump circulates non-freezing fluid to the collector when there is sufficient sun. This process is governed by a differential controller. Alternatively, a PV-driven circulation pump can be used; when there is sufficient solar energy to activate the pump, there should be enough to also heat water in the collector.

The price of this freeze protection is the inefficiency of the heat exchanger, used between the collectors' fluid and the DHW itself. Some codes require a *double* wall between any toxic nonfreezing fluid and the DHW, which further reduces efficiency.

4. *Drain-back systems* (Fig. 10.33*a*). Although these systems use water as the fluid pumped from tank to collector, this water is not the DHW itself. Instead, the DHW passes through a heat exchanger in the solar storage tank. With the DHW kept out of the collector, the solar collector can operate at lower water pressure. This, however, requires a large heat exchanger, with attendant inefficiency. It also requires care in the design and installation of piping between collector and tank, so that the collector will drain thoroughly. When the controller senses that no solar energy can be gathered, it cuts off the pump, and water drains back into the tank. Therefore, the collector will be filled with air, not water, for all nighttime and cloudy-cold daytime hours. This suggests that corrosion inhibitors should be added to the collector/tank's water because the piping is frequently exposed to air.

5. *Drain-down systems* (Fig. 10.33*b*). These are the only active systems that do not utilize heat exchangers; DHW is circulated directly through

Fig. 10.32 *Closed-loop freeze-resistant solar water heating system. (Reprinted by permission from Daniel K. Reif,* Passive Solar Water Heaters, *copyright © 1983 by Brick House Publishing Company, Inc.)*

the collector. Both higher pressure and higher efficiency result. In this system, the collector is usually filled with water that moves only when the differential controller activates the pump. Whenever the outside temperature drops near freezing, the controller activates solenoid valves and the water in the collector is drained down (dumped). In cold-winter areas, this process could result in several gallons of water wasted per winter day. Although the lack of a heat exchanger is attractive from the standpoint of cost savings and thermal efficiency, the set of electrically controlled solenoid valves is not foolproof. When malfunctions occur in pressurized systems, the loss of water can be enormous, and there is great potential for water damage to the building.

6. *Air-to-liquid systems.* These systems use air collectors and rock storage beds; a heat exchanger transfers heat from the collector's air to the DHW. Much lower efficiencies result, and the ductwork is much more space consuming than are pipes to water collectors. Worse yet, leaks in air collectors or storage beds are very difficult to detect. However, air collectors are not damaged by freezing.

7. *Phase-change systems* (Fig. 10.34). In any of the above-discussed systems, in which DHW is kept out of the collector, the fluid within the collector not only can be freeze resisting, but also can take advantage of latent heat (the considerable amount of heat stored when this fluid vaporizes and released when the fluid condenses). The primary disadvantages are the first cost, the difficulty in detecting leaks, and the resulting threat of CFC or HCFC refrigerant fluids to the environment.

Passive approaches to phase-change materials were outlined in the discussion of thermosiphoning systems. *Subambient* approaches utilize a heat pump and can glean heat even from cold-cloudy ("subambient") conditions. They thus represent a form of solar-assisted heat pump, closely related to those discussed in Section 10.6(*j*).

The *sizing* of solar water heaters usually begins with simple design guidelines. For batch heaters, this sizing guideline (Fig. 10.35) is

from 0.45 to 0.65 ft² glazing per gallon of water stored
(from 0.011 to 0.016 m² per L of water stored)

Air vent

Storage tank
90 gallons

Differential
controller

Hot

Backup
heater

Check
valve

Pump

Cold

(a)

Vacuum breaker

Air vent

Sensor

Temperature
gauge

Cold

Gate valve

Check valve

Temperature
and pressure
relief valve

Hot

Mixing valve

Normally open
solenoid valve

Normally closed
solenoid valve

Dip
tube

Auxiliary
elements

Disconnect

Drain

Ball valve

Differential
thermostat

Drain

Gate valve

Sensor

Normally closed solenoid valve

(b)

Fig. 10.33 *Comparison of* (a) *drain-back solar water-heating system with* (b) *drain-down system. The drain-back system's performance can be increased by the addition of the check valve that allows the colder water at the bottom of the storage tank to circulate into the heat exchanger. The drain-down system is best used in mild-winter climates with infrequent freezing temperatures, since the water in the collector is dumped with each freezing threat. (Adapted by permission from Russell Plante,* Solar Domestic Hot Water, *copyright © 1983 by John Wiley & Sons.)*

Cold water in

Coolant heat exchanger

PTR valve

Hot water in

Hot water out

Backup water heater

Fig. 10.34 *In a phase-change system, small heat exchangers can be built into the tops of thermosiphoning collectors; DHW is drained down when the threat of freezing arises. (Adapted by permission from* Solar Age, *November 1983.)*

Fig. 10.35 *Design guideline to the sizing of batch solar water heaters. In this example, a 40-gal storage tank should be contained within a collector that has between 18 and 27 ft² of south-facing glazing. Or a 50 ft² collector box should contain a tank of between 75 and 113 gal. (Adapted from Daniel K. Reif,* Passive Solar Water Heaters, *copyright © 1983 by Brick House Publishing Company, Inc. Reprinted by permission.)*

For the flat-plate solar collectors used in all the other solar DHW systems listed above, the sizing guidelines are:

> 12 to 25 ft² collector/person (residential)
> (1.1 to 2.3 m²/person)
> optimum tilt (up from horizontal) equal to latitude (or less)
> 1 to 1.5 gal of storage per ft² collector area
> (40.7 to 61 L per m² collector area)

This collector-sizing guideline is for *residential* hot water usage, including cooking and bathing. For warmer climates with ample insolation, use the lower figure; this will supply about half of the hot water on an annual average basis. (For *nonresidential* DHW, adjust the sizing guideline by comparing gallons of hot water per person per day used in residences to those used in the nonresidential function for which you are designing.)

A more detailed sizing procedure would consider the collector's expected heat contribution for some typical months. Such a procedure requires both weather data and assumptions about water temperatures and other values. As a result, detailed answers are best obtained from computer programs, such as those listed in Appendix H. For those interested in something more than sizing guidelines, but less than computer programs, consider the following approach to collector sizing (adapted from Brown, Reynolds, and Ubbelohde, *Inside/Out: Design Procedures for Passive Environmental Technologies,* copyright © 1982 by John Wiley & Sons).

STEP 1. *Select the collector tilt angle.* This can be done by consulting Tables B.11 to B.14, which present clear-day insolation values for various tilt angles. Appendix B, Table B.15, presents average insolation values on horizontal surfaces, which should also be checked. The tilt angle selected should be close to the optimum angle for the best month for total insolation.

STEP 2. *Check the collector efficiency.* This should be done at least twice: for the best insolation month and for the worst month. This step requires the following data:

Hourly insolation on tilted surface (Appendix B, Table B.11 to B.14).
Outdoor average temperature (Appendix B, Table B.15).

The hourly insolation values on tilted surfaces in Table B.11 to B.14, are for *clear* days. For most locations, this should be adjusted for *average* conditions, which are shown for vertical (south) and horizontal surfaces in Appendix B, Table B.15. Because for most of North America the optimum yearly DHW tilt angle will be closer to horizontal than to vertical, a simple correction to insolation must be made:

Hourly average day
insolation on
tilted surface

$$= \begin{bmatrix} \text{hourly clear-day} \\ \text{insolation on} \\ \text{tilted surface} \end{bmatrix} \begin{bmatrix} \dfrac{\text{average-day total}}{\text{horizontal insolation}} \\ \dfrac{\text{clear-day total}}{\text{horizontal insolation}} \end{bmatrix}$$

This step also requires assumptions about the input temperature of the water supplied to the collector, or T_i. In summer, this can be quite high—even higher than the thermostat set-point temperature of the hot water tank. In winter, it is likely to be lower, by perhaps 10 to 20F° (5.5 to 11C°), than the thermostate setpoint temperature. Finally, this step requires a choice of collector type, so that efficiency can be determined. Figure 10.36 shows the efficiency curves for a variety of solar collectors.

Note that the simplest of all, the unglazed flat black collectors, have the highest efficiency of all collectors at combinations of very low $T_i - T_a$ and very high I_o. These conditions are typical of summer days, on which the water is to be heated only a few degrees. This corresponds to swimming pool applications, and is one reason why such collectors are so popular for that purpose; low cost is another reason.

For the opposite conditions—those of winter space heating—more elaborate collectors are appropriate. Evacuated-tube collectors have the ad-

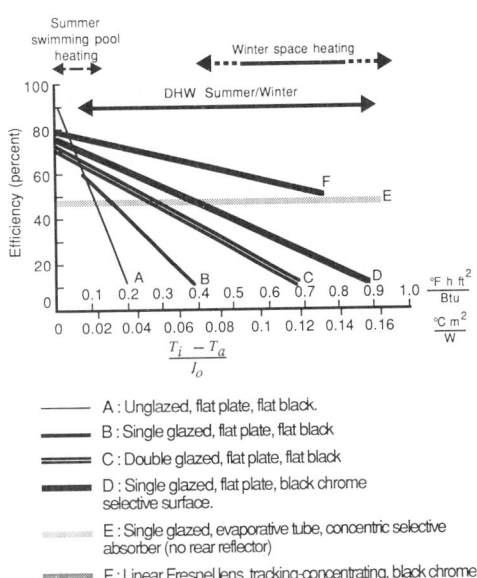

A : Unglazed, flat plate, flat black.
B : Single glazed, flat plate, flat black
C : Double glazed, flat plate, flat black
D : Single glazed, flat plate, black chrome selective surface.
E : Single glazed, evaporative tube, concentric selective absorber (no rear reflector)
F : Linear Fresnel lens, tracking-concentrating, black chrome selective surface

Fig. 10.36 *Collector efficiencies depend on insolation (I_o), water temperature entering the collector (T_i), and ambient (air) temperature (T_a). (Based on the U.S. Dept. of Housing and Urban Development's* Intermediate Minimum Property Standards Supplement, *1977.)*

vantage of nearly eliminating collector heat loss by convection. The Fresnel lens and/or tracking-concentrating collectors increase the incoming solar energy per unit area, whereas heat loss increases only slightly (due to higher ΔT). Both these approaches are expensive, and tracking collectors require added maintenance for the additional motion. Selective surfaces represent a simple, relatively cheap improvement over flat black collectors (compare D to B); they absorb solar energy just as well but emit radiant energy at a vastly lower rate, thus cutting collector heat loss by radiation.

STEP 3. *Approximate system efficiency.* Once the hot water leaves the collector, it must be led back to storage. In indirect systems, it must go through a heat exchanger. The tank will lose some heat; heat losses also occur as the cooled water is led back to the collector. Another "loss" to be considered is noncollection time. Estimates of collector performance are based on total daily insolation, yet the first few and last few hours of daylight generally will *not* bring enough insolation to warm the collector. Therefore, some reduction should be made in daily insolation totals when performance is estimated. All these factors can be accounted for by assuming a lower system efficiency. A perhaps optimistic assumption is

$$\text{system efficiency} = 0.8 \text{ (collector efficiency)}$$

STEP 4. *Determine the total heat needed for DHW.* Estimates of the quantity of hot water needed were made in Section 10.6*h*. Once the total gallons (or liters) per day are known, and the desired storage temperature determined, the heat needed for the daily supply of DHW is easy to calculate. In conventional units,

$$Q = 8.33 \text{ (gpd)} \, (T_s - T_g)$$

where Q = daily heat need, Btu
        8.33 = weight of water, pounds per gallon $\times$ Btu/lb-°F
        T_s = storage temperature, °F (see Table 10.7)
        T_g = groundwater temperature, °F (see Fig. 5.58, well-water temperatures, page 324)

In SI units,

$$Q = 1.16 \text{ (L/d)} \, (T_s - T_g)$$

where Q = daily heat need, W h
        1.16 = weight of water, $\times$ W h/L °C
        T_s = storage temperature, °C
        T_g = groundwater temperature, °C

STEP 5. *Determine the desired percentage of total heat to be solar for DHW.* There are several approaches to this problem. One common strategy is to provide 100% of the hot water needs in the best insolation month. A closely related strategy is to choose the *yearly* percent solar desired, then provide that percentage in the "average" month (such as March). Whatever strategy is chosen, a month and its percent solar must be identified.

STEP 6. *Size the collector.* This is done by combining the preceding steps, as follows.

collector area

$$= \frac{\text{daily heat need} \times \text{percent solar desired}}{\text{daily insolation} \times \text{system efficiency}}$$

This may be done for several months, as a check on optimum collector size.

EXAMPLE 10.4. Determine the approximate size of the solar collectors needed to serve the women's dormitory described in Example 10.3. Assume the location to be Springfield, Illinois.

STEP 1. *Tilt angle.* Refer to Appendix B (Table B.15). Springfield's latitude is 39.8° N. In Table B.15, average insolation is given for July. Use this month as the "best" month.

From Table B.12, the optimum tilt angle for July at 40° N latitude is horizontal. However, at 30° tilt (latitude $-$ 10°), much better performance will be obtained in "average" months, such as March and September. Choose a 30° tilt angle.

STEP 2. *Collector efficiency.* For the best month (July), the best hourly clear-day insolation is 307 Btu/h-ft², at 30° tilt, in the hour centered at noon. To adjust this for *average* hourly value, calculate

$$307 \times \frac{\begin{array}{c}2058 \text{ Btu/day average} \\ \text{horizontal (Table B.15)}\end{array}}{\begin{array}{c}2534 \text{ Btu/day clear} \\ \text{horizontal (Table B.12)}\end{array}} = 249 \text{ Btu/h}$$

Outdoor average temperature is approximately obtainable from Table B.15; TA for July is 76F°. This is quite conservative, since it is the average daily temperature; the temperature at noon should be higher. From Appendix A, the mean daily range for Springfield is 21F°. Thus 76 + 21/2 = about 86°F. The thermostat setpoint temperature of the water storage tank is assumed at 115°F. This allows adequately hot water for most dormitory usages, although additional added heat will be required for the cafeteria's dishwashing.

Assume, because it is summer, that T_i is at the setpoint of 115°F. (In winter, it would be lower). Therefore, the quantity $(T_i - T_a)/I_o$ can be calculated at

$$\frac{115 \text{ F} - 86 \text{ F}}{249 \text{ Btu/h}} = 0.116$$

Enter Fig. 10.36 with this number, and assume a single-glazed, flat plate selective surface collector (type D). The efficiency will be about 70% at these conditions.

STEP 3. *System efficiency.* Use the design guideline:

$$0.8 \ (70\% \text{ collector efficiency}) = 56\%$$

STEP 4. *Total heat needed.* The average usage in gpd can be estimated from Table 10.10. For women's dormitories, the rate is 12.3 gal/student × 300 students = 3690 gal. For the cafeteria, the average rate (type A) is 2.4 gal/meal = 2.4 × 300 = 720. Total gallons needed: 3690 + 720 = 4410.

Groundwater temperature for Springfield (from Fig. 5.58) is about 56°F. Assume that solar energy will be used to raise it to 115°F.

$$Q = 8.33 \ (4410 \text{ gpd})(115 - 56) = 2,167,380 \text{ Btu}$$

STEP 5. *Desired percent solar.* For this best hour in July, assume 100% solar contribution.

STEP 6. *Collector size.* Average daily insolation can be obtained from Appendix B, as before:

clear day total, 30° tilt, July: 2409 (Table B.12)

clear day total, horizontal, July: 2534 (Table B.12)

average day total, horizontal, July: 2058 (Table B.15)

average July daily insolation, at 30° tilt = 2409 × 2058/2534 = 1956

$$\text{collector area} = \frac{2,167,380 \text{ Btu} \times 100\%}{1956 \text{ Btu/ ft}^2 \times 56\%}$$

$$= 1979 \text{ ft}^2$$

This size should be checked against average-month performance and adjusted as desired. Note that this is a ratio of about 1979/300 = 6.6 ft^2 of collector per student—lower than the rule of thumb for typical residential DHW systems.

Note also that at the design guideline rate of 1 to 1.5 gal storage per ft^2 of collector, a tank size of 2000 to 3000 gal is indicated. In Example 10.3, the minimum recovery rate needed a storage tank of 8150 gallons, and the fastest recovery rate needed 3010 gallons. Either seems sufficient for this size of collector. ∎

Swimming pool heating is an especially attractive application for solar energy. A common de-sign guideline for collector sizing is

$$\textit{collector area} = \textit{0.5 pool surface area}$$

For summer ambient temperature operations, un-glazed collectors are both the best performing and the least expensive, as explained for Fig. 10.36. Another important consideration is a pool cover, which will not only conserve the pool's heat, but also conserve water lost by evaporation. On very hot-dry days, water losses can reach 100 gpd from a 20 × 40-ft pool (380 L/day from a 6 × 12 m pool).

(j) Heat Pump Water Heaters. The heat pump's use of the compressive refrigeration cycle was explained in Section 7.5 (also see Fig. 7.38). Air-to-water heat pumps are also used for DHW, as shown in Fig. 10.37. Because these devices re-move heat from the air, they are usually installed either in normally overheated spaces (such as res-taurant kitchens) or in unheated spaces such as garages or basements. (It can also remove heat from exhaust air, as explained in Chapter 6.) The spaces that contain these units will be cooled and dehumidified by them. Heat pumps require only a little more space than that required by the simple hot-water storage tanks they serve. Some noise is created by the compressor and the fan that moves air across the evaporator.

Heat rejected from any refrigeration unit can be used for heating water, via the heat pump. Applications include ice making machines, refrig-erated display units in grocery stores, walk-in freezers, and many others. Whenever constant refrigeration loads are present, there is the oppor-tunity to utilize waste heat.

Graywater is another opportunity for the heat pump. Where filtered graywater is collected for recycling on-site (as for irrigation systems), a wa-ter–water heat pump can heat the DHW storage tank as it lowers the graywater tank temperature. See Chapter 11 for graywater system design.

10.7 Fixtures and Water Conservation

Building design and fixture choice can affect water and energy consumption over the life of a build-ing. It is the users who turn the faucets on and off, but the designer who can encourage resource

Fig. 10.37 *Heat pump (air-water) used for heating DHW. (Adapted by permission from* Specifying Engineer, *October 1983. Copyright © 1983 by Cahners Publishing Co.)*

conservation through initial decisions, here centered largely in kitchens and bathrooms.

Visible consumption is one strategy to encourage the users of fixtures to conserve water. We see, hear, feel and sometimes taste water as it issues from the fixture. If we could see *how much* water is involved in each use, we could more readily alter our pattern of use to conserve water where appropriate. Rainwater storage tanks outside bathroom windows, used to flush toilets, might have a visible indicator of the water level; the lower this level, the more that conservation is an imperative. The water-level indicator might be as simple as visible condensation on the tank's outer surface. Small, transparent tanks of clean supply water could be placed above lavatories, filled only when empty by means of a simple float valve. The user might thus be encouraged to use only a portion of the water already in the tank. Sufficient pressure at the fixture is achieved by such a tank's elevation above the fixture. The designer might also consider making the quantity of used (waste) water visible, but such water is often unsightly and the resulting quantity is "after the fact"—too late to conserve.

Audible consumption is another strategy, calling attention to the flow of supply (or waste) water. This is achieved most simply by slightly undersizing water supply pipes so that the water velocity becomes audible (about 10 fps, or 3 m/s).

This strategy should be carefully considered; it is one thing to inform the water user, another to annoy. Audible flow is perhaps best justified on lines to exterior hose bibbs, where irrigation or car washing hoses may inadvertently be left on.

Elimination of bodily wastes accounts for a significant portion of residential water usage (see Chapter 11), and bathing accounts for a significant portion of residential hot water usage (see Table 10.7). Choice of fixtures can have an impact on water conservation, and energy conservation where hot water is involved. The bathroom is thus a site for potential resource conservation.

(a) Physiology, Psychology, and Fixtures. One of the more challenging aspects of bathroom design is the frequent conflict between physiological design criteria and such psychological influences as cultural attitudes about bathroom activities. The physiological criteria change as slowly as evolution. The cultural attitudes can change very rapidly, even within a generation. The bathroom supports two primary human activities: cleansing and elimination. Today's common attitudes about these closely linked activities are that one is "clean," the other "dirty." One might be discussed (although rarely) among friends; the other is simply not fit for conversation. This conflict in attitudes influences the design of our plumbing fixtures. The toilet is the fixture used

for elimination; it is the logical place to provide for the cleansing of the perineal zone that should immediately follow elimination. (This is especially true of public toilets, where stalls close off toilets from lavatories.) Yet very few toilets presently incorporate the cleansing feature: the mixture of cleansing and elimination in the same fixture too often seems abhorrent.

The brief summary presented in this section is based largely on a particularly revealing (and entertaining) study of the struggle between physiological and cultural criteria in fixture and bathroom design—Alexander Kira's *The Bathroom: Criteria for Design*, an expanded edition of which was published in 1976 by Viking Press.

(b) Lavatories. A key issue here is the contrast between running water and a standing water body. Lavatories are used primarily for the cleansing of hands, face, and teeth—activities done quickly with running water that is wasted directly, rather than being collected. Most lavatories are designed as collection bowls—perhaps a reflection of the days when washbasins of standing water, drawn and heated elsewhere, were brought to the bathing place. Most lavatories also have fittings that project out over the sink. Although these fittings have slowly evolved into very sleekly designed objects, most of them still dump running water directly into the drain, are hard to use as drinking fountains, and can be wounding to those who try to wash hair in the lavatory. In considering the prevalence of running as opposed to standing water, the role that running water could play in keeping lavatory surfaces clean, and the need for a drinking fountain when teeth are brushed, Kira proposed a quite different lavatory design (Fig. 10.38).

At full flow, lavatory faucets typically deliver 4 to 5 gpm (0.25 to 0.3 L/s). Newer, low-flow faucets utilize a variety of devices to function as well (or better) with less water. Such devices include aerators (which add air bubbles to the stream, making it splash less and appear larger), flow restrictors, and mixing valves to control temperature. The lower flows achieved range from ½ to 2½ gpm (0.03 to 0.16 L/s). Another promising development is the foot-operated faucet, which frees the hands from having to control water flow, thus saving a few seconds of flow during each

Fig. 10.38 *Proposed lavatory that exploits running water. The water issues from a fountain-like stream, for ease in drinking and in hair and face washing. The stream strikes the lavatory bowl in such a way as to minimize splash outside the bowl, yet sets up a self-cleaning swirling action. A small repository at the back of the bowl, over the drain, can serve for standing water when desired. (From* The Bathroom: New and Expanded Edition, © *1966, 1976 by Alexander Kira. By permission of Bantam Books, a division of Random House, Inc. All rights reserved.)*

lavatory usage. These devices could be particularly helpful at kitchen sinks, where extensive washing of objects takes place.

(c) Whole-Body Cleansing. The running/standing water contrast is even more clearly illustrated here. In cleansing, the ordinary sequence is *wet, soap, scrub, and rinse.* Standing water is good for wetting, and very good for soaping/scrubbing; running water is superior for rinsing. The psychological aspect enters here in common attitudes toward tubs and showers. Tubs are often seen as places to relax in, to spend more time in, perhaps to read in. Showers are viewed as quicker, "no-nonsense," stand-up places. Yet each could benefit from some features of the other.

Tubs should be designed so that the reclining body is supported at the back; this requires a contoured surface (Fig. 10.39) rather than the ordinary straight-line design. It also requires braces for one's feet because the body will otherwise tend to float up and away from such a backrest.

ANGLES AND LENGTHS OF TANGENTS:
102 (4) @ 80°   254 (10) @ 40°   178 (7) @ 55°   64 (2½) @ 70°
102 (4) @ 75°   229 (9) @ 32.5°   190 (7½) @ 45°   76 (3) @ 62.5°
102 (4) @ 70°   203 (8) @ 25°   203 (8) @ 35°   89 (3½) @ 55°

(a)

(b)

These tubs can be designed to accommodate persons of various leg lengths, as shown in Fig. 10.39*b*. There must also be a seat, to allow most of the body a chance to be out of the water and to facilitate safe entry/exit from the tub. Especially needed is a handheld shower for the final rinse; soapy standing water leaves a scummy film on both people and fixtures.

Showers may seem very "efficient," but cleaning would be more thorough and safe if the bather could turn off the water and sit for at least part of the soap/scrub activity, especially for the lower legs and feet. Showers with integral seats are now common.

Another consideration is the location of water controls (fittings). The user should be able to reach them easily from outside the tub or shower without wetting his or her arm, but also able to manipulate them from within the tub/shower even if temporarily blinded by soap.

Shower heads have been notorious for encouraging prodigal water usage; typical flow rates of 6 gpm (0.4 L/s) and maximum rates of 12 gpm (0.7 L/s) once were common. Even in a "short" (5-minute) shower, this rate of use could consume as much as 60 gal (227 L) of water, much of it heated. Many codes now require a limitation on showerhead flow; a fixture of 2.5 gpm (0.2 L/s) is common. These flows can be designed into the shower head, or they can be achieved by cheap, simple flow restrictors in retrofit applications. Some utilities distribute flow restrictors free of charge. Most bathers notice no difference, either in enjoyment or in cleansing, when using restricted-flow shower heads.

Fig. 10.39 *Some considerations for tub design, to facilitate relaxation during whole-body cleansing. (a) A contoured backrest allows comfortable reclining, with support to both the shoulder and lower back. Curve 2, with a median angle of 32.5°, promises comfort for most users. (b) Various tub plans, allowing long- and short-legged persons to brace their bodies against the contoured backrest. Raised seats are also provided, to facilitate safe entry/exit. (From* The Bathroom: New and Expanded Edition, © *1966, 1976 by Alexander Kira. By permission of Bantam Books, a division of Random House, Inc. All rights reserved.)*

(d) Elimination. Several conflicts arise with toilets: the already-mentioned difficulty of combining cleansing with elimination, the issue of pure (high-grade) water for impure (low-grade) purpose, and the conflict over the height of the toilet. A *lower toilet* is definitely of benefit to the average person, who will achieve far better bowel evacuation in a full squatting position. If combined with a toilet seat contoured to best support the body during defecation, the proposed toilet in Fig. 10.40a would be physiologically superior to most of today's fixtures. There are problems with low height. The typical male stands while urinating, and the lower toilet presents a more difficult target. This can have serious maintenance consequences unless males can be induced to sit or a separate urinal is provided—an unlikely option for residential bathrooms. (In our culture, urinals look "institutional"; they are also expensive.) Another problem with the lower toilet is that the elderly and some handicapped people will have difficulty getting on and off the seat (see later Fig. 10.51). Yet another problem arises from the fact that toilets are considered to be seats (for

(a)   (b)

Fig. 10.40 *Two approaches to WC (toilets) that encourage better posture during defecation and more thorough perineal cleansing.* (a) *A low toilet allows the best (full-squat) position.* (b) *The higher toilet is easier for the elderly and handicapped and better intercepts the urine from a standing male. (See Fig. 10.43 for a separate urinal design.) Both these toilets eliminate the separate toilet seat and its maintenance problems; both use small amounts of energy to warm the toilet at its point of contact with the body and to heat the water for perineal cleansing. Such cleansing is also the reason for the larger front and back openings. The pushbuttons are labeled F_T for the flush, and B for the bidet (cleansing) function. (From* The Bathroom: New and Expanded Edition, © *1966, 1976 by Alexander Kira. By permission of Bantam Books, a division of Random House, Inc. All rights reserved.)*

reading, toenail clipping, etc.). The low, squat-inducing toilet is decidedly not at a comfortable chair height.

As a result, a *higher toilet* with otherwise similar features is proposed (Fig. 10.40b). Note that both of these toilets feature openings whose shape differs from that of the conventional oval. This hourglass-shaped opening provides proper support for the body during defecation, and it allows a more generous opening, front and back, for proper perineal cleansing by hand. Another change is that there is no separate toilet seat; instead, these fixtures incorporate electric resistance heaters that warm the small portion of the toilet that contacts the body. Thus, the toilet becomes a consumer of energy—which is also used to heat the water that is provided for perineal cleansing. For those who insist on a separate seat, or for ordinary posture on conventional toilets, a physiologically sound contoured toilet seat is available (Fig. 10.41).

It is especially important that perineal cleansing be encouraged by the designers of bathrooms. The typical dry toilet tissue provided usually is not adequate for this task. In small, private bathrooms this problem is easily solved by placement of the toilet adjacent to the lavatory (or tub), where toilet tissue can be wetted with clean water. In public toilets separated by stalls, perineal cleansing requires either a clean water source built into the toilet, or a separate *bidet* within the stall—a highly unlikely provision, given the extra floor space and the cost of the bidet. Again, cultural issues arise: bidets are quite common in the private bathrooms of Europe, but a rarity in North America. Two common alternatives are (1) to build a cleansing source into the toilet itself (Fig. 10.42) or (2) to construct a toilet seat that provides for such cleansing. Several American manufacturers offer such seats, which can be easily adapted to existing toilets. Using recycled water in a toilet that also offers perineal cleansing complicates the high-grade water issue. Codes often require recycled water to be colored (blue or green), an esthetic quality unappealing for skin contact even in one's nether regions.

The *urinal* is an answer to the problem posed by lower toilets to males, who stand while urinating. Although urinals are culturally acceptable in public toilet rooms, they have never achieved a place in private bathrooms. One solution proposed by Alexander Kira is to build the home urinal into the wall (Fig. 10.43a); it could be pulled out for use, at the optimum mounting height, and at an optimum receiving shape to eliminate backsplash onto the user. When pushed back into the wall, it would flush automatically. A less-satisfactory solution would be to modify the toilet by raising its back surface (Fig. 10.43b).

Fig. 10.41 *Toilet seat for a conventional WC, which encourages better posture for defecation and somewhat better access for perineal cleansing. (Courtesy of American Standard, Inc.)*

(a)   (b)   (c)

Fig. 10.42 *Provisions for perineal cleansing built into toilets. (a) Section through the toilet before/after cleansing function. (b) At the initiation of perineal cleansing, a pipe extends and emits a controlled spray of warm water. (c) After the pipe is retracted, a warm jet of air issues from just below the seat. (From* The Bathroom: New and Expanded Edition, *© 1966, 1976 by Alexander Kira. By permission of Bantam Books, a division of Random House, Inc. All rights reserved.)*

(d)

Fig. 10.42 (d) *An adjustable-intensity spray at warm (body) temperature continues for as long as a button is depressed, in this Geberit ShowerToilet. (Courtesy of Geberit Manufacturing Inc., Michigan City, Indiana.)*

(a)   (b)

Fig. 10.43 *Designs to accommodate male urination. (a) A built-in, tilt-out urinal for the home, visible only when in use. (b) A higher back for the toilet. A bidet for cleansing is shown to the right of the toilet. (From* The Bathroom: New and Expanded Edition, © *1966, 1976 by Alexander Kira. By permission of Bantam Books, a division of Random House, Inc. All rights reserved.)*

Fig. 10.44 *Conventional flush tank toilets.* (a) *Two-piece flush tank toilet with elevated, wall-mounted tank and pull chain.* (b) *Two-piece flush tank toilet with tank bolted to bowl ("close-coupled").* (c) *One-piece flush tank toilet with integral tank. (From Milne, 1976.)*

For the toilet, or water closet (WC), there are four major categories of fixtures, depending on the amount of water used per flush. The *conventional* toilet uses 3.5 gal (13.2 L) or more per flush; the *watersaver* toilet uses from 1.7 to 3.5 gal (6.4 to 13.2 L) per flush; the *low-consumption* toilet (also called ultra-low flush) uses 1.6 gal (6 L) or less per flush, and the *waterless* toilet uses no water at all.

(e) Conventional Toilets.

[Use 3.5 gal (13.2 L) or more per flush.] These toilets may no longer be legally installed in the U.S., due to their high water consumption. They are still widely encountered in existing installations. In the common toilet, a sudden deluge of water removes human waste, and simultaneously helps cleanse the toilet. Fast-moving water requires pressure and makes noise. The older flush-tank toilet (Fig. 10.44a) stores a smaller quantity of water (about 2.5 gal, or 9.5 L) at sufficient height above the toilet bowl to achieve fast flow. Although it uses minimal amounts of water, it is noisy and its elevated tank has a maximum visual impact. The more common (in North America) version is the two-piece toilet with the tank bolted to the bowl (Fig. 10.44b). With much less pressure available, this toilet requires 5 to 7 gal (19 to 26 L) per flush, but it is quieter. Newer "shallow trap" models reduce the

quantity needed to 3.5 gal (13.2 L). The maintenance problem posed by the seam between tank and toilet is eliminated in the one-piece toilet (Fig. 10.44c). The cost of its low profile is an even greater need for water: 6 to 8 gal (23 to 30 L) per flush.

The fourth common alternative is the flush-valve toilet, which depends on the building's water pressure rather than its own stored water. A noisy system requiring very high flow rates, it is seldom found in residences. However, it requires as little as 3 gal (11.4 L) per flush.

In addition to differing tank and bowl combinations, there are varieties of types of flushing actions. Four common flush toilets are compared in Fig. 10.45.

Washdown toilets are no longer made in the United States, although many are still in use. This is the noisiest toilet, and the most likely to become plugged because it has the smallest-diameter trap. It was usually found in the two-piece flush tank toilet; with an elevated tank, the flush required was only about 2.5 gal (9.5 L).

Siphon jet toilets are in widespread use in North America, particularly in residences. A small priming jet hurries the bowl's contents along into the trap, and hastens the siphon action. With the elevated tank (two-piece) toilet, this process requires a flush of about 3.75 gal (14.2 L). With the

Washdown flushing action

(a)

Siphon jet flushing action

(b)

Siphon vortex flushing action

(c)

Blowout flushing action

(d)

more common close-coupled two-piece toilet, it requires a flush of from 5 to 7 gal (19 to 26.5 L). Siphon jet toilets are sometimes equipped with flush valves, which use less water (see the discussion of "Blowout," below).

Siphon vortex toilets are especially suitable for low-velocity water (often as a result of low pressure). They are therefore also the quietest, making them a favorite wherever bathrooms are adjacent to sleeping areas or other acoustically sensitive spaces. The water enters the bowl off-center in such a way as to form a vortex; this swirling action cleans the sides of the bowl and the trap, helping the siphon action in emptying both bowl and trap. The one-piece flush tank toilets usually have the siphon vortex flushing action, which typically requires 6 to 8 gal (22.7 to 30.3 L).

Blowout toilets combine very high velocity water and a simple trap to offer a noisy but very low maintenance toilet dependent on flush valves rather than tanks. They are very common in commercial and institutional toilet rooms, where large water supply lines and high pressures are available. The high velocity of the water lowers the quantity required from 3 to 4 gal (11.4 to 15.1 L) per flush.

Architects could easily specify toilets that require less water per flush and yet have very little impact on the user. With effort on the part of both users and architects, water can be recycled, or rainwater collected, so as to supply toilets with lower-quality water.

(f) Watersaver Toilets. [Use from 1.7 to 3.5 gal (6.4 to 13.2 L) per flush.] Although these newer toilets use much less water than conventional ones, they may not meet the more strict water limits now established in so many water-conscious states and municipalities. The U.S. requires today's toilets to not exceed 1.6 gal (6 L) per flush. Many of these toilets represent a compromise between the conventional "water waster" and the noisy operation and/or smaller water surface area that typifies so many low-con-

Fig. 10.45 *Four common flushing actions that are built into toilets. Newer flush toilets typically use either type* (b) *or type* (c); *flush valve toilets use either type* (b) *or type* (d). *(From Milne, 1976.)*

WATER AND WASTE

sumption models. Watersaver toilets tend to use a conventional flushing action, hence need enough height between the tank and the bowl to provide sufficient water pressure within the bowl during the flush.

Toilets can be designed to use less water by varying several characteristics: the pressure behind the water entering the bowl (generated either by the height of the tank above the bowl or by utilizing increasing pressure from the water supply system or another source), the shape of the bowl, the way in which the bowl fills and empties, and the trap configuration. Watersaver toilets differ somewhat in most of these characteristics from the conventional toilet; more radical changes are needed to achieve the low-consumption toilet.

(g) Low-Consumption Toilets. [Use 1.6 gal (6 L) or less per flush.] A heated controversy over these ultra-low-flush WCs developed in the late 1990s due in part to steep price increases and consumer discontent over performance. It appears that the law got ahead of the technology for a time, resulting in repeated low-consumption flushes to accomplish what one higher-consumption flush could do. However, to remove fecal matter, two 1.6 gal flushes still about equal the old 3.5 gal flush; to flush urine, one 1.6 gal flush is adequate. Manufacturers are finding ways to change the entry point of water into the toilet bowl to increase swirl and thus waste removal power.

Several of these toilets achieve water conservation by using a *flushometer tank* (Fig. 10.46). Satisfactory flushing can be achieved with much less water by flushing with a smaller amount of water entering at greater pressure. Instead of toilet tanks at ordinary air pressure, these tanks utilize water supply system pressure to compress the air trapped within the tank. Water enters the bowl with much greater force with this combination of water under system pressure and compressed air. The flush is thorough, quick, and noisy. If the water supply pressure is greater than about 65 psi (448 kPa), some problems with excessive tank pressures may be expected. A pressure-reducing valve would be helpful in such cases.

Another water-saving technique is to redesign the bowl to hold less water. The siphon jet, gravity-fed toilet whose tank is shown in Fig. 10.47

Fig. 10.46 *Example of a 1.5-gal-flush, pressure assisted, direct-fed siphon jet flush action flushometer tank, yet with a larger 10 × 12 in. (250 × 300 mm) water surface area in the bowl: the Cadet Aquameter. (Courtesy of American Standard, Inc., Piscataway, New Jersey.)*

exposes a standing water surface of about 4½ × 6 in. (114 × 152 mm), much smaller than that of Fig. 10.46. Along with a very low 1.4 gal (5.3 L) flush, its trap is wide enough to pass a 2⅛-in. (54-mm) ball.

Far lower water consumption is achieved when a central compressed air system is combined with water supply system pressure. The Microphor flush toilet (Fig. 10.48) uses compressed air and 2 qt (1.9 L) of water per flush in a two-chambered toilet. When the flush lever is pressed, the water and waste in the bowl (upper chamber) is deposited into the secondary (lower) chamber, and 2 qt of water (direct from the supply pipe) washes down the sides of the bowl to await the next user. The now-closed secondary chamber then is pressurized with compressed air, and its contents deposited into the conventional sewer line.

Air compressors are needed, along with compressed air lines. A small compressor (¼ to ½ hp) (187 to 373 W) with accompanying air tank will operate up to three toilets. Although the compressor is noisy, the toilets themselves are no more

Fig. 10.47 A very low flush is achieved with a reservoir that meters about 1.4 gal (5.3 L) of water from the tank to the bowl during flushing. The standing water surface is about 4½ by 6 in. (115 by 150 mm). The Ultra-One/G. (Courtesy of Eljer Industries, Plano, Texas.)

Fig. 10.48 (a) Compressed air is combined with a 2-quart (1.9 L) flush in the two-chambered Microphor toilet. (b) Section through the floor-discharge model. (c) Section through the wall-discharge model. (Courtesy of the Microphor Company, Willits, California.)

(a)   (b)

(c)

Series 200 Flush

Series 1000 & 2000 Flush

(d)

(e)

noisy than conventional toilets; they use the same plumbing lines. With such low flows, the designer may choose to increase the slope of waste lines; however, the 2-qt quantity has proven sufficient to carry the waste in existing installations.

In contrast, the Envirovac flush toilet (Fig. 10.49) uses a vacuum and 1.5 qt (1.4 L) per flush. Because a central sewage tank (kept under vacuum by a pump) must be used, the sewer line from the toilet to this tank may run horizontally without a slope, or even run vertically upward from the toilet. This can have significant architectural advantages when vertical clearance is tight, toilets are being added within existing structures, or toilets must be located below the level of the public sewer, as in marinas or basements. In tall buildings, significant savings in power for water supply pumping are achieved. Easier approval for building permits, reduced water hookup fees, and lower monthly sewer charges can be substantial benefits. Furthermore, the central sewage holding tank can be flushed into the sewer at off-peak times—a benefit to the treatment plant. These water-saving, architectural, and water treatment advantages must be weighed against the cost of the toilets and the tank/pump combinations, the space required for the tank/pump, the power used by the pump, and the possibility that a power failure will halt system operation.

Vacuum systems can be installed for groups of buildings, such as subdivisions. Small pipe sizes, freedom from the need for continuous-sloping sewer lines, and conservation advantages for both water supply and treatment are the benefits. In addition, "graywater" (from kitchen, laundry, and bathing fixtures) can be kept separate from toilet water ("blackwater"), and thus made easier to recycle after moderate treatment.

(h) Flushing Controls. Water conservation is encouraged by the *dual cycle* toilet, whose flushing mechanism allows a choice of fewer gallons for liquid wastes, more gallons for solid wastes. More common outside the United States, this simple mechanism's handle is pushed up for liquid flushing and down for solids.

Another development is the *automatic flush*, triggered by radiant heat from the pressure of a body at the fixture or by light reflected off the user and back to the control. This "touchless" approach seems to promise more for hygiene than for water conservation, although it does prevent the flush valve from being held open too long by a careless user.

In general, plumbing fixtures will emit less water as the supply water pressure is reduced. For this reason, *pressure-reducing valves* are becoming popular as water conservation devices (when they would not otherwise be required to protect fixtures from overpressure). Installed on the supply line to a building, they can save water throughout the structure.

Several newer low consumption toilets use vertical flush valve sleeves, where the flush handle is in the center of the top of the toilet tank. The handle is lifted to initiate the flush. An advantage is that the less time the handle is raised, the less water enters the bowl. Again, less water may be admitted to flush away liquid waste, more water for solid waste.

(i) Waterless Toilets. Because these fixtures eliminate water entirely, they are described in Chapter 11, Section 11.1.

(j) Appliances. Dishwashers, washing machines, and other appliances are big users of water—and of the energy needed to heat it. Dishwashers use 12 to 18 gal (45 to 68 L) per cycle, much of it heated well beyond the 120°F (48°C) typical of household hot water supply. Some models now allow shorter cycles, which can cut use to perhaps 7 gal (26.5 L).

Fig. 10.49 *The Envirovac system utilizes a vacuum and a 3-pt (1.4 L) flush. (a) Less than 3 pints of water remain in the bowl until toilet is used. (b) When pushbutton flush is pressed, the discharge valve opens for about 3 seconds; a rush of air into the evacuated piping carries along sewage, odor, and airborne bacteria. Simultaneously, the washdown water flow begins. The flow continues for 4 seconds after the discharge valve closes, so that most of the 3-pt flush remains in the bowl. (c) Toilets are directly connected to a central holding tank, kept under a vacuum by a pump. (d) Central tank/pump combinations can serve one house or a large building. (e) The separation of graywater and blackwater is another water-conserving opportunity. (Courtesy of Envirovac, Inc., Rockford, Illinois.)*

WATER AND WASTE

638

Clothes washing machines use from 40 to 55 gal (151 to 208 L) for full-size loads. In the past, "suds saver" features allowed soapy, hot wash water to be reused. Many newer washers allow for a wider selection of water quantities and temperatures—a feature that can save considerable water and energy. Horizontal-axis (front-loading) washers have greatly reduced the need for hot water per wash cycle.

10.8 Fixture Accessibility and Privacy

After plumbing fixtures designed for the human physique and for water conservation have been chosen, they are placed in a space that has several unusual environmental needs. Much of the information in this section is reproduced with permission from *American National Standard for Buildings and Facilities A117.1-1986, Providing Accessibility and Usability for Physically Handicapped People*, copyright © 1986 by the American National Standards Institute (ANSI). Copies of this standard may be purchased from ANSI, 1430 Broadway, New York, NY 10018. Another useful guide is the *Americans with Disabilities Act (ADA) Accessibility Guidelines for Buildings and Facilities*, available from the U.S. Architectural and Transportation Barriers Compliance Board, Washington, DC.

(a) Accessibility. Designers must allow room for wheelchair maneuvering into and within such tight spaces as toilet rooms. The minimum clear floor area should be a circle 5 ft (1525 mm) in diameter. Minimum clearances for wheelchair maneuvering are shown in Fig. 10.50. Provisions for people in wheelchairs also influence the heights at which items such as light switches, electrical receptacles, paper towels, and water controls should be installed. In general, all objects to be reached by hand in toilet rooms should be placed more than 15 in. (380 mm) and less than 48 in. (1220 mm) above the floor.

Floor space requirements and mounting-height limitations also apply to the various plumbing fixtures in bathrooms, at least one of which (in each case) should be accessible to people in wheelchairs in most buildings. These requirements for drinking fountains (which often must be

Fig. 10.50 *Wheelchair turning space, minimum requirements. (a) 60-in. (1525 mm) diameter space. (b) T-shaped for 180° turns. Dashed lines indicate minimum length of clear space required on each arm of the T-shaped space. (Reprinted by permission from American National Standard A117.1-1986, © 1986 by the American National Standards Institute.)*

located outside toilet rooms), lavatories, bathtubs, and shower stalls include clearances beside the fixture (and clearance below lavatories), and grab bar locations; see the ANSI Standard for details.

Accessible toilets (WC) have particularly large floor space requirements, as shown in Fig. 10.51. There is a conflict between the lower mounting height needed for proper defecation, the conven-

Fig. 10.51 *Clearances and grab bars for toilets. Two typical approaches* (a, b) *necessitate the clearances of the stalls shown* (c, d, e, f) *as well as arrangements for toilets and urinals not within stalls* (g, h, i). *(Reprinted by permission from American National Standard A117.1-1986, © 1986 by the American National Standards Institute.)*

tional height of 15 in. (380 mm), and the recommended height for handicapped users of 17 to 19 in. (430 to 485 mm). The doors on stalls designated for wheelchair access should swing out rather than in. Provision for a wheelchair sitting beside the outswinging door must also be made. No doors should swing into the clearance space required for any fixture.

Where six or more WC (or WC plus urinals) are located within the same toilet room, *both* the "standard stall" (Fig. 10.51*c*) and one of the "alternate stalls" (Fig. 10.51*d*) are required. This impacts the design of public toilet rooms.

Urinals for handicapped users should either be of the floor-mounted stall-type or, if wall-hung, have an elongated rim at a maximum of 17 in. (430 mm) above the floor. A clear floor space of 30 × 48 in. (760 × 1220 mm) is required in front of the urinal, and its flush controls should be no more than 44 in. (1120 mm) above the floor.

(b) Privacy. Visual privacy is easily understood, and at least minimally provided by partial-height partitions. However, *acoustic* considerations comprise another important aspect of bathroom design. Because of their nonabsorbent surfaces, bathrooms are frequently the most reverberant spaces in the house. They encourage the singer or the whistler, and may even serve as practice rooms for the family musician. Yet they can also intimidate a person who prefers to be quiet, such as a guest using the toilet while the rest of the dinner party sits quietly in the adjacent room.

Two common design responses are isolation and masking sound. Bathrooms can be separated from acoustically sensitive spaces by closets or hallways. With careful attention to sufficiently massive construction and/or other construction details, the doors, walls, ceilings, and floors of bathrooms can be constructed so as to reduce the passage of sound to acceptably low levels. (These

WATER AND WASTE

MIDDLE OF ROW

END OF ROW

(c) Standard Stalls

Fig. 10.51 *(continued)*

considerations are discussed in Chapters 26 and 27). Sound isolation also depends on attention to detail: no cracks around the bathroom door, no back-to-back electrical outlets between bathrooms and adjacent spaces, no air grilles into ducts that have other grilles nearby, and no other open window near the bathroom window. If such goals are unattainable, or if additional acoustic security is desired, masking sound can be added as needed within the bathroom. All too often, the masking noise is provided by repeated flushings of the toilet (itself an embarrassing sound for some) or the running of water in the lavatory. However, a noisy ventilating fan would serve the purpose, as would a music source, such as a radio. An elegant, expensive touch would be a recirculating fountain gracing one corner; it might provide splashing sounds only when the light or fan is on.

The final consideration in bathroom acoustics is the sound of water moving within pipes. Where such pipes are exposed or barely concealed within,

or firmly attached to, thin walls, perceptible sounds are highly likely. Although many of us may have experienced this sudden acoustic signal that a bathroom is in use, few have experienced it quite like the occupants of the Aluminaire house on Long Island, cited by Reyner Banham (1969). Designed at the height of the modernist expression of mechanical services, this house featured a dining table cantilevered from the exposed waste stack serving the bathroom above. Where water noise is undesirable, pipes should be wrapped, resiliently mounted, and/or located in a less acoustically critical wall. But remember that running water can also be a conservation signal, as well as masking sound.

10.9 Water Distribution

This section looks at ways to supply water throughout buildings at pressures sufficient to op-

(d) Alternate Stalls

(e) Rear Wall of Standard
Stalls

(f) Side Walls

Fig. 10.51 (continued)

erate plumbing fixtures. Smaller buildings may be served simply by the pressure available in water mains (or pressure tanks fed by pumped wells). This is called upfeed distribution, because the water rises directly from mains to the plumbing fixtures. For taller buildings, several other options are available: *pumped upfeed* (in which pumps supply the additional pressure needed); *hydro-pneumatic* (in which pumps force water into sealed tanks, compressing the air within; this maintains the needed water pressure); and *downfeed* (in which pumps raise the water to storage tanks at the top of the building, and water then drops down to the plumbing fixtures).

In cities with municipal water supply systems, water is distributed through street mains at pressures varying at the main from about 50 psi (about 350 kPa) to about 70 psi (about 480 kPa). For low-rise buildings of two or three stories, these pressures are adequate to act against the static pressure of water standing in the vertical piping, overcome the frictional resistance of water flow in the pipes, and still deliver water at the pressure required to operate plumbing fixtures. The flow-pressure required at the fixtures varies from 5 to 20 psi (35 to 138 kPa), depending on the type of fixture—for example, the basin faucet, showerhead and faucet, or water closet fixture. (Table

(g) Clear Floor Space at Water Closets

(h) Back Wall

(i) Side Wall

Fig. 10.51 (continued)

10.14 in Section 10.11 lists the minimum flow rates and pressures for typical plumbing fixtures).

(a) Static Pressure. The pressure exerted at the bottom of a stationary "head" of water is related directly to its height. One cubic foot of water weighs 62.4 lb. Consider a "cube" of water 1 ft square and 1 ft high. Its weight (62.4 lb) rests on a bottom area of 1 ft² (144 in.²). The static pressure at the bottom is 62.4/144 = 0.433 psi (3 kPa). Reciprocally, 1 psi of pressure will *sustain* a static (stationary) column of water 1/0.433 =

2.3 ft in height. (In SI units, 1 meter of head = 10 kPa.) When fixture pressure and pressure lost in friction-of-flow in pipes are considered, the problem becomes more complex. Example 10.6 in Section 10.11 illustrates this problem in the calculation of pipe size. For upfeed and downfeed distribution, the relationship of heights and static pressure is one controlling factor.

(b) Upfeed Distribution. In small, low buildings of moderate water demand, it is seldom difficult to achieve the proper flow-pressure at

Fig. 10.52 *Upfeed water distribution using pressure in street mains. A schematic section of the water services in a residence.*

fixtures by the use of an upfeed system. Pressure at the fixtures is usually more than required. When this causes an inconvenient splash, as at lavatory basins, a flow restrictor can be used in the faucet outlet.

Consider the typical upfeed system shown in Fig. 10.52, beginning at the point where supply water enters the building. In cold climates, water in the service entry pipe (or tube) must not freeze. The pipe must therefore be below the *frost line* of frozen ground. This could vary from 0 to 7 ft (0 to 2 m), depending on geographical location. The onset of winter in such climates requires the closing and draining of pipes supplying the hose bibbs (and other external piping) by means of a *stop-and-waste* valve. Houses left *unheated* in cold-winter weather must be entirely free of water that could freeze and burst the pipes. Note the drain valve at every low point in the system. House shutoff controls are usually located at the main, at the curb, and within the house.

Meters have recently taken on a new role:

along with measuring the water quantity for which the occupant is to be charged, they now serve a restrictive function. During water shortages, they can be used to indicate water use in excess of established limits, beyond which limit fines are imposed and in some cases water supply reduced by valves controlled by the water company.

Treatment is most often performed to reduce hardness that could clog piping and equipment, or to neutralize acidity—a source of corrosion. During the short periods when the treatment tanks are valved off for backwashing or other servicing, the *bypass shutoff valve* is opened.

From this point on, the water continues under pressure to

1. Supply makeup water to the house-heating boiler, as required.
2. Supply water to and pressurize the cold water mains and branches, including the garden hose bibbs.
3. Supply water to and pressurize the domestic

hot water system through the hot water heater, the hot water storage tank, and the mains, branches, and circulating lines.

The air-filled expansion chambers on cold water *runouts* absorb and reduce the shock of so-called "*water hammer.*" (When faucets are shut off abruptly, the force exerted by the decelerated flowing water shakes and rattles the pipes. This phenomenon is called *water hammer.*) On hot water runouts they perform the same function, *plus* they allow for the expansion of the hot water as it increases in volume with increasing temperature. Vacuum breakers prevent back-flow of polluted water into pipes carrying potable (hot *or* cold) water. Water from all fixtures and appliances, such as dishwashers, clothes washers, and boilers, is thus isolated.

The entire diagram of Fig. 10.52 shows all parts lying in a two-dimensional plane. Obviously, in a real building, such a system is three-dimensional. For instance, economy of piping would suggest that, if possible, kitchen and lavatory, as well as the two upstairs bathrooms, be placed back-to-back.

(c) Principles of Downfeed Distribution.

Water pumped directly from the street main (or from a basement "suction tank" filled by gravity from the main) is lifted to a roof-storage tank. In cold climates, the tank water in such exposed locations is kept at temperatures above freezing by heating coils in the tank. The fact that water pressure increases with distance below the tank water level is clearly shown by the construction of the tank (Fig. 10.53). The iron rings, tensioned by adjustable threaded clamps, become more closely spaced toward the bottom of the tank, where the greater water pressure makes it increasingly difficult to restrain the vertical wooden staves of this cylindrical barrel.

Tanks like these add interest, though perhaps not beauty, to a building's silhouette. A "fence" enclosing this tank would have to be about 24 ft (7.3 m) high. Architecturally, this rectangular lump on the roof would not be much of an improvement. Yet in the century following the appearance of such structures as the Goodwill Building, our technology has become more complex. Presently, for most high-rise buildings, including

Story height, floor to floor, 12' − 0"

Water level in the tank

35 ft

Static pressure at top floor fixture
35 × 0.433 = 15.2 psi

83 ft ±

Static pressure at first floor fixture
83 × 0.433 = 35.9 psi

Fig. 10.53 High-rise building of the 1870s, with downfeed distribution. Although upfeed distribution from a street main is possible in buildings higher than two or three stories, when street main pressures drop below the usual 50 psi (345 kPa), combined with heavy use, very low pressure results in fixtures in the upper stories. Therefore, water is pumped from the main to elevated wooden roof tanks high enough to assure reasonable pressure at the top story and ample pressure at the bottom of the downfeed run.

many 60 stories and more in height, an entire rooftop crowded with equipment and technical facilities is needed to serve the stories below, or the uppermost *zone.* The items could include

Water storage tanks

Two-story penthouses over elevator banks

Chimneys

Numerous plumbing vents

Exhaust blowers

Air-conditioning cooling towers

Cantilevered rolling rig to support scaffold for exterior window washing

Perimeter track for window washing rig

Photovoltaic cells and/or solar collectors for DHW

Thus, since the 1960s, a tall building commonly has a *band* or *screen* two stories (or more) high above the structural roof. The best views are thus consigned to mechanical equipment. It might be said that it all began with the need for elevated water-storage tanks.

(d) Tall Building Downfeed Distribution.

Fig. 10.54*a* shows a medium-rise building, in which one elevated tank can serve all the lower floors. For taller buildings, it is advisable to separate groups of floors into *zones* with a maximum height (for plumbing pressure limits) to about 150 ft (about 45 m). This zone-height limitation is based on the height-to-static pressure relationship. At the top of the zone [about 35 ft (10 m) below the storage tank], the minimum desirable pressure is probably at least 15 psi (103 kPa). At the bottom of the zone, the maximum desirable pressure is perhaps 80 psi (552 kPa); above this pressure, damage to fixtures might occur.

$$80 \text{ psi} - 15 \text{ psi} = 65 \text{ psi difference}$$

$$65 \text{ psi} \times 2.3 \text{ ft/psi} = \text{about 150 ft}$$

In SI units,

$$552 \text{ kPa} - 103 \text{ kPa} = 449 \text{ kPa difference}$$

$$449 \text{ kPa} \times 0.1 \text{ m/kPa} = \text{about 45 m}$$

With pressure-reducing valves at lower floors, however, these zones can be much higher, as shown in Fig. 10.54*b*.

Consider the system described in Fig. 10.54*a*, beginning with the elevated tank. The lower part of the tank often serves as a reserve space to hold a supply of water for a fire hose system. In this case, only the water in the upper part is available for use as domestic (service) water. The amount stored must be enough to supplement what the pump will deliver during the several daily hours of high demand that occur in most buildings. The pump then continues, often for several hours, to replenish the house supply that had become partially depleted during the busy period. The suction tank is a buffer between the system and the street mains. It usually holds enough reserve to allow the pumps to make up the periodic depletion in the house tank. It refills automatically by flow from the street main that, consequently, will not suffer as much of a drop in pressure as it would

if it were connected directly to the suction side of the house pumps. Neighboring water users are protected from the adverse effects of sudden demands within adjacent large buildings (Fig. 10.55).

House tanks and suction tanks are sometimes made of steel plate and divided vertically in half, each half having identical piping and controls. Hence, one half of the tank can be cleaned out at a time during hours of low demand without shutting down the entire system. One full-capacity pump is supplemented by an equivalent standby pump for alternate use. Because there is no suction lift below the pump, or any fixture pressure at the top of the house tank supply, the head against which the pump works is the sum of the distance from the suction tank water level to the top of the house tank and the feet of head equivalent to the friction loss in the tank supply pipe. For this kind of service the vertical piping is on the order of 3 or 4 in. (76 or 102 mm) in diameter for large buildings. Sizes are established by a formal engineering design.

The house supply is fed by a short pipe from the house header to the cold water header that circles the top story and connects to many downfeed cold water risers. For simplicity, Fig. 10.54*a* shows only two risers and also omits many valves and controls.

Figure 10.54*b* is even more simplified. The circulating hot water originates as cold water at the house tank header, then takes quite a long route. Descending to the bottom of the hot water heater, it rises to seek its own level at the hot water header, becoming available there for hot water downfeed on demand. All of this occurs as flow below the general pressurizing effect of the house tank. In effect, when there is a cold- and hot-water demand on a story near the top of the building, the cold water makes a short trip down to the faucet, while the hot water goes through three vertical pipes instead of one.

This arrangement, with tank above and heater below, is used in multiple forms for very tall buildings. The zones are quite independent; the only common service is that provided by the general suction tank. By this zoning method, problems of pipe expansion, excessive pipe sizes, and high pressures in lower stories are minimized. Commonly, 2½ stories, or about 35 ft (about 10 m), comprise the minimum pressure head above the

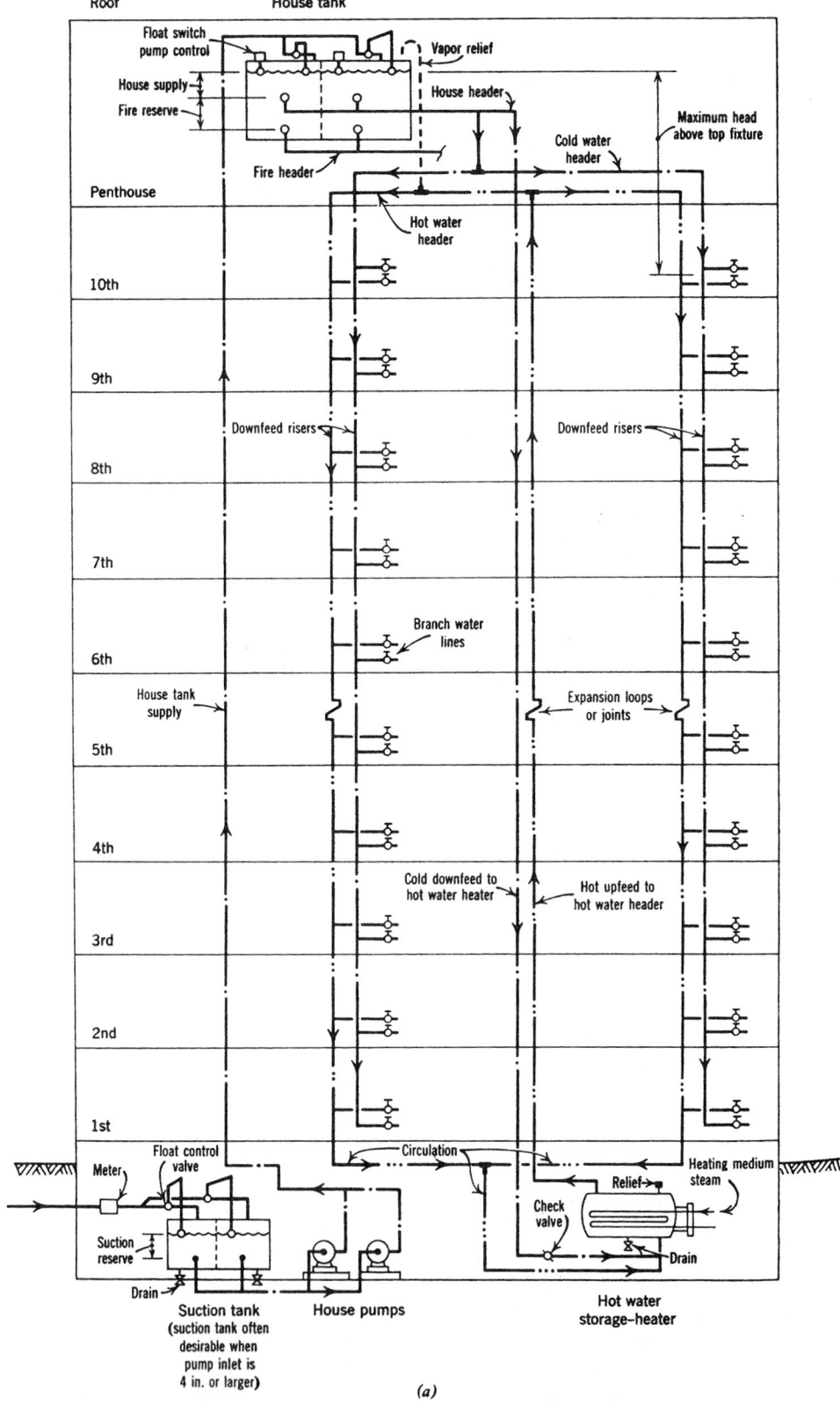

Fig. 10.54 (a) *Downfeed water distribution, schematic section, part of the water services for a 10-story building. Hot water circulation moves from the hot upfeed in two directions at the hot water header, then down to the tank through the two downfeed hot water risers. For details of one type of steel house tank, a typical centrifugal house pump, expansion joints, and expansion in hot water riser, see Figs. 10.55, 10.56.*

* In many cases a few lower floors are supplied from street mains

(b)

Fig. 10.54 (b) *Downfeed water distribution, schematic section, part of the water services for a much taller, zoned building. Zone tanks include a fire reserve but standpipes are omitted from this drawing. For detail of steam-type domestic water heater, see again Fig. 10.24.*

Fig. 10.55 (a) *House tank in elevated position for downfeed by gravity. Sediment in tank is drawn off through clean-out pipe and is prevented from entering house supply by pipe projection at A. Water for fire reserve could be provided by additional piping or by a separate tank.* (b) *Centrifugal house pump, commonly used to fill elevated house tank in top story or at intermediate mechanical floor. Electric motor responds to float switch pump control at the tank.*

top fixture served by any zone tank. The static pressure created at the fixture is thus $35 \times 0.433 = 15$ psi ($10 \times 10 = 100$ kPa). If, during flow, not too much pressure is lost in friction, flushometers (flush valves) can be placed at this level, although flush tanks, because of their lower fixture-pressure demand, must often be accepted. Quite the opposite problem occurs at the bottom of the zones, where excessive pressures must be reduced at the fixtures. In zone I of Fig. 10.54*b*, first-floor fixtures

TABLE 10.12 **Thermal Expansion of Pipe**

| Temperature Increase | | Approximate Expansion of Piping Material Inches per 100 ft (0.83 mm per m) of Length | | | | | |
F°	C°	Carbon and Carbon Moly Steel	Cast Iron	Copper	Brass and Bronze	Wrought Iron	Plastic
20	11.1	0.15	0.15	0.22	0.22	0.15	0.5
40	22.2	0.3	0.3	0.44	0.44	0.3	1.1
60	33.3	0.45	0.45	0.67	0.67	0.45	1.7
80	44.4	0.6	0.6	0.89	0.89	0.6	2.4
100	55.5	0.76	0.7	1.12	1.12	0.76	3.3
120	66.7	0.91	0.84	1.35	1.35	0.91	4.3
140	77.8	1.07	1.0	1.58	1.58	1.07	5.3
160	88.9	1.22	1.13	1.81	1.81	1.22	6.2
180	100	1.38	1.27	2.04	2.04	1.38	—
200	111.1	1.55	1.4	2.27	2.27	1.55	—

are below a head of 24½ stories, or about 149 psi (1027 kPa) of static pressure. It is obvious that pressure-reducing valves must be used and that fixture control valves must be throttled.

(e) Pipe and Tube Expansion. The range of temperature from the normal indoor air temperature of about 70°F to that of service hot water (which often exceeds 160°F) can be 90F° (50C°), with resulting expansion of hot water supply piping (Table 10.12). This longitudinal elongation of pipe, though negligibly small in houses, can be appreciable in a tall building. Two methods of allowing freedom for this longitudinal motion in long runs of expanding hot water piping are shown in Fig. 10.56a. These devices preclude the buildup

Fig. 10.56 (a) *Accommodation for the expansion of hot water piping or tubing. Left: Expansion joint of pipe and fittings. Right: A manufactured product.* (b) *(Example 10.5) Suggested scheme for location of the points of anchorage and expansion for service hot water tubing in a 20-story zone.*

of excessive stresses in the pipe material and the tendency of the pipes to buckle laterally.

EXAMPLE 10.5. A 20-story zone in a tall building has a height of 280 ft. What will be the increase in length of a copper tube carrying "service hot water" (domestic hot water), when its temperature increases from 70° to 160°F?

SOLUTION. The difference in temperature is 90F°. Approximate elongation per 100 ft for 90F° increase is 1.01 in. (interpolating in Table 10.12).

expansion = 280 ft × (1.01 in./100 ft) = 2.82 in.
∎

There are a number of ways of providing for this expansion. The one shown in Fig. 10.56*b* accepts this motion at two locations, which would make the expansion in each case 1.41 in. Equidistant anchorage to fix the tubing is provided at the bottom, the 10th floor, and the 20th floor. The support of the vertical riser at floors other than those at which it is anchored could consist of

clamps of the type illustrated in later Fig. 10.61, supported on springs.

(f) Pumped Upfeed Distribution. This distribution system (Fig. 10.57) is for medium-size buildings—those too tall to rely on street main pressure but not so tall as to necessitate heavy storage tanks on the roof. This equipment can deliver water at rates varying from what is needed for two or three faucets to full building demand, while maintaining at each outlet a pressure within 2 psi (14 kPa) of the design pressure for that outlet.

The installation shown in Fig. 10.57*a* and *b* is a triplex pump group. According to demand, one, two, or three pumps will operate. Because each pump is of the variable-speed type, virtually an infinite number of delivery rates can be achieved within the zero to maximum design rate. The pumps operate in sequence. When a very small rate of demand occurs, the smallest or "jockey" pump starts in response to a low voltage impressed on its motor. All operations are triggered

Fig. 10.57 Constant-pressure upfeed pumping (a) and (b), compared to gravity downfeed from house tank on roof (c). (By permission of Progressive Architecture.)

(a) Constant presure (upfeed schematic)

(b) Same as "a" with surge tank (when required)

(c) Conventional roof tank

and adjusted by the pressure sensor at the base of the riser. The jockey pump continues to run until it has reached its maximum delivery rate, at which time the first of the larger pumps cuts in, joined by the other larger pump when required. Sequential operation of the three pumps, each increasing in delivery as called for by the sensor, meets the requirement for increasing supply at nearly constant pressure.

Wear on the three pumps is equalized by the assigning of "lead" and "lag" positions. For a period of 24 h, one of the large pumps holds the lead position and starts after the jockey pump, giving the other large pump a smaller burden. The next day the rested pump takes over the more active assignment. All of this occurs automatically.

At full operation this triplex unit can put a suction demand on the street main that could seriously reduce the available water pressure in the neighborhood (Fig. 10.57a). Therefore, city officials sometimes require that the system feed on a surge tank, filled by casual flow from the street main, independent of the building requirements (Fig. 10.57b). This requirement is often made when the design indicates a maximum building demand in excess of 400 gpm (25 L/s).

Obvious advantages of upfeed pumping are the elimination of the house tank and the heavy structure that transmits its weight down to the footings, and of the necessary periodic cleaning of the tank. A shortcoming is the lack of reserve storage, which could cause a serious problem during an electrical power failure. However, minimum flow during this kind of emergency can be arranged if a diesel or other independent standby motor is available to drive one of the pumps.

10.10 Piping, Tubing, Fittings, and Controls

(a) Piping, Tubing, and Fittings.
The system of water supply piping or tubing should efficiently fulfill its purpose, be easily maintained, and interfere as little as possible with the interior architectural form. Except in basements, in utility rooms, and at points of access to controls, the system is usually concealed. Stud-and-joist construction provides space for concealment, but in concrete or masonry buildings, vertical and horizontal furred spaces must often be provided.

Water supply piping is subject to corrosion over time. When pipe materials corrode, they first lose some carrying capacity (due to increased wall roughness and perhaps to build-up of materials), and ultimately fail. Sediment from corrosion can adversely impact plumbing fixtures as well. Steel piping is particularly subject to corrosion. In the nonferrous group, red brass and copper tubing are effective in corrosion resistance. Copper tubing is less expensive than brass, assembles more easily, and is not subject to dezincification (attack by acids on the zinc in brass). For use in handling aggressive waters, plastic is often a good choice. Like copper, it is light in weight and assembles with great ease.

For ferrous pipes and "iron pipe size" brass, threaded connections are used. The external, tapered thread on the pipe is covered with pipe compound and screws tight against the internal tapered thread of the coupling or other fitting. The solder–joint connection in copper depends on capillary attraction that draws the solder into a cylinder of clearance between the mating surfaces of tube and fitting. This occurs after the surfaces are polished and fluxed and the parts placed in final position. They are then heated, and molten solder is applied to the circular opening where the fitting-edge surrounds the tube, with a small clearance. It is then drawn into the cylindrical connection. Solders are often tin-antimony alloys. This kind of joint permits the advantageous setting up of an entire tubing assembly without turning the parts (as in threaded installations), and before the soldering commences. For the same strength, copper tubing may have thinner walls because no threads need to be cut into it. Its smooth interior surface offers less friction to flowing water. Although threaded- and solder–joint connections are the most common in small work, there are many other types. Ferrous pipes in the larger sizes are often welded or connected by bolted flanges (Figs. 10.58 and 10.59 and Table 10.13).

(b) Plastic Pipe.
Most of the plastic pipes and fittings now produced are synthetic resins derived from such materials as coal and petroleum. These corrosion-resistant materials are widely

TABLE 10.13 **Water Supply Piping Materials**

Part A. **Water Service Only**[a]

Material	Connections[b]
ABS (Acrylonitrile butadiene styrene) plastic pipe	Mechanical with elastomeric seal (normally underground only); solvent cement; threaded joints
Asbestos–cement	Sleeve couplings of same material as pipe, sealed with elastomeric ring
Ductile iron water pipe	Depth of lead depends upon pipe size
PE (Polyethylene) plastic pipe and tubing	Flared joints (see manufacturer), heat-fusion, mechanical joints (see manufacturer)
PE-AL-PE (Polyethylene/aluminum/polyethylene) pipe	
PVC (Polyvinyl chloride) plastic pipe	Mechanical with elastomeric seal (normally underground only); solvent cement; threaded joints (may reduce pressure rating).

Part B. **Water Service and Distribution**[c]

Material	Connections
Brass pipe	Brazed, mechanical, threaded, or welded joints
CPVC (Chlorinated polyvinyl chloride) plastic pipe (and tubing, indoors)	Mechanical (see manufacturer), solvent cement, or threaded joints (may reduce pressure rating)
Copper or copper-alloy pipe and tubing	Brazed, mechanical, soldered, threaded, or welded joints
Galvanized steel pipe[d]	Threaded, or mechanical joints with an elastomeric seal
PB (Polybutylene) plastic pipe and tubing	Flared joints, heat-fusion, or mechanical joints (see manufacturer)
PEX (Cross-linked polyethylene) plastic tubing	
PEX-AL-PEX (Cross-linked polyethylene/cross-linked aluminum/polyethylene) pipe	

[a]Materials as listed in the 1997 International Plumbing Code. Water service piping is for outdoor, underground connection from building to main or supply tank.
[b]Joints between different piping materials require approved adapter fittings; dielectric or brass converter fittings required between copper (or copper alloy) and galvanized steel piping.
[c]Materials as listed in the 1997 International Plumbing Code. Water distribution piping is within the building.
[d]Sediment from corrosion over life of pipe.

used in water supply piping, fittings, and drainage systems (see Chapter 11). The National Sanitation Foundation (NSF) tests and certifies plastic pipe; the NSF seal must appear on pipes that are to carry potable water.

Most of the materials used for piping are thermoplastics and have the quality of repeated softening under the application of heat. PVDC material can carry water at 180°F (82°C), but plastic pipe should not be subjected to temperatures higher than this. Expansion (see Table 10.12) is much greater than other piping materials for the same ΔT and affects the piping design. Quite shockproof, plastic is used in the plumbing systems of most mobile homes.

(c) Valves and Controls. It is usually desirable to valve every riser, the branches that serve

Fig. 10.58 *Methods of connecting pipes and fittings, and tubes and fittings. (a) Threaded: for ferrous pipe and fittings and for "iron pipe size" (IPS) brass. (b) Soldered: for copper tubing and fittings. A sliding fit similar to that of (b) is used for the solvent weld of plastic connections.*

bathrooms or kitchens, and the runouts to individual fixtures. This facilitates repairs at any location with a minimum of shutdown within the system. Water treatment equipment will have a valved bypass (see Fig. 10.52). Pumps and other devices that may need repair should be disconnectable by unions (Fig. 10.59) after valves are closed.

The gate valve (Fig. 10.60*a*), with its retractable leaf machined to seal tightly against two sloping metal surfaces when closed, offers the least resistance to water flow when open. It is usually chosen for locations where it is left completely open most of the time. The compression-type globe valve (Fig. 10.60*b*) is usually used for the closing or throttling of flow near a point of occasional use. Faucets are usually of the compression type, as are drain valves or hose connections. They are similar to the angle valve (Fig. 10.60*d*). When it is necessary to prevent flow in a direction oppo-

Union
Used instead of coupling when future dismantling and reassembling of piping is contemplated.

This clamping element disengages by turning and then sliding away to permit uncoupling of pipes.

Fig. 10.59 *Example of threaded pipe fittings for ferrous or brass pipe. These and all common fittings are also available for solder-joint (copper) or solvent-weld (plastic) connections, and usually for transition from one system or material to another.*

site to that which is planned, a check valve (Fig. 10.60*c*) is introduced. The hinged leaf swings to permit flow in the direction of the arrow but closes against flow in the other direction.

(d) Pipe Support. Water piping systems are made even heavier because of their water content, and need regular support (Fig. 10.61). Vertical runs of piping should be supported at every story. Horizontal pipes should be supported at intervals of

> 6 ft (1829 mm) for ½-in. (12-mm) pipe;
>
> 8 ft (2438 mm) for ¾-in or 1-in. (19- or 25-mm) pipe; and
>
> 10 ft (3048 mm) for 1¼-in. (32-mm) or larger pipe.

Horizontal copper tubing should be supported at closer spacing than steel. Adequate positioning of horizontal runs is important to assure correct pitch and drainage. Hangers are adjustable for this purpose.

(e) Shock and Hot Water Expansion. Expansion chambers were shown diagrammatically in Figure 10.52, and water hammer explained in Section 10.9(*b*). Expansion chambers are often made of capped lengths of vertical pipe about 2 ft (0.6 m) long at the fixture branches (Fig. 10.62*a*). They trap air, which absorbs the impact of the water surge. A somewhat better (more controllable) device is a "rechargeable air chamber" (Fig.

WATER AND WASTE

Fig. 10.60 *Typical valves for water systems.* (a) *Gate valve.* (b) *Globe valve.* (c) *Check valve.* (d) *Angle valve.*

Fig. 10.61 *Pipe supports.* (a) *Vertical riser supported at steel beam.* (b) *Vertical riser group supported at slot in concrete slab.* (c) *Horizontal pipe hung from slab above by adjustable-length clevis hanger.* (d) *Typical metal insert in soffit of concrete to receive hanger rod.*

Fig. 10.62 *Shock relief and expansion chambers. Air chambers cushion the shock of the water hammer when the fixture faucets are shut off abruptly. They also permit hot water to expand, instead of periodically forcing open the hot-water-emergency pressure relief valve at the heater or tank. (a) Capped air chambers at each supply pipe of each fixture. (b) Rechargeable air chambers on hot and cold branch water lines. (Individual fixture chambers are omitted when these are used.) (c) Special shock absorber.*

10.62*b*). By closing the valve and draining the water through the hose bibb while the petcock above is open to admit air, the chamber may be refilled with air. Closing the petcock and hose bibb and opening the valve completes the service operation and reconnects the device with the water system. Rechargeable chambers (instead of pipe extensions) are used on branch lines adjacent to groups of fixtures. Access for service must be provided. Perhaps the best of these methods is the special shock absorber (Fig. 10.62*c*).

Air cushions also protect the water heater tank's relief valve against frequent operation, with the resultant leakage of hot water, as the hot water periodically expands and contracts in closed systems.

(f) Condensation, or "Sweating." The moisture that is always present in air often condenses on the exterior surface of cold pipes. Dropping off the pipes, it creates an unpleasantly wet condition and disfigures finished surfaces. Groundwater in some parts of North America is 50°F (10°C) and colder (see Fig. 5.58). A pipe carrying such water might have a surface temperature of about 60°F (15°C). The psychrometric chart (pages 289, 290) indicates that at a summer air temperature of 85°F (29°C), condensation will occur on this pipe when the relative humidity

exceeds 40%. To avoid condensation damage, all cold water piping and fittings should be covered. Glass fiber ½- to 1-in. (12- to 25-mm) thick is commonly chosen for this purpose. A tight vapor barrier on the exterior surface of the covering prevents the moisture-laden air from penetrating the insulation to reach the colder surface. The insulation provides another advantage of importance: it retards heat flow from the warmer air to the water, thus preventing the water from becoming disagreeably warm.

(g) Heat Conservation. Pipes carrying domestic hot water should be insulated to conserve the fuel used to heat the water and to assure a correct water temperature at the point of use. Parallel hot and cold water piping, even though insulated, should be separated to prevent heat interchange. This pipe insulation should have $k = 0.22 - 0.28$ Btu-in/h ft^2 °F, and its thickness should be a minimum of ½ in. (on pipes less than 1½ in. diameter), or 1 in. (on pipes 1½ in. diameter and above).

[In SI units, $k = 0.032 - 0.040$ W/m K, and its thickness should be a minimum of 1.3 mm (on pipes less than 25-mm diameter), or 2.5 mm (on pipes 25-mm diameter and above)].

Storage tanks and water heaters are usually manufactured with integral insulation. Older devices, however, may have less insulation than today's energy supplies warrant. As a result, many of the older water heaters are retrofitted with added insulation. Some electric utilities find the savings from conservation so attractive (relative to the cost of building new generating facilities) that they provide water-heater wrapping as a free service to customers.

10.11 Sizing of Water Pipes

There must be sufficient pressure at fixtures to assure the user of a prompt and adequate flow of water. Municipal ordinances often state that the flow must be adequate to keep the fixtures clean and sanitary. These convenience and sanitation objectives result in prescribed pressures that must be maintained at the various fixtures to assure the proper flow rates, listed in Table 10.14.

Minimum fixture pressures vary from 4 to 20 psi (28 to 138 kPa) for fixtures other than hose bibbs. Because the pressure in street mains is usually about 50 psi (345 kPa), it is possible to assure the minimum fixture pressure, provided that the water does not have to be lifted to too great a height and that not too much pressure is lost by friction in distribution piping. Friction results in

TABLE 10.14 **Flow and Pressure to Typical Plumbing Fixtures**

Fixture Served	Minimum		Maximum Flow Rate or Quantity
	Flow Rate (gpm)[a]	Pressure (psi)[b]	
Bathtub	4	8	
Bidet	2	4	
Combination fixture	4	8	
Dishwasher, residential	2.75	8	
Drinking fountain	0.75	8	
Laundry tray	4	8	
Lavatory, private	2	8	2.5 gpm at 80 psi
Lavatory, public	2	8	0.5 gpm at 80 psi
Lavatory, public, metering or self-closing	2	8	0.25 gallon per metering cycle
Shower head	3	8	2.5 gpm at 80 psi
Shower head, temperature controlled	3	20	2.5 gpm at 80 psi
Sink, residential	2.5	8	2.5 gal at 60 psi
Sink, service	3	8	2.5 gal at 60 psi
Urinal, valve	15	15	1.5 gallons per flushing cycle[a] or 1.0 gallon per flushing cycle
Water closet, blow out, flushometer valve	35	25	4 gallons per flushing cycle
Water closet, siphonic, flushometer valve	25	15	4 gallons per flushing cycle[a] or 1.6 gallons per flushing cycle
Water closet, tank, close coupled	3	8	1.6 gallons per flushing cycle
Water closet, tank, one piece	6	20	1.6 gallons per flushing cycle

Source: Copyright © 1997, International Code Council, Inc., Falls Church Virginia. *International Plumbing Code.* Reprinted with permission. All rights reserved.

For SI: 1 psi = 6.895 kPa, 1 gpm = 3.785 L/m.

[a]The higher maximum listed is for public use in places of assembly, and for patients, inmates, and residents in hospitals, nursing homes, sanitariums, prisons, asylums and reformatories.

piping that is too long in *developed length* (actual distance of water flow), or that interposes too many fittings such as elbows and tees, or is too small in diameter.

The pressure losses in an upfeed system served by street main pressure are as follows. For A below, use the highest, most remote fixture from the main.

Minimum fixture flow pressure	A
Pressure lost because of height	+B
Pressure lost by friction in piping	+C
Pressure lost by flow through meter	+D
Total street main pressure	=E

In a design, items A, B, and E are known and are reasonably constant. The value of A can be found in Table 10.14. Street main pressure, E, is a characteristic of the local water supply, and is obtained from the water utility. Item B, the pressure lost due to height, can be found by multiplying the height in feet by 0.433 (height in meters by 10) (see the discussion of static head, Section 10.9a). Item D, the pressure lost in flow through the water meter, depends on flow and pipe size (Fig. 10.63), neither of which is yet known. Therefore, the value of item D is *estimated*. [For residences and small commercial buildings, meter size

rarely exceeds 2 in. (50 mm)]. Later, it must be checked and a recalculation made if necessary. This leaves one unknown, the value of C, where

$$C = E - A - B - D$$

Pipe size is based on Fig. 10.64. Pipe diameter is determined by the point of intersection of a horizontal line representing flow and a vertical line expressing friction loss. To select a pipe size, one needs to know the probable flow and the *unit-friction loss* in the pipe and fittings. Also, the noise created by water flow must be considered. Above 10 fps (3.1 m/s) is usually too noisy; above 6 fps (1.8 m/s) may be too noisy in acoustical-critical locations.

Flow can be found for any pipe by first listing all fixtures to be served by that pipe, assigning them water supply fixture units (wsfu, listed in Table 10.15), and finding their sum. This wsfu total is then converted to the likely demand flow in gpm, from Fig. 10.65a (demand flow in L/s, from Fig. 10.65b). These curves show that flow does *not* increase in direct proportion to an increase in fixture units. The larger the installation, the less likelihood that many fixtures will be operating concurrently.

To establish the desired friction loss, divide value C (pressure lost by friction in piping) by

Fig. 10.63 *Pressure losses in disk-type water meters.* (a) *I-P units.* (b) *SI units.* (Reprinted by permission from the 1997 Uniform Plumbing Code. Copyright © by the International Association of Plumbing and Mechanical Officials, Walnut, California.)

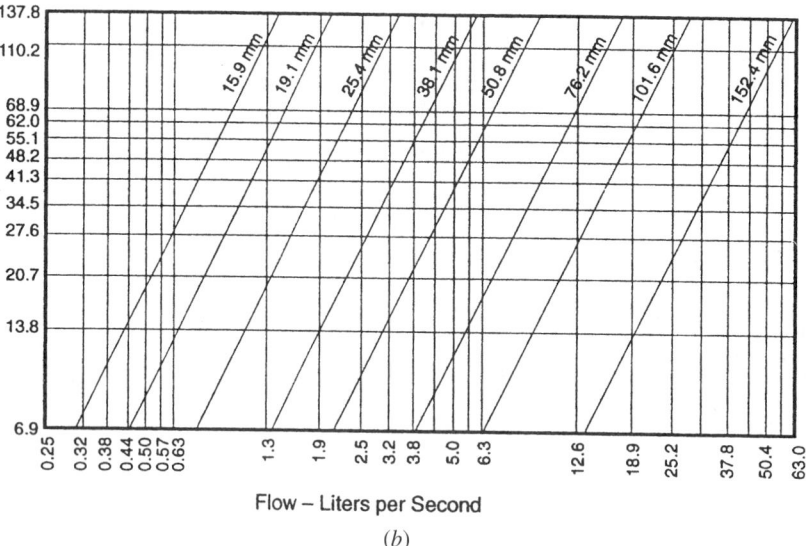

Flow – Liters per Second

(b)

Fig. 10.63 *(continued)*

the *total equivalent length* (TEL) of the piping. This length is the sum of the *developed length* (DL) (total linear distance of water travel) and the length equivalent to the fittings. For instance, Table 10.16 shows that, in a 1-in. diameter pipe run, a 90° ell causes a friction loss equivalent to that of 3 ft of pipe. The number and style of fittings is only estimated, and the size of fittings assumed. This may be puzzling but it is a common engineering procedure that sometimes requires several recalculations.

EXAMPLE 10.6. Using the following data—some of which have been arrived at by the assumptions referred to above—find the proper size for a metered water supply main.

Street main pressure (minimum)	50 psi
Height, topmost fixture above main	30 ft
Topmost fixture type	Water closet with flush valve (1.6 gal per flush)
Fixture units in the system	85 wsfu
Developed length (DL) of the piping to the highest and most remote fixture	100 ft

Pipe length equivalent to fittings (commonly estimated at 50% of the DL)	50 ft
System uses predominately	Flush valves

SOLUTION. From the minimum street main pressure, subtract the sum of the fixture pressure, the static head, and the pressure lost in the meter. This sum is

	psi
A—fixture pressure (Table 10.14)	15
B—static head 30 ft × 0.433	13
D—pressure loss in meter (estimated, Fig. 10.63)	8
subtotal	36
E—pressure in street main	50.0
E − (A + B + D)	−36.0
	14.0

The pressure lost in 100 ft (DL) of piping plus the 50 ft of piping equivalent to the pressure lost by friction in the fittings therefore can total 14 psi. Total equivalent length (TEL) is 150 ft. This procedure assures 15 psi at the critical fixture. The

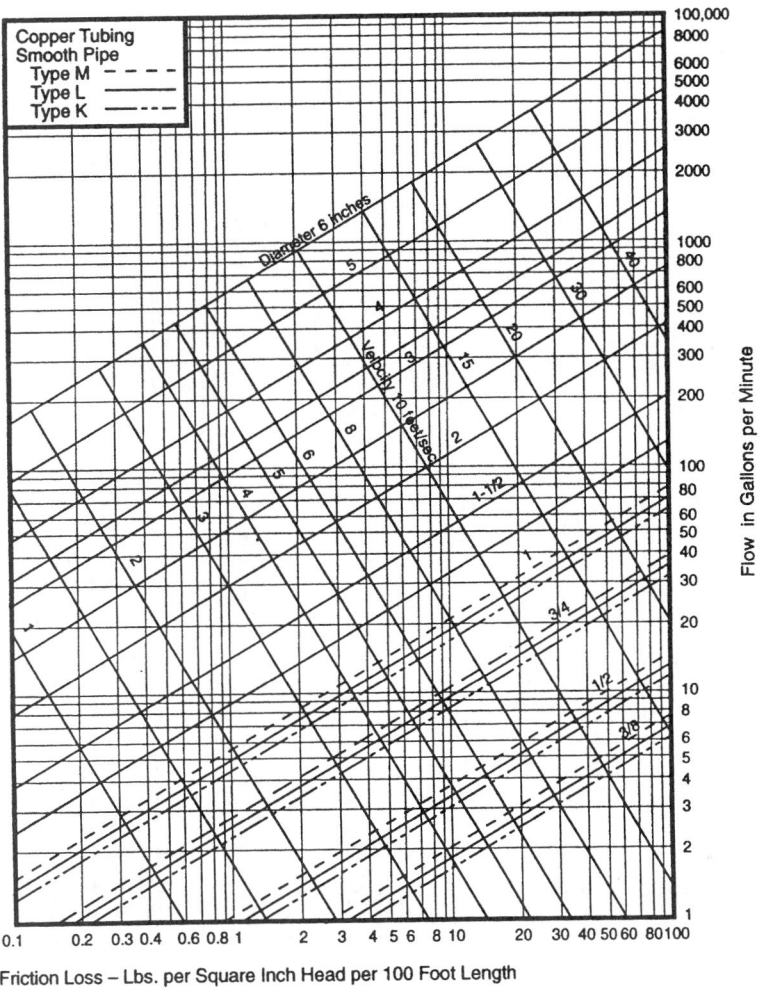

WATER AND WASTE

Friction Loss – Lbs. per Square Inch Head per 100 Foot Length

(a)

Fig. 10.64 *Flowchart for smooth pipe. Velocity is shown, as an aid in noise control: above 10 fps (3m/s), moving water can be clearly heard within pipes. (a) I-P units. (b) SI units. (Reprinted by permission from the 1997* Uniform Plumbing Code. *Copyright © by the International Association of Plumbing and Mechanical Officials, Walnut, California.)*

unit-friction loss, psi/100 ft of pipe, will be

14 psi × 100/150 TEL = 9.33 psi/100 ft

From Fig. 10.65*a*, curve 1, a flush-valve system with 85 wsfu will have a probable flow (demand) of about 64 gpm. Given this information, enter Fig. 10.64*a* horizontally at 64 gpm and vertically at 9.3 psi/100 ft. At the intersection of these lines, the pipe diameter and velocity are determined. Between 1½-in. and 2-in. diameter pipe:

velocity = 8 fps (less than 10 fps, so OK)

Therefore, a 2-in.-diameter supply pipe will be chosen, with a 2-in. meter.

Now find the actual pressure loss in the 2-in. meter for a flow of 64 gpm. Figure 10.63*a* shows that this is about 4 psi. Because this is *less* than the 8 psi estimated, the pressure at the *fixture* will be slightly higher than the minimum needed. When a final system layout is made, the fittings are tabulated and the length in piping equivalent to fittings is found. If this differs greatly from the 50 ft estimated, a recalculation is made. ■

Friction Loss – 32.9 Pa/m

(b)

Fig. 10.64 (continued)

Fixture Units

(a)

Fig. 10.65 *Estimate curves for flow, based on total wsfu (water supply fixture units). Curve 1 is for a system of predominantly flush valves. Curve 2 is for a system of predominantly flush tanks. (For fixture unit totals above 250 wsfu, see the 1997 Uniform Plumbing Code.) (a) I-P units. (b) SI units. (Reprinted by permission from the 1997 Uniform Plumbing Code. Copyright © by the International Association of Plumbing and Mechanical Officials, Walnut, California.)*

Fixture Units

(b)

Fig. 10.65 *(continued)*

TABLE 10.15 **Water Supply Fixture Units (WSFU)**

Fixture	Occupancy	Type of Supply Control	Load Values in WSFU		
			Cold	Hot	Total
Bathroom group	Private	Flush tank	2.7	1.5	3.6
Bathroom group	Private	Flush valve	6	3	8
Bathtub	Private	Faucet	1	1	1.4
Bathtub	Public	Faucet	3	3	4
Bidet	Private	Faucet	1.5	1.5	2
Combination fixture	Private	Faucet	2.25	2.25	3
Dishwashing machine	Private	Automatic		1.4	1.4
Drinking fountain	Offices, etc.	⅜″ (9.5 mm) valve	0.25		0.25
Kitchen sink	Private	Faucet	1	1	1.4
Kitchen sink	Hotel, restaurant	Faucet	3	3	4
Laundry trays (1 to 3)	Private	Faucet	1	1	1.4
Lavatory	Private	Faucet	0.5	0.5	0.7
Lavatory	Public	Faucet	1.5	1.5	2
Service sink	Offices, etc.	Faucet	2.25	2.25	3
Shower head	Public	Mixing valve	3	3	4
Shower head	Private	Mixing valve	1	1	1.4
Urinal	Public	1″ (25 mm) Flush valve	10		10
Urinal	Public	¾″ (19 mm) Flush valve	5		5
Urinal	Public	Flush tank	3		3
Washing machine (8 lbs) (3.6 kg)	Private	Automatic	1	1	1.4
Washing machine (8 lbs) (3.6 kg)	Public	Automatic	2.25	2.25	3
Washing machine (15 lbs) (6.8 kg)	Public	Automatic	3	3	4
Water closet	Private	Flush valve	6		6
Water closet	Private	Flush tank	2.2		2.2
Water closet	Public	Flush valve	10		10
Water closet	Public	Flush tank	5		5
Water closet	Public or private	Flushometer tank	2		2

Source: Copyright © 1997, International Code Council, Falls Church, Virginia. *International Plumbing Code.*
Reprinted with permission. All rights reserved.

TABLE 10.16 **Allowance in Equivalent Length of Pipe for Friction Loss in Valves and Threaded Fittings**[a]

Part A. **I-P Units**

	Equivalent Length of Pipe for Various Fittings						
Diameter of Fitting (in.)	90° Standard Ell (ft)	45° Standard Ell (ft)	Standard Tee 90° (ft)	Coupling or Straight Run of Tee (ft)	Gate Valve (ft)	Globe Valve (ft)	Angle Valve (ft)
3/8	1	0.6	1.5	0.3	0.2	8	4
1/2	2	1.2	3	0.6	0.4	15	8
3/4	2.5	1.5	4	0.8	0.5	20	12
1	3	1.8	5	0.9	0.6	25	15
1 1/4	4	2.4	6	1.2	0.8	35	18
1 1/2	5	3	7	1.5	1.0	45	22
2	7	4	10	2	1.3	55	28
2 1/2	8	5	12	2.5	1.6	65	34
3	10	6	15	3	2	80	40
3 1/2	12	7	18	3.6	2.4	100	50
4	14	8	21	4.0	2.7	125	55
5	17	10	25	5	3.3	140	70
6	20	12	30	6	4	165	80

Part B. **SI Units**

	Equivalent Length of Pipe for Various Fittings						
Diameter of Fitting (mm)	90° Standard Elbow (mm)	45° Standard Elbow (mm)	Standard Tee 90° (mm)	Coupling or Straight Run of Tee (mm)	Gate Valve (mm)	Globe Valve (mm)	Angle Valve (mm)
9.5	305	183	457	91	61	2438	1219
12.7	610	366	914	183	122	4572	2438
19.1	762	457	1219	244	152	6096	3658
25.4	914	549	1524	274	183	7620	4572
32	1219	732	1829	366	244	10668	5486
38	1524	914	2134	457	305	13716	6706
51	2134	1219	3048	610	396	16764	8534
64	2438	1524	3658	762	488	19812	10363
76	3048	1829	4572	914	610	24384	12192
102	4267	2438	6401	1219	823	38100	16764
127	5182	3048	7620	1524	1006	42672	21336
152	6096	3658	9144	1829	1219	50292	24384

Source: Reprinted by permission from the 1997 Uniform Plumbing Code. Copyright © by the International Association of Plumbing and Mechanical Officials, Walnut, California.

[a]Based on nonrecessed threaded fittings. Use one-half (1/2) these allowances for recessed threaded fittings or streamline solder fittings.

EXAMPLE 10.7. Solve the pipe sizing problem posed in the previous example now using SI units.

Street main pressure (minimum)	345 kPa
Height, topmost fixture above main	10 m
Topmost fixture type	WC with flush valve
Fixture units in the system	85 wsfu
Developed length (DL) of the piping to the highest and most remote fixture	30 m
Pipe length equivalent to fittings (commonly estimated at 50% of the DL)	15 m
System uses predominately	Flush valves

SOLUTION. From the minimum street main pressure, subtract the sum of the fixture pressure, the static head, and the pressure lost in the meter. This sum is

A—fixture pressure (Table 10.14)	103 kPa
B—static head 10m × 10 kPa/m	100 kPa
D—pressure loss in meter (estimated, Fig. 10.63)	55 kPa
subtotal	258kPa
E—pressure in street main	345 kPa
E − (A + B + D)	−258
	87 kPa

The pressure lost in 30 m (DL) of piping plus the 15 m of piping equivalent to the pressure lost by friction in the fittings therefore can total 87 kPa. Total equivalent length (TEL) is 45 m. This procedure assures 103 kPa at the critical fixture. The unit-friction loss, kPa/100 m of pipe, will be

$$87 \text{ kPa} \times 100/45 \text{ (TEL)} = 193 \text{ kPa}/100 \text{ m}$$

From Fig. 10.65*b*, curve 1, a flush-valve system with 85 wsfu will have a probable flow (demand) of about 4 L/s. Given this information, enter Fig. 10.64*b* horizontally at 4 L/s and vertically at 193 kPa/100 m. At the intersection of these lines, the pipe diameter and velocity are determined. Between 38-mm and 51-mm diameter pipe:

velocity = about 2.4 m/s (less than 3 m/s, so OK)

Therefore, a 51-mm diameter supply pipe will be chosen, with a 51-mm meter.

As in the previous Example 10.6, the actual system pressures can be found, using the exact pressure loss in the 51-mm meter. ∎

10.11 Irrigation

The highest priority for landscape watering is to design for optimum rainfall retention (see Figs. 9.13 to 9.20). In the past, provision for the watering of the landscape around buildings has frequently been limited to the installation of sill cocks (hose bibbs) at building exteriors. In many areas of North America, half the residential water usage (see Chapter 11) is for outdoor purposes. With increasing demands on a finite water supply, water-conserving irrigation equipment has become available.

Lawn sprinklers are relatively inefficient irrigating devices, as much of their water is lost to evaporation and to runoff. It is commonly estimated that lawn sprinkling will provide ½-in. of water per hour per square foot of lawn (or 0.3 gph per square foot) (0.005 L/s per m²).

Landscape sprinkling is least efficient during the daytime, when hot sun and dry air combine to increase the rate of evaporation. However, nighttime sprinkling is rarely convenient for building custodians or homeowners. One solution to this problem is the use of *sprinkler timing devices* (Fig. 10.66*a*), which control electronic valves and a network of underground supply pipes and sprinkler heads that are permanently installed within landscaped areas. These timing devices typically include controls governing the length of the watering cycle, the time at which the cycle begins (usually before sunrise, when relative humidity is highest), and the number of days between cycles. A "rainy day switch" is often included, so that the irrigation can be discontinued during rainy weather. Rain sensors can be used to shut off irrigation automatically (Fig. 10.66*b*). Tighter control over the water and plant relationship can be obtained with *tensiometers* (Fig. 10.66*c*), which monitor the moisture content of the soil at the depth of the plants' root zone. These can be installed so as to override the automatic timing device, thus watering more—or less—

(a)   (b)

(c)   (d)

Fig. 10.66 *Components for automatic landscape sprinkling systems.* (a) *Programmable controllers feature multiple stations, up to 14-day sequences, and memory provisions in case of power failure. (Courtesy of Rain Bird Sales, Inc.)* (b) *A solid-state rain sensor automatically prevents needless watering during rainfall.* (c) *A tensiometer controls irrigation by monitoring the moisture content of the soil. (Courtesy of Water Conservation Systems, Inc.)* (d) *Bubblers are low-flow substitutes for sprinklers—a step toward drip irrigation. These can be pressure-compensating, to permit constant flow. (Courtesy of Rain Bird Sales, Inc.)*

(a)   *(b)*

Fig. 10.67 *Equipment for drip irrigation. (a) On low-pressure lines, emitters are installed—one for each group of plants to be watered. The lines can be laid on the ground surface, or just under the surface. (b) Simple emitter boxes allow easy access. (Courtesy of Rain Bird Sales, Inc.)*

frequently, depending on the plants' needs. Instead of sprinklers, *bubblers* (Fig. 10.66*d*) can be installed, with very low flow rates and less evaporative water losses.

Drip irrigation takes a very different approach to that of the flooding method typical of sprinklers. From a network of plastic tubes, either just underground or on the surface, *emitters* (Fig. 10.67) slowly and steadily drip water onto the ground surface at each needy plant. Most of this water soaks into the soil at a rate that is better for most plants than the sudden, short flooding of intermittent sprinkling. Two requirements are especially important for drip systems: the water must not contain materials that can clog the small holes of the emitters, and the pressure must be low. Pressure-reducing valves at the source of the drip system are advisable; if necessary, filters should also be installed there.

Drip irrigation is not a universal solution to landscape watering; it is best for individual plants such as shrubs and small trees, but is difficult to apply to large lawns. Where appropriate, it can

achieve significant water conservation compared to sprinklers. A detailed discussion of soil texture, moisture, and plant growth is provided in Lowry (1988, Chap. 4).

Recycled or reclaimed water for irrigation, often provided by graywater systems, sometimes by stored rainwater, is gradually gaining acceptance from codes in North America. Considered nonpotable, this water must sometimes be deliberately colored, sometimes be accompanied by warning signs at each outlet. This topic is explored further in Section 11.9.

Irrigation, past and future, shaped the design of the Napa Valley Museum, California (Fig. 10.68). An adjacent creek provides water behind a small dam. A pump house carries water to a central observation tower, where the visitor sees the history of irrigation displayed in the context of the Napa Valley. The path of water in this model landscape invites the visitor to explore canal irrigation of the 18th and 19th centuries, to advanced drip irrigation and xeriscape plantings of the 20th century and beyond. (Xeriscape plant-

(a)

(b)

Fig. 10.68 *The Napa Valley Museum shows the history of irrigation in northern California. (a) Site plan. Water, piped from behind a small dam in the adjacent creek, flows through open canals, then drip irrigation, eventually discharging to a field of native plants. (b) The canals help guide the visitor through the exhibit gardens. (Courtesy of Fernau and Hartman, architects, Berkeley, California.)*

ing, as a part of the Hydrozone landscape concept, was shown in Fig. 9.22.)

REFERENCES

ANSI (1986). *American National Standard for Buildings and Facilities A117.1-1986 Providing Accessibility and Usability for Physically Handicapped People,* American National Standards Institute, New York.

Banham, R. (1969). *The Architecture of the Well Tempered Environment,* Architectural Press, London.

Kira, A. (1966, 1976). *The Bathroom: New and Expanded Edition,* Bantam Books, New York.

Lowry, William P. (1988). *Atmospheric Ecology for Designers and Planners,* Van Nostrand Reinhold, New York.

Milne, M. (1976). *Residential Water Conservation,* U.S. Office of Water Research and Technology, Department of Commerce, NTIS.

National Drinking Water Clearinghouse, *Tech Brief* Series:
(June 1996). *Disinfection*
(September 1996) *Filtration*
(February 1997) *Corrosion Control*
(May 1997) *Ion Exchange and Demineralization*
U.S. Environmental Protection Agency, Washington, DC.

U.S. Environmental Protection Agency (1975). *Manual of Individual Water Supply Systems,* U.S. EPA, Washington, DC.

U.S. Architectural and Transportation Barriers Compliance Board, *Americans with Disabilities Act (ADA) Accessibility Guidelines for Buildings and Facilities,* 1331 F Street N.W., Washington, DC 20004.

WATER AND WASTE

11

WATER AND WASTE

In both Chapters 9 and 10, we considered the waste of resources inherent in the use of potable water to flush toilets. In this chapter we examine some consequences of this conventional approach to bodily waste removal, along with alternatives that use no water at all, and indeed that offer to convert "waste" to a useful resource. For typical U.S. residences, the potential impact of such alternatives on water usage and treatment is shown in Fig. 11.1.

Almost every plumbing fixture within buildings is provided with both water supply and waste pipes. Because the toilet is usually the largest user of water (Fig. 11.1) as well as one of the worst water polluters, this chapter begins with recycling/waterless alternatives to the flush toilet and urinals that were discussed in Section 10.7. Then, more conventional systems using water to move waste, and conventional treatment systems from individual to large scale, are discussed in Sections 11.2 through 11.8. Graywater and stormwater treatment complete this chapter in Sections 11.9 and 11.10.

11.1 Waterless Toilets and Urinals

The array of waterless alternatives include toilets in which chemicals or oil are substituted for water. These devices are commonly found in airplanes, vehicles, and boats, as well as in remote and environmentally sensitive areas. The chemicals must be frequently recharged, and the waste products removed. Other waterless toilets temporarily treat the waste awaiting discharge to a sewer by freezing it, burning it or otherwise packaging it so as to remain inoffensive. Obviously, such devices can become energy-intensive solutions to waterless waste disposal. Also, they are still dependent upon public sewer systems.

(a) Composting Toilets. Once considered too unconventional to ever gain mainstream acceptance, these ecologically attractive devices are now appearing not only in homes but within institutional buildings (Fig. 11.2). Perhaps several generations needed to pass between regularly using the outhouse and using these composting toilets? They offer a self-contained and much less energy-intensive strategy for waterless toilets. Wherever water supplies are unreliable, or waste discharge difficult or prohibited, composting toilets offer an alternative. Silent and immune from potential water damage (freezing pipes, or overflowing), these systems rely upon *aerobic* digestion of waste (i.e., that which occurs in the presence of oxygen). Aerobic systems usually are essentially odor-free, and the exhaust air is rich in CO_2 and water vapor. In contrast, *anaerobic* decomposition (that which occurs without oxygen) is malodorous and produces methane gas as an important by-product. See Sections 11.7 and 11.8 for more about methane and anaerobic decomposition.

Waste material deposited in these toilets builds up into a mound, which retards the flow of air for decomposition. Wood shavings are commonly added with each use, to encourage toppling of the mound, and to add some carbon to the resulting compost. Some composting toilets include a means of regular mechanical stirring of the compost chamber, aiding aerobic decomposition.

Ventilation is important, both to reduce odors and to facilitate evaporation of excess moisture. A ventilation stack is an essential feature; it is often assisted by a very small (3 to 5 W) fan. Problems with insect hatches within the composting chamber can be lessened by keeping vegetable scraps from the chamber, as they are frequent sources of insect eggs.

The C. K. Choi Building at the University of British Columbia (Fig. 11.2) eliminated a connec-

WATER AND WASTE

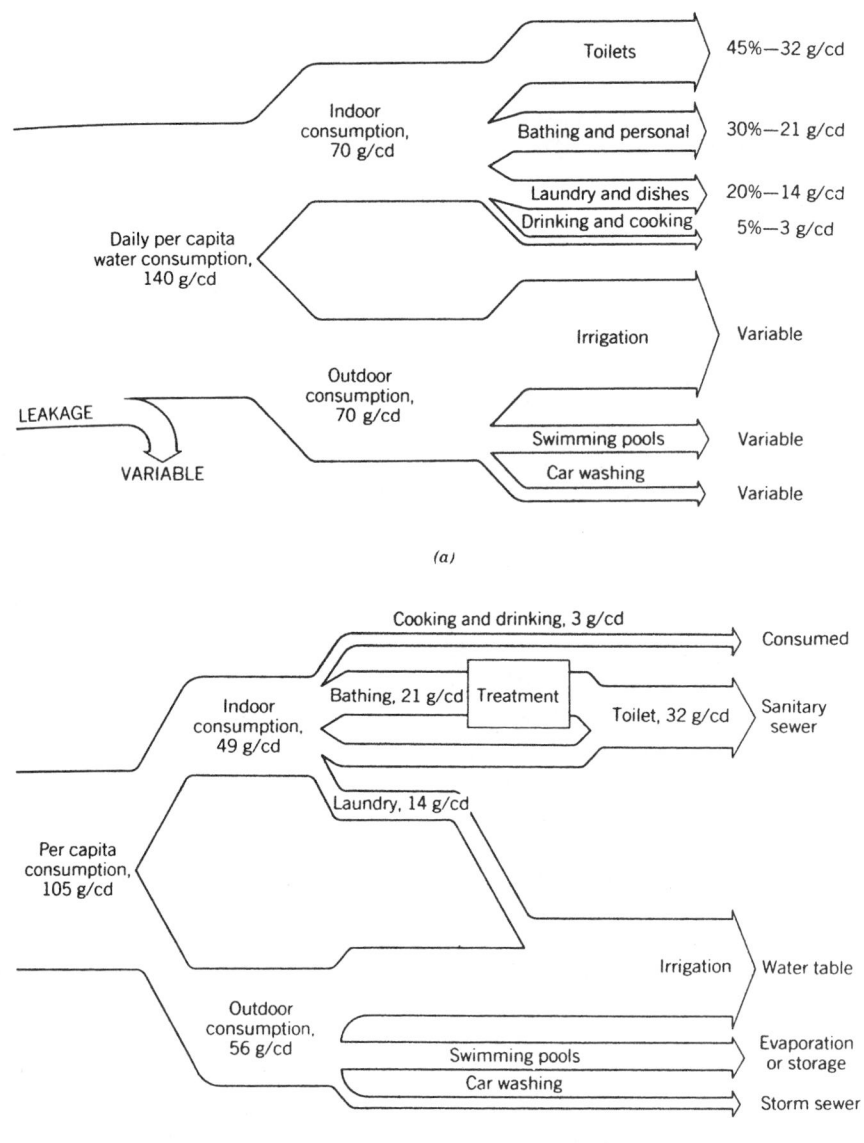

(a)

(b)

Fig. 11.1 *Opportunities for water conservation for the typical U.S. family.* (a) *At present the average U.S. residential usage is about 140 g/cd (530 L/cd)—all of it potable.* (b) *With attention to recycling and the matching of water quality to usage, potable water usage could be cut to 105 g/cd (397 L/cd)—a 25% reduction.* (c) *With further on-site treatment and recycling, potable water usage drops to 73 g/cd (276 L/cd)—a 40% reduction—and the need for public sewage treatment (for residences) disappears. (From Milne, 1976.)*

tion to the city sewer system by a combination of composting toilets and a graywater treatment system. The composting toilets in this three-story, 30,000 ft² (2,787 m²) facility for the Institute of Asian Research are estimated to save 1,500 gallons (5,680 L) of water per day. At an estimated

30,000 uses per year, these composting toilets are expected to produce an annual compost yield of about 4 ft³ (0.1 m³). In addition, their liquid end product will be added to the graywater treatment system, at the rate of about 3 to 5 gal (11 to 19 L) per day. The composting chambers are located

(c)

Fig. 11.1 *(continued)*

in a small basement below the toilet rooms at the center of the building.

The graywater includes waste from lavatories and sinks, as well as liquid from the composting toilets. This water drains into the subsurface "graywater trench," containing a variety of phragmite plant material. The roots support microbial life known for digesting and neutralizing bacteria. This graywater is then used for irrigation. Graywater systems are discussed further in Section 11.9.

Other features of this unusual institutional building include natural ventilation with night flush cooling, ceiling fans, daylighting, and rainwater collected for irrigation, instead of using city water. The 7,000 gal (26,500 L) rainwater cistern sits with the composting chambers in the basement.

(b) Vault Type Composting Toilets. The *Clivus Multrum system* (Fig. 11.3) has a large decomposition chamber that must be below the toilet and the kitchen, from which it readily accepts organic waste. It must also be accessible to remove the humus—from 3 to 10 gal (11 to 38 L) of soil per person per year.

The relatively large chamber (about the size of a Volkswagen "bug") and its low position/access requirement can have a significant impact on design. Also of significance is the device's ap-

petite for fresh air. The more air, the speedier the process of aerobic decomposition and, therefore, the less odor. In winter, this air could cause increased heat losses by infiltration in a building. The natural ventilation due to the stack effect (Section 5.6b) is at least 15 cfm (7 L/s) in the Clivus Multrum. A ventilating fan in the stack will increase this rate significantly. However, if outdoor air is brought directly to the chamber, it could be too cold for proper decomposition; 98°F (36.7°C) is optimum. Solar heating of such input air offers one alternative. However, adding too much heat to compost piles encourages oxidation rather than aerobic decomposition; inferior compost results.

Another composting toilet application is shown on the shores of Oregon's Crater Lake National Park in Fig. 11.4. Consider the challenge: this facility is at the bottom of a 1-mile trail down from the crater rim; public sewers are miles away and hundreds of feet higher in elevation; severely cold winters and deep snow close the trail for more than half the year; and Crater Lake is one of the clearest in the world, so water pollution is absolutely to be avoided. Three Phoenix composting toilet tanks sit within a solar-heated attached sunspace.

The *Phoenix* composting tanks (Fig. 11.5) feature rotatable tines that speed aerobic decomposition. They also can be fitted with an evaporation

(a)

Fig. 11.2 *Composting toilets save water—indeed, eliminate a sewer connection—at the C. K. Choi Building, University of British Columbia, Canada. (a) A long narrow plan saved a grove of trees and enhances daylight penetration. The toilet rooms are concentrated near the stairs at the center, with a small basement below containing composting chambers and a rainwater cistern for irrigation. (b) Natural ventilation and daylight join the passive strategies of water conservation. (c) The composting toilets produce both dry humus and a liquid outflow to a graywater recycling system. (Courtesy of Matsuzaki Wright, architects, Vancouver, British Columbia.)*

system, typically using a PV-driven small pump that delivers liquid to the top of a drip-tank filled with plastic balls. This presents a very large surface area, and a small fan also aids evaporation.

(c) Heater Type Composting Toilets. Yet another approach puts the composting chamber within the toilet stool itself. The more compact

the composting chamber, the more likely that heat must be added to speed decomposition, and to evaporate liquid. Although electricity is an easy and compact way to heat, this energy used in treating waste reduces the compact toilet's ecological advantage.

In the most compact of *Sun-Mar's* products (Fig. 11.6) a variable-diameter "Bio-drum™" is

SOUTH FACING ROOFS
OPTIMIZE DIRECT GAIN AT
PHOTO-VOLTAIC PANELS

HIGH NORTH FACING
WINDOWS ADMIT
DIFFUSE LIGHT

REFLECTIVE SURFACES
ENHANCE DAYLIGHTING

WINDOWS ARE OPERABLE

ATRIA ADMIT DAYLIGHT
INTO SECOND FLOOR
WORK SPACES

WINDOWS WITH LOW FLOW
VENTILATION APERATURES
FOR CONTINUOUS FLUSHING
OF BUILDING

HIGHER CEILINGS
AT GROUND FLOOR
FACILITATE DAYLIGHTING

CHIMNEY FACILITATES
GROUND FLOOR
STACK VENTILATION

(b)

CARBON DIOXIDE

WATER VAPOUR

AIR

VENT

WASTE

FAN

AIR INTAKE

COMPOSTING
WASTE

AIR CHANNEL

FINISHED
COMPOST

LIQUID OUTFLOW
TO GRAY WATER
SYSTEM

(c)

Fig. 11.2 (continued)

WATER AND WASTE

(a)

Fig. 11.3 *Clivus Multrum System, with composting of both kitchen and toilet organic wastes. (a) The lower composting chamber requires $4 \times 8 \times 7$ ft ($1.2 \times 2.4 \times 2.1$ m) of floor space, a generous supply of air (warm air speeds decomposition), and access for periodic removal of the garden-ready humus. (b) It must be arranged so that its upper end receives toilet wastes; as the bottom slopes downward, kitchen wastes are deposited on top of the decomposing toilet wastes. The vent stack assures continuous airflow, preventing odors from entering bathroom or kitchen. (c) The toilet seat and cover must be kept closed when not in use, so as to keep air flowing through the composting chamber and out the stack. The bowl (containing no water) swings open when the seat is occupied, exposing the composting chamber below. (Courtesy of Clivus Multrum USA, Inc.)*

Exhaust vent
CO_2 H_2O out

Sink (grey water)

Garbage chute

Toilet

Kitchen
waste
chamber

Toilet waste
chamber

Baffle

Optional
midsection

Hatch

Baffle

Starter layer of
peat and top soil

Air in

Humus
storage
chamber

(b)

Fig. 11.3 (continued)

(c)

Fig. 11.3 *(continued)*

turned by a fold-out hand-crank at the front of this rather conventional-looking toilet. At this size, it is rated for continuous residential use by one adult, or a family of two (for weekend/vacation use, by three adults or family of four). Sun-Mar uses a three-chamber approach, separating the quick composting of waste and toilet paper, the evaporation of liquid, and the safe storage of the finished compost.

(d) Waterless Urinals. Urine is often kept out of composting toilets for several reasons. First, excess liquid in the chamber creates a discharge issue, complicating an otherwise simple end result of occasional harvesting of garden-ready humus. Second, urine is nearly sterile and an excellent fertilizer, and its separate capture could be useful. If not captured, urine is directed either to the graywater or to the waste water system.

Waterless No-Flush™ urinals utilize a floating layer of BlueSeal® liquid that forms a barrier to

(a)   (b)

Fig. 11.4 *On the shore of Crater Lake, Oregon, (a) a building with three Phoenix composting toilets serves visitors. An attached sunspace (b) contains and warms the composting chambers, and contains an evaporator for liquid waste. The facility is closed all winter.*

Exhaust

Toilet

Fan

Spray System

Air Baffles

Rotatable Tines

Liquid Treatment

Air

(a)

(b)

Fig. 11.5 *The Phoenix composting toilet's tank serves either two toilets or one toilet and one food waste inlet. (a) Schematic section; rotatable tines mix waste and control movement of compost to access area for eventual removal. The fan exhausts air from the Phoenix tank to control odors. Passing behind the air baffles, air makes frequent contact with the compost pile assuring aerobic decomposition. After passing through the compost pile and receiving secondary treatment in the bottom of the Phoenix, liquid is periodically sprayed on top of the compost pile to keep the pile moist and inoculate fresh waste with decomposing organisms. Each tank is 39 in. (1000 mm) wide, 61 in. (1550 mm) long plus another 5 ft (1525 mm) in front for maintenance access, and is available in three heights: 55 in. (1400 mm), 73 in. (1850 mm) or 92 in. (2350 mm). Another 12 in. (300 mm) above the top of the tank is convenient for maintenance. (b) Four chambers with their rotatable tines serve the Vogelsang High Sierra Camp at Yosemite National Park, California. (Courtesy of Advanced Composting Systems, Whitefish, Montana.)*

Fig. 11.6 *The Sun-Mar "Compact" composting toilet has a "Bio-drum™" that is turned by a hand-crank. The crank folds back into the body of the toilet after use. It also uses a 200 W electric heater to speed decomposition, a 25 W fan, and a 2-in. (50-mm) vent that attaches at the rear. The self-contained unit is designed for light residential use. (Courtesy of Sun-Mar Corporation, Tonawanda, New York.)*

(a)

Cross - Section of EcoTrap®

X-Traptor®
Access Slots

EcoTrap®

BlueSeal®

Urine

To Drain

(b)

Fig. 11.7 *Using no water,* (a) *the Waterless No-Flush™ urinal employs a special liquid to allow urine to pass through* (b) *while keeping sewer gas trapped in the waste pipe. (Courtesy of the Waterless Company LLC, Del Mar, California.)*

sewer vapors, but allows urine to readily pass through (Fig. 11.7). This liquid does not dissolve, mix or react chemically with urine. It is more than 95% biodegradable, and does not evaporate at 100°F (38°C). Because one quart (0.9 L) of this liquid lasts for 15,000 to 20,000 urinal uses, it re-

places an estimated 15,000 to 60,000 gal (56,780 to 227,120 L) of water that would have been used by conventional flush urinals. The trap retains sediments and should be emptied occasionally, and the liquid is periodically replenished [3 oz (89 mL) per 1500 uses].

11.2   Principles of Drainage

Early in the history of indoor plumbing, waste drainage was a simple matter; a pipe containing the wastewater led to a sewer (Fig. 11.8a). Before long, the noxious gases that were created by the anaerobic conditions in the sewer became a threat to the health of those indoors. Thus, the *trap* was invented (Fig. 11.8b) to block the pipe so that gases could not pass. However, as moving water filled the pipe downstream from the traps, the mass of water acted as a plunger, creating higher pressures in front of it and negative pressures behind. The positive pressures might force sewer gas through the water in other traps; worse, the negative pressures could suck (or siphon) the water from the trap, leaving it open to gas passage. A way to deal with these pressures was through the installation of *vents,* so that the suction would draw air down through vents rather than water from the traps (Fig. 11.8c). A typical arrangement of fixtures, traps, and vents is shown in Fig. 11.9.

(a) Traps. The only separation between the unpleasant and dangerously unhealthy gases in sanitary drainage pipes and the air breathed by room occupants is the water caught in the fixture trap after each discharge from a fixture. Sufficient water must flow, especially in water closets, to keep this residual water clean. Traps are made of steel, cast iron, copper, plastic, or brass—except those in water closets and urinals, which are often made of vitreous china cast integrally with the fixture. The deeper the seal, the more resistance to siphonage but the greater the fouling area; therefore, a minimum depth of 2 in. (50 mm) and a maximum depth of 4 in. (100 mm) are common standards. All traps should be self-cleaning: that is, capable of being completely flushed each time the trap operates, so that no sediment will remain inside to decompose.

If A is greater than B the siphon
will empty the upper tank
as it would the trap.

Adding a vent stops
the siphon by admitting air.

Gases

Gases

Fixture vent

Crown

Trap

Dip

Cleanout

INCORRECT

INCORRECT

"Seal of the
trap 2" to 4"

The vent has broken the
siphon by admitting air.
Water at A and B find
their destination by gravity.

(a) Incorrect, no trap
Foul gases enter the room

(b) Incorrect, no vent
The water that should seal
the trap is siphoned out

(c) Correct
Air entering from the vent
breaks the siphonage

Fig. 11.8 *Function of a trap and one of the several functions of a vent (preventing siphonage).*

There are a few exceptions to the rule that each fixture should have its own trap. Common exceptions include: two laundry trays and a kitchen sink connected to a single trap; not more than three laundry trays using one trap; and three lavatories on a single trap. In the case of the laundry trays and sink, the sink is equipped with the trap and set nearest to the stack. (See stack *b* in Fig. 11.9).

Traps are usually placed within 2 ft (610 mm) of the fixture and should be accessible for cleaning through a bottom opening that is otherwise closed by a plug. Overflow pipes from fixtures are connected into the inlet side of the trap. In long runs of horizontal pipe, so-called "running traps," are used only near the drains of floors, areas, or yards and should be provided with hand-hole cleanouts. "Island" sinks pose a special problem when the

vent line cannot lead upward from such an exposed location. The sink's waste line can be taken to a distant sump, which is then itself trapped and vented (see Fig. 11.11c for a similar process).

When fixtures are used very infrequently, the water in traps can evaporate into the air, breaking the seal of the trap. In contemplating the possible frequency of use, this fact should be kept in mind by the designer. Unoccupied residences (such as weekend or vacation homes) are likely candidates for sewer gas penetration through traps emptied by evaporation. Otherwise, evaporation to a dangerous degree rarely occurs, except in the case of floor drains. Used to carry away the water used in washing floors or drained from heating/cooling equipment, floor drains may often lose the water seal between infrequent operations. Some authorities are reluctant to approve floor drains con-

Fig. 11.9 *Typical sanitary drainage system, which separates stormwater from sewage. This installation combines blackwater (from WCs) and graywater, although their separation would encourage graywater recycling. The housetrap is optional under some plumbing codes, and illegal under others (because sanitary sewers could be vented at street level through the "fresh air inlet").*

nected to the building's sewer, requiring instead that they be separately connected to a dry well. In either case the use of a special hose bibb, affording a source of water directly above the drain, is a wise precaution. It can easily be used to manually refill the trap of the drain. Another strategy is to lead overflow lines from lavatories to the floor drain trap.

(b) Vents. To admit air and discharge gases, soil and waste stacks are extended through roofs, and a system of air vents, largely paralleling the drainage system, is provided. As in the case of drainage stacks, the ventilating stacks extend through the roof or vent through the drainage stack. The functions of venting are often misunderstood. It is true that one important purpose is to ventilate the system by allowing air from the fresh-air inlet (or from the sewer, if there is no

house trap or fresh-air inlet) to rise through the system and carry away offensive gases. This provides some purification for the piping. However, several other purposes are served by the vent piping. The introduction of air near the fixture (and in the case of circuit vents, at the branch soil line) breaks the possible siphonage of water out of the trap. Under other circumstances—namely, when drainage fluids descend to a fixture group through the soil stack—the foul gases, under pressure, would bubble through the trap-seals of that group. The vent system provides a local escape for these gases. *Circuit venting*—which permits air and gases to pass in and out of the soil or waste branch instead of at each fixture, as in the case of continuous venting (individual fixture venting)—can prevent the siphonage of trap-seals or their penetration by gases (Fig. 11.10). Some codes may prohibit circuit venting.

The part of the soil stack
above the highest branch
is called the stack vent

Branch vent   Soil stack

Individual vents

Branch soil   A circuit vent at each floor
                               affords relief similar to that
                               provided by individual
                               fixture-vents

Circuit vents   Vent stack

Gases are relieved through the   Liquids descending from
vent system instead of emerging   top floor push gases ahead.
through the trap seal.

Fig. 11.10 *Sewer gas relief through vents. Gases pressurized by hydraulic action or by expansion due to putrefaction have an escape path through the vent system and will not enter the rooms. The upper floor shows* individual vents *(a vent at each fixture), the lower floor shows* circuit vents, *permitted by some codes, where one vent serves the entire branch.*

(c) Air Gaps and Vacuum Breakers. Nearly every plumbing fixture is supplied with pure water at one point, and most discharge contaminated fluids at another. The proximity of sewage to potable water at typical fixtures is inescapable; sewage could accidentally be siphoned into a pipe carrying potable water. Consider an improperly placed faucet whose outlet is below the rim of a fixture. If the fixture overflow is plugged and the fixture bowl full, the faucet can easily project into the foul drainage water. If, in this circumstance, the water piping is drained while the faucet is open, contaminated water could be drawn by suction into the water piping.

In water closets served by flushometers (flush valves), the water supply unavoidably enters the bowl below the rim. A vacuum breaker placed in

the flushometer closes with water pressure but opens to admit air if there is suction in the water pipe. This prevents siphonage in much the same way that a vent prevents trap siphonage (Fig. 11.11). The use of vacuum breakers at dishwashers and clothes washers was diagrammed in Figure 10.52. These are especially important locations because in both these appliances pumps force the wastewater into the drainline.

11.3 Piping, Fittings, and Accessories

(a) Piping and Fittings. The materials used within buildings for soil and waste piping and for venting include cast iron, copper, ABS plastic,

If faucet were below rim of a
full sink and the water supply
drained, back siphonage could
pollute potable water.

Vacuum breaker admits air
when suction occurs in the water
supply pipe to the flushometer.

Water from small
water cooled device

Connected to
potable water

Connected to
potable water
(protected by
air gap)

Potable water

Air gap
Unsanitary

Air

Air gap for
indirect waste

Unsanitary

Unsanitary

(a) **Air gap**   (b) **Vacuum breaker**   (c) **Indirect waste**

Fig. 11.11 *Backflow preventers. Unsanitary fluid wastes cannot be siphoned into the potable water piping in these examples.*

and PVC type DWV pipe (Table 11.1, Fig. 11.12). Galvanized steel is sometimes chosen for vents and for tall stacks in high-rise structures. Sometimes, different materials are used in the same system. Where dissimilar metals are connected, dielectric unions are used to prevent corrosion due to electrolysis. Materials used underground for sewage disposal (depending upon local codes) include vitrified clay tile, cast iron, copper, asbestos-cement, ABS plastic, PVC type DWV, and concrete pipe.

Cast iron. Used first in Germany around 1562 and appearing in the United States about 1813, cast iron supplanted the tubing and culverts of early epochs that employed clay, lead, bronze, and wood. Cast iron was thus the earliest of the modern materials used for piping. Its durability and resistance to corrosion makes it appropriate for a wide range of uses, from small residential work to the stacks and branches of tall buildings.

Typical fittings for sanitary drainage appear in Figs. 11.12 and 11.13. In sanitary flow systems

TABLE 11.1 **Materials for Waste Piping**

| | | Underground | |
	Above Ground DWV	Building Drainage and Vent	Building Sewer
Material			
ABS (Acrylonitrile butadiene styrene)	√	√	√
Asbestos–cement		√	√
Brass	√		
Cast-iron	√	√	√
Concrete			√
Copper (Type)	√ (K, L, M, or DWV)	√ (K or L)	√ (K or L)
Galvanized steel	√		
Glass	√		
Polyolefin	√	√	
PVC (polyvinyl chloride) type DWV	√	√	√
Vitrified clay			√

Source: Based on the *International Plumbing Code.* Copyright © 1997, International Code Council, Inc., Falls Church, Va. Reprinted with permission. All rights reserved.

Fig. 11.12 *Piping and fittings.* (a) *Connection of cast iron piping.* (b) *Coupling to connect copper tubing.* (c) *Connection of vitrified clay piping.* (d) *Detail of house trap fitting.* (e) *House drain, house trap with cleanouts and vent (fresh air inlet), and house sewer.* (f) *Cleanout showing removable threaded plug. For large buildings the terms building drain, building sewer, and so on supplant "house." Local codes differ on inclusion or omission of the house trap.* Note: *Plastic connection is similar to* (b).

composed of *any* material, changes in direction must be made with easy bends. To prevent clogging or fouling by the solid materials in the piping, right-angle connections are not used. Thus, the choices in Fig. 11.13 would be for one-eighth bend plus a 45° Y, or a one-quarter bend long sweep. The top connection of the 90° T, in the position shown, would connect *only* to a vent.

The three cast-iron soil pipe joints shown in Fig. 11.14 are semirigid, watertight, and gastight connections of two or more pieces of pipe or fittings in a sanitary system. Types *b* and *c* provide a quieter plumbing system and slightly more flexible joints than type *a*. See Fig. 11.15 for cast-iron soil pipe and fittings in bathroom groups.

Copper tube and fittings for DWV. There are several tube classifications for the copper products used in plumbing systems: types K, L, and M are for water supply systems, and type DWV for *drainage, waste,* and *vent* installations. Connections between copper tubing and its couplings or fittings are made by a sliding fit (see Fig. 11.12b). Between the mating surfaces is a cylindrical capillary space filled with solder. The process of making the joint is a simple one, utilizing a "flux," heat, and solder. Properly made, the joint will be airtight and capable of withstanding high pressure (although such pressure is not normal in waste lines). To undo the joint for repair or renovation, it is simply reheated until the solder melts. Like cast iron, copper has a history of use in ancient installations. Updated and highly developed in recent decades, it is now in widespread use.

Plastic materials for DWV. Along with copper and cast iron, plastics are suitable for sanitary

Fig. 11.13 *Cast-iron fittings, principal types, and method of flashing at roofs for steel and cast-iron.*

drainage systems. Table 11.1 lists the plastics suitable for drainage, waste, and vent, as well as building drains and sewers. Acrylonitrile-butadiene-styrene (ABS) is identified and further evaluated by the labeling shown in Fig. 11.16. One of several steps used in making a "solvent-weld" connection is seen in Fig. 11.16, as is a method of support in wood frame construction. Figure 11.17 shows assembled bathroom piping in place. One advantage of plastic piping is its relatively light weight, encouraging some preassembly under better controlled conditions, then carrying the piping to its place of installation.

(b) Accessories. Among the many special devices that can form part of a plumbing system are floor drains, backwater valves, ejectors, and interceptors.

Floor drains. When floors in buildings must be washed down after such operations as food preparation and cooking, floor drains are usually necessary. Figure 11.18 shows a typical floor drain. Because they are often connected to sanitary drainage systems (rather than dry wells) and, in long periods of disuse, might lose their trap-seals by evaporation, special precautions are necessary

to preserve the trap-seal and avoid odors and unsanitary conditions in the room.

Backwater valves. These devices (Fig. 11.19) are sometimes used when plumbing fixtures are installed at low elevations, such as in basements, or in other locations that are near the level of the sewer. They cannot be used to protect the entire plumbing system, and should only be used when necessary. They must be accessible for maintenance. An alternative to the use of backwater valves is the sewage sump.

Sewage sumps and ejectors. Whenever subsoil drainage, fixtures, or other equipment are situated below the level of public sewers, a sump pit or receptacle must be installed. Into this pit the drainage from the low fixtures may flow by gravity, and from it the contents are then lifted up into the building sewer. Sewer ejectors may be motor-driven centrifugal pumps (Fig. 11.20) or they may be operated by compressed air. The latter have no revolving parts within the receptacle. An air compressor is started when the float within the sump reaches a certain level, and air at a pressure greater than 0.433 psi for each foot

Fig. 11.14 *Various joints used to connect cast-iron soil pipe and fittings. (Courtesy of Cast Iron Soil Pipe Institute.)*

(10 kPa for each m) of lift is delivered into the space above the liquid. The air pressure closes the inlet and opens the outlet check valves, expelling the contents of the sump and elevating it to the sewer.

Interceptors. Sanitary drainage installations ultimately discharge their waste matter into private or public sewage treatment plants that attempt to digest or cope with anything that may come through the pipes. From any plumbing fixture to the end of the disposal process, all parts of systems should be openable through cleanouts and other points of access, to relieve clogging that will often occur in the piping (as well as in the septic tank or public disposal plant). Because it is impossible to control what will be discarded into the plumbing drains, trouble can occasionally be expected. The problem can be reduced somewhat by devices known as interceptors, which catch foreign matter before it travels too far into the system.

Obviously, interceptors require periodic servicing. Interceptors for as many as 25 different kinds of extraneous material are listed by some manufacturers. They include devices to catch hair,

Fig. 11.15 *Two typical piping arrangements for water closet, lavatory, and tub. (Courtesy of Cast Iron Soil Pipe Institute.)*

grease, plaster, lubricating oil, glass grindings, or troublesome unwanted material from many industrial processes. One of the few interceptors that is sometimes used in homes, and more often in institutional kitchens, is the grease interceptor, or *grease trap.*

As the waste is passed from a kitchen sink through the circuitous path within the grease interceptor, the grease floats to the top, where it is trapped between baffles, while the more fluid wastes pass through at a lower level (Fig. 11.21). If not so intercepted, grease congeals within piping and thus physically retards the sewage digestion process.

11.4   Design of Residential Waste Piping

In residential work, the piping assemblies may often be viewed as a "flag," where the mast is the

soil stack, the horizontal top of the flag is the branch vent, the bottom is the soil or waste branch, and the outer edge is the vertical pipe of the last fixture. In frame construction, the "flag" usually fits into a 6-in. (150-mm) partition. Fixture branches project from the surface of the "flag." There is considerable advantage in "back-to-back" planning of baths and kitchens; this allows the piping assembly to pick up the drainage of fixtures on both sides of it. When all the fixtures are on nearly the same level, it is unnecessary to have a separate vent stack standing beside the soil stack, as is often the case in multistory construction. Generally in one-story construction, the upper part of the soil stack forms a vent called a *stack vent,* to which the branch vents connect. A separate major vertical vent would be called a *vent stack.*

The task of fitting the two plumbing "distribution trees" (supply and waste) into available hori-

(a)

(b)

(c)

Fig. 11.16 *Details in the use of plastic pipe.* (a) *One of the steps in making a "solvent weld" of a plastic pipe to a plastic fitting.* (b) *In wood frame construction, plastic pipe assemblies can be supported by metal straps nailed to the wood joists. The supports should be more closely spaced than in the case of metal piping, because of plastic's increased flexibility.* (c) *Typical identification symbols on plastic pipe. (Courtesy of Plastic Pipe Institute.)*

ACME	The name of the manufacturer.
4 in.	Diameter of the pipe.
ABS	Acrylonitrile-butadiene-styrene, the material.
DWV	Suitable for drainage waste and vent.
SCH 40	Schedule 40. This identifies the wall thickness of the pipe.
ASTM D2661	"Standards Number" assigned by the American Society for Testing and Materials.
NSF DWV	Tested by the National Sanitation Foundation Testing Laboratory. The pipe meets or exceeds the current standards for sanitary service.

zontal and vertical chases can be difficult. This is especially true of the waste system, which has larger pipes, because it is not under pressure, and must carry solids as well as liquid. To compound the problem, because of gravity flow requirements these larger pipes must slope continuously downward, from fixtures to the sewer. In residences, the "vertical chase" is sometimes nothing more than an extra-wide wall, increasing the designer's difficulty.

In residential applications and in other relatively small buildings certain fairly standard minimum sizes [such as a 4-in. (102-mm) soil stack and

building drain] are usually adequate. Horizontal fixture branches from *individual* fixtures should be the same size as the fixture trap. Horizontal fixture branches from *groups* of fixtures are sized by the drainage fixture units of the group. For individual fixture vents, the vent size is usually the same as the size of the fixture's horizontal branch. Under many codes, vertical vents that penetrate the roof must increase to a 4-in. (102-mm) size to prevent blocking by icing in freezing weather.

Tables 11.2 through 11.5 list minimum pipe sizes to carry waste and serve for venting.

Fig. 11.17 *Plastic piping (solid lines) for water service, gas service, hot and cold water lines, and for drainage, waste, and vent. Gas service below grade can be PE, PB, or PVC. (Courtesy of Plastic Pipe Institute.)*

Fig. 11.18 *Floor drain. (Courtesy of Josam Manufacturing Company.)*

EXAMPLE 11.1. Design, lay out, and size the piping for the sanitary drainage system for the house shown in Fig. 11.22, page 692.

SOLUTION. The first step is to identify the locations where hot and cold water is needed at fixtures and where soil or waste drains must be provided. Figure 11.23*a* illustrates how this is done. A plan layout for the drains in both levels follows (see Fig. 11.23*b*, page 693).

Next comes the "plumbing section" (Fig. 11.24). The local administrative authority usually requires this to be submitted for approval. Sizes of all piping are determined from Tables 11.2 through 11.5. Drainage fixture units (dfu) for this system are summarized in Table 11.6 from data given in Table 11.2.

Although Table 11.3 permits a 3-in. (76-mm) branch for up to 20 dfu, it is common practice to

Access

Normal flow

Fig. 11.19 *Backwater valve, with access to facilitate unclogging and cleaning.*

Discharge

Submerged water-tight electric motor

Vent

Electrical control

Sewage in

Rugged bladeless centrifugal impeller, inside of housing

Float controls

Suction

Suction

Sewage pump

Sewage ejector pump

Fig. 11.20 *Sump and ejector: a submersible-type centrifugal pump for raising sewage to a higher level. Shown here for an outdoor subgrade sewer installation, it may be used in basement applications within buildings. Venting must be carried to roof. (Courtesy of Weil Pump Company.)*

JH Series Semi-Automatic Interceptor installed on floor, servicing
a double compartment sink.

JH Series Semi-Automatic Interceptor installed on floor, servicing
a sink on floor above.

(b)

JHX Series Semi-Automatic Interceptor installed recessed below
floor, in pit, servicing a double compartment sink.

(a)

Fig. 11.21 (a) *One type of institutional interceptor, a grease trap. Choice of three locations—
adjacent to sink, on floor below, or in a pit. Periodic
cleaning is both necessary and unpleasant; easy
access is helpful.* (b) *Cutaway view with identification of component parts. (Courtesy of Josam Manufacturing Company.)*

use a 4-in. (102-mm) branch for every water closet
(or every group of water closets totaling 20 dfu
or less). Because Table 11.3 allows 160 dfu for a
4-in. (102-mm) branch, this size is acceptable for
any branch to which a water closet is connected.
This size is also more than adequate for the stack
in this example.

A water closet must have a 2-in. (51-mm) vent.
Table 11.4 shows that a 2-in. (51-mm) vent, for
developed vent lengths not exceeding 150 ft (45.7
m), will serve 20 dfu. This size is acceptable for
all branch vents in this example.

The house drain should not be less than 4 in.
(102 mm). From Table 11.5, at a one-quarter-in.
fall per ft (2.1% slope), a 4-in. (102-mm) house
drain will carry 216 dfu, which is more than adequate for our 26 dfu system. ∎

There are prefabricated bathrooms in which
one manufacturer assembles the piping for preselected fixtures (Fig. 11.25). Going further, there
are a few examples of entirely one-piece bathrooms, which incorporate the maintenance advantage of having no seams between fixtures, walls,

TABLE 11.2 **Drainage Fixture Units (dfu)**

Part A. *By Type of Fixture*

Fixture(s)	Drainage Fixture Units (dfu)	Minimum Trap Size (in.)	Minimum Trap Size (mm)[a]
Automatic clothes washers: Commercial[b]	3	2	51
Residential	2	2	51
Bathroom group: Water closet, lavatory, bidet and bathtub or shower	6	—	—
Bathtub[c] (with or without overhead shower or whirlpool attachments)	2	1½	38
Bidet	2	1¼	32
Combination sink and tray	2	1½	38
Dental lavatory	1	1¼	32
Dental unit or cuspidor	1	1¼	32
Dishwashing machine,[d] domestic	2	1½	38
Drinking fountain	½	1¼	32
Emergency floor drain	0	2	51
Floor drains	2	2	51
Kitchen sink, domestic	2	1½	38
Kitchen sink, domestic, with food waste grinder and/or dishwasher	2	1½	38
Laundry tray (1 or 2 compartments)	2	1½	38
Lavatory	1	1¼	32
Shower compartment, domestic	2	2	51
Sink	2	1½	38
Urinal	4	e	
Urinal, 1 gallon (3.8 L) per flush or less	2[f]	e	
Wash sink (circular or multiple) each set of faucets	2	1½	38
Water closet, flushometer tank, public or private	4[f]	e	
Water closet, private installation	4	e	
Water closet, public installation	6	e	

Part B. *By Size of Trap*

Fixture Drain or Trap Size (in.)	Fixture Drain or Trap Size (mm)[a]	Drainage Fixture Unit (dfu) Value
1¼	32	1
1½	38	2
2	51	3
2½	64	4
3	76	5
4	102	6

Source: Based on the *International Plumbing Code.* Copyright © 1997, International Code Council, Inc., Falls Church, Va. Reprinted with permission. All rights reserved.

[a]Metric conversions by the author.

[b]For traps larger than 3 in. (76 mm), see Part B of this table.

[c]A showerhead over a bathtub or whirlpool bathtub attachments do not increase the dfu value.

[d]For ratings of fixtures not listed or for intermittent flows, see the 1997 *International Plumbing Code.*

[e]Trap size shall be consistent with fixture outlet size.

[f]For computing the loads on building drains and sewers, water closets or urinals shall not be rated at a lower dfu unless the lower values are confirmed by testing.

Fig. 11.22 (Example 11.1) Floor plans, house on Long Island, New York. Budd Mogesen, architect and planner.

TABLE 11.3 **Horizontal Fixture Branches and Stacks**[a]

Diameter of Pipe		Horizontal Branch	Maximum Total Number of dfu Allowable		
			Stacks[b]		
			One Branch Interval	Three Branch Intervals or Less	Greater than Three Branch Intervals
(in.)	(mm)[c]				
1½	38	3	2	4	8
2	51	6	6	10	24
2½	64	12	9	20	42
3	76	20	20	48	72
4	102	160	90	240	500
5	127	360	200	540	1100
6	152	620	350	960	1900
8	203	1400	600	2200	3600
10	254	2500	1000	3800	5600
12	305	3900	1500	6000	8400
15	381	7000	d	d	d

Source: Based on the *International Plumbing Code.* Copyright © 1997, International Code Council, Inc., Falls Church, Va. Reprinted with permission. All rights reserved.

[a]Does not include branches of the building drain; see Table 11.5.

[b]Stacks shall be sized based on the total accumulated connected load at each story or branch interval. As the total accumulated connected load decreases, stacks may be reduced in size. Stack diameters shall not be reduced to less than one-half the diameter of the largest stack size required.

[c]Metric conversions by the author.

[d]Sizing load based on design criteria.

Fig. 11.23 *(Example 11.1)* (a) *Plumbing requirements, water supply and partial sanitary drainage.* (b) *Sanitary drainage plan.*

TABLE 11.4 **Size and Developed Length of Stack Vents and Vent Stacks**

Diameter of Soil or Waste Stack (in.) [mm][b]	Total Fixture Units Being Vented (dfu)	Maximum Developed Length[a] of Vent (feet) [m][b]									
		Diameter of Vent (in.) [mm][b]									
		1¼ [32]	1½ [38]	2 [51]	2½ [64]	3 [76]	4 [102]	5 [127]	6 [152]	8 [203]	10 [254]
1¼ [32]	2	30									
		9.1									
1½ [38]	8	50	150								
		15.2	*45.7*								
1½ [38]	10	30	100								
		9.1	*30.5*								
2 [51]	12	30	75	200							
		9.1	*22.9*	*61.0*							
2 [51]	20	26	50	150							
		7.9	*15.2*	*45.7*							
2½ [64]	42		30	100							
			9.1	*30.5*							
3 [76]	10		42	150	360	1040					
			12.8	*45.7*	*109.7*	*317*					
3 [76]	21		32	110	270	810					
			9.8	*33.5*	*82.3*	*246.9*					
3 [76]	53		27	94	230	680					
			8.2	*28.7*	*70.1*	*207.3*					
3 [76]	102		25	86	210	620					
			7.6	*26.6*	*64.0*	*189.0*					
4 [102]	43			35	85	250	980				
				10.7	*25.9*	*76.2*	*298.7*				
4 [102]	140			27	65	200	750				
				8.2	*19.8*	*61.0*	*228.6*				
4 [102]	320			23	55	170	640				
				7.0	*16.8*	*51.8*	*195.0*				
4 [102]	540			21	50	150	580				
				6.4	*15.2*	*45.7*	*176.8*				
5 [127]	190				28	82	320	990			
					8.5	*25.0*	*97.5*	*301.8*			
5 [127]	490				21	63	250	760			
					6.4	*19.2*	*76.2*	*231.6*			
5 [127]	940				18	53	210	670			
					5.5	*16.2*	*64.0*	*204.2*			
5 [127]	1400				16	49	190	590			
					4.9	*14.9*	*57.9*	*179.8*			
6 [152]	500					33	130	400	1000		
						10.1	*39.6*	*121.9*	*304.8*		
6 [152]	1100					26	100	310	780		
						7.9	*30.5*	*94.5*	*237.7*		
6 [152]	2000					22	84	260	660		
						6.7	*25.6*	*79.2*	*201.2*		
6 [152]	2900					20	77	240	600		
						6.1	*23.5*	*73.2*	*182.9*		
8 [203]	1800						31	95	240	940	
							9.4	*29.0*	*73.2*	*286.5*	
8 [203]	3400						24	73	190	720	
							7.3	*22.3*	*57.9*	*219.5*	

TABLE 11.4 Size and Developed Length of Stack Vents and Vent Stacks (*continued*)

Diameter of Soil or Waste Stack (in.) [mm]b	Total Fixture Units Being Vented (dfu)	Maximum Developed Lengtha of Vent (feet) [m]b									
		Diameter of Vent (in.) [mm]b									
		1¼ [32]	1½ [38]	2 [51]	2½ [64]	3 [76]	4 [102]	5 [127]	6 [152]	8 [203]	10 [254]
8 [203]	5600						20 / 6.1	62 / 18.9	160 / 48.8	610 / 185.9	
8 [203]	7600						18 / 5.5	56 / 17.1	140 / 42.7	560 / 170.7	
10 [254]	4000							31 / 9.4	78 / 23.8	310 / 94.5	960 / 292.6
10 [254]	7200							24 / 7.3	60 / 18.3	240 / 73.2	740 / 225.6
10 [254]	11,000							20 / 6.1	51 / 15.5	200 / 61.0	630 / 192.0
10 [254]	15,000							18 / 5.5	46 / 14.0	180 / 54.9	570 / 173.7

Source: Based on the *International Plumbing Code.* Copyright © 1997, International Code Council, Inc., Falls Church, Va. Reprinted with permission. All rights reserved.

aThe developed length is measured from the vent connection to the open air.

bMetric conversions by the author.

TABLE 11.5 Building Drains and Sewers

Diameter of Pipe		Maximum Number of dfu Connected to Any Portion of the Building Drain or Building Sewer, Including Branches of the Building Draina			
		Fall, in. per ft (% slope)			
(in.)	(mm)b	1⁄16 (0.5%)	1⁄8 (1.04%)	1⁄4 (2.1%)	1⁄2 (4.2%)
2	51			21	26
2½	64			24	31
3	76		36	42	50
4	102		180	216	250
5	127		390	480	575
6	152		700	840	1,000
8	203	1,400	1,600	1,920	2,300
10	254	2,500	2,900	3,500	4,200
12	305	2,900	4,600	5,600	6,700
15	381	7,000	8,300	10,000	12,000

Source: Based on the *International Plumbing Code.* Copyright © 1997, International Code Council, Inc., Falls Church, Va. Reprinted with permission. All rights reserved.

aThe minimum size of any building drain serving a water closet shall be 3 in. (76 mm).

bMetric conversions by the author.

WATER AND WASTE

Fig. 11.24 (*Example 11.1*) *Plumbing section. When every fixture is vented individually, as in this example, the method is known as continuous venting. In larger systems, batteries of fixtures may be vented by a loop or circuit vent. This reduces the piping of the vent system (see Figs. 11.10 and Fig. 11.30). Note that the septic tank discharge to a leaching pit is not permitted in many locations; see Section 11.6.*

TABLE 11.6 **Drainage Fixture Units (dfu)
for Example 11.1**

Residential Fixture	Drainage Fixture Units (dfu)
Bar sink	2
Kitchen sink with dishwasher	2
Lavatory	1
Water closet	4
Automatic clothes washer	2
Master bathroom group	6
Extra lavatory	1
Shower	2
Lower floor bathroom group	6
Total	**26**

Dfu values from Table 11.2. The hose bibb drains to the ground; roof drainage is taken to dry wells.

and floors. This, of course, makes fixture replacement expensive and difficult, and tends to put the burden of access to plumbing into the adjacent rooms.

For the more ordinary bathroom, the two maintenance questions of greatest concern are (1) how easy is cleaning around and within the fixture, and (2) how accessible are those parts of the fixture most likely to need repair or replacement?

Fig. 11.25 *Pre-assembled plumbing "tree" for manufactured housing, along with the prefabricated kitchen/bath in which it is installed. (Courtesy of Wausau Homes, Inc., and the Copper Development Association.)*

Ease of cleaning is often determined more by the space around the fixture than by the design of the fixture itself. This is particularly true of toilets, where a generous space of open floor to either side makes maintenance easy and therefore likely to be more frequent. Access to fixture parts may be more difficult. An access panel in the wall of the room behind the fixtures is often provided, encouraging speedy repair and replacement for fixtures at tubs, showers, and lavatories. Ideally, accessibility to all plumbing lines—whether via access panels in walls, trenches in concrete floors, exposed basement ceilings, or adequately deep crawl spaces—should be provided. The bathroom is likely to undergo thorough "remodeling," including fixture replacement, as styles change and water/energy conservation becomes more important.

11.5 Design of Larger-Building Waste Piping

(a) Basic Planning. Multistory construction, especially in office buildings, is often designed to be flexible and free of random partitions that would interfere in the periodic renovation of interior spaces and the relocating of dividing partitions. Building "cores" contain elevators, stairs, and shafts for plumbing, mechanical and electrical equipment, and are often placed at the central section of the building, freeing the surrounding areas for access to daylight. A hole in the floor for each pipe is often chosen in preference to a slot or shaft. This method usually interferes less with the floor construction (Fig. 11.26).

Offices often need a single lavatory or a complete toilet room for executives at locations away from the central core of the building. The greater the horizontal distance from the core, the more vertical clearance will be needed to allow the drain to slope. When such vertical clearance becomes difficult, "wet" columns with a full complement of plumbing pipes offers a solution. If the pipes are to accompany a column in a steel building, structural coordination must be sought early in the planning if the pipes are to clear the structural framing of the floor (Fig. 11.27).

In some installations, the branch soil and waste

Girders frame into
column from this
direction (to avoid the
pipes)

Line of the
furring
around pipes

Vent stack
Soil or waste stack
Hot water
Cold water
H.W. circulating

Line of the
column
fireproofing

Fig. 11.27 Piping at a "wet" column. In large office buildings, there are usually several of these, remote from the core, in the general office area.

Fig. 11.26 Risers in a fireproof, multistory building. Pipes, tubes, conduits, and ducts adjoin toilet rooms and utility spaces. Ventilation ducts and a master 5-in. (127-mm) copper hot-water riser are just left of center. Soil and vent stacks with hot and cold water supplies, all of copper, are to the right of this group. At the left in lighter tone is the galvanized steel feeder conduit and distribution circuit conduits, of the same metal, for a local electrical control panel box. Note that some pipes and tubes are supported at this floor by bolted clamps. After testing and before pipes are enclosed, covering will be completed. (Courtesy of Copper Development Association.)

piping perforates a floor and crosses below the slab to join the stack. Tubing has been developed, however, that sits above the structural slab, obviating the need for hung ceilings below (Fig. 11.28). A lightweight concrete fill is cast to cover the tubing, raising the floor by 5 or 6 in. (125 or 150 mm). This can create a raised floor in the toilet room—with attendant access problems—so the higher floor level is usually carried throughout the floor of the entire story, forming a convenient space into which the electrical conduit can be placed at a time later than would have been required if it were to be placed in the structural slab. Such "raised floors" are becoming more common,

along with the HVAC strategy of displacement ventilation (see Section 8.4c).

(b) Roughing-In. This is the process of getting all pipes installed, capped, and pressure-tested before the actual fixtures are installed. An example of roughing-in for an office building was shown in Fig. 11.28. The roughing-in of supply and waste piping for school lavatories is shown in Fig. 11.29.

Institutions such as schools have extensive requirements for durability and ease of maintenance. The fixtures are made of such resistant materials as stainless steel, chrome-plated cast brass, precast stone or terrazzo, or high-impact fiberglass. The fixture controls are designed to withstand heavy use—or misuse—and the fixtures are securely tied into the structure with concealed mounting hardware designed to resist extraordinary forces. (Some schools even move the lavatories into the hallway for better visual control of at least a part of the restroom facilities.)

In prisons, extreme measures are taken to prevent plumbing fixtures from becoming weapons. Heavy-gauge stainless steel fixtures with nonremovable fittings are provided, at very high cost for both the fixture and its tamper-proof installation.

Fig. 11.28 *Horizontal waste piping* above *the structural slab, in an example of plumbing roughing for two lavatory rooms in a concrete office building. A lavatory and water closet in each room are served by soil and waste branches below and vent branches above. Hot and cold water tubing with air chambers can be seen. Although the extensions of the water tubing above the two flushometer connections appear to connect into the horizontal vent branches, they actually do not; they are capped and merely touch the bottoms of the vent branches. Note that soil branches are above the structural slab. A fill of 5 or 6 in. (125 to 150 mm) will be necessary to cover the tubing. All vertical tubing will be within the masonry block used to enclose the cubicles. (Courtesy of Copper Development Association.)*

Drainage		*Supply*	
A	Branch soil	F	Hot
B	Lavatory waste	G	Cold
C	Branch vent	H	Flushometer supply (1¼ in.)
D	WC vent	I	Capped air chamber
E	Lavatory vent	J	Capped air chamber (flushometer)

Schematic Plan
——— Soil or waste
– – – Vents

Fig. 11.29 Roughing in place for four lavatories in the toilet room of a school. The waste branches have been capped and the system tested against possible leakage. The waste branches are cast iron, the vents galvanized steel, and the water supply lines of copper with soldered fittings. At the center of the supply array are vertical capped expansion and shock tubes, one for hot, one for cold supply water.

EXAMPLE 11.2. Select sizes for drainage and vent piping for the plumbing in an office building for which the fixtures are shown in Fig. 11.30.

SOLUTION. Selected dfu from Table 11.2 are applied to each section of the piping and totaled for each branch and stack, and for the build-ing drain and the building sewer. Individual fix-ture branches shall not be less than the size indicated in Table 11.3 for the minimum size of trap for each fixture. An example of a fixture-unit summary and sample sizes of individual branches that connect into a typical branch of the men's toilet group on any floor are shown in the follow-ing table.

Features	Units per Fixture	Total Fixture Units	Diameter, Fixture Branch (in.)	Diameter, Fixture Branch (mm)
1 service sink (2-in. trap)	3	3	3	76
3 lavatories	1	3	3	76
3 urinals, washout	4	12	2	51
3 water closets, valve operated	6	18	4	102
Total fixture units, men's toilet branch		*36*		

Fig. 11.30 (Example 11.2) Plumbing section of an office building. Circuit vents serve branch soil lines, but some codes require continuous venting (individual fixture vents). The house trap and fresh-air inlet are omitted from building drain. Note that (a) Relief vent not required on top floor. (b) Men's and women's toilets on third floor are typical and would be repeated on first, second, fourth, and fifth. (c) Drinking fountains on fifth and fourth floors would be repeated on first, second, and third floors.

701

Table 11.3 indicates that a 3-in. (76-mm) horizontal fixture branch is inadequate for the above group because it will handle only 20 dfu. Therefore, a 4-in. (102-mm) pipe is selected (also a wise minimum choice whenever WC are served). Its capacity of 160 dfu will be more than enough for the 36 needed here. The same table shows that the soil stack diameter can be 4 in. (102 mm), and it is run thus for its entire height. Its capacity of 90 dfu per story (or interval) is sufficient for the 64 dfu that connect in at each T-Y connection. For three branch intervals, this size stack will carry 240 dfu (we need $3 \times 64 = 192$ dfu); for its entire height, it will carry 500 dfu (we need $5 \times 64 = 320$ dfu).

We need a vent stack serving up to 338 dfu, with a 4-in. (102-mm) soil stack, and at a maximum 70 ft (21.3 m) developed length. From Table 11.4, a 3 in. (76 mm) vent serving 540 dfu has maximum developed length of 150 ft (45.7 m), well over our minimum requirements. All vent stack diameters will increase to 4 in (102 mm) as they pass through the roof.

According to Table 11.5, the building drain and the building sewer at their pitch of one-quarter in./ft should carry at least our total of 350 dfu. A 5 in. (127 mm) building drain or building sewer has a capacity for 480 dfu. ∎

Although opinion may vary about the relative merits of continuous or circuit venting, either system, properly designed, will effectively prevent the siphoning of traps or relieve air pressures that could cause foul gases to bubble through the traps into the occupied space. Another system, especially suitable for high-rise buildings, eliminates the vent stack completely with equal effectiveness.

(c) The Sovent System. This essentially ventless system changes the nature of the effluent (discharge of wastes and sewage from the fixtures) instead of coping with the pressures and suctions that normal effluent would cause (see Figs. 11.31 and 11.32).

The "plunger" effect of a descending "slug" of water/waste within pipes was described in Section 11.2. If the effectiveness of the "plunger" can be reduced, the negative and positive pressures created by it will be also reduced. If their values can be brought down below the holding power of the several inches of water in the trap, no vents will be necessary. In the single-stack *Sovent* system illustrated in Fig. 11.32, this is done by dealing with the normal liquid effluent at each floor. Aeration there produces a *foam* that lacks the stack-filling tendency of the liquid effluent. Thus, through the creation of a *soft* plunger, pressure variations in the single stack are minimized.

Tests have shown that the positive and negative pressures produced by normal liquid effluent during its descent (and relieved by the vent piping) are often about 5 to 12 in. (127 to 305 mm) water gauge. Obviously, if the vents were not provided, the 2 to 4 in. (51 to 102 mm) of water seal in the traps would be vulnerable to penetration by gases from pipes under positive pressure or siphonage of water seals into pipes that may be under negative pressure.

Figures 11.31, 11.32, and 11.33 illustrate the components and the action of the Sovent system. Effluent, already aerated and descending from upper stories, is diverted in the stack at each lower story. The aerator fitting there affords a passage for this diverted flow and also an air space into which the effluent from the local branch soil or waste can drop. Here it spatters, mixing with the air to form a rarefied mixture of air and liquid. Tests show that this mixture does not produce pressures, positive or negative, of more than 1 in. (25 mm) water gauge. Thus a trap-seal of 2 in. (51 mm) or more is safe against siphonage or penetration.

At the foot of the single stack the aerated effluent is compacted—a process aided by a baffle in the path of the flow in the deaerator fitting (see Fig. 11.33). If not relieved, air piling up at this point could cause pressures in the stack at the first floor. An air-discharge pipe provides this relief of air from the deaerator fitting to the upper part of the building drain, above the liquid flow.

The Sovent system was invented by Fritz Sommer of Switzerland, who tested it in a 10-story drainage test tower. Since its introduction in 1962, it has been installed and used in hundreds of buildings in Europe and Africa. Canada used the Sovent method in the Habitat apartments at the 1967 Montreal exposition. Sovent was first granted U.S. code acceptance in 1968 in Rich-

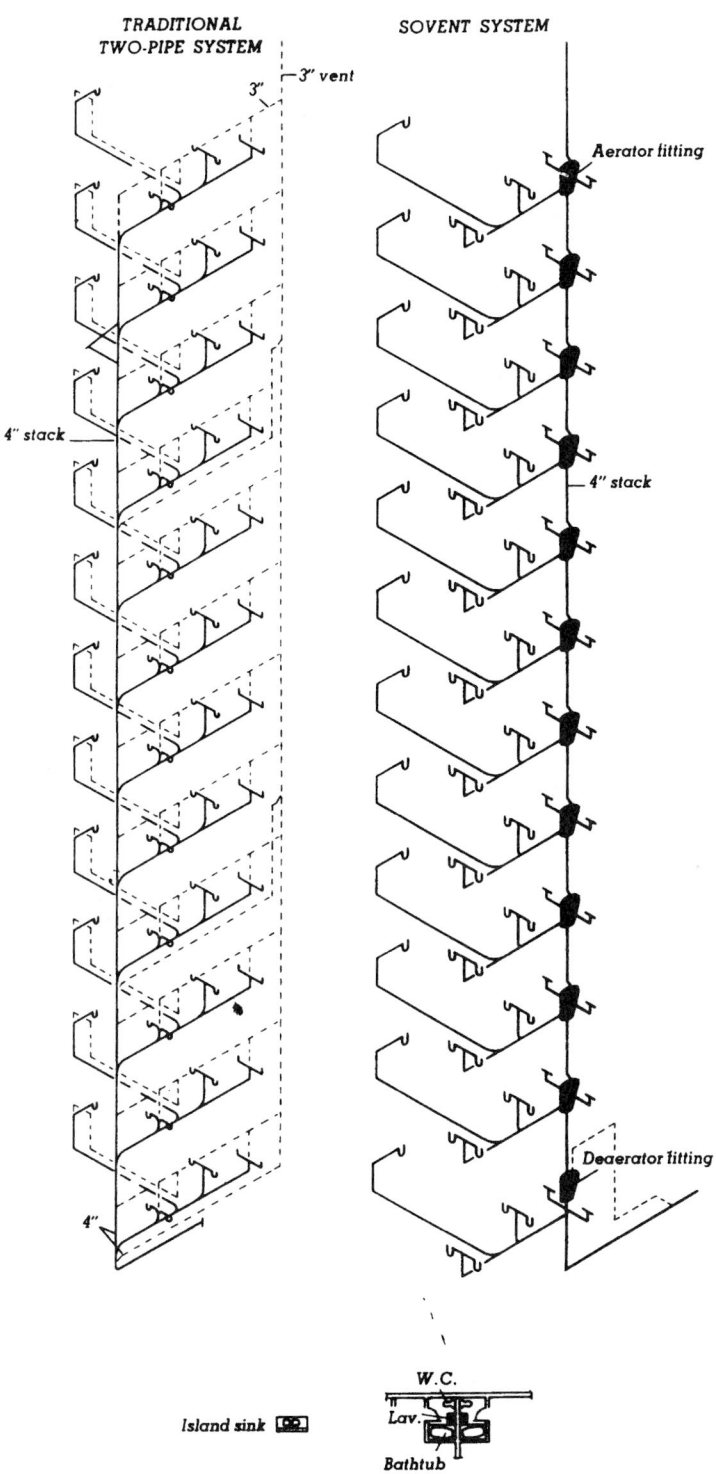

TRADITIONAL
TWO-PIPE SYSTEM

SOVENT SYSTEM

3" vent

3"

4" stack

4"

Aerator fitting

4" stack

Deaerator fitting

Island sink

W.C.
Lav.
Bathtub

Fig. 11.31 *Comparing the Sovent system to a two-pipe system for a 12-story stack serving an apartment grouping. (Courtesy of the Copper Development Association.)*

Fig. 11.33 *The deaerator consists of an air separation chamber with an internal nosepiece, a stack inlet, a pressure relief outlet at the top, and a stack outlet at the bottom. The deaerator fitting at the bottom of the stack functions in combination with the aerator fittings above to make the single stack self-venting. The deaerator is designed to overcome the tendency for the falling waste to build up excessive back pressure at the bottom of the stack when the flow decelerates at the bend into the horizontal drain. (Courtesy of the Copper Development Association.)*

11.6 On-Site Individual Building Sewage Treatment

The great majority of individual-building sewage treatment systems in North America use the septic tank (Fig. 11.34) as a *primary* treatment, where the settling of solids and anaerobic digestion take place. Subsequently, the effluent receives *secondary* treatment, which usually consists of a filtering process. Four common filtration systems are seepage pits, drainfields, mounds, and sand filters. Occasionally, a *tertiary* treatment (usually disinfection with chlorine) must be used, as when outflows from secondary treatment would flow directly into surface waterways.

(a) Primary Treatment: Septic Tanks (Fig. 11.35). These are commonly constructed of precast concrete; steel, fiberglass and polyethylene tanks are also available. The sewage enters the first chamber, where solids sink to the bottom as sludge, and scum forms on the surface. Anaerobic decomposition proceeds, with a production of methane gas. The liquid moves through the submerged opening in the middle of the tank to the second chamber, where finer solids continue to sink, and less scum forms on the surface. Fi-

Fig. 11.32 *The Sovent drainage stack consists of aerator fittings that join the horizontal branches to the stack at each floor level, and a deaerator fitting at the bottom of the stack. The stack is open to the atmosphere above the roof at the top. (Courtesy of the Copper Development Association.)*

mond, California. Following this success, its code acceptance grew rapidly during the early 1970s. However, not all jurisdictions allow Sovent installations.

Fig. 11.34 *For individual sewage treatment systems, septic tanks are commonly used for primary treatment. Four options for secondary treatment are shown here. (Tertiary treatment usually is only required for effluent discharge into waterways.) (a) Seepage pits are rarely allowed. (b) Drainfields constitute the most commonly used options. (c, d) Mounds and sand filters are more expensive to construct and are used where high water tables preclude the use of option (a) or (b).*

Fig. 11.35 *Plan and section of septic tank and seepage pit for four-bedroom house, in sandy loam soil. Pit is only suitable when the earth is absorbent and the water table low (below pit bottom). Drawing is not to scale.*

nally, the effluent, about 70% purified, leaves the septic tank for secondary treatment.

The longer that the sewage stays in the septic tank, the less polluted the effluent. This is why water conservation measures are so welcome with septic tank systems; the less the flow, the longer the water remains in the tank. The anaerobic decomposition process, while malodorous, is so thorough that the sludge needs only occasional removal—once in several years is about average for residences. The methane and its odor are kept within the tank, and access hatches are sometimes covered with soil (as in Fig. 11.35), even though this makes their occasional location far more difficult.

Most systems with septic tanks will eventually experience "failure," usually due to a breakdown in the secondary treatment rather than in the septic tank. With periodic removal of sludge, the septic tank itself is a reliable and simple primary treatment device. However, some types of domestic waste can disrupt the anaerobic process within the tank. Worst offenders are paints, varnishes, thinners, waste oil, photographic solutions, and pesticides. Also to be avoided are coffee grounds, dental floss, disposable diapers, kitty litter, sanitary napkins, tampons, cigarette butts, condoms, gauze bandages, fat and grease, and paper towels.

Septic tank sizes are commonly based on code requirements that consider the number of bed-

TABLE 11.7 **Septic Tank Capacity**[a]

Single-Family Dwellings— Number of Bedrooms	Multiple Dwelling Units or Apartments—One Bedroom Each	Other Uses: Maximum Fixture Units Served[b,c]	Minimum Septic Tank Capacity, Gal (L)[c]
1 or 2		15	750 (2,838)
3		20	1,000 (3,785)
4	2 units	25	1,200 (4,542)
5 or 6	3	33	1,500 (5,678)
	4	45	2,000 (7,570)
	5	55	2,250 (8,516)
	6	60	2,500 (9,463)
	7	70	2,750 (10,409)
	8	80	3,000 (11,355)
	9	90	3,250 (12,301)
	10	100	3,500 (13,248)

Extra bedroom: 150 gal (568 L) each.
Extra dwelling units over 10: 250 gal (946 L) each.
Extra fixture units over 100: 25 gal (95 L) per fixture unit.

Source: Reprinted by permission from the Uniform Plumbing Code, copyright © 1997 by the International Association of Plumbing and Mechanical Officials.

[a]Septic tank sizes in this table include sludge storage capacity and the connection disposal of domestic food waste units without further volume increase.

[b]See Table 11.2

[c]For larger or nonresidential installations in which sewage flow rate is known, size the septic tank as follows.

1. Flow up to 1500 gpd (5,678 L/d):

$$\text{flow} \times 1.5 = \text{septic tank capacity}$$

2. Flow over 1500 gpd (5,678 L/d)

$$(\text{flow} \times 0.75) + 1125 = \text{septic tank capacity in gallons}$$

$$[(\text{flow} \times 0.75) + 4258 = \text{liters}]$$

3. Secondary system shall be sized for total flow per 24 hours.

rooms in residences (Table 11.7) or the number of waste fixture units served (see Table 11.2). (As shown in Table 11.8, maximum size may also be related to the soil condition.) Sewage flow rates are also considered. Oversized septic tanks are more expensive to install, but they release cleaner effluent and will prolong the life of the secondary treatment process.

The size of the secondary treatment system is usually based on the size of the septic tank it serves, or on the expected total flow over a 24-h period. (Flow can be estimated from Table 9.4.) Where effluent must be raised for secondary treatment, or when the total length of secondary treatment disposal line exceeds 500 ft (154 m), a dosing tank (also called siphon or pumping chamber; see Fig. 11.34 *c, d*) is used to automatically discharge the septic tank's outflow chamber to the secondary treatment lines or sand filters.

(b) Primary Treatment: Aerobic Treatment Units (Fig. 11.36). Active rather than passive, ATUs represent an increasingly popular alternative to septic tanks for primary sewage treatment. They are frequently used to replace a septic tank in a troubled system; this "upgrade" can rejuvenate an existing drainfield, and extend its life. ATUs depend on air bubbled through the sewage to achieve aerobic digestion, which is faster than anaerobic digestion; hence, they can be

TABLE 11.8 Septic Tank and Leaching Area Design Criteria for Five Typical Soils

Type of Soil	Required ft² of Leaching Area/100 Gal (m²/L)	Maximum Absorption Capacity, Gal/ft² of Leaching Area for a 24-h Period (L/m²)	Maximum Septic Tank Size Allowable	
			Gallons	Liters
1. Coarse sand or gravel	20 (0.005)	5 (203.7)	7,500	(28,387.5)
2. Fine sand	25 (0.006)	4 (162.9)	7,500	(28,387.5)
3. Sandy loam or sandy clay	40 (0.010)	2.5 (101.9)	5,000	(18,925)
4. Clay with considerable sand or gravel	90 (0.022)	1.10 (44.8)	3,500	(13,247.5)
5. Clay with small amount of sand or gravel	120 (0.029)	0.83 (33.8)	3,000	(11,355)

Source: Reprinted by permission from the *Uniform Plumbing Code,* copyright © 1997 by the International Association of Plumbing and Mechanical Officials.

smaller in size than septic tanks. They are energy intensive and require more maintenance than will the anaerobic tank. A secondary treatment process is required, as well. The effluent typically is less polluted than is that of septic tanks; the biochemical oxygen demand (BOD) is reduced by 90% in ATUs, compared to 50% typical of septic tanks.

The sewage first enters an aeration chamber, where it is kept in turmoil so that air can continue to percolate through it. Either air is forced through the sewage by distribution lines fed by an air compressor, or the sewage can be stirred by a variety of devices, depending on the manufacturer. After about a day's retention in the aeration chamber, the aerated wastewater enters a settling chamber, allowing remaining solids to settle and to be filtered out before the effluent leaves, for further treatment in a secondary process.

(c) Secondary Treatment: Seepage Pits (Fig. 11.35). Also called cesspools, these follow treatment by either septic tanks or ATUs. They are increasingly rare, and only appropriate in very porous soil, where the water table is at least 2 ft (0.6 m) below the bottom of the pit. Because these pits are commonly 10 to 15 ft (3 to 4.6 m) below the earth's surface, a very low water table may be required. Another common usage of precast seepage pits is for "dry wells" that receive runoff from paved areas during rainstorms.

Seepage pits are sized by the square footage of wall area exposed to the earth—the "leaching area," as listed in Table 11.8. Placement of them relative to buildings, water sources, waterways, and property lines is strictly controlled—see Table 11.9 and Fig. 11.37.

(d) Disposal Fields (Fig. 11.38). These are very commonly used as a secondary treatment method because they are relatively inexpensive to build and do not require a water table so deep—or soil so permeable—as seepage pits re-

Fig. 11.36 *Aerobic treatment unit (ATU), requiring more energy and maintenance, but with cleaner effluent than the passive, anaerobic septic tank. (National Small Flows Clearinghouse.)*

Fig. 11.37 *Checking for required clearances on a suburban lot. Although the 40,000 ft² (3,716 m²) lot is more than the minimum required, space should be left for a backup disposal field (of equal size) in case of failure of the original field. A seepage pit adjacent to this septic tank would barely meet the clearance criteria. With a stream nearby, the water table is probably too high for such a pit (see Table 11.9).*

quire. Either drain lines consisting of perforated pipes (of many approved materials) are used, or square-edge agricultural tile, 4 in. (102 mm) or more in diameter [the ends of the tiles are separated by one-quarter-in. (6-mm) openings]. These lines are placed in shallow trenches, on a bed of gravel, then covered with gravel. The effluent runs out of these lines and stands in the interstices of the gravel until it seeps into the earth. In effect, the gravel provides spaces that act as a dry well to receive the fluids and accommodate them until they slowly sink into the ground.

Disposal fields are located according to regulations such as those given in Table 11.9 and sized in relation to total sewage flow and septic tank size (see Tables 11.8 and 11.10). Although Table

11.8 shows the maximum absorption capacities for soils based on gal/ft² over a 24-h period, a common alternative is to use *percolation* tests. For such tests, a "test pit" is dug and water poured into it. The absorption in gal/ft² (L/m²) in a 24-h period is determined from this test. Some codes require design sizes based on percolation test data. Codes often require that an area equal to the drainage field's required size be set aside for use in the event of failure in the original field.

Design guidelines for sizing disposal fields were presented in Section 9.3, but a more accurate method is shown here. The trench width, depth, and spacing shown in Fig. 11.38 are typical, but other combinations can be used, provided that they meet the requirements listed in Table 11.10.

TABLE 11.9 **Location of On-Site Sewage Disposal Systems**

Minimum Horizontal Distance Clear Required from:	Building Sewer		Septic Tank		Disposal Field		Seepage Pit (Cesspool)	
	(ft)	(mm)	(ft)	(mm)	(ft)	(mm)	(ft)	(mm)
Buildings or structures[a]	2	610	5	1,542	8	2,438	8	2,438
Property line adjoining private property	Clear		5	1,542	5	1,542	8	2,438
Water supply wells	50[b]	15,240	50	15,240	100	30.5 m	150	45.7 m
Streams	50	15,240	50	15,240	50	15,240	100	30.5 m
Trees	—		10	3,048	—		10	3,048
Seepage pits or cesspools	—		5	1,542	5	1,542	12	3,658
Disposal field	—		5	1,542	4[d]	1,219	5	1,542
On-site domestic water service line	1	305	5	1,542	5	1,542	5	1,542
Distribution box	—		—		5	1,542	5	1,542
Pressure public water main	10[c]	3,048	10	3,048	10	3,048	10	3,048

NOTE: When disposal fields and/or seepage pits are installed in sloping ground, the minimum horizontal distance between any part of the leaching system and ground surface shall be 15 ft (4572 mm).

Source: Reprinted by permission from the *1997 Uniform Plumbing Code.* Copyright © by the International Association of Plumbing and Mechanical Officials, Walnut, Ca.

[a]Including porches and steps, whether covered or uncovered, breezeways, roofed porte-cocheres, roofed patios, carports, covered walks, covered driveways and similar structures or appurtenances.

[b]All drainage piping shall clear domestic water supply wells by at least 50 ft (15,240 mm). This distance may be reduced to not less than 25 ft (7620 mm) when the drainage piping is constructed of materials approved for use within a building.

[c]For parallel construction—For crossings, approval by the Health Department shall be required.

[d]Plus 2 ft (610 mm) for each additional 1 ft (305 mm) of depth in excess of 1 ft (305 mm) below the bottom of the drain line.

EXAMPLE 11.3. (Fig. 11.38) Design a septic tank and drainfield system for a suburban residence under the following conditions:

Bedrooms	4
Soil	Sandy loam
Depth to water table	7 ft

SOLUTION

Septic tank capacity (Table 11.7) = 1200 gal

Fluid volume of tank (min. required) = 1200 gal ÷ 7.48 gal/ft^3 = 160 ft^3

Establish dimensions for the septic tank. Dimensions from the chosen tank (Fig. 11.35) are

$$4.5 \times 4.0 \times 9.0 = 162 \text{ ft}^3 \text{ volume}$$

OK, 162 > 160 (min. required)

Drainfield size is established from Table 11.8: sandy loam (soil type #3) requires 40 ft^2 of leaching area per 100 gal. With a septic tank of 1200 gal,

$$40 \text{ ft}^2 \times 1200 \text{ gal}/100 \text{ gal}$$

$$= 480 \text{ ft}^2 \text{ minimum absorption area}$$

The effective absorption area of the typical trench depth and spacing shown in Fig. 11.38 is

trench width	2.0 ft^2/ft of length
trench sides	2.0 ft^2/ft of length (12" each side)
total absorption	4.0 ft^2/ft

The required absorption area was determined to be 480 ft^2. Because the trench has 4.0 ft^2/ft of length, the total trench length is

$$480 \text{ ft}^2/4 \text{ ft}^2/\text{ft of length} = 120 \text{ ft}$$

A three-line disposal field is selected, so 120 ft/3 lines = 40 ft per line. Given the required clearances to the edges of the disposal field (Table

TABLE 11.10 **Disposal Field Trenches**

Part A. Dimensions

	Minimum	Maximum
Length of drain line(s)	—	100 ft (30,480 mm)
Bottom width of trench	18 in. (457 mm)	36 in. (914 mm)
Spacing of lines, o.c.[a]	6 ft. (1829 mm)	—
Depth of earth cover over lines	12 in. (305 mm)	—
	[NOTE: 18 in. (457 mm) preferred]	
Grade of lines	Level	3 in./100 ft (25 mm/m)
Filter material		
Over drain lines	2 in. (51 mm)	—
Under drain lines[a]	12 in. (305 mm)	—[d]

*Part B. **Leaching Areas***

Trench bottom[b]: minimum 150 ft^2 (14 m^2) per system
Trench side wall: minimum[c] 2 ft^2/ft of length
              maximum[d] 6 ft^2/ft of length

Source: Adapted by permission from the *Uniform Plumbing Code,* copyright © 1997 by the International Association of Plumbing and Mechanical Officials.

[a]Minimum spacing of drain lines: 4 ft (1.2 m) plus 2 ft (0.6 m) for *each* additional foot (0.3 m) of depth *beyond* 1 ft (0.3 m) below the bottom of the drain line.

[b]Exclusive of rock, clay, or other impervious formations.

[c]Based on the minimum 12-in. trench depth below drain tile.

[d]Maximum of 36-in. trench depth below drain line can be counted when calculating required absorption area.

11.9) of 5 ft to property lines, the disposal field's area can be calculated as:

width: 5 ft + 12 ft

        (two spaces of 6 ft each) + 5 ft = 22 ft

length: 5 ft + 40-ft line + 5 ft   = 50 ft

The site surface area, then, is 22 × 50 = 1100 ft^2. Note that double this area (for a second disposal field in the event of the first field's failure) is often required. ■

(e) Mounds with Leaching Beds (Fig. 11.39).

They represent a newer solution in the United States, and thus may require special approval. The guidelines for leaching-bed sizing are essentially similar to those for drainage tile disposal fields. The absorption area for leaching beds, however, must be 50% greater than that required for trenches. The bottom of the leaching bed generally must be at least 5 ft (1.5 m) above the water table, although in water-scarce areas, officials may reduce this requirement.

(f) Buried Sand Filters (Fig. 11.40).

Slow sand filters for treating supply water were discussed in Section 10.2d, and shown in Fig. 10.3. Buried sand filters work in a similar way, using primarily biological but also physical and chemical processes to clean wastewater. The media is most commonly sand, although other locally available materials such as crushed glass, mineral tailings, bottom ash, and so on, have been used. The grain size ranges from 0.3 to 3 mm in diameter.

Buried sand filters are one remedy for failed disposal fields. They are also used where high groundwater, shallow bedrock, or poor soil preclude the more simple disposal field. Although buried sand filters may require considerable site area, the ground surface can be used for lawns or other nonpaved surface activities. More detailed information on the sand filter types discussed here is available from the National Small Flows Clearinghouse.

The layout is similar to the disposal fields described in Section 11.6e, but these filters, them-

Fig. 11.38 *(Example 11.3) Tile drainfield for a four-bedroom house. Although the drawings are not to scale, the dimensions show a required area of about 20 × 50 ft (6 × 15 m) on the lot. Because paving is not permitted over drainfields, sewage treatment on a small lot demands considerable space. (a) Transverse and longitudinal sections. (b) Schematic plan.*

selves 24 to 36 in. (610 to 914 mm) deep, usually require an excavation of from 4 to 5 ft (1.2 to 1.5 m). The filter bed must be level and sited to avoid contact with groundwater and excess surface water runoff. Some authorities require the sand filter to be contained in an impermeable membrane liner. Underdrain pipes and a graded layer of washed gravel or crushed rock are placed on the bottom of the filter bed, with the finer gravel on top to keep the media grains from washing into the underdrains. Then, a layer of fine gravel first, then coarser gravel above, is placed around and

Fig. 11.39 *Mounds with leaching beds offer an option when the water table is high. The system serves a two- or three-bedroom home. (Adapted from Converse, 1978.)*

Fig. 11.40 *Buried sand filters appear to be like the common disposal fields (Fig. 11.38), but are deeper and may require an impermeable membrane liner. (National Small Flows Clearinghouse.)*

over the distribution pipes. A geotextile fabric covers the top of the entire filter bed, which is in turn covered by backfill material. A properly constructed buried sand filter that receives properly digested waste from a well maintained septic tank should last up to 20 years without maintenance.

In general, sand filters are designed for normal dosing twice a day; they need time to allow the filter media to drain between doses. The water

TABLE 11.11 **Sand Filter Design Values**[a]

Design Factor	Buried Filters	Open Filters	Recirculating Filters
Pretreatment	All filters must be preceded by settling and removal of solids		
Media			
Materials	Washed, durable granular material		
Uniformity Coefficient	Less than 4.0 between smallest and largest size particles		
Depth	24 to 36 in. (610 to 914 mm)		
Effective size	0.3 to 1 mm	0.3 to 1 mm	0.8 to 3 mm
Hydraulic loading	<1.5 gpd/ft^2	2 to 5 gpd/ft^2	3 to 5 gpd/ft$^{2\,b}$
	(<61 L/day m^2)	(82 to 204 L/day m^2)	(122 to 204 L/day m^2)
Media temperature	>41° F	>41° F	>41° F
	(>5° C)	(>5° C)	(>5° C)
Dosing frequency	<2 per day	<2 per day	5–10 min./30 min.[d]
Recirculation ratio[c]	(Not applicable)	(Not applicable)	3 : 1 to 5 : 1

Source: National Small Flows Clearinghouse, *Pipeline,* Summer 1997 Vol. 8 No. 3.

[a]These values show typical design criteria, and do not represent all possibilities.

[b]Forward flow only, not including recirculated effluent.

[c]Water recirculated: water discharged.

[d]Dosage lasts 5 to 10 minutes, every half hour.

that filters through the media is collected in under-drains that are sloped toward an outlet, and vented to the surface at the upstream end. This vent is a potential source of odors. The water is then taken either to a disinfection (usually chlorination) unit, or taken to a disposal field, for sub-surface discharge.

Water from sand filters has been cleansed of many pathogens such as harmful bacteria and viruses, and is therefore more likely to be approved for a down-line disposal field. Typical values for three types of sand filters are shown in Table 11.11. (Open and recirculating sand filters are discussed in the next section.)

11.7   On-Site Multiple-Building Sewage Treatment

The first approach involves septic tanks at each building, for primary treatment. Then the combined outflows from the individual septic tanks can enter a secondary treatment process. Either of two more types of sand filters can be employed for this purpose. There are several advantages to this approach. The effluent lines from the septic tanks carry no solids; they can be much smaller in diameter and placed in shallower trenches than conventional sewer lines. The runs are far shorter because distant centralized treatment plants are not involved. With sand filter secondary treatment, this septic tank "waste" water is more likely to find a recycling use, such as irrigation.

(a) Open Sand Filters. Also called intermittent sand filters, these are very similar in construction to the slow sand filter (Fig. 10.3). They are nearly identical in performance to the buried sand filter (Fig. 11.40); the difference is that these are at least partially above ground. A detail of the filter media is shown in Fig. 11.41. Table 11.11 shows that they are able to accept a somewhat higher flow, so they take less area than the buried filter, but the surface is no longer useable for any other purpose. (A removable cover for odor control is typical, despite their name.) They are used on systems with flows up to 120,000 gpd (454,240 L/day) of wastewater. Frequently, two such open filters are built, so that one can rest while the other is in use. Or, two filters may be

Open Sand Filter

Fig. 11.41 Open sand filter receives wastewater from above, through (1) distribution pipes (or a thin layer of flooded wastewater). Filter media (2) such as sand is on top, then (3) fine gravel, then (4) coarse gravel, surrounding the (5) underdrain. Slow sand filters (Fig. 10.3) are very similar. (National Small Flows Clearinghouse.)

used in series, with resulting even cleaner final effluent.

(b) Recirculating Sand Filters (Fig. 11.42). These eliminate odors by ensuring an adequate supply of oxygen to the wastewater. The recirculating sand filter first receives water (from the septic tank or other primary treatment), into a recirculation tank. This tank has a pump, timing mechanism, and float valves. Either in timed doses or when triggered by the float valve, doses of water are pumped to the sand filter. The treated wastewater collects in the underdrain, where the majority (from 75% to 80%) is directed back to the recirculation tank. Thus mixing with the septic tank effluent and being pumped to the sand filter, the result is weaker (cleaner) effluent, containing more oxygen, as it enters the sand filter. This eliminates odor, and also allows slightly larger filter grain size, making the system less prone to clogging. Table 11.11 compares sand filter types, showing the highest allowable flow rates per surface area for this recirculating sand filter.

At Stonehurst, a 47-lot subdivision in Contra Costa County, California, 1500-gal (5680-L) septic tanks serve each unit; each has a screened effluent vault, keeping solids out of the effluent lines. The effluent is taken through small-diameter variable grade sewers (located under the roadway) to a

Fig. 11.42 *Recirculating sand filters periodically distribute a cleaner, more oxygen-rich effluent to the sand filter. This eliminates odor and allows a larger media grain size. (1) Effluent from pretreatment (such as a septic tank). (2) To disinfection/discharge. (3) Float valve. (4) Bleed line. (5) Submersible pump. (6) Recirculated effluent line. (7) Perforated underdrain piping. (8) Layered support gravel. (9) Filter media. (10) Riser pipe. (11) Distribution piping. (12) Recirculation pump discharge to recirculating filter. (National Small Flows Clearinghouse.)*

recirculating granular medium filter. (Homes lower than the roadway use effluent pumps at the septic tank outflows.) Outflow from the recirculating filter is then taken to an ultraviolet (UV) recirculation tank, from which it is treated by ultraviolet light (see Fig. 10.6). From this tertiary treatment, it is conducted to a subsurface drip irrigation system serving a community park. In winter, a 2.5-acre (10,117-m^2) community soil absorption field is utilized.

(c) Lagoons. These basins need sun and wind and a greater land area than other methods, but are very simple to maintain and very low in energy use. They usually must be lined to prevent wastewater from polluting the groundwater below. Sometimes the outflow must undergo further treatment before release. If lagoons are not square or round, their length should not exceed three times their width. Outflow and inflow should be at opposite ends. Detailed information is available in National Small Flows Clearinghouse (1997).

Anaerobic lagoons are usually used as the first of at least two lagoons in series. Working much as does a septic tank, they hold wastewater from 20 to 50 days, and are relatively deep, 8 to 15 ft (2.4 to 4.6 m). With a surface of floating scum that blocks oxygen from the wastewater, these are unsightly and malodorous.

Aerobic lagoons are shallower than other lagoons so sunlight, and oxygen from wind and air,

can better penetrate. Aerobic bacteria and algae do the work of cleaning the wastewater. These are best in warm climates where freezing is not a threat. They hold water from 3 to 50 days. The bottoms are either paved or lined to prevent weeds growing within the lagoon.

Aerated lagoons take the aerobic lagoon a step further, actively stirring and adding oxygen to the lagoon. Because they speed aerobic action, they require shorter retention time and less land area than the passive aerobic lagoon.

Facultative lagoons (also called stabilization, oxidation, photosynthetic, or aerobic–anaerobic ponds) are the most common type used by small communities and individual households. They can work in most climates, and require no machinery. They require about 1 acre (0.4 hectare) for every 50 homes (or for every 200 people) they serve, are from 3 to 8 ft (0.9 to 2.4 m) deep, and are designed to hold wastewater anywhere from 20 to 150 days (the longer period in cold weather).

Three layers form in these lagoons; the top aerobic zone, the middle facultative zone, and at the bottom, the anaerobic zone. The depth of the surface aerobic zone depends on how much sunlight and wind and rain can contribute oxygen; the deeper this zone is, the better it controls odors rising from the anaerobic zone. Therefore, these lagoons should be sited with full access to both sun and wind.

Down in the anaerobic zone, anaerobic bacteria convert sludge to gases, including hydrogen

(A) ADVANCED FACULTATIVE PONDS WITH IN-POND DIGESTER

(B) HIGH RATE POND AND ALGAL SETTLING POND

(C) WETLANDS TREATMENT AND PARK AREA

WASTEWATER TREATMENT DIAGRAMS

Fig. 11.43

Advanced Integrated Pond System Facility™
Advanced Facultative Ponds
In Pond Digester Pits (below)
High Rate Pond
Algal Settling Ponds

Education & Laboratory Building

Wastewater Disinfection

Constructed Wetlands for wastewater treatment
Wildlife Marsh
Wastewater Treatment & Solid Waste Processing Control Buildings

Park and Recreational Areas

Water Storage Lake

Solid Waste Processing Building

Anaerobic Digester

Compost Pasteurization Buildings

(D) RESOURCE RECOVERY FACILITY LAYOUT

Fig. 11.43 *(continued)* Schematic design proposal for wastewater treatment at the Energy Efficient Resource Recovery (E2R2) Facility at Cal Poly, San Luis Obispo. (a) *Effluent enters the advanced facultative pond (AFP) through the in-pond digester where solids settle and undergo methane fermentation and are reduced to ash. Rising biogas is channeled by a proprietary submerged gas collector and passes through a highly aerobic surface layer where impurities are removed, yielding methane-rich biogas at the surface.* (b) *AFP effluent then goes to the high rate pond, where it is slowly moved along the shallow channels by an energy efficient paddle wheel providing optimum solar exposure and enabling the profusely growing algae to supply abundant oxygen for oxidation of wastewater organics and disinfection of pathogens. Subsequently, effluent moves into the quiescent Algal Settling Pond where algae settles and is periodically removed for beneficial use.* (c) *Effluent then moves into constructed wetlands where the plants in this system create an ideal environment for invertebrates and vertebrates (in free surface wetlands), and aerobic and anaerobic microbial populations to thrive and provide further treatment and effluent polishing. Finally, the effluent from the wetlands is disinfected by ozone or ultraviolet disinfection systems and stored in a lake which is part of a campus park and recreational facility.* (d) *Site plan. (Courtesy of Dr. William Oswald, Professor Emeritus, of Civil and Environmental Engineering and Public Health at University of California, Berkeley, and Principal of Oswald Engineering Associates, Inc.; Dr. Bailey Green, Research Engineer at University of California, Berkeley, and Principal of Oswald Engineering Associates, Inc.; and Prof. Daniel Panetta, Architecture Department, Cal Poly, San Luis Obispo.)*

sulfide, ammonia and methane. As these gases rise, they provide food for both the aerobic bacteria and the algae in the aerobic zone. Sludge accumulates more quickly in cold climates; but sludge removal should be expected every 5 to 10 years.

(d) Advanced Integrated Wastewater Pond System (AIWPS™).

At California Polytechnic State University, San Luis Obispo, there

is an innovative proposal to create an integrated infrastructure facility combining wastewater treatment and solid waste processing on campus. In addition to the waste treatment and resources recovery for this campus of 18,000, the schematic design for the Energy Efficient Resource Recovery (E2R2) Facility includes an educational facility, wildlife habitat, and park facilities. The wastewater and related wetlands proposals are shown in Fig. 11.43; the solid waste facility will appear in Fig. 12.1.

Three ponds are used in the AIWPS facility (Oswald, 1963, 1990). The first is the advanced facultative pond (AFP) (Fig. 11.43a), which includes an in-pond digester (IPD) or anaerobic fermentation pit at the bottom, and an oxygen-rich aerobic surface water layer. Two AFPs used in parallel allow fail-safe operation. This design enables isolation and treatment of toxic substances in the event such material is accidentally released into the community sewer system. Conventional sewage treatment plants typically cannot isolate or treat accidental toxic releases in wastewater due to their short processing time.

The raw sewage enters the system near the bottom of the IPD where anaerobic bacteria slowly reduce biosolids to ash (Oswald et al., 1994; Green et al., 1995b). Biochemical oxygen demand (BOD) is reduced to 60% to 80% and minimal accumulation of refractory solids (sludge) has been demonstrated since the late 1960s.

Earthwork construction of the ponds allows them to be quite large and deep without the substantially higher costs of concrete and steel construction typical of conventional wastewater treatment facilities. A berm surrounding the IPD prevents mixing of aerobic water with the influent below, whether by wind or thermal inversion. The aerobic upper layer oxidizes any malodorous biogases rising from the IPD. The biogas generated in the IPD is scrubbed as it passes through the water column and the aerobic surface layer. At the current campus population, between 3000 and 5000 cubic feet (85 and 142 cubic meters) per day of methane (CH_4) are estimated to be available for collection from the IPD and use as fuel (Green et al., 1995a).

The second pond is the high rate pond (HRP) designed to grow algae (Fig. 11.43b). The lined channels in this pond are approximately 1 meter or less in water depth, and include a paddle wheel (see later Fig. 11.48) to slowly mix the primary effluent. The HRP design allows maximum oxygen production by algal photosynthesis due to optimum exposure to sunlight and abundant nutrients. The pond's aerobic and facultative bacteria oxidize most of the soluble and readily biodegradable BOD.

The next pond in the series is an algal settling pond (ASP) (Fig. 11.43b). The quiet waters of this pond allow 50% to 80% of the algae to settle over several days residence time. The settled algae are periodically harvested and the algal biomass does not produce objectionable odors. When properly disinfected, the harvested algae, rich in nitrogen and phosphorus, can be used as fish or livestock feed or as a nitrogen source in compost.

Near St. Helena, California, an AIWPS facility has functioned since the late 1960s. In that time period, no sludge removal has been necessary, attesting to the efficiency of the methane fermentation process. The facility is near several large wineries and is considered a good neighbor without odor problems (Oswald et al., 1970).

While a maturation pond usually follows the ASP in an AIWPS facility, various tertiary treatment components can subsequently be used to achieve advanced tertiary treatment: the highest wastewater reclamation and reuse standards required by the state of California (Green et al., 1996). In the proposed schematic design, the ASP effluent flows into a constructed wetlands.

(e) Constructed Wetlands. Wetlands as waste treatment are shown in Fig. 11.43c. In the schematic design, two parallel constructed wetlands are proposed, allowing a comparison between their outflows.

The free surface wetlands (FSW) consist of shallow open basins or channels that are lined to prevent seepage. Soil supports a succession of wetlands vegetation, nourished by a continuing flow of effluent from the initial wastewater treatment system(s) (such as a secondary AIWPS facility, septic tank, etc.). Although the water surface is exposed to beneficial air and sunlight, proper design consideration is required to mitigate human contact with the treated effluent in FSWs. Consideration is also required to address issues such as mosquitoes or other water-borne insects.

The subsurface flow wetlands utilize a lined treatment basin filled with a coarse media such as large gravel or crushed rock. The large voids (if properly designed) allow the effluent to flow freely through rocks and plant root systems. The relationship between gravel size, slope of bed and rate of flow is complex, and explored in detail in EPA (1993). A layer of soil is used to cover the gravel bed, so subsequent human contacts with the effluent and insect problems are precluded.

In either type of constructed wetland system, saturated and nutrient-rich conditions encourage growth of both aerobic and anaerobic microbes, and certain invertebrates. The most common plants include varieties of *scripus* (bulrushes), *phragmites* (reeds), *typha* (cattails), *canna,* and *iris.* These plants not only provide large amounts of surface area for microbial activities, they also create an aerobic zone underwater by transporting air through their stems to their roots. A thin aerobic environment results among the root hairs.

Constructed wastelands become a vivid, colorful demonstration of how, in nature, there is no such thing as "waste." After passage through a slow sand filter (see again Fig. 10.3) and final disinfection, the reclaimed water from the E2R2 facility can provide campus landscape irrigation water, be released into the campus creek to aid habitat restoration, or used for aquifer recharge. Arcata, California, completed its wetlands treatment system in 1986. This system draws several hundred species of birds as its abundant vegetation polishes nutrients from and allows habitat reuse of the city's treated wastewater.

(f) Greenhouse Ecosystems. When constructed wetlands are moved indoors, greater control and less area become possible. As pioneered by marine biologist John Todd at Ocean Arks International, *Living Machines* are a series of tanks, each with its particular ecosystem, that form a "stream." The stream is followed by an indoor marsh; together they achieve a high degree of tertiary wastewater treatment.

These enclosed systems typically cost less to construct and about the same to maintain as conventional sewage treatment plants, but use less energy because photosynthesis (solar energy) and gravity flow largely replace energy-intensive pumps and large-scale aerators. These systems do not harm the environment with a final dose of chlorine, and they produce about one-quarter of the amount of sludge as do conventional systems. Because they are pleasant to look at and smell

much as do commercial greenhouses, they can be located within the neighborhoods they serve. This can save huge costs associated with large-diameter, deep-trench sewer lines and pumping stations necessary with distant, centralized treatment plants.

A diagram of one greenhouse ecosystem is shown in Fig. 11.44. This living machine begins with an anaerobic bioreactor, in effect a septic tank where solids are reduced to sludge and methane gas is produced. From here the wastewater passes to aerated tanks and then to *ecological fluidized beds,* each with a particular combination of bacteria, algae, snails, and fish. The further along the treatment process, the more "advanced" the ecosystem, with goldfish (carp) growing in the final clarifier. This process depends on the following food chain: (1) aerobic bacteria consume suspended organic matter and convert ammonia into nitrites and then nitrates; (2) algae and duckweed feed on the products of the bacterial action; (3) snails and zooplankton eat the algae, while floating duckweed provide shade that discourages algal growth in the later stages of the process; (4) fish eat the zooplankton and the snails. Plants range from water hyacinth in the early stages to an increasing variety of specialized "accumulators" that remove troublesome phosphorus, heavy metals, etc. Papyrus, canna lilies, bald cypress, willows, and eucalyptus (among other plants) each have a role. When trees outgrow the greenhouse, they are transplanted outdoors—but they are long-term storage for the heavy metals they absorb, not final solutions, because heavy metals will be returned to the earth when the trees die and decompose. Plants can be sold to nurseries or ground up as compost; the fish also enrich the compost. The smaller fish (shiners) can be sold as bait fish. These are small enough to dart back and forth in the piping between tanks, choosing their water quality. If a toxic slug of waste enters the system, they swim to later, clearer tanks until the danger is past.

In Fig. 11.45*a,* this greenhouse solar aquatics system cleans the wastewater from a mobile home park on Vancouver Island, British Columbia. In Fig. 11.45*b,* a striking envelope design occupies the center of an urban block in Kolding, Denmark. The proposed EPICenter at Montana State Uni-

WATER AND WASTE

Fig. 11.44 *The Living Machine enclosed within a greenhouse uses a food chain in a series of ecosystems to treat wastewater. (Living Technologies, Inc.)*

versity (introduced in Fig. 1.5) utilizes a Living Machine (Fig. 11.45c,d), and uses its outflow to replenish non-potable supply water within the building.

In milder climates and with specialized wastewater, these processes can be achieved without the greenhouse enclosure. The Sonoma Mountain Brewery's Living Machine (Fig. 11.46) near Sonoma, California, occupies about an acre (0.4 hectare) and treats a maximum flow of 7,800 gpd (29,525 L/day) of waste from the brewery process. Two streams move in parallel, allowing a comparison of their outflows as experiments with detailed ecosystems are conducted. A part-time operator

(a)

(b)

Fig. 11.45 (a) *A greenhouse in a mobile home park in British Columbia encloses the 12 solar tanks and marsh of a Solar Aquatics System, designed by Ecological Engineering Associates and ECO-TEK, Inc. (Photo by Charles Rusch.) (b) In the middle of an urban block in Denmark, a Living Machine is a good neighbor. (Photo © 1996 by Robert Peña.) (c) "Life of Water and Waste" cartoon for the EPICenter at Montana State University (see Fig. 1.5). (d) Section of the EPICenter, with water and waste strategies. (Courtesy of Place Architecture LLC, Bozeman, Montana, and Berkebile Nelson Immenschuh McDowell, Architects, Kansas City, Missouri.)*

WATER AND WASTE

(c)

(d)

Fig. 11.45 (continued)

(a)

(b)

Fig. 11.46 *In the mild climate of California's Sonoma valley, the waste from the Sonoma Mountain Brewery* (a) *is treated by an outdoor Living Machine* (b). *Flow is from left tanks to right tanks, then to a marsh.* (c) *Calla lilies and other plants lend beauty as well as function to this open aerobic tank. (Designed by Living Technologies, Burlington, Vermont.)*

(c)

Fig. 11.46 *(continued)*

manages this treatment plant. After an underground anaerobic holding tank, flow moves to:

1. Closed aerobic reactors, each holding 2,500 gal (9,463 L). Air is blown through the tanks from the bottom (an adjacent small building encloses the blower); wastewater spends about 15 hours in these tanks. Next are

2. Open aerobic tanks, 4 per stream. Plants, snails and organisms can migrate among the agitated tanks, choosing their preferred level of nutrients ("pollution"). Water leaving the last of these tanks is much cleaner; water and sludge collected from the tanks is taken next to the

3. Clarifier. Shaded by water hyacinth, algae cannot grow in the water at this advanced stage. In this quiet-water environment, solids settle out, and are then pumped to an adjacent

4. A. Open air reed bed ("composter"), where they are dried and then composted. The compost will serve as nutrient for the hops used in the brewing process. Excess water percolates through an underlying sand bed, and is returned to the very beginning of the

Living Machine streams. Meanwhile the clarified water passes next to

B. Ecological fluidized beds (two per stream), whose purpose is to further "polish" the wastewater. Water circulates many times through this section, pumped from the plant-shaded perimeter into a rock bed at the center. In this bed, microorganisms continue to clean the water. From the second ecological fluidized bed, water moves to an underground

5. Irrigation pond, from where it is filtered through a

6. Constructed wetlands, for further purification.

Finally, it is welcomed for irrigation of the hops and grapevines grown on site.

(g) *Pasveer Oxidation Stream.* This method of sewage treatment (Figs. 11.47 to 11.49) has been adopted at New York Institute of Technology. Serving the 450-acre campus at Old Westbury, Long Island, New York, this system provides an on-campus sewage treatment, which

Restricted Parking

Student Parking

• Connected to
  Sewage Treatment Plant

LEGEND
- •27   Harry J. Schure Center
- •37a Anna Rubin Hall (300 bldg.)
- •37b John J. Theobald Hall (400 bldg.)
- •37c "500" Building
- 38   Gerry House
- 39   French House
- 41   Art Center
- •41a Sculpture Studio
- 42   Education Hall
- •45   Student Activity Center & Gym
- 45a   Health Office
- 10   Sewage Treatment Plant
- 48   Tower House
- 49   Balding House

NEW YORK INSTITUTE OF TECHNOLOGY
Old Westbury Campus

(a)

Fig. 11.47 *Sewage treatment plant, New York Institute of Technology (NYIT). The treatment plant's location is central to all the buildings it serves, an efficient choice. (a) Surrounded by heavily wooded areas, it is not a prominent landmark, though it offers a pleasant appearance to the occasional viewer. A frequently used walkway (dotted lines on the map) passes just east of the woods that surround the plant. (b) Layout of the treatment plant. (c) Plan of the two streams. Each of these units includes stream, rotor, clarifier with adjacent sludge pump pit, and three drying beds. Because of odorless operation, the proximity of the plant to the Student Activities Building poses no problem. (Courtesy of Bogen Jenal, Engineers, P.C.)*

WATER AND WASTE

(b)   (c)

Fig. 11.47 (continued)

returns the purified effluent to the ground through 48 leaching wells located under the athletic field. The groundwater, thus restored, provides a contributing source of water for 400 ft (122 m) deep wells, distantly located, that furnish part of the water supply for the campus buildings (see Fig. 11.47).

A few of the original small buildings, later converted for administration offices and classrooms, are still served by septic tanks and leaching fields. There was no public sewer near the campus, and in the 1970s, the health authorities ruled out the use of septic tanks for the numerous additional buildings that were contemplated.

The oxidation stream process applied here was developed by the Netherlands Research Institute for Public Health Engineering (T.N.) and is now in operation in many U.S. locations as well as in Europe. It is considered to be a modified form of the *activated sludge* process (see later discussion of the Oceanside Plant, Section 11.8b), with aerobic digestion and periodic sludge removal.

The mechanical aerator (Fig. 11.48), which keeps the stream of sewage moving and provides

(a)   (b)

Fig. 11.48 (a) Rotor with plastic "greenhouse" cover. (b) The rotor induces oxidation in this oval, circulating stream. (A similar mechanism serves the high rate pond in Fig. 11.43.)

(a)

(b)

Fig. 11.49 *NYIT sewage treatment plant at completion.* (a) *The only projections above ground level are the office/laboratory and the plastic covers over the rotors.* (b) *The oxidation stream does not appear turbulent here, but it becomes so under the action of the mechanical aerator (rotor).*

the oxidation necessary for aerobic digestion, is an important feature. In this design, sludge-drying beds are placed on an island, surrounded by the continuously moving oxidation stream. Another feature of the system is its low profile (Fig. 11.49), readily screened by trees. The aerobic digestion process is not malodorous. The sounds are produced by the splashing water-wheel action of the mechanical aerators; no air compressor is required.

The plant has a full-time accredited operator; there are also two assistants and one relief operator. This design provides for an eventual population of 4330, with a 340,000 gpd (1,287,010 L/day) flow. The sludge was removed from the drying beds twice in the first five years of operation.

11.8   Larger-Scale Sewage Treatment Systems

Two examples of community-wide sewage treatment are presented here—one for a small, water-scarce community near San Diego, California, and one for an oceanside urban neighborhood serving much of San Francisco.

(a) Padre Dam Municipal Water District (Fig. 11.50).

The Santee Water Reclamation Plant represents innovations in water conservation and recycling during the 1960s and 1970s. The Padre Dam Municipal Water District is a region where rainfall is less than 15 in. (380 mm) per year. It has no local water supplies available, so water is imported 300 miles (480 km) from the Colorado River. If conventionally treated, the wastewater resulting from this flow would be discharged into the Pacific Ocean. Alternative sewage treatment (shown here) processes this fluid waste for secondary uses such as irrigation and recharging of groundwater. Many municipal supplies for potable water are taken from rivers often heavily polluted with sewage and, after treatment, are pumped into the domestic water mains. Therefore, district officials here decided to install a system that would make use of purified wastewater for many secondary uses.

However, would such a concept be acceptable to the public? An effective public relations pro-

gram concurrent with the technological advances has proved successful.

The project involved building a sewage treatment plant and utilizing seven pits left over from prior surface mining of sand and gravel. After partial purification of the sewage at the plant by the first two stages of a conventional activated sludge process, the effluent is discharged to form the lakes adjacent to the plant. This provides tertiary (oxidation) treatment after which the cleaner effluent is pumped to a filter area at the north end of the complex where it is further purified by flow through sand and gravel. Chlorination is administered, and the water then flows through a series of seven lakes Nos. 7–6–5–4–3–2–1, in that order.

The appearance of new lakes in this semiarid region created an initial interest that was rapidly augmented by a very clever program. At first the lakes were fenced in. Then the fences were removed. The seven lakes were next made available for boating. They were stocked with fish and careful studies were made which indicated that the fish were healthy and flourishing. Fishing was permitted, but for a while all fish had to be returned to the water. Anglers were later permitted to keep and consume the fish. Finally, swimming was permitted. The overflow from the last lake, No. 1, discharges into Sycamore Canyon Creek and is used for irrigation of a golf course and the recharging of groundwater.

(b) Oceanside Water Pollution Control Plant.

Imagine the challenge of locating a major new sewage treatment plant on the shore of the Pacific, within the city of San Francisco, immediately adjacent to both the city zoo and a well-established neighborhood. This plant treats all the sewage in western San Francisco watershed, an average dry weather flow of 21 million gpd (79,491,720 L/day). The treated effluent is discharged through an ocean bottom pipeline, 4.5 miles (7.2 km) offshore, with effluent and ocean water quality monitoring by an on-site laboratory. Sludge is shipped north to a serve as the required daily cover for a landfill near Novato. The plant itself is concealed by large earth berms, through which entry and exit tunnels carry plant traffic. And the odor? It is completely contained within a huge underground complex of wastewater basins

(a)

Fig. 11.50 *Santee Water Reclamation Plant and Santee Park and Recreational Facilities, Padre Dam Municipal Water District, California.* (a) *Aerial photograph of the Santee Plant, including the seven recreational lakes. Note the location of the San Diego River, at the bottom of the photograph. The water reclamation plant and the stabilization ponds do not appear in this aerial view.*

WATER AND WASTE

Fig. 11.50 (b) Raw sewage from the community of Santee enters the treatment plant, which is located at the top of this diagram. The process then proceeds southward to the point where reclaimed water is pumped to irrigation or recharges ground water. Sludge is pumped to the San Diego Metro system.

(c)

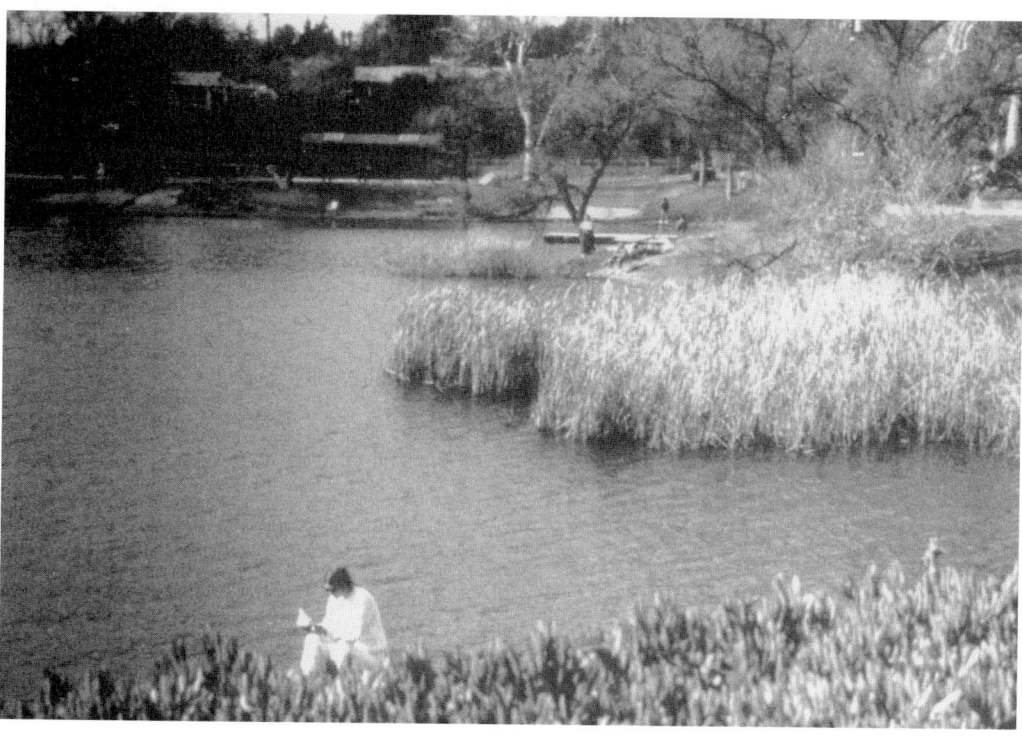

(d)

Fig. 11.50 (c) *This "laundry" process cleans up and makes reusable the valuable water that constitutes the major part of raw sewage volume.* (d) *Beside one of the recreational lakes, years after it was an open gravel pit. (Photo by Daniel Panetta.)*

Santee Water Reclamation Plant

1 — Available but not in use
2 — Not available
3 — In use

Fig. 11.50 (e) *Details of the reclamation plant. The asterisk before "Potable water study ponds" indicates continuing study and preparation for future distribution of drinking water that is reclaimed from sewage. The plant is not burdened by stormwater, which is recharged to the ground where it occurs locally in the community.*

Wet oxidation 2

Main blowers 3

Sludge drying beds 1

ary sedimentation 3

.ion

Sludge handling 2

Sludge now
being pumped to
San Diego's Metro System

N

Influente flow meter 3

Primary 3

NOTE: Can treat sludge
aerobically

en blowers 1

WATER RECLAMATION PLANT

condary 3

Design Capacities	
Main plant	4,000,000 gallons per day
Tertiary treatment	2,000,000 gallons per day
* Potable water study ponds	100,000 gallons per day
Total Construction Cost	$3,825,000
U.S. Government Construction Grants	$1,497,000
Joint Demonstration Research Project	$511,000

Santee County Water District
Federal Water Quality Administration

(e)

Fig. 11.50 (continued)

and channels, over which the zoo will build its new Mammal Conservation Center, on 5 ft (1.5 m) of topsoil cover. Exhaust air from the underground plant is filtered through activated carbon and potassium permanganate before release.

The process is simplified in Fig. 11.51a. Incoming sewage passes through a bar screen and grit chambers, removing the largest objects as well as grit, sand and gravel (these are trucked to landfill). Then the water flows to the primary clarifier, in which floating scum is skimmed off, and sludge settling on the bottom is pumped to the egg-shaped anaerobic digesters (Fig. 11.51b). The wastewater flows next to aeration tanks where bacteria grow fat, feeding off of and breaking down the remaining dissolved and suspended solids. Next is the secondary clarifier, where the fattened bacteria (activated sludge) sinks to the bottom. Part of this activated sludge is returned to the aeration tanks to bolster the bacterial culture there. The remainder is thickened and pumped to the egg-shaped digesters.

(a)

(b)

Fig. 11.51 *Oceanside Water Pollution Control Plant, San Francisco, is largely underground, as well as hidden behind high earth berms.* (a) *Flow diagram. (Courtesy CH2M Hill and the city/county of San Francisco)* (b) *View of some of the egg-shaped anaerobic digesters, from within the berm walls. Biogas produced within the four, 750,000 gal (28,375 m³) digesters supplies about 20% of the electricity used in the treatment plant.*

Thus the anaerobic digesters receive both primary sludge and thickened activated sludge. As this sludge is stabilized and reduced in volume, biogas is produced: enough to heat water for both process and facility heating, and to generate about one-fifth of the electricity used within the plant. Sludge leaving the digesters is dewatered in belt presses, then hauled to the landfill. During summer, biosolids are spread on agricultural land for soil enrichment.

Questions remain about such use of biosolids. Although a 1998 EPA national survey of sewage sludge found toxic chemicals at such low levels that regulation was deemed unnecessary, concentration of certain industries may produce local chemical or radioactive sludge pollutants at higher levels. Intercepting such wastes at their source, rather than allowing them into the sewage stream, seems an obvious imperative.

11.9 Recycling and Graywater

These chapters on water and waste have illustrated four common "grades" of water in buildings:

Potable water (usually treated, suitable for drinking)

Rainwater

Graywater (wastewater not from toilets or urinals)

Blackwater (water containing toilet or urinal waste)

Two more categories include:

Dark graywater (washing machines with dirty diaper loads, kitchen sinks, and dishwashers; usually prohibited for reuse)

Clearwater (backwash water from reverse osmosis water treatment; condensation from refrigerator or other cooling coil)

After considering the many blackwater sewage treatment complications in the previous section, we might expect that graywater's reuse would, by contrast, be a simple matter. However, there are lingering prejudices against wastewater of any kind that complicate water recycling, if not prohibit it outright. This section presents some future directions in which we might well move if water conservation is to be taken seriously, then some code-mandated approaches.

Rainwater, given an adequate supply, could meet many uses of water in and around buildings: bathing and laundry, toilet flushing, irrigation, and evaporative cooling, for example. Little or no treatment of rainwater is needed for such uses.

Blackwater requires much more extensive treatment, given the higher concentration of human waste in it. However, this may be the easiest waste to eliminate, given the waterless alternatives (Section 11.1).

Graywater (also correctly spelled "greywater") reuse opportunities are more limited than rainwater, because it carries increased threats from pathogens. For use in toilet flushing, it must undergo treatment before reuse. It is likely to contain soap, hair, and, occasionally, human waste (soiled clothes in the laundry). If it is to be used in drip irrigation systems, filtering is essential. If kitchen wastes are to be included, grease and food solids add to the problems. Graywater may well be dirtier than rainwater, but its supply is likely far more predictable and constant. However, codes tend to limit the use of graywater to underground landscape irrigation for single family houses.

The collection of these separate types of water poses a problem. In the early years of indoor plumbing, rainwater, graywater, and blackwater were mixed together, which severely overloaded sewage treatment facilities in rainy weather. Today, we usually mix graywater and blackwater. There are precedents for keeping them separate, however (Fig. 11.52).

(a) Graywater and Future Recycling. Notwithstanding present code restrictions, the potential for residential graywater reuse (given separate waste collection systems) was shown in earlier Fig. 11.1b. Bathing's 21 g/cd (gallons per capita per day) can meet much of the conventional toilet's 32 g/cd, and all of low-flush toilets' water needs. Laundry's 14 g/cd can help with irrigation; limiting contact between irrigation water and people seems prudent, hence the emphasis on underground distribution. Table 11.12 relates both quantity and quality of residential graywater to its use in irrigation. Another potential benefit of separate graywater collection is that the water's heat content might be partially reclaimed (Table 11.13). In England, the Aquasaver Company diverts and cleans water from lavatories, baths, and showers for reuse in toilet flushing, clothes washing, car washing, and irrigation. This low pressure system, installed behind panels in a house's bathroom, uses a pump to push graywater through a series of filters to remove soap, detergents, and other impurities, and then to a storage tank (in the attic, or above points of use). Here, it undergoes weir and trickle filtration and then

(a)

Legend
W.C. Water closet
Lav. Lavatory
KS Kitchen sink
TS Tub or shower

Vent

Vent

Eave gutter

Storm

Storm

Storm

Legend
B Black water
G Grey water
S Storm water

Lav.

W.C.
TS

S

Lav.

W.C.
TS

W.C.

KS

S

B

G

B G

S

Ground level

(b)

Fig. 11.52 *Residence of the 27th Vicar of the Parish Church at Bibury (established* AD *1086) in the English Cotswolds. (a) Clearly, the plumbing was added after the construction of the residence—perhaps several hundred years later. In England the drainage system of an older building often appears on the outside of the building. It seemed inappropriate to ask the Vicar's permission to inspect his indoor facilities, so we have taken the liberty of assigning probable uses. Note that (1) offsets of the vertical piping are made with "easy bends" and are pitched down in the direction of the flow; (2) vents are located at high outdoor points; and (3) stacks carrying blackwater (from toilets) are of larger pipe size than those carrying graywater. (b) In this very sparsely settled region of the Cotswolds, the dispersal of stormwater to the ground and the private septic tank treatment of blackwater plus graywater can be satisfactory. In planning new systems the possible separate treatment of B, G, and S waters is becoming an important consideration. (Photo by William McGuinness.)*

TABLE 11.12 **Residential Graywater Sources for Irrigation**

*Part A. **Sources with Pumps***

Source	Average Outflow Water Quantity	Outflow Water Relative Quality
Clothes washing machine	30–50 gal (114–190 L) per load, top loader 10 gal (38 L) per load, front loader Average 1.5 loads per week per adult 2.5 loads per week per child	Good: medium concentration of soaps, lint. With bio-compatible cleaners, quality can improve to excellent; diapers degrade quality to poor.
Automatic dishwasher	5–10 gal (19–38 L) per load	Poor: low to high quantity of solids depending on degree of prerinsing; high salt and pH from conventional cleaning compounds

*Part B. **Gravity Flow Sources***

Source	Outflow Water Quantity	Outflow Water Quality
Shower	20 gal (76 L) per day per person, low-flow 40 gal (151 L) per day per person, high-flow	Excellent: minimal concentration of soap and shampoo, but hair can clog distribution pumps and lines
Tub	40 gal (151 L) per bath	Concentration in tub water likely less than shower water
Bathroom lavatory	1–5 gal (4–19 L) per day per person	Good: concentration of soap, shaving cream and toothpaste can be high
Kitchen sink	5–15 gal (19–57 L) per day per person	Mixed: high in nutrients but also high in solids, grease and soap
Reverse-osmosis water purifier	3–5 gal per gallon of drinking water (3–5 per L)	Excellent: usually no suspended solids, contains 25% more concentration of the same dissolved solids as tap water
Water softener backwash	5% of indoor water use	Very bad: extremely high in salt
Softened water	(All graywater from softened water source)	Poor: contains salt (unless potassium chloride is used in softening process)

Source. Adapted from *Create an Oasis with Greywater*, by Art Ludwig. © 1997, Art Ludwig. Poor Richard's Press, Santa Maria, California.

TABLE 11.13 **Heat Recovery from Graywater**

Source	Volume/Use, Gal (L)	Flow Rate, Gal (L)	°F	(°C)	Quality	Coincident[a]
Kitchen sink	Up to 5 (1–20)	2.5 (10)/min	85	(30)	Poor	No
Lavatory	Up to 1.5 (1–5)	1.25 (5)/min	85	(30)	Fair	No
Bath	32 (120)	5 (20)/min	100	(37)	Good	No
Shower	15 (50)	2 (7)/min	100	(37)	Good	Yes
Washer	40 (150)	7 (25)/min	60	(15)	Moot	No

Source: Glenn Nelson, "Graywater Heat Recovery," © *Solar Age,* August 1981. Reprinted by permission.
[a]Coincident flow of warm waste water and input water to heater. If "no," a holding tank is needed for heat exchange. If "yes," a countercurrent heat exchanger can be built into the drain line. (See Fig. 11.53)

treatment with nonhazardous cleaning agents. Finally, before reuse, it passes through a carbon filter network.

Storage raises a health issue. To avoid bacteria build-up in graywater, storage should be avoided and the filtered water taken directly to irrigation. However, storage allows for controlled dosage, allows solids to settle, allows for some potential heat recovery, provides for overflow to the sewer, and is often mandated by code.

Various stages of treatment are appropriate for different kinds of graywater (Fig. 11.53). Kitchen sink waste contains grease, which should be trapped (and periodically removed) before it can clog the filters and heat exchangers that serve graywater recycling systems. Similarly, lavatory showers and laundry waste contains lint and hair that must be intercepted quickly. Devices that do so, which are often called *interceptors,* were described in Fig. 11.21. However, code approval tends to sharply limit graywater recycling, assumes no filtering, and assigns its use.

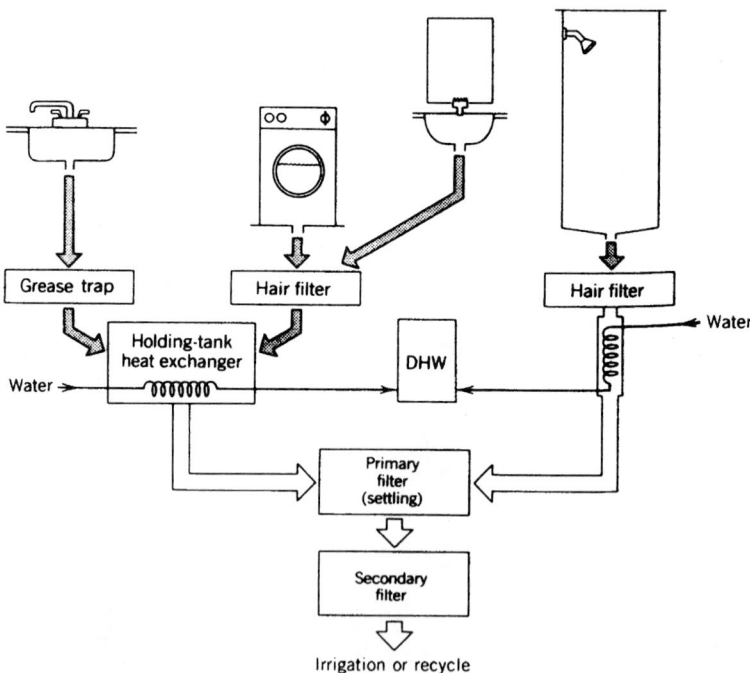

Fig. 11.53 *Sequence of water treatment and heat reclamation for domestic graywater, where such systems might be approved.*

TABLE 11.14 **Location of Residential Graywater System**

| | Minimum Clear Horizontal Distance Required | | | |
| | To Holding Tank | | To Irrigation/ Disposal Field[a] | |
From	*(ft)*	*(mm)*	*(ft)*	*(mm)*
Building structures[b]	5[c]	(1,524)	2[d]	(610)
Property line adjoining private property	5	(1,524)	5	(1,524)
Water supply wells[e]	50	(15,240)	100	(30,480)
Streams and lakes[e]	50	(15,240)	50[f]	(15,240)
Sewage pits or cesspools	5	(1,524)	5	(1,524)
Disposal field and 100% expansion area	5	(1,524)	4[g]	(1,219)
Septic tank	0	0	5	(1,524)
On-site domestic water service line	5	(1,524)	5	(1,524)
Pressurized public water main	10	(3,048)	10[h]	(3,048)

Source: Reprinted by permission from the 1997 Uniform Plumbing Code. Copyright © by the International Association of Plumbing and Mechanical Officials, Walnut, California.

[a]When irrigation/disposal fields are installed in sloping ground, the minimum horizontal distance between any part of the distribution system and the ground surface shall be fifteen (15) feet (4572 mm).

[b]Including porches and steps, whether covered or uncovered, breezeways, roofed porte-cocheres, roofed patios, carports, covered walks, covered driveways and similar structures or appurtenances.

[c]The distance may be reduced to zero feet for above ground tanks when first approved by the Administrative Authority.

[d]Assumes a 45° (0.79 rad) angle from foundation.

[e]Where special hazards are involved, the distance required shall be increased as may be directed by the Administrative Authority.

[f]These minimum clear horizontal distances shall also apply between the irrigation/disposal field and the ocean mean higher tide line.

[g]Plus two (2) feet (610 mm) for each additional foot of depth in excess of one (1) foot (305 mm) below the bottom of the drain line.

[h]For parallel construction. For crossings, approval by the Administrative Authority shall be required.

(b) Subsurface Irrigation. In the 1997 Uniform Building Code (Appendix G), water *only from* bathroom lavatories, showers and tubs, and clothes washing machines and laundry tubs, and *only in* single family residences, can *only be used for* subsurface "irrigation" on the same site as the residence. The mandated trench constructions, in their zeal to avoid any surface graywater, may deliver very little water to surface plants.

Site conditions must meet the requirements of Table 11.14 and Fig. 11.54. The bottom of the trenches must be at least 5 feet (1.5 m) *above* the highest known seasonal groundwater. Flow estimates are as follows:

Number of occupants: 2 for the first bedroom

1 for each additional bedroom

Combined showers, bathtubs, and washbasins, flow per occupant:

25 gal/day (95 L/day)

Laundry, flow per occupant: 15 gal/day (57 L/day)

A holding tank is required with a minimum 50 gal (189 L) capacity. An unvalved overflow must connect to the building sewer (ahead of any septic tank).

The irrigation/disposal field must be divided into a minimum of three valved zones (allowing the occupants to better direct flow rates). The total area of this field is the aggregate length of the perforated pipe, times the width of the proposed field. Required area is based on the estimated graywater flow (or size of holding tank, whichever

Fig. 11.54 *Example of a residence whose daily graywater flow requires an irrigation/disposal field equal to the valved zone as shown. In this example, five such valved zones are shown (minimum required are three such zones). (Reprinted by permission from the 1997* Uniform Plumbing Code. *Copyright © by the International Association of Plumbing and Mechanical Officials, Walnut, California.)*

is larger), based upon the soil types and acceptance rates, and trench design guidelines, shown in Table 11.15. If percolation tests are performed, the 24-h percolation rate results must be within these limits:

Minimum, 0.83 gal/ft² (33.8 L/m²)

Maximum, 5.12 gal/ft² (208.5 L/m²)

One of several graywater system types from the 1997 Uniform Building Code, Appendix G, is shown in Fig. 11.55.

EXAMPLE 11.4. A four-bedroom single family residence is located on sandy clay soil, more than 5 ft above the highest seasonal ground-

water level. If the bathroom and laundry facilities contribute to a graywater system, how large an irrigation/disposal field is required?

SOLUTION. The total number of occupants = 2 + 1 + 1 + 1 = 5

Estimated graywater flow is

$$[(25 + 15)\text{ gal/day}]\text{ per occupant}$$

$$\times\ 5\text{ occupants} = 200\text{ gal/day}$$

Required field size (Table 11.15) of each valved zone is

$$60\text{ ft}^2/100\text{ gal} \times 200\text{ gal/day} = 120\text{ ft}^2$$

TABLE 11.15 **Graywater Design Criteria, Six Typical Soils**

Part A. **Area of Irrigation/Leaching Field**

	Minimum Area of Irrigation/Leaching per Estimated Graywater Discharge per Day		Maximum Absorption Capacity of Irrigation/Leaching per 24-Hour Period	
Type of Soil	ft²/100 gal	m²/L	gal/ft²	L/m²
Coarse sand or gravel	20	0.005	5	203.7
Fine sand	25	0.006	4	162.9
Sandy loam	40	0.010	2.5	101.8
Sandy clay	60	0.015	1.7	69.2
Clay with considerable sand or gravel	90	0.022	1.1	44.8
Clay with small amounts of sand or gravel	120	0.030	0.8	32.6

Part B. **Irrigation/Leaching Field Construction**

	Minimum		Maximum	
	(I-P)	(SI)	(I-P)	(SI)
Number of valved zones	3			
Number of lines per valved zone	1			
Length of each perforated line			100 ft	30,840 mm
Bottom width of trench	12 in.	305 mm	18 in.	457 mm
Spacing of lines, center-to-center	4 ft	1219 mm		
Depth of earth cover of lines	10 in.	254 mm		
Depth of filter material cover of lines	2 in.	51 mm		
Depth of filter material beneath lines	3 in.	76 mm		
Grade of perforated lines	3 in./100 ft	2 mm/m		

Source: Adapted by permission from the 1997 Uniform Plumbing Code. Copyright © by the International Association of Plumbing and Mechanical Officials, Walnut, California.

If each valved zone has 2 lines at the minimum spacing of 4 ft, the minimum width of the irrigation/disposal field is 4 ft; the minimum length of the two perforated lines is therefore

$$120 \text{ ft}^2/4 \text{ ft} = 30 \text{ ft}$$

Thus, three irrigation/disposal fields, each 4 ft by 30 ft, in trenches constructed as shown in Fig. 11.54, will meet code minimums. ∎

11.10 Stormwater Treatment

Until relatively recently, stormwater runoff from buildings and paved surfaces was considered harmless and was conducted without treatment, and as rapidly as possible to receiving streams, rivers, and lakes. However, almost three-quarters of the U.S. population lives on only about 7% of the land. And as our urban watersheds experience an increase in their percentage of paved or roofed land area, a two-fold problem develops. First, a heavy rain greatly increases peaks in stream flows (because less rain is retained for later, slower release by unpaved surfaces); this accentuated peak–valley pattern of flows makes life more difficult for the aquatic flora and fauna. Second, pollutants (especially those accumulated on roadways) are washed into the stream; oil, gasoline, antifreeze, fragments of brake linings and tires, and so on, are especially unwelcome deposits. Some design strategies to minimize these impacts, such as roof retention, porous pavement, and on-site groundwater recharging, were discussed in Section 9.6.

Fig. 11.55 *Schematic diagram of a graywater system with an above-ground, gravity-emptied tank. (Reprinted by permission from the 1997* Uniform Plumbing Code. *Copyright © by the International Association of Plumbing and Mechanical Officials, Walnut, California.)*

With minimal changes in street design, storm-water receptacles can be constructed to not only provide access for maintenance, but also to screen out sediment, capture oil and other floating contaminants, and provide a more even outflow (Fig. 11.56). But much more can be done.

Phytoremediation is the science of cleaning polluted soil and water with plants. (The constructed wetlands and algal ponds of Sections 11.7 and 11.8 apply phytoremediation to sewage treatment.) Phytoremediation extracts, degrades, or contains contaminants. Plants that *extract* take up and accumulate contaminants in their shoots and leaves; when the vegetation is removed from the site, so is the contaminant. Plants that *degrade* break down contaminants (such as hydrocarbons and other organic compounds) until they are no longer toxic. Some plants degrade toxins in their

root zone, whereas others use elements of organic toxins as food. Plants that *contain* essentially immobilize toxins in long-term storage. However, when the plants die and decompose, long-lived toxins may be released.

For treating stormwater, plants are typically incorporated in drainage strips between paved parking areas. The mix of plants is similar to those cited earlier in wetland construction, with sedges and rushes especially effective at cleansing (see Fig. 11.2 for a sample trench). Shade trees could add welcome summer thermal benefits, as well as intercepting and storing some pollutants. The capacity of the planting strip should be sufficient to intercept the "first flush" of rain, which carries the heaviest load of pollutants. Underground drains intercept excess runoff, and lead to a capture chamber where floatable contaminants can

Grit/Oil chambers

Flow controls

Baffle well

(a)

(b)

Fig. 11.56 *This Vortechs™ Stormwater Treatment System (a) is designed to separate both floating oily contaminants and sediment from stormwater. (b) Rather like a septic tank, two chambers are separated by a baffle. Oil scum and sediment remain in the first chamber.*

WATER AND WASTE

be periodically removed. The screened, regulated flow from the capture basin will be much more gentle on the receiving stream.

A new community within Bellevue, Washington incorporated a stormwater treatment facility in a 20-acre (8.1-hectare) public park. Beneath tennis courts, ballparks, picnic areas, and other playing fields, a 300,000 ft³ (8,495 m³) sediment vault [surface area about 8 acres (3.2 hectares)]

slows the flow of stormwater, encouraging sediment to drop out. The water then passes to a filter basin of vegetation, sand and peat, for the primary purpose of phosphorus removal. From there it is slowly released to the receiving stream, which then drains into a lake.

Building designers rarely encounter such community-wide pollution control strategies, but parking lots and on-site roadways are encoun-

(c)   (d)

(e)   (f)

(g)

Fig. 11.56 (c) *In a "2-month" storm, water rises above the top of the inlet pipe, greatly reducing velocity and turbulence within the first chamber, encouraging sediment to precipitate to the bottom.* (d) *As flow increases, rising above the low-flow control (second chamber), oily contaminants accumulated from previous storms float upwards, and more sediment falls out.* (e) *At the height of stormwater flow, usually designed for a "10-year" storm, the high level outlet begins to discharge. Sometimes a peak flow by-pass is used. Oily scum and sediment continue to be trapped in the manhole.* (f) *As the flow subsides, water outflow is controlled, lessening the strain on the storm sewer system.* (g) *Cleanout, at lowest water levels, is best done by a vacuum truck. As the grit chamber empties, oily liquids and floating debris drain back toward the inlet and can be removed along with the sediment. Note that the bottom of the baffle between the chambers always remains submerged. (Courtesy of Vortechnics Incorporated, Portland, Maine.)*

tered regularly. Design for local pollution control, even for seemingly benign stormwater, is part of the architect's responsibilities.

REFERENCES

Converse, J., et al. (1978). *Design and Construction Manual for Mounds.* Small Scale Waste Management Project, University of Wisconsin Extension, Madison, Wis.

EPA (1993). *Subsurface Flow Constructed Wetlands for Wastewater Treatment.* U.S. Environmental Protection Agency, Office of Water, EPA 832-R-93-008, July 1993.

Green, F. B., Lundquist, T. J., and Oswald, W. J. (1995a). "Energetics of advanced integrated wastewater pond systems," *Water Science and Technology,* Vol. 31, No. 12, pp. 9–20.

Green, F. B., Bernstone, L., Lundquist, T. J., Muir, J., Tresan, R. B., and Oswald, W. J. (1995b). "Methane fermentation, submerged gas collection, and the fate of carbon in advanced integrated wastewater pond systems," *Water Science and Technology,* Vol. 31, No. 12, pp. 55–65.

Green, F. B., Bernstone, L., Lundquist, T. J., and Oswald, W. J. (1996). "Advanced integrated wastewater pond systems for nitrogen removal," *Water Science and Technology,* Vol. 33, No. 7, pp. 207–217.

Milne, M. (1976). *Residential Water Conservation,* U.S. Office of Water Research and Technology, Department of Commerce, NTIS.

National Small Flows Clearinghouse (1997). "Lagoon Systems Can Provide Low-Cost Wastewater Treatment," in *Pipeline,* Vol. 8, No. 2, Spring. National Small Flows Clearinghouse, Morgantown, West Virginia.

Oswald, W. J., Golueke, C. G., Cooper, R. C., Gee, H. K., and Bronson, J. C. (1963). "Water reclamation, algal production and methane fermentation in waste ponds," *International Journal of Air and Water Pollution,* Vol. 7, pp. 627–648.

Oswald, W. J., Meron, A., and Zabat, M. D. (1970). "Designing waste ponds to meet water quality criteria," *Proceedings of the Second International Symposium on Waste Treatment Lagoons,* Kansas City, Missouri, pp. 186–194.

Oswald, W. J. (1990). "Advanced integrated wastewater pond systems," *Supplying Water and Saving the Environment for Six Billion People: Proceedings of the 1990 American Society of Civil Engineers Convention,* Environmental Engineering Division, November 5–8, 1990 San Francisco, California, pp. 73–81.

Oswald, W. J., Green, F. B., and Lundquist, T. J. (1994). "Performance of methane fermentation pits in advanced integrated wastewater pond systems," *Water Science and Technology,* Vol. 30, No. 12, pp. 287–295.

Stoner, C., ed. (1977). *Goodbye to the Flush Toilet,* Rodale Press, Emmaus, Pa.

12

SOLID WASTE

For some designers, the last building distribution system considered is the one involving the bulkiest items: the flow of supplies in and solid waste out. Because this system usually is not seen as consuming building energy or requiring specialized equipment, it ordinarily becomes a lower-priority system in the design process. Yet provisions for delivery of supplies, and especially for the collection and storage of solid wastes, can be more space-consuming than water/waste systems, can present a fire danger, and can create severe local environmental problems. The separation of solid waste for resource recovery involves significant energy and environmental consequences. Finally, mechanical equipment associated with solid waste is now more commonly installed.

12.1 Waste and Resources

Since the late 1940s there has been a marked increase in the amount of packaging material used for consumer products. Where shoppers once refilled reusable containers for bulk supplies at the market, for example, they now buy food in bags or cans that are discarded after use. This trend has increased spatial needs in the store, where shelves of cans are required instead of a bin of bulk products, and in the home, where such packaging soon turns into waste and must be stored until garbage collection day. Energy is required to make the boxes, bags, cans, and other containers; transport them; and collect them as trash. Devices such as trash compactors add both space and energy requirements to the process. Landfills for garbage disposal fill more rapidly as solid-waste flows increase; methane and leachate from these landfills are potential environmental problems.

There may not be much that building designers can do about these increased packaging trends. But the solid wastes from buildings do contain important resources (Tables 12.1 and 12.2), and

a designer can help society to recover those resources, rather than to bury them in landfill or dump them in the ocean. Furthermore, the buildings themselves can be designed for materials recovery upon remodel or demolition, as was discussed in Sections 1.2*b* and 1.7.

The resource within solid waste can be divided into the high-grade resources represented by recyclable materials and the low-grade resources of heat and by-products obtainable from the burning or decomposing of combustible solid wastes.

(a) High-Grade Resources. These include metals such as aluminum and steel, paper and paperboard, wood, and some plastics. For buildings, they pose the problem of storage while awaiting collection. For many of these materials, re-use in their present form is desirable, which eliminates compaction as a storage strategy. Table 12.1 compares uncompacted and compacted volumes of various recyclable materials.

Glass is especially suitable for recycling when it can be reused after simple washing, as in the case of beer and soft-drink bottles. "Returnable" bottles and cans have measurably decreased roadside litter in states that now require such containers. Newspaper is so readily stored and recycled that in many communities, charitable groups or service organizations recover newspapers for profit. Recycled paperboard can easily save 50% of the energy that would be required for pulp from virgin material.

The recovery of aluminum saves 96% of the energy necessary to produce it originally. Because the aluminum production process is dependent on electricity, the recycling of aluminum is even more attractive. Energy conservation in aluminum production developed to the point that only 82% as much energy was needed to make a pound of aluminum in 1982 as in 1972. More than a third of that reduction was attributable to aluminum recycling programs; the remainder came from im-

WATER AND WASTE

TABLE 12.1 **Typical Recyclable Materials in Building Operation**

Category	Product	Recycling Label	Material Description	Conversion, Volume to Weight	
				I-P Units	SI Units
Paper	Ledger paper, white letterhead	SWL	Sorted white ledger, high-grade white paper	Uncompacted: 1 yd^3 = 500 lb Compacted: 1 yd^3 = 750 lb	Uncompacted: 1 m^3 = 297 kg Compacted: 1 m^3 = 445 kg
	Computer paper	CPO	Computer printout	Uncompacted: 1 yd^3 = 500–600 lb Compacted: 1 yd^3 = 1000–1200 lb	Uncompacted: 1 m^3 = 297–356 kg Compacted: 1 m^3 = 593–712 kg
	Colored paper	SCL	Sorted color ledger	Uncompacted: 1 yd^3 = 500 lb Compacted: 1 yd^3 = 750 lb	Uncompacted: 1 m^3 = 297 kg Compacted: 1 m^3 = 445 kg
	Newspaper	Mix	Newsprint	Uncompacted: 1 yd^3 = 350–500 lb Compacted: 1 yd^3 = 750–1000 lb	Uncompacted: 1 m^3 = 208–297 kg Compacted: 1 m^3 = 445–593 kg
	Magazines	Mix	Clay-coated paper	Not available	Not available
	Telephone books	Mix	Mixed papers–adhesives	1 book = 1–3 lb	1 book = 0.5–1.4 kg
	Cereal boxes	Mix	Coated paperboard	Not available	Not available
	Shipping boxes	OCC	Old corrugated cardboard	Uncompacted: 1 yd^3 = 285 lb Compacted: 1 yd^3 = 500 lb	Uncompacted: 1 m^3 = 169 kg Compacted: 1 m^3 = 297 kg
Glass	Food jars, beverage bottles	1 2 3	Amber glass Green glass Clear glass	Loose, whole: 1 yd^3 = 600 lb Manually crushed: 1 yd^3 = 1000 lb Mechanically crushed: 1 yd^3 = 1800 lb	Loose, whole: 1 m^3 = 356 kg Manually crushed: 1 m^3 = 593 kg Mechanically crushed: 1 m^3 = 1068 kg
Plastic	Beverage containers	PET	Polyethylene terephthalate	Whole: 1 yd^3 = 30 lb	Whole: 1 m^3 = 18 kg
	Milk containers	HDPE	High-density polyethylene	Whole: 1 yd^3 = 25 lb Crushed: 1 yd^3 = 50 lb Compacted: 1 yd^3 = 600 lb	Whole: 1 m^3 = 15 kg Crushed: 1 m^3 = 30 kg Compacted: 1 m^3 = 356 kg
	"Clamshell" containers	—	Polystyrene plastic foam	Not available	Not available
	Film plastic	LDPE	Low-density polyethylene	Not available	Not available

TABLE 12.1 **Typical Recyclable Materials in Building Operation (***continued***)**

Category	Product	Recycling Label	Material Description	Conversion, Volume to Weight	
				I-P Units	SI Units
Metals	Beverage cans		Aluminum/bi-metal	Whole: 1 yd³ = 50–70 lb	Whole: 1 m³ = 30–42 kg
				Crushed: 1 yd³ = 300–450 lb	Crushed: 1 m³ = 178–267 kg
	Food and beverage cans		Steel with tin finish	Whole: 1 yd³ = 125–150 lb	Whole: 1 m³ = 74–89 kg
				Crushed: 1 yd³ = 500–850 lb	Crushed: 1 m³ = 297–504 kg
Miscellaneous	Pallets		Wood	Not available	Not available
	Food waste		Organic solids and liquids	55-gal drum—415 lb	(208 L drum = 188 kg)
	Yard waste		Organic solids	Leaves, uncompacted: 1 yd³ = 250 lb	Leaves, uncompacted: 1 m³ = 148 kg
				Leaves, compacted: 1 yd³ = 450 lb	Leaves, compacted: 1 m³ = 267 kg
				Wood chips: 1 yd³ = 500 lb	Wood chips: 1 m³ = 297 kg
				Grass clippings: 1 yd³ = 400 lb	Grass clippings: 1 m³ = 237 kg
	Used motor oil		Petroleum product	1 gallon = 71 lb	1 L = 8.5 kg
	Tires		Rubber	1 passenger car = 20 lb	1 passenger car = 9 kg
				1 truck = 90 lb	1 truck = 41 kg

Source: Reprinted by permission from AIA, Ramsey, and Sleeper, *Architectural Graphic Standards,* 9th ed., © 1994 by John Wiley and Sons.

provements in the production process. As for steel, recycling can produce a 52% energy savings, compared to the use of virgin material.

Wood has become more valuable as fiber than as fuel. Wood chips and scraps that once were burned or buried are now recycled to become oriented strand board (OSB) used in structural insulated panels (SIPs), window frames, and many other aspects of building and furniture construction.

Plastics are more difficult to recycle. Regulations and consumer preferences discourage the use of recycled plastics in food-related items. However, the plastics from items such as soft-drink containers, margarine tubs and lids, and milk jugs can be reprocessed into plastic pellets, which are cheaper than virgin plastic pellets.

These pellets are made into nonfood items—toys, building products, sports products, and other things.

(b) Low-Grade Resources. These resources include materials for which recycling is impractical, but that are combustible. Low-grade resources include gaseous wastes, liquid and semi-liquid wastes, and solid wastes. Industrial and commercial processes can generate wastes of all types, including some with very high heat content and some that are very toxic. Although to recover heat by burning such materials seems better than to waste them altogether, air pollution regulations have severely restricted simple trash burning as a means of solid waste disposal. Incinerators must meet increasingly strict regulations and, as a re-

WATER AND WASTE

TABLE 12.2 **Solid Waste Sources in Building Operation**

| | | Types of Waste Generated | Quantities of Waste Generated | |
| | | | I-P Units | SI Units |
Classification	Occupancy	(percent, where available)		
Residential	Studio or 1-bedroom apartment	Newspaper (38/43)[a] Plastic (18/7)[a] Miscellaneous (13/18)[a]	1–1.5 yd³/unit/month (200–250 lb) or 2.5 lb/person/day	0.8–1.1 m³/unit/month (91–113 kg) or 1.1 kg/person/day
	2- or 3-bedroom apt. or single-family house	Metals (14/9)[a] Compost (10/15)[a] Glass (2/8)[a]	1.5–2 yd³/unit/month (250–400 lb) or 2.5 lb/person/day	1.1–1.5 m³/unit/month (113–181 kg) or 1.1 kg/person/day
Commercial	Office	Plastics, compost, used oil, metals, and glass (30)[b] High-grade paper (29)[b] Mixed papers (23)[b] Newspaper (10)[b] Corrugated cardboard (8)[b]	1.5 lb/employee/day or 1 yd³/10,000 ft²/day (includes 0.5 lb of high-grade paper/employee/day)	0.7 kg/employee/day, or 0.8 m³/1,000 m²/day (includes 0.2 kg of high-grade paper/employee/day)
	Department store Wholesale/retail Shopping center	Corrugated cardboard, compost, wood pallets, high-grade paper, and plastic film	1 yd³/2,500 ft²/day 70 lb/$1000 sales/day 2.5 lb/100 ft²/day	0.8 m³/250 m²/day 32 kg/$1000 sales/day 1.1 kg/100 ft²/day
	Supermarkets	Corrugated cardboard, compost, and wood pallets	1 yd³/2,500 ft²/day	0.8 m³/250 m²/day

Facility		Materials	US units	Metric units
Restaurants/entertainment		Compost (38)[b], Corrugated cardboard (11)[b], Newsprint (5)[b], High-grade paper (4)[b]	Cafeteria, 1 lb/meal; Fast food, 200 lb/$1000 sales; Restaurant, 1.5 lb/meal	Cafeteria, 0.45 kg/meal; Fast food, 91 kg/$1000 sales; Restaurant, 0.7 kg/meal
Drug stores		Corrugated cardboard and high-grade paper	1 yd³/2,000 ft²/day	0.8 m³/200 m²/day
Banks, insurance companies		High-grade paper, mixed paper, and corrugated cardboard	0.75 lb high-grade paper/person/day (survey required)	0.34 kg high-grade paper/person/day (survey required)
Hotels and Motels (not including restaurants)	High occupancy	Glass, aluminum, plastic, high-grade paper, newspaper, and corrugated cardboard	0.5 yd³/room/week; 3.2 lb/room/day	0.38 m³/room/week; 1.45 kg/room/day
	Average occupancy		0.17 yd³/room/week; 1.7 lb/room/day	0.13 m³/room/week; 0.8 kg/room/day
Institutional	Hospitals	Compost, high-grade paper, biomedical waste, corrugated cardboard, glass, and plastics	1 yd³/5 occupied beds/day	0.8 m³/5 occupied beds/day
	Nursing homes		1 yd³/15 persons/day	0.8 m³/15 persons/day
	Retirements homes		1 yd³/20 persons/day	0.8 m³/20 persons/day
Educational	Grade school	High-grade paper, mixed paper, newspaper, corrugated cardboard, compost, plastic, glass, and metals	1 yd³/8 rooms/day	0.8 m³/8 rooms/day
	High school		1 yd³/10 rooms/day	0.8 m³/10 rooms/day
	Universities		(Survey required)	(Survey required)

Source: Reprinted by permission from AIA, Ramsey, and Sleeper, *Architectural Graphic Standards*, 9th ed., © 1994 by John Wiley and Sons.

[a]Percent by volume/percent by weight.
[b]Percent by volume.

sult, are very rarely installed in buildings. A better option is to compost them in landfills. The methane generated in landfills can then be used as a high-grade fuel.

Where large quantities of mixed trash, garbage, and other refuse are collected, special resource-recovery plants can be built to recover materials, produce useful steam for electricity generation, and reduce the flow of waste to landfill. In this energy-intensive process, the mixed garbage is shredded and blown through large "air classifiers" that separate the organic (burnable) wastes from metals and glass. The burnable wastes are then used, under controlled combustion, to generate electricity. The metals are further separated magnetically into ferrous and nonferrous classes; these and the glass are then recycled.

The proposed E2R2 integrated infrastructure facility at Cal Poly State University, San Luis Obispo, combining wastewater treatment and solid waste processing, was introduced in Fig. 11.43. The solid waste facility is shown in Fig. 12.1.

The proposed schematic design for solid waste processing (SWP) in the E2R2 facility involves transporting unsorted solid waste to a central processing facility located on campus. Materials will then be sorted using mechanical equipment and hand sorting. The composition of the solid waste stream is expected to be 70% organic materials, suitable for composting, and 815 tons (739 Mg) per year of recyclable materials (i.e., metals, glass, paper, and cardboard). Roughly 10% of the solid wastes are anticipated to be unsuitable for either recycling or composting. These "inert" materials will be disposed of at the local landfill.

Once separated, the design calls for the nonrecyclable, organic materials to be uniformly ground

and combined with sludge (from the City of San Luis Obispo's wastewater treatment plant) or manure (from the University farming operations) to achieve a proper carbon to nitrogen balance. Given the dynamic nature of the recyclable materials market, it is anticipated that, at times, some potentially recyclable paper will be included in the compost operations.

After mixing, the organic materials will be loaded into a high solids, anaerobic digester and undergo methane fermentation. The biogas generated by the fermentation process will be harvested and stored on site. Prior to use, the biogas will be scrubbed (processed) to remove impurities, making it suitable for use as fuel. A portion of the methane will be used to heat the digester whereas the excess will power a micro fuel cell to generate electricity for campus use. Other potential uses include using methane in the campus's boiler as an alternative to gasoline in the university's automobile and truck fleet; and even to power an ultra–high temperature hazardous waste disposal system.

The anaerobic digestion process is estimated to generate 180,000 cubic feet (5098 cubic meters) per day of biogas. The energy content of the methane in this biogas is estimated at 35×10^9 Btu (36.9×10^9 kJ) per year.

After unloading from the high-solids digester, the remaining biosolids will be pressed to recover bacteria-rich liquids, which are useful in inoculating the next batch of materials for the digester. The dewatered biosolids will then be mixed with algae, from the adjacent AIWPS™ Facility, to create a nutrient-rich compost. This compost will be available for use on the campus' landscape, its farm lands, or for sale to local gardeners. Yearly

Fig. 12.1 *Schematic design proposal for Solid Waste Processing at the E2R2 Facility at Cal Poly State University, San Luis Obispo, California. (This works in conjunction with the proposed Wastewater Treatment facility; see Fig. 11.43.) Unsorted solid wastes are separated into recyclable, organic and inert materials (a). The organic materials are ground and combined with sludge from the city's sewage treatment plant, and loaded into the high solids digester (b) to undergo anaerobic fermentation. The biogas generated by the fermentation process is scrubbed to remove impurities from the methane, making it suitable for use as a fuel to heat the digester and to power a micro fuel cell to generate electricity for the campus. Biosolids from the digester are pressed and the liquid, rich in anaerobic bacteria, is then used to inoculate the next anaerobic fermentation cycle. Pressed biosolids are mixed with nutrient-rich algae, from the adjacent AIWPS™ Facility, and pasteurized in compost curing buildings, (c). Once cured, the compost can be used on landscaped areas, farm lands or sold to local gardeners. For site plan, see Fig. 11.43d, page 717. (Courtesy of Prof. Douglas Williams, Agricultural Engineering Department, and Prof. Daniel Panetta, Architecture Department, Cal Poly, San Luis Obispo.)*

(A) SOLID WASTE INTAKE AND PROCESSING

(B) ANAEROBIC DIGESTION & METHANE HARVESTING

(C) COMPOST PRODUCTION

SOLID WASTE PROCESSING DIAGRAMS

Fig. 12.2 *A 45,000-ft² (4,180 m²) greenhouse built over part of a 70-acre (28-hectare) Michigan landfill utilizes "landfill gas" (including about a 65% methane component) both to power its space heaters and to generate electricity for lighting. Waste heat from the generator also contributes to space heating. (Photo by George P. Graff. Courtesy of Willow Run Farms, Inc., Ypsilanti, Michigan.)*

compost production is anticipated to exceed 5000 tons (4535 Mg) air-dried weight.

As an added touch, straw bale construction is proposed for the E2R2 facility buildings. Currently, plans call for an education and laboratory building, a shared control office for wastewater treatment and solid waste processing operations, a solid waste processing building, and on-site student housing. Straw bale construction is advantageous because of its affordability as well as its exceptional insulation and noise mitigation characteristics. In addition, using rice straw for construction will help provide an alternative to burning waste straw from California's rice crop.

Conventional landfills accept unsorted garbage from a huge variety of sources. Over many years, anaerobic combustion in such enclosed landfills generates methane gas, a usable resource as described previously. Fig. 12.2 shows an architectural opportunity for such landfill sites: a greenhouse heated by burning landfill (methane) gas. Electricity generated by landfill gas supplies the lights for the greenhouse. As a design guideline, a site with 1 million tons (907,000 Mg) of refuse in place should produce enough low-heat-value "landfill gas" to support a cogeneration system that includes a greenhouse. Landfill gas, rather than natural (pipeline) gas, can also drive a fuel cell.

12.2   Resource Recovery: Central or Local?

In general, the more thoroughly mixed are the different types of solid waste, the harder it is to recover their high- and low-grade resources. From an energy conservation viewpoint, solid wastes should be kept locally as separate as possible; glass bottles should be washed and reused rather than broken and recycled, and unrecyclable but burnable solid wastes should be kept clean and dry until they can be burned or composted. The earlier that different types of metals are separated, for example, the less energy spent later on to separate aluminum from steel, and so forth. Organic food wastes could be composted for use on site, rather than ground up and added to the load on the sewage treatment system.

However, there are two disadvantages to "local" solid waste separation (at the point of their discard)—one cultural, one physical. Keeping solid wastes separate requires somewhat more effort and time on the part of the consumer. Rather than dumping everything in a garbage can, the resource-conscious consumer will wash metal cans, remove both ends and flatten them, then deposit them in a container reserved for ferrous metal. Newspaper and box cardboard are stacked separately; returnable bottles and cans are kept apart from glass; compostable kitchen wastes are kept separate from garbage. Composting in urban apartments presents a unique space-odor challenge.

This disadvantage of slightly more work and time for waste separation is compounded by the physical disadvantage of the floor or cabinet space taken up by so many separate containers. In communities fortunate enough to have recycling-oriented garbage service, these containers can be carried out and lined up to await garbage collection. The garbage trucks used for this purpose are complex assemblages of various bins, rather than simple massive caverns with a compactor. In many communities, however, each pile of recyclable materials at home must be dealt with separately, either collected by a specialized handler (nonprofit agencies are examples), or taken to a different collection point—a disadvantage in terms of both the time and energy used in transportation. Clearly, the designer's incentive to include recycling in building function will be stronger where recycling collection is well established.

The characteristics of local waste separation, then, are increased consumer time and building space requirements; see Tables 12.2 and 12.3. Central waste separation is characterized by energy-intensive (and noisy) industrial processes. Sections 12.3 and 12.4 offer a detailed look at the consequences of local waste separation on some common building types.

The massive quantities of solid waste generated in urban areas gives rise to the terms "urban mines" and "urban forests." To the extent that these solid "wastes" can be turned into resources, savings can be realized: of energy in transportation, of land area for landfills, and of virgin materials replaced by these recycled products.

The Bronx Community Paper Company (Fig. 12.3) is a proposed recycling paper mill in the midst of New York City's "urban forest." The

WATER AND WASTE

TABLE 12.3 **Space Planning for Solid Waste**

Part A. ***I-P Units***

Occupancy	Building Area (ft²)	Exterior Area Required	
		For Trash (ft²)	For Recyclable Materials (ft²)
Nonresidential Buildings	0–5000	12	12
	5001–10,000	24	24
	10,001–25,000	48	48
	25,000+	Each additional 25,000 ft² require 48 ft² each, trash and recyclables	
Multifamily Residential	2–6 Units	12	12
	7–15 Units	24	24
	16–25 Units	48	48
	25+ Units	Each additional 25 dwelling units require 48 ft² each, trash and recyclables	

Part B. ***SI Units***

Occupancy	Building Area (m²)	Exterior Area Required	
		For Trash (m²)	For Recyclable Materials (m²)
Nonresidential	0–465	1.1	1.1
	466–929	2.2	2.2
	930–2323	4.5	4.5
	2323+	Each additional 2323 m² require 4.5 m² each, trash and recyclables	
Multifamily Residential	2–6 Units	1.1	1.1
	7–15 Units	2.2	2.2
	16–25 Units	4.5	4.5
	25+ Units	Each additional 25 dwelling units require 4.5 m² each, trash and recyclables	

Source: Reprinted by permission from AIA, Ramsey, and Sleeper, *Architectural Graphic Standards,* 9th ed., © 1994 by John Wiley and Sons.

abandoned industrial site is considered a brownfield; the project aims to show that job production, social assistance, economic development, private sector profitability, and environmental protection can work in a mutually supportive way. This recycling installation includes a wastepaper de-inking plant, a newsprint paper-making machine, a wastepaper sorting plant, a sewage water clean-up system, and a steam boiler. Both the source of the raw material and the destination of the end product are potentially within short reach.

The wastepaper resource is estimated to replace (save) some 3,432,000 trees per year, representing 11,240 acres (4549 hectares). De-inking removes ink and other objectionable materials from the recycled fiber. Office paper "waste" is far more likely to undergo recycling if hauling distances to de-inking facilities are short. The water for the de-inking and paper-making processes comes from the nearby Ward's Island Wastewater Treatment Plant effluent, after treatment by reverse osmosis. (This is estimated to be one-third the price of fresh water.) Some water is lost in the paper product, much as sludge, most by evaporation. Yet of the 2.9 million gal/day supplied to the plant, 2.6 million gal/day will be sent back, after passing through a primary clarifier at the mill.

Fig. 12.3 *The proposed Bronx Community Paper Company takes advantage of the resources of New York City's "urban forest" as it de-inks waste paper, makes newsprint paper, and interacts with the nearby Ward's Island Wastewater Treatment plant. (Courtesy of Natural Resources Defense Council, and Maya Lin; © HLW International LLP/IO Media.)*

This process is made more visible by clerestory windows that will allow drivers on the Triborough Bridge to look within the plant as well as at the glass tower that supports a chlorine-free brightening process. The project was initiated by the Natural Resources Defense Council, with the Banana Kelly Community Improvement Association.

12.3 Solid Waste in Small Buildings

The choice between separation of waste or the mixing of garbage in one can is most often enjoyed by the occupants of small buildings. Where food preparation is involved, as in residences or restaurants, the collection of special containers can reach surprising complexity (Fig. 12.4).

As the point of origin for so many types of solid waste, the kitchen is often the location of waste separation and storage. There is an inherent conflict, however, between the kitchen's frequently hot and humid environment and the need for a cool, dry, well-aired place for solid wastes. This suggests that a space—pantry, air-lock entry, cabinet, or closet—be provided that opens to the kitchen on one side and to the outside on the other (Fig. 12.5). Both the daily deposits to solid waste and the weekly waste removal are made easy, and the near-outdoor conditions are better for the waste in most U.S. climates. It is also important that cleaning of the waste storage area be easy.

(a) Garbage Disposer. This common appliance (Fig. 12.6) is usually installed below the kitchen sink, often in common with a dishwasher. It grinds up organic food scraps and sends them on through the sewer. This device is a boon to central garbage collection because it lightens the weight of the garbage can and adds less moisture to the garbage (which thus can be burned more efficiently). Finely chopped organic matter has a better chance to biodegrade at the wastewater treatment plant than in a tightly packed landfill.

WATER AND WASTE

Fig. 12.4 *Local separation of solid wastes can require many containers. In communities that offer recycling garbage collection service, the consumer can recycle without transporting individual containers to many different collection points. Some recycling services ask for glass to be separated into green, brown, and clear; some offer to recycle white paper.*

However, garbage disposer units require both water and energy. Water must be kept running during the grinding process to coagulate grease for chopping, wash the blades and keep them free of debris, and cool the grinder's motor. In all, from 2 to 4 gal (7.5 to 15 L) are required for 1 minute of operation. Because more water and solid waste is deposited in the sewer system, moreover, the central sewage treatment plant requires more energy to operate.

(b) Garbage Compactor. In some small buildings, this device allows a much less bulky storage arrangement. Used selectively, it can compact several of the stacks of items shown in Fig. 12.4, such as aluminum for recycling, ferrous metals, and box cardboard. Used indiscriminately, it can make the central process of garbage separation more difficult by crushing dissimilar items

together. For a recycling-conscious single-family household, it is questionable whether the compactor will save much more storage volume than it takes for itself. However, small stores and institutions with large quantities of a bulky waste (such as cardboard) could save considerable storage space.

(c) Compost Pile. The alternative to the resource-hungry garbage disposal is the compost pile, familiar to most home gardeners as a source of excellent soil conditioners. Urban opportunities for composting seem very limited; yet raised growing beds on a balcony or window flower boxes are logical recipients of a family's compost. The problem may not be so much what to do with the final product (humus) as *where* to locate the compost pile. Several self-contained composting containers are available.

Fig. 12.5 *An entry vestibule (or airlock) to the kitchen can serve as a depository for solid wastes, as well as for coats, boots, and other items. Heavy, dirtier items belong on the floor, and many recyclables can be readily and conveniently stored on shelves. An outdoor hose bibb is useful for washing out soiled compost containers and garbage cans.*

Fig. 12.6 *The garbage disposer (or grinder) diverts solid waste (food scraps) from landfills to sewage treatment plants. Occupying less than a cubic foot, it saves garbage storage space and some of the user's time, but it requires more energy and much more water than treatment of the food scraps as garbage. Composting is another alternative. (From Milne, 1976.)*

The outdoor compost pile has several characteristics that challenge the designer. At its best, it is a frequently turned, quite warm, damp, well-aired source of rich humus (and red worms) for gardens; odors are noticeable only while the pile is turned. At its worst, it is a source of unpleasant odors, a breeding place for vermin, and a fast-food attraction for rats, dogs, and raccoons. Where odors are not objectionable, the heat generated in a frequently fed and tended compost pile might be welcome against the exterior walls of residences. Clearly, these walls must have nonorganic exterior materials!

(d) Storage Areas. As groups of residences are combined into large apartment complexes, the solid waste systems can make more significant demands on the designer. Approximate exterior storage areas are listed in Table 12.3. Where central storage compounds are provided, garbage

cans should be fenced to ward off dogs and other marauders. Different bins for recyclable materials might be provided. Where space is limited, a single bin could accept specific materials on specified days—newspapers Mondays, metals Wednesdays—if collection schedules permit. A central compost pile could provide humus for the landscaping of the complex. (An incinerator, used to recover heat from nonrecyclable burnable wastes, is almost never approved for installation today.) This combination of space and equipment has special environmental needs, including garbage truck access, noise control, and location of both incinerator stacks and compost pile with respect to prevailing winds.

12.4 Solid Waste in Large Buildings

As buildings become larger, it becomes more likely that solid wastes will be handled several times in the storage and collection process. This may inhibit the separated-waste approach be-cause not only the employees who generate it, but also the custodians who collect and store the waste, must understand what items go into which bins. However, larger buildings tend to generate large and concentrated kinds of wastes, which makes recycling more attractive, while it also in-creases the demand for storage space.

Consider the multistory office building. The office floor operations are likely to discard large quantities of white paper and smaller quantities of newspaper, box cardboard, and unrecyclable burnable trash (including floor sweepings). Much smaller quantities of food scraps (coffee grounds), metals, and glass are also generated. Given the high cost of rental space, the pressures are very high for a simple mixed-garbage can (rather than multiple separate bins). However, the cost of haul-ing trash to increasingly scarce landfills is now so high that many recycling programs pay for them-selves both by avoided landfill and by the value of recycled materials.

(a) The Collection Process. In larger build-ings, the collection of solid waste typically in-volves a three-stage process (Fig. 12.7). The first stage is the generation of the waste itself, by em-ployees who might be provided with one waste-basket each. If each employee is expected to sepa-rate wastes, this suggests a redesigned receptacle. At the typical desk in an office building, white paper, recyclables, compostables, and garbage would be deposited in separate compartments. Waste separation at the point of origin thus may not require much more floor area, just more atten-tion by the worker.

The second stage begins as custodians discon-nect these individual baskets, dump them into separate bins on a collection cart, and reconnect the individual empty baskets for the next day's deposits. At various special-purpose stations, spe-cial wastebaskets can be supplied: white paper at the computer room and the copying machine, compostables and garbage at the employee lounge. Floor sweepings are added to garbage. When the cart is full, it is wheeled to the service closet, which is probably located within the core of the building. Here would be a container for each category of waste, along with a service sink to wash out the garbage bin (and perhaps a paper shredder to be used selectively by employees).

The third stage begins at the ground floor of the service elevator where white paper and other recyclables are perhaps shredded and stored until collection; compostables are stored or sent to a roof garden compost pile; and garbage compacted and bagged. In the storage space, cool, dry, and fresh air is desirable. A sprinkler fire protection system is often required. The compactors and shredders are noisy, and must be vibration-iso-lated from the floor. At the end of the third stage, a truck or van from the recycling center collects recyclables, and a garbage truck collects gar-bage bags.

(b) Audubon House. This New York City office building (Fig. 12.8) utilizes several of the previously mentioned strategies and demon-strates some unusual design impacts of planning for recycling in offices. Built just at the end of the 19th century, this eight-story building was itself reused rather than demolished in its renovation near the end of the 20th century as the offices of the National Audubon Society. Many new materi-als made from recycled materials were installed; four new vertical chutes conduct recyclable mate-rials to a subbasement resource recovery center. Details about this project are available in Audubon/Croxton (1994).

Solid waste collection's first stage begins with two desktop recycled paper trays: one for paper for the employee's reuse, the other for recycling. Two wastebaskets are provided per employee: one for mixed paper, the other for unrecycled garbage. On each floor, central recycling points are gathered around four 20-in. (508-mm) diame-ter steel chutes that run from top floor to sub-basement. One of these chutes is for white (and computer) paper; the opening to this chute is in a vestibule for copy machines. The other three chutes openings are in a "pantry" beside a small kitchen/break room, and are for (1) organics (food wastes and soft soiled paper), (2) returnable plastic bottles and aluminum cans, and (3) mixed paper (colored paper, file folders, paperboard, Post-it notes). In addition, shelves in each pantry accept returnable glass bottles, coated papers (juice and milk cartons), magazines, and news-papers.

In Audubon House's second stage collection, custodians pick up the wastebaskets from the work areas, and the collection from the pantry

Fig. 12.7 *Hypothetical three-stage collection process with four-category waste separation for office buildings.* (a) *At each workstation, a four-compartment waste receptacle is provided. White paper is the predominant waste product.* (b) *Custodians begin the second stage by collecting waste in separate bins. Floor sweepings are added to the garbage.* (c) *At the end of the second stage, the four categories of wastes are deposited in separate bins. Service sinks, paper shredders, and other maintenance items can be incorporated in such a service closet.* (d) *The third stage begins with compactors at the base of the service core. White paper is compacted, baled, and stored; recyclables are sorted and stored; garbage is compacted, baled, and stored.* (e) *At the end of the third stage, white paper, metal, plastic, and glass are collected for recycling and garbage is collected for separation at a central plant.*

shelves. These are taken to the subbasement recycling depot and sorted. In the third stage, large movable bins await deposits at the bottom of the four chutes. Glass bottles, newspapers, and so on from the pantry shelves are collected and boxed or baled. Recyclable materials and garbage are periodically taken up to the delivery dock, to recycling and garbage collectors. The exception is the organic waste. This compostable material is first refrigerated until enough material accumulates. It is then screened (to remove the largest difficult-to-compost solids) and put into one of four composting "reactors," each with a 40-lb (18-kg) capacity. These are enclosed for odor control,

WATER AND WASTE

Fig. 12.8 *Audubon House, New York City, renovated after its first 100 years, now features four 20-in. (508-mm) diameter steel chutes that conduct recyclables to a subbasement resource recovery center. Organic materials become compost and are contributed to the roof garden. (Courtesy of the Croxton Collaborative Architects PC.)*

yet aerobic; air and heat, starter compost, and wood chips are provided to fast-start the process at about 140°F (60°C). Each reactor is vented through filters to control odors. After about 3 days, the second compost stage begins in which a steady flow of air sustains aerobic decomposition during a 3-month period. After this, the compost (greatly reduced in size and weight) has turned to humus and is taken to the roof garden.

12.5   Equipment for the Handling of Solid Waste

At Audubon House, four new steel chutes of 20-in. (508-mm) diameter were inserted in the building's vertical core. Another approach is to use but one vertical chute, with a rotating receiving facility at the base (Fig. 12.9). With up to six receiving bins on a carousel and a control panel at each floor's opening, considerable floor space is saved compared to six individual chutes. (Also, custodial workers have fewer recyclables to separate and move.) The user first checks the control panel setting (it will be at whatever bin the previous user had selected). After making the selection, the carousel below turns to position the appropriate bin. The access panel then opens (interlocked so that but one floor can be opened at one time) and the recyclable material (or garbage) can be deposited. The receiving room at the chute's base is about 12-ft (3.7-m) square, with a minimum height of 8 ft (2.4 m).

Where large amounts of solid waste are generated and little space is available in which to store it, various types of equipment can be used to change the volume or the composition of the waste so as to facilitate its transportation and storage. The first step is to determine the probable daily flow of solid waste, as is done in Tables 12.1 and 12.2. When the extent of the problem is known, equipment can be selected.

(a) Compactors (Fig. 12.10).   Of the wide variety of compacting devices, most are able to reduce the volume of solid waste to as little as 10% of the original volume. Among the many choices to be made are vertical versus horizontal compaction, automatic chute fed versus manual freestanding, whether wastes are to be bagged or baled, and the final size of each unit of compacted

waste. Compactors can be noisy; they can also be prone to fires as heat is generated in the compaction process. Many compactors have built-in sprays for both fire control and disinfecting. Access to wash water and a floor drain are highly desirable.

(b) Vacuum Systems.   The primary advantage the vacuum system is that it reduces the building volume consumed by waste chutes or storage cans on each floor. A grinder-plus-evacuated-tube system is similar to the vacuum sewage systems described in Section 10.7 (Fig. 10.49). Vacuum conveying systems are now used in several industrial applications, such as ash and wood-chip transporting. Vacuum systems that use air, not water, as the transport medium are frequently used for linens in hotels and for trash. Separate vacuum systems for trash and linens are desirable, both to lessen the soiling of the linens and to allow them to be delivered to a destination different from that for the trash.

Fig. 12.9 *One chute serves many different recyclables, as well as garbage, in this Hi-Rise Recycling System™. Interlocking access doors on each floor allow only one user at a time; the user selects the type of recyclable being deposited, using the control panel at the opening. (Courtesy of Hi-Rise Recycling Systems, Inc., Miami, Florida.)*

BALER/COMPACTOR

Fig. 12.10 A refuse baler/compactor may save storage space where particular waste stream volumes are high. Cardboard, metal cans, and plastic bottles may be more attractive to recyclable haulers when their volume is thus reduced. (Reprinted by permission from AIA, Ramsey, and Sleeper, Architectural Graphic Standards, 9th ed., © 1994 by John Wiley and Sons.)

The advantages of such pressurized systems is that lines can be small and the contents can be moved horizontally or even upward. This allows far greater flexibility in design than is possible with gravity systems.

(c) Summary. There are many options to be considered within the present waste-handling process of grind, crush, burn, and bury. As Earth's nonrenewable resources are stripped away, the recovery of materials from solid waste becomes increasingly attractive. The situation is similar to energy-and-design issues: After decades of energy-intensive building trends, designers are now giving space to daylighting atriums and allowing surfaces to act as thermal mass. A modest increase in floor space and equipment to encourage resource recovery from solid waste seems an increasingly worthwhile design investment.

12.6 The Service Core

Medium- and high-rise buildings usually concentrate many building services within a "core," from which services can be distributed as needed throughout each floor. The core typically contains stairs; passenger and service elevators; toilet rooms; service closets; mechanical, plumbing, and electrical chases; electrical/telephone closets with local switching capabilities; fire protection equipment; and supply closets. The size of such cores varies widely; for example, the taller the building, the more elevators.

Often, these service cores are identical in plan from one floor to the next. Alternatively, the vertical services can be identical (stairs, elevators, chases) and the arrangement of other elements (toilet rooms, supply closets) allowed to vary. Whether or not minor floor-plan variations occur, the cores usually depart radically from the typical plan—and ceiling height—at both roof and ground floor (Fig. 12.11). (Middle interruptions also occur where intermediate mechanical floors are provided.) Depending on the type of elevator machinery selected, the machine room ceiling height above the top floor served might be three times the ordinary floor height (see Chapter 23). On the delivery/loading floor, large trucks must be accommodated. For chillers, boilers, incinerators, and fans, higher ceilings may be required to accommodate the flues, ducts, and pipes to which they must be connected; see Section 8.3 for the

Fig. 12.11 Schematic section through a core of a multistory building. At the top and bottom, floor space and ceiling height requirements increase. The consequences on the ground (or delivery/collection) floor are particularly great.

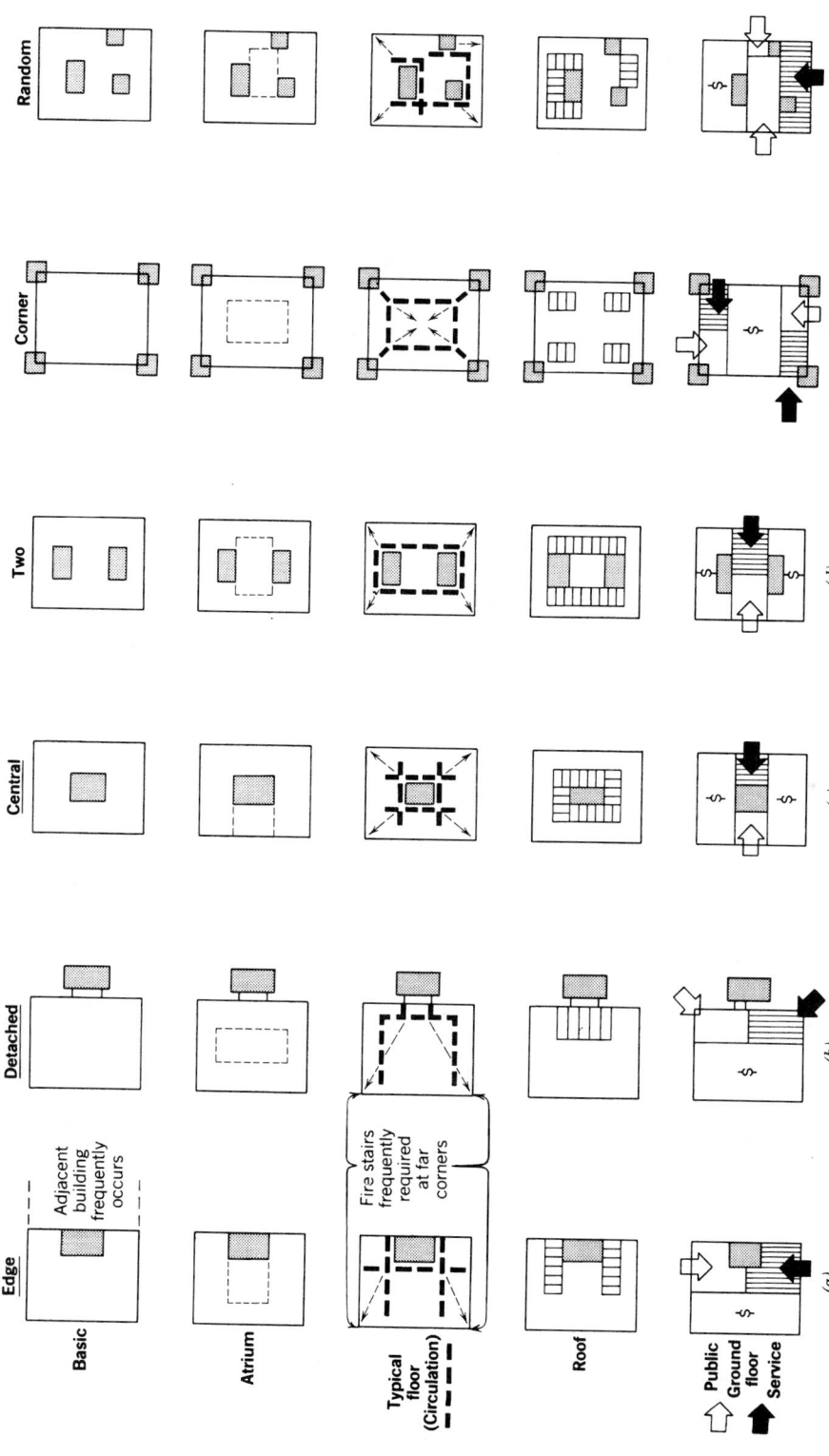

Edge | **Detached** | **Central** | **Two** | **Corner** | **Random**

Basic

Adjacent building frequently occurs

Atrium

Typical floor (Circulation)

Fire stairs frequently required at far corners

Roof

Public
Ground floor
Service

(a)   *(b)*   *(c)*   *(d)*   *(e)*   *(f)*

Fig. 12.12 *Comparison of core arrangements and building plans. (a,b) The edge and the detached cores give great flexibility of rental floor area, with light and view for core spaces. (c) The central core expands readily at the roof and the ground floor and has clear circulation and fairly flexible rental space. (d) Two-core is a popular, workable arrangement. (e) The corner cores give great flexibility to rental floors but are difficult at roof and ground floor. (f) Random cores generally occur in low-rise buildings, in which the benefits of repetitious plans are minimal. See also Table 12.4.*

TABLE 12.4 **Building Characteristics and Core Placement**

Core Designs	Edge	Detached	Central	Two	Corner	Random
Flexibility of typical rental areas	Good	Poor	Average	Fair	Good	Poor
Perimeter for rental areas	Fair	Average	Poor	Poor	Poor	Good
Ground floor high rent area	Average	Poor	Average	Fair	Good	Poor
Distance of travel from core, typical	Fair	Poor	Good	Poor	Average	Average
Clarity of circulation, typical	Average	Fair	Good	Poor	Average	Poor
Daylight, view for core spaces	Good	Poor	Excellent	Poor	Poor	Fair
Core expansion at roof	Average	Poor	Excellent	Good	Fair	Average
Core expansion at ground floor	Average	Fair	Good	Poor	Poor	Average

Excellent — Good — Average — Fair — Poor

dimensions required. Considering the entrance lobby as its extension, a building's service core can expand to nearly fill the ground floor. This core expansion typically happens on the roof and in the basement as well.

A service core can be related to the remaining service floor area in any of several ways (Fig. 12.12). A single central core is probably the most familiar arrangement (Occidental Chemical Offices, Fig. 6.11; Fox Plaza Los Angeles, Fig. 8.10; International Building, San Francisco, Fig. 8.11; CBS Tower, Fig. 8.66). Frequent variations include a core at the edge (Comstock Center, Fig. 8.72; Seeley Mudd Library, Fig. 8.74) or one that is detached, two cores symmetrically placed, corner cores (SC Johnson, Fig. 8.49), or core services dispersed somewhat randomly (Iowa Public Service Building, Fig. 8.40).

Each core location has advantages and drawbacks. From the viewpoint of rentable space, important factors are the flexibility of the served (rental) floor area; the exposure of rental area to the perimeter, and thus to light, air, and view; and the extent to which ground floor high-rent space

gets commercial exposure. For fire safety and occupant convenience, the distance of travel from the furthest rental floor area to the service core and the clarity of circulation within the rental area, are important. For the environment within the core—toilet rooms, stairs, elevator waiting areas—access to light, air, and view are desirable. For the mechanical services, ease of core area expansion on the roof, at the loading dock, and in the basement are important, as is the length of travel for ducts and other such elements within each floor. These factors are compared for various types of core designs in Table 12.4.

REFERENCES

Audubon/Croxton (1994). *Audubon House: Building the Environmentally Responsible, Energy Efficient Office.* © National Audubon Society, with Croxton Collaborative, Architects. John Wiley & Sons, New York.

AIA, Ramsey, and Sleeper (1994). *Architectural Graphic Standards,* 9th ed., © by John Wiley and Sons.

PART IV
FIRE PROTECTION

One of the most challenging aspects of building design is that of resistance to fire. Designers may enjoy considering how their buildings can be made bearably cool on a stiflingly hot day, or lighted and warm on a bitterly cold night. It is less pleasant to imagine a building burning and consider how first its occupants, then its contents—including the building itself—can be saved.

To complicate this process, there are inherent conflicts between some of the optimum features for fire resistance and some of those optimum features for passive strategies such as daylighting or passive cooling, or for forced-air systems. There are even potential design conflicts between systems for the safe evacuation of people and systems for the suppression of fire because both people and fire thrive on oxygen. Yet there are also characteristics in common between daylight, passive solar heating and cooling, acoustics, and fire protection. The challenge is to find the optimum balance.

13

FIRE PROTECTION

Just as buildings can be designed to minimize heating, cooling, and lighting equipment and thus to reduce energy consumption, they can be designed to reduce the size of fire-fighting systems and retard the spread of smoke and fire. As the average size of fires within buildings continues to decline, the emphasis is shifting to that of minimizing water and smoke damage. Whole-building conflagrations, while spectacular, are relatively rare in large buildings in North America, now that both automatic fire detection and fire-extinguishing systems are widespread.

This discussion of these topics in fire protection begins with basic design considerations for fire resistance. Smoke management (for safe evacuation and for limited smoke damage) is next, followed by fire suppression systems such as sprinklers and non–water-based approaches. Lightning protection is then discussed, along with the many fire detection and alarm systems that are keyed to the four stages of the "typical" building fire.

Throughout Part IV, the influence of the National Fire Protection Association (NFPA) is apparent. The periodically updated *Fire Protection Handbook* is an especially useful source of information for diagnosis, from which several illustrations have been reprinted in our text. Publications can be ordered from the NFPA, Batterymarch Park, Quincy, MA 02269-9109.

13.1 Design for Fire Resistance

Fire is a special kind of oxidation known as *combustion.* Oxidation, which has been discussed in the previous chapters in terms of rust within water supply equipment and aerobic digestion in waste disposal systems, is a process in which molecules of a fuel are combined with molecules of oxygen, producing a mixture of gases and energy. When this occurs rapidly, as in a fire, energy is released

as heat and light, and some gases become visible as smoke.

Fire has a triangle of needs: fuel, high temperature, and oxygen. If deprived of any of these needs, building fires will be extinguished. In general, this triangle's influence on building design is as follows. The *fuel* is the building's structure and contents; the designer controls the choice of structural and finish materials, but rarely the final contents. The *temperatures* achieved in fires are well beyond the ability of building cooling systems to control, so special water systems (in the form of sprinklers) are often installed to deprive fire of the high temperatures it needs. *Oxygen* may be denied to a fire partly by limitations on ventilation, but these can have serious safety consequences. Another design response is to install fire suppression systems that either cover the fuel (foam, dry chemicals) or displace oxygen with another gas—for example, carbon dioxide or "clean agents" that inhibit the chemical action of the flame itself.

(a) Sources of Ignition. Buildings commonly contain three basic sources of ignition: chemical, electrical, and mechanical. (Much more rarely, nuclear sources are present). In *chemical* combustion, most commonly known as "spontaneous combustion," some chemicals reach ignition at ordinary temperatures within buildings. Chemical combustion depends on the rate of heat generation (related to the degree of saturation of combustible products by the chemicals involved), the air supply (enough to supply oxygen but not enough to lower the temperature), and the insulation provided by the immediate surroundings (again—the more insulation, the easier the attainment of combustion temperatures).

Electrical heat energy is most commonly supplied by resistance heating, a familiar process in many appliances and in space-heating equipment. Less common are electric ignition by induction, dialectic process, arcing, and static electricity. Lightning is an infrequent but enormously destructive electrical source.

Mechanical heat energy is produced by friction (including sparks), by overheating of machinery, and occasionally by the heat of compression.

(b) Products of Combustion. (Fig. 13.1).

Everyone has experienced to some degree the dangers of the *thermal* products of combustion—flame and heat. These visible and tactile elements of fire can cause burns, shock, dehydration, heat exhaustion, and fluid blockage of the respiratory tract, but they are responsible for only about one-quarter of the deaths resulting from building fires. The bulk of such deaths are caused by the *nonthermal* products—smoke (including gases). Smoke can usually be seen and smelled. Made up of droplets of flammable tars and small particles of carbon suspended in gases, it irritates the eyes and nasal passages, sometimes blinding and/or choking a person. Gases are especially dangerous because, without visible smoke, they are so often difficult to detect. Some gases are directly toxic, but all are dangerous because they displace oxygen. Common gases released in building fires include carbon monoxide and carbon dioxide.

Carbon monoxide is a deadly product of combustion, and is often the most abundant. It is produced when insufficient oxygen is available to completely oxidize the burning material. It is more readily attached to hemoglobin molecules in red blood cells than is oxygen, thus depriving the brain and muscles of needed oxygen. This leads to irrational behavior and loss of consciousness, then to death.

Carbon dioxide is produced in large quantities, and rapidly overstimulates breathing, thus causing the lungs to swell. Other dangerous and common building-fire gases include hydrogen sulfide, sulfur dioxide, ammonia, oxides of nitrogen, cyanide, phosgene, and hydrogen chloride (from burning polyvinyl chloride).

In indoor fires oxygen commonly becomes insufficient because the fire consumes it so rapidly. The normal concentration of oxygen in air is about 21%. At less than 17%, muscular coordination and judgment are diminished; at 14% down to 10%, people remain conscious but become irrational and fatigue is rapid; at 10% down to 6%, collapse occurs, but revival is possible when increased oxygen is supplied. The technique of starving the fire of oxygen can therefore pose a threat to humans, both by increasing the chances of carbon monoxide production and by depriving people of oxygen.

(c) Objectives in Fire Safety. Many older

buildings were designed at a time when building fires usually resulted in the loss of several adjacent structures and, hence, the primary objective of fire fighting was to limit the conflagration to as few city blocks as possible. With the increased use of fire-resistant construction and code control of site and building planning, fires typically were confined to but one building at a time. As fire-suppression systems came into common usage within buildings, it came to be expected that fires could be confined to one or two floors within a building. Now that automatic detection/suppression systems are technically advanced, fires can usually be confined to only one room, or to even smaller areas. In the United States, most fires are now extinguished with between 1 and 5 sprinklers operating.

Three common objectives of building fire safety, in order of usual importance, are

· Protection of life

· Protection of property

· Continuity of operation

Fig. 13.1 *Although fire's thermal products—flame and heat—make strong visible and tactile impressions, fire's nonthermal products—smoke and gases—pose the greater threat to life.*

TABLE 13.1 **Major Factors Influencing Fire Growth**

Realm	Approximate Ranges of Fire Sizes	Major Factors that Influence Growth
1 (Preburning)	Overheat to ignition	Amount and duration of heat flux Surface area receiving heat from material ignitability
2 (Initial burning)	Ignition to radiation point [10-in. (254-mm) high flame]	Fuel continuity Material ignitability Thickness Surface roughness Thermal inertia of the fuel
3 (Vigorous burning)	Radiation point to enclosure point [10-in. to 5-ft high flame (254-mm to 1.5-m)]	Interior finish Fuel continuity Feedback Material ignitability Thermal inertia of the fuel Proximity of flames to walls
4 (Interactive burning)	Enclosure point to ceiling point [5-ft (1.5-m) high flame to flame touching ceiling]	Interior finish Fuel arrangement Feedback Height of fuels Proximity of flames to walls Ceiling height Room insulation Size and location of openings HVAC operation
5 (Remote burning)	Ceiling point to full room involvement	Fuel arrangement Ceiling height Length/width ratio Room insulation Size and location of openings HVAC operations

Source: Reprinted with permission from the *Fire Protection Handbook*, 18th ed., copyright © 1997. National Fire Protection Association, Quincy, Massachusetts, 02269.

Many of the elements of building fire safety are covered by building codes, but it is important to remember that codes specify the *minimum* acceptable protection. Designers can go much further than the codes require to enhance fire safety. It is also important to realize that codes *cannot* cover all aspects of fire safety, as there are too many variables to building design. Table 13.1 lists the stages of a fire and the factors that influence its growth. Many of these factors involve decisions made by building designers.

Codes typically *prescribe* design strategies that are *passive* means of limiting the spread of fire and protecting life: wall, floor, and ceiling constructions; maximum open floor areas; maximum distances to exits; and so on. Codes commonly allow some relaxing of such prescriptions when *active* fire-suppression systems (such as sprinklers) are designed into a building. Alternatively, a detailed *computer analysis* of fire spread and occupant evacuation within a given design may allow for even greater distances to exits, larger open floor areas, and alternate construction. This represents a *performance-based,* rather than a prescriptive, approach to design. It requires close cooperation between designers and fire code enforcement agents.

(d) Fire Safety and Other Environmental Control Systems.

In some instances, the optimum design for fire safety will resemble the optimum design for lighting, thermal, acoustic, and water systems. Such matching characteristics include the following.

Thermal mass, which is useful for passive heating and cooling systems, for acoustic isolation of airborne sound, and for fire barriers (most thermally massive materials will not burn easily).

High ceilings, useful for daylight distribution and displacement ventilation, but also for collecting a large quantity of smoke before it reaches the occupant level, and for allowing smoke and/or flames from a fire to be seen from a greater indoor distance.

Windows, for daylight, ventilation, and view, also allow access for firefighting and rescue, provide escape routes, relieve smoke accumulation with fresh air, and thus relieve some of the stress of trapped occupants.

Solid (noncombustible) overhangs over windows not only provide sunshading, but also discourage the vertical spread of fire over the building face and can serve as emergency exterior places of refuge.

Elevated water storage tanks provide both adequate water pressure for plumbing fixtures and water for firefighting in the first few minutes of a fire, before firefighters arrive.

In many other instances, there are potential design conflicts between building performance under ordinary conditions and building performance in the extraordinary event of a fire. Escalators invite shoppers to explore the upper levels of a department store, or connect hotel lobbies to ballrooms; they are also efficient apertures for the vertical spread of fire and smoke. (Special fire protection strategies at escalators are found in Figs. 25.9 to 25.13.) Daylighting and natural ventilation are best served by high ceilings and low partitions, which encourage light and air from the perimeter to pervade building interiors. However, if unchecked by sprinklers, fire and smoke can easily spread through such "open plans." (However, smoke and fire can build up much more rapidly in small, enclosed rooms that retain heat.) Forced-air systems that heat, ventilate, and cool are also potential pathways for smoke and fire; this hazard is especially serious when the systems penetrate floors because vertically spreading fires are harder to fight than are horizontally spreading ones. (However, carefully designed forced-air systems can aid in smoke management.) "Windowless" buildings that rely on electric light in place of daylight are especially dangerous in fires because firefighters cannot easily evacuate occupants or gain access to the building. Sunscreens that completely cover windows are disadvantageous for the same reason; nonoperable windows, although considered advantageous for tightly controlled air conditioning, must be broken for fire evacuation/access. The higher the window, the greater the danger from falling shards of glass. Many excellent insulating materials will burn readily, and some will give off toxic gases. Some of the loveliest interior finishes are both flammable and deadly sources of gases in a fire. Table 13.2 summarizes the requirements for interior finishes.

This conflict, between interior comfort throughout a building's life and safety at the moment of its impending death by fire, is the principal reason why codes cannot alone assure fire safety. The dilemma of ordinary versus extraordinary performance is one that must be faced by the designer, owner, and occupants of buildings.

(e) Protection of Life.

Section 8 of Chapter 1, *NFPA Handbook of Fire Protection,* discusses human behavior in fires in great detail. Although panic behavior drives many code requirements, such behavior has been found to be rare. Designers should consider how building occupants make decisions in a fire. In the first phase, cues are detected; the smell of smoke, sounds associated with a fire (breaking glass, sirens, alarm bells), and, more rarely, seeing flames. Open plans (with longer visible indoor distances) are more amenable to exposing such clues to a wider population. In the second phase, the occupants define the situation: Just how serious is this fire? The more numerous the cues, the more rapid the definition. How other people are reacting is influential, and in the absence of strong cues can lead to a group

TABLE 13.2 **Summary of Life Safety Code Requirements for Interior Finish**

NOTES: Class A Interior Wall and Ceiling Finish—flame spread 0–25, (new) smoke developed 0–450.
Class B Interior Wall and Ceiling Finish—flame spread 26–75, (new) smoke developed 0–450.
Class C Interior Wall and Ceiling Finish—flame spread 76–200, (new) smoke developed 0–450.
Class I Interior Floor Finish—critical radiant flux, minimum 0.45 Watts/cm^2.
Class II Interior Floor Finish—critical radiant flux, minimum 0.22 Watts/cm^2.
Automatic sprinklers—where a complete standard system of automatic sprinklers is installed, interior finish with flame spread rating not over Class C may be used in any location where Class B is normally specified, and with rating of Class B in any location where Class A is normally specified; similarly, Class II interior finish may be used in any location where Class I is normally specified, and no critical flux rating is required where Class II is normally specified. (This does not apply to new health care facilities.)
Exposed portions of structural members complying with the requirements of heavy timber construction may be permitted.

Occupancy	Exits	Access to Exits	Other Spaces
Assembly—New: >300 people	A	A or B	A or B
≤300 people	A	A or B	A, B, or C
Assembly—Existing: >300 people	A	A or B	A or B
≤300 people	A	A or B	A, B, or C
Educational—New	A	A or B	A or B; C on low partitions
Educational—Existing	A	A or B	A, B, or C
Day Care Centers—New	A I or II	A I or II	A or B NR
Day Care Centers—Existing	A or B	A or B	A or B
Group Day-Care Homes—New	A or B	A or B	A, B, or C
Group Day-Care Homes—Existing	A or B	A, B, or C	A, B, or C
Family Day-Care Homes	A or B	A, B, or C	A, B, or C
Health Care—New A. S. Mandatory	A or B	A or B C on lower portion of corridor wall	A or B C in small individal rooms
Health Care—Existing	A or B	A or B[a]	A or B[a]
Detention and Correctional—New	A I	A I	A, B, or C
Detention and Correctional—Existing	A or B I or II	A or B I or II	A, B, or C
Residential, Board and Care[b]			
Residential, Hotels and Dormitories— New	A I or II	A or B I or II	A, B, or C
Residential, Hotels and Dormitories— Existing	A or B I or II	A or B I or II	A, B, or C
Residential, Apartment Buildings—New	A I or II	A or B I or II	A, B, or C
Residential, Apartment Buildings— Existing	A or B I or II	A or B I or II	A, B, or C
Residential, 1- and 2-Family, Lodging or Rooming Houses	A, B, or C	A, B, or C	A, B, or C
Mercantile—New	A or B	A or B	A or B
Mercantile—Existing Class A or B	A or B	A or B	Ceiling A or B existing on walls A, B, or C
Mercantile—Existing Class C	A, B, or C	A, B, or C	A, B, or C

(continued)

TABLE 13.2 **Summary of Life Safety Code Requirements for Interior Finish** (*continued*)

Occupancy	Exits	Access to Exits	Other Spaces
Office—New	A or B I or II	A or B I or II	A, B, or C
Office—Existing	A or B	A or B	A, B, or C
Industrial	A or B	A, B, or C	A, B, or C
Storage	A or B	A, B, or C	A, B, or C

Source: Reprinted with permission from the *Fire Protection Handbook*, 18th edition, © 1997, National Fire Protection Association, Quincy, Massachusetts, 02269.
ªSee occupancy chapters of NFPA *101* for details.
ᵇSee Source, Chapters 22 and 23.

refusal to evacuate in the early stages of a fire. In the third phase, coping behavior begins: fight or flight. Clear exit pathways and access to fire-fighting equipment are critical in this decision.

For most low-rise buildings, a reasonable goal is the evacuation of all occupants in the time interval between the detection of a fire and the arrival of the firefighters. Designers can provide clearly defined pathways to exits ("exit access") that can be kept relatively clear of smoke (Fig. 13.2). To accommodate a wheelchair, a minimum clear width of 32 in. (813 mm) is required. "Exits" can take a variety of forms. Vertical exits (Figs. 13.3 and 13.4) include smokeproof towers, exterior and interior stairs and ramps, and escalators that meet specific requirements. *Vertical exits do not include elevators;* they are too easily stalled or, worse, opened at the floor of a fire by malfunctioning signal equipment. Exits in the horizontal plane include doors leading directly to the outside, 2-h fire-enclosed hallways, and moving walls. Special "horizontal exits" are provided by internal fire-walls penetrated by two fire doors—one swinging open in each direction. "Exit discharge" is the area outside the exits that leads to a public way, and may still need protection in a fire.

Maximum allowable distances to exits are specified in Table 13.3. These are influenced by

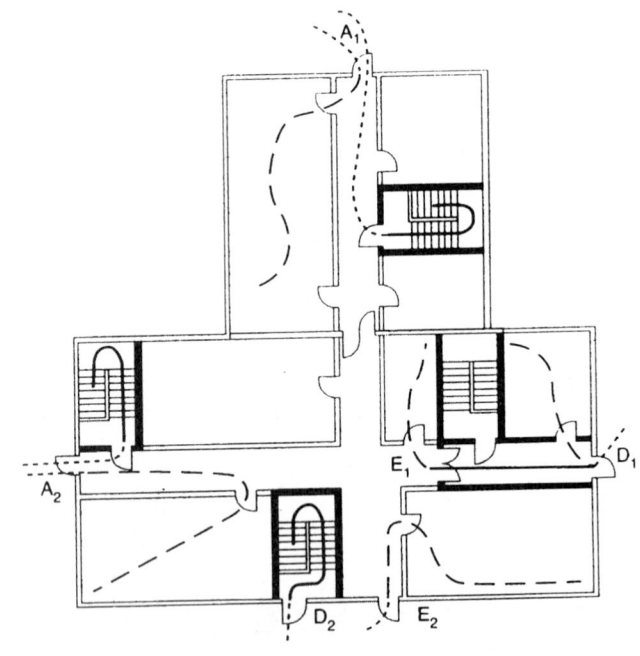

Fig. 13.2 *Exit access, exit, and exit discharge on the first floor of a multistory building. The doors A_1, A_2, E_1, and E_2 are exits and the path (dashed line) is the exit access. To the person emerging from the exit enclosure, doors A_1 and A_2 and the paths (dotted lines) are the exit discharge. Doors D_1 and D_2 are exit discharge doors. Solid line paths are within the exit. (Reprinted with permission from the* Fire Protection Handbook, *18th ed., copyright © 1997, National Fire Protection Association, Quincy, Massachusetts, 02269.)*

FIRE PROTECTION

Fig. 13.3 *Plan views of exits.* (a) *Enclosed stairway allows occupants on any floor above a fire to escape. A smokeproof tower is better, as it opens to the air at each floor, largely preventing the chance of smoke in the stairway.* (b) *Horizontal exit through an interior firewall provides a quick refuge and lessens the need of a hasty flight down the stairs. Two wall openings are needed to facilitate exit in either direction. Fire-rated doors must be arranged to be self-closing or automatic-closing by smoke detection. (Reprinted with permission from the* Fire Protection Handbook, *18th ed., copyright © 1997, National Fire Protection Association, Quincy, Massachusetts, 02269.)*

past experience with people exiting through smoke, especially when it may involve briefly traveling toward the perceived fire itself (as in the case of a dead-end corridor). When automatic fire suppression systems such as sprinklers are used, the allowable distances to exits are increased. Designers should remember, however, that at least 30% of building fire deaths result from fire cutting off the paths to exits.

Minimum egress widths per floor are found with the help of Table 13.4. First calculate the floor area (net or gross, as specified in Table 13.4). Divide the floor area by the occupant load to determine the number of occupants for whom exits must be provided for that floor. Then the exit capacity is calculated based on its clear width. Stairs are sized to meet the requirements of Table 13.5.

EXAMPLE 13.1. A multistory office building is 80 ft (24 m) wide by 300 ft (80 m) long. What exit capacity is required per floor?

SOLUTION. The gross floor area = 80 × 300 = 2400 ft². From Table 13.4, "business" categories are based on one person per gross 100 ft².

Fig. 13.4 *Four variations of smokeproof towers. Plan A has a vestibule opening from a corridor. Plan B shows an entrance by way of an outside balcony. Plan C could provide a stair tower entrance common to two areas. In plan D, smoke and gases entering the vestibule would be exhausted by natural or induced draft in the open air shaft. In each case a double entrance to the stair tower with at least one side open or vented is characteristic of this type of construction. Pressurization of the stair tower in the event of fire provides an attractive alternative for tall buildings and is a means of eliminating the entrance vestibule. (Reprinted with permission from the* Fire Protection Handbook, *16th ed., copyright © 1986. National Fire Protection Association, Quincy, Massachusetts, 02269.)*

TABLE 13.3 **Common Path, Dead End, and Travel Distance Limits**

Type of Occupancy	Common Path Limit		Dead-End Limit		Travel Distance Limit	
	Unsprinklered ft (m)	Sprinklered ft (m)	Unsprinklered ft (m)	Sprinklered ft (m)	Unsprinklered ft (m)	Sprinklered ft (m)
Assembly						
New	20/75[a,b] (6.1/23)	20/75[a,b] (6.1/23)	20[b] (6.1)	20[b] (6.1)	150[c] (45)	200[c] (60)
Existing	20/75[a,b] (6.1/23)	20/75[a,b] (6.1/23)	20[b] (6.1)	20[b] (6.1)	150[c] (45)	200[c] (60)
Educational						
New	75 (23)	75 (23)	20 (6.1)	20 (6.1)	150 (45)	200 (60)
Existing	75 (23)	75 (23)	20 (6.1)	20 (6.1)	150 (45)	200 (60)
New Day-Care Center	NR[d,e]	NR[d,e]	20 (6.1)	20 (6.1)	150[f] (45)	200[f] (60)
Existing Day-Care Center	NR[d,e]	NR[d,e]	20 (6.1)	20 (6.1)	150[f] (45)	200[f] (60)
Health Care						
New	NR[d]	NR[d]	30 (9.1)	30 (9.1)	NA[g]	200[f] (60)
Existing	NR[d]	NR[d]	NR[d]	NR[d]	150[f] (45)	200[f] (60)
New Ambulatory Care	NR[d]	NR[d]	30 (9.1)	30 (9.1)	150[f] (45)	200[f] (60)
Existing Ambulatory Care	NR[d]	NR[d]	50 (15)	50 (15)	150[f] (45)	200[f] (60)
Detention and Correctional						
New–Use Conditions						
II, III, IV	50 (15)	100 (30)	50 (15)	50 (15)	150[f] (45)	200[f] (60)
V	50 (15)	100 (30)	20 (6.1)	20 (6.1)	150[f] (45)	200[f] (60)
Existing–Use Conditions						
II, III, IV, V	50[h] (15)	100[h] (30)	NR[d]	NR[d]	150[f] (45)	200[f] (60)
Residential						
Hotels and Dormitories						
New	35[i] (10.7)	50[i] (15)	35 (10.7)	50 (15)	175[i,j] (53)	325[i,j] (99)
Existing	35[i] (10.7)	50[i] (15)	50 (15)	50 (15)	175[i,j] (53)	325[i,j] (99)
Apartments						
New	35[i] (10.7)	50[i] (15)	35 (10.7)	50 (15)	175[i,j] (53)	325[i,j] (99)
Existing	35[i] (10.7)	50[i] (15)	50 (15)	50 (15)	175[i,j] (53)	325[i,j] (99)
Board and Care						
Small, New and Existing	NR[d]	NR[d]	NR[d]	NR[d]	NR[d]	NR[d]
Large, New	NA	125k (38.1)	NA	15 (15)	NA	325[i,j] (99)
Large, Existing	110 (33.5)	160 (48.8)	50 (15)	50 (15)	175[i,j] (53)	325[i,j] (99)
Lodging and Rooming Houses	NR[d]	NR[d]	NR[d]	NR[d]	NR[d]	NR[d]
One- and Two-Family Dwellings	NR[d]	NR[d]	NR[d]	NR[d]	NR[d]	NR[d]
Mercantile						
Class A, B, C						
New	75 (23)	100 (30)	20 (6.1)	50 (15)	100 (30)	200 (60)
Existing	75 (23)	100 (30)	50 (15)	50 (15)	150 (45)	200 (60)
Open Air	NR[d]	NR[d]	0 (0)	0 (0)	NR[d]	NR[d]
Covered Mall						
New	75 (23)	100 (30)	20 (6.1)	50 (15)	100 (30)	400[i] (120)
Existing	75 (23)	100 (30)	50 (15)	50 (15)	150 (45)	400[i] (120)
Business						
New	75[m] (23)	100[m] (30)	20 (6.1)	50 (15)	200 (60)	300 (91)
Existing	75[m] (23)	100[m] (30)	50 (15)	50 (15)	200 (60)	300 (91)
Industrial						
General	50 (15)	100 (30)	50 (15)	50 (15)	200[n] (60)	250[n] (75)
Special Purpose	50 (15)	100 (30)	50 (15)	50 (15)	300[p] (91)	400[p] (122)
High Hazard	0 (0)	0 (0)	0 (0)	0 (0)	75 (23)	75 (23)
Aircraft Servicing Hangars, Ground Floor	50[p] (15)	50[p] (15)	50[p] (15)	50[p] (15)	o	o
Aircraft Servicing Hangars, Mezzanine Floor	50[p] (15)	50[p] (15)	50[p] (15)	50[p] (15)	75 (23)	75 (23)

(continued)

TABLE 13.3 **Common Path, Dead End, and Travel Distance Limits (*continued*)**

	Common Path Limit		Dead-End Limit		Travel Distance Limit	
Type of Occupancy	Unsprinklered ft (m)	Sprinklered ft (m)	Unsprinklered ft (m)	Sprinklered ft (m)	Unsprinklered ft (m)	Sprinklered ft (m)
Storage						
Low Hazard	NR[d]	NR[d]	NR[d]	NR[d]	NR[d]	NR[d]
Ordinary Hazard	50 (15)	100 (30)	50 (15)	100 (30)	200 (60)	400 (122)
High Hazard	0 (0)	0 (0)	0 (0)	0 (0)	75 (23)	75 (23)
Parking Garages, Open	50 (15)	50 (15)	50 (15)	50 (15)	200 (60)	300 (91)
Parking Garages, Enclosed	50 (15)	50 (15)	50 (15)	50 (15)	150 (45)	200 (60)
Aircraft Storage Hangars, Ground Floor	50[p] (15)	100[p] (30)	50[p] (15)	50[p] (15)	o	o
Aircraft Servicing Hangars, Mezzanine Floor	50[p] (15)	75[p] (23)	50[p] (15)	50[p] (15)	75 (23)	75 (23)
Underground Spaces in Grain Elevators	50[p] (15)	50[p] (15)	NR[d,p]	NR[d,p]	200 (60)	400 (122)

Source: Adapted with permission from the *Fire Protection Handbook,* 18th Edition, © 1997, National Fire Protection Association, Quincy, Massachusetts, 02269.

[a]20 ft (6.1 m) for common path serving >50 persons; 75 ft (23 m) for common path serving <50 persons.

[b]See Source, Chapters 8 and 9, for special considerations for aisle accessways, aisles, and mezzanines.

[c]See Source, Chapters 8 and 9, for special considerations for smoke-protected assembly seating in arenas and stadia.

[d]No requirement.

[e]See Source, Sections 10–7 and 11–7 for requirement for second exit access based on room capacity or area.

[f]This dimension is for the total travel distance, assuming incremental portions have fully utilized their allowable maximums. For travel distance within the room, and from the room exit access door to the exit, see Source for the appropriate occupancy chapter.

[g]Not applicable.

[h]See Source, Chapter 15, for special considerations for existing common paths.

[i]This dimension is from the room/corridor or suite/corridor exit access door to the exit; thus it applies to corridor common path.

[j]See Source, appropriate occupancy chapter for special travel distance considerations for exterior ways of exit access.

[k]See Source, Section 22–3, for requirement for second exit access based on room area.

[l]See Source, Sections 24–4 and 25–4, for special travel distance considerations in covered malls considered pedestrian ways.

[m]See Source, Chapters 26 and 27, for special common path considerations for single tenant spaces.

[n]See Source, Chapter 28, for industrial occupancy special travel distance considerations.

[o]See Source, Chapters 28 and 29, for special requirements on spacing of doors in aircraft hangars.

[p]See Source, Chapters 28 and 29, for special requirements if high hazard.

The population per floor is therefore:

$$2400\ ft^2/100\ ft^2/person = 240\ people.$$

Exit doors (to stairs): 240 people × 0.2 in./person = 48 in. total.

(One 34-in. clear door into each of two stairs = 2 doors × 34 = 68 in., more than the minimum.)

Stairs: 240 people × 0.3 in./person = 72 in. total

(Two stairs at 44 in. each = 88 in., more than the minimum.) ∎

The building population as estimated for fire safety is usually much greater than the population for which HVAC, water, or elevator service is designed. The stairs must be designed so as to allow those already within the stairwell to continue down without interference from access doors on any floor. Stairs with direct access to outdoor air at each floor—so-called smokeproof towers—are the safest kind. A fire stair must allow firefighters to move up while occupants are moving down. Another phenomenon is "re-entry," in which occupants who have exited decide to re-enter despite the danger. (Rescue of family members, pets, and valuables are likely reasons). This greatly complicates downward flow.

High-rise buildings present much more difficult problems because firefighting equipment can ordinarily reach no higher than seven floors

TABLE 13.4 **Summary of NFPA 101®, Life Safety Code®, Provisions for Occupant Load, and Exit Capacity**

| | Occupant Load | Level Components | |
Occupancy	ft^2 (m²) per person	Doors, Corridors, Horizontal Exits, Ramps	Stairs
Assembly			
Less concentrated use without fixed seating	15 Net (1.4)	0.2	0.3
Concentrated use without fixed seating	7 Net (0.65)	0.2	0.3
Fixed seating	Actual number of seats	0.2	0.3
Educational			
Classrooms	20 Net (1.9)	0.2	0.3
Shops and vocational	50 Net (4.6)	0.2	0.3
Care centers	35 Net (3.3)	0.2	0.3
Health care		NAS AS	NAS AS
Sleeping departments	120 Gross (11.1)	0.5 0.2	0.6 0.3
Treatment departments	240 Gross (22.3)	0.5 0.2	0.6 0.3
Residential	200 Gross (18.6)	0.2	
Board and care	200 Gross (18.6)	0.2	0.3
Mercantile			
Street floor and sales basement	30 Gross (3.7)	0.2	0.3
Multiple street floors (each)	40 Gross (3.7)	0.2	0.3
Other floors	60 Gross (5.6)	0.2	0.3
Storage—shipping	300 Gross (27.9)	0.2	0.3
Malls	See Code	0.2	0.3
Business	100 Gross (9.3)	0.2	0.3
Industrial	100 Gross (9.3)	0.2	0.3
Detention and correctional	120 Gross (11.1)	0.2	0.3

Source: Adapted with permission from the *Fire Protection Handbook,* 18th Edition, © 1997, National Fire Protection Association, Quincy, Massachusetts, 02269.

NOTES: NAS = nonsprinklered; AS = sprinklered.

[about 90 ft (27 m)] and because, typically, only two exit stairways are provided. Downward flow rates in stairs had been assumed at about 45 persons/minute/22 in. (559 mm of width), but more recently peak flows of only 24 persons/minute/22 in. (559 mm) have been observed. With a typical exit stair, a 15-story building housing 60 persons per floor per stair can be evacuated in about 9 minutes. However, with the same stair in a 50-story building housing 240 persons per floor per stair, evacuation will take at least 2 hours, 11

minutes! When doors are held open at each floor to admit occupants fleeing the fire, smoke can readily enter the stairwell. Moreover, it is increasingly common for people to refuse to evacuate a high-rise building, placing their faith in fire-extinguishing systems instead of facing the daunting descent through many floors of stairs.

Because of the impracticality of rapid evacuation, larger buildings are required to provide "refuge areas" where smoke penetration is less likely. For example, stairs can be designed to hold *all* the

TABLE 13.5 **Requirements for Exit Stairs**

	New Stairs	Existing Stairs	
		Class A[a]	Class B[a]
Minimum width clear of all obstructions[b]:			
Total occupant load[c] = 50 or more	44 in. (1.12 m)	44 in. (1.12 m)	44 in. (1.12 m)
= less than 50	36 in. (0.91 m)	36 in. (0.91 m)	36 in. (0.91 m)
Maximum height of risers	7 in. (178 mm)	7½ in. (191 mm)	8 in. (203 mm)
Minimum height of risers	4 in. (102 mm)		
Minimum tread depth	11 in. (279 mm)	10 in. (244 mm)	9 in. (229 mm)
Minimum headroom	6 ft 8 in. (2.03 m)	6 ft 8 in. (2.03 m)	6 ft 8 in. (2.03 m)
Maximum height between landings	12 ft (3.7 m)	12 ft (3.7 m)	12 ft (3.7 m)
Minimum dimension of landings in direction of travel	[d]	[e]	[e]
Doors opening immediately on stairs, without landing at least the width of door	No	No	No

Source: Adapted with permission from the *Fire Protection Handbook,* 18th Edition, © 1997, National Fire Protection Association, Quincy, Massachusetts, 02269.

[a]Class A and B stairs are defined as those having the tread and riser dimensions as listed in this table.

[b]Except projections not exceeding 3½ in. (89 mm) at and below handrail height on each side.

[c]Total occupant load includes all floors served by stairway.

[d]Every landing shall have a dimension, measured in the direction of travel, equal to the width of the stair. Such dimension need not exceed 4 ft (1.22 m) when the stair has a straight run.

[e]Stairways and intermediate landings shall continue with no decrease in width along the direction of exit travel.

occupants, in which case the design of discharge is based more on "capacity" than on "flow." At maximum crowding, about 3 ft^2 (0.28 m^2) per person is recommended. Again, smokeproof towers will be safer in such a design, although stairs can be pressurized with outdoor air. These details on smoke management are presented in the next section.

(f) Property Protection. One of the earliest design concerns in this category is that the site should permit access for firefighting equipment. Ideally, fire trucks should be able to pull alongside each exterior wall. If accessibility is limited by adjacent buildings, and roadways alongside buildings and other measures are impractical, more reliance must be placed on internal fire suppression systems. Another factor is the amount of time it will ordinarily take for firefighters to reach a site. In congested urban areas, or remote rural ones, time that elapses between the alarm and the arrival of firefighters can permit a fire to grow to unstoppable proportions. In these cases, emphasis again shifts to internal fire-suppression systems.

Another design concern is for adequate water to fight the fire. Reliance on city water mains is not always a good solution. Elevated tanks on buildings can help in the early moments of firefighting, but their capacity is soon exhausted. Some buildings therefore rely on lakes or enclosed reservoirs for firefighting supply (see the discussion in Section 13.3).

Exposure protection is becoming common in areas where highly flammable surroundings pose a serious threat of fires originating *outside* a building. Candidates for exposure protection include buildings surrounded by older wooden buildings, bordered by lumberyards or other commercial activities that utilize highly flammable materials,

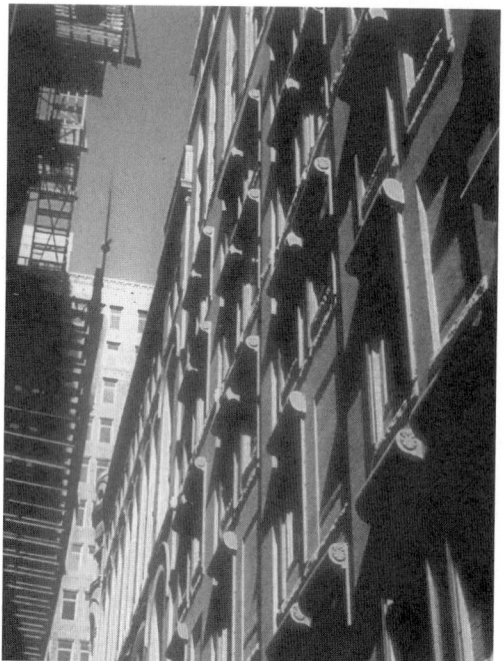

Fig. 13.5 *Rolling metal shutters, housed in valences above each window, punctuate the exterior of this downtown Pittsburgh building opening onto a narrow street. The shutters can protect the building contents from an exterior fire, but can also protect the narrow street from a fire inside the building. Note the exterior fire escapes across this narrow street. These are no longer permitted, partly because smoke plumes rising from windows can make such stairs unusable long before heat from the fire also becomes a threat.*

or even bordered by open fields of dry grass or brush. Exposure protection guards against heat transfer by radiation and convective currents and against direct fire transfer via flying embers. Exposure protection begins with the use of nonflammable materials for the building's exterior. Erecting firewalls between the building and a fire-threatening neighbor is a more drastic, but sometimes necessary, step. Exposure protection sometimes includes external water sprinkler systems, in which a sprayhead is placed over (or under) each opening such as windows or doors. Sprinkler systems that soak the roof can play a cooling role on summer days, as well as an exposure protection role. Exterior doors can be chosen for their fire-delaying characteristics. Windows can also be protected by fire-rated shutters that are designed to close automatically at high temperatures (Fig. 13.5). Sometimes exterior protec-

tion is necessary to protect the "exit discharge" from flames originating *inside* the building, as with narrow alleyways like that in Fig. 13.5 in downtown Pittsburgh.

Compartmentation has become increasingly important as buildings have become lightweight structures incorporating decreased fire resistance and open floor areas that encourage the spread of fires. Building codes establish the maximum floor areas permissible for various constructions and occupancies. If a building's floor area exceeds such limits, it must be subdivided by firewalls into areas that fall within the code limitations. Once again, sprinklers allow for increased floor area limits.

Openings in the firewalls must be protected by fire-rated doors and by fire dampers in forced-air systems (Fig. 13.6). As important as compartmentation is in buildings with large floor areas, however, the vertical spread of fire (its natural path) poses a more serious problem. For this reason, compartmentation requirements around vertical openings are often especially strict.

Concealed spaces are found in many contemporary buildings, especially over suspended ceilings but also behind walls, within pipe chases, in attics, under raised floors, and in other places. All such spaces can offer paths for the spread of fire. The designer can utilize noncombustible materials, insofar as possible, in such spaces. Another important step is to include automatic fire detection and suppression systems in these uninhabited spaces—and often the use of oxygen-deprivation approaches to fire suppression. Another design response is compartmentation; using firestops (or firewalls) to break up otherwise continuous concealed spaces. Many codes, for example, require firestopping around each 1000 ft^2 (93 m^2) of suspended ceiling area, and every 2000 ft^2 (186 m^2) of attic floor area.

Structural protection, another important requirement, allows a building to continue to stand during a fire and enables the building to be salvaged rather than demolished after the fire. Codes require various protective layers for structural materials. In order of importance, it is usually most critical to protect columns, girders, beams, and, finally, the floor slab.

(g) Continuity of Operations. For most building functions, it is desirable to minimize the

(a)

Fig. 13.6 *Openings in firewalls are protected by various devices.* (a) *Access doors to controls and other elements within firewalls can be of fire-rated construction.* (b) *Locations of fire dampers (F.D.) in a typical forced-air system—at firewalls, partitions, ceilings, floors, and shaft enclosures.* [*(Part* (b) *reprinted with permission from the* Fire Protection Handbook, *18th ed., copyright © 1997, National Fire Protection Association, Quincy, Massachusetts, 02269.)*]

(b)

disruption of operation that fire will cause. Design strategies to encourage continuity of operations include special fire alarm/suppression systems for especially critical operations areas (control rooms, for example), the design of HVAC systems to allow for 100% outside air (to aid in purging a building of smoke), and provision for the speedy removal of the water dumped on a fire from a sprinkler system. The floors in sprinkler-served buildings should be (yet rarely are) waterproof; waterproofing should also be carried up walls, columns, pipes, and other elements to a height of

4 to 6 in. (102 to 152 mm) above the floor. These provisions will help to minimize the water damage associated with fires.

13.2 Smoke Management

As it has become increasingly evident that smoke kills more people in building fires than either heat or structural collapse, it has also become more common to design for smoke management as well as for resistance to fire and fire suppression. The objectives in smoke management are the same as those for fire resistance: to reduce deaths and property damage due to smoke and provide for continuity of operations with minimal smoke interference. Smoke management systems have been used for several generations; scientific laboratory buildings equipped with fume hoods (such as the EPICenter at Montana State University, Fig. 6.25) offer a familiar example. Several options for smoke management are now available for ordinary building design.

(a) Factors in Smoke Management.

The heat of a fire produces air pressure and buoyancy that aid the spread of smoke well beyond the scene of the fire itself. In low buildings, the smoke is spread primarily by heat-induced convective air motion and by the differential air pressures caused by the expansion of gases as the temperature increases. In tall buildings, the stack effect complicates this pattern of smoke spread, encouraging the rapid vertical rise of heated air within vertical shafts. Wind forces from outside are also more likely to be a factor in tall buildings. (Wind velocities are usually higher with increasing distance from the earth's surface). Forced-air systems, so common in high-rise buildings, can also contribute to the spread of smoke. The interactions among fire, stack effect, wind, building geometry, and HVAC systems are complex. Several detailed calculation techniques are available, as explained in two publications, NFPA 92A and NFPA 92B. These methods include building and testing a scale model, or using a series of closed-form algebraic equations, or modeling compartment fire models with the help of a computer. In the last category, a more modest "zone" approach divides a fire into two zones (upper and lower), and can be done on a personal computer. The more ambitious "field" approach is to use computational fluid dynamics (CFD), requiring a large-capacity computer, but allowing the testing of wide variations in approaches to fire and smoke control; fire growth, smoke spread, structural behavior, and occupant evacuation can be examined together. Results can be surprising, leading at times to counterintuitive solutions.

(b) Confinement.

The most passive design response to smoke is to try to confine it to the

Fig. 13.7 *Smoke barriers, often called curtain boards, are useful in a fire's early stages. By confining the initial layer of heated air and smoke produced by the fire, they help to slow the spread of smoke while making more likely the early detection and suppression of the fire.*

fire area itself. Another confinement technique is to exclude smoke from specially protected areas, called *refuges*. Compartmentation is important in these approaches: where firewalls cannot be used, special smoke barriers (often called curtain boards) are suspended from the ceiling (Fig. 13.7) in an effort to trap the initial layer of hot air and smoke, and therefore to set off the fire detection/ suppression system more quickly. As useful as these partial barriers can be in the fire's early stages, they quickly lose effectiveness as the smoke layer thickens or as air pressures force the smoke below the barrier. Even in firewalls with fire door protection, small cracks around such openings provide smoke paths, aided by the air pressures and velocities that fires can quickly produce. Therefore, the typical "barrier" is not very resistant to spreading smoke, and smoke control based solely on confinement must be closely linked to an effective early detection/suppression system, such as a sprinkler system.

Compartmentation may not always be the best strategy. In the Potomac Mills Mall, Prince William County, Virginia, a 156,000 ft² (14,493 m²) addition to the shopping mall keeps the walls low and the ceiling high. The walls separating tenant spaces and public corridors are 12 ft (3.7 m) high, while the ceiling is at 17 ft (5.2 m), resulting in a large open volume of air. Smoke rising from a fire within a store can certainly spread well beyond that store, but will collect overhead in such a large space that evacuation of the mall can be completed well before the smoke reaches the occupied level. A computer model indicated that 20 minutes into a fire (and assuming *no* sprinkler operation), the smoke layer would have dropped to 12.3 ft (3.75 m) above the floor. If the walls were extended to the ceiling, this same building would experience, in the same 20 minutes, within the origin store of the fire, smoke all the way down to the floor. In such a scheme, CFD analysis may well be required before approval is granted.

(c) Dilution. For a limited time in a fire's early stages, the dilution of smoke with 100% outdoor air (provided by the HVAC system) may make conditions bearable during occupant evacuation. However, such large quantities of fresh air are needed in so short a time that dilution alone is rarely sufficient for smoke control, particularly when the smoke contains toxic fumes. When the dilution system is combined with both confinement and early detection/suppression systems, it becomes more attractive. Dilution systems that dump large quantities of outdoor air into refuge areas (including stairs) can create such high pressures within the refuge that smoke cannot enter through cracks around doors. However, when such doors are opened (as when more people enter the refuge), dilution systems that rely on the conventional HVAC system can rarely provide sufficient pressure to exclude smoke through the open doorway.

Fan-driven dilution systems raise several issues. Is the fan located where fire will not affect its performance, and where its air intake can be kept free of smoke? Given the multiple possibilities in building fires, this is a difficult promise to make. "Dedicated" systems, independent of the building's HVAC system, are often favored to enable such emergency equipment to be isolated, and because such huge quantities of air for relatively small spaces are well beyond the capability of the typical HVAC system. Where should the fresh air supply outlets be located? Most certainly, *below* the level of smoke accumulation, so as not to drive smoke downward. [This may suggest using a displacement ventilation (underfloor supply) system for such a purpose]. In a stair, how often should fresh air be supplied? Some fire engineers

say every three floors, others every ten; but supplying all fresh air from either the top or the bottom is *not* recommended; it is too likely that open doors near that source would deplete fresh air for the rest of the stair. At what velocity should fresh air be supplied? The recommended maximum is 200 fpm (1 m/s), to avoid stirring the smoke layer and inviting its descent. The designer must consider that high rates of flow at low velocities result in very large grilles, located quite low within the space.

(d) Exhaust. Special exhaust systems that function only in a fire are becoming more common. These systems employ both air velocity and air pressure to help control smoke. As with dilution systems, high velocities are avoided so that fresh air is not drawn up through the smoke layer. Such systems are particularly useful in large-volume atria spaces, as in Fig. 13.8. Smoke accumulating at the ceiling is removed, while fresh air is supplied lower in these high spaces. If smoke can be removed at the same rate that it is generated, the smoke layer can be kept from descending. Although they involve a greater initial expense, because they require special fans and special smoke-exhaust shafts, exhaust systems have several advantages over simple confinement or dilution approaches.

1. They can remove toxic gases from refuge areas (particularly when there is a dilution or outdoor air supply to such areas).
2. They help firefighters by improving the air quality in the vicinity of the fire itself.
3. They can help control the direction that a fire takes, by creating air currents that a fire will follow.
4. They remove the unburned but combustible gases from a fire before the latter can cause a "backdraft" or "flashover" (smoke-explosion).
5. By creating higher air pressures in refuge areas and lower pressures in the fire zone, they keep smoke out of refuge areas, even when doors are temporarily open.
6. With them, the tall-building phenomenon of stack effect, complicated by buoyancy and wind, is less likely to overcome the smoke management system.
7. They can help to remove postfire smoke.

(a)

(b)

Fig. 13.8 *The Rock 'n' Roll Museum on Cleveland's lakefront* (a) *features a tall, pyramidal interior space.* (b) *At the apex, a large horizontal grille marks the smoke exhaust system intake. Smoke naturally accumulates here, and can be removed while special sources of fresh air operate at ground level around the perimeter. (When hot days alternate with cool nights, such a system could be used for night ventilation of thermal mass.)*

The two Petronas Towers in Kuala Lumpur, Malaysia, are each occupied by as many as 18,000 workers during the workday. As these are the tallest buildings in the world at the turn of the 21st century, most of these workers are well beyond the range of ordinary fire rescue operations. Smoke exhaust, stair pressurization, and vestibule pressurization systems are installed, and each floor has isolation dampers, controlled by the fire alarm system, wherever air ducts serve multiple floors. However, because each floor has its own air-handling system (floor-by-floor fan rooms), vertical ductwork is minimized. There are both a main and a secondary fire command center for each of the towers and associated lower buildings.

(e) HVAC, Sprinklers, and Smoke.
Two systems within the building must be closely coordinated with smoke exhaust systems: the HVAC system and the fire detection/suppression (usually sprinkler) system. As the fire detection system activates the smoke exhaust fans, it must also override the conventional HVAC system operation, usually by switching to 100% outside supply air and simultaneously blocking the return duct system. This process pressurizes each HVAC zone to form a barrier against the smoke (Fig. 13.9), and also keeps smoke out of the return duct system. If the HVAC system is VAV, all supply control valves (dampers) must be moved to full open position. If additional protection is needed for refuge areas, additional air supply to such areas can be provided, perhaps by a separate (smoke dilution) system described earlier.

Sprinkler systems can hamper the functioning of smoke exhaust systems, both by creating a curtain of water that inhibits the movement of smoke and by cooling the smoke, thus reducing its buoyancy. As less buoyant smoke descends, visibility is reduced and dangers of smoke inhalation increase, particularly in spaces with ordinary ceiling heights. The buoyancy of smoke is the factor relied on by smoke exhaust shafts, whose intakes are located at ceiling level within each zone. Although these two systems thus appear to work at cross purposes, the suppression of the fire is obviously as important a goal as smoke management. Sprinkler systems potentially suppress fires so quickly that the size of the accompanying smoke exhaust systems can generally be reduced.

Fig. 13.9 *Smoke management by a smoke exhaust system.* (a) *The conventional HVAC system is shown operating at normal, minimum outdoor air supply conditions.* (b) *When smoke detectors trigger the smoke exhaust system, special fans begin to exhaust smoke through a special duct from the affected area only. Simultaneously, the conventional system undergoes two changes: All supply air becomes 100% outside air (for smoke dilution) and all return openings are closed off (so as to pressurize all nonfire areas, to keep smoke out of them and out of the return duct system.)*

Heavy Duty Torsion
Springs Raise Cover
to Open Position.

Hold-open Arms
Engage Automatically
to Lock Cover
in Open Position.

Fusible Link, When
Ruptured by Heat
Releases Cover Latches.

Shock Absorbers
Control Opening Speed;
Prevent Overtravel.

Outside and Inside Manual Latch
Release Controls Permit Opening
Without Disturbing Fusible Link.

Fig. 13.10 *Automatic heat-and-smoke roof hatch ventilators, primarily for one-story commercial and industrial buildings. (Courtesy of Inryco, Inc.)*

When the fire suppression system relies on oxygen displacement (such as carbon dioxide or "clean agents"), smoke exhaust systems clearly pose a threat; special attention is required if both systems are to be used. Again, oxygen displacement systems are often used in nonhabitable spaces, where smoke evacuation is less of an issue.

(f) Automatic Ventilating Hatches.

These heat-and-smoke venting devices (without fans) are often installed (Fig. 13.10) in smaller buildings, or larger one-story buildings such as shopping malls. These hatches open individually as heat or smoke from the fire trigger their control devices. As a result, indoor conditions near the fire are improved for firefighters, and firefighters on the roof are quickly alerted to the location of the fire indoors.

In summary, designers can provide several components that aid in the exhaust of smoke and the provision of access and ventilation for firefighters for almost any building: dedicated smoke exhaust and dilution systems, emergency controls on HVAC systems, operable windows and skylights (to facilitate ventilation and occupant escape), and smoke-and-heat hatches in roofs.

13.3 Water for Fire Suppression

The most popular medium for building fire suppression is water, which is readily available and relatively low in cost. Water cools, smothers, emulsifies, and dilutes. As it turns to vapor, it removes 970 Btu/lb (2256 kJ/kg) at atmospheric pressure and its volume increases 1700 times—a process that helps push away the oxygen needed by the fire. Water has several disadvantages that sometimes preclude its use for fire suppression: it damages most contents of buildings, including interior surfaces; it conducts electricity readily as a stream (less readily as a spray); many flammable oils float on the water's surface while continuing to burn; and as water vaporizes rapidly as steam, it can harm firefighters. Where these factors are major considerations, other suppression media can be considered (see Section 13.4).

Fire-suppressing systems are commonly combined with smoke management systems; Fig. 13.11 shows a typical combination of provisions in one-story buildings. Taller buildings, and especially those with multistory interior spaces (atriums), present a special challenge (Fig. 13.12). A typical approach is to construct an approximately 6-ft (1.8-m) deep curtain board (or fire spandrel) at the opening to the atrium on each floor. Smoke detectors, motorized dampers, and sprinklers can be combined in various ways. At the lobby level, the atrium floor level, and at any office floors open to the atrium, sprinklers at about 6-ft (1.8-m) o.c. will provide a water curtain at the edge of the atrium opening. Where glazing is used at the atrium's perimeter, the glazing frames sometimes can be designed for considerable thermal expansion, and sprinklers can provide a water curtain on the office side of the glazing. Fire-rated glazings and their frames are listed for both heat resistance

Fig. 13.11 *Fire protection for large-area, one-story buildings showing the fire suppression (sprinkler) system,* (A) *and the smoke management system,* (B and C). *Locations* D *and* E *are relatively safe positions for firefighters.*

and water pressure from a hose stream. Where balcony corridors adjoin the atrium, two sprinkler water curtains can be provided, one on each side of the partitions between corridor and office space. The smoke exhaust system often can provide six air changes per hour (ACH) for all spaces that open onto the atriums.

(a) Standpipes and Hoses (Fig. 13.13). Standpipes and hoses with a separate

water reserve, upfeed pumping, or fire department connections are listed in three classes and five types. The major differences are whether the system is for first-aid or for full-scale firefighting, and whether the system has an automatic water supply or a manual one.

Class I systems are for full-scale firefighting, and are typically required in both sprinklered and unsprinklered buildings more than three stories high, as well as in malls. (This is based on the

Fig. 13.12 *Fire protection in an atrium-type office building: section showing the detection/suppression system and provisions for smoke management.*

Fig. 13.13 *Standpipe system variations.* (a) *Basic flow rate requirements for Class I and Class III systems.* (b) *Example of the pipe schedule sizing method for a Class I system.* (c) *Multiple-zone system. (Reprinted with permission from the* Fire Protection Handbook, *18th ed., copyright © 1997, National Fire Protection Association, Quincy, Massachusetts, 02269.)*

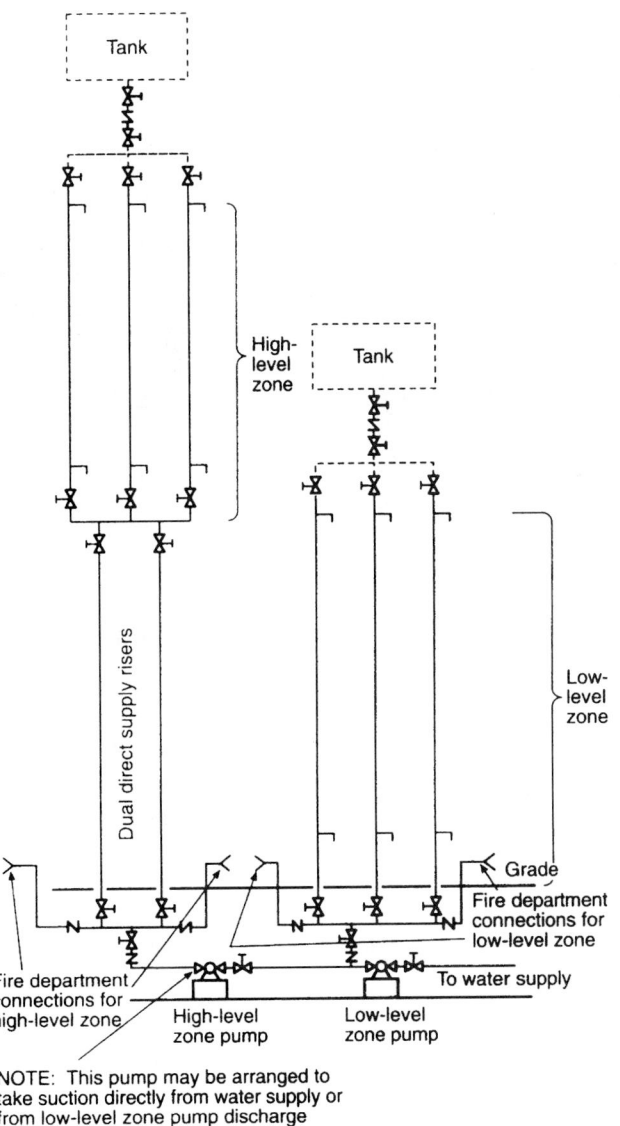

(c)

Fig. 13.13 *(continued)*

excessive time required to lay out hoses from out-door fire connections.) It is for use by trained firefighters, using 2½-in. (64-mm) hose connec-tions at designated locations. This is becoming the system of choice as use of the next two system classes declines.

Class II systems are for first-aid firefighting, before the fire trucks arrive. They use 1½-in. (38-mm) hose connections, and typically store a hose, nozzle and hose rack (Fig. 13.14) in each specified location. The difficulty for untrained people to manage a 100-ft (30.5-m) hose containing a huge flow [100 gpm (378 L/min) or more] has lead to a decline in use of this system, although it can be found in large unsprinklered buildings and in special hazard areas such as stages. Many fire de-partments would rather the building occupants evacuate than try to manage such large hoses and water quantities. Significant water damage is a potential threat when hoses cannot be controlled.

(a)

(b)

(c)

Fig. 13.14 *Components for standpipe and hose system.* (a) *Standard hose rack.* (b) *Hose rack and fire extinguisher in cabinet with glass door.* (c) *Siamese connection through which fire department pumping equipment supplies water to standpipe system. The similar sprinkler siamese is marked "sprinkler." Color codes sometimes differentiate the two.*

Class III systems combine the characteristics of classes I and II, thus serving both for first-aid and for full-scale firefighting. Both size hose connections are provided (with adapters) at each specified location. Again, the disadvantages of Class II systems also apply to Class III, and its use is declining for the same reasons.

"*Combined*" systems are either Class I or III standpipe systems that also supply water to a sprinkler system.

The standpipe system types are as follows:

Automatic-wet systems. The pipes are filled with water and are connected to a water supply capable of automatically meeting the fire-fighting demands. Water flows immediately on the opening of a hose valve.

Automatic-dry systems. The pipes are filled with pressurized air, and are connected to a water supply capable of automatically meeting

the firefighting demands. Through a device such as a dry-pipe valve, water replaces the air when a hose valve is opened.

Semiautomatic-dry systems. The pipes are filled with air and are connected to a water supply capable of automatically meeting the fire-fighting demands. Through a device such as a deluge valve, water replaces the air when both a remote sensing device at the hose station and a hose valve are opened.

Manual-dry systems. The pipes are filled with air, and there is no connection to a water sup-ply system other than that provided by the fire department (as in Fig. 13.14c).

Manual-wet systems. The pipes are filled with water, with a connection to a domestic source that is used merely to fill and test the system. Water for firefighting is provided by the fire de-partment.

Minimum flow rates, minimum pressure, and maximum pressure all are involved in the design of standpipe systems. Two methods of sizing, the pipe schedule and the hydraulic method, are used. Neither is detailed here; see NFPA 14 for the procedures. The required supply flow rates are outlined in Fig. 13.13a. As a preliminary design guideline (Fig. 13.13b), Class I and III standpipes not exceeding 100 ft (30.5 m) in height must be a minimum of 4 in. (102 mm) nominal pipe size. Class I and III standpipes more than 100 ft (30.5 m) in height must be a minimum of 6 in. (152 mm) nominal pipe size [although the topmost 100 ft (30.5 m) may be minimum of 4 in. (102 mm) nominal pipe size]. For combined standpipe and sprinkler systems, regardless of height, a minimum of 6 in. (152 mm) nominal pipe size is required.

For Class I and III systems, a minimum hose pressure of 100 psi (690 kPa) is now required, because of the widespread use of fog nozzles (rather than stream nozzles). The maximum hose pressure (also the maximum for sprinklers) is 175 psi (1207 kPa).

Water supply systems for firefighting may use roof tanks or fire pumps. It is not practical to store enough water on the roof for a protracted firefighting period, and it is usually assumed that a one-half hour supply will be more than enough to provide for the short period it takes the fire engines to arrive. When the system is used by the fire department, its pumps are attached to the street siamese (Fig. 13.14c) to deliver water from street hydrants or the building's "secondary source." The check valve closest to the siamese in use opens and the check valves at the tank close, to prevent the water from rising uselessly in the tank. After the engines are disconnected from the siamese, the water between the siamese and the adjacent check valve drains out through the ball drip so that it does not freeze.

An overhead tank is considered a most dependable source, but the height required can be architecturally undesirable and a considerable seismic disadvantage. In such a case, upfeed fire pumps (Fig. 13.13c) operating automatically to deliver water to higher stories may be used. Another option in this case is a pneumatic tank that delivers water by the power of the air that is compressed in the upper portion of the tank.

A standpipe zone in a high-rise building is determined by hydraulic calculation so that maximum hose pressures are not exceeded. Fire standpipes and their hoses (for full-scale firefighting) are now recommended to be located on the landings of fire stairs, from which personnel or firefighters can approach a fire with charged hoses. This enables the hose connection to be made in a safe (relatively smoke-free) place, half a floor below the fire. Awkwardly, it means that the door to the stair will be held open by the hose, but the risk to occupants on the stairs is considered outweighed by the advantage to the firefighters.

(b) Sprinkler System Design Impacts.

Unlike the fire hose, a sprinkler is likely to be already positioned above the point of a fire and is capable of being deployed in seconds, not minutes. Sprinkler systems are widely relied on as proven automatic fire-suppressers. (They have come a very long way from the early water-filled bucket suspended by a black powder fuse; when the fuse was lit by a fire, it blew up the keg, dispersing the water and theoretically dousing the fire.)

However, designers must remember that the use of sprinklers does *not* give one license to ignore fire code limitations, even though many codes are more lenient for sprinklered buildings. In addition, provision must be made for adequate water supply, standby power for water pumping, and water drainage during and after the fire. Sprinklers require very large supply pipes and valves, and sometimes fire pumps—all of which are frequently considered unsightly. However, sprinklers afford opportunities for integration with energy-conserving HVAC systems (Section 13.3j). Also, the cost of sprinkler system installation can usually be recovered rather quickly through reductions in fire insurance premiums.

Automatic sprinkler systems usually consist of a horizontal pattern of pipes placed just below or within the ceilings of industrial buildings, warehouses, stores, theaters, offices, homes, and other structures in which a fire hazard requires their use. These pipes are provided with outlets and sprinkler heads constructed such that abnormally high temperatures will cause them to open automatically and emit a series of fine water sprays. NFPA 13 lists the detailed requirements for sprin-

FIRE PROTECTION

kler systems (briefly referred to here) as well as numerous other regulations. In addition to pipes and sprinklers (discussed in the remainder of this section), several other system components are usually included.

Alarm gong. An alarm gong mounted on the outside of the building warns of water flow through the alarm valve on the actuating of a sprinkler head. This warning gives the building personnel an opportunity to make additional firefighting arrangements that can minimize loss and speed the termination of the fire; in this way, the sprinklers can be turned off to prevent excess water damage to building contents after the fire is out. Sprinkler alarms commonly are also connected to private regional supervisory offices that communicate promptly with municipal fire departments on the receipt of a signal. All public buildings, and other buildings as required, should be provided with fire detection and alarm systems that indicate, in the custodian's office, the location of the fire.

Siamese connections. Siamese connections permit fire engines to pump into the sprinkler system in a manner similar to that used for standpipe systems (Fig. 13.14c).

Provisions for drainage. Sprinkler heads can release a great deal of water, most of which will remain unvaporized and quickly collect at floor level. In addition to waterproofing the floors and at lower walls, columns, and other elements, provision should be made, where possible, for gravity drainage of water. Scuppers in exterior walls are preferable to floor drains, which are more easily clogged by debris. Scuppers are provided with hoods that protect against control of infiltration, birds, or insects.

Water supply. In tall buildings, sprinklers (and standpipes) can be supplied with water from elevated storage tanks for domestic water. These tanks supply a constant pressure on the distribution lines, store sufficient water to balance supply and demand, prevent excessive starting and stopping of the pump, and provide a dependable fire reserve. The last factor has been critical in the calculation of fire-insurance rates. When gravity

tanks are used with sprinkler systems, they should provide enough to operate 25% of the sprinkler heads for 20 minutes. As in the case of standpipe and hose systems, this provision gives the fire company a chance to arrive and take over.

The principal objections to the use of tanks have been their unsightliness, the problem of freezing, and—in the case of large buildings—the tanks' tremendous weight, with resulting added structural costs.

Skidmore, Owings & Merrill, in their design for General Mills' central office building in suburban Minneapolis, turned to automatically controlled pumping systems instead of elevated storage (Fig. 13.15). For the General Mills building, the fire underwriters could have required at least 50,000 gal (189,250 L) of residual water in the tank for emergency purposes, which would mean about 100,000 gal (378,500 L) of elevated storage. The alternative chosen was a reinforced-concrete structure placed underground to one side of the building and covered with from 3 to 5 ft (0.9 to 1.5 m) of earth. Small vents rising from this reservoir blend in with the lawn and landscaped shrubbery above.

The saving in structural steel tended to make the overall cost comparable to that of an elevated tank of the same capacity. The savings were made possible chiefly by refined automatic-pump control. The underground reservoir eliminates the problem of appearance and weight, but the other usual advantages of an elevated tank—reliability in case of fire, a minimum starting and stopping of motors, and the maintenance of pressure while balancing supply and demand—must be equaled in the automatic circuitry. This is not so simple as it might at first seem. Factors such as fluctuations of demand, friction within the pipes, elevations, starting surges from the pumps, and pressure-flow characteristics of the pumps themselves must be met. Various combinations of these problems undoubtedly account to a great extent for the continued use of elevated tanks. The trend is toward the use of more-sophisticated pump control.

A continuous flow from the deep-well pumps through both domestic and fire reservoirs prevents the water from becoming stale and rancid. The fire reservoir is given the necessary priority over the domestic reservoir by means of a simple

Fig. 13.15 *Diagram of upfeed pumping system for supply of sprinklers and building demands for domestic water. In the General Mills Building, a large subsurface concrete water tank provides the secondary water source that reassures fire insurance carriers. Thus, the wells and their pumps constitute a primary water source. If the well yield fails, the concrete tank supply can be tapped. An additional resource provides engine-driven pumping that operates in the event of a power outage at an electric utility company. Automatic controls and piping for the storage tank reserve system are not shown in this illustration.*

weir. Even if the domestic reservoir were completely empty, the fire reservoir would remain full. Pressure for the sprinkler is supplied by a small, 20-gpm (1.26-L/s) jockey pump. Signals from the sprinkler system bring in a 750-gpm (47.3-L/s) main pump. If this should fail, a diesel engine driven pump of equal capacity automatically takes over. Normally, the system automatically satisfies the heavy demands of air conditioning, fire control, and domestic water supply in this modern, rural isolated office building.

Valves. Valves are required to allow the sprinkler system to be shut off for maintenance, system modification, or for replacing *all sprinkler heads that have operated* after a fire. Indicating valves of various types are required so that it is always immediately obvious whether this crucial valve is open (as it nearly always should be) or closed. Table 13.6 lists common failure modes in sprinkler systems; an improperly closed valve is the leading reason.

(c) Sprinkler Construction, Orientation, and Rating. The common sprinkler head contains water by a plug or cap that is held tightly against the orifice by levers or other restraining devices. These in turn are held in place by the arms of the sprinkler body. In the past, the restraining device was usually a fusible metal link that melted at a predetermined temperature. Today, the restraining device is typically a glass bulb that contains both a colored liquid and an air bubble. As the liquid expands with the heat of a fire, it compresses the air bubble until it is absorbed, then continued expansion ruptures the bulb (again, at a predetermined temperature) and releases the water in a solid stream through the orifice.

A deflector converts this solid stream of water into a spray, whose pattern is determined by the design of the deflector. Most sprinklers direct the spray down and horizontally, rather than upward toward the ceiling. This yields better water distribution close to the head, and more effective coverage of burning material below. The spray pattern

TABLE 13.6 **Common Failure Modes for Automatic Sprinklers**

Failure Mode	Potential Causes
Water supply valves are closed when sprinkler fuses	Inadequate valve supervision Owner attitude Maintenance policies
Water does not reach sprinkler	Dry pipe accelerator or exhauster malfunctions Pre-action system malfunctions Maintenance and inspection inadequate
Nozzle fails to open when expected	Fire rate of growth too fast Response time and/or temperature of link inappropriate for the area protected Sprinkler link protected from heat Sprinkler link painted, taped, bagged, or corroded Sprinkler skipping
Water cannot contact fuel (*Note:* The intent of this failure mode is to ensure that discharge is not interrupted in a manner that will prevent fire control by a sprinkler)	Fuel is protected High piled storage is present New construction (walls, ductwork, ceilings) obstructs water spray
Water discharge density is not sufficient	Discharge needs are insufficient for the type of fire and the rate of heat release Change in combustible contents occurred Number of sprinklers open is too great for the water supply Water pressure too low Water droplet size is inappropriate for the fire size
Enough water does not continue to flow	Water supply is inadequate because of original deficiencies, changes in water supply, or changes in the combustible contents Pumps are inadequate or unreliable Power supply malfunctions System is disrupted

Source: Reprinted with permission from the *Fire Protection Handbook,* 18th ed., © 1997, by the National Fire Protection Association, Quincy, Massachusetts, 02269.

from the typical upright or pendant sprinkler is shown in Fig. 13.16.

Common types of sprinkler heads are upright (SSU), pendant (SSP), or sidewall types. Upright heads sit on top of the exposed supply piping. Pendant heads hang below piping, which thus can be concealed above suspended ceilings (Figs. 13.17*a, b*). The pendant heads themselves have a number of variations: recessed, flush, concealed, and ornamental pendant heads are available (Fig.

13.17*c*). Recessed heads have part of the sprinkler body and the deflector below the ceiling. Flush heads have only the thermosensitive element projecting below the ceiling. Concealed heads are entirely above a ceiling cover plate that falls away in a fire, exposing the thermosensitive element. Ornamental pendants are manufacturer-coated (never field-coated!) to match a desired decor.

Sidewall sprinklers (Fig. 13.17*d*) are usually located adjacent to one wall of smaller rooms—

Fig. 13.16 *Typical distribution pattern from a standard spray sprinkler. Obstructions (whether from structure or furnishings) are to be avoided within this volume. (Reprinted with permission from the* Fire Protection Handbook, *18th ed., copyright © 1997, National Fire Protection Association, Quincy, Massachusetts, 02269.)*

Heat-sensitive breakable bulb

(a)

(b)

(c)

(d)

(e)

Fig. 13.17 *Sprinkler heads.* (a) *Upright; sits above exposed piping, just below the structural deck—where the hottest gases will first accumulate.* (b) *Pendant projecting through a suspended ceiling in most installations. Both* (a) *and* (b) *use the quartzoid bulb, a transparent bulb with a colored liquid that ruptures at a preset temperature to release the water stream; heads can be manufactured as plated, polished, or colored—but never painted on site.* (c) *Pendant, styled for a more contemporary appearance.* (d) *A sidewall, front and side views.* (e) *Flow control pendant, which must be exposed within the space it serves. [Illustrations* (c), (d), *and* (e) *courtesy of Central Sprinkler Corporation.]*

TABLE 13.7 **Sprinkler Heads**

Discharge Characteristics

Nominal Orifice Size				
(in.)	(mm)	K Factor[a]	Nominal Orifice Size Marked on Frame	Percent of Nominal ½-in. Discharge
¼	6.4	1.3–1.5	Yes	25
⁵⁄₁₆	8.0	1.8–2.0	Yes	33.3
⅜	9.5	2.6–2.9	Yes	50
⁷⁄₁₆	11.0	4.0–4.4	Yes	75
½	12.7	5.3–5.8	No	100
¹⁷⁄₃₂	13.5	7.4–8.2	No	140
⅝	15.9	11.0–11.5	Yes	200
¾	19.0	13.5–14.5	Yes	250

Temperature Classifications

Max. Ceiling Temp.		Temperature Rating		Temperature Classification	Color Code	Glass Bulb Colors
°F	°C	°F	°C			
100	38	135 to 170	57 to 77	Ordinary	Uncolored or black	Orange or red
150	66	175 to 225	79 to 107	Intermediate	White	Yellow or green
225	107	250 to 300	121 to 149	High	Blue	Blue
300	149	325 to 375	163 to 191	Extra high	Red	Purple
375	191	400 to 475	204 to 246	Very extra high	Green	Black
475	246	500 to 575	260 to 302	Ultra high	Orange	Black
625	329	650	343	Ultra high	Orange	Black

Source: Reprinted with permission from the *Fire Protection Handbook,* 18th ed., copyright © 1997, National Fire Protection Association, Quincy, Massachusetts, 02269.

hotels, apartments, and so on—and throw a one-quarter–sphere spray of water entirely across such rooms. Therefore only one sidewall sprinkler head per small room often is used.

The temperature at which a sprinkler is triggered is usually specified, as in Table 13.7; it should be at least 25F° (14C°) higher than the maximum ceiling temperature ordinarily expected. "Ordinary" sprinkler heads operate at between 135° and 170°F (57° and 77°C).

Buildings to be protected by sprinkler systems fall into several hazard groups, as described in Table 13.8. Newer *quick-response* sprinkler heads are now required throughout light hazard occupancies, including office buildings. These more thermally sensitive heads open sooner than ordinary heads, and thus tend to fight a fire with even fewer heads operating, even though they may sometimes open with extraordinary heat that is not fire-associated. They are considered superior for life protection because of their earlier operation.

Other special sprinkler models include an *extra large orifice* (for delivering large water quantities where water pressures are relatively low) and *multilevel* sprinklers (for use where other sprinklers are at a higher plane within the same space). These lower sprinklers may otherwise have their action retarded by water from the higher elevation heads.

Flow control sprinklers close automatically once the ceiling temperatures are sufficiently reduced. Ordinary sprinklers, once activated, continue to operate until a main valve is manually closed; this can waste large quantities of water and result in extensive water damage. The flow control sprinkler shown in Fig. 13.17*e* is designed to operate individually and in two stages. When the ambient temperature reaches 140°F (60°C), a fusible alloy melts and the sprinkler is readied for action. Water discharge, however, does not occur until the second stage, when ambient temperatures reach 165°F (74°C). Later, when the ambient temperature falls below 95°F (35°C), the sprinkler returns to its "first stage" condition: readied for action, but no water discharge. Thus, if a fire reflash occurs, the sprinkler can again emit water, then again shut down as temperatures drop. Flow control sprinklers are not designed for use in dry-pipe systems (see Section 13.3*l*) and require special detailing when used in preaction systems (see Section 13.3*m*). They are not to be used in areas where nonfire ambient temperatures may exceed 100°F (37°C). These sprinklers are more visible because both the bottom and side elements must be exposed within a space.

(d) Sprinkler Spacing and Hazard.

The spacing of sprinkler heads and the sizing of their pipes is a complex matter, and most sprinkler systems are designed by professionals working for sprinkler manufacturers. However, some guidelines for preliminary sprinkler location can be included here. The first consideration is the degree of hazard faced by the occupants, as listed in Table 13.8. Once this is known, sprinklers and pipes can be approximately located in plan with the aid of Table 13.9. Maximum floor areas (per floor) to be protected by any single sprinkler system are shown, again by hazard, in Table 13.10. Within each space, sprinklers should be located so as to detect a fire readily and to discharge water over the greatest area (considering obstacles such as joists and beams, partial height partitions, etc.; see Fig. 13.16).

Piping for sprinkler supply can be hydraulically designed, as shown in Section 10.9. A complicating factor is the expectation that only a small percentage of the sprinklers will actually open; more than 50% of the fires studied over a 49-year period were extinguished by two or fewer sprinklers. A detailed sizing procedure would consider both the available pressure at the highest sprinkler and the expected flow rate, which can vary from 500 to 5000 gpm (31.5 to 315 L/s). The sprinklers' actual performance is then determined by

$$Q = K\sqrt{p}$$

where Q = flow rate, gpm
K = K factor, published for each sprinkler head by U.S. manufacturers (see Table 13.7)
P = pressure, psi

In SI units,

Q = flow rate, L/min
K = 14 K factor
P = pressure in bars

Sprinkler systems are usually designed for a maximum working pressure of 175 psi (1206.6 kPa). As a preliminary design guideline:

Light hazard systems need a minimum residual pressure of 15 psi (104.4 kPa) and 500–750 gpm (32–47 L/s) at the base of the system riser for 30 to 60 minutes.

Ordinary hazard systems need a minimum residual pressure of 20 psi (104.4 kPa) and 850–1500 gpm (54–95 L/s) at the base of the system riser for 60 to 90 minutes.

In the past, the "pipe schedule" design method made system layout and sizing fairly simple. Now there are requirements for more complicated hydraulically designed systems, beyond the scope of this book, for pipe sizing.

(e) Residential Sprinklers.

Now that many codes require sprinklers in all residential occupancies, some new issues arise. The residential sprinkler is a fast-response device with a tested ability to enhance survivability in the room of fire origin, and is thus listed for protection of dwelling units. It is sensitive to both smoldering and rapidly developing fires, opening quickly to fight a fire with only one or two heads operating. This is important because residences typically do not have a water supply with sufficient capacity for standard sprinkler systems. Toxic gases and smoke quickly fill the small rooms typical of residences; fast response is important for life safety.

Most codes that otherwise require residential sprinklers in all areas make an exception for bathrooms no larger than 55 ft² (5.1 m²); for closets with the least dimension not exceeding 3 ft (0.9

TABLE 13.8 Relative Fire Hazard for Various Occupancies, as Related to Sprinkler Installations

Classification of Occupancies

Occupancy classifications for this standard relate to sprinkler installations and their water supplies only. They are not intended to be a general classification of occupancy hazards. Examples in the listings below are intended to represent the norm for those occupancy types. Unusual or abnormal fuel loadings or combustible characteristics and susceptibility for changes in these characteristics, for a particular occupancy, are considerations that should be weighed in the selection and classification.

Light Hazard Occupancies

Occupancies or portions of other occupancies where the quantity and/or combustibility of contents is low, and fires with relatively low rates of heat release are expected.

The Light Hazard classification is intended to encompass residential occupancies; however, this is not intended to preclude the use of listed sprinklers in residential occupancies or residential portions of other occupancies.

Light Hazard Occupancies Include Occupancies Having Conditions Similar to:

Churches	Museums
Clubs	Nursing or convalescent homes
Eaves and overhangs[a]	Office, including data processing
Educational	Residential
Hospitals	Restaurant seating areas
Institutional	Theaters and Auditoriums, excluding stages and prosceniums
Libraries (except large stack rooms)	Unused attics

[a]If combustible construction with no combustibles beneath.

Ordinary Hazard Occupancies

Group 1: Occupancies or portions of other occupancies where combustibility is low, quantity of combustibles is moderate, stockpiles of combustibles do not exceed 8 ft (2.4 m), and fires with moderate rates of heat release are expected.

Ordinary Hazard Occupancies (Group 1) Include Occupancies Having Conditions Similar to:

Automobile parking and showrooms	Electronic plants
Bakeries	Glass and glass product manufacturing
Beverage manufacturing	Laundries
Canneries	Restaurant service areas
Dairy products manufacturing and processing	

Group 2: Occupancies or portions of other occupancies where quantity and combustibility of contents is moderate to high, stockpiles do not exceed 12 ft (3.7 m), and fires with moderate to high rates of heat release are expected.

TABLE 13.8 **Relative Fire Hazard for Various Occupancies, as Related to Sprinkler Installations** (*continued*)

Ordinary Hazard Occupancies (Group 2) Include Occupancies Having Conditions Similar to:

Cereal mills	Paper and pulp mills
Chemical plants—ordinary	Paper process plants
Confectionery products	Piers and wharves
Distilleries	Post offices
Dry cleaners	Printing and publishing
Feed mills	Repair garages
Horse stables	Stages
Leather goods manufacturing	Textile manufacturing
Libraries—large stack areas	Tire manufacturing
Machine shops	Tobacco products manufacturing
Metal working	Wood machining
Mercantile	Wood product assembly

Extra Hazard Occupancies

Occupancies or portions of other occupancies where quantity and combustibility of contents is very high and flammable and combustible liquids, dust, lint, or other materials are present, introducing the possibility of rapidly developing fires with high rates of heat release. Extra hazard occupancies involve a wide range of variables that may produce severe fires.

Extra Hazard Occupancies (Group 1) includes occupancies with little or no flammable or combustible liquids.

Extra Hazard Occupancies (Group 1) Include Occupancies Having Conditions Similar to:

Aircraft hangars[b]	Printing using inks having flash points below 100°F (37.9°C)
Combustible hydraulic fluid use areas	Rubber reclaiming, compounding, drying, milling, vulcanizing
Die casting	Saw mills
Metal extruding	Textile picking, opening, blending, garnetting, carding, combining of cotton, synthetics, wool shoddy, or burlap
Plywood and particle board manufacturing	Upholstering with plastic foams

[b]Except as governed by NFPA 409.

Extra Hazard Occupancies (Group 2) includes occupancies with moderate to substantial amounts of flammable or combustible liquids or where shielding of combustibles is extensive.

Extra Hazard Occupancies (Group 2) Include Occupancies Having Conditions Similar to:

Asphalt saturating	Open oil quenching
Flammable liquids spraying	Plastics processing
Flow coating	Solvent cleaning
Manufactured home or modular building assemblies[c]	Varnish and paint dipping

[c]Where finished enclosure is present and has combustible interiors.

Special Occupancy Hazards

See NFPA 13 for a listing of standards by occupancy.

Source: Adapted with permission from NFPA 13, *Installation of Sprinkler Systems,* © 1996, National Fire Protection Association, Quincy, Massachusetts. This reprinted material is not the complete and official position of the National Fire Protection Association on the referenced subject, which is represented only by the standard in its entirety.

FIRE PROTECTION

TABLE 13.9 Protected Area, Maximum Spacing, and Distances Below Ceiling for Sprinklers

NOTES: For SI units: 1 ft^2 = 0.0929 m^2; 1 ft = 0.305 m.

Obstructed: depth, spacing and openness of structural members impede heat flow to the sprinkler head and/or disrupt the spray pattern.

Unobstructed: structural members do not impede the operation of the sprinkler head.

Part A. Upright (SSU) and Pendant (SSP) Sprinklers

Construction Type	Light Hazard		Ordinary Hazard		Extra Hazard		High-Piled Storage	
	Protection Area, ft^2	Spacing (max.), ft	Protection Area, ft^2	Spacing (max.), ft	Protection Area, ft^2	Spacing (max.), ft	Protection Area, ft^2	Spacing (max.), ft
Noncombustible obstructed and unobstructed, and combustible unobstructed	225	15	130	15	100	12	100	12
Combustible obstructed	168	15	130	15	100	12	100	12

Part B. Standard Sidewall Spray Sprinklers

	Light Hazard		Ordinary Hazard	
	Combustible Finish	Noncombustible or Limited-Combustible Finish	Combustible Finish	Noncombustible or Limited-Combustible Finish
Maximum distance along the wall (S)	14 ft	14 ft	10 ft	10 ft
Maximum room width (L)	12 ft	14 ft	10 ft	10 ft
Maximum protection area	120 ft^2	196 ft^2	80 ft^2	100 ft^2

Part C. Extended Coverage (EC) Upright and Pendant Spray Sprinklers

Construction Type	Light Hazard		Ordinary Hazard		Extra Hazard		High-Piled Storage	
	Protection Area, ft^2	Spacing (max.), ft	Protection Area, ft^2	Spacing (max.), ft	Protection Area, ft^2	Spacing (max.), ft	Protection Area, ft^2	Spacing (max.), ft
Unobstructed	400	20	400	20	196	14	196	14
	324	18	324	18	144	12	144	12
	256	16	256	16				
			196	14				
			144	12				
Obstructed noncombustible (when specifically listed for such use)	400	20	400	20	196	14	196	14
	324	18	324	18	144	12	144	12
	256	16	256	16				
			196	14				
			144	12				
Obstructed combustible	NA	NA	NA	NA	NA	NA	NA	NA

Part D. Extended Coverage (EC) Sidewall Sprinklers

Construction Type	Light Hazard		Ordinary Hazard	
	Protection Area, ft^2	Spacing (ft)	Protection Area, ft^2	Spacing (ft)
Unobstructed, smooth, flat	400	28	400	24

TABLE 13.9 **Protected Area, Maximum Spacing, and Distances Below Ceiling for Sprinklers (*continued*)**

*Part E. **Large-Drop Sprinklers***

Construction Type	Protection Area, ft²	Spacing (ft)
Noncombustible unobstructed	130	12
Noncombustible obstructed	130	12
Combustible unobstructed	130	12
Combustible obstructed	100	10

*Part F. **Early Suppression Fast-Response (ESFR) Sprinklers***

Construction Type	ESFR Sprinkler up to 30 ft in Height		ESFR Sprinkler up to 40 ft in Height	
	Protection Area, ft²	Spacing (ft)	Protection Area, ft²	Spacing (ft)
Noncombustible unobstructed	100	12	100	10
Noncombustible obstructed	100	12	100	10
Combustible unobstructed	100	12	100	10
Combustible obstructed	NA	NA	NA	NA

*Part G. **Maximum Allowable Deflector Distances Below the Ceiling for Various Sprinklers**[a]*

Sprinkler Type	Maximum Distance (in.)
SSU/SSP	12
Standard sidewall	6
EC upright and pendant	12
EC sidewall	6
Large drop	8
ESFR	14

Source: Adapted with permission from the *Fire Protection Handbook,* 18th Edition, © 1997, National Fire Protection Association, Quincy, Massachusetts, 02269.
[a]Greater distances are permitted based on construction type, special listings, or both.

m); for open porches, garages, and carports; for uninhabited attics and crawl spaces (if not used for storage); and for entrance foyers that are not the sole means of egress.

Residential sprinklers also have a special water distribution pattern, capable of delivering water to the walls, and high enough on the walls to prevent the fire from getting above the sprinklers. This water near the ceiling also tends to cool gases at the ceiling, reducing the likelihood of excessive sprinkler openings.

The added cost of residential sprinkler systems may be recovered in several ways. As with other buildings, reductions in fire insurance rates will be helpful, although in residences the payback time is rather long. Zoning could permit smaller parcels of land for sprinkler-protected residences, since separation between buildings is less important when sprinklers protect both buildings.

*(f) **Quick-Response Sprinklers.*** All light hazard occupancies are now required to have quick-response (also called fast-response) sprinklers. These include hotels, motels, offices, and other buildings where faster sprinkler operation could enhance life safety. One measure of thermal sensitivity is the response time index (RTI), which indicates how fast the sprinkler can absorb heat

TABLE 13.10 **Maximum Floor Areas for Sprinkler Systems**

NOTE: The maximum floor area on any one floor to be protected by sprinklers supplied by any one sprinkler system riser (or combined system riser) is:

Hazard	Area (ft^2)	Area (m^2)
Light Hazard	52,000	4,831
Ordinary Hazard	52,000	4,831
Extra Hazard (hydraulically calculated)	40,000	3,716
High-Piled Storage	40,000	3,716

NOTES: Floor area occupied by mezzanines shall not be included in the above areas. Where single systems protect extra hazard or high-piled storage and ordinary or light-hazard areas, the extra hazard or storage area coverage shall not exceed the floor area specified for that hazard and the total area coverage shall not exceed 52,000 ft^2 (4,831 m^2).

Source: Based on the *Fire Protection Handbook,* 18th Edition, © 1997, National Fire Protection Association, Quincy, Massachusetts, 02269.

from its surroundings sufficient to cause activation. It is measured in the square root of meters-seconds. Quick- (fast-) response sprinklers have an RTI of 50 or less. Standard-response sprinklers have an RTI of 80 or more.

Because of the thermal lag in the glass bulb (or fusible link), the sprinkler body and the water in sprinkler pipes, air temperature around the sprinkler may reach 1000°F (538°C) before a standard sprinkler, rated at 175°F (79°C), actually opens. The fast-response sprinkler's operating element has a smaller mass, enabling it to track air temperature rise more quickly.

(g) Early Suppression Fast-Response (ESFR) Sprinklers. (See Table 13.9, part F.) These are tested for their capability to provide fire suppression of specific high-challenge fire hazards, as encountered in high-piled storage. They operate at a higher pressure and flow, so that their water droplets depend on momentum rather than gravity to penetrate to the bottom of high-velocity fire plumes. These sprinkler heads require a minimum of 50 psi (344.8 kPa) and a minimum flow of 100 gpm (6.3 L/s). They have largely replaced the large-drop sprinklers (Table 13.9, Part E) that depend on the weight of the larger water droplet to penetrate the fire plume.

(h) Extended Coverage Sprinklers. (See Table 13.9, parts C and D.) These are limited to a type of unobstructed construction consisting of flat smooth ceilings of a slope not exceeding 2 in.

per foot (158 mm/m). Note that a smooth ceiling means that luminaires and air grilles are flush or recessed, not suspended from the ceiling. However, such sprinklers can also be specifically listed for "noncombustible obstructed" construction, or for upright and pendant sprinklers within trusses or bar joists having web members not greater than 1 in. (25.4 mm) thickness, or where specifically listed for flat smooth ceilings of a slope not exceeding 4 in. per foot (316 mm/m).

(i) Future Developments. Now under development, "quick-response, early suppression" (QRES) sprinklers would use the same principles as the ESFR sprinkler, but with a smaller orifice suitable to lighter-hazard occupancies. This could be useful in a wide variety of business, mercantile, public assembly, and educational applications.

In future sprinkler systems, different types of sprays may be ejected from a single sprinkler head: one spray of larger droplets to penetrate the fire plume and thereby cool the burning surfaces, as well as adjacent surfaces; and another, finer spray to cool the ceiling itself.

(j) Wet-Pipe Systems. These are the most common and most simple systems, shown in Fig. 13.18. They are filled with water under pressure, and are limited to spaces in which the air temperature does not fall below 40°F (4.4°C). (Wet pipe systems that contain antifreeze, and admit ordinary water when a sprinkler head opens, are included in this category; the type of antifreeze is

Cross main

Inspector's test connection

Branch lines

Bulk main (riser)

Local alarm

Automatic sprinklers

Fire dept. connection

Alarm valve

OS&Y gate valve to control water supply to system

Main drain

Water supply

Fig. 13.18 *Wet-pipe sprinkler system. This most typical system is under water pressure at all times so that water will be discharged immediately when an automatic sprinkler opens. The automatic alarm value shown causes a warning signal to sound when water flows through the sprinkler piping. OS&Y is "outside stem and yoke," a valve that clearly indicates the degree to which it is open or closed. (Reprinted with permission from the* Fire Protection Handbook, *18th ed., copyright © 1997, National Fire Protection Association, Quincy, Massachusetts, 02269.)*

FIRE PROTECTION

limited when potable water supplies the sprinkler system.) In the wet-pipe system, sprinklers in the area affected are opened by sensitive elements within the sprinkler heads themselves, and immediately emit water.

A typical automatic wet-pipe sprinkler system is shown in Fig. 13.19. The building, a printing and publishing plant, is in the category of "Ordinary Hazard, Group 2" (see Table 13.8). The sprinkler design results in a nozzle spacing such that one nozzle (sprinkler head) takes care of 130 ft^2 (12 m^2) of floor area.

An urban high-rise building with a wet-pipe system is the Transamerica Building (Fig. 13.20) in San Francisco. Two fire pumps can deliver 750 gpm at 275 psi (47.3 L/s at 1896 kPa) discharge pressure; they draw from city mains at 50 psi (345 kPa) or from a 5000-gal (18,925-L) closed tank in the basement. These pumps feed into two 6-in.

(152-mm) diameter "express" risers, one in each stair tower, that rise the full height of this 48-story office building. These risers serve both the sprinkler system and fire department hose lines. In each of three 16-story zones, "local" 6-in. diameter risers branch off to feed a 2-in. (50.8-mm) looped main at each floor.

The sprinkler piping is carried above suspended ceilings; pendant sprinkler locations are coordinated with the modular grid that also locates partitions, air diffuser, and utility jacks. There are provisions for moving sprinkler heads as tenants change. At regular intervals, tees have been provided with one outlet stubbed and capped for future use. Typical office spaces are served by fully recessed, ½-in. orifice, 165°F (12.7 mm, 74°C) pendant sprinklers. Exposed pendant sprinklers are used in toilet rooms and service areas.

Fig. 13.19 *Plan of sprinklered industrial building, ordinary hazard occupancy, whose 30,000 ft² floor area is protected by 230 sprinkler heads [average 130 ft² (12 m²) per sprinkler head]. Sprinklers are closer than the maximum allowed. Sprinklers (and standpipes) may use water from street mains when pressure is adequate. Either system may use pneumatic or gravity tanks. When the latter is used to supply both systems, an independent sprinkler reserve occupies the bottom, and the fire standpipe supply occupies the top. Auxiliary fire-engine feed by siamese should be provided in all cases. Auxiliary sources and standby pressurization may be required if street main adequacy is questionable.*

Although the typical sprinkler system is served by a single riser with a main and branch lines, there are two variations that increase reliability by some redundancy in the supply lines. Figure 13.21a shows the gridded system, where each branch is served from either end, allowing each sprinkler head to receive water from either direction in the branch line. Figure 13.21b shows the loop system, where each branch line can receive water from either direction.

(k) Circulating Closed-Loop Systems. These wet-pipe systems use the rather large sprinkler piping to circulate water for HVAC water heat pumps. Water is not removed from this system, merely circulated. Water temperature in these systems must not exceed 120°F (49°C), nor fall below 40°F (4°C). Such a system used for heating and cooling in a motel is shown in Fig. 13.22. An office building with a circulating closed-loop system was shown in Fig. 8.72.

(l) Dry-Pipe Systems. As shown in Fig. 13.23, these systems are filled with compressed air (or nitrogen) rather than with water. They are used in unheated areas, including loading docks and cold-storage areas. As soon as a sprinkler head opens, the compressed air rushes out, allowing water to enter the formerly dry-pipe system through a *dry-pipe valve.* It then functions as does a wet-pipe system. The dry-pipe valve must be within a heated enclosure, since water under pressure is on one side of this valve. Also, due to the delay in delivering water throughout a previously-dry piping system, a maximum system capacity of 750 gallons (2839 L) is recommended.

Dry-pipe systems require a device to maintain design air pressure within the pipes; such air might be furnished from a compressor, from an air receiver tank, or from an existing pressurized air system (as in manufacturing operations). Dry-pipe systems also require a heated main control valve housing; and the pitching of all piping, to allow thorough drainage after usage.

(m) Preaction Systems. In this system (Fig. 13.24), the pipes are filled with air that may or may not be under pressure. In addition to the sprinkler heads, either a heat- or a smoke-detection system is installed, more sensitive than the sprinkler head. Water is held back by the *preaction valve.* When the heat- or smoke-detectors are activated, they open the preaction valve, an alarm is sounded, and the water fills the pipes. It then functions as does a wet-pipe system, with further water flow only on the opening of a sprinkler head.

A variation on this system, known as a combined dry pipe-preaction (or "double-interlock preaction") system, has the pipes filled with compressed air. Heat- or smoke-detectors release the preaction valve and sound the alarm, but air pressure keeps water out of the piping until a sprinkler head opens.

Preaction systems are popular where the building's contents are especially subject to water damage—computer rooms, retail stores, museums,

(a)

Fig. 13.20 *Transamerica Building, San Francisco, California; William Pereira & Associates, Los Angeles, California, architects. (a) The pyramid form provides rentable floor areas ranging from 22,000 to 3000 ft² (2044 to 279 m²). (b) Sixth-floor sprinkler plan, showing two risers, looped feed main, and branch lines. (Courtesy of Copper Development Association.)*

and so on—because the early alarm provided by water filling the piping often permits the fire to be found and extinguished manually, before any sprinklers open.

• ½" flush-mounted pendant sprinklers

(b)

Fig. 13.20 *(continued)*

However, designers must remember that while delaying the onset of the water spray, a fire can quickly grow in size. Typically, the 60-second delay in a preaction system could allow a fire to grow in area by at least 30%. Thus about 30% more water would be required to extinguish the fire.

(n) Deluge Systems. These systems (Fig. 13.25) have *open* sprinklers on dry pipes. As with preaction systems, a separate heat- or smoke-detection system is installed. These detectors control a *deluge valve,* which, once opened, floods the system with water and *all* heads emit water. Obviously, huge quantities of water are thus released.

Deluge systems are used where extremely rapid fire spread is expected—in aircraft hangars, for example, or other places where flammable liquid fires may break out.

(o) Mist Systems. These are considered last because they introduce the next section, fire suppression alternatives to water. With a history of success in shipboard fires, they are being studied for much wider applications. NFPA 750 is the relevant standard.

As an equivalent to a sprinkler system, a mist system offers faster initiation of an alarm as well as response to a fire. Smaller volumes of water

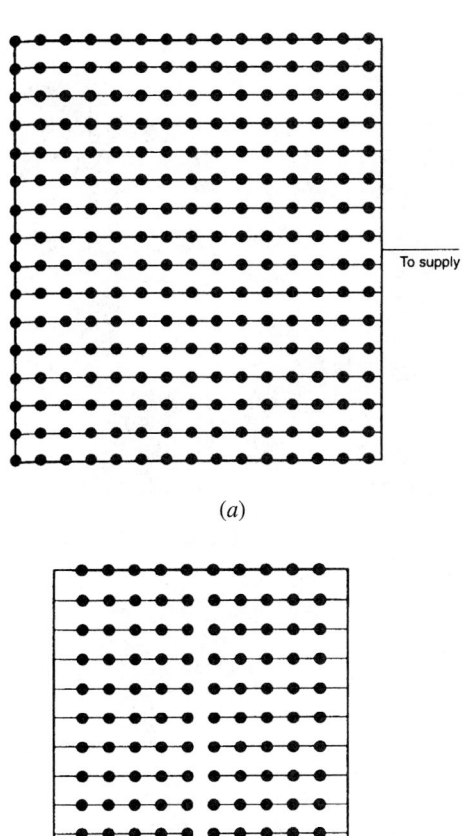

(a)

(b)

Fig. 13.21 *Alternative piping arrangements that provide more than one path for water. (a) Gridded systems supply water from either end of a branch line, offering the potential for water to come from either side of each sprinkler. (b) Looped system, with the potential for each branch line to get water from either direction. (Reprinted with permission from NFPA 13,* Installation of Sprinkler Systems, *copyright © 1996, National Fire Protection Association, Quincy, Massachusetts, 02269. This reprinted material is not the complete and official position of the National Fire Protection Association on the referenced subject, which is represented only by the standard in its entirety.)*

mean less water damage, and the mist can move more easily around obstructions.

As an alternative to halon or other clean agent gases (Section 13.4), mist systems do not pose a life-safety threat to firefighters, are more tolerant of small amounts of ventilation, reduce the radiant heat transfer from the fire, and eliminate residues associated with many clean-agent gases. They also save the expense of refilling a system with expensive clean-agent gases and allow a faster return to service after discharge.

These systems produce a much smaller water droplet, thanks to inlet pressures from 45 psi to 4100 psi (310 to 28,270 kPa), depending on the design of the sprinkler head. The heads are typically spaced closer together, and have more sensitive thermal elements.

The NFPA *Fire Protection Handbook* (1997) suggests three classes of mist systems, as follows:

Class I mists have a droplet size ≤ 200 microns, the finest mists. This is achieved at the expense of flow rate and spray velocity, and thus requires significant input of energy to produce useful quantities. They are most suitable where enclosure reduces the need for spray momentum, and fuel wetting is not critical (examples are liquid fuel fires and spray fires in enclosed spaces).

Class II mists have a droplet size range from 200 to 400 microns. With larger drops, it is easier to achieve higher mass flow rates, and considerable surface wetting is possible. They are likely to be effective on fires involving ordinary combustibles.

Class III mists have a droplet size from 400 microns up to 1000 microns. These can be generated by small-orifice sprinklers and fire hose fog nozzles, and deliver the highest mass flow rates.

The fire extinguishing mechanisms of mists are shown in Fig. 13.26. Two primary mechanisms are heat extraction (through rapid evaporation of the finely divided water droplets) and oxygen displacement. Which mechanism dominates depends on whether the fire is poorly or well-ventilated and the properties of the fuel. (Mist systems perform best in smaller enclosures with restricted ventilation, aided by heat entrapment and relatively easy oxygen displacement.) A third primary mechanism is the blocking of radiant heat. Secondary mechanisms in pool and spray fires are vapor/air dilution (mixing of water vapor and entrained air in the flammable vapor zone above the pool surface) and kinetic effects (the velocity of the flame front may be inhibited by small water droplets dispersed in a volume of flame).

FIRE PROTECTION

Fig. 13.22 *Circulating closed-loop sprinkler system, whose water is used by heating/cooling systems. This application is for a motel; a ground source heat exchanger is supplemented by a water heater when required. (Courtesy of Montana Power.) For another circulating closed-loop sprinkler system example, see Fig. 8.72, page 513.*

Tests in 1998 by the Institute for Research in Construction, National Research Council of Canada demonstrated that cycling, rather than continuous, discharge from a mist system extinguished most test fires within a shorter time, and all test fires with less water. A Class II spray with a twin-fluid (water and compressed air distributed separately, mixed at delivery) nozzle was used. The cycling discharges were "long water off" (50 seconds on, 30 seconds off) and "short water off" (30 seconds on, 20 seconds off). Compared to continuous discharge, cycling discharge produced higher room air temperatures near the ceiling and more rapid oxygen depletion—both are threats to the contents and the occupants, yet both also make the mist system more efficient at extinguishing the fire. This produced the results of less water used, and in most cases the shorter fire extinction period.

After full-scale testing of mist systems, with resulting design guidelines, they can be expected to replace both standard sprinkler and clean agent systems for many applications.

13.4   Other Fire Suppression Methods

When water poses almost as much of a threat as fire to a structure or its contents, a variety of other fire suppression methods are available. The most passive of such measures are *intumescent* materials, which expand rapidly as they are touched by fire. This process creates air pockets that insulate a surface from the fire or swell a material until it blocks openings through which fire (or smoke) could have passed. Intumescent paints, caulks, and putties are available. Some intumescent materials come in ¼-in. (6.35-mm) thick sheets, with various facing materials.

Fig. 13.23 *Dry-pipe sprinkler system for unheated occupancies. The system riser, cross mains, and branch lines are maintained at some air pressure. The dry pipe valve and water supply must be within a heated enclosure. (Reprinted with permission from the* Fire Protection Handbook, *18th ed., copyright © 1997, National Fire Protection Association, Quincy, Massachusetts, 02269.)*

(a) The Rise and Fall of Halon 1301. Halogenated hydrocarbons, known commonly as "halons," are gases (stored as liquids) in which one or more hydrogen atoms have been replaced by halogen atoms. Although the original hydrocarbons are often highly flammable gases, the substitution of halogen not only makes the gas itself inflammable, but also gives it flame-extinguishing capabilities by interrupting the chemical chain reaction in a fire.

Until the mid-1990s, the most widely used of these agents was *Halon 1301;* the name signifies that its molecule contains one carbon atom, three fluorine atoms, no chorine atoms, and one bromine atom. (When a fifth number is listed for halons, it refers to the number of iodine atoms.) Hydrogen atoms are not accounted for.

The advantages of Halon 1301 (and other halons) are that it can be released with relative safety in a flooding system in areas such as computer rooms, quickly extinguishing a fire with little harm to contents, no oxygen displacement to threaten life support, little demonstrated harm to people, and no residue left on electronic components. It is also lightweight and space-saving relative to other fire suppressants, as seen in the following subsections. Halon 1301 was used on all commercial aircraft and in many special building applications, such as computer rooms, museums and libraries, telephone exchanges, and kitchens.

However, Halon 1301 is now identified as a long-lived, significant threat to the Earth's protective stratospheric ozone layer. Its production was thus phased out as of 1994. The search for replacements for Halon 1301 (and 1211) leads to mists (discussed previously), foams, inerting gases and clean agents (discussed later), as well as special application approaches beyond the scope of this book.

Inerting gas and clean agent replacements for Halon 1301, unlike water, are really more protectors of building *contents* than of the building *struc-*

Fig. 13.24 *Preaction sprinkler system* (a). *Sprinkler piping contains air until the preaction valve allows water to enter the system piping.* (b) *In a preaction system, detectors open the valve.* (c) *In a double-interlock preaction system, both the detectors and then a sprinkler must open before the valve allows water to enter the system piping. (Reprinted with permission from the* Fire Protection Handbook, *18th ed., copyright © 1997, National Fire Protection Association, Quincy, Massachusetts, 02269.)*

Fig. 13.25 Deluge system. All sprinklers are open, therefore all operate when water is allowed to flow in the system piping. (Reprinted with permission from the Fire Protection Handbook, 18th ed., copyright © 1997, National Fire Protection Association, Quincy, Massachusetts, 02269.)

(a)

(b)

Solid fuel:
1. Low flame, harder to cool,
2. Char burns at lower O_2 concentration, and
3. Fuel wetting.

Fig. 13.26 Water mist fire extinguishing mechanisms. (a) Interaction of mist with pool flame fire. (b) Mist and flame on a solid fuel with charring. (Reprinted with permission from the Fire Protection Handbook, 18th ed., copyright © 1997, National Fire Protection Association, Quincy, Massachusetts, 02269.)

Fig. 13.27 *Detergent foam (high density) discharged from four units, two of which are seen in the illustration, after 3 minutes of operation. This illustration at the North Central Airlines hangar building at Detroit Metropolitan Airport smothers fire and prevents its spread, and will not harm airplanes or machinery. The system can handle combustible liquid fires. (Note also the supplementary grid of upright sprinklers directly below the roof surface.)*

ture. They should leave no sticky residue, such as most dry chemicals produce, and are therefore a popular choice where a clean fire-suppressing agent is required, people are present, and there are objects or processes of high value. Occasionally, they are used where water availability is low, or where space cannot be found for systems that use other, bulkier fire suppressants.

(b) Foams. Because foams (masses of gas-filled bubbles) are lighter than water and flammable liquids, they float on the surfaces of burning liquids, smothering and cooling the fire and sealing in vapors. Foams can be designed to inundate a surface or to fill cavities; they can be thin and rapidly spread, or thick, tough, and heat-resistant. Some foams can be spread by foam-water sprinkler systems. Many foaming agents are available, suited for various purposes. The NFPA *Fire Pro-*

tection Handbook chapter on foam extinguishing agents and systems presents a concise summary.

Foams are defined by their expansion ratio (final foam volume to original foam solution volume before adding air). *Low-expansion foam* has expansion up to 20:1, and is used principally to extinguish burning flammable or combustible liquid spill or tank fires. *Medium-expansion foam* has expansion from 20:1 to 200:1; *high-expansion foam* has expansion from 200:1 to 1000:1. These last two types are used for indoor fires in confined spaces: to fill enclosures such as basement room areas, or holds of ships. They are also (at about 500:1) used to control liquefied natural gas (LNG) spill fires and help disperse the resulting vapor cloud.

Because foam breaks down and vaporizes its water under attack by heat and flame, it must be applied in sufficient volume and rate to compensate for water loss, leaving enough to produce a residual foam layer. Foam may easily be broken down by a water hose stream; turbulent air or rising combustion gases may divert foam from the burning area, especially if foam distribution occurs directly above the fire. Foam solutions are conductive, therefore not recommended for use on electrical fires. Firefighter entry to a foam-filled passage requires self-contained breathing apparatus and a life line (vision and hearing are reduced).

An example of high-expansion foam is shown in Fig. 13.27, the North Central Airlines Hangar Building at the Metropolitan Airport in Detroit (Albert Kahn Associated Architects & Engineers, Inc.). The foam-generating equipment in this installation can fill the 38,400-ft^2 (3567-m^2) hangar with 1,400,000 ft^3 (39,644 m^3) of foam to a height of 36 ft (10.8 m) in less than 12 minutes. Automatic devices, on sensing abnormal heat increase, operate the foam generators, open roof vents, start the smoke control exhaust fans in the vents and transmit a fire alarm signal to the Airport Fire Department. Foam discharge is delayed 30 seconds while evacuation sirens sound to permit occupants to leave the fire area. Manual "override" controls allow personnel to start the system in the event of failure of the automatic controls, or to stop it if the fire is small and controllable by other methods.

The high-expansion foam is created by wetting a nylon net with a mixture of water and a special detergent soap concentrate. A large blower directs an air current through the net, producing an avalanche of foam. Suds blanket the fire, attacking it in several ways. The water in the suds converts to steam, absorbing the heat of the fire. The expansion of the foam into steam reduces the oxygen content to about 7%, which is insufficient to support active combustion. A cooling effect is achieved by the wetting action of the breaking bubbles. The movement of air currents toward the fire to replace the rising hot gases draws the foam, supplied from the sides (rather than directly over the fire), to the center of the fire. There it blocks the air flow and cuts off the supply of oxygen. The fire, thus contained and diminished, can be approached by firefighters for further control.

Aircraft are not harmed by the foam. Delicate machinery that might be injured by high velocity streams of water is undamaged and left clean when the foam is rinsed away. The structure—in this case, an open steel frame with a metal roof deck—is protected from excessive temperatures that might weaken it and cause it to collapse.

(c) Carbon Dioxide (CO₂). Among other inerting gases, CO_2 has long been used to prevent ignition of potentially flammable mixtures and extinguish fires involving flammable liquids or gases. Although inerting gases certainly help to extinguish fire by displacing oxygen, they are even more effective by acting as a "thermal ballast," absorbing combustion energy and reducing the temperature of the flame/vapor mixture below that necessary to sustain combustion.

Because of its oxygen displacement, CO_2 is often used in tightly confined spaces that are free of people or animals—for example, display cases, mechanical or electrical chases, and unventilated areas above suspended ceilings or below raised floors. However, it also finds applications in data centers, telecommunications equipment, and electrical equipment rooms; in short, wherever water damage is a major concern.

CO_2 is stored in cylinders under great pressure, as a liquid, requiring roughly four times as many cylinders as would Halon 1301 for the same hazard. When it is released as a gas, it absorbs about 120 Btu/lb, to provide cooling as well as smothering action. CO_2 is noncombustible and will not react with most substances. It does not conduct electricity, and will not normally damage sensitive electronic equipment. It has no residual clean-up associated with its use as a fire-suppressing agent. It spreads from discharge as a gas, but will stratify over time. When properly vented, the gas escapes to the atmosphere after a fire has been extinguished.

The main problem with CO_2 is that it must be used at concentrations of from 21% to 62% of that of the air, depending on the fire's fuel. However, at a CO_2 concentration of 9%, loss of consciousness will occur after a few minutes. This may allow enough time for an awake occupant to escape, but it does not help the firefighters' (or trapped persons') environment. Another potential prob-

Fig. 13.28 *Schematic of a CO_2 automatic fire detection/suppression system that provides total flooding of a space. Clean agent systems use similar components.*

lem is that after the initial smothering and subsequent dissipation of the CO_2, smoldering embers might reignite.

A typical CO_2 automatic detection/suppression system (Fig. 13.28) has cylinders, a detection system, and a discharge valve that releases CO_2 into a piping system with discharge nozzles (rather similar to a sprinkler system).

(d) Clean Agent Gases.

An array of future replacement gases for Halon 1301 can be expected. Initially, HCFCs and HFCs are likely to predominate. Confined and vital spaces such as control rooms, computer and communication facilities, and emergency response centers are candidates for clean agent systems; consult NFPA 2001 for system descriptions. Systems will resemble those for CO_2 (Fig. 13.28).

One of the earliest replacements to gain approval is FM-200® (heptafluoropropane, CF_3-CHF-CF_3, ASHRAE designation HFC-227ea). It works in a manner similar to halons but with zero ozone depletion potential. With a much shorter atmospheric lifetime than halons, it is also less threatening as a greenhouse gas contributing to global warming. It will typically require about 50% more cylinders as would Halon 1301 for the same hazard.

FM-200 leaves no particulates or oily residue to damage electronic instruments, is electrically nonconductive and noncorrosive, is colorless, and typically displaces only about 7% of the air in a space, with acceptable toxicity at design concentration. This makes it relatively safe for firefighters.

FM-200 decomposes to form halogen acids when exposed to open flames. Thermal decomposition products include hydrogen fluoride. With early warning and proper clean agent system installation, these by-products can be minimized.

(e) Portable Fire Extinguishers.

Most fires in buildings can be extinguished at an early stage with these common devices. Fire extinguishers are rated for the "class" of fire they are designed to fight; how many extinguishers and where they are located depends upon the hazard of the occupancy (as listed in Table 13.8). This material is a very short summary of NFPA 10, *Portable Fire Extinguishers.*

Portable fire extinguishers are labeled as follows:

Class 1A to 40A: The numerals refer to the relative extinguishing potential (40A will extinguish 40 times as much as 1A). "A" refers to the contents: water, aqueous film-forming foam (AFFF), film-forming fluoroprotein foam (FFFP), and multipurpose dry chemical (ammonium-phosphate-base). (Halogenated agents are being phased out.)

For use on "Class A" fires: ordinary combustibles such as wood, cloth, paper, rubber, and many plastics, requiring the heat-absorbing, cooling effects of water, the coating effects of dry chemicals, or the interruption of the combustion chain reaction by dry chemicals.

Class 5B to 40B: The numerals refer to the approximate square footage of deep-layer liquid fire that an inexperienced operator can extinguish. "B" refers to the contents: smothering or flame-interrupting chemicals such as CO_2, dry chemicals, AFFF, or FFFP. (Halogenated agents are being phased out.)

For use on "Class B" fires in flammable or combustible liquids, flammable gases, greases, and similar materials that are fought by excluding oxygen, by inhibiting the release of combustible vapors, or the interruption of the combustion chain reaction.

Class C: Contents are non-electrically conducting, such as CO_2 or dry chemicals. (Halogenated agents are being phased out.)

For use on "Class C" fires in live electrical equipment.

Class A:B:C: "Multipurpose" dry chemical extinguishers filled primarily with ammonium phosphate. Although indicated for all three classes of use, ammonium phosphate has the disadvantage of leaving an especially hard residue, unless promptly and thoroughly cleaned as soon as the fire is extinguished.

Class D: Contents are dry powders, such as graphite or sodium chloride. For use on a variety of combustible metals; the specific combustible metal for which the extinguisher is designed is printed on the extinguisher's nameplate.

A typical portable extinguisher was shown in Fig. 13.14*b*. The most common Class A extinguisher contains 2½ gal (9.5 L) of water, weighs about 30 lb (13.6 kg), and emits a stream from 30 to 40 ft (9 to 12 m). Fire extinguishers are rarely considered esthetically pleasing, yet should be lo-

cated in conspicuous places along ordinary paths of egress (Fig. 13.29). The requirements for maximum floor area served, and maximum actual path to extinguishers, are shown in Table 13.11.

13.5 Lightning Protection

Lightning is nature's most destructive force: the average lighting discharge is estimated at 200 million V, 30,000 A, and courses through a grounded object in less than a thousandth of a second. "Cold" lightning bolts have ample current, voltage, and duration to shatter and kill, but not to ignite combustibles. "Hot" bolts will ignite combustibles as well.

The decision of whether to protect a structure depends on an evaluation of these factors:

1. Frequency and severity of thunder storms.
2. Value and nature of building and contents.
3. Hazard to building occupants.
4. Building exposure. Buildings in open and exposed areas are more susceptible to lightning than urban buildings, although very tall towers remain vulnerable.

▲ FIRE EXTINGUISHER (LOCATED SO THAT NO OCCUPANT IS MORE THAN 75 FT. AWAY FROM AN EXTINGUISHER)

● STANDPIPE (SERVES AS AN EXTENSION OF FIRE-FIGHTERS' HOSES TO GROUND-LEVEL HYDRANTS)

Fig. 13.29 *Location of portable fire extinguishers for a typical multistory office building. Such extinguishers should be easily reached and placed in conspicuous locations along normal paths of protected egress, away from potential fire hazards. Note that three locations are shown. (Reprinted by permission from Egan,* Concepts in Building Firesafety, *© 1978 by John Wiley & Sons.)*

TABLE 13.11 **Class A Fire Extinguishers**

NOTE: For SI units, 1 ft = 0.305 m; 1 ft^2 = 0.0929 m^2

Part A. Size and Placement for Class A Hazards

	Occupancy		
	Light Hazard	Ordinary Hazard	Extra Hazard
Minimum rated single extinguisher	2-A[a]	2-A	4-A[b]
Maximum floor area per unit of A	3,000 ft^2	1,500 ft^2	1,000 ft^2
Maximum floor area for extinguisher	11,250 ft^{2c}	11,250 ft^{2c}	11,250 ft^{2c}
Maximum travel distance to extinguisher	75 ft	75 ft	75 ft

Part B. Maximum Area to be Protected per Extinguisher (ft^2)

	Occupancy		
Class A Rating on Extinguisher	Light Hazard	Ordinary Hazard	Extra Hazard
1A	—	—	—
2A	6,000	3,000	—
3A	9,000	4,500	—
4A	11,250	6,000	4,000
6A	11,250	9,000	6,000
10A	11,250	11,250	10,000
20A and more	11,250	11,250	11,250

Source: Adapted with permission from the *Fire Protection Handbook,* 18th Edition, © 1997, National Fire Protection Association, Quincy, Massachusetts, 02269.

[a]Up to two water-type extinguishers each with 1-A rating can be used to fulfill the requirements of one 2-A rated extinguisher for light hazard occupancies.
[b]Two 2½-gal. (9.5-L) water-type extinguishers can be used to fulfill the requirements of one 4-A rated extinguisher.
[c]This is the maximum area of a square that can be inscribed in a circle of 75-ft radius; where 75 ft is the maximum distance to a fire extinguisher.

5. Indirect effects. For example, loss of a water tower will seriously affect fire prevention and other services.

If a decision is reached to protect a building, it should be done completely and properly, with Underwriters Laboratories (UL) label equipment (Label A and B) and UL approved installation (Label C). A partially protected building is in reality an improperly protected building, which might be worse than one with no protection at all.

The relevant standards are NFPA 780, *Lightning Protection Code,* and Underwriters Laboratories Standard UL 96A, *Master Labeled Lightning Protection Systems.* The subject of lightning protection is a complex one, and the design of an adequate system is best left to specialists. Some of the design considerations and data on available materials are discussed briefly later.

The basic principle in lightning protection is to provide a continuous metallic path to solid (low resistance) ground for the lightning stroke because *there is no known method of protection that will prevent the occurrence of a lightning stroke.* This will prevent the stroke from passing through the nonconductive portions of a building, accompanied by intense heat and very large mechanical forces due to the high resistance of this path.

Any lightning protection system includes one (usually more) *air terminals* (pointed copper or aluminum rods projecting above a structure); *ground terminals,* terminating in highly conductive soil (or in a network of buried wires if the soil is nonconductive, as with bedrock or very dry soil); and *down conductors* that connect air and ground terminals. The conductors may be exposed on the building exterior or concealed within a structure.

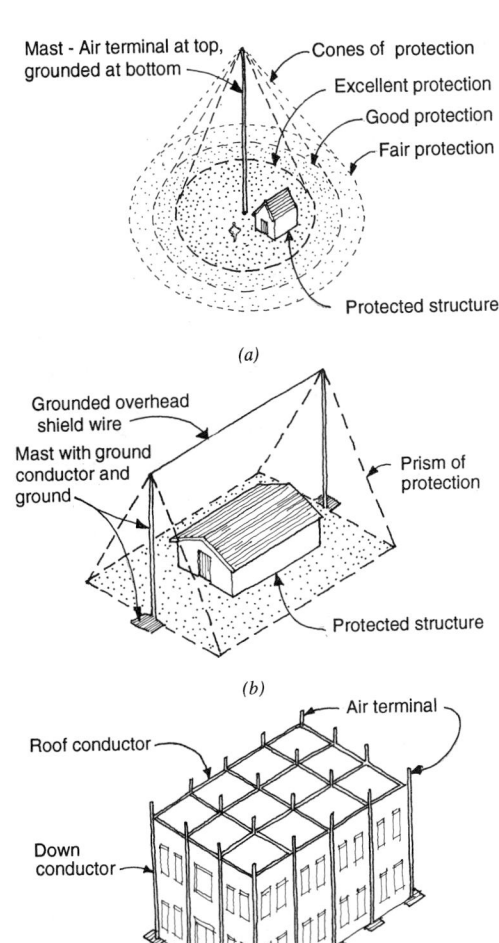

Mast - Air terminal at top, grounded at bottom

Cones of protection

Excellent protection

Good protection

Fair protection

Protected structure

(a)

Grounded overhead shield wire

Mast with ground conductor and ground

Prism of protection

Protected structure

(b)

Air terminal

Roof conductor

Down conductor

Grounding plate

(c)

Fig. 13.30 *Three approaches to lightning protection. (a) The Franklin cone is provided by a central mast that acts as air terminal, conductor and ground. (b) The overhead ground shield wire provides a triangular prism protection zone. (c) The Faraday cage wraps a protective mesh around the structure.*

Three techniques for protecting structures from lightning are shown in Fig. 13.30. All three offer the lightning stroke a good conductor, above the level of the protected nonconductive structure. The lightning stroke is drawn to the conductor, then conducted harmlessly to the earth. (Trees can be similarly protected.)

(a) Franklin Cone (Fig. 13.30a). A *Franklin cone,* named for Ben Franklin, is simply

a mast with a conductor running straight to ground. A "cone of protection" is formed that protects the objects within it from strikes by absorbing the lightning stroke at the mast and grounding it harmlessly. The closer to the ground and the mast, the better the protection: buildings within an interior solid angle of 60° get excellent protection, of 90° get good protection, and up to 126° get fair protection. (A more complex "geometric" version is described in the NFPA *Fire Protection Handbook* chapter on lightning protection systems.)

(b) Overhead Ground Shield Wire (Fig. 13.30b). A continuous air terminal, this shield is linear and horizontal, so the protected volume is a triangular prism rather than a cone. This is most commonly used to protect overhead transmission lines.

For both the single mast and the overhead ground shield, it is recommended that they project above any structure within the protected zone by at least 6 ft (1.8 m), and more is better.

(c) Faraday Cage (Fig. 13.30c). Named for Michael Faraday, this approach depends on an open interconnected mesh, covering a large nonconducting mass (a building), to draw a lightning stroke and conduct it to Earth. This leaves large areas of the roof exposed, so the higher the air terminals and the denser the mesh, the better the protection. An office tower, celebrating high air terminals at close intervals, is shown in Fig. 13.31. Because this is the most common method used to protect buildings, some details follow.

All metallic objects on the roof should be bonded to a looped conductor that joins the air terminals. These air terminals commonly are placed at both roof edges and ridges, at intervals of about 20 ft (6 m). The conductors are made usually of copper and are best housed in a plastic pipe conduit. Ground conductors, often called counterpoise conductors, are also installed to connect the earth terminals.

Metal roof and siding must not be substituted for air terminals and down conductors, unless constructed of 3/16-in. (4.8-mm) minimum sheet metal that has been made electrically continuous by bonding or an approved interlocking contact.

In *reinforced-concrete buildings,* concrete col-

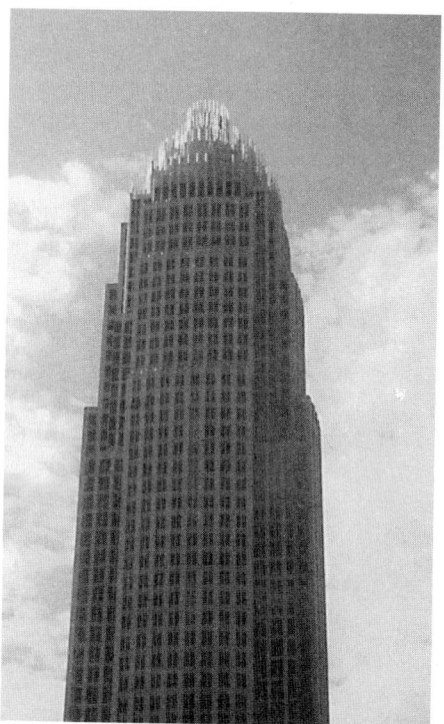

Fig. 13.31 *Lighting rods (air terminals) enliven the silhouette of the Bank of America Corporate Center in Charlotte, North Carolina. (Photo by Dale Brentrup.)*

umn steel reinforcing can be used for lightning conduction *if* the reinforcing steel is welded, rather than merely tied. Where tied, and in buildings constructed with precast concrete panels, reinforcing steel (and other metal) within a few feet of lightning conductors should be bonded to them to avoid arcing. Such arcing, as noted previously, is accompanied by heat and mechanical forces which can produce major structural damage and fire. Precast concrete buildings present a challenge because the reinforcing steel within precast panels is not typically interconnected.

In *steel structure buildings,* the columns relatively easily become lightning conductors; care must be taken to adequately bond both tops and foundations of such columns to the air and ground terminals, respectively (Fig. 13.32).

(d) Lightning Arresters. A special type of *surge arrester,* a lightning arrester is generally connected at one end to an overhead electrical line and at the other end to the ground. It thus suppresses any lightning voltage surges that appear in the electrical line, so that equipment downstream on that line is not damaged. A lightning arrester operates only when it senses a voltage surge. At that point it conducts the surge to the ground through its own low resistance, and partially dissipates the energy of the surge in the

Fig. 13.32 *Diagram of a typical lightning protection system.*

body of the arrester. Once the surge is passed the arrester returns to its high-impedance quiescent state, presenting an open-circuit connection to the normal voltage line to which it is connected.

(e) High-Rise Buildings. Lightning protection systems are particularly important for very tall buildings. Figure 13.33 shows the schematic diagram for the John Hancock Center in Chicago, a structural steel building of unusual height. Intermediate loops at the mechanical equipment floors help to overcome the problem posed by the poorer conductivity of metal piping systems, which can become additional relatively high-impedance paths for lightning, with resultant destructive effects. The intermediate loops are con-

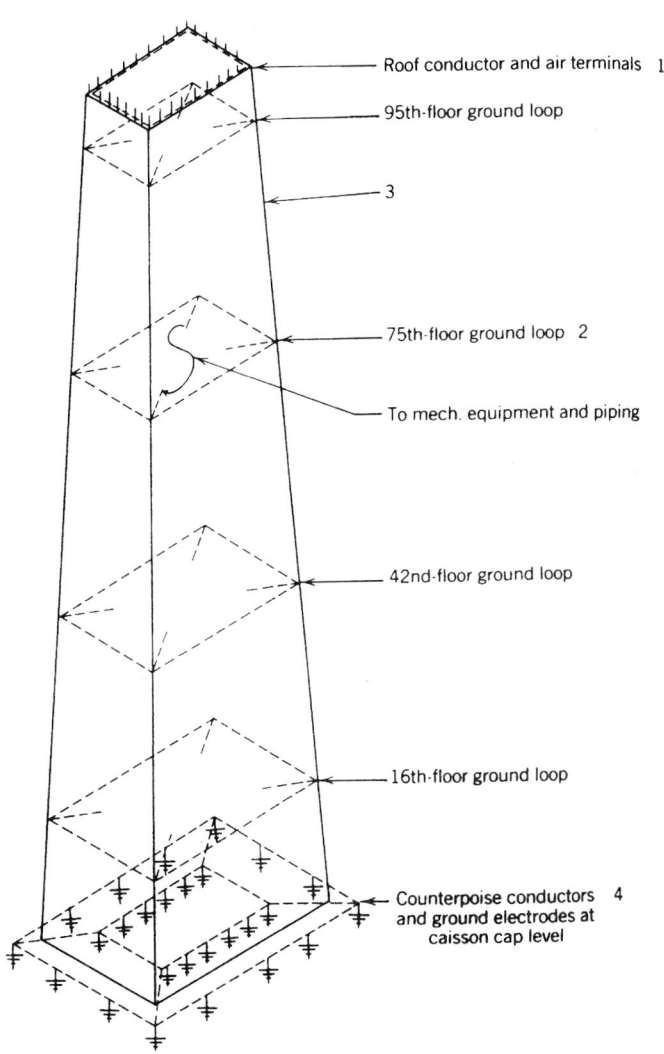

Roof conductor and air terminals 1

95th-floor ground loop

3

75th-floor ground loop 2

To mech. equipment and piping

42nd-floor ground loop

16th-floor ground loop

Counterpoise conductors 4
and ground electrodes at
caisson cap level

Fig. 13.33 *Lightning protection for the John Hancock Center, Chicago. Air terminals with roof conductors (1) are bonded to building steel and looped to assure that a single wiring break will not fault the system. Intermediate-floor ground loops (2) with connections to building mechanical equipment and piping protect against side-effect flashover to people or equipment. Building structural steel (3), properly bonded together, is used as a down conductor. Counterpoise conductors (4) cross-connect ground electrodes with ground conductors located below minimum groundwater level, to assure adequate and uniform dissipation of the charge. (Reprinted by permission from Jensen (ed.),* Fire Protection for the Design Professional, *copyright © 1975 by Van Nostrand Reinhold Company, Inc.)*

nected to each outside column, as well as to the main electrical, plumbing, air conditioning, and fire protection risers. These loops are also connected to both the roof conductor and ground counterpoise conductor.

FIRE ALARM SYSTEMS

13.6   General Considerations

A fire alarm system serves primarily to protect life and secondarily to prevent property loss. Because buildings vary in occupancy, flammability, type of construction, and value, the fire alarm system must be tailored to the needs of the specific facility. In schools, for instance, where the paramount consideration is rapid, orderly evacuation, the type of system used will generally be the same, although the means of automatic detection will vary with construction type, building height, specific area use, furnishings, and staffing. The fire alarm is part of the overall fire *protection* system design of the building. In particular, it overlaps with the design of safe egress and smoke/fire control in matters such as fan control and smoke

venting, smoke door closers, rolling shutters, elevator capture, and the like. These automatic functions are initiated by operation of the alarm system but are designed in accordance with the overall fire protection plan.

As with other alarm systems, a fire alarm system has three basic parts: signal initiation, signal processing, and alarm indication (see Fig. 13.34). The signal initiation can be manual (pull stations or telephones) or automatic (fire and smoke detectors and/or waterflow switches). The alarm signal is processed by some sort of control equipment, which in turn activates audible and visible alarms and, in some cases, alerts a central fire station or municipal authorities.

13.7   Fire Codes, Authorities, and Standards

There are probably more codes and standards in the area of fire protection than in any other area, with the possible exception of structural standards. Although these codes are devoted primarily to fire protection and life safety, they also gov-

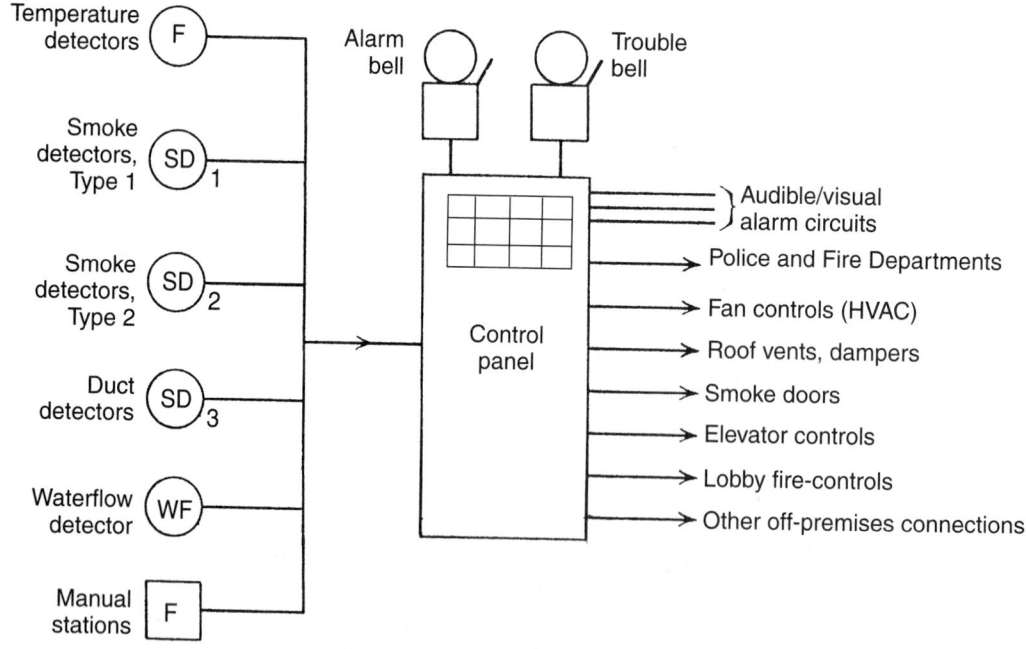

Fig. 13.34 *Typical simplified arrangement of a building fire alarm system. Multiple circuits, zoning, and other accessories such as annunciators are not shown. In a protected premises (local alarm) system, the off-premises connections to police, fire department, and other central stations do not exist.*

ern the type of fire alarm system to be installed, its components, and its installation.

NFPA codes are published by the National Fire Protection Association (NFPA), 1 Batterymarch Park, P.O. Box 9101, Quincy, MA 02269-9101. They bear directly on fire alarm system arrangements.

NFPA 101: Life Safety Code—specifies the means of egress, the type of fire alarm system, and the type and location of fire and smoke detection equipment for all types of occupancies.

NFPA 70: National Electrical Code; particularly Art. 760—Fire Alarm Systems, which was completely rewritten for the 1996 edition.

NFPA 72: National Fire Alarm Code—this integrated code replaces the various individual sections that preceded it.

NFPA 90A: Standard for the installation of Air Conditioning and Ventilating Systems.

NFPA 92A: Smoke control systems in malls, atria, and large areas. This code and 90A provide information on smoke detection in HVAC systems.

NFPA 99: Heath Care Facilities.

NFPA 110: Emergency and Standby Power Systems.

NFPA 111: Stored Electrical Energy Emergency and Standby Power Systems.

ANSI: (American National Standards Institute)—1430 Broadway, New York, NY 10018. This organization does not develop standards; it accepts for review standards that are prepared by professional and technical societies and associations. Approval and listing by ANSI validates the technical acceptability of the work.

CABO/ANSI 117.1: Accessible and Usable Buildings and Facilities. This standard specifies requirements to make buildings completely accessible and usable to people with physical disabilities of all sorts.

SBCCI: Also known as SBC (Standard Building Code), published by the Southern Building Code Congress—900 Montclair Road, Birmingham, AL 35213-1206. This code, generally used in the south and southeast areas of the United States, deals with auxiliary system controls such as door release, elevator capture, and so forth. In addition, SBC has a special section on high-rise buildings, covering fire communication, elevator control, and smoke alarms.

BOCA: Published by Building Officials and Code Administrators International, 4051 W. Flossmoor Rd., Country Club Hills, IL 60478-5795; Basic Fire Prevention Codes—cover smoke detection, alarm audibility, plus requirements for high-rise buildings.

UBC: (Uniform Building Code), published by International Conference of Building Officials (ICBO), 5360 Workman Mill Rd., Whittier, CA 90601-2298. This code, used generally in west and southwest USA covers essentially the same material as the other codes. Like them, it relies heavily on the Life Safety Code, NFPA 101.

HUD: (Department of Housing and Urban Development)—covers requirements for residential buildings and care-type facilities. FHA minimum property standards are included here.

UL: (Underwriters Laboratories) performance standards, published by Underwriters Laboratories, Inc., 333 Pfingsten Road, Northbrook, IL 60062-2096. UL laboratories test equipment for inherent safety and for performance according to manufacturer's specification. A product is "listed" if the equipment passes UL safety and performance tests.

UL Detection Standards

38: Manually Activated Signaling Boxes.

217: Smoke Detectors, Single and Multiple Stations.

268: Smoke Detectors for Fire Protective Signaling Systems.

268A: Smoke Detection for Duct Applications.

346: Water Flow Indicators for Fire Protective Signaling Systems.

521: Heat Detectors—Fire Protective Signaling Systems.

539: Single and Multiple Station Heat Detectors.

985: Household Fire Warning System Units.

1730: Smoke Detectors, Monitors and Accessories for Individual Living Units of Multiple Residences and Hotels/Motels.

UL Alarm Standards

228: Door Closers with/without Integral Smoke Detectors.
464: Audible Signal Devices.
827: Central Station Alarm Services.
1480: Speakers for Fire Protection.
1638: Visual Signaling Applications—Private mode, emergency and general utility signals.
1971: Signal Devices for Hearing Impaired.

In addition, several insurance groups (e.g., Factory Mutual) issue standards that may apply to a specific facility, in addition to local codes, standards, and fire marshal regulations. As with other aspects of construction—the architect/designer must ascertain which regulations have jurisdiction before proceeding with the design. The specific recommendations given herein are representative of current good practice. Actual design must be based on, and must meet the requirements of *current* NFPA standards plus all codes having jurisdiction in the particular locale of the project.

13.8 Fire Alarm Definitions and Terms

The following short list can serve as a reference to terminology that the reader may encounter in fire alarm work.

Addressable analog (smoke) detector (also referred to as an *intelligent detector*). An addressable (smoke) detector that continuously measures the concentration of smoke or other products of combustion in its test chamber. It continuously transmits an analog (voltage or current) signal to its (intelligent) control panel, that is proportional to this measurement. It does not transmit an alarm signal (see Fig. 13.46).

Addressable detector. A smoke or fire detector that communicates an alarm condition along with an individual identification (address), thereby immediately establishing the alarm lo-

cation. It responds to polling (testing) when it senses its unique identification (address) signal, "replying" with a signal indicating OK, trouble, or alarm.

Air sampling detector. A system consisting of aspiration equipment and piping that draws air from the protected area to the detector where analysis of the sampled air occurs.

Alarm signal. An audible and/or visual signal indicating a fire emergency, generally requiring immediate building evacuation.

Alarm verification. A technique for reducing false alarms. Common procedures include requirement for alarming for a specific minimum time and repeat alarm after initial reset.

Automatic system. A system in which an alarm-initiating device operates automatically to transmit or sound an alarm signal.

Auxiliary fire alarm system. See Section 13.9.

Breakglass. Refers to a false-alarm deterrent installed in manual fire alarm stations; a glass rod is placed across the pull-lever that breaks easily when the lever is pulled.

Coded alarm signal. An alarm signal that indicates the location of the fire alarm station operated by the pattern (code) of the audible signal.

Coded system. One in which not less than three rounds of coded alarm signals are sounded after which the system may either be silenced or set to sound a continuous alarm.

Control unit (Fire alarm panel). The controls, microprocessors, relays, switches, and associated circuits necessary to (1) furnish power to a fire alarm system, (2) receive and process signals from alarm initiating devices and transmit them to indicating devices and accessory equipment, and (3) electrically supervise the system circuitry.

Drift compensation. Adjustment function that compensates for detector sensitivity changes due to aging and environmental conditions. The adjustment may be automatic at the detector or manual/automatic at the control panel. Also referred to as *Automatic gain control* in beam-type smoke detectors.

Dual-coded system. See Section 13.11.

Intelligent control panel. Control center that receives analog signals from detectors, and based on predetermined conditions (usually in software or a microprocessor) determines whether an alarm situation exists. Sensitivity control, polling, and remote diagnostic testing of connected detectors are other functions usually found in these panels.

Listed, listing. Inclusion of a device in a list published by a recognized testing organization such as UL, indicating that the device meets the requirements of the referenced standard.

Local fire alarm systems. See Section 13.9.

Manual system. One in which the alarm initiating device is operated manually to transmit or sound an alarm signal.

Master-coded system. See Section 13.12(a).

Multiplex system. An arrangement whereby a single pair of wires can carry more than one message (signal) at a time, in either or both directions.

Noncoded system. See Section 13.12(a).

Presignal system. See Section 13.12(b).

Private mode. An alarm system installed in a location that has personnel specifically trained to act on receipt of an alarm signal.

Proprietary fire alarm system. See Section 13.9.

Public mode. An alarm system that alerts all of the protected spaces, occupants, audibly and/or visually.

Rate-of-rise detector. A temperature detector that alarms when the rate of temperature rise exceeds a specific design level, usually 15° F/ minute.

Remote-station fire alarm system. See Section 13.9.

Selective-coded system. See Section 13.12(a).

Station, fire alarm. A manually operated alarm initiating device; may be equipped to generate a continuous signal (noncoded station) or a series of coded pulses (coded station).

Supervised system. A system in which a break or ground in the wiring that prevents the transmission of an alarm signal actuates a trouble signal.

Trouble signal. A signal indicating trouble of any nature, such as a circuit break or ground, occurring in the device or circuitry associated with a fire alarm system.

Zone-coded system. See Section 13.12(a).

13.9 Types of Fire Alarm Systems

Fire alarm systems can be classified according to several different criteria including location, application, connections, coding, and the degree of automation of the detection system. Indeed, this last criterion is that most often found in manufacturer's literature. To avoid confusion, we use the classification system found in the National Fire Alarm Code (NFPA 72). This code classifies fire alarm systems essentially by location and function:

· Household fire warning systems
· Protected premises systems (local alarm)
· Off-premises systems (connections between local alarms and off-premises equipment and system). In this category are subsumed auxiliary, remote station, proprietary, central station, and municipal fire alarm systems.

(a) Household Fire Warning Systems. The term is self-explanatory. These systems are discussed at some length in Section 13.25.

(b) Protected Premises Fire Alarm Systems (Fig. 13.34). This arrangement, as the name indicates, is intended to sound an alarm only in the protected premises. Action in response to an alarm must be taken locally, either manually or automatically. Thus, notification to the fire department must be manual, although fire suppression systems can be set into operation automatically. This arrangement is applicable to privately owned facilities. When the building is unoccupied, notification to the fire department can come only incidentally, perhaps as a result of notification from a passerby. Local systems and their components are discussed in detail in Section 13.11.

(c) Auxiliary Fire Alarm Systems (Fig. 13.35). This is simply a local system equipped

FIRE PROTECTION

Fig. 13.35 *An auxiliary fire alarm system is one that is hard-wired to the municipal fire department, generally via a city box. This type of connection is usually restricted to public buildings.*

with a direct connection to a municipal fire alarm box. The received alarm signal is identical to that resulting from a manual alarm at that city box. Since the fire department is aware of all city box connections, arriving firefighters would always check the protected premises. This type of system is usually applied to public buildings such as schools, governmental offices, museums, and the like.

(d) Remote-Station Protective Signaling System (see Fig. 13.36). This system is similar to the auxiliary system, except that the alarm or

Fig. 13.36 *A remote station system is one that transmits alarm, supervisory, and trouble signals to a remote location that is continuously attended. The required actions will be taken by those attending the remote stations.*

trouble signal is transmitted via a leased telephone line to a remote location (a police facility or a telephone answering service) that is manned 24 h a day. The notice of alarm is then telephoned to the fire department. This arrangement is used in private buildings, such as stores and offices, that are unoccupied for considerable periods and for which reliance on passersby to turn in a fire signal is unacceptable. An audible alarm circuit extended from a local system to a nearby building—as, for instance, from a store to a nearby residence—does *not* constitute a remote-station system, unless all the requirements of NFPA 72 are met.

(e) Proprietary Fire Alarm Systems (see Fig. 13.37). This system, which is applicable to large, multibuilding facilities such as universities, manufacturing facilities, and the like, utilizes a dedicated central supervisory station to receive signals from all buildings. In a proprietary system, this station is on the site and is manned by persons associated with the facility. A common arrangement is for the station to be located in a guardhouse or similar supervisory location, from which point alarms are sent manually to a fire department and/or on site fire brigades. Other actions that must be taken on receipt of alarm or trouble signals are acted on by facility personnel at the central supervisory facility.

Information on the exact location or zone within each building at which an alarm occurs is transmitted to the central supervisory location. The central location has an audible alarm, some sort of visual display that indicates the alarm location, and a printer that makes a permanent record of each alarm. As mentioned, these central supervisory locations are frequently multipurpose, covering all aspects of facility security plus, frequently, control functions such as energy management.

(f) Central Station Fire Alarm System. This arrangement is similar to that described in (e) except that the system supervision and normally all of its equipment is owned and operated by a service company. In lieu of the multiple buildings of one facility shown in (e), the central station system supervises many individual unrelated local systems for a fee. Here too, opera-

Fig. 13.37 *Block diagram of a proprietary fire alarm system. The entire system, including all local devices, plus the central supervisory equipment is the property of a single owner. The central supervisory installation is attended by facility personnel who are trained to respond appropriately to alarm, trouble and supervisory signals.*

tors of the central supervisory installation receive all signals from individual users and provide the required services that include alarm verification, fire reporting, and repairs when trouble signals are received. Here also, as with proprietary systems, central station consoles often supervise access control, intrusion alarms, and related systems.

13.10 Circuit Supervision

This very common term refers to the circuit arrangements in fire alarm systems that indicates a malfunction in the wiring of alarm (and other) devices, by sounding a trouble bell. The trouble signal is separate and distinct from an alarm signal. The extent of circuit supervision required by fire codes varies with the type of facility and the specific code having jurisdiction. Minimally, an open circuit (break) or a ground in any of the detector and manual station wiring will cause a trouble signal. In large sophisticated systems all of the wiring and circuitry is supervised, including the control panel(s), alarm device circuits, and even annunciator wiring.

In conventional systems (see Section 13.11) the supervision does not indicate where the fault lies, but only that a fault exists. It is up to the system operator to troubleshoot the system in order to locate the fault. Because this may be a protracted procedure, all codes require that the system wiring be such that a single break or a single nonsimulta-

neous ground will not prevent an alarm signal from being sounded; beyond the break (Class B circuits) or on the entire circuit (Class A circuit).

13.11 Conventional Systems

Manufacturers classify fire alarm systems by the type of information transmitted to the control panel by the detectors and the alarm devices. The simplest (and oldest) system is the conventional system (Fig. 13.38). A conventional system is one that uses detectors and manual stations that transmit an alarm signal only. When they are in standby or quiescent state, they do not transmit. Such detectors and manual stations are called *conventional units;* hence the same name is used for the entire system. The signals transmitted by all detectors, and by manual stations as well, are identical, and therefore indistinguishable one from the other.

In theory, the number of alarm initiating devices, such as fire and smoke detectors, that can be wired to a single detector circuit is unlimited. (In practice there are electrical limitations.) As a result, when an alarm signal is received from such a multiple device circuit, there is no way of knowing which of the devices alarmed and whether the alarm represents an actual fire condition or whether it is a false alarm caused by dirt, moisture, a puff of smoke, a short-circuit in the wiring, or a malfunction of a detector.

Fig. 13.38 *Simplified schematic diagram of a conventional system. Alarm initiating devices are connected together on detector circuits, as are manual stations and audible/visual alarm devices. The control panel receives identical signals from all detectors, manual stations and alarm initiating circuits.*

Smoke detectors of all types (ionization, photo) are subject to false alarms activated by particulate matter in the air. Detectors are by nature threshold devices; raising sensitivity increases false alarming, decreasing sensitivity shortens the crucial early-warning period. False alarms are not merely nuisances; in facilities such as hospitals, and in places of public assembly such as theaters, public office buildings, and mass eating facilities, a false fire alarm can result in serious disruption, property loss, personal injury, and even death. As a result, conventional detectors in such buildings require continual maintenance and field sensitivity checks to minimize false alarms; an expensive and time-consuming procedure.

Because such facilities are so difficult and even dangerous to evacuate, most have fire/evacuation plans that call for some type of alarm verification before a general evacuation alarm is sounded. Because field experience has demonstrated that alarm verification can reduce false alarms appreciably without seriously degrading the early warning performance of combustion product detectors, recent issues of some fire codes now permit, or even require, alarm verification. This verification can be accomplished in conventional (hard-wired, zoned, nonaddressable) systems in a number of ways, including:

· requiring activation of at least two cross-zoned detectors in a single area

· requiring that a detector repeat its alarm after being reset (applicable only to systems with remote reset capabilities)

· requiring that a smoke detector continue to alarm for a minimum period of time—thus eliminating smoke puffs as the cause of an alarm

· physically inspecting the site protected by the detector to visually determine whether the alarm is actual or false

This last technique for alarm verification is probably the most fool-proof but is also the most difficult because it requires knowledge of the exact location of the alarming detector. This is only possible if detectors are grouped into zones, and each zone is annunciated individually (Fig. 13.39). To assist in localizing areas in a building by annunciation, zones are wired to contain as few devices as possible; the ultimate being a separate zone for each detector. As can be seen from Fig. 13.39, as the number of zones increases, so does the necessary wiring, and concomitantly, the installation costs. Thus a balance must be struck between two requirements; the need for more zones to localize and thereby simplify alarm identification and verification, and limitation of the increased cost of zoning. This balance depends on the type of facility being protected. All modern conventional systems are solid-state and all but the smallest use a wiring system called multiplexing.

This system uses a time-sharing electronic technique to transmit and receive multiple signals on a single two-wire communications circuit. This has the great advantage of reducing the primary cost item in large hard-wired systems—individually wired circuits. The alarm and audible signal devices in standard multiplexed systems are

Fig. 13.39 *Identification of detectors* (D) *can be accomplished in 3 ways:*
(a) *Each circuit, with multiple detectors, is annunciated as a zone. Applicable where detectors are close together.*
(b) *Each detector is wired individually, comprising a separate zone. Applicable for large or scattered detector spacing.*
(c) *Multiple detectors are wired on a single alarm circuit, but individually annunciated. Applicable where the number of detector circuits in the panel is limited but individual detector identification is required.*

the same as in conventional hard-wired systems. Only the system wiring and the panel architecture change. Zones, and even coding, are substantially unaltered, although the reduction in wiring costs permits the use of smaller zones. Some multiplexed systems are hybrid, using multiplexed detector circuits and hard-wired audible/visual alarm circuits. An additional advantage of multiplex panels is that they can be reprogrammed readily, which was emphatically not the case with hard-wired relay panels.

13.12(a) System Coding

(a) Noncoded Systems. Noncoded systems are *continuous* ringing evacuation types using manual and automatic alarm initiation. If desired, the devices can be zoned and, if the system is sufficiently large, annunciation can be provided. Audible devices are continuous ringing vibrating

bells and horns. Visual devices are lights and strobes.

(b) Master-Coded Systems. This system, also called *common-coded* and *fixed-coded*, generates four rounds of code that are sounded and flashed on all of the building alarm devices when any signal initiating device operates. It utilizes a single code transmitter at the panel. Normally, the system stops after four rounds of code, although it can readily be arranged to sound and flash continuously thereafter. When the code is set to ring the bells at an even 108 strokes per minute, it is known as "march time," because of the rhythmic cadence. This beat aids in the rapid panic-free evacuation of a building and therefore is frequently used in schools.

(c) Zone-Coded Systems. Identification of the alarmed zone in a system can be accomplished with zone lights, an annunciator, or by coding. In the first two cases, it is necessary to *go to the panel* *or annunciator* to determine the location of the operated device, which entails a possibly critical delay. All coded systems obviate this necessity by sounding the code on all the gongs in the building, thus immediately identifying the station and permitting the building staff to quickly investigate the cause of the alarm and take appropriate measures.

Therefore, if a *coded* system is desired, but by zone rather than by device (and this is less expensive by far), *noncoded* manual stations are used, along with automatic detectors, grouped by circuit into zones or individually zoned (Fig. 13.39). Each zone circuit trips a zone's transmitters *in the panel,* which in turn ring the zone's code on the single stroke gongs or chimes. As with all coded systems, four rounds of coded signal are sounded, after which the system is silenced. In all coded systems it is advisable to include a device that records all alarms in plain language, including time of receipt and the code sounded.

(d) Dual-Coded Systems. This arrangement is a combination of noncoded and zone-coded systems. When an alarm device operates, it initiates two separate functions—an identifying coded alarm and a continuous ringing evacuation alarm. The alarms are sounded simultaneously; the coded alarm in the building's maintenance office and the evacuation alarm on separate audible and visual devices throughout the building. A requisite to the application of this system is a continuously manned office in which the coded identifying signal can be received and acted on.

(e) Selective-Coded Systems. These are fully coded systems in which all manual devices are individually coded and all automatic devices are arranged to trip code transmitters at the panel. Each manual station can be immediately identified by its distinctive code. Automatic devices may be grouped in any fashion desired, and annunciated if desired. The combinations and circuitry are entirely in the hands of the designer. In large conventional systems, which fully selective-coded systems usually are, sprinkler transmitters and smoke detectors operate as integral subsystems of the main fire alarm panel.

(f) Presignaling. When it is desired to alert only key personnel, a system called *presignaling*

is used. Small bells or chimes are activated only at their work locations. Because these systems are always selectively coded, the personnel alerted can immediately investigate and, if necessary, manually turn in a general alarm by key operation of a station. Because of the delay involved, this type of system is used only in buildings where evacuation is difficult and sufficient staff is available to immediately investigate the cause of an alarm.

13.12(b) Signal Processing

Once a hazard has been detected, a signal is transmitted to a fire control center and appropriate action taken. In conventional systems, the alarm signal is transmitted over dedicated conductors ("hard wiring") to a control panel, consisting of either electromechanical relays in older systems or solid-state switching circuitry in modern systems. These in turn actuate audible and visible device circuits, illuminate annunciator panels, control fans and door releases, and so on, all via dedicated wiring. This arrangement has the advantages of reliability and simplicity. As the system grows, however, the wiring becomes heavy, complex, and expensive; panels become large and bulky; and changes become difficult. Also, troubleshooting of system faults becomes time consuming. Furthermore, minimizing false alarms is problematic because alarm verification in systems with large zones containing many alarm devices is very difficult within the extremely short time span permitted for this operation.

As a result of these problems, a system was developed in which every device can be individually identified and remotely checked, without separate wiring to each. It is called an *addressable system* because all the devices are individually addressable.

13.13 Addressable Fire Alarm Systems

An addressable fire alarm system is, by definition, one that uses addressable fire detection devices, both automatic and manual. These detectors are essentially identical to conventional detectors except that they are equipped with electronic circuitry, usually mounted in a special base, that makes each detector, in effect, a separate zone. Thus, in the event of an alarm, the control can require alarm confirmation from an adjacent detector or can require a repeat alarm from the same detector, after (remote) reset.

Addressable devices are continuously polled from the control panel, and each replies to the poll by reporting its condition as OK–standby, alarm, or trouble. Because each device is identified by its address, an alarm or trouble signal from a detector can be quickly and easily confirmed. Detector identification is accomplished either visually on an annunciator panel, or by alpha-numeric readout on an LCD readout panel. Wiring is simplified; up to 100 detectors can be wired on a single multiplex line. Hardware costs of addressable systems are higher than that of a conventional system because of the additional electronics in the detector bases and in the control panel. Maintenance costs, however, are somewhat lower because the panel's polling operation checks some (but not all) of a detector's electronics. Wiring costs in large systems are considerably lower than those of a conventional system. Measuring and calibrating the sensitivity of detectors is essentially the same as in a conventional system; the detector must be physically demounted and checked. This can be a major expense in a large system because periodic sensitivity checks are mandated by NFPA codes.

13.14 Addressable Analog (Intelligent) Systems

Addressable analog systems are known in the fire alarm industry as "intelligent" despite the fact that the word "intelligent" when referred to an inanimate device or system is meaningless. To further confuse the issue, there is no accord among manufacturers as to the exact meaning or use of the word when so applied. As a result, as of this writing, "very intelligent" and "extremely intelligent" devices have already appeared, and other superlatives are sure to follow. We use the accurate technical term "addressable analog" in this text and would advise readers to do the same.

As mentioned previously, conventional and addressable detectors are identical except for the

electronic identification coding of the latter. In function, both types test the adjacent space for temperature characteristics and/or products of combustion, compare the result to a preset threshold (adjustable), and latch into alarm position when the threshold is exceeded. The decision as to whether to alarm is made at the detector, and the control panel acts only to receive and translate the alarm signal into audiovisual building-wide alarm. Expensive periodic testing and adjustment of detectors in the field is essentially the same in both systems.

An analog detector gathers information from its sensor in analog form (voltage, current, or another variable characteristic proportional to the condition it senses) and transmits this data to its control panel. That is, an analog detector acts only as a sensor, and does not make the threshold comparison leading to an alarm/no-alarm decision. That decision is made by a microprocessor at the control center. In modern systems, the data received from each detector is combined with other data including reports from adjacent detectors, "historical" data on the condition of the detectors' time patterns at each detector that might falsely indicate an alarm, plus other relevant information. Analysis of all of this information leads to the decision as to whether to sound a general alarm. This arrangement has these advantages:

· False alarms are sharply reduced because of the alarm verification procedures at the panel.

· Threshold sensitivity is set at the panel, permitting adjustment without accessing the detector.

· A history of each detector's output is recorded and sensitivity degradation can be compensated automatically. This decreases maintenance costs by increasing the time period between cleanings.

· Any malfunction of a detector is noted immediately at the control panel, since all detectors are polled every few seconds.

Some manufacturers place an integral microprocessor in *each* detector that is capable of performing almost all of the functions described previously as panel functions. These detectors act as normal addressable analog detectors, transmitting analog signals to a control panel. However, in the event of a connection failure, the on-board microprocessor converts the unit to a stand-alone, self-contained detector/controller.

13.15   Automatic Fire Detection: Incipient Stage

Fire authorities agree that *most* fires pass through four stages, the last of which is the visible, flaming fire. These stages are shown in Fig. 13.40 as a function of duration and degree of hazard. Also shown is the type of automatic detector recommended for detection at each stage.

In the incipient stage the combustion products comprise a significant quantity of microscopic particles (0.01 to 1.0 μm), which are best detected by *ionization-type* detectors (Fig. 13.41). These detectors contain a small amount of radioactive material that serves to ionize the air between two charged surfaces, causing a current to flow. Combustion particles entering the detector chamber reduce air ion mobility, thus reducing current flow

and increasing voltage. These changes are sensed and the alarm is set off.

Response time of this type of detector depends on how rapidly the combustion particles can reach the detector—a factor that varies with room air currents and the type of material "burning." Once the particles reach the detector, response is essentially instantaneous. Therefore, ionization detectors are best applied indoors, in spaces with stagnant air or low air velocity (below 50 fpm) and in which little visible smoke (large particles) is expected. (Some manufacturers make a special unit whose sensitivity *increases* with air velocity for application in spaces with air velocity up to 500 fpm).

Ionization detectors should not be installed on warm or hot ceilings, or in any other location where hot air concentrates, because the hot air prevents the combustion particles from reaching the detectors. As a corollary, ionization detector sensitivity is higher at low ambient temperature.

Because ionization detectors react to minute particles of combustion, they should not be applied in, or close to, areas where normal activities

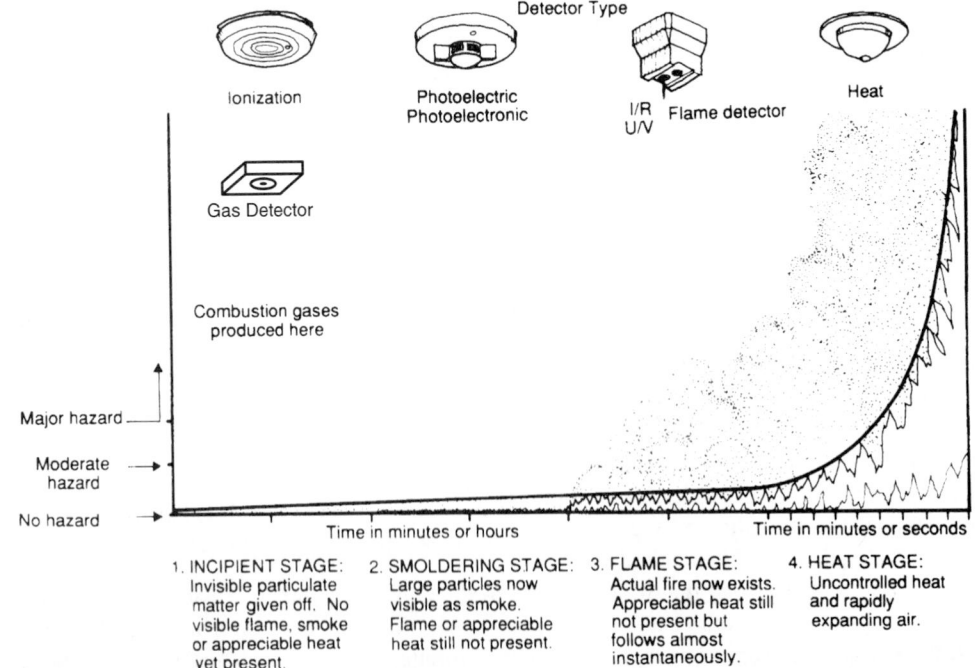

Fig. 13.40 *Four stages of a fire. Early detection of each stage requires a different type of detector that responds to the fire's particular characteristics (fire signature) at that stage.*

Fig. 13.41 *Addressable analog (intelligent) ionization type smoke detector with integral microprocessor that operates the unit's detection, diagnostic, and error-checking procedures. The detector is self-adjusting to compensate for sensitivity changes due to environmental conditions. Operating sensitivity is remotely adjustable from the control panel. (Photo courtesy of Cerberus Pyrotronics.)*

produce such particles as the obvious result will be nuisance alarms. Thus, kitchens, bakeries, welding and brazing shops, workshops using open flames and burners, and areas with concentrated engine exhaust fumes are all spaces inappropriate for their application. Ionization detectors respond best to fast-burning flaming fires. Because dust settling in the ionization chambers reduces sensitivity, these units need periodic cleaning and recalibration.

As can be seen from Fig. 13.40, the incipient stage of a fire also produces changes in the gas content of the environment. These combustion gases, which are not normally present in the air, are detectable by devices known as *gas-sensing fire detectors*. The principle of operation of these detectors depends on the type. A semiconductor type reacts to a change in conductivity due to the presence of combustion gases, whereas a catalytic element type detects a change in electrical resistance of the element caused by the presence of gases. Gas detectors are frequently used in conjunction with another type of detector, so that both gases and particulate matter are detectable.

Coverage of incipient stage detectors varies from 150 to 900 ft² per unit, depending on the unit used, the type of combustible material in the space, and ambient conditions. Once the fire has passed the incipient stage and becomes smokey, or where even in the incipient stage the combustion products are the large particles typical of

visible smoke, the ionization detector loses sensitivity and photoelectric detectors are recommended. Ionization, gas, and photoelectric detectors are classified as early-warning types.

13.16 Automatic Fire Detection: Smoldering Stage

(a) Photoelectric Smoke Detection. Refer again to Fig. 13.40. The smoldering stage of a fire is characterized by particles up to 10 μm in size. Such particles, although small, are visible to the naked eye as smoke and are best detected by photometric means. The simplest type of *photoelectric smoke detector* operates on the principle of beam obscuration, as shown schematically in Fig. 13.42. A beam of light is directed onto a photosensor and a steady-state, no-smoke circuit condition is established. The presence of smoke in sufficient concentration partially obscures the beam, changing current flow in the photocell circuit and setting off an alarm response. Sensitivity is typically set at 0.5% to 4% obscuration per foot for gray smoke and 0.5% to 10% for black smoke, to comply with UL standard 1989.

In this design, accumulation of dust and dirt and the presence of heavy fumes from industrial processes cause gradual obscuration of both the photo cell and the lamp. This, combined with lamp aging and consequent light reduction result in *in-*

Fig. 13.42 *Principle of smoke detection by beam obscuration. This design is used in two ways: as illustrated, with a projector in one location and a photocell at a remote location; or with the entire assembly in a single small housing, with light source and receiver on opposite sides of a 2- to 3-in. wide smoke chamber.*

creased sensitivity, causing false alarms that are not only a nuisance, but also can constitute a hazard. To correct this condition, continuous maintenance and periodic recalibration is necessary. As a result, this design is no longer recommended for spot-type smoke detection, but is used in beam detectors, as discussed next.

(b) Projected Beam Photoelectric Smoke Detector.

Beam-type photoelectric detectors consist of two separate units: a beam transmitter and a beam receiver, normally wall-mounted on opposite sides of a space, somewhat below the ceiling. They operate on the simple obscuration principle (Fig. 13.43). They are best applied in areas that do not lend themselves to the application of spot-type detectors. Among these are the following:

- High ceiling areas such as atria, churches, malls, auditoriums, and the like. In these spaces, ceiling mounted spot-type detectors present serious maintenance problems. This is particularly true in spaces such as theaters and auditoriums.

- Spaces with medium to high velocity air flow at the ceiling level. This condition severely dilutes smoke entering the test chamber of a spot detector. Although beam detectors are also affected negatively, their long throw and wide "vision" angle make the dispersion and dilution problem much less serious.

- Closed areas with little air flow, resulting in a hot air layer at the ceiling that prevents smoke from reaching ceiling mounted spot detectors (Fig. 13.44). Exposed ceiling beams also tend to trap hot air and reduce the effectiveness of ceiling spot detectors.

- Environments that militate against spot-detector use, such as those that are extremely dirty, corrosive, humid, very hot, or very cold. Beam-type detectors can be physically shielded against these conditions in a manner that interferes only minimally with their light-beam transmission and reception characteristics.

Beam-type detectors have an effective throw (transmitter to receiver) ranging from 30 to 330 ft (depending on ambient conditions), and can be spaced 30 to 60 ft apart. This gives a coverage range of 900 to almost 20,000 ft^2, as compared to spot detectors, whose *maximum* coverage is about 900 ft^2. Furthermore, beam detectors are wall-mounted, which reduces maintenance problems.

Finally, beam detectors are usually drift-com-

Projected Beam Smoke Detector

Fig. 13.43 *This projected beam smoke detector utilizes an infrared (IR) light beam so that visible light will not affect its operation. Two sensitivity levels are available that are selected in consonance with the distance between transmitter and receiver. Because heat attenuates or deflects the IR beam, the receiver translates this beam strength reduction as obscuration, identical to smoke obscuration. This effect makes the IR projected beam detector sensitive to high heat flaming fires as well as smokey smoldering ones. (Photo courtesy of Notifier, div. of Pittway.)*

Fig. 13.44 *Stratification of air in a closed space leads to a concentration of hot air at the ceiling that can prevent smoke from reaching a ceiling-mounted spot smoke detector. A projected beam detector mounted 1–2ft below the ceiling will alarm due to obscuration, as shown.*

pensated (thus further reducing maintenance expense) and are arranged not to alarm with full obscuration; a condition that frequently occurs during routine building maintenance and repair work.

Disadvantages of beam-type detectors as compared to spot units include:

· Response to the second (smoke) stage of a fire, and therefore not in the same early-warning category as ionization or gas detectors.

· High cost.

· Inapplicable to spaces that lack unobstructed view, such as spaces with pendant lighting fixtures, exposed ductwork, and the like.

As with all types of detectors, the use of beam-type detectors in a particular application must be consonant with the requirements of *all* fire and administrative codes having jurisdiction in that area.

(c) Scattered Light Photoelectric Smoke Detector.
A second design, usually referred to as a "photoelectronic," scattered-light, or "Tyndall-effect" detector, is illustrated in Figs. 13.45, 13.46, 13.47. In this design a beam of pulsed LED light is directed at a supervisory photocell, which serves to provide a baseline reference signal. The

alarm photocell is shielded and normally receives no light. When smoke enters the unit, light is reflected from (scattered by) the smoke particles and strikes the alarm cell. This changes the cell's resistance and the resultant signal is amplified electronically, causing an alarm. An alternative design depends on refraction of light to set off the alarm. These designs are not sensitive to nor-

1 Light beam
2 Light source
3 Lens
4 Supervisory photocell
5 Alarm photocell

Fig. 13.45 *Principle of operation of a scattered light photoelectronic spot smoke detector. A pulsed beam of light (1) from a source (2) is focused by a lens (3) on a supervisory cell (4). Smoke entering the unit reflects the light onto the alarm cell (5), changing its electrical characteristics and setting off an alarm. Photoelectric units with sensitivities up to 2% per foot are usually classified as early-warning units.*

Fig. 13.46 *Addressable analog (intelligent) scattered-light type photoelectric smoke detector with on board microprocessor that negates false-alarm signals caused by radio frequency and electro-magnetic interference (RFI/EMI). It also validates trouble signals. If required, this unit can be equipped with a resettable heat detection element. The unit its approximately 5½ in. deep and 3 in. high. (Photo courtesy of Cerberus Pyrotronics.)*

Fig. 13.47 *Smoke detector, intended for duct mounting, can utilize either a photoelectronic or ionization-type head. The unit, which is usually mounted in return-air ducts rather than supply ducts to minimize smoke dilution, uses an air sampling tube that extends across the full width of the duct. Duct smoke detectors are intended primarily to prevent smoke dispersal throughout the building by causing shutdown of the air-handling system and by control of smoke damper operation. (Courtesy of Simplex.)*

mal dust and dirt accumulation or to light source depreciation, and high sensitivity can be maintained without continual maintenance. Thus, they are usually found in commercial use and high-quality residential applications. Sensitivity is usually set at 1% obscuration per foot.

(d) Laser Beam Photoelectric Detector. A very high sensitivity laser diode source, scattered-light type photoelectric smoke detector has been developed that is classified by its manufacturer as an early warning device (Fig. 13.48). The high sensitivity results from the ability of the very sharply focused laser beam and its associated control panel software to differentiate between a smoke particle and a dust particle. It is best applied in clean environments.

(e) Air Sampling Detection System. Another system that takes advantage of the high sensitivity of laser-beam based photoelectric smoke detection is the air sampling system (Fig. 13.49). In principle the system is quite simple. Instead of waiting for air that may be carrying incipient smoke particles to reach the detector(s) by thermal air currents, this system samples air

throughout the protected space by aspiration and brings it to the detector for testing. The aspiration and air conduction system is simply a piping system with holes at sampling points, powered by a fan. The advantage of the system is that air throughout the space is sampled, thus, in effect converting each sampling opening into a highly sensitive laser-beam detector.

By zoning the piping into manageable areas that can readily be checked for alarm verification, the disadvantage of nonaddressability of the aspirating openings is avoided. Addition of a microprocessor to a multizone control unit that adjusts the alarm sensitivity of each zone based on a history of environmental smoke levels, all but eliminates false/nuisance alarms.

(f) Application of Photoelectric Detectors. Photoelectric smoke detectors of all designs will detect particles from about 0.2 μm to about 1000 μm (1.0 mm). Thus, they are useful not only for smoldering fires but also for the smokey fires that characterize the burning of certain plastics

Fig. 13.48 This unit is an addressable analog laster-beam type of scattered light photoelectric smoke detector. The sharply focused laser beam permits discrimination between smoke and dust particles, resulting in a high sensitivity unit without false alarming. The unit is shown mounted on a sounder base containing a horn assembly, making it applicable to small, enclosed rooms. (Photo courtesy of Notifier, div. of Pittway.)

and chemicals. Also, particle agglomeration, which increases with the distance the smoke travels, does not reduce their sensitivity, as it does with ionization types.

Maximum recommended spacing for photoelectric detectors, as for other types, is given by Underwriters Laboratories and Factory Mutual Lab standards. Closer spacing is often mandated by the particular application or by the structure's characteristics. Manufacturers' recommendations should be obtained for all installations. In order

to provide early-warning detection of a wider range of combustion products than is possible with either the photoelectronic or ionization-type detectors individually, several manufacturers produce a unit that combines a multichamber ionization detector with a photoelectronic detector.

13.17 Automatic Fire Detection: Flame Stage

As noted in Fig. 13.40, the appearance of flame is followed almost instantaneously by heat buildup and the rapid spread of flame, with a concomitant large increase in hazard. Detection of flame is no longer "early warning," and the prime requirement for a detector at this stage is speed. It is self-evident that the actions taken as a result of a flame detection alarm, such as fire suppression and/or evacuation, must also be very rapid. This is not the case with early-warning equipment alarms, with which, several minutes or more are normally available to investigate the authenticity of the alarm, *if desired,* before taking appropriate action.

Flame detectors are of two types: those that detect ultraviolet radiation and those that detect infrared radiation. Both types of radiation are present at the beginning of the flame stage.

Ultraviolet (UV) radiation detectors operate by optically detecting the UV radiation produced by

Fig. 13.49 In an air sampling system, air is aspirated at selected points in the protected area and piped to highly sensitive laser beam photoelectric smoke detector for testing. Large areas are zone-piped, so that trouble and alarm signals can be readily checked in the protected areas. (Drawing courtesy of Cerberus Pyrotronics.)

Fig. 13.50 *Combined ultraviolet-infrared (UV/IR) flame detector has peak sensitivity at 200 and 4300 nm, respectively, which corresponds to the maximum emission points of a typical hydrocarbon (e.g., petroleum products) fire. This unit is equipped with a microprocessor that performs diagnostic procedures and ensures proper calibration. The detector, which measures approximately $5 \times 5\frac{2}{3} \times 6$ in. deep is suitable for use in specific class hazardous areas. (Photo courtesy of Megitt Avionics.)*

flames, which is typically in the 170- to 290-nm (nanometer) range. (The entire UV range is 40 to 390 nm.) Hydrocarbon (organic material) fires in particular produce strong radiation in this range. A quartz filter in front of the UV sensor tube limits the detector's reception range to between 190 and 260 nm, which is below the visual range. This desensitizes it to sunlight and to incandescent and fluorescent sources. The detector need not "see" the flame directly; it can also detect UV radiation reflected from walls, ceilings, and the like. Obviously, the more direct the path, the stronger the radiation and the more rapidly the detector will respond. Because many devices—in particular, electric welders—produce UV radiation in sufficient quantity to activate the sensor, flame detectors are usually programmed to respond only to flickering sources at the usual flame flicker rate of 5 to 30 times per second.

In sum, UV detectors are long range, very sensitive, react in milliseconds, and respond to most types of fires. They are best applied in highly flammable or explosive storage and work areas, either indoor or outdoor. Modern units are equipped with microprocessors that perform self-test diagnostics. UV detectors are best applied to

detect hydrocarbon fueled fires such as burning alcohol, methane, propane, acetone, and all types of petroleum products—their principle disadvantage is that they are "blinded" by thick black smoke.

Infrared (IR) radiation detectors are sensitive to radiation in the IR region, between 650 and 6500 nm. Most commercial units are filtered to be sensitive to radiation in the 3800 to 4300 nm range, and to have maximum sensitivity at 4300 nm (4.3 μm), which corresponds to the radiation emitted by hot CO_2. They differ from ambient heat detectors in that they respond to radiant energy and are essentially optical detectors. IR units have about half the distance range of UV detectors, are sensitive (although not as sensitive as UV detectors), react in seconds, and must be programmed for flicker response to avoid false alarms, which can occur if they are exposed to other sources of IR, such as sunlight. Because of interference problems that cause false alarms, IR units are normally applied to enclosed spaces such as sealed storage vaults and the like. IR flame detectors are best applied to fires that result in rapid flaming combustion and the production of CO_2. Typical materials in this risk category are petroleum products, wood and paper products, coal, and plastics.

Combination UV/IR detectors such as the one illustrated in Fig. 13.50 cover a wide range of risk applications, including aircraft hangars, fueling stations of all sorts, tank farms, and flammable storage areas. The illustrated unit contains a microprocessor that performs self-diagnostics, not only of the optical system, but also of the electronic circuitry and calibration. The presence in one unit of both UV and IR detectors permits the integral microprocessor to analyze the UV/IR ratio in any radiation detected and its time signature. This analysis is compared to known fire data in the microprocessor's memory in order to discriminate between actual fires and spurious radiation, and thereby reduce false alarming.

13.18 Automatic Fire Detection: Heat Stage

Again referring to Fig. 13.40, the heat stage is the last and most hazardous stage because by this time the fire is burning openly and producing great

Fig. 13.51 *Detectors in the illustrated design are made with fixed temperature nonresetting elements that melt out at 130°F or 200°F and rate of rise elements that will alarm when the temperature rise is faster than 15°F/min. The illustrated unit is a 200°F fixed-temperature-only model. It is applicable to spaces with unusually violent temperature fluctuations and ceiling temperatures between 100° and 150°F. (Photo courtesy of Cerberus Pyrotronics.)*

heat, incandescent air, and smoke. Spread of the fire depends on fuel, air currents, and the construction of the space in which the fire is burning. Detectors intended for use at this stage respond to heat and are referred to as heat-actuated, thermal, thermostatic, or simply temperature detectors. They act much like the fusible link in a sprinkler head. Effective application is restricted to locations where the subsequent alarm permits adequate countermeasures to be taken in time to prevent injury and minimize loss. Keep in mind that the heat stage *follows* the smoke stage, and

that smoke, not heat, is responsible for most casualties in fires.

Heat detectors are made in two designs: spot units, which are mounted in the center of the area that they protect; and linear units, which sense heat along their entire length. Both types respond to the high ambient temperatures caused by hot air convection from a fire. The linear type also senses the overheating of an object or surface with which it is in contact, without the presence of fire.

(a) Spot-type Heat Detectors. Spot heat detectors are of two types: fixed-temperature units and rate-of-rise units. In the former, a set of contacts operates when a preset (nonadjustable) temperature is reached—usually 135°F or 185°F. The rate-of-rise type operates when the rate of ambient temperature change exceeds a predetermined amount (usually 15°F/min), which is indicative of the heat stage of a fire. The rate-of-rise unit is normally combined with a fixed-temperature unit in a single housing.

The fixed-temperature unit is available in either a one-time nonrenewable design that utilizes a low-melting-point alloy plug, or an automatic resetting unit that operates in the same fashion as a thermostat. For most applications, the resettable unit is preferred. Spot units are best applied in spaces that are separated from occupied areas and are subject to rapid-temperature-rise fires. Three different designs of heat detectors are shown in Figs. 13.51, 13.52, and 13.53.

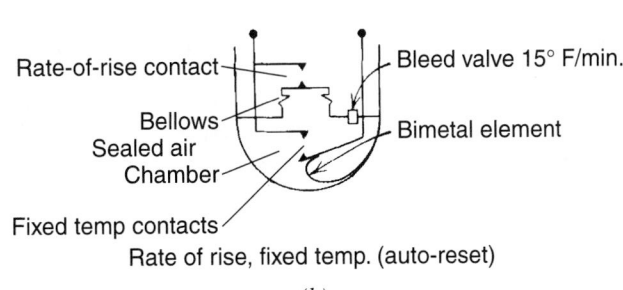

(a) (b)

Fig. 13.52 (a) *Combination rate-of-rise and fixed temperature detector. The detector and mounting plate fit on to a standard 4-in. square outlet box.*
(b) *The principles of device operation are shown schematically in this drawing. The rate-of-rise unit consists of an air chamber with a restricted bleed valve. Rapid temperature rise causes the bellows to expand before air is lost by bleeding, thereby setting off the alarm. The thermostatic element is a fixed-temperature self-restoring bimetallic unit.*

Silver
contact points
Expansion struts
High expansion
sensing shell 0.040
anodized aluminum

Cut-Away View, $\frac{1}{2}$ Actual Size

(a)   (b)

Fig. 13.53 (a) *The illustrated unit is a rate-of-temperature rise and fixed temperature (135° or 200°F) type detector. Unlike the rate-of-rise alarm design of the detector in Fig. 13.52, this unit uses the principle of differential thermal expansion for both actions. (Photo courtesy of Cerebrus Pyrotronics.)*
(b) *Cutaway of the detector with a different base than that shown in* (a). *The detector consists of an aluminum tubular shell containing two curved expansion struts under compression fitted with a pair of normally open opposed contact points that are insulated from the shell. The tubular shell and the struts have different coefficients of expansion. When subjected to a rapid heat rise the tubular shell expands and lengthens slightly. The interior struts lengthen but at a slower rate than the shell. The rapid lengthening of the shell pulls the struts together, closing the contact points and initiating the alarm. When subjected to a slow heat rise the tubular shell and the interior struts lengthen at approximately the same rate. At the detectors' set temperature point of 135°F or 200°F, the interior struts are fully extended, thereby closing the contact points and initiating an alarm. (Drawing courtesy of Notifier, div. of Pittway.)*

(b) Linear Heat Detectors. Linear heat detectors are devices in which the entire length of the linear, cable-like element is heat sensitive. Its application is to long, narrow elements, the protection of which with spot detectors is impractical. Typical applications include cable trays, cable bundles of all sorts, conveyors, and large, long items of equipment. The heat sensitive element is installed close to, or in direct continuous contact with, the protected item. As a result, any "hot-spot" along the protected equipment will be detected.

The detectors themselves are available in a number of designs. One type uses a pair of steel

Steel Wires

Thermoplastic Insulation

Outer Jacket
Approx ⅛" Diam

Fig. 13.54 *In this construction of a linear heat detector, the thermoplastic insulation on the stiff twisted steel conductors melts, allowing the two conductors to touch, and thereby register an alarm.*

wires under tension, held apart by thermoplastic insulation (Fig. 13.54). When exposed to its rated alarm temperature, the insulation melts and the wires come into contact, changing the current flow in the circuit to which they are connected and setting off an alarm. This type is available in a range of alarm temperatures [68°C (155°F), 88°C (190°F), 138°C (280°F)] and can be used in open- or closed-circuit configurations in lengths up to 2500 ft. Meters are available that indicate the exact point of a fault, so that in long runs where overheating rather than fire is detected, countermeasures can be taken rapidly, without the delay caused by having to seek out the fault point. The major disadvantage to this type of detector is that it is nonrenewable; once having faulted, the melted section must be cut out and replaced.

Another type of linear detector is basically a linear thermistor—that is, a device whose electrical resistance varies with temperature (Fig. 13.55). The resistance is monitored at the control panel, and small changes noted. Depending on the thermistor material selected the detection range can be set anywhere from 70°F (21°C) to 1200°F (650°C). This type of linear detector is more sensitive, rapid, and expensive than the preceding design. When linear detectors are installed in direct contact with the protected equipment, they detect surface temperature rather than air temperature and therefore act like early-warning devices. They are used extensively in this fashion to continuously monitor surface temperature of such devices as transformers, switchgear, generators, and all sorts of hazardous equipment.

A third design for linear heat detectors uses optical fiber with a thermoplastic cladding (coating). Optical fiber transmits light by successive internal reflection at the fiber/cladding interface throughout the entire length of the fiber. Excessive temperature anywhere along the length of the fiber causes the cladding to melt and thereby appreciably changes the intensity of light received by the light sensor at the cable's end. This change is recorded and interpreted automatically as either a trouble signal or an overheat alarm signal. This design is available in lengths up to a mile or more. Other linear sensor designs include variable electrical resistance polymers, phase change eutectic salt, and pneumatic tubing. Each type has its particular advantages, drawbacks, and specific applicability.

Conductor

Negative Resistance Coefficient Thermistor (Ceramic)

Metallic outer jacket; O.D. <⅛"

Fig. 13.55 *This linear heat detector is connected to an electrical control that monitors circuit current and thereby resistance. Any appreciable temperature rise along the entire cable length causes a sharp drop in the ceramic material's electrical resistance, which is immediately noted and pinpointed at the control point.*

FIRE PROTECTION

This control cabinet accepts samples of air through inlet tubes at top. If concentration of invisible particles exceeds preset levels, the detector initiates an alarm or activates an extinguishing system, as desired.

(a)

(b) Tamper-proof sampling head

Fig. 13.56 *This very early warning system detects microscopic combustion products in air aspirated from the protected area, using a "cloud chamber" effect.*
(a) *Sampling heads are placed in all spaces to be monitored, including inside cabinets.*
(b) *Tamperproof head housing, applicable to high-vandalism areas and correctional institutions. (Courtesy of Protec Fire Detection, plc.)*

Factors involved in the choice of a linear detector include:

- Physical characteristic and value of the protected elements
- Ease in locating the alarm area
- Requirements for resetting the system after an alarm
- Maintenance requirements
- Cost factors

13.19 Special Types of Fire Dectectors

Several other sophisticated detectors are available for early-warning detection in the incipient and smoldering fire stages. Among the most sensitive is a unit based on the operating principle of a *Wilson cloud chamber* (see Figs. 13.56, 13.57, and 13.58). When microscopic particles such as those produced by the early stages of fire are introduced into a saturated atmosphere (cloud chamber), they act as nuclei around which water vapor condenses to form visible droplets. The detector operates by continuously sampling air from the protected space and setting off an alarm when the cloud chamber indicates the presence of particulate matter.

The continuous sampling procedure and the unit calibration make the system insensitive to dust and other noncombustion particulate matter and generally free of nuisance false alarms. The system's disadvantages are the need for piping and its high price in small installations. For large installations (30 or more detection points), the price is competitive with those of comparable sys-

Fig. 13.57 *Schematic diagram of an installation in a computer facility showing aspirating heads and piping. Note that air is sampled inside cabinets and under the raised floor in addition to sampling in the room itself. (Courtesy of Protec Fire Detection, plc.)*

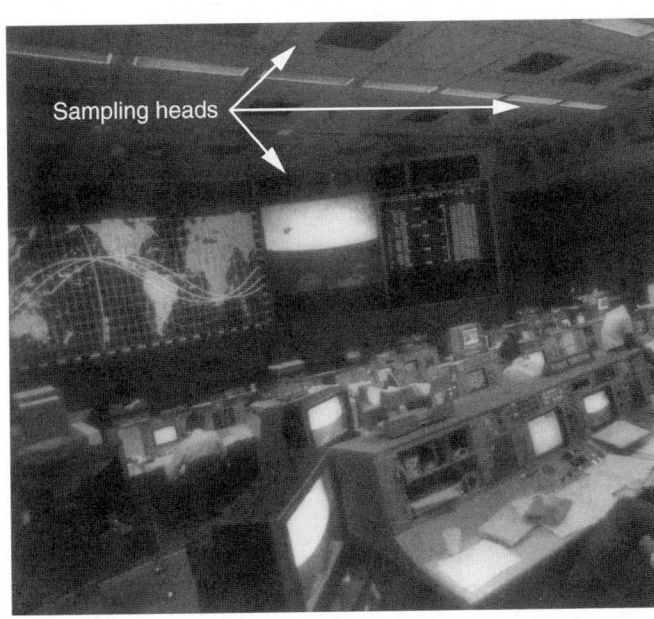

Fig. 13.58 *Typical high-tech fire detection installation, at NASA Mission Control, Houston, Texas. (Courtesy of Protec Fire Detection, plc and NASA.)*

FIRE PROTECTION

tems. This detector is particularly applicable to high-value installations, such as museums, data-processing spaces, libraries, clean rooms, and facility control rooms where its extremely high sensitivity may provide critically important advance warning of an incipient fire.

13.20   False Alarm Mitigation

Smoke detectors of all types (ionization, photoelectric) are subject to false alarms due to moisture and particulate matter in the air. Detectors are by nature threshold devices; raising sensitivity increases the false alarming, decreasing sensitivity shortens the crucial early-warning period. Much of the technological advancement since the late 1980s has been addressed to solving this intractable problem. Technology alone, however, will not entirely eliminate false alarms; the fire alarm system designer and the system user can contribute appreciably to minimizing this annoying and in certain circumstances dangerous occurrence by following some basic guidelines, including the following:

· Appropriate selection of smoke detectors. Choice of an ionization type where a photoelectric type is indicated, and vice versa, almost guarantees false alarms.

· Use of detectors with drift compensation to permit accommodating dirt/high temperature/humidity/ and air current ambient conditions, all of which increase false alarming.

· Use, where necessary, of radio-frequency interference (RFI) filters on detector circuits because line radio noise (spikes, surges) can cause alarms. This factor should be investigated in any installation with a heavy solid-state equipment load.

· Use control panels and detectors that have self-learn software and multispace-type algorithms that match sensitivity patterns to the spaces in which the detectors are installed.

· Use of detector covers in areas where any type of construction work is taking place. This prevents fouling of detectors.

· When installing conventional systems, estab-

lish a regular maintenance and testing program that includes field testing and recalibration of all detectors.

· Avoid detector placement in areas where ambient conditions can cause problems. Where such placement is unavoidable, use compensating detectors, special maintenance, and appropriate verification procedures in the event of an alarm signal. These areas include:

1. Kitchens, laundries, boiler rooms, shower rooms, and other spaces with high humidity and steam.
2. Repair shops, laboratories, and other areas where open flames are used in normal work.
3. Garages and engine test facilities where exhaust gases are present.
4. Smoking rooms and areas near spaces designated as smoking areas.
5. Areas in which heavy accumulation of dust and dirt can be expected.
6. Areas of high air movement, such as near loading docks and exit doors and near the discharge from ducts or registers.

13.21   Manual Stations

In contrast to automatic detectors, the manual station is operated by hand (Fig. 13.59). Manual stations serve to spread the alarm that has already been detected by other means, either human or automatic. In conventional systems, manual stations are either *coded* or *noncoded*. (See Section 13.12 for an explanation of coding). If identification of the exact manual noncoded station operated is desirable, an annunciation panel can be added to the system; this is equivalent to using each station as a noncoded indicating zone. Because of wiring costs, annunciated systems become expensive, and beyond 10 stations, coding should be considered (see Fig. 13.39).

When the system design is such that immediate *building-wide* aural identification of the operated station is necessary, a coded station is used. The station code is received at the control panel, processed, and then transmitted audibly on the system gongs. Not less than three rounds of code, and normally four rounds, are transmitted. The code usually comprises three or four digits (e.g.,

(a)

(b)

Fig. 13.59 (a) *Manual fire alarm station with a break-glass rod. The unit is marked "local alarm," indicating that its operation will result in building-wide evacuation signal but not a city alarm (see Fig. 13.38). Similar units are equipped with multiple sets of contacts for annunciation and other control functions, or with addressable system electronics (Fig. 13.60). Other similar units use a lock-open design that requires a key to reset; this avoids the necessity of replacing the broken glass rod. (b) The manual station, which is normally surface mounted, can be enclosed in a well-marked recessed cabinet if desired for esthetic reasons.*

cabinets because they spoil the decor of the lobby are defeating the purpose of the system. Similarly, placement of bells *inside* hung ceilings because they are unattractive is not only foolish but dangerous, and should *never* be done, regardless of the circumstances. Loss of property and even of life may result from such ill-conceived esthetic considerations.

Addressable manual stations obviate the necessity for a code wheel for each manual station but do not eliminate coding of alarm bells if building-wide identification of the operated station is required. (Obviously, addressable manual stations are only applicable to addressable systems and not conventional systems). The purpose of gong coding is to permit rapid alarm verification by persons scattered about the building. An addressable manual station is identified only at the control panel, from which location a person must be sent to verify the alarm. In a small building, that is not a problem; in a large building, the time required to make such a verification is critical, because a manual alarm is already the result of a verified emergency, unless it is a deliberate false alarm. The decision then as to whether to code the gongs must be made on the basis of data for that particular building. A typical addressable manual station is shown in Fig. 13.60.

Fig. 13.60 *This addressable manual station latches in the down (operated) position when pulled and can only be reset by key. The manual switch can be polled from the control panel to verify that it has indeed been operated, and that the alarm signal is not a circuit malfunction. The station is semi-recessed and measures approximately 4 in. wide by 5½ in. high. (Photo courtesy of Notifier, div. of Pittway.)*

2-3-2) with a pause between the ringing groups and a longer pause between the rounds. The first number may identify the building floor, the second digit the wing, and the third digit the individual station. Establishment of codes is left to the user.

Manual stations are placed in the normal path of egress from a building, so that an alarm may be turned in by a person as he or she exits. It is *imperative*, therefore, that stations be well marked and easily found. Architects who place fire alarm stations in nooks and corners and in camouflaged

Fig. 13.61 *Typical water flow indicator. The unit bolts onto a sprinkler pipe with the paddle inside the pipe. Any water motion deflects the paddle, causing a signal to be transmitted from the microswitch mounted in the box on top of the pipe. (Courtesy of Simplex/Potter.)*

13.22   Sprinkler Alarms

Water flow switches are placed in sprinkler pipelines and operate when a sprinkler head goes off (Fig. 13.61). In electrical terms, a water flow switch is a set of contacts, similar to a temperature detector. It can be used to trip a coded transmitter, setting off a sprinkler code; to show up on a sprinkler annunciator board, called a sprinkler alarm panel; or to act as a zone in a noncoded system. Wiring of water flow switches, or of switches activated by other extinguishing systems, is the same as that for a manual station.

13.23   Audible and Visual Alarm Devices

The type and placement of audible and visual alarms in a (public) building must meet the requirements of NFPA 72 (National Fire Alarm Code) and the ADA (Americans with Disabilities Act, PL 101-336). There are some disagreements between their requirements as of this writing. In such cases the fire alarm system designer in concert with the building architect should settle these differences to the satisfaction of both authorities.

(a) Audible Signals. In buildings where an evacuation signal will be sounded throughout the building (public mode) to alert all of its occupants, the requirements of NFPA 72 are normally used because they are more stringent than those of the ADA. There are several requirements, the most important of which to a building designer requires a minimum alarm sound level of 15 dBA above average ambient sound level, or 5 dB above maximum sound level (whichever is greater) for at least 1 minute. This means that the designer must actually measure or predict, with a high degree of accuracy, the (anticipated) sound levels in the various portions of the building. Average sound level estimates are available in many sources, including the NFPA code itself; maximum sound levels are not. As a result, most designers use a considerable factor of safety to ensure compliance. The upper limit permissible in any area is 120 dBA. See Chapter 26 for technical information on architectural acoustics and sound levels.

The minimum private mode requirement is the greater of 10 dBA above average ambient level or 5 dBA above maximum ambient sound for at least 1 minute. Because this requirement generally applies to an enclosed space that houses the alarm panel or an adjacent space that is continuously occupied, the acoustic design is much simpler than for the public mode alarm.

The codes also specify the minimum required sound levels for sleeping areas and mechanical equipment rooms, in addition to directives regarding the location of audible devices. No purpose would be served in duplicating that information here, particularly because these codes, as others, are subject to constant review and revision. Suffice it to say that selection and placement of audible alarm devices is a highly technical task requiring acoustic analysis of spaces, occupancy, and the characteristics of available audible devices.

(b) Visible Signals. The requirements for these signals, the primary purpose of which are to alert hearing-impaired people, are covered in NFPA 72 Chapter 6, UL 1971, ANSI 117.1, ADA, and the various building code requirements. This is one of the areas in which, at this writing, discrepancies exist between ADA requirements and those of NFPA 72, UL 1971, and ANSI 117.1,

which are in general agreement. All of the codes require that the visible alarm be a strobe light (usually a Xenon flashtube), flashing at less than 2 Hz to minimize the problems for persons with photosensitive epilepsy. This does not, however, solve the problem, because a person viewing the light from more than one strobe (as at a corridor junction) can still be exposed to a higher flash rate unless all strobes are synchronized. Careful strobe placement can also help to minimize this problem.

(c) Strobe Intensity. For nonsleeping areas, the ADA requires a strobe intensity of 75 cd and 50 ft maximum unit spacing, whereas the other standards (UL, ANSI, NFPA) require an intensity dependent on the size of the room. For sleeping areas, the ADA requirement remains the same, whereas the other standards require a 110 cd source if wall mounted and a 177 cd source if ceiling mounted. For corridors, the ADA requirement remains 75 cd at 50 ft strobe spacing, whereas the other standards have a requirement that varies with the corridor length.

Location of the strobes according to NFPA 72, Appendix A, must be such that strobe light can be seen at any point in any space, regardless of the viewers' orientation, with the additional requirement that the maximum distance between strobes will not exceed 100 ft. A typical combination audible/visible fire alarm device is shown in Fig. 13.62.

13.24   General Recommendations

The specific building system descriptions that follow are based on good present practice. Actual design must meet the requirements of the latest edition NFPA standards plus other codes having jurisdiction. Special attention is directed to Article 760 of the *NEC* (NFPA 70). This article was completely rewritten and restructured in the 1996 edition of the code. It now covers fire alarm system wiring in its broadest sense (i.e., it now refers to "all circuits controlled and powered by the fire alarm system," including "fire detection and alarm notification, guard's tour, sprinkler . . . systems, . . . safety functions, damper control and

Fig. 13.62 Combination audiovisual fire alarm signal is capable of eight different warning tones that can be field selected. The 15 cd strobe flashes once every 1.5 sec (2/3 Hz). The unit measures 4-in. square and can be mounted surface, semiflush, or flush. (Photo courtesy of Notifier, div. of Pittway Corp.)

fan shutdown"). In office buildings the fire alarm system includes elevator capture and intercom, which, if powered and controlled by the fire alarm system, fall under Article 760.

13.25   Residential Fire Alarms

Refer to NFPA 72, *Household Fire Warning Equipment,* and NFPA 101, Chapter 22, for detailed requirements. The system should provide sufficient time for the evacuation of the residents and for appropriate countermeasures to be initiated. The elements of the system are the various alarm-initiating devices, the wiring and control panel, and the audible alarm devices.

A good basic system requires the following:

· A listed smoke detector: outside and adjacent to each sleeping area, in each sleeping room, at the head of every stair, with at least one on every level, including the basement. Combined smoke/heat detectors in boiler room, kitchen, garage, and attic. Listed heat detectors in the attic, kitchen, and boiler room are frequently set at 185°F because of high ambient temperatures. Other units are set at 135°F.

· Ensure that an alarm in any detector produces an alarm in all audible and visual units.

· Control unit (central panel) annunciated to show the location of an alarmed device and arranged to shut off oil and gas lines and the attic fan (to prevent spread of smoke). Also, it should turn on lights both inside and out, operate an automatic dialer to ring a neighbor's phone or a commercial central station, and give a distinctive alarm sound when the phone is answered. An outside bell and some other device that transmits an alarm outside the residence is important.

· Backup power for the system—a supervised storage battery with trickle charger.

· Wiring of all devices connected to a system control unit to be on supervised circuits that will sound a trouble alarm in the event of a fault. Trouble alarm to be distinct from fire alarm.

Additional requirements for detectors, alarms, power supply, supervision signals, and other relevant items including special requirements for fire alarm systems that are combined with other signal systems are found in the above-referenced standard, current edition.

13.26 Multiple-Dwelling Alarm Systems

The exact requirements for location and type of detectors and their action are specified in NFPA 101 and NFPA 72, and no purpose is served in repeating them here, particularly because they are constantly updated. The title of multiple dwellings includes apartment houses, dormitories, hotels, motels, and rooming houses. The prime directive for these structures in the event of a fire emergency is early warning and orderly egress, with consideration of the fact that a fire emergency may occur when most of the building occupants are asleep. Helpful design guidelines are as follows:

· Audible/visual alarms must be positioned so that all sleeping persons, including those with hearing and/or sight impairment, will be wakened. Code requirements for audible lev-

els are minimum and are based on average spaces and furnishings. Unusually large or oddly shaped spaces should be designed individually.

· Consider the possibility of living rooms being used as sleeping areas on a regular basis.

· Provide smoke detection in corridors, service spaces, and utility and storage rooms.

· Battery-powered detectors may not be used because, unlike homeowners, apartment dwellers rely on the building management for all building services, and periodic battery checks and replacement might not be carried out.

· Provide standby power to all fire alarm circuits.

· All alarms should be identifiable, either by addressing or by annunciation. Annunciators should be located at the system control panel in the building management office, or at the lobby desk in the case of hotels and dormitories. In all buildings, a lobby annunciator for the use of firemen is advantageous.

· An alarm light over the door of each apartment or suite to indicate alarm location is desirable. This is particularly important if the central panel only provides zone annunciation.

· In high-rise buildings an emergency voice/alarm communication system should be provided.

13.27 Commercial and Institutional Building Alarm Systems

The requirement of these buildings are so varied that no specific recommendations can be made. A few suggestions, however, are in order:

· Presignaling, where permitted, is recommended for buildings that do not readily tolerate an evacuation alarm.

· In schools, particularly of the elementary grades, rapid orderly evacuation of the building is the primary requirement. Consideration must be given to the uniqueness of the sound of the fire alarm gongs to allow no possibility of confusion with program gongs, where the latter are used. Also, because regular fire-drills are mandatory in most schools, the system must be arranged to provide this facility.

· Public buildings should have an auxiliary alarm connection to the fire department.

· For medium-size buildings, an accurate cost estimate frequently indicates an advantage to addressable analog systems over addressable systems because of high detector maintenance costs of the latter. Similarly, addressable systems frequently have an economic advantage over hard-wired systems because of high labor costs for wiring.

Figs. 13.63, 13.64, and 13.65 show a few modern system control panels.

13.28   High-Rise Office Building Fire Alarm System

Sad experience has demonstrated that high-rise buildings, once thought to be "fire-proof," are emphatically not so. Indeed, due to their size, they have particularly severe fire protection problems, one of which is reliable communications during fire emergencies. As a result, fire codes now require that high-rise buildings be equipped with an emergency voice/alarm communications system. The system provides full control of transmission and building-wide distribution of all tones, alarm signals, and voice announcements on a selective or all-call basis. Specifically, it makes possible:

· Two-way active communications from the firefighters' control command post (usually in the lobby) to at least one fire station per floor, all mechanical equipment rooms, elevator machine rooms, and air-handling (fan) rooms.

· Distribution to selected areas of alert tones, signals, and prerecorded messages on independent channels. (Obviously, a complete system of loudspeakers covering all areas of the building is an integral part of the system).

· Communication with fire department or central station.

In addition to communication functions, the (lobby) fire command post provides:

Fig. 13.63 *Conventional hard-wired eight zone (max.) solid-state fire alarm control panel for supervised class A or Class B circuits. A single annunciator light indicates the alarmed zone. Alarm verification for each zone is provided. Both audible and visible alarms can be used on two alarm-signal circuits. The audible evacuation signal is a slow (20 BPM) or fast (120 BPM) march time. Note the two standby batteries at the bottom of the cabinet. The unit measures approximately 16 in. wide and 17 in. high and is designed for surface mounting on 16 in. o.c. studs. (Photo courtesy of Simplex.)*

Fig. 13.64 *Control panel for a hard-wired conventional system that can accommodate up to 36 detection zones. The panel has a 2-line, 80-character alpha-numeric display that indicates zone alarm, trouble signals and diagnostics information. The panel, which is 24-in. square, is microprocessor controlled and field programmable. A separate annunciator can be readily connected. (Photo courtesy of Simplex.)*

Fig. 13.65 *System block diagram for a compact addressable analog fire system for a medium size building. The panel, which measures approximately 15 in. wide by 16 in. high by 6 in. deep, has a capacity of 99 analog detect and 99 monitor/control modules. Integral programming provides drift compensation and sensitivity adjustment for analog detectors, detector maintenance (dirt accumulation) alert, automatic cyclic sensitivity adjustment, plus standard diagnostics. Custom graphics and the video display can be arranged to display floor plans, device locations, and the like. Relay control can be used for control of building air handling unit, elevator interlock, and security/access system interlock. Notification circuits connect to central station, remote station, proprietary, or municipal alarms. (Diagram courtesy of Notifier, div. of Pittway Corp.)*

· Visual display (annunciation) of all fire alarm devices, including sprinkler valves and waterflow indicators.

· Fire pump status indication.

· Controls for any automatic stair door unlocking system (security access system).

· Emergency generator status.

· Elevator location indicators plus operation and capture controls.

· Control of smoke control devices (doors, dampers, etc.)

· Means for testing all circuits and devices.

The exact equipment supplied depends on the building and the local fire code. The outstanding characteristics of this system are the communica-

Fig. 13.66 *Large addressable analog ("intelligent") fire alarm control panel, suitable for network connection, computer interface, and operation in conjunction with an Emergency Voice/Alarm communication system of the type used in multi-story office buildings (see Fig. 13.67). The unit can accommodate 990 analog detectors and an equal number of monitor/control modules. It is illustrated with a 96-point integral annunciator, and additional annunciation can be connected. The integral control panel provides for polling, diagnostics, sensitivity control of all detectors, automatic detector drift compensation, alarm verification, front-panel programming, and options for system connections as shown in the block diagram of Fig. 13.65. (Photo courtesy of Notifier, div. of Pittway Corp.)*

Fig. 13.67 *A multiplex emergency voice alarm/ communication system panel is illustrated that is used in large high-rise buildings, in conjunction with the fire alarm system panel (Fig. 13.66). The panel controls and feeds a multiplexed voice alarm system and, in addition, strobe light circuits and a complete firemen's telephone intercom. It provides evacuation, alert, page, and auxiliary signals to all building speakers, selectively. An integral annunciator indicates which audio devices are connected to which of the three available audio channels. All audio risers are fully supervised and power limited. (Photo courtesy of Cerberus Pyrotronics.)*

tions system, the visual display of alarm locations, and the remote control of air-handling equipment (Figs. 13.66 and 13.67).

13.29 Industrial Facilities

In addition to manual stations at points of egress, these buildings also use:

· Temperature and smoke detectors in storage areas and laboratories.

· Smoke and flame detectors in record rooms and continuous process laboratories.

· Water flow switches on all sprinklers.

The annunciators, control panel, and alarm register are best placed in the guardroom. If none is available, an auxiliary or remote-station circuit should be added to allow remote monitoring.

Because of the high ambient noise level in many plants, horns are substituted in such areas for bells and gongs, which might be inaudible.

In summary, the specific occupancy recommendations given previously are representative of good practice at the present time. However, codes are constantly being changed in this particularly sensitive field, and they vary with locale. In actual design situations, all codes and regulations having jurisdiction must be complied with.

FIRE PROTECTION

PART V
ELECTRICITY

Electricity is the most prevalent form of energy in a modern building. It supplies not only electric outlets and electric lighting, but also the motive power for ventilation, heating, and cooling equipment; traction power for elevators and material transport; and power for all signal and communications equipment.

An electric power failure can paralyze a facility. If properly designed, the facility returns to partial functioning by virtue of emergency equipment that furnishes some part of the facility's electricity needs for a limited time.

Given this complete dependence on electric power for normal functioning, it is apparent that planners must be familiar with normal electrical systems. Chapter 14 reviews basic relationships in electric circuits with emphasis on power, energy, energy costs, and methods of energy managements and electric load control. Chapter 15 introduces the reader to the concept and basis of electric equipment ratings and capacity, and continues with a description of modern wiring systems and their components. Chapter 16, essentially a continuation of Chapter 15, describes service, utilization, and emergency/standby power equipment. Also discussed are energy conservation considerations and economic factors. Chapter 17 draws on information given in the three preceding chapters to demonstrate straightforward design methods for building electrical systems.

14

PRINCIPLES OF ELECTRICITY

Historically, usable energy was most often produced by burning a fossil fuel such as coal or oil. The resultant heat energy was used directly as heat and light or converted by machines into motion. Only since the end of the 19th century, however, has this heat in turn been used to create another form of usable energy—electricity. Even the recent partial substitution of nuclear for fossil fuels has affected only the heat production portion of this process. Beyond that point, the heat is utilized in the same manner to drive generators that produce electricity. It is well to remember, therefore, that in terms of natural resources, electricity is an expensive form of energy because the efficiency of the overall heat-to-electricity conversion, on a commercial scale, rarely exceeds 40%.

14.1 Electric Energy

Electricity constitutes a form of energy itself that occurs naturally only in unusable forms such as lightning and other static discharges or in natural galvanic cells, which cause corrosion. The primary problem in the utilization of electric energy is that, unlike fuels or even heat, it cannot be stored and therefore must be generated and utilized in the same instant. This requires an entirely different concept of utilization than, for example, a heating system with its fuel source, burner, piping, and associated equipment.

The bulk of electric energy utilized today is in the form of *alternating current* (ac), produced by ac generators, commonly called *alternators*. *Direct current* (dc) generators are utilized for special applications requiring large quantities of dc. In the building field, such a requirement was once almost universal for elevator work because of the ease with which dc motors can be speed controlled. Today, however, because fine speed control of ac motors is practical, it is the technique of choice in modern elevators, with attendant energy sav-

ings and a reduction in machinery space requirements. Smaller quantities of dc, furnished either by batteries, photovoltaic (PV) equipment, or by rectifiers, are utilized for telephone and signal equipment, controls, and other specialized uses.

14.2 Unit of Electric Current— The Ampere

Electricity flowing in a conductor is called *current*, which is measured *amperes*, abbreviated *amp*, *amps*, or simply *A*. When current is used in an equation, it is usually represented by the letter I or i.

It is convenient to establish an analogy between electric systems and mechanical systems as an aid to comprehension. Current is a measure of flow, and as such would correspond to water flow in a hydraulic system (Fig. 14.1). The analogy is not complete, however, because in the hydraulic system the velocity of water flow varies, whereas in the electric system the velocity of (electric) propagation is constant and may be considered instantaneous; hence the need to utilize the electric energy the instant it is produced.

14.3 Unit of Electric Potential—The Volt

The electron movement and its concomitant energy, which constitutes electricity, is caused by creating a higher positive electric charge at one point on a conductor than exists at another point on that same conductor. This difference in charge can be created in a number of ways. The oldest and simplest method is by electrochemical action, as in the battery.

In the ordinary dry cell, or in a storage battery, chemical action causes positive charges (+) to collect on the positive terminal and electrons [i.e., negative charges (−)] to collect on the negative

Fig. 14.1 *Electric–hydraulic analogy. The circuits show that voltage is analogous to pressure, current to flow, friction to resistance, wiring to piping, and switches to valves.*

terminal. Assume for the moment that nothing is connected to the battery terminals. There is a tendency to flow between the electrified particles concentrated at the positive and negative terminals. *Potential difference* or *voltage* is the name given to this tendency, or force. It is analogous to pressure in a hydraulic or pneumatic system. Just as the pressure produced by a pump or blower causes water or air to flow in a connecting pipe, so too the potential (voltage) produced by a battery (or generator) causes current to flow in a conductor connecting the terminals between which a voltage exists. The higher the voltage (pressure), the higher the current (flow) for a given resistance (friction; Fig. 14.2). Other means of producing voltage, both direct (dc) and alter-

nating (ac), are discussed in Section 14.8. The unit of voltage is the volt, abbreviated V or v.

14.4 Unit of Electric Resistance— The Ohm

The flow of fluid in a hydraulic system is resisted by friction; the flow of current in an electric circuit is resisted by resistance, which is the electrical term for friction. In a dc circuit this unit is called *resistance* and is abbreviated R; in an ac circuit it is called *impedance* and is abbreviated Z. The unit of measurement is the *ohm*. (It is interesting to note that the scientific convention of naming units after persons whose work is closely related to the field is followed here also. Thus the units ampere, volt, and ohm are derived from André Ampere, Alessandro Volta, and George Ohm.)

Materials display different resistance to the flow of electric current. Metals generally have the least resistance and are therefore called *conductors*. The best conductors are the precious metals—silver, gold, and platinum—with copper and aluminum only slightly inferior. Conversely, materials that resist the flow of current are called *insulators*. Glass, mica, rubber, oil, distilled water, porcelain, and certain synthetics exhibit this insulating property and are therefore used to insulate electric conductors. Common examples are rubber and plastic wire coverings, porcelain lamp sockets, and oil-immersed switches.

14.5 Ohm's Law

The current I that will flow in a dc circuit is directly proportional to the voltage V and inversely proportional to the resistance R of the circuit. Expressed as an equation, we have the basic form of Ohm's law:

$$I = \frac{V}{R} \qquad (14.1)$$

In ac circuits, the same relation holds true except that instead of dc resistance we use the ac impedance. Ohm's law is frequently written in another form, that is,

$$V = IR \qquad (14.2)$$

which states the mathematical relationship that voltage is the product of current and resistance.

Fig. 14.2 *Current flows in the electric circuit as a result of the voltage (potential difference) V that exists between the terminals of the battery. By convention, current direction is from positive (+) to negative (−) in the circuit (and from − to + inside the battery).*

This form has no logical basis; therefore, we recommend that the reader remember the form of equation 14.1. It clearly states the physical situation, that is, as a result of voltage V, a current I is produced that is proportional to the electric voltage (pressure) V, and inversely proportional to the electric resistance (friction) R.

Example 14.1 illustrates the application of Ohm's law. The example chosen is applicable to both ac and dc because the load device is purely resistive. When a load is purely resistive, resistance and impedance are equal. This is more fully explained in the subsequent discussion on alternating current.

EXAMPLE 14.1. An incandescent lamp having a hot resistance of 66 ohms is put into a socket that is connected to a 115-V supply. What current flows through the lamp (after it reaches operating temperature)?

SOLUTION.

$$I = \frac{V}{R} = \frac{115 \,(\text{volts})}{66 \,(\text{ohms})} = 1.74 \, A$$

(These figures correspond to a normal 200-W general service incandescent lamp.)

Hot resistance is mentioned in the previous example because the electrical resistance of some materials changes with temperature. A typical example of this is the tungsten-filament lamp that when first turned on cold takes, for a fraction of a second, 10 to 15 times the steady-state hot filament current.

14.6   Circuit Arrangements

The two basic electric circuit arrangements are *series* and *parallel*. These concepts are the same for both dc and ac. As previously, we use purely resistive circuits so that circuit calculations are applicable to both dc and ac. In other than purely resistive circuits, ac circuit calculations are different and much more complicated than their dc counterparts.

(a) Series Circuits. In a series arrangement the elements are connected one after the other, that is, in series. Thus, resistances and voltages add. This is indicated graphically in Fig. 14.3. An

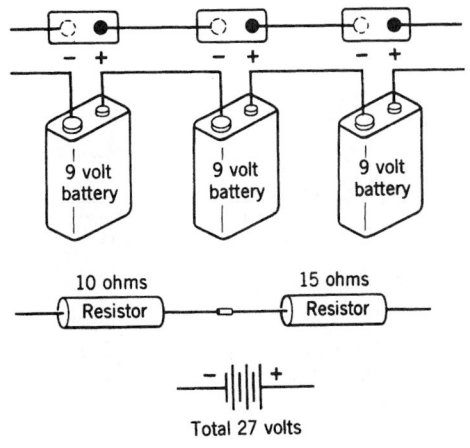

Fig. 14.3 *Physical and graphic representation of series connection of batteries and resistors.*

electric circuit may be defined as a complete conducting path that carries current from a source of electricity to and through some electrical device (or load) and back to the source. A current can never flow unless there is a complete (closed) circuit. Due to the arrangement of components in a series circuit, *the current is the same in all parts of the circuit.* A somewhat more complicated circuit is shown in Fig. 14.4 and is analyzed in Example 14.2.

It is customary to refer to connection points on such wiring diagrams by letters as *a, b, c, d,*

Fig. 14.4 *A series circuit always contains a voltage source (here, the battery), and a load (R). In this circuit we have also shown the resistance of the connecting wires, for accuracy. Such resistances are normally neglected because they usually are very small as compared to the load resistance.*

Fig. 14.5 *In a parallel connection the flow divides between the branches, but the pressure is the same in each branch.*

and so on. The battery voltage may then be called $V_{ab} = 12$ V; the voltage across the load resistance, $V_{cd} = 11.5$ V; and the resistance of the two wires $r_{bc} + r_{da} = 0.04$ ohm. The positive and negative terminals of the battery are always shown.

EXAMPLE 14.2. The battery in Fig. 14.4 is rated at 12 V; the total line resistance (both wires) is 0.04 ohm; the battery internal resistance is 0.01 ohm; and the load resistance is 1.15 ohms. Determine (1) current flowing in the circuit, and (2) the voltage across the load (V_{cd}).

SOLUTION. (a) The current flowing is

$$I = \frac{V}{R} = \frac{V_{ab}}{r_{ab} + r_{bc} + r_{cd} + r_{da}}$$

$$= \frac{12}{0.01 + 0.02 + 1.15 + 0.02} = \frac{12}{1.2} = 10\,A$$

(b) The voltage drop across the load is

$$V_{cd} = I \times R_{cd} = 10 \times 1.15 = 11.5\ V \quad \blacksquare$$

Series circuits have very limited application in building wiring.

(b) Parallel Circuits. When two or more branches or loads in a circuit are connected between the same two points, they are said to be connected in *parallel* or *multiple*. Such an arrangement and its hydraulic equivalent are shown in Fig. 14.5. From the circuit of Fig. 14.6 it should be apparent that the voltage across each load is the same, but the current in each load (branch) depends on the resistance of that load. Parallel loads, in effect, constitute separate circuits. From this we conclude that in this arrangement, the total current in the circuit is the sum of the individual

Fig. 14.6 *Note that loads connected in parallel are equivalent to separate circuits superimposed into a single circuit. Each load acts as an independent circuit unrelated to, and unaffected by, the other circuits.*

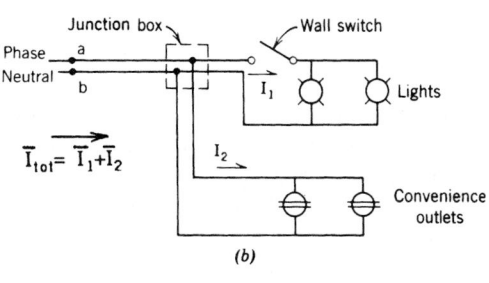

$$\overline{I}_{tot} = \overline{I}_1 + \overline{I}_2$$

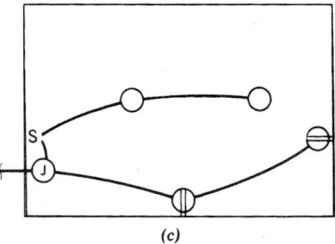

Fig. 14.7 *Parallel groupings of ceiling light outlets and wall outlets are in turn connected in parallel to each other. A circuit is shown* (a) *pictorially,* (b) *schematically, and* (c) *as on an electrical floor plan. [The horizontal bar over the currents* I_1 *and* I_2 *in* (b) *signifies the use of vectorial addition. See Example 14.6.]*

currents flowing in the branches—that is,

$$I_T = I_1 + I_2 + I_3$$

Refer to Fig. 14.6. Notice from the numbers shown there that the total current flowing in the circuit is the sum of all the branches, but that the current in each branch is determined by a separate Ohm's law calculation. Thus, in the 10-ohm load, a 12-A current flows, and so forth.

The parallel connection is the standard arrangement in all building wiring. A typical lighting and receptacle arrangement for a large room is shown in Fig. 14.7. Here the lights constitute one parallel (multiple) grouping, and the convenience wall outlets constitute a second parallel grouping. The fundamental principle to remember is that loads in parallel are additive for current and that each has the same voltage across it.

One additional point is important to appreciate. Remember from Ohm's law that current is inversely proportional to resistance. Thus, as resistance drops, current rises. Now look at the circuit of Fig. 14.7. Normally, that circuit carries the (vector) sum of the currents in the two branches. (Vectorial addition is discussed in Section 14.22, Ex. 14.6c). If by some mischance, such as deterioration of the wiring insulation, a connection appears between points *a* and *b*, the circuit is *shortened* so that there is no resistance in parallel branch *ab*. The current rises instantly to a very high level, and we have a *short circuit*. If the circuit is properly protected, the fuse or circuit breaker opens and the circuit is de-energized. If not, the heat generated by the excessive current will probably start a fire.

14.7 Direct Current and Alternating Current (dc and ac)

Whenever the flow of electric current takes place at a constant time rate, practically unvarying and in the same direction around the circuit, it is called dc. Figure 14.8 shows dc voltages of 1.5 V positive polarity, and 1.0 V negative polarity.

Whenever the flow of current is periodically varying in time and in direction, as indicated by the symmetrical positive and negative loops, or *sine waves*, in Fig. 14.9, it is called an *alternating current*. The distance along the time axis spanned by a positive and a negative ac loop is called *one cycle*. The number of such cycles occurring in one second is known as the *frequency* of the ac current. Modern ac systems in the United States operate at a frequency of 60 cycles per second, or 60 *Hertz* (after H. Hertz). This means that current at 60 hertz (abbreviated Hz) is delivered to the consumer. In Europe and much of Asia, 50 Hz is standard.

An ac circuit differs from a dc circuit in a number of important respects and, because normal current supply is 60 Hz ac, it is important to

Fig. 14.8 *Graphic representation of dc voltages with positive and negative polarity.*

Fig. 14.9 *Graphic representation of a pure, single frequency alternating current (ac). The figure shown in a sine wave. Note that a complete cycle includes both the positive and negative loops. The number of full cycles in one second is defined as the frequency of the current (or voltage).*

understand ac terminology and usage. Instead of resistance, the corresponding parameter in an ac circuit is *impedance*, which is also measured in ohms.

Depending on the circuit load, impedance can be markedly different from the dc resistance. For an ac circuit, ohm's law is

$$I = \frac{V}{Z} \qquad (14.3)$$

where Z is the symbol for impedance. We will not go into ac circuit calculations primarily because such calculations are not especially useful to the reader. What *is* useful and important is an understanding of power and energy in both dc and ac circuits. These are discussed in the following sections.

14.8   Electric Power Generation—dc

With respect to generation of large amounts of power, piezoelectric and thermoelectric effects can be ignored. Photovoltaic (PV) power using solar cells is becoming increasingly important as a practical source of electricity for isolated small uses. This subject is discussed in sections 14.9 through 14.19. This leaves the battery and the dc generator as practical sources of dc current. Because the dc generator is in reality an ac generator with a device (commutator) attached that rectifies the ac to dc, the battery is still the major direct source of direct current. (There are some special types of generators that produce direct

current *directly*, but their use to date has been extremely limited.) A discussion of the application of batteries for emergency and standby power supply can be found in section 17.21.

Another source of dc power is rectification of ac. This can be accomplished on any desired scale to provide as much dc power as there is available ac power. The principal applications of dc in older buildings is for elevator power and in all buildings, to a lesser extent, for standby power. Small amounts of dc are also used for controls and telephones.

PHOTOVOLTAIC SYSTEMS

14.9   Photovoltaic Power—General

Photovoltaic electric power systems, which until the mid- to late 1980s were little more than a curiosity in the United States because of their very limited applicability and discouraging economics, have recently begun to make a noticeable impact on the commercial electrical power scene. This has come about for several reasons, the most important of which are federal government financed research and development, and state and federal sponsored legislation in the electric utility field. The former, in the form of the photovoltaic manufacturing technology project (PV Mat) involving the Department of Energy (DOE) and 20 private companies, launched in 1990, plus the more recent (1992) building opportunities in the United States for photovoltaics (PV:bonus) cost-sharing program also sponsored by the DOE, resulted in the reduction of PV module costs from upwards of $5.00 per watt to about $1.50–2.50. In addition, it spurred the development of new PV module materials, construction techniques and product forms, increases in PV module efficiency, and an overall sharp decrease in the cost of power produced from about $0.50–$1.00 per watt to about half that amount, depending on system configuration. The latter, legislative impetus, has taken the form of utility deregulation and specific legislation requiring utilities to purchase power produced by (small) private installations.

The 1978 federal law, public utilities regulatory policy act (PURPA) required electric utilities to buy electric power at a price equal to the costs it itself avoids by not having to produce that power. The utilities generally interpret that "avoided cost" as fuel cost only, without considering other avoided costs such as additional plant construction. As a result, energy is bought under this law at about $0.03/Kwh by the same utilities that sell energy at $0.08-0.15/Kwh. This spread obviously does not encourage the construction of grid-connected PV installations (see later). Subsequently, many states adopted *net-metering* laws that ameliorated this price unbalance. In this arrangement,

the utility buys power during peak PV generation periods (see Fig. 14.10) at the same rate as it sells. At the end of a billing period (usually 1 month), the imbalance between utility purchased and customer purchased power is paid for at the above rates (i.e., if the customer used more power than he generated, it is paid for at the conventional utility rate). Conversely (and this is unusual), if the customer generated more power than he used, the utility would buy it at the "avoided cost rate."

The net metering arrangement is attractive economically to both utility and user. The utility buys power that offsets its peak afternoon load (see Fig. 14.10), which is particularly severe in warm weather, and sells power in its off-peak morning and evening periods. The residential customer sells power at the PV installation's peak output, which is also usually a period of low user consumption, and buys it back at the same price in the evening, when PV output is low but power demand is high. Thus, the arrangement obviates the necessity for an expensive battery installation that in a stand-alone system provides power during periods of low PV energy production. This advantage, combined with other advantages such as having a nonpolluting power source, have led

Fig. 14.10 *Chart representing blocks of energy produced and used by a residence equipped with a PV array. Daily energy use is shown uniform over the 24-h period as an average, to indicate overall quantity and not levels of use during the day. The bell-shaped PV energy production curve is centered on solar noon and extends from sunrise (solar 6:00 AM) to sunset (solar 6:00 PM). The area of the bell curve above the average consumption line indicates energy purchased by the utility in a grid-connected system or energy stored in batteries in a stand-alone system. [From Harris, Miller, and Thomas,* Solar Energy Systems Design *(1985) p. 715; reprinted with permission of John Wiley & Sons, Publishers.]*

to a large increase in grid-connected installations. Indeed, the attractiveness of having a "green," environmentally beneficial electric power supply is so great that a number of electric utilities have instituted a (primarily residential) PV installation program in which the utility installs and maintains a PV system on the customer's premises (usually on the roof) for which the customer pays a nominal *surcharge* to his normal electric power bill. The advantages of net-billing are available only to customers who initiate and maintain a PV system themselves.

Fig. 14.11 *Typical direct-connected PV installation feeds electric power to a positive-displacement pump that pumps water into an elevated tank.*

14.10 PV Systems—Types and Application

(a) Stand-Alone Systems. There are two basic types of PV systems: *stand-alone* and *grid-connected*. The system names well describe their essential characteristic. A stand-alone system is just that; it is isolated from any outside electrical connections and is designed to do a carefully defined specific job. Stand-alone systems are the oldest PV installations because they were essentially the only solution to the problem of electrifying a remote and/or unattended load. Typical of these installations were those supplying sign lighting, railroad crossing lights, unattended pumps, lighthouses, unattended navigational aids, microwave repeaters, motor homes, sailboats and yachts, isolated small residences, and the like. The common characteristic of all these installations is the impossibility or impracticality of feeding them from a utility grid. Where the load became larger than could be supplied by a practical solar array from the physical or economic viewpoints, and particularly when the peak load was periodic rather than continuous, a fuel-powered generator was added to the system. Such a system is known as a *hybrid stand-alone system* and is used frequently for remote residences where the connection charge levied by the local utility is so large that a PV system is an economical and environmentally friendly solution to the electrification problem. Schematic diagrams of stand-alone system types are shown in Figs. 14.11, 14.12, and 14.13.

The difference between the systems shown in Figs. 14.11 and 14.12 is the use of an energy stor-

age medium; in most cases, a battery. The *direct-connected* system of Fig. 14.11 is applicable only to loads that tolerate variable power levels ranging from zero at night to maximum at solar noon. The most common of these is water pumping, using a dc motor to drive a positive displacement (piston) pump. Such installations are used frequently to fill elevated water tanks because the slow, interruptible fill rate does not adversely affect the water system's usability.

The vast majority of stand-alone systems are assembled as shown in Fig. 14.12, using a storage

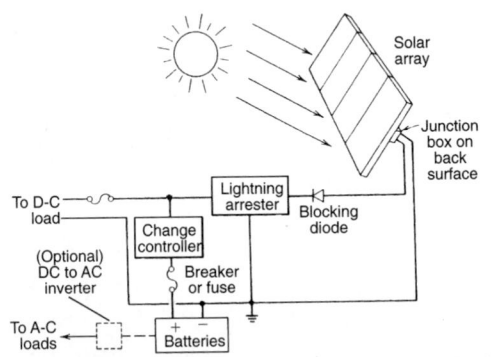

Fig. 14.12 *Schematic of the most common type of stand-alone PV installations. The blocking diode prevents reverse current flow when the battery voltage exceeds that of the PV array. The charge controller prevents overcharging of the battery and excessive current drain. If ac loads are present, a dc to ac inverter is added, sized as required. [From Harris, Miller, and Thomas,* Solar Energy Systems Design *(1985); reprinted with permission of John Wiley & Sons, Publishers.]*

Fig. 14.13 *One possible arrangement of a hybrid stand-alone PV system. A diesel-, gasoline-, or propane-driven ac generator (alternator) supplies large intermittent ac-only loads, whereas the PV array and battery supply dc loads and continuous ac loads through an inverter. During unusually long cloudy periods the alternator can charge the battery and, via an emergency connection, can also supply the regular ac loads. Switches, interlocks, and blocking diodes are omitted for clarity. Other arrangements are possible, using a larger inverter to supply an all-ac system, with the alternator supplying intermittent heavy ac loads and all loads during low battery periods.*

battery to store excess energy produced during peak insolation hours for use during periods of partial darkness (cloudy days) and full darkness (nighttime). With proper component selection, as is discussed in detail in subsequent sections, stand-alone PV systems can be used to adequately supply remote year-round residences, small mercantile establishments, and other off-grid loads.

A stand-alone system may contain a dc–ac inverter if any of the loads require ac that *cannot* be supplied by a built-in inverter. Desirable equipment such as fluorescent lighting is available with integral inverter ballasts. Here again, the choice of whether to use a single central inverter or distributed smaller inverters is, in large measure, economic. Some kitchen appliances and power tools are not easily available in dc or battery-powered format, although that situation too is rapidly changing and bears careful consideration during system design.

The hybrid stand-alone system shown in Fig. 14.13 is similar to the stand-alone system of Fig. 14.12, with the addition of a (small) ac generator. This type of system, as noted previously, is used where a heavy ac load such as a clothes washer or any other relatively large ac motor-driven appliance is in use. Where these loads are continuous, such as would be the case with refrigeration and air-conditioning equipment, a larger PV system should be considered. A continuously running engine–generator set is a source of noise and air pollution, and the necessity for fuel storage for such an installation makes it contradictory in spirit and in fact to a completely silent, pollution-free, solar-powered PV installation.

(b) Grid-Connected Systems. Prior to the advent of net-metering, the number of stand-alone PV systems in areas served by electric utility grids was small indeed, because of the overwhelming economic advantage of purchased power. Even today, with net-metering, the payback period of initial investment seldom justifies the installation on purely economic grounds, although the continuous development work in PV materials may soon change that situation.

In a stand-alone arrangement, over the life of the system battery costs can exceed that of the PV array. Lead–acid batteries are relatively cheap, but have a short life; nickel–cadmium batteries have a longer life, but are much more expensive. By replacing the batteries that furnish energy storage and carry-over in a stand-alone system with a utility tie that effectively serves the same purpose (see Fig. 14.10), and by taking advantage of the economies available from building-integrated PV elements [see subsection 14.10 (*c*) below] in new construction, an economical PV system with reasonable payback can be constructed. A schematic diagram of one such system arrangement is shown in Fig. 14.14.

The inverter shown is equipped with a safety cutoff that instantly disconnects the PV system from the grid if a utility outage is sensed. This is a requirement of the electric utilities to ensure that their maintenance personnel are not endangered by PV electric power being fed into an ostensibly de-energized line.

Fig. 14.14 *Schematic drawing of a grid connected PV array, arranged for net metering. Power from the array is converted from dc to utility grade power (constant voltage and frequency; low harmonic content) by the inverter and fed to the utility connection, through a kWh meter that measures energy produced and sold to the utility. The second (utility company) meter measures total power used by the structure. The difference in meter readings represents the positive or negative (credit) utility billing.*

The inverter creates utility-grade ac from the PV array dc and feeds it to the utility through a Kwh meter. The power company supplies house power through a second Kwh meter, as shown. In net-metering, the difference between the readings of the two meters is either billed to the consumer at the normal utility rate or is credited to the consumer at the *avoided cost* rate, depending on which meter reads higher. The consumer will notice no difference whatever in the electric service, except in the low, or possibly even negative, electric utility bills.

More recently, to encourage wider use of non-polluting PV power, some utilities have permitted the installation of small, individual, PV modules in conventional existing buildings. These units plug into convenience outlets in the building, supplying power to the building itself. Any excess electrical energy not required by the structure is fed back into the utility via their reversible energy meter. This arrangement is shown schematically in Fig. 14.15. Each PV panel is equipped with a built-in, high-quality inverter that produces utility grade power, and is equipped with the safety devices described previously. A common commercially available unit of this type is rated 100 W peak, 85 W nominal, 120 V 60 Hz, measures approximately 2 ft × 4 ft × 4 in. and weighs 25 lb (11 kg). The advantage gained with these units is that a user can begin with one or two units and then expand as desired, without a relatively high first cost, and without a centralized installation, requir-

ing professional construction and electrical personnel. Some power companies require separate disconnect switches for each PV unit rather than relying only on physical disconnection of the unit plug from the wall outlet.

(c) Economic Considerations. These can be complex if they are the deciding factor in the choice between PV and full grid connection (normal electric service). Among the cost factors are:

· Cost of grid connection.

· Cost of power from the grid.

· Life-cycle cost of each component of the system over the life of the entire installation. This is necessary because the life span of individual components varies sharply; PV modules last about 20 years, batteries about 5 years, electrical components 10 to 20 years or more, PV module mounting means 15 to 25 years, tracking mechanisms (if used) 5 to 10 years, and so on.

· Life-cycle maintenance costs (and time involved) for each item.

· Financing costs.

· Cost *credits* for PV systems including such items as use of the system battery as a UPS source for computers and peripherals, cost of grid power service interruption, and construction cost credits when PV modules are used in lieu of building elements.

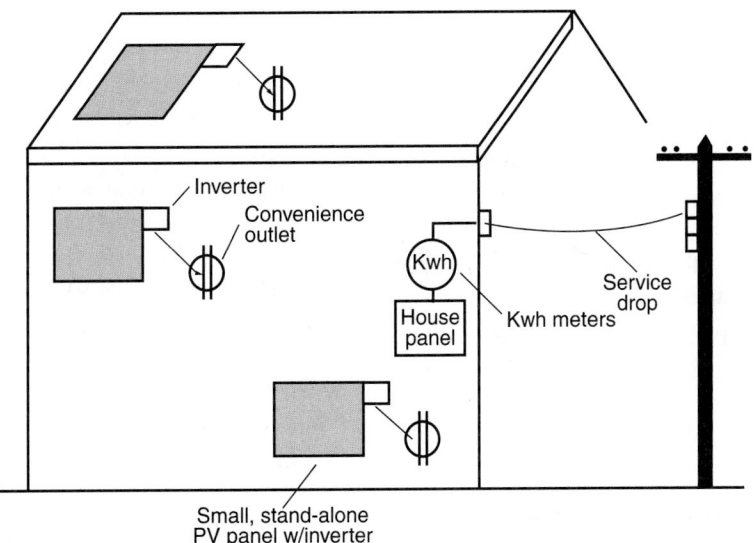

Fig. 14.15 *Schematic drawing of a grid connected PV system using small distributed PV arrays connected to the house distribution system at convenience outlets. In this arrangement, the utility meter is unratcheted (reversible) so that excess PV-produced electric energy reverses the power company's meter. A net monthly credit is calculated at the avoided-cost rate.*

The last factor has taken on considerable additional importance in recent years with the development of building-integrated PV (BIPV) elements. The traditional PV installation consisted of PV modules mounted near or on the serviced facility. In the latter case, the panels were roof mounted if a south-facing roof surface of sufficient size was available. The development of thin film amorphous silicon modules (see Section 14.13) permits incorporation of PV modules into a structure in the form of roof panels, roofing tiles, wall panels, skylights, and other elements that *replace* traditional building construction elements, fulfilling the dual function of electricity production and weatherproof building envelope materials. In such cases, the difference in cost between standard construction plus add-on PV elements and BIPV construction can be taken as a cost credit to the PV system.

14.11 Codes and Standards

PV installations are electrical systems and therefore fall in the purview of the National Electrical Code (*NEC*), NFPA 70. Article 690 of the *NEC*—Solar Photovoltaic Systems—is devoted specifically to such systems, and its requirements plus those of applicable articles in the remainder

of the code must be observed in all areas where the *NEC* is mandated by local jurisdiction. In addition, for grid-connected (interactive) systems, the specific requirements of the local utility must be followed, all equipment approved by them, and the final installation subject to their inspection.

Other codes and standards that contain important information and may be applicable if required by local authorities are:

IEC 61194 Ed. 1.0 b:1992: Characteristic Parameters of Stand-Alone Photovoltaic (PV) Systems.

IEC 61727 Ed. 1.0 b:1995: Photovoltaic (PV) Systems—Characteristics of the Utility Interface.

IEEE 1144-1996: Recommended Practice for Sizing Nickel–Cadmium Batteries for Photovoltaic (PV).

IEC 61702 Ed. 1.0 b:1995: Rating of Direct Coupled Photovoltaic (PV) Pumping Systems.

IEEE 929-1987 (R1991): Recommended Practice for Utility Interface of Residential and Intermediate Structures.

IEEE 937-1987 (R1993): Installation and Maintenance of Lead–Acid Batteries for Photovoltaic (PV).

IEC 61173 Ed. 1.0 b:1992: Overvoltage Protection for Photovoltaic (PV) Power Generating Systems.

IEC 61277 Ed. 1.0 B:1995: Terrestrial Photovoltaic (PV) Power Generating Systems.

UL 1703-1993: Flat-Plate Photovoltaic Modules and Panels.

IEEE 1145-1991: Installation and Maintenance of Nickel–Cadmium Batteries for Photovoltaic Systems.

IEEE 1013-1990: Recommended Practice for Sizing Lead–Acid Batteries for Photovoltaic Systems.

14.12 Terminology and Definitions

As in every technical field, so also in the area of solar PV, a generally accepted terminology has developed. The list that follows contains common terms and their accepted definitions. Particular attention is required to two terms that are often incorrectly used interchangeably: insolation and irradiance. *Irradiance* is the amount of solar *power* impinging on a specific area, usually measured in units of watts per square meter. *Insolation* is the amount of solar *energy* received by a given area, measured in wh/m^2, Kwh/m^2, BTU/ft^2, or MJ/m^2. The conversions are: $1\ Kwh/m^2 = 317.2\ BTU/ft^2 = 3.6\ MJ/m^2$. (See also Appendix I.)

Definitions:

Avoided cost. The minimum amount an electric utility is required to pay an independent power producer under the Public Utilities Regulatory Policy Act of 1978.

Balance of system (BOS). All components other than the PV modules themselves, including the system electronics, support structure, and storage.

BIPV. Building-integrated photovoltaic (system).

Blocking diode. A diode used to block reverse flow of current into a photovoltaic source circuit.

Interactive System. A solar PV system that operates in parallel with the electric utility lines and may be designed to deliver power to them.

Inverter. Commonly known as a power conversion system (PCS) or (incorrectly) as a power conditioning unit (PCU), an inverter is a device that changes direct-current to alternating-current.

Module. The smallest complete, environmentally protected assembly of solar cells and other components, designed to generate direct-current power under sunlight.

Panel. A collection of modules mechanically fastened together, wired, and designed to provide a field-installable unit.

Peak sun hours. The number of hours per day at an irradiance of $1\ Kw/m^2$ that is equivalent to the total daily insolation energy.

PV array. A mechanically integrated assembly of modules or panels with a support structure and other components, as required, including tracking apparatus where used, to form a power-producing unit.

Solar cell. The basic PV device that generates electricity when exposed to light.

(Solar) Photovoltaic system. All the components and subsystems that, in combination, convert solar energy into electrical energy suitable for connection to a load.

Thick-crystal photovoltaics. The most common commercial type of PV material.

Thin-film photovoltaics. PV devices made of a semiconductor material, such as copper indium diselenide or amorphous silicon, a few micrometers thick deposited on substrates of glass, ceramic, or other compatible material.

Stand-alone system. A solar PV system that supplies electrical energy without interconnection to any other power source.

14.13 Solar PV Cells

Photoelectricity, defined simply as the production of an electric charge when light strikes particular metallic surfaces, has been a recognized physical phenomenon since the middle of the 19th century. About 1905, Einstein established a mathematical basis for what has come to be known as the *photoelectric effect*. The explanation of this effect, in simplified qualitative terms, is approximately as follows: Light exhibits both wave characteristics and those of a stream of energetic particles called *photons*. When a photon strikes a photoelectric metal surface, it dislodges a single electron from its normal orbit, causing a charge to appear. When pure silicon is exposed to an intense stream of photons, as in sunlight, a large number of electrons are dislodged from their orbits and proceed to wander about the crystal lattice structure of the silicon crystals. By a process called *doping*, impurities are deliberately added to pure silicon to create a P–N (positive–negative) junction, across which the electrons flow to create an electric current, and giving the doped silicon the properties of a semiconductor. If the semiconducting doped silicon is exposed to light and the negative and positive sides of the junction connected through a load, an electron flow (current) commences from the negative side of the junction to the positive side, doing work on the load. Work, of course, is energy. The energy comes from the fast-moving photons (light) and is imparted to the dislodged electrons by impact. The overall effect, therefore, is to create a current flow proportional to the intensity of photon bombardment—that is, proportional to light intensity. In so doing, a PV cell is created. Figure 14.16 is a schematic representation of this process.

The photons in light are not uniformly energetic. It has been found experimentally that a photon must have an energy of at least 1.08 electron-volts [corresponding to a wavelength of 1150 nm (nanometers)] to dislodge an electron. This wavelength is deep into the infrared region of the sun's spectrum and considerably beyond the visible spectrum, which ranges from violet at 380 nm to red at 780 nm. Photons at longer wavelengths do not have sufficient energy to create the photoelectric effect. Because shorter wavelength photons are more energetic, it follows that the entire visible light spectrum is useful in PV action, the magnitude of current produced depending on intensity of light. Inasmuch as, according to Einstein's photovoltaic theory, a single photon can only dislodge a single electron, the photon's energy in excess of the minimum level is dissipated as heat. For this reason, PV cells generate heat as well as electricity, and some means of heat collection or dissipation is required in practical PV array construction. Commercial PV cells operate at an overall insolation-to-electric energy conversion efficiency of 8% to 12%. By layering different strata of semiconductor material, each of which is sensitive to a different limited frequency bandwidth (called *bandgap*), efficiencies as high as 35% to 40% can be achieved. Economic consid-

PHOTOVOLTAIC CELL

Fig. 14.16 *Schematic drawing of the operation of a PV cell. Photons from daylight dislodge electrons in the silicon semiconductor surface creating a positively charged "hole." The N-P junction creates an electric field that "sweeps" electrons across the junction. When the top and bottom electrodes are electrically connected through a load, an electric current flows that is capable of doing work (supplying energy). Thus, solar energy is converted to electrical energy by the PV cell. [From* Handbook of Energy Technologies, *Meyers, R.A., ed., (1983), Wiley-Interscience; reprinted with permission.]*

ELECTRICITY

erations, however, have limited today's commercial PV modules to a maximum of two layers with a conversion efficiency of about 12%.

As mentioned previously, intensive research and development work on solar cell construction has yielded a new generation of PV materials that are variously referred to as *thin-film technology* and *polycrystalline technology*, using various copper and cadmium compounds and amorphous rather than large crystal silicon modules. These materials offer higher efficiencies, can be manufactured in large elements, and, because they do not require a glass cover, can be integrated into standard building material formats as previously described (BIPV). Typical data for two commercial products using this newer technology are:

Amorphous silicon module:
Dimensions: 2 ft × 4 in. × 2 in.

Weight: 15 kg

Electrical: 50 W, 72 V, 0.7 A

Thin-film polycrystalline module:
Dimensions: 2 ft × 4 in. × 2 in.

Weight: 10 kg

Electrical: 75 W, 17 V, 4.5 A

These second-generation PV materials have high efficiency and flexible physical format; distinct advantages over the older, large crystal silicon wafer elements. However, the latter are cheaper, *extremely* hardy, reliable, and long-lived, as has been demonstrated in many many satellites and space vehicles. The newer units are initially somewhat unstable, and have a somewhat shorter life than the large crystal silicon units.

14.14 PV System Arrays

As defined previously, a PV array is a complete connected set of modules mounted and ready to deliver electricity. Two basic arrangements exist: stationary and tracking. Obviously, building mounted arrays are stationary and they have the advantage of not requiring a separate structure, although even simple roof mounting on a pitched roof can be expensive due to the labor cost involved. The materials usually used for ground-mounted arrays are concrete, galvanized steel,

and aluminum. Wood and painted iron are not recommended because of high maintenance costs and relative short life. Actual design of an array mounting depends on the array size, weight, mounting angle, and wind loads, and is therefore particular to each installation. Mounting on a wooden pole or steel pipe with a single support point is limited to relatively small arrays in low wind-speed areas because of the stress on the support arrangement. Several typical fixed mounting arrangements are shown in Figs. 14.17 through 14.19.

The angle, measured from the horizontal, at which a flat-plate collector is mounted, is known as the *tilt angle*. For a fixed (nontracking) array, total *annual* insolation is maximum for a tilt angle equal to the latitude of the location. To maximize winter insolation, when the sun is lowest, the tilt angle can be increased to latitude plus 15°. Conversely, to maximize summer insolation, when the sun is highest, the tilt angle can be reduced to latitude minus 15°. The tilt angle chosen for a fixed array depends on facility usage periods, amount of insolation for the months of the year, and economic considerations.

Tracking arrays follow the motion of the sun and thereby substantially increase the insolation per unit area of solar cells. Single axis tracking follows the sun's motion from east to west while maintaining the array tilt angle constant. This type of tracking can increase insolation by 35% to 50% throughout the year and is most effective at low latitudes. Completely passive, sealed Freon, single axis tracking drives are available for small to medium pole-mounted arrays in low wind areas. Where wind velocity is high, a motorized drive using PV-generated power may be necessary. Dual-axis tracking (altitude and azimuth) for flat-plate collectors and altitude tracking for various designs of concentrating collectors are beyond the scope of this text.

14.15 PV System Batteries

Batteries are required in a stand-alone system to store excess PV-produced energy and to return it to the user in peak energy usage hours (see Fig. 14.10). For residential use, there is usually almost no coincidence between the two time periods be-

COLLECTOR SUPPORT

(a)   (b)

Fig. 14.17 *Typical mounting techniques for PV arrays.* (a) *Arrangement for mounting large arrays on a horizontal surface.* (b) *Simple pipe rack mounting for a small array. For all installations:*

- *Investigate wind loads and anchor accordingly.*
- *Place array so that snow (if any) will not mound up around the array.*
- *Mount the array so that natural ventilation will provide necessary cooling.*

(Diagrams reprinted with permission from Architectural Graphic Standards, *8th ed. (1988), John Wiley and Sons, Publishers.)*

ELECTRICITY

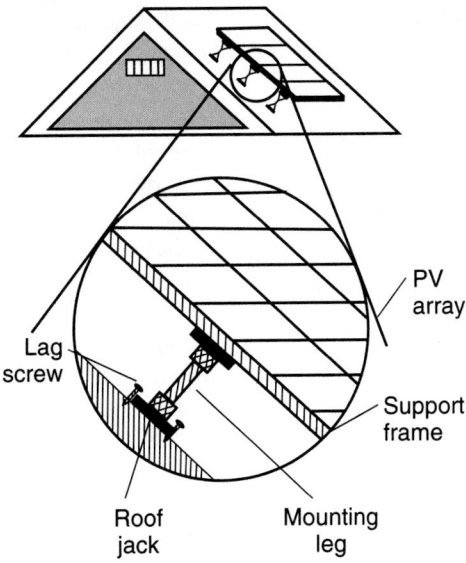

Fig. 14.18 *Typical roof mount for a PV array. The leg lengths of the supports should be designed or adjusted to give the desired collector slope. Azimuth can also be adjusted with asymmetrical legs, but appearance should be considered as well. In snow areas, mount the array close to the roof peak so that snow will not accumulate under and around the collector. Also, do not place the array above a building entrance, as snow will accumulate on the collector and subsequently "avalanche" off. [From* Stand-Alone Photovoltaic Systems, *Sandia Laboratories (1995).]*

(a)

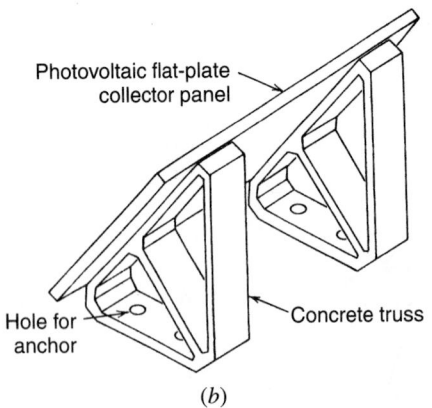

(b)

Fig. 14.19 *PV flat-plate collector support structures. (a) Wooden-pole torsion tube support. (b) Concrete truss support. [From* Handbook of Energy Technologies, *Meyers, R.A. (ed.), (1983), Wiley-Interscience; reprinted with permission.]*

cause PV production peaks at solar noon, whereas residential usage peaks in the evening and is low at noon. In comparison, commercial stand-alone PV installations such as off-grid stores, shops, cottage industries, and the like have a much higher coincidence. This means that during normal, everyday use, the drain on the battery is shallow, and a cheaper battery will frequently suffice. These considerations are expanded in the design discussion in Sections 14.17 and 14.18.

The service required of a PV system battery is that it be capable of supplying most or all of the installation's electrical requirements for a given period—usually 3 days of cloudy weather, if no other specific-use data is available. This means that the battery selected:

· must be capable of repeated discharge to 20% capacity or less, without injury.

· must be designed to accept slow, lengthy recharge.

In addition, the battery

· should have long life.

· should be suitable for the physical conditions in which it will be used (i.e., size, ambient temperature, ventilation, and maintenance availability).

Pure lead automobile batteries (also known as *lead–acid batteries*) are not suitable for PV systems because they are designed for rapid shallow discharge (engine starting) and very rapid recharge. If such a battery is repeatedly discharged more than about 20% (shallow discharge), or if it is not completely recharged after discharge, its life will be extremely short. Batteries made for use in electric vehicles such as golf carts and forklifts are constructed with heavy plates and are designed for the deep-cycle discharge needed in PV systems. Most of these units are *flooded* batteries, meaning that the plates must be completely covered with electrolyte to prevent damage. This requires frequent maintenance, making flooded batteries unsuitable for unattended sites.

Lead–calcium batteries are almost maintenance free, rarely requiring addition of water or electrolyte. As a result they are frequently used in unattended systems. They do not cycle as deeply as heavy-plate, deep-cycle units, and this must be considered in design.

Sealed batteries are completely maintenance free. They are frequently constructed with a gel-type electrolyte that does not "boil" off on recharge, and therefore needs no replacement. This makes them ideal for unattended sites provided they are specifically designed for deep-cycle service (80% discharge) and are so labeled. Most maintenance-free batteries are shallow-cycle units, and therefore unsuitable. An important characteristic of sealed units is that they require very accurate charge and discharge control to prevent either overcharge or over-discharge, or over-rapid charge or discharge. Any of these *severely* shortens battery life. As a result, gel-cell batteries require the use of a high-quality charge controller.

Lead–antimony batteries are very sturdy and tolerate deep discharge and slow or rapid charge while maintaining rated life. They require occasional water replenishment and adequate ventilation to disperse gases created during charging cycles.

Nickel–cadmium (nicad) batteries are highly desirable in PV systems because they are maintenance-free, can be discharged to essentially 100% of capacity without injury or shortening of life, are much less temperature sensitive than lead–acid batteries and have longer life. Their principal drawback is price. Nicad batteries should not be shallow discharged because they tend to develop a "memory" of the discharge cycle and after many shallow cycles may not supply sufficient current for a deep discharge. Therefore, in essentially clear-sky climates, where a 3-day period of cloudiness is rare, nickel-cadmium may not be applicable because a battery sized for a 3-day system supply will normally operate on shallow discharge.

Additional considerations in battery selection are:

Temperature. Lead-acid battery capacity declines with temperature. Nominal battery capacity rating refers to a temperature of 25°C (77°F). Batteries operating at lower temperatures should be derated as follows:

Ambient Temperature	Derating Factor
70°F (21.1°C	0.96
60°F (15.5°C)	0.90
50°F (10°C)	0.84
40°F (4.4°C)	0.77
30°F (− 1.1°C)	0.71
20°F (− 6.7°C)	0.63

It is important to note that a discharged lead–acid battery contains, essentially, water, and therefore freezes at about 30°F (−1°C). A fully charged battery will not freeze until the temperature dips below 0°F. Nickel–cadmium batteries are much less affected by temperature.

Voltage. Battery terminal voltage varies with its state of charge ranging, for a 12 V battery, at

77°F, from approximately 15 V at full charge to just over 12 V immediately after a deep discharge. If not charged immediately, voltage drops to a low of just over 11 V.

Life. A properly maintained deep-cycle lead–acid battery should last a minimum of 5 years. A nickel–cadmium battery should last 1.5 to 2 times as long.

Rating. The manufacturer's stated ampere-hour (A-H) rating of a battery refers to a single continuous low rate discharge at laboratory conditions. Actual, field use, A-H capacity of a battery depends heavily on the discharge rate, and it is this figure as calculated for the particular installation that should be used to compare batteries of the same type. All reputable manufacturers have readily available discharge curves that show voltage and duration at different discharge rates.

The basic storage battery unit is rated at 12 V, although 6-V units are available. Figure 14.20 shows typical battery connection arrangements for the most common system voltages, which are 12, 24, 48, and 72 V. Table 14.1 shows some typical physical data.

14.16 BOS

Refer to Fig. 14.13. The balance of system (BOS) consists of a charge controller, inverter if used, and accessories such as blocking diodes, lightning arrester, electrical grounding system, and other electrical specialty items. A full description of these items would take our discussion far afield into the realm of electronic controls and power system accessories. The interested reader is referred to the bibliography at the end of this chapter for a few of the many excellent references in this field.

14.17 Design of Stand-Alone PV System

There is a fundamental difference in the design procedure for array sizing and system arrangement between a stand-alone system and a grid-connected (grid-tied) system. The stand-alone system must supply all of the electric load; there is no utility backup. The grid-tied system is essen-

Prescott, AZ

WBAN NO. 23184

LATITUDE: 34.65° N
LONGITUDE: 112.43° W
ELEVATION: 1531 meters
MEAN PRESSURE: 847 millibars

STATION TYPE: Secondary

Solar Radiation for Flat-Plate Collectors Facing South at a Fixed Tilt (kWh/m²/day), Uncertainty ±9%

Tilt (°)		Jan	Feb	Mar	Apr	May	June	July	Aug	Sept	Oct	Nov	Dec	Year
0	Average	3.1	3.9	5.1	6.6	7.5	8.0	6.9	6.3	5.7	4.6	3.4	2.8	5.3
	Min/Max	2.6/3.5	3.2/4.5	4.2/6.1	5.7/7.1	6.8/8.0	7.2/8.5	6.1/7.7	5.1/7.0	5.1/6.7	3.6/5.2	2.8/3.9	2.2/3.2	5.1/5.7
Latitude −15	Average	4.4	5.1	5.9	7.0	7.5	7.7	6.7	6.5	6.5	5.8	4.8	4.1	6.0
	Min/Max	3.5/5.1	4.0/6.0	4.7/7.2	6.0/7.7	6.8/8.0	6.9/8.1	5.9/7.5	5.3/7.2	5.7/7.6	4.4/6.6	3.7/5.5	3.0/4.8	5.6/6.5
Latitude	Average	5.1	5.7	6.2	6.9	7.0	7.0	6.2	6.3	6.6	6.4	5.5	4.9	6.1
	Min/Max	3.9/6.1	4.3/6.8	4.9/7.6	5.8/7.6	6.3/7.4	6.3/7.3	5.5/6.9	5.1/7.0	5.8/7.8	4.7/7.3	4.1/6.4	3.5/5.8	5.7/6.6
Latitude +15	Average	5.5	5.9	6.1	6.4	6.1	6.0	5.4	5.7	6.4	6.5	5.9	5.4	5.9
	Min/Max	4.2/6.6	4.4/7.1	4.8/7.5	5.4/7.1	5.6/6.5	5.4/6.3	4.8/5.9	4.6/6.3	5.6/7.5	4.7/7.5	4.3/6.9	3.7/6.4	5.5/6.4
90	Average	5.0	4.9	4.3	3.5	2.6	2.3	2.3	2.9	4.0	5.0	5.2	5.0	3.9
	Min/Max	3.8/6.1	3.6/5.9	3.4/5.2	3.0/3.8	2.5/2.8	2.2/2.3	2.1/2.4	2.4/3.1	3.6/4.7	3.6/5.8	3.6/6.2	3.4/6.0	3.5/4.2

Solar Radiation for 1-Axis Tracking Flat-Plate Collectors with a North-South Axis (kWh/m²/day), Uncertainty ±9%

Axis Tilt (°)		Jan	Feb	Mar	Apr	May	June	July	Aug	Sept	Oct	Nov	Dec	Year
0	Average	4.8	5.9	7.4	9.4	10.6	11.3	9.3	8.8	8.4	7.1	5.4	4.4	7.7
	Min/Max	3.7/5.7	4.4/7.2	5.7/9.5	7.9/10.6	9.4/11.5	9.9/12.0	7.9/10.8	7.0/10.1	7.1/10.2	5.0/8.2	4.0/6.3	3.2/5.2	7.1/8.4
Latitude −15	Average	5.7	6.8	8.1	9.8	10.6	11.2	9.3	9.0	9.0	8.0	6.3	5.4	8.3
	Min/Max	4.3/6.9	5.0/8.3	6.2/10.3	8.2/11.0	9.4/11.6	9.8/11.9	7.8/10.7	7.1/10.4	7.6/10.9	5.6/9.3	4.6/7.5	3.8/6.4	7.5/9.0
Latitude	Average	6.3	7.2	8.3	9.8	10.6	10.7	8.9	8.8	9.1	8.4	6.9	6.0	8.4
	Min/Max	4.7/7.6	5.3/8.8	6.3/10.6	8.1/11.0	9.2/11.3	9.4/11.4	7.6/10.4	7.0/10.2	7.7/11.1	5.8/9.8	4.9/8.2	4.1/7.2	7.6/9.2
Latitude +15	Average	6.6	7.4	8.2	9.4	9.7	10.1	8.4	8.4	8.9	8.5	7.2	6.3	8.3
	Min/Max	4.9/8.1	5.3/9.1	6.2/10.6	7.7/10.6	8.6/10.6	8.8/10.7	7.1/9.7	6.7/9.8	7.5/10.9	5.8/9.9	5.1/8.6	4.3/7.6	7.5/9.0

Average Climatic Conditions

Element	Jan	Feb	Mar	Apr	May	June	July	Aug	Sept	Oct	Nov	Dec	Year
Temperature (°C)	2.3	3.9	5.9	9.6	14.2	19.6	22.8	21.3	18.1	12.6	6.7	2.6	11.7
Daily Minimum Temp	−5.6	−4.4	−2.2	0.8	5.1	9.9	14.4	13.2	9.3	3.4	−1.9	−5.4	3.1
Daily Maximum Temp	10.2	12.2	14.1	18.3	23.3	29.2	31.2	29.4	26.7	21.8	15.3	10.6	20.2
Record Minimum Temp	−18.3	−18.3	−13.9	−6.1	−3.3	2.2	8.3	4.4	1.1	−7.2	−13.3	−22.8	−22.8
Record Maximum Temp	23.3	25.0	27.2	30.6	35.0	39.4	38.9	37.2	35.0	32.2	26.1	21.1	39.4
HDD, Base 18.3°C	496	403	384	262	138	32	0	0	41	182	350	489	2775
CDD, Base 18.3°C	0	0	0	0	11	68	139	96	32	4	0	0	351
Relative Humidity (%)	59	53	49	38	33	27	44	51	47	46	51	59	47
Wind Speed (m/s)	3.1	3.5	4.1	4.2	4.3	4.1	3.5	3.1	3.4	3.3	3.3	3.0	3.6

Fig. 14.20 *Solar radiation data for a southwestern city.* (From Solar Radiation Data Manual, NREL, 1994.)

TABLE 14.1 **Physical Characteristics of Typical 12-V[a] Batteries**

Battery Type	Capacity at Discharge Rate 6/20/100 hr Amp-hours	Dimensions 12-V Unit L × W × H (in.)	Weight (lb)
Deep–cycle Lead–Acid	500/625/775	33 × 6.5 × 24	500
	675/840/1025	40 × 6.5 × 24	700
	850/1050/1275	50 × 6.5 × 24	800
Sealed Gel-cells	(20-h rate)		
	850	40 × 8 × 24	650
	1050	40 × 9 × 24	770
	1275	40 × 10 × 24	930
	1500	40 × 12 × 24	1100

[a]Use two units for a 24-V battery and 4 units for a 48-V battery.

tially an economic plus quality-of-life decision because utility service is available for part or all of the electric load. This and the next section therefore address these two procedures separately.

Refer to Fig. 14.12, which shows the essentials of the stand-alone system. As explained previously, because there is no facility whose energy requirement exactly matches the output of the PV array, the system operates by using a PV array to charge a battery, which then supplies the facility's load. Any coincidence between load and PV energy production simply bypasses the battery, but is only incidental to the design. The three factors that establish the design are:

· The daily energy usage.
· The period of time for which the battery must supply the electric load without recharge.
· The amount of insolation available.

The second factor is an aspect of local weather conditions and would most often represent the number of successive solid overcast or rainy days *usually* encountered. Designing for unusual weather results in a system that is oversized for normal (say 95th percentile) weather, and therefore essentially uneconomical. That in turn weighs heavily in favor of a hybrid system (see Fig. 14.13), which defeats, at least in part, the *raison d'être* of the PV system.

The design procedure is essentially straightforward and is as follows:

STEP 1. Determine the *average* daily energy usage in units of watt-hours.

STEP 2. Determine the maximum period for which the system battery must supply the entire electrical load. (In practice, this period can be extended to accommodate unusually long sunless periods by switching off nonessential loads.)

STEP 3. Determine the required system battery capacity, which is equal to the daily load (step 1) multiplied by the number of days that the battery must supply the load (step 2). System losses are taken into account in this calculation.

STEP 4. Determine the size of the solar array required. The array must supply all of the required energy; that is, the facility's daily requirements

(step 1) plus all system losses. The array size is this total, in terms of watt-hours. To determine the physical size of the array, we must first determine the number of watt-hours produced per square foot of PV module. This requires knowledge of the area's solar radiation, the type and efficiency of solar cell used, mounting type (stationary or tracking), and tilt angle (mounting angle measured from the horizontal). With these data in hand, the area of PV cells required is simply the total load in watt-hours divided by the PV array material's energy production in watt-hours per unit area (usually square feet).

STEP 5. Check that the array selected is sufficiently large to charge the battery selected in step 3. Assuming, for instance, that we had selected in step 2 a 3-day period during which the battery must supply all of the load, we will need about 3 ½ consecutive sunny days *at the same insolation* for full recharge (considering losses of approximately 20%). If the facility is to be used year-round and the PV array is sized on the basis of summer insolation, it is necessary to do a calculation for a winter month as well. Insolation data is usually available for both, and average monthly insolation accounts for the appreciable energy availability from cloudy skies as well as sunny skies. The result generally indicates a larger PV array requirement for winter months. The decision in such cases, on purely economic grounds, is often to use a small engine generator for battery-charging for winter use only. This is *not* the same as a hybrid system (Fig. 14.13), in which the generator is used year-round to pick up heavy ac loads. Some designers base the PV array size on *minimum* yearly insolation, reasoning that in other months the wasted overproduction of the array is the price paid for year-round operation. Others use a median insolation figure, reasoning that the outages will be few and of short duration. Finally, using summer insolation will probably require a small generator, as noted previously.

An illustrative example using the outlined design procedure should clarify the considerations and calculations involved.

EXAMPLE 14.3 Stand-Alone PV System for a Residence:
Design a stand-alone PV system for a 650-ft² year-round off-line cottage located near Prescott, Ari-

zona. In addition to the usual appliances, a desert cooler is used during hot summer days. The PV array can be mounted on the south-facing sloping roof at any required angle. Determine the required battery and PV array sizes. State all assumptions.

SOLUTION

STEP 1. Determine the house's average electric energy consumption. Although refrigerators that run on bottled gas are readily available, we will assume use of an electric unit. Water heating is solar, with a small gas booster heater. Cooking uses bottled gas.

24-V DC Appliances	Watts (W)	Daily Use (h)	Daily Energy (wh)
12-ft³ refrigerator/ freezer	35	24	840
12″ color TV	60	4	240
Desert cooler*	50	8	400
Hi-Fi equipment	40	2	80
Ceiling fans* (3)	25	4	300
Radios	5	10	50
Incandescent lighting	50	8	400
		Daily use dc	2310 wh

*Summer only.

120-V AC Appliances	Watts (W)	Daily Use (h)	Daily Energy (wh)
Fluorescent lighting	80	10	800
Clothes washer	500	1/2	250
Vacuum cleaner	600	1/4	150
Kitchen appliances	400	1	400
Iron	1000	1/4	250
Miscellaneous	100	3	300
		Daily use AC	2150 wh

*Maximum coincident ac load.

Total daily energy usage:
DC	2310
AC 2150 wh × 1.15 (inverter loss)	2473
	4783 wh

$$\text{Ampere-hours} = \frac{4783 \text{ wh}}{24\text{V}} = 199, \text{ say } 200 \text{ A-H}$$

STEP 2: Assume that investigation of local weather conditions indicates occasional periods throughout the year, but primarily in the winter, of 2-3 consecutive days of heavy overcast, clouds and rain.

STEP 3: Sizing the system battery. Assume use of a deep-cycle flooded cell lead battery, with appropriate charge controller. These batteries can supply about 80% of their rated ampere-hour capacity. Since the battery will be stored indoors, but in an unheated room, it should be derated 15% for conservative design.
Required battery size in ampere-hours equals:

$$\frac{3 \text{ days} \times 200 \text{ A-H daily load} \times 1.20 \text{ discharge factor}}{0.85 \text{ (temperature derating)}}$$

$$= 850 \text{ A-H}$$

One model of an 850 A-H, 24-V deep-cycle lead–acid battery (2–12V units in series) measures approximately 40 in. l × 15 in. w × 24 in. h and weighs about 1500 lbs.

STEP 4: Sizing the PV array. Because the house will be occupied all year, and because the critical period is winter when the sun is low, we opt for a tilt of latitude +15° to maximize winter PV production. Inasmuch as the array is roof mounted and will be relatively large, we use a stationary rather than tracking array. The array must supply the daily energy requirement which is 4783 wh or 4.8 kwh.

Referring now to Fig. 14.21, which gives solar radiation data for the location selected, we obtain the following data: Average solar radiation for PV cells facing south, at tilt angle of latitude +15°:

January: 5.5 kwh/m²/day; Average monthly temperature 2.3°C

June: 6.0 kwh/m²/day; Average monthly temperature 19.6°C

Figure 14.22 was drawn from the results of a computer program called PVFORM, developed at Sandia National Laboratories. It gives the daily average PV array production in wh/ft² as a function of monthly average temperature and monthly average solar radiation in kwh/m²/day. Using January and June radiation figures stated previously, we obtain from Fig. 14.22:

January: 52 wh/ft² of PV array

June: 53 wh/ft² of PV array

Fig. 14.21 *Wiring arrangements of 12-V battery packs to obtain increased capacity, increased voltage, or both.*

From this we see that PV array output is constant throughout the year and no difficulty will be encountered in using either set of data. If it is known that the PV material to be used has a higher conversion efficiency than 10%, then its output is increased proportionately.

The required area of PV array, assuming 10%

conversion efficiency, is

$$\text{PV area} = \frac{4783 \text{ wh/day required}}{52 \text{ wh/day/ft}^2} = 91 \text{ ft}^2$$

STEP 5. Because the solar radiation in Prescott, Arizona is essentially identical for summer and

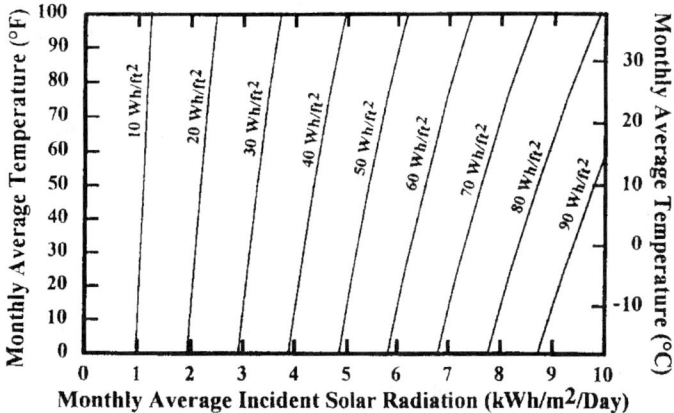

Fig. 14.22 *Daily average PV array production, Wh/ft², as a function of monthly average temperature and incident solar radiation. PV performance calculated using PVFORM for a 10% efficient solar cell array at 25°C ambient temperature. (Reprinted with permission from "Design for PV," Ass'n of Collegiate School of Architecture.)*

ELECTRICITY

winter, PV production in 3 to 3 ½ clear days will fully recharge the battery, as explained previously.

■

Summary. The suggested PV system consists of approximately 90 ft² of PV collector, an 850 A-H, 24 V battery, and a 1500 W inverter to handle the initial load plus anticipated addition of small ac appliances. The summer-only DC loads will be offset in winter loads by additional lighting and use of television, and other free-time activities.

Additional Considerations. Using accurate radiation and weather data for the site selected results in an efficient workable system. Unfortunately, accurate radiation data, particularly for tilted surfaces, are not always available, and tilt data are not easily derived from horizontal surface or normal-to-the-sun data. Areas outside of large cities are frequently specifically the places for which PV installations are planned, and it is usually for these areas that a paucity of data obtains. In such instances data for nearby areas should be used, but applied judiciously, as insolation conditions can vary sharply within a relatively small geographic area.

14.18   Design of a Grid-connected (Interactive) PV System

The design procedure for an interactive system (with feedback into the utility grid) is much different than the typical stand-alone design described in the previous section. Refer to Fig. 14.14. The principal differences are:

· The loads in an interactive system are all ac rather than a mix of ac and dc because the building is fed from the utility line or from the PV array dc to ac system inverter; dc is not available.

· The existence of a PV system has no bearing on the choice of fuel for appliances. Therefore, for instance, whether to use gas or electric cooking is a matter of fuel costs and preferences.

· Loads are generally heavier in interactive systems because one cannot run out of elec-

tric power, as with a stand-alone system. As a result, an all-electric house or a "smart" house can just as much be candidates for PV systems as any other structure.

Note: Due to unusual weather patterns in the last few years, some areas in the continental United States have been hit with uncharacteristic ice-storms that have caused unanticipated, relatively lengthy power outages. Residential and other consumers with grid-connected PV systems in these areas have taken to adding a small battery to their systems to carry their essential needs (lights, heating system, computers) through the outage.

· Load data can be obtained for a specific design, or from general utility-supplied data, similar to that shown in Figs. 14.23 and 14.24.

· Inasmuch as no battery or control equipment is used, space allocation in the structure for PV related equipment is small and system maintenance is minimal. Also, system life is as long as the PV array life.

(a) Interactive System Design Considerations. Assuming a net-metering arrangement, there is no advantage to coincidence of building loads and peak PV production. This is emphatically *not* the case where buyback is based on avoided costs. In such situations, careful scrutiny of the utility's rate schedule may yield cost-savings opportunities in matching loads to lowered-rate time periods.

· For nonresidential design, careful consideration of future electric loads should be reflected in architectural consideration for future (large) PV arrays. In this connection, the possibility of using distributed arrays as in Fig. 14.15 should be studied.

Design Procedure for an Interactive PV System.

STEP 1. Determine with the power company the metering arrangement to be used, both physical and financial. The utility will also generally supply reliable data on solar radiation, load patterns, PV array sizing and tracking feasibility, and other relevant design data.

Electrical Usage Profile for All Electric House in Southern California during Winter

Electrical Usage Profile for House in Southern California*

Electrical Usage Profile for House in Midwest

Electrical Usage Profile for House in the Pacific Northwest

Fig. 14.23 *Typical residential electrical usage profiles, based on 1500-ft² house. (Reprinted with permission from "Design for PV," Ass'n of Collegiate School of Architecture.)*

STEP 2. On the basis of a detailed cost analysis determine the optimum size of (initial) PV array.

STEP 3. Using specific data or generalized information similar to that given in Figs. 14.23 and 14.24, determine the required electrical rating and physical size of the PV array determined in step 2.

14.19 PV System—Conclusion

We have placed considerable emphasis in this section on PV design and in earlier chapters on PV use because of our belief that PV power is one of the dominant trends of the future. A recent large commercial installation (646 m² of array) in

Fig. 14.24 *Typical electric load profiles for nonresidential American buildings. (Reprinted with permission from "Design for PV," Ass'n of Collegiate School of Architecture.)*

Sunderland, United Kingdom (an area not noted for year-round sunshine), is indicative of modern trends. In this book, nonresidential examples of PV installations are to be found in:

Chapter 1

EPICenter at Montana State University; classrooms and labs.

PV Parking Canopy Competition; New York. PV shading devices cover a parking lot and the power used to recharge parked electric vehicles.

Center for Regenerative Studies at Cal Poly, Pomona; residential/labs.

Chapter 3

Band-of-sun example, showing siting for seasonal sun angles.

Chapter 5

Zion Park Visitor Center, Utah.

Cooper-Haggard House/office, California.

UK Pavilion, Seville World's Fair.

Chapter 8

Hyndburn Council Offices, England.

4 Times Square, New York; new office tower.

ELECTRIC POWER AND ENERGY

14.20 Electric Power Generation—ac

Alternating current is produced commercially by an ac generator, generally called an *alternator*. Its prime mover may be any type of engine or turbine. The process by which electricity is produced is illustrated in Fig. 14.25. It is based on the fundamental discovery in 1831 by Michael Faraday of the principle of electromagnetic induction. Put briefly, this principle states that when an electrical conductor is moved in a magnetic field, a voltage is induced in it. The direction of movement determines the polarity of the induced voltage, as shown.

If the conductor is formed into a coil and rotated in the magnetic field, a voltage of alternating polarity is produced, that is, ac. It does not matter whether the conductor or the magnet moves; the motion of the conductor and the field with respect to each other produces the voltage (see Fig.

Fig. 14.25 *The action fundamental to all generators. When a conductor of electricity moves through a magnetic field, a voltage is induced in the conductor, with polarity as shown.*

14.26). It is only one step (that of development) from this rudimentary ac generator to the large, powerful alternator that produces ac in a modern power plant. The frequency of the voltage generated is a function of the machine design and the speed at which it is driven. Normal generator frequency in the United States, as noted previously, is 60 Hz, whereas in Europe and the Mideast it is 50 Hz.

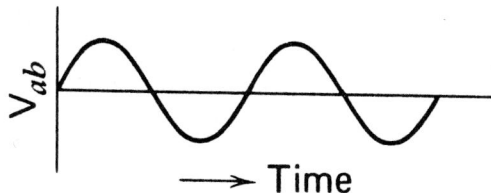

Fig. 14.26 *Rotating a coil in a magnetic field produces an alternating sinusoidal voltage at terminals a and b because of the alternating polarity (see Fig. 14.25).*

14.21 Power and Energy

It is important, indeed imperative, that the distinction between power and energy be clearly understood, because all too frequently the terms are incorrectly used interchangeably. *Energy* is the technical term for the more common expression—work. *Power* is the rate at which the energy is used. Stated otherwise and concretized, it takes a fixed amount of energy to lift a weight a given distance, but the faster it is lifted, the more power is required. Energy is the product of power and time—that is,

$$\text{energy (work)} = \text{power} \times \text{time} \quad (14.4)$$

In practical terms, energy is synonymous with fuel and, therefore, also cost. Thus, energy can be expressed as barrels (tons) of oil, cubic feet (cubic meters) of gas, tons of coal, kilowatt hours of electricity usage, and dollars of fuel cost. The concept of energy efficiency of structures can be stated in terms of annual usage of oil, gas, and electricity, or alternatively in terms of dollars of fuel cost. In technical terms, energy is expressed in units of Btu (calories), foot-pounds (joules), and kilowatt hours. Conversion factors between these (and other) units of energy are found in Appendix I. In terms of fuel, 1 kilowatt-hour (kWh) of energy is roughly equivalent to 0.5 lb of coal, 0.07 gallon of oil, 7.7 cubic feet of natural gas, and 8200 gallons of dam water.

Power is the rate at which energy is used or, alternatively, the rate at which work is done, because energy and work are synonymous. The term *power* implies continuity—that is, the use of energy at a particular rate, over a given, generally considerable, span of time. The concept of power necessarily involves the factor of time because it is, as stated previously, the *rate* at which work is done. Thus, multiplying power by time yields energy. Typical units of power in the English system are horsepower, Btu per hour, watt, and kilowatt. In the metric or SI system, the corresponding units are joule per second, calorie per second, watt, and kilowatt. In physical terms, power is also the rate at which fuel (energy) is used. Thus, power can also be expressed as gallons (liters) of oil per hour, cubic feet (cubic meters) of gas per minute, and tons (metric tons) of coal per day. (When using fuel figures, it is important to note

the assumed fuel-to-useful energy conversion efficiency. When not stated, "energy content" i.e. energy availability at 100% conversion efficiency, is assumed.)

14.22 Power in Electric Circuits

The unit of electric power is the watt (W). A larger unit of 1000 watts is the kilowatt (kW). The power input in watts to any electrical device having a resistance R and in which the current I is given by the equation

$$W = I^2R = I(IR) \quad (14.5)$$

where W is wattage. This is true for both dc and ac circuits. However, because the resistance of an item is generally not known, whereas the circuit voltage and current are known, it would be preferable to be able to calculate power using these two quantities. This can be done, but is different for dc and ac.

In dc circuits, by Ohm's law $V = IR$

and, because $W = I(IR)$ from Eq. 14.5, we obtain

$$W = VI \quad (14.6)$$

where

> W is watts
>
> R is ohms
>
> I is amperes
>
> and V is volts

In ac circuits, impedance consists of a combination of dc resistance and ac resistance (called *reactance*). Reactance causes a phase difference between voltage and current. This phase difference is represented by an angle, the cosine of which is called the *power factor* (*pf*). Power factor is extremely important in that it enables us to calculate power in an ac circuit. The ac power equation is similar to that for dc (see Eq. 14.6) with the addition of this special ac power factor term; that is

$$W = VI \times pf \quad (14.7)$$

If *pf* is not used in the equation, the product of voltage and current gives a quantity known as *volt-amperes*. In a purely resistive circuit, such as one with only electric heating elements, imped-ance equals resistance, power factor equals 1.0, and wattage equals volt-amperage. A few exam-ples here should help make applications of these equations clear.

EXAMPLE 14.4.   Referring back to Exam-ple 14.1 in Section 14.5, calculate the power drawn using Eqs. 14.5, 14.6, and 14.7.

SOLUTION.   Because an incandescent lamp is purely resistive and therefore has unity (1.0) power factor, it does not matter whether the cir-cuit is ac or dc.

From Example 14.1,

$R = 66$ ohms, $I = 1.74$ A, and $V = 115$ V

1. In a dc circuit, we would use Eq. 14.6:

$$W = VI = 115(1.74) = 200 \text{ W}$$

2. In an ac circuit, we would use Eq. 14.7:

$$W = VI \times pf = 115 \times 1.74 \times 1.0 = 200 \text{ W}$$

3. In either a dc or an ac circuit, we can use equation 14.5:

$$W = I^2R = (1.74)^2 \times 66 = 200 \text{ W} \quad \blacksquare$$

EXAMPLE 14.5.   Using the data given in Ex-ample 14.2 and Fig. 14.4, determine (a) the power lost in the wiring and (b) the power input to the load.

SOLUTION.

(a) The total power loss in the wiring is:

$$W = I^2R = I^2(r_{bc} + r_{da})$$

$$= (10)^2 \times 0.04 = 4 \text{ W}$$

(b) The power taken by the load is

$$W = I^2R = I^2R_{cd} = (10)^2 \times 1.15$$

$$= 115 \text{ W or } 0.115 \text{ kW}$$

Alternatively, we can find this power by multi-plying voltage and current. The voltage on the load is

$$IR = 10 \times 1.15 = 11.5 \text{ W}$$

and

$$W = VI = 11.5 \times 10 = 115 \text{ W} = 0.115 \text{ kW} \quad \blacksquare$$

EXAMPLE 14.6.   Refer to Fig. 14.7. Assume a 150-W incandescent lamp at each ceiling outlet. Also assume the load connected to one conve-nience outlet to be a 10-A hair dryer, with a power factor of 0.80. Calculate the current and power in the two branches of the circuit, and the total cir-cuit current, assuming a 120 V ac source.

SOLUTION.

(a) In the circuit branch feeding the lights we are given the power consumption as two 150 W lamps; that is,

$$P = 2(150) = 300 \text{ W}$$

To calculate current, we would use Eq. 14.7, which expresses power in an ac circuit:

$$W = VI \times pf$$

However, because incandescent lamps are (ef-fectively) resistive loads, their power factor is 1.0. Therefore:

$$\text{Power} = VI$$

$$300 \text{ W} = 120 \text{ V} \times I$$

$$I = \frac{300 \text{ W}}{120 \text{ V}} = 2.5 \text{ A}$$

(b) In the second branch, we have a 10-A, 0.8 *pf* load.

$$\text{Power in watts} = \text{volts} \times \text{amperes}$$

$$\times \text{power factor}$$

$$W = 120 \times 10 \times 0.8 = 960 \text{ W}$$

However, the branch volt-amperes are

$$V \times A = 120 \times 10 = 1200 \text{ VA}$$

This latter figure is important for sizing equipment.

(c) To calculate the total current flowing from the panel to both branches of the circuit, we must combine a purely resistive current (lamp circuit) with a reactive one (dryer circuit). The exact value of current is the *vectorial* sum of the two branch currents, which calculates to be 12.1 amperes. (Vec-torial addition is a technique that considers the phase angle between the two currents. Phase angle, or phase difference, was explained previously.) In normal circuit design practice, the *arithmetic* sum of 12.5 amps would be used. The error introduced is only 3.2% and because it results in slightly oversized circuit components, it is on the safe side. Only where power factors are very low would a careful designer use vectorial addition. ■

One further example at this point demonstrates the importance of power factor in normal situations.

EXAMPLE 14.7. The nameplate of a (single-phase) motor shows the following data: 3 hp, 240 V, ac, 17 A. Assume an efficiency of 90%. Calculate the motor (and therefore circuit) power factor.

SOLUTION. In Appendix I we find the conversion needed,

$$1 \text{ hp} = 746 \text{ W}$$

Therefore,

$$3 \text{ hp} = 3 \times 746 = 2238 \text{ W}$$

This, of course, represents the motor output. Because we know that for any device

$$\text{efficiency} = \frac{\text{output}}{\text{input}}$$

We have

$$\text{power input} = \frac{2238 \text{ W}}{0.9} = 2487 \text{ W}$$

However, for ac,

$$\text{power} = \text{volts} \times \text{amperes} \times \text{power factor}$$

so

$$2487 \text{ W} = 240 \text{ V} \times 17 \text{ A} \times \text{power factor}$$

and

$$\text{power factor} = \frac{2487}{240 \times 17} = 0.61$$

Note the large difference between volt-amperes and watts:

$$V \times I = 240 \times 17 = 4080 \text{ VA}$$

$$P = \text{as above} = 2487 \text{ W}$$

where P designates power. This difference is important in circuit design as we will see when we study the methods for sizing circuit components.

■

14.23   Energy in Electric Circuits

Because energy = power × time, the amount of energy used is directly proportional to both the power of the system and the length of time it is in operation. Because power is expressed in watts or kilowatts and time in hours (seconds and minutes are too small for our use), the units of energy used are watt-hours (Wh) or kilowatt-hours (kWh).

EXAMPLE 14.8

(a) Find the daily energy consumption of the appliances listed if they are used daily for the amount of time shown.

Toaster (1340 W)	15 min
Percolator (500 W)	2 h
Fryer (1560 W)	½ h
Iron (1400 W)	½ h

(b) Assuming that average cost of energy is $0.12 kilowatt-hour, find the daily operating cost.

SOLUTION.

(a) Toaster   1340 W = 1.34 kW × ¼ h
                       = 0.335 kWh

     Percolator   500 W = 0.5 kW × 2 h
                       1 = 1.00 2 kWh

     Fryer   1560 W = 1.56 kW × ½ h
                       = 0.78 kWh

     Iron   1400 W = 1.4 kW × ½ h
                       = 0.70 kWh

          Total   2.815 kWh

(b) The daily operating cost is 2.815 kWh × $0.12/kWh = $0.3378 or approximately 34 cents. ■

The power being used by a residential household varies with the time of day. If we were to graph the power in use for a typical American household during a normal weekday, the plot might look

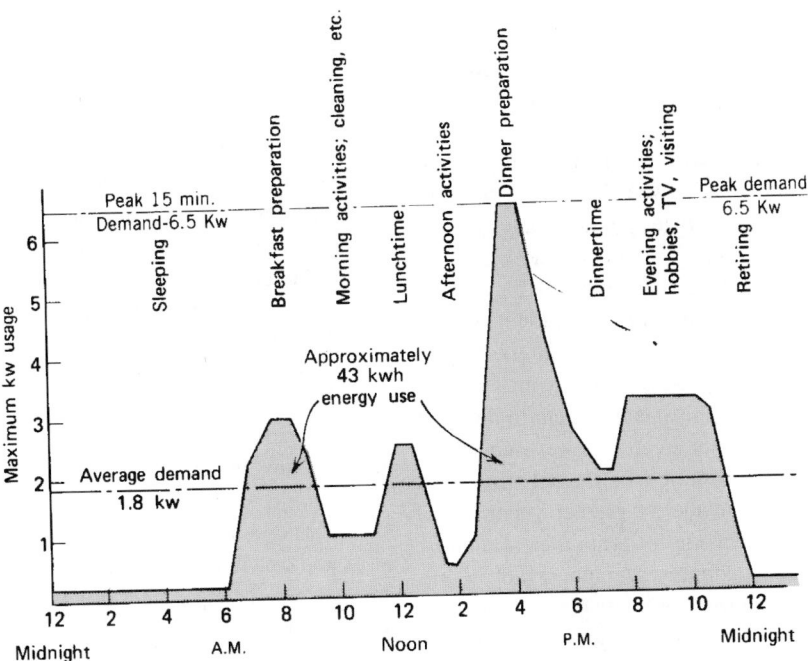

Fig. 14.27 *Hypothetical graph of power usage for a typical American household. Electric cooking is assumed. Area under the curve represents energy usage. Maximum kW demand (vertical axis) is based on a 15-min integrated demand, thus eliminating spikes in demand, such as those caused by starting a refrigeration (air-conditioning) compressor. This curve has a 24-h use of approximately 43 kWh, giving an average 24-h demand of 1.8 kW. The ratio between this average demand and the peak demand of 6.5 kW is called the* load *factor, which here is 27.5%.*

something like that in Fig. 14.27. The average power demand of the household is obviously much lower than the maximum. The ratio between the two is called the *overall load factor* and runs between 20% and 30% for a typical household. The energy used by this household for the 24-h period shown is represented by the *area* under the curve of Fig. 14.27. This can be determined by integration only, because it varies continuously. That this integration is exactly what a kilowatt-hour meter does is explained in Section 14.26, which deals with electrical measurements.

EXAMPLE 14.9. It has been estimated that the average power demand of an American household with electric cooking is 1.8 kW. Calculate the monthly electric bill of such a household, assuming a flat rate of $0.12 per kilowatt-hour.

SOLUTION. Monthly energy use

$$1.8 \, \text{kW} \times \frac{24 \, \text{h}}{\text{day}} \times \frac{30 \, \text{days}}{\text{month}} = 1296 \, \text{kWh}$$

Monthly electric power bill

$$1296 \, \text{kWh} \times \$0.12/\text{kWh} = \$155.52 \quad \blacksquare$$

We stated in Example 14.8 that the bill was based on a "flat" rate of 12 cents per kWh because flat rates are indeed very common for residential users. The actual monetary amount of this flat rate varies from a low of 3 to 4 cents per kWh in areas using hydroelectric power to as high as 17 cents per kWh in some major cities.

By terms of their franchise, electric utilities must provide for a customer's *maximum* power demand, whereas energy billing only compensates them for *average* demand, which is always lower. One technique used by electric utilities to ameliorate this condition is to levy a charge for power (kW) in addition to the normal energy (kWh) billing. This technique has long been standard in industrial and commercial user tariffs, but is unusual for residential users. This *demand charge* is primarily useful in encouraging users to reduce their peak loads. (In so doing, energy use is also somewhat reduced.)

ELECTRICITY

14.24 Electric Demand Charges

As stated, electric utility companies normally levy a kW demand charge on all but individual residential and a few special category customers. Varying with the individual utility company involved, this monthly charge runs between $2 and $15 per kilowatt of maximum average demand in any demand interval for that month. Demand intervals vary, usually being either 15 or 30 min (see Fig. 14.28).

Many utilities use a "sliding window" interval timing technique that starts a new interval every minute and updates the maximum interval demand accordingly. This enables them to find and bill for the maximum electric power demand in any 15- or 30-min period in the month. Some companies also include a "ratchet" clause that levies a demand charge for a number of months based on the maximum demand in any single month. This penalizes users with seasonal highs—that is, users with a low *yearly* load factor. The load factor is a measure of uniformity of power demand; a low load factor indicates short-time demand peaks for which the user is heavily

charged. The justification for the imposition of a demand charge and the significance of load factors can best be demonstrated by an example.

Assume that a pottery manufacturer, whose average 8-h daily load is 20 kW for lighting, pottery wheels, and the like, operates two 50-kW electric kilns twice a month for a 4-h period each time. Further assume an energy charge of $0.10 per kWh. The total monthly *energy* bill for the operation of the two kilns only would be

$$\text{cost} = 2 \times 50\,\text{kW} \times 8\,\text{h} \times \frac{\$0.10}{\text{kWh}} = \$80$$

Thus, were it not for the demand charge, the utility company, which is required by law to supply maximum customer demand, would have to provide and maintain 100 kW of generation, transmission, and distribution facilities in return for a payment that is the equivalent of 1.11 kW of average continuous load—that is,

equivalent continuous load

$$= \frac{\$80.00}{720\,\text{hr/month} \times 0.10} = 1.11\,\text{kW}$$

Fig. 14.28 *Typical instantaneous load curve for a facility. The utility demand meter records the average demand in each period (here, 15 min). The maximum interval demand—in this case between 11:15 and 11:30—is used as the basis for monthly billing. A high load factor (utilization factor) indicates that little can be gained by demand control; a low load factor indicates the reverse.*

The user's load factor can be calculated readily. By definition,

$$\text{load factor} = \frac{\text{average power demand}}{\text{maximum power demand}}$$

(14.8)

For a given time interval (e.g. day, month, year) the average power demand is equal to the energy used in this period divided by the period length.

average period power demand

$$= \frac{\text{kWh energy use}}{\text{hours of use}} \text{ for that time period.}$$ (14.9)

Therefore, the average *monthly* power demand equals the average monthly energy use divided by 720 hours. Substituting the monthly version of Eq. 14.9 as the general expression for load factor in Eq. 14.8, we obtain

LF monthly

(14.10)

$$= \frac{\text{monthly kWh energy use}/720 \text{ hours}}{\text{maximum demand}}$$

For the case under consideration, the *monthly* load factor is

$$\text{LF} = \frac{[23 \text{days} \times 8\text{h} \times 20\text{kW} + 2 \text{days} \times 4\text{h} \times 100\text{kW}] \div 720\text{h}}{120\text{kW}}$$

$$= \frac{(3680 + 800 \text{kWh/month})/720\text{h}}{120\text{kW}}$$

$$= 0.052 \text{ or } 5.2\%$$

This is obviously a very poor load factor, which results in the levying of a high demand charge. Assuming a $8.00 per kW demand tariff, this pottery manufacturer would be billed, monthly, an additional:

demand charge $= 120 \text{ kW} \times \$8.00 = \$960.00$

Inasmuch as the energy bill is only

energy cost $= 4480 \text{ kWh} \times \$0.10 = \$448.00$

this manufacturer is paying heavily for highly peaked power use.

Although the illustration selected is somewhat extreme in its type of power use, it is not uncommon to find demand charges of the same order

of magnitude as energy charges. Obviously, it is impossible to eliminate demand charges entirely, but it is certainly possible and frequently very simple to reduce them. Such a step is in the interest of the user, the utility, and the public at large: the user—for economic reasons; the utility—to permit more efficient use of its facilities; and the general public—by avoidance of unnecessary power plant construction and concomitant inefficient use of fuel at partial generator loading, and by overall reduction in fuel use. This last item is a secondary benefit of demand control.

In the next section we discuss user electric demand control, the primary function of which is to reduce electric *power* demand. This reduces demand charges and, incidentally, secondarily somewhat reduces energy consumption. Electric demand control is different from *energy* management, which is primarily concerned with reduction of all types of energy use, including electric. Electric demand control is frequently included as one part of an overall energy management system.

14.25   Electric Demand Control

Electric demand control methods vary greatly in complexity and in degree of automation, but all basically perform the same task—efficient utilization of available energy to produce a high load factor, resulting in a lowering of demand charges. An ancillary but important benefit is the improved utilization of building electrical power equipment, which normally runs underloaded. When demand control is included at the design stage, it results in smaller equipment, lower first cost, and less space utilization.

For a number of years beginning in the 1980s, many utilities offered their customers rebates that covered up to 40% of the cost of equipment and renovation that would reduce maximum demand *and* overall energy use. This program, now ended, acted as a clear financial incentive for the development of cost-effective demand control and energy conservation equipment and techniques that have become widespread in use. The result has been a considerable reduction in nationwide electric power and energy use. The rebate reduced the investment payback period to such an extent that many and perhaps most large users invested large sums in electric demand control and energy conservation and management equipment.

The advantage to the energy user is obvious; lower electric bills. The reasoning of the utility is equally simple. It costs from $3500 to $5000 per kilowatt of generating capacity for new power plant construction (depending on location and required auxiliary construction), whereas rebates ran from $150 to $1000 per kW conserved, with the larger amounts paid for peak demand reduction. Because a utility, by terms of its franchise, must supply all the power demanded, it is very much in the interest of any utility that is generating power near its equipment's maximum capacity to reduce load in general and peak loads in particular.

The oldest and simplest utility-sponsored demand control scheme, still in use today, is the time-of-day dependent, variable utility rate schedule. This scheme encourages off-peak use of electricity by offering a lower energy rate for off-peak hours use. Utilities offering off-peak rates generally install, at no charge to the user, a separate time-controlled circuit switch for use with time-deferrable loads such as water heaters, well pumps, battery chargers, and the like. The circuit is energized only during off-peak hours, which typically are mid-afternoon and after midnight. One such arrangement is shown in Fig. 14.29. Note that the switch itself has no programming buttons. All programming is done by the utility with an auxiliary programmer (shown) that can be used to program many user switches.

The basic technique of *user* demand control is a simple one; electric loads are disconnected and reconnected in such a fashion that demand peaks are leveled off and load factor is thereby improved. The extent to which a facility's electric loads can tolerate this type of switching is an indication of the potential effectiveness of a demand controller. An installation with a large *un-*interruptible load, such as computer or office lighting, benefits minimally from demand control. Most industrial and commercial installations, however, contain a large percentage of interruptible loads (interruptions may be very short) and demand control systems frequently accomplish a 15% to 20% reduction in electric bills with a resultant short payback period on equipment investment.

Fig. 14.29 *Programmable electronic time control switch designed for application with time-of-use (off-peak) utility rate schedules. Programming device is separate from the switch and can be used by a utility to program many customer switches. The illustrated unit is arranged for 365-day scheduling, which permits full coordination with utility schedules that vary with the seasons of the year. Typical controlled loads are water heaters, thermal storage units, water and air accumulators, and any other load that either inherently or by design can be delayed for several hours. (Courtesy of Paragon Electric. Co., Inc.)*

The proliferation of demand control equipment has also produced a proliferation of nomenclature, including load shedding control, automated load control, peak demand control, and programmable load control. Descriptions that include the term *energy management* refer to devices whose primary function is the control of *energy* and secondarily are equipped to accomplish electric demand control. Such devices are discussed in Chapter 5 for HVAC, Chapter 21 for lighting and Chapter 22 for complete building energy control systems.

Demand control devices are intended to control *power*, which is *timed* energy use. Demand control produces energy savings as a secondary benefit. We use the expression *electric demand control* for simplicity and avoid knowingly using a term that refers to a particular manufacturer's equipment.

We will discuss the various demand control schemes in some detail to enable the reader to differentiate between them by recognizing their specific characteristics. A manufacturer's descriptive literature usually comprises a lengthy list of its equipment's abilities, none of its shortcomings, and little of its specific comparative applicability. Furthermore, much of the equipment on the market today has additional functions not directly related to demand control. That being so, a good understanding of the essentials of demand control schemes will greatly assist the prospective user.

(a) Level 1—Load Scheduling and Duty-Cycle Control.

This level is the simplest and most obvious, and is applicable to all types of facilities. The installation's electric loads are analyzed and then scheduled to restrict demand. Thus, large loads can be shifted to off-peak hours and controlled to avoid coincident operation. The user can also take advantage of special night and weekend utility rates for loads that do not require immediate operation, such as battery charging and transfer pumping. The demand control device used is essentially a programmable time switch (see Section 16.16) for a number of circuits or loads. It is not, strictly speaking, a demand controller in that no cognizance is taken of the actual continuous electric load. Instead, it operates on a preset timed duty cycle relying entirely on a prior analysis of the loads. Typical applications of this device are control of HVAC loads, lighting loads, and process loads in small commercial, institutional, and industrial buildings.

(b) Level 2—Demand Metering Alarm.

If, in conjunction with a duty-cycle controller, some type of continuous demand metering is installed that goes into alarm when a predetermined demand level is exceeded, a basic load control system will have been established. The load analysis discussed previously would have to be extended to determine load priorities so that when the preset maximum demand load is exceeded and the alarm sounds, loads can be shed (disconnected) manually in a predetermined order of priority and subsequently reconnected, also in order of priority. This type of control is practical only for a limited size installation, inasmuch as the load switching activity is manual.

(c) Level 3—Automatic Instantaneous Demand Control.

This type of control (also called *rate control*) is, in effect, an automated version of the level-2 system described previously. The unit accepts instantaneous kilowatt load information from the utility system, compares this information to the preset kilowatt limit, and acts automatically to disconnect and reconnect loads as required. These units *do not* recognize the utility's metering interval, but act continuously on the basis of load comparison data. For this reason, these units are also referred to as *load comparator controllers*. Examination of Fig. 14.30 will make the unit's operation clear.

Typical sheddable loads might include nonessential lighting, some cooling load, domestic hot water heating, snow melting, and the like. Nonsheddable loads (that do not tolerate even short duration interruption) might include essential lighting, elevators, communications equipment, computers, process control, emergency equipment, and the like. The nonsheddable loads are

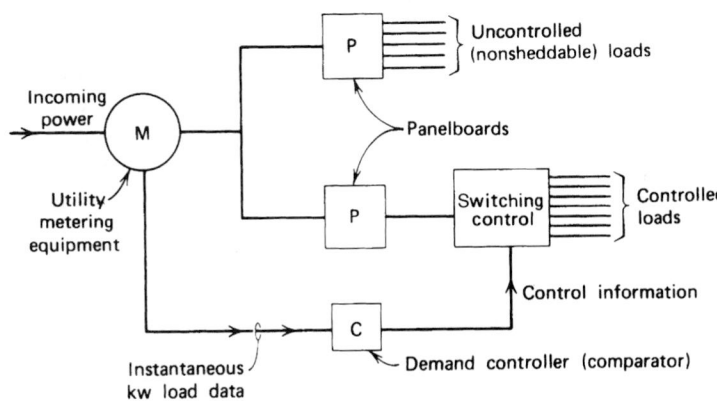

Fig. 14.30 *Block diagram of a system of automatic electric power demand control. The demand controller receives instantaneous load data from the metering equipment, compares it to preset limits, and disconnects and reconnects controllable loads automatically to keep kilowatt demand (load) within these limits.*

Fig. 14.31 *Graph of the action of an ideal rate controller. Note that toward the latter part of the period, the pattern has been established, and the actual rate of energy use corresponds closely to the ideal. This type of control is often simply a part of the software in a large centralized building control system.*

fed directly from the power line. The sheddable loads are fed through a panel of control relays that respond to on/off instructions from the demand controller. Although theoretically the energy use with or without the controller is the same, in actual practice energy savings of 15% and more are common.

The principal drawback of this system is that the load-shedding is preprogrammed. This results in an inability to readily adapt to varying load patterns, resulting from variable production schedules, time schedules, changes in weather, and so on. As a result of this limitation, this system is most useful in applications where operating modes do not change frequently and the facility is not very large. Thus stores, supermarkets, warehouses, small industrial facilities, and commercial installations are well served with this level system if they have at least 20% sheddable loads and their connected electric load is at least 150 kilovolt-amperes (kVA).

This level of demand control, as well as levels 4 and 5 described later, are most often supplied as one of many functions of a larger building energy control system. Because this is so, it is necessary to ascertain that the type of demand control to be supplied by the overall building control hardware and software is what the designer intended.

(d) Level 4—Ideal Curve Control. This control function operates by comparing the actual rate of *energy* usage to the ideal rate, and controls kilowatt demand by controlling the total *energy* used within a metering interval. The utility company determines the demand over the demand interval by integrating the kilowatt-hour energy over the interval and dividing by the interval time. Thus, the user is actually given a block of energy (kWh) that can be utilized at any desired rate, not necessarily at a constant rate. The desirable rate of energy use is shown as the "ideal curve" on the typical usage curve shown in Fig. 14.31.

Shed points can be programmed for each load independently according to a predetermined priority, and priorities can be readily adjusted and rescheduled. Loads are normally shed only toward the end of the interval, when the permissible energy total for that interval is approached, and all loads are restored at the beginning of each interval. Thus, during each interval sheddable loads are off for only a few minutes at most. Controllers operating on the "ideal curve" principle are considerably more flexible than the kilowatt rate controller described in the level-3 system and are applicable to facilities of widely divergent load size, but with at least a 300-kW connected load. As with other controllers, the principal savings will be in demand charges, but almost always with considerable economies in energy billings.

(e) Level 5—Forecasting Systems. These systems are by far the most sophisticated, the most expensive, and the most effective. They are best applied to large structures where the number of loads, load patterns, and complexity of operation preclude the effective use of the preceding systems. As a result of the large amount of load data required to be processed, these systems are usually installed as part of computerized central control facilities. Details of operation are too complex to be described here, but the basic operation can be outlined. These units operate by continuously forecasting the amount of energy remaining in the demand interval based on kilowatt-hour data received. They then examine the status and priority of each of the connected loads and decide on a course of action. Loads that in other systems are classified nonsheddable are, in this system, controlled because of the accuracy and rapidity of the control function. A pneumatic compressor, for instance, that supplies process air might, in other level control systems, be classified as noncontrollable, despite the fact that it has long off periods. With computer control, such a load would be classified as "delayable" or "inhibited" because a 30-sec or 1-min delay in activation after

the pressure switch closes its contact is normally acceptable.

The advantage of these systems is that if programmed properly, they can make small, accurate load changes throughout the interval, resulting in minimum load cycling and maximum efficiency.

14.26 Electrical Measurements

In the preceding sections we explain the fundamental electric quantities of voltage and current and define the units involved as volts and amperes, respectively. As is true of all other physical quantities that are to be used in practical application, the need existed for a simple means of measuring these quantities. This need was initially met by the development of the meter movement illustrated in Fig. 14.32(a).

Everyone at one time or another has felt the repulsion between like poles of two magnets held close together and, conversely, the attraction between opposite poles. This principle was used in the first basic meter movement. It causes a deflection of the pointer as a result of the repulsion between the field of a permanent magnet and an electromagnet. The electromagnet is formed when current flows in the coil, and its strength is proportional to the amount of measured current.

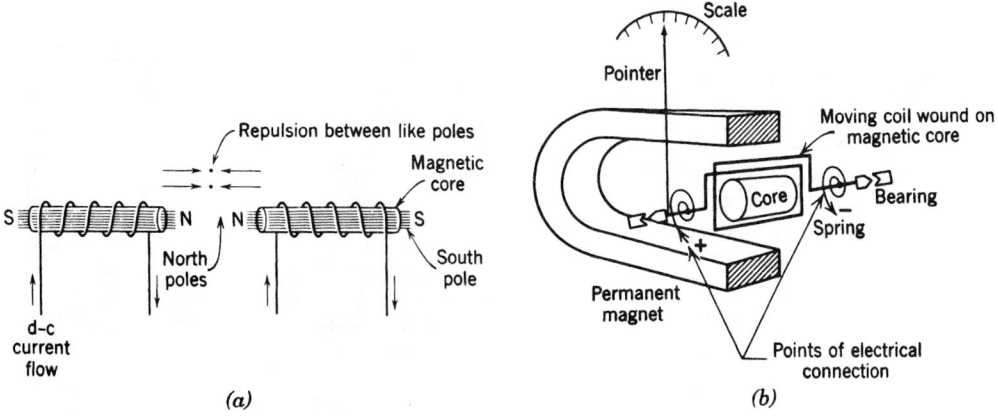

Fig. 14.32 (a) *Diagram showing basic electromagnetic principle and interaction between electromagnets. Any iron core becomes an electromagnet when current flows in a coil wound around it, as shown.* (b) *The principle of the electromagnet is used in this basic meter movement. Current flowing in the movable coil forms an electromagnet whose field interacts [see (a)] with the permanent magnet's field, causing a deflection proportional to the current flow. (Courtesy of Wechsler, Div. of Hughes Corp.)*

Thus, a strong current causes a larger deflection of the needle and therefore a higher reading on the dial. A spring [see Fig. 14.32(*b*)] provides restraining torque on the pointer. To make this very sensitive basic unit usable for large currents (it is intrinsically a microammeter, sensitive to millionths of an ampere), we simply divert, or *shunt* away, most of the current, allowing only a few microamperes to actually flow in the meter coil.

To use the same unit as a voltmeter, we put a large resistance called a *multiplier* in series with the meter, thereby again limiting the current flowing to a few microamperes. The meter scale is then calibrated in volts. All dc meters are made in this fashion. Most ac meters operate on basically the same principle, except that instead of a permanent magnet, an electromagnet is used. Thus, when the polarity reverses, the deflecting force remains in the same direction. A dc meter connected to an ac circuit simply will not read because inertia prevents the needle from bouncing up and down 60 times a second.

The meters described are conventional analog units that read electrical units in proportion to mechanical forces exerted within the unit—that is, by analogy. Modern electronics has produced a line of solid-state electrical meters (Fig. 14.33) that display the measured electrical values in analog mode (needle on a dial), digital mode, or both. They operate in a number of ways, all of which are different from that described previously, and all of which are beyond the scope of this book.

The fact that a meter reads digitally does not necessarily mean that it is highly accurate. Accuracy depends on the quality of the internal circuitry. Digital meters are obviously easier to read than analog types because no visual interpretation and interpolation is involved. This advantage, plus the constantly declining cost of sophisticated electronics, will undoubtedly lead to solid-state digital meters replacing analog types, except for special applications.

In building work, the measurement of current and voltage is generally not as important as the measurement of power and energy. To measure power, we utilize the fact learned earlier that power is proportional to the product of the voltage and current in the circuit. Although actual construction is complex, the theory of operation of a conventional coil-type wattmeter is simple. The meter has two coils; a current coil that is similar in connection to an ammeter, and a voltage coil that is similar in connection to a voltmeter (Fig. 14.34). By means of the physical coil arrangement, the meter deflection is proportional to the product of the two, and therefore to the circuit power. The meter can be calibrated as desired, depending on the size of the shunts and multipliers.

Fig. 14.33 (a) *Solid-state clamp-on type meter with digital readout and automatic ranging. This latter feature eliminates the necessity to preselect the meter range, and is particularly useful where the magnitude of current or voltage is unknown. The clamp-on feature permits use without wiring into, or otherwise disturbing the circuit being measured. The meter measures approximately 8 × 3 × 1.5 in., weighs under a pound and is battery powered. Scales are 0.1–1000 amp ac, 0.1–1000 V ac, and 0.1–1000 ohms resistance, with ± 2% accuracy.*
(b) *Solid-state auto-ranging clamp-on ac meter with analog type readout. Similar in design to the meter in (a), except with somewhat larger range scales and additional features such as peak current measurement. (Photos courtesy of TIF Instruments, Inc.)*

(a)   (b)

Fig. 14.34 *Schematic arrangement of wattmeter connections. Note that the current coil is in series with the circuit load, whereas the voltage leads are in parallel.*

To measure energy, the factor of time must be introduced because, as we know

$$\text{energy} = \text{power} \times \text{time}$$

dc energy meters are available but are not of general interest because of the rarity of dc power. ac watt-hour meters are basically small motors, whose speed is proportional to the power being used. The number of rotations is counted on the dials, which are calibrated directly in kilowatt-hours. A diagram of the basic construction of an ac kilowatt-hour meter is shown in Fig. 14.35. As can be seen, the kilowatt-hour energy consumption and the maximum interval kilowatt demand can be read directly from the dials. If the numbers involved are too large for calibration reasons, a multiplying factor is required to arrive at the proper kilowatt-hour consumption. This number is written directly on the meter nameplate, and we multiply the meter reading by it to get the actual kilowatt-hours. Several types of solid-state multifunction and energy meters are shown in Figs. 14.36, 14.37, and 14.38.

Continuous monitoring of the electrical characteristics of an entire distribution system, or of a portion fed from a specific switchboard, can be readily accomplished by use of the solid-state device shown in Fig. 14.39. This programmable analyzer can be arranged to perform power and energy power quality analysis, harmonic analysis, and the like, and to display results locally and remotely.

Because manual reading of kilowatt-hour meters of individual consumers is so labor intensive a task even without consideration of routine job difficulties such as inaccessible meters, dogs, inclement weather, and so on, utilities have long sought a more cost-effective means of determining consumer energy usage. To this end, with the benefit of modern microelectronics, a number of interesting, laborsaving kilowatt-hour meters have been developed. One type is equipped with a programmable electro-optical automatic meter reading system that can be activated locally or

Fig. 14.35 *Typical induction-type kilowatt-hour meter with kilowatt demand dial. Dials register total disc revolutions that are proportional to energy. Disc rotational speed is proportional to power. Note that the current coil is in series with the load and that the voltage coil is in parallel.*

(a)   (b)

Fig. 14.36 (a) *A modern multifunction solid-state kilowatt-hour meter. This single-phase meter includes a mechanical register similar to that shown in Fig. 14.35 plus a large liquid crystal display. It can be arranged to provide either time-of-use billing or demand/time-of-use billing in addition to its energy measurement function. It can also be configured to provide load control, demand threshold alert, end-of-interval alert, and load profile recording.*
(b) *Field programming of the meter shown in* (a) *is accomplished with a hand-held programming device, as shown. (Photos courtesy of Landis and Gyr.)*

Fig. 14.37 *Digital power factor meter, which measures power factor on single and three-phase circuits. In addition, the meter can measure 0–600 V ac and 0–2000 A ac. Connections to the circuits being measured are made with the clamp-on probes shown. A very similar unit is available that measures power in both sinusoidal and nonsinusoidal wave forms, in a range of 0.1 W through 2000 kW. (Courtesy of AEMC Corporation.)*

Fig. 14.38 *This portable programmable power analyzer unit measures approximately 13 × 7 × 12 in. and weighs 6 lb. It is capable of measuring (and computing) 19 electrical parameters, among which are current, voltage, energy, power, power factor and frequency, and of providing both instantaneous digital readout and printed records (hard copy). Connections permit all functions to be addressed by remote control. (Courtesy of AEMC Corporation.)*

IQ Analyzer

Fig. 14.39 *Power system analysis unit, designed to be mounted permanently into a system's switchboard (see Figs. 16.24 and 16.25). This programmable unit can meter circuit parameters, measure energy and power use, analyze power quality (including harmonic analysis) and log events. (Photo courtesy of Cutler-Hammer.)*

from a remote location. The meter data are transmitted electrically to a data-processing center, where they may be used by the utility to prepare subscribers' bills, prepare customer load profiles, and study, in combination with other such data, area load patterns, equipment loading, and so on. The customer can use instantaneous data to control loads, as explained at length in the preceding section.

Another meter type is equipped with a miniature radio transmitter that can be remotely activated to transmit the current kilowatt-hour reading. The meter reader moves along the street and remotely activates the meter by entering a customer code into a digital pad. A special receiver not only receives the transmitted kilowatt-hour data but also encodes and records the data automatically. These and other new kilowatt-hour meters are obviously much more costly than the traditional units, but are gradually replacing them because of the large reduction in labor costs and the quantity and quality of data made available.

15

ELECTRICAL SYSTEMS AND MATERIALS

WIRING AND RACEWAYS

The major components of a building's electrical system can be arranged in three major categories; wiring and raceways, power-handling equipment, and utilization equipment. In the first category, we include conductors and raceways of all types; in the second, transformers, switchboards, panelboards, large switches, and circuit breakers; and in the last, actual utilization equipment such as lighting, motors, controls, and wiring devices. After some discussion applicable to all electrical materials, this chapter discusses in detail the items in the first of these three categories—that is, the wiring and raceway system. Chapter 16 covers most of the remaining items in the next two categories, with the exception of lighting equipment, which is discussed in the lighting section. Signal systems, building control and automation are discussed in Chapter 22.

15.1 System Components

Referring to Fig. 15.1, note the important fact that the power distribution equipment proceeds from the service point to the utilization points in a series of steps of decreasing circuit capacity. This is analogous to the arrangement of water supply systems and heating systems; the distribution equipment is largest at the supply point, narrowing in steps to reach the last utilization points. The reverse is true in collection systems such as drainage and venting; there, piping is smallest at the initial collection points growing larger (in steps) as quantities of fluid increase.

A normal single-line diagram does not differentiate by line weight between heavy and light

(large and small) conductors (the heavier the conductor, the greater the amount of power being carried), but the single-line diagram of Fig. 15.1 does differentiate in order to show relative power levels in the system. This "size" differentiation is more clearly shown in Fig. 15.2, which is a pictorial representation of a system similar to that of Fig. 15.1 but in somewhat greater detail and omitting items beyond the panelboard. Only standard items are shown; special items, however common, such as emergency and uninterrupted power sources, energy controls, service transformers, and the like, are omitted for the sake of clarity.

15.2 National Electrical Code

The National Electrical Code (*NEC*) of the National Fire Protection Association (NFPA) defines the fundamental safety measures that must be followed in the selection, construction, and installation of all electrical equipment. This code is used by all inspectors, electrical designers, engineers, contractors, and operating personnel. Having been incorporated into Occupational Safety and Health Act (OSHA) it has, in effect, the force of law. The reader of this book should obtain for personal use the latest edition of the *NEC* from NFPA at 80 Batterymarch Street, Boston, MA 02110, or a local source. Frequent references are made to this code throughout this text.

In addition to the *NEC,* many large cities such as New York, Boston, and Washington, DC have their own electrical codes that, although similar to the *NEC,* contain numerous special requirements.

In order to ensure a minimum standard of in-

Fig. 15.1 *Single-line diagram of a typical building electrical distribution system, from the incoming service to the utilization items at the end of the system. This type of diagram is also referred to as a block diagram because the major components are shown as rectangles, or blocks. When this same type of information is presented showing the vertical spatial relationship between components, it is called a* riser diagram; *when electrical symbols are used in lieu of the blocks, it is called a* one-line *or* single-line diagram. *The connecting conductors between the major system components are drawn to reflect size and thereby power-handling capacity, for illustrative purposes only. In practice, all connecting lines are the same weight. Also, branch circuits are not usually shown; the diagram ends at the lighting and appliance panels.*

trinsic electrical safety for electrical equipment, a single agency was needed to establish standards and actually test and inspect electrical equipment. Such an organization is Underwriters Laboratories, Inc., which publishes lists of inspected and approved electrical equipment. These listings are universally accepted, and many local codes state that only electrical materials bearing the Underwriters Laboratories (UL) label (of approval) are acceptable.

15.3   Economics of Material Selection

The selection of electrical materials involves not only choosing a material or assembly that is functionally adequate and, where necessary, visually satisfactory, but also the consideration of economic factors. This is necessary because usually many brands and types of equipment that will fulfill the construction need are available. In such cases, economic factors often decide the issue.

The decision is relatively simple when the various materials differ only slightly from each other and a straightforward first-cost comparison is all that is required. Often, however, the choice is not so simple, because the materials may vary

considerably in characteristics other than functional suitability, and a more detailed cost study is required. Such economic analyses are frequently performed in comparisons between various HVAC systems. This is less so when dealing with electrical systems because, from an energy point of view, electrical systems are more passive than are HVAC systems, and it is energy costs that are often the decisive factor in economic analyses. We are speaking, of course, of life-cycle equipment costs (over the life of the *structure*) expressed in *present-value* dollars, or annual owning and operating costs, including equipment amortization costs. The type of analysis used depends on the situation. However, such comparisons are useful only when both the initial cost and the operating costs will be borne by the same individual—that is, an owner-operator. In the case of a speculative building venture, first cost is usually the deciding factor.

It is often difficult to perform such cost analyses because accurate data on life and maintenance costs for electrical equipment may not be readily available. Still, even if a formal analysis cannot be done, the principle involved should be considered in the selection of all electrical material, bearing in mind its relative importance as energy-

Typical Electrical Building Equipment

9. Plug-in switch (on duct).

8. Dry-type transformers to supply 120 volts.

7. Plug-in busway.

6. Switch unit for plug-in busway.

5. Panelboard for 120-volt circuits.

4. 277-volt panelboard.

Typical wiring closet

Plug-in bus duct or cables.

Feeder busduct or cables.

10. Centralized switching and protection for large branch circuits.

11. Power panelboard for centralized switching and protection for small power branch circuits.

12. Individual small branch circuit switching and protection.

3. Risers for carrying power up through the building from the entrance switchboard.

13. Feeders for supplying power load, elevators, and other heavy electrical loads.

12. Individual small branch circuit switching and protection.

7. Plug-in busway.

6. Switch unit for plug-in busway.

14. Centralized motor control.

1. Service feeder

2. Building switchboard to switch and protect main feeders and risers in building.

ELECTRICITY

Fig. 15.2 *Diagrammatic, pictorial representation of a typical building electrical power system with the relative power capacity being indicated by the size of the conductors. This diagram does not extend beyond the local panelboard and includes only commonly used items. Note also that the entire wiring system is, in effect, jacketed in steel. (Courtesy of General Electric.)*

utilizing equipment. Detailed economic discussions on particular items are found with the related technical discussion. See Appendix E for an explanation of cost analysis techniques.

15.4 Energy Considerations

Energy costs are a major factor in economic analysis. However, energy considerations are at least as important in and of themselves because of environmental and natural resource considerations. This aspect is examined in detail in the discussions of building energy budgets and their components, including lighting, elevators, and electric motors in all other systems.

15.5 Electrical Equipment Ratings

All electrical equipment is rated for the normal service it is intended to perform. These ratings may be in voltage, current, duty, horsepower, kW, kVA, temperatures, enclosure, and so on. Ratings related to specific equipment are discussed in the sections that follow. The ratings that are specifically and characteristically electrical are those of voltage and current.

(a) Voltage. The voltage rating of an item of electrical equipment is the maximum voltage that can safely be applied to the unit continuously. It frequently, but not always, corresponds to the voltage applied in normal use. Thus, an ordinary wall electrical receptacle is normally rated at 250-V maximum, although in normal use, only 120 V are applied to it. The rating is determined by the type and quantity of insulation used and the physical spacing between electrically energized parts.

(b) Current. The current rating of an item is determined by the maximum operating temperature at which its components can operate at full load. That in turn depends on the type of insulation used. As a case in point, consider an electric motor. The current flowing in the motor windings causes a power loss (I^2R), which generates heat. If the windings are insulated with varnished cotton

braid, with a maximum safe operating temperature of 65°C, the maximum permissible current (to which the horsepower rating of the motor is directly related) is that current which produces this operating temperature. If these same windings are insulated with a silicone or glass compound with a maximum operating temperature of 150°C, obviously more current can be carried safely, and the horsepower rating is consequently larger. Thus we see that, although a motor is rated in horsepower (or kW where SI units are used), a transformer is rated in kVA and a cable (discussed later) is rated in amperes, the actual criterion on which all these ratings are based is maximum permissible operating temperature of its insulation (and other components).

15.6 Interior Wiring Systems

At this point it is helpful to survey the different types of interior wiring systems before commencing a discussion of components. When the primary purpose of the system is to distribute electrical energy, it is referred to as an *electrical power system;* when the purpose is to transmit information, it is referred to as an *electrical signal* or *communication system.* In this chapter, we deal with electrical power systems, except that the discussion on raceways covers equipment used by communication systems as well.

Due to the nature of electricity, its distribution within a structure for power use poses basically a single problem: how to construct a distribution system that *safely* provides the energy required at the location required. The safety consideration is all-important because even the smallest interior system is connected to the utility's powerful network, and the potential for physical damage, injury, and fire is always present. The solution to this problem is to isolate the electrical conductors from the structure, except at those specific points, such as wall receptacles, where contact is desired. This isolation is generally accomplished by insulating the conductors and placing them in protective raceways.

The principal types of interior wiring systems in use today are exposed insulated cables, insulated cables in open raceways, and insulated conductors in closed raceways.

(a) Exposed Insulated Cables. In this category would be included (using the *NEC* nomenclature) cable types NM ("Romex") and AC ("BX"). (See Sections 15.11 and 15.12.) Also included are other types where the cable construction itself provides the necessary electrical insulation and mechanical protection.

(b) Insulated Cables in Open Raceways (Trays). This system is specifically intended for industrial application, and it relies on both the cable and the tray for safety.

(c) Insulated Conductors in Closed Raceways. This system is the most general type and is applicable to all types of facilities. In general, the raceway is installed first and the wiring pulled in or laid in later. The raceways themselves may be:

1. Buried in the structure; for example, conduit in the floor slab or underfloor duct. (See Sections 15.18, 15.19, 15.25, and 15.26.)
2. Attached to the structure; for example, all types of surface raceways, including conduit and wireways suspended above hung ceilings. (See Sections 15.23 and 15.30.)
3. Part of the structure; for example, cellular concrete and cellular metal floors. (See Sections 15.27 and 15.28.)

(d) Combined Conductor and Enclosure. This category is intended to cover all types of factory-prepared and factory-constructed integral assemblies of conductor and enclosure. Included here are all types of busway, busduct, and cablebus (Section 15.15); flat cable assemblies and lighting track (Section 15.16); flat cable intended for under-carpet installation (Section 15.30); and manufactured wiring systems (Section 15.31).

15.7 Conductors

Electrical conductors (wiring) are the means by which the current is conducted through the electrical system, corresponding to the piping in the hydraulic analogy. The standard of the American wire and cable industry for round cross-section conductors is the American Wire Gauge (AWG).

By convention, a single insulated conductor No. 6 AWG or larger, or several conductors of any size assembled into a single unit, are referred to as *cable*. Single conductors No. 8 AWG and smaller are called *wire*.

All wire sizes up to No. 0000 (also written No. 4/0) are expressed in AWG. The AWG numbers run in *reverse* order to the size of the wire—that is, the smaller the AWG number, the larger the size. Thus No. 10 is a heavier wire than No. 12 wire and lighter (thinner) than No. 8 wire. The No. 4/0 size is the largest AWG designation, beyond which a different designation called *kcmil* (thousand circular mil) is used. The former designation for this unit was *MCM*, a term that is still used in many sources. In this designation, wire diameter *increases* with number; thus, 500 kcmil is a heavier wire (double the area) than 250 kcmil.

Outside of the United States, where the metric system is in general use, conductor sizes are given simply as their diameter in millimeters (mm). Table 15.1 gives dimensional and stranding data for the commonest wire sizes and includes the millimeter equivalent of each size. This will prove useful in interfacing American gauges and metric sizes.

15.8 Conductor Ampacity

Conductor current–carrying capacity, or *ampacity*, is determined as explained previously, by the maximum safe operating temperature of the insulation. Heat generated as a result of the current flowing is dissipated into the environment. Thus, for a given environment (open-air, buried in earth, or enclosed), ampacity increases with increasing conductor size *and* with maximum permissible insulation temperature. These facts are clearly shown in Table 15.2. If more than three conductors are placed in a conduit, the resultant increase in temperature requires that the conductors be derated by the amount shown in Table 15.3.

Because heat dissipation from a conductor in free air is much greater than that from the same conductor enclosed in conduit or direct buried, its corresponding allowable ampacity is also greater. Conversely, if the ambient temperature around the conductor is higher than 30°C, on which all

TABLE 15.1 **Physical Properties of Bare Copper Conductors**

Size AWG Kcmil (MCM)	Area Circular Mils	Diameter				dc Resistance at 25°C, (77°F) (Uncoated) ohms/1000 ft
		(in.)		(mm)		
		Solid	Stranded	Solid	Stranded	
18 AWG	1,620	0.040	—	1.02	—	7.77
16	2,580	0.051	—	1.29	—	4.89
14	4,110	0.064	—	1.63	—	3.07
12	6,530	0.081	0.092	2.05	2.34	1.93
10	10,380	0.102	0.116	2.59	4.06	1.21
8	16,510	0.128	0.146	3.26	3.71	0.764
6	26,240	—	0.184	—	4.11	0.491
4	41,740	—	0.232	—	5.18	0.308
2	66,360	—	0.292	—	6.55	0.194
1	83,690	—	0.332	—	7.34	0.154
0 (1/0)	105,600	—	0.373	—	8.26	0.122
00 (2/0)	133,100	—	0.418	—	9.27	0.097
000 (3/0)	167,800	—	0.470	—	10.41	0.077
0000 (4/0)	211,600	—	0.528	—	11.68	0.061
250 kcmil	250,000	—	0.575	—	12.70	0.052
300 kcmil	300,000	—	0.630	—	13.92	0.043
400 kcmil	400,000	—	0.728	—	16.05	0.032
500 kcmil	500,000	—	0.813	—	19.56	0.026

Source: Except for millimeter dimensions, this table was extracted from NFPA 70-1999, the *National Electrical Code.* Copyright © 1999, National Fire Protection Association, Quincy, MA 02269.

Note: This extracted material is not the complete and official position of the National Fire Protection Association on the referenced subject, which is represented only by the standard in its entirety.

the ampacity tables are based, the permissible ampacity must be reduced.

Ampacity tables for conductors in free air, for cable types not shown in Table 15.2 and derating factors for high ambient temperatures are all found in the *NEC.* Typical ambient temperatures are given in Table 15.4.

15.9 Conductor Insulation and Jackets

Most conductors are covered with some type of insulation that provides both electrical isolation and a degree of physical protection. Additional physical shielding, where necessary, is provided by a jacket over the insulation. Insulation is rated by voltage. Ordinary building wiring is rated for 300 V or 600 V. Common types of building wire insulation are listed in Table 15.5 with their associated trade names, code letters, maximum per-

mitted operating temperatures, and special provisions.

15.10 Copper and Aluminum Conductors

Aluminum has an inherent weight advantage over copper, with concomitant lower installation costs. Economy usually lies with copper in small- and medium-size cable, because weight is not a problem and the smaller conduit required for the smaller copper conductors generally makes the combined installation cheaper. In the larger cable sizes, the aluminum weight advantage offsets the economy of smaller copper size and smaller conduit, and generally proves less expensive, particularly in areas of high labor cost such as urban areas.

Aluminum and copper both exhibit the low electrical resistivity necessary for a good electrical

TABLE 15.2 **Allowable Ampacities of Insulated Conductors Rated 0 through 2000 Volts, 60° to 90°C (140° to 194°F), Not More Than Three Current-Carrying Conductors in Raceway or Cable or Earth (Directly Buried), Based on Ambient Temperature of 30°C (86°F)**

Part A

| Size AWG, Kcmil | Temperature Rating of Conductor[a] | | |
| | 60°C (140°F) | 75°C (167°F) | 90°C (194°F) |
	Type UF	Types RHW, THW, THWN, XHHW[b]	Types THHN, XHHW[b]
14 AWG[c]	20	20	25
12[c]	25	25	30
10[c]	30	35	40
8	40	50	55
6	55	65	75
4	70	85	95
2	95	115	130
1	110	130	150
0	125	150	170
00	145	175	195
000	165	200	225
0000	195	230	260
250 Kcmil	215	255	290
300	240	285	320
350	260	310	350
400	280	335	380
500	320	380	430

Part B

| Ambient Temperature | | Correction Factors For Ambient Temperatures Other Than 30°C (86°F), Multiply the Allowable Ampacities Shown Above by the Appropriate Factor Shown Below. | | |
C	F			
21–25	70–77	1.08	1.05	1.04
26–30	78–86	1.00	1.00	1.00
31–35	87–95	.91	.94	.96
36–40	96–104	.82	.88	.91
41–45	105–113	.71	.82	.87

Source: This table was extracted from NFPA 70-1999, the *National Electrical Code.* Copyright © 1999, National Fire Protection Association, Quincy, MA 02269.

Note: This extracted material is not the complete and official position of the National Fire Protection Association on the referenced subject, which is represented only by the standard in its entirety.

[a]See Table 15.5.

[b]For dry and damp locations use 90°C rating; for wet locations use 75°C rating.

[c]Unless otherwise permitted by the *NEC,* the overcurrent protection for these conductors shall not exceed 15 A for 14 AWG, 20 A for 12 AWG, and 30 A for 10 AWG, after application of correction factors for ambient temperature and number of conductors in raceway.

ELECTRICITY

TABLE 15.3 **Current-Carrying Capacity Derating Factors**[a]

Number of Current-Carrying Conductors in Raceway[b]	Derating Factor[c]
4–6	0.80
7–9	0.70
10–20	0.50
21–30	0.45
31–40	0.40
41 and above	0.35

Source: This table was extracted from NFPA 70-1999, the *National Electrical Code.* Copyright © 1999, National Fire Protection Association, Quincy, MA 02269. *Note:* This extracted material is not the complete and official position of the National Fire Protection Association on the referenced subject, which is represented only by the standard in its entirety.

[a]These factors are to be applied after application of derating factors for ambient temperatures other than 30°C (86°F). See Table 15.2.

[b]Grounding and bonding conductors are not counted. Neutral conductors are counted in special cases. See 1999 *NEC*, Article 310-15(4).

[c]When a load diversity factor is applicable, derating factors as given in 1999 *NEC* Table B-310-11 shall be used.

conductor. There are, however, difficulties inherent in splicing and terminating aluminum. These difficulties—which can be overcome with the use of proper equipment, techniques, and workmanship—stem from aluminum's cold-flow characteristic when under pressure (causing joints to loosen) and aluminum's oxide. This oxide, which forms within minutes on any exposed aluminum surface, is an adhesive, poorly conductive film that must be removed and prevented from reforming if a successful, long-life joint or termination is to be effected. If this is not done, the oxide causes a high-resistance joint with consequent excessive heat generation and possible incendiary effects. This oxide problem can, in large measure, be overcome by the use of copper-clad aluminum wire, but the cold-flow problem remains. Furthermore, when used in branch circuits, even if properly installed initially, aluminum can create problems when wiring devices are replaced by unskilled homeowners.

As a result of a number of unfortunate incidents, some localities in the United States have banned the use of aluminum wire in branch circuitry. Heavy feeders are normally installed by experienced and skilled contractors, and the risk of a poor joint is minimized. We recommend that the use of aluminum conductors be restricted to sizes not smaller than No. 4 AWG, and that installation be permitted *only* by contractors who certify expertise in the specialized techniques involved. Also, local codes and electrical inspectors should be consulted. All references in this text, including all tables and illustrations, are to copper conductors. The following sections are devoted to a brief description of the principal building wire types.

TABLE 15.4 **Typical Ambient Temperatures**

Location	Temperature
Well-ventilated, normally heated buildings	30°C (86°F)
Buildings with major heat sources such as power stations or industrial processes	40°C (104°F)
Poorly ventilated spaces such as attics	45°C (113°F)
Furnaces and boiler rooms (min.)	40°C (104°F)
(max.)	60°C (140°F)
Outdoors in shade in air	40°C (104°F)
In thermal insulation	45°C (113°F)
Direct solar exposure	55°C (131°F)

TABLE 15.5 **Conductor Insulation and Application**

Trade Name	Type/Letter	Temperature	Application Provisions
Moisture and heat-resistant rubber	RHW	75°C (167°F)	Dry and wet locations
Single conductor, underground feeder and branch-circuit	UF	60°C (140°F) 75°C (167°F)[a]	Refer to *NEC* Article 339
Moisture-resistant thermoplastic	TW	60°C (140°F)	Dry and wet locations
Heat-resistant thermoplastic	THHN	90°C (194°F)	Dry and damp locations
Moisture and heat-resistant thermoplastic	THW	75°C (167°F) 90°C (194°F)	Dry and wet locations Special applications
Moisture and heat-resistant thermoplastic	THWN	75°C (167°F)	Dry and wet locations
Moisture and heat-resistant cross-linked synthetic polyethylene	XHHW	90°C (194°F) 75°C (167°F)	Dry and damp locations Wet locations

Source: This table was extracted from NFPA 70-1999, the *National Electrical Code.* Copyright © 1999, National Fire Protection Association, Quincy, MA 02269. This extracted material is not the complete and official position of the National Fire Protection Association on the referenced subject, which is represented only by the standard in its entirety.

[a]For ampacity limitation, see *NEC* Article 339-5.

15.11   Flexible Armored Cable

Among the most common types of exposed wiring is *NEC* type AC armored cable, commonly known in the smaller sizes by the trade name *BX*. It is an assembly of insulated wires, bound together and enclosed in a protective armor made of a spiral-wound interlocking strip of steel tape (Fig. 15.3). The cable is installed with simple U-clamps or staples holding it against beams, walls, and so on. This type of installation is frequently used in residences and in the rewiring of existing buildings. Use of type AC cable is generally restricted to dry locations. For application and installation details and restrictions, see *NEC* Article 333, "Armored Cable."

A similar construction with much broader application (covered in *NEC* Article 334) is metal-clad (MC) cable. This cable may be used exposed or concealed and in cable trays, and when covered with a moisture-impervious jacket, in wet and outdoor locations as well (Fig. 15.4).

Fig. 15.3 *Flexible armored cable, NEC type AC, trade name BX. Note especially the insulating bushing that is always installed on the end of the armor to protect the wires from damage from the sharp edges of the cut steel armor. (Courtesy of AFC Cable Systems.)*

15.12   Nonmetallic Sheathed Cable (Romex)

In application, *NEC* types NM and NMC, also known by the trade name *Romex,* are restricted to small buildings—that is, residential and other structures not exceeding three floors above grade (Fig. 15.5). The plastic outer jacket, unlike the armor on type AC, makes type NM easier to handle but more vulnerable to physical damage. For application details and restrictions, see *NEC* Article 336, "Non-metallic Sheathed Cable." Typical installation technique is shown in Fig. 15.6.

Fig. 15.5 *Construction of typical NEC type NM cable. The illustrated cable is a two-conductor, No. 12 AWG with ground, insulated for 600 V. Normally shown are the manufacturer, cable trade name, and the letters (UL), which indicate listing of this product by the Underwriters Laboratories, Inc. The ground wire is bare or covered, and the entire cable may be obtained flat (illustrated), oval, or round.*

Fig. 15.4 *(a) Metal-clad cable (NEC type MC, Article 334) with aluminum armor in lieu of the more common galvanized steel armor. Application is similar to that of the steel-armored cable, with the weight advantage of aluminum. Conductors are factory-installed, color-coded, and covered with type THHN insulation and nylon jacket. Cables of similar construction, using steel armor, are available for almost all power and control applications. (Courtesy of AFC Cable Systems.)*
(b) Jacketed-type MC cable has a wide variety of applications, including wet locations, because of its water impervious outer PVC jacket. Where specifically indicated by the manufacturer, this cable may be used for direct burial in earth or installed in concrete. (Courtesy of AFC Cable Systems.)

Fig. 15.6 *Typical wiring technique using types NM (Romex) or AC (BX) in wood stud construction. With metal stud construction, BX cables are passed through precut openings in lieu of field-drilled holes. Where exposed to damage from nails, screws, and other sources, protective metal cover plates are required. For details, see the installation limitations in the NEC.*

15.13 Conductors for General Wiring

Under this heading (Article 310), the *NEC* lists the wire types that are generally installed in raceways and are referred to by the term *building wire*. The most common types are listed in Table 15.5. These wires consist of a copper conductor covered with insulation and, in some instances, with a jacket (Fig. 15.7).

15.14 Special Cable Types

Although most building wiring is accomplished with plastic-insulated 300- and 600-V conductors of the types described in the preceding sections, some applications require the use of special cables. These include high-voltage cables, armored cables, corrosion-resistant jacketed cables, underground cables, and so on. The reader is referred to manufacturers' catalogs and the *NEC* for construction and application details. Service entrance cables and installation are discussed in Sections 16.1 and 16.4, which cover electric service.

15.15 Busway/Busduct/Cablebus

A busway (busduct) is an assembly of copper or aluminum bars in a rigid metallic housing (Fig. 15.8). Its use is almost always preferable, from an economic viewpoint, in two instances: when it is necessary to carry large amounts of current

Polyvinylchloride insulation   Glass tape wrapping

Copper bus bars

Mylar® polyester film insulation   Steel enclosure

Fig. 15.8 *Cutaway view showing construction of a typical feeder busduct. This design is highly compact and rigid, which gives desirable electrical characteristics as well as the advantage of small size. (Photo courtesy of Siemens Energy and Automation, Inc.)*

(power) and when it is necessary to tap onto an electrical power conductor at frequent intervals along its length.

In the case of a heavy current requirement, the alternatives to using busways are to use paralleled sets of round conductors or a single large conductor. Paralleled sets of conductors are almost always more expensive than a busway of similar current capacity because of the high installation cost of multiple conduits. Alternatively, using a single, large-diameter conductor becomes increasingly inefficient as cable size increases because large round conductors require more cross section per ampere of current-carrying capacity than do the flat conductors (busbars) used in busduct (Fig. 15.9).

Where many power tap-offs are required along an electric feeder run consisting of cable in conduit, costs become very high because of the large amount of expensive hand labor involved, because a connection must be made to each conductor in the run. The preferable alternative is to use "plug-in" busway to which connection can be made simply and rapidly with a plug-in device, in similar fashion to the insertion of a common plug into a wall receptacle. This has the additional advantage of convenience; connection and disconnection is simply a matter of inserting or withdrawing the plug-in device, whereas cable taps are permanent connections (Fig. 15.10).

Conductor insulation   Conductor

(a)

Conductor
Conductor insulation
Nylon jacket

(b)

Fig. 15.7 *(a) Typical construction of unjacketed building wire such as type THW (see Table 15.5). (b) The illustrated construction is typical for any nylon jacketed cable such as THWN or THHN. (The first three letters indicate the type of insulation, and the final N indicates the nylon jacket.)*

ELECTRICITY

Fig. 15.9 *Sectional view of this busduct type shows the tight assembly of insulated conductors within a metal housing. This design, unlike the ventilated type, can be mounted in any position because heat dissipation is by conduction from the busbars to the housing. The eight sets of cable shown have the same current-carrying capacity as the busduct shown. (Reproduced by permission of Square D Company.)*

A typical application of heavy-duty busduct might be a vertical feeder in a high-rise building, connecting the basement switchboard to the penthouse machine room. The same building might also use heavy-duty plug-in busduct as vertical riser(s) with taps feeding individual floors (Fig. 15.2). Typical applications for light-duty plug-in busduct (70 to 100 A) could be any machine shop or workshop. The electrical supply to individual machine tools is made very simply and flexibly with a tap-on device (see Fig. 15.10).

Busduct is specified by type, material, number of buses, current capacity, and voltage (e.g., aluminum feeder busduct, 4-wire, 1000 A, 600 V, or copper plug-in busway, 100 A, 3-wire, 600 V). Feeder busduct (no plug-in capability) is avail-

able in ratings from 400 to 4000 A. Plug-in busway is available from 30 A for lighting or light-machinery circuits (see Section 15.16) to 3000 A. A wide variety of fittings and joints are available for all busways to permit easy installation (Fig. 15.11). Designs are available for indoor and outdoor application.

Cablebus is similar to ventilated busduct, ex-

Fig. 15.10 *Construction of one type of plug-in busduct. Plug-ins are evenly spaced on alternate sides to facilitate connection of plug-in breakers, switches, transformers, or cable taps. Housing is of sheet steel with openings for ventilation. The cover plate is not shown. (Courtesy of Square D Company.)*

Fig. 15.11 *Typical installation of compact design busduct. Note that the individual busducts are supported by channels hung from the ceiling and that the same hangers support more than one level of bus. Right angle turns are easily made in the same plane (horizontal or vertical) and between vertical and horizontal planes. (Reproduced by permission of Square D Company.)*

TABLE 15.6 Life-Cycle Relative Cost Comparison of 2000-A, 208-V Feeder Installation

Feeder System Description[a]	Material Cost	Labor Cost[b]	Total First Cost	Power Loss per 100 ft[c] (kW)	Annual Energy Loss[d] (kWh)	Annual Energy Cost[e]	Life-Cycle Energy Cost[f]	Total Life-Cycle Cost[g]
Cable tray	0.63	0.37	1.00	2.90	12,536	0.086	1.081	2.08 100%
Wire and conduit	0.68	0.61	1.29	2.90	12,536	0.086	1.081	2.37 114%
Busduct	1.14	0.24	1.38	5.60	24,207	0.166	2.087	3.47 167%
Cablebus	0.69	0.45	1.14	5.91	25,547	0.174	2.203	3.34 161%

[a]Equipment rating is 600 V, cable is copper, conduit is rigid steel, cable tray is aluminum.
[b]Labor cost includes overhead.
[c]Based on published resistivity data for cable and bus—assuming 80% demand (1600 A) and all conductors in the system equally loaded.
[d]Based on 80% demand, 12 h per day, 360 days per year.
[e]Using $0.10 per kWh as the combined net rate, including demand charges.
[f]Using 20-year life cycle, 8% fixed capital cost, and 3% annual escalation in energy cost.
[g]Sum of fourth and eighth columns.

cept that it uses insulated cables instead of busbars. The cables are rigidly mounted in an open space-frame. The advantage of this construction is that it carries the ampacity rating of its cables *in free air,* which is much higher than the conduit rating, thus giving a high amperes per-dollar first-cost figure. Its principal disadvantage is bulkiness and difficulty in making tap-offs.

As an example of the type of economic analysis that should be made when considering an item as relatively simple as a heavy current electrical feeder, refer to Table 15.6, which shows the results of such a study in relative cost terms. Note that when considering first cost alone, the advantage lies with cable tray (with interlocked armor cables) and cablebus. Adding the energy-loss consideration shifts the advantage to cable tray and wire in conduit. No general conclusion should be drawn from Table 15.6 regarding costs. A change in feeder length, number of taps, hours of operation, energy rates, or any of the other factors can shift the advantage to a different system. The point of our study is to demonstrate clearly that life-cycle costs and first costs often yield entirely different results and that, therefore, this type of study is required before an engineering decision

can be made. (Life-cycle cost is taken to mean the present value of all costs over the installation's life cycle; in this case, 20 years for the system.)

Two additional items are worthy of note:

1. The very factors that yield lower first cost operate to yield higher operating cost. The smaller copper sizes in busduct and cablebus, permitted by high-temperature insulation and good ventilation, cause increased power loss because of their higher resistivity.

2. If the heat loss from the busduct or cablebus can be used to advantage, the related energy cost can be credited instead of being considered a total loss, and life-cycle costs can be changed considerably. Conversely, it can also affect the building cooling load.

15.16 Light-Duty Busway, Flat-Cable Assemblies, and Lighting Track

Special assemblies that act as light-duty (branch circuit) plug-in electrical feeders are widely used because of simplicity of installation and, more importantly, because of its plug-in mode of connection.

Fig. 15.12 Light-duty busway is rated 20 to 60 A, 300 V, and either 2 or 3 wire. Power takeoff devices twist into the bus to make contact with the circuit conductors. Within NEC *restrictions of overcurrent protection, this busway may be used for standard and heavy-duty lighting fixtures and for other electrical devices such as electrically powered tools. (Courtesy of Siemens Energy and Automation, Inc.)*

(a) Light-Duty Plug-In Busway. This construction, which may be used either for feeder or branch circuit application, is covered by the *NEC* general article on busways, with restrictions when applied as branch circuit wiring. Light-duty busways are rated from 20 to 60 A at 300 V, in 2- and 3-wire construction. A somewhat heavier design rated 60 A to 100 A at 600 V is available in 3- and 4-wire construction. Their application is principally for direct connection (with overcurrent protection) of light machinery and industrial lighting (Figs. 15.12 and 15.13).

(b) Flat-Cable Assemblies (NEC Article 363; Type FC). A specially designed cable consisting of two, three, or four No. 10 AWG conductors is field installed in a rigidly mounted standard 1⅝-in. square structural channel. Power-tap devices, installed where required, puncture the insulation of one of the phase conductors and the neutral. Electrical connection is then made to the pigtail wires that extend from the tap devices. This connection can extend directly to the device or to an outlet box with a receptacle, which then acts as a disconnecting means for the electric device being served. In this fashion, lights, small

motors, unit heaters, and other single-phase, light-duty devices can be served without the necessity of "hard" (conduit and cable) wiring. Figure 15.14 illustrates similar equipment that is specifically intended to feed industrial lighting fixtures.

(c) Lighting Track (NEC Article 410-R). This is a factory-assembled channel with conductors for one to four circuits *permanently* installed in the track. Power is taken from the track by special tap-off devices that contact the track's electrified conductors and carry the power to the attached lighting fixture, which can be positioned anywhere along the track. The tracks are generally rated at 20 A and, unlike FC cable assemblies, may feed only lighting fixtures. Taps to feed convenience receptacles are not permitted. A typical design is shown in Fig. 15.15. An application of track lighting is shown in Fig. 21.22.

15.17 Cable Tray

This system, which is covered in *NEC* Article 318, is simply a continuous open support for approved cables. When used as a general wiring system, the

Continuous hangar slot

Extruded aluminum housing

Insulating busbar support

Twist plug blade

Twist plug on plug-in unit

Copper busbars

Continuous access slot

Junction box

EBC 100 busway   Drop ceiling

(a)   (b)

Fig. 15.13 (a) *Plug-in busway rated 100 A, 3 phase, 4 wire, 600 V measures approximately 2½ in. W × 4½ in. H. The twist-in plug, which is integrally attached to a connection means (junction box in the illustration), is rated 30 to 100 A, single or 3 phase, as required. The attached junction box (or receptacle, circuit breaker, or fuse box) then feeds the utilization device (e.g., heavy-duty lighting or machinery). (Courtesy of Universal Trolley.)*
(b) *This bus can also be installed in a hung ceiling. (Courtesy of Universal Trolley.)*

Channel support
assembly

Type FC
Flat cable assembly

Cable
tap
assembly

Lighting fixture
pipe hanger with
wiring tapped off
FC cable

1⅝″ square
steel channel

Fig. 15.14 *A 3-phase, 4-wire flat cable assembly installed in a 1⅝-in. square steel channel is shown. Taps into the conductors are made by tightening the tap device. Taps can be made phase to ground to give 120 V and phase-to-phase to give 208 V. (If the cable is connected to a 277/480 V 3-phase system, then the phase-to-ground and phase-to-phase voltages will be 277 and 480 V, respectively). After a tap device is removed, the puncture made by the tap "heals" itself. (Courtesy of Chan-L-Wire/Wiremold Company.)*

cables must be self-protected. The advantages of this system are free-air rated cables, easy installation and maintenance, and relatively low cost. The disadvantages are bulkiness and the required accessibility. Cable trays are used primarily in industrial applications.

15.18 Design Considerations for Raceway Systems

The following sections deal with closed wiring raceways, which completes our discussion of raceways. The details of construction and application are not discussed here because of space limitation and because those data are readily available from manufacturers and the applicable *NEC* articles. However, enough material is provided for the reader to become familiar with the types of raceways, their common applications and limitations, and, where applicable, comparative characteristics.

Although our discussion in this chapter covers equipment used primarily in electrical power systems, empty raceways intended for signal, data,

Fig. 15.15 *Decorative lighting track. The circular housing can be obtained in a variety of finishes and hanger arrangements. The actual lighting track, shown full size, is readily available without the housing for direct surface mounting. Track is also available in a 4-conductor configuration (three 120-V conductors plus neutral). (Courtesy of Swivelier.)*

and communications wiring (by others) are normally provided under the electrical section of the construction contract. The raceway's function for such systems is in large measure the same as it is for power wiring: protection and isolation of the wiring.

Prior to the widespread use of computers in commercial establishments, the raceway space requirements for communication and signal wiring were easily established because such wiring consisted of small telephone cables plus miscellaneous signal and alarm wires. These requirements were also easily satisfied with empty conduits or cells in floor raceway systems. Today, when virtually every commercial establishment uses some type of data-processing equipment and communication networking becomes commonplace even in small facilities, raceways for communication cabling have become a major design item. They often far exceed in cross-sectional area the space required for power cabling. Their space requirements are sometimes so large that they, like ductwork, have architectural impact, and must therefore be considered early in the design process.

Sizing of raceways for power wiring is an exact process (see Section 17.16) based as it is on maximum permissible temperatures for specific materials in a given environment. Not so for communication cabling, even when a system's present requirements are known, because of the extremely rapid growth of networking and of data interchange. The problem is all the more difficult

when designing commercial space for rental to an unknown client. The advisable approach in such cases is either to provide a reasonable amount of floor-level raceway space for main cabling (see Sections 15.25 and 15.26) and to rely on add-on systems such as under-carpet wiring (see Section 15.30) and surface or ceiling raceways (Section 15.31) for additional raceway area, or, alternatively, to use a structural system that provides virtually unlimited wiring space (Figs. 15.27 through 15.29). Because the latter is a major structural/architectural decision, it must be taken at the preliminary stages of design.

Design considerations for power system raceways are discussed in the following sections and in Chapter 17. Design of raceways for data and communication system wiring includes the following considerations:

I. Number, type, and location of data-processing terminals.
II. Networking requirements:
    A. The type of local area network in large measure determines the communication media (i.e., coaxial cable, shielded and unshielded wire, and fiber optic cables), which affects the raceway space requirement. Cable type also determines the type of connectors needed (and their space requirements) and the type of floor outlets used for machine connection.

B. Cable topology (i.e., interconnection arrangements). This item is frequently not within the domain of the architectural designer, although the raceway space availability seriously affects cabling arrangement and vice versa.

C. Requirement for interconnection of networks and connection to remote networks.

III. Number, location, and characteristics of major peripheral devices, such as mass storage, printing, and plotting.

IV. Location and type of major subsystems, such as CAD/CAM design spaces.

V. Location of presentation spaces that require interconnection to computer networks.

In view of these highly technical and rapidly changing requirements, the services of a design consultant in this area is suggested.

15.19   Steel Conduit

The purpose of conduit is to:

1. Protect the enclosed wiring from mechanical injury and damage from the surrounding atmosphere.
2. Provide a grounded metal enclosure for the wiring in order to avoid shock hazard.
3. Provide a system ground path.
4. Protect surroundings against fire hazard as a result of overheating or arcing of the enclosed conductors.
5. Support the conductors.

For these reasons, the *NEC* generally requires that all power wiring be enclosed in a rigid metallic corrosion-resistant conduit. To this latter end, steel conduit is manufactured in several types, among which are hot-dip galvanized, sherardized (coated with zinc dust), enameled, and plastic covered.

There are three types of steel conduit that differ basically only in wall thickness. They are, in order of decreasing weight:

1. Heavy-wall steel conduit, also referred to simply as "rigid steel conduit"; covered by *NEC* Article 346.

2. Intermediate metal conduit, usually referred to as IMC; covered by *NEC* Article 345.
3. Electric metallic tubing, normally known as EMT or thin-wall conduit; this is covered by *NEC* Article 348.

The differences are clearly shown in Table 15.7. Several types of heavy-wall conduit plus EMT are shown in Fig. 15.16. The equivalent millimeter sizes of conduits are given in Table 15.8.

Rigid conduit and IMC use the same threaded fittings. As a result of its thin wall, EMT is not threaded; instead, it uses set-screw and pressure fittings. The thinner walls of EMT and IMC yield a larger inside diameter (ID) and, therefore, easier wire pulling. The combination of lower weight and easier wire pulling gives EMT and IMC a distinct labor cost advantage over rigid conduit, which is further enhanced in jobs with a great deal of field bending and handling of conduit. Both, however, have application restrictions, which are detailed in the *NEC*.

Generally, no conduit smaller than ½-in. nominal trade diameter is used. Ordinary steel pipe may not be used for electric purposes, and all

Fig. 15.16 *Steel conduits:* (a) *galvanized, heavy wall, rigid;* (b) *black enameled;* (c) *EMT thin wall;* (d) *plastic-coated conduit for use in highly corrosive atmospheres.*

TABLE 15.7 **Comparative Dimensions and Weights of Metallic Conduit**[a]

3/4" Nominal trade size conduit

R.S. IMC EMT AL

O.D. 1.05	1.029	0.92	1.05
I.D. 0.824	0.887	0.82	0.82
W 0.113	0.071	0.05	0.113

Nominal or Trade Size (in.)	Outside Diameter (in.)				Inside Diameter (in.)				Weight per 10 ft Length (lb)[b]			
	RS[c]	IMC[d]	EMT[e]	AL[f]	RS	IMC	EMT	AL	RS	IMC	EMT	AL
½	0.84	0.82	0.71	0.84	0.62	0.69	0.62	0.62	7.9	5.3	2.9	2.7
¾	1.05	1.03	0.92	1.05	0.82	0.89	0.82	0.82	10.9	7.2	4.4	3.6
1	1.32	1.29	1.16	1.32	1.05	1.13	1.05	1.05	16.5	10.6	6.4	5.3
1¼	1.66	1.64	1.51	1.66	1.38	1.47	1.38	1.38	21.5	14.4	9.5	7.0
1½	1.90	1.88	1.74	1.90	1.61	1.70	1.61	1.61	25.8	17.7	11.0	8.6
2	2.38	2.36	2.20	2.38	2.07	2.17	2.07	2.07	35.2	23.6	14.0	11.6
2½	2.88	2.86	2.88	2.88	2.47	2.61	2.73	2.47	56.7	38.2	20.5	18.3
3	3.50	3.48	3.50	3.50	3.07	3.23	3.36	3.07	71.4	46.9	25.0	23.9
3½	4.00	3.97	4.00	4.00	3.55	3.72	3.83	3.55	86.0	54.7	32.5	28.8
4	4.50	4.47	4.50	—	4.03	4.21	4.03	—	97.2	62.5	37.0	—

[a]Data varies slightly among manufacturers.
[b]Standard length including one coupling.
[c]Standard heavy-wall rigid steel conduit.
[d]Intermediate-weight steel conduit.
[e]Electric metallic tubing.
[f]Aluminum conduit.

TABLE 15.8 **Metric Equivalents of Rigid Steel Conduit Sizes**

Conduit Size (in.)	Dimensions (mm)		
	OD[a]	ID[b]	W[c]
½	21.3	15.7	2.8
¾	26.7	20.8	3.0
1	33.5	26.7	3.4
1¼	42.2	35.1	3.5
1½	48.3	40.9	3.6
2	60.5	52.6	4.0
2½	73.2	62.7	5.2
3	88.9	78.0	5.5
3½	101.6	90.2	5.7
4	114.3	102.4	6.0

[a]Outside diameter.
[b]Inside diameter.
[c]Wall thickness.

electric steel conduit is distinctively marked as such.

When steel conduit is installed in direct contact with the earth, it is advisable to use hot-dip galvanized type and to coat the joints with asphaltum. If the earth is very wet, the entire conduit system should be coated with an appropriate waterproofing compound. Alternatively, a plastic-jacketed conduit can be used.

Conduit is fastened to the structure in much the same way as pipe: with pipe straps and clamps. Vertical load at floor opening is taken with special support clamps. Trapeze mounting is common for conduit banks hung from the ceiling, as in Fig. 15.17.

Conduit size depends not only on the maximum permissible temperature of the contained conductors, but also on the number and diameter of the wires that may be drawn into the conduit without injuring the wire. The number and radius of bends in the conduit, as well as its total length, affect the degree of abrasion to the wiring insulation during installation. No wires should be installed until the conduit system has been inspected and approved.

For structural reasons, conduits in concrete slabs are run close to the bottom surface (in the portion of the slab in tension) or near the center. If a large number of conduits must be embedded,

it may be necessary to increase the slab thickness. In many instances, the structural slab is covered with a concrete topping, in which conduit may be installed without affecting slab integrity. In all cases, local building codes should be consulted for limitations on embedded conduits. In any event, the top of any conduit shall be at least ¾-in. below the finished floor surface in order to prevent cracking. When heavy trucking is expected, this allowance should be increased to 1½ in. minimum.

In general, the following rules should be observed and included in all specifications for conduit work in concrete slabs:

1. Conduits shall have an outside diameter (OD) no greater than one-third the slab thickness, as measured at its thinnest point.
2. Conduits running parallel to each other shall be spaced not less than three times the OD of the largest conduit center-to-center.
3. Conduit crossings shall be as near to a right angle as possible.
4. Minimum cover over conduits shall be ¾ in.

15.20 Aluminum Conduit

The use of aluminum conduit has increased greatly in recent years because of the weight advantage that aluminum has over steel, being even

ELECTRICITY

ELECTRICITY

Fig. 15.17 *Typical overhead conduit bank installation. Note that due to field conditions the insert* (a) *for hanger rods was inadequate and an additional insert* (b) *was added. This conduit bank uses EMT, which has a pipe-wall thickness approximately one-third that of heavy wall rigid conduit. The resulting weight difference in a large bank such as this is very pronounced. EMT joints are made with set screw fittings* (c). *Note how individual conduits are fixed by clamps to the trapeze channel* (d). *(Courtesy of Republic Steel Corp.)*

lighter than EMT. The savings in labor cost more than offsets the additional cost of the material itself. In addition, aluminum has better corrosion resistance in most atmospheres; it is nonmagnetic, giving lower voltage drop; it is nonsparking; and, generally, it does not require painting.

Its major drawback is its deleterious effect on many types of concrete, causing spalling and cracking when embedded. Although manufacturers can demonstrate cases of embedding in concrete without harmful effect, it is a procedure that should be avoided unless the concrete additives are rigidly controlled and the conduit is coated to prevent contact with the concrete. It is also inadvisable to bury aluminum in earth, with or without asphalt or other coating, because of the rapid corrosion often encountered. Other difficulties frequently encountered are freezing of threaded joints (because of thread deformation) and difficulty in obtaining electrical contact with

grounding straps. With the exceptions noted, aluminum conduit may be used in all locations where steel conduit is used.

15.21   Flexible Metal Conduit

This type of conduit consists of an empty, spirally wound, interlocked armor steel or aluminum raceway. It is known to the trade as *Greenfield* and is covered in *NEC* Article 350. It is used principally for motor connections and other locations where vibration is present, where movement is encountered, or where physical obstructions make its use necessary. The acoustic and vibration isolation provided by flexible conduit is one of its most important applications. It should always be used in connections to motors, transformers, ballasts, and the like. A typical application is for wiring inside metal partitions. When covered with

Fig. 15.18 *This is a particularly good application of liquid tight flexible conduit because it provides weatherproofing and acoustical isolation of the noise-producing transformer. (Courtesy of Electri-Flex Company.)*

a liquid-tight plastic jacket, it is suitable for use in wet locations (Fig. 15.18). In this construction, it is most often known by the trade name *Sealtite.*

15.22 Nonmetallic Conduit

A separate classification of rigid conduit (*NEC* Article 347) covers raceways that are formed from such materials as fiber, asbestos–cement, soapstone, rigid polyvinyl chloride (PVC), and high-density polyethylene.

For use above ground, this conduit must be flame retardant, tough, and resistant to heat distortion, sunlight, and low-temperature effects. For use underground the last two requirements are waived. Generally, nonmetallic conduit may be used without restriction in nonhazardous areas within the physical limitations of the material involved. Thus, plastic conduit has a temperature limitation, asbestos–cement has considerable physical strength limitations, and so on. As a re-

sult of these limitations, PVC conduit is the material of choice for indoor exposed use, and asbestos–cement, fiber, and PVC plastic for outdoor and underground use. A separate ground wire *must* be provided because the ground provided by a metallic conduit is absent.

15.23 Surface Metal Raceways (Metallic and Nonmetallic)

These raceways are covered in *NEC* Article 352. Surface metal raceways and multioutlet assemblies may be utilized only in dry, nonhazardous, noncorrosive locations and may generally contain only wiring operating below 300 V. Such raceways are normally installed exposed, in places not subject to physical injury.

The principal applications of surface metal raceways are:

1. Where the architecture does not permit recessing (Fig. 15.19).
2. Where economy in construction weighs very heavily in favor of surface raceways and where expansion is anticipated (Fig. 15.20).
3. Where outlets are required at frequent intervals, and where rewiring is required or anticipated (Figs. 15.21 through 15.23).
4. Where access to equipment in the raceways is required and/or where necessary due to the nature of the wiring (Figs. 15.20, 15.24, 15.25, and 15.26).
5. Where we wish to avoid the extensive and expensive cutting and patching required to "bury" a raceway during rewiring (Fig. 15.27).

15.24 Outlet and Device Boxes

These boxes are generally of galvanized stamped sheet metal. The most common sizes are the 4-in. square and 4-in. octagonal boxes used for fixtures, junctions, and devices, and the 4- × 2⅛-in. box used for single devices where no splicing is required. Box depths vary from 1½ to 3 in. Nonmetallic boxes may be used with NM and NMC

Fig. 15.19 *The exposed wood members require the use of an unobtrusive surface raceway. A small flat raceway feeds receptacle outlets into which the elaborate hanging fixtures are plugged. (Courtesy of Wiremold Company.)*

cable and with nonmetallic conduit installations. In wet locations and for outdoor work, cast-iron or cast-aluminum boxes are recommended.

As a result of an *NEC* (Article 300-21) requirement that electrical penetrations in fire-rated floors be designed to maintain fire ratings, electrical manufacturers have produced a line of poke-through fittings to meet this need. (This requirement applies also to walls, ceilings, and partitions.) One such design is shown in Fig. 15.28. These electrical penetrations have become increasingly prevalent in existing commercial spaces where the expanded need for desktop power and data wiring can be met most economically and

Fig. 15.20 *Three-section nonmetallic baseboard-type raceway measures 4 in. high and 1.5 in. deep. The three sections are intended for telephone, power, and data cabling. The junction box on the right protrudes so as not to lessen the raceway wiring space. It is equipped with vertical dividers to keep the three types of wiring completely separated. The elevated wall box on the left contains telephone and data cable outlets, and connects only to the low-voltage portions of the raceway. (Photo courtesy of Hubbell, Inc.)*

Fig. 15.21 *Multioutlet surface raceways find ready application in schools. Note the use of a surge and noise surpressor (foreground) to protect groups of computers from power line disturbances. See Sections 16.35 through 16.37 for a discussion of power conditioning equipment. (Courtesy of Walker Systems, A Wiremold Company.)*

Fig. 15.22 *Where prewired multioutlet assemblies are insufficient due to heavy loads, as in the illustrated laboratory, field-wired individual receptacles of any desired rating can be mounted in surface raceways. Spacing between outlets is governed by work positions and equipment load density. (Photo courtesy of The Wiremold Company.)*

Fig. 15.23 *Large-capacity surface metal raceways are particularly useful for wiring in full access floors (see Section 15.29) because of the heavy wiring (see wall box) and frequent rewiring. The perforated floor tile in the foreground is used to supply laminar airflow in this clean room at an integrated circuit manufacturer's laboratory facility. (Photos courtesy of Walker Systems, A Wiremold Company.)*

Fig. 15.24 *Multichannel nonmetallic surface raceway with snap-in connector modules for data network, signal, and power wiring systems. (Similar metallic raceways are also available.) The raceway itself measures 4½ in. H × 1 in. D. The internal dividers are movable or entirely removable, which permits varying the number and size of the wiring channels. Principal application of this type of wireway is in commercial occupancies using extensive desktop data processing and communication equipment. (Courtesy of Panduit.)*

Fig. 15.25 *The frequent wiring changes required for theater and exhibition lighting are easily made when the wiring is run in a suspended surface raceway. (Courtesy of Walker Systems, A Wiremold Company.)*

rapidly by through-the-floor feeds from expanded wiring in the hung ceiling below. In addition, these fittings facilitate the electrical wiring relocations so common in rental office occupancies.

Fig. 15.26 *The basic raceway illustrated is 1⅞ in. D × 3³⁄₁₆ in. W. It is shown with a divider installed, permitting use of the top section for power wiring and the bottom section for low-voltage wiring. Because data and communication cables are frequently supplied with factory-installed terminations (as in the photo), a raceway where the cable can be laid in rather than pulled in is required. Also, terminal strips and other equipment can be installed in the low-voltage section of these large raceways, making the use of separate terminal cabinets unnecessary. (Courtesy of Walker Systems, A Wiremold Company.)*

FloorTrak	2	3	4
Width	2.68"	2.985"	2.975"
Height	.465"	.610"	.885"
Ctr. Hole	.50x.312"	.75x.500"	1.0x.750"

Fig. 15.27 *Where low-voltage wire and cable cannot be concealed ("buried"), shallow, properly shaped floor track can be used. Such installations should avoid foot-traffic areas where possible to minimize trip hazards. (Courtesy of Hubbell Premise Wiring.)*

15.25   Floor Raceways

In commercial spaces with large open floor areas it is common practice to place desks and other workstations throughout the space, at considerable distances from permanent walls containing electrical services. Because in a modern office it must be assumed that each workstation requires electric power for a computer, desk lamp, and other common equipment plus a telephone line, a computer network connection, and possibly a data outlet, the problem of bringing these services to the workstations with a minimum of exposed wiring arises immediately. The required outlets can be installed on the floor adjacent to or under the workstation, or, if half-height partitions are used, in these partitions.

To bring the various electrical and communication services to the user, in the absence of any sort of overall floor raceway system, the installing contractor has one of four choices:

1. Channel the floor and install a conduit in the chase, connecting it to the nearest wall outlet. Patch the chased portion.

Power & Low Tension outlets

Fire Rated floor slab

Fire rated insert

4" hole in floor

Cable Junction Box

Fig. 15.28 *Typical poke-through electrical fitting mounts in a 4-in. hole. It is wired from underneath with the required power, telephone, signal, and data cables. Power and low-voltage cables are separated as required by code. Units are available prewired or suitable for field wiring, and adaptable for varying floor thicknesses. The floor fitting is provided with power, telephone and data cable outlets as required for the specific installation. (Courtesy of Walker, Div. of Wiremold.)*

2. Install a surface floor raceway. The usefulness of this technique is very limited because of the tripping hazard and problems in routine floor cleaning.
3. Drill through the floor twice and connect the new outlet to a nearby existing floor or wall outlet via a conduit on the underside of the floor slab. Floor penetrations must be fireproof.
4. Drill through the floor and run a conduit in or on the ceiling to outlets below. When using this technique, special "poke-through" fittings are available that restore the fire rating of the slab (see Fig. 15.28). These special "poke-through" fittings are

designed to carry all the electric services normally required at a workstation. They can then be connected to a single location multiservice floor outlet group as in Fig. 15.28, or used to wire the partitions in a work cubicle, as in Fig. 15.29.

All four of these methods have serious disadvantages; method 1 is labor intensive, method 2 is unsightly and presents a safety hazard, methods 3 and 4 disturb the occupant below, and all four methods are inflexible and therefore unsuitable for spaces where changes in wiring and workstation location are anticipated. For these reasons, overall-access in-floor and underfloor raceway systems were developed and are widely used in high-grade commercial and institutional spaces.

The *NEC* recognizes three types of in-floor raceways:

Underfloor raceways—Article 354

Cellular metal floor raceways—Article 356

Cellular concrete floor raceways—Article 358

All three types are applicable to all types of structures, and none may be used in corrosive or hazardous areas. The fundamental difference between them is that underfloor raceways are added on to the structure, whereas cellular floor raceways are part of the structure itself—and therefore have a pronounced effect on the building's architecture. (Underfloor duct systems antedate poke-through fittings, which are a relatively recent development.)

15.26 Underfloor Duct

These raceways may be installed beneath or flush with the floor. They find their widest application in office spaces because their use permits placement of power, data, and signal outlets close to desks and other furniture, regardless of furniture layout. Underfloor (UF) duct systems were widely employed until the introduction of what may be called over-the-ceiling ducts (in apposition to under-the-floor ducts) and flat-cable under-carpet wiring. These systems are discussed in Sections 15.30 and 15.31. The reason that alternate systems were developed is simply economic: underfloor

POWER CABLE

COMPUTER CABLE

TELEPHONE CABLE

POWER CABLE

POKE-THROUGH FITTING
ACCESS THROUGH
CONCRETE SLAB BY
CORE DRILLING OR
BY PRESET SLEEVES
INSTALLED IN SLAB

FLEXIBLE CONDUIT
FOR POWER,
ELECTRONICS
(COMPUTER), AND
COMMUNICATIONS

JUNCTION BOX
FOR POWER

JUNCTION BOX
FOR COMMUNICATIONS

JUNCTION BOX
FOR ELECTRONICS

CONDUIT TO NEXT
JUNCTION BOX
OR UTILITY CLOSET

CEILING GRID BELOW

POKE THROUGH SYSTEM

NOTES

Poke-through systems are used in conjunction with overhead branch distribution systems run in accessible suspended ceiling cavities to serve outlets in partitions. When services are required at floor locations where adjacent partitions or columns are not available, as in open office planning, they must either be brought down from a wireway assembly (known as a power pole) or up through a floor penetration containing a fire-rated insert fitting and above-floor outlet assembly. To install a poke-through assembly, the floor slab must either be drilled or contain preset sleeves arranged in a modular array. Poke-through assemblies are used in conjunction with cellular deck and underfloor duct systems when precise service location required does not fall directly above its associated system raceway.

With one floor penetration, the single poke-through assembly can serve all the power, communications, and data requirements of a work station. Distribution wiring in the ceiling cavity can be run in raceways. A more cost-effective method is to use armored cable (bx) for

cost-effective method is to use armored cable (bx) for power and approved plenum rated cable for communications and data when the ceiling cavity is used for return air. To minimize disturbance to the office space below when a poke-through assembly needs to be relocated or added, a modular system of prewired junction boxes for each service can be provided, although it is more common to elect this option for power only. A different type of wiring system must be selected for a floor slab on grade, above lobby or retail space, above mechanical equipment space, or above space exposed to atmosphere.

Low initial cost of a poke-through system makes it both viable and attractive for investor-owned buildings where tenants are responsible for future changes and for corporate buildings where construction budget is limited. It is effective when office planning includes interconnecting work station panels containing provisions to extend wiring above the floor, reducing the number of floor penetrations needed for services.

Fig. 15.29 *Typical application of a poke-through fitting to provide power, telephone, and data service to a modern work station. This drawing shows the electrical services being tapped at junction boxes in a hung ceiling conduit system on the floor below. The ceiling wiring system can also be a raceway network (as in Fig. 15.50) in lieu of the hard wiring shown here. (From Ramsey and Sleeper,* Architectural Graphic Standards, *8th ed., 1988, reprinted by permission of John Wiley & Sons.)*

duct systems are expensive and, because they are inflexible, being literally cast in concrete, they are therefore frequently underutilized in one area while inadequate in another. However, before discussing the relative merits of systems, an under-

standing of what UF duct systems are and how they are assembled and utilized is necessary.

An underfloor duct system is simply an arrangement of parallel rectangular metal or heavy plastic raceways laid on the structural slab and covered with concrete fill. Access to the wiring in these *distribution ducts* is made via inserts that connect to openings in the ducts and terminate in floor fittings, for both power and signal/data wiring. Cable feeds to the distribution ducts are supplied by a second set of rectangular raceways called *feeder ducts,* usually laid at right angles to the distribution ducts.

In a *single-level* underfloor duct system, the distribution and feeder ducts are on the same level, and the interwiring between them is accomplished in junction boxes. The advantage of

a single-level system is shallow concrete fill; normally 2½ to 3 in. The limiting factor of a single-level system is the junction box, which becomes more complex and multisectioned with increasing number of ducts and wires. Newer systems utilize a one-piece triple-cell duct for both distribution and feeder ducts, with factory set inserts every 24 in. that straddle all three cells at once (Fig. 15.30). By placing distribution ducts on 5-ft centers with adequate crosswise feeder ducts and utilizing large flat junction boxes, a cost-effective installation adequate for all but the heaviest wiring de-

ELECTRICITY

(a) Portion of floor plan

(b) Distribution (and feeder) duct

Fig. 15.30 *Details of single-level underfloor duct system utilizing three-section cell duct for both distribution and feeder ducts.*
(a) *Section of typical large open floor space in commercial facility. Distribution ducts may be placed as close as 5-ft centers to satisfy dense desk spacing.*
(b) *Three-cell distribution duct utilizes a 4-in. wide center cell (4.9 in.²) for power and either 3-in. wide (3.7 in.²) or 6-in. wide (7.4 in.²) outer cells for signal and data cabling. Minimum concrete fill depth is 2½ in., resulting in a minimum 1-in. cover over the distribution ducts. Service fittings are flush with the floor. (Courtesy of Square D Company.)*

mands can be assembled. For such areas, distribution ducts can be arranged to feed under-carpet cables (see Sections 15.18 and 15.30).

Because the initial cost of a full UF system is high, an alternative arrangement utilizes *only* feeder ducts on approximately 25-ft centers, with flat (under-carpet) cable box connectors spaced approximately every 20 ft along the feeder ducts. The low-tension portion of this system relies completely on flat telephone and data transmission cables, including fiber optic cables. Because these cables are generally precut and factory terminated, the system requirements must be carefully analyzed (see Section 15.18) before committing to a complete under-carpet wiring system.

A two-level underfloor duct system is essentially the same as a single-level system except that the distribution ducts and feeder ducts are on different levels (Fig. 15.31). This arrangement has the advantages of simplifying junction boxes and of giving the system unlimited feeder capacity but the distinct disadvantage of requiring a minimum of 3⅝ in. of concrete fill. This additional slab thickness can frequently be avoided by depressing part of the slab to accommodate feeder ducts run *under* the distribution ducts, as shown in Fig. 15.32.

A modern two-level system is illustrated in Fig. 15.33. Here, the feeder ducts run above the distribution ducts and intersect at an especially constructed junction fitting into which the distribution ducts partially recess, in order to reduce overall system height. The required concrete fill is either 3 or 4 in., depending on the depth of the distribution cells (2 or 3 in.).

In all underfloor duct systems, the principal

Fig. 15.32 *Setting a two-level underfloor duct system. To avoid thickening fill, a depression in the slab can accept feeder ducts. Ducts would be run near the bay center to avoid the negative steel of joists near columns.*

cable capacity bottleneck is usually the supply point to the *feeder* ducts. One solution to this problem uses a special feed arrangement at panels (Fig. 15.34). Another possibility is to subdivide large floor areas, supplying each via a system of multiple feed points arranged in closets or at wall panels. In such systems care must be taken to ensure sufficient interconnection capacity between feed points because data networks are not only floor-wide but frequently building-wide.

Underfloor ducts may be cast into the structural slab in lieu of being installed in fill or topping, but the slab must be designed to accommodate them. The use of a fill or topping on the structural slab for underfloor duct has these advantages:

1. Ducts can be run in any direction, without conflict to structural elements.
2. Finishing is simplified.
3. Coordination is simplified.
4. Formwork and construction sequence are simplified.

The disadvantages are:

1. Additional concrete increases costs directly by increasing weight. This is particularly expensive in seismic designs.
2. Height of building may be increased.

In retrofit jobs where underfloor duct is decided on rather than one of the other floor or ceiling raceway systems, the ducts will obviously be placed in a new floor fill.

In conclusion, some general comments on the application of underfloor duct systems are in order. Underfloor duct systems are *expensive*. They can add 50% to the building's electric system cost,

Fig. 15.31 *Typical two-level junction box demonstrates the simplicity of the two-level system. (Courtesy of Square D Company.)*

Detail of preset insert

Junction fittings

Preset inserts

22″

Data/signal
Power
Data/signal

Fig. 15.33 *Modified two-level underfloor duct system requires 3 in. of concrete fill for 2-in. high ducts and 4 in. of fill for 3-in. ducts. Preset inserts straddle all three cells and provide two duplex power receptacles plus connection to both low-tension cells. Distribution ducts recess into feeder-header ducts at junction boxes to reduce overall height. See insert. (Courtesy of Walker Systems, A Wiremold Company.)*

Electric panel

Signal cable feed connection

Standard duct
Super duct

6 4 6

6

6 4 6

Distribution or Feeder duct

Feeder duct

Feeder details

Fig. 15.34 *Due to the large capacity of both distribution ducts and feeder ducts, central cable feed points such as at electric closets can cause bottlenecks. Illustrated is one possible solution consisting of a double-duct feed arrangement. Signal cable would feed in from cable boxes (not shown). (Diagrams courtesy of Square D Company.)*

ELECTRICITY

without consideration of the construction costs involved. To justify their use, therefore, the building should meet these criteria:

1. Open floor areas, with a requirement for outlets at locations removed from walls and partitions.
2. Under-carpet wiring system inapplicable.
3. Outlets from ceiling systems unacceptable.
4. Frequent rearrangement of furniture and other items requiring electrical and signal service anticipated.

The facilities that may meet these criteria are prestige office buildings, museums, galleries, and other display-case spaces, high-cost merchandising areas, and selected areas in industrial facilities. Bear in mind that even in high-cost office construction, underfloor duct systems are difficult to justify economically unless furniture layout will change. In doubtful cases, alternate arrangements can be planned and an intelligent choice made after costs and impact on the building structure are studied.

15.27 Cellular Metal Floor Raceway

The underfloor duct system described previously is best applied to rectilinear arrangements. Random arrangements, such as those found in office landscaping, require a fully accessible floor—if indeed the floor is to be used for electrification. This may be provided by a cellular (metal) floor that is an integrated structural/electrical system. The floor can be partially or completely electrified. One of the many structural element designs available is shown in Fig. 15.35.

The cellular floor is part of the structural system and is designed accordingly. Electrical wiring is fed into the cells from header ducts and/or trenches that run perpendicular to the floor cells and constitute a system of underfloor ducts in themselves. The header ducts in turn are fed from electric panels and signal data-transmission and telephone cabinets, in much the same manner as underfloor ducts are fed.

Three types of wiring systems generally run in separate floor cells and header ducts; electric power, data-transmission wiring, and telephone

and signal systems. The last two may be combined in a single cell only if the signal system voltage and power level are low and the local telephone company approves. A complete range of outlets and fittings is available.

15.28 Precast Cellular Concrete Floor Raceways

This structural concrete system is similar to a cellular metal floor in application and has the same advantages: large capacity, versatility in that each cell is a potential raceway, and flexibility in outlet placement and movement. Here too, as with the metal cell constructions, first cost is higher than that of standard underfloor duct installation, although life-cycle cost is frequently lower, depending on space use.

A cell is defined in *NEC* Article 358 as a "single, enclosed, tubular space in a floor made of precast cellular concrete slabs, the direction of the cell being parallel to the direction of the floor member." Feed for these cells is provided, as with metal cellular floor construction, by header ducts. Although header ducts are normally installed in concrete fill above the hollow core structural slab, a header arrangement with feed from the ceiling below is also entirely practical. Like the metallic cellular floor, the cells can be used for air distribution and even for piping, although these items are generally installed in a hung ceiling.

15.29 Full Access Floor

This construction is applicable to spaces with very heavy cabling requirements, particularly if frequent recabling and reconnection is required. It provides rapid and complete access to an underfloor plenum. The system was originally developed for data-processing areas that have a requirement for large, fully accessible cable spaces and for large quantities of conditioned air. The solution to both of these requirements is an infinite access floor, usually constructed of lightweight die-cast aluminum panels supported on a network of adjustable steel or aluminum pedestals. Panels are available from 18 × 18 in. to

Trench duct section

(c)

Flush carpeted or monument service fittings provide access to recessed power and outlets.

Cellular floor deck acts with concrete in composite design.

Individual cells serve as secondary raceways carrying power, telephone and data transmission lines to the preset inserts.

Large 22 sq. in. cells handle data and signal wiring.

Preset inserts are installed over prepunched holes in Celluflor before concrete is placed.

Trench duct, the primary raceway, runs across floor deck, carrying power, phone and data transmission lines from the service closet. Steel cover plates can be readily removed for access.

(a)

Signal, data cable outlet
Power receptacle
Alternate feed to under-carpet wiring

Alternate arrangement with above floor "monument" outlet

Preset insert gives access to all 3 cells

Signal cables
16 sq. in. cell

Power cables
5.6 sq. in. cell

Data cables
22 sq. in. cell

(b)

Fig. 15.35 (a) *One of many designs for electrified cellular floors. The floor cells are available in many designs, depending primarily on the structural requirements. The trench (c) that straddles the cells provides the electrical feeds through precut holes in the cells. The trench itself is completely accessible from the top and, when opened, exposes all the wiring and the cells below.*
(b) *Activated preset insert. Note that the insert straddles the center (power) cell and provides access to the two adjoining low-tension wiring cells. Power and signal wiring are completely separated at all times by metal barriers. If desired, a standard surface "monument" fitting can be mounted on the floor or connection can be made to undercarpet cables in lieu of the flush plate shown. When an insert is to be deactivated, the flush cover plate is simply replaced with a blank plate.*
(c) *Section through trench duct, which acts as feeder for distribution ducts. Trench is available with or without bottom, in any required height, widths from 9 to 36 in., and one, two, or three compartments, depending on floor cell design and cabling requirements. (Courtesy of Walker Systems, A Wiremold Company.)*

36 × 36 in., and floor depth is normally 12 to 24 in., although taller pedestals are available. The subfloor space thus created can be used for cabling and also to carry conditioned air either in ducts or by using the entire space as an air plenum. (In the latter case, the wire and cable must be suitable for air plenum use; see Fig. 15.23). The construction is usually fireproof. Obviously, sufficient ceiling height is necessary to accommodate the raised floor.

Where air requirements are limited or nonexistent and the floor is intended primarily for cabling, pedestals as short as 6 in. can be used, thus relieving the ceiling height problem (Fig. 15.36). In such access floor spaces, use of multiservice distribution modules and modular wiring avoids cable tangles and reduces labor costs significantly (Figs. 15.36 through 15.39).

15.30 Under-Carpet Wiring System

This system, which is covered in *NEC* Article 328, was originally developed as both an inexpensive alternative to an underfloor or cellular floor system and as a means for providing a flexible floor-level branch circuit wiring system. Essentially, the system consists of a factory-assembled flat cable (*NEC* type FCC), approved for floor installation *only* under carpet squares, plus the accessories necessary for connection to 120-V power outlets. The cable itself consists of three or more flat copper conductors, placed edge to edge and enclosed in an insulating material (Fig. 15.40). The entire assembly is covered with a grounded metal shield, which, like a metal conduit, provides both physical protection and a continuous electrical ground path. In addition, a bottom shield is required, which is usually heavy PVC or metal.

The cable, when properly installed on a hard, flat surface, is approximately 0.03 in. high, and thus is essentially undetectable when covered with carpet. Because the carpet squares are designed to be readily removable, the entire system can be repositioned to meet changing furniture layout requirements with a minimum of disruption and no structural work. The cable is designed to carry normal physical loads such as office traffic and furniture placement without affecting its electrical performance.

The attractiveness and simplicity of the system led to the development of similarly designed flat low-tension cables for signal and communication wiring, and more recently both electrical and fiber optic cables and accessories for data transmission (Fig. 15.41). Manufacturers also offer a complete line of junction fittings, connectors, adapters, and receptacles (Fig. 15.42).

The problems inherent in this type of on-the-floor wiring system, such as cable crossings, splicing, interfacing with round cable systems, interconnections at floor boxes and fittings, feed connections from cabinets, underfloor ducts, floor cells, and through-the-floor fittings, have all been solved by a full line of manufactured devices designed for specific usages.

Because under-carpet wiring systems are separate and distinct from wire and conduit systems, they, like underfloor duct systems, are usually shown on a separate electrical plan. A small typical plan of this type is shown in Fig. 15.43. Figures 15.44 and 15.45 are photographs of essential portions of such an installation. Note that a complex system such as that shown in Fig. 15.45 requires recessing a floor box into the slab, which to an extent, defeats the essential simplicity and flexibility of an under-carpet system. Although these systems, at least in simplest form, are particularly applicable to retrofit work, their low cost, combined with the inherent advantages of a flexible *floor-level* wiring system, particularly in open office areas, has made them a widely used first choice in new construction as well.

15.31 Ceiling Raceways and Manufactured Wiring Systems

The need for flexibility in a facility's electrical system coupled with the high cost of underfloor electrical raceway systems encouraged the development of equivalent over-the-ceiling systems. These systems are actually more flexible than their underfloor counterparts because they energize lighting, provide power and telephone facilities, and even supply outlets for the floor above (see Figs. 15.28, 15.29, and 15.36), in addition to permitting very rapid layout changes at low cost. This last characteristic is particularly desirable in stores where frequent display changes necessitate corre-

POWER CABLE

COMPUTER CABLE

TELEPHONE CABLE

POWER CABLE

FLUSH FLOOR OUTLET FOR POWER, ELECTRONICS, AND COMMUNICATIONS

PLUG-IN JUMPER CABLE (POWER)

REMOVABLE ACCESS FLOOR PANELS

ELECTRONICS (COMPUTER) MODULAR CABLE TO ZONE JUNCTION BOX (NO CONDUIT)

ZONE PLUG-IN JUNCTION BOX (POWER)

JUMPER CABLE TO NEXT WORK STATION

ACCESS FLOOR SUPPORT SYSTEM FRAMEWORK

MODULAR COMMUNICATIONS (TELEPHONE) CABLE TO ZONE JUNCTION BOX (NO CONDUIT)

POWER CONDUIT TO NEXT JUNCTION BOX

RAISED ACCESS FLOOR SYSTEM / MODULAR PLUG-IN DISTRIBUTION

NOTES

Where projected frequency of future changes is relatively high, a raised access floor system will provide the maximum flexibility and lowest cost to relocate or add services for work stations. When used in conjunction with a modular system of power, communication and data wiring plug-in receptacles, and cable connector sets, changes can be made without the need of an electrician or wiring technician. Advantages come at a premium, as access floor is the highest in initial cost of all systems described in this section.

Raised access floor is essentially a basic computer floor that is restricted in application to distribute only power, communication, and data services to work stations. The absence of air distribution and high density of cabling associated with computers permits raised floor height to be reduced to nominal 6 in. As the depth of standard 2 ft sq formed steel floor panel is less than 2 in., over 4 in. clear height under the panels provides sufficient clearance to accommodate most hardware associated with distributing services. Virtually any variety of above-floor or flush outlets

can be mounted on a floor panel and connected to lengths of cable with plug or connector fitting at the other end.

Access floor can be provided with or without stringers, which are used to minimize "creep" effect. Laser beam equipment speeds up accurate leveling of pedestal heights. For a custom installation without ramps or steps, the base floor is structurally designed to be depressed below permanent building elements such as lobbies, stairs, and toilets. Panels can be ordered with factory-installed carpeting or, alternatively, magnetic-backed carpet squares can be added after installation.

The introduction of an access floor does not necessarily require an increase in floor-to-floor height, and if so, the cubage added is at a much lower per unit cost than for the rest of the building. When special attention is given to coordinating lighting with other elements in the suspended ceiling or when lighting is provided below as from the work stations, the cavity can be compressed to compensate for the raised floor.

Fig. 15.36 *Pictorial representation of a full access floor designed to provide complete electrical, data, and signal services to a modern workstation. The infinite access and unlimited space are ideal for heavily wired, rapidly changing work stations. (From Ramsey and Sleeper,* Architectural Graphic Standards, *8th ed., 1988, reprinted by permission of John Wiley and Sons.)*

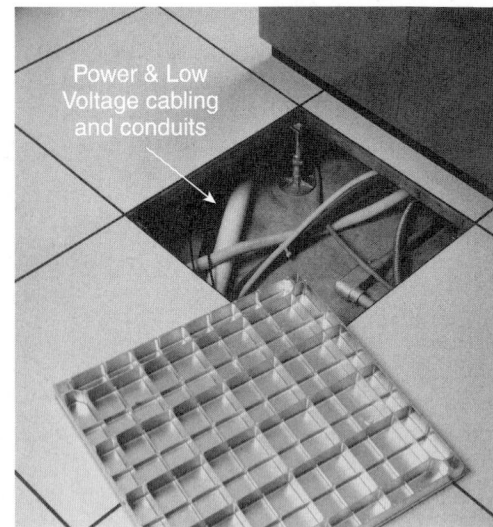

Fig. 15.38 *Reinforced nonferrous (aluminum) floors are used frequently in hi-tech applications requiring the large wiring capacity and convenience of full-access floors. Floor construction can be stringerless, as illustrated, or with stringers as in Fig. 15.39 (b). (Photo courtesy of Tate Access Floors, Inc.)*

Fig. 15.37 *Full-access floors provide infinite access to the myriad cables used in a modern commercial office. The modular wiring, using snap-on connectors and preassembled junction boxes, drastically reduces the labor cost of installation and the frequent cabling changes [see Fig. 15.39 (a)]. (Photo courtesy of Tate Access Floors, Inc.)*

(a)

Bolted Stringer System

(b)

Fig. 15.39 (a) *Typical power/low-voltage floor box for use with a full-access floor. See under-desk box in Fig. 25.37. (Photo courtesy of Tate Access Floors, Inc.)*
(b) *Steel-bolted stringer support for access floor panels. Many stringer and stringerless designs are available (see Fig. 15.38). (Photo courtesy of Tate Access Floors, Inc.)*

Conductor .009
Metallic shield — PVC
Polyester   Total thickness .031"

Fig. 15.40 Schematic section through one design of NEC type FCC under-carpet cable. The copper conductors illustrated are the equivalent of #12 AWG. The PVC acts as insulation, and the polyester as both insulation and physical protection. All designs require a metallic top shield and a metallic or nonmetallic bottom shield for physical protection.

sponding electrical facility changes. Beyond the extreme flexibility made possible by the ceiling raceway system, it has the additional advantage that it itself, not being cast in concrete like its underfloor counterpart, can be altered at will. Thus, not only layout changes (as mentioned previously) but also changes in the function of spaces can readily be accommodated. This is particularly important characteristic in merchandising and educational facilities, where spaces can have their function repeatedly changed during the course of the building's life.

Fig. 15.41 Preterminated 25-pair under-carpet telephone cable. These cables are commercially available in lengths from 5 to 50 ft. in 5-ft. increments. (Reprinted with permission of AMP.)

Fig. 15.42 Typical components of an under-carpet wiring system. The under-carpet FCC power cable is shown without the metallic top shield required in actual installation. It is a color-coded five-conductor cable (neutral, equipment ground, and three circuit conductors or two circuit conducts and an isolated ground conductor). The floor outlets shown are: front, single power outlet; rear (left to right), duplex power outlet (one of which has isolated ground), standard duplex power outlet, and combination data cable, communications, and telephone outlet. (Courtesy of Hubbell, Inc.)

System details vary among manufacturers but the systems are essentially the same and, in principle, resemble underfloor systems. A typical system is constructed of metallic or nonmetallic surface type raceways arranged in a tree formation [i.e., large trunk (header) raceways feed multiple smaller branch (distribution) raceways, and so on]. The raceways are hung in the ceiling plenum from the concrete slab above. The hung ceiling must consist of lift-out panels because this type of wiring system is not permitted in spaces rendered inaccessible by the building structure. Header ducts (large area raceways) are fed from electrical panels and signal, data, and telephone cabinets in the electrical and low-tension wiring service closets, respectively. Data headers are normally of larger size than the power header and can carry other low-voltage, low-power signal wiring as well. Distribution ducts (laterals) tap onto the headers. These laterals act as subdistribution raceways, feeding lighting fixtures and data, signal, telephone, and power outlets on the same floor and, via poke-through fittings, outlets on the floor above.

Undercarpet System Legend

⊖ Single duplex floor fitting

▲ Combination data and phone floor fitting

⊖◉▶ Combination fitting (power, data and phone)

⎯⎯⎯ Power cable

● Tap (step down, 5 cond to 3 cond)

▣ Equal tap (3 cond to 3 cond)

- - - - Data phone cables (low count, 4 pr)

--25-·- Data and phone cables, multiple

▨ Flush wall box-power

▨▲ Flush wall box-phone and data

▨ Surface wall box-power

▨ Floor box-power, phone, data, or combo

TYPICAL UNDER-CARPET WIRING FLOOR PLAN LAYOUT

Fig. 15.43 *Typical layout of under-carpet power and low-tension/data cabling for a small office. A power, phone, and data cable connection on the floor is provided under or immediately adjacent to each desk (see Figs. 15.41, 15.42, 15.44, and 15.45 for typical details). (Reproduced with permission of AMP.)*

928

Flush wall box

Carpet square

Power cable

Fiber optic data cables

Floor fitting

Power connection

Data connection

Fig. 15.44 Under-carpet flat power cables connect to round supply cables in a flush wall box and then extend to a combination power/low-tension/data floor fitting. Data cables are connected to the combination floor fitting from their system boxes, either individually or via other floor fittings. (Photo courtesy of Walker Systems, A Wiremold Company.)

Coaxial computer data cable

Carpet squares

Power cable

Fiber optic cables

Telephone cables

Power cable

Floor juction box

Fig. 15.45 Four service floor-junction boxes handle under-carpet power, telephone, low-tension electric, and fiber-optic cables. The floor box measures approximately 14 in. square and 2 in. deep. It can accommodate up to four 3-phase power circuits or eight 25-pair telephone/data cables or 24 coaxial cables or, as illustrated, a mixture of cable types. (Reprinted with permission of AMP.)

Fig. 15.46 *Typical floor-to-ceiling electrical/communication raceway poles. Units are available in a wide variety of shapes, sizes, and cross-sectional configurations, in aluminum and steel (see Fig. 15.47). (Courtesy of Hubbell Premise Wiring.)*

The standard method for extending wiring from the ceiling plenum raceways to floor-level or desk-level signal and power outlets is by use of vertical multisection raceways fed from the top (see Fig. 15.46). These service poles are available in a large variety of designs, finishes, and cross-sectional raceway areas and are easily installed in almost any location. They may be prewired and usually contain several power outlets, a telephone connection, and possibly data cable outlets. The result in a hung-ceiling office area or an exposed ceiling slab area (Figs. 15.47, 15.48, and 15.49) is certainly less elegant than a floor-level wiring system, but for most users it is satisfactory, and its low cost as compared to any type of floor-duct system is a prime redeeming feature.

When electrical connections to poles, lighting fixtures, receptacles, and communication/data outlets are made with "hard" wiring, considerable field labor is required with corresponding high cost. Furthermore, the relative permanence of such wiring lessens the inherent flexibility of the raceway system. To solve both problems, a number of manufacturers have developed a line of modular branch-circuit wiring elements. These, covered in *NEC* Article 604 under the very logical

Fig. 15.47 *(a) Poles are fed at the top in the hung ceiling or from exposed ceiling raceways. These feeds can be made either conventional hard-wired or modular (plug-in) as shown. Modular connectors are used for power and low-tension (telephone, communication, data) wiring.*
(b) Two of the many cross-sectional configurations available are illustrated. Other designs divide the pole into three sections to suit the specific application. (Courtesy of Hubbell Premise Wiring.)

Fig. 15.48 *Power poles extend down from the ceiling to any desired height. In this library, power is required above the base cabinets, and the power pole is easily arranged to supply it. (Courtesy of Wiremold Company.)*

Fig. 15.49 *The archetypical application of a power/signal pole is to provide electrical service to an island desk, as illustrated. This pole is equipped with telephone, data, and power outlets, all of which are in use. (Courtesy of Hubbell, Inc.)*

ELECTRICITY

Fig. 15.50 *Manufactured modular wiring assemblies are used for tap connections to feed ceiling lighting fixtures or any other ceiling connection, and a complete range of junction, switching, tap, and poke-through units. (Courtesy of Walker Systems, A Wiremold Company.)*

931

Fig. 15.51 *Manufactured modular wiring connectors simplify installation and reduce labor costs initially and in any subsequent rearrangements. (Courtesy of AFC Cable System.)*

name of "manufactured wiring systems," consist of metal-clad or armored cable sets terminating in polarized plugs. The polarization prevents accidental interconnection of low-voltage, 120- and 277-V systems. Ceiling raceways can be equipped with matching receptacles, and connection to fix-tures, poles, and other devices becomes a simple matter of plug insertion (see Figs. 15.46, 15.50, and 15.51).

The result is a wiring system of extreme flexibility in which even extensive changes can be made very rapidly with minimal disruption and virtually no mess. Manufactured wiring systems are permitted in accessible areas only, for obvious reasons. They are therefore also applicable to access floor spaces as seen in Figs. 15.36 through 15.39. The additional cost of manufactured wiring elements is frequently offset by labor savings even on initial installation and certainly after one or two field changes. Cable sets are available for power (120 V and 277 V), telephone, and all types of low-voltage signal equipment. The cables must be approved for use in conditioned air plenums and hung ceilings.

To take full advantage of the potential labor-cost savings inherent in the system, field labor must be minimized. This is accomplished by factory equipping all utilization equipment, including lighting fixtures, with appropriate plug-in connectors.

16

ELECTRICAL SYSTEMS AND MATERIALS

Service and Utilization

Having discussed the first of three major categories of electrical equipment, conductors and raceways, in Chapter 15, we turn now to the two remaining categories: power-handling and utilization equipment. The first step is to examine the means by which electric service is brought into a structure.

16.1 Electric Service

The codes and standards that apply to electric service include:

1. National Electrical Code ©; in particular, Section 230.
2. National Electrical Safety Code, published by the Institute of Electrical and Electronic Engineers (IEEE), 345 E. 47th Street, New York, N.Y. 10017. This code is recognized as an American National Standard (ANSI C-2). It deals with clearances for overhead lines, grounding methods, underground construction, and related matters.
3. Standards of the utility supplying electric service.

Public utility franchises require only that service be made available at the private property line. Thus, service is normally tapped onto the utility lines at a mutually agreeable point at or beyond the property line. The service tap may be a connection on a pole with an *overhead service drop* or an *underground service lateral* to the building, or a connection to an underground utility line with a service lateral to the building. Electrical construction work on private property is usually at the owner's expense.

Under certain conditions the owner can influence the type of construction utilized by the electric utility company in conveying the electric service to the site. This is often the case in large tract developments and in places where owners are willing to share some of the cost of better grade construction. Also, in many areas, the utilities themselves have instituted "beautification" programs in an effort to decrease the objectionable appearance of much of their equipment.

Service from a utility line to a building may be run overhead or underground, depending on the following factors:

- Length of service run.
- Type of terrain.
- Customer participation in the cost of service installation.
- Service voltage.
- Size and nature of electric load.
- Importance of appearance.
- Local practices and ordinances.
- Maintenance and service reliability.
- Weather conditions.
- Type of interbuilding distribution, if applicable.

16.2 Overhead Service

The principal advantage (to the utility) of overhead electric lines is low cost. Depending on terrain and other factors, the cost of overhead as compared to underground installation has been in the range of 10% to 50% (the latter when compared to direct burial cable installation). This accounts for the majority of installations being overhead. In recent years special techniques in underground installation have lowered that cost, making it a reasonable economic alternative when its advantages are considered.

Where the length of service run is several hundred feet or more, voltages higher than the facility's utilization level may be involved. This weighs heavily in favor of overhead lines, particularly with voltages exceeding 5000 V. Similarly, when terrain is rocky and the electrical load is heavy, the cost of underground installation rises sharply. Because overhead lines are easily maintained and repaired and faults easily located, service continuity with overhead lines is generally acceptable. In areas with severe weather conditions, called *heavy loading areas*, where combinations of snow, wind, and ice increase the possibility of outages on overhead lines, underground service is preferable. This is particularly true when even short service outages cause hardship or financial loss. Reliability of overhead service can be improved markedly by taking service from two separate, and preferably separated, overhead lines. Final decision on the type and location of electrical service will be taken after meetings among the building's architect, electrical engineer, and the electric utility's technical personnel.

Overhead cables are of several types: bare, weatherproof, or preassembled aerial cable. Bare copper cable supported on porcelain or glass insulators on crossarms are normally used for high-voltage lines (2.4 kV and higher). Low-voltage circuits (600 V and below) are generally run on porcelain spool secondary racks using single conductor (1/c) weatherproof cable. Preassembled aerial cable consists of three or four insulated cables wrapped together with a metallic tape and suspended by hooks from poles. This type of construction may be used up to 15 kV (Fig. 16.1). It often proves to be more economical than crossarm or rack installation and more resistant to damage from severe weather conditions.

A typical detail of an overhead electric service entrance to a multiple residence is given in Fig. 16.2.

Fig. 16.1 *Preassembled aerial messenger cables are carried by steel cables clamped to the poles.*

16.3 Underground Service

The advantages of underground electric service are attractiveness (lack of overhead physical and visual clutter), service reliability, and long life. The principal disadvantage is high cost. To overcome this, utilities frequently use direct burial techniques that, by eliminating the raceway, reduce costs considerably. Because direct buried cable cannot be pulled out if it faults, as is the case with a raceway-installed cable, restoration of service after a cable fault is time consuming. It is recommended that the decision of which technique is to be used be based on the consideration of these data:

· Cost premium for underground raceway installation, including handholes if required (see Section 16.4).

· History of outages for direct burial installation by this installer, in this immediate area.

· Cost and availability of repair service (utilities frequently will repair customer-owned underground service laterals *for a fee*).

· Impact of electric service outage in terms of time delays, inconvenience, necessity to dig up lawns and paved areas, and cost impact in the case of a commercial facility.

16.4 Underground Wiring

The methods available for underground wiring are:

· Direct burial (Fig. 16.3)

· Installation in Type I, concrete-encased duct [Fig. 16.4(a)].

· Installation in Type II, direct burial duct [Fig. 16.4(b)].

Item	Description
1	WEATHERHEAD FOR SERVICE ENTRANCE CABLE
2	WEATHERHEAD FOR CONDUIT
3	3 CONDUCTOR SERVICE ENTRANCE CABLE, #8 MINIMUM
4	PORCELAIN WIRE HOLDER
5	GALVANIZED CONDUIT MINIMUM 1"
6	3/4" GROUND ROD
7	1/2" GALVANIZED IRON PIPE W/GROUND WIRE
8	GROUND CLAMP
9	CONDUIT OR CABLE STRAPS
10	WATERTIGHT CONNECTOR
11	GALVANIZED FITTING
12	METER
13	METER SOCKET
14	SERVICE ENTRANCE SWITCH
15	#8 AWG, INSULATED, MINIMUM
16	CIRCUIT TO LOAD
17	SOLDERLES CONNECTORS

ELECTRICITY

First floor

Apt. 1   Apt. 2

Wiring trough   Note 3

Apt. 1   Apt. 2

Notes:
1. Omit item #10 if conduit is used.
2. Cold water pipe ground may be used in lieu of ground rod.
3. Meters may alternatively be placed inside the building.

Fig. 16.2 *Detail of typical overhead electric service to a four-family residence. Note that meters in small residential buildings are frequently placed on the exterior of the building. They can alternatively be installed inside, provided that ready access is available or that some type of remote meter reading system is installed (see Section 14.14).*

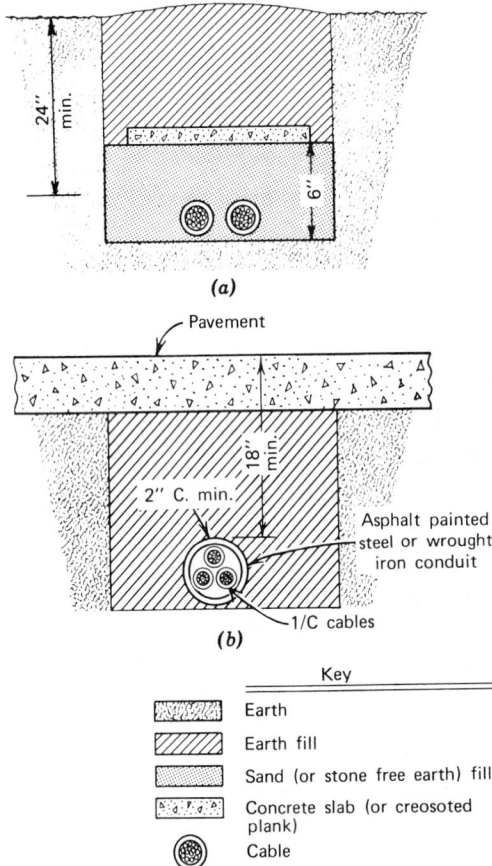

Key

▨	Earth
▨	Earth fill
▨	Sand (or stone free earth) fill
▨	Concrete slab (or creosoted plank)
◉	Cable

Fig. 16.3 (a) *Technique for installation of direct-burial cable.* (b) *Under highways, streets, and other high-load areas, cable should be installed in metal conduit.*

The first alternative offers low cost and ease of installation, with the disadvantage stated previously. The second offers high strength and permanence, but at the highest price of the three. The last offers median cost, but little strength. It is applicable only for installations on undisturbed earth and/or under light paving.

Nonmetallic duct (conduit) intended for underground electrical use is commercially available in two wall thicknesses. NEMA (National Electrical Manufacturers Association) type II with a heavy wall provides the physical protection required for direct burial installation with no concrete encasement. Type I is manufactured with a thinner wall and is intended for encasement in a minimum of 2 in. of concrete. Common trade

Installing Type I Nonmetallic
Underground Duct

Duct tiers
with separators

Tier by tier method
in concrete

(a)

Installing Type II Nonmetallic
Underground Duct

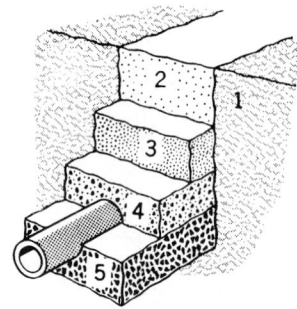

1. Trench wall
2. Ordinary backfill
3. Selected backfill
4. Selected backfill
5. Bedding

(b)

Fig. 16.4 *Underground duct installation.* (a) *Concrete duct bank should have minimum 6 in. earth cover.* (b) *Heavy-wall duct should be buried at least 24 in. in ordinary traffic areas and 36 in. where subject to heavy traffic. Each layer is about 8 in. thick.*

names for asbestos–cement and fiber ducts are Transite and Orangeburg. Plastic conduit is referred to as PVC, or simply as plastic.

Nonmetallic conduit is most frequently used without concrete encasement for low-voltage and signal wiring and with encasement for high-voltage wiring. It offers several advantages over steel conduit for underground work, such as lower cost and freedom from corrosion.

When underground electric wiring is duct installed and the run extends over several hundred feet (the exact distance depending on pulling tension), a pulling handhole or manhole is necessary. Handholes are used for low-voltage power and signal cables, and runs with a small number of cables. Manholes are used for high-voltage cables and where large duct banks must be accommodated. Precast handholes and manholes are readily available in most standard sizes and are usually cheaper than field-formed and poured units.

Cable used in underground wiring must be specially manufactured and approved for that purpose. Type SE is the basic service entrance cable, constructed with moisture- and flame-resistant covering. When provided with moisture proofing for underground use, the designation is SE type U, or simply USE. Underground cable for other than service runs is classified type UF (underground feeder).

16.5   Service Equipment

Referring back to Fig. 15.1, note that interposed between the high-voltage incoming utility lines and the secondary service conductors is a block labeled "transformer." This equipment item is required whenever the building utilization voltage is different from the service voltage. It may be pole or pad mounted outside the building, or installed in a room or vault inside the building, as discussed later.

16.6   Transformers

A transformer is a device that changes or *transforms* alternating current of one voltage to alternating current of another voltage. Transformers used in building work consist essentially of an iron core on to which are wound at least two coils: a primary (coil) winding and a secondary (coil) winding. A voltage impressed on the primary winding induces (through the iron core) a voltage in the secondary winding in proportion to the ratio of turns in the two coils. Thus a *step-down* transformer has a larger number of turns in its primary winding than in its secondary winding, and a *step-up* transformer has the reverse. In theory transformers are reversible, although in practice they are rarely used as such. Transformers cannot be used on dc.

A transformer would typically be used to step down an incoming 4160 V service to 480 V distribution within a building. Another transformer would then be used to step down the 480 V to 120 V for use on receptacle circuits. It is well to remember that ordinarily 120, 208, 240, 277, and 480 V are called *low* or *secondary voltages,* and 2400, 4160, 7200, 12,470 and 13,200 V are *high* or *primary voltages.*

Transformers are available in single-phase or three-phase construction. (See Section 17.3 for an explanation of these terms.) Transformer power capacity is rated in kilovolt-amperes (kVA). For a single-phase unit, this figure is the product of the full load current and the voltage. Because the voltages on primary and secondary are different, so are the primary and secondary currents—because the kVA remains constant.

For example, a single-phase 100 kVA, 2400/120 V transformer will carry at full load:

$$\text{primary current} = \frac{100{,}000 \text{ VA}}{2400 \text{ V}} = 41.6 \text{ A}$$

$$\text{secondary current} = \frac{100{,}000 \text{ VA}}{120 \text{ V}} = 832 \text{ A}$$

That is, the ratio of primary to secondary current is exactly the reverse of the ratio of primary to secondary voltage, because their product, which is the transformer's kVA rating, remains constant. Expressed in an equation,

primary current × primary voltage =

secondary current × secondary voltage

or

$$V_p I_p = V_s I_s \qquad (16.1)$$

where V_p is primary voltage
        I_p is primary current
        V_s is secondary voltage
        I_s is secondary current

Heat is generated by the passage of current through the transformer coils due to the winding cable resistance. This heat is transferred to the unit's *cooling medium,* where it is radiated or otherwise disposed of. The unit's cooling medium is a characteristic of major importance. Transformers are either dry (air cooled) or liquid filled. The choice depends on required electrical characteristics, proposed physical location, and cost factors. Although detailed considerations are beyond the scope of this book, some selection criteria are presented in the following sections. In general, units rated above 5 kV are liquid filled and units in the 600-V class are dry. Units installed indoors, except in vaults, are normally dry type, and are intended for general purpose light and power circuits (Table 16.1). *Load center* transformers are installed in unit substations, both indoor and outdoor. *Distribution* transformers are mounted on a pole or on a concrete pad outdoors. *Substation* transformers are large and are always concrete-pad mounted.

The *insulation class* of a transformer affects its permissible temperature rise and operating temperature, and as a direct result, its physical size, electrical power losses, overload capacity, and life.

The physical size of a transformer of given kVA rating and voltage depends on the type of insulation used. In order of decreasing physical size and increasing operating temperature we have, for dry transformers, 105, 150, 185, and 220°C systems that represent organic, inorganic (two types), and silicone insulating materials, respectively (Table 16.2).

The dimensional data in Table 16.3 vary considerably from one manufacturer to another, and therefore the table is useful only for a concept of bulk volume and weight.

Although 220°C system insulation transformers can withstand 150°C rise (see Table 16.2) some users specify 220°C system insulation with 115°C rise or even 80°C rise; that is, a better grade of insulation is used with an *underrated* transformer. The advantages of this design are fourfold:

· Longer life.

· Higher overload capacity.

· Lower operating cost.

· Increased capability to withstand heating effects of harmonics commonly found in commercial and industrial electrical systems (see Section 17.18).

TABLE 16.1 **Typical Transformer Data**

Transformer Type (Application)	Maximum Capacity[a] (kVA)	Insulating (Cooling) Medium	Voltage Range	
			Primary	Secondary
General purpose, dry type	1000	Air	120–600 V	120–600 V
Load center pad-mounted unit substation	3000	Air	2.5–15 kV	120–600 V
Load center pad-mounted unit substation	5000	Oil Silicone Fluid[b] Epoxy/ resin[c]	2.5–34.5 kV	120–4800 V
Distribution system	10,000	Oil Silicone	2.5–67 kV	120–4800 V

[a]Three-phase bank.
[b]High-fire-point paraffinic hydrocarbon fluid, manufactured expressly as a transformer dielectric-coolant.
[c]Solid dielectrics used for special installation, such as high hazard areas.

TABLE 16.2 **Air-Cooled Transformer Electrical Insulation Temperature Ratings (Based on 40°C Ambient)**

Insulation Class (System)[a]	Insulation Type	Average Conductor Temperature Rise	Ambient Temperature	Hot-Spot Temperature Gradient	Total Maximum Temperature
	Organic (A)				
105°C	Mica, glass, resins	55°C	40°C	10°C	105°C
150°C	(B)	80°C	40°C	30°C	150°C
185°C	Asbestos (F)	115°C	40°C	30°C	185°C
220°C	Silicones (H)	150°C	40°C	30°C	220°C

[a]The modern terminology for insulation class uses "system" in lieu of "class."

TABLE 16.3 **Typical Dry-Type Transformer: Dimensions and Weights**[a]

kVA Output Continuous	Temp. Rise (°C)	Approximate Dimensions (in.) $H \times W \times D$	Approx. Weight (lb)
Primary: 480 V Secondary: 120/240 V			
Single-phase			
3	115	$13 \times 8 \times 8$	60
5	115	$14 \times 12 \times 12$	120
10	115	$15 \times 12 \times 12$	150
15	150	$24 \times 16 \times 16$	200
25	150	$24 \times 16 \times 16$	230
50	150	$30 \times 20 \times 20$	430
100	150	$40 \times 24 \times 24$	650
Primary: 480 V Secondary: 208Y/120 V			
Three-phase			
45	150	$26 \times 26 \times 15$	380
75	150	$30 \times 30 \times 20$	600
150	150	$40 \times 36 \times 26$	1100
225	150	$44 \times 36 \times 26$	1400

[a]Dimensions and weights vary among manufacturers. All figures increase with higher primary voltage and with lower temperature rise.

ELECTRICITY

A 220°C system transformer operated at full load *continuously* (an unusual situation) has a short life—estimated on the basis of accelerated aging tests to be between 3 and 10 years. The same class insulation transformer (220°C system) rated at 80°C has a life expectancy of more than 100 years. With respect to operating cost, the same situations obtain here as with the high-temperature insulation discussed in Section 15.15 and studied in Table 15.5. The truism that one gets nothing for nothing can be expressed here as: In return for the smaller, lighter, cheaper, and hotter transformer (220°C system), we have higher losses, and therefore higher operating cost and shorter life.

ASHRAE/IES Standard 90.1-1989, *Energy Efficient Design of New Buildings,* which applies to most new construction, requires that for buildings with a total transformer capacity in excess of 300 kVA a calculation be made comparing the operating costs of various types of transformers and the results used to "balance energy costs with necessary operating flexibility, reliability and safety."

The author performed a typical calculation using the ASHRAE form and assuming a daily/weekly operating cycle and loading that is representative of commercial use. The two transformers studied were dry type, rated 750 kVA; one designed for 80°C rise and the second for 150°C temperature rise. Based on an energy cost of 7 cents per kWh, the lower energy waste of the more expensive 80°C unit repays the first-cost differential in about 4 years. Obviously, a higher electric energy cost reduces the payback time, and vice versa. Beyond the payback period the advantage is entirely on the side of the 80°C unit. Using a 30-year life, 8% fixed capital cost, and 3% annual cost escalation, the life-cycle cost of the 80°C unit is $10,000 less than that of the 150°C unit. (At the assumed loading, even the 150°C unit will probably last 30 years; the 80°C unit much longer.)

Furthermore, with a total permissible hot-spot temperature of 220°C (424°F), the location of the 150°C rise unit must be very carefully chosen because it can create a serious heat generation and radiation problem.

To summarize, a transformer is specified by type, phase, voltages, kVA rating, sound level, and insulation class. Thus, 112.5 kVA, 3-phase,

480/120-208 V, air-cooled, indoor, dry-type transformer with 220°C insulation system and 115°C rise, 45 db (decibel) maximum sound level, is an adequate transformer description. Sound ratings of transformers, as well as installation techniques and acoustical treatment, are discussed in Part IX, "Acoustics."

16.7   Transformers Outdoors

A service transformer bank is necessary, as explained above, when the facility utilization voltage is different from the utility voltage [Fig. 16.5(a)]. The designer occasionally opts for a step-up, step-down arrangement when the service run is so long that the conductor cost when run at a low voltage would be excessive because of large-sized cables. In such instances, the cost of the double transformer installation must be more than offset by the savings in feeder cost to be economically justifiable [Fig. 16.5(b)].

The advantages of an outdoor transformer installation are:

· No building space required.

· Reduced noise problem within building.

Fig. 16.5 Service transformer arrangements. (a) *High-voltage service with step-down service transformer at the facility, and* (b) *low-voltage service, with transformation at both ends of a long service run.*

· Lower first cost.

· Ease of maintenance and replacement.

· No interior heat problem.

· Opportunity to use low-cost, long-life oil-filled units.

However, it is frequently easier to find space indoors (preferably in a basement) than to find a suitable exterior spot, and noise may be more disturbing from the available exterior spaces, such as courtyards, than from a basement. Costs can run high if long secondary voltage runs are required; heat can often be handled by louvers or areaways adjoining a basement or even made use of profitably if the transformer load is fairly constant. Also, because exposure to direct sunlight decreases a transformer's rating by increasing its temperature, a shaded spot may be difficult to find. Furthermore, an exterior transformer, other than pole mounted, is a questionable choice in an area with a high incidence of vandalism, regardless of sturdiness of construction.

Finally, appearance of an exterior unit may be objectionable. This point has received much attention from manufacturers, and numerous designs have been developed that minimize the appearance problem (Fig. 16.6). The most popular type of exterior transformer installation is the pad mount. It has all of the advantages listed previously in addition to extreme simplicity of installation—it is simply set on a concrete pad. Consult manufacturers' catalogs for dimensional data.

16.8 Transformers Indoors: Heat Loss

When an indoor transformer installation is indicated, special consideration must be given to the transformer's heat-generating properties. Between 1% and 1½% of a transformer's rating, depending on type, is converted to heat at full load. For a 750 kVA, dry-type, 150°C rise unit, 12 kW, or 41,000 Btu/h, of heat loss is generated at full load! Losses are lower for 80°C rise units. Liquid-filled units have approximately the same losses as 80°C rise dry units. Unless the heat can be used, sufficient ventilation—either natural or forced—must usually be provided to keep the room ambient temperature from exceeding 40°C.

(a)

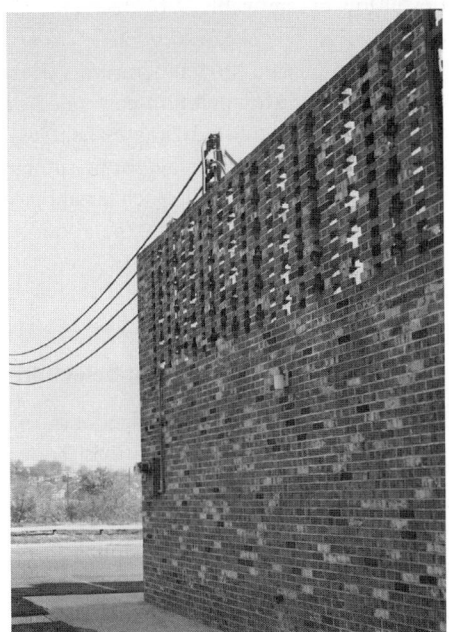

(b)

Fig. 16.6 (a) *Pad-mounted exterior transformers are neat, compact, and, if sited properly, unobtrusive. Large units can be partially screened by shrubbery.*

(b) *When the size is such that visual screening becomes a problem, consideration should be given to a structural screen such as a decorative brick wall, which also provides some screening. (Courtesy of General Electric, U.S.A.)*

ELECTRICITY

Ventilation by natural convection is most desirable, necessitating the location of the transformer room on an exterior wall (with an areaway if the room is located below grade). The size of the *free* area required for a ventilation opening is 3 in.^{2}/kVA of capacity, plus an additional 1 in.^{2}/kVA for switchgear heat losses, if any. If a louver is used in the openings, the size of the opening must usually be doubled (total: 8 in.^{2}/kVA) because most louvers have a 50% free area. For good convection, it is desirable to divide the louvered areas in half, placing one-half near the ceiling and the remaining half near the floor. To provide for equipment removal, if the areaway is large enough, it may be useful to add louver area between the upper and lower openings and make the full louver removable. A bird screen is also desirable.

Because outside air temperature varies, it is advisable to use a temperature-controlled, adjustable louver. In severe cold climates, heat loss from the electrical equipment may not be sufficient in winter to warm the room. In such instances, a unit heater should be installed in the room with a thermostat set at 55°F.

16.9 Transformers Indoors: Selection

When transformers are installed indoors, they are subject to stringent *NEC* regulations that are designed to make the installations intrinsically safe. These regulations are detailed in *NEC* Article 450, and no purpose is served in duplicating them here. Instead, we present the reader with the essential considerations involved.

(a) Oil-Insulated Transformers. These present a fire hazard indoors because flammable oil can spread from a tank leak or rupture. To prevent this, most oil-filled transformers must be installed in a fire-resistant vault, the construction of which involves a heavy cost. (See *NEC* Art. 450 for exceptions.) Advantages offsetting this cost are the oil-filled transformer's small size, low first cost, low losses, long life, excellent electrical characteristics, low noise level, and high overload capacity. Despite these, the vault requirement has had the effect of restricting oil-filled units to industrial facilities and other structures where electrical considerations favor its use.

(b) "Less-Flammable" Liquid-Insulated Transformers. Transformers rated 35 kV or less that are insulated with a liquid whose fire point is not less than 300°C ("less-flammable" liquid) may be installed indoors without a vault. A liquid confinement area is required, plus either a fire extinguishing system or restrictions on the combustibility of the building's contents. The insulating liquids used are special types of silicones and hydrocarbons, and the transformer installation must meet all requirements for the specific liquid involved. Use of these units achieves the electrical and physical advantages of oil-insulated units without the cost penalty of vault construction and installation. First cost of the units depends on the fluid type and the specific electrical characteristics required. In general, they are somewhat less expensive than nonflammable fluid–insulated units with similar electrical ratings.

(c) Nonflammable Fluid–Filled Transformers. Certain nonflammable liquid coolants, called *askarels,* generally containing polychlorinated biphenyl (PCB) were once very widely used. However, since 1979, use of PCB coolants in new equipment has been banned by federal regulation due to negative ecological impact, and existing units are gradually being phased out. Newer nonflammable fluid coolants have increased the price of these transformers considerably (Table 16.4). Such units have most of the advantages of oil-filled units and do *not* require a vault unless voltage is very high. They do, however, require a sump or catch basin of sufficient capacity for all of the contained liquid. This and the high first cost are the negative aspects of this design.

(d) Dry-Type Transformers. These are the units of choice in the majority of indoor installations, despite shorter life, higher losses, high noise level, greater weight, and larger size than the liquid-filled units. The principal advantage is ease of installation and almost unrestricted choice of location. As explained previously, by using an underrated transformer (220°C system, 80°C rise),

TABLE 16.4 **Relative Installation Costs for a 300 to 1000 kVA Transformer**

Transformer Type	Temperature Rise[a] (°C)	Relative Transformer Cost	Construction Cost	Total Relative First Cost
Oil filled	65	1.00	50–100[b]	1.50–2.00
Less-flammable and nonflammable fluids	65	1.15–1.30	20–40[c]	1.35–1.70
Dry, ventilated	80	1.65	—	1.65
	80	1.50	—	1.50
	150	1.35	—	1.35
	150	1.20	—	1.20

[a]Transformers of equal temperature rise have approximately equal life and equal losses.
[b]Cost of vault depends on local labor costs and size of transformer. The relative cost decreases with increasing transformer size.
[c]Cost of catch basin. As in the case of a vault, relative cost depends on labor rates and transformer size.

losses (heat) can be reduced and life extended. Also, for a price premium, noise level can be reduced. Table 16.4 gives a comparison of installed costs for these four classes of transformers.

16.10 Transformer Vaults

A transformer vault is basically a fire-rated enclosure provided because of the possibility of fire due to rupture of an oil-filled transformer case. However, this must not be construed as implying that transformers are hazardous or delicate devices prone to faults. On the contrary, transformers are extremely tough, sturdy, long-lived, capable of sustaining large and prolonged overloads, and, indeed, among the most reliable elements of an electrical system. However, faults do occur, and an oil-filled transformer is a potential fire hazard.

Transformer vaults should be located, if practicable, where they can be ventilated to the outside air without flues or ducts. The combined net area of all ventilating openings (gross area less screens, louvers, etc.) should be as explained in the preceding section: not less than 3 in.^{2}/kVA of transformer, but in no case less than 1 ft^2. Further ventilation recommendations plus details of enclosure construction materials, fire rating, door and sill details, and other relevant and important information are provided by *NEC* Article 450.

16.11 Service Equipment Arrangements and Metering

Metering must be provided ahead (electrically) of the building's service entrance switch(es). The metering is either at the utility or the facility voltage, and either at the service point or inside the building. The choice is generally left to the owner with the understanding that interior meter equipment must be readily accessible to utility personnel. Although in increasingly large numbers of facilities actual meter reading is accomplished remotely, the meter equipment must still be available for inspection and service.

If high-voltage service is purchased, then the transformers and all equipment beyond the service connection must be furnished by the owner. Conversely, if low-voltage service is purchased, all equipment necessary to provide low voltage is furnished by the utility. Obviously, the electric service rates for low-voltage service are *higher* than for high-voltage service in order to compensate the utility for the cost of providing and maintaining the step-down transformer and associated equipment. Therefore, it is often advisable for the owner of a large facility to investigate the economics of purchasing power at high voltage. Because many owners are not equipped to maintain high-voltage equipment, arrangements can sometimes be made to pay the utility to provide and maintain transformers while taking the cost advantage of high-voltage service.

For a single-occupant building or a building in which electric energy is included in the rental charge, only a single meter is necessary. Provision for such metering may be made in the main switchboard or the meter may be independently mounted. In both cases, the meter is furnished and installed by the utility company. Where submetering is required, such as in apartment houses, banks of meter sockets are installed to accommodate the multiple meters. Federal regulations forbid master metering in new multiple-dwelling constructions because it encourages energy waste (Fig. 16.7).

A low-voltage underground service detail as it would appear on a set of contract drawings, including relevant details, is given in Fig. 16.8. Note that here the service switch and meters are separately mounted.

Fig. 16.7 *Three-section commercial submetering switchboard comprising a center service-entrance section rated 4000 A, 3 phase, 480 V maximum, equipped (usually) with a main switch or circuit breaker. Each of the end cubicles will accept six tenant meters, each on a 320 A 3-phase circuit, maximum, protected by an integral circuit breaker or fused switch. Tenant wiring enters the top or bottom of the board. Switchboards of this type can be installed on every floor of a large, multitenant commercial office building. (Photo courtesy of Cutler-Hammer.)*

16.12   Service Switch(es)

The purpose of the electric service switch(es) is to disconnect the normal service to the building. (See Section 17.20 for a discussion of emergency electric service.) Thus, in the event of fire, no electrical hazard will face firefighters. It is therefore obvious that this disconnecting apparatus must be located at a readily accessible spot near the point at which the service conductors enter the building. If such a location is not feasible, service conductors may be run in concrete encasement under the building and are considered "outside the building" up to the point at which they emerge from the floor in the building (see Fig. 16.8). At that point, the service disconnect must be installed. The service switch or, more accurately, the "service disconnecting means," may comprise one to six properly rated switches. These are frequently assembled into a switchboard. However, before discussing switchboards, a description of switches, circuit breakers, and fuses, the components of which switchboards are constructed, is in order.

16.13   Switches

In this and the following two sections we discuss "traditional" electrical switching devices, which close and open an electric circuit by physically moving two electrical conductors into contact with each other to close the circuit, and physically separate them to open the circuit. The motive force can be supplied by hand, an electrical coil, a spring, or a motor, as discussed later, and the device name may change accordingly, but the action is the same. Solid-state switches perform the same electrical switching function but by a completely different process, without moving parts. They are discussed in Section 16.16.

An electrical switch is rated by current and voltage, duty, poles and throw, fusibility, and enclosure. The current rating is the amount of current that the switch can carry continuously and interrupt safely. Switches intended for motor control are also rated in horsepower. The voltage rating of a switch is, as for other electrical equipment, by voltage class. Thus a switch may be rated

Fig. 16.8 *Typical electrical plan and details of an underground low-voltage service (600-V class) to an industrial facility. Note that the portion of UG service beneath the building must be encased in concrete. A service transformer (if required) could be mounted on the utility pole or on a pad at the property line.*

250 V, 600 V, or 5 kV, as required. Switches intended for normal use in lighting and power circuits are called *general duty safety switches*. Switches intended for frequent interrupting, high-fault currents, and ease of maintenance are rated HD for *heavy duty*.

An examination of Fig. 16.9(a) should clarify what is meant by the number of poles and throws of a switch. Unless otherwise noted, a switch is assumed to be single throw. Because the *NEC* states generally that the grounded neutral conductor of a circuit should not be broken, most switches carry the neutral through unbroken, by means of a solid link within the switch. This gives rise to the term *solid neutral* (SN) *switch.* Switches are available in 1-, 2-, 3-, 4-, and 5-pole construc-

(a) Single-pole single-throw switch.

(b) Two-pole single-throw switch.

(c) Three-pole and solid-neutral (3P and SN) switch.

(d) Single-pole double-throw switch (also called, in small sizes, a 3-way switch).

(e) Single-pole double-throw switch with center "off" position (in control work called a hand-off-automatic switch).

(f) Use of two single-pole double-throw (3-way) switches for switching of a lighting circuit from two locations.

(a)

Fig. 16.9 (a) *Typical switch configurations. Note that switches are always shown in the open position.*
(b) *Graphic representation of switch (contact) positions in a contactor or relay. Note that a latching contactor shows the switch in one of two positions to indicate the latching nature of the contactor.*

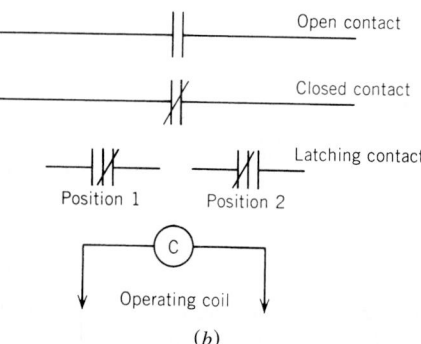

TABLE 16.5 **Control Equipment Enclosures**

NEMA Designation: Type	Description	Application
1	General purpose	Dry, indoor use
2	Drip-proof	Indoor, subject to dripping
3	Dust-tight, rain-tight, and sleet-resistant	Indoor/outdoor, where subject to windblown dust and water
3R	Rain-proof and sleet-resistant	Outdoor, subject to falling rain, snow, and sleet
3S	Dust-tight, rain-tight, and sleet-proof	Outdoor, subject to windblown water, dust, and sleet; most severe exterior duty
4	Watertight and dust-tight	Indoor/outdoor, subject to water from all directions; not sleet-proof
4X	Watertight, dust-tight, corrosion-resistant	Same as type 4 with added corrosion protection
7–9	Hazardous	Differing in application by class and group of hazardous use; see NEC
12	Industrial use, dust-tight, and drip-tight	Indoor only, general use, industrial and other "dirty" environments

tion. Poles are indicated by a "P"; thus, 3-pole is written "3P," and so on.

A switch may be constructed with or without provision for fusing. If provided, the switch is fusible; if not, the switch is nonfusible. All individual enclosed switches are mounted in an appropriate cabinet. The National Electrical Manufacturers Association (NEMA) standardized the nomenclature and application of enclosures for all electrical control equipment, of which switches are only one item. These are detailed in Section 16.15. Summarizing then, we could adequately describe a switch thus: switch, HD, 3P & SN, 200A/150AF, 600 V, in NEMA 12 enclosure. This "translates" as a heavy duty switch, 3 poles and solid neutral, 200 amp rating with 150 amp fuses, 600 volt rating, in an industrial use enclosure (Table 16.5). Such a switch is illustrated in Fig. 16.10.

16.14 Contactors

A contactor is a switch. Instead of a handle-operated, movable blade, a contactor uses *contact* blocks of silver-coated copper, which are forced together to *make* (close) or are separated to *break* (open) the circuit. The common wall light switch is a small, mechanically operated contactor. A relay is an electrically operated contactor. Most contactors are operated by means of an electromagnet that causes the contacts to close. They open either by spring action or by gravity.

Contactor terminology is somewhat different from that of a switch. Its condition when deenergized is its *normal* state. Thus, a contactor whose contacts are open when the coil is not energized is referred to as *normally open* (NO), whereas a contactor whose contacts are closed when deenergized is referred to as *normally closed* (NC). Units intended for motor connection are called *motor starters* and are discussed in Section 16.23. Current, voltage, and number of poles have the same nomenclature for contactors and relays as for switches.

The great advantage of contactors over switches is their facility for remote control. Switches must be manually thrown—or, in very special cases, by a motor. The magnetic contactor by contrast is inherently a remotely controlled device, making it ideal for a myriad of control functions. It can be controlled by manual or remote pushbutton or by automatic devices such as timers, float switches, thermostats, pressure switches, and so on. Because control can be both remote and automatic, the application of contactors is universal in control of lighting, heating,

ELECTRICITY

Fig. 16.10 *The internal construction of a heavy-duty, fused, 3-pole 3-wire switch in an industrial NEMA 12 enclosure. Note the gasketing on the inside of the cover, which seals the sheet steel box and makes it usable in atmospheres containing lint, dust, dirt, sawdust, and the like. The external handle is a double lockable indicating type, for safety. (Courtesy of Siemens Energy & Automation, Inc.)*

air conditioning, motors, and the like. Graphic representation of contactors is shown in Fig. 16.9(b).

16.15 Special Switches

Many special types of switches are available. Most of these types are beyond the scope of this book, except for the following, which are used extensively in building work.

(a) Remote-Control (RC) Switches. A contactor that latches mechanically after being operated is known as a *remote-control switch.* It differs from a relay in that the latching operation performs the dual function of latching the contacts and disconnecting the electric circuit. To unlatch

the contacts, the electric circuit must be reenergized. It is therefore also known as an *electrically operated mechanically held contactor.* This characteristic is advantageous where the position of the contacts will be maintained for a long period (as in a lighting control circuit), and it is undesirable to hold the contacts by continuous energizing of a coil, as would be required with an ordinary relay. A typical unit is shown in Fig. 16.11 and a typical application is shown in Fig. 17.13. A common application of these switches is for switching of a "block" load, such as exterior lighting, whole floor or whole building lighting, and the like.

(b) Automatic Transfer Switch. This device, which is an essential part of all emergency and standby power arrangements, is basically a double throw switch—generally 3-pole—so arranged that on failure of normal service it *automatically transfers* to the emergency service.

Fig. 16.11 *This remote-control switch (mechanically held, electrically operated contactor) is a 12-pole, 20-A, 600-V unit measuring approximately 7 in. H × 6 in. W × 4 in. D, making it suitable for stud-wall mounting. Compact solid-state control modules are available that mount directly onto the unit and convert it to a flush, wall-mounted multicircuit programmable controller (see Section 16.16). (Courtesy of Automatic Switch Co.)*

When normal service is restored, it automatically retransfers to it. (Retransfer can also be arranged to have a minimum delay, or it can be entirely manual). The switch control devices are voltage sensors that sense the condition of the service and operate the switch accordingly. Auxiliary devices can be built on to the basic switch, the most common of which are emergency generator starting equipment. Figure 16.12 illustrates a typical unit. Transfer switches used in uninterrupted power supplies are normally solid-state because they involve no switching time lapse.

(c) Time-Controlled Switches. In this category are included all switches whose operation is time based. The timer can be either the familiar electromechanical device consisting of a low-speed miniature drive motor, to which some type of contact making device is physically connected, or it can be solid-state electronic timer, which in turn controls either a relay or a solid-state switch. The latter arrangement is the basis of all modern programmable time controls (see Section 16.16), which find wide application in lighting control (Sections 21.3 and 22.27), energy management and automated building control (Sections 14.13 and 22.28), and clock and program systems (Section 22.14).

Motor-operated timers (other than hand-wound types) depend for their accuracy on power-line frequency, have moving parts that wear, must be reset after power outages (some units have

spring-wound reserve power motors), and become cumbersome and expensive with increase in the number of controlled events. Electronic units are small, cool, quiet, independent of line frequency for accuracy (in many designs), carry through outages when standby-battery equipped, and have virtually unlimited event control capacity. They have in large measure now replaced the older design except for the simplest applications (Figs. 16.13, 16.14, and 16.15).

16.16 Solid-State Switches, Programmable Switches, Microprocessors, and Programmable Controllers

A solid-state switch is an electronic device with a conducting state and a nonconducting state, corresponding to a conventional switch in its closed position and its open position. The change between the two states is accomplished by the application of a control signal (voltage). The change of state is instantaneous, noiseless, and occurs without arcing. The current ratings of solid switches are approximately the same as those of their mechanical counterparts, ranging from fractions of an ampere to many hundreds of amperes. Voltage ratings are usually limited to secondary systems ratings (0 to 600 V) due to the nature of the semiconductors comprising the active elements of these switches, although higher-voltage equipment is available for special application.

Load Connections

Source 1
Line Connections

Solid State Logic
Panel

Manual Operating
Handle

Time Delay
Engine Start Timer

Engine Start
Connection

Quick Disconnect
Plug

Source 2
Line Connections

Fig. 16.12 Automatic transfer switch with solid state logic. The unit can be operated manually if required. It is equipped with engine starting contacts, auxiliary contacts for monitoring, electrical safety interlocks, and automatic differential voltage sensing on the normal source. An 800-A 3-phase unit of this design, in a NEMA indoor sheet metal enclosure, measures 73 in. H × 26 in. W × 21 in. D and weighs approximately 600 lbs. (Photo courtesy of Cutler-Hammer.)

Fig. 16.13 *Time switch arranged for three different types of control functions that are applicable to different types of load: (1) full photocell control—security lighting; (2) dusk ON, preset OFF—signs, parking lots; (3) preset ON, OFF—interiors. This unit's dial is a 7-day calendar type that permits setting a different schedule for every day of the week. The dial is not astronomical (solar) because all the turn-on functions are either photocell or time-of-day controlled. Solar dials are useful when photocell control is not used because sunrise and sunset times vary with the season. (Courtesy of Tork, Inc.)*

Fig. 16.14 *Programmable time switch. This unit, which measures 9.5 in. H × 8 in. W × 4.5 in. D in its cabinet, permits independent programming of eight 20-A, 240-V circuits. Each circuit (channel) can be arranged for multiple ON–OFF daily operations and can be programmed on a 7-day a week basis, with special 365-day functions such as automatic daylight savings time adjustment. (Courtesy of E.Z. Controls.)*

If an electronic timing device is added to a solid-state switch, a time-controlled electronic switch is created that functions exactly as the electromechanical time-based switches described in Section 16.15. These switches have the advantage that their time control is independent of the utility line frequency and there are no moving parts to wear out. If a small programmable memory circuit in the form of an erasable programmable read-only memory (EPROM) chip is added, we then have a programmable time switch. The simplest of these are single-circuit 15-A devices intended to replace common wall switches (see Fig. 16.47). A more complex unit, corresponding roughly to the electromechanical unit shown in Fig. 16.13 but much more flexible, is shown in Fig. 16.14. Addition of a read only memory (ROM) chip preprogrammed with astronomical information

Fig. 16.15 *This unit controls two 15-A, 240-V circuits. The built-in astronomical feature automatically adjusts the ON–OFF setpoint according to sunset and sunrise throughout the year. (Courtesy of Paragon Electric Co., Inc.)*

Fig. 16.16 *Block diagram of a programmable controller. These devices are designed to supervise field conditions continuously, and are therefore used in industrial applications as well as for lighting and environmental system control.*

(sunrise–sunset) for the particular latitude involved converts the switch to a "sun-tracker." Such a switch is particularly applicable to control circuits with year-round dusk-to-dawn control (see Fig. 16.15).

If, in lieu of a programming device, which, until reprogrammed, gives fixed invariable instructions to a (multichannel) switch, as is the case with a programmable time switch, we install a *microprocessor* [i.e., a logic/memory device programmed to *respond* according to a particular algorithm (control plan) to given input signals], we have constructed a programmable controller (Figs. 16.16, 16.17, and 16.18). By definition, a *programmable controller* is an electronic device that uses a programmable memory for internal storage of

Fig. 16.17 *Programmable controller (center) designed for use as an energy monitoring and building system control device. The basic unit accepts up to 16 analog inputs, such as illuminance, temperature, kW demand load, time signals, central control signals, and the like. It processes data according to its current program (note programming keypad) and can signal and control up to 16 outputs. Additional functions available include time-of-day switching including astronomical considerations and duty-cycle load switching for electric demand control. The controller is approximately 14 in. W × 16 in. H × 6 in. D, and is arranged for surface mounting. On either side are accessory communication and relay modules. In the foreground are application software and typical sensors. (Photo courtesy of Paragon Electric Co., Inc.)*

Fig. 16.18 *General purpose programmable controller. The processor unit contains the central processor (CPU), random access memory (RAM), and power supply. Peripherals include the illustrated hand-held programming unit and EEPROM (electrically erasable programmable read-only memory) module, and various computer/software interfaces (not shown). The unit illustrated which measures approximately 10 × 9 × 3 in., has 16 I/0 circuits and a 885-word memory. It is readily adaptable to a variety of automated control processes in manufacture, building control, and material handling. (Courtesy of Allen-Bradley Co.)*

instructions that in turn implement specific functions such as logic, timing, counting, and so on.

The internal control/memory is frequently referred to as a *microprocessor,* whereas with respect to the overall programmable controller, it acts as a central processing unit (CPU). The programmable controller is therefore a special type of computer. It differs from a conventional computer in that its program is relatively short and exists in hardware (microprocessor), whereas a computer reads its program, which can be very lengthy, into its memory from software. Furthermore, the programmable controller is designed for a specific type of function, whereas the conventional computer is a general-use calculating (computing) device that processes almost any type of software.

Programmable controllers have all but replaced the ubiquitous *hard-wired* relay panel in applications such as industrial control, elevator control, process control, and the like, because of the ease with which such a controller can be reprogrammed (in lieu of rewiring a relay panel), the speed and accuracy of control, and the vastly increased complexity possible. See Chapter 22 for definitions and for an explanation of application of these devices to building control.

16.17   Equipment Enclosures

Proper NEMA nonmenclature, descriptions, and applications for the more common enclosures are found in Table 16.5. It is important to note that there is no enclosure described as WP or weatherproof. Equipment intended for outdoor use should be specified:

1. In a type 3R enclosure to protect against rain (Fig. 16.19).

Raintight NEMA 3R circuit breaker enclosure.

Fig. 16.19 *Type 3R outdoor enclosure. This is the type usually intended when "weatherproof" specified. (Courtesy of General Electric U.S.A.)*

2. In a type 3S enclosure to protect against wind-driven rain and sleet.

3. In a type 4 enclosure to protect against the above plus splashing and condensation.

Note also that the type 12 industrial enclosure is similar to type 1 except that it is gasketed for dust and drip resistance and therefore is well applied in *all* "dirty" indoor environments, including commercial and institutional spaces (see Fig. 16.10).

16.18 Circuit-Protective Devices

To protect insulation, wiring, switches, and other apparatus from the destructive effects of overload and short-circuit currents, an automatic means for opening the circuit is required. The two most common devices employed to fulfill this function are the *fuse* and the *circuit breaker*, the latter frequently abbreviated *c/b*.

(a) Fuses. The fuse is a simple device consisting of a *fusible* link or wire of low melting temperature that, when enclosed in an insulating fiber tube, is called a *cartridge fuse,* and when in a porcelain cup is known as a *plug fuse.* Figure 16.20 shows common types of fuses. Plug fuses, such as those in residential use, are rated 5 to 30 A, 150 V to ground, maximum. Cartridge fuses of various designs are made up to 6000 A and 600 V.

(b) Circuit Breakers. A circuit breaker is an electromechanical device that performs the same protective function as a fuse and, in addition, acts as a switch. Thus it can be used in lieu of a switch-and-fuse combination to both protect and disconnect a circuit. Most circuit breakers are equipped with both thermal and magnetic trips. The thermal trip acts on overload, whereas the magnetic trip acts on short circuit. Both the thermal and the

Fig. 16.20 *Plug-type fuses are made in two physical types;* (a) *nonrenewable type with a standard (Edison) base that screws directly into a standard socket.*
(b) *Nonrenewable NEC type S fuses have a smaller base than type* (a) *and therefore require an adapter to fit into a standard socket. The adapter is current rated and nonremovable. This prevents deliberate or accidental use of a type S fuse of incorrect rating. Type S is required in new construction where plug fuses are used. Both fuses shown are nonrenewable and must be replaced after "blowing." The type S fuse shown in* (b) *is a dual-element time-delay fuse.*
Cartridge fuses are available in a variety of designs. Illustrated are:
(c) *Nonrenewable, single-element type and* (d) *nonrenewable, dual-element, time-delay type. Because fuses are inherently very fast-acting devices, time delay must be built into a fuse to prevent "blowing" on short-time overloads such as those caused by motor starting. A dual-element fuse as shown in* (b) *allows the heat generated by temporary overloads to be dissipated in the larger center metal element, preventing fuse blowing. If the overload reaches dangerous proportions, the metal will melt, releasing the spring and opening the circuit. The notched metal portions of the fuse element, at both ends of the dual center element, provide short-circuit protection. The time required to clear (blow) a fuse is generally inversely proportional to the amount of current. In renewable cartridge fuses, the spent (blown) element is replaceable, reusing the original outer cartridge and blade connectors.*

ELECTRICITY

magnetic action have inverse-time characteristics: that is, the heavier the overload, the faster the trip action. Modern circuit breakers in commercial and industrial applications are frequently equipped with solid-state electronic tripping control units, which provide fully adjustable overload, short-circuit, and ground-fault protection.

Air circuit breakers are available in two types: the molded-case breaker and the "large air breaker." Molded-case breakers consist of a complete mechanism encased in a molded phenolic case. A light-duty molded case 50-A frame, plug-in circuit breaker is illustrated in Fig. 16.21(a). The molded case breaker shown in Fig. 16.21(b) has a 400 A frame, and can be equipped with an adjustable electronic tripping unit, ground fault trip, plus other features formerly available only on large air-circuit breakers. The large air-circuit breaker is a more complicated and highly adjustable device that can be used in applications that preclude use of molded case breakers. A modern solid-state adjustable trip controller for a large air circuit breaker is shown in Fig. 16.22, along with a circuit breaker with a similar solid-state trip unit.

All breakers can be equipped with remote trip and auxiliary contacts, and all good breakers have trip-indicating handles and are *trip-free* (i.e., will trip out harmlessly if closed in on a short-circuited line). Low-voltage (600-V class) circuit breakers are available in frame sizes from 50 A to 4000 A and one to three poles. Characteristics vary widely and are beyond the scope of this work.

(c) Characteristics of Fuses and Circuit Breakers. Although both are circuit-protective

Position indication; On-Trip-Off

Molded phenolic case

Load connection

Thermal/magnetic trip element

Quick make/break trip-free mechanism

Line connection

Mounting spring

(a)

(b)

Fig. 16.21 (a) *Essential elements of a plug-in–type molded case circuit breaker are shown in cutaway. Details vary with breakers designed for specific applications. The unit shown is typical of breakers in the 50–100 ampere size. (Photo courtesy of The Square D Company.)*
(b) *Heavy-duty molded case circuit breaker, 400 A frame, 2–4 pole, 240–600 volts. This c/b is available with either a standard thermal-magnetic trip or an adjustable electronic trip with ground fault protection. Among the accessories available are auxiliary contacts, shunt trip, undervoltage release, and an electrical solenoid operator that permits remote control of the breaker. The unit is suitable for individual or panelboard mounting. A typical 3-pole unit measures 5½ in. W × 10 in. H × 4 in. D and weighs about 12 lbs. (Photo courtesy of Cutler-Hammer.)*

(a)   (b)

Fig. 16.22 (a) *This device is a multifunction microprocessor-based trip control unit for a large circuit breaker. It provides true-root-mean-square (rms) sensing of phase and ground currents, selectable tripping characteristics, and trip alarm contacts. In addition, the unit can be used to display and transmit information on phase and ground currents. (Photo courtesy of Cutler-Hammer.)*
(b) *Electrically operated large air circuit breaker, with electronic microprocessor-controlled trip unit and additional remote-control accessories. The illustrated unit, rated 2000 A, is shown before stationary (nondrawout) mounting in a switchboard. (Courtesy of General Electric U.S.A.)*

ELECTRICITY

devices, their characteristics differ markedly, as can be seen from the following tabulations.

Fuses—Switch and Fuse Combination

Advantages
Simple and foolproof
Constant characteristics (no ageing)
Initial economy
Very high interrupting capacity (IC)
No maintenance
Instantaneous; energy limiting

Disadvantages
Fuses are single pole only
Necessity for storage of replacement fuses
Nonrenewable (one-time operation)
Nonadjustable
Nonindicating[a]

No electric or remote control[a]
Not trip free

Circuit Breakers
Advantages
Usable as switch
Multipole
No replacement storage
Resettable
Indicates trip
Trip free
Remote control
Adjustable

Disadvantages
Low to medium IC, except for special units
Periodic maintenance
High initial cost
Complex construction changes with age

[a]Can be accomplished with accessories.

Fig. 16.23 *One-line diagram of a typical switchboard. Switches are normally shown in the open position. Switches must be on the line (supply) side of fuses. Each line in a single-line diagram represents a 3-phase circuit. If circuit breakers were used, the entire board would be composed of units as illustrated in circuit 6.*

These lists demonstrate that there is no relevance to the oft-posed question, "Which are preferable—fuses or breakers?" The answer depends on the specific application involved and is often based on highly technical factors, beyond this book's scope. Circuit breakers are generally used in lighting and appliance panelboards.

16.19 Switchboards and Switchgear

A switchboard is a large, free-standing assembly of switches, fuses, and/or circuit breakers, which normally provide switching and overcurrent protection to a number of circuits connected to a single source. Metering and instruments are also often included. A switchboard may be represented in a single-line diagram, as in Fig. 16.23. In an electrical system, it serves to distribute, with adequate protection, bulk power into smaller "packages." Thus, by hydraulic analogy, the main buswork of the switchboard is equivalent to a main header, the switches to on/off valves, the fuses to flow-limiting devices, and the feeders to subheaders connected to the main header.

Modern switchboards (Figs. 16.24–16.27) are

Fig. 16.24 *Free-standing, indoor service, low-voltage (600 V) switchboard with group-mounted fixed molded-case circuit breakers. The deeper section contains the service entrance main circuit breaker (maximum 3000 A) plus instrumentation. Section width varies from 30 to 45 in. and depth from 18 to 36 in., depending on main c/b rating. The smaller section contains branch circuit breakers. All units are 90 in. high. (Photo courtesy of Cutler-Hammer.)*

Metering unit

Electronic
trip unit
(see Fig 16.22)

Electronic trip unit
(see Fig 16.22)

Fig. 16.25 *Metal-enclosed, indoor, low-voltage power switchboard with fixed or drawout large circuit breakers. Maximum bus capacity is 5000 A. Section dimensions vary with the size and type of contained equipment. All units are 90 in. high. Each circuit breaker is equipped with a microprocessor-controlled trip unit, similar to the unit shown in Fig. 16.22 (b). It can be seen clearly on the front of each unit. In addition, another programmable solid-state metering/analysis unit, similar to that shown in Fig. 14.24, can be seen at the left of the top compartment of the entry (left) cubicle of both this switchgear and the switchboard shown in Fig. 16.24. (Photo courtesy of Cutler-Hammer.)*

ELECTRICITY

Typical indoor assembly with a
breaker withdrawn on rails

Fig. 16.26 *Three sections of medium-voltage metal-clad drawout switchgear, suitable for indoor use as shown, and outdoor use with cabinet modification. Voltage ratings of 5–27 kV; current ratings 1200–3000 A. Each indoor section measures 36 in. W × 95 in. H × 96 in. D and contains two compartments for circuit breakers and/or instrumentation. Due to the drawout characteristic (shown), a 72 in. front aisle is required. Similar outdoor aisle-less switchgear measures 36 in. W × 115 in. H × 101 in. D. (Photo courtesy of Cutler-Hammer.)*

One line diagram

Fig. 16.27 *Four-bay (section) power-operated high-voltage service entrance switchgear rated 13.8 kV, consisting of (see single-line diagram inset, and reading left to right) a fused feeder section, two bays with parallel incoming lines and automatic transfer in the event of power failure, and a second switch and fuse feeder section. Fuses are electronic, with circuitry designed to provide a trip signal when excessive current is sensed rather than simply relying on the thermal characteristic of a fusible element. This gear, which measures 93 in. H × 172 in. W × 44 in. D and weighs 8000 lb, is suitable for interior or exterior installation. (Courtesy of S&C Electric Co.)*

all deadfront; that is, they have all circuit breakers, switches, fuses, and live parts completely enclosed in a metal structure. The operator controls all devices by means of push buttons and insulated handles on the front panel. When a switchboard has circuit breakers equipped with bayonet-type contacts and each is mounted in a movable drawer (like the drawers of a standard letter file), they are described as the *drawout* type. This drawout arrangement facilitates emergency replacements, inspection, and repairs, and is illustrated in Fig. 16.26.

There is no clear distinction made between the terms *switchboard* and *switchgear,* although generally, low-voltage switchboards with large circuit breakers and all high-voltage equipment (above 600 V) are referred to as *switchgear.* When molded case circuit breakers are utilized in a switchboard, it is often referred to as a *building-type switchboard.*

Recommended minimum space requirements for various types of switchgear are shown in Fig. 16.28. Working space around all types of electrical equipment must meet the requirements of *NEC* Article 110. Main metal-clad switchgear for commercial, industrial, and public buildings is almost invariably located in a basement and housed in separate well-ventilated electrical switchgear rooms. Smaller subdistribution switchboards require no special room. A wire screen enclosure to bar tampering or vandalism plus a large DAN-GER—HIGH VOLTAGE sign is usually adequate. The architect must provide adequate exits, hallways, and hatches for the installation and removal of all equipment. The specifications for switchgear should state the maximum overall dimensions of sections that can be accommodated in one piece.

When switchgear is to be installed outdoors, one of three methods is employed: build a small

(a) High-voltage metal clad switchgear

(b) Low-voltage drawout switchgear with large air breakers

(c) Fixed low-voltage switchgear with large air breakers

(d) Building-type switchboard with molded case breakers

Fig. 16.28 *Typical* nominal *switchgear space requirements. Each room should have two doors when switchgear is connected to high-capacity systems or operates at high voltage. Switchgear is shown here in plan view. For detailed space requirements, see* NEC *Article 110.*

structure to enclosed normal indoor gear, utilize weatherproof outdoor gear, or utilize switchgear that is built into its own exterior enclosing structure, as seen in Fig. 16.27. These housings are equipped with heat and light, and often prove the most economical choice.

16.20 Unit Substations (Transformer Load Centers)

An assembly, comprising a primary voltage switch-and-fuse or circuit breaker, a step-down transformer, meters, controls, buswork, and secondary (low-voltage) switchgear, is known as a *unit substation* or a *load-center substation.* It is available for indoor or outdoor use installation. Its function is to accept an incoming high-voltage power line, transform the high voltage to a level that can be utilized in the facility, and distribute the low-voltage power through low-voltage (secondary) switchgear. A dimensional physical sketch of a typical unit substation, along with its electrical single-line diagram is shown in Fig. 16.29.

The location in the building of a unit substation is governed by the type of transformer utilized, as explained in the discussion on indoor transformer installations. For this reason, almost all indoor

unit substations utilize dry-type (air-filled) transformers. Unit substations are utilized to effect the economies inherent in prefabricated construction with coordinated components. A basement location is most often selected, with ventilation requirements as detailed previously. Access should be restricted to authorized persons only.

16.21 Panelboards

An electrical panel, or *panelboard,* serves essentially the same function as a switchboard except on a smaller scale; that is, it accepts a relatively large block of power and distributes it in smaller blocks. Like the switchboard, it comprises main buses to which are connected circuit-protective devices (breakers or fuses) that feed smaller circuits. The panelboard level of the system is usually the final distribution point, feeding out to the branch circuits that contain the electrical utilization apparatus and devices, such as lighting, motors, and so on. Small panels, particularly in residential work, are frequently referred to as *load centers.*

The panelboard components—that is, the buses, breakers, and so on—are mounted inside an open metal cabinet called a *backbox.* The backbox is prefabricated with knockouts at the top,

ELECTRICITY

Fig. 16.29 *Approximate sizes and weights of a typical large single-ended unit substation. Such a unit would supply a building with a maximum demand of 750 kVA. The incoming 13,800-V cables enter cubicle A and connect to the primary switch and fuses. The load side of the fuses connects to the transformer, which in turn connects to the secondary switchgear. The main secondary switchgear then feeds various switchboards and panelboards distributed through the building.*

bottom, and sides to permit connection of conduits carrying circuit conductors (Fig. 16.30). The main feeders that supply power to the panel's busbars enter through the large knockouts in the metal backbox, as can be seen in Fig. 16.30. Details of panelboard construction can be seen in Fig. 16.31. A panel may be equipped with a main circuit breaker whose function it is to disconnect the entire panel in the event of a major fault. Figures 16.30 and 16.31 show panelboards with only branch circuit devices and no main breaker or switch. These panels are described as "lugs in mains only," which means that the panel has only connectors (lugs) on the main busbars for connection of the main feeder cables, and no main protective device. Three basic panelboard types are shown in Fig. 16.32.

The line terminal of each circuit protective device (breaker or fused switch) is connected to the busbars of the panelboard. The load terminal of the device then feeds the outgoing branch circuit. This is shown schematically in Fig. 16.33, which is a line-drawing representation of an electrical panelboard. Notice that the circuit breakers are arranged in two vertical rows, corresponding to actual construction. In the illustrated line drawing, a 3-phase panel is shown (i.e., 3-phase busbars and a neutral bar). The busbars of the panelboard are energized by a feeder from a switchboard or large power distribution panel.

Panelboards are described and specified by type, bus arrangement, branch breakers, main breaker, voltage, and mounting—although not necessarily in that order. A typical description might be: lighting and appliance panel, 3-phase, 4-wire; 200A mains; main c/b, 225A frame, 150A trip. Branch breakers—all 100A frame; 8 ea. SP-20A, 4 ea. 2P-20A, 4 ea. 3P-20A; flush with hinged locked door.

16.22 Principles of Electric Load Control

The key to energy conservation and electric demand limitation is load control. In the recent past, the extent to which both of these desirable op-

Single pole
C/B

Gutter space

2-Pole C/B

Neutral Bar

Locknuts

Circuit
wiring

Backbox

Knockouts

Feeder

Neutral
wiring

Circuit
conduits

Circuit
numbers

Phase
bars

Circuit
directory

Door
lock

Fig. 16.30 *A wired panelboard with the cover removed is shown. This panel is single-phase, 3-wire, meaning that it has 2 phase bars and a neutral (note the three heavy feeder cables). Lighting and appliance panels such as this are 4½–6 in. deep, 16–20 in. wide, and of sufficient height to accommodate the devices. The top device may be no higher than 78 in. AFF and the bottom one no lower than 18 in. AFF to meet NEC requirements.*

Fig. 16.31 *Cutaway of a typical lighting and appliance panelboard. Branch circuit protective devices are normally arranged in two vertical rows, even in 3-phase panels. The circuit directory on the inside of the panelboard door gives a brief description of the loads connected to each branch circuit breaker.*

erating functions were put into practice was essentially an economic decision: Would the electric billing savings justify the expense of installing the required control? Today, and increasingly in the future, legislation resulting from environmental pressures mandates energy-use limitation. Already many states have enacted legislation requiring lighting control in certain nonresidential buildings, and the trend toward additional and more stringent requirements is quite clear.

Lighting control strategies are discussed in Chapter 21. The essential control principle that applies to all loads, not only to lighting, is straightforward; the load is switched or modulated in response to a control signal [Fig. 16.34(a)]. In terms of hardware, the process is somewhat more complex. Figure 16.34(b) shows in principal the equipment and wiring required in a common application; time control of exterior lighting, with photocell override. Figure 16.35 shows a more complex arrangement for area lighting control

with local override, as might be installed in an office area of a commercial office building. In each case, there are signal-initiation devices, one or more relays that performs the actual electrical switching, and the control and power wiring, in addition, of course, to the main power wiring from the panelboard.

Many users found that the payback from reduced electric billing was insufficient to cover the control equipment cost, particularly where local and/or remote override is required, and as a result relied on inefficient and generally ineffective manual control. That option, however, is limited today, and will become even more so in the future. As a result, a more economical way to accomplish the required control functions was developed, thanks in large measure to modern miniaturized programmable control elements. These can be installed directly in the lighting panelboard. Such a panel is known in the industry by the generic term *intelligent panelboard* and, of course, by various trade names that identify specific manufacturers.

ELECTRICITY

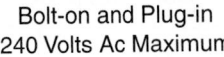

Bolt-on and Plug-in   Circuit Breaker Panelboard   Fusible Panelboards
240 Volts Ac Maximum

(a)   (b)   (c)

Fig. 16.32 *Three different designs of panelboards.* (a) *Circuit breaker–type panelboard, using bolt-on and plug-in molded case circuit breakers 1- to 3-pole, 15–100 A. A main device up to 400 A can be installed. Panel is rated for 240 V ac. The illustrated panel, which is arranged for surface mounting, measures 30 in. H × 20 in. W × 5¾ in. D.*
(b) *Circuit breaker distribution–type panelboard with maximum bus ratings of 1200 A and 600 V ac. Branch circuit breakers can be 15–100 A, single-pole and 15–1200 A, 2- or 3-pole. These panels are free-standing but can be wall-mounted, provided the highest device does not exceed 6 ft. 6 in. AFF. Similar units are available with a front door. Depending on contents, cabinets can be 24–36 in. W × 60–90 in. H × 6–8 in. D.*
(c) *Switch- and fuse-type distribution panelboard, free-standing, with similar ratings to the c/b type in* (b). *Cabinets are 36–44 in. W, 74–90 in. H, and 10½ in. D. (Photos courtesy of Cutler-Hammer.)*

16.23 Intelligent Panelboards

The idea behind the intelligent paperboard is straightforward. If much of the necessary electric load control and switching functions are incorporated into the panelboard by use of a compact centralized programmable microprocessor, the corresponding external devices and the associated wiring are eliminated. The switching function is made possible by the use of motor- or solenoid-operated panel circuit breakers, thus taking advantage of their switching capability and eliminating the need for relays, contactors, and switches. The internal central controller permits local programmable control of each panel circuit breaker *individually*.

This central panel-mounted controller can also accept signal data from individual remote or network sources and can provide status reports, alarm signals, operational logs and local by-pass

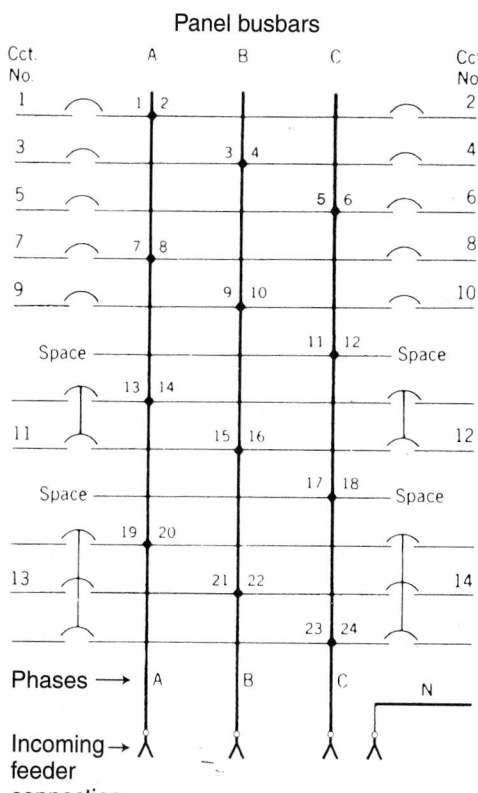

Fig. 16.33 *Typical schematic diagram for a panel. Note that single-pole, 2-pole, and 3-pole circuit breakers are connected to one-, two-, and three-phase buses, respectively. They supply 120 V two-wire, 120/208 V 3-wire, and 120/208 V 3-phase 4-wire, respectively. This panelboard would be described as a 3-phase 4-wire lighting and appliance panel, with 10 SP, 2-2P and 2-3P c/b, lugs in mains only. The voltage of the panel and the ampere ratings of buses and all circuit breakers would then be specified.*

Fig. 16.34 (a) *A low-voltage lighting control scheme always includes the elements shown. Signal initiation can come from a sensor such as a photocell or from a control device such as a timer, an override signal, a central computer signal, and the like. The signal activates the power switching device, which in turn switches or modulates the load. The power source is always the associated lighting panelboard.*
(b) *Schematic arrangement of control scheme for exterior lighting. If the system is compact, a switching timer would be used. For a spread system, timer-controlled switching relays at each light would be used with attendant additional control wiring. Grouping of the lights onto different control sequences (road, parking lot, building entrance, etc.) further complicates the wiring.*

and override functions. Control cards in an adjacent cabinet, connected to the controller, can provide local switch override, daylight dimming and on–off control, telephone override energy control, electric demand control, and networking. With careful planning, the high initial cost of the intelligent panelboard can be offset by elimination of remote relays, relay panels, programmable time switches, remote-control switches, and all of their associated wiring. Simplified and improved facility operation plus reduced maintenance costs and electric billings are additional benefits that can be achieved with proper application and care-

ful design. Figure 16.36 shows the essential elements of an intelligent panelboard and a typical network arrangement using the remote-control capabilities of such panels.

16.24 Electric Motors

Motors are very frequently supplied with its driven equipment, such as fans, blowers, and so

ELECTRICITY

Fig. 16.35 *Schematic diagram showing equipment and connections that might be used for control of a block of lighting in a large commercial building. The wiring can be even more complex and extensive if control of smaller lighting groups is desired, or if control is desired from more than one location.*

on, as a complete package. The actual choice of motor is left to the driven-equipment supplier, it being maintained that the supplier is best qualified to select a motor that will optimally match the driven-equipment requirements for whose proper operation the supplier is responsible. In practice, however, the supplier is frequently guided primarily by the price motive. The specifier, therefore, should be sufficiently knowledgeable so that the particular motor desired can be specified. The following paragraphs are written with that purpose in mind, and are therefore primarily concerned with application.

(a) Direct-Current Motors. These motors are not normally used in building work. The fine speed control for which they were primarily used previously is now available, more economically, with ac motors, as explained in Section 16.26.

(b) Alternating-Current Motors. These motors fall into three general classifications: polyphase induction motors, polyphase synchronous motors, and single-phase motors. Within these categories there are further subdivisions. Of these many types, the vast majority in use in building equipment are squirrel-cage induction

Local control module; see (*b*)

Remotely controlled motor operated circuit breakers see (*c*)

Space for power-supply, power and control cable connection and accessories

Fig. 16.36 (a) *This "intelligent" panelboard contains (from top to bottom) microprocessor control module, 26 single-pole, remote controllable (motor-operated) circuit breakers, space for 10 additional poles, and a compartment containing a dc power supply (for c/b operation) and accessories (continued on next page).*

(*a*) "Intelligent" panelboard

(b) Single-pole motor operated C/B

(c) Control Module

(d)

Fig. 16.36 *(continued from preceding page)* (b) *This motorized single-pole circuit breaker can be manually or remotely controlled. Electric operation latches the c/b so that in the event of power failure, all breakers remain in their pre-outage condition. The illustrated unit is a 100-A frame, 20-A trip unit. A status window on each unit displays its condition, while electronic sensors in each unit sense voltages and indicate on the control panel the status of each unit.*

(c) *The control module, mounted at the top of the panelboard, is equipped with programming buttons that allow control of each c/b individually. The illustrated basic control module programs and controls essential functions including configuration, c/b status, resetting, zoning, settings for hardware characteristics, override modes, screen selection, and self-diagnostics. Additional control modules (not shown) supply time-based programming, logging functions, and timed functions. The illustrated breaker status screen is one of five displays. The others are setup, program, override, and diagnostic screens.*

(d) *Intelligent panels can be incorporated into a large control network by use of interface connections of various types (not shown). The illustration is a block diagram showing one of the networking possibilities. (All photos courtesy of Square D.)*

machines; therefore, this type is studied in some detail.

(c) Squirrel-Cage Induction Motors.

This motor owes its interesting name to an early design in which the rotor consisted of a group of bars welded together into a cylindrical cage-type shape. The design, invented by Nikola Tesla, is basically unchanged today except for refinements. Squirrel-cage motors are manufactured in four different NEMA designs to meet different application requirements. Of these the most common are:

Type B: standard design, high efficiency and power factor, normal torque. Applicable to fans, blowers, and pumps.

Type C: high starting torque, fair efficiency and power factor. Applicable to compressors, conveyors, and other devices that start under load.

A motor nameplate gives important data on the motor that are not self-evident:

1. *Type:* This is the manufacturer's designation and indicates primarily the enclosure. Common enclosures are open drip-proof, totally enclosed, fan-cooled, and weather-protected.

2. *Duty* (Time rating): Continuous or intermittent.

3. *Service factor:* Permissible overload; generally 15%.

4. *kVA code:* Indicates by a letter the maximum starting current per horsepower. This is useful in selecting motor protective devices. More recently, actual locked rotor amperes are given.

5. *Frame:* A NEMA standard number that indicates the motor's physical dimensions.

6. *Motor voltage:* The standard motor voltages are 208, 230/460, and 575 V. Induction motors generally operate satisfactorily ±10% voltage. Only 208 V motors should be used on 208 V systems because actual line voltage may be as low as 200 V. Using a 230 V at 200 V motor will result in sharply reduced torque, increased temperature rise, and poor overload capacity.

7. *Motor nominal full-load efficiency:* A requirement of ASHRAE Standard 90.1-1989 and EPACT.

8. *Design letter:* Indicates inherent motor design characteristics.

(d) Electric Motor Energy Considerations.

The Federal Energy Policy Act of 1992 (EPACT) requires that, effective October 1997, the most common designs of induction motors, accounting for 70% to 80% of all polyphase induction motors sold in the United States meet new high-efficiency and power-factor standards. Specifically, the new efficiency standards apply to single-speed, polyphase squirrel-cage induction motors, designs A and B, continuous rating, and operating at 230/460 V, 60 Hz. These motors are known as *premium efficiency* or *PE* motors.

The decision as to whether to use a PE motor in applications not mandated by law is essentially an economic one. Many motor manufacturers make available straightforward software that enables a specifier to rapidly make a detailed life-cycle cost comparison among motor types, with operating time, loading, first cost, power and maintenance costs, and so on as the variables. As a guideline, motor applications involving low starting torque and frequent motor operation at partial load are likely candidates for short payback periods on the price premium of PE motors.

16.25 Motor Control Standards

Until recently, American designers and equipment manufacturers utilized motor control equipment manufactured to NEMA exclusively. It was also common practice in the construction industry to split the motor and motor-control package by having equipment manufacturers supply the motorized equipment while the electrical construction contractor supplied the motor-control equipment. Few coordination problems between controller and motor drive were encountered because control equipment sized and built to NEMA standards is large, heavy duty, and applicable to a wide range of motor sizes and characteristics.

The International Electrotechnical Commission (IEC) is a technical organization with 42 member nations, including the United States. It develops recommended standards for electrical equipment, including motor controls. The IEC design concept is one of close coordination between a motor and its controller. This application sensitivity, coupled with flexible ratings for a single piece of equipment depending on its application, makes equipment built to IEC standards smaller, lighter, and cheaper than similarly rated equipment built to NEMA standards. As a result, many U.S. manufacturers utilize IEC motor controls in conjunction with their motive equipment, particularly on international jobs. Because American architects, engineers, and contractors are involved in a large number of construction projects outside territorial United States, it behooves those responsible for designing, specifying, and purchasing electrical equipment in general and motor-control equipment in particular to be aware of the comparative characteristics of NEMA-based and IEC-based equipment.

16.26 Motor Control

(a) Fundamentals. A conventional ac motor controller is basically a contactor (see Section 16.14) designed to handle the heavy inrush currents encountered in starting an ac motor. Its function is twofold: to start and stop the motor and to protect it from overload. These two separate and distinct functions are accomplished by combining a set of contacts for on/off control with a set of thermal overload elements for overload protection in a single unit. When the contacts are operated by hand, the controller is called a *manual starter;* when the contacts are operated by a magnetic coil controlled by push buttons, thermostats, or other devices, the unit is known as *magnetic controller* or, simply and more commonly, a *magnetic starter.*

Motors 1 hp or less are generally controlled by a manual switch that contains an overload protection device. It is advisable to utilize such a device for all fractional horsepower motors.

Most starters are of the full-voltage across-the-line (ATL) type; that is, the contacts place the motor directly onto the line and the motor starts up immediately. When such a procedure is undesirable because of voltage dip and flicker caused by the large inrush current or because of utility company limitations, a reduced voltage starter, sometimes called a *compensator,* is used. These units initially apply reduced voltage to the motor,

thus reducing starting inrush current and line voltage drop. This in turn reduces dimming of lights, flicker, and other undesirable effects.

Older reduced-voltage starters use a stepping arrangement whereby voltage is increased incrementally, thus limiting inrush current. Unfortunately, starting torque is also limited, thus restricting application of these starters. Solid-state starters have been available since the late 1980s that provide continuous (stepless) controlled motor starting. These units not only limit inrush current, but by adjusting acceleration time, the required starting torque of the motor can be maintained. Such starters provide *soft* starts (i.e., starts that minimize the mechanical stresses caused by rapid application of accelerating torque). Additional advantages of these units are reduced size and weight, long life, more sophisticated motor protection, and additional operating functions such as jogging and reversal. Disadvantages are higher cost and possible radio frequency noise problems (Fig. 16.37).

(b) Motor Speed Control. Many applications in HVAC, fluid piping, and industrial systems require speed control of electric motors. Until recently, this constituted a problem whose solution was expensive, often space consuming, and rarely energy efficient. The reason for this is that the common squirrel-cage induction motor is essentially a constant speed device, where speed is determined by power line frequency and motor design. Speed drops slightly (slips) as load increases. Until the advent of cheap electronic power equipment, speed control was usually accomplished by using a wound-rotor motor in lieu of a squirrel-cage unit. This drastically increases both motor and controller costs while providing only limited speed control.

Advances in power electronics have made practical a variable voltage, variable frequency (VVVF) controller that gives smooth continuous speed control over a range exceeding 30 to 1 while maintaining motor torque. These highly reliable controllers [also known as variable frequency drives (VFD)] provide considerable energy economies that usually result in rapid payback of the relatively high first cost of equipment. The essentials of the VVVF control scheme are shown in Fig. 16.38. Voltage control, in addition to fre-

Fig. 16.37 "Soft" starter type for 3-phase induction motors up to 1400 hp, shown without enclosure. This microprocessor-based unit provides two selectable starting characteristics with stepless acceleration, plus additional control and protection functions including reversing control, over and under voltage and current protection, and two normal motor operating speeds. (Photo courtesy of MagneTek.)

Fig. 16.38 Schematic representation of the action of a variable voltage and frequency motor speed controller. Line voltage at 60 Hz is rectified to dc and the inverted to ac. Frequency and voltage control are built into the inverters, designed to maintain a constant voltage-to-frequency ratio. This constancy is necessary to prevent malfunction of the motor as its speed is varied.

quency control, is necessary in order to maintain a constant voltage to frequency ratio which is necessary for efficient safe operation. An increase in this ratio results in overheating of the motor whereas a decrease results in insufficient torque. Additional advances in power electronics have produced motor speed controllers with controlled speed ratios approaching 1000:1.

One disadvantage of these speed controllers has been the production of line harmonics and radio noise of sufficient strength to constitute a serious engineering problem. At this writing, these problems have been mitigated, and will undoubtedly be pursued to the point that, except for highly sensitive areas, the interference will be negligible. At this point, however, it behooves the careful designer to investigate these effects for the particular application. Typical adjustable frequency controllers are shown in Fig. 16.39.

16.27 Motor Control Equipment

All the starters discussed previously are available in a wide range of sizes, voltages, and types of enclosure. Every motor controller is required by the *NEC* to have a disconnecting means for safety reasons. Where convenient, this disconnect switch may be combined with the starter into a single unit, known as a *combination starter*. A circuit breaker or fused switch is often used in such an arrangement, which then constitutes the branch circuit protection and disconnecting means (Fig. 16.40).

A typical, brief description of a conventional motor controller would be similar to the following: Combination circuit-breaker type, across-the-line motor controller, NEMA size 2, three O.L. elements, 208 V, in a NEMA 1 enclosure. Starter shall contain integral on/off push buttons.

Table 16.6 gives the approximate dimensions of one particular type of combination starter. As with other electrical equipment, a current manufacturer's catalog must be consulted for reliable data for *that* manufacturer. Dimensions vary somewhat among manufacturers producing a product that meets NEMA specifications. As noted previously, physical characteristics can vary greatly between items made to meet IEC specifi-

Fig. 16.39 *Three modern programmable microprocessor-controlled motor controllers.*
(a) *Integrated motor and controller for motors up to 1 hp. The controller provides soft start, a speed control range of 40:1, high efficiency, multiple speed settings, and full motor protection.*
(b) *Controller for motors up to 3-phase, 400 hp; this type unit provides programmable functions including speed control from standstill to double rated speed, reversing, adjustable acceleration, and full motor protection.*
(c) *Controller that can be programmed for three different control modes. Speed control range, regulation, and torque curves vary in each mode to suit the specific controlled load. The controller has low harmonic distortion plus fault diagnostic software, in addition to the normal overload and protective characteristics. It also has terminals for remote monitoring and control plus network control capabilities. A 20 hp 230-V controller in a NEMA 1 (indoor) enclosure measures approximately 16 in. × 10 in. × 9 in. and weighs 24 lbs. (Photos courtesy of MagneTek.)*

Fig. 16.40 *Interior of a conventional (non–solid-state) combination fused switch type, across-the-line motor controller. Note that the unit is essentially a switch and a starter wired together and installed in a single cabinet. Approximate dimensions of the illustrated unit are 20 in. W × 48 in. H × 9 in. D, with a 75 hp capacity. (Courtesy of Allen-Bradley Co.)*

TABLE 16.6 **Rating and Approximate Dimensions**[a] **of ac Full-Voltage Conventional Single-Speed Motor Controllers, 3-Phase Combination Circuit Breaker Type**[b]

NEMA Size[c] Designation	Maximum Horsepower[c]	Width	Height	Depth
			(in.)	
0	3	10	24	7
1	7½	10	24	7
2	15	10	24	7
3	30	20	24	9
4	50	20	48	9
5	100	20	56	11

[a]Dimensions vary among manufacturers.
[b]All starters are housed in a NEMA 1 indoor ventilated enclosure.
[c]Maximum hp that can be controlled at 208–230 V. Generally, when operating at 460 V, a starter one size smaller can be used.

cations and those made to NEMA specifications for control equipment of similar rating.

In spaces where a group of motors are installed in close proximity to one another, it is often convenient from the control aspect, as well as economical, to combine the motor starters, disconnect switches, motor controls, and indicating devices into a single large assembly. Such an assembly is called a *motor control center.* Typical motor control center types are shown in Fig. 16.41.

16.28 Wiring Devices: General Description

The general term *wiring devices* includes all devices that are normally installed in wall outlet boxes, including receptacles, switches, dimmers, fan controls, and so on. Attachment plugs, also called *caps,* and wall plates are also included in wiring devices.

The three grades of device quality, using NEMA and UL grading nomenclature, are hospital grade, federal specification grade, and UL general purpose grade, in descending order of price and quality. Inasmuch as these grades correspond to published specification requirements, they are the only reliable gage of construction quality. Hospital-grade devices are identified by a green dot on the device face. These devices are built to withstand severe abuse while maintaining reliable operation, and must meet UL requirements for this grade.

Manufacturers usually grade their devices as hospital grade, premium or industrial specification grade, commercial specification grade, and residential grade, in descending order of quality and price. Hospital grade is approximately the same among manufacturers because industry standards must be met. Industrial and commercial specification grades correspond roughly to federal specification grade, and residential grade to UL general purpose. Manufacturers' grades are qualitative, although reputable makers list the NEMA and UL specifications that each of their grades meet. It is only on this basis that a technical quality comparison can be made.

In application, industrial specification grade equipment is usually used in industrial and high-grade commercial construction; commercial specification grade in most educational and good residential buildings; and standard or residential grade in low cost construction of all types. The grade of wiring devices should, as with all electrical equipment, be consistent with the quality of construction in the entire facility.

Although historically the term *wiring device* referred to devices that operate at line voltage (120, 208, 240, 277 V) in a branch circuit, the term today also includes low-voltage lighting control devices, fan controls, and the like. The accepted criteria seems to be that only devices rated 30 amps or less, that can be mounted in a small wall box, are termed wiring devices. Communication attachments are not usually considered wiring devices even when they are supplied by the electrical

Vertical wiring trough

Horizontal wiring trough

Metering

Typical motor controllers

Control compartment

Control compartment

Control pushbuttons

Switch handle

(a)

(c)

20" Back to back
Top view

(b-1)

(b-2)

Side 2 Side 1

20" Back to back
Side view

a. Control Unit
b. Vertical Wireway
d. Top Horizontal Wireway
e. Bottom Horizontal Wireway

Fig. 16.41 *Motor control centers (MCC) of various constructions.* (a) *Conventional MCC construction with magnetic motor starters, manual local control, analog metering, and conventional overload and protective equipment.*

(b) *Back-to-back construction of conventional MCC adds only 5 in. to the basic 15 in. depth. Units are 90 in. high and 20 in. wide per section. Solid-state controllers can be installed in this unit, as well as in the MCC shown in* (a), *along with conventional magnetic contactors.*

(c) *Modern MCC that uses solid-state programmable controllers with heaterless overload protection. The unit also provides protection against phase loss, single phasing, and ground faults. Single side units in this construction are 90 in. H × 20 in. W × 16–20 in. D (depending on starter size). Back-to-back construction is 21 in. D. [Photos courtesy of Allen-Bradley* (a), *Siemens-Furnas* (b), *and Cutler-Hammer* (c).]

ELECTRICITY

contractor. They are part of *premise wiring,* which is the term commonly used to describe the data and communication system wiring, raceways, and devices (see Fig. 15.24).

16.29 Wiring Devices: Receptacles

A receptacle is, by *NEC* definition, "a contact device installed at the outlet for the connection of single attachment plug." This usually takes the form of the common wall outlet or, as illustrated later, larger and more complex devices. A comment here about terminology is required. The common wall outlet is properly called a *convenience receptacle outlet, a receptacle outlet,* or a *convenience outlet.* The term *wall plug,* which is heard so often, is incorrect. A *plug* is another name for the attachment device (cap) on the wire that carries electricity to the appliance. The device at the end of the line cord is plugged into the wall; hence "plug."

Because by *NEC* definition a receptacle is a contact device for the connection of a single attachment plug, and because the normal wall convenience receptacle takes two attachment plugs, it is properly called a *duplex convenience receptacle* or *duplex convenience outlet.* Most people shorten this to *duplex receptacle* or *duplex outlet.*

Receptacles are identified by number of poles, number of wires, and whether the device is designed for connection of a separate equipment ground. (This ground may connect to a separate equipment grounding wire or to the metallic conduit system, according to system design.) The number of receptacle *poles* equals the number of current carrying contacts, including the neutral. The equipment ground connection (pole) is not counted because it does not (normally) carry current. The number of *wires* includes the equipment ground connection because it is wired. The equipment ground connection (grounding pole) should not be confused with the system ground (usually the neutral pole) nor may the wiring for the two be interchanged (see Section 17.3).

In a typical application, a receptacle for an electric dryer with 4800 W, 208 V heating element, and ⅙ hp, 115 V motor would be NEMA 14-50R (Fig. 16.42). The motor would connect across *W* and *X,* the heater across *X* and *Y,* and the appli-

ance case to *G.* Receptacles installed on standard 15- or 20-A branch circuits must be of the grounding type. Receptacles connected to different voltages, frequencies, or current type (ac or dc) on the same premises must be polarized so that attachment plugs are not interchangeable. Figure 16.42 shows some of the standard receptacle configurations and their ratings. Figure 16.43 shows typical receptacle construction plus enclosures for use in damp and wet locations.

Receptacles are readily available from 10 to 400 A, 2 to 4 poles, and 125 to 600 V. In addition, special types such as locking, explosion proof, tamper proof, and decorative design units are made. Also, specific usage units such as range receptacles are available. All receptacles other than the normal 15/20 A, 3-wire, parallel-slot type should be specified to be furnished with at least two matching caps (plugs). A typical receptacle specification would be: receptacle, duplex, 2-pole, 3-wire, grounding type, 20 A, 250 V, federal specification grade, for indoor use. Receptacles are normally mounted vertically between 12 and 18 in. above the finished floor (AFF), except that in shops, labs, and other areas where tables are placed against the walls, 42 in. is the usual mounting height.

In addition to these types, several manufacturers produce a receptacle with built-in ground-fault protection (Fig. 16.44). For a discussion of these ground-fault circuit interrupter (GFCI) receptacles, see Section 17.4.

The great sensitivity of modern electronic equipment to voltage surges and to electrical *noise* (random, spurious electrical voltages) has led to the development of two special receptacles. Receptacles with built-in surge suppression protect the connected equipment from overvoltage spikes. Receptacles with an insulated equipment grounding terminal separate the device ground terminal from the system (raceway) ground because it has been determined that much of the unwanted electrical noise can be eliminated by such disconnection. The receptacle's insulated ground terminal is connected to an isolated green ground conductor that carries through the entire system but is connected to the system ground at the service entrance only; hence the name *isolated ground.* Isolated ground receptacles are identified by an orange triangle on their face (see Fig. 16.45).

Note: "W" denotes white system ground or neutral
     "G" denotes green equipment ground.

Fig. 16.42 *Receptacle configuration chart of selected common general-purpose, nonlocking devices with associated NEMA designations.*

A receptacle with a built-in surge suppression (see Section 16.37) is shown in Fig. 16.45.

16.30 Wiring Devices: Switches

Switches up to 30 A that can be outlet-box mounted fall into this category. Generally "ac only" switches are preferable to the ac/dc type because of better construction. The usual ac switch rating is 15, 20, or 30 A at 120 or 120/277 V. Usual constructions are single-pole, 2-pole, 3-way, 4-way, momentary-contact, 2-circuit, maintained-contact SPDT and DPDT. Operating handles are toggle-type, key, push, touch, rocker, rotary, and tap-plate types. Mercury and ac quiet types are relatively noiseless; toggle, tumbler, and ac/dc types are generally not. A typical switch specification might be: switch, single-pole, ac, quiet type, federal specification grade, 15 A, 120 V, with press handles lighted when OFF, for side wiring only.

ELECTRICITY

Commercial Specification Grade Receptacle

Steel mounting yoke.

Brass grounding assembly.

Mounting and terminal screws.

Brass line contacts.

Green hex head grounding screw.

Grounding clip provides ground continuity between strap and metal enclosure.

(a)

Gasket seals out moisture for electrical safety.

Self-closing covers allow use in wet locations.

Polycarbonate material resists abuse and temperature extremes. UV-resistant, non-corrosive, non-conductive.

Stainless steel mounting screw.

(b)

Fig. 16.43 (a) *Cutaway of a commercial, specification-grade duplex convenience receptacle, showing its construction.*
(b) *Receptacles installed in damp locations, outdoors but not exposed to the rain, or in wet locations but used to energize portable equipment and are continuously attended should have a cover that is weatherproof when the attachment plug cap is removed, of which one type is illustrated.* (NEC Article 410-57b(2).)
(c) *Receptacles in wet locations (that may be in use unattended) must have a weatherproof enclosure the integrity of which is not affected by the attachment of a plug. One such enclosure is illustrated.* (NEC Article 410-57b(1).) *(All photos courtesy of Eagle Electric Manufacturing Co., Inc.)*

Polycarbonate resists abuse and temperature extremes. UV-resistant, non-corrosive, non-conductive.

Self-closing tinted transparent cover.

Base with cord ports 2" from wall.

Lockable

(c)

A switch incorporating a solid-state rectifier is readily available that gives high/off/low control for *incandescent lamps* and costs very little more than an ordinary switch. Typical applications are areas where a lower illumination level is often acceptable and always desirable as an energy-conserving measure. In high-security areas where the easily defeated normal key switch is inadequate, a tumbler lock–controlled unit can be used. Loads that can be timed-out, such as bathroom heaters and ventilating fans, can be controlled by a spring-wound timer switch as illustrated in Fig. 16.46 that also illustrates a few common switch types. Figure 16.47 shows two common switch designs in cutaway revealing their construction and operating mechanisms. A solid-state switch with adjustable time delay is shown in Fig. 16.48.

Thanks to miniature electronics, a programmable switch is available that fits into a wall-outlet box in lieu of an ordinary switch. The unit acts as a solid-state 15-A switch and can be readily programmed to switch the controlled circuit or device at preset times (see Fig. 16.48).

Ground connection

Test and reset buttons

Indicator light (optional) glows when unit has tripped.

GFCI

Fig. 16.44 *Where it is desired to provide ground-fault protection at a single outlet only, and thereby localize power interruption (rather than using a GFCI circuit breaker on the entire circuit), a receptacle containing a ground-fault interrupter, as illustrated, may be used. A test button is provided to permit periodic testing, and a reset button reenergizes the receptacle after it has tripped on ground fault and the fault has been cleared. (Courtesy of Eagle Electric Manufacturing Co., Inc.)*

Fig. 16.45 *Hospital grade, insulated (isolated) ground receptacle used where heavy duty is expected, and both reliability and freedom from electronic "noise" are required. Hospital grade is denoted by the green dot on the upper left of the receptacle face, and the insulated ground by the orange triangle on the upper right. (Some manufacturers make the entire receptacle orange.) (Photo Courtesy of Leviton Manufacturing Co.)*

16.31 Wiring Devices: Specialties

The most common device in this group, which also includes pilot lights, fan controls, and other small motor controls, is the lighting dimmer. The original lighting dimmers were variable auto transformers. They were large, heavy, and expensive, and therefore found application in theater lighting, displays, and the like. The advent of small electronic units (SCR) made the modern wall-box dimmer possible. The simplest units are rotary, with limited capacity and applicable only to incandescent loads. One of their disadvantages, still found in inexpensive units, is the production of annoying radio frequency interference (RFI) and line harmonics. Newer models are available that provide preset control, combine dimming and switching, are usable for smooth fan speed control, can be remotely controlled (in addition to local control), provide automatic fadeout, and dim fluorescent lamps (provided with dimming ballasts). A few types are shown in Figs. 16.49 and 16.50.

ELECTRICITY

Toggle
Key
Specification Grade switches
15, 20 and 30 ampere

Key

Maintained or momentary contact switch
15 and 20 ampere
(Double throw, center off)

Rotary switch
15 ampere
120 V

Lighted button in "Off" position
Press switch
15 and 20 ampere

60 min spring-wound timer 20 A 120 V

Fig. 16.46 *A few of the most common wiring device switches; generally installed in a small wall box and used for control of lighting circuits.*

Industrial Specification Grade Switch

Terminal and mounting screws.

Copper alloy terminal contact arms hold silver alloy contacts.

Green hex head grounding terminal screw.

Faces color coded for amperage identification. (Not an industry-wide standard)

Steel mounting strap.

Grounding clip provides continuity between mounting strap and grounded metal enclosure.

(a)

Ground terminal screw.

Long wide strap provides good stable, contact.

Cam action switch mechanism for smooth switch movement.

Mounting and terminal screws.

(b)

Fig. 16.47 *Cutaway drawing of two common wall-switch designs showing operating mechanisms and construction details.* (a) *Conventional toggle design.* (b) *Rocker plate design. (Drawings courtesy of Eagle Electric Manufacturing Co.)*

Fig. 16.48 *Solid-state adjustable electronic switch timer that mounts in a common wall-switch box. The unit switches OFF, with delay adjustable from zero to 15 min. in five discrete steps. Units such as these can control incandescent and fluorescent lighting and small motors. They are applicable where either direct manual control or occupancy sensors are inapplicable. (Photo courtesy of Leviton Manufacturing Co.)*

16.32 Low-Voltage Switching

The switches illustrated and discussed in Section 16.30 are all full-voltage types; that is, they are wired directly into the load circuit and operate at line voltage and full current. A control scheme that uses light-duty, low-voltage (24-V) switches to control line voltage relays, which in turn do the actual circuit switching, is illustrated in Figs. 16.51 and 16.52. The system is variously referred to as *low-voltage switching, remote control switching,* and *low-voltage control.*

Because the loads in this system are relay con-

trolled, any number of switches and control devices can be wired in parallel to operate the relay and thereby effect the switching. Thus, the control can be local, remote, and master, with individual override by local control devices such as occupancy sensors, and central control or override at a central controller by timers, daylight controllers, and so on. The advantages of this system of control over full-voltage switching are, therefore:

· Flexibility of control location.

· Individual load control override by local control devices (occupancy sensors, photocells; see Section 21.4).

· Group load override by central control devices such as timers, daylight controllers, and energy management systems.

· Low cost of low-voltage, low-current wiring as compared to full-voltage wire and conduit.

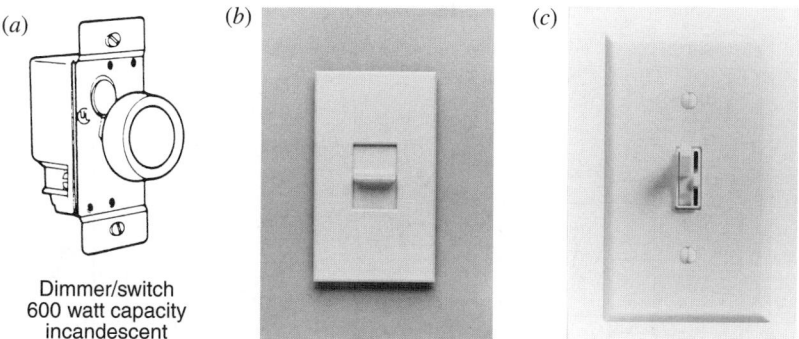

(a)

Dimmer/switch
600 watt capacity
incandescent

(b)

(c)

Fig. 16.49 (a) *Simple solid-state dimmer, suitable for a maximum 600-W incandescent load only. The switch permits level preset in pushbutton design. Inexpensive units may cause considerable radio frequency interference (RFI).*
(b) *Slide dimmer for a 600-W incandescent load, including ON–OFF and level preset. This dimmer is available with a wireless receiver that responds to a hand-held infrared remote control.*
(c) *Combined ON–OFF switch and slide dimmer provides preset level control. Units similar to (b) and (c) are available for loads to 1000 W, fluorescent lamp control (with appropriate ballast), and fan speed control.*
(Photos (b) and (c) courtesy of LUTRON Electronics Co., Inc.)

(a)

(b)

Fig. 16.50 (a) *Combined tap switch and slide dimmer provides preset level control. A built-in night light indicates this design's residential application.*
(b) *Multifunction tap switch in the center permits switching to full brightness, gradual brightening to preset, 10-second fade-out, and OFF. Rocker switch at the right controls levels. LED lights behind pinholes at the left provide an indication of the preset level. Both illustrated unit designs are available for incandescent and fluorescent lamp control (with special ballast) and fan speed control. Both are designed with RFI suppression. (Photos courtesy of LUTRON Electronics Co., Inc.)*

Fig. 16.51 *Low-voltage switching control. Multipoint control and central control are illustrated. The diagram shows the relays located at the load. Central relay cabinets are used in dense load areas.*

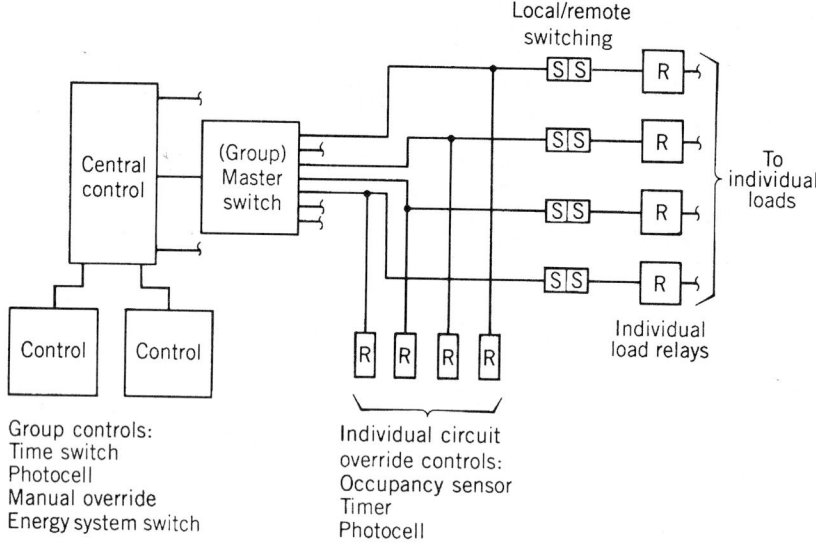

Fig. 16.52 *Schematic arrangement of a large low-voltage switching system. Individual loads can be switch-controlled from multiple locations, including a group (master) control point. In addition, local control devices such as photocells and occupancy sensors can override the local switches, to maximize energy conservation. Central (group) control devices can perform the same override function at the group level.*

Fig. 16.53 *Components of low-voltage switching and control systems for lighting and other electric loads.*

(a) *The basic switching device is a single-pole double-throw latching relay with a 24 V coil and 20 A, 120 to 277 V contacts. It is shown with a plug-in connector on its leads, making it suitable for rapid mounting in a relay panel (see Fig. 16.54). The same relay without the connector is suitable for hard-wiring at a controlled device.*

(b) *Low-voltage (24 V) wall switches can be mounted single or ganged in groups of two to eight switches. Switches are available in button, rocker, key-operated, and lighted designs [as in (b), (c), and (d)].*

(c) *Switching station with push-button low-voltage switches, each with an LED to indicate switch position (ON–OFF).*

(d) *Switch bank with illuminated push-button switches identified with room number and floor. The ON–OFF switches at the left of the bank are override switches for the corresponding horizontal line. (Photos courtesy of Touchplate Technologies, Inc.)*

· Inherent system flexibility and simplification of alterations.

· Monitoring of the status of individual loads at a centralized control panel by use of relays with auxiliary contacts (see Fig. 16.51).

The second and third points are particularly important from the energy conservation point of view, as it has been amply demonstrated that reliance on manual control for energy usage limitation yields disappointing results.

The basic components of the entire system are the control relay and the switches, shown in Fig. 16.53. Small systems use hard wiring (physically wired connections), with the hard-wired relays installed either dispersed at each controlled light (or lighting groups) or grouped in a relay panel. When the number of relays exceeds about 20, these hard-wired panels become so densely wired that control changes become difficult and tedious. As a result, prewired computer- or microprocessor-controlled panels are used in large systems, where connections can be made and changed easily. The interior of one such panel is shown in Fig. 16.54. The architecture of a medium-size computer-controlled system is shown in Fig. 16.55. Manufacturers have made software available that permits rapid accurate design of complex large systems. A typical computer screen in such a system is shown in Fig. 16.56.

Relay Cabinet

Terminal strip for connection of switches and sensors.

Relay; see Fig. 16.53(a)

Relay load terminals

Fig. 16.54 Prewired relay cabinet (partial). Field wiring consists of control wiring to switches and sensors and power wiring to the controlled loads. (Photo courtesy of Touchplate Technologies, Inc.)

16.33 Wireless Switching and Control

Full-voltage local switching is described in Section 16.30, and low-voltage switching, both local and remote, is described in Section 16.32. The next logical step is remote wireless control and switching. These systems have undergone extensive development. They impinge on the design considerations of architects and building engineers to the extent that facilities for their accommodation must be provided if such systems are to be used in a facility.

These system aspects as well as system functioning and application to lighting control are discussed in detail in Chapter 21.

16.34 Power Line Carrier Systems

Although the advantages inherent in flexible, programmable load switching systems such as those described are undeniable, particularly in connection with energy management systems, their high first cost frequently makes the entire project uneconomical. This is particularly true in retrofitting large, complex facilities with energy management controls, where the cost of wiring, even when using low-voltage control wires, can amount to more than half the total cost of the project.

As a result of the desire to minimize wiring costs, a system of control signal transmission that uses the electric power wiring (either new or existing) as its conductors was developed, thus obviating the need and expense of dedicated control wiring. The system is called a power line carrier (PLC) system because the power line carries the control signal. An ancillary benefit of such an arrangement is that not only can it be added to an existing system, but it can also be removed, thus making possible and economical the leasing of energy management control systems.

The system operates by injecting into the power wiring a series of low-voltage, high-frequency, binary-coded control signals, which then disperse over the entire power network. Only a receiver that is "tuned" to receive a particular code reacts, generally by operating a local device, which in turn connects, disconnects, dims, or brightens a particular load. In practice, the control signal generator can be a small manually programmed controller as in a residence, or a com-

ELECTRICITY

Input devices

Switching devices
see Fig 16.53

Occupancy sensors,
photocells, timers,
other remote controls

Dimmed
fluorescent
lighting (with
electronic
ballasts)

Switched loads

Dimmed
incandescent
lighting

Control wiring

See
Fig. 16.56

Data
cable

Load
wiring

Relay
panel see
Fig. 16.54

Dimming
panel

Remote
panel

Power wiring

PC with Windows™
software by
Touch-plate.
Computer not
required to be on-line
for daily operation

Fig. 16.55 *Block diagram of a large, computer-controlled, low-voltage control system. The relay panel (Fig. 16.54) contains the incoming and outgoing power and control wiring, the switching relays, and the computer interface card. System arrangement is controlled by computer software (see Fig. 16.56).*

puter-operated energy management or lighting controller in a commercial facility. The receivers are normally of four types, most of which are designed to fit into an ordinary wiring device box:

1. A wall-switch module, which acts as a normal manual wall light switch but also contains a coded receiver and a fully rated relay (20 A, 125 V, or 277 V).
2. A wall receptacle module, which contains a 20-A receptacle along with a receiver and relay that serve to switch power on and off.
3. A switching module, which can be connected to activate contactors, small motors, and the like.

4. A dimming module, which provides both local and remote control.

Some of these components are shown in Fig. 16.57. Larger, multiple-circuit switching modules are also available for more extensive systems. Each miniature receiver can be manually tuned in the field (i.e., its code can be set and reset readily), giving the entire system extreme flexibility.

The system operates in the same manner as described for a low-voltage switching system or for an energy management or lighting control system, with the essential difference that no dedicated wiring is required. In retrofit jobs, existing

Fig. 16.56 *Typical computer screen of the low-voltage control software, showing facility floorplans (scanned in) and controlled loads. The icons at the top of the screen are customized and are used to control relays, make connections, log, check status, print, and log events by time and date. Other screens enable group and event definition, labeling, establishment of priorities, and other global sequential and individual control functions. (Courtesy of Touchplate Technologies, Inc.)*

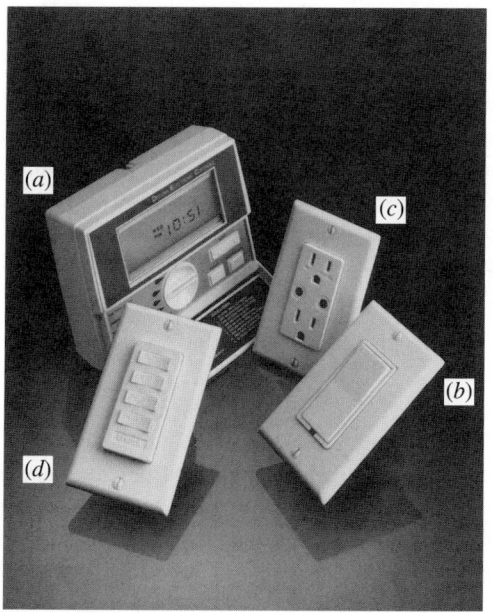

Fig. 16.57 *Components of a residential-type PLC system.*
(a) *Control unit, at which both manual and automatic (programmed) control of a maximum of 256 address codes is possible.*
(b) *Wall-switch module has local dimming and switching functions in addition to being remotely controlled from control center* (a).
(c) *Duplex wall receptacle either split wired with local programming of one-half of the outlet or complete remote control.*
(d) *Outlet box wall-mounted control switch that provides local group control (switching and/or dimming) for one to four devices (address codes). (Photo courtesy of Levington Manufacturing Co.)*

wall switches and receptacle outlets are replaced with receiver-controlled units; in new installations, these devices are installed initially.

Due to the fact that the control signals travel over the power wiring and are therefore attenuated by high-resistance connections, poor grounds, and faulty wire insulation, PLC installations require a high-quality power wiring installation to operate properly. In retrofit work, preliminary tests occasionally indicate the impracticability of a PLC installation because of the condition of the power wiring system. Another problem encountered in PLC systems is interference with the PLC control signals from radio noise generated by faulty power equipment and by improperly shielded or grounded electronic equipment. These problems can frequently be overcome by careful testing, installation of filters and jumpers, and by repair of improperly operating equipment. In addition to energy management and normal lighting control, PLC control can be applied to group switching of other loads, such as office machines, in addition to simplifying control of temporary light and power for special events.

16.35   Power Conditioning

(a) General. *Power conditioning* is a relatively new electrical term that describes the process of converting utility supplied electrical power, which is frequently characterized by transient surges, spikes, radio frequency noise, and voltage fluctuation, to a pure (noiseless), accurately voltage-regulated sinusoidal waveform, normally referred to as *computer-grade power.*

Some form of power conditioning is required by all data processing and telecommunication equipment, and by many other types of electronic equipment. This is due to their extreme sensitivity to fluctuations in the power supply voltage and particularly to random high-frequency voltages superimposed on the power supply voltage, known as *radio frequency interference* (RFI) or simply *radio noise.* In the case of data-processing equipment (computers and peripherals), radio noise can cause data errors; slow voltage fluctuations can cause overheating, data loss, and premature equipment failures; and large rapid voltage

fluctuations (spikes, transients) can cause equipment burnout and system collapse. The economic ramifications of such effects on large commercial data-handling systems are so severe that power conditioning has become almost universal. [Power conditioners must not be confused with uninterrupted power supplies (UPS), which function to maintain power during utility *failure,* although in many commercial installations, both functions are served by a single equipment package.]

(b) Source of Disturbance. Utility power systems are designed to maintain voltage and frequency within certain limits on an average, long-term basis. For good technical reasons utility equipment cannot react to the short-time disturbances that are so inimical to electronic equipment. Furthermore, much of the noise in utility lines is introduced by users and is therefore beyond the utility's point of control. The disturbances on the electrical power line can roughly be classified into three types: voltage variations, electrical noise, and transients.

1. *Voltage Variations.* These can be caused by short-time current drain, such as during motor starts, deliberate voltage reduction (brownouts), load switching of EMS systems, excessive line voltage drop due to heavy loading or current leakage, and similar problems. These variations are *relatively* slow and long lasting.

2. *Electrical Noise.* This is simply low-amplitude higher-frequency voltage that, when superimposed on the power line, results in a distorted waveform. Sources of such extraneous voltages are electronic equipment power supplies, lighting dimmers, solid-state motor controls, PLC systems, arc welding, switching transients caused by branch circuit switches and relays, and voltages induced by local magnetic fields.

3. *Transients.* These are large short-duration voltage variations, also known as *surges* or *spikes.* They are the principal culprits in major equipment failures. They are caused by lightning strikes to overhead lines (even many miles from the affected installation), by electrical system faults, and by utility switching of high-voltage lines, transformers, capacitor banks, generators, and circuit reclosers.

16.36   Power Conditioning Equipment

Each of the three types of problems described in the preceding section requires its own type of correction. Voltage variations are corrected with voltage regulators; noise problems by electrical isolation, filtering, and noise suppression; and transients are treated with surge suppressors. Each corrective device is a separate entity that can be applied individually to suit the particular problem encountered or in combination with another conditioning device. The choice in specific installations is a highly technical one and should be made by the facility's electrical engineer after careful analysis of the quality of the utility service. A few general recommendations can be made.

1. All computer installations, including the smallest, should be protected from line transients with an appropriate surge suppressor. Multitap plug-in strips with integral surge suppressors are unsatisfactory unless they have surge current, clamping voltage, and surge-energy specifications, and these meet the installation requirements.

2. Major data-processing installations normally require all three types of line treatment. The most economical way to provide them is a single integrated power conditioning unit.

3. Considerable improvement in the quality of electrical power can frequently be achieved by the simple expedient of running separate electrical feeders for sensitive loads.

4. Physical isolation of sensitive equipment areas is frequently helpful, particularly when disturbances are being induced by proximity to switching, arcing, and rectifying equipment. Discharge-type lighting, including fluorescent, mercury, sodium, and metal–halide, can cause interference, especially when used with electronic ballasts.

5. It was found some years ago that much of the "noise" in electrical power systems was introduced through the (equipment) grounding pole of an electrical receptacle. Because this connection is required for electrical safety, it cannot be eliminated. It can, however, be separated from the wiring system ground and still maintain its function. Receptacles so constructed are color coded either entirely orange or have an orange triangle on the face (see Section 16.29). This "fix" is particularly effective where electronic dimmers, ballasts, and switching devices are present. Because the insulated ground receptacle (sometimes referred to as an *isolated* ground receptacle) is a special item, it must be so indicated on the construction contract documents. Furthermore, because these receptacles require special isolated wiring, the entire matter must be studied carefully during the electrical design stage and before issuance of construction contracts.

6. A problem related to equipment that produces radio noise is that this same equipment also produces harmonics that appear on the system neutral and can cause overloading. This subject is covered in Section 17.18.

The issue of power quality is so important in modern commercial installations that the following steps should be taken in each such design:

· A detailed report should be obtained from the electric utility that will supply electric power showing power outages, voltage constancy, frequency control, and, if possible, oscillograms of wave forms at random intervals over the past year. Particularly important are periods of peak loading and periods of weather extremes.

· A power quality study should be made in the area in which the facility will be (or is) located. Sophisticated instrumentation for this purpose as well as specialists in this field are readily available (see Section 14.23).

· An in-depth study of all of the electrical equipment to be used in the building should be made for the purpose of

(a) Reducing the sources of electrical disturbances

(b) Determining the tolerance or sensitivity of the data-processing equipment to be used.

In the absence of reliable data for (b), as is the case in speculative construction, a high degree of sensitivity must be assumed. In any case, for equipment that is known to produce radio noise, such as electronic lighting ballasts and dimmers, the specifications should include limitations plus a requirement for submission of laboratory tests on random samples.

· Many large power companies now offer "computer grade power" that meets stated limitations to voltage fluctuation and the like. This type of service is desirable for all new office space or full building renovations. For individual office spaces, separate power conditioning equipment is required.

Table 16.7 and Fig. 16.58 give the reader some idea of the physical characteristics of power conditioning equipment. Dimensions are typical and are indicative only of equipment bulk. Designs vary widely among manufacturers, as do, therefore, individual dimensions. Weights are not given for the same reasons. Voltage-regulating equipment containing regulating transformers is very heavy. Power conditioning equipment is frequently combined with UPS equipment (see Section 16.38).

16.37 Surge Suppression

The full term for this aspect of power conditioning is *transient voltage surge suppression* (TVSS). Because it is transient voltages, also called *surges* or *spikes,* that can cause major physical damage to computer systems and other types of electronic equipment, it is this aspect of system protection that has received most attention (including a fair amount of highly imaginative advertising). Surge suppressors are available in a huge variety of designs and ratings, ranging from the ubiquitous cord-connected multiple-outlet strip to large, 3-phase, service-entrance units. However, they all have a single purpose: to suppress (limit) a voltage surge to a level that the protected equipment can withstand without damage. This is done in two ways:

1. By placing one or more devices in series with the incoming voltage transient that presents a high impedance, thereby limiting the let-through current.

TABLE 16.7 **Power Conditioning Equipment**

Equipment Functions				Dimensions (in.)		
Voltage Regulation	Noise Suppression	Surge Suppression	kVA[a]	H	W	D
x			5	15	20	20
			10	35	40	20
			20	40	40	30
x	x		0.25	6	6	10
			0.5	7	7	12
			1.0	8	10	15
			3.0	10	12	20
x	x		5	10	10	15
			10	20	30	15
			20	30	30	20
x	x	x	5	40	30	30
			10	50	35	30
			20	70	40	35
		x	NA[a]			
			15 A	3	3	16
			100 A	10	12	12
			500 A	15	15	12

[a]Surge suppressors are rated by transient energy-handling capability, and therefore a kVA rating is not applicable. However, because transient energy increases with kVA size of protected equipment, a service ampere rating is given.

Fig. 16.58 *This single-phase power conditioner supplies noise filtering and surge suppression by use of an isolation transformer, ground noise filters, and category A and B surge suppression (see Section 16.37(c)). It is rated 5 A, 120 V, and 6 in. H × 7 in. W × 11 in. D, and weighs 18 lb. A similar unit rated 20 A, 120 V measures 8 in. H × 8½ in. W × 14 in. D. (Photo courtesy of Leviton Manufacturing Co.)*

2. By placing one or more devices across the incoming power line, in parallel with the protected load, that presents a low impedance to the high transient voltage, and thereby bypass (shunt away) the incoming wave's current. Many TVSS units use both of these techniques in a single "hybrid" unit.

(a) Terminology. To be able to select and compare TVSS units and their ratings, a basic technical vocabulary is necessary.

Avalanche diode. A solid-state device placed in parallel with the protected load. It responds to overvoltage in nanoseconds by conducting (i.e., placing) its very low impedance across the incoming voltage. This results in a large ("avalanche") bypass current flow that serves to reduce the incoming voltage. It has a low clamping voltage (definition follows) and low energy absorption capacity. This latter characteristic causes "wear" and eventual failure.

Clamping voltage. The voltage at which shunting devices (such as the avalanche diode) begin to conduct and thereby "clamp" the incoming voltage surge. Therefore, the lower a device's clamping voltage, the better is its suppression action.

EMI/RFI rejection. A measure of the attenuation by the TVSS device of electromagnetic interference (EMI) and radio frequency interference. A TVSS device, due to its internal filters, always provides some radio noise attenuation. This characteristic is not of primary significance to a device used primarily for surge suppression.

Gas tube surge protector. A high-energy shunt device, typically found in high-capacity, good-quality TVSS units that are subject to large surges, such as those caused by lightning. It consists of a gas-filled tube with two electrodes. The tube conducts when a high-voltage surge causes its electrodes to arc over. Because of low arc impedance, the current drain and energy absorption are very high, thus clamping the incoming surge. Its principal drawback is its relatively slow response, which necessitates its use in conjunction with other faster devices.

Isolation transformer. A transformer with two separate windings and no conductive electrical connection between them. By virtue of the inductive coupling of the windings, considerable attenuation of sharp voltage spikes is achieved. It can be used in high-capacity TVSS units in lieu of multiple solid-state components in parallel. It is also useful in attenuating high-frequency noise at any voltage level.

Joule rating. A *joule,* equal to 1 watt-second, is the SI unit of energy. A device's joule rating is a measure of its capacity to absorb and dissipate heat generated by its action. Although joule rating is important where large amounts of energy are to be absorbed, as in (hard-wired) units at a service entrance or at a power distribution point, published ratings can be misleading because there is, at this writing, no industry-wide standard calculation procedure for joule rating. As a result, the industry standard specification form (NEMA-LS-1) for TVSS devices does not include this rating.

ELECTRICITY

Maximum surge current. A measure of a device's ability to divert and dissipate surge current without failing. Useful metric for TVSS devices used at service entrances, and for reasons explained previously, more reliable than published joule ratings.

Metal-oxide variable resistor (MOV). A shunt device somewhat slower than an avalanche diode but with much higher energy absorption capability and higher clamping voltage than an avalanche diode. MOV is very widely used in TVSS designs. Like avalanche diodes, MOVs suffer degradation with use, thus limiting the useful life of the entire TVSS unit.

Response time. A measure of the rapidity with which a TVSS device begins to clamp a voltage surge, usually measured in microseconds or nanoseconds. Not of great significance because both avalanche diodes and MOVs (found in all good-grade TVSS designs) are faster than the voltage build-up time of an incoming surge. Response time is not included in the NEMA standard specification, apparently because it is not considered to be a very important characteristic.

Sinewave tracking. A term of uncertain meaning found frequently in manufacturer's literature.

(b) TVSS Operation. Figure 16.59 is intended to be self-explanatory. Each element acts to reduce the magnitude of the incoming voltage surge by either bypassing (shunting away) current or presenting a high impedance to the voltage surge. Both actions reduce the voltage. A commercial TVSS may contain any or all of the illustrated components, single or in multiple, depending on the current and voltage it is expected to handle.

(c) Standards. Standards applicable to surge suppression equipment are:

ANSI/IEEE C62.41: Guide for Surge Voltages in Low-Voltage AC Power Outlets.

ANSI/IEEE C62.45: Guide for Surge Testing for Equipment Connected to Low-Voltage AC Power Circuits.

UL 1449: Standard for Safety—Transient Voltage Surge Suppressors.

NEMA LS.1: Low Voltage Surge Protective Devices.

Surge suppression equipment can be subjected to widely varied voltage stresses, depending on where in an electrical system a unit is installed. Standard C62.41 categorizes installation points according to the severity of service expected; that

Fig. 16.59 All of the elements usually found in a surge suppression device are shown. Each acts to reduce the voltage surge; parallel elements by shunting away (bypassing) surge current, and series elements by increasing the electrical impedance to the voltage surge.

is, the (electrical) environment in which the surge suppression equipment operates.

C62.41, Category C represents the most severe environment, corresponding to locations on the line side of the facility service switches. This environment is subject to the full force of lightning and switching surges, and therefore requires robust equipment. As a voltage surge travels through a building's electrical system, it is attenuated by cables, transformers, and other electrical equipment. Therefore, the further along one travels in the building's electrical distribution system, the less severe the voltage surges become. All TVSS devices used in category C environments are hard-wired (Fig. 16.60).

C62.41, Category B covers environments on the load side of the service equipment, including heavy feeder distribution panels and short branch circuits. Most SSTV devices in category B areas are hard-wired (Fig. 16.61).

C62.41, Category A, which represents the least severe service, includes utilization outlets and long branch circuits (because the long cables attenuate the voltage surge). Most category A SSTV devices are arranged for loads to be

Fig. 16.61 *TVSS unit designed to be used at service and branch panelboards in residences and small commercial installations. The illustrated unit is rated single-phase, 120/240 V, 3-wire service, category B3. The unit measures 7 in. W × 7 in. H × 4 in. D and is equipped with indicating lights that indicate power and suppression status. (Photo courtesy of Leviton Manufacturing Co.)*

plugged in directly. Devices used in category A environments include the common cord connected multioutlet strip and surge-suppression equipped receptacles (Fig. 16.62).

Fig. 16.60 *Surge suppressor designed for use on the incoming service of a facility. Rated for category C3 application; shunt type, 120/208 V, 3-phase, unlimited current. Provided with RFI/EMI filtering, replaceable surge suppressor modules and arranged to indicate power and protection status of each phase. (Photo courtesy of Surge Control Ltd.)*

Lights indicate working condition and replacement requirement

Fig. 16.62 *Typical category A utilization TVSS outlet, equipped with indicating lights that show the condition of the surge suppressor circuitry. (Photo courtesy of Eagle Electric Co.)*

ELECTRICITY

Each of these three major categories is in turn subdivided into three minor categories: C3, C2, C1; B3, B2, B1; A3, A2, and A1, where 3 is the most severe and 1 the least. The differentiation depends on the electrical capacity of the facility's service and distribution system. Large systems have higher surges and less attenuation in the system, thus requiring larger and heavier TVSS equipment than smaller installations. The three major categories are shown graphically and in text in Fig. 16.63.

Standard C62.45 established the test procedures and test voltage levels to be used by manufacturers in testing and classifying their TVSS products in order to simulate the voltage stresses found in A, B, and C environments.

Underwriters Laboratories (UL) Standard 1449 establishes standards for surge suppressor safety and performance. An important aspect of this standard is that it establishes clamping voltage ratings. The ones of interest to building designers are the lowest four, which are 330, 400, 500, and

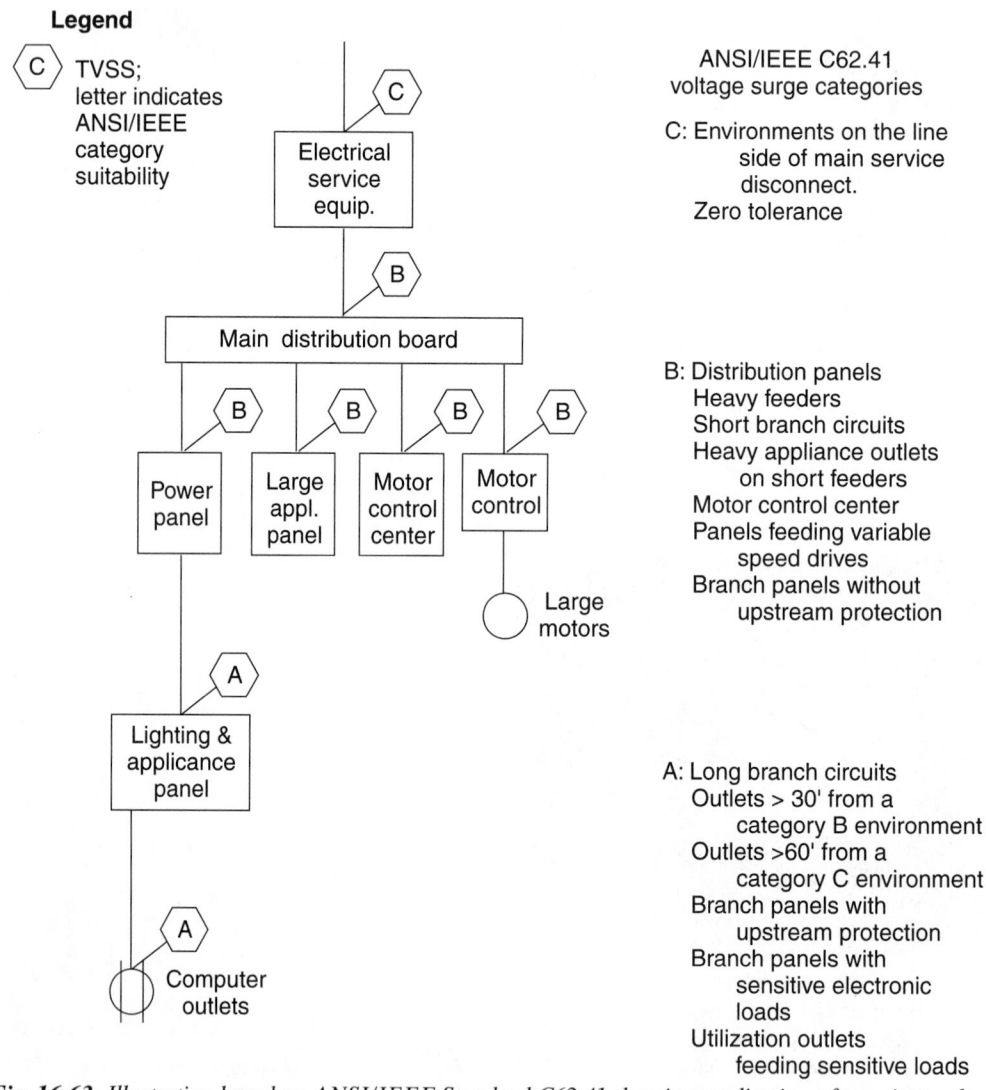

Fig. 16.63 *Illustration based on ANSI/IEEE Standard C62.41 showing application of transient voltage suppression devices in a typical electrical facility, based on the three categories of electrical environments (C, B, A) defined by the standard.*

600 V. The UL standard requires testing for actual clamping levels, which are then rounded *upward* to one of the standard ratings. Thus, a TVSS that clamps at 240 V and one that clamps at 320 V would both be rated at 330 V, despite the fact that the 240-V unit gives better protection. This is particularly important for protecting computers and other electronic equipment equipped with switching power supplies, and requires comparison of *actual* clamping voltage of units rather than reliance on the voltage classification.

NEMA LS-1 standardizes the specification format for suppression units, and in so doing establishes a basis for comparison among various manufacturers' products. Important characteristics that must be specified include line voltage, maximum operating current and voltage, maximum surge current, and clamping voltage rating. (Actual clamping voltage is not supplied unless specifically requested, as it involves specific testing.)

All of the mentioned standards are voluntary industry and technical society publications. Recently (1997), the U.S. government developed a Performance Verification Test specification that establishes a quality grading system for cord-connected surge suppressors. This standard does not correspond readily to the C62.41 classifications, except in the B category. It is worth noting that the highest classification in this system is Class 1, Grade A, Mode 1, which corresponds to ANSI category B3, UL 1449 clamping voltage rating of 330 V, and line to neutral protection.

(d) Application of TVSS Devices. As should be evident at this point in our discussion, selection and application of surge suppression devices for even the simplest application requires considerable technical background, for which the foregoing material is intended. Important guidelines that assist in this area are:

- Every power level should have its own TVSS. Small systems can function with C62.41 categories B and A only. Large systems may require protection for two or three B levels (B3, B2, B1) and more than one A level.

- In the absence of accurate data on the surge tolerance of electronic equipment, use protection for the most sensitive level. At the

utilization level, this is category A3 and U.S. government classification Grade A, Class 1, Mode 1.

- Mode 1 (also called Mode A) indicates protection line-to-line or line-to-neutral. This mode should be specified for stand-alone items. It is preferable to Mode 2 (or B), which is connected line-to-ground, because this can introduce random signals (contaminates) into the equipment grounding system.

- The primary criteria for selection and comparison are *actual* clamping voltage and maximum surge current capacity. Other important characteristics are those listed in NEMA LS-1.

- TVSS units should be installed as close as possible to the loads being protected.

- Telephone and data lines must also be protected, particularly if the service is overhead. Most telephone companies install an appropriate TVSS at the main distribution frame near the service entrance. This may not be sufficient to protect electronic equipment from disturbance that can cause distortion of data. A specialist should be consulted to determine the required protection levels and select appropriate equipment (Figs. 16.64 and 16.65).

16.38 Uninterruptible Power Supply

As explained previously, power conditioning equipment can supply clean utility power; it cannot, however, supply any power during a utility outage. That eventuality is covered by an alternate supply of power. Facilities that contain computers and/or data-processing equipment cannot tolerate power outages in excess of about 8 to 50 milliseconds (ms) without serious risk of data loss, with all the negative ramifications of such a loss.

A UPS is an arrangement of normal and backup power supplies that transfer a facility's critical load from normal to backup supply in so short a time that no computer malfunction results. This transfer time varies somewhat among different schemes and manufacturers, but is always less than 8.3 ms, which is the minimum period of

Fig. 16.64 *Surge suppressor designed for use with standard snap-in low-voltage telephone and industry jacks. The illustrated unit is designed to protect telephone/telecommunication lines, both analog and digital. The units are rated 185 V, with a maximum surge current of 5000 A and test clamping voltage of 240 V. The units can be used with telephones, modems, fax machines, and the like. The unit measures approximately 4 in. W × 2½ in. H × 1 in. D. (Photo courtesy of Surge Control Ltd.)*

Fig. 16.65 *TVSS device designed to protect video lines from voltage transients. The unit plugs into a standard outlet for the purpose of establishing a ground connection for current bypass. This unit is equipped with BNC connections; others have telephone, PL, and RS connectors for use on telephone, communication, and data lines. The illustrated unit has a clamping voltage of 15 V and a maximum surge current capacity of 5000 A. The unit measures approximately 2 in. W × 4 in. H × 1 in. D. (Photo courtesy of Leviton Manufacturing Co.)*

power outage that computers must tolerate without disturbance to meet the computer industry's manufacturing guidelines. This time period is double the maximum transfer time required by IEEE Standard 446, "Emergency and Standby Power," of ¼ cycle at 60 Hz, or 4.16 ms. Thus, all computer systems fed from a UPS that meets this standard have a safety factor of at least 2 in transfer time.

The period of time after transfer that the equipment will run on the standby source depends on system design. In most cases it is 5 to 10 minutes, which is usually enough time to permit an orderly manual or automatic shutdown. Where shutdown is not a viable alternative, as with computer-controlled manufacturing processes, the standby power system can readily be designed to supply power indefinitely.

The actual selection of a UPS system for a facility is a highly technical and complex process, beyond our scope here. However, an informed architect or project manager should have sufficient familiarity with these systems to permit preliminary selection and planning. The material that follows is addressed to that end.

(a) Alternate power source. For most systems, battery backup is sufficient because, as noted above, only an orderly shutdown is required. Where this is not the case, a two-stage transfer can be used; the first transfer to a battery backup that can carry the load for up to 1.5 h (or more) depending on load magnitude, and a second transfer to a long-term standby generator set. Thereafter the availability of generated standby power is limited only by fuel supply. Alternatively, the generator can be on-line, and the load picked up directly on the first transfer.

In large industrial installations, the standby source may also be a second utility line. In such cases, only a single transfer is required. Questions of service reliability and equipment redundancy are both technical and economic, and must be studied carefully for each individual installation.

(b) Equipment Arrangement: Classic Standby and On-Line Topologies. Most
UPS systems today are described by their manufacturers as on-line or simply as UPS rather than standby, primarily because there is no generally accepted, industry-wide definition of "on-line" and also because many systems are hybrid and do not fall easily into either category. Figure 16.66 shows the usual static (nonrotary) UPS equipment arrangement that applies, in principle, to both standby and on-line modes.

In the standby mode (Path A), utility power is normally passed through some power-conditioning equipment to shield against surges and remove random noise, and then, via position A in the static transfer switch, to the load. The ac line also provides a small current to the small battery charger that keeps the battery fully charged. The dc to ac inverter is open-circuited at the transfer switch and delivers no power. In the event of a utility power failure, the transfer switch moves to position B, and power flows from the battery through the inverter and transfer switch to the load (Path B). With proper equipment design, the transfer is usually accomplished smoothly, although an instantaneous change from no-load to full load on the inverter can sometimes cause a transfer voltage loss. For this reason, on-line schemes are preferred.

In the straightforward on-line arrangement shown in Fig. 16.67, power path A (normal) is through the rectifier and inverter to the load via position B in the transfer switch. The battery floats on the dc line. Failure of utility power causes the power path to change to path B (i.e., battery to inverter to load). The transfer switch remains in the B position. As a result, voltage to the load is undisturbed. If one of the Path A components fails, the transfer switch moves

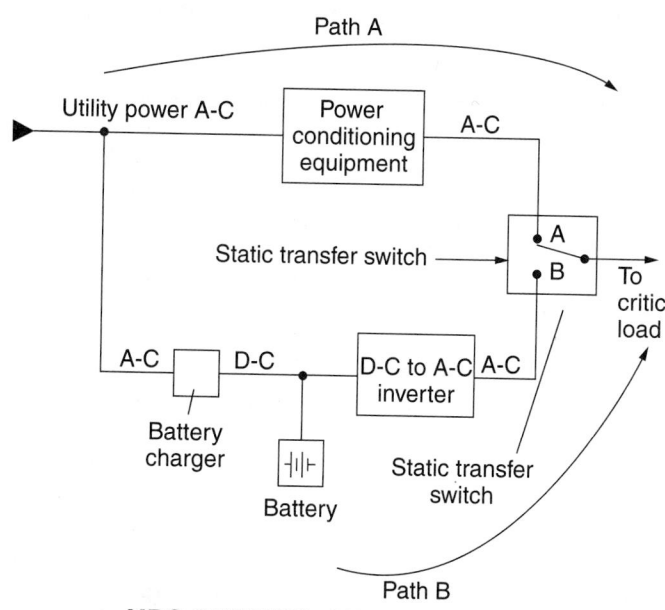

UPS SYSTEM; STANDBY MODE

Fig. 16.66 In this UPS standby arrangement, power normally flows from the utility to the critical (computer) loads through the A position of the transfer switch (PATH A). The battery floats until a failure of utility power puts it on line through the inverter and position B in the solid-state, effectively instantaneous, transfer switch (PATH B). The inverter is normally inactive. The battery charger is small because it carries only charging current.

ELECTRICITY

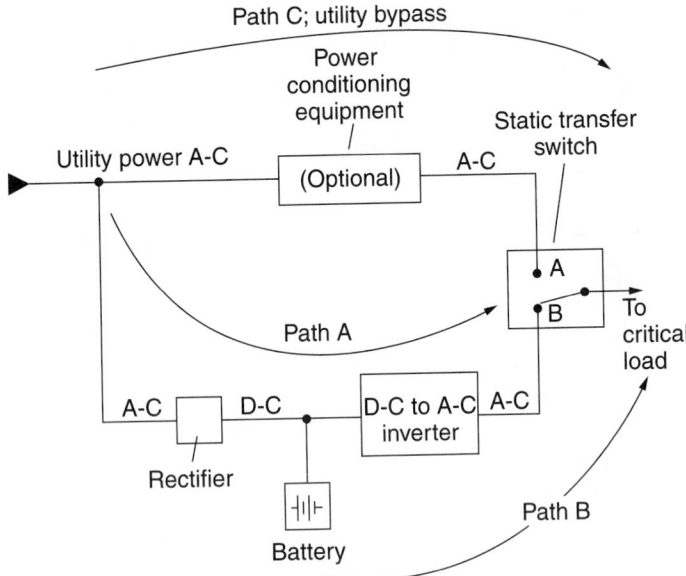

Fig. 16.67 In this UPS on-line mode the normal power path A is from the utility through a full-size rectifier to the load, via an inverter. The battery floats, taking a small charging current from the rectifier. Utility failure brings the battery line through the inverter (PATH B). In both instances, the transfer switch remains in B position and the load senses no change. Only in the event of UPS equipment failure is power taken directly from the utility (PATH C).

UPS SYSTEM; ON-LINE MODE

to position A and power is supplied directly from the utility lines via path C, which is referred to as a *utility bypass*. The advantage of the on-line arrangement is that it eliminates reliance on the quality of utility power. Power conditioning and surge suppression are provided by the solid-state equipment.

The disadvantages of this classic on-line topology are high cost due to the full load capacity size of the charger/rectifier and an overall efficiency in path A of about 75% to 80%. The resultant heat production can be problematic, and is certainly expensive in terms of power costs.

(c) Additional UPS topologies.

1. A money-saving arrangement called "on-line" without bypass, is simply the bottom portion of Fig. 16.67. First cost is reduced by elimination of the transfer switch and considerable cabling. Because failure of any of the components disables the system completely, as there is no back-up power path, this system is not recommended for critical loads.

2. A system variously known as line-interactive, standby line-interactive, and hybrid interactive is shown in Fig. 16.68. In its normal mode (Path A), power flows from the utility lines

through power conditioning equipment and position A of the transfer switch to the load. (This is identical to the normal mode of the standby system shown in Fig. 16.66.) In addition, however, utility ac is tapped at the power conditioning equipment to feed the inverter that, operating in reverse as a rectifier (ac to dc), serves as a battery charger.

In the event of a utility power failure, power is drawn from the battery, and passing through the inverter it is converted to ac. It then connects to the load via the B connection in the transfer switch. The power path is indicated on Fig. 16.68 as Path B. The advantage of this topology is the elimination of a large rectifier by dual use of the inverter, and elimination of transients caused by switching-on the inverter at full load. The system is efficient, producing little heat, and is highly reliable.

(d) System Selection and Comparison.

The systems described previously in some detail are intended to give the reader a topical familiarity with UPS equipment arrangements. Actual selection depends on the types and magnitudes of the critical loads, detailed outage histories, space considerations, cost factors, equipment and service redundancy, and overall system efficiency. This latter characteristic can be a major

UPS SYSTEM; LINE INTERACTIVE MODE

Fig. 16.68 In this UPS hybrid mode, the normal power path A takes power directly from the utility but keeps the inverter on-line, working in reverse to furnish dc charging current to the battery. Utility failure causes a change to power path B (i.e., from battery through the inverter now operating normally to invert dc to ac). The static switch changes to position B.

decision factor where large loads are being supplied because overall system efficiencies vary from a high of about 95+% for classic standby (Fig. 16.66) to as low as 65% for some arrangements. Because all of the lost power turns to heat, a low efficiency 100 kVA/80 kW system would produce about 27 kW or about 92,000 Btu/hr! Unless this heat can be used effectively, power costs and ventilation arrangements should be factored into the selection process.

(e) Ancillary Characteristics of UPS Systems. Programmable microprocessors and effec-

tive sensors and transducers have been added to basic UPS systems to produce what is known as intelligent UPS equipment. These additional devices now make available operating and diagnostic data including battery status, equipment operating temperatures, load data, waveform displays, on-line power quality analyses, operational and event logs, and the like. These data help an informed operator optimize system functioning, anticipate problems, and keep an accurate printed operational record. Table 16.8 gives some typical dimensional and run-time data for UPS units of different topologies.

ELECTRICITY

TABLE 16.8 **Typical Dimensional Data for UPS Equipment**

Topology	Capacity (kVA)	Dimensions $H \times W \times D$ (in.)	Weight[a] (lb)	Full Load Run Time (min.)
Standby	0.3	$6 \times 6 \times 12$	15	10
Single phase	0.6	$7 \times 6 \times 14$	25	5
On-Line	3	$30 \times 12 \times 24$	250	20
Single Phase	7.5	$36 \times 12 \times 36$	500	10
3 Phase	10	$60 \times 24 \times 36$	1200	15
Interactive	0.4	$6 \times 6 \times 12$	20	5
Single phase	1.0	$8 \times 8 \times 15$	40	5
	2.0	$16 \times 8 \times 20$	105	20
	3.5	$18 \times 10 \times 24$	140	10

[a]With batteries.

ELECTRICITY

16.39 Emergency/Standby Power Equipment

The *NEC* makes a clear distinction between emergency systems and standby systems, covering the former in Article 700 and the latter in articles 701 and 702. The equipment, circuitry, and arrangement of both is similar; the purpose is somewhat different. The reader is also referred to NFPA Standard 110, *Emergency and Standby Power Systems,* which covers equipment requirements for these systems.

Emergency systems are intended to supply electric power to equipment essential for *safety* to human life on interruption of normal supply. Included in this classification would be illumination in areas of assembly, to permit safe exiting and panic prevention, and such other vital functions as fire detection and alarm systems, elevators, fire pumps, public address and communication, and orderly shutdown or maintenance of hazardous processes.

Standby systems are divided into two categories: those legally required (Article 701) and optional systems (Article 702). The former are intended to power processes and systems (other than those classified as emergency systems) whose stoppage might create hazards or hamper firefighting operations. This classification is broader than the emergency system and could include HVAC systems, water supply equipment, and industrial processes whose interruption could cause a safety or health hazard. It is intended primarily as a safety measure.

Optional standby systems can cover any or all loads in a facility at the discretion of the owner, and are normally intended to protect property and prevent financial loss in the event of a normal service interruption. A typical application would be a critical industrial process or an ongoing research project.

Health care facilities are covered by a separate set of regulations: *NEC* Article 517 and NFPA Standard 99, both of which are referenced as legally binding in the vast majority of jurisdictional codes. It is important to note that for both emergency and legally required standby systems, the *provision* of the system must be mandated by the authority having jurisdiction over construction,

whether it be a local, state, or federal agency, or a combination of these. The *NEC* dictates how the system is to be designed and constructed; its existence depends on another authority.

A case in point is the fundamental item of exit lights. These are mandated by the NFPA Life Safety Code 101, and Subpart E of OSHA regulations. They are not required by the *NEC* and reference to *NEC* will not assure their provision. The code(s) having authority and jurisdiction must specifically require an emergency and/or standby system by that nomenclature in order for *NEC* provisions to apply. The exact items of equipment to be powered are selected by the designer, keeping in mind the specific and general requirements of the codes having jurisdiction. The majority of codes make the provisions of the *NEC* and the relevant NFPA standards legally binding. Most codes require emergency systems; far fewer require standby systems, and then only for essential water and water-treatment systems and a few other essential uses. The designer must thoroughly investigate the matter of jurisdictional codes before proceeding with consideration of emergency electric power systems and equipment.

System design arrangements are discussed in Sections 17.3 and 17.6. Emergency lighting equipment and system design is covered in Chapter 21. System equipment, which falls into two principal categories—that is, generator and battery installations—is discussed later. Optional standby systems normally use a fueled prime mover and not batteries.

(a) Engine–Generator Sets.

An engine–generator set installation comprises basically three components: the fuel system, including storage, if necessary; the set itself, plus exhaust facilities; and the space housing the unit (Fig. 16.69). The principal advantages of the engine–generator set are unlimited kVA capacity, duration of power limited only by size of fuel tank, use for peak-load shaving, and, if properly maintained, indefinite life. Disadvantages are noise, vibration, nuisance of exhaust piping, need for constant maintenance and regular testing, and difficulties with fuel storage. Gasoline can be stored for only a year at most, and subsequent disposal is difficult. Diesel fuel keeps somewhat longer, but disposal

ELECTRICITY

AC POWER
WIRING

CONDENSATION
DRAIN TRAP

MUFFLER

ENGINE
EXHAUST

THIMBLE through
outside wall, for
exhaust pipe

FLEXIBLE
SECTION

DC CONTROL
WIRING

AIR IN

AIR OUTLET DUCT
connects to louvered
opening in outside wall

VIBRATION
ISOLATORS

CONCRETE BASE

IMPORTANT!

COOLING AIR INLET MUST BE AT
LEAST 1-1/2 TIMES LARGER THAN
RADIATOR DUCT OUTLET AREA.

FLOW OF COOLING AIR AND
HEATED AIR CAN BE CONTROLLED
BY AUTOMATICALLY OPERATED
LOUVRES

Fig. 16.69 *Typical installation diagram of a radiator-cooled diesel engine–generator set. The set should be oriented so that the hot radiator cooling air vented through an outside wall flows in the same direction as prevailing winds. A solid masonry wind and noise barrier not less than 6 ft from the radiator air outlet is recommended, as well as an elevated exhaust pipe outlet. (Illustration courtesy of ONAN Corp.)*

is also difficult. Use of gas for a diesel or gas engine obviates the fuel storage problem, but poses the alternate problem of availability of gas service during emergencies. In some large cities, steam is commercially available as an energy source. Here too, service reliability, particularly in the event of a widespread electric service failure, must be carefully investigated.

(b) Battery Equipment. Storage batteries are often used to supply limited amounts of emergency power, for lighting (see Section 21.36) and for UPS systems as detailed previously. Such units are mounted in individual cabinets or in racks for large installations, and are always provided with automatic charging equipment.

Battery types are undergoing intensive development. At this writing, the types principally in use are lead–acid, nickel–cadmium, lead–antimony, lead–calcium cells, and several alkaline types. The choice depends on the application. Installation requirements such as ventilation, gas detection, isolation, and the like depend entirely on the type of battery, size of the installation, and battery voltage. That being so, no general guidelines can be given except to follow *NEC* Article 480 requirements and consult a battery specialist for other data.

Batteries have the distinct advantage that they can be installed either in a central system with distribution of the battery power throughout the facility or in small package units around the building. Central systems are 24 to 125 V, dc or ac, and normally feed emergency lighting only. Individual packs are used often to supply ac power via built-in inverters. The great disadvantage of battery systems is limited duration of power. The *NEC* requires that batteries maintain loads for a 1½ h *minimum,* but larger capacity is frequently installed.

16.40   System Inspection

Each electric wiring system is inspected at least twice by the local inspection authorities: once after raceways (roughing) have been installed and before the wiring and closing-in of walls and once after the entire job is complete. The purpose of these inspections is to determine whether design, material, and installation techniques are meeting the national and local code requirements. Excellence of installation is the responsibility of the contractor. The designer, however, must be completely familiar with installation work and the equipment's physical characteristics in order to properly design an electrical system. Such a design will not present the contractor with unwarranted difficulties. The designer must also be very wary of equipment substitutions by the contractor who, having submitted a bid on the basis of plans and specifications, should be required to supply the specified equipment.

17

ELECTRIC WIRING DESIGN

In wiring design, there are numerous possible solutions to each problem—some good, some fair, and some poor. Experience guides designers to a solution that best suits the job because it is their responsibility to establish the most economical design within the framework of the design criteria. This chapter opens with a discussion of these criteria and continues with wiring design. Actual electrical design intended for construction must meet all the requirements of local and national building codes. To our knowledge, all building codes include NFPA 70, the *National Electrical Code* (*NEC*), as the minimum standard for all electrical design, and its requirements are thereby made legally binding. The electrical design data and procedures adduced in this chapter and in other sections of this book are based on *NEC* requirements and are, to the best of our knowledge, current. However, they are intended only as didactic materials. Code requirements change, as the *NEC* is revised every 3 years. The designer is therefore advised to obtain and study the latest edition of all codes having jurisdiction, including the *NEC*, to ensure that the electrical design meets the requirements of these codes.

17.1 General Considerations

(a) Flexibility. Every wiring system should incorporate sufficient flexibility in branch circuitry, feeders, and panels to accommodate all probable patterns, arrangements, and locations of electric loads. The degree of flexibility to be incorporated depends in large measure on the type of facility. Thus laboratories, research facilities, and small educational buildings require a great deal more flexibility than do residential, office, and fixed-purpose industrial installations. As part of the design for flexibility, provision for expansion must be provided, as experience has demonstrated that

most facilities will grow, both physically and in electrical demand. Overdesign, however, is as bad as underdesign, being wasteful of money and resources, both initially and in operation.

(b) Reliability. The reliability of electrical power within a facility is determined by two factors: the utility's service and the building's electrical system. The service record of the utility should be studied along with the economic impact of a power outage to determine whether and to what extent standby power equipment is justified (see Sections 16.39 and 17.21). Emergency equipment required for the safety of a building's occupants is determined by local, state, and national building codes.

Beyond the electrical service point, the reliability of power is entirely dependent on the wiring system. Here, too, economic studies must be made to determine the quality of equipment and the amount of redundancy (duplicate equipment) to be installed. The subject of reliability is a complex one, and we can state only a few general principles here.

1. The reliability of an electric system is only as good as that of its weakest element. Therefore, it may be necessary to provide redundancy at anticipated weak points in the system.
2. The electrical service and the building's distribution system act together. An extremely reliable (and expensive) service is of little use if the power cannot reach the desired points.
3. Critical loads within the facility should be pinpointed to determine how best to reliably serve them. This is done by establishing reliable power paths to them or by furnishing individual standby power packages for them. The latter course is often chosen for health care and other critical loads.

(c) Safety. Although rigid adherence to the requirements of the *NEC* and other applicable NFPA codes ensures an initially safe electrical installation, the designer must constantly be alert to such factors as electric hazards caused by misuse or abuse of equipment or by equipment failure. Also, attention to the physical size of equipment eliminates the oft-encountered hazard caused by obstruction of access spaces, passages, closets, and walls with electric equipment. Finally, lightning protection can be subsumed under the heading of safety; this topic is discussed in Chapter 13.

(d) Economic Factors. This item can readily be divided into two frequently interrelated items: first cost and operating cost. All other factors being equal, first cost depends in large measure on whether the constructor is interested in minimum first cost or minimum owning cost. We have demonstrated that these two costs frequently stand in inverse relationship to one another (exceptions are mentioned later). Low first-cost equipment generally results in higher energy cost, higher maintenance cost, and shorter life.

The decision, however, is not purely an economic one, inasmuch as the electrical energy cost factor in the operating-cost equation is directly related to energy consumption, with one exception. That exception is the utility's demand charge, which is discussed in Section 14.12. Means for minimizing this cost are covered in detail in Section 14.13. Here, there occurs a coincidence of reduction to both first cost and operating costs. Load-leveling equipment permits the electrical distribution system to be sized without consideration of coincident load peaks, thus resulting in smaller equipment operating more efficiently—near its full-load capacity. All other reductions in electric energy cost flow directly from the corresponding reduction in energy consumption.

(e) Energy Considerations. This factor is a complex one, involving considerations of energy codes and budgets, energy conservation techniques (see Section 17.5), and energy control.

Buildings constructed with governmental participation may be subject to energy budget limitations expressed in Btu/square foot/year. Although the lion's share of this budget is taken by heating/cooling and lighting systems, the electrical distribution system is also subject to conformity to stated codes. Most important among these codes is ASHRAE Standard 90, a current copy of which is *de rigeur* in all design offices. Detailed energy conservation considerations are discussed in detail in Section 17.5. A discussion of the IES Recommended Procedure for Lighting Power Limit Determination is found in Section 20.7.

(f) Space Allocations. The general impression that electrical equipment is small and easily concealed is accurate only for wire and conduit. Panels, motor control centers, busduct, distribution centers, switchboards, transformers, and so on can be large, bulky, noisy, and highly sensitive to tampering and vandalism. Thus, space allocations must be concerned with maintenance ease, ventilation, expandability, centrality (to limit length of runs), limitation of access to authorized personnel, and noise limitation in addition to the fundamental consideration of space adequacy.

(g) Special Considerations. These depend on the nature of the facility and may include items such as security, central and/or remote controls, interconnection with other facilities, and the like.

17.2 Load Estimating

When initiating the electrical design of a facility, it is necessary to estimate the total building load in order to plan such spaces as transformer rooms, conduit chases, and electrical closets. This information is also required by the local power company well in advance of the start of construction. An exact load total can be made only after completing the design but, because this is often several months later, a good preliminary estimate is required. Such an estimate can be made from the figures given in Table 17.1. These figures are average. When it appears that the building will have heavier or lighter loads because of lighting levels, energy codes, load management equipment, or other factors, the figures should be modified accordingly. The figures in the individual categories should be added without application of demand or diversity factors in order to obtain the maxi-

TABLE 17.1 **Electric Load Estimating**[a]

I	II	III	IV	V	VI
	Volt-Amperes per Square Foot[b]				Ten Year Percent Load Growth
		Misc.	Air Conditioning[f]		
Type of Occupancy	Lighting[c,d]	Power[e]	Electric	Nonelectric	
Auditorium					
General	1.0–2.0	0	12–20	5–8	20–40
Stage	20–40	0.5			
Art gallery	2.0–4.0	0.5	5–7	2.0–3.2	20–40
Bank	1.5–2.5		5–7	2.0–3.2	30–50
Cafeteria	1.0–1.6	0.5	6–10	2.5–4.5	20–40
Church and synagogue	1.0–3.0	0.5	5–7	2.0–3.2	10–30
Computer area	1.2–2.1	2.5[g]	12–20	5–8	50–200
Department store					
Basement	3–5	1.5	5–7	2.0–3.2	50–100
Main floor	2.0–3.5	1.5			
Upper floor	2.0–3.5	1.0			
Dwelling (not hotel) 0–3000 ft²	3.0	0.5	—	—	50–100
3000–120,000	0.4	0.15	—	—	
above 120,000	1.5–2.5	2.0	—	—	
Garage (commercial)			—	—	10–30
Hospital	1.0–3.5	1.5	5–7	2.0–3.2	40–80
Hotel	1.0–2.0	0.5			
Lobby	1.0–1.5	1.0	5–8	2.0–3.5	30–60
Rooms (no cooking)	2.0–3.0	5–20	3–5	1.5–2.5	
Industrial loft building	1.2–2.2	0.5	—	—	50–100
Laboratories	1.5–3.0	1.5	6–10	2.5–4.5	100–300
Library	1.0–2.0	0.5	5–7	2.2–3.2	30–40
Medical center	1.5–3.5	2.5	4–7	1.5–3.2	50–80
Motel	1.2–2.5	0.5	—	—	30–60
Office building	1.5–2.8	2.5	4–7	1.5–3.2	40–80
Restaurant			6–10	2.5–4.5	20–40
School	2.0–2.5	2.0	3.5–5.0	1.5–2.2	50–80
Shops	2.0–3.5	0.5			
Barber & beauty	2.0–3.0	0.5	5–9	2–4	40–80
Dress	2.0–3.0	0.5			
Drug	2.0–3.5	0.25			
Five & ten	0.25–1.0	—	4–7	1.5 to 3.2	
Hat, shoe, specialty	0.3	—			
Warehouse (storage)	0.25	—	—	—	10–30
In the above except single dwellings:					
Halls, closets, corridors,	0.5	—	—	—	
storage spaces	0.25	—	—	—	

[a]Figures assume energy-conservation techniques applied.

[b]These figures do not include allowance for future loads.

[c]The figures given in Article 220 of the *NEC* are minimum figures for calculation of electric feeder sizes.

[d]See Section 20.5 for a discussion of power limits.

[e]These figures are based on experience and must be applied judiciously.

[f]Includes the loads of air-handling equipment and pumps.

[g]This figure does not include the power used by the computer.

ELECTRICITY

mum load for which the building service equipment must be sized, in the absence of electric load-control (load-leveling) equipment. (Obviously, noncoincident loads such as heating and cooling in the same space are not combined.)

At this point an analysis can be made to determine the feasibility of incorporating electric load-control equipment into the facility (see Section 14.13). Input to this study includes the utility's complete rate schedule, including all penalty clauses, a detailed analysis of the building's equipment load patterns, and any external constraints such as maximum loads imposed by power and energy budgets. Equipment load patterns must be carefully analyzed because they determine a load's "sheddability." Thus, for kitchen equipment, load interruption may be undesirable, but shifting of cooking time by half an hour is usually feasible. For HVAC equipment, building thermal inertia and "stretching" maximum and minimum temperatures and humidities permit considerable latitude without adverse effects. Also, as explained in Section 14.13, the degree and duration of load shedding is a function of the type of control equipment utilized. It is well to repeat here what is stated there—load control affects maximum demand with only minor effect on total energy consumption. The external constraints referred to are the energy budgets recommended or required by codes, legislation, and funding bodies.

After the load control analysis is complete, or simultaneously with it, a building energy consumption analysis must be performed. This may be done manually, although numerous computer programs are available that not only increase accuracy, but also permit consideration of more factors. The results of this analysis indicate whether the target electrical energy budget is being met. If not, loads must be modified by reconsideration of projected systems and system criteria, by incorporating energy-conservation devices and techniques into the electrical system, and by drawing up energy-use guidelines that will be applied when the building is occupied. Because this last item depends for its success on the day-to-day voluntary actions of the building's occupants, experience has shown that it should not be considered as a major conservation source. Conservation measures are covered in Section 17.5.

The electrical loads in any facility can be categorized as follows:

1. Lighting.
2. Miscellaneous power, which includes data-processing equipment, convenience outlets, and small motors.
3. Heating, ventilating, and air conditioning (HVAC).
4. Plumbing and other piping systems.
5. Vertical transportation equipment and fixed material handling equipment.
6. Kitchen equipment.
7. Special equipment.

Category 1 is self-explanatory and is covered by column II of Table 17.1. Note, however, that lighting loads are carefully prescribed in ASHRAE/IES 90.1 and that these ASHRAE/IES prescribed loads should be used where this standard has jurisdiction. Contradictions between recommendations in standards having jurisdiction must be adjudicated by local authorities before design is established. Consult the latest current edition of ASHRAE/IES 90.1 for details of prescribed loads.

Category 2, column III includes, in addition to receptacles and small motors, such items as desktop computers and data-processing terminals including all peripheral equipment, plus plug-in heaters, water fountains, and so on.

Category 3, column IV includes all loads imposed by the HVAC equipment. Included therefore are fuel pumps, boiler motors, exhaust fans, and so forth. Also included in column IV, for air conditioning loads, are refrigeration compressors. This item is omitted in column V because the air conditioning utilizes absorption machines that do not use electricity for primary power. When air conditioning is not anticipated, the HVAC load is still appreciable because of heating and ventilating (H and V) requirements. A rough estimate for this load would be two-thirds of the loads in column V.

Category 4 includes all loads associated with the water and sanitary system, including house water pumps, air compressors and vacuum pumps, sump pumps and ejectors, well pumps and fire pumps, water heaters and pneumatic tubes, plus such special items as display fountain pumps. Also included in this category are electric loads connected with fixed piped systems such as cooking gas, medical gas piping, distilled water systems, and so on.

Because these loads vary widely with local conditions and with facility design as much as with type of facility, an estimate cannot be made on a volt-amperes per square foot basis, by type of building. For this reason, no figure is included in Table 17.1. If actual data cannot be used, it is helpful to remember that plumbing and piping electrical loads are relatively small, rarely exceeding 20% of the HVAC system electric load, although, for the most part, they are unrelated to it.

Category 5, vertical transportation, is related to square footage in some types of buildings, but not in others. A close estimate of power and energy requirements can be made by the job's elevator consultant. Lacking this, a fair estimate can be made from the data given in the vertical transportation chapters of this book. In addition to elevators, escalators, moving walkways, and ramps, these loads should include dumbwaiters, horizontal and vertical conveyors, trash and linen transport systems, automated container delivery, and fixed conveyors. These loads are readily available from the material transport consultant, who is frequently also the elevator consultant.

Category 6, kitchen equipment, is also not included in Table 17.1, although obviously present in all restaurants, most hospitals, and some office and religious-use buildings. The reason for this omission is that the primary power for the major load, the cooking equipment, may be either gas or electric. Other large energy-use equipment such as dishwashers can be electric, gas, or steam fed. Furthermore, no correlation can be made between the facility type, area, and load, even if electrically powered, because population and schedule are also major factors. When kitchens are planned, a kitchen consultant or other experienced planner can supply an estimate of the electric power requirement.

Category 7, special equipment, is so variegated that no figures can be listed. Under this title are subsumed such items as laboratory equipment, shop loads, display area loads, floodlighting, canopy heaters, display window lighting, industrial loads and so on. This load data must be gathered for individual items of equipment and added to the foregoing totals.

Table 17.2 gives a tabulation of service entrance size in amperes, based on single- and 3-phase service for typical occupancies. These figures are intended for quick estimate purposes and should be adjusted after the design is completed.

17.3 System Voltage

There are several systems of voltage commonly available in the United States and Canada. (See

TABLE 17.2 **Nominal Service Size in Amperes**
Nominal service sizes are 100 A, 150 A, 200 A, 400 A, 600 A.

Facility	1,000	2,000	5,000	10,000	Remarks
Single-phase, 120/240 V, 3-wire					
Residence	100 A	100 A	150 A	—	Minimum 100 A
Store[a]	100 A	150 A	—	—	
School	100 A	100 A	150 A	—	
Church[a]	100 A	150 A	—	—	
3-phase, 120/208 V, 4-wire					
Apartment House	—	—	100 A	150 A	
Hospital[a]	—	—	200 A	400 A	
Office[a]	—	100 A	400 A	600 A	
Store[a]	—	100 A	400 A	600 A	
School	—	100 A	100 A	200 A	

[a]Fully air conditioned using electric-driven compressors. Based on figures in Table 17.1.

Fig. 17.1 *120-V, single-phase, 2-wire service. This is also the arrangement of the usual branch circuit.*

NEC Art. 230-SERVICES, for Code requirements.)

(a) 120 V, Single-Phase, 2-Wire (Fig. 17.1). This is used for the smallest of facilities, such as out-buildings, and isolated small loads up to 6 kVA. Load is calculated by multiplying current and voltage. For 60-A service, which is the normal limit for this type of service, no more than 50 A are usually drawn. Thus

$$VA = 120 \times 50 = 6000 \text{ VA} = 6 \text{ kVA}$$

The nominal system voltage is 120 V, although it is also referred to as 110 V and 115 V.

(b) 120/240 V, Single-Phase, 3-Wire (Fig. 17.2). This system is for somewhat heavier loads. The code requires that all one-family residences with six or more 2-wire circuits or a net computed load (by code calculation) of 10 kVA or more have a minimum of 100 A, 3-wire service. Service disconnect for 100-A service would be a 100A, 2-pole, solid neutral switch, fused at no more than 80% of rating, or 80 A. This is usually written 100A, 2P & SN, 80AF. This service is used principally for residences, small stores, and other occupancies where the load does not exceed 80 A or 19.2 kVA. Load is calculated thus:

$$kVA = \frac{V \times I}{1000} = \frac{240 \times 80}{1000} = 19.2$$

Although it may appear otherwise, the neutral carries no more than full-load current. Note that each "hot leg" of the 3-wire system carries line current. Thus, total load can also be calculated,

load kVA = twice load on each line

Fig. 17.2 *120/240 V, single-phase, 3-wire service. The single-phase transformer, which is normally located on the power company's pole in the street or road, is center-tapped to establish a neutral. The neutral connection is always grounded.*

Assuming a balanced 80-A load,

$$\text{total kVA} = 2 \times 80\,\text{A} \times 120\,\text{V}$$

$$= 2 \times 9600\,\text{VA} = 19{,}200\,\text{VA}$$

$$= 19.2\,\text{kVA}$$

If the loads were unbalanced with, say, 30 A in one line and 50 A in the second, total load would be:

$$120 \times 30 + 120 \times 50 = 3600 + 6000 = 9600\,\text{VA}$$

$$= 9.6\,\text{kVA}$$

Actual system voltages can be 120/240, 115/230, or 110/220, although 120/240 is the accepted industry standard. 120 V loads are connected between line and neutral; heavier loads such as a clothes dryer or an electric stove are connected between phase lines at 240 V.

For example, to find the line currents caused by the three loads shown on Fig. 17.3:

1. 120-V, 1200-W iron, line A to neutral
2. 120-V, 1440-W hair dryer, line B to neutral
3. 240-V, 4800-W dryer, line A to line B

we calculate

$$I_1 = \frac{1200}{120} = 10\,\text{A}$$

$$I_2 = \frac{1440}{120} = 12\,\text{A}$$

$$I_3 = \frac{4800}{240} = 20\,\text{A}$$

Note that the neutral only carries the unbalance of 2 A (see Fig. 17.3).

Total current in line A = 20 + 10 = 30 A

Total current in line B = 20 + 12 = 32 A

Total current in neutral N = 2 A

Total load = 120(30) + 120(32) = 3600 + 3480

$$= 7440\,\text{VA, or}$$

Loads are 1200 + 1440 + 4800 = 7440 VA

Because the loads are almost entirely resistive heating loads (small hair blower motor and approximately a ⅙-hp dryer motor), the entire load has a power factor of almost 1.0. This, however, affects only energy calculations because equipment is sized, in general, by kVA capacity. The 120/240-V, single-phase system is derived from a center-tapped single-phase transformer.

(c) 120/208 V, Single-Phase, 3-Wire (Fig. 17.4). Although this system appears similar to the one described in subsection (b), it is really part of a 3-phase system. It is most often found *within* a building that takes 3-phase service rather than constituting a service voltages arrangement. It is used to serve a load that does not require 3-phase, 4-wire, but does require a voltage higher than 120 V to feed its load. Calculation of loads and line currents is considerably more complex than in subsection (b) because of the 120° phase displacement between phases A and B. Here, as before, the neutral carries no more than line current regardless of whether the system is balanced.

(d) 120/208 V 3-Phase, 4-Wire (Fig. 17.5). This system is a widely used 3-phase arrangement, applicable to all facilities except very large ones. In those, lengths of feeders and sizes of loads become so great that a higher system voltage is required. In this system, 120 V loads such as lighting, computers and accessories, recep-

Fig. 17.3 Note that the neutral carries the difference in current between the A and B legs and therefore a maximum that is equal to the current in one of the legs (when the other is zero).

Fig. 17.4 *120/208 V, single-phase, 3-wire service. This arrangement comprises two-thirds of the full 120/208 V, 3-phase, 4-wire connection shown in Fig. 17.5.*

tacles, and so on, are fed at 120 V by connection between each phase leg (Fig. 17.6) and neutral. Motors larger than ½ hp and all 3-phase loads are fed at 208 V by connection to the 3-phase legs. Single-phase, 208 V loads such as heaters are accommodated as described in subsection (c), by connection between two phase legs. Such loads are often referred to as "2-pole" loads, alluding to the 2-pole current breakers used to feed them.

(e) **277/480 V 3-Phase, 4-Wire (Figs. 17.7 and 17.8).** This system is applicable to large buildings (either horizontally or vertically) where lighting is principally fluorescent and/or HID and the 120 V load does not exceed one-third of the total load. It provides 277 V for fluorescent and HID lighting, 480 V for machinery. Small (3–25 kVA) dry-type transformers are used to step down from 480 V, for 120 V loads. This system is ideally suited to multistory office build-

Fig. 17.5 *120/208 V, 3-phase, 4-wire system. The neutral connection is connected to the system ground and is not broken by the service switch.*

Fig. 17.6 *The flexibility of the 120/208 V system is illustrated and accounts for its wide application. Loads are shown schematically. In practice, the loads are fed via protective devices in the panel. These are omitted here for clarity. A, B, C, and N represent the panel buses.*

ings and large single-level or multilevel industrial buildings. Cost savings are generated by the smaller feeder and conduit sizes and smaller switchgear, which more than offset the additional cost of step-down transformers for the 120 V load.

As an example of the economies possible with this system, let us consider the wiring required for a 15 kW heater. At 3-phase, 208 V:

$$I = \frac{15{,}000 \text{ W}}{\sqrt{3} \times 208 \text{ V}} = 42 \text{ A}$$

requiring No. 8 RHW wire (45-A capacity). At 480 V:

$$I = \frac{15{,}000 \text{ W}}{\sqrt{3} \times 480 \text{ V}} = 18 \text{ A}$$

requiring only No. 12 RHW wire (20-A capacity). This voltage system is also referred to as 265/460 V and 255/440 V.

Voltages above 150 V to ground are generally avoided in residential branch circuits, but may be used in commercial and industrial facilities within the guidelines established by the *NEC*.

(f) 2400/4160 V, 3-Phase, 4-Wire. This system is only used in very large commercial buildings or in industrial buildings with machinery requiring these voltages. The cost of running these voltages within a building is high because of *NEC* requirements and the inherent higher cost of 5-kV equipment. A detailed cost and engineering

Fig. 17.7 *277/480 V, 3-phase, 4-wire service system. The system is identical to the 120/208 V system shown in Fig. 17.5 except for voltages.*

Fig. 17.8 *Normal load arrangement for this system is illustrated. The lighting can be fluorescent or HID (mercury, metal–halide, sodium). Transformers, either single-phase or 3-phase, supply 120 V for receptacles and 208 V for loads requiring that voltage.*

TABLE 17.3 **System and Utilization Voltages**[a]

Standard Voltages[b]			
System Voltage (Transformers)		Utilization Voltage[c] (Motors)	
Nominal	With 4% Drop[c]	Current Standard[d]	Obsolete Standard
120	115.2	115	110
208[d]	199.7	200	208
240[d]	230.4	230	220
480	460.8	460	440
600	576.0	575	550

[a]To eliminate any confusion between system and utilization voltages, the current NEMA standards are tabulated above.

[b]When specifying transformers, use system voltages; for motors, use utilization voltage.

[c]Note that utilization voltage corresponds to a 4% drop from system voltage, well within the normal motor tolerance.

[d]Motors for 208-V systems are rated 200 V. Motors for 240-V systems are rated 230 V. They cannot be used interchangeably without seriously affecting motor performance.

analysis by a competent engineer is required for each case. Voltages above this level are widely used in large industrial plants and are beyond the purposes of this book.

Reference was made earlier to the varied voltages assigned to the same voltage system. Thus, at the lowest level, we have 110, 115, and 120 V; at the next level, 200, 208, 220, 230, and 240 V; at the next, 255, 265, and 277 V; and finally, 440, 465, and 480 V. These voltage differences arise because of the historical difference between transformer voltage standards, which establish the *system* voltage, and motor voltage standards, which govern *utilization* voltage level (Table 17.3). Present motor voltage standards are established at a level that is consonant with the system voltage. Note the close correspondence between motor standard voltage and system voltage with a (normal) 4% feeder voltage drop. Thus we see that on 240- and 480-V systems, 230- and 460-V motors are suitable.

Difficulties arise in application of 230- and 240-V motors to a 208-V system. Although motors will operate at plus or minus 10% voltage, 230- and 240-V motors should *not* be used on 208-V

TABLE 17.4 **Effects of Undervoltage on Utilization Equipment**[a]

Load	10% Undervoltage
Lighting	
Incandescent	Output reduced 30%
Fluorescent	Output reduced, poor start
Mercury	Low output, poor start
Motors	20% lower torque, hotter operation, reduced life, overloading
Heaters	20% reduction in output
Small tools	Stalling, low power

[a]Computers and peripherals are generally supplied with voltage-regulating power supplies which make them tolerant of ±10% variation in steady-state supply voltage. Conversely, they are highly intolerant of rapid supply voltage fluctuation.

systems. Instead, motors specially wound for 200 V should be specified. A brief summary of the effects of undervoltage is given in Table 17.4.

17.4 Grounding and Ground-Fault Protection

The vast majority of secondary wiring systems are solidly grounded. The reasons for this arrangement are several and varied. Among them are:

1. To prevent sustained contact between the low-voltage secondary system and the high-voltage primary system in the event of an insulation failure. Such contact could cause a breakdown of the secondary system insulation and severely endanger the system's users.
2. To prevent single grounds from going unnoticed until a second ground occurs, which would extensively disable the secondary system.
3. To permit locating ground faults with ease.
4. To protect against voltage surges.
5. To establish a neutral at zero potential for safety and for reference.

Points 2 and 4 are highly technical and a full explanation is beyond the purpose of this book.

Point 5 requires that the neutral in a single secondary system is:

1. Never interrupted by switches or other devices.
2. Connected to ground only at one point—the service entrance.
3. Color-coded white, natural grey, or by 3 continuous white stripes on any insulation color other than green, along the entire conductor length, (in the United States) for easy recognition.

A typical service-grounding diagram is given in Fig. 17.9. Universal acceptance and use of grounded secondary 120-V systems *introduces* another shock hazard while eliminating the dangers described previously. This is shown in Fig. 17.10(*a*). An accidental fault within an appliance could connect the metal case of the appliance to the line. This can occur with such common devices as an electric saw, a clothes washer or dryer, or a food mixer. A person contacting the appliance housing and simultaneously a ground, such as a water pipe, would receive a nasty 120-V shock. If the contact were made with wet hands, the shock could be fatal. Unfortunately, however, until such an incident occurred, the internal fault would remain an unnoticed but incipient source of danger.

To eliminate this hazard, appliance manufacturers have always recommended that appliance housings be grounded to a cold-water pipe. In addition, the appliances are supplied with 3-wire plugs: two wires connected to the appliance and the third wire to the housing. To accommodate such plugs and to provide a ground path, the *National Electric Code* requires all receptacles to be of the grounding type and all wiring systems to provide a ground path, separate and distinct from the neutral conductor (see Fig. 16.42). The result of such wiring is shown in Fig. 17.10(*b*), where the ground current passes harmlessly through the internal fault, along the ground-wire path and back to the panel. A person contacting the appliance housing establishes a parallel ground-path. However, because this path is usually of much higher resistance than the ground-conductor path, only a very small current will flow. Wet hands materially reduce contact resistance, and shock current can increase to a dangerous level. If the

Fig. 17.9 *Typical service-grounding arrangement. Note that the grounded neutral is unbroken through-out. The ground bus, if present, is separate and distinct from the neutral bus, and both are grounded at the service entrance point. For the sake of clarity, not all bonding connections, which are required to ground all non–current-carrying metal parts of the electrical system (boxes, enclosures, conduits, etc.) are shown.*

ground current is sufficiently high, the branch circuit breaker or fuse opens, disconnecting the circuit.

When wiring systems are installed in metallic conduit, the conduit itself or the conduit plus a separate conductor within the conduit may be used as the grounding path. This latter method with the additional ground wire is very much preferable, as explained in Fig. 17.10. When nonmetallic or flexible metallic wiring is used (Romex or BX), a separate grounding conductor run with the regular circuit conductors *must* be used.

All insulated ground conductors must have their covering colored green for identification as a grounding conductor. Many industrial installations install complete "green-ground" systems in an attempt to eliminate shock hazard and reduce insulation failures. This has not been entirely successful for the reason alluded to previously; that

is, in order to clear the ground fault, its current must be high enough to trip the branch circuit protective device. Otherwise, the ground fault continues to "leak," unnoticed by the system's protective devices. Unfortunately, ground faults are by nature low-current, leak-type faults, because they result from weak spots in insulation, dirt accumulation, and so on. Therefore, although the shock hazard is greatly reduced by the green-ground path, the fault continues to leak and arc until it becomes hot enough to cause a major breakdown, frequently accompanied by fire.

To eliminate this potentially dangerous situation, which occurs anytime there is a leak of current to ground in an electric circuit, the ground-fault circuit interrupter (GFCI or GFI) was developed [Fig. 17.10(*c*)]. This device compares, with extreme precision, the current flowing in the hot and neutral legs of a circuit; if there is

ELECTRICITY

Fig. 17.10 *Three types of circuit arrangements.*
(a) *Shows a 2-wire, grounded-neutral circuit, with no means of preventing shocks from ground faults.*
(b) *Illustrates a similar arrangement but includes a green ground wire, which considerably lessens the danger of ground-fault shocks. Note, however, that as long as the ground-fault current is below the rating of the branch circuit protective device (i.e., below 15 to 20 A), it will continue to flow within the appliance, causing overheating, arcing, and the eventual destruction of the appliance.*
(c) *Illustrates the use of ground-fault circuit interrupter devices, now mandatory for bathrooms, outdoor locations, swimming pool circuits, and other areas where ground faults are common and particularly hazardous.*

a difference, it indicates a ground fault and the device trips out. The rapidity of this action—approximately one-half second—eliminates the possibility of a potentially dangerous shock hazard, which exists even in a properly grounded circuit, as in Fig. 17.10(*b*), and all the more so for the circuit arrangement of Fig. 17.10(*a*). The separate ground path required by the *NEC* (bonded metallic conduit system, green-ground,

or both) serves to minimize the shock current taken by a person before the GFCI operates. The device can be applied at the panel to replace a normal circuit breaker (see Fig. 16.8b) or at an individual outlet to replace a normal receptacle device. It is advisable to use GFI devices on all appliance circuits (see the *NEC* for locations where GFI use is mandatory, such as outdoors and in kitchens and bathrooms). Application to lighting circuits is generally not required because fixtures are usually out of reach and are switch controlled. In mixed lighting and receptacle circuits, the GFCI is best applied at the outlet that is to be protected.

17.5   Energy Conservation Considerations

Before proceeding with a detailed description of design procedure, in this section many of the energy conservation ideas and techniques applicable to electrical distribution systems are surveyed. This is done for ease of reference and cross-reference because the individual items appear throughout the lengthy design procedure. ASHRAE Standard 90.1, latest issue, which is mandated legally in many states, calls for a design of energy distribution systems "to conserve energy." The following design procedures are intended to fulfill that broad directive.

1. After establishing an energy budget based on projected loads and normal operation, set an energy reduction figure of 10% to 20% and accomplish it with the techniques that follow. Annual energy consumption estimates are best made with the aid of one of the many computer programs available for that purpose.

2. Learn to recognize the energy-use characteristics of all materials and systems specified (see Table 15.6). In general, select high-efficiency equipment (motors, transformers, etc.). If not available as such, use materials and equipment with the lowest temperature rise because these have lowest losses. Economic justification can be established with life-cycle cost analysis. To avoid making a detailed cost analysis on every item, utilize one of the many available shortcut calculations for payback time (see Appendix E). This generally indicates the material of choice. When comparisons are close, a detailed analysis may be necessary.

3. Provide electric load control equipment (demand control) either as part of an overall building control system or separately (see Sections 14.13 and 21.2).

4. In any multitenant residential building, provide individual user metering. (Metering of heating and cooling energy should be part of the HVAC contract.) All tenants should be made financially responsible for the energy they use. Exceptions to this rule would be made in the case of hotels, dormitories, and other transient facilities.

5. Where a choice of service voltages is available, consider the highest available. Similarly, consider the highest voltage in each class for interior distribution systems. This means 480 V in the 600-V class, 4 kV in the 5-kV class, and 13 kV in the 15-kV class. The result will be low-line losses, small panelboards at the branch circuit level, and generally lower electrical contract cost (see Section 17.3).

6. In any building, the maximum total voltage drop *shall not* exceed 3% in branch circuits or feeders, for a total of 5% to the farthest outlet, based on steady-state design load conditions.

7. Avoid the use of electric heating elements if alternatives are available. Electric heat is an inefficient use of natural resources because of the low overall efficiency of fuel-to-electricity conversion.

8. Provide metering points (for fixed or plug-in meters) throughout the system to permit accurate analysis of power and energy use. Meters, both instantaneous reading and recording types, provide essential data on equipment loading, load patterns, load coincidence, power-factor, load voltage, power demand, and energy consumption. Analysis of these data indicates how to program and shift loads for maximum operational efficiency. A flexible design that permits this load shifting is assumed (see item 10).

9. Include provisions for power-factor correction in the system, both at devices and at the feeder level. Then, if metering (see item 8) indicates the necessity, add capacitors as required. High power factor reduces line losses, permits maximum utilization of equipment capacity, and avoids utility penalty charges. Utilization equipment rated greater than 1000 W and lighting equipment greater than 15 W, with an inductive reactance load component, should have a power factor of not less than 85% under rated load conditions. Utilization equipment with a power factor of less than 85% should be corrected to at least 90% under rated load conditions.

10. Size equipment as close as possible to the load. This normally results in maximum efficiency and high power factor. Where load varies considerably, for example, between day and night or weekend, consider splitting the loads so that part of the equipment can be switched off when load is low, and the remaining load can be fed from a "night and weekend" feeder. Such a design permits one to shut down whole systems rather than operate at very low load with concomitant high losses and low power factor. The design must be sufficiently flexible to permit shifting of loads between feeders if measurements on the operation facility indicate that this is desirable (see item 8). The aim is to operate equipment as near rated load as possible and deenergize lightly loaded sections by shifting load to other partly loaded equipment.

11. Use the most efficient type of control. This means solid-state control for motors. Variable voltage variable frequency (VVVF) control for variable speed equipment, remote switch control for blocks of lighting, electronic control systems for elevators, and so on.

12. Arrange automatic time controls for 24-h loads, such as vent fans, water cooler, vending machines, and calculators.

13. Seal all electric riser shafts to avoid heat loss by stack action.

14. Generally select the coolest possible locations for electric equipment. Low ambient temperature (below 40°C) permits use of smaller equipment for the same load with concomitant lower cost and losses. If below-grade space is available, it is well suited for this purpose.

15. Provision for future expansion should be made by means of a design that accommodates additional equipment in lieu of oversizing equipment initially. Here again, the higher cost of two pieces of equipment can normally be justified by a detailed owning-and-operating cost analysis using realistic cost escalation and capital cost figures.

16. Energy-conservation techniques in lighting and lighting control are found in Sections 20.7 and 21.2 to 21.4.

17. Energy conservation techniques in vertical transportation are found in Sections 23.39, 23.40, 23.41, and 25.12.

17.6 Design Procedure

The steps involved in the electrical wiring design of any facility are outlined later in this section. These may in some instances be performed in different order, or two or more steps may be combined, but the procedure normally used is that listed as follows.

1. Make an electrical load estimate based on areas involved, building data, and any other pertinent data (see Section 17.2).

2. In cooperation with the local electric utility, decide on the point of service entrance, type of service run, service voltage, metering location, and building utilization voltage (see Sections 16.1 to 16.11 and Section 17.5, item 5).

3. Determine with the client the proposed usage of all areas and type and rating of all client-furnished equipment, including their specific electric ratings and service connection requirements.

4. Determine from other consultants such as HVAC, plumbing, elevators, kitchen, and the like, the exact electrical rating of all the equipment in their designs. This determination is often made after conferences during which the electrical consultant makes valuable recommendations to these other specialists about the comparative characteristics and cost of equipment (see Section 17.5, items 2, 7, 9, and 10).

5. Determine the location and estimate the size of all required electric equipment spaces including switchboard rooms, emergency equipment spaces, electric closets, and so forth. Panelboards are normally located in closets but may be located in corridor walls or elsewhere. These decisions are necessary at this point to enable the architect to reserve the spaces for electrical equipment. Once the design is accomplished in detail, the estimated space requirements can be checked and necessary adjustments made.

6. Design the lighting for the facility. This step, as discussed in detail in Chapters 18 through

21, is complex and involves a continued interaction between the architect and the lighting designer.

7. Depending on the type of facility, it may be necessary to separate the lighting layout from the layout plan(s) for other devices such as signalling, low tension systems, and receptacles. When an underfloor wiring system is used, it is customary to show it on a separate plan to avoid clutter and confusion. This decision is made at the preliminary stage by the project engineer. Once the decision has been made as to how this is to be handled, the lighting fixture layout can be made.

8. On the plan(s) locate all electrical apparatus including receptacles, switches, motors, and other power-consuming apparatus. Underfloor, under-carpet, and "over-ceiling" wiring and raceway systems would be shown at this stage, in general on a separate plan.

9. On the plans, locate data-processing and signal apparatus such as telecommunication outlets, network connections, phone outlets, speakers, microphones, TV outlets, fire and smoke detectors, and so on. At this stage, provision is also made for load control wiring, building automatic control wiring, computer control, and the like. At this stage the decision is also made as to whether "intelligent panels" will be used (see Section 16.23). Because some of these systems may be covered by separate contracts, the division of responsibility must be clearly defined. Often, only an empty conduit system and power outlets or sources are required in the electrical work contract.

The material that follows deals with wiring design only; all further discussion of signal equipment is reserved for Chapter 22.

10. Circuit all lighting, devices, and power equipment to the appropriate panels, and prepare the panel schedules. Include in this step the circuitry for emergency equipment.

11. Compute panel loads.

12. Prepare the riser diagram. This includes design of distribution panels, switchboards, and service equipment.

13. Compute feeder sizes and all protective equipment ratings.

14. Check the preceding work.

15. Coordinate the electrical work with the other trades and with the architectural plans. This is not really a separate step, but a continuing process starting at point 9 and covering all subsequent stages of the work.

The material for items 1, 2, 3, 4, 6, 7, and 9 is covered elsewhere in this book. The remaining steps; that is, 5, 8, 10, 12, 13, and 15, are discussed in order below.

17.7   Electric Spaces

The spaces required for electric equipment in a facility vary greatly, depending on the design and nature of the building. The working spaces required around major pieces of electrical switchgear and transformers were discussed previously (see Fig. 16.28). The *NEC* in Article 110 further specifies the minimum working spaces required in front of electrical equipment.

(a) Residences. In private residences, the service equipment and the building panelboard are generally incorporated into a single unit. The main disconnect(s) is usually installed as the main switch/breaker of the panel. A number of typical residential service-panel arrangements are shown in Fig. 17.11. The panel is normally placed in the garage, utility room, or basement. To minimize voltage drop, the panel should be placed as close to the major electrical loads as practicable, without sacrificing valuable space or making the panel inaccessible. Frequently a smaller panel can be subfed from the main panel to feed the kitchen and laundry loads. In apartments, panels are normally placed in the kitchen or the corridor immediately adjoining the kitchen. This location is chosen so that the panel circuit breaker may act as the required disconnecting means for most fixed appliances (see *NEC* Article 422C).

(b) Commercial Spaces. The location of the required panelboards depends on their type and number and on availability of space. In the

Fig. 17.11 Typical 150-A service arrangements, applicable to residences and small commercial buildings. The service switch can be a separate unit (a) or combined with the branch circuit panelboard (b–d). The panel may be a single unit (a,b), two units in a single enclosure, fed by separate service switches (c), or a central panel and a subfed single use panel (d)—in this case, a kitchen subpanel.

research building of which Fig. 17.12 is a part plan, lighting panels are recessed into the corridor wall because the building is only two stories high and the panels can be vertically stacked and fed by a single conduit. If this building were six or more stories high, an electric closet (see Section 17.8) would be advisable to accommodate the panel and riser conduits. Of course, when panels are installed in finished areas such as corridors, flush mounting is required.

To limit the voltage drop on a branch circuit in accordance with the code requirements, panelboards should be located so that no circuit exceeds 100 ft in length. If 15-A or 20-A branch circuits longer than this are unavoidable, No. 10 AWG wire should be used for runs of 100 to 150 ft and No. 8 AWG for longer circuits. These circuits are normally wired with No. 12 AWG wire.

The laboratory between the two offices of Fig. 17.12 is intended to function as a self-contained unit and is therefore equipped with its own panel.

Multioutlet assemblies, all wiring within the room, and the panel itself are surface-mounted to allow ready access to all components for the frequent rewiring encountered in laboratories. A main circuit breaker should be provided in such a panel to act as a main disconnect, whether required by Code or not. Where panels are convenient to the load controlled, the panel circuit breakers may be used for switching.

Panels supplying large blocks of load simultaneously switched, such as auditorium house lights, lobby lights, large single-use office areas, store lighting, and the like, can be constructed with built-in electrically operated, mechanically held contactors to switch the entire load, with control at any desired remote location. These remote-control (RC) switches are discussed in Section 16.15. If only part of the panel's circuits is so arranged (i.e., for remotely controlled block switching), a split bus panel can be used, with the RC switch controlling only that part of the panel's load (Fig. 17.13).

Fig. 17.12 *Typical floor plans for lighting and power for a section of an office–laboratory building. Separate lighting and power plans are drawn for the sake of clarity. For circuited plan, see Fig. 17.23. Data outlets handle data processing and telecommunications cabling.*

Small offices, stores, and other small buildings frequently have lighting panels mounted in a convenient finished area and utilize the panels' circuit breakers for load switching. In large buildings, strategically located electric closets are provided to house all electrical supply equipment. Power panels and distribution panels are located as required by the loads fed through them.

In general, branch circuit panels, distribution panels, and switchboards are best located near the electrical load center. This minimizes feeder length and reduces voltage drop, making it the most economical arrangement.

Every completely enclosed switchgear room, emergency generator room, or transformer vault should be equipped with an emergency light source (see Section 21.36). In generator rooms these should be battery operated to give illumination for generator repairs in the event of generator failure during a power outage.

(b)

Fig. 17.13 (a) *Single-line diagram of a split bus panel. This design is used when a block of load is to be controlled as a unit. Control point is either remote, at the panel, or both. Contactor (RC switch) is electrically operated, mechanically latched.*
(b) *RC switches (mechanically held contactors) are shown to the left of a bank of panels. The contactors were installed to accomplish photocell and timer control of a department store's lighting. (Photo courtesy of Automatic Switch Co.)*

17.8 Electric Closets

In the design of building electric system, particularly in multistory construction, it is often advantageous and convenient to group the electrical equipment in a small room called an *electric closet* (Fig. 17.14). The shape of this space can be varied

to fit the architectural requirements, and it should provide the following:

1. One or more locking doors.
2. Vertical stacking, above and below other electric closets and located so as not to block conduits entering or leaving horizontally. Thus locations on outside walls and adjoining shafts, columns, and stairs are poor.
3. Space free of other utilities such as piping or duct passing through the closet, either horizontally or vertically.
4. Sufficient wall space to mount all requisite and future panels, switches, transformers, telephone cabinets, and communication equipment. Wall cabinet space must be coordinated with raceway connections to underfloor ducts and over-the-ceiling raceways systems (see Sections 15.25 to 15.31).
5. Floor slots or sleeves of sufficient size for all present and future conduit or bus risers.
6. Sufficient floor space so that an electrician can work comfortably and safely on initial installation and on repair.
7. Adequate illumination and ventilation.

17.9 Equipment Layout

Wiring devices, principally comprising receptacles and switches, are located as required by the equipment to be served and by the anticipated area use.

Switches for control of lighting or receptacles are normally placed on the strike side of the door. Other devices such as plug-in strips on walls and special-purpose receptacles are shown and identified. Signal outlet locations are often noted but generally remain uncircuited on floor plans, a riser or floor raceway plan being utilized to show interconnections. These include fire-alarm equipment, telephone and intercom equipment, data and communications, radio and TV outlets, and so on. These devices may be identified by a special symbol or note where a standard symbol is not available.

As mentioned previously, lighting fixture outlets are normally placed on the same drawing as wiring devices unless the large number of the latter precludes showing the lighting, without undue cluttering of the drawings. In such an event, the lighting is shown on one drawing and receptacles

ELECTRICITY

Fig. 17.14 *Typical electric closets with some usual equipment. If warranted by amount of equipment, separate closets may be used for signal and telephone conduits and cabinets. Smaller spaces may be used, depending on types of equipment cabinets and their arrangement (see NEC Article 110). In all cases, sufficient space must be provided to allow full opening of all cabinet doors. Empty wall space for future cabinets and equipment should also be considered.*

on another, with signals shown on the one least occupied. A ceiling or underfloor wiring system would necessitate such separation (see Figs. 15.43, 15.44, and 15.45). Motors, heaters, and other fixed and permanently wired equipment are shown and identified on the receptacle drawings (also called *power drawings,* in contradistinction to lighting drawings). Equipment furnished with a cord and plug is not normally shown. However, the receptacle intended for supplying such a device is shown and identified. A typical device layout is shown in Fig. 17.12. An abbreviated symbol list is given in Fig. 17.15. For a complete electrical symbol list, see Stein (1997).

17.10 Application of Overcurrent Equipment

Before beginning an explanation of circuiting, it is necessary to explain the principles underlying

overcurrent protection. As outlined in Chapter 16, the function of an overcurrent device is to open (interrupt) a circuit when the current rating of the equipment being protected is exceeded. The following general rules govern the application of overcurrent protection:

1. Overcurrent devices must be placed on the line or supply side of the equipment being protected (Fig. 17.16).
2. Overcurrent devices must be placed in all *ungrounded* conductors of the protected circuit (Fig. 16.33).
3. All equipment should be protected in accordance with its current-carrying capacity (Fig. 17.17).
4. In general, conductor sizes shall not be reduced in a circuit or tap, unless the smallest-size wire is protected by the circuit overcurrent devices (Fig. 17.18).
5. Overcurrent devices shall be located so as to be readily accessible.

ELECTRICAL SYMBOL LIST

RACEWAYS

CONDUIT AND WIRING CONCEALED IN CEILING OR WALLS TICS INDICATE NO. OF CONDUCTORS EXCLUDING GROUNDS; 2 #12, ¾" CONDUIT UON.

CONDUIT AND WIRING CONCEALED IN OR UNDER FLOOR

CONDUIT AND WIRING EXPOSED

FEEDER F-6, SEE SCHEDULE, DWG NO.

BX WIRING

NONMETALLIC CABLE (ROMEX) WIRING

HOME RUN TO PANEL 2PLA—NUMERALS INDICATE CIRCUITS, 3 #12 AWG, ¾" RIGID STEEL CONDUIT

HOME RUN TO TELEPHONE CABINET TC2A

EMPTY CONDUIT, SUBSCRIPT INDICATES INTENDED USE T—TELEPHONE, IC—INTERCOM, FA—FIRE ALARM, ETC.

SURFACE METAL RACEWAY, SEE NOTE _____, DWG _____ SIZE AND RECEPTACLES AS SHOWN

OUTLETS

CEILING   WALL

OUTLET AND LIGHTING FIXTURE: LETTER IN CIRCLE INDICATES TYPE—SEE SCHEDULE. SUPERSCRIPT NO. INDICATES CIRCUIT. SUBSCRIPT LETTER INDICATES SWITCH CONTROL. RECTANGLE INDICATES FLUORESCENT

OUTLET AND EXIT SIGN FIXTURE, UPPER CASE LETTER INDICATES TYPE, ARROWS INDICATE REQUIRED SIGN ARROWS

OUTLET BOX, BLANK COVER—NOTE 2

JUNCTION BOX, BLANK COVER

WIRING DEVICES

DUPLEX CONVENIENCE RECEPTACLE OUTLET 15 AMP 2 P 3 W 125 V, GROUNDING, WALL MTD VERTICAL, ℄ 12" AFF.

RECEPTACLE, RATED AS SHOWN. EQUIPPED WITH GROUND FAULT CIRCUIT INTERRUPTER

SPECIAL RECEPTACLE, LETTER DESIGNATES TYPE, SEE SCHED. DWG. NO. ___ WALL MOUNTED

FLOOR OUTLET TYPE B, SEE DWG. NO. _____

MULTI-OUTLET ASSEMBLY, SEE DWG. NO. ___ FOR SCHEDULE AND DETAILS (SEE SPEC.)

or

SINGLE POLE SWITCH, 15A 125 V, 50" AFF UON. SUBSCRIPT LETTER INDICATES OUTLETS CONTROLLED

SWITCH, 3 WAY, 15A 125 V, SEE SPEC.; CONTROLLING OUTLETS 'a'

SWITCH, KEY OPERATED, 15A 125 V

SWITCH, SPECIAL PURPOSE, TYPE A, SEE SPEC.; SEE DWG. NO. _____

ABBREVIATIONS RELEVANT TO SWITCHES:

SP   —SINGLE POLE
DP   —DOUBLE POLE
SPDT —SINGLE POLE DOUBLE THROW
DPDT —DOUBLE POLE DOUBLE THROW
RC   —REMOTE CONTROL

OUTLET-BOX-MOUNTED DIMMER

Fig. 17.15 *Abbreviated electrical symbol list. For complete list, see Stein (1997).*

ELECTRICITY

A,a	AMPERES	GFCI	GROUND FAULT CCT-INTERRUPTER	OH	OVERHEAD
AFF	ABOVE FINISHED FLOOR	HP	HORSEPOWER	OC	ON CENTER
C/B	CIRCUIT BREAKER	LTG	LIGHTING	SW	SWITCH
CCT	CIRCUIT	MH	MOUNTING HEIGHT, MANHOLE	T	THERMOSTAT
EM	EMERGENCY			TEL	TELEPHONE
EC	EMPTY CONDUIT	N	NEUTRAL	TV	TELEVISION
F	FUSE	NC	NORMALLY CLOSED	UON	UNLESS OTHERWISE NOTED
FA	FIRE ALARM	NO	NORMALLY OPEN	UG	UNDERGROUND
GND	GROUND	NIC	NOT IN CONTRACT	WP	WEATHERPROOF

EQUIPMENT

SW., 3 POLE AND SN., 60A/30 AF. NEMA 3R ENCL., ND (NORMAL DUTY) 250 V

CONTACTOR, ENCLOSED, ELECTRICALLY OPERATED, MECHANICALLY HELD, 3 POLE 60 A

ITEM OF ELECTRICAL EQUIPMENT, AS INDICATED

ELECTRIC PANELBOARD LP-1, RECESSED, SEE SCHEDULE ON DWG. _____

ELECTRIC PANEL (BOARD) P-2, SURFACE MTD

CABINET, MOUNTING, SIZE, AND PURPOSE AS SHOWN

SIGNALING DEVICES

BELL OR GONG, INSCRIBED LETTER INDICATES SYSTEM (SEE BELOW) AND SUBSCRIPT LETTER OR NUMBER INDICATES TYPE e.g. A—8" VIBRATING BELL, 12 V. DC; B—12" WEATHERPROOF SINGLE STROKE GONG, 120 V. DC, etc.

FIRE DETECTOR, TYPE 1

INTRUSION DETECTOR, TYPE 1   } ETC.

SMOKE DETECTOR, TYPE 2

MANUAL STATION—WATCHMEN TOUR, FIRE ALARM, ETC. LETTER INDICATES SYSTEM (SEE BELOW)

COMMUNICATIONS OUTLET, LETTER INDICATES TYPE CABINET OR CONTROL PANEL, SMOKE DETECTION, USE IDENTIFYING TYPE LETTER IF MORE THAN ONE TYPE IS USED ON THE PROJECT

AUXILIARY DEVICE, AS INDICATED

HORN OR LOUDSPEAKER, TYPE A, TYPE 2

TELEPHONE OUTLET, TYPE B

COMMUNICATIONS OUTLET, LETTER INDICATES TYPE

CLOCK SYSTEM OUTLET, TYPE A

TV ANTENNA OUTLET

OUTLET, LETTER INDICATES TYPE

Fig. 17.15 (continued)

ELECTRICITY

SYSTEM TYPES		AUXILIARY DEVICES	
SYSTEM TYPES		**AUXILIARY DEVICES**	
F, FA	FIRE ALARM	BATT	BATTERY
S, SD	SMOKE DETECTION	S, SP	SPEAKER, LOUDSPEAKER
I, IA	INTRUSION ALARM	TC	TELEPHONE CABINET
T, TEL	TELEPHONE	WF	WATERFLOW
TV	TELEVISION		
IC	INTERCOM		
D	DATA, TELECOMMUNICATION		

MOTORS AND MOTOR CONTROL

COMBINATION TYPE MOTOR CONTROLLER; ATL STARTER PLUS FUSED DISCONNECT SWITCH, NEMA SIZE II, SEE SCHEDULE DWG. ___ ___ ___

MOTOR #1, 5 HP, 3ϕ SQUIRREL CAGE UON

DEVICE 'T' (SEE LIST OF ABBREVIATIONS)

S_T MANUAL MOTOR CONTROLLER WITH THERMAL ELEMENT.

ATL   ACROSS THE LINE STARTER—MAGNETIC

FS   FUSED SWITCH

CB   CIRCUIT BREAKER

FV   FULL VOLTAGE

MCC   MOTOR CONTROL CENTER

S   START BUTTON—MOMENTARY CONTACT

ST   STOP BUTTON—MOMENTARY CONTACT

CONTROL DIAGRAMS AND WIRING DIAGRAMS

MOMENTARY CONTACT PUSH BUTTON—NO—(START)

MOMENTARY CONTACT PUSH BUTTON—NC—(STOP)

NORMALLY OPEN CONTACT—NO

NORMALLY CLOSED CONTACT—NC

OPERATING COIL FOR RELAY OR OTHER MAGNETIC CONTROL DEVICE. WITH ONE NO AND ONE NC CONTACT. LETTERS NORMALLY USED ARE C, R FOR CONTROL COIL AND RELAY

POWER WIRING

CONTROL WIRING

ONE LINE DIAGRAMS

CONDUCTOR, SIZE, AND TYPE INDICATED

CONDUCTORS CROSSING, CONNECTED

GROUND CONNECTION

GROUNDED WYE

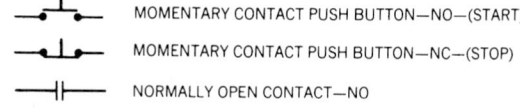

TRANSFORMER, 2 WINDING, 1ϕ INDICATES SINGLE PHASE; Δ-Y INDICATES 3 PHASE AND TYPE OF CONNECTION; NUMBERS INDICATE VOLTAGES.

FUSE, 30 A TYPE - - -

CIRCUIT BREAKER, MOLDED CASE, 3 POLE, 225 A FRAME/125 A TRIP

FUSED DISCONNECT SWITCH, 3 POLE, 100 A, 60 A FUSE

MOTOR, 3 PHASE SQUIRREL CAGE UON, 20 HP, MOTOR #3

Fig. 17.15 (continued)

Fig. 17.17 *The overcurrent protection must correspond to the rating of the protected equipment. Where ratings do not correspond exactly, the next larger standard size for protective devices up to 800 A may be used. Above this, use the next lower rating. See* NEC *Article 240-3 for further restrictions.*

Fig. 17.16 *Location of overcurrent protective equipment. Protective equipment should always be located at the point where the conductor receives its source of supply so that when it operates, the current supply is cut off.*

17.11 Branch Circuit Design

A branch circuit by *NEC* definition refers only to the circuit conductors, although for our purposes and in trade parlance it includes the protective device and the outlets served. Such circuits may be multioutlet general-purpose type [Fig. 17.19(*a*)], multioutlet appliance type [Fig. 17.19(*b*)], or single-outlet type intended for a specific piece of equipment [Fig. 17.19(*c*)]. The multioutlet types are limited to 50 A in capacity, whereas the single-outlet type is governed in size only by the requirements of the item being served, and may be 200 A or 300 A in size.

In its simplest form, a branch circuit comprises only two circuit wires. However, multiwire branch circuits carrying 2- to 3-phase wires plus a neutral are also widely used. Generally, each branch circuit should be sized for the load connected to it

plus the load expansion that is expected. These general rules of good practice should be followed:

1. In all but the smallest installations, connect lighting, convenience receptacles, and appliances on separate (groups of) circuits, although this is not an *NEC* requirement.

2. General-purpose branch circuits should be 20 A and wired with No. 12 AWG wire. Switch legs may be No. 14 AWG if the lighting load permits and the wiring meets the tap requirements of *NEC* Article 240-21.

3. Limit the circuit load on 15-A and 20-A circuits to the values shown in Table 17.5. This provides the required building load expansion capability in the branch circuitry; that is, by bringing the loads on the branch circuits up to maximum, the additional building loads can be absorbed. However, because it is not always economical or feasible to expand existing circuits, panels are always equipped with spare breakers. These can also be utilized to pick up building load expansion, or the user may use a combination of the two techniques, expanding some existing circuits and adding some new ones. (See Section 17.15 for a discussion of this subject.)

Because lighting and specific devices are circuited according to their nameplate rating, the only circuitry item left to the judgment of the designer is the number of convenience receptacles per circuit. The *NEC* specifies that plug outlets (convenience receptacles) be counted, in totaling loads, at 1.5 A each unless included in the load for general lighting. This point requires some clar-

Main supply

) 100 amp

#2 AWG, THW (115 amp capacity)

Main feeder, #1 AWG, THW (130 amp capacity)

25' max.

$\frac{1}{3}$ capacity $= \frac{100}{3} = 33$ amp

Use #8 AWG, THW, 50 A. capacity minimum

40 amp fuse

10' max.

No. 10 THW in conduit

2–15 amp loads or 30 A. device

Size to protect feeder

Taps may be made if smaller conductor is protected by main feeder protection (100 A C/B protects the No. 2).

Taps up to 25' long are permitted if tap conductor has min. 1/3 ampacity of main feeder O/C device and terminates in a single C/B or set of fuses, which limits the current on the tap to the wire capacity. (40 A fuse limits current on 50 A capacity wire).

Taps up to 10' long are permitted provided they have sufficient capacity for the loads or device they feed; eg, if the wiring feeds 2–15 amp loads, it must have 30 amp capacity and should be No. 10 AWG, THW or equal.

Taps of any length may be made if the conductor is protected at the tap point by an appropriately sized overcurrent device.

Fig. 17.18 *Some permissible tap arrangements. See* NEC *Article 240-21 for additional restrictions on taps to those given in the figure.*

Generally 15, 20 A

LTG   A

(a) General purpose branch circuit. Supplies outlets for lighting and appliances, including convenience receptacles.

15, 20, 30, or 50A

A   A   A

(b) Appliance branch circuit. Supplies outlets intended for feeding appliances. Fixed lighting not supplied.

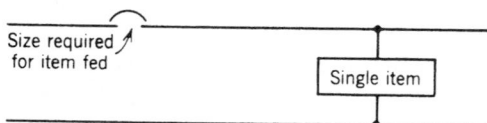

Size required for item fed

Single item

(c) Individual branch circuit, designed to supply a single, specific item.

Fig. 17.19 *Branch circuit types. See* NEC *Article 210 for detailed descriptions.*

TABLE 17.5 **Recommended Branch Circuit Loads**[a]

	0% Expansion[b,c]		
Size Circuit	Circuit[d] Amperes	Volt-Amperes[d] at 120 V	Volt-Amperes at 277 V
15 A	12	1440	—
20 A	16	1920	4440
	25% Expansion[e]		
15 A	9.6	1150	—
20 A	12.8	1520	3600
	50% Expansion[f]		
15 A	8	960	—
20 A	11	1300	3000

[a]The loading shown will provide the specified expansion in the branch circuits. Where branch circuit expansion is not practical, the required expansion can be obtained from panel spares.

[b]See Table 17.1 for anticipated load growth.

[c]For 0% expansion, initial load = 80% of circuit rating, which is the maximum permissible for continuous loads. See Table 17.6, note 6.

[d]For branch circuits feeding utilization equipment such as air-conditioners in addition to convenience outlets, see *NEC* Article 210-23.

[e]To accomplish 25% expansion, the circuit is derated to 80%, i.e., 80% of 12 A = 9.6 A, etc.

[f]To accomplish 50% expansion, the circuit is derated by ⅓, i.e., ⅔ of 12 A = 8 A, etc. (Then 50% expansion on 8 A yields 12 A.)

ification. The receptacle load for general illumination in dwellings is included in the 3 v-a/ft² specified by *NEC* Table 220-3(a) and found above in Table 17.1. This is because receptacle outlets on general illumination circuits are assumed to supply illumination (i.e., lamps), and are therefore necessarily included in the lighting load. Application of the 1.5 A (180 v-a) per outlet criterion to such outlets would, in the opinion of the *NEC*, unnecessarily limit the number of outlets permitted on such circuits.

The same is true for appliance circuits. On both these types, according to the *NEC*, the number of receptacle outlets is not limited. On 15- and 20-A receptacle circuits, which are neither general illumination nor small appliance branch circuits, the number of outlets is limited to 10 (1800 v-a) on a 15-A circuit and 13 (2340 v-a) on a 20-A circuit, by applying the 1.5 A per outlet criterion.

It is our opinion that the distinction between circuit types is not completely clear, and that good practice dictates the limitation of convenience receptacles per circuit to a considerably smaller number. Thus, following the guidelines of Table 17.5 of 9- and 12-A loading on 15- and 20-A circuits, respectively, we would have by this method

$$15 \text{ A circuit: } \frac{9 \text{ A}}{1.5 \text{ A/outlet}} = 6 \text{ outlets per cct}$$

$$20 \text{ A circuit: } \frac{12 \text{ A}}{1.5 \text{ A/outlet}} = 8 \text{ outlets per cct}$$

These figures must be used judiciously. If the devices to be energized are small, these quantities may be used. Such would be the case in a business office, where only an electric pencil sharpener or a desk lamp (100 VA) would be plugged in. However, for modern computer-equipped offices, service and repair facilities, and the like, no more than four receptacles on a 20-A circuit and prefer-

TABLE 17.6 **Branch Circuit Requirements**

	Branch Circuit Size				
	15 A	*20 A*	*30 A*	*40 A*	*50 A*
Minimum size copper conductors	No. 14	12	10	8	6
Minimum size taps	No. 14	14	14	12	12
Overcurrent device rating	15 A	20	30	40	50
Lampholders permitted	Any type	Any type	Heavy duty	Heavy duty	Heavy duty
Receptable rating permitted (see note 7)	15 A	15 or 20	30	40 or 50	50
Maximum load (see notes 6 & 8)	15	20	30	40	50

NOTES:
1. Wiring shall be types RHW, RHH, T, THW, TW, THWN, THHN, XHHW in raceway or cable.
2. On 15-A circuit, maximum single appliance shall draw 12 A. On 20-A circuit, maximum single appliance shall draw 16 A. If combined with lighting or portable appliances, any fixed appliance shall not draw more than 7.5 A on a 15-A circuit, and 10 A on a 20-A circuit.
3. On a 30-A circuit, maximum single appliance draw shall be 24 A.
4. Heavy-duty lampholders are units rated not less than 750 W.
5. 30, 40, and 50-A circuits shall not be used for fixed lighting in residences.
6. When loads are connected for long periods (3 hours or more), actual load shall not exceed 80% of the branch circuit rating. Conversely, continuous type loads shall be figured at 125% of actual load in all load calculations.
7. A single receptacle on an individual branch circuit shall have a rating not less than the circuit, for example, 15 A on a 15-A circuit, etc. 15-A receptacles on 20-A circuit shall not supply a load greater than 12-A for appliances. 20-A receptacles on 20-A circuit shall be limited to a 16-A load.
8. Rating of a single cord connected utilization equipment shall not exceed 24 A when connected to a 30 A circuit.
9. For additional restrictions and data on branch circuit use and application, see *NEC* Table 210-24.

ably two or three receptacles should be used on a 15-A circuit. Of course, diversity of use is an all-important factor, and the more accurately it can be estimated, the better the design result.

A further note of caution is in order: Receptacles should be arranged, if at all possible, so that the loss of a single circuit does not deprive an entire area of power. That is, for the sake of reliability, circuitry should be alternated to give each space parts of different circuits.

NEC Table 210-24 specifies certain requirements for conductors, devices, and loads permissible on general-purpose branch circuits. These are excerpted and summarized in Table 17.6.

17.12   Branch Circuit Design Guidelines: Residential

1. The *NEC* (1999 edition) requires for residences sufficient circuitry to supply a load of 3 VA/ft^2 in the building, excluding unfinished

spaces such as porches, garages, and basements. Using Table 17.5, which already includes a 20% derating of circuits, to provide for continuous loading, this works out to 480 ft^2 on a 15-A circuit (1440 VA) and 640 ft^2 on a 20 A cct (1920 VA). Allowing for some expansion such as a finished basement, enclosing a porch to expand living area, and similar, results in a good practice recommendation of 400–480 ft^2 per 15-A cct and 530–640 ft^2 per 20 A cct.

These good-practice figures normally provide enough circuits for all but the heaviest-loaded residences. However, if an actual design indicates the need for additional circuits, they obviously must be provided. The illustrative examples in *NEC* Chapter 9 are based on absolute minimum requirements.

2. The *NEC* (Article 210-11) requires a minimum of two 20-A appliance branch circuits [Fig. 17.19(*b*)] to feed all the receptacle outlets in the kitchen, pantry, breakfast, and/or dining room and similar areas, and *only* these outlets. This is

because any receptacle in these areas is a potential appliance outlet and must be fed and circuited as such. Permanently installed appliances such as food disposal, dishwashers, fan hood, and the like may *not* be connected to these appliance circuits. An exception is made to permit clock outlets to be fed from these circuits. Furthermore, the *NEC* requires that all kitchen outlets intended to serve countertop areas must be fed from at least two of these appliance circuits (which circuits may also feed appliance outlets in the other spaces specified previously). Thus, not all countertop workspace will be deenergized by the failure of a single circuit.

Although the code does not limit the number of appliance outlets to be wired into each circuit, good practice dictates that these receptacles be circuited with no more than four such outlets per 20-A circuit. This in turn usually requires more than the code minimum of two appliance circuits. Among other requirements, the *NEC* states that for kitchen and dining areas:

· No point on a wall behind a countertop be more than 24 in. from an outlet.
· All countertop convenience receptacles shall be of the ground-fault interrupter (GFCI) type.

3. Locations utilized primarily or frequently for workshop-type activities, such as garages, utility rooms, and basements, should be provided with receptacles wired in appliance-type circuits (i.e., 20-A receptacles on 20-A circuits), with no more than four such receptacles to a circuit. Receptacles in garages, sheds, crawlspaces, below-grade finished or unfinished basements, and outdoors must be of the GFCI type. (For exceptions to this GFCI rule, see *NEC* Article 210-8.)

4. Additional circuits similar to appliance circuits (no fixed lighting outlets) should be furnished to supply one outlet in each bedroom of a house that is not centrally air conditioned. Such outlets are intended for window air conditioners. (Good architectural and HVAC design provides window arrangement, attic ventilation, insulation, sun-screening, and the like to obviate the necessity for these noisy energy users.)

5. The *NEC* requires that at least one 20-A

appliance circuit supply the laundry outlets only. This requirement satisfies good practice. If an electric clothes dryer is anticipated (and it should be unless it is definitely known that a gas dryer will be used), an individual branch circuit (distinct from the laundry outlets circuits and rated for the load) should be supplied to serve this load via a heavy-duty receptacle. (Obviously, facilities for hanging clothes must be provided for those who prefer not to waste energy.)

6. Lay out convenience receptacles so that no point on a wall is more than 6 ft from an outlet. Use 20-A, grounding-type receptacles only. Do not combine receptacles and switches into a single outlet except where convenience of use dictates high mounting of receptacles.

7. Circuit the lighting and receptacles so that each room has parts of at least two circuits. This includes basements and garages.

8. Avoid placing all the lighting in a building on a single circuit.

9. Supply at least one 20-A wall-mounted receptacle adjacent to each bathroom basin location. Such receptacles must have GFCI protection. They shall be fed from a 20-A circuit that energizes only such bathroom receptacles. Bathroom lighting, exhaust fan, heaters, or other outlets shall not be connected to the bathroom receptacle circuit.

10. Provide at least two GFCI-protected and weatherproof receptacles on the outside of the house; one in front and one at the rear. Switch control of the outside receptacles from *inside* the house is an additional convenience.

11. In rooms without overhead lights, provide switch control for one-half of a strategically located receptacle that is intended to supply a lamp (see Fig. 17.20 for the wiring arrangement in such a case).

12. Provide switch control for closet lights. Pull chains are a nuisance (but are considerably cheaper).

13. In bedrooms, supply two duplex outlets at each side of the bed location to accommodate electric blankets, clocks, radios, lamps, and other such appliances.

Fig. 17.20 *Split wiring of a duplex receptacle. Upper half is switch controlled; lower half is "hot" all the time. This allows wall switch control of a lamp or other device while maintaining part of receptacle live, for independent use. Notice that the receptacle is mounted with the grounding pole at the top. This is the safest way to install receptacles because a metallic item such as a paper clip falling on to an inserted cap will contact the ground pole only.*

14. A disconnecting means, readily accessible and within sight of the controlled item, must be provided for electric ranges, cooktops, and ovens. It is good practice to utilize a small kitchen panel recessed into a kitchen wall to control the large kitchen appliances and provide the required disconnecting means.

15. Perimeter lighting, sensor control, and manual override can do much to lessen vandalism and discourage prowlers.

16. To accommodate the very rapidly increasing number of persons who work at home, at least partially, every study and work room or, in their absence, every large master bedroom should be equipped electrically to double as a home office. This room or area should be supplied *minimally* with these electrical services:

- Six duplex 15-A or 20-A receptacles connected to at least two different circuits, one of which should serve no other outlets. All of these receptacles should be equipped with appropriate surge suppression (see Section 16.37).
- An additional separate insulated and isolated ground wire, connected only at the service point, shall be run to the box(es) containing two of the receptacles required in the preceeding point, and there terminated, clearly marked and labelled. These grounds are intended for possible use with 1G receptacles, if it is found that the normal receptacles are too electrically "noisy."

- Two telephone jacks in recessed boxes should be provided in the area. In addition, an empty 3/4″ conduit from the telephone entry service point to an empty 4″ square box in the area should be provided. An appropriate surge suppressor must be provided for the incoming telephone service lines.
- For lighting design of this area, refer to the discussion on VDT lighting in Chapter 21.

A tabulation of residential electrical equipment, including recommended circuit and receptacle descriptions, is shown in Table 17.7. A complete residential wiring plan for a small house is shown in Fig. 17.21. Although residential plans are frequently left uncircuited, a completely circuited design is shown here for didactic purposes. The plan does not show the electrical equipment recommended for a home office that are described in item 16 for two reasons:

1. The plan is so compact that there is no single preferable location for the numerous outlets required.
2. Adding the required outlets to any selected area would clutter the plan to the point of illegibility. We would suggest that a student reader select a location and, on a larger scale plan or a large-scale part plan, add the required equipment, wiring, and notes.

17.13 Branch Circuit Design Guidelines: Nonresidential

(a) Schools. Because schools comprise an assembly of varied-use spaces, including instruction, lab, shop, assembly, office, gymnasium, plus special areas such as swimming pools, photographic labs, and so on, it is not possible to generalize on branch circuit design considerations, except for the following:

1. To accommodate the opaque, slide, and video projectors frequently used in the classroom, 20-A outlets wired two to a circuit are placed at the front and back of each such room. A similar receptacle, wired six or eight to a circuit is placed on each remaining wall.

TABLE 17.7 **Load, Circuit and Receptacle Chart for Residential Electrical Equipment**

Appliance	Typical Connected Volt-Amperes[a]	Volts	Wires[b]	Circuit Breaker or Fuse, amp	Outlets on Circuits	NEMA Device[d] and Configuration (see Fig. 16.42)
Kitchen						
Range[e,c,i]	12,000	115/230	3 #6	60	1	14–60R
Oven (built-in)[c,i]	4,500	115/230	3 #10	30	1	14–30R
Range tops[c,i]	6,000	115/230	3 #10	30	1	14–30R
Dishwasher[c]	1,200	115	2 #12	20	1	5–20R
Waste disposer[c]	300	115	2 #12	20	1	5–20R
Microwave oven	1,000	115	2 #12	20	1 or more	5–20R
Refrigerator[f]	300	115	2 #12	20	1 or more	5–20R
Freezer[f]	350	115	2 #12	20	1 or more	5–20R
Laundry						
Washing machine	1,200	115	2 #12	20	1	5–20R
Dryer[c,i]	5,000	115/230	3 #10	30	1	14–30R
Hand iron[e]	1,650	115	2 #12	20	1	5–20R
Living Areas						
Workshops[e,j]	1,500	115	2 #12	20	1 or more	5–20R
Portable heater[e]	1,600	115	2 #12	20	1	5–20R
Television	300	115	2 #12	20	1 or more	5–20R
Audio center[g]	350	115	2 #12	20	1 or more	5–20R
VCR[g]	150	115	2 #12	20	1 or more	5–20R
Personal computer and peripherals[g,h]	1,000	115	2 #12	20	1 or more	5–20R
Fixed Utilities						
Fixed lighting	1,200	115	2 #12	20	1 or more	—
Air conditioner ¾ hp[i,j]	1,200	115	2 #12	20 or 30	1	5–20R 14–30R
Central air conditioner[c,i,j]	5,000	115/230	3 #10	40	1	—
Sump pump[j]	300	115	2 #12	20	1 or more	—
Heating plant (i.e., forced-air furnace)[i,k]	600	115	2 #12	20	1	—
Attic fan[j]	300	115	2 #12	20	1 or more	5–20R

[a]Wherever possible, use the actual equipment rating.

[b]Number of wires does not include equipment grounding wires. Ground wire is No. 12 AWG for 20-amp circuit and No. 10 AWG for 30- and 50-amp circuits.

[c]For a discussion of disconnect requirements, see *NEC* Article 422.

[d]Equipment ground is provided in each receptacle.

[e]Heavy-duty appliances regularly used at one location should have a separate circuit. Only one such unit should be attached to a single circuit at the same time.

[f]Separate circuit serving only one other outlet is recommended.

[g]Surge protection recommended.

[h]Isolated ground may be required.

[i]Separate circuit recommended.

[j]Recommended that all motor-driven devices be protected by a local motor-protection element unless motor protection is built into the device.

[k]Connect through disconnect switch equipped with motor-protection element.

BASIC PLAN HOUSE - FIRST FLOOR
ELECTRIC PLAN; HOT WATER HEATING
EQUIPMENT AND DEVICE LAYOUT
(For Hot Water Heating Plan, See Fig. 3.33)

Fixture Schedule See Fig. 17.21 (c)
Symbol List See Fig. 17.21 (d)
Circuited Plan, See Fig. 17.21 (e),(f)
Plan, Schedule See Fig. 17.21 (g)

Notes:

1. Switch and outlet for exhaust fan. Switch wall mtd. above counter-blacksplash. Outlet with blank cover mounted adjacent to fan wall opening. Separate switch may be omitted if fan is supplied with integral switch.

2. Dishwasher receptacle wall mtd. behind unit, 6" AFF.

3. Range and oven outlet boxes wall mtd., 36" AFF. Flexible connection to units.

4. Receptacles at countertop locations to be wall mounted 2" above backsplash.

5. Max. ht. of top c/b to be 78" AFF.

6. Wiring shown as run exposed indicates absence of finished ceiling in basement level. All BX to be run through framing members. Attachment below ceiling joists not permitted. See Section 12.4.

7. Connect to 2–Type G fixtures ceiling mounted at $1/3$ points of crawl space.

8. Connect to 1–Type G fixture at center of crawl space.

9. Connect to shut–down switch at top of stairs. Boiler control wiring by others. See Note 10.

10. Boiler wiring safety disconnect. Provide RED wall plate, clearly marked "BOILER ON–OFF".

11. Equipped with self-closing gasket WP cover.

(a)

Fig. 17.21 *Electrical plan of a small house.*
(a) *Layout of lighting and electrical devices for main floor. (continues)*

BASIC HOUSE PLAN – BASEMENT
ELECTRIC PLAN; H.W. HEATING
EQUIPMENT AND DEVICE LAYOUT

(b)

LIGHTING FIXTURE SCHEDULE			
TYPE	DESCRIPTION	MANUFACTURER	REMARKS
A	48" L X 12" W X 4" DEEP NOMINAL, 2 LAMP/FLUORES-CENT, WRAP–AROUND ACRYLIC LENS, 40 W (NOM.) T8 LAMPS. SURFACE MTD.	BRITE–LITE CO. CAT. #2740/KFF OR EQUAL	4" DEPTH MAXIMUM
B	24" L, 1 LAMP 20W FLUOR. FIXTURE, WRAP–AROUND WHITE DIFFUSER, MOUNT ABOVE MEDICINE CABINET.	BRITE–LITE CO. CAT. #1/20/BFF OR EQUAL	MAX. MTG. HT. 78" TO ℄.
C	ADJUSTABLE HEIGHT PENDANT INCANDESCENT, 3–75W MAX., BUILT–IN 3–POSITION SWITCH.	HOMELAMP CO. CAT. #3/75/DRP OR EQUAL	————
D	10" D. DRUM–TYPE FIXTURE, WHITE GLASS DIFFUSER. CENTER LOCK–UP, 2–60W INCAND. MAX., SURF. MTD.	BRITELITE CO. CAT. #2/60/HF OR EQUAL	6" MAX. DEPTH.
F	12" D. DRUM FIXTURE, CONCEALED HINGE ON OPAL GLASS DIFFUSER FOR RELAMPING WITHOUT GLASS REMOVAL, 2–75W INCAND. MAX. SURFACE MTD.	DENMARK LIGHTING SPECIAL UNIT #374821	NO SUBSTITUTION WILL BE ACCEPTED.
G	PORCELAIN LAMPHOLDER, PULL CHAIN WITH WIRE GUARD, 100 W. INCAND. SURF. MTD.	————	————
H	SAME AS TYPE G, EXCEPT W/O GUARD.	————	————
K	DECORATIVE OUTDOOR LANTERN, MAX. 150W INCAND., WALL MTD. 84" AFF TO ℄.	TO BE CHOSEN BY OWNER	————
L	UTILITY OUTDOOR LIGHT, ANODIZED ALUMINUM BODY AND CYLINDRICAL OPAL GLASS DIFFUSER. 1–100W INCAND. MAX. 84" AFF TO ℄.	UTIL–LITE CO. CAT. #1/100/BP OR EQUAL	IF VANDALISM IS OF CONCERN, SUBST. PLASTIC DIFFUSER.
IR	RECESSED FIXTURE WITH 150 W INFRARED HEAT LAMP	HEAT-LIGHT CO.	————

(c)

Fig. 17.21 (b) *Layout of lighting and electrical devices for basement.*
(c) *Lighting fixture schedule. (continues)*

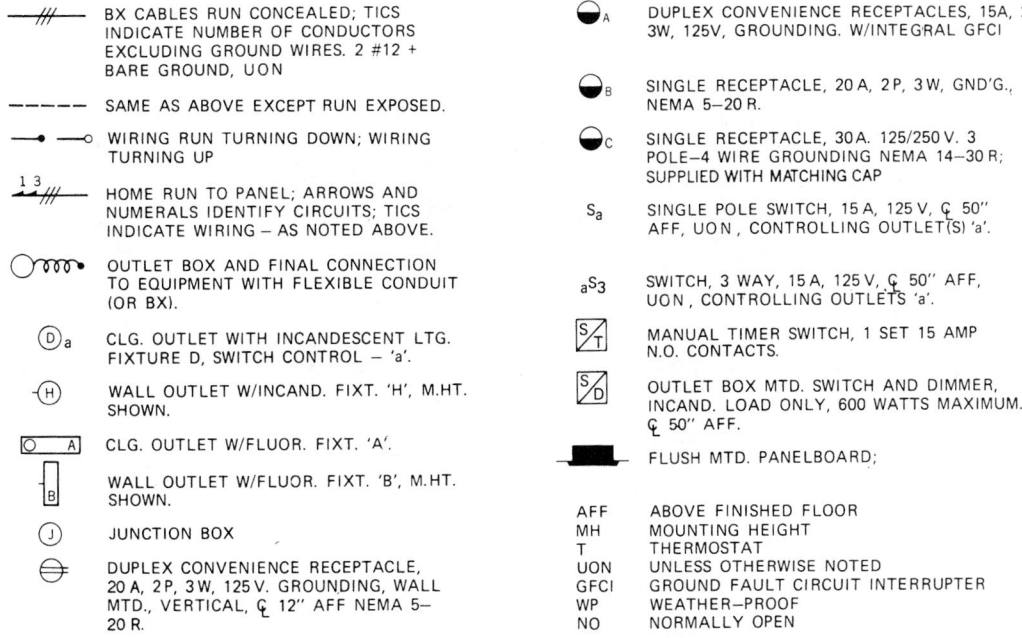

—///— BX CABLES RUN CONCEALED; TICS INDICATE NUMBER OF CONDUCTORS EXCLUDING GROUND WIRES. 2 #12 + BARE GROUND, UON

——— SAME AS ABOVE EXCEPT RUN EXPOSED.

—•—○ WIRING RUN TURNING DOWN; WIRING TURNING UP

1 3 —///— HOME RUN TO PANEL; ARROWS AND NUMERALS IDENTIFY CIRCUITS; TICS INDICATE WIRING – AS NOTED ABOVE.

OUTLET BOX AND FINAL CONNECTION TO EQUIPMENT WITH FLEXIBLE CONDUIT (OR BX).

D_a CLG. OUTLET WITH INCANDESCENT LTG. FIXTURE D, SWITCH CONTROL – 'a'.

H WALL OUTLET W/INCAND. FIXT. 'H', M.HT. SHOWN.

A CLG. OUTLET W/FLUOR. FIXT. 'A'.

B WALL OUTLET W/FLUOR. FIXT. 'B', M.HT. SHOWN.

J JUNCTION BOX

DUPLEX CONVENIENCE RECEPTACLE, 20 A, 2P, 3W, 125 V. GROUNDING, WALL MTD., VERTICAL, ℄ 12″ AFF NEMA 5–20 R.

A DUPLEX CONVENIENCE RECEPTACLES, 15A, 2P, 3W, 125V, GROUNDING. W/INTEGRAL GFCI

B SINGLE RECEPTACLE, 20 A, 2P, 3W, GND'G., NEMA 5–20 R.

C SINGLE RECEPTACLE, 30A. 125/250 V. 3 POLE–4 WIRE GROUNDING NEMA 14–30 R; SUPPLIED WITH MATCHING CAP

S_a SINGLE POLE SWITCH, 15 A, 125 V, ℄ 50″ AFF, UON, CONTROLLING OUTLET(S) 'a'.

$_aS_3$ SWITCH, 3 WAY, 15 A, 125 V, ℄ 50″ AFF, UON, CONTROLLING OUTLETS 'a'.

S/T MANUAL TIMER SWITCH, 1 SET 15 AMP N.O. CONTACTS.

S/D OUTLET BOX MTD. SWITCH AND DIMMER, INCAND. LOAD ONLY, 600 WATTS MAXIMUM. ℄ 50″ AFF.

FLUSH MTD. PANELBOARD;

AFF ABOVE FINISHED FLOOR
MH MOUNTING HEIGHT
T THERMOSTAT
UON UNLESS OTHERWISE NOTED
GFCI GROUND FAULT CIRCUIT INTERRUPTER
WP WEATHER–PROOF
NO NORMALLY OPEN

(d)

BASIC HOUSE PLAN – FIRST FLOOR
CIRCUITED ELECTRIC PLAN; H.W. HTG.

(e)

Fig. 17.21 (d) *Symbols and abbreviations.*
(e) *Circuited plan, main floor of house. (continues)*

ELECTRICITY

1029

BASIC HOUSE PLAN – BASEMENT
CIRCUITED ELECTRIC PLAN,
H.W. HTG.

(f)

Circ. no.	Description	LOAD V-A.	Circuit beakers		Load V-A	Description	Circ. no.
						Panel schedule for Basic Plan of Fig. 17.21	
1	LTG – { Kit. Dr., Br. #1, Hall, Outside, bath + Φ }	1045 1 R	20	1 2 20	30 6 R	Outlets – LR. & corridor + Exh. fan	2
3	LTG – { Outside Lr., Stair, Br. #2, Bsmt. + 6 Φ }	935 6 R	20	3 4 20	6 R	Outlets – BR. 1 & 2	4
5	Dishwasher	1500	20	5 6 20	1500	Outlets – Kit., Dr.	6
7	Outlets – BR 1, Bsmt	2 R	20	7 8 20	1500	Outlets – Kit., Dr.	8
9	Outlets – Bsmt., BR 2	2 R	20	9 10 20	1500	Laundry outlet – Bsmt.	10
11	Bathroom outlets	2 R	20	11 12 20	1300	Boiler or furnace	12
13	Spare	—	20	13 14 30 A 2 P	6Kw	Range	14
15	Spare	—	20	15 16			
17	Clothes dryer	5Kw	30 A 2 P	17 18 30 A 2 P 19 20	4800w	Oven	16
	Space for 2 – 1P or 1 – 2 pole		21 22 23 24			Space for 1 – 2P or 2 – 1 pole	
	Load total – phase A					Load total, phase B	

Panel Data
Mains, GND. BUS: 150 A MNS., 60 A GND. BUS
Main C/B or ~~SW/F~~   150/100
Branch C/B INT. CAP. AMP. 10,000
Mounting – ~~Surf~~/recess
Remarks: Front suitable for painting

Voltage   120/208
1 PH. 3 wire

(g)

Fig. 17.21 (f) *Circuited plan, basement.*
(g) *Panel schedule of the house.*

2. To accommodate computers, a detailed layout of the space is required showing the proposed computer area. A two-section surface mounted or recessed raceway on the wall behind the row of computers should be provided. The power section will have two duplex 20-A receptacles at each computer station, wired on alternate circuits. The second section of the raceway is intended for low-tension wiring, such as network cabling, and wiring to computer peripherals.

3. Light switching should provide:

· High–low levels for energy conservation and to permit low-level lighting for screen viewing. With fluorescent lighting, this can be accomplished by alternate ballast wiring and switching, thus avoiding the high cost of dimming equipment.

· Separate switching of the lights on the window side of the room, which is often lighted sufficiently by daylight. Control should be initiated automatically by photocell (see Section 21.3).

4. Provide appropriate outlets for all special equipment in labs, shops, cooking rooms, and the like.

5. Use heavy-duty devices and key-operated switches for public area lighting (corridors, etc.), plastic instead of glass in fixtures, and vandal-proof equipment wherever possible. All panels *must* be locked and should be in locked closets.

6. The *NEC* requires sufficient branch circuitry to provide a minimum of 3 V-A/ft^2 for general lighting in schools (refer to *NEC* Article 220). Unlike the figure for residential occupancy, this figure does *not* include receptacles. Receptacles are calculated separately at 180 V-A each for ordinary convenience outlets.

(b) Office Space.

1. In small office spaces (less than 400 ft^2), provide either one convenience outlet for every 40 ft^2, or one outlet for every 10 linear ft of wall space, whichever is greater. In larger office spaces, provide one outlet every 100 to 125 ft^2 beyond the initial 400 ft^2 (10 outlets). These outlets are intended for miscellaneous electrical devices, in addition to anticipating the constantly increasing number of data system accessories. In addition, provide at *every* desk (on an adjacent wall, in a "power pole," or in the floor) at least one duplex 20-A receptacle for a computer terminal. These receptacles should be circuited at no more than six to 20-A branch circuit, and less if the equipment to be fed so dictates. Figure 17.22 shows one possible circuiting arrangement for the room layouts shown in Fig. 17.12. Although other arrangements are possible, the net result is the same (see also Fig. 17.23).

Fig. 17.22 *Alternative methods of circuiting are shown. Room 205 shows the actual junction box location with flexible connections to the box at each fixture. Room 207 shows circuit numbers and switch designations only; the placement of junction boxes is understood, and conduit runs are omitted for clarity and because they most often are not representative of actual installation. Room 209 shows an outlet box at each fixture, with schematic conduit connections. All of these systems are in common use.*

Fig. 17.23 *Typical receptacle circuiting of several rooms in an office–lab building. Lighting (Fig. 17.22) and power (receptacles) are shown on separate plans to avoid crowding (see Fig. 17.15 for symbols and Fig. 17.12 for notes). Lighting in offices is recessed; lighting in labs is surface mounted for flexibility. Note the double circuiting of the type-D receptacles.*

2. Corridors should have a 20-A, 120-V outlet every 50 ft to supply cleaning and waxing machines.

3. As with all nonresidential buildings, convenience receptacles are figured at 180 V-A each.

4. Only specification grade equipment should be used.

(c) Industrial Spaces. These areas are so specialized that no meaningful guidelines can be given.

(d) Stores. In stores, good practice requires at least one convenience outlet receptacle for every 300 ft² in addition to outlets required for loads such as lamps, show windows, and demonstration appliances.

17.14 Load Tabulation

While circuiting the loads, a panel schedule is drawn up that lists the circuit numbers, load description, wattage (actually volt-amperes), and the current rating and number of poles of the circuit-protective device feeding each circuit. Spare circuits are included to the extent that the designer considers them necessary and consonant with economy, but normally no less than 20% of the number of active circuits. Finally, spaces are left for future circuit breakers in approximately the

same quantity as the number of spare circuits, but always to round off the total number of circuits.

Panelboards are normally manufactured with an even number of poles. Thus if a panel had, with spares, 21 poles, the designer would ask for (at least) three spaces, to give a 24-circuit box. A typical panel schedule is shown in Fig. 17.24, which serves the laboratory of Figs. 17.22 and 17.23.

In calculating panel loads, the following rules apply:

1. Each specific appliance, device, lighting fixture, or other load is taken at its nameplate rating, except certain kitchen and laundry appliances for which the *NEC* allows a demand factor (see *NEC* Article 220).

2. Each convenience outlet, in other than residential spaces, is counted as 1.5 A (180 V-A).

3. Loads for special areas and devices such as show window lighting, heavy-duty lampholders, and multioutlet assemblies are taken at the figures given in *NEC* Article 220.

4. Spare circuits are figured at approximately the same load as the average active circuits (1200 to 1500 V-A).

5. Spaces are not counted in the load.

6. Continuous loads such as lighting are calculated at 125% of their actual value. See *NEC* Articles 210-19(a) and 210-20(a), which refer to load

ELEC. PANEL-LP-1			
120/208V 3φ 4W		LOAD IN WATTS	
No.	LOAD	φA	φB	φC
---	---	---	---	---
1	Lighting	1200		
2	Lighting	1200		
3	Lighting		1450	
4	Lighting		1050	
5	Lighting			1100
6	Lighting			1200
7	Lighting	800		
8	Lighting	1100		
9	Lighting		700	
10	Lighting		1050	
11	Lighting			1000
12	Lighting			1200
13	Receptacle 2	900		
14	Receptacle 3	900		
15	Receptacle		900	
16	Receptacle		900	
17	Spare			1200
18	Spare			1200
19	Spare	1200		
20	Receptacle 4		1000	
				1000
21	Spare	1200		
			1200	
22-26	Spaces only			
	25% addition for continuos loads 5	1025	1063	1125
	Phase totals	9525	9313	9025
	Panel total		27863	V-A
	Max φ current		77 amp	
	25% spare capacity	~	20 amp	(Future)
	Max. Phase I		97 amps	

Main breaker _225A 3pole (see text example 17.1, section 17.18)_

Trip _100 A_

Feeder size _4 #2 THW in 1¼"C._

NOTES:
1. All C/B 1P; 50A frame, 20A trip except ccts 20 and 21, which are 2P, 50AF/20AT.
2. 5 receptacles @ 1.5A each.
3. Corridor receptacle; 120 V section.
4. Corridor receptacle; 208 V section.
5. See Section 17.15

Fig. 17.24 *Schedule for lighting panel LP-1.*

such that average load will be lower than the maximum demand (see Section 17.17 for multipanel feeders). If it is known that certain loads will not or cannot be used simultaneously, the load total should reflect only the larger of the two. Thus, heating and cooling loads are *generally* not concurrent. Nor is building night floodlighting concurrent with the business office load, but it may be with general interior lighting. Note in Fig. 17.24 that 2-pole loads (208 single-phase) appear in two columns. Similarly, 3-phase loads would appear in three columns. Also note that the phase loads are *not equal*. It is the responsibility of the designer (or contractor) to circuit the loads so that the phases are as closely balanced in load as possible. If this is not done, one phase will carry considerably more current than the others. Because the panel feeder must be sized for the maximum phase current, this may lead to an oversized feeder and therefore a waste of money.

Having tabulated and balanced the loads and totaled them by phase, the maximum current is calculated. A portion of the spare capacity available in the branch circuits is added to the total as the basis for the calculation of the feeder load. This spare capacity, shown in Table 17.5, is something between 25% and 50%. The exact amount to be added initially in feeder sizing is developed in the ensuing discussion.

17.15 Spare Capacity

Load calculations for dwelling occupancies and other spaces are detailed in *NEC* Article 220 and examples given in Chapter 9 of the *NEC*. Because these calculations are specialized but routine and are covered there in detail, they are not repeated herein. Again, it is emphasized that possession of a current copy of the *National Electrical Code* (NFPA Bulletin 70) is a *sine qua non* for a proper electrical design. Code requirements, it must be remembered, are minimum; quality design often exceeds these minimum requirements.

Having arrived at the panel load totals as detailed previously, the next step is to size the conductors feeding the panel. To do this, an examination of the spare capacity of the panel and of the feeders is necessary, in order that the system design be consistent, giving equal capacity for fu-

calculations for conductor sizing and feeder overcurrent protection, respectively, for combinations of continuous and noncontinuous loads. These calculations apply to panel load calculations as well and are best accomplished by adding 25% to the panelboard load as shown in Fig. 17.24.

In calculating total panel loads, as shown in Fig. 17.24, no demand factor may be applied except as specifically stated in the *NEC* (see *NEC* Articles 220-11 and 220-13). This is true despite the knowledge that most often the usage will be

ELECTRICITY

TABLE 17.8 **Panel Initial and Expanded Loads**[a]

	Panels in Facilities Expecting Limited Expansion; Circuits Initially Loaded to Give 25% Expandability (See Table 17.5)	Panels in Facilities Expecting Extensive Expansion; Circuits Initially Loaded to Give 50% Expandability (See Table 17.5)
Initial load	100%	100%
Initial plus spares	120%	120%
Load after all circuits including spares are loaded to maximum allowable	150%	180%
Load after utilizing 20% spaces also	175%	210%

[a]For development of these figures, see McGuinness et al. (1980), pp. 690–691.

ture growth in all its components. Considering the panel circuitry first, let us examine the effect of load expansion, including spares and spaces (Table 17.8).

As noted in Section 17.11, spare capacity is built into the branch circuitry *and* into the panels. Most often, expansion is accomplished by additional loading on some circuits and by adding new circuits via spare circuit breakers in the panel. Table 17.8 gives the *ultimate* capacity of the panel; that is, fully loaded circuits and fully utilized spares and spaces. Because this ultimate capacity is rarely achieved, panel feeders need only be sized for initial loads as detailed in Section 17.14, and provision made for rewiring to meet anticipated expansion by one of the techniques listed in Section 17.16.

These results can be summarized as follows: For panels in buildings expecting limited expansion, for which branch circuits are loaded to 80% of capacity (25% *branch circuit* expansion, see Table 17.5, note *e*), the ultimate panel load without new conduit work (i.e., merely by filling out circuits), is 1.5 L, or 50% beyond the initial load. By adding breakers in the spaces, this load can be expanded to 75% beyond the initial load. The corresponding figures for panels that are lightly loaded (66% capacity, i.e., 50% branch circuit expansion as in Table 17.5 note *f*), in anticipation of considerable load growth, are 80% to 110%, respectively. These results are summarized in Table 17.8.

17.16 Feeder Capacity

To achieve economy, the panel feeder must accommodate the initial load plus some portion of the future load. Spare capacity in feeders (to accommodate a considerable portion of the panel spare capacity as shown in Table 17.8) is provided by one or more of the following procedures:

1. Provide feeder (and conduit) capacity initially to handle the entire eventual load. This method is most expensive, requiring initial outlay for no return, and is rarely used.

2. Provide feeder for initial load plus spare, with properly sized conduit. Conduit is sized for type THW or RHW without covering. This method, as we shall see, yields limited spare capacity.

3. Provide feeder for initial load plus spare, with conduit oversized by one size. Size conduit for type THW wire, which is very widely used because of attractive price and excellent electrical properties. Some additional costs are entailed here.

If the initial wiring is done with type TW wire, the effect is approximately the same as having oversized the conduit by one size and wiring initially with THW wire. This is caused by the lower ampacity rating of TW as compared to THW, resulting in a larger conduit size for the same initial ampacity. The reader can work out exact

TABLE 17.9 **Maximum Number of Conductors in Trade Sizes of Rigid Metal Conduit**

Type Letters	Conductor Size AWG, kcmil	Conduit Trade Size (in.)									
		½	¾	1	1¼	1½	2	2½	3	3½	4
TW, THW, RHW, and RHH (without outer covering) THHW	6	1	3	5	8	11	18				
	4	1	1	3	6	8	14	20			
	2	1	1	2	4	6	10	14			
	1	1	1	1	3	4	7	10			
	0		1	1	2	3	6	8	13		
	00		1	1	2	3	5	7	11		
	000		1	1	1	2	4	6	9	13	
	0000			1	1	1	3	5	8	10	
	250			1	1	1	3	4	6	8	11
	300			1	1	1	2	3	5	7	9
	350				1	1	1	3	5	6	8
	400				1	1	1	3	4	6	7
	500				1	1	1	2	3	5	6
	600				1	1	1	1	3	4	5
THHN	6	2	4	7	12	16	27				
	4	1	2	4	7	10	16	24			
	2	1	1	3	5	7	12	17	26		
	1	1	1	1	4	5	9	12	19	26	
XHHW (4–500 kcmil)	0	1	1	1	3	4	7	10	16		
	00		1	1	2	3	6	9	13		
	000		1	1	1	3	5	7	11	15	
	0000		1	1	1	2	4	6	9	12	16
	250			1	1	1	3	5	7	10	13
	300			1	1	1	3	4	6	9	11
	350			1	1	1	2	3	6	7	10
	400			1	1	1	2	3	5	7	9
	500				1	1	1	2	4	5	7
	600				1	1	1	1	3	4	

Source: Extracted from the *NEC*.

figures in each with the help of Tables 15.2a, and 15.2b, and conduit capacity Table 17.9, or by using the *NEC* tables in Article 310 and Chapter 9.

4. Provide feeder for initial load plus spare and oversize conduit by *two* sizes. This yields most of the capacity necessary in facilities anticipating large expansion.

5. Provide for initial load plus spare, with an empty conduit for future. This method is expensive because of high conduit cost, and is advisable only infrequently.

In procedures 2, 3, and 4, the future capacity beyond that initially supplied is handled by the use of larger gauge wire in the existing conduit. To examine exactly what these alternatives provide in spare capacity, we have tabulated in Table 17.10 the maximum ampacity of various size conduits and the future capacity obtainable. Table 17.9 is taken directly from the *NEC*. Note from Table 17.10 that simply by rewiring we can obtain up to 30% additional capacity, whereas if the conduit had been oversized the additional capacity would be from 35% to 146%.

TABLE 17.10 **Maximum Wire and Ampacity of a Conduit, and Ampacity Gain on Rewiring**

| Initial Conduit Size (in.) | Initial Installation THW Cable | | Rewiring with XHHW | | | | |
| | Max.[a] Wire AWG/ kcmil | Max. Amperes | Using Original Conduit | | | Capacity Increase Having Oversized Conduit | |
			Maximum Wire	Maximum Amperes	Capacity Increase	One Size	Two Sizes
1½	1	130	1/0	170	30%	100%	146%
2	3/0	200	4/0	260	30%	60%	115%
2½	250	255	300	320	25%	69%	100%
3	400	335	500	430	28%	35%	81%
3½	600	420	600	475	13%	64%	—

[a]Assuming four single conductors in conduit: 3-phase conductors plus neutral.

Returning now to the question of how large to make the feeder for a given panel load, we must balance future panel load, initial cost of feeder, and future capacity of existing conduit. It is best to avoid the installation of empty conduits because this is expensive. Rewiring, however, is relatively inexpensive and oversizing conduit is the method of choice if rapid expansion is anticipated.

Referring to Table 17.10, note that normal design uses THW cable. Design with T or TW is, in effect, a first step in oversizing conduit and is generally not economical. The second step is a deliberate oversizing of conduit that results in much increased conduit ampacity, as reflected by the figures. Using these figures in actual practice requires that the designer juggle cable and conduit cost against anticipated load growth to arrive at the most economical long-term solution. Applying these numbers to concepts previously developed, we have:

1. Buildings designed for 25% branch circuit expansion (see Table 17.5) have a panel capacity of 1.75 times the load (Table 17.8). If it is desired to design the panel feeder to carry the full expansion possible, then calculate the feeder on the basis of panel load plus 20% spares and oversize the feeder by 20%. This gives a feeder capacity of

$$1.20 \times 1.2 = 1.44 \times \text{initial load}$$

Table 17.10 indicates that rewiring adds another 20% on the average. This gives

$$1.20 \times 1.44 = 1.73 \times \text{initial load}$$

which corresponds closely to the 1.75 desired.

2. For a building with 50% branch circuit expansion, utilizing full panel space capacity gives us an ultimate panel capacity of 2.1 times the initial load (see Table 17.8). This is accomplished as follows: Oversize the feeder by 15% and oversize the conduit by one size. The latter step gives an average expansion of 65%. Therefore,

$$\text{feeder capacity} = 1.15 \times 1.2 \times 1.65 = 2.3$$

which is approximately the desired figure.

If, as in the case of laboratories, more than 100% expansion is anticipated (see Table 17.1), conduit should be oversized by two sizes and initial wiring oversized by approximately 25%. Feeders thus arranged will handle the new panels required to meet the anticipated expansion.

Two factors should be carefully noted here. First, note that the smaller conduits offer the largest expandability, although, in dollars per amperes, they are more expensive. Second, in order to take advantage of spaces in a panel, conduit stubs should be taken from the panel and extended into hung ceilings or another procedure used to make the panel circuitry easily accessible in the future.

17.17   Panel Feeder Load Calculation

EXAMPLE 17.1. Refer to Fig. 17.24. The panel is for a laboratory/office area. Because large expansion is anticipated, circuitry follows the bottom section of Table 17.5. Ultimate panel load would be 26 cct at 1920 V-A = 50 kVA = 138 A. Thus the initial feeder is sized for 115 A (4 #2 THW, 1¼ in. C), but rewiring with XHHW will allow as much as 150 A in the same conduit. A 225-A frame c/b is chosen initially because eventually the trip will be raised to 150 A.

The *NEC* in Article 220 specifies minimum volt-amperes per square foot (V-A/ft^2) figures for various occupancies, for lighting, and for miscellaneous power loads. Proper design procedure therefore requires that after detailed design of an area, the actual loading be compared to these minima, and the larger of the figures, as regards number of circuits and feeder load, be used. An example will help to make this clear.

EXAMPLE 17.2. Assume a single floor of an office building 100 × 200 ft. Assume also that 15% of the area is corridor and storage, equally divided between the two. Calculate the load and feeder size. Assume a good grade speculative construction venture.

SOLUTION.

$$\text{Office space} = 85\% \text{ of } 20{,}000 \text{ ft}^2$$
$$= 17{,}000 \text{ ft}^2$$
$$\text{Corridor and storage} = 15\% \text{ of } 20{,}000 \text{ ft}^2$$
$$= 3{,}000 \text{ ft}^2$$

The *NEC* specifies a minimum of 3½ V-A/ft^2 for lighting and 1 V-A/ft^2 for miscellaneous receptacles in office space. It further specifies ½ V-A/ft^2 minimum for corridors and ¼ V-A/ft^2 for storage. Therefore:

Lighting

Office load:	17,000 × 3½	= 59.5 kVA
Corridor:	1,500 × ½	= 0.74 kVA
Storage:	1,500 × ¼	= 0.38 kVA
Minimum lighting load		= 60.6 kVA

Receptacles

$$17{,}000 \times 1 \text{ V-A/ft}^2 = 17 \text{ kVA}$$

These figures would then be compared to the actual design loads. Receptacles are counted at 180 V-A each, as noted in Section 17.14, item 2. If the code minima exceed the design load (as it well may if lighting is properly designed), panels must be equipped with sufficient additional circuits to make up the difference. The number of such circuits is up to the designer, because circuit loading is not specified.

For instance, suppose that the lighting design was accomplished at 2 V-A/ft^2. Panel circuits would have to be provided for the additional 1½ V-A/ft^2, as follows:

$$17{,}000 \text{ ft}^2 \times 1.14 \text{ V-A/ft}^2 = 25.5/\text{kVA}$$

Assuming a 30% to 50% future load expansion, we would circuit the lighting loads at approximately 1300 V-A per circuit (see Table 17.5). Therefore, we would provide

$$\frac{25{,}000 \text{ V-A}}{1{,}300 \text{ V-A}} = 20 \text{ additional circuits}$$

With respect to minimum feeder load, the *NEC* specifies that it be increased by 25% if loads are continuous (3 or more hours). This requirement allows for breakers to heat up in panels while carrying continuous load, and is waived for circuit breakers that are ambient compensated— that is, are rated to carry 100% load. Because we have established 80% of the breaker rating as maximum load (see Table 17.5), *we have already accounted for this factor in circuitry but must utilize it in feeder calculation.* Assuming that the code minima are the design loads, then, for feeder calculation:

$$\text{Lighting load} = 60.6 \text{ kVA} \times 125\% = 75.75 \text{ kVA}$$

$$\text{Receptacle load as above} = 17.0 \text{ kVA}$$

$$\text{Minimum feeder load} = 92.75 \text{ kVA}$$

$$25\% \text{ future load} = 23.2 \text{ kVA}$$

$$\text{Design feeder load} = 116 \text{ kVA}$$

Because this load would be divided between several panels, the building electrical design might be such that the panels were not all fed by one feeder (see Fig. 17.25). However, assuming they

TABLE 17.11 **Current and Volt-Ampere Relationships**

120-V Single-Phase	120/240-V 3-Wire	120/208-V Single-Phase, 3-Wire	120/208-V 3-Phase	277/480-V 3-Phase	277-V Single-Phase
$I = \dfrac{V\text{-}A}{120}$	$I = \dfrac{V\text{-}A}{240}$	$I = \dfrac{V\text{-}A}{208}$	$I = \dfrac{V\text{-}A}{360}$	$I = \dfrac{V\text{-}A}{830}$	$I = \dfrac{V\text{-}A}{277}$

were, the feeder would be calculated in terms of 3-phase current thus:

$$I = \frac{kVA}{0.360} = \frac{116}{0.360} = 322 \text{ A}$$

Using THW cable, a minimum of 400 MCM would be required. Conduit would be a minimum of 3 in., but might be increased to 3 ½ or 4 in., according to the considerations of spare capacity discussed in the previous section. ■

As an aid in computing currents, Table 17.11 lists the relevant relations. A note of caution is required regarding the use of computer programs to perform the previous calculations. The reader who has followed the discussion carefully understands that considerations of economy and anticipated expansion are combined in designing a feeder system. None of the programs that we have seen will make the necessary comparisons and calculations unless specifically directed, and this frequently requires a multistage calculation. Most programs are arranged to select feeders on the basis of current loads, plus a specified spare capacity, without regard to expansion or economic considerations. It therefore is important to understand the methods used by any particular computer program in order to apply it intelligently.

17.18 Harmonic Currents

A phenomenon of recent vintage, that of large harmonic currents, has been the cause of considerable difficulty in modern electrical installations. Without going into detail in this highly technical subject, a brief description of the problem and its causes can be given.

Conventional electrical loads such as lighting, resistive devices (heaters), motors, and the like are linear (i.e., the load impedance remains essentially constant regardless of instantaneous voltage). This is not the case with most electronic equipment. Computers, modems, printers, electronic lighting ballasts, variable-speed motor drives, and solid-state equipment of all types are essentially nonlinear loads. As such they produce harmonic currents, of which the odd-order ones are additive in the power system neutral conductor. The most troublesome of these are the third harmonic and its odd multiples (9th, 15th, 21st, . . .). These currents can become so large in a modern computerized office installation with electronic lighting ballasts that instead of the neutral conductors carrying the unbalanced current in a 3-phase system (zero in a balanced system), it actually carries *more* current than the phase wires.

Other serious negative effects of harmonic currents are:

· Deterioration of electronic equipment performance; continuous or sporadic computer malfunctions

· Overheating of neutral—possibly causing neutral burnout and resulting in equipment being subjected to severe voltage variations

· Overheating and premature failure of transformers—even when the transformer nameplate rating seems adequate

· Overheating of motors because of operation with a distorted voltage waveform

· Nuisance tripping of circuit breakers and adjustable speed drives

· Telephone interference

· Capacitor fuse blowing

As the amount of electronic equipment in use increases (as it does continuously), the problem of destructive harmonic currents becomes progressively more severe.

In view of the fact that today, at least half of the electric load in a modern office-type facility is composed of nonlinear, harmonic producing equipment, it follows that all such facilities, existing and in-planning, must take necessary corrective measures. In the past, these measures consisted of oversizing equipment to avoid overload burnout; adding passive harmonic filters (which act to reduce harmonic content) in the electric distribution system; using isolation transformers at sensitive loads; selecting power sources with low output impedance to minimize voltage distortion; using controls that are relatively insensitive to harmonic distortion; adding meters throughout the system that measure true rms voltage and current rather than the average figures shown by conventional meters; and other expensive and essentially passive power line conditioning (see Sections 16.35 and 16.36). In view of the increasing severity of the problem, computer-controlled variable power conditioning equipment (called "active" line conditioning) has become available. This power-conditioning equipment operates in a fashion similar to that described for active noise cancellation in Section 28.28. The conditioner instantaneously and continuously analyzes the harmonic content of the line voltage and injects an equal but exactly out-of-phase voltage, to cancel the harmonics and produce a pure sinusoidal voltage supply. The harmonic currents that are required by non-linear loads are supplied by a digital signal generator. Other techniques are also used, a description of which is not within our scope.

In any rehab work, the electrical designer must obtain a detailed electrical system analysis of the existing system, performed by competent engineers experienced in the field of power quality. Many existing systems carry as much as 70% to 80% harmonic current and constitute a major system failure waiting to happen. A proper power quality study, performed with such instruments as true rms meters, harmonic analyzers, frequency selective voltmeters, and spectrum analyzers, will yield a true picture of the existing system and permits the electrical rehab work to be engineered with harmonic limitations as one of the important design parameters.

17.19 Riser Diagrams

When all devices are circuited and panels are located and scheduled, we are ready to prepare a riser diagram. A typical diagram, shown in Fig. 27.25, represents a block version of a single-line diagram except that, as the name implies, vertical relationships are shown. All panels, feeders, switches, switchboards, and major components are shown up to, but not including, branch circuiting. This diagram is an electrical version of a vertical section taken through the buildings.

EXAMPLE 17.3. Feeder F10 of Fig. 17.25 serves lighting panels 1A, 2A, and 3A. Calculate the required feeder size, considering loads, future expansion, and voltage drop.

SOLUTION. The loads on these panels have been computed in accordance with the previous considerations and are:

Connected load: LP-1A---110 A
                   LP-2A---125 A
                   LP-3A---<u>100 A</u>
                            335 A

These figures include connected load, spares, and a 25% future factor (the *NEC* requires that feeders be sized minimally for loads calculated by using the given *NEC* square foot loads plus any demand factors listed; see Article 220). If actual panel circuit loads are larger than the *NEC* minima, then they are the loads used in feeder calculations and any reasonable demand may be used, provided that at no time do the totals drop below the minima specified in the *NEC*. Therefore, we may apply diversity factors between panel loads in a judicious manner.

In office building work, typical diversity factors are

Lighting Panels Fed from a Single Feeder	Diversity Factors
1, 2	1.00
3, 4	1.09
5, 6, or 7	1.18
8, 9, or 10	1.33

Thus the load on feeder F-10, using 100% demand per panel and 1.09 diversity between panels, would be $335 \times 1.0/1.09 = 307$ A.

ELECTRICITY

Fig. 17.25 *Typical power riser diagram. Ordinarily, the main switchboard would be shown as a large rectangle with the feeders emanating from it, and a switchboard schedule would detail the contents. Here, because of the unusual bus arrangement, the main switchboard appears as it would on a single-line diagram.*

Methods for handling future expansion were discussed previously. In this case, the feeder, before voltage drop considerations, would be (from Tables 15.2, 15.3, and 17.9) 4-350 MCM THW in 3-in. C. In this case, the figures in Table 17.10 can be misleading if not read carefully. A 3-in. conduit will take a maximum of 400 kcmil THW. However, because we are initially using only 350 kcmil that also requires a 3-in. conduit, the expansion possible by rewiring with XHHW is

$$\frac{\text{ampacity of 500 kcmil, XHHW in 3'' C}}{\text{ampacity of 350 kcmil, THW in 3'' C}} = \frac{430 \text{ A}}{310 \text{ A}}$$

$$= 1.39\text{—i.e., 39\% expansion}$$

This should be sufficient. ∎

The final consideration in sizing a feeder is voltage drop. The *NEC* suggests (does not re-

Fig. 17.26 *Curves for determining voltage drop in copper cables. Curves show maximum one-way circuit length for 1% voltage drop.*

quire) that sizing branch circuit conductors for a maximum of 3% voltage drop to the farthest outlet, and sizing feeders so that the maximum *combined* feeder and branch circuit voltage drop to the farthest outlet does not exceed 5% will give reasonably efficient operation. These figures apply to all types of loads.

We have always found that it makes economic sense to restrict branch circuit voltage drop to 2% to the farthest outlet while maintaining the overall 5% limitation. This allows a minimum 3% drop in feeders, which are frequently long and, because of size and ampacity inefficiency, relatively more expensive than small branch circuit wiring.

Many tables and curves are published by manufacturers from which voltage drop can be obtained. Such a set of curves is shown in Fig. 17.26, which shows maximum length of run for a 1% drop. Applying these curves to our last example; and assuming an 80-ft run gives the following:

From the curves, 307 A on 350 MCM cable gives a 1% drop in 50 ft. Therefore, the drop in 80 ft will be 1.6%, which is well within our criteria,

assuming that branch circuits were limited to 2% voltage drop.

In summary, then, feeders are sized in accordance with load (actual or square feet, whichever is larger) and voltage drop. Conduit may be oversized for large future load expansion.

17.20   Service Equipment and Switchboard Design

The main switchboard shown in Fig. 17.25 constitutes a combination of service equipment and feeder switchboard. The service equipment portion of the board comprises the metering and the four main switches feeding risers, motor control center (MCC), roof machine room, and elevators. The feeder board comprises switches 5 through 12. Such an arrangement is permissible inasmuch as the *NEC* allows up to six fused switches or circuit breakers to serve as the service disconnect means. This arrangement was chosen in order to separate to the largest extent possible the motor loads (elevators, air-conditioning equipment, basement power, etc.) from the lighting. Such a procedure minimizes lighting fluctuations resulting from motor starting and yields simpler maintenance. Also, the size of the main switch is reduced. This switchboard would be of the metal-clad dead-front type with switches or circuit breakers, as desired.

Other recommendations affecting service equipment are as follows:

· All equipment used for service including cable, switches, meters, and so on, shall be approved for that purpose.

· It is recommended that a minimum of 100-A, 3-wire, 120/240 V service be provided for all individual residences.

· No service switch smaller than 60 A or circuit breaker frame smaller than 100 A should be used.

· In multiple-occupancy buildings, tenants must have access to their own disconnect means.

· All building equipment shall be connected on the load side of the service equipment except that service fuses, metering, fire alarm and signal equipment, and equipment serving emergency systems may be connected ahead of the main disconnect (see Fig. 17.25).

· For additional information on electric service, see *NEC* Article 230.

17.21 Emergency Systems

(a) General. Some of the considerations relevant to power reliability were discussed in Section 17.1(*b*), and a brief review of the equipment available to supply emergency power was presented in Section 16.39. Emergency lighting equipment is covered in Section 21.36. In this section, we discuss possible arrangements of emergency power supply. The choice of arrangement and the size and type of equipment depends in large measure on the requirements of local codes, which determine the loads that must be supplied from the emergency system. The reader should note that although we are using the term

emergency, the concepts involved are equally applicable to standby systems. Where differences occur, they are pointed out.

The three classes of electrical power supply systems that are subsumed under the general title of "emergency" electric supply systems are:

· Emergency Systems; covered by *NEC* Art. 700

· Legally Required Standby Systems; covered by *NEC* Art. 701

· Optional Standby Systems; covered by *NEC* Art. 702

The essential differences between the three classifications are concerned with the purpose, application, and duration of the specific non-normal power supply. All three types are activated when normal power fails, but only the first two types are legally required when so specified by any governmental agency having jurisdiction. Optional standby systems are, as their name states, optional; they are entirely dependent on the requirements of the facility's owner or operator.

Emergency systems are those essential to the safety of human life. As such, and depending on the type of facility involved, they may supply power for fire detection, lighting, alarm and communication systems, elevators, fire pumps, and such loads as ventilation, refrigeration, and industrial processes where power interruption would imperil human safety. Generally, emergency systems must pick up the loads automatically within 10 seconds of power interruption. Where the emergency power supply consists of batteries, they must have a full-load capacity of 90 minutes. Where the power supply is an engine–generator set, an on-site fuel supply that will suffice for a minimum of two hours of full load is required. The entire wiring system for the emergency power system must be separate and distinct from the normal power system, except of course, where both systems are connected to the same item of equipment.

Legally required standby systems are those required to provide power for essentially the same types of loads as emergency systems, except that their absence does not involve immediate danger to human life. Such purposes as fire fighting, control of health hazards, long-term rescue operations and industrial hazard prevention would be among the electrical loads served. For these systems, loads must be picked up automatically with 60 seconds of failure of normal power, and maintained for the same minimum periods of time as specified for emergency power systems. Wiring may be run in the same enclosures as that of the normal power system.

Optional standby power systems are intended, essentially, to minimize economic loss, and therefore can supply any load selected by the building's user. Connection of the selected standby source can be manual or automatic and duration of service is entirely at the user's discretion. No strictures are placed on the wiring system other than those applied to the normal electrical wiring system.

(b) NFPA Codes. The *NEC* does not determine whether an emergency system is required; that is the function of the jurisdictional authorities, and is generally covered by NFPA 101-1997 (see following list). Once the determination has been made, the equipment and installation must comply with the requirements of the *NEC*. Relevant NFPA codes included (latest issue):

NFPA 70 National Electrical Code
NFPA 99 Standard for Health Care Facilities
NFPA 101 Life Safety Code
NFPA 110 Standard for Emergency Light and Power
NFPA 111 Standard on Stored Electric Energy Emergency and Standby Power Systems

(c) Technical Considerations. In general, when emergency power is discussed, it is assumed to be replacing "normal" power; that is, the assumption underlying most codes and ordinances is that power must be supplied to selected loads within the building because of utility power outages. In good practice design, cognizance must also be taken of situations in which normal power has not failed and the outage is localized because of an equipment failure. That aspect of design—reliability—is left to the designer. Some of the arrangements discussed here differentiate between the nature of outages—that is, a utility or general outage versus an equipment or local

outage. An exception to this generalization occurs with health-care facilities, where *NEC* in Article 517 specifies an internal electrical design that in large measure covers both types of outages, down to the distribution level. The interested reader should refer to the referenced *NEC* article for further reading.

An emergency system includes all devices, wiring, raceways, and other electrical equipment, including the emergency source that is intended to supply electric power to the selected load. An important aspect of code requirements permits the use of a single power source for (1) emergency, (2) legally required standby, and (3) optional standby systems provided it is equipped with automatic selective load pickup and load-shedding equipment that will assure adequate power to the three types of systems in the order of priority stated (1, 2, 3).

1. Where emergency loads are light, a storage battery arranged to be connected automatically on power outage may be used. Where all emergency loads can be operated on direct current, the arrangement of Fig. 17.27(*a*) is used. (Note that ac lighting can accept dc emergency power if equipped with a local inverter, as discussed in Section 21.36). If ac is required, the arrangement of Fig. 17.27(*b*) can be utilized.

2. Where emergency loads are larger than can be practically supplied by a battery installation, and where start-up and power transfer time of up to 10 seconds is tolerable, a generator set is employed. A combination of sources may be used in a single building. For instance, a generator can supply bulk power loads and a battery installation selected lighting loads.

The system can be arranged with a single transfer switch that senses normal power loss, as in Fig. 17.28(*a*), or it can use multiple switches, each one of which will sense power loss *at its downstream location,* as in Fig. 17.28(*b*). The latter system provides greater power reliability, provided that the design is such that the emergency power uses an independent power path to the transfer switches. Otherwise, a fault in a junction box or an item of equipment that interrupts normal power downstream will also prevent emergency power from reaching that point.

Central battery bank to supply d-c

(a)

Central battery bank to supply a-c and d-c loads

Emergency loads

Fig. 17.27 *Use of a central battery for emergency power supply. Where all loads can be energized with dc, the arrangement of* (a) *is satisfactory. When ac as well as dc must be distributed, a central inverter is added as in* (b), *and the ac and dc emergency loads fed separately.*

3. Many codes permit the use of two separate electric services in lieu of a normal service plus an emergency source, provided that the two sources are independent—that is, come from different utility transformers or feeders; enter the building at different points and, preferably, from different directions; and use separate service drops or laterals. The point is, of course, that the type of reliability desired can be obtained only by minimizing the possibility of a single event interrupting both services. The usual arrangement is for one such utility service to be "normal" and the other "standby."

4. The least reliable arrangement is one in which the emergency loads are connected ahead of the main disconnects and are so arranged that a downstream fault within the building will not affect these items. This situation is illustrated in the riser diagram of Fig. 17.25, where the stair and exit panel, which supplies egress lighting, and the fire alarm panel are connected ahead of the building main disconnect and protected with their own fuses. This arrangement is permitted only for Legally Required Standby System (*NEC* Article 701) and not for Emergency Systems as defined in *NEC* Article 700. The items that may be con-

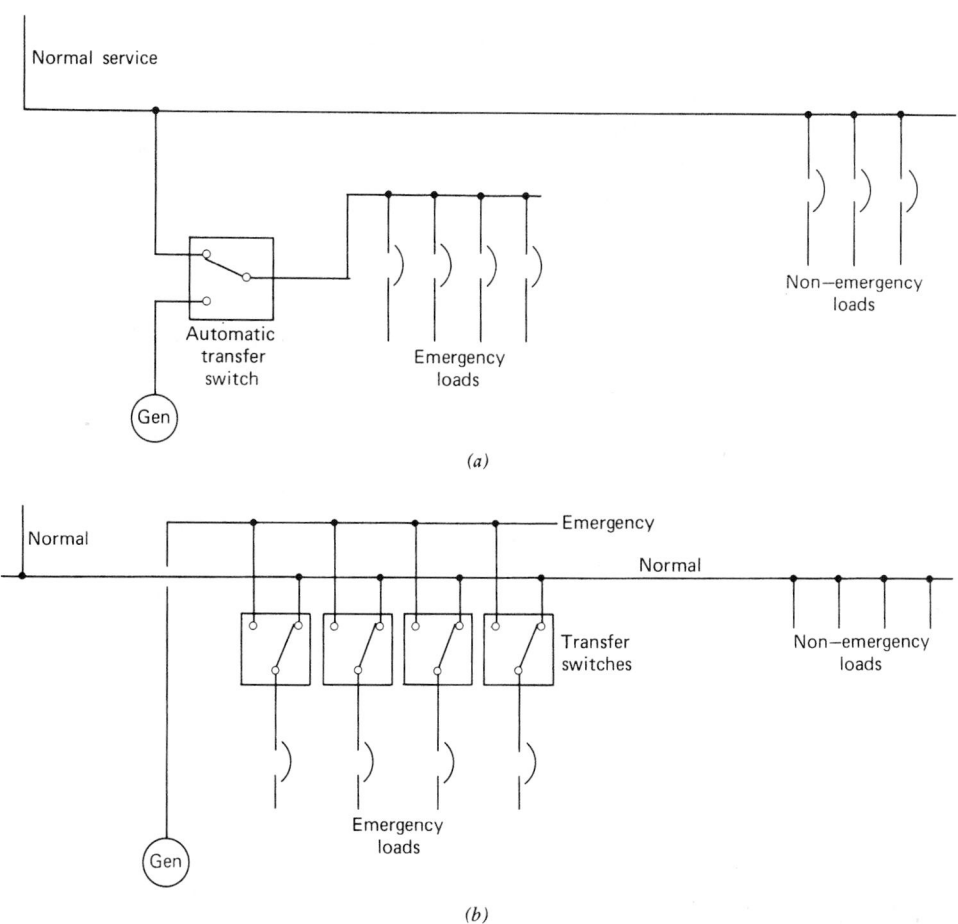

Fig. 17.28 *Alternative arrangements of emergency/normal power feed. In* (a) *a single transfer switch serves the normal power and transfers to the generator on power failure. In* (b) *the transfer switches are smaller, thus reducing the chance of a single equipment failure faulting out the entire emergency power system.*

nected ahead of the main service equipment are generally limited to emergency lighting, fire alarm, fire pumps, standby power equipment and other alarm and protective equipment, each of which must be provided with a separate disconnect means and overcurrent protection (see *NEC* Article 230-82).

This power arrangement obviously can do nothing in the event of a power outage. It was once very common, but is now falling into disuse as a result of more stringent codes and its obvious shortcomings.

5. The *NEC* recognizes a category of equipment for emergency illumination called "unit equipment." These devices, discussed and illustrated in Section 21.36, consist of individual self-contained packages with battery, charger, and light source *permanently* mounted and wired at required locations. The panel device feeding these units should be clearly identified.

REFERENCES

McGuinness, W. J., Stein, B., and Reynolds, J. S. (1980). *Mechanical Equipment for Buildings,* Wiley, New York.

Stein, B. (1997). *Building Technology, Mechanical and Electrical Systems,* Wiley, New York.

PART VI
ILLUMINATION

It has been estimated that 90% of the information we obtain from our senses is received via our sense of sight. Vision, in turn, is made possible by the presence of light, the proper provision of which is discussed in this part of the book. Because architecture is a uniquely visually oriented profession, its practitioners must thoroughly understand the art and science of illumination in order to be able to integrate structure and light into a working whole that will appear, and function, according to the design intent.

Chapter 18, Lighting Fundamentals, introduces the subject with terminology, definitions, basic characteristics, and measurements. A clear distinction is made between the study of light as a physical phenomenon and its study for visual photometric effect. The former considers light as a form of energy, whereas the latter views it as a visual phenomenon (i.e. , the physiological response to a form of energy). Pursuing the subject of human response to light, the chapter continues with a discussion of factors in visual acuity. This subject is concluded with an exposition of the development of current American and European lighting-level recommendations.

The discussion continues with consideration of factors in lighting quality (as distinct from quantity), with particular emphasis on analysis and control of reflected glare. The chapter concludes with material on color vision, color in light, and color temperatures of light sources.

Chapter 19, Light Sources: Their Characteristics and Application, covers the two principal light sources: daylight and electric light. The daylighting section begins with a discussion of the characteristics of natural light and then proceeds to explain and demonstrate, with illustrative examples, a number of daylighting design methods. These are both graphic and analytic, American and international. Here too, as throughout the book, full cognizance is taken of computerized calculation procedures.

The second subject, electric light sources, covers in detail incandescent, fluorescent, mercury, metal–halide, and sodium lamps, plus a few of the more recently developed electric light sources of interest to architects and engineers.

The discussions include a detailed description of construction and operating characteristics, including lamp accessories such as ballasts and lamp holders. Emphasis is placed on luminous efficiency and overall life-cycle costs. The chapter concludes with a discussion of chromaticity of light sources, their spectral distribution, and color rendering index.

Chapter 20, Lighting Design, leads the reader step-by-step through the design procedure methodology. The first section, on preliminary design, covers cost factors, power budgets, energy considerations, and illumination methods. This is followed by a section on detailed design that discusses lighting fixture characteristics and illumination calculation techniques. The chapter concludes with a discussion of lighting design evaluation. Illustrative examples are used to demonstrate and compare calculation techniques.

Chapter 21, Lighting Application, applies the knowledge and techniques explained in the three preceding chapters to specific occupancies, each of which has its special needs, namely: residential, educational, commercial, and industrial facilities. In addition, important special topics are discussed in detail, including emergency lighting, exterior lighting, and lighting for computer use. Finally, lighting control strategies and techniques are covered in detail from the points of view of energy conservation, economics, and controls.

Throughout the four chapters of this section, prime emphasis is placed on principles and logical procedures and only secondarily on "hardware."

The lighting field is one that changes constantly, not only with new sources, fixtures, lenses and lightpipes, but also with reconsideration of very basic and long-held beliefs relating to visibility, visual acuity, light quality, and spectrum.

A competent lighting designer must have suf-ficient background, both scientific and design, to be able to evaluate, adapt and adopt these developments. This background can only be acquired by a thorough understanding of the principles of lighting and their application, which are the burden of the following four chapters.

18

LIGHTING FUNDAMENTALS

Architecture is the masterly, correct and magnificent play of masses brought together in light. Our eyes are made to see forms in light; light and shade reveal these forms; cubes, cones, spheres, cylinders, or pyramids are the great primary forms which light reveals to advantage. *Le Corbusier*

18.1 Introductory Remarks

For many years an artificial division existed in the field of lighting design, dividing it into two disciplines: architectural lighting and utilitarian design. The former found expression in design that took little cognizance of visual task needs and displayed an inordinate penchant for incandescent wall washers, architectural lighting elements, and form-giving shadows. The latter saw all spaces in terms of illuminance levels and cavity ratios, and performed its design function with foot-candles (lux) and dollars as the ruling considerations. That both these trends have in large measure been eliminated is due largely to the efforts of thoughtful architects, engineers, and lighting designers, assisted in part by the energy consciousness that followed the 1973 Arab oil embargo. The latter spurred research into satisfying real vision needs within a framework of minimal energy use. That research and its resulting energy codes and continuing development of higher efficiency sources is today motivated by environmental considerations.

Another positive factor in the rationalization of lighting design has been the work of the Illuminating Engineering Society of North America (IESNA). Its activities in research, standardization, and publication have done much to place lighting design on a stable scientific basis while taking full cognizance of its essential artistic aspects. It is precisely this combination of science and art that makes lighting design an architectural-type discipline. For each project, a responsible lighting designer will consider *quantitatively:*

1. Daylight—its introduction and integration with electric light.
2. The interrelationship between the energy aspects of electric and natural lighting, heating, and cooling.
3. The effect of lighting on interior space arrangement and vice versa.
4. The characteristics, means of generation, and utilization techniques of electric lighting.
5. Visual needs of specific occupants and of specific tasks.
6. The effects of brightness patterns on visual acuity

and *qualitatively:*

7. The location, interrelationship, and psychological effects of light and shadow—that is, brightness patterns.
8. The use of color, both of light and of surfaces, and the effect of illuminant source on object color and sometimes the reverse.
9. The artistic effects possible with patterns of light and shadow including the changes inherent in daylighting, and so on.
10. Physiological and psychological effects of the lighting design, particularly in spaces occupied for extended periods.

The list is almost endless because so much of the information we receive from our senses comes via our eyes, and what we see is a direct consequence of scene lighting.

As a result of the need to consider these and other interrelated factors, many of which are mutually incompatible, the lighting designer is faced with many difficult decisions. The purpose of the lighting chapters in this book is then twofold: to provide the background that will help lighting designers make these decisions correctly, and to make them proficient in the use of lighting as a design material.

PHYSICS OF LIGHT

18.2 Light as Radiant Energy

The IESNA defines light as visually evaluated radiant energy or, more simply, a form of energy that permits us to see. If light is considered as a wave, similar to a radio wave or an alternating current wave, it has a frequency and a wavelength. Figure 18.1 shows the position of light in the wave spectrum with relation to other wave phenomena of various frequencies.

From the chart we see that even the longest wavelength light (red) is a much higher frequency than radio and radar, and that visible light constitutes only a very small part of the wave energy spectrum. Color is determined by wavelength. Starting with the longest wavelengths (red), we proceed through the spectrum of orange, yellow, green, blue, indigo, and violet to arrive at the shortest visible wavelengths (highest frequency). Bordering the visible spectrum are infra-red at the low-frequency (long wavelength) and ultra-violet at the high-frequency end. Both are invisible to humans but are not so to some animals.

When a light source produces energy over the entire visible spectrum in *approximately* equal quantities, the combination appears "white," whereas a source producing energy over only a small section of the spectrum produces its characteristic colored light. Examples are the blue–green clear mercury lamp and the yellow sodium lamp.

For our purposes, all light will be considered "white" unless specifically noted otherwise. This position is scientifically tenable because light sources with large differences in chromatic content all appear white after a short accommodation period, and all standard commercial sources permit colors to be easily and correctly identified. It is only when sources differing widely in chromaticity are viewed side by side that the variation in "whiteness" can be noticed, by the effect on colored objects and on neutral surfaces. This subject is discussed in several places in the following chapters, and in depth in Chapter 19, in the explanations of source chromaticity.

18.3 Transmittance and Reflectance

Lighting design is possible because light is predictable; that is, it obeys certain laws and exhibits certain fixed characteristics. Although some of these are so well known as to appear self-evident, a review is in order.

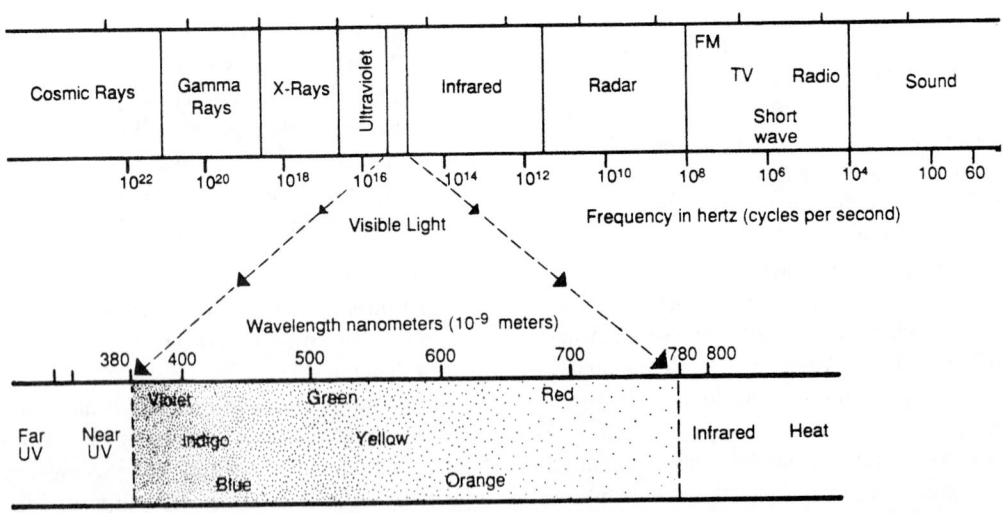

Fig. 18.1 *Electromagnetic spectrum. See Chapter 19 for chromaticity diagrams of various light sources.*

The *luminous transmittance* of a material such as a luminaire lens or diffuser is a measure of its capability to transmit incident light. By definition, this quantity, known variously as *transmittance, transmission factor,* and *coefficient of transmission,* is the ratio of the total transmitted light to the total incident light. In the case of incident light containing several spectral components passing through a material that displays *selective* absorption, this factor becomes an average of the individual transmittances for the various components and must be used cautiously. A piece of frosted glass and a piece of red glass may both have a 70% transmission factor, but obviously affect the incident light differently. In general then, transmission factors should be used only when referring to materials displaying nonselective absorption—that is, those that transmit the various component colors equally. Clear glass, for instance, displays a transmittance between 80% and 90%, frosted glass between 70% and 85%, and solid opal glass between 15% and 40%. The remainder is absorbed and reflected. See Table 19.6 for typical transmission factors.

Similarly, the ratio of reflected to incident light is variously called *reflectance, reflectance factor,* and *reflectance coefficient.* Thus, if half the amount of light incident on a surface is bounced back, the reflectance is 50%, or 0.50. The remainder is absorbed, transmitted, or both. The amount of absorption and reflection depends on the type of material and the angle of light incidence because light impinging on a surface at grazing angles tends to be reflected rather than absorbed or transmitted (Fig. 18.2). An example of almost perfect reflection from an opaque surface would be that from a well-silvered mirror, whereas almost complete absorption takes place on an object covered with lamp black or matte-finish black paint. The effect of the material's surface finish on reflection is shown in Fig. 18.3. See Tables 19.14 and 20.3 for typical reflectance values. Reflectance measurement is discussed in Section 18.11.

The reflection that occurs on a smooth surface such as polished glass or stone is called *specular reflection,* as in Fig. 18.3(*a*). If the surface is rough, multiple reflections take place on the many small surface projections and the light is diffused, as in Fig. 18.3(*b*). Reflectance is a measure of total light reflected; it may be specular or diffuse, or a combination of both, as shown in Fig. 18.3(*c*).

Diffuse transmission takes place through any translucent material such as frosted glass, white glass, milky Plexiglas, tissue paper, and so on. This diffusing principle is widely employed in luminaires to spread the light generated by the source within the fixture. Diffuse and nondiffuse transmission are illustrated in Fig. 18.4(*a*) and (*b*).

18.4   Terminology and Definitions

Before beginning any discussion of lighting studies, techniques, and effects, it is important to have a basic understanding of the physical concepts and terminology involved and their interrelations. We use and explain both the conventional American Standard (AS) and the SI systems of units in this text. The latter, also called the *metric system,* is used as the basic system by the IESNA and in this book, whereas the lighting industry uses both systems. To familiarize the reader with the SI system, we frequently use duplicate units, with the second system's unit enclosed in square brackets [ ].

Fig. 18.2 *Relation between angle of incidence and percentage reflectance. This effect is important when considering the penetration of sunlight into interior spaces and, conversely, the exterior glare produced by reflection of the sun from building windows.*

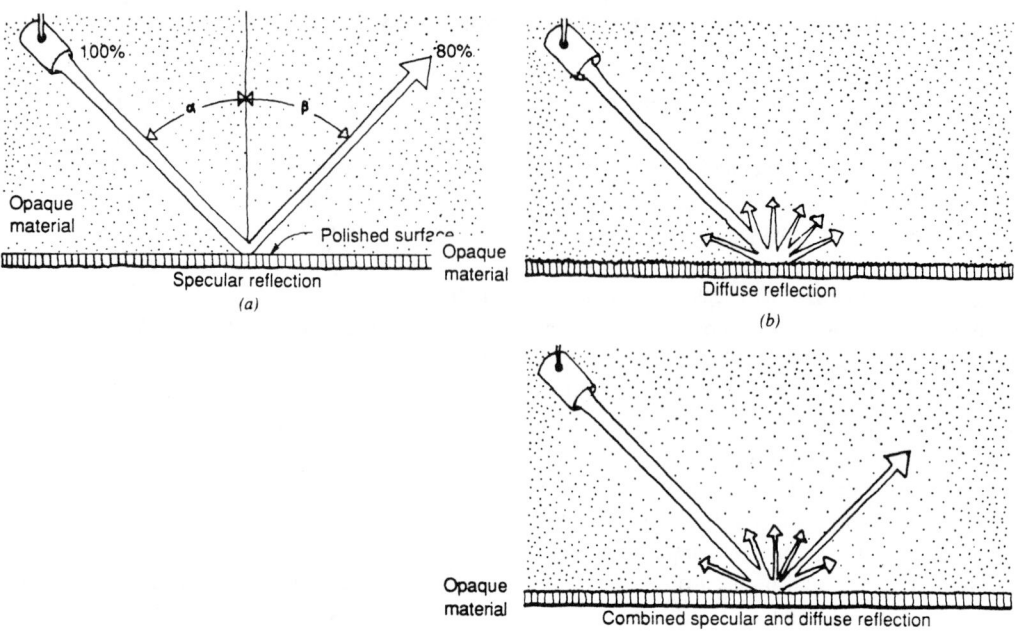

Fig. 18.3 *Reflection characteristics*
(a) *In specular reflection angle of incidence equals angle of reflection ($\alpha = \beta$). Because 80% of light is reflected, reflectance is 80%; 20% of light is absorbed.*
(b) *In diffuse reflection, incident light is spread in all directions by multiple reflections on the unpolished surface. Such surfaces appear equally bright from all viewing angles.*
(c) *Most materials exhibit a combination of specular and diffuse reflection. Such a surface mirrors the source while producing a bright background.*

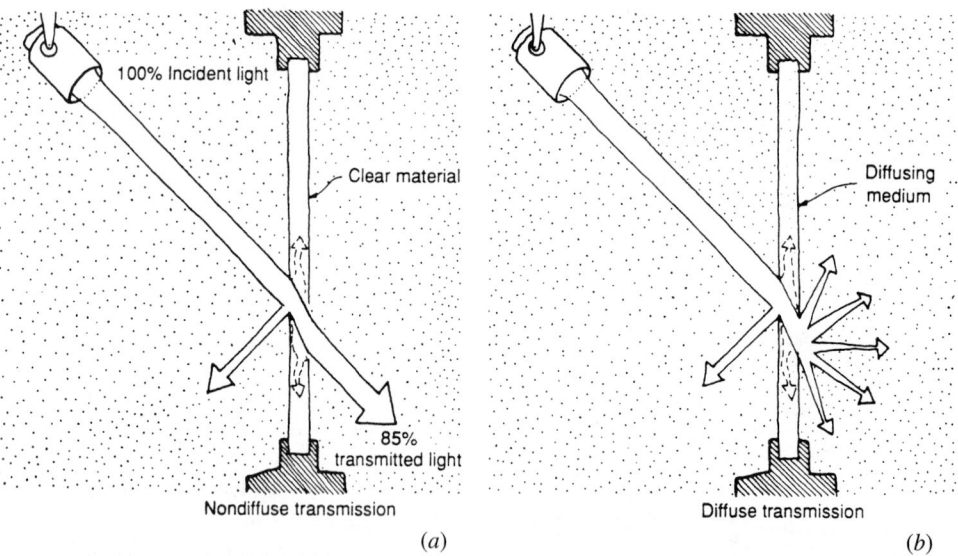

Fig. 18.4 *Transmission characteristics.*
(a) *In nondiffuse transmission, the light is refracted (bent) but emerges in the same beam as it enters. Clear materials such as glass, water, and certain plastics exhibit this type of transmission. In the instance illustrated, the transmittance is 85% (the remaining 15% is reflected and absorbed). The source of light is clearly visible through the transmitting medium.*
(b) *With diffuse transmission, the source of light is not visible and, in the case of multiple sources, the diffusing surface exhibits generally uniform brightness if the spacing between the light sources does not exceed approximately 1½ times their distance from the material.*

18.5 Luminous Intensity

The SI unit of *luminous intensity* is the candela [candlepower] abbreviated cd [cp], and normally represented by the letter *I*. It is analogous to pressure in a hydraulic system and voltage in an electric system, and represents the force that generates the light that we see. An ordinary wax candle has a luminous intensity horizontally of approximately one candela, hence the name. The candela and candlepower have the same magnitude. Luminous intensity is a characteristic of the source only; it is independent of the visual sense.

18.6 Luminous Flux

The unit of luminous flux, in both SI and conventional (AS) units, is the lumen (lm). If we take a 1 candela [candlepower] source that radiates light equally in all directions and surround it with a transparent sphere of 1 meter [foot] radius (Fig. 18.5), then *by definition* the amount of luminous energy (flux) emanating from 1 square meter [square foot] of surface on the sphere [lumen] is 1 lm. Because there are 4π m² [ft²] surface area in such a sphere, it follows that a source of 1

candela [candlepower] intensity produces 4π, or or 12.57 lm. The lumen, as luminous flux, or quantity of light, is analogous to flow in hydraulic systems and current in electric systems and is normally represented by the letter ϕ.

In physical terms, the lumen is a unit of power, like the watt. However, unlike the watt, which is a radiometric unit directly convertible to other power units such as Btu/h, the lumen is a measure of *photometric* power. This means light power as perceived by the human eye, and therefore a function of human physiology. Put another way, lumens or luminous flux is the time rate of flow of *perceived* luminous energy. Because the visual response of the eye is frequency dependent, the apprehended light power is therefore also frequency dependent, varying with the spectral content of the impinging light and the spectral sensitivity of the eye. Figure 18.6(*a*) shows the spectral content of the visible energy produced by a 500-W incandescent lamp. Measured radiometrically, it amounts to 45 W. However, when passed through a selective filter [Fig. 18.6(*b*)], which is effectively what happens when the light enters the eye, the resultant "understood" light power appears as in Fig. 18.6(*c*), and therefore can no longer be measured in watts. Instead, we use a

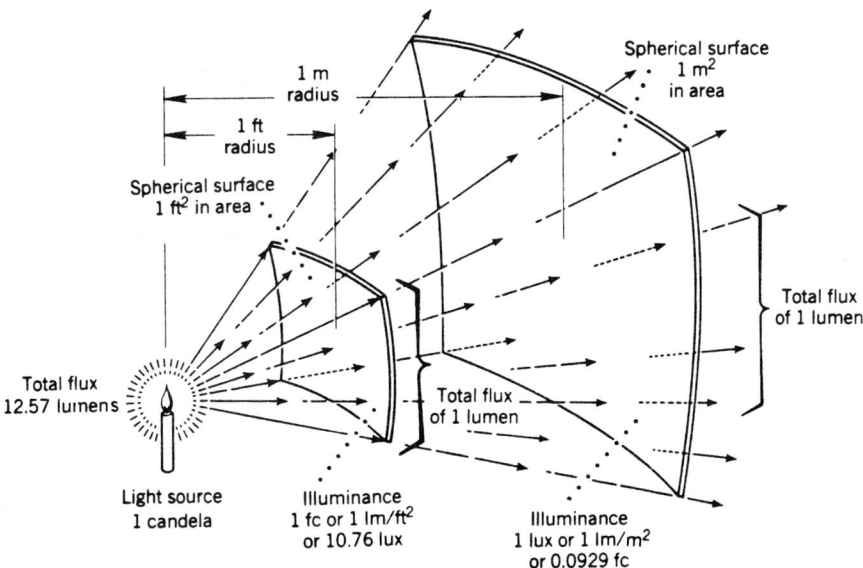

Fig. 18.5 *A source of one candela intensity produces 4 π (12.57) lumens of light flux. Thus, each square foot [square meter] of (spherical) surface surrounding such a source receives 1 lumen of light flux. This quantity of light flux produces an illuminance of 1 footcandle [lux] on the spherical surface.*

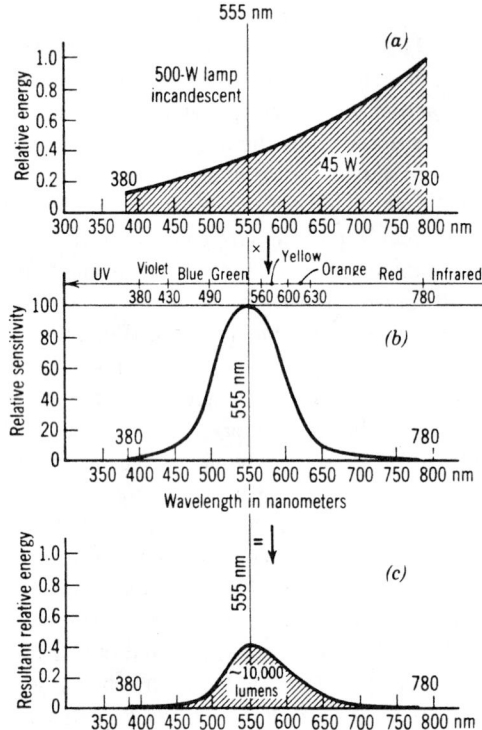

Fig. 18.6 *Graphical demonstration of the method by which the unit of light flux is defined.*
(a) *Shows the spectrum of the light produced by a 500-W incandescent lamp. It amounts to approximately 45 W, measured radiometrically. When filtered by the human eye, whose spectral sensitivity curve is given in* (b), *this light power is perceived as shown in* (c). *The new light power curve is expressed in lumens and indicates the quantity of light as perceived by the eye.*

unit of eye-perceived, or photometric, power called the *lumen*. It should be obvious that if the spectral content curve in Fig. 18.6(*a*) were differently shaped, *even if the total radiometrically measured power were the same,* the resultant perceived power in Fig. 18.6(*c*) would be different.

Refer to Fig. 18.6(*b*). A correlation can be made between photometric and radiometric power at the point of maximum response of the eye, which occurs at 555 nanometers wavelength. [A nanometer is 10^{-9}m. The older unit, now obsolete, is the angstrom (Å), which is 10^{-8}cm, or 10^{-10}m (i.e., ¹⁄₁₀ of a nanometer)]. One watt of monochromatic light at that wavelength produces 683 lumens. However, because common light

sources such as incandescent, fluorescent, mercury, and so on, are not monochromatic but produce light in many parts of the spectrum (see Figs. 19.68, 19.69, and 19.70), no single conversion factor between watts and lumens exists. Each source has its own luminous efficiency (lumens/watt), determined by its spectrum. For the 500-W lamp used as an illustration in Fig. 18.6, its luminous efficiency (efficacy) is 10,000 lm/500 watts, or 20 lumens per watt (lm/w or lpw).

18.7 Illuminance

One lumen of luminous flux, uniformly incident on 1 meter [square foot] of area, produces an *illuminance* of 1 lux (lx) [*footcandle* (fc)]. Illuminance is normally represented by the letter E. Restated, illuminance is the density of luminous power, expressed in terms of lumens per unit area. If we were to consider a light bulb to be analogous to a sprinkler head, then the rate of water flow would be the lumens and the amount of water per unit time per m² [ft²] of floor area would be the lux [footcandles]. Thus the SI unit, lux, is smaller than the corresponding AS unit, footcandles, by the ratio of square meters to square feet. That is,

$$10.764 \text{ lux} = 1 \text{ footcandle}$$

or, multiply footcandles by 10.764 to obtain lux. These relations are shown in Fig. 18.5. Restating mathematically yields

$$lux = \frac{lumens}{square\ meters\ area}$$
$$lx = \frac{lm}{m^2} \tag{18.1}$$

and

$$footcandles = \frac{lumens}{square\ foot\ area}$$
$$fc = \frac{lm}{ft^2} \tag{18.2}$$

As an *approximation* (8% error)

$$10 \text{ lx} \cong 1 \text{ fc} \tag{18.3}$$

EXAMPLE 18.1. A 34-W, 425-mA (milliampere), 48-in. [122-cm] fluorescent tube produces 3200 lm. What is the illuminance on the floor of a 3-meter square room assuming 60% overall efficiency and uniform illumination?

SOLUTION.

Useful lumens = $0.6 \times 3200 = 1920$

$$lx = \frac{1920}{3 \times 3} = 213.3 \ lx$$

$$fc = 213.3 \times 10.76 = 19.8 \ fc$$

Calculating footcandles directly, we obtain

$$fc = \frac{1920}{3 \times 3} \times \frac{m^2}{(3.28 \ ft/m^2)} = 19.8 \ fc$$

By approximation:

$$fc \cong lx/10 = 21.3$$

Note that this calculation gives *average* illumination in the space. Illuminance at a *point* can be computed from intensity, as explained in Section 18.12. ■

18.8 Luminance, Exitance, and Brightness

An object is perceived because light coming from it enters the eye. The impression received is one of object *brightness*. This brightness sensation, however, is subjective, and depends not only on the object *luminance* (L), but also on the state of adaptation of the eye (see Sections 18.18 and 18.19). For this reason, the physiological sensation is generally referred to in the literature as *subjective* or *apparent brightness,* or simply *brightness,* whereas the measurable, reproducible state of object luminosity is its *luminance* (formerly "photometric brightness") Luminance is normally defined in terms of intensity; it is the luminous intensity per unit of *apparent* (projected) area of a primary (emitting) or secondary (reflecting) light source. Thus its units are candela per area. Specifically, the SI unit of luminance is candela per square meter (cd/m^2), sometimes referred to as the *nit.* Another unit formerly in common use in the conventional (AS) system is the footlambert. Conversion factors for SI and AS units (plus other, obsolete units, for the convenience of readers using older sources) are given in Table 18.1. Other luminance terms such as *stilb, apostilb, blondel, millilambert,* and *candela per square in.* are best avoided. In this book, the term *luminance* is used except where it is specifically intended to refer to the physiological sensation involved, in which case the terms *brightness, subjective brightness,* or *apparent brightness* will be used. Luminance has no readily conceivable mechanical or electrical analogy.

A word of caution at this point is in order. Although definitions and terminology are established for the specific purpose of accurate information exchange, the lighting literature is replete with articles, comments, and rebuttals that exist only because of the looseness of definitions and terminology. Some authors insist on applying "brightness" only to self-luminous surfaces, using

ILLUMINATION

TABLE 18.1 **Lighting Units—Conversion Factors**

Unit	Multiply	By	To Obtain
Illuminance (E)	Lux	0.0929	Footcandle
	Footcandle	10.764	Lux
Luminance (L)	cd/m^2	0.2919	Footlambert
	cd/cm^2	10000	cd/m^2
	$cd/in.^2$	1550	cd/m^2
	cd/ft^2	10.76	cd/m^2
	millilambert	3.183	cd/m^2
	Footlambert	3.4263	cd/m^2
Intensity (I)	Candela	1.0	Candlepower

ILLUMINATION

"lightness" as the equivalent term for objects deriving their luminance from reflection. Thus, the sun has brightness; the moon, lightness. In this book, we use the term *brightness* for the subjective reaction to either source type. Other authors (constantly) point out that the luminance–brightness relation breaks down when light other than "white" is used. Although this is demonstrable, it is of real interest only in theatrical lighting where colored light is very frequently used. For our purposes we assume white light, and as pointed out previously, the color accommodation characteristic of our eyes recognizes as white (colorless) light of a large chromatic range. Within that range, and for a very large range of intensities, object color is readily recognizable and the fixed luminance–brightness ratio is maintained. Contrasting word usages such as *dim* and *dark, light* and *bright, clear* and *muddy, shallow* and *deep,* and so on as applied to lighting are best left to experienced lighting designers because the terms are almost entirely subjective, and therefore unhelpful to novice designers.

Another concept that the lighting designer will encounter is know as *luminous exitance,* or simply as *exitance,* and, as the name implies, describes the total luminous flux density leaving (exiting) a surface, irrespective of directivity or viewer position. For instance, if a surface 1 meter square emits 1 lumen, its luminous exitance is 1 lumen per square meter (1 lm/m²) or 0.093 lm/ft². A surface that is a perfect diffuser, whether by emitting light diffusely or reflecting light diffusely, is known as a *Lambertian surface.* It is fairly simple to demonstrate mathematically that the luminance of such a surface equals $1/\pi$ times its exitance. The importance of this relationship is its usefulness as an approximation. Although very few surfaces are truly Lambertian, many are approximately so, and this relationship can be used as an engineering-accuracy approximation in many such cases.

The concept of exitance is important in detailed photometric calculations such as those involved in determining coefficients of utilization (Section 20.28), surface luminance coefficients (Section 20.43), and in detailed point illuminance calculations. All of these are beyond the scope of this book because they are not usually performed by the lighting designer. Use of the derived coefficients is demonstrated in the referenced sections.

Detailed point calculations are today almost universally performed by computer, and the necessary mathematics is built into the computer program. Readers interested in further background on luminous exitance are referred to Murdoch (1985). References are found at the end of Chapter 21.

Because object luminance is that which is visually perceived and is a prime factor in visibility (and glare), it is important that the reader be able to perform basic luminance calculations. Although the eye does not differentiate between primary sources that generate and emit light, and secondary sources that derive their luminance from reflection or transmission, the differentiation is important in calculation procedures. See Fig. 18.7 for a graphic representation of the basic relationships.

EXAMPLE 18.2. Luminance of a light-emitting surface.

(a) Calculate the luminance of an A-19 standard inside-frosted, 100-W incandescent light bulb with a maintained output of 1700 lm. Assume (for simplicity's sake) that the bulb is spherical.

Illuminance
 100 fc or
 100 lm/ft²
Diffuse Reflectance
 0.6 or 60%
 Reflected luminance
 100 fc × 0.6 reflectance
 = 60 fL or
 60 lm/ft² reflected

1 ft

1 m

Light source
 100 candelas or
 1257 lumens output

Illuminance (inside surface)
100 lumens/m² or 100 lux
Transmittance 30%
Diffuse luminous existance =
.3(100) = 301 m/m²
Luminance (transmitted) =
$\dfrac{\text{Transmitted intensity}}{\text{Projected area of sphere}} =$
$\dfrac{(0.3)\,(100\ \text{cd})}{\pi\ \text{m}^2} = \dfrac{30}{\pi}\ \text{cd/m}^2$

Fig. 18.7 *Luminance may be either reflected or transmitted. In the former case, it is calculated as the product of the incident lumens and the reflectance; in the latter case, as the transmitted intensity divided by projected area.*

(b) Assume that an opal glass globe of 8-in. [20-cm] diameter and a transmittance of 35% surrounds the above bulb. Calculate the luminance of the globe. Use SI units throughout.

SOLUTION. (a) Assume that the filament is a point source (an essentially valid assumption) and that the inside frosting of the glass does not reduce the output (it does by about 1%). The inside frosting serves to convert the point source filament to a uniformly emitting globe. The definition of a point source tells us that 1 candela produces 4π lumens, distributed spherically. Therefore:

$$\frac{1 \text{ candela}}{4\pi \text{ lm}} = \frac{I \text{ candela}}{1700 \text{ lm}}$$

The intensity, I, of the filament, and therefore also of the bulb (because the frosting is assumed not to reduce output) is

$$I = 1700 \text{ lumens} \times \frac{1 \text{ candela}}{4\pi \text{ lm}} = 135 \text{ candela}$$

An A-19 bulb has a diameter of 19/8″ (6 cm). The definition of luminance is

$$L = \frac{I}{A} \frac{(\text{intensity})}{(\text{projected area})}$$

The projected area becomes approximately a circle with a diameter of 6 cm (assuming a spherical bulb), whose area equals πr^2 or π (0.03 m)^2. So

$$L = \frac{135 \text{ candela}}{\pi (0.03 \text{ m})^2} = 47,750 \frac{\text{candela}}{\text{m}^2}$$

This luminance is a potentially severe glare source, if it is in the field of vision (depending on its distance from the eye; see Section 18.28).

(b) We have already calculated the intensity of the source, but it is reduced to 35% by the globe. The projected area of the globe is larger than that of the bulb; that is,

$$A \text{ proj.} = \pi r^2 = \pi \frac{(0.2 \text{ m})^2}{(2)}$$

So the expression for luminance becomes

$$L = \frac{135 \text{ candela } (0.35)}{\pi (0.10 \text{m})^2} = 1500 \frac{\text{candela}}{\text{m}^2}$$

This is no longer a potential source of direct glare. ∎

Note two important principles demonstrated by this example:

1. Intensity is not area or distance dependent; it varies only with transmission factors.
2. Luminance varies with both intensity and area.

For example, if a white lamp whose inside coating reduces lamp output by about 10% had been used in lieu of an inside frost, both intensity and luminance would have been reduced by 10%. However, if a clear glass lamp had been used, intensity would remain the same, but luminance would increase inversely with the ratio of filament area to bulb area. Therefore, a clear 100-W lamp, whose filament has an area of perhaps of 0.3 cm^2, would have a luminance of

$$\frac{135 \text{ candela}}{0.3(E\text{-}4)\text{cm}^2} = 4,500,000 \text{ cd/m}^2$$

which is so severe a glare source as to be disabling when in the near field of vision.

EXAMPLE 18.3. Calculate the luminance of a 34-W, T12, 4-ft white fluorescent lamp. Assume a viewing angle normal to the long axis of the lamp and that the lamp is a diffuse (Lambertian) emitter. Use SI units.

SOLUTION. This problem can be solved in two ways: using the relationship between exitance and luminance or by calculating intensity.

By exitance/luminance: Use the 2770 lm at 40% life as an average condition of lamp output. The luminous length is 4 ft [120 cm] and the diameter is 12/8 in. [3.8 cm]. The luminous surface area of the tube is then

$$A = L \times d \times \pi = (120 \text{ cm}) \times (3.8 \text{ cm}) \times \pi$$

$$= (1.2 \text{ m}) (.038 \text{ m})$$

$$= 0.0456 \pi \text{ m}^2$$

$$\text{exitance} = \frac{\text{luminous flux}}{\text{area}} = \frac{2770 \text{ lm}}{0.0456\pi \text{ m}^2}$$

$$\text{luminance} = \frac{I}{\pi}(\text{exitance})$$

$$= \frac{2770}{0.0456\pi^2} \text{ cd/m}^2 = 6155 \text{ cd/m}^2$$

ILLUMINATION

By intensity: The equivalent spherical intensity of the lamp can be calculated by using the relationship that 1 candela produces 4π lumens. Therefore,

$$\text{equivalent intensity } I = \frac{2770/m}{4\pi} \text{ candela}$$

The radius of a sphere of equivalent surface area to the fluorescent tube would be

$$4\pi r^2 = 0.0456\pi \text{ m}^2$$

$$r^2 = \frac{0.0456}{4} \text{ m}^2 = 0.0114 \text{ m}$$

$$r = 0.1067 \text{ m}$$

The equivalent projected area of such a sphere is πr^2, or

$$A = 0.0114\pi$$

and its luminance is

$$L = \frac{\text{intensity}}{\text{area}} \frac{\text{cd}}{\text{m}^2} = \frac{2770/4\pi}{0.0114\pi} \frac{\text{cd}}{\text{m}^2}$$

$$= \frac{2770}{0.0456\pi^2} \frac{\text{cd}}{\text{m}^2}$$

$$= 6155 \text{ cd/m}^2$$

This luminance can constitute a glare problem, depending on its position in the field of view. If we had used a 40-W lamp with 3200 lm output, its luminance would be higher in direct proportion to the ratio of output; that is,

$$L_{3200} = L_{2770} \times \frac{3200}{2770} = 6155 \frac{\text{cd}}{\text{m}^2} \frac{(3200)}{(2770)}$$

$$= 7110 \text{ cd/m}^2$$

This is the origin of the frequently used average figure of 7000 cd/m^2 as the luminance of a standard 48-in. fluorescent tube. It is a borderline glare source.

Because of their energy saving characteristic, the 32-W, 48-in. T8 lamp with 3000 lm output has come into wide use. That it should not be used unshielded can be readily demonstrated by a similar luminance calculation:

$$\text{Tube area} = L \times \pi d$$

$$= (120 \text{ cm})(\pi)8/8''\left(2.54 \frac{\text{cm}}{\text{in.}}\right)$$

$$= 958 \text{ cm}^2 = .0958 \text{ m}^2$$

$$\text{exitance} = \frac{\text{lumens}}{\text{area}} = \frac{3000 \text{ lm}}{.0958 \text{ m}^2} = 31,330 \text{ lumens/m}^2$$

$$\text{luminance} = \frac{1}{\pi} (\text{exitance}) = \frac{31,330}{\pi} = 9972$$

or approximately 10,000 cd/m^2. Because this level of luminance is potentially a mild glare problem. These lamps should not be used in bare-bulb fixtures. ■

EXAMPLE 18.4. To demonstrate the usefulness of the luminance/exitance approximation, calculate the luminance, in SI units, of the page that you are now reading. Assume a uniform illuminance of 500 lux, a diffuse reflectance of 0.77, and a viewing angle normal to the page.

SOLUTION. From the definition of illuminance, 1 lux is produced by 1 lumen falling on 1 square meter. Therefore, the exitance, or density of reflected lumens from the page, is

$$\text{exitance} = 500 \frac{\text{lm}}{\text{m}^2} \times 0.77 = 385 \frac{\text{lm}}{\text{m}^2}$$

and

$$L = \frac{1}{\pi} \times 385 \frac{\text{lm}}{\text{m}^2} = 122.5 \text{ cd/m}^2$$

(Typical luminances are given in Table 18.2; see Section 18.19) ■

18.9 Illuminance Measurement

Field measurements of illuminance levels are most commonly made with a portable illuminance meter, three of which are illustrated in Figs. 18.8 through 18.10. These devices contain a photoelectric material connected to a microammeter via electronic control circuitry and are calibrated in lux, footcandles, or both.

As explained in Section 18.6 and as shown in Fig. 18.6, the human eye is not equally sensitive to the various wavelengths (colors). Maximum sensitivity at high illuminance levels is in the yellow–green area (wavelength of 555 nm), whereas sensitivity at the red and blue ends of the spectrum is quite low. This effect is so pronounced that 10 units of blue energy are required to produce the same visual effect as 1 unit of yellow–green.

Fig. 18.8 Color- and cosine-corrected analog illuminance meter calibrated in lux and equipped with a remote cable–connected photocell. The unit has nine ranges covering 0 to 200,000 lux with a maximum ± 3.5% error. When equipped with the adapter shown in the lower left of Fig 18.9 the unit becomes a wide-angle (20°) aperture luminance meter with a range of 0 to 2,000,000 cd/m² in nine steps. The unit is battery powered, measures approximately 3 × 4 × 1½ in., and weighs 12 oz. (Photo courtesy of Gossen.)

Fig. 18.9 Electronic, digital, color- and cosine-corrected light meter that measures illuminance (0 to 200,000 lux) and luminance (0 to 2,000,000 cd/m²) in four ranges with a maximum of 2½% error. The unit also measures solar radiation in W/m². It is equipped with a recorder output for use in extended-time monitoring applications. The meter measures 3½ × 6 × 1 in. thick and weighs under 1 lb. (Photo courtesy of GOSSEN.)

Fig. 18.10 Autoranging digital light meter that displays the intensity of photometric excitation of the sensor to which it is connected in appropriate units—lux, klux, or W/m². Sensors that measure illuminance, solar irradiance, or photosynthetic radiation are cable connected to the input terminal at the top of the unit. The sensors are small, measuring 1¼ in. D × 1¾ in. H. The illuminance sensor is frequently used to measure light in architectural models—for instance, as in Fig. 19.26—as, due to its small size, it has minimal effect on light distribution in any but the smallest models. The meter has a range of 0 to 199,900 lux with better than 1% accuracy, is battery powered, measures 5½ × 3 × 1½ in., and weights 9 oz. (Courtesy of Li-Cor, Inc.)

Therefore, if a meter is to be useful, its inherent response, which is quite different from that of the human eye, must be corrected to correspond to the eye. For this reason, meters are "color corrected."

The cells (meters) must also be corrected for light incident at oblique angles that does not reach the cell due to reflection from the surface glass and shielding of the light-sensitive cell by the meter housing. This correction is known as *cosine correction*. A good meter must therefore be (and it will plainly so indicate) color and cosine corrected.

Modern photometers may have considerable electronic circuitry, which provides such functions as automatic ranging, integration for flickering or time-varying sources, and connection facilities for data storage and transmission. For determining average room illuminance when using a conventional nonintegrating meter a number of readings should be taken and an average computed. Where no definite heights is specified, readings are taken at 30 in. (75 cm) above the floor, a level known as the *working plane* because it is approximately normal desk height. The meter must always be held with the cell parallel to the plane of the test. Thus, to measure wall illuminance, the meter must be held with the cell parallel to the wall. If electric lighting readings are desired and the test is being conducted during daylight hours, readings should be taken with and without the electrical illumination and the results subtracted. Detailed instructions for conducting field surveys are contained in the IESNA publication, "How to Make a Lighting Survey." Briefly, a survey of an existing indoor lighting installation should establish:

1. Type, rating, and age of sources.
2. Type, design, and model of luminaires.
3. Maintenance schedule.

It should also measure:

1. Mounting height of luminaires.
2. Spacing and pattern of luminaires.
3. Reflectances of walls, floor, ceiling, and major items of furniture and equipment.
4. Illuminance levels throughout the area plus levels at all working plane elevations. In addition, vertical plane illuminance at walls and other major vertical planes should be measured. The significance of vertical surface luminance is discussed in detail in subsequent chapters.

18.10   Luminance Measurement

From the aspect of appreciation of the visual scene, including particularly considerations of glare, the measurement of luminance is more important and meaningful than that of illuminance (lux) [footcandles]. This is so because it is luminance or, more accurately, subjective brightness and brightness contrasts caused by photometric luminance that we see, not illuminance. Light, as such, is invisible. That lux measurements are still more widely taken than luminance measurements and utilized as a gauge of the adequacy of a lighting installation is due to two factors:

1. Lux meters are cheaper and simpler to use than luminance meters.
2. Design recommendations for lighting levels are given in terms of illuminance (Section 18.24). Lux measurements can therefore be used as a rapid, simple method of determining whether a particular lighting installation meets these design requirements (assuming that material reflectances and reflectance ratios are correctly chosen).

Luminance meters are available in a number of configurations, one of which is shown in Fig. 18.11. An approximation of the luminance of a reflecting or luminous source can be obtained using an illuminance (footcandle) meter of the type shown in Fig. 18.8. For diffuse *reflecting* surfaces, the cell of the meter is placed against the surface and then slowly retracted 2 to 4 in. (5–10 cm) until a constant reading is obtained. The luminance, in

Fig. 18.11 *Direct-reading, narrow-angle, spot-type luminance meter has an acceptance angle of 1°, a range of 0.001 to 299,000 cd/m² (0.001–87000fL), variable response speed to permit measurement of flickering sources, and a comparison mode that permits direct luminance comparison of two sources. Results are displayed digitally. (Courtesy of Minolta Corp.)*

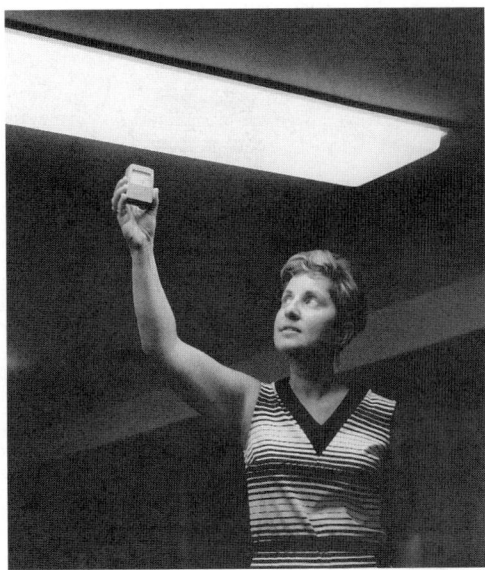

Fig. 18.12 *When the cell of a direct-reading illuminance meter is held in contact with a luminous source, the surface luminance can be read directly or simply calculated (see text).*

footlamberts, is then 1.25 times the reading in footcandles, the 1.25 factor compensating for wide-angle losses.

For a diffuse luminous source, the cell of an illuminance meter is placed directly against the surface (Fig. 18.12); the source luminance in footlamberts is equal to the reading on the meter in footcandles because footlamberts = lumens per area = footcandles. When using a meter calibrated in lux, the readings must be divided by π to obtain the diffuse source luminance in cd/m².

18.11 Reflectance Measurements

It is often desirable to know the reflectance of a given surface because luminance can then be readily computed (see Fig. 18.7). Two methods of measuring diffuse (nonspecular) reflectance are shown in Fig. 18.13: the known-sample method and the light–ratio method. If a sample of known reflectance factor (RF) is available, this method should be used because it yields more accurate results than the ratio method. The sample should be no smaller than 8 in. × 8 in.

It is a good idea for an inexperienced lighting designer to determine the reflectances, illumi-

ILLUMINATION

Fig. 18.13 *Two simple methods of measuring diffuse reflectance of a surface.*

nance, and luminance levels of spaces and surfaces familiar to him or her, such as an office desk, adjoining wall, and the like—even to the extent of marking these figures on the respective surfaces in order to develop an appreciation of and a memory for these parameters. This enables the designer to visualize the result of a lighting design and should be of considerable assistance. See Table 19.14 for typical reflectance values.

18.12   Inverse Square Law

We have already seen that, by definition, a point source of 1 cd intensity produces an illumination of 1 lux on the inside surface of a surrounding sphere of 1 m radius (r). Because the surface area of this sphere is 4π m^2, a 1-cd source produces 4π lm of luminous flux. Now, assume a sphere of 2-m radius surrounding this same source (Fig. 18.14). Because the same amount of flux is spread over a larger area, the illumination on the larger sphere

is inversely proportional to the ratio of the sphere areas; that is,

$$\frac{lux_2}{lux_1} = \frac{area_1}{area_2} \qquad (18.4)$$

or

$$lux_2 = lux_1 \times \frac{area_1}{area_2}$$

Therefore,

$$lux_2 = lux_1 \times \frac{4\pi r_1^2}{4\pi r_2^2}$$

$$= lux_1 \times \frac{r_1^2}{r_2^2} \qquad (18.5)$$

In other words, the illumination is inversely proportional to the square of the distance from the source. In general terms,

$$lux = \frac{cd\ intensity}{distance^2}$$

Fig. 18.14 *Relations between candela, lumens, and lux defined with reference to a standard light source of 1 mean spherical candlepower (1 candela) located at the center of a sphere of 1 meter radius.*

where distance is expressed in meters (feet). (This holds true for surfaces normal to a source. For other situations, see Section 20.38.)

This relationship can also readily be derived by using any solid angle and the area it intercepts, as in Fig. 18.15. A glance at this figure shows clearly that the area intercepted is proportional to the square of the distance from the source and therefore the illumination is inversely proportional, as stated previously.

Fig. 18.15 *Demonstration of inverse square law properties using a solid angle of unit size. Note that the surfaces are necessarily spherical because points on a planar surface are not equidistant from the source.*

18.13 Luminous Intensity: Candela Measurements

Luminous intensity (candela) [candlepower] cannot be measured directly but must be computed from its illumination effects. The simplest way of doing this is to use the inverse square relationship developed in the preceding section. Measure the illumination produced on a plane at right angles to the source at a known distance, and apply Eq. 18.6. For accurate measurement, the distance should be at least 5 and preferably 10 times the maximum dimension of the source because for anything other than a point source the equation is an approximation. The candela [candlepower] thus calculated is the luminous intensity in the direction being viewed. Because luminous intensity is not uniform in all directions for anything except an ideal point source, and because a single intensity figure for a source is desirable for calculation purposes, the average of a number of intensity figures taken from several directions is used. This average figure is called the *mean spherical candlepower (mscp)*, and represents an equivalent point source that produces 4π lm for every candela. Thus a 10-cd lamp exhibits an average intensity of $\pm$ 10 cd in all directions and produces 40π lumens.

18.14 Intensity Distribution Curves

If the luminous intensity figures calculated in the preceding section are plotted on polar coordinate axes, the resultant figure is called (using the AS unit) a *candlepower distribution curve* (CDC) for the particular source involved. The procedure for making this curve is straightforward. A photo cell is rotated around the source in a single plane, illuminance measured, and intensity (cd) calculated. Alternatively, the photo cell can be fixed and the source rotated. If the source's distribution is symmetrical, as shown in Fig. 18.16, then only a single set of values is required, and the resultant plot is valid in all vertical planes through the source. Thus for incandescent lamps, downlights, open circular reflectors, and the like, only a single CDC is required. For a nonsymmetric source such as a fluorescent luminaire, CDC curves in several planes are required to define the fixture's distribu-

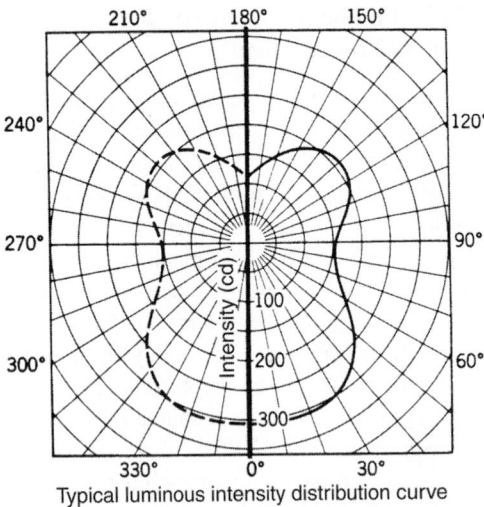

Typical luminous intensity distribution curve
for a general diffuse type of luminaire

Fig. 18.16 *Typical luminous intensity (cd) distribution curve for a general diffuse-type luminaire. Because the unit is symmetrical about its vertical axis, only one curve need be shown. Furthermore, only the right side of this curve need by shown, due to symmetry, as in Fig. 18.17.*

tion characteristic. Normally, manufacturers will provide longitudinal and crosswise curves, plus a diagonal (45° plane) curve on request. This is illustrated in Fig. 18.17, where the three planes and typical resultant curves are shown.

Most CDC plots are made on polar coordinates because such a plot clearly shows directions and magnitudes. Nevertheless, polar plots tend to crowd near the nadir and accurate magnitude readings at the cutoff angle are difficult to make. For this reason it is occasionally desirable to obtain a plot on rectangular coordinates. One such plot is shown in Fig. 18.18. The usefulness of intensity distribution curves will become clear in our subsequent discussions on lighting fixture diffusers (Section 20.19), point-by-point calculations (Section 20.39), and direct and reflected glare (Sections 18.28 to 18.30). It should be noted that the area of the CDC curve is *not* a measure of the lumen output.

Fig. 18.17 *Due to the asymmetry of a fluorescent luminaire, intensity distribution curves in (at least) three planes are required. However, because there is symmetry—for this fixture—in each individual plane, only one side of a curve is required. By convention, the right side is used.*

Fig. 18.18 *Luminous intensity distribution curves plotted in rectangular coordinates. Note that candela values near the cutoff angles are easily read, which is not the case in polar plots.*

LIGHT AND SIGHT

18.15 The Eye

Because any discussion of light and lighting techniques is irrelevant to our purposes unless ultimately related to vision, we turn to a cursory examination of the human eye before proceeding further with discussions of lighting.

Light impinging on the eye enters through the pupil, the size of which is controlled by the iris, thereby controlling the amount of light entering the eye. The lens focuses the image on the retina, from which the optic nerve conveys the visual message by electric impulse to the brain. Figure 18.19 shows the structure of the eye and the parallel structure of a camera.

The central portion of the eye, near the fovea, contains light-sensitive cells called *cones* because of their shape. The cones are responsible for the ability to discriminate detail and also give us our sensation of color. Proceeding outward from the

ILLUMINATION

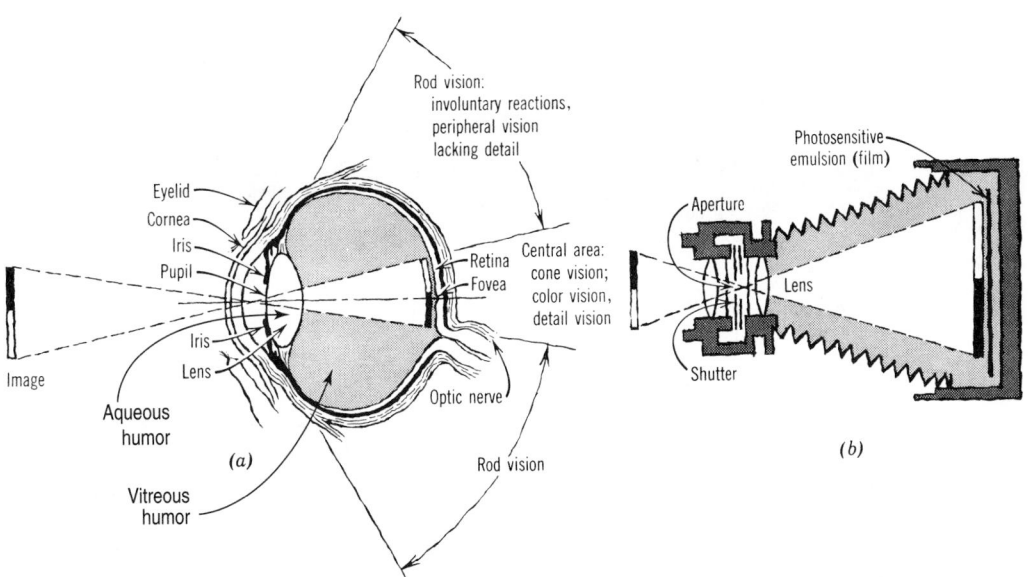

Fig. 18.19 *The human eye and the camera operate on similar optic principles. The cornea acts as an outer refracting lens that introduces light into the iris. The iris and pupil control the f-stop, or opening of the eye, and correspond roughly to a range of f2.1 to f11. The lens, which acts as a perfectly smooth automatic zoom lens, can focus from about 2 in. to infinity. Light is focused on the retina, which contains in all some 150 million light-sensitive cells of two types: rod cells, which can detect luminances from 1/1000 cd/m^2 to approximately 120 cd/m^2 in black-and-white mode; and cone cells, which see luminances in the range 3 to 1,000,000 cd/m^2 in full-color mode. The fovea, which is the center of focus, is an area of pinhead size containing about 100,000 cone cells, which accounts for the extreme precision of foveal (center-focus) vision.*

fovea, a second type of cell is encountered called a *rod* cell, also named for its shape. These cells are extremely light sensitive, giving response to light 1/10,000 as bright as that required by cone cells. However, rod cells lack color sensitivity, thus accounting for the fact that in dim light (rod vision), we have no color perception and all colors appear as varying shades of gray. Rod cells also lack detail discrimination, making "night vision" quite coarse. Finally, rod cells are slower acting than cone cells and therefore have a low degree of flicker fusion, or, stated conversely, they are highly motion sensitive. Because these cells occur at the outer portions of the retina, their motion sensitivity results in our being best able to detect movement when looking out of the "corner of

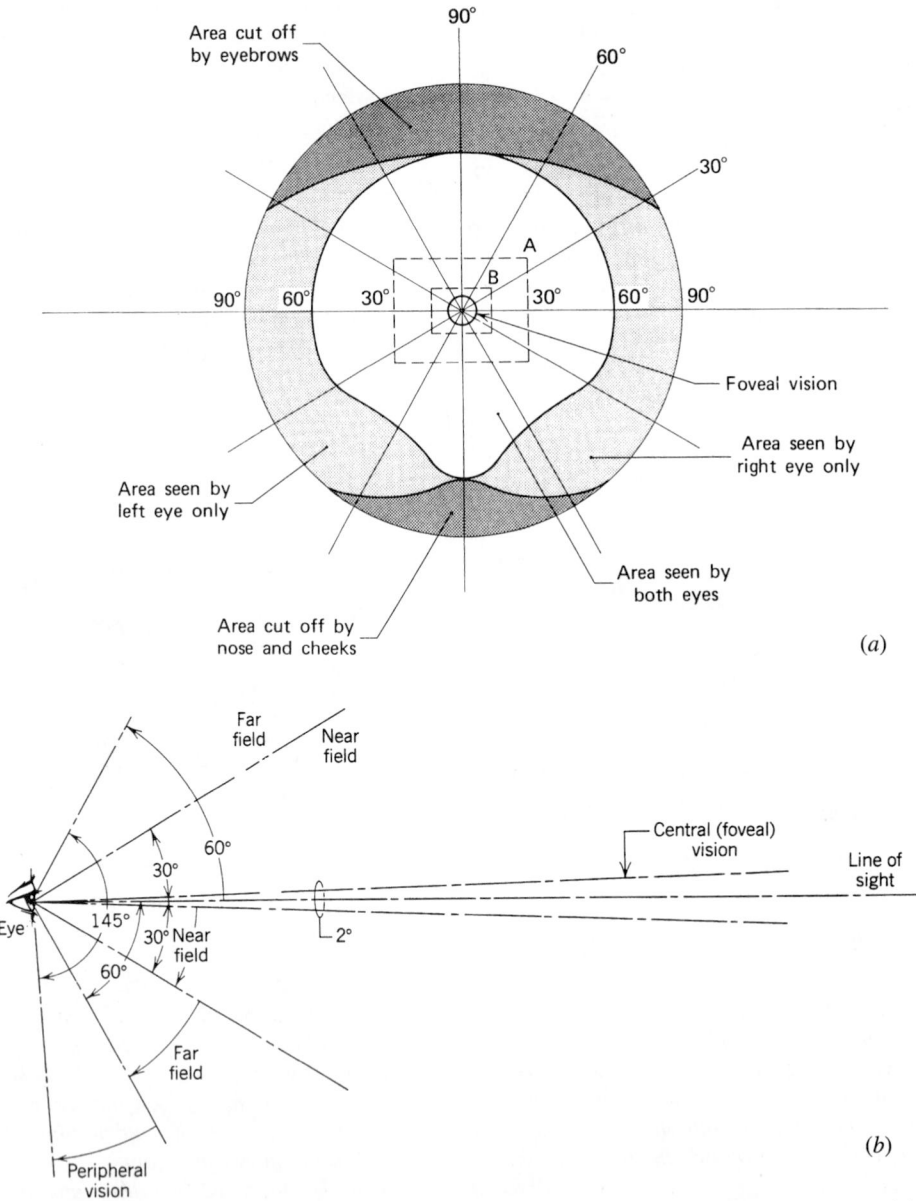

Fig. 18.20 *The fields of vision of a normal pair of human eyes* (a) *and the subtended angles* (b). *The rectangles A and B superimposed on the field of vision in* (a) *represent a large magazine and a small book, respectively.*

the eye." Looking at a fluorescent tube directly and then obliquely demonstrates this effect.

Figure 18.20 is a sketch illustrating the angles involved in the field of vision. Of particular interest is the extreme narrowness of the cone of central (foveal) vision, in which acute perception of detail takes place. This area is so small that the eye must refocus on each dot in a colon (:) if you wish to examine each individually. Surrounding this central area is a cone of binocular vision of 30° half-angle, called the *near field* or *surround,* in which area most of coarser sight information is gathered. Beyond this cone we have far field and peripheral, primarily horizontal, monocular vision. It is the far field and peripheral areas that largely give us our subjective, ambience-type reactions.

18.16   Factors in Visual Acuity

The three components of any seeing task are obviously the object or task, the lighting conditions, and the observer. Listed below are the variables affecting each of these three components. Based on the results of many investigations they can be categorized as of primary or secondary importance.

I. The Task
   Primary Factors
   (*a*) Size
   (*b*) Luminance (brightness)
   (*c*) Contrast, including color contrast
   (*d*) Exposure time—needed or given

   Secondary Factors
   (*e*) Type of object—required mental activity; familiarity with the object (in reading, familiarity is so important as to become the primary factor)
   (*f*) Degree of accuracy required
   (*g*) Task; moving or stationary
   (*h*) Peripheral patterns

II. The Lighting Condition
   Primary Factors
   (*a*) Illumination level
   (*b*) Disability glare
   (*c*) Discomfort glare

   Secondary Factors
   (*d*) Luminance ratios
   (*e*) Brightness patterns
   (*f*) Chromaticity

III. The Observer
   Primary Factors
   (*a*) Condition of the eyes (both health and age)
   (*b*) Adaptation level
   (*c*) Fatigue level

   Secondary Factors
   (*d*) Subjective impressions; psychological reactions

Although in the following discussions these factors are considered individually, many are interrelated. Thus luminance I(*b*) and adaptation III(*b*) result from the presence of illumination II(*a*); subjective impressions III(*d*) are dependent on brightness patterns II(*e*) and chromaticity II(*f*); fatigue III(*c*) results from a combination of many of the factors, and so on.

In the literature, it is common to find reference to the *quantity* and *quality* of the lighting environment. In terms of these factors, the quantity of light has reference to item II(*a*) and the quality to items II(*b*) through II(*f*).

The basic visual tasks are the perception of low contrast, fine detail, and brightness gradient. Assuming a good lighting environment—that is, low glare, acceptable luminance ratios, and white light plus a normal pair of unfatigued eyes—visual acuity is primarily dependent on items I(*a*) to I(*d*), the interrelated effects of which have been determined by a large number of field tests. Remember that the seeing task under discussion involves foveal vision (i.e., focusing and concentrating on small-area detail). This is a vastly different task than normal reading, where the eye rapidly scans familiar images without focusing on details and the brain immediately understands even when much of the information is missing, as in poor reading copy. The task discussed next is quite different and could be compared to studying mathematical equations or reading an unfamiliar language or even proofreading spelling. All of these tasks require detailed examination of each symbol individually.

ILLUMINATION

Fig. 18.21 *Relationship between object size and visibility is demonstrated by comparison of subtended angles* (a) *and* (b).

18.17 Size of Visual Object

Visual acuity is generally proportional to the physical size of the object being viewed given fixed brightness, contrast, and exposure time. Because the actual parameter is not physical size but subtended visual angle, visual ability can be increased by bringing the object nearer the eye (Fig. 18.21). It is assumed that we are dealing with a pair of *young eyes,* because at ages above 40, the accommodation ability of the eye becomes limited and bringing the object closer blurs the focus.

18.18 Subjective Brightness

The sensation of vision, as explained previously, is caused by light entering the eye. This light may be thought of as a group of convergent rays, each ray coming from a different point in space and therefore carrying different visual information. The composite of these rays comprises the entire visual picture that the eye sees and the brain comprehends. The individual rays differ from each other in intensity and chromaticity depending on the part of the viewed object from which they were reflected. The intensity of these cones of light determines and describes the perceived brightness of the object being viewed (Fig. 18.22).

The human eye detects luminance over an astonishing range of more than 100 million to 1, the lower levels being accomplished after an adjustment period called *adaptation time.* This period varies from 2 minutes for cone vision to up to 40 minutes for rod vision for dark adaptation, but is much faster for both types for light adaptation (going from dark to light). The effects of adaptation on apparent (photometric) brightness are discussed in the following section. Tables 18.2 and 18.3 list some measured luminances of everyday visual tasks.

An interesting characteristic of light-level adaptation is a shift in the sensitivity curve of the eye [Fig. 18.6(*b*)]. Whereas for the light-adapted eye (photopic vision) maximum sensitivity occurs

(a)   (b)

Fig. 18.22 (a) *Composition of the visual image.* (b) *Luminance of a nonuniform surface.*

TABLE 18.2 **Typical Luminance Values**[a]

Object	Luminance	
	cd/m²[b]	Footlamberts
Black glove on cloudy night	0.0003	0.0001
Wall brightness in a well-lighted office	100	30
This sheet of paper in an office	120	35
Green electroluminescent lamp	150	45
Asphalt paving—overcast day	1,300	380
North sky	3,500	1,000
Moon, candle flame	4,000–5,000	1,300
Fluorescent tube	6,000–8,000	2,200
Kerosene flame	8,500	2,500
Hazy sky or fog	15,000	4,400
Snow in sunlight	25,000	7,300
100-W inside frost incandescent lamp	50,000	14,600
Sun	2.3 E9	0.67 E9

[a]Values are rounded off.

[b]To obtain footlamberts, divide by 3.42.

at 555 nm in the yellow–green region, the dark-adapted eye (scotopic vision) peaks at 520 nm in the blue–green region. This means that as the light dims, the warm colors—yellow, orange, red—become grayed and the blues and violets stand out. This phenomenon can be important in lighting design of restaurants, where light levels generally vary inversely with restaurant quality. Very few foods are blue or violet.

Returning then to the primary consideration of visual acuity as affected by luminance, we can state that, in general, visual performance increases with object luminance. However, a great deal depends on the background against which an object is viewed and the consequent contrast in brightness between the object being viewed and its surroundings.

18.19 Contrast and Adaptation

The discussion that follows assumes full-spectrum "white" light and ignores the effects of chromaticity, which is considered separately. Many researchers in the area of visibility have concluded that contrast is the single most important factor in visual acuity. This is actually self-evident when we realize that in point of fact, the eye sees only

TABLE 18.3 **Preferred and Permissible Luminances**

Item	Luminance in cd/m²
Recommended road luminance	1–2
Minimum discernible, chromatic	2–3
Clearly discernible human features	15–20
Preferred wall luminance	25–150
Preferred ceiling luminance	50–250
Preferred task luminance	100–500
Permissible luminaire luminance (depending on position in field of vision)	1000–7000

ILLUMINATION

contrast. This can readily be demonstrated by viewing a large, evenly lighted, monochromatic, diffuse-finish surface (preferably white) that encompasses the entire visual field. The eye is unable to focus on such a surface because it sees no contrast, but only a single luminance. Therefore, the eye itself attempts to provide the missing contrast by seeing an internally reflected view of the retina, in yellow.

To properly evaluate the effect of contrast (luminance ratio) on visibility, we must first determine the nature of the visual task, or, more simply, exactly what it is that we are trying to see. As stated before, the basic visual tasks are detail discrimination and detection of low contrast. Examples of the former are drafting, industrial product inspection, or something as simple as discriminating between the numbers 3, 6, and 8, which are similarly shaped. Detection of low contrast include reading faint copy, sewing a black fabric with black thread, and the like.

Contrast is a dimensionless ratio, defined as

$$C = \frac{L_T - L_B}{L_B} \text{ or } \frac{L_B - L_T}{L_B} \text{ or } \left| \frac{L_B - L_T}{L_B} \right| \qquad (18.7)$$

where L_T and L_B are the luminance of the task and background, respectively, in any units. Thus, C varies from 0 for no contrast to 1.0 for maximum contrast. In most situations, the illumination on the task and background is the same. Therefore, because luminance is the product of illuminance (lux) and reflectance, contrast can also be expressed as

$$C = \left| \frac{R_B - R_T}{R_B} \right| \qquad (18.8)$$

where R_T and R_B are the reflectances of the task and the background, respectively. From this equation, we can conclude that *contrast is generally independent of illumination* (ignoring specularity). Thus, for the black-on-white lettering that you are now reading,

$$C = \frac{0.77 - 0.045}{0.77} = 0.94$$

which accounts for the excellent legibility. (We are assuming no specularity.) Reflectance figures are taken from Table 18.9.

It is obvious that high contrast is the critical factor in visual appreciation of outline, silhouette,

and size, which are the factors involved in the task of reading. Thus, black-on-white print can be read with ease even in moonlight, which is at best 0.1 lux illuminance because the contrast is so high (94%). An important conclusion can then be drawn about a *reading task:* with high contrast (clear, legible print), visibility is essentially *independent* of illumination above a certain minimum. Indeed, high illuminance values can be detrimental because they generally go hand-in-hand with high luminance sources, and these in turn can cause veiling reflections (see Section 18.29).

Now refer to Fig. 18.23. Note that as the contrast between the letters in the word "performance" and the background diminishes, the individual letters become harder to read. The end letters of the word require an illumination of up to 1000 lux, and that suffices only because we expect the letter *"e"* at the end. Were it an unknown sign, illumination of the magnitude of 10,000 lux or more would be needed. These latter letters are an example of the second type of visual task mentioned previously—low-contrast poor copy requiring surface detail study.

The *"e"* at the end of "performance" in Fig. 18.23 is printed with the same density as the *"C"* in "Contrast." It exists, and with enough lighting, the negative effect of lack of contrast can be overcome. This is not so with copy from a used-up printer ribbon or a washed-out photocopy. There, the data simply do not exist, and increased lighting only makes this fact more evident.

High background luminance makes an object look darker, and therefore assists in outline detail discrimination, which is precisely the visual task involved in reading. For this reason, black-on-white is desirable for reading. (This is a special case of lateral adaptation, which is discussed in Section 18.36.) Conversely, high background lu-

Fig. 18.23 *High contrast is helpful when the seeing task involves detection of silhouette detail.*

minance makes surface examination more difficult. A simple experiment demonstrates this effect. Stand near a window and hold your hand in front of you with the floor as background. The skin surface detail is perfectly clear—in rough proportion to its luminance. Now hold your hand up against the window with the daytime sky as background. The hand outline is clear, but the skin surface appears dark—the brighter the sky, the darker the skin surface. The reason for this is that the eye automatically adapts to the average brightness of the entire scene.

It is well known that when using an automatic exposure control camera to photograph a dark object on a light background such as a person in snow scene, it is necessary to manually increase the camera aperture in order to obtain additional light to photograph the detail of the darker object. (In doing this, we overexpose the rest of the scene.) Because we cannot easily control the aperture of our eyes, we must compensate for the detrimental effect of high background luminance in another way; for example, by increasing the surface luminance of the visual task. Indeed, this method is frequently employed [see Section 18.31(*b*)]. Limited visual compensation can be made by squinting; this reduces the field of vision and the overall scene brightness. For maximum visual acuity, *the luminance of a surface-type task*

should be the same as, or slightly higher than, that of the background, but ratios of 3 : 1 are acceptable in most circumstances.

Another way of understanding this is to consider the adaptation characteristic of the human eye (Fig. 18.24). As stated, the eye adapts to the brightness level of the *overall* scene and sees each object in the scene in the framework of that adaptation level. Thus, at an adaptation level of 1 fL [3.4 cd/m^2], a measured luminance ratio of 1 : 10 (horizontal scale on Fig. 18.24) appears to be only approximately 1 : 4 (vertical scale); that is, the apparent ratio is *smaller* than the actual. Put another way, the low level of eye adaptation causes the eye to diminish the difference between high brightnesses. This effect becomes smaller as the adaptation level rises, until at an adaptation level of 1000 fL [3400 cd/m^2], (daylight conditions), the apparent and actual *ratios* correspond; that is, smaller ratios are recognizable. Because visual acuity is, by definition, the ability to distinguish between different levels of luminance, we have in effect demonstrated that *visual acuity increases with increased adaptation level.*

The second important conclusion that can be drawn is that at high adaptation levels, apparent brightness is lower than actual brightness, and vice versa. Thus, a shadowed object near a window looks *darker* than it actually is; contrary to

Fig. 18.24 *The effect of an eye's adaptation level on perceived (subjective) brightness is clearly shown. See text for full discussion. (Adapted from A. Cotton, Principles of Illumination, Wiley, New York, 1960.)*

first expectation, it must be better lighted than a similar object further inside the room for equal visibility. That this effect (high-level adaptation) is primarily important in daylight situations is also apparent from the curves. At 100 fL [340 cd/m²] adaptation level, which is approximately that of a brightly lighted interior space apparent and actual luminance levels coincide. The reverse effect, resulting from low adaptation levels, can be very important in design situations where low lighting levels are found, such as theaters, lecture halls, restaurants, and storage spaces. Sources of light that would be entirely acceptable at a higher adaptation level can easily become an annoying glare at a low level. Good examples are a theater usher's flashlight or the blinding glare of an oncoming car's headlights. Reduction of contrast due to veiling reflections and the measurement of contrast and contrast reduction are discussed in Section 18.29.

The foregoing discussion of contrast deliberately avoided any discussion of *object* colors and the effect of color contrast on visual acuity for several reasons:

- Office-type tasks (paperwork) are most often black-on-white tasks.

- Most work tasks involving colored objects deal with unsaturated colors, where the pronounced effects of color contrast are minimal.

- The effect of object color on visual acuity is very complex because it involves the color characteristic not only of the object, but also of the background and the surround plus the chromaticity of the illuminant.

That being said, we must, however, at least mention a number of important object color phenomena that bear on visual acuity.

- The subjective brightness of a colored (heterochromatic) object is greater than that of an achromatic object for the same photometric luminance. This effect varies with hue and saturation. It is more pronounced with saturated colors than with those of low chroma and more in the blue–purple–red area than the yellow–green area.

- Colored objects on a dark-to-black background appear light and desaturated. Con-

versely, colors on a light-to-white background appear darker and more saturated.

- Adjacent complementary colors produce a pale-to-white border between them. The effect also appears when the task and background colors are complementary, and is most pronounced with saturated (high chroma) colors.

These remarks deal with *object* color only. As stated at the beginning of this section, we are assuming white, full-spectrum light, so that neither object color nor visual acuity are affected. Effects on visual acuity resulting from unbalanced illuminants such as those from high-pressure sodium lamps and, to a lesser extent, from halide and fluorescent sources are the subject of intensive and ongoing research. Some of the results are adduced in the discussion of illuminant chromaticity at the end of this chapter. These effects are most pronounced and therefore most important in relation to elderly persons and persons with visual defects. We must again, therefore, emphasize that unless otherwise specifically stated, our discussions of vision, light, and lighting assumes full-spectrum white light and young, healthy eyes.

18.20 Exposure Time

Registering a meaningful visual image is not an instantaneous process, but one that requires finite amounts of time. Just as a photograph can be taken in dim light by using a longer exposure, so can the human eye better distinguish and discriminate fine detail in poor light given time (and neglecting eyestrain). Of course, the time needed depends on the type of task, but the principle of shorter time at higher illumination, within limits, remains the same. This is particularly true when the object being viewed is not static, but is in motion.

The phenomenon, however, is not linear. For one specific task tested, increasing the luminance by a factor of 6 halved the seeing time, whereas a further sixfold increase in luminance reduced the time only another 20%. Thus, as in the case of improved contrast with increasing background brightness, we have a case of diminishing returns.

With the parameter of time, as with other parameters of visual acuity, the same qualification applies. When dealing with material that does not require detail discrimination, improved performance does not necessarily result from improved illumination. It has been amply demonstrated that speed of reading and comprehension are substantially independent of illumination levels above a minimum, but are very much dependent on the contrast quality of the material.

18.21 Secondary Task-Related Factors

Refer again to the list of factors in visual acuity in Section 18.16. We have discussed at some length the primary task-oriented factors I(*a*)–(*d*), and at this point wish to consider secondary factors (*e*)–(*h*) of that list. These items refer essentially to the level of concentration required. Thus, spray painting a large metal object or packing fruit are very different from inspecting the painted object for defects or the fruit for bruises. The former tasks are largely mechanical and repetitious, whereas the latter tasks require continuous judgmental decisions based on visual information. Because both the latter tasks are frequently moving (assembly-line type of work) and both involve penalties for inaccuracy (rejection at a later inspection state or negative feedback from the purchaser), the lighting required for these tasks is several orders of magnitude better than that for related but largely mechanical tasks. Indeed, extrapolation from laboratory condition tests yields only a range of lighting recommendations. Thereafter, considerable field testing and adjustment is required.

An observer performing a lab test has a different level of concentration and performance than that of a person at an 8-hour-a-day task. The latter compensates for unsatisfactory seeing condition by:

1. Moving the work to a better viewing angle.
2. Moving head and eyes to a more comfortable position.
3. Reducing the distance between eyes and task to the extent that the eyes can accommodate.

4. Complaining about poor contrast task so that something is done about it (such as fixing the photocopying machine).
5. Taking more time to perform the seeing task involved. (This item, if it affects production, frequently spurs management to make appropriate alterations in the work environment.)

The last item in the task list—peripheral patterns—deals with the visual surround rather than the immediate area of the work. Other than glare sources, which are discussed separately, there are many items that, although outside the central field of vision, can disturb the viewer's concentration and therefore the task performance of a worker. These include movement (vehicles, machines, persons), to which peripheral vision is particularly sensitive; large variations in the brightness pattern of the background caused by such activities as periodic opening of an outside door or welding; and even nonvarying patterns, which are disturbing because of their very nature, such as checkerboard light–dark patterns, or devices on which it is difficult to focus the eyes, such as crossed patterns of wires and bars. None of these items is strictly lighting oriented, but they are noted here to demonstrate that an adequate lighting design necessarily includes adjustments for particular field conditions.

18.22 Observer-Related Visibility Factors

Refer to Section 18.16, category III. It is a well-documented fact that the visual performance of *healthy* eyes decreases with age. This reduction evinces itself principally in two areas: an increase in minimum focusing distance caused by increasing lens rigidity, and a decrease in sensitivity caused by clouding of the cornea, lens, and vitreous humor. Both these effects can be compensated, the former with external lensing (eyeglasses) and the latter by increased task size, luminance, contrast, and exposure time, as explained previously. The point is, of course, that the age of the worker is an important parameter in the vision equation.

Furthermore, maximum performance may not be synonymous with maximum comfort or minimum fatigue. Indeed, the reverse may sometimes be true. Most experts agree that what is normally referred to as "eyestrain" is a condition of the eye muscles resulting from extensive and intensive eye use (see Section 18.16, items I[*e, f*]). Thus, excellent performance under excellent lighting conditions can still produce fatigue because of the demanding nature of the task. In addition, as discussed later, discomfort glare or even excessive lighting can cause fatigue without affecting performance.

The lighting designer must be concerned not only with providing adequate uniform lighting levels, but also with all of the factors involved. Many, indeed, are beyond his or her control. Some, such as task contrast, previously thought to be outside the lighting designer's province, should be examined by the designer in the framework of an overall lighting plus task plus observer problem, and recommendations made. Acceptance of these recommendations is a management decision.

Recent work in field testing of visibility situations in real-work conditions considers all of these parameters in arriving at a visibility judgment and, in addition, goes far toward identifying any difficulty as lying with the task, the lighting condition, or the observer. The entire field of visibility and visual performance is undergoing very active continuing research, and definitive answers to the elusive question of how to design for optimal viewing have not yet been found, nor indeed is there any assurance that they will be. What has emerged over the years have been various parameters and criteria for judging aspects of visibility, such as ESI, which is a contrast-related visibility criterion; RVP, which rates a vision situation in terms of speed and accuracy of performance; VCP, which judges the visual comfort of an overall scene; and a somewhat ephemeral and entirely subjective impression called *visual clarity*. Other research has concentrated on the relationship between eye pupil size and visual acuity, and based on results from this research, the relationship between pupil size and various aspects of light and lighting. What has been confirmed to date is the complexity of human work–oriented vision plus our inability thus far to adequately quantify the interrelationships between the very many factors involved—quantification being the basis of reliable, duplicatable design.

18.23 The Aging Eye

The last few decades have seen a remarkable increase in life expectancy in modern Western countries with a resultant sharp increase in the "aged" population. In the United States at this writing, about 15% of the population is above 65 years of age, a proportion that is expected to reach 20% by the year 2020. As a result, lighting design must take cognizance not only of special requirements in buildings specifically intended for use by the aged, but also, increasingly, of general use public buildings. To this end a brief review of these special requirements is presented here.

Refer to Fig. 18.19. Light enters the eye through the cornea, passes through the aqueous humor, and enters the lens through the pupil. After being focused by the lens, it continues through the vitreous humor and finally projects the viewed image, reversed, on the retina. As the eye ages, a whole spectrum of physiological changes may occur; some are usual and are therefore classified as normal; others, such as cataracts, are less common but are still considered by ophthalmologists to be an expected development. The unusual developments are classified as *pathologies,* only indirectly or partially related to aging, and therefore outside the purview of "normal" lighting design. The normal developments and their influence on lighting design are briefly described in the following subsections.

(a) Cornea. This perfectly clear outer lens tends to become cloudy, with corresponding reduction of visual clarity and acuity. This results in a requirement for more light to overcome the reduction in light intensity on the retina. The overall effect is very similar to that of a neutral density filter on a camera lens, the difference being that such a filter is most often used to reduce excessive ambient light without excessively reducing the shutter opening (increasing the *f*-stop). The need for additional light for aging eyes is recognized in most modern systems of illumination specification. See Table 18.6 for the factors used in the IESNA system.

(b) Lens. The lens, which begins life as a very lightly yellow–tinted flexible crystalline body, gradually thickens and yellows. As a result of the thickening, flexibility is reduced, resulting in the well-known inability to close-focus. The yellowing both reduces the overall light intensity in the eye and selectively filters the blue portion of the spectrum. Research seems to indicate improved visual acuity through pupil size control when the incident light is rich in the blue area of the spectrum. Because lens yellowing *reduces* the blue frequencies, the overall effect is again to require additional light.

A second and more important degenerative phenomenon of the lens is its gradual clouding. When the opacity is confined to the perimeter, its effect is negligible because vision is unaffected. When small, opaque areas appear within the visual axis through the lens, vision is affected in two ways:

1. The viewed image is dimmed and blurred due to opacities in the field of view.
2. Light entering the lens is scattered by inter-reflections from the opaque particles, resulting in a subjective impression of glare. This effect is particularly severe outdoors where light enters the lens from all angles, and both indoors and outdoors as a result of concentrated, high luminance (glare) sources.

The net result of these reactions is a requirement for more light, but an even more pressing requirement that sources of glare and peripheral light be eliminated. (People with this condition frequently wear eyeglasses and sunglasses with large opaque ear pieces to block peripheral light and thus reduce glare.) Also, because short-wavelength (blue) light interreflects and scatters more readily than does long-wavelength (yellow–red) light, these people are more comfortable with incandescent sources and low-color temperature fluorescent lamps (2700–3100° K) than with sources rich in the blue–green spectrum. (Ophthalmologists frequently prescribe yellow-tinted eyeglass for people with this lens condition to filter out blue light.)

Finally, a less common but still prevalent condition of the aging lens is the development of fluorescent particles, called *fluorigens,* in the vision path. In the presence of UV light, such as exists in daylight, fluorescent, and HID sources, these particles fluoresce, causing scatter, blur, and glare. The solution to this problem is a combination of yellow-tinted eyeglass lenses and a reduction of light sources containing appreciable quantities of UV.

(c) Pupil. The pupil controls the amount of light entering the eye and is therefore intimately involved in the constantly changing accommodation level of the eye. The pupil muscles react more slowly as they age, thus lengthening accommodation time. Dark-to-light accommodation is very rapid in the young eye, and is barely noticed except for extreme changes such as exiting a cinema into sunlight. With an aging eye, the slower pupil results in severe glare sensations with even much smaller brightness changes.

The net result of all of the "normal" conditions of the aging eye is a heightened sensitivity to glare, an intolerance to the blue–UV end of the spectrum, and an overall requirement for higher illuminance levels. For the lighting designer, these needs translate as requirements for very careful selection and placement of luminaires, increased use of indirect lighting, and particular attention to the spectrum of the light sources used. Because some of these requirements are not only mutually incompatible but are also contrary to energy-efficient design practice, it may be particularly difficult to satisfy all the requirements in spaces occupied by a wide range of occupant ages. In work areas of this type, it may be wise to provide for the possibility of readily changing lighting conditions in a limited area to accommodate older occupants rather than attempt an overall design, keeping in mind constantly the glare and color factors discussed previously.

QUANTITY OF LIGHT

18.24 Illuminance Levels

Returning to the list of factors in Section 18.16, and having discussed the task-oriented and observed-oriented items [except for item III(d), psychological reactions, which is covered in Section 18.36], we turn now to item II, the *lighting condition.* This is frequently, if somewhat inaccurately, divided into two groups—quantity and quality of lighting—with item II(*a*) representing quantity and items II(*b*) to II(*f*) representing quality. That such a division is not accurate becomes clear in our discussion of glare in Sections 18.28 to 18.31.

An understanding of the factors involved in visual acuity as discussed previously does not answer the most basic lighting design question, which is, "How much light must I provide for the specific visual task at hand?" That this question is extremely difficult to answer is evidenced by the fact that even today, recommendations for similar tasks vary among countries with highly developed technologies by ratios as high as 10 to 1. Because this is obviously an unsatisfactory situation in an era of global markets, international construction, European Common Market, and international cooperative lighting research, the trend since the late 1980s has been to attempt a degree of standardization.

The American (IESNA) recommendations were originally developed analytically by extrapolation from extensive laboratory tests. The function of these tests was to determine the conditions under which small differences in contrast could be detected for specific degrees of accuracy, with variable parameters of task luminance, size, and exposure time. The idea behind the tests was that visual acuity could be defined as the ability to distinguish differences in contrast.

The British and, to a large extent, the European approach was to study specific tasks in actual and simulated field conditions. To these results, modifying factors of size, contrast, accuracy requirement, speed, and duration were added. The disadvantage of this approach was the necessity to study a large number of visual tasks individually. In the absence of a very specific visual task description, a detailed listing of visual categories was developed, as is discussed in the next section. Recently, American IESNA recommendations have been modified by a graded system more in line with British and CIE recommendations that establishes a median or average requirement for a task, within a range, and then modifies the median up or down to specific conditions of speed, accuracy, error importance, task duration, background reflectances, and viewer visual capability.

In addition, because of the moral and legal pressure (ASHRAE/IES Standard 90.1 is mandated by many codes) for energy conservation, the IESNA has taken additional salutary steps toward rationalization of its very influential illumination standards. They include recognition of fatigue and task familiarity (e.g., reading) as factors in determining illuminance levels, establishment of lighting power budgets, and energy standards (see Section 20.5) that encourage use of daylight as a normal component of a space's illumination and use of task/ambient lighting design as the preferred technique where high levels of task lighting are required.

18.25 Illuminance Category

Before discussing illuminance (lux) recommendations it is necessary to understand the basis of their derivation, applicability, and shortcomings. As noted above, most of the IESNA task illuminance recommendations are derived by extrapolation from threshold contrast visibility tests which yield a required task luminance. Assuming uniform, diffuse task reflectance and uniform illuminance, it is then a simple step to calculate required illuminance since luminance is simply the product of illuminance and reflectance. In SI units,

$$L(\text{cd/m}^2) = \frac{E(\text{lux}) \times \text{RF}}{\pi}$$

and in AS units,

$$\text{fL} = \text{fc} \times \text{RF}$$

where   L is luminance
        E is illuminance
        RF is reflection factor
        fL is luminance in foot-lamberts
        fc is illuminance in footcandles (see Example 18.4 and Fig. 18.7).

This simple relationship is used to derive much of the current IESNA tables, with modifying factors as explained previously (see Table 18.6).

A number of reservations on this method have been voiced by respected authorities. One is based on research that indicates that suprathreshold visibility requirements are more readily related to eye brightness adaptation levels than to threshold contrast luminance levels (see Fig. 18.24 and the associated text discussion). Another objection indicates that deriving suprathreshold luminances from threshold values depends on applied criteria, and can therefore vary considerably. Still another objection is based on the readily demonstrable fact that the sensation of vision is not mathematically related to photometric luminance. Thus, a black surface of 10% reflectance illuminated with 900 lux and a white surface of 90% reflectance illuminated with 100 lux both have exactly the same luminance, yet the eye always sees the white surface as lighter than the black by a large margin. All of these reservations add up to a caveat to the lighting designer to use the IESNA illuminance recommendations as one aspect of the overall design and not dogmatically.

Returning to recommendations, visual task studies indicate that assuming good contrast, the required luminances, categorized by type of task, are roughly as follows:

Category of Visual Task	Required Luminance (cd/m²)
Causal	10–20
Ordinary	20–100
Moderate	100–200
Difficult	200–400
Severe	Above 200

Dependance of required illumination on task reflectance (RF) can be seen by a glance at the following tabulation, which shows quantitatively the illumination requirements in the previous categories for tasks of radically different reflectance.

Category of Visual Task	Required E (lux)*	
	RF = 50%	RF = 10%
Casual	62–125	300–625
Ordinary	125–625	625–3,125
Moderate	625–1,250	3,125–6,250
Difficult	1,250–2,500	6,250–12,500
Severe	>2,500	>12,500

* Lux figures rounded.

This illustrates that a single illumination scheme is often inadequate for an area containing widely differing visual tasks. Note that a 10% RF makes all tasks difficult and that casual seeing would comprise only outline recognition.

18.26 Illuminance Recommendations

Because American architects and engineers are involved in many construction projects outside of the United States, we present in this section the essentials of British illuminance recommendations, which are similar to those of the CIE and many EC countries, in addition to the American IESNA recommendations. In any specific design, the standards of that country should be consulted. Most developed nations, including Australia, Brazil, China, France, Germany, Japan, and others have their own published lighting standards.

(a) British Lighting Standards. These standards are published by the Chartered Institute of Building Service Engineers (CIBSE), Delta House, 222 Balham High Rd., London, SW129BS, UK. The particular publication in which illuminance recommendations appear is the *Code of Interior Lighting,* 1994, amended 1997. The method of use is to determine the recommended average illuminance level, called the *standard maintained illuminance,* from either the very detailed, extensive listing of specific tasks in the previously mentioned publication (a small sample of which is reproduced in Table 18.4); or, if only the representative type of task is known, then from the task category chart in Table 18.5. Having established this recommendation, the designer then modifies it (if necessary) by using the flow chart shown in Table 18.6, which either increases or decreases the recommended illuminance to suit the task size, contrast, duration, and error risk. Explanatory usage notes accompany Table 18.6.

(b) American Illuminance Recommendations (IESNA). This (1981) system is patterned, to an extent, after CIE recommendations. In lieu of the single illuminance recommendation that characterized the early IES tables, this system provides a luminance range determined by task difficulty (contrast, size), within which range a specific illuminance is selected based on three weighting factors—age of the observer, the importance of speed and/or accuracy, and the reflectance of the background on which the task is seen. The procedure operates as follows:

1. An illuminance category (A through I) is initially selected, based on either a general de-

ILLUMINATION

TABLE 18.4 **Typical Illuminance Recommendations, CIBSE (UK)**

Offices and Shops		
	Standard Maintained Illuminance (lux)	Notes
Offices		
General offices	500	Local lighting may be appropriate
Computer work stations	300–500	See ref. 25
Conference rooms, executive offices	300–500	Dimming or switching to permit use of visual aids may be necessary
Computer and data preparation rooms	500	See ref. 25
Filing rooms	300	Vertical surfaces may be especially important
Drawing Offices		
General	500	
Drawing boards	750	Local lighting may be appropriate
Computer aided design and drafting	300–500	Special lighting is required; see ref. 25
Print rooms	300	

Reference:

(25) *CIBSE Lighting Guide LG3: Areas for Visual Display Terminals.*

General Notes

The information in the Table will be influenced by reference to the "core recommendations" in Sections 2.4, 2.5, and 2.6.1–2.6.3.

Check that installations designed to meet the needs of visual display screen tasks also satisfy the task-to-wall and task-to-ceiling ratios recommended in Section 2.4.4. See also Sections 3.3.2.9–3.3.2.11.

Where air conditioning or mechanical ventilation is required, air-handling luminaires may be appropriate. For information on the lighting of atria, refer to General building areas in the schedule.

Source: Reproduced with permission from CIBSE publication *Code for Interior Lighting* (1994).

scription of the activity involved (Table 18.7) or, if known, on a specific activity in a specific setting. [The extensive tables for this latter selection are found in the IESNA *Lighting Handbook,* 8th ed. (1993), and are not reproduced here.] The category selection gives a three-number range of illuminances that corresponds to minimum significant subjective illuminance changes.

2. In the next step, three weighing factors are introduced. Age under 40 reduces the light requirement, unusual demand for speed or accuracy increases it, and particularly low or high background reflectance (below 30% or above 70%) increases or reduces it, respectively. This step is not a mechanical one in that at the higher luminances (categories D through I), the designer is expected to become thoroughly familiar with the importance, duration, and visual difficulty of the specific task involved, because the design aims at *task* lighting and not simply general room illumination (see Table 18.5). Having determined the weighting factors, the designer can then select the recommended (target) illuminance from Tables 18.8 and 18.9.

The resultant recommended illuminance is "raw" or conventional illumination; that is, average maintained lux [fc], *on the task* for categories D through I, and *in the room* for categories A through C. Therefore, the calculation procedure for the target illuminance is a lumen method for categories A through C and a point-by-point method for the *task* illumination of categories D through I.

TABLE 18.5 **Examples of Activities/Interiors Appropriate for Each Maintained Illuminance**

Standard Maintained Illuminance (lux)	Characteristics of Activity/Interior	Representative Activities/Interiors
50	Interiors used rarely with visual tasks confined to movement and casual seeing without perception of detail	Cable tunnels, indoor storage tanks, walkways
100	Interiors used occasionally with visual tasks confined to movement and casual seeing calling for only limited perception of detail	Corridors, changing rooms, bulk stores, auditoria
150	Interiors used occasionally or with visual tasks not requiring perception of detail but involving some risk to people, plant, or product	Loading bays, medical stores, plant rooms
200	Interiors occupied for long periods or for visual tasks requiring some perception of detail	Foyers and entrances, monitoring automatic processes, casting concrete, turbine halls, dining rooms
300[a]	Interiors occupied for long periods, or when visual tasks are moderately easy [i.e., large details (>10 min arc) and/or high contrast]	Libraries, sports and assembly halls, teaching spaces, lecture theaters, packing
500[a]	Visual tasks moderately difficult [i.e., details to be seen are of moderate size (5–10 min arc) and may be of low contrast]; also, color judgement may be required	General offices, engine assembly, painting and spraying, kitchens, laboratories, retail shops
750[a]	Visual tasks difficult [i.e., details to be seen are small (3–5 min arc) and of low contrast]; also, good color judgements or the creation of a well-lit, inviting interior may be required	Drawing offices, ceramic decoration, meat inspection, chain stores
1000[a]	Visual tasks very difficult [i.e., details to be seen are very small (2–3 min arc) and can be of very low contrast]; also, accurate color judgements or the creation of a well-lit, inviting interior may be required	General inspection, electronic assembly, gauge and tool rooms, retouching paintwork, cabinet making, supermarkets

(continued)

ILLUMINATION

TABLE 18.5 **Examples of Activities/Interiors Appropriate for Each Maintained Illuminance** (*continued*)

Standard Maintained Illuminance (lux)	Characteristics of Activity/Interior	Representative Activities/Interiors
1500[a]	Visual tasks extremely difficult [i.e., details to be seen are extremely small (1–2 min arc) and of low contrast]; optical aids and local lighting may be of advantage	Fine work and inspection, hand tailoring, precision assembly
2000[a]	Visual tasks exceptionally difficult [i.e., details to be seen exceptionally small (<1 min arc) with very low contrasts]; optical aids and local lighting will be of advantage	Assembly of minute mechanisms, finished fabric inspection

[a] 1 minute of arc (min arc) is one sixtieth of a degree. This is the angle of which the tangent is given by the dimension of the task detailed to be seen divided by the viewing distance.

Source: Reproduced with permission from CIBSE *Code for Interior Lighting,* 1994.

For a space with several tasks of varying visual difficulty, the designer is expected to so design the lighting and controls that task requirements are met without overlighting. A uniform layout keyed to the most severe task is tedious and energy wasteful, and is strongly to be discouraged.

The IES-recommended illuminance values are *not* applicable to installations where a visual task is not the deciding factor. Such installations include merchandising spaces, displays of all sorts, theatrical and artistic lighting, lighting for mood, lighting for safety, light used as part of an industrial process, and so on. Recommended illuminances for exterior spaces are found in the same group of tables in the IESNA handbook referred to previously.

(c) Proposed IESNA Recommendations.
Changes in the IESNA system have been proposed that will most probably amend the above described illuminance selection system as follows:

1. The range of illuminances for categories A through I will consist of only two values rather than the present three in the ratio of 2:1; the higher illuminance being used for tasks judged to be unusually difficult. The new values will be (in lux):

Occasional Tasks (lux)		Common Tasks (lux)		Special Tasks (lux)	
A	30–60	D	300–600	G	1,500–3,000
B	60–120	E	500–1,000	H	3,000–6,000
C	120–240	F	750–1,500	I	5,000–10,000

2. The method for determining whether a task in a specific category is to be judged difficult (and therefore require the larger of the two category values) will depend on six basic factors: requirements of speed and accuracy; effect of errors on safety and health; visual capacity of viewer; background reflectances; and duration of viewing. (Note that these factors correspond closely to the flow chart decision factors in the British CIBSE system shown in Table 18.6.)

3. Additional factors related to the lighting condition will be factored into the choice of lighting category. These include scene geometry, shadowing and modeling, luminance ratios, presence of daylight, glare considerations, and others. Inasmuch as, of this writing, the proposed changes have not yet been adopted; they are presented here for information only. We strongly advise all persons concerned with lighting design to obtain a copy of the current IESNA *Lighting Handbook* and to use the current illuminance recommendation system presented therein.

TABLE 18.6 **Design Maintained Illuminance Flow Chart**[a]

Standard Maintained Illuminance (lux)[b]	Task Size and Contrast[c]		Task Duration[d]		Error Risk[e]	Design Maintained Illuminance (lux)[f]
	Unusually Difficult to See?	Unusually Easy to See?	Unusually Long Time?	Unusually Short Time?	Serious for People, Plant, or Product?	
200	Yes—200	200	Yes—200	200	Yes—200	200
	250	250	Yes—250	250	Yes—250	250
300	Yes—300	Yes—300	Yes—300	Yes—300	Yes—300	300
	400	400	Yes—400	Yes—400	Yes—400	400
500	Yes—500	Yes—500	Yes—500	Yes—500	Yes—500	500
	600	600	Yes—600	Yes—600	Yes—600	600
750	Yes—750	Yes—750	Yes—750	Yes—750	Yes—750	750
	900	900	Yes—900	Yes—900	Yes—900	900
	1000	1000	Yes—1000	Yes—1000	Yes—1000	1000
			1300		1300	1300
			1500		1500	1500

[a] To use the chart, follow the horizontal path from the "Standard" maintained illuminance in the schedule until the answer to a question is "Yes." If the "Yes" is strong, follow the solid arrow; if moderate, follow the dashed arrow.

[b] The flow chart should be used for all standard maintained illuminance recommendations from 200–750 lux for general activities and interiors in the lighting schedule. For recommendations of 150 lux or less, the modifying factors are not relevant (see Table 18.5). Where a "standard" maintained illuminance of more than 750 lux is recommended, this always applies to a stated task for specific industries or activities where the modifying factors have usually been taken into account.

[c] The "standard" maintained illuminance given in the schedule assumes the task is representative of its type. If the task is much more visually difficult than usual (e.g., smaller size, lower contrast), then an increase in the maintained illuminance is appropriate. Reduced contrast may arise from the use of safety lenses or safety screens because they reduce the transmission of light. Increase illuminance to take account of the age or eyesight of the operator. Conversely, if the task detail is such that the task is easier to see than usual (e.g., larger size, higher contrast), a reduction in maintained illuminance can be made.

[d] The "standard" maintained illuminance given in the schedule assumes that the task is to be undertaken over a conventional working period. If the work is to be undertaken continually for a much longer period than usual, the maintained illuminance should be increased in order to diminish the risk of visual fatigue. Conversely, if the work is to be carried out over a much shorter period than usual, the maintained illuminance may be reduced.

[e] The schedule assumes that the consequences of any errors are typical of the activity. However, if errors have unusually serious consequences for people, plant, or product, an increase in illuminance may be appropriate.

[f] If the design maintained illuminance is more than two full steps (e.g., 300–750 lux) on the illuminance scale above the standard maintained illuminance, consideration should be given to whether changes in the task details or organization of the work are more appropriate than substantial increases in the maintained illuminance.

Source: Reproduced with permission from CIBSE publication *Code for Interior Lighting* (1994).

QUALITY OF LIGHTING

18.27 Considerations of Lighting Quality

Quality of lighting is a term used to describe all of the factors in a lighting installation not directly connected with quantity of illumination. Certainly it is obvious that if two identical rooms are lighted to the same *average* illuminance, one with a single bare bulb and the second with a luminous ceiling, there is a vast difference in the two lighting systems. This difference is in the *quality* of the lighting, a term that describes the overall scene—that is, the luminances, diffusion, uniformity, and chromaticity of the lighting.

ILLUMINATION

TABLE 18.7 Illuminance Categories and Illuminance Values for Generic Types of Activities in Interiors

Type of Activity	Illuminance Category	Ranges of Illuminances	
		Lux	Footcandles
General lighting throughout spaces			
Public spaces with dark surroundings	A	20–30–50	2–3–5
Simple orientation for short temporary visits	B	50–75–100	5–7.5–10
Working spaces where visual tasks are only occasionally performed	C	100–150–200	10–15–20
Illuminance on task			
Performance of visual tasks of high contrast or large size	D	200–300–500	20–30–50
Performance of visual tasks of medium contrast or small size	E	500–750–1000	50–75–100
Performance of visual tasks of low contrast or very small size	F	1000–1500–2000	100–150–200
Illuminance on task, obtained by a combination of general and local (supplementary) lighting			
Performance of visual tasks of low contrast and very small size over a prolonged period	G	2000–3000–5000	200–300–500
Performance of very prolonged and exacting visual tasks	H	5000–7500–10000	500–750–1000
Performance of very special visual tasks of extremely low contrast and small size	I	10000–15000–20000	1000–1500–2000

Source: Courtesy of Illuminating Engineering Society of North America.

TABLE 18.8 Effect of Weighting Factors on General Lighting Illuminance Values, for Illuminance Categories A, B, and C[a]

Weighting Factors		Illuminance (lux) Categories		
Average Age of Occupants	Average Room Surface Reflectance (%)[b]	A	B	C
Under 40	Over 70	20	50	100
	30–70	30	75	150
	Under 30	30	75	150
40–50	Over 70	30	75	150
	30–70	30	75	150
	Under 30	30	75	150
Over 55	Over 70	30	75	150
	30–70	30	75	150
	Under 30	50	100	200

[a] Based on IESNA recommendations.
[b] Average weighted surface reflectances, including wall, floor, and ceiling reflectances, if they encompass a large portion of the task area or visual surround. For instance, in an elevator lobby, where the ceiling height is 25 ft, neither the task nor the visual surround encompasses the ceiling, so only the floor and wall reflectances would be considered.

TABLE 18.9 **Effect of Weighting Factors on Task Lighting Illuminance Value, for Illuminance Categories D through I**[a]

Illuminance on Task

Weighting Factors			Illuminance Categories					
Average of Workers' Ages	*Demand for Speed and/ or Accuracy*[1]	*Task Background*[2] *Reflectance (%)*	*D*	*E*	*F*	*G*[b]	*H*[b]	*I*[b]
Under 40	NI[c]	Over 70	200	500	1000	2000	5000	10000
		30–70	200	500	1000	2000	5000	10000
		Under 30	300	750	1500	3000	7500	15000
	I[c]	Over 70	200	500	1000	2000	5000	10000
		30–70	300	750	1500	3000	7500	15000
		Under 30	300	750	1500	3000	7500	15000
	C[c]	Over 70	300	750	1500	3000	7500	15000
		30–70	300	750	1500	3000	7500	15000
		Under 30	300	750	1500	3000	7500	15000
40–55	NI	Over 70	200	500	1000	2000	5000	10000
		30–70	300	750	1500	3000	7500	15000
		Under 30	300	750	1500	3000	7500	15000
	I	Over 70	300	750	1500	3000	7500	15000
		30–70	300	750	1500	3000	7500	15000
		Under 30	300	750	1500	3000	7500	15000
	C	Over 70	300	750	1500	3000	7500	15000
		30–70	300	750	1500	3000	7500	15000
		Under 30	500	1000	2000	5000	10000	20000
Over 55	NI	Over 70	300	750	1500	3000	7500	15000
		30–70	300	750	1500	3000	7500	15000
		Under 30	300	750	1500	3000	7500	15000
	I	Over 70	300	750	1500	3000	7500	15000
		30–70	300	750	1500	3000	7500	15000
		Under 30	500	1000	2000	5000	10000	20000
	C	Over 70	300	750	1500	3000	7500	15000
		30–70	500	1000	2000	5000	10000	20000
		Under 30	500	1000	2000	5000	10000	20000

NOTES:

1. In determining whether speed and/or accuracy is not important, important, or critical, the following questions need to be answered: What are the time limitations? How important is it to perform the task rapidly? Will errors produce an unsafe condition or product? Will errors reduce productivity and be costly? For example, in reading for leisure there are no time limitations and it is not important to read rapidly. Errors will not be costly and will not be related to safety; thus, speed and/or accuracy is not important. If, however, prescription notes are to be read by a pharmacist, accuracy is critical because errors could produce an unsafe condition and time is important for customer relations.
2. The task background is that portion of the task upon which the meaningful visual display is exhibited. For example, on this page the meaningful visual display includes each letter, which combines with other letters to form words and phrases. The display medium, or task background, is the paper, which has a reflectance of approximately 85%.

[a] Based on data from the *IESNA Handbook,* 8th ed, 1993.

[b] Obtained by a combination of general and supplementary lighting.

[c] NI = not important; I = important; C = critical.

Fig. 18.25 *Glare zones. The direct and reflected glare light paths are delineated on the diagram. Direct glare presupposes a head-up position, whereas reflected glare assumes eyes down at a reading angle. Factors that affect the severity of glare are: adaptation level of the eyes, the apprehended size of the glare source, luminance ratios, room size and surface finishes, and size and position of lighting fixtures and windows. Light sources in the far-field and peripheral-vision areas beyond the central 90° cone are less troublesome as glare sources (see Fig. 18.20).*

Excessive luminances and/or excessive luminance ratios in the field of vision are commonly referred to as *glare*. The quality of the lighting system must also include the visual comfort of the system; that is, the absence of glare. When the glare is caused by light sources in the field of vision, it is known as *direct* or *discomfort glare*. When the glare is caused by reflection of a light source in a viewed surface, it is known as *reflected glare* or *veiling reflection* (see Fig. 18.25).

18.28 Direct (Discomfort) Glare

The factors involved in producing discomfort glare are the luminance, size, and position of each light source in the vision field plus the adaptation level of the eye. The discomfort of direct glare stems from two facts: first, the eye adapts (rapidly) to the average brightness of the overall visual scene; and second, the eye is attracted to the highest luminance in that scene. (The latter fact is used effectively in merchandising displays.) Thus, if an area of high brightness, such as a window or a lighting fixture exists in the visual scene and we are looking at an area of lower brightness, such as a work task, three visually disturbing things occur:

1. The eye adapts to a higher luminance level, thus effectively reducing the subjective brightness of the task—or, put more simply, making it harder to see what we are looking at (see Fig. 18.24). This is readily demonstrated by alternately blocking and unblocking a direct glare source with one's hand while trying to perform a moderately difficult visual task, and noting the immediate improvement in visibility when the glare source is obscured.

2. The eye is drawn simultaneously in two directions: involuntarily to the source of high luminance, and volitionally to the object we are looking at. The resultant tension causes considerable visual discomfort.

3. The adaptation level is continuously varying as the eye is drawn to the glare source and away again.

Glare is proportional to a source's luminance and its apprehended *solid* angle. Therefore, a small, bright source is usually not a problem, whereas a large, low brightness source (such as a luminous ceiling) may be. Indeed, a small, bright source adds sparkle to the field of vision, and many observers find it a pleasant addition in a monotonous lighting environment. Although discomfort glare from a scene is cumulative, source luminance is more important than the number of sources. For instance, if the luminance of a number of sources is halved, the reduction in glare is greater than is achieved by reducing the number of such sources by half. Indeed the latter procedure has little effect on discomfort glare.

The remaining two factors are less self-evident. Glare decreases rapidly as the brightness source is moved away from the direct line of vision, and thus the glare produced depends on the source's position in the field of view. The amount of discomfort glare produced by a source is inversely proportional to the background luminance (eye adaptation level). Thus, a ceiling fixture with a luminance of 4000 cd/m² at 65° might easily constitute a source of discomfort glare in a space with an eye adaptation level of 150 cd/m². The same fixture would not be objectionable in a daylight condition, where the eye adaptation level might be 1500 cd/m². A more striking example is that of an automobile's headlights, which at night are so severe a source of glare as to constitute "disabling glare," whereas in daylight, with its concomitant high eye adaptation level, these lights are very noticeable, but not usually disturbing.

Keeping in mind the dependence of direct glare on eye adaptation level, a useful rule of thumb is that luminances of large sources should not exceed 2500 cd/m², and that of small sources should not exceed 7500 cd/m². The former is roughly the luminance of blue sky; the latter approximates that of a fluorescent lamp. The terms "large" and "small" depend not only on actual physical dimensions of the source but also on the distance from the observer. That is, the actual criterion is apprehended size, or subtended visual angle, as shown in Fig. 18.21.

The glare effect of a number of individual glare source contributions in an interior space can be quantified by a criterion called *visual comfort probability* (VCP), which is defined as the percentage of normal-vision observers who will be *comfortable* in that specific visual environment. The IESNA has established a set of standard conditions for which VCP of sources can be calculated. These include a 1000-lux illuminance, repre-

sentative room dimensions, fixture height and observer position, and a head-up field of view limited to 53° above and directly forward from the observer (Fig. 18.26).

With these conditions, direct glare will not be a problem if all three of the following conditions are satisfied:

1. The VCP is 70 or more.
2. The ratio of maximum-to-average luminaire luminance does not exceed 5:1 (preferably 3:1) at 45°, 55°, 65°, 75°, and 85° from the nadir, crosswise and lengthwise.
3. Maximum luminaire luminances crosswise and lengthwise do not exceed:

Angle Above Nadir (degrees)	Maximum Luminance (cd/m²)
45	7710
55	5500
65	3860
75	2570
85	1695

A typical set of manufacturer's luminance and VCP data is shown in Fig. 18.27 for a ceiling-mounted fluorescent fixture with 4-40WT12 lamps. Note that all VCP values are considerably above the 70 minimum criteria. If full VCP data of this type are not available, they can be calculated with almost any lighting calculation program, given the luminaire luminance data. Despite the usefulness of the VCP criterion, it is

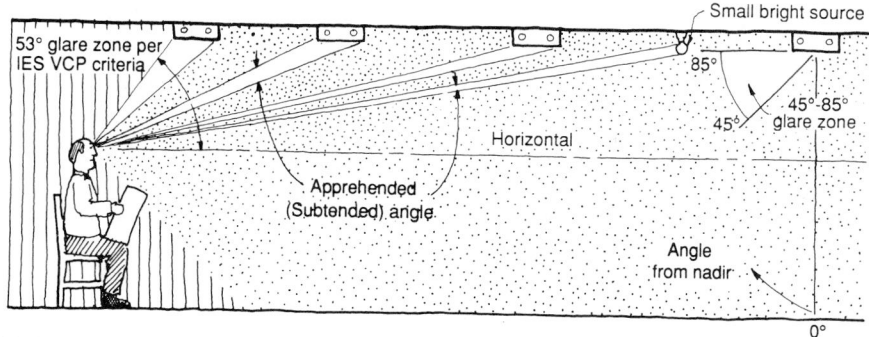

Fig. 18.26 *Determination of direct glare. The glare contribution of each source depends on its size (subtended or apprehended solid angle), luminance, and location in the field of view. Note that the apprehended solid angle of a small source is such that even with high luminance, it is not objectionable. Such sources are normally called* sparkle. *Glare is much more objectionable with a dark background than with a light one; therefore, light-colored paints on ceilings and upper walls are recommended.*

4–40-W Lamps
Prismatic Lens Diffuser

Average Luminance Data cd/m²

Vertical Angles	Across Axes	Along Axes
60°	2000	1750
65°	1060	1075
70°	500	560
75°	410	380
80°	480	343
85°	560	420

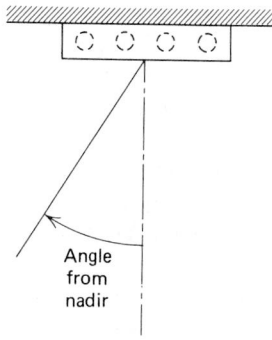

Angle from nadir

Reflectances:
  Wall 50%
  Ceiling cavity 80%
  Floor cavity 20%
Work plane illumination: 1000 lux

IES Visual Comfort Probability Data

Room Size		Luminaires Lengthwise			Luminaires Crosswise		
W (ft)	L (ft)	Ceiling Height (in ft)					
		8.5	10.0	13.0	8.5	10.0	13.0
20 ×	20	80	77	76	79	75	72
	30	80	78	76	79	76	73
	40	82	79	77	79	77	74
	60	80	80	78	79	77	75
30 ×	20	84	80	77	82	78	74
	30	83	80	77	81	79	74
	40	82	80	78	81	79	75
	60	82	80	78	80	79	75
	80	82	80	78	80	78	76

Fig. 18.27 A typical set of manufacturer's published VCP and luminance data.

inherently limited by its own standard conditions, which are not easily applied to other situations:

1. In small spaces, VCP has little significance.
2. Tabulated VCP figures are given for the *worst* viewing position in the room. Because VCP varies dramatically with observer position, the VCP values given are always lower than the space's *average* VCP.

In view of these (and other) reservations, most of which tend to make the actual direct glare situation better than the VCP calculation would indicate, it is recommended that layouts giving a VCP somewhat below 70 not be discarded out of hand. Instead, they should be examined carefully and, if possible, several observer positions calculated using one of the many readily available PC-based computer programs. With these as a guide, the designer can usually rearrange and substitute equipment to obtain the desired condition.

See Section 20.21 for a comparison of the direct glare characteristics of lighting fixture diffusers.

18.29 Veiling Reflections and Reflected Glare

Although there is no generally accepted convention with respect to nomenclature, many people refer to *reflected glare* when dealing with specular (polished or mirror) surfaces and to *veiling reflections* when considering source reflections in dull or semimatte finish surfaces, which always exhibit some degree of specularity. We use the terms interchangeably.

(a) Nature of the Problem. The problem of veiling reflections is much more complex than that of direct glare because it involves both the source *and* the task and is inherent in the act of

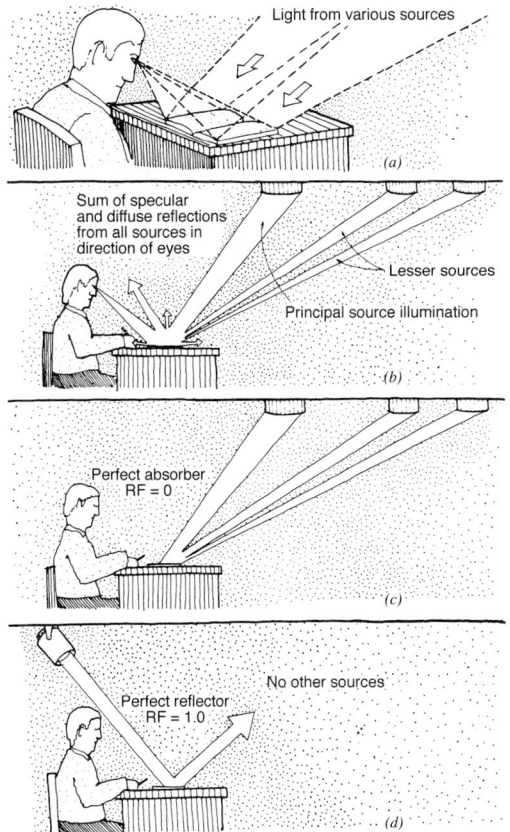

Fig. 18.28 (a) *The nature of the seeing process requires that light from the source(s) be reflected by the task into the eye.*
(b) *The light entering the eye is the sum of all of the reflected light, specular and diffuse, from all sources, in the direction of the eye. If the task is specular, all of the sources will be seen reflected in the task.*
(c) *A perfectly absorptive object is jet black because it reflects nothing.*
(d) *A perfectly reflective object positioned as shown is also black because geometrically it cannot reflect light into the eyes.*

seeing (Fig. 18.28). Vision is produced by light being reflected from the object seen. Thus if the object being viewed were replaced by a mirror, we would see the source(s) of light clearly [see Fig. 18.28(a)]. In commercial spaces there are usually one or more lighting fixtures near the observer that furnish most of the light by which to see. These *principal* sources are the main contributors to reflected glare. Other, more remote fixtures in

the room are lesser sources of veiling reflections [see Fig. 18.28(b)].

To the extent that the sources can be seen in the vision task, glare exists. It is imperative to an understanding of this problem to appreciate the importance of the reflection characteristic of the object being viewed. If the object were perfectly absorbent—that is, if it had a reflection coefficient of 0%—it would appear completely black, as no light would be reflected into the eye [see Fig. 18.28(c)]. Conversely, if the object were perfectly specular, like a clean mirror, and no light source were within the geometry of reflection, it too would appear black [see Fig. 18.28(d)]. Thus, if we took a clean mirror out on a moonless night and shined a light on if from over our shoulder, it would be practically invisible because no light would be reflected back into the eye.

The reader might try this experiment: In an inside space with a single overhead luminaire, try to examine the surface of a very clean, dustfree mirror. You will find that the best angle to hold it is *almost* at the angle at which the light source is seen. This is because the mirror is *almost* completely specular, and it is the slight diffuse reflection near the viewing angle that permits us to see the surface. This means that reflected glare is due to task surface specularity, whereas object definition, that is, the ability to see the task itself, is due to task surface diffuseness. A corollary of this conclusion is that veiling reflections, which are caused by mirroring of a source in the task, are proportional to source luminance and substantially independent of illuminance level. The brighter the source, the more troublesome its reflection.

Glare sources within the geometry of reflected vision are shown in Fig. 18.29 and the effects are shown in Fig. 18.30. Figure 18.29 clearly shows that although large sources are difficult to avoid, small sources can usually be easily avoided by a small change in the source-task-eye geometry, as by moving the head or tilting the task. Table 18.10 lists a few sample reflectance figures to demonstrate that most materials exhibit both a specular and a diffuse reflectance. In studying Fig. 18.29, it is important to note that a majority of visual work is done in the zone of 20° to 40° from the vertical, below the eye, with a maximum at the 25° reading angle (Fig. 18.31).

Fig. 18.29 *Geometry of reflected glare.*
(a) *Because normal desktop, head-down viewing angles vary from 20° to 40° from the vertical, the offending zone is the area on the ceiling corresponding to specular reflection between these two angles. (In specular reflection, angle of incidence equals angle of reflection). Note that the higher the ceiling, the larger this area becomes.*
(b) *In an office situation, the draftsman would see ceiling fixtures in the offending zone reflected in his instruments, parallel straightedge, and work. Note the important fact that the offending zone moves back and becomes smaller as the table tilts up (see Fig. 18.35).*

Fig. 18.30 *Note the effect of light in the area of the ceiling that causes reflected glare; contrast is reduced, and the print washes out. The usual viewing angle to a horizontal surface is between 20° and 40° from the vertical; we show 25° because it is the most common viewing angle. With vertical incident light on a diffuse surface (a), such as the pages of a textbook, the print is dark and clear. When the angle of light incidence is equal to the viewing angle (b), we have a mirror reflection situation. Even with diffuse paper, the print is light at best and almost invisible at worst. As the angle of incidence becomes larger (c), reflected glare decreases. When the incident light is at a very low angle (d), there is little reflected glare but all the print appears lighter. (Photos by Stein.)*

TABLE 18.10 **Typical Reflectances**

Material	Reflectance	
	Specular	Diffuse
Matte black paper	0.0005	0.04
Matte white paper	0.0030	0.77
Newspaper	0.0065	0.68
Very glossy white photo paper	0.048	0.83
Metallic paper—copper	0.11	0.28
Dull black ink	0.006	0.045
Super gloss black ink	0.039	0.016

Source: Courtesy of the IESNA.

(b) Contrast Reduction. The principal effect of the reflection of a light source in a visual object is to reduce contrast between the object and its background, thus reducing visibility. It is as if a bright veil were spread over the object being viewed, which accounts for the nomenclature "veiling reflection." As the angle of the incident light approaches the viewing angle, the specularly reflected component of this light becomes more and more pronounced, and task contrast drops. This is clearly visible in Figs. 18.30 and 18.31. The worst situation occurs when the incident angle equals the viewing angle. When the specular reflectance of the task and background is high, as with the glass screen of a visual display terminal, for instance, an image of the source is superimposed on the object, making viewing impossible (see Fig. 18.31). However, even with the highly specular finish of "slick" magazine paper, vision is still possible because of the very high contrast between black ink and white paper, although with much reduced clarity and considerable annoyance.

When considering specular *and* diffuse reflectance, the equation for contrast given in Section 18.19 must be rewritten as

$$C = \frac{(L_{BD} + L_{BS}) - (L_{TD} + L_{TS})}{L_{BD} + L_{BS}} \quad (18.9)$$

where L_B and L_T are background and task luminances caused by diffuse (D) and specular (S) reflectance; that is, L_{BS} is the background luminance due to its specular reflectance, and so forth. If we were to rework the calculation of contrast of Section 18.19, including specularity, the result would be much different.

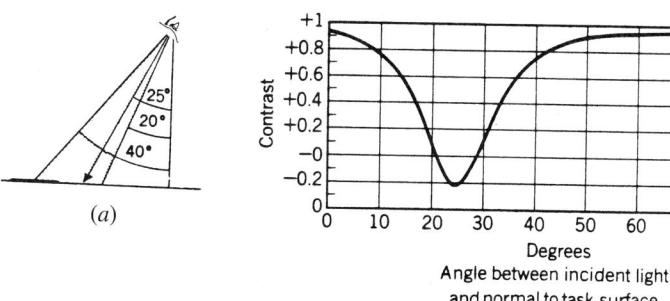

(a)

(b)

Fig. 18.31 (a) *Normal viewing angle of a task on a horizontal surface ranges from 20° to 40° from the vertical. Most common viewing angle is 25°.*
(b) *Graph showing contrast reduction of a task with a specular background (such as a sheet of glossy paper) as a function of the angle between incident light and the normal to the task surface. Viewing angle is assumed to be 25° from normal (see Fig. 18.30). Note that between 22° and 27° the contrast is negative. This indicates that background luminance exceeds that of the task, making the task essentially invisible. What is visible, clearly, is a reflected (mirrored) image of the source.*

EXAMPLE 18.5. Assume an interior space lighted to an average illuminance of 75 fc (750 lux) using bare bulb fluorescent fixtures (luminance [=] 7000 cd/m² or 2000 fL). The task is drafting with India ink on vellum. Reflectances are:

	Specular	Diffuse
Ink	0.021	0.038
Paper	0.018	0.71

Calculate the task contrast, without and with reflection of the 2000 fL source on the work.

SOLUTION.

(a) Without specularity, using Eq. 18.6 for diffuse reflection only:

$$C = \frac{R_B - R_T}{R_B} = \frac{0.71 - 0.038}{0.71} = 0.947$$

(b) With specularity, using equation 18.9,

$$C = \frac{(75 \text{ fc} \times 0.71 + 2000 \text{ fL} \times 0.018)}{L_{BD} + L_{BS}}$$

$$= \frac{(75 + 0.038) + (2000 \times 0.021)}{L_{BD} + L_{BS}}$$

$$= \frac{(53.25 + 36) - (2.85 + 42)}{53.25 + 36} = 0.497$$

which is just over half of the previous contrast! If the contrast is normalized to the maximum contrast (as is usually done), the contrast reduction R can be expressed as

$$R = 1 - \frac{C}{C_{max}} \qquad (18.10)$$

Thus, in this case, contrast reduction would be

$$R = 1 - \frac{0.497}{0.946} = 0.47$$

That is, contrast reduction would be 47%. ∎

A similar calculation for clear, black typewritten material on good white bond paper yields a contrast reduction from 94% to 77%, or $R = 17\%$. This 77% figure simply serves to emphasize the fact that with high task to background contrast, effective seeing is possible almost regardless of the lighting condition. However, this most emphatically does not relieve the lighting designer of the responsibility to provide a comfortable and efficient lighting environment in which pronounced veiling reflections do not exist. In general, any contrast reduction of more than 15% is undesirable.

Because both specular and diffuse reflectances frequently vary with the angle of view and exact figures are rarely available, accurate calculation is difficult. If a lighting system exists, or a mock-up can be made, measurements of contrast reduction can be made accurately with a contrast/luminance meter of the type shown in Fig. 18.32. A standard contrast device that is designed to correspond to a normal office task (black typeface and white paper background) is positioned on the work surface and exposed to the ambient illumination. The task contrast is then measured at the same angle at which it would normally be viewed. Contrast reduction is automatically calculated and displayed.

A contrast reduction map of the work surface can thus easily be made [Fig. 18.32(c)]. Thereafter, changes can be made in position of the work, viewer, or illumination sources to minimize con-

---

Fig. 18.32 (a) *This unit is a high-precision electronic digital instrument that measures luminance, luminance ratio, contrast, and contrast reduction. Use of the unit is shown in (b) (c) and (d). Luminance measurement range is 0 to 200,000 cd/m² (56,400 fL) at high accuracy.*
(b) *The meter measures contrast reduction at a specific viewing angle. Because the sample task can be positioned at any point on the table, a chart of contrast reduction can be made for any viewing position and any task location.*
(c) *Note that with a lighting fixture directly in the "offending zone"— that is, above and in front of the viewer—a severe loss of contrast occurs over much of the work surface (see Fig. 18.29).*
(d) *By shifting the relative position of the viewer and the source so that no source exists in the offending zone, contrast reduction is held to 3% to 4% over most of the work surface. Contrast reduction in the normal work area should not exceed 15%. (Courtesy of Brüel & Kjaer.)*

Fig. 18.32 *See caption at the bottom of page 1090.*

trast reduction. Note the pronounced effect of simply shifting the source out of the ceiling glare source zone (offending zone) [see Figs. 18.32(*d*) and 18.29]. Contrast reduction in the primary work area should not exceed 15% for good task visibility with specular work items such as glossy papers.

18.30   Equivalent Spherical Illumination (ESI) and Relative Visual Performance (RVP)

(a) Equivalent Spherical Illumination. Another way of approaching the problem of contrast reduction is to define a reference lighting system that is effectively free of veiling reflections, and then relate actual lighting systems to it with a figure of merit. Conversely, one can measure the effectiveness of a given lighting system in terms of the equivalent glare-free system. Both of these ideas, which are essentially the same, are the basis of the concept of equivalent spherical illumination (ESI).

In order to achieve a lighting system almost free of reflected glare, it is necessary to construct an enclosed volume whose surfaces are uniformly diffusely reflective and whose primary source is obscured to the maximum extent possible. As illustrated (Fig. 18.33), the integrating sphere is such a device. Light is introduced from the outside, split by a deflector, and evenly distributed throughout the sphere by the multiple reflections from the white-painted walls. The result is an evenly illuminated volume. When a task is intro-

Fig. 18.33 *Spherical illumination is produced by illuminating an object by diffuse reflection from the inside walls of an integrating sphere. The light source and observer are normally external.*

duced, the illumination falling on it is *entirely* uniform; that is, there are no high luminance sources reflected in it. It is therefore termed *spherically* illuminated. (Note the parallel to sky illumination.) The extent to which any other illumination system can duplicate this glare-free environment is that system's *equivalent* spherical illumination (ESI), representing the portion of its total illumination that is spherical—that is, diffuse and glare-free. Determination of ESI is accomplished by comparing contrast rendition in the spherical and test systems.

A study of school lighting (Fig. 18.34) gave the illustrated results for four viewing positions in a classroom lighted with ceiling-mounted continuous rows of 2 ft × 4 ft, 4-lamp, 40-W fluorescent fixtures with lens-type wraparound diffusers, on 10 ft centers. Carefully note that:

1. ESI depends entirely on viewing position and viewing angle, other factors in the space being equal.

2. In an ostensibly very well-lighted (215 fc) position (M1), the glare-free illumination is only 28 fc! This does not mean that visual work in this position is impossible. It does mean, emphatically, that in position M1, a pronounced veiling reflection exists on all specular objects. (Because of the size, orientation, and location of the glare source, this reflected glare is difficult to avoid.) It further means that a large amount of energy is being utilized (effectively wasted) to produce essentially negative results.

The results could have been anticipated, at least qualitatively by examination of the observer positions vis-à-vis the layout. Positions M1 and M3 have bright sources in the offending zone—M1 more so than M3—as is borne out by the results. M2 is an excellent position in that it receives light contributions from the two sides, its illuminance value being lower than the others due to wide row spacing. M4 is ideally placed; no glare sources are in the offending zone and a row of fixtures is positioned *behind* it, which makes it geometrically impossible to act as a glare source. The ESI analysis gives quantitative expression to our qualitative judgment and, as such, is a valuable design tool. Note particularly that the ESI results shown in Fig. 18.34 clearly correspond to the re-

		Observer Position			
		M1	M2	M3	M4
TI	2L	108	92	125	118
	4L	215	185	250	235
CRF	2L	.75	1.00	.82	1.01
	4L	.76	1.00	.83	1.03
ESI	2L	17.8	91.9	31.5	127.8
	4L	28.4	185.3	58.1	308.3

TI–Task illumination
2L–2 lamps (inside pair)
4L–4 lamps
CRF–Contrast Rendition Factors

Fig. 18.34 *A test classroom illuminated by three widely spaced rows of four-lamp fixtures with lens-type wraparound diffusers. Observer positions are shown by arrows. The row of fixtures in front of position M4 is too far forward to be in the offending zone. (Reproduced from Sampson, 1970).*

sults of a similar test made with the contrast meter, as shown in the charts of Fig. 18.32.

As with VCP criteria for direct glare, so with ESI; there are ameliorating factors that generally make a given lighting system better than these criteria figures would indicate. Some of these factors have already been mentioned but bear repetition.

1. ESI is critically dependent on observer position and viewing angle. Although position is generally fixed by chair location, observers can and do change viewing angle and head aspect to correct for glare situations.

2. The nature of the task (i.e., its specularity) is assumed to be fixed and unique. In some situations the task nature varies, and thus also the contrast. When tasks are constant, severe veiling reflections frequently lead to measures being taken that improve the task, the lighting, or both.

3. The lighting distribution characteristics of the fixture involved is a critical factor in glare production. The characteristic of a wrap-around lens diffuser is such (See Section 18.31(c[4]), Fig. 20.23, and Section 20.19] that considerable light falls in the glare zone. Other diffuser characteristics yield different results.

The concept and use of ESI has come under considerable criticism in the professional lighting literature because it addresses contrast, which is not identical with visibility, because it requires recalculation for even the small geometry changes that can change contrast dramatically, it tracks raw footcandles, and it is based on extrapolation from threshold conditions, the efficacy of which has been called into doubt. These published reservations and criticisms of ESI have disturbed the lighting profession sufficiently to result in wide abandonment of its use and its disappearance from most modern computer lighting programs.

A textbook is not a proper forum for polemics. The author simply wishes to state that he has used the ESI concept and calculations many times since 1970 with excellent results, and will continue to do so. The procedure does exactly what it is designed to do: to point out locations of poor lighting geometry immediately and quantitatively, and to flag luminaires with unsuitable distribution characteristics for the proposed use. That is, as stated in an earlier *IESNA Handbook* note, it is "used as a tool in determining the effectiveness of controlling veiling reflections and as a part of the evaluation of lighting systems."

(b) Relative Visual Performance. In more recent years a metric called *relative visual performance* (RVP) has appeared in the literature and has merited considerable acceptance. It tests (also via computer calculation) the effectiveness

ILLUMINATION

[i.e., the relative (to perfection) visual performance of a given visual environment] of task accomplishment as regards speed and accuracy. Like ESI, it is based on luminance and contrast, but unlike ESI, it judges relative performance of a task rather than simply contrast reduction. It seems to ignore the discomfort of veiling reflections if the task can be performed efficiently. The author has computed RVP for a number of common lighting layouts with various types of lighting fixtures. The results give uniformly high values for RVP varying between 0.95 and 0.99, which to our mind makes effective judgment of glare situations very difficult. The reader is encouraged to use the available lighting programs to calculate ESI and RVP for proposed lighting layouts and, where possible, to compare results with completed construction.

18.31 Control of Reflected Glare

Because the causes of veiling reflections are well understood, it would seem that a solution to the problem should long since have been adduced. Unfortunately, this is not the case. Although there is no known lighting method or material that completely eliminates veiling reflections, there are a number of techniques that minimize contrast loss due to veiling reflections while maintaining adequate illumination. These are:

- Physical arrangement of sources, task, and observer so that reflected glare is minimal [see Section 18.31(a)].

- Adjusting brightness (eye adaptation level) so that objectionable brightness is minimized [see Section 18.31(b)].

- Design of the light source so that it causes minimal reflected glare [see Section 18.31(c)].

- Changing the task quality [see Section 18.31(d)].

(a) Physical Arrangement of System Elements. Arrange the lighting geometry so as to avoid sources of high luminance at reflection angles when dealing with specular tasks. This is often difficult to accomplish in modern offices,

which frequently have both horizontal and vertical work surfaces, the latter being the specular screen surface of a visual display terminal (VDT). The latter problem is so widespread that today, it effectively governs the design of office lighting. It is discussed in detail in Section 21.24. As should be clear from Figs.18.32 and 18.34 and the related discussion, in a space using multiple sources, particularly in continuous rows, placing the work between rows with the line of sight parallel to the long axis of the units is an effective technique (see Fig. 18.34, position M2). Position M4 is "dangerous" in that the center row can be source of reflections. In this case it is not, due to the 10-ft row spacing. Note that the offending zone for horizontal tasks depends on the tilt of the desk. Thus for a horizontal desk, the offending zone is forward of the desk, as in Fig. 18.35(a); with an elevated table, the ceiling glare source zone may well be behind the source, as in Fig. 18.35(b).

All of the geometric solutions mentioned presuppose a detailed, fixed furniture layout, a situation that obtains in many but certainly not all cases. In the absence of such data, two alternatives are possible: a uniform layout and adjusting furniture to it, or vice versa. In practice, a combination of both is the most practical approach. Because low watts per square foot budgets have made ducted lighting fixture heat removal systems (air troffers) much less prevalent, fixtures are easily shifted. This mobility is further enhanced by the extreme flexibility of lighting fixtures fed from ceiling plug-in raceways, as discussed in Section 15.31. Figure 18.36 shows such a rearrangement, which results in saving five fixtures, a load reduction of 800 W, and an *improvement* in visibility.

(b) Control of Area Brightness and Eye Adaptation Level. As discussed in Section 18.19, loss of contrast can be compensated for (and glare reduced) by increased overall nonglare illumination. In so doing we are simply making the task brighter to override the detrimental veiling reflection. The problem with this technique, however, is that a large increase in illuminance is required to overcome the glare. This increase can, in many instances, be most practically accomplished not by increasing overall room illumination with the associated extremely high energy consump-

Fig. 18.35 (a) *If luminaires are kept out of the trapezoidal offending zone, contrast will be excellent. If the bulk of one or more luminaires projects into this zone, and in particular into the critical zone, contrast will drop sharply. The dimensions shown are for a flat desk 3 ft × 5 ft and a 9 ft ceiling height. (b) The dependence of glare zone on table tilt is illustrated. The offending zone becomes smaller as the table is raised, so that with a table near vertical position, glare is all but eliminated. See also Fig. 18.29.* [(a) *from Ross and Barruzini, 1974.*]

Fig. 18.36 *The original uniform fixture layout utilized three rows of six 2 × 4, four-lamp fixtures, giving a total load of 2880 W, a load density of 2.6 W/ft², and a uniform illumination level of approximately 90 (raw) fc [900 lux]. The original layout is shown dotted and numbered. The rearranged layout uses 13 fixtures (shown shaded) for a total 2080 W, a load density of 1.9 W/ft², and better than 100 ESI fc [1000 lux] on each work surface. In addition five fixtures are saved.*
Note: This level of illuminance is only justified for difficult visual tasks. See Table 18.7.

tion, but by adding a supplementary task-lighting source so arranged as to be free of reflected glare. By making this supplementary source's position adjustable [as in Fig. 18.29(*b*)], we accomplish three things.

1. Veiling reflection is overcome.
2. The high level of illumination needed for exacting tasks is provided with minimum energy expenditure.
3. The observer is granted complete control with resultant optimum lamp placement plus psychological satisfaction that generally prevents worker complaint. (Optimum position is generally to the left and slightly forward of the task).

We can demonstrate the effectiveness of a supplemental desk lamp by returning to Example 18.5.

EXAMPLE 18.6. Recalculate the contrast reduction of the ink on vellum visual task of Example 18.5, assuming that a desk lamp raises the illuminance to 200 fc (2000 lux) and is positioned so as to be glare-free. (*Note:* An adjustable lamp with 2–15 W fluorescent tubes produces about that luminance on the desk.)

ILLUMINATION

SOLUTION. Contrast from Eq. 18.9:

$$C = \frac{(L_{BD} + L_{BS}) - (L_{TD} + L_{TS})}{L_{BD} + L_{BS}}$$

$$= 1 - \frac{200 \times 0.038 + 2000 \times 0.021}{200 \times 0.71 + 2000 \times 0.018} = 0.72$$

With this contrast (0.72), the contrast reduction from the original no-glare situation has been improved from the original 47% reduction (0.946 to 0.497) to 24% reduction (0.946 to 0.72). However, because even a contrast reduction of 24% is undesirable, a change in task-source geometry or a change in source luminance would be required, assuming that the task itself must remain unchanged. ∎

(c) Control of Source Characteristics.

The reflected luminance that causes loss of contrast is proportional to the luminaire's luminance at that viewing angle, and therefore may be reduced by reducing luminaire luminance at that angle. This can be accomplished in four ways:

Dimming or switching lamps (see number 1 following).

Using luminaires with lower overall luminance (see number 2 following).

Using the luminaire as a primary source to illuminate a large, low-brightness secondary source (see number 3 following).

Reduce the luminaire luminance *only at the offending angles* (see number 4 following).

1. Reducing the total output of a fixture also reduces its output in the critical portion of the ceiling glare zone and can actually *increase* the ESI illuminance (i.e., improve task contrast).

2. In lieu of using a few small high-output sources, utilize larger-area, low-output sources (Fig. 18.37). This has the effect of reducing the source luminance in the ceiling glare zone while increasing the illumination contribution from outside the glare zone, resulting in better contrast for the same or lower illumination level (lux). The disadvantage of this technique is increased lighting fixture cost.

3. To overcome the economic disadvantage of multiple low-output, low-luminance sources, the ceiling can be used as a secondary source illuminated from high-output indirect or semi-indirect fixtures (see Section 20.11). These sources, which can be fluorescent or HID (e.g., metal–halide), have the advantage of high efficiency. The space's ceiling height must be sufficient to permit suspending the unit while avoiding "hot spots" on the ceiling. The minimum suspension length depends on the luminaire characteristics and is normally provided by the manufacturer. To assure high efficiency, the ceiling should be painted with a high-reflectivity matte white paint and kept clean. Results obtained from a semi-indirect installation using 1500-mA, very high-output lamps are shown in Fig. 18.38.

4. Because most horizontal task vision takes place between 20° and 40° from the vertical (see Figs. 18.29 and 18.30), any fixture that emits little or no light below 40° from the horizontal *cannot* produce veiling reflection regardless of its position in the field of view (Fig. 18.39). As a result, diffuser manufacturers produce a prismatic diffuser

(a)   (b)   (c)

Fig. 18.37 *A concentration of light in the glare zone* (a) *produces the largest amount of reflected glare. As the number of light sources is increased* (b) *in the glare zone and luminance is decreased, reflected glare is decreased. Least glare is from an all-luminous ceiling, which also has the lowest luminance* (c).

At observer position:
"Raw" illumination = 59 fc
ESI = 68.5 fc

Fig. 18.38 *With a high-reflectance diffuse finish ceiling, this semidirect perimeter lighting installation yields a higher ESI illuminance than raw illuminance at the viewing location illustrated, indicating excellent contrast rendering. The plastic lens at the bottom of the fixture serves to provide perceived light source luminance and to avoid an impression of gloominess, despite the satisfactory overall luminance level. (Reproduced from Sampson, 1970.)*

the output of which is diminished below 30° and above 60° in order to minimize both reflected and direct glare. Due to the characteristic shape of the distribution curve, they are known industry-wide as "batwing" diffusers. For observers positioned so that their sight lines are parallel to the longitudinal axis of the ceiling fixtures, lenses with linear (side-to-side) batwing characteristics perform well. If the observing position varies in aspect with respect to the fixture, a radial batwing curve (in all directions) is required. Note carefully

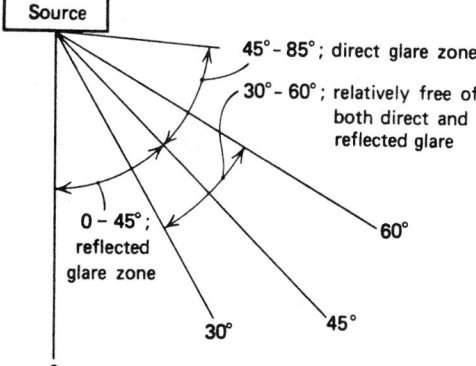

Fig. 18.39 *Glare zones are 0° to 45° and 45° to 85° for reflected and direct glare, respectively. Therefore, a diffuser that emphasizes the 30° to 60° zone will be least objectionable on both counts.*

that these diffusers have only limited usefulness in reducing reflected glare in specular vertical surfaces (VDT screens). All types of diffusers and their characteristics are discussed in Section 20.21.

(d) Changing the Task Quality. At this point it should be abundantly clear that reducing the task specularity is at least as effective a means of reducing veiling reflections as changing the lighting system characteristics, if not more so. It is therefore recommended that task contrast and specularity be actively considered and recommendations made in a framework of energy and cost effectiveness. Thus, to produce adequate visibility it is often cheaper and always more energy economical to upgrade the task (in the visibility sense) than to change the lighting system.

18.32 Luminance Ratios

As explained previously, other factors being equal, visual performance increases with contrast—that is, with difference in luminance between the object being viewed and its immediate surroundings. Conversely, however, the difference between the average luminance of the visual field (task) and the remainder of the field of vision should be low to avoid the discomfort of large

ILLUMINATION

TABLE 18.11 **Recommended Maximum Luminance Ratios**[a]

NOTE: To achieve a comfortable brightness balance, it is desirable to limit luminance ratios between areas *of appreciable size* as seen from normal viewing positions as follows:

1 to one third	Between task and adjacent surroundings
1 to one tenth	Between task and more remote darker surfaces
1 to 10	Between task and more remote lighter surfaces
20 to 1	Between luminaires (or fenestration) and surfaces adjacent to them
40 to 1	Anywhere within the normal field of view

[a] These ratios are recommended as maximums; reductions are generally beneficial.

Fig. 18.40 *The reflected glare from luminaires disappears when a piece of light, diffuse linoleum is placed over the dark, polished desktop. Light-colored desktops with a 35% to 50% reflectance result in task-to-background ratios within the 3:1 recommended range. Before the linoleum was placed, a reflection similar to the one seen on the right also existed on the left of the desk due to another luminaire. (Courtesy of IESNA.)*

rapid changes in eye adaptation level. Restated, *contrast is desirable in the object of view but undesirable in the wider surrounding field of view.*

Reflectances of 50%, 30%, and 80% for walls, floor, and ceiling, respectively, and 35% for furniture, establishes a fairly high eye adaptation level so that direct glare (which results from excessive luminances in the field of view) is minimized. Recommendations for maximum luminance ratios to achieve a *comfortable* environment are presented in Table 18.11. Effective visual performance is entirely possible in environments with much higher ratios. It is simply not as visually comfortable and may be fatiguing.

To achieve the recommended luminance ratios, it is obviously necessary to carefully control the reflectances of the major surfaces in a room. The reflectance figures given are averages for work-type commercial and educational spaces. The marked difference between a background with proper reflectance and one with excessive brightness ratios caused by the low surrounding reflectances is shown in Fig. 18.40.

18.33 Patterns of Luminance: Subjective Reactions to Lighting

In the list of characteristics in Section 18.16, we included among the secondary factors in illumina-

tion *patterns of luminance*—that is, the patterns of light and shadow in a space resulting from the illumination. Thus, a single source may produce sharp shadows whereas a luminous ceiling or a completely indirect illumination system produces almost completely diffuse light (Fig. 18.41). *Diffusion* is the degree to which light is shadowless and is therefore a function of the number of directions from which light impinges on a particular point and their relative intensities.

Perfect diffusion, rarely obtainable (or desirable), would have equal intensities of light impinging from all directions, therefore yielding no shadows. The only naturally occurring example of perfectly diffuse lighting is a daytime fog, which we know to be extremely disturbing to the eye, demonstrating that some directivity is desirable. Diffusion can be judged by the depth and sharpness of shadows. A room with well-diffused illumination resulting from multiple sources and high room surface reflectances yields soft multiple shadows that do not obscure the visual task. Because purely diffuse lighting is monotonous and not entirely conducive to extended periods of work effectiveness, some directional lighting is often introduced as an adjunct to diffuse general lighting to lend interest by producing shadows

Fig. 18.41 *Diffuse illumination. The luminous ceiling installation provides shadowless, almost perfectly diffuse lighting. By comparison, the single ceiling bulb produces sharp shadows and very little light diffusion.*

and brightness variations. Where texture must be examined or surface imperfections detected by grazing angle reflections, highly directional lighting is required. Indeed, as seen in Fig. 18.42, directional light is what creates shape and is precisely the characteristic best used to influence architectural space and form.

Sections 20.11 through 20.16, which deal with systems of lighting, illustrate a few of the light/dark patterns produced by different lighting arrangements. The combinations of uplighting and downlighting are legion; each produces its own shadows and modeling, and each has a quality of its own. It is very much in the interest of the lighting designer to be familiar with these effects so that he or she can mentally visualize them as the design progresses. Indeed, it would be well for a designer to prepare a reference sketchbook of such shadow diagrams. It is these patterns of light and darkness that give the ambience and the subjective reactions of sociability/isolation, clarity/fuzziness, spaciousness/crampedness, simplicity/clutter, formality/informality, boredom/excitement, definition/shapelessness, and so on. Indeed, many lighting designers *begin* a lighting design by sketching pictorially the area to be lighted, showing the patterns of brightness and shadow desired to achieve their objective. This technique

ILLUMINATION

Fig. 18.42 *Totally diffuse lighting (a) destroys texture, whereas a combination of diffuse and directional lighting (b) produces the required modeling shadows. (Courtesy of Holophane.)*

is obviously most useful in nonoffice areas such as lobbies, waiting areas, all type of recreational spaces, restaurants, merchandising spaces, and so on. In office areas the technique can also be applied to provide the points of visual interest referred to in our discussion.

Color has a great deal to do with subjective reactions and is discussed separately. The subject of psychological reactions to lighting environment is extensive and complex and can be only touched on here to the extent of mentioning a few of the salient lighting techniques and their usual subjective responses.

In addition to modeling and texture accent, small high-brightness sources, usually called *sparkle,* create points of interest and visual excitement. Lighting installations generally yield a sense of vividness or activity proportional to the level of illumination. This is not the case with very diffuse lighted areas, which, even at high footcandle levels, are tedious. This is particularly noticeable in large, luminous ceiling installations that are particularly oppressive when the ceiling is low. Small exposed incandescent lamps, a brightly lighted, rough-textured wall, and pendant fixtures with pierced reflectors are some of the techniques used to create visual interest.

Visual attention can be drawn by high brightness. This well-known fact is used constantly in displaying merchandise. Note the following usual reactions:

- A 3:1 luminance ratio between object and surround will be noticed, but usually will not affect behavior or draw attention.
- A 10:1 luminance ratio will attract attention and, if interesting, hold it.
- A 50:1 luminance ratio or higher will highlight the object thus illuminated, practically to the exclusion of all else in the field of view.

Because areas of high luminance draw the eye's attention, all of the individual brightness sources in the field of view produce an overall impression. If there is some form or order or pattern to them (as a pattern of lighting fixtures), then the overall impression is not disturbing—it can be thought of as visually harmonious. If, however, they are in disarray, they produce a discordancy in the eye precisely as sound produces discordancy in the

ear. This visual "noise" is frequently referred to as *visual clutter* and can be very disturbing. The designer is well advised to keep this important fact in mind when arranging light sources that are the primary sources of luminance in an enclosed space. Aspects of fixture patterning are shown in Section 20.16.

Other subjective reactions to lighting on which there is wide consensus are:

1. Bright walls (about 25% of horizontal light level) increase the impression of spaciousness. Conversely, dark walls diminish a space. As a corollary, high fixture luminance attracts the eye away from the walls and diminishes spaciousness.
2. Worker-adjustable task lights increase the feeling of control and therefore comfort.
3. Downlights (and color highlights) increase feelings of relaxation and comfort.
4. Hidden source indirect lighting and very low brightness lighting fixtures cause discomfort because of the inability to locate the source of light.

FUNDAMENTALS OF COLOR

The subject of color is vast. We can relate here only to a few aspects of color that it is imperative that the lighting designer understand. Furthermore, because it is difficult to discuss color without actually using it, the coverage is brief.

18.34 Color Temperature

A light source is often designated with a *color temperature,* such as 3400 K for halogen lamps, 4200 K for certain fluorescent tubes, and so on. This nomenclature derives from the fact that when a light-absorbing body (called a *black body*) is heated, it first glows deep red, then cherry red, then orange, until it finally becomes blue–white hot. The color of the light radiated is thus related to its temperature. Therefore, by developing a black-body color temperature scale, we can compare the color of a light source to this scale and assign to it a "color temperature"—that is, the temperature to which a black body must be heated

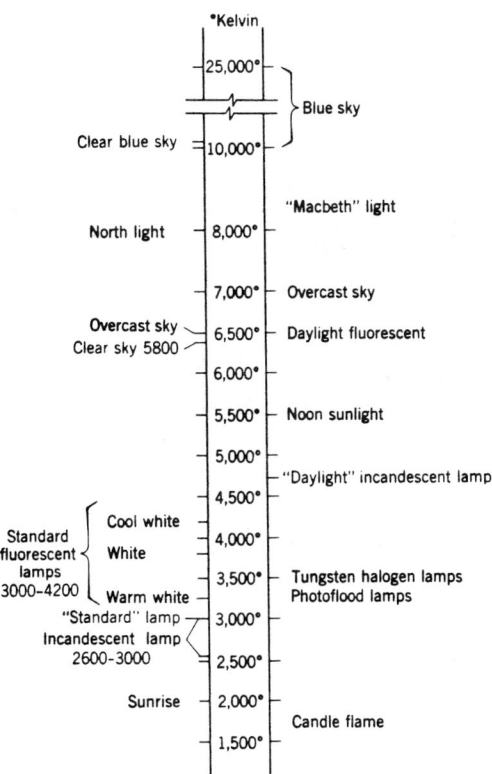

Fig. 18.43 *Approximate color temperatures of common illuminants.*

white, selective absorption occurs. The component colors are absorbed in different proportions so that the light reflected or transmitted is composed of a new combination of the same colors as had impinged on the surface. Thus, a white light reflected from a red wall acquires a red tint because the component colors of the white light other than red were absorbed in greater proportion than the red. When reflected, the red light takes prominence, thus giving the reflected light a red tint. This is illustrated in Fig. 18.44(*a*). Similarly, a white light when passed through a piece of red glass emerges as a reddish light because the other components are absorbed in much greater proportions than the red. This well-known phenomenon is illustrated in Fig. 18.44(*b*).

It is this phenomenon that allows us to see color at all; the individual object pigmentation absorbs other colors of light and reflects or transmits to the eye only its own hue, in greater concentration than in the incident light.

to radiate a light similar in color to the color of the source in question. Temperature is measured in Kelvin, which is a scale that has its zero point at −460°F. Figure 18.43 shows the assigned color temperature of some common light sources.

Strictly speaking, a color temperature can be assigned only to a light source that produces light by heating, such as the incandescent lamp. Other sources, such as fluorescent lamps, produce light by processes that are detailed in the Chapter 19. Such sources are assigned a *correlated color temperature* (CCT), which is the temperature of a black body whose chromaticity most nearly matches that of the light source. For such sources there is *no relation whatever* between their operating temperature and the color of the light produced.

Any nonspectral color illuminant is composed of two or more component color illuminants. When such a composite light—for example, white—falls on a surface other than black or

18.35 Object Color

The color of the illuminant (light) and, correspondingly, the coloration of the objects within a space constitute an important facet of the lighting quality. The two factors, however, must not be considered separately because by definition the color of an object is its ability to modify the color of light incident on it by selective absorption. The color reflected or transmitted is apprehended by the eye as the color of the object. An object is technically said to be "colorless" (not transparent) when it does not exhibit selective absorption, reflecting and absorbing the various components of the incident light nonselectively. Thus, white, black, and all shades of gray are colorless, neutral, achromatic, or, more precisely, lack hue.

Hue is defined as that attribute by which we recognize and therefore describes colors as red, yellow, green, blue, and so on. Just as it is possible to form a series from white to black with the intermediate grays, so it is possible to do the same with a hue.

The difference between the resultant colors of the same hue so arranged is called *brilliance* or *value*. White is the most brilliant of the neutral colors and black the least; pink is a more brilliant

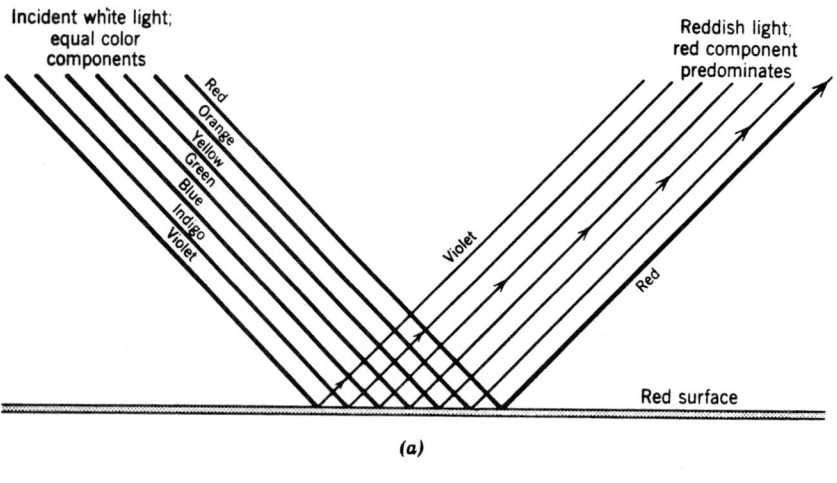

Incident white light;
equal color
components

Reddish light;
red component
predominates

Red
Orange
Yellow
Green
Blue
Indigo
Violet

Violet

Red

Red surface

(a)

Incident white light;
equal color
components

Violet
Indigo
Blue
Green
Yellow
Orange
Red

Red glass

Reddish light;
red component
predominates

(b)

Fig. 18.44 *Selective absorption of* (a) *reflected light and* (b) *transmitted light.*

red hue than ruby; and golden yellow a more brilliant (lighter) yellow hue than raw umber.

Colors of the same hue and brilliance may still differ from each other in saturation, which is an indication of the vividness of hue or the difference of the color from gray. Thus, pure gray (black plus white) has no hue; as we add color, we change the saturation without changing the brilliance. The three characteristics then that define a particular coloration are *hue, brilliance.* and *saturation.* Using these terms, we may define "bay" as a color

red–yellow in hue of low brilliance and low saturation, whereas carmine is a color red in hue, of low brilliance, and very high saturation.

Various systems of color classification have been devised, including the ISCC–NBC color system, the Munsell Color System, the Ostwald Color System, C.I.E., and the Chromaticity Diagram. In the well-known and widely used Munsell Color System (Fig. 18.45), brilliance is referred to as *value* and saturation as *chroma;* thus, a color is defined by hue, value, and chroma. The bril-

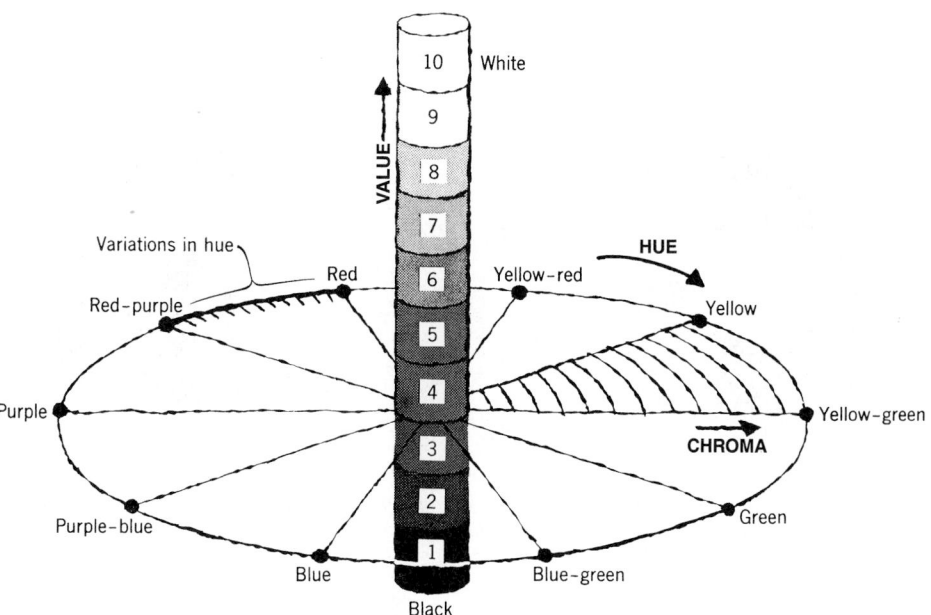

Fig. 18.45 *The Munsell Color System defines a color by three characteristics: hue (color), chroma (saturation), and value (grayness).*

liance (value) of a pigment or coloration is related to its reflectance to white light. The higher the brilliance or value, the higher the reflectance factor, as might be expected when one considers that white and black are the poles of brilliance. Chroma or saturation may be thought of as either the difference from gray or the purity of the color. Spectral colors have 100% purity, and therefore maximum chroma.

When white is added to a pigment, it produces a *tint;* adding black produces a *shade.* When pigments are mixed to produce a particular color, we create this color by a subtractive process. That is, each pigment absorbs certain proportions of full-spectrum white light; when mixed, the absorptions combine to subtract (absorb) various colors of the white spectra and leave only those colors that finally constitute the hue, value, and chroma of the pigment. This subtractive effect is also utilized when producing colors by filtering white light. Each filter selectively absorbs component colors, transmitting only the component desired. Thus, a blue filter transmits only blue, and so on [see Fig. 18.44(b)]. Conversely, when light of the three primary colors of red, green, and blue are combined, they form white by an additive process (Fig. 18.46).

The additive and subtractive primary colors are complementary; they combine to give a white or neutral gray, respectively. Thus, red and blue–green, blue and yellow, and green and magenta are complementary. Therefore, if a red object is illuminated with blue–green light, the object color appears gray because the red pigment absorbs the blue–green and reflects nothing; hence the gray. This accounts for the once common "lost red car" in parking lots illuminated with clear mercury lamps with their characteristic blue–green color. This phenomenon is today rare, as clear mercury lamps have fallen into disuse in favor of metal–halide and sodium lamps.

18.36 Reactions to Color

Light of a particular hue (other than white) is rarely used for general illumination except to create a special atmosphere. When a space is lighted with colored light, the eye adapts by a phenomenon known as *color constancy* so that it can, to a considerable degree (depending on the chromaticity of the light), recognize colors of objects despite the spectral quality of the illuminant. Thus, even when wearing heavily tinted sunglasses, we

ILLUMINATION

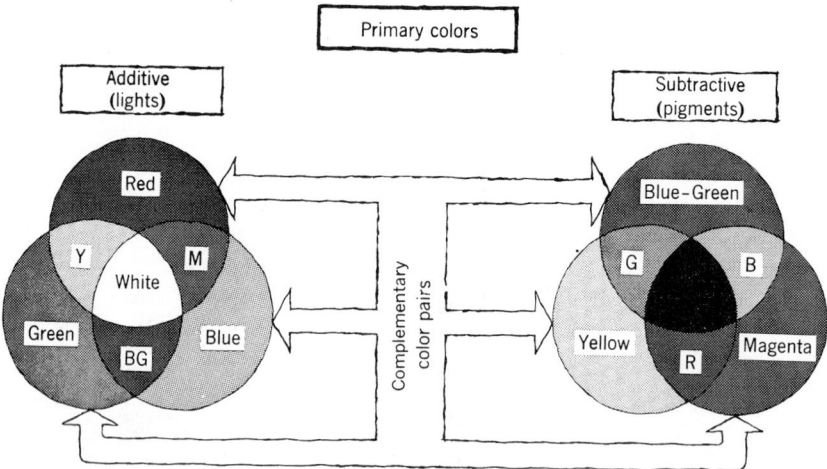

Fig. 18.46 *Primary and complementary colors. Complementary color pairs are shown by arrows. Pigments from color by an absorptive (subtractive) process; colored lights from colors by an additive process.*

can still distinguish the color of objects quite easily. Indeed, after only a very short while we no longer notice the green, yellow, blue, amber, or other color cast caused by the tinted lenses. However, the eyes become more sensitive to the missing colors that would make up white light. This phenomenon can be used to make meat look redder on a butcher's counter by using blue-rich, red-poor, cool white lighting in the remainder of the store.

A similar phenomenon occurs when the eye is exposed to a monochromatic scene, where the chromaticity is due to coloration of the objects rather than the illumination. The eye in such a situation becomes sensitized to the complementary color; thus, if after looking at a green surface one shifts the gaze to a white surface, one sees the complementary red color. Returning to our meat market, the use of green paint on the walls also enhances the redness of the meat. This effect in reverse also partly accounts for the extensive use of green for paints, linens, and gowns, and so on, in operating rooms. The eyes of the surgeons and nurses when diverted from the redness of the surgical area are more comfortable seeing green on a green background than on a white one.

By a process known as *lateral adaptation,* the apparent color of an object changes when background color is changed. Thus, a green object looks somewhat blue-green on a yellow background because the eye is supplying the complementary color to yellow—that is, blue. Similarly, the same green object looks slightly yellow-green when on a blue background, the eye supplying the yellow.

Apparent brightness of a color is a function of its hue, in that light colors appear lighter than dark colors even when measured brightness is the same. Thus, spaces may be defined by color within an area of equal illumination. Also, all colors tend to appear less saturated; that is, they appear "washed out" when illumination is high. Thus, pigments of high saturation (chroma) must be used in well-lit spaces if they are to be effective, although extensive use of saturated colors is generally best avoided.

Other well-known psychological effects of colors are the coolness of blues and greens and the warmth of reds and yellows. Thus, cool colors might well be used in a fur salon and warm colors in a display of summer wear. Red and yellow are "advancing" colors because objects lit with them tend to "advance" toward the observer, giving the appearance of becoming larger. The opposite effect is noted with blue and green, accounting for their being known as "receding" colors.

A practical, energy-saving application of these color phenomena would be to use warm colors to compensate somewhat for lowered thermostats in the winter, and cool colors for the opposite effect in summer. How to accomplish this without the expense of repainting twice a year is left to the ingenuity of the architect and interior designer. In an atmosphere designed to be calm and restful, greens should generally predominate either in illuminant color, object color, or both, except in eating areas, which should be lighted with reds and yellows because cool colors are generally unappetizing. Yellows and browns emphasize motion sickness, whereas blues and greens tend toward the reverse. Warm and saturated colors produce activity; conversely, cool, unsaturated colors are conducive to meditation. Cool colors also seem to shorten time passage and are well applied in areas of dull, repetitive work.

A further discussion of color control, illuminant colors, color measurement, and color matching is found in Sections 19.34, 19.35 and 19.36, dealing with spectral energy distribution of sources and color rendering index.

BIBLIOGRAPHY

See listing at the end of Chapter 21.

ILLUMINATION

19

LIGHT SOURCES:
THEIR CHARACTERISTICS
AND APPLICATION

19.1 General Remarks

Long before the dawn of recorded history, the double blessing of fire, heat and light, was discovered. Even today, in our sophisticated space age, fire is still used almost universally as the source of heat and, in not inconsiderable proportion of the world's dwellings, as the source of light. Electrical lighting had its real beginning in about 1870 with the development of commercially usable arc lamps, and was given greater impetus nine years later by Edison's first practical incandescent lamp. Today's practical electric light sources fall generally into two generic classifications: the incandescent lamp, including tungsten–halogen types; and the gaseous discharge lamp, which includes fluorescent, mercury, metal–halide, sodium lamps, and the more recent induction lamp. The microwave-powered sulphur lamp seems to be in a separate classification of plasma-type lamps.

The efficiency of a light source is termed its *efficacy* and is measured in lumens per watt (lm/W or lpw). Table 19.1 lists efficacies of modern light sources, including ballast losses where applicable. It is misleading to use the efficacy of the light sources alone, as is often done in the literature, because the ballast/power supply is an integral part of the lamp circuit. Note that Table 19.1 gives a range of efficacies for each lamp type. In general, efficacy increases with wattage. It is, therefore, energy-economical to use a small number of high-wattage lamps rather than the reverse. (It is also usually more economical with respect to luminaires.)

Because electric lighting in American nonresidential buildings consumes 25% to 60% of the electric energy utilized, any attempt to reduce this must necessarily include integration of the cheapest (insofar as energy is concerned), most abundant, and, in many ways, most desirable form of lighting available—daylight.

DAYLIGHTING

19.2 Daylighting as a Lighting Design Factor

The provision of daylight in structures in the United States had largely been considered an amenity rather than a necessity before the 1973 energy crisis. As such, its provision had been the province of architecture rather than lighting design. The reasons for this are clear. Daylight is indeed an amenity. Windows provide visual contact with the outside, and the resultant daylight provides a bright, pleasant, airy ambience. When daylight enters through windows (sidelighting as opposed to toplighting), its horizontal directivity provides good modeling shadows, minimal veiling reflections, and excellent vertical surface illumination. Furthermore, the continual variation of daylight, which is one of its prominent characteristics, provides a constantly changing pattern of space illumination—one that is not easily attainable with artificial light. Because these changes are gradual, the eyes adapt easily and the effect is one of visual interest. Undoubtedly as a result of these effects, numerous studies have conclusively demonstrated a marked preference for daylight over any other form of lighting.

However, although concern has been expressed about possible physical ill effects, none have been demonstrated *conclusively* to have

ILLUMINATION

TABLE 19.1 **Efficacy of Various Light Sources**

Source	Efficacy (lpw)
Candle	0.1
Oil lamp	0.3
Original Edison lamp	1.4
1910 Edison lamp	4.5
Modern incandescent lamp	8–22
Tungsten halogen lamp	16–20
Standard fluorescent lamp[a, b, c]	35–80
Compact fluorescent lamp[d]	55–75
Mercury lamp[a, c, e]	25–55
Metal halide lamp[a, c, e]	75–105
High pressure sodium[a, c]	55–115
Induction lamp[f]	48–70
Sulphur lamp[f]	90–100
Maximum predicted by year 2010	150
Maximum theoretical limit	250 (approximate)

[a] Includes ballast losses.

[b] With electronic ballasts, these figures become 40–100 lpw.

[c] With standard core-and-coil ballasts.

[d] With electronic ballasts.

[e] Approximately 10% higher with electronic ballast.

[f] With power supply.

been caused by lack of daylight; that is, by working in an electrically lighted space. Inasmuch as an electric lighting system had to be installed in any event, to furnish interior illumination during periods when daylight is insufficient, the practice arose in the United States of ignoring daylight and even of shutting it out deliberately! Careful design of an electric-lighting system can provide a satisfactory visual atmosphere, as these chapters explain. Furthermore, unlike daylight, control of such systems is relatively simple. Perhaps most important, an interior electric-lighting system has minimal impact on the building architecture, least of all on the all-important building facade. Finally, the energy to power electric-lighting systems *was* cheap.

The option of ignoring daylight in our high-energy-cost and rapidly diminishing natural resource world is no longer available. That being so, the designer must learn to cope with the special problems that daylight use presents in order to reap its benefits. Because daylight is variable, it creates special problems of glare control, direct sunlight control, and heat-gain limitation. In large measure, the science (and art) of daylighting is

not so much how to provide daylight as how to do so without the attendant undesirable effects. Put otherwise, the lighting designer must adapt and adopt the technique of Permanent Supplementary Artificial Lighting in Interiors (PSALI), which is almost universally applied in Europe.

One caveat must be introduced immediately. Windows are a two-way portal. HVAC designers have long been aware of the major influence of windows on the thermal balance of a building and of the need for window controls. The same awareness is necessary for daylight design. Windows can be a daytime glare source and a nighttime light-loss source unless appropriate controls are used. Furthermore, windows, particularly skylights, are a source of nighttime light pollution, which affects not only astronomers but also ordinary citizens. That subject is discussed briefly in Section 21.39.

The material that follows, which discusses PSALI, sky luminance, outdoor illumination, and calculation of interior daylight levels, is necessarily brief because of space considerations. Probably no subject bearing on architectural engineering has received more attention since the late 1970s than daylight—both its source and its ramifications. A bibliography of books and papers bearing on the subject would alone cover more space than is allotted to the subject itself in these pages. The interested reader is referred to the bibliography at the end of Chapter 21, which lists a few of the important sources in this field.

19.3   Permanent Supplementary Artificial Lighting Interiors (PSALI)

This technique, which is really a design approach, views electric lighting as supplementary to daylighting, and not vice versa. The PSALI technique recognizes that nonresidential buildings are used principally during daylight hours and that sufficient daylight is generally available during these hours to provide much of a structure's lighting needs. The success of PSALI design approach is predicated on three assumptions:

1. The large variation of interior daylight illuminance levels during the course of the day will not adversely affect visual performance, even when the actual levels fall below accepted recommendations. Stated otherwise, the same visual performance can be achieved with less daylight than electric light, when compared on a lux [footcandle] basis. The rationale for this conclusion is based on two well-known phenomena of human vision: lightness constancy and Weber's law of visual contrast. Lightness constancy is the ability of the eye to distinguish between light and dark objects over an extremely wide range of illumination. Thus, even though a lump of coal in sunlight is brighter than this piece of paper in moonlight, we will correctly identify the one as black and the other as white. This is because the eye appreciates lightness or brightness not only by actual luminance of the object being viewed, but also by the surrounding area. Thus, in a space with *slowly* changing illumination levels, as is the case with daylighted spaces, the eye adapts to the altered level and sees basically the same scene, in terms of lightness (and darkness) of objects. That is, the eye's "understanding" of the pattern of brightnesses in the room remains constant.

Weber's Law states that within the very wide illuminance range of lightness constancy (above 10 lux), contrast ratios remain constant. Because visual performance is closely related to contrast discrimination, it is also largely unaffected over the wide absolute range of daylight illuminance.

2. Illuminance recommendations to be used are the current ones (see Section 18.24).

3. Daylight and artificial light can be readily and successfully combined; that is, artificial light can supplement daylight when the latter is insufficient (Fig. 19.1). We made particular reference to nonresidential buildings because residences are occupied and extensively utilized at night, when no daylight is available and PSALI is inapplicable.

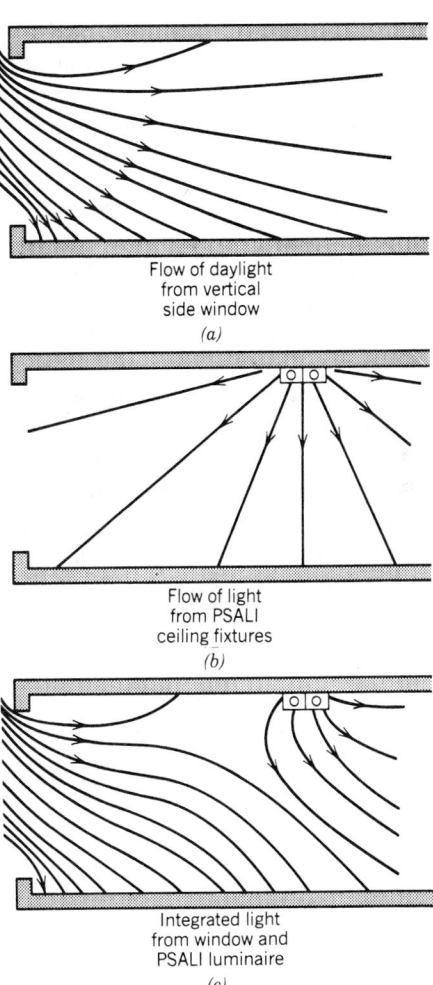

Flow of daylight
from vertical
side window
(a)

Flow of light
from PSALI
ceiling fixtures
(b)

Integrated light
from window and
PSALI luminaire
(c)

Fig. 19.1 *Integration of daylight and artificial light, showing flow of each and of the combination. (Drawings adapted from Lynes et al., 1966, and Ne'eman and Longmore, 1973; reproduced with permission of J. A. Lynes)*

ILLUMINATION

Obviously, residences are also used during daylight hours, but the conditions of fixed work location and long-duration visual tasks generally do not apply. As detailed in various sections of this book, home offices are becoming increasingly common. These spaces should be treated as office spaces and designed as such. Nonresidential buildings that are regularly used at night as well as during daylight hours should be designed for both conditions, with appropriate controls.

Because interior daylighting is obviously directly dependent on exterior lighting levels, an understanding of the latter is a clear prerequisite to designing for the former. However, before beginning a discussion of exterior illuminance and the various methods by which the resultant interior daylight illuminance can be calculated, it is appropriate to ask several basic questions regarding daylight calculations in general:

1. Inasmuch as the majority of buildings must be designed for nighttime use as well—that is, with a complete system of electric lighting—what purpose does a knowledge of daylight distribution in such buildings serve?
2. Given the great variability of daylight with time of day, season, and weather, how can any meaningful calculations be made?
3. Furthermore, if daylight distribution is calculated for a variety of conditions, what use can be made of the results? What are the results we want, the definition of which will establish the exterior conditions that we wish to study?

The answers to these questions are fairly clear and will define the direction that our study of daylight will take. The purpose of daylight analysis is manifold.

1. Granted the variability of daylight, it is possible to calculate maximum, minimum, and, if enough data are available, statistical averages. Furthermore, there are extensive areas of the United States where, for long periods of time, the weather is constant—clear blue skies in the Southwest, solid overcast in the Northwest—and for those areas, fairly accurate calculations of interior daylight($\pm$ 25%) can be made.

2. Maximum-daylight data, including direct sunlight, are needed for design of glare controls, shading devices, light shelves, and other types of reflectors, and for zoning of automatic daylight compensation controls (see Section 21.4). These data are also important in establishing maximum cooling load.

3. Minimum-daylight data are necessary to establish maximum PSALI (Fig. 19.2) and to design the controls that differentiate PSALI from nighttime lighting. This also defines maximum electric lighting loads, which are needed for heating, cooling, and energy budget calculations.

4. Average-daylight data are necessary to establish percentiles on the basis of which annual energy budgets can be determined. Furthermore, average data are important in designing fenestration because use of minimums probably result in excessive glass area, with the concomitant thermal problems discussed in Chapter 5. Thus, it is reasonable to design fenestration on the basis of a 50% to 60% figure, which states that a given daylight level will be maintained for 50% to 60% of the daylight hours. In the analyses that follow, it is assumed that fenestration has been established. Obviously, the procedures presented can also be used to establish fenestration, by starting with rules of thumb and utilizing a modified trial-and-error procedure to improve the initial design.

5. The problem of inaccuracy in interior lighting calculations can be divided into two: that caused by inaccurate input data, and that which is inherent in all calculations based on statistical averages. Referring to the first, we will see in the discussion that follows that hand-type, pencil-and-paper methods and the use of sky luminance data based on season, latitude, and solar altitude may introduce an error of 25% or more, depending on the type of sky. Furthermore, the calculation procedure itself involves approximations of considerable magnitude, so that results may be 30% to 50% from actual measured levels. (That being so, the pointlessness of working to three decimal places should obvious.) However, these *hand* calculation procedures are intended only to give the designer a "feel" for the daylight levels in the space and, more important, for the gradation, or

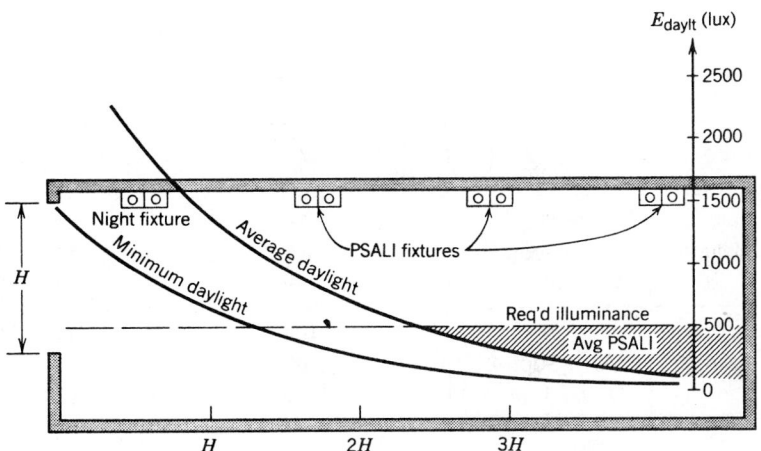

Fig. 19.2 *Section through a room with daylighting from one side window. The typical average daylight curve shows that for this building, to a depth of about 2½ times window height* H, *daylight provides the necessary illuminance. Beyond that, PSALI is required to maintain the required level. Even with minimum daylight conditions, a depth of* H *into the room is daylighted, and PSALI lighting is required only beyond that depth. Thus, the first (nearest the window) lighting fixture is used only in nighttime lighting, the second and third would be dimmed or multilevel switched, and the fourth (back wall) would be lighted during all hours of building use. Average daylight would represent the 85th percentile—that is, the level that (statistically) is maintained for 85% of the daylight hours. Setting the minimum is flexible and depends on the type of control system. It usually represents the 95th percentile—that is, the level that is maintained for 95% of daylight hours.*

contours, of daylight. Greater accuracy is possible only with computer-aided design procedures, which input more specific sky luminance data and reduce approximations.

For annual energy budget calculations, hour-by-hour daylight calculation is required, and, of course, this is done only by computer.

The second "inaccuracy" is a statistical deviation. Because of the infinite variability of weather in *temperate* climates, it is not possible to calculate the daylight condition for a specific time on a specific day, *nor is it necessary*. For this reason, the following procedures use single figures for whole seasons and rely on proper design of controls of both daylight and electric light, with PSALI, to furnish a well-lighted, low-energy-use interior.

It may be helpful at this point for the reader to turn ahead to Section 21.4e and read through the section of daylight compensation to understand how PSALI with proper controls results in low energy use while maintaining the necessary lighting levels.

19.4 Characteristics of Outdoor Illumination

(a) Factors. The most prominent characteristic of daylight is its variability. Obviously, the source of all daylight is the sun. The level of exterior illumination, at a particular place and time, depends on:

1. Solar altitude, which can be determined if latitude, date, and time of day are given (see Table 19.2).
2. Weather conditions (e.g., cloud cover, smog).
3. Effects of local terrain (natural and man made obstructions and reflections; see Fig. 3.7).

The position of the sun in the sky is expressed in terms of its altitude above the horizon and its azimuth angle. Azimuth angle is defined as its horizontal position angle, measured from the *south*. Both are normally expressed in degrees (Fig. 19.3*a*). Because the sun is due south at noon,

TABLE 19.2 **Typical Solar Altitude and Azimuth Data as a Function of Date and Time of Day**

Latitude (°N)		Date	Solar Time [a]						
			AM: 6 / PM: 6	7 / 5	8 / 4	9 / 3	10 / 2	11 / 1	Noon
30	Altitude	June 21	12	24	37	50	63	75	83
		Mar.–Sept. 21	—	13	26	38	49	57	60
		Dec. 21	—	—	12	21	29	35	37
	Azimuth	June 21	111	104	99	92	84	67	0
		Mar.–Sept. 21	90	83	74	64	49	28	0
		Dec. 21	—	60	54	44	32	17	0
34	Altitude	June 21	13	25	37	50	62	74	79
		Mar.–Sept. 21	—	12	25	36	46	53	56
		Dec. 21	—	—	9	18	26	31	33
	Azimuth	June 21	110	103	95	90	78	58	0
		Mar.–Sept. 21	90	82	72	61	46	26	0
		Dec. 21	—	—	54	43	30	16	0
38	Altitude	June 21	14	26	37	49	61	71	75
		Mar.–Sept. 21	—	12	23	34	43	50	52
		Dec. 21	—	—	7	16	23	27	28
	Azimuth	June 21	109	101	90	83	70	46	0
		Mar.–Sept. 21	90	81	71	58	43	24	0
		Dec. 21	—	—	54	43	30	16	0
42	Altitude	June 21	16	26	38	49	60	68	71
		Mar.–Sept. 21	—	11	22	32	40	46	48
		Dec. 21	—	—	4	13	19	23	25
	Azimuth	June 21	108	99	89	78	63	39	0
		Mar.–Sept. 21	90	80	69	56	41	22	0
		Dec. 21	—	—	53	42	29	15	0
46	Altitude	June 21	17	27	37	48	57	65	67
		Mar.–Sept. 21	—	10	20	30	37	42	44
		Dec. 21	—	—	2	10	15	20	21
	Azimuth	June 21	107	97	88	74	58	34	0
		Mar.–Sept. 21	90	79	67	54	39	21	0
		Dec. 21	—	—	52	41	28	14	0
48	Altitude	June 21	17	27	37	47	56	63	65
		Mar.–Sept. 21	—	10	20	29	36	40	42
		Dec. 21	—	—	1	8	14	17	19
	Azimuth	June 21	106	95	85	72	55	31	0
		Mar.–Sept. 21	90	79	67	53	38	20	0
		Dec. 21	—	—	52	41	28	14	0

[a] Solar time and clock time do not usually coincide. They are related by the equation of time. See, for instance, IESNA Recommended Practice RP-23, 1989.

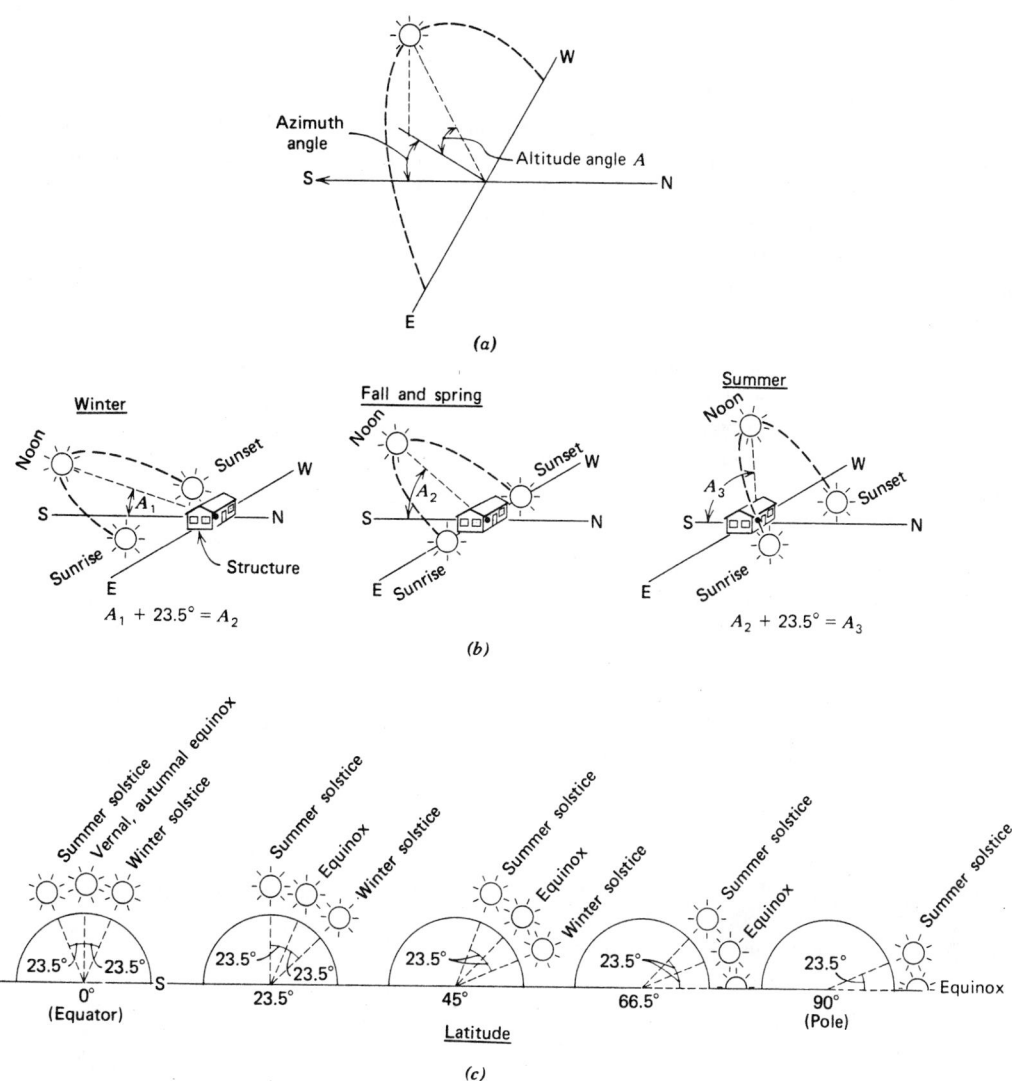

Fig. 19.3 (a) *The position of the sun is expressed in terms of vertical angle above the horizon (altitude) and horizontal angle (azimuth), measured from the south.* (b) *Approximate position of the sun in each of the seasons, at a mid-northern latitude (approximately 45°). Note that altitude angle is maximum in summer, minimum in winter, and in between in spring and fall. Note, too, the length of daylight hours: maximum in summer, minimum in winter, and in between in spring and fall.* (c) *Maximum sun altitude at various latitudes for both solstices and equinoxes. Maximum summer sun altitude is 90° minus latitude plus 23.5°. Minimum winter sun altitude is 90° minus latitude minus 23.5°. Thus, for all latitudes the yearly difference between maximum and minimum altitudes is twice 23.5°, or 47°, as shown.*

azimuth angles are east of south before noon and west of south after noon. It is assumed that the reader is familiar with the basic astronomical phenomena governing the motion of the earth that produce seasonal and latitudinal variations in the position of the sun. For our purpose, we simply state the important facts, which are:

1. For all latitudes, the sun's altitude is highest in summer, lowest in winter, and in between in spring and fall (Fig. 19.3b and Table 19.2). This fact, coupled with the shorter winter day, leads to the apparent rapid motion of the sun across the horizon in the winter and the apparent slow motion in the summer.

2. As a location approaches the equator (low latitude, either north or south), the sun's daily maximum altitude increases. The seasonal altitude *variation,* however, is the same for all latitudes (except at those extreme north and south latitudes, where the sun is above or below the horizon for extended time periods; see Fig. 19.3c). This factor not only affects exterior illumination levels but also has a pronounced effect on the design and efficacy of sun-shading devices.

3. The sun's azimuth angle is entirely dictated by the time of day because the sun by definition traverses the sky between sunrise and sunset. The principal significance of the azimuth angle is encountered when discussing building orientation, exposures, and shading angles (see Table 19.2).

The factor of cloud cover, unlike that of solar position, is predictable only statistically on the basis of extensive U.S. Weather Service observations at numerous weather stations throughout the United States. At locations other than those for which these data are available, an educated guess is necessary. Outside the United States, the designer must rely on locally available data, which are frequently meager or entirely unavailable. The third factor—that of local terrain and construction conditions that either reduce illumination by shadowing or increase it by reflection—is so particular and individual that it can be considered only on a case-by-case basis (see Fig. 3.7). Most of the calculation methods consider the effects of shadowing and obstruction and, to a lesser extent, the effects of reflections (see Sections 19.5–19.7).

For manual calculation procedures, it is sufficient to establish four basic sky conditions. These are:

1. Solid overcast sky.
2. Clear sky, without sun (in the shade).
3. Clear sky, with sun.
4. Partly cloudy sky.

19.5 Solid Overcast Sky

This condition, which obtains for much of the year in northerly climates such as England, Scandinavia, and the Pacific Northwest, is also called the *CIE sky,* because it was adopted by the Commis-

sion Internationale de l'Eclairage (CIE) as the standard design sky for daylighting calculations (*Daylight,* 1970). This sky as defined by the CIE, has a nonuniform brightness distribution, increasing from horizon to zenith in approximately the ratio of 1:3. Sky luminance at any altitude angle above the horizon is defined as

$$L_A = L_Z \frac{1 + 2 \sin A}{3} \qquad (19.1)$$

where L_A = luminance at $A°$ above horizon (in any direction)
L_Z = luminance at zenith

Fig. 19.4 (a) *The completely overcast sky has a zenith luminance* L_Z, *which is three times the horizon luminance, according to the CIE formulation. With such a sky, illumination on unobstructed exterior horizontal surfaces* (E_H) *is about 2½ times that on a similar vertical surface* (E_V). *This latter relation is also shown in Fig. 19.5.* (b) *One widely accepted formulation for clear sky luminance is simply an inversion of the overcast distribution* (a); *horizon luminance is three times that of the zenith. Obviously, the area around the sun is brightest. The area opposite the sun is darkest and can be considered as essentially uniform at approximately 3500 cd/m² (1000 fL). Actual luminance varies from about 1000 cd/m² (300 fL) at the center to 7000 cd/m² (2000 fL) at the sides, averaging 3500 cd/m² (1000 fL) over the whole area.*

Unobstructed exterior surface illuminance, solid overcast sky

Fig. 19.5 *Curves giving unobstructed exterior surface illumination from an overcast sky, directly. To obtain illuminance in lux, multiply footcandles by 10.76 (or 10 as an approximation). The ratio between horizontal and vertical surface illumination is 2.5:1, as shown in Fig. 19.4a. (Data based on U.S. Weather Service observations; courtesy of Libbey-Owens-Ford.)*

Thus at the horizon, where $A = 0\,°$,

$$L_A = \frac{L_Z}{3}$$

as stated previously. The illuminance (lux) produced by this distribution on unobstructed exterior horizontal and vertical surfaces stands in the approximate ratio of 2.5:1 (Fig. 19.4a). This results from an integration of Eq. 19.1 over the whole sky.

There is, however, agreement among all sources that with an overcast sky, exterior horizontal illuminance varies directly with the sun's altitude, *irrespective of azimuth,* despite the overcast. Various formulations for the relation have been put forward. One formulation that gives good agreement with observations is given by Krochman (1963):

$$E_H = 300 + 21,000 \sin A \text{ (lux)} \qquad (19.2)$$

where E_H is exterior horizontal illuminance and A is the solar altitude, in degrees. Solar altitude (and azimuth) for various times of day can be obtained from Table 19.2. Figure 19.5 is a plot of year-round averages for both vertical and horizontal illuminance from an overcast sky as a function of solar altitude, based on U.S. Weather Service observations.

To manually perform accurate calculations of indoor daylight levels using a nonuniform sky luminance distribution, one requires some sort of sky protractor with which to measure sky angles from the interior observation point. With these angles, the sky luminance of each small patch of sky can be determined and, in turn, the resultant interior daylight. Protractors are available from the Building Research Station, London, that simplify the work considerably, but it is still a time-consuming and laborious procedure. Today, the ready availability of computer programs that cal-

ILLUMINATION

culate interior daylight for all types of sky conditions makes such procedures substantially obsolete.

It is interesting to compare exterior horizontal illumination obtained from the two sources given: Krochman's formula (Eq. 19.2) and the observation-based data of Fig. 19.5 for a few typical conditions. Solar altitude is obtained from Table 19.2.

Latitude:	38°	
Solar Time:	10 AM	
Dates:	Dec.21, March/Sept 21, June 21	
	Eq. 19.2	Fig. 19.5
Dec 21	790 fc	800 fc
Mar/Sept 21	1359 fc	1480 fc
June 21	1735 fc	2150 fc

The degree of agreement is generally satisfactory and either source will yield suitable results.

19.6   Clear Sky

(a) Horizontal Illumination. Exterior horizontal illumination on a cloudless day consists of two components: diffuse illumination from the entire sky plus the much larger component of direct sunlight. As with overcast sky, various empirical formulas for both components have been proposed, and here too, all sources agree that the total illumination, diffuse plus direct, varies directly with solar altitude. [The most widely accepted empirical formulation for clear sky luminance distribution is that of Kittler (see *IESNA Handbook,* 8th ed. 1993, p. 364)].

Figure 19.6 gives values for both components of exterior horizontal illumination, based on observations. The *sky only* values are used to determine shaded skylight illumination or daylong ground illumination outside a shaded window— that is, a north-facing window, or an east/west window when the sun is on the opposite side of the building. In determining ground illumination, the values given in Fig. 19.6 must be reduced somewhat, as they represent unobstructed horizontal illumination, whereas the area outside a building window is partially obstructed from sky light by the building itself. If the building is so large that the ground outside the shaded window effectively receives diffuse radiation only from the half of the sky away from the sun, an average figure for E_H of 1000 fc ($\approx$10,000 lux) can be used. This is because the luminance of the half of sky

Fig. 19.6 *Components of the exterior horizontal illumination on an unobstructed surface, from a clear sky, as a function of solar altitude. Total illumination E_H is the sum of the two components. (From data in Rennhackkamp, 1967.)*

away from the sun varies from a minimum of approximately 300 fL for the deep-blue patch directly opposite the sun to about 2000 fL at the sides, giving an average half-sky luminance of about 1000 fL. This, is turn, gives a horizontal illuminance E_H, diffuse, of abut 1000 fc (see Fig. 19.4*b*).

Figure 19.6 also gives horizontal illuminance from the sun only, as a function of solar altitude. This value, when combined with the proper portion of diffuse illumination as discussed previously, is useful in determining ground illumination outside a sunny building exposure or illumination on an unshaded skylight. The light incident on an external reflector or light shelf at a window can also be determined from these figures.

(b) Vertical Surface Illumination. Inasmuch as most daylighting is accomplished via vertical fenestration, vertical surface illumination is the major component of interior daylight. It is also important for determining the daylight contribution of vertical elements in skylights. There is no simple relationship between horizontal and vertical illumination from a clear sky as there was for an overcast sky because the illumination on a vertical surface depends on solar azimuth as well as altitude. More specifically, it depends on the *bearing angle,* which is defined as the horizontal angle between vertical planes containing the sun and a perpendicular to the vertical surface

in question. A bearing angle of 0° indicates that the sun plane is perpendicular to the vertical surface (Fig. 19.7). As with E_H, E_V (vertical illuminance) is divided into two components: sky only and direct sun only, which are plotted in Figs. 19.8 and 19.9 as a function of solar altitude and bearing angle. The *sky only,* component is effectively for the half-sky because a vertical surface can be exposed to a maximum of only half of the full sky.

The data are used as follows: Surfaces exposed to the sun are illuminated by the sun component at the relevant bearing angle (Fig. 19.8) plus the sky component at the bearing angle (Figs. 19.9*a* and 19.9*b*). Bearing angle for the sky component is important because of the nonuniform luminance of the clear sky; brightest near the sun and dullest opposite it. A vertical surface shaded from the sun (bearing angle >90°) is illuminated by the sky component, at the bearing angle, plus any reflected light from objects facing the window, at its bearing angle. For instance, a north-facing window at noon solar time has a bearing angle of 180°. It is illuminated by the sky component at 180° plus reflected light from south facing buildings *at 0° bearing angle.* These buildings (at 0°) are illuminated by both sun and sky (see Fig. 19.7).

The data presented in the curves of Fig. 19.8 are based on year-long averages of observations taken by the U.S. Weather Service, modified by later observations of Rennhackkamp (1967). When more accurate observation data are available, they obviously should be used in preference.

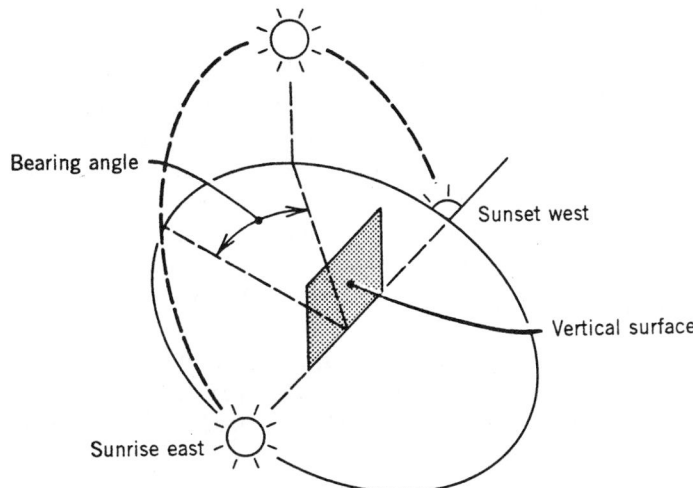

Fig. 19.7 *Diagram showing the geometry of a vertical surface in relation to the sun position. The angle between a vertical plane perpendicular to the vertical surface (window) and a vertical plane containing the sun is defined as the bearing angle. (Other sources refer to this angle as the* window-to-sun azimuth angle, *or simply* azimuth angle).

ILLUMINATION

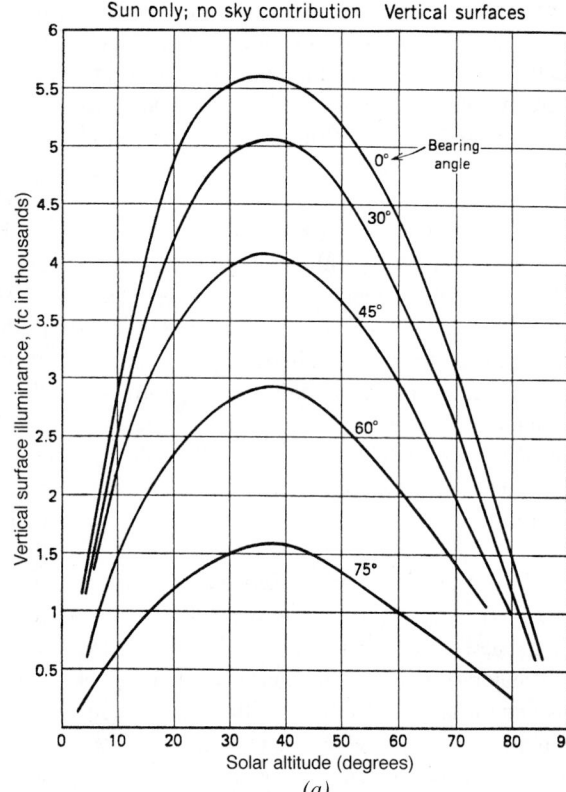

Sun only; no sky contribution   Vertical surfaces

Fig. 19.8 Vertical surface illuminance, year-long average, sun only, no sky contribution. (Courtesy of Libby-Owens-Ford.)

In many areas of the world, data on solar radiation are available, but illumination data are not (see Treado and Gillette, 1983). To translate these radiation data to illumination, an "efficiency" figure is required for solar energy in units of lumens per watt of received radiation. Reliable average figures are:

> Total radiation, sun and sky = 119 lm/W
>
> Diffuse (sky) radiation only $\sim=$ 14 lm/W
>
> Direct solar radiation only $\sim=$ 105 lm/W

Therefore, a surface receiving a total radiation of 750 W/m² is illuminated to

$$E = 750\,\text{W}\,(119\,\text{lm/W}) = 89{,}250\,\text{lux, or }8272\,\text{fc}$$

Obviously, the angle between the sun and the surface affects the amount of received radiation. Therefore, if only horizontal or perpendicular-to-the-sun radiation data are available, somewhat complex trigonometric calculations will give the radiation on a surface at any bearing angle. The relevant equations are found in many sources; see, for instance, Wieder (1982, pp. 35–37).

19.7   Partly-Cloudy Sky

There is no way of expressing mathematically the sky luminance of a partly cloudy sky because of its infinite variability. However, statistical data on cloud cover are available from observations at many weather stations, and these data should be used in computer-calculated, hour-by-hour energy programs. For the purpose of illumination, it is important to note that the illumination from a partly cloudy sky is *higher* than that from a clear sky by 10% to 15% because of the additional reflected sunlight from cloud edges. Several attempts have been made to account for this type

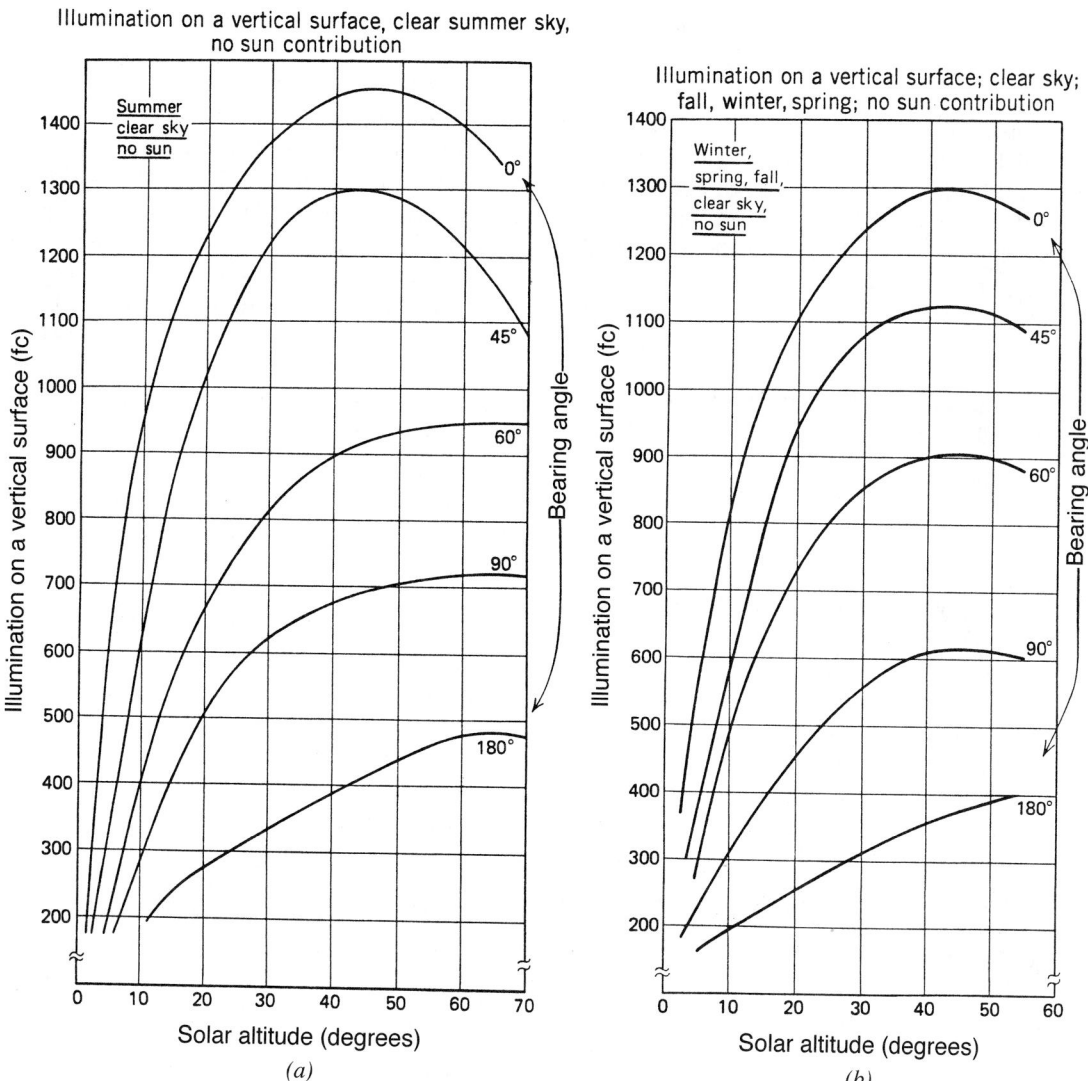

Fig. 19.9 *Vertical surface illumination from a clear sky with no sun contribution for various seasons of the year. (Courtesy of Libby-Owens-Ford.)*

of sky in terms of the effect on daylight factor within the room, but none have received general acceptance. As we see in Section 19.8, the concept of daylight factor was originally intended for overcast sky only, but, because of its great usefulness, it has been extended to clear sky.

In the following sections, several interior daylight analysis procedures are presented. We specify "analysis" rather than design because, unlike the standard design procedure, we start with a daylight inlet (window, skylight, clerestory) and determine the daylight produced.

As discussed in detail in Section 19.3, hand-calculation methods (including small programmable calculators) address only maximum, minimum, and (possible) average conditions. Computer-aided daylight analyses can produce much more detailed and accurate results in both numerical and graphical form (two- and three-dimensional contours). These analysis techniques are also demonstrated. It is our feeling that interior daylight contours (isolux lines) are frequently more meaningful than isolated individual absolute numbers, and for this reason we also review "hand" graphic methods that produce such contours. Finally, because lighting design has as its final result a visual image, we discuss briefly the use of physical models, which are particularly appropriate to the study of daylight.

19.8 Concept and Characteristics of Daylight Factor

Daylight factor (DF) is defined as

$$D_F = \frac{\text{indoor illuminance, at a given point}}{E_H} \quad (19.3)$$

where E_H is the unobstructed horizontal illuminance. D_F is expressed as a percentage. The concept is obviously applicable only where E_H can be calculated or measured; that is, where the sky luminance distribution is known or can reasonably be estimated. Thus, it can be used with the CIE-defined overcast sky (Eq. 19.1) and clear sky whose distribution is fixed for the purpose of calculations. *Direct sun is excluded.* DF cannot readily be used for skies with constantly changing luminance (partly cloudy) because under such conditions the DF at a point also varies continuously, making the concept useless as a calculation tool *for absolute daylight values.* However, even with such a sky, plots of DF contours show the *relative* daylight levels in an interior space. This is a great value to the lighting designer in appreciating the patterns of brightness within a room. As becomes clear from an explanation of its components, a daylight factor associated with one type of sky luminance distribution cannot be used directly with another sky without introducing considerable inaccuracy.

The reason for the primary reliance on overcast sky conditions is historical. The method was developed in England (see Hopkinson et al., 1966), where the solid overcast sky condition is predominant for much of the year. Considerable additional pioneer work on daylighting was done in Scandinavia, where the same sky condition prevails. As a result, the overcast sky became the basis for DF recommendations and remains so today.

The validity of the DF concept as a method for expressing interior daylight conditions, both absolute and relative, should be obvious. With a given sky luminance distribution (which varies with solar altitude), variation in daylight *inside* corresponds exactly to variations *outside* [i.e., the ratio (DF) remains the same]. This assumes minimal effect from obstructions and ground reflections, as is indeed the case with a dull overcast sky. Then, as explained in Section 19.3, the phenomenon of lightness constancy and Weber's law of contrast constancy smooths out the variations caused by changing solar altitude, making the *apparent* brightness due to daylight appear fairly uniform. Thus, the daylight factor method permits study of interior daylight distribution for varying fenestration, architectural arrangement, and building orientation.

The overcast sky in the DF method, as stated, represents minimum exterior illumination because the sky is frequently brighter. Hence, actual interior daylight levels frequently exceed the design minima by as much as 30%. Because the overcast sky is *not* a glare source, the large window areas utilized in overcast sky design locales are not a glare source even when exterior levels are considerably above the design minima. However, such windows *are* a source of severe glare in clear sky conditions.

Just as daylight levels in a sidelighted space vary, dropping as we move away from the window, so too DF varies because it is the ratio between that interior daylight level to exterior daylight level. This variation, however, is constant for a given architectural and window configuration. Therefore, knowing the DF variation for a given space and the exterior illumination as derived from sky luminance data, the actual interior daylight levels throughout the space can easily be calculated. Going one step further, minimum DF levels (as well as minimum artificial illumination levels) can be *specified* for different occupancies. These are based on a minimum average daylight level in a given geographic location. The architecture is then designed to meet these requirements. Thus a design based on a combination of minimum exterior illuminance and corresponding

$$DF = SC + ERC + IRC_1 + IRC_2$$

Fig. 19.10 *Total daylight factor (DF) is composed of sky component (SC), externally reflected component (ERC), and internally reflected component (IRC). IRC, in turn, is subdivided into reflected skylight and reflected groundlight. Note that surfaces deep in the room are lighted in large measure with re-reflected light.*

minimum DF requirements results in sufficient daylight for a large part of the working day under almost all exterior conditions.

The daylight at any point within an enclosed space is comprised of three components (see Fig. 19.10):

1. Sky component (SC)
2. Externally reflected component (ERC)
3. Internally reflected components (IRC$_1$ + IRC$_2$)

The *sky component* (SC) is that portion of total daylight at a point received directly from the area of the sky *visible through* the window. As sky component is *received* light, it takes into account light reduction due to window obstructions (mullions, etc.) and losses in transmission; that is,

SC = incident skylight − window losses

The *externally reflected component* (ERC) is light reflected from exterior obstructions onto the point under consideration. This *does not include ground-reflected light.* ERC is of significance only in built-up areas (where there are structures opposite the window) and can be estimated as the portion of the sky component for that area of obstructed sky, reduced by the *reflectance factor* (RF) of the obstruction; that is,

ERC = sky component × RD (of obstruction)

Thus, if 25% of the sky is obstructed by a building with a 20% RF, we have

$$ERC = SC \times 0.25 \times 0.20$$

or

ERC = 5% of SC,
to be added to the remaining 75% SC

The *internally reflected component* (IRC) is the light received at the point under consideration that has been reflected from interior and exterior surfaces. Nevertheless, as IRC$_2$ is generally small, IRC $\simeq$ IRC$_1$ (see Fig. 19.10). IRC is, therefore, entirely dependent on surface reflectances and on the amount of window glazing, and becomes a large portion of DF deep within the interior space (Table 19.3, and Fig. 19.11). IRC is normally calculated using published interreflectance tables, as direct calculation is extremely complex. In summary, the DF then is the sum of the three components:

$$DF = SC + ERC + IRC$$

all calculated individually for each location being considered.

Typical curves for both horizontal and vertical daylight factors, for a room with single side-lighting (windows on one side) are shown in Fig. 19.12. Note that DF$_V$ is larger then DF$_H$, reversing

ILLUMINATION

TABLE 19.3 **Effect of Wall Reflectance Factor on Proportion of Internally Reflected Component (IRC) in Daylight Factor**

Distance from Window (ft)	30% Wall Reflectance		60% Wall Reflectance	
	Total DF	$\frac{IRC}{DF}$ (%)	Total DF	$\frac{IRC}{DF}$ (%)
0	30	1	31	3.5
5	16	1.9	17	6.5
10	5.5	5.5	6.3	16.9
15	2.1	14.3	2.9	37.9
20	1.3	23	2.1	52.4

ROOM DATA:
Room 24 × 28 ft   70% ceiling RF
Window on 28-ft wall—one side only   20% floor RF
Window area = 20% of floor area

the condition that exists outside (see Fig. 19.4a). This is due to the light coming only from the side. The reverse would be true for a toplighted space.

These curves are produced by a longhand daylight-protractor-aided technique (Building Research Station, London). The three components—SC, IRC, and ERC—are calculated separately and added to produce the total daylight factor. Any change in parameters, such as window dimensions or height above the working plane, ceiling height, surface reflectance, ground reflection, and obstructions, alters these curves and re-

quires recalculation and replotting. Obviously, exact calculation of even a few variants in a space is a tedious and time-consuming procedure when done point by point, longhand. For this reason several alternative approaches are available:

1. Use of computer program (see Section 19.12).
2. Use of simplifications, such as standard curves, tabular data, or the CIE method described in Section 19.9.
3. Use of a library of graphic light distribution plots with varying parameters (see Section 19.10).
4. Use of a less laborious manual calculation procedure. One such technique, known commonly as the lumen method or the IES method, is demonstrated in Section 19.11.

Space limitation prevents us from presenting complete data on each method. This is especially true with tabular data (item 2) and the library of graphic plots. As a result, we only survey those methods while demonstrating in some detail the CIE and IES sidelighting methods and computer procedures. All use the same input data, to permit comparison of results.

All the methods are primarily applicable to analysis rather than design; that is, calculating daylight given the openings, not vice versa. The architectural designer, however, must establish an initial design, and this is done either intuitively or, lacking experience, by following some rules

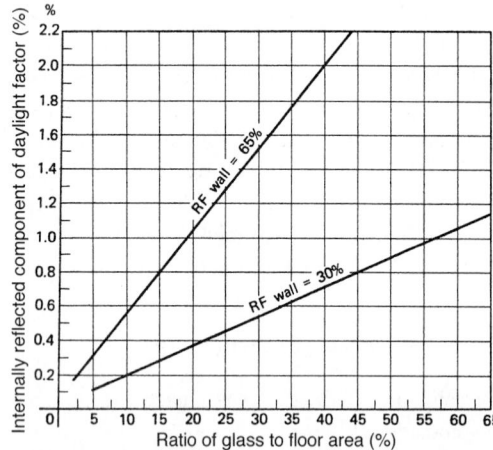

Fig. 19.11 *Plot of IRC of DF as a function of amount of glazing, expressed as DF, (i.e., as a percentage of exterior illuminance). As expected, the effect of lighter wall finish becomes more pronounced as the fenestration area increases.*

Distance from window (multiples of window height above work plane)

Fig. 19.12 *Typical daylight factor curves for horizontal (DF_H) and vertical (DF_V) illumination for a room with large windows on one side only. Note that the sky component represents almost the entire DF near the window, but reduces its proportion at greater depths. There, interreflected light constitutes 50% of the available daylight (see also Fig. 19.27).*

of thumb. A few such rules are given in Table 5.3 for overcast sky. Despite the fact the DF is a ratio and not an absolute value, such as lux, it is applied as an absolute. That is, recommendations are made of minimum daylight factor for specific seeing tasks. Recommended minimum levels of daylight factor vary, but those listed in Table 5.3 are average.

To demonstrate the value of starting-point DF design figures such as those given in Table 5.3, we can translate them into actual illuminances in lux and compare the results to the recommendations in Tables 18.4, 18.5, 18.7, and 18.8. We select two cities in the continental United States that have overcast skies for appreciable portions of the year—Columbus and Seattle. Their approximate latitudes are 40° N and 48° N, respectively. Using Table 19.2 for solar altitude, Fig. 19.6 for overcast sky horizontal illuminance, and taking 10 AM/3 PM for the average daily illuminance figures, we have tabulated the results in Table 19.4.

Note that only for midwinter (i.e., 21 December) are the actual values consistently below the recommendations of Tables 18.4, 18.5, 18.7, and 18.8. For most of the year, daylight provides all the light necessary for the tasks at hand if the recommendations of Table 5.3 are followed. Obviously, then, where exterior illuminance is indeed as low as 5000 to 7000 lux, PSALI would be required for all areas beyond H feet; that is, one window height from the window (Fig. 19.12). In addition to the recommendations of Table 5.3, the designer should also see to it that the ratio between minimum and average DF, which relates to the contrast ratios in the space, should not be less than 30%; that is,

$$\frac{DF_{min}}{DF_{avg}} \geq 0.3$$

Finally, minimum DF in any portion of a space should not drop below 0.5%, which is sufficient for circulation.

ILLUMINATION

TABLE 19.4 Horizontal Illuminances (E_H) from Overcast Sky, at Selected Times, in Columbus, Ohio, and Seattle, Washington, from Overcast Sky, Corresponding to Recommended Daylight Factor

Location	10 AM Solar Altitude[a]	Horizontal Illumination E_H[b] (fc/lux)	DF Recommendation (%)[c]	DF Recommendation (lux)	Illuminance Recommendation[d] (lux)
Columbus, Ohio	21 June 60°	2100/22,500	1.5–2.5 2.5–4 4–8	338–563 563–900 900–1800	300–500 500–750 750–1000
40° N latitude	21 Mar./Sept. 41%	1400/15,500	1.5–2.5 2.5–4 4–8	225–375 375–600 600–1200	300–500 500–750 750–1000
	21 Dec. 21°	700/7500	1.5–2.5 2.5–4 4–8	113–188 188–300 300–600	300–500 500–750 750–1000
Seattle, Washington	21 June 56°	1950/21,000	1.5–2.5 2.5–4 4–8	315–525 525–840 840–1680	300–500 500–750 750–1000
48° N latitude	21 Mar./Sept. 36°	1220/13,000	1.5–2.5 2.5–4 4–8	195–325 325–520 520–1040	300–500 500–750 750–1000
	21 Dec. 14°	500/5400	1.5–2.5 2.5–4 4–8	81–135 135–216 216–532	300–500 500–750 750–1000

[a] From Table 19.2.

[b] From Fig. 19.5.

[c] From Table 5.3.

[d] From Tables 18.4, 18.5, 18.7, and 18.8.

19.9   Daylight Analysis: CIE Method

(a) General. This method resulted from a search for a simple, rapid, straightforward, and reasonably accurate daylight calculation method that would yield reliable results without the time-consuming constructions and calculations necessitated by other manual methods. After a study of considerable length and intensity, the CIE adapted and adopted a system developed in Australia by Dr. A. Dresler (*Daylight Design Diagrams,* 1963). The current CIE recommendations were published in 1970 (*Daylight,* 1970).

This system is based on the daylight factor described previously as applied to the *standard overcast CIE sky.* Dr. Dresler developed a set of more than 100 curves covering rooms of varying proportions and fenestration. One such typical curve is shown in Fig. 19.13. The curves relate minimum DF (at a point 2 ft from the wall opposite the window) to the maximum permissible room depth, for given reflectances and a standard

window design, thus establishing the room's proportions. Depth, or width, is the dimension at right angles to the window wall.

The curves were calculated using techniques developed in England as described in the preceding section. The published CIE recommendations make a clear, strong case for the minimalist approach in selecting sky illumination and in design. They state simply that the number of design variables is so large and daylight itself so variable that a simple routine method can be based only on minimal conditions for a given (selected) daylight duration percentile. Therefore, the diagrams give the *lowest* level of daylight that can reliably be expected for a given percentage (percentile) of normal working hours in sidelighted rooms, and the *average* level in toplighted spaces. Thus, the designer can select the 95th percentile for minimum conditions, the 80th to 85th percentile for average conditions, and the 60th percentile for maximum conditions. On the basis of these, maximum design PSALI, energy consumption esti-

Fig. 19.13 *Maximum room depth required to maintain minimum DF is proportional to window size. Thus, for a room less than 25-ft long with a 5-ft-high window for 60% of the room's length, depth cannot exceed 12 ft if 2% DF is to be maintained at a point 2 ft from the rear wall. (From Daylight Design Diagrams, 1963.)*

ILLUMINATION

mates, and selection of lighting control zones can be made.

(b) Characteristics of the Method. Advantages of the system are:

1. Consideration of obstructions, reflections, and interior reflections.
2. Use of overcast sky luminance that varies with latitude.
3. Applicability to a very wide range of side and top fenestration designs.
4. Establishment of required room proportions is architecturally more useful than solving for specific dimensions.

Limitations of the system are:

1. Inapplicable to clear sky and sun conditions.
2. Inapplicable to other than rectangular rooms.
3. Unusable with sun-shading or high-reflectance ground.

4. Results give points of minimum, twice minimum, and four times minimum daylight only. Other points must be interpolated or extrapolated.
5. Window proportions and position in a wall are fixed.

Overall, the system accomplishes what it intended. The limitations listed are inherent in any rapid, simplified daylight calculation technique. They are listed here for information and not as criticism. The standard itself, which should be in the library of every daylight designer, also clearly states these limiting characteristics.

(c) Calculation Procedure. The CIE system is usable in two modes:

1. Given complete architectural data, find resultant daylight.
2. Given incomplete architectural data and required daylighting, find maximum room

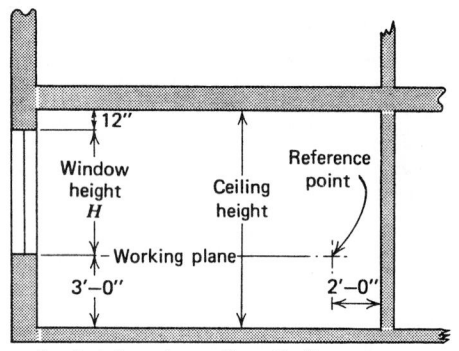

Section through a unilaterally lit room showing the assumed dimensions. These dimensions are the same for bilateral lighting except that the reference point is midway between the window walls.

(a)

Plane of window walls showing window width expressed as a percentage of total room length.

(b)

Fig. 19.14 *Sketches indicating the parameters of the CIE calculation systems.*
(a) *A vertical section through a room with dimensional data relevant to this system. Note that the sill height has been selected to coincide with a working plane at 90 cm (3 ft). Height of working plane varies between 30 and 36 in., the former being more common in American use, the latter in European use. A lower sill contributes only ground-reflected light at the working plane. Where window sill is much above the working plane (i.e., short windows, high on the wall), this design system is inapplicable.*
(b) *Calculation size (length) of windows with respect to overall room length. (From* Daylight, International Recommendations for the Calculation of Natural Daylight, *1970, Commission Internationale de l'Eclairage; reproduced with permission.)*

depth and/or other room proportions that satisfy the daylighting requirement.

Obviously the former mode is simpler because it leads directly to an answer, as demonstrated by the illustrative example that follows. Inasmuch as the latter mode, has many answers, it is more complex. That is, various combinations of fenestration and room dimensions will yield adequate daylight. For this reason, the designer should set room length (parallel to the window wall) and percent fenestration of the window wall, leaving depth (perpendicular to the window wall) as a variable or, alternatively, set room length and depth, with percentage of fenestration as the variable. Ceiling height is usually fixed by other considerations. See Fig. 19.14 for sketches showing room parameters.

The procedure in mode 1 is as follows (the reader should work through the mode 2 procedure and apply it to a design project):

1. Express room depth in terms of window height. From Fig. 19.15 determine *design* daylight factor—that is, the daylight factor 2 ft from the wall opposite the window—on the window centerline.

2. The design daylight factor is larger than the actual service daylight factor because it is based on clear, clean glass and unobstructed horizon. As this is rarely the case, the actual (service) daylight factor must be calculated using the correction factors of Fig. 19.16 and Table 19.5; that is,

$$DF_{service} = DF_{design} \times \text{correction factors}$$

3. Using the service daylight factor in Eq. 19.3,

$$\text{required exterior illumination} = \frac{\text{required interior ollumination}}{\text{service DF}}$$

obtain required exterior illumination.

4. From the curves in Fig. 19.17, obtain the percentage of hours between 0900 and 1700 that the required illumination is maintained. For time periods other than this, see Table 19.5c.

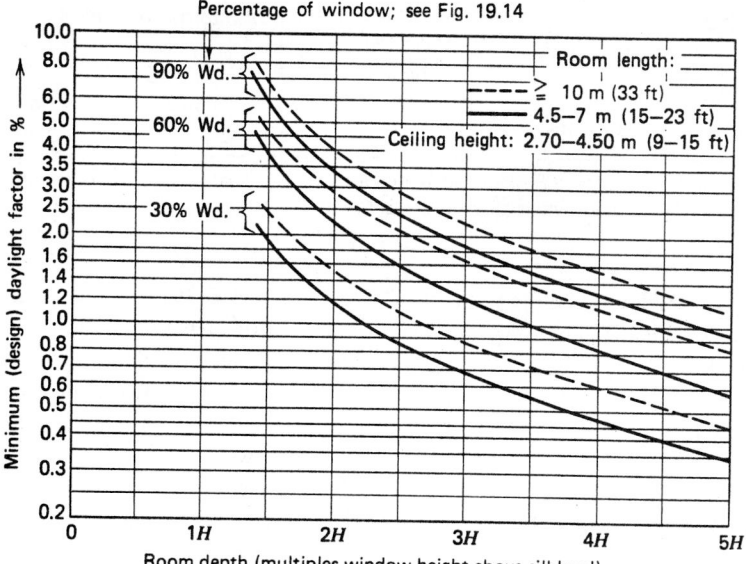

Relation between room depth and minimum (design) daylight factor
(For various room lengths and window widths)

Fig. 19.15 Basic design diagram that relates minimum daylight factor to room depth. Inasmuch as room depth is expressed in terms of window height, the curves effectively relate minimum daylight factor (2 ft from back wall) to room proportion. (From Daylight, International Recommendations for the Calculation of Natural Daylight, 1970, Commission Internationale de l'Eclairage; reproduced with permission.)

TABLE 19.5 **Correction Factors to Be Used in CIE Daylight Calculations**

(a)

Diffuse Transmittance of Glass (%)	Correction Factor
80	0.95
70	0.8
60	0.7
50	0.6
40	0.45
30	0.35

(b) Correction Factors to Allow for Dirt Accumulation on Glass

Locality	Class of Industry	Angle of Slope (Measured to the Horizontal)		
		90–75°	60–45°	30–0°
Country or outer-suburban area	Clean	0.9	0.85	0.8
	Dirty	0.7	0.6	0.55
Built-up residential area	Clean	0.8	0.75	0.7
	Dirty	0.6	0.5	0.4
Built-up industrial area	Clean	0.7	0.6	0.55
	Dirty	0.5	0.35	0.25

(c) Percentages to Use When Figure 19.17 Curves Are Applied to Periods Other Than 09.00–1700

Curve in Figure 19.17	95%	90%	85%	80%	70%	60%
Alternative period	Percentage of alternative period					
07.00–15.00	95	90	85	80	70	60
08.00–16.00	100	100	95	85	70	60
07.00–17.00	95	85	75	65	55	45
06.00–18.00	75	70	65	60	50	40

Source: From Daylight, International Recommendations for the Calculation of Natural Daylight, Publication CIE No. 16 (E-3.2), 1970; Commission Internationale de l'Éclairage.

5. From Fig. 19.18, determine the room positions at which illumination is twice of four times minimum.

The curves given here for unilateral sidelighting are typical of those found in the standard (*Daylight,* 1970), which covers, in addition, bilateral sidelighting and toplighting of the three principal varieties—that is, skylights, sawtooth roofs, and monitor roof.

An example of this simple, rapid technique should make its use clear. We have selected a geographic location appropriate to this method.

EXAMPLE 19.1. A classroom in a single-story Seattle elementary school is 25-ft long and 18-ft deep and has a 9-ft, 6-in. ceiling. It receives daylight unilaterally from windows totalling 18 ft in length (see room sketch, Fig. 19.19). Window glazing is wired glass having a transmittance of 80%. The school is situated in a built-up residential area with a satisfactory cleaning schedule. Determine the portion of the year during which tasks requiring a minimum illumination of 150 lux can be carried out by daylight throughout the room. Also, determine what levels can be maintained for 85% of daylight hours, and at what distances

External obstructions and the angle of obstruction. The angle of obstruction α is measured from the center of the window to the outline of the opposite building.

(a)

ILLUMINATION

(b)

Fig. 19.16 (a) *Angle of obstruction α of an external object.* (b) *Correlation factors to account for the influence of external obstructions on minimum daylight factor. (From* Daylight, International Recommendations for the Calculation of Natural Daylight, *1970, Commission Internationale de l'Eclairage; reproduced with permission.)*

Fig. 19.17 *Minimum maintained external illumination as a function of latitude for a given percentage of the normal working day. (From* Daylight, International Recommendations for the Calculation of Natural Daylight, *1970, Commission Internationale de l'Eclairage; reproduced with permission.)*

Fig. 19.18 (a) *Distance from window at which daylight factor is twice minimum daylight factor.* (b) *Distance from window at which daylight factor is four times minimum daylight factor. (From* Daylight, International Recommendations for the Calculation of Natural Daylight, *1970, Commission Internationale de 1'Eclairage; reproduced with permission.)*

Fig. 19.19 *Plan and window wall elevation of room of Example 19.1. The three daylight contours are estimated based on the calculated centerpoint. They represent the levels maintained for 85% of daylight hours. Levels twice as high are maintained for 60% of the daylight hours.*

from the window. (Note that 150 lux corresponds to a DF of 2% applied to an E_H of 7500 lux.)

SOLUTION. *General.* The latitude of Seattle is 47.6°N. The design condition for Seattle is solid overcast sky for 85% of the hours between 0900 and 1700. From Fig. 19.17, we see that the minimum unobstructed horizontal E_H during these hours is 7200 lux. (No azimuth for the window is specified because exterior horizontal illuminance is independent of azimuth according to the CIE formulation for overcast sky.)

STEP 1. Window height H is 5 ft (see Fig. 19.19). Room depth in terms of window height:

$$\frac{18\text{-ft depth}}{5\text{-ft window height}} = 3.6\,H$$

Window coverage:

$$\frac{3 \times 6\,\text{ft}}{25\,\text{ft}} = 72\%$$

From Fig. 19.15, for room length of 25 ft, ceiling height 9 ft 6 in., window coverage 72%:

$$\text{design DF at } 3.6\,H = 1.3$$

STEP 2. From Table 19.5 and Fig. 19.16b, we obtain correction factors:

Glass transmission 0.95

Glass cleanliness 0.8

Therefore,

$$\text{service DF} = 1.3 \times 0.95 \times 0.8 \simeq 1.00$$

STEP 3. Required exterior illumination:

$$E_H = \frac{150\,\text{lux min}}{1.00\,\text{DF}_{\min}} \times 100 = 15{,}000\,\text{lux}$$

STEP 4. From Fig. 19.17 we see that with the given conditions, an illuminance of 150 lux will be maintained for less than 60% of the hours between 0900 and 1700. The level that *is* maintained for 85% of the hours is

$$E_{\min} = 7200\,\text{lux} \times 1.0\,\text{DF} \simeq 72\,\text{lux}$$

STEP 5. To complete the picture, we have from Fig. 19.18

(a) DF doubles at 1.8H corresponding to 144 lux at depth of 9 ft (1.8 × 5 ft).

(b) DF quadruples at 1.2H, corresponding to 288 lux at a depth of 6 ft (1.2 × 5 ft). These values (*a,b*) are maintained for 85% of daylight hours.

STEP 6. The calculated illuminance levels are accurate only at the centerline of the window wall. By visual estimate and extrapolation, rough contours can be drawn. They are shown on Fig. 19.19, and also in the next section on Fig. 19.21c for the purpose of comparing results with the graphic method discussed there. ∎

In summary, then, the CIE method is simple, rapid, and flexible, but provides only limited data. It is usable in both the design and analysis mode. Its exterior illuminance data (Fig. 19.17) seem to agree well with actual measurement (7250 lux on the figure, as compared to 7200 lux measured average for Seattle). If contours are roughed in as shown on Fig. 19.19, guidelines for PSALI design and lighting control strategy can be drawn.

19.10   Graphic Daylighting Design Method (GDDM)

A method that applies the principle of daylight factor to overcast sky and shows results as DF (isolux) contours within a room (rather than individual DF factors at specific points) was developed by M.S. Millet and J.R. Bedrick (1980). Its primary advantage over the CIE method is that its results are a family of DF contours that are more meaningful to a lighting designer than is numerical output. Its shortcomings are that it, too, is not readily applied to clear-sky conditions and that it requires that the designer acquire a "library" of 200 or so patterns that cover most design situations. For this latter reason, only the outline of the method is presented here. We suggest that the interested reader acquire access to the catalog of patterns to permit application of this useful and relatively accurate method. (Application to clear sky is possible but not practical because of the size of the required pattern library.)

The authors of the method utilized a computer simulation program (UWLIGHT) to develop daylight distribution patterns in a room, resulting from either sidelights or skylights. To generalize the system, windows are identified by height to

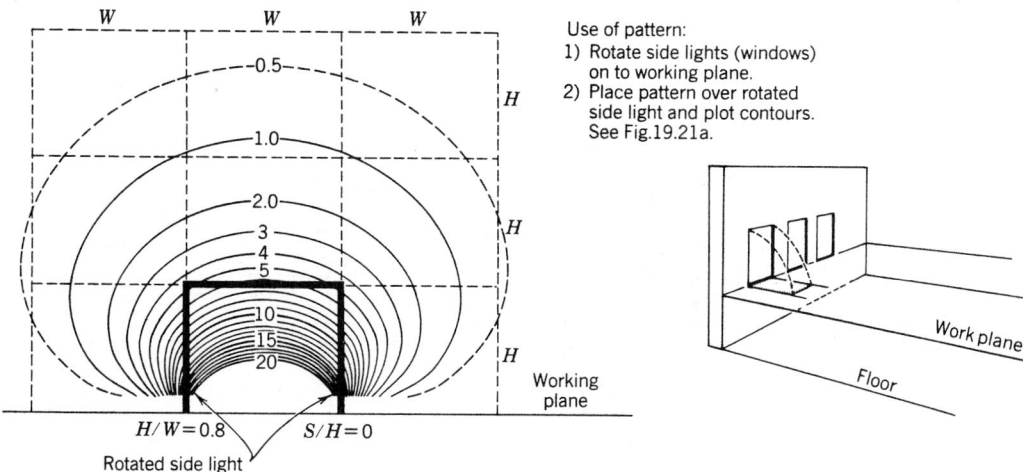

Fig. 19.20 *Typical isolux contour map for a window with height-to-width ratio of 0.8, and sill at the working plane elevation. Numbers represent sky component of daylight factors for overcast sky condition. The rectangles are the window outlines rotated: (i. e., projected) onto the work plane. See insert. (From Millet and Bedrick, 1980, p. 191.)*

width proportion (*H/W*), and the position of the isolux contours on the plan is determined by the ratio of the elevation of the sill over the work plane to the window height. The latter factor is a considerable advantage over the CIE system, which is restricted to a sill height *at* the work plane. Thus, the GDDM system can account for high windows, clerestories, and other designs intended to introduce daylight deep into the space—something that the CIE method cannot do.

Figure 19.20 shows a typical isolux pattern for a window whose *H/W* ratio is 0.8 and whose sill is at the work plane. This particular window pattern was selected because it corresponds to the window of Example 19.1, Fig. 19.19, and enables us to run through the same problem and compare results.

EXAMPLE 19.2.   Repeat the design problem of Example 19.1 using the GDDM method.

STEP 1. Select the appropriate window pattern. Referring to Fig. 19.19, window proportion is

$$\frac{H}{W} = \frac{5}{6} = 0.83$$

We therefore select the pattern shown in Fig. 19.20 as the closest to our requirements.

STEP 2. On a plan of the room, trace the isolux pattern for each window (Fig. 19.21*a*). *S/H* = 0

indicates that the pattern begins at the window wall, as shown. The patterns overlap because the windows are close together. Where contours meet, the DF of the contours are added, producing points for new combined contours. The combined contours and their values, in DF, are shown in Fig. 19.21*b*.

STEP 3. In the final step, the value of the combined contours is corrected to account for internally reflected components of daylight plus light reduction due to glazing. The final contours are shown on Fig. 19.21*c*. Note that this diagram gives the designer a much more complete picture of the daylight contours than the results of the CIE method. For the purpose of comparison, the three calculated DF factors of the CIE method are shown on Fig. 19.21*c*. ■

19.11   Daylight Design: IESNA Lumen Method

A hand calculation procedure that addresses all types of sky conditions was developed in the early 1950s at Southern Methodist University (Dallas) by J. W. Griffith and associates (Griffith, Arner, and Wenzler, 1957). The work was sponsored and published by the Libby-Owens-Ford (LOF) Glass Co. ("How to Predict. . .," 1976). The procedure

ILLUMINATION

Fig. 19.21 (a) *Daylight contours for each window of Fig. 19.19 are plotted on the floor plan of the room being studied. Numbers in parentheses are combined sky components.*
(b) *The isolux contours of (a) are combined to form new isolux contours that represent the total sky component of daylight within the room.*
(c) *The final isolux contours are calculated, including correction factors. The numbers represent daylight factors. Note the variance between these contours and the points calculated by the CIE method. The five design points calculated by the IESNA method (discussed in Section 19.11) are also shown. A comparison of the results of the three methods described indicates that on the room centerline for which location comparison of all methods is possible, agreement is within engineering accuracy (see text discussion).*

was then adopted by the IES of North America for its recommended Practice of Daylighting in 1962. The current IESNA recommended practice (RP-23-1989), also frequently called the *lumen method*, is a modification of that method.

(a) General Remarks. This method is in large measure parallel to the IESNA method for calculating artificial lighting in that it treats the window or toplight as a large area lighting source; essentially, a transilluminated lighting fixture. It then applies coefficients to the light output of this "daylight fixture," similar in usage to the familiar coefficient of utilization and light-loss factor of a lighting fixture, and arrives at an interior daylight level. For sidelighting, the lighting levels are calculated at five predetermined points in the room. For toplighting, it is an average overall level, assuming proper spacing of toplights. The method is described completely in the relevant current IESNA publication (IES-RP23-1989). The reader is urged to obtain a copy because only selected portions are reproduced in the discussion that follows as required for the illustrative examples presented.

(b) Characteristics of the Method. Before beginning a description of the technique and demonstrating it with illustrative examples, a few general remarks are in order. The IESNA method is probably the most flexible manual technique available. It has these major characteristics for sidelighting:

1. It takes into account reflected light from the ground and adjacent structures as well as the reduction in sky light due to such structures.

2. It cannot accommodate direct sunlight, but conversely, it readily accommodates the shading devices normally used to block direct insolation.

3. Provision is made for various types of glazing as well as common window controls such as horizontal and vertical blinds.

4. The principles of zonal cavity calculation of interior lighting (see Section 20.32) are applied. The window height determines the cavities (i.e., the floor cavity extends to the window sill height and the ceiling cavity from the top of the window to the ceiling). The room cavity is therefore the window height.

5. The working plane is always at the sill height of the window. Where this is decidedly not the case (a difference of up to a foot is usually negligible), such as when using a clerestory or a floor-to-ceiling window, working plane illuminance can be calculated by superposition. For instance, with a clerestory, we would subtract a working plane-to-celerestory sill-height window from a working plane-to-top of clerestory window to obtain the desired result. Note, however, that a degree of inaccuracy is unavoidable when the working plane is *above* the sill.

6. Cavity reflectances are fixed (Fig. 19.22) at 70-50-30 for ceiling, room, and floor cavities, respectively.

7. The system calculates only five points in a room, on the window centerline. As stated in

Fig. 19.22 *Standard conditions in a room for daylighting calculation: sidelighting. (From IESNA RP-23-1989; reprinted with permission.)*

reference to the three points calculated by the CIE method, this is not normally sufficient to give a picture of the interior daylight distribution.

8. The method is usable in only one mode; that is, given location and full architectural data, daylighting can be calculated. It cannot readily be used to determine room proportions, given the other data, as can the CIE method. Thus, it is an *analysis* tool, not a *design* tool, except by trial and error.

(c) Calculation Procedure.

The calculation procedure for sidelighting is discussed here as it is more frequently encountered than toplighting. The reader is referred to the current *IESNA Recommended Practice* (RP-23-1989) for a discussion and presentation of the latter. It should present no difficulties. It is similar to, and simpler than, the sidelighting calculation procedure. In the discussion that follows, we use the same notation and terms as found in RP-23-89 *except* for *bearing angle* [see Fig. 19.8 and Section 19.2*b*(4)], which is referred to in the IESNA procedure as *solar window azimuth.* (We use the term *bearing angle* because that is the term in common use in most international sources.)

The method, as described in the referenced *IESNA Recommended Practice,* consists of four steps:

STEP 1. Determine the vertical illuminance on the exterior of the window.

STEP 2. Reduce this illuminance by the necessary factors (glazing, window controls, dirt) to obtain the transmitted illuminance.

STEP 3. Select the coefficients of utilization for the five calculation points on the basis of room dimensions and portion of sky seen by the window. The latter is determined by the ratio of vertical to horizontal illuminance at the window.

STEP 4. Calculate the illuminances at the five interior points by simple multiplication.

The terminology of the IESNA Recommended Practice is:

E_{xvk} = exterior vertical illuminance at window from half sky (an unobstructed vertical window sees only half the sky)

E_{xHk} = exterior horizontal illuminance at window from full sky

E_{xhk} = exterior horizontal illuminance at window from half sky

E_{xvg} = exterior vertical illuminance at window from ground

τ = net transmittance of the window

CU_k = coefficient of utilization for sky light

CU_g = coefficient of utilization for ground light

E_i = interior illuminance at a specific reference point

In detail, then, the four-step procedure just described is as follows:

STEP 1. Determine vertical and horizontal surface illuminances from Figs. 19.6, 19.7, and 19.9. Calculate half-sky illuminance.

STEP 2. Determine net transmittance of window. This factor is the product of the glazing factor (Table 19.6), factors representing loss from glass, plus any complete uniform diffuser (*not* blinds, which are accounted for in the coefficient of utilization tables) and a light-loss factor (LLF; Table 19.7), which represents the cleanliness status of the window. This step also accounts for net glazing (i.e., gross window area less mullions, glazing bars, and so on).

STEP 3. Select coefficients of utilization from Tables 19.8 to 19.13. These tables represent window without blinds for clear and overcast sky conditions.

STEP 4. Calculate interior illuminance for each of the five reference points. The basic equation for each point is

$$E_i = \tau (E_{xvk} \times CU_k + E_{xvg} \times CU_g)$$
(19.4)

The ground component vertical illuminance E_{xvg} is the product of the ground reflectance (Table 19.14) and half the horizontal illuminance:

$$E_{xvg} = RF_g \times \frac{E_{xHk}}{2}$$
(19.5)

(For the derivation of this expression, see RP-23, p. 8.) Due to space constraints, the coefficient tables for windows with venetian blinds are not presented here. For these as well as all toplighting

TABLE 19.6 **Transmittance Data of Glass and Plastic Materials**

Material	Approximate Transmittance (%)
Polished plate/float glass	80–90
Sheet glass	85–91
Heat-absorbing plate glass	70–80
Heat-absorbing sheet glass	70–85
Tinted polished plate	40–50
Figure glass	70–90
Corrugated glass	80–85
Glass block	60–80
Clear plastic sheet	80–92
Tinted plastic sheet	9–42
Colorless patterned plastic	80–90
White translucent plastic	10–80
Glass-fiber–reinforced plastic	5–80
Double glazed—two lights clear glass	77
Tinted plus clear	37–45
Reflective glass[a]	5–60

Source: IES RP-23-1989; reprinted with permission.

[a] Includes single glass, double-glazed units, and laminated assemblies. Consult manufacturer's material for specific values.

data, refer to *IESNA Recommended Practices,* RP-23-1989.

To demonstrate the use of the IESNA method, we apply it to the same example used in the CIE and GDDM methods and compare results.

EXAMPLE 19.3. Using the building of Example 19.1, find illuminance E_i at the five points shown in Fig. 18.22 for a spring day and a winter day, at 10 AM (2 PM). Assume the sky to be overcast so that a direct comparison with Examples 19.1 and 19.2 can be made. Reflectances are assumed to be 70% ceiling, 50% wall, 30% floor, to correspond to the IESNA method standard conditions.

SOLUTION. For the sake of completeness, we repeat here the conditions of our problem:

Location: Seattle, Washington

Latitude: 47.6° N

Room: 25-ft long × 18-ft deep (wide) × 9-ft, 6-in. high

Window: height 5 ft above 30 in. working plane; total length, 18 ft

Transmittance: 85%; net glass area 92%

Ground Reflectance: not previously specified. Assume an extensive area of mixed grass and concrete walkways, with an average overall reflectance of 20% (see Table 19.4). (For more accurate calculation, see the method described in the *IESNA Recommended Practice*).

TABLE 19.7 **Typical Light-Loss Factors for Daylighting Design**

Location	Light-Loss Factor Glazing Position		
	Vertical	Sloped	Horizontal
Clean areas	0.9	0.8	0.7
Industrial areas	0.8	0.7	0.6
Very dirty areas	0.7	0.6	0.5

Source: IES RP-23-1989; reprinted with permission.

TABLE 19.8 **Coefficient of Utilization from Window Without Blinds; Sky Component** $E_{xvk}/E_{xhk} = 0.75$

Room Depth/ Window Height	Depth	Window Width/Window Height							
		.5	1	2	3	4	6	8	Infinite
1	10	.824	.864	.870	.873	.875	.879	.880	.883
	30	.547	.711	.777	.789	.793	.798	.799	.801
	50	.355	.526	.635	.659	.666	.669	.670	.672
	70	.243	.386	.505	.538	.548	.544	.545	.547
	90	.185	.304	.418	.451	.464	.444	.446	.447
2	10	.667	.781	.809	.812	.813	.815	.816	.824
	30	.269	.416	.519	.544	.551	.556	.557	.563
	50	.122	.204	.287	.319	.331	.339	.341	.345
	70	.068	.116	.173	.201	.214	.223	.226	.229
	90	.050	.084	.127	.151	.164	.167	.171	.172
3	10	.522	.681	.739	.746	.747	.749	.747	.766
	30	.139	.232	.320	.350	.360	.366	.364	.373
	50	.053	.092	.139	.163	.174	.183	.182	.187
	70	.031	.053	.081	.097	.106	.116	.116	.119
	90	.025	.041	.061	.074	.082	.089	.090	.092
4	10	.405	.576	.658	.670	.673	.675	.674	.707
	30	.075	.134	.197	.224	.235	.243	.243	.255
	50	.028	.050	.078	.094	.104	.112	.114	.119
	70	.018	.031	.048	.059	.065	.073	.074	.078
	90	.016	.026	.040	.048	.053	.059	.061	.064
6	10	.242	.392	.494	.516	.521	.524	.523	.588
	30	.027	.054	.086	.102	.111	.119	.120	.135
	50	.011	.023	.036	.044	.049	.055	.056	.063
	70	.009	.018	.027	.032	.035	.040	.041	.046
	90	.008	.016	.023	.028	.031	.034	.035	.040
8	10	.147	.257	.352	.380	.387	.391	.392	.482
	30	.012	.026	.043	.054	.060	.067	.070	.086
	50	.006	.013	.021	.026	.029	.033	.035	.043
	70	.005	.011	.017	.021	.023	.026	.027	.034
	90	.004	.010	.015	.019	.021	.023	.025	.030
10	10	.092	.168	.248	.275	.284	.290	.291	.395
	30	.006	.014	.026	.032	.036	.041	.044	.059
	50	.003	.008	.014	.017	.019	.022	.024	.032
	70	.003	.007	.012	.014	.016	.018	.019	.026
	90	.003	.006	.011	.013	.015	.016	.017	.024

Source: IES RP-23-1989; reprinted with permission.

TABLE 19.9 **Coefficient of Utilization from Window Without Blinds; Sky Component**
$E_{xvk}/E_{xhk} = 1.00$

Room Depth/ Window Height	Depth	Window Width/Window Height							
		.5	1	2	3	4	6	8	Infinite
1	10	.671	.704	.711	.715	.717	.726	.726	.728
	30	.458	.595	.654	.668	.672	.682	.683	.685
	50	.313	.462	.563	.589	.598	.607	.608	.610
	70	.227	.362	.478	.515	.527	.530	.532	.534
	90	.186	.306	.424	.465	.481	.468	.471	.472
2	10	.545	.636	.658	.660	.661	.665	.666	.672
	30	.239	.367	.459	.484	.491	.499	.501	.506
	50	.121	.203	.286	.320	.335	.348	.351	.355
	70	.074	.128	.192	.226	.243	.259	.264	.267
	90	.058	.101	.156	.188	.207	.215	.221	.223
3	10	.431	.561	.607	.613	.614	.616	.615	.631
	30	.133	.223	.306	.337	.348	.357	.357	.366
	50	.058	.103	.155	.183	.197	.211	.213	.218
	70	.037	.064	.098	.119	.132	.147	.150	.154
	90	.030	.051	.079	.098	.110	.122	.126	.129
4	10	.339	.482	.549	.560	.563	.566	.565	.593
	30	.078	.139	.204	.234	.247	.258	.260	.272
	50	.033	.060	.094	.114	.126	.139	.143	.150
	70	.022	.039	.061	.074	.083	.095	.099	.104
	90	.019	.032	.050	.061	.070	.080	.084	.089
6	10	.211	.343	.433	.453	.458	.461	.461	.518
	30	.033	.065	.103	.123	.135	.145	.148	.167
	50	.015	.029	.047	.057	.064	.073	.077	.086
	70	.011	.021	.033	.040	.045	.051	.054	.060
	90	.010	.019	.028	.034	.038	.044	.046	.052
8	10	.135	.238	.326	.353	.362	.366	.367	.452
	30	.016	.034	.058	.072	.080	.090	.094	.116
	50	.008	.017	.027	.034	.039	.045	.048	.059
	70	.006	.013	.021	.026	.028	.032	.035	.043
	90	.005	.012	.019	.023	.025	.029	.031	.038
10	10	.090	.165	.244	.272	.283	.290	.291	.395
	30	.009	.020	.036	.045	.052	.060	.064	.087
	50	.005	.010	.019	.023	.026	.030	.033	.044
	70	.004	.009	.015	.018	.020	.023	.025	.033
	90	.003	.008	.014	.016	.018	.020	.022	.030

Source: IES RP-23-1989; reprinted with permission.

ILLUMINATION

TABLE 19.10 **Coefficient of Utilization from Window Without Blinds; Sky Component**
$E_{xvk}/E_{xhk} = 1.25$

Room Depth/ Window Height	Depth	Window Width/Window Height							
		.5	1	2	3	4	6	8	Infinite
1	10	.578	.607	.614	.619	.621	.633	.634	.635
	30	.405	.525	.580	.594	.599	.612	.614	.615
	50	.287	.423	.519	.547	.556	.569	.571	.573
	70	.218	.347	.461	.501	.515	.522	.525	.526
	90	.186	.307	.428	.473	.491	.483	.486	.487
2	10	.472	.549	.566	.569	.570	.574	.575	.581
	30	.221	.337	.422	.447	.456	.465	.467	.472
	50	.120	.202	.285	.321	.337	.353	.357	.361
	70	.078	.136	.204	.242	.261	.281	.287	.290
	90	.064	.112	.174	.211	.233	.244	.251	.253
3	10	.377	.488	.527	.533	.534	.536	.536	.549
	30	.130	.217	.298	.329	.341	.352	.353	.362
	50	.062	.110	.165	.195	.211	.228	.231	.237
	70	.040	.070	.109	.132	.147	.166	.171	.175
	90	.033	.057	.090	.112	.127	.142	.148	.152
4	10	.300	.424	.484	.494	.497	.499	.499	.524
	30	.080	.143	.209	.240	.255	.267	.269	.283
	50	.036	.066	.104	.126	.140	.156	.160	.168
	70	.024	.043	.068	.083	.094	.109	.115	.120
	90	.021	.036	.056	.070	.080	.092	.099	.103
6	10	.193	.314	.395	.415	.420	.423	.423	.476
	30	.036	.071	.113	.136	.149	.161	.165	.186
	50	.017	.033	.053	.065	.074	.084	.089	.100
	70	.012	.024	.037	.045	.050	.058	.061	.069
	90	.011	.021	.031	.038	.043	.049	.053	.060
8	10	.128	.226	.310	.337	.346	.351	.352	.433
	30	.019	.039	.066	.082	.092	.104	.109	.134
	50	.009	.019	.031	.040	.045	.052	.056	.069
	70	.007	.015	.023	.029	.032	.037	.040	.049
	90	.006	.013	.021	.025	.028	.032	.035	.043
10	10	.088	.164	.241	.270	.282	.290	.291	.396
	30	.011	.024	.043	.054	.062	.071	.076	.103
	50	.005	.012	.022	.026	.030	.035	.038	.052
	70	.004	.010	.017	.020	.023	.026	.028	.038
	90	.004	.009	.016	.018	.020	.023	.025	.034

Source: IES RP-23-1989; reprinted with permission.

TABLE 19.11 **Coefficient of Utilization from Window Without Blinds; Sky Component**
$E_{xvk}/E_{xhk} = 1.50$

Room Depth/ Window Height	Depth	Window Width/Window Height							
		.5	1	2	3	4	6	8	Infinite
1	10	.503	.528	.536	.541	.544	.557	.558	.559
	30	.359	.464	.514	.528	.534	.549	.550	.552
	50	.261	.384	.471	.499	.508	.524	.526	.527
	70	.204	.325	.432	.470	.485	.497	.499	.500
	90	.179	.295	.412	.456	.475	.474	.477	.478
2	10	.412	.477	.490	.492	.493	.498	.499	.505
	30	.201	.304	.379	.402	.410	.422	.424	.429
	50	.115	.192	.269	.304	.320	.339	.343	.347
	70	.078	.136	.204	.241	.261	.286	.292	.295
	90	.066	.117	.183	.221	.246	.262	.271	.273
3	10	.331	.426	.458	.461	.462	.465	.465	.477
	30	.121	.202	.275	.304	.316	.327	.329	.337
	50	.062	.109	.164	.193	.209	.228	.232	.238
	70	.041	.073	.114	.138	.154	.176	.183	.188
	90	.035	.062	.099	.123	.141	.159	.169	.173
4	10	.265	.372	.422	.430	.433	.435	.435	.456
	30	.077	.137	.199	.229	.243	.256	.259	.272
	50	.037	.069	.107	.130	.144	.161	.167	.175
	70	.026	.046	.073	.089	.101	.119	.126	.132
	90	.022	.039	.063	.078	.090	.106	.114	.120
6	10	.173	.281	.351	.368	.373	.375	.375	.422
	30	.037	.073	.115	.137	.151	.164	.168	.189
	50	.018	.036	.058	.071	.080	.092	.098	.110
	70	.013	.026	.040	.049	.056	.064	.069	.078
	90	.012	.023	.035	.043	.048	.057	.062	.070
8	10	.117	.207	.282	.305	.314	.319	.320	.393
	30	.020	.042	.071	.087	.098	.111	.116	.143
	50	.010	.021	.035	.044	.050	.058	.063	.078
	70	.007	.016	.026	.032	.036	.041	.045	.055
	90	.076	.014	.023	.028	.031	.036	.040	.049
10	10	.082	.153	.224	.250	.262	.269	.271	.368
	30	.012	.026	.047	.059	.068	.078	.084	.114
	50	.006	.014	.024	.030	.034	.040	.044	.060
	70	.005	.011	.019	.022	.025	.029	.032	.043
	90	.004	.010	.017	.020	.023	.026	.028	.038

Source: IES RP-23-1989; reprinted with permission.

ILLUMINATION

TABLE 19.12 **Coefficient of Utilization from Window Without Blinds; Sky Component** $E_{xvk}/E_{xhk} = 1.75$

Room Depth/ Window Height	Depth	Window Width/Window Height							
		.5	1	2	3	4	6	8	Infinite
1	10	.435	.457	.465	.471	.474	.486	.488	.489
	30	.317	.407	.452	.466	.471	.486	.488	.489
	50	.234	.343	.422	.447	.456	.472	.475	.476
	70	.187	.297	.395	.430	.445	.458	.461	.462
	90	.168	.276	.384	.426	.444	.447	.450	.451
2	10	.357	.412	.422	.424	.424	.430	.431	.436
	30	.180	.271	.335	.356	.363	.375	.378	.381
	50	.106	.177	.246	.278	.293	.313	.318	.321
	70	.074	.130	.194	.229	.249	.274	.282	.284
	90	.065	.116	.181	.219	.244	.264	.273	.276
3	10	.288	.369	.394	.397	.397	.400	.401	.411
	30	.110	.183	.247	.272	.282	.294	.296	.304
	50	.058	.104	.154	.181	.196	.215	.221	.226
	70	.040	.072	.112	.136	.152	.176	.184	.188
	90	.035	.063	.101	.126	.144	.166	.177	.182
4	10	.232	.324	.365	.371	.373	.375	.375	.394
	30	.071	.127	.183	.209	.222	.235	.238	.250
	50	.036	.067	.104	.125	.139	.157	.163	.171
	70	.025	.046	.072	.089	.101	.119	.127	.134
	90	.022	.041	.065	.082	.095	.114	.124	.130
6	10	.153	.247	.307	.320	.324	.326	.327	.367
	30	.035	.070	.109	.130	.143	.155	.160	.180
	50	.018	.036	.058	.071	.080	.091	.098	.110
	70	.013	.026	.041	.051	.058	.067	.073	.082
	90	.012	.023	.037	.046	.052	.062	.069	.078
8	10	.104	.184	.249	.269	.276	.281	.282	.346
	30	.020	.042	.070	.086	.096	.109	.115	.141
	50	.010	.022	.036	.046	.052	.060	.066	.081
	70	.008	.017	.027	.033	.038	.044	.048	.059
	90	.007	.015	.024	.030	.034	.040	.044	.054
10	10	.074	.138	.201	.223	.233	.240	.242	.328
	30	.012	.027	.048	.059	.067	.078	.084	.114
	50	.006	.014	.026	.032	.036	.043	.047	.064
	70	.005	.011	.020	.024	.027	.031	.034	.046
	90	.004	.010	.018	.022	.024	.028	.031	.042

Source: IES RP-23-1989; reprinted with permission.

TABLE 19.13 **Coefficient of Utilization from Window Without Blinds; Ground Component**

Room Depth/ Window Height	Depth	Window Width/Window Height							
		.5	1	2	3	4	6	8	Infinite
1	10	.105	.137	.177	.197	.207	.208	.210	.211
	30	.116	.157	.203	.225	.235	.241	.243	.244
	50	.110	.165	.217	.241	.252	.267	.269	.270
	70	.101	.162	.217	.243	.253	.283	.285	.286
	90	.091	.146	.199	.230	.239	.290	.292	.293
2	10	.095	.124	.160	.178	.186	.186	.189	.191
	30	.082	.132	.179	.201	.212	.219	.222	.225
	50	.062	.113	.165	.189	.202	.214	.218	.220
	70	.051	.093	.141	.165	.179	.194	.198	.200
	90	.045	.079	.118	.140	.153	.179	.183	.185
3	10	.088	.120	.157	.175	.183	.185	.163	.167
	30	.059	.107	.154	.176	.187	.198	.193	.198
	50	.039	.074	.114	.134	.146	.157	.163	.170
	70	.031	.055	.085	.101	.111	.122	.127	.130
	90	.028	.047	.070	.083	.092	.107	.113	.115
4	10	.073	.113	.154	.174	.183	.187	.176	.184
	30	.040	.082	.127	.148	.159	.170	.177	.185
	50	.025	.049	.078	.094	.103	.113	.117	.123
	70	.020	.036	.054	.065	.071	.079	.083	.087
	90	.019	.032	.046	.054	.060	.069	.073	.076
6	10	.056	.106	.143	.164	.175	.184	.173	.194
	30	.021	.050	.081	.098	.107	.117	.123	.138
	50	.013	.027	.041	.049	.054	.060	.064	.072
	70	.011	.021	.029	.033	.035	.039	.041	.046
	90	.011	.020	.026	.030	.032	.035	.037	.042
8	10	.036	.082	.122	.143	.156	.166	.170	.208
	30	.011	.029	.050	.062	.070	.078	.082	.101
	50	.007	.016	.024	.028	.031	.035	.038	.046
	70	.006	.013	.018	.020	.021	.023	.025	.030
	90	.006	.013	.017	.019	.020	.022	.023	.028
10	10	.024	.061	.109	.120	.131	.144	.147	.200
	30	.006	.017	.034	.040	.046	.053	.056	.076
	50	.004	.010	.016	.018	.020	.023	.024	.033
	70	.004	.009	.013	.014	.015	.016	.016	.022
	90	.004	.009	.013	.013	.014	.015	.016	.021

Source: IES RP-23-1989; reprinted with permission.

ILLUMINATION

TABLE 19.14 **Reflectances of Building Materials and Outside Surfaces**

Material	Reflectance (%)	Material	Reflectance (%)
Bluestone, sandstone	18	Asphalt (free from dirt)	7
Brick		Earth (moist cultivated)	7
Light buff	48	Granolite pavement	17
Dark buff	40		
Dark red glazed	30		
Cement	27	Grass (dark green)	6
Concrete	55	Gravel	13
Granite	40	Macadam	18
Marble (white)	45	Slate (dark clay)	8
Paint (white)		Snow	
New	75	New	74
Old	55	Old	64
		Vegetation (mean)	25

Source: IES RP-23-1989; reprinted with permission.

STEP 1. Illuminances:
Solar altitude (Table 19.2)

$$\text{Dec. 21: } -14°$$

$$\text{Mar. 21: } -36°$$

Vertical window illuminance from Fig. 19.6.

$$\text{Dec. 21: } E_{xvk} = 200 \text{ fc} = 2{,}150 \text{ lux}$$

$$\text{Mar. 21: } E_{xvk} = 500 \text{ fc} = 5{,}380 \text{ lux}$$

Horizontal illuminance from full sky from Fig. 19.6.

$$\text{Dec. 21: } E_{xHk} = 500 \text{ fc} = 5{,}380 \text{ lux}$$

$$\text{Mar. 21: } E_{xHk} = 1{,}240 \text{ fc} = 13{,}340 \text{ lux}$$

Horizontal illuminance from half sky.

$$\text{Dec. 21: } E_{xhk} = 2{,}690 \text{ lux}$$

$$\text{Mar. 21: } E_{xhk} = 6{,}670 \text{ lux}$$

From Eq. 19.5, the vertical illuminance from the ground is

$$\text{Dec. 21: } E_{xvg} = RF_g \times \frac{E_{xHk}}{2} = 0.2\,(5380) = 1076 \text{ lux}$$

$$\text{Mar. 21: } E_{xvg} = 0.2(13{,}340) = 2670 \text{ lux}$$

STEP 2. Window transmittance (Tables 19.6 and 19.7). Glass transmittance 85%, LLF 0.9, net glass area 92%.

$$\tau = 0.85(0.9)(0.92) = 0.70$$

STEP 3. Selection of coefficients of utilization (Table 19.8 to 19.13) for the five calculation points (10%, 30%, 50%, 70%, 90%)
Sky component E_{xvk}/E_{xhk}:

$$\text{Dec.21: } 2150/2690 = 0.8; \text{ use Table 19.8}$$

$$\text{Mar. 21: } 5380/6670 = 0.8; \text{ use Table 19.8}$$

$$\frac{\text{room depth}}{\text{window height}} = 18/5 = 3.6 \text{ use } 4.0$$

$$\frac{\text{window length}}{\text{window height}} = 18/5 = 3.6 \text{ use } 4.0$$

Ground component: use Table 19.13

Point (%)	CU_k (sky)	CU_g (ground)
10	0.673	0.183
30	0.235	0.159
50	0.104	0.103
70	0.065	0.071
90	0.053	0.060

STEP 4. Calculation of illuminance (Eq. 19.4)

$$E_i = \tau (E_{xvk} \times CU_k + E_{xvg} \times CU_g)$$

E_i point (%)	Dec. 21 (lux)	Mar. 21 (lux)
10	1151	2877
30	474	1182
50	234	585
70	151	378
90	125	312

■

To compare the results with those of the CIE and GDDM methods, the values calculated for December 21 are plotted on the room plan of Fig. 19.21c. The December values are selected because they are minimum and would correspond most closely to the 85th percentile figures of both previous methods. The agreement is excellent for the half of the room depth nearest the window. For the deeper half, where ground contribution is significant, the IESNA method yields higher figures because it considers the rereflected ground light contribution. Overall, the agreement among the three methods is excellent. The figures for March 21 would correspond most closely to the 50th to 60th percentile of Fig. 19.17; that is, somewhat more than double the minimum figures—and here again the agreement is good.

To demonstrate the use of the IESNA method for clear-sky conditions, we can work through a short example.

EXAMPLE 19.4. Find the five IESNA point illuminances for the same room as Example 19.3, on 21 June at 10 AM for clear-sky conditions. Assume that the window faces southwest (azimuth angle −45°).

SOLUTION. Solar azimuth at 10 AM, 21 June, 48° N latitude is 56° (Table 19.2). Bearing angle (Fig. 19.8) is therefore 45° + 56° or 101° (i.e., no direct sunlight enters the window, which is a necessary condition of the IESNA clear-sky method). Solar altitude is 55°.

STEP 1. Horizontal illuminances (Fig. 19.7)

Sun only: 80,000 lux
Sky only: 18,000 lux, full sky
             9,000 lux, half sky (E_{xhk})
Vertical illuminance (Fig. 19.9b):
                         Bearing angle 101°;
700 fc = 7500 lux (E_{xvk})

From Eq. 19.5, vertical illuminance on the exterior of the window, resulting from ground light is

$$E_{xvg} = 0.2 \times \frac{80{,}000 + 9{,}000}{2} = 8900 \text{ lux}$$

STEP 2. $\tau = 0.70$, as previously.

STEP 3. Selection of CU from Tables 19.8 to 19.13:

$$\frac{E_{xvk}}{E_{xhk}} = \frac{7{,}500}{9{,}000} = 0.83; \text{ Table 19.8}$$

Coefficients are the same as in Example 19.3.

STEP 4. Calculation of illuminance.

$$E_i = \tau(E_{xvk} \times CU_k + E_{xvg} \times CU_g)$$
$$= 0.7\,(7500 \times CU_k + 8900 \times CU_g)$$

E_i at Point (%)	June 21 (lux)
10	4674
30	2225
50	1188
70	784
90	652

■

Note that these results indicate that on a clear summer day, with no direct sunlight on the window (bearing angle >90°), there is enough daylight throughout the space for routine work of moderate difficulty (500 lux requirement), and, in most of the space, for difficult and exacting work.

19.12 Daylight Analysis: Computer Use

The most effective (and painless) method of performing the calculations necessary to produce meaningful numerical data is by use of one of the many computer programs available. The advantage gained is that depending on the software used, data can be produced in any detail desired, for the entire room, with interior and exterior obstructions, for any configuration of windows, and with varying room finishes. We used LUMEN MICRO 7[c] from LIGHTING TECHNOLOGIES, Boulder, Colorado, to produce the results shown in Figs. 19.23 and 19.24. We ran the program with the same input data as in Examples 19.1 through 19.3. Agreement with the results of the other three methods is well within engineering accuracy. This is to be expected because all of the methods utilize the CIE overcast sky luminance distribution formulation. In summary, then, it is our recommendation that each daylighting prob-

Room 1-1 (Floor 1)

Grid Summary

Grid Properties
Grid Type: Horizontal Illuminance
Grid Height: 2.50 ft
Units: English

Calculation Grid

	0.5	1.5	2.5	3.5	4.5	5.5	6.5	7.5	8.5	Dis 9
24.4	63.4	78.9	74.7	65.1	55.6	48.9	43.2	38.3	33.8	30
23.1	130.7	98.2	78.4	63.9	52.8	44.8	38.2	33.4	29.4	26
21.9	136.7	109.6	85.2	67.3	54.2	45.0	37.9	32.6	28.3	25
20.6	135.1	109.7	86.9	69.3	56.0	46.2	38.6	33.0	28.7	25
4.4	132.2	103.7	79.8	63.0	50.9	42.4	36.0	31.2	27.5	24
3.1	122.9	89.6	70.7	57.5	47.7	40.6	35.0	30.7	27.2	24
1.9	59.6	62.1	57.1	51.0	44.8	39.7	35.0	31.5	28.2	25
0.6	19.0	40.0	48.8	49.4	46.0	42.9	39.6	36.2	32.7	30

Distance from Grid Origin Point

Output Generated by Lumen Micro 7 for Windows. Copyright 1996 Lighting Technologies, Inc., Boulder, Colorado

Fig. 19.23 Partial printout of the results of computer calculation of daylight in the room shown in Fig. 19.19, for the conditions of Example 19.1. Figures are given in footcandles. Due to the calculation grid density (1 × 1.3 ft.), only a partial printout is shown. (Lumen-Micro 7© program, courtesy of Lighting Technologies.)

lem be worked through assuming three different sky conditions:

1. *Completely overcast sky.* Use sky luminances that are maintained for 85% of the time; that is, minimum luminance. Use a computer program or the GDDM method if facilities are available. If not, use the CIE or IESNA methods and sketch in contours intuitively. As stated previously, these results establish maximum PSALI needed and extent/type of lighting control strategy.

2. *Average sky on March 21.* That is, overcast or clear sky, whichever is more applicable with no direct insolation either due to geometry or fixed shading. This condition in the design mode assists in determining glass area and room proportions. In the analysis mode, the results are usable for HVAC design and energy calculations. Hand calculation is not recommended; computer use is. For hour-by-hour energy calculations, computer calculation is required.

3. *Clear sky with sun.* These results indicate the need for shading and sun control devices and give maximum HVAC loads. In all studies, use the most accurate sky luminance data available; that is, if local data based on field observations are available, they are preferable to the generalized data presented here. Spaces lighted with multiple sources such as bilateral sidelighting, windows, and toplighting, or daylight plus artificial light are calculated by superposition (calculating the effect of each source separately and adding the results).

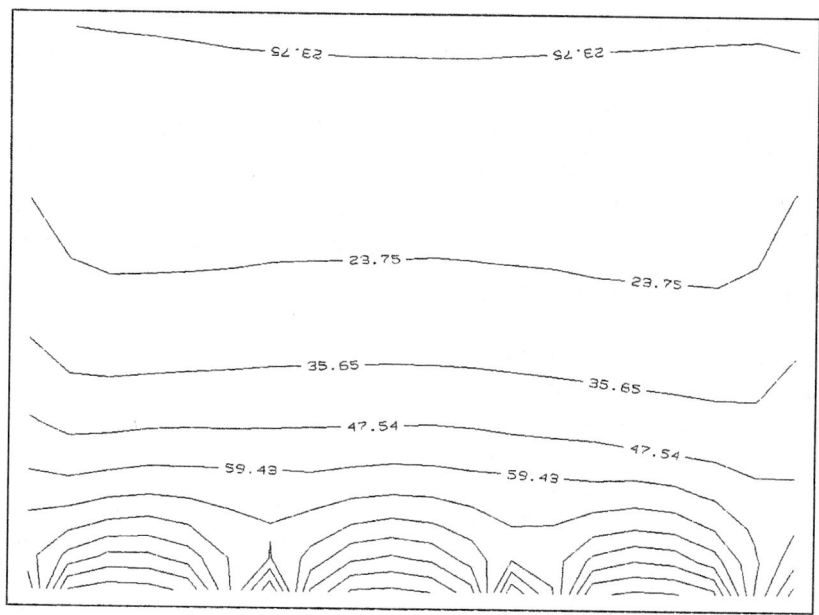

Horizontal Grid #2 Calculation Results (Room 1-1)

Horizontal Grid #2 Calculation Results (Room 1-1)

Fig. 19.24 *Graphical output results of Example 19.1 as produced by Lumen-Micro© software.*
(a) *Footcandle contour plot of the entire room. Where contour lines are densely drawn, footcandle illuminance values are omitted for clarity. The contour lines are ≈12 fc apart.*
(b) *Enlarged scale plot of the footcandle contours in front of the room's center window, with grid of calculated results superimposed. (Lumen-Micro© program, courtesy of Lighting Technologies.)*

ILLUMINATION

19.13 Physical Modeling in Daylight Design

As every architectural and engineering student is aware, physical models are a useful and occasionally an indispensable tool for the investigation of complex phenomena. The variables in daylight design are so numerous and the ranges so wide that even simple physical models can give the lighting designer a visual understanding of a daylighted space that only the most experienced designer can mentally visualize, imperfectly, on the basis of numerical results. Physical models are an adjunct to calculation procedures and not a substitute (or vice versa) as each serves a different purpose. Modeling gives the designer a three-dimensional visual depiction of the space with the brightnesses and shadows produced by daylight. These can then be altered by changing window design and orientation, and addition of light shelves, reflectors, and window controls, to produce the visual image that corresponds to the intended design throughout the day and also throughout the year.

The principal problem in the use of models is how to expose them to the desired sky conditions. Thus, if it is desired to view a particular space under both overcast and clear sky conditions, and the latter for different seasons of the year (different times of day are obviously not a major problem), waiting for a real sky condition that corresponds to the one desired is obviously not a practical alternative. (Use of models in partly cloudy sky conditions is not useful because of the infinite variability of this condition; computer simulation is more practical.) Overcast skies are common in temperate climates, and therefore a model can be tested in real sky conditions without undue delay. In arid areas, where overcast skies are rare, the overcast sky condition would not constitute a design condition for local construction. The reasonable alternatives to real sky exposure are, in descending order of desirability:

Artificial sky

Heliodon (tilt-table)

Narrow-beam electric lights and diffuse light boxes

1. *Artificial sky.* Carefully designed and controlled artificial sky domes can duplicate overcast and clear-sky conditions with a high degree of accuracy, and thus are ideal for testing physical models. Unfortunately, to achieve an acceptable degree of accuracy, these domes are large, complex, and very expensive to construct. A number of units exist in major universities and test facilities around the country, and designers with access to them would do well to utilize them. Small domes and mirror boxes have much more limited usefulness and are applicable primarily to overcast sky conditions. For construction details, see Moore (1958).

2. *Heliodon.* The heliodon, or tilt-table, as shown in Fig. 19.25 and used to produce the photos shown in Fig. 19.26, is a simple but very effective device that operates by rotating and tilting the building model with respect to the

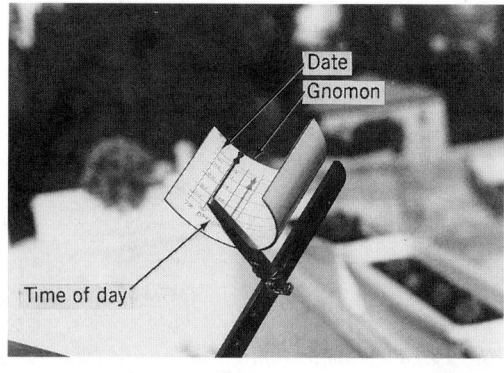

| (a) | (b) |

Fig. 19.25 (a) *Table-type heliodon with adjustments for latitude, elevation, and tilt permit exposing the model to desired clear sky conditions.* (b) *Date and time of day are set by adjusting the position of the gnomon's shadow.*

(a) Room with windows as in Example 19.1

(b) Large high window, center of wall

(c) Two small high horizontal windows

(d) Two small low vertical windows .....

(e) ...... with small skylight

Fig. 19.26 *These photographs illustrate the effectiveness of even a crude model in daylight studies. The model was constructed out of a cardbord box, at a scale of 1:20, to represent a room of the same horizontal dimensions as the classroom of Example 19.1 but with a higher ceiling. A Munsell Gray Scale was mounted along the back wall on gray paper corresponding to the center chips of the gray scale. The upper-left back wall has chips of varying chroma, and the right wall has chips representing typical interior finishes. The bottom row shows accent colors. A reflex camera with an f2, 28 mm-wide angle was used for all photos.*
(a) *Overcast sky, 21 June, 6* PM. *This photo represents the room of Example 19.1, with windows along one entire wall.*
(b–e) *Overcast sky, 21 June, 6* PM. *These photos show the effect of window placement. The square-foot area of fenestration is equal in all the photos.*

(f) 21 Jun

(g) 21 Mar/Sept

(h) 21 Dec

(i) South

(j) Southeast

(k) Northeast

Fig. 19.26 (f–h) *Clear sky, SW exposure, bright sun, 3* PM. *The heliodon permits viewing afternoon sun penetration for the four annual seasons. Note the relative darkening of room surfaces because of eye adaptation to the extremely high luminance of the sun on the floor.*
(i–k) *Clear sky, bright sun, 21 June, 3* PM. *At the same time that window* (f) *is receiving direct sunlight, other exposures receive sky light only* (i, j), *and one exposure* (k) *receives reflected sunlight from an opposing structure.*

real sky until the desired solar altitude and azimuth is reached. This technique is most effective with clear sky conditions. The objection often raised to this procedure is that in tilting the model, a different portion of the sky and ground is "seen" through the window than would be the case in real sky conditions. Al-

though this is obviously true, tests have shown the error to be minimal because:

(a) The solar component is much larger than the clear sky component when it is diffused by a window diffuser or even when reflected from facing structures and/or high reflectance ground (see Fig. 19.9).

(b) The clear sky component is roughly equal to the reflected component of low reflectance ground, so that a change in the portion of each that is "viewed" through the window is not significant.

3. Use of a *narrow-beam electric light source* to simulate the sun and back-lighted translucent panels to simulate an overcast sky are crude and highly inaccurate methods that, at best, can be used to obtain some idea of interior light and shadow patterns.Light levels in scale models can be accomplished quite readily by using small photocells intended specifically for the purpose. The scale of the model must be large enough that the photocell(s) does not have a significant influence on the light distribution with the space. For further information on physical modeling, the reader is referred to Robbins (1986) and Moore (1985).

19.14 Additional Factors in Interior Daylighting

(a) Horizontal and Vertical Surfaces. Inasmuch as the sky component of daylight enters side fenestration at an angle, it can be resolved into horizontal and vertical components, as shown in Fig. 19.27. The vertical component that illuminates horizontal surfaces is proportional to the sine of the angle of incidence, and the horizontal component that illuminates vertical surface is proportional to the cosine of this angle. Therefore,

for horizontal tasks, windows should be *as high as possible* and, for vertical tasks, *as low as possible.* As most tasks are horizontal, high windows give better lighting and deeper penetration than low windows of the same area.

(b) Window Details. The effect of window construction on total fenestration area reduction is often neglected. Even windows with narrow mullions and light metal frames have 8% to 10% obscuration; heavy window supports and small glass lights can result in 12% to 15% obscuration, with proportional daylight reduction. Further obscuration results from dirt accumulation, wired glass, and mechanical system items such as pipes and ducts inside the room, adjacent to windows. The effect of window placement on interior light distribution is shown in Fig. 19.28.

(c) Surface Reflectances. Interior reflectances are very important in daylight design. In addition to determining the magnitude of the IRC, in large measure they determine the eye adaptation level. A high adaptation level is desirable to avoid a sensation of glare when the window and its immediate surround are in the field of vision. Furthermore, the IRC contributes largely to the diffuseness of the room light. With low IRC, the sky component of daylight is the essential illuminant, and diffuseness and room penetration are reduced. Floors receive the sky component directly and should have at least 20% reflectance. Ceilings receive the ground compo-

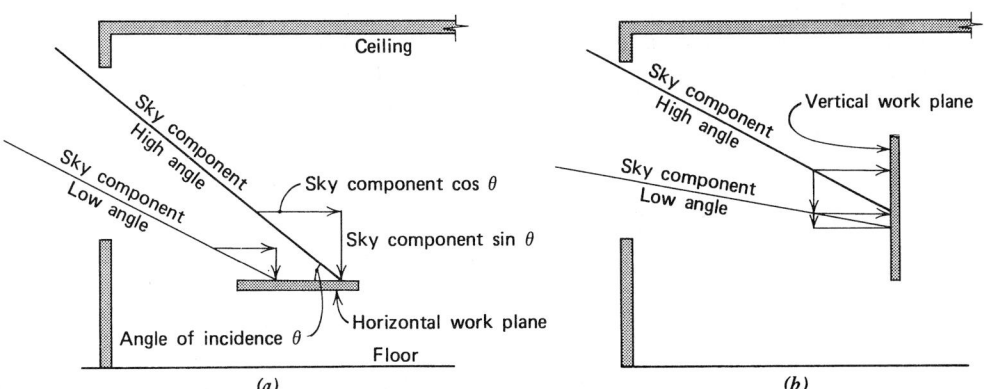

Fig. 19.27 *Effect of daylight incidence angle on illumination components. High-angle skylight is more effective for illumination on the horizontal plane* (a), *whereas low-angle skylight is more effective on vertical surfaces* (b). *See also Fig. 19.12 for variation of horizontal and vertical daylight factor with depth.*

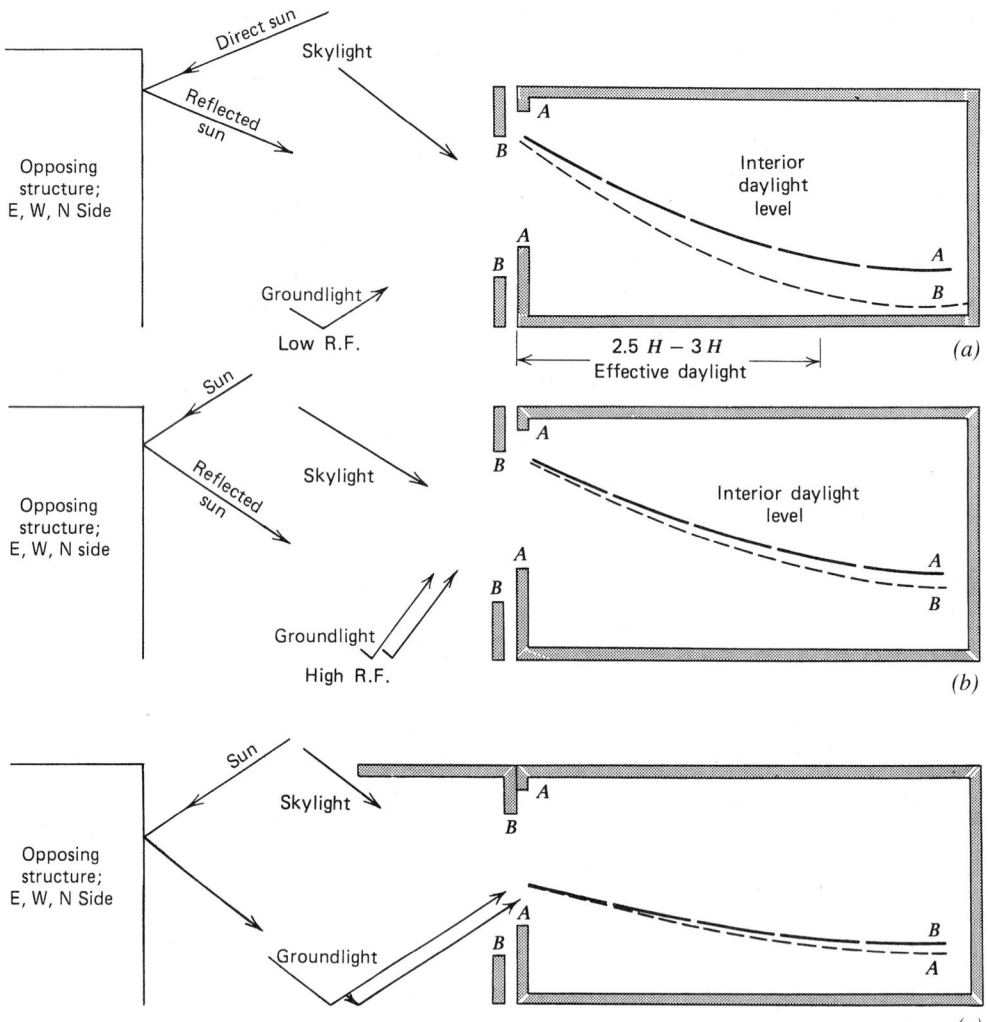

Fig. 19.28 *The effect of window placement on interior daylight. Where skylight predominates* (a) *because of high window/high sill/low ground reflectance/strong reflected component from opposite structure, illuminance drops off rapidly as depth increases. When reflected ground light is a major component, due to high ground reflectance* (b) *and/or strong re-reflected sun from opposite structure, the window position on wall has minimum effect. Very low sill height brings curve B above curve A, as in* (c). *Note that where sky light is blocked* (c), *lower sill height obviously results in better penetration. Re-reflected sun can be added to both skylight and groundlight, as shown.*

nent and should have at least 70% reflectance. Walls receive reflected light. As they are the surfaces seen at normal vision angles, they are responsible for eye adaptation levels and should have at least 50% reflectance.

Exterior surface reflection can provide the deep daylight penetration that is required for effective daylighting. Thus, a concrete or light-painted ground surface (RF of 50% to 70%) will furnish 25% to 65% of the light incident on a window, depending on shading and orientation. When combined with a high-reflectance ceiling, good interior distribution can be achieved.

(d) Glare and Heat Control (Sunshading).

These are among the most difficult problems to overcome in daylighting. All exposures except north receive direct sun during clear sky weather, with the attendant glare and heat problem. Although a northern exposure can receive direct sun during the summer, the early (and late) hour, oblique angle, and low sun altitude combine to essentially eliminate all problems, except possibly glare. This subject is discussed at length in Chapters 5 and 6, primarily from the viewpoint of comfort. The reader is referred to Section 6.3*a* and to Table 6.6 for a description of adjustable sunshading devices.

It is understood that furniture should be oriented so that windows are to the left side or, if not possible, to the rear. Windows in the line of sight must be sufficiently distant as not to be a glare source and must be provided with some type of glare control, such as blinds, curtains, shades, or the like.

(e) Fixed Sunshades and Lightshelves.
The three principal problems in the use of daylight as a practical lighting tool, as explained previously, are glare and heat control, particularly in areas near windows, and provision of sufficient light at greater distances from windows. The third would provide more even light distribution in sidelighted spaces, thus reducing energy use and simultaneously permitting use of deeper outside-wall spaces. This in turn has the ancillary benefit of buildings designed with a lower length-to-depth ratio, which in turn results in further energy use economies.

The design of fixed exterior sunshading devices that provide the desired window shading for specific dates and seasons is covered in detail in Appendix D. The technique discussed there is based on the use of shading masks. (Readers desiring to refresh their knowledge on the use of shading masks are referred to Olgyay and Olgyay, 1957; Ramsey and Sleeper, 1981; Lim et al., 1979; and Harkness and Mehta, 1978). In an interesting study, Millet, Lakin, and Moor (1981) compared various fixed shading devices that have the same shading mask (Fig. 19.29) as to their effect on interior daylight distribution, for both direct sun and overcast sky conditions. The study shows that devices with identical shading masks can yield vastly different interior daylight results. We would, therefore, suggest that the lighting designer/architect would be well advised to test a model (either physically or mathematically) of

any intended shading device under the anticipated sky conditions before fixing the design in concrete.

Fixed exterior sunshades, particularly horizontal overhangs, can be arranged to function simultaneously as a light shelf—that is, as a reflector—to increase the flow of light deep into the space. Several examples of sunshade–light shelves are shown in Figs. 19.30, 19.31 and 19.32.

The variations in shade/light-shelf design are legion. That being so, attempts to rationalize light shelf design almost always foundered on the large number of variables involved. One such attempt (Saraiji and Mistrick, 1993) developed a series of utilization coefficient tables for specific shelf designs and sky conditions. Although the continual improvements in computer software and hardware will eventually place readily usable design tools at the designers' disposal, physical modeling will always retain its great usefulness.

(f) Suggested Daylight Techniques.
Some of the principles involved in the use of daylight in construction are illustrated in Figs. 19.33 through 19.36. The reader is referred to the references at the end of Chapter 21 for full coverage of the many aspects of daylighting that space limitations prevented covering here. Among these are toplighting, treatment of light-shelves and overhangs, shading devices, daylight color, and similar topics.

19.15 Advanced Daylighting Techniques

The use of computers with appropriate software has effectively solved the problem of interior daylight distribution given the sky condition "viewed" by the window or other daylight admitting surface, and has made large strides towards defining, categorizing, and analyzing the infinite number of different sky conditions. As a result, the major emphasis in daylighting development today is concentrated in three areas:

1. Control of the incoming energy to limit glare and heat gain.
2. Collection and concentration of solar energy so as to increase the amount available beyond that passively received by openings on the building envelope, and concentration of the energy to make redirection and con-

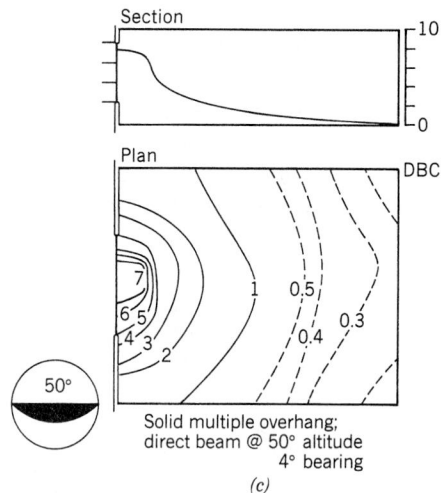

Fig. 19.29 Three horizontal shading device designs with identical shading masks but resulting in entirely different interior daylight distributions under (simulated) direct sunlight conditions (labeled DBC). *The solid overhand (a) provides complete (50°) window shading and results in a rapid drop-off of light in depth. The slanted slatted shade (b) acts both as shade and reflector, whereas the small stacked overhangs (c) act as lightshelves, shading below and reflecting above. (Reproduced with permission from Millet, Lakin, and Moore, 1981.)*

Fig. 19.30 Lightshelf acts to reduce daylight near the window and increases it at greater depth. Shelf material (opaque, translucent) and angle of installation (horizontal, sloped up) markedly affect performance.

duction feasible. Many of the devices, arrangements, and systems perform several functions that overlap among the three areas listed. A full discussion is far beyond the purposes and scope of this book. Instead, a survey of some of the most successful and promising techniques is given. The interested reader can explore the extensive literature for design and performance details.

3. Redirection and conduction of the incoming energy in order to distribute it within a structure.

The systems can best be categorized into stationary and moving (sun-tracking) systems. In the

Fig. 19.31 *Arrows illustrate principal light paths into room with light-directing glass blocks above and clear glass below. Clear section permits unobstructed view outdoors. These blocks are particularly effective in low-ceilinged rooms because the lower band of blocks near the 6-ft sill delivers most of the light to the deeper portions of the room. The use of a lightshelf below the blocks acts both as a reflector to increase the light transmission above and as a fixed horizontal sunshade for the window below.*

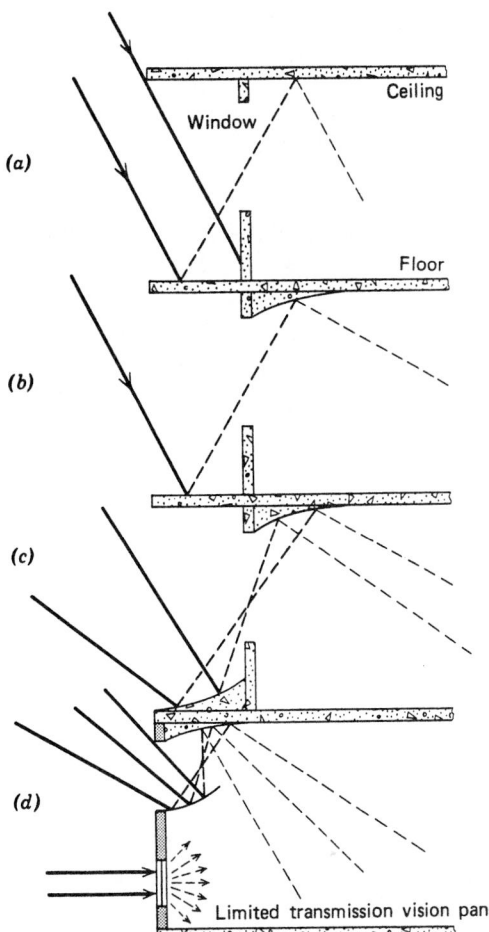

Fig. 19.32 *Window treatments that permit reflected daylight to enter the room but block direct sunlight.* (a) *Simple overhang acts as shield and reflector. Penetration is poor.*
(b) *Addition of high-reflectance, curved section at ceiling increases room penetration.*
(c) *Using a curved collector and reflector increases daylight deep into the room.*
(d) *Reflector can be placed inside the building wall, using building wall for sun shield. Daylight penetration can be very deep because sky component angle is large.*

stationary category, the most commonly used devices are light shelves of various designs (Figs. 19.31 and 19.32; and Beltrain, 1997), and prismatic materials in various configurations (see Moeck, 1998, Caddet, 1997-1). These are designed to reflect and redirect sunlight so that none enters the space directly while using as much of the impinging energy as possible. A recently developed variable transmittance (electrochromic) glass is slated for use in solar heat gain control. The material's transmittance is controlled by the application of electric potential (see Ehrlich, 1998).

Movable computer-controlled sun-tracking reflectors were originally developed as a means for the collection and concentration of solar radiation. The same principal can also be applied to the collection of visible solar radiation for use as lighting, with the important difference that in such applications the very large infrared component of solar radiation must be diverted. In systems that use the collected light in the same immediate loca-

tion, such as any of several commercial "sun fixtures," tracking reflectors are used to illuminate a diffuser of some sort, and heat is rejected by sealed air chambers or other transparent thermal buffers in the light path. These "sun fixtures" are by nature large and bulky, and are best applied as rooftop skylights. Depending on the installation's

Fig. 19.33 *When oriented east or west, expanses of glass must be provided with sun-control devices. This unshielded, west-facing, 15-ft, glass-door facade creates intolerable disabling glare, the more so for the dimly lit interior corridor and associated low eye-adaptation level. (Photo by Stein.)*

location, cleaning maintenance can be a major labor cost factor.

More advanced systems use concentrating reflectors and lenses to introduce concentrated energy in the visual range into some type of light-conducting device. These may be fiber-optic bundles, prismatic light pipes, or some type of mirrored channel. The problem of heat rejection becomes more severe as the degree of solar energy concentration increases. Thus, light-pipe designs are much less critical in this regard than are optical-fiber systems. Efficiency and economic feasibility of these systems are interdependent because of the materials used; the farther the light is transmitted, the lower is the system's overall efficiency and the higher the cost to make the system technically feasible. As a result, light-conducting designs, which are discussed at some length in Sections 21.40–21.46, Remote Source Lighting, are at present applied only in special areas such as those requiring explosion-proof lighting, or to produce special lighting effects.

Fig. 19.34 *Daylight factor here has a large ground component due to cutoff of the sky component by deep overhang, which also acts as a reflector for the ground light. Low-reflectance surfaces are suitable for a corridor that has no demanding visual task. (Photo courtesy of Libbey-Owens-Ford.)*

Fig. 19.35 *Full-height clear glazing with negligible direct sun shading is applicable only on north elevations or climates with continual overcast sky, as in this Pacific Northwest installation. Note that work positions are not placed facing the glass to avoid excessive brightness ratios in the field of view. Exterior planting reduces any incipient glare. (Photo courtesy of Libbey-Owens-Ford.)*

Fig. 19.36 *Unusual application of skylighting provides daylighting to an inner court of the International Monetary Fund Headquarters, Washington, DC. Translucent acrylic panels give sun control while limiting heating and glare. (Courtesy of Roper IBG.)*

INCANDESCENT LAMPS

19.16   The Incandescent Filament Lamp

(a) Construction. The standard lamp consists simply of a tungsten filament inside a gas-filled, sealed glass envelope (Fig. 19.37). Current passing through the high-resistance filament heats it to incandescence, producing light. Gradual evaporation of the filament causes the familiar blackening of the bulbs and eventual filament rupture and lamp failure. Incandescent lamps are available in many bulb and base types, with special designs for particular application (Figs. 19.38 and 19.39). To diffuse the light, most bulbs are coated inside with white silica. This provides almost complete light diffusion at a cost of approximately 2% to 3% of the light output. Colored light is also readily available from either coated bulbs or bulbs of colored glass.

The lamp base is the means by which connection is made to the socket and thereby to the source of electric current. Most lamps are made with screw bases of various sizes, the most common being the medium screw base. General service lamps of 300 W and larger use the mogul screw base. Where exact positioning of the filament is important, as it is when lamps are placed in precise reflectors or in lens systems, one of the special bases illustrated in Fig. 19.38 is used.

(b) Operating Characteristics. These are critically dependent on the voltage at the lamps; therefore, the life, output, and efficiency of a lamp can be markedly altered by even a small change in operating voltage, as illustrated by Fig. 19.40.

For example, operating a 120-V lamp at 125 V (104.2%) means approximately:

16% more light (lumens)

7% more power consumption (watts)

8% higher efficacy (lumens per watt)

42% less life (hours)

ILLUMINATION

Fig. 19.37 *Construction of a typical general-service incandescent lamp.*

Operating a 120-V lamp at 115 V (95.8%) means approximately:

15% less light (lumens)

7% less power consumption (watts)

8% lower efficacy (lumens per watt)

72% more life (hours)

Particular note should be taken of the effect of voltage on lamp life. In installations where lamp replacement is difficult and/or expensive, and use of an incandescent lamp is indicated, lamps may be operated slightly undervoltage and life prolonged, thereby decreasing the frequency of replacement. However, because efficiency is decreased by this procedure and recognizing that energy cost is normally a major factor over the life of any lighting installation, a detailed cost analysis should be made by the consulting engineer involved. Conversely, where lamps are replaced before burnout on a group replacement

system and initial installation cost per lux and/or energy costs are high, lamps may be operated overvoltage, thereby increasing output and efficiency but shortening life. This latter procedure is normal in sports-lighting installations because of the high installation cost of tower-mounted floodlights, making it mandatory to extract the maximum light from each unit. In stadium installations that have yearly lamp operation schedules averaging less than 200 h, 10% overvoltage operation doubles the light output but still allows a once-a-year, off-season relamping and is therefore a highly economical procedure.

In general, however, it is advisable to operate incandescent lamps at rated voltage, accepting balanced efficiency, output, and life.

(c) Other Characteristics.

1. *Lumen maintenance.* Light output decreases slowly with lamp life as the bulb blackens.

Bulb shapes

A—Standard shape
B—Flame shape
C—Cone shape
ER—Elliptical reflector

G—Globe
GA—Combination of G and A
P—Pear shape
K—Arbitrary designation

PS—Pear shape
      straight neck
PAR—Parabolic aluminized
        reflector

R—Reflector
S—Straight
T—Tubular

ILLUMINATION

Base types

 Miniature

Miniature bayonet

Candelabra

Intermediate

Single contact bayonet candelabra

Double contact bayonet candelabra

Disc

Recessed single contact

Medium

Three contact medium

Medium prefocus

Medium side prong

Medium skirted

Mogul

Three contact mogul

Mogul prefocus

Mogul end prong

L.C.L.—Light center length

Fig. 19.38 *Common incandescent lamp bulb and base types with nomenclature. The bulb nomenclature indicates type and size; the letter being an abbreviation of the shape and the number equal to the maximum diameter in eighths of an inch. Thus, a PS-52 is a pear-shaped bulb, 52/8 (6½) in. in diameter, and an PAR-38 is a parabolic reflector lamp 38/8 (4¾) in. in diameter.*

Bulb diameter is given in 1/8 in. Example: An A-19 bulb has a diameter of 19/8 in. or 2 3/8 in.

MOL: Maximum overall length: This figure refers to the maximum length of the bulb.

LCL: Light center length: This dimension, important when designing reflectors, is measured from the filament to a point that varies with base type. See Fig. 19.31.

	A – STANDARD SHAPE									PS – PEAR SHAPE							
WATTS	15	25	40	60	75	100	100	150	150	150	200	300	300	500	750	1000	1500
BULB	A-15	$A\text{-}19_1$	$A\text{-}19_2$	$A\text{-}19_3$	$A\text{-}19_3$	$A\text{-}19_3$	$A\text{-}21_1$	$A\text{-}21_2$	A-23	PS-25	PS-30	PS-30	PS-35	PS-40	PS-52	PS-52	PS-52
DIAMETER"	$1^{7}/_{8}$	$2^{3}/_{8}$	$2^{3}/_{8}$	$2^{3}/_{8}$	$2^{3}/_{8}$	$2^{3}/_{8}$	$2^{5}/_{8}$	$2^{5}/_{8}$	$2^{7}/_{8}$	$3^{1}/_{8}$	$3^{3}/_{4}$	$3^{3}/_{4}$	$4^{3}/_{8}$	5	$6^{1}/_{2}$	$6^{1}/_{2}$	$6^{1}/_{2}$
M.O.L."	$3^{1}/_{2}$	$3^{7}/_{8}$	$4^{1}/_{4}$	$4^{7}/_{16}$	$4^{7}/_{16}$	$4^{7}/_{16}$	$5^{1}/_{4}$	$5^{1}/_{2}$	$6^{3}/_{16}$	$6^{15}/_{16}$	$8^{1}/_{16}$	$8^{1}/_{16}$	$9^{3}/_{8}$	$9^{3}/_{4}$	13	13	13
L.C.L."	$2^{3}/_{8}$	$2^{1}/_{2}$	$2^{15}/_{16}$	$3^{1}/_{8}$	$3^{1}/_{8}$	$3^{1}/_{8}$	$3^{7}/_{8}$	4	$4^{5}/_{8}$	$5^{3}/_{16}$	6	6	7	7	$9^{1}/_{2}$	$9^{1}/_{2}$	$9^{1}/_{2}$
BASE	MED	MED	MED	MED	MED	MED	MED	MED	MED	MED	MED	MED	MOG	MOG	MOG	MOG	MOG
STANDARD FINISH	IF	IF	IF W	IF	IF	IF	IF	IF	CL IF	CL IF	IF	IF	IF	IF	CL IF	CL IF	CL IF

CL – CLEAR   IF – INSIDE FROSTED

Fig. 19.39 *Typical dimensional data for common general-service incandescent lamps.*

Lamp position (vertical or horizontal) during operation and the closely related bulb temperatures affect this characteristic.

2. *Color.* The light has a large yellow–red component and is therefore highly flattering to the skin. As is explained in detail in Section 19.35, the spectral content of the light produced by a heated (incandescent) source depends on its temperature: high-wattage lamps are bluer, low-wattage lamps are yellower. Dimmed lamps give yellow–red light.

3. *Surroundings.* Generally, incandescent lamps are impervious to external heat, cold, or humidity. Starting is completely unaffected by ambient conditions.

4. *Lamp efficiency.* Incandescent lamps produce light as a by-product of heat; as a result they are inherently inefficient. Luminous efficiency (efficacy) increases with wattage. Thus a 60-W general-service lamp produces 890 initial lumens, or 14.8 lm/W, whereas an A-21 100-W lamp produces slightly less output than two 60-W lamps (1750 lm); put otherwise, it results in a 18% energy savings. Refer back to

Fig. 19.40 *Characteristics of a standard 120-V general-service incandescent lamp as a function of voltage.*

Section 18.6 for a discussion of photometric efficiency, or efficacy.

(d) Summary. The principal advantages of incandescent lamps are low cost; instant start and restart; simple, inexpensive dimming; simple, compact installation requiring no accessories; cheap fixtures; focusability as a point source; high power factor; lamp life independent of number of starts; and skin-flattering, full-spectrum color.

The principal disadvantages are low efficacy (see Fig. 19.1), short lamp life, and critical voltage sensitivity. Low efficacy means more fixtures and large heat gain. Short lamp life results in high lamp replacement labor cost. Voltage sensitivity may require careful and expensive circuit design. Also, light concentration of the filament (point source) requires careful fixture design in order to avoid glare and, if undesirable, sharp shadows.

Because of the poor energy characteristics, incandescent lamp use should be limited to the following applications:

· Infrequent use

· Frequent short-duration use

· Where low-cost dimming is required

· Where the point source characteristic of the lamp is important, as in focusing fixtures

· Where minimum initial cost is essential

· Where its characteristic color rendering is desired

A brief list of conventional incandescent lamps and their physical and operating characteristics is given in Table 19.15. Lamp data for use in design should be taken from current manufacturers' literature. Data presented here are typical.

ILLUMINATION

TABLE 19.15 **Typical Incandescent Lamp Data (Listing a Few of Many Sizes and Types of 115-, 120-, and 125-V Lamps)**

Lamp Watts[a]	Average Rated Life (h)	Approximate Color Temperature[b] (°K)	Initial Lumens	Lumens per Watt[c]	Shape of Bulb[d]	Base	Description
15	2,500	—	126	8.4	A-15	Med	Long life
25	1,000	—	240	9.6	A-19	Med	Rough service
25	2,500	2,500	232	9.3	A-19	Med	—
40	1,500	—	495	12.4	A-19	Med	—
50	2,000	—	525	10.5	ER30	Med	—
60	1,000	2,800	890	14.8	A-19	Med	—
60	2,500	—	800	13.3	A-19	Med	Long life
75	750	—	1,220	16.3	A-19	Med	—
100	750	2,870	1,650	16.5	A-19	Med	—
100	750	—	1,750	17.5	A-21	Med	—
100	2,500	—	1,500	15.0	A-19	Med	Long life
135	750	—	2,425	18.0	A-21	Med	Economy
150	750	2,900	2,810	18.7	A-21	Med	—
150	750	—	2,600	17.3	PS-25	Med	—
200	750	2,930	4,000	20.0	A-23	Med	—
300	750	2,940	6,300	21.0	PS-25	Med	—
500	1,000	3,000	10,850	21.7	PS-35	Mogul	—

[a] Figures in this column designate the input watts thus: 60 = 60 W.
[b] See Section 19.25.
[c] Efficacy (luminous efficiency), in lumens per watt, increases with filament temperature; therefore, with wattage.
[d] See Fig. 19.38.

19.17 Special Incandescent Lamps, Including Energy-Saving Lamps

In the field of incandescent lamps other than the tungsten–halogen lamp, which is discussed separately, numerous special types are available. Some of the more important types are covered briefly in the following pages.

Rough service and *vibration* lamps are built to withstand rough handling and continuous vibration, respectively. Both conditions are extremely hard on general-service lamp filaments. Neither type is intended for general use, and both types have lower luminous efficacy than a general-service lamp (see Table 19.15).

Extended-service lamps are designed for 2500-h life. They are useful in locations where maintenance is irregular and/or relamping is difficult. The lamp is really designed for slightly higher voltage than that at which it is applied, and therefore efficacy is reduced (see Table 19.15 and Fig. 19.40). So-called long-life lamps, which are guaranteed to burn for 2, 3, or 5 years, are actually lamps designed for higher voltages than those at which they operate. As they normally sell at a high cost and are very inefficient, their use is seldom advisable. Before using such lamps, a cost comparison, including cost of lamps, energy, and relamping, should be made (see Appendix E).

(a) Reflector Lamps. These are made in "R," "BR," "ER," and "PAR" shapes (see Fig. 19.38). They contain a reflective coating on the inside of the glass envelope that gives the entire lamp accurate light-beam control. Many reflector lamp types are available in narrow or wide beam design, commonly called *spot* and *flood,* respectively. R lamps are generally made in soft glass envelopes for indoor use, whereas PAR lamps are hard glass, suitable for exterior application. When using R and PAR lamps, the fixture acts principally as a lamp-holder inasmuch as beam control is built into the lamp. They have an improved reflector design that increases the efficiency of the lamps.

The Energy Policy Act (EPACT) of 1992 made a number of incandescent lamps in these designs obsolete because of the act's minimum efficacy requirement. These requirements state that R and PAR lamps rated 115 V to 130 V with medium screw base, bulb diameter greater than 2.75 in., and nominal wattages between 40 and 205, shall have minimum efficacies as shown in Table 19.16. Among the formerly popular lamps no longer manufactured due to low efficacy are the 75-W R30, R40, and PAR38; the 150-W R40 and PAR38; and the 200-W R40. All major manufacturers now produce incandescent reflector lamps that meet EPACT requirements. Among these are the elliptical reflector (ER) and bulge reflector (BR) lamps that use a more efficient reflector design. These lamps are normally catalog listed as "energy-saving" lamps.

(b) Energy-Saving Lamps. The requirement to conserve lighting energy has resulted in a line of energy-saving lamps being produced by every major manufacturer. These lamps, of all types, are frequently known by trademarked names. They are rated at a wattage lower than the standard lamps they are intended to replace, and are generally more efficient. However, not all reduced-wattage lamps have higher efficacy than the corresponding standard lamps. Some simply reduce light output proportionately to the wattage. In such instances, if light levels must be maintained, a *loss* is incurred because more fixture bodies are required to hold the lower output lamps for the same total output. Finally, because

ILLUMINATION

TABLE 19.16 **Minimum Required Efficacy and R and PAR Lamps**

Nominal Lamp Wattage (w)	Minimum Efficacy (lpw)
40–50	10.5
51–66	11.0
67–85	12.5
86–115	14.0
116–155	14.5
156–205	15.0

lamp life for energy-saving lamps is often different from that of a standard lamp and first cost is usually higher, a simple comparison is not possible.

As has been repeatedly stated, any additional first cost should be analyzed by a life-cycle cost analysis and the payback period calculated. The designer will find that a proper control system is often more economically attractive than low-wattage lamps, and that energy-saving lamps are primarily useful in retrofit work.

19.18 Tungsten–Halogen (Quartz–Iodine) Lamps

This lamp type, illustrated in Fig. 19.41 is similar to the standard incandescent lamp in that it produces light by heating a filament. It differs in that a small amount of halogen gas (iodine or bromine) is added to the inert gas mixture that fills the bulb. This addition results in retardation of filament evaporation, which is the usual cause of incandescent lamp failure and thereby extends lamp life (Fig. 19.42).

Although the lamp has only slightly higher efficacy than an equivalent common incandescent, it has the advantages of longer life, lower lumen depreciation (98% output at 90% life), and a smaller envelope for a given wattage (see Fig. 19.41). The latter characteristic is due to the high temperature required by the halogen cycle, which in turn requires a compact, high-temperature filament. As a result, the lamp is effectively a point source, making it ideal for use in precision reflectors. Indeed, it is in this area of compact precision reflectors, as discussed later, that most of the more recent developments in tungsten–halogen lamp technology have occurred.

Due to the lamp's high filament temperature, the bulb envelope is generally made of quartz, which can withstand the temperature better than glass; this in turn gave rise to the name—quartz–iodine—which is sometimes applied to this lamp type. (Recently, special high-temperature glass has also been used.) Another result of the high filament temperature is that the gas pressure inside the quartz envelope is elevated and the lamps have been known to rupture violently, spraying

Double-ended tungsten halogen lamp 120 volts, 50-1500 watts
(*a*)

Single ended pin terminal lamps; 12 volts, 120 volts 20-65 watts

Single ended screw base lamps; 12 volts, 120 volts 50-500 watts

Fig. 19.41 *Tungsten–halogen lamps are available in a variety of designs, the most common of which are shown.*

hot quartz fragments over a wide area. As a consequence, all manufacturers now provide a cautionary notice with their lamps. The wording varies, but essentially states that due to the possibility of rupture, lamps should be handled carefully, guarded against abrasion and overvoltage operation, and, most important, adequately shielded or screened. The shield can be a reflector cover, a fixture lens, a screen, or other devices that will contain hot flying fragments in the event the lamp shatters. Another caveat that has gained wide publicity stems from the high lamp temperature and the attendant fire hazard of installations using

Fig. 19.42 *The self-regenerative halogen cycle results in slow evaporation of the tungsten filament with consequent lower light depreciation and longer life than those of a standard incandescent lamp.*

exposed lamps, as in torchiere lamps, inverted reflector indirect lighting fixtures, and the like.

A number of standards organizations and professional societies have adopted and published cautionary notices, including the *IESNA Lighting Handbook,* ANSI Standard C78,1451.1988, Underwriters' Laboratories UL1591 (1990), Canadian Standards Association, and the International Electrotechnical Commission (IEC). Exceptions to these precautions obtain when a lamp is inherently protected by encapsulation inside a sealed envelope. This encapsulated construction is now manufactured in R, PAR, MR, and modified A-lamp shape. These encapsulated lamps are intended for direct replacement of standard incandescent lamps of the same bulb shape.

Other halogen lamp characteristics are similar in all respects to the incandescent lamp. Color temperature ranges between 2000 and 3400 K; spectral energy distribution is typical of a black-body radiation (see Section 19.25), and dimming characteristics are similar to those of standard incandescent lamps.

19.19 Halogen Lamp Types

The basic lamp is a small gas-filled quartz tube, as shown in Fig. 19.41. Because the lamp must be used with some sort of reflector, it is manufactured with different terminations to suit the fixture reflector or secondary lamp envelope into which it is placed. All the lamp types shown in Fig. 19.41 can be used where lamp-only replacement is in-

tended, such as in floodlights or reflector fixtures, by using appropriate bases—slide contacts for double-ended lamps, screw bases for screw-base lamps, and special ceramic pinhole bases for pin-type lamps.

(a) Encapsulated Lamps (Figures 19.43 and 19.44). These lamps are sealed units intended for direct replacement of either the corresponding distribution-type incandescent lamp in the case of reflector units, or general service incandescent lamps for reflectorless units. Being sealed units, the halogen lamp is not replaceable, and the entire unit is discarded on burnout. Reflector units are available in a wide variety of beam patterns that designers will find detailed in manufacturers' catalogs. Typical data is given in Fig. 19.45.

Reflector lamps are also available with a variety of filters; so-called cool lamps that direct much of the heat out through the back of the lamp; high-efficiency units that reflect and that concentrate the heat on the lamp filament; lamps with ultraviolet (UV) filters for use in displays of UV-sensitive objects; and others. For this type of information, complete current manufacturers' catalogs are a necessity.

(b) MR-16 Precision Reflector Units. Miniature single-ended 12-V lamps in 2-in. diameter (and smaller) multifaceted dichroic (heat-ejecting) reflectors with a bipin base have found very wide acceptance in all types of display and accent lighting applications. These units, illustrated in Fig. 19.46, are essentially an entire lighting fixture, like R and PAR lamps, requiring only a base for electrification. The lamps are known by the generic name of MR-16, after an early 2-in. diameter model, and although each major manufacturer utilizes its own trade name, equivalence can be established by referring to the ANSI Code. The multi-mirror-faceted dichroic reflectors produces a "cool" precision light beam by ejecting approximately two-thirds of the lamp heat through the back of the reflector. Means to dissipate this heat must be adequate, as in all installations, but particularly with tungsten–halogen lamps, whose life is sensitive to operating temperature. Beam characteristics of a few MR-16 lamps are given in Fig. 19.46.

(a) Lamps with reflectors

12,120 volts, 40-150 watts
Spots; 9-15° beam spread
Floods; 30-55° beam spread

120 V 50, 75W
Narrow spot
Narrow flood

(b) General service encapsulated lamps

Fig. 19.43 *Tungsten–halogen lamps can be mounted in a variety of ways inside an enclosing glass envelope. (a) Lamp is used either horizontally or vertically to suite reflector design or (b) used without a reflector in a protective glass envelope with Edison medium screw base.*

Fig. 19.44 *PAR halogen lamp with standard medium screw bases, in sizes PAR 16, 20, 30, and 38, with short or elongated lamp neck. Lamp wattages range from 45 to 90 W. (Courtesy of Philips Lighting Co.)*

FLUORESCENT LAMPS

19.20 Fluorescent Lamp Construction

The second major category of light sources is that of electric discharge lamps, of which the fluorescent lamp is the best-known and most widely used type. Since its introduction in 1937, it has almost completely supplanted the incandescent lamp in all fields except specialty lighting and residential use. The typical linear fluorescent lamp comprises a cylindrical glass tube sealed at both ends and containing a mixture of an inert gas, generally argon, and *low-pressure* mercury vapor. Built into each end is a cathode that supplies the electrons to start and maintain the mercury arc, or gaseous discharge. The short-wave UV radiation, which is produced by the mercury arc, is absorbed by

Light Distribution – Candlepower Distribution

(a)   *(b)*

(c)   *(d)*

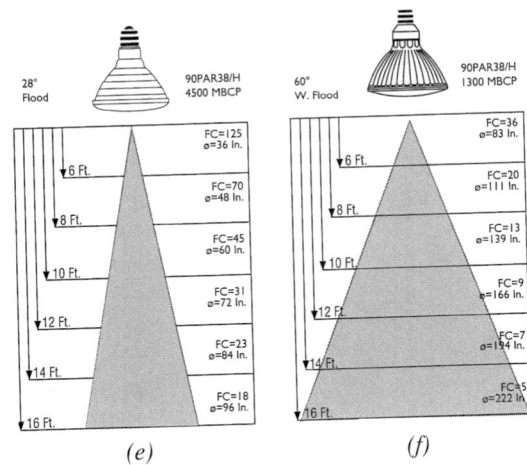

(e)   *(f)*

MBCP = Maximum Beam Candlepower
ø = Diameter of beam spread in inches
FC = Footcandle measured at 0°

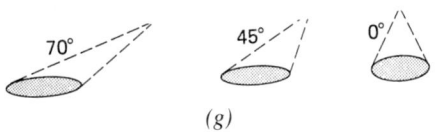

(g)

the phosphors with which the inside of the tube is coated and is reradiated in the visible range. The fluorescent lamp is so called because its phosphors fluoresce, or radiate light, when exposed to UV radiation. The particular mixture of phosphors used governs the quantity and spectral quality of the light output.

The descriptions that follow cover *standard* lamps and circuits—that is, lamps operating at power-line frequency (60 Hz), with standard loading (200, 430, 800, 1500 mA), using *core-and-coil magnetic* ballasts, in common starting and operating arrangements. Special lamps, accessories, and circuits, including low-wattage lamps and bal-

---

Fig. *19.45* *Typical PAR halogen lamp data. (a–f) The beam of the PAR lamp is conical in shape. Each type of PAR lamp has a distinct illumination pattern that varies in size and light intensity, depending on the angle at which the lamp is aimed and on its distance from the area illuminated. When centered directly on the surface to be lighted and at right angles to it, the small PAR sizes illustrated give a round lighting pattern. The horizontal illuminance (fc) figures shown are at the* center *of the beam and at the specified perpendicular distance from the lamp. Thus, for the 45-W, PAR 16, 10° narrow spot lamp shown in (a), the horizontal illuminance produced at the beam center, on a surface perpendicular to the long axis of the lamp, at a distance of 4 ft is 313 fc. The beam diameter is 8.4 in. measured to the point where beam intensity (candela) is 50% of the intensity at the center of the beam. On a flat surface, as shown, the horizontal illuminance is equal to ½ of beam center illuminance* $\times \cos^3 \phi/2$*. Thus, at 4 ft, the horizontal luminance at beam edge is*

$$I = \frac{313}{2} \times \cos^3\left(\frac{10°}{2}\right) = \frac{313}{2} \times (.996)^3 = 155 \, fc$$

(see Section 20.38, Calculating Illuminance from a Point Source). Similarly, the horizontal illuminance at beam edge of the 90-W, 60°-wide flood (item f) *at 10 ft from the lamp is*

$$I = 13/2 \, fc \times \cos^3\left(\frac{60}{2}\right) = 4.2 \, fc$$

(g) *The round lighting pattern changes to oval or elliptical when the lamp is aimed at an angle, making illuminance calculations much more complex. (Data in* a-f *courtesy of Philips Lighting Co.)*

ILLUMINATION

ILLUMINATION

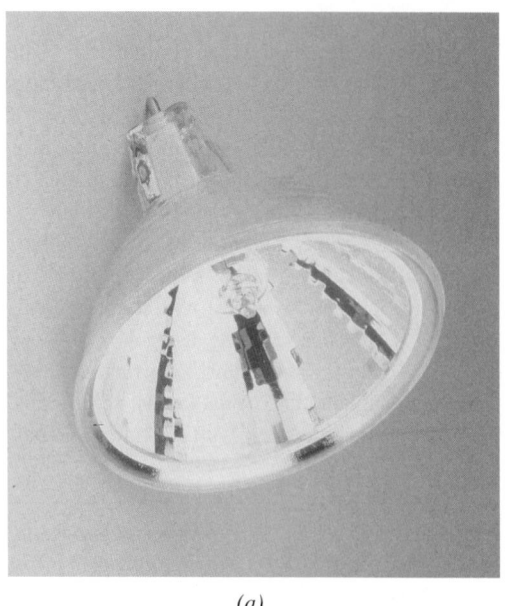

(a)

Fig. 19.46 (a) *Photo of a particular design of MR16 type lamp. The illustrated unit is designed to reject about two-thirds of its heat from the back of the lamp (for heat sensitive display applications), and the special quartz filament envelope reduces UV in the beam more than normal quartz. The lamps, rated 20 to 65 W and 10° to 60° beam angle are are all the same dimensions.*
(b) *Typical "illumination cones" for the lamp shown in (a). The figures have the same significance as those in Fig. 29.45. The abbreviations are: NSP, narrow spot; NFL, narrow flood; Fl, flood; VWFL, very wide flood. (Courtesy of OSRAM-SYLVANIA Products, Inc.)*

FOOT CANDLE CONES

TRU-AIM TITAN™

10° NSP

Distance from Source (in ft.)	Diameter (in ft.)	Center FC			
		20W	35W	50W	65W
0'					
3'	0.5	556	922	1278	1556
6'	1.0	139	231	319	389
9'	1.6	62	102	142	173
12'	2.1	35	58	80	97
15'	2.6	22	37	51	62

25° NFL

Distance from Source (in ft.)	Diameter (in ft.)	Center FC	
		50W	65W
0'			
3'	1.3	356	444
6'	2.7	89	111
9'	4.0	40	49
12'	5.3	22	28
15'	6.7	14	18

40° FL

Distance from Source (in ft.)	Diameter (in ft.)	Center FC			
		20W	35W	50W	65W
0'					
3'	2.2	78	139	222	233
6'	4.4	19	35	56	58
9'	6.6	9	15	25	26
12'	8.7	5	9	14	15
15'	10.9	3	6	9	9

60° VWFL

Distance from Source (in ft.)	Diameter (in ft.)	Center FC			
		20W	35W	50W	65W
0'					
3'	3.5	39	72	111	117
6'	6.9	10	18	28	29
9'	10.4	4	8	12	13
12'	13.9	2	5	7	7
15'	17.3	2	3	4	5

lasts, high-frequency operation with electronic ballasts, 26-mm (T-8) tubes operating at 265 mA, modified waveform ballasts, triphospher lamps, and special-shape lamps are all discussed separately.

(a) Preheat Lamps. The original fluorescent lamp was a preheat design. Construction of a typical hot cathode lamp (preheat and rapid start) is shown in Fig. 19.47a; the basic preheat circuit is shown in Fig. 19.48a. The circuit utilizes a separate starter, which is a small cylindrical device that plugs into a preheat fixture. When the lamp circuit is closed, the starter energizes the cathodes; after a 2- to 5-second delay, it initiates a high-voltage arc across the lamp, causing it to start. Most starters are automatic, although in desk lamps the preheating is accomplished by depressing the start button for a few seconds and then releasing it. This closes the circuit and allows

BULB
Usually straight glass tube. May also be circular or U—shaped.

PHOSPHOR
Coating inside the bulb transforms ultraviolet radiation into visible light. Color of light produced depends on composition of phosphor.

CATHODE
"Hot cathode" at each end of lamp is coated with emissive material which emits electrons.

(a)

BASE
Different types, as shown below.

MERCURY
A minute quantity of liquid mercury is placed in the bulb to furnish mercury vapor.

GAS
Usually argon or a mix— ture of inert gasses at low pressure. Krypton is sometimes used.

Base types

T—12
Med. bi-pin

(b)

T—12
Recessed
double contact

(d)

Instant start
cathode

(f)

T—12
Single pin

(g)

Preheat and
rapid start
lamps (both ends)
(c)

Recessed double contact:
high-output lamps
and very-high output lamps
(e)

High-voltage end   Low voltage end

Instant - start
(Slimline) lamps
(h)

Fig. 19.47 *Details of typical fluorescent lamps and associated lampholders. (a) Construction of preheat–rapid-start bipin base lamp. (Courtesy of GTE/Sylvania, Inc.) This type of lamp has type (b) base and is held in type (c) lampholder. High-output HO and VHO rapid-start lamps use recessed dc base (d) and lampholders (e). Instant-start lamps are similar in construction to (a) except with cathode construction (f), have a single pin base (g), and use single-pin lampholders (h), which are different for each end.*

the heating current to flow; releasing the button causes the arc to strike.

Standard preheat lamps operate at 430 mA. All preheat lamps have bipin bases (see Fig. 19.47b). They range in wattage from 4 to 90 W and in length from 6 to 90 in. A typical ordering abbreviation for a preheat lamp would be F15T12WW. This translates: fluorescent lamp, 15 W, tubular-shaped bulb, 12/8-in. diameter (number represents diameter in eighths of an inch), warm white

(a)

(b)

(c)

Fig. 19.48 *Basic circuits for preheat, rapid-start, and instant-start (Slimline). For the sake of clarity, only single-lamp circuits are shown, and power-factor correcting capacitors, auto-transformers, and compensators are omitted. (a) Basic preheat circuit. Starter may be any of several types, manual or automatic. Most preheat lamps are T-12 and operate at 430 mA.*
(b) Basic rapid-start circuit. Note the special end windings used to supply voltage to heat the cathode continuously. To assure proper starting, all standard RS lamps must be mounted within ½ in. of a grounded metal strip extending the full length of the lamp (1 in. for HO and VHO lamps). Normal output lamps operate at 430 mA; HO at 800 mA and VHO at 1500 mA.
(c) Basic instant-start circuit. Voltage from ballast transformer is high enough to strike an arc directly. Note that unlike preheat and rapid-start lamps, these are single pin because cathodes are not pre-heated. T-6 and T-8 lamps normally operate at 200 mA; T-12 lamps at 430 mA. Because of the high voltage involved, the lampholder at one end is a disconnecting device that opens the circuit when the lamp is removed.

color (3000° K). Preheat lamps are common in sizes up to 30 W; in larger sizes they have been supplanted by rapid-start and instant-start types.

(b) Rapid-Start (RS) Lamps.

These are similar in construction to the preheat lamps; the essential difference is in the circuitry (see Fig. 19.48*b*). The circuit eliminates the delay inherent in preheat circuits by keeping the lamp cathodes constantly energized (heated). When the lamp circuit is energized, the arc is struck immediately. No external starter is required. Because of this similarity of operation, rapid-start lamps will operate satisfactorily in a preheat circuit. The re-

verse is not true, because the preheat requires more current to heat the cathode than the rapid-start ballast provides.

The 40-W T12 rapid-start lamp that for years was the industry standard for commercial and institutional work is today rapidly disappearing because of federal standards (see Section 19.21) and because of new higher efficiency lamps with better color. Indeed, the number of different types of lamps (and ballasts) is so large and changes so rapidly that careful study of current technical catalogs is required for each application. A survey of the major types, as of this writing, is found in Section 19.23.

Standard rapid-start lamps operate at 425 to 430 mA. If this current is increased, the output of the lamp also increases. Two generic types of higher output rapid-start lamps are available. One operates at 800 mA and is called simply high output (HO). The second, which operates at 1500 mA (1.5 A), is called by different manufacturers very high output (VHO), superhigh output, or simply 1500-mA, rapid-start lamp. There is also a 1500-mA lamp that uses what looks like a dented or grooved glass tube. This lamp has somewhat higher output than the standard VHO tube. All high-output lamps use double-contact bases and special ballasts (see Fig. 19.47d and e). High-output lamps are used in applications such as outdoor sign lighting, street lighting, and merchandise displays where high output is required from a limited size source. Because of the serious heat problems involved, VHO lamps are frequently operated without enclosing fixtures. Conversely, HO and VHO lamps are frequently used in cold environments that would prevent proper operation of standard output 430-mA lamps. Most HO and VHO lamps are slightly less efficient than the standard 430-mA, rapid-start lamp and have considerably shorter life.

A circuit known as trigger start uses a preheat lamp in a rapid-start type of circuit, but with higher strike voltage. It is used for lamps up to 20 W and does not use a starter. After the arc is struck, lamp voltage is reduced.

(c) Instant-Start Fluorescent Lamps.
Slimline lamps are the best known type of instant-start fluorescent lamps. They use a high-voltage transformer to strike the arc without any cathode preheating. These lamps have only a single pin at each end that also acts as a switch to break the ballast circuit when the lamp is removed, thus lessening shock hazard (see Figs. 19.45c and 19.47f, g, and h). The lamps are generally operated in two-lamp circuits at various currents; normal currents are 200 and 430 mA, and standard lengths are 24, 36, 42, 48, 60, 64, 72, 84, and 96 in. The high-voltage starting characteristic of instant-start circuits lowers lamp life to about half that of the corresponding rapid-start lamp.

Slimline lamps and ballasts are more expensive than rapid-start and are somewhat less efficient. However, they are manufactured in certain sizes

and currents not made in rapid-start (e.g., 96 in., 430 mA), and they have the additional advantage of being able to start in much lower ambient temperatures (below 50° F) than rapid-start circuits. This starting characteristic makes the instant-start circuit particularly applicable to outdoor use. Typical characteristics of a few of the many types of *standard* preheat, rapid-start, and instant-start lamps are given in Table 19.17.

19.21 Federal Standards for Fluorescent Lamps

The National Energy Policy Act of 1992 (EPACT) mandates minimum efficacy in lumens per watt (lpw) and minimum color rendering index (CRI) for the four fluorescent lamp types listed in Table 19.18. These minima made many popular lamps obsolete. They included the F-40 T-12 lamp in cool white, warm white, white, and daylight; the F-40 T-12 cool white U-shaped tube with 6-in. leg spacing; the 96-in. T-12 cool white Slimline and high-output lamps, among others. More efficient lamps in both T-12 and T-8 tubes, with much improved CRI are now readily available to replace these obsolete types (see Section 19.23).

19.22 The Fluorescent Lamp: Characteristics of Operation

(a) Lamp Life.
The lamp life of a standard fluorescent tube is dependent on the burning hours per start. The figures listed in lamp catalogs for lamp life are based on a burning cycle of 3 h per start and represents the average life of a group of lamps; that is, half the lamps of any group will have burned out at this time. Typical lamp mortality curves are shown in Fig. 19.49, and the effect of burning hours per start is shown in Fig. 19.50.

This item is significant in connection with energy costs and utilization. From an energy source viewpoint, if an area is not utilized for periods of 10 min or more, fluorescent lamps should be shut off. This takes into account the resource energy required to replace a tube as a result of shortening its life. From a cost viewpoint, the break-even point depends on these factors:

TABLE 19.17 **Typical Fluorescent Lamp Data: Standard Lamps, 60 Hz, Conventional Core-and-Coil Ballasts**

Lamp Abbreviations[a]	Lamp Data		Lamp Life (h)[b]	Initial Output (lm)[c]	Initial Efficacy (lpw)[d]	CRI[e]
	Lamp (W)	Length (in.)				
Preheat lamp,[f] 430 ma						
F15T8CW	15	18	7,500	870	38	62
F20T12 3000°K	20	24	9,000	1,300	43	85
Rapid-start, preheat lamps,[g] 430 ma						
F40T12 3000°K	40	48	20,000+	3,300	75	85
F40T12 3500°K	40	48	20,000+	3,300	75	73
Rapid-start, high output, 800 ma						
F48T12CW/HO	60	48	12,000	4,050	58	62
F60T12CW/HO	70	60	12,000	5,150	64	62
F72T12CW/HO	85	72	12,000	6,350	65	62
Rapid-start, very high output, 1,500 ma						
F48T12CW/VHO	110	48	12,000	7,050	57	62
F72T12CW/VHO	160	72	12,000	11,000	63	62
F96T12CW/VHO	215	96	12,000	15,200	65	62
Instant-start (Slimline) lamps, 430 ma						
F24T12CW	20	24	7,500	1,100	40	62
F48T12CW	39	48	9,000	2,900	67	62
F72T12CW	55	72	12,000	4,550	69	62
F96T12 4100°K	75	96	12,000	6,500	79	70

[a] Lamp symbol: CW, cool white; 3000°K, color temperature; lamp abbreviations vary among manufacturers.
[b] Life figures are for 3 h burning per start.
[c] After 100 h burning.
[d] Includes average ballast loss.
[e] Color rendering index.
[f] Data given for a preheat circuit.
[g] Data given for lamps in a rapid-start circuit.

TABLE 19.18 **EPACT Minimum Standards for Fluorescent Lamps**

Lamp Type	Wattage (W)	Minimum CRI	Minimum Efficacy (lpw)
F40	>35	69	75
	28–35	45	75
F40/U	>35	69	68
	28–35	45	64
F96T12	>65	69	80
Slimline	52–65	45	80
F96T12/HO	>100	69	80
	80–100	45	80

Fig. 19.49 *Typical mortality curves of standard fluorescent lamps.*

1. Lamp life reduction as a function of burning hours per cycle.
2. Cost of energy.
3. Cost of lamp and lamp replacement.
4. Amount of time lamp remains off when shut off.
5. Cost of switching equipment (if any).
6. Life of the building.

With this number of variables it is not possible to give general solutions, and an individual analysis is required. However, several analyses by the author have shown that assuming ordinary office conditions, using lamp life data as given in Fig. 19.50, 20-year fixture life, $0.085/kWh energy cost escalating 3% annually, $1.25 lamp cost, 15-min relamping time, and $8 per lamp to provide the

Fig. 19.50 *Effect of burning hours on fluorescent lamp life. Note that at 3 burning hours per start, the average lamp life is 100% of the nominal catalog figure.*

necessary switch (one switch per two 2-lamp fixtures), lamps should be switched off any time they are not in use for 5 to 8 min or more. (The spread is caused primarily by variation in local labor rates.) This figure should be used in setting timing for occupancy sensor switching (see Sections 21.2–21.4.) It is thus clearly an economic fallacy to leave lamps burning to achieve longer lamp life.

(b) Effect of Temperature. The temperature of the coolest point on the lamp bulb wall determines the lamp's mercury vapor pressure, which in turn determines the lamp lumen output, wattage, and color. Maximum output for standard lamps occurs at a *bulb* temperature of 40°C (104°F). The bulb wall temperature itself is affected by room ambient temperature, airflow over the lamp (as in air-return fixture), and the temperature of adjacent surfaces such as the ballast enclosure. Thus, catalog data on lamp output and wattage, which are based on laboratory tests of bare bulbs at 25°C ambient temperature in still air may be very far from field measurements. For this reason, lamp (and ballast) manufacturers list lamp wattage in the catalog as "nominal." Typical actual wattage values for a standard nominal 40-W lamp with standard ballast would be 40, 39, and 34, respectively; that is, highest for laboratory conditions, 3% less for an open lamp fixture, and 14% less for an enclosed fixture because of its elevated lamp temperature. The well-documented effect of increased light output in air-return luminaires is due to the cooling action

Fig. 19.51 *Recommended operating ranges of circuit voltages for standard fluorescent lamps. The curves indicate the percentage changes in output lumens, efficiency, total watts, and current for line-voltage changes from the rated value. The nominal circuit voltages are 120, 208, and 277 V.*

of the air on overheated fluorescent tubes. However, over-cooling, which may occur with outdoor fixtures or open suspended fixtures in air-conditioned rooms, also reduces output.

Special all-weather and jacketed lamps are available that maintain fairly constant lumen output over a wide ambient temperature range. For outdoor use where starting below 50°F is necessary, rapid-start lamps require special low-temperature ballasts. Slimline lamps with normal ballasts start readily down to 20°F and, by using the next higher voltage ballast, starting down to −20°F is possible.

(c) Voltage Effects. Voltage either above or below rating adversely affects life (Fig. 19.51). Normal operating voltage range for core-and-coil ballasts is 110 to 125 V on 120-V circuits, 200 to 215 V on 208-V circuits, and 250 to 290 V on 277-V circuits.

(d) Lumen Maintenance. Lumen output of a fluorescent tube decreases rapidly during the

first 100 h of burning and thereafter much more slowly. Most catalogs list "initial lumens," which is the lamp output at *laboratory conditions* after 100 h burning, and "mean" or "design" lumens, which is lamp output at 40% of life. Lumen depreciation is generally unaffected by burning hours per start.

(e) Efficacy. The design efficacy (lumens per watt) of a fluorescent lamp depends on operating current and the phosphors utilized. Figure 19.52 shows the energy distribution of a typical standard fluorescent lamp alone, not including ballast losses. The range of efficacy for standard lamps is 40 to 85 lm/W, *including* ballast losses in the wattage figure (see Table 19.17). This is important because discharge lamps are inoperative without ballasts, and neglecting ballast losses yields an artificially high and therefore misleading efficacy.

It is important that the designer appreciate the significance of efficacy as compared to lamp wattage. In the usual commercial design situation, an interior space requires a calculated quantity

Energy distribution:
typical standard fluorescent lamp

Fig. 19.52 *Fluorescent lamps with efficacies of up to 85 lm/W are among the most efficient light sources available, yet still convert less than one-fourth of their energy to useful light. Ballast losses are not included in the percentages shown.*

of light, in lumens. The designer usually attempts to deliver that light at minimum overall cost; that is, a combination of first cost and operating cost. Minimum first cost usually means *maximum light output* per fixture, consistent with good lighting design. Minimum operating costs means maximum *lumens per watt*. These two demands are not normally consonant. Over the life of the building, operating cost dominates, as life-cycle analyses show, and the lamp-ballast combination of choice is normally the one with highest efficacy.

Wattage, in and of itself, is a meaningless quantity unless associated with a lumen output figure. Thus, energy saving low-wattage or high-output lamps with their special ballasts are seldom the indicated choice in new design because their efficacy does not usually justify the cost premium. These special lamps are useful in retrofit work, in which case, as noted previously, field measurements of illuminance and lamp temperature are required before selecting the replacement lamp-ballast combination. Table 19.19 in Section 19.23 lists comparative efficacy values of standard and specialty lamps under similar installation conditions.

Conventional (nonelectronic) ballast losses, which constitute 5% to 12% of lamp wattage, de-

pend on ballast type, circuit, manufacturer, type of fixture, ambient temperature, and number of lamps connected. Figures used in Table 19.17 are average. The amendment to Public Law 100-357, which mandates high-efficiency ballasts for certain lamp types, is discussed in Section 19.24.

To make a proper comparison between lamp types, it is not sufficient to compare cost to produce a given quantity of lumens as high-output lamp installations use a smaller number of fixtures. A meaningful cost comparison requires a full life-cycle or annual owning cost analysis.

(f) Dimming and Reduced Output. It is often desirable to reduce the output of a fluorescent lamp in order to reduce energy consumption, correct overlighting, compensate for daylight, change an area's function, and so forth. If it is anticipated that full output will again be required, dimming is indicated. If the change is long-term or permanent, a stepped or fixed reduced-output arrangement is probably indicated. Output reduction is accomplished by reducing effective lamp current. This can be accomplished by lowering primary voltage, adding impedance to the circuit, or shortening the time of current flow in each half cycle. The last method describes the gating action of a modern solid-state thyristor dimmer (SCR, triac).

1. *Partial Dimming.* The thyristor action shown in Fig. 19.53 can be applied to control the effective input voltage to a fluorescent luminaire with *conventional magnetic* ballasts, and by so doing dim the lamp. This is commonly known as "2-wire" dimming because the controlled

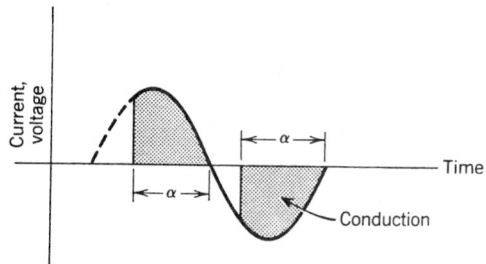

Fig. 19.53 *Gating action of a solid-state dimmer. By controlling the gating, the conduction angle α can be adjusted, thus varying the effective voltage over the entire cycle.*

ILLUMINATION

voltage is applied to the two input wires of the luminaire. However, due to the absence of the high-strike voltage and cathode heating circuit necessary for full-range dimming this arrangement is only useful down to 40% to 50% of full output. Below this level, light output drops sharply, lamp color changes, starting becomes unstable, and lamp life is affected. However, within the 100% to 50% range, the energy-output curve is linear; that is, efficacy is maintained, making this arrangement very desirable as an economic lighting-energy management system. See Sections 21.2 to 21.4, which cover lighting control.

Low-energy lamps are generally *not* suitable for dimming service. This is because they operate with special ballasts that reduce lamp current in various ways. Some ballast designs change current waveform, others disconnect the cathode circuit after the lamp starts, and still others change the circuit impedance and current-voltage characteristics. As most of these circuit alterations would interfere with dimmer functioning their use is inadvisable without a prior detailed engineering analysis. Also, low-energy lamps themselves are usually not operable with dimming ballasts.

2. *Full-Range Dimming.* When dimming is required to levels as low as 5% of full output, each individual lamp must be provided with a special (magnetic) dimming ballast. (Electronic ballasts are discussed in Section 19.26.) Full-range dimming ballasts use "3-wire" control to provide high voltage for starting and restrike (even at low output levels) and continuous cathode heating. This latter is important to assure that even at low lamp arc current the lamp does not flicker and that lamp life is unaffected, because only arc current is reduced while cathodes remain fully heated. Finally, the 3-wire control arrangement of individual dimming ballasts permits use of a central controller, which makes presetting, zoning, and timing of local dimming functions possible.

The lamp of choice in dimming circuits is the standard 430-mA rapid start, although 800- and 1500-mA lamps with magnetic ballasts can also be dimmed. An important characteristic of this dimming system is the linear relation between light output and power input, such that lamp efficacy is maintained down to about a third of output, after which efficacy drops and color shifts toward the blue–green portion of the spectrum because of bulb cooling.

3. *Output Reduction.* The cost of dimming equipment is usually justified only if smooth continuous-level changes are required (see Sections 21.3 and 21.4). Where discrete-level switching is acceptable, one solution is the use of multilevel ballasts. Such ballasts, once quite popular, are used less today because many local codes prohibit use of wall switches or other simple switching devices to change levels, requiring instead that an electrician perform the function. Obviously, this is inconvenient, time-consuming, and expensive. The reason for the prohibition is that the level change involves switching of ballast capacitors, with the possibility of high voltages (about 300 V) and current surges (up to 50 A). However, circuiting can be arranged so that switching is performed when the unit is de-energized, and if that is acceptable to local code authorities, these ballasts can be very useful.

Where it is desired to reduce lamp output on a semipermanent basis while maintaining efficacy, impedance can be added to the circuit in various forms. When it is desired to restore full output, the device is simply removed. Alternatively, special lamps are available that contain integral impedances and operate at lowered output with standard magnetic ballasts. Both these solutions are best applied in enclosed fixtures because the reduced lamp current may cause excessive lamp cooling in open fixtures with consequent drop in efficacy and change in lamp color.

(g) Flashing. Fluorescent lamp flashing ballasts are similar to dimming ballasts in that the cathode circuit remains energized continuously and only the arc circuit is turned on and off. This permits flashing the lamps without affecting lamp life. Cold cathode, preheat, and rapid-start lamps are suitable for flashing service.

(h) Other Characteristics. Fluorescent light color is discussed in Sections 19.34 and 19.35.

Linear fluorescent lamps are large and therefore necessitate a relatively expensive luminaire to both hold the lamps and control the light. Inasmuch as the tubes emit light throughout their considerable length, accurate beam control is difficult, making fluorescent units best applicable to area lighting.

In sum, then, the advantages of fluorescent lamps are long life; low cost; high output and efficacy; availability in an extremely wide range of sizes, colors, and brightnesses; and relative insensitivity to voltage fluctuation (important in brownout areas). Disadvantages are large size, which creates storage, handling, and relamping problems; and the luminaire situation previously referred to.

19.23   Special Fluorescent Lamps

(a) Low-Energy Lamps. The need for energy conservation, the discontinuance of certain lamp types due to EPACT regulations, and the desire to reduce lighting levels in existing overlighted spaces has resulted in a complete line of low-energy lamps. Their wattage ratings are lower than those of standard lamps because they are intended primarily as lower energy direct replacements for existing lamps. All such lamps are clearly marked by the manufacturer. They operate on lower current than standard lamps, achieved by one of the techniques discussed in Section 19.22*f*(3). They require special matching ballasts for maximum effectiveness, and have an efficacy equal to, or somewhat higher than, standard lamps and ballasts. They have the disadvantage of higher cost, need for special ballasts where maximum energy reduction is desired, generally shorter life, unusability in most dimming circuits, and problems of inventory and proper lamp replacement. To our mind, their use is indicated only where other light and wattage-reduction schemes, such as circuit dimming or reduced wattage ballasts, are inapplicable.

(b) Triphosphor Fluorescent Lamp. This is a high-efficiency, 1-in. (26-mm) diameter (T8) lamp whose high-output rare earth phosphors have the additional desirable characteristic of excellent color rendering and correspondingly high

color rendering index (CRI). (See Section 19.36 for an explanation of CRI.) The lamps are called triphosphor lamps because in addition to standard phosphors, they use an additional inner coating of rare earth phosphors that produce light primarily at three wavelengths; about 450 nm (blue), 540 nm (green), and 610 nm (red). Three well-known commercial lamps of this type are OSRAM-SYLVANIA's OCTRON®, GE's TRIMLINE T8™, and PHILIP's TL70/TL80™. These lamps operate at 265 mA as compared to 430 mA for "standard" T12 lamps. As a result, they require special ballasts and are not interchangeable with T12 lamps. This fact is important in retrofit work. The lamps are operable on magnetic ballasts, and low-frequency (60 Hz) electronic ballasts, but give their best performance when used with matching high-frequency electronic ballasts. See Table 19.19 for a performance comparison. The lamps have a better light depreciation characteristic than T12 lamps and essentially the same lamp life. Although the original design of triphosphor lamp was highly temperative sensitive, making it unsuitable for use in open fixtures or in dimming operations, modern units perform extremely well in open parabolic and louvered direct/indirect luminaires. When used with high-frequency dimming ballasts, smooth, flicker-free dimming is possible down to 20% of full output and lower, depending on the ballast. The effect of dimming on lamp color should be checked with the lamp manufacturer. When used with h-f electronic ballasts, up to four 32-W T8 lamps can be operated from a single ballast.

The lamps are manufactured (at present) in straight lengths of 2, 3, 4, 5, and 8 ft, with nominal wattages of 17, 25, 32, 40, and 59 W, respectively. They are also available in U shape with 6-in. leg spacing and 22.5 in. overall length and 32-W nominal rating. At least one manufacturer also makes the lamps in a high-output version, usable with electronic ballasts only. All straight lamps are made in three color temperatures—3000°K, 3500°K, and 4100°K—and in two versions: one with a CRI of approximately 75 and the second with a CRI of about 85 (and lower output). Selected sizes are also made with cooler color temperatures of 5000°K and 6500°K.

Although triphosphor lamps are more expensive than the corresponding T12 lamps, as are

TABLE 19.19 **Comparative Characteristics[a] of 48 in. Tubular Fluorescent Lamps**

Characteristic	F40 T12[b] Standard		F40 T12 Low Energy		F32 T8 Triphosphor	
CRI	73	85	62	84	75	85
Initial lumens	3,200	2,300	2,600	1,900	2,800	2,900
Mean lumens	2,800	1,900	2,300	1,600	2,550	2,650
Wattage						
Nominal	40	40	34	34	32	32
w/Magnetic ballast[c]	40	40	34	34	32	32
w/HF electronic ballast[d]	35	35	30	30	29	29
Overall efficiency (lpw)						
w/Magnetic ballast[c]	80	58	76	56	88	91
w/HF electronic ballast[d]	92	66	87	63	97	100
Lamp diameter (in.)	1½		1½		1	
Surface luminance (cd/m²)	6,000–8,000		4,050–6,500		9,000–11,000	
Lamp life (h)	20,000+		20,000+		20,000+	

[a] Figures are representative; for exact figures, consult current catalogs.
[b] Lamps meet EPACT requirements.
[c] One low-energy magnetic ballast for two lamps.
[d] One high-frequency electronic ballast for four lamps.

their electronic ballasts, the increased efficiency and corresponding lower operating cost makes the payback period for the first cost differential very short. This fact coupled with the much improved color characteristic makes these lamps the economic and engineering choice for most retrofit and new installations, with the single exception of purely speculative construction, where first cost considerations are dominant.

(c) U-Shaped Lamps. U-shaped lamps were developed to answer the need for a high-efficacy fluorescent source that could be utilized in a square fixture because the linear fluorescent lamp shape is frequently not architecturally suitable. The U lamp is basically a standard fluorescent tube bent into a U shape and available with 3⅝- or 6-in. leg spacing; the former can be accommodated three to a 2-ft square fixture. (The narrower T8 envelope of triphosphor lamps permits a tighter bend; those U lamps have 1⅝-in. leg-to-leg spacing.) U lamps operate on standard ballasts and have slightly lower output than the corresponding straight tube. In all other respects the

U lamp has the same characteristics as the straight lamp of similar type.

(d) T5 Lamps. Linear T5 (⅝-in. diameter) lamps are a relatively recent product line, produced primarily to conform to metric ceiling grid dimensions. The lamps are generally available in four designs:

14 W, 549 mm (21.6 in.) long

21 W, 849 mm (33.4 in.) long

28 W, 1149 mm (45.2 in.) long

35 W, 1449 mm (57.0 in.) long

That is, they are 2½ to 3 in. shorter than corresponding wattage T12 and T8 lamps. The lamps are manufactured with rare earth (RE) phosphors that give them slightly better overall efficacy (lpw) than T8 lamps, whether operated with magnetic or electronic ballasts. Due to the small diameter, the lamp base is miniature bipin, requiring special compatible lampholders. Ballasts should be chosen specifically to operate T5 lamps. The lamps are available in the same color temperatures as

T8 and T12 lamps, with a CRI between 80 and 85. They are also available in high-output design, which requires special ballasts.

These lamps, in addition to metric ceiling application, are useful for architectural lighting applications where their small diameter can be used to advantage and the lamp is concealed. This latter consideration is very important because lamp surface luminance is approximately 17,000 cd/m^2 for standard output lamps and 25,000 to 35,000 cd/m^2 for T5HO lamps. Such luminances can produce severe glare and must therefore be handled with great care in luminaire design and application.

19.24 Compact Fluorescent Lamps

The unwieldy linear shape of the straight tubular fluorescent lamp and its double-ended electrical connections prevented its wide use in other than large open spaces, despite its great energy efficiency advantage over incandescent lamps. The advent of various types of folded fluorescent lamps, commonly called *compact fluorescent lamps* (CFL), plus the vast improvement in output color accomplished by the use of rare earth (RE) phosphors has opened areas of lighting application to CFLs previously held exclusively by the notoriously inefficient, yet simple, reliable and cheap incandescent lamp.

CFLs are simply folded fluorescent tubes with both ends terminated in a common base. Two basic designs exist. In one the lamp plugs into a base containing a compact electronic ballast, which in turn screws into a medium screw (Edison) socket. In the second design, the electronic ballast and lamp are a single, inseparable unit terminating in a screw base. Both designs are therefore usable as direct replacements for an incandescent lamp. The advantage of the first design lies in the fact that ballast life is at least four

and as much as eight times as long as lamp life. As a result, an exhausted lamp is simply replaced in the existing ballast, resulting in a considerable economy. In the second design the entire lamp/ballast assembly must be replaced, although this design has the advantage of lower first cost and simplified replacement storage. Both designs are shown in Fig. 19.54, which also illustrates some of the major designs of CFL. Because all major American manufacturers produce similar products but identify them with specific trade names, we have provided for the reader's assistance a short cross-reference table of these names in Table 19.20. Although the products are similar, they differ in various aspects. Specific data for each lamp is available in manufacturers' technical publications.

CFLs are manufactured in an ever-increasing number of designs to fit almost every application in which incandescent lamps are used. Efficacy of the lamp/ballast combination ranges from 55 to 75 lpw, assuming an electronic ballast. Lamps with magnetic ballasts are available but are not favored because of excessive heat, weight, and reduced efficiency. Wattages range from 5 to 55 W. Lamp colors are similar to straight lamps (i.e., 3000°K, 3500°L, 4200°K, and 5000°K). All lamps have a CRI of 82. Life is 10,000 to 12,000 h, based on 3 h/start.

In connection with lamp life, a large caveat is in order. These lamps are fluorescent, and like all fluorescents, life is proportional to burning hours per start. They are therefore most definitely not applicable to such usages as closets, pantries, storage facilities, and the like where lamp on-time is short and on–off cycles are frequent. They are applicable to ceiling down lights, desk lamps, decorative fixtures, outdoor lighting in appropriate fixtures, and all applications where burning time per start is at least 3 h.

TABLE 19.20 **Cross-Reference of Trade Names for Major CFL Products**

OSRAM/SYLVANIA	GE	PHILIPS
Dulux® (twin tube)	BIAX®	PL
Dulux® EL	COMPAX®	Earthlight®, PL
Dulux® (double twin)	Double BIAX®	PL-C
Dulux® L	High Lumen BIAX®	PL-L
Dulux® T/E	Triple BIAX®	PL-T

Fig. 19.54 *Family portrait of compact fluorescent lamp designs. Pin base lamps designed for use with a separable ballast are shown in the foreground, the same lamps mounted in their screw base ballasts are in the left background, and the one-piece combined lamp/ballast design stand in the right background. Globe and reflector type replacements for incandescent lamps are in the center of the photo. (Courtesy of OSRAM-SYLVANIA.)*

One of the favored applications of CFL is to replace an incandescent reflector flood lamp such as a 75-W R30FL with an 18-W CFL in a specially designed reflector. Designers are cautioned to carefully inspect the candlepower distribution curve of the lamp reflector combination because it is never identical to that of an incandescent reflector lamp and may be vastly different, yielding disappointing results. Another caution in the use of CFLs in enclosed luminaires refers to heat buildup inside the luminaire that can reduce the CFL's output by as much as 50%. To avoid this problem, designers should insist on obtaining photometric reports from independent laboratories, particularly for ceiling downlight luminaires.

19.25   Magnetic Core-and-Coil Ballasts

The function of a ballast in a fluorescent lamp circuit is threefold:

- To supply controlled voltage to heat the lamp filaments in preheat and rapid start circuits

- To supply sufficient voltage to start the lamp by striking an arc through the tube

- To limit the lamp current once the lamp is started. The simplest ballast is a choke coil, so called because it serves to "choke" (limit) the circuit current with its high inductive impedance

A modern magnetic core-and-coil ballast consists of magnetic steel core on which is wound a coil, usually with copper wire. Older ballasts were constructed with an iron core and aluminum wire, which resulted in high energy losses. In addition, a modern ballast has various capacitors and switches that serve to increase power factor and decrease losses.

The organizations involved with ballast standards and testing are:

CBM—Certified Ballast Manufacturers Association

ANSI—American National Standards Institute; originates standards on a national level

ETL—Electrical Testing Laboratories, Inc.; a private, independent organization and a recognized authority in measurements and testing of lamps and lighting equipment.

UL—Underwriters Laboratories, Inc.; an independent, nonprofit organization that tests for public safety

In the United States, ballasts should be UL labeled and CBM/ETL certified. The UL label assures intrinsic safety. CBM establishes high-quality design criteria, and ETL tests the ballasts to determine that the design standards have been met. The important characteristics of magnetic ballasts are discussed next.

(a) Ballast Efficiency.
Because of the considerable energy loss in inefficient ballasts, manufacturers and standards organizations have established two figures of merit by which ballast energy efficiency can be judged. *Ballast factor* (BF) is the ratio of light output of a lamp when operated on a test ballast to its output when operated on a standard reference ballast, using an ANSI test procedure. Minimum BF required for an F40 rapid-start lamp ballast to carry ANSI/CBM certification is 92.5%. (As pointed out previously, only ballasts carrying CBM certification should be used where such certification is applicable.) The second figure of merit is the *ballast efficacy factor* (also called *ballast efficiency factor;* BEF). This rating is the ratio of ballast BF to its input watts. It is thus a measure of ballast efficiency. An amendment to the National Appliance Energy Conservation Law of 1988 mandates that after 1 January 1990, all ballasts for the three most common fluorescent lamp circuits—one- and two-lamp 40-W rapid start, two-lamp 96-in. instant start (Slimline), and two-lamp 96-in. high output (HO)—must have the minimum BEF specified in the law. Such ballasts are marked with a capital letter E in a circle. Exceptions to the law include special ballasts for these lamps (e.g., dimming), ballasts for other lamps, and ballasts used in residential buildings no higher than three stories above grade. The required minimum ballast efficacy factors are listed in Table 19.21.

(b) Ballast Protection.
The *National Electrical Code* (*NEC*) requires that all ballasts for *indoor* fixtures except simple reactance ballast (choke coils) and ballasts for exit and emergency egress lighting be protected by an integral thermal-sensing device that disconnects the ballast in the event of overheating. Overheating is caused by excessive voltage, excessive ambient temperature, or failure of a ballast component. These devices are either thermostatic (self-resetting) or fuse-type (self-destructive). Inasmuch as two of the three conditions that cause overheating are usually correctable, our recommendation is to specify the self-resetting type of protector in installations employing electrical maintenance personnel. Thermally protected ballasts are designated type "P" by the UL.

(c) Ballast Heat.
Ballast heat is transferred to the luminaire body by direct metal-to-metal contact (which must be unimpeded) and is then dissipated by radiation and convection from the fixture. Obviously, therefore, the location and method of fixture installation affect the heat trans-

ILLUMINATION

TABLE 19.21 **Minimum Required Ballast Efficacy Factors (BEF)**

Ballast for Operation of	Nominal Input Voltage	Total Nominal Lamp Watts	Minimum BEF
1—F40T12 lamp	120	40	1.805
	277	40	1.805
2—F40T12 lamps	120	80	1.060
	277	80	1.050
2—F96T12 Slimline lamps	120	150	0.570
	277	150	0.570
2—F96T12HO lamps	120	220	0.390
	277	220	0.390

fer from the fixture and, consequently, the ballast temperature. Pendant fixtures (more than 6 in. below the ceiling) and fixtures recessed into ventilated suspended ceilings do not generally present a temperature problem. Fixtures mounted on insulating surfaces, such as low-density acoustic tile, or into unventilated or heated ceiling spaces, or when boxed by a fire-rated enclosure and recessed into a fire-rated ceiling, *do* present serious heat-dissipation problems. When the material adjacent to the fixture is combustible, the installation may constitute a fire hazard (see *NEC* Article 410-76). As each installation situation represents almost a unique case because of the variables of luminaire ballast, ceiling material, and ambient temperature, the designer should require a temperature test of the specific fixture in the installation situation involved before acceptance of the unit. This can be specified along with shop drawings, photometric data, and a sample unit.

Conventional magnetic ballasts are designed to operate in the luminaire at a maximum temperature of 90°C, at 25°C ambient. Ballast life is directly affected by its operating temperature. At normal operating temperature, a ballast life of 12 to 15 years can be expected. A rule of thumb is that ballast life is halved for every 10°C above 90°C operation and, conversely, is doubled for every 10°C reduction in operating temperature below 90°C. Ballast temperature rises 0.9°C per 1°C of ambient temperature above 25°C. Where ambient conditions above 25°C (77°F) are anticipated, special cool operation ballasts should be specified.

Normal ballasts are designed to start fluorescent lamps in an ambient temperature range of 50° to 105°F. If conditions outside these limits are expected, special ballasts must be specified. For outdoor installations, ballasts suitable for starting at +20°, 0°, and −20°F are available.

(d) Ballast Power Factor (pf). This factor is determined by ballast design and is either high (above 0.9) or low (0.5–0.6). High pf ballasts are more expensive, but the additional cost is readily repaid by lower line losses, smaller circuit conductors in long runs, and larger number of fixtures per circuit. Energy conservation and economic considerations dictate the use of power-factor cor-

rected (high power factor) ballasts. Ballasts listed in Table 19.21 must have a *minimum* power factor of 90%. Such ballasts are identified by an E in a circle marked on the ballast, indicating compliance with National Energy Conservation Amendments (NECA) of 1988 to the Energy Policy and Conservation Act (EPACT).

(e) Radio Noise. Frequently referred to as *radio frequency interference* (RFI), this noise is *not* produced by the ballast, but by the arc discharge in the fluorescent tube. (Occasionally, a defective ballast does cause RFI.) To minimize RFI, ballasts are available with integral RF noise suppressors. In extreme cases, additional suppression can be obtained by installation of RF noise attenuators in the fixture.

(f) Stroboscopic Effect. This is no longer a problem with single-lamp ballasts due to the use of long-persistence phosphors in all lamps.

(g) Flicker. Flicker is noticeable at the ends of fluorescent tubes when viewed in peripheral vision. It is caused by extinguishing and reignition of the arc, and is only visible (just barely) where the long persistence phosphors are thin or entirely absent (i.e., at the extreme ends of the lamp). It is not normally disturbing as one rarely looks at a bare tube, even peripherally, for an extended period.

(h) Noise (see Section 28.30). Magnetic ballasts produce a hum or buzz when operating that is transmitted to the fixture because of the integral contact required for heat transfer. This contact amplifies the 120 Hz noise caused by the ballast's core vibration. The room noise level produced by a ballast depends on sound rating of the ballast, fixture design, and acoustical characteristics of the room.

Ballasts are sound rated by a letter that indicates not actual sound developed, but performance in a space. The ballast selected should be suitable for the lowest sound level likely to be encountered in the subject space. Generally, A-rated ballasts are selected for offices and other normally quiet areas where ballast noise would be objectionable. Where ballast noise (or heat buildup) may present a problem, ballasts can be mounted remotely if provision is made for heat dissipation and noise control.

(i) Special Ballasts. Lamps operating at other than 430 mA, including low current and high current units, require matching ballasts to supply the required current, waveform, and circuitry. Use of one maker's low-current lamp with another's low-current ballast is not suggested without prior testing or specific manufacturer's recommendation. The principal varieties of non-standard ballast are:

1. *Low-current ballasts* are intended to match specific low-current lamps, including T8 triphosphor units, slimline lamps and others.

2. *High-current units* are intended to be used with high-output lamps. The purpose of this combination is either to increase output in an existing installation or to reduce the number of fixtures in a new installation.

3. *Energy saving ballasts* are designed to reduce the total wattage of the lamp/ballast combination. Part of this power reduction is produced by more efficient design of the ballast itself. Another part is due to a lower current rating for the lower wattage of the lamp itself. A third portion is frequently a switching arrangement in the ballast that disconnects the lamp filaments after an arc is struck (the lamp ignites). This technique can save 4 to 8 W per lamp/ballast pair.

4. *Multilevel ballasts* are useful when it is desired to change lighting level evenly. The usual unit is two-level—that is, full output and 50%—but three-level units are available for full, two-thirds, and one-third output. As explained previously, switching may be a problem.

5. *Dimming ballasts* are made for standard lamps only and are normally applied one per lamp.

In addition to these types there are special units for low or high ambient temperature, weather-proof units, and low leakage-to-ground units for hospital applications.

19.26 Electronic Ballasts for Fluorescent Lamps

It has long been known that operation of fluorescent lamps at frequencies above 60 Hz has many beneficial effects. Efficacy increases dramatically, the heavy and expensive magnetic core-and-coil ballast is replaced by small, light, quiet, and inexpensive solid-state components, heat (energy) loss is reduced by as much as 90%, maintenance is reduced, and dimming is greatly simplified. However, until about the late 1980s, the drawback to all these desirable characteristics had been the problem of providing the requisite high-frequency power reliably and economically. With the development of compact, *reliable,* and inexpensive electronic ballasts, the problem has been solved. These devices are actually small solid-state rectifier-inverter ballasts that change the incoming 60 Hz ac to a frequency between 20 and 30 kHz. This local at-the-fixture conversion makes high-frequency lamp operation, with all its advantages, economically attractive; so much so that core-and-coil ballasts have been largely replaced for operating T8, T5, and CFL lamps, and may very soon become the industry standard as their price becomes competitive with the magnetic ballast.

The following is a brief comparison of the characteristics of fluorescent lamp operation with an electronic ballast at 25 to 30 kHz versus operation with an *energy-saving* core-and-coil magnetic ballast at 60 Hz. For the purpose of comparison, and where applicable, two 40-W T12 RS lamps are assumed *and light output equalized* for the two situations.

(a) Circuit Wattage, Losses, and Component Temperatures. All these factors are intimately interrelated. In a typical open two-lamp fixture, average power consumption with standard ballast is 80 W; with electronic ballast, it drops to 70 W. This 10-W reduction is primarily due to reduction in ballast heating. As is characteristic of solid-state equipment, the electronic ballast is almost lossless, and as a result runs at 50° to 60°C, as compared to 85° to 90°C for the magnetic unit. This has several additional desirable effects: the bulb runs cooler because the en-

ILLUMINATION

tire luminaire temperature is lowered, and the fixture mounting constraints that are directed at controlling fixture ambient temperature may be almost completely eliminated. See NEC Art. 410-76.

(b) Flicker. Conventional ballast circuits have a flicker, which is visible in peripheral vision and may be annoying. Flicker is effectively eliminated with all high-frequency electronic ballasts.

(c) Dimming and Level Control. This capability can be much more easily and cheaply incorporated into the electronic design without the appreciable increase in flicker that is characteristic of conventional circuits. Local control at the electronic ballast permits individual fixture level adjustment—a "tuning" aspect absent from conventional circuits. Also, advanced design electronic (EL) ballasts using integrated circuits can be photocell or otherwise remotely controlled in dimming, two-level switching modes, and on–off control modes.

(d) Noise. Solid-state ballasts are almost completely silent and readily meet A-rating requirements.

(e) Other Characteristics. Both standard and electronic ballasts are available with power factor and ballast factor greater than 90%.

Standard ballasts will usually start a lamp at 50°F minimum. A special ballast is required for starting down to 0°F. Electronic ballasts normally start down to 0°F.

Standard circuits are voltage sensitive. Electronic ballasts can easily be provided with a voltage-regulating circuit that can handles ±10% line voltage variation with no change in lamp performance, thus accommodating "brownouts" with no difficulty.

Failure of one lamp in a two-lamp conventional circuit usually has negative effects depending on circuitry. In electronic circuits, including Slimline, lamps are operated in parallel. Failure of one lamp causes the second either to be unaffected or to increase its output by 15% to 25%, depending on design.

Neither EMI nor RFI is a problem with properly designed, shielded, and grounded equipment.

Conventional circuits have 5% to 15% third-harmonic distortion, which appears as an undesirable current in the neutral of a 3-phase system. EL ballast circuits generally have higher third-harmonic currents, although stringent specifications can be met by proper design. Recently developed EL ballasts exhibit a third-harmonic distortion of less than 1%, thus not only eliminating the problem but actually improving on magnetic ballasts.

EL ballasts readily meet BF and BEF standards because of their inherently low losses. However, care must be exercised in applying EL ballasts to low-energy lamps. Integrated-circuit–controlled EL ballasts are available with integral wattage sensing, making them applicable to a range of different lamps. The normally higher price of EL ballasts can frequently be offset by the use of a single ballast for three or four lamps as opposed to one- and two-lamp magnetic (EM) ballasts. This advantage is particularly marked where remote modulating lighting control is utilized (see Sections 21.2–21.4).

EL ballasts are much lighter than their standard counterparts because of the absence of the heavy, magnetic, iron core. As a result, handling, maintenance, and storage are easier. For example, a standard EM two-lamp 40-W rapid-start ballast is approximately $10 \times 1\frac{3}{4} \times 2\frac{1}{2}$ in. and weighs $3\frac{1}{2}$ lb. Recent designs of equivalent EL ballasts for two *or three* lamps have less than half the physical bulk and one-quarter the weight. The ongoing miniaturization of electronics has produced (in Europe) an electronic ballast for a CFL lamp that fits *inside* the medium screw base. Although electronic ballasts still carry a price premium, an economic analysis almost always indicates not only a life-cycle cost advantage to the EL ballast, but also a short investment payback period due to sharply reduced energy, demand (where applicable), and maintenance costs.

HIGH-INTENSITY DISCHARGE (HID) LAMPS

Subsumed under this heading are mercury, metal–halide, and sodium-vapor lamps. These lamps have inherently high efficacy and, with appropriate color correction, can be utilized in most applications, indoor or outdoor, that do not have

Fig. 19.55 Bulb shapes for most HID lamps, with their maximum overall length (MOL).

exacting color criteria. Metal–halide lamps have excellent color and therefore almost unlimited applicabily. Bulb shapes for most HID lamps are shown in Figure 19.55

19.27   Mercury-Vapor Lamps

The mercury-vapor lamp was the first HID lamp to be developed, and for many years it was the only HID lamp commercially available. It has been supplanted in large measure by the metal–halide lamp because of the latter's better color. It is still in use today, principally in outdoor applications, and in relatively inaccessible locations because of its reliability, extremely long life, acceptable efficacy, and low price.

The lamp operates by passing an arc through a *high-pressure* mercury vapor contained in a quartz arc tube (Fig. 19.56). This produces radiation in both the UV region (as in the low-pressure fluorescent lamp tube) and in the visible region, principally in the blue–green band. This color is characteristic of the clear mercury lamp. Details of spectral distribution are given in Section 19.35.

(a) UV Radiation. A considerable portion of a mercury lamps' energy spectrum is in the UV range. This does not normally constitute a hazard to persons exposed to the light from even a clear lamp because the outer glass bulb absorbs most of the UV while transmitting the light in the visible range. However, if the outer bulb is broken, the quartz arc tube will continue to burn and the UV emitted constitutes a safety hazard, particularly to the skin and eyes of persons exposed to the radiation. As a result, all manufacturers include

a warning to this effect with all mercury lamps sold. Users are also generally informed that a safety-type mercury lamp is readily available which will self-extinguish if the outer glass envelope is broken.

An alternative is to use mercury lamps in an enclosing fixture designed to both prevent lamp breakage from external sources, such as vandalism, and protect users of the lamp in the (unlikely) event of a spontaneous lamp fracture.

In connection with protection from the injurious effects of UV radiation, it is well to note two facts:

1. The shorter the UV wavelength, the more potentially irritating to skin and eyes. Germicidal UV is in the short-wave range (200 to 300 nm).

Fig. 19.56 Typical construction of a clear mercury lamp.

2. White plaster and polished metal are good reflectors of UV. As a result, UV reflected from such surfaces is almost as dangerous as UV from line-of-sight exposure to the UV source.

(b) Lamp Life. Lamp life is extremely long, averaging 24,000 h or more based on 10 burning hours per start. Mercury lamps are not suitable for applications that are subject to constant switching; therefore, a relatively long period of burning per start was selected. Life is affected by ambient temperature, line voltage, and ballast design. (In the interest of safety, it is suggested that mercury lamps be shut off at least once a week for at least 30 minutes to allow them to cool completely.)

(c) Lumen Maintenance. This depends on the type of lamp and its burning position. Manufacturers publish data on *each* of their lamp types. In general, clear lamps have the best lumen maintenance, followed by color-improved and phosphor-coated units.

(d) Color Correction and Efficacy. Color correction is normally required because the blue–green light from a clear lamp distorts almost all object colors. (Mercury lamps are frequently used to illuminate outdoor gardens because the blue–green light enhances the green of trees and vegetation.) Color correction is achieved by adding phosphor to the inside of the outer bulb and stains on the bulb itself. The phosphors convert UV to visible light exactly as in fluorescent lamps, and the stain on the glass acts as a filter to some of the blue–green radiation. The phosphors reradiate generally in the red band, which is entirely absent in the basic lamp color. Depending on the arc tube design and the phosphors used, the color of the emitted light can be corrected to make it acceptable for general indoor use. Lamps are available in clear, white, color-corrected, and white-deluxe, in ascending order of color improvement. Efficacy, including ballast loss, ranges from 25 lm/W for a 50-W lamp to a maximum of 55 lm/W for a 1000-W color-corrected lamp. Note that, in general, efficacy is lower than for fluorescent lamps. CRI ranges from a low of 20 for a clear lamp to a high of (only) 50 for a deluxe white lamp. A short listing of representative lamp data is given in Table 19.22.

TABLE 19.22 **Typical Data for Mercury-Vapor Lamps**

Lamps		Bulb[b]	Initial Lumens	Efficacy[c] (lpw)
Watts	Type[a]			
100	DX	A-23	4,300	39
100	DX	ED-23 1/2	4,400	40
100	DX	R-40	2,800	25
250	DX	ED-28	13,000	47
250	DX/SB	E-28	6,000	24
400	DX	R-57	23,000	54
450	W/SB	BT-37	9,700	28
700	DX	BT-46	43,000	55
750	W/SB	R-57	14,000	19

NOTE. All lamps using an external ballast have a life of 24,000+ h, based on 10 h burning per start. Life of self-ballasted lamps: 250W—12,000 h; 400W and 700W—16,000 h.
All deluxe lamps with external ballast have a CRI of 45 and a CCT of 3700°K. All self-ballasted lamps have a CRI of 50 and a CCT of 3300°K.
[a] Type abbreviations: DX = deluxe; w = white; SB = self-ballasted.
[b] For bulb shape and dimensions, see Fig. 19.56.
[c] Efficacy includes an estimated loss in a magnetic ballast. For self-ballasted lamps, efficacy is as shown.

ILLUMINATION

(e) Ballasts and Lamp Starting. Ballasts are required, as with all arc discharge lamps, to start the lamp and thereafter to control the arc. The basic conventional (magnetic) ballast is simply a reactor that controls the arc after the discharge has been initiated. Three to six minutes are required for the lamp to reach full output because heat must be generated by electron flow to vaporize the mercury in the arc tube before the arc will strike. Once extinguished, the lamp must cool before restrike is possible. This restart delay amounts to 3 to 8 minutes, depending on the ballast type, and is an important consideration in design, as a momentary outage will extinguish all lamps, leaving an interior area in the dark. Mercury-lamp luminaires are available that utilize small halogen lamps to supply light during such outages. Alternately, some incandescent lighting can be utilized that will maintain minimum illumination.

Principal ballast types are reactor, regulating, and electronic. Because lamp operating characteristics depend heavily on the type of ballast, the choice of which involves highly technical electrical considerations, we recommend that it be left to a electrical consultant. Magnetic mercury ballasts are large, heavy, and quite noisy. Where this may be a problem, remote mounting should be considered, or lighter, quieter, and more expensive electronic ballasts used.

(f) Dimming. Dimming of mercury lamps is practicable with the use of a dimming ballast and solid-state dimming control. These are available for 175-W through 1000-W units. Cost analyses performed by the author indicate the economic feasibility of such control, including sensing equipment that automatically maintains illumination levels at preset levels. Where dimming is not required, two-level high–low operation is obtainable with a number of commercially available control systems.

(g) Self-Ballasted Lamps. These have been available for some years and consist of a screw-base color-corrected lamp with an internal resistive/reactive ballast. They have a CRI of 50, an efficacy of 20 to 25 lpw, and a correlated color temperature (CCT; see Section 19.34) of 3500° to 4000°K. Their great advantage is long life, which can be used to advantage in applications involving burning periods of 8 to 10 minimum, relative inaccessibility, limited space that precludes ballast installation, and indifferent color requirements.

(h) Application. Mercury-vapor lamps are applicable to indoor and outdoor use with proper attention to color and fixture luminance. The most common exterior application is for parking lots. Indoor application is generally limited to mounting 10-ft AFF or higher to avoid direct glare problems and permit adequate area coverage. Use in industrial spaces and stores was common, but today retrofitting with metal–halide lamps is common. Warehouses and non–color-sensitive industrial areas continue to use mercury lamps.

19.28 Metal–Halide Lamps

This lamp began its life in the early 1960s as a modified mercury-vapor lamp. Addition of halides such as thallium, indium, and sodium to the arc tube results in changes in the output, efficacy, color, and life of the lamp. Because of the high efficacy and good color of this lamp, it has undergone intensive development; probably more than any other lamp type. As a result, the metal–halide lamp today is available in an extremely varied range of designs and sizes. The number of types and sizes is so large and changes so rapidly that any abbreviated tabulation presented here would be inadequate at best, and misleading at worst. As with all lamps, but more so with lamps under intensive development, a current manufacturer's catalog should be consulted for accurate lamp data. Typical metal–halide lamp characteristics and types are discussed in the following subsections.

(a) Lamp Configurations. Construction details of the basic metal–halide lamp are shown in Fig. 19.57. Its similarity to its "parent," mercury vapor lamp, illustrated in Fig. 19.56, is apparent. In addition to this design in a BT-shaped bulb envelope (Fig. 19.57), metal–halide (MH) lamps are manufactured in elliptical bulbs, PAR reflector lamps, and single-and double-ended tubular shapes (Figs. 19.55, 19.58 and 19.59).

(b) Safety. Being essentially a modified mercury lamp, MH lamps carry the same safety warning as mercury lamps (Section 19.27a). An additional warning, however, refers to the fact that MH lamp arc tubes have a tendency to rupture violently (explode), and therefore the lamp must be used in an approved enclosing luminaire. Because this tendency to rupture is strengthened by continuous burning, users are requested to turn off MH lamps once daily for a period of time sufficient to allow the lamp to completely cool and stabilize. If daily shut-off is impractical, a once weekly shut-off is required as an absolute minimum, and a circuit timer can be used to perform this safety action automatically. All major manufacturers also make lamps with internal shields that will contain the flying pieces of ruptured arc-tube without damage to the outer bulb. Such lamps may be used in open lighting fixtures. One such lamp is illustrated in Fig. 19.59.

As with mercury-vapor lamps, manufacturers also produce a line of safety MH lamps that self-extinguish within 15 minutes of an outer bulb fracture, thus limiting exposure to harmful UV radiation to 15 minutes. Both of these safety designs are clearly noted on the lamps by trade name and sometimes by description.

(c) Designs, Shapes, and Ratings. The lamp designs available as of this writing are:

- Standard lamps, available in ED, BT, and PAR shapes, in wattage ratings from 50 to 1500 W, efficacies of 75 to 105 lumens per watt with magnetic ballasts, and slightly higher with electronic ballast (efficacy increases with wattage). Lamps are clear or phosphor coated, with CCT from 3000°K to 4200°K, and CRI from 65 to 85. Life, at 10 burning hours per start, varies from 10,000 to 20,000 h depending on lamp type, size, and burning position.

- Safety-shielded lamps with integral shields to contain ruptures arc tube. These lamps are made in essentially the same sizes and shapes as standard lamps with approximately the same characteristics.

- Self-extinguishing lamps designed to shut down automatically within 15 minutes of a break in the outer glass envelope. The lamps

are generally made only in high wattages and are specifically recommended for installation in spaces prone to lamp breakage, such as sport areas. They are similar in operating characteristics to standard lamps.

- High-output lamps are designed specifically for a specific burning position that is specified on each lamp. Output is 5% to 8% higher than standard lamps, but color is somewhat poorer, with a CRI range of 65 to 70. Shapes are essentially the same as standard lamps.

- Single-ended and double-ended tubular lamps. These lamps are characterized by very high CRI (80–93); somewhat shorter life than standard lamps, particularly for the single-ended units (6,000–10,000 h); and slightly lower efficacy (70–85 lpw). They are intended for applications requiring very high color rendering accuracy. These lamps are generally used with electronic ballasts.

(d) Operating Characteristics. As with mercury lamps, MH lamps are not instant starting, requiring approximately 2 to 3 minutes on initial startup and 8 to 10 minutes for restrike. (Tubular lamps require only about half these times.) As a result, when used for indoor installations, a secondary instant-start source must be available. A number of manufacturers produce special hot-restrike ballasts that provide immediate restrike of lamps on restoration of power after an outage. Lamp output on restrike is inversely proportional to the duration of the power outage.

It is important to note here that the spectrum of light produced by MH lamps *changes* as the lamp ages. The change is gradual, definite, and usually unnoticed, and depends on the particular design of lamp. Where color rendering is important, or where mixed with other light sources, the user should choose MH lamps that are especially made for color stability. These lamps are designed not to vary in CCT more than 200° K over the lamp life. Finally, dimming or reduced output operation of MH lamps is not normally recommended because of the very noticeable color shift that occurs when the lamp is dimmed.

One characteristic of MH lamps that is peculiar to them is their great sensitivity to burning posi-

Fig. 19.57 *Construction details of a 400-W standard metal–halide lamp, which can be mounted either horizontally or vertically. (Courtesy of OSRAM-SYLVANIA.)*

tion—so much so that lamps are designed for base-up, base-down, horizontal, and universal burning positions. If the burning position of a lamp is altered, its lumen output, lumen mainte-

Fig. 19.58 *Various configurations of metal–halide lamps. Clockwise from the bottom left: phosphor-coated and clear elliptical bulbs; PAR 30 and 38 reflector lamps; single-ended and double-ended tubular lamps. All of the illustrated lamps have ceramic arc tubes and exhibit a CRI in excess of 80. (Courtesy of GE Lighting.)*

Fig. 19.59 *Construction details of a PAR enclosure for a metal–halide lamp. The lamp itself is tubular (see Fig. 19.58) and is constructed with a protective shield around the arc tube, designed to contain arc tube fragments in the event of a violent rupture [see text, Section 19.28(b)]. (Courtesy of OSRAM-SYLVANIA.)*

nance, and life are all reduced by factors up to 20% at best, and for some lamps much more severely. For this reason, all MH lamps are marked with their proper burning position.

(e) Lamp Ballasts. MH lamps operate satisfactorily on a simple reactor ballast, although a separate ignitor is usually required to start the lamp. These ballasts have a low power factor of about 50%, which is undesirable from aspects of energy conservation, wiring economy, electrical losses, and component heating. High power factor magnetic ballasts (pf > 90%) are readily available. Magnetic ballasts are large, heavy, and tend to be noisy. The latter characteristic can be improved by the use of a potted (epoxy-filled) ballast. Electronic ballasts are also readily available. Their drawback has been price and harmonic distortion. Both of these factors are constantly being improved and should not constitute a barrier to their use. Dimming and multilevel ballasts are also readily available. However, as noted previously, they are not frequently used because of the large shift that they cause in lamp color.

(f) Application. MH lamps are being applied increasingly in commercial, institutional, and mercantile interiors because of the excellent efficacy, color, and life. Traditional exterior applications have been expanded, and now include all

ILLUMINATION

types of floodlighting, sports lighting, building exterior illumination, and area lighting, particularly in high pole-mounted fixture groups.

19.29   High-Pressure Sodium (HPS) Lamps

The highest efficacy general-purpose HID source available is the high-pressure sodium lamp (HPS), marketed by its developer, General Electric, under the trade name *Lucalox®*, by Philips as Ceramalux™ and by OSRAM-Sylvania under the name *Lumalux®*. Basic construction of the lamp is shown in Figs. 19.60 and 19.61, which show schematic drawings and photographs of the same designs. Typical performance data of various types of HPS lamps are given in Table 19.23. Construction is quite different from that of mercury and MH lamps and, although it operates as an arc discharge unit, its excellent characteristics stem from the spectral absorption phenomenon of the sodium contained under high pressure in the arc tube. The resultant light has a pronounced yellow tint.

(a)   Primary Characteristics. Following are typical characteristics of standard HPS lamps:

Wattages	50 to 1000
Lamp efficacy	65 to 135 lm/W
Efficacy, including ballast losses	55 to 115 lm/W
Life	24,000+ h (10 h/start)
Lumen maintenance	80% to 90% at 50% life
Warm-up time	3 to 4 min
Restrike time	½ to 1½ min
CCT	1900° to 2100°K
CRI	20 to 22

The extremely low CRI (20–22) is not acceptable where any degree of color rendition is required, thus limiting the standard HPS lamp to exterior areas and road lighting. Color can be improved

Fig. 19.60 *Simplified drawings of the internal construction of two high-pressure sodium (SON) lamp designs.*
(a) *Noncycling lamp (in an E-shape glass bulb) designed to indicate by a special color when the lamp has reached the replacement stage. Unlike standard HPS lamps, this lamp will not cycle on and off at the end of its useful life.*
(b) *Retrofit HPS (SON) lamp in a BT bulb, intended for direct replacement of an existing mercury lamp. This lamp operates efficiently on a mercury lamp ballast. (Drawings courtesy of OSRAM-SYLVANIA.)*

TABLE 19.23 Typical Data for Clear^a High-Pressure Sodium (SON) Lamps

Watts	Bulb^b	Life (h)	Initial Lumens^c	Lamp Efficacy (lpw)^d
"White" Lamps: CRI—85, CCT—2700° K				
35	T10	10,000	1,250	36
50	ED-17	10,000	2,000	40
100	ED-17	10,000	4,200	42
Color-Corrected Lamps: CRI—60, CCT—2200° K				
100	ED-17	15,000	7,300	73
250	ED-18	15,000	22,000	88
400	ED-18	15,000	37,000	93
Standard HPS Lamps: CRI—22, CCT—2100° K				
50	ED-17	24,000+	4,000	80
100	ED-17	24,000+	9,500	95
250	ET-18	24,000+	28,000	112
400	ET-18	24,000+	48,000	120
750	BT-37	24,000+	110,000	147
1,000	E-25	24,000+	133,000	133

NOTE: Data extracted from current manufacturers' catalogs.

^{a} Data is identical for coated lamps except for bulb shapes and lumen output.

^{b} Other bulb shapes are available in some sizes. See Fig. 19.55 for bulb data.

^{c} Initial lumens for coated lamps are 6% to 9% lower.

^{d} Based on initial lumens. Efficacy with ballast is approximately 10% lower.

considerably by increasing the pressure inside the arc tube. This causes some of the sodium in the arc to be reabsorbed, and the radiated light widens its spectrum in the red range at the expense of efficacy and lamp life. Color improved lamps have the following ratings:

Wattages	70 to 400
Lamp efficacy	60 to 92 lpw
Life	15,000 h
CRI	60
CCT	2200° to 2300°K

A still further increase in lamp pressure improves the lamp color again and yields a "white" HPS in a limited wattage range. General characteristics are as follows:

Wattages	35 to 100
Lamp efficacy	35 to 52 lpw
Life	10,000 h
CRI	85
CCT	2700°K

Non-cycling lamp

Retrofit Lamp

Fig. 19.61 *The two types of HPS lamps illustrated in the drawing of Fig. 19.61. (Courtesy of OSRAM-SYLVANIA.)*

ILLUMINATION

These low wattage, reduced-efficacy, and shortened-life lamps are normally operated with small, lightweight electronic ballasts. Their principal application is in areas where color rendering is of prime importance, such as in merchandising displays.

Because of its high output and narrow linear arc tube, any HPS lamp is a potential glare source, and wattages 150 W and larger must be either completely shielded or mounted at sufficient height (in an open reflector) to be above the near field of vision. One way to reduce the glare problem is to mount the lamp vertically, thus presenting a smaller apprehended (subtended) angle to the eye. Unfortunately, it was found that using a vertically mounted clear lamp in an open (overhead) reflector designed for a large luminous source such as a phosphor-coated mercury lamp produced an extremely narrow distribution. Reflectors designed for horizontally mounted HPS lamps produced excessive glare (Fig. 19.62a). This obviously made an economical area lighting layout almost impossible.

By adding a diffusing coating to the lamp, the entire glass envelope becomes the light-emitting source. This reduces lamp luminance and therefore glare problems drastically, but also reduces output (and efficacy) by 6% to 8%. Light distribution from a coated lamp in an open reflector is vastly improved, as can be seen from Fig. 19.62(b).

(b) Other Operating Characteristics. In contrast to both mercury-vapor and MH lamps, HPS lamps do not emit any appreciable UV radiation and therefore do not present a health hazard to persons nearby if the glass bulb breaks. Also, unlike MH lamps, HPS lamps have no tendency to rupture violently, and therefore can be used safely in open reflectors (see Fig. 19.62). Again, unlike MH lamps, HPS lamps can be operated in any position without affecting any of its burning characteristics.

As with all discharge lamps, the HPS lamp requires a ballast for ignition and arc control. Due to the extremely high voltage required for lamp ignition, both magnetic and electronic ballasts contain an electronic ignition circuit. Because of this, an HPS lamp *must* be used with a compatible ballast that carries the same ANSI designation as found on the outer glass bulb. If used with an incompatible ballast (and fixture), the lamp may

(a) Clear HPS lamp operating at 12,000 lumens in open-bottomed fixture.

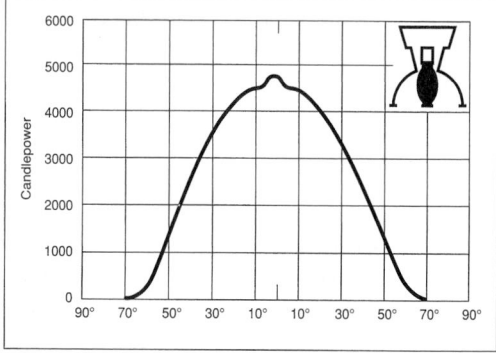

(b) Coated HPS lamp operating at 12,000 lumens in open-bottomed fixture.

Fig. 19.62 (a) *The narrow linear arc-tube of an HPS lamp is not suitable for use in an open reflector designed for a larger source such as a phosphor-coated mercury lamp.*
(b) *Use of a coated HPS lamp creates a large diffuse source with a highly improved light distribution. (Drawings courtesy of OSRAM-SYLVANIA.)*

rupture and constitute a serious safety hazard. As with halide lamps, ballasts are available for HPS lamps that provide instantaneous restrike after a power interruption. Light output on restrike is inversely proportional to the length of the outage; for example, after a 10-second outage, restrike light output level will be 85% of full capacity, whereas after a 2-min outage, lamp output will be only 10%. It then takes another minute ± to regain full output.

(d) Lamp Design Types. In addition to the standard HPS lamps, including the color-corrected types, there are three additional special designs available at this writing that were developed to solve a specific problem. They are:

1. *Noncycling lamp.* As an HPS lamp ages, its arc voltage rises. Eventually, the ballast is unable to sustain the arc and the lamp extinguishes. After cooling, the lamp lights to full brightness and soon thereafter extinguishes again. This on–off cycling is characteristic of an HPS lamp at the end of its life. To eliminate it, a special noncycling lamp was developed that uses very little sodium amalgam in the arc tube. The lamp therefore becomes, automatically, more environment friendly because of the much-reduced mercury content. To enhance this latter aspect, these lamps are made with a lead-free base and lead-free solder. Photometric characteristics are similar to those of standard lamps. The E-shape lamp in Figs. 19.60(*a*) and 19.61 is of this design.

2. *Standby lamps.* As noted previously, HPS lamps require a minute or more to restrike after being extinguished. A crowded public area plunged into complete darkness is a recipe for disaster, and therefore such areas must always be furnished with instant-on emergency lighting. Standby lamps have two arc tubes. When the lighted one is extinguished due to a momentary power loss, the second (cool) arc tube immediately begins to glow—assuming voltage has returned. This arrangement *may* be acceptable to some code authorities as a substitute for instant-on replacement lighting following a momentary power outage. It is *not* acceptable as emergency lighting as required by NFPA 101-Life Safety Code.

3. *Retrofit lamps.* These are HPS lamps designed as a direct replacement for mercury lamps of the same wattage. They are enclosed in BT-shape envelopes of the same size as the mercury lamp being replaced, and they operate properly on mercury lamp ballasts. A retrofit lamp is illustrated in Figs. 19.60(*b*) and 19.61. High-pressure sodium lamps are also referred to as SON lamps in addition to the HPS abbreviation.

19.30 Low-Pressure Sodium (SOX) Lamps

This lamp, also referred to as SOX, produces light characteristic of sodium's monochromatic saturated yellow color, making it inapplicable for general lighting. Because of its very high efficacy of more than 150 lm/W *including* ballast loss, it can be applied wherever color is not an important criterion. Thus, SOX is used for street, road, and area lighting, as well as for emergency or after-hours indoor lighting. Another desirable aspect of SOX lamps is their 100% lumen maintenance. This, coupled with the discharge lamp's typically long life (18,000+ h), makes SOX lamps the most economical source available today in terms of cost per million lumens produced. Research, however, has raised a question about visual acuity in colored light, and seems to indicate that visual acuity and effectiveness are not necessarily linearly related to lumens (illuminance), and that other factors, including chromaticity, are involved. At this writing, the results of these research areas are not yet conclusive.

19.31 Induction Lamps

During the last few years, several major American lamp manufacturers have introduced what is essentially a fluorescent lamp, except that it does not use electrodes. Two such designs are shown in Figs. 19.63 and 19.64. The lamp is filled with low-pressure mercury vapor. When ionized by a high-frequency induction coil inside the lamp, the mercury vapor produces UV light. This, in turn, strikes the phosphor coating on the inside of the lamp, producing visible light. This is exactly the light-producing process used by standard fluorescent lamps, the difference being that in this lamp, the gas is ionized by an induction coil. As a result, the lamp is known as an *induction lamp*.

Characteristics of the design shown in Fig. 19.63 are as follows:

Wattage	85 W total, including external devices
Initial lumens	6000±
System efficacy	70 lpw
CCT	3000° or 4000°K
CRI	80+
Ignition time	Under ½ second
Time to 75% output	Up to 1 min
Hot restrike time after outage	Less than ½ second
Life (50% survival)	100,000 h

Lumen maintenance	70% at 60,000 h
Burning position	Any

The extremely long life and excellent lumen maintenance of this lamp plus the essentially instantaneous restrike time make it suitable for illuminating public areas and architectural lighting. As with

(a)

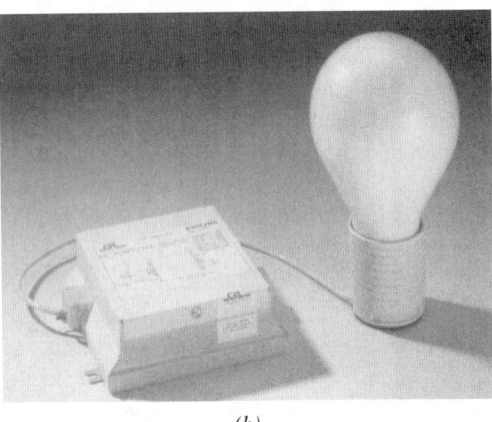

(b)

Fig. 19.63 (a) *A schematic cutaway of the induction lamp shows its operating principles. The high-frequency generator (C) produces a high-frequency current, which circulates in the coil on the power coupler (B). This ionizes the mercury vapor inside the lamp (A), producing UV light. The UV strikes the fluorescent coating inside the lamp, producing visible light.*
(b) *The lamp and its high frequency generator. The lamp, which is rated 85 W including all losses, is 4.3 in. in diameter and 7.5 in. high overall. The generator measures 5.5 in. × 5.1 in. × 1.77 in. high. The lamp and generator together weigh 2.2 lb. (Illustration and photo courtesy of Philips Lighting Company.)*

all lighting systems, a comparison with other lighting sources can only be made for each proposed usage, based on specific individual considerations.

A self-contained design of lower wattage is illustrated in Fig. 19.64. This lamp is built into a modified R-shaped envelope with a standard Edison base, and is therefore readily applied as a direct replacement for a 100-W incandescent reflector lamp. The lamps' published performance figures are:

Total wattage	23
Initial lumens	1100
Efficacy	48 lpw
CCT	3000°K
CRI	82
Life	10,000 h
Lumen maintenance at 70% life	75%

One of the special problems involved in fixture design for this lamp is concerned with shielding the lamp's electromagnetic radiation and minimizing harmonic conduction. Both of these items are highly technical and beyond our scope here. It is sufficient to say that careful fixture design can minimize these effects to the point that they meet all required standards, making the induction lamp a commercially acceptable and highly desirable addition to the lamp market.

19.32 Sulphur Lamps

The principal of microwave energy excitation has been applied successfully in another recently developed lamp that consists of a golf-ball–size globe filled with an inert gas and a few milligrams of sulphur. In this lamp, in contrast to the induction lamp described previously, no mercury vapor is used, and the radiation is full-spectrum light with very little UV or infrared. The originally developed lamp was a 6-kW unit that emitted more than 400,000 lumens. The most recent (1998) version of the lamp has the following characteristics:

Total power input	1320 W
Initial lumens	130,000

Mercury vapor fill   Induction coil   Electronic ballast

Phosphor coating   Plastic housing   Medium screw base

Fig. 19.64 *Cutaway of the GE induction lamp showing the essential elements: induction coil, phosphor-coated bulb with mercury-vapor fill, and electronic ballast. This 23-W lamp in a modified R (reflector)-shape bulb has a height just under 5 in. overall and a 3 in. maximum bulb diameter. (Photo courtesy of GE Lighting.)*

Efficacy	101 lpw
Diameter of globe	29 mm
CCT	5600°K
CRI	80
Life of lamp (estimated)	60,000 h
Life of exciting magnetron	15,000 h
Lumen maintenance	Very high
Color constancy with aging	Excellent

From these data we see that the lamp has approximately the same output as a 1000-W HPS lamp but much better color characteristics. It also compares to about 1200 W of MH lamp(s) with similar efficacy, but again, because of its full-spectrum radiation, it has superior color characteristics. Initial trials of the lamp as the driver of a "light pipe" have been highly successful. Because the lamp is in intensive development at this time, its eventual commercial usage is difficult to predict. It is certainly suitable at this stage for large area indoor and outdoor lighting, and because of its small size, it can be used in focusing reflectors. Remote source lighting for hazardous areas and clean rooms seems another viable application.

19.33 Special Purpose Lamps

Among the many special purpose lamps of interest to lighting designers are those that are applied as high-output, high-intensity sources to light conductors in remote-source installations (see Section 21.40). Of increasing interest are sources to "feed" light pipes of all shapes prismatic channels, mirrored channels, and fiber-optic bundles. Sources include miniature arc lamps, short arc MH, and compact Xenon arc lamps. All of these sources produce white light of good color, high CRI, and good efficacy. Space limitations prevent our including data here on these sources that, in any case, would be academic as lighting designers are interested in the performance of entire systems, of which these lamps are only a part. Interested readers are referred to manufacturers' data sheets for lamp characteristics.

COLOR OF LIGHT

19.34 Chromaticity

The Commission Internationale de l'Eclairage (CIE) color system is the internationally accepted

ILLUMINATION

standard for designating illuminant color. In this system, the relative proportions of each of the three primary colors (red, green, and blue) required to produce a given illuminant color are calculated. These values are called the *tristimulus values* for that color, and are designated by capital letters: X (red), Y (green), and Z (blue).

19.35 Spectral Distribution of Light Sources

In addition to providing sufficient light of adequate quality, the lighting designer must also be concerned with the spectral content of the selected illuminant, because perceived object color depends heavily on the illuminant. As shown earlier in Chapter 18, perceived object color is the result of selective absorption and reflection of components of the illuminating light by the pigments of the object being viewed. It is therefore necessary that the *illuminant* contain the color of the *object* in order for us to see the object's color. It is not so obvious that the relative energy of an illuminant at a particular wavelength determines the saturation and brilliance with which we see a color. To understand this, refer to Fig. 19.65. In graphic form, the relative spectral energy of a few common light sources have been plotted as a function of wavelength—that is, color. If we compare the graphs in Figs. 19.65(*a*) and (*c*), which show the spectral content of two of the

Fig. 19.65 *Spectral energy distribution of specific fluorescent lamp types with their correlated color temperatures (CCT) (see Section 18.34) and color rendering indices (CRI). In actual fact, the curves are not continuous, but consist of individual color lines. They are shown connected for simplicity. Because only a radiating blackbody has a true color temperature, a source with mixed color illuminants is assigned a* correlated *color temperature, which is the temperature of a blackbody radiator whose chromaticity most nearly matches that of the light source.*

High Intensity Discharge (HID) Lamps

Fig. 19.66 *Spectral energy distribution of typical high-intensity discharge (HID) lamps, with their CCT and CRI.*

most common light sources, cool white and warm white fluorescent lamps, respectively, we note that the principal difference lies in the amount of blue in their spectrum. As a result, a blue object will be bright under cool white light and dull (grayed) under warm white light. The situation is more pronounced with the standard HPS [Fig. 19.66(c)] as compared to the clear MH lamp [Fig. 19.66(b)]. A blue object will be gray under the sodium lamp and unrecognizable as blue, whereas under the MH lamp, its blue color will show clearly, and so on.

This concern for perceived object color, which relates not only to furnishings but also to paints and prefinished construction materials such as carpets or floor tiles, is quite properly the province of the architect and lighting designer, who in turn must possess the necessary knowledge and infor-

mation to make the appropriate choices of both illuminant *and* object color. Spectral composition graphs of the types shown in Figs. 19.65 through 19.67 are available from manufacturers for all light sources, and they should be examined when considering the characteristics of a particular light source.

One of the best ways to compare illuminants is first to expose a dull white surface to the illuminants, side by side but separated by an opaque divider, to get an impression of the illuminant color, and then expose a series of colored chips—again, side by side—to see which colors are brightened and which are grayed. The intensity of illumination also influences the appearance of colors, and it must be considered in choosing object colors. As intensity is increased, reflection increases, particularly with pale tints (high value) that con-

ILLUMINATION

Incandesent Lamps

Fig. 19.67 (a) *Standard photopic (cones) eye sensitivity curve. Note that maximum sensitivity occurs in the daylight range of 500–750 nm.*

(b) *Spectral energy distribution of 2 specific types of daylight. North light, with a color temperature of 8,000–10,000° K peaks in the blue range, whereas noon daylight contains all spectral colors in roughly equal proportions.*

(c) *Tungsten–halogen lamps are incandescent light sources, and therefore contain all spectral colors. As wattages increase, the color changes from orange–red to white and the CRI drops slightly.*

(d) *A simple filament-type incandescent lamp is very close to being a blackbody radiator (i.e., its actual temperature and CCT are almost the same). This is indicated by its high CRI (97).*

tain much white pigment and thus tend to wash out color. Therefore, with high-intensity lighting, saturation of colors should be high for true, brilliant color rendition.

Refer now to Fig. 19.67. Note from diagrams *b, c,* and *d* that the spectrum of a light source that produces light as a result of heating is continuous. Sunlight is equal in spectrum to a blackbody radiator at 5,500 K; north light equal to one at about 8,000 to 10,000 K; a 500-W incandescent lamp approximately equal to one at 2,850 K; and so on.

If the spectrum of a blackbody radiator is plotted on a chromaticity diagram, its locus is a continuous curved line, as seen in Fig. 19.68. The chromaticity of all true blackbody radiators falls exactly on this line, with the location depending on temperature. Daylight, for most purposes, falls on this locus, although because of selective atmospheric absorption and other phenomena, it is actually slightly off. Incandescent lamp chromaticity is very close to this locus because it is also a heat–light radiator.

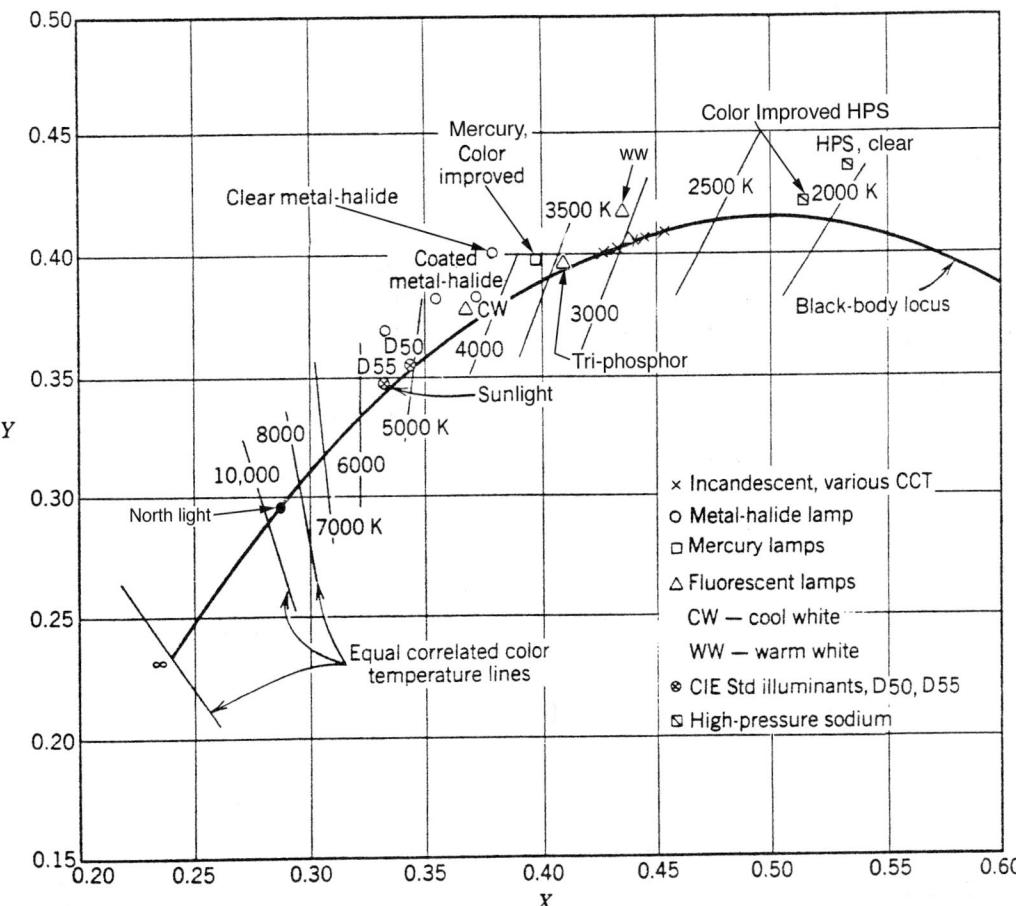

Fig. 19.68 *Portion of a chromaticity diagram showing relation of common illuminant chromaticities to that of the blackbody locus. Illuminants whose coordinates fall on the same line crossing the blackbody locus have the same CCT (see Section 18.34), but may have entirely different component colors.*

A source that produces light by means of individual phosphors can also have chromaticity on this locus if the phosphors are selected to produce a continuous spectrum similar to that of a blackbody radiator. Thus, we see on Fig. 19.68 that trophosphor fluorescent lamps [Fig. 19.65(*d*)] and MH lamps [Fig. 19.66(*d*)] have spectral components over the entire spectrum, yielding chromaticities fairly close to the blackbody locus. For other sources, the CCT (correlated color temperature) is established by their chromaticity locus in relation to the diagonal lines crossing the blackbody locus, as seen on Fig. 19.68. Each of these lines is isothermal—that is, all chromaticities on it have the *same CCT*. Thus, the reader can see readily that two sources with widely differing spectral content and therefore object color rendering can have the same CCT.

19.36 Color Rendering (Index) CRI

Color rendering is defined as the degree to which perceived colors of objects illuminated by a test source conform to the colors of the same objects as illuminated by a reference source. The *color rendering index* (CRI) of a source is a two-part concept, comprising a color temperature that establishes the reference standard and a number that indicates how closely the illuminant approaches the standard. *The standard is always daylight at that color temperature.* Therefore, the CRI

ILLUMINATION

TABLE 19.24 **Effect of Illuminant on Object Colors**

Lamp	CRI (approximate)	CCT (°K)	Whiteness	Colors Enhanced	Colors Grayed	Notes
FLUORESCENT						
Warm white	52	3050	Yellowish	Orange, yellow	Red, blue, green	
Cool white	62	4200	White	Yellow Orange Blue	Red	
Cool white deluxe	77	4050	White	Green Orange Yellow	Red	
Triphosphor	75	2800	Yellowish	Red Orange	Deep red, blue	
	80	3000	Pale yellowish	Red Orange Green	Deep red	
MERCURY						
Clear	20	7000	Blue-Green	Blue Green	Red Orange	Poor overall color rendering
Deluxe	45	3700	Pale-purplish	Deep-blue, red	Blue-green	Shift over life to greenish
METAL–HALIDE						
Clear	65	4000	White	Blue Green Yellow	Red	May shift to pinkish over life
Phosphor coated	80	4200	White	Blue Green Yellow	None	Shifts to pinkish over life
HPS						
Standard	21	2100	Yellowish	Yellow Green	Red, blue	
Color-corrected	60	2200	Yellowish-White	Red Green Yellow	Blue	CRI decreases slightly over life
Incandescent	99+	2900	Yellowish	Red, orange, yellow	Blue, green	

of a light is really a measure of how closely it approximates daylight of the same color temperature. Two lights cannot be compared unless their color temperatures are equal or quite close. A CRI of 100 indicates an illuminant whose spectral content is equal to daylight of that temperature. CRIs for typical common lamps are given in Figs. 19.65 through 19.67.

Fig. 19.69 *A hand-held lightweight ($\simeq$ 8oz) chroma meter for determining chromaticity and color temperature of a light source. The unit reads X, Y coordinates, color temperature in degrees Kelvin, and illuminance in lux. (Courtesy of Minolta Corp).*

Table 19.24 lists the color characteristics of a few of the major sources. An illuminant's own color appearance on a neutral surface depends on its own spectral content, but if the observer is placed in a space illuminated with this source, after a short exposure time the eye becomes adapted to the source color and detects only a degree of whiteness rather than an actual tint.

Where it is necessary to detect small color differences between two objects, a light poor in object color or complementary to the object color should be used at a relatively high illumination level, followed by a light high in object color at the same illumination level. If this is not possible, two widely different but broad-spectrum illuminants should be used, preferably at the same illumination level. Another technique is the use of a special, fixed-color source. For a full discussion, see the *IES Handbook.*

It should be remembered in all considerations of color, comparison, matching, and rendering that object color depends on the spectral energy distribution of the light source (illuminant), and therefore any change in the spectral content changes the object's appearance. Two sources of the same color temperature and, therefore, apparent whiteness can have quite different spectral content and therefore render object colors differently. A case in point would be a 3000°K warm white fluorescent tube and an incandescent lamp (500-W photoflood) of approximately the same color temperature. Color temperature is an expression of dominant color, not spectral distribution.

A convenient hand-held chromaticity meter is illustrated in Fig. 19.69. This unit measures the *X, Y* coordinates of an illuminant, which can then be plotted on a standard CIE color diagram to determine absolute chromaticity. This is very useful in comparing illuminants to predict color response and avoid color metamerisms. The meter also reads color temperature in kelvin and illuminance in lux.

REFERENCE

See the complete lighting bibliography at the end of Chapter 21.

ILLUMINATION

20

LIGHTING DESIGN

20.1 General

Lighting design is a combination of applied art and applied science. There can be many solutions to the same lighting problem, all of which will satisfy the minimum requirements, yet some will be dull and pedestrian whereas others will display ingenuity and resourcefulness. The competent lighting designer approaches each problem afresh, bringing to it a knowledge of current technology and years of background and experience, yet rarely being satisfied with a carbon copy of a previous design. It is these years of background with their successful and not-so-successful designs coupled with a constant striving for improvements that are the characteristics differentiating the competent lighting consultant, designer, or engineer from the person who attempts to force each new job into the unwilling mold of a previous design.

Because of the large number of interrelated factors in lighting, no single design is the correct one, and for this very reason it is not entirely desirable to solve a lighting problem with a step-by-step technique. However, since experience has shown this technique to be good avenue of approach for the uninitiate who lack the experience necessary to view an entire solution, we have adopted it.

20.2 Goals of a Lighting Design

Simply stated, the goal of lighting is to create an efficient and pleasing interior. These two requirements—that is, the utilitarian and esthetic—are not antithetical, as is demonstrated by every good lighting design. Light can and should be used as a primary architectural material.

1. Lighting levels should be adequate for efficient seeing of the particular task involved. Variations within acceptable luminance ratios in a given field of view are desirable to avoid monotony and to create perspective effects.

2. Lighting equipment should be unobtrusive, but not necessarily invisible. Fixtures (luminaires) can be chosen and arranged in various ways to complement the architecture or to create dominant or minor architectural features or patterns. Fixtures may also be decorative and thus enhance the interior design.

3. Lighting must have the proper quality, as discussed previously. Accent lighting, directional lighting, and other highlighting techniques increase the utilitarian as well as architectural quality of a space.

4. The entire lighting design must be accomplished efficiently in terms of capital and energy resources, the former determined principally by life-cycle costs and the latter by operating-energy costs and resource-energy usage. Both the capital and energy limitations are, to a large extent, outside the control of the designer, who works within constraints in these areas. Obviously, these constraints are maxima.

With these goals before us we can write a lighting design procedure, keeping in mind that the order of steps shown is not necessarily the same in each lighting problem and that all factors being closely interrelated, it is often necessary to apprehend several of the stages simultaneously before arriving at a decision.

It is appropriate to note at this point that the lighting design approach and procedure that we explain in this chapter is primarily an analytic one; that is, the design procedure establishes requirements primarily in numerical form and then manipulates the variables of sources, fixtures, placement of units, and so on, to arrive at a design conclusion. There exists an alternative approach, frequently referred to as *brightness design,* in which the designer labels surfaces on a perspective plan of a space with desired luminances as established by a mental picture (or by other means) and designs the lighting accordingly. This approach, which can be very effective, is highly intuitive (i.e., requires considerable prior experience on the part of the designer). For this reason, and because that approach requires a good deal of hands-on, field trial-and-error work, we feel that it is not appropriate to a textbook and therefore is not used here. The only exception to this statement occurs in our discussion of daylight models in Chapter 19. There, the entire purpose of the model is to give the designer a visual/mental picture of the space's brightness patterns on the basis of which windows, window treatments, and possible exposures can be varied to achieve the desired light and shadow patterns (luminances and luminance ratios).

20.3   Lighting Design Procedure

(a) Project Constraints (Fig. 20.1). This flowchart, which represents the design procedure and its interactions, should be referred to throughout the necessarily lengthy discussion that follows in order to maintain perspective. It is important that the reader be aware of job constraints and of the interactions between the lighting designer and the remainder of the design group. We deliberately emphasize this to demonstrate the interdisciplinary nature of lighting design in general and its connection with HVAC and daylight (fenestration) in particular. This approach, which is most often referred to as the *systems design approach,* is followed throughout the discussion.

Item 4 of the list in Section 20.2 referred to constraints. These can alternatively be classified as the owner-designer-user team and the jurisdictional authorities. In some detail these are:

1. *Owner-designer-user group.* The owner establishes the cost framework, both initial and operating. A part of both of these may be a rent structure, which in turn determines and is determined by the space usage. If the owner is also the occupant, the cost factors change somewhat but remain in force. The architect determines the amount and quality of daylighting and the architectural nature of the space to be lighted. Many of these data are

detailed in the building program. Obviously, the architect and lighting designer (who may be the same person) should interact in this aspect of building design.

2. The jurisdictional authorities *may* include:

DOE—U.S. Department of Energy

GSA—General Services Administration

NFPA—National Fire Protection Association (codes)

ASHRAE—American Society of Heating, Refrigerating and Air Condition Engineers

IESNA—Illuminating Engineering Society of North American

NIST—National Institute of Science and Technology (formerly National Bureau of Standards)

Most of these are jurisdictional by reference; that is, the actual authorities may specify that the lighting system meet the requirements of ASHRAE, IESNA, and so on. If federal funds are involved, DOE/GSA standards will probably be involved. The principal area of involvement is that of energy budgets and lighting levels, both of which affect every aspect of lighting design including source type, fixture selection, lighting system, fixture placement, and even maintenance schedules. For this reason, the first step in the lighting design procedure is to establish the *project lighting cost framework and the project energy budget.*

(b) Task Analysis. As shown in Fig. 20.1, this step essentially determines the needs of the task. Factors to be considered in addition to the nature of the task are its repetitiveness, variability, who is performing the task (i.e., physical condition of the occupant), task duration, cost of errors, and special requirements. Several of these factors have been discussed in the preceding sections dealing with quality of light. These are cross-referenced in the appropriate sections in the following analysis.

(c) Design Stage. This is the active consideration stage during which detailed suggestions are raised, considered, modified, accepted, or rejected. This is also the most interactive stage, as is clearly seen from Fig. 20.1. At its completion, a detailed, workable design is in hand. The critical

Lighting Design Procedure

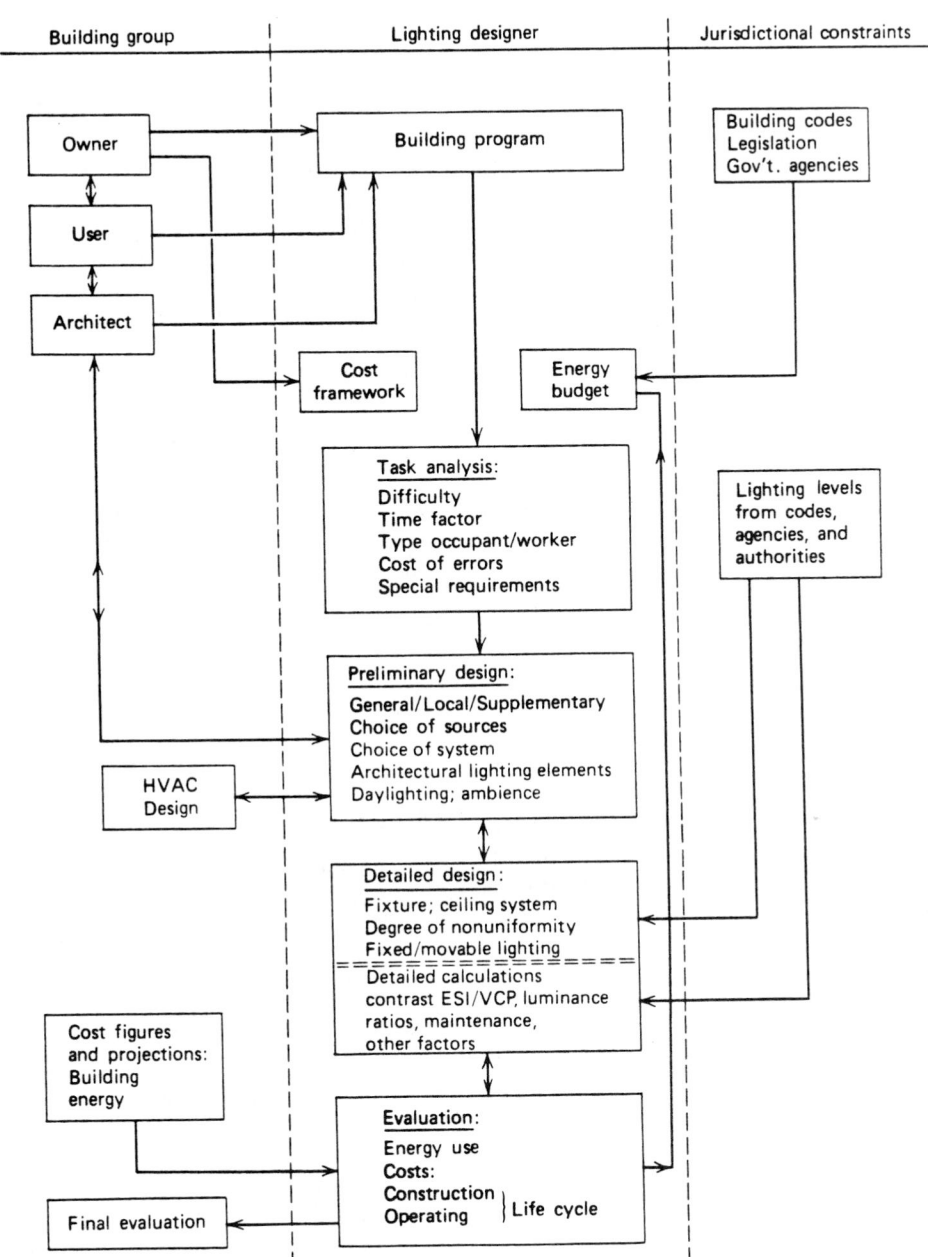

Fig. 20.1 Lighting design procedure chart.

interactions here are with the architect in daylighting and with the HVAC group in power loads. The former may result in relocating a space within the building, the latter in making a change in a lighting system or HVAC system. In brief, this stage consists of the following steps:

1. Select the lighting system. Select type of light source and distribution characteristic of fixture(s) or area source and consider effects of daylighting, economics, and electric loads.

2. Calculate the lighting requirements. Use the

applicable calculation method and establish the fixture pattern, considering the architectural effects.

3. Design the supplemental decorative and architectural (built-in) lighting.

4. Review the resultant design. Check the design for quality, quantity, esthetic effect, and originality.

(d) Evaluation Stage. With the design on paper, it can now be analyzed for conformance to the principal constraints of cost and energy. If the design stage has been carefully accomplished, with due attention to these factors, the result of the final evaluation should be gratifying. The results of this stage are fed to the architectural group for use in the final overall project evaluation. In the following sections, we consider in detail each of the steps in the design procedure.

20.4 Cost Factors

This is a particularly difficult item for a novice lighting designer because it requires experience in the field and an acquaintance with commercially available equipment. Also, the inevitable trade-offs between first cost and operating cost cannot be made intelligently unless the cost structure is clearly understood. The following guidelines should be of considerable assistance both in avoiding unpleasant surprises when a job is estimated and in preparing cost analyses:

1. Decide at the outset what cost criteria will be applied—that is, the relative importance of first cost, operating costs, annual owning costs, and life-cycle costs.

2. Trade-off decisions are required between first cost and operating costs. For example, incandescent lamps and fixtures are low in first cost and high in operating cost, and so on. Dimming and control equipment falls into this area of decisions.

3. Manufacturers' catalog items are *always* cheaper than specials and can be priced more readily.

4. Compare annual owning costs of two systems or methods. Conversion of these data to life-cycle cost comparisons is straightforward.

5. The impact of lighting energy on the operating cost of the entire building must be studied and the apportionment of costs determined. The only practical means of accomplishing this is by computer program. Programs can be readily adjusted to reflect the effect of the lighting system on building costs, and in particular on HVAC first cost and operating costs. It is incorrect to artificially separate the lighting system from the HVAC system with which it intimately interacts.

The lower the lighting system's energy level, the lower the building's overall operating cost. The argument that heat from a lighting system is fully utilized to heat the building and is therefore not wasted is a specious one, which has been refuted on many counts:

HVAC system first cost is higher.

HVAC year-round cost is higher.

Lighting energy cost is higher.

Life-cycle costs are higher.

Energy resource use is higher.

This subject is discussed in detail in Section 21.23.

20.5 Power Budgets

The requirement to establish a project's lighting power budget in accordance with a specified procedure has now been incorporated into the building codes of many states.

The purpose of this budget determination procedure is not to dictate design procedure. Indeed, all standards explicitly so state. Instead, the purpose is to develop an overall maximum power budget *within* which the designer is free to do as he or she wishes. Obviously, prodigality in one area is necessarily at the expense of another area as maximum power is inflexible, and the entire power budget is built on reasonable design techniques. Still, there is enough leeway in the budget and enough exceptions so that the designer is not overly restricted.

The nationally accepted standard that defines the establishment of a lighting power budget is: ASHRAE/IES 90.1 (latest edition date)—*Energy Efficient Design of New Buildings Except Low Rise Residential Buildings,* published by ASHRAE, 1791 Tullie Circle, NE, Atlanta, GA 30329.

ASHRAE 90.1 sets forth design requirements for the efficient use of energy in new buildings intended for human occupancy. Specifically excepted in the current edition are single- and multi-family residences of three or less stories above grade and buildings whose primary function is directed to a specific purpose, for which human occupancy may be required, but is secondary. These include industrial buildings, buildings with very low energy use, and very small buildings ($<100\,\text{ft}^2$ area). The standard encompasses energy use requirements for all of a building's environmental systems (i.e., HVAC, lighting, electrical, power, water, envelope, and energy management).

The standard recognizes the advances that have taken place in the field of lighting controls (see Sections 21.2–21.4) as a result of which it is able to focus on *energy use* limitation (which is its specific purpose) without necessarily limiting *power use*. For details as to how this is accomplished, the reader is referred to the current issue of Standard 90.1.

20.6 Task Analysis

Refer to Fig. 20.1. This is the stage at which the quantity and quality of lighting required for the tasks are decided. The factors affecting this choice as shown in Fig. 20.1 are difficulty, time factor, occupant, cost of errors, and special requirements.

(a) Difficulty. The components of visual difficulty were discussed at length in Sections 18.16 to 18.22, and the results in terms of lighting levels appear in Tables 18.4 to 18.9. Essentially, the designer examines the type of task involved, and after determining the applicable authority he selects the required illuminance.

In the absence of specific instructions or reasons to the contrary, the designer uses IESNA recommendations. If there are several tasks to be performed at the same point and the most difficult occurs infrequently, it may be reasonable to provide supplementary portable lighting or even to suggest moving to another brighter location. If it is the major task, lighting should be based on it and provision made for intensity reduction for less demanding work.

Variation in task difficulty is particularly common in spaces in public buildings. Thus, a school gym can be used for athletics, band concerts (despite the acoustics), and town meetings—three totally disparate lighting requirements. In these and similar instances it is common practice to treat the space as essentially three different spaces and design lighting for each with a careful eye to maximum common equipment usage. Similar problems are encountered in basements, multipurpose rooms, and conference/meeting/lecture/exhibition rooms. Fortunately, most such spaces do not have severe seeing tasks.

The task variation referred to here is the variation that occurs in one very specific location and is not to be confused with task variation in an area, however restricted. Thus, a small private office of, say, 8×8 ft, has a desk, file cabinet, and circulation space, three tasks of differing but constant difficulty in one small space. The corresponding lighting for these is also fixed and varies with the task severity. The values listed in Table 18.4 to 18.9 represent the required illumination on the surface in question whether horizontal, vertical, or in between. Inasmuch as the flux method of calculating illumination normally yields the 30-in., horizontal-plane illumination level, it is helpful in the early design stages to know the ratio of horizontal to vertical illumination for various lighting systems. This ratio is approximately:

Narrow distribution direct and semidirect	3:1
Wide distribution direct and semi-direct	2.5:1
General diffuse indirect	1.5:1

Once the design is advanced, computer calculation will yield exact illuminance data on any desired surface, including important vertical surfaces.

As the illumination values listed assume adherence to both recommended luminance ratios and reflectances (see Section 18.32), it is necessary to select, in conjunction with the interior designer, finishes and reflectances for surfaces within the area. If, for instance, in a private office a dark wall finish of 10% reflectance is chosen, it will be necessary for the lighting designer to compensate

TABLE 20.1 **Approximate Reflection Factors**

Medium Value Colors	
White	80–85
Light gray	45–70
Dark gray	20–25
Ivory white	70–80
Ivory	60–70
Pearl gray	70–75
Buff	40–70
Tan	30–50
Brown	20–40
Green	25–50
Olive	20–30
Azure blue	50–60
Sky blue	35–40
Pink	50–70
Cardinal red	20–25
Red	20–40

for this by additional wall lighting to maintain the recommended maximum 10:1 brightness ratio (see Table 18.11 and the discussion of point *g* in Section 20.7). The atmosphere created by vertical surface luminances is discussed later. Table 20.1 lists reflectances of some common interior paint finishes.

(b) Time Factor. As discussed in Section 18.20, the length of time in which the task must be accomplished is important in exacting work. Beginning with moderately difficult tasks—that is, luminance of about 170 cd/m² (50 fL)—prolonged intensive application or rapidly changing tasks would require illumination to be raised one (or more) levels. Alternatively, quality could be improved by increasing daylight or task contrasts.

(c) Occupant. Inasmuch as the age and other specific characteristics of the worker are usually not known, a standard distribution is assumed and the recommendations as tabulated take account of this. However, if there is a high percentage of older workers, as is the case in certain industries, lighting should be raised one level.

This compensates for inability of aging eyes to accommodate and for the tendency to tire easily (see Section 18.23).

(d) Cost of Errors. This involves an economic tradeoff between savings resulting from improving seeing accuracy against the cost of the improved lighting. Performance can be brought close to perfection, but the cost of so doing increases much more rapidly than the proportional increase in performance. Lighting is usually designed to provide 90%± work accuracy. Thus, this step is basically an economic calculation, the criteria for which must come from the owner or user. Tasks in which this problem is encountered include inspection, proofreading, textile matching, very fine machining, and jewelry manufacture.

(e) Special Requirement. These include any nonstandard task lighting requirements. Some of these are specific illuminant color, directionality for shadowing, reflections as required for inspection, polarization, and controlled variations, as required in a space with varied tasks or varying daylight factor. In addition to these, the physical dimensions of the task often create special requirements of their own. We tend to assume a small object in the horizontal plane because that is the normal office task. However, there are exceptions, such as a drafting board, a large machine, and inspection bench, or a cutting table. Consequently, these special requirements arise:

1. *Large tasks.* In large tasks, the angle of seeing varies from 20° to 70° from the vertical, resulting in radically changing glare angles and reflections from the task.

2. *Three-dimensional tasks.* These tasks shadow themselves, particularly when containing undercuts and reveals. An architect's model shop presents such tasks. When it is necessary to see into an opening, an intense narrow beam is required.

3. *Tools.* Tools cast shadows below and in front when lighted from above and behind. A fabric cutter must see ahead of and below the cutting machine.

4. *Nonhorizontal tasks.* These must be calculated for the plane in which they stand. As

noted in subsection 20.6a, the ratio between horizontal and vertical illumination varies between 1.5:1 to 3:1, depending on the system. Task lighting requirements are stated in the plane of the task. This can have a pronounced effect on the lighting system selected and its arrangement.

5. *Task observed from various positions.* There are instances where a fixed task is observed from several angles, such as a drawing in a conference room or a wall display. Illumination must be adequate for all viewing angles.

PRELIMINARY DESIGN

20.7 Energy Considerations

Energy considerations must pervade every aspect of the design process. Some background material is in order here, to place the lighting energy subject in proper perspective. Best current estimates indicate that lighting consumes approximately 25% of the electric power generated in the United States. In terms of *resources,* this amounts to approximately 4 million barrel (bbl) of oil per day. The same sources indicate usage by occupancy as approximately:

Residential	20%
Industrial	20%
Stores	20%
Schools and offices	15%
Outdoor and other	25%

In commercial buildings, lighting consumes about 20% to 30% of the building's electric energy, more in residences and less in industrial facilities. By judicious design a reduction of 40% to 50% in lighting energy is attainable. Translated into resources, this reduction can readily amount to more than 1 million bbl of oil per day. Few will disagree that such a goal is well worth the effort. Every watt per square foot reduction in lighting energy results in at least 1.25 W/ft² (13.5 w/m²) savings in air-conditioned buildings. It has been demonstrated by actual designs that offices and schools can be *well* lighted with less than 1.5 W/ft² (27 W/m²). The question to be answered then is: What design guidelines can be followed to effect this energy-conscious design?

The two standards that provide the most authoritative and reliable guidelines at this writing are IESNA Standard LEM-3 (1987) and ASHRAE/IESNA Standard 90.1-1989. They are so indicated where referenced. Both standards are in the process of extensive revision. The material that follows is basic and will most probably remain unchanged in any conceivable revision.

1. IESNA Standard LEM-3 (1987). *Lighting Design Criteria for Effective Energy Utilization* provides the reader with appropriate criteria for energy-conscious lighting design.

(a) *Design lighting for expected activity.* This is the task lighting approach. It is wasteful of energy to light any surface to a higher level than it requires. Nonuniform lighting is recommended where high illuminance levels are required for selected tasks in multitask spaces. One way to accomplish this for areas where exact furniture layout is not available is to use readily movable fixtures. Providing overall high-level illumination with provision for switching to reduce levels is not advisable because of increased first cost and the psychological impetus to operate at maximum levels. Another solution is to use fixed luminaires for general low-level lighting and supplementary task lighting. Other factors and techniques to be borne in mind are:

(1) Group tasks with similar lighting requirements.

(2) Place the most severe seeing tasks at the best daylight locations.

(3) Improve the quality of difficult visual tasks. This is more energy-economical than providing additional light (see Section 18.21).

(4) The advantages of nonuniform lighting increase as the space between work stations increases.

(5) When using the task-ambient design approach, keep in mind that a nonuniform ceiling layout may give a chaotic appearance to a space. Therefore, the preferred approach is uniform ambient lighting and local task lighting.

(b) *Design with effective, high-quality, efficient, low-maintenance, thermally controlled luminaires. Effective* means providing use-

ILLUMINATION

ful light and minimum direct glare. In cases where much of the viewer's time is spent in a head-up position, as in schools, or where the viewer can compensate for veiling reflections, the decision should lean toward high VCP (see Section 18.28). Where work and viewing position are fixed and most of the viewer's time is spent head down, the decision should lean toward low reflected glare (see Sections 18.29–18.31).

(1) *A high-quality* luminaire is made with permanent finishes such as Alzac or multicoat baked enamel or any of the high-quality permanent aluminum finishes. This ensures that its performance after 8 to 10 years of service will be comparable to the original.

(2) An *energy efficient* luminaire is one with a high luminaire efficacy rating (LER). This (new) metric is explained in detail in Section 20.29. It combines photometric efficiency (high coefficient of utilization; CU; see Section 20.27) with energy efficiency of the luminaire's components to obtain a high overall *efficacy* for the luminaire, expressed in lumens per watt.

(3) A *low-maintenance* luminaire remains clean for extended periods and is designed so that all reflecting surfaces can be easily and rapidly cleaned without demounting. Enclosed fixtures should be gasketed. Nongasketed units collect and retain dust and cause rapid output depreciation. Relamping should be simple and rapid, to encourage group relamping programs that are energy efficient and cost effective. A 20% increase in maintained light is possible if lamps are replaced at the end of their *useful* life—that is, when output is down to 70% of initial maintained lumens—and if fixtures are cleaned and maintained on a fixed schedule. No cost tradeoff is generally involved because periodic maintenance and relamping are normally cheaper than one-at-a-time maintenance and burnout replacement, *and* yields 20% higher average lumens. Fixtures in relatively inaccessi-

ble locations such as high ceilings must be designed for low maintenance, and maintenance should be on a fixed schedule.

(4) A *thermally controlled* luminaire is one that controls the heat generated by the light source. This item depends to a large extent on the type of HVAC system, the lighting heat load, and the type of fixtures employed. Detailed analysis of this point involves HVAC considerations and the overall impact of lighting energy on the building. In the late 1980s, the thermal problem was principally directed at removing fixture heat. Today, the use of electronic ballasts and low-wattage lamps has reduced this seriousness of this problem (Fig. 20.2).

(c) *Use efficient light sources and accessories.* This point is self-explanatory. The ready availability of high efficacy, high color rendering index (CRI), compact sources has made this item far less problematic than previously. The only cost tradeoff involved is between relatively expensive high CRI sources and cheaper low CRI units. The cost differential is not large and the choice is frequently made on other than an economic basis.

Spill light and borrowed light are often neglected sources. Glass in upper wall sections can provide sufficient corridor lighting from borrowed office lighting. Sources with high lumen *maintenance* should be given preference.

(d) *Select the appropriate lighting system.* As detailed in Sections 20.10 to 20.15, there are five types of lighting systems, each with its particular distribution characteristics and, therefore, applicability. In addition, all luminaires must be carefully located to provide uniformity of ambient lighting and task lighting, where applicable. When using indirect luminaires, correct spacing and hanging-stem length will avoid ceiling "hot spots," which can cause direct and reflected glare. Properly designed indirect lighting allows fully effective use of a space in that users are not restricted to facing any partic-

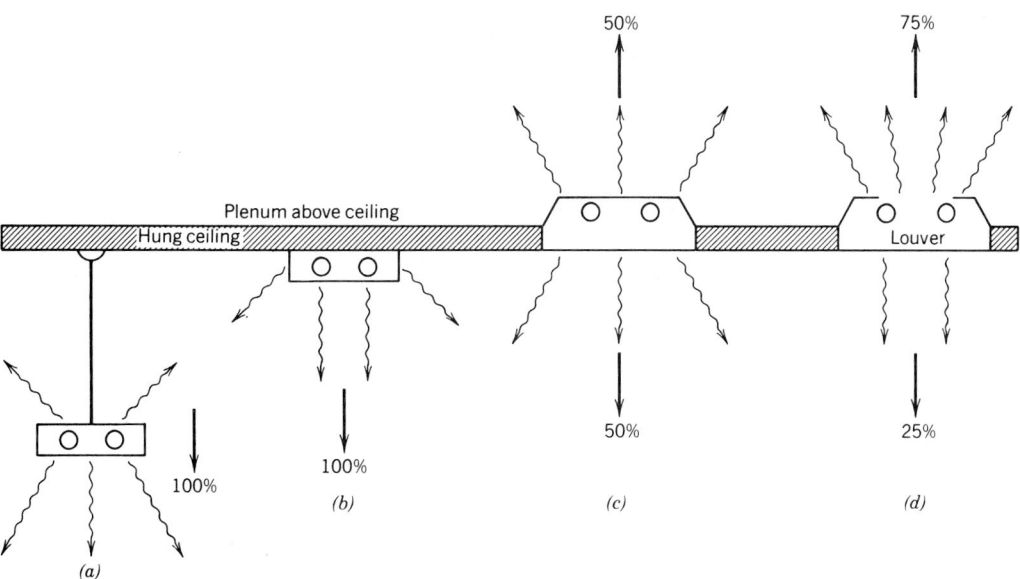

Fig. 20.2 *Method of fixture installation controls the transfer of luminaire's heat.*
(a) *Suspended units contribute all their heat to the space, and themselves remain fairly cool.*
(b) *Surface-mounted fixtures also place all their heat in the space, but because of blocked transfer upward, run hot.*
(c) *Completely enclosed recessed units transfer about 50% of their heat to the plenum.*
(d) *Open louvered, baffled units transfer about 75% of their heat. When ducted, heat transfer up can be as high as 85%.*

ular direction to avoid direct or reflected glare from fixtures. Mounting height of suspended fixtures must be coordinated with cavity sizes and finishes to decrease light loss and maximize coefficient of utilization.
(e) *Use daylight, properly.* Daylight must be considered as a normal light source subject to weather variations and time of building use. Obviously, a three-shift industrial plant cannot use daylight on all shifts, but it can for at least one shift, and design should reflect this fact. Part of proper daylight design is control of window luminance, which can cause severe and even disabling glare. A corollary of excessive luminance is excessive heat gain. Both are manageable with common control devices, manual and automatic. Window control devices must also be designed to reflect light back into the space at night, to avoid light loss via windows.
(f) *Use energy-efficient lighting control strategies.* The subject of lighting control, including manual and automatic switching,

dimming, sensing, and intensity control, is covered in Sections 21.2 to 21.4 for both new installations and retrofit work. Proper design of controls can reduce energy consumption over a noncontrolled installation by as much as 60% without reducing lighting effectiveness. These controls are mandated in the revised issue of ASHRAE/IESNA 90.1.
(g) *Use light finishes on ceilings, walls, floors, and furnishings.* This point is self-explanatory and is examined in a number of sections. A brief summary of recommended ranges would be:

Ceilings	80–92%
Walls	49–60%
Furniture, office machines, and equipment	25–45%
Floors	20–40%

In addition to producing higher illumination levels in the room, high reflectances minimize uncomfortable luminance ratios, as between luminaire and upper wall or between task and background. (see Table 18.11.) For luminance ratios:

(1) Between task and near surround—aim for 3:1.

(2) Between task and immediate area—aim for 10:1.

(3) Between luminaires and their background—aim for 20:1.

(4) Anywhere in the normal field of view—aim for 40:1.

Note that the targets themselves are *maximums* and values above maximum should be accepted only with excellent justification.

2. *ASHRAE/IESNA Standard 90.1.* The 1989 edition (with minor revisions) is in force at the time of this writing, with a revised edition due for publication in the near future. The standard is large and the lighting energy design requirements are extensive and detailed. A copy of the standard current at the time of design is required; a copy of the ASHRAE user's manual is strongly recommended.

(a) *Mandatory requirements* (present and anticipated)

(1) All interior and exterior lighting must conform to the stated energy limitations. Trade-offs between the two are not permitted.

(2) All interior and exterior lighting systems in buildings larger than 5000 ft^2, except emergency and exit lighting, must be equipped with some form of automatic shut-off control.

(3) Where applicable, occupancy sensors and automatic daylight compensation control should be used.

(4) Separate tasks or spaces must have separate controls.

(5) Areas with lighting power densities above 1.1 W/sf should have facilities for at least two lighting levels.

(b) Recommendations

(1) Where using task/ambient lighting, the ambient level should not be lower than a third of the task level to avoid uncomfortable luminance ratios.

(2) Accent lighting should not exceed five times the ambient level. Therefore, in merchandising areas where the contrast between ambient and task levels is critical, reduce ambient levels as much as practicable.

(3) When specifying super-reflective aluminum in fluorescent (or other) luminaires, determine that the material's high reflectance will be maintained in the specific application and that the required luminaire maintenance procedures will be available.

(4) Utilize self-luminous exit lights or those utilizing LED displays where permitted by local codes.

20.8   Preliminary Design

Again referring to Fig. 20.1, the preliminary design phase is the time during which ideas crystalize, but in terms of areas and patterns as well as light and shadow, and not yet in terms of hardware. At this stage the quality of the system is decided on; that is, the luminances, diffuseness, chromaticity, and proportion of vertical to horizontal lighting are determined. The last factor establishes in large measure the room "mood" or lighting ambience. In preceding sections these items were discussed in some detail. In the sections that follow on lighting systems (direct, indirect, etc.), the quality of each is considered and applications suggested. In the overall view, however, the ultimate quality of the lighting system, its visual pleasantness, centers of visual attention, highlights and shadows, as well as texture and forms, are a deft and perhaps artistic combination of the previously mentioned considerations and establish, as the term implies, the quality of the lighting design. A few observations, not covered elsewhere, are mentioned in the following paragraphs.

Planes other than the "working plane" must always be considered. The ratio of vertical to horizontal illumination of the chosen lighting system determines wall luminance, which in turn greatly influences the overall impression of the space's brightness. (see Fig. 20.12). The floor finish has a pronounced effect on the ceiling illumination for direct lighting systems because in direct systems, ceiling illumination derives only from room inter-reflectances, with floor reflectance being particularly important in large spaces (see Fig. 20.45).

The chromaticity of the room lighting depends primarily on the source but secondarily on the luminaire and surface finishes. A "white" source can be tinted slightly by the use of a colored reflector in the luminaire. Of course, the effect on luminaire output of such a change must be considered. In the case of semi-indirect and indirect lighting, this same effect can be accomplished by the use of colored ceiling and upper wall surfaces, which serve as secondary reflectors and become the actual luminous source for the room. Recommendations in Table 19.24 cover choice of source as affected by its chromaticity.

20.9   Illumination Methods

There are three methods of illumination: general, local/supplementary, and combined general and local.

(a) General Lighting. This is a system designed to give uniform and, generally, although not necessarily, diffuse lighting throughout the area under consideration. The method of accomplishing this result varies from the use of luminous ceiling to properly spaced and chosen downlights, but the resultant lighting on the *horizontal working plane* must be the same; that is, reasonably uniform. It may be, but is not necessarily, task lighting.

(b) Local/Supplementary Lighting. These are two terms that are used interchangeably. By definition, both *local lighting* and *supplementary lighting* provide a restricted area of relatively high intensity. A desk lamp, a high-intensity down-light on a merchandising display, and a track light illuminating wall displays in practice are all referred to as local, supplementary, or local–supplementary lights. Typical of this genre are the units illustrated in Fig. 20.3.

(c) Combined General and Local Lighting. This illumination method is used in spaces where the general visual task is low, but supplementary lighting is required in a limited area for a particular task.

These three *methods* of illumination can be accomplished in many ways by the use of luminaires and luminous sources of different types, because the illumination method is a function of *both* luminaire arrangement and the luminaire's inherent lighting distribution. The term used to describe the effect of the combination of a particular fixture type applied in a particular way is the *lighting system.* Thus, a reflector-type fixture, when aimed down, gives *direct* light. The same fixture beamed up at the ceiling gives *indirect* light. The following section describes the systems that constitute the vast majority of lighting installations.

20.10   Types of Lighting Systems

No single lighting system can be said to be the only choice in a given instance; on the contrary, the designer normally has a choice of at least two systems that, if utilized properly, yield illumination of adequate quantity and good quality. However, other factors, such as harmonization with the architecture and economics, usually tip the balance in favor of one or the other.

The five generic types of lighting systems are indirect, semi-indirect, diffuse or direct-indirect, semi-direct, and direct.

20.11   Indirect Lighting

See Fig. 20.4*a.* Between 90% and 100% of the light output of the luminaires is directed to the ceiling and upper walls of the room. The system is called *indirect* because practically all the light reaches the horizontal working plane indirectly; that is, via reflection from the ceiling and upper walls. *Therefore, the ceiling and upper walls in effect become the light source,* and if these surfaces have a high-reflectance finish, the room illumination is highly diffuse (shadowless). Because the source must be suspended at least 12 in. (30 cm) and preferably 18 in. (45 cm) or more from the ceiling

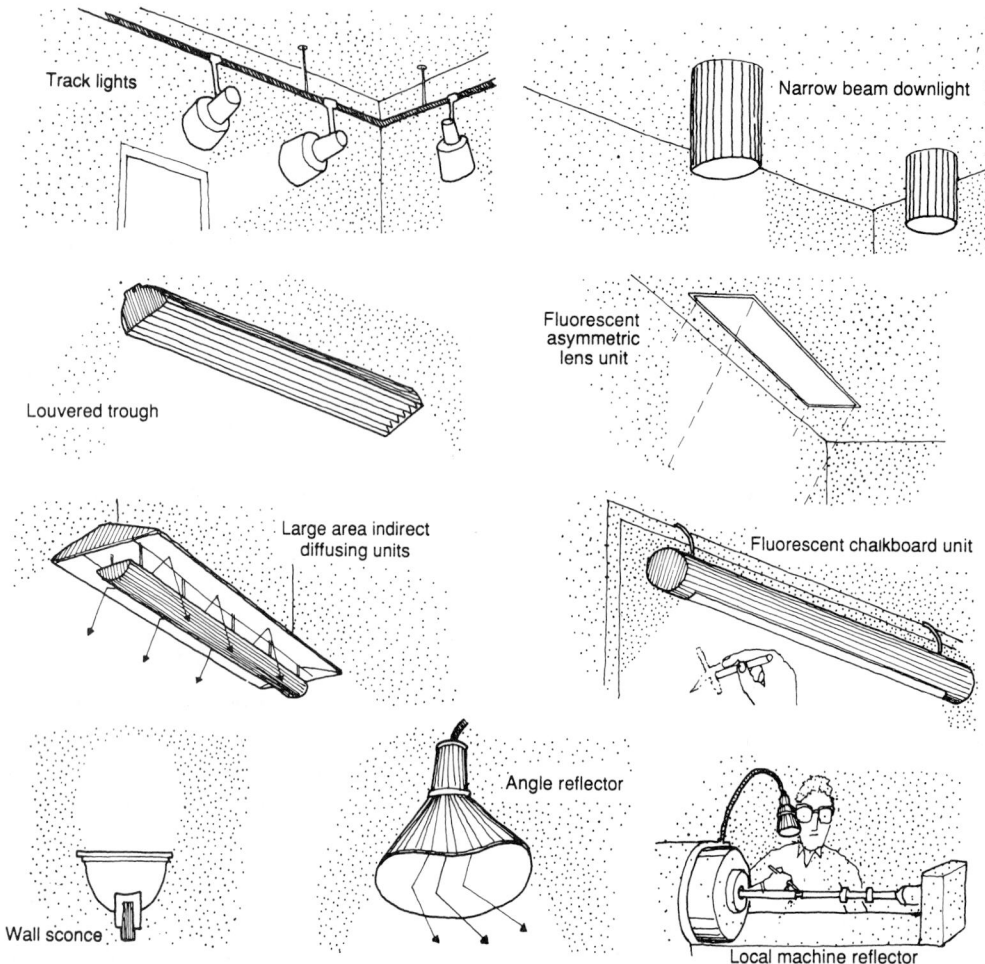

Fig. 20.3 *Typical supplementary lighting units for incandescent, CFL, and linear fluorescent sources.*

(depending on the unit's output) to avoid ceiling "hot spots," this system requires a minimum ceiling height of 9 ft, 6 in. (~3 m).

If luminaires are correctly spaced, the resultant illumination is uniform, and direct and reflected glare are both low.

To avoid an excessive luminance ratio between the luminaire and its surrounding field, the luminaire can be made translucent on the bottom, the sides, or both. Approximately 750 lux (75 fc) is the maximum horizontal-plane illumination attainable without exceeding *overall* ceiling luminance of about 2500 cd/m^2 (730 fL) (see Section 18.28). With practically no veiling reflections, this illuminance level is sufficient for all but the most difficult tasks. The lack of shadow, low source brightness, and highly diffuse quality created by indirect lighting give a very quiet, cool ambience to the space, suitable for private offices, lounges, and plush waiting areas. Areas having specular visual tasks such as an office with visual display terminals (VDTs) use this system to advantage. In such spaces, indirect fixtures *without* luminous bottoms or sides should be specified.

When properly designed, particularly when the source of light is architectural coves (see Fig. 20.4*c*), the ceiling has a floating, sky quality, which is pleasant and can be used to give an impression of height in a low-ceilinged room. (This system is not to be confused with a transilluminated ceiling, which is really a direct lighting system of entirely different quality and effect.) A further character-

Indirect lighting

Indirect fluorescent

(a)

(b)

(c)

Fig. 20.4 *Indirect lighting.* (a) *The luminaires deliver 90–100% of their output above their own horizontal plane.*
(b) *The ceiling and upper wall surfaces of the space are directly illuminated, and by reflection become large secondary sources that illuminate the space below. When properly designed, this type of installation yields a substantially uniformly bright ceiling.*
(c) *Use of architectural coves gives an acceptable luminance gradient on the ceiling, and, if properly designed, nearly uniform, glareless illumination in the room. This system of illumination is particularly useful in VDT spaces.*

istic of the indirect lighting system is loss of texture on vertical surfaces, as is common to all fully diffuse lighting.

Indirect lighting is inherently inefficient as much of the useful light reaches the working plane only after double reflection—within the luminaire and off the ceiling. Although to a considerable extent this inefficiency is offset by the glare-free lighting, applications to difficult seeing tasks frequently require an additional light source. Thus,

an indirectly lighted architect's drafting room, having tables equipped with supplementary lamps would take advantage of both systems—the local high-intensity light at a maximum of 200 fc ($\sim$2000 lux) for the restricted area being worked on, and overall table lighting of 400 to 500 lux of high-quality lighting free of veiling reflections. The latter would also solve any reflected glare problems arising from the many viewing angles required by large tasks such as drawings.

ILLUMINATION

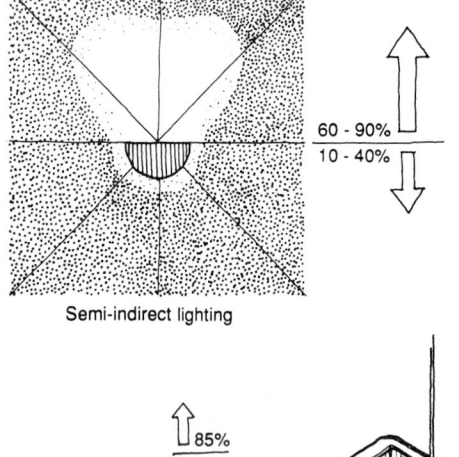

Semi-indirect lighting

60 - 90%
10 - 40%

85%
15%

Ballast   Tube

Diffusing or   Metal
louvered bottom   housing

Semi-indirect fluorescent

Fig. 20.5 *Semi-indirect lighting.*

20.12   Semi-Indirect Lighting

Between 60% and 90% of the light is directed upward to the ceiling and upper walls. This distribution is similar to that of indirect lighting, except that it is somewhat more efficient and allows higher levels of illumination without undesirable brightness contrast between the luminaire and its background, along with lower ceiling brightness. A typical fixture, illustrated in Fig. 20.5, employs a translucent diffusing element through which the downward light component passes. The ceiling remains the principal radiating source and the diffuse character of room lighting remains. Direct and reflected glare are both very low, as they are with indirect lighting (see Fig. 18.38). In both indirect and semi-indirect systems, it is often desirable to add accent lighting or downlighting to break the monotony inherent in these systems

and establish visual points of interest, or to create required modeling shadows.

In both indirect and semi-indirect lighting systems, the light undergoes a number of ceiling and wall reflections before reaching the horizontal working plane. The use of colored paints, particularly on the ceiling, can serve to tint the room illumination slightly by selective absorption.

20.13   Direct–Indirect and General Diffuse Lighting

Direct–indirect lighting provides approximately equal distribution of light upward and downward, resulting in a bright ceiling and upper wall (Fig. 20.6). For this reason, luminance ratios in the upper-vision zone are usually not a problem. As the ceiling is an important although secondary source of room illumination, diffuseness is good, with resultant satisfactory vertical-plane illumination.

Diffusing fixtures (Fig. 20.7) gives light in all directions, whereas direct–indirect fixtures have little horizontal component. Stems for both types should be of sufficient length to avoid excessive ceiling brightness; generally not less than 12 in. (30 cm).

Because the impression of illumination depends to a large extent on *wall luminance* (as this is the surface we see most often), a space with general diffuse illumination *appears* lighter than one with direct–indirect illumination due to the darker walls in the latter [see Figs. 20.7(*b*) and 20.11].

By avoiding excessively bright luminaires and giving attention to positioning of sources and viewing angles, direct and reflected glare can both be kept low. Furthermore, because the luminaire (as any other luminous source in the field of view whose luminance is higher than the average scene) draws the eyes' attention, particular care must be taken to limit its luminance and to avoid disturbing fixture patterns (see Fig. 20.14).

Efficiency of these two systems is good. Both are well applied in spaces requiring overall uniform lighting at moderate levels such as classrooms, standard office work spaces, and merchandising areas.

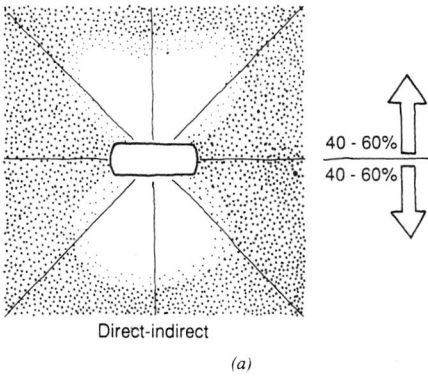

Direct-indirect

(a)

40 - 60%

40 - 60%

(b)

Direct-indirect fluorescent

40 - 60%

40 - 60%

(c)

Fig. 20.6 (a) *Direct–indirect lighting. Upper and lower room surfaces are luminous* (b), *but center of walls is not because of the lack of horizontal light from fixtures* (c). *Principal light on working plane comes directly from the luminaire.*

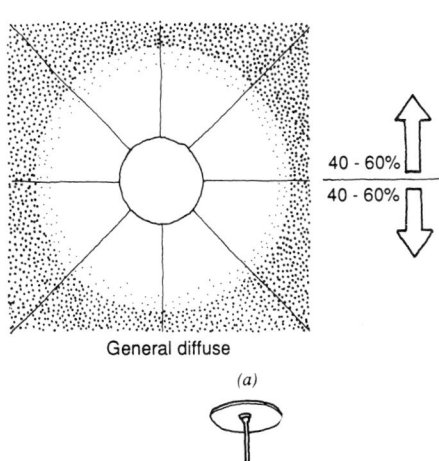

General diffuse

(a)

40 - 60%

40 - 60%

(b)

General diffuse incandescent

Fig. 20.7 (a) *General diffuse lighting.* (b) *Note that all room surfaces are illuminated and become secondary sources, although those closest to the fixtures (ceiling, upper wall) are the brightest secondary sources. The primary source of illumination is the direct radiation from the fixture. The floor contribution is low due to its normally low reflectance.*

ILLUMINATION

Fig. 20.8 *Semidirect lighting provides its own ceiling brightness* (a), *with surface-mounted fixtures* (b) *or pendant/surface units* (c). *Other characteristics are similar to direct lighting.*

20.14 Semi-Direct Lighting

With this type of lighting system, 60% to 90% of the luminaire output is directed downward and the remaining upward component serves to illuminate the ceiling (Fig. 20.8). If the ceiling has a high reflectance, this upward component is normally sufficient to minimize direct glare from the luminaires, depending on eye adaptation level. The degree of diffuseness depends in large measure on the reflectances of room furnishings and of the floor. Shadowing should not be a problem when upward components are at least 25% and ceiling reflectance not less than 70%. With smaller upward components, the system is essentially direct lighting (see Section 20.15). The system is inherently efficient. Reflected glare can be controlled by the methods discussed in Section 18.29. With adequate wall illumination, the quality of the lighting gives a pleasant working atmosphere. It is applicable to offices, classrooms, shops, and other working areas.

20.15 Direct Lighting

In this system essentially all the light is directed downward. As a result ceiling illumination is entirely due to light reflected from floor and room furnishings. This system, then, more than any other, requires a light, high-reflectance, diffuse floor unless a dark ceiling is desired from an architectural or decorative viewpoint. Occasionally, the ceilings are deliberately painted a dark color and pendant direct fixtures used to lower the apparent ceiling of a poorly proportioned room or hide unsightly piping, ductwork, and so on.

The effect of direct lighting depends greatly on whether the luminaire light distribution patterns are spread or concentrating (Figs. 20.9 and 20.10). In the former cases, considerable diffusion of light results from reflections on floor, furniture, and walls. The result is a working atmosphere with slightly darkened walls and ceiling. This type of lighting, which is most widely represented by the recessed fluorescent troffer in a suspended ceiling, is common for general office lighting. The luminaires themselves form a ceiling surface of light and dark areas, and the quality of the entire system is not unpleasant. Difficulties associated with direct glare and veiling reflections can be controlled by proper use of reflectances, use of low-brightness units, and judicious arrangement of viewing positions. When direct lighting units are used in a uniform pattern, this latter option disappears, and the need for particularly low-brightness units and high ceiling reflectivity, or specialty diffusers such as those with batwing distribution, is increased.

Fig. 20.9 *Spread-type direct lighting* (a) *illuminates all room surfaces except the ceiling* (b), *which is only illuminated by reflection from the floor. Some diffuseness is evident. The most common type of unit in this category is the direct fluorescent, either surface-mounted* (b) *or troffer type, recessed in ceiling* (c).

Fig. 20.10 *With concentrating direct distribution* (a), *the floor is the only luminous surface* (b) *other than the ceiling fixture. Diffuseness is absent. Walls are dark. Incandescent downlights* (c) *and, to a lesser extent, CFL downlights are of this type unless equipped with spread-type lenses.*

(a)   *(b)*

Fig. 20.11 *Large-dimension lighting fixtures may be used in a low-ceiling room if the apparent size of the unit is reduced. Here at a mounting height of 7 ft 6 in., 4 × 4 ft units are acceptable because the lattice on the face of each unit gives the impression of reduced fixture size. Note also that the apparent illumination in (b) is greater than in (a) (although both are exactly equal on the table surface) due to the wall wash in the background. The eye perceives vertical surface illumination more readily than horizontal and retains the impression for the entire space.*

Direct lighting gives little vertical surface illumination, requiring the addition of perimeter lighting in business atmospheres (Fig. 20.11).

Concentrating downlights create sharp shadows and a theatrical atmosphere, not appropriate to a working commercial space. They can be used in restaurants and other areas where the privacy type of atmosphere generated by limited-area horizontal illumination and minimal vertical-surface illumination is desired. When a lighting fixture is designed with a black cone or baffle or other device that is nonreflecting at the viewing angle, it appears dark even when lighted. It is our opinion that installations providing high-horizontal surface illumination, with no apparent source of brightness, such as those using black-cone downlights, are disturbing to our normal bright-sun-and-sky orientation, and should be used cautiously and only in limited areas. This same comment, but to a lesser extent, is applicable to very low brightness diffusers such as the parabolic wedge type (see Fig. 21.33). (There, however, the unit has the redeeming characteristic of low reflected glare, which is not the case with downlights.) Stated otherwise, the psychological impression of a space without a visible source of light is gloomy and cave like. When using very low brightness sources, as in VDT areas, this negative impression can be alleviated by the addition of luminous surfaces in nonreflecting areas or points of light sparkle.

In summary then, spread-type direct lighting is suitable for general lighting, whereas concentrated direct lighting, which reduces vertical illumination, is appropriate for highlights, local and supplementary lighting, and specialized privacy-atmosphere installations.

20.16   Size and Pattern of Luminaires

Because of its luminance, each luminaire or other luminous source is a point of visual attention. To the extent that luminaires are numerous, large, very bright, or arranged in striking patterns, attention is drawn to them and away from other surfaces. Furthermore, color elements or accent lighting can be added deliberately to draw attention. Rigid rules cannot be established to cover these criteria, but examples can demonstrate the principles involved.

Luminaire size should correlate with room size and ceiling height. Fluorescent fixtures larger than 2 × 4 ft (60 × 120 cm) should not be used in ceilings lower than 10 ft unless their size is minimized by some sort of surface pattern (see Fig.

(a) (b)

Fig. 20.12 (a) *Transilluminated (luminous) ceiling. When properly designed, piping and ductwork above the suspended diffusing material are not seen and do not affect the installation's light distribution. As a rule of thumb, the strip fluorescent fixture spacing S should not exceed 1.5 times the fixture height above the diffusing element for uniform illumination.*

(b) *A luminous ceiling provides low brightness and highly diffuse, uniform illuminance, generally at levels exceeding 500 lux. They are particularly useful for specular tasks where supplemental lighting is impractical. To relieve the monotony of large unbroken expanses of diffusers as in* (a), *designers frequently use clearly defined diffuser panels as shown.*

(c)

Fig. 20.13 *Coffer type light sources can be manufactured* (a) *or architectural* (b).

(a) *Manufactured coffers are large direct lighting fixtures generally available in 48-in. (120-cm) width with standard length 4-ft fluorescent lamps installed across the width, as shown. Length is variable, as required.*

(b) *Architectural coffers can be constructed in any shape or size using hidden linear fluorescent lamps or CFL units to illuminate a white (plaster) flat or domed surface, which in turn illuminates the space beneath. Both types of coffer can give an illusion of great depth and of a floating illuminated surface.*

(c) *Multiple coffers create a dominant architectural effect and when designed in conjunction with skylights can furnish soft, glare-free illumination throughout the day and night.*

ILLUMINATION

Fig. 20.14 (a) *Longitudinal lines in the direction of the sight line increase apparent length, direct traffic flow, and decrease direct glare.*
(b) *Lines perpendicular to the line of sight shorten and widen a space, but also increase direct glare.*
(c) *Diagonal lines minimize shadows and break rectangular patterns. They are architecturally dominant.*
(d) *Rectangular pattern is architecturally dominant. It is a poor choice in stores where attention downward is desired.*

20.11). Transilluminated ceilings (Fig. 20.12) are *totally* lighting fixture and require a minimum of 12-ft (3.66-m) mounting height. When installed below this level, particularly in large rooms, the effect is oppressive, as if the sky were lowered on us. To offset this effect, the use of colored, shaped, or dark panels is of some help. In place of a luminous ceiling, large-area, coffer-type fixtures can be utilized that give the impression of depth (Fig. 20.13).

To achieve the uniformity of illumination desirable for general lighting, regular spacing is required. However, various effects may be obtained within the regularity to accomplish an architectural purpose, as shown in Fig. 20.14(a–d). The pattern of lights must never be at cross-purposes with any dominant architectural pattern; rather, it should either reinforce an architectural form or be neutral. If a strong architectural element is absent, a dominant lighting pattern may be desirable. Conversely, a strong architectural element can either be reinforced [Fig. 20.15(a)] or utilized to carry a neutral lighting pattern [Fig. 20.15(b,c)].

Continuous row installations eliminate the dominant checkerboard effect of (closely spaced) individual luminaires and are cheaper. Coves and cornices give the ceiling a floating or lightness effect. Geometric patterns can be used to add interest or break the monotony of large areas, such as are found in department stores. Generally, downlights are not dominant, and regularity of placement is not essential (Fig. 20.16). However, when downlights are surface mounted (a generally inadvisable procedure), the result can be very far from visually neutral (Fig. 20.17). Nonuniform layouts with large sources create a distinct pattern problem inasmuch as they are too large to be neutral, and the nonuniformity can create visual

(a)

(b)

(c)

Fig. 20.15 (a) *Treatment of dominant architectural patterns. In each case the lighting designer was faced with essentially the same problem: a low-level seeing task in a large space with a dominant architectural ceiling. Different solutions were arrived at, each of which is consonant with the architectural ceiling. Designs (a) and (b) accomplish this by following the dominant line; (c) follows it with a neutral pattern. (Photo by M. B. Warren.)*
(b) *Floor reflection and daylight provide the required ceiling and wall illumination. (Photo by L. Reens.)*
(c) *The lighting in this extremely strong architectural pattern was harmonized deftly by recessing fixtures deeply into the lattice motif. Metal–halide HID units and tungsten–halogen units were used. (Courtesy of GTE/Sylvania, Inc.)*

(a)

Fig. 20.16 (a) *Downlights are architecturally neutral and may therefore be spaced evenly*

(b)

. . . . *or unevenly.* **Fig. 20.16** (b)

confusion (Fig. 20.18). The only cure for this problem is to minimize the source brightness by using low-brightness luminaires (see Fig. 21.29 and Section 20.28).

In spaces where circulation is the primary "seeing task," yet containing isolated areas requiring greater illumination, such as waiting areas in transportation terminals, a perimeter lighting system with additional local lighting as required is a viable solution (Fig. 20.19). In addition to their illumination function, lighting patterns are often used as directional -markers. This is particularly useful in transportation terminals, where signs only partially serve this purpose. Figure

20.20 illustrates this principle, as well as use of a lighting pattern to emphasize an architectural aspect of the space.

A frequently neglected consideration is the appearance of a luminaire when de-energized. With proper daylight and energy-conserving design, many sources are unlighted during the normal-use hours of the space. (Obviously, low-brightness sources change least in appearance.) Cognizance of the visibility and appearance of luminaires in daylight, regardless of whether illuminated, from inside or outside a building, can be used to advantage, as shown in Figs. 20.21 and 20.22.

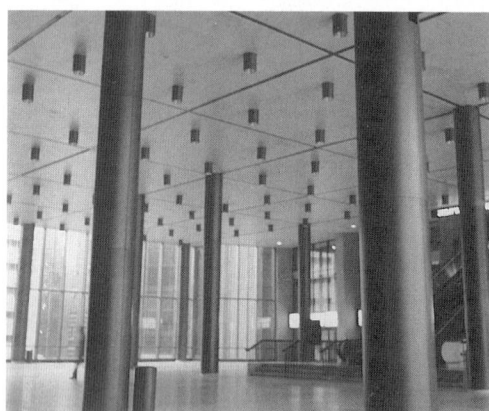

Fig. 20.17 *The large, can-shaped, surface-mounted downlights dominate the area's appearance despite the high ceiling. The pimpled look could be eliminated by recessing the units. Or, taking advantage of the prominence of these units, interest could be added by altering the regular pattern (see Fig. 20.20).*

20.17   Other Design Considerations

Because the phenomenon of vision as affected by lighting is the core of our discussion, it is appropriate to present here some miscellaneous yet important observations that are not readily subsumed under any of the major headings in our discussion.

1. As shown in Fig. 20.14*a* and *b*, the impression of room length and width can be emphasized by the direction of lines of lighting. An even wall wash of light can also be used to shorten and widen a hallway or corridor.

2. As can be seen in Fig. 20.11, a light wall in the line of sight increases the room's apparent size. Therefore, both light and low-chroma high-value paint hues can be used to expand a space (lighted walls, light colors) or contract it (unlighted walls, dark color).

Fig. 20.18 (a) *The uniform layout of Fig. 18.36, shown here in perspective, is neutral in that it does not dominate the space or draw the eye.* (b) *The nonuniform layout can be dominant in the pejorative sense if the eye is drawn by the lack of pattern or symmetry.*

3. Vertical surface illuminance should be approximately 25% to 35% of horizontal for a space to appear dimensionally undistorted. Because high luminance attracts the eye, fixtures with sparkle draw the eye away from the walls and thereby shrink the space's horizontal dimensions.

4. As shown in Figs. 20.13(*c*), 20.14(*c,d*), 20.16, and 20.17, luminaire patterns can be dominant (i.e., can become a focus of attention) by virtue of their size or arrangement. The same is true of wall-lighting patterns to an even greater extent, as walls are always in the direct line of sight. Therefore, wall lighting that creates meaningless scallops, spots, irregular gradients, and points of sparkle become a dominant visual element. In most cases this is neither intended nor desired. Unscalloped, even-gradient wall illumination is readily accomplished with linear sources, with elliptical reflector lamps or luminaire reflectors, utilizing proper luminaire spacing.

5. Concentrated pools of light in an overall low-ambient-light space effectively isolate the il-

Fig. 20.19 *Cornices, valances, and coves are luminous ceiling borders. In large rooms, suspended coves achieve a uniform ceiling brightness gradient and when designed with a downward component or combined with local lighting, as illustrated, give a pleasant, intimate atmosphere.*

Fig. 20.20 *The lines of lights in the center converge optically to produce a directional flow of traffic toward the escalator. The remaining floor lighting is provided by visually pleasing circular patterns around columns. (Penn Station, New York—photo by Stein.)*

(a)   *(b)*

Fig. 20.21 (a) *Lighting can be utilized as a medium to connect the inside and outside of a building. The simple expedient of continuing the lighting pattern beyond the window or wall glass provides visibility from inside out as well as outside in. Care must be exercised to avoid fixture placement that reflects in the glass.* (b) *As fixtures are readily visible even when unlit during daylight hours, their outline can be accentuated and the resultant pattern utilized as an architectural motif. (Courtesy of Welton Becket & Assoc.)*

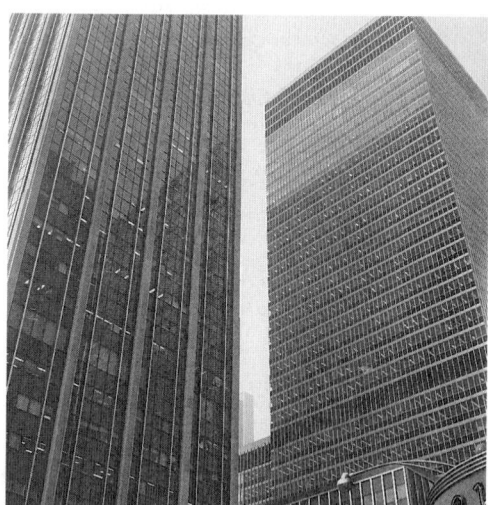

Fig. 20.22 *That patterns of lighting are plainly visible even during the daytime is apparent from this photo. The attractiveness of uniformity of fixture pattern can readily be seen. (Photo by Stein.)*

luminated areas from each other (i.e., define a limited specific geographic space). This can be used to advantage in restaurants and work or school areas where it is desired to define individual "territories" in a single large space.

LUMINAIRES

20.18   Design Considerations

(a) General. Refer again to Fig. 20.1. At this point in the design process, the lighting hardware is chosen on the basis of the considerations adduced in the preliminary design stage and then appropriate calculations performed. Some spaces require overall, uniform illumination. These spaces are calculated by the lumen method, which yields average illumination. Other spaces utilize local lighting alone, or local lighting in addition to general lighting, requiring point-by-point illumination calculations or some other method for restricted-area calculation. Additional considerations at this design stage are the control strategy

(see Sections 21.2–21.4); type of ceiling system (for example, modular, movable fixture, and integrated service); and ancillary considerations of ballast noise, luminaire heat distribution, and maintenance. Also decided here is whether to utilize work-station-mounted or built-in lighting, both of which are principally applicable to open-plan spaces.

(b) Luminaire Characteristics. The purpose of a luminaire or lighting fixture (the terms are synonymous) are twofold: physically, to hold, protect, and electrify the light source; and photometrically to control the lamp output (i.e., to redirect the light produced) because most common light sources emit light in substantially all directions. The means by which this beam shaping is accomplished are well known. The characteristics of reflectors, baffles, lenses, and louvres that perform these functions are discussed in some detail in the following sections. However, the problem in luminaire selection is that the requirements are, in many respects, incompatible, and therefore a trade-off between various luminaire characteristics must be made. Thus, for instance, high efficiency frequently entails high fixture luminance with resultant glare. Desirable wall lighting means high-angle luminaire output with resultant direct glare (see Figs. 18.25 and 18.26). Low-angle light means minimum direct glare but possible veiling reflections. High shielding angle (>35°) means

good visual comfort but reduced efficiency, and so on. It is therefore obvious that

· No single luminaire design is ideal for even a majority of applications, and that

· To make an intelligent selection among the hundreds of lighting fixtures commercially available, it is absolutely necessary that the specifier understand both the specific requirements of the application and the light control characteristics of the luminaire being considered.

20.19 Lighting Fixture Distribution Characteristics

At this point, review of Section 18.14 by the reader would be useful. The two distribution curves shown in Fig. 20.23(a) are actual test results of two 2-lamp, 1 ft wide by 4 ft long, semi-direct fluorescent fixtures with prismatic enclosure. The flat bottom of the curve in Fig. 20.23(*a*) indicates even illumination over a wide area, therefore permitting a high spacing-to-mounting height ratio (1.4) for uniform illumination. The rounded bottom of the curve in Fig. 20.23(*b*) indicates uneven illumination and closer required spacing for horizontal uniformity as defined in Section 20.22.

The straight sides of the curve in Fig. 20.23(*a*) show a fairly sharp cutoff, and the small amount

ILLUMINATION

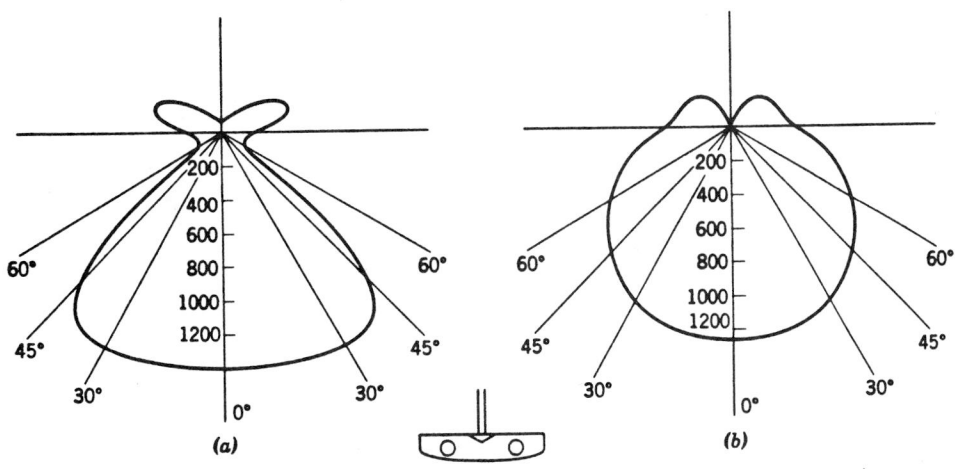

Fig. 20.23 *Semidirect fluorescent fixture crosswise distribution (two lamps, 32 W each, prismatic enclosure). Note the sharp cutoff and wide horizontally even distribution of* (a) *in contrast to the diffuse, broad, and horizontally uneven distribution of* (b).

of light above 45° means high efficiency, probably insufficient wall lighting, barely adequate diffuseness, and very little direct glare problem, but a distinct possibility of veiling reflections. Conversely, the curve in Fig. 20.23(*b*) shows a large amount of horizontal illumination (above 45°) with resultant direct glare, diffuseness, and relative inefficiency, because horizontal light is attenuated by multiple reflections before reaching the horizontal working plane. Here, however, low output below 45° minimizes reflected glare. The uplight component of luminaire (*a*) is directed outward to cover the ceiling and will not cause hot spots; the corresponding light from fixture (*b*) is concentrated above the fixture and gives uneven illumination of the ceiling.

These conclusions were reached based on the following observations:

1. *Uniformity of illumination* requires that the intensity at angles above the nadir (0° from vertical) be greater than those at 0° so that points distant from the fixture centerline obtain the same illumination as those below the fixture [because illuminance varies inversely with (the square of) distance]. This is exactly the case with the flat-bottom characteristics of Fig. 20.23(*a*). Therefore, such fixtures can be spaced more widely than the units of Fig. 20.23(*b*).

2. *High efficiency* is achieved by directing the luminaire output to the work plane (i.e., from 0° to 45° from the vertical). Light above 45° is directed to the walls and reaches the working plane only after multiple attenuating interreflections.

3. *Diffuseness* exists when light reaches the work plane from multiple directions. This requires that light be reflected from walls and ceilings to the work plane, which in turn requires luminaire light output above 45° from the vertical.

4. *Direct glare* is caused by light output at high angles (i.e., above 45° from the vertical). Direct glare from linear fluorescent fixtures can be minimized by placing their long axis parallel to the line of sight, because such fixtures normally have low *endwise* high-angle output.

5. *Reflected glare* is caused by reflection of low-angle output from the task. Therefore, fixtures with control means that limit output between 0° and 45° minimize the potential for veiling reflections [Section 20.21(*b*)].

6. *Shielding* is a function of the shape of the fixture housing plus any additional lamp concealment means, such as louvres or baffles. The shielding angle is defined as the angle between a horizontal plane through the louvres or baffles

Baffled downlights (1), (2), (3) control unwanted higth-angle light by cutoff as illustrated. Black baffles aid by absorbing and appearing dark. Other colors give a ring of light at the baffled edge.

Cones (4), (5) control brightness by cutoff and by redirection of light due to shape. They are either parabolic or elliptical. A light specular finish such as aluminum appears dull; a black specular finish appears unlighted.

Fig. 20.24 *Methods for shielding downlights using circular shields for vertically symmetrical sources such as incandescent and HID lamps. Halogen lamps mounted vertically in* R *lamp envelopes are symmetrical; horizontally mounted units in PAR or other reflectors are not (see Fig. 19.43). CFL lamps used in reflectors are not normally a serious source of direct glare, and are considered to be vertically symmetrical. A shielding angle of 45° minimum is recommended for high-luminance lamps. Black finishes require high-quality maintenance because dust shows as a bright reflection.*

Lamp holder
FLUORESCENT LUMINAIRE
Fluorescent lamp
Shielding angle
Shielding angle
45°
30°
(1)
(2)
45° Transverse (crosswise) shielding
30° Longitudinal (endwise) shielding

Fig. 20.25 *Shielding of fluorescent lamps is less critical due to lower lamp luminance. For T12 lamps, 45° × 35° crosswise/lengthwise shielding as shown is excellent and 35° × 30° is satisfactory. For T8 lamps, 45° × 35° should be used. Because fixture luminance is higher in the transverse direction than lengthwise, a better cutoff angle is required. The shielding elements may be louvres or baffles. Opaque shielding elements have a higher visual comfort rating than translucent plastic units.*

and the inclined plane at which the lamp first becomes visible as one approaches the fixture (Figs. 20.24 and 20.25). *Cutoff angle* is usually defined as being synonymous with shielding angle. However, because some sources define it as the complement of shielding angle, it is best to avoid the term and use only *shielding angle*.

7. *Ceiling illumination* is produced by light above the horizontal. As with light below the horizontal, a spread characteristic [Fig. 20.23(a)] means good ceiling coverage, no hot spots, and, ultimately, good diffuseness. Concentrated uplight means a potential hot spot if the fixture suspension hanger is too short, and in any event it yields uneven ceiling illumination.

Thus, we see that a rapid inspection of a fixture curve performed by an informed person can yield a large amount of information on the fixture's performance. The reader is encouraged at this juncture to review our comments on the two distribution curves of Fig. 20.23 and then analyze similarly other distribution curves in manufacturers' catalogs.

20.20 Luminaire Light Control

(a) Lamp Shielding. Except where it is desired to use a bare lamp as a source of sparkle, such as in chandeliers and other decorative fixtures, all lamps in interior fixtures should be shielded from normal sight lines (i.e., sight lines in a head-up, eyes straight-ahead position; see Fig. 18.20). The reason is obvious; bare lamps are so bright (see Table 18.2) that they usually constitute a source of direct or even disabling glare, depending on apprehended angle (closeness to the eye and size of lamp) and eye adaptation level. The range of permissible luminaire luminances (listed in Table 18.3) of 1000 to 7000 cd/m² depends on these two variables (apprehended angle and adaptation level).

As a general rule, exposed incandescent lamps, 6 W and larger, are sources of direct glare and should be avoided. Note that the upper direct-glare limit corresponds to the luminance of a bare 34-W T12 fluorescent tube, which accounts for the fact that bare-lamp fluorescent fixtures are well tolerated. However, when such a fixture is relamped (and re ballasted) with a more efficient, better CRI, T-8 lamp whose luminance *exceeds* 10,000 cd/m², it becomes a source of annoying direct glare that actively impairs visual ability.

Shielding of lamps is accomplished with the fixture housing/reflector or with baffles and louvres, as mentioned previously (see Figs. 20.24, 20.25, and 20.26). Fluorescent fixtures require shielding most when placed crosswise to the line of sight, thus exposing the entire length of lamp to

(b)

(a)

Fig. 20.26 *Shielding elements.*
(a) *The most basic shielding element is the lamp reflector which may double as the fixture body.*
(b) *Shields perpendicular to the long axis of a linear fluorescent lamp are normally called baffles. They are less important than lengthwise shields (louvres) because endwise lamp luminance is lower than crosswise luminance.*
(c) *Two-way shielding is most effective but seriously lessens luminaire efficiency. (From Stein, B.,* Building Technology, Mechanical and Electrical Systems, *2nd ed, 1997; reproduced by permission of John Wiley & Sons.)*

(c)

Fig. 20.27 *Parabolic reflector action: (a) with source at focal point rays are parallel; (b) with source below focal point they converge; (c) with source above focal point they diverge. This focusing action is illustrated by fixtures correspondingly designated. Note that type (c) requires a large ceiling opening to achieve even minimal efficiency.*

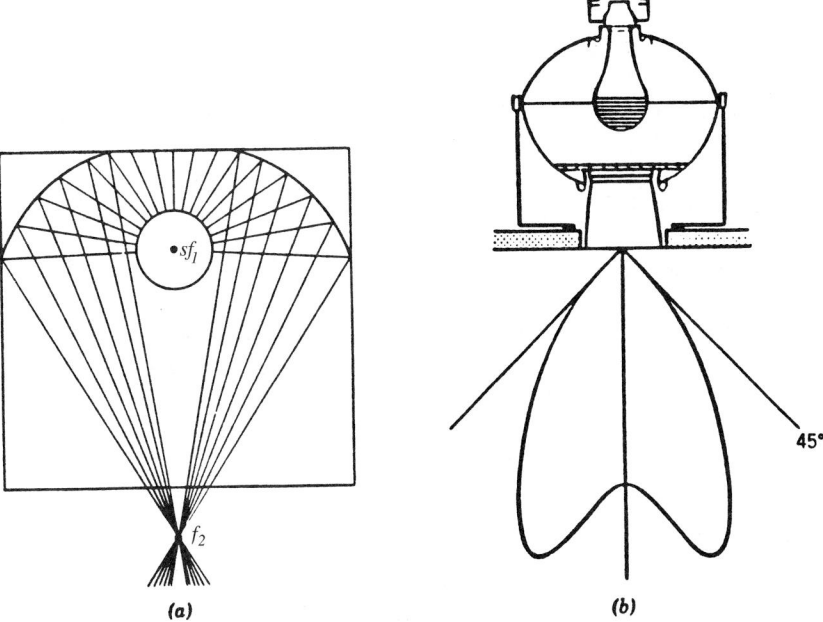

(a)   **(b)**

Fig. 20.28 (a) *Action of elliptical reflector section. With the light source at focal point* f1 *the light converges at the other focal point* f2. *This effect is useful in fixture design as in* (b). *By projecting light up only (through use of silvered bowl lamp) the output light can be redirected through a constricted aperture at the other focal point, with little loss. Beyond the aperture, the light diverges. This design is the basis of high efficiency "pinhole" downlights. Without the (large) elliptical reflector most of the lamps output is trapped inside the fixture, yielding a very low efficiency unit.*

the field of view. Such fixtures require longitudinal baffles, deep housings, or louvres to provide the necessary shielding. Alternatively, when at all possible, place fixtures with their long axis in the direction of sight lines, as stated previously.

(b) Reflectors. It is important to understand the action of luminaire reflectors. The basic shapes and beam patterns are illustrated in Figs. 20.27 through 20.29. Note from Fig. 20.28 that the so-called pinhole downlight requires an elliptic reflector to focus the light through this hole at point *f2* in order to maintain even minimal fixture efficiency. Elliptic reflectors are large, and frequently space above the ceiling is too restricted for their use. Lamps with integral elliptic reflectors can be utilized with a standard baffled reflector to achieve roughly the same effect.

(c) Reflector Materials. Until fairly recently, reflector materials were of two types: white gloss paint for portions of fixture body interiors that acted as reflectors, and formed anodized alu-

minum sheet for the shaped reflectors of the types shown in Figs. 20.27 through 20.29. The reflectances (reflection factors) of both these materials are approximately the same, varying between 0.84 and 0.88 *when new and clean*. Neither, however, is specular; the paint finish is actually primarily

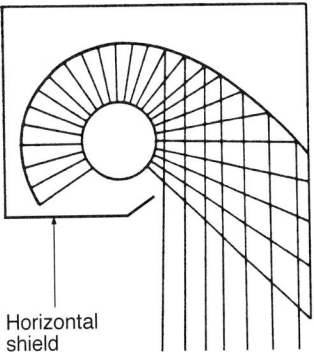

Horizontal shield

Fig. 20.29 *The extended section reflector allows the source to be concealed (shielded) while projecting its light directly downward but horizontally displaced from the source.*

Fig. 20.30 (a) *Approximately 40% of lamp output in an open linear fluorescent fixture is unrestricted. The remainder leaves the fixture after one or more reflections.* (b) *A mirrored reflector narrows the distribution pattern of the luminaire by specular reflection, and increases output somewhat (see text).*

diffuse whereas the aluminum is principally specular. Where shaped reflectors are not used, as in the case of a fluorescent troffer, the lack of specularity is essentially immaterial because at worst, the diffuseness will reduce luminaire output slightly by increasing the number of interreflections within the fixture body. [Fig. 20.30(*a*)]. The idealized specular reflections for shaped reflectors shown in Figs. 20.27 through 20.29 are just that; in reality, the reflectances are considerably more diffuse and become increasingly so with reflector aging and dirt accumulation.

Painted fixture body interiors lose their high reflectance by rapid aging due to elevated temperatures and accumulation of dust and dirt. This causes a decrease in overall reflectance, which is compensated by initial overdesign, as explained in Section 20.32. However, because a rectangular fixture body is not an accurately shaped reflector, the result of overall reflectance reduction is simply an overall reduction in output, yet maintaining the same photometric *distribution* characteristics.

Energy conservation programs and utility rebates (now terminated) led to the introduction in retrofit of very high-reflectance auxiliary reflectors that were added to existing fluorescent troffers approximately as shown in Fig. 20.30(*b*). Unfortunately, many of the claims of highly increased efficiency were based on retrofitting aged, dirty luminaires, and their results can be very misleading. A reasonable estimate of the possible improvement in luminaire efficiency can be made by considering these facts:

· Approximately 40% of a lamp's output in an open luminaire is directed downwards and

is therefore completely independent of any reflector action.

· The difference in reflectance between a new, clean painted surface and that of an old, dirty surface is, *at most,* 50%. That means that the *maximum* light loss of an open fixture, due to poor maintenance, is 50% of 60% (the maximum reflected light component), or 30% of the overall light output. Reference is always to an open-bottom fixture.

· The maximum reflectance of the best (and most expensive) silver reflectors is about 95%, comprising 93% specular and 2% diffuse. This is only 10% higher than the original *minimum* 85% paint reflectance. Therefore, at most, retrofit of a very dirty old fixture with a high-quality (expensive) silver reflector improves performance by the 30% lost to dirt plus 10% of the 60% reflected light for a total of 36% *maximum.* Relamping, of course, also improves output, but that is not connected with fixture body reflectance.

· Simple cleaning of a very dirty fixture body restores 20% to 25% of the light loss. The remaining loss is due to aging of the paint. A cost analysis is required to determine whether the 10% to 15% differential in light output between simple cleaning and silvered reflector addition is economically feasible.

A very important consideration in retrofit and new construction is the photometric characteristic of high-reflectance linear fluorescent luminaires. In general, they are shaped like the curve of Fig. 20.23(*b*). This means, as explained in Section 20.16

that light is concentrated downward, resulting in a requirement for closer luminaire spacing to obtain uniform illumination. In new construction, this can be considered in design, although it can be a serious economic penalty. In retrofit work, it can result in unacceptable lack of illuminance uniformity, requiring additional luminaires and expensive relocation of existing units.

Another factor to be considered is the degree of maintenance required to keep "super" reflectors in pristine condition in order to achieve the 15%± maximum output differential. To determine this, designers should request an aging test and inspect a previously retrofited installation with similar ambient conditions and cleaning schedules to those of the area under consideration.

20.21 Luminaire Diffusers

Subsumed under this heading, in trade parlance, are all of the devices placed between the lamp(s) and the illuminated space, whose function it is to diffuse the light, control fixture brightness, redirect the light, obscure (hide), and shield the lamps. As most of these devices perform multiple functions, they are discussed individually.

(a) Translucent Diffusers. Because these do not redirect the light, but merely diffuse it, the distribution characteristic is circular, as seen in

Fig. 20.31(*a*). Typical examples of this type are white opal glass, frosted glass, and white plastics such as Plexiglas, polystyrene, vinyl, and polycarbonates. The distribution is basically the same as it would be for bare lamps. Lamp hiding power is good. Depending on the material, direct glare can be a problem (VCP is poor). Veiling reflections are high. Spacing to mounting-height ratio (S/MH) does not exceed 1.5. The fixture is generally inefficient. Wall illumination is good because of a large component of high-angle light (which reduces VCP). The net result of using this type of diffuser is to lower lamp luminance by distributing the lamp output over a larger diffusing area. Applications include corridors, stairwells, high ceiling spaces, and other areas without demanding visual tasks.

A special type of flat plastic panel that polarizes the transmitted light was introduced some years ago. These panels held great promise because they produce a marked decrease of veiling reflection at an angle of 60°, but much less at other angles. Because most viewing is in the range 20° to 40°, using these panels does not result in any appreciable reduction in reflected glare in normal office-work situations. From experience in a drafting room equipped with luminaires utilizing high-efficiency multilayer polarizers, it can be stated that visual discomfort from reflected glare, as personally experienced and as reported by a large staff, is not noticeably reduced.

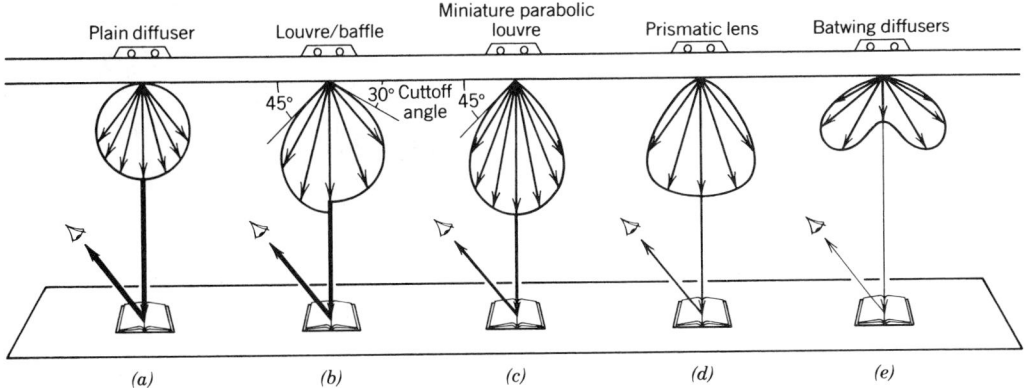

Fig. 20.31 *Comparison to typical candlepower distribution curves for common linear or PL lamp fluorescent luminaire diffuser elements (for a full description, see text). Note that for a given geometry of viewer and luminaire, the severity of veiling reflections depends entirely on the fixture's photometric characteristic. In the individual figures, the potential to produce reflected glare is indicated by weight of the line representing fixture output and reflectance from the work task. The batwing distribution (e) concentrates its output in the 30° to 60° range, which minimizes both direct and reflected glare.*

Fig. 20.32 (a) *Section through a conventional miniature parabolic wedge eggcrate type of louvre. These units give exceptionally low brightness when seen at normal viewing angle (see Fig. 20.33) and are therefore useful in VDT areas. Most such units are made of aluminized plastic. Fixtures equipped with these units exhibit low overall efficiency due to the large amount of light trapped by the broad top of each parabolic wedge.*

(b) *A modified wedge design uses a curved top on each wedge to redirect and utilize light striking the top. Solid lines represent light rays redirected by the bottom curve, whereas dotted lines show light redirected by the top curve, which was lost in the design of* (a). *Typical louvre cell dimensions are ½-in. cube for design* (a) *with a consequent 45° shielding angle and ⅝ - to ¾-in. square by ½-in. high for design* (b), *giving a 35° to 45° shielding angle.*

(b-1)

(a)

(b-2)

(b) Louvres and Baffles.

These are generally rectangular section, metal or plastic, and serve primarily to shield the source (see Figs. 20.18, 20.24, and 20.25) and also diffuse the output, particularly when using plastic translucent louvres. Candlepower distribution curves are shown in Fig. 20.31(*b*). The exact curve shape depends on the shielding angle, design of the louvre, and its finish. Louvres finished in specular aluminum or dark colors exhibit low direct glare. The large downward light component can cause serious veiling reflections. Overall fixture efficiency is average.

The S/MH, a luminaire metric that indicates the maximum spacing permissible for a given luminaire mounting height that yields uniform illumination and is given as a dimensionless ratio, is fully explained in the next section. For this diffuser type it does not exceed 1.5 and varies inversely with shielding angle. This is because the basic circular distribution is changed to egg shape by cutoff and redirection, reducing the high-angle light. Thus, a 45° shielding angle has lower direct glare but requires closer spacing.

A special design of this category is the miniature egg-crate parabolic wedge louvre shown in Figs. 20.32 and 20.33. These units redirect a large portion of the light directly downward, and because of this redirection and their specular finish,

they appear completely dark; darker indeed than the unlighted portion of the ceiling when viewed obliquely (see Figs. 20.33 and 21.00). Fixtures using these louvres have low efficiency due to

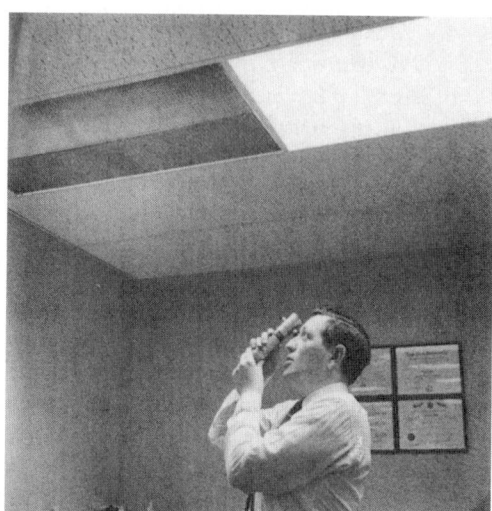

Fig. 20.33 *In the illustration the author is checking the luminance of a fluorescent fixture equipped with a miniature parabolic wedge louvre, which has very low luminance above 45° (see Fig. 20.32). This characteristic is readily apparent in comparison to the adjacent fixture, which utilizes a prismatic lens diffuser.*

trapped light, with a maximum coefficient of utilization of about 0.5 (see Section 20.32). VCP is very high, but veiling reflections can be troublesome; S/MH varies between 1.0 and 1.5. Shielding angle is usually 45°. A typical candlepower distribution curve is shown in Fig. 20.31(c). When using these units, additional wall lighting is almost always required.

Because of the low efficiency of the parabolic louvre design shown in Fig. 20.32(a) caused by the wide light-trapping tops of the louvres, an improved version was developed with the tops shaped to reflect incident light efficiently. This improved design, which is shown in Fig. 20.32(b), increases luminaire efficiency by about 20%.

(c) Prismatic Lens.

There are many designs available with varying distribution characteristics. Figures. 20.23(a) and 20.31(d) can be taken as typical of this genre. They produce an efficient fixture (high coefficient of utilization), good diffusion, wide permissible spacing—S/MH as high as 2.0—and low direct glare (high VCP). Veiling reflections can be troublesome, depending on viewing angles and position (see Fig. 18.34).

(d) Fresnel Lens.

The action of this lens is similar to that of a reflector. Lamp hiding power is poor, but efficiency is high and visual comfort is good. S/MH is seldom more than 1.5 (Fig. 20.34).

(e) Batwing Diffusers.

The theory of this type of diffuser is covered in paragraph 4 of Section 18.31(c). A typical characteristic is shown in Fig. 20.31(e). There are two fluorescent luminaire designs that produce the batwing shape distribution characteristic—a prismatic lens and parabolic reflectors and baffles.

1. *Prismatic batwing diffusers.* These are either linear or radial; that is, they produce the batwing distribution either in one direction or in all directions. Typical characteristics of both types are shown in Fig. 20.35. Note that the characteristic shape is more pronounced in the linear diffuser, which indicates better control of veiling reflections in that direction (usually crosswise). Fixtures equipped with these diffusers have good efficiency, low direct and reflected glare, and good diffusion. As with all enclosed ungasketed fix-

tures, the lens acts as a dust trap, necessitating frequent cleaning to maintain high output.

2. *Deep parabolic reflectors.* These luminaires (see Fig. 20.36) produce modified versions of the characteristic batwing distribution in the normal (crosswise) direction. Distribution in the parallel or lengthwise direction is circular (diffuse), indicating minimum beam control in that direction. These fixtures, like the batwing lens–type diffuser units, have high efficiency, high S/MH, low reflected glare, and low to very low surface brightness, making them usable in VDT areas. They

By utilizing a Fresnel lens fixture, a smaller housing without a reflector can be used, while still maintaining beam control

Fig. 20.34 *Action of a Fresnel lens. The lens performs the same function as a reflector, controlling the beam as a function of source placement. By utilizing a lens fixture, the curved reflector can be largely eliminated, yielding a smaller fixture while maintaining accurate beam control. Common design* (c) *uses regressed lens to provide shielding, although lens brightness is not normally objectionable.*

ILLUMINATION

(a)

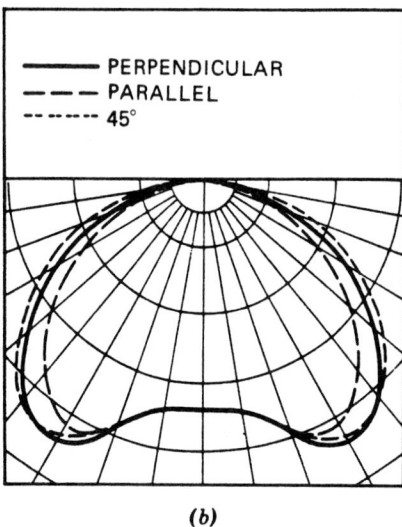

——— PERPENDICULAR
– – – PARALLEL
– – - – - 45°

(b)

Fig. 20.35 (a) *Linear batwing distribution with extremely sharp cutoff in the upper and lower ranges. Curve taken across lamp axis for single-lamp unit. (b) Distribution curves for a radial batwing distribution lens. Note that the perpendicular, parallel, and diagonal curves are almost identical. Zonal flux is maximum in the 30° to 60° range and drops off at both extremes, as desired.*

are normally applied with the long axis in the direction of sight lines.

20.22   Uniformity of Illumination

In any space intended to be lighted uniformly with multiple, discrete, ceiling-mounted direct-lighting system light sources, it is necessary to establish a fixture spacing that gives acceptable uniformity of illumination. A ratio of maximum to minimum illumination on the working plane of 1.2 to 1.3 is readily acceptable because lesser ratios are not easily noticed. For general background or circulation lighting, up to 1.5 is acceptable. The recommended spacing-to-mounting-height (above the working plane) ratio given by manufacturers (see figures immediately above the distribution curves for each fixture in Table 20.2) are generally based on a 1.0 illuminance ratio. (Fig. 20.37). Therefore, the recommendation may be exceeded somewhat without seriously affecting uniformity.

When the luminaire's distribution characteristic is symmetrical in all directions, as is generally the case with small source lamps such as incandescent, CFL, and HID, only a single S/MH figure is required. However, with the asymmetrical distribution of most fluorescent fixtures, a S/MH ratio is required both crosswise and lengthwise. Due to the characteristic of the lamp itself, the transverse (crosswise, perpendicular) ratio is almost always considerably higher than the longitudinal (parallel, endwise, lengthwise) ratio. See, for instance, fixtures 26, 28, and 42 of Table 20.2. The S/MH (also called *spacing criteria;* SC) for a specific luminaire is determined by measuring the distance between two test luminaires that yields the same illuminance on the working plane midway between them as directly under each one. This ignores (deliberately) any contributions from other fixtures in a multifixture installation and from interreflections, accounting only for the direct component of illuminance from the two test fixtures. Therefore, in an actual installation with several rows of fixtures, the illuminance at point P_1 in Fig. 20.38 is *higher* than the average by 20% to 30% because of the other fixtures and interreflections, and the illuminance at point P_2 is approximately equal to the room average. The illuminance levels along the walls, assuming a distance between the last row and the wall equal to one-half of the side-to-side spacing, range from 60% of average at point P_3 down to 50% at point P_4, assuming light-colored walls.

When walls are dark due to paint or aging, bookshelves, dark wood paneling, and the like, the illuminance levels drop to less than 50% of average, which is obviously insufficient as task lighting. To counteract this effect, particularly when placement of furniture is such that visual tasks *will* occur near walls, the designer has three choices:

TOTAL LUMINAIRE EFFICIENCY   =   66.0 %
TOTAL REFLECTANCE OF PAINT   =   88.3 %
CIE TYPE – DIRECT

PLANE	:	0-DEG	90-DEG
SPACING CRITERIA	:	1.2	1.6
SHIELDING ANGLES	:	36	43
PLANE	:	0-DEG	90-DEG
LUMINOUS LENGTH	:	21.250	21.250

LUMINANCE DATA IN FOOTLAMBERTS

ANGLE IN DEG	AVERAGE 0-DEG	AVERAGE 45-DEG	AVERAGE 90-DEG
45	2451.	3101.	1927.
55	494.	639.	55.
65	0.	0.	0.
75	0.	0.	0.
85	0.	0.	0.

Recessed

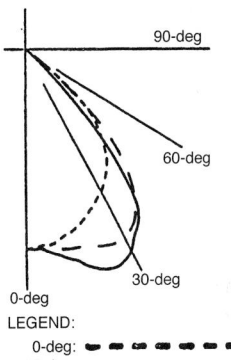

LEGEND:
0-deg:
45-deg:
90-deg:

ILLUMINATION

Visual Comfort Probability

Reflectance - 80, 50, 20
Work Plane Illumination - 100 FC @ 2.5 ft.   **VCP**

Room Room W L	Ht.	Luminaires Lengthwise 8.5	10.0	13.0	Ht.	Luminaires Crosswise 8.5	10.0	13.0
20 20		68	81	77		90	90	87
20 40		69	81	77		91	90	84
30 30		69	81	77		91	90	84
30 60		69	82	78		91	90	84
40 40		69	82	78		91	90	84
40 60		70	82	78		91	90	84
60 30		69	82	78		91	90	84
60 40		70	82	78		91	90	84
60 60		70	82	78		91	91	84

Coefficients of Utilization

Coefficients of Utilization
Zonal Cavity Method   Floor Reflectance - .20

RC RW	80 70	50	30	10	50 50	30	10
1	75	73	71	69	69	67	66
2	71	67	65	62	64	62	60
3	67	62	59	56	59	57	54
4	63	57	54	50	55	52	50
5	59	53	49	46	51	48	45
6	55	49	44	41	47	43	41
7	51	44	40	37	43	39	36
8	48	41	36	33	39	35	33
9	44	37	32	29	36	32	29
10	41	34	29	26	33	29	26

Energy Data

LER: FP-42   **Energy Cost: $5.71***
Input Watts: 134   **BF: .90**

The above energy calculations were conducted using a specific lamp/ballast combination. Actual results may vary depending upon the lamp and ballast used. Lamp and ballast specifications are subject to change without notice.

*Comparative annual lighting energy cost per 1000 lumens based on 3000 hours and $0.08 per KWH.

Fig. 20.36 *Modern 2-ft square deep parabolic reflector luminaire designed for three F40, 22.5-in. twin-tube CFL lamps, with a rated output of 3150 lm each. Total power input is 134 W, and BF is 0.9, giving an luminaire efficacy rating (LER) of 42. The typical modified batwing crosswise distribution and circular lengthwise distribution are clearly shown. Photometric data of interest to lighting designers are also shown. (Courtesy of Columbia Lighting.)*

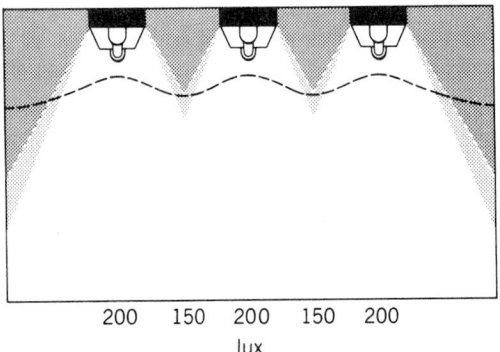

200   150   200   150   200

lux

Fig. 20.37 *The ratio of maximum to minimum illumination should not exceed 1.30 in areas requiring uniform illumination (see text).*

1. Reduce the distance between the last row of lights and the wall to a third or less of the row-to-row spacing. This also provides required wall lighting.
2. Provide some type of continuous perimeter lighting or wall-wash units, both of which increase illuminance levels at the walls.
3. A combination of the methods 1 and 2.

Because endwise illuminance from linear fluorescent fixtures is considerably lower than crosswise, end walls have lower illuminance than side walls and greater illuminance variation. It is, therefore, particularly important to provide some additional illumination as discussed previously, and terminate fixture rows no more than 1 ft from an end

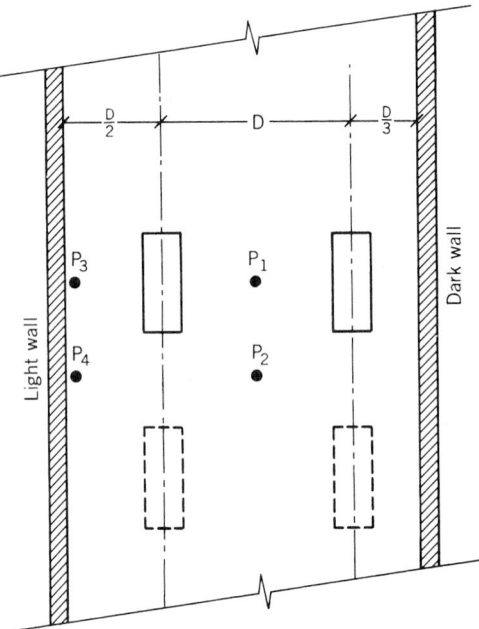

Fig. 20.38 *The diagram shows lighting fixtures installed according to the manufacturer's recommended spacing criteria (spacing to mounting above the workplane ratio), with a row-to-row spacing D and a row-to-wall spacing, D/2, as shown at the left wall. Assuming a high reflectance finish on the wall (light color), illuminances P_3 and P_4 are at least one-half of the illuminance directly below a fixture. At a dark wall, as on the right, illuminance would fall below this value. As a consequence, the designer would move the row of luminaires closer to the wall, as shown (for further discussion, see text).*

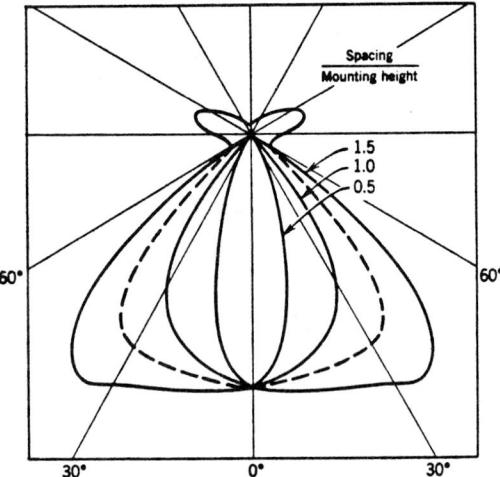

Fig. 20.39 *Typical distribution curves for approximating ratio of fixture spacing to the mounting height (S/MH) above the working plane, for direct lighting luminaires. To determine S/MH for a specific unit, compare its curve to those shown and interpolate. Although these curves were developed for direct lighting point sources such as incandescent (and HID), they can also be used with asymmetric distribution luminaires such as fluorescent. In such cases, the curves give the S/MH ratio in the plane whose characteristic is used—that is, lengthwise or crosswise. The dashed curve was transferred from Fig. 20.23 (a). Its permissible S/MH is somewhat higher than the curves indicate because it is a semidirect distribution and its ceiling light component permits wider spacing between units. (After Odle and Smith, from IESNA/Journal, January 1963.)*

wall. This is all the more important where visual tasks without supplementary task lighting occur at these walls.

We mentioned prevlously that the fixture of Fig. 20.23(a) had a high S/MH because of its flat-bottomed curve. This ratio, when not given by the manufacturer, may be approximated from Fig. 20.39. An accurate method of calculating maximum to minimum illumination ratios is available (see *IESNA Handbook,* 1993).

The foregoing discussion of illumination uniformity concerned itself with uniformity on a horizontal work plane. Occasionally, it is necessary to know the degree of uniformity vertically—that is, on horizontal planes at different elevations *directly below the fixtures.* Four different lighting situations are normally encountered. They are point sources, such as point source downlights; line sources, such as continuous-row florescent fixtures; infinite sources, such as luminous ceilings—whether transilluminated or indirect; and parabolic reflector beams, such as from PAR lamps. The vertical uniformity of each type is shown graphically in Fig. 20.40.

20.23 Luminaire Mounting Height

The mounting height of luminaires is normally established before their spacing. In arriving at a mounting height for fixtures with an upward component, a balance must be struck between low mounting, which controls ceiling brightness and gives good utilization of light, and the reticence to dominate an area, particularly a large room, by using such a low mounting height that the apparent ceiling height is affected (Fig. 20.41). Rules of thumb for mounting height are:

1. Indirect and semi-indirect luminaires should normally be suspended no less than 18 in. from ceiling, and preferably 24 to 36 in. Single lamp luminaires with a very wide distribution (inverted batwing) may be suspended as little as 12 in. from the ceiling. Manufacturers' recommendations should be sought on this point.
2. Direct–indirect and semi-direct fluorescent fixtures should be suspended not less than 12 in. for two-lamp units and 18 in. for three- and four-lamp units.

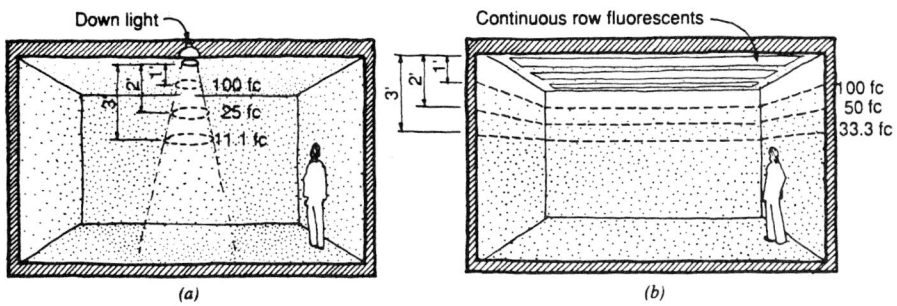

(a), (b) Illumination directly below the fixture varies inversely with the square of the distance for a point source and inversely as distance for a line source

(c), (d) Illumination remains constant at all distances from either an infinite (or nearly) source or a parabolic reflector

Fig. 20.40 *Variation of illuminance vertically, directly below the fixtures, for different source types.*

(a)

Fig. 20.41 *Mounting height of fixtures may be lower in a small room than in a large room because of the illusion of lowness created in a large room.*

(b)

The effect of pendant length on coefficient of utilization (efficiency; see Section 20.28) is given in Fig. 20.42.

Fig. 20.42 *Coefficient of utilization (CU; lighting system efficiency) as a function of pendant length for various distributions. With a substantial downward component as in direct–indirect or general diffuse (a), system efficiency rises slowly as the fixture descends (pendant length increases). Maximum differentials occur in small rooms and can reach 20%. Where the ceiling is the light source as in indirect and semi-indirect systems (b), the pendant length does not change the room illumination (see also Fig. 20.40). This curve can be used to estimate CU for indirect and semi-indirect, where not found in Table 20.2 or manufacturers' catalogs.*

20.24 Lighting Fixtures

Before proceeding further with design we will discuss the principal item of lighting hardware: the luminaire itself. This section and the sections that follow cover luminaire construction, installation, and appraisal. The architect need simply stop to consider that lighting fixtures constitute 25% to 30% of the electrical budget or 4% to 5% of the overall building budget to appreciate their importance. Because the difference between a quality unit and an inferior one is often not readily visible to the casual observer, particular care must be taken in the specification of lighting fixtures and in examination of shop drawings *and* samples. All fixtures, if applied properly, give a sufficient quantity of light, but only a good unit combines quantity with good quality, ease of installation, facility of maintenance, and indefinite life. In addi-

tion, installation must be proper to ensure mechanical rigidity and safety, electrical safety, freedom from excessive temperatures, and requisite accessibility of component parts and of the fixture outlet box. The following material is a combination of *NEC* minimum requirements plus factors that the author has found important beyond these minima.

20.25 Lighting Fixture Construction

1. All fixtures should be wired and constructed to comply with local codes, *NEC* (Article 410), and Underwriters Laboratories(UL) Standard for Lighting Fixtures, and shall bear the UL label, where label service is available. Reflector Luminaire Manufacturers (RLM) standards should be adhered to for all porcelain-enameled fixtures.

2. Fixtures should generally be constructed of 20-gauge (0.0359-in.) thick steel minimum. Cast portions of fixtures should be no less than ¹⁄₁₆ in. thick.

3. All metals should be coated. The final coat should be a baked-enamel white paint of minimum 85% reflectance, except for anodized, aluminum, or silvered surfaces.

4. No point on the outside surface of any fixture should exceed 90°C after installation and on continuous operation. For an exception, see *NEC* Article 410 M.

5. Each fixture should be identified by a label carrying the manufacturer's name and address and the fixture catalog number.

6. Glass diffuser panels in fluorescent fixtures should be mounted in a metal frame. Plastic diffusers should be suitably hinged. "Lay-in" plastic diffusers should not be used.

7. Plastic diffusers should be of the slow-burning or self-extinguishing type with low smoke-density rating and low heat-distortion temperatures. This latter should be low enough so that the plastic diffuser distorts sufficiently to drop out of the fixture before reaching ignition temperature.

8. It is *imperative* that plastics used in air-handling fixtures be of the noncombustible, low smoke-density type. These requirements also apply to other nonmetallic components of such fixtures.

9. All plastic diffusers should be clearly marked with their composition material, trade name, and manufacturer's name and identification number. Results of ASTM combustion tests should be submitted with fixture shop drawings.

The characteristics of many plastic diffusers change radically with age and exposure to ultraviolet (UV) light.

A brief survey of the most common standard transparent and translucent lighting fixture diffusers follows:

Glass. Transparent to translucent, scratch resistant, easily cleaned, nonflammable, available in all grades of impact resistance, non-yellowing, usable with all sources, unaffected by and effective in blocking UV, readily formed into desired patterns. Heavy and expensive.

Acrylic (Plexiglass). Clear to translucent, easily scratched, slow burning, low impact resistance, resistant to yellowing and available in special low-yellowing composition (UV grade). Usable indoors and outdoors. Not usable at the elevated temperatures (>90°C) found in some HID applications. Does not readily embrittle, warp, or craze. Molds well.

Polycarbonate. Initially very clear and highly impact resistant, but with a tendency to opacity and strength loss with age. Good scratch resistance and excellent thermal resistance. Usable indoors and out, with all sources. Readily molded to prismatic forms. Self-extinguishing (burning rate). Expensive.

Polystyrene. Usable only indoors because of rapid yellowing when exposed to exterior UV. Slow burning, but smoke generation properties problematic with some fire codes. Not usable in the long term because of discoloration, particularly with UV producing sources. Good thermal resistance. Readily molded. Cheap. Not scratch resistant.

As can readily be seen, no ideal diffuser material exists. The designer must select the material that best suits the use and budget, considering both initial and replacement costs, as well as long-term optical properties.

10. Ballasts should be mounted in fixtures with captive screws on the fixture body, to allow ballast replacement without fixture removal.

11. All fixtures mounted outdoors, whether under canopies or directly exposed to the weather,

ILLUMINATION

should be constructed of appropriate weather-resistant materials and finishes, including gasketing to prevent entrance of water into wiring, and should be marked by the manufacturer, "Suitable for Outdoor Use".

20.26   Lighting Fixture Structural Support

Although some codes allow fluorescent fixtures weighing less than 40 lb to be mounted directly on the horizontal metal members of hung-ceiling systems, experience has shown that vibration, member deflection, routine maintenance operation on equipment in hung ceilings, and poor workmanship can cause such fixtures to fall, endangering life. It is therefore strongly recommended that all fixtures, surface, pendant, or recessed, whether mounted individually or in rows, be supported from the black channel iron supporting the ceiling system (purlins) or directly from the building structure, but in no case by the ceiling system itself. This is particularly important in the case of an exposed "Z" spline ceiling system.

20.27   Lighting Fixture Appraisal

The intense competition in the lighting field necessitates close scrutiny of the characteristics of luminaires and all accessories. To compare the relative merits of similar lighting fixtures as manufactured by different companies, complete test data plus a sample in a regular shipping carton from a normal manufacturing run are needed.

The following list should be used as a basic guide, with additional items added according to job requirements:

1. *Photometric and Design Data.* Manufacturers should furnish complete test data, including candlepower distribution curve(s), coefficients of utilization (CU), wall and ceiling luminance coefficients, luminance data from 45° to 85°, a table of visual comfort probability (VCP), energy data including LER (see Section 20.29), and recommended S/MH (SC). These data should come from a reliable, independent testing laboratory, and not from the manufacturer's test facilities. In addition, many

manufacturers either regularly publish or make available on request various design aids such as isolux (isocandle) curves and point-by-point computer printouts for different layouts. These very useful data and their application are covered in Section 20.28.

2. *Construction and Installation.* The designer should check the sample for workmanship; rigidity; quality of materials and finish; and ease of installation, wiring, and leveling. Installation instruction sheets should be sufficiently detailed. Results of actual operating temperature tests in various installation modes should be included. Air-handling fixtures should be furnished with heat-removal data, pressure-drop curves, air-diffusion data, and noise-criteria (NC) data for different airflow rates.

3. *Maintenance.* Luminaires should be simply and quickly relampable, resistant to dirt collection, and simple to clean. Replacement parts must be readily available.

20.28   Luminaire–Room System Efficiency: Coefficient of Utilization

Because of internal reflections inside a luminaire, some of the generated lumen output of the lamp is lost. The ratio of output lumens to lamp (input) lumens, expressed as a percentage, represents the luminous efficiency of the fixture. This characteristic has little meaning by itself, however, as the efficiency of a luminaire in doing a particular lighting job depends on the space in which it is used.

To illustrate, let us consider the case of a large room with a high, dark ceiling. If we were to use a high-efficiency (say, 80%) indirect lighting unit in such a room, most of the light directed upward would be lost (absorbed) and the actual lighting on the working plane would be very low. If, however, the same room were illuminated with 50% efficiency direct-lighting units utilizing the same wattage, the illumination on the working plane would be considerably higher than in the first case.

Similarly, if we consider a small room with dark walls and ceiling, lighted alternatively by diffuse lighting and direct lighting units of the same wattage and unit efficiency, the horizontal-plane illumination is higher in the case of the direct units because of the large loss of the horizontal and upward components of the diffuse lighting on the walls and ceiling. It should be obvious then that fixture efficiency *alone* is not sufficient but *the overall luminous efficiency of a particular unit in a particular space* is the required figure of merit. This number, inasmuch as it describes the utilization of the fixture output in a specific space, is known as the *coefficient of utilization* (CU). It is defined as the ratio between the lumens reaching the horizontal workplane and the generated lumens. As each luminaire has a different coefficient for every different space in which it is used, a system of standardization has been evolved utilizing room cavities (explained later) of certain proportions and various surface reflectances. The fixture coefficients are then computed and tabulated as shown in Table 20.2. It should be emphasized that the figures given in Table 20.2 are for the generic fixture type only; in an actual job, luminaire data as found in manufacturers' catalogs should be used. To summarize, CU is a factor that combines fixture efficiency and distribution with room proportions, mounting height, and surface reflectances.

20.29 Luminaire Efficacy Rating (LER)

As a result of an EPACT mandate calling for an industry-wide testing and information program designed to improve lighting fixture energy, a collaborative effort produced NEMA standard *LE5, Procedure for Determining Efficacy of Luminaires.* Unlike the coefficient of utilization discussed in the previous section, which defines the illumination efficiency of a particular luminaire in a particular space, standard LE5 determines the energy efficiency of the luminaire *alone.* Because this efficiency is expressed in lumens per watt (lumens output per watt input), it uses the same descriptive term as is used for light sources (i.e., efficacy). This metric takes into account all power used by a luminaire, including ballast, and includes ballast factor, which is itself a ballast energy efficiency metric. The expression used to calculate LER is:

$$LER = \frac{\text{photometric efficiency} \times \text{ballast factor}}{\text{luminaire input watts}}$$

An LER metric applies to a specific type of fluorescent luminaire, and is identified by an abbreviation as:

FL = fluorescent lensed

FP = fluorescent parabolic

FW = fluorescent wraparound

FI = fluorescent industrial

FS = fluorescent strip light

Thus, the energy figures shown in Fig. 20.36 are specifically labeled "FP" to denote a fluorescent parabolic luminaire. This enables a designer to compare LER figures of different fixtures on a common basis. This commonality also extends to ballast type (i.e., magnetic or electronic). In addition, standard LE5 lists benchmark LER figures that are considered to represent an acceptable luminaire. At this writing, LER data is available only for the five types listed here. The intention is to add HID luminaires and eventually cover most common luminaire types.

An additional item of data useful in comparisons that is included in the NEMA standard is economic, in the form of yearly cost per 1000/lm, based on 3000 burning hours and $0.08 per kWh (see Fig. 20.36.) Actual cost is easily calculated from this figure.

DETAILED DESIGN PROCEDURES

20.30 Calculation of Average Illuminance

Having selected a luminaire on the basis of all the foregoing criteria, it remains only to calculate the number of such units required in each space, for uniform *general* illumination, and to arrange them properly. Although a number of calculation methods are available, the lumen (flux) method is simplest and most applicable to our needs for area-lighting calculations. Illuminance calculation from point, line, or area sources is covered in Sections 20.39 to 20.42. Luminance (photometric brightness) calculations are covered in Section 20.44.

Before beginning a detailed description of the zonal cavity calculation method, a general comment on precision is in order. The precision of any calculation need not exceed either the accuracy

ILLUMINATION

TABLE 20.2 Coefficients of Utilization for Typical Luminaires with Suggested Maximum Spacing Ratios

To obtain a coefficient of utilization:

1. Determine cavity ratios for the room, ceiling, and floor.
2. Determine the effective ceiling and floor cavity reflectances from Table 20.3. Use initial ceiling, floor, and wall reflectances.
3. Obtain coefficient of utilization (CU) for 20% effective floor cavity reflectance from appropriate table below for luminaire type to be used. Interpolate, when necessary, to obtain CU for exact room cavity ratio for nearest effective ceiling cavity reflectances above and below reflectance obtained in step 2; interpolate between these CUs to obtain CU for step 2 ceiling cavity reflectance.
4. If effective floor cavity reflectance differs significantly from 20%, obtain multiplier from Table 20.4 and apply this to the CU obtained in step 3.
5. To obtain CU for a ceiling cavity reflectance (ρ_{CC}) of 30 or 10%, multiply the figure for ρ_{CC} = 50% by 0.85 and 0.70, respectively. This is an approximation. For exact figures see *IES Handbook* (1987).
6. Use the figure in the last column (ρ_{CC} = 0; ρ_w = 0 for outdoor lighting, i.e., no walls or ceiling.
7. Legend:

$$\rho_{CC} = \text{percent effective ceiling cavity reflectance}$$

$$\rho_w = \text{percent wall reflectance}$$

$$RCR = \text{room cavity ratio}$$

$$\text{Maximum } S/MH \text{ guide} = \text{ratio of maximum luminaire spacing to mounting above work plane.}$$

NOTE: In some cases, luminaire data in this table are based on an actual typical luminaire; in other cases, the data represent a composite of generic luminaire types. Therefore, whenever possible, specific luminaire data should be used in preference to this table of typical luminaires.

The polar intensity sketch (candlepower distribution curve) and the corresponding spacing-to-mounting height guide are representative of many luminaires of each type shown.

Typical Luminaire	Typical Distribution and Percent Lamp Lumens	Maintenance Category	Maximum S/MH	RCR	ρ_{CC}: 80			70			50			0
				ρ_w:	50	30	10	50	30	10	50	30	10	0
					Coefficients of Utilization for 20% Effective Floor Cavity Reflectance (ρ_{FC} = 20)									
1 Pendant diffusing sphere with incandescent lamp	5% / 45%	V	1.5	0	.87	.87	.87	.81	.81	.81	.69	.69	.69	.44
				1	.71	.67	.63	.66	.62	.59	.56	.53	.50	.31
				2	.61	.54	.49	.56	.50	.46	.47	.43	.39	.23
				3	.52	.45	.39	.48	.42	.37	.41	.36	.31	.18
				4	.46	.38	.33	.42	.36	.30	.36	.30	.26	.15
				5	.40	.33	.27	.37	.30	.25	.32	.26	.22	.12
				6	.36	.28	.23	.33	.26	.21	.28	.23	.19	.10
				7	.32	.25	.20	.29	.23	.18	.25	.20	.16	.09
				8	.29	.22	.17	.27	.20	.16	.23	.17	.14	.07
				9	.26	.19	.15	.24	.18	.14	.20	.15	.12	.06
3 Porcelain-enameled ventilated standard dome with incandescent lamp	10% / 85%	IV	1.3	0	.99	.99	.99	.97	.97	.97	.92	.93	.93	.83
				1	.88	.85	.82	.86	.83	.81	.83	.80	.78	.72
				2	.78	.73	.68	.76	.72	.67	.73	.69	.66	.61
				3	.69	.62	.57	.67	.61	.57	.65	.60	.56	.52
				4	.61	.54	.49	.60	.53	.48	.58	.52	.48	.45
				5	.54	.47	.41	.53	.46	.41	.51	.45	.41	.38
				6	.48	.41	.35	.47	.40	.35	.46	.39	.35	.32
				7	.43	.35	.30	.42	.35	.30	.41	.34	.30	.28
				8	.38	.31	.26	.38	.31	.26	.37	.30	.26	.24
				9	.35	.28	.23	.34	.27	.23	.33	.27	.23	.21
7 EAR-38 lamp above 51-mm (2 in.)-diameter aperture (increase efficiency to 54½% for 76-mm (3 in.)-diameter aperture)	0% / 43½%	IV	0.7	0	.52	.52	.52	.51	.51	.51	.48	.48	.48	.44
				1	.49	.48	.48	.48	.48	.47	.47	.46	.46	.42
				2	.47	.46	.45	.46	.45	.44	.45	.44	.43	.41
				3	.45	.44	.43	.45	.43	.42	.44	.42	.42	.40
				4	.43	.42	.41	.43	.41	.40	.42	.41	.40	.38
				5	.42	.40	.39	.41	.40	.38	.41	.39	.38	.37
				6	.40	.39	.37	.40	.38	.37	.39	.38	.37	.36
				7	.39	.37	.36	.39	.37	.36	.38	.37	.35	.35
				8	.37	.36	.34	.37	.35	.34	.37	.35	.34	.33
				9	.36	.34	.33	.36	.34	.33	.35	.34	.33	.32
18 "High-bay" wide distribution ventilated reflector with clear HID lamp	½% / 77½%	III	1.5	0	.93	.93	.93	.91	.91	.91	.87	.87	.87	.78
				1	.85	.82	.80	.83	.81	.79	.79	.78	.76	.70
				2	.77	.73	.70	.76	.72	.69	.73	.70	.67	.63
				3	.70	.65	.61	.68	.64	.60	.66	.62	.59	.56
				4	.63	.58	.53	.62	.57	.53	.60	.56	.52	.49
				5	.57	.51	.47	.56	.51	.47	.55	.50	.46	.44
				6	.51	.45	.41	.51	.45	.41	.49	.44	.40	.38
				7	.46	.40	.35	.45	.39	.35	.44	.39	.35	.33
				8	.41	.35	.31	.41	.35	.31	.40	.34	.31	.29
				9	.37	.31	.27	.37	.31	.27	.36	.30	.27	.25

(continued)

TABLE 20.2 Coefficients of Utilization for Typical Luminaires with Suggested Maximum Spacing Ratios (*continued*)

Typical Luminaire	Maintenance Category	Maximum S/MH	RCR	ρcc → 80			70			50			0
				ρw → 50	30	10	50	30	10	50	30	10	0
				Coefficients of Utilization for 20% Effective Floor Cavity Reflectance (ρFC = 20)									
26 Diffuse aluminum reflector with 35° crosswise shielding 17% / 66%	II	1.5/1.3	0	.95	.95	.95	.91	.91	.91	.83	.83	.83	.66
			1	.85	.82	.80	.82	.79	.77	.75	.73	.72	.59
			2	.76	.72	.68	.74	.70	.66	.68	.65	.62	.52
			3	.69	.63	.59	.66	.61	.57	.62	.58	.54	.46
			4	.62	.56	.51	.60	.54	.50	.56	.51	.47	.41
			5	.55	.49	.44	.53	.48	.43	.50	.45	.41	.36
			6	.50	.43	.39	.48	.42	.38	.45	.40	.36	.31
			7	.45	.38	.34	.43	.37	.33	.41	.36	.32	.27
			8	.40	.34	.29	.39	.33	.29	.37	.31	.28	.24
			9	.36	.30	.25	.35	.29	.25	.33	.28	.24	.20
28 Diffuse aluminum reflector with 35° crosswise × 35° lengthwise shielding 17% / 56½%	II	1.5/1.1	0	.83	.83	.83	.79	.79	.79	.72	.72	.72	.56
			1	.75	.72	.70	.72	.69	.67	.65	.64	.62	.50
			2	.67	.63	.60	.65	.61	.58	.59	.57	.54	.45
			3	.61	.56	.52	.58	.54	.51	.54	.50	.48	.40
			4	.55	.49	.45	.53	.48	.44	.49	.45	.42	.36
			5	.49	.44	.40	.47	.42	.39	.44	.40	.37	.31
			6	.45	.39	.35	.43	.38	.34	.40	.36	.33	.28
			7	.40	.35	.31	.39	.34	.30	.36	.32	.29	.25
			8	.36	.31	.27	.35	.30	.26	.33	.28	.25	.21
			9	.33	.27	.23	.32	.26	.23	.29	.25	.22	.19
33 Luminous bottom suspended unit with very-high-output lamp 66% / 12%	VI	N.A.	0	.77	.77	.77	.68	.68	.68	.50	.50	.50	.12
			1	.67	.64	.62	.59	.57	.54	.44	.42	.41	.10
			2	.59	.54	.50	.52	.48	.45	.38	.36	.34	.09
			3	.51	.46	.42	.45	.41	.37	.34	.31	.28	.07
			4	.45	.40	.35	.40	.35	.31	.30	.27	.24	.06
			5	.40	.34	.30	.35	.30	.27	.26	.23	.20	.05
			6	.36	.30	.26	.32	.27	.23	.24	.20	.18	.05
			7	.32	.26	.22	.28	.23	.20	.21	.18	.15	.04
			8	.29	.23	.19	.25	.21	.17	.19	.16	.13	.03
			9	.26	.20	.17	.23	.18	.15	.17	.14	.12	.03
35 Two-lamp prismatic wraparound; multiply by 0.95 for four lamps 11½% / 58½%	V	1.5/1.2	0	.81	.81	.81	.78	.78	.78	.72	.72	.72	.59
			1	.71	.69	.66	.69	.66	.64	.64	.62	.60	.50
			2	.64	.59	.56	.61	.58	.54	.57	.54	.51	.44
			3	.57	.52	.48	.55	.50	.47	.51	.48	.45	.38
			4	.51	.46	.41	.49	.44	.41	.46	.42	.39	.34
			5	.46	.40	.36	.44	.39	.35	.41	.37	.34	.29
			6	.41	.35	.31	.40	.35	.31	.38	.33	.30	.26
			7	.37	.31	.27	.36	.31	.27	.34	.29	.26	.23
			8	.33	.28	.24	.32	.27	.23	.30	.26	.22	.19
			9	.30	.24	.20	.29	.24	.20	.27	.23	.19	.17
38 Four-lamp, 610-mm (2 ft)-wide troffer with 45° plastic louvre 0% / 50%	IV	1.0	0	.60	.60	.60	.58	.58	.58	.56	.56	.56	.50
			1	.54	.52	.50	.52	.51	.49	.50	.49	.48	.44
			2	.48	.45	.43	.47	.44	.42	.45	.43	.41	.39
			3	.43	.40	.37	.42	.39	.37	.41	.38	.36	.34
			4	.39	.35	.32	.38	.35	.32	.37	.34	.32	.30
			5	.35	.31	.28	.35	.31	.28	.34	.30	.28	.26
			6	.32	.28	.25	.32	.28	.25	.31	.27	.25	.23
			7	.29	.25	.22	.29	.25	.22	.28	.25	.22	.21
			8	.26	.22	.20	.26	.22	.20	.25	.22	.20	.18
			9	.24	.20	.17	.24	.20	.17	.23	.20	.17	.16
42 Fluorescent unit with flat prismatic lens, four-lamp, 610 mm (2 ft) wide 0% / 63% / 60°	V	1.4/1.2	0	.75	.75	.75	.73	.73	.73	.70	.70	.70	.63
			1	.67	.65	.63	.66	.64	.62	.63	.62	.60	.55
			2	.60	.57	.54	.59	.56	.53	.57	.54	.52	.49
			3	.54	.50	.47	.53	.49	.46	.52	.48	.45	.43
			4	.49	.44	.40	.48	.44	.40	.47	.43	.40	.37
			5	.44	.39	.35	.43	.38	.35	.42	.38	.34	.33
			6	.40	.34	.31	.39	.34	.31	.38	.34	.30	.29
			7	.36	.30	.27	.35	.30	.27	.34	.30	.27	.25
			8	.32	.27	.23	.32	.27	.23	.31	.26	.23	.22
			9	.29	.24	.20	.28	.23	.20	.28	.23	.20	.19

(continued)

Typical Luminaire	Maintenance Category	Maximum S/MH	RCR	80 ρw: 50	30	10	70 50	30	10	50 50	30	10	0 0
				Coefficients of Utilization for 20% Effective Floor Cavity Reflectance (ρFC = 20)									
44 Radial batwing distribution—louvred fluorescent unit	IV	N.A.	0	.71	.71	.71	.70	.70	.70	.66	.66	.66	.60
			1	.65	.63	.61	.63	.62	.60	.61	.59	.58	.54
			2	.59	.55	.53	.58	.55	.52	.55	.53	.51	.48
			3	.53	.49	.46	.52	.48	.45	.50	.47	.45	.42
			4	.47	.43	.40	.47	.43	.40	.45	.42	.39	.37
			5	.42	.38	.34	.42	.37	.34	.41	.37	.34	.32
			6	.38	.33	.30	.38	.33	.30	.37	.33	.30	.28
			7	.34	.29	.26	.33	.29	.26	.33	.28	.25	.24
			8	.30	.25	.22	.30	.25	.22	.29	.25	.22	.20
			9	.27	.22	.18	.26	.22	.18	.26	.21	.18	.17
45 Radial batwing distribution—four-lamp, 610-mm (2 ft)-wide fluorescent unit with flat prismatic lens and overlay	V	N.A.	0	.57	.57	.57	.56	.56	.56	.53	.53	.53	.48
			1	.50	.48	.47	.49	.47	.46	.47	.46	.44	.41
			2	.44	.41	.38	.43	.40	.38	.41	.39	.37	.34
			3	.39	.35	.32	.38	.34	.31	.37	.33	.31	.29
			4	.34	.30	.27	.33	.29	.26	.32	.29	.26	.24
			5	.30	.25	.22	.29	.25	.22	.28	.24	.22	.20
			6	.26	.22	.19	.26	.22	.18	.25	.21	.18	.17
			7	.23	.19	.16	.23	.19	.16	.22	.18	.16	.14
			8	.21	.16	.13	.20	.16	.13	.19	.16	.13	.12
			9	.18	.14	.11	.18	.14	.11	.17	.14	.11	.10
46 Bilateral batwin distribution—one lamp, surface-mounted fluorescent with prismatic wraparound lens	V	N.A.	0	.87	.87	.87	.84	.84	.84	.77	.77	.77	.64
			1	.76	.73	.70	.73	.70	.67	.67	.65	.63	.53
			2	.66	.61	.57	.64	.59	.56	.59	.56	.52	.44
			3	.59	.53	.48	.56	.51	.47	.53	.48	.44	.38
			4	.52	.45	.40	.50	.44	.40	.47	.42	.38	.32
			5	.46	.39	.34	.44	.38	.33	.41	.36	.32	.27
			6	.41	.34	.29	.39	.33	.29	.37	.31	.27	.23
			7	.36	.30	.25	.35	.29	.24	.33	.27	.23	.20
			8	.32	.26	.21	.31	.25	.21	.29	.24	.20	.17
			9	.29	.22	.18	.28	.22	.18	.26	.21	.17	.14
47 Radial batwing distribution—four-lamp, 610-mm (2 ft)-wide fluorescent unit with flat prismatic lens	V	1.7	0	.71	.71	.71	.69	.69	.69	.66	.66	.66	.60
			1	.62	.60	.58	.61	.59	.57	.59	.57	.55	.51
			2	.55	.51	.47	.53	.50	.47	.51	.48	.46	.42
			3	.48	.43	.39	.47	.43	.39	.45	.41	.38	.36
			4	.42	.37	.33	.41	.37	.33	.40	.36	.32	.30
			5	.37	.32	.27	.36	.31	.27	.35	.30	.27	.25
			6	.33	.27	.23	.32	.27	.23	.31	.26	.23	.21
			7	.29	.24	.20	.29	.24	.20	.28	.23	.20	.18
			8	.26	.21	.17	.25	.20	.17	.25	.20	.17	.15
			9	.23	.18	.14	.23	.18	.14	.22	.17	.14	.13
48 Two-lamp fluorescent strip unit	I	1.6/1.2	0	1.01	1.01	1.01	.96	.96	.96	.87	.87	.87	.68
			1	.85	.81	.77	.81	.77	.73	.73	.70	.67	.53
			2	.73	.66	.61	.69	.63	.58	.63	.58	.54	.42
			3	.63	.56	.50	.60	.53	.48	.55	.49	.44	.35
			4	.56	.47	.41	.53	.46	.40	.48	.42	.37	.29
			5	.49	.40	.34	.46	.39	.33	.42	.36	.31	.24
			6	.43	.35	.29	.41	.34	.28	.38	.31	.26	.20
			7	.39	.31	.25	.37	.29	.24	.34	.27	.23	.17
			8	.34	.27	.21	.33	.26	.21	.30	.24	.19	.15
			9	.31	.23	.18	.30	.23	.18	.27	.21	.17	.12
49 Two-lamp fluorescent strip unit with 235° reflector fluorescent lamps	I	1.4/1.2	0	1.13	1.13	1.13	1.09	1.09	1.09	1.01	1.01	1.01	.85
			1	.96	.92	.88	.93	.89	.85	.87	.83	.80	.68
			2	.83	.76	.70	.80	.74	.68	.75	.69	.65	.55
			3	.73	.65	.58	.70	.63	.57	.66	.59	.54	.46
			4	.64	.55	.49	.62	.54	.48	.58	.51	.46	.39
			5	.56	.47	.41	.55	.46	.40	.51	.44	.38	.33
			6	.50	.41	.35	.49	.40	.34	.46	.38	.33	.28
			7	.45	.36	.30	.44	.35	.30	.41	.34	.28	.24
			8	.40	.32	.26	.39	.31	.25	.37	.30	.25	.21
			9	.36	.28	.22	.35	.27	.22	.33	.26	.21	.18

ILLUMINATION

(continued)

TABLE 20.2 **Coefficients of Utilization for Typical Luminaires with Suggested Maximum Spacing Ratios** (*continued*)

Typical Luminaire	Typical Distribution and Percent Lamp Lumens	Maintenance Category	Maximum S/MH	RCR	ρ_{CC}: 80			70			50			0
					ρ_W: 50	30	10	50	30	10	50	30	10	0
					Coefficients of Utilization for 20% Effective Floor Cavity Reflectance ($\rho_{FC} = 20$)									
50				1	.42	.40	.39	.36	.35	.33	.25	.24	.23	Coves are
				2	.37	.34	.32	.32	.29	.27	.22	.20	.19	not recom-
				3	.32	.29	.26	.28	.25	.23	.19	.17	.16	mended-
				4	.29	.25	.22	.25	.22	.19	.17	.15	.13	for lighting
				5	.25	.21	.18	.22	.19	.16	.15	.13	.11	areas
				6	.23	.19	.16	.20	.16	.14	.14	.12	.10	having
Single-row fluorescent lamp cove without				7	.20	.17	.14	.17	.14	.12	.12	.10	.09	low reflec-
reflector, multiplied by 0.93 for two rows and				8	.18	.15	.12	.16	.13	.10	.11	.09	.08	tances
by 0.85 for three rows.				9	.17	.13	.10	.15	.11	.09	.10	.08	.07	

For luminaire 53 — additional block with $\rho_{CC} = 10\%$ ($\rho_W = 50, 30, 10$):

Typical Luminaire		Maintenance Category	Maximum S/MH	RCR	50 (ρ_W: 50)	30	10	$\rho_{CC} = 10\%$ ($\rho_W = 50$)	30	10
53 ρ_{CC} from below ~45%				1	.51	.49	.48	.47	.46	.45
				2	.46	.44	.42	.43	.42	.40
				3	.42	.39	.37	.39	.38	.36
				4	.38	.35	.33	.36	.34	.32
				5	.35	.32	.29	.33	.31	.29
				6	.32	.29	.26	.30	.28	.26
Louvred ceiling. Ceiling efficiency ~50; 45°				7	.29	.26	.23	.28	.25	.23
shielding opaque louvres of 80% reflectance.				8	.27	.23	.21	.26	.23	.21
Cavity with minimum obstructions and painted				9	.24	.21	.19	.24	.21	.19
with 80% reflectance paint—use $\rho_{CC} = 50$.				10	.22	.19	.17	.22	.19	.17

NOTES:
1. Data extracted from *IES Handbook* (1981), Reference Volume, with permission.
2. Multiply by 1.05 for three lamps and 1.1 for two lamps.

required or the precision of data available. Thus, there is no point in working to three decimal places if a ±10% rounding of the result is common and acceptable, or if the data available are only accurate to one decimal place. As the reader will see, the lumen (lighting flux) method of *average* illuminance calculation is replete with assumptions and estimates. Among these are:

1. It is assumed that the space is empty. This is not normally the case.
2. It is assumed that all surfaces are perfect diffusers. This is not the case.
3. All surface reflectances are estimates, ±10%.
4. Maintenance conditions are estimates, at best ±10%.
5. No allowance is made for deviation of the performance of an individual product from its specification.

Any attempt to account accurately for these approximations would enormously complicate the calculation and would serve no useful purpose for this type of *average* illuminance calculation. For this reason, the procedure presented here introduces some approximations (which we feel are well justified) in the interest of simplication. These approximations are noted wherever used.

20.31   Calculation of Horizontal Illuminance by the Lumen (Flux) Method

The lumen method of calculation is a procedure for determining the *average* maintained illuminance (footcandles, lux) on the working plane in a room. The method presupposes that luminaires will be spaced so that uniformity of illumination is provided in order that an *average* calculation will have validity. The method is based on the definition of 1 lux [foot-candle] of illuminance as 1 lumen incident on 1 meter [square foot] of area; that is,

$$\text{lux [fc]} = \frac{\text{lumens}}{\text{area m}^2 \text{ [ft}^2]}$$

As explained previously, the ratio between the lumens reaching the working plane in a specific space and the lumens generated is the coefficient of utilization, CU.
Or,

lumens on the working plane =
$$\text{lamp lumens} \times \text{CU}$$

Therefore,

$$\text{illuminance } E = \frac{\text{lamp lumens} \times \text{CU}}{\text{area}}$$

The coefficient CU is selected from tables provided by the manufacturer of the selected luminaire by a technique known as the *zonal cavity method* (explained in Section 20.33). In the absence of specific CU data, an approximation can be made by using the generic fixture types in Table 20.2.

The illuminance figure so calculated is the *initial average* illumination. This initial level is reduced by the effect of temperature and voltage variations, dirt accumulation on luminaires and room surfaces, lamp output depreciation, and maintenance conditions. All of these effects are cumulatively referred to as the *light-loss factor* (LLF):

$$\text{maintained } E = \text{initial } E \times \text{LLF}$$

(This factor was previously known as the maintenance factor, MF.) The procedure required to arrive at this factor is explained in the following section.

Our final expression for maintained illuminance E as calculated by the lumen method is, therefore,

$$E = \frac{\text{lamp lumens} \times \text{CU} \times \text{LLF}}{\text{area}} \qquad (20.1)$$

where E is lux if area is expressed in square meters, or footcandles where the area is in square foot.

Lamp lumens is the total within the space, and equal to

No. of fixtures × lamps per fixture
× initial lumens per lamp

Our formula than becomes

$$E = \frac{\begin{array}{c}\text{No. of luminaires} \times \text{lamps/luminaire} \times \\ \text{lumens/lamp} \times \text{CU} \times \text{LLF}\end{array}}{\text{area}} \qquad (20.2)$$

or, conversely, solving for the number of luminaires required to achieve a target-maintained illuminance E:

$$\text{No. of luminaires} = \frac{E \times \text{area}}{\begin{array}{c}\text{lamps per luminaire} \times \\ \text{lumens per lamp} \times \\ \text{CU} \times \text{LLF}\end{array}} \qquad (20.3)$$

For large areas, a much more useful figure is the area illuminated per luminaire; that is,

$$\text{area per luminaire} = \frac{\begin{array}{c}\text{lamps per luminaire x lumen} \\ \text{per lamp} \times \text{CU} \times \text{LLF}\end{array}}{E}$$

$$(20.4)$$

For instance, it is much more convenient to know that to maintain, say, 60 fc with a given luminaire, 70 ft^2 per unit is required than it is to know that for an 18,000-ft^2 floor, 257 fixtures are necessary. The former figure allows us to establish a pattern, say 7 × 10; the latter figure is too large to be immediately useful. Therefore, for rooms requiring more than a small number of luminaires, the latter calculation should always be used.

20.32   Calculation of Light-Loss Factor (LLF)

The LLF is composed of elements that can be categorized as *recoverable* and *nonrecoverable*. The former can be improved by maintenance; the latter cannot. The total LLF is the product of all the individual factors. The survey that follows includes approximations. For more precise data, see *IESNA Handbook* (1993).

Among the nonrecoverable loss factors are the following.

(a) Luminaire Ambient Temperature.
Light output changes when a *fluorescent* fixture

operates at other than its design temperature. With normal indoor installation use 1.0—that is, no depreciation. For other conditions, refer to technical data on the luminaire involved.

(b) Voltage. When the lamp operates at rated voltage, use 1.0. Details of source sensitivity to voltage are given in Chapter 19.

(c) Luminaire Surface Depreciation. This factor is proportional to age and depends on type of surface involved. The designer must estimate this factor based on knowledge of the materials used.

(d) Components. Losses due to components include ballast factor, ballast-lamp photometric factor, equipment operating factor, and lamp position (tilt) factor. Air troffers also introduce a thermal factor. In the absence of specific data for component factors, use a total of 0.92.

In the absence of reliable data for any of the foregoing nonrecoverable factors, use an overall factor representing the product $a \times b \times c \times d$ of 0.88.

The factors below are *recoverable*—that is, can be returned to their initial state by maintenance.

(e) Room Surface Dirt. This factor is self explanatory. Obviously, lighting systems that depend heavily on surface reflections, such as indirect, are more seriously affected than systems that deliver most of their useful light directly. Assuming a 24-month cleaning cycle and normal conditions of cleanliness, use the appropriate factor in the following list. Alter it for other conditions such as infrequent maintenance and unusual cleanliness or dirtiness.

Direct lighting: $0.92 \pm 5\%$

Semi-direct lighting: $0.87 \pm 8\%$

Direct–indirect lighting: $0.82 \pm 10\%$

Semi-indirect lighting: $0.77 \pm 12\%$

Indirect lighting: $0.72 \pm 17\%$

(f) Lamp Lumen Depreciation. This factor depends on the type of lamp and the replace-ment schedule. Use the following when exact data are unavailable:

	Group Replacement	Replacement on Burnout
Incandescent	0.94	0.88
Tungsten–halogen	0.98	0.94
Fluorescent	0.90	0.85
Mercury	0.82	0.74
Metal–halide	0.87	0.80
High-pressure sodium	0.94	0.88

(g) Burnouts. This self-explanatory factor depends on maintenance schedules and method of replacement. Use the following:

Group replacement procedures: 1.0

Individual replacement on burnout: 0.95

(h) Luminaire Dirt Depreciation (LDD). This factor depends on luminaire design, atmosphere conditions in the space, and maintenance schedule. The maintenance category is obtained from manufacturer's data or from Table 20.2. The type of atmosphere is determined by considering the space involved. Assuming a 12-month cleaning schedule and normal room cleanliness, use the base number in Fig. 20.43 and change it to match conditions of dirt and maintenance. The categories correspond to those used by the IESNA.

Total LLF is the product of all the depreciation factors, that is,

$$\mathrm{LLF} = a \times b \times c \times d \times e \times f \times g \times h$$

For example, a fluorescent air troffer in a regularly maintained group lamp–replacement, air-conditioned office might typically have an LLF of 0.8. The same fixture in the same office, but with walls and fixture cleaned only when replacing burned-out lamps, would typically have an LLF of 0.55. Thus, if in the first case the maintained illumination is E fc, in the second case it is 0.55/0.80 or 0.69 E fc; that is, a reduction of 31% as a result of poor maintenance. When a detailed determination of LLF is not possible, use the factors given in Section 20.34. They are somewhat more conservative than those given in Section 20.5c.

Fig. 20.43 *The LDD factor is determined from the category of luminaire (which is an indication of its proneness to dirt accumulation) plus a knowledge of room ambient conditions.*

20.33 Determination of Coefficient of Utilization (CU) by the Zonal Cavity Method

The CU connects a particular fixture to a particular space, by relating the luminaire's light distribution characteristic to the room size and its surface reflectances. To account for the luminaire's mounting height and its relationship to the working plane, the space is divided into three cavities: the ceiling cavity above the fixture, the floor cavity below the working plane, and the room cavity between the two (Fig. 20.44). Given the surface reflectances, the effective reflectances of the floor and ceiling cavities can be obtained. With these, the CU can be selected from the tables (either Table 20.2 or manufacturer's data) and the lumen formula (Eq. 20.3) applied to arrive at average illuminance. A step-by-step explanation of the method plus illustrative examples demonstrates the procedure. The reader should follow the steps with the flow chart in Fig. 20.45 and the calculation form in Fig. 20.46, in hand.

STEP 1. First, dimensional data are recorded. In offices, schools, and many other occupancies, the work plane is 30 in. above finished floor

(AFF). In drafting rooms it is 36 to 38 in.; in shops, 42 to 48 in.; in carpet stores and sail-cutting rooms the work plane is at the floor level. The three "h" terms are the heights of the various cavities. In this step also, identify the initial reflectance of the room surfaces and fill in the sketch in Fig. 20.46. Similarly, establish wall and floor reflectances. Utilize the nearest reflectance to those given in Table 20.3.

STEP 2. See Fig. 20.46. This step involves determining the cavity ratios of the room by calculation. The basic expression for a cavity ratio (CR) is

$$CR = 2.5 \times \frac{\text{area of cavity wall}}{\text{area of work plane}} \quad (20.5)$$

In a rectangular space, the area of the cavity wall is $h \times (2l \times 2w)$ or $2h(l + w)$; therefore

$$CR = \frac{2.5 \times 2h \times (l \times w)}{\text{area of work plane}}$$

or

$$CR = 5h \times \frac{l + w}{l \times w} \quad (20.6)$$

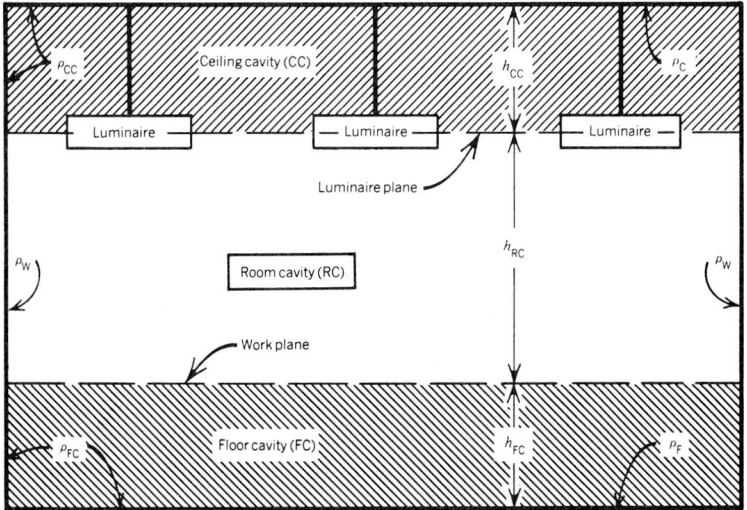

Notes:

1. Obtain effective ceiling cavity reflectance (ρ_{CC}) for combination of ceiling and wall reflectance from Table 20.3. Note that for surface-mounted or recessed luminaires, CCR = 0 and the ceiling reflectance may be used as the effective cavity reflectance. Since LLF accounts for room surface depreciation, use initial reflectances.

2. Obtain effective floor cavity reflectance (ρ_{FC}) from combination of floor and wall reflectances from Table 20.3.

3. Obtain coefficient of utilization for 20% effective floor cavity reflectance condition from table for luminaire, interpolating between tabulated values as required to match room size and ceiling and wall reflectance combinations. If effective floor cavity reflectance (ρ_{FC}) obtained in step 3 differs significantly from 20%, obtain the multiplier from Table 20.3. Multiply the coefficient of utilization by this multiplier.

4. If *initial illumination* is desired, CU selection should be obtained by above procedure and the light loss factor LLF omitted from the formula.

5. Legend:

ρ_C = ceiling reflectance
ρ_{CC} = ceiling cavity reflectance
ρ_W = wall reflectance
ρ_F = floor reflectance
ρ_{FC} = floor cavity reflectance
  h = height in feet or meters
h_{RC} = height of room cavity

Fig. 20.44 *Room cavities as used in the zonal cavity method.*

ILLUMINATION

TABLE 20.3 **Percent Effective Ceiling or Floor Cavity Reflectance (ρ_{CC}, ρ_{FC}) for Various Reflectance Combinations**

Percent Ceiling ρ_C or Floor Reflectance ρ_F:	90				80				70			50			30				10		
Percent Wall Reflectance ρ_W:	90	70	50	30	80	70	50	30	70	50	30	70	50	30	65	50	30	10	50	30	10
Ceiling or Floor Cavity Ratios CCR or FCR																					
0	90	90	90	90	80	80	80	80	70	70	70	50	50	50	30	30	30	30	10	10	10
0.2	89	88	86	85	79	78	77	76	68	67	66	49	48	47	30	29	29	28	10	10	9
0.4	88	86	83	81	78	76	74	72	67	65	63	48	46	45	30	29	27	26	11	10	9
0.6	88	84	80	76	77	75	71	68	65	62	59	47	45	43	29	28	26	25	11	10	9
0.8	87	82	77	73	75	73	69	65	64	60	56	47	43	41	29	27	25	23	11	10	8
1.0	86	80	74	69	74	71	66	61	63	58	53	46	42	39	29	27	24	22	11	9	8
1.2	86	78	72	65	73	70	64	58	61	56	50	45	41	37	29	26	23	20	12	9	7
1.4	85	77	69	62	72	68	62	55	60	54	48	45	40	35	28	26	22	19	12	9	7
1.6	85	75	66	59	71	67	60	53	59	52	45	44	39	33	28	25	21	18	12	9	7
1.8	84	73	64	56	70	65	58	50	57	50	43	43	37	32	28	25	21	17	12	9	6
2.0	83	72	62	53	69	64	56	48	56	48	41	43	37	30	28	24	20	16	12	9	6
2.2	83	70	60	51	68	63	54	45	55	46	39	42	36	29	28	24	19	15	13	9	6
2.4	82	68	58	48	67	61	52	43	54	45	37	42	35	27	28	24	19	14	13	9	6
2.6	82	67	56	46	66	60	50	41	53	43	35	41	34	26	27	23	18	13	13	9	5
2.8	81	66	54	44	66	59	48	39	52	42	33	41	33	25	27	23	18	13	13	9	5
3.0	81	64	52	42	65	58	47	38	51	40	32	40	32	24	27	22	17	12	13	8	5
3.5	79	61	48	37	63	55	43	33	48	38	29	39	30	22	26	22	16	11	13	8	5
4.0	78	58	44	33	61	52	40	30	46	35	26	38	29	20	26	21	15	9	13	8	4
4.5	77	55	41	30	59	50	37	27	45	33	24	37	27	19	25	20	14	8	14	8	4
5.0	76	53	38	27	57	48	35	25	43	32	22	36	26	17	25	19	13	7	14	8	4

Extracted from *IES Handbook 8 ed.*, reprinted with permission; for more complete data see *IES Handbook* 1993.

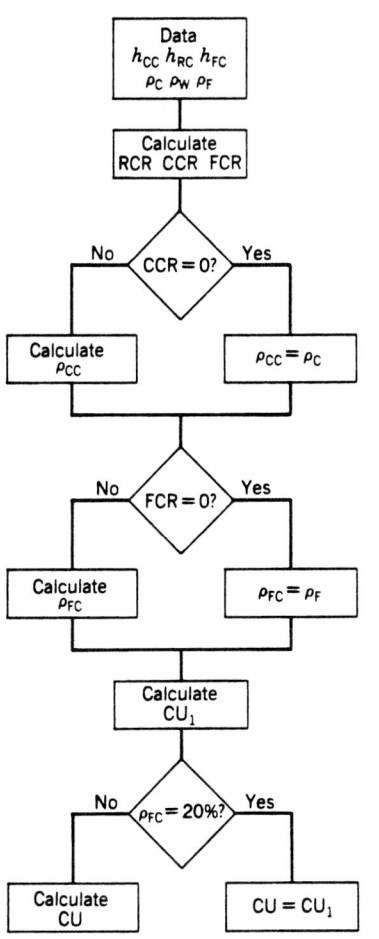

Fig. 20.45 *Zonal cavity method flowchart.*

For other than rectangular rooms, the area can be calculated as required. For instance, in a circular room, the cavity wall area $= h \times 2\pi r$ and the work plane area is πr^2. Thus,

$$CR = \frac{2.5 \times h \times 2\pi r}{\pi r^2} = \frac{5h}{r} \qquad (20.7)$$

For each of the cavities in a rectangular room we have:

Room cavity ratio

$$RCR = 5h_{RC}\frac{l+w}{l \times w} \qquad (20.8)$$

Ceiling cavity ratio

$$CCR = 5h_{CC}\frac{l+w}{l \times w} \qquad (20.9)$$

Floor cavity ratio

$$FCR = 5h_{FC}\frac{l+w}{l \times w} \qquad (20.10)$$

For reference, because all cavity ratio figures are related, having determined one, the others are

$$CCR = RCR\frac{h_{CC}}{h_{RC}} \qquad (20.11)$$

$$FCR = RCR\frac{h_{FC}}{h_{RC}} \qquad (20.12)$$

and

$$CCR = FCR\frac{h_{CC}}{h_{FC}} \qquad (20.13)$$

STEP 3. See Table 20.3 and Figs. 20.45 and 20.46. This step involves obtaining the effective ceiling cavity reflectance (ρ_{FC}) from Table 20.3. Note that the wall reflectance remains as selected in step 1. If the fixtures are surface mounted or recessed, then CCR = 0 and ρ_{CC} = selected ceiling reflectance.

STEP 4. See Table 20.3 and Figs. 20.45 and 20.46. This step involves obtaining the effective floor cavity reflectance ρ_{FC} as in step 3, for ρ_{CC}. If the floor is the working plane, FCR = 0 and ρ_{FC} = selected floor reflectance.

STEP 5. Select CU from manufacturer's data. Note that interpolation may be necessary for ceiling cavity reflectance (ρ_{CC}), which is between the figures in the CU table. See Example 20.1 in the next section. Also, factors for CU correction for ρ_{FC} other than 20% (standard in CU tables) are given in Table 20.4.

STEP 6. Calculate illuminance and number of fixtures or area per luminaire in the usual fashion.

Illustrative examples and shortcut methods are demonstrated below. CU coefficients are listed in Table 20.2 for generic luminaire types.

20.34   Zonal Cavity Calculations: Illustrative Examples

EXAMPLE 20.1. It is suggested that the reader photocopy Fig. 20.46 and fill it in as the solution progresses.

GENERAL INFORMATION

1. Project identification: _____
   (Give name of area and/or building and room number)

2. Average maintained illumination for design: _____ lux [footcandles]

Luminaire data:
3. Manufacturer: _____
4. Catalog number: _____

Lamp data:
5. Type and color: _____
6. Number per luminaire: _____
7. Total lumens per luminaire: _____

SELECTION OF COEFFICIENT OF UTILIZATION

8. Step 1: Fill in sketch at right.

9. Step 2: Determine Cavity Ratios by formulas.

9a. Room cavity ratio, RCR   = _____
9b. Ceiling cavity ratio, CCR = _____
9c. Floor cavity ratio, FCR   = _____

10. Step 3: Obtain effective ceiling cavity reflectance (ρ_{CC}) from Table 20.3. ρ_{CC} = _____

11. Step 4: Obtain effective floor cavity reflectance (ρ_{FC}) from Table 20.3. ρ_{FC} = _____

12. Step 5: Obtain coefficient of utilization (CU) from manufacturer's data.   CU = _____

SELECTION OF LIGHT LOSS FACTORS

Unrecoverable
13. Luminaire ambient temperature   _____
14. Voltage to luminaire   _____
15. Luminaire surface depreciation   _____
16. Other factors **(components)**   _____

Recoverable
17. Room surface dirt depreciation   _____
18. Lamp lumen depreciation   _____
19. Lamp burnouts factor   _____
20. Luminaire dirt depreciation LDD _____

21. Total light loss factor, LLF (product of individual factors above): _____

CALCULATIONS

(Average maintained illumination level)

$$\text{Number of luminaires} = \frac{(\text{Illuminance}) \times (\text{Area})}{(\text{Lumens per luminaire}) \times (\text{CU}) \times (\text{LLF})}$$

22.   = _____ =

$$\text{Lux [footcandles]} = \frac{(\text{number of luminaires}) \times (\text{lumens per luminaire}) \times (\text{CU}) \times (\text{LLF})}{(\text{area})}$$

23.   = _____ =

24. Calculated by: _____ Date: _____

Fig. 20.46 Zonal cavity method calculation form. (Courtesy of IESNA.)

TABLE 20.4ᵃ **Factors for Effective Floor Cavity Reflectances Other Than 20%**
(Any Wall Reflectance)ᵇ

For 30% effective floor cavity reflectance, *multiply* by appropriate factor below.
For 10% effective floor cavity reflectance, *divide* by appropriate factor below.

Room Cavity Ratio	Percent Effective Ceiling Cavity Reflectance, ρ_{CC}			
	80	70	50	10
1	1.08	1.06	1.04	1.01
2	1.06	1.05	1.03	1.01
3	1.04	1.04	1.03	1.01
4	1.03	1.03	1.02	1.01
5	1.03	1.02	1.02	1.01
6	1.02	1.02	1.02	1.01
7	1.02	1.02	1.01	1.01
8	1.02	1.02	1.01	1.01
9	1.01	1.01	1.01	1.01
10	1.01	1.01	1.01	1.01

ᵃ Extracted from *IESNA Handbook* (1993), and reprinted with permission.
ᵇ For more precise data, for varying ρ_w, see *IESNA Handbook* (1993).

Given. Classroom: 6 m ×8 m ×3.70 m height, elementary school. Initial reflectances: ceiling 80%, entire wall 50%, floor 20%. (Note that the sketch on Fig. 20.46 can accommodate different reflectances for the upper, center, and lower wall sections.) Provide adequate lighting using fluorescent fixtures. Assume yearly maintenance, lamp replacement at burnout, proper voltage and ballasts, medium clean atmosphere.

SOLUTION. *(a) Illuminance Level.* Refer to Table 18.7. Depending on the class grade, the illuminance category could be either D or E. We use E, which gives an illuminance range of 500–750–1000 lux. In Table 18.9 under category E we find: age—under 40, speed/accuracy are important, and background is more than 70% (white paper). This yields a recommendation of 500 lux. The daylight contribution in much of the space is usually considerable because of the hours of use. It frequently exceeds this 500-lux level, but to demonstrate the calculation procedure, we assume an absence of daylight. Lines 1, 2, and 8 of Fig. 20.46 and the room sketch can be filled in.

(b) Luminaire Selection. The criteria we have previously developed for a classroom situation re-

quires an installation that yields

1. Low direct glare (high VCP) because school children spend a large proportion of their time in a heads-up position.
2. Low veiling reflections because much of the seeing task involves high reflectance materials, occasionally specular.
3. High efficiency and low energy use to meet IESNA/ASHRAE and most governmental requirements.
4. Minimum required maintenance, in view of the poor cleaning and maintenance situation that exists in many schools.

Although ceiling height is sufficient to permit use of indirect lighting (e.g., luminaire 33 in Table 20.2) with all its distinct advantages, it is not chosen, because

5. The luminaire maintenance category is bad, and given the type of maintenance expected, light reduction would be serious.
6. Indirect lighting depends on a highly reflective ceiling, requiring yearly cleaning and repainting at intervals not exceeding 5 years. This is not generally the case in public schools.

ILLUMINATION

As a result, we select a *two*-lamp version of luminaire no. 44 from Table 20.2. This unit is a parabolic aluminum reflector with louvres and exhibits a 45° cutoff and crosswise batwing distribution. (Fig. 20.36 is a unit of this design.) This unit meets requirements 1 through 4.

Although its distribution curve (see Table 20.2) shows no upward component, most commercial units of this basic design do have slots in the top of the reflector and show 5% to 10% uplight. In practice, we would select such a unit. This avoids an excessively dark ceiling and undesirable luminance ratios between fixture and background. Furthermore, fixtures with top slots stay cleaner because of the upward air movement through the fixture caused by the warm lamps and ballast. Mounting height should be about 270 cm (~9 ft) to permit easy maintenance and good row spacing. For this luminaire the recommended maximum spacing criterion (SC) can be estimated at between 1.5 and 2.0. Work plane height is 75 cm (30 in.).

(c) Calculation.

STEP 1. The required data that should appear on the sketch in Fig. 20.46 are

h_{CC} = 1.0 m $\qquad$ h_{RC} = 1.95 m

h_{FC} = 0.75 m $\qquad$ l = 8 m

ρ_C = 80% $\qquad$ ρ_w = 50%

ρ_F = 20% $\qquad$ w = 6 m

STEP 2. From Eqs. 20.8, 20.9, and 20.10

$$RCR = 5h_{RC}\frac{l+w}{l \times w} = 2.84$$

$$CCR = 5\,(1)(0.29) = 1.46$$

$$FCR = 5\,(0.75)(0.29) = 1.09$$

STEP 3. From Table 20.3 obtain effective reflectances: For ρ_{CC} use ρ_C = 0.8, ρ_w = 0.5, and CCR = 1.46.
Therefore,

$$\rho_{CC} = 0.61$$

STEP 4. For ρ_{FC} use ρ_F = 0.2, ρ_w = 0.5, and FCR = 1.09. Therefore,

$$\rho_{FC} = 0.18 \text{ by interpolation}$$

STEP 5. Interpolation between RCR of 2.8 and 3.0 is necessary. The coefficient of utilization for the selected luminaire—no. 44 of Table 20.2—can now be obtained by double interpolation.

ρ_w = 0.50	CU		
ρ_{cc} →	0.70	0.61	0.50
RCR			
2.	0.58		0.55
2.84		?	
3.	0.52		0.50

Required CU = 0.52. No correction from Table 20.4 is required, as ρ_{FC} is close to 20%. At this stage, lines 9 through 12 of Fig. 20.46 can be filled in.

STEP 6. The LLF (see Section 20.32) results from establishing items 13 to 20 of Fig. 20.46. These are:

Items 13–16	0.88
Item 17	0.95
Item 18	0.85
Item 19	0.95
Item 20	0.80

Item 21: LLF = (0.88)(0.95)(0.85)(0.95)(0.80) = 0.54

STEP 7. *Lumen calculation.* A typical classroom is normally divided into a large student seating area and a teacher's area. The illuminance requirement is approximately the same for both so that the room can be treated as a single unit for visual task purposes. A good lighting design includes chalkboard lighting, the spill light from which is usually sufficient to illuminate the very front of the room. Treating the entire space as one visual task area, and assuming two nominal 40-W, 4100° K fluorescent lamps per fixture, the number N of luminaires required is, from Eq. 20.3,

$$N = \frac{(500 \text{ lux})(6 \times 8 \text{ m})}{2(3200 \text{ lm})(0.52)(0.54)} = 13.35$$

The remaining lines of Fig. 20.46 can now be completed.

Fig. 20.47 *Layout of pendant parabolic aluminum reflector luminaires in a typical classroom. The units have a modified batwing distribution in the crosswise direction, which mandates their being hung with their long axis parallel to the line of sight, as shown. Furthermore, the fixtures have lower brightness in that direction ie VCP for students in the head-up position is excellent.*

Note that the distance between the outside fixture row and the window wall is more than one-half of the side-to-side spacing because of daylight contribution during school hours, whereas the inside fixture row to inside wall spacing is less than one-half of the side-to-side spacing in order to maintain sufficient illuminance near that wall. The single fixture to the right of the teacher's desk (to the teacher's left) was so placed to avoid the veiling reflections that would result from a fixture directly above the desk, and also to take advantage of the fixture's transverse batwing distribution characteristic.

Refer now to Fig. 20.47 for the layout. Two lengthwise rows of fixtures, spaced 300 cm apart, give excellent coverage, as shown. The resultant spacing-to-mounting height ratio:

$$\frac{300 \text{ cm spacing}}{270 \text{ cm mounting height}} = 1.1$$

is well below the maximum of 1.5 to 2.0 estimated by inspection of the candlepower distribution curve. The sixth fixture in the outside (window) row provides illumination for the teacher's desk. Two *single-lamp* batwing distribution or asymmetric distribution luminaires provide chalkboard

and front-of-room lighting. The window wall row is separately switched and is further from the wall than the inside row, due to the usual presence of daylight. Point-by-point illuminance can be computed by one of the available computer programs. Daylight contribution can also be included. Actual *average* illuminance in the room would be simply the ratio of the number of fixtures used to the number required (calculated) times the target illuminance:

$$E = \frac{11 + \frac{1}{2} + \frac{1}{2}}{13.35} (500 \text{ lux}) \approx 450 \text{ lux}$$

assuming similar characteristics for the one- and two-lamp fixtures. ∎

EXAMPLE 20.2. Large modern business office, 60 ×100 × 8 ft hung ceiling height. Initial reflectances are 0.80, 0.50, 0.30. Provide general lighting using standard 4-ft fluorescent lamps with high-efficiency magnetic ballasts in recessed troffer luminaires. The space is fully air conditioned. Lamps are replaced on a burnout basis and the fixture is then cleaned.

SOLUTION. From Table 18.7 we obtain a recommended target illuminance for routine office tasks of 500 lux. As almost every desk in a modern business office is equipped with a visual display terminal (VDT), we must select a fixture that is designed to minimize the extremely disturbing reflection of luminaires in VDT screens. (See Section 21.20 for a full discussion of this ubiquitous problem.) Luminaires with parabolic diffusers, reflectors, and/or baffles, which exhibit very low high-angle luminance, meet this requirement, as explained in Section 20.21.

A second consideration is that in an office of this size, which may well be "landscaped" using half-height partitions, viewing directions vary greatly, obviating any imperatives inherent in a luminaire's directionality. As a result, we selected a specular parabolic baffled 2-ft square unit, which is directionally and architecturally neutral and displays the required low levels of luminance at high angles. The unit's characteristics are given in Fig. 20.48, when equipped with two 32-W, T8, U-shaped lamps, with 2800 initial lumens. Note that the luminaire exhibits a modified batwing intensity distribution in the transverse direction.

REK D/RIK-D 2/31-U

Recessed mounted luminaire with parabolic mirrored louver
(without mirrored reflector)

Shielding angle : Parl.: 60 - Norm.: 60

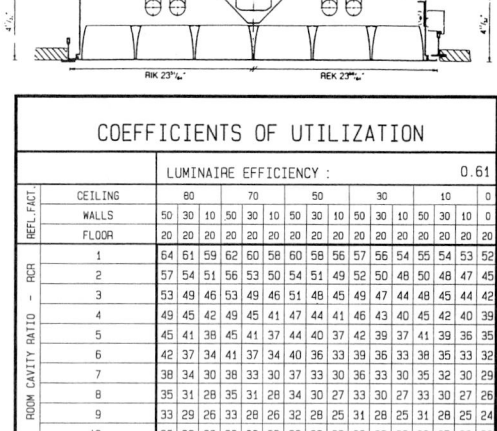

CANDLEPOWER	DEG.	Parl.	22.5	45	67.5	Norm.	Flux
	0	1609	1609	1609	1609	1609	
	5	1603	1603	1607	1608	1605	177
	10	1577	1579	1589	1598	1600	
	15	1534	1534	1551	1565	1564	506
	20	1473	1480	1490	1522	1529	
	25	1400	1412	1431	1481	1496	773
	30	1312	1327	1365	1464	1502	
	35	1212	1235	1311	1448	1501	964
	40	1102	1141	1263	1328	1352	
	45	970	1034	1138	1135	1149	892
	50	539	572	617	621	635	
	55	122	129	139	140	144	223
	60	14	14	15	15	16	
	65	12	6	11	7	10	10
	70	3	3	2	3	4	
	75	2	2	1	1	2	2
	80	2	1	1	1	2	
	85	1	1	1	2	2	2
	90	2	2	2	2	2	
	95	0	0	0	0	0	0
	100-180	0	0	0	0	0	

2500 cd
(based on 2 × 2900 lm)

COEFFICIENTS OF UTILIZATION

LUMINAIRE EFFICIENCY : 　　　　0.61

REFL. FACT.		CEILING	80			70			50			30			10			0
		WALLS	50	30	10	50	30	10	50	30	10	50	30	10	50	30	10	0
		FLOOR	20	20	20	20	20	20	20	20	20	20	20	20	20	20	20	20
ROOM CAVITY RATIO – RCR	1		64	61	59	62	60	58	60	58	56	57	56	54	55	54	53	52
	2		57	54	51	56	53	50	54	51	49	52	50	48	50	48	47	45
	3		53	49	46	53	49	46	51	48	45	49	47	44	48	45	44	42
	4		49	45	42	49	45	41	47	44	41	46	43	40	45	42	40	39
	5		45	41	38	45	41	37	44	40	37	42	39	37	41	39	36	35
	6		42	37	34	42	37	34	40	36	33	39	36	33	38	35	33	32
	7		38	34	30	38	33	30	37	33	30	36	33	30	35	32	30	29
	8		35	31	28	35	31	28	34	30	27	33	30	27	33	30	27	26
	9		33	29	26	33	28	26	32	28	25	31	28	25	31	28	25	24
	10		30	26	23	30	26	23	30	26	23	29	26	23	29	25	23	22

ZONAL SUMMARY				AVG　[cd/m²]			
ZONE	LUMENS	LAMP	FIXT	DEG.	NORM.	45°	PARL.
0 - 30	1456	25.1	41.0	45	5091	5041	4296
0 - 40	2420	41.7	68.2	50	3092	3009	2624
0 - 50	3312	57.1	93.3	55	786	759	665
0 - 60	3535	61.0	99.6	60	98	94	85
0 - 70	3545	61.1	99.9	65	75	81	66
0 - 90	3549	61.2	100.0	70	35	16	24
90 - 180	0	0.0	0.0	75	28	14	21
0 - 180	3550	61.2	100.0	80	31	21	31

Fig. 20.48 *Complete photometric data on a 2-ft² fixture with 36-cell mirrored parabolic louvre. The unit uses two T8 U-shaped 32-W fluorescent lamps. Note that the transverse distribution characteristic has a modified batwing shape typical of this type of louvre. The louvre's specular finish yields a very low brightness at high angles, as can be seen from both the distribution curve and the luminance (cd/m²) data. (Courtesy of Zumtobel-STAFF.)*

Calculations. The reader should fill in a copy of Fig. 20.46 as we proceed. Working plane is taken as desktop height (i.e., 30 in.).

STEP 1. $h_{CC} = 0$; $h_{RC} = 5.5$; $h_{FC} = 2.5$

STEP 2. CCR = 0; RCR = 0.73; FCR = 0.33

STEP 3. $\rho_W = 50\%$; $\rho_{CC} = 80\%$ (for a recessed fixture we use the ceiling reflectance).

STEP 4. From Table 20.3, $\rho_{FC} = 29\%$

STEP 5. We find CU for $\rho_{FC} = 20\%$ by extrapolation,

RCR	CU	
0.73	?	
1.0	0.64	CU = 0.66
2.0	0.57	

STEP 6. From Table 20.4, multiplier for $\rho_{FC} = 30\%$ (or 29%) is 1.085.

STEP 7. Final CU = 1.085(0.66) = 0.72

STEP 8. LLF per Fig. 20.46:

1.0	0.92
1.0	0.85
0.9	0.95
0.92	0.88

LLF = 0.54

STEP 9.

$$\text{Area/luminaire} = \frac{\text{lamp} \times \text{lumens/lamp} \times \text{CU} \times \text{LLF}}{\text{illuminance}}$$

$$= \frac{2(2800)(0.66)(0.54)}{500/10.76 \text{ lux per fc}} = 43 \text{ ft}^2$$

With a mounting of 5.5 ft above the working plane, an estimated spacing criteria of 1.8 transverse and 1.2 parallel yields maxima of 9.9 ft transverse and 6.6 ft parallel on fixture centerlines. We would test grids of 9 × 5 ft and 8 × 6 ft on a computer to see which yielded better results for uniformity of illumination and lower reflected glare in *all* directions because, as noted previously, viewing directions will probably vary. In the absence of a computer program, we would select the 9 × 6 grid to take advantage of the batwing characteristic. Note the important fact that a maintained level of 500 lux (50 fc of high-quality illumination) for VDT areas is achieved with a power density of less than 1.6 W/ft² (17.54 W/m²). ∎

20.35 Zonal Cavity Calculation by Approximation*

Although the foregoing zonal cavity calculations are straightforward and essentially simple, they can become tedious if more than a few areas are involved. Two alternatives exist: to utilize one of the many computer programs available, or simply to shorten the calculations with reasonable approximations. Computer assistance is discussed in Section 20.38; an effective computational method using approximations is demonstrated in this section.

Fill in the calculation sheet for illuminance calculation by approximation, shown in Fig. 20.49. Assume all rooms are square. To do this for a rectangle, take one-third the difference in dimensions and add to the smaller dimension to obtain equivalent width *w*. Then, for square rooms

$$RCR = \frac{10h_{RC}}{w}$$

then, using the calculated side dimension of the square equivalent room, assume:

1. $\rho_{CC} = 0.80$ for a large room, i.e., equal to or larger than 10 × 10 m (30 × 30 ft)
   $\rho_{CC} = 0.70$ for a medium room, i.e., between 10 × 10 m (30 × 30 ft) and 4 × 4 m (12 × 12 ft)
   $\rho_{CC} = 0.60$ for a small room, i.e., equal to or smaller than 4 × 4 m (12 × 12 ft)
2. Assume $\rho_{FC} = 0.20$
3. Assume LLF = 0.65 for good conditions, 0.55 for average, and 0.45 for poor.

* Based on a method developed by B.F. Jones.

EXAMPLE 20.3 (by approximation). Classroom as in Example 20.1.

1. "Square" the room

$$w_{SQ} = 6 + \frac{8-6}{3} = 6\tfrac{2}{3}$$

2. $\rho_{CC} = 70 \; \rho_W = 50 \; \rho_{FC} = 20$

$$RCR = \frac{10 \times 1.95}{6\tfrac{2}{3}} = \frac{19.5}{6.66} = 2.93$$

Obtain CU from Table 20.2, fixture no. 44.

CU = 0.52 by visual inspection

$$N = \frac{500(6 \times 8 \text{ m})}{2(3200)(0.52)(0.55)}$$

$$= 13.11 \text{ fixtures}$$

Thus, the result is substantially the same as the accurate calculation. Let us check Example 20.2 ∎

EXAMPLE 20.4 (by approximation)

1. $w_{SQ} = 60 + \frac{100-60}{3} = 73 \text{ ft}$

2. $\rho_{CC} = 80 \; \rho_W = 50 \; \rho_{FC} = 20$

$$RCR = \frac{10 \times 5.5}{73} = \frac{55}{73} = 0.75$$

CU = 0.66 by inspection
(mental interpolation)

LLF = 0.55

$$\text{area/luminaire} = \frac{2 \times 2800 \times 0.66 \times 0.55}{50 \text{ fc}}$$

$$= 40.6 \text{ sq ft}$$

This result is within 5.5% of the accurate calculation. Thus, we see that these simple approximations give answers sufficiently accurate for most uses and are therefore recommended.

In conclusion, then, with respect to zonal cavity calculations, we can make the following statements:

1. For preliminary and routine calculations of rectangular rooms, with assumed reflectances, use the assumptions listed previously. A modified calculation form is provided in Fig. 20.49 to assist with this method.

ILLUMINATION CALCULATION BY APPROXIMATION

Step 1:   **GENERAL INFORMATION**

(a) Project identification: _____
                              (Give name of area and / or building and room number)

(b) Average maintained illumination for design: _____ lux [footcandles]

Luminaire data: _____   Lamp data: _____
(c) Manufacturer: _____   (e) Type and color: _____
(d) Catalog number: _____   (f) Number per luminaire: _____
                                             (g) Total lumens per luminaire: _____

SELECTION OF COEFFICIENT OF UTILIZATION

Step 2: Determine the
        equivalent square-room
        RCR (room cavity ratio):

$$W_{sq} = W + \frac{L-W}{3}$$

$$RCR = \frac{10 \ hrc}{W_{sq}}$$

Step 3: Determine ceiling cavity
        equivalent reflectance by
        equivalent square room size.

$\rho_{cc} = \_\_ \%$

$\rho_w = \_\_ \%$

Work plane

$\rho_{fc} = 20\%$

h_{cc}

hrc

h_{fc}

L = _____
W = _____

30' (10m)

30' (10m)   W   12' (4m)

12' (4m)

W

$\rho_{cc} =$ 0.80   0.70   0.60

Step 4: Fill in sketch of room above.
Step 5: Assume floor cavity reflectance $\rho_{fc} = 20\%$.
Step 6: Obtain coefficient of utilization CU from manufacturers' data: CU = _____

Step 7: Light loss factor LLF   Good conditions: 0.65
                                     Average conditions: 0.55
                                     Poor conditions: 0.45

Step 8:   **CALCULATIONS**

$$\text{Number of luminaires} = \frac{\text{Recommended illuminance x area}}{\text{(Lumens per luminaire) x (CU) x (LLF)}}$$

$$\text{OR} \quad \text{Lux [footcandles]} = \frac{\text{(Number of luminaires) x (lumens per luminaire) x (CU) x (LLF)}}{\text{area m}^2 \text{ (ft}^2)}$$

Fig. 20.49 *Calculation sheet for illuminance calculation using an approximate zonal cavity method (based on a method developed by B. F. Jones).*

2. For rooms where a high degree of accuracy is desired and actual reflectances are known, use the long method, with visual interpolation.

3. For rooms of unusual shape or rooms with special conditions, such as coffered ceilings, mixed-material walls, and partial height partitions, use computer assistance.

4. For spaces in which a number of different solutions are to be tried, use a computer. ∎

20.36 Effect of Cavity Reflectances on Illuminance

It should be obvious to every reader at this point in our study that the reflectances of the various room cavities have a marked effect on the CU because of light reflections within the room. To demonstrate this graphically, we have plotted in Figs. 20.50, 20.51, and 20.52 the effect of varying cavity reflectances on the three principal types of fixture distribution: semi-indirect, direct-indirect, and direct-spread. Note that as expected, the ceiling cavity reflectance has the most pronounced effect with indirect fixtures, and the floor reflectance with direct units. Because lighting costs amount to 3% to 5% of *total construction* cost for many types of buildings such as offices, a 20% differential in lighting units can amount to as much as 1% of the total cost of the facility. This amount would not only pay for the increased cost of higher reflectance finishes and materials but also would reduce both initial and operating costs. These data clearly indicate the necessity for the lighting designer to have considerable influence on the choice of room materials and finishes, a situation that, unfortunately, does not usually occur.

20.37 Modular Lighting Design

An increasingly large number of buildings are being designed on a modular system, resulting in a need for flexible lighting to fit the module utilized. In such buildings, once the general lighting scheme and the fixture involved are established, it is convenient to draw a family of curves for the

Fig. 20.50 Effect of surface reflectances on the CU of a luminaire with semi-indirect distribution. As expected, because the ceiling becomes the light source, its reflectance has the most pronounced effect. With this particular unit having a 25% downward component, floor finish also has an appreciable effect, increasing CU by an average 10% for a 30% reflectance floor. The effect of wall reflectance naturally increases as rooms become smaller and the proportion of wall surface is larger. The change in CU between a 30% and a 50% reflectance wall varies from 15% for a 400-ft² room to 5% for a 4000-ft² room.

fixture chosen, thereby facilitating the utilization of the modular unit in various spaces. "Area" may readily be replaced with multiples of modular areas, as shown in Fig. 20.53.

20.38 Calculating Illuminance at a Point

The lumen (flux) method of horizontal illuminance calculation explained previously is appropriate for spaces in which illumination is essentially uniform throughout. However, even in such a space, illuminance varies at least ±10%, and near columns, walls, windows, bookcases, and the like, considerably more. Therefore, to answer the often asked question "How much light will I have on my desk?", the designer must turn to other methods. Three are available:

1. Calculation of illuminance at selected points by computer, as explained in Section 20.43.
2. Utilization of one of the design aids explained in the Section 20.39.
3. Longhand calculation by one of the methods presented in Sections 20.40 through 20.42.

Fig. 20.51 *Effect of surface reflectances on the CU of a luminaire with direct–indirect distribution. With this distribution, the effects of ceiling and floor are most pronounced, with an appreciable wall effect only in small rooms.*

Fig. 20.52 *Effect of surface reflectances on the CU of a luminaire with direct (spread) distribution. Floor finish is most important, with wall reflectance important only in small rooms. As these fixtures have no upward component, all ceiling illumination is derived from reflection. Thus, in a room with floor reflectance less than 20%, ceiling finish has no effect on room illumination.*

ILLUMINATION

Fig. 20.53 *Luminaire design chart. For frequently used fixtures, this type of chart gives a rapid design figure for various size rooms. As seen from the ordinates, the figures can be translated into number of modules and watts per square foot.*

These methods also yield results where the lumen method is simply inapplicable. Among such situations are layouts that are intentionally nonuniform; calculation of illumination on planes other than horizontal (e.g., wallwashers); calculation of illuminance resulting from architectural lighting elements such as coves, valances, and the like; and illuminance calculations for nonstandard light sources for which coefficient data of the type given in Table 20.2 are not available.

Luminaire	Contribution (Lux)
F2, F3	75 × 2
F5, F8	50 × 2
F6, F7	225 × 2
F9, F12	75 × 2
F10, F11	400 × 2
F13, F16	60 × 2
F14, F15	300 × 2
F18, F19	100 × 2
	2570 lux total

Fig. 20.54 *The ellipses represent isolux lines for a single luminaire at a given height above the work plane. They are centered on the point (the work area of a desk) for which it is desired to determine the illuminance. The total illuminance at that point is the sum of the individual luminaire contributions. The center of the luminaire is the point of references. Therefore, when two or more isolux lines pass through a fixture, its contribution is determined by the interpolated isolux line passing through its center. Note the symmetry around the vertical axis, necessitating a plot of only half the ellipses.*

ILLUMINATION

20.39   Design Aids

By *design aids* we mean any of the various curves, charts, plots, or tables either prepared by the designer or made available by luminaire manufacturers, the purpose of which is to simplify and speed lighting design when using that particular lighting fixture. The reliability of data so obtained depends entirely on the manufacturer involved, and their use should be governed accordingly. More recently, it has become customary for major manufacturers to provide computer output charts and tables based on the designers' proposed layout(s). When using these, it behooves the designer to study the data carefully in the computer program being used as the fixture supplier is certainly not a disinterested party, and the program may be "weighted" accordingly. A brief description of common design aids follows.

(a) Isolux Charts. These charts, also called *isofootcandle charts,* are based on the type traditionally supplied by manufacturers of outdoor lighting equipment such as street lights and floodlights, but equally applicable to interior lighting. Their use is illustrated in Fig. 20.54. The basic tool is an isolux diagram for a single luminaire. This is calculated (see Sections 20.40–20.43), measured from a full-scale mockup (the most accurate if not usually the most practical method), or obtained from the manufacturer. Inasmuch as the relative positions of source and illuminated point are reversible—that is, if a source at *A* causes illuminance *E* at point *B*, then the source at *B* will cause the same illuminance *E* at point *A*—placing the center of the isolux chart at the point in question permits direct reading of the illumination contribution of every other luminaire. It then remains simply to sum the individual contributions to obtain the (scalar) illumination at the desired point. An example is shown on Fig. 20.54.

(b) Illuminance "Cone" Charts. See Fig. 20.55. When the light distribution of a direct downlight is symmetrical, as is generally the case, a cone can be drawn showing illuminance directly under the fixture at various distances. The projected circles are defined by maximum illuminance at the center, and half of this illuminance at the edge. This projected circle can be used in the

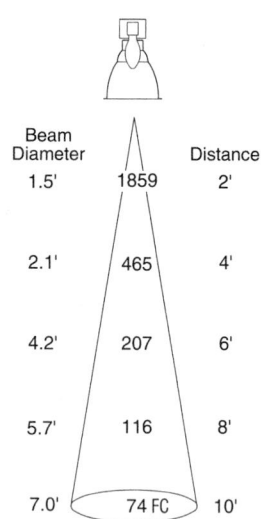

Fig. 20.55 *For downlights with symmetrical circular distribution, a "cone of light," as shown, can be drawn. The illuminance at varying distances on the beam centerline directly below the luminaire are given in the center column. A circle at the circumference of which illuminance is half this maximum is drawn at each distance from the downlight (2 ft, 4 ft, etc.). The numbers in the left column show the diameter of this (beam) circle. A detail of the downlight, which uses a 100-W, ED-17, metal–halide lamp is shown above the chart, for information only. (Courtesy of Zumtobel-STAFF.)*

same fashion as the isolux chart in the preceding paragraph, except that only two values are given—that at the center and that on the circumference.

Wallwash Lighting Data Chart

ceiling		0'	1'	2'	3'
1'		38	38	38	38
2'		44	44	44	44
3'		51	51	51	51
4'		56	58	58	56
5'		59	61	61	59
6'		58	57	57	58
7'		52	49	49	52
8'		42	42	42	42
9'		18	18	18	18
floor		33	33	33	33

Fig. 20.56 Addition of a interior reflector to the downlight of Fig. 20.55 converts it to a dual-purpose downlight and wallwash unit. The wall illuminances produced by multiple units spaced 3 ft apart and ceiling mounted 3 ft from the wall are given in the chart. Similar charts are available for other luminaire spacings. (Courtesy of Zumtobel-STAFF.)

(c) Illuminance Tables and Charts.

These take various forms but all give specific illuminance data, in numerical form, for specific points. The values are obtained from computer printout or actual test. Figure 20.56 shows the illuminance pattern on a wall produced by the wallwash version of the downlight shown in Fig. 20.55. The difference between the two fixtures is the addition of an interior reflector in the wallwasher version.

20.40 Calculating Illuminance from a Point Source

It is well to note at the outset that all of the following methods calculate illumination at a *point*. The answer to the constant query, "How much light will I have on my desk from a luminaire at this location?" is arrived at by taking several points on the desk and calculating illumination at each one. Shortcuts can be made for symmetry and so on. The reader should also understand that most of these techniques are laborious and are generally performed by the consulting engineer or lighting specialist rather than the architect. They are presented here as source material for the technically oriented reader.

The basis of point source calculations is the inverse square law developed in Section 18.12:

$$fc = \frac{cp}{D^2} \qquad (18.6)$$

where *fc*, *cp*, and *d* are footcandle illumination, candlepower intensity, and distance, respectively. Refer to Fig. 20.57. The horizontal illumination at a point *P* as shown in Fig. 20.57 (*a*) is

$$\text{horizontal } fc_P = \frac{cp}{D^2} \cos\theta \qquad (20.14)$$

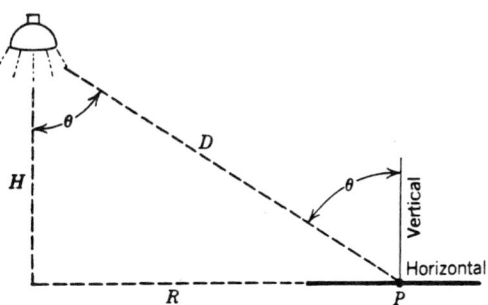

Horizontal footcandles at point *P*
$$E_H = \frac{CP \times \cos\theta}{D^2} = \frac{CP}{H^2} \cos^3\theta$$

Vertical footcandles at point *P*
$$E_V = \frac{CP \times \sin\theta}{D^2} = \frac{CP}{R^2} \sin^3\theta$$

Fig. 20.57 Relationship between intensity in candlepower (CP) and illuminance, when source can be considered a point source—that is, when inverse square law applies. Source major dimension must not exceed 0.2D to be considered a point source. Measurement in feet yields fc; distances in meters yield lux.

ILLUMINATION

and the vertical illumination at that same point is

$$\text{vertical } fc_P = \frac{cp}{D^2} \sin\theta \qquad (20.15)$$

However, because

$$\cos\theta = \frac{H}{D} \text{ and } \sin\theta = \frac{R}{D}$$

we have then at point P

$$\text{horizontal illumination} = \frac{cp}{H^2} \cos^3\theta \qquad (20.16)$$

and

$$\text{vertical illumination} = \frac{cp}{R^2} \sin^3\theta \qquad (20.17)$$

Inasmuch as the candlepower intensity in the direction of point P is taken from a candlepower distribution curve, and θ is known, these expressions are readily usable. Very few commercial sources are actually point sources. However, *when the maximum dimension of the source is less than five times the distance to point P,* the equations give satisfactory results. Note that these equations can be used to calculate and plot isolux diagrams for point sources, of the type shown in Figs. 20.54 and 20.55.

EXAMPLE 20.5. Referring to Fig. 20.57 and the candlepower distribution curve of Fig. 20.58, find the horizontal and vertical illumination at point P, which is 10 ft below and 12 ft horizontally distant from the source.

SOLUTION

$$H = 10 \text{ ft} \qquad R = 12 \text{ ft}$$

$$\theta = \tan^{-1}\frac{12}{10} = 50°$$

$$\sin\theta = 0.766 \qquad \cos\theta = 0.643$$

$$cp \text{ at } 50° = 6600 \text{ (from Fig. 20.58)}$$

$$\text{horizontal illumination} = \frac{6600}{10^2}(0.643)^3 = 17.5 \text{ fc}$$

$$\text{vertical illumination} = \frac{6600}{12^2}(0.766)^3 = 20.8 \text{ fc}$$

■

Fig. 20.58 *Typical candlepower distribution plot for use in inverse square law calculation.*

20.41 Calculating Illuminance from Linear and Area Sources

When the source is too large to be considered a point source (the definition is relative and depends on distance to the illuminated surface), it is referred to either as a *linear source* or an *area source.* The *direct component* of the illuminance at a point, resulting from such sources, can be calculated by manual graphical or analytical means, both of which are based on an assumed distribution, generally lambertian (diffuse). Inasmuch as most lighting fixtures being applied today do *not* have lambertian characteristics (e.g., parabolic reflectors, prismatic diffusers), the results of such calculations are necessarily approximate. (Skylights and luminous ceiling *do* have a lambertian distribution, and there these calculation methods do give reliable results.) In addition to the direct component of illuminance, a *reflected component* must be added that depends on the point's location in the room and the room's characteristics. This calculation involves charts, diagrams, and tables. The interested reader will find a full description of these manual methods in Section 9, *Reference Volume, 1987 IESNA Lighting Handbook.*

However, because these manual methods are laborious and frequently less than reliable, they are not presented here. The ready availability of desktop computers and computer programs that can handle detailed input for the specific light source involved, without broad approximations, has made these manual procedures obsolete. We recommend that when point-by-point illuminance calculation is desired, such a program be used. Alternatively, one of the design aids described previously, based on a specific light source, can be used.

20.42 Computer-Aided Design

As pointed out repeatedly in previous sections, the use of a computer in lighting design is a practical necessity if really useful results are to be obtained. Once the desired luminance patterns and illuminance levels have been established, and luminaires selected and located by using average illuminance (zonal cavity) calculations or by use of one of the design aids previously demonstrated and/or manufacturer's assistance, the responsible designer will confirm the preliminary design with accurate calculations. Furthermore, only a computer analysis will give useful point-by-point illuminance figures plus the necessary data on VCP and reflected glare for selected work locations and viewing directions.

In addition, the computer gives the designer a degree of flexibility not otherwise possible in that:

1. It performs the calculations accurately and rapidly.
2. It frees the designer for other, less routine work.
3. It gives the designer the ability to change parameters repeatedly, without making the amount of calculation excessive, as would be the case in hand calculation.

It is this last characteristic that gives the designer greatest flexibility. The ability to run a series of calculations for a pendant fixture with pendant lengths varying from 12 to 36 in. in 3-in. intervals, or to change paint colors and reflectance on various surfaces and note the effect, or to test different layout patterns as mentioned in Example 20.2, gives the designer a very powerful and extremely useful design tool. In addition, the computer can consider related items, such as first costs, energy use, operating costs, and impact on HVAC; items whose complexity because of interrelations put them well beyond the pencil-and-hand-calculator's ability. A few of these useful programs currently available are listed in Appendix H.

20.43 Computer-Aided Design: Illustrative Example

In order to demonstrate the huge amount of accurately calculated data available at a designer's fingertips when using any of the well-written, comprehensive lighting design programs, we have chosen to run Example 20.2 with a very powerful modern program: Lumen-Micro 7 © (Lighting Technologies, Boulder, Co.) The first step, of course, is to input the data and the target illuminance. For convenience it is repeated here:

· Room: $60' \times 100' \times 8'$

· Reflectances: 0.80 ceiling, 0.50 wall, 0.30 floor

· Required uniform illuminance: 50 fc (500 lux)

  (Note: To avoid decimal dimensions, the program was configured in English units. It could very simply have been configured in SI units.)

· Luminaire: Recessed parabolic troffer

Some modern programs (as this one) have a large library of luminaires that is constantly updated, plus an efficient search program. This obviates the tiresome task of physically searching through several dozen catalogs, and the even more tiresome task of keeping the catalogs current. We decided to use a 2-ft square luminare in order to reduce the possibility of both direct and reflected glare to persons viewing the fixture crosswise, and further chose to use U-shaped lamps. A very rapid search in the computer's extensive library of lighting fixtures resulted in the choice of the luminaire shown in Fig. 20.59 as printed from the computer's library. On the basis

ILLUMINATION

Fig. 20.59 *Photometric report obtained from the computer library of Lumen-Micro 7© as a result of a search for a suitable 2-ft square, recessed parabolic luminaire for the example in Section 20.43 (see text). (Courtesy of Lighting Technologies, Boulder, Co.)*

Photometric Summary:

Luminous size: 2.0 x 2.0 x 0.0 ft
No. of Lamps: 3
Lamp Lumens: 2800
Input Wattage: 97
Efficiency: 60.0%
Max Candela: 2916

No. of Lamps:	3	Lamp Lumens:	2800
Multiplier:	1	Input Wattage:	97
Ballast Factor:	1	Ball/Lamp Factor:	1
Horz. Angles:	5		
Vert. Angles:	11		

	0°	22.5°	45°	67.5°	90°
0°	2888	2888	2888	2888	2888
5°	2881	2880	2882	2889	2893
15°	2738	2771	2837	2898	2916
25°	2515	2580	2685	2750	2745
35°	2188	2300	2297	2186	2241
45°	1494	1653	1578	1272	1089
55°	247	351	296	48	17
65°	3	8	1	10	1
75°	0	10	0	0	1
85°	0	0	0	0	1
90°	0	0	0	0	0

COEFFICIENTS OF UTILIZATION - Zonal Cavity Method:

Floor Refl(%) = 20

Ceiling Refl(%)	80				70				50			30			10			0
Wall Refl(%)	70	50	30	10	70	50	30	10	50	30	10	50	30	10	50	30	10	0
RCR																		
0	71	71	71	71	70	70	70	70	67	67	67	64	64	64	61	61	61	60
1	68	66	64	62	66	64	63	61	62	61	59	60	59	58	58	57	56	55
2	63	60	57	55	62	59	56	54	57	55	53	55	53	52	53	52	51	50
3	59	55	51	48	58	54	51	48	52	50	47	51	48	47	49	47	46	45
4	56	50	46	43	54	49	46	43	48	45	42	47	44	42	46	43	41	40
5	52	46	42	39	51	45	41	39	44	41	38	43	40	38	42	40	37	36
6	49	42	38	35	48	42	38	35	41	37	34	40	37	34	39	36	34	33
7	45	39	35	32	45	38	34	31	38	34	31	37	34	31	36	33	31	30
8	43	36	32	29	42	36	31	29	35	31	29	34	31	28	33	30	28	27
9	40	33	29	26	39	33	29	26	32	29	26	32	28	26	31	28	26	25
10	38	31	27	24	37	31	27	24	30	26	24	30	26	24	29	26	24	23

$$RCR = \frac{5 * H * (Room\ Length + Room\ Width)}{(Room\ Length * Room\ Width)}$$

H = distance between workplane and luminaire plane

Fig. 20.59 *(See caption at the bottom of page 1267.)*

of a preliminary calculation, an area of approximately 75 ft² per fixture seemed appropriate (because in the original long-hand calculation an area of about 50 ft² was used for a 2 lamp fixture). A row spacing of 7.5 ft and a column spacing of 10 ft was "drawn" on the computer, and complete point-by-point calculation run for horizontal and vertical illuminance, VCP and RVP. The numerical results are shown in Figs. 20.60–20.62. As the room is very large and symmetrical and is being

calculated with no daylight contribution, a calculation for one quarter of the space is sufficient. Figure 20.60 gives a summary of the results. Were there daylight calculations, their results would show as well. The *average* illuminance in the space is 48 fc, as compared to the 50 fc target illuminance, and is therefore well within the desired range. Unit power density (UPD) is only 1.29 wsf, which is within the requirements of the major energy codes. The Luminaire Summary shows the

Output Generated by Lumen Micro 7 for Windows. Copyright 1996 Lighting Technologies. Inc., Boulder, Colorado

Room / Site Summary

Room 10   (Floor 1)

Calculation Summary

	Attained	
Zonal Cavity Illum.*:	48	fc
Unit Power Density:	1.29	W/sq. ft.

*Zonal cavity illuminance does not take into account objects within the space or daylighting.

Room Summary

Overall Size:	100 ft. x 60 ft. x 8 ft.
Reflectances:	Ceiling: .8
	Walls: .5
	Floor: .3

Daylighting Information

Latitude:	0
Longitude:	0
Ground Refl.:	0
Bldg. Rotation:	0

Luminaire Summary

Type	Catalog Number
2PMO 3U	Lithonia "Optimax" 2PMO 3U31 12LS

Luminaire Positions

Type	X	Y	Z	Rotate
2PMO 3U	5.00	3.75	8.00	0.00
2PMO 3U	15.00	3.75	8.00	0.00

Fig. 20.60 *Summary of the results of a single layout plan. (Courtesy of Lighting Technologies, Boulder, Co.)*

ILLUMINATION

Calculation Grid: Horizontal Grid #1 (1 of 1)

Room 10 (Floor 1)

Grid Summary

Grid Properties
Grid Type: Horizontal Illuminance
Grid Height: 2.50 ft
Units: English

Calculation Grid

	1.3	3.8	6.3	8.8	11.3	13.8	16.3	18.8	21.3	Dis 23
28.8	35.9	59.2	59.2	38.6	38.7	59.4	59.5	38.7	38.9	59
26.3	38.2	65.0	64.9	41.5	41.6	65.1	65.2	41.6	41.8	65
23.8	36.0	59.3	59.2	38.6	38.8	59.5	59.5	38.7	38.9	59
21.3	36.1	59.4	59.4	38.8	38.9	59.6	59.7	38.9	39.0	59
18.8	38.1	64.9	64.8	41.4	41.6	65.1	65.1	41.5	41.7	64
16.3	36.1	59.4	59.3	38.8	38.9	59.6	59.6	38.9	39.0	59
13.8	35.8	59.1	59.1	38.5	38.6	59.3	59.4	38.6	38.7	59
11.3	38.2	65.1	64.9	41.5	41.7	65.2	65.2	41.6	41.7	65
8.8	35.8	59.1	59.1	38.5	38.7	59.4	59.4	38.6	38.7	59
6.3	36.5	59.5	59.5	38.9	39.1	59.8	59.8	39.0	39.2	59
3.8	37.1	63.5	63.3	40.6	40.9	63.8	63.8	40.7	40.9	63
1.3	30.0	47.0	46.8	32.1	32.9	47.4	47.6	32.1	32.5	47

Distance from Grid Origin Point

Fig. 20.61 *Point by point horizontal illuminance results. (Courtesy of Lighting Technology, Boulder, Co.)*

fixture chosen; had there been several units used, all would be listed. Similarly, a list of luminaires in their positions in the space appears at the bottom of the figure. Inasmuch as only one type was used in this space, the long, repetitive list was truncated. Figure 20.61 is a printout of the point-by-point horizontal illuminance on the working plane (30 in. AFF), and it immediately demonstrates the value of a point-by-point calculation. Although the *average* illuminance is 48 fc and therefore satisfies the target requirement, the illuminance drops as low as 38.5 fc and rises to 65 fc in the body of the space, away from the walls. (The calculation grid is 2.5 ft 2.5 ft.) In practice, this situation might cause the designer to select a closer fixture spacing or perhaps change the

luminaire, because 38.5 fc is 23% below the target illuminance.

On another printout of horizontal illuminance [Fig. 20.62] we have plotted luminaire locations, desk locations, and one drafting table. (The plan is not scalable because the printout row and column spacing are different.) Note that on desks against the wall, the *average* illumination is near 50 fc whereas on the drafting table it is about 45 because of its location. A staggered arrangement might improve uniformity, and therein lies one of the great advantages of computer calculation. With an additional ½ hour of work, an experienced designer who is accustomed to working with a specific program could try three luminaires, each in three different arrays, and compare the

Calculation Grid: Horizontal Grid #1 (1 of 1)

Room 10 (Floor 1)

Grid Summary

Grid Properties
Grid Type: Horizontal Illuminance
Grid Height: 2.50 ft
Units: English

Calculation Grid

Fig. 20.62 Illuminance chart with luminaires and proposed furniture layout plotted on the chart. (Chart courtesy of Lighting Technologies, Boulder, Co.)

results to arrive at an optimum lighting/economic solution.

Figure 20.63 shows partial printouts of three other metrics of the lighting design. Vertical illumination [Fig. 20.63(a)] is useful for seeing tasks involving other than horizontal surfaces. In conventional offices where the majority of work is on the horizontal plane, vertical illuminance is not critical. As pointed out repeatedly in previous sections, diffuse lighting is required to achieve good vertical surface illumination from overhead sources. In this space, with direct lighting only and (effectively) no walls to reflect light and increase diffusion, vertical illumination comes primarily as a component of the direct lighting, with a small contribution from light re-reflected from the floor (see Fig. 21.44).

Figure 20.63 (b) indicates uniformly high VCP when viewing north (directly ahead). This is to be expected from a parabolic reflector luminaire, which is noted for low luminance, and therefore low direct glare. Relative visual performances (RVP) is the metric that replaced ESI as an indication of supra-thereshold visual response, considering task size, contrast, and background luminance. The task we selected for these calculations is a difficult one: HB pencil on vellum, on a white table. Nevertheless, the calculations indicate perfect RVP. In our discussion of reflected glare, ESI and RVP in Sections 18.29 and 18.30 we expressed serious reservations as to the usefulness of RVP as a visual performance metric when considering veiling reflections; the results shown in Fig. 20.63 (c) seem to reinforce these reservations.

ILLUMINATION

ILLUMINATION

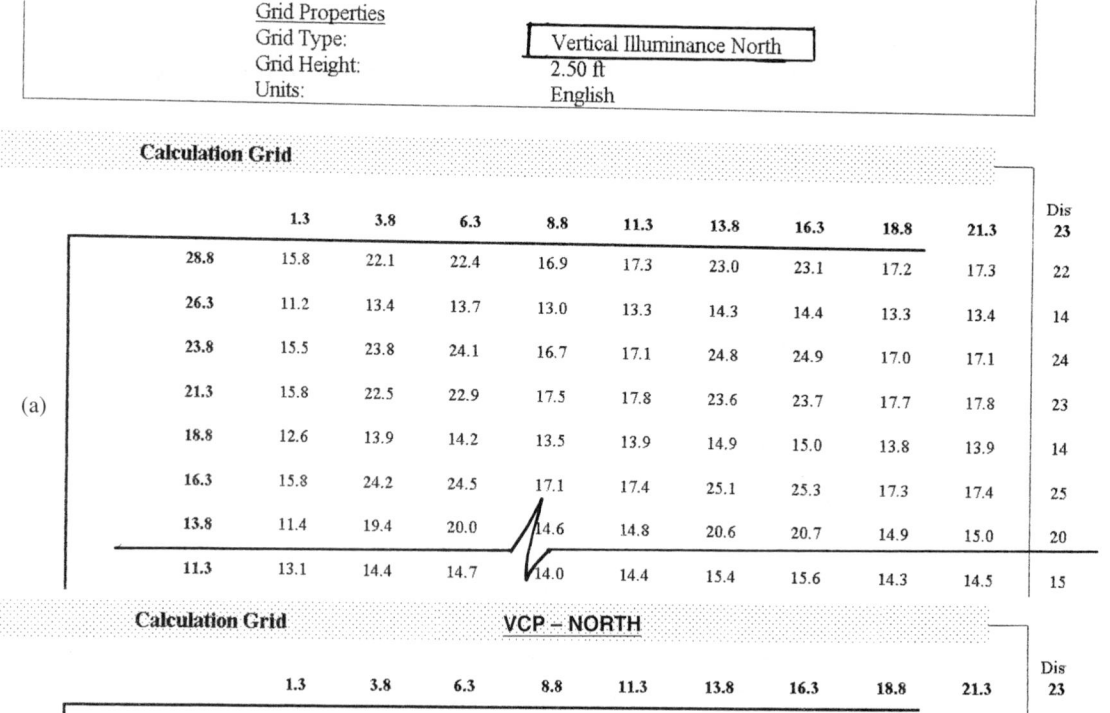

Grid Summary

Grid Properties
Grid Type: Vertical Illuminance North
Grid Height: 2.50 ft
Units: English

Calculation Grid

	1.3	3.8	6.3	8.8	11.3	13.8	16.3	18.8	21.3	Dis 23
28.8	15.8	22.1	22.4	16.9	17.3	23.0	23.1	17.2	17.3	22
26.3	11.2	13.4	13.7	13.0	13.3	14.3	14.4	13.3	13.4	14
23.8	15.5	23.8	24.1	16.7	17.1	24.8	24.9	17.0	17.1	24
21.3	15.8	22.5	22.9	17.5	17.8	23.6	23.7	17.7	17.8	23
18.8	12.6	13.9	14.2	13.5	13.9	14.9	15.0	13.8	13.9	14
16.3	15.8	24.2	24.5	17.1	17.4	25.1	25.3	17.3	17.4	25
13.8	11.4	19.4	20.0	14.6	14.8	20.6	20.7	14.9	15.0	20
11.3	13.1	14.4	14.7	14.0	14.4	15.4	15.6	14.3	14.5	15

(a)

Calculation Grid   VCP – NORTH

	1.3	3.8	6.3	8.8	11.3	13.8	16.3	18.8	21.3	Dis 23
28.8	91.1	86.0	82.4	92.2	93.3	89.3	89.3	93.3	93.3	89
26.3	97.8	94.5	96.7	97.4	97.5	97.2	97.2	97.5	97.5	97
23.8	100.0	100.0	100.0	100.0	100.0	100.0	100.0	100.0	100.0	100
21.3	91.1	86.0	82.4	92.2	93.3	89.3	89.3	93.4	93.4	89
18.8	97.8	94.5	96.7	97.4	97.5	97.2	97.2	97.5	97.5	97
16.3	100.0	100.0	100.0	100.0	100.0	100.0	100.0	100.0	100.0	100
13.8	91.1	86.0	82.4	92.2	93.3	89.3	89.3	93.4	93.4	89
11.3	97.8	94.5	96.7	97.5	97.5	97.2	97.2	97.5	97.5	97

(b)

Calculation Grid   RVP – NORTH

	1.3	3.8	6.3	8.8	11.3	13.8	16.3	18.8	21.3	Dis 23
28.8	1.00	1.00	1.00	1.00	1.00	1.00	1.00	1.00	1.00	1.0
26.3	1.00	1.00	1.00	1.00	1.00	1.00	1.00	1.00	1.00	1.0
23.8	1.00	1.00	1.00	1.00	1.00	1.00	1.00	1.00	1.00	1.0
21.3	1.00	1.00	1.00	1.00	1.00	1.00	1.00	1.00	1.00	1.0
18.8	1.00	1.00	1.00	1.00	1.00	1.00	1.00	1.00	1.00	1.0
16.3	1.00	1.00	1.00	1.00	1.00	1.00	1.00	1.00	1.00	1.0
13.8	1.00	1.00	1.00	1.00	1.00	1.00	1.00	1.00	1.00	1.0
11.3	1.00	1.00	1.00	1.00	1.00	1.00	1.00	1.00	1.00	1.0

(c)

Fig. 20.63 *Partial printout report for the computer analysis of problem 20.2.* (a) *Point-by-point calculation of vertical illuminance at the working plane level. Note that it averages less than half of horizontal illuminance.* (b) *VCP chart of the room.* (c) *RVP chart indicating perfect seeing conditions in the entire area looking North (see text discussion). (Courtesy of Lighting Technologies, Boulder, Co.)*

Fig. 20.64 *A combined graphic (contours) and point-by-point printout of horizontal illuminance for lighting design. The almost rectangular figures represent closed contours at 40 fc. (Courtesy of Lighting Technologies, Boulder, Co.)*

Graphic output is often easier to grasp than a raft of numbers. Using the same calculation grid as previously, we printed out horizontal illuminance contours overlaid with the point-by-point calculations in Fig. 20.64. (They can just as easily be printed separately.) The rectangular boxes on the figure are closed 40 fc contour lines.

20.44 Average Luminance Calculations

The basic equations relating luminance to candela intensity and to illuminance are covered in Section

18.8, which also deals with the calculation of source luminance and reflected luminance when the illuminance (lux) level is known. Thus, once horizontal illuminance has been calculated by any of the methods described previously and knowing the reflectance of an object, its horizontal luminance can readily be calculated (see Section 18.8). However, as explained in detail in Sections 18.32 and 18.33, which discuss lighting quality, and in Sections 20.10 through 20.16, which deal with lighting systems and patterns, the luminance impression of a visual environment is affected more

TABLE 20.5 Wall Luminance Coefficients and Ceiling Cavity Luminance Coefficients for Typical Luminaires

To obtain a luminance coefficient, follow the procedure detailed at the head of Table 20.2 to find a coefficient of utilization. Data should be obtained from manufacturers of actual luminaires used.

Typical Luminaire	RCR	ρ_{cc}: 80			50			80		50	
		ρ_w: 50	30	10	50	30	10	50	30	50	30
		Wall Luminance Coefficients for ρ_{FC} = 20						Ceiling Cavity Luminance Coefficient ρ_{FC} = 20			
1								.42	.42	.25	.25
	1	.32	.18	.06	.27	.15	.05	.42	.40	.25	.23
	2	.27	.15	.05	.23	.13	.04	.42	.38	.24	.23
	3	.24	.13	.04	.20	.11	.03	.41	.37	.24	.22
	4	.21	.11	.03	.17	.09	.03	.41	.36	.24	.22
	5	.19	.10	.03	.16	.08	.02	.40	.35	.24	.21
	6	.18	.09	.03	.14	.07	.02	.39	.34	.23	.21
	7	.16	.08	.02	.13	.07	.02	.39	.34	.23	.21
	8	.15	.07	.02	.12	.06	.02	.38	.34	.23	.20
Pendant diffusing sphere	9	.14	.07	.02	.12	.06	.02	.38	.33	.23	.20
with incandescent lamp	10	.13	.06	.02	.11	.05	.01	.37	.33	.22	.20
3								.15	.15	.09	.09
	1	.23	.13	.04	.21	.12	.04	.15	.13	.09	.08
	2	.22	.12	.04	.21	.11	.04	.14	.11	.08	.07
	3	.21	.11	.03	.20	.11	.03	.13	.10	.08	.06
	4	.20	.10	.03	.19	.10	.03	.13	.08	.08	.05
	5	.19	.09	.03	.18	.09	.03	.12	.08	.07	.05
	6	.18	.09	.03	.17	.09	.02	.12	.07	.07	.04
	7	.17	.08	.02	.16	.08	.02	.11	.06	.07	.04
Porcelain-enamaled	8	.16	.08	.02	.15	.07	.02	.11	.06	.06	.04
ventilated standard dome	9	.15	.07	.02	.14	.07	.02	.10	.05	.06	.03
with incandescent lamp	10	.14	.07	.02	.13	.07	.02	.10	.05	.06	.03
7								.08	.08	.04	.04
	1	.06	.03	.01	.05	.03	.01	.08	.07	.04	.04
	2	.05	.03	.01	.05	.03	.01	.07	.06	.04	.04
	3	.05	.03	.01	.04	.02	.01	.06	.05	.04	.03
	4	.05	.02	.01	.04	.02	.01	.05	.04	.03	.03
	5	.05	.02	.01	.04	.02	.01	.05	.04	.03	.02
	6	.04	.02	.01	.04	.02	.01	.05	.03	.03	.02
Reflector downlight with	7	.04	.02	.01	.04	.02	.01	.04	.03	.03	.02
battles and inside-frosted	8	.04	.02	.01	.04	.02	.01	.04	.03	.02	.02
lamp (see note on this unit	9	.04	.02	.01	.04	.02	.01	.04	.02	.02	.02
in Table 20.2)	10	.04	.02	.00	.04	.02	.01	.04	.02	.02	.01
28								.27	.27	.16	.16
	1	.17	.10	.03	.14	.08	.03	.26	.24	.15	.14
	2	.16	.09	.03	.14	.08	.02	.25	.23	.15	.14
	3	.15	.08	.02	.13	.07	.02	.24	.21	.14	.13
	4	.15	.08	.02	.13	.07	.02	.24	.21	.14	.12
	5	.14	.07	.02	.12	.06	.02	.23	.20	.14	.12
	6	.13	.07	.02	.11	.06	.02	.23	.19	.14	.12
	7	.13	.06	.02	.11	.06	.02	.23	.19	.13	.11
Diffuse aluminum reflector	8	.12	.06	.02	.11	.05	.02	.22	.18	.13	.11
with 35° crosswise and 35°	9	.12	.06	.02	.10	.05	.01	.22	.18	.13	.11
lengthwise shielding	10	.11	.05	.01	.10	.05	.01	.22	.18	.13	.11
33								.65	.65	.38	.38
	1	.20	.12	.04	.13	.08	.02	.65	.63	.38	.37
	2	.19	.10	.03	.12	.07	.02	.64	.61	.38	.37
	3	.17	.09	.03	.11	.06	.02	.64	.60	.37	.36
	4	.16	.08	.02	.11	.06	.02	.63	.59	.37	.36
	5	.15	.08	.02	.10	.05	.02	.63	.59	.37	.36
	6	.14	.07	.02	.09	.05	.01	.60	.58	.37	.35
	7	.13	.07	.02	.09	.04	.01	.62	.58	.37	.35
Luminous bottom	8	.12	.06	.02	.08	.04	.01	.61	.57	.37	.35
suspended unit with extra-	9	.12	.06	.02	.08	.04	.01	.61	.57	.36	.35
high-output lamp	10	.11	.05	.01	.07	.04	.01	.61	.57	.36	.35

ILLUMINATION

(continued)

TABLE 20.5 **Wall Luminance Coefficients and Ceiling Cavity Luminance Coefficients for Typical Luminaires** (*continued*)

Typical Luminaire	RCR	ρ_{CC}: 80			50			80		50	
		ρ_w: 50	30	10	50	30	10	50	30	50	30
		Wall Luminance Coefficients for $\rho_{FC} = 20$						Ceiling Cavity Luminance Coefficient $\rho_{FC} = 20$			
44 Radial batwing distribution—louvred fluorescent, unit	0							.114	.114	.066	.066
	1	.137	.078	.025	.125	.072	.023	.105	.094	.061	.055
	2	.131	.072	.022	.121	.067	.021	.097	.079	.057	.047
	3	.127	.068	.020	.118	.064	.019	.092	.068	.054	.041
	4	.123	.064	.019	.115	.061	.018	.087	.060	.052	.036
	5	.119	.060	.018	.112	.058	.017	.084	.053	.050	.032
	6	.114	.057	.016	.108	.055	.016	.080	.048	.048	.029
	7	.110	.054	.015	.104	.053	.015	.078	.044	.046	.027
	8	.106	.052	.015	.101	.050	.014	.075	.041	.045	.025
	9	.102	.049	.014	.097	.048	.014	.073	.038	.043	.023
	10	.097	.047	.013	.093	.046	.013	.070	.036	.042	.022
46 Bilateral batwing distribution—one lamp, surface-mounted fluorescent, with prismatic wraparound lens	0							.236	.236	.138	.138
	1	.234	.133	.042	.208	.119	.038	.229	.210	.134	.124
	2	.213	.117	.036	.190	.106	.033	.222	.193	.130	.115
	3	.195	.104	.031	.175	.095	.029	.216	.180	.127	.108
	4	.181	.094	.028	.162	.086	.026	.211	.170	.124	.102
	5	.170	.087	.025	.153	.080	.023	.206	.163	.122	.098
	6	.159	.080	.023	.143	.073	.021	.201	.157	.119	.095
	7	.149	.074	.021	.135	.068	.020	.197	.152	.117	.092
	8	.141	.069	.019	.128	.064	.018	.193	.148	.115	.090
	9	.134	.065	.018	.121	.060	.017	.189	.144	.113	.088
	10	.126	.061	.017	.115	.056	.016	.185	.141	.111	.086
35 Two-lamp prismatic wraparound—multiply by 0.95 for four lamps								.22	.22	.13	.13
	1	.19	.11	.03	.17	.10	.03	.21	.20	.12	.12
	2	.18	.10	.03	.15	.09	.03	.21	.18	.12	.11
	3	.16	.09	.03	.14	.08	.02	.20	.17	.12	.10
	4	.15	.08	.02	.14	.07	.02	.19	.16	.11	.10
	5	.14	.07	.02	.13	.07	.02	.19	.15	.11	.09
	6	.14	.07	.02	.12	.06	.02	.18	.15	.11	.09
	7	.13	.06	.02	.12	.06	.02	.18	.14	.11.	.09
	8	.12	.06	.02	.11	.05	.02	.18	.14	.11	.08
	9	.12	.06	.02	.11	.05	.01	.17	.13	.10	.08
	10	.11	.05	.01	.10	.05	.01	.17	.13	.10	.08
42 Fluorescent unit with flat prismatic lens; four-lamp, 2 ft wide—multiply by 1.05 for three lamps and 1.10 for two lamps								.12	.12	.07	.07
	1	.16	.09	.03	.15	.09	.03	.11	.10	.06	.06
	2	.15	.08	.03	.14	.08	.02	.10	.08	.06	.05
	3	.15	.08	.02	.14	.07	.02	.10	.07	.06	.04
	4	.14	.07	.02	.13	.07	.02	.09	.06	.05	.04
	5	.13	.07	.02	.12	.06	.02	.09	.06	.05	.03
	6	.12	.06	.02	.12	.06	.02	.09	.05	.05	.03
	7	.12	.06	.02	.11	.06	.02	.08	.05	.05	.03
	8	.11	.05	.02	.11	.05	.02	.08	.04	.05	.03
	9	.11	.05	.01	.10	.05	.01	.08	.04	.05	.02
	10	.10	.05	.01	.10	.05	.01	.07	.04	.04	.02
47 Radial batwing distribution—four-lamp, 610-mm (2 ft)-wide fluorescent unit with flat prismatic lens—see note 2	0							.114	.114	.066	.066
	1	.175	.100	.032	.163	.094	.030	.107	.093	.063	.055
	2	.168	.092	.028	.157	.087	.027	.102	.079	.060	.047
	3	.157	.083	.025	.147	.080	.024	.097	.068	.057	.041
	4	.147	.076	.022	.138	.073	.022	.093	.060	.055	.036
	5	.139	.071	.021	.131	.068	.020	.090	.054	.053	.033
	6	.130	.065	.019	.123	.063	.018	.086	.049	.051	.030
	7	.122	.060	.017	.116	.059	.017	.082	.045	.049	.027
	8	.115	.056	.016	.110	.055	.016	.079	.042	.047	.025
	9	.109	.053	.015	.104	.052	.015	.076	.039	.045	.024
	10	.103	.049	.014	.099	.048	.014	.072	.037	.043	.022

NOTES:

1. Data extracted, with permission, from *IESNA Handbook* (1993), Reference Volume. Coefficients for fixtures 1, 3, 7, 28, 33, 35, and 42 have been rounded to two decimal places.

2. Multiply coefficients by 1.05 for three lamps and 1.1 for two lamps.

by vertical surface luminance than by horizontal. (See also Table 18.2 and Fig. 20.11.) For this reason it is important to be able to calculate *average* vertical surface (wall) luminance in the same simple, straightforward fashion that *average* horizontal illuminance is calculated. In addition, it is useful to know the average luminance of the ceiling cavity in a space in order to judge the contrast between all luminous objects, including luminaires, that have the ceiling cavity as background.

Straightforward calculation of both wall and ceiling cavity luminance (L_W and L_{CC}) is possible through the use of luminance coefficients that are similar in application and calculation to utilization coefficients. These coefficients are listed in Table 20.5 for some of the generic fixture types listed in Table 20.2. Others are listed in the *IESNA Lighting Handbook*. For actual installation calculations, it is preferable to obtain coefficients from luminaire manufacturers. The average luminance calculations are parallel to those for illuminance.

Average initial wall luminance (cd/m²):

$$L_W = \frac{\text{lamp lumens} \times \text{wall luminance coefficient}}{\pi \times \text{floor area in m}^2} \quad (20.18)$$

and average initial ceiling cavity luminance in cd/m²:

$$L_{CC} = \frac{\text{lamp lumens} \times \text{ceiling cavity luminance coefficient}}{\pi \times \text{floor area in m}^2} \quad (20.19)$$

If area is expressed in square feet and π omitted from the equations, L is expressed in foot-lamberts.

To obtain *maintained* values, an LLF similar to that explained in Section 20.32 is introduced. It is calculated similarly, except that item 17 (see Fig. 20.46), room surface dirt, is calculated using the following figures:

Lighting System	Wall Luminance	Ceiling Luminance
Direct	0.82 ± 10%	0.75 ± 10%
Semidirect	0.87 ± 7%	0.82 ± 10%
Direct–indirect	0.92 ± 5%	0.85 ± 8%
Semi-indirect	0.87 ± 7%	0.88 ± 7%
Indirect	0.82 ± 10%	0.90 ± 5%

For ceiling-mounted or recessed luminaires, L_{CC} is the average luminance of the ceiling between luminaires. For pendant luminaires, the calculated L_{CC} is that of an imaginary plane at the height of the luminaires. It is useful in determining the brightness ratios when compared to luminaire luminance at the seeing angle involved. The ceiling cavity, as the wall, is assumed to have a lambertian characteristic—that is, perfect diffuseness, making luminance independent of viewing angle.

It would be instructive to calculate the wall luminance of the office of Example 20.2. The photometric data in Fig. 20.48 do not include the wall luminance coefficient because luminance coefficients are not normally published by luminaire manufacturers. However, based on other data available, a figure of 0.22 for wall luminance is a good estimate given RCR, ρ_{CC}, and ρ_w of 0.66, 80%, and 30%, respectively. Initial wall luminance is then

$$L_W = \frac{3200 \times 0.22}{32 \text{ ft}^2} = 22 \text{ fL} = 75.5 \text{ cd/m}^2$$

This is within the preferred range of 50 to 200 cd/m² (see Table 18.3). In actuality the average wall luminance would probably be higher because of the practice of placing the last row of luminaires quite close to the wall.

EVALUATION STAGE

20.45 Design Evaluation

The design stage having been completed, the final step is evaluation of the design from the three aspects—lighting, costs, and energy. The lighting aspects include quantity, quality, luminance ratios, mood, ambience, texture, color, variation, psychological impressions, orientations, and daylight use—in short, a review of all the lighting factors previously discussed in detail. A good deal of experience is required to visualize the actual results from the drawings. The novice designer would do well to be assisted in doing the review by someone having such experience. The other two aspects of evaluation, cost and energy, can be performed readily with the aid of contractor's estimating figures for cost and a straightforward calculation for energy. The results are compared to the budget figures developed at the preliminary design stage.

As we have repeatedly stressed, the useful cost figures are life cycle, annual operating, and first cost for economic comparisons, operating budgets, and construction budgets, respectively. In Chapter 21 we present recommendations for specific occupancies accompanied by actual cost studies and energy analyses. Detailed cost studies including the impact of lighting on air conditioning, the proportional cost of the wiring system, and the proper apportionment of costs involves the entire building and can be accurately performed only by computer. Studies of this type are generally made by consulting engineers rather than architects, and then only after initial, operating, and total costs have been set in proper perspective for the particular job, by the architect and client. This is necessary because often, as in the case of speculative construction, the client's overriding consideration is first cost, thereby rendering a complete cost analysis unnecessary. Any attempt to completely separate costs for lighting, HVAC, structure, and so on is arbitrary because of the intimate interaction between these factors. Lighting designers are well advised to keep themselves and the construction team aware of this, if they are to fulfill their responsibility.

REFERENCES

See complete listing for all lighting sections at the end of Chapter 21.

ILLUMINATION

21

LIGHTING APPLICATION

21.1 Introduction

In Chapters *18* through *20,* we examined lighting fundamentals, sources, and design procedures. In this final lighting chapter, we consider application of lighting principles to specific situations. The facilities covered in some detail include residential, educational, commercial, institutional, and industrial occupancies. Each is examined from the viewpoint of its special requirements, and design approaches are suggested. The latter include lighting materials and sources as well as comparative economics and energy considerations. The chapter concludes with consideration of special types of indoor lighting, plus a short section on exterior lighting.

LIGHTING CONTROL

21.2 Requirement for Lighting Control

The term *lighting control* means all the techniques by which a lighting system can be operated, and covers both manual and automatic controls. The control strategy must be decided on simultaneously with the lighting design because the control scheme must be appropriate to the light source. In turn, the system accessories and arrangement depend on the control scheme. For instance, if dimming is decided on, using a fluorescent light source, then the range of dimming determines the type of ballast, the ballast switching points, and the degree of dimming flexibility.

The primary purposes of lighting control are flexibility and economy: flexibility to provide the modifications of luminances and patterns desired by the designer, and economy of both energy resources and monetary resources (see Appendix E for treatment of economic analyses). A properly designed lighting control system can reduce energy usage up to 60% over a simple ON/OFF system installation. In addition, financial operating economies result from:

- Reduced energy use.
- Reduced-air conditioning costs as a result of lower lighting waste heat.
- Longer lamp and ballast life due to lower operating temperatures and lower output.
- Lower labor costs due to control automation.

As explained previously in Chapter 20, lighting in most new nonresidential construction is designed within the energy constraints of ASHRAE/ IESNA Standard 90.1, *Energy Efficient Design of New Buildings.* The 1989 issue of this standard has a system of lighting power credits for lighting control systems designed with automatic energy-conserving controls. These credits permit the effective connected load to be reduced by factors that increase with energy conservation effectiveness. Thus, for example, circuits with a simple on–off mode initiated by a daylight sensor have a smaller power credit than daylight sensing with continuous dimming because the latter is more energy economical (but much more expensive initially). To avoid confusion, particularly in view of overlapping and sometimes inaccurate terminology, it is necessary to differentiate between control functions, control devices, and control systems.

For lighting, the only control *functions* are switching and dimming. The control *devices* are the means by which the switching and dimming functions are accomplished. They are numerous, ranging from a simple wall switch through time switches and dimmers of all sorts. Generally also included in this category are control *initiation* devices, such as occupancy sensors and photocells. The control *system* is the entire assembly of control and signal initiating equipment together with their interconnections. Included here also are microprocessors and programmable controllers. The system can be a stand-alone arrangement or, alternatively, as is the case in large facilities, it may be part of an energy management system (EMS) or a building automation system (BAS) or both. The difference in operation in these instances lies in the control algorithm, which is primarily energy oriented for an EMS and overall building function oriented for a BAS. In the discussion that follows, the control criteria are energy conservation, cost reduction, and operating flexibility.

21.3　Lighting Control: Switching

There are two basic control functions—switching and dimming. Switching is an on–off function. By selecting the number of lighting elements to be switched in each switching action, the designer can establish the number of control levels. The more levels, the finer the control. Thus, in a space requiring several levels of uniform illumination for different functions, the designer has many control alternatives. He or she can switch entire fixtures, but this adversely affects uniformity. Taking three-lamp fluorescent fixtures as an example, the designer can obtain better uniformity and four levels of illumination by switching the ballasts (assuming one two-lamp and half of a two-lamp ballast per fixture):

All ballasts on	100% illumination
Two-lamp ballast on	66% illumination
½ of two-lamp ballast on	33% illumination
All ballasts off	0% illumination

(Magnetic ballasts are usually preferred in this arrangement because a single electronic ballast is often used for up to four lamps as one means of obtaining its benefits while reducing its cost per lamp.)

This type of switching has the advantage of light reduction in relatively small steps, at low cost. A typical arrangement is shown in Fig. 21.1. Use of split-wired two-lamp units is advantageous from the cost and energy viewpoints. Even more uniform light reduction and finer control are possible with two-level ballasts (at increased cost). There, each lamp remains lighted but at either full or half output. Thus, the designer could have 0%, 50%, and 100% levels or, by combining alternate ballast switching, with two-level ballasts could have 0%, 17%, 33%, 50%, 67%, 83%, and 100%. However, if that degree of control is desirable, dimming is probably preferable. The choice depends on the type of space and on the situation economics, as discussed later. An alternative method of achieving lower lighting levels in discrete steps, by switching, is to introduce impedance into the lighting circuit. This acts to reduce circuit current and light output. Such devices are readily available for control of fluorescent lamps (see Fig. 21.6).

Recognition of the fact that an increased number of control (switch) points makes possible finer control and therefore energy conservation led to requirements in ASHRAE/IESNA 90.1 relating

Fig. 21.1 *Schematic diagram of switching arrangements to achieve multiple discreet lighting levels with three-lamp fluorescent lighting fixtures. Two-lamp ballasts are used in the interest of energy conservation and financial economy. In scheme* (a)*, ballasts are switched, thus removing either one or two lamps from service. Finer control is achieved by using two-level ballasts or by introducing impedance* (b) *into the circuit either in a block for an entire circuit or distributed in each fixture.*

to the number of control points in a space and their types. Many field studies have demonstrated conclusively that reliance on manual switching is not an effective long-term energy conservation strategy, regardless of the good intentions of the space occupants. Indeed, studies have shown that space "ownership" affects even the low level of conservation possible with manual switching (i.e., private offices, conference rooms and small storage spaces may be switched off, whereas multioccupancy and common-use spaces such as libraries, large office spaces, smoking areas, and the like may not). As a result, Standard 90.1 therefore first specifies the minimum number of control points required in a space and then awards automatic switching or dimming with a higher number of equivalent control points.

The basic requirement is for one control point for every 450 ft^2 or fraction thereof of enclosed lighted space plus one control point for each task (or group of tasks) located in the space. Thus, the classroom of Example 20.1 would require two control points for its 517 ft^2 of area plus one control point for all similar tasks grouped in the room, for a total of three control points. This requirement could be met with three wall switches or one wall switch and an occupancy sensor, or an automatic dimming system, probably initiated by daylight sensors. The point is, of course, to encourage use of automatic controls, which, as has been pointed out repeatedly, is the only proven method of attaining significant energy conservation.

21.4 Lighting Control: Dimming

The techniques and equipment required for dimming each of the different light sources, as well as the effect on the color of the light produced and on the lamp, are discussed in Chapter 19. Figure 21.2 shows typical lumen output versus power input curves for common light sources. Note that for fluorescent lamps, *even with conventional magnetic ballasts,* dimming down to approximately 40% of output is possible without reducing efficacy. This desirable characteristic can be exploited in control schemes where it is desired to change light output gradually without sacrificing efficiency. Due to efficacy reduction below 40%

Fig. 21.2 *Typical dimming curves for generic light source types. Note that fluorescent is efficient and approximately linear, down to 40% output. All other sources have reduced efficacy when dimmed.*

output, an economical and efficient control scheme combines dimming and switching of multilamp fluorescent fixtures to yield an almost stepless output range of 13% to 100% output. Continuous dimming over a 10% to 100% range is practicable with special individual magnetic dimming ballasts or with electronic ballasts. As discussed in Chapter 19, electronic ballasts are much more energy efficient than conventional ones, and must be considered for all *new* installations, dimmed or not. For retrofit work, silicon-controlled rectifier (SCR) or triac dimmers give excellent results with existing conventional core-and-coil ballasts, down to 40% output, as noted previously.

21.5 Lighting Control: Control Initiation

Control initiation is either manual or automatic, or a combined manual–automatic function. This last is usually in the form of an override function; manual override of an automated procedure to assure adequate control for special or unusual situations, and automatic reset of manual function to reestablish normal or steady-state operation.

(a) Manual Control. Numerous studies dating back at least fifty years have indicated in-

creased employee satisfaction when at least a degree of control of the working environment is in his or her hands. This satisfaction is frequently accompanied by increased output, at least in the short term. Unfortunately, manual control of lighting levels has been demonstrated to be wasteful of energy due to the tendency to leave lights on at maximum level even when daylight is supplying more than enough light, or when leaving a room for an extended period. A modicum of energy conservation in this latter instance is possible, with the installation of "time-out" switches in spaces normally used for short periods, such as supply closets (see Fig. 16.48). However, for normal working spaces, even installation of manual dimmers in private offices is not effective because of the nuisance of having to go to the dimmer control location on the wall to readjust. Obviously, manual dimming in multiple-occupancy spaces is effective only in creating personnel dissatisfaction and friction.

Modern electronics has made practical a remote-control dimming system that can control single or multiple luminaires, making it applicable to all occupancies. Figure 21.3 illustrates the use of the system in an open multiple-occupancy space. This enables individual workers to adjust the output of luminaire(s) closest to their workstation without disturbing other employees. This adjustment, which is simply accomplished [see Fig. 21.3(a)] by remote control, can alleviate direct and/or reflected glare, or can be a temporary expedient suitable to the task at hand, as for instance work with a visual display terminal (VDT). Figure 21.3(b–d) shows the system components. Wiring of the system can be arranged so that a single receiver–dimmer can control up to 20 ballasts or, conversely, multiple dimmers are connected on a single circuit. This latter arrangement would be the case in wiring a group of small single- or double-occupancy rooms. Such rooms are frequently wired with a wall-mounted dimmer, which

(a-1)

Fig. 21.3 (a-1, a-2) *Ceiling-mounted fluorescent fixtures within an 8-ft radius of the optical sensor can be dimmed by remote control, and restored almost instantaneously.*
(Fig. cont. on next page), (b) *The hand-held infrared remote control device (shown with permanently mounted holster and a keeper-cord) arranged for continuous dimming and for dimming override for maximum or minimum light.*
(c) *The IR activated dimmer/receiver, which fits into a 4-in. square standard outlet box, can be mounted inside the luminaire as in* (a-1) *or adjacent to it. The optical tube that carries the IR signal must be exposed on the ceiling. It may be mounted on the underside of an open luminaire or close by.*
(d) *High-frequency electronic dimming ballast for 1–4 lamps can provide full-range dimming down to 1–5% of output or down to 10% output, depending on type. (Courtesy of Lutron Electronics Co.)*

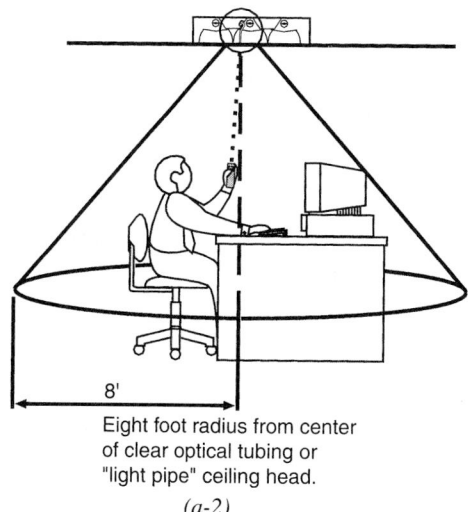

8'

Eight foot radius from center
of clear optical tubing or
"light pipe" ceiling head.

(a-2)

(b)

Wiring to
electronic
dimming
ballast

Optical tubing
terminating
in end cap

(c)

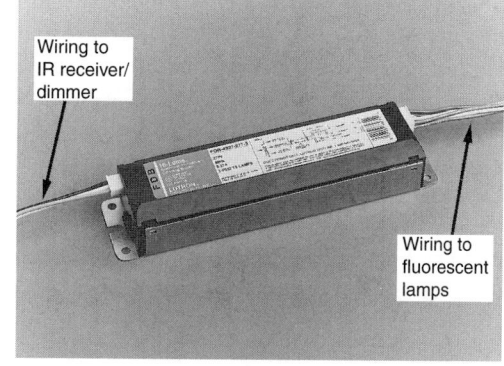

Wiring to
IR receiver/
dimmer

Wiring to
fluorescent
lamps

(d)

Fig. 21.3 *(continued)*

ILLUMINATION

can then be activated by the remote control, as shown in Fig. 21.4.

(b) Automatic Control Initiation. Automatic controls are of two types: an open-circuit type and a closed-loop feedback type, also known as static and dynamic control, respectively. The former initiates a control function that is independent of the actual lighting situation, whereas the latter reacts to the condition of the lighting situation it controls via a feedback loop.

1. Static control. The most common type of open-circuit lighting control is the programmable time-base controller. These vary from the small,

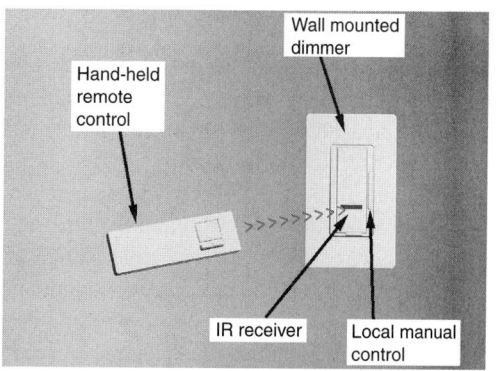

Wall mounted
dimmer

Hand-held
remote
control

IR receiver

Local manual
control

Fig. 21.4 *Remote controller may be used to operate a wall-mounted dimmer. (Courtesy of Lutron Electronics Co.)*

relatively simple units that replace wall switches and fit into a common device box to the more sophisticated units shown earlier in Figs. 16.13 through 16.15. These devices are available in a myriad of designs and capacities, but all perform the same basic function—(remote) control of loads and circuits on a preprogrammed time basis. The programming in turn is determined after analysis of operating schedules, task requirements, and field conditions. With "tight" programming, energy savings of up to 50% over an uncontrolled installation are possible.

Because these devices act only on a time-base and are insensitive to actual field conditions, an override feature must be incorporated to permit accommodation of special conditions. Thus, if the timer is set to shut off the row of lights adjacent to the windows between 10 AM and 3 PM, local override must be provided to accommodate dark rainy days and the like. Similarly, if lights are shut off during nonworking hours, provision must be made for persons working overtime. The override arrangement can be entirely local, in which case it may lead to energy waste because it depends on local cancellation; it can be local with time-out, which can be a nuisance to a person working for an extended period; or it can incorporate an override feedback link to the controller, usually operated by telephone lines. In general, programmable time controls are best applied to facilities with regular, repetitive schedules and few exceptional situations.

2. Dynamic control. The second type of automatic control initiation responds to sensor-indicated field conditions via an information feedback loop. It is frequently referred to as dynamic control because the initiation of a control function depends not on a fixed programmed parameter such as time, but on real-time field parameters (i.e., as measured at that instant). In the case of lighting control, these parameters may be ambient illuminance, time, system kW demand, kWh usage in a time period, and space occupancy, singly or in a combination, depending on the programming algorithm (i.e., how the controller's microprocessor has been programmed). The control device in its entirety is called a *programmable controller* (see Section 16.16), which, in combination with the field sensors and the interconnecting wiring, constitute the control system. Some systems are

"wireless," using high-frequency signals impressed on the power wiring system to transmit control signals. This arrangement is known as a power line carrier (PLC) system and is discussed in Section 16.34. Other systems are completely wireless, using radio frequencies and a system of wireless transmitters and receivers.

In addition to its CPU (microprocessor), the programmable controller contains input/output (I/O) interfaces, memory, and means for programming (and reprogramming). An operational block diagram is shown in Fig. 16.16. The controller accepts not only the usual time-based signal function, but also information (feedback) from field devices via its I/O device. It then "processes" the signal in its CPU, which consists of logic and storage memory, and sends out a resultant processed control signal.

Large lighting control systems use a computer in lieu of a programmable controller and usually have additional control facilities, such as telephone interfaces and local control/relay/switching centers, which process local sensor input and control local lighting blocks (see Figs. 16.56 and 21.5). Identical systems are used for building automation (Chapter 22), HVAC control, energy management, and the like, the difference being the type of sensing devices and the control algorithm.

21.6 Lighting Control Strategy

It is apparent that a good lighting control system varies the lighting supplied to match the lighting required, *as the requirement varies,* and thus overlighting and underlighting are avoided. In addition, the control system must be capable of permitting initial adjustments and of accommodating external, non–lighting-connected constraints such as commands from a peak-demand controller. The common lighting system situations addressed by the control system follow.

(a) System "Tuning." In every lighting installation there is a difference between the design intent and the field result. This is due to assumptions and imprecision in calculation, differences between specified and installed equipment, equipment location changes, and so on. The responsible lighting designer "tunes" the lighting system in the field to attain the intended design. This usually

Fig. 21.5 *Schematic arrangement of a large lighting control system. The controller can schedule and supervise thousands of control points via the local control panels. The local control panel accepts coded commands from the controller and operates individual devices and circuits. It also accepts local control signals from manual (switches) and automatic devices. Override is provided via telephone command to the controller, and all functions, including overrides, appear as "hard copy" on the printer.*

means *reducing levels* in nontask areas because spill light is frequently sufficient for circulation, rough material handling, and the like. This tuning can result in energy reduction of 20% to 30%, depending on the control technique.

The smaller the group of light sources controlled, the more accurate the tuning and the larger the energy saving, proportionally. Lighting system retuning is also required when the function of an entire space is changed, or when a single area is altered by furniture move or by a task change. An ancillary benefit of field tuning is glare reduction, which frequently improves task visibility dramatically. Tuning is a one-step function in the sense that once accomplished, it does not require change unless the space function changes. Therefore, it should also be reversible to accommodate such changes. Tuning should not be confused with *lumen maintenance,* described in (*d*), which is a control strategy designed to compensate for normal system output decline and its corollary–initial system overdesign.

As stated, the action most often required is a reduction in illuminance. This can be accomplished by:

1. Making appropriate field modifications to adjustable fixtures (aiming, lamp position, etc.).
2. Replacing lamps with others of lower watt-

ages, and replacing fluorescent tubes with low-wattage tubes. These changes reduce light output, reversibly.
3. Replacing fluorescent ballasts with low-current ballasts, thereby lowering output.
4. Adding current-limiting impedances to luminaires or lighting circuits as mentioned in Section 21.3a above (Fig. 21.6).
5. Ballast switching or use of multilevel ballasts (see Section 21.3*a*).

Fig. 21.6 *This type of compact solid-state electronic circuit reduces ballast power input by approximately 30% and lamp output by about 28% (see the curve in Fig. 21.2), resulting in a net gain in efficacy. The power factor remains above 90%. Similar units are available for larger power decrease. They can be mounted on the lamp end or in the fixture channel, as shown. An ancillary benefit is cooler ballast operation, resulting in extended life. (Photo courtesy of Remtec Systems.)*

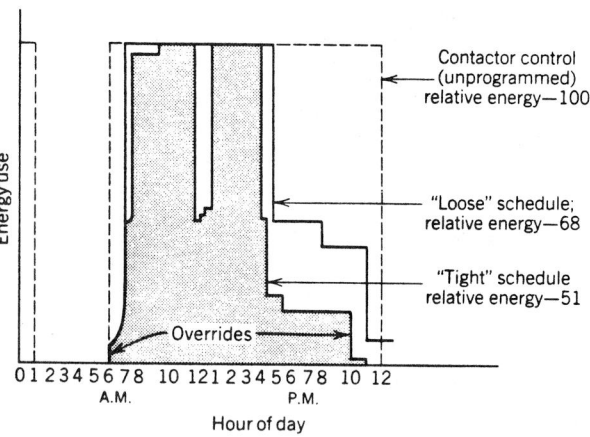

Fig. 21.7 *Actual plot of energy usage on the 58th floor of the World Trade Center in New York, with three degrees of lighting control. Note the importance of override by occupants when using "tight" scheduling. Override was provided on a 1000-ft² zone basis. (Data extracted from Peterson and Rubenstein, 1982.)*

6. Dimming by adjustment of potentiometer at individual fixtures so equipped.
7. Replacing standard wall switches with time-out units (see Fig. 16.48), programmable units, or dimmer units.

(b) Variable Time Schedule. No normal task area has a constant 24-h, 365-day lighting requirement. In commercial and industrial spaces, work areas have regularly scheduled periods during which task lighting is not required. These include coffee and lunch breaks, cleaning periods, shift changes, and unoccupied periods. Programmed time controls can readily save 10% to 25% of the energy use as compared to relying on occupants to manually operate the controls. The action of such a controller in an actual installation is shown in Fig. 21.7. "Tight" scheduling took account of lunch hour and provided lighting only in restricted areas being cleaned rather than whole floors. Payback period for the investment in control equipment varies between 1½ and 5 years. Note that static control that is insensitive to actual field conditions has only limited ability to conserve energy.

(c) Occupancy Sensing. Within a normal 9 AM to 5 PM working schedule, offices in commercial spaces are unoccupied for 30% to 60% of the time. The reasons are manifold—coffee breaks, conference, work assignment, illness, vacation, and reassignment to a different work location are a few. Occupancy sensors can operate relays to

turn off lights after a preset minimum period of about 10 minutes or can dim the light level to a minimum in areas such as corridors, which always require some light. (They can also turn off other energy consumers such as fan coil units, air-conditioners, and fans.) Reestablishment of the original lighting level can be instantaneous, delayed, or manual on the action of the occupant. Another useful function of an occupancy sensor is to provide automatic override in schedule systems, thus both relieving the occupant of the necessity of using a manual override and limiting the energy use to actual occupancy time. It can also light the occupant's way into a space, and shut off the system after he or she has left.

Occupancy sensors that react to presence are of three types—passive infrared (IR), ultrasonic, and a hybrid of both technologies. Infrared sensors (Fig. 21.8) react to the motion of a heat source within its range. It operates by creating a pattern of beams, and alarms when a heat source (such as a person) moves from one beam to another. It will *not* alarm to a stationary heat source. Although the IR sensor is quite sensitive, it has several disadvantages:

· Small movements may not be detected, as they may not cross from one beam (zone) to another. As a result a person sitting quietly may not be detected and the sensor may shut down/off the lights.

· Very slow movement may not be detected even when they cross zones.

· The IR detector must "see" the heat source.

(a)

(b)

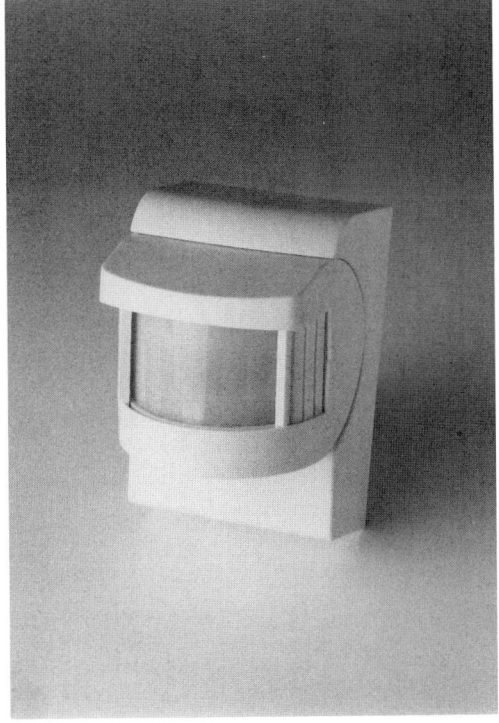

(c)

Fig. 21.8 *Passive infrared (PIR) occupancy sensors. All sensors can be equipped with an adjustable override to prevent turning on lights when ambient light is sufficient. All units have adjustable delayed-off timing and flashing LED that indicates sensor operation.*
(a) *Flush-mount ceiling unit. Note the 360° circular pattern of the lens, which indicates omnidirectional coverage. Unit is approximately 4-in. in diameter.*
(b) *Flush-mount combination wall switch and PIR occupancy sensor. Operation of the switch overrides the sensor function.*
(c) *Surface-mount wall PIR occupancy sensor. Note the semicircular shape of the lens, which indicates wide horizontal coverage. (Photos courtesy of Leviton Manufacturing Co.)*

ILLUMINATION

Therefore, a heat source blocked by furniture will not be detected.

· The beams have a discrete width and depth. Therefore, there may be "dead" spots under the beams if the units are not carefully selected to have adequate multilevel beams, or properly located to give the desired coverage.

Ultrasonic sensors emit energy in the 25 to 40 kHz range, which is well above the range of human hearing. The waves immediately fill a space by reflecting and re-reflecting off all hard surfaces, establishing a pattern that is detected by the sensor. Any movement within the space disturbs this pattern and is immediately detected by the sensor. They have distinct advantages over IR sensors in that they do not require a direct line-of-sight exposure to the movement and they detect small movements. This latter characteristic, however, is also a disadvantage inasmuch as movement of curtains and even air movement can trigger a sensor. It is frequently necessary to adjust (reduce) the sensor's sensitivity to avoid false sensing. Infortunately, this also decreases its coverage.

Hybrid (dual-technology) sensors combine the characteristics of both sensors by usually requiring both sensors to react to turn lights on, but once ON, a reaction in either sensor keeps the lights on. In addition, sophisticated electronic circuity "learns" a space's occupancy patterns and is programmed to react accordingly.

Placement of sensors is very important, and should be tested before final installation. Studies have shown that reducing the minimum ON period below 10 minutes is counterproductive, and frequently causes space occupants to shut off the sensors. Depending on type and mounting position, sensors cover a maximum area of 250 to 1000 ft^2 per unit. Payback for this equipment runs between 6 months and 3 years, depending on the type of space and the degree of control already existing. They are best applied in areas that are divided into individual rooms and work spaces. Sensors can be wall or ceiling mounted, or mounted on a wall-outlet box in a combined sensor/wall-switch configuration. A few designs are shown in Fig. 21.9. (See Section 22.2, which discusses the security application of motion detectors and illustrates their operation.)

ASHRAE/IESNA 90.1 recognizes the great effectiveness of these devices by giving both connected power credit and a high control point equivalency.

(d) Lumen Maintenance. Referring to Section 20.32, we see that in order to maintain a minimum lighting level to the end of a maintenance period, we must deliberately overdesign initially. The extent of the overdesign is the reciprocal of the light-loss factor (LLF). With an average LLF of 0.60, this initial overdesign amounts to (1/0.60), or 66%. Assuming a linear light falloff over a 2-year maintenance period, this overdesign results in an average of 33% annual energy waste. (In the next 2-year maintenance cycle, the savings are slightly less due to a small amount of unrecoverable loss.) Because the light depreciation is a continuous and very gradual process over the maintenance period, the most appropriate control strategy is one that reduces the initial overlighting by the required amount (as measured in the field) and gradually restores it as the system ages. This control strategy, known as *lumen maintenance,* is accomplished by a dimming system operating in conjunction with local light sensors (photocells). The photocells measure ambient light, and in response to their signals the controller(s) operates the dimming units to raise (or lower) the light output. Depending on the size of the installation, the dimmers can either be dispersed or centralized. The modulating action of such a system over the maintenance period is shown in Fig. 21.10. This strategy, in a purely manual mode (periodic maintenance, initial light reduction in accordance with the length of the maintenance period), gives only about a 10% energy reduction, whereas in an automated system almost all of the 33% annual energy waste can be eliminated.

In a new installation, the choice of whether to use electronic ballasts and full-range dimming, conventional ballasts and partial dimming, or a system of multilevel switching is an economic one and depends on many factors, one of the most important of which is the cost of energy. Often a combined system is advisable. In interior zones initial lighting reduction does not exceed 30% to 40%, and full-range dimming is not required. In perimeter zones, daylight often provides all of the required light, and either full-range dimming, dimming plus switching, or a multilevel switching

(a)

(b)

ILLUMINATION

Fig. 21.9 *Ultrasonic occupancy detectors.*
(a) *Ceiling-mount sensors; bidirectional on the left and unidirectional on the right. (Courtesy of Leviton Manufacturing Co.)*
(b) *Sensors in various designs, intended for differing applications. Left to right:*
  • *Ceiling-mount two-way sensor intended for large rooms up to 2800 ft². Unit measures 6 in. × 3¾ in. × 1 in.*
  • *Dual-function sensor and wall switch; designed for rooms up to 300 ft²; wall mount on single- or double-gang wall box; measures 4½-in. square × 1 in.*
  • *One-way sensor designed for use in rooms up to approximately 1250 ft². Unit measures 3¼ in. × 4 in. × 1 in. and is intended for ceiling mount. (Courtesy of Novitas Inc.)*

system is required (see Section 21.3a and the discussion of daylight compensation). Payback period for a lumen maintenance installation of this type varies from 1 to 5 years. Shorter payback periods can be obtained by using multilevel switching rather than dimming because of its lower first cost.

An additional favorable effect of initial light reduction is the lengthening of effective lamp life and a reduction in the rate of its lumen deprecia-

Fig. 21.10 *Graph of energy use by a system that reduces initial lighting level to compensate for initial overdesign and gradually increases level as system output depreciates. In subsequent cycles the energy savings are reduced slightly because of unrecoverable output loss.*

tion. When a lamp is operated at rated voltage, its lumen output drops during its life, according to the type of lamp. However, if lamps are operated at reduced output, as is the case if lamps are dimmed to compensate for initial overlighting, the lamp life cycle is greatly extended, lumen depreciation is reduced, and lamp energy consumption is linearly reduced.

Typical life extension (to economic replacement) figures are:

Fluorescent	80%
Metal–halide	40%
High-pressure sodium	20%

These figures are for interior zones; for perimeter zones with ambient daylight compensation, they are higher.

(e) Daylight Compensation.

It should be apparent that a control system arranged for continuous ambient light compensation, as described in the preceding subsection, is automatically also arranged to compensate for ambient daylight. The difference is that ambient compensation for lumen maintenance due to light loss factors is a very gradual process of *increasing* output, whereas daylight compensation can be a minute-by-minute variation, and generally in the direction of *de-*

creased electric lighting. Because of these possible rapid variations, switching systems are undesirable, as the constant on–off or level switching of lamps can be very annoying to occupants and deleterious to lamps.

Automatic dimming is the system of choice. Recognizing that in perimeter areas daylight often supplies all the required light, the system for fluorescent installations must be either full-range dimming with electronic ballasts or partial dimming with either magnetic or electronic ballasts. Figure 21.11 shows the action of both types of dimming systems in a fluorescent installation with daylight compensation. The crucial design element in a daylight-compensating system is the establishment of zone areas. Depending on latitude and climate, the southern and possibly the east and west exposures can have an interior (second) perimeter zone that receives sufficient daylight for a large enough portion of the year to economically warrant dimming. The northern exposure has only a narrow perimeter zone (Fig. 21.12). As a rule of thumb, the size of the zones is established by determining the maximum room depth that receives at least half its illuminance from daylight for several hours a day. A computer daylight and dimming cost study with perimeter zone depth as a variable is an effective approach. Several such programs are available.

Placement of the control photocells depends on the control system. Where daylight compensation is desired in conjunction with lumen maintenance (see the preceding subsection), area photocells are desirable as they give a feedback control signal for the specific area involved. Alternatively, a daylight-factor (DF) map of a space can be made, preferably after the installation is complete and the furniture in place, and photocells located at one point, based on this map (Fig. 21.13). We know from our study of daylight in Chapter 19 that the DF at any inside point is constant. Therefore, by measuring actual daylight level at a point either immediately inside or outside the window, shielded from direct sun and from inside electric lighting, we can relate it to any area in the room by the ratio of daylight factors and establish a switching level at which the lighting for that inside area is switched, partially or fully.

An example should make this clear. In Fig. 21.13, the photocells are mounted at point *A*. The

ILLUMINATION

Lighting energy saved with dimming/switching system

Additional energy saved by full-range dimming

Fig. 21.11 *Typical graph of energy savings with daylight compensating lighting control. A full-range dimming system is more effective than one that dims down only to 40% because daylight often supplies most of the required lighting. An economic analysis is required to determine whether the additional cost of such a system is justified.*

ratio of daylight between points A and B is the ratio of their daylight factors—that is, 20/10, or 2.0. Therefore, if 500 lux of daylight is required at point B before switching lights in that area, twice that amount (2.0 × 500 lux) or 1000 lux is required at point A. Similarly, a 500-lux requirement at point C corresponds to (20/4) (500) or 2500 lux at point A. Therefore, switching can be initiated by two single-level photocells or a multilevel unit at point A, with settings at 1000 and 2500 lux. Other switching arrangements can be made on the same principle. Dimming initiation can also be arranged in this fashion.

Daylight compensation can reduce energy use in perimeter areas by up to 60%, depending on latitude, climate, depth of perimeter zone, hours of building use, initial power density, and so on. The amount saved for this entire building obviously depends on the building configuration—that is, the ratio of perimeter to total area. Payback time is usually in the range of 3 months to 3 years.

In summary, a well-designed lighting control system can provide energy conservation of up to

Fig. 21.12 *Typical building section showing approximate daylight perimeter zones. Exact delineation of zones depends on latitude, climate, window design, and cost of electric energy.*

Fig. 21.13 *Typical daylight factor (DF) curve plotted on a room section with one side window. The photocell control technique for daylight compensation described in the text assumes a constant DF distribution indoors—that is, fixed luminance sky and no direct sun.*

ILLUMINATION

ILLUMINATION

60%, extremely long lamp life, reduced cooling costs, extended ballast life, and reduced maintenance costs. Investment payback period (see Appendix E), when considering all these factors, is always short and therefore financially attractive. An aspect of a centralized lighting control system not discussed previously, because it is not directly concerned with lighting, is its use in connection with peak demand reduction (see Section 14.13). When interconnected with a demand controller, this approach acts to reduce electric lighting load in accordance with a predetermined preferential-load schedule, and can achieve significant savings.

RESIDENTIAL OCCUPANCIES

21.7   Residential Lighting: General

The buzzwords in modern residential construction are automation (convenience), environmental considerations (energy conservation, "green" lights), and the home as a working area (home office, special communications considerations). These underlying considerations color decisions on sources, control, energy use, and budget, as is discussed later.

Residential lighting offers to the lighting designer great opportunity for originality and ingenuity because a residence combines more diverse functions and needs than almost any other building. Furthermore, it often requires that all work be done at minimal cost, and that the result please a range of tastes. The designer approaches the problems with a list of requirements, a perception of the space, and two basic tools: the lighting fixture and the architectural lighting element. The former was discussed at length in Sections 20.18 through 20.29.

There are numerous books and countless periodical articles devoted to residential lighting design. We would recommend using IESNA publication RP-11 (1995), *Design Criteria for Lighting Interior Living Spaces,* as a reliable reference and guide.

21.8   Residential Lighting: Energy Factors

Although residences are excluded from the requirements of ASHRAE/IESNA 90.1, (1989), the reader is cautioned that subsequent revisions may change that and that there may be local jurisdictional standards whose requirements would bear on energy use. General recommendation include:

1. Provide means for multiple light levels in all areas. A kitchen during food preparation does not have the same lighting requirements as the kitchen being entered for a "refrigerator raid." Low-level lighting provision should be made in *all* rooms, including bathrooms. To accomplish this, use high–low switches, simple dimmers, multilevel ballasts, and multilevel switching. An ancillary benefit is that ambience can be changed thereby in multiuse rooms such as dining rooms, family and recreation rooms, and finished basements.
2. Provide local task lighting for relatively difficult visual tasks such as the kitchen location at which menus are planned and accounts are handled.
3. Provide dimming and switching for accent lighting.
4. Use programmable timers with photocell override for exterior lights.
5. In large residences, consider low-voltage or wireless control for ease of remote control and energy savings (see Section 21.12).
6. Use daylight in areas normally occupied during daylight hours such as kitchens and living rooms. Consider skylights with built-in artificial lighting for these areas.

Although automatic daylight compensation is generally not justified economically, area switching is and should be considered in all spaces, including work areas of children's rooms and home offices.

21.9   Residential Lighting Sources

Incandescent sources have traditionally found very wide use in residences despite their inefficiency because of their desirable characteristics, which include flattering skin color, low first cost, small size, focusability, and simple, economical dimmability. Modern sources with at least some of these characteristics and much higher efficiency should be considered:

- Fluorescent sources with color temperatures between 3000° and 3500° K can be used in kitchens and other work areas. Where linear fluorescents are not desirable, PL lamps and folded CFL units can readily be used. Keep in mind however that the lamp life of *all* fluorescent sources is shortened by switching.

- Architectural elements such as coves and cornices can readily use CFL lamps.

- Tungsten–halogen lamps should be restricted to highlighting and specialty requirements. They are incandescent lamps, and in addition to low efficiency, must be carefully applied because of their concentrated heat.

- Consider use of low wattage (9–12 W) PL fluorescent lamps in frequently used dark corridors and stairwells rather than incandescents. These small lamps can be left on for extended periods to provide the required low-level lighting without the constant switching and lamp replacement required for incandescents. This suggestion is even more practical in difficult access spots such as stairwells.

- For closets, pantries, and other frequently switched lighting, incandescent remains the source of choice.

- For home offices and dual-purpose areas that serve both work and recreation, design the lighting for each use individually, with maximum common use. Fluorescent sources are recommended for work purpose.

- Bathrooms can use 3000° to 3500°K fluorescent sources for general lighting, a separate incandescent source for short-time use, and low brightness globe-shape incandescents for mirror lighting.

- Although many manufacturers produce downlights and wall washers for CFL sources, none that we have seen deliver the "punch" of incandescents. Where this brightness is not required, CFL downlights are a good choice when burned for at least 3 hours per use.

- HID sources are appropriate for all exterior lighting.

TABLE 21.1 **Illuminance Recommendations for General Lighting**

Activity or Area	Average Lux
Conversation and relaxation	50–100[a]
Passage areas	50–100[a]
Areas other than kitchen	200–500
Kitchen	500–1000

[a]General lighting in these areas need not be uniform.

21.10   Residential Lighting: Design Suggestions

The following are a few general design suggestions:

- Use general/task-lighting concept with recommended levels as in Tables 21.1 and 21.2.
- Provide luminance ratios as in Fig. 21.14.
- Provide general lighting in all spaces sufficient for movement and casual seeing. Hallways require little lighting; stairs require more. Light stairs from directly above or ahead to create a shadow directly below the

Zone 2	The immediate surroundings (area adjacent to the visual task)	
Desirable ratio		1/3 to equal to task*
Minimum acceptable ratio		1/5 to equal to task*

Zone 3	The general surroundings (not immediately adjacent to task)	
Desirable ratio		1/5 to 5 times task*
Minimum acceptable ratio		1/10 to 10 times task*

*Typical task luminance range is 40 to 120 cd/m² [12–35 fL] and seldom exceeds 200 cd/m² [60 fL].

Fig. 21.14 *Seeing zones and recommended luminance ratios for residential visual tasks.*

ILLUMINATION

TABLE 21.2 **Current Illuminance Recommendations for Specific Residential Visual Tasks**[a]

Seeing Task	Typical American Recommendation: Average Lux[b]	Other Authorities: Average Lux[b]
Dining	100–200	100–150
Grooming, makeup	200–500	500
Handcraft		
Ordinary seeing tasks	200–500	200–500
Difficult seeing tasks	500–1000	500–750
Critical seeing tasks	1000–2000	>1250
Kitchen duties		
Food preparation and cleaning involving difficult seeing tasks	500–1000	750–1000
Serving and other noncritical tasks	200–500	200–300
Laundry tasks	200–500	100–300
Reading and writing		
Handwriting, reproductions, and poor copies	500–1000	750
Books, magazines, and newspapers	200–500	300
Sewing, hand or machine		
Dark fabrics	1000–2000	>1250
Medium fabrics	500–1000	700–1000
Light fabrics	200–500	300–500
Table games	200–300	300

[a]Selection of illuminance within the given range is made on the basis of criteria given in Section 18.24.
[b]Divide by 10 to get footcandles. Due to the range of values, use of the exact 10.76 figure is unnecessary.

Fig. 21.15 *Track lighting is available in a very wide variety of luminaire bodies. Those illustrated include (left to right) spheres, step cylinders, swivel reflector-lamp holders, flat-back cylinders, external transformer low-voltage spots, and gimbal-ring cylinders. Not illustrated are wall washers, adjustable and filtered spots, framing projectors, barndoor shutter units, and so on. In addition, a variety of single and multicircuit track designs are made (see Fig. 15.15). (Photos courtesy of Ruud Lighting.)*

tread front. Lighting from the front eliminates shadows and can create a safety hazard.

· Do *not* avoid ceiling lights as is so frequently done. Wide-profile ceiling fixtures provide general lighting; switch-controlled table lamps do not.

· Lighting in areas specifically intended for use by older occupants such as "grandma apartments" that are parts of larger residences and in residential buildings intended for older occupants should take cognizance of the special requirements listed in Section 18.23.

21.11 Residential Lighting: Luminaires and Architectural Lighting Elements

Guidelines that can assist the designer in selecting luminaires from the huge variety available commercially are as follows:

1. Utilize diffuse distribution for general lighting; narrow-distribution downlight for area and furniture accents; and narrow-distribution, ceiling-recessed incandescent wallwashers for accenting surface texture such as that of brick walls.
2. Use built-in lighting to the extent possible, including architectural lighting elements. We believe that this demonstrates integrity of concept. For this reason, we recommend that the flexibility of track lighting be utilized for accent and task lighting but not for general lighting (Fig. 21.15).
3. Private residences are the exception to the rule of selecting off-the-shelf items in preference to specials. The lighting should complement the architecture and furnishings, and frequently this can best be accomplished by original designs.

Architectural lighting elements are coves, cornices, valances, coffers, skylights, and other luminous constructions not normally comprising a lighting fixture. Although such units are normally less efficient than lighting fixtures, their use is often indicated by architectural considerations. Empirical design data are given in Figs. 21.16 and 21.17.

When using fluorescent lamps in architectural lighting elements, it is possible to avoid dark spots between lamps by placing lamps at a slight angle rather than end-to-end, thus enabling ends to overlap. Similar overlapping is readily accomplished with PL lamps as well. Reflectors increase installation efficiency. When used in coves, reflectors should be aimed 15° to 25° above the horizontal and field-adjusted for best ceiling coverage. When using two-lamp strips, they should be arranged vertically, as shown in Fig. 21.17. Light output for double-lamp installation rarely exceeds 1.75 times the single-lamp output.

21.12 Residential Lighting: Control

In large residences, remote control of lighting becomes more of a necessity than a convenience from the viewpoints of control, safety, and status information. We have already discussed two wiring and control systems that perform the necessary functions satisfactorily; low-voltage switching in Section 16.32 and power line carrier systems in Section 16.34. Completely wireless, radio-wave controlled lighting is particularly applicable to retrofit work because it requires only that control modules be substituted for existing switches. A conventional system might use a wall-mounted master at each entrance in the kitchen and in the home office (Fig. 21.18), a bedside tabletop master in the master bedroom (Fig. 21.19), and radio-operated wall switches and dimmers in all rooms.

EDUCATIONAL FACILITIES

21.13 Institutional and Educational Buildings

The lighting requirements for some spaces in educational facilities coincide with requirements for commercial (office) and institutional buildings. To that extent, the remarks herein are applicable there also. Generally, educational buildings (excluding private colleges and universities) are maintained from operating funds obtained from taxes, and the budget is *always* tight. Therefore,

(a) Lighted Cornices

Cornices direct all their light downward to give dramatic interest to wall coverings, draperies, murals, etc. May also be used over windows where space above window does not permit valance lighting. Good for low-ceilinged rooms.

(b) Lighted Valances

Valances are always used at windows, usually with draperies They provide up-light which reflects off ceiling for general room lighting and down-light for drapery accent. When closer to ceiling than 10 inches use closed top to eliminate annoying ceiling brightness.

(c) Lighted Coves

Coves direct all light to the ceiling. Should be used only with white or near-white ceilings. Cove lighting is soft and uniform but lacks punch or emphasis. Best used to supplement other lighting. Suitable for high-ceilinged rooms and for places where ceiling heights abruptly change.

(g) Lighted Canopies

The canopy overhang is most applicable to bath or dressing room. It provides excellent general room illumination as well as light to the user's face.

(d) Lighted High Wall Brackets

High wall brackets provide both up and down light for general room lighting. Used on interior walls to balance window valance both architecturally and in lighting distribution. Mounting height determined by window or door height.

(e) Lighted Low Wall Brackets

Low brackets are used for special wall emphasis or for lighting specific tasks such as sink, range, reading in bed, etc. Mounting height is determined by eye height of users, from both seated and standing positions. Length should relate to nearby furniture groupings and room scale.

(f) Lighted Soffits

Soffits over work rease are designed to provide higher level of light directly below. Usually they are easily installed in furred-down area over sink in kitchen. Also are excellent for niches over sofas, pianos, built-in desks, etc.

Bath or dressing room soffits are designed to light user's face. They are almost always used with large mirrors and counter-top lavatories. Length usually tied to size of mirror. Add luxury touch with attractively decorated bottom diffuser.

(h) Luminous Wall Panels

Luminous wall panels create pleasant vistas; are comfortable background for seeing tasks; add luxury touch in dining areas, family rooms and as room dividers. Wide variety of decorative materials available for diffusing covers.

Fig. 21.16 *Residential architectural lighting elements. (Courtesy of IESNA.)*

ILLUMINATION

(a) Typical Valance

This "typical" dimensional drawing applies only to commonly encountered window valance situations. Obviously, other window treatments could necessitate modifications in these critical dimensions; i.e., vertical blinds, double track situations, curved bay windows, etc.

The same "job-tailored" variations can occur in the design of any type of structural lighting device. Therefore no other dimensional drawings have been included here.

Wood blocking locates lamp out from wall to minimize upper wall brightness, and approximately 4" in front of drapery track. For good spread of light down draperies.

For good spread of light on ceiling . . keep shielding in line with top of channel approximately 12" below ceiling and bevel top inside edge 45"

Lamp approximately 2" behind shielding for easy removal.

4" Min — 2"

Inside flat white to redirect light.

Shielding size determined by proportions of interior space may vary from 6" to 10" for good light spread and adequate shielding.

(b) and (c) Cove Installations

Proper cove proportions: Height of front lip of cove should shield cove from the eye yet expose entire ceiling to the lamp. Orientation of fluorescent strip as shown is preferable. Cove interiors should be painted with high reflectance matte-finish paint. *Westinghouse Lighting Handbook. (Out of print)*

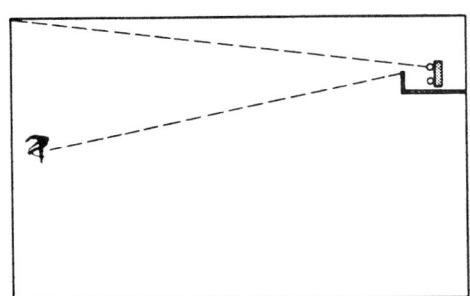

(d) Typical Cornices

Wall washing equipment mounted in valances and cornices provide improved brightness ratios and may be used for lighting desks against walls, or vertical illumination of walls and objects mounted thereon. *Westinghouse Lighting Handbook. (Out of print)*

Wood block

Facing–wood, plastic or metal. Size depends on height of cornice and required shielding angle

Single lamp fluorescent strips, install canted to overlap ends and avoid dark spots. Lamps selected to complement drape or wall color. Use dimming ballasts.

Paint all surfaces matte white.

***Fig. 21.17** Selected architectural lighting elements: construction details.*

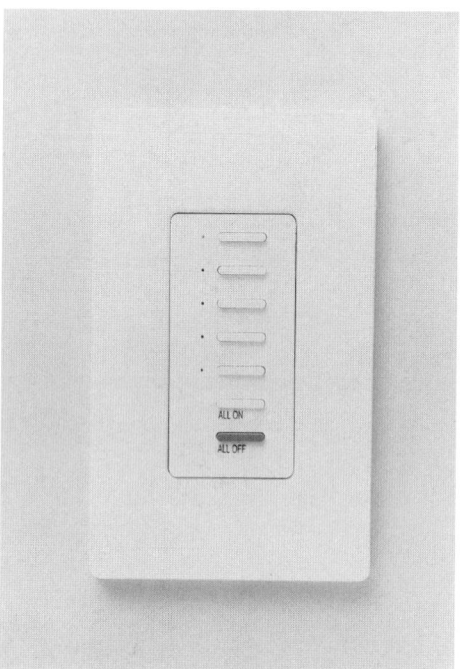

Fig. 21.18 *Typical wall-mounted master station has five control buttons in addition to the ALL ON and ALL OFF on every master. The control buttons can be arranged to control a single outlet or a group, so as to establish lighting of a scene. Thus, the entry unit might have homecoming, extended absence, pathlighting, nightlighting, and garage, whereas the kitchen master may have lighting scenes for breakfast, dinner, formal and informal entertaining, and night. Each scene can be arranged to turn on, brighten, or dim, and shut off selected lighting. (Courtesy of Lutron Electronics Co., Inc.)*

on the lighting fixture to ensure proper relamping.

3. Long-life sources should always be given preference because of lower maintenance. Thus, corridor and stair lighting should be fluorescent of HID. This is also important in locations where relamping is difficult, as in high-ceiling rooms such as gyms and assembly rooms.

4. In calculating levels, low figures for LLF should be used to allow for aging of paints and dirt accumulation. Cleaning of lighting fixtures in schools is virtually unknown. An LLF figure of 0.5 is reasonable. This being so, provision should be made to reduce initial overlighting (see Section 21.4).

5. Most schools are *not* air conditioned. The masking air noise being absent, careful control must be exercised on noise and vibration from ballasts, diffusers, and so on. Electronic ballasts are preferred if the construction budget tolerates the additional cost. Ballast noise increases with current rating. Therefore, 800-mA, high-output and 1500-mA, very-high-output lamps must be used with caution, particularly in locations that amplify sounds, or where low noise criteria (NC) obtains.

all equipment in these public buildings must be extremely hardy, vandal-proof, as maintenance-free as possible, and low in energy consumption. Maintenance in such buildings is generally poor and on a repair rather than preventive basis. With these overall criteria in mind, the following remarks apply to lighting equipment:

1. Use sources with highest possible efficacy. Remember that daylight has the highest efficacy, followed by HPS, fluorescent, and metal–halide sources.

2. Where specific color lamps are called for, such a 3500°K T8 and the like, the requirement should be permanently stenciled in large letters

Fig. 21.19 *Typical components of a wireless control system. Left to right:*
- *Bedside master station has five scene buttons and 10 individual outlet controls in addition to ALL ON and ALL OFF buttons.*
- *Repeater, which is used in large buildings to strengthen signals.*
- *Typical radio-controlled wall dimmer with local manual control. (Courtesy of Lutron Electronics Co., Inc.)*

6. Lighting equipment must be designed for an absolute minimum of maintenance, and those within easy reach should be vandal-proof. This means captive screws, rust-preventive plated parts, captive-hinged diffusers whose cleaning requires only one person, ballast replacement without demounting fixtures (plug-in ballasts are available), non-yellowing plastics, and high-quality finish and assembly.

For recommended illuminances for the various visual tasks in an educational facility, see Sections 18.24 to 18.26 and ANSI/IESNA RP-3 1988, *Guide for Educational Facilities Lighting*. In the following sections, we discuss the lighting requirements of specific school building areas.

21.14 General Classrooms

The classroom is the basic space in a school. Unlike the classic schoolroom with fixed seats, a single viewing direction, and a fixed teacher location, many modern classrooms, at all grade levels, utilize multiple-student groups, teacher mobility, multiple tasks in the same overall space, and movable seating arrangements. Such spaces require:

1. Controls that permit subdivision of lighting and level control within the subdivision (see Sections 21.2–21.4).
2. A lighting system with an appreciable indirect component and good diffusion to minimize the problem of veiling reflections due to viewing direction (see Sections 18.27–18.32).
3. Low-brightness luminaires with high VCP in all viewing directions inasmuch as a considerable portion of the students' time is spent in a head-up position (see Section 18.28).

In addition to these special recommendations, some lighting recommendations applicable to all classrooms are:

4. Classrooms using VDT units require special lighting (see Section 21.24).
5. Generally, fluorescent luminaires should use T8 tri-phosphor lamps with CCT of 3500° to 4100°K and high CRI. Electronic ballasts are preferred. Incandescent sources should not be used. Daylight, to the maximum extent, is desirable. Some form of daylight compensation is mandatory.
6. In comparing costs of alternative appropriate lighting systems, use life-cycle costing techniques only, because these facilities are nonprofit and long term. See Appendix E for an explanation of the principles of economic analysis. Use a computer program to perform the economic analysis where possible. Do *not* compare costs on the basis of footcandles per dollar (i.e, by dividing maintained illumination by cost), because this leads to preference for higher illumination levels and consequently, higher energy usage.
7. Illuminance category (see Section 18.26) and other lighting recommendations for the various types of classrooms and their activities are:

(a) *Reading.* C, D, E, depending on the visibility characteristics of the reading material.

(b) *Mechanical Drawing and Drafting.* E, F. Drafting tables should be equipped with an adjustable fluorescent lamp. Incandescents are not recommended because of heat, low efficacy, and both direct and reflected glare. Proper positioning of the drawing lamp provide glareless illumination even when working with pencil on vellum.

(c) *Typing.* See the special considerations of lighting for VDT units in Section 21.24.

(d) *Sight Saving.* These are general classrooms for visually impaired students. Illuminance category F is recommended, which indicates the need for supplemental lighting. These units should be individually adjustable by the user for both psychological and physiological reasons.

Figure 21.20 shows a perimeter classroom lighting layout using a two-lamp T8 louvered luminaire, and calculated initial horizontal and vertical illuminance values. A single lamp version of the luminaire and its photometry is shown in Fig. 21.21. Note the crosswise batwing distribution that assists in limiting reflected glare.

ILLUMINATION

Fig. 21.20 *Computer-generated image of a 24 ft W × 28 ft L classroom, using a perimeter layout of 16 4-ft luminaires in a 14 ft × 22 ft rectangle, suspended 18-in. from the ceiling and 9 ft AFF. The luminaires use T8 triphosphor lamps and high frequency electronic ballasts. Average horizontal luminance on the desks is 74 fc (800 lux), and average vertical luminance on the chalkboards is 39 fc (420 lux). Room power density is 1.8 W/ft². (Courtesy of Zumtobel-STAFF Lighting Co.)*

TABLE 21.3 **Illuminance Levels for Safety**[a]

Hazard Requiring Visual Detection:	Slight	Slight	High	High
Normal Activity Level:	Low	High	Low	High
Areas	Normal classrooms Lounges Small offices Dorm rooms Washrooms	Shops Business classrooms Large offices Corridors Drafting rooms Lecture rooms Large classrooms Parking area Exterior walkways	Stairs Libraries Reading rooms Cafeteria Swimming pools Locker rooms Interior sports Bleachers	Boiler rooms Auditoriums Exits Exitways Laboratories Kitchens
Lux (Footcandles)	5.4 (0.5)	11 (1.0)	22 (2.0)	5.4 (0.5)

Source: ANSI/IES RP3-1988; reproduced with permission.

[a]Minimum illumination for safety or personnel, absolute minimum at any time and at any location where safety is related to seeing conditions.

ZX-ST 1/32W T8
BIVERGENT STEPPED VANE LOUVER

Total Luminaire Efficiency 65%

0% Uplight	100% Downlight

Spacing Criteria

Lateral Plane	0	90
	1.2	1.5

TOTAL LAMP LUMENS = 2900
INPUT WATTS = 32

Candela Distribution

Vertical Angle	Horizontal Angle			Zonal Lumens
	0	45	90	
0	797	797	797	
5	790	846	861	82.2
15	748	929	1043	256.4
25	670	917	1059	412.3
35	541	851	861	486.6
45	363	590	425	400.2
55	170	245	29	184.6
65	53	42	4	45.9
75	20	13	2	15.5
85	6	0	1	3.6
90	1	0	0	

Luminance Data in Candela / Sq. Meter

Angle in Vertical	Average 0	Average 45	Average 90
45	2763	4491	3235
55	1595	2299	272
65	675	535	51
75	416	270	42
85	371	0	62

(a)

Fig. 21.21 (a) *This luminaire uses a very high reflectance silvered reflector and an unusual stepped mirrored baffle that provide low brightness for direct glare control, a batwing crosswise distribution for reflected glare control, and a sparkle on each baffle step for interest. The luminaire snap mounts to a suspended electrified track for ease of maintenance and installation.* (b) *Photometrics of the luminaire shown in* (a). *(Courtesy of Zumtobel-STAFF Lighting Co.)*

Coefficients of Utilization

Effective Floor Cavity Reflectance = 20%

pcc	0.8				0.7				0.5			0.3		
pw	0.7	0.5	0.3	0.1	0.7	0.5	0.3	0.1	0.5	0.3	0.1	0.5	0.3	0.1
0	78	78	78	78	76	76	76	76	73	73	73	69	69	69
1	73	71	69	67	71	69	67	66	67	65	64	64	63	62
2	68	64	61	58	67	63	60	57	61	58	56	59	56	55
3	63	58	54	50	62	57	53	50	55	52	49	53	51	48
4	59	52	48	44	57	52	47	44	50	46	43	49	45	43
5	55	48	43	39	53	47	42	39	46	42	39	45	41	38
6	51	44	39	35	50	43	38	35	42	38	35	41	37	34
7	48	40	35	31	46	39	35	31	38	34	31	38	34	31
8	44	37	32	28	43	36	31	28	35	31	28	35	31	28
9	42	34	29	26	41	33	29	26	33	28	26	32	28	25

(b)

21.15 Special-Purpose Classrooms

(a) Shops. Illuminance categories for all types of bench and machine work would be

Rough	D
Medium	E
Fine	F, G
Extra fine	G, H

As for other tasks requiring levels above 500 lux, a task-ambient design involving an adjustable supplemental task light is recommended. As with drafting, adjustability of the task light is particularly important because of the frequently specular quality of the visual task, such as in a metal machining process. Another aspect of lighting design in shops that does not exist to the same extent in other school environments is that of safety lighting (Table 21.3).

(b) Music Rooms. The illuminance category here is *D* for clear printed music and *E* or *F* for older copy or handwritten scores. These requirements are for vertical surface illumination, as music scores are normally held in that position. Direct–indirect luminaires provide a high degree of diffusion that is helpful in this regard. For all spaces with vertical tasks, a computer analysis of the design is recommended to determine compliance with recommendations.

Fig. 21.22 *Art exhibition room, illustrating good and bad lighting techniques. Upper wall fenestration is excellent for deep daylight penetration. Track lighting is ideal for display of art. The mixture of incandescent downlights for general lighting is excessive and an eyesore. Also, the positioning of the track lights can create both direct and reflected glare problems and annoying shadows unless the sources are selected properly and ceiling height is at least 10 ft.*

(c) Art Rooms. The primary requirement here is for constant color daylight. Thus, north windows and skylights are highly desirable. For electric lighting, because color is so important, high-CRI fluorescent lamps are required. General illumination should be augmented by user-adjustable supplementary lighting. If use of models is anticipated, adjustable accent lights are advisable. For display of artwork, adjustable wall illumination is required. Ceiling track-mounted units are an excellent choice (see Figs. 21.15 and 21.22).

21.16 Assembly Rooms, Auditoriums, and Multipurpose Spaces

The varied activities in these rooms make flexible lighting imperative. For performances, low-level dimmed incandescent lighting is required. Here,

incandescent is the recommended source because of the lower cost of dimming and the short burning periods. For assembly, this can be augmented by architectural elements along walls and draperies and in the ceiling. For study, additional ceiling fluorescents or HID units can be switched on. The combinations are legion; the different usages are the critical consideration (Figs. 21.23 and 21.24). Acoustic considerations are acute because of the low NC criteria. Thus, electronic ballasts for HID sources as well as fluorescent should be used.

An additional consideration is step lighting. These units should be mounted to the side, or in risers, to illuminate the tread, and particularly its leading edge. Stage lighting is too highly specialized to be discussed here. Obviously, however, some form of stage lighting is required and the building designer is well advised to consult an expert.

ILLUMINATION

Fig. 21.23 *Schools frequently utilize spaces for multiple functions. This space, normally used as a dining area, doubles as an assembly room. The architecture did not lend itself to conventional fixtures. High-intensity, indirect tungsten–halogen units, tucked into concrete beam junctures, provide sufficient light for both uses.*

21.17   Gymnasium Lighting

Gyms present a situation similar to auditoriums in that they have widely varying usages. All fixtures should be sturdy and guarded. Phosphor-coated mercury, HPS, and high CRI metal–halide are excellent choices for color, life, control, and efficiency. Multiple levels should be available by

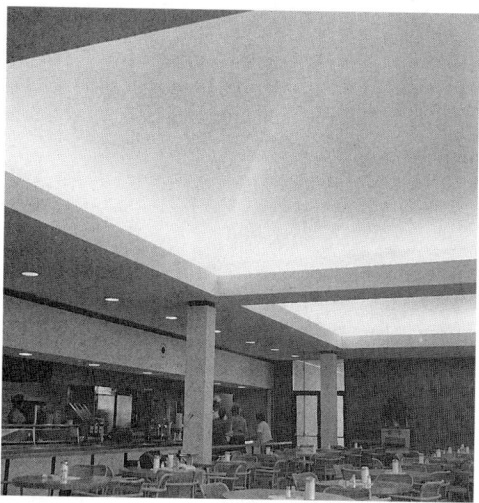

Fig. 21.24 *Institutional cafeteria illuminated by cove lighting in deep pyramidal coffer. Lighting is even, glare-free, soft in quality, and pleasant, yet of sufficient intensity to permit using the cafeteria as a working–meeting space. (Photo by Stein.)*

switching or dimming. For dance and social events requiring low-level general lighting, other fixtures can be lamped with long-life incandescent or tungsten–halogen, which provide good color for low-intensity lighting and also provide illumination during HID startup or restart after an outage. All fixtures should be designed for relamping from the floor. Locker rooms should use guarded-strip fluorescents.

21.18   Lecture Hall Lighting

Lecture hall lighting is similar, with respect to sources and other considerations, to illumination for classrooms. Adjustable level fluorescent lighting is necessary for demonstrations, video, and the like. Auxiliary lighting for demonstration table and chalkboard completes the design. High-ceiling installations can utilize metal–halide for general lighting. Controls for lighting should be at the demonstration table (Fig. 21.25).

21.19   Laboratory Lighting

Laboratories differ from classrooms in that tables are fixed, bench surfaces are frequently very dark, many of the items used exhibit specular reflection, vertical surface illumination is important, and visual tasks are not normally prolonged or severe. With low ceilings, use direct fixtures with uplight component, run crosswise to the tables. Luminaires with batwing distribution minimize reflected glare from specular equipment. If ceiling height is sufficient, indirect lighting is highly desirable for the same reason. Indirect lighting also provides a high degree of diffuseness necessary for vertical surface illumination. See Fig. 21.26 for suggested layouts.

21.20   Library Lighting

Libraries comprise several different seeing tasks, each of which requires its own lighting solution.

(a) General Reading Room.   Here two solutions are possible, and both are in common use. In the first, general lighting is supplied over the

Fig. 21.25 *Typical lecture room lighting utilizes 45° cutoff baffled parabolic reflector troffers for minimum direct and reflected glare, adjustable track lights for demonstration table illumination, and an asymmetric reflector for chalkboard lighting. The large visual display unit in the room will not have veiling reflections with the illustrated lighting arrangement. See Section 21.24 for a discussion of lighting in areas using visual display devices.*

Fig. 21.26 *Laboratory lighting schemes. Running lights across tables or in aisles is preferable to the transverse direction from the aspect of reflected glare. (a) Pendant direct–indirect units. (b) and (c) variations of the single, semidirect HO design illustrated in Table 20.2, fixture 33. The relatively high noise level of ballasts is generally not objectionable in labs, although quiet electronic ballasts are recommended for their acoustic and energy characteristics.*

Single rapid-start or high output
lamp with special lens
or parabolic reflector

24″ max

45°   45°

Fixture
characteristics

Floor

Fig. 21.27 Stack lighting is best accomplished by fixtures with lenses specifically designed for the purpose. Fixtures with baffles and plastic diffusers generally do not give adequate vertical surface illumination.

entire area, which is sufficient for reading tasks. For this purpose, fluorescent or metal–halide sources are normally applicable, the latter with ceiling heights of at least 10 ft. The long life and high efficacy of these sources are suited to the long burning hours found in libraries. The second and more energy-efficient solution involves low-level general lighting supplemented by local reading lighting on the tables or in carrels. This solution is consonant with task-lighting orientation and is to be preferred. Reading lights should be fluorescent, user adjustable if possible, and arranged to avoid veiling reflections when not user adjustable.

Wherever HID sources are used, an instant restart source must be available to supply minimal lighting after an outage. Many commercial HID luminaires contain a small tungsten-halogen source for this purpose. Ballast noise can be a problem in low-NC-criteria spaces such as libraries. Electronic ballasts for fluorescent and HID sources are available and should be employed.

(b) Stack Areas. In stack areas, the required vertical surface illumination is best supplied by one of the special fluorescent units designed for this purpose. These are mounted between racks and no higher than 24 in. above them for best results (Fig. 21.27).

21.21 Special Areas

Most schools contain areas devoted to functions not covered previously. Some lighting recommendations for a few of these areas are as follows:

1. *Spaces where color rendering and color matching are important,* such as sewing rooms and textile and art work spaces, must be lighted with particular attention to illuminant color. Consistent color rendering without disturbing metamerisms requires a continuous spectrum source such as incandescent (including tungsten–halogen) and, of course, daylight. For lighting color matching, see Section 19.36.

2. *Food service areas* must be well lighted to emphasize cleanliness and food attractiveness. Color rendering of food is particularly important so as to enhance its appetizing appearance. Use 3100 K fluorescent or metal–halide in preparation and serving areas, and eating areas, but in the latter at lower illuminance levels.

3. *Medical attention spaces* require high-level, good color rendering light for diagnosis, and conventional office lighting for records and desk tasks.

4. *Offices, storage spaces, industrial spaces, and outdoor facilities* have the same requirements as similar spaces in other buildings, and are covered in the discussion that follows.

5. *Corridors and stairways in all types of buildings* require special lighting. Corridors intended only for circulation need only be lighted to ±10 fc unless a specific seeing task requires higher levels—for example, bulletin boards and lockers (Fig. 21.28). Lighting can also be used to give direction by longitudinal arrangement. Wall-mounted or recessed wall lighting is particularly effective in corridors, giving walk illumination

ILLUMINATION

Fig. 21.28 Lighting of school corridors. High-reflectance walls, floor, and ceiling improve utilization of light and increase the feeling of cheerfulness. The lighting technique illustrated is appropriate for school corridors. The rows of luminaires at each sidewall illuminate bulletin boards, special displays, and the faces and interiors of lockers more effectively than do units centered in the ceiling.

plus lighting for posters, bulletins, and so on. Fluorescent luminaires mounted across corridors, particularly when corridors are long, are effective in reducing the tunnel impression. Incandescent sources are not recommended because of low efficacy, high maintenance, and frequency of relamping. Fluorescent and HID sources are also suggested for stairwells. Care must be exercised here, however, to avoid direct glare, which causes attention to shift from the stairs to the light and may thereby cause a hazard.

21.22   Other Considerations in School Lighting

(a) Controls. See Sections 21.2 to 21.6. Because schools operate for the most part on fixed time schedules and during daylight hours, lighting controls of the preprogrammed time-base type and daylight compensation controls are readily applicable. Other energy-conserving strategies, as discussed at length in the previously referenced sections, can be utilized, as applicable.

(b) Safety Lighting. See Table 21.3 for recommended illuminance levels. As pointed out,

designers must be aware of requirements of all jurisdictional codes.

(c) Emergency Lighting. See Sections 16.39, 17.21, and 21.36. Here, too, the local, state, and other NFPA codes establish minimum requirements. Exits must be clearly identified with lighted signs, and a lighted path to these exits provided. As with safety lighting, it is not sufficient that average illuminance meet requirements. Illuminance in any given area must be free of large level differences, which can cause disabling glare, particularly in view of the relatively high adaptation levels of occupants' eyes immediately before a lighting outage.

COMMERCIAL INTERIORS

21.23   Office Lighting: General

The following information applies primarily to offices in commercial buildings and secondarily to similar spaces in other occupancies, such as in educational and industrial buildings. In the latter cases, the general remarks applicable to facilities of those types take precedence. The special problems associated with VDTs (CRTs, computer screens, word processors) are discussed in Section 21.24. Task-ambient (nonuniform) lighting is covered in Sections 21.25 and 21.26. The reader is referred to ANSI/IESNA RP-1 (1993), *American National Standard Practice for Office Lighting,* for a full discussion of office lighting.

(a) Light Sources. In the interest of energy economy and good color, T8 3500° to 4000°K triphosphor linear lamps or equivalent CFL units are recommended, along with high-frequency electronic ballasts. HID can be used in indirect installations with sufficient ceiling height (min. 9 ft, 6 in. clear). Color-corrected HPS and metal–halide of high CRI are both suitable.

Source color must be coordinated with the color scheme of room surfaces and furnishings. In areas with large daylight contribution, source-correlated color temperature should be at least 4000 K. Incandescents may be used for storage areas, closets, and other short burning-period uses. Incandescent and tungsten–halogen track lighting is used to advantage to illuminate displays of all sorts.

(b) Illumination Levels. These are to be found in Section 18.26 for the particular type of activity involved. See also standard RP-1 (1993). Recommended reflectances for room surfaces are:

Ceiling	80% minimum
Walls	50–70%
Partitions	50–70%
Floor	20–40%
Desktops, furniture	25–45%
Window blinds	40–60%

In landscaped offices, the half-height partitions block 30% to 80% of the light from ceiling lighting fixtures, depending on furniture arrangement. It is therefore all the more important that partition finish (including fabrics) be light colored. It is also desirable to have upper-wall sections painted to match the ceiling—that is, with a lighter finish than the remainder of the wall. This serves the dual function of increasing ceiling cavity brightness, particularly with suspended fixtures, and increasing vertical illumination due to reflection from this surface.

(c) Vertical Surface Illumination. This is required for visual tasks in offices, such as files, desk drawers, card files, and copy stands. Large area luminaires and a high degree of diffusness are desirable. This is especially true in large offices where wall reflections are absent. Light-finish furniture surfaces, fixtures that yield an illuminated ceiling, and high-reflectance floors also assist in this.

21.24 Lighting for Areas with Visual Display Terminals (VDTs)

One result of the emergence of the computer era has been the explosive proliferation of VDTs in business offices. These devices, variously called VDTs, monitors or monitor screens, video displays, word processors, or simply computers, have become standard desktop items. Because of its specular face, the VDT creates a special problem in office lighting that becomes the deciding consideration in selecting the lighting system. Simply stated, the primary problem is to avoid reflection in the screen of any luminous source in the area including luminaires, windows, illuminated walls, and even light-colored clothing. Any such reflection makes reading of the data on the screen always difficult, and sometimes impossible.

The problem of reflections in the VDT screen is not only one of work difficulty. With a *sharply defined* reflection in the screen, as is that of a lighting fixture, the problem becomes physiological (Fig. 21.29). As we already know, the eye is naturally and ineluctably drawn to any bright areas in the field of vision. On a VDT screen with a clear reflected luminaire image, there are two bright items in the vision field: the luminaire reflection and the VDT text or graphics. These two images are separated in depth by the thickness of the glass on the face of the VDT. The eyes are drawn to and attempt to focus on both images, but because of the difference in depth, the eye focus changes continuously, causing eyestrain and severe fatigue. This effect is pronounced only with a clear reflected image; a fuzzy or blurred image appears only as a bright veiling reflection. As a result of this and associated problems, a number of governmental agencies have already issued guidelines for employees using VDTs that limit

ILLUMINATION

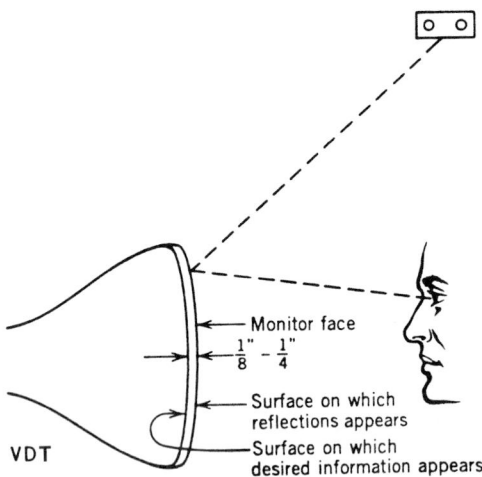

Fig. 21.29 *A clear reflection on the face of a VDT is physiologically disturbing because it causes the eye's focus to move continuously between the plane of the monitor image (inside) and the plane of the reflection (monitor face). This constant refocusing causes eye fatigue.*

work periods and monitor physical effects. The health department of the state of New Jersey, for instance, has issued guidelines for all state employees that call for 15-minute breaks every 2 h and periodic eye examinations. In addition, these guidelines call for all VDTs to have tiltable swivel mounts, operator adjustable, and *separate* adjustable-angle keyboards. Fixed-position key-

boards are not permitted. The reasoning here is obviously that each operator's situation is unique because of differences in work position, posture, and the body's physical dimensions. Therefore, the VDT operator must be given the wherewithal to adjust and thereby maximize the visual (and physical) comfort at his or her working position.

A secondary problem is avoidance of reflections on the usually specular keyboard and other specular objects in the vicinity. Matte-finish keyboards are now available, as is semispecular glass. The problem with the latter is that it tends to dull and blur the monitor information, which, especially when working with complex graphics, is highly undesirable. This is even more true of retrofit glare filters placed over the VDT screen.

It is obvious (Fig. 21.30) that the problem is one of geometry, complicated by the fact that the screen is often up to 20° from the vertical and the keyboard is up to 30° from the horizontal. A further difficulty lies in the contradictory nature of adjacent visual tasks. Viewing of the screen calls for a low (vertical) ambient light level (75 to 125 lux), whereas the reading and writing work done in conjunction with the screen data requires a much higher (horizontal) level (300 to 700 lux). A full solution to the problem therefore involves careful attention to the selection and location of the VDT equipment and control of room surface reflectances in addition to proper lighting design. The following recommendations refer to all these

Fig. 21.30 *Primary sources of reflection in the VDT screen and keyboard are luminous areas such as lighting fixtures and windows. Specular surfaces such as window glass can become secondary reflection sources via re-reflection of light from luminaires (or other light sources), as shown.*

considerations, with the knowledge that some may not be in the purview of the architect/lighting designer.

(a) Equipment.

1. The screen should be recessed as deeply as possible into the VDT. This reduces the ambient light level at the screen and make reading easier. It also reduces glare by restricting the room surface area that is reflected in the screen. (If the VDT is not constructed in that fashion, a simple handmade shade does almost as well.)
2. Convex screens reflect a larger ceiling area than flat screens, and therefore should be avoided.
3. All parts of the VDT, including keys, should have a matte finish.
4. The VDT should be mounted on an adjustable tilting swivel that permits changing the geometry of reflection.
5. The keyboard should be separate from the monitor, with adjustable tilt angle.

(b) Location.

1. Avoid locations where the VDT *viewer* faces a window or other large bright area.
2. Equip windows within the geometry of reflection with low-reflectance vertical blinds. Blackout curtains are not required. Remember that windows can become a secondary source of reflection—that is, luminaire to window to VDT screen (see Fig. 21.30).

(c) Luminance Ratios. Throughout the course of the lighting chapters we repeatedly emphasize the need to limit the luminance ratios in the visual field, with the stringency diminishing as we recede from the center of the vision field. Thus, in Section 18.32, Table 18.11, we recommend a maximum luminance ratio of 3:1 between task and immediate surround and 1:10 between task and remote lighter surfaces. These ratios become essentially impossible to maintain when the "task" involved is a VDT because of its low luminance.

The luminance of a modern color monitor with text or graphic/text display varies from 20 to 85 cd/m² and averages about 50 cd/m² (~20fL). It should therefore be obvious that a 3:1 task-to-surround luminance ratio would be impossible to achieve without almost completely darkening the room. A sheet of white paper with 85% reflectance would have to be illuminated with approximately 60 lux to achieve this ratio, and that level of illuminance, in almost any conceivable office environment, is unacceptably low. As a result, with respect to VDT area lighting design, the ANSI/IESNA standard RP-1 recommends these maximum luminance ratios between a VDT screen and its surroundings:

VDT to lighter immediate surround	1:3
VDT to darker immediate surround	3:1
VDT to lighter remote surround	*
VDT to darker remote surround	10:1

*The permissible ratio of screen luminance to higher remote luminance depends on the size and location of the luminance, as explained in Section 18.28. The recommendation is that no point on the ceiling of the room containing a VDT exceed 850 cd/m² on any 60-cm square (17 times the average screen luminance), and that lighting fixture luminance preferably be limited to:

850 cd/m² at 55° from the vertical
350 cd/m² at 65° from the vertical
175 cd/m² at 75° or more from from the vertical

The limitation becomes more severe as the angle increases because high-angle luminance is reflected in the VDT screen from luminaires behind the VDT.

(d) Lighting.

1. Indirect lighting. With a ceiling height of 10 ft or more, an indirect lighting system with the 850 cd/m² luminance constraint can provide a uniform ambient illumination level of about 30 fc (300 lux) horizontal and approximately half that vertically. If these task levels are not sufficient, portable desktop task lights can be utilized. It is advisable to design for somewhat higher levels and incorporate switching or dimming controls for field "tuning" of the installation. A ceiling height of less than 10 ft would yield unacceptably high ceiling luminance in a totally indirect installation.

2. Semi-indirect and direct–indirect lighting. Spaces with a ceiling height of 9 to 11 ft can utilize these systems to advantage, using semi-indirect in the taller rooms and direct–indirect in lower ceiling areas. Uplight can vary between 40% and 80%, depending on the pendant stem length and the degree of spread of the upward component. Ceiling and fixture luminance should meet the criteria given previously. The ratio between fixture luminance and the background ceiling luminance and between areas of ceiling of varying luminance should not exceed 5 : 1 at any viewing angle within the reflection geometry of the screen. Inasmuch as manufacturers provide only fixture luminance data, the only way to determine ceiling luminances with different pendant lengths is with a full-scale mock-up. A few manufacturers provide *near-field* candela intensity data with which ceiling luminances can be calculated using a sophisticated computer program. However, because the majority of manufacturers use only far-field intensity measurement techniques, luminances of surfaces close to the luminaire, such as the ceiling, cannot be calculated accurately and only a full-scale mock-up gives reliable data. When designed properly, a direct–indirect system with approximately half the luminaire output in each direction can provide a very effective VDT environment. Figures 21.31 and 21.32 show a well-designed direct–indirect unit with its characteristics and a typical installation. Figure 21.33 shows application of a direct–indirect luminaire in an open-plan office with a relatively low ceiling. The batwing uplight characteristic prevents excessive ceiling brightness.

3. Direct lighting. In spaces with a ceiling height under 9 ft and where otherwise required, a low-brightness direct-lighting luminaire can be used. The CIE standard for direct luminaires in areas with heavy VDT use calls for a maximum luminance of 200 cd/m^2 (58 fL) at 50° from the vertical and above, and the same luminance at 60° and above in areas with moderate VDT use. Such a low luminance assuredly prevents both annoying reflections on monitor screens and also direct glare, which can be problematic in VDT

areas because the working posture is in a head-up rather than the traditional head-down position used for horizontal tasks. A glance at American manufacturers' catalogs shows that very few local manufacturers meet such stringent criteria, adhering instead to the preferred maxima given in the IESNA Standard RP-1: 850 cd/m^2 (~250 fL) at 55°, 350 cd/m^2 (~100 fL) at 65° and 175 cd/m^2 (~50 fL) at 75°, all measured from the vertical.

Furthermore, these luminances, which can be met by many one- and two-lamp low-brightness fixtures, avoid the dingy, cavelike atmosphere created by very-low-brightness luminaires. The reader should note, however, that these luminances refer to the direction causing the reflection. Therefore, if viewing direction is not fixed, it is necessary to obtain and check parallel, normal, and 45° figures from the manufacturer.

To attain the required low luminance, fluorescent troffers can be equipped with miniature parabolic wedge louvres [Fig. 20.32(*a*)] or the more efficient large-cell specular parabolic louvres (Fig. 21.34). One problem with the latter is that although a good shielding angle (minimum 35°, preferably 45°) avoids lamp image reflection in the screen and yields high VCP (minimum of 80 recommended), the lamps are reflected in the specular louvres, causing bright spots, which in turn can reflect in a monitor screen. To avoid this, some designers place a diffusing material (translucent overlay) *above* the louvres, which makes the louvre luminance uniform but reduces efficiency. Another alternative is to use semispecular louvres, which increase overall louvre luminance but without the highlights of specular louvres. A third alternative is to use specially made high-grade (and expensive) louvres whose curvature is carefully controlled to eliminate the ripple reflections seen in Fig. 21.34. Finally, specular louvres must be very well maintained as every speck of dust shows up as a bright spot on a dark background.

(d) Finishes. Walls, floors, and furniture should be finished in low-chroma, low-value colors, with a maximum reflectance of 50%. Very dark and very light desktops are to be avoided because of excessive luminance ratios and the latter also because of reflections. Similarly, operators should be advised to avoid light-colored clothing and specular clothing accessories. In gen-

— Hanger

— Hanger

7 5/16"

3 15/16"

1 31/32"

Lens/refractor panels

3 3/8"

Deep cell parabolic baffle matte silver finish

2'- 4"

1 1/2"

1 19/32"

3 5/32"

48 13/16"

51 3/16"

Photometrics

(a)

ID-VM 1/32W T8
4' DIRECT/INDIRECT

Total Luminaire Efficiency 79%

76% Uplight   24% Downlight

Spacing Criteria
Lateral Plane   0   90
                   1.2   0.7

TOTAL LAMP LUMENS = 2900
INPUT WATTS = 31

Coefficients of Utilization

pcc	0.8				0.7				0.5			0.3		
pw	0.7	0.5	0.3	0.1	0.7	0.5	0.3	0.1	0.5	0.3	0.1	0.5	0.3	0.1
0	80	80	80	80	71	71	71	71	55	55	55	40	40	40
1	74	70	68	65	65	63	61	59	49	47	46	36	35	34
2	67	62	58	54	60	56	52	49	43	41	39	32	31	29
3	62	55	50	46	55	50	45	42	39	36	34	29	27	26
4	57	49	44	39	51	44	40	36	35	32	29	26	24	23
5	52	44	38	34	47	40	35	31	32	28	26	24	22	20
6	48	40	34	30	43	36	31	27	29	25	23	22	20	18
7	45	36	30	26	40	33	28	24	26	23	20	20	18	16
8	42	33	27	23	37	30	25	22	24	20	18	18	16	14
9	39	30	25	21	35	27	23	19	22	19	16	17	15	13

Effective Floor Cavity Reflectance = 20%

(b)

Candela Distribution

Vertical Angle	Horizontal Angle 0	45	90	Zonal Lumens
0	459	459	459	
5	457	454	454	42.6
15	433	410	395	112.9
25	389	281	186	129.7
35	324	116	100	109.0
45	243	73	58	82.2
55	141	29	26	44.1
65	19	15	17	17.3
75	4	9	16	10.6
85	3	6	11	7.4
95	8	30	19	39.1
105	80	221	187	179.1
115	166	287	382	272.9
125	251	325	402	291.9
135	327	373	417	288.8
145	390	421	444	263.4
155	442	471	471	213.5
165	477	499	503	139.2
175	497	500	506	47.7
180	499	499	499	

180.0° 600 500 400 300 200 100
157.5°
135.0°
112.5°
90.0°
67.5°
45.0°
22.5°
0.0°

Luminance Data in Candela / Sq. Meter

Angle in Vertical	Average 0	Average 45	Average 90
45	1516	455	362
55	1084	223	200
65	198	157	177
75	68	153	273
85	152	304	557

(b)

(c)

Fig. 21.31 (a) *Semi-indirect, low brightness, deep-cell parabolic baffle fluorescent fixture. The parabolic baffle provides the required low luminance in the parallel direction (see text) as does the fixture body in the normal plane so that the fixtures should not constitute a glare problem. (See luminance figures in the photometric table.)*
(b) *Photometry. Maximum ceiling luminance and ceiling luminance ratios depend on hanger length, ceiling reflectance, and fixture spacing.*
(c) *Photograph of semi-indirect luminaire showing construction and suspension hardware. (Courtesy of Zumtobel-STAFF Lighting Co.)*

Fig. 21.32 *Semi-indirect lighting installation in a New York stock brokerage office, utilizing the fixture shown in Fig. 21.31. Every desk is equipped with a VDT. The translucent sides of the fixture are within the acceptable 5:1 luminance ratio to the ceiling recess luminance, and therefore should not create a reflected glare problem. (Courtesy of Zumtobel-STAFF Lighting Co.)*

Fig. 21.33 *A direct–indirect luminaire (see insert) with batwing uplight distribution can be used in a space with ceilings as low as 9 ft and even less, without excessive ceiling brightness, as shown in this open-plan landscaped office. (Courtesy of Zumtobel-STAFF Lighting Co.)*

Fig. 21.34 The illustrated installation shows a modern newspaper office with a VDT on every desk and a variety of viewing directions. The lighting fixtures utilized are deep-cell specular parabolic louvre units, which typically have low surface brightness and crosswise batwing distribution. Note particularly that the VDT screens are deeply recessed into the monitor case to minimize glare and that each VDT is mounted with tilt and swivel capability. (Courtesy of Zumtobel-STAFF Lighting Co.)

eral, the standard recommended maximum luminance ratios of 1:3 in the near field and 1:10 in the far field should be the design goal here as well.

21.25 Office Lighting Guidelines

(a) Private Offices. In these spaces, a task-ambient approach is frequently indicated because there is usually only one primary visual task location, with the remainder of the space devoted to circulation and storage. Often, sufficient illumination for the latter is supplied by spill light from the task area, particularly when the task light is a pendant unit with uplight component, as seen in Fig. 21.35. In large rooms, provide downlighting in sitting areas and some type of wall illumination using wallwashers, sconces, or a recessed perime-

ter unit to brighten the often dark wood-paneled walls. Accent lights are required for pictures and other displays.

(b) General Offices. The two basic approaches to general office lighting are a uniform layout that provides task-level lighting in the entire area or a task-ambient design. The former, discussed in this section, is most appropriate to speculative-type construction, where the furniture layout is unknown but a complete job is required. (Frequently, however, the construction contractor provides only sufficient electric power for lighting and miscellaneous power; the lighting of the space is designed *after* the space has been rented, as "tenant" work. In such cases, a task-ambient design is possible because tenant work is tailored to the user's needs.)

Fig. 21.35 Geometry of a pendant task-ambient fixture. This unit is intended not only to supply task illumination, but also a measure of ambient lighting via its 20% uplight component. In a small room, this component is sufficient for general circulation. When installed as shown, veiling reflections are minimized and glare on the VDT screen is essentially eliminated.

As has been pointed out repeatedly, a task-level overall layout is wasteful of energy and creates problems with building energy budgets. Therefore, a uniform layout, when used, should be designed so that levels can be lowered easily in areas not requiring task lighting (see Sections 21.2–21.6). Furthermore, because no control of the geometry of direct and reflected glare is possible in such layouts, the luminaire selected must give minimum average glare in *all* viewing directions. The system that best meets these requirements is one that uses a large number of low-brightness luminaires in a dense layout. This is a very expensive solution, however, unless the ceiling is either modular or coffered.

Modular ceilings, with a single-lamp luminaire in each module, are frequently a feasible and economical approach. A structurally coffered ceiling with a luminaire in alternate coffers (Fig. 21.36) is another possibly feasible alternative depending on coffer size. The higher cost of luminaires may be offset by the absence of a hung ceiling. For a flat ceiling, a low-brightness troffer would be selected. An unusual design luminaire that creates a direct–indirect low-brightness unit in a troffer enclosure is shown in Fig. 21.37. Batwing distribution is generally not sought because of high direct glare (low VCP) and strong veiling reflections at work locations whose line of vision is crosswise to a lighting fixture. Where the viewing direction is established but exact furniture layout is unknown, a uniform layout with low-brightness parabolic reflectors with batwing characteristic can

be used to advantage. Luminaire selection criteria for VDT areas were given in the preceding section. Where VDTs are not the overriding consideration, the following criteria are helpful:

1. In terms of appearance, a 2 × 4 fixture is appropriate in a 2-ft^2 tile ceiling; a 1 × 4 fixture is suitable in a 1 × 1 ceiling, and a square 2 × 2 fixture is suitable in all tile patterns.
2. Two-in. deep parabolic louvres permit a shallow fixture, but lamp shielding is poor. A 3-in. minimum louvre depth is suggested for all large-cell parabolic louvre luminaires.
3. Three-lamp parabolic louvre units, which are usually too bright for VDT areas, can be used to advantage. Two-lamp units give better uniformity and VCP, at a price premium.
4. Two-foot square deep cell parabolics with 31-W T8 U lamps or biaxial compact lamps are an excellent choice from the aspects of VCP, uniformity of illumination, and appearance.
5. Premium-quality specular parabolic louvres are available that eliminate hot (bright) spots.
6. In relatively small offices with 1 × 1 ceiling tiles, 1-ft^2 troffers with deep miniature parabolic wedge louvres and 16-W U lamps or compact fluorescents are an excellent all-purpose lighting solution (Fig. 21.38).

Fig. 21.36 *A coffer ceilinged room (a) is illuminated with 1 × 4 two-lamp units in alternate coffers. The appearance both unlighted (a) and lighted (b) is symmetrical and pleasing. A nonuniform layout would be less objectionable here than in a flat-hung ceiling. (Photos by Stein.)*

(a)

(b)

Fig. 21.37 *Novel luminaire construction giving a direct–indirect appearance* (a) *in a direct troffer construction* (b). *The medium brightness luminaire interior reflects the light, giving a deep, airy appearance. The translucent lamp enclosure maintains an acceptable luminance against the background (interior) reflector. The photometric characteristic is essentially circular in both longitudinal and transverse planes. (Courtesy of Zumtobel-STAFF Lighting Co.)*

7. Direct–indirect shallow luminaires with a wide, inverted batwing uplight distribution can be pendant mounted as low as 7 ft, 6-in. AFF (Fig. 21.33).

(c) Office Lighting Equipment. Office lighting equipment is generally not handled roughly. Fixtures may have touch latches, light hinges, and adjustable devices without fear of breakage or vandalism.

(d) Maintenance. In most offices maintenance is provided on a trouble-call basis. Lamps are replaced on burnout, and the fixture is then cleaned. Because of the long life of fluorescents and HID sources, this generally means a 3- to 5-year cleaning cycle. An LLF of 0.65 is reasonable in air-conditioned spaces, lower in open-window offices.

(e) Fenestration. When fenestration is absent, a lighted valance is recommended around the room. This removes the wall-ceiling line and partially compensates for the lack of windows. It

Fig. 21.38 *One-foot square, 9-cell luminaire, with deep parabolic louvres. This unit is applicable to small offices and/or areas with 1 ft × 1 ft ceiling-tile arrangements. (Courtesy of Zumtobel-STAFF Lighting Co.)*

ILLUMINATION

also brightens the walls and increases illumination on desks placed adjacent to the walls.

(f) Control. Control strategy (see Sections 21.2–21.6) should *minimally* provide (after tuning):

1. Daylight compensation.
2. Possibility to operate individual small groups of lights while the remainder are off to permit off-hours work.
3. Path lighting through large spaces to permit traverse without turning on all lights.
4. Careful scheduling with supervised local override.With these general guidelines in mind, the following sections discuss specific topics in office lighting.

21.26 Task-Ambient Office Lighting Design Using Ceiling-Mounted Units

Efficient modern office lighting, like other work-area lighting, is often predicated on a task-ambient design. This approach has the advantages of minimum contrast reduction at the task and minimum power use, the latter resulting in low operating cost. The method involves designing a uniform ceiling layout that provides low-level lighting for circulation and miscellaneous easy visual tasks, plus work station–mounted task lighting. This latter can either be integral with the furniture (see the following section) or a separate, generally adjustable unit on or adjacent to the desk. The former is particularly useful for very severe seeing tasks when adjustability of angle and distance between light and work is vital. Integral furniture units are usually not adjustable, and are therefore limited to a maximum of about 750 lux on the task. Higher levels would generate excessive heat and, almost certainly, glare.

21.27 Task-Ambient Office Lighting Using Furniture-Integrated Luminaires

In lieu of ceiling-mounted luminaires to provide ambient lighting, indirect HID lamps can be mounted on the top of furniture or can be free-standing. Task lighting would be integral with the

office furniture. Advantages of this arrangement include the following:

1. The problem of furniture layout and layout changes is eliminated.
2. Initial construction cost is reduced.
3. Energy requirements are lowered because of short distances between light source and task.
4. Each occupant has ON–OFF control of his or her task lighting, including, in some designs, positioning control.
5. Maintenance is very much simplified because fixtures are readily accessible from the floor.
6. Floor-to-floor height can frequently be reduced.
7. Tax advantages normally accrue due to higher depreciation rates on furniture than on the buildings.

Disadvantages include the following:

1. Difficulty in dissipating heat and minimizing magnetic ballast noise due to proximity of sources to user. For this reason, electronic ballasts should be used.
2. Veiling reflections are *always* present, and can be severe.
3. Luminance ratios in the near and far surround may exceed recommended levels.
4. Difficulty in lighting a free-standing open desk, because most of the fixture types are undercounter or sidewall mounted.
5. Difficulty in evenly lighting large table or L-shape desk areas because of the concentrating nature of the lighting units.
6. Not readily applicable to automatic switching and dimming schemes.
7. For satisfactory operation with fixed task lighting, the desk must be useable by both left- and right-handed people. This requires a considerable degree of duplication, as shown by lamps on both sides of the drawing in Fig. 21.39.

Figure 21.39 graphically shows the problem of local desk lighting. The author's own experience has been that only systems that permit user positioning adjustment of the light source have a high degree of user acceptability.

Fig. 21.39 *Furniture-mounted task lighting can create severe reflected glare or hand shadows if the luminaire location is not selected carefully. Further constraints are: location in elevation—for shielding (needs to be low) and for good distribution (needs to be high), appearance considerations, and compatibility with the worker's physical movements. Light source position adjustability removes most user objections to this type of task lighting.*

21.28   Integrated and Modular Ceilings

The cost, appearance, and design-flexibility advantages of an integrated ceiling design over field-assembled and coordinated systems have long been known. As a result, ceiling systems with integrated lighting, acoustic control, and air-handling capabilities are commercially available in modular sizes, among which are 60-in. square, 48-in. square, and 30 × 60 in. Modules are made in flat and pyramidal shapes, the latter having several distinct advantages over the flat:

1. More interesting and esthetically pleasing.
2. More acoustic absorbency due to ceiling angles and large surface area.
3. Recessed center provides visual baffling, permitting use of higher brightness sources while maintaining high VCP.

Possible luminaire arrangements for both flat and pyramidal shapes are given in Fig. 21.40, and an example is shown in Fig. 21.41. In addition to the design flexibility available, electrified track can be integrated into the system runners to supply both the lighting fixtures and power poles.

21.29   Lighting and Air Conditioning

The reduction of lighting power density levels to below 2 W/ft^2 in all but special areas has considerably reduced the impact of lighting-generated

heat on a building's HVAC system. In non–air-conditioned buildings, the lighting heat contribution is partially applicable to building heating. Fixture efficiency is directly affected by its temperature. Fluorescent units operate at an optimum temperature of 77° F. Temperatures above and below this decrease output and fixture efficiency. Thus, heat removal from units is desirable even at low lighting-energy levels. The most effective method of fixture heat removal is by duct connection to the unit itself. This method, however, is relatively expensive and immobilizes the fixture. Alternatively, the plenum can be exhausted with air passing over the fixtures, picking up excess heat. These details are essentially part of the HVAC design and are covered in Part II.

INDUSTRIAL LIGHTING

21.30   General

In industrial lighting, the primary consideration is economic. Given acceptable standards of comfort and safety for the working staff, additional costs for lighting must be self-justifying economically. In one case, a good lighting installation was improved at considerable cost. Production jumped 15%, of which 3% was sufficient to amortize the cost of the lighting alteration. In another case, an outlay for new inspection lighting reduced product failures and proved economically sound. In a

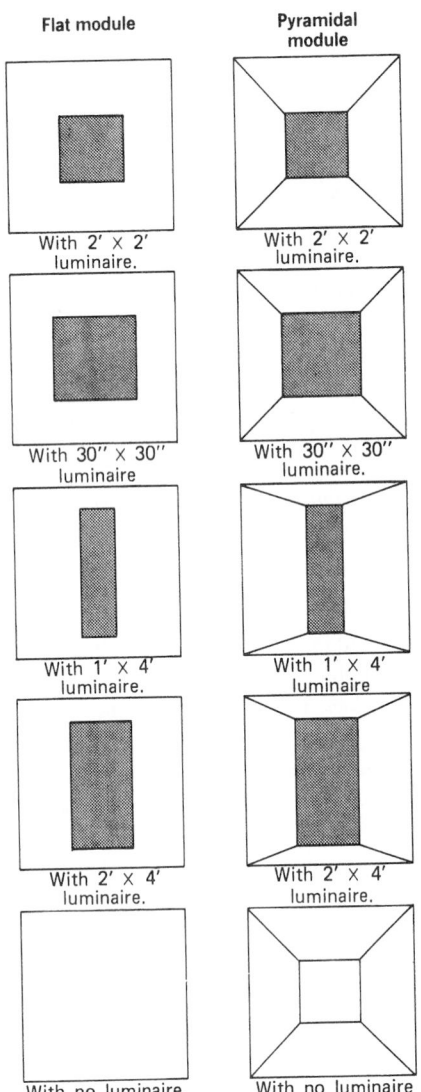

Flat module | Pyramidal module

With 2' × 2' luminaire.

With 30" × 30" luminaire

With 1' × 4' luminaire.

With 2' × 4' luminaire.

With no luminaire.

Fig. 21.40 *Various configurations of 5-ft²
modules.*

Fig. 21.41 *A square modular ceiling utilizing
square lighting units demonstrates architectural
symmetry, blending, and, incidentally, that lighting
is easily visible from the outside, even during the
day. (Photos by Stein.)*

third, improved lighting reduced accidents, improved employee morale, and consequently improved production. The cases studied are far too numerous to mention; general principles are adduced instead. For control strategies, see Sections 21.2 to 21.6.

21.31   Levels and Sources

Levels of illumination are detailed in ANSI/
IESNA RP-1991, *The American National Standard Practice for Industrial Lighting.* In addition,
the IESNA publishes Annex A to this standard,
which describes in detail a method for prescribing
illuminance for specific industrial occupancies and
tasks. Where levels higher than 50 to 75 fc are
required, general illumination must be supplemented by local illumination. Industrial facilities
lend themselves readily to daylighting because
many are one-story structures. Thus roof monitors, skylights, and clerestories are readily applicable and extremely desirable. However, inasmuch
as industrial facilities are frequently sited in industrial areas with attendant heavy atmospheric soot
and dirt, a frequent cleaning and maintenance
program is necessary if the LLF is to be kept at
reasonable levels. This observation is obviously
also applicable to indoor lighting facilities.

Sources for industrial applications should exhibit high-efficacy, good lumen maintenance, and
long life. Of the sources available today, fluorescent and HID meet these criteria. Induction and
sulphur lamps may join this group after additional
development and field testing. Where color is not
critical, HPS is the recommended source. Adaptation to its warm yellow color is rapid, and if it is
mixed with metal–halides or mercury sources, no
problem should be encountered. HID sources are
easier to maintain, store, clean, and relamp than
fluorescent and have equal or better efficacy, but
have the disadvantage of delay and lower output
on restrike. Because of their relatively small size,

HID sources are used in focusing reflectors that produce intensity (cp) distribution characteristics designed for specific illumination objectives. Thus, HID sources are generally used for high-bay (>25 ft) and medium-bay (15–25 ft) installations. For low-bay lighting, industrial reflector fluorescent luminaires and low-bay HID reflectors are both applicable.

One of the most common industrial lighting tasks is warehouse aisle lighting that must provide adequate vertical surface illumination on racks on both sides of the aisle. The required cp distribution to perform this task efficiently is a modified batwing curve. This is more easily accomplished in low-bay lighting with a continuous row fluorescent installation than with discrete HID units. A typical fluorescent unit of this type and its photometric characteristics are shown in Fig. 21.42.

21.32 Industrial Luminance Ratios

For reasons explained at length in preceding sections, luminance ratios must be controlled. Recommendations are given in Table 18.11. In many situations it is difficult to control the surrounding brightness. Ceilings, which so frequently are covered with piping, ducts, and other equipment, should be *light*. Therefore, this mechanical equipment must be painted with matte, light unsaturated colors; maintenance and cleaning must be good; and fixtures should have an upward component of light to avoid more than a 20:1 ratio of task-to-ceiling luminance.

Use of bright saturated colors for general surface painting should be avoided because they draw attention and frequently have special significance. In addition to color-coded piping (banding is preferable), red frequently means fire equipment; green, first aid; orange, danger; and so on. White is also to be avoided, being excessively bright and susceptible to dirt. Recommended minimum reflectances are:

Ceiling	75–85%
Walls	40–60%
Equipment	25–45%
Floors	20%

21.33 Industrial Lighting Glare

The problem of direct glare can be acute in low-bay installations, and that of reflected glare in high-bay designs, when either uses a point source. One method of reducing direct glare is the use of low-brightness prismatic lens units with a black reflector behind the prismatic lens. Methods of minimizing veiling reflection from all sources have been discussed previously.

21.34 Industrial Lighting Equipment

The cost of maintenance increases with labor rates. For this reason, high-quality lighting equipment yields lowest owning and life-cycle costs. For instance, the cost of replacing a ballast for HID lighting units frequently *exceeds* the cost of the ballast. It is thus obvious that it is more economical to utilize long-life, high-quality ballasts, particularly where luminaires are not readily accessible.

Other suggestions for lowering costs, both initial and operating, include using ventilated luminaires that tend to be self-cleaning by convection (Fig. 21.43) in addition to giving the needed upward light component, using bus-mounted fixtures for rapid installation and repair (see Fig. 15.12), using lowering mechanisms on high-bay units to avoid catwalk or platform relamping with concomitant *extremely* high cost, using fixtures arranged for "stick" relamping from the floor in medium- and low-bay work, and generally incorporating modern equipment into the plant.

Proper maintenance is of paramount importance in industrial facilities because of the prevalence of dirt, vibration, and rough service. Under "maintenance" is subsumed cleaning, relamping, inspection, and preventive maintenance. Relamping on a burnout basis is extremely uneconomical because of disruption of production and lowered production due to lumen depreciation before burnout. Relamping should be done on a planned group basis. Similarly, if the specific facility has a high dirt accumulation rate, cleaning must also be done on a planned group basis rather than only at relamping time.

Ballast noise, including the high levels of magnetic HID ballasts, is not usually a factor in indus-

(a)

ZX-SRV W2/32 2/32W T8
VERTICAL SURFACE SEMI-SPECULAR REFLECTOR

Total Luminaire Efficiency 82%

0% Uplight   100% Downlight

Spacing Criteria

Lateral Plane	0	90
	1.2	1.5

TOTAL LAMP LUMENS = 5800
INPUT WATTS = 58

Candela Distribution

Vertical Angle	Horizontal Angle			Zonal Lumens
	0	45	90	
0	1759	1759	1759	
5	1753	1758	1754	167.8
15	1692	1787	1905	511.2
25	1567	1918	2130	859.6
35	1383	1894	1813	1063.3
45	1138	1425	1004	962.9
55	846	704	487	675.8
65	539	284	227	378.4
75	252	79	34	128.8
85	44	5	2	16.1
90	6	2	2	

Luminance Data in Candela / Sq. Meter

Angle in Vertical	Average 0	Average 45	Average 90
45	4331	5423	3821
55	3969	3303	2285
65	3432	1808	1445
75	2620	821	354
85	1359	154	62

Coefficients of Utilization

Effective Floor Cavity Reflectance = 20%

pcc	0.8				0.7				0.5			0.3		
pw	0.7	0.5	0.3	0.1	0.7	0.5	0.3	0.1	0.5	0.3	0.1	0.5	0.3	0.1
0	98	98	98	98	96	96	96	96	92	92	92	88	88	88
1	91	88	85	82	89	86	83	81	83	80	78	79	78	76
2	84	78	73	69	82	77	72	68	74	70	67	71	68	65
3	77	70	64	59	75	68	63	58	66	61	57	64	60	56
4	71	62	56	51	69	61	55	50	59	54	50	57	53	49
5	66	56	49	44	64	55	49	44	53	48	43	52	47	43
6	61	51	44	39	59	50	43	39	48	43	38	47	42	38
7	57	46	39	34	55	45	39	34	44	38	34	43	38	34
8	53	42	35	31	51	41	35	31	40	35	31	39	34	30
9	49	39	32	28	48	38	32	28	37	32	28	36	31	27

(b)

Fig. 21.42 (a) *Two-lamp, 48 in., industrial fluorescent luminaire designed to provide good vertical illumination for low-bay warehouse aisle lighting. The interior reflector is semispecular aluminumshaped to provide the modified batwing candlepower curve shown in* (b). (b) *Photometric data. (Courtesy of Zumtobel-STAFF Lighting Co.)*

Fig. 21.43 *Graph demonstrating the advantage of ventilated fixtures.*

trial facilities because of high ambient noise. In relatively quiet installations and/or where fluorescent fixtures are mounted a short distance above the work bench, as in inspection and fine assembly, this is not true and electronic ballasts are needed.

21.35 Vertical-Surface Illumination

In industrial facilities more than any other, the illumination of vertical surfaces is crucial. This is a result of the nature of work: machines, storage, gauges, and so on all require high-level vertical-surface illumination (Fig. 21.44). The illumination on the vertical surface is the result of the horizontal component of the lighting. This is

$$\text{fc} = \frac{\text{cp}}{D^2}\sin\theta = \frac{\text{cp} \times R}{D^3} = \frac{\text{cp} \times \cos^2\theta\sin\theta}{H^2}$$

To maximize the horizontal component, we set to zero the derivative of fc with respect to θ. Thus

$$\frac{d\text{fc}}{d\theta} = \frac{\text{cp}}{H^2}(-2\cos\theta\sin^2\theta + \cos^3\theta) = 0$$

or

$$2\sin^2\theta = \cos^2\theta$$

$$\tan^2\theta = \tfrac{1}{2}$$

$$\tan\theta = 0.707 = \frac{R}{H}$$

$$\theta \cong 35°$$

Therefore, maximum vertical illumination (illumination resulting from the horizontal-lighting

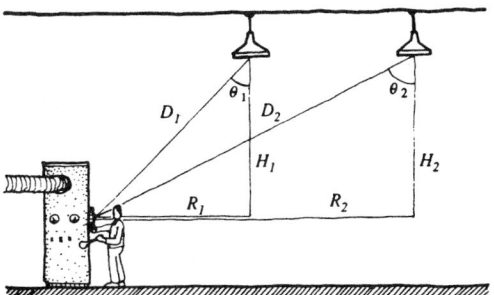

Fig. 21.44 *Vertical surface illumination is maximum where theta is approximately 35° (see derivation in text).*

component) is obtained when the angle between the fixture's vertical axis and the work is approximately 35°. Hence, we should select a fixture whose candlepower distribution curve demonstrates a high value of that angle. Of course, the derivation is for a single location and fixture. For good vertical and angular illumination over a large area, arrange fixtures with considerable overlap.

SPECIAL LIGHTING: APPLICATION TOPICS

21.36 Emergency Lighting

Emergency lighting is required when the normal lighting is extinguished, which can occur for any of three reasons:

1. General power failure.
2. Failure of the building's electrical system.
3. Interruption of current flow to a lighting unit, even as a result of inadvertent or accidental operation of a switch or circuit disconnect.

As a result of the third requirement, sensors must be installed at the most localized level—that is, at the lighting fixture (voltage sensor) or in the lighted space (photocell sensor).

(a) Codes and Standards. Because emergency lighting is a safety-related item, it is covered by various codes, all of which have jurisdiction. In addition, there are widely accepted technical society and industry standards whose recommendations normally exceed the minima required by the codes.

1. Life Safety Code, ANSI/NFPA 101, current issue—1997. This code defines the locations within specific types of structure requiring emergency lighting and specifies the level and duration of the lighting.

2. National Electric Code, NFPA 70, current issue—1999. This code deals with system arrangements for emergency light (and power) circuits, including egress and exit lighting. It discusses power sources and system design.

ILLUMINATION

3. "Health Care Facilities," NFPA 99, current issue—1996. This code deals with special emergency light and power arrangements for these facilities.

4. OSHA regulations. These are primarily safety oriented and, in the area of emergency lighting, discuss primarily exit and egress lighting requirements.

5. Industry standards. These include the publications of the IESNA and the IEEE; in particular, *Recommended Practice for Emergency and Standby Power Systems, ANSI/IEEE Standard 446, current issue—1996.* Because codes and standards are constantly being revised and updated, the designer in any actual construction project must determine which codes have jurisdiction, obtain current editions, and design to fulfill their requirements. The following material is in the nature of general information and good practice, but is not intended to take the place of construction and safety codes.

(b) Minimum Illumination Levels and Duration of Emergency Lighting.

Most codes and authorities require a *minimum average illuminance at floor level,* throughout the means of egress, of 1.0 fc (10 lux). This is considered sufficient, after eyes accommodate, to permit orderly egress. No point along the path of egress should have an illuminance less than 0.1 fc (1 lux), and the maximum-to-minimum ratio of illuminances along the egress path shall not exceed 40:1. The language of the codes is precise and its consequences must be completely understood in order to properly design an emergency lighting system.

Life Safety Code 101 relates to life safety—that is, safe egress during an emergency that often involves fire. In such instances smoke normally obstructs vision at eye level, and it is therefore extremely important that the required illumination be available at floor level. Ceiling-mounted emergency lighting may not illuminate the floor in smoky areas, and may even worsen a situation by creating a bright, foglike condition. It is therefore widely recommended that adequate egress lighting be provided at *baseboard level.* Some codes already mandate baseboard egress lighting.

The requirement that the *average minimum* illumination be 1 fc, along with the mandated minimum of 0.1 fc at any point and a maximum 40:1 ratio of maximum to minimum illumination should effectively eliminate faulty design that utilizes several widely separated bright sources.

Other authorities recommend that escape routes be lighted to not less than 1% of their normal illuminance, but in no case less than 5 lux (0.5 fc). Where this 1% of normal requirement recommendation falls below 1 fc (normal illumination less than 100 fc), we would abide by the 1 fc minimum required by the Life Safety Code.

Duration of the 1 fc level of emergency lighting is normally specified to be a minimum of 90 minutes (for egress) and 0.6 fc thereafter. Facilities that cannot be evacuated quickly require higher levels for indefinite periods.

(c) Central Battery Emergency Lighting Systems.

Sections 16.39 and 17.21 cover emergency power systems, including generators and central battery systems. Generators are normally arranged to replace the utility during outages, using the building's normal electrical distribution system, with the difference that only essential loads remain connected. This obviously does not meet the requirement that emergency lighting be supplied on power failure, *even if the fault is local* (i.e., within the building).

Central battery systems are of three types:

1. Those supplying low-voltage dc (6 to 48 V) to dedicated emergency lighting units (Fig. 21.45*a*). This central battery installation consists of a battery, charger, and switching equipment. The emergency lights are generally not part of the normal lighting system, consisting instead of dedicated incandescent fixtures. The system has the same limitation as that of a generator, and therefore is not applicable to most emergency lighting situations except when equipped with multiple downstream sensors. As such sensing would be highly uneconomical, this system is limited to a single (large) space with common outage sensing. This arrangement has the further disadvantage that the emergency lighting units may obtrude on the building's decor and architecture inasmuch as they are not part of the normal lighting system. Also, because the lighting units are

Fig. 21.45 *Schematic diagrams of emergency power system using central batteries. In* (a) *the emergency system is completely separate from the normal lighting system and utilizes dedicated lighting units.*
(b) *This arrangement furnishes dc to the luminaire, where it is inverted to ac (except for incandescent lamps). The fixtures are utilized during both normal and emergency operation. As switching is instantaneous, HID lamps are not extinguished.*
(c) *Arrangement similar to* (b) *except that a central inverter supplies ac continuously. Here, too, the lamps are used in both normal and emergency modes.*

isolated devices, it is difficult to obtain a satisfactory degree of emergency illumination uniformity as explained previously [see also (e)].

2. Systems supplying 120-V dc [Fig. 21.45 (b)]. This arrangement supplies line voltage dc to emergency lighting fixtures, which are also used as part of the normal system. HID and fluorescent luminaires are furnished with integral inverters to change the incoming dc to ac. This arrangement has the advantage of a central, well-maintained battery supply, but the disadvantage of requiring a full dc distribution system.

3. Systems supplying line voltage ac [Fig. 21.45 (c)]. Here the central unit contains a battery, charger, central dc/ac inverter and switching equipment. Such a system can supply power to incandescent, fluorescent, and HID systems in spaces where a single voltage-loss sensing point is sufficient. The emergency lights are part of the normal lighting system. This arrangement is most frequently used where a distributed system with local batteries (discussed later) would not be readily

applicable due to the size of the load, as would be the case with high-wattage HID luminaires. A typical 10-kVA on-line UPS cabinet designed to supply emergency ac power to HID lamps measures approximately 40 in. W × 50 in. H × 20 in. D and weighs about 1000 lb.

(d) Distributed (Local) Emergency Lighting Arrangements. As a result of the inability of a central system to respond to a localized outage a frequently used emergency lighting arrangement is one where the source of power as well as all the voltage sensing and switching equipment are installed at the emergency lighting fixture, which is usually part of the normal lighting system. The emergency pack consists of a rechargeable battery and charger, voltage sensing and switching equipment, and in the case of fluorescent lights, an electronic package (ballast) that operates the lamp at high frequency and generally at reduced output (Fig. 21.46). These packages, which are designed to supply the code required 90 min of emergency lighting, are maintenance

ILLUMINATION

Fig. 21.46 *The emergency lighting ballast package contains a 90-min capacity high-temperature battery, inverter, sensing and switching equipment, and electronic ballast circuitry to operate the lamps(s) at full or reduced output. The emergency ballast is energized on loss of normal power. The entire package measures 9½ × 2⅜ × 1½ in. (Courtesy of Big Beam Emergency System, Inc.)*

free for periods of up to 5 years. A second type of distributed emergency lighting unit is the familiar packaged unit with integral incandescent lights. Several designs are shown in Figs. 21.47 and 21.48.

(e) Emergency Lighting Design Considerations.

The levels and duration discussed in (b) refer only to egress lighting. However, general emergency lighting is required to avoid distress and even panic that unfortunately may rapidly ensue. The level of required emergency illumination in specific areas should be related to the area's level of normal illumination and the degree of hazard in the area. Therefore, we would provide:

(a)

(b)

Fig. 21.47 *Commercial emergency lighting units.*
(a) *This decorative unit contains a sealed, maintenance-free lead–acid battery, an automatic charger and an automatic, low battery voltage disconnect. The battery is designed to supply a minimum of 90 minutes operation of fully adjustable 5-W sealed beam lamps. The unit measures 14 in. W × 5 in. H × 5 in. D.*
(b) *Conventional design emergency unit contains a lead–acid or lead–calcium battery, as specified, which is housed with its required electronic control and safety equipment in a polycarbonate housing. The lamps are rated 7.2 W each. The entire unit measures approximately 11 in. L × 12 in. H × 10 in. D; exact dimensions varying with battery type, voltage, and capacity. (Photos courtesy of Dual-Lite, a unit of General Signal.)*

(a)　　　　　　　　　　　　　　　　(b)

Fig. 21.48 *Industrial-type emergency lighting units.*
(a) *This traditional-design unit is available in a variety of battery voltages and uses a pure lead or a lead–calcium maintenance-free battery, as specified. The dark-brown cabinet conforms to a NEMA 1 enclosure. The unit can be arranged to supply additional remote 7.2 W sealed-beam heads. The unit measures 14 in. W × 8 in. D × 19 in. H (over the lamps), and is provided with an external charge indicator light and test switch.*
(b) *This compact polyester enclosure unit is sealed and gasketed for rugged industrial application. It measure approximately 10 in. W × 11 in. D × 14 in. H and contains the usual electronics for control and charging of its lead battery. A charge indicator and a test switch (to simulate power failure) are mounted on the side. (Photos courtesy of Dual-Lite, a unit of General Signal.)*

ILLUMINATION

Exit area	5 fc (50 lux)
Stair	3.5–5 fc (35–50 lux)
Hazard areas, such as machinery room	2–5 fc (20–50 lux)
Other spaces	1.0 fc (10 lux)

These levels should be essentially uniform. When the illumination level in an interior space drops instantaneously from a level of 30 to 100 fc to 1 to 5 fc, the eyes require up to 5 min to fully accommodate. During this long period, the space's occupants are partially sightless—a condition that lends itself readily to panic. For this reason, bright, spotlight-type heads must be *very carefully arranged.* Otherwise they can create disabling glare and distorting shadows and impede eye accommodation.

Although it is customary to furnish the required emergency illumination from ceiling- or wall-mounted fixtures, consideration must be given to the code requirement as noted in (b) that specified emergency lighting levels be maintained at floor level. Because heavy smoke can readily obscure light from overhead fixtures, it is advisable to install some (preferably directional) emergency lighting fixtures near the floor level.

(f) Exit Lighting. Most codes require 5 fc (50 lux) on nonilluminated exit signs, internally illuminated signs with the same visibility as an externally illuminated sign, or self-luminous signs with a specified luminance (usually at least 0.21 cd/m^2). Some exit signs are equipped with a battery and controls that provide 1½ h of illumination on loss of utility power. Others are arranged to illuminate the exit area beneath the sign (Fig. 21.49), and still others are equipped with a flasher and/or an audible beeper that assists people in finding the exit in a light-obscuring, smoke-filled room. Finally, some exit signs are nonelectrical

Fig. 21.49 *Typical application of exit light with built-in battery, charger, and controls. Note that the bottom of the unit is designed to illuminate the area immediately in front of the exit. (Photo by Stein.)*

self-illuminating, requiring continuous illumination and are part of the emergency lighting system.

21.37 Floodlighting

Floodlighting, both interior and exterior, is used extensively for such diverse locations as are listed in Table 21.4, in addition to the more common sports lighting, which is not listed. At the designer's disposal are a variety of sources with respect to output, color, life, efficiency, and wattage (see Chapter 19).

Although a detailed floodlighting design involves complex calculations beyond the scope of this work, it is often sufficient for the designer to utilize a watts per square foot table such as Table 21.4 to determine the approximate floodlighting

requirements. The designer should also consult the unit power allowances for exterior lighting listed in ASHRAE/IESNA 90.1.

Thus, if one is concerned with lighting a self-service parking lot at a neighborhood shopping center, and metal–halide is selected, Table 21.4 tells us that approximately 0.055 W/ft^2 will suffice. This is well within the limits of ASHRAE/IESNA 90.1 (1989). If the lot is 200 × 500 ft, or 100,000 ft^2, then 0.055 × 100,000 = 5500 W is required. This figure is a good first estimate. Obviously, the actual power level depends on the specific equipment and the actual design. This may vary from the initial estimate by ±20%.

Arrangement and choice of equipment remains before the problem can be considered solved. Considerable assistance on this score can be obtained from either the lighting engineer involved or from representatives of the equipment manufacturers.

Although most floodlight installations use a single type, the installation of Fig. 21.50(*b*) used a combination of metal–halide and mercury to obtain the desired effect.

21.38 Street Lighting

Although detailed street-lighting calculations and design considerations are beyond our scope (see appropriate IESNA standards), a few remarks are in order. New installations now use HID source almost exclusively. The low efficacy and short life of incandescent sources and the bulkiness of linear fluorescents make them obsolete. Furthermore, high street-lighting levels reduce vandalism and crime, improve night merchandising, and add to an area's attractiveness. A typical walkway luminaire is shown in Fig. 21.51.

21.39 Light Pollution

An oft-neglected corollary to the principle of placing light where it is required is not to place light where it is not required. *Light pollution* is frequently defined as unwanted light in public places, whereas *light trespass* is the intrusion of unwanted light onto private property. The latter would include light intrusion into windows and onto pri-

TABLE 21.4 **Lighting Application Guide**

	Minimum Footcandles Maintained[a]	Watts per Square Foot Generally Required			
Application		Tungsten–Halogen	Mercury Units	Metal–Halide	High-Pressure Sodium
Automobile Parking					
Attendant parking	2	0.38	0.17	0.11	0.075
Industrial lots	1	0.13–0.15	0.06–0.07	0.037–0.044	0.026–0.03
Self-parking lots	1	0.13–0.15	0.06–0.07	0.037–0.044	0.026–0.03
Shopping Centers					
Neighborhood	1	0.13–0.19	0.06–0.09	0.037–0.055	0.026–0.038
Average commercial	2	0.26–0.3	0.12–0.135	0.075–0.087	0.052–0.06
Heavy traffic	5	0.65	0.29	0.19	0.13
Automobile Sales Lots					
Front row (Front 20 ft)	50	10.	4.5	2.9	2.0
Remainder	10	1.5–1.8	0.68–0.81	0.44–0.52	0.3–0.36
Building					
Construction	10	1.5–1.8	0.68–0.81	0.44–0.52	0.3–0.36
Excavation	2	0.26–0.3	0.12–0.14	0.075–0.09	0.052–0.06

	Adj. Area		Watts per Square Foot Generally Required						
Buildings up to 50 ft High	**Light**	**Dark**							
Light surfaces	15	5	3.3	1.2	1.5	0.54	0.96	0.35	0.66	0.24
Medium light	20	10	4.3	2.2	1.94	1.0	1.25	0.64	0.86	0.44
Dark surfaces	50	20	10.0	4.3	4.5	1.94	2.9	1.25	2.0	0.86
	Adj. Area								
Billboards and Signs	**Light**	**Dark**							
Good contrast	50	20	10.0	4.3	4.5	1.94	2.9	1.25	2.0	0.86
Poor contrast	100	50	20.0	10.0	9.0	4.5	5.8	2.9	4.0	2.0

Application	Minimum Footcandles Maintained[a]	Tungsten–Halogen	Mercury Units	Metal–Halide	High-Pressure Sodium
Protective Lighting					
Gates and vital area	5	1.2	0.54	0.35	0.24
Building surrounds	1	0.15–0.19	0.07–0.09	0.044–0.055	0.03–0.04
Roadways					
Along buildings	1	0.24	0.11	0.07	0.05
Open areas	0.5	0.08–0.1	0.036–0.045	0.023–0.029	0.02
Storage yards	20	3.6–4.3	1.6–1.94	1.04–1.25	0.72–0.86
Storage yards (inactive)	1	0.15–0.19	0.07–0.09	0.044–0.055	0.03–0.04
Shopping Centers					
Parking areas (attraction)	5	0.65	0.29	0.19	0.13
Buildings (attraction)		(See Buildings)			
Used Car Lots		(See Automobile Parking)			

[a]All footcandle levels for ground area applications are *horizontal* values.

vate property, whereas the former would cover excessive brightnesses everywhere, plus stray light that finds its way into the night sky to the distress not only of astronomers, but also of anyone who simply wishes to enjoy the beauty of a star-filled sky. At this writing, no standards exist that define the light levels that constitute an intrusion or a nuisance glare. However, a few simple guidelines assist the designer in avoiding creation of a nuisance:

1. Light all exterior vertical surfaces from above, not below, wherever possible. This reduces sky-light pollution.
2. Use luminaires with sharp cutoff beyond the illuminated area. Shields can be added to standard luminaires to accomplish this.

ILLUMINATION

(a)

(b)

Fig. 21.50 (a) *Floodlighted section of wall surrounding the Old City of Jerusalem, Israel, adjacent to the Jaffa Gate. Light sources are 400-W HPS units, giving an average illumination level of 50 lux. Sodium source was chosen to enhance the yellow–red color of the stone.*
(b) *Church of All Nations, Mount of Olives, Jerusalem, Israel. Floodlight sources are 250- and 400-W mercury and metal–halide units, giving an average illumination of 70 lux. Sources were selected to complement the colors in the mosaic at the top of the facade. (Photos courtesy of City of Jerusalem and J. Stroumsa, Chief Engineer.)*

Fig. 21.51 *The "lollypop" fixture, even if esthetically pleasing, gives poor illumination downward (note the large collar). The narrow pole is weakened by the large handhole. (Photo by Stein.)*

3. After an installation is complete, inspect it at night to determine whether any nuisance has been created.

Further information on this subject can be found in IESNA publication CP-46-84, *Astronomical Light Pollution and Light Trespass.*

21.40 Remote Source Lighting

Every experienced lighting designer has at some time felt the need for a remote source luminaire to fill a specific lighting need. Among the most commonly encountered of these situations are:

· Display lighting for light and heat-sensitive objects such as old books, fabrics, drawings and paintings, and, in general, objects containing organic materials, dyes, and coloring. Of particular importance in this category is the acute sensitivity of such objects to UV light, which exists in daylight and most electrical light sources.

· Installations where relamping is a major logistic and financial problem. These include high-ceiling auditoriums and public assembly spaces of all sorts, high bay lighting areas and other difficult-access locations, clean rooms, spaces with security entry limitations, and rooms that cannot tolerate disturbances or interruptions, such as air-traffic control rooms, continuous process manufacturing control areas, and the like.

· Installations where lamp heat is highly undesirable and its removal is difficult, expensive, or both. Among these are store show-windows, refrigerated showcases, and instances where lamp heat is felt by the space occupant(s), such as halogen and other incandescent lighting at workstations and low ceiling spaces.

· Installations where the presence of electrical wiring is undesirable such as patient-controlled hospital bed lighting or light sources in devices used by children.

· Spaces classified by the *NEC* as electrically hazardous. These areas use large, heavy, expensive, and inefficient explosion-proof lighting.

· Installations where the electric and magnetic fields produced by fluorescent and HID lighting are unacceptable.

· Applications where the light source must be very small and effectively invisible.

Additional applications undoubtedly come to the mind of the reader.

Two basic designs have been developed over the years to fill these lighting requirements; arrangements using optical fibers and those using light guides of various types. Applications of these two techniques do not generally overlap, and they are therefore discussed separately in the following sections.

21.41   Fiber-Optic (FO) Lighting

Fiber optics have been in use for years, principally in instruments designed to permit illumination and observation of essentially inaccessible locations. In medical instruments designed for intrusive applications, FO cable has the additional advantages of absence of heat and elimination of electrical wiring, both of which could be at the very least uncomfortable and frequently dangerous when inserted into the body.

The physical principle by which an optical fiber conducts light, is that of total internal reflection, illustrated in Fig. 21.52. The fiber is constructed of an inner core of transparent material (silica, glass, optical plastics) and an outer coating (cladding) of another material of *lower* refractive index. Because of this difference in refractive indices, light rays within the acceptance cone angle (2θ) are reflected at the interface of the two materials and thus proceed almost unimpeded down the core by a series of 100% reflections. Although there is (theoretically) no loss at the reflection, the core material itself does exhibit an optical impedance that varies with the type of material. Ultraclear glass fibers, used in communications, have losses as low as 1 dB per kilometer, whereas the best plastic optical fiber in use today has losses in the area of ⅓ dB per *meter*—that is, several hundred times as high. However, because of the short distances involved these plastic fibers are satisfactory for lighting work (1 dB loss is the equivalent of 8.6% loss per meter, or 2.6% loss per foot).

Air, with a refractive index of 1.0, can also be used as a ''cladding'' material (unclad core), resulting in a fiber that emits light over its entire length. This light is shown in Fig. 21.52 as light ''lost'' in the cladding. In practice, a bare core is not used because it would be damaged at supports and dirt accumulation on the surface would result in high losses. Sidelight fibers (lateral mode) are constructed commercially of a core, cladding, and sheathing, all of which are transparent plastic materials with decreasing refractive index from core to sheath. Note also from the figure that the angle θ, and therefore the size of the acceptance cone, increases with the difference in the refractive indices between cone and cladding. It is desirable to have an acceptance cone as large as possible (high numerical aperture) so that maximum light from the source (illuminator) can be introduced (coupled) into the fiber. Glass fibers are generally not used in lighting applications because of high cost, small light-carrying capacity and large bending radius. Table 21.5 shows the construction and major characteristics of typical lighting-use optical fiber cables.

Core Index N_1	Cladding Index N_2	NA*	Cone Half Angle θ	Acceptance Cone (2θ)
1.4	1.3	0.52	31°	62°
1.4	1.0	0.98	78.5°	157°

*NA (Numerical Aperture) = $\sqrt{N_1^2 - N_2^2}$ = sin θ

Fig. 21.52 *Schematic representation of the phenomenon of total internal reflection (TIR) in optical fibers (OF). Light rays entering the end of an OF at an angle to the fiber centerline no greater than θ are totally reflected at the interface between core and cladding. Angle θ is determined by the difference between the refractive indices of the core and cladding materials and increases as that difference increases. Light rays entering within the acceptance core (solid angle 2θ) travel along the fiber; all other light rays are lost.*

21.42 Fiber-Optic Terminology

Because FO cable is becoming common in lighting work, it behooves a lighting designer to be familiar with FO terminology. A brief glossary of the most important terms follows.

Acceptance Angle (also *Acceptance Cone*). The maximum solid interior angle of a cone that defines the spread of light rays that will enter a single fiber. As this angle is determined only by the fiber and the cladding materials, it is also the cone of light ray acceptance of a fiber bundle.

Attenuation. The degree of light reduction as it travels along the fiber. Because attenuation varies with frequency, the light color exiting an optical fiber is different from that entering, and the change becomes more pronounced with increasing fiber length.

Axial Mode (also *End-Light*), refers to the mode of conduction from one end of the fiber (cable) to the other, with no deliberate loss by emission over the cable length.

Core. The center of an optical fiber that carries the light.

Fiber-Optic Port. A factory- or field-applied terminating connector that enables coupling an FO cable to a light source or other item in the system. Glass FO is factory terminated; plastic fibers can be either.

Glass Optical Fiber. The type always used in communication work and seldom in lighting application. Expensive, requires a large bending radius, and exhibits very low attenuation.

Illuminator. The term applied to the light source in an FO lighting installation. Usually a metal box containing the light source (usually

TABLE 21.5 **Typical Characteristics of Plastic Optical Fiber Cables**

Type	Construction	Diameter Fiber/Core (mm)	OD (mm)	NA	Cone Angle (degrees)	Attenuation Percent per Meter
End-Light Fiber Bundle	See Illustration (a)	0.15–2.0	6–18	0.5	60	4.7
Side-Light Fiber Bundle		0.15–2.0	6–14	0.5	60	4.7
End-Light Solid Core	See Illustration (b)	3–13	5–16	0.62	76	8.2
Side-Light Solid Core		7–13	9–16	0.57	70	23.7

[1]Sheathing is opaque for end-light cables and transparent for side-light cables.

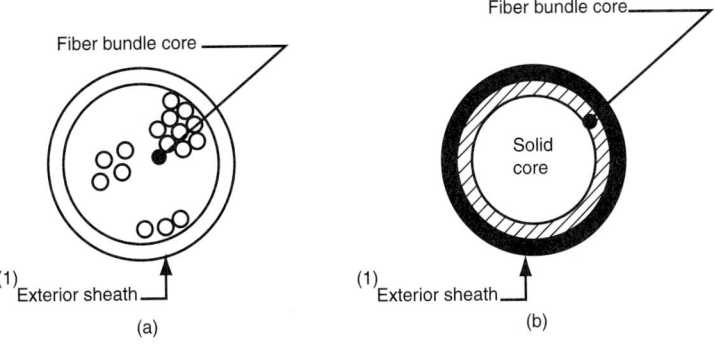

(a)   (b)

metal–halide or halogen) and its accessories, plus required optical devices that collect and concentrate the light from the source and couple it optically to an optical port to which an optical cable is connected. The illuminator may also contain color filters, local and remote-control connections, and almost always a fan or blower for forced air cooling of the source lamp.

Lateral Mode (also *Sidelight(ing)*). It refers to a light conduction and distribution system whereby light is emitted over the entire length of the fiber, simulating neon tubing. It is accomplished by using transparent cladding and sheathing, with refractive indices selected to increase light rays entering (and leaving) the cladding.

Numerical Aperture. A numerical metric based on the refractive indices of the core and cladding materials, indicating the angle of the acceptance cone (i.e., that portion of the light that is conducted by the fiber core).

Plastic Optic Fiber. The fiber material normally used in lighting work; usually a clear acrylic-type compound as the core, with various plastic compounds as cladding or sheathing, depending on the design and mode of operation. The fibers are classified by size as small core, up to 2 mm in diameter; and large core, up to 20 mm in diameter. Small core fibers are manufactured in continuous and essentially unlimited lengths. Large diameter fibers and sheathed fiber-bundle cables do not usually exceed 150 ft in length.

ILLUMINATION

Refractive Index. The characteristic of a light conducting (non-opaque) material to bend a light ray entering from another medium, expressed numerically.

Tail. A single optical fiber or a bundle of fibers extending from an illuminator to an output point. A single fiber separated from a bundle for a specific small isolated illumination purpose (often decorative) is also referred to as a tail.

21.43   Fiber Optic Lighting—Arrangments and Applications

The development of large, efficient plastic optical fibers coupled with a concomitant price reduction has produced a lighting tool whose applications are only limited by the imagination of designers and manufacturers. In one form or another, optical fibers are applicable where:

- A single remote source can supply a large number of relatively small point-source lights.

- Burial of the light carrier (fiber bundle) in almost any substrate is desired or required.

- The heat, UV content, and electrical fields associated with most sources are absent.

- The presence of electrical wiring and its associated hazards are undesirable or unacceptable.

A few of the many current uses are described in the following subsections.

(a) Axial-Mode Linear Devices. In this arrangement, a bundle of fibers is placed into a longitudinal enclosure and individual fibers or small groups are separated from the bundle(s) and brought out of the enclosure as a light emitting point [Fig. 21.53(a)]. Most of these *light bars* are custom made for a specific application, enabling the designer to specify all of the parameters including spacing of points, intensity of each point, dimensions of the bar, individual fiber size and bundle size, color characteristics, and so on. Light bars are used in retail display lighting, accent lighting, directional devices, decorative applications, and other applications requiring a low-

Fig. 21.53 *Linear constructions using end-light fibers.*
(a) *A light-bar is simply a box containing multiple tails that are brought out of the linear box-type container at intervals, usually equal. By selecting fiber size, exit spacing, and illuminator size, a low-intensity linear lighting fixture is created.*
(b) *By using large fibers and or bundles brought out in adjustable groups, a linear bar with adjustable, multihead spotlights can be readily constructed.*

intensity, linear, "sparkling" and/or lighting device.

The essential design of a light bar can be adapted to the use of large core fibers or multiple fiber bundles to produce a linear string of focusable, medium-intensity spot lights [Fig. 21.53(b)]. In lieu of the close-spaced tiny light points of Fig. 21.53(a), large fibers or bundles can be extracted. They can be used as light sources or may be coupled to optical plastic-fiber rods or other devices. These can then be "aimed" as desired to illuminate individual areas in a cabinet, store window, and the like. Here, too, the units are commonly custom designed to perform a specific lighting task.

(b) Axial-Mode Discrete Sources (Fig. 21.54). By using large groups of end-light fiber bundles combined to produce a point light source, semiconventional lighting fixtures can be con-

Fig. 21.54 *Relatively large end-light FO bundles can be utilized as point sources in conventional types of lighting fixtures.*

structed with a variety of common diffusing elements. The advantage of these discrete sources as compared to common electrically powered lighting fixtures are the same as listed previously: absence of heat and electricity, and low maintenance and higher efficiency of high output lamps in the illuminators. One disadvantage common to all plastic-fiber FO lighting is the necessity of keeping runs short due to the attenuation of the fibers. Lengths in excess of 20 ft are inefficient and therefore uncommon.

(c) Lateral-Mode FO Lighting. As explained previously, lateral mode fibers emit light throughout their length. This makes them particularly suitable for linear lighting tasks such as illuminating stair nosings, path lighting, and all sorts of decorative trim lighting. Figure 21.55 shows a possible arrangement for under-shelf lighting, illumination under stair nosings, and the like. The practicability of burying parts of the cable in concrete is useful in giving the remaining lighted cable a disconnected, floating appearance. Color filters in the illuminator, when applied to multiple cable runs as in Fig. 21.55, can produce dramatic outlining effects.

21.44 Hollow Light Guides

The idea of conducting light from one place to another apparently arose several millennia ago from architects' desire to provide daylight deep

into interior spaces. The interior court is one example of an architectural light guide that is only recognized as such in multistorey buildings. Figure 21.56 illustrates use of interior light shaft in a six-storey building. Notice the rapid attenuation as daylight descends by multiple reflection, accounting for the increasing size of windows at lower levels. Indeed, attenuation is the principle problem with light pipes and guides that depend on multiple internal reflections to conduct light any appreciable distance. The best metallic mirrors available today specularly reflect only about

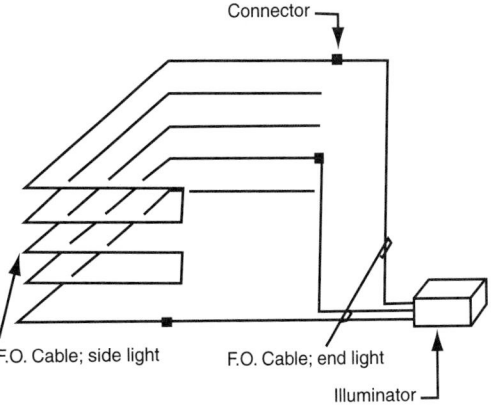

Fig. 21.55 *Side-light–emitting FO cables are ideal for linear lighting tasks such as stair-edge illumination, under-shelf lighting, and outline lighting of all sorts. The illustrated 3-tail arrangement is representative of this type of FO lighting.*

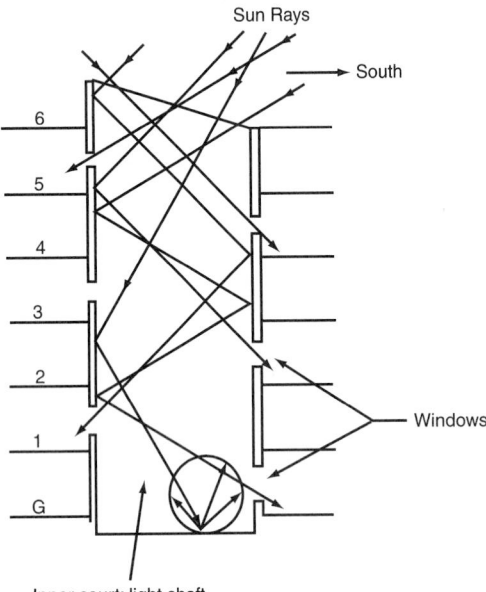

Sun Rays

South

6

5

4

3

2

1

G

Windows

Inner court; light shaft

Fig. 21.56 An interior court or light well (shaft) with reflective wall surface acts as a light guide to introduce daylight at lower floors. Buildings in the northern hemisphere are oriented as shown. Due to attenuation of re-reflected sunlight beams traveling down the shaft, lower floor windows are larger than those at upper floors to capture more (of the attenuated) light. (Reprinted from NASA Tech. Brief LAR-12333.)

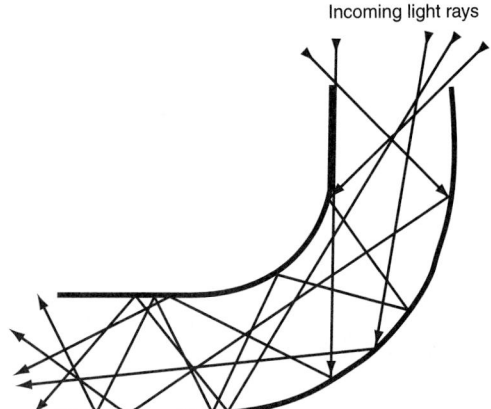

Incoming light rays

Fig. 21.57 Curvature in a light guide increases the number of reflections that a light ray makes over the guide length, thereby increasing attenuation. (Reprinted from NASA Tech. Brief LAR-12333.)

95% of the impinging light. It is therefore relatively easy to calculate the number of 5% loss reflections that reduce the output light from a light pipe with random nondirectional input, to an uneconomical level (where the "free" light received is more expensive due to construction and maintenance costs than locally produced electric light). Furthermore, open-ended light pipes of this sort (Fig. 21.57) are notoriously difficult to keep clean of dust and surface-reflectance degeneration, so that the 95% initial specular reflectance rapidly decreases to 85% semidiffuse reflection, thereby increasing ray re-reflection and consequently the overall light attenuation. (Sealing the upper end of the pipe with a transparent medium is not advisable due to the rapid accumulation of dust and condensation.) In order to capture an appreciable amount of sunlight (sky-light is insufficient), an open-ended pipe at roof level

must be very large. This in turn increases construction and maintenance costs and decreases (in the case of a commercial building) usable/rental area, making the light-guide idea generally impractical except for very short guides. These (commercially available) units are essentially small sky-lights with an elongated, reflective collar.

One solution to the rapid attenuation in a light guide caused by multiple interior wall reflections is to collimate the incoming light. This has the additional advantage of permitting reduction of the light guide's cross-sectional area. Such a system, however, requires a sun-tracking arrangement plus mirrors and lenses in a collimating optical train. A typical arrangement is shown schematically in Fig. 21.58. The increased efficiency of arrangements of this sort is offset by losses in the optical train and increased equipment cost (although the light pipe itself is smaller and therefore cheaper). Systems of this type are in use today, with generally satisfactory results, although system economics is marginal except in arid sunny climates.

21.45   Prismatic Light Guides

The first major improvement in hollow light guide design occurred in 1981, with the development

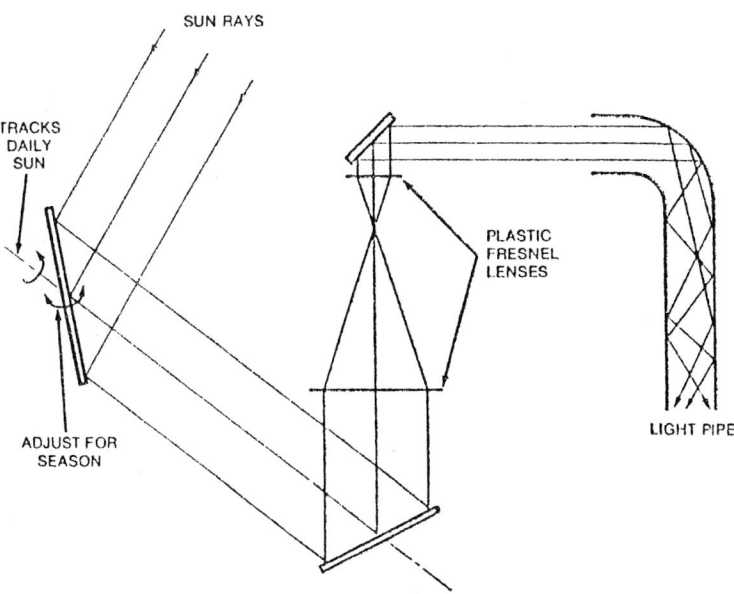

Fig. 21.58 *Concentration and collimation of sunlight can be accomplished with a sun-tracking mirror and an optical train of mirrors and lenses. Full tracking of the sun requires altitude and azimuth drives on the collection mirror. To reduce costs without severely reducing collection efficiency, a full azimuth drive tracks the sun from east to west daily while the mirror tilt around a horizontal axis (altitude tracking) is adjusted only seasonally to the mid-season position. (The sun's maximum altitude varies 23.5° from equinox to solstice.) (Reprinted from NASA Tech. Brief LAR-12333.)*

and patenting of a prismatic rectangular hollow acrylic light guide (Whitehead, 1981, 1982). The patentable novelty of this device is that the facets of the prismatic exterior transparent acrylic walls of the pipe act as total internal reflection "mirrors," thus preventing light from escaping from the transparent box guide. As a result, light is guided along the length of the device with very low losses, because reflection at the prismatic walls is theoretically perfect and losses in the air space within the guide are very small. In actual practice it was found that at each reflection about 2% of the light was lost by absorption and, because of microscopic imperfections in the prisms, about 6% escaped through the wall. This 6% "loss" converted the light guide into a long rectangular lighting fixture emitting light *uniformly* over its entire length. By placing a mirror at the end of the guide, all the light reaching the end is reflected back into the guide, thereby increasing the light output over its length. Overall efficiency of this extended "lighting fixture" is high, which fact, combined

with very low maintenance, produced a highly desirable lighting product. Its first application was in an electrically hazardous area previously illuminated by explosion-proof lighting fixtures (Fig. 21.59).

Fig. 21.59 *Views of the hollow, clear acrylic, prismatic light guide developed by L. A. Whitehead. The cross-section measures 13-cm square, and the prism angle α is 90°. (Reprinted from L. A. Whitehead et al.,* A New Efficient Light Guide for Interior Illumination, *1981.)*

21.46   Prismatic Film Light Guide

In 1988, the 3M company developed a thin plastic prismatic film that utilizes the principle of total internal reflection at the prism face. Tests have shown that this material loses only about 1% at each reflection. As a result, tubular hollow light guides using this material can extend for lengths of several hundred feet if the input light is collimated (Figs. 21.60 and 21.61). When used as an extended lighting fixture it exhibits good photometric characteristics.

Prismatic light pipes have all of the advantages of FO lighting but fewer limitations, although in practice the two systems do not compete because they address entirely different needs. In addition to all the advantages of remote-source FO lighting including absence of heat, minimum UV, absence of electrical wiring and the associated electromagnetic fields and electrical wiring hazards, and usability at any temperature, interior or exterior, prismatic light guides (and lighting fixtures) are

Circular hollow light guide

Fig. 21.60 *If the prismatic plastic film is formed into a cylindrical shape as shown, it will act as a light guide with very low losses. (Reprinted from a paper presented at Globalcon '96, by K. G. Kniepp of the 3M Company.)*

constructed to handle very large quantities of light and do not produce color distortion (as do optical fibers). As a result, a luminous light pipe (light fixture) can be coupled to a very high output source such as an arc lamp, Xenon lamp, or the relatively new sulphur lamp, and the pipe used to illuminate large areas, either interior and exterior. The economies in wiring, luminaires, installation, and maintenance, in addition to the high efficacy of these large sources, more than offset the cost of the prismatic material.

Because of the very high light conduction efficiency of optical lighting film (OLF), using it as a lighting fixture requires special techniques and materials to extract light from the pipe. One method is to apply another prismatic material to the inside of the light pipe with a different prism angle that causes incident light to change direction and exit the pipe. By placing this material judiciously, selected areas of the pipe circumference emit light, in effect producing a longitudinal lighting fixture with the desired directionality. Typical lighting applications of prismatic film light-pipe fixtures are high bay industrial and commercial installations, large exterior signs, tunnel illumination, highway signs, continuous rail-type guidance illumination, and exterior architectural building lighting. Figure 21.62 shows a circular prismatic light guide designed for exterior use, with several configurations and the associated photometric data. Figure 21.63 shows an application of the prismatic light-guide fixture.

21.47   Standards and Nomenclature

Due to the relative newness of long, remote-source tubular lighting, industry-wide standards do not exist and even nomenclature varies. Clear differentiation between devices intended primarily to carry light from one location to another and devices intended to illuminate along their length does not exist. Nor is terminology differentiated between light guides operating on the principal of internal reflection from specular surfaces and prismatic light guides. All these devices are variously and interchangeably referred to as

- hollow light guide
- light pipe

Reflecting Endcap

(a)

Light Guide

Input Endcap

Luminaire

LIGHTPIPE System Assembly

White PVC Extrusion
Housing Panel – 3 sides
Diffuse Reflecting Film
Prismatic Light Guide Film

7¾" ± ⁰⁄₁₆" sq.

⅝" Typ.

Pattern 12 Lens

LIGHTPIPE Light Guide – Cross Section   A-A

(b)

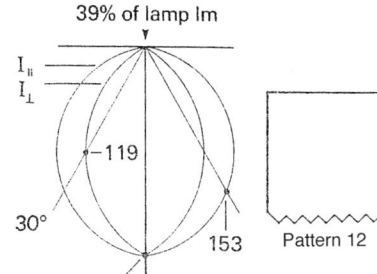

39% of lamp lm

$I_{\parallel}$
$I_{\perp}$

–119

30°

153   Pattern 12

200 cd / 1000 lamp lm

ρcc	80		50		30	
ρw	50	30	50	30	50	30
RCR						
0	.46	.46	.43	.43	.41	.41
1	.41	.40	.39	.38	.37	.37
2	.37	.35	.35	.33	.34	.32
3	.33	.31	.32	.30	.31	.29
4	.30	.27	.29	.26	.28	.26
5	.27	.24	.26	.23	.25	.23
6	.25	.22	.24	.21	.23	.21
7	.22	.19	.21	.19	.21	.19
8	.20	.17	.19	.17	.19	.17
9	.18	.15	.18	.15	.17	.15
10	.17	.14	.16	.14	.16	.13

CU Table — One Side Emitting

(c)

Fig. 21.61 *Details of a rectangular light-guide fixture.*
(a) *The entire lighting unit comprises a "luminaire" at one end containing the light source (250 or 400 W metal–halide lamp) and its accessories connected to the hollow prismatic light guide. The end of the guide is sealed with a mirror that reflects the light back into the light guide.*
(b) *The light guide itself is enclosed on three sides and equipped with a prismatic lens diffuser on the open (bottom) surface, through which light is emitted. This specific design is usable to a length of 40 ft. Longer units are equipped with light sources at both ends.*
(c) *Photometric characteristics of the luminaire–light guide assembly. (Courtesy of TIR Systems, LTD.)*

ILLUMINATION

ILLUMINATION

Luminaire/Light Guide Connector

Coupling Endcaps

Mirror Endcap

Luminaire

Input Endcap

Light Guide Bracket

Light Guide

Conduit to Remote Ballast

Luminaire Bracket

The End-Feed Light Pipe System

(a)

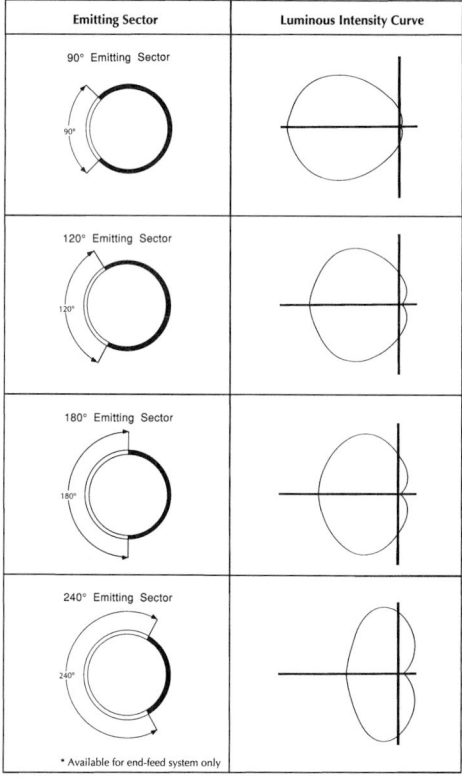

Emitting Sector	Luminous Intensity Curve
90° Emitting Sector	
120° Emitting Sector	
180° Emitting Sector	
240° Emitting Sector	

* Available for end-feed system only

(b)

Fig. 21.62 *Details of a 7-in. D hollow prismatic end-feed light-guide fixture.*
(a) *The entire unit consists of an opaque "luminaire" section containing a 250 W metal–halide lamp, connected to a 7-in. D acrylic tube up to 44 ft in length containing prismatic optical film, and terminating in a reflective (mirrored) end cap.*
(b) *The portion of the tube circumference that emits light (uniformly) is controlled by placement of a diffuse reflecting film between the inner prismatic film and the outer transparent acrylic envelope [see Fig. 21.61 (b)]. The reflective film is shown as the black portion of the tube circumference.*
Figure 21.62 continued on the next page.

CU Tables — End-Feed System

ρ_{cc}	70		50	
ρ_w	50	30	50	30
RCR				
0	.34	.34	.32	.32
1	.30	.29	.29	.28
2	.26	.25	.25	.24
3	.24	.21	.23	.21
4	.21	.19	.20	.18
5	.19	.16	.18	.16
6	.17	.14	.16	.14
7	.15	.13	.15	.12

CU Table - 90° Emitting

ρ_{cc}	70		50	
ρ_w	50	30	50	30
RCR				
0	.36	.36	.34	.34
1	.30	.29	.28	.27
2	.26	.24	.24	.23
3	.23	.20	.21	.19
4	.20	.17	.19	.16
5	.17	.15	.17	.14
6	.15	.13	.15	.12
7	.14	.11	.13	.11

CU Table - 120° Emitting

ρ_{cc}	70		50	
ρ_w	50	30	50	30
RCR				
0	.39	.39	.36	.36
1	.33	.31	.30	.29
2	.28	.25	.26	.24
3	.24	.22	.23	.20
4	.21	.18	.20	.17
5	.19	.16	.18	.15
6	.17	.14	.16	.13
7	.15	.12	.14	.11

CU Table - 180° Emitting

ρ_{cc}	70			50	
ρ_w	50	30	10	50	30
RCR					
0	.39	.39	.39	.35	.35
1	.32	.30	.28	.29	.27
2	.27	.24	.22	.24	.22
3	.23	.20	.18	.21	.18
4	.20	.17	.15	.18	.16
5	.18	.14	.12	.16	.13
6	.16	.13	.10	.14	.11
7	.14	.11	.09	.13	.10

CU Table - 240° Emitting

Length of Run (feet)		10 ft.	20 ft.	30 ft.	40 ft.
Lumens/foot		500 lm/ft.	270 lm/ft.	165 lm/ft.	110 lm/ft.
Peak Luminous Intensity (cd)\ (Mean Exitance (lm/ft^2))	90° emitting	1569\(1070)	1708\(583)	1569\(357)	1435\(245)
	120° emit.	1143\(850)	1243\(462)	1143\(283)	1043\(194)
	180° emit.	1091\(631)	1187\(343)	1091\(210)	998\(144)
	240° emit.	696\(497)	757\(270)	696\(166)	N/A

Note: All values listed are based on the maintained output of a single T250 luminaire.

Photometric Data - End-Feed System

Fig. 21.62 (continued from previous page) (c) *Photometric data for an end-fed system with a single 250 W metal–halide source. (Courtesy of TIR Systems, LTD.)*

ILLUMINATION

Fig. 21.63 *Application of prismatic tubular light-guide to tunnel lighting (see Fig. 21.61).*
(a) *Perspective drawing showing tubular lights on both sides of the tunnel and the resultant uniform tunnel illuminance levels.*
(b) *Drawing of one section of center-feed light guide. (Courtesy of TIR Systems, LTD.)*

· hollow light pipe

· prismatic light guide

· prismatic light pipe

· hollow prismatic light guides

· remote source hollow light guide

and various other names. NEMA is developing a standard for this class of device under the somewhat misleading name of Remote Illumination Systems. We can only suggest that the reader who has the opportunity to work with these devices select an accurate descriptive name, such as 6-in. diameter hollow prismatic light guide, or the like.

REFERENCES AND BIBLIOGRAPHY

ANSI/IEEE Standard 446. (1996). *Recommended Practice for Emergency and Standby Power*, IEEE/Wiley, New York.

Beltron, L. O., et al. (1997). *Advanced Optical Daylight Systems,* "Light Shelves and Light Pipes," *Journal of the IESNA,* Winter.

Brown, G. Z., Reynolds, J., et al. (1992). *InsideOut,* 2nd ed., Wiley, New York.

Bryan, H. J. (1980). "A Simplified Procedure for Calculating the Effects of Daylight from Clear Skies," *JIES,* April, pp. 142–151.

Caddet Newsletter #1 (1997). "Innovative Skylight Offers Natural Light Without Excess Heat."

Daylight Design Diagram (1963). Service Division, Commonwealth of Labor and National Service, Melbourne, Australia.

Daylight, International Recommendations for the Calculation of Natural Daylight (1970). Publication CIE No. 16 (E-3.2), Commission Internationale de l'Eclairage, Paris.

Dordai, P. C. (1997). "Integrated Solutions for Daylight," Architectural Lighting, April/May.

Ehrlich, C. (1998). "Simulating the Visual Performance of Electrochromic Glazing for Solar Control," *Journal of the IESNA,* Winter.

Elmer, W. B. (1980). *The Optical Design of Reflectors,* Wiley, New York.

Griffith, J. W., Arner, W. J., and Wenzler, (1957). "Practical Daylighting Prediction," IES National Technical Conference, Atlanta, Ga.

Harkness, E., and Mehta, M. (1978). *Solar Radiation Control in Buildings,* Applied Science Publishers, London.

Hopkinson, R. G., Pethebridge, R., and Longmore, J. (1966). *Daylighting,* Heinemann, London.

"How to Predict Interior Daylight Illumination" (1976). Pamphlet, Libbey-Owens-Ford Co.

Hughes, P. C., and Neer, R. M. (1981). *Lighting for the Elderly: A Psychobiological Approach to Lighting,* The Human Factors Society, Inc.

IES *Lighting Handbook,* 8th ed. (1993). IESNA, New York.

IES RP1-1993. *American National Standard Practice for Office Lighting,* IESNA, New York.

IES RP2-1985. *Recommended Practice for Lighting Merchandising Areas,* IESNA, New York.

IES RP3-1988. *American National Standard Guide for Educational Facilities Lighting,* IESNA, New York.

IES RP7-1991. *American National Standard Practice for Industrial Lighting,* IESNA, New York.

IESNA RP-21 (1983). *Recommended Practice for the Calculation of Daylight Availability,* IESNA, New York.

IESNA (1989). *Recommended Practice for the Lumen Method of Daylight Calculations,* IESNA, New York.

IESNA LEM-3-1987. *IES Design Considerations for Effective Building Lighting Energy Use,* IESNA, New York.

Jones, B. F. (1983). "Very Very Simple Hand Calculations for Daylighting," *Proceedings, 1983 International Daylight Conference,* Phoenix, Az. pp. 87–92.

Kambich, D. G., and McCloskey, N. M. (1984). "Practical Consequences of Using Relative Visual Performance to Analyze Lighting Systems," *JIES,* October.

Kimball, H. H., and Associates (1923). "Daylight Illumination on Horizontal, Vertical and Sloping Surfaces," *U.S. Weather Review,* Vol. 50, p. 615.

Lim, R., et al. (1979). *Environmental Factors in the Design of Building Fenestration,* Applied Science Publishers, London.

Lynes, J. A., Burt, W., Jackson, G. D., and Cuttle, C. (1966). "The Flow of Light into Buildings," *Trans. IES (London),* Vol. 31, No. 3, pp. 65–91.

Matsuura, K. (1979). "Turning-Off Line in Perimeter Areas for Saving Lighting Energy in Side-Lit Offices," *Energy and Buildings,* Vol. 2.

Millet, M. S., a Bedrick, J. R. (1980). *Graphic Daylighting Design Method.* LBL/DOE, Washington, DC.

Millet, M. S., and Bedrick, J. R., and Adams, C. (1980). "Graphic Daylight Design Method," paper presented at International Symposium on Daylighting, Berlin, July.

Millet, M. S., Lakin, J. E., and Moore, J. (1981). "Rainy Day Shading: The Effect of Shading Devices on Daylighting and Thermal Performance," Department of Architecture, University of Washington, Seattle.

Mistrick, R. G., and Thongtipaya, J. (1997). "Analysis of Daylight Photocell Placement and View in a Small Office," *Journal of IESNA,* Summer.

Moech, M. (1998). "On Daylight Quality and Quantity and its Application to Advanced Daylight Systems," *Journal of the IESNA,* Winter.

Moore, F. (1985). *Concepts and Practice of Architectural Daylighting,* Van Nostrand Reinhold, New York.

Murdoch, J. B. (1985). *Illumination Engineering,* Macmillan, New York.

Ne'eman, E., and Longmore, J. (1973). "The Integration of Daylight with Artificial Light," paper presented at CIE Symposium, Instanbul.

Olgyay, A., and Olgyay, V. (1957). *Solar Control and Shading Devices,* Princeton University Press, Princeton, NJ.

Pierpoint, W. (1983). "A Simple Sky Model for Daylighting Calculations," *Proceedings, International Daylighting Conference,* Phoenix, Ariz., pp. 47–51.

Ramsey, C. G., Sleeper, H. R., and AIA (1988). *Architectural Graphic Standards,* 8th ed., Wiley, New York, pp. 76–86.

Rennhackkamp, W. M. H. (1967). "Sky Luminance Distribution on Warm Arid Climates," *Proceedings, 16th International Conference on Illumination,* Washington, DC.

Robbins, C. L. (1986). *Daylighting: Design and Analysis,* Van Nostrand Reinhold, New York.

Ross and Barruzini, (1975). "Energy Conservation Applied to Office Lighting," Federal Energy Administration.

Rubenstein, F., Verderber, R., and Karayel, M. (1985). "Determination of the Optimum Sector Size for Automatic Lighting Controls," *JIES,* Fall.

ILLUMINATION

ILLUMINATION

Sampson, F. K. (1970). *Contrast Rendition in School Lighting,* Educational Facilities Laboratory, New York.

Saraiji, R. M. N., and Mistrick, R. G. (1993). "Development of Coefficient of Utilization for Light Shelves," *Journal of the IESNA,* Winter.

Selkowitz, S., Navvab, M., and Mathews, S. (1983). "Design and Performance of Light Shelves," *Proceedings, International Daylighting Conference,* Phoenix, Ariz.

Selkowitz, S. (1982). *Daylight Design Overlays for Equidistant Sum-Path Projections,* LBL, Berkeley, Ca.

Smith, S. W., and Rea, M. S. (1982). "Performance of a Reading Test Under Different Levels of Illumination," *JIES,* October, p. 29.

Stein, B., and Reynolds, J. (1992). *Mechanical and Electrical Equipment for Buildings,* 8th ed., Wiley, New York.

Stein, B. (1997). *Building Technology Mechanical and Electrical Systems,* 2nd ed., Wiley, New York.

Treado, S., and Gillete, F. (1983). "Measurement of Sky Luminance, Sky Illuminance and Horizontal Solar Radiation," *JIES,* April, pp. 130–135

Whitehead, L. A. (1981). *Prism Light-Guide Having Surface Which Are in Octature.* U.S. Patent 4260220.

Whitehead, L. A. (1982). "Illumination of a Hazardous Area with Piped Light," *LD&A* Oct.

Wieder, S. (1982). *An Introduction to Solar Energy for Scientists and Engineers,* Wiley, New York.

Yorks, P., and Ginthner, D. (1987). "Wall Lighting Placement: Effect on Behavior in the Work Environment," *LD&A,* July.

PART VII
SIGNAL SYSTEMS

Modern buildings, from the simplest residence to the most complex industrial facility, depend completely on electrical, signal, alarm control, and communication systems for normal functioning. The systems discussed in this part are security, music/sound, intercom, clock and program, paging, and building automation. Fire and smoke detection and alarms are covered completely in Chapter 13 (Part IV), Fire Protection. The discussion of individual, single-purpose systems is from the point of view of a user—that is, we emphasize system application rather than design. Operating principles are adduced and application data are related to available equipment. Thus, although equipment is constantly being improved, op-erating and application principles remain substantially constant, thereby permitting the designer to readily adapt new hardware to existing system arrangements.

Where the discussion turns to building automation, which is essentially complex control systems with extremely sophisticated hardware, we limit our discussion to system architecture and application data. Any discussion of hardware, protocols, and software would take us too far afield and would necessarily lag behind the times in a field that doubles its computing power, and therefore control ability, roughly every 18 months.

22
SIGNAL SYSTEMS

22.1 Introduction

No area of equipment design and application to buildings has seen such rapid changes as the field of signal equipment. Under this title is subsumed all signal, communication, and control equipment, the function of which is to assist in effecting proper building operation. Included are surveillance equipment such as fire and access control; audio and visual communication equipment such as telephone, intercom, and television (both public and closed circuit); and timing equipment such as clock and program and all types of time-based controls. Specifically excepted is the entire area of data processing, which is not within the province of the building designer, except for equipment space allocation.

Clock and program equipment, which once were the exclusive interest of schools and some industrial facilities, are now incorporated into building mechanical equipment control systems. Closed-circuit TV, which was once limited to classroom and college use, is commonplace in mercantile areas as part of surveillance systems. The hundreds of signals generated throughout a large facility are logged, channeled, and controlled by means of specially programmed computers and microprocessors. Signal systems that once were separate and distinct are now frequently combined and serve multiple purposes.

Obviously, a detailed study of the design and application of such equipment is beyond the scope of this book or, for that matter, of any single book. We attempt, however, to discuss the basic operation of the various systems, some of the equipment available, its application to different types of facilities, and the impact of these systems on the spaces within a structure. The types of facility considered are single and multiple residences, schools, stores, office buildings, and industrial facilities. Hospitals and laboratories are combinations of these area types plus facilities too highly specialized to be discussed here.

22.2 Principles of Intrusion Detection

To understand the design of intrusion detection systems, it is necessary first to understand the characteristics of the commonly available intrusion detectors (sensors) on which these systems are based. Once the intrusion alarm has been given by a sensor (as with fire detection), the signal must be processed and appropriate measures taken. This may include sounding loud alarms, turning on lights, sending signals to central proprietary or private surveillance services or police, and so on.

(a) Sensors with Normally Open (NC) Contacts. These devices are no longer used in any reasonable system because the circuits are unsupervised. As a result, a defective circuit does not indicate trouble, and the system is thereby rendered ineffective.

(b) Simple Normally Closed (NC) Contact Sensors. These come in a variety of designs, the most common of which are magnetic contacts for doors and windows, spring-loaded plunger contacts for doors and windows, window foil, and pressure/tension devices. They operate to transmit a alarm signal, as noted previously. Being used in closed, supervised circuits provides a trouble signal when the sensors, or circuits, are deranged.

(c) Mechanical Motion Detectors. Where window foil or fixed contacts are impractical, a mechanical motion detector can be used. This device is basically a spring-mounted contact suspended inside a second contact surface. Any appreciable motion of the surface on which the device is placed causes the contacts to "make" *momentarily,* turning in an alarm. These devices are very sensitive and can be activated by sonic booms, wind, and even a heavy truck passing by. Because of this, most such units are provided with sensitivity adjustment.

(d) Photoelectric Devices. These devices operate on the simple principle of beam interruption. When the beam is received, a contact in the receiver is closed. Interruption of the beam causes the contact to open, setting off the alarm. Older devices of this design use a visible light beam and rely, for concealment, on the fact that light is invisible except when reflected from an intervening object. This is quite effective indoors, but outside dust, insects, birds, and so on, show the location of the beam, permitting it to be circumvented. Birds and small animals set it off, too. Dispersion of visible light also limits the throw of the devices when used outside. Modern units use lasers or infrared (IR) beams, which are less easily detected and can be arranged to distinguish between intruder and other disturbance. These latter devices have a longer effective range for exterior use. When using a laser beam, the signal can be picked up, amplified, and retransmitted in a different direction, thus establishing a perimeter security "fence" from a single source.

(e) Passive Infrared (PIR) "Presence" Detector. [Fig. 22.1(*a*)]. This device acts on the principle that all objects emit IR radiation, or heat. The amount radiated depends primarily on the object's temperature and secondarily on its material, color, and texture. The PIR sensor uses a lens or mirror that focuses on a small area and concentrates the PIR radiation collected from the conic volume of space between the device and that area into an IR sensor, forming, in effect, a conical beam of coverage, called a *zone*. IR radiation in an area that is undisturbed (not necessarily unoccupied) changes very slowly (because object temperature changes very slowly). As a result any rapid change in the IR reading of that zone indicates an object entering (or leaving) the space, and this triggers an alarm. (We have already seen how these detectors can be used as occupancy sensors to turn off lights when a space's occupant leaves.) See Section 21.6.

The ability to focus on a particular area is utilized to cover areas both horizontally and vertically by the simple expedient of using multiple lenses or mirrors to create multiple zones that cover any desired volume of space (see Fig. 21.8). Also, inasmuch as the IR detector is not intrinsically sensitive to motion, but only to heat, it is usable where motion in the monitored area is

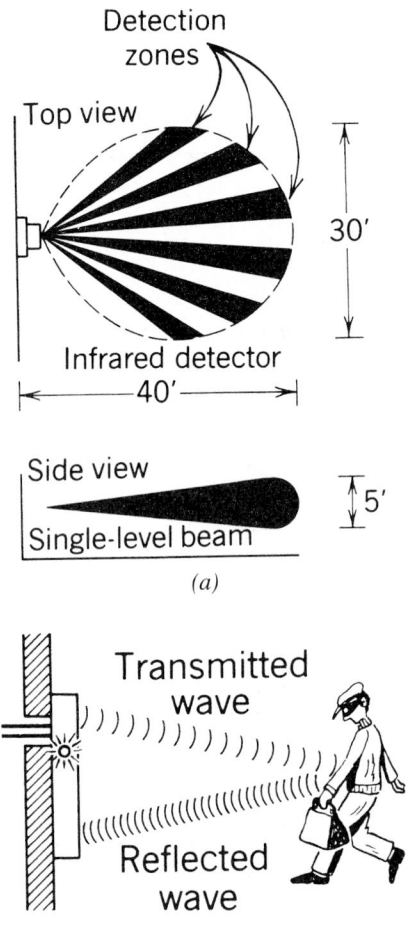

(a)

(b)

Fig. 22.1 (a) *Passive infrared (IR) detectors give basically a 30 × 40 ft oval protective zone, starting as a narrow beam and widening with distance. Focusability permits exact coverage of any area in a space. Units are also readily available with multiple vertical sensitivity stacked beams that give the same type of high-level interbeam vertical sensitivity as is available from the multiple horizontal beams (zones).*
(b) *Motion detectors depend on the Doppler effect. They detect changes in the frequency of a signal reflected from a moving object. Sensitivity is highest when relative motion is greatest; that is, when the intruder is moving directly toward (or away) from the detector.*

unavoidable. The principal disadvantage to PIR detectors is that rapid temperature changes caused by direct insolation, a cold breeze, a heater turning on, and the like can cause false alarms. PIR detectors can be, and are, applied as sensitive motion detectors by using a multi-beam (zone) unit. Motion is detected as changes in the IR radiations of adjacent zones, or in the radiation of a zone (target) with respect to the background, both of which characterize motion.

(f) Motion Detectors [Fig. 22.1(b)].

These devices, which operate either at microwave frequencies or at ultrasonic frequencies, detect motion in the protected area by the Doppler effect. (This is the same effect that changes the perceived sound of a car horn or train whistle as the vehicle passes.) Any moving body changes the received frequency of the signal it reflects, and an alarm sounds. However, because the Doppler effect depends on relative motion between the source and the moving body, an intruder moving laterally may go undetected, if sensitivity has been reduced to avoid false alarms. Therefore, units should be located so that the path of an intruder is as nearly as possible directly toward or away from the detector. Ultrasonic units are cheaper than microwave, but can be disturbed by strong air turbulence and very loud noises. Microwave units are undisturbed by air or noise, but because they penetrate solids (like ordinary TV signals), they can be affected by motion outside the protected area.

(g) Acoustic Detectors.

These units alarm when the noise level exceeds a preset maximum. Alternatively, they can be arranged to respond to a particular range of frequencies, corresponding to the noise of breaking glass, forced entry, or whatever is desired. Although applied principally in security systems, they can also be used as occupancy sensors for switching of lighting.

A typical selection of the previously described devices above shown in Fig. 22.2.

As with fire detectors, a balance must be struck between sensitivity of detectors and the nuisance of false alarms as, unfortunately, increasing the former increases the latter as well. One very effective method of reducing nuisance alarms is to use multiple detectors of different technologies that, in effect, verify each other. Thus, as with occupancy detectors, units are available that combine PIR and ultrasonic detectors in a single housing. Such a unit does not transmit an alarm until both detectors indicate intrusion. This dual technique is applicable to area and perimeter protection as well as portal surveillance.

The previous discussion and that immediately following refer only to intrusion detection (i.e., indication of a human presence where it should not be). *Access control,* which is another area of security design entirely, deals with controlled entry to an area. It is generally more complex than intrusion control, uses different technologies, and a different philosophy of design (see Section 22.24.)

PRIVATE RESIDENTIAL SYSTEMS

22.3 General

Modern private residences utilize a variety of signal devices that greatly enhances their functional value. Indeed, automation in residences, which includes control of and by a host of signal systems, is today a prime factor in valuation (see Sections 16.32–16.34 and 22.32). Figure 22.3 shows a conventional nonautomated residential plan, providing what would be considered *minimally* adequate sound and signal equipment.

In general, all signal systems require a source of signal; equipment to process the signal, including transmitting it; and finally, a means of indicating the signal, whether it be audibly, visually, or on permanent record "hard copy." A complex system still falls into this threefold category except that the individual items of equipment and their functions become more sophisticated. In connection with complexity of equipment, the reader is undoubtedly familiar with the sophisticated automated systems now on the residential market that handle security, fire alarm, time functions, lighting, and so on. They function on dedicated wiring, control bus, or on power line carrier (PLC) signals (high-frequency signals impressed on electric power wiring). As it is not our purpose to discuss automation or central control at this point, but rather the basic systems, the following descriptions consider the systems separately, despite the clear trend to consolidated control.

Residential fire alarm systems are covered in Chapter 13. The components and design of such a system, including detectors, circuit arrangement, and the like are also discussed there.

In Table 22.1, we list the systems and equipment found in the residence shown in Fig. 22.3 by the threefold functions of source, processing,

Fig. 22.2 *Typical intrusion detection equipment:*
(1) Passive infrared detector. Maximum sensitivity is 2°C differential between target and background, and target motion of 2½ in./s. Unit is approximately 6-in. deep. They are available with different coverage patterns and a maximum coverage of 2000 ft².
(2) High-sensitivity multizone wide-angle passive IR detector gives coverage of 17 zones horizontally and vertically.
(3) Ultrasonic detectors arranged to give broad coverage of approximately 200 × 50 ft. Each unit is 11 × 5½ × 3½ in.
(4) High-sensitivity balanced signal type of ultrasonic detector differentiates between intrusion signal and random environmental signals, thus reducing false alarms.
(5) Unobtrusive, button-type ultrasonic sensor intended for ceiling mounting.
(6–8) System control panel, annunciator, and mechanical alarm interface units. (Photos courtesy of Sentrol, an SLC Technologies Company.)

TABLE 22.1 **Elements of Residential Signal Systems**

System Type	Signal Generator	Signal Processor[a]	Signal Transducer
Fire alarm	Temperature and smoke detectors	Control cabinet(s)	Bells, annunciator, buzzer, siren
Intrusion alarm	Door and window switches, motion detectors	Control cabinet	Bells, buzzer, annunciator, siren
Emergency call system	Pull, push button	Control cabinet	Bells, annunciator, corridor lights
Intercom	Microphone, speaker–microphone	Amplifier	Speakers in various stations

[a]The proper wiring and switching is included under this title in all cases.

Fig. 22.3 (a) *Electrical signal equipment plan—lower level.*

SIGNAL SYSTEMS

Fig. 22.3 (b) *Electrical signal equipment plan—upper level.*

(b) Upper level

Living Room

DN

Dining Room

Foyer

ENTRY

Bath

Bath

Master B.R.

Control panel

To OLCP

Study

Note 4

Kitchen

Note 1

Note 2

Note 3

3/4" EC

Notes:
1. The fire detection, smoke detection, and intrusion alarm devices all operate from a single control panel, see (d). The annunciator indicates the device operated and its location.
2. The connection between the signal control panel and OLCP (outside lighting control panel) activates all outside lights when a signal device trips. Selected lights inside the house can also be connected to go on. See riser diagram (d).
3. Two 3/4-in. empty plastic conduits, extending from 4-in. boxes in living room wall down to family room and terminating in 4-in. flush boxes. Boxes to be 18 in. AFF and fitted with blank covers. Also, extend a 3/4-in. plastic EC from one 4-in. box in living room to 12-in. speaker backbox recessed in dining room ceiling. Locate in the field. From the second 4-in. box in living room extend a 3/4-in. empty plastic conduit to an empty 4-in. box in the master bedroom, 18-in. AFF. Finish with blank cover.
4. Coordinate the location and size of the telephone entrance service cabinet (or box) with the 'phone company.
   Extend a 3/4-in. empty plastic conduit from the telephone entrance cabinet to each of the signal raceways in the study. Extend one 1 1/4-in. or two 3/4-in. empty plastic conduits from the cabinet to the empty signal raceway around the perimeter of the family room.
5. Provide a 3/4-in. EC through the wall and capped at both ends, for entry of cables from an exterior satellite dish. Coordinate location with TV/CATV/satellite dish contractor.
6. Boiler room to contain smoke detector, fixed 190 F heat detector, and remote station intercom outlet.

1350

Symbols for signal equipment

F   6 in a-c vibrating bell, concealed in recessed box, with grill cloth cover, 84 in. AFF.

WP F   8 in. weatherproof bell or siren

F   Buzzer, a-c, similar installation to above

F₁   Temp. detector; rate-of-rise and fixed temp., resettable

2   Temp. detector; fixed temp., 190°C

SD   Smoke detector with resettable fixed temp. detector.

①₁   Intrusion detector; magnetic door switch.

2   Intrusion detector; magnetic window switch.

3   Intrusion detector; electronic, motion detector.

◇   Annunciator, custom design

CP   Central panel for fire alarm, smoke detector, and intrusion

DB   Door bell

CH   Chimes signal

▼₁   Prewired phone outlet; jack 12 in. AFF.

2   Prewired phone outlet; fixed, 12 in. AFF.

3   Prewired phone outlet; fixed wall outlet 60 in. AFF.

IC₁   Intercom outlet, outdoor, W.P. 60 in. AFF.

2   Intercom outlet, master station 60 in. AFF.

3   Intercom outlet, remote station 60 in. AFF.

TV   Prewired TV antenna outlet, 12 in. AFF.

TC   Telephone cabinet

(c) Symbol list

Riser diagram

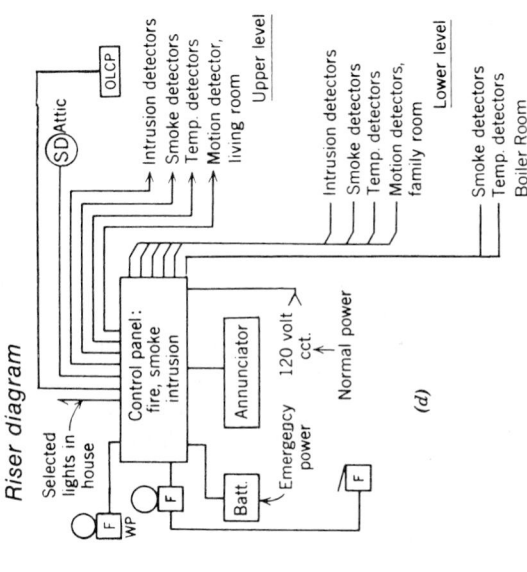

(d)

Fig. 22.3 (c) *Symbol list for signal equipment.*
Fig. 22.3 (d) *Signal equipment riser diagram.*

and indication. Note that the fire alarm, smoke detection, and intrusion alarm systems have been combined into a single system. This simplifies operation and avoids unnecessary equipment duplication. As we discuss the more complex systems, it will be seen that the basic functions remain unchanged.

As shown in Fig. 22.3, a single control panel can serve a multiplicity of residential systems. An annunciator, either integral with the panel or in a separate adjacent enclosure, displays the nature and location of the alarm device that has "tripped." A riser diagram for this residence is shown in Fig. 22.3(d). The alarm devices themselves are not shown on the riser because they appear on the plans and duplication serves no useful purpose.

22.4   Residential Intrusion Alarm Systems

Although any or all of the devices described in Section 22.2 may be used, residences normally utilize door and window magnetic switches, and PIR and/or motion detectors as shown in Fig. 22.3. A manual switch at the end of a long cord is also often provided so that the resident may set off the alarm at will if an intruder is heard. If the system employs the same audible signal devices as the fire system, the sound should be distinctive so that the nature of the alarm can be discerned aurally. Intrusion alarm systems can be continuously supervised by connection with central stations of companies whose business is such supervision and who either respond directly to an alarm call or notify local police authorities of any illegal entry.

22.5   Residential Intercom Systems

The public demand for step-saving conveniences has resulted in the wide acceptance of the home "intercom." Although available with various features, the basic system comprises one or more masters and several remote stations, one of which monitors the front door, allowing it to be answered from various points within the home. Where desired, a closed-circuit TV system can be added so that visual identification at entrances, in addition to voice communication, is effected.

In general, master stations allow selective calling, whereas remote stations operating through the masters are nonselective. The systems are particularly useful when left in the open (monitor) position for remote "baby-sitting." The applicability of such systems to residences with outbuildings should be immediately apparent. As the wiring is low voltage and low power, multiconductor color-coded intercom cable is generally used, run concealed within walls, attics, and basements.

Systems are also available that impose the voice signals onto the house power wiring. This has the advantage of eliminating separate wiring and making remote stations portable—they are connected simply by plugging into a power outlet.

22.6   Residential Telecommunication and Data System

Prior to court decisions permitting users to install their own telephone equipment, the actual wiring within a structure was done only by the utility in the user's raceway system. For some years now, work beyond the service entrance may be done by the owner, in fashion similar to other signal work. In residential work, the telephone company normally follows the route of the electric service, entering the building overhead or underground as desired. In both cases, a separate service entrance means must be provided: if aerial, a sleeve through the wall; if underground, a separate entrance conduit. Unless a residence has many entering lines, no source of power is required for the telephone equipment.

Wiring of telephone instruments when installed *after* completion of the residence consists of a single surface-mounted cable that, even if skillfully installed, is unsightly at best and completely objectionable at worst. Prewiring consists of running the cables on the wall framing and into empty device boxes to which instruments are later connected. The huge increase in the number of private residences with multiple phone lines, dedicated fax lines, and special high-speed data transfer lines connected to home office outlets has made telephone planning as much of a necessity as planning for any other system. This is not only true where residences are designed with a dedicated working-office space. The ease with which modern telecommunications has made it possible to work at home is resulting in an exponential increase in work-at-home situations, at least part-time. Failure to provide for multiple lines with adequate raceways is improper planning. In this regard the locally franchised telephone company's technical representative can offer valuable planning advice. The system of raceways, boxes, and outlets dedicated to communications systems of all sorts, generally excluding audio signals, has come to be known as "premise wiring."

22.7 Premise Wiring

As noted previously, premise wiring refers generally to communication system wiring, including raceways and outlets, and paradoxically, fre-

quently *not* including wiring. The "wiring" is installed by the various communication contractors and includes fiber optic (FO) cabling under the title of wiring, despite the fact that wiring is traditionally understood to be metallic. The use of FO cables for most data cabling will undoubtedly increase in the near future because of their proven advantages in bandwidth, freedom from interference, and high-level security.

Most of the premise wiring raceways are surface mounted because access is frequently required and because data cables are often preterminated, making it difficult to pull them into recessed raceways. Finally, premise wiring raceways are usually large, and as such are much easier and more economical to install as surface-mounted units. Figures 22.4 and 22.5 show a few

(a)   *(b)*   *(c)*

(d)   *(e)*

Fig. 22.4 *Typical boxes, devices, and plates used in data and telecommunication wiring systems.*
(a) *Single and double gang flush wall plates with various jacks and connectors.*
(b) *Communication outlets styled to match standard rectangular electrical wiring device plates.*
(c) *Angled faceplates with a variety of jacks and connectors.*
(d) *Multi-media outlet that accepts both copper and fiber optic (FO) cable connections.*
(e) *Multi-media fiberoptic outlet with built-in storage spool for storage of spare FO cable.*
 [Photos (a–d) courtesy of Hubbell Premise Wiring; photo (e) courtesy of Panduit Corp.]

SIGNAL SYSTEMS

Fig. 22.5 *A range of raceway sizes and shapes is available to meet all cabling requirements. (Photo courtesy of Panduit Corp.; see also Fig. 15.24).*

of the hundreds of devices and raceways used in premise wiring, also referred to as "wire management." See also Fig. 15.24 for a cutaway view of a typical premise wiring raceway.

MULTIPLE-DWELLING SYSTEMS

See Chapter 13 for a description of multiple-occupancy residential fire detection and alarm systems.

22.8 Multiple-Dwelling Entry and Security Systems

Apartment houses, residences, and hotels combine the functions of the access control doorbell systems in the familiar lobby-to-apartment communication system. The most basic system is a series of pushbuttons in the lobby and an intercom speaker or telephone with which to communicate with residents. At the other end, the tenant has a speaker microphone plus a lobby door-opener button. This system can also be arranged to utilize the tenants' regular telephones. When the number

of occupants is large, an alphabetical roster is added to the apartment-button panel to avoid the nuisance of scanning all the apartment names when the sought party's apartment number is not known. When the number is larger yet, a simple pushbutton per apartment arrangement becomes cumbersome, and is usually replaced by an alphabetical tenant register plus a dial or button phone. Closed-circuit television is frequently added to the lobby system, enabling the occupant not only to converse with, but also see, the caller. Such a system increases the electrical contract cost for an average apartment house by 5% to 7%.

In addition to the security provisions provided by the apartment-to-lobby audio and video connections, additional security and alarm devices have been used, such as emergency call buttons within the apartment. These perform any alarm functions required to cover the situation of an intruder who manages to bypass the lobby security check. In geriatric designs, these buttons serve to *unlock* the apartment door to allow aid to enter, if summoned by lights and alarms. In luxury apartments, apartment doors can be monitored from a central security desk and any unscheduled door movement subjected to immediate investigation. These systems are custom designed to the needs and requirements of the owner.

A security problem applicable to all facilities, including residential ones, deals with the problem of limiting entry in unsupervised areas to authorized persons (i.e., unsupervised access control). The problem of keys and locks is well known, despite advances in that field. More sophisticated means include magnetic cards and electronic combination locks, which, because of ease of code change, are particularly useful in residential facilities that cater to transients.

Another aspect of security that is particularly appropriate to housing for the elderly and handicapped, although it is applicable to all housing installations, is the emergency call system. The purpose of this system is to alert *outsiders* to an emergency situation *inside* a closed apartment. This alarm system is essentially a way to call for help in time of illness or other distress. Many construction and housing codes include descriptions of the required equipment. Most often they prescribe a call initiation button in each bedroom *and* bathroom, that registers an audible (alarm) and visible (annunciated) signal at a location that is monitored, locally or remotely, 24 hours a day. Additional signals are required in the floor corridor and at the apartment, the purpose of which is to alert immediate neighbors to the distress call.

22.9 Multiple-Dwelling Television Systems

All modern multiple residences supply each room with one or more TV/FM jack outlets. The TV jack may supply "house" signals from a rooftop satellite dish plus a house VCR or cable TV with pay TV as an option. As these systems are always subcontracted, the design requirements in new construction include a system of empty conduits connecting to cable pulling points in cabinets. Raceways should be sized liberally because of the constant expansion in the electronic entertainment field. As pointed out previously, the ideal raceway for signal cabling is 100% accessible (i.e., a surface-mounted raceway with removable cover). However, the unsightliness of this solution, plus the fact that for most cabling concealed raceways are adequate, lead to their use in new work. In economy-type construction, only floor and wall sleeves are supplied and coaxial cables are run exposed in residential spaces. The cable contractor arranges for the electric power needed for any local amplification. Normally, a 15-A dedicated circuit is sufficient.

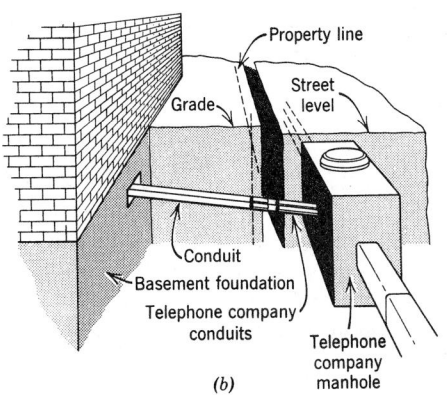

Fig. 22.6 *Telephone cable may enter a building underground, originating on overhead lines* (a) *or in manholes* (b).

22.10 Multiple-Dwelling Telephone Systems

As in the small residence, the telephone service normally follows the same entrance path and method of entrance as the electric power service. For the sake of economy in underground construction, the two services often share the same trench, albeit in different raceways, and utilize twin manholes where such are required. Typical entrance arrangements for any large building, residential or other, are shown in Fig. 22.6.

The service entrance space requirements vary with the size of the building and telephone capacity. For a small apartment house of the garden or three-story type, a clear wall space of 4 to 6 ft is

sufficient. A terminal (equipment) room is required only in large buildings (see Section 22.21). The telephone system is part of the overall telecommunications system utilized in modern residences, preparations for which fall into the premise wiring area discussed previously. As such, generalizations are not useful, and planning must consider the type of occupancy and future requirements.

Rental apartment buildings and dormitories differ from commercial structures in that the floor plans of all floors are similar, so that the arrangement of risers is relatively simple. It is common practice to utilize cable only in risers that extend through vertically aligned closets in apartments.

Fig. 22.7 *Punched-card coded type of "key" in common use in hotels. The punched hole coding is changed after each room use. The drawing of a key on the card is the size of an ordinary door key. (Photo by Stein.)*

To accommodate these cables, sleeves through the floor between closets is necessary. If a riser is located in a shaft other than a closet, conduit is normally utilized to allow for easy installation, protection, and repair. If the location is accessible, as in an alcove, only a sleeve is provided. When the riser is located outside the apartment, each dwelling unit is connected to the riser by a conduit with a junction box at either end. Beyond the apartment service point, the individual rooms can normally be prewired entirely without conduit, or with only a few short sleeves. In condominium and co-op structures, large raceways are run between the service entrance point to cabinets on each floor of the building. From terminal devices in these cabinets, the required service can be extended into each apartment. Prewiring of the apartment avoids the unsightliness of exposed cabling or of cabling within surface raceways.

22.11 Hotels and Motels

(a) Security. Because of the transient population of these facilities and the need to provide maximum service and security to the guests during their stay, the principal problem in addition to room access security is hotel equipment security, (i.e., prevention of what is euphemistically known as *shrinkage*).

1. *Room Access Security.* The ineffectiveness of key locks in preventing undesired entry even in private homes is well known; in hotels, the key is little more than a psychological barrier. As a result, most modern hotels have installed electronic room locks whose opening device code is changed with every new guest. These locks may be of the coded pushbutton type, whose coding is changeable only from a central lock–security console; a magnetically or punched-hole coded-card type (Fig. 22.7); or a programmable electronic lock and coded-key type. (Fig. 22.8). All of these systems have relative advantages and disadvantages, depending on the number of rooms, whether the installation is new construction or retrofit, and average length of guest stay.

2. *Equipment Security.* The need to provide service requires that guest rooms be equipped with television and possibly a VCR; meeting rooms with TV, VCR, projectors, and computer terminals; and that these and other devices be made available for guest use either as part of the room fee or on a rental basis. Theft control is a consultant specialty on its own and far beyond the scope of this book. One such system (equally applicable to equipment in hotels, schools, office buildings, and industrial facilities) senses the disconnection of the equipment from its power connection (wall plug) and transmits an alarm over the power lines to an annunciator at a selected control location. This arrangement has the advantage of alarming immediately on equipment removal, thus normally permitting appropriate retrieval action to be taken in time.

(b) Telecommunication, Data. The telephone and data communication system in large modern hotels is important and complex. Hotels catering to business persons must make available computer terminals and modems that can be used

(a)   *(b)*

Fig. 22.8 *This electronic door-locking system uses a nonduplicatable fixed-code metal key* (b) *to operate a programmable electronic lock* (a). *The lock, which can be reprogrammed by the use of a portable programmer with upward of 10 million different codes, reads the key code with an internal infrared reader as shown diagrammatically in the drawing* (b) *of a cutaway lock The keys, which can be programmed with up to 1.5 million codes, are discarded after use. Each lock accepts up to 3000 different codes and can be programmed in two modes: passage mode to admit user in a specific time slot, and time mode that is programmed to accept specific keys in specific time slots. In addition, the locks can be* audited *to reveal access records, invalid attempts, activation/deactivation, and mechanical key override. (Photo courtesy of ILCO UNICAN Inc.)*

at the very least in rooms dedicated to that purpose, and in high-quality hotels, in (selected) guest rooms as well. This requires that the building designer must provide adequate raceway (and cabling) facilities, as discussed in Section 15.18. The current practice of holding business meetings and technical conferences in hotels also requires that conference and meeting rooms be arranged for very heavy concentrations of electronic equipment that can be installed and rearranged in short order. This normally calls for some sort of access floor and modular cabling, as described in Sections 15.25 to 15.30.

SCHOOL SYSTEMS

22.12 General

The proper operation of a modern school requires that flexible and efficient signal and communications equipment be available to the administrative and teaching staff. Such equipment, engineered to meet the needs of the individual institutions, does much toward optimum utilization of staff and student time. School fire detection and alarm systems are discussed in Chapter 13.

22.13 School Security Systems

Although intrusion alarms and security systems were not historically normal school requirements, this situation has unfortunately changed. Sensing devices on doors and windows can be arranged both to trip local alarm devices and, via auxiliary circuits, to notify police headquarters. Often, vandals can be frightened off by having the alarm system actuate a protective lighting system that illuminates the building exterior and any interior areas desired, such as record rooms. A perimeter alarm detection system of the types described in Section 22.2 can be installed in particularly

Fig. 22.9 *Delayed egress exit alarm. When the bar is pushed an alarm sounds immediately, but the egress is delayed 15 seconds to permit security personnel to investigate before the door release operates. In the event of fire or other emergency, the door releases instantly. The alarm can be bypassed and reset by key only. (Courtesy of NAPCO.)*

vandalism-prone areas to assist in preventing entry to the school premises after hours. Although expensive, they are very frequently cost-effective.

An exit-control alarm is a type of exterior door security device applicable to schools (and other facilities) with doors that are locked from the outside but that must be openable from the inside in the event of emergency. This device (Fig. 22.9) alarms as desired (i.e., audibly or visually, locally or remotely) when the controlled door is opened. It is normally equipped with a timed bypass that allows authorized personnel to key-operate the bypass but prevents the door being held open

without alarming. One model of door-opener alarm requires a continuous pressure on the door opening bar of 10 to 20 seconds before the door opens, with the alarm being activated immediately on application of pressure. This arrangement (usable only where fire codes permit) allows sufficient time for the facility's staff to investigate the attempted exit before the door opens and is applicable only where exit control is the overriding consideration. As mentioned in Chapter 13, stair and exit door exit locks can be arranged to be centrally released from the fire control center.

22.14   School Clock and Program Systems

The clock system and the program system were at one time separate and distinct, sharing only the time-keeping facilities provided by the master clock. Now that electromechanical programming devices are effectively obsolete, the two systems are actually one, but the traditional two-system name remains. Referring to Fig. 22.10, we see that the heart of the system is the time base (electronic clock). This is precisely the same device that provides timing for all programmable switches and controllers. In a clock and program device it controls clock signals, audible devices, and, if desired, other switching functions.

The clock system may use analog clocks (with hands), digital units or both. The former are usually locally powered, and a correction signal is periodically transmitted via dedicated wiring from

Fig. 22.10 The central clock and program device uses its electronic time base to control clocks of all types, time-based audible signals, and event controllers.

Fig. 22.11 *Typical microprocessor-based master time clock and program device. This unit, which measures 14 × 12 × 5 in. deep, provides time correction signals to synchronous wired clocks of the digital or analog type and can be adapted to control minute impulse clocks. Programming of four output circuits can be daily or weekly, and can accommodate daylight savings, holidays, and leap-year update with a total program capacity in excess of 600 events. Programming is done locally or via a modem from a remote computer. This type of device is also applicable to facilities other than schools requiring a clock and/or program system. (Courtesy of Midwest Time Control Inc.)*

the master clock at the controller. This same controller continuously transmits a binary-coded signal to digital "slave" clocks, which can be either of the self-illuminated light-emitted display (LED) type or of the liquid-crystal display (LCD) type. LCD units are easily visible only in high ambient illumination and when viewed directly (not at an acute angle). LED units are best applied in areas of low ambient illuminance. Large-face analog clocks are easily visible in all ambient light situations.

The programming function of the controller serves to delineate audibly the various time periods into which the school day and week are divided. A single-circuit unit is utilized in the instance of an institution that operates entirely on one schedule, such as an elementary school on a morning period–lunch–afternoon period regimen. For a school employing different schedules for its various parts, the controller can provide multiple program schedules on different circuits, depending on its design. Controllers are user programmable and are normally provided with a crystal-controlled master clock to ensure accurate time keeping regardless of line frequency variation, backup power source to maintain user programming and master clock local display, and various conveniences such as daylight saving time correction, security arrangements, event timers, and so on. If desired, the clock and program controllers can also be used for mechanical system control by the simple expedient of adding relays and switching devices, as shown in Fig. 22.10. Typical clock and program controllers are shown in Figs. 22.11 and 22.12.

The audible devices in a program system may be bells, gongs, buzzers, horns, or a tone reproduced on a classroom loudspeaker. This last system has the following advantages:

1. Clear audibility in each classroom, with adjustable volume to accommodate both quiet and noisy areas.
2. No possibility of confusion between program tone and other signals such as fire alarm gongs.
3. Multiple use of the speaker unit for classroom sound as well as program tones.
4. Complete flexibility of programming that is not possible with hall gongs. This is particularly desirable in schools with special programs for special groups of students.

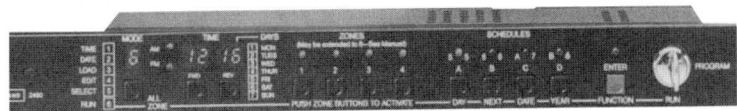

Master Control Clock System

Fig. 22.12 *User-programmable master clock and program controller. This sophisticated unit has nonvolatile memory, eight program zones, automatic daylight savings and holiday adjustment, a maximum of four schedules, and is usable with most types of secondary clocks. (Courtesy of Rauland-Borg.)*

SIGNAL SYSTEMS

Intercom Controller

Fig. 22.13 Intercom controller. The illustrated microcomputer design interfaces with the existing telephone system. In addition, it provides paging, tone distribution, and program distribution, with control from a telephone. In addition, the unit, which measures 19 in. × 9.5 in. × 3.4 in. high, weighs 9 lbs, and has a built-in master clock. (Courtesy of Rauland-Borg.)

22.15   School Intercom Systems

Various types of intercom systems are available depending on the needs of the school building involved. In small schools, a simple wired intercom system connecting the various offices is usually sufficient. This is supplemented by outside telephones in the administrative offices and a functional paging system that is normally part of the school's sound system arrangement. With larger buildings and correspondingly larger numbers of extensions and multiple-function demands, more sophisticated equipment is required. The unit illustrated in Fig. 22.13 is typical of modern school intercom equipment, and is actually a private telephone system of considerable flexibility. Such a system is generally interfaced with the school sound system, and may provide these functions:

1. Intercom between staff members and offices.
2. Direct communication with classrooms, including selective and all-call capability.
3. Paging zone, group call, and conference call functions.
4. Interconnection with the outside phone system.
5. Class-change signals via interface with master clock system.
6. School intercomputer communication via modems.

A combination program/intercom controller is shown in Fig. 22.14. These systems use direct pushbutton dialing and programming, thus eliminating the necessity for switchboards and operators. All stations are coded with three-digit codes, and all switching is solid state, minimizing maintenance problems. Such systems are adequate for all but the largest institutions.

22.16   School Sound Systems

An integrated sound-paging-radio system designed for school use offers several modes of operation and considerable flexibility. Its function is to provide a means for distributing recordings (CD disks, tapes), broadcast (AM/FM), or live sound to preselected areas of the school. Thus, a simple system might provide a CD player and single microphone input and a single channel to all the speakers in the school, whereas a complex system can be arranged to operate with three simultaneous input signals distributed to six different areas of the school (see Fig. 22.14).

A conventional system consists of a control console containing most of the input units, amplifiers, switching devices, and connections to the remote loudspeakers. The input units may comprise one or more AM/FM tuners, VCR, CD player, tape deck, and microphones. One microphone is normally located at the console with others in the principal's office, auditorium, school office, or other selected locations. If desired, microphone outlets can be spotted around the school

Fig. 22.14 Compact program/intercom control center. This unit, which measures 20.5 × 11 × 8.5 in. high, has a separate power amplifier for intercom and program material. It distributes program material or provides intercom with 25 loudspeaker locations, and has additional capabilities, such as emergency/all-call paging, tone calling, time signal, night bell, and telephone paging. Program material derives from separate external sources. (Courtesy of Bogen Communications Inc.)

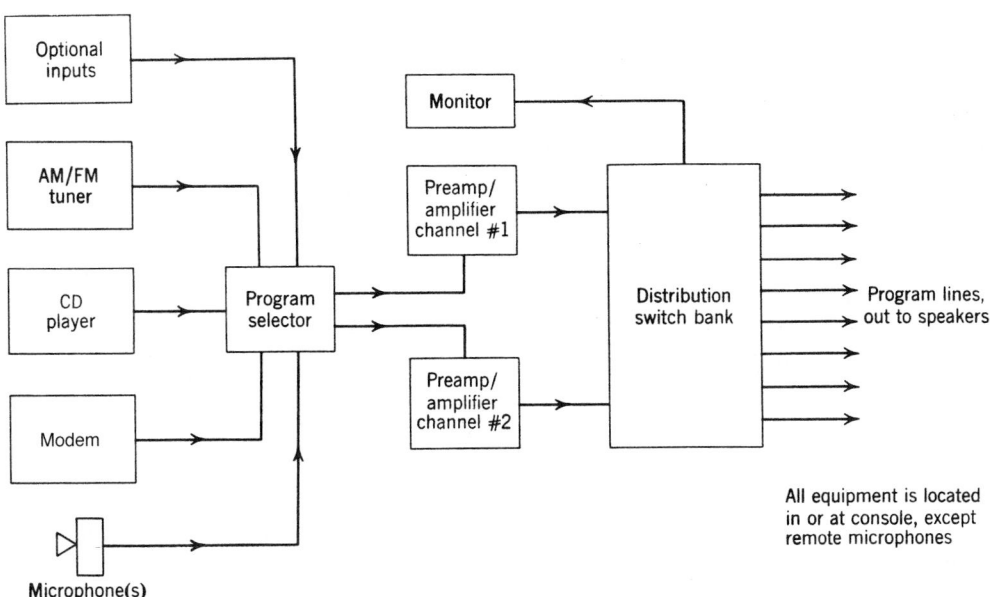

Fig. 22.15 *Block diagram of a two-channel sound system. Optional facilities that can be added to the console are tape deck, private telephone communications, modem connection, and equipment for rebroadcast of signals between areas. Intercom and master clock and program can also be incorporated or interconnected.*

and a spare microphone and stand supplied that can be plugged in at any of these points (Fig. 22.15).

Loudspeakers, located in classrooms, gymnasium, auditorium, cafeteria, and outdoors receive the amplified signal through the switching mechanisms located in the console. It is the function of these switches to deliver the program material to the various loudspeaker circuits, called *program lines.* Thus, using a system with multiple amplifiers, music can be piped to the cafeteria, an educational broadcast program to selected classrooms, and instructions to an outdoor gym class or team practice. An all-call feature also allows announcements to reach all speakers in the system simultaneously. The intercom system discussed previously can be interconnected with the sound system to allow conversation between classrooms and the console or other points. A small system can be contained in a compact desktop console (see Fig. 22.14). A large system frequently requires a console. A console is usually built in a desk arrangement, and it is advisable to provide sufficient space for it and for the person who operates it. Often an alcove of 30 to 50 ft² is reserved for it and a library of recordings.

Loudspeakers may be in flush or surface baffles, at the discretion of the designer. Gymnasium, cafeteria, and auditorium units are normally flush-mounted in the ceiling. For large areas such as these it is well to provide a volume control, enclosed in a recessed wall box with a locking cover. A common variation of this system uses separate subsystems for the cafeteria, auditorium, ball field, or other areas that utilize sound equipment frequently. These smaller systems have their own input, amplification, and control devices but utilize speakers in common with the central console. Normally, the console has an override feature that allows it to override local systems.

For a discussion of high-quality sound systems required for recital halls in music schools and the like, refer to Sections 27.16 to 27.18.

22.17 School Electronic Teaching Equipment

This area, like so many others that utilize computers, is growing and changing so rapidly that almost anything written on it is immediately out of date. We therefore simply outline the present and pro-

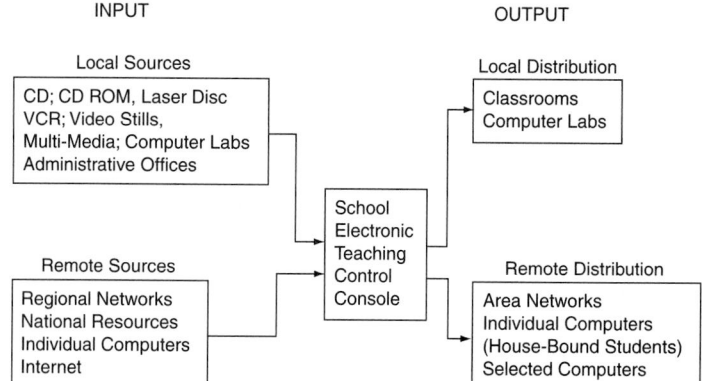

INPUT   OUTPUT

Fig. 22.16 *Block diagram of connections to an electronic teaching console, using present technology.*

jected near-future uses of electronic teaching equipment. Figure 22.16 shows in block diagram form the arrangements possible with current technology.

(a) Passive-Mode Usage. In this category we can include all recorded material, in whatever form, that is available to the student via some form of information retrieval technique. This would include printed, audio, and video material, in conventional and electronic library forms.

(b) Interactive Mode. Here students use a computer teaching terminal interactively to study at their own pace on a one-to-one basis, with the computer acting as tutor. Modern teaching programs sense the students' weak points (as do good teachers) and emphasize them in teaching. The building designer must be aware of the rapid developments in this field at all educational levels, including elementary school, in order to make adequate electric power, cable raceway, lighting, and HVAC provisions.

OFFICE BUILDING SYSTEMS

22.18 General

In this category we include systems found in all office-, professional-, and sales-type buildings. Such buildings house tenants with varying schedules and requirements. This must be considered in the design of the signal and communications systems for such buildings.

Although in many of the medium-size and large buildings control, alarm, and security func-

tions are combined in multiuse apparatus and consoles (discussed later in the chapter), the basic systems are essentially separate despite shared-use equipment. We discuss them individually to demonstrate function and equipment, and combined to show economies and modern practice.

22.19 Office Building Security Systems

Although automatic surveillance systems *are* applicable to office and mercantile occupancies, they are more frequently found in industrial facilities and are discussed under that heading. Office buildings normally utilize some type of manual watchman's tour system so that surveillance of unoccupied areas is conducted on some regular basis.

The simplest type is nonelectric and comprises a number of small cabinets, each containing a key, placed at intervals around the interior and exterior of the building. The watchman uses these keys to operate a special clock that he carries about, thus recording the exact time at which he "clocked in" at any specific location. Alternatively, the clock is wall-mounted and the guard carries only a key (Fig. 22.17). A computerized version of this system is available that simplifies station check-in, automatically records guard visit data, and provides hard-copy printout.

Guard tour systems are also available that permit constant supervision and are particularly effective where more than one person is on duty. Such systems show on a panel the location and progress of the watchman, by means of lights that glow when the device at each location is operated.

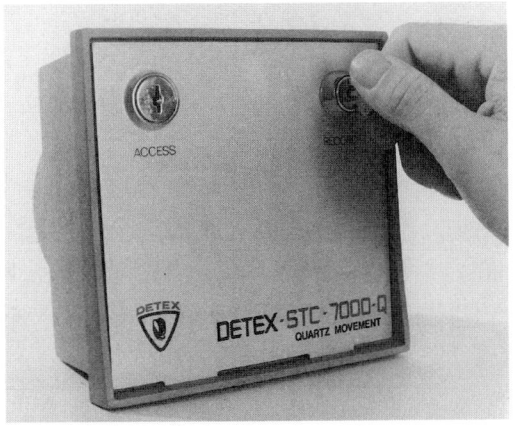

Fig. 22.17 *Watchman's tour station. Here, the clock and recording tape are inside. The station is operated by simply inserting a key as shown and giving it a quarter-turn. An alternative version of this system, commonly used, consists of a portable clock and tape, operated by a key in a wall box at each station. (Courtesy of Detex Corp.)*

Fig. 22.18 *This desktop multifunction intercom unit is usable in either handset or hands-free mode. In addition to standard intercom functions, each handset, which is connected to its central exchange with a four-conductor cable, can be used for such additional facilities as automatic redial, call conferencing, searching, all-call, group call, call transfer, call registration (store incoming call numbers), privacy mode, special meaning tones, and networking possibility. Also, the unit can be used for voice and "beeper" paging. Exchanges, which can handle up to 4000 stations, are microcomputer-controlled solid-state assemblies that vary in physical size according to the number of stations served. A typical large exchange might measure 60 in. W $\times$ 24 in. H $\times$ 20 in. D. (Courtesy of Dukane Corp., Communications Systems.)*

Because part of the effectiveness of these systems lies in the timing of the tour, a system can be arranged to sound an alarm if a particular station is not operated within a specific time period. Telephone jacks spaced at points along the guard's route allow the guard to communicate with the supervising office or other point without interrupting the scheduled tour. For protection of areas housing extremely valuable items or documents, an intrusion alarm system may be employed.

22.20 Office Building Communications Systems

This planning item is composed of four parts, which are frequently melded into a single network:

1. Intraoffice voice communication, or intercom (Fig. 22.18).
2. Interoffice and intraoffice data communication using telephone and communication cabling.
3. Outside-the-building communication via telephone company lines.
4. Paging function.

Today, in most locations, the user has the choice of purchasing or leasing as much of the system as is desired. In other words, the communication system, including cables, instruments, switching equipment, and so on, can be all privately owned, all supplied by the local telephone company, or almost any other arrangement desired. However, except for a small interoffice intercom, duplication is eliminated—that is, the same instruments and switching equipment are used for both intercom and outside connection.

What to buy, rent, or lease constitutes an economic decision because the required functions are satisfied by either private or telephone company equipment.

22.21 Office Building Communications Planning

Planning for the telephone and other communications equipment in an office building is of prime importance because of the large amounts and critical locations of required space. Therefore, it must be done simultaneously with other space planning. Exact requirements for office space are generally unknown at design time, and even if they were known, planning would have to account for changes in space usage as well as increased communications and data transmission services. For this reason, all planning is based on usable office area. Planning is essentially for spaces only, from incoming service to final instrument, because cabling and equipment are furnished and installed either by a private telephone equipment supplier or by the telephone company.

The planning information that follows is applicable to office buildings with average telephone and data transmission loads. Buildings whose tenants include brokerage houses, insurance companies, headquarters of multibranch operations, and other heavy telephone and data transmission users may need more space. However advances in technology, including the use of FO cables, act to reduce equipment size and space requirements. As a result, the best course of action is to design for current requirements with a reasonable estimate for the future based on advice from experts in the field.

Space is required for:

1. Service entrance room that houses incoming cable (network cable), terminated empty conduits for expansion and data cables, a network cable splice box, connection (network interface) cabinets and equipment required to interface the building equipment including telephones, modems, fax machines, and the like, to the main connection cabinets. This space is usually referred to as the *equipment room.*

2. Riser spaces (shafts) and riser closets. These are stacked vertically, and carry main cables.

3. Satellite closets where required.

4. Horizontal distribution between closets and devices, including conduit, boxes and cabinets, underfloor raceways, and over-the-ceiling systems.

A block diagram of a typical system in shown in Fig. 22.19.

(a) Service Entrance. At least two 4- or 5-in. conduits extend from the exterior service connection to the basement or ground floor service equipment room. One conduit will contain the service (network) cable; the other(s) is spare. Spare conduits should be terminated 12 in. AFF of the equipment room, in a threaded fitting, and sealed against entrance of dirt, moisture, or rodents. The service entrance cable terminates in a splice box or cabinet, from which, in large buildings, additional cables extend to one or more wall cabinets, each measuring about 3-ft wide, 6-ft high, and 2-ft deep. Each cabinet can serve a maximum of 30,000 to 100,000 ft^2 of tenant area, depending on the density of service required. To accommodate these cabinets from which risers extend, a clear wall space of 6 ft for every cabinet should be provided. This wall area should be covered with ¾-in. marine plywood for cabinet mounting. A 1-in. PVC conduit should extend between cabinets from a suitable electrical system ground point, with a ground wire sized according to telephone company requirements. The room should be dry, ventilated, well lighted for close work in wiring and color recognition, and supplied with at least two 20-A duplex convenience outlets on a separate circuit.

In the event that additional connection, switching, or communication equipment is required, the project electrical engineer should obtain the equipment size and electrical requirements, supply the necessary space requirements to the architect, and design the electrical facility. In large equipment rooms, circulation and egress must be considered, as well as emergency light and power.

(b) Riser Shafts. These accept the cables extending beyond the equipment room and carry them vertically through the building. Connection from the wall cabinets in the equipment room up to the riser shaft should preferably be in conduit.

The riser shafts provide means for the cables to extend vertically and terminate at each floor.

Fig. 22.19 *Riser diagram of the telecommunications system raceway and space arrangements for a typical commercial office building (see text for details).*

Ideally, the riser spaces comprise a series of vertically aligned closets connected by 3½- or 4-in. sleeves set in the floors and extending 1 in. above the floor. Sleeves are preferable to slots because they can easily be sealed and fireproofed; slots cannot. Install a minimum of four 4- in. conduit sleeves between closets, vertically, plus an additional 4-in. sleeve for every additional 50,000 ft^2 of office space to be served. It is preferable to separate communications closets from electrical power closets. Where multiple risers are used, shafts should be interconnected by several 2-in. conduits to allow for interconnection of systems. Riser cables are also known as *backbone* cables.

(c) Riser Closets. Here, cables from the riser system are interconnected to switching and power equipment, as well as to the cables that radiate from the closet to station locations throughout the floor area. These closets may also be called *zone closets* or *apparatus closets,* particularly if they function with an underfloor raceway system. Walls of the closet should be lined with

plywood at least ¾-in. thick to support the weight of switching and connection panels, power equipment, terminals, connecting blocks, and other hardware. Each riser (apparatus) closet must be provided with a switched ceiling light, and a separate 20-A, 120-V circuit with two duplex receptacles. A source of emergency power is desirable to avoid curtailment of telephone service during outages.

Riser closets should preferably have a minimum net area of 20 ft^2 (2 m^2) and a minimum clear wall of 5 ft for cabinet mounting. Each closet can supply telecommunication service for 8,000 to 10,000 ft^2 (740–930 m^2) of floor area. Utilization telephone equipment should not be more than 250 ft from a closet.

(d) Satellite Closets. Unlike riser (apparatus) closets, satellite closets do not contain switching and power equipment. Their primary use is to provide cable-connecting and terminating facilities in large complex facilities, where riser closet space is insufficient.

(e) Auxiliary Equipment Rooms. Where extensive cross connection is required or tenants utilize private switch-board (PBX) equipment, the required equipment is placed in a relatively small, auxiliary equipment room. These spaces, which are actually small closets or alcoves, should contain a 20- to 30-A, 2P, 120/208-V circuit, a 20-A, 120-V outlet, a grounding point, good lighting, and, obviously, sufficient equipment space. Space requirements vary with each installation and are obtainable from the equipment supplier. Ventilation is essential, as is absorptive acoustic material on the ceiling and at least one wall. Connections between these spaces and other communication equipment closets should be via floor or ceiling ducts or multiple 3½-in. conduits.

(f) Horizontal Distribution. Cabling from riser, satellite, apparatus, and equipment closets to individual outlets and instruments can be underfloor (Sections 15.25–15.29), in ceilings and plenum spaces (Section 15.31), under carpet (Section 15.30), or in surface raceways (Sections 15.23 and 22.7). Because of the large raceway volumes required, conduit is infrequently used. When using underfloor raceways, a header capacity of 2 in.^{2} per workstation is reasonable, based on one multiline telephone and one video display unit per station.

(g) FO Cables. In installations with very heavy data transmission loads, in those using video systems, and/or in applications for which the high-security, low-noise, and broad-bandwidth characteristics of FO cables are desirable, they are used in lieu of copper cabling. Space requirements and connection accessories are obtainable from manufacturers. Because cables are often supplied with factory-applied terminations, raceways must be of the large-capacity full-access type.

22.22   Office Building Control and Automation Systems

As the modern office building's mechanical and electrical systems increased in complexity, the need arose for a central point of supervision, con-

trol, and data collection from which to survey and control an entire building's functioning, plus increased automation at the utilization levels. From the supervisory location, water, air conditioning, heating, ventilating, electrical, and other systems could be controlled with great accuracy and convenience. Data on temperatures, pressures, flow, current, voltage, and all of the many parameters of mechanical systems could be made available instantly so that operational decisions could be made and automated procedures programmed. Also, all systems could be monitored here, and all alarms instantly acted on automatically.

Such *supervisory control centers* are now installed as a matter of course in office buildings. They are equipped with computers that process the data received to arrive at operational decisions intended to optimize system performance. As such, they result in a considerable savings in operating and maintenance costs. These systems in their most generalized form are referred to as *building automation systems* (BAS) and are discussed at length in Sections 22.26 to 22.31.

Architecturally, the BAS location requires good lighting and ventilation, extensive raceway space, but little area. Because systems are tailored to the building, no guidelines can be stated for space requirements.

INDUSTRIAL BUILDING STSTEMS

22.23   General

All industrial facilities ranging from the taxpayer loft, housing a small hand-assembly plant, to the immense steel manufacturing plant, require a variety of signal and communication equipment. Although, as in the case of commercial structures, a detailed analysis of the equipment is out of place here, it behooves the building designer to know generally the function, operation, and availability of this type of equipment.

Fire alarm systems for industrial buildings are discussed in Chapter 13. Audible alarms in industrial facilities for any of the building security systems must be selected with the usually high ambient noise level in mind. See Figs. 22.20 and 22.21 for recommendations.

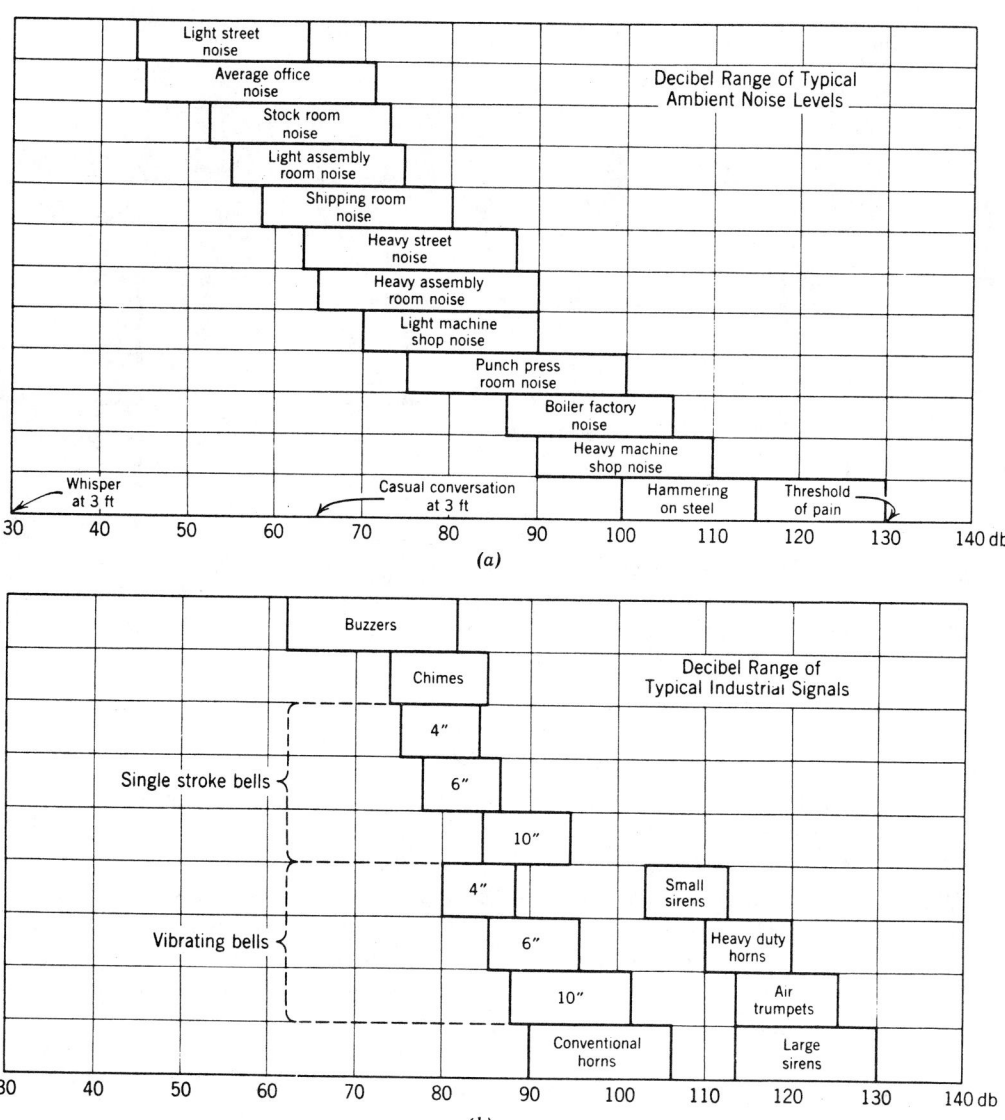

Fig. 22.20 (a) *Decibel range of typical industrial noise levels.* (b) *Decibel rating of typical industrial signals may be compared with ambient noise level* (a) *to facilitate selection.*

22.24 Industrial Building Personnel Access Control

The design and engineering of personnel access control systems is a specialty that has burgeoned since the late 1980s to the proportions of a independent profession. The large variety of identification technologies plus the control and supervisory functions of computers, has greatly enhanced the access control capabilities of even a relatively small system. As a result, we do not attempt to discuss system design, but address our discussion to the operation and application of the subsystems that comprise the whole.

The traditional access control question was always, "Who is permitted to pass through a specific portal?" The means by which the identification is made that answers this question varies with the importance of access control at that portal. Thus, entrance to a large, multipurpose space used by many people must be rapid and must avoid the

SIGNAL SYSTEMS

Heavy duty horn	Double gong bell	2-way horn	Megaphone horn	Underdome bell	Buzzer	Chime

Fig. 22.21 *Plan of a small industrial facilitiy, showing suggested locations of typical signals.*

delay engendered by a physical barrier. The means commonly used is one or more guards who visually and in relatively cursory fashion inspects a badge with an ID photo. The next level of supervised, barrier-free access control may be a barrier-free "turnstile" that requires presentation of an access card to an electronic proximity reader. One such unit is illustrated in Fig. 22.22. This type of barrier is limited to a maximum pass-through rate of about one person per second. Because these two control methods are barrier-free, the action taken on alarm is to set up physical barriers and sound the appropriate alarms.

When a physical access barrier is involved, the passage is slow, depending only partly on the identification process. Even with the most rapid electronic identification, a door or other portal closure must be physically released and operated, which can be very time-consuming, In the particular instance, shown in Fig. 22.23 the identification process is very time-consuming as well, because it involves human intervention.

Unattended physical barriers can be released by a multitude of differing identification technologies, depending on their level of importance. At the lower end of the security scale are barriers that require some knowledge and/or inspect an object presented by the user, such as a magnetic, bar-coded, or proximity-reader card (Fig. 22.24). Because locks can be controlled electronically and programmed remotely, they can be programmed to give access at certain times only, to specific groups or individual cards only, or to prevent access to specific cards (e.g., cards reported lost), or for any other reason. The element of time-controlled access is relatively new; it integrates extremely well with intrusion security systems because portals can be easily coordinated with changing work schedules, both general and specific. Thus, a person can be barred entrance to an area with which that person should not be concerned at that time of day (Fig. 22.25).

The most sophisticated identification methods in use today are biometric: examining a specific physical characteristic that is unique to each person. This is considered a higher level of entry control because it identifies a person, and not an object. Several of these methods are illustrated in Figs. 22.26 through 22.28.

In view of the plethora of available access-control technologies, access is being redefined beyond its original physical admittance meaning. Access control is today used to limit access to copy machines, fax machines, phone lines, and other office facilities frequently used by employees for other than purely business purposes.

Fig. 22.22 *Optical turnstile in an office lobby. These barrier-free access controls require presentation of a magnetic card to the proximity reader in the stanchion. Unauthorized access (or egress) causes an alarm that triggers audible and visual alarms and other measures as designed. This type of barrier-free access control is normally monitored by a guard positioned adjacent to the turnstile. See desk at left. (Photos courtesy of SMARTER Security Systems.)*

22.25 Industrial Building Sound and Paging Systems

The ubiquitous belt beepers and cellular phones are frequently ineffective in industrial facilities because of ambient noise levels. In such areas, voice paging at high audio levels (see Figs. 22.20 and 22.21), radio-frequency paging calls on earphones or speakers built into sound-reducing ear protection (see Fig. 28.53), and paging calls using paging lights are used. Similar methods are employed in noisy areas for announcements and special voice messages. In lower noise areas such as offices adjoining manufacturing spaces, sound-

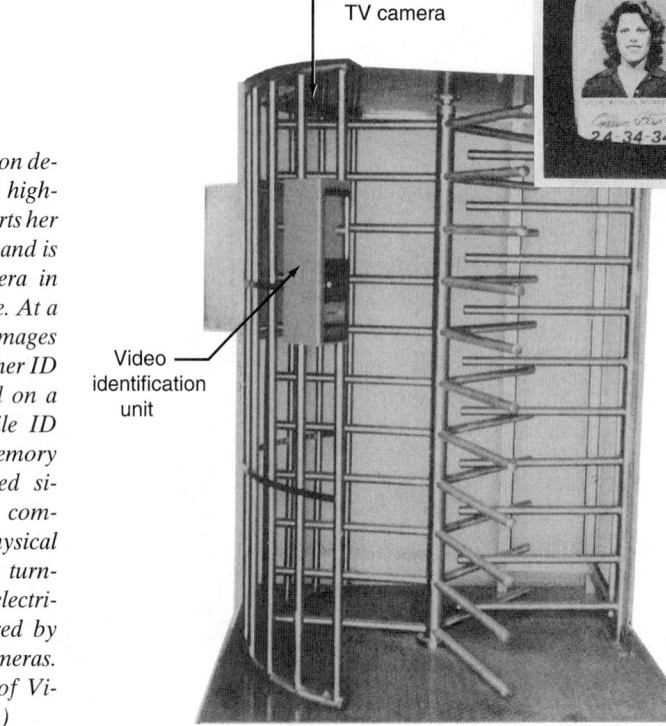

Overhead
TV camera

Fig. 22.23 A person desiring entry at this high-security portal inserts her identification card and is viewed by a camera in the entrance device. At a remote location, images of the subject and her ID card are displayed on a screen (inset) while ID data from a memory bank are displayed simultaneously for comparison. Actual physical entry through the turnstile is controlled electrically and monitored by an overhead cameras. (Photos courtesy of Visual Methods, Inc.)

Video
identification
unit

Fig. 22.24 This digital access lock has a programmable keypad and is arranged for remote or key override. Individuals or groups can be barred access by special local program codes. Multilevel user codes are available that act in the same fashion as multilevel master keys in cylinder locks. (Courtesy of NAPCO Security Systems.)

(b)

Fig. 22.25 *Stand-alone electronic locking system. (a) The arrangement and interconnection of equipment at each portal is shown. The card reader (swipe or insert types) are built into the door mechanism that also contains the electric strike mechanism.*

(b) Each lock (swipe style on the left and insert type on the right) is individually programmed using the laptop computer seen in the backgroud. Each lock can be programmed for up to 1500 users, each on a time-schedule basis, convering not only working hours, holidays, weekends, and the like, but also individual time schedules. All programming is performed locally, and the passage record of each lock can be downloaded easily. (Courtesy of Ilco-Unican Corp., Electronic Access Control Div.)

Fig. 22.26 *This foolproof identification procedure relies on the fact that no two persons have identical retinal vascular patterns. The illustrated device, usually placed at an unmanned controlled access portal, scans the entrant's retina, compares it to the previously recorded pattern, and then either permits or denies entry. (Photo courtesy of Eyedentify Inc.)*

reinforced conventional office-type systems are usually employed.

AUTOMATION

22.26   General

It is important that the distinction between remote control and automation be clearly understood. *Remote control* is simply a technique by which an action that can be performed manually locally is performed from a remote location by some intermediate means. The means may be low-voltage wiring (Section 16.32), a high-frequency wired signal (Section 16.34), a wireless signal such as the ubiquitous household TV/CD/VCR remote control, a similar IR lighting control unit (Fig. 21.4), or a radio-frequency wireless control (Section 21.12, Fig. 21.19). In each case, however, we have a single-stage action, manually initiated. When the signal initiation is nonmanual—that is, *automatic,* from a timing device (Section 16.16), a sensor such as a daylight or an occupancy sensor (Section 21.6) or a programmable device such as a microprocessor or computer (Fig. 21.5)—we have an *automated* system. The system can be elementary as just described, in which an automated signal controls a single action. It can also be very complex with many levels, yet treat only a single function, as an automated lighting system with sensors activating or overriding scene-presets

SIGNAL SYSTEMS

(a)

(b)

Fig. 22.27 *Iris scanning system.*
(a) *When used as an access control device at a portal, the entrant stands about 10 in. to 12 in. away from the scanner as the iris is scanned. The technology relies on the fact that each person's iris has unique and randomly formed features.*
(b) *After scanning, the image is digitized as shown and compared to a computer record. The entire procedure is very rapid; entry access or denial is automatic. (Courtesy of IriScan Inc.)*

Fig. 22.28 *This biometric identification device takes more than 90 hand measurements that are then converted into a unique identification "template" that is stored in the system computer. Once identification is made it can be used for access control. Access points can be programmed to permit entry to authorized personnel on any time/date schedule desired. (Courtesy of Recognition Systems Inc.)*

that activate dimmers and switches. Such a system is referred to as a *stand-alone* (automated) system. When several of these stand-alone systems are interconnected and supervised by a higher-level controller, then each is referred to as a *subsystem* and the overall system is referred to as an *integrated control system*. Such an integrated system, when applied to the individual systems in a building, is referred to as a *building automated system* (BAS).

22.27 Stand-Alone Lighting Control Systems

A straightforward stand-alone lighting control system is shown in Fig. 22.29. The initial program-

Fig. 22.29 Block diagram of a typical stand-alone automated lighting control system (see text for description of functioning). A connection to another stand-alone system (fire alarm), shown dotted, would exist at the overall building control level. (Photos courtesy of Lutron Electronics.)

SIGNAL SYSTEMS

ming is entered from the laptop computer (*a*) into the control panel (*b*) using system (manufacturer's) software. The control panel contains the required microprocessors, timers, and output control devices to establish lighting levels and settings in the entire structure via dimming and switching panels. In addition, through interfaces (connection devices), the control panel reacts to signals from daylight and occupancy sensors in a preprogrammed fashion to switch and dim certain lights. Windowshade control is two-way: they may close automatically when sunlight is detected in a specific area, causing lights to brighten, or they may open when direct sunlight is absent. In addition, a conference room arranged for audiovisual display may control lights and shades via a local lectern-mounted scene-preset controller, and so on. When another stand-alone system such as fire alarm (which itself may be part of an overall security/safety system) sends an emergency signal, the control panel reacts as programmed. It may turn on all lights to full capacity, or it may brighten exit paths and dim other lights, and so forth. This interconnection between two stand-alone systems is characteristic of a BAS.

22.28 Building Automation System (BAS)

The clear trend in modern construction, except for small or simple structures, is to use integrated system design plus centralized monitoring and control of building systems. The subsystems almost always included in a BAS are HVAC, energy management, and lighting control. Inclusion of security, life safety (fire alarm, fire control and surpression, plus emergency aspects of vertical transportation), material handling, and some aspects of communications depends on the specific needs of the building. This trend toward building automation, which previously had been economically justifiable only in large owner-user facilities, is today not only economically feasible but also very nearly an economic necessity because of high labor costs and the relatively low cost of computer and microprocessor controls. What the rapid advances in microelectronics and computer technology have done is to make possible and practical detailed multipoint monitoring and control in real

time, the result of which (assuming proper programming and BAS operation) is a highly cost-efficient and environmentally appropriate facility. Indeed, the advantages of such an arrangement are so great that retrofitting of existing buildings, which is obviously more cost intensive than new construction, has become a major industry.

Because the technique of building automation has grown so rapidly in the past few years, the terminology surrounding it and its equipment suffers from a lack of standardization and uniformity. Thus, terms such as *building automation, intelligent buildings, computerized building control, integrated computer control, supervisory data systems, integrated building control, facilities management systems,* and so on, are used almost interchangeably. We use the expression *building automation system* (BAS) in the discussion that follows. Inasmuch as every BAS is by definition programmed, any discussion necessarily involves basic control and computer terminology in addition to that which specifically applies to BAS. A short glossary should prove helpful in this regard.

22.29 Glossary of Computer and Control Terminology

Address. A number used to identify an I/O channel or a module.

Algorithm. Detailed description, often in flow-diagram form, of the method for solving a problem. The outline used for computer or microprocessor programming.

Alphanumeric. Characters, which are alphabet letters, symbols, or numbers.

Analog data. Numerical information about physical variables given in representative, continuously variable form (e.g., dial meters, temperature gauges, etc.).

Artificial intelligence. Branch of computer science concerned with computer programs that "learn" on the basis of experience and modify their own behavior accordingly.

ASCII. A 7-bit standard alphanumeric code established by ANSI. The acronym stands for American Standard Code for Information Interchange.

Baud. Rate at which information is transmitted in terms of bits per second.

Binary. A numeric system using only 1 and 0, normally used to represent circuit conditions ON/OFF.

Bit. The number 1 or 0. Contraction of binary digit.

Building automation systems (BAS). Building-wide computer-based system that monitors and controls selected aspects of building operations. The BAS is understood to include all of the equipment involved and their interconnections.

Bus. An electrical data channel.

Byte. A string of bits; 8 bits unless otherwise specified.

Central processing unit (CPU). That portion of a computer, microprocessor, or programmable controller that processes data and executes programs. The "brains" of the computer.

Converter. General term for a device that converts data or energy from one form to another. Data converters are analog to digital (ADC) or digital to analog (DAC). Energy-form converters (e.g., temperature or illuminance to voltage) are properly called *transducers*.

Digital data. Data in digital form, usually binary.

Direct digital control (DDC). Control arrangement that processes input data according to a programmed algorithm and yields a control function (e.g., the action of a local microprocessor controller or a computer control function).

Duplex, duplex transmission. Simultaneous bi-directional data transmission.

EIA. Electronic Industries Association. Establishes data communication standards (e.g., RS-232-C).

Energy management system (EMS). Computer-based system for monitoring and controlling facility energy use. Frequently part of a BAS.

Hardware. All physical equipment associated with a computer system, as opposed to software.

I/O. Input/output.

LCD. Acronym for *Liquid Crystal Display*.

LED. Acronym for *Light Emitting Diode*.

Load control, load management. Arrangement whereby energy-using devices are monitored and controlled. Part of EMS.

Local area network (LAN). Signal network providing intercomputer communications.

Logic. Factors involved in the design of a CPU.

Microprocessor. A CPU, usually in the form of a single integrated circuit (IC), preprogrammed to perform specific functions in response to specific input.

Modem. Device that connects a data-originating device to a communication line; from "modulator–demodulator."

Multiplex(er) (MX, MUX). Use of a single communications line for simultaneous transmission of multiple signals.

Nonvolatile memory. Memory that remains intact when electric power is shut off.

Operating system. Software, that controls input to, output from, and operation of the CPU.

Parity. Method of verifying the accuracy of recorded data.

Point. General term to describe either a sensor/monitor location or a specific control signal in a BAS.

Port. An I/O connection point.

Programmable controller. Electronic device containing a microprocessor, a programming means or device, and I/O interfaces (see Section 16.16).

Protocol. Conventions governing format and timing of signaling between communication devices (e.g., between computers in area networks).

RAM, ROM, PROM, EPROM. Random access memory, read-only memory, programmable ROM, erasable PROM.

Real time. Arrangement in which a computer receives and processes data without introduction of any deliberate time delay (i.e., effectively an immediate response).

RS-232. Standard for interconnecting digital communications devices, such as computers, printers, and monitors.

Software. Computer program; as opposed to hardware.

Stand-alone. Descriptive term for a control device or system that can perform independently and usually can also interconnect as a subsystem (controller) in a larger system.

UPS. Uninterrupted power supply.

Volatile memory. Memory that loses its information on loss of power.

Word. A group of bits that is treated as a single unit.

22.30   BAS System Arrangement

The BAS described in this section is that in use today (1999). A block diagram is given in Fig. 22.30. Notice that information is transferred *vertically* between levels within a stand-alone system, whereas *horizontal* intersystem information transfer occurs only at the highest level (level 1), because that is the level at which systems are inter-

connected. (In practice, some interconnection also occurs at level 2.) This means that a smoke alarm from level 5 of the fire alarm system travels to level 1, where the central controller tells the other stand-alone systems to take the required action. These systems in turn then signal downwards, essentially to level 4, where appropriate commands are given to controls for supply fans, elevators, security, and so on. The stand-alone systems are essentially independent, and in practice cannot easily be interconnected because of the absence of standardization in wiring and computer protocols. The ostensibly simple problem of connector standardization has yet to be resolved, as anyone who has worked hands-on in the communication field can attest, although the situation is much improved over the near chaos that once reigned. Fully or even partially integrated system buildings are at this writing in the planning stages only. These "intelligent buildings" are discussed in some detail in the next section.

Returning to the block diagram in Fig. 22.30 of today's BAS: any specific BAS may contain more or fewer monitors, sensors, dependent and stand-alone local controllers, and interconnected

Fig. 22.30 *Overall block diagram of a building automation system (BAS). In an actual system, the various levels and devices can be combined and telescoped by use of equipment that combines several level functions in a single housing.*

systems, but the overall arrangement remains the same. Actually, depending on the equipment manufacturer and the system complexity, some of the functions and levels can be telescoped or grouped. Our discussion deals with the general case. To concretize the discussion, we use fire control and evacuation system terminology as a relatively simple and straightforward example of overall system design, function, and interconnections.

Level 5 in the diagram is the lowest system level. It contains automatic operation passive sensors such as flame, smoke, ionization, temperature, and water-flow detectors; physical condition indicators such as smoke and fire-door position sensors; and manual fire alarm pull stations. All yield a digital signal (ON–OFF),which is transmitted to level 3. As level 5 is purely a sensor/monitor level, it does not receive any control or operating signals.

Level 4 contains sensor/operators or actuators (i.e., devices that both transmit a condition signal and receive an operation signal). In our example, this level could contain smoke and fire-door actuators, smoke vent controls, sprinkler system valves, and fan control (override) switches. This last item can either reside at this level or can be controlled through the HVAC system via level 1, 2, or 3. Control at this level would be hard-wiring of a single device or block.

Level 3 could represent a floor, zone, or area controller, with either relay circuitry or microprocessor control (programmable controller), depending on system complexity. This level receives information from levels 4 and 5, processes it, returns control information to level 4, operates specific local audible and visible alarms directly, and sends alarm and condition information to the level 2 building controller and to the master controller (computer) at level 1. It also receives and processes signals from levels 1 and 2, containing such information as alarm reset, alarm device operation for building emergency evacuation, and any reprogram or reset function.

Level 2 contains the building's central fire and evacuation system (life safety system) controls. This level can contain some of the level 1 programming logic relating to firemen's control of elevator operation and HVAC fan functioning, or it can simply act as an I/O device to level 1 for these

functions, again depending on system design and complexity. In some facilities, these life safety system functions may be controlled from a separate console that is interconnected with the BAS. In a multibuilding facility, this level contains the individual building fire system central control, interconnections to other building fire system controllers, as well as external connections for city alarm and external supervisory services.

Level 1 is the central computer console. This level receives status reports on the individual devices from level 4 or 5; on areas, zones, and subsystems from level 3; and on the entire fire/evacuation system from level 2. Hard-copy printout of all alarms and periodic status reports are generally made at this level, although they can also be accomplished at level 2. Visual display terminals (VDTs) at level 1 can be used to view the status of any area and all the devices in the fire/evacuation system (Fig. 22.31 or, for any other system, Figs.

Fig. 22.31 *A workstation screen shows a selected building area, its zoning (Johnson Plaza East, third floor, zones 1–6), all of the area's fire/evacuation system devices and their status, and the control system's present functioning. The illustrated screen shown an active alarm condition with messages: Elevators being recalled to lobby; Floor is now pressurized; Audio evacuation messages active; CALL FIRE DEPARTMENT. (Courtesy of Johnson Controls.)*

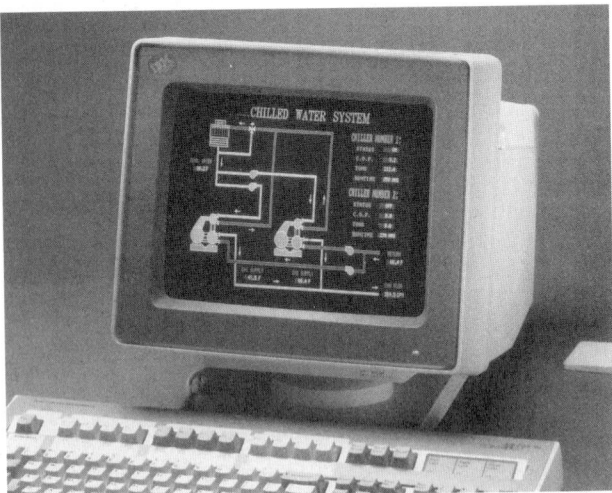

Fig. 22.32 *Workstation screen shows any section of any building system and the status of all components. Here we see chiller status (ON–OFF, COP, tons, run time), water temperature and flow in supply and return lines, and condenser water temperature. (Courtesy of Johnson Controls.)*

22.32 and 22.33) graphically or tabularly (Fig. 22.34). Pictorial graphics for nontechnical end users are also possible (Fig. 22.35). At level 1, information is transferred to HVAC, elevator, security, and lighting systems in preprogrammed alarm modes, with manual override possible. A fire emergency scenario (Fig. 22.31) might interconnect with the HVAC, elevator, security, water, electrical, and lighting systems to activate exhaust and pressurizing fans and deactivate supply fans, place elevators in fire-evacuation mode, override access barriers to permit unhindered facility evacuation, connect fire pumps to emergency power feeders, disconnect high-voltage feeders, and activate emergency lighting, all via preprogrammed intersystem functions. The system operator at level 1 could then view any part of any system and reactivate or override preprogrammed functions as desired. With this system architecture, this is the level at which human supervisory intervention is planned. At other levels it only occurs if trouble arises, periodic maintenance is scheduled, or an alteration is required.

Returning to Fig. 22.30, we can consider a much more complex, nonemergency system: energy. Here, level 5 senses and continuously monitors inside and outside temperature and humidity, duct air temperature, space pressurization, airflow, and the operating and load status of all fans, compressors, and refrigeration devices. Level 4 contains monitors and DDC controls for valves, boilers, chillers, fans, humidifiers, dampers, cooling towers, and compressors (Fig. 22.36). Levels 2 and 3 contain electric load managers for duty cycling and demand limiting, HVAC controls for economizers, run-time optimizers, supply air and supply water reset, chilled and condenser water reset, enthalpy switching, load sequencing, and the like. These functions are performed either by a stand-alone EMS controller (see Fig. 16.15) or by individual demand, HVAC, and chiller load managers controlled by level 1 logic. Finally, level 1 contains the programming for all of the foregoing functions, any of which can be displayed at a level 1 workstation terminal at any time. The system operator can then reprogram any function

Fig. 22.33 *Any workstation screen can display any portion of any system in the BAS. Here an air handling unit (AHU) is displayed with data on air flow, air temperatures, and RH in each duct section. In addition, cumulative fan run time, economizer status, and area occupancy status is displayed. (Courtesy of Johnson Controls.)*

Fig. 22.34 *Global workstation screen is arranged to display four separate data screens. Clockwise from upper left:*
1. *Comfort chart showing a plot of room psychrometric conditions on a graph of room temperature vs relative humidity.*
2. *Graphic screen showing the layout of an air handling unit with relevant data.*
3. *Status chart of items in AHU-1.*
4. *Temperature readout of return air for AHU-1.*
(Courtesy of Johnson Controls.)

Fig. 22.35 *Workstation screen showing a floor plan, indicating room status and temperature by color and in general descriptive terms—hot, cold, normal occupied, and normal unoccupied. A similar floor plan screen can be displayed for status of fire alarm and security systems and the lighting condition. (Courtesy of Johnson Controls.)*

Fig. 22.36 *Digital controller programmable to a variety of digital control applications, such as chiller control, boiler control, or area environmental control. (Courtesy of Johnson Controls.)*

after analyzing system performance (Fig. 22.31) at the level 1 computer and obtain immediate visual and hard-copy readout of the changes.

22.31 Intelligent Buildings

Although the term *intelligent building* seems to attribute a human characteristic to an inanimate object, the term's actual meaning, at least as defined by the Intelligent Buildings Institute (*Guideline,* 1987), is, "a building which provides a productive and cost-effective environment through optimization of its four basic elements—structure, systems, services, and management—and the interrelationships between them. . . . Optimal building intelligence is the matching of solutions to occupant need." In the framework of such a definition, the intelligence of a building depends on the ability of the design team to understand the occupants' present needs and, based on study and interaction with the client, make an educated forecast of the occupants' future needs. Because a building's useful life depends in large measure on the accuracy of this forecast, building systems are being designed with "open architecture" that lends itself easily and economically to changes, both minor and major. The term *open architecture* is generally understood to mean, in terms of BAS, a system design, that at the very least, can do the following:

· Utilize equipment from any manufacturer whose equipment meets industry-wide specifications.

· Readily reprogram equipment to serve a variety of functions.

· Supply required system information at any I/O port.

· Permit access to databases at any workstation.

· Utilizes networks for information transfer at one or more levels.

It is apparent that these requirements demand open protocols, standardized transmission media and connectors, and virtually unlimited cabling space. (This last requirement is easily fulfilled by using any of the underfloor or cellular floor raceway systems described in Chapter 15.) At this writing, proprietary interests and the prevalence of stand-alone system design is retarding the move to open architecture in the United States, whereas it is being adopted rapidly in Asian mercantile centers.

Referring to Fig. 22.30, which shows the prevalent present BAS architecture, an open system would eliminate at least two of the levels, reducing the system to a three-level structure, that *might* be structured as follows:

1. *Level 1,* the uppermost level, contains workstations connected to a main data bus. This level, which might be called the *management* or *information level,* has direct access to databases and information networks. It serves to establish operational programming and procedures. All points are accessible from any workstation. A typical workstation might appear as shown in Fig. 22.34.

2. *Level 2,* the intermediate level, might be called the *supervisory* or *network level.* Network controllers serve the dual function of constituting a network platform supervising stand-alone controllers from various systems at the third level, and transferring information to other network controllers supervising other areas of the building via the LAN. In addition, the network controllers exchange information with one or more worksta-

Fig. 22.37 *Network control unit acts as a LAN platform, performing supervisory functions for a portion of the building. It connects to operator workstation(s) at the upper management level to other network controllers at its level via the LAN and to specific stand-alone controllers at the lower, control level. It can also be programmed to act as a high-quality industrial process controller. (Courtesy of Johnson Controls.)*

Fig. 22.38 Fire control unit. This device controls "intelligent" addressable analog devices in an "intelligent" fire control system (see Section 13.14 for a detailed description of this type of system). (Courtesy of Johnson Controls)

tions. A network controller can also act as a process controller. A typical unit might appear as in Fig. 22.37.

3. *Level 3,* the control level, contains dedicated controllers in stand-alone systems. The units are highly flexible and can be locally reprogrammed to control the same function in a wide variety of different systems. Two such controllers are shown in Figs. 22.38 and 22.39.

the systems achieve a level of user–system integration not yet available in a complex BAS, in that the user's telephone can control any portion of the system directly. This is possible because only one level or, at most, two levels of control are involved i.e., the commands are ON/OFF/SET, where the latter refers to a particular electronic contact. This contact in turn establishes a new preprogrammed condition such as night-setback,

22.32 Intelligent Residences

The operational principles of remote control and automation discussed previously in this chapter and also in Chapters 16 and 21 have found ready application in residential automation and control equipment. Residences so equipped and variously referred to as intelligent, smart, or automated homes appear increasingly in the real estate market news. The degree of automation in these houses varies considerably, the simplest being no more than a low-voltage control system of the type described in Section 16.32 with time-based programming using a relatively simple microprocessor. More complex systems use a touch-screen computer for control or a portable plug-in programmable microprocessor to set various portions of the system. The best (and most expensive) of

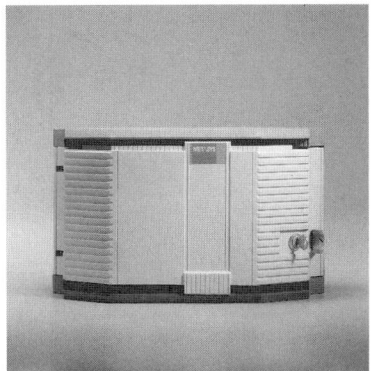

Fig. 22.39 Unitary controller is designed to control single zone HVAC subsystems such as heat pumps, fan coils, and the like. It is preset from the network level but can also be locally monitored and adjusted to satisfy local requirements. These adjustments are made with a portable plug-in control unit. (Courtesy of Johnson Controls.)

lighting preset for extended absence, reset upper level temperatures to daytime levels, and so forth. These systems also allow relatively straightforward reprogramming to meet the occupant's specific needs.

Because all new residential construction includes a degree of automation initially that will surely increase in the future, all residences, single and multiple, should be provided with the cabling and/or raceway system that is the backbone of any automation system. At this writing, the cost of plastic optical cable (POC) is approximately competitive with copper cabling, in large quantities. The decision as to what cabling and distribution system to install is technical and economic. At the very least, every room in a residence should be equipped with audio/video/control point outlets. In addition, a study, recreation room, and at least one bedroom should be wired as an incipient fully equipped home office. This applies, of course, not only to anticipated automation wiring, but also to power outlets and lighting.

BIBLIOGRAPHY

Heating, Piping, Air Conditioning (1997). Jan., Supplement, pp. 23–31, 37–48, 59–63.

IEEE, *Communications* (1993), Oct. Vol. 31, No. 10, pp. 22–61.

Grandguillaume, A., et al. (1996). " The Future of Intelligent Buildings." *ARUP Journal 4,* pp. 8–10.

REFERENCES

Fire Protection Handbook, 18th ed. (1997).

National Fire Protection Association Codes (1999). *NFPA,* Boston.

Life Safety Code, *NFPA* 101 (1997).

PART VIII
TRANSPORTATION

No facilities in a multistory structure are taken for granted more consistently than elevators and escalators, which are relied on to move people quickly and safely under all conditions, including emergencies. This movement should be rapid, trouble-free, and economical. Furthermore, because vertical transportation accounts for 10% to 15% of a construction budget, somewhat less of the building area and somewhat more of the operating cost, and is a determining factor in building shape, core layout, and lobby design, we see that elevator and escalator selection is a major task for the architectural designer.

Computers and power electronics have changed the face of elevator engineering. Electromechanical relays have given way to solid-state switches, circuit boards, and microprocessors; lobby control stations and human elevator "starters" have been replaced by sophisticated self-learning computerized controllers; transport time has been reduced and comfort has been improved by an order of magnitude. These great improvements, however, have not significantly changed the fundamentals of "elevatoring" i.e., the selection of one or more elevators that give the required passenger handling capacity and interval with minimum trip time and cost. We therefore, discuss in considerable detail the principles governing elevator motion and control which then lead to an understanding of elevator selection to meet building requirements.

Chapter 23 Vertical Transportation: Passenger Elevators introduces the subject with a description of components including traction equipment, cars, safety devices, and systems of control and supervision. Special topics, such as codes and standards and requirements for the handicapped, are also covered. The discussion then turns to actual selection of units. Criteria are established for elevator performance, including interval, waiting time, handling capacity, average trip time, and round-trip time. Extensive car performance data (curves) are presented that can be used with design criteria to select the size, number, and speed of elevators that fulfill the requirements. The discussion continues with principles of elevator zoning. The chapter continues with an analysis of spatial requirements for elevators, including shafts, lobbies, and core arrangement and concludes with special considerations such as energy use, safety, security and noise. Throughout the chapter, application of principles and data is demonstrated through the use of examples.

Chapter 24, Vertical Transportation: Special Topics, addresses important passenger elevator considerations other than elevator selection. These include power and energy requirements, with emphasis on energy conservation, fire–emergency operation, emergency power requirements, and passenger security. The next part of the chapter is devoted to special cases: freight elevators, hydraulic elevators, nonvertical designs, residential elevators, and chair lifts, with discussion of equipment, capacities, selection, economy, and comparison (where applicable) to conventional design. Cognizance is taken of environmental considerations that discourage the use of plunger hydraulic elevators. This has led to increased use of "hole-less" hydraulic elevator designs, including the very popular roped-hydraulic design. The material also covers interesting and novel designs including double-deck cars and observation cars.

The chapter concludes with a section on material-handling devices: dumbwaiters, horizontal and vertical conveyors, pneumatic systems, container delivery systems, and automated self-propelled vehicles. This section answers the need to plan for material handling in commercial and institutional buildings, rather than treating the subject as an afterthought, with resultant inefficiency, overloading of freight facilities, and misappropriation of passenger elevators.

Chapter 25, Moving Stairways and Walks, treats these subjects with a discussion similar to the coverage of passenger elevators. Descriptions of equipment components, capacities, sizes, speeds, and methods of selection are followed by special topics including lighting, fire protection, power and energy requirements, and budget estimating. Information on inclined walks (ramps) is also presented.

23

VERTICAL TRANSPORTATION: PASSENGER ELEVATORS

GENERAL INFORMATION

23.1 Introduction

Among the many decisions that must be reached by the designer of a multistory building, probably none is more important than the selection of the vertical transportation equipment—that is, the passenger, service, and freight elevators and the escalators. Not only do these items represent a major building expense, being in the case of a 25-story office building as much as 10% of the construction cost, but the quality of elevator service is also an important factor in a tenant's choice of space in competing buildings.

Although the final decision as to the type of equipment rests with the architect, the factors affecting it are so numerous that it behooves the building designer to consult with an elevator expert. This service is available from consultants in the field and from the major elevator and escalator manufacturers. It is the function of this chapter to familiarize the architect and engineer with the nature and application of vertical transportation equipment, to enable them to make preliminary design decisions and interact efficiently with consultants.

23.2 Passenger Elevators

This chapter is concerned primarily with the general-purpose traction elevator. Hydraulic elevators are covered in Chapter 24.

Ideal performance of an elevator installation provides minimum waiting time for a car at any floor level; comfortable acceleration; rapid transportation; smooth and rapid braking; accurate automatic leveling at landings; and rapid loading and unloading at all stops. The system must also provide quick and quiet power operation of doors; good floor and travel direction indication, both in the cars and at landings; easily operated car and landing call buttons (or other devices); smooth, quiet, and safe operation of all mechanical equipment for all conditions of loading; comfortable lighting; reliable emergency and security equipment; and generally pleasant car atmosphere.

In addition to passenger-oriented service considerations, elevators have architectural aspects as well. Cars and shaftway doors must be treated in a manner consonant with the architectural unity of the building. More important, however, is the fact that shaftways are major space elements whose integration into the building is a prime factor in composition, as is the design of the elevator lobby.

23.3 Codes and Standards

Perhaps more than any other item of construction, elevators are governed by strict installation codes. The "bible" of the industry in the United States is the American National Standards Institute (ANSI/ASME) Code A17.1, *Safety Code for Elevators, Dumbwaiters, Escalators and Moving Walks,* an up-to-date copy of which should be an integral part of every architect's and engineer's working library. The code has legal force in most parts of the United States. In addition, ANSI/ASME Standard A17.3-1996 covers existing elevators and escalators and standard A17.4-1991 covers emergency evacuation of passengers from elevators. As with other items, some states and municipalities have their own elevator codes (Massachusetts, Wisconsin, Pennsylvania, New York City, Seattle, and Boston, among others) that are generally based on, and more stringent than, the ANSI code.

In addition to the elevator code, other construction and installation codes have an influence on elevator work. Thus, NFPA 101, *Life Safety Code,* states certain fire safety requirements, NFPA 70 (the National Electric Code) governs some of the electrical aspects of elevator construction, and state and local laws add a multitude of requirements and restrictions bearing on fire safety, emergency power, security regulations, and special accommodations for handicapped persons. Provisions for the handicapped are covered by ANSI A117.1 (Barrier-free), a special industry code, the requirement of Americans with Disabilities Act (ADA), and in most locations by local law. As with most large industries, the elevator industry is self-regulating and standardized. The National Elevator Industry, Inc. (NEII) publishes standard elevator layouts for traction and hydraulic installations. Elevator consultants and elevator company representatives are normally knowledgeable as to all the codes and standards in force, but this does not relieve the architect–engineer of legal responsibility for the installation. Therefore, we strongly recommend that in the preliminary planning stage all pertinent regulations concerning vertical transportation be acquired and studied.

ELEVATOR EQUIPMENT

23.4   Principal Components

The car, cables, elevator machine, control equipment, counterweights, hoistway, rails, penthouse, and pit make up the principal parts of a traction elevator installation. An idea of the functioning and orientation of these items of equipment can be obtained from Fig. 23.1.

The cars are the only items with which the average passenger is familiar. Indeed, some of the building's prestige depends on proper design of the car. Essentially, the cab is a cage of some fire-resistant material supported on a structural frame, to the top member of which the lifting cables are fastened. By means of guide shoes on the side members, the car is guided in its vertical travel in the shaft. The car is provided with safety doors, operating-control equipment, floor-level indicators, illumination, emergency exits, and ventilation. It is designed for long life, quiet operation, and low maintenance.

The cables (ropes) that are connected to the cross-head (top beam of the elevator) and carry the weight of the car and its live load are made of groups of steel wires especially designed for this application. Four to eight cables, depending on car speed and capacity, are placed in parallel. Although multiple ropes are used primarily to increase traction area on the drive sheaves, they also increase the elevator safety factor inasmuch as each rope is normally capable of supporting the entire load. The minimum factor of safety varies from 7.6 to 12 for passenger elevators and 6.6 to 11 for freight elevators. The cables from the top of the car pass over a motor-driven cylindrical sheave at the traction machine (grooved for the cables) and then downward to the counterweight.

The counterweight is made up of cut steel plates stacked in a frame attached to the opposite ends of the cables to which the car is fastened. It is guided in its travel up and down the shaft by two guide rails typically installed on the back wall off the shaft. Its weight equals that of the empty car plus 40% of the rated live load. It serves several purposes: to provide adequate traction at the sheave for car lifting, reduce the size of the traction machine, and reduce power demand and energy cost. (These advantages come at the price of strengthening the overhead machine room floor, which must carry the additional structural load of the counterweight.)

Approximately 75% of the energy expended in lifting a car is returned to the system by regeneration when the car is lowered. Regeneration is the process in which the traction motor becomes a *generator* when the car is lowered and feeds power back into the electrical system. The "lost" energy appears as heat, primarily in the machine room. See Section 23.40 for a discussion of system energy requirements.

To compensate for hoist rope weight, which becomes an important factor in high-rise elevators, cables are attached to the bottom of the car and the counterweight, thus equalizing loads regardless of the cab position. These cables can be seen in Fig. 23.1.

The elevator machine turns the sheave and lifts or lowers the car. It consists of a heavy structural

GEARED

Fig. 23.1 *Components of a geared installation with solid-state control and motor drive. (Courtesy of Otis Elevator Co.)*

frame on which are mounted the sheave and driving motor, the gears (if any), the brakes, the magnetic safety brake, and certain other auxiliaries. In many existing installations the elevator driving (traction) motor receives its energy from a separate motor–generator (m-g) set, which is in operation during the period that the particular elevator is available for handling traffic. This m-g set is properly considered a part of the elevator machine, although it may be located some distance from it. In modern installations, solid-state power and control equipment replaces the m-g set, as discussed in Section 23.17. A governor, which limits the car to safe speeds, is mounted on or near the elevator machine.

The control equipment is usually divided into three groups:

1. *Drive (motion) control* is concerned with the velocity, acceleration, position determination, and leveling of the car.
2. *Operating control* covers car door operation and functioning of car signals, including floor call buttons and all indicating devices.
3. *Supervisory control* is concerned with group operation of multiple-car installations.

The actual physical devices in these control systems were electromechanical in the past but are solid-state in modern installations. The indicating and control devices that are seen and used by the elevator user, including car and hallway buttons, lanterns, and audible devices, are all coordinated into the overall operational control scheme, which produces rapid, safe, and comfortable vertical transportation.

The *shaft,* or *hoistway,* is the vertical passageway for the car and counterweights. On its sidewalls are the car guide rails and certain mechanical and electrical auxiliaries of the control apparatus. At the bottom of the shaft are the car and counterweight buffers. At the top is the structural platform on which the elevator machine rests. The elevator machine room (which may occupy one or two levels) is usually directly above the hoistway. It contains the traction machine and the m-g set or solid-state control that supplies energy to the elevator machine and control equipment. All machinery and control equipment are designed for quiet, vibration-free operation.

23.5 Gearless Traction Machines

A gearless traction machine consists of a dc or ac motor, the shaft of which is directly connected to a brake wheel and driving sheave. The elevator hoist ropes are placed around this sheave. The absence of gears means that the motor must run at the same relatively low speed as the driving sheave. As it is not economically practical to build motors for operation at very low speeds, a gearless machine is utilized for medium and high-speed elevators—that is, 500 fpm and above. The motors range from 20 to 400 hp.

Gearless machines are generally utilized for passenger service, with car capacities of 2000 to 4000 lb, although specials of up to 10,000 lb, such as at the World Trade Center, have been built. Below 500 fpm, geared machines are used. At this writing, maximum car speeds are 2000 fpm. Faster drives have already been developed and will undoubtedly be an important factor in the development of ever taller, practical, workable buildings.

A gearless traction machine is considered superior to a geared machine because it is more efficient, quieter in operation, requires less maintenance, and has longer life. The decision as to whether these advantages are worth the additional cost involved is made only after a careful analysis. Generally, a gearless machine is chosen where rise is more than 250 ft and very smooth, high-speed operation is desired. In the intermediate range of rise and speeds (i.e., 150–250 ft height and 400–500 fpm), excellent equipment, both geared and gearless, is available.

Because virtually every major elevator company operates internationally and outside the United States elevator size and speed are specified in kilograms and meters per second, it is useful to have approximate conversion factors in mind. As the accurate equivalant for 1 kg is 2.2 lb and for 1 m/s is 196.86 fpm, the factors of 2 and 200 respectively give approximate American equivalents (Table 23.1).

23.6 Geared Traction Machines

A geared traction machine has a worm and gear interposed between the driving motor and the hoisting sheave (Fig. 23.2). The driving motor can

TABLE 23.1 **Elevator Capacity and Speed; Approximate Metric Equivalents**

Weight		Speed	
Standard Metric (kg)	Approximate U.S. (lb)	Standard Metric (m/s)	Approximate U.S. (fpm)
1000	2000	1.0	200
1250	2500–3000	1.6	300
1600	3500	2.0	400
2000	4000	2.5	500
		3.15	600
		4.0	800
		5.0	1000

therefore be a smaller, cheaper, high-speed unit rather than the large, low-speed unit that a gearless installation would require.

The motor used in a geared installation, as in gearless, depends on the type of drive system and may be either dc or ac. Characteristics of the various drive systems and their application are discussed in detail in Sections 23.12 to 23.15. Geared machines are used for car speeds up to 450 fpm and a maximum rise of about 300 ft. With an appropriate drive and control system, a geared traction machine can give almost the same high-quality, accurate, smooth ride as is available from a gearless installation.

Fig. 23.2 *Cutaway of a typical dc geared traction machine. Note the grooves for multiple ropes in the traction sheave. (Photo courtesy of Schindler Elevator Corp.)*

Labels on figure: Traction sheave (ropes not shown); Automatic position brake solenoid spring and arm; D–c motor; Worm and gear; Heavy steel unit frame

23.7 Arrangement of Elevator Machines, Sheaves, and Ropes

The simplest method of arranging vertical travel of a car is to pass a rope over a sheave and counterbalance the weight of the car by a counterweight. Then, by rotating the sheave, the car moves up or down and requires very little energy to do it. This is essentially the scheme that is used on a majority of high-speed passenger elevators, as illustrated in Fig. 23.3(*a*).

When the four or more supporting ropes merely pass over the sheave *T* and connect directly to the counterweights, the lifting power is exerted by the sheave through the traction of the ropes in the parallel grooves on the sheave. This system is referred to as the *single-wrap traction elevator machine*. The function of sheave *S* is merely that of a guide pulley; it is called the *deflector sheave*.

The arrangement shown in Fig. 23.3(*b*) is called *double-wrap 1:1 roping*. It provides greater traction than the single-wrap arrangement and is used in many automatic high-speed installations.

A 1:1 roping arrangement [Fig. 23.3(*a, b*, and *d*)] gives no mechanical advantage. The 2:1 roping [Fig. 23.3(*c*)] has a mechanical advantage of 2, which permits use of a high-speed, low-power, and, therefore, lower-cost traction machine. This arrangement is used for a wide variety of installations varying from medium-speed (500–700 fpm) gearless passenger elevators to low-speed, heavy-duty freight units.

In types *a, b,* and *c* in Fig. 23.3, the elevator machines are located in a machine room pent-

Fig. 23.3 *Elevator roping and sheave arrangements. In* (b) *and* (c), *the rope passes over traction sheave* T *and sheave* S, *doubles back over* T *and then extends past* S *to the counterweight* CW. *This double-wrap arrangement provides additional traction at the drive sheave.*

house at the top of the hoistway. If, for architectural or other reasons, it is desired to eliminate this penthouse, a basement machine room can be used with the roping shown in Fig. 23.3(*d*). This arrangement uses geared traction equipment, with speeds up to 400 fpm. All the illustrated ropings are applicable to the full range of car capacities.

23.8   Safety Devices

The main brake of an elevator is mounted directly on the shaft of the elevator machine (see Fig. 23.2). The elevator is first slowed by dynamic braking of the motor, and the brake then operates to clamp the brake drum, thus holding the car still at the floor.

A dual safety system, designed to stop an elevator car automatically before the cars speed becomes excessive, is normally used. The device that acts first is a centrifugal governor or an electronic speed sensor that cuts off the power to the traction motor and sets the brake in the event of a limited overspeed. This usually stops the car. Should the speed still increase, the governor actuates two safety rail clamps, which are mounted at the bottom of the car, one on each side. They clamp the guide rails by wedging action, bringing the car to a smooth stop (Fig. 23.4).

Oil or spring buffers are usually placed in the elevator pit. Their purpose is not to stop a falling car but to bring it to a partially cushioned stop if it should overtravel the lower terminal. If a car overtravels (down or up), travel sensors de-energize the traction motor and set the main brake. Safety arrangements under emergency condition of fire or power failure are discussed in Section 23.43.

23.9   Elevator Doors

The choice of car and hoistway door affects the speed and quality of elevator service considerably. Doors for passenger elevators are power operated and are synchronized with the leveling controls so that the doors are fully opened by the time a cab comes to a complete stop at the landing. The closing time, however, varies with the type of door and size of opening. For safety reasons, the kinetic energy of an automatic door is limited to 7 ft-lb, and its closing pressure to 30 lb. To provide fastest closing within this energy limitation, a center-opening door is used. Also, to reduce passenger transfer time and avoid discomfort, a clear opening of 3 ft, 6 in. (106.7 cm) is used in most commercial installations, which permits simultaneous loading and unloading without undue passenger contact. (Some consultants feel that simultaneous

Fig. 23.4 *Elevator safety devices. The governor or velocity transducer senses over-speed, clamping the safety trip rope and releasing the safety jaws, which exert a constant retarding force on the car rails, thus bringing the car to a gradual and safe stop. (Photo courtesy of Schindler Elevator Corp.)*

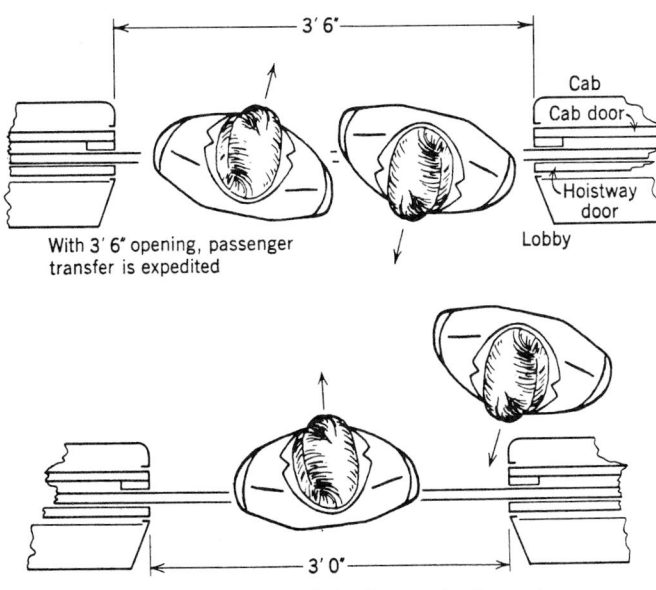

With 3′ 6″ opening, passenger transfer is expedited

With opening smaller than 3′ 6″ simultaneous loading and unloading is difficult and transfer time is lengthened

Fig. 23.5 *Transfer of passengers with door openings of different widths. With openings smaller than 3 ft, 6 in. (42 in.), simultaneous loading and unloading is difficult and transfer time is lengthened. With a 42-in. opening, large people or people with bulky outerwear may brush against each other in passing. For complete isolation of passengers, a 48-in. opening may be necessary.*

TRANSPORTATION

(a)
Single slide: 24"-36"

(b)
Center opening: 42"-60"

(c)
Two speed: 42"

(d)
2-speed, center open: 60"

Fig. 23.6 *Typical hoistway doors and applications.*
(a) *Single slide door, 24 to 36 in. wide, for small commercial building or residential use.*
(b) *Standard commercial 42-in. center opening door for office buildings use, or 48- to 60-in. center opening for hospital or service car.*
(c) *Two-speed 42-in. general commercial use door.*
(d) *Two-speed, center-opening, 60-in., department store door, freight and passenger, and nonautomatic service.*

passenger transfer is practical only with a 48-in. clear opening; Fig. 23.5). When an opening narrower than 36 in. is used, loading is delayed until unloading is complete, and speed and quality of service are thereby markedly reduced. Such small doors are applicable only in residential or small, light-traffic buildings. The available door types are shown in Fig. 23.6.

A two-speed door design is used where space conditions dictate or where a wide opening is required. The two-speed nomenclature reflects the fact that the two halves of the door must travel at different speeds to complete their travel simultaneously [see Fig. 23.6(*c*)].

Installations can be equipped with an electronic sensing device that detects passengers in a wide area on the landing in front of the car door, rather than only directly in the door path. Such detection, often accompanied by an audible signal, causes the car door to remain open for a predetermined length of time, or a closing door to reverse. These devices are particularly useful in installations where passengers cannot approach the entrance or cannot enter the car quickly—for example, riders with baggage or holding children, people in wheelchairs, and employees moving bulky objects such as beds or carts in hospitals or wheeled objects in office and industrial facilities.

All automatic elevators, regardless of whether equipped with detection beams, are required by ANSI to have a safety edge device on the car doors that causes the car and hoistway doors, which operate in synchronism, to reopen when the safety edge meets any obstruction. Some cars doors are arranged to "nudge" when almost closed and/or after a specific time period. ADA requirements for the disabled are discussed in Section 23.11.

23.10   Cabs and Signals

Possibly the only area in which the architect has a free hand in selection of equipment is the decor of the cabs and the styling of hallway and cab signals. A normal elevator specification is a functional one that describes the intended operation of the equipment and normally includes an amount to cover optional decor of the cabs. The type and functioning of signal equipment is also specified, but finish and styling are optional. Cab interiors may be finished in wood paneling, plastic (Micarta or Formica), stainless steel, or almost any material desired. Floors may be tile, wood,

or carpeting, as selected. Illumination may be from ceiling fixtures, coves, or completely illuminated luminous ceiling, of standard or special design. For each bank of elevators, it is wise to furnish at least one set of wall mats, to protect wall finishes when cars are being used to move tenant furniture. This is especially important where no separate service car has been provided.

Car and hallway signals and lanterns should be designed to fulfill their basic functions, take proper consideration of the needs of the handicapped, and be consonant with the decor of the cabs and the corridors. (For a discussion of cab control panel, see Section 23.26.) The hall buttons should indicate desired direction of travel and by visual means confirm that a call has been placed. The hall lantern located at each car entrance must visually indicate the direction of travel of an arriving elevator and preferably its present location. An audible signal should signal its imminent arrival. The latter feature allows waiting passengers to move to the next arriving car in a bank and thereby speeds service. Hall stations can be equipped with special switches for fire, priority, and limited access service, as required (Figs. 23.7 and 23.8).

(a) Hall Station   *(b)* Basic hall station   *(c)* Expanded hall station

Fig. 23.7 *Typical hall stations.*
(a) *Simple pushbutton station with illuminated ring on each button to indicate call registration.*
(b) *Large buttons show travel direction in addition to lighting to indicate call registration.*
(c) *Special hall stations may incorporate priority, emergency, or secure access controls as shown.*
(Photos courtesy of Otis Elevator Co.)

(a) Hall Lantern   (b) Combination hall lantern and car position display

Fig. 23.8 *Typical hall lanterns.* (a) *The top section is green and lights when a car approaches, indicating UP travel. The bottom section is red, and when lighted indicates an approaching DOWN car.* (b) *The lantern shows travel direction and present car location. (Photos courtesy of Otis Elevator Co.)*

Within the cab, indication of travel direction and present location can be accomplished either with separate fixtures or by indicators built into the car panel. Most manufacturers can provide voice synthesizers built into the car panel that announce the floor, direction of travel, and any other desired message such as safety or emergency messages.

23.11 Requirements for the Handicapped

Elevator manufacturers follow ADA requirements as a minimum, and may add additional conveniences for the handicapped as the specific facility or local codes require. As with all aspects of design for construction, no planning should begin until the project architect has verified and assembled all applicable codes and standards.

The basic physical limitations addressed are those of ambulation, sight, and hearing. Thus, to ease access for passengers in wheelchairs (or with walking aids), the ADA requires excellent car leveling, 36-in. minimum clear door opening (42-in. recommended) delayed door closing, detection beams that reopen the door *without contact* on sensing a passenger, inside car dimensions that

permit turning a wheelchair, buttons and emergency controls within easy reach, and appropriate car furnishings.

For those with sight impairment, the ADA requires audible signals in addition to easily seen and recognized visual ones, both in the car and at landings, to indicate call registration, car approach, car landing, direction of travel, floor, car position, and so on. In this connection, a voice synthesizer is of invaluable assistance. In addition, Braille plates adjacent to car floor buttons and large, easily recognizable symbols adjacent to passenger-controlled emergency controls are to be used. Hearing-impaired passengers are assisted by the large visual signals required for sight-impaired passengers. In addition, call buttons must visually indicate that a call has been placed, and the indication extinguished when the call has been answered. Many of these ideas also make life easier for passengers with full faculties. A note of caution is in order, however. Delayed door closing may increase travel time appreciably. In buildings with traffic peaks, it may be necessary to designate one or more specific elevators for use by the handicapped during these periods.

For the convenience of the reader, the actual requirements of the ADA as stated in the *Federal*

Register, Vol. 56, No. 144, July 26, 1991, follow, with the accompanying illustrations as they appear in the Federal Register.

ADA Accessibility Guidelines for Buildings and Facilities

4.10 Elevators

4.10.1 General. *Accessible* elevators shall be on an accessible route and shall comply with 4.10 and with the *ASME A*17.1—1990. Safety Code for Elevators and Escalators. *Freight elevators shall not be considered as meeting the requirements of this section unless the only elevators provided are used as combination passenger and freight elevators for the public and employees.*

4.10.2 Automatic Operation. Elevator operation shall be automatic. Each car shall be equipped with a self-leveling feature that will automatically bring the car to floor landings within a tolerance of ½ in. (13 mm) under rated loading to zero loading conditions. This self-leveling feature shall be automatic and independent of the operating device and shall correct the overtravel or undertravel.

4.10.3 Hall Call Buttons. Call button in elevator lobbies and halls shall be centered at 42 in. (1065 mm) above the floor. Such call buttons shall have visual signals to indicate when each call is registered and when each call is answered. Call buttons shall be a minimum of ¾ in. (19 mm) in the smallest dimension. The button designating the up direction shall be on top. (See Fig. 20). *Buttons shall be raised or flush. Objects mounted beneath hall call buttons shall not project into the elevator lobby more than 4 in. (100 mm).*

4.10.4 Hall Lanterns. A visible and audible signal shall be provided at each hoistway entrance to indicate which car is answering a call. Audible signals shall sound once for the up direction and twice for the down direction or shall have verbal annunciators that say "up" or "down". Visible signals shall have the following features:

(1) Hall lantern fixtures shall be mounted so that their centerline is at least 72 in. (1830 mm) above the lobby floor. (See Fig. 20)

(2) Visual elements shall be at least 2½ in. (64 mm) in the smallest dimension.

(3) Signals shall be visible from the vicinity of the hall call button (see Fig. 20). In-car

NOTE: The automatic door reopening device is activated if an object passes through either line A or line B. Line A and line B represent the vertical locations of the door reopening device not requiring contact.

Fig. 20
Hoistway and Elevator Entrances

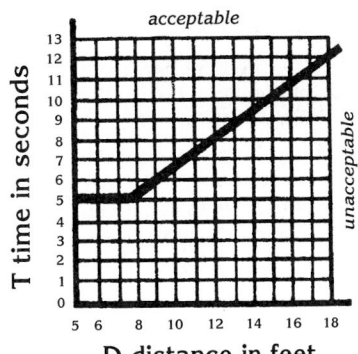

Fig. 21
Graph of Timing Equation

lanterns located in cars, visible from the vicinity of hall call buttons, and conforming to the above requirements, shall be acceptable.

4.10.5 *Raised and Braille* **Characters on Hoistway Entrances.** All elevator hoistway entrances hall have *raised* and *Braille* floor

TRANSPORTATION

designations provided on both jambs. The centerline of the characters shall be 60 in. (1525 mm) above *finish* floor. Such characters shall be 2 in. (50 mm) high and shall comply with 4.30.4. Permanently applied plates are acceptable if they are permanently fixed to the jambs (see Fig. 20).

4.10.6 Door Protective and Reopening Device. Elevator doors shall open and close automatically. They shall be provided with a reopening device that will stop and reopen a car door and hoistway door automatically if the door becomes obstructed by an object or person. The device shall be capable of completing these operations without requiring contact for an obstruction passing through the opening at heights of 5 in. and 29 in. (125 mm and 735 mm) above finish floor (see Fig. 20). Door reopening devices shall remain effective for at least 20 seconds. After such an interval, doors may close in accordance with the requirements of *ASME A*17.1—1990.

4.10.7 Door and Signal Timing for Hall Calls. The minimum acceptable time from notification that a car is answering a call until the doors of that car start to close shall be calculated from the following equation:

$$T = D/(1.5 \text{ ft/s}) \text{ or } T = D/(445 \text{ mm/s})$$

where T total time in seconds and D distance (in feet or millimeters) from a point in the lobby or corridor 60 in. (1525 mm) directly in front of the farthest call button controlling that car to the centerline of its hoistway door (see Fig. 21). For cars with in-car lanterns, T begins when the lantern is visible from the vicinity of hall call buttons and an audible signal is sounded. *The minimum acceptable notification time shall be 5 seconds.*

4.10.8 Door Delay for Car Calls. The minimum time for elevator doors to remain fully open in response to a car call shall be 3 seconds.

4.10.9 Floor Plan of Elevator Cars. The floor area of elevator cars shall provide space for wheelchair users to enter the car, maneuver within reach of controls, and exit from the car. Acceptable door opening and inside dimensions shall be as shown in Fig. 22. The clearance between the car platform sill and the edge of any hoistway landing shall be no greater than 1¼ in. (32 mm).

4.10.10 Floor Surfaces. Floor surfaces shall comply with 4.5.

4.10.11 Illumination Levels. The level of illumi-

(a)

(b)

Fig. 22
Minimum Dimensions of Elevator Cars

nation at the car controls, platform, and car threshold and landing sill shall be at least 5 footcandles (53.8 lux).

4.10.12 Car Controls. Elevator control panels shall have the following features:

(1) Buttons. All control buttons shall be at least ¾ in. (19 mm) in their smallest dimension. They *shall* be *raised* or flush.

(2) Tactile. *Braille,* and Visual Control Indicators. All control buttons shall be designated by *Braille* and by *raised* standard alphabet characters for letters, Arabic characters for numerals, or standard symbols as shown in Fig. 23.(a), and as required in *ASME A*17.1—1990. *Raised* and

(a)
Panel Detail

(b)
Car Control Height

(c)
**Alternate Locations of Panel
with Center Opening Door**

(d)
**Alternate Locations of Panel
with Side Opening Door**

**Fig. 23
CAR CONTROLS**

Braille characters and symbols shall comply with 4.30. The call button for the main entry floor shall be designated by a *raised* star at the left of the floor designation [see Fig. 23(a)], and as required in *ASME A17.1—1990. Raised* and *Braille* characters and symbols shall comply with 4.30. The call button for the main entry floor shall be designated by a raised star at the left of the floor designation [see Fig. 23(a)].

All raised designations for control buttons shall be placed immediately to the left of the button to which they apply. Applied plates, permanently attached, are an acceptable means to provide raised control designations. Floor buttons shall be provided with visual indicators to show when each call is registered. The visual indicators shall be extinguished when each call is answered.

(3) Height. All floor buttons shall be no higher than 54 in. (1370 mm) above the *finish floor for side approach and 48 in. (1220 mm) for front approach.* Emergency controls, including the emergency alarm and emergency stop, shall be grouped at the bottom of the panel and shall have their centerlines no less than 35 in. (890 mm) above the finish floor [see Fig. 23(a) and (b)].

4.10.13 Car Position Indicators. In elevator cars, a visual car position indicator shall be provided above the car control panel or over the door to show the position of the elevator in the hoistway. As the car passes or stops at a floor served by the elevators, the corresponding numerals shall illuminate, and an audible signal shall sound. Numerals shall be a minimum of ½ in. (13 mm) high. The audible signal shall be no less than 20 decibels with a frequency no higher than 1500 Hz. An automatic verbal announcement of the floor number at which a car stops or which a car passes may be substituted for the audible signal.

4.10.14 Emergency Communications. If provided, emergency two-way communication systems between the elevator and a point outside the hoistway shall comply with *ASME A*17.1—1990. The highest operable part of a two-way communication system shall be a maximum of 48 in. (1220 mm) from the floor of the car. It shall be identified by a raised symbol and lettering complying with 4.30 and located adjacent to the device. If the system uses a handset then the length of the cord from the panel to the handset shall be at least 29 in. (735 mm). *If the system is located in a closed compartment the compartment door hardware shall conform to 4.27, Controls and Operating Mechanisms. The emergency inter-communication system shall not require voice communication.*

[End of extract from Federal Register]

ELEVATOR CAR CONTROL

23.12 Drive Control

The movement of an elevator car and all of its parts is controlled by three different systems that combine and interact to provide a unified control system. The *supervisory system* controls a bank of elevators as a group and dictates which car answers which call. The *operational control* system determines when and where physical motion of a car and its doors should occur. It deals with the operation of the car doors and the integration of car buttons, lanterns, and passenger-operated devices into the overall control and indicating system. Operating control passes its information about car and door control to the *motion control* system (also known as drive control). Motion control determines the car's acceleration, velocity, braking, leveling, and regenerative braking, plus all aspects of door motion.

Elevator *car* control is separate and distinct from the control system that governs the functioning of a group of cars acting as a system. That arrangement is generally designated as the *supervisory system,* and is discussed separately.

Elevator car acceleration and deceleration are accomplished by controlling the speed of the motor that drives the elevator traction machine. This speed control can be accomplished in a number of ways, all of which are in use in elevator installations. They are:

- Thyristor control of an asynchronous (squirrel cage) ac traction motor.

- Thyristor control of a dc traction motor.

- Motor–generator set control of a dc traction motor (Ward–Leonard system).

- Variable-voltage, variable-frequency control of an asynchronous ac traction motor.

(Rheostatic control of single and multispeed ac motors is essentially obsolete, and therefore is not discussed).

We describe each system briefly and note its applicability, advantages, and disadvantages.

23.13 Thyristor Control, ac and dc

(a) ac. High-power transistors make accurate speed control of inexpensive ac squirrel-cage motors practical by the simple expedient of supplying carefully regulated variable voltage to the motor. The resultant speed control is smooth and stepless, making it applicable to passenger elevator installations. However, the high motor slip caused by the constant frequency of the variable-voltage

Fig. 23.9 *Arrangement of electric speed-control equipment for elevator drives.*
(a) *Solid-state silicon-controlled rectifier (SCR) thyristor control produces variable-voltage dc and finds application in good-quality new and retrofit medium- and high-speed installations.*
(b) *The traditional Ward–Leonard system, which produces finely controlled, variable-voltage dc. This is the system in use today in the vast majority of existing high-quality installations.*
(c) *Variable-frequency ac is energy efficient, highly accurate, and applicable to all types of installations.*

ac supply causes large thermal losses with resultant low operating efficiency. Too, the "chopper," which is used to accomplish the necessary voltage control, introduces undesirable harmonics into the power system in considerable quantity. These harmonics cause undesirable radio noise and can cause system component overheating. Finally, the system's low power factor increases line losses and necessitates increased feeder sizes. As a result, this option, which is applicable to low and midrise passenger service (maximum rise 250 ft

and maximum car speed of 350 fpm), using geared machines has been largely replaced by VVVF control (see Section 23.16).

(b) dc [see Fig. 23.9(a)]. Excellent ride quality can be had by utilizing the dc version of thyristor control to supply variable voltage to a dc traction motor. This arrangement provides the high-quality ride and leveling characteristic of dc drives and removes the rise and speed limitations of its ac counterpart, but not the system's inherent low power factor, heavy line harmonics, and high machine room thermal losses. Line harmonics can be sharply reduced by the use of power line filters. This control system is widely used today, with both geared and gearless traction machines as applicable, for installations of all rises and car speeds. Its extremely smooth ride makes it one of the three systems of choice in high-quality construction. The other two systems are the traditional Ward-Leonard variable-voltage dc system and the variable-voltage, variable-frequency (VVVF) ac system.

23.14   Variable-Voltage dc Motor Control

See Fig. 23.9(*b*). Before the development of electronic motor control, the only practical method of obtaining the precise motor speed control necessary for smooth stepless elevator operation was to impress a variable dc voltage on a dc traction motor. This variable dc voltage was obtained from an auxiliary m-g set comprising an ac motor and a dc generator. This arrangement is known as a *Ward–Leonard system* or as a *unit multivoltage* (UMV) drive. Although this arrangement has disadvantages, it is the classic high-quality elevator drive arrangement and is found in the vast majority of high-quality geared and gearless installations built before 1990. The disadvantages of this system—low overall efficiency, expensive machines, high thermal losses in machine rooms, high maintenance costs for three rotating machines, and high noise levels were, until the development of a practical VVVF system, accepted as the price of a system that operates faultlessly, with supreme accuracy at all speeds, for long periods of time given proper maintenance. As noted previously, it is today still one of the three principal drive control systems being used in new installations. Its principal advantage over VVVF control is that it is "forgiving" of a less than ideal building power supply.

In modernization work, buildings with existing Ward–Leonard systems are frequently converted to solid-state dc thyristor control. This change maintains the excellent ride characteristics of variable-voltage dc traction machine installations while improving system performance by replacing the auxiliary m-g set with a low-loss, low-maintenance, quiet solid-state thyristor drive control.

23.15   Variable-Voltage, Variable-Frequency (VVVF) ac Motor Control

See Fig. 23.9(*c*). This system, which is entirely solid state, is in the opinion of many elevator professionals the drive system of choice for all high-quality new installations of any speed or rise. The system consists of a rectifier, which changes the incoming ac to dc, and an inverter, which creates variable-voltage, infinitely variable frequency, three-phase ac from the (rectified) dc. This ac is then applied to a standard squirrel-cage ac motor, which operates essentially at the speed corresponding to the frequency of the input. By maintaining a constant voltage-to-frequency ratio input to the traction motor, it is possible to provide continuously variable, highly accurate speed control at extremely high efficiencies throughout the full speed range of the motor. This is precisely what is required for high-quality geared and gearless installations, without the energy losses associated with the auxiliary m-g set of UMV control. The VVVF speed control system eliminates most of the disadvantages of the solid-state thyristor control system and has the following characteristics:

- Overall system efficiency is high at all motor speeds.
- Traction motors are economical single-speed squirrel-cage ac.
- Motors are 10% to 15% smaller than other drive systems.
- System power factor is close to unity.
- Line harmonics are lower than with thyristor controls.
- Speed control and leveling are equal to dc traction motor control.
- Equipment is 98% solid-state, thus requiring minimum maintenance.
- Electric energy use and peak loads are reduced, thus reducing electric billings up to 35%.

TABLE 23.2 **Comparative Energy Use of Various Motion Control Arrangements**

Energy Use	Relative Amount
Hydraulic elevators (discussed in Section 24.10)	10
UMV motion control; geared dc traction motor	7–7.5
UMV motion control; gearless dc traction motor	6–6.5
Thyristor control; geared dc traction motor	5–5.5
Thyristor control; gearless dc traction motor	4–4.5
Variable-voltage variable-frequency control, ac traction motor	2.5–3.0

· System is applicable to all rises and speeds.

· Machine rooms are smaller, cooler, and quieter than those using Ward–Leonard UMV controls.

At this point we can summarize the preceding discussion with a tabular comparisons of the various drive systems in use today (Tables 23.2 and 23.3). The exact degree of improvement possible when changing to solid-state equipment in a retrofit job depends on the quality of both the old and the new systems. It is greatest when a microcomputer group control system replaces a relay-logic terminal dispatch system.

23.16 Elevator Operating Control

Assuming the system to be energized and at rest, registration of a call from a station in the lobby or an upper-floor corridor activates the system. The particular elevator that answers the call is selected by the supervisory system. In a UMV system the car's m-g set is started, whereas with solid-state motion control power is available immediately. Switching devices in the car control panel release the brake, energize the elevator motor, and accelerate the car to its rated speed. Reverse operations are initiated when decelerating and stopping (landing) the car. When the car stops, the brake holds the sheave and elevator stationary.

The motion of a single car is determined by the action of three principal items of equipment: the car controller, the motion controls, and the system supervisory equipment. The action of this latter equipment is discussed in Sections 23.17 to 23.21. The function of the car controller is to provide information on the car's exact location, car panel calls, and hall calls. This information is fed into the supervisory system and the motion control equipment, which in turn act to initiate all the procedures necessary to answer all calls via the individual car controller panels. The car controller panel also supplies the necessary signals to car and hall lanterns that indicate car position and direction of travel.

TABLE 23.3 **Comparative Characteristics of Elevator Drive Systems**

Type	Rise (ft)	Speed (fpm)	Control	Initial Cost	Operating Cost	Performance
Geared ac	250	150–350	Thyristor	Medium	Medium	Fair
	300	150–450	VVVF	High	Low	Excellent
Geared dc	175	50–450	UMV	High	High	Excellent
	250	50–450	Thyristor	Medium	Low	Very good
Gearless dc	Unlimited	400–1200	UMV	High	High	Excellent
			Thyristor	Medium	Low	Very good
Gearless ac	Unlimited	400–2000	VVVF	Medium	Low	Excellent

NOTE: Life of equipment is generally indefinite except for worm and gear of geared unit, which has a 30- to 40-year life.

23.17 System Control Requirements

An operating system is one that provides the automatic response of a group of cars to calls for service. An effective system must take cognizance of all hall calls and car calls, car travel directions, and car position, in relation to each other and in relation to the call requirements, plus the trends of traffic. This last is required in order that the system anticipate demand rather than react to it. Because traffic and calls are never static, the control system that can satisfy all these demands in a large elevator system is necessarily an extremely sophisticated one. On small systems, the operating control is much simpler, as described in later sections. Throughout the discussion that follows, solid-state systems are assumed.

23.18 Single Automatic Pushbutton Control

This system is the simplest of the passenger-operated automatic control schemes. It handles only one call at a time, providing an uninterrupted trip for each call. A single corridor button at each level can register a call only when the car is not in motion. This system is used only in private residences and for light-use freight elevators.

23.19 Collective Control

Cars stop at each floor that registers a car irrespective of direction; hence the "collective" nomenclature. This leads to slow and annoying service. As a result, this system is no longer used in new installations in the United States, although it is common in other countries.

23.20 Selective Collective Operation

This type of collective operation is "selective" in that it is arranged to collect all waiting "up" calls on the trip up and all hall "down" calls on the trip down. The control system stores all calls until they are answered, and automatically reverses the direction of travel at the highest and lowest calls. When all calls have been cleared, the car will remain at the floor of its last stop awaiting the next call. Any hall button call will set the car into operation.

Selective collective control is standard in locations where service requirements are moderate, such as in apartment houses, small offices, and professional buildings. Because these locations often require more than one car, a group control scheme for up to three cars automatically assigns each hall call to the car best situated to answer it, prevents more than one car from answering a call, allows one car to be detached for freight duty, and automatically parks cars at the ground floor when they are not required.

Although selective collective control is in common use for residential and other light-to-moderate service requirement buildings, its inherent strong tendency toward bunching of cars can result in long waiting periods. This characteristic is particularly annoying in groups of three cars. Frequently, a passenger arrives at a landing to find that all three cars have just passed, going in the same direction. The result is that service is only slightly better than that which would be rendered by a single car, except that load (handling) capacity is greater. For this reason, operation of more than two cars with this system is not recommended, and operation of more than three cars is not feasible.

23.21 Computerized System Control

Prior to the advent of computer supervisory control, large banks of elevators were controlled by lobby-based human "starters" who attempted, with limited success, to recognize and anticipate traffic patterns and thus speed service. However, due to the huge amount of information that must be processed, service, particularly during heavy traffic, was frequently less than satisfactory. This changed with the advent of computerized, microprocessor-controlled operating systems.

A satisfactory control system must continuously monitor demand, and control each car's motion in response to demand only; that is, it must analyze all the possibilities and answer each call in optimum fashion. The definition of "optimum" depends, of course, on the system design strategy, and this varies among manufacturers. Such a system is possible only with the aid of a central computer combined with programmable microprocessor controlled peripherals because the amount of data that must be collected and instantaneously processed is enormous.

All manufacturers attempt to optimize the parameters by which system quality is measured; specifically, to minimize interval, hall waiting time, and average trip time. How this is accomplished depends on the relative weight assigned to each item of input data in an extremely complex computer algorithm. One manufacturer uses an algorithm that calculates a figure of merit for *each* car to answer *each* waiting hall call. The number arrived at represents the weighted sum of prospective passenger hall waiting time (interval and hall wait time) and traveling passenger delay (average trip time). The car with the best figure of merit (minimum time) is assigned to the landing call. Another manufacturer uses a somewhat simpler dynamic call allocation algorithm that relies on the high-speed calculation capabilities of a central computer to make a last moment decision as to which car answers which call. Another algorithm, in addition to analyzing each car's capability to answer a call, calculates the effect of any decision on the overall elevator service quality for the building and uses this factor as well in the final decision.

Still another algorithm uses a car dispatch program based on preprogrammed and learned traffic patterns modified by the history of the previous few minutes of operation. In the basic program, each car computes its own response time to a waiting hall call, considering its own position, velocity, and car calls, and compares it to that of all other cars in the group. The car that will provide optimum hall call service time (waiting time plus trip time) answers the hall call. To facilitate this type of car "bidding" for calls, some systems are arranged so that each car controller can act as a master group controller. At any one time, one car in the group acts as master, but if taken out of service, another car controller becomes the master.

In addition, some programs use an artificial intelligence module to learn traffic data and history and continuously change the car dispatching mode. These changes consist in part of sectoring or zoning the building so that cars are grouped

to provide optimal service. This type of system is particularly useful in buildings with single- or multifloor-use occupants, with repetitive traffic patterns. Most algorithms permit overriding the program mode in response to crowd sensors or to particularly heavy, concentrated, service demand such as might be the case if occupants of an entire floor left the building at an unusual hour.

The actual program logic for even a small group supervisory traffic control system is beyond the scope of this book, but its guidelines are not. An adequate system should:

· Be programmed initially for the anticipated service needs. These needs can be analyzed in existing buildings with computerized traffic analyzers (Fig. 23.10) and in planned buildings by detailed occupancy analysis using computerized elevator system simulation (Fig. 23.11).

· Be reprogrammable to meet changes in building needs at nominal cost and with minimum shutdown time. It must be possible to detach cars from the system during testing, reprogramming, and routine maintenance and to provide minimal service even during offhours.

· Provide for priority calls (based on landing waiting times), statistical analysis of traffic in order to anticipate patterns, adaptive (zoned) car parking to meet specific needs, adjustable door timing based on type of call (lobby, landing, car), backup dispatch means (in case of dispatch system failure), and automatic call cutout for constant (stuck) signals.

· Provide means for lobby and management office viewing (monitor) (Fig. 23.12) and obtaining hard copy (printer output) of elevator traffic information both stored and in real time (see Fig. 23.12). This data-handling equipment must provide a fault mode that stores, displays, and diagnoses system operating faults. The owner/manager unit may have on-site control and reprogramming capabilities, if so required.

· Provide additional functions such as emergency power elevator selection and control (see Section 23.42), priority service, selective hall/car call cutout, swing (separate) car oper-

Fig. 23.10 Building elevator traffic studies can be accomplished quickly and accurately using a traffic analyzer with multiple inputs to feed data via appropriate software into a conventional laptop computer (a). The unit illustrated has 64 inputs, all of which are scanned once per second, and is capable of unattended monitoring for a week. The accumulated data on call registration, response time, waiting time, maximum interval, cars in service, and so on, can be viewed visually on the computer's LCD screen or printed out as in (b). (Courtesy of UNITEC Parts Co.)

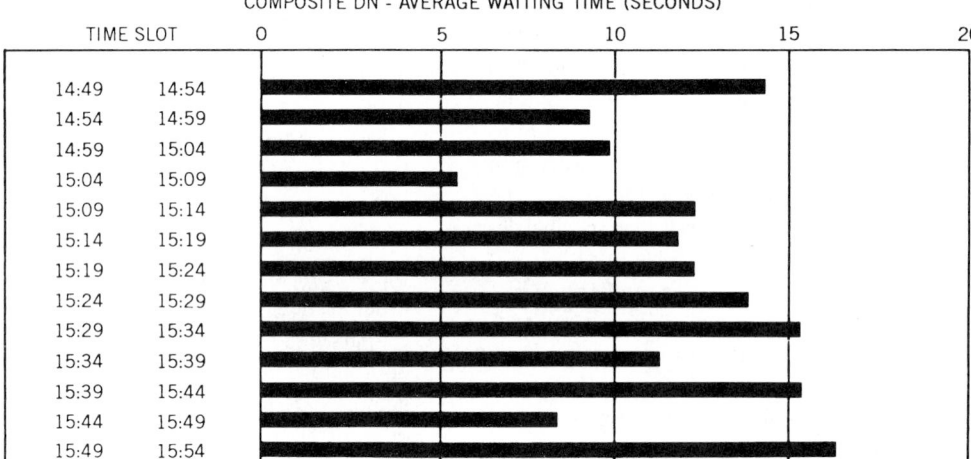

COMPOSITE DN - AVERAGE WAITING TIME (SECONDS)

ation, and hoistway access controls for maintenance.

- Provide, where specifically required by the building management, a riot control feature (access limitation at entrance levels), crossover floor operation in a zoned system, convention service (intense short-time usage at selected levels), and controlled access at given floors and for specific occupants.

- Be fully coordinated with the fire protection system in accordance with the local fire regulations (see Section 23.43).

- Act in consonance with the elevator security equipment so that operation of security/alarm devices initiates automatic elevator motion control procedures. This too must be coordinated with the local security authorities, and the automatic procedures must be subject to manual override (see Section 23.44).

Proper operation of the system should also result in:

1. All floors getting equal service, including the basement, if required.

Fig. 23.11 *Elevator system simulation. A computerized mathematic model of an automated elevator system permits the operator to enter building data and vary elevator control strategy. The results, displayed graphically, show the system response. (Courtesy of Otis Elevator Co.)*

2. Proper handling of multifloor tenants in office buildings, to permit efficient interfloor traffic.
3. Appropriate action in emergencies, such as general or local circuit power failure or any type of abnormal car to signal operation.

23.22 Rehabilitation Work: Performance Prediction

The primary reason for rehabilitating an existing elevator system is to improve its operating performance. Traction machinery has extremely long life, particularly in the gearless configuration. Thus, the usual rehabilitation project consists of replacing the m-g set, car control panel, and group controller with solid-state, microprocessor-controlled programmable equipment, while retaining the original traction equipment. The car door operator, which controls door action, is usually also replaced because door opening and closing characteristics are an important factor in overall trip time, and therefore system performance.

Using computers, major manufacturers have developed elevator system simulators. These devices enable architects and owners to input pro-

posed (or existing) building data, and to receive visual and printed readout on the performance of proposed systems (see Fig. 23.11). Because such programs are interactive, the operator can change the input data and the characteristics of the proposed equipment until the desired performance level is reached. These devices are particularly useful in modernization work because the owner sees in advance the operation of a proposed system *in his building* and can make the required decisions on this basis.

23.23 Lobby Elevator Panel

The traditional lobby elevator control and information panel for each elevator bank that was usually wall-mounted adjacent to the related elevators, has become one or more computer monitor screens, positioned at a lobby desk (Fig. 23.12), and/or in the building maintenance office. In addition, an information-only screen is frequently wall-mounted adjacent to the related elevators for the edification of waiting passengers.

The information displayed on the screen includes car locations, movement direction, waiting corridor calls, and any special status data. The control functions available at the computer terminal permit intervention to establish special types of operation including:

· Car movement without operating the usual audible and visual signals (*inconspicuous riser*).
· One or more cars removed from supervisory control and operated manually (*attendant or independent service.*)
· Cars elected for night or weekend service while the other cars are shut down.
· Car(s) assigned to a particular floor on a fixed- or priority-basis call (*convention feature or priority*).

Among other control functions are those concerned with emergency service, including the "fireman's return" feature required by ANSI and many local fire codes (Section 23.43) and the controls related to switching of power between cars in

Fig. 23.12 *Dedicated elevator display computer terminals are part of the extensive control and communication equipment installed at a lobby control desk in a modern New York City skyscraper. (Photo by Stein.)*

Elevator bank displays

the event that operation on emergency generator power is necessary (Section 23.42).

In addition, means of two-way communication with each car and other selected locations are provided at the control center.

23.24   Car Operating Panel

A typical modern car operating panel is illustrated in Fig. 23.13. Every car panel (station) is equipped with full-access buttons for call registry, door-open, alarm, emergency stop, and fireman control. Also always provided is a intercom device that permits communication with the building control office. A door-close button is sometimes provided if extensive hand operation of the car is antici-pated. It is activated only when the car is under manual control. Controls that do not concern the normal passenger are grouped in a locked com-partment in the car station. These include hand-operation switch, light, fan, and power switches, and any special controls such as security and emer-gency devices. Finally, a compartment accessible only to technicians contains the devices control-ling door motion, car signals, door and car posi-tion transducers, load-weighing control, door and platform detection beam equipment, speech syn-thesizer (if used), and visual displays.

ELEVATOR SELECTION

23.25   General Considerations

The selection of elevators for any but the simplest buildings requires the simultaneous consideration of several factors: adequate elevator service for the intended building usage, the economics of ele-vator selection, and the architectural integration of spaces assigned to elevators, including lobbies, shafts, and machine rooms. In large buildings many combinations are possible because these factors are interdependent. The selection of an optimum system for such cases is most practical and accurate with the aid of a computer or simula-tor, and their use has become standard practice in the industry. Hand computation, following certain guidelines, can yield good results for small straightforward buildings and reliable preliminary data for almost all buildings.

The criteria usually used in determining eleva-tor service quality are:

· Interval and average waiting time.

· Handling capacity.

· Travel time.

The guidelines that are followed in computation of elevator requirements, by hand or computer, result in a determination of these three criteria,

- Car position indicator.
- Continuous floor location reading.

- Car direction indicator.

- Special indication

- Overload warning light

- Capacity Plate

Capacity 550 kg.
8 PERSONS
NO SMOKING

- Communication module
- Hands-free intercom (connected to building control center and emergency fireman service).
- Voice synthesizer

- Floor Buttons.
- Illumination and tactile registration when pushed.
- Floor designation of contrasting colour.
- Braille floor number symbols

- Basic operating buttons
- Door close.
- Door open.
- Emergency stop
- Alarm

- Special services cabinet
- Independent service key switch.
- Fan switch.
- Attendant service key switch.
- Light switch.
- Emergency light/fan test switch.
- Stop/Run switch (parking).

Typical Car Operating Panel (COP)

Fig. 23.13 *Typical car operating panel. Designs of the panel vary widely, but the essential components are as shown.*

and their comparison with predetermined target values.

23.26 Definitions

A clear definition of important terms, including variant usages, follows:

Average lobby time or average lobby waiting time. Average time spent by a passenger between arriving in the lobby and leaving the lobby in a car.

Handling capacity (HC). Figure of merit for an elevator system, indicating the maximum number of passengers that can be handled in a given period—usually 5 min; thus the term "5-min handling capacity." When expressed as a percentage of the building population it is called *percent handling capacity* (PHC).

Interval (I) or lobby dispatch time. Average time between departure of cars from the lobby.

Registration time. Waiting time at an upper floor after registering a call.

Round-trip time (RT). Average time required for a car to make a round trip, starting from the lower terminal and returning to it. The time includes a statistically determined number of upper-floor stops in one direction and, *when calculating elevator requirements based on up-peak traffic,* an express return trip.

Travel time or average trip time (AVTRP). Average time spent by passengers from the moment they arrive in the lobby to leaving the car at an upper floor.

Zone. Group of floors in a building that is considered as a unit, with respect to elevator service. It may consist of a physical entity—a group of upper floors above and below which are blind shafts—or it may be a product of the elevator group control system, changing with system needs.

23.27 Interval or Lobby Dispatch Time and Average Lobby Waiting Time

In an ideal installation, at least from the riding public's point of view, a car would be waiting at the lower terminal on the rider's arrival or would

TABLE 23.4 **Recommended Elevator Intervals and Related[a] Lobby Waiting Time**

Facility Type	Interval (sec)	Waiting Time[a] (sec)
Office Buildings		
Excellent service	15–24	9–14
Good service	25–29	15–17
Fair service	30–39	18–23
Poor service	40–49	24–29
Unacceptable service	50+	30+
Residential		
Prestige apartments	50–70	30–42
Middle-income apartments	60–80	36–48
Low-income apartments	80–120	48–72
Dormitories	60–80	36–48
Hotels—1st quality	30–50	18–30
Hotels—2nd quality	50–70	30–42

[a]Based on the relationship: waiting time = 0.6 × interval.

be available after a short wait. Because cars leave the lobby separated in time by the *interval (I)* and passengers arrive at the lobby in random fashion, the average waiting time in the lobby should be half the interval. Field measurements show, however, that it is actually higher than this. The figure most often used in the industry is 60%—that is,

$$\text{average lobby waiting time} = 0.6I$$

Table 23.4 lists intervals and their relative quality for office buildings and the related average waiting time, based on the foregoing relationship. Because some control systems zone the building in such a way that some cars do not return to the lobby, the interval as a figure of merit may be somewhat misleading in such buildings. The table also lists recommended intervals for other types of buildings.

With intervals in the recommended range, riders are not conscious of any irksome delay in elevator service. Consciousness of delay is considered a major drawback in rental desirability and should be sedulously avoided for all conditions of traffic except during morning and evening peaks, when a certain delay is expected and therefore tolerated, however grudgingly. Even in peak periods any modern group supervisory system will recognize a *timed-out* call—that is, a call with registration time exceeding 50 sec, as a priority call. Priority calls are answered by the first available car, usually within 15 sec. If a considerable amount of interfloor traffic is expected during peak periods, as may be the case when a large company occupies several upper floors, elevator capacity should be increased by 20% to 40% over that calculated to maintain proper intervals.

23.28 Handling Capacity

The frequency, or interval, with which a car appears at the lobby is one of the two factors that determine the passenger capacity of an elevator system. The other is obviously the size of the elevator car. The system's *handling capacity* is completely determined by these two factors—car size and interval—and is independent of the number of cars. This can be best understood by visualizing the system as a single set of doors that open periodically (interval) to remove a given number of passengers (car capacity) from the waiting group. Whether that set of doors represents a single car or many cars that take turns is immaterial. The only factors that fix handling capacity are passenger load (car capacity) and frequency of loading (interval) (Table 23.5).

Note that cognizance is taken of the fact that during peak traffic periods, cars are not loaded to maximum capacity, but only to about 80%—a figure determined by actual count in many existing installations.

TABLE 23.5 **Car Passenger Capacity (*p*)**

Elevator Capacity (lb)	Maximum Passenger Capacity	Normal Passenger[a] Load per Trip
2000	12	10
2500	17	13
3000	20	16
3500	23	19
4000	28	22

[a]The number of passengers carried on a trip during peak conditions is approximately 80% of the car capacity.

TABLE 23.6 **Minimum Percentage Handling Capacities (*PHC*)**

Facility	Percent of Population to Be Carried in 5 Min
Office Buildings	
Center city	12–14
Investment	11.5–13
Single purpose	14–16
Residential	
Prestige	5–7
Other	6–8[a]
Dormitories	10–11
Hotels—1st quality	12–15
Hotels—2nd quality	10–12

[a]Due to more urgent traffic demands, particularly at the school and work exodus.

As a convenient measure of capacity, the handling capacity of a system for 5-min is taken as a standard. This is because a 5-min rush period is used as a measure of a system's ability to handle traffic. This may be expressed thus:

handling capacity (HC) = passengers/car

$$\times \text{ cars/sec} \times 5 \text{ min}$$

$$\times 60 \text{ sec/min}$$

Because the number of cars per second is the reciprocal of the interval (e.g., 30 sec between cars is the same as 1/30 of a car per second), this equation reduces to:

$$HC = \frac{\text{Passengers/car}}{\text{Interval}} \times 300$$

or

$$HC = \frac{300p}{I}$$

where *p* is car loading (number of passengers/car). When the interval is 30 sec, the system handling capacity is 10*p*, a convenient figure to remember.

To establish a figure of merit for building service, system *HC* must be related to building size. This is normally done by establishing a minimum percentage of building population that the system will handle in 5-min, called PHC. A good system for a diversified office building will handle no less than 12% of the building population. Similar figures are shown in Table 23.6 for various types of facilities.

In planning a building's elevator requirements, its population must be estimated. This is particularly difficult in speculative-type, diversified-use buildings. However, based on rental cost, area, and building type, a fair estimate can be made. Population estimates for office buildings are based on net area—that is, actual available area for tenancy, Table 23.7 gives suggested density figures, whereas Table 23.8 gives average office building efficiency figures for use in calculating net area.

23.29   Travel Time or Average Trip Time

The average trip time or time to destination is the sum of the lobby waiting time plus travel time to the median floor stop. Car round-trip time is also used as a criterion but it is not as meaningful as trip time. In a commercial atmosphere, a trip of less than 1 min is highly desirable, a 75-sec trip is acceptable, a 90-sec trip is annoying, and a 120-sec trip is the limit of toleration. Obviously, in the more relaxed atmosphere of a residence, where interval alone can account for a minute or more of trip time, these maxima are revised upward.

From Fig. 23.14 we see that the 2000- and 2500-lb cars used in residential buildings can have a 17-story rise, even with 60-sec interval, without excessive trip time. However, the 3500-lb car, which is almost universal in office buildings (Fig. 23.15), is limited to a maximum of 16 floors local run before exceeding the 90-sec limit, and about 6 to 8 floors to stay within the 75-sec criterion.

An important reservation on the foregoing statements must be noted. The curves of Figs.

TABLE 23.7 **Population of Typical Buildings for Estimating Elevator and Escalator Requirements**

Building Type	Net Area
Office Buildings	Square feet per person
Diversified (Multiple Tenancy)	
Normal	110–130[a]
Prestige	150–250
Single Tenancy	
Normal	90–110
Prestige	130–200
Hotels	Persons per sleeping room
Normal use	1.3
Conventions	1.9
Hospitals	Visitors and staff per bed[b]
General private	3
General public (large wards)	3–4
Apartment Houses	Persons per bedroom
High-rental housing	1.5
Moderate-rental housing	2.0
Low-cost housing	2.5–3.0

[a]Density may vary for different floors. Clerical and stenographic area may have a population density as high as 70 square feet per person.
[b]If visiting hours are restricted, visitor population will determine elevator requirements. If visiting is not restricted to a certain few hours, staff requirements may determine elevator design. Where traffic is heavy, a combination of passenger cars and larger "hospital" cars should be used to provide optimum service.

23.14 and 23.15 are based on statistical calculations, empirical data, and field observations, as discussed in the next section. This being so, the average values that these curves give should be considered to be ±15%, and borderline cases can be shifted either way. Designs that show high travel time on paper frequently work out well in the field, because lobby loading is often less than 80%, upper floor stops are less than the statistical figure due to groups of people going to the same floor, and staggered working hours relieve traffic

TABLE 23.8 **Office Building Efficiency**

Building Height	Net Usable Area as Percentage of Gross Area
0–10 floors	Approximately 80%
0–20 floors	Floors 1–10 approximately 75%
	11–20 approximately 80%
0–30 floors	Floors 1–10 approximately 70%
	11–20 approximately 75%
	21–30 approximately 80%
0–40 floors	Floors 1–10 approximately 70%
	11–20 approximately 75%
	21–30 approximately 80%
	31–40 approximately 85%

NOTE: Applicable to buildings with 15,000 to 20,000 gross square feet per floor.
Source: Reprinted from G. R. Strakosch, *Vertical Transportation, Elevators and Escalators,* 2nd ed. New York: Wiley, 1983.

peaks. Also, a feature called high-call reversal takes account of the fact that cars do not travel to the top on each trip, but reverse at the top call. This can reduce average trip time by 5% to 10%. Finally, sophisticated solid-state traffic controls allow more rapid acceleration and deceleration without discomfort, variable door closing time, and more efficient selection of landing call answers, all of which can further reduce trip time another 5%.

23.30   Round-Trip Time

The figure for round-trip time during up-peak traffic conditions, which are those used for calculating elevator requirements, is composed of the sum of four factors: (1) time to accelerate and decelerate; (2) time to open and close doors at all stops; (3) time to load and unload; and (4) running time (Figs. 23.16–23.18). It is physically the elapsed time from door opening at the lower terminal to door opening at the same terminal at the end of a round trip. Because the actual number of stops made by a car is unknown, a statistical probability figure is used, based on the passenger capacity of the car and number of local floors above the lower terminal. In calculating this

Fig. 23.14 *Plots of average trip time for various car speeds and capacities with 9-ft, 6-in. floor height and 30-s interval.*

TRANSPORTATION

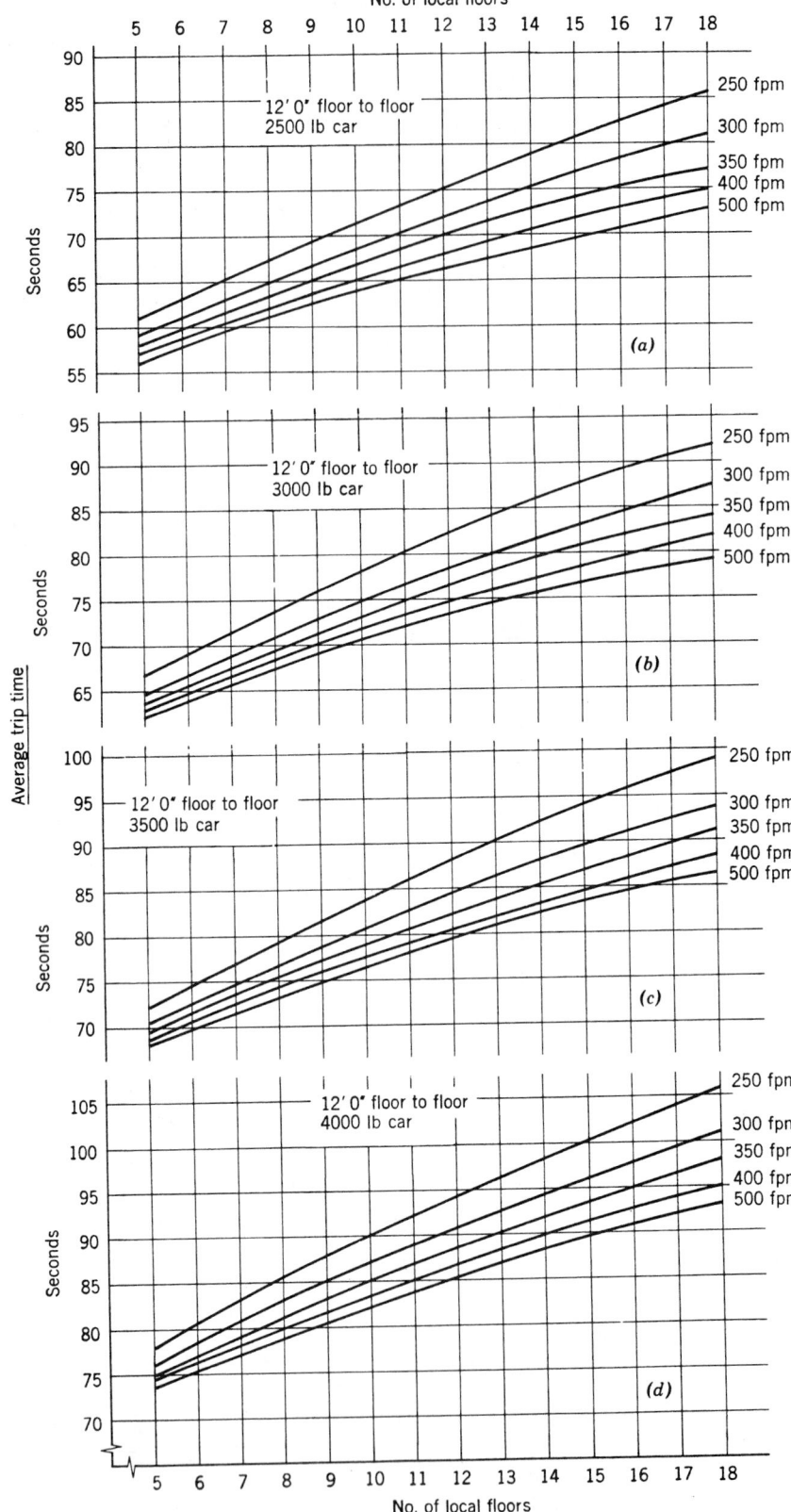

Fig. 23.15 *Plots of average trip time for various car speeds and capacities for 12-ft floor height and 30-s interval.*

Fig. 23.16 *Plots of round-trip time for various car speeds and capacities with 9-ft, 6-in. floor heights and 30-s interval.*

Fig. 23.16 (continued)

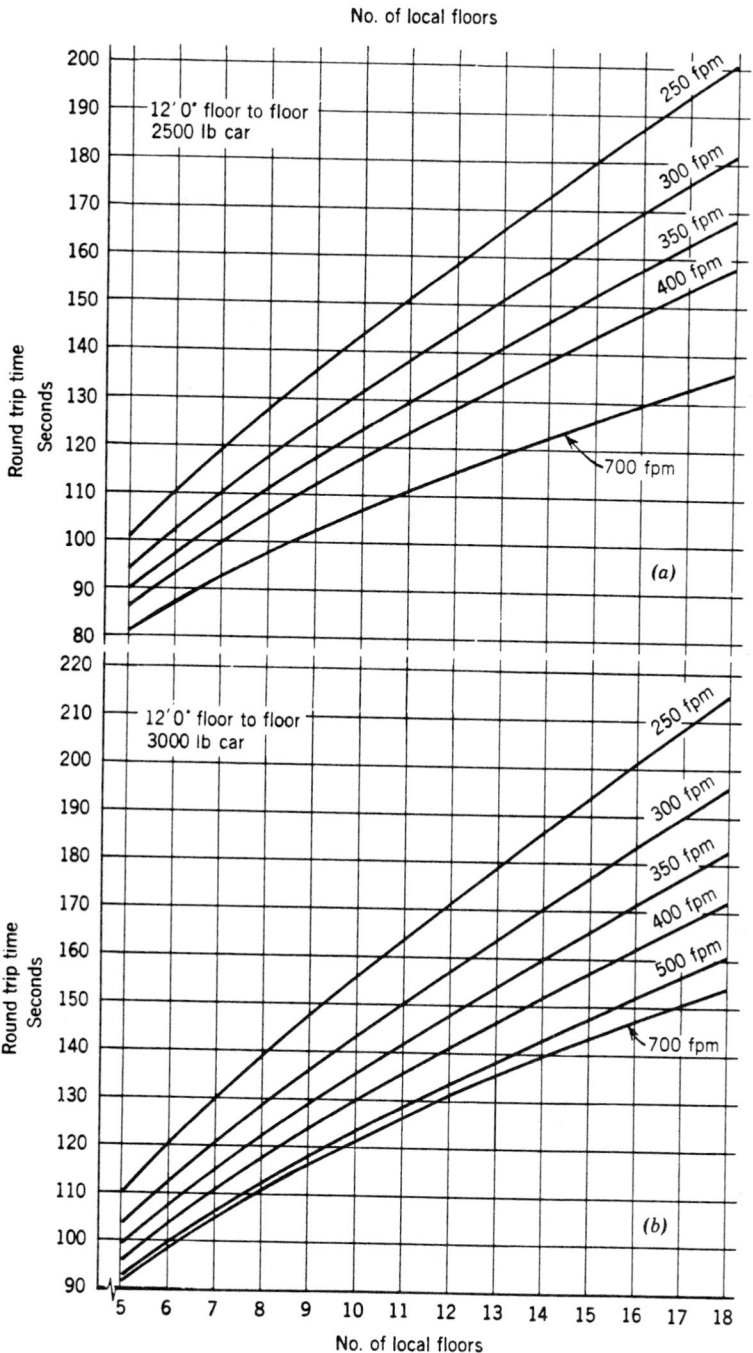

Fig. 23.17 *Plots of round-trip time for various car speeds and capacities with 12-ft floor height and 30-s interval.*

Fig. 23.17 (continued)

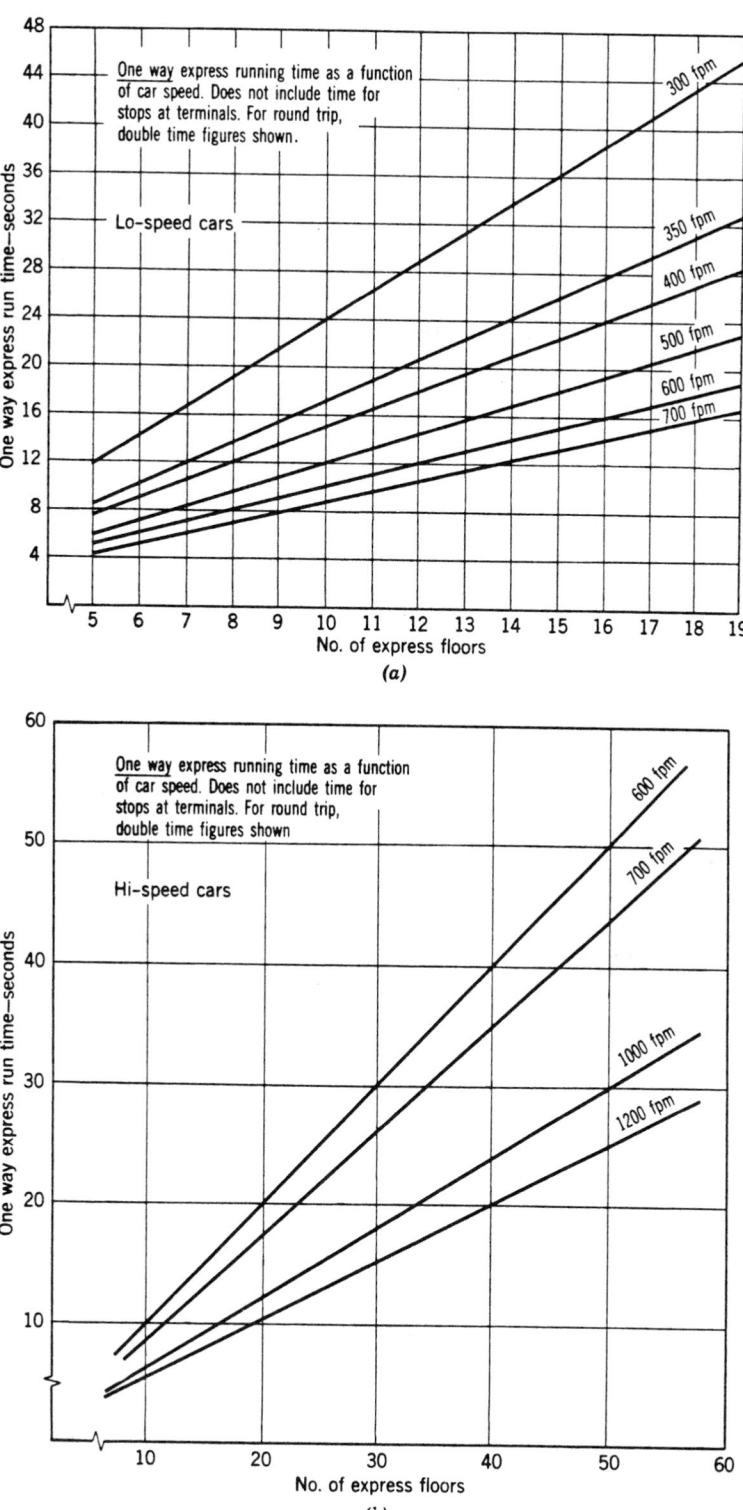

Fig. 23.18 *One-way express running time, not including terminal time.*

round-trip time (RT), it is assumed that a car will depart the lower terminal when loaded. No intentional delay is included at either lower or upper terminal. The RT thus calculated is a median figure, with any single actual round trip taking more or less time. In detail, round-trip time consists of the time expended in:

1. Loading at the lobby.
2. Door closing at the lobby.
3. Accelerating from the terminal and from each stop.
4. Decelerating at each stop.
5. Passenger transfer at each stop.
6. Door operations at each stop.
7. Running time at rated speed between stops.
8. Return express run from the last stop.

These figures are obtained as follows:

1. *Field observations:* Items 1 and 5 are based on a 3-ft, 6-in. door opening. A smaller door opening increases passenger transfer time.
2. *Calculations:* Items 2, 3, 4, 6, 7, and 8. *Door-closing time* is based on a 3-ft, 6-in. center-opening door with adjustable speed.

Acceleration and deceleration times are calculated with a maximum of 4 ft/s/s because anything beyond that results in physical discomfort to the passengers.

Running time at rated speed takes place after the car has accelerated and before it begins to decelerate. If we consider that it takes between 20 and 30 ft to accelerate to 700 fpm, depending on rate of acceleration, we see that in local runs the car never gets to rated speed. It simply accelerates and decelerates. The higher-speed equipment with its larger motor accelerates more quickly and gives some time advantage on the return express run, but has no great time advantage overall. This accounts for the small reduction above 500 fpm in Figs. 23.16 and 23.17.

In calculating RT for cars in upper zones, it is necessary to know the time required to traverse the express floors. This may be obtained from Fig. 23.18. The times given therein are for *one-way* express runs. Thus, to calculate RT for an upper zone car, take the RT corresponding to the upper local floors and add *twice* the figure obtained for express run time from Fig. 23.18.

23.31 System Relationships

The symbols that will be used in all elevator calculations are:

p	individual car capacity, equal to 80% of maximum during peak hours
h	5-min capacity of a single car
N	number of cars in a system
HC	system 5-min handling capacity, expressed in number of persons
RT	Round-trip time, in seconds
$AVTRP$	Average trip time, in seconds
I	interval, in seconds
D	population density, in square feet per person
PHC	percent of population to be moved in 5-min, expressed as a percentage figure

Having considered the definition of interval, handling capacity, average time and round trip time, we can demonstrate at this point the interrelationships between these quantities and the other equations governing the remaining factors that define elevator systems.

We previously established that handling capacity is determined by car capacity p and interval I:

$$HC = \frac{300p}{I} \qquad (23.1)$$

In a system comprising a single car, the interval (I) is obviously equal to the round trip time (RT). In a system with more than one car, the interval is reduced in proportion to the number of cars. Thus,

$$I = \frac{RT}{N} \qquad (23.2)$$

The 5-min handling capacity (h) of a single car is then

$$h = \frac{300p}{RT} \qquad (24.3)$$

if we remember that for a single car, its interval is its round trip time. It follows then that if the handling capacity of a single car is h, then the handling capacity of N cars is N times as much. Thus,

$$HC = N \times h$$

or

$$N = \frac{HC}{h} \qquad (24.4)$$

23.32   Car Speed

The selection of the car speed to be used is a matter of trial and error, the final selection being that required to give an RT that in turn gives an acceptable interval. However, in order to establish a starting point, it has been found that a minimum car speed corresponding to a given building height—or, in elevator parlance, "rise"—can be established. Similarly, although car size can be selected at any value, it has been shown that for certain facility types, specific-size cars are indicated. These recommendations are tabulated in Table 23.9.

It should also be borne in mind that elevator equipment falls into distinct speed categories. Thus, most manufacturers use geared equipment through 400 fpm [2 m/s] and gearless thereafter. The next category is 500 fpm [2.5 m/s] gearless followed by 600 fpm [3.0 m/s] to 700 fpm [3.5 m/s], and so on. Therefore, it is wise to avoid moving into the next higher and more expensive equipment category, if possible. This may mean exceeding recommended interval or dropping slightly low in handling capacity. It will be found, however, that this can be done without injury to the elevator system, provided a good group supervisory control system is employed.

23.33   Single-Zone Systems

Having established the relationships that govern the design and performances of an elevator system

TABLE 23.9   Elevator Equipment Recommendations

Building Type	Car Capacity[a] (lb)	[kg]	Rise (ft)	[m]	Minimum[a] Car Speed (fpm)	[m/s]
Office Building	2500 3000 3500	1250 1250 1600	0–125 126–225 226–275 276–375 Above 375	0–40 41–70 71–85 86–115 >115	350–400 500–600 700 800 1000	2.0 2.5 3.15 4.0 5.0
Hotel	2500 3000	1250 1250	As above		As above	
Hospital	3500 4000	1600 2000	0–60 61–100 101–125 126–175 176–250 >250	0–20 21–30 31–40 41–55 56–75 >75	150 200–250 250–300 350–400 500–600 700	0.63 1.0 1.6 2.0 3.15 4.0
Apartments	2000 2500	1000 1250	0–75 76–125 126–200 >200	0–25 26–40 41–60 >60	100 200 250–300 350–400	0.63 1.0 1.6 2.0
Stores	3500 4000 5000	1600 2000 2500	0–100 101–150 151–200 >200	0–30 31–45 46–60 >60	200 250–300 350–400 500	1.0 1.6 2.0 2.5

[a]Car capacity is determined by building size; car speed by rise.

TRANSPORTATION

comprising a single zone, it would be helpful to follow through an illustrative example.

EXAMPLE 23.1. Office building, downtown, diversified use, 14 rentable floors above lobby, each 12,000 sq ft net. Floor-to-floor height—12 ft. Determine a workable elevator system.

SOLUTION. From Table 23.6, recommended average HC is 13%. From Table 23.4, maximum recommended interval is 25 sec. From Table 23.7, average population density is 120 ft² per person.

Trial #1.
Building population:

$$\frac{14 \text{ floors at } 12,000 \text{ ft}^2}{120 \text{ ft}^2 \text{ per person}} = 1400 \text{ persons}$$

Suggested minimum handling capacity:

$PHC = 13\%$

$HC = 0.13 \times 1400 = 182$ persons
rise = 14 floors at 12 ft = 168 ft

From Table 23.9, we select a car size of 3500 lb at 500 fpm.

3500 lb

500 fpm

Then, from Figs. 23.17(c) and 23.15(c):

$RT = 155$ s $AVTRP = 82$ s

Single-car capacity: $h = 300p/RT$ (see Table 23.5 for p):

$$h = \frac{300\,(19)}{155} = 36.8 \text{ persons}$$

$$N = \frac{HC}{h} = \frac{182}{36.8} = 4.9, \text{ say 5 cars}$$

$$I = \frac{RT}{5} = \frac{155}{5} = 31 \text{ sec}$$

$$\text{actual } PHC = \frac{5(13\%)}{4.9} = 13\%$$

These figures are acceptable, but we should try faster cars, to reduce the interval. We select 700 fpm.

Trial #2.

3500 lb

700 fpm

$RT = 151$ s $AVTRP = 81$ s (by extrapolation)

$$h = \frac{300\,(19)}{151} = 37.7 \text{ persons}$$

$$N = \frac{182}{37.7} = 4.8, \text{ say 5 cars}$$

$$I = \frac{RT}{N} = \frac{151}{5} = 30 \text{ sec}$$

$$\text{actual } PHC = \frac{5\,(13\%)}{4.8} = 13.5\%$$

This solution is only marginally better than the previous 500 fpm solution, and the increased cost would not be justified. A trial using smaller cars with shorter RT is called for.

Trial #3.

3000 lb cars

500 fpm

$RT = 143$ s; $AVTRP = 76$ s

$$h = \frac{300\,(16)}{143} = 33.6 \text{ persons}$$

$$N = \frac{HC}{h} = \frac{182}{33.6} = 5.4 \text{ cars}$$

Using 5 cars:

$$I = \frac{RT}{N} = \frac{142}{5} = 28.4 \text{ sec}$$

$$\text{actual } PHC = \frac{5}{5.4}(13\%) = 12\%$$

Trial #4.
Using six 3000 lb, 500 fpm cars:

$$I = \frac{RT}{N} = \frac{143}{6} = 23.8 \text{ sec}$$

and

$$PHC = \frac{6}{5.4}(13) = 14.4\%$$

Both solutions are acceptable.

Tabulating the calculation results, we have

Solution	Cars (lb)	Speed (fpm)	RT (s)	AVTRP (s)	I (s)	PHC (%)
1	5 (3500)	500	155	82	31	13
2	5 (3500)	700	151	81	30	13.5
3	5 (3000)	500	143	76	28.4	12
4	6 (3000)	500	143	76	23.8	14.4

Solutions 1, 3, and 4 are acceptable. Solution 2 was discounted due to cost. Interestingly, solution 3, using smaller cars than the corresponding solution 1, and therefore being more economical, gives better results except for handling capacity. Although the best solution is number 4, which gives excellent interval and *HC*, the additional cost of a sixth car plus the revenue loss from the rentable area occupied by the sixth shaft and the high additional cost of maintenance weigh heavily against this option. A trial with five 3000 lb cars at 700 fpm is in order, with the foreknowledge that a considerable cost increase would result because 700 fpm cars require gearless equipment, whereas 500 fpm is available in either geared or gearless format, both giving excellent service.

Trial #5.

3000-lb cars

700 fpm

$RT = 140 \qquad AVTRP = 73 \text{ sec}$

$$h = \frac{300\,(16)}{140} = 34.3 \text{ persons}$$

$$N = \frac{182}{34.3} = 5.3; \text{ use 5 cars}$$

$$I = \frac{RT}{N} = \frac{140}{5} = 28 \text{ sec}$$

$$PHC = \frac{5}{5.3}\,(13\%) = 12.3\%$$

As expected, the improvement over the performance at 500 fpm is very slight: an interval of 28 sec in lieu of 28.4 and a handling capacity of 12.3% in lieu of 12%. The large increase in first cost for gearless equipment would not be justified. ■

The final selection would at this point be made on the basis of cost. When considering cost, note that first cost is the governing factor only in a speculative venture. With an owner–operator building, the cost comparison should be on a life-cycle basis. Cost figures must reflect the impact of elevator space requirements on net rentable area in the building. Comparative cost figures are given in Table 23.10.

As mentioned earlier and shown in Fig. 23.11, planners of new buildings today generally take advantage of elevator selection software provided by consultants, manufacturers, or by in-house capabilities. The results of one such program as applied to the problem just solved manually are shown in Fig. 23.19. This particular program was prepared by Otis Elevator Co. Note that the results in Fig. 23.19(*a*) correspond exactly to the final results obtained manually.

The correspondence is not surprising because the round-trip curves in Fig. 23.17 are based on a 3.3-sec door time and a 4. 0 ft/s^2 car acceleration, as shown in the figure. Note that high call reversal occurs at the top floor [13.5 (i.e., 14)] and that the number of up stops is 10 (9.7 from statistical calculations). The up-peak calculation assumes no counterflow traffic (i.e., an express down run and no interfloor traffic), as shown. These can be added, however, yielding much different results, as shown in Fig. 23.19(*b*). We see there that counterflow traffic (down stops) of only 3% and interfloor traffic of 1%, both of which are reasonable, change the interval to only fair quality and handling capacity to unacceptable.

23.34 Multizone Systems

In general, buildings with less than 15 stories are elevatored with a single zone (i.e., all cars serve all floors), and buildings with more than 20 stores are split into two or more zones. The area in between—16 to 19 stories—can go either way, depending on population density and required interval. Actually, a modern group supervisory system can automatically zone a building when

TABLE 23.10 **Relative First Cost**[a] **Figures for Passenger Elevators at Various Speeds (fpm) (for a Given Drive System)**

Car Size (lb)	Hydraulic (fpm)	Geared Traction (fpm)			Gearless Traction (fpm)			
	100	200	350	500	500	700	1000	1200
2000	40	80	100	130	165	170	220	235
2500	43	85	115	145	175	180	235	250
3000	50	90	120	150	180	185	250	265
3500	58	95	125	155	190	195	265	275
4000	60	100	135	165	200	205	280	300
4500[b]	70	120	150	185	225	230	300	325
5000[b]	75	130	160	200	240	250	330	350

[a]Costs are ±10%; based on standard fixtures, cabs, and entrances, and average rise for the speed indicated.
[b]Service elevator or hospital elevator.
NOTE: See Table 23.9 for speed/rise recommendation.

traffic requires it. However, such an arrangement, although efficient, is expensive both as to equipment and from a construction viewpoint because it does not take advantage of the considerable savings engendered by blind lower shaftways for upper-zone elevators. Analysis of multizone systems is complex and today is rarely done by hand. The interested reader is referred to Stein et al. (1986) for a detailed explanation of the technique involved. Most designers and consultants use one of the many extant computerized simulation and selection programs. These have the advantage that in addition to using basic criteria and building parameters, they can also consider the effect of variations in traffic control. Furthermore, the best of these programs can evaluate the engineering and economic impact of such factors as varying rental rates for different floors, rental space, machine room and hoistway space, core layout, and the structural ramifications of the elevator system.

23.35 Other Elevator Selection Recommendations

(a) Office Buildings. All necessary design criteria can be selected from Tables 23.4 to 23.7. Supervisory group control is normally microprocessor based. Approximately one service car per 10 passenger cars should be provided or, alternatively, one car for every 300,000 ft² of net area.

Service cars should be 5000 lb or larger without dropped ceiling and, if also used for passenger service, equipped with wall pads. An oversized door (e.g., 4 ft, 0 in. or 4 ft, 6 in.) is particularly useful in handling furniture. Service elevators should have a shaftway door at every level plus easy access to the truck dock or other freight entrance as well as the lobby. These cars operate as service cars normally but can serve as passenger cars in peak periods to alleviate congestion and delay. This fact is particularly useful in marginal service designs. See also Table 23.11 for approximate building costs.

(b) Apartment Buildings. Studies indicate that apartment building traffic depends not only on population but also on location and type of tenant. Buildings with many children experience a school-hour peak; buildings in midtown with predominantly adult tenancy exhibit evening peaks due to the homecoming working group and out-going dinner traffic. Where two cars are re-

TABLE 23.11 **Office Buildings: Cost of Elevator and Electric Work**

Item	Number of Stories		
	20	35	60
Elevator work	10.9%	11.9%	12.2%
Electric work	13.3%	12.6%	12.2%

OtisPlan Single Group Performance

Single Deck Up Peak

name: Sample Problem
floors above lobby: 14
lobby height: 12.0ft
average floor height: 12.0ft
express zone height: 0.0ft
door time: 3.3sec
12ft flight time: 4.3sec
car acceleration: 4.00ft/sec/sec
high call reversal: 0.0
up probable stops: 0.0

population per floor: 100people
% counterflow: 0.0
% interfloor: 0.0
number of cars: 5
car speed: 500.0fpm
car capacity: 3000lbs
up car loading: 16people/car

added trip time: .0sec

(Use 0 to calculate high call
 reversal and probable stops)

*** RESULTS ***

round trip time: 144.0sec
interval: 28.8sec
up handling capacity: 11.9%/5 min

high call reversal: 13.5
up probable stops: 9.7

(a)

OtisPlan Single Group Performance

Single Deck Up Peak

name: Sample Problem
floors above lobby: 14
lobby height: 12.0ft
average floor height: 12.0ft
express zone height: 0.0ft
door time: 3.3sec
12ft flight time: 4.3sec
car acceleration: 4.00ft/sec/sec
high call reversal: 0.0
up probable stops: 0.0

population per floor: 100people
% counterflow: 3.0
% interfloor: 1.0
number of cars: 5
car speed: 500.0fpm
car capacity: 3000lbs
up car loading: 16people/car

added trip time: 0.0sec

(Use 0 to calculate high call
 reversal and probable stops)

*** RESULTS ***

round trip time: 183.1sec
interval: 36.6sec
up handling capacity: 9.4%/5 min

high call reversal: 13.5
up probable stops: 9.7

(b)

Fig. 23.19 *Printout of the results of a computerized elevator selection program.*
(a) Up-peak conditions, assuming no down stops (0% counterflow) and no interfloor traffic. The results
correspond very closely to those obtained manually in Example 23.1 (see text).
(b) Addition of even light counterflow and interfloor traffic can seriously affect the system's performance,
as can be seen from the round trip, interval, and handling capacity figures. (Courtesy of Otis Elevator Co.)

quired, the second car should function both as a service car and a passenger car. The cars may be banked or separated, as desired. If a single car is used, it should be service elevator size.

Self-service collective control is the general choice, with provision for attendant control in prestige buildings. With small cars and short rise, a swing-type manual corridor door is acceptable; in larger installations both the cab and corridor door should be the power-operated sliding type.

Service elevators must be large enough to handle bulky furniture and should therefore be at least 4000 lb, with a 48-in. door and high ceiling. Hoistways must be isolated from sleeping rooms by lobbies or other space. Similarly, machine rooms must be isolated because the starting and stopping of motors and other machine room noises are an effective barrier to sound sleep. Security arrangements are discussed in Section 23.44.

(c) Hospitals. As mentioned in Table 23.6, the governing factor in the determination of elevator requirements may be either normal hospital traffic or visitor traffic, depending on the visiting-hour schedule. Due to the large volume of vehicular traffic such as stretcher carts, wheelchairs, beds, linen carts, and laundry trucks, the elevator cabs are much deeper than the normal passenger type. This type of car, when used for passenger service, holds more than 20 persons and therefore gives slow service. For this reason, it is occasionally advisable to utilize some normal passenger cars in addition to hospital-size cars, particularly in large hospitals.

The use of tray and bulk carts in food service imposes a considerable load on the elevators before, during, and after meals, and passenger service is seriously disrupted. To alleviate this congestion and delay, many architects and hospital administrators prefer the use of dumbwaiter cars or another of the many types of materials-handling system that can handle a 15½ × 20 in. food tray. These systems can also be used for transporting pharmaceuticals and other items and are discussed in Sections 24.18 and 24.19.

Elevators should be grouped centrally, although separated by type of use. Car control is normally self-service collective.

Population of the hospital may be estimated from Table 23.7. Experience has shown that a carrying capacity of 45 passengers in a 5-min period is adequate (estimating each vehicle as equivalent to 9 passengers).

Intervals should not exceed 1 min. All recommendations for service to the handicapped should be adopted (see Section [23.11]).

(d) Retail Stores. Retail stores present a unique problem in vertical transportation inasmuch as the objective is partially to transport persons to specific levels and partially to expose the passengers to displayed merchandise. For this reason, modern stores rely heavily on escalators, with one or two elevators intended for use by staff and handicapped persons. When, for some reason, it is desired to equip a store with elevators, use the recommendations shown in Table 23.9, calculated for a load of between 10% and 20% of store population. Control should be automatic, selective collective. Cars are arranged in a straight line to facilitate loading and waiting.

THE PHYSICAL PROPERTIES AND SPATIAL REQUIREMENTS OF ELEVATORS

23.36　Shafts and Lobbies

The elevator lobbies and shafts form one of the major space factors with which the architect is concerned. The elevator lobby on each floor is the focal point from which the corridors radiate for access to all rooms, stairways, service rooms, and so forth. Such lobbies obviously must be located one above the other. The ground-floor elevator lobby (also called the lower terminal) must be conveniently located with respect to main entrances. The equipment within or adjacent to this area should include public telephones, building directory, elevator indicators, and possibly a control desk (see Fig. 23.12).

All lobbies should be adequate in area for the peak-load gathering of passengers to ensure rapid and comfortable service to all. The number of people contributing to the period of peak load (15- to 20-min peak) determines the required lobby area on the floor.

Not less than 5 ft^2 (0.5m^2) of floor space per person should be provided at peak periods for waiting passengers at a given elevator or bank of elevators. The hallways leading to such lobbies should also provide at least 5 ft^2 (0.5m^2) per person, approaching the lobby. Under self-adjusting relaxed conditions, density is about 7 ft^2 [0.65m^2] per person. However, in peak periods crowding occurs, reducing this 3 to 4 ft^2 per person [3–4 persons/m^2]. An acceptable compromise is 5 ft^2 per person.

Car size	D	W
2500 lb	70"	8' 6"
3000 lb	7' 6"	8' 6"
3500 lb	8' 2"	8' 6"
4000 lb	8' 2"	9' 6"

(a)

Car size kg	speed m/s	D mm	W mm
1000	2.5	2200	2400
1250	2.5	2200	2600
1600	2.5	2500	2600
2000	2.5	2500	2600
1000	4.0	2200	2400
1250	4.0	2200	2600
1600	4.0	2500	2600
1250	6.0	2250	2500

(b)

Fig. 23.20 *Rough hoistway dimensional data to be used in architectural single-line planning stage. (a) American elevator sizes and dimensions. (b) European elevator sizes and dimensions.*

The main lower terminal of elevator banks is generally on the street-floor level, although it may be on the mezzanine level, when the elevations of the street entrances vary so that one side of the building is at mezzanine level whereas another entrance is lower. Such a situation is ideal for the use of escalators, which economically and rapidly carry large numbers of persons between levels, thus making practical and efficient a single main lower elevator terminal. The upper terminal is usually the top floor of the building. Typical dimensional data and lobby arrangements are shown in Figs. 23.20 to 23.22.

23.37 Dimensions and Weights

Manufacturers and elevator consultants will supply standard layouts for elevators including dimensions, weights, and structural loads on request. Furthermore, to assist in preliminary design, major manufacturers have agreed on and publish a set of Standard Elevator Layouts via

their trade organization, the National Elevator Industry, Inc. One such standard is reproduced in Fig. 23.23 for 500- to 700-fpm gearless units in the full range of car capacities. These standards are available from the NEII at 185 Bridge Plaza North, Rm. 310, Fort Lee, NJ 07024.

As may be seen from Fig. 23.23, it is necessary, in providing for an elevator installation, to consider such factors as the depth of the pit, the dimensions of the hoistway, the clearance from the top of the hoistway to the floor of the penthouse, the size of the penthouse, and the loads that must be carried by the supporting beams.

The penthouse floor (and secondary level floor where required) are located above the shaft of each elevator and need approximately 1½ stories of additional height above the top of the support beam of a given elevator when it is standing at its top-floor location. The actual floor area required by the elevator traction machine and its controls is roughly two times the area of the elevator shaft itself. The machine room contains the bulk of the elevator machinery. Because some of this equipment must be moved for maintenance, it is advisable to furnish an overhead trolley beam that can be used during installation as well. The maximum beam load is supplied by the elevator manufacturer.

Some typical machine room dimensional data are listed in Table 23.12, taken from actual installations. Because of the multiple drive options and the flexibility in equipment arrangement, no general conclusions can be drawn from these figures;

TABLE 23.12 Typical Machine Room Dimensions

Bank Description		Machine Room Dimensions (Rounded to Nearest Foot)
2—2500 lb	125 fpm	16 × 18
3—2500 lb	350 fpm	19 × 26
3—2500 lb	700 fpm	21 × 26
6—3500 lb	500 fpm	26 × 29
6—3500 lb	700 fpm	27 × 27
6—3500 lb	1000 fpm	27 × 27
6—5000 lb	450 fpm	31 × 33

Fig. 23.21 *Lobby groupings for single zones: three-, four-, six-, and eight-car groups.*

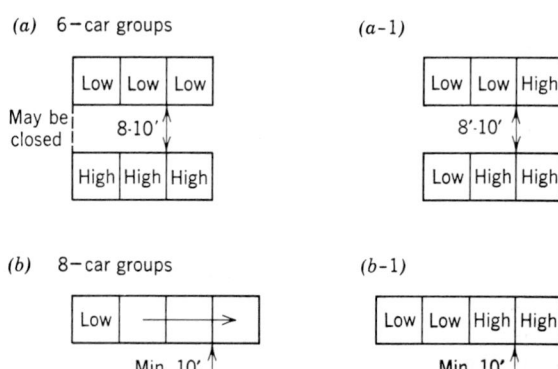

Fig. 23.22 *Lobby groupings for multiple zones. Arrangement* (a) *is preferable to* (a-1), *and* (b) *to* (b-1). *Groups with more than four cars in a row are not used because end-to-end walking time would excessively lengthen landing stops and hence total travel time.*

ELECTRIC PASSENGER ELEVATORS

RATED SPEEDS 500-700 fpm

Fig. 23.23 *Typical elevator installation dimensional data. (Reproduced with permission of National Elevator Industry, Inc., 185 Bridge Plaza North, Fort Lee, NJ 07024. Copyrighted 1992,* Vertical Transportation Standards, *7th edition.)*

(Fig. 23.23 continued on p. 1428)

TRANSPORTATION

"OH" BOTTOM OF BEAM (Note 1) ft-in.					
SPEED (fpm)	RATED LOAD (lb)				
	2000	2500	3000	3500	4000
500	17-6	17-6	18-3	18-3	17-6
600	18-6	18-6	18-6	18-6	18-6
700	20-6	20-6	20-6	20-6	20-6

CAR & HOISTWAY								
RATED LOAD (lb)	AREA ■	ft-in. (Note 11)					ENTRANCE (Note 10)	
		A	B	C	D	E	TYPE	
2000	24.2	5-8	4-3	7-4	6-11	3-0	SSSO ▲	SSCO
2500	29.1	6-8	4-3	8-4	6-11	3-6	SSSO ✱	SSCO ▲
3000	33.7	6-8	4-7	8-4	7-5	3-6	SSSO ✱	SSCO ▲
3500	38.0	6-8	5-3	8-4	8-1	3-6	SSSO ✱	SSCO ▲
4000	42.2	7-8	5-3	9-6	8-1	4-0	SSCO ✱	

"P" PIT DEPTH (Note 3) ft-in.					
SPEED (fpm)	RATED LOAD (lb)				
	2000	2500	3000	3500	4000
500	10-1	10-1	10-1	10-1	10-1
600	11-5	11-5	11-5	11-5	11-5
700	11-5	11-5	11-5	11-5	11-5

MACHINE ROOM ft-in.					
RATED LOAD (lb)	W1 WIDTH	W2 WIDTH	"Y" DEPTH FOR RATED SPEED (fpm)		
			500	600	700
2000	7-4	15-0	18-6	18-6	18-6
2500	8-4	17-0	18-6	18-6	18-6
3000	8-4	17-1	18-6	18-6	18-6
3500	8-4	17-1	18-6	18-6	18-6
4000	9-6	19-5	18-6	18-6	18-6

Notes
1. "OH" Dimensions are based on a 8'-4" overall car height.

2. Supports for elevator machine beams not by elevator supplier.

3. For travel greater than 400'-0" increased pit depth may be required. Consult elevator supplier.

4. 3'-6" x 7'-0" Recommended.

5. Pit ladder not by elevator supplier.

6. Dividing beams, not by elevator supplier, to be designed to sustain rail forces. Consult elevator supplier.

7. When compliance with seismic risk zone 2 or greater requirements is anticipated, provide additional hoistway space.

Notes
■ Maximum allowable inside car area in ft^2 per ASME A17.1, Rule 207.1.

✱ These car dimensions and entrance types provide wheelchair accessibility and accommodate an ambulance type stretcher (76 in. x 24 in.) in the horizontal position.

▲ These car dimensions and entrance types provide wheelchair accessibility.

Fig. 23.23 (continued)

they are listed simply to give a conceptual picture. A manufacturer's layout giving dimensional data for hoistway and machine room is shown in Fig. 23.24.

When penthouse space is not available and a hydraulic unit is not desired, a basement traction unit, also referred to as an *underslung arrangement,* can be used. These units are always low speed (100–350 fpm) and are therefore applicable where rise is limited and traffic is light to medium. Figure 23.25 shows a typical shaft section for this design with car and dimensional data.

23.38 Structural Stresses

For the purpose of structural design, it is necessary to know the overhead load that must be supported by the foundations, by structural columns extending upward to the penthouse, and by the main beams that support the penthouse floor and subfloor. These loads (reactions) are supplied by manufacturers and usually include the actual dead weights of equipment when the elevator is not in motion, plus the added weight caused by the momentum of all moving parts and passengers

Fig. 23.24 *Manufacturers layout data for a bank of six 3500-lb (1588-kg), 700-fpm (3.45-m/s) gearless passenger elevators. Equipment shown in the machine room is thyristor control for dc traction machines. Because each controller has the capability of group supervisory control in this design (Otis Elevonic 411), no separate group supervisory equipment is shown. No additional space would be required if a UMV drive with m-g sets were selected rather than thyristor control. (Courtesy of Otis Elevator Co.)*

TRANSPORTATION

Basement Type–Moderate Speed

Building height restrictions, aesthetics, and the possibility of future floors are three reasons to utilize a basement type application. This type of elevator system eliminates the use of an elevator penthouse. 1:1 or 2:1 roping arrangements are available to minimize the space required for your building design.

NOTE: Machine may be located at any suitable elevation other than the basement.

CENTER OPENING ENTRANCE APPLICATION
(N/A FOR 2000 LB. CAP.)
(SINGLE SLIDE OPTIONAL)

2:1 ROPING

2:1 ROPING ARRANGEMENT

Overhead dimension is reduced to a minimum. Greater pit depth is required due to the sheave arrangement as compared to a 1:1 installation.

SIZES & CAPACITIES

CAPACITY	2000 lbs. (907 kg)		2500 lbs. (1134 kg)		3000 lbs. (1361 kg)		3500 lbs. (1588 kg)	
A	6'-0"	(1829)	7'-0"	(2134)	7'-0"	(2134)	7'-0"	(2134)
B	5'-0"	(1524)	5'-0"	(1524)	5'-6"	(1676)	6'-2"	(1880)
C	3'-0"	(914)	3'-6"	(1067)	3'-6"	(1067)	3'-6"	(1067)
D	8'-0"	(2438)	9'-0"	(2743)	9'-0"	(2743)	9'-0"	(2743)
E	6'-10"	(2083)	6'-10"	(2083)	7'-4"	(2235)	8'-0"	(2438)
F	5'-4"	(1626)	5'-4"	(1626)	5'-10"	(1778)	6'-6"	(1981)
G	1'-0"	(305)	1'-0"	(305)	1'-0"	(305)	1'-0"	(305)
H	1'-0"	(305)	1'-0"	(305)	1'-0"	(305)	1'-0"	(305)

MINIMUM MACHINE ROOM OVERHEAD & PIT DIMENSIONS

SPEED	100 FPM (0.51 m/s)		200 FPM (1.02 m/s)		250 FPM (1.27 m/s)	
L	10'-6"	(3200)	10'-6"	(3200)	10'-6"	(3200)
O	13'-0"	(3962)	13'-2"	(4013)	13'-7"	(4140)
P	5'-6"	(1676)	6'-6"	(1981)	6'-11"	(2108)

NOTES:
1. If a counterweight safety is applied, consult Montgomery KONE
2. Layouts and dimensions shown are for machine located at rear of hoistway.
3. Add 4" (102) to dimension D in seismic zones.
4. Other capacities, speeds, and arrangements are available.

Fig. 23.25 *Typical data for basement traction machine (underslung) arrangement, used where penthouse is unavailable or undesirable. (Courtesy of Montgomery KONE.)*

when the elevator is at top speed and is suddenly stopped rapidly by the safety devices.

POWER AND ENERGY

23.39 Power Requirements

The power required by an elevator drive is that required to perform the necessary traction and to overcome friction. Because power is equal to the *rate* at which work is done, the elevator motor size is directly proportional to the speed of the system. In other words, it requires proportionately more power to lift a 3000-lb car at 700 fpm than at 200 fpm. This relationship is shown in Fig. 23.26, which shows the minimum size of a dc elevator traction motor as a function of speed, for different capacity cars. (For power data on hydraulic elevators, see Section 25.11.) As friction is higher in geared machines than in gearless units, its traction motor must be larger for the same car speed. The size of the traction machine shown in Fig. 23.26 is independent of the power supply design (m-g set, VVVF, thyristor control) because it is determined purely on traction system requirements. (In actual practice, however, traction motors with VVVF control are frequently smaller because they operate more efficiently.)

An elevator moves only about 50% of the time, the remainder being spent standing at various landings. As the number of cars in a bank increases, the probability of *all* the cars being in operation simultaneously decreases, resulting in a system demand factor less than 1.0. The factor for different group sizes is shown directly in Fig. 23.26.

As an example of the use of the curves, consider a bank of five 3500-lb, 600-fpm units. From Fig. 23.26, each car requires 48 hp:

group demand factor = 0.67

total instantaneous power required = 5 × 48 × 0.67 = 160 hp

Note that this is the *traction motor* power requirements. If an m-g set with an overall efficiency of 80% were used to drive the traction motor, the

Fig. 23.26 *Elevator traction motor power requirements per car. An m-g set motor (if used) is approximately 20% larger than the traction machine.*

elevator system power requirement is

$$\text{system power} = \frac{160\,\text{hp}}{80\%\,\text{eff}} = 200\,\text{hp}$$

which would have to be provided by the building power system. If a solid-state power supply system with a (typical) efficiency of 92% were used, the system power requirement would be only

$$\text{system power} = \frac{160\,\text{hp}}{92\%\,\text{eff}} = 174\,\text{hp}$$

which is a 13% reduction from the previously calculated 200 hp requirement.

23.40   Energy Requirements

The energy used by an elevator is essentially the system friction, including the heat generated by the brakes plus the electrical losses in the traction motor and power supply equipment—rotary or solid state. The energy expended in raising the car and its passengers is simply stored as potential energy. It is *returned to the power system* when the car and passengers descend, via the system of regenerative braking in use in almost all elevator systems. Refer to Fig. 23.27, which shows the approximate efficiencies of the components of a typical system. With these data, we are able to calculate a system's energy consumption.

EXAMPLE 23.2. Given a system of five 3500-lb, 600-fpm (gearless) cars, calculate:

(a) Heat generated in the machine room during peak periods. Assume solid-state control.

(b) Approximate monthly energy cost using a combined demand/energy rate of $0.08/kWh.

SOLUTION.

(a) During peak periods, the traction motor operates approximately 50% of the time and is at standstill the other half. Assume that while operating it draws 90% of full load (with VVVF power supply this figure is reduced considerably). Therefore, for one car, from Fig. 23.26,

$$\text{traction motor} = 48\,\text{hp}$$

Total loss per machine:
In controls:

$$\frac{48\,\text{hp}}{0.9\,\text{eff}} \times 90\%\,\text{load} \times 50\%\,\text{operation}$$
$$\times\,10\%\,\text{loss} = 2.4\,\text{hp}$$

In traction motor:

48 hp × 90% load × 50% operation × 20% loss = 4.32 hp

Total = 6.72 hp = 17.1000 Btu/h

Because there are five elevators operating, the total heat generated is

$$5 \times 17,000\,\text{Btu/h} = 85.500\,\text{Btu/h}$$

This is the rating of a home furnace. As a result, machine room temperature in warm climates fre-

Fig. 23.27 *Block diagram showing losses in the system per horsepower delivered to the elevator car, and the equivalent wattages. Note that the losses in a geared system are almost double those of a gearless one. Figures shown are for solid-state thyristor controls.*

quently reach 120° F! (No diversity is taken because all the machines are operating and the heating is additive; diversity is applicable only in calculating instantaneous load.)

Inasmuch as solid-state elevator equipment is much less tolerant of high ambient temperatures than were the old electromechanical switches and relays that preceded them, the machine room should be held to a maximum dry bulb temperature of 90°F. (Temperatures above 90°F can result in unreliable elevator system performance.) This can sometimes be accomplished by thermostatically controlled forced ventilation, particularly if spill air from an air-conditioning system is present. However, because machine rooms are frequently on the building roof and exposed on all surfaces to direct solar radiation, air conditioning may be necessary. It is also important to prevent machine room temperature from dropping below 55°F. This can usually be done with one or more unit heaters, which will normally operate only during the winter, and then only nights and weekends.

In actual design situations, accurate heat loss figures, which are available from manufacturers, would be used, along with accurate heat gain and heat loss calculations for the specific machine room being designed. Frequently, use of thermal insulation, thermostatically controlled louvers, and sunshading can ease the thermal load and result in appreciable financial and energy savings. Because of the high initial and operating cost of air conditioning, some manufacturers use control components that are tolerant of high temperatures. This point should be carefully examined with proposed elevator equipment manufacturers and the conclusions reflected in the project elevator specifications.

(b) To calculate the monthly energy cost, an estimate must be made of the total usage of the system. Assuming the system to be in an office building, a reasonable breakdown of operation during a 24-h day would be

2 h peak use

2 h 70% of peak

6 h 50% of peak

14 h 10% of peak

This gives an average of 30% of peak load for the bank. Therefore, per car

$$\text{energy} = 30\% \times \text{total losses} \times 24\,\text{h}$$

$$= 0.3 \times 6.72\,\text{hp} \times 24\,\text{h}$$

$$= 48\,\text{hp-h} = 36\,\text{kWh/day/car}$$

Monthly cost would be

$$36\,\frac{\text{kWh}}{\text{day}} \times 25\,\text{days} \times \$0.08$$

$$= \$72/\text{month/car}$$

$$= \$360/\text{month for the bank}$$

This figure would be lower with VVVF (variable-frequency, variable-voltage) power supply and higher for a Ward–Leonard (m-g set) arrangement. ■

23.41 Energy Conservation

A reduction in energy consumption can be accomplished by implementing the following recommendations:

For Existing Elevators

1. Increase interval during nonpeak hours.
2. Replace m-g sets with solid-state dc power supply or ac traction motors with VVVF power supply. This conserves energy not only due to higher efficiency of the power supply, but also because energy consumption of idling machines is eliminated.
3. Recycle machine room waste heat.
4. Shut down some units completely during off hours.

For Building in the Planning Stage

1. Base design on maximum recommended trip time.
2. Use the lowest speeds possible, within a type—that is, geared or gearless.
3. Use gearless equipment whenever possible.
4. After construction, implement the recommendations for existing elevators previously detailed.

Because elevator shafts have a powerful stack effect, measures should be taken to counteract this during the heating season.

TRANSPORTATION

23.42 Emergency Power

Major power failures and local "brownouts" have demonstrated forcefully the need for a standby or emergency power source of adequate size to operate a building's elevators. Few experiences are so harrowing as being trapped in the crowded confines of a small box suspended in a long vertical shaft, with little or no light, and complete strangers for companions.

A common misconception relative to elevators is that on failure of power, the cars will automatically descend to the nearest landing where exit is then possible. In actuality, the brake is set immediately on power outage and the car remains stationary. Hydraulic cars can be lowered by operation of a manual valve; *small* traction cars can be cranked to a landing by hand, but large cars are fixed in position. This is particularly bad for cars in blind shafts—that is, express shafts with no shaftway doors. In such cases escape from the cars via hatchway is not practicable and, when emergency power is not available, breaking through the shaftway walls is the only recourse.

In addition to simple inconvenience, loss of elevator service in facilities such as hospitals and mental and penal institutions constitutes a danger to life. For this reason most codes require that emergency power be available in specific building types to operate at least one elevator at a time, and for elevator lighting and communications. Many installations separate the emergency power functions, providing a generator for elevator traction power and separate individual elevator battery packs for communications, lighting, and preferably, also the car fan. The last two items can be furnished by the elevator manufacturers with the cars as an option.

The generator is normally sized to supply one elevator motor at a time, with manual or automatic switching arranged between unit controllers. Thus, each car in turn can be brought to a landing and thereafter a single car retained in service. Obviously, if it is desired to operate more than one car, a larger generator can be installed. This might well be the case in a multiwing building with critical service requirements, such as a hospital.

The amount of power required, the size of the emergency generator, and the equipment size necessary to absorb regenerative power are all data that can be furnished by the consulting engineer and the elevator manufacturer.

SPECIAL CONSIDERATIONS

23.43 Fire Safety

Most fire codes specify the procedures that the elevator control equipment must implement once a fire emergency has been initiated. Details vary somewhat, but in general the actions are these:

1. All cars close their doors and return nonstop to the lobby or other designated floor, where they park with doors open. Thereafter, they are operable in manual mode only, by use of the firemen's key in the car panel.
2. All car and hall calls are canceled and call registered lights and directional arrows extinguished.
3. The fire emergency light or message panel in each car is activated to inform passengers of the nature of the alert and that cars are returning to a designated terminal.
4. Door sensors and in-car emergency stop switches are deactivated.
5. Traveling cars stop at the next landing without opening doors and then proceed to the designated terminal.

The cars can then be used by trained personnel to transport fire personnel and equipment, and for evacuation. In the event of a false alarm, the emergency procedure can be overridden at the (lobby) control point and the system returned to normal while the source of the alarm is located. (This is a particularly important feature in large buildings with automatic fire alarm systems containing hundreds of fire, smoke, and water-flow detectors.)

23.44 Elevator Security

Elevator security has two aspects: physical security of riders, and consideration of the elevator as a portal in a building-access security system.

(a) Rider Security. This problem is particularly difficult inasmuch as the traveling elevator

Man Trying To Hide In Corner Under Camera

23.28 Wide-angle TV camera intended for elevator cab surveillance. A prominent, printed warning in the cab is an integral part of the system's effectiveness. (Photo courtesy of Visual Methods, Inc.)

is an enclosed space that can be rendered inaccessible simply by pressing the emergency stop button. Thereafter, an attacker can escape at the floor of his choice. To ameliorate this danger to an extent, elevators are equipped with alarm buttons that alert residents and security personnel, if any. Every elevator, by code, must be equipped with communication equipment. Two-way communication system with "no-hands" operation in the car is particularly effective for security. When a closed-circuit TV monitor is added, utilizing a wide-angle camera in each car (Fig. 23.28), the security problem will have been addressed to a considerable extent. Obviously, using a communication and TV system presupposes continuous monitoring of the building security desk.

At least one major manufacturer now markets a device that alarms automatically on detecting sudden violent motions or a sharp, pointed instrument. The matter of handling the alarm is problematic because an automatically locked door can be forced open manually and, furthermore, the advisability of locking a violent person in an elevator with potential victims is questionable.

(b) Access Control. This is often a matter of restricting access to (and from) a floor or car. This can be accomplished by pushbutton combination locks or coded cards, the proper use of which permit access (see Chapter 22). However,

if a second person happens to accompany the authorized person, the effectiveness of this type of access barrier is seriously compromised. In sum, the most effective security system is a combination of automatic monitoring and access devices coupled with continuous supervision by persons who know the appropriate action to take in an emergency.

23.45 Elevator Noise

As already noted, elevator operation, with its rotating, sliding, and vibrating masses, can be a cause of serious noise disturbance to quiet areas such as sleeping rooms, libraries, and certain types of office space. Noise can be reduced by the appropriate application of vibration isolators (e.g., between guide rails and the structure) and by proper control, but primarily by placing noise-sensitive areas away from shafts and machine rooms. Furthermore, the clatter and whirring sound of the older machine room, caused by relays, step switches, m-g sets, and sliding contacts, can be entirely eliminated by the use of solid-state equipment.

23.46 Elevator Specifications

Two basic types of specification for elevator equipment, as for other types of equipment, are utilized. The performance specifications describe job conditions and invite contractors to submit detailed proposals including full engineering. The burden of comparing proposals then falls on the owner, who—if competent to properly perform such an evaluation—would most probably do better to utilize an equipment-type specification in the first place.

In recent years the use of performance specifications has increased because of the advent of preengineered, premanufactured systems. These are supplied by the major manufacturers and have the following advantages:

1. Approximately 10% lower cost than a custom-designed system.
2. A complete engineered and tested system whose performance and cost are known exactly.

TRANSPORTATION

3. Rapid delivery.
4. Minimum supervision required by the owner and architect.

If architects decide to use a custom-designed system, they must prepare detailed drawings and specifications. The specifications must include:

· Elevator type, rated load, and speed.
· Maximum travel.
· Number of landings and openings.
· Type of control and supervisory system.
· Details of car and shaft doors.
· Signal equipment.
· Characteristics of power supply.
· Finishes.

This last item can be left as a dollar allowance for architectural treatment of the car interior. Because the selection of, and technical specifications for elevators are specialized and complex, the services of an elevator consultant are usually required.

In addition to the technical portions of the specifications, it is imperative that the following items be covered in detail.

(a) Owner's Responsibility. The *construction* contractor normally provides the following:

1. The hoistway, including properly designed, lighted, drained, waterproofed, and ventilated machine room and pit.
2. Access doors, ladders, and required guards.
3. Guide rail bracket supports, and support for machine and sheave beams.
4. Electric feeder terminating in a switch in machine room.
5. Hoistway outlets for light, power, and telephone.
6. Temporary light and power during construction.
7. Concrete machine foundations.
8. Vents, holes, and other work to satisfy fire codes.
9. All cutting, patching, and chasing of walls, beams, masonry, and so on.
10. Coordination of all work.
11. Any special work, as negotiated.

(b) Elevator Contractor's Responsibility. Provide complete, working, tested, and approved system in accordance with specification, plus any special work such as painting, special tests, work scheduling, and temporary elevator service.

(c) Special Job Conditions. These include work restrictions, scheduling, penalties or bonuses, test reports, and the like.

In alteration and modernization work, the problems of coordination are more complex, and an elevator contractor experienced in this type of work must be selected. To this end, in all elevator contract work, bids should be solicited from parties named on qualified bidder lists. Part of an elevator contract comprises maintenance of the installation for a specific period after completion. Contractors with poor maintenance facilities should be avoided.

23.47   Novel Designs

The elevator industry is constantly developing new equipment to improve operation and safety for standard system designs. In addition, novel designs that are essentially different from standard traction arrangements are always being developed in an attempt to increase the efficiency of space use and to decrease the high cost of standard traction machinery. Among interesting designs being developed in the first category is one that permits a car to travel horizontally in addition to its normal vertical motion, the purpose of which is to increase the number of cars using a single shaft. The second category includes a design using a linear motor (as opposed to a rotating unit) to supply traction power. These and other special designs are discussed in detail in Chapter 24.

A recently developed interesting variation of the conventional traction design, that effects a considerable space reduction is shown in Figs. 23.29–23.32. At this writing its principal applications are in low speed, low rise installations now generally serviced by hydraulic elevators, but with higher speeds and rises in development. The novelty of the design lies in the use of flat (disc shaped) synchronous a-c gearless hoisting motor, which, due to its flat disc shape, can be mounted

Top of main
car guide rail

Disk shape
gearless
a-c hoist motor

Controls

Rigid attachment
of motor to rail

Cable sheave

Hoisting ropes (cables)

Fig.   23.29 *Disc-shaped hoisting motor rigidly mounted on the elevator guide rail. The a-c synchronous motor is connected directly to the hoisting cable drive sheave with no intervening gears. Brakes and controls are built into the assembly. (Courtesy of Montgomery-KONE)*

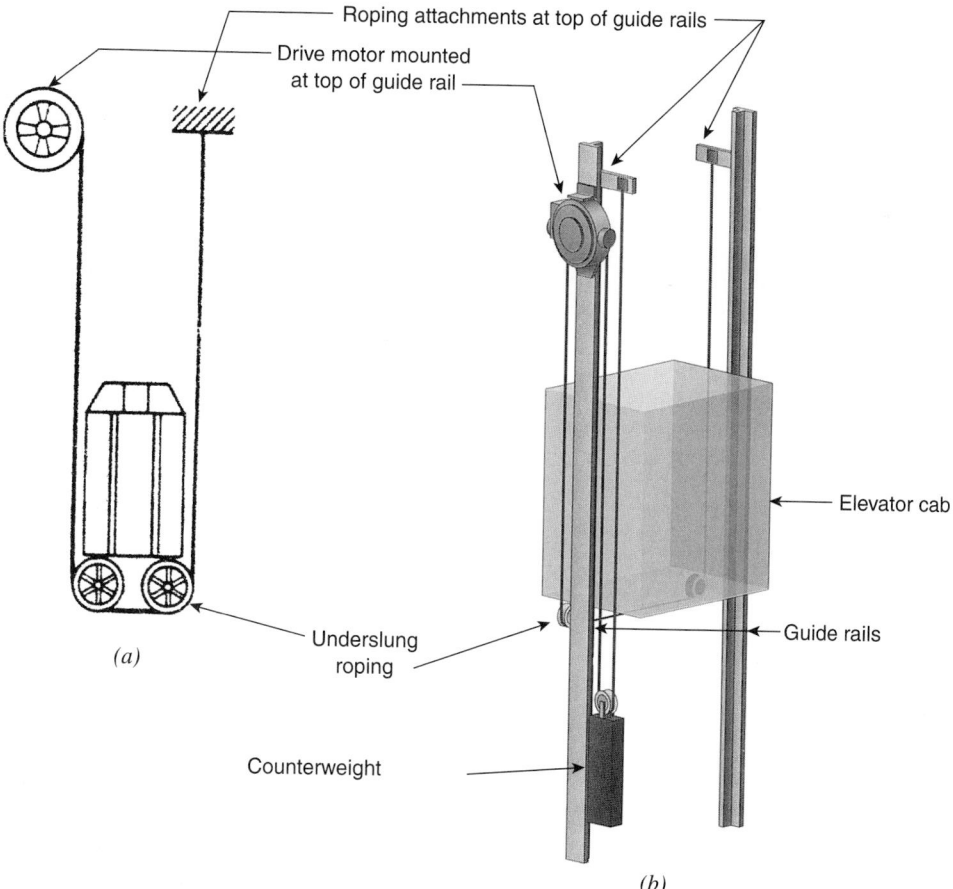

Roping attachments at top of guide rails

Drive motor mounted
at top of guide rail

Elevator cab

Guide rails

Underslung
roping

(a)

Counterweight

(b)

TRANSPORTATION

Fig. 23.30 *Schematic (a) and pictorial (b) representations of the disc-motor-driven elevator arrangement. [Part (b) Courtesy of Montgomery-KONE]*

SIMPLEX

(a)

Fig. 23.31 *(a) Section through top of hoistway showing dimensional data for a single 2500 lb, 200 fpm installation with rise up to 80 ft. Note that the drive motor occupies less than 2 ft in the width of the hoistway, and that the elevator motion and operating controls (Section 23.4) are installed in a closet 42 in. wide and approximately 20 in. deep, at the top landing.*

(b) Elevator system basic data for the simplex (single) unit shown in (a). (Courtesy of Montgomery-KONE)

(b)

MonoSpace Elevator System	
Speed	200 FPM (1.0 m/s)
Capacity	2500 lb. (1134 kg)
A	7'-0" (2134)
B	5'-0" (1524)
D	9'-0" (2744)
E	6'-4" (1931)
Entrance Types	Single Slide
Overhead (O)	12'-9" (3887) min
Minimum Travel	8'-4" (2540)
Maximum Travel	80'-0" (24384)
Landings Served	2 - 10
Pit (P)	5'-6" (1677) min
Machine Room (W x D x H)	3'-9" x 16" x 8'-0" (1143 x 407 x 2439)

directly on to the main car guide rail at one side of the shaft. See Fig. 23.29. This essentially obviates the necessity for a penthouse and large machine room above the hoistway. Due to the traction motor's position at the side of the hoistway, the car is roped in an underslung arrangement as shown in Fig. 23.30. An additional space economy is achieved by use of a small drive controller built into an alcove at the top landing. See Fig. 23.31. A pictorial hoistway representation

Fig. 23.32 *Pictorial representation of a disc type traction motor hoistway showing the system essentials. (Courtesy of Montgomery-KONE)*

in Fig. 23.32 shows the equipment arrangement, demonstrating the absence of a penthouse and the limited machine room space requirement. An additional advantage of this arrangement is that the elevator loads and reactions are borne by the (stiffened) guide rail and transferred directly to the concrete pit below the bottom landing. This reduces the reactions borne by the machine room level in conventional traction design and results in reduced structural loads. When compared to hydraulic elevators this design exhibits consider-able energy economy due to its use of a gearless traction machine.

REFERENCES

Stein, B., Reynolds, J., and McGuinness. W. (1986). *Mechanical and Electrical Equipment for Buildings,* 7th ed., Wiley, New York.
Strakosch, G. R. (1983). *Vertical Transportation, Elevators and Escalators,* 2nd ed., Wiley, New York.

24

VERTICAL TRANSPORTATION: SPECIAL TOPICS

Chapter 23 dealt primarily with the construction, control, and selection of traction passenger elevators (i.e., units that are raised and lowered by cables attached to a crosshead beam above the cab). These elevators move in dedicated shafts: one per car, with the traction machine usually placed at the top of the shaft. This chapter discusses elevators with special shaft arrangements, special cars, lifting arrangements other than cable traction, and elevators designed primarily to carry freight rather than passengers. The chapter closes with a fairly extensive section devoted to material handling and movement in buildings using elevator-like traction cars and other means.

SPECIAL SHAFT ARRANGEMENTS

The fact that each traction elevator in a building has a full-height basement-to-penthouse shaft that occupies valuable building space, and that the elevator only uses part of the space only part of the time, has always disturbed finance-conscious building operators. Development of high-speed drives and very sophisticated control systems has been spurred, in part, by the desire to increase the efficiency of elevator use of building space. An alternate approach to this problem lies in innovative use of shaft space so that more than one car can use a single shaft. A detailed analysis of these solutions is beyond our scope here; however, a rapid review is of interest.

24.1 Sky Lobby Elevator System

For skyscraper buildings such as the World Trade Center in New York and multiple-use buildings

such as the John Hancock Center in Chicago—both of which are, in effect, stacked multiple buildings—the elevator solution involves transporting large groups of people from the street lobby to an upper lobby called a *sky lobby* or *plaza*. At this point, the passengers transfer to another elevator to continue their upward journey.

The traditional approach in tall, single-purpose buildings is to zone the structure, and have banks of elevators serving groups of floors. The difficulty with this solution is that the upper zone cars travel through a long, expensive blind shaft to reach the floors being served. The sky plaza approach, stacks (figuratively) two (or more) shafts vertically with a lobby in between. The effective result is to have *two* cars operating in the equivalent of a single full height shaft. To mitigate the annoyance felt by passengers resulting from a lengthened trip, the sky lobby is attractively appointed and may serve as an upper observation floor. Also, an elevator trip broken up by a lobby stop is thought to be less annoying than a somewhat shorter but still long, uninterrupted one. Finally, this plan is used principally where a clear differentiation in building use occurs; usually consisting of office areas below and residential above, as in the John Hancock building. Thus, most lower-section occupants never use the sky lobby. Passengers headed for upper zones can use a sky lobby "shuttle" that travels from the entry level to the sky lobby and then continues the trip in an upper section car.

Some designs also provide for a single-shaft upper zone elevator group that serves upper zone passengers who prefer not to use the sky lobby. A schematic drawing of the elevator arrangement in the Hancock building in Chicago should make the advantages of the system clear (Fig. 24.1).

John Hancock Center, Chicago.

Fig. 24.1 The use of a sky plaza in this dual-purpose building eliminates the blind shafts below the upper section (residential) elevators. The lengthened trip time is well tolerated by riders destined for the upper residential section, whereas the lower office section remains unaffected and retains its short interval and trip time. (G.R. Strakosch, Vertical Transportation, 2nd ed., John Wiley & Sons; reprinted with permission.)

24.2 Double-Deck Elevators

This is an old technique, recently revived and revised to answer the needs of tall buildings such as the Sears and Citicorp towers. Its principal purpose is to limit the otherwise prohibitively large amount of space occupied by elevator shafts (Fig. 24.2). The double-deck car increases shaft capacity, decreases the number of local stops, and increases the rental area available. This technique can also be combined with sky lobbies for further space economy, as was done in the Sears Tower.

HYDRAULIC ELEVATORS

24.3 Conventional Plunger-Type Hydraulic Elevators

The elevators discussed thus far are traction types; that is, they are raised and lowered as a result of the tractive force of cables attached to, or passing under, the car. These cables, in turn, are raised and lowered by a motor-driven traction sheave. In contrast to these, the conventional hydraulic or plunger elevator is raised and lowered quite simply, by means of a movable rod (plunger) rigidly fixed to the bottom of the elevator car. The absence of cables, drums, traction motors, elaborate controllers, safety devices, and penthouse equipment makes this system inherently inexpensive and often the indicated choice for low-speed (up to 200 fpm), low-rise (up to 65 ft) applications, where construction of the plunger pit does not present difficulties and/or the absence of a penthouse is desirable.

As a matter of historical interest, the first hydraulic elevators used water as the system fluid, supplied at sufficient pressure from roof water tanks. The tanks in turn were kept full by building water pumps. All hydraulic elevators today use oil and obtain their motive power from a sealed oil-piping circuit powered by an oil pump.

The components of a typical hydraulic unit are shown in Fig. 24.3. This system operates very much the same way as a hydraulic automobile jack. Oil from a reservoir is pumped under the plunger thereby raising it and the car. The pump is stopped during downward motion, the car being lowered by gravity and controlled by the action of bypass valves, which also control the positioning of the car during upward motion. Control systems normally used are similar to those for traction types—for example, collective, and selective collective. Similarly, door arrangements are the same as in traction types—that is, single slide, center opening, and two speed. Automatic level-

Fig. 24.2 (a) *The double-deck car serves to increase cab capacity and decrease shaft space. Coincidence of calls in the upper and lower cabs reduces the number of local stops made by the double car.* (b) *Graphical representation of the space saved in a 40-story building by the use of double-deck elevators. (Courtesy of Otis Elevator Co.)*

ing is readily available and is standard on all automatic units.

From the point of view of architecture and construction, the major inherent advantage of hydraulic units is the absence of overhead machine room, penthouse, and traction equipment. In Fig. 24.3, we see that only the guide rails project above the car and, if these are camouflaged, the impression of a freestanding elevator car is given. This effect can be used to good advantage *inside* large, open spaces such as exist in shopping malls and stores; when combined with glass-enclosed, observation-type cabs, the effect is striking (see Figs. 24.12 and 24.13).

Additional advantages of hydraulic elevators over traction units include:

· The elevator load is carried by the ground and not the structure. By contrast, traction units place a large structural load on the penthouse and machine room floors and overhead steel as well.

· The hoistway is smaller due to the absence of a counterweight and its guide rails.

HOLED HYDRAULIC

Plunger

Fluid Tank

Controller

Two Car Buffers

In-Ground Cylinder

Fig. 24.3 *Phantom view of a conventional "holed" hydraulic elevator, so-called because the elevator car rests on a hydraulically activated steel plunger that descends into a "hole" in the ground. The hole is actually a buried hollow steel cylinder, into which oil is pumped, under pressure, to raise the car. The oil pump is inside the fluid tank. (Illustration courtesy of Otis Elevator Co.)*

· Cars can be lowered manually by operation of oil valves. This is particularly useful and important in the event of control equipment failure or power failure.

· There is essentially no limit to the load that can be lifted.

The major inherent *disadvantage* of the standard hydraulic elevator is its operating expense. Because it is not counterweighted, it requires a relatively large motor to drive the oil pump, and *all* the energy is lost in heat (see Table 23.2). As an example of the operating cost, consider a 3500-lb, 125-fpm hydraulic unit in a department store. Such a unit requires a 40-hp motor. Assuming the unit to be in operation 10 h/day, 6 days a week, and assuming a normal 60% time-in-operation figure, we have (remembering that the motor operates only in the up direction)

$$\text{energy used/day} = \frac{40\,\text{hp}}{0.82\,\text{eff}} \times \frac{0.746\,\text{kW}}{\text{hp}}$$

$$\times\,60\% \times 10\,\text{h} \times \tfrac{1}{2}$$

At \$0.08/kWh, we have

$$\text{monthly energy cost} = 110 \times \frac{6\,\text{days}}{\text{week}}$$

$$\times \frac{4.33\,\text{weeks}}{\text{month}} \times \$0.08$$

$$= \$229/\text{month}$$

Compare this to the previously calculated (Section 23.40) monthly energy cost of \$72 when using solid-state equipment and \$126 with m-g set power supply, for a 3500-lb, 600-fpm traction car, for an appreciation of the value of a counterweight.

Other *disadvantages* of hydraulic units as compared to traction units include

· Limited to low rise, low speed applications

· Ride quality is inferior to that of a good traction unit, although it is entirely acceptable for residential, mercantile, and industrial applications.

· Because oil viscosity changes with temperature, the ambient temperature of the space containing the pump and oil storage tank must be controlled to maintain ride quality and performance.

· The high inrush current taken by the pump each time it starts, which is every time the elevator travels upward, requires a "stiff" power supply to avoid problems of light flicker and other undesirable line-voltage fluctuations.

· Noise from the pump and motor plus piping

noise can be disturbing. This problem can be ameliorated by moving the pump mechanism (up to 50 ft from the elevator shaft).

Plunger-type hydraulic units have fallen into some disfavor recently because of problems caused by oil seepage into the ground. The plunger travels in a buried steel casing that, despite excellent butyl or other liquid-tight coatings, corrodes after an extended period and leaks oil. In many locations this violates stringent EPA regulations dealing with groundwater pollution. Because repair of such an oil leak is expensive and entails extended elevator outage, other arrangements as detailed in the next two sections have come into increasing use.

Hydraulic elevators are best applied to low-speed, low-rise applications such as office buildings and residential buildings up to four stories in height, low-rise department stores, malls, basement and garage shuttles, theater elevators, and stage lifts and freight applications of all sorts; in particular, those intended for very heavy loads. Another very useful application is for use of handicapped persons who cannot use escalators or negotiate stairs. Typical layout and dimensional data for standard plunger units are given in Fig. 24.4, along with capacities and application recommendations.

24.4 Hole-less Hydraulic Elevators

Where drilling a plunger hole presents difficulties, a hydraulic installation using a telescoping plunger or a roping arrangement can be installed. The telescoping jack design is shown schematically in Fig. 24.5 in a single jack arrangement. As this causes a lateral stress in the building due to the cantilevered car, a dual-jack arrangement is used more frequently. When supported on both sides as shown in Fig. 24.6, all of the vertical elevator load is transferred directly to the ground. The telescoping jack, in which all sections move simultaneously, is more complex than a simple plunger unit and requires more maintenance, although maintenance is simplified by the fact that the entire length of the jack is readily accessible. The ride in the telescoping jack arrangement is not as smooth as that on a straight plunger elevator due to the simultaneous movement of the jack's telescoping sections, causing a degree of jerk.

24.5 Roped Hydraulic Elevators

The roped hydraulic arrangement is simpler than a telescoping plunger unit because it uses only a single moving jack section as compared to two or even three in the telescoping unit, for the same rise. It accomplishes this by using 2:1 roping, which means that the car travels twice as far as the piston. This is quite simply accomplished by passing the rope over a pulley in the piston crosshead. One end of the rope is attached to a fixed point in the pit below the car, and the other end is attached to the base of the car (Fig. 24.7). The piston (plunger, jack) lifts the crosshead, which in turn lifts the car twice as far.

The arrangement shown in Fig. 24.7 uses a single jack and a cantilevered car. Other arrangements use two jacks to eliminate lateral building load (Fig. 24.8). A 2:1 roping is standard in the United States. The simplicity and reliability of the single- or double-jack roped arrangement has made it in recent years by far the most common choice for low-rise, light- to medium-duty hydraulic elevators. Because it is a roped unit, it is equipped with a slack-rope safety in addition to the other safeties used on direct-connected hydraulics.

FREIGHT ELEVATORS

24.6 General

The preceding material, which dealt with passenger traffic, had as its prime consideration the most effective solution to the problem of transporting a given number of persons vertically. The problem with respect to freight elevators is similar: to transport a given tonnage of freight efficiently, economically, and quickly. The service car in a facility can be considered to be a freight car but, if utilized for passenger duty at all, it must meet passenger service requirements. If passenger duty is not required, or if much freight is to be handled, a car designed specifically for freight is used.

Factors to be considered in freight elevator selection, in addition to tonnage movement per hour, are size of load, method of loading, travel, type of load, type of doors, and speed and capacity of cars. Due to the interrelation of these factors the actual process of selection involves making

TRANSPORTATION

CAPACITY (LBS)	STD. SPEED (FPM)	MAXIMUM STANDARD		CLEAR OPENING (ii)	PLATFORM SIZE		MINIMUM CLEAR CAB INSIDE		HOISTWAY					
		OPENINGS	TRAVEL (i)		A WIDTH	B DEPTH	CA WIDTH	CB DEPTH	C CLEAR WIDTH	D(iv) WALL TO WALL	R	S	E	Q
2000	100	8	70'-0"	3'-0"	6'0"	5'-1"	5'-8"	4'-3"	7'-4"	5'-9"	3'-5"	11"	2'-1"	8"
	125		77'-0"											
2500	100	8	70'-0"	3'-6"	7'0"	5'-1"	6'-8"	4'-3"	8'-4"	5'-9"	3'-11"	11"	2'-11/2"	8"
	125		76'0"											
	150		76'-0"											
3000	100	8	70'-0"	3'-6"	7'0"	5'-7"	6'-8"	4'-9"	8'-4"	6'-3"	3'-11"	11"	2'-41/8"	8"
	125		72'0"											
	150		73'-0"											
3500	100	8	69'-0"	3'-6"	7'0"	6'-3"	6'-8"	5'-5"	8'-4"	6'-11"	3'-11"	11"	2'-81/8"	8"
	125		69'0"											
	150		69'-0"											
4000	100	8	66'-0"	4'-0"	8'0"	6'-3"	7'-8"	5'-5"	9'-4"	6'-11"	—	—	2'-81/2"	8"
	125		66'-0"											
	150		66'-0"											

Fig. 24.4 *Typical dimensional, capacity, and layout data for conventional plunger-type hydraulic elevator. Door systems are single slide (SS) or center opening (CO). The car can be used as a viewing type unit by utilizing a glass wall at the back of the car. (Extracted with permission from published data of Schindler Elevator Corp.)*

Cantileved car

Car support rails

3-section
telescoping jack

To oil pump
and reservoir

Fig. 24.5 *Hole-less hydraulic elevator driven by a single telescoping jack. The cantilevered car exerts a lateral structural load on the building. This arrangement is used less than the two-jack mechanism shown in Fig. 24.6. (Courtesy of OTIS Elevator.)*

Car Panel

Electronic Detectors

Dual Lifting Pistons

Microprocessor Control

Submersible Pump
and Motor

No Drilling Required

Fig. 24.6 *Dual-jack hole-less hydraulic elevator. The balanced vertical load is borne by the ground only, with no building component. The unit illustrated is suitable only for a low rise (two-stop) installation, with no greater than a 2500 lb car. (Courtesy of OTIS Elevator.)*

assumptions on the basis of recommendation and then arriving at a solution, very much as was done for passenger elevators.

It is beyond the scope of this book to discuss in detail the selection of material-handling elevators because of the large number of considerations involved. Therefore, we restrict the following paragraphs to descriptive material and recommendations. Also, inasmuch as freight elevators form such an important link in industrial processes, a careful and detailed material-flow study should be made before freight elevators are selected. Elevator manufacturers' representatives and materials-handling consultants can be very helpful in this regard.

24.7 Freight Car Capacity

Figure 24.9 is a section through a typical traction type freight car shaft. Capacities corresponding to a specific platform size are due to the varying square-foot loads that are permissible. Cognizance of this is taken by the ANSI Code (A17.1) for Elevators, which has established five load classifications for freight elevators.

Class A. General Freight Loading, by hand truck. Single items may not exceed 25% of the car-rated load. Rated load is based on 50 pounds per square foot (psf) of net inside platform area.

TRANSPORTATION

Pulley & Crosshead

2:1 Roping

To Pump, Reservoir
& Controls

Car

Cantilevered frame
supports car

Base of car
attachment

Pit attachment

Fig. 24.7 *Low-rise residential-type elevator of the roped hydraulic type. The cantilevered car is lifted by cables from the cable crosshead, which is in turn lifted (and lowered) by the single-section telescoping piston (jack). The 2:1 roping arrangement lifts the car twice as far as the piston travels. The power unit, which includes the oil tank, pumps, and control, is usually mounted at the lower level. Control is automatic, including automatic leveling. Depending on the specific design, the car is a 700- to 750-lb, 30- to 36-fpm unit, normally with a single-story rise. The hatchway is 16 to 26 ft², with the larger size used for a car intended to accommodate a wheelchair. (Courtesy of CEMCO, Corbett Elevator Co.)*

2:1 ROPED-HYDRAULIC HOLELESS ELEVATOR
Isometric View

Class B. Motor Vehicle Loading. Car will carry automobiles or automobile trucks. Rating is based on a load of 30 psf of net inside platform area.

Class C1. Industrial truck loading; truck carried.

Class C2. Industrial truck loading; truck not carried.

Class C3. Concentrated loading; no truck used; increments greater than 25% rated capacity.

For classes C1, C2, and C3—rated load is based on 50 psf. Cars have automatic leveling.

24.8 Freight Elevator Description

As speeds are generally between 50 and 200 fpm, a geared-type traction machine or a hydraulic unit is used. The preferred system of control is collective, with variable-voltage, dc supply either umv or VVVF. However, if the car is used infrequently (fewer than five trips a day), economy is very important, accurate leveling is not essential, and a rougher ride is tolerable, then a two-speed ac rheostatic control may be used.

For low-rise installations, a hydraulic unit is most often employed. These, as with the variable-voltage dc traction units, provide accurate control, smooth operation, and accurate automatic level-

ROPED HYDRAULIC

Governor

Sheave

Plunger and
Plunger Guides

Above Ground
Cylinders

Ropes

Fluid Tank

Controller

Car Buffer

Jack Stands

Fig. 24.8 *For heavier loads than can be readily handled by the cantilevered car in a single-jack design (Fig. 24.7), a balanced dual jack (cylinder) is used, as seen in this phantom view. This design, which can accomplish somewhat higher rise than a single-jack design, in addition to much heavier loads, also requires a larger shaftway to accommodate the second cylinder and rails. (The second pulley and crosshead are ommitted for clarity in the drawing.) (Courtesy of Otis Elevator Co.)*

ing. Hydraulic units rarely exceed 60 ft in height and operate at speeds up to 125 fpm. Accessories such as governors, safeties, and brakes are similar to those for passenger elevators previously described.

Motion control panel

M-G set in UMV
system only

Geared
machine

Governor

Secondary
sheave

Limit switches

Counterweight

Guide rails

Staps and
cams

Hoist cables

Solid guide
shoes

Landing
switches

Sling

Cab

Platform

Safety

Governor
cables

Buffers

Governor tension

Fig. 24.9 *Components of a typical freight elevator installation utilizing a variable-voltage–controlled geared traction machine. The sling that lifts the car is frequently arranged with double sheaves over which the hoisting ropes pass. This roping arrangement increases the mechanical advantage of the lifting ropes.*

General-purpose freight elevators, whether traction (Fig. 24.10) or hydraulic (Fig. 24.11), in load ranges up to 20,000 lb, are standard design items, applicable to all types of commercial and industrial buildings. Heavier units are individually engineered. Units 20,000 lb and more require special safeties. As with passenger elevators, structural reactions for traction units are supplied by the manufacturer to the architect, who is responsible for providing adequate structural supports. This item is of great importance in larger car installations, because traction unit rails must be sup-

TRANSPORTATION

TRACTION FREIGHT ELEVATORS

	LIGHT AND MEDIUM DUTY FREIGHT ELEVATORS				
CAPACITY	2500 lbs. (1134 kg)	3000 lbs. (1361 kg)	4000 lbs. (1814 kg)	6000 lbs. (2722 kg)	8000 lbs. (3629 kg)
A	5'-4" (1626)	6'-4" (1930)	6'-4" (1930)	8'-4" (2540)	8'-4" (2540)
B	7'-0" (2134)	8'-0" (2438)	8'-0" (2438)	10'-0" (3048)	10'-0" (3048)
C	5'-0" (1524)	6'-0" (1829)	6'-0" (1829)	8'-0" (2438)	8'-0" (2438)
D	7'-10"(2388)	8'-10"(2692)	8'-10"(2692)	10'-10"(3302)	10'-10"(3302)
E	1'-7" (483)	1'-7" (483)	1'-7" (483)	1'-7" (483)	1'-7" (483)
F	0'-11" (279)	0'-11" (279)	0'-11" (279)	0'-11" (279)	0'-11" (279)
L	13'-0" (3962)	14'-0" (4267)	14'-0" (4267)	14'-0" (4267)	14'-0" (4267)

	HEAVY DUTY POWER TRUCK LOADING FREIGHT ELEVATORS				
CAPACITY	10,000 lbs. (4536 kg)	12,000 lbs. (5443 kg)	16,000 lbs. (7258 kg)	18,000 lbs. (8165 kg)	20,000 lbs. (9072 kg)
A	8'-4" (2540)	10'-4" (3150)	10'-4" (3150)	10'-4" (3150)	12'-4" (3759)
B	12'-0" (3658)	14'-0" (4267)	14'-0" (4267)	16'-0" (4877)	20'-4" (6198)
C	8'-0" (2438)	10'-0" (3048)	10'-0" (3048)	10'-0" (3048)	12'-0" (3658)
*D	11'-4" (3454)	13'-6" (4115)	14'-0" (4267)	14'-2" (4318)	16'-6" (5029)
E	1'-7" (483)	1'-7" (483)	1'-7" (483)	1'-7" (483)	1'-7" (483)
F	0'-11" (279)	0'-11" (279)	0'-11" (279)	0'-11" (279)	0'-11" (279)
L	14'-0" (4267)	15'-0" (4572)	15'-0" (4572)	17'-0" (5182)	21'-0" (6400)

MINIMUM OVERHEAD & PIT DIMENSIONS FOR LIGHT & MEDIUM DUTY FREIGHT ELEVATORS				
SPEED	50 FPM (0.25 m/s)	75 FPM (0.38 m/s)	100 FPM (0.51 m/s)	200 FPM (1.02 m/s)
O	16'-0"(4877)	16'-0" (4877)	16'-0"(4877)	16'-0"(4877)
P	5'-6"(1676)	5'-6" (1676)	5'-6"(1676)	6'-0"(1829)

NOTES:
2. Dimension G = 5" (127) for regular type counterbalanced hoistway doors and 6 ¾" (172) for pass type counterbalanced hoistway doors.
3. Pass type hoistway doors are required when floor heights are less than 11'-0" (3353) for 7'-0" (2134) openings and less than 12'-6" (3810) for 8'-0" (2438) openings.
4. Dimension H = 7'-0" (2134) on light and medium duty and 8'-0" (2438) (or as required) for heavy-duty doors. Doors higher than 8'-0" (2438) require additional overhead.
8. 2:1 Roping recommended for heavy duty power truck loading freight elevators
* D dimension includes space required when vertical columns are added (inside hoistway) for rail support. If no columns required, D=A+E+F.

Fig. 24.10 *Typical dimensional data for traction-type freight elevators. (Courtesy of Montgomery-KONE.)*

ported every few feet and additional steel provided to accomplish this.

24.9   Freight Elevator Cabs, Gates, and Doors

Cabs for freight service are normally built of heavy-gauage steel with a multilayer wooden floor, the entire unit being designed for hard service. Guarded ceiling light fixtures are required.

Cab gates slide up vertically and are a minimum of 6 ft high. Hoistway doors are normally vertical lift, center-opening, manual or power operated. Both cab gate and hoistway doors are counterweighted and open fully to give complete floor and head clearance.

24.10   Freight Elevator Cost Data

The cost of a freight elevator installation, as with passenger elevator installation, is dependent on

HYDRAULIC FREIGHT ELEVATORS

LIGHT AND MEDIUM DUTY FREIGHT ELEVATORS						
CAPACITY	2000 lbs. (907 kg)	3000 lbs. (1361 kg)	4000 lbs. (1814 kg)	5000 lbs. (2268 kg)	6000 lbs. (2722kg)	8000 lbs. (3629 kg)
A	5'-0" (1524)	5'-6" (1676)	6'-6" (1981)	8'-6" (2591)	8'-6" (2591)	8'-6" (2591)
B	6'-0" (1829)	7'-0" (2134)	8'-0" (2438)	10'-0" (3048)	12'-0" (3658)	12'-0" (3658)
C	4'-8" (1422)	5'-2" (1575)	6'-2" (1880)	8'-2" (2490)	8'-2" (2490)	8'-2" (2490)
D: manual doors	6'-4" (1930)	6'-10" (2083)	7'-10" (2388)	9'-10" (2997)	10'-0" (3048)	10'-6" (3200)
D: power doors	6'-10" (2083)	7'-4" (2235)	8'-4" (2540)	10'-4" (3150)	10'-6" (3200)	10'-6" (3200)
O: 7'-0" door ht. (2134)	13'-2" (4013)	13'-2" (4013)	13'-2" (4013)	13'-2" (4013)	13'-2" (4013)	13'-2" (4013)
P: 7'-0" door ht. (2134)	4'-0" (1219)	4'-0" (1219)	4'-0" (1219)	4'-6" (1372)	4'-6" (1372)	5'-0" (1524)

HEAVY DUTY POWER TRUCK LOADING FREIGHT ELEVATORS					
CAPACITY	10,000 lbs. (4536 kg)	12,000 lbs. (5443 kg)	16,000 lbs. (7258 kg)	18,000 lbs. (8165 kg)	20,000 lbs. (9072 kg)
A	10'-6" (3200)	10'-6" (3200)	10'-6" (3200)	10'-6" (3200)	12'-6" (3810)
B	14'-0" (4267)	14'-0" (4267)	16'-0" (4877)	16'-0" (4877)	20'-0" (6096)
C	10'-2" (3098)	10'-2" (3098)	10'-2" (3098)	10'-2" (3098)	12'-2" (3708)
D: manual doors	12'-6" (3810)	12'-6" (3810)	12'-6" (3810)	12'-6" (3810)	14'-6" (4420)
D: power doors	12'-6" (3810)	12'-6" (3810)	12'-6" (3810)	12'-6" (3810)	14'-6" (4420)
O: 7'-0" door ht. (2134)	13'-2" (4013)	13'-2" (4013)	13'-2" (4013)	13'-2" (4013)	13'-2" (4013)
O: 8'-0" door ht. (2438)	14'-2" (4318)	14'-2" (4318)	14'-2" (4318)	14'-2" (4318)	14'-2" (4318)
P: 7'-0" door ht. (2134)	5'-0" (1524)	6'-0" (1829)	6'-0" (1829)	6'-0" (1829)	6'-0" (1829)
P: 8'-0" door ht. (2438)	5'-0" (1524)	6'-0" (1829)	6'-0" (1829)	6'-0" (1829)	6'-0" (1829)

NOTES:
2. Dimension E = 5" (127) for regular type counterbalanced hoistway doors and 6³/₄" (172) for pass type counterbalanced hoistway doors.
3. Pass type hoistway doors are required when floor heights are less than 11'-0" (3353) for 7'-0" (2134) openings and less than 12'-6" (3810) for 8'-0" (2438) openings.

4. Layout and dimensions shown for freight elevators based on bi-parting counterbalanced type hoistway doors – dimension F.
5. Dimensions O and P are minimums based on car speeds up to 150 FPM (0.76 m/s).

TRANSPORTATION

Fig. 24.11 *Typical dimensional data for a hydraulic freight elevator. (Courtesy of Montgomery-KONE.)*

many factors, principally capacity, type of control, use, and type of door operation.

Inasmuch as exact pricing, like actual selection, is outside our scope, we recommend that a reputable manufacturer or elevator consultant be consulted for such information. We can, however, make some general remarks on pricing, as follows:

1. Variable-voltage controlled equipment, depending on type, costs 20% to 50% more than rheostatically controlled equipment.
2. Above a basic two-floor rise, the cost increases linearly with rise.
3. Electric door operation can increase the cost of a car installation 10% to 25%.

As an example of *comparative* pricing, using a nominal 100 for an 8000-lb, 75-fpm, four-floor, manual door car with ac rheostat control and automatic leveling, the same car with variable-voltage control, 150 fpm, and electrically operated doors would cost approximately 180.

SPECIAL ELEVATOR DESIGNS

24.11   Observation Cars

By placing a traction lifting mechanism *behind* the car, attaching the car at the back, and using a glass-enclosed, observation-style car, a spectacular unit can be constructed that becomes an attraction in itself. Basic construction and examples are seen in Figs. 24.12 and 24.13. If the back screen is treated properly, the car gives the impression of movement without any apparent motive force or machinery.

The same effect can be accomplished by using a single-jack hydraulic lift mechanism and a cantilevered car, as in Figs. 24.9 and 24.11. Observation cars, which are placed on the *outside* of a wall, do not require shaft space. Thus, in addition to an interesting and attractive appearance, they increase the amount of usable interior space, thereby effecting a considerable cost economy.

24.12   Inclined Elevators

Although elevators are normally conceived as traveling vertically, this is not necessarily so. Slant or inclined elevators have been constructed in numerous locations where a building is built onto an inclined surface, usually a hillside or mountainside. One such elevator is shown in Fig. 24.14. Another well known example is shown in Fig. 24.15. The motive mechanism used varies with the angle of inclination. In most instances the car rides on inclined rails and is pulled up by a traction cable. It is counterweighted either by a weight riding on another set of rails (Fig. 24.14) in the case of a single car, or by the weight of another car in a two-car installation.

24.13   Rack and Pinion Elevators

These operate on the straightforward principle of a rotating cogged-wheel pinion attached to a vertical rack. Rotation of the pinion forces the attached cab up and down along the rack (see Fig. 24.16). The advantages of this system are its inherent simplicity and safety, unlimited rise, low maintenance and operating costs, plus minimum space requirements. This last characteristic was primarily responsible for the selection of a rack and pinion design for a 210-ft-rise emergency elevator installed as part of the renovation and rehabilitation of the Statue of Liberty in 1986.

That elevator is used primarily to evacuate victims of heart attacks, which average one a month. Carrying them down the 171-step spiral staircase that connects the main upper landing to the observation platform in the crown in obviously not practical. The tight headroom available—less than 9 ft—ruled out a traction unit, as did the extremely tight shaftway, which measures only 2 ft, ½ in. × 4 ft, 10 in. The car itself (see Fig. 24.17) measures only 1 ft, 11 in. × 3 ft, 10 in. outside dimensions. Space was so tight that the drive was divided between two motors, both of which are mounted under the passenger compartment. Rack and pinion lifts are used both indoors and outdoors primarily in industrial environments for both passengers and material vertical transport.

25.14   Residential Elevators and Chair Lifts

Although the special needs of the handicapped have only been widely recognized officially in legislated requirements since the late 1970s, the elevator industry has been providing for the handicapped for years, on a voluntary basis. Small private-residence elevators can double as wheelchair lifts, and are available in a wide range of designs, including winding-drum units (Figs. 24.18 and 24.19), roped hydraulics (Figs. 24.7 and 24.8), and worm and screw units. Standard traction designs are used infrequently due to the overhead space requirement. Similarly, standard hydraulics requiring a plunger bore hole are rarely used.

Geared Traction Observation Elevators

(b)

Notes (Hydraulic and Traction)
State and local code requirements may vary. Data is for general application.

All layout details are based upon the use of center opening type entrances.

Elevator machine room temperature should be maintained between 65°F and 100°F.

Please consult Montgomery KONE for exact information for working drawings and for designs which may vary from those depicted.

(d)

(a)

(c)

TRANSPORTATION

Fig. 24.12 *Details of a geared, traction-type observation elevator.*
(a) *The glass-enclosed observation portion of this elevator faces the rear. The shaftway doors are seen in the wall behind the elevator. The traction ropes are also clearly seen above each elevator.*
(b) *Section drawing of the unit. Note that the counterweight is mounted at the side of the car, so as to be essentially invisible and not detract from the overall relatively uncluttered view of the car in motion.*
(c) *Vertical section of the installation, including the overhead machine room.*
(d) *Notes applicable to (b) and (c).*
Illustration continued on page 1454.

BASIC DIMENSIONS: TRACTION OBSERVATION ELEVATORS						
DIMENSION	CAPACITY					
	3000 lbs.		3500 lbs.		4500 lbs.	
	Option 2		Option 2		Option 2	
"DBG"	25"		31"		37"	
A	7' - 0"		7' - 0"		7' - 0"	
B	5' - 8"		6' - 6"		7' - 6"	
C	3' - 6"		3' - 6"		3' - 6"	
D	9' - 5"		9' - 5"		9' - 5"	
E	1' - 8"		1' - 8"		1' - 8"	
F	9"		9"		9"	
G	4' - 7$1/2$"		5' - 1$1/2$"		5' - 7$1/2$"	
H	5' - 3$1/2$"		5' - 9$1/2$		6' - 3$1/2$"	
J	10$3/4$"		1' - 2$3/4$"		1' - 8$3/4$"	
K	NA		NA		NA	
L	NA		NA		NA	
M	3' - 4$1/2$"		3' - 10$1/2$"		4' - 4$1/2$"	
N	7' - 6"		7' - 6"		7' - 6"	
S	13' - 8"		13' - 8"		13' - 8"	
SPEED	200 fpm	350 fpm	200 fpm	350 fpm	200 fpm	350 fpm
*O[1]	14' - 8"	15' - 4"	14' - 8"	15' - 4"	15' - 2"	15' - 6"
**O[1]	15' - 2"	15' - 10"	15' - 2"	15' - 10"	15' - 2"	15' - 6"
*P	5' - 0"	5' - 6"	5' - 0"	5' - 6"	5' - 4"	5' - 10"
**P	5' - 4"	5' - 10"	5' - 4"	5' - 10"	5' - 4"	5' - 10"

NOTE: * Dimension when no shroud is used.
** Dimension when 2' - 0" shroud is used.
[1] Based on standard height cab (8' - 0" to underside of canopy).
→ The relationships between dimensions G, H, J and K are based on an assumed thickness of 8" for the rear wall of the hoistway.
→ The hoistway dimensions shown are based on a counterweight not requiring safeties.

(e)

VERTICAL REACTIONS	TRACTION ELEVATOR CAPACITY			
	2500 lbs.	3000 lbs.	3500 lbs.	4500 lbs.
R1	17500 lbs.	18900 lbs	20900 lbs.	24400 lbs.
R2	13900 lbs.	15000 lbs.	16300 lbs.	19100 lbs.

NOTE: Reactions include allowance for impact but DO NOT include weight of concrete slab. Reactions are for preliminary use only. Exact reactions will be provided when exact conditions are known.

(f)

Fig. 24.12 *(continued from page 1453.)*
Details of a geared, traction-type observation elevator.
(e) Dimensions applicable to drawings (b) and (c).
(f) Vertical reactions required for structural design.
(Courtesy of Montgomery-KONE Inc.)

TRANSPORTATION

(a)

HYDRAULIC
GLASS BACK OPTION 4

(b)

WALL MUST CLEAR SHROUD

OPTIONAL TOP SHROUD

OPTIONAL BOTTOM SHROUD

(c)

BASIC DIMENSIONS: HYDRAULIC OBSERVATION ELEVATORS				
	CAPACITY			
	2500 lbs.	3000 lbs.	3500 lbs.	
Dimension	OPTION 1,2,3,4	OPTION 1,2,3,4	OPTION 1,2,3,4	OPTION 5
A	7' - 0"	7' - 0"	7' - 0"	7' - 0"
B	5' - 3"	5' - 11"	6' - 6"	7' - 2"
C	3' - 6"	3' - 6"	3' - 6"	3' - 6"
D	8' - 4"	8' - 4"	8' - 4"	8' - 4"
E	8"	8"	8"	8"
F	8"	8"	8"	8"
G	4' - 3"	4' - 7"	4' - 10$^{1/2}$"	4' - 10$^{1/2}$"
H	2' - 8$^{3/4}$"	3' - 0$^{3/4}$"	3' - 4$^{1/4}$"	3' - 4$^{1/4}$"
M	3' - 3"	3' - 7"	3' - 10$^{1/2}$"	3' - 10$^{1/2}$"

Basic dimensions outlined above are based upon conventional inground jack application.

(d)

HYDRAULIC MACHINE ROOM SIZES (Typically recommended)		
	CAPACITY	
SIZE	2500 lbs. & 3000 lbs.	3500lbs.
WIDTH	9' - 0"	9' - 6"
LENGTH	6' - 8"	6' - 9"
DOOR	3' - 6" × 7' - 0"	

NOTE: Hydraulic machine room location should be at the lowest landing adjacent to the hoistway. Consult your Montgomery KONE Professional for alternative locations and optimum sizes. Material above is typical for 150 FPM speed.

(e)

Fig. 24.13 *Hydraulic observation elevator.*
(a) The hydraulic plunger is clearly visible below each car, on a background of shaftway doors.
(b) Dimensioned sectional drawing of the car and its enclosure.
(c) Vertical section of the elevator travel.
(d) Dimensions applicable to drawing (b) and (c).
(e) Dimensional data for the required machine room. Note that in drawing (c), the machine room is not shown because it can be remote from the elevators, connected only by piping. See Figs. 24.4 through 24.7. (Courtesy of Montgomery-KONE.)

TRANSPORTATION

Acrylic dome enclosure

(a)

Acrylic weatherproof roof

Shaft doors

Counterweight rails

Traction rope

Car rails

(b)

Fig. 24.14 (a) *Exterior inclined elevator enclosure, seen from inside of the 14-story hotel it serves in Zichron Yaakov, Israel. The shaft is inclined at 26° and contains a single 10-passenger car. (Photo by Stein.)*
(b) *Inclined elevator shaft containing rails (tracks) for the elevator, its counterweight, and the traction ropes. Shaft doors are swinging type, interlocked with the car. (Photo by Stein.)*

Fig. 24.15 *The St. Louis Gateway Arch has a 10-passenger, inclined elevator in each leg. Placement of doors, arrangement of counterweight, and size of shaft all depend on the angle of incline. The car moves 82 ft horizontally during its 386 ft of total travel, at an incline of approximately 12°. (Photo courtesy of Bethlehem Steel Corporation, which supplied the elevator rope for this installation.)*

In recognition of the fact that most residential elevators (and elevators intended primarily for use of the disabled, in public and private buildings) are low-speed, limited load units, the ASME developed a new section for the elevator code (A17.1b part XXV) to cover such units. The section applies to Limited Use/Limited Application (LU/LA) elevators, defined as "a power passenger elevator where the use and application is limited by size, capacity, speed and rise, intended primarily to provide vertical transportation for people with physical disabilities." The new code section goes on to limit LU/LA elevators to a maximum size of 18 ft², load of 1400 lbs, rise of 25 ft, and speed of 30 fpm. Due to these limitations, safety requirements are different from those of standard traction elevators, drives may be different (e.g., winding drum cable lifts), space requirements are smaller, and overall costs are considerably reduced. Elevators that exceed the limitations must conform to the requirements of the standard elevator code ANSI/ASME 17.1.

Chair lifts and wheelchair platform lifts also come in a variety of designs. The chair lift shown

TRANSPORTATION

Fig. 24.16 The basis of the rack and pinion drive is simply a driven pinion on a stationary rack; rotation of the pinion forces the structure attached to the pinion to move along the rack. (Courtesy of Alimak Elevator Co.)

High-speed emergency elevator in Statue of Liberty.

Fig. 24.17 The rack and pinion principle is used in the very small Statue of Liberty emergency elevator. The rack extends the entire length of the shaft and the car moves along the rack, driven by two motorized pinions below the passenger compartment. (Courtesy of Alimak Elevator Co.)

in Fig. 24.20 uses a rack and pinion drive that permits it to negotiate a turn as illustrated. For straight stair lifts, a winding-drum design similar to that used for residential elevators is often used. Other drives in use include worm gear and chain and screw worm and cog. The drive selection depends on load, rise, duty, and, of course, price. The wheelchair platform lift illustrated in Fig. 24.21 uses a ball screw drive. Other designs use some of the drives listed previously, including a roped-hydraulic drive for units requiring a large rise. The hydraulic scissor-jack type of wheelchair platform lift is most frequently found in public and institutional buildings because it is particularly applicable to short-rise, heavy-load, frequent-duty applications, and requires minimal maintenance.

All of these units are covered by various sections of the elevator code and must be installed in accordance with code requirements, including safeties and controls. Residential units are normally arranged to operate on 120 V ac, although heavy-duty units may require 240 V service.

24.15 Linear Elevator Motor Drive

A revolutionary elevator drive design developed by OTIS utilizes a linear induction motor built into the counterweight frame. Motive power is therefore supplied at the counterweight, thus entirely eliminating the overhead traction machine and machine room. At this writing, the only installations using this design are in Japan. The inherent architectural, construction, energy usage, maintenance, and cost advantages of this elegant design make it an extremely attractive option for American buildings designers once it has achieved the required code and jurisdictional approvals. However, because the code is predicated on the use of a traction machine, which in this design is entirely absent, an essentially new section of code must be prepared to cover this design, and in particular its safety aspects. It should not be expected that

Various cab sizes are available to allow for more headroom or more platform area, as needed.

28" Rail centers

1"

Clear

7½"   1½" 2"

A
Cab width

7½"   5"

Including safety-lock clearance

Hatch width

2"

B
Cab depth

Hatch depth

9'-3" overhead (8'-0" w/Motor at back)

*9'-0" Travel (STD)

6'-3" Clear height

18'-11" Overall height (STD)

8" Pit or ramp

500 LB, 30 FPM
3' X 3' CAR

TOP SHEAVE BRACKET

TRUCK ASSEMBLY

MACHINE
(LOW MOUNT)

WINDING DRUM
RESIDENTIAL ELEVATOR

Cab sizes		
	A (width)	B (depth)
3 x 3	3 ft.	3 ft.
3 x 4	3 ft.	4 ft.
Special	To Suit (12 sq. ft. max. platform area)	

Fig. 24.18 *Isometric drawing of a typical residential elevator of the winding drum design. Units of this type are usually limited to 500-lb car capacity, 40-ft rise, and 30-fpm speed. The cab is rigidly attached in cantilever fashion to a rolling truck, which is raised and lowered by cables attached to the winding drum. The driving motor and winding drum shown here at the base of the assembly can be installed at any floor stop along the hoistway, or overhead. Control is normally pushbutton automatic with limit-switch leveling. A 6-in. pit is required at the bottom of the hoistway. Door interlocks, cable failure, and overrun safeties are standard items in such installations, which must meet elevator code requirements. (Courtesy of Waupaca Elevator Co., Inc.)*

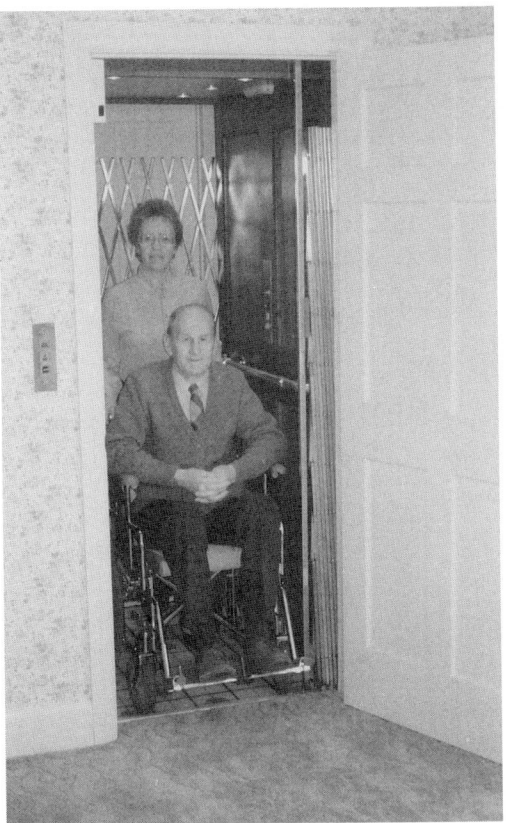

Fig. 24.19 Winding-drum residential elevator specifically designed for use by handicapped persons. Car floor area is 12 ft², which accommodates a wheelchair plus another passenger. Note that the car is arranged with front and back access for maximum flexibility. (Courtesy of Waupaca Elevator Co., Inc.)

so extensive a code revision will be a short-term project. A schematic diagram of the design is shown in Fig. 24.22.

MATERIAL-HANDLING

24.16 General

The material-handling equipment discussed briefly here is that which finds application in commercial and institutional buildings. Industrial materials handling is an entirely separate subject not germane to our purpose.

The need to transport material within a building has always existed, and until about the late 1970s was done largely manually, with mechanical assistance. Thus, offices used messengers; and hospitals used dumbwaiters, service elevators, conveyors, and chutes. The single exception to this situation was (and still is) the extensive use of pneumatic tube systems in large stores, although in new installations their use is declining. Today's systems accomplish the same end—that is, the transfer of materials—but automatically and, in general, much more rapidly. First cost of these systems is frequently high, but the reduction in labor and increase in speed generally yield a short payback period combined with a marked rise in efficiency.

Modern, commercial material-handling systems can be grouped roughly into four categories:

1. *Elevator-type systems.* These are vertical lift, car-type systems including the common dumbwaiter and ejection lifts, which are basically automated dumbwaiters.
2. *Conveyor-type systems.* These include horizontal and vertical conveyors.
3. *Pneumatic systems.* These include sophisticated pneumatic tube systems and pneumatic trash and linen systems.
4. *Other systems.* Systems that do not fit easily into any of the previously mentioned categories include automated messenger carts and automatic track–type container delivery systems.

24.17 Manual Load/Unload Dumbwaiters

Dumbwaiters often provide the most convenient and economical means of transporting relatively small articles between levels. In department stores, such units transport merchandise from stock areas to selling or pickup counters; in hospitals, dumbwaiters often transport food, drugs, linens, and other necessary small items. In multilevel restaurants and the like, dumbwaiters are almost always used for delivery of food from the kitchen and for return of soiled dishes.

Dumbwaiter cars are limited to a platform area of 9 ft² and a maximum height of 4 ft. The car may be, and frequently is, compartmented by shelves. Normal speed ratings are 50 to 150 fpm, with a capacity of up to 500 lb. Cars may be of the traction (counterweighted) or drum (direct pickup)

TRANSPORTATION

(a)

2⁵/₈"

18"

17¹/₈"

6⁷/₈"

23"

13¹/₂"
(Folded)

8¹/₂"   SIDE VIEW

END VIEW

(b)

Fig. 24.20 (a) *Chair lift with a rack and pinion drive permits the lift to negotiate a turn in the stairs. This unit is limited to a live load of 300 lb (135 kg), runs at 25 fpm (0.13 m/s), climbs at any incline up to 51°, and is driven by a ½ hp motor. Maximum rise is 60 ft. (b) Dimensional data for the chair lift shown. (Courtesy of Access Industries, Inc.)*

(a)

38 5/8"

30-1/2" TO
CENTER OF DOOR
OR GATE

36"   48-1/2"

48"

3/8" to 3/4"
STD. 48" PLATFORM

PLAN VIEW

1 1/2"
Typ.

52" Min.
53" Max.

2" to 3"

3/8" to 3/4"

48 3/4" Min.--49 1/2" Max.

HOISTWAY LAYOUT
(REF. ASME A17.1 PART XX)

(b)

TRANSPORTATION

Fig. 24.21 *The wheelchair platform lift shown in (a) is designed for a maximum load of 750 lbs (340 kg) and a maximum lifting height of 171 in. (4.34 m). The drive is a motor-driven recirculating ball screw unit for low rise and 1:2 roped-hydraulic lift for high rise. Controls are mounted at the top of the right post of the gate frame. The unit is a straight-through design with gates at both ends. Dimensions are shown in (b). (Courtesy of Access Industries, Inc.)*

LINEAR INDUCTION MOTOR ELEVATOR SYSTEM

Fig. 24.22 *Schematic drawing of OTIS Japan's linear induction motor (LIM) elevator design. Because the motor is built into the counterweight and supplies motive power linearly (vertically), the conventional overhead rotary traction machine is entirely eliminated, as is also the machine room. The system operates more efficiently, quietly, and smoothly than the conventional design. (Courtesy of OTIS Elevator.)*

type. Control is normally "call and send" between two floors, although multibutton selector switch or central dispatching arrangements are available for applications with more than two floors. Loading may be floor, counter, or any other specified height (see Fig. 24.23 for typical layouts).

24.18   Automated Dumbwaiters

These units are also known as *ejection lifts* because of the method of delivery (Fig. 24.24). They find their best application in institutions and other facilities that require rapid vertical movement of relatively large items. Thus, this device is ideally suited for delivery of food carts, linens, dishes, bulk-liquid containers, and so on. The load can be a cart (Fig. 24.25) or a basket (Fig. 24.28) containing the items being transported. At the delivery terminal, the item must be picked up and transferred horizontally to its final destination if remote from the delivery point. Loading can be manual or automatic. Sophisticated ejection lifts use programmable controllers for automated loading, dispatch, and ejection; electronic sensors to determine whether space is available for a load; and automated return of the unloaded cart.

Payload capacity for cart systems is available up to 1000 lb and car speeds up to 350 fpm. Maximum cart size is approximately 32 in. W × 68 in. L × 70 in. H. Round-trip time for a 200 fpm unit with five loading stations is approximately 2 min. Major considerations for these units are their relatively high cost and the large shaft area required.

24.19   Horizontal Conveyors

Although horizontal conveyors find their best application in industrial facilities, they are also usable in commercial buildings such as mail-order houses, which require a continuous flow of material. Restrictions in application stem from inflexible right-of-way requirements, noise generation, and a degree of danger if left unprotected or exposed to unauthorized persons. Cost is relatively low, and capacity is virtually unlimited.

24.20   Selective Vertical Conveyors

The action of this system is similar to the automated dumbwaiter in that the system transfers vertically and automatically loads and unloads, but the similarity ends there. Vertical conveyors are constructed with a moving continuous-loop chain to which are attached carriages that pick up and deliver tote boxes (called *trays*). At sending and receiving stations, the operator places the items to be moved (up to 60 lb) in the tote box, "addresses" the box in one of several ways depending on the system, and places it at a pickup point (Fig. 24.26). The first empty carriage on the

ELECTRIC DUMBWAITERS WITH MANUAL DOORS
RATED SPEED 50–150 fpm

(a) PLAN 2

(b) ELEVATION COUNTER HEIGHT LOADING

(c)

RATED LOAD (lb)	VOL. ■ ft³	COUNTER HEIGHT LOADING DIMENSIONS in.										
		CAR			DOORS ✱		HOISTWAY					
		A WIDTH	B DEPTH	C HEIGHT	D WIDTH	E HEIGHT	F	G	H	J	K ▲	L	M ✱✱
PLAN 1 OR 2 (MACHINE OVERHEAD OR BELOW) RATED SPEED 50–150 fpm												
APPLICATION – BULK SUPPLY, FOOD SERVICE (TRAY) & MATERIAL HANDLING												
100	8.33	20	24	30	20	30	6 1/2	7	5	3	30 ▲	48	6
200	12	24	24	36	36	24	6 1/2	7	5	3	30 ▲	48	12
300	17.5	30	24	42	30	42	6 1/2	7	5	3	30 ▲	48	18
400	30	36	30	48	48	48	6 1/2	7	5	3	30 ▲	48	24
500	36	36	36	48	48	48	6 1/2	7	5	3	30 ▲	48	24

   ✱ ONE OPENING MUST BE LARGER THAN THE CAR SIZE SO THE ASSEMBLED CAR CAN
     BE PLACED INTO, OR REMOVED FROM THE DUMBWAITER HOISTWAY.

  ✱✱ DEPTH OF PIT REQUIRED IF SLIDE DOWN TYPE DOORS ARE USED.

   ● 48 in. IF BI-PARTING HOISTWAY DOORS ARE USED.

   ▲ IF BASEMENT MACHINE (DRUM OR TRACTION) 42 in.

   ■ CAR SIZES ARE OPTIONAL UP TO 9 ft² OF FLOOR AREA AND UP TO 48 in. HIGH.

NOTES
    1. 2'-0" X 2'-0" ACCESS DOOR IF MACHINE ABOVE.

    2. 2'-0" X 2'-0" ACCESS DOOR IF MACHINE BELOW.

(d)

Fig. 24.23 *Typical layout of counter-height dumbwaiter. The shaft openings may be single (one side) or dual access. Doors can be biparting, slide up, or slide down. In addition to the sizes shown in the table, light-load cars rated 25 and 50 lb at 50 fpm are available in standard designs. (Reproduced with permission of National Elevator Industry, Inc. 185 Bridge Plaza N., Fort Lee, NJ 07024; copyrighted 1992, Vertical Transportation Standards.)*

TRANSPORTATION

Lift Model	Capacity	Inside Car Dimensions			Inside Shaft Dimensions		Hoistway Door Frame Dimensions		Pit (a)	Overhead (a)	Suggested Vertical Transportation Speed	
	Lbs.	WC	Dc	Hc	Ws	Ds	Wd	Hd			Landings	Feet per Minute
Tote-Matic CC-2	100	20"	31"	37"	40"	41"	20"	34"	12"	13'4"	7-10	150
											4-6	100
											2-3	50

(a) These figures must meet the requirements of ANSI/ASME 17.1 and local codes.

SECTION

TOTE-MATIC

PLAN

Fig. 24.24 *This automated dumbwaiter or vertical ejection lift is designed to eject (unload) a container automatically at a preselected station. (Note the dispatching station to the right of the doors, in the photo.) The tote box is carried on an ejection conveyor, which electrically senses arrival at its destination and effects ejection. Station doors are automatically electrically operated. The dumbwaiter car drive mechanism is a counterweighted traction drive. Car capacity is 100 lb, although containers are normally limited to 50 lb for ease of handling. Car speeds are normally 50 to 150 fpm (see table). A typical hoistway section is shown. Vertical travel is not limited. (Courtesy of Courion Industries, Inc.)*

PLAN AND SECTION

CART-MATIC

Machine Room Above

PLAN

SECTION

TABLE OF STANDARDS

Capacity	Inside Car Dimensions			Inside Shaft Dimensions		Hoistway Door Frame Dimensions		Pit **	Overhead **
Lbs.	Wc	Dc	Hc	Ws	Ds	Wd	Hd		
500	28″	46″	48″	48″	56″	28″	50″	46″	14′4″
1000	30″	55″	67″	56″	65″	30″	69″	60″	16′8″
1000	34″	72″	72″	60″	82″	34″	74″ *	60″	17′4″

** May vary with local codes

(a)

(b)

(c)

Fig. 24.25 (a) *Plan, section, and dimensions for vertical ejection lift system (automated dumbwaiter).* (b) *Open ejection-lift unit showing cart ejection mechanism. Shaftway doors are vertical biparting.* (c) *The same unit being loaded with food carts and dispatched to the various floors of the hospital. At the upper floors the carts are rolled away by attendants. Later, the lifts are used to return soiled dishes and trays. (Courtesy of Courion Industries, Inc.)*

Selective Vertical Conveyor Mechanical Components

Control Center

Drive
Stand

Drive
Sprocket

Station

Firedoor

Tray

Gravity
Runoff

Operator
Display
Panel

Rails (4)

Upper
Tie
Channel

Floor
Channel

Car

Chain

Take-Up
(In Pit)

(a)

(b)

Fig. 24.26 (a) *System diagram showing operation of a selective vertical conveyor. Tote box addressing and dispatching can be accomplished remotely from the control center.*
(b) *Typical sending and receiving terminal. Stations are arranged vertically in a common shaftway. Each station contains a keypad for addressing and dispatching boxes. (Courtesy of TransLogic Corp.)*

chain picks up the box and delivers it to its address. Functions in modern vertical conveyors are monitored by a microprocessor, which tracks system operation and furnishes maintenance data and operational diagnostics. Drawbacks of this system are the large shaft required, noise, and cumbersome arrangements when interfacing with horizontal conveyors. Cost is moderate.

24.21   Pneumatic Tubes

This well-tried system will undoubtedly continue in use where physical transfer of an item is required. Where information must be moved, electronic data reproduction has in large measure replaced the transfer of pieces of paper between two points. Pneumatic tube systems are available with 2¼- to 6-in. range of tube diameters (special shapes are also used) and with single or multiple loops.

Whereas older systems were generally pressurized using a single large and noisy compressor, newer systems are computer controlled, utilize a small blower in each of the zones, operate basically on vacuum but also on pressure, are relatively quiet, and are capable of being constructed in unlimited system length. Carriers travel at 25 fps. The computerized control center provides information and control of all system components, including status, traffic data, station assignments, scheduling, and the like. Overall, pneumatic tube systems perform their task reliably, rapidly, and efficiently at relatively low cost if installed during initial building construction. Typical system components are shown in Fig. 24.27.

24.22   Pneumatic Trash and Linen Systems

Rapid movement of bagged or packaged trash and linen from numerous outlying stations to a central collecting point is usually handled by this system. (Health codes require using separate tubes for trash and linen.) Linen systems are found generally in hospitals; trash systems in various facilities, frequently in conjunction with compactors. The system is basically a network of large

Control Panel

Dispatcher

Storage Racks

Receiver

(a)

(b)

Fig. 24.27 (a) *The recessed tube station illustrated is a full-facility combined sending and receiving point. Dispatching and receiving share the same pipe but operate independently automatically. Arriving carriers* (b) *are decelerated by an air cushion and drop into the front bin on arrival.*
(b) *Carriers (3¾ in. D × 15½ in. L and 5¾ in. D × 16 in L) are made of impact-resistant transparent polycarbonate plastic. Liners are used with fragile items. The bands around the carriers that maintain the air seal in the transport tubes are replaceable. (Courtesy of TransLogic Corp.)*

pipes, negatively pressurized, with numerous loading stations throughout the building. Pipe sizes are 16-, 18-, or 20-in., operating at high static pressure. A system normally can handle only one unit load at a time, but moves it so quickly (20 to 30 fps) that system capacity is large and delays are not encountered. Material placed into a load-

TRANSPORTATION

ing station is picked up as soon as the previous load clears. Compressors are large and very noisy, requiring considerable space allocation and acoustical isolation. In addition to the main vacuum system, a high-pressure air line is required to operate the doors, and sprinkler heads must be installed every few floors. Overall costs are low to moderate. For the specific task performed, a cheaper and more efficient transfer technique is difficult to find.

24.23　Automated Container Delivery Systems

This useful arrangement employs captive and secure containers locked onto a motorized carriage that, in turn, is locked onto the track system. Power for the motor in the carriage is picked off a third rail at 24 V dc. The entire assembly moves horizontally or vertically with equal ease. Containers move at a constant 120 fpm horizontally but slower on rises, depending on container load and the steepness of the rise. Containers are available in a number of shapes and volumes to suit the particular installation. Two standard sizes are approximately 18 in. × 6 in. × 13 in. narrow profile and 18 in. × 12 in. × 8 in. low profile. Normal payload is 20 lbs, although for horizontal runs

only, considerably heavier loads can be carried.

In simple single- or dual-track point-to-point or loop systems, right-of-way conflicts cannot occur and routing is simply a matter of addressing the car. For complex systems involving loops and branches, routing is decided by a central computerized controller that finds the shortest route for each car (up to 250 cars per system), side-tracks, parks cars, and, in effect, operates a miniature railroad type system. The system is easily added onto a structure as a retrofit operation because of track flexibility and small size. Its major drawback is it high cost. Components are shown in Fig. 24.28.

24.24　Automated Self-Propelled Vehicles

These robot battery-powered vehicles follow a route determined by a passive guidance floor tape. They can be arranged to interface automatically with vertical transport means (elevators) so that the route can cover various levels in a facility. The floor tape, which can be installed below carpets, is entirely passive. It determines only the path that the vehicle will follow. All sensing, instruction, motive power, and vehicle control are located on the vehicle itself, which can carry approximately a 300-lb payload and can operate a full 8-h day

(a)

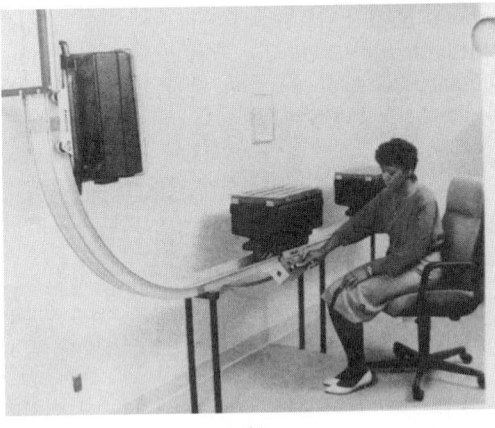
(b)

Fig. 24.28 *(a) One of many designs of carriage and container for an electric track vehicle system. This design uses a pivoted container that is maintained in a vertical position regardless of the carriage position, which may be horizontal, vertical sloped, or even inverted.*
(b) Dispatching and receiving station for a track vehicle system. (Courtesy of TransLogic Corp.)

1   Pneumatic Safety Bumper

2   Sensor Panel
    Contains metal detection sensors
    which confirm presence of the guide
    tape to the vehicle guidance system.

3   Steering System
    Tricycle wheel design.

4   Batteries

5   Coupling Unit
    Transcar tows trolleys. This device is
    activated to couple Transcar with
    trolley for transport. When trolley is
    delivered to destination, coupling unit
    retracts and vehicle drives from
    underneath trolley.

6   Drive Motors
    Each drive wheel is powered by a 24
    VDC motor.

7   Control Unit
    Includes all electronic equipment

8   Charging System

(a)

(b)

(c)

Fig. 24.29 *Components of the automated vehicle are shown in this block diagram (a). The vehicle fits underneath its load, as can be seen in (b) and (d), which show two typical applications: automated food cart delivery in a hospital and automated mail service in a large office. The external programming device is seen in (c). (Courtesy of TransLogic Corp.)*

(d)

TRANSPORTATION

without recharging its on-board batteries. Vehicle speeds are variable, ranging from 20 to 120 fpm. Programming of route, stops, vertical interfaces, changing of cycle, and other functions is accomplished with an external programming device and an on-board programmable controller. Applications for an automated vehicle of this type are limitless: parts delivery and pickup in industrial facilities, food and supplies distribution in hospitals, and mail and document pickup and delivery in offices are among its basic functions. Typical units are shown in Fig. 24.29.

24.25   Conclusion

The foregoing very brief summary simply describes the types of equipment available. For each facility being planned the architect must study the material transfer problems, remembering that buildings not only handle and process but also generate material; an office building generates about a pound of waste per 100 ft^2 per day—a prodigious amount in today's large office structures. This type of dry waste can be compacted, bailed, and sold, unlike garbage and wet waste. In addition to considerations of the type of material being handled, there are factors of speed, scheduling, location of stations, labor and material costs, space requirements, noise generation, and energy requirements. To consider and evaluate all these factors in a large and complex facility is generally beyond the ability of the architect alone. Thus, expert advice from consultants who specialize in materials handling and from manufacturers' representatives should be sought.

25
MOVING STAIRWAYS AND WALKS

MOVING ELECTRIC STAIRWAYS

25.1 General

The moving stairway, also referred to as an *escalator* or as an *electric stairway,* was first operated at the Paris Exposition in 1900. Its modern successors deliver passengers comfortably, rapidly, safely, and continuously at constant speed and usually with no delay at the boarding level. The annoyance of waiting for elevators is eliminated. Also, no time is lost by acceleration, retardation, leveling, and door operation, or by passenger interferences in getting in or out of the cars, and so on. Instead of formal lobbies and hallways leading to a bank of elevators on each floor and a ride in a small, enclosed box, the electric stairway is always in motion, inviting passenger to ride on an open, airy, observation-type conveyance, that cannot ever trap passengers due to equipment or power failure. In contrast to the purely utilitarian function of an elevator, an escalator also has a decorative/design function, and its open, observation characteristic is frequently used to expose the rider to very specific visual panoramas.

25.2 Parallel and Crisscross Arrangements

Moving stairs can be constructed in three ways: two parallel arrangements and one crisscross arrangement. The crisscross arrangment, however, can be operated in two modes—spiral and walk-around—whereas the parallel arrangement defines usage by the physical arrangement of the stairs. The aptness of these descriptive terms become evident with the diagrams and photos that follow immediately.

The essential difference between the two plans is that in the crisscross arrangement, the upper and lower terminal entrances and exits to the up and down escalators, are separated by the horizontal length of an escalator, whereas in either of the parallel arrangements the two escalators face in the same direction.

(a) The Crisscross Arrangement. This is simpler to visualize and also more common. We therefore examine it first. Refer to Fig. 25.1 (*a*) and (*b*). Notice that the stair construction in both is *identical*; the difference occurs in the direction of operation of the second level of stairs. In the spiral crisscross arrangement of Fig. 25.1 (*a*), the rider begins an upward trip on stair L1A and by means of a 180° turn continues the trip uninterruptedly on the stair L2A, travelling, effectively, in an upward spiral. The downward-travelling rider performs the same spiral on L2B and L1B. This arrangement is rapid, pleasant, and very economical of space because the stairs nest into each other. It can be used for as much as five floors without excessive annoyance to the rider.

Now let us examine the arrangement of Fig. 25.1 (*b*). By reversing the direction of the second level stairs L2A and L2B, the up passenger must leave L1A at the first upper landing and walk around the entire length of the stair, to continue the upward trip on L2B. Similarly, the down passenger begins the trip on L2A, but at the first landing, must traverse the escalator length to reach the next down elevator L1B; hence the descriptive name, walk-around crisscross. This arrangement obviously requires floor construction around the escalators, which is used in stores to display special sale merchandise. Indeed, this display purpose is the reason that stores force passengers to endure the extremely annoying forced walk-around.

A similar, but generally much shorter and therefore less objectionable walk-around exists in

(a) Criss-Cross; Spiral Operation

Fig. 25.1 *Escalators constructed in criss-cross arrangement. (a) By operating stairs L1A and L2A up, a passenger from the lower terminal travels upward in a spiral, with only a turn-around at each level. The same is true for a down passenger. This system is called "spiral" operation, and the entire installation is referred to as crisscross spiral (operation). (b) By reversing the stair direction of upper level stairs L2A and L2B, passengers at each intermediate landing (only one shown) must traverse the entire horizontal stair length to reach the next stair traveling in the desired direction. (See arrows on the drawing). As a result, this arrangement is known as crisscross walk-around operation. (Drawings courtesy of Otis Elevator Co.)*

(b) Criss-Cross; Walk-Around Operation

Fig. 25.2 *Side and end elevations, and plan views of the crisscross escalator arrangements. In the spiral operation mode (see text), separating the escalators forces the rider to walk distance* D *to continue a trip. In the walk-around operation mode, escalators are frequently separated to provide a walk-around path.*

the case of the spiral crisscross plan, where the escalators are separated by distance D [Fig. 25.2 (*a*)]. This distance is usually limited to about 10 ft because any greater distance places the trip-continuation escalator out of sight and causes not only annoyance at the enforced walk but also confusion and consequently resentment. Separation of the escalators is frequently an architectural consideration and does have the advantage of easier mixing of riders entering at the various levels with riders making a continuous trip. Designers must, however, be continuously aware of the possibility of a negative reaction to separation of escalators, which can be further reinforced when:

1. Insufficient floor space is provided for the transit between escalators, causing crowding, pushing, and delay.
2. Insufficient elevator service is provided for passengers wishing to travel at least three floors. This forces people to make a multistory escalator trip, which in itself can be wearying, particularly when carrying parcels. If such a trip is further lengthened by an enforced walk-around at each floor, it becomes a source of severe irritation, often sufficient to keep customers away from the store.

Figure 25.2 shows plan and elevation of the crisscross arrangement, for both spiral and walk-around operation, since, as shown, construction is identical for both. Figures 25.3 and 25.4 illustrate the crisscross plan.

(b) Parallel Escalators. This configuration requires more floor space than the crisscross arrangment and is therefore used less. As noted, it is constructed in either of two designs: the parallel spiral arrangement shown in Fig. 25.5 (*a*) and stacked parallel arrangement, with forced walk-around, shown in Fig. 25.5 (*b*). The principal advantage of the parallel arrangment is its impressive appearance, as can be seen in Figs. 25.6 through 25.9. The stacked arrangement must be used with caution due to the inconvenience to the rider of an enforced long walk-around to continue the trip. This arrangement is found most often in mass purchase–type facilities and in malls. In department stores or malls this arrangement is much less objectionable as many of the people are there to browse and window-shop rather than to purchase and leave, as is the case in a single purpose store. In a multistory store, the inconvenience of this walk, which is frequently compounded by the crowds of people that normally

Traffic guide at turn-around

Fig. 25.3 *Bank of glass balustrade escalators with mirrored sides, in a spiral operation, crisscross plan. Note the dividers at the turnaround at the intermediate level whose function is to guide traffic smoothly, either away from the escalator or to the turn point. This effectively eliminates undesirable bunching and crowding. (Photo courtesy of O&K Rolltreppen, Germany.)*

Fig. 25.4 *Crisscross design, applicable to either spiral or walk-around operation. The horizontal area in the right foreground can be used to display impulse-buying merchandise to walk-around riders during holiday seasons, after which operation can be returned to the spiral mode. (Photo courtesy of O&K Rolltreppen, Germany.)*

Fig. 25.5 *Side and end elevations, and plan views of escalators in* (a) *spiral parallel and* (b) *stacked parallel arrangements. In the spiral parallel arrangement* (a) *the separation has no effect on the rider, who simply makes a turn to continue in the same up or down direction. In the stacked arrangement* (b), *the rider must traverse the entire length of the escalator at each level in order to continue since the escalators are stacked vertically on each floor.*

congregate at special sale merchandise displays and counters, rapidly engenders annoyance. Thus, the stacked arrangement is seldom used above two floors (one such walk-around) and when it is used above two floors, riders can be expected to gravitate to elevators.

Escalators between two contiguous levels do not present the continued trip problem and therefore are frequently used in the parallel arrangement. This configuration is particularly common

in public buildings, transportation terminals, and other heavy traffic areas, where the advantage of a single location for the entrance to the bank of escalators eliminates confusion and the safety hazard engendered by hesitant, confused riders in heavy traffic.

The consideration of division of rider traffic between elevators and escalators in a store is an important one. The general philosophy of store owners is to make escalators the primary means

Fig. 25.6 *An unusual design of stacked parallel escalators with a forced walkaround in a continuous trip. The exposed truss and drive elements, combined with a transparent balustrade, add considerable visual interest to the installation. (Photo courtesy of O&K Rolltreppen, Germany.)*

Fig. 25.7 *An unusual installation of parallel escalators adapted to a long four-story building. In the more common instance of a building that is taller than it is long, the escalators are stacked as in Fig. 25.6. Here, the design is adapted to the large central court/atrium, with offset stepped landings and entrances to the buildings. (Photo courtesy of O&K Rolltreppen, Germany.)*

of vertical transportation, for the obvious reason of merchandise exposure. As a result, elevators are frequently placed at one or both ends of a store whereas the escalator banks are central. However, care must be taken to provide sufficient elevatoring in stores exceeding three floors, particularly in stores using the stacked or walkaround crisscross plans, because of the inconvenience of multifloor escalator trips discussed previously.

25.3   Location

Because escalators are constantly moving and are generally part of a horizontal and vertical trip, they must be placed directly in the main line of traffic. This is in contrast to the elevator bank, which, being a vertical transportation unit, can be set off as an element on its own for people to approach and utilize. Escalators must therefore be placed in the area served, and with a total and even dominating view of it. This allows potential riders to immediately:

1. Locate the escalators.
2. Recognize the individual escalator's destination.
3. Easily and comfortably move toward the escalator.

One of the most effective ways to disorient traffic movement is to inadequately or poorly mark escalator destinations. The resultant milling about, false starts, and constant inquiries can be observed in numerous otherwise well-designed buildings, and often in stores where large displays block the original open line-of-sight.

Sufficient lobby space must be provided at the base for queuing where anticipated, and most particularly at discharge points. A restricted or poorly

Fig. 25.8 Escalators in transportation terminals are subject to periods of very heavy use. To avoid crowding at the entrance to the up (left) moving stair, the balustrade is effectively extended with a glass-sided divider. Passengers on the down (right) escalator are guided by large overhead signs and a standing sign at the base and balustrade extension, to avoid hesitation and bunching. Note also the large open area in front of both units. (Photo courtesy of O&K Rolltreppen, Germany.)

marked discharge area causes passenger hesitation and traffic backup. Because the escalator discharges continuously, backup of traffic is dangerous and therefore intolerable. This is particularly crucial in theaters and stadiums, where even momentary hesitation during peak traffic periods can be disastrous. To avoid this, four design steps, in descending order of importance, are taken:

1. Provide well-marked escalators with sufficient traffic-carrying capacity.
2. Provide collecting space at intermediate landings so that pressure can be relieved.
3. Provide a physical divider at intermediate landing turnaround points that guides riders away from the discharge point and provide adequate space (and time) for riders either to leave at that level or to follow the guide around and continue the trip (see Fig. 25.2).
4. Provide a slight setback for the next escalator so that the necessary 180° turn can readily be negotiated.

At the exit terminal, an escalator should discharge into an open area with no turns or choice of direction necessary. Where such is absolutely unavoidable, *large* clear signs should make hesitation unnecessary. (See Fig. 25.8). Landing space beyond

Fig. 25.9 Among the longest moving stairways in Europe are those installed in the Stockholm subway system. This parallel bank of three escalators is 70-m (230-ft) long with a 33-m (108-ft) rise. Unlike American standards of 30° incline, 100-fpm speed, and 32- or 48-in. size, these units are at 27.3° incline, 45-m/min (147.6-fpm) speed, and 1000-m (40-in.) width. (Photo courtesy of O&K Rolltreppen, Germany.)

TRANSPORTATION

the escalator newels should be a minimum of 8 ft for 32-in. units and 10 ft for 48-in. units, for a standard 100-fpm speed. For escalators that will be reversed to accommodate change in traffic direction, this landing space must be provided at top and bottom.

The crisscross arrangement has the advantage of lower cost, minimum floor space occupied, and lowest structural requirements. The parallel arrangement, being less efficient and more expensive, has as a compensating virtue a very impressive appearance that strongly draws people to it. For this reason it is frequently employed, particularly in banks of three or four units in transportation terminals (see Figs. 25.8 and 25.9). In such large installations, flexibility is maintained by operating all but one escalator in the direction of heaviest traffic. Reversibility of escalators provides this most desirable feature.

25.4   Size, Speed, Capacity, and Rise

Moving stairs are built according to manufacturers' and industry standards and are therefore available in standard designs. However, all major manufacturers also produce special designs for particular applications. The data that follow refers to standard designs.

All escalators in the United States are installed at an angle of 30° from the horizontal, with a minimum vertical clearance of 7 ft for escalator passengers. The 30° inclination means that the rise is equal to 57% of the unit's projected floor area for its inclined portion. The length of the horizontal portions of the stairway depends on the specific design. To meet ADA requirements,

elongated newels with at least two horizontal treads before the landing plate are needed. Today, although the maximum linear speed permitted by the safety code (ANSI/ASME 17.1) is 125 fpm, the industry has standardized on a single speed of 100 fpm.

The elevator code now defines the *width* of an escalator as the width of the stair tread, in inches. The previous width designation referred to the distance between balustrades; a figure difficult to define because of design variations. That measurement is now called *size*. The standard sizes and widths are:

Escalator Size (in.)	Tread Width (in.)
32	24
48	40

Table 25.1 lists theoretical, nominal (design), and observed average escalator passenger capacities. Maximum loads assume approximately 1¼ persons per tread for a 32-in. unit, and almost 2 persons per tread for a 48-in. unit. In actuality, maximum capacity is approached only during peak-loads periods in transportation terminals and stadiums. At other times, full (heavy) load is represented by the nominal capacity figure and average (long period) load by the observed capacity figure. Although a 40-in. tread can indeed carry two persons, psychological factors plus physical ones, such as bulky clothing, packages, purses, and briefcases, mitigate against such loading. As a result, on a 40-in. wide tread one person uses each step in a diagonal pattern, and on 24-in. stairs one person occupies every other tread.

Some major manufacturers make two standard models (not styles) of moving stairs; a standard-

TABLE 25.1 **Escalator Passenger Capacity**

Size (in.)	Tread Width (in.)	Speed (fpm)	Passengers per Hour		
			Maximum[a]	Nominal[b]	Observed[c]
32	24	100	5200	4000	2300
48	40	100	9000	6750	4500

[a]Theoretical maximum (see text).

[b]Heavy loading (see text).

[c]Average long-period loading.

duty unit intended for general indoor use, which provides low to medium rise, and a heavy-duty, sturdier unit intended for all weather heavy-traffic use such as at transport terminals. The latter, because of heavier construction, can provide a greater maximum rise than that of standard units. Maximum rise for off-the-shelf design units varies among manufacturers. Approximate maximum figures are given in Table 25.2.

Specially designed units are available with rises of up to about 60 ft. In escalator design all the motive power is delivered at one point; that is, the drive motor drives the main chain, which drives the top sprocket, which drives the step chain, which pulls up the steps, causing the entire assembly to move. This arrangement is suitable for moderate rises of up to approximately 25 ft; beyond that, the design becomes increasingly inefficient. As the rise increases, the loads on all the drive components, including chains and sprockets, increase sharply. Furthermore, to accommodate the heavier equipment, truss width increases, as does wellway size and balustrade decks. For rises above 25 to 35 ft (depending on unit width) the drive motor is too large to fit inside the truss and requires a separate machine room below the truss, with attendant cost. All these factors combine to limit standard escalators to a maximum rise of 60 ft (varies slightly between manufacturers).

TABLE 25.2 **Approximate Maximum Escalator Rise**

Unit Size (in.)	Type	Supports	Maximum Rise (ft)
32	Standard	Ends	24
48			16
32	Standard	Ends and center	30
48			20
32	Heavy	Ends	24
48			18
32	Heavy	Ends and center	40
48			20

25.5 Components

The major components of a standard escalator installation are shown in Fig. 25.10. Safety devices are discussed in Section 25.6.

The truss is a welded steel frame that supports the entire apparatus. (See Fig. 25.4). The tracks are steel angles attached to the truss on which the step rollers are guided, thus controlling the motion of the steps. The sprocket assemblies, chains, and machine provide the motive power for the unit, somewhat similar to the chain drive of a bicycle.

The handrail is driven by sheaves powered from the top sprocket assembly. It is synchronized with the tread motion to provide stability to riding passengers and a support for entering and leaving passengers. Handrails disappear at inaccessible points at newels. The balustrade assembly is designed for maximum safety of persons stepping on or off the escalators.

Transparent balustrades are made of tempered glass, and are frequently referred to as *crystal balustrades*. In these units, the handrail is pinch-driven within the truss. In addition to metal and glass as balustrade materials, fiberglass, wood, and various plastic materials are also used.

The control cabinet, which is normally located near the drive machine, contains malfunction indicators in addition to the drive controls. The cabinet may also contain a microprocessor malfunction analyzer and communication means for transmitting escalator operating conditions to a central control point. Operation of an emergency stop button which is wired to the controller and placed near or on the escalator housing at both ends stops the drive machine and applies the brake. Key-operated control switches at the top and bottom newels start, stop, and reverse the stairway.

25.6 Safety Features

Protection to passengers during normal operation is assured by a number of safety features associated with moving stairways:

· Handrails and steps travel at exactly the same speed (100 fpm) to assure steadiness and bal-

TRANSPORTATION

Machine space
Controller
Drive Machine
Handrail

Upper landing

Left
ballustrade

Left handrail

Rise

Handrail drive

Right ballustrade

Right handrail

Steps

Truss
*Step chain
(located beneath the
cladding, cannot be
seen in this illustration)*

Comb
plate

Lower
landing

Floor
plate

Machine space
*Tension carriage
main drive*

Fig. 25.10 *View of a standard escalator showing principal parts. (Drawing courtesy of Otis Elevator Co.)*

ance and to aid stepping on or off the combplates.

· The steps are large and steady, and are designed to prevent slipping.

· Step design and step leveling with the combplates at each landing prevent tripping upon entering or leaving the escalator. This is accomplished with two or three (depending on manufacturer) horizontal steps at each end of the escalator.

· The balustrade is designed to prevent the catching of passengers' clothing or packages. Close clearances provide safety near the combplates and step treads.

· Adequate illumination is provided by the building at all landings, at the combplates, and completely down all stairways. Some escalator designs provide built-in lighting, as discussed in Section 25.8.

· An automatic service brake will bring the stairway to a smooth stop if:

(a) The drive chain or the step chain is broken or abnormally stretched.

(b) A foreign object is jammed into the handrail inlet, between the skirt guard and step or between steps, causing them to separate.

(c) A power failure occurs.

(d) The emergency stop button is operated (one is located at each end of the escalator).

(e) Any of the fire safety system devices operates (see Section 25.7).

(f) A tread sags, rises, or breaks.

(g) A drive motor malfunction occurs.

In case of overspeed or underspeed, an automatic governor shuts down the escalator, prevents reversal of direction (up or down), and operates the service brake.

If the escalator is stopped by operation of a safety device, passengers can then walk the steps as they would any stationary stairway.

25.7 Fire Protection

Four methods of providing protection in case of fire near escalators are available: the rolling shutter, the smoke guard, the spray-nozzle curtain, and the sprinkler vent. One of these methods is required by code when more than two floors are pierced. Figure 25.11 illustrates clearly how the wellway at a given floor level may be entirely closed off by the fire shutter, thus preventing draft and the spread of fire upward through escalator wells. Temperature and smoke detectors automatically actuate the motor-driven shutters. The shutter in Fig. 25.11 is shown at the third-floor level, but other shutters may be installed at the tops of horizontal wellway openings at any floor.

Figure 25.12 illustrates the smoke-guard method of protection. It consists of fireproof baffles surrounding the wellway and extending downward about 20 in. below the ceiling level. Smoke and flames rising upward to the escalator floor opening meet a curtain of water automatically released from sprinkler heads of the usual type, shown at the ceiling level. The baffle is a smoke and flame deflector. The vertical shields between adjacent sprinklers ensure that the spray from one will not cool the nearby thermal fuses, preventing the opening of adjacent sprinklers.

The spray-nozzle curtain of water (not shown) is similar to the smoke-guard protection. Here, closely spaced, high-velocity water nozzles form a compact water curtain to prevent smoke and flames from rising through the wellways. Automatic thermal or smoke relays open all nozzles simultaneously.

The sprinkler-vent fire control system is shown in Fig. 25.13. The fresh air intake housed on the roof contains a blower to drive air downward through escalator floor openings, while the exhaust fan on the roof creates a strong draft upward through an exhaust duct; this duct in turn draws air from the separate ducts just under the ceiling of each moving stairway floor opening. Three such separate wellway ducts are shown. Each duct has a number of smoke-pickup relays that automatically start the fresh air fans. The usual spray nozzles on the ceiling near the stairways aid in quenching the fire.

25.8 Lighting

Adequate illumination of a moving stairway, particularly at the landings, is important from decorative as well as safety standpoints. In a stairwell-type installation, where general-area lighting does not provide sufficient illumination for the escalator, lighting consonant with the adjacent illumination is installed on the ceiling above the stairway, with special emphasis on lighting the combplate. In Fig. 25.9, banks of fluorescent lights are placed across the escalator bank at frequent intervals along the rise. Note the additional concentrated light at the combplate. Two different lighting treatments of similar installations are shown in Figs. 25.14 and 25.15 (see also the balustrade lighting in Fig. 25.16.)

25.9 Application

1. Main floor locations should be chosen in the direct flow of traffic to assure maximum use.
2. Vertical arrangements should be made to accomplish specific purposes, such as expo-

Fig. 25.11 *Rolling-shutter method of wellway fire protection.*

Fig. 25.12 (a) *Smoke-guard method of fire protection for a 32-in. moving stairway, crisscross type. The escalator floor opening (per floor) is approximately 4 ft, 4 in. ×14 ft, 8 in.*
(b) *Reflected ceiling plan and section showing baffle and sprinkler layout. (Courtesy of Otis Elevator Co.)*

Fig. 25.13 *Sprinkler vent fire protection for escalator openings. An exception (with control) to the rule against perforation in floors.*

Fig. 25.14 *Stair lighting is supplied by a continuous fluorescent strip at the base of the balustrade in this California shopping mall. (Photo courtesy of Montgomery KONE.)*

Fig. 25.15 A continuous fluorescent source is placed under the handrail of the crystal balustrade in this covered mall in West Germany. Note the additional light at the base for illuminating the combplate. (Photo courtesy of O&K Rolltreppen, Germany.)

sure of merchandise, maximum passenger capacity, and maximum accessibility to various areas.

3. The aspect of reversibility of an electric stairway should be considered in applications where major traffic flow is unidirectional. Light traffic in the reverse direction

can be handled by a normal fixed stair, adjacent to the escalator (see Fig. 25.16). Similarly, a bank of two escalators can operate either both up, both down, or one up and one down to handle variable traffic conditions in such areas as office buildings and transportation terminals.

4. Exterior escalators can provide an attractive, interesting, and economical solution to transporting people to elevated entry points in a building (see Fig. 25.16.)

25.10   Elevators and Escalators

The use of elevators and escalators should be considered together, as a single problem in vertical transportation for the particular facility being designed. In this connection, and particularly in case of modernization, Fig. 25.17 provides an interesting comparison.

In certain facilities there are times during which demand for vertical transportation is so large that elevators are not a feasible solution. A prime example is the school building. During class change, virtually the entire buildings population moves, with as much as 80% moving between floors. Since class change time is at most 10 min., the only reasonable solution is the combined use

Fig. 25.16 An all-weather exterior electric stairway is often the best solution to an elevated building entrance or exit. Note the elongated left balustrade that prevents bunching at the escalator entrance, and the fluorescent fixture built into the balustrade that provides nighttime step lighting. (Photo courtesy of O&K Rolltreppen, Germany.)

Elevators

Electric stairways

Fig. 25.17 *Comparative space requirements for equivalent passenger-handling capacity. Note the marked space savings offered by escalators.*

of fixed and moving stairs. In other buildings such as multifloor stores, the escalator provides for short trips of one or two floors, and the elevator generally transports passengers traveling three or more stories.

A comparison of travel time between escalator and elevator is of interest. Using a normal speed of 100 fpm, a 12-ft floor requires 14.4 s for travel plus about 5 to 6 s to transfer to the next section, for a total of approximately 20 s. Thus a four-story trip would take approximately 75 s. A similar elevator trip would take at most 60 s. The additional escalator time is not noticed in the activity of boarding, turning, and riding. However, a trip of more than four stories becomes tiresome, and all the more so when there is an enforced "walk-around" at each floor.

25.11 Electric Power Requirements

Standard American electric stairways are driven by three-phase, 60-Hz, ac induction motors at standard voltages (208, 230, and 460 V). Approximate horsepowers of driving motors are shown in Table 25.3.

It is recommended that no more than four escalators be served by a single electric feeder, and further that not all the escalators of an installation, whatever the number, be served from the same feeder.

Since obviously one cannot be trapped on an escalator, emergency power is rarely required. Ventilation for the machinery should be supplied for approximately 40% of the power, to be dissi-

pated as heat. Thus a 10-hp motor would require the dissipation of $0.40 \times 10 \times 2500$ Btu/hp, or approximately 10,000 Btu/h.

25.12 Special-Design Escalators

As with elevators, escalators of nonstandard design are available on special order. These include units with slopes other than 30°, speeds other than 100 fpm, and rises beyond normal. The most unusual of these special designs, is the curved escalator, which made its debut overseas and is now available in the United States. This design, which solves the very common problem of directing passengers to make a 90° turn as they leave the escalator, is an engineering tour-de-force. The entire drive mechanism had to be designed anew because tread speed along the inside curve must be lower than that on the outside curve. Handrail drive was less of a problem since the handrail is

TABLE 25.3 Typical Standard-Duty Escalator Motor Sizes

Escalator Size (in.)	Maximum Rise (ft)	Size of Motor (hp)
32	14	5
	22	7½
	30	10
48	10	5
	15	10
	20	15

TRANSPORTATION

narrow and therefore speed variation within it is minimal.

25.13 Preliminary Design Data and Installation Drawings

At the preliminary design stage, the architect requires rough dimensional and structural data for escalators. Since at this stage of design a specific manufacturer for the stairs has not yet been selected, and because these data vary from one supplier to another, the information in Fig. 25.18 is given as a range that covers most of the major manufacturers. Similar data are available from the NEII. Once construction contracts have been awarded and an escalator supplier selected, exact data on space requirements and structural reactions are made available by the vendor. On these, two "working points" are identified. From these

		Range	
		From	To
A		13'-2"	16'-10"
B		7'-5"	9'-4"
C		5'-9"	7'-6"
D		3'-3"	4'-2"
E		3'-3"	3'-9"
F		12'-5"	14'-10"
G		2'-11"	3'-1"
H		2'-7"	2'-9"
I	32"	3'-11"	4'-6"
	48"	5'-3"	5'-8"
J		0	1"
K		2'-11"	3'-2"
L	32"	4'	4'-4"
	48"	5'-3"	5'-10"
M	32"	2'	2'
	48"	3'-4"	3'-4"
N		1'-9"	2'-2"
O		3'-6"	4'-0"

PRELIMINARY ESCALATOR SPACE REQUIREMENTS

Fig. 25.18 *Escalator dimensional data in a range representing most of the major manufacturers. These data can be used for preliminary planning only.*

two points all other measurements are made, that is, locating the centerline of the truss sections, placing the lower and upper landing truss support beams, and so on.

25.14 Budget Estimating for Escalators

The cost of an escalator includes the cost of the associated mechanical and electrical equipment, plus the shipping and installation charges. The manufacturer provides expert engineering and a field erector who supervises the installation.

A tabulation of *relative* escalator prices on a base of 100, is given in Table 25.4. Prices for units with rise above 35 ft rise very rapidly and depend on the type of unit used. The designer is referred to the suppliers for quotes on all units. To these figures must be added the cost of builder's work, wellway protection, lighting, outside balustrades, and plaster.

MOVING WALKS AND RAMPS

25.15 General

Moving walks and ramps are different from moving stairways in application, function, construction, and capacity. Escalators have as their primary function the movement of large numbers of

people vertically, when such vertical distance does not exceed approximately five stories, as noted above. The moving stair performs this specific transportation function extremely well with minimum cost, space, and maintenance.

When vertical transportation of wheeled vehicles and large parcels is required, the use of an electric stairway is at best awkward, if not entirely impossible. For such functions and others discussed below, the moving ramp serves very well.

Unlike the elevator and escalator, the moving walk or ramp serves a dual function, that is, horizontal transportation only, or a combined function of horizontal and vertical transportation. For the purpose of our discussions, we will define a moving *walk* as one with an incline not exceeding 5° where the principal function is horizontal motion and inclined motion is incidental to the horizontal. A moving *ramp* is a device with an incline limited by code to 15°, where vertical motion is generally more important than the horizontal component. It should be understood that the walk and ramp are physically the same device, differently applied.

25.16 Application of Moving Walks

The principal uses of moving walks, also known as autowalks, are to:

1. Eliminate and/or accelerate burdensome walking.
2. Eliminate congestion.
3. Force movement.
4. Easily transport large, bulky objects.

Anyone who has walked the seemingly endless distances in a major airport, carry or dragging a heavy suitcase, can appreciate the near absolute necessity for a moving walkway. It is for this reason that transportation terminals have become major users of this item (Fig. 25.19). Other transportation terminals, such as rail and ship terminals, also can often find excellent applications for the moving walk, since much heavy and bulky luggage is moved in these areas.

Application of the apparent distance compression that moving walks provide permits placement of parking areas remote from the pedestrians' destination. Thus, a store can extend its parking

TABLE 25.4 Relative[a] Escalator Prices

Rise (ft)	Size	
	32 in.	48 in.
14	100[a,b]	118
16	103	121
18	105	124
20	108	127
22	111	130
24	113	133
26	115	135
28	118	138
30	120	141

[a]Base price figure is 100 for the shortest, narrowest, slowest unit (i.e., 14-ft rise, 32 in. wide). All units are 100 fpm. Add 5% for glass balustrade for any unit.

TRANSPORTATION

TRANSPORTATION

Fig. 25.19 *Twin auto-walks at a transportation terminal. Of particular interest are the sloped entry and exit and the continuous fluorescent lighting beneath the handrails of the balustrades. (Photo courtesy of O&K Roll-treppen, Germany.)*

area with no annoyance to patrons who must make the long trip to their cars with bulky packages or shopping carts. These advantages are all the more appreciated by persons with a walking impediment.

A second application of walks, as noted previously, is the routing of traffic to avoid congestion, milling about, and lost time and motion. This is particularly applicable in transportation terminals, where persons are always traveling in opposite directions through the same and often restricted area, such as in "fingers" leading from airplanes to the main air terminal.

Moving walks are also useful to move people past a display window or some other point where congestion caused by stopping, is undesirable. This "movement of objects" application demonstrates clearly that the moving walk is in function simply a large conveyor belt, regardless of its construction.

25.17 Application of Moving Ramps

The moving ramp that combines horizontal and vertical movement is principally applicable:

  1. To move persons and wheeled vehicles vertically.

Fig. 25.20 *Unusual angular design of multiple ramps in a shopping mall in Brisbane, Australia. The use of ramps rather than escalators permits shopping carts and bulky packages to be moved with ease between the building levels. (Photo courtesy of O&K Rolltreppen, Germany.)*

Fig. 25.21 *Multiple inclined ramp installation in a European department store (3 up, 2 down). Of interest in this installation are the level entries, lighted balustrades, the ease with which shopping carts are handled on the wide pallet, and the use of the space between the transparent balustrades for merchandising. (Photo courtesy of O&K Rolltreppen, Germany.)*

2. To move persons who lack the agility required to use an escalator.
3. To vertically move large, bulky objects.

Ramps have found a fertile field of application in multilevel stores where escalators are not feasible for shopping-cart users. Such stores may also utilize rooftop parking that is made accessible to cart users via a moving ramp. Since luggage carriers are not easily used on escalators, transportation terminals, which are almost always multilevel, also find extensive application for the moving ramps (Figs. 25.20 and 25.21).

25.18   Size, Capacity, and Speed

The speed, physical dimensions, and therefore passenger capacity of walks and ramps are not as extensively standardized as is the case with escalators. Manufacturers utilize several different tread widths, combined with various speeds and ramp angles of incline. The combinations are designed to suit the application. Furthermore, since the maximum permissible walk ramp speed varies

TABLE 25.5 **Maximum Permissible Operating Speeds of Moving Ramps**

	Maximum Speed [fpm (m/s)]	
Angle of Incline	Level Entrance	Sloping Entrance
0–3°	180 (0.9)	180 (0.9)
3–5°	180 (0.9)	160 (0.8)
5–8°	180 (0.9)	140 (0.7)
8–12°	140 (0.7)	130 (0.65)
12–15°	140 (0.7)	125 (0.625)

TABLE 25.6 **Moving Walks/Ramps: Commercial Ratings**

Angle (deg)	Pallet Width [in. (mm)]	Ramp Speed [fpm (m/s)]
0–3	32 (800)	100 (0.5), 130 (0.65), 150 (0.75)
Walk	40 (1000)	100 (0.5), 130 (0.65), 150 (0.75)
	62 (1400)	90 (0.45), 100 (0.5), 130 (0.65)
10, 11, 12	32 (800)	100 (0.5), 130 (0.65)
Ramp	40 (1000)	90 (0.45), 100 (0.5), 130 (0.65)

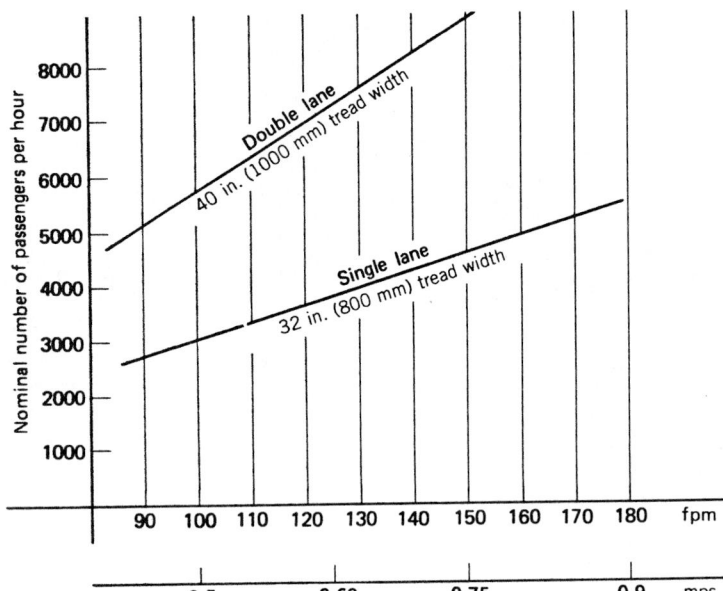

Fig. 25.22 The capacity of moving walks varies with speed, angle of incline, and tread width. Capacity shown is for maximum incline permitted at that walk speed. Because of the requirement for handrail support, tread widths greater than double lane are not often utilized.

with angle of slope, and with design of entering point, passenger capacity ratings vary with each design. Higher speeds are allowed by code for level entrance than with sloping entrance, for the obvious reason that the level entrance is easier to board. Tables 25.5 and 25.6 and Fig. 25.22 give data for typical units.

Since capacity varies with width, speed, and type of entrance, exact capacity figures must be obtained for each specific design. Maximum practical lengths at present are approximately 1000 ft, with longer units in design. Wide autowalk units (1400 mm) are now extensively used in transportation terminals since they permit passengers with luggage trolleys who walk on the moving walkway to easily pass other passengers with luggage carts, who prefer to stand still and move at the relatively

slow rate of 100 to 130 fpm (about one-half normal walking speed). Recent installations of such wide autowalks are found at Heathrow, Gatwick, and Munich airports, among others.

25.19 Components

Moving walks are manufactured in only one design: a derivative of the escalator, which uses a flattened pallet in place of a step. In all other respects—the drive mechanism, safeties, brake, handrails, and so on—the unit is similar to an escalator. Preliminary layout data and detailed plans are, as with moving stairs, available from vendors, who frequently are the same vendors as supply escalators.

PART IX

ACOUSTICS

The best (and simplest) distinction between the terms "sound" and "noise" is a subjective one—sound is desirable, noise is not. This definition does not consider the content of the sound. For example, speech, which is a desirable sound in most instances, can become a "noise" when emanating from a neighbor's apartment or an adjoining office. An air conditioner, which is most often considered to be a "noise," can provide a desired "sound" by masking the intruding speech. The function of architectural acoustics is simply to reinforce this definition; that is, to enhance desired sound and to maximally attenuate noise.

Chapter 26, Fundamentals of Architectural Acoustics, introduces the subject with a discussion of physical sound theory and physiological hearing phenomena. The latter include the negative effects of noise, which are primarily psychological (annoyance) at low levels but become physical and can result in hearing damage at high levels. Criteria of the U.S. Occupational Safety and Health Administration (OSHA) are given. The discussion continues with a description of the types of sounds encountered in architectural acoustics study, in order to establish criteria for acceptable background noise in enclosed spaces. These are developed into noise criteria (NC) and room criteria (RC) curves.

Chapter 27 moves to the core of the subject—room acoustics—with an explanation of absorption and reverberation, and develops acoustic design criteria for various indoor activities. The chapter concludes with a description of high-grade sound reinforcement systems.

Chapter 28 is devoted to noise control. Noise within a space is controlled by absorption, whereas control of noise transference between spaces is a matter of isolation. The chapter begins with a discussion of the application of absorptive materials for room noise reduction. It treats the problem of interspace noise conduction, dividing the problem into two parts (airborne noise and structure-borne noise), because the solutions are different. Relevant criteria, including sound transmission class (STC) and impact isolation class (IIC), are adduced and explained, and solutions to noise transference problems of different types are suggested. The problem of speech privacy for both enclosed spaces and open (office) areas is treated in detail. The chapter offers sections on mechanical system noise control and on acoustic recommendations and criteria, ending with a section on exterior acoustics. An extensive section containing reference material closes the acoustic portion of the book.

26

FUNDAMENTALS OF ARCHITECTURAL ACOUSTICS

SOUND THEORY AND HEARING PHENOMENA

26.1 General

Architectural acoustics may be defined as the technology of designing spaces, structures, and mechanical systems to meet hearing needs. With proper design, "wanted" sounds can be heard properly and "unwanted" sounds, or "noise," can be attenuated or masked to the point where it does not cause annoyance. However, achieving good acoustics has become increasingly more difficult for a variety of reasons. To cut costs, the weight of construction materials used in many of today's buildings is reduced. Since light structures generally transmit sound more readily than heavy ones, this practice poses major acoustical problems. Population density of office spaces has increased, thus raising the amount of noise generated. Worse yet—from the acoustics point of view—most offices are today designed as open areas with, at best, only thin partial height dividers separating workers. Forty percent or more of a building budget may be allocated for mechanical systems—most of which make noise. Outside noise sources such as cars, trucks, trains, and airplanes present problems in isolating interior spaces from exterior sound.

Building owners and tenants are aware that good acoustic environments in buildings are required for high productivity and hence rental or purchased space desirability, and the architect is expected to provide them. A clear understanding of the principles explained in this and the following chapters will assist the architect in accomplishing straightforward designs alone and, in more complex instances, cooperating knowledgeably with the project's acoustic consultant. The importance of proper acoustic design is all the more

critical, since after-the-fact acoustic "repair" is often difficult and frequently impossible without substantial structural alterations.

All acoustic systems have three common elements—source, transmission path, and receiver. The source can be made louder or quieter and the path can be made to transmit more or less sound. The listener's reception of sound also may be influenced. This chapter presents essential aspects of architectural acoustics to assist a designer in defining acoustic goals. Moreover, it describes the achievable goals as well as various methods for reaching them through design.

26.2 Sound: Definition and Generation

Sound can be defined in a number of different ways depending on the aspect we desire to study. Thus sound is a physical wave, or a mechanical vibration, or simply a series of pressure variations, in an elastic medium. For airborne sound, the medium is air. For structure-borne sound, the medium is concrete, steel, wood, glass, and combinations of all these materials. A much more limited definition, more appropriate to our study, is to define sound simply as an audible signal. This means that the science of architectural acoustics is concerned with the building occupant; sounds that he or she cannot detect are generally not our concern.

It is assumed that the hearer has a pair of healthy young ears with a detection range of 20 to 20,000 hertz (Hz). With these givens, it is probably best to view sound as a series of pressure variations. In air, these pressure variations take the form of periodic compressions and rarefactions. In Fig. 26.1, the bell radiates a pure tone in all directions equally—that is, it creates a circular

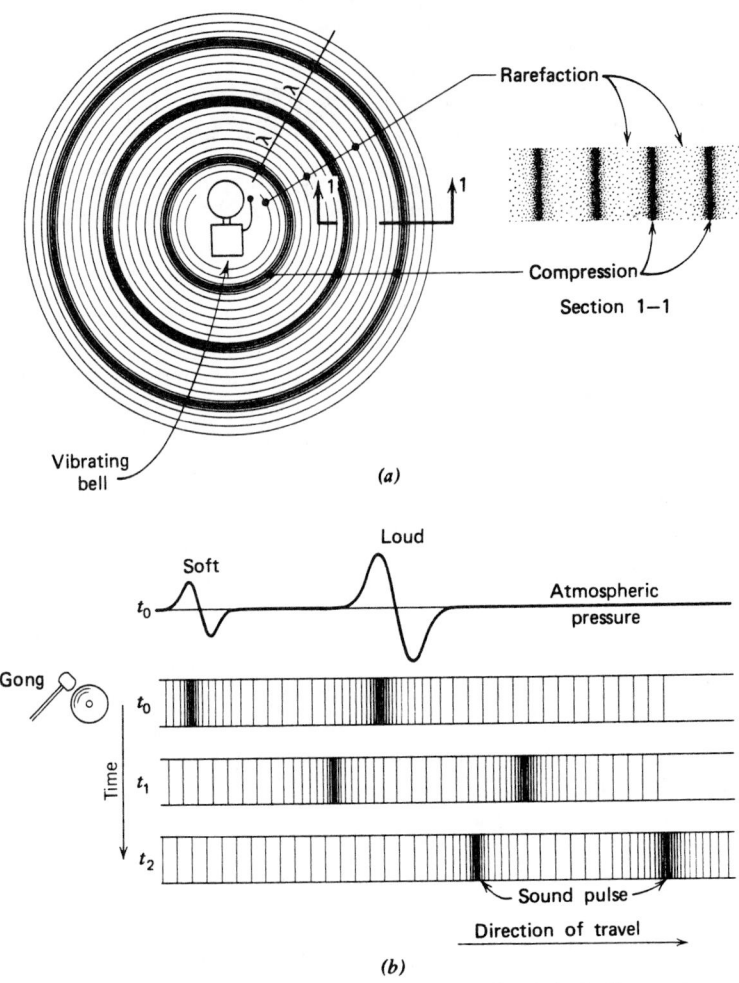

Fig. 26.1 *Sound pressure waves. (a) The continuous vibration from the bell causes a series of compressions and rarefractions of the air to travel outward longitudinally from the source. Amplitude information is carried by pressure; that is, greater amplitude means greater compression and greater rarefraction. (Compression and rarefraction are shown diagrammatically as line density although they are actually molecular phenomena, as shown in the upper drawing).*
(b) Two single impulses of different magnitude (amplitude) traveling away from the source. Note how amplitude information is carried by difference in pressure.

wave front. As the bell vibrates, it sets up vibrations in the air, of the same frequency, which can best be visualized in the form of a sectional view. Notice that the pressure changes containing the sound information travel in the same direction as the wavefront—longitudinally. This is unlike a radio signal, for instance, in which the wave travels longitudinally but the information—that is, the modulation—is transverse. Sound is therefore *longitudinal* mechanical wave motion.

26.3 Frequency

The number of times the cycle of compression and rarefaction of air occurs in a given unit of time is described as the frequency of a sound. For example, if there are 1000 such cycles in one second, the frequency of the sound is 1000 cps or 1000 Hz in the standard nomenclature. Thus in Fig. 26.1(a), higher frequencies would be shown by compressions and rarefactions that are closer together and lower frequencies by those that are farther apart. In sound, the concept of frequency is often referred to by a term borrowed from music—*pitch*. The higher the frequency, the higher the pitch, and vice versa.

As stated, the approximate frequency range of a healthy young person's hearing is 20 to 20,000 Hz. The upper limit decreases with age as a result of a process called *presbycusis*. Recognition of this phenomenon can be of importance in schools, since very high-pitched sounds that are inaudible to most adults (the loss is more pronounced in men than women) can be a source of extreme annoyance to students. For example, dentists report that high-speed turbine drills and teeth-cleaning devices cause extreme auditory discomfort to many young patients. These devices produce sounds in the 15- to 20-kilohertz (kHz) range (Fig. 26.2).

The human speaking voice has a range of approximately 100 to 600 Hz in *fundamental frequencies*, but harmonics (overtones) reach to ap-

proximately 7500 Hz. Most speech information is carried in the upper frequencies, whereas most *acoustic energy* exists in the lower frequencies. The critical frequency range for speech communication is 300 to 4000 Hz. Overtones outside these frequencies give the voice its characteristic sound and specific identity. Telephone and radio communication are accomplished in a considerably narrower frequency band by sacrificing some voice quality and intelligibility. (See Fig. 26.14.)

A sound composed of only one frequency is a pure tone. Except for the sound generated by a tuning fork, few sounds are truly pure. Musical sounds (tones) are composed of a fundamental frequency and integral multiples of the fundamental frequency (harmonics). Most common sounds are complex combinations of frequencies. Figure 26.3 shows examples of pure tones, musical notes, and common sounds: Fig. 26.4 shows the frequency ranges of some common devices and phenomena. The frequencies in the scale of Fig. 26.4 all stand in the ratio of 2 : 1 to each other—that

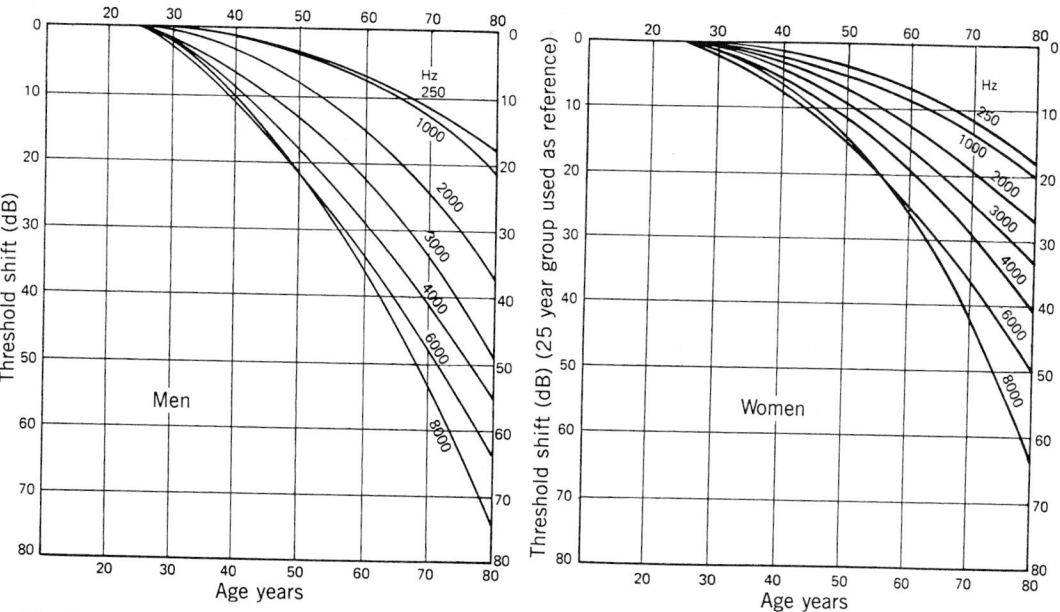

Fig. 26.2 *The curves illustrate the average hearing threshold shift for men and women, with increasing age relative to 25 year old reference group. Note that at age 60 the male threshold shift at upper speech frequencies (4000 Hz) is 30 dB as compared to 20 dB for women, a difference of 10 dB or one-half the subjective loudness (see Table 26.3). For men older than 50 and women older than 60, frequencies above 10 kHz are effectively inaudible, and even at normal speech frequencies of 1000 to 2000 Hz subjective loudness is reduced to one-half of that of a 25-year-old listener. (Reproduced with permission from F. A. White,* Our Acoustic Environment, *Wiley, New York, 1975).*

Fig. 26.3 *Schematic representations of a pure tone, a musical note, and more complex sounds (speech, music, and noise), showing the variation of sound pressure with time and frequency.*

is, 16:32:63:125:250, and so on. Borrowing again from musical terminology, they are one *octave* apart. These particular frequencies are also accepted internationally as the center frequencies of octave bands used for the purpose of sound specification. For technical reasons, a geometric mean is used. Thus 250 is the center frequency of an octave band ranging from $250/\sqrt{2}$ to $250\sqrt{2}$ and that particular octave is known as the 250-Hz octave. If finer division is required, ½-octave and ⅓-octave bands are used.

26.4   Velocity of Propagation

Sound travels at different velocities depending on the medium. In air, at sea level, sound velocity is 344 m/s, or 1130 fps. This corresponds to 770 miles per hour, or 1239 kilometers per hour—slow indeed when compared to light, which has a velocity of 186,000 miles per *second*. Since sound travels not only in air but also through parts of a structure, it is of interest to know the velocities in other media (Table 26.1). For our purposes, velocity changes due to temperature and altitude (atmospheric pressure) may be ignored and, for rough calculation, 1100 fps and 350 m/s may be used as velocity in air, since both are within 3% error.

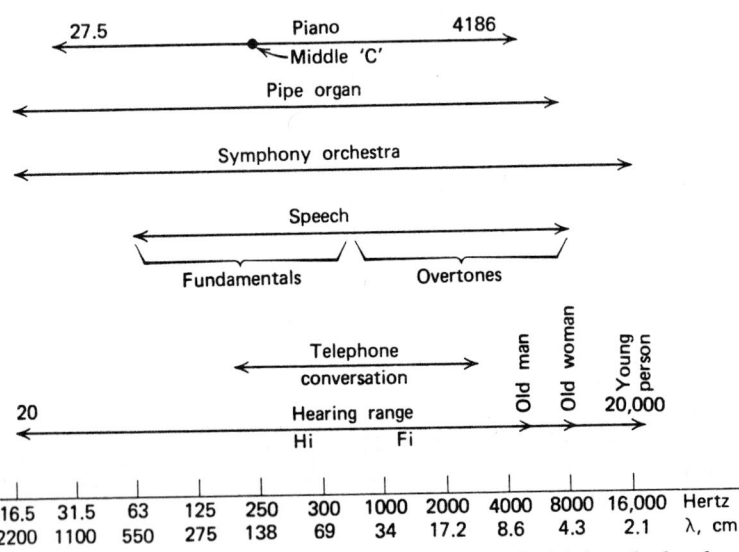

Fig. 26.4 *Frequency ranges of common instruments. Wavelength (λ) is calculated on an assumed propagation velocity of 344 m/s.*

TABLE 26.1 **Sound Propagation Velocity in Various Media**

Medium	Velocity	
	Meters per Second	*Feet per Second*
Air	344	1130
Water	1410	4625
Wood	3300	10,825
Brick	3600	11,800
Concrete	3700	12,100
Steel	4900	16,000
Glass	5000	16,400
Aluminum	5800	19,000

NOTE: These figures are approximate, since the listed materials vary in density. Average frequency is used.

26.5 Wavelength and Types of Propagation

The wavelength of a sound may be defined as the distance between similar points on successive waves or the distance the sound travels in one cycle. The relationships between wavelength, frequency, and velocity of sound is expressed as

$$\lambda = \frac{c}{f} \qquad (26.1)$$

where λ = wavelength, ft (m)
c = velocity of sound, fps (m/s)
f = frequency of sound, Hz

Low-frequency sounds are characterized by long wavelengths and high-frequency sounds by short wavelengths. Sounds with wavelengths ranging from ½ in. to 50 ft can be heard by humans. A simple nomograph (see Fig. 27.18) permits rapid determination of wavelength, given the frequency, and vice versa.

26.6 Sound Magnitude

The magnitude of a sound signal is a relatively complex subject because of the different terms in use and because of the great range of numbers involved. When we speak of sound magnitude, we think of loudness, which is a subjective, ear-oriented reaction not linearly related to the sound power (watts). The strength of sound is described variously as sound power, sound pressure, sound pressure level (SPL), sound intensity, and sound intensity level (IL), all of which differ from each other, and from subjective loudness. To clearly understand these concepts, and it is imperative that they be so understood, a comprehension of how we hear and how sound is propagated in free space is necessary.

26.7 Sound Intensity: Free Field Propagation

A point sound source of constant power P (watts) radiating in free space—that is, at a distance far from the effects of any reflecting surface, is represented in the drawing of Fig. 26.5. The *sound intensity* at any distance from the source is ex-

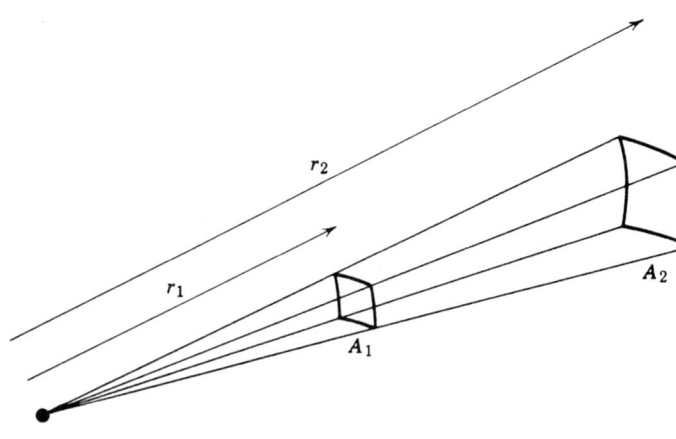

Point source

Fig. 26.5 The same total energy passes through areas A_1 and A_2. Since A is proportional to the square of r, the energy density or intensity is inversely proportion to r^2.

pressed as

$$I = \frac{P}{A} \qquad (26.2)$$

where I = sound (power) intensity, W/cm^{2a}
       P = acoustic power, watts
       A = area, cm^{2a}

^{a}It is traditional in architectural acoustics to express area in square centimeters although the MKS (SI) system would require area to be stated in square meters. Conversion data are supplied in Table 28.15. See also Table 28.16 for a listing of symbols and abbreviations.

Since the sound radiates freely in all directions.

$$I = \frac{P}{4\pi r^2} \, \text{W/cm}^2 \qquad (26.3)$$

where r is the radius of an imaginary enclosing sphere. This is a form of the classic *inverse square law*, stating that *intensity is inversely proportional to the square of the distance from the source.* [In English units, this is

$$I = \frac{P}{930 \times 4\pi r^2} \, \text{W/ft}^2 \qquad (26.4)$$

since 1 ft^2 = 930 cm^2.] Using Eq. 26.3 to determine the intensities I_1 and I_2 at distances r_1 and r_2 from point source P, and dividing the equations, we find that the intensities at distance r_1 and r_2 from the source stand in the ratio

$$\frac{I_1}{I_2} = \frac{r_2^2}{r_1^2} \qquad (26.5)$$

Note the exact correspondence of these relations to the derivations for illumination from a point source, found in Section 18.12 and Fig. 18.15. Figure 26.6 shows graphically how a sound pulse is attenuated in *strength* (but not in waveform) as it travels outward from the source by action of distance.

The preceding derivation is based on a *point source*—that is, a source that is small relative to the wavelength produced. This type of source produces spherical waves. Line sources such as strings produce cylindrical waves. Large vibrating surfaces such as walls produce plane waves. The importance of these facts will become clear in the discussions on barriers and diffraction in Chapter 28.

The threshold of hearing—-that is, the minimum sound power intensity (I) that a normal ear can detect—is 10^{-16} W/cm^2. (Actually, the ear responds to pressure, as explained below.) The maximum sound intensity that the ear can accept without damage is approximately 10^{-3} W/cm2. This gives a range of 10^{13}, or 10 million to 1 (10,000,000,000,000 : 1). Table 26.2 gives an idea of the physical significance of these numbers. As an interesting comparison, note that the maximum (painful) acoustic intensity level in Table 26.2 is 0.001 W/cm^2, or 10 watts/m^2; this is only 1% of average clear sky solar radiation of 1000 W per square meter.

Two problems arise immediately when dealing with quantities of this type; the numbers themselves are very small and the ratios are very large. Furthermore, the human ear responds logarithmically, not arithmetically to sound pressure (and intensity); that is, doubling the intensity of a sound does not double its loudness—the change is barely perceptible. To solve these problems it would be much more convenient if we were to construct a scale that:

1. Started at zero for the minimum sound (intensity or pressure) we can hear.

Fig. 26.6 *Attenuation of a sound signal in air as it travels away from the source. The shape (information) remains constant when traveling in a nondispersive medium such as air. This is not the case with travel in solids, where different frequencies travel at different velocities, and therefore cause a wave-shape change with time and distance. Since velocity of propagation is constant, time and distance are linearly related and can therefore be plotted on the same axis.*

TABLE 26.2 **Comparison of Decimal, Exponential, and Logarithmic Statements of Various Acoustic Intensities**

Intensity (W/cm²)		Intensity Level—Logarithmic Notation	Examples
Decimal Notation	Exponential Notation		
0.001	10^{-3}	130 db	Painful
0.0001	10^{-4}	120 db	
0.00001	10^{-5}	110 db	75-piece orchestra
0.000001	10^{-6}	100 db	
0.0000001	10^{-7}	90 db	Shouting at 5 ft
0.000000001	10^{-9}	70 db	Speech at 3 ft
0.00000000001	10^{-11}	50 db	Average office
0.0000000000001	10^{-13}	30 db	Quiet unoccupied office
0.00000000000001	10^{-14}	20 db	Rural ambient
0.000000000000001	10^{-15}	10 db	
0.0000000000000001	10^{-16}	0 db	Threshold of hearing

2. Used whole numbers rather than negative powers of 10.
3. Had some fixed relationships between an arithmetic difference and a loudness change; say, 10 units equals a doubling (or halving) of loudness. Thus, on such a scale, the difference between 20 and 30, 60 and 70, would always be a doubling of loudness.

Such a scale exists. It is the decibel scale.

26.8 Intensity Level (IL) and the Decibel (dB)

The word "level" indicates a quantity relative to a base quantity. *Intensity level* is the ratio between a given intensity and a base intensity. If we express intensity level as

$$IL = 10 \log \frac{I}{I_0} \qquad (26.6)$$

where IL = intensity level, dB
  I = intensity, W/cm²
  I_0 = base (i.e., 10^{-16} W/cm², the threshold of hearing)
  log = logarithm to base 10

then we have established a scale that satisfies the three conditions set forth above. The quan-

tity *IL*, is dimensionless, since it indicates simply a ratio between two numbers. It is measured in decibels for convenience in expressing the large numbers involved. Table 26.2 shows the great convenience of using the logarithmic decibel scale as compared to either decimal notation or exponential notation. Table 26.3 gives a short listing of subjective loudness changes expressed in decibels. Note that *10 dB indicates a doubling of loudness,* and 20 dB is loudness doubled twice—that is, four times as loud. The *difference* (Δ) between any two intensity levels can be

TABLE 26.3 **Subjective Loudness Changes and Corresponding Intensity Level Changes**

Change in Level (db)	Subjective Change in Loudness
3	Barely perceptible
6[a]	Perceptible
7	Clearly perceptible
10	Twice or half as loud
20	Four times or one-quarter as loud

[a]Six decibels corresponds to the change encountered when distance to the source in a free field is doubled (halved).

expressed as:

$$\Delta IL = IL_2 - IL_1 = 10 \log \frac{I_2}{I_0} - 10 \log \frac{I_1}{I_0}$$

Therefore,

$$\Delta IL = 10 \log \frac{I_2}{I_1} \text{ dB} \qquad (26.7)$$

A few examples using decibel notation and logarithmic calculations should help the reader become familiar with this useful system.

EXAMPLE 26.1. Two sound sources (I_1 and I_2) produces intensity levels of 60 and 50 dB, respectively, at a point. When these are functioning simultaneously, what is the total sound intensity level? (We assume identical frequency content and random phase; — that is, the phase relationship between the two sources changes in a random manner.)

SOLUTION. Note that we are dealing with intensity level, not intensity, since intensity itself has little significance for us. The technique involved in adding two sound intensity levels has three steps:

1. Convert both to actual intensity.

$$IL = 10 \log \frac{I}{I_0}$$

so

$$60 = 10 \log \frac{I_1}{10^{-16}}$$

or

$$6.0 = \log \frac{I_1}{10^{-16}}$$

Then, using the definition of a base 10 logarithm:

$$10^6 = \frac{I_1}{10^{-16}}$$

$$I_1 = (10^{-16}), 10^6 = 10^{-10} \text{ W/cm}^2$$

By similar calculation,

$$I_2 = 10^{-11} \text{ W/cm}^2$$

2. Add arithmetically.

$$I_1 + I_2 = 10^{-10} + 10^{-11}$$

$$= (10 \times 10^{-11}) + 10^{-11}$$

$$I_{tot} = 11 \times 10^{-11} \text{ W/cm}^2$$

3. Reconvert to decibels. To find the intensity level IL corresponding to the new total intensity $I_1 + I_2$, we simply apply Eq. 26.6:

$$IL_{tot} = 10 \log \frac{I_{tot}}{I_0}$$

$$= 10 \log \frac{11 \times 10^{-11}}{10^{-16}}$$

$$= 10(\log 11 + \log 10^5)$$

$$= 10(1.04 + 5)$$

$$= 60.4 \text{ dB}$$

which is only a fraction larger than the original 60 dB of the stronger sound. ∎

EXAMPLE 26.2. Assume two noise signals of 60 dB each. What is the combined strength in decibels?

SOLUTION. One method would be to calculate levels as in Example 26.1. A shorter method is to find the difference between the sum and either signal, and to add it to either original signal. Using Eq. 26.7:

$$\Delta IL = IL_{comb} - IL_1 = 10 \log \frac{I_{comb}}{I_1}$$

$$= 10 \log \frac{2I_1}{I_1}$$

$$= 10 \log 2$$

$$= 10(0.30)$$

$$= 3 \text{ dB}$$

This answer (which is independent of any particular level) gives us the extremely important fact that *doubling a signal intensity raises the intensity level by 3 dB*. (In our case, the combined intensity level would obviously be 60 dB + 3 dB, or 63 db). Similarly, quadrupling a signal's intensity raises the received level by 6 dB. This is because

$$\Delta IL = 10 \log \frac{4I}{I}$$

$$= 10 \log 4$$

$$= 10(0.60)$$

$$= 6 \text{ dB}$$

Therefore, quadrupling 60 dB gives 66 dB, or 60 dB + 60 dB = 63 dB and 63 dB + 63 dB = 66 dB. This technique is very useful when combining a large number of identical levels, as in the following example. ■

EXAMPLE 26.3.   A factory will contain 20 identical machines, each of which generates a noise level of 80 dB. What will be the combined noise level? (Ignore problems of frequency content, phase, and fields.)

SOLUTION.

$$\Delta IL = IL_{\text{tot}} - IL_{\text{single}}$$

$$= 10 \log \frac{I_{\text{tot}}}{I_{\text{single}}}$$

$$= 10 \log \frac{20 \, I_{\text{single}}}{I_{\text{single}}}$$

$$= 10 \log 20$$

$$= 10(1.3) = 13 \text{ dB}$$

Therefore the total noise level will be

$$IL_{\text{tot}} = 80 \text{ dB} + 13 \text{ dB} = 93 \text{ dB} ■$$

A chart for combining decibel levels of two sources is given in Fig. 26.7 that eliminates the somewhat lengthy procedure detailed in Example 26.1. Referring to Table 26.3, we note that the human ear is not responsive to fractional dB changes; indeed, even a 3 dB change is barely perceptible. This being so, we recommend that the detailed calculation and the chart be reserved for situations where a high degree of accuracy is required.

For everyday calculations, the following rules of thumb may be used to combine dB levels of two sources:

· When the difference between two sources is 1 dB or less, add 3 dB to the higher level to obtain the total.

· When the difference is 2 to 3 dB, add 2 dB.

· When the difference is 4 to 8 dB, add 1 dB.

· When the difference is 9 dB or more, ignore the lower source.

A comparison in Table 26.4 of addition by these rules of thumb and by an accurate method shows that at usual levels, the error involved is always less than 1%.

Returning to the inverse square law expressed in Eq. 26.5, we are now in a position to determine

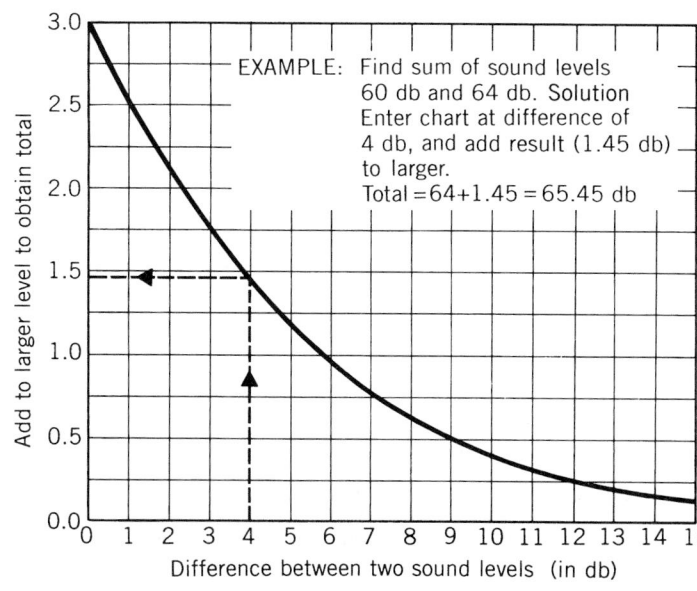

Fig. 26.7 Chart for adding two uncorrelated sound pressure levels when both are expressed in decibels. (Reprinted by permission, from E. B. Magrab, Environmental Noise Control, Wiley, New York 1975).

ACOUSTICS

TABLE 26.4 **Addition of Uncorrelated Sound Pressure Levels**

dB Level		Sum	
Lower	Higher	Approximate[a]	Accurate
60	60	63	63
60	62	64	64.4
60	64	65	65.5
60	66	67	67
60	68	69	68.7
60	70	70	70.5

[a]See text for rules of thumb.

the effect on sound intensity level of moving away from the sound source.

EXAMPLE 26.4.   Given a sound source that produces intensity IL at a distance d_1 from the source (the reader can substitute any numbers desired, or follow the problem with symbols), what are the intensities at twice the distance? Four times?

Note: Sound intensity that obeys the inverse-square law on which these calculations are based is developed from a source in an open (unenclosed), obstruction-free space (e.g., outdoors). Also, intensity measurement I_1 must be taken at a sufficient distance d_1 from the source that the free field has developed. See the discussion in Section 27.7 for an explanation of sound fields.

SOLUTION.   We know from Eq. 26.7 that

$$\Delta IL = \frac{I_2}{I_1}$$

and we also know from Eq. 26.5 that

$$\frac{I_1}{I_2} = \frac{d_2^2}{d_1^2}$$

therefore, since $d_2 = 2d_1$

$$\frac{I_2}{I_1} = \frac{(d_1)^2}{(2d_1)^2} = \frac{1}{4}$$

Substituting in Eq. 26.7, we have

$$\Delta IL = 10 \log \frac{I_2}{I_1}$$

$$= 10 \log \tfrac{1}{4}$$

$$= 10(-0.6)$$

$$= -6\,\text{dB}$$

which tells us that sound intensity level (not pressure) is reduced by 6 dB. Similarly, when distance is quadrupled, it is reduced by 12 dB.   ∎

To summarize, then, *intensity level changes 3 dB with every doubling or halving of power and changes 6 dB with every doubling or halving of distance from a point source.* Figures 26.8 and 26.9 illustrate the latter relationship.

The ear responds to sound pressure, not intensity. As explained below, *sound pressure level* (SPL) is equal numerically to *intensity level* (IL)

Fig. 26.8 *Decibel level increase as a function of power (intensity) increase.*

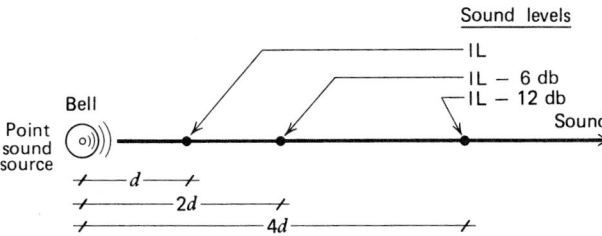

Fig. 26.9 *Sound energy levels at varying distances from the source. Each doubling of distance reduces the level by 6 dB. These relationships hold true only in a free field.*

(at least for normal temperature and pressure, i.e., for our use in architectural acoustics). Therefore, the foregoing examples and manipulations of intensity level are equally applicable to sound pressure level.

As explained in some detail in the following section, the combined effect of two sounds depends on their frequency content. In the examples above we assumed signals either of identical frequency and random phase or of very-wide-frequency spectrum—so wide that phase phenomena are not significant. In architectural acoustics work, such an assumption is generally valid.

26.9 Human Hearing

Refer to Fig. 26.10. The outer ear is funnel shaped and serves as a sound- gathering input terminal to the auditory system. Sound energy travels through the auditory canal (outer ear) and sets in motion the components of the middle ear, comprising the eardrum, hammer, anvil, and stirrup. The stirrup acts as a piston to transmit vibrations into the fluid of the inner ear. This fluid motion causes movement of hair cells in the cochlea, which in turn stimulates nerves at their bases. They in turn transmit electrical impulses along the eighth cranial nerve, to the brain. These impulses we understand as sound.

Frequency recognition is accomplished in the cochlea, by the basilar membrane. This membrane resonates at one end at about 20 Hz and at the other at 20 kHz, giving the ear its frequency range. The ear hears and recognizes distinct frequencies, yet the hearing mechanism has the ability, apparently as directed by the brain, either to hear individual frequencies or to combine them into a sin-

gle sound. Thus when we hear a string quartet we can, *at will,* hear either the entire quartet or each instrument individually. With concentration (vision helps in this), a "trained" ear can pick out a single instrument in an orchestra of 120 pieces, even if there is more than one such instrument in the group. Conductors do this regularly. Similarly, the ear can perform the selection known as the "cocktail party effect," that is, pick out one voice in background noise that may be 20 dB louder than the wanted signal. In effect, the ear is attenuating the unwanted signals. Normally, however, the ear does precisely the opposite. It combines sounds that are clearly distinct from each other in frequency and phase. The three tones in a chord struck on a piano are different in frequency and, if played as a very rapid triplet, out of phase. Yet the ear combines them and hears a single sound, despite the fact that the maxima of the three tones do not occur simultaneously.

In the preceding decibel calculations (Section 26.8), as well as those that will follow, it is assumed that the ear ignores phase differences and combines frequencies. This may not always be the case, particularly when the frequencies are very far apart. For this reason the type of information found in Table 26.5 (i.e., single-number representations of complex sounds) can be misleading and must be used with caution. This problem is discussed in the next section and in Section 26.12, which explains measurements and weighting networks.

At the threshold of hearing (approximately 0 dB), the displacement of air molecules impinging on the eardrum, and the eardrum excursion, are approximately one angstrom unit (1 Å = 10^{-8} cm), which is approximately the diameter of an *atom.* Were the ear an order of magnitude more sensitive, it would hear thermal noise. The

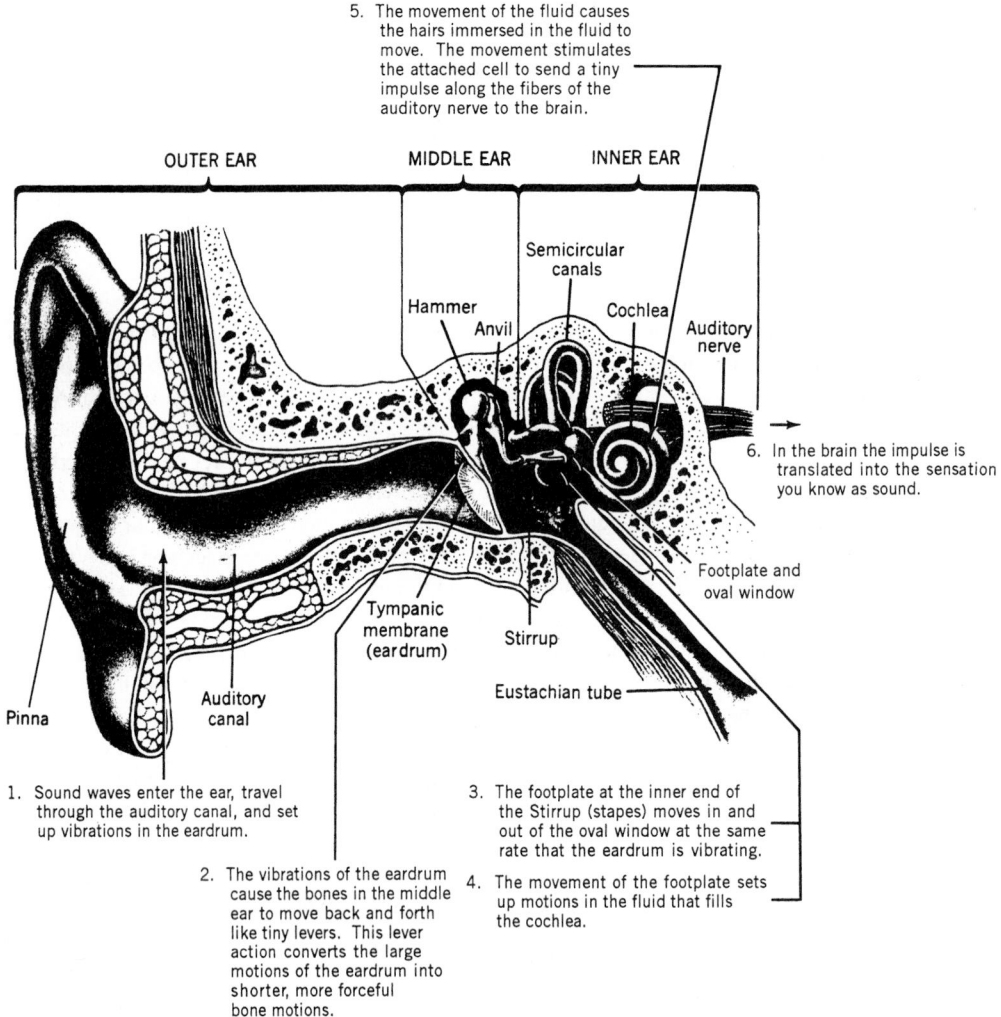

5. The movement of the fluid causes the hairs immersed in the fluid to move. The movement stimulates the attached cell to send a tiny impulse along the fibers of the auditory nerve to the brain.

OUTER EAR   MIDDLE EAR   INNER EAR

Semicircular canals

Hammer   Cochlea
Anvil   Auditory nerve

6. In the brain the impulse is translated into the sensation you know as sound.

Footplate and oval window

Tympanic membrane (eardrum)   Stirrup

Eustachian tube

Auditory canal

Pinna

1. Sound waves enter the ear, travel through the auditory canal, and set up vibrations in the eardrum.

2. The vibrations of the eardrum cause the bones in the middle ear to move back and forth like tiny levers. This lever action converts the large motions of the eardrum into shorter, more forceful bone motions.

3. The footplate at the inner end of the Stirrup (stapes) moves in and out of the oval window at the same rate that the eardrum is vibrating.

4. The movement of the footplate sets up motions in the fluid that fills the cochlea.

Fig. 26.10 *Drawing of the human ear showing principal parts and functioning as sound receptor. [Drawing reproduced with permission from F. A. White,* Our Acoustic Environment, *Wiley, New York, 1975. Notes are reproduced from* Quieting: A Practical Guide to Noise Control *(1976).]*

human ear is thus close to the practical limit of sensitivity. At the other end of the noise spectrum, the threshold of pain corresponds to a sound level of 130 dB, and to an eardrum motion of approximately 0.25 mm. An astonishing range indeed.

As is obvious from the brief description of the hearing process above, movement of the eardrum is caused by air pressure. Therefore the quantities that interest us more than intensity (*power density*) are pressure, which is *force density,* and pressure level, which is the ratio of sound pressure to a base level, expressed in dB.

26.10 Sound Pressure and Sound Pressure Level (SPL)

The usual base or reference pressure corresponds to the threshold of hearing and is taken to be 20 μ Pa or 2×10^{-4} microbar (μ bar) (see Table 28.15). As intensity, this pressure is established as 0 dB for the purpose of sound pressure *level* calculation. Since the ear responds logarithmically to intensity and since pressure varies as the square root of intensity, we can write the expression

$$SPL = 10 \log \frac{p^2}{p_0^2}$$

TABLE 26.5 **Common Noise Levels**

Sound Pressure Level, SPL (dBA)	Typical Sound	Subjective Impression
150		(Short exposure can cause hearing loss)
140	Jet plane takeoff	
130	Artillery fire, riveting, machine gun	(Threshold of pain)
120	Siren at 100 ft, jet plane (passenger ramp), thunder, sonic boom	Deafening
110	Woodworking shop, hard-rock band, accelerating motorcycle	Sound can be felt (Threshold of discomfort)
100	Subway (steel wheels), loud street noise, power lawnmower, outboard motor	
90	Noisy factory, truck unmuffled, train whistle, machine shop, kitchen blender, pneumatic jackhammer	Very loud, conversation difficult; ear protection required for sustained occupancy
80	Printing press, subway (rubber wheels), noisy office, supermarket, average factory	(Intolerable for phone use)
70	Average street noise, quiet typewriter, freight train at 100 ft, average radio, department store	Loud, noisy; voice must be raised to be understood
60	Noisy home, hotel lobby, average office, restaurant, normal conversation	
50	General office, hospital, quiet radio, average home, bank, quiet street	Usual background; normal conversation easily understood
40	Private office, quiet home	
30	Quiet conversation, broadcast studio	Noticeably quiet
20	Empty auditorium, whisper	
10	Rustling leaves, soundproof room, human breathing	Very quiet
0 db		Intolerably quiet. Threshold of audibility

or

$$SPL = 20 \log \frac{p}{p_0} \qquad (26.8)$$

where SPL = sound pressure level, dB
p = pressure, Pa or μ bar
p_0 = reference base pressure, (20 μ Pa or $2E-4$ μ bar)

Since for both intensity level and sound pressure level, the 0-dB base corresponds to the hearing threshold, we have equalized the dB scales for sound pressure level (SPL) and intensity level (IL) *and the dB values of the two can be used interchangeably.* Obviously, though, the actual intensity and the actual pressure corresponding to a particular decibel level are different—completely different, in magnitude and units. For instance, 70 dB equals 10^{-9}W/cm intensity and 0.063 Pa pressure. However, the important fact is that 70 dB corresponds approximately to a particular noise level. We say "approximately" because assigning a single-number decibel level to a sound present two difficulties:

1. The sound pressure level varies with time, except for a pure steady tone.
2. The different components of most common sounds vary in pressure level.

To overcome this problem two techniques are used. If a sound has a dominant frequency, that frequency's level can be used (Fig. 26.11). This would be the case for a relatively constant sound such as a motor, or a fan or blower. Other sounds that vary widely in level and frequency can be plotted on an octave-band chart using maximum

level for minimum percent of time (Fig. 26.12). Where the position of the listener is not specified in the table, it is assumed to be at normal *close* distances: that is, 10 to 20 ft from a train, 3 to 5 ft from a radio, and the like.

26.11 Loudness Level: Fletcher–Munson Contours

The human ear is not uniformly sensitive over its entire frequency range of 20 Hz to 20 kHz. The 120- to 130-dB upper limit pain threshold occurs at all frequencies. However, at the lower limit, the 0-dB threshold occurs only at 1000 Hz.

Fig. 26.11 *Common sound sources plotted at their dominant frequencies and levels as typically heard by the observer. The equal loudness curves (see Section 26.11) show why certain sounds seem louder than others, despite the pressure levels that would indicate the contrary [Reprinted with modification from F. A. White,* Quieting: A Practical Guide to Noise Control *(1976).]*

Fig. 26.12 *Curves of SPL of common noise sources plotted according to octave band frequency content. Values shown are averages of multiple measurements. [Reprinted from* A Guide to Air-borne, Impact, and Structure-Borne Noise Control in Multifamily Dwellings *(1968).]*

The ear is in fact most sensitive at 3000 to 4000 Hz, at which frequencies the threshold is about −5 dB. This nonlinear response exists throughout the ear's range. To determine the nature of this nonlinearity, a great number of tests were conducted with pure tones at different frequencies, in which listeners were asked to equate subjective loudness of signals. These tests resulted in a family of curves called *equal loudness level contours,* or, alternatively, *Fletcher–Munson equal loudness contours,* (after two of the principal researchers). These curves (Fig. 26.13) are internationally rec-

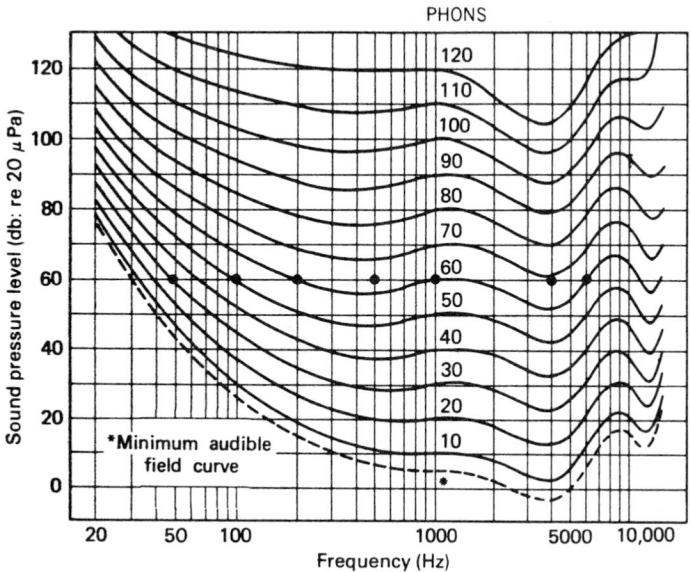

Fig. 26.13 *Standard equal loudness contours. These curves are accurate for a listener with normal binaural hearing, situated in the near field of a source producing pure tones directly ahead of the listener. The subjective loudness of the perceived sound of each contour is quantified in a unit called the phon. The phon value of each curve is shown in the center of the figure. Note that an SPL of 60 dB corresponds to 30 phons at 50 Hz, 50 phons at 100 Hz, 63 phons at 500 Hz, 60 phons at 1 kHz, 68 phons at 4 kHz, and 60 phons at 6 kHz. This indicates the relative flatness of the ear's response in the central frequency loudness range, and the sharp drop at low frequencies.*

ACOUSTICS

ognized and standardized, and are used as the reference for normal hearing response. They are also used to "weight" measuring devices, as is explained in Section 26.12. Note that *by definition:*

1. All points on a single contour have the same *subjective* sensation of loudness.
2. The loudness level in *phons* (number shown in the center of each curve in Fig. 26.13) of the *entire* contour is defined by the dB level of that contour at 1000 Hz.

The phon scale was constructed as a way of defining perceived sound (subjective loudness impression) in terms of the ear's nonlinear response. Assigning a single number—in a unit called the *phon*—to an equal loudness contour makes it possible to compare subjective loudness impressions of two sounds regardless of frequency or actual sound pressure level. Thus in Fig. 26.11, the subjective loudness of a whisper in the ear (20 phons) is the same as that of the distant pounding surf, despite the fact that the SPL of the former is 20 dB and of the latter, 80 dB. The 60-dB differential is due to the ear's sharp drop in sensitivity at low frequencies. However, because the phon scale is nonlinear in terms of perceived loudness (a given phon differential does not correspond to the same perceived loudness change throughout the phon scale) and because phons cannot be combined arithmetically to give a resultant subjective loudness (60 phons plus 50 phons is not 110 phons), the scale has found only limited use.

The Fletcher–Munson contours demonstrate some interesting phenomena:

1. Sensitivity drops off sharply at low frequencies, particularly at low-dB levels. (For this reason, most high-fidelity amplifiers provide automatic bass boost at low-volume levels).
2. Maximum sensitivity occurs between 3 and 4 kHz—precisely the frequencies that contain most information in human speech. (See Section 26.14 and Fig. 26.14).
3. In the normal listening range of 45 to 85 dB, and the most used frequency range of 150 Hz to 6 kHz, the contour is substantially flat; that is, the ear's response is effectively linear. It is only at extremes of sound level and frequency that nonlinearity occurs.

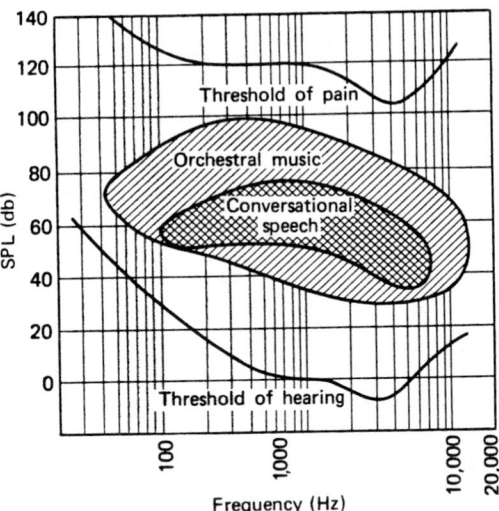

Fig. 26.14 *The positions of speech and wide range music in the human ear's aural field are illustrated. Speech is in the nominally linear response area of the ear as is most music (see Fig. 26.13). Beyond these frequencies, the ear's action is effectively to attenuate the signal.*

26.12 Measurement of Sound Pressure Level

The need for a means of measuring sound levels is obvious. One very useful instrument is the integrating sound-level meter illustrated in Fig. 26.15. To correlate meter readings with subjective loudness impressions, most such single-reading instruments are furnished with weighting networks, the characteristics of which are given in Fig. 26.16. The A network corresponds to an inverted 40-phon contour and discriminates against low frequencies (see Fig. 26.13). The B and C networks correspond to the 70-phon and 100-phon contours, respectively. In addition, a completely linear response is available. The reasoning behind the use of these weighting network is evident when they are compared to the equal loudness curves of Fig. 26.13. The original intent was to use the A network at levels up to 55 phons, the B network to 85 phons, and C network at higher levels. However, in practice it was found that only the A level corresponded fairly well to subjective loudness reports. As a result, the B network has fallen into disuse and the A network is today used for all measurements regardless of loudness. The

Fig. 26.15 *Modern, general purpose (type 2) integrating, data logging sound level meter. This unit alone will measure and record isolated, nonrepetitive noise events in addition to continuous, fluctuating and impulsive sounds, and will give both instantaneous and equivalent sound levels (L$_{eq}$). The latter is used when an equivalent noise level over a selected measurement time of varying sound is required. The meter will measure dBA, dBC, and linear, with fast, slow, peak or impulse response. When equipped with a filter set, the meter can be used for octave or ⅓-octave analysis. (Photo courtesy of Quest Technologies, Inc.)*

Fig. 26.16 *Internationally standardized A, B, and C weighting curves.*

ACOUSTICS

discrepancy at high loudness levels is apparent when the 40 phon–A weighting network curve is compared to the equal loudness curves above 80 phon. All measurements should be identified with the weighting network used, as 50 dBA or 100 dBC and the like.

It is known that the ear "averages" sounds over a minimum period, and that sound impulses shorter than this period sound quieter than they would sound as steady-state noise. This minimum sound length (i.e., the time required for the sound to achieve full loudness) varies in the range of 50 and 200 milliseconds ($\frac{1}{20}$–$\frac{1}{5}$ s) and depends to an extent on the frequency content of the sound. A figure of 70 ms ($\frac{1}{14}$ s) is frequently found in the literature. This figure becomes important when considering the effect of echoes (see Sections 27.11 through 27.14).

More accurate measurements of complex sounds than are possible with a standard meter are accomplished with sophisticated instruments that measure intensity in octave bands and plot the results, as in Fig. 26.12. Such measurements are necessary for accurate application of sound absorption and attenuation materials whose characteristics are also nonlinear over the frequency spectrum. Single-number dBA readings are known as overall levels and are useful as preliminary data and for broad-spectrum design.

Table 26.5 shows a list of common noise levels as measured by the dBA scale. Such single-value numbers are useful to establish a mental-aural comparison base, and to use in maximum exposure calculations, as discussed in Section 26.20.

26.13   Other Factors in Hearing

(a) Masking. When two separate sources of sound are perceived simultaneously, the perception of each is made more difficult by the presence of the other. This effect is known as masking, and is defined as the number of decibels by which the threshold of audibility of a sound is raised by the presence of another sound. Masking is greatest when two sounds are close in frequency or frequency content, since the ear has greater difficulty separating them. Also, a low frequency will more effectively mask a high frequency than the reverse, for the same decibel levels. With broad-

frequency sounds, the masking effect is difficult to predict, since it depends in part on how "hard" the listener is listening. Masking is an extremely important technique in noise control where background noise levels are deliberately manipulated to mask unwanted sounds (see Section 28.21). The background noises used for this purpose are of the broadband continuous variety, such as "pink" noise (see Glossary) that are themselves noninformation-bearing. They serve to destroy the intelligibility of lower level, information bearing sounds that would cause annoyance, and are therefore particularly useful in open (landscaped) office plans. Frequently, mechanical equipment noise such as that from air diffusers and fans can be utilized to good advantage as masking noise.

(b) Directivity. The exact mechanism by which the binaural aspect of hearing detects direction is not entirely understood. It is clear, however, that in enclosed space, echoes from surface reflection (reverberation) will blur most directivity and "stereo" information will be almost completely dependent on high frequencies in the near field (the sound field that is close to the source and therefore minimally affected by reverberation). Since high frequencies with short wavelengths travel in a relatively straight line, they will reach the hearer before their reflections and in so doing will cue the hearer as to their origin. Low frequencies, with long wavelengths, reach a room's perimeter rapidly and even in the near field are already mixed with their reflections, thus disguising the location of their origin. That is the reason that an annoying low frequency noise in a room is so difficult to locate; it seems to come from everywhere.

SOUND SOURCES

26.14   Speech

As can be seen from Fig. 26.14, the ear's sensitivity is maximum in the speech frequency and normal energy range. Individual speech sounds vary in time length between 30 and 300 ms and the ear normally perceives them individually and clearly. Speech is comprised of *phonemes,* which are individual and distinctive sounds that, to an extent, vary from language to language. That is, certain ones exist in one language and not in another. Since certain phonemes carry more information than others, it is these that good architectural acoustics must be particularly careful to preserve intact, to maintain intelligibility. In English, consonants carry much more information than vowels, as can readily be demonstrated by writing a sentence first without consonants and then without vowels:

Most speech energy is concentrated in the 100- to 600-Hz range.

o ee ee i oeae i e 100 o 600 e ae

Mst spch nrgy s cncntrtd n th 100 t 600 Hrtz rng.

The male voice centers its energy around 500 Hz; the female about 900 Hz. It is, however, in the high frequencies that consonants have most of their energy. Phonemes such as *s* and *sh* have most of their energy above 2 kHz and both are particularly important in conveying intelligence.

Normal speech averages between 40 and 50 dBA sound pressure level at 3 to 4 ft with a dynamic range of from about 30 dBA for soft speech to about 65 dBA for loud speech at the same distance. Extremes of speech are 10 dBA for a soft whisper and 80 dBA for a shout, but in both these instances intelligibility is sharply reduced because of lack of consonant power. Indeed, in shouting, emphasis is necessarily on vowels, so that it is generally accepted that 70 dBA SPL is about the upper limit of fully intelligible human speech. (Singers who frequently exceed 90 dBA do so at great loss of intelligibility).

Another result of the high-frequency content of consonants, and therefore intelligibility, is its directiveness. The higher the frequency, the greater the sound's directivity and the less its diffraction (ability to be heard beyond a barrier). Therefore, intelligibility of speech is greatest directly in front of the speaker and least behind him. High-frequency tones are most easily absorbed and least diffracted.

26.15 Other Sounds

Instrumental music is much broader in dynamic range and more complex in frequency than speech. It has no direct parallel to intelligibility.

"Reception" of music is a combination of physiological and psychological phenomena. As such, it is beyond most of the purposes of this study but is briefly examined in our discussion of room acoustics, auditoriums, and halls. *Noise* is variously defined as unwanted sound, sound with no intelligence content, and broadband sound, depending on the listener and the situation. All three definitions are correct at various times and situations. For our purpose, we must assume that any sound can be referred to as noise.

NOISE CRITERIA

26.16 Negative Effects of Noise

Although noise effects and their control are the subject of Chapter 28, noise criteria and their background are discussed here as part of our overall study of hearing and sound sources. There are two basic approaches to the negative effects of noise; a psychological-practical one and a purely physiological one. The latter is concerned with the physical impact of noise on the body including hearing loss and other deleterious conditions (see Section 26.19). The former is concerned with noise levels that cause annoyance and disturbance to daily activities including work, relaxation, and rest, and is discussed in the following sections.

26.17 Noise and Annoyance: Articulation Index and Speech Interference Level

Research has developed accurate data on loudness. The concept of annoyance, however, being primarily subjective and psychological, is much more elusive. Tests have shown that *in general,* annoyance as a result of noise is:

1. Proportional to the loudness of the noise.
2. Greater for high-frequency than low-frequency noise.
3. Greater for intermittent than continuous noise.
4. Greater for pure-tone than for broadband noise.
5. Greater for moving or unlocatable (reverberant) noise than for a fixed-location sound.

ACOUSTICS

6. Much greater for an information bearing noise (neighbor's radio) than for nonsense noise (foreign language on neighbor's radio).

To establish criteria for *acceptable background noise* (i.e., noise whose annoyance level is acceptable), certain of these effects must be neglected for the sake of simplicity. [They can be and are considered in construction design and in establishing levels of masking noise (see Chapter 28)]. We thus ignore:

Factor 3, since design is based on continuous sounds.

Factor 4, since broadband noise is assumed.

Factor 5, since the noise source is assumed to be fixed in location.

Factor 6, since we can only consider noise level, not content.

Thus the particularity and special characteristics of noises such as a barking dog (3), a whistle (4), a single passing vehicle (5), and intelligible sounds (6) are not considered.

In order to quantify the concept of acceptable background noise, it was necessary to remove it from the purely psychological arena and relate it to a physical phenomenon. This was done by studying the effect of the two remaining annoyance factors (1 and 2) on speech communication. This study resulted in two concepts: the Articulation Index (AI) and the Speech Interference Level (SIL). They are determined by reading a carefully selected set of phonetically balanced nonsense syllables to a test audience in the presence of different levels and compositions of background noise. The ratio of correct answers to total syllables is the Articulation Index. An Articulation Index (Beranek, 1971) in excess of 0.5 indicated a condition in which acceptable intelligibility could be expected for *male* voices. A simplified version of the AI called Speech Interference Level (SIL) was devised by Beranek. The SIL consists simply of the arithmetic average in decibels of the background noise sound levels in four octave bands centered on 500, 1000, 2000, and 4000 Hz, for which acceptable intelligibility could be expected, for a given voice effort, at a given distance between speaker and listener.

26.18　Noise Criterion Curves

Beranek also developed the well-known and widely accepted noise criteria (NC) curves shown in Fig. 26.17. The NC curves take cognizance of the field-determined fact that most people prefer to speak at a level no greater than 22 phons above the background noise. By combining the SIL levels in decibels with this fact, the NC curves are derived; they represent a loudness level 22 phons higher than the SIL in dB. These contours represent then the maximum *continuous* background noise that will be considered acceptable in the environment specified and correspond fairly accurately to background noise level in commercial environments. A similar set of curves called noise rating (NR) curves find considerable application outside the United States. They are less stringent then NC curves in low frequencies but more stringent in high frequencies.

In application, the octave-band spectrum of the specific noise being studied, over the range of 63 to 8000 Hz, is measured and plotted on an NC curve sheet. The lowest NC curve that is not exceeded by any portion of the plot is the NC rating of the noise. Thus specifying a maximum noise level of NC-30 for a space means that no portion of the sound pressure level curve of any continuous background noise in the area may cross the NC-30 contour. Conversely, a piece of equipment rated NC-35 has an octave-band spectrum completely below NC-35. A fan rating of NC-53 would indicate that at some point on its frequency spectrum plot the fan exceeded NC-50 by 3 dB (Fig. 26.17).

The NC number of a noise varies between 5 and 10 dB below the measured dBA. The virtue of the NC curve system is that it is a single number specification for an entire frequency spectrum. Its disadvantage is that it was derived for, and is primarily accurate for speech conversation against a continuous, non-intelligence-bearing noise. This is not the situation that is most troublesome in an office; indeed, continuous equipment noise may be *helpful* in masking unwanted speech interference, as mentioned above (see Sections 28.16 through 28.21). Nevertheless, NC curves remain the commonly used criteria for continuous *nonintentional nonintelligible* background noise acceptability.

26.19   Room Criterion Curves

Due to the recognized shortcomings of the NC curve criteria, specifically that they are undefined in the very low frequency (16 and 31.5 octave bands) and are not sufficiently stringent at frequencies above 2 kHz, a similar criteria was developed called room criteria (RC) curves. These curves, shown in Fig. 26.18, have been adopted by ASHRAE as the suggested noise limitation criterion in preference to NC curves. The curves are used to evaluate the acceptability of background mechanical system noise for listed unoccu-pied spaces. The RC curves differ from NC curves in a number of aspects:

· they are straight lines

· their slope is constant at −5 dB, as a result of extensive tests, mostly in the range of 40–50 dB.

· Regions labelled A and B in the diagram (Fig. 26.18) address the problem of very low frequencies and high sound pressure levels; an area that was ignored in the NC criteria. This is the area of rumble and vibration that

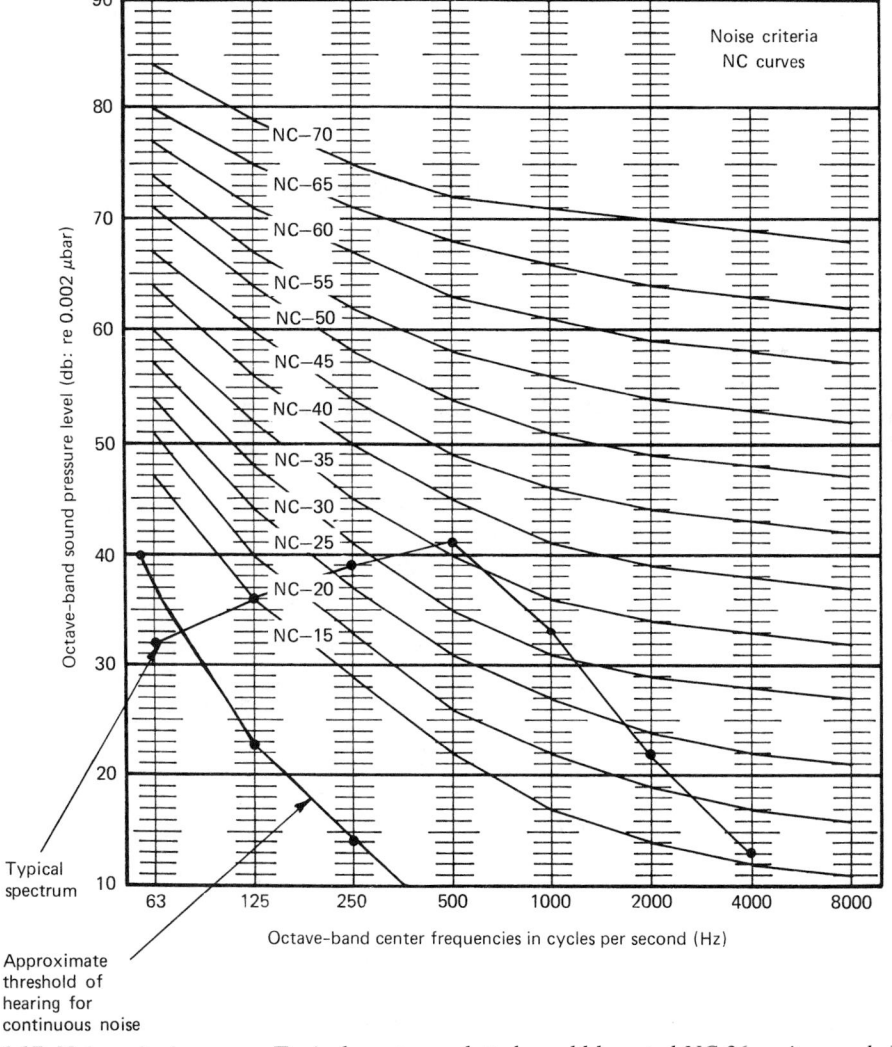

Fig. 26.17 *Noise criteria curves. Typical spectrum plotted would be rated NC-36, as it exceeds NC-35 at 500 Hz by 1 dB. (See Table 28.7 for specific recommndations for interior spaces).*

can cause extreme annoyance in many occupancies.

Refer again to Fig. 26.18. The procedure for determining the RC value of a specific item of equipment whose noise spectrum is known, is as follows: 1.

1. Calculate the arithmetic average of the SPL values in the 500, 1000, and 2000 Hz octave bands. This number is the RC value of that noise spectrum.
2. Draw a straight line at −5 dB through this value of RC, at 1000 Hz.
3. Plot the SPL values of the octave band center frequency on the RC curve sheet and compare it to the RC line drawn in Step 2.
4. Classify the equipment sound quality from this comparison, as follows:

a. *Neutral.* If the octave band data plotted in step 3 do not exceed the RC line drawn in step 2 by
   (1) more than 5 dB at or below 500 Hz and
   (2) more than 3 dB at or above 1000 Hz

then the sound is considered neutral (bland, uncharacteristic) and the designator letter N is placed after the RC level. A piece of mechanical equipment with an N designation is then understood to have a neutral tone quality that most people would classify as unobtrusive and lacking specific character. ASHRAE design guidelines for HVAC system noise levels are listed as RC (N).

Room Criterion Curves

Fig. 26.18 *Room criterion (RC) curves. These curves have been adopted by ASHRAE in lieu of NC curve criteria. See text for an explanation of their use [Reprinted with permission from* ASHRAE Applications Handbook *(1995).]*

b. *Rumble.* If the octave band plot *does* exceed the RC line by more than 5 dB at 500 Hz or below, the spectrum is classified as "rumbly" and the descriptor letter *R* is appended to its RC number.

c. *Hiss.* If the octave band plot *does* exceed the RC line by more than 3 dB at or above 1000 Hz it is classified as "hissy" and the descriptor letter H is appended to its RC

level. The result of this classification is to give the acoustic designer not only an average SPL for an item of equipment, but also a subjective sound quality that should be of considerable assistance in determining the noise abatement measures to be taken.

d. *Vibration.* The shaded areas in Fig. 26.18 labelled A and B represent high sound

---

§ 1910.95 Occupational noise exposure.

(a) **Protection against the effects of noise exposure shall be provided when the sound levels exceed those shown in Table G-16 when measured on the A scale of a standard sound level meter at slow response.** When noise levels are determined by octave band analysis, the equivalent A-weighted sound level may be determined as follows:

Equivalent sound level contours. Octave band sound pressure levels may be converted to the equivalent A-weighted sound level by plotting them on this graph and noting the A-weighted sound level corresponding to the point of highest penetration into the sound level contours. This equivalent A-weighted sound level, which may differ from the actual A-weighted sound level of the noise, is used to determine exposure limits from Table G-16. [1910.95 amended at 39 FR 19468, June 3, 1974]

(b)(1) When employees are subjected to sound exceeding those listed in Table G-16, feasible administrative or engineering controls shall be utilized. If such controls fail to reduce sound levels within the

levels of Table G-16, personal protective equipment shall be provided and used to reduce sound levels within the levels of the table.

(2) If the variations in noise level involve maxima at intervals of 1 second or less, it is to be considered continuous.

(3) In all cases where the sound levels exceed the values shown herein, a continuing, effective hearing conservation program shall be administered.

TABLE G-16—PERMISSIBLE NOISE EXPOSURES[1]

Duration per day, hours	Sound level dBA slow response
8	90
6	92
4	95
3	97
2	100
1½	102
1	105
½	110
¼ or less	115

[1] When the daily noise exposure is composed of two or more periods of noise exposure of different levels, their combined effect should be considered, rather than the individual effect of each. If the sum of the following fractions: $C_1/T_1 + C_2/T_2 + \cdots C_n/T_n$ exceeds unity, then, the mixed exposure should be considered to exceed the limit value. Cn indicates the total time of exposure at a specified noise level, and Tn indicates the total time of exposure permitted at that level.

[1910.95 Table G-16 amended at 39 FR 19468, June 3, 1974]

Exposure to impulsive or impact noise should not exceed 140 dB peak sound pressure level.

Fig. 26.19 *The standard for exposure to noise in the workplace. (From OSHA Rev., July 1988.)*

pressure levels in the 31.5 and 63 Hz octave bands. In the A area vibration will be felt in light construction, and furniture, and rattling may occur in loosely constructed devices, cabinets, glassware and the like. In the lower energy B area, rattling will be less likely but vibration may still be felt.

The original NC curves developed by Beranek were modified by him in 1989, and issued as balanced noise criteria (NCB) curves. They amended a failing of the NC curves by adding very low frequency coverage and the NCB curves were

straightened, so that they resemble RC curves above 125 Hz, except that the slope angle is −3.33 dB as compared to −5 dB of the RC curves. The slope at the higher frequencies has also eliminated the hiss that characterized equipment noise conforming to the NC criteria. These adjustments have yielded a "balanced" neutral sound, which reflects the chart's name.

Application of these curves to spectra of mechanical equipment is similar to that used for RC curves, except that letter descriptors are not used. For a full exposition of the construction and application of BNC curves, see Beranek, 1998.

26.20 High Noise Levels: Hearing Protection: OSHA

It has long been recognized that continuous exposure to high noise levels causes a degree of tempo-

Fig. 26.20 *A modern noise dosimeter. The unit is carried by the subject whose noise exposure is being tested, with the clip mount microphone at shoulder (ear) level. The meter will display, store and calculate maximum, minimum, and peak levels of SPL; L_{eq}; TWA; exposure levels; and dosage. Noise level histories are stored and can be printed out as desired. The unit shown measure 5.5 in. × 2.8 in. × 1.4 in. × (140 × 70 × 40 mm) and weighs 15.5 oz (440 g). (Photo courtesy of Quest Technologies.)*

TABLE 26.6 **Typical Industrial Noise Levels**[a]

Equipment	dBA
Printing press plant (medium size automatic)	86
Heavy diesel-propelled vehicle (about 25 ft away)	92
Heavy-duty grinder	93
Air compressor	94
Plastic chipper	96
Cutoff saw	97
Multiple spot welder	98
Turbine condenser	98
15 cu ft air compressor	100
Drive gear	103
Banging of steel plate	104
Magnetic drill press	106
Air chisel	106
Positive displacement blower	107
Air hammer	107
Vacuum pump	108
Jolt squeeze hammer	122

[a]These are approximate values for typical generic equipment types and should not be used as design values.

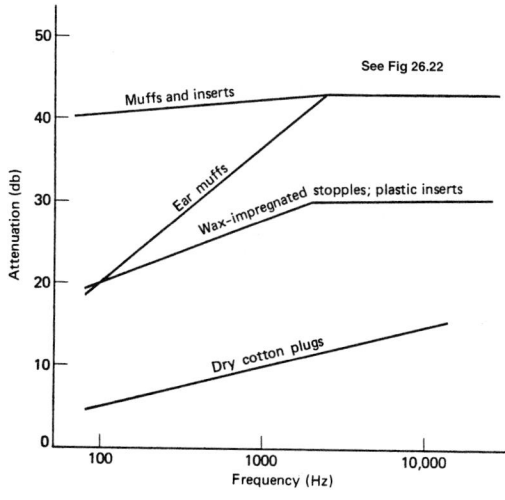

Fig. 26.21 *Typical sound attenuation characteristics of various types of ear protectors. For design purpose, specific attenuation characteristics must be used, and the attenuation figures modified to reflect field usage. Three common types of ear protectors are shown in Fig. 26.22. [From* Quieting: A Practical Guide to Noise Control *(1976).]*

(a)

(b)

(c)

Fig. 26.22 *Three common types of ear protection devices:*
(a) PVC foam ear plugs are rolled and squeezed to reduce their diameter before placing into the ear canal, where they re-expand to their previous size. The devices have an EPA noise reduction rating of 20, when tested to ANSI 53.19-1974. They are generally discarded after a single use.
(b) Ear protector for intermittent use consists of (replaceable) foam pads mounted on a flexible neck band. The foam does not require rolling. The unit is rated NRR 20.
(c) Spring-type head band and foam-filled ear cushions provide a rating of NRR 20 for this earmuff design. Mean attenuation characteristics for the three types of ear protectors are given in Table 26.7. (Photos courtesy of E-A-R Specialty Composites, a Div. of AEARO Company.)

TABLE 26.7 **Mean Attenuation Data (dBA) for Ear-Protectors at Test Frequencies (Hz) Shown**

Foam Earplugs [Fig. 26.22(a)]									
125	250	500	1000	2000	3150	4000	6300	8000	NRR
37.4	40.9	44.8	43.8	36.3	41.9	42.6	46.1	47.3	29
Semi-Aural Hearing Protector [Fig. 26.22(b)]									
28.7	28.3	28.0	28.6	32.2	42.1	44.3	47.2	44.6	20
Earmuff [Fig. 26.22(c)]									
11.7	16.2	26.5	31.6	32.5	35.0	38.1	41.8	41.8	20

Data extracted with permission from published AEARO catalog material.

rary deafness in a majority of people and that long periods of such exposure, even on an intermittent 8-h workday basis, can produce permanent hearing impairment. Most experts place the safe 8-h upper limit at 85 dB. In addition, studies have indicated that continual exposure to noise levels as low as 75 to 85 dBA level can produce or contribute to numerous physical and psychological ailments including headache, digestive problems, tachycardia, high blood pressure, anxiety, and nervousness—an extensive catalog of human illnesses. Since continuous noises are most severe in industry, regulatory legislation in the United States has been most stringent and effective in this area.

In 1969 the Walsh–Healy Public Contracts Act was passed, and thereafter its provision for maximum permissible exposure to noise levels was incorporated into the Occupational Safety and Health Act. Both the act and the associated regulatory agents, the Occupational Safety and Health Administration, are known as OSHA. The relevant provisions of this act are reproduced in Fig. 26.19. To avoid complex regulations, limitations of exposure are given in terms of single number dBA values. Since workers rarely remain in a single acoustic environment for 8-h, their total daily exposure, or *time weighted average* (TWA) exposure can be calculated from timed dBA measurements using formulas and tables given in the OSHA code, and then compared to permissible levels. Alternatively, a dose meter (Fig. 26.20)

can be used, which automatically integrates the noise to which a person is exposed over a given time period and reads out the permissible TWA exposure directly.

Typical industrial noise levels are given in Table 26.6. When levels are exceeded, management must take steps to reduce the exposure, either by reducing the noise or by providing hearing protectors. Typical characteristics of a few types of ear protectors are given in Fig. 26.21. Photographs of three of the most common ear protectors in industrial use are shown in Fig. 26.22, and their laboratory-measured mean attenuation spectra are given in Table 26.7.

OSHA does not deal extensively with impulse noises, except to state that they shall not exceed 140-dB peak sound pressure level. Impulse noise is quite different from continuous noise, since apprehended levels depend on duration, and specification is difficult. Much work has been done in this area by the military for obvious reasons. The interested reader is referred to the literature since this subject is substantially outside the area of architectural acoustics.

REFERENCES

See the complete list of architectural acoustics reference, as well as a Glossary at the end of Chapter 28.

27

SOUND IN ENCLOSED SPACES

In Chapter 26 we learned the fundamentals of sound theory, some characteristics of human hearing and speech, and information on the negative effects of unwanted and excessive sounds. This chapter will discuss in some detail the interreaction between an enclosing space and a sound generated within it, which is essentially the definition of architectural acoustics.

field, making the space "noisy". Conversely, if much energy is absorbed and little reflected, the room will sound quiet to speech and "dead" to music. Quantification of these two primary characteristics of an enclosed space—absorption and reverberation (echo)—is the subject of the next six Sections.

27.1 Sound in Enclosures

When a continuous sound is generated in an enclosure, fields are set up as described in Section 27.7 below. When the sound is not a continuous tone or noise but a series of discrete sounds, following one upon the other and containing information as in speech or music, the room must be designed to maintain and enhance the information's intelligibility. That is what is meant by design of room acoustics.

Generated sound radiates out from the source until it strikes a room boundary or other large surface. Before reaching this surface the sound intensity is attenuated by distance (inverse square law—Section 26.7) and by absorption in the air. This latter is appreciable only in large rooms, and at frequencies above 2000 Hz. When the sound reaches the wall, it is partially reflected and partially absorbed, and a small portion is transmitted into adjoining spaces. The energy transmitted is has little effect on the space within which the sound originates although, as discussed in Chapter 28, it may be very important in the surrounding spaces.

The ratio between the energy absorbed and the energy reflected by a surface will significantly affect what one hears within a space. Specifically, if little energy is absorbed and much is reflected, two effects will be noticeable. Intermittent sounds will be mixed together (which may make speech *less* intelligible or music more pleasant), and steady sounds will accumulate into a reverberant

ABSORPTION

27.2 Sound Absorption

When sound energy impinges on a material, part is reflected and the remainder is absorbed, in the sense that it is not reflected. (Some of the energy is transmitted although it is so small that for our discussion here it will be ignored.) Materials are neither perfect reflectors nor perfect absorbers. The term used to define a material's sound absorption characteristic is called its *coefficient of absorption* and it is usually represented by the lowercase Greek letter alpha (α). This sound absorption coefficient is defined as

$$\alpha = \frac{I_a}{I_i} \qquad (27.1)$$

where
I_a = sound power density (intensity) absorbed by the material, W/cm^2

I_i = intensity impinging on the material, W/cm^2

α = absorption coefficient, (no units since it represents a ratio)

This absorption coefficient α can be also be thought of as a measure of absorption efficiency; the larger the absorption coefficient, the more efficient a sound absorber the material. Thus an absorption coefficient of 1.0 would indicate 100% absorption and zero reflection of the impinging sound energy. We have already said that no material is a perfect absorber, but an absence of material, that is, an open space, transmits (absorbs) all the impinging energy and therefore can be considered as a perfect absorbing "material."

Since open space has this characteristic, α has also been defined as the ratio between the absorption of a given material and that of an *open window* of the same area. By definition then, for an opening (open window, open door, etc.)

$$\alpha = 1.0$$

The total absorption (A) of a given quantity of material is proportional to its area and its absorption coefficient, that is,

$$A = S\alpha \qquad (27.2)$$

where A = total absorption, *sabins*
  S = surface area, square feet or square meters
  α = coefficient of absorption

Since α is a ratio and thus unitless, and S is a unit of area, $S\alpha$ should be in units of area as well. Instead, sound absorption units are called *sabins* in honor of W. C. Sabine, a pioneer in architectural acoustics. One sabin (m^2) is the sound absorption equivalent to an open window 1 square meter in area. Similarly, 1 sabin (ft^2) is equivalent to 1 square foot of open window. Obviously, 1 sabin (m^2) equals 10.76 sabin (ft^2). In this text, when not specified, a sabin (ft^2) is intended.

All rooms are constructed of several materials, each having a different absorption coefficient. In addition, most rooms contain furnishings, that have their own individual coefficients of absorption. Thus to determine the total absorption of the room, it is necessary to sum the component absorptions, that is,

$$\Sigma S\alpha = S_1\alpha_1 + S_2\alpha_2 + \ldots + S_n\alpha_n$$

or

$$\Sigma A = A_1 + A_2 + \ldots + A_n \qquad (27.3)$$

where $\Sigma S\alpha$ = total absorption in the room, sabins
  S_1, S_2, etc. = area of each material
  α_1, α_2, etc. = coefficient of each material
  A_1, A_2, etc. = total absorption of each different material

If S is expressed in square feet, than A will be in sabins (ft^2); if S is in square meters then A is sabins (m^2).

Absorption coefficients for some common materials, acoustic materials, and auditorium furnishings are tabulated in Table 27.1. It is important to note that for most common materials, absorption (and therefore α) varies with frequency. Thus accurate absorption calculations must be made individually for the frequencies being studied.

27.3 Mechanics of Absorption

At this point, it is appropriate to examine absorption as an acoustic phenomenon so that we may understand the application of absorptive material. Refer to Fig. 27.1(*a*). In an untreated room of normal construction, when the sound waves strike the walls or ceiling, a small portion is transmitted, a small portion is absorbed, and most of the sound is reflected. The exact proportions obviously depend on the nature of construction. When acoustical treatment is applied to the room surfaces as in Fig. 27.1(*b*), some of the energy in the sound waves is dissipated before the sound reaches the wall. The transmitted portion is slightly reduced, but the reflection is greatly reduced.

It is important to understand however, that the principal effect of absorptive material is on the reflected sound. The transmitted sound energy is essentially determined by the mass of the solid airtight barrier between the two spaces. This is graphically represented in Fig. 27.1(*c*). The effect of added acoustic absorption in a space is shown graphically in Fig. 27.2 and is calculated in Section 27.9.

27.4 Absorptive Materials

We will now examine the acoustic materials themselves and the effect of varying type, quantity, thickness, and installation methods. There are three families of devices for sound absorption—fibrous materials, panel resonators, and volume resonators. All three types absorb sound by changing sound energy into heat energy. Only fibrous materials and panel resonators are used commonly in buildings. Volume resonators, also known as Helmholtz resonators, after their originator, are used principally as devices for absorbing a narrow band of frequencies. The discus-

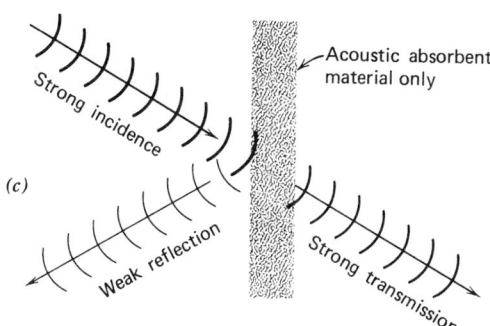

Fig. 27.1 (a) *Action of an incoming sound wave striking a heavy barrier. Much of the energy is reflected, some is absorbed, and a little is transmitted.*
(b) *When absorbent material is applied to the heavy wall, it "traps" sound, preventing reflection, while wall mass acts to reduce transmission.*
(c) *Action of acoustic absorbent material alone. Very little energy is reflected, some is absorbed, and most is transmitted.*

sion in this Section refers to fibrous absorbers; the other two types are discussed separately in Section 28.2.

Fibrous materials absorb acoustic energy by the frictional drag of air being moved in the tiny spaces between the fibers. The absorption provided by a specific material depends on its thickness, density, porosity, and resistance to airflow. Since the action depends on absorbing energy by "pumping" air through the material, the air paths *must extend from one side to the other.* A fibrous material with sealed pores is almost useless as an acoustic absorbent. (Therefore, painting will generally ruin a porous absorber.) A simple test is to blow smoke through the material. If the smoke passes through freely and the material is porous, fibrous, and thick, it should be a good sound absorbent. Absorbency increases with increasing material porosity up to approximately 70% porosity; above that figure absorbency remains fairly constant.

Table 27.1 gives absorption coefficients for fibrous absorbent materials and for some building materials and furnishings. Several important conclusions can be drawn from examination of this table and Fig. 27.3:

1. For absorbent materials, absorption is normally higher at high frequencies than at low.

2. Absorption is not always proportional to thickness, but depends on the type material being used and the method of installation (see Fig. 27.3). It is clear from this figure that beyond a nominal thickness, little is to be gained by additional thickness except at very low frequencies, or when installed discontinuously, as described in the next item.

3. It is possible to obtain an α greater than 1.0 by using very thick blocks. See "Sound blocks" in Table 27.1. As they are installed at a distance from each other, and their edge absorption is very large (particularly at high frequencies), they exhibit an α greater than 1.0.

4. Installation methods have a pronounced effect as discussed in the following section.

5. All other factors remaining constant, the thicker an absorbent material, the better its low frequency absorption characteristic.

TABLE 27.1 **Octave Band Average Sound Absorption Coefficients**[a] (α)

	Absorption Coefficients						
General Building Materials and Furnishings[b]	125 Hz	250 Hz	500 Hz	1000 Hz	2000 Hz	4000 Hz	NRC[c]
Brick, unglazed	0.03	0.03	0.03	0.04	0.05	0.07	0.005
Brick, unglazed, painted	0.01	0.01	0.02	0.02	0.02	0.03	0.00
Carpet, heavy, on concrete	0.02	0.06	0.14	0.37	0.60	0.65	0.29
Carpet, heavy, on 40-oz hairfelt or foam rubber	0.08	0.24	0.57	0.69	0.71	0.73	0.55
Concrete block, coarse	0.36	0.44	0.31	0.29	0.39	0.25	0.35
Concrete block, painted	0.10	0.05	0.06	0.07	0.09	0.08	0.05
Fabrics							
Light velour, 10 oz/yd², hung straight, in contact with wall	0.03	0.04	0.11	0.17	0.24	0.35	0.15
Medium velour, 14 oz/yd², draped to half area	0.07	0.31	0.49	0.75	0.70	0.60	0.55
Heavy velour, 18 oz/yd², draped to half area	0.14	0.35	0.55	0.72	0.70	0.65	0.60
Floors							
Concrete or terrazzo	0.01	0.01	0.015	0.02	0.02	0.02	0.00
Linoleum, asphalt, rubber, or cork tile on concrete	0.02	0.03	0.03	0.03	0.03	0.02	0.05
Wood	0.15	0.11	0.10	0.07	0.06	0.07	0.10
Glass							
Large panes of heavy plate glass	0.18	0.06	0.04	0.03	0.02	0.02	0.05
Ordinary window glass	0.35	0.25	0.18	0.12	0.07	0.04	0.15
Gypsum board, ½ in. nailed to 2 × 4's 16 in. o.c.	0.10	0.08	0.05	0.03	0.03	0.03	0.05
Marble or glazed tile	0.01	0.01	0.01	0.01	0.02	0.02	0.00
Openings							
Stage, depending on furnishings				0.25–0.75			
Deep balcony, upholstered seats				0.50–1.00			
Grilles, ventilating				0.15–0.50			
Plaster, gypsum or lime, smooth finish on tile or brick	0.013	0.015	0.02	0.03	0.04	0.05	0.05
Plaster, gypsum or lime, on lath	0.14	0.10	0.06	0.05	0.04	0.03	0.05
Plywood paneling, ⅜ in. thick	0.28	0.22	0.17	0.09	0.10	0.11	0.15
Rough wood as tongue and groove cedar	0.24	0.19	0.14	0.08	0.13	0.10	0.14
Slightly vibrating surface (e.g., hollow core door)	0.02	0.02	0.03	0.03	0.04	0.05	0.03
Readily vibrating surface (e.g., thin wood paneling on 16-in. studs)	0.10	0.07	0.05	0.04	0.04	0.05	0.05
Water surface, as in a swimming pool	0.008	0.008	0.013	0.015	0.020	0.025	0.00

Absorption of Seats and Audience[d]	125 Hz	250 Hz	500 Hz	1000 Hz	2000 Hz	4000 Hz	NRC[c]
Audience, in upholstered seats, per ft² of floor area	0.60	0.74	0.88	0.96	0.93	0.85	—
Unoccupied cloth upholstered seats, per ft² of floor area	0.49	0.66	0.80	0.88	0.82	0.70	—
Wooden pews, occupied, per ft² of floor area	0.57	0.61	0.75	0.86	0.91	0.86	—
Students in tablet-arm chairs, per ft² of floor area	0.30	0.42	0.50	0.85	0.85	0.84	—

Acoustic Absorptive Materials	*Mtg*[e]			*Absorption Coefficients*			
		125 Hz	*250 Hz*	*500 Hz*	*1000 Hz*	*2000 Hz*	*4000 Hz*	*NRC*[c]

	Mtg[e]	*125 Hz*	*250 Hz*	*500 Hz*	*1000 Hz*	*2000 Hz*	*4000 Hz*	*NRC*[c]
High-performance vinyl-faced fiberglass ceiling panels:								
1 in. thick	E405	0.73	0.88	0.71	0.98	0.96	0.77	0.90
1.5 in. thick	E405	0.79	0.98	0.83	1.03	0.98	0.80	0.95
Painted nubby glass cloth panels								
3/4 in. thick	E405	0.81	0.94	0.65	0.87	1.00	0.96	0.85
1 in. thick	E405	0.78	0.92	0.79	1.00	1.03	1.10	0.95
Random fissured 3/4-in. thick panels	E405	0.52	0.58	0.60	0.80	0.92	0.80	0.70
Perforated metal panel with infill 1 in. thick	E405	0.70	0.86	0.74	0.88	0.95	0.86	0.85
Typical averages, mineral fiber tiles and panels								
³⁄₄ in. fissured	E405	0.47	0.50	0.52	0.76	0.86	0.81	0.65
³⁄₄ in. textured	E405	0.49	0.55	0.53	0.80	0.94	0.83	0.70
⁵⁄₈ in. fissured	E405	0.28	0.33	0.66	0.73	0.74	0.75	0.60
⁵⁄₈ textured	E405	0.29	0.35	0.66	0.63	0.44	0.34	0.50
⁵⁄₈ in. perforated	E405	0.27	0.29	0.55	0.78	0.69	0.53	0.60
3 in. thick × 16 in. square on 24-in. centers	A	0.40	0.61	1.92	2.54	2.62	2.60	

[a] Complete tables of coefficients of the various materials that normally constitute the interior finish of rooms may be found in books on architectural acoustics. This table will be useful in making simple calculations.

[b] Selected data courtesy of Owens-Corning Fiberglass.

[c] Noise reduction coefficient: the arithmetic average of the α values at 250, 500, 1000, and 2000 Hz.

[d] When the audience is randomly spaced, use an average of 5.0 sabins (ft²) per person.

[e] See Fig. 27.4.

ACOUSTICS

1523

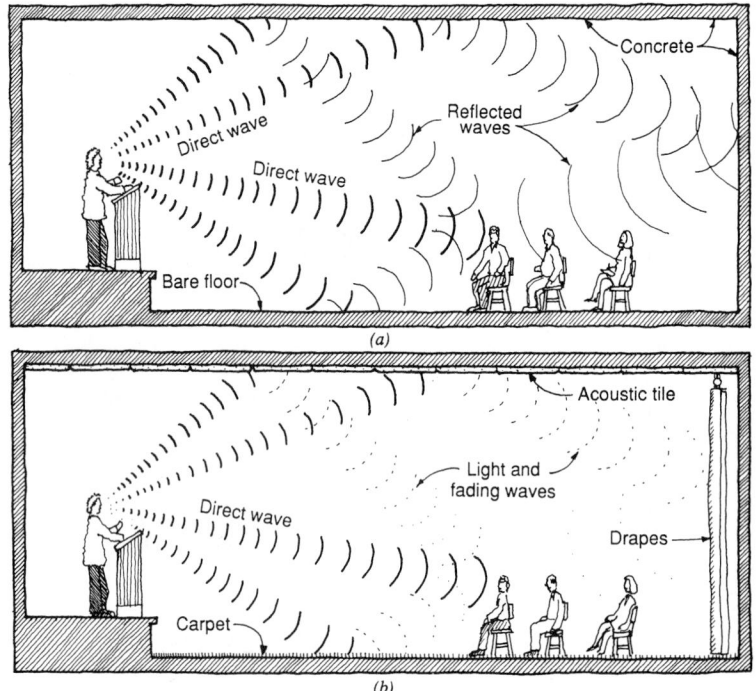

Fig. 27.2 In the untreated space (a) reverberant (reflected) sound constitutes a large portion of received sound in much of the room. These reflections are largely eliminated in (b) by wall and ceiling absorption. Note that direct wave is completely unaffected.

Fig. 27.3 Variation of absorption coefficient with thickness of felt absorbent. Note particularly that beyond 1 kHz, all thicknesses give the same α, whereas at low frequencies the absorption is proportional to thickness. Note also that it requires a very heavy layer to give appreciable absorption at low frequency. (Courtesy of Brüel and Kjaer.)

Fig. 27.4 *Standard test-mounting methods for absorptive material, in accordance with which absorptive coefficients are given by manufacturers. Number following the letter E represents the distance in millimeters from the mounting surface.*

27.5 Installation of Absorptive Materials

Coefficient ratings for absorptive materials are always given with mountings corresponding to ASTM requirements (see References 1, 2). The most common standardized mounting methods are shown in Fig. 27.4. Installation of absorbent material directly on a wall or ceiling is the least effective means, since exposure to sound energy is minimal. When an air gap is left between the porous layer and the rigid surface, the combination acts almost as well in midfrequencies, as an absorbent layer equivalent in thickness to the air plus porous material (Fig. 27.5). One problem with this technique, however, is that at the $\lambda/2$ node of a standing wave there is a severe drop in absorption, as can be seen from Fig. 27.5(c). Thus, at 1000 Hz, one-half wavelength is approximately 7 in. At that distance, α drops severely, but is a maximum at $\lambda/4$ or $\frac{1}{2}$ in.

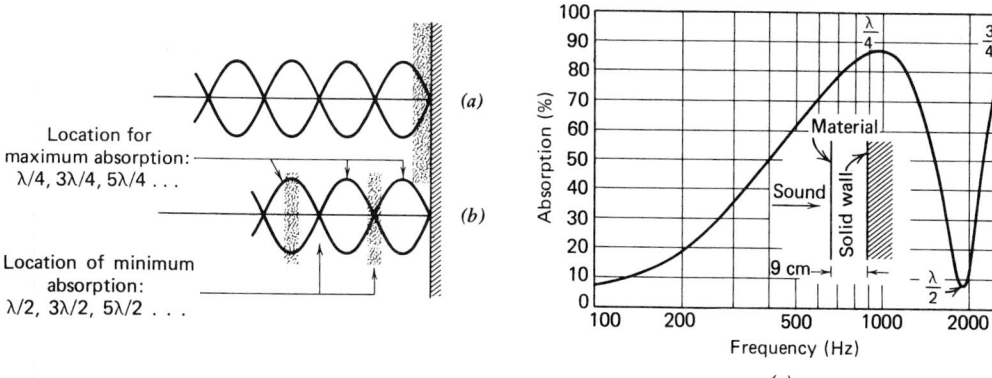

Fig. 27.5 *Sound waves striking a surface of a large mass will create standing waves at certain frequencies, depending on room dimension (a). Only insulation placed at the peaks of these waves ($\lambda/4$, $3\lambda/4$....) is effective as seen in (c). [Illustration (c) reproduced with permission of Brüel and Kjaer.]*

For ceiling tile hung at 16 in. below the slab (Fig. 27.4, method E405), the drop in absorption occurs at

$$\lambda/2 = 16 \text{ in.}$$

$$\lambda = 32 \text{ in.} = 2.67 \text{ ft}$$

$$f = \frac{1128}{2.67} = 422 \text{ Hz}$$

which is midfrequency. This factor should be considered in applying absorptive material. Avoid a spacing corresponding to a drop in absorption at a sensitive frequency. To obtain good low-frequency absorption, it is essential that a deep air space be provided behind the absorbent material, and that walls be treated in addition to the ceiling.

In increasing order of effectiveness, absorbent material can be applied:

1. Directly to room surface.
2. Hung below ceiling and supported away from the walls.
3. Hung from the ceiling as louvres or baffles.
4. Made up into shapes such as cubes, or tetrahedrons, and suspended from the ceiling.

The last two techniques are extremely effective because they expose a very large surface of porous material—much larger than could be accomplished with wall or ceiling covering. Of course, these suspended objects become architectural elements and must be handled accordingly. In contrast, surface coverings are relatively neutral. In general, treatment should not be limited to one room surface such as the ceiling. All three principal surfaces in the direction of sound propagation, that is, ceiling, floor, and back wall, should be treated *approximately equally* for best results. The common practice of treating the ceiling only is generally inadvisable, since high frequencies are highly directive and may not reach the ceiling until the third reflection.

In order to fully understand the effects of absorbent materials in an enclosed space, it is necessary to introduce two new concepts to the reader: reverberation and sound fields. This will be done in the next two sections, after which we will return to our discussion of the use and effects of acoustic absorptive material in enclosed spaces.

ROOM ACOUSTICS

27.6   Reverberation

Reverberation is the persistence of sound after the sound source has ceased, as a result of repeated reflections in an enclosed space. Reverberation time (T_R) is defined as the time required for the sound level to decrease 60 dB after the sound source has stopped producing sound. For rooms of usual size and shape, the reverberation time at a specific frequency may be found by the formula:

$$T_R = K \times \frac{V}{\Sigma A} \text{ seconds} \qquad (27.4)$$

where K = a constant, equal to 0.05 when measurements are in feet and 0.16 when in meters
 V = room volume, ft^3 or m^3
 ΣA = total room absorption, sabins (ft^2 or m^2) at that frequency

(For spaces of unusual shapes, see Beranek, 1988). Reverberation is one of the most pronounced hearing reactions in an enclosed space. It is the ear's reaction to echoes, giving a subjective impression of "liveness" or "deadness" to the space.

A space with highly reflective surfaces, and therefore low average absorption coefficient ($\overline{\alpha} <$ 0.2), sounds live. Conversely, a highly absorptive nonreflective environment ($\overline{\alpha} > 0.4$) sounds dead. ($\overline{\alpha}$ is the average absorption coefficient of the entire space, as explained in detail in the Section 27.8). Since room absorption is related to total surface area, which in turn is related by room proportions to room volume, it is possible to relate all three factors in a single diagram. Figure 27.6 is drawn using room proportions of 2 : 1.5 : 1, for L : W : H, based on Fig. 27.22 for preferred room proportions. These proportions (2 : 1.5 : 1) represent the average of the extremes of Fig. 27.22. The maximum differential introduced by using average proportions is 2½%—well within engineering accuracy.

The ultimate dead space is one where there is no reflection at all—as is the case outdoors in a flat open area. Indoors, we receive an auditory cue in the form of reflected sound (feedback) which helps us regulate our sound power output (voice level). Outdoors, the absence of this cue,

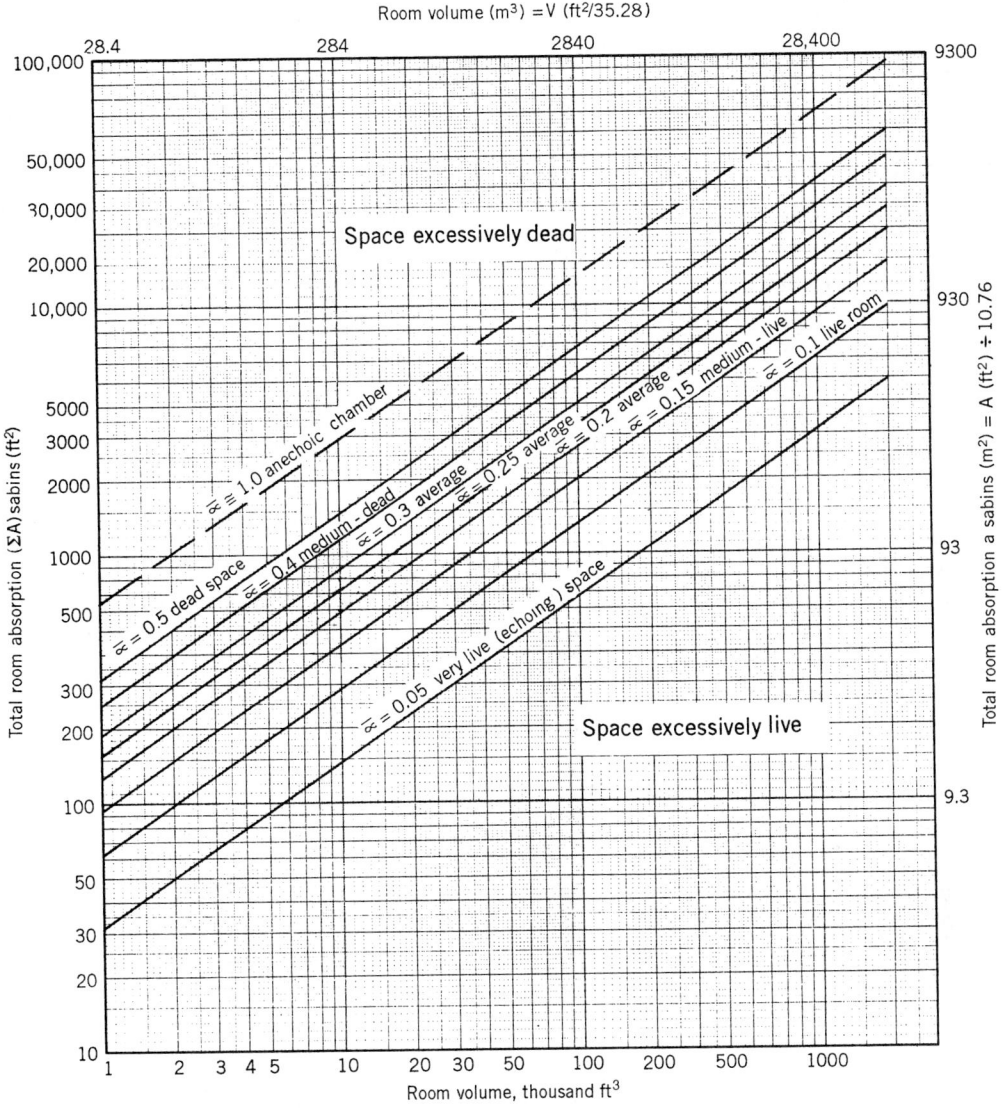

Fig. 27.6 *Chart indicating a room's "liveness" as a function of its volume and total Sabin absorption. The* L-W-H *room proportions chosen—2H:1.5H:H—represent the average of the three extreme points of the recommended room proportion triangle in Fig. 27.22. For these proportions,* S = 6.25V$^{2/3}$ *and therefore* A = S $\bar{\alpha}$ = 6.25 $\bar{\alpha}$ V$^{2/3}$.

to which we are so accustomed by our largely indoor existence, automatically causes us to raise our output. For this reason most people tend to speak excessively loudly outdoors. Conversely, a highly reflective indoor condition gives a large feedback signal, which usually results in a lowered vocal output.

In common room acoustics studies, reverberation times are calculated at 125, 500, 1000, and 2000 Hz. The midfrequency (500 to 1000 Hz)

range is generally the reference used in specifying the reverberation time of a room when studying the speech characteristic of the space.

Reverberation can be considered as a mixture of previous and more recent sounds. The converse of *reverberation* or *reverberance* is *articulation*. An articulate environment keeps each sound event separate rather than running them together. Spaces for speech activities should be less reverberant—more articulate—than those designed for performance of music. See Sections 27.11 and 27.12 for reverberation criteria for speech and music rooms, respectively.

27.7 Sound Fields in an Enclosed Space

The inverse square law described in Section 26.7 holds true for the acoustic *far field,* which is the sound field sufficiently far from the source that intensity is proportional to power, and inversely to distance. This type of acoustic field is developed in open, obstruction-free space. Propagation in an enclosed space is quite different. There, when a sound reaches a wall or other large (with respect to wavelength) obstruction, part of the sound energy is reflected and part absorbed. As a result, the sound at any point in the room is a combination of direct sound from the source plus reflected sound from walls and other obstructions. If the reflections are so large that the sound level becomes uniform throughout the room, the acoustic field within the room is termed a *diffuse* one (no shadows), and intensity measurements with respect to a specific source are meaningless. Of course, if it is our intention to measure sound pressure level at a specific point, such as a seat in an auditorium, the type of acoustic field in the room is irrelevant.

Most indoor spaces do not have such a high level of reflection that a diffuse field is created. Instead, there is a *near field* near the source, a *free field* beyond the near field, and a *reverberant field* near the walls (Fig. 27.7). They can be recognized as follows:

1. The *near field* is generally within one wavelength of the lowest frequency of sound produced by the source. Within this distance SPL measurements vary widely and are not meaningful. (The maximum wavelength for the human male voice is about 11 ft.)

2. Close to large obstructions such as walls, the *reverberant field* is dominant and approaches a diffuse condition. In well-designed music auditoriums the reverberant (diffuse) field predominates and sound pressure level remains relatively constant beyond the free field area.

3. The *free* (far) *field* exists between the near and reverberant fields, and there intensity varies as pressure and inversely with distance squared. In this field, sound pressure level drops 6 dB with

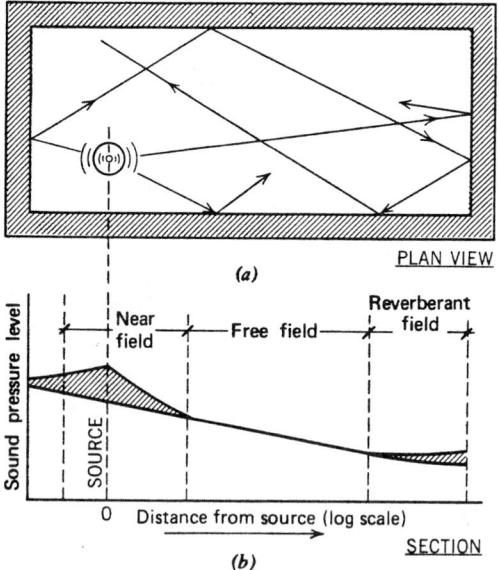

Fig. 27.7 *Type of sound field in an enclosed space depends in large measure on reflections (reverberation) and absorption. In a typical room there is a near field adjacent to the sound, a free field beyond that, and a reverberant field adjacent to the walls. In a large hall or auditorium the reverberant field dominates and sound pressure level remains approximately constant.*

each doubling of distance from the source, and it is in this field that meaningful SPL measurements can be made, with respect to a specific small source.

27.8 Sound Power Level (PWL) and Sound Pressure Level (SPL)

Sound *power level* (PWL) is a measure of the amount of sound generated by a source, independent of its environment. Sound *pressure level* (SPL) can be thought of as the resultant "noise" or sound in an enclosed space, resulting from a source in that space, as affected by the characteristics of the space and the position of the listener. It is thus an end effect.

Since it is desirable to be able to predict the sound pressure level (SPL) in a space during the design stage, *before* construction, and also because of the difficulty of SPL measurement *after* construction in rooms with various types of sound fields (as explained above) it is useful to have equations and or a graphic means of relating PWL to SPL. Sound power level is supplied by manufacturers with each item of equipment, at either octave or one-third octave points. In free space, SPL and PWL are simply related by the inverse square law, whereas in enclosed spaces the room characteristics come into play. Roughly speaking, an analogy to lighting can be drawn: SPL corresponds to room illumination, that is, footcandles or lux and PWL corresponds to the lumen output of the source.

The basic relationship between sound pressure level and sound power level for a single, small source, in a room large enough to have both free and reverberant fields is:

English units:

$$SPL = PWL + 10 \log \left(\frac{Q}{4\pi r^2} + \frac{4}{R} \right) + 10.5 \quad (27.5)$$

Metric units:

$$SPL = PWL + 10 \log \left(\frac{Q}{4\pi r^2} + \frac{4}{R} \right) + 0.2 \quad (27.6)$$

where Q is a directivity constant
  SPL = sound pressure level, dB
  PWL = sound power level, dB
  r = distance from source, ft[m]
  R = room factor, ft²[m²]

The factor R can be calculated from

$$R = \frac{\Sigma S \overline{\alpha}}{1 - \overline{\alpha}}$$

where ΣS = total room surface area, ft²(m₂)
  $\overline{\alpha}$ = average absorption coefficient of all materials in the room

that is,

$$\overline{\alpha} = \frac{\Sigma A \text{ (total room absorption)}}{\Sigma S \text{ (total room surface area)}} \quad (27.7)$$

or

$$\overline{\alpha} = \frac{S_1\alpha_1 + S_2\alpha_2 + \ldots + S_n\alpha_n}{S_1 + S_2 + \ldots + S_n} \quad (27.8)$$

The directional constant Q is either inherent in the sound source, and as such will be part of the given data, or can be obtained from Fig. 27.8 for a nondirectional source made directional by adjacent reflecting surfaces.

Thus for a source suspended or supported at least ½ wavelength (at its lowest frequency) from the ceiling or floor, $Q = 1$; a source near the floor and distant from the walls uses $Q = 2$; and so on. In a space containing more than one sound source, the sound pressure levels can be combined as explained in Section 26.8. In the two extreme cases [i.e., rooms with only far (direct) fields and room with only reverberant (diffuse) fields, corresponding to dead and live rooms, respectively] for a nondirectional source located on the floor and away from walls ($Q = 2$), Eqs. 27.5 and 27.6 reduce to:

Dead room, direct field:

  English units:

$$SPL = PWL - 20 \log r + (Q - 0.6)$$
$$= PWL - 20 \log r + 1.4 \quad (27.9)$$

  Metric units:

$$SPL = PWL - 20 \log r + (Q - 10.9)$$
$$= PWL - 20 \log r - 8.9 \quad (27.10)$$

Live room, diffuse-reverberant field:

  English units:

$$SPL = PWL - 10 \log \Sigma A + 16.3 \quad (27.11)$$

  Metric units:

$$SPL = PWL - 10 \log \Sigma A + 6.0 \quad (27.12)$$

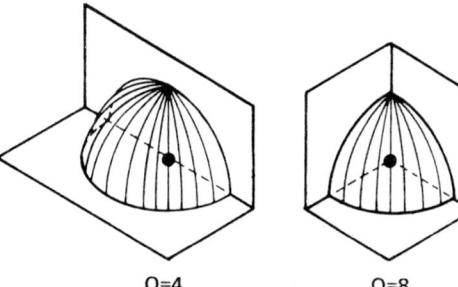

Q=1   Q=2   Q=4   Q=8

Fig. 27.8 Diagrams illustrating directivity factors for either inherently directive sources or nondirective sources placed adjacent to large reflecting surfaces. (Courtesy of Barry Blower Co.)

Fig. 27.9 *Calculated curves show that the development of a reverberant field (constant SPL, or flat curve) is dependent on the room's absorption characteristics. On each curve the point at which a reverberant field begins is indicated by a dot on the curve. In a live room ($\overline{\alpha}$ = 0.1) the reverberant field begins 20 ft from the source. In a dead room ($\overline{\alpha}$ = 0.5). It begins at 120 ft from the source, that is, at the back wall. Effectively, then, there is no reverberant field in such a room (or in any room with $\overline{\alpha}$ = 0.5). Slope of asymptote is 6 dB per octave, that is, inverse square attenuation.*

To understand the effect of room absorption on the development of the reverberant field in a large space, refer to Fig. 27.9. The field is plotted from Eq. 27.5, for a room 80 ft × 120 ft × 30 ft high, using average absorptions $\overline{\alpha}$ ranging from 0.05 to 0.7, that is, from very live to very dead. Note particularly that the addition of sufficient absorbent material can, by drastically reducing sound reflections from room surfaces, entirely prevent the development of a reverberant (diffuse) field. This is particularly important in noisy industrial interiors where the building materials commonly employed are generally nonabsorbent, and the noise sources are both numerous and loud. An example of the use of absorptive material to effect a considerable degree of in-room noise reduction is demonstrated in the next Section. We emphasize that the effect is almost exclusively in the room where the absorptive material is used since, as pointed out several times previously, the acoustic energy transmitted to the adjacent spaces is only minimally affected. Sound transmission from one enclosed space to an adjoining space through a common barrier depends almost entirely on the type of barrier. This is discussed in detail in Chapter 28.

27.9 Noise Reduction by Absorption

Equation 27.6 relates SPL and PWL as a function of distance from the source and room absorption, using metric units. The following example shows the application of this equation and the result of adding absorption to the space.

EXAMPLE 27.1 An open blower is installed on the floor, away from the walls, in a large enclosed space that is 6 m long, 12 m wide, and 4 m high. The floor is concrete (α = 0.01); walls and ceiling painted block (α = 0.07). The PWLs supplied by the manufacturer are 90 dB at 500 Hz and 87 dB at 2000 Hz. Calculate the noise level SPL at distances of 5 m and 10 m from the blower outlet: (a) in the original room; (b) with double the absorption; and (c) with quadruple the absorption.

SOLUTION.

1. From Eq. 27.6, we have the expression

$$SPL = PWL + 10\log_{10}\left(\frac{Q}{4\pi r^2} + \frac{4}{R}\right) + 0.2$$

In our example, Q = 2 (see Fig. 27.8).

2. The first step is to calculate room factor R for the three situations of absorption.

$$R = \frac{S\bar{\alpha}}{1 - \bar{\alpha}}$$

$$S = 2(48) + 2(24) + 2(72) = 288 \text{ m}^2$$

from Eq. 27.8

$$\bar{\alpha} \text{ original} = \frac{3(48)(0.07) + 1(72)(0.07) + (72)(0.01)}{288}$$

$$= 0.055$$

$\bar{\alpha}_a = 0.055$		$R_a = 16.76$
$\bar{\alpha}_b = 2(0.055) = 0.11$		$R_b = 35.6$
$\bar{\alpha}_c = 4(0.055) = 0.22$		$R_c = 81.23$

3. Calculating SPL values and tabulating, we obtain

	500 Hz		2000 Hz	
SPL for:	*5 m*	*10 m*	*5 m*	*10 m*
Original room	84.1	84.0	81.1	81.0
Double $\bar{\alpha}$	81.0	80.8	78.0	77.8
Quadruple $\bar{\alpha}$	77.7	77.3	74.7	74.3

∎

The results indicate the very important fact that *doubling the absorption decreases the noise level by only 3 dB*. Therefore it requires a quadrupling of absorption to make the decrease noticeable (see Table 26.3). This is obviously an expensive procedure with diminishing returns.

If the entire space is considered to be a reverberant field, then from Eq. 27.11 the noise (sound) intensity level (IL) can be expressed as follows:

$$IL = PWL - 10 \log \Sigma A + 16.3 \text{ dB} \quad (27.11)$$

where

ΣA = total absorption in room, sabins (ft²)
IL = intensity level, dB
PWL = sound power level, dB

Although increasing absorption decreases the noise level, the level cannot be reduced below the free field level for that distance from the source because the free field situation corresponds to outdoors, where $\bar{\alpha} = 1.0$.

The amount of noise reduction provided by additional absorption may be determined by

$$\text{noise reduction } (NR) = IL_1 - IL_2$$
$$= 10 \log \Sigma A_2 - 10 \log \Sigma A_1$$

Therefore,

$$NR = 10 \log \frac{\Sigma A_2}{\Sigma A_1} \quad (27.13)$$

where

NR = noise reduction in dB
ΣA_2 = total absorption, initial condition
ΣA_1 = total absorption, final condition

From Eq. 27.13 it is obvious that doubling the absorption results in a noise reduction of 3 dB, since $10 \log_{10} 2 = 3$ (dB).

It is of interest to work out a practical example using Eqs. 27.11 and 27.13 and comparing the results with the more precise relation given in Eq. 27.5.

EXAMPLE 27.2   Referring to Fig. 27.10, calculate the original noise level and the subsequent

(a)   *(b)*   *(c)*

Fig. 27.10 *Quieting a room (a) by addition of absorptive material is cost effective only through step 2 (b). Further quieting would be accomplished locally (c), which might obviate the necessity for wall treatment (step 2). Ceiling treatment should remain.*

ACOUSTICS

noise reduction by three steps of sound absorption treatment, assuming a completely reverberant field in the space, that is, SPL independent of location. Data:

Original Condition. Painted concrete block chamber, 10 × 10 × 10 ft.
Fan power level:

At 500 Hz = 88 dB

At 2000 Hz = 78 dB

SOLUTION.

Absorption (Hz)	Area (ft²)	α	Total Absorption ($\Sigma S\alpha$)
500	600	0.06	36 sabins (ft²)
2000	600	0.09	54 sabins (ft²)

Sound intensity level before treatment:
At 500 Hz:

$$IL = \text{sound power} - 10 \log \Sigma A + 16.3 \text{ dB}$$

$$= 88 \text{ dB} - 10 \log 36 + 16.3 \text{ dB}$$

$$= 88 \text{ dB} - 15.6 \text{ dB} + 16.3 \text{ dB}$$

$$= 88.7 \text{ dB}$$

At 2000 Hz:

$$IL = 78 \text{ dB} - 10 \log 54 + 16.3 \text{ dB}$$

$$= 78 \text{ dB} - 17.3 \text{ dB} + 16.3 \text{ dB}$$

$$= 77 \text{ dB}$$

(b) *Ceiling Treatment Only*

At 500 Hz:

$$\alpha = 0.82$$

$$\text{additional absorption} = 100 (0.82 - 0.06)$$

$$= 76 \text{ sabins}$$

$$NR = 10 \log \frac{76 + 36}{36}$$

$$= 4.9 \text{ dB}$$

At 2000 Hz:

$$\alpha = 0.94$$

$$\Delta A = 100 (0.94 - 0.09) = 85$$

$$NR = 10 \log \frac{85 + 54}{54}$$

$$= 4.1 \text{ dB}$$

(c) *Ceiling and Half-Wall Treated*

At 500 Hz:

$$\text{added absorption} = 300 (0.82 - 0.06)$$

$$= 228 \text{ sabins}$$

$$NR = 10 \log \frac{228 + 36}{36}$$

$$= 8.7 \text{ dB}$$

At 2000 Hz:

$$\text{added absorption} = 300 (0.94 - 0.09)$$

$$= 255 \text{ sabins}$$

$$NR = 10 \log \frac{255 + 54}{54}$$

$$= 7.5 \text{ dB}$$

(d) *Ceiling and Full-Wall Treatment*

At 500 Hz:

$$\Delta A = 500 (0.82 - 0.06)$$

$$= 380 \text{ sabins}$$

$$NR = 10 \log \frac{380 + 36}{36}$$

$$= 10.6 \text{ dB}$$

At 2000 Hz:

$$\Delta A = 500 (0.94 - 0.09)$$

$$= 425 \text{ sabins}$$

$$NR = 10 \log \frac{425 + 54}{54}$$

$$= 9.5 \text{ dB}$$

Summary

	IL (SPL)[a]	
	500 Hz	*2000 Hz*
Bare room	88.7	77
Ceiling treated	−4.9 dB	−4.1 dB
	(−4.9)[a]	(−4.2)[a]
Half-wall treatment	−8.7 dB	−7.5 dB
	(−8.7)[a]	(−7.7)[a]
Full-wall treatment	−10.6 dB	−9.5 dB
	(−10.7[a]	(−9.6)[a]

[a] The numbers in parentheses are the SPL differences as calculated from Eq. 27.5, using $Q = 2$ and $r = 5$ ft.

∎

Two important conclusions can be drawn from these results:

1. The third step of adding absorptive material is not worthwhile since negligible additional room quieting is accomplished. This clearly demonstrates the law of diminishing returns as applied to room quieting with absorptive materials. Starting with a live room, the initial application is effective. Beyond that, additional quieting by absorption is not economical, and the same outlay would be better applied in quieting the machine itself, probably with a machine enclosure as indicated in Fig. 27.10.

2. If we compare the figures in parentheses that resulted from a full calculation (Eq. 27.5) to those arrived at using the much simpler equations applicable only to a fully diffuse field (Eq. 27.11) we see that the differences are negligible. This means that the short cut method can always be used in preliminary calculations and frequently for final results as well. The diffuse (reverberant) field calculation also lends itself nicely to a graphical solution.

Using the diffuse field Eq. 27.11

$$SPL = PWL - 10 \log \Sigma A + 16.3$$

and remembering that reverberation time is expressed in terms of total absorption and room volume; that is

$$T_R = \frac{kv}{\Sigma A} \qquad (27.14)$$

where K is 0.05 when room volume is in ft³, and 0.16 when volume is cubic meters. Using these equations, we can then rewrite Eqs. 27.11 and 27.12 in terms of reverberation time and room volume:

for room volume in ft³

$$SPL - PWL = -10 \log \frac{.05V}{T_R} + 16.3 \qquad (27.13)$$

for room volume in m³

$$SPL - PWL = -10 \log \frac{.16V}{T_R} + 6.0 \qquad (27.14)$$

These two equations are plotted graphically in Figs. 27.11 and 27.12. The charts are accurate for

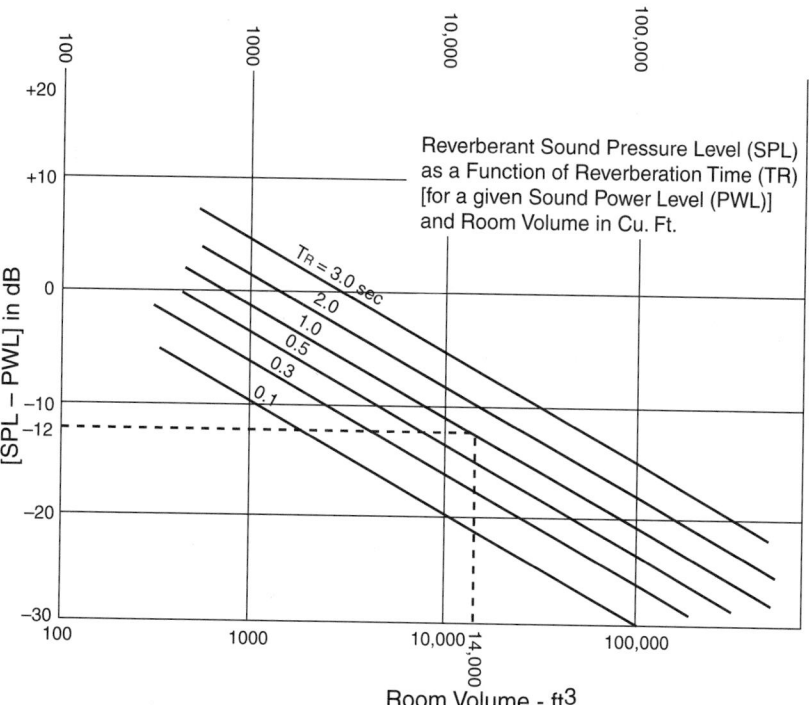

Fig. 27.11 *Chart for determining reverberant sound pressure level (SPL) when the power level (PWL) of an item of equipment is known. Room volume in cubic feet and reverberation time in seconds are the variables.*

ACOUSTICS

Fig. 27.12 *Chart for determining reverberant sound pressure level (SPL) when the power level (PWL) of an item of equipment is known. Room volume in cubic meters and reverberation time in seconds are the variables.*

live room, diffuse (reverberant) fields, and a close approximation for other situations. An illustrative example will demonstrate use of the charts.

EXAMPLE 27.3   A generator in a 14000 ft³ room has a PWL of 95 dB (re: 10^{-12} W) at 400 Hz.

(a) Using the chart in Fig. 27.11, find the SPL at that frequency. Assume that the sound field in the room is reverberant. The room's reverberation time is 1.0 s at 400 Hz.

(b) Check the result using the chart in Fig. 27.12

SOLUTION.

(a) Enter the chart in Fig. 27.11 at 14000 ft³ and draw a vertical line until it intersects the 1.0-s sloping line. Extend a line from that intersection point to the X-axis, and read − 12 dB. Then, since

$$SPL - PWL = -12 \text{ dB}$$

and PWL is given as 95 dB,

then

$$SPL = 95 - 12 = 83 \text{ dB}$$

(b) 14000 ft³ is equal to 400 m³, within engineering accuracy (it is actually 396.7 m³.) Enter the chart in Fig. 27.12 at 400 m³, and extend a line upward to T_R of 1 s, and left to the X scale. Read − 12 dB. The two answers check, as they should. ■

27.10   Noise Reduction Coefficient

The last column in Table 27.1 is labeled NRC—noise reduction coefficient. This figure is the arithmetic average of the absorption coefficients at 250, 500, 1000, and 2000 Hz. The name is ill chosen inasmuch as it cannot be used as the words seem to imply. It is useful as a single-number criterion for the midband effectiveness of a porous absorber. Obviously, if careful design is required or high- and low-end frequencies are of interest, NRC is nearly useless, and detailed calculations over the entire frequency range must be made. Similarly, two materials with the same NRC can perform quite differently, since the NRC is an average and few materials have a flat characteristic.

ROOM DESIGN

27.11 Reverberation Criteria for Speech Rooms

The overriding criterion for speech is intelligibility. Since speech consists of short disconnected sounds 30 to 300 ms in length (see Section 26.14), among which are high-frequency, low-energy phonemes, the ideal room must ensure the ear's undistorted reception of these phonemes. This requires keeping reverberation to a minimum. We can obtain a good approximation of the subjective feeling of liveness of a room, for purposes of speech, from the relation

$$T_R \text{ (speech)} = 0.3 \log \frac{V}{10} \qquad (27.16)$$

where

T_R (speech) = optimum reverberation time in seconds, for speech

V = room volume, m^3

For instance, a typical classroom might have a volume of 150 m^3 (5300 ft^3). Optimum reverberation time is

$$T = 0.3 \log 15 = 0.35 \text{ s}$$

Reverberation times longer than this would sound live, shorter ones dead and flat. Indeed an increase of 20% in reverberation time would make the room excessively live and boomy and would negatively affect speech intelligibility. Figure 27.13 gives optimum *midfrequency* reverberation times as a function of room size and use. Figure 27.14 gives *maximum* reverberation time figures for speech, in large rooms. For good speech intelligibility, reverberation time should remain essentially flat down to 100 Hz.

The reflected sounds associated with reverberation can have either a salutary or a deleterious effect. The ear cannot distinguish between sounds

Fig. 27.13 *Optimum reverberation times for the frequency range 500 to 1000 Hz, as a function of room volume. (The cubic foot volume figures are based on the approximate conversion factor of 35 ft^3/m^3). (Reprinted by permission from E. B. Magrab,* Environmental Noise Control, *Wiley, New York, 1975, p. 206.)*

Fig. 27.14 *Maximum recommended reverberation time for speech in auditoriums and lecture halls. (From P. V. Brüel and Kjaer,* Sound Insulation and Room Acoustics. *Chapman & Hall, London, 1951, as reprinted in Brüel and Kjaer, 1963.)*

that arrive within a maximum of 50 ms (1/20 s) of each other. (Some authorities use 40 ms, i.e., 1/25 s.) Sounds arriving within this time *reinforce* the direct path signal and appear to come from the source. Sounds arriving after this time are apprehended as a fuzzy echo or elongation of the sound, reducing intelligibility and directivity.

Since the range of 40 to 50 ms corresponds at 344 m/s to 13.76 to 17.2 m, a speech room should be so arranged that the difference between the first reflection path and the direct path is no greater than 17 m and preferably 14 m or less at mid-frequency (Fig. 27.15). For more details concerning this and related factors on reflected paths, refer to Section 27.13.

Too *low* a reverberation time (very high absorption, minimum reflection) is also undesirable because:

1. It limits the size of the room to that which can be covered by direct sound only.

2. It is disturbing to the speaker, since absence of reflection prevents him or her from gauging proper voice level and tends to cause excessive effort (shouting, as when outdoors).

Thus proper design of a room for speech is a compromise between the need for some reflection and the desire to minimize reflection to preserve intelligibility.

27.12   Criteria for Music Performance

Adequate design for a music space requires recognition of the following:

1. Large-volume spaces require direct-path sound reinforcement by reflection.

2. Relatively long reverberation time is needed to enhance the music—the exact amount

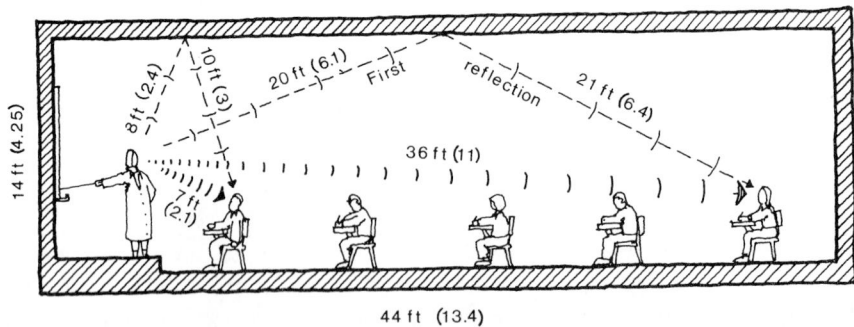

Fig. 27.15 *Sound paths in a typical medium sized lecture room. Note that for both extremes of listener position, the maximum path-length difference between direct and first reflection is 11 ft. Thus signal is reinforced, and intelligibility should be excellent if room absorption is provided to limit reverberation time to about 1/2 s maximum (see Fig. 27.14). Numbers in parentheses are dimensions in meters.*

depending on the type of music (Figs. 27.13 and 27.16). Designers should keep in mind that reverberation time recommendations vary as much as 100% between respected sources.

3. It is generally agreed that reverberation time should vary inversely with frequency [i.e., T_R should be longer at lower frequencies (than the midfrequency recommendation) and shorter at higher frequencies]. The longer T_R at low frequencies adds fullness to music and "body" to speech. Thus T_R at 100 Hz should be, according to most researchers, 35% to 75% longer than T_R at the center frequencies.

4. Short T_R at upper frequencies adds directivity to the music. With large ensembles, directivity gives the sense of depth and instrument location necessary for proper appreciation. This is often referred to as clarity or definition in music. With a solo instrument, this problem is diminished.

5. Brilliance of tone is primarily a function of high-frequency content. Since these frequencies are most readily absorbed, a good direct path must exist between sound source and listener. Since our eyes and ears are close together, a good sound path exists when a good vision path exists. At the other end of the spectrum, lack of sufficient bass expresses itself as a loss of "fullness," which is often caused by resonant absorption.

The actual design of music performance space is a very complex procedure involving extensive calculations of absorption, reverberation time and ray diagramming, and juggling of materials, dimensions, and wall angles. Simulation techniques and acoustic models are also employed.

Recent research and simulation studies of concert and recital halls have demonstrated that the sensation of "fullness" of music, or what is today referred to as sound envelopment, is enhanced by lateral reflections that reinforce the direct signal. It has also been found that the subjective judgement of reverberance is more strongly af-

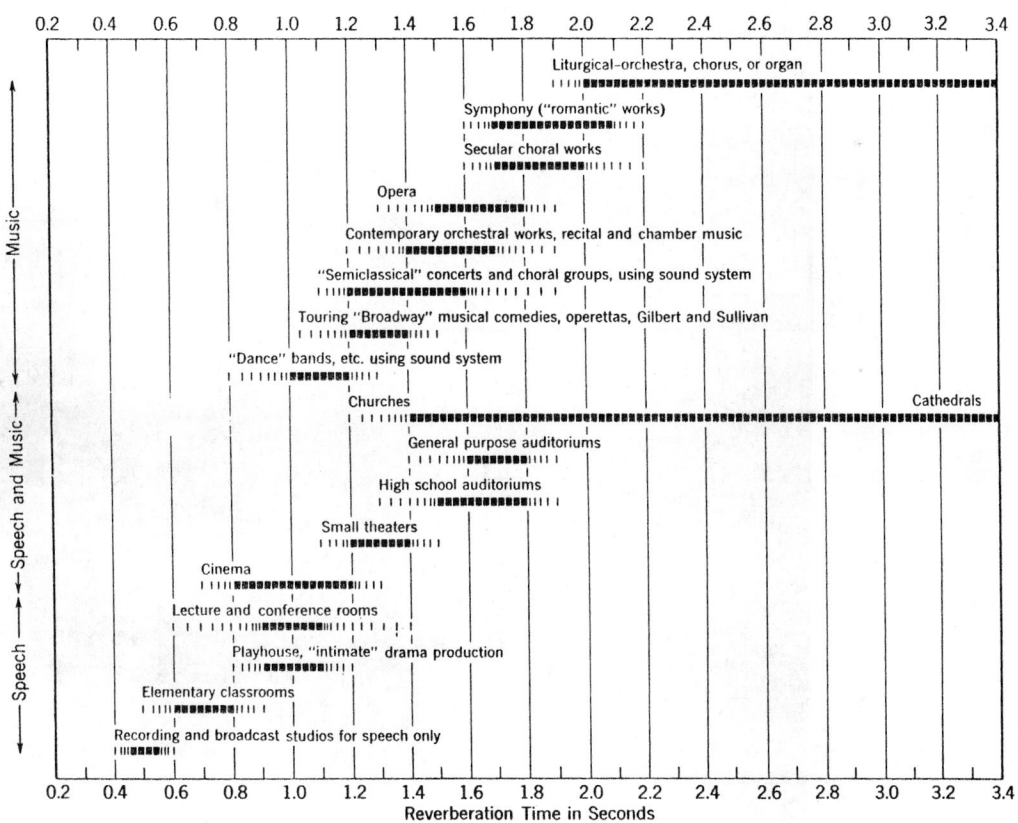

Fig. 27.16 *Optimum reverberation times at midfrequencies (500–1000 Hz) for various types of facilities. The wide range shown reflects the effect of room proportions, shape and volume on the optimum T_R, as well as differences of opinion among designers.*

ACOUSTICS

fected by "early decay time," (the time required for a 10 dB decrease in signal strength) than the conventional 60 dB decay time. Finally, crispness or clarity of the music (particularly important in recital halls and for chamber music) depends on the reflections arriving within 40 to 70 ms. All of these factors are considered both in the original design and in the often lengthy 'tuning' process of a space intended for music performance. Most modern design also uses movable reflector panels and other active variables. After construction is completed, extensive tests are conducted and field adjustments are made.

27.13   Sound Paths

Ideally, every listener in a lecture hall, theater, or concert hall should hear the speaker or performer with the same degree of loudness and clarity.

Since this is obviously impossible by direct-path sound, the essential design task is to plan methods for reinforcing desirable reflections and minimizing and controlling undesirable ones. Normally only the first reflection is considered in ray diagraming (discussed in the next section) since it is strongest. Second and subsequent reflections are usually attenuated to the point that they need not be considered except for the special situations of flutter, echoes, and standing waves discussed below.

(a) Specular Reflection. Specular reflection occurs when sound reflects off a hard polished surface. This characteristic can be used to good advantage to create an effective image source. In ancient Greek and Roman theaters, seats were arranged on a steep, conical surface around the performers. The virtue of this arrangement [Fig. 27.17(*a*)] is that the sound power travels to each

Fig. 27.17 *Use of an angled reflector panel* (b) *creates an image source that stands in approximately the same relation to the audience as the performer in the classic Greek theater* (a).

Wavelength(ft, in.)	Frequency (Hz)	Wavelength (m)
113	10	34.4
	15	
56 — 50	20	20 — 17.2
	30	10
22.6 — 20	50	6.88
	70	5.0
11.3 — 10	100	4.0 — 3.44
	150	3.0
5.6 — 5	200	2.0 — 1.72
— 4	300	
— 3		1.0 — 1 meter
2.3 — 2	500	69 cm
	700	0.5
1.1 — 1	1 kHz	0.4 — 34 cm
0.56 — 8 in.	2 kHz	0.2 — 17 cm
— 6 in.	3 kHz	
— 4 in.		0.1
0.23	5 kHz	7 cm
— 2 in.	7 kHz	0.05
0.11	10 kHz	3.4 cm

Fig. 27.18 *Nomograph for determining wavelength in feet or meters, given frequency in hertz, or vice versa. Speed of sound is taken as 344 m/s and 1128 ft/s. To use, hold a straightedge horizontally across the nomograph, and read the figures directly.*

location with minimal attenuation. The same effect can be accomplished by placing the sound source above the seats. This is not practical physically, but it can be accomplished effectively by the use of a reflecting panel [Fig. 27.17(*b*)]. The panel dimension must be at least one wavelength at the lowest frequency under consideration. Fig-

ure 27.18 is a conversion chart from frequency to wavelength in feet and meters.

(b) Echoes. As explained in Section 27.11, a clear echo is caused when reflected sound *at sufficient intensity* reaches a listener more than 50 ms after he has heard the direct sound. (Some authorities place this figure as high as 80 ms.) Echoes, even if not distinctly discernible, are undesirable. They make speech less intelligible and music sound "mushy." The relative undesirability depends on the time delay and loudness relative to the direct sound, which, in turn, are dependent on the size, position, shape, and absorption of the reflecting surface.

Typical echo-producing surfaces in an auditorium are the back wall and the ceiling above the proscenium. Figure 27.19 shows these problems and suggests remedies. Note that the energy that produced the echoes can be redirected to places where it becomes useful reinforcement. If echo control by absorption alone were used on the ceiling and back wall, that energy would be wasted. The rear wall, since its area cannot be reduced too far, may have to be made more sound absorptive to reduce the loudness of the reflected sound.

(c) Flutter. A flutter, perceived as a buzzing or clicking sound, is comprised of repeated echoes traversing back and forth between two nonabsorbing parallel (flat or concave) surfaces. Flutters often occur between shallow domes and hard, flat floors. The remedy for a flutter is either to change the shape of the reflectors or their parallel relationship, or to add absorption. The solution chosen will depend on reverberation requirements, cost, and esthetics.

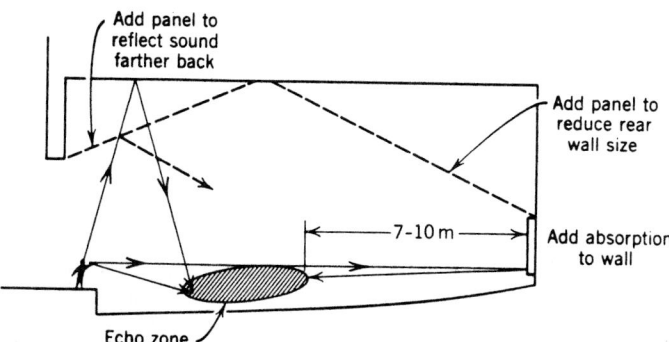

Fig. 27.19 *Auditorium section showing the causes and remedies for two typical echoes.*

ACOUSTICS

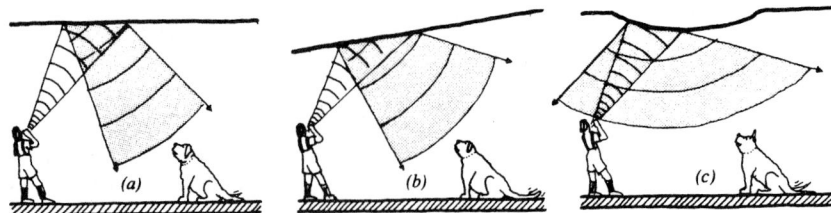

Fig. 27.20 *Sound diffusion can be created with reflectors of different shapes, ranging from the horizontal flat* (a), *inclined flat* (b), *or convex* (c). *Diffusion improves from* (a) *to* (c).

(d) Diffusion. This is the converse of focusing and occurs primarily when sound is reflected from convex surfaces. A degree of diffusion is also provided by flat horizontal and inclined reflectors (Fig. 27.20). In a diffuse sound field the sound level remains relatively constant throughout the space; an extremely desirable property for musical performances.

(e) Focusing. Concave domes, vaults, or walls will focus reflected sound into certain areas of rooms. This has several disadvantages. For example, it will deprive some listeners of useful sound reflections and cause hot spots at other audience positions [Fig. 27.21(a)].

(f) Creep. This describes the reflection of sound along a curved surface from a source near the surface. Although the sound can be heard at points along the surface, it is unaudible away from the surface. Creep is illustrated in Fig. 27.21(b).

(g) Standing Waves. Standing waves and flutters are very similar in principle and cause, but are heard quite differently. When an impulse (such as a hand clap) is the energy source, a flutter will occur between two highly reflective parallel walls. It is perceived as a slowly decaying buzz. When a steady, pure tone is the source, a standing wave will occur, but only when the parallel walls are spaced apart at some integral multiple of a half-wavelength.

When the parallel walls are exactly one-half wavelength apart, the tone will sound very loud near the walls and very quiet halfway between them. This is because at the center, the reflected waves traveling in one direction are exactly one-half wavelength away from those traveling in the other, and thus equal *but opposite in pressure,* which results in total cancellation. In other rooms standing waves are noted as points of quiet and maximum sound in the room. Standing waves are important only in rooms small with respect to the wavelengths generated (smallest room dimension <30 ft for music or <15 ft for speech).

Another effect of standing waves called resonance, is the accentuation of a particular frequency that will cause a standing wave. Thus, if one speaks (or plays a musical instrument) standing near a wall of a room, about 8 ft × 8 ft, one will notice an abnormal and sometimes unpleasant loudness in the sound at about 280 Hz.

Focusing

(a)

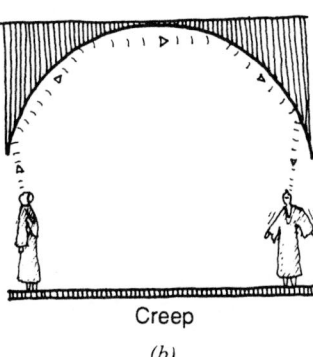

Creep

(b)

Fig. 27.21 *Two undesirable phenomena in room acoustics.*

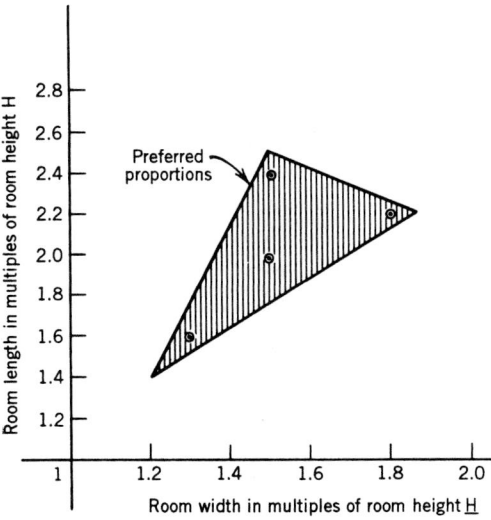

Fig. 27.22 *Triangular figure contains the room proportions recommended, for avoidance of disturbing low-frequency acoustic phenomena such as flutter, standing waves, and resonances, particularly at frequencies below 100 Hz. The points shown at the three apexes were averaged to obtain the center, average proportion of* 2H : 1.5H : H. *See also Fig. 27.6 for application of these proportions.*

Similarly, when a musician plays a scale, one note may seem far louder than the adjacent ones, and listeners in one section of the room will hear a quality of sound different from that heard by those in other sections. This effect *must* be avoided for music performance, but is merely an annoyance in rooms designed for speech use. This is one of the reasons that one finds music rehearsal rooms, broadcast studios, and so on with nonparallel walls and undulating ceilings. These irregularities prevent most of the undesirable effects described above from occurring. However, since rooms with nonparallel walls and undulating ceilings are for most applications unacceptable, and since standing waves and other resonant phenomena are related to room geometry, it is possible to calculate room proportions that will minimize these effects. Figure 27.22 shows such room proportions, which are applicable to bass frequencies (i.e., below 100 Hz). These are the problematic frequencies, since at higher frequencies (and in large rooms) many standing waves occur and the total effect is much less disturbing and frequently hardly noticeable.

27.14 Ray Diagrams

Ray diagramming is a design procedure for analyzing reflected sound distribution throughout a hall, using the first reflection only. Figure 27.23 shows a ray diagram. The rays are drawn normal (perpendicular) to the spherically propagating sound waves. Specular reflection is assumed, that is, the angles between the reflecting panel and the incident and reflected rays are always equal. Thus, in addition to the direct sound, each listener is receiving reflected sound energy. It is as though there were additional sound sources, the real one and numerous image sources. Figure 27.23 shows the application of a ray diagram to the design of a lecture hall. In Fig. 27.23(*a*), the stage height and seating slope are arranged to provide good sight lines, and the ceiling height is established by reverberation requirements, esthetics, cost, and so on. It can be seen that less than half the ceiling is providing useful reflection. By dividing the ceiling into two panels [Fig. 27.23(*b*)], people in the rear of the room perceive the direct source plus two image sources, and the useful reflecting area is increased by 50%. In Fig. 27.23(*c*), the shape has been further refined to include a lighting slot and a loudspeaker grille.

As mentioned previously, the sensation of musical envelopment that is so satisfying to a listener, particularly when dealing with large groups playing symphonic music, is due in large measure to reflections received from the side (lateral reflections). This being so, ray diagramming that was originally performed on room *sections* to determine ceiling shapes and the proper placement of ceiling-hung reflector panels, is now done in plan as well. This is particularly important in fan shaped auditoriums, where the canted wall does not provide useful lateral reflections. The solution to this problem is to build a saw tooth wall or to provide reflector panels along the wall that will provide the desired lateral reflection. Nonhorizontal ceiling reflector panels can also provide a measure of lateral reflection, particularly in balconies.

Although they are a useful design tool, ray diagrams have certain restrictions. Also, design must always be a compromise between ray diagrams for various "speaking positions" on the stage. Thus a paraboloid may be a perfect shape

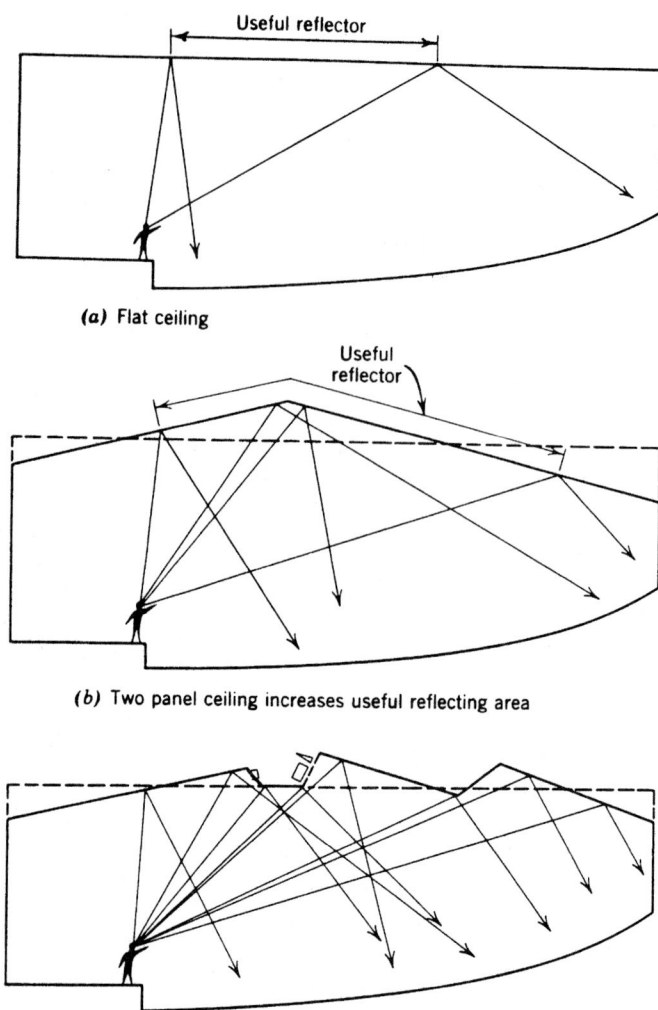

(a) Flat ceiling

(b) Two panel ceiling increases useful reflecting area

(c) Multifaceted ceiling incorporates lights and loudspeakers

Fig. 27.23 Section through a typical lecture room showing use of ray diagrams.

for one source position but will be very poor for other positions.

27.15 Auditorium Design

Auditorium is a general term used to describe a space where people sit and listen to speech or music. Acoustical design of an auditorium includes room acoustics, noise control, and sound system design. Noise control is covered in Chapter 28, and sound systems are discussed in Sections 27.16 through 27.18. If the program for the auditorium includes activities that require different acoustical environments (as is frequently the case), then it must be decided early whether the acoustics will be a compromise between the program extremes or adjustable for various activities. Acoustical environments can be altered by changing volume, moving reflecting surfaces, and adding or subtracting sound-absorbing treatment. Figure 27.24 illustrates several examples of acoustical adjustability.

Factors that influence the acoustical design include audience size, range of performance activities, and sophistication of the potential audience. A small school auditorium and a professional theater will have widely divergent demands from

Fig. 27.24 Adjustable acoustic elements in an auditorium.

both audience and performers. The audience size determines the basic floor area of an auditorium, assuming no balconies. Once this area has been fixed, the volume of the room is developed according to reverberation requirements of the space.

Figure 27.25 shows a typical auditorium in plan and section. The shape of wall and ceiling surfaces is developed to give proper distribution of sound and eliminate focusing or echoes. Essential characteristics of the design include:

1. Ceiling and side walls at the front of the auditorium distribute sound to the audience. These surfaces must be close enough to the performers to minimize time delays between direct sound and reflected sound.
2. Ceiling and side walls provide diffusion.

Acoustics must be considered in selection of materials used in an auditorium. All auditoriums use both sound-reflecting and sound-absorbing materials. Since the largest area of sound-absorbing material in any auditorium is the audience, the difference in acoustical characteristics that occur without an audience may be minimized by using fully upholstered seating.

Chairs with fully upholstered seats and backs, covered in an open-weave material, will have absorption characteristics approximating an audience. Using the auditorium in Fig. 27.25 as an example, the reverberation characteristics of an auditorium with various materials may be examined. In the first example, the room use is assumed to be for music performance. The only sound absorption is that provided by the audience and seating. In the second set of calculations absorptive curtains were installed along the rear wall and a portion of the side wall. This configuration might by used for lectures in a room that is adjustable between speech and music configurations. A third configuration might use permanent sound-absorbing treatment installed on the ceiling and rear and side walls. Because of its low reverberation time, this configuration would be appropriate only for movies and lectures, not for music activities.

These simple examples indicate the effect of changes in the amount of absorption on the characteristics of a room. Adjustable treatments permit the characteristics of the room to be modified to any point between the extremes to meet the program acoustic requirements of a multipurpose hall.

Existing spaces may require remedial treatment to eliminate unwanted phenomena such as focusing and echoes, as shown in Fig. 27.26. In the first example, the surface of the dome was covered with sound-absorbing material to eliminate focusing; in the second, sound-absorbing treatment was applied to a curved rear wall to eliminate an echo. Such treatment also will affect the reverberation characteristics.

ACOUSTICS

Section   Plan

Simplified Calculations of Midfrequency (500 and 1000 Hz) Average Reverberation Times

$$\text{Reverberation time } (T_R) = \frac{0.05 \times \text{volume (cu ft)}}{\text{total absorption (sabins)}} \text{ seconds}$$

$$\text{volume} = 155{,}500 \text{ cu ft}$$

More Reverberant Condition (curtains retracted)				*Less Reverberant Condition* (curtains exposed)			
	Area	∝	Absorption		Area	∝	Absorption
Seating and stage (with audience and performers)	3323	.92	3060	Seating and stage (with audience and performers)	3323	.92	3060
Wall area Concrete block	8000	.2	1600	Wall area Concrete block (balance covered by curtains)	3600	.2	720
				Curtains	4400	.45	1970
Lower rear wall Permanent sound-absorbing treatment	450	.88	396	Lower rear wall Permanent sound-absorbing treatment	450	.88	396
Total absorption			5056	Total absorption			6146
More reverberant condition				Less reverberant condition			

$$T_R = \frac{0.05 \times 155{,}500}{5056} = 1.5 \text{ s} \qquad T_R = \frac{0.05 \times 155{,}500}{6146} = 1.2 \text{ s}$$

Fig. 27.25 *Auditorium with surface treatments for control of reflections and reverberation.*

SOUND REINFORCEMENT SYSTEMS

27.16 Objectives and Criteria

The purpose of a sound reinforcement system is just what the name indicates—to reinforce the sound, which would otherwise be inadequate. Thus an ideal sound system will give the listener the same loudness, quality, directivity, and intelligibility as if the source of sound were immediately adjacent—a distance of 2 to 3 ft for speech and farther for music, depending on type and number of instruments. This situation must obtain for ev-

Fig. 27.26 Sound-absorbing treatment used to eliminate focusing from dome and curved auditorium wall.

ery position in the space within ±3 dB. The other factors should remain constant. Of these factors, loudness and intelligibility have been discussed at length and should be well understood. By quality we mean that frequency response should be linear so that reproduced sound bears the same relation between its frequency components as the original sound. (Quality is then field-adjusted by "voicing" or "equalization," as discussed below.)

Directivity is the characteristic whereby the sound appears to be coming from the originating source, that is, the loudspeakers should be directionally "invisible," and the listener must have the impression of actually hearing the source. It should be emphasized that sound systems cannot

correct poor acoustic design completely, although they can improve a bad situation.

Generally, sound systems will be required in spaces larger than 50,000 ft³ ($\approx$1400 m³). In terms of population, this volume translates as 550 persons in lecture rooms (15 ft average ceiling height and 6 ft² per person) and 325 persons in theaters (20 ft average ceiling height and 7.5 ft² per person). In such a room (50,000 ft³) a normal speaking voice can maintain a volume level of only 55 to 60 dB, depending on room design and voice strength. With background noise at NC 30 (see Table 28.7), the speaker will be heard; at higher noise levels intelligibility will suffer.

27.17 Components and Specifications

All sound systems consist of three basic elements: input devices, amplifier(s), and loudspeaker systems.

(a) Input. Input usually means a microphone, a source of commercial broadcast material of various types and means for reproducing recorded material in all common commercial formats. Connections to local computers and computer networks would be available in sophisticated systems.

(b) Amplifier and Controls. Amplifiers must be rated to deliver sufficient power to produce intensity levels of 80 dB for speech, 95 dB for light music, and 105 dB for symphonic music. This assumes a *maximum* background noise level of 60 dBA. Thus 80 dB speech intensity will be 20 dB higher, or four times as loud as the noise level. If noise level is *known* to be below 60 dB maximum, amplifier and loudspeaker power ratings can be reduced accordingly. The amplifier should carry technical specifications for signal-to-noise ratio, linearity, and distortion. Exact figures here depend on application and are left to the acoustics specialist or sound engineer to supply.

In addition to the usual volume, tone mixing and input-output selector controls, the amplifier *must* contain special equalization controls for signal shaping. These are highly critical filter networks which by selective amplification and attenuation of portions of the overall audio frequency spectrum, *voice* or *equalize* a system after installa-

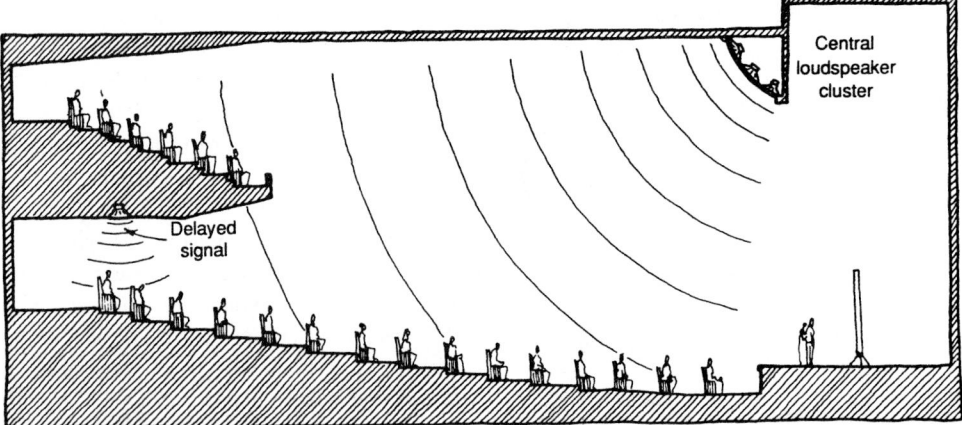

Fig. 27.27 *Loudspeaker system using delayed signal to under-balcony area.*

tion. Equalization is a sine qua non of a good sound system; without it the system will howl, sound rough, give insufficient and poorly distributed gain and sound level, and generally bad sound. Essentially, voicing tailors the system to the acoustic properties of the space. A system not equipped for equalization is not a professional system, and results will verify it. Furthermore, the specification must provide for the services of a competent sound engineer to perform the equalization after installation and construction is complete.

Another control frequently required in theater systems is a delay mechanism or circuit that can introduce a time delay into a signal being fed to a loudspeaker. Figure 27.27 shows a sound system that covers a majority of an auditorium from a central loudspeaker cluster. The under-balcony seating areas are hidden from the central cluster and receive the reinforced sound from distributed loudspeakers in the under-balcony soffit. To provide directional realism, the signal to the under-balcony loudspeakers must be delayed to allow the weaker signal from the central speakers to arrive first. Delay is necessary because electrical signals travel at the speed of light, whereas sound is much slower (one-millionth the speed, approximately). With this arrangement, sound will seem to come from the source, and the directivity so necessary to realism is maintained.

(c) Loudspeakers. These are the heart of any sound system and obviously must be of the same high quality as the remainder of the system.

Indeed, system economies will show up much more quickly in loudspeaker performance than in any other component. Selection of speakers is a complex, technical task not within our scope. Nevertheless, a few general remarks are in order. The best systems with traditional components use central-speaker arrays consisting of high-quality, sectional (multicell), directional, high-frequency horns, and large-cone woofers. These assemblies are frequently very large, and the architect should be aware of the dimensions that must be accommodated. Smaller units with folded horns can be used, at a sacrifice in low-frequency response. If only speech is to be reproduced, these will perform adequately. Distributed systems use small (4–12 in. diameter) low-level speakers, ceiling mounted, and firing directly down.

Recent developments in loudspeaker design have produced units much smaller than previously required for high-power high-quality low-frequency sound reproduction. Here again the considerations are highly technical and speakers should be supplied under a performance specification that guarantees user satisfaction subject to specified test procedures.

27.18 Loudspeaker Considerations

Loudspeaker system design and placement must be coordinated with the architectural design. The two principal types of loudspeaker systems are central and distributed. The loudspeakers in a conventional central system are a carefully designed array of directional high-frequency units combined with less directional low-frequency units, placed above and slightly in front of the primary speaking position. In most theaters, this location is just above the proscenium on the centerline of the room. Located in this position, the system provides directional realism and is simple in its design.

A distributed loudspeaker system consists of a series of low-level loudspeakers located overhead throughout the space. Each loudspeaker covers a small area, in a manner similar to downlights. This type of system is used in low-ceiling areas where a central loudspeaker cluster could not provide proper coverage. It also can be used for public address functions if directional realism is not essential, in spaces such as exhibition areas, airline terminals, and offices. In public areas such as

transportation terminals, where the absence of absorptive material makes such spaces highly reverberant, particular care must be taken in speaker positioning and volume levels. Failure to do so will result in the unfortunately very common condition of extremely loud yet unintelligible speech.

Distributed loudspeaker systems provide flexibility for use in spaces where source and listener locations vary according to the use of the space, since loudspeakers can easily be switched to provide proper coverage. In general, a listening position should receive sound from only one loudspeaker. Systems that cover seating areas with signals from several scattered loudspeakers will increase the loudness of the sound but tend to produce garbled speech. This rule is the principal reason that the arrangements shown in Fig.

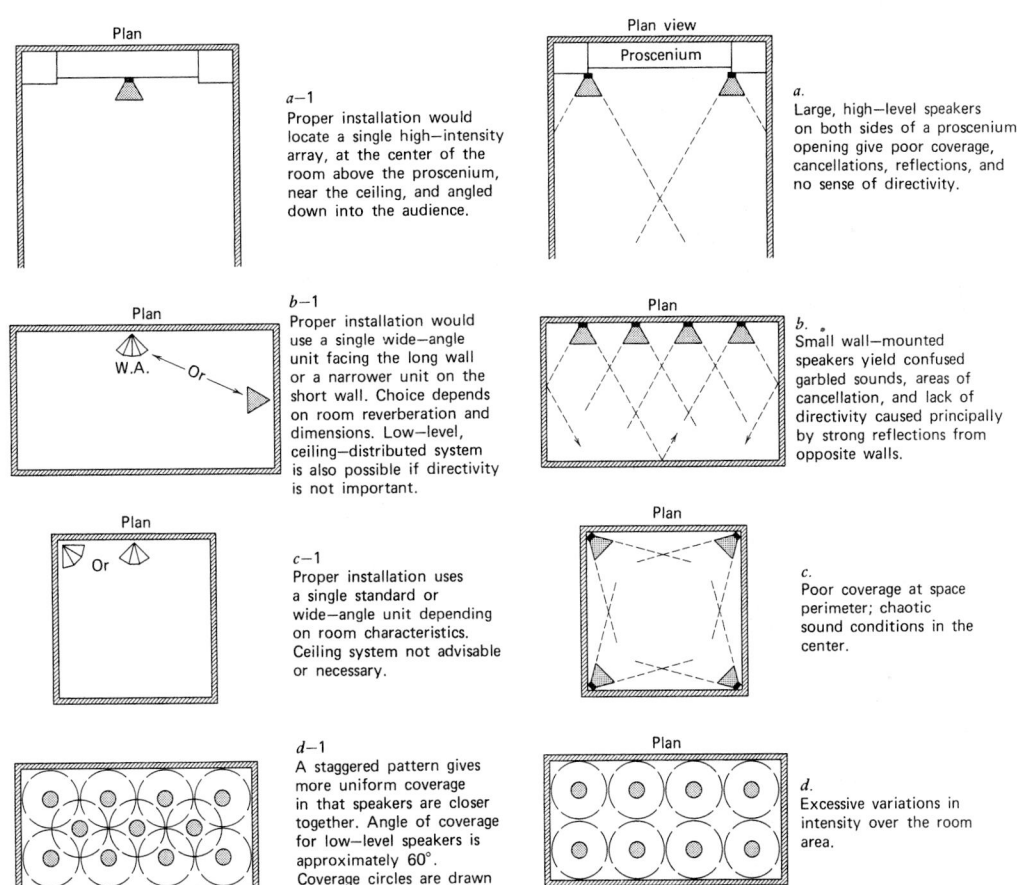

Fig. 27.28 *Poor layouts* (a) *to* (d) *and good layouts* (a-1) *to* (d-1). *Wide-angle speakers are available to fit most needs. Ceiling speakers give effective coverage on a 60° cone. Working plane is taken to be 4 ft AFF for seated audience and 6 ft AFF for standing listeners.*

ACOUSTICS

27.28(*a–d*) will guarantee a bad job. The common practice of placing one loudspeaker on either side of a proscenium opening [Fig. 27.28(*a*)], or rows of speakers on one or both sides of a room [Fig. 27.28(*b*)], is particularly to be deplored.

Essential parameters for location and design of the sound system control position creates problems for the architect. The sound system operator must be within the coverage pattern of the loudspeakers. For proper operation, he or she should be able to hear the sound as it is heard by the audience. Some current auditorium designs locate sound system controls within the audience seating pattern (see Fig. 27.24). Others place a control room with a completely open wall or a large window at the rear of the auditorium. Monitor loudspeakers and earphones are inadequate substitutes for actual listening within the auditorium. In churches the control equipment can usually be located at the rear of the congregation area.

REFERENCES

See complete list of acoustics references at the end of Chapter 28.

28

BUILDING NOISE CONTROL

NOISE REDUCTION

28.1 Principles of Noise Reduction

Noise control in buildings is comprised of three components:

1. Reduction of noise generation at the source by proper selection and installation of equipment.
2. Reduction of noise transmission from point to point (along the transmission path) by proper selection of construction materials and appropriate construction techniques.
3. Reduction of noise at the receiver through acoustical treatment of the relevant spaces to meet the noise criterion criteria developed in Chapter 26.

Speech privacy is achieved by manipulation of all of the above plus the use of masking noise where necessary.

Noise reduction is essentially the science of converting acoustical energy into another, less disturbing form of energy—heat. Since the amounts of energy involved are minute—130 dB corresponds to 1/1000 of a watt, or 0.003 Btu/h—the heat produced is completely negligible. This energy conversion is accomplished by absorption of the sound energy by the room contents and wall coverings, and by the structure itself. The former controls noise levels *within* a space and the latter noise transmission between spaces. The reasons for this will become clear as our discussion proceeds.

ABSORPTION

Absorptive noise control treatment of a room will affect the reverberant noise level *within* that room but will have minimal effect on the noise level in adjoining spaces. Refer to Fig. 28.1 for a graphic presentation of this fundamental fact. The best

that can be accomplished with acoustic room treatment is elimination of the reverberant field, that is, to make the intensity at the room boundaries what it would have been in free space, as in Fig. 28.1 (*d*). (Even this is extremely difficult; the actual field at the wall would be above 72 dB, except in a completely anechoic chamber.) Adding further wall or other acoustic absorbent as in Fig. 28.1 (*e*) does nothing in the room itself and has minimal effect on the overall transmission loss, since the transmission loss in the acoustic material itself is very low, as can be seen in Fig. 27.1.

The subject of acoustic energy absorption and absorptive materials, and their effect on room acoustics including noise reduction, is treated extensively in Sections 27.1 through 27.10, for porous absorptive materials. As stated there, two other types of absorptive material are in use, although much less common. They are panel resonators and cavity resonators.

28.2 Panel and Cavity Resonators

Panel resonators are built with a membrane such as thin plywood or linoleum in front of a sealed air space generally containing absorbent material. The panel is set in motion by the alternating pressure of the impinging sound wave. The sound energy is converted into heat through internal viscous damping. Panel resonators are used where efficient low-frequency absorption is required and middle- and high-frequency absorption is unwanted or provided by another treatment (Fig. 28.2). Panel resonators are often used in recording studios.

A *volume* or *cavity resonator* (Helmholtz resonator) is an air cavity within a massive enclosure, connected to the surroundings by a narrow neck opening. The impinging sound causes the air in the neck to vibrate, and the air mass behind causes

(a) TV set in free space produces 75-db sound level, which drops 6 db for each doubling of distance. Attenuation by inverse square law (see Section 26.7).

(b) TV still produces 75 db. In the free field, sound drops to 72 db but builds up to 74 db at the wall due to reverberant field reinforcement (see Figs. 27.6 and 27.8). Wall attenuation is 30 db. Sound on other side of the wall is 74 − 30 = 44 db.

(c) Acoustic tile ceiling acts to reduce room reverberant field. Free field is extended. Level at wall is 73 db. Level in second space is 73 − 30 = 43 db.

(d) Entire room is acoustically treated, effectively eliminating reverberant field. Room is "dead." Level on second side of wall is 72 db less acoustic tile loss, less wall loss (that is, 72 − 2 − 30 = 40 db).

(e) Add another 2¼ in. of acoustic wall treatment. Room is "dead." Level at wall 72 db. Level in second space = 72 − 4 − 30 = 38 db.

Fig. 28.1 *Graphic description of the effect of porous absorptive material on fields and SPL in adjoining spaces.*

the entire construction to resonate at a particular frequency. At that frequency absorption approaches unity, and drops fairly sharply above and below this frequency (see Fig. 28.2). By adjusting neck opening and cavity dimensions, the unit can be tuned to resonate at different frequencies. This makes it extremely useful when a major single frequency is present, as with 120-Hz transformer hum. Concrete blocks can be used as resonators by tuning their cavities. Their absorption characteristic over the entire frequency band is improved by adding absorptive material in the cavities. The

blocks also serve as standard concrete construction blocks. See Fig. 28.3 for typical block details and Fig. 28.4 for a common application. Fibrous filler can be used in the block to increase high-frequency absorption.

28.3 Acoustically Transparent Surfaces

The soft porous material of which acoustic absorbents are constructed may be covered with perforated metal or other materials, to act as physical

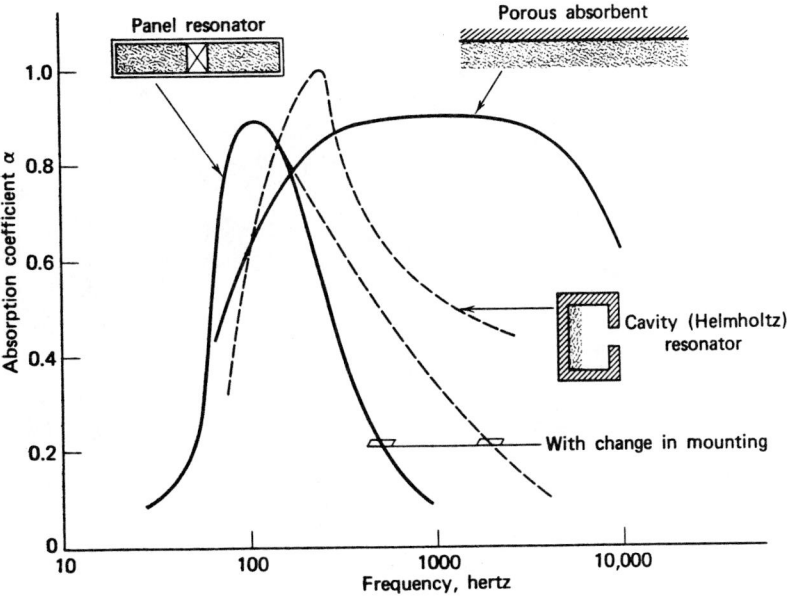

Fig. 28.2 *Typical absorption curves for the three major types of absorbers. Absorption characteristic of each type can be changed by varying design, as discussed in the text.*

Sound Absorption Coefficients — Type RSC

Size	Type	Surface	Exposed Slots/ Cavities	FREQUENCY – Hertz																	NRC
				125	160	200	250	315	400	500	630	800	1000	1250	1600	2000	2500	3150	4000	5000	
6"	RSC	PAINTED	2/3	.48	.70	.93	1.14	1.05	.97	.91	.84	.75	.76	.77	.70	.67	.68	.56	.51	.59	**.85**
8"	RSC	PAINTED	2/4	.48	.85	1.17	.99	.90	.88	.98	.79	.62	.58	.60	.61	.70	.69	.70	.64	.51	**.80**
12"	RSC	PAINTED	2/4	.57	*	*	.76	*	*	1.09	*	*	.94	*	*	.54	*	*	.59	*	**.85**

(c)

Fig. 28.3 *Details of a Helmholtz resonator design using concrete blocks, with tuned block cavities as the resonating chambers. (a) Six-in. block with three sequential cavities, tuned with a metal divider (septum). (b) Twelve-in. block with four cavities and added absorptive filler. (c) Sound absorption coefficients. (Courtesy of The Proudfoot Co., Inc.)*

Fig. 28.4 Typical application of concrete block resonators of the type shown in Fig. 28.3, in a school gymnasium. (Courtesy of The Proudfoot Co., Inc.)

protection and as stiffeners. These coverings are generally acoustically transparent except at higher frequencies. The frequency at which a noticeable reduction in absorption occurs for a perforated metal cover with circular holes can be estimated as

$$f = \frac{40p}{d} \tag{28.1}$$

where f = frequency, Hz
         p = percentage of open area
         d = diameter of holes, in.

Thus for ¼-in. holes and 60% open area, which is a typical commercial construction,

$$f = \frac{40(6)}{0.25} = 9500 \text{ Hz}$$

which is very high and generally not of major concern. It is always preferable, given a fixed percentage of open area, to use covers with small holes rather than large ones since, as is obvious from the formula, this raises the frequency at which absorption drops. It is also desirable to stagger the holes as this improves absorption. An

open weave fabric is almost completely transparent to sound and is often used as a decorative cover on absorbent wall coverings.

28.4   Absorption Recommendations

To summarize, absorption techniques are useful and effective:

1. To change room reverberation characteristics.
2. In spaces with distributed noise sources such as offices, schools, restaurants, and machine shops.
3. In spaces with hard surfaces and little absorptive content.
4. Where listeners are in the reverberant field. (No amount of absorptive material can reduce intensity levels in the free field.)

Concentrated noise sources are better handled by individual equipment enclosures than be room treatment, since enclosures reduce the amount of noise emitted into the room, which room surface treatment cannot do. In addition to ceiling tiles and wall panels, acoustic absorbent material can be applied in the form of a sprayed-on finish (acoustic plaster), suspended unit absorbers, carpets, draperies, and the like.

28.5 Characteristics of Absorptive Materials

a. *Acoustic tile* is available in size multiples of 12 in., from 12 in. × 12 in. up to 48 in. × 96 in., in a huge variety of patterns and finishes, including units with fire ratings. Installation methods include lay-in, nailing to furring strips and gluing. Tile materials are generally mineral fiber or faced fiberglass, with NRC absorption ratings in the range of 0.45 to 0.75 for mineral fiber tiles and up to 0.95 for fiberglass. The latter, which are frequently used in open-office applications have

articulation class (AC) ratings of 170 to 210 (see Section 28.20(d). Tiles for application in high humidity areas should be certified by the manufacturer for that use (see Fig. 28.4).

b. *Perforated metal-faced units* are usually installed in a lay-in suspension ceiling although nailing to furring strips is possible. Units range in size from 12 in. × 24 in. through 24 in. × 96 in. and larger on special order. The fill material is either wrapped mineral wool or fiberglass, with NRC ratings somewhat lower than acoustic tile of the same material. Metal finish is generally baked enamel in a range of colors. The units are applicable to all spaces and have the advantage of easy cleaning, high luminous reflectivity and incombustibility. With the acoustic backing removed, the perforated unit can be used for air return (Fig. 28.5).

c. *Acoustic panels* (boards) are made of treated wood fibers, bonded with an inorganic

(a)   *(b)*   *(c)*   *(d)*

Fig. 28.5 *Four of the many types of acoustic ceiling tiles available are illustrated.* (a) *and* (b) *are 2 ft × 4 ft × 1½ in. fiberglass ceiling tiles with an acoustically transparent, plastic, humidity shield facing, and a foil back for good sound dispersion in the plenum of open-plan offices.*
(a) *Tile has 1.0 NRC, 200 AC, 0.89 light reflectance and a thermal* R *factor of 4.0.*
(b) *Tile has similar acoustic and thermal characteristics, but due to its coarser surface, its light reflectance factor is 0.85. Both tiles are applicable to open plan offices and are usable with indirect lighting systems.*
(c) *Perforated metal cassette panels are used in lobbies, corridors, entries, outdoor canopies and soffits and areas with high humidity. They are usable in HVAC systems for air returns, and when supplied with an acoustic infill (mineral fiber or fiberglass), have good acoustic properties.*
(d) *General purpose, coarse texture, mineral fiber ceiling tile is rated 0.70 NRC, 25 CAC, and 0.73 light reflectance. It can be used acoustically in offices, libraries and restaurants and public spaces. (Four photos courtesy of Armstrong World Industries.)*

(a)

(b)

Fig. 28.6 *Acoustic panel material made of pressed organic fiber with an inorganic binder can be used as ceiling tile (a) or wall panel (b). The latter has cloth covering as an outer finish. (Photos courtesy of Tectum.)*

cement binder, in sizes ranging from 12 in. × 24 in. to 24 in. × 120 in., thicknesses from 1 in. to 3 in. and smooth or "shredded" finish. Panels are used in ceiling suspension systems, or nailed or glued when applied to walls and structural ceilings. Their principal advantage is high structural strength which makes them applicable to installations requiring both acoustic treatment, strength and abuse-resistance. A second advantage is an excellent flame-spread rating. Typical applica-

tions include full span corridor ceilings, long span direct-attached ceiling finish, wall panels in school gyms and corridors and the like. NRC ratings range from 0.40 to 0.70. Acoustic panels are usually resistant to humidity but installations in high humidity spaces should be confirmed with the product manufacturer. This is particularly true for panels with "reveal" edges (Fig. 28.6).

d. *Acoustic plaster,* a material comprising a plaster-type base into which is introduced fibrous or light aggregate, is useful for application to curved and other non-linear surfaces, in thickness up to 1.5 in. Its advantage is ease of application and a fire rating; its disadvantage is its inability to resist even mild abuse and its inapplicability to humid atmospheres. The noise absorption characteristics of acoustic plaster vary widely with composition, thickness, and application technique, and are generally below those of acoustic tile and panels.

e. *Sound blocks,* baffles and hanging panels are simply masses of absorptive acoustic material that achieve absorption coefficients in excess of 1.0 by exposing more than a single absorptive surface to the impinging sound. In all cases, because of their prominent appearance, necessitated by their shape, they obtrude into the space and frequently constitute the space's major architectural element. This is especially true of hanging baffles (Fig. 28.7).

f. *Wall panels* consist of a wood or metal backing on which is mounted a mineral fiber or fiberglass substrate and a fabric covering. NRC coefficients vary from 0.5 for direct-mounted 1 in. mineral fiber substrate to as high as 0.85 for strip-mounted 1½ fiberglass substrate panels. Wall panels are available in widths from 18 in. to 48 in. and lengths up to 120 in. Fabric coverings generally carry a fire-spread rating. These panels are frequently used in offices, conference rooms, auditoriums, theaters, teleconferencing centers and educational facilities (see Fig. 28.8).

g. *Resonator sound absorbers* are available in a wide variety of sizes and shapes. Although some designs are available "off the shelf" most are tailored to the specific acoustic needs of the job, using one of several standard designs. One fairly common type is shown in Fig. 28.9. In general

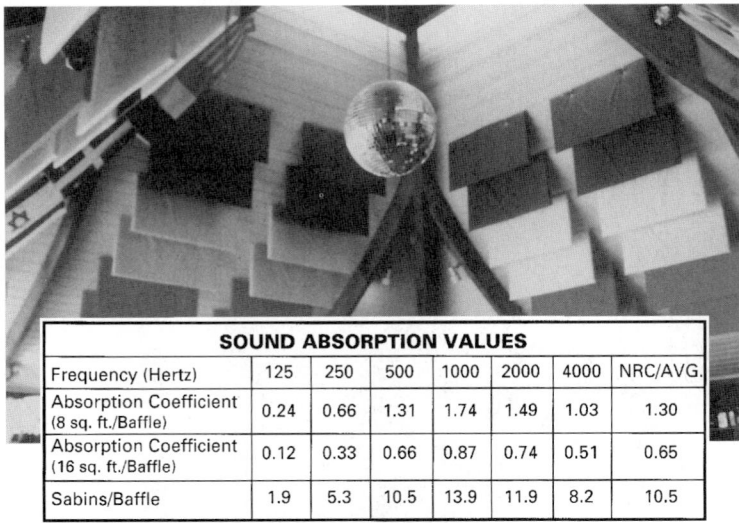

SOUND ABSORPTION VALUES							
Frequency (Hertz)	125	250	500	1000	2000	4000	NRC/AVG.
Absorption Coefficient (8 sq. ft./Baffle)	0.24	0.66	1.31	1.74	1.49	1.03	1.30
Absorption Coefficient (16 sq. ft./Baffle)	0.12	0.33	0.66	0.87	0.74	0.51	0.65
Sabins/Baffle	1.9	5.3	10.5	13.9	11.9	8.2	10.5

Fig. 28.7 *Application of hanging acoustic baffles to reduce reverberation in high pyramid-shaped ceiling of a restaurant. The baffles are 1½-in. thick fiberglass, with a tough fired rated cover. Sound absorption values are given in the accompanying table. (Courtesy of The Proudfoot Co., Inc.)*

Fig. 28.8 *Acoustic wall panel consists of a rigid backing covered with either mineral fiber or fiberglass and finished with any of a wide selection of fabrics. NRC averages 0.6 for mineral fiber substrate and 0.8 for fiberglass, for contact mounting (A mounting) and 0.7/0.9 when the panels are mounted on 2 cm furring strips (D-20 mounting). The latter mounting is particularly effective in increasing absorption below 500 Hz. (Photo courtesy of Armstrong World Industries.)*

resonators are large and must therefore be integrated into the architectural design of a space. Exterior shapes can be altered and the units can be installed less obtrusively to fulfill this architectural requirement.

h. *Carpeting and drapery* are used to cover large acoustic reflective surfaces in a space. Carpeting can be selected in almost any degree of density, looping and depth, plus an additional depth of padding, to produce a high degree of absorption in mid and high frequencies (see Table 27.1). In general, absorption is proportional to pile height and density and increases when the carpet is installed on a thick fibrous pad. Where drapery is not feasible and wall panels are impractical, carpeting can be installed on walls. In such instances, installation on furring strips with an enclosed air space behind will increase absorption over the entire acoustic spectrum and especially at low frequencies, where direct contact installation exhibits poor absorption.

Draperies are essentially acoustically transparent and only provide appreciable absorption in the middle and upper frequencies with heavy, dense "fuzzy" fabrics, particularly when draped with a high degree of fold. As with carpeting, absorption increases over the entire spectrum when a heavy folded drapery encloses an air space between itself and the wall. Approximate absorption figures are given in Table 27.1.

ACOUSTICS

SOUND ABSORPTION VALUES							
Frequency (Hertz)	125	250	500	1000	2000	4000	NRC/AVG.
Absorption Coefficient (7.87 sq. ft./Absorber)	0.46	0.74	1.08	1.21	1.27	1.26	1.08
Sabins/Absorber	3.65	5.84	8.48	9.51	9.98	9.90	8.45

(c)

Fig. 28.9 (a) *Resonator/absorber unit consists of a molded fiberglass cylinder, 12-in. diameter and 24-in. length, containing a metal resonator at each end for low-frequency absorption and a fill of acoustic fiberglass for full spectrum absorption.*
(b) *The unit which weighs about 6 lb is suspended on thin wires or chains that are essentially invisible when the unit is mounted 10 ft AFF or higher.)*
(c) *A typical application is a school gymnasium. Because of size, the units constitute a major architectural element. By color selection and hanging patterns they can be effectively used as such. The absorption characteristics of these units are listed in the table, as a function of frequency. (Courtesy of The Proudfoot Co., Inc.)*

SOUND ISOLATION

28.6 Airborne and Structure-Borne Sound

In contradistinction to the preceding material which was concerned with the phenomenon of in-room sound reduction by absorption, the following sections will discuss the characteristics of sound *transmission* between enclosed spaces. A distinction is often made between airborne and structure-borne sound, although in reality they differ only in the origin of the sounds. *Airborne*

sound originates in a space with any sound-producing source and although it changes to *structure-borne* sound when the sound wave strikes the room boundaries it is still referred to as airborne because it originated in the air. *Structure-borne* sound is generally understood as energy delivered by a vibrating or impacting source directly contacting the structure. Hence a child crying in the adjoining apartment is contributing airborne sound; the same child bouncing a ball on the floor is creating structure-borne sound, in this case by impact. The building pumps that were installed without proper damping mounts create structure-borne sound by vibration.

In reality, all sound transmission is both airborne and structure-borne since, once having entered the structure, the sound travels along the structure and causes the structure to vibrate, in turn generating airborne sound. Figure 28.10 should assist in understanding this action. In Fig. 28.10 (*a*), the sound is airborne, originating in the air on one side of the partition. The incidence of sound energy *causes the partition to vibrate,* generating sound on the other side. Sound does not "pass through" unless an air path exists. If the partition is air tight then the sound energy causes the structure to become a secondary source. The partition vibrates primarily in the direction of the sound, that is, in the vertical plane. It also vibrates in other modes causing some sound energy to pass into the floor and ceiling, depending on the mode of attachment. This energy becomes structure-borne sound.

In Fig. 28.10 (*b*), the process is similar, but reversed. Energy is introduced into the structure directly and efficiently by mechanical contact, that is, by vibration and impact. Sound travels along the structure, as shown, and by causing the structure to vibrate, creates *airborne sound.* In a structure with rigid wall-to-floor connections, these sounds are clearly heard throughout the structure. It is a common misconception that in a (relatively) massive concrete structure with masonry walls, such as a multi-story residential building, light impact such as footfalls will not be transmitted. On the contrary, the rigidity of the structure, and in particular the rigid airtight connections between partitions, floors and ceilings provide an excellent path for structure-borne sound. Only impact absorption by heavy carpeting, will attenuate the sound of footsteps, and resilient floor-wall

connections will attenuate structure-borne sound.

Airborne sound (originating in the air) is generally much less disturbing than structure borne, since its initial energy is very small and it attenuates rapidly at boundaries. Structure-borne sound is generally at a much higher energy level initially and attenuates slowly as it travels through the structure, thereby causing disturbance over large sections of a building. This disturbance is magnified by the "sounding board effect."

We are all familiar with the fact that a tuning fork must be held up to the ear to be heard directly, but if its handle is placed on a table the sound is amplified. This action is not really amplification but an increase in the efficiency of energy transfer. In general, the efficiency of a radiator is proportional to the ratio of its surface dimensions to the sound wavelength. A tuning fork vibrating at concert A (440 Hz) of wavelength $2\frac{1}{2}$ *ft* cannot efficiently couple its energy into the air. It is simply too small. By placing the instrument on a table whose dimensions are approximately one wavelength, we permit it to transfer its energy efficiently; hence the "amplification." The same effect can be extremely troublesome in structure-borne sound. A vibrating pump itself makes little sound. However, it transfers a large amount of energy into the structure, which will appear as audible sound at each partition, floor, and wall that is rigidly coupled to the structure. Soft (damping) connections prevent energy transfer, thereby greatly attenuating the transmission of sound energy into connecting efficient radiating surfaces; hence the desirability of such flexible connections.

Airborne sound changes direction easily (diffracts), with low frequencies being most flexible in this regard. Structure-borne sound travels much more rapidly than airborne sound (see Table 26.1) and with attenuation as low as 1 dB per kilometer. A sound traveling along a massive structure will radiate outward from the structure only minimally (although enough to be very annoying) because the large mass minimizes vibration in that direction. Thus in Fig. 28.10 (*b*), noise from impact on the floor above will be louder (for equal impact energy) than noise from machines below because the former generates sound directly downward while the latter introduces energy into the entire network of parallel paths.

The sections immediately following deal with airborne sound and the means for controlling it

Transmission of Impact and Structure - Borne Noise.

(b)

Fig. 28.10 (a) *Airborne sound (noise) is so called because it originates in air; its energy level is low. Here the loudspeaker is the source of acoustic energy which is converted to a vibration of the partition (shown much exaggerated), which in turn becomes a secondary source of airborne sound for the adjoining space. A small amount of energy is reradiated by the ceiling and floor of both spaces directly, and indirectly via the rigid partition-to-floor and partition-to-ceiling connections.*
(b) *Structure-borne sound originates from mechanical contact between structure and vibrating or impacting sources. As such, its energy level is usually much higher than airborne sound. This energy is transmitted with little attenuation throughout the structure via the rigid partition-to-floor and partition-to-ceiling connections. The entire structure is then set into vibration as shown, converting the (large) structure borne sound energy to noise throughout the building.*

Fig. 28.11 *A simple case of airborne sound transmission between adjacent rooms through a common barrier. With a sound source in one room, the transmitted sound level is dependent not only on the transmission loss of the barrier but also on the area of the barrier and the receiving-room absorption. The background noise level in the receiving room determines whether the transmitted sound will be noticed. See Section 28.18 and Figs. 28.36 and 28.37.*

(Fig. 28.11). Impact noise (structure-borne sound) is covered in Sections 28.22 to 28.34.

AIRBORNE SOUND

28.7 Transmission Loss (TL) and Noise Reduction (NR)

The transmission loss (TL) of a barrier is the ratio, expressed in decibels, of the acoustic energy reradiated by the barrier to the acoustic energy incident on it. This number is a figure of merit for the sound-isolating quality of the wall itself and is obtained from controlled laboratory tests. (In Europe, transmission loss is referred to as *sound reduction index, R.*) However, the number that is of greater importance to the building designer is the actual noise reduction between two spaces separated by a barrier, that is, the action of the barrier in situ. That noise reduction is defined as the difference between the intensity levels in the two rooms, that is,

$$NR = IL_{room\,1} - IL_{room\,2} \qquad (28.2)$$

and is related to the TL of the barrier by the expression

$$NR = TL - 10\log\frac{S}{A_R} \qquad (28.3)$$

where NR = noise reduction, dB
TL = barrier transmission loss, dB
S = area of barrier wall, ft^2 (m^2)
A_R = total absorption of *receiving* room, sabins, ft^2 (m^2)

We see, therefore, that noise reduction and transmission loss are not equal but are related by the size of the dividing wall S and the absorption characteristic of the receiving room, A_R. A moment's thought will confirm the logic of this relation. When sound energy impinges on the wall, the wall in turn becomes the sound source, radiating into the receiving room. Therefore, the amount of sound energy transferred is proportional to the (log of) area S of the common wall between the two spaces.

The sound level in the receiving room is related to its own reverberance (absorption characteristic A_R) as we have seen repeatedly. Thus, if the receiving room is a reverberant, live space, A_R is low and NR is less than TL. Conversely, if the receiving room is dead, A_R is large and noise reduction can be greater than transmission loss, depending on the ratio of barrier wall size to room area (see Fig. 28.11). In lieu of calculating, the following rule of thumb can be use:

1. For a live receiving room,
$$NR = TL - 1 \text{ dB}$$

2. For a medium receiving room,
$$NR = TL + 4 \text{ dB}$$
3. For a dead receiving room,
$$NR = TL + 7 \text{ dB}$$

The extreme case of "deadness" of the receiving room is one with no walls, that is, sound transmission from inside to the exterior. In such cases, NR exceeds TL by 10 to 15 dB, depending on the size of the exterior opening and the point outside where IL is measured. To acquire facility with sound-isolation techniques, the reader must become familiar with the relationship of transmission loss to the barrier walls' physical characteristics, its mass, rigidity, material of construction, and method of construction and attachment. These considerations are the subject of the following sections.

Note: The reader is cautioned to be careful with the term *noise reduction* (NR), since a similar term—*noise reduction coefficient* (NRC)—also

exists that is completely unrelated. The latter (Section 27.10) is very poorly named.

28.8 Barrier Mass

Sound transmission between spaces requires that the barrier be set into vibration by the incident sound energy. Although this was stated above, we repeat it here to emphasize the fundamental importance of this simple statement. (We are, of course, referring to a barrier that is impervious to air—i.e., a solid barrier. Otherwise, the moving air molecules carrying the sound will simply pass through with minimal transmission loss.) The impinging sound energy acts as a force on the wall. Since $F = MA$, the larger the mass, the less it will vibrate. When other factors (particularly angle of incidence) are taken into account, the resultant relationship is known in acoustics as the *mass law*. It states that for a nonporous, homogeneous structure of low *stiffness*, the sound transmission

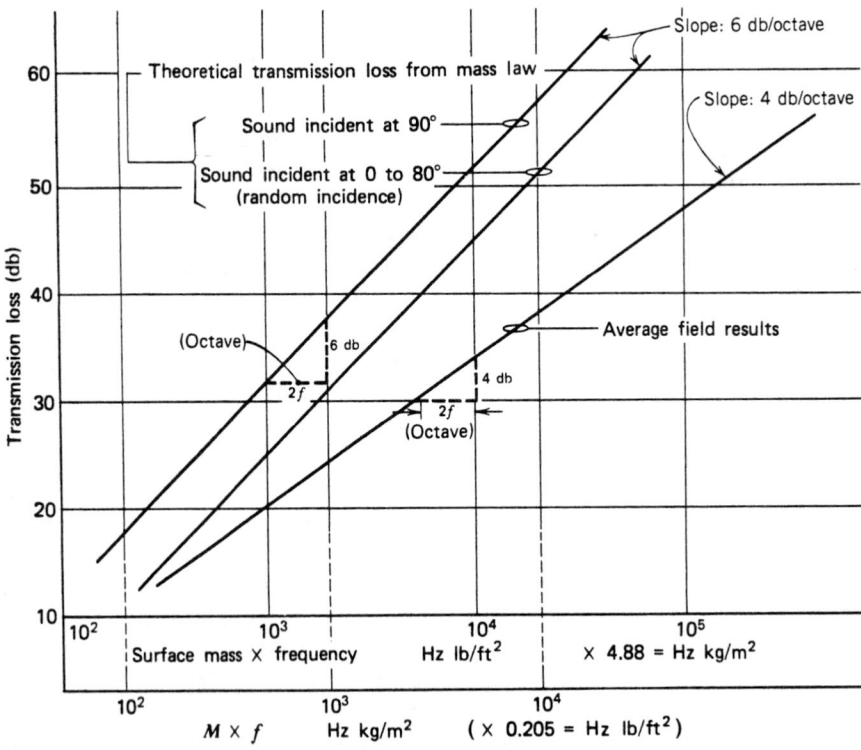

Fig. 28.12 *Graphic representation of mass law action in attenuation of transmitted sound. Perpendicular (90°) sound incidence results in maximum transmission loss. Field or random incidence (0° to 80°) is approximately 6 dB lower, but maintains the 6 dB per octave slope. Field results are lower due to flanking and stiffness effects but the slope of the curve averages only 4 dB per octave (frequency doubling).*

loss is proportional to the logarithm of surface mass (the weight of the wall per unit of surface area) and to the frequency of vibration. Thus doubling the mass or frequency will, theoretically, cause an increase of 6 dB in the transmission loss or, stated otherwise, the slope of TL versus the frequency times mass (fM) curve is 6 dB. Figure 28.12 is a graphic representation of mass law operation. With sound incident at 9°, maximum energy is imparted to the barrier and the entire mass resists, resulting in maximum transmission loss. In practice, however, sound is incident from 0° to 80° (called *field* or *random incidence*) reducing the mass effect, but keeping the slope at 6 dB per octave. Due to nonhomogeneity, porousness, and stiffness, actual field results indicate transmission losses nearer 4 dB per octave, as shown by the lower curve in the figure.

28.9   Stiffness and Resonance

The stiffness of a barrier is a function of its material composition and the rigidity of its mounting. The former depends on its internal cohesiveness—that is, its modulus of elasticity—and the latter depends on its boundary restraints—whether the barrier is tightly or loosely held. A homogeneous material of high Young's modulus (such as steel) has great cohesiveness between its molecules. As soon as one molecule is set in motion by the incident sound energy the motion is passed to the next, and so on, making the material an excellent sound conductor. Homogeneous materials with low modulus of elasticity have high internal damping (motion of molecules is not transmitted well), and they are good sound insulators. Composite materials such as concrete and organic materials such as wood do not conform to these general rules.

Rigidity of mounting can be likened to a drumhead—the tighter it is stretched, the better it resounds. Rigidity (stiffness) in a panel barrier resists damping and assists vibrations, making it a good transmitter, and conversely a poor noise transmission insulator. As a result, a material such as lead, which has high mass and low modulus of elasticity, and resists rigid mounting, is an excellent sound attenuator.

The effects of stiffness and mass both vary with frequency, unfortunately in opposite directions. Stiffness acts to reduce transmission loss as frequency increases while, as we have seen, mass acts to increase it; the combined effect is shown in Fig. 28.13. Therefore, stiffness is most effective at low frequencies and mass at high frequencies. At very low frequencies the mass and stiffness effects negate each other, giving the resonance dips shown. Beyond approximately 200 Hz most common wall construction enters the mass law range and continues with it until the critical frequency. Deviations from a smooth 4 to 6 dB per octave slope are due to the nonhomogeneous nature of most wall constructions. At the critical frequency the phase of incident sound waves corresponds or *coincides* with the phase of vibration (shear wave) of the barrier in such a way as to pass a large portion of the incident energy. See insert on Fig. 28.13. This shows in Fig. 28.13 as the "coincidence dip." This effect is most pronounced in thin homogeneous partitions and light stiff ones.

Critical frequency, f_c, as a function of panel thickness for common materials, is plotted in Fig. 28.14. To avoid a coincidence dip in the audible range, partitions can be either very heavy and/or very stiff, which greatly decreases the critical frequency, or heavy and limp (resilient), which greatly increases the critical frequency. In practical terms, cost effectiveness is heavily in favor of the latter alternative. Thus, for instance, the TL of a wood partition can be improved by grooving it to increase flexibility thereby increasing the critical frequency. The dramatic improvement in TL, resonances, and coincidence dip achieved by use of resilient mounting of a simple masonry partition is shown in Fig. 28.15. Both walls have the same weight—21+lb/ft². The solid wall A has better attenuation below 200 Hz in the stiffness-controlled range. Above that frequency the resilient-mounted partition is 10 dB better, which means that the transmitted noise is only one-half as loud. The sound transmission class (STC) (which is a figure of merit for partition sound transmission, as explained in Section 28.11) of wall A is 40; that of wall B is 51. Both show a coincidence dip at approximately 250 Hz (see Fig. 28.14), but that of wall B is shallow, that of A deep and wide. Furthermore, wall B is consis-

ACOUSTICS

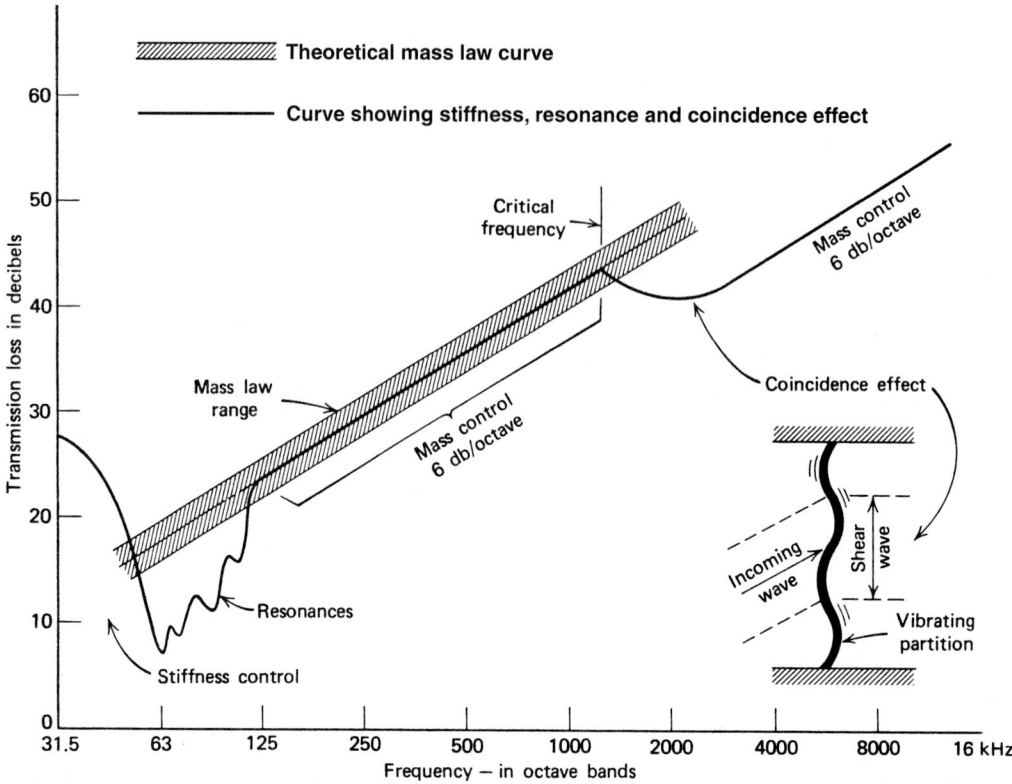

Fig. 28.13 *Partition transmission loss as a function of frequency, showing the effects of stiffness, resonance and coincidence. At frequencies below resonance, control is almost purely a stiffness function; at frequencies above critical, control is almost purely a surface mass function.*

tently better than mass law attenuation; wall A consistently worse (due to stiffness).

28.10 Compound Barriers (Cavity Walls)

Since the maximum theoretical increase in transmission loss with mass increase is 6 dB per doubling of mass, it is apparent that this method of transmission loss improvement rapidly reaches the limits of practicality. Indeed as we have seen the transmission loss of actual single homogeneous walls fall below the mass law curve. This is because mass increase brings with it stiffness increase, which as we have seen acts to *reduce* transmission loss. If, however, a barrier is constructed of two separate layers without rigid interconnection, its performance is better than the calculated transmission loss based on mass alone. Note that even the nonrigid wire ties of wall B in Fig. 28.16 lower its STC by five points. At low

frequencies, where stiffness controls the transmission loss (see Fig. 28.13), the cavity in Fig. 28.16 C acts as a rigid connection between the layers, adding stiffness and increasing transmission loss. At higher frequencies, in the mass law range, the air in the cavity acts as a damping coupling to reduce stiffness. The net result is an improvement in performance throughout the frequency range.

Transmission loss for the entire cavity wall increases with the width of the air space at the rate of approximately 5 dB per doubling. Performance can be improved still further by filling the void with porous, sound-absorbent material. This acts to further decrease the stiffness of the compound structure *and* to absorb sound energy that reflects back and forth between the two inside surfaces. The performance of cavity walls is reduced by any rigid interconnections between leaves. Thus a common stud wall with frequent rigid interconnections acts little better than a single homogeneous wall. However, a stud wall with staggered

Fig. 28.14 *Critical frequency as a function of thickness for several common materials. (Reprinted with permission from E. B. Magrab,* Environmental Noise Control, *Wiley, New York, 1975.)*

ACOUSTICS

Fig. 28.15 *Transmission loss characteristics of two equal-weight partitions with similar boundary constraints. The solid partition A is worse than the mass law due to stiffness. The resilient-mounted wall B performs better than the mass law and much better than wall A, except at the lowest frequencies. (Data extracted from* A Guide to Airborne, Impact, and Structure-Borne Noise Control in Multifamily Dwellings, *1968.)*

A: 100 lb/sq ft, STC 52

B: 100 lb/sq ft, STC 49

C: 120 lb/sq ft, STC 62

Wires ties*

Double brick wall
2 in. air cavity

Double brick wall,
6 in. cavity

A: Single 9 in. brick wall

B: 12 in. total thickness

C: 18 in. total thickness

*Without wire ties, STC rises to 54

Fig. 28.16 *Transmission loss curves showing the effect of air space on heavy wall construction. All three walls are approximately the same mass. The 2-in. airspace in wall B is not significant until the higher frequencies, whereas the large, 6-in. airspace of wall C is effective throughout the frequency spectrum. (Data extracted from* A Guide to Airborne, Impact, and Structure-Borne Noise Control in Multifamily Dwellings, *1968.)*

studs exhibits much improved performance over a single-material wall or a common stud wall. The above effects are illustrated in Figs. 28.16 and 28.17 (see also Appendix F). The effects of mass (Section 28.8), stiffness (Section 28.9) and compound barriers with and without filler are shown qualitatively and graphically in Fig. 28.18.

28.11 Sound Transmission Class (STC)

Various attempts at using a single-number average transmission loss to describe a barrier's characteristics have been made with only limited success. Indeed these averages can be misleading, since they ignore both deficiencies and proficiencies at particular frequencies. Their use,

A: 6 lb/sq ft; STC 34

$5\frac{1}{8}$"

2 × 4 Wood studs 16 in. OC

$\frac{1}{2}$" Gypsum wallboard both sides,

A ———— · — · — · — · — · ——

B: 6.2 lb/sq ft; STC 44

5"

2 × 3 Wood studs 8 in. OC

$\frac{1}{2}$" Gypsum board both sides,

B ———————————————————

C: 13.8 lb/sq ft, STC 46

$5\frac{3}{4}$"

2 × 4 Wood studs 8 in. OC

$\frac{1}{2}$" Gypsum board both sides;

0.9 in. Fiber blanket

C————————————————

Fig. 28.17 *Transmission loss curves illustrating the effect on lightweight walls of stiffness reduction and addition of absorptive material in the cavity. Curve A is a standard stud wall as found in frame construction. Curve B shows the advantage of staggered studs over the entire frequency range. The dip at 3 kHz is a coincidence dip for a single leaf. Addition of absorptive material (Curve C) improves the attenuation characteristic at both ends and is particularly useful for its low end improvement. (Data extracted from* A Guide to Airborne, Impact, and Structure-Borne Noise Control in Multifamily Dwellings, *1968.)*

ACOUSTICS

INCIDENT NOISE

TRANSMITTED NOISE

1. MASSIVE WALL   2. ISOLATED SURFACE   3. DOUBLE WALL   4. SOUND ABSORBER IN CAVITY   5. HOLES AND CRACKS

Constructions 1-4 reduce transmitted noise; holes and cracks (5) can drastically reduce a wall's transmission loss and thereby its effectiveness as a sound barrier

Fig. 28.18 *The various techniques used to increase the transmission loss of a partition are shown in illustrations 1–4 (see also text, Sections 28.8, 28.9, and 28.10). In contrast, even a very small air path through the partition as in illustration 5, can effectively destroy its effectiveness as a sound barrier. See also Fig. 28.31 (Data extracted from* A Guide to Airborne, Impact, and Structure-Borne Noise Control in Multifamily Dwellings, *1968.)*

therefore, in all but rough work is to be discouraged.

To avoid the shortcomings of averages and yet to benefit from the indisputable convenience of single-number ratings, a system of standard contours was developed in the United States called sound transmission class (STC) contours. (A similar system, conforming to ISO 717-1, 1982 is used in Europe, that uses a weighted sound reduction index R_W; see References.) In practice, the STC number for a particular barrier construction is derived by comparing actual test results measured in a series of sixteen $\frac{1}{3}$-octave bands, to the standard STC contours according to a fixed procedure. The technique is illustrated in Fig. 28.19. Figure 28.20 shows two TL curves and STC ratings of

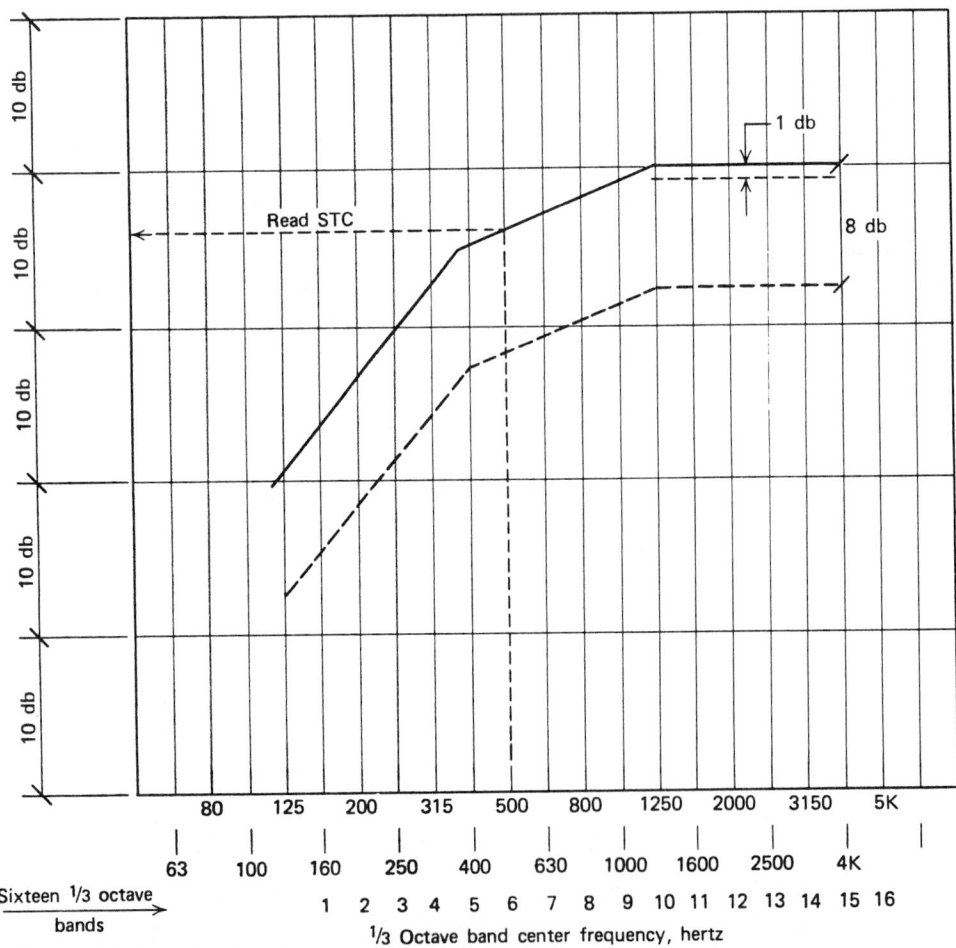

The STC is determined by comparison with a transparent overlay of Figure 27.19 on which the STC contour is drawn. The STC contour is shifted vertically, relative to the test curve, until some of the measured TL values for the test specimen fall below those of the STC contour (the solid line) and the following conditions are fulfilled:

1. The sum of the deficiencies (i.e., the deviations below the contour) shall not be greater than 32 db.
2. The maximum deficiency at a single test point shall not exceed 8 db [the broken (dashed) line beneath the STC contour].

When the contour is adjusted to the highest value (in integral db) that meets the above requirements, the sound transmission class for the specimen is the TL value corresponding to the intersection of the contour and the 500-Hz ordinate.

Fig. 28.19 *Overlay from which sound transmission class (STC) is determined graphically.*

Fig. 28.20 *Curves A and B are two different construction types with the same average TL. However, application of the STC curve criteria given in Fig. 28.19 yields an STC of only 33 for construction B because of its deep center dip, as compared to STC of 42 for construction A. (From E. B. Magrab,* Environmental Noise Control, *Wiley, New York, 1975.)*

TABLE 28.1a **Improvements in STC Rating of Stud[a] Partitions[b]**

Description	STC[c]
Basic partition: single wood studs, 16 in. on centers, ½-in. gypsum board each side, air cavity	35
Add to basic partition	
Double gypsum board, one side	+2
Double gypsum board, both sides	+4
Single-thickness absorbent material in air cavity	+3
Double-thickness insulation	+6
Resilient channel supports for gypsum board	+5
Staggered studs	+9
Double studs	+13

[a] For application to metal stud partitions, use adders as in note *b,* but begin with STC = 40 for 3⅝-in. basic partition.

[b] When using two improvements add additional +2; for three improvements add +3.

EXAMPLE: Improvements to 35 STC basic partition:

Staggered wood studs	+9
Double gypsum board, one side	+2
Single insulation	+3
Adder	+3
Total	+17
Total STC	35 + 17 = 52

[c] The STC figures are conservative. Other sources list the same constructions with 1 to 5 points higher STC.

each. Because STC fails to give credit for performance *above* the established requirements, octave-band transmission-loss data, rather than STC ratings, should be used in all critical work such as music rooms or mechanical rooms where certain particular frequencies may be dominant.

Figure 28.21 gives three standard STC contours that are of interest because they are used by the FHA to specify grades of construction. The criteria for their application are found Section 28.33. An appreciation of the degree of speech sound isolation provided by walls with different STC ratings is given in Section 28.17 (see Table 28.5). Since the subjective reaction on the quiet side depends on the background noise level there, the table gives this reaction for two NC curve levels. To assist the designer, extensive sound transmission testing has been performed on most types of standard wall and partition construction and the results published. Table 28.1 and Appendix F give descriptions of construction with typical details, transmission loss data, STC ratings, and other pertinent data.

28.12   Composite Walls and Leaks

It is frequently necessary to determine the transmission loss of a composite wall, that is, a wall with a window, door, vent opening, and the like. It should be clearly appreciated that the two materials are "in parallel," to borrow an electrical concept, and the behavior is similar to that situation. That is, the overall performance will be strongly affected by the poorer of the two, with some tem-

ACOUSTICS

TABLE 28.1b **STC Ratings of Masonry Walls**

Description	STC[a]
4-in. lightweight[b] hollow block	36
4 in. dense hollow block	38
6-in. lightweight hollow block	41
6-in. dense hollow block	43
8-in. lightweight hollow block	46
8-in. dense hollow block	48
12-in. lightweight hollow block	51
12-in. dense hollow block	53
4-in. brick	41
6-in. brick	45
8-in. brick	49
12-in. brick	54
6-in. solid concrete	47
8-in. solid concrete	50
10-in. solid concrete	53
12-in. solid concrete	56

[a] See note c, Table 28.1a.
[b] All ratings of lightweight block assume sealing with paint. Note that this reduces absorption.

MODIFICATIONS

Add sand to cores of hollow blocks	+3
Add plaster to one side	+2
Add plaster to both sides	+4
Add furring strips, lath and plaster:	
1 side	+6
2 sides	+10
Add plaster via resilient mounting:	
1 side	+10
2 sides	+15

pering of the degradation when the poorer barrier is much smaller in area than the other material. Fig. 28.22 enables us to calculate situations of this type.

Since an opening in a wall is effectively a second material of TL = 0, the curves of Fig. 28.22 can be replotted for this situation as in Fig. 28.23. Note carefully that the curves very rapidly flatten out; thus any wall with 1% open area will have a *maximum* transmission loss of 20 dB, which is all but useless as sound insulation. For this reason it is imperative that all openings be completely sealed, particularly those around doors and windows. A hairline crack degrades a wall 6 dB; a keyhole will degrade a door 3 dB, and so on. Special considerations for doors and windows are

discussed below. Care must also be taken with such common acoustic leaks as back-to-back electric outlets, pipes passing through walls, and medicine cabinets—in fact any break in the integrity of a partition. All such openings must be caulked to make an airtight joint if any appreciable degree of sound isolation is to be maintained.

Examples 28.1 and 28.2 show how barriers of high *TL* are degraded by standard openings. To maintain the integrity of a barrier, special care must be taken in the design of windows and doors, as explained below.

EXAMPLE 28.1. Given a wall 9 ft × 18 ft with a transmission loss of 52 dB at 1000 Hz, containing a 3 ft × 7 ft, 6-in. hollow core door of 22 dB transmission loss at that frequency. Find the overall *TL* of the composite wall.

SOLUTION. Refer to Fig. 28.22.

$$TL_1 - TL_2 = 30\,\text{dB}$$

$$\frac{S_2}{S_1} = \frac{3 \times 7\frac{1}{2}}{9 \times 18} \times 100 = 13.9\%$$

From curves in Fig. 28.22:

$$TL_1 - TL_c = 21.5$$

$$TL_c = 52 - 21.5 = 30.5\,\text{dB}$$

That is, a door with an area of only 14% of the entire wall reduces the *TL* of the structure from 52 to 30.5 dB—that is, from excellent to *very poor*. ∎

EXAMPLE 28.2. An exterior brick/frame wall having a *TL* of 54 at 1000 Hz, measuring 8 ft × 16 ft, is pierced by two wood frame windows, each of area 3 ft × 4 ft, with single ⅛-in. glass, with a *TL* of 34 at 1000 Hz. Find the combined transmission loss.

SOLUTION.

$$TL_1 - TL_2 = 54 - 34 = 20\,\text{dB}$$

$$\frac{S_2}{S_1} = \frac{2 \times 3 \times 4}{8 \times 16} \times 100 = 18.75\%$$

From Fig. 28.22:

$$TL_1 - TL_c = 12.5\,\text{dB}$$

$$TL_c = 54 - 12.5 = 41.5$$

Again, the result is a reduction from an excellent wall to a poor one. ∎

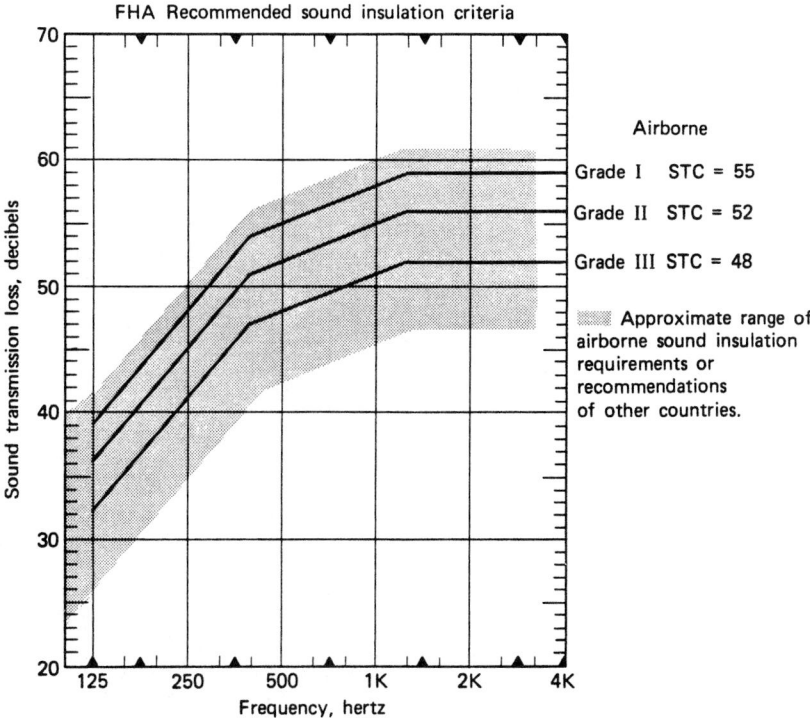

Fig. 28.21 *Sound insulation criteria recommended by the FHA. (From* A guide to Airborne, Impact, and Structure-Borne Noise Control in Multifamily Dwellings, *1968.)*

Fig. 28.22 *Transmission loss of a two-element composite barrier as a function of the relative transmission loss of the components. (From E. B. Magrab,* Environmental Noise Control, *Wiley, New York, 1975.)*

Fig. 28.23 *Effect of a hole of a given percentage of the total area on a partition of TL dB. (From E. B. Magrab,* Environmental Noise Control, *Wiley, New York, 1975.)*

28.13 Doors and Windows

As can be appreciated from the preceding section, doors and windows can in large measure determine the overall transmission loss of a wall. Since in almost every instance doors and windows have a lower acoustic transmission loss than that of the wall in which they are mounted, particular care must be taken not to further degrade performance with air leaks.

(a) Doors. Figure 28.24 gives nominal TL data for the most common types of doors as a function of frequency. Average TL for doors, that is, the arithmetic average of the octave band TLs in the range of 150 Hz to 3000 Hz, are not useful for two reasons:

· The very important low frequency attenuation data is absent.

· Sharp peaks and valleys in the curves [see for instance the 6 dB peaks at 500 Hz in Fig.

28.24 (*a*)] are unrecognized. As a result, a particularly troublesome frequency may not be sufficiently attenuated. In the absence of a complete frequency analysis, the STC rating of a door is a better indication than an average TL figure. Typical STC values are given in Table 28.2.

Conclusions that can be drawn from inspection of Fig. 28.24 (a) are:

1. Louvered doors (and doors undercut to permit air movement) are useless as sound barriers.
2. The most important step in sound-proofing doors is complete sealing around the opening. The door in the closed position should exert pressure on these gaskets making the joints airtight [see Fig. 28.24 (*b*)].

When a single door does not provide sufficient attenuation, and specially constructed high attenuation commercial acoustic doors are not practi-

---- 1 3/4" solid wood core door with
gaskets and drop closure

—— 1 3/4" hollow wood core door with
gaskets and drop closure

- — — Same hollow door, no gaskets or
closure, 1/4" airgap at sill

········ Louvered door, 25-30% open area

(a) Sound Transmission Loss of Doors

(1) DOOR JAMB GASKETS

(2) DOOR SILL CLOSURES

(b)

Fig. 28.24 (a) *Chart of typical transmission loss values for representative door constructions, as a function of frequency.*
(b) *Method for gasketing a door edge enclosure (1), and sealing the gap between the bottom of the door and the door saddle (2). (Chart and drawings extracted from* A Guide to Airborne, Impact, and Structure-Borne Noise Control in Multifamily Dwellings, *1968.)*

TABLE 28.2 **Typical STC Values for Doors**

Construction	STC
Louvered door	15
Any door, 2-in. undercut	17
1½ in. hollow core door, no gasketing	22
1½ in. hollow core door, gaskets and drop closure	25
1¾ in. solid wood door, no gasketing	30
1¾ in. solid wood door, gaskets and drop closure	35
2 hollow core doors, gasketed all around with sound lock	45
2 solid core doors, gasketed all around, with sound lock	55
Special commercial construction, with lead lining and full sealing	45–65

ACOUSTICS

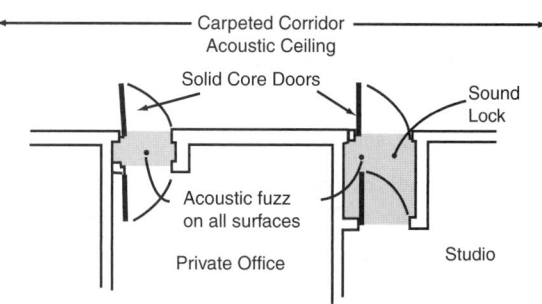

Fig. 28.25 *A two-door sound lock should increase the transmission loss into the room by a minimum of 10 dB for the small lock and 15 dB for the larger one, depending on the type, thickness and mounting technique of the absorptive material in the lock. The solid core doors must be sealed by one of the techniques shown in Fig. 28.24* (b).

cal, a simple and very effective technique is the construction of a sound lock consisting of two doors, preferably with sufficient space between to permit door swing. See Fig. 28.25. All surfaces in the sound lock should be covered completely in absorbent material, and the floor carpeted. Such an arrangement will increase attenuation across

the board by at least 10 dB and as much as 20 dB at some frequencies, depending on the shape of the sound lock and the type, amount and mounting of absorptive material in the sound lock. The two doors of the sound lock must be gasketed as explained above.

Another important consideration with respect

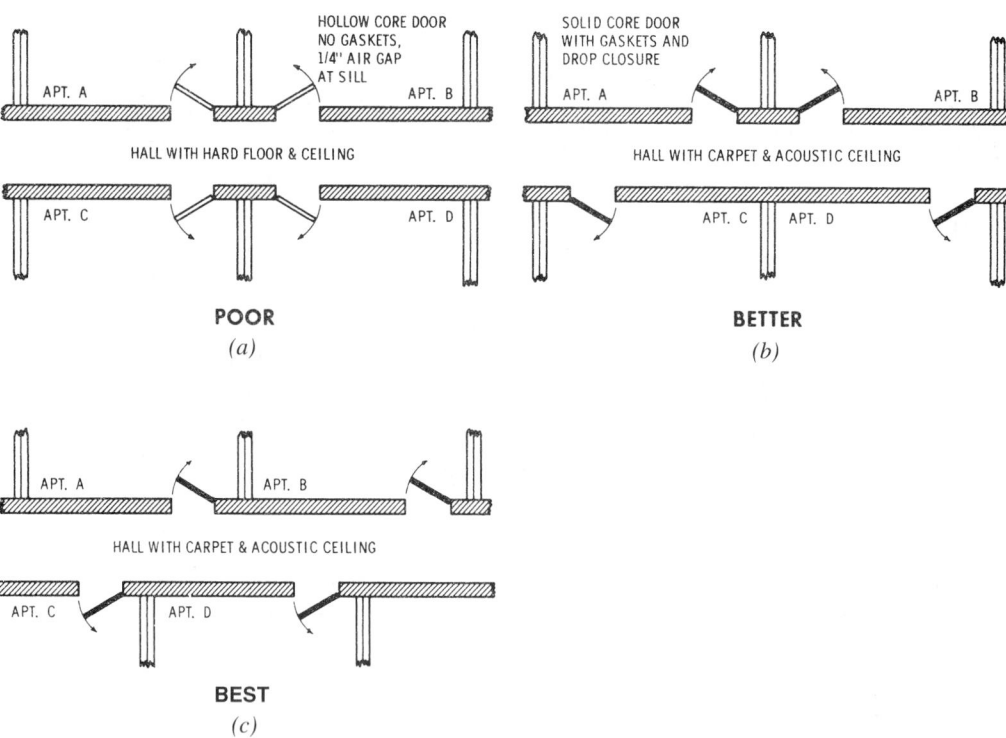

Fig. 28.26 *Proper arrangement of doors to rooms on a common corridor can diminish noise transfer in the area.*
(a) Poor arrangement because any noise emanating from one of the rooms or from the corridor has a very short and unattenuated path into the remaining rooms.
(b) Better arrangement than plan (a) because noise from any source must travel a minimum of a room width along an absorbent corridor to reach any other room. Noise to the remaining rooms is further attenuated. The weak point of this plan is a noise short circuit from Apt. A to Apt. B.
(c) This plan is best because there are no short circuits for sound travel. Although the A–C and B–D paths are slightly shorter than in plan (b), the difference would be unnoticeable. (Extracted from A Guide to Airborne, Impact, and Structure-Borne Noise Control in Multifamily Dwellings, *1968.)*

———— 1/8-in. plate glass
– – – – 1/4-in. plate glass
———— 0.45-in., 3 ply, laminated glass panel
– – – – 0.62-in., 4 ply, laminated glass panel
–··–·– 0.80-in., 5 ply, laminated glass panel

(a)

Aluminum framed windows with glass panes
isolated with neoprene gaskets;
———— two 1/4-in. glass panes, 1/2-in.
       air space.
– – – – 1/4-in. and 3/16-in. glass panes,
       2 1/2-in. air space.
–··–·– 1/4-in. and 7/32-in. glass panes,
       3 3/4-in. air space.

(b)

Fig. 28.27 (a) *Sound transmission loss frequency spectrum shows the effect of thickening glass. The ¼-in. plate shows a very sharp (10 dB) coincidence dip at 2K, making it less effective than ⅛″ plate between 1500 Hz and 2500 Hz. Further thickening with laminated glass eliminates the coincidence drop, but reaches a practical limit at about ½-in. thickness.*
(b) *Note that two ¼-in. panes with a ½-in. air space acts as a stiff thick pane and exhibits the sharp coincidence drop at 2 kHz. Larger sealed inter-pane air spaces markedly improve the acoustic isolation performance.*

to sound intrusion via doors, is the location of a door with respect to sources of unwanted sound. This is particularly important in multiple residence buildings of all types including private homes, apartment houses, dormitories, hotels and rooming houses. The same principal applies to commercial spaces where numerous private spaces such as offices open on to a common lobby or foyer Fig. 28.26 shows the effect of improvements in door placement.

(b) Windows. Windows are critically important to block exterior noise, and all the more so, since exterior wall construction is generally of high STC, making the window the deciding factor in the composite exterior wall transmission loss.

Sound leaks through cracks in closures of operable windows will normally establish a window's rating regardless of type of glazing. Fortunately, the attention now given to the sealing of windows for thermal purposes has had a salutary effect on their acoustic properties. As with doors, the importance of proper gasketing and sealing cannot be over emphasized. Double glazing is effective only when the two panes are separated by a wide air gap (Fig. 28.27). A narrow air gap acts as a stiff spring between the panes and transmits the sound energy almost unattenuated. The result is approximately that of a single pane of double weight. Note that here too—as with absorptive material—that the requirements of acoustic and thermal insulation are opposed. A small sealed

ACOUSTICS

air space between panes is desirable for thermal insulation, because a large space allows convection currents to transfer heat. For acoustical purposes a small sealed space is not very useful as explained above, whereas a large space traps acoustic energy and is an effective noise barrier as is clearly seen in Fig. 28.27.

In addition to a window's sound transmission characteristic when closed, it is important to consider the *TL* when open, because of ventilation and natural cooling requirements. The sound attenuation between the center of a room with a clear-through open window and a point some distance outside, is 5 to 15 dB. This drops to about 5 dB as the receiver–observer approaches the open element. By making the path from inside to outside indirect, the "open" window attenuation can be increased to as much as 25 dB, but with considerable reduction of airflow, hence ventilation. Several possible arrangements with approximate midfrequency TL figures are given in Fig. 28.28. This principle can be applied advantageously when exterior noise reduction is important, but sealed windows are undesirable. Window opening style and placement can also have an effect on the amount of exterior noise admitted, as shown in Figs. 28.29 and 28.30. Typical STC ratings of common window constructions are given in Table 28.3.

28.14 Diffraction: Barriers

The physical process by which sound passes around obstructions and through very small openings is called diffraction. Simply stated, diffraction is a process whereby any point on a sound wave establishes a new wave when passing an obstacle. Thus, although much of a sound wave is blocked by a small opening, the portion that does get through establishes a new wave front [see Fig. 28.18 (5)]. The *amplitude* of the diffracted wave is determined by the relationship between the size of the opening and the wavelengths of the signal components. For a small hole, short wavelengths (high frequencies) are attenuated less than long wavelengths (low frequencies). See Fig. 28.31.

When sound encounters a finite-length barrier, it diffracts around and over it, approximately as shown in Fig. 28.32. The attenuation of the diffracted sound depends on frequency, type of source, and the dimensions of the barrier. For a *point source,* with only a single *practical* path around the obstruction (barrier) the noise reduction in dB can be calculated from Maekawa's empirical equation:

$$NR = 20 \log \left[\frac{\sqrt{2\pi N}}{\tanh \sqrt{2\pi N}} \right] + 5 \text{ dB} \qquad (28.4)$$

where
$$N = (F/565)\,(A + B - d)$$
$$NR = \text{noise reduction, dB}$$
$$F = \text{frequency, Hz}$$
$$A + B = \text{shortest path length around the barrier, ft (over or around)}$$
$$d = \text{straight-line distance, source-to-receiver, ft}$$

Note that this equation:

1. Is applicable only to exterior barriers where sound passing over the barrier is partially diffracted and partially attenuated by distance. In an interior situation, sound passing over a partial height barrier (see Fig. 28.40) strikes the ceiling and is reflected down, increasing the received

TABLE 28.3 **Sound Transmission Class (STC) of Window Construction**

Window Construction	STC
Operable wood sash, ⅛-in. glass, unsealed	23
Operable wood sash, ¼-in. glass, unsealed	25
Operable wood sash, ¼-in. glass, gasketed	30
Operable wood sash, laminated glass, unsealed	28
Operable wood sash, double-glazed, ⅛-in. panes, ⅜-in. air space, gasketed	29
Fixed sash, double ⅛-in. panes, 3-in. air space, gasketed	44
Fixed sash, double ¼-in. panes, 4-in. air space, gasketed	48

Fig. 28.28 *The degree of attenuation of external noise can be regulated when using pairs of double-hung (or horizontally sliding) windows with acoustic sealant and absorbent materials. Ventilation airflow varies inversely with transmission loss.*

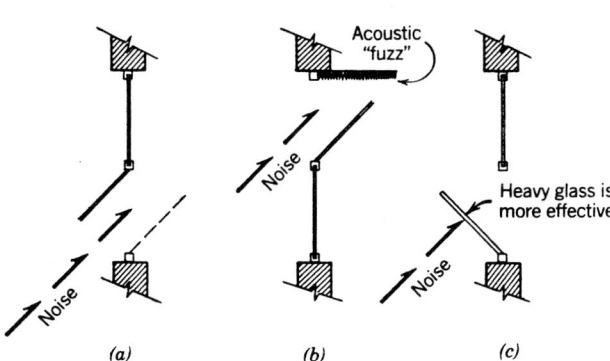

Fig. 28.29 *Alternative arrangements of the same basic "hopper" window design can yield results differing by as much as 10 dB. Design (a) is entirely open, and the noise path is unobstructed deep into the room. Design (b) is about 5 dB better than (a) at frequencies above 1 kHz because of higher absorption and less diffraction. Lower frequencies diffract readily around the window leaf and are less affected by absorptive material. Design (c) can be 10 dB better than (a) particularly at high frequencies because it interposes a rigid barrier into the noise path. In this arrangement, the glass thickness is important.*

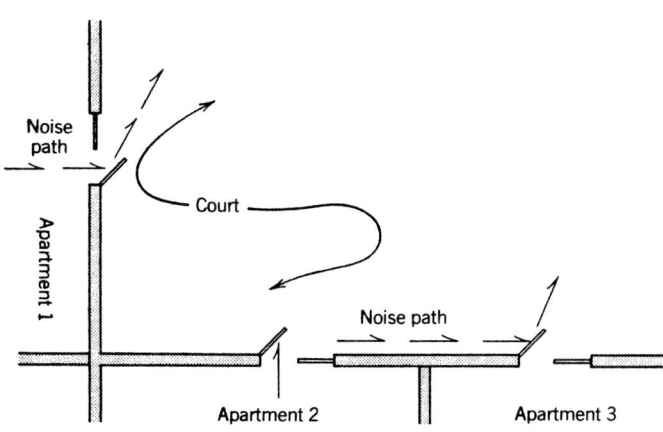

Fig. 28.30 *Noise transfer between contiguous corner spaces, as in apartments 1 and 2, can be particularly severe if windows are improperly designed. Swinging windows, as shown, are preferable to double-hung or hopper, because they act to reflect sound away from the adjacent space. Similarly, adjacent spaces on the same wall, as apartments 2 and 3, can benefit from this type of swinging arrangement, which is preferable to sliding or double-hung designs.*

ACOUSTICS

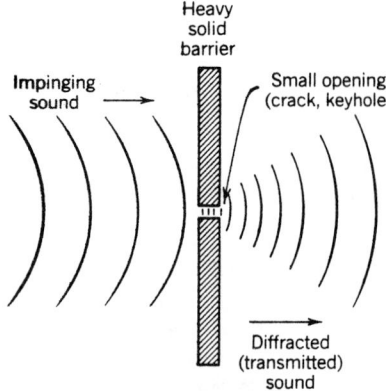

Fig. 28.31 *Sound passes through small openings by diffraction. The intensity of the transmitted sound is proportional both to frequency and to the size of the opening. It is always less than the intensity of the impinging sound. See also Fig. 28.18, illustration 5.*

sound and effectively reducing barrier attenuation. Maximum exterior barrier attenuation is 24 dB as compared to about 15 dB for a partial height interior partition.

2. Assumes that the barrier is very long, (or very high) so that only one sound path exists. In practice, a barrier whose length (height) is at least four times the distance between the source and the wall is sufficient, if *the barrier is close to the source.* If the barrier is close to the receiver, it must be longer (higher) still.

3. Assumes a point source. Line sources (such as traffic) show 20 to 25% less attenuation for the same barrier.

The equation will, however, give reliable, usable results when the dimensions of the source are small with respect to the barrier, as is the case for speech; individual motors, fans, engines, and other mechanical devices; and individual motor vehicles. The chart in Fig. 3.19 relates barrier dimensions and position to traffic noise reduction. Note that there, frequency is not a variable, since the chart has been plotted for an average attenuation at 220 Hz, which is center frequency for random car and truck traffic.

It should be apparent that the best location for a barrier is either very close to the source or very close to the receiver. The worst position for attenuation is halfway between. All effective barriers are assumed to be opaque and to have a minimum surface density of 5 lb/ft² (~25 kg/m²). The inherent TL of the barrier need not be very high; a massively thick barrier has only marginally higher attenuation than one with the minimum

Fig. 28.32 *Comparison of the effect of a barrier on sources of different frequencies. The low-frequency sound (a) diffracts more readily over the barrier than the high-frequency sound (b) because of its longer wavelength. Thus the lower the frequency, the smaller the "acoustic shadow" and the lower the barrier attenuation. Of course, the "shadow" is not sharply defined; it represents increasing attenuation as the observer approaches the barrier. See Fig. 28.33 for evaluation of barrier attenuation.*

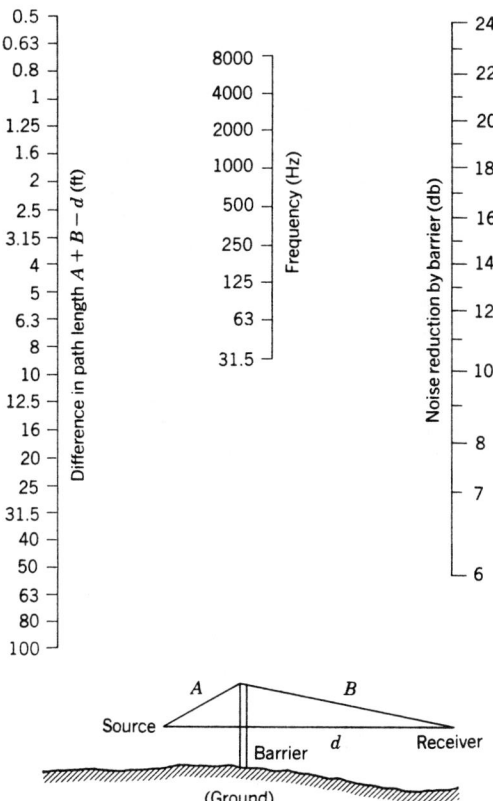

Fig. 28.33 *This nomograph to estimate the noise reduction afforded by a barrier is based on Eq. 27.4 and assumes a point (small) source and only a single path around the barrier. The dimensions A, B, and d are taken from the insert sketch. Dimensions A plus B represent the shortest path around the barrier—which may be over or around it. (Reprinted with permission from B. Fader, Industrial Noise Control, Wiley, New York, 1981.)*

surface weight specified above. Absorptive material placed on the source side of a barrier will reduce the noise reflected back toward the source but will not effectively increase the barrier's attenuation with respect to the receiver. Although maximum theoretical noise reduction of an *exterior* barrier is about 24 dB, in practice it rarely exceeds 20 dB. Figure 28.33 is a nomograph based on Eq. 28.4.

28.15 Flanking

Just as sound will pass through the acoustically weakest part of a composite wall, so it will also find parallel or flanking paths, that is, an acoustic short-circuit. Proper design of window locations to avoid flanking paths has already been shown in Fig. 28.30. The same situation obtains with respect to doors as shown in Fig. 28.26 and any other openings between spaces. Thus in Fig. 28.34 a high STC wall between the two spaces is in large measure defeated by flanking paths F5, F6, and F7. In other spaces the most common flanking path is via the plenum, as in Fig. 28.34, path F1, and Fig. 28.35 (*b*) and (*d*). Ductwork with registers or grilles in various rooms acts as an excellent intercom system unless completely lined with sound-absorptive material (see Section 28.27). Even then, low-frequency sound is only minimally attenuated, and special measures must be employed if good transmission loss is required. This subject is discussed further below, under Mechanical System Noise Control.

SPEECH PRIVACY

28.16 Principles of Speech Privacy between Enclosed Spaces

The subject of speech privacy has always been of paramount importance in office design. It has been demonstrated by numerous studies that productivity and noise are related inversely, when the noise is information bearing. When the noise does not carry information, it can be annoying and therefore counterproductive or it can be useful as a masking sound, depending on its frequency content, intensity level and constancy. If we refer back to Section 28.7 that discusses the noise reduction of an *airtight* barrier between two spaces, we saw that the sound intensity levels in the source room (1) and the receiving room (2) are related by the expression

$$IL_2 = IL_1 - NR$$

where *NR* is noise reduction and IL_2 and IL_1 are sound intensity levels in the receiving and source rooms, respectively. If the receiving room is completely quiet and has no sound source other than the transmitted sound IL_2, then that sound will always be a potential source of annoyance to the occupant of the receiving room as long as its intensity level is above the hearing threshold. If however there is an *constant* ambient sound level in

ACOUSTICS

FLANKING NOISE PATHS

F1 OPEN PLENUMS OVER WALLS, FALSE CEILINGS
F2 UNBAFFLED DUCT RUNS
F3 OUTDOOR PATH, WINDOW TO WINDOW
F4 CONTINUOUS UNBAFFLED INDUCTOR UNITS
F5 HALL PATH, OPEN VENTS
F6 HALL PATH, LOUVERED DOORS
F7 HALL PATH, OPENINGS UNDER DOORS
F8 OPEN TROUGHS IN FLOOR-CEILING STRUCTURE

NOISE LEAKS

L1 POOR SEAL AT CEILING EDGES
L2 POOR SEAL AROUND DUCT PENETRATIONS
L3 POOR MORTAR JOINTS, POROUS MASONRY BLK
L4 POOR SEAL AT SIDEWALL, FILLER PANEL ETC.
L5 BACK TO BACK CABINETS, POOR WORKMANSHIP
L6 HOLES, GAPS AT WALL PENETRATIONS
L7 POOR SEAL AT FLOOR EDGES
L8 BACK TO BACK ELECTRICAL OUTLETS
L9 HOLES, GAPS AT FLOOR PENETRATIONS

OTHER POINTS TO CONSIDER, RE: LEAKS ARE (A) BATTEN STRIP A/O POST CONNECTIONS OF PREFABRICATED
WALLS, (B) UNDER FLOOR PIPE OR SERVICE CHASES, (C) RECESSED, SPANNING LIGHT FIXTURES, (D) CEILING
& FLOOR COVER PLATES OF MOVABLE WALLS, (E) UNSUPPORTED A/O UNBACKED WALL BOARD JOINTS (F) EDGES
& BACKING OF BUILT-IN CABINETS & APPLIANCES, (G) PREFABRICATED, HOLLOW METAL, EXTERIOR CURTAIN
WALLS.

Fig. 28.34 *Flanking transmission of airborne noise. (Reprinted from* A Guide to Airborne, Impact, and Structure-Borne Noise Control in Multifamily Dwellings, *1968.)*

the receiving room, then, depending on its characteristics, it may mask the transmitted sound IL_2, even to the extent of making it completely inaudible. In most instances however, it simply reduces or eliminates the annoyance without completely masking its existence. As has been pointed out previously, what we hear (and therefore what can potentially be a source of disturbance) depends on our level of attention to both what we are doing and conversely to the—intrusive sound. (A remarkable exception is the ability of some, generally young, students to study in the presence of very loud, familiar—and therefore information bearing—music. Indeed, some claim that they can *only* study that way.)

In general however, tests have shown that a majority of adults will not consider the intruding noise level IL_2 to be annoying if the intensity of a properly designed background noise is either greater, or no more than 2 dB less than IL_2. Thus, a transmitted IL_2 of 40 dBA will not be considered annoying by most people if the level of the background noise is at least 38 dBA. The upper level of usable background masking noise is usually taken to be about 50 dBA. Any higher intensity level will itself become a source of annoyance. Figure 28.36 gives a graphic representation of the relation between transmitted and background noise levels in a receiving room. In summary then of the above discussion we can simply state that

DO

DON'T

(a)

CEILING SLAB

PARTITION WALLS BETWEEN
ROOMS SHOULD EXTEND
FROM FLOOR SLAB TO
CEILING SLAB

FLOOR SLAB

ROOM A   ROOM B

(b)

GYPSUM BD.

NOISE PATH   AVOID EXTENDING
WALLS TO UNDERSIDE
OF SUSPENDED CEILING

APT. A   APT. B

(c)

ROOF

ATTIC SPACE   SOUND BARRIER

GYPSUM BD.
CEILING

EXTEND WALL TO ROOF
OR DIVIDE ATTIC SPACE
WITH FULL—HEIGHT BARRIER

FLOOR—CEILING
ASSEMBLY   PARTITION
APT. A   WALL   APT. B

APT. C   APT. D

(d)

ATTIC OR PLENUM

AVOID OPEN
ATTIC SPACES
OR PLENUMS

APT. A   APT. B

APT. C   APT. D

Fig. 28.35 *Construction techniques to avoid flanking paths. (Reprinted from* A Guide to Airborne, Impact, and Structure-Borne Noise Control in Multifamily Dwellings, *1968.)*

Noise reduction, $NR \begin{cases} = IL_1 - IL_2 \\ = TL - 10 \end{cases} \log \dfrac{S}{A_2} \begin{cases} S = \text{area of barrier} \\ A = \text{absorption of room 2} \end{cases}$

IL_1

Source room level
Room 1

IL_2

Transmitted level
Receiving room
Room 2

Background
noise level

ACOUSTICS

Fig. 28.36 *The background noise level determines whether the transmitted sound will actually be heard. In this case, because the background noise level is considerably higher than the transmitted level from the adjacent room, the transmitted noise will not be noticeable (i.e., it will be masked).*

the degree of *speech privacy* in a space is a function of two factors:

· the degree of sound isolation provided by the barriers between rooms, and

· the ambient sound level in the receiving room.

28.17   Sound Isolation Descriptors

Since the degree of intrusiveness of extraneous noise in a space varies, depending as it does on both transmitted and ambient sound, a descriptive scale of some sort is obviously necessary. If we restrict our discussion to speech sounds, since ultimately we are interested in office design where speech is the primary sound source, then a descriptive scale as shown in Table 28.4 can be established. With these absolute descriptions in mind, and remembering that the hearing condition in a receiving room can be altered by changing either the barrier characteristics, or the background sound level, *or both,* we can express the effectiveness of a barrier as a speech sound barrier in terms of its STC for a given ambient noise level. Since the ambient noise level is frequency dependent, it can be approximated by an NC value. This is particularly useful when the ambient noise level is generated by machinery or by the sound of the air-conditioning system, rather than by a shaped signal from an electronic masking sound system.

Table 28.5 shows the hearing conditions in a receiving room with an NC-25 background noise as a function of barrier STC rating. If the background noise level were raised to NC-30, then each descriptor would roughly increase one level in quality (i.e., poor would become fair, fair would become good, and so forth). Stated otherwise, the *apparent* isolation provided by a barrier may be increased by raising the background (masking) noise level in the receiving room. Figure 28.37 shows two conditions of adjacent spaces. Although the source room level is uniform and partitions on both sides of the source room are identical, the background noise in the two receiving rooms, is different. In *A,* the background is NC-35; in *B,* it is NC-25. The occupant of room *A* is not disturbed by the little heard from the source room. The occupant of room *B* hears clearly. Occupant *A* will probably praise the partition, whereas occupant *B* will complain. Although the levels of reradiated sound are identical in the two receiving rooms, the intruding signal is masked by the background noise in *A* but it is clearly audible in *B.* Thus the apparent noise reduction is substantially higher in *A* than in *B.* This clearly demonstrates the effectiveness of masking sound in providing *apparent* sound isolation and *speech privacy.*

Sound isolation can also be improved by careful planning. Storage and circulation can serve as buffers for noise-sensitive areas. Physical separation of noisy areas from quiet ones often eliminates the need for complicated and expensive compound barriers.

TABLE 28.4 **Relative Quality of Sound Isolation**

Ranking	Descriptor	Hearing Condition[a]
6	Total privacy	Shouting is barely audible.
5	Excellent	Normal voice levels not audible. Raised voices barely audible but not intelligible.
4	Very good	Normal voice levels barely audible. Raised voices are audible but largely unintelligible.
3	Good	Normal voice levels are audible but generally unintelligible. Raised voices are partially intelligible.
2	Fair	Normal voice levels are audible and intelligible some of the time. Raised voices are generally intelligible.
1	Poor	Normal voice audible and intelligible most of the time.
0	None	Normal voice levels always intelligible.

[a] Hearing condition in the presence of ambient noise, if any.

TABLE 28.5 **Relation Between Barrier STC and Hearing Condition on the Receiving Side, Background Noise Level at NC-25**

Barrier STC	Hearing Condition	Descriptor and Ranking[a]	Application
25	Normal speech can be understood quite easily and distinctly through the wall.	Poor/1	Space divider
30	Loud speech can be understood fairly well. Normal speech can be heard but not easily understood.	Fair/2	Room divider where concentration not essential
35	Loud speech can be heard, but is not easily intelligible. Normal speech can be heard only faintly, if at all.	Very Good/4	Suitable for offices next to quiet spaces
42–45	Loud speech can be faintly heard but not understood. Normal speech is inaudible.	Excellent/5	For dividing noisy and quiet areas; party wall between apartments
46–50	Very loud sounds, such as loud singing, brass musical instruments, or a radio at full volume can be heard only faintly or not at all.	Total Privacy/6	Music room, practice room, sound studio, bedrooms adjacent to noisy areas

[a] See Table 28.4.

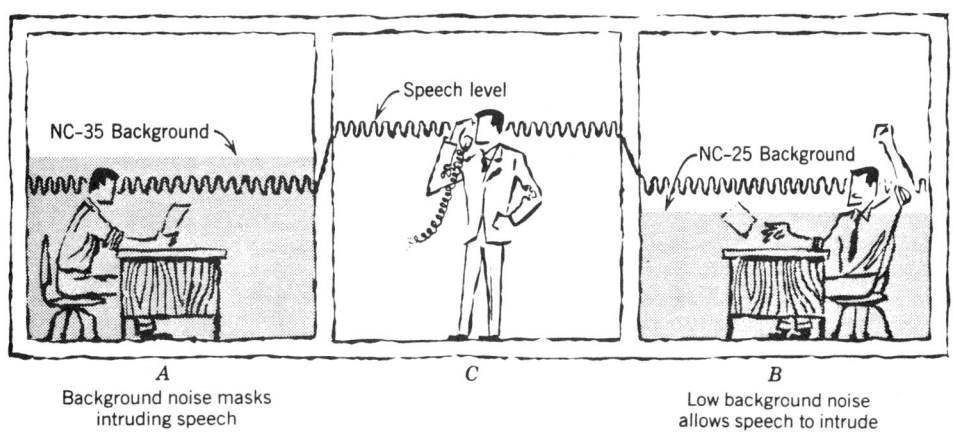

Fig. 28.37 *The occupant in room A with background noise NC-35 (≈45 dBA) is unaware of the noise that is so disturbing to occupant B, whose NC-25 (≈36 dBA) is insufficient to mask C's loud speech.*

ACOUSTICS

a. Procedure for determining speech privacy rating

Speech rating

Step 1. Speech effort — from source room

Loud   Raised   Conversational

72   66   60

Step 2. Source room floor area (A_1) —
effect of source room absorption

125   250 500 1000 (sq ft)

10   6   3   0

Step 3. Privacy allowance —
degree of privacy required

Confidential   Normal

15   9

Isolation rating

■ Speech rating total   ____

Step 4. Sound transmission class (STC) —
common barrier

Step 5. Noise reduction factor (A_2/S) —
effect of receiving room
absorption and barrier size

1   5   10 (Unitless)

A_2/S -2 0 2 3 4 5 6 7 8

Step 6. Adjacent room background noise level (dBA) —
masking sound available

■ Isolation rating total   ____

Speech privacy rating number

Find speech privacy rating number by subtracting isolation rating total from speech rating total. ____
Then use graph at bottom of sheet to predict degree of satisfaction.

b. Anticipated response to privacy situation

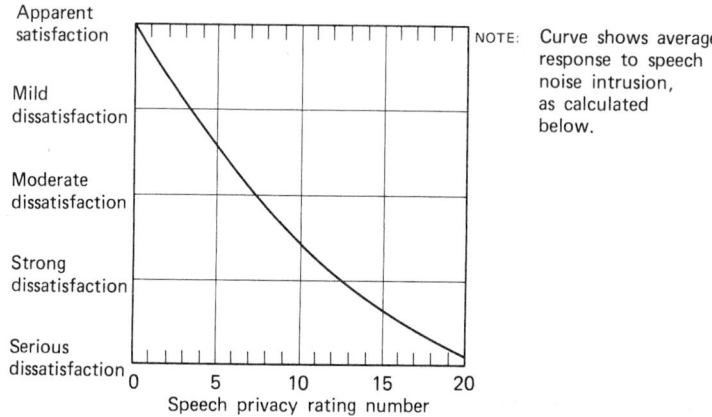

NOTE: Curve shows average
response to speech
noise intrusion,
as calculated
below.

Apparent satisfaction — Mild dissatisfaction — Moderate dissatisfaction — Strong dissatisfaction — Serious dissatisfaction

Speech privacy rating number (0 5 10 15 20)

Fig. 28.38 Speech privacy analysis sheet. (Based on Cavanaugh et al., 1962.)

28.18 Speech Privacy Design for Enclosed Spaces

The study of speech privacy received considerable emphasis with the advent of open-plan offices (office landscaping), although the same problem prevails in both instances. Essentially the problem was to determine the factors affecting speech privacy and to quantify them with a degree of accuracy sufficient for design purposes. It rapidly became evident that although the physical principles are the same, the solutions to speech privacy design problems are radically different for closed spaces and for open-plan (landscaped) offices. In the former, the acoustic characteristics of the air-tight barrier between two spaces is the crucial element in the design, because all of the inter-space sound is transmitted through this barrier. In contrast, sound transmission between adjacent "cubicles" in an open office are primarily the sum of *reflected* and *refracted* sound, with the direct component passing through the barrier of second-ary importance. As a result the acoustic transmis-sion characteristics of the barrier also became sec-ondary in importance.

The above mentioned studies indicate that there are six factors involved in enclosed space speech privacy, which can be subsumed under two headings (Fig. 28.38):

1. Speech rating of the source room (Room No. 1)
   a. Speech effort—a measure of the loud-ness of speech.
   b. Source room factor—gives the approxi-mate effect of room absorption on speech level in the source room. The scale in Fig. 28.38 is drawn for average absorption. For live rooms raise the fac-tor by two points; for dead rooms lower it by two points. Factors *a* + *b* give the approximate source room voice level.
   c. Privacy allowance—What is the measure of privacy required?

The privacy criteria definitions that were used in this step (Fig. 28.38, step 3) are:

Normal privacy—such that the receiving room occupant can understand a small portion of normal voice conversation in the source room by listening intently. This corresponds to "Good" and ranking 3 in Table 28.4. It was

found that most occupants can work normally with this level of speech intrusion.

Confidential privacy—assumes that only a few words will occasionally be intelligible. This pri-vacy level corresponds to "Excellent" and ranking 5 in Table 28.4. The 6 speech-rating points between Normal Privacy (9 points) and Confidential (15 points) in Fig. 28.38 corre-spond roughly to a 5-dB difference. If we re-member that a sound intensity differential of 3 dB is barely perceptible, 6 dB is clearly no-ticeable and 10 dB is a doubling of perceived sound (Table 26.3), we can appreciate that the difference between *normal* and *confidential* privacy is small and calls for accurate design, plus a measure of field adjustment of mask-ing sound.

2. Isolation rating of the receiving room (Room No. 2)
   d. STC rating of the barrier. Table 28.6 gives some typical STC ratings for of-fice partitions.
   e. Noise reduction factor A_2/S is an indica-tion of receiving-room absorption, that is, the difference between *NR* and *TL,* where A_2 is the area of receiving room and *S* the area of the barrier between rooms. Absorption is assumed to be av-erage. For live rooms lower this factor two points; for a dead room raise two points.
   f. Recommended background noise level in receiving room, use Table 28.7.

An analysis sheet for enclosed spaces is provided in Fig. 28.38 (see also Cavanaugh et al., 1962, and Young, 1965.) The two examples of this analysis that follow should clarify its use. The reader

TABLE 28.6 **Typical STC Ratings of Interior Partitions**

Type of Partition	STC
Demountable partition	STC 20–30
Drywall partition up to acoustical ceiling	STC 30
Drywall partition extending 12 in. above acoustical ceiling tile system, into ceiling plenum	STC 35
Drywall partition with cavity insulation, full height to the underside of slab above	STC 40–45
Two-layer drywall partition with insulation, erected full height to the underside of slab above	STC 50

TABLE 28.7 **Recommended Category Classification and Suggested Noise Criteria Range for Steady Background Noise**

Type of Space (and Acoustical Requirements)	NC Curve	Equivalent[a] dBA
Concert halls, opera houses, and recital halls (for listening to faint musical sounds).	10–20	20–30
Broadcast and recording studios (distant microphone pickup used).	15–20	25–30
Large auditoriums, large drama theatres and houses of worship (for excellent listening conditions).	20–25	30–35
Broadcast, television, and recording studios (close microphone pickup only).	20–25	30–35
Small auditoriums, small theatres, small churches, music rehearsal rooms, large meeting and conference rooms (for good listening), or executive offices and conference rooms for 50 people (no amplification).	25–30	35–40
Bedrooms, sleeping quarters, hospitals, residences, apartments, hotels, motels, and so forth (for sleeping, resting, relaxing).	25–35	35–45
Private or semiprivate offices, small conference rooms, classrooms, libraries, and so forth (for good listening conditions).	30–35	40–45
Living rooms and similar spaces in dwellings (for conversing or listening to radio and TV).	35–45	45–55
Large offices, reception areas, retail shops and stores, cafeterias, restaurants, and so forth (for moderately good listening conditions).	35–50	45–60
Lobbies, laboratory work spaces, drafting and engineering rooms, general secretarial areas (for fair listening conditions).	40–45	50–55
Light maintenance shops, office and computer equipment rooms, kitchens, and laundries (for moderately fair listening conditions).	45–60	55–70
Shops, garages, power-plant control rooms, and so forth (for just acceptable speech and telephone communication). Levels above PNC-60 are not recommended for any office or communication situation.	—	—
For work spaces where speech or telephone communication is not required, but where there must be no risk of hearing damage.	—	—

[a] For information only. This data is not part of the NC information and does not appear in the source.
Source: Extracted with permission from E. B. Magrab, *Environmental Noise Control*, Wiley, New York, 1975.

should follow the analysis with Fig. 28.38 in hand. The numbered steps in the examples correspond to the numbers in the figure.

EXAMPLE 28.3 Source room:

General clerical office, 40 × 60 × 9 ft, average $\bar{\alpha}$, 16-ft long full height partition, STC 40

Receiving room:

Conference room, 16 × 24 ft, medium-dead room

Background noise level, 40 dBA (NC-30) (from Table 28.7)

Privacy analysis (see Fig. 28.38)

(a) 1. Speech effort: raised   66
2. Area $A_1 > 1000$ ft^2 0
3. Privacy—normal   9
     Speech rating $a = 75$
(b) 4. STC (given)   40
5. A_2/S 3
     $(16 × 24)/(16 × 9) = 2.6$, corresponding to 2.0 on the scale; add 1.0 for higher than average $\bar{\alpha}$
6. Background noise level   40
     Isolation rating $b = 83$
     Speech privacy rating $a - b = -8$

Therefore the STC rating of the partition can be reduced to 32 without affecting speech privacy.

EXAMPLE 28.4 Source room:

Drafting room 20 × 30 ft, medium-live

Common wall 12 × 8 ft high

STC: 26 (half glass, with door)

Receiving room:

Supervisor's office, 12 × 14 × 8 ft, average absorption

Background noise level, 35 dBA

Privacy analysis:

1. Speech effort—conversation   60
2. Source room factor:
   for area   +2
   for liveness   +1   Total   +3
3. Privacy—(almost) confidental   13
                         Speech rating $a = 76$
4. STC (given)   26
5. A_2/S
   [(12 × 14)/(12 × 8) = 1.8, corresponds to a reduction factor of 1.6; gives 2.0 on the scale since system uses whole numbers only.] No adder required for average absorption.
                         Therefore, 2 + 0 = 2
6. Background noise level   35
                    Isolation rating: $b = 63$
              Speech privacy rating $a - b = 13$
           which indicates strong dissatisfaction

The suggested corrections here are to increase the wall's STC to 36 by gasketing the door and increasing the background noise level in the receiving room to 40 dBA (NC 30). This would give a speech privacy rating of −2:

STEP 4.	STC of barrier	36
STEP 5.	A_2/S	2
STEP 6.	Background noise	40
	Isolation rating	78

Speech privacy rating = $a - b = 76 - 78 = -2$
This result should be satisfactory according to the chart of Fig. 28.38. ∎

28.19 Principles of Speech Privacy in Open Area Offices

The huge increase in the service sectors of world economy, has brought with it a corresponding increase in desk jobs, each of which will very soon be equipped with a computer console. This increase has also necessitated an increased space density for office workers, made possible by the general elimination of paper storage (files), and the corresponding elimination of the necessity for employees to circulate. The increased density problem has been in large measure solved by open-office plans with ever smaller "cubicles" usually with single occupancy, but recently also with dual occupancy. This situation has obviously aggravated the serious problem of annoyance due to the intrusion of speech sounds from neighboring workers, that is, the lack of *speech privacy.*

Since production is adversely affected by the inability of a worker to concentrate because of annoyance with speech intrusion—an annoyance that usually *increases* with passing time—the proper design of open office plans can have major economic benefits.

(a) Sound Paths in Open Offices. In contrast to the single, or at most dual sound paths that exist with full height enclosures (Fig. 28.39), the sound paths in an open plan arrangement, including first reflections, are direct, diffracted, ceiling reflected and laterally reflected (Figs. 28.40 and 28.41). Careful study of these illustrations shows a number of important facts affecting speech privacy in open offices.

1. The angles of reflection of sound waves from the ceiling depend on the location and height of the source, and the ceiling height [Fig. 28.40 (*a*)]. Measurements have shown that these angles vary from a minimum of 30° for a standing speaker in the center of a cubicle to a maximum of 60° for a speaker close to, and facing a partition, for ceiling heights up to 9 ft. Since much of the sound energy reaching an adjacent occupant does so after reflecting off the ceiling at these angles (30°–60°), a ceiling material with high absorption at these angles of incidence is required for speech privacy. This effectively negates the use of noise reduction coefficient (NRC) as a useful factor in describing a ceiling material's absorption charac-

ACOUSTICS

Fig. 28.39 *Sound paths between enclosed spaces are determined by the type of barrier separating them. In the case of a full height barrier reaching to the underside of the ceiling slab, the only path sound path is through the barrier and its STC determines the noise level in the receiving room. In the instance of a ceiling height partition with overhead plenum, most of the sound energy will travel the upper, less attenuating path. Factors affecting the level of received sound are the ceiling CAC rating (transmission characteristic) and the acoustic characteristics of the plenum, including all its contents. In all cases, sound within a space attenuates with increasing distance, as schematically shown.*

teristic in open office design, since NRC averages absorption at all angles. All major ceiling material manufacturers have tested, and will supply accurate angular absorption data on their products. A single number descriptor that relates to this characteristic, called Articulation Class, is discussed in detail in Section 28.20.

2. Since absorption at the ceiling increases with the angle of incidence of a sound wave, it is always desirable for speech privacy to have maximum ceiling height. Most authorities recommend a *minimum* ceiling height of 9 ft.

3. Absorption efficiency of ceiling tiles varies inversely with STC, which is to be expected, since sound transmission depends on mass and absorption at air pockets. Light fiberglass ceiling tiles will typically have an absorption coefficient α of 0.95 at voice frequencies, with a CAC (STC) of 22–24 as compared to more massive mineral fiber tiles with a maximum α of 0.8–0.85, but an STC of 34–36. Since we are interested in maximum absorption at the ceiling, the tile of choice is always that with the highest α at voice frequencies. The fact that more energy will pass through a tile of lower CAC (STC) does not affect the acoustic result since most of the sound energy enters the plenum above the ceiling through openings in the far from air-tight suspension system and is then largely absorbed and dissipated by the spray fire-

proofing, sound absorbing material, ductwork and structural members in the plenum.

4. Since sound will always find the path of least resistance to travel, very little sound energy will pass through partitions between cubicles; the paths over the partition by diffraction and reflection are much less resistant. That being so, STC of these partitions need not be high. Where the source is close to the partition as is the case with a seated speaker facing the partition, and delivering sound energy directly at the partition at a height of approximately 44–48 in. and from a distance of about 3 ft, the STC of the partition should be 25–26. For speakers at greater distances and heights an STC of 20–22 is usually sufficient.

An exception to this rule of thumb *may* occur when a large (17″–19″) computer console is interposed between the seated speaker and an absorbent partition. Large video display terminals (VDTs) are coming increasingly into use. Their smooth, highly reflective glass surface creates a strong sound path to the rear of the speaker and decreases the sound energy absorbed by the partition behind the VDT. The variables however are so numerous that conservative design will use the higher STC rating. (Contrast these values with those required of a full height fixed partition construction typical of enclosed spaces, as listed in Table 28.6.)

Principal Sound Paths

① Direct path through 5'6" high partition
② Diffracted path over partition
③ Reflected path over partition
④ Sound absorbed and diffused

Theoretical attenuation

Actual attenuation

Fig. 28.40 (a) *The principal sound paths between occupancies in an open office plan are: (1) direct, (2) diffracted, (3) ceiling reflected and (4) reflected from the slab above the plenum. Only paths (2) and (3) are problematic from the speech privacy viewpoint, requiring masking sound. Note that the angle of incidence of the sound wave at the ceiling varies from 30° for a standing speaker to 60° for a seated speaker, requiring a ceiling material of high absorption between these angles. Large VDT screens (17–19"), now in common use, create another path of strongly reflected sound, indicated by dashed lines. (b) Sound is attenuated by distance, dropping 4–6 dB for each doubling of distance from the source (6 dB in the open; 4 dB in enclosed spaces due to interreflections). Average SPL of conversational speech is 60 dBA at 3 ft from the speaker. Attenuation of a diffracted signal at a partial height partition is 4–8 dB. Note that the diffracted sound (without a contribution from ceiling-reflected sound) drops to 41–45 dB in the adjacent cubicle. In a two-person open office, the received signal from a standing speaker is at least 54–56 dBA [i.e., perfectly intelligible, even in the presence of maximum (50 dB) masking noise].*

The absorption coefficient α of the partial height partition at voice frequencies should be a minimum of 0.8 and preferably 0.85–0.95. Some manufacturers have assigned an Articulation Class rating to their partition products although that descriptor is usually reserved for ceiling tiles as explained above. The recommended AC ratings for partitions range from 180–220.

5. Partitions must be tall enough to block direct line-of-sight voice transmission, since such a path is unattenuated except by distance. Median mouth height of a standing American male is 63 in. This is the basis of the widely accepted recommendation that partitions should not be lower than 65 in. and preferably 66–72 in. between adjoining cubicles. Since a 72-in. high partition

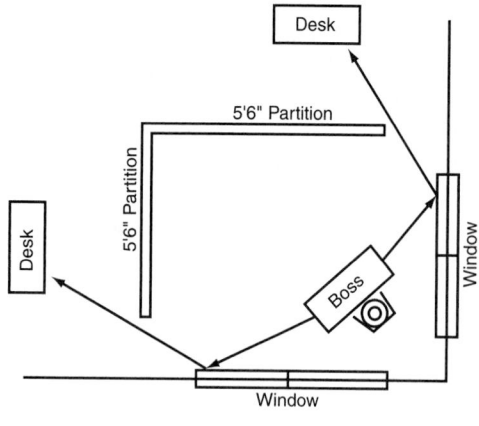

Plan View

Fig. 28.41 *Confidentiality in offices on exterior building walls is difficult to maintain because of the lack of attenuation of sound reflected from windows. The problem is compounded by the custom of placing management personnel in such offices (see text).*

blocks vision for all but the tallest people, giving a subjective closed-in sensation to the occupant of a (small) cubicle, this height is normally used only between departments, with intra-department partitions at 63–66 in. Increasing the height of a partition from 65 in. to 72 in. will increase path attenuation to an adjacent cubicle by 1–3 dB, depending on ceiling height, and speaker height and location.

6. Refer to Fig. 28.40 (*b*). This figure indicates signal attenuation due to distance, in three different paths. Speech intensity at conversational level is approximately 60 dBA at 3 ft from the speaker. Using the fact that sound in free field attenuates 6 dB for every doubling of distance, and making the assumption that the sound field in a cubicle approaches that of a free field because of the large amount of highly absorptive material in the area, we obtain a sound intensity level of 54 dBA at 6 ft. from the speaker and 48 dBA at 12 ft. In practice, the received sound level would be several dB higher because, despite the high α of the space, there exist intraspace reflections that increase the sound level.

Referring now to Fig. 28.40 (*b*), we see that the minimum sound level at a receiver *within* the cubicle would be 55–56 dB; a level that no practical amount of background noise can mask. Thus

two occupants of a single, large cubicle, will always hear each other quite clearly. The attenuation of a partition in the diffracted paths (over and around the partition) depends on the location and height of the speaker and varies from 4–8 dB. The attenuation of the signal in the transmitted path (through the partition) will be at least 10 dB. Based on these figures, it is recommended that the *minimum* horizontal distance between occupants of adjoining cubicles when seated at their work stations be 10 ft for minimum speech privacy. (Degrees of speech privacy are discussed in the next section.) Speech levels in team work areas readily reach 66 dB. This necessitates either locating such areas away from normal working spaces or the use of full height fixed or demountable partitions to completely enclosed such areas. The spaces that require careful siting or complete enclosure because of raised voice levels (64–66 dB) include video conferencing rooms, telecommunications spaces, and areas where workers use speakerphones or voice-activated computers. The latter two devices are usually forbidden in densely populated work areas where a reasonable degree of speech privacy is required for the conduct of regular business tasks.

7. Refer to Fig. 28.41. As pointed out previously, sound will be received via paths of least resistance. These are often flanking paths that do not become evident except in plan view. In (*a*) of the figure, the flanking paths are particularly important because the first reflection occurs at a window. Glass has negligible absorption and because of its smoothness, specular reflection. As a result, the corner office occupant's voice will be heard clearly via the flanking paths shown, thus destroying the confidentiality of conversation in that office. Since offices on the building perimeter are usually reserved for middle and upper management, and the large windows in such offices act to minimize the speech privacy so important to managerial personnel, the space designer has several options to ameliorate this condition:

(a) Use full-height fixed partitions, with fixed glass vision panels if required, and doors rather than openings.

(b) Use heavy drapes over the "offending" glass windows, although this option defeats the visual and daylighting purpose of the windows.

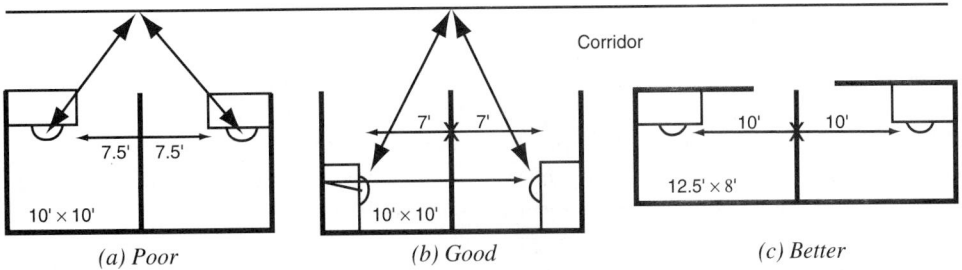

Fig. 28.42 *Open-office shapes and furniture arrangement have a marked effect on speech privacy.*
(a) *A strong unattenuated signal reflects from the opposite corridor wall, directly into the adjoining office. The side-to-side 15 ft spacing with an intervening partition is satisfactory.*
(b) *The sound reflected from the corridor wall is much weaker than in (a) because of increased distance, and decreased voice intensity from the side. Lateral sound is also weak due to the attenuation at the first reflection off the absorbent partition.*
(c) *Reflection from the corridor is eliminated by use of a front closure partition with glass at the upper 18 in. for visual purposes. Side-to-side sound is weak due to wide spacing and the intervening partition.*

(c) Locate spaces requiring confidentiality in groups, sound-buffered from open-office spaces by unoccupied areas such as storage rooms.

It is also important to note that although the arrow signifying a sound path in Fig. 28.41 shows reflection from the windows, the sound energy will also strike the exterior walls which are usually plastered and therefore highly reflective. Here, the placement of absorbent acoustic insulation on all walls, to ceiling height is not impractical as it is with windows, but does entail considerable expense.

8. Refer to Fig. 28.42. Furniture arrangement establishes the source location of speech energy and consequently all of the sound paths that contribute to speech privacy—or, more accurately, the lack of speech privacy. In layout (a), sound power, unattenuated except by distance, reflects off the *back* of the opposing aisle partition and travels to the occupant of the neighboring cubicle. Since the back of an acoustic partition is usually metallic and uninsulated, this arrangement would entail the additional expense of an insulated rear surface on the corridor panels, to maintain a degree of speech privacy between cubicles.

Changing the desk location in the same shaped cubicle to that shown in Fig. 28.42 (b) improves speech privacy considerably by reducing the SPL of both the reflected and flanking paths, since the speakers face a highly absorptive surface. It may be unnecessary to insulate the rear of the corridor panels. A disadvantage of this arrangement lies in the short distance between speakers and neighbors if the same pattern of cubicles is continued longitudinally. A second possible disadvantage may be employee dissatisfaction with a working position facing a blank wall.

The arrangement in Fig. 28.42 (c) uses the same area per cubicle. By changing its shape and adding a 6-ft high acoustic partition whose top 18 in. is glass, the designer has reduced the sound energy levels in all paths while overcoming any employee resentment related to working in a blank unexposed corner. Furthermore, the problematic flanking path in (a) has been eliminated as has probably any requirement for exterior acoustic insulation on the corridor partitions.

28.20 Open Office Speech Privacy Levels and Descriptors

(a) Factors. Although the factors involved in determining the level of acoustic privacy that can be expected at a specific open-office location, as a result of neighboring speech sources should be understood at this juncture of our discussion, it may be helpful to state them clearly. They are:

1. Loudness of the source (s).

2. Acoustic characteristics of the source(s). This includes location, height above floor, directivity, frequency spectrum, and information content. The last factor refers to the ability of the listener to make sense of what is being said or

heard. Therefore, most types of non-information–bearing noise would not be considered as sources of intrusive sound. A completely foreign language should also fall into this category although in practice that is so only when the words and syllables are muffled. Clearly heard words in a foreign language do constitute a source of speech annoyance although to a far lesser extent than a comprehended language. Speech privacy calculation procedures ignore foreign languages and assume that all speech is a possible source of annoyance.

3. Signal attenuation along each path between the source and the receiver. This factor is different for every normal position of even a single source, as, for instance, for a seated or standing source person.

4. Level and frequency content of background noise (deliberate masking, HVAC noise, continuous machinery noise and the like).

5. Degree of privacy required. These are discussed in detail in the next subsection. It is important to keep in mind however that the three classifications—confidential, normal and transitional (minimal, marginal) are assumed to remain constant as long as the physical factors involved do not change. This ignores the oft-demonstrated fact that intrusive noise causes physchological and physiological reactions in some people that tend toward aggravation and increased severity with passage of time. A speech privacy situation originally classified as Confidential may deteriorate to Normal as the hearer's increasing sensitivity to noise causes him or her to strain to hear, and thereby to actually hear noise that ordinarily is masked. This is, in effect, another aspect of the "cocktail party effect" discussed in Chapter 26.

(b) Levels (Degrees) of Speech Privacy.
Speech privacy is accomplished by masking the intruding speech with background noise. By consensus among acousticians, the definitions of the three levels of speech privacy in an open office are:

1. Confidential privacy. Normal voice levels are audible but generally unintelligible. Raised voices are partially intelligible. Noise level is minimal. To accomplish this level of privacy, it has been found that the background noise level must be no more than 2 dB less than the intruding sound, and no more than about 3 dB more than the intruding sound, to satisfy the "minimal noise level" requirement. In this acoustic situation approximately 95% of people will not sense any sound intrusive disturbance, and will be able to concentrate on most types of work.

2. Normal privacy. At this level, normal voice levels from adjacent spaces are heard but are not intelligible without concentration (i.e., straining to hear and catch every syllable). Raised voices are generally intelligible. Overall noise level is low. This level of speech privacy is achieved when the background noise level is within 6 dB (less than) the intruding speech level. This would correspond roughly to an intruding speech level of 50 to 54 dB, and background noise level of 44 to 45 dB, that together would give a range of 51 to 55 dB; levels that can be considered to meet the "low noise" requirement.

3. Transitional (minimal, marginal) privacy. At this speech privacy level, speech at normal voice levels in adjacent open offices is readily understood most of the time, and the overall noise level is average. This noise level obtains when the intruding speech level is 10 dB or more than the background noise level. Since background noise is limited to about 50 dB, this privacy level would mean an intruding speech intensity of 60 dB or more. Since 60 dB is approximately normal speech at 3 ft, this "privacy" level would occur with two occupants in a single office, or a single occupant receiving intruding noise from at least three neighboring offices. This speech intrusion level would be considered intolerable by about 40% of people, and would negatively affect the work efficiency of a higher percentage. In summary, it is interesting to compare these three open-office speech privacy levels with the "absolute" grading given in Table 28.4:

Open Office Class	Table 28.4 Rating
Confidential	Good
Normal	Fair
Marginal	Poor

It should therefore be apparent that by its very nature, an open office cannot achieve the top three grades of privacy listed in Table 28.4. If those levels of privacy are desired or required fully enclosed spaces are required.

(c) Articulation Index (AI).

In order to quantify speech privacy, for open office design, a single number metric called articulation index (AI) was developed in the 1970s by the acoustics consulting firm of Bolt, Beranek and Newman, based in part on studies of speech intelligibility by Bell Labs in the 1940s. Essentially, the AI relates speech intelligibility, speech intensity and background noise level at the center of five octave-band frequencies, that encompass the spectrum of the human voice: 250, 500, 1000, 2000, and 4000 Hz. AI is determined by measuring the percentage of individual words that can be understood with specific speech and background sound levels. An AI of zero percent indicates zero intelligibility and therefore ideal speech privacy. This, of course does not mean silence; it means that with a specific combination of speech intensity level and background noise level, the intelligibility was nil, and therefore speech privacy, defined as a lack of intelligible intruding speech, is ideal. At the other end of the scale, a high percentage of intelligible words yield a high AI and a correspondingly low speech privacy rating. In practice, the resultant AI figures are related to listener satisfaction and speech privacy descriptors, as shown in Table 28.8.

The calculation procedure for AI involves use of weighting factors that are applied to intensity level differences between speech and background noise levels at different frequencies in order to reflect the connection between intelligibility and frequency. As we pointed out in Chapter 26, most of the information in English words is carried in consonants, whose frequencies are generally above 2 khz. Thus the problem frequently encountered in telephones conversations, of distinguishing between "f" and "s," "b" and "v," "t" and "d," and so on are due to excessive attenuation of the high frequencies that distinguish these letters, one from another. The AI calculation emphasizes the importance of high frequencies to intelligibility by using the following weighting factors:

Octave Band Center Frequency (Hz)	Relative Weighting Factor
250	1
500	2.5
1000	3.5
2000	5.0
4000	4

The calculation of AI values for an actual open office design is complex and laborious, since it considers all sound paths between each source and each receiver, including the acoustic characteristics of all reflective and absorptive surfaces in the path. The results are a specific AI factor for every receiver location. Because of the very large number of calculations involved, the calculation is always done by computer. Once the results are obtained, changes in materials, plan arrangements and dimensions can be made where the calculated AI does not satisfy the space's speech privacy requirement.

These changes can be predicted fairly accurately, since it has been demonstrated that a 3-dB change in the relative level of the intruding speech signal with respect to the background noise level will result in a 0.1 level change in AI. Thus, an increase of 3 dB in the background noise level or a decrease of 3 dB in the intruding signal (by increased absorption or path length) will have the

TABLE 28.8 **Articulation Index (AI) and Speech Privacy**

AI	Persons Satisfied with Speech Privacy (%)	Open Office Speech Privacy Descriptor
0–0.05	95–92%	Confidential
0.06–0.2	90–80%	Normal
0.21–0.3	79–65%	Minimal
>0.3	<65%	Unacceptable

effect of increasing the AI between the involved work stations by 0.10, which is the difference between Normal and Poor speech privacy. Since 3 dB is a barely perceptible change in intensity (see Table 26.3), we can appreciate how sensitive speech privacy is to small changes in intensities.

(d) Articulation Class. Numerous measurements in actual open-office installations have indicated that the absorption characteristics of the ceiling are the most important factors in speech privacy design. As noted in the previous section, the angles of incidence of speech sound on the ceiling range between 30° and 60°, with the majority of sound energy at the top of this range. A figure of merit, for absorption, called the articulation class (AC) was established that indicates absorption efficiency at angles of incidence between 45° and 55°. The usual range of AC is between 180 and 220 (no units) with higher numbers representing better absorption.

28.21   Design Recommendations for Speech Privacy in Open Offices

(a) General Factors. The architectural arrangement of spaces in an open office design has a marked influence on speech privacy. Areas should be grouped according to their speech privacy requirements. Spaces rated as *Confidential* should be placed on the perimeter of the open area to limit their exposure to speech intrusion, with the caveats relating to exterior windows and walls discussed above, observed. The design emphasis for these area should be not only an AI between 0.0 and 0.05, but also a low overall sound level, *including* background noise. Similarly, high noise producing areas should be grouped and placed on the perimeter, at maximum distance from *confidential* speech privacy areas. Use of demountable full height partitions for such spaces should be considered.

Because of reflection from perimeter walls, open-area spaces should be as large as practical, with insulated perimeter walls. Ceiling height should not be less than 9 ft clear, with a 3 ft plenum above. Extreme care must be taken with air-conditioning ductwork which, if untreated, will act as an excellent speech and noise conduit,

with multiple ceiling outlets. Furthermore, these outlets, that were once relied upon to produce an even level of background noise, no longer do so. Most HVAC systems today use variable air volume (*VAV*) designs, whose noise levels vary ±10 dB, making them useless as reliable masking sound.

(b) Individual Office (Cubicle) Design. Offices should be designed for maximum closure, and maximum partition length. Separation between occupants of adjacent offices should never drop below 10 ft, with 12-ft minimum as a design target for *normal* privacy and 16 ft for *confidential* privacy. Minimum office area should be 80 ft², with a design target of 100 to 120 ft² for normal privacy and 200 ft² for confidential privacy. Desk arrangements should be checked for optimum speech paths (for privacy), recognizing that office furniture arrangements need not be uniform in all offices (see Section 28.19).

(c) Ceilings. As the most important design element in speech privacy, care must be taken to avoid unintentional strongly reflective speech paths, as from metal pan air diffusers, flat lighting fixture diffusers and the like. If the use of such or similar items is unavoidable, highly absorptive vertical baffle strips may be placed on their perimeter to block sound paths. In general, ceiling tiles should have an articulation class rating of 220 minimum, and minimum absorption coefficients at incidence angles of 30°–60° as follows:

Frequency (Hz)	α
250	0.65
500	0.65–0.75
1000	0.85
2000	0.90
4000	0.90

(c) Partitions. As explained above, minimum height of partitions should be 65 in., with 72 in. height units separating offices from aisles and dividing departmental groups. AC rating, if available, should range from 200 to 220. STC ratings as explained above, depend to an extent on speaker locations and vary from STC 20 to STC 26. Joints between partitions should be carefully

sealed as even small openings can seriously compromise a partition's already limited efficiency. All partitions should reach the floor, although the lower portion is not always insulated in low speech privacy areas.

(d) Floors. Although carpeted floors do not seriously affect overall sound absorption, they do drastically reduce the sounds of chair movement and footfall sounds. For this reason all floors in open office areas should be carpeted. The difference in effectiveness of shallow pile carpet as opposed to deep pile is minimal, and the same differential can be achieved by using a polyurethane cushion backing in lieu of the more common jute pad. The principal purpose of carpeting is to cushion footfall impact so that its energy is not introduced into the structure. This subject is covered in detail in the discussion of structure-borne sound that follows.

(e) Lighting Fixtures. Flat-bottom lighting fixtures must never be used. Fixtures should not be placed directly over partitions so as not to create an inter-office speech reflection path. Experience has shown that the best lighting fixture (from the speech privacy point of view) is one with deep parabolic reflector cells, and overall dimensions of 1 ft × 4 in. or 2 ft × 4 in.

(f) Masking Sound. It is imperative that the level of masking sound be uniform throughout the open office area, and at as low a level as will yield the desired speech privacy. Nonuniformity will immediately be noticed as people move about, and the masking sound itself will become a source of auditory annoyance. For a similar reason, loudspeakers should not be visible. Visible units become themselves a source of interest initially, and then a source of annoyance. Speakers should be placed *in* the plenum, preferably facing upward to increase dispersion and improve uniformity. Speakers mounted in the ceiling and facing down should be avoided. Most ceiling tiles in open-office spaces have a low CAC so that sound will easily penetrate into the office area below the ceiling.

The masking noise system should comprise a signal (noise) generator, a sophisticated equalizer for shaping the signal, an amplifier with appro-

priate controls and a distribution system for the speakers to be fed. Speakers are normally 12 in., and are installed in a grid on 12- to 16-in. centers. The amplifier should be arranged so that volume levels can be remotely controlled. This permits time control so that background noise volume can be reduced automatically after working hours. This is necessary so that the few people working late are not annoyed by relatively loud background noise in the absence of intruding speech.

The noise produced by a masking sound system is variously described as white noise, noise of air rushing through an opening (whoosh sound), noise of water in piping, and the like. The actual sound can be tailored to the user's preference by adjusting the filters in the system's equalizer. Generally, the masking noise emphasizes low frequencies, because higher frequencies are immediately noticed as an annoying hiss. As noted above, background noise level should not exceed 48 to 50 dBA. In some installations the masking sound installation doubles as a PA system, although this practice is not recommended because the noise is disconnected during an announcement and when it returns, it is noticed. Background noise must be designed so as to blend into the ambience of the background, and anything that disturbs the hidden quality of masking sound is sedulously to be avoided.

(g) Design Procedure. Unfortunately, due to the large number of variables involved in open-office speech privacy design, a straightforward manual design method that will yield reliable results does not exist. However, a number of computer programs are available that will calculate AI for any location, as a result of specified speech source intrusion. On the basis of these calculations, changes can be made to the design and the program rerun to achieve improvements where the calculated AI is excessive. The most effective way to perform a complete design is then, on the basis of the above calculations, to actually construct a full scale mockup which can then be field tested and "tuned." Although this is expensive and time-consuming, it is frequently far cheaper than making the requisite changes after actual construction.

One of the distinct advantages of this two-step design procedure is that it enables the designer

to equalize the acoustic insulation "strength" of various paths, so that the attenuation of major paths are equal. There is no economic or engineering sense in a system that is much more effective in one path than in another, since sound will always choose the path of least acoustic resistance. To accomplish this balancing, the designer has many variables to "juggle." They include barrier height and material, ceiling material, baffle sizes and positions (if used), distances between source and receiver, position and directions of sources, and level of background noise.

(h) Standards. The acoustic design and testing of open-offices is covered by a group of *American Society for Testing and Materials* (ASTM) standards that should be available to anyone engaged in open-office design. The standards are available from ASTM, 100 Barr Harbor Drive, West Conshohocken, PA 19428-2959.

ASTM E1041-85(1994)e1—Standard Guide for Measurement of Masking Sound in Open Offices. This guide describes the measurement of masking sound in an open-office environment. The objectives of the measurements are to determine the uniformity of masking sound both in time and from position to position, and how closely the masking sound matches the specified spectrum.

ASTM E1110-86(1994)e1—Standard Classification for Determination of Articulation Class. This classification provides a single figure rating that can be used for comparing buildings systems and subsystems for speech privacy purposes. Excluded from this classification are applications involving female speakers or children, languages other than English, and sound spectra other than speech.

ASTM E1111-92(1996)e1—Standard Test Method for Measuring the Interzone Attenuation of Ceiling Systems. This test method is intended to provide measurements of the sound reflective characteristics of ceiling systems when used in conjunction with partial-height space dividers.

ASTM E1130-90(1994)e1—Standard Test Method for Objective Measurement of Speech Privacy in Open Offices Using Articulation Index. This test method describes a field test method

for measuring speech privacy objectively between locations in open offices. This test method relies upon acoustical measurement, published information on speech levels, and standard methods for assessing speech communication. This test method does not measure the performance of individual open office components which affect speech privacy; it measures the privacy which results from a particular configuration of components. This test method relies upon the AI, which predicts the intelligibility of speech for a group of talkers and listeners.

ASTM E1179-87(1993)e1—Standard Specification for Sound Sources Used for Testing Open Office Components and Systems. This specification states the requirements for sound sources used for measuring the speech privacy between open-offices or for measuring the laboratory performance of acoustical components. The sound source is a loudspeaker located in an enclosure driven with an appropriate test signal.

ASTM E1264-96—Standard Classification for Acoustical Ceiling Products. This classification covers ceiling products that provide acoustical performance and interior finish in buildings. It classifies acoustical ceilings by type, pattern, and certain ratings for acoustical performance, light reflectance, and fire safety.

ASTM E1374-93— Standard Guide for Open Office Acoustics and Applicable ASTM Standards. This guide discusses the acoustical principles and interactions that affect the acoustical environment and acoustical privacy in the open-office. In this context, it describes the application and use of the series of ASTM standards that apply to the open-office.

ASTM E1375-90(1994)e1—Standard Test Method for Measuring the Interzone Attenuation of Furniture Panels Used as Acoustical Barriers. This test method covers the measurement of the interzone attenuation of furniture panels used as acoustical barriers in open-plan spaces to provide speech privacy or sound isolation between working positions.

ASTM E1376-90(1994)e2—Standard Test Method for Measuring the Interzone Attenuation of Sound Reflected by Wall Finishes and Furniture Panels. This laboratory test method measures the degree to which reflected sound is attenuated by the most commonly found vertical surfaces in open-plan spaces. The vertical surfaces covered by this test method include wall finishes such as sound-absorbent panels, and furniture panels or screens. It does not cover such items as window finishes or furniture other than panels.

ASTM E1573-93—Standard Test Method for Evaluating Masking Sound in Open Offices Using A-Weighted and One-Third Octave Band Sound Pressure Levels. This test method specifies the procedures that can be used to evaluate the spatial and temporal uniformity of masking sound in open offices using A-weighted sound levels. It also specifies the procedure for evaluating the masking sound spectrum and level using ⅓-octave band sound pressure levels.

STRUCTURE-BORNE SOUND

28.22   Structure-Borne Impact Noise

Structure-borne noise is at least as serious a problem as airborne noise for the following reasons:

1. There is no air cushion between the source and the structure; thus high-intensity energy is introduced into the structure, through which it travels with minimum attenuation and at great speed.

2. Sound, once introduced into the structure, is attenuated well only by discontinuities in the structure. Since the structure must have structural integrity to carry the loads, discontinuities of the type that will stop noise are complex and expensive.

3. The entire structure constitutes a network of parallel paths for sound. Therefore, partial solutions are useless, since sound will find flanking paths. The entire structure must be sound-proofed to yield good results.

4. Unlike the case of airborne noise, additional mass does not usually block structure-borne noise, particularly in long spans where a floor can act as a diaphragm, thereby improving the structure-to-air noise transfer efficiency (like a drum).

5. The increasing use of exposed structural ceilings eliminates the attenuation that can be introduced by a plenum above a hung ceiling. This is particularly bad, since most structure-borne noise is carried by floor structures (rather than walls), which radiate sound up and down.

The discussion that follows will be limited to impact noise. Refer to Section 28.26 for a brief treatment of vibration that is felt rather than heard, and is in effect a very low-frequency noise. Many of the recommendations and techniques that will minimize impact noise will also reduce vibration.

28.23   Control of Impact Noise

Impact noise problems can be controlled in two ways—by preventing or minimizing the impact, and by attenuating it once it has occurred. Since we are concerned with structure, the latter problem is discussed first. The former problem is covered in Section 28.25. Impact on floors is more serious than wall impact because the latter will be partially attenuated at the wall/floor joint, whereas the former is introduced directly into the building framework.

The discussion below addresses each of the solutions shown in Fig. 28.43.

(a) Cushion the Impact. See Fig. 28.43(*a*). This obvious solution will frequently eliminate all but severe problems. The resilient materials in common use are floor tile of rubber and cork, or carpeting on pads, in ascending order of impact insulation. See Section 28.24 and Appendix G for quantitative data on impact isolation.

(b) Float the Floor. See Fig. 28.43(*b*). Since the key to elimination of structure-borne sound is *isolation*, separating the impacted floor from the structure floor by a resilient element is extremely effective. This element can be rubber or mineral wool pads, or blankets, or special spring metal sleepers. The effectiveness depends on the mass of the floating floor, compliance of the resilient support, and degree of isolation of the floating floor. This last element is extremely important, since flanking paths via end contacts with walls can short-circuit the floating element's sound im-

ACOUSTICS

Cushion impact   Float floor   Suspend ceiling   Sound absorber in cavity   Isolate and seal piping, etc.

(a)   (b)   (c)   (d)   (e)

Fig. 28.43 *Methods of controlling impact sound transmission through floors. (Reprinted from* Quieting: A Practical Guide to Noise Control, 1986.)

pedance and defeat the system. With floating floors it is important that:

1. Mass of the floating floor be large enough to properly spread the loads. Otherwise, the pad will compress and deform sufficiently to transmit the impact.

2. Total construction must be airtight. Airtight is soundtight.
3. Particular care be exercised where partitions rest on the floating floor [see Fig. 28.44(a)].
4. Short-circuits at walls or by penetrations be avoided; see Fig. 28.10(b). Details of proper

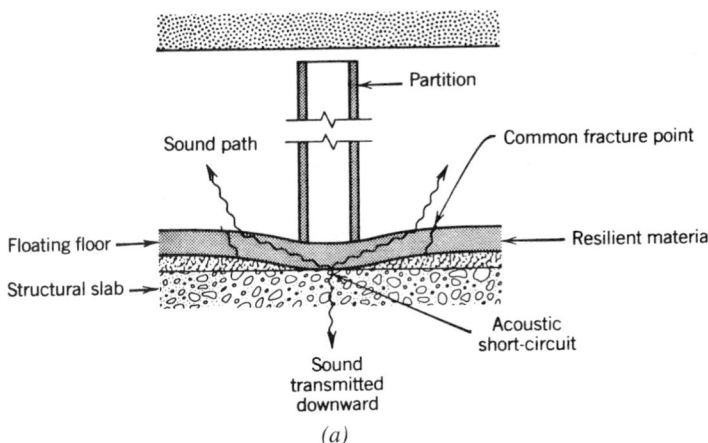

(a)

Fig. 28.44 (a) *Caution must be exercised when supporting partitions on floating floors to prevent structural failures or short-circuiting of the floating element, as illustrated. (Reprinted from* A Guide to Airborne, Impact, and Structure-Borne Noise Control in Multifamily Dwellings, 1968.)
(b) *Flanking paths in mixed-construction type floors. FHA does not recommend mixing construction types unless provisions have been made to prevent flanking (e.g., expansion joints or breaks in all structural paths between each space). (Reprinted from* A Guide to Airborne, Impact, and Structure-Borne Noise Control in Multifamily Dwellings, 1968.)

(b)

construction techniques are given in *A Guide to Airborne, Impact and Structure-Borne Noise Control in Multi-family Dwellings* (see References).

5. Construction throughout be consistent. Mixed construction types invite flanking noise paths [see Fig. 28.44(*b*)].

(c) Suspend the Ceiling—and Use Absorber in Cavity. See Fig. 28.43(*c,d*). As stated, the most disturbing noise is radiated down from the ceiling. A flexibly suspended ceiling with an acoustic absorbent layer suspended in it can be very effective if not flanked by paths leading into the walls and from there reradiating into the space below. It is imperative that the entire floor slab above be decoupled from the walls below by resilient separators.

(d) Isolate All Piping. See Fig. 28.43(*e*). All rigid structures such as piping must be isolated so as not to form a flanking path, and caulked with resilient sealant so as not to constitute an air–sound leak.

28.24 Impact Isolation Class (IIC)

This is a single-number, impact isolation rating for floor construction, similar in intent and derivation to STC wall ratings. Tests are made with a standard tapping machine and noise levels measured in 1/3-octave bands. These are plotted and compared to a standard contour, approximately as with sound transmission class. Details of typical floor constructions along with IIC ratings are given in Appendix G. Resilient floor finishes on any of the floor constructions not provided with them will add to the IIC ratings approximately as follows:

1/16-in. vinyl tile	0
1/8-in. linoleum or rubber tile	4 ± 1
1/4-in. cork tile	10 ± 2
Low-pile carpet on fiber pad	12 ± 2
Low-pile carpet on foam rubber pad	18 ± 3
High-pile carpet on foam rubber pad	24 ± 3

MECHANICAL SYSTEM NOISE CONTROL

28.25 Mechanical Noise Sources

Mechanical devices obviously make noise. And generally, the more power they consume, the more noise they make. In many of today's buildings, 40% of the total construction budget is spent on mechanical systems located throughout a building.

In most buildings, the primary sources of mechanical noise are the components of the air conditioning and air-handling systems such as fans, compressors, cooling towers, condensers, ductwork, dampers, mixing boxes, induction units, and diffusers. The curve of Fig. 28.45 depicts typical air-handling system noise and indicates the portions of the spectrum produced by each group of components.

Pumps are another source of mechanical noise whose noise, along with the noise of flowing liquid is transmitted along pipes to locations throughout the building.

Elevators, escalators, and freight elevators also introduce mechanical noise into buildings. Escalators and freight elevators pose few problems, since they are localized in a specific area and have low operation speeds. Elevator car operation however, is rapid, and it affects large areas. In addition, the motors and controls are located on or above the prime upper floors of a building. Motor, shaftway, and other equipment noise must be properly controlled to prevent annoyance to building tenants located near the shaftways or

Fig. 28.45 *Frequency spectrum of noise from HVAC system components.*

elevator penthouses and mechanical equipment rooms. Vibration isolation of these major components is a specialized problem beyond the scope of this text.

28.26   Quieting of Machines

Machines cause noise by vibration. This noise is imparted directly to the surrounding air, and by vibrational contact to the surrounding structure. Therefore, there are three ways to reduce this noise:

1. Reduce the vibration itself.
2. Reduce the airborne noise by decoupling the vibration from efficient radiating sources.
3. Decouple the vibrating source from the structure.

Refer to Fig. 28.46. Items 1, 3, and 4 reduce vibration; items 4, 5, 6, and 7 reduce and decouple the vibration from the radiating cabinet; and items 2 and 8 decouple the vibrating source from the structure. Once the noise becomes airborne or

structure borne, the isolation techniques studied above are employed.

Vibration reduction takes two forms, damping and isolation. One form of damping is accomplished by rigidly coupling the vibrating source to a large mass, frequently called an inertia block. Much of the energy is absorbed and dissipated as friction; the remainder results in lower-amplitude vibration (Fig. 28.47). *Isolation* is accomplished by supporting the vibrating mass on resilient supports. These take many forms and can be used in tandem. Thus machines are supported on fibrous, rubber, or spring steel vibration isolators, and the entire mass can be supported on a floating floor, which in turn rests on resilient vibration isolators as in Fig. 28.47. Large machines are supported on special commercial "sandwiches" of asbestos, lead, cork, and other strong resilient materials. Piping is supported on cork pads and hung on resilient hangers.

Use of a diesel driven electric generator for load peaking or cogeneration can cause very serious noise and vibration problems. The ideal solution to this situation is to completely isolate the unit in a separate out-building that is designed

1. Install motors, pumps, fans, etc. on most massive part of the machine.
2. Install such components on resilient mounts or vibration isolators.
3. Use belt drive or roller drive systems in place of gear trains.
4. Use flexible hoses and wiring instead of rigid piping and stiff wiring.
5. Apply vibration damping materials to surfaces undergoing most vibration.
6. Install acoustical lining to reduce noise buildup inside machine.
7. Minimize mechanical contact between the cabinet and the machine chassis.
8. Seal openings at the base and other parts of the cabinet to prevent noise leakage.

Fig. 28.46 *Techniques to reduce the generation of airborne and structure-borne noise in machines and appliances. (Reprinted from* A Guide to Airborne, Impact, and Structure-Borne Noise Control in Multifamily Dwellings, *1968.)*

Fig. 28.47 *Soundproofing a mechanical equipment room. Additional noise reduction to the space below can be achieved by inserting a layer of highly absorbent material in the space above the suspended acoustical ceiling. (Reprinted from* A Guide to Airborne, Impact, and Structure-Borne Noise Control in Multifamily Dwellings, *1968.)*

specifically to contain the very high noise level produced. If that is not practical and an inside location is necessary, a complete enclosure may be required to ameliorate the noise problem.

Vibration damping can be an even more serious problem, which can be solved satisfactorily with sufficient mass and proper vibration isolation. In the case of vibrating sheet metal, soft foam-type damping material glued directly to the metal is effective in damping. Flexible joints in all pipes and ducts connected to vibrating machines are mandatory. This includes flexible conduit connectors to all motors, transformers, and lighting fixtures using magnetic ballasts.

28.27 Duct System Noise Reduction

Design of a quiet duct system entails more than specifying an absorptive duct lining. Air turbulence generates noise. Turbulence increases as the velocity of airflow increases, and anywhere in the

duct system where smooth laminar flow is disturbed, such as at sharp bends. The "sharpness" of the bend permitted depends in turn on air velocity; the higher the air velocity the more aerodynamic the duct system must be to prevent turbulence, and therefore, noise. Table 28.9 demonstrates this principle clearly. The further one proceeds upstream from a point of turbulence, the higher air velocity may be, to produce the same noise level, since truly laminar flow is essentially noiseless. In principle, velocities should be as low as practical, since air turbulence noise increases exponentially with velocity.

Sound travels as easily against as with the airflow in ductwork. Therefore, both supply and return systems must be lined to control transmission of fan noise. Maximum fan noise reduction occurs at bends in the ducts. For maximum fan noise reduction in short runs, a pair of 90° bends is sometimes deliberately inserted. However, because 90° bends also introduce turbulence that generates noise, introducing bends as fan noise

TABLE 28.9 **Maximum Air Speeds in Ducts to Yield NC-15 or NC-25 Background Levels**[a]

	Supply		Return	
Location	NC-15	NC-25	NC-15	NC-25
Slot speed at min. ½-in. opening	250 fpm	350 fpm	300 fpm	420 fpm
10 ft of duct before opening	300	420	350	490
Next 20 ft	400	560	450	630
Next 20 ft	500	700	570	800
Next 20 ft	640	900	700	980
Next 20 ft	800	1120	900	1260
Next 20 ft	1000	1400	1100	1540
Next 20 ft	1300	1820	1450	2030
Next 20 ft	1600	2240	1800	2520

[a] Ducts with 1- to 2-in. inside duct lining, all duct sizes.

attenuators can be counter-productive at air velocities above 600 fpm. Another disadvantage of added system friction is the additional energy required to move the air, and its cost. This point will be discussed further, in the section on active noise cancellation.

Other design approaches that make a quiet system include smooth transitions at changes of duct size and large radius bends with turning vanes, the purpose of which is to reduce turbulence. Attenuation drops rapidly with increasing size of duct, and therefore ducts should not be deliberately oversized. Cross-talk between rooms and between ducts can be minimized by using lined ducts, separating adjacent ducts as much as possible, and gluing damping material on the

Fig. 28.48 *Since ducts are efficient sound transmission paths, precautions must be taken to avoid cross-talk, ventilation air noise and equipment noise. Avoid running ducts as a common supply or return between rooms unless they are properly baffled and lined with sound-absorbing material. The common practice, in wood frame structures, of using troughs between joists as a common return duct between rooms and between separate dwelling units results in serious noise transmission problems. Caulk or seal around ducts at all points of penetrations through partitions. Use double-wall ducts, acoustical lining, flexible boots, and resilient hangers where required. Dwelling units should be serviced by separate supply and return ducts, which branch off a main duct system. (Reprinted from* A Guide to Airborne, Impact, and Structure-Borne Noise Control in Multifamily Dwellings, *1968.)*

outside and lining on the inside. Damping material is particularly effective in preventing the thin metal walls of the ducts from resonating. Mufflers and silencers are effective in reducing the high frequency components of fan noise but much less so with low frequencies. The pressure drop they introduce, which can be considerable, must be compensated for in the fan.

The *ASHRAE Handbook*'s material on noise control in fans, plenums, housings, and ducts should be consulted for their recommendations. Figure 28.48 shows some of the ways by which cross-talk and flanking noises can be reduced. Figure 28.49 shows some of the techniques employed for quieting duct noise. Active noise canceliation

is particularly useful in duct systems since it does not reduce air flow as do liners, baffles, and other mechanical silencing devices and is effective at low frequencies which these devices are not (see Section 28.28).

The increased use of variable air volume (VAV) systems has introduced some noise problems which should not be neglected. VAV system noise can be minimized by following a few basic design rules. Maintain minimum system static pressure since fan noise increases exponentially with static pressure. Select the air volume modulating device at the fan with care as it can be a noise source. Since outlet air volume control involves duct area restriction with attendant ve-

Fig. 28.49 *Unlined duct has negligible sound attenuation and acts as an excellent speaking tube. Inside lining (a) gives 2–3 dB attenuation per foot in the range 1–2 kHz, dropping rapidly above and below those frequencies, and giving negligible low-frequencies attenuation. (b) Double lining gives higher attenuation and reduces cross-talk between ducts. Duct silencers and baffles (c–e) give high broadband attenuation: a maximum of 10–12 dB/ft in the range 1–2 kHz, and lower above and below. They are useful to reduce fan noise in short runs but cause considerable pressure drop. (Reprinted from* A Guide to Airborne, Impact, and Structure-Borne Noise Control in Multifamily Dwellings, *1968.)*

ACOUSTICS

locity increase and resultant noise, the design must include some sort of downstream silencing equipment. Ceiling diffuser acoustic characteristics must be coordinated with design air velocity and with any requirement for masking noise. Finally, avoid the use of throttling dampers on ceiling diffusers since a partially closed damper can generate very high noise levels.

28.28   Active Noise Cancellation

Since noise is a phenomenon consisting of acoustic wave energy being transmitted at certain frequencies, it is theoretically possible to *eliminate* (not mask) noise by simultaneous transmission of identical wave energy exactly out of phase with the noise. The sum of the two energy waves is zero—hence silence (Fig. 28.50). The phase matching however must be exact; if the injected out-of-phase noise is not precisely a negative image of the original noise, then not only will the signals not cancel but they may even increase the noise level. The technique for accomplishing noise cancellation is straightforward in theory but much less so in practice. A microphone samples the noise source, and feeds that signal into an analysis/ synthesis device. This, as the name implies, analyses the frequency and amplitude content of the noise and synthesizes the anti-noise, which is then fed to a loudspeaker in the original noise path. The resultant residual noise is detected by a downstream microphone, and fed back into the controller as a feedback correction signal. This entire procedure is shown in block diagram form, in Fig. 28.51. The technical problems that must be overcome are formidable. Without delving deeply into the physics involved, we can describe them qualitatively.

(a) It takes a finite amount of time to analyze the frequency content of the noise and to synthesize the anti-noise. If the noise is random it will have changed by the time the anti-noise signal is injected and will therefore not be cancelled. As a result, *the only type of noise that can effectively be attenuated by anti-noise is one that is continuous and/or predictable.* Random noise, such as that of a barking dog, cannot be actively attenuated with present technology. Candidates for noise cancellation are sounds like those produced by operating machinery, that is, continuous and repetitive. This includes very low frequency noise, commonly called vibration.

(b) The analysis and synthesis process is achieved by a device called an *adaptive digital filter.* For technical reasons, the higher the frequencies (pitch) involved, the more complex and expensive are the required digital electronics, microphones and loudspeakers. This further narrows the range of practical noise-cancellation candidates to those producing low frequencies, such as blowers, fans, rotors, internal combustion engines and their exhausts, transformers (hum), air movement in ducts, fluid movement in pipes and the like.

(c) The wave form of the noise, its modes and its dispersion in space inpact heavily on the type and amount of equipment required to

Fig. 28.50 Silencing of noise by introduction of a synthesized noise signal exactly out of phase with the original signal. The resultant sound is effectively zero—that is, silence.

Fig. 28.51 *Block diagram of active noise cancellation. The noise signal is detected, its periodicity determined and its waveform analyzed, and an out-of-phase noise is synthesized and injected into the acoustic environment. A residual noise detector provides a feedback signal, which acts to improve noise cancellation.*

effect attenuation economically (i.e., commercially). The simplest type of noise to treat is one which is confined by some sort of waveguide, at low frequency, and exhibits constant sound pressure and phase. Such a wave is known as a plane wave, and it can be "treated" with a single sampling microphone, a single loudspeaker and a single processor.

Considering the above three criteria, one of the ideal candidates for economic noise cancellation treatment is duct noise, which is low frequency, continuous, and a plane wave. Figure 28.52 shows the construction of a typical commercial duct noise suppression unit, plus application photographs. Practical duct noise cancellation equipment in use today can reduce levels from NC-50 to NC-35, and are particularly effective at the "rumble" frequencies below 200 Hz.

Important auxiliary advantages of duct noise cancellation are energy conservation and economic benefits. Passive duct noise reduction, particularly at low frequency, requires massive noise absorbers, requiring 7 to 10 ft of duct for installation. See Fig. 28.49(*c,d,e*). In addition to high first cost for these devices and their installation, they introduce a static pressure loss of ½ in. to 1½ in. w.g., depending on air flow and speed. This in turn requires a higher power (and noisier) fan. Economic analysis of such designs generally shows an advantage to active noise cancellation equipment. Such an analysis should be performed in every design where duct silencers are desired or required.

Another important application of active noise cancellation, already in wide use, is in the area of hearing conservation for people exposed to high noise levels at work. Here the out-of-phase noise

is introduced into miniature loudspeakers (earphones) in an acoustically transparent headset. This allows the wearer to hear random noise such as speech, clearly, while repetitive cyclic noise from engines and the like is attenuated (Fig. 28.53).

Other applications of active noise cancellation already in use include engine exhausts, heavy machine vibration, interiors of luxury automobiles, military and space vehicles and selected shipboard spaces. Applications will undoubtedly increase rapidly with advances in digital signal processing technology, equipment miniaturization and reduction in equipment costs.

28.29 Piping System Noise Reduction

As with airflow, noise increases exponentially with liquid flow velocity. Piping is not a major noise source normally, since the radiating diameter is small, except for flow velocities much in excess of 8 fps where the pipe is in contact with the structure. This is of course most serious where the pipe passes through NC-15 to NC-25 areas (see Table 28.7). Domestic water system mains should be limited to 50 psi in other than tall buildings and pressure in branches to 35 psi. In highrise structures pressure-reducing valves will be required in high-pressure mains to meet these recommendations. Obviously piping must be designed to prevent water hammer, and noise sources must be located away from quiet areas.

Pumps, as with all rotating equipment, are sources of vibration and noise and should be treated as described in Section 28.26. Figure 28.54 shows a typical pump installation with appro-

ACOUSTICS

Fig. 28.52 *Active duct noise cancellation equipment.*
(a) *The system components, consisting of the sensing and error-signal microphones, controller and a loudspeaker are built into a duct section of the required size, that is then coupled into the duct system.*
(b) *A mock-up of an office installation showing the relative size and positions of a dual-channel system.*
(c) *Photograph of a multichannel duct-noise cancellation system in a new office building, designed with large unlined HVAC ducts. (Courtesy of Digisonix.)*

Fig. 28.53 Schematic diagram of a headset with active noise cancellation. Because of physical proximity of all the elements, only a single microphone is used. (Courtesy of Noise Cancellation Technologies.)

priate noise reduction measures. For at least a distance of 100 pipe diameters beyond the pump, resilient pipe hangers should be used. With centrifugal pumps as with fans and blowers, machines sound concentrates in narrow bands and, if extremely disturbing, can be attenuated with resonant filters. Reciprocating pumps are more diffi-

cult to control as the pulsations are more vibration than noise. Flexible connections in the piping and U-joints in the piping will absorb much of this vibration.

28.30 Electrical Equipment Noise

Electrical equipment is generally overlooked as a noise source and this is unwise. Most electrical noise is 120-Hz hum. This can be very disturbing because it is so low a frequency and, as we have noted repeatedly, low-frequency noise is difficult to attenuate passively. Transformer noise levels are dictated by NEMA and ANSI standards. For a premium price lower noise units are obtainable. Table 28.10 lists maximum noise levels for dry-type units. Most manufacturers warranty noise levels below these. Oil- and silicone-filled units are normally quieter than dry-type transformers, as are units designed for lower temperature rise. Transformer noise can be minimized by these steps:

1. Mount unit on vibration isolators.
2. If transformer is wall hung, use resilient

Fig. 28.54 Typical pump installation with appropriate isolation and damping measures.

TABLE 28.10 **Maximum Sound Levels: Dry-Type Transformers**

kVA	Decibels (NEMA Standard)
0–9	40
10–50	45
51–150	50
151–300	55
301–500	60

hanging. If it is floor mounted, place on a massive slab as possible.

3. Locate the unit so that reflections do not amplify the sound. Sound-absorbent material on the walls behind the units is not useful at 120 Hz. Only cavity resonators will absorb appreciable amounts of sound at that frequency.

4. Use only flexible conduit connections.

5. Avoid locating transformers adjacent to, or immediately outside, quiet areas. A common error in this regard is placing a transformer pad immediately below the window of an NC 15–25 area.

The second major source of 120-Hz hum is conventional, core-and-coil discharge light fixture ballasts. This includes fluorescent plus all the HID sources. Fortunately, electronic ballasts, which are practically noiseless, are rapidly replacing core-and-coils ballasts in fluorescent fixtures, and to a lesser extent in HID units. Table 28.11 lists the recommended application of non-electronic fluorescent ballasts. Conventional coil type HID ballasts can be very noisy, and care must be exercised in their placement. With all ballasts, the method of mounting has a marked effect on the radiated noise. As pointed out earlier, when a small vibrating source is coupled rigidly to a larger body, noise is amplified because of increased source-to-air coupling. Since core-and-coil fluorescent ballasts for linear fluorescent lamps are necessarily closely coupled to large metal fixtures for heat dissipation purposes, the sound radiation is much amplified. A large number of such fluorescent fixtures mounted in a plenum can create a serious noise problem. Solution of the problem lies either in ballast replacement or in the use of absorptive material in plenums, flexible conduit connection to fixtures, and resilient fixture hanging. In severe cases ballasts can be remote-mounted. Coil-type HID ballasts are inherently noisier than fluorescent but like ballasts for compact fluorescent lamps are generally less troublesome, being coupled to small radiating bodies.

28.31 Noise Problems Due to Equipment Location

Roof-mounted HVAC units have proven themselves very economical and very noisy. The vibration, short duct runs, and sound reflections are serious problems that can be solved with vibration isolators, sound mufflers, and careful location of equipment. Roof-mounted cooling towers are a particular problem when located adjacent to a taller building. That problem has led to a spate of lawsuits and noise control legislation in many cities. For this reason particular attention should be paid in design to all exterior equipment.

TABLE 28.11 **Acoustic Criteria for Selection of Conventional Core-and-Coil Fluorescent Lamp Ballasts**

For an Installation in	Use of Ballasts with This Rating Will Usually Be Satisfactory
TV or radio station, church, synagogue	A
Offices, residence, library, reception or reading rooms, school study hall	B
Noisy office, doctor's or dentist's office, classroom	C
Industrial applications	D

In high-rise buildings, problems are cause by the conflicts between the stringent noise requirements of the prime upper floor space and the near presence of elevator machine rooms, mechanical equipment rooms, and cooling towers. These problems are almost impossible to solve after construction and require the services of an acoustics expert during design.

28.32 Sound Isolation Enclosures, Barriers, and Damping

In buildings with concentrated high-level noise sources such as certain types of machinery, it is always more desirable to reduce the noise at its source rather than attempt to treat the larger enclosing space. This is most effectively accomplished by enclosing the noise source with materials that will furnish a combination of reverberant noise reduction by absorption as explained in Section 27.9 and Fig. 27.10, and blocking of airborne sound with high transmission loss as detailed in Sections 28.7 to 28.12. These materials are available in the form of curtains, panels, and prefabricated partial and full enclosures, tailored to the specific characteristic of the noise source (Fig. 28.55). These enclosures are not normally the responsibility of the building designer. It is, however, important to know that they will be used, as well as their characteristics, so that appropriate isolation can be designed into the building for the residual sound that is radiated from the enclosures.

Where the noise is at least partially the result of vibrating (sheet) metal enclosures as for instance on laundry machines, mixers, bins, chutes, polishing drums and the like, a very effective noise control technique is to damp the vibration. This can be done by permanent attachment of a layer of foam to the vibrating metal which, by minute flexing of the foam mass, coverts the vibration energy to heat. The noise reduction can be further enhanced by adding a heavy limp barrier material to the outside of the foam. Such combinations are called composite damping barrier material, several of which are illustrated in Fig. 28.56. The foam layer also acts to decouple the barrier and thereby increase the noise attenuation.

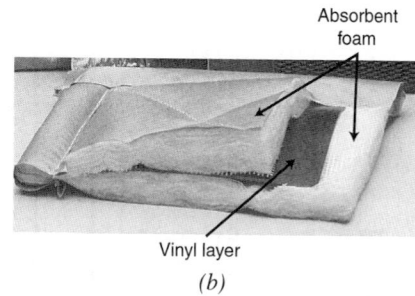

Fig. 28.55 (a) *Typical flexible nonmetallic composite acoustical barrier material which can be used as a free-hanging curtain or applied to a rigid surface. It consists of a layer of flexible vinyl sandwiched between two layers of fiberglass or other acoustic foam. The construction is seen clearly in the cutaway*
(b) *The material is most effective when applied with an airspace between it and any rigid support barrier. See Fig. 27.5 and Section 27.5. (Courtesy of E-A-R Specialty Composites, div. of AEARO Co.)*

STC AND IIC RECOMMENDATIONS AND CRITERIA

Recommendations for background noise levels (NC criteria) are given in Table 28.7. Criteria for partition isolation (STC) and impact isolation (IIC) are given in the following sections and in Table 28.12.

28.33 Multiple Occupancy Residential STC/IIC Criteria

The most important and binding criteria for residential work in the United States are issued by

Fig. 28.56 (a,b,c) *Composite damped, decoupled barrier material comprising a layer of damping foam plus an integral layer of loaded vinyl barrier material. The exterior may be covered with a decorative cover layer. The foam layer acts to damp the vibrating surface to which it adheres, and to decouple the barrier layer. Preliminary attenuation test data for a composite consisting of ¼ in. foam and 1 lb/ft² is shown in (c). (Courtesy of E-A-R Specialty Composites, div. of AEARO Co.)*

ACOUSTICS

TABLE 28.12 **Recommended STC for Partitions; Specific Occupancies**

Type of Occupancy	Wall, Partition, or Panel Between		Sound Isolation Requirement: Background Level in Room Being Considered	
	Room Being Considered AND	Adjacent Area	Quiet	Normal
Normal school buildings without extraordinary or unusual activities or requirements	Classrooms	Adjacent classrooms	STC 42	STC 40
		Corridor or public areas	STC 40	STC 38
		Kitchen and dining areas	STC 50	STC 47
		Shops	STC 50	STC 47
		Recreation areas	STC 45	STC 42
		Music rooms	STC 55	STC 50
		Mechanical equipment rooms	STC 50	STC 45
		Toilet areas	STC 45	STC 42
	Music practice rooms	Adjacent practice rooms	STC 55	STC 50
		Corridor and public areas	STC 45	STC 42
Executive areas, doctors' suites; confidential privacy requirements	Office	Adjacent offices	STC 50	STC 45
		General office areas	STC 48	STC 45
		Corridor or lobby	STC 45	STC 42
		Washrooms and toilet areas	STC 50	STC 47
Normal office; normal privacy requirements	Office	Adjacent offices	STC 40	STC 38
		Corridor, lobby, exterior	STC 40	STC 38
		Washrooms, kitchen, dining	STC 42	STC 40
Any occupancy, using rooms for group meetings	Conference rooms	Other conference rooms	STC 45	STC 42
		Adjacent offices	STC 45	STC 42
		Corridor or lobby	STC 42	STC 40
		Exterior of building	STC 40	STC 38
		Kitchen and dining areas	STC 45	STC 42
Large offices, drafting areas, banking floors, etc.	Large general office areas	Corridors, lobby, exterior	STC 38	STC 35
		Data-processing area	STC 40	STC 38
		Kitchen and dining areas	STC 40	STC 38
Motels and urban hotels, hospitals and dormitories	Bedrooms	Adjacent bedrooms[a]	STC 52	STC 50
		Bathroom[a]	STC 50	STC 45
		Living rooms[a]	STC 45	STC 42
		Dining areas	STC 45	STC 42
		Corridor, lobby, or public spaces	STC 45	STC 42

[a] Separate occupancy.

Source: Courtesy of U.S. Gypsum.

ACOUSTICS

the Department of Housing and Urban Development in conjunction with the Federal Housing Administration (HUD/FHA). The reader is referred to the latest issue of *A Guide to Airborne, Impact and Structure-Borne Noise Control* and to any subsequent HUD/FHA publications. Tables 28.13 and 28.14 give the essential data presented there in the current edition.

The recommendations are divided into grades I, II, and III. Grade II is the most important category and is applicable primarily in residential urban and suburban areas considered to have the "average" noise environment. The nighttime exterior noise levels might be about 40 to 45 dBA, and the permissible interior noise environment should not exceed NC 25–30 characteristics. Grade I is suburban, with "quiet" noise environment characterized by a nighttime exterior noise level of about 35 to 40 dBA. *Grade I STC/IIC criteria are three points higher than those of grade II.*

The fundamental criteria of airborne sound insulation between dwelling units are, for grade II:

Wall partitions	STC > 52
Floor-ceiling assemblies	IIC > 52

These apply where similar function spaces are contiguous, as bedroom to bedroom and living room to living room. Where this is not the case, the isolation must be revised up to meet the higher sensitivity requirement.

Grade III recommendations are minimal and can be characterized as "noisy," with an average nighttime exterior noise level of about 55 dBA or higher. *Grade III STC/IIC recommendations are four points lower than those of grade II.*

28.34 Specific Occupancies

(a) Schools. School buildings house spaces of many kinds—classrooms, auditoriums, gymnasiums, cafeterias, shop areas, and music suites—that pose acoustics problems.

1. *Auditoriums.* All auditoriums require a sound system for some of the activities accommodated (see Figs. 27.24 and 27.27). The most difficult aspect, architecturally, is integration of the loudspeaker system into the design. To provide

TABLE 28.13 Criteria for Airborne Sound Insulation of Wall Partitions Between Dwelling Units

Partition Function Between Dwellings			Grade II STC
Apt. A		Apt. B	
Bedroom	to	Bedroom	52
Living room	to	Bedroom[a]	54
Kitchen[b]	to	Bedroom[a]	55
Bathroom	to	Bedroom[a]	56
Corridor	to	Bedroom[a,c]	52
Living room	to	Living room	52
Kitchen[b]	to	Living room[a]	52
Bathroom	to	Living room	54
Corridor	to	Living room[a,c,d]	52
Kitchen	to	Kitchen[e]	50
Bathroom	to	Kitchen	52
Corridor	to	Kitchen[a,c,d]	52
Bathroom	to	Bathroom	50
Corridor	to	Bathroom[a,c]	48

[a] Whenever a partition wall might serve to separate several functional spaces, the highest criterion must prevail.

[b] Or dining or family or recreation room.

[c] It is assumed that there is no entrance door leading from corridor to living unit.

[d] Criterion applies to the partition. Doors in corridor partitions must have the rating of the partition, not vice versa.

[e] Double wall construction is recommended to minimize kitchen impact noises.

Source: Reprinted from *A Guide to Airborne, Impact, and Structure-borne Noise Control in Multifamily Dwellings* (1968). For Grade I, add 3 points; for Grade III, subtract 4 points.

proper sound reinforcement, loudspeakers must be located properly without large obstructions. To accomplish this, the loudspeaker system should be incorporated in the earliest design stages.

In general, a school auditorium must be a multipurpose facility. It should be designed to meet speech requirements and also should be suitable for the school's music activities. Often a modified gymnasium (gymnatorium) or cafeteria (cafetorium) functions as an auditorium. Obviously, acoustics compromises occur in such facilities.

TABLE 28.14 **Criteria for Airborne and Impact Sound Insulation of Floor–Ceiling Assemblies Between Dwelling Units**

Partition Function Between Dwellings			Grade II	
Apt. A		Apt. B	STC	IIC
Bedroom	above	Bedroom	52	52
Living room	above	Bedroom[a]	54	57
Kitchen[b]	above	Bedroom[a,c]	55	62
Family room	above	Bedroom[a,d]	56	62
Corridor	above	Bedroom[a]	52	62
Bedroom	above	Living room[e]	54	52
Living room	above	Living room	52	52
Kitchen	above	Living room[a,c]	52	57
Family room	above	Living room[a,d]	54	60
Corridor	above	Living room[a]	52	57
Bedroom	above	Kitchen[c,e]	55	50
Living room	above	Kitchen[c,e]	52	52
Kitchen	above	Kitchen[c]	50	52
Bathroom	above	Kitchen[a,c]	52	52
Family room	above	Kitchen[a,c,d]	52	58
Corridor	above	Kitchen[a,c]	48	52
Bedroom	above	Family room[e]	56	48
Living room	above	Family room[e]	54	50
Kitchen	above	Family room[e]	52	52
Bathroom	above	Bathroom[c]	50	50
Corridor	above	Corridor	48	48

[a] This arrangement requires greater impact sound insulation than the converse, where a sensitive area is above a less sensitive area.

[b] Or dining or family or recreation room.

[c] It is assumed that plumbing fixtures, appliances, and piping are installed with proper vibration isolation.

[d] The airborne STC criteria in this table apply as well to vertical partitions between these two spaces.

[e] This arrangement requires equivalent airborne sound insulation and perhaps less impact sound insulation than the converse.

Source: Reprinted from *A Guide to Airborne, Impact, and Structure-borne Noise Control in Multifamily Dwellings* (1968). For Grade I, add 3 points; for Grade III, subtract 4 points.

The large areas of sound-absorbing treatment in either kind of space make them unsuitable as auditoriums and most events that require speech amplification.

2. *Classrooms.* Typical classrooms are approximately 30-ft square with 10-ft ceilings. Adequate speech communication is easily achieved in a room of this size. Classroom acoustic design usually involves:

a. Locating sound-absorbing treatment to reduce classroom noise levels.

b. Ensuring adequate privacy between adjacent spaces.

c. Control of air-handling system noise.

Acoustic tile ceilings yield adequate sound absorption for most classrooms. An NRC of 0.7 is recommended (see Section 27.10).

Partition systems must produce sufficient isolation to prevent disturbance from activities in other classrooms and corridors. Such partitions should run full height from floor to ceiling slab or roof construction. If return air transfer ducts are needed, their noise reduction characteristics must be as good as the walls or doors that they penetrate (for NC data, see Table 28.7). Unit ventilators commonly used for classrooms produce approximately the required level of background noise.

3. *Music Suites.* School music programs usually range from individual instruction to band and choral concerts. The teaching spaces required for such a program include practice rooms, ensemble rooms, and large rehearsal spaces. Both room acoustics design and sound isolation are important in music suites. Privacy between adjacent spaces is critical, since simultaneous use is necessary.

4. *Dining Areas.* The activity in cafeterias or lunchrooms usually generates a great deal of noise. The kitchen and serving areas should be separated from the eating spaces. Treat ceilings and wall areas in the cafeteria with sound-absorbing material. Unless the ceiling is completely treated with a highly efficient sound absorbing material, the environment will be unsatisfactory due to its high noise level. Minimum NRC of this material should be 0.8.

5. *Gymnasiums.* Activities in gymnasiums create so much noise that even extensive treatment will not quiet these spaces. A quiet gymnasium probably would be unsatisfactory, since spectators are conditioned to consider the noise as an enjoy-

ACOUSTICS

able aspect of athletic events. However, to provide a proper environment for normal sports activities, the ceiling area should be sound absorbing. In addition, if a sound amplification system is to be used, sound-absorbing wall treatment may be required to eliminate echoes that would reduce intelligibility of announcements. An NRC of 0.7 is suggested with sound-absorbent material to be ceiling mounted.

If a gymnasium will also serve as an auditorium, loudspeaker system placement needs special consideration. For example, the loudspeakers should be located above the source location for speeches and plays.

6. *Swimming Pools.* The acoustic environment of swimming pools is often chaotic. Most sound-absorbing materials disintegrate in the high-humidity conditions prevalent in pool areas. Use special sound-absorbing units that have moisture-resistant properties.

7. *Shops.* Metal, woodworking, and scenery shops in schools contain many noise sources—saws, planers, drill presses, and manual tools. Each generates high airborne and structure-borne noise levels. Consolidating noisy areas and maximizing the distance between them and quiet spaces are essential. Ceiling and wall absorptive treatment with an NRC of at least 0.75 is recommended.

(b) Houses of Worship. The basic activities of these buildings usually combine speech and music. Thus, the worship environment must be acoustically hospitable to both. The architectural plan also must respond to religious requirements including the relative positioning of pulpits, lecterns, the altar (if any) and choir.

Successful acoustics can be achieved by designing the overall environment for music and providing special assistance for speakers. Large congregational spaces frequently include a sound-reflecting canopy over the pulpit to direct the speaker's voice to the congregation. In some large buildings, a loudspeaker located above the canopy further reinforces speech from the pulpit. The choir and organ communicate with the entire volume of the building and, therefore, benefit from a reverberant environment.

(c) Offices. Although office buildings contain public spaces, auditoriums, and restaurants, prime occupancy is in office areas. Most acoustics problems in office buildings relate to privacy—either between spaces within a single firm or between adjacent firms. Speech privacy has been discussed at length in Sections 28.16 through 28.21, including consideration of open-plan offices. Mechanical equipment noise problems are discussed in Sections 28.25 through 28.32.

(d) Apartment Buildings. Large apartment buildings house hundreds and even thousands of residents. Privacy and freedom from annoyance are high on the list of tenant requirements. See HUD/FHA criteria in Section 28.34.

The performance of partitions is compromised in many designs by careless planning of convenience outlets, medicine cabinets, and mechanical services. Moreover, direct-exhaust duct connections between apartments and back-to-back placement of medicine cabinets result in loss of privacy. Back-to-back convenience outlets must be avoided.

Installation of rugs or carpeting provides the best protection against footfall noise. Many leases now require that a tenant provide such impact-reducing floor covering over most of the floor area in the apartment. Good design also dictates that similar spaces in adjacent apartments be grouped—bedrooms next to bedrooms, for example. Absorptive material in bedrooms should be ceiling mounted. A minimum NRC of 0.6 is recommended.

Apartment house site selection seldom includes consideration of acoustics. Nevertheless, truck routes, superhighways, and airports can be annoying "neighbors." As mentioned above, cooling towers serving adjacent buildings must be considered during planning stages.

OUTDOOR ACOUSIC CONSIDERATIONS

28.35 Sound Power and Pressure Levels in Free Space (Outdoors)

The equations of Section 27.8 are not applicable to outdoor propagation, in which the large reflective component of the indoor condition is absent. Although the propagation of sound outdoors may not appear to be of immediate importance in architectural acoustics, outdoor noise sources such as traffic, cooling towers, and aircraft are frequently loud enough to disturb activities within or immediately adjacent to a building. Conversely, the noise made by building equipment such as cooling towers, heat pumps, and even window air conditioners may be loud enough to disturb neighbors in a nearby building. For this reason it is desirable to have some basic understanding of outdoor sound propagation.

For preliminary evaluation of an outdoor noise problem, assuming a small nondirectional source on the ground, SPL can be determined from Eqs. 27.5 and 27.6. For large sources such as cooling towers and traffic, which do not exhibit inverse square properties, sound level estimates are best made on the basis of experience and empirical data beyond the scope of this book (see Magrab, 1975; Schaudinischky, 1976). For small outdoor sources, the equipment power level can be estimated by measuring the sound power level at 5 ft and adding 15 dB. Other factors, such as moisture in the air, the presence of trees, wind, and temperature gradients, will affect outdoor sound propagation to some extent, but they can be ignored except when great distances (i.e., over 1000 ft) are involved. Barriers, which are most effective outdoors, were discussed in Section 28.14.

28.36 Building Siting

As important as interior structural design is building siting vis-a-vis exterior noise sources. Since this subject is somewhat beyond our scope, the discussion is brief. Buildings should be sited, with respect to noise sources:

1. To use natural terrain noise barriers (Fig. 28.57).
2. With respect to trees as noise barriers, rely only on thick wooded areas (Fig. 28.57).
3. To avoid naturally poor sites (Fig. 28.57).
4. To avoid sound reflection from other buildings (Fig. 28.57).

Factor 4 is also important in a multiwing building; in avoiding U-shapes or other configurations where a central court becomes an echo chamber.

Where avoidance of an exterior noise source is impossible, quiet zones can be buffered from the noise by placing higher noise areas on the noisy side of the building. Thus, in a school, classrooms and offices can be buffered by a cafeteria and gym; in a residence, bedrooms by living rooms and corridors; in an office building, private offices by noisier clerical offices; and so on.

REFERENCE MATERIAL

28.37 Glossary

A-Scale. Identifies the filtering system that has characteristics that roughly match the response characteristics of the human ear. Referred to as dBA.

Absorption Coefficient α. The ratio of the sound absorbed to the sound incident on the material or device.

Anechoic Room. An anechoic room provides a free field acoustic testing environment like the out-of-doors. All the sound emanating from a source is essentially absorbed at the surfaces of the room.

ANSI. American National Standards Institute, a nonprofit, national technical association that publishes standards covering definitions, test methods, recommended practices, and specifications of materials. Formerly American Standards Association (ASA) and United States of American Standards Institute (USASU).

Articulation Class (AC). A figure of merit for acoustic absorption, indicating absorption efficiency of a material for angles of incidence of the sound wave between 45° and 55°. The range is 150–250 (no units), with higher numbers indicating better absorption.

Articulation Index (AI). A numerical index in the range of 0.0–1.0 indicating the degree of speech intelligibility in an open office design, measured with background noise, including masking sound, if any. Also a measure of speech privacy. An index of 0.0 indicates no speech intelligibility, hence ideal speech privacy; an index of 1.0 indicates perfect speech intelligibility and therefore no speech privacy whatsoever.

ACOUSTICS

THIS   NOT THIS

(a)

(b)

|← 100 FT →|

A THICK GROWTH OF LEAFY TREES AND
UNDERBRUSH REDUCES NOISE ABOUT
6 to 7 DB/100 FT (AVERAGE OVER
AUDIBLE FREQ. RANGE)

LOW—FREQ. LOSS: 3–4 dB
HIGH—FREQ. LOSS: 10–12 dB

HIGH FREQ. REDUCTION 3–4 dB

SINGLE ROW OF TREES IS WORTHLESS
AS NOISE BARRIER. DUE TO INTER—REFLECTION
MULTI—ROWS OF TREES ARE MORE EFFECTIVE

(c)

AVOID HOLLOWS OR DEPRESSIONS.
THEY ARE GENERALLY NOISIER THAN
FLAT OPEN LAND.

(d)

BUILDING SITES IN OPEN
AREAS ARE LESS NOISY
THAN SITES IN CONGESTED
BUILDING AREAS

TRAFFIC ARTERIES BETWEEN
TALL BUILDINGS ARE
QUITE NOISY.

Fig. 28.57 (a) *Use of natural noise barriers.*
(b) *Effectiveness of wooded areas as noise barriers. Noise reduction of trees.*
(c) *An example of a poor building site.*
(d) *Building sites near traffic arteries and other buildings. (Reprinted from* A Guide to Airborne,
Impact, and Structure-Borne Noise Control in Multifamily Dwellings, *1968.)*

ACOUSTICS

ASA. Formerly American Standards Association, now American National Standards Institute (ANSI).

ASTM. Formerly American Society for Testing and Materials, now ASTM, a nonprofit, national technical society that publishes definitions, standards, test methods, recommended installation practices and specifications for materials. ASTM is a consensus group of the entire building materials industry. ASTM sets standards for products and establishes methods for testing.

Baffle or Barrier, Sound. A shielding structure or partition used to increase the effective length of a sound transmission path between two locations.

Ceiling Attenuation Class (CAC). A single-number or range for evaluating the effectiveness of an acoustical ceiling construction in isolating audible airborne sound transmission, tested at 16 one third-octave frequencies. Higher numbers indicate more effectiveness in preventing noise between rooms. Tested in accordance with ASTM E1414. Previously indicated as Ceiling Sound Transmission Class (CSTC) or Sound Transmission Class (STC).

Critical Frequency. The lowest frequency at which the wavelength of a bending wave, traveling in a structure, is the same as the wavelength in air at that frequency.

Damping. Dissipation of structure-borne noise. For the most part, this is accomplished by using a material with a high internal energy-absorbing capacity (i.e., high internal damping).

Decibel (dB). Adopted for convenience in representing vastly different sound pressures. The sound pressure level (SPL) in decibels is 10 times the logarithm to the base 10 of the squared ratio of the sound pressure to a reference pressure of 20 micropascals. This reference pressure is considered the lowest value at 100 Hz that the ear can detect.

Diffraction. The tendency of sound waves to flow readily around obstacles which are small in comparison to the wave length of the sound.

Diffuse Sound Field. A sound field that at any given point is made up of sound waves of all angles of incidence.

Direct Sound Field. The sound in a region in which all or most of the sound arrives directly from the source without reflection.

FIIC. Field Impact Insulation Class, is determined by actual field test. Also see FSTC, Field Sound Transmission Class, also determined by actual field test.

Flanking Sound Path. The transmission of sound or noise from one room to another by indirect paths, rather than directly through an intervening partition.

Flutter. A multiple echo set up between parallel reflecting surfaces.

Free Sound Field (Free Field). A field in a homogeneous medium, free from boundaries. In the free field, the sound pressure level decreases 6 dB for a doubling of distance, from a point source.

FSTC. Field STC, or Sound Transmission Class, is determined by actual field test, performed per ASTM E336. Also see FIIC, which is Field Impact Insulation Class, also determined by actual field test.

Impact Isolation (Insulation) Class (IIC). A single-figure rating that provides an estimate of the impact sound-isolating performance of a floor-ceiling assembly.

Intensity. The amount of sound energy per second that is carried across a unit area.

Intensity Level (IL). A measure of the acoustic power passing through a unit area expressed on a decibel scale referenced to some standard (usually 10^{-12} W/cm^2).

Interference. The destructive or reinforcing action of two or more waves arriving at the same position simultaneously.

Loudness. The subjective human definition of the intensity of a sound.

Masking. The presence of a background noise increases the level to which a sound signal must be raised in order to be heard or distinguished. If the level of background noise is significantly higher than that of the sound signal, the signal cannot be heard. This effect is known as masking.

Mass Law. States that the transmission loss of walls (in part of the frequency range) is controlled entirely by the mass per unit area

ACOUSTICS

of the panel. Also that the transmission loss increases 6 dB for each doubling of frequency or each doubling of the panel mass per unit area.

Noise. Any undesired sounds, usually of different frequencies, resulting in an objectionable or irritating sensation.

Noise Reduction (NR). (1) The reduction in sound pressure level caused by making some alteration to a sound source. (2) The difference in sound pressure level measured between two adjacent rooms caused by the transmission loss of the intervening wall.

Noise Reduction Coefficient (NRC). The average sound absorption coefficient to the nearest .05 measured at the four one-third octave band center frequencies of 250, 500, 1000, and 2000 Hz.

Octave Band (OB). A range of frequency where the highest frequency of the band is double the lowest frequency of the band. The band is usually specified by the center frequency.

Phon. Loudness level, at a particular frequency, equal to the 1000-Hz decibel level of that equal-loudness contour.

Pink Noise. Wide frequency spectrum noise, whose amplitude drops 3 dB per octave with increasing frequency (equal energy per octave). Useful for masking.

Random Noise. A noise whose magnitude and/or frequency cannot be predicted precisely at any given time. A rough approximation of random noise is the static heard on a radio between stations (see *Noise, White Noise, Pink Noise*).

Reverberation. The persistence or echoing of previously generated sound caused by reflection of acoustic waves from the surfaces of enclosed spaces.

Reverberation Time. The time required for a sound to decay to a value one-millionth of its original intensity or to reduce sixty decibels after the sound source has stopped.

Sabin. The unit of acoustic absorption. One sabin (ft^2) (m^2) is the absorption of 1 square foot (meter) of perfect sound-absorbing material or open space with no reflecting surfaces.

Sound Absorption Coefficient. The fraction of the incident energy absorbed (not reflected) by a material when a sound wave strikes it is called the sound absorption coefficient of that material. Usually represented by the Greek letter alpha (α).

Sound Barrier. A material installed in a plenum or a partition to prevent the passage of sound from one area to another. Sound-deadening board and lead sheet or special insulation make good sound barriers.

Sound-Level Meter. An instrument for the direct measurement of sound pressure level. Sound-level meters may also incorporate octave-band filters for measuring sound directly in octave bands.

Sound Power Level (PWL). A measure of the total airborne acoustic power generated by a noise source, expressed on a decibel scale referenced to some standard (usually 10^{-12} W).

Sound Pressure Level (SPL). A measure of the air pressure change caused by a sound wave. Expressed on a decibel scale referenced to some standard (usually 0.0002 μ bar).

Sound Transmission Class (STC). A single-number rating of a building element's efficacy in blocking the transmission of sound, as compared to a standard transmission attenuation/frequency curve. See also *Ceiling Attenuation Class.*

Transmission Loss (TL). The reduction of airborne sound power that is caused by placing a wall or barrier between the reverberant sound field of a source and its receiver. Transmission loss is a property of the wall or barrier.

White Noise. Noise of wide frequency range in which the amplitude of the noise is essentially the same in all frequency bands (equal energy per frequency band).

28.38   Reference Standards

See Section 28.11(h) for a description of the contents of these standards.

ASTM C423-90. Test Methods for Sound Absorption and Sound Absorption Coefficients by the Reverberation Room Method.

ASTM E90. Method for Laboratory Measurement of Airborne-Sound Transmission Loss of Building Partitions.

ASTM E413 (1994). Classification for Rating Sound Insulation.

ASTM E795-93. Practices for Mounting Test Specimens During Sound Absorption Tests.

ASTM E1041-85 (1994)e1. Guide for Measurement of Masking Sound in Open Offices.

ASTM E1110-86 (1994)e1. Classification for Determination of Articulation Class.

ASTM E1111-92 (1996)e1. Test Method for Measuring the Interzone Attenuation of Ceiling Systems.

ASTM E1130-90 (1994)e1. Test Method for Objective Measurement of Speech Privacy in Open Offices Using Articulation Index.

ASTM E1179-87 (1993)e1. Specification for Sound Sources Used for Testing Open Office Components and Systems.

ASTM E1264-96. Classification of Acoustical Ceiling products.

ASTM E1374-93. Guide for Open Office Acoustics and Applicable ASTM Standards.

ASTM E1375-90 (1994)e1. Test Method for Measuring the Interzone Attenuation of Furniture Panels Used as Acoustical Barriers.

ASTM E1376-90 (1994)e2. Test Method for Measuring the Interzone Attenuation of Sound Reflected by Wall Finishes and Furniture Panels.

ASTM E1414. Method for Measuring Sound Attenuation Between Rooms sharing Common Ceiling Plenum.

ASTM E1573-93. Test Method for Evaluating Masking Sound in Open Offices Using A-Weighted and One-third Octave Band Sound Pressure Levels.

International Standardization Organization. "Acoustics; Rating of Sound Insulation in Buildings and of Building Elements; Airborne Sound Insulation in Buildings and of Interior Building Elements" IS0 717-1, 1982.

American Society for Testing and Materials. Classification for Rating Sound Insulation, ASTM

28.39 Units and Conversions

See Table 28.15.

TABLE 28.15 Acoustic Units and Conversions

Units	MKS	CGS
Force	kilogram-meter/s^2 = newton	gram-cm/s^2 = dyne
Intensity	watts/meter^2	watts/cm^2
Pressure	newtons/meter^2 = pascals	dynes/cm^2 = microbars

	In Conversion:		
Quantity	Multiply	By	To Obtain
Force	newtons	10^5	dynes
	dynes	10^{-5}	newtons
Intensity	watts/cm^2	10^4	watts/m^2
	watts/m^2	10^{-4}	watts/cm^2
Pressure	pascals	10	microbars
	microbar	10^{-1}	pascals

NOTE: One atmosphere = 1 bar = 10^6 μbar.

ACOUSTICS

28.40   Symbols

See Table 28.16.

TABLE 28.16 **Symbols and Abbreviations Used Commonly in Acoustics**

$A =$	Total absorption, sabins; area in unit being used
$A_R =$	Absorption in receiving room, sabins
$A_{1,2\cdots} =$	Total absorption of each different material in a space, sabins
AC	Articulation Class
$c =$	Velocity of sound, feet per second
CAC	Ceiling attenuation class
$d =$	Distance from source, meters or feet
$dB =$	Decibel
$f =$	Frequency of sound, hertz (hz)
$I =$	Intensity, watts per square centimeter
$I_a =$	Absorbed energy, watts per square centimeter
$I_i =$	Incident energy, watts per square centimeter
$I_0 =$	Reference intensity, 10^{-16} watt per square centimeter
IIC =	Impact insulation class, no units
IL =	Intensity level, decibels
NC =	Noise criterion, no units
NRC =	Noise reduction coefficient, no units
$NR =$	Noise reduction, decibels
$p =$	Pressure, pascals or microbars
$p_0 =$	Reference base pressure 2×10^{-5} Pa
$P =$	Acoustic power, watts
Pa =	Pascal, unit of pressure (SI)
$PWL =$	Sound power level, decibels
$R =$	Room constant, square feet (meters)
$r =$	Distance from source, meters or feet
$S =$	Surface area, in unit being used
$SPL =$	Sound pressure level, decibels
STC =	Sound transmission class, no units
$T_R =$	Reverberation time, seconds
$TL =$	Transmission loss, decibels
$V =$	Volume
$W,P =$	Sound power, watts
$W_0 =$	Reference base sound power 10^{-12} watt
$a, \alpha =$	Absorption coefficient (no units)
$\overline{a}, \overline{\alpha} =$	Average absorption coefficient (no units)
$\lambda =$	Wavelength, feet or meters
$\Sigma =$	Sum of, or total (no units)
$\Sigma S\alpha = \Sigma A =$	Total absorption, sabins
$\Delta =$	Change in a quantity or difference between two quantities

NOTE. Where definitions are expressed in feet, centimeters or meters is also understood, with proper conversion factors, and vice versa.

28.41   REFERENCES

A Guide to Airborne, Impact, and Structure-Borne Noise Control in Multifamily Dwellings (1968). U.S. Department of Housing and Urban Development, Washington, DC.

Application of Sound Power Level Ratings (1973). Publication 303, AMCA, Arlington Heights, IL.

Beranek, L. L. (1988). *Noise and Vibration Control,* rev. ed. INCE, Washington, DC.

Brüel, P. V. and Kjaer. *Measuring Sound,* Denmark.

Cavanaugh, W. J., Farrell, W. R., Hirtle, P. W., and Watters, B. G. (1962). "Speech Privacy in Buildings," *Journal of the Acoustical Society of America,* Vol. 34, No. 4.

Crocker, M. J. (Ed.). (1998). *Handbook of Acoustics,* Wiley, New York.

Dupree, R. B. (19XX). *Catalog of STC and IIC Ratings for Wall and Floor/Ceiling Assemblies,* Office of Noise Control, California Department of Health Services, Berkeley, CA.

Fader, B. (1981). *Industrial Noise Control,* Wiley, New York.

Harris, C. (1979). *Handbook of Noise Control,* 2nd ed., McGraw-Hill, New York.

Hedeen, R. A. *NIOSH Compendium on Materials for Noise Control,* DHEW Publ. #80-116, U.S. Department of Health, Education and Welfare, Cincinnati, OH.

Information on Levels of Environmental Noise (1974). Report 550/9-74-004, U.S. Environmental Protection Agency, Washington, DC.

Magrab, E. B. (1975). *Environmental Noise Control,* Wiley, New York.

Pearsons, K. S., and Bennet, R. L. (1974). *Handbook of Noise Rating,* National Aeronautics and Space Administration Report, NASA CR-2376, Washington, DC. pp. 55–56.

Practical Building Acoustics (1976). Sound Research Labs, Holstead, England.

Quieting: A Practical Guide to Noise Control (1976). National Bureau of Standards Handbook 119, NBS, Washington, DC.

Schaudinischky, L. H. (1976). *Sound, Man and Building,* Applied Science Publishers, London.

Schiff, M. L. (1975). "Noise Control for Draft Fans." *Specifying Engineer,* pp. 62–70.

Terminology, Laws and Application of Fan Sound Data (1972). Publication SD100-1972, Barry Blower Co.

White, F. A. (1975). *Our Acoustic Environment,* Wiley, New York.

Young, R. W. (1965). "Revision of the Speech Privacy Calculation," *Journal of the Acoustical Society of America,* Vol. 38, No. 4.

ACOUSTICS

PART X
APPENDICES

The purpose of this part is to present extensive important reference and ancillary data in a separate, easily accessed section. the appendices are entitled:

A   Climate Conditions for the United States, Canada, and Mexico
B   Solar Data
C   Annual Solar Performance
D   Sunshading
E   Economic Analysis
F   Sound Transmission Data for Walls
G   Sound Transmission and Impact Insulation Data for Floor/Ceiling Construction
H   Options for Computation Assistance
I   Metrication; SI Units; Conversions
J   ASHRAE Standards for Energy Efficient Building Design

A

CLIMATIC CONDITIONS FOR THE UNITED STATES, CANADA, AND MEXICO

These data are taken from two ASHRAE publications: the 1997 *Handbook of Fundamentals,* and ANSI/ASHRAE Standard 90.2-1993, *Energy Efficient Design of New Low-Rise Residential Buildings,* reprinted with permission of the American Society of Heating, Refrigerating, and Air-Conditioning Engineers, Inc., Atlanta, Ga.

Station Designations

AP = airport

CO = urban areas, influenced by surroundings

No designation = semirural, similar to airport locations

Winter Design Conditions

HDD65 are annual heating degree-days base 65° F. They are from ANSI/ASHRAE Standard 90.2-1993, *Energy Efficient Design of New Low-Rise Residential Buildings,* and should be used to enter the figures in Appendix J that specify maximum *U*-values, etc., for residential buildings. (Note that in Appendix B, HDD65 are also given, and may vary slightly from these values because they are based upon the average from a different set of years.)

Design dry-bulb (97.5%) are outdoor temperatures, °F, that will, in a normal year, be the lowest experienced during 97.5% of the hours in a winter season defined as

United States, December through February

Canada, January

Summer Design Conditions

Design dry-bulb and mean coincident wet-bulb (2.5%) are outdoor temperatures, °F, that will, in a normal year, be the highest experienced during all but 2.5% of the hours in a summer season defined as

United States, June through September

Canada, July

Mean daily range is the average difference, in F°, between daily maximum and minimum dry-bulb temperatures during the above-defined summer season.

Design wet-bulb (2.5%) is the outdoor temperature, °F, that will, in a normal year, be the highest experienced during all but 2.5% of the hours in a summer season as defined above.

CDH74 are normal cooling degree-hours base 74°F. They are from ANSI/ASHRAE Standard 90.2-1993, *Energy Efficient Design of New Low-Rise Residential Buildings,* and should be used to enter the figures in Appendix J that specify maximum *U*-values, etc., for residential buildings.

Other stations: many more stations in each state or province, and for other countries, are found both in the ASHRAE 1993 *Handbook of Fundamentals,* Chapter 24, and ANSI/ASHRAE Standard 90.2-1993, *Energy Efficient Design of New Low-Rise Residential Buildings,* Tables 9.1 and 9.2.

Interpretations Between Stations.

As a general rule, weather data from listed stations can be adjusted for nearby locations in these ways:

*Adjustment for elevation.** For lower elevations, which tend to be warmer, *increase* the design temperatures:

*Elevations for many locations are given in Appendix B, Table B.15.

Dry-bulb temperature: 1F° per 200 feet of elevation

Wet-bulb temperature: 1F° per 500 feet of elevation

Adjustment for air mass (at coastlines). Along the west coast, both dry-bulb and wet-bulb design temperatures increase with distance from the ocean. Along the Gulf Coast, dry-bulb temperatures increase for the first 200 to 300 miles inland, while wet-bulb temperatures decrease slightly. (Beyond this 200- to 300-mile belt, both dry-bulb and wet-bulb values decrease.)

TABLE A.1 **Outside Design Conditions: United States and Canada**

State and Station[a]	HDD65	Winter Design Dry-Bulb (97.5%)	Summer Design Dry-Bulb and Mean Coincident Wet-Bulb (2.5%)	Mean Daily Range	Design Wet-Bulb (2.5%)	CDH74
ALABAMA						
Auburn	2614	22	93/76	21	78	19189
Birmingham AP	2876	21	94/75	21	77	20324
Huntsville AP	3289	16	93/74	23	77	18603
Mobile AP	1694	29	93/77	18	79	28912
Montgomery AP	2283	25	95/76	21	79	24564
ALASKA						
Anchorage AP	10825	−18	68/58	15	59	35
Fairbanks AP	14280	−47	78/60	24	62	752
Juneau AP	9113	1	70/58	15	59	34
Nome AP	14380	−27	62/55	10	56	272
ARIZONA						
Flagstaff AP	7256	4	82/55	31	60	373
Phoenix AP	1444	34	107/71	27	75	54404
Prescott AP	4951	9	94/60	30	65	3828
Tucson AP	1737	32	102/66	26	71	35946
Winslow AP	4844	10	95/60	32	65	10636
Yuma AP	983	39	109/72	27	78	62507
ARKANSAS						
Fayetteville AP	4179	12	94/73	23	76	14956
Fort Smith AP	3482	17	98/76	24	79	23485
Little Rock AP	3161	20	96/77	22	79	23839
Texarkana AP	2506	23	96/77	21	79	27118
CALIFORNIA						
Bakersfield AP	2127	32	101/69	32	71	29954
Barstow AP	2585	29	104/68	37	71	27584
Burbank AP	1681	39	91/68	25	70	11400
Eureka/Arcata AP	4729	33	65/59	11	60	19
Fresno AP	2649	30	100/69	34	71	19366
Long Beach AP	1485	43	80/68	22	69	7755
Los Angeles AP	1595	43	80/68	15	69	4306
Los Angeles CO	1210	40	89/70	20	71	10575
Needles AP	1394	33	110/71	27	75	65218
Oakland AP	2880	36	80/63	19	64	435
Pomona CO	1973	30	99/69	36	72	10240
Redding AP	2548	31	102/67	32	69	27881
Sacramento AP	2775	32	98/70	36	71	10464

TABLE A.1 **Outside Design Conditions: United States and Canada** (*continued*)

State and Station[a]	HDD65	Winter Design Dry-Bulb (97.5%)	Summer Design Dry-Bulb and Mean Coincident Wet-Bulb (2.5%)	Mean Daily Range	Design Wet-Bulb (2.5%)	CDH74
San Diego AP	1284	44	80/69	12	70	4643
San Francisco AP	3164	38	77/63	20	64	265
San Francisco CO	3078	40	71/62	14	62	216
San Luis Obispo	2503	35	88/70	26	71	1085
Santa Barbara AP	2493	36	77/66	24	67	894
Santa Maria AP	3064	33	76/63	23	64	131
Santa Monica CO	1873	43	80/68	16	69	1908
COLORADO						
Alamosa AP	8718	−16	82/57	35	61	46
Boulder	5466	8	91/59	27	63	7684
Colorado Springs AP	6353	2	88/57	30	62	3658
Denver AP	6023	1	91/59	28	63	5908
Fort Collins	6488	−4	91/59	28	63	2946
Grand Junction AP	5684	7	94/59	29	63	12097
Leadville	—	−4	81/51	30	55	—
Pueblo AP	5471	0	95/61	31	66	10983
CONNECTICUT						
Bridgeport AP	5501	9	84/71	18	74	4977
Hartford	6197	7	88/73	22	75	4840
New Haven AP	—	7	84/73	17	75	—
DELAWARE						
Wilmington AP	4989	14	89/74	20	76	8167
DISTRICT OF COLUMBIA						
Washington National AP	4125	17	91/74	18	77	12389
FLORIDA						
Daytona Beach AP	904	35	90/77	15	79	27107
Fort Myers AP	440	44	92/78	18	79	37444
Gainesville AP	1069	31	93/77	18	79	27679
Jacksonville AP	1402	32	94/77	19	79	24148
Key West AP	114	57	90/78	9	79	50236
Miami AP	198	47	90/77	15	79	39401
Miami Beach CO	132	48	89/77	10	79	39007
Orlando AP	660	38	93/76	17	78	33985
Panama City, Tyndall AFB	—	33	90/77	14	80	—
Pensacola CO	1570	29	93/77	14	79	29023
Tallahassee AP	1651	30	92/76	19	78	25185
Tampa AP	741	40	91/77	17	79	33677
West Palm Beach AP	262	45	91/78	16	79	35175
GEORGIA						
Athens	2980	22	92/74	21	77	16147
Atlanta AP	3025	22	92/74	19	76	16803
Augusta AP	2571	23	95/76	19	79	19510
Columbus, Lawson AFB	2361	24	93/76	21	78	22115
Macon AP	2282	25	93/76	22	78	24443
Rome AP	3136	22	93/76	23	78	17745
Savannah-Travis AP	1921	27	93/77	20	79	22773
Valdosta-Moody AFB	—	31	94/77	20	79	—

TABLE A.1 **Outside Design Conditions: United States and Canada (*continued*)**

State and Station[a]	HDD65	Winter Design Dry-Bulb (97.5%)	Summer Design Dry-Bulb and Mean Coincident Wet-Bulb (2.5%)	Mean Daily Range	Design Wet-Bulb (2.5%)	CDH74
HAWAII						
Hilo AP	0	62	83/72	15	74	14770
Honolulu AP	0	63	86/73	12	75	36213
Wahiawa	—	59	85/72	14	74	—
IDAHO						
Boise AP	5808	10	94/64	31	66	7979
Lewiston AP	5435	6	93/64	32	66	7923
Moscow	6691	0	87/62	32	64	1986
Pocatello AP	7131	−1	91/60	35	63	3286
ILLINOIS						
Carbondale	4568	7	93/77	21	79	14114
Champaign/Urbana	5766	2	92/74	21	77	9949
Chicago, O'Hare AP	6459	−4	89/74	20	76	6606
Chicago CO	6013	2	91/74	15	77	8196
Moline AP	6504	−4	91/75	23	77	8616
Peoria AP	6233	−4	89/74	22	76	9503
Rockford	6955	−4	89/73	24	76	6523
Springfield AP	5662	2	92/74	21	77	12438
INDIANA						
Evansville AP	4734	9	93/75	22	78	14797
Fort Wayne AP	6324	1	89/72	24	75	6816
Indianapolis AP	5653	2	90/74	22	76	9082
Lafayette	6229	3	91/73	22	76	7728
Muncie	—	2	90/73	22	76	—
South Bend AP	6381	1	89/73	22	75	6606
Terre Haute AP	5527	4	92/74	22	77	9499
IOWA						
Ames	6882	−6	90/74	23	76	7453
Burlington AP	6094	−3	91/75	22	77	8277
Des Moines AP	6563	−5	91/74	23	77	10424
Dubuque	7379	−7	88/73	22	75	4661
Iowa City	6378	−6	89/76	22	78	10098
Mason City AP	7886	−11	88/74	24	75	5584
Sioux City AP	6953	−7	92/74	24	77	10117
Waterloo	7542	−10	89/75	23	77	6641
KANSAS						
Dodge City AP	5069	5	97/69	25	73	18470
Goodland AP	6103	0	96/65	31	70	10079
Manhattan, Fort Riley	5122	3	95/75	24	77	18826
Topeka AP	5323	4	96/75	24	78	16640
Wichita AP	4791	7	98/73	23	76	21208
KENTUCKY						
Bowling Green AP	4314	10	92/75	21	77	14745
Lexington AP	4819	8	91/73	22	76	11208
Louisville AP	4529	10	93/74	23	77	13268

TABLE A.1 **Outside Design Conditions: United States and Canada** (*continued*)

State and Station[a]	HDD65	Winter Design Dry-Bulb (97.5%)	Summer Design Dry-Bulb and Mean Coincident Wet-Bulb (2.5%)	Mean Daily Range	Design Wet-Bulb (2.5%)	CDH74
LOUISIANA						
Alexandria AP	1960	27	94/77	20	79	27315
Baton Rouge AP	1673	29	93/77	19	80	26938
Lafayette AP	1560	30	94/78	18	80	28543
Lake Charles AP	1580	31	93/77	17	79	28573
Monroe AP	2406	25	96/76	20	79	26604
New Orleans AP	1311	33	92/78	16	80	32758
Shreveport AP	2271	25	96/76	20	79	28295
MAINE						
Bangor, Dow AFB	7951	−6	83/68	22	71	1232
Caribou AP	9621	−13	81/67	21	69	920
Portland	7502	−1	84/71	22	72	1104
MARYLAND						
Baltimore AP	4714	13	91/75	21	77	9504
Baltimore CO	4086	17	89/76	17	78	15162
Cumberland	5108	10	89/74	22	76	7088
Hagerstown	5092	12	91/74	22	76	7346
Salisbury	4415	16	91/75	18	77	9166
MASSACHUSETTS						
Boston AP	5596	9	88/71	16	74	5358
Springfield	5955	0	87/71	19	73	5205
Worcester AP	6951	4	84/70	18	72	1504
MICHIGAN						
Alpena AP	8413	−6	85/70	27	72	837
Detroit	6564	6	88/72	20	74	4889
Escanaba	8549	−7	83/69	17	71	805
Flint AP	7067	1	87/72	25	74	2929
Grand Rapids AP	6927	5	88/72	24	74	4555
Muskegon AP	6926	6	84/70	21	73	2869
Sault Ste. Marie AP	9309	−8	81/69	23	70	844
Traverse City AP	7800	1	86/71	22	73	2968
MINNESOTA						
Bemidji AP	10207	−26	85/69	24	71	2191
Duluth AP	9906	−16	82/68	22	70	849
International Falls AP	10607	−25	83/68	26	70	1614
Minneapolis/St. Paul AP	8010	−12	89/73	22	75	6809
Rochester AP	8279	−12	87/72	24	75	3901
MISSISSIPPI						
Columbus AFB	2868	20	93/77	22	79	21829
Jackson AP	2390	25	95/76	21	78	25152
Meridian AP	2478	23	95/76	22	79	23827
Vicksburg CO	2200	26	95/78	21	80	24595
MISSOURI						
Columbia AP	5214	4	94/74	22	77	14475
Kansas City AP	4814	6	96/74	20	77	20256
St Louis AP	4948	6	94/75	21	77	17843

TABLE A.1 **Outside Design Conditions: United States and Canada** (*continued*)

State and Station[a]	HDD65	Winter Design Dry-Bulb (97.5%)	Summer Design Dry-Bulb and Mean Coincident Wet-Bulb (2.5%)	Mean Daily Range	Design Wet-Bulb (2.5%)	CDH74
St Louis CO	—	8	94/75	18	77	—
Springfield AP	4665	9	93/74	23	77	16262
MONTANA						
Billings AP	7220	−10	91/64	31	66	6167
Bozeman	9876	−14	87/60	32	62	407
Cut Bank AP	9161	−20	85/61	35	62	1378
Glasgow AP	8952	−18	89/63	29	66	4762
Great Falls AP	7784	−15	88/60	28	62	3574
Helena AP	8186	−16	88/60	32	62	2547
Kalispell AP	8363	−7	87/61	34	63	1210
Lewistown AP	8623	−16	87/61	30	63	2127
Miles City AP	7907	−15	95/66	30	68	10094
Missoula AP	7844	−6	88/61	36	63	1068
NEBRASKA						
Grand Island AP	6491	−3	94/71	28	74	11957
Lincoln CO	5974	−2	95/74	24	77	14607
North Platte AP	6914	−4	94/69	28	72	8492
Omaha AP	6201	−3	91/75	22	77	13180
Scottsbluff AP	6711	−3	92/65	31	68	7339
NEVADA						
Elko AP	7252	−2	92/59	42	62	3840
Ely AP	7705	−4	87/56	39	59	659
Las Vegas AP	2535	28	106/65	30	70	43153
Lovelock AP	5843	12	96/63	42	65	7845
Reno AP	6032	10	92/60	45	62	2192
Tonopah AP	5754	10	92/59	40	62	5888
Winnemucca AP	6417	3	94/60	42	62	4284
NEW HAMPSHIRE						
Berlin	—	−9	84/69	22	71	—
Concord AP	7483	−3	87/70	26	73	1993
Portsmouth, Pease AFB	6919	2	85/71	22	74	2239
NEW JERSEY						
Atlantic City CO	4766	13	89/74	18	77	5880
Newark AP	4977	14	91/73	20	76	9107
Trenton CO	4953	14	88/74	19	76	7393
NEW MEXICO						
Albuquerque AP	4415	16	94/61	27	65	11012
Farmington AP	5734	6	93/62	30	65	5047
Las Cruces	3120	20	96/64	30	68	14507
Los Alamos	6389	9	87/60	32	61	1235
Raton AP	—	1	89/60	34	64	—
Roswell, Walker AFB	3131	18	98/66	33	70	20216
Silver City AP	—	10	94/60	30	64	—
Tucumcari AP	3938	13	97/66	28	69	14754
NEW YORK						
Albany AP	6929	−1	88/72	23	74	2998
Binghamton AP	7346	1	83/69	20	72	1646

TABLE A.1 **Outside Design Conditions: United States and Canada** (*continued*)

State and Station[a]	HDD65	Winter Design Dry-Bulb (97.5%)	Summer Design Dry-Bulb and Mean Coincident Wet-Bulb (2.5%)	Mean Daily Range	Design Wet-Bulb (2.5%)	CDH74
Buffalo AP	6799	6	85/70	21	73	3044
Ithaca	7182	0	85/71	24	73	1583
Massena AP	8184	−8	83/69	20	72	2012
NYC—Central Park	4869	15	89/73	17	75	9537
NYC—Kennedy AP	5171	15	87/72	16	75	7634
Rochester AP	6718	5	88/71	22	73	3764
Syracuse AP	6789	2	87/71	20	73	3516
NORTH CAROLINA						
Asheville AP	4300	14	87/72	21	74	6292
Charlotte AP	3348	22	93/74	20	76	15159
Greensboro AP	3877	18	91/73	21	76	11020
Raleigh/Durham AP	3538	20	92/75	20	77	11845
Wilmington AP	2473	26	91/78	18	80	17648
NORTH DAKOTA						
Bismarck AP	9080	−19	91/68	27	71	4554
Fargo AP	9349	−18	89/71	25	74	4284
Minot AP	9425	−20	89/67	25	70	4157
Williston	9252	−21	88/67	25	70	4034
OHIO						
Akron–Canton AP	6248	6	86/71	21	73	4808
Cincinnati CO	4958	6	90/72	21	75	10698
Cleveland AP	6179	5	88/72	22	74	4772
Columbus AP	5692	5	90/73	24	75	7490
Dayton AP	5260	4	89/72	20	75	11400
Mansfield AP	6254	5	87/72	22	74	4948
Toledo AP	6576	1	88/73	25	75	5081
Youngstown AP	6562	4	86/71	23	73	2975
OKLAHOMA						
Norman	—	13	96/74	24	76	—
Oklahoma City AP	3742	13	97/74	23	77	22978
Stillwater	3802	13	96/74	24	76	23621
Tulsa AP	3741	13	98/75	22	78	26468
OREGON						
Astoria AP	5250	29	71/62	16	63	30
Bend	7082	4	87/60	33	62	647
Eugene AP	4803	22	89/66	31	67	1296
Medford AP	4803	23	94/67	35	68	6151
Pendleton AP	5275	5	93/64	29	65	8121
Portland AP	4693	23	85/67	23	67	1851
Portland CO	4417	24	86/67	21	67	2105
Salem AP	4978	23	88/66	31	68	1081
PENNSYLVANIA						
Allentown AP	5815	9	88/72	22	75	5804
Erie AP	6768	9	85/72	18	74	2222
Harrisburg AP	5339	11	91/74	21	76	9071
Philadelphia AP	4954	14	90/74	21	76	8896

TABLE A.1 **Outside Design Conditions: United States and Canada** (*continued*)

State and Station[a]	HDD65	Winter Design Dry-Bulb (97.5%)	Summer Design Dry-Bulb and Mean Coincident Wet-Bulb (2.5%)	Summer Mean Daily Range	Summer Design Wet-Bulb (2.5%)	CDH74
Pittsburgh AP	5957	5	86/71	22	73	5009
Pittsburgh CO	—	7	88/71	19	73	—
Scranton/Wilkes-Barre	6332	5	87/71	19	73	3774
State College	6252	7	87/71	23	73	3484
RHODE ISLAND						
Newport	6122	9	85/72	16	75	1682
Providence AP	5909	9	86/72	19	74	3631
SOUTH CAROLINA						
Anderson	2960	23	92/74	21	76	16014
Charleston AFB	2146	27	91/78	18	80	20315
Charleston CO	1868	28	92/78	13	80	23283
Columbia AP	2632	24	95/75	22	78	21805
Spartanburg AP	3246	22	91/74	20	76	14069
SOUTH DAKOTA						
Huron AP	8109	−14	93/72	28	75	8677
Pierre AP	7580	−10	95/71	29	74	10444
Rapid City AP	7308	−7	92/65	28	69	8286
Sioux Falls AP	7890	−11	91/72	24	75	8638
TENNESSEE						
Bristol–Tri City AP	4367	14	89/72	22	75	8954
Chattanooga AP	3591	18	93/74	22	77	17017
Knoxville AP	3666	19	92/73	21	76	14214
Memphis AP	3214	18	95/76	21	79	24504
Nashville AP	3767	14	94/74	21	77	18543
TEXAS						
Abilene AP	2628	20	99/71	22	74	31904
Amarillo AP	4240	11	95/67	26	70	15718
Austin AP	1759	28	98/74	22	77	35180
Brownsville AP	607	39	93/77	18	79	42529
Bryan AP	1784	29	96/76	20	78	33178
Corpus Christi AP	970	35	94/78	19	80	42515
Dallas AP	2303	22	100/75	20	78	36697
Del Rio, Laughlin AFB	1511	31	98/73	24	77	42032
El Paso AP	2672	24	98/64	27	68	22966
Fort Worth AP	2420	22	99/74	22	77	36294
Galveston AP	1252	36	89/79	10	80	31911
Houston AP	1548	32	94/77	18	79	30474
Houston CO	—	33	95/77	18	79	—
Laredo AFB	927	36	101/73	23	78	52560
Lubbock AP	3522	15	96/69	26	72	18218
Lufkin AP	1929	29	97/76	20	79	30370
Midland AP	2664	21	98/69	26	72	24201
Port Arthur AP	1476	31	93/78	19	80	31667
San Angelo, Goodfellow AFB	2325	22	99/71	24	74	32735
San Antonio AP	1605	30	97/73	19	76	36179
Sherman Perrin AFB	2943	20	98/75	22	77	29686

TABLE A.1 **Outside Design Conditions: United States and Canada** (*continued*)

State and Station[a]	HDD65	Winter Design Dry-Bulb (97.5%)	Summer Design Dry-Bulb and Mean Coincident Wet-Bulb (2.5%)	Mean Daily Range	Design Wet-Bulb (2.5%)	CDH74
Waco AP	2128	26	99/75	22	78	36671
Wichita Falls AP	3017	18	101/73	24	76	34510
UTAH						
Cedar City AP	6002	5	91/60	32	63	5036
Salt Lake City AP	5805	8	95/62	32	65	9898
Vernal AP	7671	0	89/60	32	63	2678
VERMONT						
Barre	8529	−11	81/69	23	71	1377
Burlington AP	7956	−7	85/70	23	72	2562
VIRGINIA						
Charlottesville	4195	18	91/74	23	76	10287
Norfolk AP	3458	22	91/76	18	78	13677
Richmond AP	3969	17	92/76	21	78	12260
Roanoke AP	4321	16	91/72	23	74	9275
WASHINGTON						
Bellingham AP	5726	15	77/65	19	65	185
Olympia AP	5710	22	83/65	32	66	341
Seattle CO	4684	27	82/66	19	67	897
Seattle–Tacoma AP	5122	26	80/64	22	64	1050
Spokane AP	6885	2	90/63	28	64	3453
Walla Walla AP	5057	7	94/66	27	67	9511
Yakima AP	6035	5	93/65	36	66	4118
WEST VIRGINIA						
Beckley	5585	4	81/69	22	71	2408
Charleston AP	4705	11	90/73	20	75	8790
Huntington CO	4687	10	91/74	22	77	11185
Wheeling	5456	5	86/71	21	73	6793
WISCONSIN						
Ashland	9067	−16	82/68	23	70	1288
Eau Claire AP	8466	−11	89/73	23	75	3886
Green Bay AP	8146	−9	85/72	23	74	2453
La Cross AP	7544	−9	88/73	22	75	6795
Madison AP	7643	−7	88/73	22	75	3343
Milwaukee AP	7327	−4	87/73	21	74	3313
WYOMING						
Casper AP	7649	−5	90/57	31	61	4370
Cheyenne AP	7315	−1	86/58	30	62	2087
Lander AP	7915	−11	88/61	32	63	3940
Laramie AP	8933	−6	81/56	28	60	428
Rock Springs AP	8358	−3	84/55	32	58	1043
Sheridan AP	7852	−8	91/62	32	65	4115
Province and Station						
ALBERTA						
Calgary AP	9657	−23	81/61	25	63	1176
Edmonton AP	10782	−25	82/65	23	66	1137
Medicine Hat AP	8762	−24	90/65	28	68	3291

TABLE A.1 **Outside Design Conditions: United States and Canada (*continued*)**

State and Station[a]	HDD65	Winter Design Dry-Bulb (97.5%)	Summer Design Dry-Bulb and Mean Coincident Wet-Bulb (2.5%)	Mean Daily Range	Design Wet-Bulb (2.5%)	CDH74
BRITISH COLUMBIA						
Dawson Creek	11212	−33	79/63	26	64	427
Nanaimo	5834	20	80/65	21	66	560
Prince George AP	9677	−28	80/62	26	64	431
Prince Rupert CO	7398	2	63/57	12	58	3
Vancouver AP	5454	19	77/66	17	67	221
Victoria CO	5607	23	73/62	16	62	42
MANITOBA						
Flin Flon	12551	−37	81/66	19	68	1424
Winnipeg AP	10661	−27	86/71	22	73	2615
NEW BRUNSWICK						
Fredericton AP	8530	−11	85/69	23	71	1000
Saint John AP	8582	−8	77/65	19	68	49
NEWFOUNDLAND						
Corner Brook	8550	0	73/63	17	66	505
St. John's AP	8683	7	75/65	18	67	1163
NORTHWEST TERR.						
Fort Smith AP	14013	−45	81/64	24	66	724
Inuvik	18182	−53	77/60	21	62	1055
Yellowknife AP	15352	−46	77/61	16	63	655
NOVA SCOTIA						
Halifax AP	7963	5	76/65	16	67	332
Sydney AP	8172	3	80/68	19	70	668
Yarmouth AP	7315	9	72/64	15	66	14
ONTARIO						
Kitchener	7463	−2	85/72	23	74	2447
Ottawa AP	8442	−13	87/71	21	73	2051
Sudbury AP	9812	−19	83/67	22	70	1222
Toronto AP	7457	−1	87/72	20	74	2676
PRINCE EDWARD ISLAND						
Charlottetown AP	8438	−4	78/68	16	70	409
QUEBEC						
Chicoutimi	9781	−22	83/68	20	70	1135
Montréal AP	8167	−10	85/72	17	74	2035
Québec AP	9297	−14	84/70	20	72	882
Sept Iles AP	11077	−21	73/61	17	65	225
Val d'Or AP	11158	−27	83/68	22	70	1269
SASKATCHEWAN						
Prince Albert AP	11806	−35	84/66	25	68	502
Regina AP	10577	−29	88/68	26	70	1973
Saskatoon AP	10912	−31	86/66	26	68	1439
YUKON TERRITORY						
Whitehorse AP	12578	−43	77/58	22	59	454

[a] U.S. solar data available from the National Climatic Data Center, Federal Building, Asheville, NC 28801. Canadian solar data available from The Canadian Climate Center, Atmospheric Environment Service, 4905 Dufferin St., Downsview, Ontario M3H 5T 4.

TABLE A.2 **Outdoor Design Conditions: Mexico and Puerto Rico**

Country and Station	HDD65	Winter[a] Design Dry-Bulb 97.5%	Summer[b] Design Dry-Bulb 2.5%	Mean Daily Range	Design Wet-Bulb 2.5%	CDH74
MEXICO						
Guadalajara	—	42	91	29	67	—
Mérida	—	61	95	21	79	—
Mexico City	—	39	81	25	60	—
Monterrey	—	41	95	20	78	—
Vera Cruz	—	62	89	20	83	—
PUERTO RICO						
San Juan	0	68	88	11	80	55224

[a] For the 3-month winter period.
[b] For the 4-month summer period.

APPENDICES

B

SOLAR DATA

This appendix presents both *clear day* and *average day* data.

Solar Intensity and Solar Heat Gain Factors (Tables B.1–B.10).

These are listed for five latitudes; solar heat gain factors occur through one layer of double-strength sheet (DSA) glass. The data are based on an "average cloudless" day at each latitude. Tables B.1 through B.10 are copyrighted by the American Society of Heating, Refrigerating, and Air-Conditioning Engineers, Inc., Atlanta, Ga., and are reprinted by permission from *ASHRAE 1989 Handbook of Fundamentals.*

Solar Position and Clear Day Global Irradiance (Insolation) (Tables B.11–B.14).

These are listed for four latitudes. The data are presented in conventional units for the following tilt angles (degrees above horizontal): latitude $-10°$, latitude, latitude $+10°$, latitude $+20°$, and vertical. To convert to SI units: Btu/h-ft^2 $\times$ 3.152 = W/m^2.

Elevation, Latitude, Average Horizontal Insolation, Average Vertical Insolation, Average Temperature, and Heating Degree Days (Table B.15).

These are listed in conventional units for January and July, as well as yearly totals, for a number of locations (Fig. B.1). To convert horizontal surface (HS) and vertical surface (VS) to SI units: Btu/h ft^2 $\times$ 3.152 = W/m^2. To convert average temperature (TA) to Celsius: (°F $-$ 32)/1.8 = °C. To convert DD(F) to DD(C): DD(F) $\times$ 5/9 = DD(C) or DD(K).

TABLE B.1 Solar Intensity and Solar Heat Gain Factors[a] for 16° N Latitude (Conventional Units)[b]

Date	Solar Time am	Direct Normal Btuh ft²	N	NNE	NE	ENE	E	ESE	SE	SSE	S	SSW	SW	WSW	W	WNW	NW	NNW	HOR	Solar Time pm
Jan 21	7	141	5	6	44	92	124	134	126	96	49	6	5	5	5	5	5	5	14	5
	8	262	14	15	55	147	210	240	233	189	114	25	14	14	14	14	14	14	79	4
	9	300	21	21	32	122	200	244	251	219	152	58	22	21	21	21	21	21	150	3
	10	317	26	26	27	66	150	209	233	223	178	102	31	26	26	26	26	26	203	2
	11	325	29	29	29	31	77	148	195	210	194	146	75	31	29	29	29	29	236	1
	12	327	30	30	30	30	32	73	139	184	199	184	138	72	32	30	30	30	248	12
	HALF DAY TOTALS		110	112	196	461	760	1000	1096	1020	781	426	211	127	111	110	110	110	805	
Feb 21	7	182	8	17	84	138	169	172	150	103	36	8	8	8	8	8	8	8	25	5
	8	273	17	19	96	180	231	247	224	166	77	18	17	17	17	17	17	17	101	4
	9	305	23	24	64	153	214	242	233	188	110	30	23	23	23	23	23	23	174	3
	10	319	28	29	33	92	161	202	211	188	134	61	30	28	28	28	28	28	229	2
	11	326	32	32	32	37	83	136	167	172	149	102	49	33	32	32	32	32	263	1
	12	328	33	33	33	33	34	60	107	142	154	142	106	60	34	33	33	33	275	12
	HALF DAY TOTALS		124	137	321	609	865	1023	1034	885	582	287	174	132	124	124	124	124	930	
Mar 21	7	201	11	53	124	172	192	183	145	82	15	10	10	10	10	10	10	10	40	5
	8	272	20	50	140	205	239	235	195	123	35	19	19	19	19	19	19	19	120	4
	9	299	26	35	109	179	218	225	197	138	57	27	26	26	26	26	26	26	192	3
	10	312	31	33	61	120	165	182	172	134	76	34	32	31	31	31	31	31	247	2
	11	318	34	35	36	53	87	114	125	116	89	55	36	35	34	34	34	34	280	1
	12	320	35	35	36	36	37	47	69	87	93	86	68	47	37	36	36	35	291	12
	HALF DAY TOTALS		141	226	494	755	928	975	879	643	319	187	153	142	139	139	139	139	1025	
Apr 21	6	14	2	8	12	14	14	12	8	2	1	1	1	1	1	1	1	1	1	6
	7	197	24	94	153	187	191	167	117	45	14	13	13	13	13	13	13	13	53	5
	8	256	27	99	172	216	227	204	150	69	24	22	22	22	22	22	22	22	131	4
	9	280	31	79	149	193	208	193	147	77	31	29	29	29	29	29	29	29	197	3
	10	293	35	54	102	141	158	151	120	73	37	34	33	33	33	33	33	33	249	2
	11	299	38	40	54	72	86	88	78	60	43	38	38	36	36	36	36	37	279	1
	12	301	39	39	39	40	41	43	45	45	45	43	41	40	39	39	39	39	289	12
	HALF DAY TOTALS		179	403	674	859	922	851	653	352	174	159	157	156	155	155	155	156	1057	
May 21	6	44	14	30	41	45	43	34	19	4	3	3	3	3	3	3	3	3	5	6
	7	193	50	120	168	191	185	150	92	24	16	16	16	16	16	16	16	17	62	5
	8	244	52	132	189	218	215	179	115	38	25	24	24	24	24	24	24	25	135	4
	9	268	49	116	171	198	197	167	109	45	32	30	30	30	30	30	30	32	197	3
	10	280	47	89	130	151	150	126	84	44	37	35	35	35	35	35	35	37	245	2
	11	286	47	63	79	87	83	70	52	41	40	39	38	38	38	39	39	41	273	1
	12	288	46	46	44	43	42	41	41	41	41	41	41	41	42	43	44	46	282	12
	HALF DAY TOTALS		283	575	804	916	897	748	493	217	172	167	167	167	167	168	169	176	1058	
Jun 21	6	53	20	39	52	55	51	39	20	4	4	4	4	4	4	4	4	4	7	6
	7	188	62	128	172	190	179	141	80	20	16	16	16	16	16	16	16	18	64	5
	8	238	66	142	194	217	207	167	99	31	25	25	25	25	25	25	25	27	135	4
	9	261	63	130	178	198	190	154	93	37	31	31	31	31	31	31	31	33	194	3
	10	273	59	104	140	154	145	115	70	39	37	36	36	36	36	36	36	38	241	2
	11	279	57	76	90	92	82	63	46	41	40	39	39	39	40	41	43	43	268	1
	12	281	57	55	50	45	43	42	41	41	41	41	41	42	42	45	50	55	277	12
	HALF DAY TOTALS		356	648	850	929	876	700	430	194	174	171	171	171	172	173	176	190	1049	
Jul 21	6	41	14	29	39	42	40	31	18	4	3	3	3	3	3	3	3	3	6	6
	7	184	51	118	164	185	179	145	88	23	16	16	16	16	16	16	16	17	62	5
	8	236	55	132	187	214	210	174	111	37	25	25	25	25	25	25	25	26	133	4
	9	259	52	117	170	196	193	163	106	44	32	31	31	31	31	31	31	33	194	3
	10	272	50	92	131	151	148	123	81	44	38	36	36	36	36	36	36	38	241	2
	11	278	49	66	81	88	83	69	52	42	41	40	39	39	39	40	40	42	269	1
	12	279	49	48	46	44	43	42	42	42	42	42	42	42	43	44	46	48	277	12
	HALF DAY TOTALS		296	580	799	903	878	729	478	215	176	172	171	171	171	172	173	182	1043	
Aug 21	6	11	2	7	10	12	12	10	6	2	1	1	1	1	1	1	1	1	1	6
	7	180	26	92	145	176	180	156	109	42	15	14	14	14	14	14	14	14	53	5
	8	240	30	100	168	209	219	196	143	65	25	23	23	23	23	23	23	23	128	4
	9	266	33	82	148	190	203	187	142	74	33	30	30	30	30	30	30	30	193	3
	10	279	37	58	104	140	155	147	117	71	39	36	35	35	35	35	35	35	243	2
	11	285	40	43	57	75	86	87	76	59	44	40	39	38	38	38	38	39	273	1
	12	287	41	41	41	42	42	43	44	45	46	45	44	43	42	41	41	42	282	12
	HALF DAY TOTALS		191	410	666	837	891	817	624	339	180	167	165	164	163	163	163	164	1033	
Sep 21	7	179	12	50	114	158	176	168	133	76	15	11	11	11	11	11	11	11	39	5
	8	253	21	49	134	196	227	224	186	119	36	20	20	20	20	20	20	20	116	4
	9	281	28	36	106	173	211	217	191	134	57	28	27	27	27	27	27	27	185	3
	10	295	32	34	61	118	161	178	168	132	76	35	33	32	32	32	32	32	238	2
	11	302	35	36	37	54	86	113	123	114	88	56	38	36	35	35	35	35	271	1
	12	304	36	36	37	38	39	49	69	86	93	86	69	48	39	38	37	36	282	12
	HALF DAY TOTALS		146	226	475	722	885	931	842	622	319	192	159	148	145	144	144	144	991	
Oct 21	7	166	8	18	79	128	156	159	139	95	33	9	8	8	8	8	8	8	25	5
	8	259	17	20	95	174	223	237	215	159	74	19	17	17	17	17	17	17	99	4
	9	292	24	25	65	150	209	235	225	182	106	31	24	24	24	24	24	24	170	3
	10	307	29	30	34	92	158	197	205	183	130	60	31	29	29	29	29	29	224	2
	11	314	32	32	33	39	83	133	163	167	145	100	49	34	32	32	32	32	258	1
	12	316	33	33	33	34	35	60	105	139	150	138	104	60	35	34	33	33	270	12
	HALF DAY TOTALS		127	141	318	592	836	986	996	852	563	283	175	136	128	127	127	127	911	
Nov 21	7	134	5	6	43	89	119	129	120	92	47	6	5	5	5	5	5	5	14	5
	8	255	15	15	55	145	206	235	228	185	111	25	15	15	15	15	15	15	78	4
	9	295	21	21	33	121	197	241	247	215	150	57	22	21	21	21	21	21	149	3
	10	312	26	26	28	67	147	206	230	220	176	100	31	26	26	26	26	26	201	2
	11	320	29	29	29	31	77	146	192	207	191	144	74	31	29	29	29	29	234	1
	12	322	30	30	30	30	32	72	137	181	196	181	137	72	32	30	30	30	246	12
	HALF DAY TOTALS		112	113	197	456	749	983	1077	1001	767	420	210	128	112	112	112	112	799	
Dec 21	7	118	4	5	30	72	101	112	107	85	48	7	4	4	4	4	4	4	10	5
	8	255	13	14	41	132	198	233	231	193	124	33	13	13	13	13	13	13	69	4
	9	297	20	20	25	108	191	241	254	227	165	72	21	20	20	20	20	20	138	3
	10	315	25	25	26	56	144	208	239	233	192	117	35	25	25	25	25	25	191	2
	11	323	28	28	28	29	73	150	202	221	207	161	86	30	28	28	28	28	223	1
	12	325	29	29	29	29	30	77	149	197	212	196	149	76	30	29	29	29	234	12
	HALF DAY TOTALS		104	105	159	402	710	975	1099	1050	836	484	228	125	105	104	104	104	748	
			N	NNW	NW	WNW	W	WSW	SW	SSW	S	SSE	SE	ESE	E	ENE	NE	NNE	HOR	PM

[a]Total solar heat gains for DSA glass (based on a ground reflectance of 0.20).

[b]Half-day totals computed by Simpson's rule, time interval = 10 min.

TABLE B.2 Solar Intensity and Solar Heat Gain Factors[a] for 32° N Latitude (Conventional Units)[b]

Date	Solar Time am	Direct Normal Btuh ft²	N	NNE	NE	ENE	E	ESE	SE	SSE	S	SSW	SW	WSW	W	WNW	NW	NNW	HOR	Solar Time pm
Jan 21	7	1	0	0	0	1	1	1	1	1	0	0	0	0	0	0	0	0	0	5
	8	203	9	9	29	105	160	189	189	159	103	28	9	9	9	9	9	9	32	4
	9	269	15	15	17	91	175	229	246	225	169	82	17	15	15	15	15	15	88	3
	10	295	20	20	20	41	135	209	249	250	212	141	46	20	20	20	20	20	136	2
	11	306	23	23	23	24	68	159	221	249	238	191	110	29	23	23	23	23	166	1
	12	310	24	24	24	24	25	88	174	228	246	228	174	88	25	24	24	24	176	12
	HALF DAY TOTALS		79	79	107	284	570	856	1015	1014	853	553	264	112	80	79	79	79	512	
Feb 21	7	112	4	7	47	82	102	106	95	67	26	4	4	4	4	4	4	4	9	5
	8	245	13	14	65	149	205	228	216	170	95	17	13	13	13	13	13	13	64	4
	9	287	19	19	32	122	199	242	248	216	149	55	20	19	19	19	19	19	127	3
	10	305	24	24	25	62	151	213	241	232	189	112	31	24	24	24	24	24	176	2
	11	314	26	26	26	28	76	156	208	227	212	165	87	28	26	26	26	26	207	1
	12	316	27	27	27	27	29	79	155	204	221	204	155	79	29	27	27	27	217	12
	HALF DAY TOTALS		100	103	201	445	735	978	1080	1010	780	452	228	122	100	100	100	100	691	
Mar 21	7	185	10	37	105	153	176	173	142	88	20	9	9	9	9	9	9	9	32	5
	8	260	17	25	107	183	227	237	209	150	62	18	17	17	17	17	17	17	100	4
	9	290	23	25	64	151	210	237	227	183	107	30	23	23	23	23	23	23	164	3
	10	304	28	28	30	87	158	202	215	195	144	70	29	28	28	28	28	28	211	2
	11	311	31	31	31	34	82	142	179	188	168	120	59	32	31	31	31	31	242	1
	12	313	32	32	32	32	33	66	122	162	176	162	122	66	33	32	32	32	252	12
	HALF DAY TOTALS		124	162	359	629	875	1033	1041	888	589	326	193	136	125	124	124	124	874	
Apr 21	6	66	9	35	54	65	66	56	38	12	4	3	3	3	3	3	3	3	7	6
	7	206	17	80	146	188	200	182	136	65	16	14	14	14	14	14	14	14	61	5
	8	255	23	61	144	200	227	219	177	107	30	22	22	22	22	22	22	22	129	4
	9	278	28	36	103	168	206	212	187	133	58	29	28	28	28	28	28	28	188	3
	10	290	32	34	52	108	155	177	172	141	87	39	33	32	32	32	32	32	233	2
	11	295	35	35	36	47	83	118	135	132	108	70	40	36	35	35	35	35	262	1
	12	297	36	36	36	37	38	53	82	106	115	106	82	53	38	37	36	36	271	12
	HALF DAY TOTALS		161	296	550	792	952	992	889	645	360	228	177	157	153	152	152	152	1015	
May 21	6	119	33	77	108	121	116	94	56	13	8	8	8	8	8	8	8	9	21	6
	7	211	36	111	170	202	204	174	118	42	19	18	18	18	18	18	18	19	81	5
	8	250	29	94	165	208	220	199	149	73	27	25	25	25	25	25	25	25	146	4
	9	269	33	61	128	177	198	190	155	93	37	32	31	31	31	31	31	31	201	3
	10	280	36	40	76	121	150	156	138	99	54	37	35	35	35	35	35	35	243	2
	11	285	38	39	42	59	83	99	102	90	68	47	40	39	37	37	37	37	269	1
	12	286	38	39	40	40	41	47	59	70	74	70	59	47	41	40	40	39	277	12
	HALF DAY TOTALS		222	438	702	900	985	933	747	447	250	199	183	177	175	174	174	175	1098	
Jun 21	6	131	44	92	123	135	127	99	51	12	10	10	10	10	10	10	10	11	28	6
	7	210	47	122	176	204	201	168	108	35	20	20	20	20	20	20	20	21	88	5
	8	245	36	106	171	208	214	189	135	60	28	27	27	27	27	27	27	27	151	4
	9	264	35	74	137	178	193	180	139	77	35	32	32	32	32	32	32	32	204	3
	10	274	38	47	86	125	146	145	123	83	45	38	36	36	36	36	36	36	244	2
	11	279	40	41	47	64	82	91	89	75	56	43	41	40	39	39	39	39	269	1
	12	280	41	41	41	42	42	46	52	58	60	58	52	46	42	42	41	41	276	12
	HALF DAY TOTALS		261	504	762	935	985	897	678	372	225	197	189	185	184	184	183	186	1122	
Jul 21	6	113	34	76	105	117	113	90	53	12	9	9	9	9	9	9	9	9	22	6
	7	203	38	111	167	198	198	169	114	41	20	19	19	19	19	19	19	19	81	5
	8	241	31	95	163	204	215	194	145	70	28	26	26	26	26	26	26	26	145	4
	9	261	34	64	129	175	195	186	150	90	37	32	32	32	32	32	32	32	198	3
	10	271	37	42	78	121	148	153	134	96	53	38	36	36	36	36	36	36	240	2
	11	277	39	40	43	60	83	98	99	88	66	47	41	40	38	38	38	38	265	1
	12	279	40	40	41	41	42	48	58	68	72	68	58	48	42	41	41	40	273	12
	HALF DAY TOTALS		231	444	701	890	967	912	726	433	248	202	187	182	180	179	179	180	1088	
Aug 21	6	59	10	33	50	60	60	51	34	11	4	4	4	4	4	4	4	4	8	6
	7	190	19	79	141	179	190	172	128	61	17	15	15	15	15	15	15	15	61	5
	8	240	25	63	141	195	219	210	170	102	31	23	23	23	23	23	23	23	128	4
	9	263	30	39	104	166	200	206	181	127	57	31	29	29	29	29	29	29	185	3
	10	276	34	36	55	109	153	173	167	136	84	40	35	34	34	34	34	34	229	2
	11	282	36	37	39	50	84	116	131	127	104	69	41	38	36	36	36	36	256	1
	12	284	37	37	37	39	40	54	81	103	111	103	81	54	40	39	37	37	265	12
	HALF DAY TOTALS		171	303	546	774	922	955	854	618	352	231	184	166	162	161	160	160	999	
Sep 21	7	163	10	35	96	139	159	156	128	80	20	10	10	10	10	10	10	10	31	5
	8	240	18	26	103	173	215	224	198	143	60	19	18	18	18	18	18	18	96	4
	9	272	24	26	64	146	202	227	218	177	105	31	24	24	24	24	24	24	158	3
	10	287	29	29	32	86	154	196	208	189	141	70	31	29	29	29	29	29	204	2
	11	294	32	32	32	36	81	139	174	182	163	118	59	34	32	32	32	32	234	1
	12	296	33	33	33	33	35	66	120	158	171	158	120	66	35	33	33	33	244	12
	HALF DAY TOTALS		130	164	345	598	831	982	993	852	574	325	197	142	130	129	129	129	845	
Oct 21	7	99	4	7	43	74	92	96	85	60	24	5	4	4	4	4	4	4	10	5
	8	229	13	15	63	143	195	217	206	162	90	17	13	13	13	13	13	13	63	4
	9	273	20	20	33	120	193	234	239	208	144	54	21	20	20	20	20	20	125	3
	10	293	24	24	26	62	147	207	234	225	183	109	32	24	24	24	24	24	173	2
	11	302	27	27	27	29	76	152	203	221	207	160	85	29	27	27	27	27	203	1
	12	304	28	28	28	28	30	78	151	199	215	199	151	78	30	28	28	28	213	12
	HALF DAY TOTALS		103	106	200	433	708	941	1038	972	753	441	226	125	104	103	103	103	679	
Nov 21	7	2	0	0	0	1	1	1	1	1	0	0	0	0	0	0	0	0	0	5
	8	196	9	9	29	103	156	184	184	155	100	27	9	9	9	9	9	9	32	4
	9	263	16	16	17	90	173	225	241	221	166	80	17	16	16	16	16	16	88	3
	10	289	20	20	21	41	134	206	245	246	209	138	45	21	20	20	20	20	136	2
	11	301	23	23	23	24	67	157	218	245	234	188	109	29	23	23	23	23	165	1
	12	304	24	24	24	24	25	87	171	224	243	224	171	87	25	24	24	24	175	12
	HALF DAY TOTALS		80	81	108	282	561	841	996	995	838	544	261	113	81	80	80	80	509	
Dec 21	8	176	7	7	19	84	135	163	166	143	97	31	7	7	7	7	7	7	22	4
	9	257	14	14	15	77	162	218	238	222	171	89	15	14	14	14	14	14	72	3
	10	288	18	18	18	34	127	204	246	251	216	148	52	19	18	18	18	18	119	2
	11	301	21	21	21	22	63	157	222	252	243	197	116	29	21	21	21	21	148	1
	12	304	22	22	22	22	23	89	177	232	252	232	177	89	23	22	22	22	158	12
	HALF DAY TOTALS		71	71	84	227	500	792	965	986	852	578	275	107	71	71	71	71	440	
			N	NNW	NW	WNW	W	WSW	SW	SSW	S	SSE	SE	ESE	E	ENE	NE	NNE	HOR	PM

[a]Total solar heat gains for DSA glass (based on a ground reflectance of 0.20).

[b]Half day totals computed by Simpson's rule, time interval = 10 min.

TABLE B.3 Solar Intensity and Solar Heat Gain Factors[a] for 40° N Latitude (Conventional Units)[b]

Date	Solar Time am	Direct Normal Btuh/ft²	N	NNE	NE	ENE	E	ESE	SE	SSE	S	SSW	SW	WSW	W	WNW	NW	NNW	HOR	Solar Time pm
Jan 21	8	142	5	5	17	71	111	132	133	114	75	22	6	5	5	5	5	5	14	4
	9	239	12	12	13	74	154	205	224	209	160	82	13	12	12	12	12	12	55	3
	10	274	16	16	16	31	124	199	241	246	213	146	51	17	16	16	16	16	96	2
	11	289	19	19	19	20	61	156	222	252	244	198	118	28	19	19	19	19	124	1
	12	294	20	20	20	20	21	90	179	234	254	234	179	90	21	20	20	20	133	12
HALF DAY TOTALS			61	61	73	199	452	734	904	932	813	561	273	101	62	61	61	61	354	
Feb 21	7	55	2	3	23	40	51	53	47	34	14	2	2	2	2	2	2	2	4	5
	8	219	10	11	50	129	183	206	199	160	94	18	10	10	10	10	10	10	43	4
	9	271	16	16	22	107	186	234	245	218	157	66	17	16	16	16	16	16	98	3
	10	294	21	21	21	49	143	211	246	243	203	129	38	21	21	21	21	21	143	2
	11	304	23	23	23	24	71	160	219	244	231	184	103	27	23	23	23	23	171	1
	12	307	24	24	24	24	25	86	170	222	241	222	170	86	25	24	24	24	180	12
HALF DAY TOTALS			84	86	152	361	648	916	1049	1015	821	508	250	114	85	84	84	84	548	
Mar 21	7	171	9	29	93	140	163	161	135	86	22	8	8	8	8	8	8	8	26	5
	8	250	16	18	91	169	218	232	211	157	74	17	16	16	16	16	16	16	85	4
	9	282	21	22	47	136	203	238	236	198	128	40	22	21	21	21	21	21	143	3
	10	297	25	25	27	72	153	207	229	216	171	95	29	25	25	25	25	25	186	2
	11	305	28	28	28	30	78	151	198	213	197	150	77	30	28	28	28	28	213	1
	12	307	29	29	29	29	31	75	145	191	206	191	145	75	31	29	29	29	223	12
HALF DAY TOTALS			114	139	302	563	832	1035	1087	968	694	403	220	132	114	113	113	113	764	
Apr 21	6	89	11	46	72	87	88	76	52	18	5	5	5	5	5	5	5	5	11	6
	7	206	16	71	140	185	201	186	143	75	16	14	14	14	14	14	14	14	61	5
	8	252	22	44	128	190	224	223	188	124	41	22	21	21	21	21	21	21	123	4
	9	274	27	29	80	155	202	219	203	156	83	29	27	27	27	27	27	27	177	3
	10	286	31	31	37	92	152	187	193	170	121	56	32	31	31	31	31	31	217	2
	11	292	33	33	34	39	81	130	160	166	146	102	52	35	33	33	33	33	243	1
	12	293	34	34	34	34	36	62	108	142	154	142	108	62	36	34	34	34	252	12
HALF DAY TOTALS			154	265	501	758	957	1051	994	782	488	296	199	157	148	147	147	147	957	
May 21	5	1	0	1	1	1	1	1	0	0	0	0	0	0	0	0	0	0	0	7
	6	144	36	90	128	145	141	115	71	18	10	10	10	10	10	10	10	11	31	6
	7	216	28	102	165	202	209	184	131	54	20	19	19	19	19	19	19	19	87	5
	8	250	27	73	149	199	220	208	164	93	29	25	25	25	25	25	25	25	146	4
	9	267	31	42	105	164	197	200	175	121	53	32	30	30	30	30	30	30	195	3
	10	277	34	36	54	105	148	168	163	133	83	40	35	34	34	34	34	34	234	2
	11	283	36	36	38	48	81	113	130	127	105	70	42	38	36	36	36	36	257	1
	12	284	37	37	37	38	40	54	82	114	113	104	82	54	40	38	37	37	265	12
HALF DAY TOTALS			215	404	666	893	1024	1025	881	601	358	247	200	180	176	175	174	175	1083	
Jun 21	5	22	10	17	21	22	20	14	6	2	1	1	1	1	1	1	1	2	3	7
	6	155	48	104	143	159	151	121	70	17	13	13	13	13	13	13	13	14	40	6
	7	216	37	113	172	205	207	178	122	46	22	21	21	21	21	21	21	27	97	5
	8	246	30	85	156	201	216	199	152	80	29	27	27	27	27	27	27	27	153	4
	9	263	33	51	114	166	192	190	161	105	45	33	32	32	32	32	32	32	201	3
	10	272	35	38	63	109	145	158	148	116	69	39	36	35	35	35	35	35	238	2
	11	277	38	39	40	52	81	105	116	110	88	60	41	39	38	38	38	38	260	1
	12	279	38	38	40	41	52	72	89	95	89	72	52	41	40	38	38	38	267	12
HALF DAY TOTALS			253	470	734	941	1038	999	818	523	315	236	204	191	188	187	186	188	1126	
Jul 21	5	2	1	2	2	2	2	1	0	0	0	0	0	0	0	0	0	0	0	7
	6	138	37	89	125	142	137	112	68	18	11	11	11	11	11	11	11	12	32	6
	7	208	30	102	163	198	204	179	127	53	21	20	20	20	20	20	20	20	88	5
	8	241	28	75	148	196	216	203	160	90	30	26	26	26	26	26	26	26	145	4
	9	259	32	44	106	163	193	196	170	118	52	33	31	31	31	31	31	31	194	3
	10	269	35	37	56	106	146	165	159	129	81	41	36	35	35	35	35	35	231	2
	11	275	37	38	40	50	81	111	127	123	102	69	43	39	37	37	37	37	254	1
	12	276	38	38	38	40	41	55	80	101	109	101	80	55	41	40	38	38	262	12
HALF DAY TOTALS			223	411	666	885	1008	1003	858	584	352	248	204	186	181	180	180	181	1076	
Aug 21	6	81	12	44	68	81	82	71	48	17	6	5	5	5	5	5	5	5	12	6
	7	191	17	71	135	177	191	177	135	70	17	16	16	16	16	16	16	16	62	5
	8	237	24	47	126	185	216	214	180	118	41	23	23	23	23	23	23	23	122	4
	9	260	28	31	82	153	197	212	196	151	80	31	28	28	28	28	28	28	174	3
	10	272	32	33	40	93	150	182	187	165	116	56	34	32	32	32	32	32	214	2
	11	278	35	35	36	41	81	128	156	160	141	99	52	37	35	35	35	35	239	1
	12	280	35	35	35	36	38	63	106	138	149	138	106	63	38	36	35	35	247	12
HALF DAY TOTALS			164	273	498	741	928	1013	956	751	474	296	205	166	157	156	156	156	946	
Sep 21	7	149	9	27	84	125	146	144	121	77	21	9	9	9	9	9	9	9	25	5
	8	230	17	19	87	160	205	218	199	148	71	18	17	17	17	17	17	17	82	4
	9	263	22	23	47	131	194	227	226	190	124	41	23	22	22	22	22	22	138	3
	10	280	27	27	28	71	148	200	221	209	165	93	30	27	27	27	27	27	180	2
	11	287	29	29	29	31	78	147	192	207	191	146	77	31	29	29	29	29	206	1
	12	290	30	30	30	30	32	75	142	185	200	185	142	75	32	30	30	30	215	12
HALF DAY TOTALS			119	142	291	534	787	980	1033	925	672	396	222	137	119	118	118	118	738	
Oct 21	7	48	2	3	20	36	45	47	42	30	12	2	2	2	2	2	2	2	4	5
	8	204	11	12	49	123	173	195	188	151	89	18	11	11	11	11	11	11	43	4
	9	257	17	17	23	104	180	225	235	209	151	64	18	17	17	17	17	17	97	3
	10	280	21	21	22	50	139	205	238	235	196	125	38	22	21	21	21	21	140	2
	11	291	24	24	24	25	71	156	212	236	224	178	101	28	24	24	24	24	168	1
	12	294	25	25	25	25	27	85	165	216	234	216	165	85	27	25	25	25	177	12
HALF DAY TOTALS			88	89	152	351	623	878	1006	974	791	493	247	117	89	88	88	88	540	
Nov 21	8	136	5	5	18	69	108	128	129	110	72	21	6	5	5	5	5	5	14	4
	9	232	12	12	13	73	151	201	219	204	156	80	13	12	12	12	12	12	55	3
	10	268	16	16	16	31	122	196	237	242	209	143	50	17	16	16	16	16	96	2
	11	283	19	19	19	20	61	154	218	248	240	194	116	28	19	19	19	19	123	1
	12	288	20	20	20	20	21	89	176	231	250	231	176	89	21	20	20	20	132	12
HALF DAY TOTALS			63	63	75	198	445	721	887	914	798	551	269	101	63	63	63	63	354	
Dec 21	8	89	3	3	8	41	67	82	84	73	50	17	3	3	3	3	3	3	6	4
	9	217	10	10	11	60	135	185	205	194	151	83	13	10	10	10	10	10	39	3
	10	261	14	14	14	25	113	188	232	239	210	146	55	15	14	14	14	14	77	2
	11	280	17	17	17	17	56	151	217	249	242	198	120	28	17	17	17	17	104	1
	12	285	18	18	18	18	19	89	178	233	253	233	178	89	19	18	18	18	113	12
HALF DAY TOTALS			52	52	56	146	374	649	822	867	775	557	276	94	53	52	52	52	282	
			N	NNW	NW	WNW	W	WSW	SW	SSW	S	SSE	SE	ESE	E	ENE	NE	NNE	HOR	PM

[a] Total solar heat gains for DSA glass (based on a ground reflectance of 0.20).

[b] Half day totals computed by Simpson's rule, time interval = 10 min.

TABLE B.4 Solar Intensity and Solar Heat Gain Factors[a] for 48° N Latitude (Conventional Units)[b]

Date	Solar Time am	Direct Normal Btuh/ft²	N	NNE	NE	ENE	E	ESE	SE	SSE	S	SSW	SW	WSW	W	WNW	NW	NNW	HOR	Solar Time pm
Jan 21	8	37	1	1	1	4	18	29	34	35	30	20	6	1	1	1	1	1	2	4
	9	185	8	8	8	53	118	160	176	166	129	69	10	8	8	8	8	8	25	3
	10	239	12	12	12	22	106	175	216	223	195	136	50	12	12	12	12	12	55	2
	11	261	14	14	14	15	53	144	208	239	233	190	116	26	14	14	14	14	77	1
	12	267	15	15	15	15	16	86	171	226	245	226	171	86	16	15	15	15	85	12
	HALF DAY TOTALS		43	43	46	117	316	567	729	776	701	512	259	85	43	43	43	43	203	
Feb 21	7	4	0	0	1	3	3	3	3	2	1	0	0	0	0	0	0	0	0	5
	8	180	8	8	36	103	149	170	166	136	82	17	8	8	8	8	8	8	25	4
	9	247	13	13	16	90	168	216	230	209	155	71	14	13	13	13	13	13	66	3
	10	275	17	17	17	38	131	203	242	244	207	138	44	18	17	17	17	17	105	2
	11	288	19	19	19	20	65	158	221	249	239	192	113	27	19	19	19	19	130	1
	12	292	20	20	20	20	22	89	176	231	250	231	176	89	22	20	20	20	138	12
	HALF DAY TOTALS		68	68	107	274	541	816	968	967	813	531	261	104	68	68	68	68	395	
Mar 21	7	153	7	22	80	123	145	145	123	80	23	7	7	7	7	7	7	7	20	5
	8	236	14	15	76	154	204	222	206	158	82	15	14	14	14	14	14	14	68	4
	9	270	19	19	3	121	193	234	239	207	142	52	20	19	19	19	19	19	118	3
	10	287	23	23	24	58	146	208	237	231	189	115	33	23	23	23	23	23	156	2
	11	295	25	25	25	26	74	156	210	232	218	172	94	28	25	25	25	25	180	1
	12	298	26	26	26	26	27	83	161	211	228	211	161	83	27	26	26	26	188	12
	HALF DAY TOTALS		100	118	250	494	775	1012	1100	1014	767	465	244	126	101	100	100	100	636	
Apr 21	6	108	12	53	86	105	107	93	64	23	6	6	6	6	6	6	6	6	15	6
	7	205	15	61	132	180	199	189	148	84	18	14	14	14	14	14	14	14	60	5
	8	247	20	32	111	179	219	225	196	138	55	21	20	20	20	20	20	20	114	4
	9	268	25	26	60	141	197	223	215	176	106	33	25	25	25	25	25	25	161	3
	10	280	28	28	31	77	148	193	209	194	150	80	31	28	28	28	28	28	196	2
	11	286	31	31	31	33	78	140	181	193	177	133	69	33	31	31	31	31	218	1
	12	288	31	31	31	31	34	71	131	172	186	172	131	71	34	31	31	31	226	12
	HALF DAY TOTALS		147	242	461	724	957	1098	1081	895	605	370	226	156	141	140	140	140	875	
May 21	5	41	17	31	40	42	39	29	14	3	3	3	3	3	3	3	3	3	5	7
	6	162	35	97	141	162	160	133	85	24	12	12	12	12	12	12	12	13	40	6
	7	219	23	90	158	200	212	191	142	68	21	19	19	19	19	19	19	19	91	5
	8	248	26	54	132	190	218	214	178	113	38	25	25	25	25	25	25	25	142	4
	9	264	29	32	82	151	194	208	192	147	77	32	29	29	29	29	29	29	185	3
	10	274	33	34	39	90	145	178	184	163	116	57	35	33	33	33	33	33	219	2
	11	279	35	35	36	40	79	126	155	160	142	101	54	37	35	35	35	35	240	1
	12	280	35	35	35	36	38	63	107	139	150	139	107	63	38	36	35	35	247	12
	HALF DAY TOTALS		215	388	645	893	1065	1114	1007	749	483	316	225	184	174	173	173	174	1045	
Jun 21	5	77	35	61	76	80	72	53	24	6	5	5	5	5	5	5	5	8	12	7
	6	172	46	92	155	175	169	138	84	22	14	14	14	14	14	14	14	16	51	6
	7	220	29	101	165	204	211	187	135	60	23	21	21	21	21	21	21	21	103	5
	8	246	29	64	139	191	215	206	168	101	34	27	27	27	27	27	27	27	152	4
	9	261	31	36	91	153	190	199	180	133	66	33	31	31	31	31	31	31	193	3
	10	269	34	36	45	94	143	169	171	148	101	50	36	34	34	34	34	34	225	2
	11	274	36	36	38	44	79	118	142	145	126	88	49	38	36	36	36	36	246	1
	12	275	37	37	37	38	40	60	96	124	134	124	96	60	40	38	37	37	252	12
	HALF DAY TOTALS		257	459	722	955	1095	1102	955	678	436	299	228	197	189	188	188	191	1108	
Jul 21	5	43	18	33	42	45	41	30	15	3	3	3	3	3	3	3	4	6	7	7
	6	156	37	96	138	159	156	129	82	24	13	13	13	13	13	13	14	41	6
	7	211	25	90	156	196	207	186	138	66	22	20	20	20	20	20	20	20	92	5
	8	240	27	56	132	187	214	209	174	110	38	26	26	26	26	26	26	26	142	4
	9	256	30	34	83	149	191	204	187	143	75	33	30	30	30	30	30	30	184	3
	10	266	34	35	41	90	143	174	180	158	113	56	36	34	34	34	34	34	217	2
	11	271	36	36	37	42	79	124	151	156	138	99	54	38	36	36	36	36	237	1
	12	272	36	36	36	37	39	63	104	136	146	136	104	63	39	37	36	36	244	12
	HALF DAY TOTALS		223	448	646	886	1050	1092	983	730	474	315	229	190	181	179	179	180	1042	
Aug 21	6	99	13	51	81	98	100	87	60	22	7	7	7	7	7	7	7	7	16	6
	7	190	17	61	128	172	190	179	141	79	19	15	15	15	15	15	15	15	61	5
	8	232	22	34	110	174	211	216	188	132	53	23	22	22	22	22	22	22	114	4
	9	254	27	28	63	139	192	216	208	169	102	34	27	27	27	27	27	27	159	3
	10	266	30	30	33	78	145	188	203	188	144	78	33	30	30	30	30	30	193	2
	11	272	32	32	32	36	78	137	175	187	171	129	68	35	32	32	32	32	215	1
	12	274	33	33	33	33	36	71	128	167	189	167	128	71	36	33	33	33	223	12
	HALF DAY TOTALS		157	251	459	709	929	1060	1040	862	587	366	231	165	151	149	149	149	869	
Sep 21	7	131	8	21	71	108	128	128	108	71	21	8	7	7	7	7	7	7	20	5
	8	215	15	16	72	144	191	207	193	148	77	16	15	15	15	15	15	15	65	4
	9	251	20	20	34	116	184	223	227	197	136	52	21	20	20	20	20	20	114	3
	10	269	24	24	25	58	141	200	228	221	182	112	34	24	24	24	24	24	151	2
	11	278	26	26	26	28	73	151	203	223	210	166	92	29	26	26	26	26	174	1
	12	280	27	27	27	27	29	82	156	204	220	204	156	82	29	27	27	27	182	12
	HALF DAY TOTALS		105	121	240	465	729	953	1040	963	737	453	243	131	106	105	105	105	614	
Oct 21	7	4	0	0	2	3	4	4	3	2	1	0	0	0	0	0	0	0	0	5
	8	165	8	9	35	96	139	159	155	126	77	16	8	8	8	8	8	8	25	4
	9	233	14	14	16	88	161	207	220	199	148	68	15	14	14	14	14	14	66	3
	10	262	18	18	18	39	128	196	233	234	199	133	43	18	18	18	18	18	104	2
	11	274	20	20	20	21	64	153	213	241	231	186	109	27	20	20	20	20	128	1
	12	278	21	21	21	21	23	87	171	223	242	223	171	87	23	21	21	21	136	12
	HALF DAY TOTALS		71	71	108	266	519	780	925	925	779	513	256	106	72	71	71	71	391	
Nov 21	8	36	1	1	4	18	29	34	35	30	20	6	1	1	1	1	1	1	2	4
	9	179	8	8	9	52	115	156	171	161	125	67	10	8	8	8	8	8	26	3
	10	233	12	12	12	22	104	172	212	218	191	133	49	13	12	12	12	12	55	2
	11	255	15	15	15	15	52	142	204	234	228	186	114	26	15	15	15	15	77	1
	12	261	15	15	15	15	17	85	168	222	240	222	168	85	17	15	15	15	85	12
	HALF DAY TOTALS		44	44	47	117	310	555	713	760	686	502	255	86	44	44	44	44	204	
Dec 21	9	140	5	5	6	36	86	120	133	127	100	56	8	5	5	5	5	5	13	3
	10	214	10	10	10	16	91	156	194	201	179	126	49	10	10	10	10	10	38	2
	11	242	12	12	12	13	46	134	195	225	220	180	111	25	12	12	12	12	57	1
	12	250	13	13	13	13	14	81	163	215	233	215	163	81	14	13	13	13	67	12
	HALF DAY TOTALS		33	33	34	73	233	458	610	665	616	468	247	76	34	33	33	33	141	

N	NNW	NW	WNW	W	WSW	SW	SSW	S	SSE	SE	ESE	E	ENE	NE	NNE	HOR	PM

[a]Total solar heat gains for DSA glass (based on a ground reflectance of 0.20).

[b]Half day totals computed by Simpson's rule, time interval = 10 min.

TABLE B.5 **Solar Intensity and Solar Heat Gain Factors[a] for 64° N Latitude (Conventional Units)[b]**

Date	Solar Time am	Direct Normal Btuh ft²	N	NNE	NE	ENE	E	ESE	SE	SSE	S	SSW	SW	WSW	W	WNW	NW	NNW	HOR	Solar Time pm
Jan 21	10	22	1	1	1	1	9	16	20	21	19	13	5	1	1	1	1	1	1	2
	11	81	3	3	3	3	15	45	67	77	75	62	38	8	3	3	3	3	6	1
	12	100	3	3	3	3	4	33	67	89	96	89	67	33	4	3	3	3	8	12
HALF DAY TOTALS			5	5	5	6	25	79	121	142	141	119	75	23	5	5	5	5	11	
Feb 21	8	18	1	1	3	10	15	17	17	14	9	2	1	1	1	1	1	1	1	4
	9	134	5	5	6	43	89	118	128	119	90	45	6	5	5	5	5	5	13	3
	10	190	8	8	8	18	87	144	176	180	157	108	38	9	8	8	8	8	28	2
	11	215	10	10	10	11	44	122	177	202	197	160	97	20	10	10	10	10	41	1
	12	222	11	11	11	11	12	73	147	194	210	194	147	73	12	11	11	11	45	12
HALF DAY TOTALS			29	30	33	89	244	446	578	617	560	411	212	66	30	29	29	29	106	
Mar 21	7	95	4	11	47	74	90	91	79	53	17	4	4	4	4	4	4	4	9	5
	8	185	9	10	46	113	158	177	170	135	78	14	9	9	9	9	9	9	32	4
	9	227	13	13	16	88	159	203	215	194	143	64	14	13	13	13	13	13	59	3
	10	249	16	16	16	35	122	190	226	228	194	130	42	16	16	16	16	16	84	2
	11	260	17	17	17	18	60	148	209	236	228	184	109	25	17	17	17	17	99	1
	12	263	18	18	18	18	19	85	168	221	239	221	168	85	19	18	18	18	105	12
HALF DAY TOTALS			68	74	150	334	596	854	984	958	779	504	257	104	68	68	68	68	335	
Apr 21	5	27	8	18	24	27	26	20	12	2	1	1	1	1	1	1	1	1	2	7
	6	133	12	59	102	127	132	118	84	35	8	8	8	8	8	8	8	8	21	6
	7	194	14	41	113	163	189	185	153	96	25	13	13	13	13	13	13	13	51	5
	8	228	17	19	79	153	201	217	201	153	79	19	17	17	17	17	17	17	85	4
	9	248	21	21	32	111	180	219	225	197	138	55	22	21	21	21	21	21	116	3
	10	260	23	23	24	51	134	194	225	221	185	118	38	24	23	23	23	23	140	2
	11	266	24	24	24	26	68	148	202	225	214	171	99	29	24	24	24	24	155	1
	12	268	25	25	25	25	27	83	159	208	224	208	159	83	27	25	25	25	160	12
HALF DAY TOTALS			131	218	410	671	943	1150	1186	1036	763	487	273	149	121	120	120	120	651	
May 21	4	51	30	44	51	51	43	28	8	3	3	3	3	3	3	3	3	10	6	8
	5	132	48	95	125	135	125	96	50	11	9	9	9	9	9	9	9	11	26	7
	6	185	28	97	150	181	183	158	109	40	15	15	15	15	15	15	15	15	55	6
	7	218	21	63	138	189	211	201	161	94	24	19	19	19	19	19	19	19	90	5
	8	239	23	28	97	167	209	220	198	146	68	25	23	23	23	23	23	23	124	4
	9	252	26	27	45	122	183	215	215	184	123	46	27	26	26	26	26	26	152	3
	10	261	28	28	30	61	135	188	212	205	167	102	36	28	28	28	28	28	174	2
	11	265	30	30	30	32	72	141	188	207	195	154	87	33	30	30	30	30	188	1
	12	267	30	30	30	30	33	78	146	189	204	189	146	78	33	30	30	30	192	12
HALF DAY TOTALS			247	425	680	950	1177	1291	1218	985	708	465	288	191	169	168	168	176	911	
Jun 21	4	93	53	83	96	94	78	50	14	7	7	7	7	7	7	7	7	21	16	8
	5	154	62	114	148	158	145	110	55	14	12	12	12	12	12	12	12	14	39	7
	6	194	36	107	162	191	192	163	110	39	18	17	17	17	17	17	17	18	71	6
	7	221	24	71	145	193	213	200	158	89	25	22	22	22	22	22	22	22	105	5
	8	239	25	33	104	170	208	216	192	139	62	27	25	25	25	25	25	25	137	4
	9	251	28	29	51	124	181	210	208	175	115	43	29	28	28	28	28	28	165	3
	10	258	30	30	32	65	134	183	204	195	157	94	36	30	30	30	30	30	186	2
	11	262	32	32	32	34	72	137	180	196	184	144	82	35	32	32	32	32	199	1
	12	263	32	32	32	32	35	76	138	179	193	179	138	76	35	32	32	32	203	12
HALF DAY TOTALS			322	533	801	1061	1253	1317	1195	946	679	455	296	211	192	190	191	213	1021	
Jul 21	4	53	32	47	55	54	46	29	9	4	4	4	4	4	4	4	4	11	8	8
	5	128	49	94	123	133	124	95	50	11	10	10	10	10	10	10	10	11	28	7
	6	179	30	96	148	177	180	155	106	39	16	15	15	15	15	15	15	15	57	6
	7	211	22	64	137	186	207	197	157	92	25	20	20	20	20	20	20	20	92	5
	8	231	24	30	97	165	205	215	193	142	67	26	24	24	24	24	24	24	124	4
	9	245	27	28	47	121	180	211	211	179	120	46	28	27	27	27	27	27	152	3
	10	253	29	29	31	62	134	185	208	200	164	100	37	29	29	29	29	29	174	2
	11	257	31	31	31	33	72	139	185	202	191	151	86	34	31	31	31	31	187	1
	12	259	31	31	31	31	34	78	143	185	200	185	143	78	34	31	31	31	192	12
HALF DAY TOTALS			258	434	684	946	1163	1269	1193	965	697	462	292	198	177	175	175	180	918	
Aug 21	5	29	9	20	27	30	28	22	13	2	2	2	2	2	2	2	2	2	3	7
	6	123	13	58	97	121	125	111	80	34	9	9	9	9	9	9	9	9	23	6
	7	181	15	42	109	157	180	176	145	92	26	14	14	14	14	14	14	14	53	5
	8	214	19	21	78	148	193	208	192	147	76	21	19	19	19	19	19	19	87	4
	9	234	22	22	34	109	174	211	217	189	133	55	23	22	22	22	22	22	117	3
	10	246	25	25	26	52	131	188	217	214	178	114	39	25	25	25	25	25	140	2
	11	252	26	26	26	28	69	144	196	217	207	166	97	31	26	26	26	26	154	1
	12	254	27	27	27	27	29	82	155	201	217	201	155	82	29	27	27	27	159	12
HALF DAY TOTALS			142	226	410	657	914	1109	1141	997	740	478	275	158	131	130	130	130	656	
Sep 21	7	77	4	10	39	62	74	75	65	44	15	4	4	4	4	4	4	4	8	5
	8	163	10	10	43	103	143	160	154	123	71	14	10	10	10	10	10	10	31	4
	9	206	14	14	17	83	148	189	200	181	133	61	15	14	14	14	14	14	57	3
	10	229	16	16	17	35	116	179	213	214	183	123	41	17	16	16	16	16	81	2
	11	240	18	18	18	19	59	141	198	224	216	174	104	26	18	18	18	18	96	1
	12	244	19	19	19	19	21	82	160	209	227	209	160	82	21	19	19	19	101	12
HALF DAY TOTALS			71	77	142	307	547	787	910	891	731	480	249	106	72	71	71	71	324	
Oct 21	8	17	1	1	3	10	14	16	16	13	8	2	1	1	1	1	1	1	1	4
	9	122	5	5	6	40	82	109	118	110	83	42	6	5	5	5	5	5	13	3
	10	176	9	9	9	18	83	135	165	169	147	102	36	9	9	9	9	9	29	2
	11	201	11	11	11	11	43	116	167	191	186	152	92	20	11	11	11	11	41	1
	12	208	11	11	11	11	13	70	140	184	199	184	140	70	13	11	11	11	46	12
HALF DAY TOTALS			31	31	34	86	231	420	542	580	527	388	202	66	32	31	31	31	108	
Nov 21	10	23	1	1	1	1	10	17	21	22	20	14	5	1	1	1	1	1	2	2
	11	79	3	3	3	3	15	44	65	75	74	61	37	8	3	3	3	3	6	1
	12	97	4	4	4	4	4	32	66	87	93	87	66	32	4	4	4	4	8	12
HALF DAY TOTALS			5	5	5	6	26	79	120	141	140	117	74	23	6	5	5	5	11	
Dec 21	11	4	0	0	0	0	1	2	3	4	4	3	2	0	0	0	0	0	0	1
	12	16	0	0	0	0	1	5	11	14	15	14	11	5	1	0	0	0	1	12
HALF DAY TOTALS			0	0	0	0	1	5	9	11	11	10	7	3	0	0	0	0	1	
			N	NNW	NW	WNW	W	WSW	SW	SSW	S	SSE	SE	ESE	E	ENE	NE	NNE	HOR	PM

[a]Total solar heat gains for DSA glass (based on a ground reflectance of 0.20).

[b]Half day totals computed by Simpson's rule, time interval = 10 min.

TABLE B.6 Solar Intensity and Solar Heat Gain Factors[a] for 16° N Latitude (SI Units)[b]

Date	Solar Time am	Direct Normal W/m²	N	NNE	NE	ENE	E	ESE	SE	SSE	S	SSW	SW	WSW	W	WNW	NW	NNW	HOR	Solar Time pm
Jan 21	7	445	17	19	138	291	390	424	397	303	155	19	17	17	17	17	17	17	43	5
	8	827	45	48	174	463	662	757	734	596	359	79	45	45	45	45	45	45	249	4
	9	948	67	67	102	384	630	770	791	690	481	183	69	67	67	67	67	67	472	3
	10	1001	82	82	86	209	474	658	737	704	563	321	97	82	82	82	82	82	640	2
	11	1025	92	92	92	96	242	467	614	663	612	462	236	96	92	92	92	92	745	1
	12	1032	95	95	95	95	100	228	438	580	628	580	438	228	100	95	95	95	782	12
HALF DAY TOTALS			348	352	618	1453	2398	3153	3458	3217	2465	1344	666	401	350	348	348	348	2539	
Feb 21	7	575	24	55	265	435	532	544	474	326	113	26	24	24	24	24	24	24	80	5
	8	862	53	60	304	567	729	778	706	525	244	56	53	53	53	53	53	53	319	4
	9	961	74	77	202	482	676	763	733	592	347	96	74	74	74	74	74	74	549	3
	10	1006	90	91	104	292	508	636	665	593	423	193	94	90	90	90	90	90	722	2
	11	1027	99	99	102	118	262	428	527	542	471	323	154	103	99	99	99	99	831	1
	12	1034	103	103	103	103	108	189	336	448	487	448	336	189	108	105	103	103	868	12
HALF DAY TOTALS			390	431	1013	1922	2730	3228	3263	2792	1836	906	547	417	393	390	390	390	2933	
Mar 21	7	634	36	167	393	544	606	578	458	260	47	33	33	33	33	33	33	33	126	5
	8	857	63	157	442	648	752	741	615	390	111	61	61	61	61	61	61	61	380	4
	9	943	84	110	343	565	689	709	622	435	180	86	82	82	82	82	82	82	606	3
	10	983	98	103	191	379	519	575	543	424	240	107	100	98	98	98	98	98	778	2
	11	1003	108	110	113	166	273	361	395	366	281	173	114	110	108	108	108	108	885	1
	12	1008	111	111	112	114	117	149	216	273	295	273	216	149	117	114	112	111	919	12
HALF DAY TOTALS			443	712	1558	2383	2926	3076	2773	2028	1006	588	483	448	440	483	437	437	3233	
Apr 21	6	44	7	24	37	43	43	37	24	7	2	2	2	2	2	2	12	2	4	6
	7	622	75	298	482	589	604	528	369	141	45	42	42	42	42	42	42	42	169	5
	8	807	85	312	543	682	718	644	473	217	74	70	70	70	70	70	70	70	413	4
	9	885	97	248	469	610	657	608	465	244	97	90	90	90	90	90	90	90	623	3
	10	924	112	171	321	444	499	476	380	231	118	109	106	106	106	106	106	106	784	2
	11	942	120	127	169	228	270	276	245	189	136	121	118	115	115	115	115	115	882	1
	12	948	123	123	124	125	126	129	135	141	143	141	135	129	126	125	124	123	911	12
HALF DAY TOTALS			565	1272	2127	2711	2909	2684	2059	1111	547	503	494	491	490	490	492	492	1333	
May 21	6	138	43	94	128	141	134	106	59	11	9	9	9	9	9	9	9	9	10	6
	7	608	157	378	531	603	583	474	290	76	49	49	49	49	49	49	49	55	17	5
	8	771	165	415	598	689	677	564	362	121	78	76	76	76	76	76	76	80	425	4
	9	845	156	366	539	626	622	526	344	141	100	96	96	96	96	96	96	100	621	3
	10	883	149	281	410	478	474	398	264	139	116	111	111	111	111	111	111	116	772	2
	11	902	147	198	248	273	262	220	166	130	126	123	120	120	120	122	124	128	862	1
	12	907	146	144	140	134	132	131	130	129	129	129	130	131	132	134	140	144	890	12
HALF DAY TOTALS			893	1813	2537	2891	2829	2360	1555	685	541	528	527	526	527	529	533	557	3338	
Jun 21	6	168	64	124	163	175	162	123	64	13	12	12	12	12	12	12	12	13	24	6
	7	593	195	404	543	598	565	445	252	63	52	52	52	52	52	52	52	57	203	5
	8	750	209	449	612	684	653	526	313	98	79	79	79	79	79	79	79	84	425	4
	9	823	199	409	560	626	601	487	294	118	98	98	98	98	98	98	98	105	613	3
	10	861	187	329	441	486	459	363	222	125	116	113	113	113	113	113	113	121	759	2
	11	879	181	241	283	290	258	200	146	130	126	122	122	122	122	125	128	136	847	1
	12	885	179	173	158	142	134	132	130	129	129	129	130	132	134	142	158	173	873	12
HALF DAY TOTALS			1122	2043	2683	2930	2763	2207	1357	612	540	540	540	542	545	555	599	599	3308	
Jul 21	6	128	43	90	122	134	127	99	55	11	9	9	9	9	9	9	9	10	17	6
	7	579	161	373	518	585	564	456	277	74	51	51	51	51	51	51	51	55	194	5
	8	743	173	415	590	676	661	549	349	110	78	78	78	78	78	78	78	83	419	4
	9	818	165	371	537	618	610	513	334	138	102	99	99	99	99	99	99	104	611	3
	10	857	157	289	413	475	467	389	257	138	118	113	113	113	113	113	113	120	759	2
	11	876	154	207	255	277	262	218	164	133	128	125	122	122	122	125	128	131	848	1
	12	882	154	151	145	138	135	134	133	132	132	132	133	134	135	138	145	151	875	12
HALF DAY TOTALS			933	1829	2521	2848	2770	2298	1507	680	554	541	539	539	540	542	547	575	3289	
Aug 21	6	136	7	21	31	37	36	31	20	6	2	2	2	2	2	2	2	2	4	6
	7	569	81	289	458	555	567	493	343	131	48	45	45	45	45	45	45	45	167	5
	8	757	94	315	531	660	691	617	451	206	79	74	74	74	74	74	74	74	404	4
	9	838	104	258	467	598	640	589	448	235	103	95	95	95	95	95	95	95	608	3
	10	879	118	183	328	443	490	464	368	224	122	114	110	110	110	110	110	110	766	2
	11	899	127	136	180	235	271	273	240	185	138	127	124	120	120	120	120	124	860	1
	12	905	129	130	130	131	132	134	139	143	145	143	139	134	132	131	130	129	889	12
HALF DAY TOTALS			603	1294	2099	2640	2810	2578	1969	1069	568	528	519	516	515	514	515	518	3258	
Sep 21	7	565	38	157	360	497	554	529	419	240	48	35	35	35	35	35	35	35	122	5
	8	797	67	156	424	618	716	705	587	374	113	64	64	64	64	64	64	64	367	4
	9	887	87	114	335	547	665	684	602	424	181	90	86	86	86	86	86	86	585	3
	10	931	101	107	193	372	507	560	529	415	240	111	103	101	101	101	101	101	752	2
	11	952	111	114	118	169	272	356	389	361	279	176	118	114	111	111	111	111	856	1
	12	958	114	114	116	118	121	153	217	272	293	272	217	153	121	118	116	114	889	12
HALF DAY TOTALS			461	712	1500	2276	2791	2937	2658	1963	1007	605	501	466	457	455	454	454	3126	
Oct 21	7	524	25	56	249	404	492	502	437	300	105	27	25	25	25	25	25	25	79	5
	8	816	55	64	299	548	702	747	677	502	234	59	55	55	55	55	55	55	313	4
	9	920	76	80	204	473	659	741	711	573	336	97	76	76	76	76	76	76	537	3
	10	968	92	94	108	291	499	621	647	577	412	189	97	92	92	92	92	92	707	2
	11	991	102	105	122	261	420	515	528	459	315	154	106	102	102	102	102	102	814	1
	12	997	105	105	107	111	189	227	432	474	572	474	432	227	189	111	107	105	850	12
HALF DAY TOTALS			402	444	1002	1869	2637	3110	3141	2689	1776	891	553	428	404	402	402	402	2872	
Nov 21	7	423	17	19	134	280	375	406	379	289	147	19	17	17	17	17	17	17	43	5
	8	806	46	49	174	456	651	742	719	583	350	78	46	46	46	46	46	46	247	4
	9	929	67	67	105	382	623	779	779	679	472	180	70	67	67	67	67	67	468	3
	10	984	83	83	87	210	470	651	727	694	554	316	98	83	83	83	83	83	635	2
	11	1009	92	92	92	98	242	462	607	654	603	455	234	98	92	92	92	92	740	1
	12	1016	96	96	96	96	101	227	432	572	619	572	432	227	101	96	96	96	775	12
HALF DAY TOTALS			352	357	620	1438	2363	3101	3396	3159	2421	1324	664	404	355	352	352	352	2520	
Dec 21	7	372	13	14	93	228	318	354	339	268	150	21	13	13	13	13	13	13	31	5
	8	803	42	43	129	416	625	734	728	609	390	105	42	42	42	42	42	42	219	4
	9	936	63	63	78	341	604	761	800	716	520	226	66	63	63	63	63	63	436	3
	10	993	78	78	81	178	445	657	753	735	605	369	112	79	78	78	78	78	602	2
	11	1019	87	87	87	92	231	474	638	698	654	506	270	94	87	87	87	87	704	1
	12	1026	91	91	91	91	95	241	469	620	670	620	469	241	95	91	91	91	739	12
HALF DAY TOTALS			328	330	501	1269	2241	3076	3468	3312	2638	1527	721	393	331	328	328	328	2361	

N	NNW	NW	WNW	W	WSW	SW	SSW	S	SSE	SE	ESE	E	ENE	NE	NNE	HOR	PM

[a] Total solar heat gains for DSA glass (based on a ground reflectance of 0.20).

[b] Half day totals computed by Simpson's rule, time interval = 10 min.

TABLE B.7 Solar Intensity and Solar Heat Gain Factors[a] for 32° N Latitude (SI Units)[b]

Date	Solar Time am	Direct Normal W/m²	N	NNE	NE	ENE	E	ESE	SE	SSE	S	SSW	SW	WSW	W	WNW	NW	NNW	HOR	Solar Time pm
Jan 21	7	4	0	0	1	3	4	4	4	3	2	0	0	0	0	0	0	0	0	5
	8	640	28	29	93	330	505	597	596	502	326	88	29	28	28	28	28	28	102	4
	9	849	48	48	53	286	553	721	775	711	534	258	52	48	48	48	48	48	278	3
	10	931	63	63	64	129	427	659	784	788	670	444	144	64	63	63	63	63	430	2
	11	967	71	71	71	75	213	502	698	784	750	602	347	90	71	71	71	71	524	1
	12	977	74	74	74	74	79	277	548	718	777	718	548	277	79	74	74	74	556	12
	HALF DAY TOTALS		249	250	338	897	1797	2700	3201	3198	2692	1746	831	252	249	249	249	249	1614	
Feb 21	7	352	13	23	148	258	323	336	299	212	83	14	13	13	13	13	13	13	30	5
	8	771	41	43	204	469	646	719	683	538	300	53	41	41	41	41	41	41	200	4
	9	905	60	60	101	386	627	764	783	680	471	174	63	60	60	60	60	60	402	3
	10	964	75	75	78	195	475	670	760	733	595	352	99	75	75	75	75	75	556	2
	11	990	83	83	83	88	240	491	656	717	670	519	275	89	83	83	83	83	652	1
	12	998	86	86	86	86	91	250	489	644	696	643	489	250	91	86	86	86	684	12
	HALF DAY TOTALS		314	324	635	1405	2318	3086	3408	3188	2461	1426	718	316	314	314	314	314	2181	
Mar 21	7	583	30	116	332	483	556	545	447	276	63	29	29	29	29	29	29	29	100	5
	8	821	54	78	339	576	717	746	661	473	195	57	54	54	54	54	54	54	315	4
	9	914	74	77	203	475	662	746	716	578	339	95	74	74	74	74	74	74	516	3
	10	959	88	88	96	273	499	637	677	615	456	221	93	88	88	88	88	88	667	2
	11	980	96	96	98	108	258	447	564	592	529	380	185	101	96	96	96	96	762	1
	12	987	99	99	99	99	105	209	386	511	554	511	386	209	105	99	99	99	795	12
	HALF DAY TOTALS		393	512	1131	1985	2761	3258	3284	2801	1858	1030	609	429	394	391	391	391	2757	
Apr 21	6	210	29	110	172	205	207	177	119	38	11	11	11	11	11	11	11	11	23	6
	7	649	53	253	462	593	631	575	428	204	49	45	45	45	45	45	45	45	191	5
	8	804	73	192	453	632	715	689	559	337	95	69	69	69	69	69	69	69	408	4
	9	876	89	113	324	532	649	669	591	418	183	92	87	87	87	87	87	87	593	3
	10	913	101	106	165	342	490	599	543	445	275	123	104	101	101	101	101	101	736	2
	11	932	109	109	115	149	263	371	426	415	340	222	126	113	109	109	109	109	825	1
	12	937	112	112	112	116	120	167	260	335	363	335	260	167	120	116	112	112	854	12
	HALF DAY TOTALS		508	932	1734	2500	3003	3129	2806	2033	1135	721	559	496	482	479	479	479	3203	
May 21	6	374	104	244	340	381	367	295	175	39	26	26	26	26	26	26	26	28	67	6
	7	666	112	350	535	638	643	550	374	134	60	57	57	57	57	57	57	59	256	5
	8	787	93	295	519	655	694	629	470	229	86	80	80	80	80	80	80	97	461	4
	9	849	104	194	404	558	625	601	488	294	117	100	97	97	97	97	97	110	633	3
	10	882	114	127	240	383	473	491	435	313	170	117	110	110	110	110	110	118	766	2
	11	898	121	124	132	186	260	312	322	285	215	147	126	122	118	118	118	118	847	1
	12	904	121	123	125	127	130	148	186	220	223	220	186	148	130	127	125	123	873	12
	HALF DAY TOTALS		700	1382	2214	2840	3106	2943	2356	1409	788	627	577	551	550	548	551	551	3464	
Jun 21	6	412	140	289	388	425	401	314	174	39	32	32	32	32	32	32	32	35	89	6
	7	662	148	384	556	644	634	529	342	109	62	62	62	62	62	62	62	65	279	5
	8	773	115	335	540	656	677	597	427	188	89	84	84	84	84	84	84	84	476	4
	9	831	110	234	431	563	609	567	440	244	111	101	101	101	101	101	101	101	642	3
	10	863	120	150	272	395	461	458	387	261	143	120	114	114	114	114	114	114	770	2
	11	880	126	130	148	202	258	288	280	237	177	135	128	125	122	122	122	122	847	1
	12	885	128	129	130	132	134	144	164	183	191	183	164	144	134	132	130	129	871	12
	HALF DAY TOTALS		824	1589	2403	2950	3108	2829	2137	1174	710	621	595	585	582	579	579	586	3538	
Jul 21	6	358	106	240	332	370	355	285	168	39	27	27	27	27	27	27	27	30	70	6
	7	640	118	349	526	623	626	534	361	129	62	59	59	59	59	59	59	61	257	5
	8	761	98	300	515	645	680	613	456	221	89	82	82	82	82	82	82	82	457	4
	9	823	107	202	405	553	615	588	475	284	117	100	100	100	100	100	100	100	626	3
	10	856	118	133	247	383	467	481	424	303	167	120	113	113	113	113	113	113	757	2
	11	873	124	128	137	191	261	308	314	277	210	147	129	125	121	121	121	121	836	1
	12	879	126	127	128	130	133	150	184	214	226	214	184	150	133	130	128	127	861	12
	HALF DAY TOTALS		728	1402	2210	2809	3052	2877	2290	1365	783	637	591	574	568	566	565	568	3431	
Aug 21	6	187	30	103	159	188	189	162	108	35	12	12	12	12	12	12	12	12	25	6
	7	599	59	250	444	565	598	542	402	192	53	49	49	49	49	49	49	49	192	5
	8	756	79	200	446	614	690	662	536	321	96	74	74	74	74	74	74	74	403	4
	9	830	95	124	328	523	632	648	570	402	179	98	93	93	93	93	93	93	583	3
	10	869	106	113	175	344	482	544	526	429	266	126	110	106	106	106	106	106	722	2
	11	889	115	117	122	157	264	365	414	401	328	217	131	119	115	115	115	115	808	1
	12	894	117	117	117	122	126	170	255	324	350	324	255	170	126	122	117	117	837	12
	HALF DAY TOTALS		539	957	1721	2443	2910	3014	2694	1950	1109	729	581	524	510	507	506	506	3152	
Sep 21	7	514	32	109	302	437	502	492	405	252	62	30	30	30	30	30	30	30	97	5
	8	758	57	81	324	546	679	706	626	450	190	61	57	57	57	57	57	57	304	4
	9	857	77	82	201	459	637	717	688	557	330	99	77	77	77	77	77	77	498	3
	10	905	91	91	101	270	485	617	655	596	444	220	97	91	91	91	91	91	645	2
	11	928	100	100	102	113	257	437	549	576	516	373	187	106	100	100	100	100	737	1
	12	935	103	103	103	103	110	210	379	499	540	499	379	210	110	103	103	103	769	12
	HALF DAY TOTALS		409	518	1089	1888	2621	3097	3132	2689	1812	1025	620	446	411	408	408	408	2664	
Oct 21	7	312	13	24	136	233	291	301	268	190	75	14	13	13	13	13	13	13	30	5
	8	724	42	46	200	450	616	684	649	511	285	54	42	42	42	42	42	42	198	4
	9	862	63	63	104	378	608	738	755	655	454	170	65	63	63	63	63	63	395	3
	10	923	77	77	81	197	465	652	737	711	577	342	101	77	77	77	77	77	546	2
	11	952	86	86	86	91	239	481	639	697	651	505	269	86	86	86	86	86	639	1
	12	960	89	89	89	89	95	247	478	626	677	626	478	247	95	89	89	89	671	12
	HALF DAY TOTALS		325	335	632	1365	2234	2968	3275	3067	2376	1390	714	394	328	325	325	325	2143	
Nov 21	7	5	0	0	1	3	4	5	4	3	2	0	0	0	0	0	0	0	0	5
	8	619	29	29	93	323	493	581	579	488	316	86	29	29	29	29	29	29	102	4
	9	829	49	49	55	284	545	709	761	697	523	253	49	49	49	49	49	49	277	3
	10	912	64	64	65	131	423	650	772	775	659	437	142	65	64	64	64	64	428	2
	11	949	72	72	72	76	213	496	688	773	739	593	342	91	72	72	72	72	521	1
	12	959	75	75	75	75	80	274	541	708	766	708	541	274	80	75	75	75	553	12
	HALF DAY TOTALS		253	254	342	890	1769	2653	3143	3138	2643	1717	824	355	256	253	253	253	1607	
Dec 21	8	556	22	22	59	265	426	515	523	452	305	98	23	22	22	22	22	22	69	4
	9	812	43	43	46	244	512	686	751	701	539	282	48	43	43	43	43	43	228	3
	10	908	57	57	57	106	402	642	777	792	683	466	163	59	57	57	57	57	375	2
	11	949	66	66	66	69	200	497	700	795	766	620	367	92	66	66	66	66	468	1
	12	960	69	69	69	69	73	281	559	733	794	733	559	281	73	69	69	69	500	12
	HALF DAY TOTALS		223	223	265	717	1577	2499	3043	3111	2687	1823	868	339	225	223	223	223	1389	
			N	NNW	NW	WNW	W	WSW	SW	SSW	S	SSE	SE	ESE	E	ENE	NE	NNE	HOR	PM

[a]Total solar heat gains for DSA glass (based on a ground reflectance of 0.20).

[b]Half day totals computed by Simpson's rule, time interval = 10 min.

TABLE B.8 Solar Intensity and Solar Heat Gain Factors[a] for 40° N Latitude (SI Units)[b]

Date	Solar Time am	Direct Normal W/m²	N	NNE	NE	ENE	E	ESE	SE	SSE	S	SSW	SW	WSW	W	WNW	NW	NNW	HOR	Solar Time pm
Jan 21	8	446	17	17	55	223	350	417	420	358	60	17	17	17	17	17	17	17	44	4
	9	753	37	37	41	233	485	648	706	658	504	260	42	37	37	37	37	37	173	3
	10	865	51	51	51	97	390	627	761	776	671	460	161	53	51	51	51	51	303	2
	11	913	59	59	59	62	193	493	699	796	769	623	372	89	59	59	59	59	390	1
	12	926	62	62	62	62	66	293	563	740	802	740	563	283	62	62	62	62	419	12
	HALF DAY TOTALS		194	194	231	628	1426	2316	2852	2941	2566	1770	860	318	196	194	194	194	1117	
Feb 21	7	175	6	10	71	127	160	167	150	107	43	6	6	6	6	6	6	6	11	5
	8	692	33	35	158	407	576	651	628	505	296	56	33	33	33	33	33	33	136	4
	9	854	52	52	70	337	587	738	773	689	496	209	55	52	52	52	52	52	309	3
	10	926	65	65	67	155	450	666	777	768	641	408	120	66	65	65	65	65	451	2
	11	958	73	73	73	77	224	504	690	769	730	579	325	86	73	73	73	73	538	1
	12	967	76	76	76	76	80	271	536	702	760	702	536	271	80	76	76	76	568	12
	HALF DAY TOTALS		267	271	478	1140	2043	2888	3308	3202	2591	1602	790	361	269	267	267	267	1730	
Mar 21	7	540	27	92	295	441	514	509	425	271	69	26	26	26	26	26	26	26	83	5
	8	789	50	57	288	534	686	732	665	494	232	54	50	50	50	50	50	50	268	4
	9	889	67	70	147	429	640	749	744	625	404	127	69	67	67	67	67	67	450	3
	10	938	80	80	85	226	482	653	722	682	539	299	91	80	80	80	80	80	587	2
	11	961	88	88	88	94	247	476	623	673	622	473	244	94	88	88	88	88	673	1
	12	968	91	91	91	91	97	238	458	602	650	602	458	238	97	91	91	91	702	12
	HALF DAY TOTALS		358	440	954	1777	2626	3265	3429	3055	1539	1270	694	417	360	357	357	357	2411	
Apr 21	6	282	36	144	228	275	279	241	164	56	16	15	15	15	15	15	15	15	34	6
	7	651	50	223	442	584	633	588	451	255	51	45	45	45	45	45	45	45	193	5
	8	795	69	140	402	601	706	703	594	391	130	69	67	67	67	67	67	67	389	4
	9	865	84	91	254	488	638	691	640	494	260	91	84	84	84	84	84	84	557	3
	10	901	96	99	117	291	480	589	608	538	381	177	101	96	96	96	96	96	685	2
	11	920	104	107	107	122	255	411	506	522	459	323	164	109	104	104	104	104	766	1
	12	926	106	106	106	109	114	196	341	448	486	448	341	196	114	109	106	106	794	12
	HALF DAY TOTALS		487	835	1580	2390	3020	3314	3135	2466	1539	935	628	495	467	464	464	464	3020	
May 21	5	3	1	2	3	3	3	2	1	0	0	0	0	0	0	0	0	0	0	7
	6	453	113	284	403	458	446	364	223	56	33	33	33	33	33	33	33	35	96	6
	7	681	90	320	520	638	659	580	412	172	63	59	59	59	59	59	59	59	276	5
	8	787	86	230	471	629	694	655	519	295	92	80	80	80	80	80	80	80	461	4
	9	844	99	131	330	518	620	632	551	382	168	101	96	96	96	96	96	96	616	3
	10	875	107	114	171	332	467	529	513	419	262	127	111	107	107	107	107	107	737	2
	11	891	115	115	121	152	256	357	409	400	331	222	133	120	115	115	115	115	812	1
	12	896	117	117	117	121	126	171	258	329	355	329	258	171	126	121	117	117	836	12
	HALF DAY TOTALS		679	1275	2102	2818	3231	3232	2778	1897	1129	780	630	568	554	550	550	552	3418	
Jun 21	5	68	32	54	68	70	63	46	20	5	4	4	4	4	4	4	4	7	8	7
	6	488	150	329	450	501	478	380	222	53	39	39	39	39	39	39	39	43	126	6
	7	681	118	355	543	648	654	562	385	145	68	65	65	65	65	65	65	67	306	5
	8	776	94	268	492	633	680	626	480	252	93	85	85	85	85	85	85	85	484	4
	9	829	105	160	359	524	607	601	507	332	142	104	100	100	100	100	100	100	633	3
	10	859	112	121	197	345	457	500	468	366	218	122	115	112	112	112	112	112	750	2
	11	874	119	123	128	166	254	332	367	347	279	188	130	124	119	119	119	119	821	1
	12	879	121	121	121	127	131	163	227	281	301	281	227	163	131	127	121	121	844	12
	HALF DAY TOTALS		799	1483	2314	2968	3275	3151	2580	1649	995	743	642	602	592	588	587	594	3551	
Jul 21	5	7	3	5	7	7	6	5	2	0	0	0	0	0	0	0	0	1	1	7
	6	435	116	281	395	447	433	352	216	55	34	34	34	34	34	34	34	37	100	6
	7	656	95	321	513	625	643	564	400	166	66	62	62	62	62	62	62	62	278	5
	8	762	90	236	468	620	680	639	505	285	94	83	83	83	83	83	83	83	459	4
	9	818	102	138	333	513	610	618	537	371	165	104	99	99	99	99	99	99	611	3
	10	850	110	118	177	333	462	519	501	407	255	129	114	110	110	110	110	110	729	2
	11	866	117	120	125	157	256	352	400	389	321	217	135	123	117	117	117	117	802	1
	12	871	120	120	120	125	130	172	253	320	345	320	253	172	130	125	120	120	826	12
	HALF DAY TOTALS		705	1296	2102	2792	3180	3164	2707	1842	1110	783	643	586	572	568	567	570	3395	
Aug 21	6	255	38	137	214	256	259	223	151	52	18	17	17	17	17	17	17	17	38	6
	7	603	55	223	426	557	602	557	426	222	55	49	49	49	49	49	49	49	196	5
	8	747	75	149	397	584	681	676	569	374	128	74	72	72	72	72	72	72	386	4
	9	819	89	97	259	481	621	669	618	475	251	97	89	89	89	89	89	89	549	3
	10	857	102	105	126	294	472	574	590	519	567	173	107	102	102	102	102	102	674	2
	11	876	109	109	113	130	257	403	492	505	443	313	165	116	109	109	109	109	753	1
	12	882	112	112	112	115	120	197	333	434	470	434	333	197	120	115	112	112	780	12
	HALF DAY TOTALS		518	861	1571	2338	2929	3196	3015	2370	1496	932	647	524	496	492	492	492	2983	
Sep 21	7	472	28	87	265	395	460	456	381	244	66	27	27	27	27	27	27	27	80	5
	8	725	52	61	275	504	646	689	626	467	224	57	52	52	52	52	52	52	258	4
	9	830	71	73	148	413	613	717	712	599	391	129	73	71	71	71	71	71	434	3
	10	882	84	84	89	224	468	631	697	659	522	293	96	84	84	84	84	84	567	2
	11	906	92	92	92	99	245	463	604	652	603	461	242	99	92	92	92	92	651	1
	12	914	95	95	95	95	101	237	446	584	631	584	446	237	101	95	95	95	679	12
	HALF DAY TOTALS		374	447	917	1683	2484	3092	3257	2918	2119	1250	699	432	376	373	373	373	2329	
Oct 21	7	153	6	10	64	113	142	148	132	94	38	7	6	6	6	6	6	38	12	5
	8	644	35	37	155	387	545	615	592	476	280	56	35	35	35	35	35	35	136	4
	9	811	54	54	73	329	567	710	743	661	476	202	57	54	54	54	54	54	305	3
	10	884	67	67	70	157	439	646	752	742	619	395	120	69	67	67	67	67	443	2
	11	917	76	76	76	80	223	491	670	745	707	562	317	89	76	76	76	76	529	1
	12	927	78	78	78	78	84	267	521	681	737	681	521	267	84	78	78	78	558	12
	HALF DAY TOTALS		277	282	479	1106	1965	2771	3173	3074	2516	1555	780	369	280	277	277	277	1704	
Nov 21	8	430	17	17	55	217	339	404	406	346	228	66	18	17	17	17	17	17	44	4
	9	733	38	38	42	231	476	634	691	643	492	254	43	38	38	38	38	38	173	3
	10	846	52	52	52	99	385	617	748	763	659	145	159	54	52	52	52	52	303	2
	11	894	60	60	60	63	192	486	688	783	757	613	367	89	60	60	60	60	388	1
	12	908	63	63	63	63	67	280	555	728	789	728	555	280	67	63	63	63	418	12
	HALF DAY TOTALS		197	198	235	623	1403	2273	2797	2884	2516	1738	850	319	199	197	197	197	1115	
Dec 21	8	279	9	9	25	129	212	259	264	230	157	52	10	9	9	9	9	9	20	4
	9	685	31	31	33	188	427	584	646	611	477	260	40	31	31	31	31	31	124	3
	10	825	45	45	45	78	358	594	732	755	661	462	173	47	45	45	45	45	244	2
	11	882	53	53	53	55	177	477	685	786	765	624	379	90	53	53	53	53	327	1
	12	898	56	56	56	56	60	281	560	736	798	736	560	281	60	56	56	56	356	12
	HALF DAY TOTALS		165	165	178	461	1180	2046	2594	2736	2446	1757	870	298	167	165	165	165	891	
			N	NNW	NW	WNW	W	WSW	SW	SSW	S	SSE	SE	ESE	E	ENE	NE	NNE	HOR	PM

[a]Total solar heat gains for DSA glass (based on a ground reflectance of 0.20).

[b]Half day totals computed by Simpson's rule, time interval = 10 min.

TABLE B.9 Solar Intensity and Solar Heat Gain Factors[a] for 48° N Latitude (SI Units)[b]

Date	Solar Time am	Direct Normal W/m²	N	NNE	NE	ENE	E	ESE	SE	SSE	S	SSW	SW	WSW	W	WNW	NW	NNW	HOR	Solar Time pm
Jan 21	8	116	4	4	13	57	90	109	110	94	63	18	4	4	4	4	4	4	7	4
	9	584	24	24	26	168	371	505	555	523	406	217	30	24	24	24	24	24	79	3
	10	754	37	37	37	69	333	554	682	702	615	429	159	39	37	37	37	37	174	2
	11	823	45	45	45	47	166	455	656	753	734	598	365	83	45	45	45	45	244	1
	12	842	48	48	48	48	51	271	541	713	772	713	541	271	51	48	48	48	269	12
	HALF DAY TOTALS		134	134	144	368	996	1788	2298	2448	2212	1616	817	267	136	134	134	134	639	
Feb 21	7	11	0	1	5	8	10	11	10	7	0	0	0	0	0	0	0	0	1	5
	8	568	24	25	114	324	470	537	524	428	259	52	24	24	24	24	24	24	78	4
	9	780	42	42	49	284	530	683	727	660	488	225	45	42	42	42	42	42	210	3
	10	869	54	54	55	120	415	641	764	768	653	434	137	55	54	54	54	54	331	2
	11	908	61	61	61	64	205	498	696	787	755	607	355	84	61	61	61	61	409	1
	12	920	63	63	63	63	68	280	555	728	790	728	555	280	68	63	63	63	435	12
	HALF DAY TOTALS		214	216	337	864	1708	2575	3054	3051	2565	1677	825	328	216	214	214	214	1246	
Mar 21	7	482	23	71	253	387	458	458	388	253	71	23	22	22	22	22	22	22	64	5
	8	744	44	48	239	486	644	701	651	498	257	49	44	44	44	44	44	44	214	4
	9	853	60	60	104	381	609	738	753	652	449	164	62	60	60	60	60	60	371	3
	10	906	72	72	75	183	460	656	749	728	596	363	104	72	72	72	72	72	493	2
	11	932	79	79	79	83	232	492	663	731	689	541	297	88	79	79	79	79	568	1
	12	939	81	81	81	81	86	261	509	666	720	666	509	261	86	81	81	81	594	12
	HALF DAY TOTALS		317	372	790	1558	2446	3193	3471	3199	2420	1466	769	399	319	316	316	316	2006	
Apr 21	6	340	39	167	271	330	337	294	203	74	20	19	19	19	19	19	19	19	46	6
	7	646	49	191	417	567	628	595	468	264	56	45	45	45	45	45	45	45	189	5
	8	779	64	99	350	566	690	709	619	435	173	67	64	64	64	64	64	64	359	4
	9	847	79	83	191	444	621	703	679	554	335	105	79	79	79	79	79	79	507	3
	10	884	90	90	97	244	466	610	660	613	472	251	97	90	90	90	90	90	618	2
	11	902	97	97	97	105	245	443	570	609	558	419	217	103	97	97	97	97	689	1
	12	908	99	99	99	99	106	224	414	543	587	543	414	224	106	99	99	99	713	12
	HALF DAY TOTALS		463	765	1454	2285	3019	3465	3411	2825	1908	1169	713	492	445	442	442	442	2761	
May 21	5	129	52	97	125	133	122	91	44	9	8	8	8	8	8	8	8	10	16	7
	6	511	112	305	443	512	504	418	267	75	38	38	38	38	38	38	38	40	125	6
	7	690	73	283	498	631	668	604	448	214	66	61	61	61	61	61	61	61	287	5
	8	782	83	170	418	599	689	675	562	358	120	79	79	79	79	79	79	79	449	4
	9	834	93	101	259	475	611	656	605	463	243	100	93	93	93	93	93	93	585	3
	10	864	103	106	124	283	458	561	579	513	366	179	109	103	103	103	103	103	690	2
	11	879	109	109	113	127	249	396	488	505	447	320	170	116	109	109	109	109	756	1
	12	884	111	111	111	114	120	198	336	439	474	439	336	198	120	114	111	111	778	12
	HALF DAY TOTALS		679	1224	2035	2817	3360	3515	3176	2363	1525	997	709	579	550	546	546	549	3297	
Jun 21	5	243	111	192	241	252	228	166	75	19	17	17	17	17	17	17	17	24	38	7
	6	544	146	348	488	552	534	434	266	70	46	46	46	46	46	46	46	49	162	6
	7	693	93	317	521	642	467	591	427	188	72	67	67	67	67	67	67	67	324	5
	8	775	90	203	440	604	678	651	529	320	107	84	84	84	84	84	84	84	479	4
	9	822	98	114	286	482	600	629	567	418	208	105	98	98	98	98	98	98	610	3
	10	849	108	113	142	296	450	534	540	465	319	157	114	108	108	108	108	108	711	2
	11	864	114	114	120	138	248	373	449	456	396	279	156	121	114	114	114	114	775	1
	12	869	116	116	116	120	126	189	303	391	423	391	303	189	126	120	116	116	796	12
	HALF DAY TOTALS		811	1446	2279	3013	3454	3477	3013	2140	1376	942	718	620	597	593	592	602	3495	
Jul 21	5	135	57	104	133	141	129	96	46	11	9	9	9	9	9	9	9	12	18	7
	6	492	116	302	436	501	492	407	259	75	40	40	40	40	40	40	40	43	130	6
	7	666	78	285	492	619	653	588	436	207	69	63	63	63	63	63	63	63	290	5
	8	757	86	176	416	590	675	660	547	348	119	81	81	81	81	81	81	81	448	4
	9	809	96	106	263	471	601	643	591	450	237	104	96	96	96	96	96	96	582	3
	10	839	106	109	130	285	453	550	566	500	356	177	112	106	106	106	106	106	684	2
	11	855	112	112	117	132	249	390	477	492	435	312	169	119	112	112	112	112	749	1
	12	859	114	114	114	117	124	198	329	428	462	428	329	198	124	117	114	114	771	12
	HALF DAY TOTALS		705	1247	2039	2795	3311	3446	3101	2303	1496	993	721	598	570	565	565	569	3287	
Aug 21	6	311	42	161	256	310	316	273	190	70	22	21	21	21	21	21	21	21	51	6
	7	599	53	193	403	543	598	565	444	250	59	49	49	49	49	49	49	49	193	5
	8	732	69	108	347	549	665	681	593	417	168	73	69	69	69	69	69	69	358	4
	9	801	84	90	198	437	605	681	655	534	323	108	84	84	84	84	84	84	502	3
	10	839	95	95	104	247	453	593	639	592	456	245	104	95	95	95	95	95	610	2
	11	858	102	102	102	112	247	433	553	589	539	406	215	110	102	102	102	102	679	1
	12	864	104	104	104	104	113	224	404	526	568	526	404	224	113	104	104	104	702	12
	HALF DAY TOTALS		496	790	1449	2237	2929	3343	3282	2720	1852	1156	728	521	475	471	471	471	2741	
Sep 21	7	414	24	66	224	342	403	403	342	224	66	24	23	23	23	23	23	23	62	5
	8	678	46	51	228	455	602	654	608	467	244	52	46	46	46	46	46	46	206	4
	9	792	63	63	107	366	581	702	716	621	430	163	66	63	63	63	63	63	358	3
	10	848	75	75	79	182	444	630	719	699	573	353	107	75	75	75	75	75	476	2
	11	876	82	82	82	88	230	476	639	704	664	524	292	93	82	82	82	82	549	1
	12	884	84	84	84	84	91	257	497	643	695	643	494	257	91	84	84	84	574	12
	HALF DAY TOTALS		332	381	758	1467	2300	3007	3280	3039	2324	1429	766	334	331	331	331	1937		
Oct 21	7	12	0	1	5	9	11	12	11	8	3	0	0	0	0	0	0	0	1	5
	8	522	25	27	111	304	439	501	489	398	242	51	25	25	25	25	25	25	79	4
	9	734	44	44	52	276	508	652	693	629	466	216	47	44	44	44	44	44	208	3
	10	825	56	56	58	122	403	618	736	739	628	418	136	58	56	56	56	56	327	2
	11	866	64	64	64	67	203	483	673	760	729	587	345	86	64	64	64	64	403	1
	12	878	66	66	66	66	71	274	538	704	763	704	538	274	71	66	66	66	429	12
	HALF DAY TOTALS		223	225	340	838	1637	2461	2918	2918	2459	1619	808	334	226	223	223	223	1233	
Nov 21	8	115	4	4	13	57	90	108	109	93	62	18	4	4	4	4	4	4	7	4
	9	565	25	25	27	165	363	492	540	509	394	211	31	25	25	25	25	25	81	3
	10	735	38	38	38	70	328	543	668	688	602	420	156	40	38	38	38	38	175	2
	11	804	46	46	46	48	165	448	645	739	720	587	358	83	46	46	46	46	244	1
	12	823	48	48	48	48	53	267	531	700	758	700	531	267	53	48	48	48	269	12
	HALF DAY TOTALS		138	138	147	368	979	1752	2250	2396	2165	1583	804	268	140	138	138	138	642	
Dec 21	9	442	16	16	18	113	272	378	420	401	316	176	25	16	16	16	16	16	42	3
	10	676	30	30	30	51	286	491	613	636	563	398	156	32	30	30	30	30	119	2
	11	765	38	38	38	40	146	421	615	708	694	569	351	80	38	38	38	38	181	1
	12	789	41	41	41	41	44	256	514	679	734	679	514	256	44	41	41	41	204	12
	HALF DAY TOTALS		105	105	107	231	735	1444	1924	2098	1944	1477	778	239	107	105	105	105	444	
			N	NNW	NW	WNW	W	WSW	SW	SSW	S	SSE	SE	ESE	E	ENE	NE	NNE	HOR	PM

[a] Total solar heat gains for DSA glass (based on a ground reflectance of 0.20).

[b] Half day totals computed by Simpson's rule, time interval = 10 min.

TABLE B.10 Solar Intensity and Solar Heat Gain Factors[a] for 64° N Latitude (SI Units)[b]

Date	Solar Time am	Direct Normal W/m²	N	NNE	NE	ENE	E	ESE	SE	SSE	S	SSW	SW	WSW	W	WNW	NW	NNW	HOR	Solar Time pm
Jan 21	10	69	2	2	2	4	29	51	63	66	58	41	16	2	2	2	2	2	4	2
	11	256	9	9	9	9	46	143	210	242	238	196	120	24	9	9	9	9	18	1
	12	316	11	11	11	12	104	211	280	302	280	211	104	12	11	11	11	11	24	12
HALF DAY TOTALS			16	16	16	18	80	250	382	449	446	374	237	72	17	16	16	16	33	
Feb 21	8	56	2	2	9	31	46	53	53	44	27	6	2	2	2	2	2	2	3	4
	9	422	16	16	18	136	280	373	403	375	285	143	18	16	16	16	16	16	41	3
	10	601	26	26	26	57	276	453	554	567	495	341	119	28	26	26	26	26	90	2
	11	679	32	32	32	34	139	386	558	638	622	506	307	64	32	32	32	32	128	1
	12	701	34	34	34	34	37	231	464	613	662	613	464	231	37	34	34	34	143	12
HALF DAY TOTALS			93	93	103	279	770	1408	1822	1947	1767	1298	668	210	95	93	93	93	334	
Mar 21	7	300	12	34	147	235	284	288	249	168	54	13	12	12	12	12	12	12	27	5
	8	582	29	31	147	357	499	559	536	427	246	44	29	29	29	29	29	29	100	4
	9	717	41	41	51	277	502	642	679	613	450	203	44	41	41	41	41	41	187	3
	10	786	49	49	50	111	386	598	714	719	613	410	132	51	49	49	49	49	264	2
	11	821	54	54	54	57	190	468	658	746	718	579	344	80	54	54	54	54	314	1
	12	831	56	56	56	56	61	269	530	696	754	696	530	269	61	56	56	56	331	12
HALF DAY TOTALS			213	234	473	1052	1879	2695	3103	3021	2457	1591	811	330	215	213	213	213	1057	
Apr 21	5	85	24	56	77	85	81	64	36	6	4	4	4	4	4	4	4	5	8	7
	6	419	38	187	320	400	416	371	266	111	26	24	24	24	24	24	24	24	67	6
	7	613	43	129	355	516	595	584	482	303	80	41	41	41	41	41	41	41	162	5
	8	720	54	60	249	483	633	685	634	484	250	60	54	54	54	54	54	54	269	4
	9	783	65	65	100	351	567	691	710	620	436	174	68	65	65	65	65	65	367	3
	10	820	72	72	76	161	421	613	709	698	582	372	119	74	72	72	72	72	441	2
	11	839	77	77	77	82	215	468	637	710	676	540	312	93	77	77	77	77	448	1
	12	846	79	79	79	79	85	262	503	655	708	655	503	262	85	79	79	79	504	12
HALF DAY TOTALS			413	687	1294	2116	2974	3628	3741	3268	2408	1537	861	471	381	378	378	378	2053	
May 21	4	160	94	139	162	161	135	87	25	10	10	10	10	10	10	10	11	31	20	8
	5	416	152	298	393	425	395	302	158	34	29	29	29	29	29	29	29	33	81	7
	6	584	90	304	474	570	579	499	343	126	49	46	46	46	46	46	46	46	175	6
	7	688	65	199	436	596	665	634	507	297	78	60	60	60	60	60	60	60	284	5
	8	754	72	89	307	527	660	694	623	459	215	78	72	72	72	72	72	72	390	4
	9	796	82	85	143	384	577	678	679	579	389	147	85	82	82	82	82	82	481	3
	10	822	89	89	94	193	427	593	668	645	528	322	114	89	89	89	89	89	549	2
	11	836	93	93	93	100	226	446	594	652	614	485	275	105	93	93	93	93	592	1
	12	841	95	95	95	95	103	248	460	597	644	597	460	248	103	95	95	95	606	12
HALF DAY TOTALS			780	1341	2145	2998	3712	4073	3841	3108	2234	1467	909	603	533	529	530	554	2875	
Jun 21	4	294	181	263	302	297	247	156	43	21	21	21	21	21	21	21	23	67	50	8
	5	485	195	360	466	498	458	346	175	44	39	39	39	39	39	39	39	45	124	7
	6	614	113	338	510	603	605	516	347	122	57	55	55	55	55	55	55	57	223	6
	7	698	74	225	457	610	672	632	498	282	80	68	68	68	68	68	68	68	331	5
	8	754	79	105	327	535	657	682	605	438	197	85	79	79	79	79	79	79	433	4
	9	791	89	93	161	393	572	662	655	552	362	137	92	89	89	89	89	89	520	3
	10	814	96	96	102	205	423	577	642	615	497	297	114	96	96	96	96	96	585	2
	11	827	100	100	100	108	228	431	568	619	581	456	258	110	100	100	100	100	627	1
	12	831	101	101	101	101	110	240	436	566	610	566	436	240	110	101	101	101	641	12
HALF DAY TOTALS			1016	1080	2526	3347	3953	4154	3769	2985	2142	1436	935	667	605	601	603	673	3219	
Jul 21	4	168	101	149	173	172	144	93	27	12	12	12	12	12	12	12	12	35	24	8
	5	405	154	297	389	420	390	298	156	36	32	32	32	32	32	32	32	36	88	7
	6	564	94	303	467	560	567	488	335	124	51	49	49	49	49	49	49	49	181	6
	7	665	69	201	431	585	652	620	495	290	80	63	63	63	63	63	63	63	289	5
	8	730	75	94	307	520	648	680	609	449	211	81	75	75	75	75	75	75	393	4
	9	771	85	88	148	382	567	665	664	566	380	146	88	85	85	85	85	85	481	3
	10	797	92	92	98	196	422	583	655	631	516	315	116	92	92	92	92	92	548	2
	11	812	96	96	96	104	226	439	582	638	601	475	271	108	96	96	96	96	590	1
	12	816	98	98	98	98	107	246	452	585	630	585	452	246	107	98	98	98	604	12
HALF DAY TOTALS			814	1370	2157	2985	3669	4004	3763	3046	2198	1457	920	626	557	553	554	582	2985	
Aug 21	5	92	28	62	85	94	89	71	40	8	6	6	6	6	6	6	6	6	11	7
	6	388	42	182	306	380	395	352	252	107	30	27	27	27	27	27	27	27	73	6
	7	570	48	132	344	494	567	555	458	289	81	45	45	45	45	45	45	45	168	5
	8	675	59	66	247	468	609	657	607	464	241	66	59	59	59	59	59	59	273	4
	9	737	70	70	107	345	549	666	683	597	420	172	74	70	70	70	70	70	368	3
	10	775	78	78	82	165	412	594	685	674	562	361	122	80	78	78	78	78	440	2
	11	794	82	82	82	89	216	456	617	686	653	523	305	99	82	82	82	82	485	1
	12	801	84	84	84	84	92	260	469	635	684	633	489	260	92	84	84	84	501	12
HALF DAY TOTALS			447	714	1293	2073	2884	3498	3600	3147	2333	1509	869	500	413	409	409	410	2069	
Sep 21	7	242	12	30	122	194	234	238	206	139	47	13	12	12	12	12	12	12	26	5
	8	513	30	33	136	324	451	505	484	387	225	45	30	30	30	30	30	30	97	4
	9	651	43	43	54	261	468	596	631	570	420	193	47	43	43	43	43	43	181	3
	10	722	52	52	53	111	366	563	672	677	577	388	131	54	52	52	52	52	255	2
	11	758	57	57	57	61	185	445	623	706	680	549	329	83	57	57	57	57	303	1
	12	769	59	59	59	59	65	260	504	660	715	660	504	260	65	59	59	59	320	12
HALF DAY TOTALS			224	241	448	968	1727	2484	2871	2305	1513	787	336	227	224	224	224	224	1021	
Oct 21	8	54	2	2	10	30	45	52	51	42	26	6	2	2	2	2	2	2	4	4
	9	383	17	17	19	127	259	345	372	346	263	133	19	17	17	17	17	17	42	3
	10	556	28	28	28	58	261	426	520	532	465	321	115	29	28	28	28	28	91	2
	11	633	34	34	34	36	135	367	528	603	588	479	292	64	34	34	34	34	130	1
	12	655	36	36	36	36	40	222	441	581	628	581	441	222	40	36	36	36	144	12
HALF DAY TOTALS			98	98	109	273	728	1324	1711	1829	1661	1225	638	209	100	98	98	98	339	
Nov 21	10	72	2	2	2	5	31	53	67	69	62	43	17	3	2	2	2	2	4	2
	11	250	9	9	9	10	46	140	207	238	234	192	118	24	9	9	9	9	19	1
	12	307	11	11	11	11	13	102	207	274	295	274	207	102	13	11	11	11	25	12
HALF DAY TOTALS			17	17	17	19	81	248	378	444	440	369	234	71	18	17	17	17	35	
Dec 21	11	12	0	0	0	0	2	7	10	12	11	9	6	1	0	0	0	0	1	1
	12	51	1	1	1	1	2	16	34	45	48	45	34	16	2	1	1	1	3	12
HALF DAY TOTALS			1	1	1	1	3	16	28	35	36	32	22	8	1	1	1	1	2	
			N	NNW	NW	WNW	W	WSW	SW	SSW	S	SSE	SE	ESE	E	ENE	NE	NNE	HOR	PM

[a]Total solar heat gains for DSA glass (based on a ground reflectance of 0.20).

[b]Half day totals computed by Simpson's rule, time interval = 10 min.

APPENDICES

TABLE B.11 **Solar Position and Clear Day Insolation, 32° N Latitude**

Date	Solar Time AM	Solar Time PM	Solar Position Alt	Solar Position Azm	Direct Normal	Horiz	22	32	42	52	90
Jan 21	7	5	1.4	65.2	1	0	0	0	0	1	1
	8	4	12.5	56.5	203	56	93	106	116	123	115
	9	3	22.5	46.0	269	118	175	193	206	212	181
	10	2	30.6	33.1	295	167	235	256	269	274	221
	11	1	36.1	17.5	306	198	273	295	308	312	245
	12		38.0	0.0	310	209	285	308	321	324	253
	Surface Daily Totals				2458	1288	1839	2008	2118	2166	1779
Feb 21	7	5	7.1	73.5	121	22	34	37	40	42	38
	8	4	19.0	64.4	247	95	127	136	140	141	108
	9	3	29.9	53.4	288	161	206	217	222	220	158
	10	2	39.1	39.4	306	212	266	278	283	279	193
	11	1	45.6	21.4	315	244	304	317	321	315	214
	12		48.0	0.0	317	255	316	330	334	328	222
	Surface Daily Totals				2872	1724	2188	2300	2345	2322	1644
Mar 21	7	5	12.7	81.9	185	54	60	60	59	56	32
	8	4	25.1	73.0	260	129	146	147	144	137	78
	9	3	36.8	62.1	290	194	222	224	220	209	119
	10	2	47.3	47.5	304	245	280	283	278	265	150
	11	1	55.0	26.8	311	277	317	321	315	300	170
	12		58.0	0.0	313	287	329	333	327	312	177
	Surface Daily Totals				3012	2084	2378	2403	2358	2246	1276
Apr 21	6	6	6.1	99.9	66	14	9	6	6	5	3
	7	5	18.8	92.2	206	86	78	71	62	51	10
	8	4	31.5	84.0	255	158	156	148	136	120	35
	9	3	43.9	74.2	278	220	225	217	203	183	68
	10	2	55.7	60.3	290	267	279	272	256	234	95
	11	1	65.4	37.5	295	297	313	306	290	265	112
	12		69.6	0.0	297	307	325	318	301	276	118
	Surface Daily Totals				3076	2390	2444	2356	2206	1994	764
May 21	6	6	10.4	107.2	119	36	21	13	13	12	7
	7	5	22.8	100.1	211	107	88	75	60	44	13
	8	4	35.4	92.9	250	175	159	145	127	105	15
	9	3	48.1	84.7	269	233	223	209	188	163	33
	10	2	60.6	73.3	280	277	273	259	237	208	56
	11	1	72.0	51.9	285	305	305	290	268	237	72
	12		78.0	0.0	286	315	315	301	278	247	77
	Surface Daily Totals				3112	2582	2454	2284	2064	1788	469
Jun 21	6	6	12.2	110.2	131	45	26	16	15	14	9
	7	5	24.3	103.4	210	115	91	76	59	41	14
	8	4	36.9	96.8	245	180	159	143	122	99	16
	9	3	49.6	89.4	264	236	221	204	181	153	19
	10	2	62.2	79.7	274	279	268	251	227	197	41
	11	1	74.2	60.9	279	306	299	282	257	224	56
	12		81.5	0.0	280	315	309	292	267	234	60
	Surface Daily Totals				3084	2634	2436	2234	1990	1690	370

TABLE B.11 **Solar Position and Clear Day Insolation, 32° N Latitude (*continued*)**

Date	Solar Time AM	Solar Time PM	Solar Position Alt	Solar Position Azm	Direct Normal	Global Irradiance (Btu/h ft²) Horiz	South-Facing Elevation Angle 22	32	42	52	90
Jul 21	6	6	10.7	107.7	113	37	22	14	13	12	8
	7	5	23.1	100.6	203	107	87	75	60	44	14
	8	4	35.7	93.6	241	174	158	143	125	104	16
	9	3	48.4	85.5	261	230	220	205	185	159	31
	10	2	60.9	74.3	271	274	269	254	232	204	54
	11	1	72.4	53.3	277	302	300	285	262	232	69
	12		78.6	0.0	279	311	310	296	273	242	74
	Surface Daily Totals				3012	2558	2422	2250	2030	1754	458
Aug 21	6	6	6.5	100.5	59	14	9	7	6	6	4
	7	5	19.1	92.8	190	85	77	69	60	50	12
	8	4	31.8	84.7	240	156	152	144	132	116	33
	9	3	44.3	75.0	263	216	220	212	197	178	65
	10	2	56.1	61.3	276	262	272	264	249	226	91
	11	1	66.0	38.4	282	292	305	298	281	257	107
	12		70.3	0.0	284	302	317	309	292	268	113
	Surface Daily Totals				2902	2352	2388	2296	2144	1934	736
Sep 21	7	5	12.7	81.9	163	51	56	56	55	52	30
	8	4	25.1	73.0	240	124	140	141	138	131	75
	9	3	36.8	62.1	272	188	213	215	211	201	114
	10	2	47.3	47.5	287	237	270	273	268	255	145
	11	1	55.0	26.8	294	268	306	309	303	289	164
	12		58.0	0.0	296	278	318	321	315	300	171
	Surface Daily Totals				2808	2014	2288	2308	2264	2154	1226
Oct 21	7	5	6.8	73.1	99	19	29	32	34	36	32
	8	4	18.7	64.0	229	90	120	128	133	134	104
	9	3	29.5	53.0	273	155	198	208	213	212	153
	10	2	38.7	39.1	293	204	257	269	273	270	188
	11	1	45.1	21.1	302	236	294	307	311	306	209
	12		47.5	0.0	304	247	306	320	324	318	217
	Surface Daily Totals				2696	1654	2100	2208	2252	2232	1588
Nov 21	7	5	1.5	65.4	2	0	0	0	1	1	1
	8	4	12.7	56.6	196	55	91	104	113	119	111
	9	3	22.6	46.1	263	118	173	190	202	208	176
	10	2	30.8	33.2	289	166	233	252	265	270	217
	11	1	36.2	17.6	301	197	270	291	303	307	241
	12		38.2	0.0	304	207	282	304	316	320	249
	Surface Daily Totals				2406	1280	1816	1980	2084	2130	1742
Dec 21	8	4	10.3	53.8	176	41	77	90	101	108	107
	9	3	19.8	43.6	257	102	161	180	195	204	183
	10	2	27.6	31.2	288	150	221	244	259	267	226
	11	1	32.7	16.4	301	180	258	282	298	305	251
	12		34.6	0.0	304	190	271	295	311	318	259
	Surface Daily Totals				2348	1136	1704	1888	2016	2086	1794

NOTE: 1 Btu/h ft² × 3.152 = W/m².

Source: Reprinted by permission from Peter J. Lunde, *Solar Thermal Engineering,* © 1980, John Wiley & Sons Inc., New York.

APPENDICES

TABLE B.12 **Solar Position and Clear Day Insolation, 40° N Latitude**

Date	Solar Time AM	Solar Time PM	Solar Position Alt	Solar Position Azm	Direct Normal	Global Irradiance (Btu/h ft²) Horiz	South-Facing Elevation Angle 30	40	50	60	90
Jan 21	8	4	8.1	55.3	142	28	65	74	81	85	84
	9	3	16.8	44.0	239	83	155	171	182	187	171
	10	2	23.8	30.9	274	127	218	237	249	254	223
	11	1	28.4	16.0	289	154	257	277	290	293	253
	12		30.0	0.0	294	164	270	291	303	306	263
	Surface Daily Totals				2182	948	1660	1810	1906	1944	1726
Feb 21	7	5	4.8	72.7	69	10	19	21	23	24	22
	8	4	15.4	62.2	224	73	114	122	126	127	107
	9	3	25.0	50.2	274	132	195	205	209	208	167
	10	2	32.8	35.9	295	178	256	267	271	267	210
	11	1	38.1	18.9	305	206	293	306	310	304	236
	12		40.0	0.0	308	216	306	319	323	317	245
	Surface Daily Totals				2640	1414	2060	2162	2202	2176	1730
Mar 21	7	5	11.4	80.2	171	46	55	55	54	51	35
	8	4	22.5	69.6	250	114	140	141	138	131	89
	9	3	32.8	57.3	282	173	215	217	213	202	138
	10	2	41.6	41.9	297	218	273	276	271	258	176
	11	1	47.7	22.6	305	247	310	313	307	293	200
	12		50.0	0.0	307	257	322	326	320	305	208
	Surface Daily Totals				2916	1852	2308	2330	2284	2174	1484
Apr 21	6	6	7.4	98.9	89	20	11	8	7	7	4
	7	5	18.9	89.5	206	87	77	70	61	50	12
	8	4	30.3	79.3	252	152	153	145	133	117	53
	9	3	41.3	67.2	274	207	221	213	199	179	93
	10	2	51.2	51.4	286	250	275	267	252	229	126
	11	1	58.7	29.2	292	277	308	301	285	260	147
	12		61.6	0.0	293	287	320	313	296	271	154
	Surface Daily Totals				3092	2274	2412	2320	2168	1956	1022
May 21	5	7	1.9	114.7	1	0	0	0	0	0	0
	6	6	12.7	105.6	144	49	25	15	14	13	9
	7	5	24.0	96.6	216	114	89	76	60	44	13
	8	4	35.4	87.2	250	175	158	144	125	104	25
	9	3	46.8	76.0	267	227	221	206	186	160	60
	10	2	57.5	60.9	277	267	270	255	233	205	89
	11	1	66.2	37.1	283	293	301	287	264	234	108
	12		70.0	0.0	284	301	312	297	274	243	114
	Surface Daily Totals				3160	2552	2442	2264	2040	1760	724
Jun 21	5	7	4.2	117.3	22	4	3	3	2	2	1
	6	6	14.8	108.4	155	60	30	18	17	16	10
	7	5	26.0	99.7	216	123	92	77	59	40	14
	8	4	37.4	90.7	246	182	159	142	121	97	16
	9	3	48.8	80.2	263	233	219	202	179	151	47
	10	2	59.8	65.8	272	272	266	248	224	193	74
	11	1	69.2	41.9	277	296	296	278	253	221	92
	12		73.5	0.0	279	304	306	289	263	230	98
	Surface Daily Totals				3180	2648	2434	2224	1974	1670	610

TABLE B.12 **Solar Position and Clear Day Insolation, 40° N Latitude (*continued*)**

Date	Solar Time AM	Solar Time PM	Solar Position Alt	Solar Position Azm	Direct Normal	Horiz	30	40	50	60	90
							Global Irradiance (Btu/h ft²) — South-Facing Elevation Angle					
Jul 21	5	7	2.3	115.2	2	0	0	0	0	0	0
	6	6	13.1	106.1	138	50	26	17	15	14	9
	7	5	24.3	97.2	208	114	89	75	60	44	14
	8	4	35.8	87.8	241	174	157	142	124	102	24
	9	3	47.2	76.7	259	225	218	203	182	157	58
	10	2	57.9	61.7	269	265	266	251	229	200	86
	11	1	66.7	37.9	275	290	296	281	258	228	104
	12		70.6	0.0	276	298	307	292	269	238	111
	Surface Daily Totals				3062	2534	2409	2230	2006	1728	702
Aug 21	6	6	7.9	99.5	81	21	12	9	8	7	5
	7	5	19.3	90.0	191	87	76	69	60	49	12
	8	4	30.7	79.9	237	150	150	141	129	113	50
	9	3	41.8	67.9	260	205	216	207	193	173	89
	10	2	51.7	52.1	272	246	267	259	244	221	120
	11	1	59.3	29.7	278	273	300	292	276	252	140
	12		62.3	0.0	280	282	311	303	287	262	147
	Surface Daily Totals				2916	2244	2354	2258	2104	1894	978
Sep 21	7	5	11.4	80.2	149	43	51	51	49	47	32
	8	4	22.5	69.6	230	109	133	134	131	124	84
	9	3	32.8	57.3	263	167	206	208	203	193	132
	10	2	41.6	41.9	280	211	262	265	260	247	168
	11	1	47.7	22.6	287	239	298	301	295	281	192
	12		50.0	0.0	290	249	310	313	307	292	200
	Surface Daily Totals				2708	1788	2210	2228	2182	2074	1416
Oct 21	7	5	4.5	72.3	48	7	14	15	17	17	16
	8	4	15.0	61.9	204	68	106	113	117	118	100
	9	3	24.5	49.8	257	126	185	195	200	198	160
	10	2	32.4	35.6	280	170	245	257	261	257	203
	11	1	37.6	18.7	291	199	283	295	299	294	229
	12		39.5	0.0	294	208	295	308	312	306	238
	Surface Daily Totals				2454	1348	1962	2060	2098	2074	1654
Nov 21	8	4	8.2	55.4	136	28	63	72	78	82	81
	9	3	17.0	44.1	232	82	152	167	178	183	167
	10	2	24.0	31.0	268	126	215	233	245	249	219
	11	1	28.6	16.1	283	153	254	273	285	288	248
	12		30.2	0.0	288	163	267	287	298	301	258
	Surface Daily Totals				2128	942	1636	1778	1870	1908	1686
Dec 21	8	4	5.5	53.0	89	14	39	45	50	54	56
	9	3	14.0	41.9	217	65	135	152	164	171	163
	10	2	20.7	29.4	261	107	200	221	235	242	221
	11	1	25.0	15.2	280	134	239	262	276	283	252
	12		26.6	0.0	285	143	253	275	290	296	263
	Surface Daily Totals				1978	782	1480	1634	1740	1796	1646

NOTE: 1 Btu/h ft² × 3.152 = W/m².

Source: Reprinted by permission from Peter J. Lunde, *Solar Thermal Engineering,* © 1980, John Wiley & Sons Inc., New York.

APPENDICES

TABLE B.13 **Solar Position and Clear Day Insolation, 48° N Latitude**

Date	Solar Time AM	Solar Time PM	Solar Position Alt	Solar Position Azm	Direct Normal	Global Irradiance (Btu/h ft²) Horiz	South-Facing Elevation Angle 38	48	58	68	90
Jan 21	8	4	3.5	54.6	37	4	17	19	21	22	22
	9	3	11.0	42.6	185	46	120	132	140	145	139
	10	2	16.9	29.4	239	83	190	206	216	220	206
	11	1	20.7	15.1	261	107	231	249	260	263	243
	12		22.0	0.0	267	115	245	264	275	278	255
	Surface Daily Totals				1710	596	1360	1478	1550	1578	1478
Feb 21	7	5	2.4	72.2	12	1	3	4	4	4	4
	8	4	11.6	60.5	188	49	95	102	105	106	96
	9	3	19.7	47.7	251	100	178	187	191	190	167
	10	2	26.2	33.3	278	139	240	251	255	251	217
	11	1	30.5	17.2	290	165	278	290	294	288	247
	12		32.0	0.0	293	173	291	304	307	301	258
	Surface Daily Totals				2330	1080	1880	1972	2024	1978	1720
Mar 21	7	5	10.0	78.7	153	37	49	49	47	45	35
	8	4	19.5	66.8	236	96	131	132	129	122	96
	9	3	28.2	53.4	270	147	205	207	203	193	152
	10	2	35.4	37.8	287	187	263	266	261	248	195
	11	1	40.3	19.8	295	212	300	303	297	283	223
	12		42.0	0.0	298	220	312	315	309	294	232
	Surface Daily Totals				2780	1578	2208	2228	2182	2074	1632
Apr 21	6	6	8.6	97.8	108	27	13	9	8	7	5
	7	5	18.6	86.7	205	85	76	68	59	48	21
	8	4	28.5	74.9	247	142	149	141	129	113	69
	9	3	37.8	61.2	268	191	216	208	194	174	115
	10	2	45.8	44.6	280	228	268	260	245	223	152
	11	1	51.5	24.0	286	252	301	294	278	254	177
	12		53.6	0.0	288	260	313	305	289	264	185
	Surface Daily Totals				3076	2106	2358	2266	2114	1902	1262
May 21	5	7	5.2	114.3	41	9	4	4	4	3	2
	6	6	14.7	103.7	162	61	27	16	15	13	10
	7	5	24.6	93.0	219	118	89	75	60	43	13
	8	4	34.7	81.6	248	171	156	142	123	101	45
	9	3	44.3	68.3	264	217	217	202	182	156	86
	10	2	53.0	51.3	274	252	265	251	229	200	120
	11	1	59.5	28.6	279	274	296	281	258	228	141
	12		62.0	0.0	280	281	306	292	269	238	149
	Surface Daily Totals				3254	2482	2418	2234	2010	1728	982
Jun 21	5	7	7.9	116.5	77	21	9	9	8	7	5
	6	6	17.2	106.2	172	74	33	19	18	16	12
	7	5	27.0	95.8	220	129	93	77	59	39	15
	8	4	37.1	84.6	246	181	157	140	119	95	35
	9	3	46.9	71.6	261	225	216	198	175	147	74
	10	2	55.8	54.8	269	259	262	244	220	189	105
	11	1	62.7	31.2	274	280	291	273	248	216	126
	12		65.5	0.0	275	287	301	283	258	225	133
	Surface Daily Totals				3312	2626	2420	2204	1950	1644	874

TABLE B.13 **Solar Position and Clear Day Insolation, 48° N Latitude** (*continued*)

Date	Solar Time AM	Solar Time PM	Solar Position Alt	Solar Position Azm	Direct Normal	Global Irradiance (Btu/h ft²) Horiz	South-Facing Elevation Angle 38	48	58	68	90
Jul 21	5	7	5.7	114.7	43	10	5	5	4	4	3
	6	6	15.2	104.1	156	62	28	18	16	15	11
	7	5	25.1	93.5	211	118	89	75	59	42	14
	8	4	35.1	82.1	240	171	154	140	121	99	43
	9	3	44.8	68.8	256	215	214	199	178	153	83
	10	2	53.5	51.9	266	250	261	246	224	195	116
	11	1	60.1	29.0	271	272	291	276	253	223	137
	12		62.6	0.0	272	279	301	286	263	232	144
	Surface Daily Totals				3158	2474	2386	2200	1974	1694	956
Aug 21	6	6	9.1	98.3	99	28	14	10	9	8	6
	7	5	19.1	87.2	190	85	75	67	58	47	20
	8	4	29.0	75.4	232	141	145	137	125	109	65
	9	3	38.4	61.8	254	189	210	201	187	168	110
	10	2	46.4	45.1	266	225	260	252	237	214	146
	11	1	52.2	24.3	272	248	293	285	268	244	169
	12		54.3	0.0	274	256	304	296	279	255	177
	Surface Daily Totals				2898	2086	2300	2200	2046	1836	1208
Sep 21	7	5	10.0	78.7	131	35	44	44	43	40	31
	8	4	19.5	66.8	215	92	124	124	121	115	90
	9	3	28.2	53.4	251	142	196	197	193	183	143
	10	2	35.4	37.8	269	181	251	254	248	236	185
	11	1	40.3	19.8	278	205	287	289	284	269	212
	12		42.0	0.0	280	213	299	302	296	281	221
	Surface Daily Totals				2568	1522	2102	2118	2070	1966	1546
Oct 21	7	5	2.0	71.9	4	0	1	1	1	1	1
	8	4	11.2	60.2	165	44	86	91	95	95	87
	9	3	19.3	47.4	233	94	167	176	180	178	157
	10	2	25.7	33.1	262	133	228	239	242	239	207
	11	1	30.0	17.1	274	157	266	277	281	276	237
	12		31.5	0.0	278	166	279	291	294	288	247
	Surface Daily Totals				2154	1022	1774	1860	1890	1866	1626
Nov 21	8	4	3.6	54.7	36	5	17	19	21	22	22
	9	3	11.2	42.7	179	46	117	129	137	141	135
	10	2	17.1	29.5	233	83	186	202	212	215	201
	11	1	20.9	15.1	255	107	227	245	255	258	238
	12		22.2	0.0	261	115	241	259	270	272	250
	Surface Daily Totals				1668	596	1336	1448	1518	1544	1442
Dec 21	9	3	8.0	40.9	140	27	87	98	105	110	109
	10	2	13.6	28.2	214	63	164	180	192	197	190
	11	1	17.3	14.4	242	86	207	226	239	244	231
	12		18.6	0.0	250	94	222	241	254	260	244
	Surface Daily Totals				1444	446	1136	1250	1326	1364	1304

NOTE: 1 Btu/h ft² × 3.152 = W/m².

Source: Reprinted by permission from Peter J Lunde, *Solar Thermal Engineering,* © 1980, John Wiley & Sons Inc., New York.

APPENDICES

TABLE B.14 Solar Position and Clear Day Insolation, 64° N Latitude

Date	Solar Time AM	Solar Time PM	Solar Position Alt	Solar Position Azm	Direct Normal	Global Irradiance (Btu/h ft^2) Horiz	South-Facing Elevation Angle 54	64	74	84	90
Jan 21	10	2	2.8	28.1	22	2	17	19	20	20	20
	11	1	5.2	14.1	81	12	72	77	80	81	81
	12		6.0	0.0	100	16	91	98	102	103	103
	Surface Daily Totals				306	45	268	290	302	306	304
Feb 21	8	4	3.4	58.7	35	4	17	19	19	19	19
	9	3	8.6	44.8	147	31	103	108	111	110	107
	10	2	12.6	30.3	199	55	170	178	181	178	173
	11	1	15.1	15.3	222	71	212	220	223	219	213
	12		16.0	0.0	228	77	225	235	237	232	226
	Surface Daily Totals				1432	400	1230	1286	1302	1282	1252
Mar 21	7	5	6.5	76.5	95	18	30	29	29	27	25
	8	4	12.7	62.6	185	54	101	102	99	94	89
	9	3	18.1	48.1	227	87	171	172	169	160	153
	10	2	22.3	32.7	249	112	227	229	224	213	203
	11	1	25.1	16.6	260	129	262	265	259	246	235
	12		26.0	0.0	263	134	274	277	271	258	246
	Surface Daily Totals				2296	932	1856	1870	1830	1736	1656
Apr 21	5	7	4.0	108.5	27	5	2	2	2	1	1
	6	6	10.4	95.1	133	37	15	9	8	7	6
	7	5	17.0	81.6	194	76	70	63	54	43	37
	8	4	23.3	67.5	228	112	136	128	116	102	91
	9	3	29.0	52.3	248	144	197	189	176	158	145
	10	2	33.5	36.0	260	169	246	239	224	203	188
	11	1	36.5	18.4	266	184	278	270	255	233	216
	12		37.6	0.0	268	190	289	281	266	243	225
	Surface Daily Totals				2982	1644	2176	2082	1936	1736	1594
May 21	4	8	5.8	125.1	51	11	5	4	4	3	3
	5	7	11.6	112.1	132	42	13	11	10	9	8
	6	6	17.9	99.1	185	79	29	16	14	12	11
	7	5	24.5	85.7	218	117	86	72	56	39	28
	8	4	30.9	71.5	239	152	148	133	115	94	80
	9	3	36.8	56.1	252	182	204	190	170	145	128
	10	2	41.6	38.9	261	205	249	235	213	186	167
	11	1	44.9	20.1	265	219	278	264	242	213	193
	12		46.0	0.0	267	224	288	274	251	222	201
	Surface Daily Totals				3470	2236	2312	2124	1898	1624	1436
Jun 21	3	9	4.2	139.4	21	4	2	2	2	2	1
	4	8	9.0	126.4	93	27	10	9	8	7	6
	5	7	14.7	113.6	154	60	16	15	13	11	10
	6	6	21.0	100.8	194	96	34	19	17	14	13
	7	5	27.5	87.5	221	132	91	74	55	36	23
	8	4	34.0	73.3	239	166	150	133	112	88	73
	9	3	39.9	57.8	251	195	204	187	164	137	119
	10	2	44.9	40.4	258	217	247	230	206	177	157
	11	1	48.3	20.9	262	231	275	258	233	202	181
	12		49.5	0.0	263	235	284	267	242	211	189
	Surface Daily Totals				3650	2488	2342	2118	1862	1558	1356

TABLE B.14 **Solar Position and Clear Day Insolation, 64° N Latitude (*continued*)**

Date	Solar Time AM	Solar Time PM	Solar Position Alt	Solar Position Azm	Direct Normal	Global Irradiance (Btu/h ft²) Horiz	South-Facing Elevation Angle 54	64	74	84	90
Jul 21	4	8	6.4	125.3	53	13	6	5	5	4	4
	5	7	12.1	112.4	128	44	14	13	11	10	9
	6	6	18.4	99.4	179	81	30	17	16	13	12
	7	5	25.0	86.0	211	118	86	72	56	38	28
	8	4	31.4	71.8	231	152	146	131	113	91	77
	9	3	37.3	56.3	245	182	201	186	166	141	124
	10	2	42.2	39.2	253	204	245	230	208	181	162
	11	1	45.4	20.2	257	218	273	258	236	207	187
		12	46.6	0.0	259	223	282	267	245	216	195
	Surface Daily Totals				3372	2248	2280	2090	1864	1588	1400
Aug 21	5	7	4.6	108.8	29	6	3	3	2	2	2
	6	6	11.0	95.5	123	39	16	11	10	8	7
	7	5	17.6	81.9	181	77	69	61	52	42	35
	8	4	23.9	67.8	214	113	131	123	112	97	87
	9	3	29.6	52.6	234	144	190	182	169	150	138
	10	2	34.2	36.2	246	168	237	229	215	194	179
	11	1	37.2	18.5	252	183	268	260	244	222	205
		12	38.3	0.0	254	188	278	270	255	232	215
	Surface Daily Totals				2808	1646	2108	2008	1860	1662	1522
Sep 21	7	5	6.5	76.5	77	16	25	24	24	23	21
	8	4	12.7	62.6	163	51	92	92	90	85	81
	9	3	18.1	48.1	206	83	159	159	156	147	141
	10	2	22.3	32.7	229	108	212	213	209	198	189
	11	1	25.1	16.6	240	124	246	248	243	230	220
		12	26.0	0.0	244	129	258	260	254	241	230
	Surface Daily Totals				2074	892	1726	1736	1696	1608	1532
Oct 21	8	4	3.0	58.5	17	2	9	9	10	10	10
	9	3	8.1	44.6	122	26	86	91	93	92	90
	10	2	12.1	30.2	176	50	152	159	161	159	155
	11	1	14.6	15.2	201	65	193	201	203	200	195
		12	15.5	0.0	208	71	207	215	217	213	208
	Surface Daily Totals				1238	358	1088	1136	1152	1134	1106
Nov 21	10	2	3.0	28.1	23	3	18	20	21	21	21
	11	1	5.4	14.2	79	12	70	76	79	80	79
		12	6.2	0.0	97	17	89	96	100	101	100
	Surface Daily Totals				302	46	266	286	298	302	300
Dec 21	11	1	1.8	13.7	4	0	3	4	4	4	4
		12	2.6	0.0	16	2	14	15	16	17	17
	Surface Daily Totals				24	2	20	22	24	24	24

NOTE: 1 Btu/h ft² × 3.152 = W/m².

Source: Reprinted by permission from Peter J. Lunde, *Solar Thermal Engineering,* © 1980, John Wiley & Sons Inc., New York.

*Fig. **B.1*** *U.S. and Canadian locations for which solar data are published in this text. (Basic temperature data for many more locations are found in Appendix A.) (Adapted by permission from J. D. Balcomb et al., Passive Solar Design Handbook, Vol 3, © 1983, American Solar Energy Society, Inc., Boulder, Colo.)*

TABLE B.15 Average Insolation, Temperature, and DD Data

Elevation in feet, latitude in degrees north latitude, HS (horizontal surface) and VS (vertical surface) in Btu/day ft², TA in degrees F, D50 through D65 in degree days F.

United States

MONTGOMERY, ALABAMA   Elev 203   Lat 32.3

	HS	VS	TA	D50	D55	D60	D65
Jan	752	896	48	148	256	394	556
Jul	1841	820	81	0	0	0	0
Yr	1390	946	65	445	866	1474	2269

PHOENIX, ARIZONA   Elev 1112   Lat 33.4

	HS	VS	TA	D50	D55	D60	D65
Jan	1021	1462	51	78	162	285	428
Jul	2486	964	91	0	0	0	0
Yr	1371	1326	70	187	459	919	1552

TUCSON, ARIZONA   Elev 2556   Lat 32.1

	HS	VS	TA	D50	D55	D60	D65
Jan	1099	1539	51	80	166	292	442
Jul	2341	922	86	0	0	0	0
Yr	1874	1307	68	214	525	1036	1752

FORT SMITH, ARKANSAS   Elev 463   Lat 35.3

	HS	VS	TA	D50	D55	D60	D65
Jan	744	996	39	346	497	651	806
Jul	2065	908	82	0	0	0	0
Yr	1406	1013	61	996	1622	2405	3336

FRESNO, CALIFORNIA   Elev 328   Lat 36.8

	HS	VS	TA	D50	D55	D60	D65
Jan	657	886	45	176	308	457	611
Jul	2685	1076	81	0	0	0	0
Yr	1714	1210	62	507	1021	1741	2650

LOS ANGELES, CALIFORNIA Elev 105   Lat 33.9

	HS	VS	TA	D50	D55	D60	D65
Jan	926	1293	55	21	83	186	331
Jul	2307	942	69	0	1	5	19
Yr	1596	1157	62	64	299	849	1819

SACRAMENTO, CALIFORNIA Elev 26   Lat 38.5

	HS	VS	TA	D50	D55	D60	D65
Jan	597	829	45	183	315	464	617
Jul	2688	1131	75	0	0	0	0
Yr	1646	1198	60	554	1097	1871	2843

SAN DIEGO, CALIFORNIA   Elev 30   Lat 32.7

	HS	VS	TA	D50	D55	D60	D65
Jan	976	1325	55	9	58	160	314
Jul	2186	902	70	0	0	1	6
Yr	1600	1151	63	23	170	623	1507

SAN FRANCISCO, CALIFORNIA Elev 16   Lat 37.6

	HS	VS	TA	D50	D55	D60	D65
Jan	708	1023	48	82	210	363	518
Jul	2392	1034	63	0	2	21	93
Yr	1556	1156	57	202	705	1643	3042

SANTA MARIA, CALIFORNIA Elev 236   Lat 34.9

	HS	VS	TA	D50	D55	D60	D65
Jan	854	1198	51	51	150	296	450
Jul	2341	965	62	0	3	30	112
Yr	1610	1172	57	155	624	1604	3053

DENVER, COLORADO   Elev 5331   Lat 39.7

	HS	VS	TA	D50	D55	D60	D65
Jan	840	1465	30	623	778	933	1088
Jul	2273	1053	73	0	0	0	0
Yr	1570	1334	50	2592	3588	4733	6016

GRAND JUNCTION, COLO   Elev 4839   Lat 39.1

	HS	VS	TA	D50	D55	D60	D65
Jan	791	1296	27	726	880	1035	1190
Jul	2465	1094	79	0	0	0	0
Yr	1661	1346	53	2514	3412	4434	5605

HARTFORD, CONNECTICUT   Elev 180   Lat 41.9

	HS	VS	TA	D50	D55	D60	D65
Jan	477	694	25	781	936	1091	1246
Jul	1649	861	73	0	0	0	0
Yr	1060	835	49	2971	3948	5075	6350

WASHINGTON, DC   Elev 289   Lat 38.9

	HS	VS	TA	D50	D55	D60	D65
Jan	572	793	32	555	710	865	1020
Jul	1817	883	75	0	0	0	0
Yr	1210	912	54	2004	2869	3864	5010

MIAMI, FLORIDA   Elev 7   Lat 25.8

	HS	VS	TA	D50	D55	D60	D65
Jan	1057	1121	67	1	4	18	53
Jul	1763	787	82	0	0	0	0
Yr	1474	941	76	3	14	55	206

ORLANDO, FLORIDA   Elev 118   Lat 28.5

	HS	VS	TA	D50	D55	D60	D65
Jan	999	1151	60	13	42	105	197
Jul	1801	795	81	0	0	0	0
Yr	1488	984	72	39	126	348	733

TALLAHASSEE, FLORIDA   Elev 69   Lat 30.4

	HS	VS	TA	D50	D55	D60	D65
Jan	877	1033	53	73	150	256	408
Jul	1748	786	81	0	0	0	0
Yr	1434	969	68	215	501	951	1563

ATLANTA, GEORGIA   Elev 1033   Lat 33.6

	HS	VS	TA	D50	D55	D60	D65
Jan	718	884	42	246	393	546	701
Jul	1812	821	78	0	0	0	0
Yr	1347	941	61	758	1362	2150	3095

BOISE, IDAHO   Elev 2867   Lat 43.6

	HS	VS	TA	D50	D55	D60	D65
Jan	485	770	29	651	806	961	1116
Jul	2613	1309	75	0	0	0	0
Yr	1499	1255	51	2420	3395	4536	5833

CHICAGO, ILLINOIS   Elev 623   Lat 41.8

	HS	VS	TA	D50	D55	D60	D65
Jan	507	756	24	797	952	1107	1262
Jul	1944	984	75	0	0	0	0
Yr	1217	960	51	2954	3881	4940	6127

MOLINE, ILLINOIS   Elev 594   Lat 41.4

	HS	VS	TA	D50	D55	D60	D65
Jan	535	803	22	884	1039	1194	1349
Jul	1939	974	75	0	0	0	0
Yr	1226	973	50	3191	4117	5178	6395

SPRINGFIELD, ILLINOIS   Elev 614   Lat 39.8

	HS	VS	TA	D50	D55	D60	D65
Jan	585	852	27	723	877	1032	1187
Jul	2058	984	76	0	0	0	0
Yr	1304	1003	53	2558	3434	4425	5558

TABLE B.15 Average Insolation, Temperature, and DD Data (*continued*)

INDIANAPOLIS, INDIANA   Elev 807   Lat 39.7

	HS	VS	TA	D50	D55	D60	D65
Jan	496	668	28	685	840	995	1150
Jul	1806	891	75	0	0	0	0
Yr	1167	873	52	2511	3403	4421	5577

SOUTH BEND, INDIANA   Elev 774   Lat 41.7

	HS	VS	TA	D50	D55	D60	D65
Jan	416	566	24	806	961	1116	1271
Jul	1852	944	72	0	1	4	6
Yr	1140	864	49	3112	4084	5199	6462

DES MOINES, IOWA   Elev 965   Lat 41.5

	HS	VS	TA	D50	D55	D60	D65
Jan	581	912	19	949	1104	1259	1414
Jul	2097	1037	75	0	0	0	0
Yr	1314	1065	49	3491	4435	5510	6710

DODGE CITY, KANSAS   Elev 2582   Lat 37.8

	HS	VS	TA	D50	D55	D60	D65
Jan	827	1303	31	596	750	905	1060
Jul	2295	1013	79	0	0	0	0
Yr	1562	1232	55	2131	2980	3945	5046

TOPEKA, KANSAS   Elev 886   Lat 39.1

	HS	VS	TA	D50	D55	D60	D65
Jan	681	1033	28	683	837	992	1147
Jul	2128	993	78	0	0	0	0
Yr	1387	1036	54	2325	3175	4137	5243

LEXINGTON, KENTUCKY   Elev 988   Lat 38.0

	HS	VS	TA	D50	D55	D60	D65
Jan	546	714	33	531	685	840	995
Jul	1850	881	76	0	0	0	0
Yr	1221	892	55	1865	2686	3632	4729

BATON ROUGE, LOUISIANA   Elev 75   Lat 30.5

	HS	VS	TA	D50	D55	D60	D65
Jan	785	889	51	90	174	294	451
Jul	1746	786	82	0	0	0	0
Yr	1380	913	67	232	530	1006	1670

NEW ORLEANS, LOUISIANA   Elev 10   Lat 30.0

	HS	VS	TA	D50	D55	D60	D65
Jan	835	950	53	73	150	252	403
Jul	1813	801	82	0	0	0	0
Yr	1438	943	68	197	465	887	1465

SHREVEPORT, LOUISIANA   Elev 259   Lat 32.5

	HS	VS	TA	D50	D55	D60	D65
Jan	762	920	47	154	264	403	552
Jul	2014	864	83	0	0	0	0
Yr	1428	973	66	428	832	1415	2167

PORTLAND, MAINE   Elev 62   Lat 43.6

	HS	VS	TA	D50	D55	D60	D65
Jan	450	689	22	884	1039	1194	1349
Jul	1659	894	68	0	1	5	27
Yr	1052	857	45	3648	4758	6039	7498

DETROIT, MICHIGAN   Elev 627   Lat 42.4

	HS	VS	TA	D50	D55	D60	D65
Jan	417	585	26	760	915	1070	1225
Jul	1835	951	73	0	0	0	0
Yr	1122	869	50	2931	3890	4986	6228

SAULT STE. MARIE, MICHIGAN   Elev 725   Lat 46.5

	HS	VS	TA	D50	D55	D60	D65
Jan	325	492	14	1110	1265	1420	1575
Jul	1835	1045	64	1	7	33	96
Yr	1044	861	40	4969	6198	7607	9193

MINNEAPOLIS, MINNESOTA   Elev 837   Lat 44.9

	HS	VS	TA	D50	D55	D60	D65
Jan	464	768	12	1172	1327	1482	1637
Jul	1970	1071	72	0	1	5	11
Yr	1172	996	44	4584	5631	6824	8159

MERIDIAN, MISSISSIPPI   Elev 308   Lat 32.3

	HS	VS	TA	D50	D55	D60	D65
Jan	744	883	47	163	274	413	575
Jul	1823	815	81	0	0	0	0
Yr	1371	933	65	510	955	1582	2388

SAINT LOUIS, MISSOURI   Elev 564   Lat 38.7

	HS	VS	TA	D50	D55	D60	D65
Jan	627	898	31	581	735	890	1045
Jul	2049	959	79	0	0	0	0
Yr	1329	1006	56	1961	2762	3686	4750

SPRINGFIELD, MISSOURI   Elev 1270   Lat 37.2

	HS	VS	TA	D50	D55	D60	D65
Jan	684	956	33	534	686	840	995
Jul	2063	936	78	0	0	0	0
Yr	1364	1011	56	1848	2618	3522	4571

HELENA, MONTANA   Elev 3898   Lat 46.6

	HS	VS	TA	D50	D55	D60	D65
Jan	419	719	18	989	1144	1299	1454
Jul	2334	1312	68	1	3	12	33
Yr	1266	1134	43	4151	5342	6689	8190

NORTH OMAHA, NEBRASKA   Elev 1325   Lat 41.4

	HS	VS	TA	D50	D55	D60	D65
Jan	634	1034	20	924	1079	1234	1389
Jul	2106	1038	75	0	1	3	7
Yr	1323	1078	49	3369	4309	5381	6601

ELY, NEVADA   Elev 6253   Lat 39.3

	HS	VS	TA	D50	D55	D60	D65
Jan	819	1380	24	818	973	1128	1283
Jul	2447	1094	67	0	2	11	23
Yr	1675	1391	44	3716	4922	6291	7814

LAS VEGAS, NEVADA   Elev 2178   Lat 36.1

	HS	VS	TA	D50	D55	D60	D65
Jan	978	1553	44	216	346	493	645
Jul	2588	1039	90	0	0	0	0
Yr	1866	1431	66	631	1129	1788	2601

RENO, NEVADA   Elev 4400   Lat 39.5

	HS	VS	TA	D50	D55	D60	D65
Jan	800	1345	32	561	716	871	1026
Jul	2692	1167	69	0	1	5	17
Yr	1764	1439	49	2292	3345	4590	6022

ALBUQUERQUE, NEW MEXICO   Elev 5312   Lat 35.0

	HS	VS	TA	D50	D55	D60	D65
Jan	1016	1562	35	459	614	769	924
Jul	2489	995	79	0	0	0	0
Yr	1830	1379	57	1497	2292	3216	4292

LOS ALAMOS, NEW MEXICO   Elev 7380   Lat 35.9

	HS	VS	TA	D50	D55	D60	D65
Jan	893	1340	29	650	804	958	1113
Jul	2051	913	68	0	0	0	0
Yr	1579	1244	48	2822	3851	5043	6437

ALBANY, NEW YORK   Elev 292   Lat 42.7

	HS	VS	TA	D50	D55	D60	D65
Jan	456	674	22	884	1039	1194	1349
Jul	1725	908	72	0	1	3	9
Yr	1068	843	48	3424	4428	5586	6888

TABLE B.15 **Average Insolation, Temperature, and DD Data** (*continued*)

BUFFALO, NEW YORK				Elev 705		Lat 42.9
	HS	VS	TA	D50	D55	D60	D65
Jan	349	465	24	815	970	1125	1280
Jul	1776	935	70	0	0	3	12
Yr	1037	780	47	3322	4363	5551	6927

NEW YORK, NEW YORK				Elev 187		Lat 40.8
	HS	VS	TA	D50	D55	D60	D65
Jan	500	708	32	552	707	862	1017
Jul	1688	861	77	0	0	0	0
Yr	1101	849	55	1931	2759	3737	4848

SYRACUSE, NEW YORK				Elev 407		Lat 43.1
	HS	VS	TA	D50	D55	D60	D65
Jan	385	538	24	818	973	1128	1283
Jul	1758	931	72	0	1	3	11
Yr	1037	791	48	3215	4218	5366	6678

CHARLOTTE, NORTH CAROLINA				Elev 768		Lat 35.2
	HS	VS	TA	D50	D55	D60	D65
Jan	719	944	42	255	402	555	710
Jul	1831	841	79	0	0	0	0
Yr	1346	981	61	828	1451	2257	3218

RALEIGH-DURHAM, NORTH CAROLINA				Elev 440		Lat 35.9
	HS	VS	TA	D50	D55	D60	D65
Jan	694	924	41	300	451	605	760
Jul	1776	832	78	0	0	0	0
Yr	1297	955	59	990	1659	2509	3514

BISMARCK, NORTH DAKOTA				Elev 1647		Lat 46.8
	HS	VS	TA	D50	D55	D60	D65
Jan	467	847	8	1296	1451	1606	1761
Jul	2184	1241	71	0	2	8	18
Yr	1251	1145	41	5235	6364	7627	9044

FARGO, NORTH DAKOTA				Elev 899		Lat 46.9
	HS	VS	TA	D50	D55	D60	D65
Jan	415	720	6	1367	1522	1677	1832
Jul	2120	1210	71	0	1	5	13
Yr	1206	1075	41	5485	6607	7858	9271

CINCINNATI, OHIO				Elev 889		Lat 39.1
	HS	VS	TA	D50	D55	D60	D65
Jan	500	659	31	587	741	896	1051
Jul	1771	869	76	0	0	0	0
Yr	1160	858	54	2117	2973	3951	5070

CLEVELAND, OHIO				Elev 804		Lat 41.4
	HS	VS	TA	D50	D55	D60	D65
Jan	388	507	27	716	871	1026	1181
Jul	1828	929	71	0	1	4	9
Yr	1093	808	50	2804	3768	4879	6154

COLUMBUS, OHIO				Elev 833		Lat 40.0
	HS	VS	TA	D50	D55	D60	D65
Jan	459	606	28	670	825	980	1135
Jul	1755	876	74	0	0	0	0
Yr	1128	834	52	2524	3438	4491	5702

OKLAHOMA CITY, OKLAHOMA				Elev 1302		Lat 35.4
	HS	VS	TA	D50	D55	D60	D65
Jan	801	1114	37	413	565	719	874
Jul	2128	925	82	0	0	0	0
Yr	1463	1073	60	1232	1903	2734	3695

MEDFORD, OREGON				Elev 1299		Lat 42.4
	HS	VS	TA	D50	D55	D60	D65
Jan	407	565	37	417	571	725	880
Jul	2475	1207	72	0	1	3	11
Yr	1356	1033	53	1576	2505	3621	4930

SALEM, OREGON				Elev 200		Lat 44.9
	HS	VS	TA	D50	D55	D60	D65
Jan	332	471	39	348	502	657	812
Jul	2142	1154	67	0	1	7	43
Yr	1130	897	52	1265	2220	3411	4852

HARRISBURG, PENNSYLVANIA				Elev 348		Lat 40.2
	HS	VS	TA	D50	D55	D60	D65
Jan	536	763	30	617	772	927	1082
Jul	1764	883	76	0	0	0	0
Yr	1152	887	53	2221	3093	4086	5224

PHILADELPHIA, PENNSYLVANIA				Elev 30		Lat 39.9
	HS	VS	TA	D50	D55	D60	D65
Jan	555	792	32	549	704	859	1014
Jul	1758	876	77	0	0	0	0
Yr	1170	905	55	1935	2775	3749	4865

PITTSBURGH, PENNSYLVANIA				Elev 1224		Lat 40.5
	HS	VS	TA	D50	D55	D60	D65
Jan	424	553	28	679	834	989	1144
Jul	1689	857	72	0	1	3	7
Yr	1071	793	50	2635	3574	4669	5930

PROVIDENCE, RHODE ISLAND				Elev 62		Lat 41.7
	HS	VS	TA	D50	D55	D60	D65
Jan	506	750	28	670	825	980	1135
Jul	1695	878	72	0	0	0	0
Yr	1114	884	50	2566	3543	4669	5972

CHARLESTON, SOUTH CAROLINA				Elev 39		Lat 32.9
	HS	VS	TA	D50	D55	D60	D65
Jan	744	904	49	120	222	360	521
Jul	1799	813	80	0	0	0	0
Yr	1346	940	65	360	756	1355	2146

RAPID CITY, SOUTH DAKOTA				Elev 3169		Lat 44.0
	HS	VS	TA	D50	D55	D60	D65
Jan	542	928	22	871	1026	1181	1336
Jul	2223	1161	73	1	2	8	13
Yr	1344	1177	47	3681	4749	5965	7324

KNOXVILLE, TENNESSEE				Elev 981		Lat 35.8
	HS	VS	TA	D50	D55	D60	D65
Jan	621	785	41	302	449	602	756
Jul	1804	839	78	0	0	0	0
Yr	1275	909	60	1018	1671	2504	3478

MEMPHIS, TENNESSEE				Elev 285		Lat 35.0
	HS	VS	TA	D50	D55	D60	D65
Jan	683	870	41	312	455	606	760
Jul	1972	879	82	0	0	0	0
Yr	1368	965	62	988	1588	2357	3227

NASHVILLE, TENNESSEE				Elev 591		Lat 36.1
	HS	VS	TA	D50	D55	D60	D65
Jan	580	721	38	369	519	673	828
Jul	1891	869	80	0	0	0	0
Yr	1272	891	59	1195	1874	2720	3696

APPENDICES

TABLE B.15 **Average Insolation, Temperature, and DD Data** (*continued*)

AUSTIN, TEXAS				Elev 620		Lat 30.3
	HS	VS	TA	D50	D55	D60	D65
Jan	864	1008	50	116	207	333	483
Jul	2105	865	85	0	0	0	0
Yr	1478	974	68	289	602	1088	1737

BROWNSVILLE, TEXAS				Elev 20		Lat 25.9
	HS	VS	TA	D50	D55	D60	D65
Jan	913	923	60	18	51	116	225
Jul	2212	867	84	0	0	0	0
Yr	1550	917	74	44	127	325	650

DALLAS, TEXAS				Elev 489		Lat 32.8
	HS	VS	TA	D50	D55	D60	D65
Jan	821	1035	45	189	312	457	608
Jul	2122	890	86	0	0	0	0
Yr	1470	1014	66	505	943	1543	2290

EL PASO, TEXAS				Elev 3917		Lat 31.8
	HS	VS	TA	D50	D55	D60	D65
Jan	1125	1572	44	210	355	509	663
Jul	2450	934	82	0	0	0	0
Yr	1901	1327	63	561	1102	1810	2678

HOUSTON, TEXAS				Elev 108		Lat 30.0
	HS	VS	TA	D50	D55	D60	D65
Jan	772	852	52	71	150	263	416
Jul	1828	805	83	0	0	0	0
Yr	1353	884	69	161	409	825	1434

LUBBOCK, TEXAS				Elev 3241		Lat 33.6
	HS	VS	TA	D50	D55	D60	D65
Jan	1031	1497	39	343	494	648	803
Jul	2412	956	80	0	0	0	0
Yr	1768	1279	60	1069	1739	2582	3545

SAN ANTONIO, TEXAS				Elev 794		Lat 29.5
	HS	VS	TA	D50	D55	D60	D65
Jan	895	1026	51	93	179	302	451
Jul	2121	863	85	0	0	0	0
Yr	1501	973	69	213	490	941	1570

BRYCE CANYON, UTAH				Elev 7588		Lat 37.7
	HS	VS	TA	D50	D55	D60	D65
Jan	914	1511	20	936	1091	1246	1401
Jul	2424	1044	62	1	8	47	128
Yr	1742	1404	40	4693	6106	7675	9133

SALT LAKE CITY, UTAH				Elev 4226		Lat 40.8
	HS	VS	TA	D50	D55	D60	D65
Jan	639	1017	28	683	837	992	1147
Jul	2590	1186	77	0	0	0	0
Yr	1606	1301	51	2648	3612	4725	5983

BURLINGTON, VERMONT				Elev 341		Lat 44.5
	HS	VS	TA	D50	D55	D60	D65
Jan	385	572	17	1029	1184	1339	1494
Jul	1721	941	70	0	1	4	20
Yr	1023	815	44	4142	5230	6464	7876

NORFOLK, VIRGINIA				Elev 30		Lat 36.9
	HS	VS	TA	D50	D55	D60	D65
Jan	678	932	41	300	450	605	760
Jul	1853	868	78	0	0	0	0
Yr	1327	990	59	974	1646	2489	3488

RICHMOND, VIRGINIA				Elev 164		Lat 37.5
	HS	VS	TA	D50	D55	D60	D65
Jan	632	863	38	390	543	698	853
Jul	1774	849	78	0	0	0	0
Yr	1250	936	58	1296	2021	2909	3939

ROANOKE, VIRGINIA				Elev 1175		Lat 37.3
	HS	VS	TA	D50	D55	D60	D65
Jan	660	911	36	423	577	832	887
Jul	1796	854	75	0	0	0	0
Yr	1271	958	56	1486	2277	3211	4307

SEATTLE–TACOMA, WASHINGTON						
				Elev 400		Lat 47.4
	HS	VS	TA	D50	D55	D60	D65
Jan	262	378	38	367	521	676	831
Jul	2248	1299	65	0	2	16	80
Yr	1056	857	51	1386	2393	3662	5185

SPOKANE, WASHINGTON				Elev 2365		Lat 47.6
	HS	VS	TA	D50	D55	D60	D65
Jan	315	496	25	763	918	1073	1228
Jul	2357	1368	70	0	2	7	21
Yr	1227	1068	47	3061	4150	5411	6835

CHARLESTON, WEST VIRGINIA						
				Elev 951		Lat 38.4
	HS	VS	TA	D50	D55	D60	D65
Jan	498	638	35	483	636	791	946
Jul	1682	827	75	0	0	0	0
Yr	1125	822	55	1726	2540	3488	4590

MADISON, WISCONSIN				Elev 860		Lat 43.1
	HS	VS	TA	D50	D55	D60	D65
Jan	515	822	17	1029	1184	1339	1494
Jul	1934	1009	70	0	1	5	14
Yr	1193	978	45	4086	5143	6352	7730

MILWAUKEE, WISCONSIN				Elev 692		Lat 42.9
	HS	VS	TA	D50	D55	D60	D65
Jan	479	731	19	949	1104	1259	1414
Jul	1962	1017	70	0	1	4	15
Yr	1194	957	46	3774	4833	6045	7444

SHERIDAN, WYOMING				Elev 3966		Lat 44.8
	HS	VS	TA	D50	D55	D60	D65
Jan	517	900	21	899	1054	1209	1364
Jul	2329	1237	70	1	2	9	28
Yr	1333	1170	45	3860	4991	6279	7708

Canada

EDMONTON, ALBERTA				Elev 2220		Lat 53.6
	HS	VS	TA	D50	D55	D60	D65
Jan	324	746	4	1421	1574	1728	1883
Jul	1977	1378	62	1	7	38	117
Yr	1114	1205	36	6317	7563	9016	10650

SUFFIELD, ALBERTA				Elev 2549		Lat 50.3
	HS	VS	TA	D50	D55	D60	D65
Jan	433	937	7	1333	1486	1640	1794
Jul	2173	1377	67	0	2	11	49
Yr	1239	1269	40	5500	6637	7923	9393

VANCOUVER, BRITISH COLUMBIA						
				Elev 310		Lat 37.5
	HS	VS	TA	D50	D55	D60	D65
Jan	254	395	37	425	572	724	878
Jul	2021	1239	63	0	1	13	82
Yr	1060	916	50	1791	2781	4041	5588

WINNIPEG, MANITOBA				Elev 820		Lat 49.95
	HS	VS	TA	D50	D55	D60	D65
Jan	461	1011	0	1588	1740	1893	2047
Jul	2025	1264	67	0	2	9	45
Yr	1190	1199	36	6925	8062	9338	10790

TABLE B.15 **Average Insolation, Temperature, and DD Data** (*continued*)

HALIFAX, NOVA SCOTIA				Elev 136	Lat 44.6		TORONTO, ONTARIO				Elev 443	Lat 43.7	
	HS	VS	TA	D50	D55	D60	D65	HS	VS	TA	D50	D55	D60	D65
Jan	456	737	26	752	900	1051	1204	487	777	22	891	1041	1194	1348
Jul	1694	929	65	0	1	8	57	1958	1035	70	0	0	3	18
Yr	1076	907	46	3457	4500	5746	7211	1171	948	46	3842	4853	6013	7343

OTTAWA, ONTARIO				Elev 377	Lat 45.5		NORMANDIN, QUEBEC				Elev 450	Lat 48.8	
	HS	VS	TA	D50	D55	D60	D65	HS	VS	TA	D50	D55	D60	D65
Jan	510	914	13	1169	1320	1473	1627	454	921	0	1564	1719	1873	2028
Jul	1875	1040	69	0	1	4	23	1707	1031	62	1	7	40	118
Yr	1158	1015	43	4912	5965	7158	8529	1092	1053	34	7037	8308	9762	11376

Source: Reprinted by permission from J. D. Balcomb et al., *Passive Solar Design Handbook,* Vol. 3, © 1983, American Solar Energy Society, Inc., Boulder.

APPENDICES

C

ANNUAL SOLAR PERFORMANCE

This material is abridged from J. D. Balcomb et al., *Passive Solar Heating Analysis,* © 1984, American Society of Heating, Refrigerating, and Air-Conditioning Engineers, Inc., Atlanta, Ga.

This appendix lists 30 passive solar heating systems (3 water wall, 9 Trombé wall, 9 direct gain, and 9 sunspace) for the U.S. and Canadian locations shown in Fig. B.1. The *Passive Solar Heating Analysis* lists 94 systems, for about twice as many locations. The specifications for all 94 systems are listed here, so that the original source may be consulted if your passive system is not represented by one of those in this appendix.

TABLE C.1 **Characteristics of Selected Passive Solar Heating Systems**

Part A. **Water Wall Systems**

Designation	Thermal Storage Capacity[a] (Btu/ft² F)	Wall Thickness (in.)	No. of Glazings	Wall Surface	Night Insulation
WW-A1	15.6	3	2	Normal	No
WW-A2	31.2	6	2	Normal	No
WW-A3[b]	46.8	9	2	Normal	No
WW-A4	62.4	12	2	Normal	No
WW-A5	93.6	18	2	Normal	No
WW-A6	124.8	24	2	Normal	No
WW-B1	46.8	9	1	Normal	No
WW-B2	46.8	9	3	Normal	No
WW-B3	46.8	9	1	Normal	Yes
WW-B4[b]	46.8	9	2	Normal	Yes
WW-B5	46.8	9	3	Normal	Yes
WW-C1	46.8	9	1	Selective	No
WW-C2[b]	46.8	9	2	Selective	No
WW-C3	46.8	9	1	Selective	Yes
WW-C4	46.8	9	2	Selective	Yes

Part B. **Trombe Wall Systems: Vented**

Designation	Thermal Storage Capacity[c] (Btu/ft² F)	Wall Thickness[c] (in.)	ρck[d] (Btu²/h-ft⁴-F²)	No. of Glazings	Wall Surface	Night Insulation
TW-A1[b]	15	6	30	2	Normal	No
TW-A2[b]	22.5	9	30	2	Normal	No
TW-A3[b]	30	12	30	2	Normal	No
TW-A4[b]	45	18	30	2	Normal	No
TW-B1	15	6	15	2	Normal	No
TW-B2	22.5	9	15	2	Normal	No
TW-B3[b]	30	12	15	2	Normal	No
TW-B4	45	18	15	2	Normal	No
TW-C1	15	6	7.5	2	Normal	No
TW-C2	22.5	9	7.5	2	Normal	No
TW-C3	30	12	7.5	2	Normal	No
TW-C4	45	18	7.5	2	Normal	No
TW-D1	30	12	30	1	Normal	No
TW-D2	30	12	30	3	Normal	No
TW-D3	30	12	30	1	Normal	Yes
TW-D4[b]	30	12	30	2	Normal	Yes
TW-D5	30	12	30	3	Normal	Yes
TW-E1	30	12	30	1	Selective	No
TW-E2[b]	30	12	30	2	Selective	No
TW-E3	30	12	30	1	Selective	Yes
TW-E4	30	12	30	2	Selective	Yes

Note: Vented systems may be made to perform as unvented systems by sealing both top and bottom vent openings.

TABLE C.1 **Characteristics of Selected Passive Solar Heating Systems** (*continued*)

Part C. **Trombe Wall Systems: Unvented**

Designation	Thermal Storage Capacity[c] (Btu/ft^2 F)	Wall Thickness[c] (in.)	ρck[d] (Btu^2/h-ft^4-F^2)	No. of Glazings	Wall Surface	Night Insulation
TW-F1	15	6	30	2	Normal	No
TW-F2	22.5	9	30	2	Normal	No
TW-F3[b]	30	12	30	2	Normal	No
TW-F4	45	18	30	2	Normal	No
TW-G1	15	6	15	2	Normal	No
TW-G2	22.5	9	15	2	Normal	No
TW-G3	30	12	15	2	Normal	No
TW-G4	45	18	15	2	Normal	No
TW-H1	15	6	7.5	2	Normal	No
TW-H2	22.5	9	7.5	2	Normal	No
TW-H3	30	12	7.5	2	Normal	No
TW-H4	45	18	7.5	2	Normal	No
TW-I1	30	12	30	1	Normal	No
TW-I2	30	12	30	3	Normal	No
TW-I3	30	12	30	1	Normal	Yes
TW-I4	30	12	30	2	Normal	Yes
TW-I5	30	12	30	3	Normal	Yes
TW-J1	30	12	30	1	Selective	No
TW-J2[b]	30	12	30	2	Selective	No
TW-J3	30	12	30	1	Selective	Yes
TW-J4	30	12	30	2	Selective	Yes

Part D. **Direct-Gain Systems**

Designation	Thermal Storage Capacity[c] (Btu/ft^2 F)	Mass Thickness[c] (in.)	Ratio of Mass to Glazing Area	No. of Glazings	Night Insulation
DG-A1[b]	30	2	6	2	No
DG-A2[b]	30	2	6	3	No
DG-A3[b]	30	2	6	2	Yes
DG-B1[b]	45	6	3	2	No
DG-B2[b]	45	6	3	3	No
DG-B3[b]	45	6	3	2	Yes
DG-C1[b]	60	4	6	2	No
DG-C2[b]	60	4	6	3	No
DG-C3[b]	60	4	6	2	Yes

TABLE C.1 **Characteristics of Selected Passive Solar Heating Systems** (*continued*)

Part E. Sunspace Systems

Designation	Type	Tilt (Degrees)	Common Wall[f]	End Walls[g]	Night Insulation
SS-A1[b]	Attached	50	Masonry	Opaque	No
SS-A2	Attached	50	Masonry	Opaque	Yes
SS-A3	Attached	50	Masonry	Glazed	No
SS-A4	Attached	50	Masonry	Glazed	Yes
SS-A5	Attached	50	Insulated	Opaque	No
SS-A6	Attached	50	Insulated	Opaque	Yes
SS-A7	Attached	50	Insulated	Glazed	No
SS-A8	Attached	50	Insulated	Glazed	Yes
SS-B1[b]	Attached	90/30	Masonry	Opaque	No
SS-B2[b]	Attached	90/30	Masonry	Opaque	Yes
SS-B3[b]	Attached	90/30	Masonry	Glazed	No
SS-B4	Attached	90/30	Masonry	Glazed	Yes
SS-B5	Attached	90/30	Insulated	Opaque	No
SS-B6	Attached	90/30	Insulated	Opaque	Yes
SS-B7	Attached	90/30	Insulated	Glazed	No
SS-B8	Attached	90/30	Insulated	Glazed	Yes
SS-C1[b]	Semi-enclosed	90	Masonry	Common	No
SS-C2[b]	Semi-enclosed	90	Masonry	Common	Yes
SS-C3	Semi-enclosed	90	Insulated	Common	No
SS-C4	Semi-enclosed	90	Insulated	Common	Yes
SS-D1	Semi-enclosed	50	Masonry	Common	No
SS-D2	Semi-enclosed	50	Masonry	Common	Yes
SS-D3	Semi-enclosed	50	Insulated	Common	No
SS-D4	Semi-enclosed	50	Insulated	Common	Yes
SS-E1[b]	Semi-enclosed	90/30	Masonry	Common	No
SS-E2[b]	Semi-enclosed	90/30	Masonry	Common	Yes
SS-E3[b]	Semi-enclosed	90/30	Insulated	Common	No
SS-E4	Semi-enclosed	90/30	Insulated	Common	Yes

Source: Reprinted by permission of American Society of Heating, Refrigerating, and Air-Conditioning Engineers, Inc., from *Passive Solar Heating Analysis.*

[a] Per unit of projected area.

[b] Listed in this text.

[c] The thermal storage capacity is per projected area of south aperture (ft^2 of glazing), or, equivalently, the quantity ρct. The wall thickness (t) is listed only as an approximate guide by assuming that ρc = 30 Btu/ft^3°F (ρ of 150 lb/ft^3 and c of 0.2 Btu/lb°F), typical of ordinary concrete. Example: for DGA1 thermal storage capacity,

$$(30 \text{ Btu/ft}^3\text{°F}) \times (2 \text{ in./12 in/ft}) \times (6 \text{ ft}^2 \text{ of mass/1 ft}^2 \text{ of glazing}) = 30 \text{ Btu/ft}^2\text{°F}$$

[d] ρck is the product of the density (lb/ft^3), specific heat (Btu/lb F), and thermal conductivity (Btu/h ft F): see Table C.2.

[e] See Fig. C.1 for additional description.

[f] Common walls should provide closable openings that total at minimum 6% (maximum 15%) of the wall area. Insulated common walls should have minimum R-10 insulation, and thermal mass in water at a minimum of 3.8 gal/ft^2 (0.5 ft^3/ft^2) of projected area. Masonry common walls should have minimum 3:1 mass:glass area ratio.

[g] Opaque end walls are strongly recommended; glazed end walls perform more poorly in most winter climates, and are subject to severe summer overheating.

TABLE C.2 Thermal Properties of Various Materials

Material	Source[a]	Density, ρ (lb/ft^3)	Specific Heat, c (Btu/lb-°F)	Heat Capacity, ρc (Btu/ft^3-°F)	Conductivity, k (Btu/h-°F-ft)	Diffusivity, $k/\rho c$ (ft^2/h)	$\rho c k$ (Btu^2/h-ft^4-°F^2)
Water	1	62.3	1.0	62.3			
Concrete	2, 3	144	0.21	30.3	0.89	0.029	27.0
Concrete block	2, 3						
Heavy weight		135	0.21	28.4	0.74	0.026	21.0
Medium weight		105	0.22	23.1	0.41	0.018	9.5
Light weight		85	0.23	19.6	0.27	0.014	5.3
Brick	1, 4						
Paving		135	0.19	25.7	0.75	0.029	19.3
Face		130	0.19	24.7	0.75	0.030	18.5
Building		120	0.19	22.8	0.42	0.018	9.6
Mortar or grout	1	116	0.20	23.2	0.42	0.018	9.7
Adobe	5	80	0.20	16.0	0.38	0.024	6.1
		100		20.0	0.75	0.038	15.0
Gypsum or plasterboard	1	50	0.26	13.0	0.097	0.058	9.8
Douglas fir plywood	1	34	0.29	9.9	0.067	0.007	0.7
Hardwood	1	45	0.30	13.5	0.092	0.007	1.2

Source: Reprinted by permission from *Passive Solar Heating Analysis*, © 1984, American Society of Heating, Refrigerating, and Air-Conditioning Engineers, Inc., Atlanta, Ga.

Note: Concrete and masonry products can be manufactured in a variety of densities. The values listed are for the most commonly available products. The thermal properties of concrete and concrete block vary with density. Empirical correlations between these properties are[2] $k = 0.05e^{0.02\rho}$ and [3] $c = 1/(3.934 + 0.006\rho)$.

[a] 1. *ASHRAE Handbook: 1981 Fundamentals* (American Society of Heating, Refrigerating, and Air-Conditioning Engineers, Inc., Atlanta, Ga., 1981).

2. "Calculating the Steady State *U*-Value of Concrete Masonry Walls," Expanded Shale, Clay, and Slate Institute, 4905 Del Ray Avenue, Bethesda, Md.

3. David Whiting, Albert Litvin, and Stanley E. Goodwin. "Specific Heat of Selected Concretes," *Research and Development Bulletin RD058.01B*, Portland Cement Association, 5420 Old Orchard Road, Skokie, Ill.

4. "Brick Passive Solar Heating Systems. Material Properties—Part IV," *Technical Notes on Brick Construction 43D*, Brick Institute of America, 1750 Old Meadow Road, McLean, Va. September/October 1980.

5. Benjamin T. Rogers, private communication; and W. L. Sibbitt, measurements at Los Alamos National Laboratory. Note that the thermal conductivity of adobe varies greatly with moisture content, which is part of the reason for the wide range quoted.

Fig. C.1 *Types of sunspaces, described in Table C.1, Part E.* (a) *and* (b) *Considered* attached *to the building;* (c)*,* (d)*, and* (e) *considered* semienclosed *by the building. The architectural detail at the sides of type* (d) *is insignificant; no shading of the sunspace by the building was accounted for in the performance estimates of Table C.2. (Reprinted by permission from J. D. Balcomb et al.,* Passive Solar Heating Analysis, © *1984, American Society of Heating, Refrigerating, and Air-Conditioning Engineers, Inc., Atlanta, Ga.)*

TABLE C.3 Annual Passive Heating Performance: SSF

For each location, DD listed are DD65. For a description of SSF (Solar Savings Fraction) see Fig. 5.8, page 211; LCR (load–collector ratio) is described in Section 5.10. LCR units are Btu/DD-ft^2.

United States

MONTGOMERY, ALABAMA SSF (%) — 2272 DD65

Type	LCR = 200	100	70	50	40	30	25	20
WWA3	17	29	38	48	55	64	70	77
WWB4	18	36	48	60	68	78	83	89
WWC2	18	36	47	59	67	76	82	88
TWA1	17	24	29	35	40	46	51	57
TWA2	16	26	33	41	46	54	60	67
TWA3	15	26	34	43	49	58	63	70
TWA4	14	26	34	43	49	58	64	72
TWB3	14	24	31	39	45	54	60	67
TWD4	17	33	43	54	62	72	78	84
TWE2	18	34	44	55	63	72	78	84
TWF3	12	23	30	39	45	54	60	67
TWJ2	15	31	41	52	60	70	76	82
DGA1	12	21	28	34	39	45	49	53
DGA2	13	23	31	39	45	53	58	64
DGA3	15	27	37	47	55	64	70	77
DGB1	12	22	29	37	43	51	56	62
DGB2	13	24	32	41	49	58	65	71
DGB3	15	28	38	49	57	68	75	82
DGC1	14	25	34	43	50	60	66	72
DGC2	15	27	36	47	55	66	72	79
DGC3	18	32	42	55	64	75	81	87
SSA1	21	32	40	48	55	63	68	75
SSB1	17	28	35	42	48	56	62	68
SSB2	21	35	44	54	61	70	76	82
SSB3	17	26	33	40	46	54	59	66
SSC1	15	26	34	43	49	58	64	71
SSC2	17	31	41	51	58	68	74	80
SSE1	20	33	42	51	58	67	72	78
SSE2	23	40	51	62	69	78	83	88
SSE3	18	28	35	43	48	56	62	68

TUCSON, ARIZONA SSF (%) — 1758 DD65

Type	LCR = 200	100	70	50	40	30	25	20
WWA3	31	52	64	75	82	89	93	96
WWB4	37	62	75	85	91	96	98	99
WWC2	37	61	74	84	90	95	97	99
TWA1	25	38	47	56	63	72	78	84
TWA2	27	44	55	65	73	81	86	91
TWA3	28	46	58	69	76	84	89	93
TWA4	27	46	58	69	77	85	89	94
TWB3	25	42	53	64	72	81	86	91
TWD4	33	56	69	80	86	93	96	98
TWE2	35	57	70	81	87	93	96	98
TWF3	24	42	53	65	73	81	86	91
TWJ2	31	54	67	78	85	92	95	97
DGA1	24	41	52	62	68	76	80	84
DGA2	24	43	54	65	72	81	85	89
DGA3	27	49	62	74	80	87	90	93
DGB1	24	44	58	69	76	83	87	91
DGB2	25	45	59	72	79	87	91	94
DGB3	27	50	65	78	85	91	94	96
DGC1	28	51	66	78	84	90	93	95
DGC2	28	51	66	79	85	92	95	97
DGC3	31	56	72	84	90	95	96	98
SSA1	33	51	62	72	79	86	90	94
SSB1	29	45	55	66	73	81	86	91
SSB2	35	55	67	77	84	90	94	97
SSB3	27	43	53	64	71	79	84	89
SSC1	28	47	59	70	77	86	90	94
SSC2	32	54	66	78	84	91	94	97
SSE1	35	55	67	77	83	90	93	96
SSE2	41	64	76	85	91	95	97	99
SSE3	30	46	56	67	73	82	86	91

PHOENIX, ARIZONA SSF (%) — 1556 DD65

Type	LCR = 200	100	70	50	40	30	25	20
WWA3	33	54	66	77	83	90	93	96
WWB4	39	64	76	86	91	96	98	99
WWC2	39	63	75	85	90	95	97	99
TWA1	26	40	49	58	65	73	79	84
TWA2	29	46	56	67	74	82	87	91
TWA3	29	48	59	70	77	85	89	93
TWA4	29	48	60	71	78	86	90	94
TWB3	26	44	55	66	73	82	86	91
TWD4	35	58	70	81	87	93	96	98
TWE2	36	59	71	82	88	93	96	98
TWF3	26	44	55	67	74	82	87	92
TWJ2	33	56	68	79	86	92	95	97
DGA1	26	43	54	64	70	77	81	85
DGA2	26	45	56	67	74	82	86	90
DGA3	29	51	64	75	81	87	91	93
DGB1	27	47	60	71	77	84	87	91
DGB2	27	48	61	73	80	88	91	94
DGB3	29	53	67	79	85	91	94	96
DGC1	31	54	58	79	85	90	93	95
DGC2	30	54	68	80	86	92	95	97
DGC3	33	59	74	85	90	95	96	98
SSA1	34	52	62	73	79	86	90	94
SSB1	29	46	56	67	73	81	86	90
SSB2	36	56	67	78	84	90	93	96
SSB3	28	44	54	64	71	79	84	89
SSC1	30	49	61	72	78	86	90	94
SSC2	34	56	68	79	85	91	94	97
SSE1	36	56	67	77	83	90	93	96
SSE2	42	64	76	85	90	95	97	99
SSE3	30	47	57	67	74	82	86	91

FORT SMITH, ARKANSAS SSF (%) — 3338 DD65

Type	LCR = 200	100	70	50	40	30	25	20
WWA3	11	21	29	37	43	51	57	64
WWB4	12	27	37	49	57	67	73	80
WWC2	12	27	37	48	55	65	72	79
TWA1	14	19	23	27	31	36	40	45
TWA2	12	20	25	31	36	43	48	54
TWA3	11	19	26	33	38	46	51	58
TWA4	10	19	25	33	38	46	52	59
TWB3	10	18	23	30	35	43	48	54
TWD4	12	25	34	44	51	61	67	75
TWE2	13	26	35	44	52	61	67	75
TWF3	8	16	22	29	34	42	47	54
TWJ2	10	23	32	41	48	58	65	72
DGA1	8	14	19	24	28	33	36	39
DGA2	9	17	23	29	35	42	46	52
DGA3	11	21	28	37	44	53	59	66
DGB1	8	15	19	25	30	36	40	46
DGB2	9	17	23	31	36	45	51	58
DGB3	11	21	29	38	45	56	63	71
DGC1	10	18	24	30	36	44	50	56
DGC2	11	20	27	35	42	52	59	66
DGC3	13	24	32	43	51	63	70	78
SSA1	16	25	31	38	43	50	55	62
SSB1	13	21	26	33	37	44	49	55
SSB2	15	27	35	43	50	59	64	71
SSB3	12	19	25	30	35	41	46	52
SSC1	10	19	25	32	37	45	51	57
SSC2	11	23	31	40	47	56	62	69
SSE1	14	24	31	39	45	53	58	65
SSE2	16	30	39	50	57	66	72	79
SSE3	14	21	27	33	37	44	49	55

TABLE C.3 **Annual Passive Heating Performance: SSF (***continued***)**

LOS ANGELES, CALIFORNIA **SSF (%)** 1793 DD65										SAN DIEGO, CALIFORNIA **SSF (%)** 1512 DD65									
Type	LCR =	200	100	70	50	40	30	25	20	Type	LCR =	200	100	70	50	40	30	25	20
WWA3		37	58	70	81	87	93	95	98	WWA3		39	61	73	83	89	94	97	98
WWB4		43	68	80	89	93	97	99	99	WWB4		46	71	83	91	95	98	99	100
WWC2		42	67	79	88	93	97	98	99	WWC2		45	70	82	91	95	98	99	100
TWA1		29	43	52	62	69	77	82	88	TWA1		30	45	55	65	72	80	85	90
TWA2		32	49	60	71	78	86	90	94	TWA2		34	52	64	74	81	88	92	95
TWA3		33	52	63	74	81	88	92	96	TWA3		35	55	67	78	84	91	94	97
TWA4		32	52	64	75	82	89	93	96	TWA4		34	55	67	78	85	91	94	97
TWB3		29	48	59	70	77	85	90	94	TWB3		31	51	62	73	80	88	92	95
TWD4		39	62	74	85	90	95	97	99	TWD4		41	65	77	87	92	96	98	99
TWE2		40	63	75	85	91	95	97	99	TWE2		43	66	78	88	92	97	98	99
TWF3		29	48	59	71	78	86	90	94	TWF3		31	51	63	74	81	89	92	96
TWJ2		37	60	72	83	89	94	97	98	TWJ2		39	63	76	86	91	96	98	99
DGA1		29	47	58	68	74	81	85	89	DGA1		31	51	62	72	78	84	88	91
DGA2		29	49	60	71	78	85	89	93	DGA2		31	52	64	74	81	88	91	95
DGA3		32	55	68	79	84	90	93	95	DGA3		34	58	71	81	87	92	94	95
DGB1		30	52	65	76	82	88	91	94	DGB1		32	56	69	79	85	90	93	95
DGB2		30	52	66	78	84	91	94	96	DGB2		32	56	70	81	87	93	95	97
DGB3		33	57	72	83	89	94	96	97	DGB3		35	61	75	86	91	95	96	97
DGC1		34	59	73	84	89	93	95	96	DGC1		37	63	77	87	91	95	96	97
DGC2		34	58	73	84	90	95	97	98	DGC2		36	63	77	87	92	96	98	99
DGC3		37	63	78	89	93	96	97	98	DGC3		39	68	82	91	94	97	98	98
SSA1		41	60	71	80	86	92	95	97	SSA1		43	62	73	82	88	93	95	98
SSB1		35	53	64	74	81	88	91	95	SSB1		37	56	67	77	83	89	93	96
SSB2		42	64	75	84	89	94	97	98	SSB2		45	66	77	86	91	95	97	99
SSB3		34	52	63	73	79	87	90	94	SSB3		36	54	65	75	81	88	92	95
SSC1		33	53	65	76	83	90	93	96	SSC1		36	56	68	79	85	92	95	97
SSC2		38	60	72	83	88	94	96	98	SSC2		40	63	75	85	91	95	97	99
SSE1		43	64	75	84	90	94	97	98	SSE1		45	67	78	86	91	95	97	99
SSE2		50	73	83	91	95	98	99	100	SSE2		52	75	85	92	96	98	99	100
SSE3		36	54	65	75	81	88	92	95	SSE3		38	57	67	77	83	90	93	96

SACRAMENTO, CALIFORNIA **SSF (%)** 2845 DD65										FRESNO, CALIFORNIA **SSF (%)** 2657 DD65									
Type	LCR =	200	100	70	50	40	30	25	20	Type	LCR =	200	100	70	50	40	30	25	20
WWA3		18	31	39	48	55	63	68	74	WWA3		18	31	40	49	56	64	69	75
WWB4		19	37	49	60	67	76	81	87	WWB4		20	38	49	61	68	77	82	87
WWC2		20	37	48	59	66	75	80	85	WWC2		20	38	49	60	67	76	81	86
TWA1		18	25	30	36	41	47	51	57	TWA1		18	26	31	37	41	48	52	58
TWA2		17	27	34	42	47	55	60	66	TWA2		18	28	35	42	48	55	60	67
TWA3		16	28	35	44	50	58	63	69	TWA3		17	28	36	44	50	58	64	70
TWA4		15	27	35	44	50	58	64	70	TWA4		16	28	36	44	51	59	64	71
TWB3		15	25	32	40	46	54	60	66	TWB3		15	26	33	41	47	55	60	67
TWD4		18	34	44	55	62	71	76	82	TWD4		19	35	45	56	63	72	77	83
TWE2		20	35	45	56	63	71	77	82	TWE2		20	36	46	56	63	72	77	83
TWF3		13	24	32	40	46	54	59	66	TWF3		14	25	32	40	47	55	60	66
TWJ2		17	32	42	53	60	69	74	80	TWJ2		17	33	43	53	60	69	75	81
DGA1		13	23	29	36	40	46	49	53	DGA1		14	23	30	36	41	47	50	54
DGA2		14	25	32	40	46	54	58	64	DGA2		14	25	33	41	47	54	59	65
DGA3		16	29	38	48	55	64	69	75	DGA3		17	30	39	49	56	65	70	76
DGB1		13	23	31	39	44	51	55	60	DGB1		14	24	31	39	45	52	56	61
DGB2		14	25	34	43	50	58	64	70	DGB2		15	26	35	44	50	59	64	70
DGB3		16	30	39	50	58	67	73	79	DGB3		17	30	40	51	58	68	74	80
DGC1		16	27	36	45	51	59	63	69	DGC1		16	28	37	46	52	60	64	70
DGC2		16	29	39	49	56	65	70	76	DGC2		17	30	39	49	56	65	71	77
DGC3		19	33	44	56	63	73	79	85	DGC3		19	34	45	56	64	74	79	85
SSA1		22	33	41	48	54	61	65	71	SSA1		22	34	41	49	54	61	66	72
SSB1		19	29	36	43	48	55	60	66	SSB1		19	29	36	43	49	56	61	66
SSB2		22	36	45	54	60	68	73	79	SSB2		22	36	45	54	60	69	74	79
SSB3		18	28	34	41	46	53	57	63	SSB3		18	28	34	41	46	53	58	64
SSC1		16	27	35	44	50	58	63	69	SSC1		16	28	36	44	50	59	64	70
SSC2		18	32	42	51	58	67	72	78	SSC2		18	33	42	52	59	68	73	79
SSE1		21	34	42	50	56	64	68	74	SSE1		22	34	42	51	57	64	69	75
SSE2		24	41	50	60	67	75	79	84	SSE2		25	41	51	61	67	75	80	85
SSE3		20	30	36	43	48	55	60	65	SSE3		20	30	36	44	49	56	60	66

TABLE C.3 **Annual Passive Heating Performance: SSF** (*continued*)

SAN FRANCISCO, CALIFORNIA **SSF (%)** 3050 DD65

Type	LCR = 200	100	70	50	40	30	25	20
WWA3	24	41	51	62	69	77	82	88
WWB4	27	49	61	73	80	87	91	95
WWC2	27	48	61	72	79	87	91	94
TWA1	21	31	38	46	51	59	64	71
TWA2	22	35	44	53	59	68	73	80
TWA3	21	36	46	56	63	71	77	83
TWA4	20	36	46	56	63	72	78	84
TWB3	19	33	42	52	59	68	73	79
TWD4	25	44	56	67	75	83	87	92
TWE2	26	46	57	68	75	83	88	92
TWF3	18	32	42	52	59	68	73	80
TWJ2	23	42	54	65	73	81	86	91
DGA1	18	31	40	48	54	61	65	69
DGA2	18	33	42	52	59	67	72	78
DGA3	21	38	49	60	68	76	81	86
DGB1	18	33	43	53	60	68	72	77
DGB2	19	34	45	56	64	73	79	84
DGB3	21	39	51	63	72	81	85	90
DGC1	21	38	49	61	68	76	81	85
DGC2	22	39	51	63	71	80	85	89
DGC3	24	43	57	70	78	86	90	94
SSA1	30	45	54	63	68	76	80	85
SSB1	25	39	48	57	63	70	75	81
SSB2	30	47	58	67	74	81	86	90
SSB3	24	38	46	55	61	68	73	79
SSC1	21	37	47	57	64	72	78	84
SSC2	24	42	54	64	72	80	85	90
SSE1	30	47	57	66	72	79	84	88
SSE2	35	54	65	75	81	87	91	94
SSE3	26	40	48	57	63	71	75	81

DENVER, COLORADO **SSF (%)** 6022 DD65

Type	LCR = 100	70	50	40	30	25	20	15
WWA3	21	29	37	43	52	58	66	75
WWB4	27	38	50	58	68	75	82	89
WWC2	27	37	48	56	67	73	80	88
TWA1	19	23	27	31	37	41	46	53
TWA2	19	25	31	36	44	49	55	64
TWA3	19	26	33	39	47	52	59	69
TWA4	18	25	33	39	47	53	61	70
TWB3	17	23	30	36	43	49	56	65
TWD4	25	34	44	52	62	68	76	85
TWE2	26	35	45	52	62	69	76	85
TWF3	16	22	29	35	43	48	55	65
TWJ2	23	32	42	49	59	66	74	83
DGA1	14	19	25	28	33	37	40	45
DGA2	16	22	30	35	43	47	53	61
DGA3	20	28	37	44	54	61	68	76
DGB1	14	19	25	30	37	42	48	54
DGB2	17	23	31	37	46	52	60	68
DGB3	21	29	38	46	57	65	73	82
DGC1	17	24	31	37	46	52	59	66
DGC2	20	27	36	43	53	60	68	77
DGC3	24	33	43	52	64	72	80	88
SSA1	25	31	38	44	51	56	63	71
SSB1	21	26	33	38	45	50	56	65
SSB2	27	35	44	50	59	65	72	81
SSB3	19	25	31	35	42	46	52	61
SSC1	19	25	33	38	46	52	59	68
SSC2	23	31	41	48	57	63	71	80
SSE1	24	31	39	45	54	59	66	74
SSE2	30	40	50	57	67	73	80	87
SSE3	21	27	33	38	45	49	56	64

SANTA MARIA, CALIFORNIA **SSF (%)** 3061 DD65

Type	LCR = 200	100	70	50	40	30	25	20
WWA3	25	44	55	67	74	82	87	92
WWB4	29	53	66	78	84	91	95	97
WWC2	29	52	65	77	84	91	94	97
TWA1	21	32	40	49	55	64	69	76
TWA2	22	37	47	57	64	73	79	85
TWA3	22	39	49	60	67	77	82	88
TWA4	22	38	49	61	68	77	83	88
TWB3	20	35	45	56	63	73	78	84
TWD4	27	48	60	72	79	87	91	95
TWE2	28	49	61	73	80	88	92	95
TWF3	19	35	45	56	64	73	79	85
TWJ2	25	46	58	70	77	86	90	94
DGA1	19	34	43	52	59	66	71	75
DGA2	19	35	46	56	64	72	77	83
DGA3	22	40	53	65	73	81	85	89
DGB1	19	35	47	59	66	74	79	83
DGB2	20	37	49	62	70	79	84	89
DGB3	22	41	55	69	77	86	90	93
DGC1	22	41	54	67	75	83	87	90
DGC2	23	42	55	69	77	85	90	93
DGC3	25	47	62	76	83	91	94	96
SSA1	32	49	60	70	76	83	87	92
SSB1	27	43	53	63	70	78	83	88
SSB2	33	52	64	74	80	88	91	95
SSB3	26	41	51	61	68	76	81	86
SSC1	23	40	50	61	69	78	83	89
SSC2	26	46	58	69	77	85	89	94
SSE1	33	52	63	73	80	87	91	94
SSE2	38	61	72	82	88	93	96	98
SSE3	28	43	53	63	70	76	83	88

GRAND JUNCTION, COLORADO **SSF (%)** 5607 DD65

Type	LCR = 100	70	50	40	30	25	20	15
WWA3	20	27	35	41	49	55	62	71
WWB4	26	36	47	55	65	71	79	87
WWC2	26	35	46	54	64	70	77	85
TWA1	19	22	26	30	35	38	43	50
TWA2	19	24	30	35	41	46	52	61
TWA3	19	25	31	37	44	49	56	65
TWA4	18	24	31	37	45	50	57	66
TWB3	17	22	29	34	41	46	52	61
TWD4	24	32	42	49	59	65	73	81
TWE2	25	33	43	50	59	66	73	82
TWF3	15	21	28	33	40	45	52	61
TWJ2	22	30	40	47	57	63	70	79
DGA1	14	18	23	26	31	34	37	41
DGA2	16	22	28	33	40	45	50	57
DGA3	20	27	36	42	52	58	65	73
DGB1	14	18	24	28	34	38	43	49
DGB2	16	22	29	35	43	49	56	64
DGB3	20	28	37	44	54	61	69	78
DGC1	17	22	29	34	42	47	53	61
DGC2	19	26	34	41	50	56	64	73
DGC3	23	31	41	49	61	68	76	85
SSA1	24	29	36	41	48	52	58	66
SSB1	20	25	31	35	42	46	52	60
SSB2	25	33	41	47	56	62	68	77
SSB3	18	23	29	33	39	43	48	56
SSC1	18	24	31	36	43	49	55	64
SSC2	22	30	38	45	54	60	67	76
SSE1	23	29	37	42	49	54	61	69
SSE2	28	37	47	54	63	69	75	83
SSE3	20	25	31	35	41	46	51	59

APPENDICES

TABLE C.3 **Annual Passive Heating Performance: SSF** (*continued*)

HARTFORD, CONNECTICUT **SSF (%)** 6354 DD65

Type	LCR = 100	70	50	40	30	25	20	15
WWA3	7	10	13	16	19	22	25	30
WWB4	9	16	23	28	36	41	49	58
WWC2	10	15	22	27	35	40	46	55
TWA1	12	13	14	15	16	17	18	20
TWA2	9	11	13	15	18	20	22	25
TWA3	8	10	13	15	18	21	24	28
TWA4	6	9	12	15	18	21	24	29
TWB3	7	10	12	15	18	20	23	28
TWD4	10	15	21	25	32	37	43	52
TWE2	10	15	21	26	32	37	43	51
TWF3	5	7	10	12	15	17	20	24
TWJ2	8	13	19	23	30	34	40	49
DGA1	4	5	6	6	7	7	6	5
DGA2	7	9	12	14	17	19	22	26
DGA3	10	14	18	22	28	32	37	45
DGB1	4	5	6	6	7	7	7	7
DGB2	7	9	12	15	18	20	23	28
DGB3	10	14	19	23	29	33	39	48
DGC1	5	7	9	10	12	13	14	15
DGC2	9	12	15	18	23	26	30	36
DGC3	12	17	22	27	33	38	45	54
SSA1	13	16	19	21	25	27	30	34
SSB1	11	13	15	17	20	22	25	28
SSB2	14	18	24	27	33	37	43	50
SSB3	10	11	14	15	17	19	21	23
SSC1	6	8	11	13	15	17	20	23
SSC2	8	12	17	21	26	30	35	42
SSE1	10	13	16	18	21	23	26	29
SSE2	13	18	24	29	35	40	45	53
SSE3	11	13	15	17	19	21	23	26

MIAMI, FLORIDA **SSF (%)** 206 DD65

Type	LCR = 200	100	70	50	40	30	25	20
WWA3	82	96	99	100	100	100	100	100
WWB4	89	98	100	100	100	100	100	100
WWC2	88	98	100	100	100	100	100	100
TWA1	64	84	92	97	98	100	100	100
TWA2	73	91	96	99	100	100	100	100
TWA3	76	93	97	99	100	100	100	100
TWA4	76	93	98	99	100	100	100	100
TWB3	71	90	96	99	100	100	100	100
TWD4	84	97	99	100	100	100	100	100
TWE2	85	97	99	100	100	100	100	100
TWF3	72	91	96	99	100	100	100	100
TWJ2	83	96	99	100	100	100	100	100
DGA1	75	92	97	99	100	100	100	100
DGA2	74	92	97	99	100	100	100	100
DGA3	79	93	96	97	97	97	97	97
DGB1	81	95	98	99	99	99	99	99
DGB2	80	96	99	100	100	100	100	100
DGB3	83	96	98	98	98	98	98	98
DGC1	88	97	98	99	99	99	98	98
DGC2	86	98	99	100	100	100	100	100
DGC3	88	98	98	99	99	98	98	98
SSA1	81	95	99	100	100	100	100	100
SSB1	75	92	97	99	100	100	100	100
SSB2	84	97	99	100	100	100	100	100
SSB3	74	92	97	99	100	100	100	100
SSC1	78	94	98	99	100	100	100	100
SSC2	83	96	99	100	100	100	100	100
SSE1	85	97	99	100	100	100	100	100
SSE2	91	99	100	100	100	100	100	100
SSE3	76	93	97	99	100	100	100	100

WASHINGTON, DC **SSF (%)** 5010 DD65

Type	LCR = 100	70	50	40	30	25	20	15
WWA3	12	16	22	25	31	35	40	48
WWB4	15	23	32	39	48	54	62	72
WWC2	15	23	31	38	46	52	60	69
TWA1	14	16	18	20	23	25	28	32
TWA2	13	16	20	22	27	30	34	40
TWA3	12	15	20	23	28	32	37	43
TWA4	10	14	19	23	29	33	38	45
TWB3	11	14	18	22	27	30	35	42
TWD4	15	21	29	35	43	49	56	65
TWE2	15	22	29	35	43	49	56	65
TWF3	9	12	16	20	25	28	33	40
TWJ2	13	19	27	32	40	46	53	62
DGA1	7	9	12	14	16	17	18	19
DGA2	10	13	18	21	26	29	33	39
DGA3	13	18	25	29	37	42	48	57
DGB1	7	9	12	14	17	19	21	23
DGB2	10	14	18	22	27	31	36	42
DGB3	14	19	25	31	38	44	51	61
DGC1	9	12	16	19	23	25	29	33
DGC2	12	17	22	26	33	37	43	51
DGC3	16	22	29	35	43	50	58	68
SSA1	17	21	26	29	34	38	42	49
SSB1	14	17	21	25	29	32	36	42
SSB2	18	24	31	36	43	48	54	63
SSB3	13	16	20	22	26	29	33	38
SSC1	10	14	18	21	26	30	34	40
SSC2	13	19	25	30	37	42	48	57
SSE1	15	19	24	28	33	37	41	48
SSE2	19	26	34	39	47	53	59	68
SSE3	15	18	22	24	29	31	35	41

ORLANDO, FLORIDA **SSF (%)** 733 DD65

Type	LCR = 200	100	70	50	40	30	25	20
WWA3	45	68	80	89	93	97	98	99
WWB4	52	78	88	95	97	99	100	100
WWC2	52	77	87	94	97	99	100	100
TWA1	34	51	61	72	78	86	90	94
TWA2	38	59	70	80	86	92	95	98
TWA3	40	62	73	83	89	94	97	98
TWA4	39	62	74	84	90	95	97	99
TWB3	36	57	69	80	86	92	95	97
TWD4	47	72	83	91	95	98	99	100
TWE2	49	73	84	92	96	98	99	100
TWF3	36	58	70	80	87	93	95	98
TWJ2	45	70	82	90	94	98	99	100
DGA1	37	58	69	78	84	90	93	95
DGA2	36	59	71	81	86	92	95	97
DGA3	40	66	77	86	90	94	95	96
DGB1	38	64	76	85	90	94	96	97
DGB2	38	64	77	87	92	96	98	99
DGB3	41	69	82	90	94	97	97	98
DGC1	44	72	84	91	94	97	97	98
DGC2	42	71	83	92	95	98	99	100
DGC3	46	76	88	94	96	98	98	98
SSA1	46	67	78	87	92	96	98	99
SSB1	40	61	72	82	88	93	96	98
SSB2	48	71	82	90	94	98	99	100
SSB3	39	59	71	81	86	92	95	98
SSC1	41	63	75	85	90	95	97	99
SSC2	46	70	81	90	94	98	99	100
SSE1	50	72	83	91	95	98	98	100
SSE2	57	80	90	95	98	99	100	100
SSE3	41	62	73	82	88	93	96	98

TABLE C.3 **Annual Passive Heating Performance: SSF** (*continued*)

TALLAHASSEE, FLORIDA **SSF** (%)							1563 DD65		BOISE, IDAHO **SSF** (%)							5837 DD65
Type	LCR = 200	100	70	50	40	30	25	20	Type	LCR = 100	70	50	40	30	25	20	15
WWA3	24	41	52	63	70	79	84	90	WWA3	18	23	29	34	40	45	50	57
WWB4	27	49	62	74	82	89	93	96	WWB4	22	31	40	47	56	62	69	77
WWC2	27	49	61	73	81	88	92	96	WWC2	22	30	39	46	54	60	67	75
TWA1	21	31	38	46	52	60	65	72	TWA1	17	20	24	26	30	33	36	41
TWA2	22	35	44	53	60	70	75	82	TWA2	17	21	26	30	35	38	43	49
TWA3	22	36	46	56	64	73	79	85	TWA3	16	21	27	31	37	41	46	53
TWA4	21	36	46	57	64	74	79	86	TWA4	15	21	26	31	37	41	47	54
TWB3	19	33	42	52	59	69	75	81	TWB3	15	20	25	29	35	39	44	50
TWD4	25	45	57	69	76	85	89	93	TWD4	20	28	36	43	51	56	63	71
TWE2	26	46	58	69	77	85	90	94	TWE2	21	29	37	43	51	56	63	71
TWF3	18	32	42	52	60	69	75	82	TWF3	13	18	23	27	33	37	42	49
TWJ2	23	42	54	66	74	83	88	92	TWJ2	19	26	34	40	48	54	61	69
DGA1	18	31	40	48	54	62	66	71	DGA1	12	15	19	21	25	26	28	30
DGA2	18	33	42	53	60	68	74	79	DGA2	14	19	24	28	34	38	42	48
DGA3	21	37	49	62	69	78	83	87	DGA3	18	24	32	37	45	50	56	64
DGB1	18	32	43	54	61	70	75	80	DGB1	12	16	20	23	27	29	32	35
DGB2	19	34	45	58	66	75	81	86	DGB2	15	20	26	30	36	40	46	52
DGB3	21	38	51	65	73	83	87	92	DGB3	19	25	33	39	47	52	59	68
DGC1	21	38	49	62	71	79	83	88	DGC1	15	19	25	28	33	37	41	45
DGC2	21	38	51	65	73	82	87	91	DGC2	17	23	30	35	42	47	53	60
DGC3	24	43	57	72	80	88	92	95	DGC3	21	28	37	43	52	58	65	74
SSA1	27	42	52	62	69	77	82	88	SSA1	22	26	32	35	41	44	48	54
SSB1	23	36	46	55	62	71	77	83	SSB1	18	22	27	31	36	39	43	49
SSB2	28	45	56	67	74	83	87	92	SSB2	23	30	37	42	49	54	59	67
SSB3	22	35	44	53	60	69	74	81	SSB3	17	21	25	28	33	36	39	44
SSC1	21	37	47	57	65	74	80	86	SSC1	15	20	26	30	35	39	44	50
SSC2	24	42	54	66	73	82	87	92	SSC2	18	25	33	38	46	51	57	65
SSE1	28	45	55	66	73	81	86	91	SSE1	20	25	31	35	40	44	48	54
SSE2	32	53	65	76	83	90	93	96	SSE2	25	32	40	46	53	58	64	71
SSE3	24	37	46	56	63	71	77	83	SSE3	19	23	27	31	35	38	42	47

ATLANTA, GEORGIA **SSF** (%)							3105 DD65		CHICAGO, ILLINOIS **SSF** (%)							6125 DD65
Type	LCR = 200	100	70	50	40	30	25	20	Type	LCR = 100	70	50	40	30	25	20	15
WWA3	12	22	29	37	43	52	57	65	WWA3	9	13	16	19	23	26	30	35
WWB4	12	28	38	49	57	67	74	81	WWB4	12	19	26	32	40	46	53	62
WWC2	13	28	38	48	56	66	72	79	WWC2	12	18	25	31	38	44	51	60
TWA1	15	20	24	28	32	37	41	46	TWA1	13	14	16	17	19	20	22	24
TWA2	13	20	26	32	37	44	49	55	TWA2	11	13	16	18	21	23	26	30
TWA3	12	20	26	33	39	47	52	59	TWA3	9	12	16	18	22	25	28	33
TWA4	10	19	26	33	39	47	53	60	TWA4	8	11	15	18	22	25	29	34
TWB3	10	18	24	31	36	43	49	55	TWB3	9	11	15	17	21	24	27	32
TWD4	12	25	35	45	52	61	68	75	TWD4	12	17	24	29	36	41	47	56
TWE2	13	26	35	45	52	62	68	75	TWE2	12	18	24	29	36	40	47	55
TWF3	9	17	23	30	35	43	48	55	TWF3	7	9	13	15	19	21	25	29
TWJ2	11	23	32	42	49	59	65	73	TWJ2	10	15	22	26	33	38	44	53
DGA1	9	15	20	25	29	34	37	40	DGA1	5	7	8	9	10	10	11	10
DGA2	10	17	23	30	35	43	47	53	DGA2	8	11	14	17	20	23	26	30
DGA3	12	21	29	38	44	54	60	67	DGA3	12	16	21	25	31	35	41	48
DGB1	9	15	20	26	31	37	42	47	DGB1	5	7	8	9	11	11	12	12
DGB2	10	18	24	32	37	46	52	59	DGB2	8	11	15	17	21	24	28	33
DGB3	12	22	30	39	46	57	64	72	DGB3	12	16	22	26	32	37	43	52
DGC1	10	19	25	32	37	45	50	57	DGC1	7	9	12	13	16	17	19	21
DGC2	11	21	28	36	43	53	59	67	DGC2	10	14	18	21	26	30	34	41
DGC3	13	25	33	44	52	63	71	79	DGC3	14	19	25	30	37	42	49	58
SSA1	17	26	32	40	45	52	57	64	SSA1	15	18	21	24	27	30	33	38
SSB1	14	22	28	34	39	46	51	57	SSB1	12	15	18	20	23	25	28	32
SSB2	16	28	36	45	52	60	66	73	SSB2	16	20	26	30	36	40	46	53
SSB3	13	21	26	32	37	43	48	54	SSB3	11	13	16	17	20	22	24	27
SSC1	11	20	26	33	38	46	51	58	SSC1	8	11	14	16	19	21	24	28
SSC2	12	24	32	41	47	57	63	70	SSC2	10	15	20	24	30	34	39	46
SSE1	15	26	33	41	47	55	60	67	SSE1	12	15	18	21	25	27	30	34
SSE2	17	32	41	51	58	68	74	80	SSE2	15	21	27	32	39	43	49	57
SSE3	15	23	28	34	39	46	51	57	SSE3	12	15	17	19	22	24	27	30

TABLE C.3 **Annual Passive Heating Performance: SSF** (*continued*)

MOLINE, ILLINOIS **SSF** (%) — 6399 DD65

Type	LCR = 100	70	50	40	30	25	20	15
WWA3	9	13	16	19	24	26	30	36
WWB4	12	19	26	32	40	46	53	63
WWC2	12	18	25	31	39	44	51	60
TWA1	13	14	16	17	19	20	22	24
TWA2	11	13	16	18	21	23	26	30
TWA3	9	12	16	18	22	25	28	33
TWA4	8	11	15	18	22	25	29	34
TWB3	9	11	15	17	21	24	27	32
TWD4	12	17	24	29	36	41	47	56
TWE2	12	18	24	29	36	41	47	55
TWF3	7	9	13	15	19	21	25	29
TWJ2	10	15	22	26	33	38	44	53
DGA1	5	7	8	9	10	11	11	10
DGA2	8	11	14	17	20	23	26	30
DGA3	12	16	21	25	31	35	41	49
DGB1	5	7	8	9	11	11	12	12
DGB2	8	11	15	17	21	24	28	33
DGB3	12	16	22	26	32	37	43	52
DGC1	7	9	12	13	16	17	19	21
DGC2	10	14	18	21	26	29	34	41
DGC3	14	19	25	30	37	42	49	59
SSA1	15	18	21	24	27	30	33	38
SSB1	12	14	17	20	23	25	28	32
SSB2	15	20	26	30	36	40	46	53
SSB3	11	13	15	17	20	22	24	27
SSC1	8	11	14	16	19	21	24	29
SSC2	10	15	20	24	30	34	39	47
SSE1	12	15	18	21	24	27	30	34
SSE2	15	21	27	32	38	43	49	57
SSE3	12	15	17	19	22	24	27	30

INDIANAPOLIS, INDIANA **SSF** (%) — 5585 DD65

Type	LCR = 100	70	50	40	30	25	20	15
WWA3	9	12	16	19	22	25	29	34
WWB4	11	18	25	31	39	45	52	61
WWC2	11	18	25	30	37	43	49	58
TWA1	12	14	15	17	18	19	21	23
TWA2	10	13	15	17	20	22	25	29
TWA3	9	12	15	18	21	24	27	31
TWA4	8	11	14	17	21	24	28	33
TWB3	8	11	14	17	20	23	26	31
TWD4	11	17	23	28	35	40	46	55
TWE2	12	17	23	28	35	39	46	54
TWF3	6	9	12	14	18	20	23	28
TWJ2	10	15	21	26	32	37	43	52
DGA1	5	6	8	8	9	10	10	9
DGA2	8	11	14	16	20	22	25	29
DGA3	11	15	20	24	30	34	40	47
DGB1	5	6	8	9	10	10	11	11
DGB2	8	11	14	17	20	23	26	31
DGB3	12	16	21	25	31	36	42	51
DGC1	7	9	11	13	15	16	17	19
DGC2	10	13	17	21	25	28	33	39
DGC3	13	18	24	29	36	41	48	57
SSA1	15	18	21	24	27	30	33	38
SSB1	12	15	17	20	23	25	28	32
SSB2	15	20	26	30	36	40	46	53
SSB3	11	13	16	17	20	22	24	27
SSC1	8	10	13	15	18	20	23	27
SSC2	10	14	19	23	29	33	38	45
SSE1	12	15	18	21	24	27	30	34
SSE2	15	21	27	32	38	43	49	56
SSE3	12	15	17	19	22	24	26	30

SPRINGFIELD, ILLINOIS **SSF** (%) — 5563 DD65

Type	LCR = 100	70	50	40	30	25	20	15
WWA3	12	16	21	24	30	33	38	45
WWB4	15	23	31	38	46	53	60	70
WWC2	15	22	30	36	45	50	58	67
TWA1	14	16	18	20	22	24	27	30
TWA2	13	16	19	22	26	29	32	38
TWA3	12	15	19	22	27	30	35	41
TWA4	10	14	19	22	27	31	36	43
TWB3	11	14	18	21	26	29	33	39
TWD4	14	21	28	34	42	47	54	63
TWE2	15	21	28	34	42	47	54	63
TWF3	9	12	16	19	24	27	31	37
TWJ2	13	19	26	31	39	44	51	60
DGA1	7	9	11	13	15	16	17	17
DGA2	10	13	17	20	25	28	32	37
DGA3	14	18	24	28	35	40	47	55
DGB1	7	9	12	13	16	17	19	21
DGB2	10	14	18	21	26	29	34	40
DGB3	14	19	25	30	37	42	49	59
DGC1	9	12	15	18	21	23	26	30
DGC2	12	16	21	25	31	35	41	49
DGC3	16	22	28	34	42	48	56	66
SSA1	17	20	25	28	32	36	40	46
SSB1	14	17	21	23	28	30	34	40
SSB2	18	23	29	34	41	46	52	60
SSB3	13	15	19	21	25	27	30	35
SSC1	10	14	17	20	25	28	32	38
SSC2	13	18	24	29	36	40	46	55
SSE1	14	18	23	26	31	34	38	44
SSE2	18	25	32	37	45	50	57	65
SSE3	14	17	21	23	27	29	33	38

SOUTH BEND, INDIANA **SSF** (%) — 6465 DD65

Type	LCR = 100	70	50	40	30	25	20	15
WWA3	7	9	12	14	16	18	20	23
WWB4	9	14	21	26	33	38	44	52
WWC2	9	14	20	25	31	36	41	49
TWA1	12	12	13	14	15	16	16	17
TWA2	9	11	13	14	16	17	19	21
TWA3	8	10	12	14	16	18	20	23
TWA4	6	9	11	13	16	18	21	24
TWB3	7	9	12	13	16	18	20	23
TWD4	9	14	19	23	29	34	39	47
TWE2	10	14	19	23	29	33	38	45
TWF3	5	7	9	11	13	15	17	20
TWJ2	8	12	17	21	27	31	36	43
DGA1	4	4	5	5	5	5	4	1
DGA2	7	9	11	13	16	17	20	22
DGA3	10	14	18	21	26	29	34	41
DGB1	4	4	5	5	5	5	5	3
DGB2	7	9	12	14	16	18	21	24
DGB3	10	14	19	22	27	31	36	44
DGC1	5	7	8	9	10	10	10	10
DGC2	9	11	15	17	21	23	26	31
DGC3	12	16	21	25	31	35	41	49
SSA1	13	15	18	20	23	25	27	29
SSB1	11	13	15	17	19	20	22	25
SSB2	14	18	23	26	31	35	39	46
SSB3	10	11	13	14	16	17	18	19
SSC1	6	8	10	11	13	14	15	17
SSC2	8	11	15	18	23	26	30	36
SSE1	10	12	15	16	18	20	21	23
SSE2	13	17	23	26	32	36	40	47
SSE3	11	13	15	16	18	19	20	22

TABLE C.3 **Annual Passive Heating Performance: SSF** (*continued*)

DES MOINES, IOWA SSF (%)								6718 DD65
Type LCR =	100	70	50	40	30	25	20	15
WWA3	11	14	19	22	27	30	34	41
WWB4	13	21	29	35	43	49	57	66
WWC2	14	20	28	34	42	47	54	64
TWA1	13	15	17	18	21	22	24	27
TWA2	12	14	17	20	23	26	29	34
TWA3	11	14	17	20	24	27	31	37
TWA4	9	13	17	20	25	28	32	39
TWB3	10	13	16	19	23	26	30	36
TWD4	13	19	26	31	39	44	51	60
TWE2	14	20	26	31	39	44	50	59
TWF3	8	11	14	17	21	24	28	33
TWJ2	11	17	24	29	36	41	48	57
DGA1	6	8	10	11	12	13	14	14
DGA2	9	12	16	18	23	25	29	34
DGA3	13	17	22	27	33	38	44	52
DGB1	6	8	10	11	13	14	15	17
DGB2	9	13	16	19	23	27	31	36
DGB3	13	18	23	28	34	39	46	56
DGC1	8	11	14	15	18	20	22	26
DGC2	11	15	20	23	29	32	38	45
DGC3	15	20	26	31	39	45	52	62
SSA1	16	19	23	25	29	32	36	41
SSB1	13	16	19	21	25	27	31	35
SSB2	16	22	27	32	38	43	48	56
SSB3	12	14	17	19	22	24	27	30
SSC1	9	12	16	18	22	25	28	33
SSC2	12	16	22	26	33	37	43	51
SSE1	13	16	20	23	27	30	34	39
SSE2	16	22	29	34	41	46	52	61
SSE3	13	16	19	21	24	26	29	33

TOPEKA, KANSAS SSF (%)								5243 DD65
Type LCR =	100	70	50	40	30	25	20	15
WWA3	16	21	27	32	38	43	49	57
WWB4	20	28	38	45	55	61	69	78
WWC2	20	28	37	44	53	60	67	76
TWA1	16	19	22	24	28	31	34	39
TWA2	15	19	24	27	33	36	41	48
TWA3	15	19	24	29	35	39	44	52
TWA4	14	18	24	29	35	40	45	54
TWB3	13	18	23	27	32	36	42	50
TWD4	18	26	34	41	49	55	63	72
TWE2	19	27	35	41	50	55	63	72
TWF3	12	16	21	25	31	35	40	48
TWJ2	17	24	32	38	47	53	60	70
DGA1	10	13	16	19	22	24	26	28
DGA2	13	17	22	26	32	35	40	46
DGA3	16	22	29	34	42	48	55	64
DGB1	10	13	17	19	23	26	29	34
DGB2	13	17	23	27	33	38	43	51
DGB3	17	23	30	36	44	50	58	69
DGC1	13	17	21	25	30	34	38	44
DGC2	15	21	27	32	39	44	51	60
DGC3	19	26	34	40	50	57	65	76
SSA1	20	24	30	33	39	43	48	55
SSB1	16	20	25	29	34	38	42	49
SSB2	21	28	35	40	48	53	60	68
SSB3	15	19	23	26	31	34	38	44
SSC1	14	18	23	27	33	37	43	50
SSC2	17	23	30	36	44	49	56	65
SSE1	18	23	29	33	39	43	49	56
SSE2	23	30	39	45	53	59	66	74
SSE3	17	21	25	29	33	37	41	48

DODGE CITY, KANSAS SSF (%)								5053 DD65
Type LCR =	100	70	50	40	30	25	20	15
WWA3	21	27	35	41	50	55	63	72
WWB4	26	36	47	55	66	72	80	88
WWC2	26	36	46	54	64	71	78	86
TWA1	19	22	27	30	35	39	44	51
TWA2	19	24	30	35	42	47	53	61
TWA3	19	25	32	37	44	50	57	66
TWA4	18	24	31	37	45	51	58	67
TWB3	17	23	29	34	41	46	53	62
TWD4	24	33	43	50	60	66	74	82
TWE2	25	33	43	50	60	66	74	83
TWF3	16	21	28	33	40	46	52	62
TWJ2	22	30	40	47	57	63	71	80
DGA1	14	18	23	27	31	34	38	42
DGA2	16	22	28	33	41	45	51	58
DGA3	20	27	36	42	52	58	65	74
DGB1	14	18	24	28	35	39	44	50
DGB2	17	22	29	35	44	49	57	65
DGB3	20	28	37	44	55	62	70	79
DGC1	17	22	29	34	42	48	55	62
DGC2	19	26	34	41	51	57	65	74
DGC3	23	31	41	50	61	69	77	86
SSA1	24	30	37	42	49	54	60	68
SSB1	20	25	31	36	43	47	53	62
SSB2	26	33	42	48	57	63	70	78
SSB3	19	24	29	33	40	44	50	57
SSC1	18	24	31	36	44	49	56	65
SSC2	22	30	39	45	55	61	68	77
SSE1	23	30	37	43	51	56	63	71
SSE2	29	38	48	55	64	70	77	85
SSE3	21	26	32	36	42	47	53	61

LEXINGTON, KENTUCKY SSF (%)								4732 DD65
Type LCR =	100	70	50	40	30	25	20	15
WWA3	12	16	21	25	30	34	39	45
WWB4	15	23	31	38	47	53	60	70
WWC2	15	22	30	37	45	51	58	67
TWA1	14	16	18	20	23	25	27	31
TWA2	13	16	19	22	26	29	33	38
TWA3	12	15	19	23	27	31	35	41
TWA4	11	14	19	23	28	31	36	43
TWB3	11	14	18	21	26	29	34	40
TWD4	15	21	28	34	42	47	54	64
TWE2	15	22	29	34	42	47	54	63
TWF3	9	12	16	19	24	27	32	38
TWJ2	13	19	26	32	39	45	51	61
DGA1	7	9	12	13	15	16	17	18
DGA2	10	14	17	21	25	28	32	37
DGA3	14	19	24	29	36	41	47	55
DGB1	7	10	12	14	16	17	19	21
DGB2	10	14	18	21	26	30	34	40
DGB3	14	19	25	30	37	43	50	59
DGC1	10	13	16	18	21	24	27	31
DGC2	12	17	22	26	31	36	41	49
DGC3	16	22	29	34	42	48	56	66
SSA1	17	21	25	29	34	37	41	47
SSB1	14	18	21	24	29	32	35	41
SSB2	18	24	30	35	42	47	53	61
SSB3	13	16	20	22	26	28	32	36
SSC1	10	14	18	21	25	28	32	38
SSC2	13	18	25	29	36	41	47	55
SSE1	15	19	24	27	32	35	40	46
SSE2	19	26	33	38	46	51	58	66
SSE3	15	18	21	24	28	31	34	39

APPENDICES

TABLE C.3 Annual Passive Heating Performance: SSF (*continued*)

BATON ROUGE, LOUISIANA						**SSF (%)**		1670 DD65
Type	LCR = 200	100	70	50	40	30	25	20
WWA3	21	35	45	56	63	72	78	84
WWB4	23	43	56	68	75	84	89	93
WWC2	23	43	55	67	74	83	88	92
TWA1	19	28	34	41	46	54	59	65
TWA2	19	31	39	47	54	63	68	75
TWA3	19	32	40	50	57	66	72	79
TWA4	18	31	40	50	57	67	73	79
TWB3	17	29	37	46	53	62	68	75
TWD4	22	39	51	62	70	79	84	90
TWE2	23	40	52	63	70	80	85	90
TWF3	16	28	37	46	53	62	68	75
TWJ2	20	37	48	60	68	77	82	88
DGA1	15	27	34	42	47	54	58	63
DGA2	16	28	37	46	53	61	67	73
DGA3	18	33	43	55	63	72	77	83
DGB1	15	27	36	46	53	62	67	72
DGB2	16	29	39	50	58	68	74	80
DGB3	19	33	45	57	66	76	82	87
DGC1	18	32	42	53	61	71	76	81
DGC2	19	33	44	56	65	75	80	86
DGC3	21	38	50	64	73	83	88	92
SSA1	24	38	47	56	63	71	76	82
SSB1	21	33	41	50	56	65	70	77
SSB2	25	41	51	62	69	77	83	88
SSB3	20	31	39	48	54	63	68	75
SSC1	19	32	41	51	58	67	73	79
SSC2	21	37	48	59	66	76	81	87
SSE1	24	40	49	60	67	75	80	86
SSE2	28	47	59	70	77	85	89	93
SSE3	22	33	41	50	56	65	70	77

SHREVEPORT, LOUISIANA						**SSF (%)**		2175 DD65
Type	LCR = 200	100	70	50	40	30	25	20
WWA3	17	30	39	49	56	65	71	78
WWB4	19	38	49	61	69	79	84	90
WWC2	19	37	49	60	68	78	83	89
TWA1	17	25	30	36	41	48	52	58
TWA2	17	27	34	42	48	56	61	68
TWA3	16	27	35	44	50	59	65	72
TWA4	15	27	35	44	51	60	66	73
TWB3	14	25	32	40	47	55	61	68
TWD4	18	34	45	56	63	73	79	85
TWE2	19	35	46	57	64	74	79	86
TWF3	13	24	31	40	46	55	61	68
TWJ2	16	32	42	53	61	71	77	84
DGA1	13	22	29	36	41	47	51	55
DGA2	13	24	32	40	47	55	60	66
DGA3	16	28	38	48	56	66	72	78
DGB1	13	22	30	38	45	53	58	64
DGB2	14	25	33	43	50	60	66	73
DGB3	16	29	39	50	59	70	76	83
DGC1	15	26	35	45	52	62	68	74
DGC2	16	28	38	49	57	68	74	80
DGC3	18	32	44	56	66	77	83	88
SSA1	21	33	41	49	56	64	70	76
SSB1	18	28	35	43	49	58	63	70
SSB2	21	36	45	55	62	71	77	83
SSB3	17	27	34	41	47	55	60	67
SSC1	15	27	35	44	51	60	66	73
SSC2	17	32	42	52	60	69	75	82
SSE1	21	34	43	52	59	68	73	80
SSE2	24	41	52	63	70	79	84	89
SSE3	19	29	36	44	50	58	63	70

NEW ORLEANS, LOUISIANA						**SSF (%)**		1465 DD65
Type	LCR = 200	100	70	50	40	30	25	20
WWA3	24	41	52	63	70	79	84	89
WWB4	28	50	63	74	81	89	93	96
WWC2	28	49	62	73	80	88	92	95
TWA1	21	31	38	46	52	60	65	72
TWA2	22	35	44	54	60	70	75	81
TWA3	22	36	46	57	64	73	79	85
TWA4	21	36	46	57	64	74	79	85
TWB3	20	33	42	52	60	69	75	81
TWD4	25	45	57	69	76	85	89	93
TWE2	27	46	58	69	77	85	89	94
TWF3	19	33	42	53	60	69	75	82
TWJ2	24	43	55	66	74	83	88	92
DGA1	18	32	40	49	55	62	66	71
DGA2	19	33	43	53	60	68	73	79
DGA3	21	38	50	62	69	78	83	87
DGB1	18	33	43	54	61	70	75	80
DGB2	19	34	45	58	66	75	80	86
DGB3	22	39	51	65	73	83	87	91
DGC1	21	38	50	62	70	79	83	87
DGC2	22	39	51	65	73	82	87	91
DGC3	24	44	57	72	80	88	92	95
SSA1	28	43	52	62	69	78	82	88
SSB1	24	37	46	56	63	72	77	83
SSB2	28	46	57	68	75	83	87	92
SSB3	23	36	44	54	61	69	75	81
SSC1	22	37	47	58	65	74	79	85
SSC2	25	43	54	66	73	82	87	91
SSE1	28	45	55	67	74	82	86	91
SSE2	33	53	66	76	83	90	93	96
SSE3	25	38	47	56	63	72	77	83

PORTLAND, MAINE						**SSF (%)**		7499 DD65
Type	LCR = 100	70	50	40	30	25	20	15
WWA3	7	10	13	15	18	21	24	28
WWB4	9	15	22	28	35	41	47	57
WWC2	9	15	21	26	33	38	45	53
TWA1	12	13	14	15	16	17	18	19
TWA2	9	11	13	15	17	19	21	24
TWA3	8	10	13	15	18	20	22	26
TWA4	6	9	12	14	18	20	23	28
TWB3	7	10	12	14	17	19	22	26
TWD4	10	15	20	25	31	36	42	50
TWE2	10	15	21	25	31	35	41	49
TWF3	5	7	10	12	15	17	19	23
TWJ2	8	13	18	22	29	33	39	47
DGA1	4	5	5	6	6	6	6	4
DGA2	7	9	12	14	17	19	21	25
DGA3	10	14	18	22	27	31	36	43
DGB1	4	5	5	6	6	7	6	5
DGB2	7	9	12	14	17	20	22	26
DGB3	11	14	19	23	28	33	38	46
DGC1	5	7	8	10	11	12	13	13
DGC2	9	12	15	18	22	25	29	34
DGC3	12	16	22	26	32	37	43	53
SSA1	13	16	19	21	24	26	29	32
SSB1	11	13	15	17	20	22	24	27
SSB2	14	18	23	27	33	37	42	49
SSB3	10	11	13	15	17	18	20	22
SSC1	6	8	10	12	14	16	18	21
SSC2	8	12	16	20	25	28	33	40
SSE1	10	13	15	17	20	22	24	27
SSE2	13	18	24	28	34	38	44	51
SSE3	11	13	15	17	19	20	22	25

TABLE C.3 **Annual Passive Heating Performance: SSF** (*continued*)

BALTIMORE, MARYLAND SSF (%) 4731 DD65

Type	LCR = 100	70	50	40	30	25	20	15
WWA3	13	18	24	28	35	39	45	53
WWB4	17	26	35	42	52	58	66	75
WWC2	17	25	34	41	50	56	64	73
TWA1	15	17	20	22	25	28	31	36
TWA2	14	17	21	25	30	33	38	44
TWA3	13	17	22	26	31	35	41	48
TWA4	12	16	22	26	32	36	42	50
TWB3	12	16	20	24	29	33	38	46
TWD4	16	23	31	38	46	52	59	69
TWE2	17	24	32	38	46	52	59	69
TWF3	10	14	19	22	28	32	37	44
TWJ2	14	21	29	35	44	49	57	66
DGA1	8	11	14	16	19	20	22	24
DGA2	11	15	19	23	28	32	37	42
DGA3	15	20	26	32	39	45	52	60
DGB1	8	11	14	17	20	22	25	29
DGB2	11	15	20	24	30	34	40	47
DGB3	15	21	27	33	41	47	55	65
DGC1	11	14	18	22	26	29	34	39
DGC2	13	18	24	29	36	41	47	56
DGC3	17	24	31	37	47	53	62	72
SSA1	18	22	27	31	37	41	46	52
SSB1	15	19	23	27	31	35	40	46
SSB2	19	26	33	38	45	51	57	66
SSB3	14	17	21	24	29	32	36	41
SSC1	12	16	20	24	30	33	38	46
SSC2	15	21	28	33	41	46	52	61
SSE1	16	21	26	31	36	40	46	53
SSE2	21	28	36	42	51	56	63	72
SSE3	16	19	23	26	31	34	38	44

DETROIT, MICHIGAN SSF (%) 6234 DD65

Type	LCR = 100	70	50	40	30	25	20	15
WWA3	7	10	13	15	18	20	23	26
WWB4	9	15	22	27	35	40	46	55
WWC2	9	15	21	26	33	38	44	52
TWA1	12	13	14	15	16	16	17	18
TWA2	9	11	13	15	17	19	21	23
TWA3	8	10	13	15	17	19	22	25
TWA4	7	9	12	14	17	20	23	27
TWB3	7	10	12	14	17	19	22	26
TWD4	10	15	20	25	31	35	41	49
TWE2	10	15	20	25	31	35	40	48
TWF3	5	7	10	12	14	16	19	22
TWJ2	8	13	18	22	28	33	38	46
DGA1	4	5	6	6	6	6	5	4
DGA2	7	9	12	14	17	19	21	24
DGA3	10	14	18	22	27	31	36	43
DGB1	4	5	6	6	7	7	6	5
DGB2	7	9	12	14	17	20	22	26
DGB3	11	14	19	23	28	32	38	46
DGC1	5	7	9	10	11	12	12	13
DGC2	9	12	15	18	22	25	28	33
DGC3	12	17	22	26	32	37	43	52
SSA1	13	16	19	21	24	26	28	32
SSB1	11	13	15	17	20	21	24	26
SSB2	14	18	23	27	32	36	41	48
SSB3	10	11	13	15	17	18	20	21
SSC1	6	8	10	12	14	16	18	20
SSC2	8	12	16	20	25	28	32	39
SSE1	10	13	15	17	20	21	23	25
SSE2	13	18	24	28	33	37	43	49
SSE3	11	13	15	17	19	20	22	24

BOSTON, MASSACHUSETTS SSF (%) 5622 DD65

Type	LCR = 100	70	50	40	30	25	20	15
WWA3	10	14	18	21	26	29	33	40
WWB4	13	20	28	34	43	49	56	66
WWC2	13	20	27	33	41	47	54	63
TWA1	13	15	16	18	20	21	23	26
TWA2	11	14	17	19	23	25	28	33
TWA3	10	13	17	20	24	27	31	36
TWA4	9	12	16	19	24	27	32	38
TWB3	9	12	16	18	23	26	29	35
TWD4	13	18	25	31	38	43	50	59
TWE2	13	19	26	31	38	43	50	58
TWF3	7	10	14	16	20	23	27	33
TWJ2	11	17	23	28	35	41	47	56
DGA1	6	7	9	10	12	12	13	13
DGA2	9	12	15	18	22	25	28	33
DGA3	12	17	22	26	32	37	43	51
DGB1	6	7	9	11	12	13	14	16
DGB2	9	12	16	18	23	26	30	36
DGB3	13	17	23	27	34	39	46	55
DGC1	8	10	13	15	17	19	22	25
DGC2	11	15	19	23	28	32	37	44
DGC3	14	20	26	31	38	44	51	62
SSA1	15	19	22	25	29	32	36	41
SSB1	13	15	19	21	25	27	31	36
SSB2	16	21	27	32	38	43	48	56
SSB3	11	14	17	19	22	24	27	31
SSC1	9	12	15	17	21	24	27	32
SSC2	11	16	21	26	32	36	42	50
SSE1	13	16	20	23	27	30	34	39
SSE2	16	22	29	34	41	46	52	61
SSE3	13	16	19	21	24	26	29	33

SAULT STE. MARIE, MICH. SSF (%) 9201 DD65

Type	LCR = 100	70	50	40	30	25	20	15
WWA3	5	7	8	10	11	12	13	13
WWB4	6	11	17	21	27	31	37	44
WWC2	7	11	16	20	25	29	34	40
TWA1	10	11	11	12	12	12	12	11
TWA2	8	9	10	11	12	13	13	14
TWA3	6	7	9	10	12	13	14	15
TWA4	4	6	8	10	12	13	14	16
TWB3	6	7	9	10	12	13	15	16
TWD4	7	11	16	19	25	28	33	39
TWE2	8	12	16	19	24	27	31	37
TWF3	3	5	6	8	9	10	11	12
TWJ2	6	9	14	17	22	25	30	36
DGA1	2	3	3	2	2	1	0	-4
DGA2	5	7	9	11	13	14	15	17
DGA3	9	12	15	18	23	26	30	35
DGB1	2	2	3	2	2	1	0	-2
DGB2	5	7	9	11	13	14	16	18
DGB3	9	12	16	19	24	27	32	38
DGC1	4	5	5	6	6	6	5	3
DGC2	7	9	12	14	17	19	22	25
DGC3	10	14	19	22	28	31	36	43
SSA1	12	13	15	17	19	20	21	22
SSB1	9	11	12	14	15	16	17	18
SSB2	12	16	20	23	27	30	34	39
SSB3	8	9	11	11	12	13	13	14
SSC1	4	5	7	7	8	8	8	7
SSC2	6	9	12	14	18	20	23	27
SSE1	8	9	11	12	13	13	13	11
SSE2	10	14	18	22	26	29	32	36
SSE3	10	11	12	13	14	15	15	14

APPENDICES

TABLE C.3 Annual Passive Heating Performance: SSF (*continued*)

MINNEAPOLIS, MINNESOTA SSF (%) 8165 DD65								
Type LCR =	100	70	50	40	30	25	20	15
WWA3	7	9	12	14	17	19	21	24
WWB4	9	14	21	26	33	38	45	54
WWC2	9	14	20	25	32	36	42	50
TWA1	11	12	13	14	15	16	16	17
TWA2	9	11	13	14	16	18	19	21
TWA3	7	10	12	14	16	18	20	23
TWA4	6	8	11	13	16	19	21	25
TWB3	7	9	12	13	16	18	21	24
TWD4	9	14	19	24	30	34	40	48
TWE2	10	14	20	24	29	34	39	46
TWF3	5	7	9	11	13	15	17	20
TWJ2	7	12	17	21	27	31	37	44
DGA1	4	4	5	5	5	5	4	2
DGA2	7	9	11	13	16	18	20	23
DGA3	10	13	18	21	26	30	35	41
DGB1	3	4	5	5	6	5	5	3
DGB2	7	9	12	14	17	19	21	25
DGB3	10	14	19	22	28	31	37	44
DGC1	5	6	8	9	10	11	11	11
DGC2	8	11	15	17	21	24	27	32
DGC3	12	16	21	25	31	36	42	50
SSA1	12	15	17	19	21	23	25	28
SSB1	10	12	14	16	18	19	21	23
SSB2	13	17	22	25	30	34	38	45
SSB3	9	10	12	13	15	16	17	18
SSC1	6	8	10	11	13	14	16	18
SSC2	8	11	15	19	23	27	31	37
SSE1	9	11	13	15	17	18	20	21
SSE2	12	16	21	25	31	34	39	46
SSE3	10	12	14	15	17	18	19	20

SAINT LOUIS, MISSOURI SSF (%) 4754 DD65								
Type LCR =	100	70	50	40	30	25	20	15
WWA3	15	20	26	30	36	41	47	55
WWB4	19	27	37	44	53	59	67	76
WWC2	19	27	36	42	51	57	65	74
TWA1	16	18	21	23	27	29	32	37
TWA2	15	19	23	26	31	35	39	46
TWA3	14	18	23	27	33	37	42	50
TWA4	13	18	23	27	33	38	43	51
TWB3	13	17	22	25	31	35	40	47
TWD4	18	25	33	39	48	53	61	70
TWE2	18	26	33	39	48	53	61	70
TWF3	11	15	20	24	29	33	38	46
TWJ2	16	23	31	37	45	51	58	67
DGA1	9	12	15	18	20	22	24	26
DGA2	12	16	21	25	30	34	38	44
DGA3	16	21	28	33	41	46	53	62
DGB1	10	13	16	18	22	24	27	31
DGB2	12	17	22	26	32	36	41	49
DGB3	16	22	29	34	43	48	56	66
DGC1	12	16	20	23	28	31	35	41
DGC2	15	20	26	30	37	42	49	57
DGC3	19	25	33	39	48	54	63	73
SSA1	19	24	29	32	38	42	47	53
SSB1	16	20	24	28	33	36	41	47
SSB2	21	27	34	39	46	52	58	67
SSB3	15	18	22	25	30	33	37	43
SSC1	13	17	22	26	31	35	40	47
SSC2	16	22	29	34	42	47	54	63
SSE1	18	22	28	32	38	42	47	54
SSE2	22	29	37	43	52	57	64	73
SSE3	17	20	24	28	32	35	40	46

MERIDIAN, MISSISSIPPI SSF (%) 2393 DD65								
Type LCR =	200	100	70	50	40	30	25	20
WWA3	15	28	36	45	52	61	67	74
WWB4	16	34	46	58	66	75	81	87
WWC2	17	34	45	56	64	74	80	86
TWA1	16	23	28	33	38	44	49	54
TWA2	15	24	31	38	44	52	57	64
TWA3	14	25	32	41	47	55	61	68
TWA4	13	24	32	41	47	56	62	69
TWB3	13	22	29	37	43	52	57	64
TWD4	16	31	41	52	60	70	76	82
TWE2	17	32	42	53	60	70	76	83
TWF3	11	21	29	37	43	51	57	64
TWJ2	14	29	39	50	57	67	73	80
DGA1	11	20	26	32	37	43	46	50
DGA2	12	22	29	37	43	51	56	62
DGA3	14	26	35	45	52	62	68	75
DGB1	11	20	27	34	40	48	53	59
DGB2	12	22	30	39	46	56	62	69
DGB3	14	26	36	46	55	66	72	80
DGC1	13	24	31	41	47	57	63	69
DGC2	14	26	34	45	52	63	69	76
DGC3	16	30	40	52	61	73	79	86
SSA1	20	31	38	46	52	61	66	72
SSB1	16	26	33	41	46	54	59	66
SSB2	20	33	42	52	59	68	74	80
SSB3	16	25	31	39	44	52	57	63
SSC1	14	24	32	41	47	55	61	68
SSC2	15	29	38	49	56	65	71	78
SSE1	19	31	40	49	55	64	70	76
SSE2	22	38	49	60	67	76	81	87
SSE3	17	27	33	41	46	54	59	66

SPRINGFIELD, MISSOURI SSF (%) 4571 DD65								
Type LCR =	100	70	50	40	30	25	20	15
WWA3	16	22	28	33	40	45	52	60
WWB4	21	30	40	47	57	63	71	80
WWC2	21	29	39	46	55	62	69	78
TWA1	17	19	23	25	29	32	36	41
TWA2	16	20	25	29	34	38	43	51
TWA3	15	20	26	30	36	41	46	55
TWA4	14	19	25	30	37	41	48	56
TWB3	14	18	24	28	34	38	44	52
TWD4	19	27	36	42	51	57	65	74
TWE2	20	28	36	43	51	57	65	74
TWF3	12	17	22	26	33	37	43	51
TWJ2	18	25	33	40	49	55	62	72
DGA1	11	14	18	20	24	26	28	30
DGA2	13	18	23	27	33	37	42	48
DGA3	17	23	30	36	44	50	57	66
DGB1	11	14	18	21	25	28	32	37
DGB2	14	18	24	28	35	39	46	54
DGB3	17	24	31	37	46	52	60	71
DGC1	13	18	22	26	32	36	41	47
DGC2	10	21	28	33	41	46	54	63
DGC3	20	27	35	42	52	59	67	78
SSA1	21	26	31	36	42	46	51	59
SSB1	17	22	27	30	36	40	45	52
SSB2	22	29	37	42	50	55	62	71
SSB3	16	20	25	28	33	37	41	48
SSC1	14	19	25	29	35	39	45	53
SSC2	18	24	32	38	46	51	58	68
SSE1	19	25	31	36	42	47	52	60
SSE2	24	32	41	47	56	62	69	77
SSE3	18	22	27	30	36	39	44	51

TABLE C.3 Annual Passive Heating Performance: SSF (*continued*)

HELENA, MONTANA SSF (%) — 8197 DD65

Type	LCR = 100	70	50	40	30	25	20	15
WWA3	11	14	19	22	27	30	34	39
WWB4	14	21	29	35	43	49	55	64
WWC2	14	20	28	34	41	47	53	62
TWA1	13	15	17	19	21	22	24	27
TWA2	12	14	18	20	24	26	29	33
TWA3	11	14	18	20	25	28	31	36
TWA4	9	13	17	20	25	28	32	38
TWB3	10	13	16	19	23	26	30	35
TWD4	13	19	26	31	39	44	50	58
TWE2	14	20	26	31	39	43	49	57
TWF3	8	11	14	17	21	24	28	33
TWJ2	12	17	24	29	36	41	47	55
DGA1	6	8	10	11	13	13	14	14
DGA2	9	12	16	19	23	26	29	33
DGA3	13	17	23	27	34	38	44	51
DGB1	6	8	10	12	14	15	16	17
DGB2	9	13	17	20	24	27	31	37
DGB3	13	18	24	28	35	40	46	55
DGC1	8	11	14	16	19	21	23	26
DGC2	11	15	20	24	30	33	38	44
DGC3	15	20	27	32	40	45	52	61
SSA1	16	19	23	25	29	32	35	39
SSB1	13	16	19	21	25	27	30	34
SSB2	17	22	27	32	38	42	47	54
SSB3	12	14	17	19	22	24	26	29
SSC1	9	12	16	18	22	25	28	32
SSC2	12	17	22	27	33	37	42	49
SSE1	13	17	20	23	27	29	32	35
SSE2	17	23	29	34	40	45	50	57
SSE3	13	16	19	21	24	26	29	32

ELY, NEVADA SSF (%) — 7818 DD65

Type	LCR = 100	70	50	40	30	25	20	15
WWA3	18	24	32	37	46	51	58	67
WWB4	23	33	44	52	62	69	76	85
WWC2	23	32	43	50	60	67	75	83
TWA1	17	20	24	29	32	35	40	47
TWA2	17	21	27	31	38	43	49	57
TWA3	16	22	28	33	41	46	52	61
TWA4	15	21	28	33	41	46	53	63
TWB3	15	20	26	31	38	43	49	58
TWD4	21	29	39	46	56	62	70	79
TWE2	22	30	40	47	56	63	70	79
TWF3	13	18	25	29	37	42	48	57
TWJ2	19	27	37	44	53	60	68	77
DGA1	11	15	20	23	27	30	33	37
DGA2	14	19	25	30	37	41	47	54
DGA3	18	24	32	39	48	55	62	71
DGB1	11	15	20	24	30	34	39	45
DGB2	14	19	26	31	39	45	52	61
DGB3	18	25	33	40	51	58	66	76
DGC1	14	19	25	30	38	43	49	57
DGC2	17	23	31	37	46	53	60	70
DGC3	21	29	38	46	57	65	74	83
SSA1	23	29	36	40	48	52	58	66
SSB1	19	24	30	34	41	46	51	59
SSB2	25	32	41	47	55	61	68	76
SSB3	18	22	28	32	38	42	48	55
SSC1	16	21	27	32	40	45	51	60
SSC2	19	27	35	42	51	57	64	74
SSE1	22	28	36	41	49	54	60	68
SSE2	27	37	46	53	62	68	75	83
SSE3	20	24	30	34	41	45	51	58

OMAHA, NEBRASKA SSF (%) — 6606 DD65

Type	LCR = 100	70	50	40	30	25	20	15
WWA3	12	16	21	24	30	33	38	45
WWB4	15	22	31	37	47	53	60	70
WWC2	15	22	30	36	45	51	58	68
TWA1	14	16	18	20	22	24	27	30
TWA2	13	15	19	22	26	29	32	38
TWA3	11	15	19	22	27	30	35	41
TWA4	10	14	19	22	27	31	36	43
TWB3	11	14	18	21	25	29	33	40
TWD4	14	21	28	34	42	47	54	64
TWE2	15	21	28	34	42	47	54	63
TWF3	9	12	16	19	24	27	31	37
TWJ2	13	19	26	31	39	44	51	61
DGA1	7	9	11	13	15	16	17	17
DGA2	10	13	17	20	25	28	32	37
DGA3	13	18	24	28	35	40	47	55
DGB1	7	9	11	13	15	17	19	21
DGB2	10	14	18	21	26	29	34	40
DGB3	14	19	25	29	37	42	49	59
DGC1	9	12	15	18	21	23	26	30
DGC2	12	16	21	25	31	35	41	49
DGC3	16	21	28	33	42	48	56	66
SSA1	16	20	24	27	32	35	39	45
SSB1	14	17	20	23	27	30	34	39
SSB2	18	23	29	34	40	45	51	59
SSB3	12	15	18	21	24	26	29	34
SSC1	10	13	17	20	25	28	32	38
SSC2	13	18	23	29	36	40	47	55
SSE1	14	18	22	25	30	33	37	40
SSE2	18	24	31	37	44	49	56	65
SSE3	14	17	20	23	26	29	32	37

LAS VEGAS, NEVADA SSF (%) — 2602 DD65

Type	LCR = 200	100	70	50	40	30	25	20
WWA3	24	41	52	63	71	80	85	90
WWB4	28	50	63	75	82	89	93	96
WWC2	28	49	62	74	81	88	92	96
TWA1	21	31	38	46	52	61	66	73
TWA2	22	35	44	54	61	70	76	82
TWA3	22	37	47	57	64	74	79	85
TWA4	21	36	47	58	64	74	80	86
TWB3	19	33	43	53	60	69	75	82
TWD4	25	45	57	69	77	85	89	93
TWE2	27	46	58	70	77	85	90	94
TWF3	18	33	43	53	60	70	76	82
TWJ2	24	43	55	67	75	83	88	92
DGA1	18	32	40	49	55	63	67	72
DGA2	18	33	43	53	60	69	74	80
DGA3	21	38	50	62	70	78	83	88
DGB1	18	33	44	55	62	71	75	80
DGB2	19	35	46	58	67	76	81	86
DGB3	21	39	52	65	74	83	87	92
DGC1	21	38	51	63	71	79	84	88
DGC2	21	39	52	65	74	83	87	91
DGC3	24	44	58	72	80	88	92	95
SSA1	26	40	50	59	66	74	79	85
SSB1	22	35	44	53	60	69	74	80
SSB2	27	44	54	65	72	81	85	90
SSB3	21	34	42	51	57	66	71	78
SSC1	22	37	47	58	65	75	80	86
SSC2	24	43	55	66	73	82	87	92
SSE1	27	43	53	64	71	79	84	89
SSE2	31	51	63	74	80	88	91	95
SSE3	23	36	45	54	60	69	74	80

APPENDICES

TABLE C.3 **Annual Passive Heating Performance: SSF** (*continued*)

RENO, NEVADA **SSF (%)**							6027 DD65		LOS ALAMOS, NEW MEXICO							6437 DD65	
Type	LCR = 100	70	50	40	30	25	20	15	Type	LCR = 100	70	50	40	30	25	20	15
WWA3	25	33	41	48	57	63	70	78	WWA3	20	27	35	41	49	55	62	72
WWB4	31	42	54	62	72	78	84	91	WWB4	25	36	47	55	66	72	80	88
WWC2	31	41	53	60	70	76	83	90	WWC2	25	35	46	54	64	71	78	86
TWA1	21	25	30	35	41	45	50	58	TWA1	18	22	26	29	34	38	43	50
TWA2	22	28	35	40	48	53	60	69	TWA2	18	23	29	34	41	46	52	61
TWA3	22	29	37	43	51	57	64	73	TWA3	18	24	31	36	44	49	56	66
TWA4	21	29	37	43	52	58	65	74	TWA4	17	23	31	36	45	50	57	67
TWB3	20	26	34	40	48	53	60	69	TWB3	16	22	28	33	41	46	53	62
TWD4	28	38	48	56	66	72	79	87	TWD4	23	32	42	49	59	66	74	83
TWE2	29	39	49	57	66	72	79	87	TWE2	24	33	43	50	60	66	74	83
TWF3	19	25	33	39	47	53	60	69	TWF3	15	20	27	32	40	45	52	62
TWJ2	26	36	46	54	64	70	77	85	TWJ2	21	30	40	47	57	63	71	80
DGA1	17	23	29	33	38	42	46	50	DGA1	13	17	22	26	31	34	37	41
DGA2	19	26	34	39	47	52	58	65	DGA2	15	21	28	33	40	45	50	58
DGA3	23	32	41	49	58	64	71	79	DGA3	19	26	35	42	52	58	65	74
DGB1	17	23	30	36	43	48	53	60	DGB1	13	18	23	27	34	38	44	50
DGB2	20	27	35	42	51	57	64	72	DGB2	16	21	28	34	43	49	56	65
DGB3	24	32	43	51	62	69	76	84	DGB3	20	27	36	43	54	62	70	80
DGC1	21	28	36	43	52	57	64	71	DGC1	16	22	28	34	42	48	55	63
DGC2	23	31	41	48	59	65	72	80	DGC2	18	25	33	40	50	57	65	74
DGC3	27	37	48	57	69	75	83	89	DGC3	22	31	41	49	61	69	77	86
SSA1	29	36	43	49	56	61	67	75	SSA1	25	31	38	44	51	56	63	71
SSB1	24	30	37	43	50	55	61	69	SSB1	21	26	32	37	44	49	56	64
SSB2	31	39	48	55	64	69	76	83	SSB2	27	34	43	50	59	65	72	80
SSB3	23	29	35	40	47	52	58	65	SSB3	19	24	30	35	42	46	52	60
SSC1	22	29	37	43	51	57	64	73	SSC1	17	23	30	36	43	49	56	65
SSC2	26	35	45	52	61	67	75	83	SSC2	21	29	38	45	54	61	68	78
SSE1	29	36	45	51	59	64	70	78	SSE1	24	31	39	45	53	59	65	74
SSE2	35	45	55	62	71	76	82	89	SSE2	30	39	50	57	67	73	79	87
SSE3	25	31	38	43	50	55	61	68	SSE3	21	27	33	37	44	49	55	63

ALBUQUERQUE, N. MEXICO **SSF (%)**							4293 DD65		ALBANY, NEW YORK **SSF (%)**							6891 DD65	
Type	LCR = 100	70	50	40	30	25	20	15	Type	LCR = 100	70	50	40	30	25	20	15
WWA3	28	37	46	54	63	69	76	85	WWA3	7	9	12	14	17	18	21	24
WWB4	35	47	59	67	77	83	89	94	WWB4	8	14	21	26	33	38	45	54
WWC2	34	46	58	66	76	82	88	94	WWC2	9	14	20	25	31	36	42	51
TWA1	22	27	33	38	45	50	56	64	TWA1	11	12	13	14	15	15	16	17
TWA2	24	31	39	45	53	59	66	75	TWA2	9	10	12	14	16	17	19	21
TWA3	25	32	41	48	57	63	70	79	TWA3	7	9	12	13	16	18	20	23
TWA4	24	32	41	48	58	64	71	80	TWA4	6	8	11	13	16	18	21	25
TWB3	22	29	38	44	53	59	66	75	TWB3	7	9	11	13	16	18	20	24
TWD4	31	42	53	61	71	77	84	91	TWD4	9	13	19	23	30	34	40	48
TWE2	32	43	54	62	72	78	84	91	TWE2	9	14	19	23	29	33	39	46
TWF3	21	29	37	44	53	59	66	76	TWF3	5	7	9	11	13	15	17	20
TWJ2	29	40	51	59	69	75	82	90	TWJ2	7	12	17	21	27	31	37	45
DGA1	19	26	33	37	44	48	52	58	DGA1	3	4	5	5	5	4	4	2
DGA2	21	29	38	44	52	57	64	71	DGA2	6	8	11	13	15	17	20	23
DGA3	25	35	46	54	64	70	76	84	DGA3	10	13	17	21	26	29	34	41
DGB1	20	27	35	41	50	55	61	68	DGB1	3	4	4	5	5	5	4	3
DGB2	22	30	40	47	58	64	71	79	DGB2	6	9	11	13	16	18	21	24
DGB3	26	36	47	56	68	75	82	89	DGB3	10	14	18	22	27	31	36	44
DGC1	23	32	42	49	60	66	72	79	DGC1	5	6	7	8	10	10	10	10
DGC2	25	34	46	54	65	72	79	86	DGC2	8	11	14	17	20	23	27	32
DGC3	30	40	53	63	75	82	88	93	DGC3	12	16	21	25	31	35	41	50
SSA1	30	38	46	52	61	66	73	81	SSA1	12	15	17	19	22	24	27	30
SSB1	25	32	40	46	54	60	67	75	SSB1	10	12	14	16	18	20	22	25
SSB2	32	42	52	59	68	74	81	88	SSB2	13	17	22	26	31	35	39	46
SSB3	24	30	38	43	51	57	63	72	SSB3	9	11	12	14	15	16	18	19
SSC1	24	32	41	48	57	63	71	79	SSC1	6	7	9	11	13	14	16	18
SSC2	29	39	50	57	67	73	80	88	SSC2	7	11	15	18	23	26	31	37
SSE1	31	39	49	55	64	70	77	85	SSE1	9	11	14	16	18	20	21	23
SSE2	38	49	60	67	76	82	88	93	SSE2	12	17	22	26	32	36	41	48
SSE3	26	33	40	46	54	60	66	75	SSE3	11	12	14	16	17	19	20	22

TABLE C.3 Annual Passive Heating Performance: SSF (*continued*)

BUFFALO, NEW YORK — 6931 DD65

Type	LCR = 100	70	50	40	30	25	20	15
WWA3	5	6	8	9	11	11	12	13
WWB4	6	11	16	21	27	31	37	44
WWC2	6	11	16	19	25	29	34	41
TWA1	10	11	11	12	12	12	11	11
TWA2	8	9	10	11	12	12	13	13
TWA3	6	7	9	10	11	12	13	14
TWA4	4	6	8	9	11	12	14	16
TWB3	6	7	9	10	12	13	14	16
TWD4	7	11	15	19	24	28	32	39
TWE2	8	11	15	19	23	27	31	37
TWF3	3	5	6	7	9	10	11	12
TWJ2	5	9	13	17	22	25	29	36
DGA1	2	2	2	2	1	0	−1	−4
DGA2	5	7	9	10	12	13	14	16
DGA3	9	12	15	18	22	25	29	34
DGB1	2	2	2	2	1	0	0	−4
DGB2	5	7	9	10	12	14	15	17
DGB3	9	12	16	19	23	27	31	37
DGC1	4	4	5	5	5	5	4	2
DGC2	7	9	12	14	16	18	21	24
DGC3	10	14	18	22	27	31	35	42
SSA1	11	13	15	17	19	20	21	22
SSB1	9	11	12	13	15	16	17	18
SSB2	12	15	19	23	27	30	34	40
SSB3	8	9	11	11	12	13	13	13
SSC1	4	5	6	7	8	8	8	7
SSC2	6	8	12	14	17	20	23	27
SSE1	8	9	11	12	13	13	14	13
SSE2	10	14	18	22	26	29	33	38
SSE3	10	11	12	13	14	15	15	15

SYRACUSE, NEW YORK SSF (%) — 6681 DD65

Type	LCR = 100	70	50	40	30	25	20	15
WWA3	5	7	9	10	12	13	14	15
WWB4	6	12	17	22	28	33	38	46
WWC2	7	11	17	21	26	30	36	43
TWA1	11	11	12	12	12	13	12	12
TWA2	8	9	10	11	13	13	14	15
TWA3	6	8	9	11	12	14	15	16
TWA4	5	7	9	10	12	14	15	18
TWB3	6	7	9	11	13	14	16	18
TWD4	7	11	16	20	25	29	34	41
TWE2	8	12	16	20	25	28	33	39
TWF3	4	5	7	8	10	11	12	13
TWJ2	6	10	14	18	23	26	31	38
DGA1	2	3	3	3	2	1	0	−3
DGA2	6	7	9	11	13	14	16	17
DGA3	9	12	16	18	23	26	30	36
DGB1	2	3	3	3	2	1	0	−2
DGB2	6	7	9	11	13	15	16	19
DGB3	9	12	16	19	24	28	32	38
DGC1	4	5	5	6	6	6	5	4
DGC2	7	9	12	14	17	19	22	25
DGC3	11	14	19	22	28	31	36	44
SSA1	12	14	16	17	19	20	22	24
SSB1	9	11	13	14	16	17	18	19
SSB2	12	16	20	23	28	31	35	41
SSB3	8	10	11	12	13	13	14	14
SSC1	4	6	7	8	9	9	9	9
SSC2	6	9	12	15	19	21	24	29
SSE1	8	10	11	13	14	14	15	15
SSE2	10	14	19	22	27	30	34	40
SSE3	10	11	12	13	15	15	16	16

NEW YORK, NEW YORK — 4851 DD65

Type	LCR = 100	70	50	40	30	25	20	15
WWA3	10	14	19	22	27	30	35	41
WWB4	13	21	29	35	44	50	57	67
WWC2	13	20	28	34	42	48	55	64
TWA1	13	15	17	18	21	22	24	27
TWA2	11	14	17	20	23	26	30	35
TWA3	10	14	17	20	25	28	32	38
TWA4	9	13	17	20	25	28	33	39
TWB3	9	13	16	19	23	26	30	36
TWD4	13	19	26	31	39	44	51	60
TWE2	14	20	26	32	39	44	51	60
TWF3	7	11	14	17	21	24	28	34
TWJ2	11	17	24	29	36	42	48	57
DGA1	6	8	10	11	12	13	14	14
DGA2	9	12	16	18	23	25	29	34
DGA3	12	17	22	27	33	38	44	52
DGB1	6	8	10	11	13	14	16	17
DGB2	9	12	16	19	23	27	31	37
DGB3	13	17	23	28	35	40	47	56
DGC1	8	11	13	15	18	20	23	26
DGC2	11	15	20	23	29	33	38	45
DGC3	15	20	26	31	39	45	53	63
SSA1	16	19	23	26	30	33	37	43
SSB1	13	16	19	22	26	28	32	37
SSB2	17	22	28	33	39	44	50	58
SSB3	12	14	17	20	23	25	28	32
SSC1	9	12	15	18	22	25	29	34
SSC2	11	16	22	27	33	37	43	52
SSE1	13	17	21	24	28	31	35	41
SSE2	17	23	30	35	43	48	54	62
SSE3	13	16	19	22	25	27	31	35

CHARLOTTE, N. CAROLINA SSF (%) — 3226 DD65

Type	LCR = 200	100	70	50	40	30	25	20
WWA3	13	23	31	39	45	54	59	66
WWB4	13	29	40	51	59	69	75	82
WWC2	13	29	39	50	58	67	74	81
TWA1	15	20	24	29	33	38	42	47
TWA2	13	21	27	33	38	45	50	57
TWA3	12	21	27	35	40	48	54	60
TWA4	11	20	27	35	41	49	54	61
TWB3	11	19	25	32	37	45	50	57
TWD4	13	26	36	46	53	63	69	76
TWE2	14	27	37	47	54	63	69	77
TWF3	9	18	24	31	36	44	50	56
TWJ2	11	24	34	44	51	61	67	74
DGA1	9	16	21	26	30	35	39	42
DGA2	10	18	24	31	37	44	49	55
DGA3	12	22	30	39	46	55	61	68
DGB1	9	16	21	28	32	39	44	49
DGB2	10	19	25	33	39	48	54	61
DGB3	12	23	31	40	48	58	65	73
DGC1	11	20	26	33	39	47	53	59
DGC2	12	22	29	38	45	55	61	69
DGC3	14	26	35	45	53	65	72	80
SSA1	17	26	33	40	45	53	58	64
SSB1	14	22	28	35	40	47	52	58
SSB2	17	29	37	46	52	61	67	74
SSB3	14	21	26	33	37	44	49	55
SSC1	11	20	27	34	40	48	53	60
SSC2	12	25	33	42	49	58	64	72
SSE1	16	26	34	42	47	55	61	68
SSE2	18	32	42	52	59	69	74	81
SSE3	15	23	29	35	40	47	51	57

APPENDICES

TABLE C.3 **Annual Passive Heating Performance: SSF** (*continued*)

RALEIGH-DURHAM, N. CAROL. SSF (%)　3520 DD65

Type	LCR = 200	100	70	50	40	30	25	20
WWA3	11	21	28	36	41	50	55	62
WWB4	11	26	37	48	55	66	72	79
WWC2	12	26	36	46	54	64	70	77
TWA1	14	19	22	27	30	35	39	44
TWA2	12	19	24	30	35	42	47	53
TWA3	11	19	25	32	37	45	50	56
TWA4	9	18	24	32	37	45	51	57
TWB3	10	17	23	29	34	41	46	53
TWD4	11	24	33	43	50	60	66	73
TWE2	12	25	34	43	50	60	66	73
TWF3	8	16	21	28	33	41	46	52
TWJ2	10	22	31	40	47	57	63	71
DGA1	8	14	19	23	27	32	35	38
DGA2	9	16	22	29	34	41	45	51
DGA3	11	20	27	36	43	52	58	65
DGB1	8	14	19	24	29	35	39	44
DGB2	9	17	23	30	35	44	49	56
DGB3	11	21	28	37	44	55	61	70
DGC1	9	17	23	30	35	42	48	54
DGC2	10	20	27	35	41	50	57	64
DGC3	13	24	32	42	50	61	68	77
SSA1	16	24	30	37	42	50	54	61
SSB1	13	21	26	32	37	44	48	54
SSB2	15	26	34	43	49	58	63	70
SSB3	12	19	24	30	34	41	45	51
SSC1	10	18	24	31	36	44	49	56
SSC2	11	22	30	39	46	55	61	68
SSE1	14	24	31	38	44	52	57	63
SSE2	16	30	39	49	56	65	71	77
SSE3	14	21	26	32	37	43	48	54

FARGO, NORTH DAKOTA SSF (%)　9278 DD65

Type	LCR = 100	70	50	40	30	25	20	15
WWA3	6	8	11	12	15	16	18	20
WWB4	8	13	20	24	31	36	42	50
WWC2	8	13	19	23	29	33	39	46
TWA1	11	12	13	13	14	14	15	15
TWA2	9	10	12	13	15	16	17	18
TWA3	7	9	11	13	15	16	18	20
TWA4	6	8	10	12	15	16	19	21
TWB3	7	8	11	12	15	16	18	21
TWD4	8	13	18	22	28	32	37	44
TWE2	9	13	18	22	27	31	36	43
TWF3	4	6	8	10	12	13	15	17
TWJ2	7	11	16	20	25	29	34	41
DGA1	3	4	4	4	4	3	2	0
DGA2	6	8	11	12	15	16	18	20
DGA3	10	13	17	20	25	28	33	39
DGB1	3	4	4	4	4	4	3	1
DGB2	6	8	11	13	15	17	19	22
DGB3	10	13	18	21	26	30	35	41
DGC1	5	6	7	8	9	9	9	8
DGC2	8	11	14	16	20	22	25	29
DGC3	11	15	20	24	30	34	39	47
SSA1	12	14	16	17	19	21	22	24
SSB1	9	11	13	14	16	17	18	20
SSB2	12	16	20	23	28	31	36	41
SSB3	8	10	11	12	13	14	14	14
SSC1	5	7	8	10	11	12	13	14
SSC2	7	10	14	17	21	24	28	33
SSE1	8	10	12	13	14	15	16	15
SSE2	11	15	20	23	28	31	35	41
SSE3	10	11	13	14	15	16	16	17

BISMARCK, NORTH DAKOTA SSF (%)　9057 DD65

Type	LCR = 100	70	50	40	30	25	20	15
WWA3	8	11	15	17	21	23	26	30
WWB4	11	17	24	30	37	43	49	58
WWC2	11	17	23	28	36	40	47	55
TWA1	12	13	15	16	17	18	19	21
TWA2	10	12	15	16	19	21	23	26
TWA3	9	11	14	16	20	22	25	29
TWA4	7	10	14	16	20	22	26	30
TWB3	8	10	13	16	19	21	24	29
TWD4	11	16	22	27	33	38	44	52
TWE2	11	16	22	27	33	37	43	51
TWF3	6	8	11	13	17	19	21	25
TWJ2	9	14	20	24	31	35	41	49
DGA1	5	6	7	8	8	8	8	7
DGA2	8	10	13	15	19	21	24	27
DGA3	11	15	20	23	29	33	38	45
DGB1	4	6	7	8	9	9	9	9
DGB2	8	10	14	16	20	22	25	29
DGB3	11	15	21	25	30	35	40	48
DGC1	6	8	10	12	14	15	16	17
DGC2	10	13	17	20	24	27	31	37
DGC3	13	18	24	28	35	39	46	55
SSA1	13	16	19	21	24	26	28	31
SSB1	11	13	15	17	20	22	24	27
SSB2	14	18	23	27	33	36	41	48
SSB3	10	11	13	15	17	18	20	21
SSC1	7	9	12	14	17	18	21	24
SSC2	9	13	18	22	27	31	35	42
SSE1	10	13	16	18	20	22	24	26
SSE2	13	18	24	28	34	38	43	50
SSE3	11	13	15	17	19	20	22	24

CINCINNATI, OHIO SSF (%)　5079 DD65

Type	LCR = 100	70	50	40	30	25	20	15
WWA3	10	13	17	21	25	28	32	38
WWB4	13	20	27	33	42	47	55	64
WWC2	13	19	27	32	40	45	52	62
TWA1	13	14	16	18	20	21	23	25
TWA2	11	14	17	19	22	25	28	32
TWA3	10	13	16	19	23	26	30	35
TWA4	9	12	16	19	23	26	31	36
TWB3	9	12	15	18	22	25	29	34
TWD4	12	18	25	30	37	42	49	58
TWE2	13	19	25	30	37	42	48	57
TWF3	7	10	13	16	20	23	26	31
TWJ2	11	16	23	28	35	40	46	55
DGA1	6	7	9	10	11	12	12	12
DGA2	9	12	15	18	21	24	27	32
DGA3	12	16	22	26	32	36	42	50
DGB1	6	7	9	10	12	13	14	14
DGB2	9	12	15	18	22	25	29	34
DGB3	12	17	23	27	33	38	44	53
DGC1	8	10	13	15	17	18	20	23
DGC2	11	15	19	22	27	31	36	42
DGC3	14	20	26	31	38	43	50	60
SSA1	15	19	23	25	29	32	36	41
SSB1	13	16	19	21	25	27	31	35
SSB2	16	22	27	32	38	42	48	56
SSB3	12	14	17	19	22	24	27	30
SSC1	8	11	15	17	21	23	26	31
SSC2	11	16	21	25	31	35	41	49
SSE1	13	16	20	23	27	30	33	38
SSE2	16	22	29	34	41	46	52	60
SSE3	13	16	19	21	24	26	29	33

TABLE C.3 **Annual Passive Heating Performance: SSF** (*continued*)

CLEVELAND, OHIO SSF (%) 6160 DD65

Type	LCR = 100	70	50	40	30	25	20	15
WWA3	6	9	11	13	15	16	18	20
WWB4	8	13	20	24	31	36	42	50
WWC2	8	13	19	23	29	34	39	47
TWA1	11	12	13	14	14	15	15	15
TWA2	9	10	12	13	15	16	17	19
TWA3	7	9	11	13	15	16	18	20
TWA4	6	8	10	12	15	17	19	22
TWB3	7	9	11	13	15	16	19	21
TWD4	9	13	18	22	28	32	37	45
TWE2	9	14	18	22	28	31	36	43
TWF3	5	6	8	10	12	13	15	17
TWJ2	7	11	16	20	25	29	34	41
DGA1	3	4	4	4	4	4	2	0
DGA2	7	8	11	12	15	16	18	21
DGA3	10	13	17	20	25	28	33	39
DGB1	3	4	4	5	4	4	3	1
DGB2	7	9	11	13	15	17	19	22
DGB3	10	14	18	21	26	30	35	42
DGC1	5	6	7	8	9	9	9	8
DGC2	8	11	14	16	20	22	25	29
DGC3	12	16	21	24	30	34	39	47
SSA1	13	15	18	19	22	24	26	28
SSB1	10	12	14	16	18	19	21	23
SSB2	13	17	22	25	30	34	38	44
SSB3	9	11	13	14	15	16	17	18
SSC1	5	7	9	10	11	12	13	14
SSC2	7	11	14	17	21	24	28	33
SSE1	9	12	14	15	17	18	19	20
SSE2	12	17	22	25	31	34	39	45
SSE3	11	13	14	16	17	18	19	20

OKLAHOMA CITY, OKLAH. SSF (%) 3699 DD65

Type	LCR = 200	100	70	50	40	30	25	20
WWA3	12	22	29	37	44	52	58	65
WWB4	12	28	38	50	58	68	74	81
WWC2	13	27	38	49	56	66	73	80
TWA1	14	20	23	28	32	37	41	46
TWA2	13	20	26	32	37	44	49	55
TWA3	12	20	26	33	39	47	52	59
TWA4	10	19	26	33	39	47	53	60
TWB3	10	18	24	31	36	44	49	56
TWD4	12	25	34	45	52	62	68	76
TWE2	13	26	35	45	53	62	68	76
TWF3	9	17	23	30	35	43	48	55
TWJ2	11	23	32	42	49	59	66	73
DGA1	8	15	20	25	29	34	37	41
DGA2	9	17	23	30	35	43	47	53
DGA3	11	21	29	38	45	54	60	67
DGB1	8	15	20	26	31	37	42	47
DGB2	10	18	24	31	37	46	52	59
DGB3	12	22	29	39	46	57	64	72
DGC1	10	18	24	32	37	46	51	58
DGC2	11	21	28	36	43	53	60	68
DGC3	13	25	33	44	52	64	71	79
SSA1	16	25	32	39	44	51	57	63
SSB1	14	21	27	33	38	45	50	56
SSB2	16	27	35	44	51	60	65	72
SSB3	13	20	25	31	36	42	47	53
SSC1	11	19	26	33	38	46	52	59
SSC2	12	23	32	41	48	57	63	71
SSE1	15	25	32	40	46	54	59	66
SSE2	17	31	40	51	58	67	73	80
SSE3	15	22	27	34	38	45	50	56

COLUMBUS, OHIO SSF (%) 5705 DD65

Type	LCR = 100	70	50	40	30	25	20	15
WWA3	8	11	14	16	20	22	25	29
WWB4	10	16	23	29	36	41	48	57
WWC2	10	16	22	27	34	39	46	54
TWA1	12	13	14	15	17	17	18	20
TWA2	10	12	14	16	18	20	22	25
TWA3	8	11	13	16	19	21	24	27
TWA4	7	10	13	15	19	21	24	29
TWB3	8	10	13	15	18	20	23	27
TWD4	10	15	21	26	32	37	43	51
TWE2	11	16	21	26	32	36	42	50
TWF3	6	8	11	13	16	18	20	24
TWJ2	9	13	19	23	30	34	40	48
DGA1	4	5	6	7	7	7	7	5
DGA2	7	10	12	15	18	20	22	26
DGA3	11	14	19	23	28	32	37	44
DGB1	4	5	6	7	8	8	8	7
DGB2	7	10	13	15	18	21	24	28
DGB3	11	15	20	24	29	33	39	47
DGC1	6	8	10	11	12	13	14	15
DGC2	9	12	16	19	23	26	30	35
DGC3	13	17	23	27	33	38	44	53
SSA1	14	17	20	22	25	27	30	34
SSB1	11	13	16	18	21	23	25	29
SSB2	15	19	24	28	34	38	43	50
SSB3	10	12	14	16	18	20	22	24
SSC1	7	9	11	13	16	17	19	22
SSC2	9	13	17	21	26	30	34	41
SSE1	11	14	17	19	22	24	26	29
SSE2	14	19	25	29	36	40	45	53
SSE3	12	14	16	18	20	22	24	26

MEDFORD, OREGON SSF (%) 4935 DD65

Type	LCR = 100	70	50	40	30	25	20	15
WWA3	16	21	26	31	36	40	45	51
WWB4	20	28	37	43	52	58	64	72
WWC2	20	28	36	42	50	56	62	70
TWA1	17	19	22	24	28	30	33	37
TWA2	16	20	24	27	32	35	39	44
TWA3	15	19	24	28	33	37	41	47
TWA4	14	19	24	28	34	37	42	49
TWB3	14	18	23	26	31	35	40	46
TWD4	19	26	34	39	47	52	59	67
TWE2	20	26	34	40	47	52	59	66
TWF3	12	16	21	25	30	33	38	44
TWJ2	17	24	31	37	45	50	56	64
DGA1	11	14	17	19	21	23	24	25
DGA2	13	17	22	26	31	34	38	43
DGA3	17	23	29	34	41	46	52	59
DGB1	11	14	18	20	23	25	28	30
DGB2	14	18	23	27	33	37	41	47
DGB3	17	23	30	36	44	49	55	63
DGC1	13	17	22	25	30	32	36	39
DGC2	16	21	27	32	39	43	48	55
DGC3	20	26	34	40	49	54	61	69
SSA1	21	26	30	34	39	42	46	51
SSB1	18	22	26	29	34	37	41	46
SSB2	22	28	35	40	47	51	57	64
SSB3	17	20	24	27	31	34	37	42
SSC1	14	18	23	27	32	35	39	44
SSC2	17	23	30	35	42	46	52	59
SSE1	19	24	29	33	38	41	45	50
SSE2	24	31	38	44	51	55	61	68
SSE3	18	22	26	29	33	36	40	44

APPENDICES

TABLE C.3 **Annual Passive Heating Performance: SSF** (*continued*)

SALEM, OREGON SSF (%)							4854 DD65	
Type LCR = 100	70	50	40	30	25	20	15	
WWA3 15	20	25	29	35	38	43	48	
WWB4 19	27	36	42	50	56	63	71	
WWC2 19	27	35	41	49	54	60	68	
TWA1 16	19	22	24	27	29	31	35	
TWA2 16	19	23	26	30	33	37	42	
TWA3 15	19	23	27	32	35	40	45	
TWA4 14	18	23	27	32	36	40	47	
TWB3 14	18	22	25	30	34	38	44	
TWD4 18	25	32	38	46	51	57	65	
TWE2 19	26	33	38	46	51	57	65	
TWF3 12	16	20	24	28	32	36	42	
TWJ2 16	23	30	36	43	48	55	63	
DGA1 10	13	16	18	20	21	23	23	
DGA2 13	17	22	25	30	33	37	41	
DGA3 17	22	28	33	40	45	51	58	
DGB1 11	13	17	19	22	24	26	28	
DGB2 13	18	22	26	32	35	40	45	
DGB3 17	23	29	35	42	47	54	62	
DGC1 13	16	21	24	28	30	33	37	
DGC2 16	21	26	31	37	41	46	53	
DGC3 19	26	33	39	47	53	59	68	
SSA1 21	25	29	33	37	40	44	49	
SSB1 17	21	25	28	33	36	39	44	
SSB2 22	28	34	39	46	50	55	62	
SSB3 16	20	23	26	30	33	36	40	
SSC1 14	18	22	25	30	33	37	42	
SSC2 17	22	29	33	40	45	50	57	
SSE1 19	23	28	32	36	39	43	47	
SSE2 23	30	37	42	49	54	59	66	
SSE3 18	22	25	28	32	35	38	42	

PHILADELPHIA, PENN. SSF (%)							4875 DD65	
Type LCR = 100	70	50	40	30	25	20	15	
WWA3 12	17	22	26	32	36	41	49	
WWB4 16	24	33	40	49	55	63	72	
WWC2 16	23	32	38	47	53	61	70	
TWA1 14	16	19	21	24	26	29	33	
TWA2 13	16	20	23	27	31	35	41	
TWA3 12	16	20	24	29	33	38	44	
TWA4 11	15	20	24	29	33	39	46	
TWB3 11	15	19	22	27	31	36	42	
TWD4 15	22	30	35	44	49	57	66	
TWE2 16	22	30	36	44	49	56	66	
TWF3 9	13	17	20	25	29	34	40	
TWJ2 13	20	27	33	41	47	54	63	
DGA1 8	10	12	14	17	18	19	20	
DGA2 10	14	18	22	26	30	34	39	
DGA3 14	19	25	30	37	42	49	58	
DGB1 7	10	13	15	17	19	22	24	
DGB2 11	14	19	22	28	31	37	43	
DGB3 14	20	26	31	39	44	52	62	
DGC1 10	13	17	19	23	26	30	34	
DGC2 13	17	23	27	33	38	44	52	
DGC3 16	22	29	35	44	50	58	69	
SSA1 17	21	26	29	34	38	43	49	
SSB1 14	18	22	25	29	33	37	43	
SSB2 19	24	31	36	43	48	54	63	
SSB3 13	16	20	22	26	29	33	38	
SSC1 11	14	19	22	27	30	35	42	
SSC2 14	19	26	31	38	43	49	58	
SSE1 15	20	25	28	34	37	42	48	
SSE2 19	26	34	40	48	53	60	68	
SSE3 15	18	22	25	29	32	36	41	

HARRISBURG, PENN. SSF (%)							5226 DD65	
Type LCR = 100	70	50	40	30	25	20	15	
WWA3 11	14	19	22	28	31	36	42	
WWB4 14	21	29	36	45	51	58	68	
WWC2 14	20	28	34	43	48	56	65	
TWA1 13	15	17	19	21	23	25	28	
TWA2 12	14	18	20	24	27	30	35	
TWA3 10	14	18	21	25	28	33	39	
TWA4 9	13	17	20	25	29	34	40	
TWB3 10	13	16	19	24	27	31	37	
TWD4 13	19	26	32	40	45	52	61	
TWE2 14	20	27	32	40	45	52	61	
TWF3 8	11	14	17	22	25	29	35	
TWJ2 11	17	24	29	37	42	49	58	
DGA1 6	8	10	11	13	14	15	15	
DGA2 9	12	16	19	23	26	30	35	
DGA3 12	17	23	27	34	38	45	53	
DGB1 6	8	10	12	14	15	16	18	
DGB2 9	12	16	19	24	27	32	38	
DGB3 13	18	23	28	35	40	47	57	
DGC1 8	11	14	16	19	21	24	27	
DGC2 11	15	20	24	29	33	39	46	
DGC3 15	20	27	32	40	46	53	64	
SSA1 16	19	23	26	31	34	38	44	
SSB1 13	16	19	22	26	29	32	38	
SSB2 17	22	28	33	40	44	50	58	
SSB3 12	14	17	20	23	25	28	33	
SSC1 9	12	16	19	23	26	30	35	
SSC2 12	17	22	27	34	38	44	52	
SSE1 13	17	21	24	29	32	36	42	
SSE2 17	23	30	36	43	48	55	63	
SSE3 13	16	19	22	25	28	31	36	

PITTSBURGH, PENN. SSF (%)							5937 DD65	
Type LCR = 100	70	50	40	30	25	20	15	
WWA3 7	9	12	14	17	19	21	24	
WWB4 9	15	21	26	33	38	45	53	
WWC2 9	14	20	25	32	36	42	50	
TWA1 12	12	13	14	15	16	16	17	
TWA2 9	11	13	14	16	18	19	22	
TWA3 8	10	12	14	17	18	21	23	
TWA4 6	9	11	14	17	19	21	25	
TWB3 7	9	12	14	16	18	21	24	
TWD4 9	14	20	24	30	34	40	48	
TWE2 10	14	20	24	30	34	39	46	
TWF3 5	7	9	11	14	15	17	20	
TWJ2 8	12	17	21	27	31	37	44	
DGA1 4	5	5	5	5	5	4	2	
DGA2 7	9	11	13	16	18	20	23	
DGA3 10	14	18	21	26	30	35	41	
DGB1 4	4	5	6	6	6	5	3	
DGB2 7	9	12	14	17	19	21	25	
DGB3 10	14	19	22	28	32	37	44	
DGC1 5	7	8	9	10	11	11	11	
DGC2 9	11	15	17	21	24	27	32	
DGC3 12	16	21	26	32	36	42	50	
SSA1 13	16	19	21	23	25	28	31	
SSB1 11	13	15	17	19	21	23	26	
SSB2 14	18	23	27	32	36	40	47	
SSB3 10	11	13	15	17	18	19	21	
SSC1 6	8	10	11	13	15	16	18	
SSC2 8	12	16	19	23	27	31	37	
SSE1 10	12	15	17	19	21	23	24	
SSE2 13	18	23	27	33	37	42	48	
SSE3 11	13	15	17	19	20	21	23	

TABLE C.3 **Annual Passive Heating Performance: SSF** (*continued*)

PROVIDENCE, RHODE IS. SSF (%)						5974 DD65		
Type	LCR = 100	70	50	40	30	25	20	15
WWA3	10	13	18	21	26	29	33	40
WWB4	13	20	28	34	43	49	56	66
WWC2	13	19	27	33	41	47	54	63
TWA1	13	14	16	18	20	21	23	26
TWA2	11	14	17	19	22	25	28	33
TWA3	10	13	16	19	24	27	31	36
TWA4	9	12	16	19	24	27	31	38
TWB3	9	12	15	18	22	25	29	35
TWD4	12	18	25	30	38	43	50	59
TWE2	13	19	25	30	38	43	50	59
TWF3	7	10	13	16	20	23	27	32
TWJ2	11	16	23	28	35	40	47	56
DGA1	6	7	9	10	11	12	13	13
DGA2	8	11	15	18	22	24	28	33
DGA3	12	16	21	26	32	37	43	51
DGB1	5	7	9	10	12	13	14	15
DGB2	9	12	15	18	22	26	30	36
DGB3	12	17	22	27	34	39	45	55
DGC1	7	10	12	14	17	19	21	25
DGC2	10	14	19	22	28	31	37	44
DGC3	14	19	26	31	38	44	52	62
SSA1	15	19	22	25	30	33	36	42
SSB1	12	15	19	21	25	28	31	36
SSB2	16	21	27	32	38	43	49	57
SSB3	11	14	17	19	22	24	27	31
SSC1	8	11	15	17	21	24	27	32
SSC2	11	16	21	25	32	36	42	50
SSE1	13	16	20	23	27	30	34	39
SSE2	16	22	29	34	42	47	53	61
SSE3	13	16	19	21	24	27	30	34

RAPID CITY, SOUTH DAKOTA SSF (%)						7332 DD65		
Type	LCR = 100	70	50	40	30	25	20	15
WWA3	13	18	23	27	33	37	42	50
WWB4	17	25	34	40	50	56	63	73
WWC2	17	24	33	39	48	54	61	70
TWA1	15	17	19	21	24	27	30	34
TWA2	13	17	21	24	28	32	36	42
TWA3	13	17	21	25	30	34	38	45
TWA4	11	16	21	25	30	34	39	47
TWB3	11	15	20	23	28	32	37	43
TWD4	16	23	30	36	45	50	57	66
TWE2	17	23	31	37	45	50	57	66
TWF3	10	13	18	21	26	30	35	41
TWJ2	14	21	28	34	42	47	54	64
DGA1	8	11	13	15	18	19	20	21
DGA2	11	15	19	22	27	31	35	40
DGA3	14	20	26	31	38	43	50	58
DGB1	8	11	14	16	19	21	23	26
DGB2	11	15	20	32	29	33	38	44
DGB3	15	20	27	32	40	45	53	62
DGC1	10	14	18	21	25	27	31	35
DGC2	13	18	24	28	34	39	45	53
DGC3	17	23	31	36	45	51	59	69
SSA1	18	22	26	30	34	38	42	48
SSB1	15	19	22	25	30	33	37	42
SSB2	19	25	31	36	43	48	54	62
SSB3	13	16	20	23	26	29	33	37
SSC1	11	15	20	23	28	31	36	42
SSC2	14	20	27	32	39	44	50	59
SSE1	16	20	25	28	33	37	41	47
SSE2	20	27	34	40	47	52	59	67
SSE3	15	18	22	25	29	32	35	40

CHARLESTON, S. CAROLINA SSF (%)						2150 DD65		
Type	LCR = 200	100	70	50	40	30	25	20
WWA3	18	31	40	50	57	66	72	79
WWB4	19	38	50	62	70	79	85	90
WWC2	20	38	49	61	69	78	84	89
TWA1	17	25	30	37	41	48	53	59
TWA2	17	27	34	42	48	57	62	69
TWA3	16	28	36	45	51	60	66	73
TWA4	15	27	36	45	52	61	67	74
TWB3	14	25	33	41	47	56	62	69
TWD4	18	35	45	56	64	74	79	86
TWE2	19	36	46	57	65	74	80	86
TWF3	13	24	32	41	47	56	62	69
TWJ2	17	32	43	54	62	72	77	84
DGA1	13	23	29	36	41	48	51	56
DGA2	13	24	32	41	47	55	60	67
DGA3	16	29	38	49	57	66	72	78
DGB1	13	23	31	39	45	54	59	65
DGB2	14	25	34	43	51	61	67	74
DGB3	16	29	39	51	60	71	77	83
DGC1	15	27	36	46	53	63	69	75
DGC2	16	29	38	49	58	68	75	81
DGC3	18	33	44	57	66	77	83	89
SSA1	22	33	41	50	56	65	70	76
SSB1	18	29	36	44	50	58	64	70
SSB2	22	36	46	56	63	72	77	83
SSB3	17	27	34	42	48	56	61	68
SSC1	16	28	36	45	51	60	66	73
SSC2	18	32	42	53	60	70	76	82
SSE1	21	34	43	53	60	69	74	80
SSE2	24	42	53	64	71	80	85	90
SSE3	19	29	36	44	50	58	64	70

KNOXVILLE, TENNESSEE SSF (%)						3486 DD65		
Type	LCR = 200	100	70	50	40	30	25	20
WWA3	10	18	24	31	36	44	49	55
WWB4	9	23	32	43	50	60	66	74
WWC2	10	23	32	42	49	58	65	72
TWA1	13	17	21	24	27	32	35	39
TWA2	11	17	22	27	31	37	41	47
TWA3	10	17	22	28	33	39	44	50
TWA4	8	16	21	28	33	40	45	51
TWB3	9	15	20	26	30	37	41	47
TWD4	10	21	29	38	45	54	60	68
TWE2	11	22	30	39	46	55	60	68
TWF3	7	14	19	25	29	36	40	46
TWJ2	8	19	27	36	43	52	58	65
DGA1	7	12	16	20	23	27	29	32
DGA2	8	15	19	25	30	36	40	45
DGA3	10	18	25	32	38	47	53	60
DGB1	7	12	16	21	24	29	32	36
DGB2	8	15	20	26	31	38	43	49
DGB3	10	19	26	33	40	49	55	64
DGC1	8	15	20	25	29	36	40	45
DGC2	9	18	24	31	36	44	50	57
DGC3	12	22	29	38	45	55	62	71
SSA1	15	22	28	34	38	45	49	55
SSB1	12	19	24	29	33	39	43	49
SSB2	14	24	31	39	45	53	59	65
SSB3	12	18	22	27	31	36	40	45
SSC1	9	16	21	27	32	38	43	49
SSC2	9	19	27	35	41	49	55	62
SSE1	13	21	27	34	39	48	51	57
SSE2	14	27	35	44	51	60	65	72
SSE3	13	19	24	29	33	39	43	48

TABLE C.3 Annual Passive Heating Performance: SSF (*continued*)

MEMPHIS, TENNESSEE SSF (%) — 3237 DD65

Type	LCR = 200	100	70	50	40	30	25	20
WWA3	11	21	27	35	41	49	54	61
WWB4	11	26	36	47	54	65	71	78
WWC2	12	26	35	46	53	63	69	77
TWA1	14	19	22	27	30	35	38	43
TWA2	12	19	24	30	35	41	46	52
TWA3	11	19	25	31	36	44	49	55
TWA4	9	18	24	31	37	44	50	56
TWB3	10	17	23	29	34	41	46	52
TWD4	11	24	32	42	49	59	65	72
TWE2	12	25	33	43	50	59	65	72
TWF3	8	16	21	28	33	40	45	51
TWJ2	10	22	30	40	47	56	62	70
DGA1	8	14	18	23	26	31	34	37
DGA2	9	16	22	28	33	40	44	50
DGA3	11	20	27	36	42	51	57	64
DGB1	8	14	19	24	28	34	38	43
DGB2	9	17	22	29	35	43	48	55
DGB3	11	21	28	37	43	54	60	69
DGC1	10	17	23	29	34	41	46	52
DGC2	11	20	26	34	40	49	56	63
DGC3	13	24	31	41	49	60	67	76
SSA1	16	24	30	37	42	49	54	60
SSB1	13	21	26	32	36	43	47	53
SSB2	15	26	34	42	49	57	63	70
SSB3	12	19	24	30	34	40	44	50
SSC1	10	18	24	31	36	43	48	54
SSC2	11	22	30	38	45	54	59	67
SSE1	14	24	30	38	43	51	56	62
SSE2	16	29	38	48	55	64	70	77
SSE3	14	21	26	32	36	43	47	53

AUSTIN, TEXAS SSF (%) — 1737 DD65

Type	LCR = 200	100	70	50	40	30	25	20
WWA3	21	37	47	58	65	75	80	86
WWB4	24	45	58	70	77	86	90	94
WWC2	24	44	57	69	76	85	89	94
TWA1	20	29	35	42	48	56	61	67
TWA2	20	32	40	49	56	65	70	77
TWA3	19	33	42	52	59	68	74	81
TWA4	18	32	42	52	60	69	75	82
TWB3	17	30	38	48	55	64	70	77
TWD4	22	41	52	64	72	81	86	91
TWE2	24	42	53	65	73	81	86	91
TWF3	16	29	38	48	55	64	70	77
TWJ2	21	39	50	62	70	79	84	90
DGA1	16	28	36	44	49	57	61	66
DGA2	16	29	38	48	55	64	69	75
DGA3	19	34	45	57	65	74	79	84
DGB1	16	29	38	48	56	64	69	75
DGB2	17	30	41	52	60	70	76	82
DGB3	19	35	46	60	68	79	84	89
DGC1	19	33	44	56	64	74	79	84
DGC2	19	34	46	59	68	77	83	88
DGC3	22	39	52	66	75	85	89	93
SSA1	25	39	48	58	65	73	78	84
SSB1	21	34	42	51	58	67	72	79
SSB2	26	42	53	63	71	79	84	89
SSB3	20	32	40	49	56	65	70	77
SSC1	19	33	42	53	60	69	75	81
SSC2	22	39	50	61	68	78	83	89
SSE1	25	41	51	62	69	77	82	88
SSE2	29	49	61	72	79	87	91	94
SSE3	22	34	43	52	58	67	72	79

NASHVILLE, TENNESSEE SSF (%) — 3704 DD65

Type	LCR = 200	100	70	50	40	30	25	20
WWA3	8	15	20	26	31	37	42	47
WWB4	7	19	28	37	44	54	60	68
WWC2	8	19	27	36	43	52	58	66
TWA1	12	16	18	21	24	27	30	33
TWA2	10	15	19	23	27	32	35	40
TWA3	8	14	19	24	28	34	38	43
TWA4	7	13	18	24	28	34	38	44
TWB3	7	13	17	22	26	32	35	41
TWD4	8	18	25	34	40	48	54	61
TWE2	9	19	26	34	40	48	54	61
TWF3	6	11	16	21	24	30	34	39
TWJ2	6	16	23	31	37	46	51	59
DGA1	6	10	13	16	18	21	23	25
DGA2	7	12	17	21	25	31	34	39
DGA3	9	16	22	28	34	41	47	54
DGB1	6	10	13	16	19	22	25	28
DGB2	7	13	17	22	26	32	36	42
DGB3	9	17	22	29	35	43	49	57
DGC1	7	12	16	20	24	29	32	36
DGC2	8	15	20	26	31	38	43	49
DGC3	10	19	25	33	39	49	55	64
SSA1	13	20	25	30	34	39	43	49
SSB1	11	17	21	25	29	34	38	42
SSB2	12	21	28	35	40	48	53	60
SSB3	10	15	19	23	27	31	35	39
SSC1	7	13	18	23	26	32	36	41
SSC2	8	16	23	30	35	43	48	55
SSE1	11	18	23	29	33	39	44	49
SSE2	12	23	31	39	45	53	59	66
SSE3	12	17	21	25	29	34	37	41

BROWNSVILLE, TEXAS SSF (%) — 650 DD65

Type	LCR = 200	100	70	50	40	30	25	20
WWA3	39	60	72	82	88	94	96	98
WWB4	45	70	82	90	94	98	99	100
WWC2	45	69	81	90	94	97	99	99
TWA1	30	45	54	64	71	79	84	89
TWA2	33	51	63	73	80	87	91	95
TWA3	34	54	66	76	83	90	93	96
TWA4	34	54	66	77	84	90	94	97
TWB3	31	50	61	72	79	87	91	95
TWD4	41	64	76	86	91	96	98	99
TWE2	42	65	77	87	92	96	98	99
TWF3	30	50	62	73	80	88	91	95
TWJ2	39	62	75	85	90	95	97	99
DGA1	31	50	61	70	76	83	87	90
DGA2	31	51	63	73	80	87	90	94
DGA3	34	57	70	80	86	91	93	95
DGB1	32	54	67	78	84	89	92	94
DGB2	32	55	69	80	86	92	95	97
DGB3	34	60	74	85	90	94	96	97
DGC1	36	62	76	85	90	94	95	97
DGC2	35	61	75	86	91	95	97	99
DGC3	39	66	80	90	94	97	98	98
SSA1	42	62	73	82	87	93	96	98
SSB1	36	55	66	76	83	89	93	96
SSB2	44	65	77	86	91	95	97	99
SSB3	35	54	65	75	81	88	92	95
SSC1	35	55	67	78	84	91	94	97
SSC2	39	62	74	84	90	95	97	99
SSE1	45	66	77	86	91	95	97	99
SSE2	51	74	85	92	96	98	99	100
SSE3	37	56	67	77	83	90	93	96

TABLE C.3 **Annual Passive Heating Performance: SSF** (*continued*)

DALLAS, TEXAS SSF (%)							2297 DD65			HOUSTON, TEXAS SSF (%)							1434 DD65		
Type	LCR = 200	100	70	50	40	30	25	20		Type	LCR = 200	100	70	50	40	30	25	20
WWA3	17	30	39	49	56	66	72	79		WWA3	21	37	47	58	65	74	80	86
WWB4	19	37	49	62	70	79	85	90		WWB4	24	45	58	70	77	86	90	94
WWC2	19	37	49	61	69	78	84	89		WWC2	24	44	57	69	76	85	89	93
TWA1	17	24	30	36	41	48	52	59		TWA1	20	29	35	42	48	55	61	67
TWA2	16	26	33	42	48	56	62	69		TWA2	20	32	40	49	56	65	70	77
TWA3	16	27	35	44	50	60	65	73		TWA3	20	33	42	52	59	68	74	80
TWA4	14	26	35	44	51	60	66	74		TWA4	18	32	42	52	59	69	75	81
TWB3	14	24	32	40	47	56	62	69		TWB3	18	30	38	48	55	64	70	77
TWD4	18	34	45	56	64	74	79	86		TWD4	22	41	52	64	72	81	86	91
TWE2	19	35	46	57	65	74	80	86		TWE2	24	42	53	65	72	81	86	91
TWF3	13	23	31	40	46	56	62	69		TWF3	16	29	38	48	55	64	70	77
TWJ2	16	32	42	54	61	71	77	84		TWJ2	21	38	50	62	69	79	84	89
DGA1	12	22	29	36	41	47	51	56		DGA1	16	28	36	44	49	56	61	65
DGA2	13	24	31	40	47	55	60	66		DGA2	16	29	38	48	55	63	69	74
DGA3	15	28	38	49	56	66	72	78		DGA3	19	34	45	57	64	74	79	84
DGB1	12	22	30	38	45	54	59	64		DGB1	16	28	38	48	55	64	69	74
DGB2	13	24	33	43	51	61	67	74		DGB2	17	30	40	52	60	70	75	81
DGB3	16	28	38	50	59	71	77	83		DGB3	19	35	46	59	68	78	83	89
DGC1	15	26	35	45	53	63	69	75		DGC1	18	33	44	56	64	73	78	83
DGC2	15	28	38	49	57	68	74	81		DGC2	19	34	46	59	67	77	82	88
DGC3	18	32	43	57	66	77	83	89		DGC3	21	39	52	66	75	84	89	93
SSA1	21	32	40	49	56	64	70	76		SSA1	25	39	48	58	65	73	78	84
SSB1	17	28	35	43	49	58	63	70		SSB1	21	34	42	52	58	67	72	79
SSB2	21	35	45	55	62	71	77	83		SSB2	26	42	53	64	71	79	84	89
SSB3	17	26	33	41	47	55	60	67		SSB3	20	32	41	50	56	65	70	77
SSC1	15	27	35	44	51	60	66	73		SSC1	19	33	42	52	60	69	75	81
SSC2	17	32	42	52	60	70	76	82		SSC2	22	38	49	61	68	77	83	88
SSE1	20	33	42	52	59	68	74	80		SSE1	25	41	51	62	69	77	82	88
SSE2	23	41	52	63	70	79	84	90		SSE2	29	49	61	72	79	87	90	94
SSE3	18	28	35	43	49	58	63	70		SSE3	22	34	43	52	59	67	73	79

EL PASO, TEXAS SSF (%)							2685 DD65			LUBBOCK, TEXAS SSF (%)							3551 DD65		
Type	LCR = 200	100	70	50	40	30	25	20		Type	LCR = 200	100	70	50	40	30	25	20
WWA3	22	38	49	60	68	77	83	88		WWA3	17	30	40	50	57	67	73	80
WWB4	25	47	60	72	80	88	92	95		WWB4	18	38	50	63	71	80	86	91
WWC2	25	46	59	71	79	87	91	95		WWC2	19	37	49	61	69	79	85	90
TWA1	20	29	36	44	49	58	63	70		TWA1	17	24	30	36	41	48	53	60
TWA2	20	33	41	51	58	67	73	80		TWA2	16	26	34	42	48	57	63	70
TWA3	20	34	44	54	61	71	77	83		TWA3	15	27	35	44	51	60	67	74
TWA4	19	34	44	54	62	72	77	84		TWA4	14	26	35	45	52	61	67	75
TWB3	18	31	40	50	57	67	73	79		TWB3	14	24	32	41	47	56	62	70
TWD4	23	42	54	66	74	83	88	93		TWD4	17	34	45	57	65	75	80	87
TWE2	24	43	55	67	75	84	88	93		TWE2	19	35	46	58	65	75	81	87
TWF3	16	30	40	50	57	67	73	80		TWF3	12	23	31	40	47	56	63	70
TWJ2	21	40	52	64	72	81	86	91		TWJ2	16	32	43	54	62	72	78	85
DGA1	16	29	37	46	52	59	64	69		DGA1	12	22	29	36	41	48	52	57
DGA2	17	30	40	50	57	66	71	77		DGA2	13	23	32	41	47	56	61	67
DGA3	19	35	47	59	67	76	81	86		DGA3	15	28	38	49	57	67	73	79
DGB1	16	30	40	51	58	67	72	78		DGB1	12	22	30	39	46	55	60	66
DGB2	17	31	42	55	63	73	79	84		DGB2	13	24	33	43	51	62	68	75
DGB3	19	36	48	62	71	81	86	91		DGB3	15	28	39	51	60	72	78	84
DGC1	19	35	47	59	68	77	82	86		DGC1	14	26	35	46	54	65	70	76
DGC2	20	36	48	62	70	80	85	90		DGC2	15	28	38	50	59	70	76	82
DGC3	22	40	55	69	78	87	91	94		DGC3	17	32	44	57	67	79	84	90
SSA1	25	39	49	58	65	74	79	85		SSA1	21	33	41	50	56	65	70	77
SSB1	21	34	43	52	59	68	73	80		SSB1	17	28	35	43	50	58	64	71
SSB2	26	42	53	64	71	80	85	90		SSB2	21	35	45	55	63	72	77	84
SSB3	20	32	41	50	56	65	71	77		SSB3	16	26	33	41	47	55	61	68
SSC1	20	34	44	55	62	72	77	84		SSC1	15	27	35	45	52	61	67	74
SSC2	22	40	51	63	71	80	85	90		SSC2	17	32	42	53	61	71	77	83
SSE1	25	41	52	62	69	78	83	89		SSE1	20	34	43	53	60	69	74	81
SSE2	29	49	61	73	80	87	91	95		SSE2	23	41	52	63	71	80	85	90
SSE3	22	35	43	52	59	68	73	80		SSE3	18	28	36	44	50	58	64	70

APPENDICES

TABLE C.3 **Annual Passive Heating Performance: SSF** (*continued*)

SAN ANTONIO, TEXAS **SSF (%)**   1570 DD65

Type	LCR = 200	100	70	50	40	30	25	20
WWA3	23	40	50	61	69	78	83	88
WWB4	26	48	61	73	80	88	92	95
WWC2	26	47	60	72	79	87	91	95
TWA1	21	30	37	45	50	59	64	71
TWA2	21	34	42	52	59	68	74	80
TWA3	21	35	45	55	62	71	77	83
TWA4	20	35	45	55	63	72	78	84
TWB3	19	32	41	51	58	67	73	80
TWD4	24	43	55	67	75	83	88	93
TWE2	26	44	56	68	75	84	88	93
TWF3	18	31	41	51	58	68	74	80
TWJ2	22	41	53	65	73	82	87	91
DGA1	17	30	38	47	53	60	64	69
DGA2	18	32	41	51	58	67	72	78
DGA3	20	36	48	60	68	77	81	86
DGB1	17	31	41	52	59	68	73	78
DGB2	18	33	44	56	64	74	79	85
DGB3	20	37	50	63	72	81	86	91
DGC1	20	36	48	60	68	77	82	86
DGC2	21	37	49	63	71	80	85	90
DGC3	23	42	56	70	78	87	91	94
SSA1	27	41	51	61	68	76	81	87
SSB1	23	36	45	54	61	70	75	82
SSB2	27	45	55	66	73	82	86	91
SSB3	22	34	43	52	59	68	73	80
SSC1	21	35	45	56	63	72	78	84
SSC2	23	41	53	64	71	80	85	91
SSE1	27	44	54	65	72	80	85	90
SSE2	31	52	64	75	82	89	92	96
SSE3	24	37	45	55	62	70	76	82

SALT LAKE CITY, UTAH **SSF (%)**   5988 DD65

Type	LCR = 100	70	50	40	30	25	20	15
WWA3	19	25	32	37	44	49	55	64
WWB4	23	33	43	50	60	66	73	82
WWC2	23	32	42	49	59	65	72	80
TWA1	18	21	25	28	32	35	39	45
TWA2	18	22	28	32	38	42	47	54
TWA3	17	22	29	33	40	44	50	58
TWA4	16	22	28	33	40	45	51	60
TWB3	16	21	26	31	37	42	48	55
TWD4	22	30	39	45	55	60	68	76
TWE2	23	31	39	46	55	61	68	76
TWF3	14	19	25	30	36	41	46	54
TWJ2	20	28	37	43	52	58	65	74
DGA1	12	16	20	23	27	30	32	35
DGA2	15	20	26	30	36	41	46	52
DGA3	19	25	33	39	47	53	60	68
DGB1	13	17	21	25	30	33	37	41
DGB2	15	21	27	32	39	44	50	58
DGB3	19	26	34	40	50	56	64	73
DGC1	15	20	26	31	37	41	46	52
DGC2	18	24	31	37	45	51	58	66
DGC3	22	29	38	46	56	62	70	79
SSA1	23	28	34	38	44	48	54	60
SSB1	19	24	29	33	39	43	48	55
SSB2	24	31	39	45	52	58	64	72
SSB3	18	22	27	31	36	39	44	50
SSC1	16	22	28	32	39	43	49	57
SSC2	20	27	35	41	49	55	62	70
SSE1	22	27	34	38	45	49	55	62
SSE2	27	35	44	50	58	63	70	78
SSE3	20	24	29	33	38	42	47	53

BRYCE CANYON, UTAH **SSF (%)**   9136 DD65

Type	LCR = 100	70	50	40	30	25	20	15
WWA3	16	22	30	35	43	49	56	65
WWB4	21	31	42	49	60	67	75	84
WWC2	21	30	40	48	58	65	73	82
TWA1	16	19	22	25	30	33	38	44
TWA2	15	20	25	29	36	40	46	55
TWA3	15	20	26	31	38	43	50	59
TWA4	14	19	26	31	39	44	51	60
TWB3	13	18	24	29	35	40	47	55
TWD4	19	28	37	44	54	60	68	78
TWE2	20	28	38	45	54	60	68	78
TWF3	12	17	23	27	34	39	46	55
TWJ2	17	25	35	42	51	58	65	75
DGA1	10	14	18	21	25	28	30	34
DGA2	12	17	23	28	35	39	44	51
DGA3	16	23	31	37	46	52	60	69
DGB1	10	14	18	22	27	31	36	41
DGB2	13	18	24	29	37	42	49	58
DGB3	17	23	32	38	48	55	64	74
DGC1	13	17	23	28	35	39	46	54
DGC2	15	21	29	35	43	50	58	67
DGC3	19	27	36	43	55	62	72	81
SSA1	23	29	35	40	47	52	58	66
SSB1	18	23	29	34	40	45	51	59
SSB2	24	32	40	46	55	61	68	76
SSB3	17	22	27	32	38	42	47	55
SSC1	14	19	25	30	37	42	49	58
SSC2	18	25	33	40	49	55	62	72
SSE1	21	28	35	41	48	53	60	68
SSE2	27	36	46	53	62	68	75	83
SSE3	19	24	30	34	40	45	50	58

BURLINGTON, VERMONT **SSF (%)**   7878 DD65

Type	LCR = 100	70	50	40	30	25	20	15
WWA3	4	6	8	9	10	11	12	13
WWB4	6	10	16	20	26	31	36	44
WWC2	6	10	15	19	25	29	34	41
TWA1	10	11	11	11	11	11	11	10
TWA2	7	8	9	10	11	12	12	13
TWA3	6	7	8	10	11	12	13	14
TWA4	4	6	8	9	11	12	14	15
TWB3	5	7	8	10	11	12	14	16
TWD4	7	10	15	19	24	27	32	39
TWE2	7	11	15	18	23	26	31	37
TWF3	3	4	6	7	8	9	10	11
TWJ2	5	9	13	16	21	25	29	35
DGA1	2	2	2	1	1	0	−1	−5
DGA2	5	7	8	10	11	13	14	16
DGA3	8	11	15	17	22	25	29	34
DGB1	2	2	2	1	1	0	−1	−4
DGB2	5	7	9	10	12	13	15	17
DGB3	9	12	15	18	23	26	31	37
DGC1	3	4	4	4	5	4	3	2
DGC2	7	9	11	13	16	18	20	23
DGC3	10	13	18	21	26	30	35	42
SSA1	11	13	14	16	17	18	20	21
SSB1	9	10	11	13	14	15	16	17
SSB2	11	15	19	22	26	29	33	38
SSB3	8	9	10	10	11	11	12	11
SSC1	4	5	6	6	7	7	7	7
SSC2	5	8	11	14	17	19	22	27
SSE1	7	8	10	11	12	12	12	11
SSE2	9	13	17	21	25	28	32	37
SSE3	9	10	11	12	13	13	14	14

TABLE C.3 **Annual Passive Heating Performance: SSF** (*continued*)

NORFOLK, VIRGINIA **SSF** (%)							3493 DD65	
Type	LCR = 200	100	70	50	40	30	25	20
WWA3	12	22	29	37	42	51	56	63
WWB4	12	27	37	49	56	67	73	80
WWC2	12	27	37	47	55	65	71	78
TWA1	14	19	23	27	31	36	40	45
TWA2	12	20	25	31	36	43	48	54
TWA3	11	20	26	33	38	46	51	57
TWA4	10	19	25	33	38	46	52	59
TWB3	10	18	23	30	35	42	48	54
TWD4	12	25	34	44	51	61	67	74
TWE2	13	26	35	44	51	61	67	74
TWF3	8	16	22	29	34	42	47	53
TWJ2	10	23	32	41	48	58	64	72
DGA1	8	15	19	24	28	33	36	39
DGA2	9	17	23	29	35	42	46	52
DGA3	11	21	28	37	44	53	59	66
DGB1	8	15	20	25	30	36	40	45
DGB2	9	17	23	31	36	45	50	58
DGB3	11	21	29	38	45	56	63	71
DGC1	10	18	24	31	36	44	49	56
DGC2	11	20	27	36	42	52	58	66
DGC3	13	24	33	43	51	62	70	78
SSA1	16	25	31	38	43	50	55	62
SSB1	13	21	26	33	38	44	49	55
SSB2	16	27	35	44	50	59	64	71
SSB3	13	20	25	31	35	42	46	52
SSC1	10	19	25	32	37	45	50	57
SSC2	11	23	31	40	46	56	62	69
SSE1	15	25	31	39	45	53	58	64
SSE2	16	30	40	50	57	66	72	79
SSE3	14	22	27	33	38	44	49	55

ROANOKE, VIRGINIA **SSF** (%)							4314 DD65	
Type	LCR = 100	70	50	40	30	25	20	15
WWA3	17	23	30	35	42	47	54	63
WWB4	22	31	41	49	59	65	73	82
WWC2	21	30	40	47	57	64	71	80
TWA1	17	20	23	26	30	33	37	43
TWA2	16	21	26	30	36	40	45	53
TWA3	16	21	27	31	38	43	49	57
TWA4	15	20	26	31	38	43	50	59
TWB3	14	19	25	29	35	40	46	54
TWD4	20	28	37	44	53	59	67	76
TWE2	21	29	38	44	53	59	67	76
TWF3	13	17	23	28	34	39	45	53
TWJ2	18	26	35	41	50	57	64	74
DGA1	11	14	18	21	25	27	30	33
DGA2	13	18	24	28	35	39	44	50
DGA3	17	23	31	37	46	52	59	67
DGB1	11	15	19	22	27	30	35	40
DGB2	14	19	25	30	37	42	48	56
DGB3	18	24	32	38	48	54	63	73
DGC1	14	18	24	28	34	38	44	51
DGC2	16	22	29	35	43	49	56	65
DGC3	20	28	36	43	54	61	70	79
SSA1	21	27	32	37	43	48	54	61
SSB1	18	22	28	32	38	42	47	55
SSB2	23	30	38	44	52	57	64	73
SSB3	17	21	26	29	35	39	44	50
SSC1	15	20	26	30	37	41	47	56
SSC2	18	25	33	39	48	53	61	70
SSE1	20	26	33	37	44	49	55	63
SSE2	25	34	43	49	58	64	71	79
SSE3	18	23	28	32	37	41	46	53

RICHMOND, VIRGINIA **SSF** (%)							3941 DD65	
Type	LCR = 200	100	70	50	40	30	25	20
WWA3	9	17	23	30	35	43	48	54
WWB4	8	22	31	42	49	59	65	73
WWC2	9	22	31	41	48	57	64	71
TWA1	13	17	20	24	26	31	34	38
TWA2	11	17	21	26	30	36	40	46
TWA3	9	16	21	27	32	38	43	49
TWA4	8	15	21	27	32	39	44	50
TWB3	8	15	19	25	29	36	40	46
TWD4	9	20	28	37	44	53	59	67
TWE2	10	21	29	38	45	54	60	67
TWF3	6	13	18	24	28	35	39	45
TWJ2	8	19	26	35	42	51	57	64
DGA1	6	11	15	19	22	26	28	30
DGA2	7	14	19	24	29	35	39	44
DGA3	9	18	24	31	37	46	52	59
DGB1	6	11	15	20	23	28	31	35
DGB2	8	14	19	25	30	37	42	48
DGB3	10	18	25	33	39	48	55	63
DGC1	8	14	19	24	29	35	39	44
DGC2	9	17	23	30	35	43	49	56
DGC3	11	21	28	37	44	54	61	70
SSA1	14	21	27	33	37	43	48	53
SSB1	12	18	23	28	32	38	42	47
SSB2	13	23	30	38	44	52	57	64
SSB3	11	17	21	26	29	35	39	44
SSC1	8	15	20	26	31	37	42	48
SSC2	9	19	26	34	40	48	54	61
SSE1	12	20	26	33	37	44	49	55
SSE2	13	25	34	43	49	58	64	71
SSE3	13	19	23	28	32	37	41	46

SEATTLE-TACOMA, WASH. **SSF** (%)							5188 DD65	
Type	LCR = 100	70	50	40	30	25	20	15
WWA3	14	18	23	26	30	33	36	41
WWB4	17	25	32	38	46	51	57	64
WWC2	17	24	32	37	44	49	55	62
TWA1	16	18	20	22	25	26	28	31
TWA2	14	18	21	24	27	30	33	36
TWA3	14	17	21	24	29	31	35	39
TWA4	12	17	21	24	29	32	35	40
TWB3	13	16	20	23	27	30	34	38
TWD4	17	23	30	35	42	46	52	59
TWE2	17	24	30	35	42	46	51	58
TWF3	11	14	18	21	25	28	31	36
TWJ2	15	21	28	33	39	44	49	57
DGA1	9	12	14	16	17	18	18	18
DGA2	12	16	20	23	27	30	33	37
DGA3	16	21	27	31	37	41	46	53
DGB1	10	12	15	17	19	20	21	22
DGB2	13	16	21	24	29	32	35	40
DGB3	16	22	28	33	39	44	49	57
DGC1	12	15	19	21	24	26	28	30
DGC2	15	19	25	28	34	37	42	47
DGC3	19	24	31	36	44	49	54	62
SSA1	19	23	27	30	34	36	39	43
SSB1	16	20	23	26	30	32	35	38
SSB2	21	26	32	36	42	46	51	57
SSB3	15	18	22	24	27	29	31	34
SSC1	12	16	20	22	26	28	31	35
SSC2	15	20	26	30	36	40	45	51
SSE1	17	21	25	28	32	34	38	39
SSE2	21	27	34	38	44	48	53	58
SSE3	17	20	23	26	29	31	33	36

APPENDICES

TABLE C.3 **Annual Passive Heating Performance: SSF** (*continued*)

SPOKANE, WASHINGTON SSF (%) — 6839 DD65

Type	LCR = 100	70	50	40	30	25	20	15
WWA3	11	14	18	21	24	27	29	32
WWB4	14	20	28	33	40	45	51	58
WWC2	14	20	27	32	38	43	48	55
TWA1	14	15	17	19	21	22	23	25
TWA2	12	15	18	20	23	24	27	29
TWA3	11	14	17	20	23	25	28	31
TWA4	10	13	17	20	23	26	29	33
TWB3	10	13	17	19	22	25	28	32
TWD4	13	19	25	30	36	41	46	53
TWE2	14	20	26	30	36	40	45	52
TWF3	8	11	14	17	20	22	25	28
TWJ2	12	17	23	28	34	38	43	50
DGA1	7	9	10	11	12	12	12	11
DGA2	10	13	16	19	22	25	27	30
DGA3	13	18	23	27	33	36	41	47
DGB1	7	9	11	12	14	14	14	14
DGB2	10	13	17	20	24	26	29	33
DGB3	14	18	24	28	34	39	44	51
DGC1	9	12	14	16	19	20	21	21
DGC2	12	16	20	24	29	32	35	40
DGC3	16	21	27	32	39	43	49	56
SSA1	16	19	22	25	28	30	32	35
SSB1	13	16	19	21	24	26	28	30
SSB2	17	22	27	31	36	40	44	50
SSB3	12	15	17	19	21	22	24	25
SSC1	10	12	15	18	20	22	24	26
SSC2	12	17	22	25	30	34	38	43
SSE1	13	17	20	22	24	26	27	29
SSE2	17	22	28	32	37	41	45	50
SSE3	14	16	19	21	23	25	26	28

MADISON, WISCONSIN SSF (%) — 7731 DD65

Type	LCR = 100	70	50	40	30	25	20	15
WWA3	8	11	14	17	20	23	26	30
WWB4	10	16	24	29	37	42	49	58
WWC2	10	16	23	28	35	40	47	55
TWA1	12	13	14	15	17	18	19	20
TWA2	10	12	14	16	19	20	23	26
TWA3	8	11	14	16	19	21	24	28
TWA4	7	10	13	16	19	22	25	30
TWB3	8	10	13	15	18	21	24	28
TWD4	10	16	22	26	33	38	44	52
TWE2	11	16	22	26	33	37	43	51
TWF3	6	8	11	13	16	18	21	25
TWJ2	9	14	19	24	30	35	41	49
DGA1	4	5	6	7	8	8	7	6
DGA2	7	10	13	15	18	20	23	26
DGA3	11	15	19	23	28	32	38	45
DGB1	4	5	6	7	8	8	8	8
DGB2	8	10	13	15	19	21	24	29
DGB3	11	15	20	24	30	34	40	48
DGC1	6	8	10	11	13	14	15	16
DGC2	9	12	16	19	23	26	30	36
DGC3	13	17	23	27	34	39	45	55
SSA1	14	16	19	22	25	27	30	33
SSB1	11	13	16	18	21	22	25	28
SSB2	14	19	24	28	33	37	42	50
SSB3	10	12	14	16	18	19	21	23
SSC1	7	9	12	13	16	18	20	23
SSC2	9	13	18	21	27	30	35	42
SSE1	11	13	16	18	21	23	25	28
SSE2	14	19	25	29	35	39	45	52
SSE3	12	14	16	17	20	21	23	26

CHARLESTON, W. VIRGINIA SSF (%) — 4594 DD65

Type	LCR = 100	70	50	40	30	25	20	15
WWA3	11	15	20	23	28	31	36	42
WWB4	14	21	30	36	45	50	58	67
WWC2	14	21	29	35	43	48	55	65
TWA1	14	15	18	19	21	23	25	28
TWA2	12	15	18	21	24	27	30	35
TWA3	11	14	18	21	26	29	33	38
TWA4	10	13	18	21	26	29	34	40
TWB3	10	13	17	20	24	27	31	37
TWD4	14	20	27	32	40	45	52	61
TWE2	15	20	27	32	40	45	51	60
TWF3	8	11	15	18	22	25	29	35
TWJ2	12	18	25	30	37	42	49	58
DGA1	7	9	11	12	14	14	15	15
DGA2	10	13	16	19	24	26	30	35
DGA3	13	18	23	28	34	39	45	53
DGB1	7	9	11	12	14	15	17	18
DGB2	10	13	17	20	25	28	32	38
DGB3	13	18	24	29	36	41	47	56
DGC1	9	12	15	17	20	22	24	27
DGC2	12	16	21	24	30	34	39	46
DGC3	15	21	28	33	40	46	53	63
SSA1	17	20	24	27	32	35	39	44
SSB1	14	17	20	23	27	30	33	38
SSB2	18	23	29	34	40	45	51	59
SSB3	13	15	19	21	24	27	30	34
SSC1	10	13	16	19	23	26	30	35
SSC2	12	17	23	27	34	38	44	52
SSE1	14	18	22	26	30	33	37	42
SSE2	18	24	31	37	44	49	55	63
SSE3	14	17	20	23	26	29	32	36

MILWAUKEE, WISCONSIN SSF (%) — 7445 DD65

Type	LCR = 100	70	50	40	30	25	20	15
WWA3	8	11	14	16	20	22	25	30
WWB4	10	16	23	29	37	42	49	58
WWC2	10	16	23	28	35	40	46	55
TWA1	12	13	15	15	17	18	19	20
TWA2	10	12	14	16	18	20	22	26
TWA3	9	11	14	16	19	21	24	28
TWA4	7	10	13	15	19	22	25	29
TWB3	8	10	13	15	18	21	24	28
TWD4	10	16	21	26	33	37	43	52
TWE2	11	16	22	26	32	37	43	51
TWF3	6	8	11	13	16	18	21	25
TWJ2	9	14	19	24	30	35	40	49
DGA1	4	6	6	7	8	8	7	6
DGA2	7	10	13	15	18	20	23	26
DGA3	11	15	19	23	28	32	37	45
DGB1	4	5	6	7	8	8	8	8
DGB2	8	10	13	15	19	21	24	28
DGB3	11	15	20	24	30	34	40	48
DGC1	6	8	10	11	13	14	15	16
DGC2	9	12	16	19	23	26	30	36
DGC3	13	17	23	27	34	39	45	54
SSA1	14	17	20	22	25	27	30	34
SSB1	11	14	16	18	21	23	25	29
SSB2	15	19	24	28	34	38	43	50
SSB3	10	12	14	16	18	19	21	23
SSC1	7	9	12	13	16	18	20	23
SSC2	9	13	18	21	26	30	35	41
SSE1	11	14	16	19	21	23	26	28
SSE2	14	19	25	29	35	39	45	52
SSE3	12	14	16	18	20	22	24	26

TABLE C.3 Annual Passive Heating Performance: SSF (*continued*)

SHERIDAN, WYOMING SSF (%) — 7717 DD65

Type	LCR = 100	70	50	40	30	25	20	15
WWA3	12	17	22	26	31	35	40	47
WWB4	16	24	32	39	48	54	61	71
WWC2	16	23	31	38	46	52	59	68
TWA1	14	16	19	21	23	25	28	32
TWA2	13	16	20	23	27	30	34	40
TWA3	12	16	20	23	28	32	36	43
TWA4	11	15	20	23	29	33	37	44
TWB3	11	14	19	22	27	30	35	41
TWD4	15	22	29	35	43	48	55	65
TWE2	16	22	30	35	43	48	55	64
TWF3	9	13	17	20	25	28	33	39
TWJ2	13	20	27	32	40	46	53	62
DGA1	7	10	12	14	16	17	19	19
DGA2	10	14	18	21	26	29	33	38
DGA3	14	19	25	30	37	42	48	56
DGB1	7	10	13	15	17	19	21	23
DGB2	10	14	19	22	27	31	36	42
DGB3	14	19	26	31	38	44	51	60
DGC1	10	13	17	19	23	26	29	33
DGC2	13	17	23	27	33	37	43	51
DGC3	16	22	30	35	43	49	57	67
SSA1	17	21	25	28	33	36	40	46
SSB1	14	17	21	24	28	31	35	40
SSB2	18	24	30	35	42	46	52	60
SSB3	13	16	19	22	25	28	31	35
SSC1	11	14	18	22	26	30	34	40
SSC2	13	19	25	30	37	42	48	56
SSE1	15	19	24	27	32	35	39	44
SSE2	19	25	33	38	46	50	57	65
SSE3	15	18	21	24	28	30	34	38

SUFFIELD, ALBERTA — 9423 DD65

Type	LCR = 100	70	50	40	30	25	20	15
WWA3	11	15	19	23	27	30	34	39
WWB4	14	21	29	35	44	49	56	65
WWC2	14	21	28	34	42	47	53	62
TWA1	14	15	17	19	21	23	25	28
TWA2	12	15	18	21	24	27	30	34
TWA3	11	14	18	21	25	28	32	37
TWA4	10	13	18	21	25	29	33	38
TWB3	10	13	17	20	24	27	31	36
TWD4	14	20	27	32	39	44	50	59
TWE2	14	20	27	32	39	44	50	58
TWF3	8	11	15	18	22	25	28	33
TWJ2	12	18	24	29	37	41	48	56
DGA1	7	9	11	12	13	14	15	14
DGA2	10	13	17	19	24	26	30	34
DGA3	13	18	23	28	34	38	44	51
DGB1	7	9	11	13	15	16	17	18
DGB2	10	13	17	20	25	28	32	37
DGB3	13	18	24	29	36	41	47	55
DGC1	9	12	15	17	20	22	24	26
DGC2	12	16	21	25	30	34	39	45
DGC3	15	21	28	33	40	46	52	61
SSA1	15	18	22	24	28	30	33	37
SSB1	12	15	18	21	24	26	29	32
SSB2	16	21	27	31	37	41	46	53
SSB3	11	14	16	18	21	22	24	27
SSC1	10	13	16	19	23	25	28	33
SSC2	12	17	23	27	33	37	43	50
SSE1	13	16	19	22	25	27	30	33
SSE2	16	22	28	32	39	43	48	55
SSE3	13	16	18	20	23	25	27	30

Canada

EDMONTON, ALBERTA SSF (%) — 10671 DD65

Type	LCR = 100	70	50	40	30	25	20	15
WWA3	9	12	16	18	21	23	26	28
WWB4	11	18	25	30	37	42	48	56
WWC2	12	17	24	29	36	40	46	53
TWA1	12	14	15	17	18	19	20	22
TWA2	10	13	15	17	20	22	24	26
TWA3	9	12	15	17	20	23	25	28
TWA4	8	11	14	17	21	23	26	29
TWB3	8	11	14	17	20	22	25	28
TWD4	11	17	23	27	34	38	43	50
TWE2	12	17	23	27	33	37	42	49
TWF3	6	9	12	14	17	19	22	25
TWJ2	10	15	21	25	31	35	41	47
DGA1	5	7	8	9	9	9	9	7
DGA2	8	11	14	17	20	22	25	28
DGA3	12	16	21	25	30	34	39	45
DGB1	5	7	8	9	10	11	11	10
DGB2	8	11	15	17	21	24	27	30
DGB3	12	16	22	26	32	36	41	48
DGC1	7	9	12	13	15	17	18	18
DGC2	10	14	18	21	26	29	33	37
DGC3	14	19	25	29	36	41	46	53
SSA1	13	16	19	21	23	25	27	28
SSB1	11	13	16	17	20	21	23	25
SSB2	14	19	23	27	32	35	40	45
SSB3	10	12	14	15	17	18	18	19
SSC1	8	10	13	15	17	19	21	22
SSC2	10	14	19	23	28	31	35	40
SSE1	10	13	15	17	19	20	21	21
SSE2	13	18	23	27	32	36	40	45
SSE3	11	13	16	17	19	20	21	22

VANCOUVER, BRITISH COL. SSF (%) — 5607 DD65

Type	LCR = 100	70	50	40	30	25	20	15
WWA3	15	19	24	27	31	34	37	41
WWB4	18	25	33	39	46	51	57	65
WWC2	18	25	32	38	45	49	55	62
TWA1	16	18	21	23	25	27	29	32
TWA2	15	18	22	25	28	31	34	37
TWA3	14	18	22	25	29	32	35	40
TWA4	13	17	22	25	30	33	36	41
TWB3	13	17	21	24	28	31	35	39
TWD4	17	24	31	36	43	47	53	60
TWE2	18	24	31	36	42	47	52	59
TWF3	11	15	19	22	26	29	32	37
TWJ2	16	22	29	33	40	45	50	57
DGA1	10	12	15	17	18	19	20	19
DGA2	12	16	21	24	28	31	34	37
DGA3	16	22	28	32	38	42	47	53
DGB1	10	13	16	18	20	21	22	23
DGB2	13	17	22	25	30	33	36	41
DGB3	17	22	29	33	40	45	50	57
DGC1	13	16	20	22	26	27	29	31
DGC2	15	20	25	29	35	38	42	48
DGC3	19	25	32	37	45	49	55	62
SSA1	20	23	27	30	34	36	39	42
SSB1	16	20	24	26	30	32	35	38
SSB2	21	26	32	36	42	46	51	57
SSB3	15	19	22	24	27	29	31	34
SSC1	13	17	21	23	27	29	32	35
SSC2	16	21	27	31	37	41	45	51
SSE1	18	22	26	28	32	34	36	38
SSE2	21	28	34	38	44	48	52	58
SSE3	17	20	24	26	29	31	33	36

TABLE C.3 **Annual Passive Heating Performance: SSF** (*continued*)

WINNIPEG, MANITOBA **SSF** (%)							10776 DD65	
Type LCR = 100	70	50	40	30	25	20	15	
WWA3	8	10	13	16	19	21	24	28
WWB4	10	16	23	28	36	41	48	57
WWC2	10	16	22	27	34	39	45	54
TWA1	12	13	14	15	16	17	18	19
TWA2	10	11	14	15	18	20	22	25
TWA3	8	10	13	15	18	20	23	27
TWA4	7	9	13	15	18	21	24	28
TWB3	8	10	13	15	18	20	23	27
TWD4	10	15	21	25	32	37	43	51
TWE2	11	16	21	25	32	36	42	49
TWF3	5	8	10	12	15	17	20	24
TWJ2	8	13	19	23	29	34	40	47
DGA1	4	5	6	7	7	7	6	5
DGA2	7	9	12	14	18	20	22	26
DGA3	10	14	19	22	28	32	37	44
DGB1	4	5	6	7	7	8	8	7
DGB2	7	10	13	15	18	21	24	28
DGB3	11	15	19	23	29	34	39	47
DGC1	6	7	9	10	12	13	14	15
DGC2	9	12	16	19	23	26	30	35
DGC3	12	17	22	27	34	38	45	53
SSA1	12	15	17	19	22	23	26	28
SSB1	10	12	14	16	18	20	22	24
SSB2	13	17	22	25	31	34	39	46
SSB3	9	10	12	13	15	16	17	18
SSC1	6	9	11	13	15	17	19	21
SSC2	9	12	17	20	26	29	34	40
SSE1	9	11	14	15	18	19	21	22
SSE2	12	17	22	26	31	35	40	46
SSE3	11	12	14	15	17	18	20	21

OTTAWA, ONTARIO **SSF** (%)							8461 DD65	
Type LCR = 100	70	50	40	30	25	20	15	
WWA3	8	11	14	17	20	23	26	31
WWB4	10	16	24	29	37	43	50	59
WWC2	10	16	23	28	36	41	47	56
TWA1	12	13	14	15	17	18	19	21
TWA2	10	12	14	16	19	20	23	26
TWA3	8	11	14	16	19	22	25	29
TWA4	7	10	13	16	19	22	26	31
TWB3	8	10	13	15	19	21	24	29
TWD4	10	15	22	26	33	38	44	53
TWE2	11	16	22	26	33	38	44	52
TWF3	6	8	11	13	16	18	21	26
TWJ2	9	14	19	24	31	35	41	50
DGA1	4	5	6	7	8	8	7	6
DGA2	7	10	13	15	18	20	23	27
DGA3	11	14	19	23	28	33	38	46
DGB1	4	5	6	7	8	8	9	8
DGB2	7	10	13	15	19	21	25	29
DGB3	11	15	20	24	30	34	40	49
DGC1	6	7	9	11	13	14	15	17
DGC2	9	12	16	19	24	27	31	37
DGC3	13	17	23	27	34	39	46	55
SSA1	13	16	19	21	24	26	29	33
SSB1	11	13	15	17	20	22	24	28
SSB2	14	18	23	27	33	37	42	49
SSB3	10	11	13	15	17	18	20	22
SSC1	7	9	11	13	16	18	21	24
SSC2	9	13	18	21	27	31	36	43
SSE1	10	13	16	18	21	23	25	28
SSE2	13	18	24	28	35	39	44	52
SSE3	11	13	15	17	19	21	23	25

HALIFAX, NOVA SCOTIA **SSF** (%)							7167 DD65	
Type LCR = 100	70	50	40	30	25	20	15	
WWA3	10	13	17	20	25	28	32	38
WWB4	12	19	27	33	42	47	55	64
WWC2	12	19	26	32	40	45	52	61
TWA1	13	14	16	17	19	21	23	25
TWA2	11	13	16	19	22	24	27	32
TWA3	10	13	16	19	23	26	29	35
TWA4	8	12	16	19	23	26	30	36
TWB3	9	12	15	18	22	24	28	34
TWD4	12	18	24	30	37	42	49	58
TWE2	13	18	25	30	37	42	48	57
TWF3	7	10	13	16	19	22	26	31
TWJ2	10	16	22	27	34	39	46	55
DGA1	6	7	9	10	11	11	12	11
DGA2	8	11	15	17	21	24	27	31
DGA3	12	16	21	25	32	36	42	50
DGB1	5	7	9	10	11	12	13	14
DGB2	9	12	15	18	22	25	29	34
DGB3	12	17	22	26	33	38	45	54
DGC1	7	10	12	14	16	18	20	23
DGC2	10	14	18	22	27	31	36	43
DGC3	14	19	25	30	38	43	51	60
SSA1	15	19	22	25	29	32	36	41
SSB1	13	15	19	21	25	27	30	35
SSB2	16	21	27	31	38	42	48	56
SSB3	11	14	17	19	22	24	26	30
SSC1	8	11	14	17	20	23	26	30
SSC2	11	15	21	25	31	35	41	48
SSE1	13	16	20	23	27	29	33	37
SSE2	16	22	29	34	41	46	52	59
SSE3	13	16	19	21	24	26	29	33

TORONTO, ONTARIO **SSF** (%)							7280 DD65	
Type LCR = 100	70	50	40	30	25	20	15	
WWA3	8	11	14	17	21	23	27	31
WWB4	10	17	24	30	38	43	50	59
WWC2	11	16	23	28	36	41	47	56
TWA1	12	13	15	16	17	18	19	21
TWA2	10	12	14	16	19	21	23	27
TWA3	9	11	14	16	20	22	25	29
TWA4	7	10	13	16	20	22	26	31
TWB3	8	10	13	16	19	21	24	29
TWD4	11	16	22	27	33	38	44	53
TWE2	11	16	22	27	33	38	44	52
TWF3	6	8	11	13	16	19	22	26
TWJ2	9	14	20	24	31	35	41	50
DGA1	5	6	7	7	8	8	8	7
DGA2	8	10	13	15	18	21	23	27
DGA3	11	15	19	23	29	33	38	46
DGB1	4	5	7	7	8	9	9	9
DGB2	8	10	13	16	19	22	25	29
DGB3	11	15	20	24	30	35	41	49
DGC1	6	8	10	11	13	14	16	17
DGC2	9	12	16	19	24	27	31	37
DGC3	13	17	23	28	34	39	46	55
SSA1	14	17	20	22	25	28	31	35
SSB1	11	14	16	18	21	23	26	29
SSB2	15	19	24	28	34	38	43	51
SSB3	10	12	14	16	18	20	22	24
SSC1	7	9	12	14	17	18	21	24
SSC2	9	13	18	22	27	31	36	43
SSE1	11	14	17	19	22	24	27	30
SSE2	14	19	25	30	36	40	46	53
SSE3	12	14	16	18	20	22	24	27

TABLE C.3 **Annual Passive Heating Performance: SSF** (*continued*)

NORMANDIN, QUEBEC **SSF (%)**   11364 DD65

Type	LCR = 100	70	50	40	30	25	20	15
WWA3	6	9	11	14	17	19	21	25
WWB4	8	14	21	26	33	39	45	54
WWC2	8	14	20	25	32	36	43	51
TWA1	11	12	13	13	14	15	16	17
TWA2	8	10	12	13	16	17	19	22
TWA3	7	9	11	13	16	18	20	24
TWA4	5	8	11	13	16	18	21	25
TWB3	6	8	11	13	16	18	20	24
TWD4	8	13	19	23	30	34	40	48
TWE2	9	14	19	23	29	34	39	47
TWF3	4	6	8	10	13	15	17	20
TWJ2	7	11	17	21	27	31	37	45
DGA1	3	3	4	5	5	4	4	2
DGA2	6	8	10	13	15	17	20	23
DGA3	9	13	17	20	26	30	35	42
DGB1	3	3	4	4	5	5	5	4
DGB2	6	8	11	13	16	18	21	25
DGB3	10	13	18	21	27	31	37	45
DGC1	4	6	7	8	9	10	11	12
DGC2	8	10	14	16	20	23	27	33
DGC3	11	15	20	25	31	36	42	51
SSA1	12	14	16	18	21	23	25	27
SSB1	9	11	13	15	17	19	20	23
SSB2	12	16	21	24	30	33	38	44
SSB3	8	10	11	12	14	15	16	17
SSC1	5	7	9	11	13	14	16	18
SSC2	7	11	15	18	23	27	31	37
SSE1	8	10	13	14	16	18	19	20
SSE2	11	16	21	25	30	34	39	45
SSE3	10	11	13	14	16	17	19	20

D

SUNSHADING

D.1 Introduction

This appendix presents a graphic-analytic method for the design of fixed sunshading devices. Although it is assumed that the reader is to an extent familiar with the principles of sunshading and shading mask construction, a brief review of these topics is presented, to develop the necessary background for the design methods that will be demonstrated. It is further assumed that the reader recognizes the importance of sunshading and the importance of the effect of insolation on thermal gain (and loss), daylighting, glare, and so on. A full discussion of these topics is found in Chapter 1 and 6 of this volume.

The shadow and dimensions of a projection from a wall depends on:

1. The shape and dimensions of the projection.
2. The direction in which the wall faces.
3. The position of the sun.

The first two factors are fixed; the third varies with the hour of the day and the date, that is, with the apparent motion of the sun.

D.2 Solar Position Altitude and Azimuth

Refer to Fig. 19.2(a) and Tables B.11 to B.14. This position of the sun at any instant with respect to an observer on the ground is defined by its *altitude angle A* and its *azimuth angle Az*. Note that in our discussion *azimuth is measured from south.* (In other texts it is measured from north; there is no consensus on this point.) The altitude of the sun varies during the day, starting and ending at 0° at sunrise and sunset, and reaching a daily maximum at *solar* noon. (See Section D.3 for explanation of solar and clock time.) The altitude angle at noon varies from day to day, reaching a yearly maximum at the summer solstice on June 21, a minimum at the winter solstice on December 21 and a point halfway between on the vernal and autumnal equinoxes of March 21 and September 21. These seasonal changes are shown graphically on Fig. 19.2(b) and Figure D.1. Note that the altitude change is approximately 23½° every 3 months, making the daily change approximately a quarter-degree and the maximum difference in altitude angle about 47°, between the summer and winter solstices. See Tables 19.2 and B.11 to B.14 for accurate altitude/azimuth figures.

Referring to the ecliptic (Fig. D.1a), we can see that the height of the sun (its altitude angle), depends on the observer's position on the earth; specifically, his latitude. This can be expressed as follows:

maximum solar altitude = 113.5°—latitude

equinox solar altitude = 90°—latitude

minimum solar altitude = 66.5°—latitude

These relations are shown graphically in Fig. 19.2(c).

On the equinoxes (March 21 and September 21), the sun rises due east and sets due west. In the northern hemisphere, the sun rises north of east between March 22 and September 20 and sets north of west. This is shown graphically on Fig. 19.2(b) and is projected, in Fig. D.2, for one specific latitude. Note the positions of sunrise and sunset with respect to the E–W line, to understand the explanation above.

D.3 Sun Time and Clock Time

Although for the purpose of sunshading it is not normally important to differentiate between sun time and clock time, it is important that the reader know that such a difference exists, and can be as much as an hour. The following is a brief discussion of timekeeping. The interested reader can refer to the References for a more detailed explanation. (In our discussion we will refer to the sun's motion rather than apparent motion, since

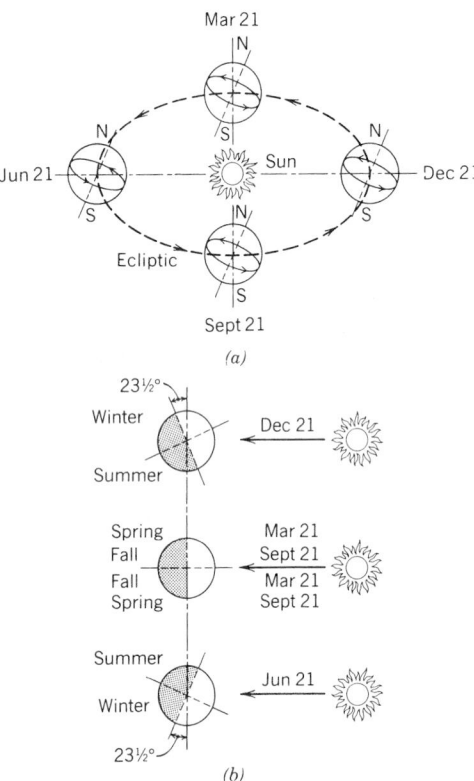

Fig. D.1 *The ecliptic is the annual path of the earth around the sun. The tilt of the earth's axis to the plane of the ecliptic results in the seasonal variations. Note in* (a) *and* (b) *that the northern hemisphere tilts away from the sun in December, resulting in winter's low sun altitude and cold weather. In June the effect is reversed.*

(clock) and apparent (actual) solar time is called the *equation of time* and can be determined from a curve called the *analemma* or from tabulations in various sources. Furthermore, the 360° earth circumference is divided into 24 one-hour time zones, each therefore separated from the adjacent zone by 15° of longitude, which corresponds to approximately 1000 miles. Therefore an observer at any point other than directly on a time meridian (which is normally in the center of the time zone) must make a time correction of up to 30 min to correspond to actual solar time. (In some countries a reference meridian is more than 30 min from the time zone extremes. In such places, the longitude correction plus the equation of time correction can total well over an hour, and one is no longer able to neglect it. This situation does *not* exist in the continental United States.) Although not vital, because the length of shaded time is the same, it is good practice to make a table of mean time versus apparent solar time for any project where shading and insolation are important. In any case, the designer must remember that unless otherwise specifically noted (and this is rare), all sun charts show sun time, that is,

relative to the observer, the sun does indeed move.)

The basis of all timekeeping is the length of the solar day, that is, one full rotation of the earth on its axis. More rigidly defined, it is the interval between two sun passages across the observer's meridian. However, because the sun moves about 1° per day on the ecliptic (360° in 365 days), it must move approximately 361° in a solar day. Furthermore, the sun's motion on the ecliptic is not uniform—it moves more rapidly when farther from the earth. As a result, the actual length of the solar day varies. Since it is impractical for us to use timepieces that need daily adjustment, we use an average, or *mean solar day,* as the basis for timekeeping. The difference between mean

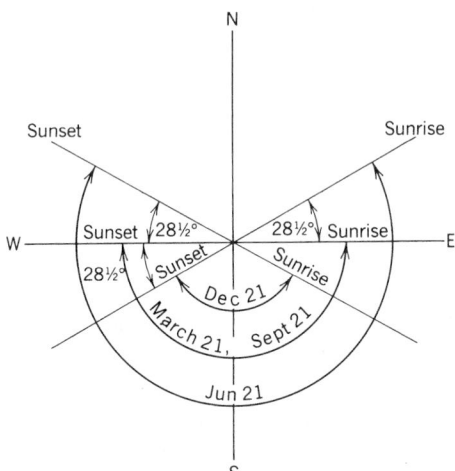

Fig. D.2 *Projected sunpaths for solstices and equinoxes at latitude 32°N (Jerusalem, Savannah, Tucson). The sun's azimuth angle, like its altitude, varies with time of day and date. The sun's azimuth path from sunrise to sunset, for this latitude is shown in horizontal projection. Sunrise on June 21 is 28½° North of East, and sunset is correspondingly North of West. On December 21 sunrise and sunset are 28½ South of East and West.*

APPENDICES

actual solar time. This includes the well-known Libbey-Owens-Ford (LOF) *Sun Angle Calculator* (1975).

D.4   Sunpath Projection

The design of sunshading devices necessitates projecting the three-dimensional solar path onto a two-dimensional surface. There are basically two types of projection: vertical and horizontal. A version of the former is familiar to users of Waldram diagrams in daylight design. Horizontal projec-

tions are more common. The basic idea behind the horizontal projection is shown in Fig. D.3. The sky vault is represented as a hemisphere that projects as a circle, the center of which is the zenith. The observer looks down onto the hemispherical sky vault and projects the sunpaths onto the two-dimensional, projected circle. The four methods of projection are:

1. Gnomonic
2. Orthographic
3. Stereographic
4. Equidistant

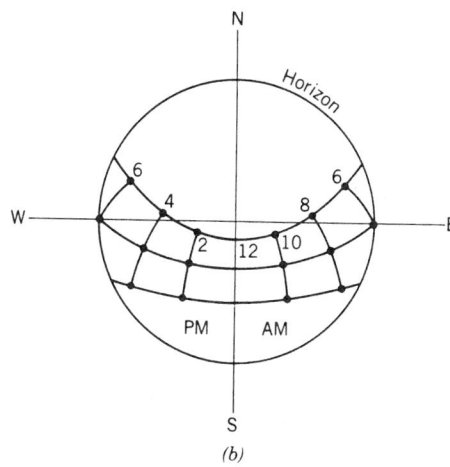

Fig. D.3 *Basis of horizontal projection of the sunpaths. The sky vault* (a) *is represented as a hemisphere, which projects as a circle* (b) *on the projection plane. The sunpaths in* (a) *project as curved lines on the projection plane* (b), *the exact shape of which depends on the projection method (Fig. D.4). The numbers on the sunpaths in* (a) *are hours: 6 AM, 8 AM, 10 AM, 12 noon, 2 PM, 4 PM, 6 PM. When hour points are connected between sunpaths* (b), *they form hour lines, which are useful for interpolation in completed diagrams.*

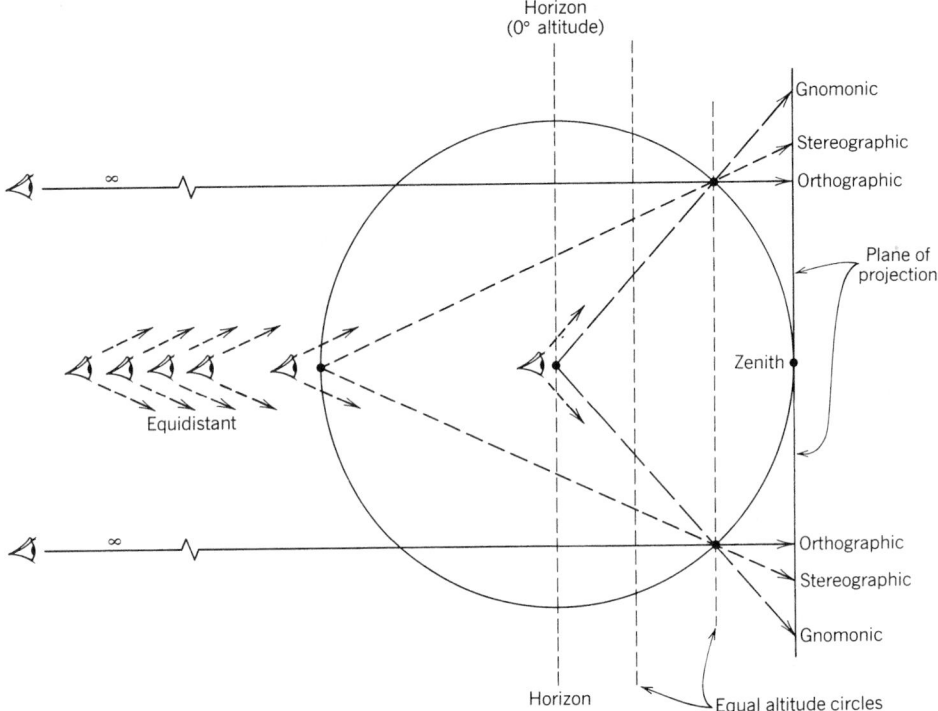

Fig. D.4 *The four standard horizontal projection methods, with observer position and projected point. Note that in the equidistant projection the observer position varies and therefore the projected point also varies. The projected point is therefore not shown. Note also that as the altitude drops and approaches the horizon, the gnomonic projection extends out to infinity.*

The position of the observer's eye for each of these projections is shown in Fig. D.4. Spacing between circles of equal altitude is the basic difference between the four methods. In all of them, lines of equal azimuth project as straight lines emanating from the center of the circle.

1. The *gnomonic projection* is derived from the sundial. The observer is at the center of the projection; therefore low sun angles (sunrise, sunset) extend to infinity. For this reason, this method is rarely used for solar charts but is used frequently for building shadow studies. The method is illustrated in the sun peg shadow charts of Fig. 3.13.

2. The *orthographic projection* views the sky vault from infinity and projects as in a standard architectural plan, that is, with all projection lines parallel.

3. The *stereographic projection* is used widely throughout the world, except in the United States. The observer is fixed at the top of the

hemisphere, resulting in wide-spaced, low-angle altitudes. This makes it particularly useful for northern latitudes, that is, above 40°N, since the sun at these latitudes is usually low in the sky. A special protractor is required (see Section D.5). The principal advantage of this projection is that it is very simple to draw sunpaths at any scale and for any latitude without having to rely on published sunpaths. Thus for any serious sunshading design, the designer can prepare a large-scale sunpath chart at the exact latitude of the project and do accurate graphical analysis. For a detailed description of the method of preparing the charts, and protractor, see Szokolay (1980, pp. 314–317).

4. The *equidistant projection* is used almost exclusively in the United States because of the ready availability and wide acceptance of the LOF sunpath charts (*Sun Angle Calculator,* 1975). These charts, eight of which are illustrated in Fig. D.5, show the latitude circles as spaced equidistantly, a projection that requires

24°N LATITUDE

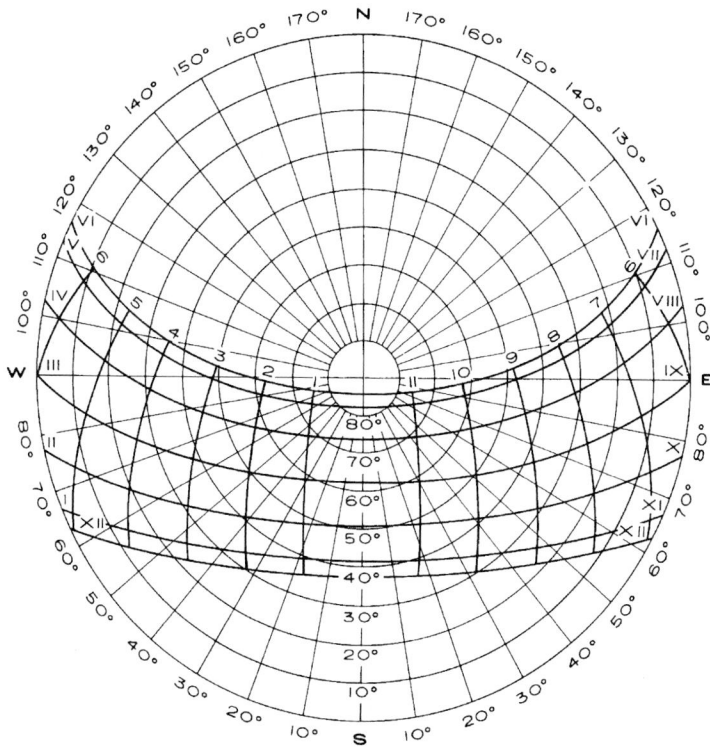

28°N LATITUDE

Fig. D.5 *Set of equidistant sunpath projections at 4° latitude intervals. The projection for 32°N latitude is approximately that of Savannah, Georgia, Tucson, Arizona, San Diego, California, and Jerusalem, Israel. In all of these locations sun shading is desirable at least from the vernal to the autumnal equinox. (Reprinted with permission from* Architectural Graphic Standards, *8th ed., 1988, John Wiley, Publishers.)*

32°N LATITUDE

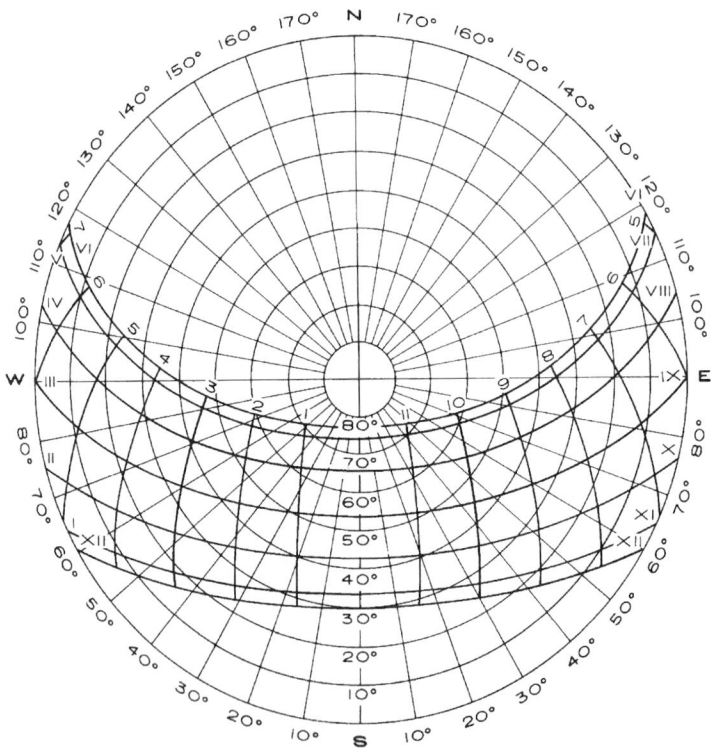

36°N LATITUDE

Fig. D.5 (continued)

40°N LATITUDE

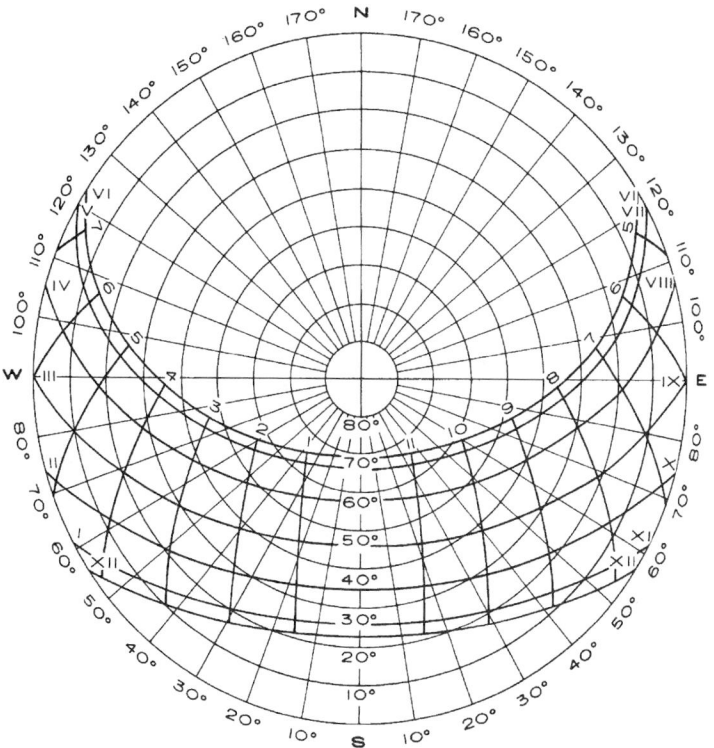

44°N LATITUDE

Fig. D.5 *(continued)*

48°N LATITUDE

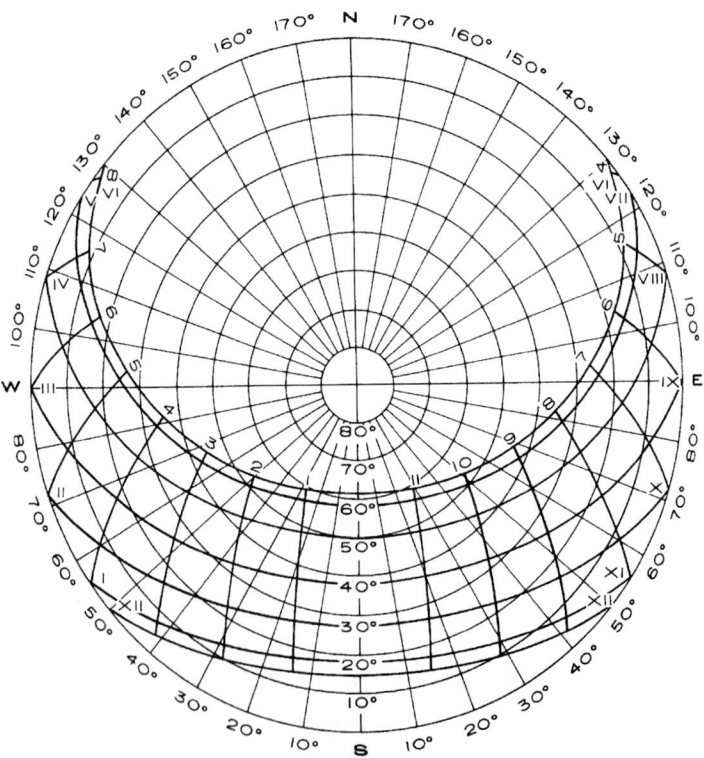

52°N LATITUDE

Fig. D.5 *(continued)*

moving the observer during projection, as shown in Fig. D.4. This method is equally applicable to all latitudes. Its disadvantage is the difficulty of preparing sunpaths and the required shadow angle protractor. Thus the user is essentially limited to published material, with its limitations of size and latitude. The LOF diagrams are available only in steps of 4° of latitude.

D.5 Shadow Angles and Shading Masks

The sun's altitude and azimuth angles are very useful in developing the sunpath diagrams but are much less so in defining the shadow angles cast by a projection, on a wall exposed to the sun. Since there is no universal agreement on angle nomenclature, we will very carefully define the angles involved. Refer to Fig. D.6, which shows the shadow cast by a horizontal overhang on a wall exposed to sunlight. Note that the shadow is defined by two angles; the *vertical shadow angle* (VSA), which indicates the position of the leading edge of the shadow as defined from the leading edge of the overhang, and the *horizontal shadow angle* (HSA), which defines the leading edge of a shadow cast by a vertical element (absent here, and indicated by a dotted line), as defined with respect to that element's leading edge. The terms VSA and HSA are in use throughout the world,

although in the United States, the vertical shadow angle is also known widely as the *profile angle* because it is so designated on the LOF sun angle calculator. Figure D.7 shows the same information in slightly different form; line *DA* is the shadow cast by line *DE*, which is the intersection between a horizontal projection and a vertical projection. This line is particularly important in determining the required extent of a shading element, as will be shown below.

Refer now to Fig. D.8. The locus of the leading edge of the shadow cast by all horizontal elements with the same VSA (profile angle) [Fig. D.8a-1] projects as a segmental line (Fig. D.8a-2), and when plotted on the projection plane (similar to the sunpath projection in Fig. D.3), shows as a segmental mask (Fig. D.8a-3). To draw this segment, a protractor is required, as shown in Fig. D.8c. Note that the angular notation on the segmental section of the protractor corresponds to VSA as shown in Fig. D.8a-3. Users of the LOF sun angle calculator can use the profile angle overlay to draw the required segment. It is important not to confuse the VSA with its complement. A simple way to remember this is to consider the limits of VSA. With no shading, VSA is 90° and the segment on Fig. D.8a-3 disappears. With a very deep overhang, VSA approaches 0° and the *unshaded* area in Fig. D.8a-3 shrinks. This action shows clearly on the protractor of Figure D.8c.

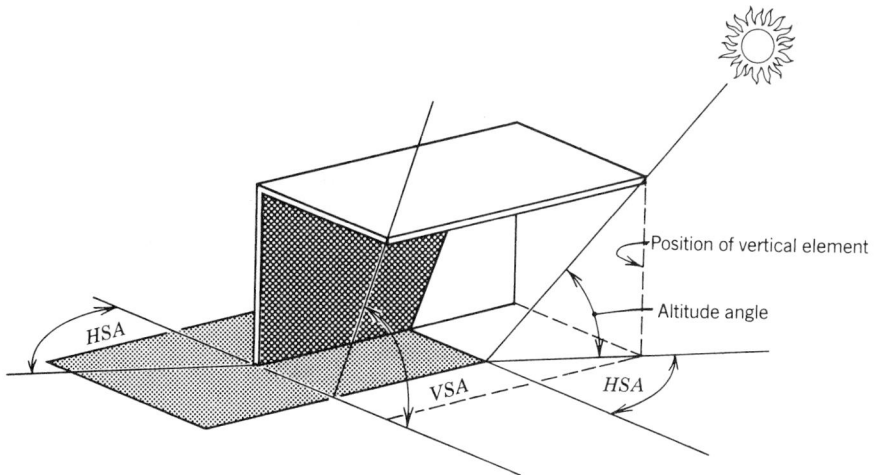

Fig. D.6 *The shadow cast by a horizontal overhang is best defined by angles VSA and HSA, the vertical shading angle and horizontal shading angles, respectively. VSA is also known in the United States as the profile angle.*

Fig. D.7 *The shadow DA cast by the intersection line DE, between a horizontal and vertical shading element, defines the edge of each of these shadows. DE can also be thought of as a pin, normal to, and extending from the wall, casting a shadow DA on the wall. The location and size of line DA are best defined in terms of angles VSA and HSA.*

VSA = vertical shadow angle
HSA = horizontal shadow angle
AL = altitude angle
DA = shadow of line DE

$$\tan HSA = \frac{AB}{BC} = \frac{AB}{DE}$$

$$\tan VSA = \frac{DB}{DE} = \frac{\tan AL}{\cos HSA}$$

Refer now to Fig. D.8*b*. Vertical shading elements Fig. D.8*b*-1 with the same HSA show as radii (*b*-2) and project as radial lines from the center, on the radial mask, at angle HSA from normal to the wall (*b*-3). These radial lines can be drawn with the assistance of an ordinary protractor, or by use of the protractor in Fig. D.8*c*. The reader must bear in mind that full segments, and full radial lines are the shading masks of *infinitely long elements,* which, of course, do not actually exist. Shading masks that appear in the literature (Olgyay, 1975; Ramsey and Sleeper, 1981) are frequently drawn as if for infinite elements, and the reader must know how to truncate them if accurate shading design is desired.

D.6   Use of Shading Masks

Once a shading mask has been drawn, it is applicable to any latitude and exposure direction. The use of the mask is straightforward. It is drawn on some transparent medium (e.g., paper or plastic sheet), laid on top of a sunpath diagram of the proper latitude, *and drawn to the same scale.* Its centerpoint is placed on the centerpoint of the sunpath diagram, and it is rotated until its facing direction (the direction of a normal to the wall) is aligned with the appropriate azimuth line on the sunpath diagram. Assuming that the shading mask has been correctly drawn to provide 100% shading for the *entire window,* the shaded hours for the window in question can then be read directly from the sunpath diagram underneath. (This is why the shading mask must be drawn on a transparent medium.) See Fig. D.9.

It is extremely important to keep in mind that a horizontal overhang shading element the same width as the window can provide full shade for the window only when the sun is exactly opposite the window, that is, when the bearing angle is 0° (see Fig. 19.7). This situation obtains for only an instant (Fig. D.10). At all other times, some part of the window will be exposed to the sun. There-

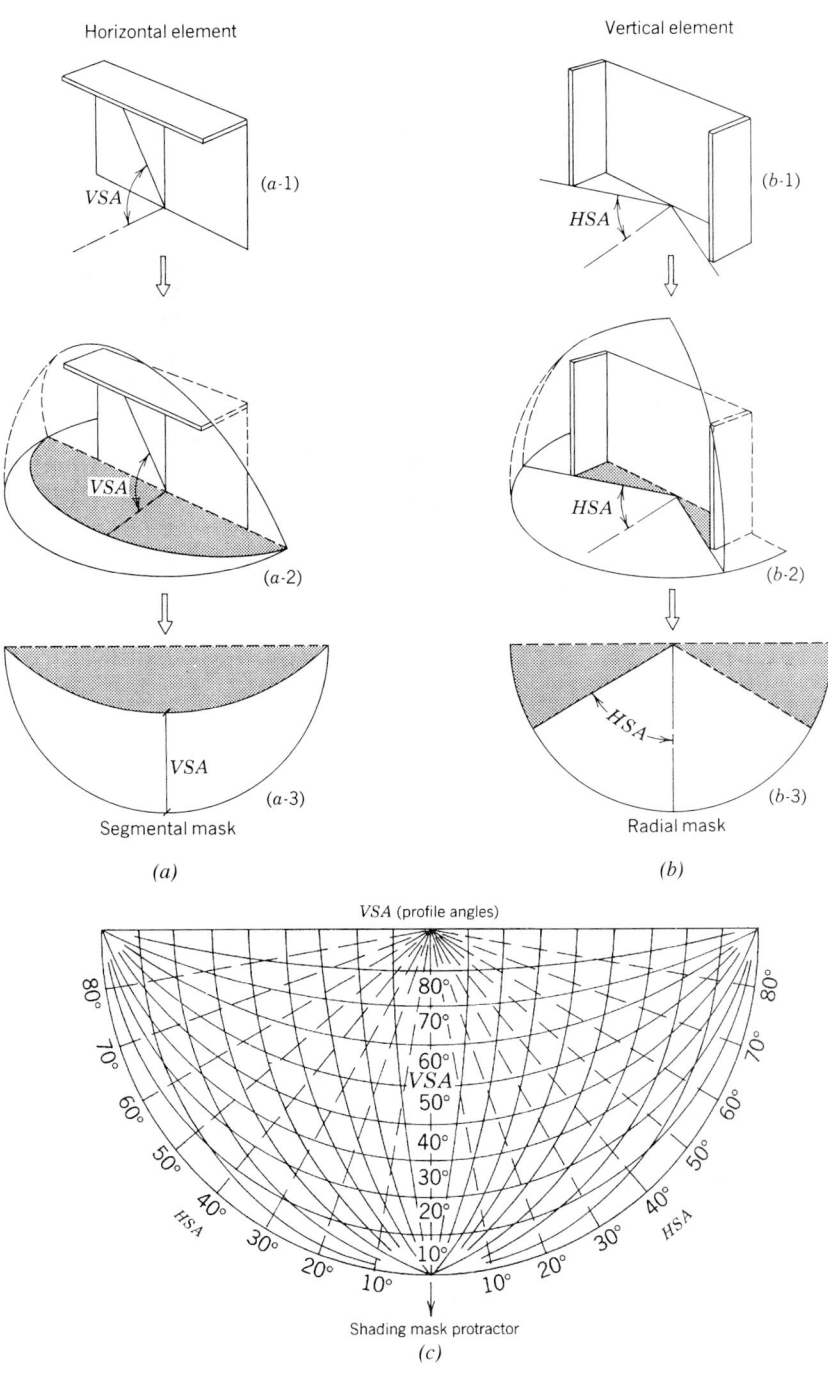

Fig. D.8 *A shading mask is the horizontal projection of the shadow cast by elements with the same VSA or HSA. Thus any infinitely long* horizontal element *with the VSA shown in (a-1) will have a shading mask as shown in (a-3). Similarly,* any infinitely high *vertical elements with the HSA shown in (b-1) will have the shadow mask shown in (b-3). A protractor of the type shown in (c) is required to draw the segmental horizontal element mask properly. This device must be drawn to the same scale and with the same projection method* (equidistant, stereographic) *as the sunpath diagram on which it will be used. The protractor of (c) matches the sunpath diagrams of Fig. D.5 and is drawn with equidistant projection.*

APPENDICES

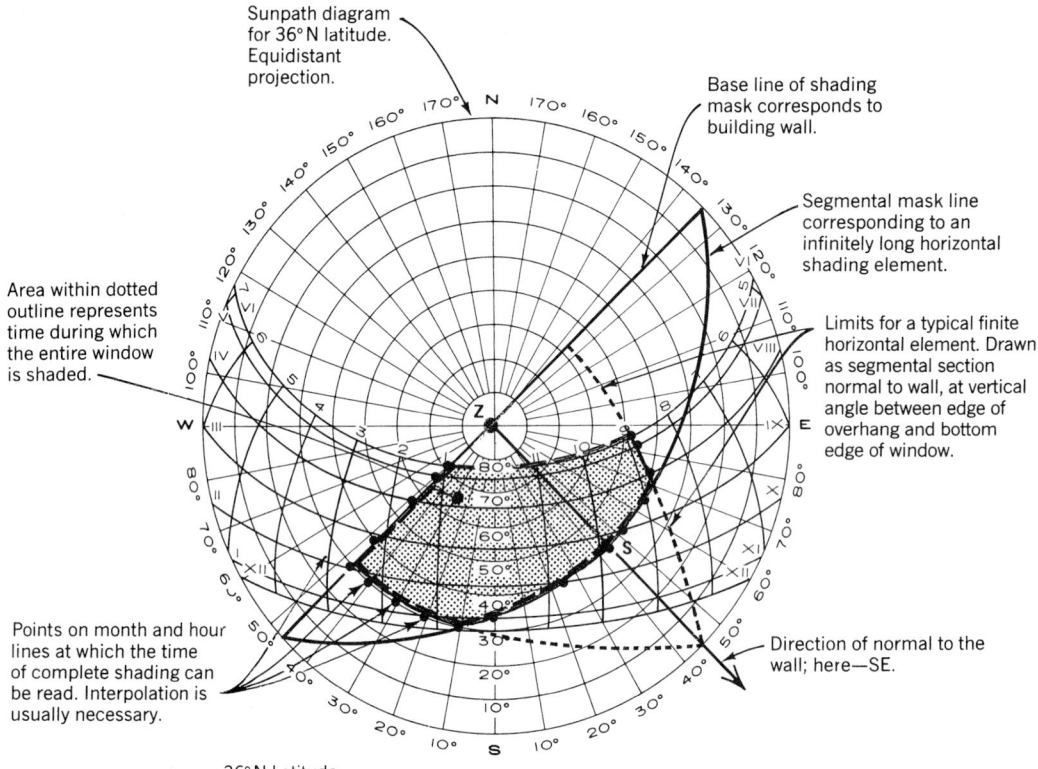

Sunpath diagram for 36°N latitude. Equidistant projection.

Base line of shading mask corresponds to building wall.

Segmental mask line corresponding to an infinitely long horizontal shading element.

Limits for a typical finite horizontal element. Drawn as segmental section normal to wall, at vertical angle between edge of overhang and bottom edge of window.

Area within dotted outline represents time during which the entire window is shaded.

Points on month and hour lines at which the time of complete shading can be read. Interpolation is usually necessary.

Direction of normal to the wall; here—SE.

36°N Latitude

Fig. D.9 *A typical shading mask for a horizontal overhang on a window is laid over a sunpath diagram, to permit determination of the shaded hours. The mask is drawn to provide full shading for the entire window. The hours during which the entire window is shaded are determined by the intersection of the mask perimeter with the date/hour lines of the sunpath diagram, and are tabulated on the figure. However, only at the hours falling along the normal-to-the-wall line ZS (where the bearing angle is 0°) and within the dotted area is a window fully shaded by an overhang element only as wide as the window. Indeed, line ZS is the shading mask of such an element with respect to 100% shading of the entire window (see Fig. D.10). Note that the element shown here is not symmetrical about the center, indicating an extension to the left to provide a larger shade period after noon. This is typical with east-facing windows, and is reversed for western exposures.*

Time of Day When Window is Fully Shaded

With Required Overhang Extensions			With Element Same Width as Window
Date	From	To	Time
21 June	0900	1240	1115
21 July/May	0850	1250	1100
21 Aug./Apr.	0845	1330	1035
21 Sept./Mar.	0940	1400	0955
21 Oct./Feb.	1040	1410	—
21 Nov./Jan.	12N	1300	—
21 Dec.	1230		—

fore, to provide full shading for more than an instant with only a horizontal element and no vertical (side) elements, *the overhang must extend beyond the sides of the window.* The amount of

such extension can be determined both graphically and analytically. Since graphic solutions are amply treated in the literature (Harkness and Mehta, 1978; Lim et al., 1979; Cowan, 1980), we will restrict ourselves to the analytic solution, leaving to the reader the parallel graphic solution. In doing this work, the reader should keep in mind the requirement that shading masks for 100% coverage must be prepared in order that the extremities of the opening be shaded.

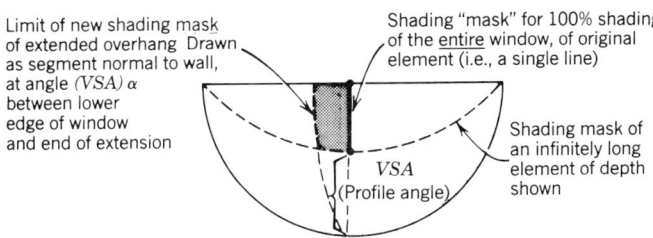

Fig. D.10 *A horizontal overhang with sufficient depth to provide 100% shading (i.e., to cast a shadow reaching to the window sill line, but only as wide as the window) can provide full shading only when the sun is exactly opposite the window. At any other time (bearing angle ≠ 0), part of the window will be exposed to the sun. To compensate for this, if full shading is desired, an extension is required, as shown. For south-facing windows the extension is symmetrical on both sides. For east-facing windows an extension to the left is usually used to provide full shading for hot afternoon hours; for west-facing windows, the reverse is customary. Full shading from the low early morning or late afternoon sun is not practical with horizontal elements.*

D.7 Finite Horizontal Elements

The first design step is to establish the required *depth* of the unit, that is, the amount it projects from the wall. Although any percent coverage desired is possible, most elements are designed to give either 50 or 100% coverage. The reason for this is explained in the illustrative design example below. Figure D.11 shows how the required depth and the corresponding segmental mask are determined. The next step is to determine the required side extensions beyond the window's sides. This can best be illustrated by a practical example, using the analytic technique.

First, however, a few remarks about the example presented are in order.

Due to the symmetry of solar motion about the ecliptic, the position of the sun is symmetrical on both sides of its maximum/minimum positions, that is, the solstices. Thus the sun's position on May 21 is the same as on July 21, since both dates

are one month from the summer solstice. As a result, any *fixed* shading device will give the same shade in the spring (before 21 Jun) as in summer (after 21 Jun). Since in many locations spring is cool, the desired summer shading will produce spring shading that *may not* be desirable. The solutions to this apparent dilemma are either to use a movable or variable size fixed shading device, or to compromise with the amount of shading, that is, to design for 50% shading for late summer, giving 50% insolation in early spring and 100% shading in early summer (and late spring).

The example that follows requires 100% shading for most of the summer (spring) for two reasons:

1. At the selected latitude of 32°N, sunshading is frequently desirable in the spring.
2. We wish to show that 100% shading using only a horizontal overhang can result in a very large element and that the shading

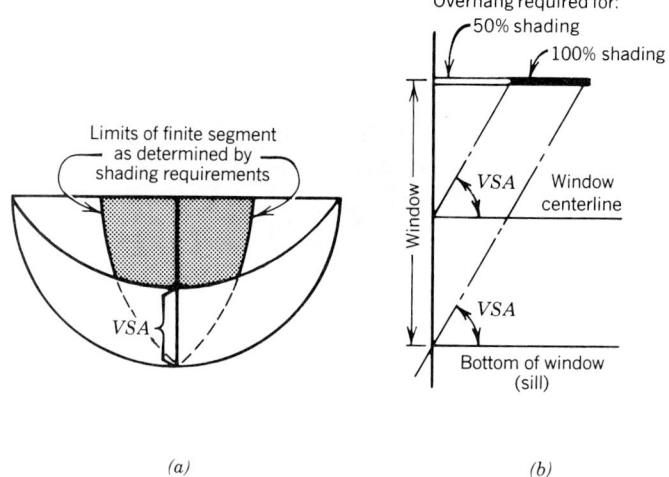

(a)   *(b)*

Fig. D.11 *To find the required depth of a horizontal overhang, having established the required segmental shading mask (a) it is only necessary to read VSA (profile angle) off the protractor in Fig. D.8c that corresponds to the segment in (a) and plot this angle as shown in (b) (wall section). Angle VSA can be plotted for 50% coverage, 100% coverage, or any other figure. The depth of the overhang can then be measured directly from the section, which is drawn to scale. Alternatively, if the overhang depth is known, the corresponding segment can be drawn.*

characteristic of such an element can be duplicated with a much smaller combined horizontal-vertical element.

EXAMPLE D.1. Design a simple horizontal over-the-window shading element that will provide 100% window shading from 1 PM to 4 PM, from April 21 to September 21 for a window 100 cm (39 in.) high and 150 cm (59 in.) wide, in a wall facing 20° west of south. The building is at latitude 32°N.

ANALYTIC SOLUTION.

STEP 1. The first step is to draw a sketch (Fig. D.12a) showing the orientation of the building. All the sketches connected with this example are shown in Fig. D.12.

STEP 2. Using either the LOF sun angle calculator, or Fig. D.5, tabulate the sun positions for the four times that bracket the shading period. Tabulate them as shown in Fig. D.12b and show sun directions as in Fig. D.12a. The purpose of establishing sun direction is to be able to immediately determine the required shading element end treatment, by sketching the direction in which the

shade falls. This is shown in the fourth column of the tabulation of Fig. D.12b.

STEP 3. By inspection of the solar azimuth and the bearing angle (the angle between a perpendicular to the wall and the vertical plane through the sun), it is apparent that all shadows fall to the right, and therefore the horizontal element must be lengthened to the left. Maximum extension is required for shading at 1600 on April 21. The next step is to establish the shadow angles VSA and HSA for each condition. These can be read directly from the LOF sun chart (profile angle = VSA; bearing from normal line = HSA) or by overlaying the protractor of Fig. D.8c on Fig. D.5. HSA need not be read off; it is simply the difference between sun azimuth and building bearing (20°) and so can be calculated directly. Fill in the required angles in the VSA and HSA columns of the table in Fig. D.12b.

STEP 4. At this point it is advisable to draw sketches of the window in elevation, section, and plan to understand the calculations. See Fig. D.12c. It is necessary for us first to establish the depth of the overhang, and with that known, to then calculate the required extension of the left.

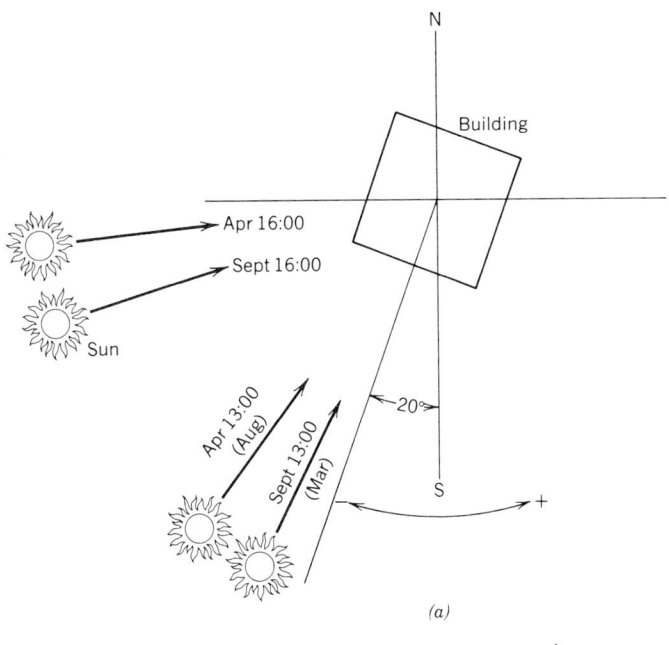

(a)

Date	Time	Sun Azm	Shade direction	VSA	HSA[a]	DE[b] (depth, cm)	AB[c] (extension, cm)
Apr 21	13:00	−37°	▨	67°	−17°	42	−40
	16:00	−84°	◨	54.5°	−64°	71	−266
Sept 21	13:00	−27°	▨	55°	−7°	70	−16
	16:00	−73°	▨	37.5°	−53°	130*	−172

[a] $HSA = |\text{Sun azm}| - |\text{Bldg. bearing}|$, i.e., $|\text{Azm}| - |20°|$
[b] $DE = \text{Window height}/\tan VSA = 100/VSA$
[c] $AB = DE \tan HSA = \text{overhang depth} \times \tan HSA = 130 \tan HSA$;
    the negative sign indicates an extension to the left.

(b)

Fig. D.12 *Sketches for solution of Example D.1 (see text).*

Refer now to Fig. D.7. Note carefully that the first item that we wish to calculate is the depth of the overhang, DE. Since

$$\tan VSA = \frac{DB}{DE}$$

and we know the window height, DB, to be 100 cm (39 in.), we have

$$DE = \frac{100 \text{ cm}}{\tan VSA}$$

We perform this simple calculation for every VSA, and we will select the deepest overhang as the one required. The figures are listed under column DE (depth) of Fig. D.12b.

STEP 5. Having found the overhang depth required, we can calculate the end treatment needed, that is, the amount that the overhang must be extended to cover the window. Study Fig. D.7. The extension needed is obviously AB, since the area ABD is unshaded. Therefore, if we move

130 cm

Fig. D.12 *(continued)*

D a distance of *AB* (on the figure—to the right; in our example—to the left), point *A* will move to *B*, and shading will be complete. Since

$$\tan HSA = \frac{AB}{DE}$$

and we have already established that *DE* is 130 cm (52 in.) we have

$$AB = 130 \tan HSA$$

These four values are calculated and tabulated in Fig. D.12*b*. We find that the required extension to the left, beyond the window, is 255 cm (105 in.). At this point we can complete the sketch in Fig. D.12*c* with dimensions.

STEP 6. This step is the evaluation stage that follows every design. Our evaluation would be that if, indeed, the stated shading hours must be maintained, an overhang 130 cm (51 in.) deep and 416 cm (164 in.) wide (266 + 150) is excessively large, and we would recommend a vertical element on the left side. The type of vertical element recommended is shown in Fig. D.12*d;* it is a sloping element that reaches from the front edge of the overhang to the window. To understand how such a unit shades, refer again to Fig. D.7. Note very carefully that because *DA* is the shadow of *DE*, triangle *DAB* is the shadow area corresponding to a vertical side element *DEB*. That is, if *DEB* were solid, *DAB* would be its shade. We thus see the extremely important and useful fact that a *sloping side element like DEB is equivalent to an infinitely long extension on that side.* Thus the full segmental areas shown in shading masks can be established not by an impossible infinitely long horizontal element, but by a horizontal element the width of the window and two sloping elements. In our case, such an element would eliminate the necessity for a massive 266 cm (105 in.) side extension.

REFERENCES

Cowan, H. J. (ed.) (1980). *Solar Energy Applications in the Design of Buildings,* Applied Science Publishers, London.

Harkness, E. and Mehta, M. (1978). *Solar Radiation Control in Buildings,* Applied Science Publishers, London.

Lim, B. P., Rao, K. R., Tharmaratnam, K., and Mattar, A. M. (1979). *Environmental Factors in the Design of Building Fenestration,* Applied Science Publishers, London.

Olgyay, A., and Olgyay, V. (1951). *Solar Control and Shading Devices,* Princeton University Press, Princeton, New Jersey.

Ramsey, C. G., and Sleeper, H. R. (1981). *Architectural Graphic Standards,* 8th ed., AIA/Wiley, New York.

Sun Angle Calculator (1975). Libbey-Owens-Ford, Available from LOF, 811 Madison Ave., Toledo, OH 43695.

Szokolay, S. V. (1980). *Environmental Science Handbook.* Halsted/Wiley, New York.

APPENDICES

E

ECONOMIC ANALYSIS

E.1 Economic Decision

As has been pointed out repeatedly in this text, economic factors are a major consideration in almost every decision in the building design process. The discussion below is necessarily brief but covers the essential principles. For in-depth study of economic analyses in the design and construction process, the reader is referred to the references at the end of this Appendix. Publications of the National Institute of Science and Technology, (formerly The National Bureau of Standards; Marshall and Ruegg, 1980a,b; Ruegg et al., 1978) are the sources of some of the principles explained below, and are so referenced.

The most common economic questions are:

1. How can the cost of different systems *that produce the same result* be compared? This may apply to the purchase of a single item such as a motor, to the lighting of a room, or to the choice of the type of HVAC installation for an entire system. The principle involved is the same.

2. Assuming that improving a system, existing or proposed, will reduce operating costs and that many different types of improvement are possible, what preliminary economic guidelines can be established to determine whether any proposed investment appears cost effective? Stated otherwise, how can the initial, simple rate of return of a proposed investment be determined so that it can be compared to the *minimum expected rate of return* on investment?

3. Having determined in answer to question 2 that a number of different proposals, each with its own investment and payback (savings) figures, are apparently cost effective, how can these be compared to determine which is *most* cost effective. Put another way, which proposal has the maximum *net benefit or savings*. This is the type of decision that must be made when comparing

energy conservation projects, such as the lighting control systems described in Section 21.4.

4. Having established which of the proposals is most cost effective, how can the *actual* rate of return of the investment be determined? This figure, often referred to as *internal rate of return* (IRR), can then be compared to the required rate of return, to determine whether the investment is economically desirable.

5. What is the *payback period* of the investment? These questions are considered individually in the discussion below.

E.2 Life-Cycle Cost (LCC)

Cost comparisons made on the basis of first cost are legitimate only when operating costs do not bear consideration, either because there are none or because the builder is not the owner. Proposals for pouring a concrete slab might represent the former situation, and speculative construction for immediate sale the latter. In such cases the technique of cost comparison is simple. However, when operating costs are involved that vary from system to system, the only reasonable way to determine and compare system costs is by *life-cycle costing*. The basic theory and calculation techniques are well known (see AIA, 1977; Marshal and Ruegg, 1980a,b; Ruegg et al., 1978), so they will be reviewed only briefly here.

Stated simply, a life-cycle cost represents the total cost of an item or system over its entire life cycle, that is, the sum of first cost and all future costs, less salvage. When the life cycle of the system being studied does not correspond to that of the building in which it is installed (as is usually the case), the designer must decide which life cycle to use. For instance, a lighting system has an estimated life of 15 to 20 years, whereas the building life is usually at least double that. It is usually wiser to make the study over the shorter period, since almost certainly, changes in technology will present an entirely different picture when the time comes to replace the system. This is true of most mechanical/electrical systems being designed today.

Life-cycle cost, in equation form, is

$$LCC = IC + MC + AC + OC \quad (E.1)$$

where LCC = life-cycle cost
         IC = investment cost, (i.e., first cost minus salvage)
       MC = maintenance and repair cost
       AC = amortization (replacement) cost (not always included in life-cycle calculation)
       OC = operating costs, including labor and energy

All costs are expressed in terms of present value, which is explained below. Costs can also be annualized. We use the present value format because we feel it gives a clearer answer to the economic decision questions posed above. When calculating costs, the designer must include related ancillary costs. For instance, in comparing lighting systems, the cost figures must reflect the impact on the HVAC system and the wiring system. (For lighting system cost comparisons, see IESNA 1993.)

When the life cycle is short—say, 2 years maximum—the last three components of Eq. E.1 can be assumed to be constant without introducing an intolerable error. However, for longer periods this is obviously not a valid assumption because of escalating labor and energy costs. Since these two factors weigh heavily in all economic comparisons, and since they generally exist in different proportions in the systems being considered, escalation in their costs will obviously change the economic balance when costs are projected over the equipment life. It is therefore necessary to change the usual present value formula to reflect this escalation.

The present value of future payments of 1.00, made at the *end* of each of a series of n periods, is

$$PV = \frac{(1 + i)^n - 1}{i(1 + i)^n} \quad (E.2)$$

where PV = present value
         i = interest (discount) rate expressed as a decimal (e.g., 8% = 0.08)
        n = number of periods

The factor i is more accurately the discount rate rather than the simple interest rate, the difference being that the latter reflects inflation, either actual or anticipated. If, however, the future payments escalate, the expression becomes:

$$PV = \frac{1 - \left(\dfrac{K}{1 + i}\right)^n}{\dfrac{1 + i}{K} - 1} \quad (E.3)$$

where K is the escalation rate per period, and the remaining items are as in Eq. E.2. Thus for a 5% annual escalation, $K = 1.05$, and so on. The values of PV are tabulated in Table E.1 for interest (discount) rates of 8%, 10%, and 12%, and escalation rates of 3%, 5%, 8%, and 10%. Other values can be calculated from the equations.

It is apparent that the result of a life-cycle cost analysis depends heavily on accurate forecasting when using escalation and that comparative results can readily shift not only with escalation estimates of the various components but also with the length of the life cycle.

E.3   Initial (Simple) Rate of Return

Addressing the second question in Section E.1, we wish to determine whether to examine a proposed cost-saving investment in detail. That is, is the initial (simple) rate of return sufficient? The difference between initial and *actual* rates of return is that the former is simply

$$\text{initial rate of return} = \frac{\text{annual savings}}{\text{investment}}$$

whereas the actual (internal) rate of return depends on the life of the project and the *total* savings accrued during this life-cycle. The calculation method for the latter is given in Section E.5.

Initial rate of return is a useful criterion in that it can be very easily calculated and then compared to the *minimum* desired rate of return on investment. We emphasize *minimum,* since actual rate of return is always lower than the simple, initial rate of return, except where savings escalate during the project life cycle. If this preliminary comparison indicates that the simple rate of return exceeds the minimum by a margin of a few percent, it is likely that the project is economically

TABLE E.1*a* **Present Value of n Future Payments Beginning at the End of the First Period**

	Interest (Discount) Rate per Period						
n	6%	8%	10%	12%	15%	20%	25%
1	0.94	0.93	0.91	0.89	0.87	0.83	0.80
2	1.83	1.78	1.74	1.69	1.63	1.53	1.44
3	2.67	2.58	2.49	2.40	2.28	2.11	1.95
4	3.47	3.31	3.17	3.04	2.86	2.59	2.36
5	4.21	3.99	3.79	3.61	3.35	2.99	2.69
6	4.92	4.62	4.36	4.11	3.78	3.33	2.95
7	5.58	5.21	4.87	4.56	4.16	3.61	3.16
8	6.21	5.75	5.34	4.97	4.49	3.84	3.33
9	6.80	6.25	5.76	5.33	4.77	4.03	3.46
10	7.36	6.71	6.15	5.65	5.02	4.19	3.57
11	7.89	7.14	6.50	5.94	5.23	4.33	3.66
12	8.83	7.54	6.81	6.19	5.42	4.44	3.73
13	8.85	7.90	7.10	6.42	5.58	4.53	3.78
14	9.30	8.24	7.37	6.63	5.72	4.61	3.82
15	9.71	8.56	7.61	6.81	5.85	4.68	3.86
16	10.11	8.85	7.82	6.97	5.95	4.73	3.89
17	10.48	9.12	8.02	7.12	6.05	4.78	3.91
18	10.83	9.37	8.20	7.25	6.13	4.81	3.93
19	11.16	9.60	8.37	7.37	6.20	4.84	3.94
20	11.47	9.82	8.51	7.47	6.26	4.87	3.95
21	11.76	10.02	8.65	7.56	6.31	4.89	3.96
22	12.04	10.20	8.77	7.65	6.36	4.91	3.97
23	12.30	10.37	8.88	7.72	6.40	4.93	3.98
24	12.55	10.53	8.99	7.78	6.43	4.94	3.98
25	12.78	10.68	9.08	7.84	6.46	4.95	3.99

feasible, and a detailed analysis can be undertaken.

E.4   Cost-Effectiveness Comparison

This analysis is undertaken in answer to question 3, that is, several projects are being considered, all of which meet the stated criterion. Which is most desirable from a cost-effectiveness viewpoint? (If only one project is under consideration, the analysis proceeds directly to the IRR calculation described in the next section).

The difference between life-cycle project cost and life-cycle project savings, that is, net lifetime savings, is a measure of cost effectiveness. Therefore, to compare the cost effectiveness of two projects, it is necessary simply to calculate this differential. To express this as an equation, we write:

net savings = life-cycle savings − life-cycle cost

Since this is true for all projects, a comparative differential can be calculated by using the differential in each term of the life-cycle cost equation, that is,

$$\Delta_{net} = \Delta_{savings} - \Delta_{costs}$$

Taking a very simple example, assume the following conditions.

	Project A	Project B
Investment	$10,000	$15,000
Life cycle	10 years	8 years
Annual savings	$2,000	$3,200
Initial rate of return	20%	32.3%
Discount rate	12%	12%

Since project B has an apparently higher rate of return, we would formulate the differential as B minus A. Using Table E.1*a*, the present worth of lifetime savings for project B (without escalation)

TABLE E.1b **Present Value of** *n* **Future Payments Beginning at the End of the First Period and Escalating at** *K* **per Period**

Interest (Discount) Rate per Period 8%

	Annual (Periodic) Escalation Rate K			
n	1.03	1.05	1.08	1.10
1	0.95	0.97	1.00	1.02
2	1.86	1.92	2.00	2.06
3	2.73	2.84	3.00	3.11
4	3.56	3.73	4.00	4.19
5	4.35	4.60	5.00	5.28
6	5.10	5.44	6.00	6.40
7	5.82	6.26	7.00	7.54
8	6.50	7.06	8.00	8.70
9	7.15	7.84	9.00	9.88
10	7.78	8.59	10.00	11.08
11	8.37	9.33	11.00	12.30
12	8.94	10.04	12.00	13.55
13	9.48	10.73	13.00	14.82
14	9.99	11.41	14.00	16.11
15	10.48	12.06	15.00	17.43
16	10.95	12.70	16.00	18.77
17	11.40	13.32	17.00	20.13
18	11.82	13.92	18.00	21.53
19	12.23	14.51	19.00	22.94
20	12.62	15.08	20.00	24.39
21	12.99	15.63	21.00	25.86
22	13.34	16.17	22.00	27.35
23	13.68	16.69	23.00	28.88
24	14.00	17.20	24.00	30.43
25	14.30	17.69	25.00	32.01

TABLE E.1c **Present Value of** *n* **Future Payments Beginning at the End of the First Period and Escalating at** *K* **per Period**

Interest (Discount) Rate per Period 10%

	Annual (Periodic) Escalation Rate K			
n	1.03	1.05	1.08	1.10
1	0.94	0.96	0.98	1.00
2	1.81	1.87	1.95	2.00
3	2.63	2.74	2.89	3.00
4	3.40	3.57	3.82	4.00
5	4.12	4.36	4.73	5.00
6	4.80	5.12	5.63	6.00
7	5.43	5.84	6.51	7.00
8	6.02	6.53	7.37	8.00
9	6.57	7.18	8.22	9.00
10	7.09	7.81	9.05	10.00
11	7.58	8.41	9.87	11.00
12	8.03	8.98	10.67	12.00
13	8.46	9.53	11.46	13.00
14	8.85	10.05	12.23	14.00
15	9.23	10.55	12.99	15.00
16	9.58	11.27	13.74	16.00
17	9.90	11.48	14.47	17.00
18	10.21	11.91	15.19	18.00
19	10.50	12.32	15.90	19.00
20	10.76	12.72	16.59	20.00
21	11.02	13.09	17.27	21.00
22	11.25	13.45	17.94	22.00
23	11.47	13.80	18.59	23.00
24	11.68	14.12	19.24	24.00
25	11.87	14.44	19.87	25.00

TABLE E.1d **Present Value of** *n* **Future Payments Beginning at the End of First Period and Escalating at** *K* **per Period**

Interest (Discount) Rate per Period 12%

	Annual (Periodic) Escalation Rate K			
n	1.03	1.05	1.08	1.10
1	.92	.94	.96	.98
2	1.77	1.82	1.89	1.95
3	2.54	2.64	2.79	2.89
4	3.26	3.41	3.66	3.82
5	3.92	4.14	4.49	4.74
6	4.52	4.82	5.29	5.64
7	5.08	5.45	6.07	6.52
8	5.59	6.05	6.82	7.38
9	6.06	6.61	7.54	8.23
10	6.49	7.13	8.23	9.07
11	6.89	7.63	8.90	9.89
12	7.26	8.09	9.55	10.69
13	7.59	8.52	10.17	11.49
14	7.90	8.92	10.77	12.26
15	8.19	9.30	11.35	13.03
16	8.45	9.66	11.91	13.78
17	8.69	9.99	12.45	14.51
18	8.91	10.31	12.97	15.23
19	9.11	10.60	13.47	15.95
20	9.30	10.87	13.95	16.64
21	9.47	11.13	14.42	17.34
22	9.63	11.37	14.87	18.00
23	9.78	11.60	15.30	18.66
24	9.91	11.81	15.72	19.31
25	10.04	12.01	16.12	19.95

is $3200 × (4.97) or $15,904, and that of project A is $2000 × (5.65) or $11,120. Therefore, the differential cost benefit (CB) of B minus A is $\Delta_{\text{life savings}}$ minus $\Delta_{\text{project costs}}$, or

$$\Delta_{CB} = (\$15,904 - \$11,120) - (\$15,000 - \$10,000)$$
$$= \$4784 - 5000 = -\$216$$

indicating that project A is more cost effective despite the apparently higher rate of return of project B. If we include escalation in costs, obviously the life savings of the longer life project (A) will shrink more than the savings of shorter life project B. Using Table E.1d, we find that the projects are about equal, with 3% annual escalation. With higher rates of escalation, project B is more cost effective.

E.5 Internal Rate of Return (IRR)

This figure represents the rate of return at which lifetime savings are exactly equal to lifetime costs. It cannot be calculated directly, but only by trial and error. We must assume a rate of return, calculate present value of savings, and compare PV to lifetime costs. Because of this difficulty, IRR is frequently neglected, and initial rate of return used instead, although as we saw above, this can readily give misleading results. To illustrate calculation of IRR, we will return to our very simple example and calculate IRR for projects A and B.

For project A, we need a present worth factor of $10,000/$2000 = 5, in 10 years. Referring to Table E.1a, we see that at a discount (interest) rate of 15%, the present worth factor is 5.019, indicating that IRR is slightly above this. (It is actually 15.1%.) For project B we need an 8-year present worth factor of $15,000/$3200, or 4.69. Referring again to Table E.1a, we see that this factor falls between 12% and 15%. By a series of trials with Eq. E.2, we arrive at 13.7%, which is 1.4% less than the same factor for project A, confirming the conclusion reached in the Section E.4.

E.6 Payback Period

The payback period referred to in most proposals is the reciprocal of the initial rate of return and is usually referred to as the *simple* payback period.

However, just as we have seen that the actual rate of return differs from the simple rate (and is usually *lower*), so also the actual or *discount payback period* is different, and usually *longer* than the simple one. This period is, logically, the period of time required for the accumulated net savings to equal the initial investment, with all figures expressed in present value dollars. To illustrate, let us return to the example given above.

1. *Proposal A.* As in the IRR calculation, we need a present worth factor of $10,000/ $2000, or 5, at a 12% discount rate. From Table E.1a we find this to be about 8.1 years. This compares to a simple payback period of 5 years. Note that this is *not* the reciprocal of internal rate of return.

2. *Proposal B.* Here we need a present worth factor of $15,000/$3200 or 4.7. At 12% discount rate, from Table E.1a, this is about 7.3 years. This compares to a simple payback period of 4.7 years.

Note that from these results the very important fact that a shorter payback period does not necessarily indicate a more cost-effective investment. Here, proposal A has a better return precisely because it yields good savings for a longer period. Summing up the results of our simple study, we have:

	Project A	Project B
Initial rate of return	20%	21.3%
Internal rate of return	15.1%	13.7%
Simple payback period	5 years	4.7 years
Discounted payback period	8.1 years	7.3 years

This shows very clearly that although initial rate of return and simple payback are indications of a proposal's value, they are useless in comparative studies.

REFERENCES

American Institute of Architects (1977). *Life Cycle Cost Analysis—A Guide for Architects,* AIA, Washington, DC.

Edison Electric Institute. *Interest Tables, 0–25%*, EEI, New York.

Illuminating Engineering Society of North America (1993). *Lighting Handbook, Application* Volume, 8th ed., Section 3, IESNA, New York.

Marshall, H. E., and Ruegg, R. T. (1980a). *Energy Conservation in Buildings: An Economics Guidebook for Investment Decision.* National Bureau of Standards Handbook 132, NBS, Washington, DC.

Marshall, H. E., and Ruegg, R. T. (1980b). *Simplified Energy Design Economics,* National Bureau of Standards Special Publication 544, NBS, Washington, DC.

Ruegg, R. T. et al. (1978). *Life Cycle Costing,* National Bureau of Standards Publication 113, NBS, Washington, DC.

ASHRAE Handbook, Applications Volume, SI ed., (1995) Chapter 33, Owning and Operating Costs.

F

SOUND TRANSMISSION DATA FOR WALLS

To use Appendix F:

1. Find desired construction Type Code in Index F.1.
2. Find desired STC corresponding to the selected construction Type Code.
3. Refer to Table F.1 for details of wall construction.
4. Refer to Index F.2 for wall thickness, weight, STC, fire rating, and the transmission losses at standard octave midpoints.

EXAMPLE. An interior masonry partition with an STC between 50 and 55 is desired. Of particular interest is the TL at 1000 HZ.

1. From Index F.1, we find Type Code "c."
2. From Index F.1, we note that constructions W6, W12b, W15b and W22b are suitable for the proposed use.
3. From Table F.1 we decide that construction W15b is most suitable for the proposed use.
4. From Index F.2 we find that construction W15b has a transmission loss of 55 dB at 1000 Hz.

INDEX F.1 Sound Transmission Class: Walls

Type Code:
a. *Wooden stud* **f.** *Plaster* **j.** *Fiber board*
b. *Metal stud* **g.** *Gypsum wallboard* **k.** *Lead*
c. *Masonry* **h.** *With resilient element* **l.** *Gypsum core board*
d. *Concrete* **i.** *Absorbent blankets or fill* **m.** *Double wall*
e. *Staggered stud*

STC	Type	Item No.	STC	Type	Item No.
63	d,f	W4	46	a,f	W30a
62[a]	c,f,j,m	W23	46	a,f	W31a
56	c	W7	46	a,f,k	W31c
56[a]	c,f	W8	46	a,e,g,i	W37
55[a]	b,g,i	W63	45	c,d	W10b
54[a]	c,f,m	W22b	45	a,f,i	W38
54	b,f,h	W52	45	b,f,h	W50a
53[a]	d,f	W2	45	g,l,m	W85a
53	c,f,h	W15b	44	c	W12a
52[a]	c,f	W6	44	a,e,g	W34a
51	b,f	W44a	44	a,e,g	W35a
51	b,g,h,i	W67	44	a,f,h	W40a
50	c,k	W12b	43	c,d	W10a
50	b,g,i	W60	43[a]	d,j,m	W21
50	g,i,l,m	W85b	43	a,e,f	W35a
49[a]	c,f,m	W22a	43	a,e,f	W36a
48	c,d	W9	43	b,f,k	W43b
48	a,e,f,i	W36b	42[a]	c,f	W5
48	b,f,k	W43c	42	a,f	W14a
47	d	W1	41	b,f	W43a
47	a,g,k	W28b	41	b,g	W55a
47	a,f,k	W31b	40	c,f	W13
47	a,g,h	W39a	40	a,g	W32a
46	c,f	W11a	39	a,g	W28a
46	c,f,h	W15a	36	g,l	W80

[a]Field measurement.

INDEX F.2 Sound Transmission Loss: Walls

Designation	Thickness (in.)	Weight (lb/ft²)	Transmission Loss (db)							Fire Rating hr
			125	250	500	1K	2K	4K	STC	
W1	3	39	35	40	44	52	58	64	47	½
W2	7	80	39	42	50	58	64	—	53	3
W4	Approx. 16	184	50	54	59	65	71	68	63	4+
W5	5½	55	34	34	41	50	66	—	42	2.5
W6	10	100	41	43	49	55	57	—	52	4+
W7	12	121	45	45	53	58	60	61	56	4+
W8	25	280	50	53	52	58	61	—	56	4+
W9	12	79	46.5	44	46	52	54	56	48	4
W10a	6	34	32	33	40	47	51	48	43	1
W10b	6	34	37	36	42	49	55	58	45	1
W11a	5¼	35.8	36	37	44	51	55	62	46	2
W12a	3¾	26.1	40	40	40	48	55	56	44	1.5
W12b	5	31	41	46	46	56	63	67	50	1.5
W13	4	21.5	39	34	38	43	48	46	40	3
W14a	5	23.4	37	42	39	44	49	49	42	4
W15a	5	27	38	37	44	51	56	59	46	3
W15b	6	31	45	44	50	55	56	59	53	4
W20	7	22.9	32	46	49	53	58	66	52	3
W21	10¼	37	41	42	46	51	52	—	43	Not available
W22a	12	100	37	41	48	60	60.5	—	49	4+
W22b	12	100	40	44	55	67.5	70	—	54	4+
W23	18	120	48	54	58	64	69	—	62	4+
W28a	5	6	21	28	35	42	45	41	39	0.5
W28b	Approx. 5⅛	12	27	37	43	52	56	—	47	0.5
W30a	5¾	13.4 to 15.7	32	37	42	47	47	63	46	0.75
W31a	5¾	13.4 to 15.7	32	37	42	48	48	63	46	0.75
W31b	Approx. 5⅞		33	41	45	52	55	65	47	0.75
W31c	Approx. 5⅞	17–19	32.5	40	43	47	50	62	46	0.75
W32a	5½	8.2	27	31	39	45	52.5	48	40	1
W34a	5	6.2	36	36	40	47	52	45	44	0.5
W35a	6½	13.4	41	41	46	49	41	54	44	1.5
W35b	5¾	15.6	48	46	48	48	48	59	43	1
W36a	6¼	11.1	36	33	42	42	41	51	43	0.75
W36b	6¼	12.8	37	37	49	50	52	66	48	1
W37	5¾	13.8	39	40	42	47.5	55	51.5	46	0.5
W38	5⅜	14.2	39	45	48	50	44	54	45	1
W39a	6¼	6.7	30	40	46	50	49	49	47	1
W40a	Approx. 6½	14.4	43	41	48	50	42	56	44	1
W43a	3⅜	12.3	27	37	43	46	39	47	41	0.75
W43b	3½	15.2	35	43	45	47	48	58	43	0.75
W43c	Approx. 3½	18.2	36	45	47	50	53	61	48	0.75
W44a	5	15.7	34	38	47	50	52	58	51	1
W50a	5¼	13	35	46	48	51	48	43	45	0.75
W52	5	19	50	52	55	56	52	60	54	1
W55a	4⅞	6	29	36	40	46	40	46	41	1
W60	3½	5.4	34	40	47	50	53	54	50	1
W63	6⅛	11.5	36	47	51	57	57	62	55	2
W67	6½	11.3	41	46	49	51	50	60	51	2
W80	2¼	10.2	34	34	37	38	39	45	36	1
W85a	5⅛	14.6	36	35	45	51	53	57	45	3
W85b	6	12.8	37	37	54	56	56	62	50	3

TABLE F.1 **Acoustic Characteristics of Walls: For Other Data See Index F.2**

Designation	Description	Section Sketch

Solid Concrete

W1 3-in.-thick solid concrete wall poured in situ in test opening. All surface cavities were sealed with thin mortar mix.

W2 6-in.-thick concrete wall with ½-in.-thick layer of plaster on both sides.

W4 Wall of 4, 6, and 8 × 8 × 16 in. sand and gravel aggregate solid concrete blocks; on each side, ¼- to ½-in.-thick layer of cement gypsum plaster and sand.

W5 4½-in.-thick brick wall with ½-in.-thick layer of plaster on each side.

W6 9-in.-thick brick wall with ½-in.-thick layer of plaster on each side

W7 12-in.-thick brick wall.

W8 24-in.-thick stone wall with ½-in.-thick layer of plaster on both sides.

TABLE F.1 **Acoustic Characteristics of Walls: For Other Data See Index F.2 (*continued*)**

Designation	*Description*	*Section Sketch*

Hollow Concrete Block

W9   12-in. wall made of hollow 8 × 8 × 12 in. and 8 × 4 × 16 in. concrete blocks.

W10a   6-in. hollow concrete blocks constructed with vertical mortar joints staggered.

W10b   Similar to W10a except wall painted.

Cinder Block

W11a   4 × 8 × 16 in. hollow cinder blocks; on each side, ⅝ in. of sanded gypsum plaster.

Cement Block

W12a   3⅝ × 7¾ × 13½ in. lightweight-aggregate cement blocks with ½-in. mortar joints; three coats of masonry paint applied to each side of partition.

W12b   Same as W12a except 1 × 2-in. furring strips were nailed vertically to partition on one side; 1/16-in. layer of lead, 3.94 lb/ft^2, nailed to furring strips, ¼-in. plywood-covered lead with joints caulked.

Hollow Gypsum Block

W13   3-in. hollow gypsum blocks cemented together with ⅜-in. mortar joints; on each side, ½-in. sanded gypsum plaster.

W14a   4-in. hollow gypsum blocks cemented together with ⅜-in. mortar joints; on each side, ½-in. sanded gypsum plaster.

APPENDICES

(*continued*)

TABLE F.1 **Acoustic Characteristics of Walls: For Other Data See Index F.2** (*continued*)

Designation	Description	Section Sketch

Hollow Gypsum Block, Resilient One Side,
Plaster Both Sides

W15a 3 × 12 × 30 in. hollow gypsum blocks with ½-in. mortar joints. On one side ⁷/₁₆-in. sanded gypsum plaster; on the other side resilient clips, spaced 18 in. on centers vertically and 16 in. on centers horizontally, held to ¾-in. metal channels 16 in. on centers, to which expanded metal lath was wire-tied; 1¹/₁₆-in. sanded gypsum plaster. ¹/₁₆-in. white-coat finish applied to both sides.

W15b Similar to W15a except 4 × 12 × 30 in. gypsum blocks were used.

Hollow Concrete

W21 Precast concrete hollow wall panels with in situ concrete posts and beams. Panels have 1½-in.-thick concrete shells with 6¼-in. air space between them. Layer of fiberboard ½-in. thick is adhered to the exposed surfaces of the panel.

Double Walls

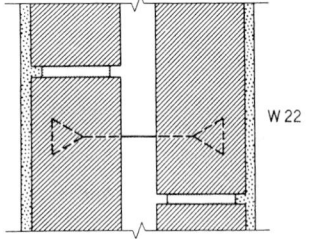

W22a Double wall with 4½-in.-thick brick leaves separated by a 2-in. cavity (wire ties between leaves); ½-in. plaster on exposed sides.

W22b Similar to W22a but no wire ties between the leaves.

W23 Double wall with 4½-in.-thick brick leaves, 6-in. cavity (no ties); on exposed sides, ½-in. plaster on 1-in.-thick wood-wool slabs mortared to the brick walls.

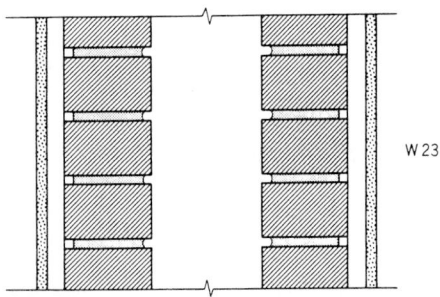

Wood Stud Walls

W28a 2 × 4 in. wooden studs, 16 in. on centers, ½-in. gypsum wallboard nailed to each side. All joints taped and finished.

W28b Similar to W28a except a layer of lead, 2.95 lb/ft², was laminated to each side of panel.

TABLE F.1 Acoustic Characteristics of Walls: For Other Data See Index F.2 (*continued*)

Designation	Description	Section Sketch
W30a	2 × 4 in. wooden studs, 16 in. on centers, attached to 2 × 4 in. wooden floor and ceiling plates, 3/8-in. gypsum lath nailed to studs on both sides, 1/2-in. sanded plaster with white-coat finish.	W 30 W 31 W 32
W31a	2 × 4 in. wooden studs, 16 in. on centers, 3/8-in. gypsum lath nailed to studs on both sides, 1/2-in. sanded plaster with white-coat finish.	
W31b	Similar to W31a except a 0.065-in.-thick layer of lead weighing 3.85 lb/ft^2 was laminated to each side of panel.	
W31c	Similar to W31a except a 0.13-in.-thick layer of lead weighing 7.9 lb/ft^2 was laminated to one side of panel.	
W32a	2 × 4 in. wooden studs, 16 in. on centers; on each side two layers of 3/8-in. gypsum wallboard cemented together; joints in exposed surfaces taped and finished.	

Staggered Wood Stud Walls

W34a	2 × 3 in. wooden studs, 16 in. on centers, staggered 8 in. on centers, attached to 2 × 4 in. wooden plates at ceiling and floor; 1/2-in. gypsum wallboard nailed 7 in. on centers on both sides to studs. All joints taped and finished.	W 34a
W35a	2 × 3 in. wooden studs, 16 in. on centers, staggered 8 in. on centers (attached to 2 × 4 in. wooden plates at floor and ceiling) two layers of 5/8-in. tapered-edge gypsum wallboard, first layer nailed 7 in. on centers, second layer nailed 16 in. on centers. All exposed joints taped and finished.	W 35a W 35b
W35b	Similar to W35a except the wall was constructed with 3/8-in. perforated gypsum lath and 1/2-in. sanded gypsum plaster with white-coat finish.	

(*continued*)

APPENDICES

TABLE F.1 **Acoustic Characteristics of Walls: For Other Data See Index F.2 (*continued*)**

Designation	Description	Section Sketch
W36a	2 × 4 in. wooden studs, 16 in. on centers, staggered 8 in. on centers, and offset ½ in. On each side ⅜-in. gypsum lath nailed to studs, ½-in. gypsum vermiculite plaster, machine applied, and a hand-applied white-coat finish.	W 36a
W36b	Same as W36a except the space between the studs contained vermiculite fill with a density of 6.3 lb/ft³.	W 36b
W37	2 × 4 in. wooden studs, 16 in. on centers, staggered 8 in. on centers, attached to 2 × 4¾ in. wooden floor and ceiling plates; ½-in. gypsum wallboard nailed on both sides to studs, 0.9-in. wood-fiber wool blanket stapled on the inside of one side of the wall. All joints taped and finished.	W 37

Slotted Wood Studs

W38	2 × 4 in. slotted wooden studs, 16 in. on centers, attached to 2 × 4 in. wooden floor and ceiling plates; ⅜-in. gypsum lath nailed 7 in. on centers to studs, ½-in. gypsum plaster with white-coat finish applied to both sides. 3-in. mineral fiber batts stapled between studs.	W 38

TABLE F.1 Acoustic Characteristics of Walls: For Other Data See Index F.2 (*continued*)

Designation	Description	Section Sketch

Wood Studs; Resilient Mounting

W39a 2 × 4 in. wooden studs, 16 in. on centers, attached to 2 × 4 in. wooden floor and ceiling plates; resilient channels nailed horizontally to both sides of studs 24 in. on centers, ⅝-in. gypsum wallboard screwed 12 in. on centers to channels. All joints taped and finished.

W40a 2 × 4 in. wooden studs, 16 in. on centers; resilient clips, nailed to studs on both sides, held ⅜-in. gypsum lath, ½-in. sanded gypsum plaster with white-coat finish.

Steel Truss Stud Wall

W43a 1⅝-in. steel truss studs; ⅜-in. gypsum lath, ½-in. plaster on both sides.

W43b Similar to W43a except a layer of lead, 2.95 lb/ft^2, was laminated to one side of partition.

W43c Similar to W43a except a layer of lead, 2.95 lb/ft^2, was laminated to each side of partition.

W44a 3¼-in. steel truss studs, 24 in. on centers, attached to metal floor and ceiling tracks; on both sides ⅜-in. perforated gypsum lath attached with wire clips wire-tied to studs, ½-in. sanded gypsum plaster.

Steel Truss Studs; Resilient Mountings

W50a 2½-in. steel truss studs 16 in. on centers, ⅜-in. gypsum lath attached with resilient clips to studs, ½-in. plaster applied to both sides.

APPENDICES

(*continued*)

TABLE F.1 **Acoustic Characteristics of Walls: For Other Data See Index F.2 (*continued*)**

Designation	Description	Section Sketch
W52	3¼-in. steel truss studs, 16 in. on center; on each side resilient clips fastened 16 in. on centers to studs, ¼-in. metal rod wire-tied to clips, diamond mesh metal lath wire-tied to metal rods, ¾-in. sanded gypsum plaster.	W 52

Metal Channel Stud Wall

W55a	3⅝-in. metal channel studs, 24 in. on centers, set into 3⅝-in. metal floor and ceiling runners; ⅝-in. gypsum wallboard screwed to studs on both sides. All joints taped and finished.	W 55a
W60	2½-in. metal channel studs, 24 in. on centers, set in 2½-in. metal floor and ceiling runners; ½-in. vinyl-coated gypsum wallboard adhesively attached and screwed to studs on both sides. All joints sealed with caulking compound. Aluminum batten strips screwed 12 in. on centers to gypsum board at joints; top and bottom finished with aluminum ceiling and base trim. 2-in. mineral fiber blankets hung between studs.	W 60
W63	3⅝-in. metal channel studs, 24 in. on centers, set into 3⅝-in. metal runners, which were attached through continuous beads of nonsetting resilient caulking compound to floor and ceiling, respectively. Two layers of ⅝-in. gypsum wallboard attached to both sides of studs. First layer screwed 8 in. on centers at joints and 12 in. on centers in field; second layer laminated and screwed 24 in. on centers to first layer, with joints staggered 24 in. 1½-in.-thick mineral fiber felt, 3 lb/ft³, stapled between studs. All exposed joints taped and finished. The ¼-in. clearance around the perimeter closed with a nonsetting resilient caulking compound.	W63
W67	3⅝-in. metal channel studs, 24 in. on centers, set in 3⅝-in. metal floor and ceiling runners; ⅝-in. gypsum wallboard screwed to studs on both sides. On one side, resilient channels screwed horizontally, 24 in. on centers to inner layer; ⅝-in. gypsum wallboard screwed to channels. On the other side, ⅝-in. gypsum wallboard laminated directly to inner layer. 3-in. mineral fiber blankets hung between studs. All exposed joints taped and finished.	W 67

TABLE F.1 **Acoustic Characteristics of Walls: For Other Data See Index F.2 (*continued*)**

Designation	Description	Section Sketch

Gypsum Partitions

W80 24-in.-wide panels constructed of 1 × 24 in. gypsum core board offset 1½ in. at edges to form tongue-and-groove edge; ⅝-in. vinyl-faced gypsum wallboard laminated to both sides of core board. Panels inserted into two-piece metal floor and ceiling tracks. Gypsum to gypsum screws at ¼ and ½ points along vertical edges of face boards.

W 80

W85a Double wall with 1⅜-in. air space. Each leaf consisted of 24-in.-wide panels of ⅝-in. gypsum core board strips, 7½ and 4⅜ in. wide, offset 1½ in. at edges to form tongue and groove; ⅝-in., vinyl-faced, gypsum wallboard laminated to both sides of core board strips. Panels screwed 12 in. on centers to 1¼ × 1 in. angle floor and ceiling runners.

W 85a   W 85b

W85b Similar to W85a except space between leaves was 2⅛ in. and contained 2-in. mineral fiber blankets stapled to one leaf. ¼-in. perimeter clearance closed with a nonsetting resilient caulking compound. Vertical face layer joints sealed with joint compound.

Source: All data extracted from A guide to Airborne, Impact and Structure-Borne Noise Control in Multi-Family Dwellings, HUD/FHA/NBS 1971.

APPENDICES

G

SOUND TRANSMISSION AND IMPACT INSULATION DATA FOR FLOOR/CEILING CONSTRUCTION

To use Appendix G:

1. Find desired construction Type Code in Index G.1 or G.2. Index G.1 lists the constructions by STC. Index G.2 lists the constructions by IIC.
2. Find desired STC/IIC ratings corresponding to selected Type Code.
3. Refer to Table G.1 for details of construction.
4. Refer to Index G.3 for thickness, weight, STC, IIC, fire rating, and transmission loss at standard octave midpoints.

INDEX G.1 Sound Transmission Class: Floor/Ceiling Constructions

Type Code:

a. *Wooden joist*
b. *Metal joist*
c. *Concrete or masonry*
d. *Plaster ceiling*
e. *Gypsum board ceiling*

f. *With resilient elements*
h. *With carpeting*
i. *With absorbent blankets*
j. *With separate ceiling joists*

STC	IIC	Type	Item No.	STC	IIC	Type	Item No.
55[a]	57[a]	c,d,f	F14	46[a]	42[a]	c,d	F23
54[a]	64[a]	c,d,f	F17b	45	44	a,e,f	F44
52	80	a,e,i,j	F48	44	42	c,f	F3-2d
51[a]	53[a]	c,d,f	F10	44	41	c	F3-1a
51[a]	48[a]	c,d	F7a	44	80	c,h	F2-1a
50[a]	53[a]	c,e,f	F25	44	29	c	F1-c
50[a]	51[a]	c,e	F27	44	25	c	F1a
50[a]	48[a]	c,d	F16	43[a]	43[a]	a,d	F32b
49[a]	48[a]	c,d	F9	42[a]	32[a]	c	F22
48[a]	47[a]	c,d	F12	40	32	a,e,i	F40a
48	33	b,c,d	F60a	39[a]	37[a]	a,e	F34
47	62	b,c,e	F58	37	33	a,e	F38a
47[a]	42[a]	c,d	F24	37	32	a,e	F39a
47	59	b,c,d,h	F57b	34[a]	32[a]	a,e	F30
47	37	b,c,d	F57a	29[a]	32[a]	a,e	F35a
46	74	b,c,d	F60c	29	56	a,e,h	F35b

[a]Field measurement.

EXAMPLE. A standard (simple) wooden joint floor-ceiling construction is required with a minimum IIC of 55. Any STC above 35 is acceptable.

1. Since IIC is the determining factor, refer to Index G.2. Basic Type Code is "a."
2. From index G.2 we note that none of the listed "a" constructions given an IIC of 55. However, from Section 28.24 we note that the addition of carpeting will add 10 to 27 points to the IIC, making items F34, F38a, F39a, and F30 all suitable. We have discounted special and resilient constructions, because of the requirement for standard (simple) construction.
3. From Table G.1 we select construction F30 as being most appropriate for application of carpeting.
4. From Index G.3 we have (without carpet): thickness 9½ in., STC 34, IIC 32 (+ carpet), fire rating ¼ h. With high pile carpeting on foam rubber pad, this construction will have IIC > 55, STC > 35.

INDEX G.2 Impact Insulation Class: Floor/Ceiling Constructions

Type Code:

a. *Wooden joist*
b. *Metal joist*
c. *Concrete or masonry*
d. *Plaster ceiling*
e. *Gypsum board ceiling*

f. *With resilient ceiling element*
g. *With resilient floor element*
h. *With carpeting*
i. *With absorbent blankets*
j. *With separate ceiling joists*

IIC	STC	Type	Item No.	IIC	STC	Type	Item No.
80	52	a,e,h,i,j	F48	44	45	a,e,f	F44
80	44[b]	c,h	F2-1a	43[a]	43[a]	a,d,g	F32b
74	46	b,c,d,h	F60c	42[a]	47[a]	c,d	F24
64[a]	54[a]	c,d,g	F17b	42[a]	46[a]	c,d	F23
62	47	b,c,e,h	F58	42	44[b]	c,g	F3-2(d)
59	47[b]	b,c,d,h	F57b	41	44[b]	c	F3-1(a)
57[a]	55[a]	c,d,g	F14	37	47	b,c,d	F57a
56[a]	29[b]	a,e,h	F35b	37[a]	39[a]	a,e	F34
53[a]	51[a]	c,d,g	F10	33	48	b,c,d	F60a
53[a]	50[a]	c,e,g	F25	33	37	a,e	F38a
51[a]	50[a]	c,e	F27	32[a]	42[a]	c	F22
48[a]	51[a]	c,d	F7a	32	40	a,e,i	F40a
48[a]	50[a]	c,d	F16	32	37	a,e	F39a
48[a]	49[a]	c,d,g	F9	32[a]	34[a]	a,e	F30
47[a]	48[a]	c,d	F12	32[a]	29[a]	a,e	F35a
				29	44[b]	c	F1c
				25	44	c	F1a

[a]Field measurement.
[b]Estimated on the basis of similar structures.

INDEX G.3 **Floor/Ceiling Sound Transmission and Construction Data**[a]

Designation	Thickness (in.)	Weight (lb/ft²)	Transmission Loss (dB)								Fire Rating (h)
			125	250	500	1K	2K	4K	STC	IIC	
F1a	4	53	47	42	45	56	58	66	44	25	1
F7a	5¼	61	41	42	47	54	59	63	51	48	2
F9	6⅝	65	40	42	46	48	57	62	49	48	2
F10	8¼	90	38	43	47	53	60	—	51	53	2½
F12	10	62	38	40	44	51	56	59	48	47	3
F14	9½	83	38	44	52	55	60	—	55	57	3
F16	8⅛	65	40	42	46	52	58	—	50	48	2
F17b	9¼	57	38	46	52	59	64	—	54	64	2
F22	6¼	28	34	34	38	45	55	61	42	32	¾
F23	9½	45	33	37	43	52	58	62	46	42	¾
F24	10¼	65	34	37	43	52	57	—	47	42	¾
F25	10	45	30	38	46	58	64	—	50	53	¾
F27	7⅝	50	36	40	47	54	58	—	50	51	¾
F30	9½	7	19	24	31	35	45	—	34	32	¼
F32b	11	12	30	31	41	47	52	—	43	43	¾
F34	10¼	9.9	15	32	44	48	54	53	39	37	1
F35a	10	9.2	14	17	30	44	47	52	29	32	—
F38a	11¾	9	30	38	36	43	48	49	37	33	½
F39a	11⅞	9.5	22	32	36	45	49	56	37	32	1
F40a	11⅞	10	25	36	38	46	51	57	40	32	1
F44	10½	10.1	43	41	41	52	50	60	45	44	¾
F48	12⅜	10.7	35	42	52	56	69	74	52	80	¾
F57a	18⁹⁄₁₆	23.2	33	44	45	46	58	62	47	37	3
F58	21½	20.4	27	37	45	54	60	65	47	62	1
F60a	11	38.2	42	44	44	51	51	61	48	33	1½
F60c	11⅝	39	39	43	44	52	52	65	46	74	—

[a]Material extracted from *A Guide to Airborne, Impact and Structure-Borne Noise Control in Multifamily dwellings*, HUD/FHA/NBS, 1971.

TABLE G.1 **Acoustic Characteristics of Floors: For Acoustic Data see Index G.3**

Code[a]	Description	Section Sketch

Reinforced Concrete Slab

F1a   4-in.-thick reinforced concrete slab, isolated from support structure. Concrete was reinforced with 6 × 6 in. No. 6 AWG reinforcing mesh placed at the centerline horizontal plane of the slab. All surface cavities were sealed with a thin mortar mix.

F1c   Same as F1a except ⅛-in.-thick vinyl tile was adhered to concrete.

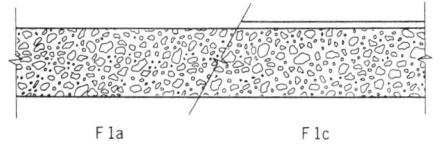

Reinforced Concrete with Floor Coverings
See also F1c above.

F2-1(a)   4-in.-thick reinforced concrete slab with carpeting and pad. The carpeting was of ¼-in. wool loop pile with ⅛-in. woven jute backing, 0.49 lb/ft²; the foam rubber pad was ¼ in. thick and weighed 0.53 lb/ft².

F3-1(a)   4-in. reinforced concrete slab with ½ × 9 × 9 in. oak blocks, 1.8 lb/ft², set in mastic.

F3-2(d)   4 in. concrete slab with ⅛-in. cork.

F7a   4⅜-in.-thick reinforced concrete slab. On the floor side, ¾-in.-thick, sand-cement screed with ⅛-in. linoleum floor covering. On the ceiling side, ⅜-in. layer of plaster.

F9   4⅜-in.-thick reinforced concrete slab. On the floor side, ½-in.-thick layer of bitumen with ½-in.-thick soft wood fiberboard, which was covered with a thin layer of bitumen with sand and a ¾-in.-thick sand-cement screed. On the ceiling side ⅜-in. layer of plaster.

Reinforced Concrete Slab, Floating Floor

F10   5-in.-thick reinforced concrete. On the floor side, 1½-in.-thick wire mesh reinforced sand-cement screed floating on ½-in.-thick bitumen-bonded, glass-wool quilt covered with building paper. On the screed, ½-in.-thick pitch-mastic with a linoleum floor covering. On the ceiling side, ½-in. layer of plaster.

F12   4⅜-in.-thick reinforced concrete slab. On the floor side, ¾-in.-thick sand-cement screed. On the ceiling side, brick wire mesh, suspended 4 in. with wire hangers, held ⅞-in. gypsum plaster.

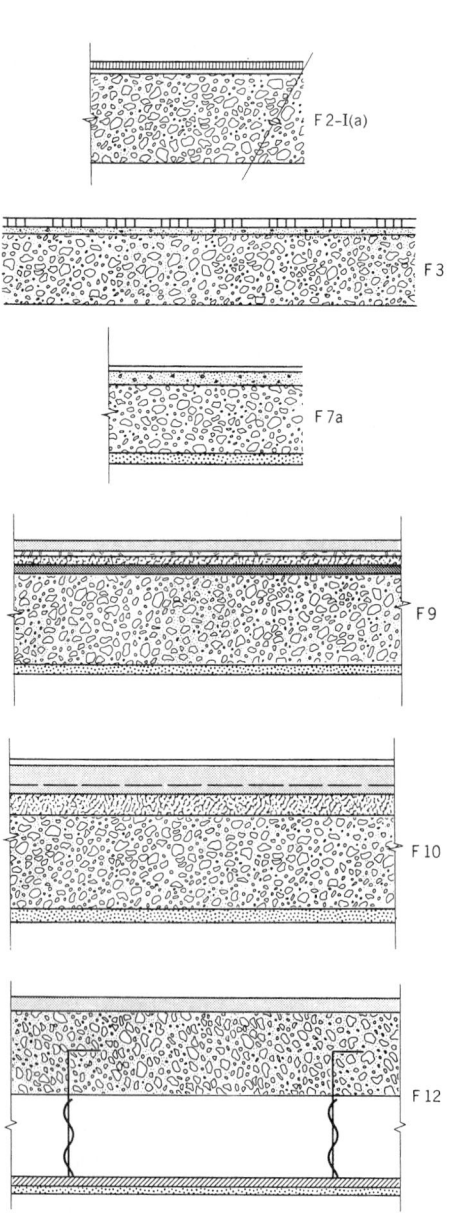

TABLE G.1 **Acoustic Characteristics of Floors: For Acoustic Data see Index G.3** (*continued*)

Code[a]	Description	Section Sketch

F14 6-in.-thick reinforced concrete slab. On the floor side, ¾-in.-thick tongue-and-groove wood flooring nailed to 1½ × 2 in. wooden battens, 16 in. on centers, floating on 1-in.-thick, glass-wool quilt. On the ceiling side, ½-in. layer of plaster.

Concrete with Hollow Blocks

F16 5 × 10 in. hollow masonry blocks, 14 in. on centers, with spaces between blocks filled with 5-in.-thick reinforced concrete. On the floor side, ⅞-in.-thick wood blocks adhered to 1½-in.-thick, sand-cement screed. On the ceiling side, ¾-in. layer of plaster.

F17b 4 × 12½ in. hollow masonry blocks, 15½ in. on centers, with spaces between blocks filled with 4-in.-thick reinforced concrete. On the floor side, 2-in.-thick, sand-cement screed; linoleum on 1-in.-thick wood flooring nailed to 1 × 2 in. wooden battens, spaced 15½ in. on centers, floating on glass-wool quilt, approximately 1 in. thick. On the ceiling side, ¾-in. layer of plaster.

Concrete Channel Slab

F22 Prefabricated concrete channel slabs mortared together 20 in. on centers. Each slab had a 3-in.-deep trapezoidal channel with bases of 11 and 14¾ in. On the floor side, ¾-in.-thick, sand-cement finish.

Ribbed Concrete

F23 7¼-in. ribbed concrete floor. Ribs were 5¼ × 3¾ in., spaced 21 in. on centers, with 1 × 2 in. wooden nailing strips cast into ends. On the floor side, the slab was 2 in. thick with a ¾-in.-thick, sand-cement screed. On the ceiling side, ⅝-in.-thick wooden laths nailed to nailing strips, held ⅝-in.-thick plaster.

Concrete Channel Beam

F24 7-in. precast trapezoidal concrete channel beams, 14 in. on centers, with spaces between beams filled with sand-cement mix. On the floor side, 1½-in.-thick, sand-cement screed with 1-in.-thick, wood-block floor

APPENDICES

TABLE G.1 **Acoustic Characteristics of Floors: For Acoustic Data see Index G.3** (*continued*)

Code[a]	Description	Section Sketch

covering. On the ceiling side, approximately ¾-in.-thick layer of plaster on expanded metal lath.

Precast Concrete Beam, Floating Floor

F25   5-in. precast concrete channel beams, 14½ in. on centers, with spaces between beams filled with a sand-cement mix. On the floor side, ⅞-in.-thick, tongue-and-groove wood flooring nailed to 1 × 2 in. wooden battens, 20 in. on centers, on approximately 1-in.-thick, glass-wool quilt on ¾-in.-thick, sand-cement screed. On the ceiling side, ⅛-in. layer of plaster on ⅜-in. gypsum wallboard nailed to 1 × 2 in. wooden battens spaced 14½ in. on centers.

F 25

Hollow Concrete Beam

F27   5-in. precast trapezoidal hollow concrete beams, 14½ in. on centers, with bases of 14 and 12½ in. Spaces between beams filled with sand-cement mix. On the floor side, 1-in.-thick, sand-cement screed with 3/16-in. cork tile floor covering. On the ceiling side, ⅜-in.-thick gypsum wallboard attached to 1 × 2 in. wooden battens held by metal clips.

F 27

Wooden Joist

F30   2 × 8 in. wooden joists 16 in. on centers. On the floor side, ⅞-in., tongue-and-groove flooring nailed to joists; on the ceiling side, ⅜-in. gypsum wallboard nailed to joists with joints sealed.

F 30

F32b   2 × 8 in. wooden joists 18 in. on centers. On the floor side, ⅞-in., tongue-and-groove wood flooring nailed to joists. On the ceiling side, 1-in. battens nailed through glass-wool quilt, approximately 1 in. thick; ½-in. layer plaster on ¼-in.-thick wood lath.

F 32b

F34   2 × 8 in. wooden joists, 16 in. on centers. On the floor side, ½-in.-thick C-D plywood nailed 8 in. on centers to joists, 25/32-in.-thick hard wood flooring on plywood. On the ceiling side, ½-in.-thick gypsum wallboard nailed 6 in. on centers to joists. All joints taped and finished; ceiling tile adhered to gypsum board.

F 34

TABLE G.1 Acoustic Characteristics of Floors: For Acoustic Data see Index G.3 (*continued*)

Code[a]	Description	Section Sketch
F35a	2 × 8 in. wooden joists, 16 in. on centers. On the floor side, 1½-in.-thick, tongue-and-groove wood fiberboard nailed to joists, vinyl tile floor covering. On the ceiling side, ½-in.-thick gypsum wallboard nailed 6 in. on centers to joists. All joints taped and finished.	
F35b	Similar to F35a except fiberboard covered with carpet and pad.	
F38a	2 × 10 in. wooden floor joists spaced 16 in. on centers. ⅝-in. fir plywood subfloor nailed to joists 8 in. on centers; ½-in. plywood underlayment nailed to subfloor with joints staggered to miss joints of the subfloor; ⅛ × 9 × 9 in. vinyl asbestos tile glued to underlayment. On the ceiling side, ½-in. gypsum wallboard nailed 12 in. on centers with all joints and nailheads taped and finished.	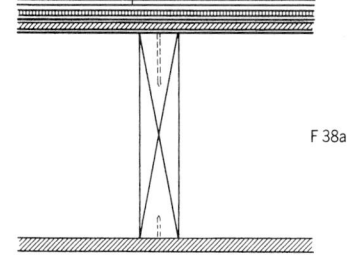
F39a	2 × 10 in. wooden joists 16 in. on centers. On the floor side, ½-in.-thick plywood subfloor nailed 6 in. on centers along edges and 10 in. on centers in field, building paper underlayment, ²⁵⁄₃₂ × 2¼ in. oak wood flooring nailed at each joist intersection and midway between joists. On the ceiling side, ⅝-in.-thick gypsum wallboard, nailed 6 in. on centers to joists. All joints taped and finished.	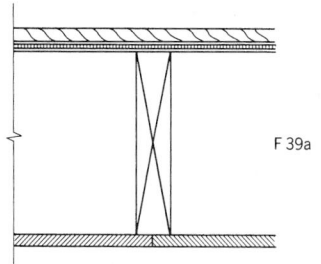

Wooden Joist, Resilient Ceiling

F44 2 × 8 in. wooden joists 16 in. on centers. On the floor side, ¾-in.-thick wood subfloor, layer of building paper, and ¾-in.-thick tongue-and-groove fir finish flooring. On the ceiling side, resilient runners bridged across joists and nailed 12 in. on centers to joists; ⅝-in.-thick gypsum wallboard screwed to resilient runners. All joints taped and finished.

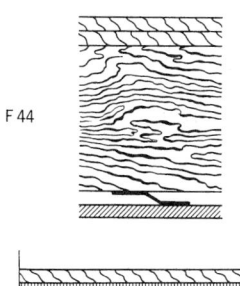

Wooden Joist with Insulation

F40a 2 × 10 in. wooden joists 16 in. on centers with 3-in.-thick mineral fiber batts stapled between joists. On the floor side, ½-in.-thick plywood subfloor nailed 6 in. on centers along edges and 10 in. on centers in field, building paper underlayment, ²⁵⁄₃₂ ×

(*continued*)

TABLE G.1 **Acoustic Characteristics of Floors: For Acoustic Data see Index G.3 (*continued*)**

Code[a]	Description	Section Sketch

2¼ in. oak wood flooring nailed at each joist intersection and midway between joists. On the ceiling side, ⅝-in.-thick gypsum wallboard nailed 6 in. on centers to joists. All joints taped and finished.

F48 2 × 8 in. wooden joists, 16 in. on centers. On the floor side, 1⅛-in.-thick regular C-D rough plywood nailed 6 in. on centers along periphery and 16 in. on centers at other bearings, plywood covered with an all-hair pad (40 oz/yd²) and all-wool pile (44 oz/yd²) carpet. The total weight of the carpet was 4.14 lb/yd² and the total thickness was ⅜ in. On the ceiling side, 2 × 4 in. wooden joists, 16 in. on centers, staggered 8 in. on centers relative to the floor joists, 3-in.-thick fibered glass blankets stapled between ceiling joists, ⅝-in.-thick gypsum wallboard nailed to ceiling joists. All joints taped and finished; entire periphery of panel caulked and sealed. The ceiling was supported independently from the floor structure.

F 48

Steel Joist with Concrete Floor

F57a 2½-in.-thick perlite concrete, 72 lb/ft³ on 28-gauge corrugated steel units supported by 14-in. steel bar joists; ⅛-in.-thick asphalt tile cemented to concrete. On the ceiling side, ¾-in. furring channels, 13½ in. on centers, wire-tied to joists, 3.4 lb/yd² diamond mesh metal lath wire-tied to furring channels, 9⁄16-in. coat of plaster with 1⁄16-in., white-coat finish.

F57b Same as F57a except carpet and pad in lieu of asphalt tile.

F58 18-in. steel joists 16 in. on centers. On the floor side, ⅝-in.-thick C-D rough plywood nailed to joists, 1⅝-in.-thick foamed concrete, 100 lb/ft³, slab constructed on the plywood; concrete covered with an all-hair pad (40 oz/yd²) and an all-wool pile (44 oz/yd²) carpet. Total weight of the carpet, 4.14 lb/yd², total thickness, ⅜ in. On the ceiling side, ⅝-in.-thick gypsum wallboard nailed to joists. All joints taped and finished; entire periphery of panel caulked and sealed.

F 57

F 58

TABLE G.1 **Acoustic Characteristics of Floors: For Acoustic Data see Index G.3** (*continued*)

Code[a]	Description	Section Sketch
F60a	7-in. steel bar joists spaced 27 in. on centers. On the floor side, ⅜-in. metal rib lath attached to top of joists, and 2-in.-thick poured concrete floor. On the ceiling side, ¾-in. metal furring channels wire-tied to joists 16 in. on centers; ⅜ × 16 × 48 in. plain gypsum lath held with wire clips and sheet metal end joint clips; ⁷⁄₁₆-in. sanded gypsum plaster and ¹⁄₁₆-in., white-coat finish.	F 60
F60c	Structure F60a with nylon carpeting and foam rubber pad placed on the floor. The carpet pad had an uncompressed thickness of ¼ in., backed with a woven jute fiber cloth. The carpet had ⅛-in. woven backing and ¼-in. looped pile spaced 7 loops per inch with a total thickness of ⅜ in.	

[a]Material extracted from *A Guide to Airborne, Impact and Structure-Borne Noise Control in Multi-Family Dwellings,* HUD/FHA/NBS, 1971.

H

OPTIONS FOR COMPUTATION ASSISTANCE

This book presents "manual" calculation procedures, those simple enough to require only a hand-held calculator and an informed (and patient) user. In this section we present a very short list from a huge and rapidly expanding array of software programs for personal computers (and occasionally, larger mainframes).

Updated, improved versions of existing programs appear as frequently as do new programs. Thus, any list prepared in 1 year quickly becomes obsolete by the next. The descriptions here thus avoid version numbers, website addresses, even telephone numbers due to soon-outdated area codes, in order to describe the basics of the program. Mailing addresses are included; latest websites should be readily obtainable by web search engines.

Excellent programs have been developed at many universities and research labs. We apologize to those whose programs do not appear here; the following list attempts to name the programs most widely used (or available) in North America at the turn of the century.

(a) Building Products Life-Cycle Assessment (Chapters 1, 4, 12).

BEES. Aids in the selection of building products by generating an overall score. It weighs the environmental and economic life-cycle performance scores, with relative importance scales specified by the user. Global warming, acidification, nutrification, natural resource depletion, IAQ, and solid waste impacts are included.
National Institute of Standards and Technology (NIST), Green Buildings Program, 100 Bureau Drive, Gaithersburg, Maryland 20899-0001.

(b) Design Strategies and Climate Analysis (Chapter 2).

Climate Consultant. A detailed data base which can be queried by users, this elaborates on the climate design strategy method described in Figs. 2.13–2.15, plotting every hour of the year on a variety of charts, and providing commentary.
Research Professor Murray Milne, UCLA, Department of Architecture and Urban Design, Los Angeles, California 90095.

Green Building Advisor. You provide the project location, building type and size, and this program responds with "moderately" and "strongly" suggested design strategies, relevant case studies, information library, and products directory. This is a tool for brainstorming, rather than detailed design decisions.
Environmental Building News, 28 Birge St., Brattleboro, Vermont 05301.

Solar 2. Graphically displays the performance of a window with any combination of fins and overhang, as well as remote objects such as walls of an exterior courtyard, or a distant building; for Windows 95.
Research Professor Murray Milne, UCLA, Department of Architecture and Urban Design, Los Angeles, California 90095.

(c) Heating and Cooling Performance (Chapters 5, 6, 7).

Building Design Advisor. Provides energy modeling in a CAD-like environment, simulating energy and daylighting performance from CAD drawings.
Lawrence Berkeley Laboratory, 1 Cyclotron Road, Berkeley, California 94720.

EnergyPlus. Expected to be available in 2000, this mainframe program aims to combine the best features of BLAST and DOE-2, providing a detailed HVAC system performance analysis that includes passive energy sources and strategies. Considerable operator training required. BLAST Support Office, University of Illinois, 140 Mechanical Engineering Building MC-244, 1206 W. Green St., Urbana, Illinois 60801.

Energy Scheming. From schematic drawings generated on-screen (Mac), rather than from numbers, this program calculates thermal performance for four months (representing seasons), with passive strategies included. Design advice accompanies the numbers.
Prof. G. Z. Brown, Energy Studies in Buildings Laboratory, Department of Architecture, University of Oregon, Eugene, Oregon 97403-1206.

Energy 10. Building energy simulations for smaller [≤10,000 ft^2 (929 m^2)] commercial and institutional projects. Input can be extremely simple (floor area, location, function), and a "shoebox" generic building is then generated with design alternatives and energy comparisons. A detailed "weathermaker" climate data feature allows for adjustments to serve quite local climates.
Passive Solar Industries Council, 1511 K St. NW, Suite 600, Washington, DC, 20005.

ENVSTD. Determines compliance with ASHRAE/IESNA 90.1, *Energy Efficient Design of New Buildings Except Low-Rise Residential Buildings.* The version of the software is updated as the standard is updated.
ASHRAE, 1791 Tullie Circle NE, Atlanta, Georgia 30329-5478.

Opaque. Graphically displays the performance of a wall or roof of any composition, any color, any orientation, and at any latitude. Research Professor Murray Milne, UCLA, Department of Architecture and Urban Design, Los Angeles, California 90095.

SOLAR 5. Using information about climate, orientation, geometry and shading configurations, users get a three-dimensional plot of building energy performance using hourly TMY data for over 230 sites. Almost all passive heating and cooling strategies can be simulated. Most HVAC system parameters can also be modeled.

Research Professor Murray Milne, UCLA, Department of Architecture and Urban Design, Los Angeles, California 90095.

(d) Environmental Assessment (Chapters 1 through 8).

BREEAM (Building Research Establishment Environmental Assessment Method). This well-tested UK program, in use since 1990, rates designs and building management policies, awarding certificates of overall performance (pass, good, very good, or excellent). Priority areas can be targeted early in design stages, making this potentially useful as a design tool.
Building Research Establishment, Marston, Watford, WD2 7JR, United Kingdom.

Green Building Assessment Tool (GBA Tool). Developed for use in the 1998 Green Buildings Challenge, this is being further developed at various levels of detail, and includes a weighting system with defaults that can be modified to reflect regional priorities. The links between building operation practices and environmental design strategies will be strengthened.
U.S. Green Buildings Council, 110 Sutter St., Suite 906, San Francisco, California 94104.

(e) Solar Water Heating (Chapter 10).

F-Chart. These programs allow analysis of various collector and system types (including some passive heating strategies).
F-Chart Software, 4406 Fox Bluff Road, Middleton, Wisconsin 53562.

(f) Fire Safety (Chapter 13).

Fire-Safe Building Design for Architects and Designers. A self-paced, interactive CD-ROM combining graphics, text, narration, animation and video, allows the user to manipulate graphics, make materials selections, consult supporting fire safety tables, review video, and explore a library of static and animated appendices. Feedback is provided, encouraging improvements to fire safety in a design.
National Fire Academy, USFA/FEMA, 16825 South Seton Ave., Emmitsburg, Maryland 21727.

I

METRICATION;
SI UNITS; CONVERSIONS

I.1 General Comments on Metrication

The building profession has been slower than most in adopting the metric (more accurately, the Systéme International, or SI) system of units for many reasons, a discussion of which is not relevant here. The change will come. Many of the major professional societies use a mixture of both systems because certain units are so entrenched that changing them is an almost insurmountable task. However, the advantages of the SI system are well known to all and need no justification here. In this book we have followed current industry practice—a mixture of the two systems. For this reason tables of conversions and approximations are presented below to enable the reader to live with both systems. Also given below are some useful facts that should make use of the SI (metric) system a bit easier.

I.2 SI Nomenclature, Symbols

For a full discussion of the SI system the reader is referred to *AIA Metric Building and Construction Guide*, edited by S. Braybrooke (Wiley, New York, 1980). The units in common use include the basic SI units; meter, kilogram, and second (MKS), plus a host of derived, supplementary, and non-SI units such as pascal (pressure), radian (solid angle), and kilowatt-hour (energy). Also, multiple and submultiple units such as liter, metric ton, and millibar are so common that they stand as separate units instead of being expressed as 10^{-3} m^3, 10^3 kg and 10^{-2} Pa. For this reason we omit a detailed analysis of the subject and simply supply data that by experience we have found useful.

Table I.1 lists prefixes with their accepted symbols. Symbols do not change in plural. That is, 6 millimeters is written 6 mm and 20 kilograms is

TABLE I.1 **These Prefixes May Be Applied to All SI Units**

Multiples and Submultiples	Prefixes	Symbols
1 000 000 000 = 10^9	giga	G
1 000 000 = 10^6	mega	M[a]
1 000 = 10^3	kilo	k[a]
100 = 10^2	hecto[b]	h
10 = 10	deka	da
0.1 = 10^{-1}	deci	d
0.01 = 10^{-2}	centi	c[a]
0.001 = 10^{-3}	milli	m[a]
0.000 001 = 10^{-6}	micro	μ[a]
0.000 000 001 = 10^{-9}	nano	n
0.000 000 000 001 = 10^{-12}	pico	p

[a]Most commonly used.

[b]A hectare is a square hectometer (i.e., 10^4 m^2).

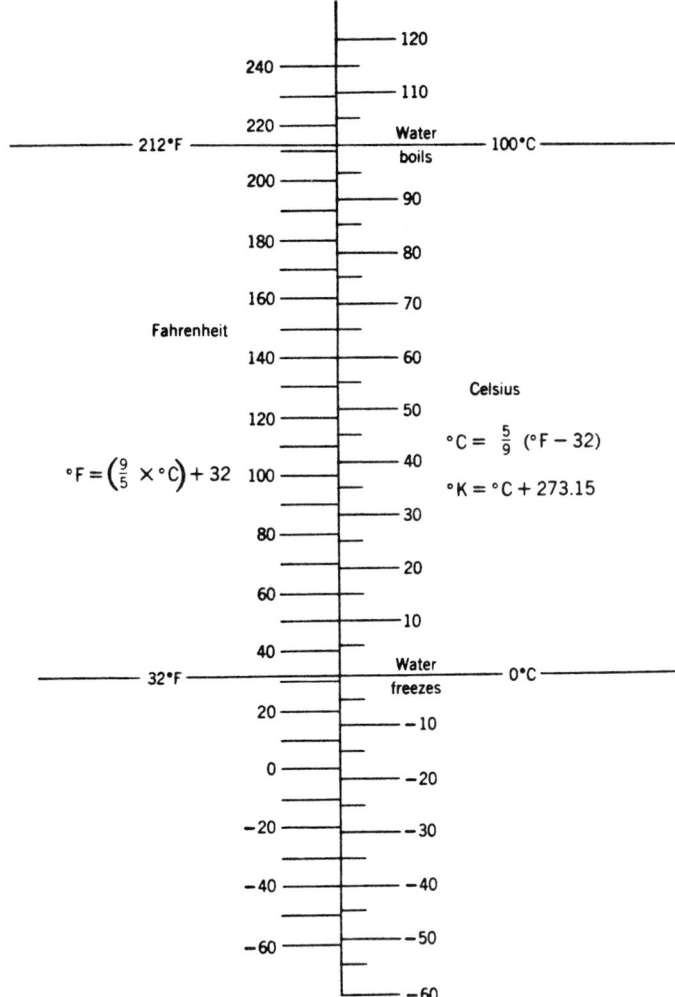

Fig. I.1 *Conversion: Fahrenheit degrees–Celsius degrees.*

written 20 kg. All units and prefixes except Fahrenheit and Celsius are uncapitalized when written out, as in megaton or meter.

I.3 Common Usage

1. *Length:* meter (m), kilometer (km), millimeter (mm), micrometer (μm), nanometer (nm).
2. *Area:* square meter (m²), hectar (ha).
3. *Volume:* cubic meter (m³), liter (L).
4. *Flow:* cubic meters per second m³/s.
5. *Velocity, Airflow:* meters per second (m/s).
6. *Weight:* kilogram (kg), gram (g).

The SI system clearly differentiates between mass (kg) and force (kg·m/s²), the latter being given a separate name and symbol, newton (N). Weight is not used, because it is a force that depends on acceleration and is therefore variable. However, the construction industry in large measure continues to use the terms mass, weight, and force interchangeably.

7. *Force:* newton (N), kilonewton (kN). A newton is the force required to accelerate 1 kg at 1.0 m/s².
8. *Pressure:* pascal (Pa), kilopascal (kPA). A pascal is a newton per square meter (N/m²).
9. *Energy, Work, Quantity of Heat:* joule (J),

kilojoule (kJ), megajoule (MJ). A joule is a watt-second (W·s).

10. *Temperature:* degree Celsius (°C), degree kelvin (K).

The SI units (kelvins) is not used as commonly as °C. (*Celsius* is the accepted term; *centigrade* is obsolete.) The Celsius and Kelvin scales are subdivided equally but start at different points— that is, 0 K is −273.15°C. Therefore, to determine Kelvin from Celsius, simply add 273.15. Increments are equal because of equal subdivision; that is, a change of 10 K is the same as a 10°C change. Because of its special importance, a separate Fahrenheit/Celsius conversion chart is given in Fig. I.1.

11. *Illumination:* see Table 18.1
12. *Acoustics:* see Table 28.15
13. *CGS/MKS Conversions:* see Table 28.15
14. *Abbreviations:* See Table I.2
15. *Approximations:* See Table I.3

I.4 Conversion Factors

Table I.4 is an alphabeticized list of useful conversion factors. Because normal use involves a hand calculator, we have avoided scientific notation and used decimal notation instead; that is, we write 0.00378, not 3.78×10^{-3} or 3.78 E-3.

TABLE I.2 **Typical Abbreviations: All Systems of Units**

atmospheres	atm	kilopascals	kPa
British thermal units	Btu	kilowatts	kW
British thermal units per hour	Btu/h	kilowatt-hours	kWh
calorie	cal	liters	L
cubic feet	cf, ft³	liters per second	L/x
cubic feet per minute	cfm, ft³/min	megajoules	MJ
cubic feet per second	cfs, ft³/s	meganewtons	MN
cubic meters	m³	megapascals	MPa
feet	ft	meters	m
feet per second	fps, ft/s	meters per second	m/s
gallons	gal	miles per hour	mph
gallons per hour	gph, gal/h	millimeters	mm
gallons per minute	gpm, gal/min	millimeters of mercury	mm Hg
grams	g	newtons	N
hectares	ha	ounces	oz
horsepower	hp	pounds	lb
inches	in.	pounds of force	lbf
inches of mercury	in. Hg	pounds per cubic foot	lb/ft³
joules	J	second	sec, s
kilocalories	kcal	square feet	ft²
kilograms	kg	square inches	in.²
kilograms per second	kg/s	square meters	m²
kilojoules	kJ	watts	W
kilometers	km	watts per square meter	W/m²
kilometers per hour	kph, km/h	yards	yd
kilonewtons	kN		

APPENDICES

TABLE I.3 **Common Approximations**

Approximate Common Equivalents			
1 inch	= 25 millimeters	1 millimeter	= 0.04 inch
1 foot	= 0.3 meter	1 liter	= 61 cubic inches
1 yard	= 0.9 meter	1 meter	= 3.3 feet
1 mile	= 1.6 kilometers	1 meter	= 1.1 yards
1 square inch	= 6.5 square centimeters	1 kilometer	= 0.6 mile
1 square foot	= 0.09 square meter	1 square centimeter	= 0.16 square inch
1 square yard	= 0.8 square meter	1 square meter	= 11 square feet
1 acre	= 0.4 hectare	1 square meter	= 1.2 square yards
1 cubic inch	= 16 cubic centimeters	1 hectare	= 2.5 acres
1 cubic foot	= 0.03 cubic meter	1 cubic centimeter	= 0.06 cubic inch
1 cubic yard	= 0.8 cubic meter	1 cubic meter	= 35 cubic feet
1 quart	= 1 liter	1 cubic meter	= 1.3 cubic yards
1 gallon	= 0.004 cubic meter	1 liter	= 1 quart
1 ounce	= 28 grams	1 cubic meter	= 250 gallons
1 pound	= 0.45 kilogram	1 kilogram	= 2.2 pounds
1 horsepower	= 0.75 kilowatt	1 kilowatt	= 1.3 horsepower

TABLE I.4 **Useful Conversion Factors: Alphabetized**

Multiply	By	To Get
acres	4047	square meters
atmospheres	33.93	feet of water
atmospheres	29.92	inches of mercury
atmospheres	760.0	millimeters of mercury
Btu (energy)	0.252	kilocalories
	1.055	kilojoules
Btu/h (power)	0.2928	watts
Btu/h/ft^2 (energy transfer)	3.152	watts per square meter
BtuF (heat capacity)	1.897	kilojoules per kelvin[a]
Btu/lb°F (specific heat)	4.182	kilojoules per kilogram per kelvin[a]
Btu/h/Fft (thermal conductivity[b])	1.729	watts per kelvin[a] per meter
Btu/hFft^2 (conductance[c])	5.673	watts per kelvin[a] per square meter
Btu/Fday (building load coefficient, BLC)	0.022	watts per kelvin[a]
Btu/Fdayft^2 (load-collector ratio, LCR)	0.236	watts per kelvin[a] per square meter
cubic feet	0.028	cubic meters
cubic feet	7.481	gallons
cubic feet	28.32	liters
cubic feet per minute	0.472	liters per second
cubic feet per second	2.832	liters per second
cubic inches	16.39	cubic centimeters
cubic meters	35.32	cubic feet
cubic meters	1.308	cubic yards
cubic meters	264.2	gallons
cubic yards	0.765	cubic meters
feet	0.305	meters

TABLE I.4 **Useful Conversion Factors: Alphabetized (*continued*)**

Multiply	By	To Get
feet	304.8	millimeters
feet per second	0.3048	meters per second
foot-pounds of force per second	1.356	watts
gallons	3.785	liters
gallons her hour	0.00152	liters per second
gallons per minute	0.0022	cubic feet per second
gallons per minute	0.06308	liters per second
grams	0.035	ounces (avoirdupois)
hectares	2.471	acres
horsepower	0.746	kilowatts
horsepower	746	watts
inches	25.4	millimeters
inches of mercury	0.033	atmospheres
inches of mercury	1.133	feet of water
inches of mercury (60°F)	3377	newtons per square meter
inches of mercury	0.491	pounds per square inch
inches of water	0.002458	atmospheres
inches of water	0.036	pounds per square inch
inches of water (60°F)	248.8	newtons per square meter
kilocalories	3.968	British thermal units
kilocalories	4190	joules
kilograms	2.205	pounds
kilograms per cubic meter	1.686	pounds per cubic yard
kilograms per square meter	0.0033	feet of water
kilograms per square meter	0.0029	inches of mercury
kilograms per square meter	0.205	pounds per square foot
kilograms per square meter	0.001422	pounds per square inch
kilojoules	0.948	British thermal units
kilojoules per kilogram	0.430	British thermal units per pound
kilometers	0.621	miles
kilometers per hour	0.621	miles per hour
kilonewtons	0.1004	tons of force
kilonewtons	224.8	pounds of force
kilopascals	20.89	pounds of force per square foot
kilowatts	1.341	horsepower
kilowatt-hours	3.6	megajoules
liters	0.03532	cubic feet
liters	61.02	cubic inches
liters	0.2642	gallons
liters	1.057	quarts
liters per second	2.119	cubic feet per minute
liters per second	951.0	gallons per hour
liters per second	15.85	gallons per minute
megajoules	0.278	kilowatt-hours
meganewtons	100.36	tons of force
megapascals	145.04	pounds of force per square inch
megapascals	9.324	tons of force per square foot
meters	3.281	feet

(*continued*)

TABLE I.4 **Useful Conversion Factors: Alphabetized** (*continued*)

Multiply	By	To Get
meters	1094	yards
meters per second	196.86	feet per minute
meters per second	2.237	miles per hour
miles	1.609	kilometers
miles per hour	1.609	kilometers per hour
miles per hour	0.447	meters per second
milliliters	0.061	cubic inches
milliliters	0.035	fluid ounces
millimeters	0.039	inches
millimeters of mercury	133.3	newtons per square meter
million gallons per day	18.94	cubic meters per hour
newtons	0.225	pounds of force
ounces (avoirdupois)	28.35	grams
ounces (fluid)	28.41	milliliters
pounds	0.454	kilograms
pounds of force	4.448	newtons
pounds of force per square foot	47.88	pascals
pounds of force per square inch	6.895	kilopascals
pounds per cubic foot	16.02	kilograms per cubic meter
pounds per cubic yard	0.593	kilograms per cubic meter
pounds per square foot	4.882	kilograms per square meter
square feet	0.0929	square meters
square inches	645.2	square millimeters
square kilometers	0.386	square miles
square meters	10.76	square feet
square meters	1.196	square yards
square miles	2.590	square kilometers
square yards	0.836	square meters
tons of force	9.964	kilonewtons
tons of force per square foot	107.25	kilopascals
tons of force per square inch	15.44	megapascals
torr (millimeters of mercury at $0°C$)	133.3	newtons per square meter
watts	3.412	British thermal units per hour
watts	0.738	foot-pounds of force per second
watts per square meter	0.317	British thermal units per square foot
yards	0.914	meters

Add your own conversion factors here.

Multiply	By	To Get
Multiply	By	To Get
Multiply	By	To Get

[a] $°K$ or $°C$.

[b] Thermal conductivity (K).

[c] Thermal conductance (C) or transmittance (U).

J

ASHRAE STANDARDS FOR ENERGY EFFICIENT BUILDING DESIGN

J.1 Criteria for Components of the Envelope: New Low-Rise Residential Building Design

The design guidelines in this section are taken from ANSI/ASHRAE Standard 90.2-1993, *Energy-Efficient Design of New Low-Rise Residential Buildings,* with subsequent addenda. These are published by the American Society of Heating, Refrigerating, and Air-Conditioning Engineers, Inc., Atlanta, Ga. This standard offers two procedures for residential building design: a *prescriptive path* in which minimum standards of insulation and maximum fenestration and ventilation rates are specified, and an *annual energy cost method,* in which a proposed design is compared to a similar but prescriptive-path design, in order to justify departures from the prescriptive path.

The annual energy cost method is not discussed further here, but it allows for the benefits of innovative designs, materials and equipment when they cannot be evaluated adequately under the prescriptive path procedures. Such approaches include active and passive solar heating and cooling, heat recovery, air-to-air heat exchangers, high-efficiency equipment, radiant barriers, thermal mass, operable shading and window insulation, and thermal storage. In this procedure, specified equations are used to evaluate the annual energy cost of proposed design with its prescriptive-path counterpart. If the annual energy cost of the proposed design is less than or equal to that of its prescriptive-path counterpart, then the proposed design is freed of further need to adhere to this Standard.

The prescriptive path has two parts: single family houses and multifamily structures. Single family houses are further divided into part (a) with ducts outside the conditioned space, and part (b) with ducts inside the conditioned space. Based upon a location's combination of heating degree-days (base 65°F) and cooling degree hours (base 74°F), this procedure specifies maximum overall U-values for roofs and above grade walls, floors, doors and windows; maximum conductances (C) for the wall assembly on below-grade basement and crawl-space walls; minimum resistances (R) for insulation of perimeters of slab-on-grade floors, and minimum shading coefficients (SC) for windows. These are presented as graphs, reproduced here in Figs. J.1 through J.16 for single family houses, and J.17 through J.35 for multifamily structures. The U value and SC for fenestration varies with the ratio of fenestration area to floor area, as shown in Table J.1. Note that a maximum fenestration area = 21% of floor area is allowed under this prescriptive path, although larger areas may be allowed with a trade-off calculation procedure, described in the following paragraph. For locations with greater than 10,000 HDD base 65, see Table J.2 for all of the $U, C, R,$ and SC requirements for both single-family and multifamily structures.

An "envelope trade-off procedure" allows for fenestration areas greater than allowed in Table J.1: a maximum 21% fenestration area/floor area; or a maximum of 15% fenestration area/floor area, with 5% of that fenestration area in skylight area. However, an absolute maximum of 32% fenestration area/floor area is allowed, using this trade-off procedure. If greater fenestration areas are desired, the more complex annual energy cost procedure must be used. This envelope trade-off

(text continued on page 1762)

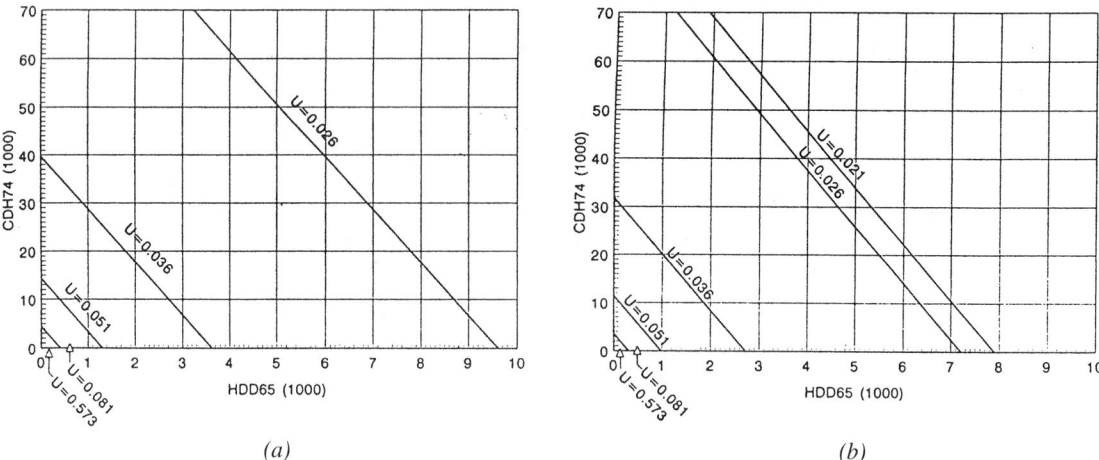

(a)   *(b)*

Fig. J.1 *Single family houses:* (a) *Maximum U, ceilings with attics (ducts within conditioned space).*
(b) *Maximum U, ceilings with attics (ducts outside conditioned space). (Reprinted with permission of
the American Society of Heating, Refrigerating, and Air-Conditioning Engineers, Inc. from* ANSI/
ASHRAE Standard 90.2-1993, Energy Efficient Design of New Low-Rise Residential Buildings.*)*

TABLE J.1 **Residential Fenestration *U* Values and SC**

Part A. ***U-Value Requirements***

	Btu/h ft² °F	
Required U-value from Fig. J.15 or Fig. J.34	*If Fenestration Area >15% and ≤18% of Conditioned Floor Area*	*If Fenestration Area >18% and ≤21% of Conditioned Floor Area*
U = 1.31	*U* = 1.31	*U* = 0.87
U = 0.87	*U* = 0.81	*U* = 0.65
U = 0.49	*U* = 0.44	*U* = 0.41
U = 0.36	*U* = 0.33	*U* = 0.29

Part B. ***Shading Coefficient (SC) Requirements***

Required SC from Fig. J.16 or Fig. J.35	*If Fenestration Area >15% and ≤18% of Conditioned Floor Area*	*If Fenestration Area >18% and ≤21% of Conditioned Floor Area*
SC = 0.70	SC = 0.70[a]	SC = 0.70[a]
SC = 0.50	SC = 0.50[b]	SC = 0.50[b]

Source: From ANSI/ASHRAE Standard 90.2-1993, *Energy-Efficient Design of New Low-Rise Residential Buildings,* with subsequent addenda. © by the American Society of Heating, Refrigerating, and Air-Conditioning
Engineers, Inc., Atlanta, Ga.

[a]If the required U = 1.31 (based on Fig. J.15 or Fig. J.34) then SC = 0.63.
[b]If the required U = 1.31 (based on Fig. J.15 or Fig. J.34) then SC = 0.43.

TABLE J.2 Residential Envelope Requirements Above 10,000 HDD65

Component	Single Family Housing HDD65				Multifamily Structures HDD65			
	10,001–12,000	12,001–15,000	15,001–18,000	Above 18,000	10,001–12,000	12,001–15,000	15,001–18,000	Above 18,000
Ceilings with attic: max. U =	0.021	0.021	0.018	0.018	0.021	0.021	0.018	0.018
without attic: max. U =	0.033	0.033	0.033	0.033	0.033	0.033	0.033	0.033
Walls above grade: max. U = (except metal frame)	0.042	0.042	0.038	0.038	0.042	0.038	0.038	0.038
metal frame: max. U =	—	—	—	—	0.067	0.067	0.067	0.067
below grade: max. C =	0.063	0.063	0.063	0.063	0.063	0.063	0.063	0.063
crawl space: max. C =	0.056	0.056	0.056	0.056	0.056	0.056	0.056	0.056
wood frame adjacent to unconditioned space, max. U =	0.059	0.059	0.059	0.059	0.059	0.059	0.059	0.059
metal frame adjacent to unconditioned space, max. U =	—	—	—	—	0.123	0.123	0.079	0.079
mass adjacent to unconditioned space, max. U =	0.172	0.172	0.080	0.080	0.172	0.172	0.080	0.080
Floors, slab on grade, perimeter min. R =	8	10	12	14	8	10	12	14
frame over ambient temperature, max. U =	0.033	0.033	0.033	0.033	0.033	0.033	0.033	0.033
frame over unconditioned space, max. U =	0.033	0.033	0.033	0.033	0.033	0.033	0.033	0.033
concrete slab, max. U =	—	—	—	—	0.058	0.058	0.058	0.058
Doors, max. U =	0.19	0.19	0.19	0.19	0.19	0.19	0.19	0.19
Fenestration,[a] max. U =	0.36	0.36	0.36	0.36	0.36	0.36	0.36	0.36
min. SC =	0.7	0.7	0.7	0.7	0.7	0.7	0.7	0.7

Source: From ANSI/ASHRAE Standard 90.2-1993, *Energy-Efficient Design of New Low-Rise Residential Buildings*, with subsequent addenda. © by the American Society of Heating, Refrigerating, and Air-Conditioning Engineers, Inc., Atlanta, Ga.

Note: units are as follows: *U*: Btu/h ft² °F (whole component)
C: Btu/h ft² °F (wall insulation only)
R: h ft² °F/Btu (perimeter insulation only)

[a]For fenestration areas in excess of 12% of floor area, see Table J.1 for *U* and SC values.

APPENDICES

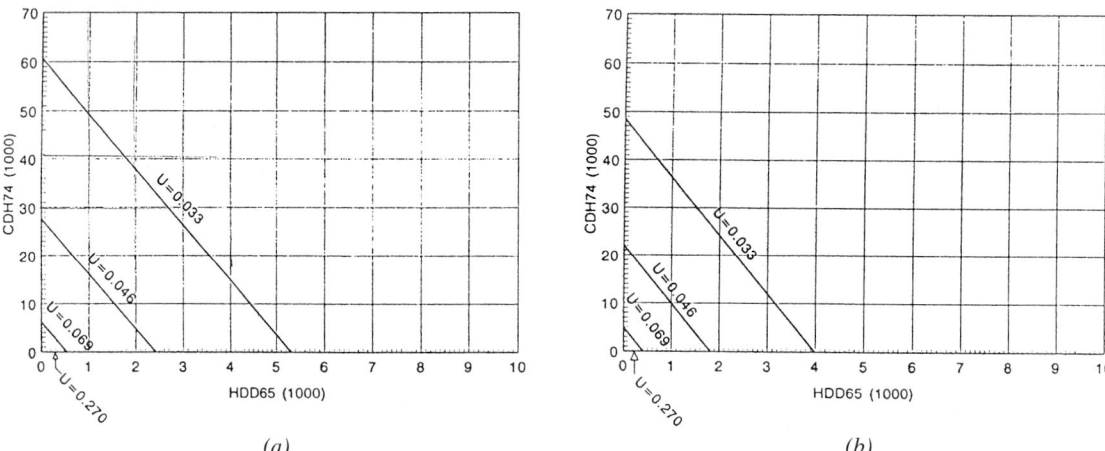

Fig. J.2 *Single family houses:* (a) *Maximum* U, *ceilings without attics (ducts within conditioned space).* (b) *Maximum* U, *ceilings without attics (ducts outside conditioned space). (Reprinted with permission of the American Society of Heating, Refrigerating, and Air-Conditioning Engineers, Inc. from* ANSI/ ASHRAE Standard 90.2-1993, Energy Efficient Design of New Low-Rise Residential Buildings.*)*

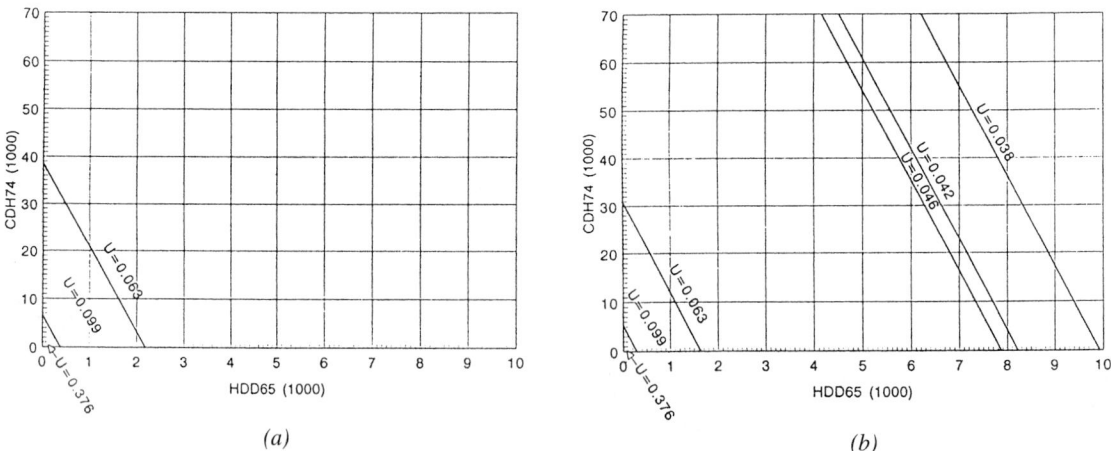

Fig. J.3 *Single family houses:* (a) *Maximum* U, *above-grade frame walls and band joists (ducts within conditioned space).* (b) *Maximum* U, *above-grade frame walls and band joists (ducts outside conditioned space). (Reprinted with permission of the American Society of Heating, Refrigerating, and Air-Conditioning Engineers, Inc. from* ANSI/ASHRAE Standard 90.2-1993, Energy Efficient Design of New Low-Rise Residential Buildings.*)*

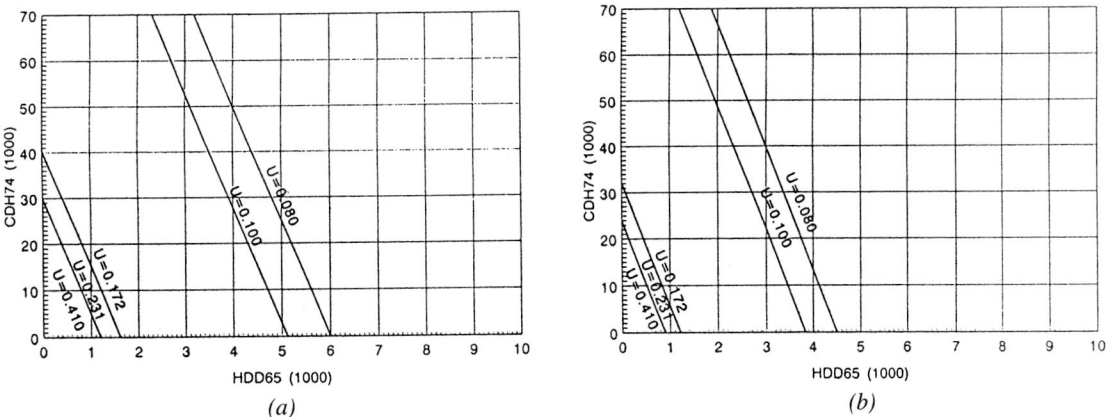

Fig. J.4 *Single family houses:* (a) *Maximum* U, *above-grade concrete, masonry or log walls with exterior or integral insulation* (*ducts within conditioned space*). (b) *Maximum* U, *above-grade concrete, masonry or log walls with exterior or integral insulation* (*ducts outside conditioned space*). *(Reprinted with permission of the American Society of Heating, Refrigerating, and Air-Conditioning Engineers, Inc. from* ANSI/ASHRAE Standard 90.2-1993, Energy Efficient Design of New Low-Rise Residential Buildings.*)*

Fig. J.5 *Single family houses:* (a) *Maximum* U, *above-grade concrete, masonry or log walls with interior insulation* (*ducts within conditioned space*). (b) *Maximum* U, *above-grade concrete, masonry or log walls with interior insulation* (*ducts outside conditioned space*). *(Reprinted with permission of the American Society of Heating, Refrigerating, and Air-Conditioning Engineers, Inc. from* ANSI/ASHRAE Standard 90.2-1993, Energy Efficient Design of New Low-Rise Residential Buildings.*)*

APPENDICES

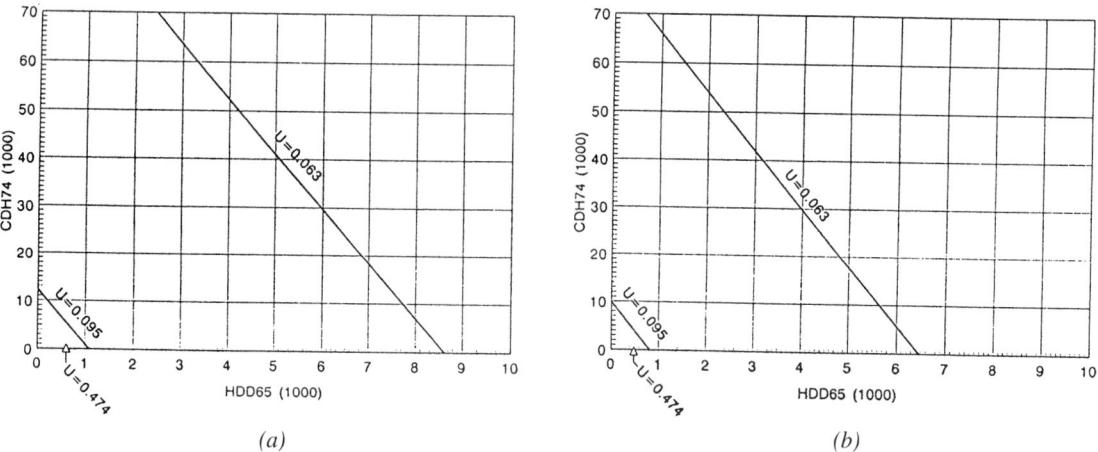

Fig. J.6 *Single family houses:* (a) *Maximum* U, *wood frame walls adjacent to unconditioned space (ducts within conditioned space).* (b) *Maximum* U, *wood frame walls adjacent to unconditioned space (ducts outside conditioned space). (Reprinted with permission of the American Society of Heating, Refrigerating, and Air-Conditioning Engineers, Inc. from* ANSI/ASHRAE Standard 90.2-1993, Energy Efficient Design of New Low-Rise Residential Buildings.*)*

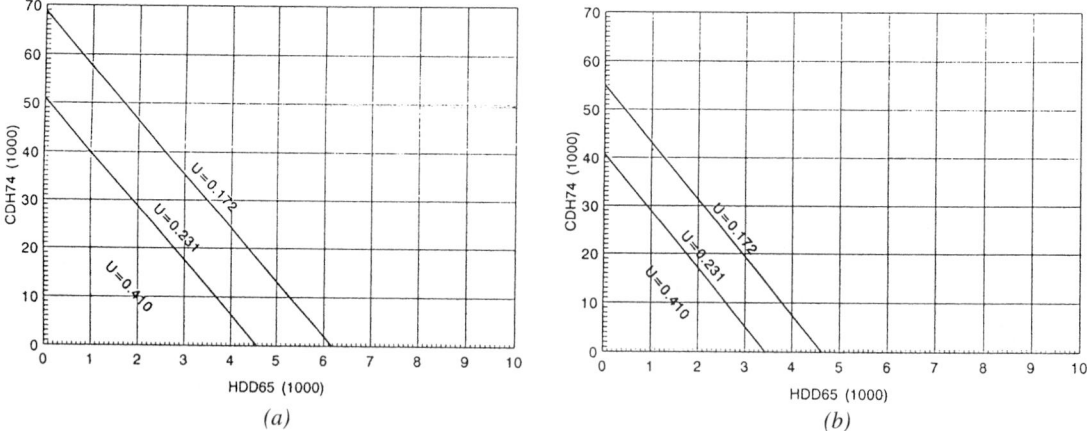

Fig. J.7 *Single family houses:* (a) *Maximum* U, *concrete or masonry walls adjacent to unconditioned space (ducts within conditioned space).* (b) *Maximum* U, *concrete or masonry walls adjacent to uncondi-tioned space (ducts outside conditioned space). (Reprinted with permission of the American Society of Heating, Refrigerating, and Air-Conditioning Engineers, Inc. from* ANSI/ASHRAE Standard 90.2-1993, Energy Efficient Design of New Low-Rise Residential Buildings.*)*

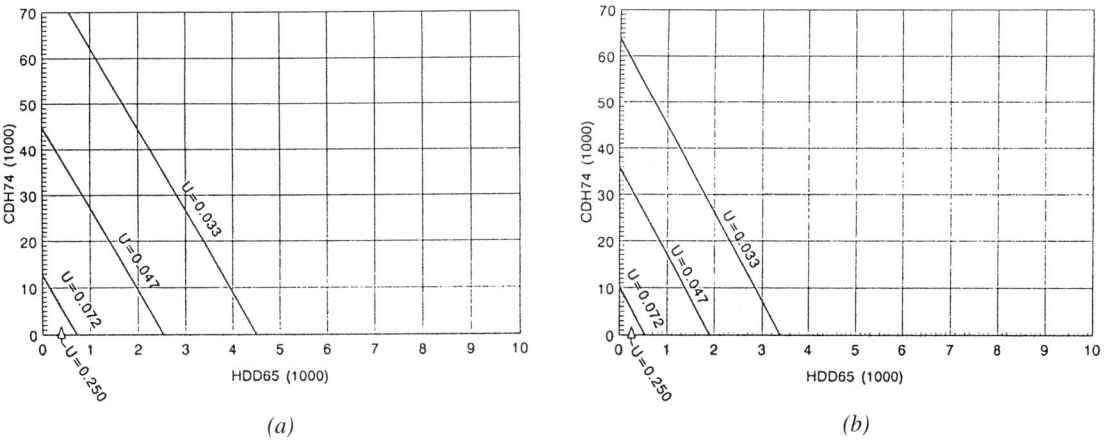

Fig. J.8 *Single family houses:* (a) *Maximum* U, *wood frame floors over exterior ambient conditions—overhangs, carports, porches (ducts within conditioned space).* (b) *Maximum* U, *wood frame floors over exterior ambient conditions—overhangs, carports, porches (ducts outside conditioned space).* (*Reprinted with permission of the American Society of Heating, Refrigerating, and Air-Conditioning Engineers, Inc. from* ANSI/ASHRAE Standard 90.2-1993, Energy Efficient Design of New Low-Rise Residential Buildings.*)*

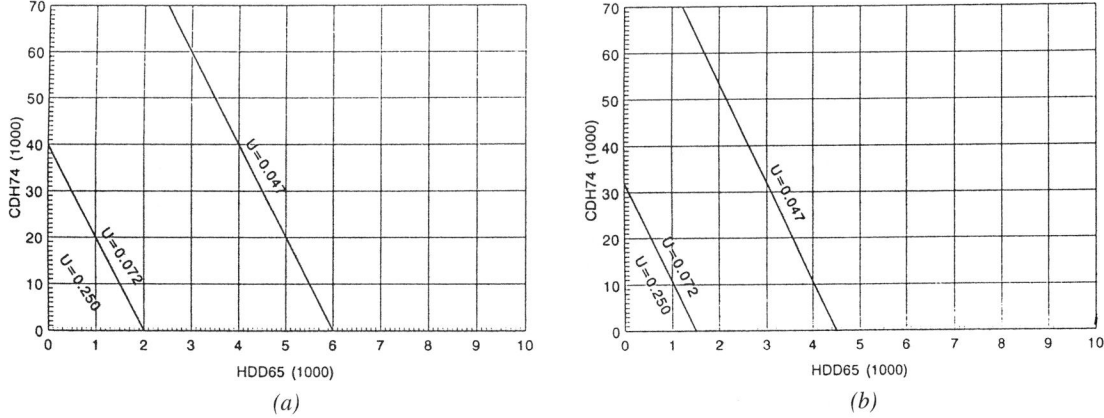

Fig. J.9 *Single family houses:* (a) *Maximum* U, *wood frame floors over unconditioned space—vented crawl spaces, basements, enclosed garages or porches (ducts within conditioned space).* (b) *Maximum* U, *wood frame floors over unconditioned space—vented crawl spaces, basements, enclosed garages or porches (ducts outside conditioned space).* (*Reprinted with permission of the American Society of Heating, Refrigerating, and Air-Conditioning Engineers, Inc. from* ANSI/ASHRAE Standard 90.2-1993, Energy Efficient Design of New Low-Rise Residential Buildings.*)*

APPENDICES

Fig. J.10 *Single family houses:* (a) *Maximum C-value of wall assembly, below-grade basement walls with exterior or integral insulation (ducts within conditioned space).* (b) *Maximum C-value of wall assembly, below-grade basement walls with exterior or integral insulation (ducts outside conditioned space).* (Reprinted with permission of the American Society of Heating, Refrigerating, and Air-Conditioning Engineers, Inc. from* ANSI/ASHRAE Standard 90.2-1993, Energy Efficient Design of New Low-Rise Residential Buildings.)

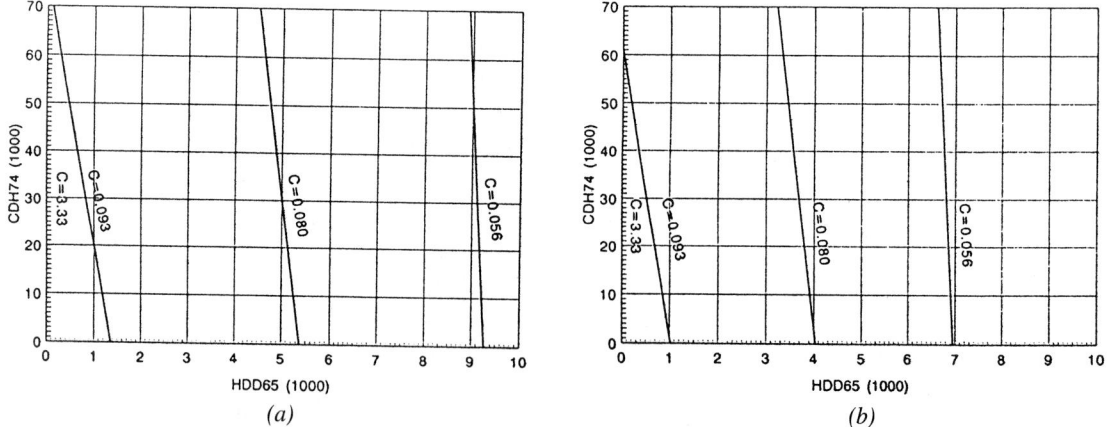

Fig. J.11 *Single family houses:* (a) *Maximum C-value of wall assembly, below-grade basement walls with interior insulation (ducts within conditioned space).* (b) *Maximum C-value of wall assembly, below-grade basement walls with interior insulation (ducts outside conditioned space).* (Reprinted with permission of the American Society of Heating, Refrigerating, and Air-Conditioning Engineers, Inc. from* ANSI/ASHRAE Standard 90.2-1993, Energy Efficient Design of New Low-Rise Residential Buildings.)

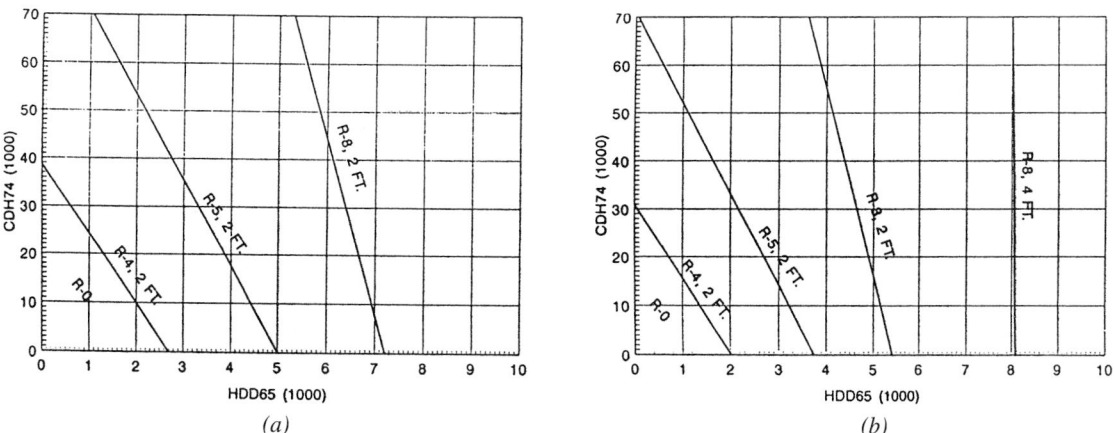

(a)   *(b)*

Fig. J.12 *Single family houses:* (a) *Minimum* R *(insulation only) and minimum total vertical +*
horizontal depth of continuous insulation, slab-on-grade floors (ducts within conditioned space). (b)
Minimum R *(insulation only) and minimum total vertical + horizontal depth of continuous insulation,*
slab-on-grade floors (ducts outside conditioned space). (Reprinted with permission of the American
Society of Heating, Refrigerating, and Air-Conditioning Engineers, Inc. from ANSI/ASHRAE Standard
90.2-1993, Energy Efficient Design of New Low-Rise Residential Buildings.)

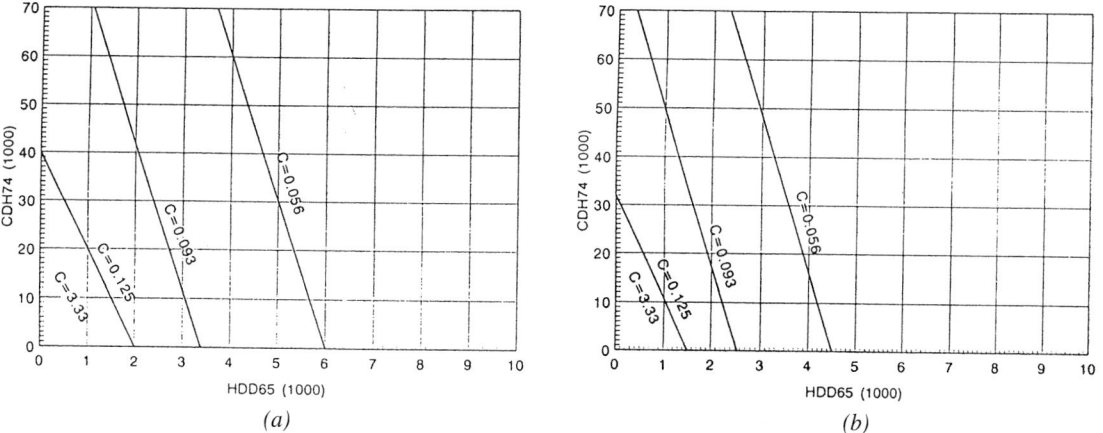

(a)   *(b)*

Fig. J.13 *Single family houses:* (a) *Maximum* C-*value of wall assembly, crawl space walls (ducts within*
conditioned space). (b) *Maximum* C-*value of wall assembly, crawl space walls (ducts outside conditioned*
space). (Reprinted with permission of the American Society of Heating, Refrigerating, and Air-Condition-
ing Engineers, Inc. from ANSI/ASHRAE Standard 90.2-1993, Energy Efficient Design of New Low-
Rise Residential Buildings.)

APPENDICES

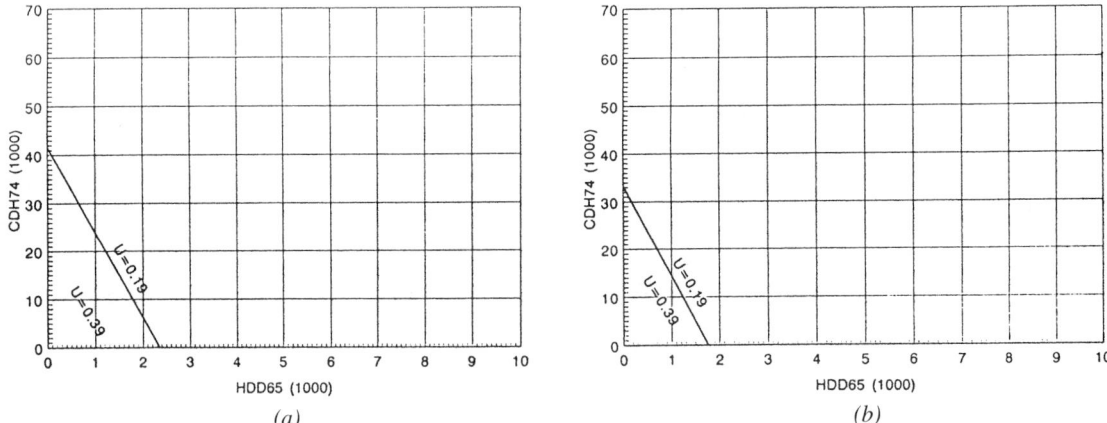

(a)   *(b)*

Fig. J.14 *Single family houses:* (a) *Maximum* U, *nonwood doors (ducts within conditioned space).* (b) *Maximum* U, *nonwood doors (ducts outside conditioned space). (Reprinted with permission of the American Society of Heating, Refrigerating, and Air-Conditioning Engineers, Inc. from* ANSI/ ASHRAE Standard 90.2-1993, Energy Efficient Design of New Low-Rise Residential Buildings.*)*

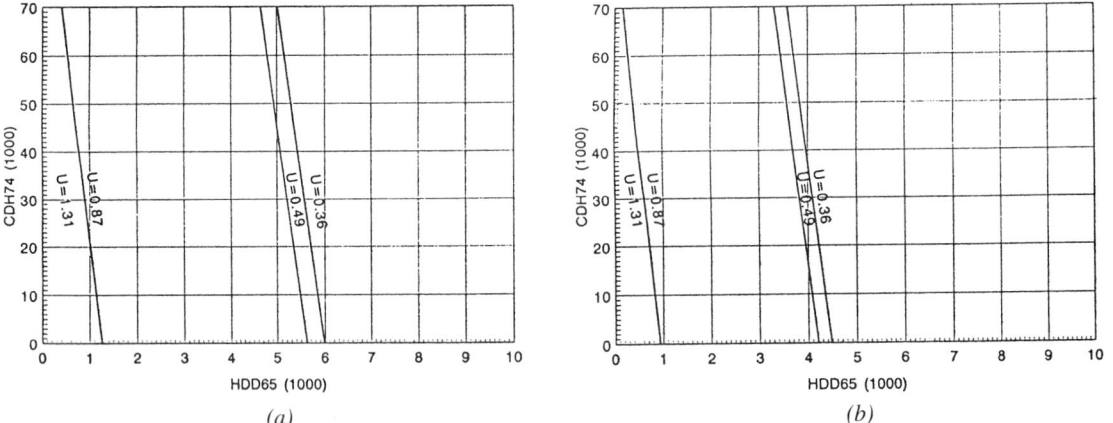

(a)   *(b)*

Fig. J.15 *Single family houses:* (a) *Maximum* U, *fenestration including framing (ducts within conditioned space).* (b) *Maximum* U, *fenestration including framing (ducts outside conditioned space). (Reprinted with permission of the American Society of Heating, Refrigerating, and Air-Conditioning Engineers, Inc. from* ANSI/ASHRAE Standard 90.2-1993, Energy Efficient Design of New Low-Rise Residential Buildings.*)*

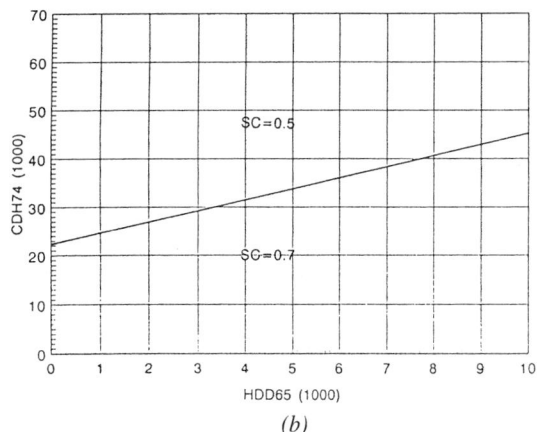

(a)   *(b)*

Fig. J.16 *Single family houses:* (a) *Fenestration maximum shading coefficients (ducts within conditioned space).* (b) *Fenestration maximum shading coefficients (ducts outside conditioned space).* *(Reprinted with permission of the American Society of Heating, Refrigerating, and Air-Conditioning Engineers, Inc. from* ANSI/ASHRAE Standard 90.2-1993, Energy Efficient Design of New Low-Rise Residential Buildings.*)*

Fig. J.17 *Multifamily structures: Maximum U, ceilings with attics. (Reprinted with permission of the American Society of Heating, Refrigerating, and Air-Conditioning Engineers, Inc. from* ANSI/ASHRAE Standard 90.2-1993, Energy Efficient Design of New Low-Rise Residential Buildings.*)*

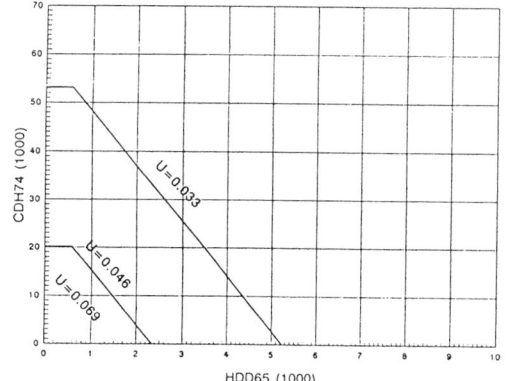

Fig. J.18 *Multifamily structures: Maximum U, ceilings without attics. (Reprinted with permission of the American Society of Heating, Refrigerating, and Air-Conditioning Engineers, Inc. from* ANSI/ASHRAE Standard 90.2-1993, Energy Efficient Design of New Low-Rise Residential Buildings.*)*

APPENDICES

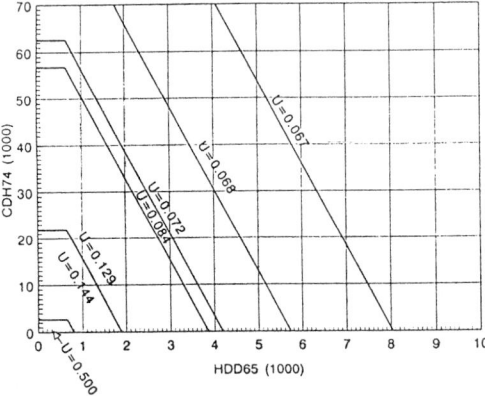

Fig. J.19 *Multifamily structures: Maximum* U, *above-grade frame walls and band joists. (Reprinted with permission of the American Society of Heating, Refrigerating, and Air-Conditioning Engineers, Inc. from* ANSI/ASHRAE Standard 90.2-1993, Energy Efficient Design of New Low-Rise Residential Buildings.*)*

Fig. J.20 *Multifamily structures: Maximum* U, *above-grade metal frame walls. (Reprinted with permission of the American Society of Heating, Refrigerating, and Air-Conditioning Engineers, Inc. from* ANSI/ASHRAE Standard 90.2-1993, Energy Efficient Design of New Low-Rise Residential Buildings.*)*

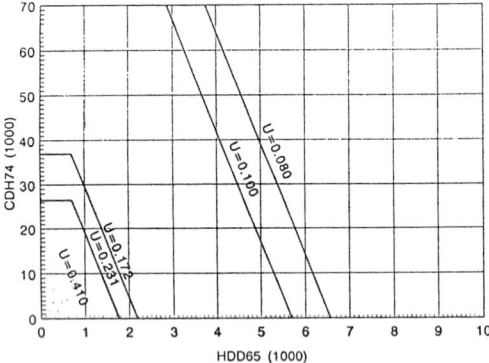

Fig. J.21 *Multifamily structures: Maximum* U, *above-grade concrete, masonry or log walls with exterior or integral insulation. (Reprinted with permission of the American Society of Heating, Refrigerating, and Air-Conditioning Engineers, Inc. from* ANSI/ASHRAE Standard 90.2-1993, Energy Efficient Design of New Low-Rise Residential Buildings.*)*

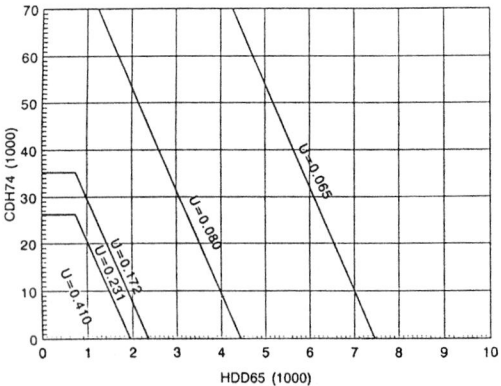

Fig. J.22 *Multifamily structures: Maximum* U, *above-grade concrete, masonry or log walls with interior insulation. (Reprinted with permission of the American Society of Heating, Refrigerating, and Air-Conditioning Engineers, Inc. from* ANSI/ASHRAE Standard 90.2-1993, Energy Efficient Design of New Low-Rise Residential Buildings.*)*

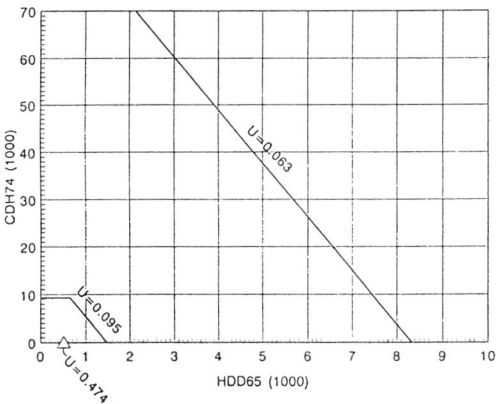

Fig. J.23 *Multifamily structures: Maximum U, wood frame walls adjacent to unconditioned space. (Reprinted with permission of the American Society of Heating, Refrigerating, and Air-Conditioning Engineers, Inc. from* ANSI/ASHRAE Standard 90.2-1993, Energy Efficient Design of New Low-Rise Residential Buildings.*)*

Fig. J.24 *Multifamily structures: Maximum U, metal frame walls adjacent to unconditioned space. (Reprinted with permission of the American Society of Heating, Refrigerating, and Air-Conditioning Engineers, Inc. from* ANSI/ASHRAE Standard 90.2-1993, Energy Efficient Design of New Low-Rise Residential Buildings.*)*

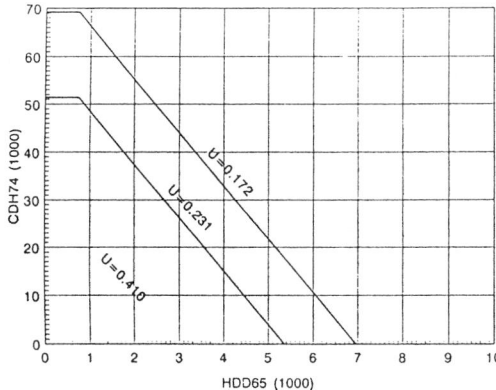

Fig. J.25 *Multifamily structures: Maximum U, concrete or masonry walls adjacent to unconditioned space. (Reprinted with permission of the American Society of Heating, Refrigerating, and Air-Conditioning Engineers, Inc. from* ANSI/ASHRAE Standard 90.2-1993, Energy Efficient Design of New Low-Rise Residential Buildings.*)*

Fig. J.26 *Multifamily structures: Maximum U, wood frame floors over exterior ambient conditions—overhangs, carports, porches. (Reprinted with permission of the American Society of Heating, Refrigerating, and Air-Conditioning Engineers, Inc. from* ANSI/ASHRAE Standard 90.2-1993, Energy Efficient Design of New Low-Rise Residential Buildings.*)*

APPENDICES

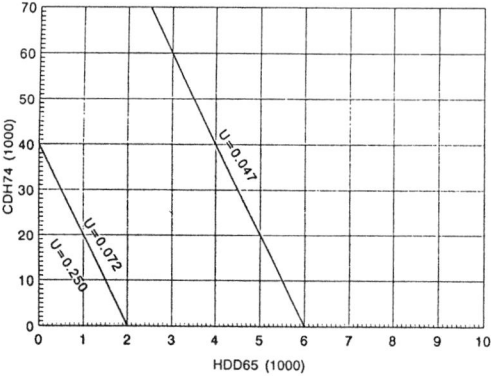

Fig. J.27 *Multifamily structures: Maximum U, wood frame floors over unconditioned space—vented crawl spaces, basements, enclosed garages or porches. (Reprinted with permission of the American Society of Heating, Refrigerating, and Air-Conditioning Engineers, Inc. from ANSI/ASHRAE Standard 90.2-1993, Energy Efficient Design of New Low-Rise Residential Buildings.)*

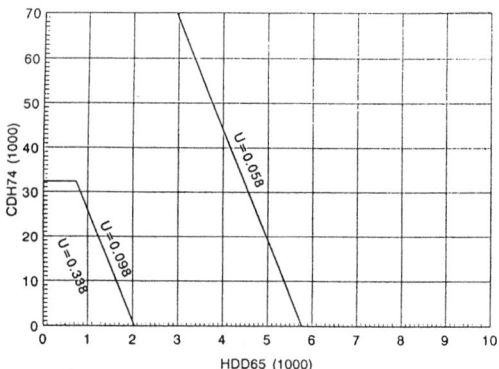

Fig. J.28 *Multifamily structures: Maximum U, concrete floors over exterior ambient conditions—garages. (Reprinted with permission of the American Society of Heating, Refrigerating, and Air-Conditioning Engineers, Inc. from ANSI/ASHRAE Standard 90.2-1993, Energy Efficient Design of New Low-Rise Residential Buildings.)*

Fig. J.29 *Multifamily structures: Maximum C-value of wall assembly, below-grade basement walls with exterior or integral insulation. (Reprinted with permission of the American Society of Heating, Refrigerating, and Air-Conditioning Engineers, Inc. from ANSI/ASHRAE Standard 90.2-1993, Energy Efficient Design of New Low-Rise Residential Buildings.)*

Fig. J.30 *Multifamily structures: Maximum C-value of wall assembly, below-grade basement walls with interior insulation. (Reprinted with permission of the American Society of Heating, Refrigerating, and Air-Conditioning Engineers, Inc. from ANSI/ASHRAE Standard 90.2-1993, Energy Efficient Design of New Low-Rise Residential Buildings.)*

Fig. J.31 *Multifamily structures: Minimum* R *(insulation only) and minimum total vertical + horizontal depth of continuous insulation, slab-on-grade floors. (Reprinted with permission of the American Society of Heating, Refrigerating, and Air-Conditioning Engineers, Inc. from* ANSI/ ASHRAE Standard 90.2-1993, Energy Efficient Design of New Low-Rise Residential Buildings.*)*

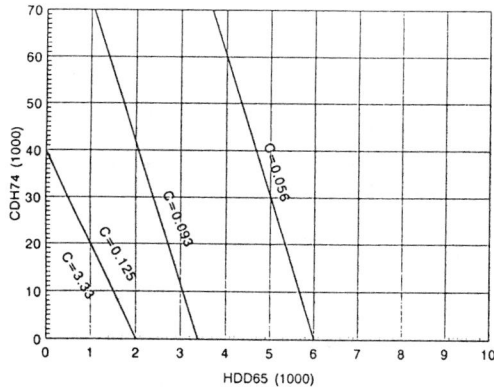

Fig. J.32 *Multifamily structures: Maximum C-value of wall assembly, crawl space walls. (Reprinted with permission of the American Society of Heating, Refrigerating, and Air-Conditioning Engineers, Inc. from* ANSI/ASHRAE Standard 90.2-1993, Energy Efficient Design of New Low-Rise Residential Buildings.*)*

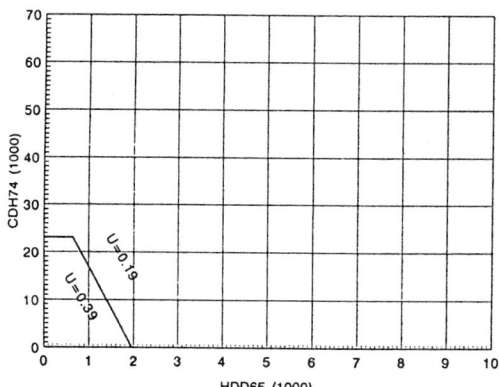

Fig. J.33 *Multifamily structures: Maximum* U, *nonwood doors. (Reprinted with permission of the American Society of Heating, Refrigerating, and Air-Conditioning Engineers, Inc. from* ANSI/ ASHRAE Standard 90.2-1993, Energy Efficient Design of New Low-Rise Residential Buildings.*)*

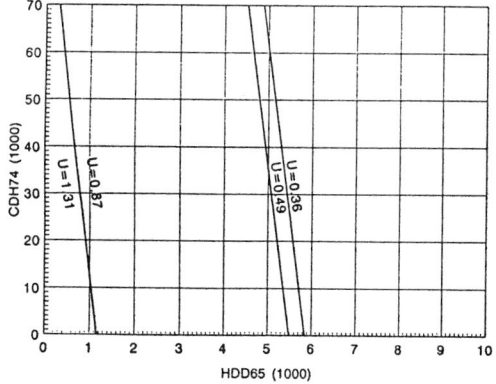

Fig. J.34 *Multifamily structures:* (a) *Maximum* U, *fenestration including framing. (Reprinted with permission of the American Society of Heating, Refrigerating, and Air-Conditioning Engineers, Inc. from* ANSI/ASHRAE Standard 90.2-1993, Energy Efficient Design of New Low-Rise Residential Buildings.*)*

APPENDICES

Fig. J.35 *Multifamily structures: Fenestration maximum shading coefficients. (Reprinted with permission of the American Society of Heating, Refrigerating, and Air-Conditioning Engineers, Inc. from ANSI/ASHRAE Standard 90.2-1993, Energy Efficient Design of New Low-Rise Residential Buildings.)*

procedure also allows for fenestration *U*-values in excess of those prescribed, by lowering the *U*-values of other envelope components, primarily opaque walls. This trade-off considers both heat-ing and cooling load factors, and is detailed in ANSI/ASHRAE Standard 90.2-1993, *Energy-Efficient Design of New Low-Rise Residential Buildings,* Section 5.9.

J.2   Criteria for Components of the Envelope: Buildings Except for Low-Rise Residential Building Design

At the time of this book's completion, ASHRAE had not yet approved BSR/ASHRAE/IESNA 90.1, their proposed revision of ASHRAE/IES 90.1-1989, *Energy Efficient Design of New Build-ings Except New Low-Rise Residential Buildings.* Therefore, no information from the proposed re-vision could be printed. Rather than reprint soon-to-be-outdated information from 90.1-1989, we therefore omit detailed references to this impor-tant standard. The latest version of BSR/ASH-RAE/IESNA 90.1, *Energy Standard for Buildings Except Low-Rise Residential Buildings,* is highly recommended as a reference.

INDEX

A.O. Smith Water Products Company, 460, 608
Aalto, Alvar, architect, 31
Abrams, Donald W., 247, 324, 329
ABS (acrylonitrile butadiene styrene), 652, 682
Absorptance, solar, 135–136
Absorption, *see* Sound absorption, Noise reduction
Absorption chillers, 379, 462–463
Absorption dehumidification, 359–362
Absorption refrigeration cycle, 377–379
ac, 857
Access, *see* Solar envelopes
Access control, detectors, 1346
  hotels and motels, 1356
  industrial buildings, 1367
  multiple dwellings, 1354
  offices, 1362
  principles, 1345
  residential systems, 1352
  schools, 1357
Access control and security, *see specific item*
Accessibility, in bathrooms, 639–640
ACH (air changes per hour):
  fire protection, 787
  IAQ, 338–340, 346
ACHE (adaptive control of home environments), 375
Acid rain, 95
Acidity, *see* pH
Acoustic criteria:
  auditoriums, 1542
  building siting, 1614
  fluorescent lamp ballasts, 1606
  music rooms, 1535
  recommendations for specific occupancies, 1609
  speech rooms, 1535
  see also Room acoustics
Acoustic definitions and glossary, 1163
Acoustic material, ASTM standards, 1594
Acoustic materials, 1553
  absorptive, 1553
  articulation class, 1592
  resonators, 1551
  transparent surfaces, 1550
Acoustic privacy, bathrooms, 639–640, *see also*
  Speech privacy
Acoustic references, 1623
Acoustic standards, 1621
Acoustic symbols and abbreviations, 1622
Acoustics, 1321, *see also* Acoustics criteria, Acoustic
  materials, Noise, Noise reduction, room acous-
  tics, Sound
Activated carbon filters, 579, 733
Activated charcoal filters, *see* Filters, air
Activated sludge process, 706–708, 726, 728, 733
Active solar heating, design guidelines, 224–225, *see*
  also Passive solar heating; Solar heating
Activity level:
  body temperature and, 37–39

comfort and, 40, 42–44
  heat gain calculations, 282–283
  thermal zoning and, 37–39
ADA, elevator requirements, 1394
Adaptive behavior, 47
Adiabatic processes, *see* Psychrometry
Adjustable sunshades, *see* Sunshades
Adsorption filters, *see* Filters, air
Advanced Composting Systems, 677
Advanced integrated wastewater pond systems, *see*
  AIWPS™
Aeration, 582–583
Aerobic decomposition:
  activated sludge, 707–708, 726, 728, 733
  composting toilets, 669
  greenhouses, 719–720
  lagoons, 715, 718
  Living Machines, 719–724
Aerobic treatment units (ATU), 707–708
Aerogel, 129–130
African Place (Expo '67 Montreal, Canada), 99, 102
AFUE (annual fuel utilization efficiency), 254–255,
  391, 404
Agua Blanca Ranch (Tucson, Arizona), 313–315
Air:
  sites and, 83, 91–106
  temperature, 39–42
  thermal properties of, 158–162
  see also Interior air quality; Pollution; Psychrome-
    try; Ventilation
Air & water HVAC systems, *see* HVAC systems
Air chambers, *see* Expansion chambers
Air change method (infiltration), 145
Air classifiers, 752
Air compressors, *see* Compressed air
Air conditioners (unit), 379, 381
Air diffusers, *see* Diffusers; Registers
Air ducts, *see* Ducts
Air-extract windows, 492–493, 509, 511, 513–514
Air filtration, *see* Filters, air
Air gaps:
  plumbing fixtures, 681–682
  windows, 139
Air motion, 41–42, 377–380
Air pollution, *see* Pollution
Air pressures, 495
Air quantities, HVAC systems, 487, 492
Airspaces, *see* Air gaps
Air supply (HVAC), 484–495
Air terminals (lightning), 816–819
Air-to-air heat exchangers, *see* Heat exchangers
Air velocities, recommended, 48, 52, 408–409
Air washers, *see* Filters, air
Air Xchange, Inc., 356, 360–361
Airborne sound, *see* Sound
Air-handling equipment: 475, 478, 480–481
  space required, 452, 480

Aitken, Donald, 492
AIWPS™ (advanced integrated wastewater pond systems), 716–719, 752
Alarm gong (fire sprinklers), 792
Albany County airport (New York), 18–21
Albedo, 379–380
Albert Kahn Associated Architects, 813
Alberta Agriculture, 356
Algae growth, 584
    sewage treatment and, 717–718, 724
    see also Cooling towers
Alkalinity, 574–575
All-air HVAC systems, see HVAC systems
Allen, Edward, architect, 41, 430, 451–452
All-water HVAC systems, see HVAC systems
Alternate Sources of Energy (magazine), 387–388
Alternative refrigerants, see Refrigerants
Aluminaire House (Long Island, New York), 640
Aluminum, recycling of, 5, 747, 749
Aluminum conduit, 911
Aluminum wire, see Conductors, electric
Amana Refrigeration, Inc., 407
American Furnace Division, Singer Company, 413
American Gas Association, 393
American Gas Cooling Center, 425
American Institute of Architects (AIA), 98, 112, 369, 425, 527
American National Standards Institute (ANSI), 638–642, 667, see also ANSI/ASHRAE
American Society of Heating, Refrigerating, and Air-Conditioning Engineers, Inc., 146, 329, 369, 425, 484, 527, Appendix H
American Society of Landscape Architects, 99, 112, 561
American Solar Energy Society, 73, 76, 100, 258, 314, 316, 318, 324, 329, 348, 492
American Standard, Inc., 630, 634
American Wire Gauge (AWG), see Conductors, electric
Ammeter, see Electrical measurements
Ampacity (current-carrying capacity), 897
Ampere, 853
Ammonia, 377, 420
Amplitude of earth temperature, 133–134
Amsler Hagenah MacLean, architects, 403
Anaerobic decomposition, 669, 704, 706–708
    digesters, 733–734, 752–753
    greenhouses, 719–720
    lagoons, 715–718
    Living Machines, 719–724
Andrews, John, architect, 102, 547, 549–550
Anemostat Corporation, 503
Angle valve, 654
Annual fuel utilization efficiency, see AFUE
Anshen and Allen, architects, 443–444
ANSI/ASHRAE Standard 55a-1995, 43–48, 66
ANSI/ASHRAE Standard 90.2–1993, 132–136, 138, 143, 146, 164, 351, 375, 387, 391, 395, 401, 404, 425, Appendix J
Antelope Valley California Poppy Reserve (Palmdale, California), 325–329, 618
Appliances:
    approximate heat gains, 229
    hourly heat gains, 282, 285
    see also Washing machines

Aqua-Flo, Inc., 581
Aquasaver Company, 735, 738
Aquifers, 576
Arcata wetlands (California), 719
Architectural acoustics, Chapter 26
Arens, Edward, 48–51, 66
Argon, 129, 139
Arup, see Ove Arup Partnership
Asbestos, 334–336
Asbestos-cement pipe, 682
Ashley, Robert, 308, 329
ASHRAE Standard 62–1989, 146, 180–187, 363
ASHRAE/IES Standard 90.1–1989, 130, 132, 134, 146, 296–297, 425, 527
Atlantic Ultraviolet Corporation, 582
Atmosphere, 95–97, see also Global warming
Atocha Railroad Station (Madrid, Spain), 379, 381
Atrium buildings:
    daylight in, 201, 205
    fire protection for, 782–784, 786–787
Attics, thermal flow through, 135
Auditorium acoustic design, see Room acoustics
Audubon House (New York City), 760–763, 766
Automation, see Building automation systems
Awnings, 121–122, 175, 177, 221, see also Shading coefficients

Backwater prevention, 684, 689
BACnet, 484
Bacteria:
    in air, 334, 337
    in water, 546, 552, 733, 738
Balance point temperature, 252–254
Balcomb, J. Douglas, 163, 209, 216–217, 263, 271, 276, 329, Appendix C
Balcomb residence, see Unit 1 First Village
Ballasts:
    fluorescent lamps, 1180–1184
    see also individual types
Banana Kelly Community Improvement Association, 757
Band of sun, 77
Banham, Reyner, 332, 369, 640, 667
Bank of America Corporate Center (Charlotte, North Carolina), 818
Banwell White Arnold Hemberger & Partners, architects, 221, 264, 274, 346
Barden, A. A., 389, 425
Barrier-free design, 1394
    elevators, 1394
    see also Bathroom design
Barriers, 115–116, see also Acoustic barriers; Building envelope, connectors/filters/barriers
Barriers, acoustic, exterior, 1574
BART (Bay Area Rapid Transit) station (Pleasanton, California), 569
Barton Hall, University of New Hampshire, 346
Baseboard convectors, 390, 393–395, 398–399
Basement walls and floors, 133–134, 167
Batch solar water heaters, 615–618, 620
Bathroom design:
    accessibility, privacy, 639–640
    cleansing functions, 531–532
    elimination functions, 626–627, 629–631
    history of, 531–532

installation/maintenance, 696, 698
prefabricated, 690, 697
public toilet rooms, 698–699
required fixtures, 540, 542–543
water conservation and, 632, 669–671
Bathtubs, 531–532, 627–628, 661, 691, 737–738
required number of, 542–543
Battery equipment, emergency power, 996–1044
Batwing diffusers, *see* Lighting fixtures
Beaufort scale, 100
Bedales School, Olivier Theatre (Hampshire, England), 346, 349
Bedrick, James, 200, 329
BEES (software), Appendix H
Bellevue stormwater treatment (Washington), 743
Bentel & Bentel, architects, 594–597
Bentley, D., 318, 329
Beranek, W. J., 100, 112
Berkebile Nelson Immenschuh McDowell, architects, *see* EPICenter
Berm walls, 73–75, 325–329, *see also* Basement walls
Berry, B., 301, 329
Beth Israel Chapel and Memorial Garden (Houston, Texas), 99, 102–104
Bibury, Residence of the 27th Vicar of the Parish Church, 736
Bidets, 630–631, 661, 691, *see also* Perineal cleansing
Biochemical oxygen demand, *see* BOD
Bioclimatic charts, 48–51
design concepts, 55–59
Biogas, *see* Methane
Biological contaminants:
air, 334, 337
water, 577
Blackwater, 735–736
BLAST (software), Appendix H
BLC (building load coefficient), defined, 267
Bleshman Regional Day School (Paramus, New Jersey), 509
Blood:
temperature and, 37–39
CO and, 770
Blowout toilets, 633
BMS (building management systems), 22, 344, 375, 482–484, 492, 495
Bobenhausen, Robert, 465–466, 527
BOCA, 821
BOD (biochemical oxygen demand), 541, 708, 718
Body temperature and heat, 37–40
comfort and, 49–50
heat gain calculations and, 282–283
Bogen Jenal, Engineers, P.C., 725
Boiler:
hot water heating systems, 391–393
large HVAC systems, 453–462
space requirements, 458
Boiler flue economizer, 466, 471
Boody, Harry John, 357–358
Bored wells, 586
Boston City Hall, 72–73
Bourne, Richard, 382, 425
Bowen, Arthur, 324, 329
Bragdon, Clifford, 93, 112
Branch circuit design, *see* Electric wiring design
Brass pipe, 651–652, 682
Breathable walls, 348–349

BREEAM (Building Research Establishment Environmental Assessment Method) (software), Appendix H
Breeching, *see* Boilers
Brightness, *see* Luminance
Brightness, subjective, 1068
British Research Establishment, 201, 329, Appendix H
British Research Establishment, Building 16 (Garston, England), 483–484
Bronx Community Paper Company (New York City), 755–757
Brower, David, 36
Brown, G. Z., 84, 112, 201, 211, 324, 329, 432, 439, 551, 622, Appendix H
Brown, Robert, 98, 111–112
Brownfield, 756
BSR/ASHRAE/IESNA 90.1 Energy Standard, 391, 404, Appendix J
Btu (British thermal unit):
defined, 124
heat flow and, 124–127
Buchanan/Watson, architects, 223
Building automation systems, (BAS), 1371, 1374, 1376
intelligent buildings, 1380
offices, 1366
stand alone, 1372
terminology, 1374
Building construction, demolition, renovation, 5
Building Design Advisor (software), Appendix H
Building drains and sewers, 690, 695
Building envelope:
climate and, 75–77, 115–120
components summary, 196–197
connectors/filters/barriers/switches, 115–120
conservation criteria for, 250
form considerations, 115–120, 192–195
heat flow through, 120–143
heat gain calculations, example, 286–288
thermal zoning and, 195–196
Building management systems, *see* BMS
Building Science Engineering, 473
Bullock's Department Store (San Mateo, California), 385–386
Burr Lawrence Rising & Bates Architects, 493
Burt Hill Kosar Rittelmann Associates, architects, 513–515
Busduct, 903
Busway, 903
BX cable, 901
Bypass shutoff valves, 643

Cable, *see* Conductors, electric
Cable tray, 906
Cablebus, 903
Caborn, J. M., 99
Calcium, 574–575, *see also* Softening of water
California Energy Commission, 320, 329
California Highway Patrol Offices (Gilroy, California), 308–311
California State Polytechnic University at Pomona:
Center for Regenerative Studies, 21–25
California State Polytechnic University at San Luis Obispo:
E2R2 Facility, 716–719, 752–755
Experimental Residence, 245–246

Calm, James E., 377
Cambridge CoHousing (Cambridge, Massachusetts), 469, 473
Canadian Research Council, 348–349, 354, 808, 850
Candela, 1053, 1064
Candlepower, 1053
Candlepower distribution curves (CD), 1064
Cannon Design, Inc., architects, 350–351
Carbon dioxide:
    climate and, 95, 376–377
    combustion, 391
    concentrations, IAQ, 333–335, 369
    fire hazards of, 770
    fire protection systems with, 786, 813–814
Carbon monoxide, 333–335, 770
Carmody, John, 141, 146, 172
Carrier Corporation, 378–379, 505
Cartridge filtration, 579
Cartwright, Virginia, 203, 205, 207
Cascade Manor Apartments (Eugene, Oregon), 117
Cast iron pipe, 682–685
Cast Iron Soil Pipe Institute, 685–686
Catalytic combustors, 386–389
Catchment areas, *see* Cisterns
CBS Tower (New York City), 504, 506
Ceiling fans, 377–380
Ceiling height:
    daylight factors, 197–203
    fire protection, 772, 782–783
Ceiling raceway systems, 924
Cellular concrete floor, 922
Cellular metal floor, 922–923
Center for Regenerative Studies (Cal Poly at Pomona), 21–25
Center Moriches Free Public Library (Long Island, New York), 271–274
Central HVAC systems, 373, 428–429, 453–484, *see also* HVAC systems
Central Sprinkler Corporation, 795
Centre Georges Pompidou (Paris, France), 436, 438
Centrifugal chillers, 463–464
Centrifugal pumps, 587, 590–591
Cermak, J. E., 100, 112
CFCs (chlorofluorocarbons), 95–96, 129, 335, 376–377, 462, 509
CFD (computational flow dynamics), 782–783
CH2M Hill (engineers), 734
Chalfoun, Nader, 329
Champion Cooler Corporation, 384
Chandra, S., 146, 300, 329, 344, 369
Check valves, 654, 685
Chemicals:
    dehumidification with, 291
    spontaneous combustion, 769
    waterless toilets, 669–678
    water quality, 574–577
Chilled water:
    district systems, 518–519
    temperatures, 463
Chiller-heaters, 520–522
Chillers, 460, 462–467
    space requirements, 451, 463, 467
Chimney, 10, 401, 404
Chloramine, 580
Chlorides, 576
Chlorination, 579–580
Chromaticity, 1195, 1199

Chronomite Laboratories, Inc., 600
Circuit breakers, 953
Circuit venting, 680–681, 696, 701–702
Cisterns:
    bacteria checks of, 546
    examples of, 549–550, 672, 722
    filtering for, 546–548
    irrigation and, 558, 560
    sizing of, 537–540, 546–551
    soft water and, 583
    see also Rainwater
C. K. Choi Building, University of British Columbia, Canada, 669–673
Clark, Gene, 318, 329
Clark-Ditton Architects, 391
Class of 1959 Chapel, Harvard Business School (Cambridge, Massachusetts), 271–274
Clean agent gases, 786, 814
Cleansing, of body, 626–627
Climate Consultant (software), Appendix H
Climate:
    body temperature and, 37–40
    building envelope and, 195
    classification of, 52–55
    comfort and, 40–53
    cooling data, 295–297, 317
    data, 56
    design strategies and, 55–65
    passive cooling and, 55–58, 63–65
    passive solar heating and, 58–65
    site and, 67
    switches and, 116–122
    timetables, 53–55, 199, 212
    see also Site
Climate window, *see* Air-extract window
Clivus Multrum USA, Inc., 671, 674
Clock and program systems, 1358
Closed-circuit evaporative cooler, 466, 471
Closed shell building envelope, 115–118
Closed-loop solar DHW systems, 619–620
Clothing, 40, 43–44
CLTD/CLF (cooling load temperature differential/cooling load factor), 294–295
Coagulation, 578
Coefficient of performance, *see* COP
Coal, *see* Energy sources
Cogeneration, 519–527
Cohen, S., 93, 112
Cold climate moisture control, 144
Collectors, *see* Solar collectors
Color, 1100
    effect of illuminant, 1200
    hue, value, chroma, 1103
    Munsell system, 1102
    of light sources, 1195–1201
    object, 1072
    reactions to, 1103
    temperature, 1100
Color, object, 1101
Color rendering index (CRI), 1199
Color systems, 1101
Colyer/Freeman Group, architects, 309–311, 326–328, 618
Combined sewers, *see* Storm sewers
Combustion:
    fire, 769–770
    heating systems, 401–404

Comfort, 41–55
  adaptive model of, 46–47
  air motion and, 49–52, 377–380
  asymmetry and, 51–52
  body temperature regulation and, 37–40, 49–53
  cleansing and elimination and, 626–627
  clothing and, 40
  combined radiant slab and warm air heat, 412–413
  cooling and air motion, 48–52, 98–104
  DHW temperatures, 597, 599
  displacement ventilation and, 487
  double-duct HVAC systems and, 446–447
  historical criteria for, 42–52
  HVAC systems and, 431–433
  individual control in offices, 489–490
  Interior air quality (IAQ) and, 338
  psychological implications, 113
  radiant floor systems and, 395
  relation of air and radiant temperature changes, 48–52
  standards, 42–52
  VAV HVAC systems and, 498
  warm air heating systems and, 405–406
  window radiant temperature and, 492–493
  zones, 42–52
Commerzbank, Frankfurt, Germany, 344–345, 509
Community Media Center (Santa Rosa, California), 487
Compactor (trash), 747–749, 763–764
Compartmentation for fire protection, 780, 782–783
Compost, 752–753, 755–759, 761–763
Composting toilets, 669–677
Compressed air:
  ATUs, 708
  sewage ejectors, 684–685
  toilets, 634–637
Compressive refrigeration, 375–376, 413
Computational flow dynamics (CFD), 782–783
Computer:
  annual energy consumption prediction, 113, 294–295, Appendix H
  DHW system control, 604–605
  energy conservation controls (HVAC), 480–484
  fire safety, 771–773
  see also BMS; Intelligent buildings
Computer terminals, lighting for, 1307
Comstock Center (Pittsburgh, Pennsylvania), 509, 511, 513–515
Concealed spaces (fire), 780
Concrete pipe, 682
Condensation:
  heat flow and, 143–144
  passive cooling and, 220, 320
  psychrometry, 288–293
  rainwater systems, 561, 569
  water supply systems, 655
Condensing water equipment, see Cooling towers
Conductance (thermal), 125–127
Conduction, see Heat flow
Conductivity, 50, 124–125
Conductors, electric, 897
  American wire gauge, (AWG), kcmil, MCM, 898
  ampacity, 897
  ampacity tables, 899, 900
  busduct, busway, cablebus, 903, 905
  cable types, 901–905
  copper and aluminum, 898

current-carrying capacity, 899
flexible armored cable, 901
insulation and jackets, 898, 901
metric sizes, 898
nonmetallic sheathed (ROMEX), 901
physical properties (tables), 898
temperature ratings, 899
underground wiring, 934
Conductors (thermal properties), 127
Conduit wire capacity (table), 1035
Connectors, see Building envelope
Consentini Associates, 431, 525–526
Conservation Center, Society for the Protection of New Hampshire Forests (Concord, New Hampshire), 220–222
Conservation Energy Systems, 354–355
Constructed wetlands, see Wetlands
Contractors, electric, 947
Continuity of operations (fire), 780–781
Contrast (in lighting), 1069
  measurement, 1091
Controls:
  DHW systems, 604–605, 614
  HVAC systems, 375, 480–484
  hydronic heating systems, 395–397
  hydro-pneumatic tanks (fire protection), 792–793
  IAQ, 369
  irrigation, 663–664
  lavatories, 627
  solar gain, 18–21
  toilet flushing, 637
  user-operated, 31–32, 116–120, 343–345
  warm air systems, 410
  water supply, 652–654
  see also BMS
Convection, see Heat flow
Converse, J., 712, 745
Conversion factors, Appendix I
  acoustic units, 1621
  general, tables, 1745
  lighting, 1055
Converters, steam to hot water, 459
Cooling:
  approximate equipment sizing, 373–374
  building envelope design criteria, Appendix J
  design considerations, 191–195
  enthalpy and, 423–425
  influences on HVAC choice, 431–433
  sites and, 98–104
  smaller building systems, 377–386
  storage of, 472, 478–479
  ventilation and, 98–104
  see also Heating and cooling equipment; HVAC systems; Passive cooling
Cooling coils, added to furnaces, 412–415
Cooling equipment, see Heating and cooling equipment; HVAC systems
Cooling load, see Heat gain
Cooling systems, small buildings, 377–386, see also Heating and cooling equipment; HVAC systems
Cooling towers, 465–466, 468–471, 581, 584–585
Coolth, 34, 471
Cooltowers (passive), 240–244, 312–316
COP (Coefficient of performance), 404, 417–421
Copper:
  contamination in water, 576

Copper (*Continued*)
DWV pipes, 682–683
water supply pipes, 651–652
Copper Development Association, 698–699, 703–704, 805–806
Corbett, Michael, 559–560
Core arrangements, high rise buildings, 764–766
Corrosion:
control, in water treatment, 577, 582–583
plumbing materials, 651
Cottage Restaurant (Cottage Grove, Oregon), 28–32
Courtyards, 41, 58, 534
Coves, lighting, 1297
Crack method infiltration calculations, 145–146
Crater Lake composting toilet, (Oregon), 671, 676
Crawl spaces, 132
Cross ventilation:
calculations, 300
cooling strategy, 57
design guidelines, 230–232
see also Passive cooling; Ventilation
Crowther, Karen, 303–304
Croxton Collaborative Architects PC, 760, 762, 766
CRS Industries Inc., 368
Crutzen, Paul, 95, 112
Cryptosporidium, 573, 581–582
Crystal Tower (Osaka, Japan), 475
Curtain boards, 782, 786–787
Custodian activities and spaces, *see* Core arrangements; Solid waste systems

Dampers:
air ducts, 407, 410
fire, *see* Fire dampers
Daniel Boone High School (Johnson City, Tennessee), 422
Daylight compensation, *see* Lighting control
Daylight factor (DF), 197–203
Daylighting: 63, 191–194, 1107–1156
artificial sky, 1148
characteristics of, 1111
clear sky, 1116
computer calculations, 1145
daylight factor (DF), 1120
design, CIE method, 1125
design, graphic method (GDDM), 1132
design, LOF, IESNA, and lumen method, 1133
design guidelines, 197–203
fire protection and, 772
glare and heat control, 1152
graphic method (GDDM), 1132
heliodon, 1148
overcast sky, *see* Daylighting, Solid overcast sky
partly cloudy sky, 1118
passive cooling and, 248–249
physical modeling, 1148
PSALI, 1109
side window, 1111, 1152
sites and, 75–83
solar motion and position, 1113
solid overcast sky, 1114, 1124
sunshades, light shelves, 1153
vertical surface illuminance, 1117, 1151
windows in, 1152
DCLF (design cooling load factors), 277, 280–281
DDC (direct digital control), 482–484

Deadband (thermostats), 481
Decibel in acoustics, 1499, 1501
Deciduous trees, 111, *see also* Plants
Decipol, 338
Defecation, 629–630, 638
Degree days, Appendices A, C
heating season calculations, 254
passive cooling, 296–297
Dehumidification, 291–293
DeKay, Mark, 76–77, 201, 205, 329
Delta-T solar, 272–276
Deluge sprinkler systems, 806, 811
Demand charge electric, 883
Demand control, electric, 881
Demolition, 9
Density:
air, 145
building materials, 147–157
Desert coolers, *see* Evaporative coolers
Desiccant cooling, 359–362
Desiccant dehumidification, 291, 359–362
Design conditions (thermal), 56, 113, Appendix A
Design guidelines:
air handling equipment, 480–481
building envelope, 208–211
cisterns, 537–540
cooling (refrigeration), 453
cooling towers, 466
daylighting, 197–203, 248–249
furnace floor area, 405
heat gains, 225–229
HVAC space allocation, 450–453
passive cooling, 225, 230–249
passive solar heating, 205–224, 248–249
phase change materials, 220
rock beds, 220
roof ponds (heating), 224
sewage treatment systems, 541
solar (active) space heating, 224–225
solar DHW:
active, 224–225, 622
passive, 620, 622
solid waste, 756
swimming pool solar heating, 625
tons of refrigeration, 453
water supply, 536–539
water thermal storage, 472
Design strategies, 35, 55–65
climate, 55–65
fire protection, 769–781, 791–793
HVAC systems, 427–444, 764–766
IAQ, 332, 341–343
site, 73–75
see also Heating and cooling, design strategies
DETD (design equivalent temperature differences), defined, 277
Detergents:
water treatment, 574
fire fighting systems, 812–813
Developed length, piping, 657–659, 694–695
Dewpoint, defined, 288
DF, *see* Daylight factor
DFU (drainage fixture units), 688, 690–691
DG, *see* Direct gain solar heating
DHW (domestic hot water) systems:
central *vs.* distributed, 601–603

conventional heat sources, 599–600
demand estimation, 604–612
distribution trees, 603–605
expansion of piping, 649–650
forced circulation, 603–604
heat conservation, 600–605
heat pump water heaters, 625–626
instantaneous heaters, 600
minimum performance standards (EF), 600–601
recirculating, 601–604
sizing, 604–612
solar heating, *see* Solar water heating
space heating and, 597–598
tankless heaters, 598–600
temperature considerations, 597, 599, 604–605
thermosiphon, 603
types of, 599–600
Diatomaceous earth filters, 579
Diffuse lighting, 1216
Diffusers, 485–489
lighting fixtures, *see* Lighting fixtures
Dilution, smoke management, 783
Dimming, dimmers (lighting), 977, 1281
Direct burial wiring, 934, 936
Direct current (dc), 857
Direct gain solar heating:
characteristics, 259
compared to SS, TW, 58, 256–257
design guidelines, 205–224
system performance, Appendix C
Direct lighting, 1218
Direct–indirect lighting, 1216
Direct refrigeration (DX) systems, 412–413, 445
Discount rate, 1712
Disease:
waterborne, 573
see also IAQ
Dishwashers, 637, 661, 691, 737
Disinfection (water), 579–582
Displacement ventilation, 486–489, 509, 783
Disposal field trenches, *see* Drainage fields
Disposer (garbage), 757–759
Distillation, 585
Distribution tree:
design considerations, 21, 374–375, 439
DHW system, 603–604, 645–647, 649
fire sprinkler system, 791
HVAC systems, 374–375, 433, 436, 438–444
hydronic heating, 393–394
waste removal, 686–688
District heating and cooling systems, 516–519
DOE-2 (software), Appendix H
Doelle, Leslie, 94, 112
Domestic hot water systems, *see* DHW
Doors:
fire exits, 774–775
hourly heat gains, 279
infiltration rates, 189–190
U values, 136, 169–170
see also Building envelope
Dosing tank (or chamber), 706–707, 712
Double-duct HVAC systems, 446–447
examples of, 440, 443–444, 499–500, 502–503
Dove Publications (Pecos, New Mexico), 261
Down conductors (lightning), 816–819
Downfeed water distribution, 641, 644–650

Downspouts, *see* Gutters and leaders
Draft:
comfort and, 50–51
flues and, 401
see also Boilers
Drainage:
DWV (drainage, waste, vent) piping, 681–690
fixture units, *see* DFU
principles of, 678–681
sprinkler systems and, 792
Drainage fields, 705, 708–712, 739–741
Drainage plane, 144
Drain-back and drain-down solar hot water systems, 619–621
Drilled wells, 586
Drinking fountains, 661, 691
accessibility, 638
lavatories as, 627
required number of, 542–543
social opportunities of, 531
Drip irrigation, 665
Driven wells, 586
Dry-bulb temperatures, *see* Temperatures
Dry pipe sprinkler systems, 805, 809
Dry rot, 144
Dry wells, 557, 680, 708
Dual-condenser chillers, 463, 465
Ducts:
air controls in, 409–410
approximate sizing of, 406–410
detailed sizing of, 410, 485
night ventilation of mass and, 303
warm air heating systems, 404–412
see also Distribution trees
Duct heaters, *see* Single duct reheat HVAC systems
Dumbwaiters, 1462–1463
Durst Organization, 525–526
DWV (drainage, waste, vent) piping, 681–690
DX, *see* Direct expansion refrigeration

Eagelson, Engineers, 443–444
Early suppression fast-response (ESFR) sprinklers, 801–802
Earth contact tempering, 58
Earth temperature amplitude, 133–134
Earth tubes, 247–248, 320, 322–329
Eaton, K. J., 101–102, 112
E. coli bacteria, 577
Echoes, *see* Room acoustics
Ecological Engineering Associates and ECO-TEK, Inc., 721–722
Ecological fluidized beds, 719–720, 724
Ecology House (San Rafael, California), 337
Economic analysis, Appendix E
life cycle cost, 1711
payback period, 1715
present value, 1713
rate of return, 1712, 1715
Economics, Appendix E
electrical material selection, 894
installation costs, transformers, 943
life cycle costs, 905
Economizer cycle, 469, 472, 482
Eco-roof Pavilion (Portland, Oregon), 553
Ecotope Group, 166
Educational Facilities Laboratories, 521

EER (energy efficiency ratio), 404
EF, *see* Energy factors
Efficacy, 1108
  minimum required, 1163
Egan, M. David, 815, 850
Egress, *see* Exits
Einhorn Yaffee Prescott, architects, 20
Electric circuits parallel, 856
Electric circuits series, 855
Electric closets, 1015, *see also* Core arrangements
Electric current, *see* Ampere
Electric demand control, 883–887, *see also* Energy
  management
Electric energy, *see* Energy, electric
Electric feeder design, 1034
Electric furnace, 404
Electric load calculations, 1033, 1037
Electric meters, 887–892
Electric panel schedules, 1030, 1033
Electric potential, 853
Electric power, 877
  ac circuits, 878–879
  dc circuits, 878
  generation, 858, 877
Electric power and energy, 878
Electric power quality, 982
Electric raceway systems, design considerations, 907
Electric resistance, 854
Electric riser diagrams, *see* Electric wiring design
Electric service, 933–934
  amperage, 1001
  arrangements, 943
  metering, 943
  overhead, 933
  site plan, 945
  underground, 934
Electric shock hazards, 1009
Electric stairs, *see* Moving stairways
Electric usage profiles, 875–876
Electric wiring design, Chapter 17
  branch circuit design, 1020
  design procedure, 1011
  emergency systems, 1042
  energy conservation, 1010
  equipment layout, 1014
  feeders, 1034
  general considerations, 997
  grounding, ground fault protection, 1007–1010
  load estimating, 998–1001
  load tabulation, 1026, 1032
  nonresidential wiring, 1025
  overcurrent protection, 1016
  residential wiring, 1023
  riser diagrams, 1039
  service equipment design, 1013
  service size, 1001
  shock hazards, 1009
  single-line diagrams, 894
  spaces for equipment, 1012
  spare capacity, 1033
  system and utilization voltages, 1006
  system voltages, 1001–1006
Electrical conductors, *see* Conductors, electric
Electrical conduit, 909–913
  metallic, 909–912
  nonmetallic, 913

Electrical equipment, Chapters 15, 16
  busways, 903
  circuit protective, devices, 953
  emergency power, 994
  enclosures, 947, 952
  motors and motor controllers, 963–972
  poke-through fittings, 918
  pole-type raceways, 930
  power conditioning, 982–994
  programmable controllers, 949
  raceways, 909
  ratings, 896
  receptacles, 972
  surge suppression, 984
  switchboards and panel boards, 956–965
  switches, 944–950
  transformers, 938
  UPS, 991
  wiring, 897
  wiring devices, 970
Electrical load estimating, 999
Electrical materials, economics of, *see* Economics
Electrical measurements, 887–892
Electrical noise, 982
Electrical plans, 1014, 1027–1032
Electrical power transients, 982
Electrical raceways, 909–922
  aluminum conduit, 909
  ceiling, 924, 931
  cellular metal floors, 922
  conduit, 909
  flexible metal conduit, 912
  floor raceways, 917
  full access floor, 922, 925
  nonmetallic conduit, 913
  precast cellular concrete floor, 922
  steel conduit, 909
  surface raceways, 913
  underfloor raceways, 917
Electrical single-line diagrams, *see* Electric wiring
  design
Electrical symbols list, 1017–1019
Electrical wiring and raceways, Chapter 15
Electrical wiring design, typical floor plans, 1014,
  1027–1032
Electricity, 851
  cogeneration, 519–527
  consumption trends, 10–11, 15–16
  fire hazards of, 769–770
Electric lighting:
  VAV HVAC systems and, 498
  approximate heat gains from, 226–229
  environmental impact, 10–11
  integral with structure, daylighting and acoustics,
    198–199
  see also Daylighting
Electric power generation:
  on site, 15–16, 519–527
  solid waste, 752
  water use and, 536
Electric resistance heaters, 389–390
  compared to heat pump, 28
Electrodialysis, 584
Electrolysis, 521
Electromagnetic spectrum, 1050
Electronic teaching equipment, 1361

Electrostatic filtering, *see* Filters, air
Elevator machines, geared and gearless, 1387–1389
Elevator system design:
  average trip time, 1409, 1411, 1412
  building population, 1410
  car capacities, 1409
  computer-aided design, 1423
  costs, 1422
  definitions, 1407
  handling capacity, 1408–1409
  interval, waiting time, 1407
  lobby arrangements, 1424, 1426
  machine rooms, 1425, 1427, 1429
  recommendations, 1408, 1422
  round-trip time, curves, 1410, 1414–1416
  typical calculations, 1418
  zoning, 1421
  see also specific item
Elevators, passenger, Chapter 23
  cabs and signals, 1393
  car panel, 1406
  codes and standards, 1385, 1457
  comparative drive characteristics, 1401
  components, 1386
  computer-aided design, 1423
  control, operating, 1398, 1401
  doors, 1390
  drive (motion) control, 1398
  energy use, 1401, 1431
  fire protection and, 774
  handicapped persons, accommodation, 1394
  linear elevator motor drive, 1461
  lobby panel, 1405
  machines (geared and gearless), 1388
  metric sizes, 1389
  operation and supervision systems, 1402
  physical properties, space requirements, 1424
  recommendations, 1408, 1419
  rehabilitation, 1405
  roping, 1389
  safety devices, 1390
  system design, selection, 1406
  system simulation, 1405
  traffic studies, 1404
  zoning, 1421
Elevators, special topics:
  dimensions and weights, 1424
  emergency power, 1434
  energy conservation, 1433
  energy requirements, 1432
  fire safety, 1434
  locations in service cores, 764–766
  National Elevator Industry Inc. (NEII) standards, 1425
  noise, 1435
  novel designs, 1436
  power and energy requirements, 1431
  roughing dimensions, 1425
  security, 1434
  shafts and lobbies, 1426
  sky-plaza system, 1441
  specifications, 1435
  structural data, 1428
  typical layouts, 1427
  see also specific item

Elevators, special types:
  chair lifts, 1452
  double-deck, 1442
  freight, 1445
  hydraulic, 1442, 1444–1445
  observation cars, 1452
  rack and pinion, 1452
  residential, 1452
  slant, 1452
  see also specific type
Elimination of body wastes, 626–627, 629–631
Eljer Industries, 635
Embodied energy in materials, 5–9
Emerald People's Utility District (Eugene, Oregon),
  116, 198–210, 219, 225, 232, 237–238, 253–254,
  267–268, 275–276, 306–308, 450–453
  (design data, 209)
Emergency lighting, 1306, 1321–1326
Emergency power equipment, 994
Emergency systems, electrical, 994, 1042
Emissivity (sky), 379–380
Emittance, 128
Energy, electric, 853, 878
  costs, 880
  demand charges, 881–882
Energy Concepts Company, 420
Energy conservation:
  elevators, 1433
  equipment for central HVAC systems, 466–472
  in motors, 963
  lighting, 1206, 1209, 1279, 1292
  residential lighting, 1292
  trends in, 96
  see also ANSI/ASHRAE Standard 90.2;
    ASHRAE/IES Standard 90.1; *specific type*
Energy consumption:
  annual calculations, 113, 254–255
  end use in buildings, 12
  trends, 9–11
Energy efficiency ratio, *see* EER
Energy Efficient Resource Recovery Facility (Cal
  Poly, San Luis Obispo, California), 716–719,
  752–755
Energy embodied in materials, *see* Embodied energy
  in materials
Energy factors (EF), 600–601
Energy management, 883
EnergyPlus (software), Appendix H
Energy recovery ventilators (ERV), 357–358, 507–509
Energy Scheming (software), Appendix H
Energy sources:
  architectural history and, 9–11
  building construction, operation and demolition, 5
  heat values of fuels, 251
  trends in, 9–11, 35
  see also Nonrenewable energy; Renewable energy
Energy storage (HVAC), 471–472, 475–479
Energy Studies in Buildings Laboratory, University of
  Oregon, 553, Appendix H
Energy transfer wheels, 359–362
Energy 10 (software), Appendix H
Energy-conserving lamps, 1163
Enermodal Engineering, Ltd., 507–508
Engineer (mechanical), design role of, 372–373
Enthalpy, 291–293

Envelope, *see* Building envelope

Environment and energy production, 10–11, *see also* Sites

Environmental Building News, 350, 359, Appendix H

Environmental illness, *see* Multiple chemical sensitivity

Environmental Planning and Research, Inc., architects, 386

Environmental Research Laboratory, University of Arizona, 240–241, 384

Environmental Resource Guide, 8, 36

Envirovac, Inc., 636–637

ENVSTD (software), Appendix H

EPA, *see* U.S. Environmental Protection Agency

EPACT (Energy Policy Act), 1163, 1171

EPICenter, Montana State University, 11–17, 21–23, 366–367, 719–722

Equinox Design, architects, 28–32, 202–203

Equipment:
  basic components of HVAC, 427–429
  heat gains from, 284–285
  heating efficiency factors, 401–404
  need for, 371–372
  space allocations for, 450–453
  see also Heating and cooling equipment

Equivalent (total) length of piping, 658, 662

ERV, *see* Energy recovery ventilators

Escalators:
  fire protection and, 772
  see also Moving stairways

ESFR sprinklers, 801–802

ESI (Equivalent Spherical Illumination), 1092

Ethernet, 484

Eugowra, NSW (Australia) Ranch of John Andrews, 547, 549–550

Eutectic salts, *see* Phase change materials

Evaporation, *see* Distillation; Heat flow; Hydrologic cycle

Evaporative coolers, 382, 384

Evaporative cooling, 55–58
  closed circuit, 471
  cooltowers, 240–244, 312–316
  design guidelines for, 238–244
  design strategy, 55–58
  detailed design procedures for, 308–313, 328–329
  fire suppression (misting) and, 806–808, 811
  indirect, 356, 382, 384–386
  mist cooling systems, 379, 381, 494–495
  psychrometry and, 291, 312
  site design and, 104
  urban-rural contrasts, *see* Heat island
  water and, 316
  see also Cooling towers

Eutectic salts, *see* Phase change materials

Exhaust fans (IAQ), 351–354

Exhaust systems, smoke, 783–787

Exit lighting, 1325

Existance (luminous), 1055

Exits, 773–779
  maximum distances to, 776–777

Expansion:
  air (hydronic systems), 393–396
  pipes, 393, 649–650, 653–655

Expansion chambers, 643–644, 653–655

Exposure protection, fire, 779–780

Extended coverage sprinklers, 800, 802

Facultative lagoons, *see* Sewage

Famolare Warehouse (Brattleboro, Vermont), 263–264

Fan-Assisted Evaporative Cooling, *see* Evaporative Cooling

Fan coil HVAC systems:
  all-water, 449–450, 511–512, 516
  with supplementary air, 448–449, 504, 507

Fanger, P. O., 43–44, 338, 369

Fans:
  cooling with, 377–380
  exhaust, 351–354
  HVAC systems, 478
  locations, 428, 430
  smoke management, 783–787
  see also Air-handling equipment; Filters

Fan-powered variable-volume (VAV) systems, 446–447, 498

Faraday cage (lightning protection), 817

Farnsworth house (Fox River, Illinois), 401

Faucets (lavatory), 627

F-Chart (software), Appendix H

FCR, *see* Fenestration cooling rating

Fenestration, *see* Windows

Fenestration cooling rating (FCR), 139

Fenestration heating rating (FHR), 138–139

Ferguson, Bruce, 537, 552, 554, 572

Fernau & Hartman, architects, 545, 666

Fetzer Winery Administration Building (Hopland, California), 238–239

FFU, *see* Floor fan units

FHR, *see* Fenestration heating rating

Fielden Clegg Architects, 349, 483

Fibers, in air, 334

Fiber-optic lighting, 1329

Fiber-optic terminology, 1330

Filters, air, 336, 343, 362–368
  adsorption, 364
  air washers, 364–365
  efficiencies, 364
  electronic, 365, 367–369
  HEPA, 362, 364
  location of, 365–367
  particulate, 363–364
  see also Building envelope, connectors/filters/barriers

Filtration (water), 577–579, 581
  cisterns, 546–548
  see also Sand filters; Sewage

Finned tube convectors, 394

Finnish contraflow masonry heater, 389

Fire alarm signals, 844
  audible, 844
  visible, 844–845

Fire alarm systems, 820
  addressable, 828
  addressable analog, 829
  air sampling, 834
  alarm verification, 826
  apartment house, 846
  auxiliary type, 823
  central stations, 824
  coding, 827–828
  commercial and institutional, 846
  conventional, 825
  definitions, 822–825

dormitories, 846
false alarm mitigation, 841
hotel, 846
industrial building, 849
"intelligent," 829
local type, 823
manual stations, 842–843
multiple dwelling, 846
office buildings, 847
presignaling, 828
proprietary type, 824
protected premises, 823
remote-station type, 824
residential, 823, 845
schools, 846–847
signal processing, 828
sprinkler flow alarms, 792, 844
Fire and smoke detection, 830
application guidelines, 834
in ducts, 834
Fire and smoke detectors:
addressable, 831, 834, 835
air-sampling type, 834–835
cloud chamber type, 840
flame stage, 835
gas sensing, 830
heat stage, 836
incipient stage, 830
infrared (IR) type, 835–836
ionization type, 830
laser-beam type, 834
linear heat detector, 838–840
photoelectric, 831–833
smoldering stage, 831
spot heat detector, 837–838
ultraviolet (UV) type, 835–836
see also specific type
Fire codes, 770–772, 820
Fire dampers, 780–781
Fire extinguishers, 790, 814–816
Fire growth stages, 771
Fire hoses, 787–791
Fireplaces, 10, 41, 386–389
Fire protection:
clean agent systems, 786, 814
carbon dioxide systems, 813–814
design considerations, 769–781, 791–793
environmental control systems and, 772
foam systems, 812–813
halogenated hydrocarbon systems, 809–812
hazards for various occupancies, 798–799
ignition sources, 769–770
objectives, 770–772
products of combustion, 770
protection of life, 772–779
protection of property, 779–781
smoke management, 781–786
water for suppression, 786–811
water supply and, 788–789, 791–793
see also Sprinkler systems
Fire pumps, 788–789, 791–793, 803
Fire-rated glazing, 786–787
Fire spandrels, see Curtain boards
Fires, stages of, 771, 830–836
Fire-Safe Building Design for Architects and Designers (software), Appendix H

Firestops (firewalls), 780, 782–783
Fire (water) tanks, 645–648, 772, 779, 791–793
Fishbeck, Thompson, Carr & Huber, engineers, 494–495
Fisher, J., 36
Fixed sunshades, see Sunshades
Fixture units:
drainage, 691
water supply, 661
Flat cable assemblies, 905
Fleischhacker, P., 315, 317–318, 329
Fletcher–Munson equal loudness contours, 1506
Flocculation, 578
Floodlighting, 1326
Floor drains, 679–680, 684, 688, 691, 792
Floor fan units (FFU), 489
Floors:
calculating slab edge heat loss, 165–166
conservation criteria, Appendix J
heat flow through, 132–133
hourly heat gains, 278–279
sprinkler systems and, 792
U-values, 163
see also Building envelope
Florida Solar Energy Center, 300
Flowforms, 582–583
Flues, see Chimneys, see also Boilers
Fluidmaster, Inc., 605
Fluorescent lamps, 1166–1184
ballasts, conventional, 1180
ballasts, electronic, 1183
compact types, 1179
dimming, 1175
efficacy, 1174
energy, distribution, 1175
EPACT requirements, 1171
high-frequency operation, see Electronic ballasts
life, 1173
operating characteristics, 1171
special types, 1177
Fluoridation, 575, 585
Flush valve toilets, 632, 648, 681, 691
Flushometer tanks, 634
FM-200®, clean agent gas, 814
Foam fire protection systems, 812–813
Foamability, in water, 574
Fogging, see Evaporative cooling
Footcandle, 1054
Footing drains, 557
Forced-air registers, see Registers
Forced circulation hot water, 603–604
Forced ventilation, see Exhaust fans
Formaldehyde, see VOC
Foss, Engelstad, Heil Associates, architects, 479
Fossil fuels, see Fuel sources, Energy sources
Fountains, 531–534, 640, see also Drinking fountains
Four-pipe all-water HVAC systems, 449–450, 504
Four Times Square (New York City), 525–526
Fox and Fowle, architects, 525–526
Fox Plaza Building (Los Angeles), 440–442
Fox Plaza Building (San Francisco), 433, 436–437
Francis, C. E., 322
Franklin cone (lightning protection), 817
Fraunhofer Institute, 258
Frederick Meijer Conservatory (Grand Rapids, Michigan), 492, 494–495

Freight elevators, 1445
Freon, 376
Fresh air, *see* Ventilation
Friction allowance (ducts), 485
Friction (fire), 770
Friction loss in piping, 656–663
Fuel cells, 521–527, 752
Fuel consumption, 294–295
Fuel sources, *see* Energy sources; Nonrenewable energy; Renewable energy
Fuel storage, 403–404
Full access floors, 922, 925–926
Furnaces, 404–407, *see also* HVAC systems
Fuses, 953–955

Gandemer, J., 101–102, 112
Garbage, *see* Solid waste systems
Garbage compactors, 758–764
Garbage disposers, 757–759
Gardner, I., 299, 329
Gases:
    fire hazards, 770
    greenhouse, *see* Global warming
    smoke management, 772–779
Gas-fired cast iron boiler, *see* Boilers
Gas-fired water heaters, 389–390
Gate valve, 654
GAX (generator-absorber heat exchange), 420
Geberit Manufacturing Inc., 631
General Mills Building (Minneapolis, Minnesota), 792–793
Generator, emergency power, 994
GeoExchange Systems, *see* Ground source heat pumps
Geotextile fabric, 712–713
Geothermal energy, *see* Energy sources
Geyser pumping solar hot water systems, 616–617
Giardia cysts, 578–579, 581–582
Giampolo and Associates, Inc., engineers, 386
Gillespie, Terry, 98, 11, 112
Gillies Stransky Brems Smith Architects, 497
Givoni, Baruch, 45–46, 55, 66, 242, 291, 312–313, 322–323, 329
Glare, direct, 1084
    discomfort, disabling, j 1084
    VCP, 1085
Glare, reflected:
    control of, 1094
    see also Veiling reflections
Glass pipe, 682
Glass recycling, 747–751
Glazing, *see* Windows
Gleick, Peter, 534, 536, 572
Global warming, 3, 95–96
Globe thermometer, 43–44
Globe valve, 654
Grab bars, 639–642
Graedel, Thomas, 95, 112
Graham, John & Associates, architects, 117
Graywater, 636–637, 735–742
    examples, 669–673, 722
    flow estimates, 739
    heat exchange, 625, 722, 735, 738
    location constraints, 739
Grease traps, 686, 690
Green, Dr. Bailey, 717–718, 745

Green Building Advisor (software), Appendix H
Green Building Assessment Tool (GBA Tool) (software), Appendix H
Greenheck Fan Corporation, 354
Greenhouse ecosystems, 719–723
Greenhouse effect, *see* Global warming
Greenhouses (solar heating), *see* Sunspace systems
Green on the Grand office building (Kitchener, Ontario), 507–509
Greensboro, N.C., student housing, 357–358
Greenways in cities, 70–71
Gridded sprinkler piping, 807
Grilles, *see* Diffusers; Registers
Ground fault (circuit) interrupter (GFI, GFCI), 975
Ground fault protection, 975, 1007
Grounding, 1007
Ground-source heat pumps, 420–422, 469–471, 473–475
Ground terminals (lightning), 816–819
Groundwater:
    chemical characteristics, 575
    contamination, 575–576
    porous pavement and, 553–555
    sites and, 104–197, 553–561
    water quality and, 535–536
    see also Rainwater
Grove, Sir William, 521
Gruen, *see* Victor Gruen Associates
Guise, David, 506
Gutters and leaders, 555–557, 561–572

Hadden, D., 299, 329
Haggard, Ken, 320, 329
Haglund, Bruce, 329
Hahn, Philip, 348
Halide lamps, *see* Metal halide lamps
Halogen lamps, *see* Tungsten halogen lamps
Halogenated fire protection systems, 809, 812, 814–815
Halon 1301, 809, 812–814
Hancock, *see* John Hancock
Handicapped persons, bathroom design for, 638–642, *see also* Barrier free design
Hannon, Bruce, 8, 36
Harbortown apartment and townhouse complex (Detroit, Michigan), 520–523
Hard water, 574–575, 583–584
Hardin, Garrett, 71
Harmonic currents, 1038
Hastings, S. Robert, 228, 329
Hay, Harold, 245, 320
Hayward, Dr. John, 331–332
HC (natural hydrocarbons), 377
HCFC (hydrochlorofluorocarbons), 335, 377, 462, 509, 814
Head, water supply, 594, *see also* Static pressure
Hearing, 1503
    equal loudness contours, 1506
    Fletcher–Munson curves, 1506
    masking noise, 1510
    presbycusis, 1495
    threshold, 1498
Hearing protection, 1514–1518
Heat exchangers, 146, 353–362, 469
Heat flow:
    air, 39–40, 127–128

building envelope, 120–143
convection/conduction/radiation, 39–42, 120
evaporation, 39–42
moisture/infiltration, 143–146
production of body heat, 37–40
solids, 124–127
ventilation and, 144–146
see also Psychrometry
Heat gain:
comparison of methods, 295
design guidelines, 225–229
hourly calculation, approximate, 276–288
hourly calculation, detailed, 293–295
internal, 252–253, 282–285
lighting and, 282
people and, 282–283
Heating:
building envelope design criteria, 209, Appendix J
design considerations, 191–195, 387
influences on HVAC choice, 431–433
see also Cooling; HVAC systems
Heating and cooling design strategies, 55–65, *see also* Passive cooling; Passive solar heating
Heating and cooling equipment, *see* HVAC systems
Heating and cooling systems, smaller buildings, 412–423
Heating systems:
approximate sizing, 373–374
combustion and storage, 401–414
cooling system addition, 412–413
hot water, 391–403
hydronic system design, 397–399
rooms, 387
small systems, 385–386
warm air systems, 404–412
wood heating, 386–389
Heating, ventilating and air-conditioning, *see* HVAC systems
Heat island, urban, 67–71
Heat loss:
heating season fuel consumption, 251–255
hourly fuel consumption rates, 250–251
hourly heat loss calculations, 249–251
Heat Mirror, 140, 142
Heat pipes, 358–359
Heat Pipe Technologies, Inc., 359
Heat pumps, 28, 413–423
defined, 413
ground source, 107, 420
heating systems and, 418
hot water systems and, 625–626
refrigeration cycle and, 376
solar energy and, 416–420
Heat source/sink, 55, 474
Heat recovery systems, *see* Heat exchangers; Energy conservation, equipment for central HVAC systems
Heat values of fuels, 251
Heavy metals, 719
Helical screw compressors, 463, 465
Hellmuth, Obata & Kassabaum, architects, 339–340, 343
Hemicycle House, *see* Solar Hemicycle
HEPA filters, *see* Filters, air
Heschong, Lisa, 42, 141, 146, 172
HFC (hydrofluorocarbons), 335, 377, 814

HID (high-intensity discharge) lamps, 1184
High-grade energy, 27–28
High-grade resources, in waste, 747
High limit switch, 410
High mass cooling, 55–58
High-pressure sodium (HPS) lamps, 1190
High-temperature water, 518–519
Hi-Rise Recycling Systems, Inc., 763
History of:
cogeneration, 520
distribution trees, 439–440
energy sources, 9–11
fire protection, 770
HVAC-structure relationship, 438
irrigation, 666
sewage systems, 533
solar energy and architecture, 9–18
sprinklers, 791
storm sewers, 533, 551–552
traps and vents, 678
ventilation, 331
window technology, 129–130
warm air heating, 404–405
waterborne disease, 573
water use, 531–534, 536
Hitachi chiller-heaters, 522
HLW International, 757
Hoeschele, Marc, 382, 425
Honeywell, Inc., 484
Hopkinson, R. G., 201, 329
Horizontal exits, 773–779
Horizontal fixture branches and stacks, 693
Horizontal rainwater piping, 570–571
Horizontal shadow angle, *see* Sunshading
Horsepower (hp), 124
boilers, 453–455
Hose bibb, 643
Hoses, *see* Fire hoses
Hot humid climate moisture control, 144–145
Hot water convectors (radiators), 402–403
Hot water systems, *see* DHW; Heating systems
Hotel HVAC controls, 483
House tanks, *see* Water storage tanks
HSPF (heating season performance factor), 418
Human behavior in fires, 772–774, 777–778
Humidification, 291, 294
Humidity, 42, *see also* Relative humidity
Humidity ratio, 288, 291
Humphreys, M., 46–47, 66
Humus, *see* Compost; Composting toilets
Hunter Douglas, Inc., 258
Hurd Millwork Company, 142
HVAC (heating, ventilating, air-conditioning) systems:
air & water systems, 447–449, 500, 503–515
air distribution within spaces, 484–495
all-air systems, 445–447, 495–503
all-water systems, 449–450, 511–512, 515–518
building organization and, 427–444, 764–766
central equipment (large buildings), 453–484
central *vs.* local systems, 373, 428
cogeneration, 519–527
concealed *vs.* exposed, 436–438
design process, 372–373
direct refrigerant (DX) systems, 412–413, 445
distribution trees, 433–436, 438–444

HVAC (heating, ventilating, air-conditioning) (*Continued*)
  district heating and cooling, 516–519
  fire protection and, 772, 785
  mechanical-structural integration, 203–209, 438–440
  single-zone systems, 495–496
  smoke management and, 772, 781–787
  space requirements, 450–453
  smaller buildings, Chapter 7
  system types (introduction), 444–453
  trends, 427, 440
  typical applications, 434–435
  uniformity *vs.* diversity, 428–430
  zoning, 427–428, 430–435
  see also Heating and cooling systems; Heating systems
Hydraulic elevators, 1442
Hydraulic Institute, 586
Hydrogen, 11, 18, 35, 523, *see also* Energy sources, Fuel cells
Hydrologic cycle, 534–536
Hydronic heating systems, 391–403, 412–413, 415
Hydronics Institute Division of GAMA, 391, 398, 425
Hydro-pneumatic water supply systems, 641
Hydrozone concept (landscaping), 558, 561
Hyndburn Borough Council Offices (England), 470–471, 474–475
Hypochlorinators, 580
Hypothalamus, 37
Hyytiainen, H., 389, 425

IAQ (interior air quality):
  design strategies, 341–343
  energy conservation and, 331–332
  equipment, 351–369
  heat exchangers, 353–361
  history of, 331–332
  HVAC and, 486–487, 491, 495, 525
  predicting levels of, 337–341
  sources and impacts, 332–337
  summary, 332
  testing, 340–341
  trends, 331
  wood heat and, 386–387
  zoning, 341–344
I = B = R rating, boilers, 391
Ice storage tanks, 472, 478–479
Ignition (fire), 769–770
Illuminance, 1054, 1076
  measurement, 1058
Illuminance levels, 1075
  American IES recommendations, 1077, 1080
  British IES recommendations, 1077, 1079
  residential, 1294
Illumination, *see* Illuminance
I. M. Pei and Partners, architects, 121
Impact insulation data for floor, ceiling construction, Appendix G
Impact Isolation Class (IIC), 1597
  specific occupancy recommendations, 1609
Impact noise, 1595
  control of, 1595
Incandescent lamps, 1157–1165
  EPACT rrequirements, 1163
  special types, 1163

Incinerators, 749, 751, 759
Incremental paving, 553
Indirect evaporative cooling, 352–357
Indirect gain, 59–65, *see also* Trombe wall, Water wall
Indirect lighting, 1213
Induction HVAC systems, 448–449, 503–506
Induction lamps, 1193
Industrial lighting, 1317
Inert gas (window insulation), 139
Infiltration:
  defined, 144
  heat flow and, 143–147
  heat flow calculations, 145–146
  heat gain calculations, 282–283
  rates of, 185–190
  vapor barriers and, 144
  windows, 138
Information systems, 18, 21
Infrared heaters (gas), 389–391
Inland Revenue Centre (Nottingham, England), 301–303
Inryco, Inc., 786
Insolation, *see* Solar Radiation
Instantaneous water heaters, 600
Institute for Research in Construction (Canada), 808
Insulation, 126–128
  embodied energy, 8
  fire protection and, 772
  glazing films, 140
  thermal transmission, 126–128
  see also Passive solar heating; Roof ponds
Insulators (thermal properties), 126–127
Intelligent buildings, 18, 1380
Interceptors, 685–686, 690, 738
Intercom and communication systems, 1352, 1360
Interior air quality, *see* IAQ
Interior finish and fire protection, 772–774
Interior zones, *see* HVAC systems
Internal-load-dominated buildings, 192–194
International Building (San Francisco), 440, 443–444, 562–564
International Plumbing Code, 542–543, 652, 656, 661, 682, 691, 693–695
Intrusion detection, *see* Access control
Intumescent materials, 808
Inverse square law, lighting, 1062
Invisible Structures, Inc., 555
Ion-exchange process, 584
Iowa Public Service offices (Sioux City), 472, 475, 479
IPLV (integrated part load values), 404
Iron, in water, 576
Irrigation systems, 663–667
  graywater and, 724, 728, 735–742
Irritants (IAQ), 333–335
Irurah, Daniel, 8
Island sinks, 679
Isolated combustion systems, 406
Isolated Gain, 59–65, *see also* Sunspaces
ITT Bell and Gossett, 459

Jacobs House II, *see* Solar Hemicycle
Jacuzzi Bros., Inc., 588, 592, 595
Jensen, Rolf, 819, 850
Jestico & Whiles & Associates, architects, 474–475
Jet pumps, 587, 590–593
John Hancock Center (Chicago, Illinois), 819–820

John Hancock Headquarters (Boston, Massachusetts), 121
Johnson Controls, Inc., 489
Johnson, Tim, 220, 329
Johnson Fain Pereira Associates, architects, 441–442
Josam Manufacturing Co., 688, 690
Joslin, Jeff, 9, 36
Joule, 124

Kallman, McKinnell and Knowles, architects, 72
Karen Terry house (Santa Fe, New Mexico), 261
Keen Engineering, 488
Kelbaugh house (Princeton, New Jersey), 260–261, 263
K factor (sprinkler heads), 797
Kim, Casey and Harase, Inc., engineers, 441–442
Kinloch Estate (Virginia), 594–597
Kira, Alexander, 627–631, 667
Kiss Cathcart Anders Architects, 19
Knowles, Ralph, 76–77, 112
Kolding (Denmark) greenhouse water treatment, 719, 721
Krypton, 129, 139
Kwok, Alison, 45, 66

Laboratory HVAC, 482–483, *see also* EPICenter
Laboratory lighting, 1303
Labs, Kenneth, 73, 112, 320, 329
Lagoons, *see* Sewage
Land and Place, landscape design, 583
Landfills, 747, 752–755, *see also* Methane
Landsberg, Helmut, 68, 112
Landscaping:
  graywater and, 724, 728, 735–742
  irrigation, 662–667
  rainwater and, 546–547
  shading, solar access and, 107–112
LAS (linear alkylate sulfonate), 574
Latent heat:
  defined, 277
  hourly heat gains, 282–283
  sizing equipment for, 423–424
Latitude (map, North America), 79
Laubin, Reginald and Gladys, 119, 146
Lavatories, 661, 691, 737–738
  accessibility, 638
  design considerations, 627
  faucets and conservation, 627
  required number of, 542–543
Law and Education Building, Boston University, 121
Lawn sprinklers, 663
Lawrence Berkeley Laboratory, 382, 510, Appendix H
Layseca, Marta, 313–316, 329
LCR (load collector ratio), 59–61, 263, 266–271, Appendix C
Leaching beds, 708–710
Lead:
  in air, 333
  in water, 575–576
Leaders, *see* Gutters and leaders
Lechner, Norbert, 63, 66
Legionnaire's disease, 337, 465, 581, 584, 597
Levitt and Sons (Levittown), 412–413, 415
Library Square (Vancouver British Columbia), 488–489
Lieberg, Owen S., 519

Liebl, John, 415
Life cycle costs, 1711
Light guides, 1333–1340
Light pollution, 1326
Light sources, Chapter 19
  chromaticity, 1195, 1199
  color of light, 1195
  color temperature, 1100
  color-rendering index, 1199
  daylight, 1106
  efficacies, table, 1108
  fluorescent lamps, 1166
  HID sources, 1184
  high-pressure sodium (HPS) lamps, 1190
  incandescent lamps, 1157
  low-pressure sodium (SOX) lamps, 1193
  mercury lamps, 1185
  metal halide lamps, 1187
  tungsten–halogen lamps, 1164
Light, physics of, 1050
Lighting:
  and air-conditioning, 1317
  architectural elements, 1296
  architectural lighting elements, 1296–1297
  bibliography, 1340
  brightness, 1055
  candlepower distribution curves, 1063
  computer terminal spaces, 1307
  controls, 1279
  conversion factors, 1055
  emergency, 1306, 1321
  energy conservation and, 10–11
  ESI, 1082
  floodlighting, street lighting, 1326
  glare, direct, 1084
  glare, reflected (veiling reflections), 1086
  illuminance, lux, footcandle, 1054
  indirect, 1213
  inverse square law, 1062
  luminance, 1055
  luminous flux, lumen, 1054
  luminous intensity, candela, 1053
  luminous intensity measurements, 1063
  modeling shadows, 1099
  office, 1306, 1313
  quality, 1041
  quantity, 1075, 1082
  references, 1340
  reflectance, reflectance factor, 1050
  reflectance measurements, 1061
  reflectances, 1089
  residential, 1292
  RVP, 1082
  safety, 1300, 1306, 1321
  school, 1295
  semi-indirect, 1216
  spectrum, 1050
  task-ambient, 1313, 1316
  terminology and definitions, 1051
  transmittance, transmission factor, 1050
  VDT areas, 1307
  see also Lighting design, Lighting calculation, Light sources, Luminaires
Lighting application, Chapter 21
Lighting calculation:
  illuminance, 1247
  illuminance, average, 1247

Lighting calculation (*Continued*)
  illuminance, computer-aided, 1267
  illuminance, effects of cavity reflectances, 1261
  illuminance, from linear and area sources, 1266
  illuminance, light loss factor (LLF), 1248
  illuminance, using design charts, 1263
  illuminance, zonal cavity methods, 1250, 1259
  illuminance at a point, 1262
  luminance, 1273
  from point source, 1265
  see also Zonal cavity
Lighting control, 1279
  daylight compensation, 1290
  dimming, 1281
  dynamic control, 1284
  initiation, 1281
  lumen maintenance, 1288
  occupancy sensors, 1286
  static control, 1283
  strategies, 1284
  switching, 1280
  time scheduling, 1286
  wireless, 1282–1283, 1298
Lighting design, Chapter 20
  architectural lighting elements, 1225
  auditorium, 1302
  classrooms, 1299, 1301
  commercial interiors, 1306
  computer-aided, 1267
  corridors and stairways, 1305
  cost factors, 1206
  with design aids, 1263
  educational, institutional buildings, 1295
  emergency lighting, 1306, 1321
  energy considerations, *see* Energy conservation
  evaulation, 1276
  floodlighting, 1326
  furniture-integrated luminaires, 1316
  goals, 1203
  illumination methods, 1213
  industrial occupancies, 1317
  integral with acoustics, structure, HVAC, 203–209
  integrated and modular ceilings, 1317
  laboratories, 1303–1304
  library, 1303, 1305
  lighting systems, 1213
  luminance calculations, 1273
  modular, 1261
  office, 1306, 1313
  power budgets, 1206
  preliminary design, 1212
  procedure, 1204, 1247
  residential, 1292
  task analysis, 1207
  task-ambient (nonuniform), 1313, 1316
  uniformity of illumination, 1236, 1241
  VDT areas, 1307–1313
  vertical surfaces, 1321
  see also Lighting designs, Lighting calculation,
    Light sources, Luminaires
Lighting fixtures, *see* Luminaires
Lighting fundamentals, Chapter 18
Lighting power budgets, *see* Energy conservation
Lighting systems, 1213
  diffuse lighting, 1216
  direct lighting, 1218
  direct–indirect lighting, 1216

  indirect lighting, 1213
  semidirect lighting, 1218
  semiindirect lighting, 1216
Lighting track, 905
Lightness constancy, 1109
Lightning arresters, 818–820
Lightning protection systems, 815–820
Lightshelves, 203–207, 490–491, 497, 1153
Lima Register Company, 412
Lin, Maya, 757
Liquefied natural gas (LNG), 812
Lithium chloride, 360
Living Machines, 719–724
Living Technologies, Inc., 720, 723
LNG, *see* Liquefied natural gas
Load collector ratio, *see* LCR
Load factor, electric, *see* Electric demand control
Load shedding, electric, *see* Electric demand control
Local HVAC systems, *see* HVAC systems
LonMark, 484
Loop (sprinkler) piping, 807
Lorsch, Harold, 369, 425, 453, 500, 527
Loudspeakers, *see* Sound reinforcement systems
Loveland, Joel, 329
Low-consumption toilets, 632, 634
Low-emittance glass, 130, 139
Low-grade energy, 27–28
Low-grade resources in waste, 747–752
Low-pressure sodium (SOX) lamps, 1193
Low-voltage switching, 976
Lowry, William, 18, 69–71, 112, 665, 667
LSG (light to solar gain ratio), 138, 171–172, 175
LU–LA (limited use, limited application) elevator
  code, 1457
Ludwig, Art, 737
Lumen, 1053
Luminaires, 1209, 1226, 1243
  batwing diffusers, 1235
  coefficient of utilization, 1238, 1246
  construction, installation, appraisal, 1244
  diffusers, 1233
  efficiency, 1210
  efficiency rating (LER), 1246
  furniture-integrated, 1316
  heat transfer, 1210
  lenses, 1235
  louvres and baffles, 1228, 1234
  mounting height, 1242
  parabolic louvres, 1235
  photometric characteristics, 1227
  reflectors, 1230–1232
  residential, 1295
  shielding, 1228–1230
  size and pattern of, 1220
Luminance, existance and brightness, 1055
Luminance, measurement, 1060
  patterns, 1098
  preferred, permissible levels, 1069
  recommended, ratios, 1098
  typical values, 1069
Luminance calculation, 1273
Luminance coefficients (table), 1274
Luminous ceilings, 1221
Luminous efficiency of lamps, *see* Efficacy
Luminous existance, 1055
Luminous flux, 1053
Luminous intensity, 1053, 1063

Lux, 1054
Lyle, John Tillman, 21, 24, 36

Make-up air, 352–355
Make-up water, 643
Manganese, 576
Marsh, William, 95, 112
Masking noise, 1584
Masking sound, 91–92, 489–490, 639–640
Masonry heaters (wood), 389
Masonry walls, *U* values, 131–132
Mass heat capacity, 304
Material safety data sheets, 340
Material-handling equipment, 1459
    automated vehicles, 1470
    container delivery systems, 1468
    dumbwaiters, 1462
    horizontal conveyors, 1465
    pneumatic systems, 1465, 1468
    vertical conveyors, 1465
Materials:
    energy consumption in, *see* Embodied energy
    thermal properties of, 147–157
    recycled, 4–9, 747–755
    virgin, 4–9
Matsuzaki Wright, architects, 672–673
Max Fordham and Partners, engineers, 349
Mayer Art Center, Philips Exeter Academy (Exeter, New Hampshire), 401, 403
Mazria, Edward, 78, 82, 220, 329
Mazria/Schiff & Associates, architects, 417–418
McCracken Solar Co., 585
McPherson, E. G., 99, 112, 561
Mean radiant temperature (MRT), 42
Mean spherical candlepower (MSCP), 1063
Measurements, electric, *see* Electrical measurements
Mechanical equipment noise, 92, *see also* Noise control
Membrane filtration, 579, *see also* Reverse osmosis
Mercury lamps, 1185
Metabolism, 37–40
Metal halide lamps, 1187
Metals recycling, 749
Methane gas, 334–335, 337, 706, 718, 720, 747, 752–755, *see also* Anaerobic decomposition, Global warming
Metric conversions, Appendix I
Meyers, R., 36
Microclimate, 71–73, 465, *see also* Sites
Microphor Company, 634–635
Mies Van Der Rohe, architect, 401
Miller, H., 303–304, 329
Millet, Marietta, 200, 329
Milne, Murray, 55, 66, 291, 537, 546, 572, 632–633, 667, 670–671, 745, 759, Appendix H
Mist fire protection systems, 806–808, 811
Mist cooling systems, 379, 381, 494–495
Mixing boxes (terminals), 497, 503
Modular boilers, 458–462
Modular wiring systems, 924, 930–932
Mogesen, Bud, architect and planner, 692–693
Moisture and heat flow, 143–146, *see also* Psychrometry
Montana Power, 808
Montgomery, D., 111–112
Moore, Fuller, 345
Moreland-Unruh-Smith, architects, 122

Morgan, Christopher C., AIA, architect, 380
Moshe Safdi and Associates, Inc., architects, 272–274, 488
Motion detection, 1345
    mechanical, 1345
    passive linfrared (PIR), 1346
    photoelectric, 1346
    see also Occupancy sensors
Motor controllers, control centers, 967
Motors, electric, 963
    high-efficiency, 966
Mounds (sewage treatment), 705, 711–712
Moveable insulation, *see* Passive solar heating; Roof ponds
Moving stairways, Chapter 25
    comparison with elevators, 1484
    components, 1479
    criss-cross and parallel, 1471
    fire protection, 1481
    first costs, 1487
    lighting, 1481
    location, application, 1476, 1481
    passenger capacity, 1478
    power requirements, 1485
    rise, 1479
    safety features, 1479
    size, capacity, speed, 1478
    special designs, 1485
    structural, dimensional data, 1486
    see also Escalators
Moving walks and ramps, 1487
Mozer, Michael, 375, 425
MRT, *see* Mean radiant temperature
Mt. Airy Library, (Mount Airy, North Carolina), 413–414, 417–418
Mt. Angel Abbey Library (Oregon), 28, 31–32
Multiple chemical sensitivity, 337
Multipurpose buildings, thermal zoning of, 430–435
Multizone HVAC systems, 445, 447, 498–499, 501
Munsell Color System, 1102

Nanofiltration (water), 581
Napa Valley Museum (Napa County, California), 665–666
National Drinking Water Clearinghouse, 573, 578, 580, 583–584, 667
National Electrical Code (NEC), 893
National Electrical Safety Code, 933
National Fenestration Ratings Council (NFRC), 137
National Fire Academy, Appendix H
National Fire Protection Association (NFPA), 769, 771–779, 781–782, 791, 794–796, 798–803, 806–807, 809–812, 814, 816–817, 850
National Institute of Standards and Technology (NIST), Appendix H
National Oceanic and Atmospheric Administration (NOAA), 56, 547
National Renewable Energy Laboratory, 351–352, 362, 367, 369
National Sanitation Foundation, 652, 687
National Small Flows Clearinghouse, 708, 711, 713–715
National Water Well Association, 324
National Wildflower Research Center (Austin, Texas), 558, 560
Natural gas, *see* Energy sources
Natural Gas Cooling Equipment Guide, 425

Natural Resources Defense Council, 757
Natural ventilation, *see* Ventilation
NEC, *see* National Electrical Code
Nelson, Glenn, 738
Netherlands Research Institute for Public Health Engineering, 726
New Canaan Nature Center (New Canaan, Connecticut), 220, 223
New England Merchants' Bank (Boston, Massachusetts), 121
New Orleans Convention Center, 552
New York Institute of Technology (Old Westbury, New York), 724–728
New York Power Authority, 19
NFPA *see* National Fire Protection Association
NFPA codes, 821
Nicol, J. F., 46–47, 66
Night insulation, *see* Passive solar heating
Night roof spray thermal storage (NRSTS), 380–383
Night ventilation of mass, 55–59
    compared to conventional cooling, 308
    design guidelines, 233–239
    detailed calculations, 301–303
Niles, Phillip, 320, 322, 329
Nitrates, 576
Nogales (Ariz.) Border Station, 380–382
Noise:
    air supply, 485–487, 515
    and annoyance, 1511
    control, *see* Noise reduction
    definition, 1491
    dosage, 1514
    duct system, 1599
    electrical equipment, 1606
    exposure measurement, 1515
    hearing protection, 1514–1518
    high levels, 1514
    masking, 92, 1584
    mechanical equipment, 1597
    NC curves, 1512
    OSHA high-level exposure criteria, 1514, 1518
    piping, 1605
    room criteria curves (RC), 1513, 1515
    side planning, 91–95
    transformers, 1606
    typical levels, 91–93, 1515
    water supply systems, 657
    see also Noise reduction, Masking sound, Noise criteria
Noise cancellation, active, 1602
Noise control and reduction:
    barriers, exterior, 92–95
    building siting for, 91–95
Noise criteria:
    curves, 1512, *see also* Room criteria curves
    recommendations for specific occupancies, 1584
    see also Acoustic criteria
Noise exposure, 1514–1518
    measurement, dosage, 1515
    OSHA criteria, 1514, 1518
Noise reduction:
    absorption techniques, 1530, 1550
    absorptive, 1549
    acoustic tiles, 1553, *see also* Sound absorption
    active noise cancellation, 1602
    airborne sound, 1556

background noise, masking, 1584
barrier critical frequency, 1561
barrier mass, Mass Law attenuation, 1560
barriers, exterior, 1574
barriers, sound transmission characteristics, 1560, 1607
coefficient, 1535
compound barriers (cavity walls), 1562
diffraction barriers, 1574
duct system noise, 1599
electrical equipment, 1606
equipment enclosures, 1607
FHA criteria for barriers, 1609, 1611
flanking paths, 1577
Impact Isolation Class (IIC), 1406
impact noise, 1404
machine quieting, 1598
mechanical system noise, 1597
noise reduction (NR) of barrier, 1559
NRC (noise reduction coefficient), 1535
panel and cavity (Helmholtz) resonators, 1549
piping, 1605
principles, 1549
speech privacy, 1577
STC (sound transmission class) of barriers, 1564
structure borne sound, 1595
transmission loss, composite barriers, 1562
transmission loss, diffraction barriers, 1576
transmission loss, doors and windows, 1570
transmission loss (TL) of barrier, 1559
    see also Acoustic criteria, Room acoustics, Sound absorption, Speech privacy
Noise reduction coefficient (NRC), 1535
Nonrenewable energy sources, 3, 9–12
Norberg-Schulz, Christian, 115, 146
North Central Airlines hangar building (Detroit Metropolitan Airport), 812–813
NRSTS, *see* Night roof spray thermal storage
Nuclear energy, *see* Energy sources
Nuisance control (water supply), 584–585

Oak Ridge National Laboratory, 164
Occidental Chemical Corporate Office Building (Niagara Falls, New York), 350–351
Occupancy sensors, 1286
Occupant loads (fire), 778
Occupational Safety and Health Act (OSHA), 1514, 1518
Ocean Arks International, 719
Oceanside Water Pollution Control Plant (San Francisco), 728, 733–735
Ocosta Junior/Senior High School (Westport, Washington), 492–493
Octagon residence (Liverpool, England), 331–332
Odors:
    air, 332–335
    water, 574–577
Office air supply systems, 484–493
Office arrangement types, 105, 429–430
Office equipment heat gains, 284
Office HVAC controls, 483–484
Office lighting, integral with structure, daylighting, acoustics, 198–199
Office of Building Technologies, 12, 36
Office solid waste recycling, 760–766
Ohm, Ohms Law, 854

Oil, *see* Energy sources
Oil-fired boiler, *see* Boilers
Olf, 338
Olin Water Service, 575
Olgyay, Victor, 48, 53, 66, 175, 275
ONSI Corporation, 523
Open-celled pavers, 553, 555
Open frame building envelope, 115–118
Open plan offices, 429–430
Operative temperature, 42–48
Orange Bowl (Miami, Florida), 555
Orientation of buildings, 106, 195–196, 220–224
Ornamentation, water and, 533–534
OSB (oriented strand board), 130
OSHA noise exposure criteria, 1514
Oswald, Dr. William, 717–718, 745
OS&Y (outside stem and yoke) valve, 803
Outdoor air standards, 340, *see also* Infiltration; Ventilation
Outgassing, *see* VOC
Outlet boxes, electric, 913
Overall thermal transmittance, *see* U values
Ove Arup Partnership, 298–299
Overcurrent protection, 1016, *see also* Circuit breakers, Fuses
Overhead ground shield wire (lightning protection), 817
Overhead raceway systems, *see* Ceiling raceway systems
Overland Partners, architects, 560
Oxidation, 582, 769
Oxygen (fire protection), 769–770
Ozone, 333, 335, 465, 580–581
  depletion of, 95–96, 809

Pacific Gas and Electric, 378–379
Package heat pump systems, 411, 423
Padre Dam Municipal Water District (California), 728–733
Panelboard, electric, 959–965, 1030, 1033
Panetta, Daniel, 717, 731, 752
Paper recycling, 747–751, 755–757
Park Plaza office building (Boston, Massachusetts), 476–477
Parking lot water treatment, 555, 742
Particulate filters, *see* Filters, air
Particulates (IAQ), 333–336
Passive building comfort standards, 48–55
Passive cooling:
  calculations for, 295–329
  comparing design procedures, 232–233
  daylighting and, 248–249
  design guidelines, 225, 230–249
  design strategies, 55–59
Passive solar heating, 26, 29–30, 32–34
  backup heating systems for, 250
  calculations, 255–276
  comparing systems, 256–257, 260–261
  daylighting and, 248–249
  design guidelines, 205–224, 248–249
  design strategies, 58–65
  glazing insulation, 163, 255–259
  internal temperature range, 272–276
  multiple system calculations, 270–271
  site planning, 73–92, 108–110
  variations, 271–272

Passive Solar Industries Council, Appendix H
Passive subslab depressurization, 346–348
Pasteur filters, 579
Pasveer oxidation stream, 724–728
Payback period, simple and discounted, 1715
Pei, *see* I. M. Pei
Pellet stoves, 388
PME (proton membrane exchange) fuel cells, 523–524
Percolation tests, 709, 740
Pereira, *see* William Pereira
Perimeter all-water systems with local fresh air, 511–512, 515–518
Perimeter zones, *see* HVAC systems, zoning
Perineal cleansing, 629–631
Personal Environments®, 489–490
Pesticides:
  IAQ, 334, 337
  water, 576
Petronas Towers (Kuala Lampur), 785
pH, 574–575, 580, 582
Phase change solar DHW systems, 617, 620–621
Phase change materials, 130–131, 219–223
Phoenix composting toilets, 671–672, 676–677
Phosphoric acid fuel cells, 523–527
Photocatalytic oxidation, 366–367
Photovoltaic power systems, 858
  array tilt angle, 866
  arrays, 866–868
  batteries, 866–870, 873
  building-integrated (BIPV) elements, 863
  codes and standards, 863
  design of grid-connected system, 874
  design of stand-alone system, 869–874
  direct connected systems, 860
  distributed PV arrays, 862
  economic considerations, 862
  grid-connected system, 862
  hybrid stand-alone system, 860
  interactive PV system, 874
  net metering arrangement, 859
  regulatory statutes, 858
  solar cells, 864
  solar insolation, 866, 870, 873
  stand-alone systems, 860
  terminology and definitions, 864
Photovoltaics (PV), 11, 18–19, 33–35, 70, 81, 265–266, 297–299, 376, 525–526, 583, 619, 672–673, *see also* Energy sources
Phragmites, 719
Physiological factors:
  cleansing and elimination, 626–627
  thermal, 41–43, 113, 120
Phytoremediation, 742
Piano & Rogers, architects, 438
Pipes:
  expansion (hot water), 393, 649–650, 653–655
  noise in, 639–640
  supports for, 653–654
  waste removal systems, 681–685
  water supply, 651–663
Pipe station columns, *see* Wet columns
Place Architecture LLC, *see* EPICenter
Plante, Russell, 601, 616–617
Plants, 107–112
Plastic Pipe Institute, 687–688

Plastic piping, 651–652
Plastic recycling, 748–751
Platzer, W., 260
Plenum (under-floor) return air, 492
Plenum supply air, *see* Displacement ventilation
Plumbing:
    drainage fixture units, 691
    drainage principles, 678–681
    fixture pressures, flow rates, 656
    fixtures required, 540, 542–543
    fixture trap sizes, 691
    large buildings waste systems, 687–704
    pipe sizing, supply systems, 656–663
    residential waste systems, 686–697
    water supply fixture units, 661
    well water, pumps, 594
    see also Waste water removal; Water supply
Pneumatic material handling systems, 1465
Poke-through (electrical wiring) system, 918
Pollution:
    air, 95–97, 332–337
    water, 533–536, 552, 573–577, 584–585, 741–745
Population Action International, 532
Population estimates in buildings, 777–778
Population growth, 3, 5
Porous pavement, 553–555
Positive displacement pumps, 586–587, 590–591
Potable water, 531, 573, 735
Potomac Mills Mall (Virginia), 782–783
Power conditioning, 982–995
Power line carrier (PLC) system, 979
Power line disturbances, 982
Power quality, *see* Power conditioning
Preaction sprinkler systems, 805, 810
Precipitation, *see* Rain; Rainwater
Preheating ventilation air, 348–352
Premise wiring, 1353
Present value of future payments, *see* Economics
Pressure reducing valves, 637, 645–649
Pressure tanks, 587–589, 593–595
Pressure-temperature (P/T) relief valves, 612
Primary treatment of waste water, 704–708, 733
Princeton Energy Group, 431
Prismatic light guides, *see* Light guides
Prison plumbing fixtures, 698
Privacy, bathrooms, 639–640
Profile angle, 1701
Programmable controllers, 604–605, 949
Programmable time switches, 949
Programmable timers for irrigation, 663–663
Projected roof area, 557
PSALI (permanent supplementary artificial lighting in
    interiors), 1109
Psychological factors:
    cleansing and elimination, 626–627
    thermal, 41–43, 113, 120
Psychrometry, 288–294, 423–424
P/T, *see* Pressure-temperature relief valves
Pulse boilers, 459–460
Pumped upfeed distribution (water), 641, 650–651
Pumping chambers, *see* Dosing tanks
Pumps:
    fire protection systems, 788–789, 791–793
    hot water heating, 398
    selection, 587, 590–591
    sewage ejector, 684–685

water supply in buildings, 645–648, 650–651
wells and, 586–597
PV, *see* Photovoltaics
PVC (polyvinyl chloride) pipes, 652, 682–684,
    687–688
PV Parking Canopy (New York), 19

Quartz lamps, *see* Tungsten halogen lamps
Quick-response sprinklers, 796, 801

Raceways, electric, *see* Electrical raceways
Radiant asymmetry, 51–52
Radiant barriers, 127, 378–379
Radiant floors, 395–399
Radiant heaters, 386–403
Radiant panels, 143, 398–399
Radiant panels with supplementary air HVAC sys-
    tems, 448–449, 504–510
Radiation, *see* Heat flow; Radon
Radiation contamination in water, 577
Radiators, *see* Heating systems, hot water
Radon, 144, 334, 337, 342, 346–348
    map of risk, 350
Rain:
    annual rainfall, 546
    hydrologic cycle, 534–536
    maximum rainfall, 562–564
    runoff, 104, 107, 534–536
    urban heat island and, 67–71
Rain Bird Sales, Inc., 664–665
Rainwater, 533, 541, 545–572
    collection and storage, 533, 546–551
    components, 561–572
    flushing toilets with, 471, 474–475
    quality, 534–536, 546
    recharge, 555–561
    recycled water and, 735
    roof retention, 552–554
    site planning and, 99, 103–107, 551–561
    solar energy and, 534, 541, 545
    stormwater systems, 551–552, 741–745
    see also Cisterns; Water
Raised floor systems, 487, 509–512, 698–699
Rate of return, *see* Economic analysis
RC (N) criteria, 407–409
Real Goods Solar Living Center (Hopland, Califor-
    nia), 582–583
Receptacles, electric, 972
    configurations chart, 973
    ground-fault type, 975
Recharge basins, *see* Rainwater
Rechargeable air chambers, *see* Expansion chambers
Reciprocating chillers, 463, 466
Reciprocating engine cogeneration, 520–521
Recovery rate, DHW systems, 604–612
Recycling:
    buildings, 35
    building materials, 5–9
    solid waste systems and, 752–757
    see also Graywater
Reflectance, in lighting, 1050–1051
Reflectance measurement, 1061
Reflectances, building materials, surfaces (table),
    379–389, 1144
    colors, 1208
    paper and ink, 1089

Reflected glare, 1086:
  geometry, 1088
  viewing angles, 1088
  see also Veiling reflection
Reflective glazing, 91–92
Refrigerants, 95–96, 376–377
Refrigerant dehumidifiers, 362–363
Refrigeration:
  air conditioners, 379, 381
  compressive, 375–376
  equipment space needs, 451
  heating systems and, 412–415
  psychrometry and, 423–424
Refrigeration cycles, 375–379, 454
Refuge areas, 778–779, 782–783
Registers, 407–412, 486
Reiach & Hall and GMW Partnership, architects, 431
Reif, Daniel, 618–620, 622
Relative humidity, defined, 44
Remote source lighting, 1328
  hollow light guides, 1333
  prismatic light guides, 1334
Renewable fuel sources, 3, 9–12, 95–97
Replacance (air quality), 338–339
Residential electrical loads (table), 1026
Residential electrical usage, 875
Residential lighting, 1292
Residential signal systems, 1347
Residential sprinklers, 797–798, 801
Residential wiring, 1023
Resistance (heat flow), defined, 126–127
Resonators, panel and cavity (Helmholtz), 1549, *see also* Sound absorption
Response time index (RTI), 801–802
Reverberation, *see* Room acoustics
Reverberation time, acoustic, *see* Room acoustics
Reverse osmosis (RO), 584, 737
Reverse return (hydronic systems), 393–394
Reynolds, John, 84, 109–110, 112, 211, 308, 329, 551, 622
Risers (plumbing), 645–651
RO, *see* Reverse osmosis
Robie house (Chicago, Illinois), 401
Rock beds, 220, 382, 384
Rock 'n Roll Museum (Cleveland, Ohio), 784
Rocky Mountain Institute, 378–379
Rolling metal shutters, 780
Rommell, M., 260
Roof:
  conservation criteria, Appendix J
  heat gain calculations, 225–229
  hourly heat gain calculations, 277–279
  rainfall retention, 552–554
  rainwater collection and, 546
  solar performance, 382
  thermal transmission and, 134–136
  U values, 163, 168
  see also Building envelope
Roof drains, 553, 567–569
Roof hatch ventilators, 786
Roof ponds:
  auxiliary cooling for, 318
  calculations (cooling), 313, 315–329
  compared to DG, SS, TW, 256–257
  design guidelines, cooling, 241, 244–246
  design guidelines, heating, 224
  detailing for, 320–322
  strategy, passive heating/cooling, 58
  see also Passive cooling; Passive solar heating
Roofscapes, 4, 644
Roof sprays, 379–383
Room acoustics, 1526
  articulation, 1527
  auditorium design, 1542
  echoes, creep, diffusion, flutter, focusing, 1539
  liveness chart, 1527
  ray diagrams, 1541
  reflections, 1538
  reverberation criteria, 1535, 1537
  reverberation time, 1526, 1537
  room proportions, 1541
  sound paths, 1536, 1538
  standing waves, 1540
  see also Acoustic criteria
Room criteria (RC) curves, 1513, 1515
Rossetti Associates, architects, 479
Roth and Moore, architects, 517–518
Rothe-Johnson, architects, 511–512
Roughing-in (plumbing), 698–702
Rousseau, David, 333–334, 364, 369
RTI, *see* Response time index
Rules of thumb for design, *see* Design guidelines
Runaround coils, 466, 469, 471
Runoff, *see* Rain
Runtal/North American Energy Systems, 402–403
RVP (relative visual performance), 1092

Saarinen, Eero, architect, 504, 506
Safdi, Moshe, *see* Moshe Safdi
Sand filters:
  buried, 711–714
  cisterns and, 546–548
  compared, 713–714
  open (intermittent), 548, 705, 713
  recirculating, 713–715
  slow, 578–579
Sandia National Laboratory, 245, 262
Sanitary sewers, 533
San Francisco City and County, 734
San Luis Solar Group Complex (Santa Margarita, California), 263, 265–266
Santee Water Reclamation Plant, *see* Padre Dam
Saturation, of air, 288
SBS, *see* Sick building syndrome
SC (shading coefficients), 138, 174–176, 178–179, Appendix J
Scale control, water, *see* Cooling towers
Schedule, in thermal zoning, 195
Schneider, S. F., 96, 112
Schubert, R. P., 348
S. C. Johnson Commercial Markets, Inc. building (Sturtevant, Wisconsin), 490–491
Screw compressor, 463, 465
Scroll compressor, 463, 466
Scuppers, 792
Secondary waste water treatment, 705, 708–714, 733
Sediment, 552
Sedimentation, 578
Seeley G. Mudd Library at Yale University, 515–518
Seepage pits, 705–706, 708
SEER (seasonal energy efficiency rating), 404, 418
Selective transmitter films, 140

Selkowitz, Stephen, 141, 146, 172
Semidirect lighting, 1218
Semi-indirect lighting, 1216
Sensible heat, defined, 277
Seppanen, O., 492
Septic tanks:
    approximate sizing, 541
    capacity, 707
    drainage fields, 541, 705, 708–710
    individual systems, 704–711
    leaching beds, 708–710
    location constraints, 709–711
    primary, for multibuilding systems, 714
    water and, 706
Sert, Jackson and Gourley, architects, 121
Service cores, *see* Core arrangements
Sewage:
    AIWPS™, 716–718
    estimated flow, 540–541, 544
    garbage disposers and, 757–759
    greenhouse ecosystems, 719–723
    lagoons, 541, 715–719
    mounds, 541, 711–712
    large scale treatment, 728–735
    on-site treatment, individual buildings, 704–714
    on-site treatment, multiple buildings, 714–728
    package plants, 541
    sumps, ejectors, 684–685, 689
    see also Sand filters; Septic tanks
Sewer gas, *see* Methane; Vents; Traps
Sewers, building, 695, *see also* Sanitary sewers, Storm
        sewers
Shading, *see* Sunshades
Shading coefficients, *see* SC
Shading masks, 1113, 1701
Shadow angles, 1701
Sheet Metal and Air Conditioning Contractors Na-
        tional Association, Inc. (SMACNA), 369
Shepley Rutan & Coolidge, architects, 121
SHF (sensible heat factor), 423
SHGC (solar heat gain coefficients), 138, 171–172,
        175, Appendix B
Shock (water hammer), 644
Shooshanian Engineering Associates, Inc., 476–477
Shower heads, 628
Showers, 627–628, 661, 691, 737–738
    required number of, 542–543
Shurcliff, William A., 339, 369
SI nomenclature, 1741
Siamese connection, 790–792
Sick building syndrome (SBS), 120, 331, 337, *see also*
        IAQ
Sidelighting, *see* Daylighting
Signal equipment systems, Chapter 22
Signal systems, hotels and motels, 1356
    industrial building, 1366
    multiple dwelling, 1354
    office building, 1362
    residential, 1347
    schools, 1357
Sill cock, *see* Hose bibb
Silica gel, 359, 362
Singer Company, 413
Single-duct reheat HVAC systems, 446–447, 499, 502
Single-duct Variable Volume HVAC systems, *see*
        VAV
Single-line diagram, electric, 894

Single package systems, 416
Single-zone HVAC systems, 445, 447, 495–496
Sinks, 542–543, 661, 691, 737–738, *see also* Lavatories
SIP, *see* Structural Insulated Panels
Siphonage from traps, 678–680
Siphon chamber, *see* Dosing tank
Siphon-jet and -vortex toilets, 632–634
Sites:
    analysis, 3, 71–73
    climate and, 67
    compost and, 755
    design strategies, 73–75
    fire protection and, 779
    IAQ and, 341–342
    plants and, 107–112
    rain/groundwater and, 104–107
    septic systems and, 709–711
    solid waste collection and, 755, 758–759
    sound/air and, 83, 91–106
    sun and daylight on, 75–92
    thermal zoning, 71–73
Skidmore, Owings & Merrill, architects, 498–499,
        522–523, 792–793
Skin-load-dominated buildings, 192–194
Skydome, 82
Skylights:
    design guidelines, 197–203
    thermal transmission, 137–142
    U-values, 173–174
Skyline solar charts, 77, 82–83
Skytherm system, 320
Skyview factor, 71
Slab floors, 132–133, 165–166
Slant/Fin Ltd., 399
Sling psychrometer, 44
Sludge, *see* Sewage, large scale treatment
SMACNA, *see* Sheet Metal and Air Conditioning
        Contractors
Smart houses, 18, 375
Smart windows, 130
Smoke:
    confinement, 782
    dilution, 783
    exhaust, 783–787
    factors in management, 782
    fire hazards, 770
    management, 772–779
    sprinklers and, 785
Smoke barriers, 782
Smokeproof towers, 774–775
Snow, 107
Society for the Protection of New Hampshire Forests,
        see Conservation Center
Sodium, 576, 584
Sodium hypochlorite, 580
Sodium lamps, *see* High-pressure sodium lamps
Softening of water, 581, 583–584, 737
Soil conductivity, 322
Soil gases, *see* Radon
Soil moisture, 536
Soil types, drainage fields, 708, 741
Sol-air temperature, 135–136, 277
Solar access, *see* Solar envelopes
Solar altitude, azimuth, 77–88, 1112, Appendix D
    diagrams, 1112
Solar Aquatics System, 721
Solar-assisted heat pumps, 416

Solar chimney effect, 241, 302–303, 515
Solar collectors:
  access to sun, 36, 108, 112
  active collector, 615
  air-extract windows as, 492
  band-of-sun analysis, 77
  calculations, 620–625
  controls, 614
  design guidelines:
    active, 224–225, 622
    passive, 620, 622
  DHW systems, 612–625
  efficiency comparisons, 623
  heat pumps and, 416
Solar energy:
  architectural history and, 9–18
  availability *vs.* task, 613
  DHW systems, 612–625
  electricity production by, *see* Photovoltaics
  rainwater and, 541, 545
  urban heat island and, 67–71
  water distillation, 585
  see also Energy sources, Nonrenewable energy, Renewable energy
Solar Energy Research Institute, *see* National Renewable Energy Laboratory
Solar envelopes, 35, 75–77
Solar 5 (software), Appendix H
Solar heat gain coefficients (SHGC), 138, 171–172, 175, Appendix B
Solar heating, *see* Passive solar heating
Solar Hemicycle House (Middleton, Wisconsin), 73–75
Solaronics, Inc., 391
Solar radiation (insolation), 48–51, 252–254
Solar reflectance index (SRI), 379–389
Solar reflection, 83, 91–92, 105, 107, 380
Solar savings fraction (SSF), 60–61, 211–219, Appendix C
Solar stills, 585
Solar water heating:
  active systems, 613–625
  compared to conventional, 602
  passive systems, 613–618
  sizing, 620–625
  supply and usage patterns, 613
  system types, 616–617
Solder (pipe connections), 651
Solenoid valves, 620
Solids, heat flow through, 124–127
Solid waste systems, 747–766
  flow rates, 750–751
  equipment for, 763–766
  large buildings, 759–763
  material quantities, 748–749
  resource recovery, 755–757
  resources and, 747–755
  small buildings, 757–759
  space planning, 756
  storage areas, 756, 759
Solomon Inc., Architecture and Urban Design, 99, 103–104
Sommer, Fritz, 702
Sonoma Mountain Brewery (Sonoma, California), 720, 723–724
Sound:
  absorption, *see* Sound absorption

  airborne, 1556, 1559
  attenuation with distance, 1503
  decibel addition and subtraction, 1501
  decibel levels, 1499, 1505
  definition, 1423
  directivity, 1529
  fields in enclosed space, 1524, 1528
  Fletcher–Munson contours, 1506
  free-field propagation, 1497
  frequency, 1494, 1496
  generation, 1493
  intensity and intensity level, 1497, 1499, 1505
  inverse square law, 1498
  isolation (bathrooms), 639–640
  loudness, subjective, 1499
  loudness level, Phon scale, 1506
  measurement, 1508
  noise reduction (NR), 1559
  outdoors, 93, 1613
  power and power level, 1528, 1534
  pressure and pressure level, 1494, 1504, 1528, 1534
  reinforcement systems, 1544
  reverberant field, 1528–1530
  reverberation, *see* Room acoustics
  sites and, 91–94
  transmission loss (TL), 1559
  typical levels, 1506
  velocity of propagation, 1496
  vibration, 1517
  wavelength, 1497
  see also Noise, Acoustics, Room acoustics, Acoustic criteria
Sound absorption, 1519
  articulation class, 1592
  coefficient (table), 1519, 1522
  definition, 1519
  material installation methods, 1525
  materials, 1520, 1553
  mechanics, 1520
  noise reduction, 1530, 1549
  panel and cavity resonators, 1549
  recommendations for noise reduction, 1552
  resonators, 1551
  see also Noise control
Sound isolation, *see* Speech privacy
Sound reinforcement systems, 1544
Sound systems, 1360
Sound transmission:
  doors, 1570
  windows, 1573
  see also Sound transmission class
Sound transmission, floor, ceiling constructions, *see* Appendix G
Sound transmission, wall constructions, 1560, *see also* Appendix F
Sound transmission class (STC), 1564
  graphical determination, 1566
  specific occupancy recommendations, 1609
Sound transmission data for floor, ceiling construction, Appendix G
Southwall Corporation, 140
Sovent system, 702–704
Space requirements, HVAC equipment, *see* Heating/cooling equipment
Specific heat, materials, 147–157
Specific volume, air, 288–291
Spectral distribution of light sources, 1196

Speech, 1510
    articulation index (AI), 1511
Speech interference level (SIL), 1511
Speech privacy, 1577
    analysis sheet, 1582
    articulation class (AC), 1592
    articulation index (AI), 1591
    and barrier STC, 1581, 1583
    between enclosed spaces, 1577, 1582
    design recommendations, open office, 1592
    open area offices, 1585–1595
    sound isolation descriptors, terms, 1580, 1589
Split-systems, 413, 416
Spontaneous combustion, 769
Spray polyurethane foam (SPF), 135
Sprinkler alarms, 792, 844
Sprinkler systems:
    circulating closed-loop systems, 804–805, 808
    deluge systems, 806, 811
    design considerations, 791–793
    drainage provisions, 792
    dry-pipe systems, 805, 809
    ESFR (early suppression fast response), 801–802
    example system, 804
    extended coverage, 800, 802
    exterior protection with, 780
    failure modes, 794
    fire codes and, 791
    flow control, 796–797
    hazard and type, 797–799
    head types, 793–796
    HVAC system integration, 785, 804–805, 808
    maximum floor areas, 802
    mist systems, 806–808, 811
    preaction systems, 805, 810
    quick-response sprinklers, 796, 801
    residential, 797–798, 801
    smoke exhaust systems and, 785
    solid waste collection and, 760
    spacing, 797–801
    spray pattern, 795
    system types, 802–811
    upfeed pumping and, 792–793
    water curtains and, 786–787
    wet-pipe systems, 802–804
Sprinkler waterflow switch, 844
SRI, *see* Solar reflectance index
SRI International, 336
SS (Sunspace solar heating systems), *see* Sunspace
    systems
Stack effect, *see* Stack ventilation
Stacks for boilers, *see* Boilers
Stack vent (plumbing), 686, 694–695
Stack ventilation:
    calculations, 300–303
    composting toilets and, 669–676
    cooling strategy, 57
    design guidelines, 230, 232–233
    fire protection and, 782–784
    HVAC system and, 440–442, 483–484
    IAQ and, 331–332, 346–350
    see also Passive cooling; Ventilation
Stairs, 773–778
Standby power equipment, 1042
Standpipes, 787–791
Static head (pressure):
    duct systems, 407, 411

fire sprinkler systems, 797
hot water heating, 398
waste systems, 702
water supply, 642, 645, 648, 657–663
Steel pipe, 651–652, 682
Stein, Richard, 36
St. Enoch's Square (Glasgow, Scotland), 430–431
St. George's School (Wallasey, England), 26–27
Stonehurst subdivision (California), 714–715
Stoner, Carol H., 745
Stop-and-waste valves, 643
Storage, strategies, 34–35
Storm sewers, 533
Stormwater retention, 507–509
Stormwater treatment systems, 741–745
Straw bale construction, 755
Stream water, 552
Street lighting, 1326
Structural fire protection, 780
Structural insulated panels (SIPs), 129–131, 164
Structural-mechanical integration, 438–440
Structure, daylighting, solar and acoustic integration,
    198–199
Structure-borne sound, 1595, *see also* Impact noise,
    Sound
Submersible pumps, 587, 589–591
Subsurface irrigation (graywater), 739–742
Suction tanks, *see* Water storage tanks
Sulfates, 576
Sulphur lamps, 1194
Sun position charts, 77–80
Sun, sites and, 75–92
Sun-Mar Corporation, 672, 676–677
Sunpaths, 77–92, 1694, 1697–1700
    projections, 1695
Sun peg shadow plots, 83–91
Sunshades and sunscreens, 121–122, 350–351
    adjustable characteristics, 177
    fire protection and, 772
    moveable *vs.* fixed, 108–110, 117, 142
    passive cooling and, 140–142, 297–299
    reflection control, 83, 91–92
    shading coefficients, 175–176
Sunshading, Appendix D
    fixed sunshade design, 1705
    shading mask, 1701
    shadow angles, 1701
Sunspace (SS) systems, 59–62, 271–274
    characteristics, 260–263
    compared to DG, TW, 58, 256–257
    composting toilet and, 676
    system performance, Appendix C
    see also Passive solar heating
Suntime and clock time, 1693
Sunwater Systems, Inc., 720
Supervisory control systems, *see* Building automation
    systems
Superwindows, 140–142
Supply registers, *see* Registers
Surface temperature, 42, 136–137, 143–144
Surge suppression:
    application, 989
    clamping voltage, 985
    operation, 986
    standards, 986
    terminology, 985
    see also TVSS

Sustainability, design strategies for, 21–35
Swales, 104, 107
Swamp coolers, *see* Evaporative coolers
Sweat, 37–39
Swimming pools:
  solar heating for, 613, 625
  water supply for, 594–597
Switchable windows, 27, 130, 344
Switchboard, switchgear, 956–959
Switches, electric, 944–950
  automatic transfer, 948
  configuration, 946
  device type, 973
  low voltage, 976–979
  programmable, 946, 976
  remote control (RC), 948
  service, 944
  time-controlled, 949–951
  see also Building envelope
Switching, low-voltage, 976
Switching, wireless, 979
Symbols, acoustics, 1622
Symbols, electrical, 1017

Tanks, *see* Water storage tanks
Tankless water heaters, 598–600
Tanner Leddy Maytum Stacy Architects, 233
Task air cleansing, 367–369
Task dehumidification and humidification, 362–363
Task lighting, 31–32
Task ventilating, 341
Tectronic Products Co. Inc., 368
Telecommunication and Data Systems:
  hotel and motels, 1356
  industrial, 1369
  offices, 1363
  residential, 1352
  see also specific type
Telephone systems, 1355
Temperature, 37
  air, 39–42
  daily range (MDR), 234–238, Appendix A
  design conditions, 56, Appendix A
  DHW systems, 597–599
  dry-bulb, 45
  extreme, 56
  hydronic systems, 398
  MRT, 42
  normal (average), 56
  operative temperature, 42–48
  supply drinking water, 574
  swing (in solar heated buildings), 272–276
  typical ambient, 900
  well water (deep earth), 324
  wet-bulb, 44, 317
  see also Surface temperature
Tensiometers, 663–664
Termodeck, 471, 474–475
Tertiary waste water treatment, 704–705, 715, 718–720, 728, 732
TETD/TA (temperature differential time averaging method), 294–295
TFM (transfer function method), 294–295
Thermal bridging, 131–132, 134–135, 164, 168
Thermal expansion (pipes), 649–650
Thermal gradient, 136–137
Thermal lag, 272, 275, 294

Thermal mass:
  common heat capacities, 304
  design considerations, 211, 217–219
  fire protection and, 772
  night ventilation of, 55–59
Thermal sailing, 31–32, 119–120
Thermal storage:
  design strategies and, 34–35
  heat gain calculations and, 294–295
  HVAC and, 472, 478–479
  passive solar heating design guidelines, 217
  phase change materials, 219–223
  roof ponds, 241, 244–246, 315–329
  wall systems with, 298–299
Thermal transfer wheels, *see* Heat exchangers
Thermal zoning:
  design considerations, 195–196, 297–299, 432
  HVAC systems, 427–428, 430–435
  hydronic heating systems, 399–401
Thermosiphon hot water systems, 603, 615–618
Thermostat, *see* Controls
Thirty-three West Monroe office building (Chicago), 498–501
Thompson, T., 312, 329
Thoreau Center for Sustainability (San Francisco), 233–235
Three Rivers Aluminum Company, Inc., 344
Tile drainage fields, *see* Drainage fields
TIM *see* Transparent insulating materials
Tipis, *see* Vernacular buildings
TLCD Architecture, 487
Todd, John, 719
Toilets:
  accessibility issues, 638–642
  cleansing (body) provisions, 629–631
  conventional, 632–633
  design considerations, 629–631
  drainage fixture units, 691
  flushing controls, 637
  history of, 533
  low-consumption, 632, 634–637
  required number of, 542–543
  water conservation and, 632, 669–671
  waterless, 669–677
  water saver, 632–634
  water supply fixture units, 661
  see also Bathroom design
Tons, cooling, 374
Toplighting, *see* Daylighting
Toronto District Heating Corporation, 519
Total energy systems, *see* Cogeneration
Total equivalent length (TEL), 658, 662
Toxic substances:
  air, 335–337, *see also* IAQ
  water, 575–577
Trane Company, 410
Transamerica Building (San Francisco, California), 803–804
Transformer load center (unit substation), 959
Transformers, 937
  indoor installation, vaults, 941–943
  installation costs, 943
  insulation class, 938–940
  noise (sound) levels, 1606
  outdoor installation, 940
  types, 942
  typical data, 938–939

Transmittance (light), glass and plastic materials (table), 1137
Transparent insulating materials (TIM), 132, 258–260
Transpired collector (air preheat), 348, 351–352
Transportation, energy use in, 17–18
Traps (drainage), 678–680, 691
Trash, *see* Solid waste systems
Trash compactors, *see* Compactors, trash
Trees, *see* Deciduous trees; Plants
Trenches, *see* Disposal fields
Trends in:
    electricity consumption, 9–11, 15–16
    energy conservation, 96
    energy sources, 9–11, 35
    HVAC systems, 427, 440
    IAQ, 331
    U and *R* values, 129–130
Tri-water HVAC system, 513–515
Trombe walls (TW), 59–62
    characteristics, 263–264
    compared to DG, SS, 58, 256–257
    performance, Appendix C
    see also Passive solar heating, Thermal storage
Tubs, *see* Bathtubs
Tuned mass damper, 475
Tungsten halogen lamps, 1164
    precision reflector type (MR-16), 1165
Turbidity (water), 573–577, 581
Turbine cogeneration, 520–521
Turbine ventilators, 28–32, 346–348
Turbine well pumps, 587–591
Turning vanes, 409–410
TVSS (transient voltage surge suppression), 984
TW (Trombe wall) solar heating systems, *see* Trombe walls
Two-pipe all-water HVAC systems, 450, 504

Ubbelohde, Susan, 84, 112, 211, 329, 551, 622
UBC (uniform building code), 821
UL (underwriters laboratories), 894
UL fire alarm standards, 821
Ultra-low-flush, *see* Low-consumption toilets
Ultraviolet Light (UV):
    air treatment and, 337, 362, 367
    water treatment and, 581–582
    waste water tertiary treatment, 715
Ultraviolet radiation, in mercury lamps, 1185
Undercarpet wiring system, 924–929
Underfloor duct, 917–922, *see also* Displacement ventilation; Raised floors
Underground wiring, 934
Underslab ventilation, 346–348
Uniform Plumbing Code, 544, 565–566, 568, 570–571, 657–662, 707–708, 710–711, 739–742
Uninterruptible power supply (UPS), 989
    arrangements, 991–992
    dimensional data, 993
    system selection, 992
Union of Concerned Scientists, 524
Unit 1, First Village Solar Residence (Los Alamos, New Mexico), 260–262
Unit substation, 959–960
Unitary systems, 411–413
United Kingdom Pavilion (Seville, Spain World Expo), 297–299
University of Arizona Experimental Residence (Tucson, Arizona), 241

University of Nebraska, roof ponds, 320
Unthank Poticha Waterbury, architects, 556
Unthank Seder Poticha, architects, 106, 555
Upfeed water distribution systems, 641–644, 650–651
Urban forests, 755–757
Urban mines, 755
Urbanization and climate, 67–71
Urinals, 629–631, 639, 661, 691
    required number of, 542–543
    waterless, 676, 678
U.S. Architectural and Transportation Barriers Compliance Board, 638, 667
U.S. Department of Energy, 600, *see also* National Renewable Energy Laboratory
U.S. Environmental Protection Agency (EPA), 5, 36, 350, 540, 548, 572, 576, 589–591, 594, 667, 719, 735, 745, *see also* National Drinking Water Clearinghouse
U.S. EPA office/laboratory buildings (Research Triangle Park, North Carolina), 339–344
U.S. Green Buildings Council, Appendix H
U.S. Department of Housing and Urban Development (HUD), 606–607, 623
Utah Department of Natural Resources building (Salt Lake City), 497–498
UV, *see* Ultraviolet light
U values:
    average, calculation, 131
    building components, Chapter 4
    defined, 128
    heat flow and, 128–129
    trends in, 129–130

Vacuum breakers, 643–644, 681–682
Vacuum solid waste systems, 763–764
Vacuum system toilets, 636–637
Valley Architects, 238
Valves:
    drainage systems, 684–685
    water supply, 652–654
    fire protection systems, 793–794, 803, 805–806
Vancouver Island (B. C.) greenhouse treatment system, 719, 721
Van Geem, Martha, P.E., 8
Vapor barriers, 144
Vapor pressure, 143
Variable air volume, *see* VAV systems
Vasoconstriction, 37
Vasodilatation, 39
VAV (single duct, variable air volume) systems, 446–447, 489, 496–501, 504
    controls for, 482
    smoke management and, 785
Vegetation, *see* Plants
Veiling reflections, 1086
    contrast reduction, 1089
    control of, 1094
    ESI, 1082
    geometry of, 1088
Ventilation:
    calculations, cross ventilation, 300
    calculations, stack ventilation, 300–301
    composting toilets, 669
    design guidelines, 230–233
    design strategies, 332
    exhaust fans, 351–354
    fire protection and, 783, 785–786

forced, 146
heat exchangers and, 353–361
heat flow and, 144–146
history, 331
infiltration and, 144
hourly heat gains, 282
passive and low energy approaches, 341–345
passive cooling, 55–57
required rates of, 144, 180–187, 337–340
site design and, 83, 91–104
smoke management and, 783, 785–786
turbine ventilators, 28–32, 346–348
window design for, 26–27
see also HVAC systems; IAQ
Vents (plumbing), 678–681, 694–696
Vent stack, 686, 694–695
Vernacular buildings, 118–119
Vertical shadow angle, *see* Sunshading
Victor Gruen Associates, Inc., architects, 437
Village Homes (Davis, California), 557–559
Visible Transmittance (VT), 138, 171–172, 175
Vision, 1065
  adaptation periods in, 1068
  aging eyes, 1074
  color, 1072
  contrast and adaptation, 1071
  eye adaptation levels, 1071
  eye physiology, 1065
  factors in acuity, 1067
  fields of, 1066
Visual comfort probability (VCP), 1085
Visual display terminals (VDT), lighting, 1307
Vitrified clay tile pipe, 682
VOC (volatile organic compounds), 333–335, 344, 367, 369
Vogelsang High Sierra Camp (Yosemite National Park, California), 677
Volatile organic compounds, *see* VOC
Volt, voltage, 853
Voltage, systems, single- and three-phase, 1001–1006
Voltage, undervoltage effects, 1007
Voltage drop, 1041
Voltage system and utilization, 1006
Voltmeter, *see* Electrical measurements
Vortechnics Incorporated, 743–744
VT, *see* Visible Transmittance
Vulcan Radiators, 402
VVVF (Variable voltage, variable frequency), 1400

Walls:
  approximate heat gains, 226–229
  conservation criteria for, Appendix J
  hourly heat gain calculations, 277–279
  thermal transmission, 130–132
  U values for, 130–132, 163–164
  see also Building envelope
Ward's Island wastewater treatment plant (New York), 756
Warm air heating systems, 404–412
Washing machines, 637–638, 661, 691, 737–738
Wasley, James, 333–334, 364, 369
Waste water removal:
  drainage fixture units, 691
  drainage principles, 678–681
  history of, 678
  large buildings, 697–704
  piping, fittings, accessories, 681–690

planning for, 697–698
residential systems, 686–697
waste treatment, 21–25
water and, 669–671, 706
see also Sewage; Solid waste systems; Toilets
Water:
  architecture and, 531–534
  building materials and, 536
  ceremonial purposes, 103–104, 532
  characteristics, 534, 573–577
  cleansing purposes, 531–532
  cooling purposes, 298, 533
  electrical power generation and, 536
  fire protection and, 534, 786
  fountains, 41, 531–534, 640
  hydrologic cycle, 534–536
  nourishment purposes, 531
  ornamental purposes, 533–534
  planning and, 536–544
  quality of, 531–536, 573–577
  recycling (graywater), 636–637, 735–742
  temperature (wells), 324
  transportation purposes, 532–533
  waste removal and, 532–533
  see also Cooling towers; Evaporative cooling; Rainwater; Water supply
Water closets (WC), *see* Toilets
Water Conservation Systems, Inc., 664
Water curtain, fire suppression, 786–787
Water filters, *see* Filtration
Water flow switch, 844
Water hammer, 644
Water heaters, *see* Boilers; DHW systems
Waterless Company LLC, 676, 678
Waterless toilets, urinals, 669–678
Water loop heat pump systems, 422–423, 448–449, 509, 511, 513–515, 804–805, 808
Water mains (pressures), 641, 657–658
Water meters, 643, 657–658
Water pipes, *see* Pipes
Watersaver toilets, 632–634
Water softening, 581, 583–584, 737
Water source heat pumps, 422–423
Water storage tanks:
  fire, 772, 779, 791–793
  HVAC, 472, 476–478
  water supply, 594, 644–645
Water supply:
  bathroom fixtures, 625–642
  condensation and, 655
  conservation strategies, 625–626, 670–671
  disinfection, 579–582
  distribution systems, 640–651
  filtration, 577–579
  fire protection, 788–789, 791–793
  heat conservation and, 655–656
  history of, 536
  hot water systems, 597–626
  irrigation systems, 662–667, 724, 728, 735–742
  measurement terms, 537
  pipe sizing, 656–663
  pipe support for, 653–654
  pressure tanks, 587–597
  pumps, 586–597
  quality considerations, 573–577
  rates of usage, 531–533, 536
  shock in, 644, 653–655

Water supply (*Continued*)
  sources, 585–597
  storage tanks, 472, 476–478, 594, 644–645, 772, 779, 791–793
  treatment, 577–585
  valves and controls, 652–654
  wells, 585–586
  zoning, 645–649
  see also Cisterns; Rainwater
Water table, 536, 708–712
Water walls (WW), *see also* Passive solar heating
  characteristics, 263–266
  compared to DG, SS, TW, 58, 256–257
  system performance, Appendix C
Watson, Donald, architect, 28–30, 73, 112, 223, 320, 329
Watson House (Connecticut), 28–30
Watt, defined, 124
Wausau Homes, Inc., 697
WC (Water closets), *see* Toilets
WE Group, PC, architects, 202–203
Weber's law of visual contrast, 1009
Weil Pump Company, 689
Wells, 585–586, 588–594
Western Ventilating Equipment, Inc., 347
Wet-bulb temperature, *see* Temperature, wet-bulb
Wet columns, 697–698
Wetlands, constructed, 717–719, 724
Wet-pipe sprinkler systems, 802–804
Wheelchairs:
  bathroom design for, 638–642
  fire exits and, 774
WhiteCap™ Roof Spray Cooling System, 383
Whole-building heat loss criteria, 208–211
Whole-house exhaust, 352–354, 378–379
Wildlife Center of Virginia (Waynesboro), 422
Wilkinson, Glen, 127, 146
William Pereira & Associates, architects, 805
Williams, Douglas, 752
Willow Run Farms, Inc., (Ypsilanti, Michigan), 754
Wind:
  building design examples, 102–104, 106
  cross ventilation calculations, 300
  design guidelines, cooling, 230–232
  site design and, 97–104
  urban heat island and, 67–71
  windbreaks, 97–99
Windows:
  air leakage rate, 138

approximate heat gain calculations, 226–229
comparing performance of, 141, 218
conservation criteria and, 191–192
daylighting and, 198–201
fire protection and, 772
hourly heat gain calculations, 277, 280–282
infiltration, 189–190
passive cooling and, 230–232
passive solar heating and, 218
shading coefficients, 138, Appendix J
history, technology of, 129–130
thermal transmission, 137–142
U values for, 129–130, 137–138, 171–172
ventilation, 27, 102, 343–345
see also Building envelope
Wind-gravity ventilators, *see* Turbine ventilators
Winitsky, David, 78, 82
Wire and cable, *see* Conductors, electric
Wireways, *see* Conduit, Electric raceways
Wiring design, *see* Electric wiring design
Wiring devices, 970
Wiring systems, interior, 896
Wittwer, V., 260
Wood recycling, 749
Wood stoves, 10, 386–389
Workstation delivery systems (air supply), 489–491
Worldwatch, 532
Wright, Frank Lloyd, 73–74, 401
WSFU (water supply fixture units), 657, 660–661
WW, *see* Water walls

Xenon, 129
Xeriscape, 561, 665–667

Yoklic, Martin, 313–316, 329

Zimmerman Design Group, architects, 490–491
Zinc, in water, 576
Zion National Park Visitor Center (Utah), 243–244
Zonal cavity illuminance calculations, 1250, 1259
Zoning:
  body heat, 37–39
  building design, 195–196
  graywater systems, 739–741
  HVAC larger buildings, 427–428, 430–435
  hydronic heating, 399–401
  water supply, 645–649
  see also IAQ; Comfort zone; Thermal zoning